A Member of the International Code Fam

INTERNATIONAL CODE COUNCIL®

ORG 1873 INTERNATIONAL FIRE CHIEFS ASSOCIATION

INTERNATIONAL FIRE CODE®

COMMENTARY

2003

First Printing: June 2004

ISBN # 1-58001-131-4

PRINTED IN THE U.S.A.

PREFACE

The principal purpose of the Commentary is to provide a basic volume of knowledge and facts relating to building construction as it pertains to the regulations set forth in the 2003 *International Fire Code*. The person who is serious about effectively designing, constructing and regulating buildings and structures will find the Commentary to be a reliable data source and reference to almost all components of the built environment.

As a follow-up to the *International Fire Code*, we offer a companion document, the *International Fire Code Commentary.* The basic appeal of the Commentary is thus: it provides in a small package and at reasonable cost thorough coverage of many issues likely to be dealt with when using the *International Fire Code* — and then supplements that coverage with historical and technical background. Reference lists, information sources and bibliographies are also included.

Throughout all of this, strenuous effort has been made to keep the vast quantity of material accessible and its method of presentation useful. With a comprehensive yet concise summary of each section, the Commentary provides a convenient reference for regulations applicable to the construction of buildings and structures. In the chapters that follow, discussions focus on the full meaning and implications of the code text. Guidelines suggest the most effective method of application, and the consequences of not adhering to the code text. Illustrations are provided to aid understanding; they do not necessarily illustrate the only methods of achieving code compliance.

The format of the Commentary includes the full text of each section, table and figure in the code, followed immediately by the commentary applicable to that text. At the time of printing, the Commentary reflects the most up-to-date text of the 2003 *International Fire Code.* As stated in the preface to the *International Fire Code,* the content of sections in the code which begin with a letter designation (i.e., Section 307) are maintained by another code development committee. Each section's narrative includes a statement of its objective and intent, and usually includes a discussion about why the requirement commands the conditions set forth. Code text and commentary text are easily distinguished from each other. All code text is shown as it appears in the *International Fire Code*, and all commentary is indented below the code text and begins with the symbol ❖.

Readers should note that the Commentary is to be used in conjunction with the *International Fire Code* and not as a substitute for the code. **The Commentary is advisory only;** the code official alone possesses the authority and responsibility for interpreting the code.

Comments and recommendations are encouraged, for through your input, we can improve future editions. Please direct your comments to the Codes and Standards Development Department at the Chicago District Office.

TABLE OF CONTENTS

Chapter 1: Administration

General Comments

This chapter addresses the administration and enforcement of the code. The objectives and mandate for enforcement are beyond the scope of this chapter. Before adopting the code, a state or local government must establish and designate an agency having staff trained to administer and enforce the code. The administrative relationships, designation of the enforcement authority (fire code official), funding, training and certification of inspectors and scope of the enforcement program are determined by the adopting body.

Management personnel generally perform functions such as planning, organizing, directing, controlling, analyzing and budgeting. Though the code administrator's duties may include all of these functions, this chapter takes a much narrower view of the code administrative function, dealing mainly with technical and legal areas. Fire prevention code administration must be considered in the context of a complex environment containing political, social, economic, technical and legal dimensions. Enforcement, too, is a broad, all-inclusive term that includes a range of activities aimed at identifying and eliminating hazards; in this case, hazards causing or contributing to a fire or impairing life safety.

Four functions are commonly associated with enforcement: inspecting, detecting, notifying and reporting [see Figure 1(1)]. Chapter 1 serves as the basis for administering a code enforcement program consisting of these functions. This chapter describes the technical and legal requirements associated with administering a code enforcement program to achieve these functions. The examination of these concepts specifically provides a better understanding of the fire code official's authority, duties and liabilities.

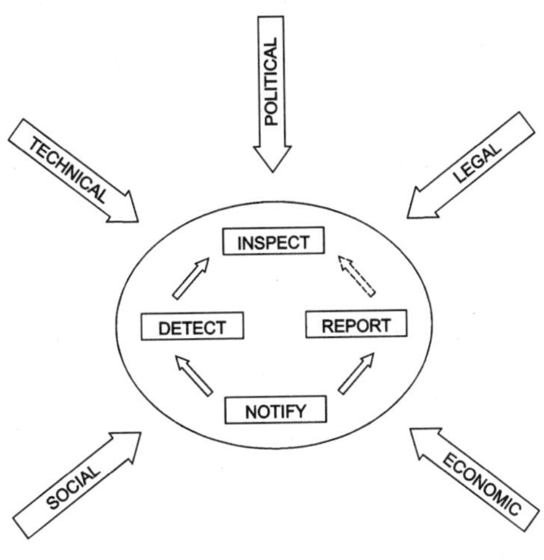

Figure 1(1)
CODE ENFORCEMENT ENVIRONMENT

Two main duties of the fire code official are administration and enforcement. In administration, the following concepts are most important:

Code Administrative Environment

Many administrative or management functions are not addressed in the code. Before provisions of this document can be of any use, many basic questions must be answered. Jurisdictions adopting a code enforcement program are using discretionary powers to fulfill a community need. The need in the community must be clearly identified, the program mission clearly established and the most appropriate delivery system selected. To address the technical and legal demands of the code administrative environment, the code assumes that jurisdictions adopting the document are interested in protecting the health, safety and welfare of its citizens from the effects of fires and explosions. Additionally, the code assumes that these jurisdictions are authorized to use the police power of the state to receive these benefits. Finally, the code assigns principal responsibility for enforcing this document to the department or agency (fire department or fire prevention bureau) most frequently available to perform this mission.

The particular objectives and social or political mandate of a code enforcement program are not considered in the context of this document. These items, however, are often cited as the most frustrating problems faced by code administrators. Code enforcers often complain of being overwhelmed by demands for leniency or special consideration based on the economic, social or political effects of their decisions.

As stated, this chapter establishes ground rules for enforcing the code; however, these ground rules are only the technical and legal requirements binding both fire code officials and the general public. For guidance on the political, social and economic considerations associated with code enforcement activities, adopting authorities must turn elsewhere; however, none of this is intended to imply that these considerations are absent from the code process. To the contrary, by establishing these requirements as "minimums," the membership has, through a democratic process of public hearings and debate, attempted to weigh these considerations carefully when deliberating, modifying and adopting the provisions appearing in this document (voting members are fire code officials representing jurisdictions). In the end, each jurisdiction must give careful consideration to how these requirements should be adopted, who should be responsible for enforcing them, how this personnel should be trained, how the operation will be financed and when and how to modify or change operations if necessary. These considerations deserve careful, thorough public attention before a decision is made to adopt and enforce the code.

Scope and Applicability of the Code

The code applies to all structures and premises, both new and existing, in all matters related to occupancy and maintenance for the protection of lives and property from fire. Conditions possibly causing or contributing to the start or spread of fire or protection of life from hazards incident to occupancy and maintenance are regulated as follows:

Retroactivity: Because the code applies to both new and existing structures and premises, the existing building provisions may be considered retroactive. Existing structures and premises built in compliance with the codes and standards in effect at the time of their original construction or alteration are not exempt from code compliance.

Other codes and standards: The code relies heavily on other codes and standards to specify a means of complying with its provisions, including the *International Building Code*® (IBC®), the *International Mechanical Code*® (IMC®), the *International Fuel Gas Code*® (IFGC®) and the standards referenced in the text. Additionally, other federal, state and local codes and ordinances may establish certain requirements related to fire protection and life safety. Code requirements are intended to complement other regulations. When conflicts arise between code provisions and the referenced standards, the code provisions apply. Where a standard provides additional technical detail or guidance beyond that provided in the related code text, the fire code official must use judgement when applying these provisions to prevent conflicts with the code provisions. If a conflict arises, it is the fire code official's duty to determine which provisions secure the code's intent. When a conflict between codes or other legal action causes a portion of this document to be "struck down," such action is not intended to invalidate the remaining code provisions. The severability of code provisions, however, does not imply that these same provisions should be considered or applied outside of their context as a part of the code.

Fire Code Official's Judgement

The code relies heavily not only on other codes and standards but also on the judgement and experience of the fire code official.

Approval: The code details occupancy and maintenance requirements; however, it relies heavily on performance criteria, as opposed to detailed specifications, to accomplish this task. The fire code official, therefore, must exercise judgement when approving or permitting operations, processes and procedures required by the code. Proof of compliance may include certification or labeling by independent testing laboratories; however, regardless of the conclusions of these external agencies and authorities, the fire code official remains the sole judge of what fulfills the intent of the code. This becomes particularly important when the fire code official is asked to evaluate equivalent methods and materials. Piles of data may be impressive, but they may be meaningless when considered in the context of the code's intent. Data in support of alternative methods and materials must demonstrate not only compliance with the code's intent but also relevance to the issues at hand. Evidence, such as a label or an independent laboratory test report, may be used inappropriately to support an application for recognition of equivalency. The fire code official must evaluate all submitted evidence to make sure it applies to its intended use, as well as to the code's intent. In an increasingly technical and litigious society, learning how to make such decisions and judgements may be the biggest challenge facing fire code officials. Relying on strict interpretations of intent or the "letter" of the code may be the conservative way, but conservative approaches may simply increase the social and political pressures confronting fire code officials. Computers have become desktop fixtures in today's professional offices. Decision aids taking advantage of contemporary computer technology have become increasingly popular as well. These models permit designers to quickly and easily evaluate the relationships and performance of a variety of complex variables.

Another model that does not rely on a computer is NFPA 550 [see Figure 1(2)]. This model requires little training to use or understand and is an all-inclusive representation of the variables contributing to fire safety. The model may, therefore, serve as a useful tool for qualitatively evaluating the contribution of various approaches to an overall fire-safety system. Once equivalent alternative methods have been identified using the Fire Safety Concepts Tree, (see "General Comments" in Chapter 3) quantitative (cost/benefit) analyses may be applied. These decision aids permit a designer to propose more innovative and creative responses to complex problems. Fire code officials must begin to recognize, use and interpret these tools and data to maintain effective protection.

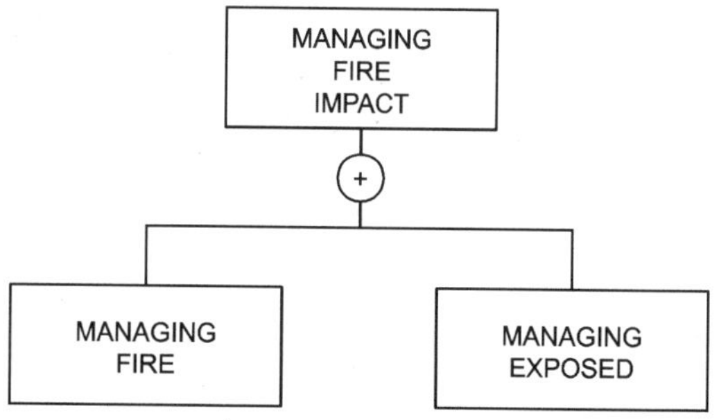

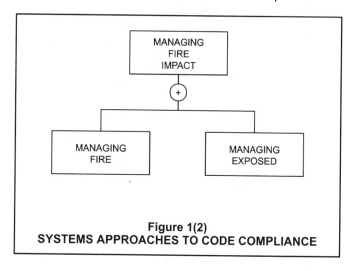

Figure 1(2)
SYSTEMS APPROACHES TO CODE COMPLIANCE

Fire Code Officials and Liability

Like all professionals, fire code officials are subject to legal action. The two most common legal actions that may be pursued against fire code officials are breach of contract lawsuits and tort claims. Tort claims, by far, are the most common lawsuits. These lawsuits allege that some damage, injury or harm (a tort) resulted from the actions of the fire code official. A successful tort claim must prove that the plaintiff was injured or harmed; that the fire code official had a legal duty or obligation to perform with respect to the plaintiff and that the cause of the plaintiff's injury was the fire code official's actions or inactions while performing these duties.

The Law of Torts includes the following:

The tort: Damages arising from the acts of fire code officials fall into two broad categories; property and personal [see Figure 1(3)]. Property torts involve the control, use, operation or ownership of personal and real property by private individuals. Personal torts involve physical, verbal or written assaults on the character, person, psyche or privacy of individuals. Such assaults or invasions may involve actual contact or threat of harm. For example, fire code officials' acts of commission may restrain business or trade activity, while acts of omission may fail to recognize that hazards need to be corrected, thus resulting in life or property losses.

Property	Personal
Trespass	Assault and Battery
Conversion	False Arrest or Imprisonment
Nuisance	Defamation, Slander and Libel

Source: Rosenbauer, D.L., *Introduction to Fire Protection Law.*

Figure 1(3)
TYPES OF TORTS

Two actions dominate lawsuits filed against enforcement authorities: Most lawsuits either allege improper acts by the fire code official (acts of commission) or failure to fulfill specified or implied legal obligations (acts of omission). In the former, plaintiffs usually seek temporary or permanent relief from a fire code official's decision. In these actions, plaintiffs usually allege improper interpretation or application of the code or its intent. Other lawsuits usually allege failure to exercise a reasonable standard of care in the performance of duties of the fire code official. In either type of lawsuit, and often in the case of omissions, plaintiffs seek compensatory and even punitive damages. Infringements on constitutional protections may be, though occurring infrequently, the basis for lawsuits against fire code officials. Common constitutional issues raised in lawsuits against fire code officials include violations of the Fourth Amendment's protection against unreasonable searches and seizures, the Sixth Amendment's due process protections and the Fourteenth Amendment's equal protection provisions. First Amendment rights guaranteed under the freedom of association protections may be raised in cases involving public assembly occupancies, especially churches.

Condition of negligence: To prevail in a tort claim action, a plaintiff must demonstrate negligence on the part of the defendant. Negligence may be simple—a failure to exercise reasonable or adequate care when performing assigned duties (commonly known as misfeasance)—or it may be gross—represented by wanton, willful, reckless or malicious disregard for public safety. Criminal activities, including dereliction (nonfeasance) or the failure to perform required assigned duties, may be cause for claims of gross negligence. Likewise, malfeasance, the willful or malicious violation of a legal duty, may constitute grossly negligent behavior. The following three elements must be proven to sustain a claim of negligence: the defendant had a duty to act, the defendant failed to exercise the required standard of care in the performance of that duty and, as a result of that failure, damage or harm was incurred by the plaintiff.

Duty to act: The code establishes few duties of the fire code official. Instead, it places greatest emphasis on the responsibility of structure or premises owners and operators to perform their duties with adequate regard for public health, safety and welfare. The duties owed the public by the fire code official fall under the following categories: approvals, enforcement, personnel, inspections, investigations, reports and record keeping. Other duties may be assumed by fire code officials through the performance of their official duties. Recently, some courts have ruled that failure to perform timely reinspections or exhaust legal remedies against violators in fire code cases creates a special relationship between the fire code official and the occupants of properties in violation of the code, especially when the occupants do not own the property and are not responsible for code compliance. Some court rulings have even implied that conducting inspections not otherwise required by the code constitutes an ultra vires (beyond the authority of) liability. Fire code officials should consult their jurisdiction's legal counsel to determine how these decisions, the jurisdiction's enforcement policies and the code provisions combined affect their enforcement program and jurisdictional and personal liabilities.

Standard of care: Taken together, the fire code official's duties are the basis for determining his or her standard of care. When assessing whether fire code officials have met this standard, judges and juries must determine whether he or she performed the required duties as a reasonable, comparably trained and experienced fire code official. Failure to meet the appropriate standard of care may be classified in three ways: nonfeasance, misfeasance and malfeasance. Nonfeasance is the failure to perform a required duty. Improper performance of a required duty constitutes misfeasance, and malicious or willful violation of a required duty is malfeasance. Of the three, misfeasance or simple negligence is the most common cause of action. The code and most tort claims either hold the government immune from specific claims of misfeasance or severely limit damage awards in such cases.

For all purposes, sovereign immunity—the doctrine inherited from British common law mandating that "the King can do no wrong" — is obsolete. Similarly, courts in many states have abandoned the public duty doctrine, which states that a duty to all is a duty to no one. Holding that most code provisions and governmental regulations secure benefits for select groups, some state courts recognize that specific enforcement activities secure greater benefits for some members of the public than others. Such judicial reasoning holds that the inspector's duty applies to the individual who may be injured as a result of failure to detect a hazard or diligently pursue compliance. Moreover, this duty may include acts of omission, such as failure to perform required inspections. With courts today recognizing only limited immunity for government officials, fire code officials must become more aware of their duties and liabilities. Though tort claim acts limit damage awards, they still permit lawsuits to proceed against governmental officials and agencies to determine their responsibility for negligent acts. Claims of gross negligence arising from nonfeasance or malfeasance are less common than misfeasance actions but are predictably harder to defend. The code provides no relief from liability where the fire code official either fails to perform a required duty or acts ultra vires; that is, beyond his or her authority. The jurisdiction is generally immune from claims when its agents perform acts beyond the scope of their authority, unless such acts were implicitly endorsed by the government (explicit endorsement may constitute a discretionary governmental act and, similarly, immunize the government). Nonfeasance is considered a criminal offense in many jurisdictions. An employee's dereliction of duty exempts the jurisdiction from immunity under most circumstances, unless the employee's failure to perform was the direct result of explicit instructions from governmental superiors, however, the employee may be held criminally liable.

In addition to the Law of Torts, the following have an impact on fire code officials and liability:

Awards: Lawsuits may seek declarative judgements in favor of the plaintiff—injunctive relief or monetary awards. Monetary awards fall into four categories: nominal, special, compensatory and punitive. The first purpose of monetary awards should be to the claimant or plaintiff for real losses. This is the purpose of compensatory and special damages. Compensatory awards reimburse the claimant or plaintiff for the direct costs resulting from the defendant's negligence or carelessness. Many times, a plaintiff will also seek additional compensation for the indirect results of the defendant's acts. Such special damage claims may result in additional compensation beyond that provided by compensatory damages. Punitive awards are intended to punish the defendant for the misdeed and discourage future unlawful activity by the defendant. These awards are often held up as examples to the community as a whole and are a way to discourage unlawful activities by others. Nominal damage awards serve to assign blame in intentional tort cases when the facts of the case do not merit a substantial settlement.

Protection: The best protection against a lawsuit is professional conduct and preparation; that is, training, education and research. Lawsuits filed against public officials have become commonplace and are probably inevitable. In 1983, H. M. Markman suggested six rules to manage legal liability [see Figure 1(4)].

- **You cannot prevent someone from filing a lawsuit against you.**
- **Do not take the lawsuit personally.**
- **Understand your risk exposure or exposures.**
- **Be professional.**
- **You are not an insurer.**
- **Do not make stupid mistakes.**

Figure 1(4)
MARKMAN'S SIX RULES

Though no single rule should be considered more important than another, the last one is perhaps the best to remember. Everyone makes mistakes, so strive to learn from the mistakes rather than repeat them. Nonetheless, every mistake may be potential exposure. Acting professionally helps minimize exposure to error, especially when training and common sense are encouraged. Using common sense, exercising reasonable care and acting professionally are no insurance against a lawsuit, but they all may provide considerable protection in the event a lawsuit is filed. No matter how hard everyone tries, someone may sue. When a lawsuit is filed, the most important things to remember are not to take it personally and not to forget the other five rules.

Enforcement

The enforcement duty of the fire code official's position is composed of four distinct functions: inspection, detection, notification and reporting. All four functions define phases in the enforcement process duties of fire code officials.

During the code enforcement process, structures or premises requiring inspections are identified. Inspectors are assigned and inspections are performed. During these inspections, any code violations found are usually noted. Then, the owner or occupant is verbally advised or notified that the deficiencies noted are code violations. To promote code compliance, the inspector may suggest remedial actions that may be taken to establish compliance. Finally, a written violation notice serving as further notice to the owner or occupant is issued. The written notice also serves as a permanent record or report of the inspection.

Inspection

Inspections are careful examinations of plans or premises for the presence of fire and life safety hazards. Upon observing a hazardous condition, the fire code official begins a process directed at correcting the situation. This may be accomplished by removing or eliminating the hazardous condition or providing some countermeasure designed to lessen its effects on the property, occupants or neighbors. In the case of inspections, care should imply a systematic method that keeps the inspection process in a proper perspective and recognizes that code enforcement is limited to legal and technical means of pursuing fire safety. Achieving fire safety objectives means using a balanced approach composed of some elements seeking to prevent ignitions and others attempting to control fire effects. Fire safety objectives are not defined by the code but rather by the users. Each jurisdiction must establish what risks and costs are reasonable while pursuing fire safety.

There may be as many different methods of conducting inspections as there are inspectors, and no single method is necessarily the correct one; however, each method probably has some strong and weak points. The following three approaches can form the basis for any number of different inspection techniques:

Outside to inside: Beginning outside is not only logical but necessary. Inspectors too often neglect hazards and clues outside the building that suggest significant danger to the occupants. An inspector must ask the following question: "Do the things I see outside match those I see inside?" For example:

- Do doors identified as exits inside actually discharge outside to acceptable refuge areas or the public way?
- Are trash receptacles or other obstructions outside located so the effectiveness of exits is not reduced, or do receptacles alone pose a fire hazard?
- Does the site permit sufficient access for fire-fighter rescue operations and fire suppression?

Top to bottom: Once inside a structure, deciding where to start is more than a matter of preference. By beginning at the top, an inspector's job becomes easier so that any violations are searched for in all areas. One question inspectors may ask is: "If completeness is the principal criterion, why not start at the bottom and work up?" The answer to this question is that walking down stairs is easier than walking up stairs. This easier path of travel allows the inspector to concentrate more completely on the inspection itself. After performing many inspections, there will not be a need for additional exercise obtained from beginning at the bottom.

General to specific: Without constructing a detailed inspection framework, many fire code officials find it helpful to move from the general to the specific when evaluating occupancies and hazards. This helps keep the whole problem in focus while preserving attention to detail. The inspector can focus on a specific problem without losing sight of the "big picture."

Detection

Systematic inspection procedures, like those described, should aid in the detection of code violations. By keeping the premises, processes or objectives in clear focus, the inspector keeps the task in context. A systematic inspection process implies not only organization but an understanding of the whole process. Achieving fire safety objectives means understanding how structures, premises, occupants and fire interact [see Figure 1(5)]. To keep the system in balance and prevent uncontrolled fires means understanding how people use structures and premises to achieve useful and productive purposes.

In such a context, a fire hazard is anything that either fails to prevent an uncontrolled fire or permits a fire to spread unchecked. Similarly, hazardous conditions are those preventing occupants from escaping or fire fighters from entering a structure and premises to control a fire.

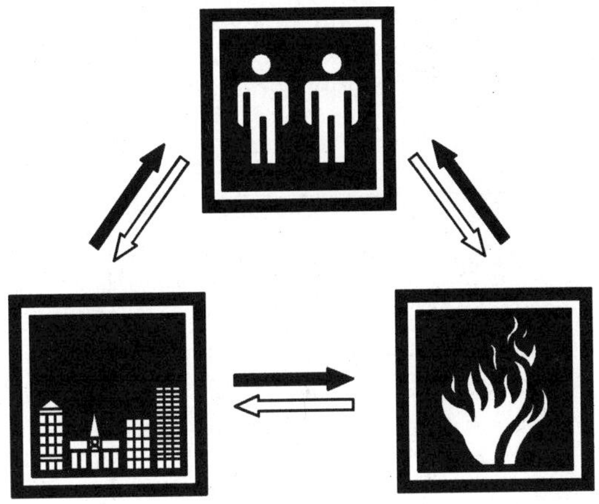

Figure 1(5)
HOW BUILDINGS, PEOPLE AND FIRE INTERACT

Notification

Inspection programs cannot identify and abate all hazards. Code enforcement alone cannot secure absolute protection for people and property. Furthermore, many code requirements maintain or reinforce features not intended to prevent a fire but rather minimize a fire's effects should one occur. Every inspection program, therefore, should consider the benefits of educating building owners and occupants about the hazards endangering their lives and property. Not only do such efforts help secure compliance with code requirements but they are likely to secure long-term commitments to fire safety as well. Another equally apt metaphor describes the fire prevention process as the "Three E's:" engineering, education and enforcement. A balanced approach comprised of these three elements can be an especially effective way of achieving desired fire safety objectives.

Reporting

The first three elements of the code enforcement process are directed at identifying and eliminating hazards at their source. Reporting is intended to help document and reinforce the lessons learned from the previous three phases. The words, "If it's not written down, it didn't happen!" reinforce the message that reporting is just as important as any of the other three elements of the code enforcement system. Few people enjoy paperwork. Without documentation, however, prosecuting an effective code enforcement program becomes nearly impossible. Accurate, concise and timely records are essential for both legal and historical reasons. Documenting the inspection and violation history of a particular premises or owner is essential when prosecuting criminal actions under the code provisions.

Figure 1(6) illustrates a systems approach using data generated by fire incidents and inspections to direct code enforcement, public education activities or code development. This approach is equivalent to the one typically used to make ordinary decisions about problems with many competing solutions, plan for the future and consider the cost and benefits of these decisions.

Understanding the code administration process and the environment influencing it allows the fire code official to be more effective. Adhering to the provisions of Chapter 1 not only minimizes the fire code official's liability, but also provides an effective code enforcement program. Just as owners and occupants have obligations under the code, so does the fire code official. Following these procedures enables him or her to identify and respond to the community's needs; thus reducing the community's fire risk.

Purpose

Chapter 1 establishes provisions to make sure that code administration and enforcement is reasonable, appropriate and fair. This chapter outlines the duties and powers of the fire code official; the scope of the fire code official's authority to enforce the code; the applicability of the document and proofs of compliance; the means of securing compliance with its provisions and procedures for protecting due process rights of applicants, owners, occupants and others affected by the code provisions and the enforcement activities of the fire code official.

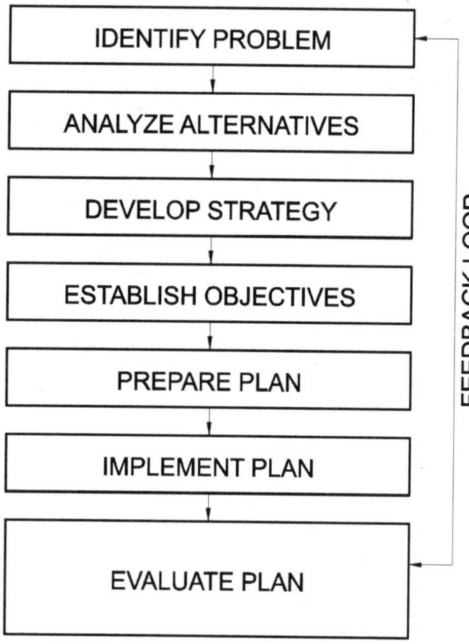

❖ This system is used mostly to solve everyday problems, weigh costs and benefits of alternatives and plan responses to problems. It is also a helpful tool for responding to fire protection problems. As an example, fire data suggest the emergence of a trend toward more fires caused by misuse of auxiliary heating appliances. Alternative solutions include:

- **A public education campaign on safe use of heaters,**
- **A targeted inspection program to identify and correct improper heating practices in homes and businesses or**
- **Doing nothing but hoping for warm weather.**

The evaluation must consider a variety of complex factors and methods of measuring the success of each approach. Once an approach is selected and implemented, it must be evaluated to examine whether the desired reduction in fires occurs. If not, perhaps one of the other alternatives should be reexamined, or the problem should be redefined.

Figure 1(6)
NORMAL PLANNING APPROACH

SECTION 101 GENERAL

101.1 Title. These regulations shall be known as the *Fire Code* of [NAME OF JURISDICTION], hereinafter referred to as "this code."

❖ This section identifies the jurisdictional applicability in legal terms. The local jurisdiction is to insert its name into this section by including a modification to the code in the adopting ordinance. This will make the code applicable to the local jurisdiction. See page v of the code for a sample ordinance for adoption.

101.2 Scope. This code establishes regulations affecting or relating to structures, processes, premises and safeguards regarding:

1. The hazard of fire and explosion arising from the storage, handling or use of structures, materials or devices;
2. Conditions hazardous to life, property or public welfare in the occupancy of structures or premises;
3. Fire hazards in the structure or on the premises from occupancy or operation;
4. Matters related to the construction, extension, repair, alteration or removal of fire suppression or alarm systems.

❖ The code does not attempt to achieve perfection by requiring every conceivable or available safeguard for every structure, premises or operation within the scope of the code. Rather, the code seeks to establish a minimum acceptable safety level to balance the many factors that must be considered, including loss statistics, relative hazard and the economic and social impact. The code is maintained through the use of a democratic code change process so that everyone affected by these minimum requirements has an equal opportunity to present their concern, both for and against any of the requirements.

101.2.1 Appendices. Provisions in the appendices shall not apply unless specifically adopted.

❖ The code has several appendices, which provide additional information regarding the provisions in the code and additional regulations that are available for adoption if desired by the adopting jurisdiction. If the jurisdiction decides to include any of the appendices as part of the code, each of the appendices to be adopted must be specifically listed in the adoption ordinance for the code. A sample adoption ordinance is on page v of the code.

101.3 Intent. The purpose of this code is to establish the minimum requirements consistent with nationally recognized good practice for providing a reasonable level of life safety and property protection from the hazards of fire, explosion or dangerous conditions in new and existing buildings, structures and premises and to provide safety to fire fighters and emergency responders during emergency operations.

❖ Code requirements regulate conditions that are likely to cause or contribute to fires or explosions; endanger life or property if a fire occurs or contribute to the spread of a fire. The intent of the code is to regulate conditions related to the health, safety and welfare of the public; the fire fighters and other emergency responders called upon to conduct emergency operations in or on any building, structure or premises. Note that the code requirements are a minimum (see commentary, Section 101.2 for a discussion on minimum requirements).

101.4 Severability. If a section, subsection, sentence, clause or phrase of this code is, for any reason, held to be unconstitutional, such decision shall not affect the validity of the remaining portions of this code.

❖ All sections of the code not invalidated by legal action remain in effect. While a dispute over a particular issue (such as hazardous materials quantity limitation) may have caused litigation that resulted in the provision being found unconstitutional, the remainder of the code is still applicable.

101.5 Validity. In the event any part or provision of this code is held to be illegal or void, this shall not have the effect of making void or illegal any of the other parts or provisions hereof, which are determined to be legal; and it shall be presumed that this code would have been adopted without such illegal or invalid parts or provisions.

❖ The code provisions are intended to be construed as severable. If any part of the code is ruled invalid by a court of competent jurisdiction, the remaining sections of the code are intended to stand as though the invalid section never existed. Fire code officials and adopting bodies should carefully and promptly evaluate the impact of any such ruling on ongoing enforcement activities and the remaining code provisions. Such changes that are necessary to preserve and protect the enforcement authority of the jurisdiction and the public should be instituted through legislative action as soon as practical. Additionally, the International Code Council® (ICC®) offices should be advised of court actions invalidating any code provisions. For the same reason local officials must evaluate the effects of court decisions, the influence of court decisions on the remainder of the code must be evaluated for national impact as well.

SECTION 102 APPLICABILITY

102.1 Construction and design provisions. The construction and design provisions of this code shall apply to:

1. Structures, facilities and conditions arising after the adoption of this code.
2. Existing structures, facilities and conditions not legally in existence at the time of adoption of this code.
3. Existing structures, facilities and conditions when identified in specific sections of this code.

4. Existing structures, facilities and conditions which, in the opinion of the code official, constitute a distinct hazard to life or property.

❖ This section establishes the scope of application of the code provisions that regulate construction and design.

Item 1 specifies that the code requirements apply to new construction that occurs following the adoption of the code.

Item 2 means that the code requirements are to apply to existing structures that did not have a certificate of occupancy at the time the code was adopted. An example would be a building that was built when there was no adopted construction code in the jurisdiction.

Item 3 refers to those sections in the code that specifically target existing structures, facilities and conditions for retroactive application of certain code requirements (for example, Sections 505.1, 701.4, 806.1, 903.6 and 1010.1, among others).

Item 4 generally requires the fire code official to determine that a "distinct hazard to life or property" exists prior to enforcing a code provision retroactively. Simply claiming that a violation exists because a building does not comply with the most recent edition of the code does not necessarily establish that a hazard actually exists. In cases where enforcement would result in substantial expense to the property owner, such a determination of hazard must be supported by adequate evidence that would be defensible in a court of law. This is true when a building has remained in compliance with the edition of the code under which it was originally constructed.

102.2 Administrative, operational and maintenance provisions. The administrative, operational and maintenance provisions of this code shall apply to:

1. Conditions and operations arising after the adoption of this code.
2. Existing conditions and operations.

❖ This section specifies that the administrative, operational and maintenance requirements of the code apply to conditions and operations that exist when the code is adopted and new conditions and operations that begin after the code is adopted. For example, a dry-cleaning operation that existed when the code was adopted or one that began after the code was adopted would be required to meet the requirements in Chapter 12 regarding dry cleaning.

[EB] 102.3 Change of use or occupancy. The provisions of the *International Existing Building Code* shall apply to all buildings undergoing a change of occupancy.

❖ A change in occupancy in an existing structure may change the level of inherent hazards that the code was initially intended to address.

Regardless of whether the change is to an occupancy considered to be more or less hazardous, this section applies the provisions of the *International Existing Building Code*® (IEBC™) to the structure with the new occupancy to match the specific requirements of the code to the specific hazards of the new occupancy.

102.4 Application of building code. The design and construction of new structures shall comply with the *International Building Code.* Repairs, alterations and additions to existing structures shall comply with the *International Existing Building Code.*

❖ The code is the companion fire and life safety maintenance code to the IBC. Maintenance of other building features is governed by other *International Codes*®. When existing buildings change, are altered or increased in area or when compliance with the code requires alterations, additions or modifications, the IEBC regulations and the appropriate fire code official's authority must prevail. This makes it essential that the fire code officials responsible for enforcing the building, existing building and fire codes establish a sound working relationship. Communication is essential to achieve compliance with the fire code official's orders.

[EB] 102.5 Historic buildings. The construction, alteration, repair, enlargement, restoration, relocation or movement of existing buildings or structures that are designated as historic buildings when such buildings or structures do not constitute a distinct hazard to life or property shall be in accordance with the provisions of the *International Existing Building Code.*

❖ This section provides a blanket exception from code requirements when the building in question has historic value. The most important criterion for application of this section is that the building must be recognized by a qualified party or agency as having historic significance. Usually this is done by a state or local authority after considerable scrutiny of the historical value of the building. Most, if not all, states have such authorities, as do many local jurisdictions. The agencies with such authority can be located at the state or local government level or through the local chapter of the American Institute of Architects (AIA).

As long as the building is not a distinct hazard to life or property, the provisions of the IEBC are to be used for the operations listed in this section of the code.

102.6 Referenced codes and standards. The codes and standards referenced in this code shall be those that are listed in Chapter 45 and such codes and standards shall be considered part of the requirements of this code to the prescribed extent of each such reference. Where differences occur between the provisions of this code and the referenced standards, the provisions of this code shall apply.

❖ The application of referenced standards is limited to those portions of the standards that are specifically identified. The code is intended to be in harmony with the referenced standards. If conflicts occur because of scope or purpose, the code text governs.

For example, Section 903.3.1.1.1 includes a list of exemptions from the NFPA 13 requirements; those which override the requirements in NFPA 13.

102.7 Subjects not regulated by this code. Where no applicable standards or requirements are set forth in this code, or are contained within other laws, codes, regulations, ordinances or bylaws adopted by the jurisdiction, compliance with applicable standards of the National Fire Protection Association or other nationally recognized fire safety standards, as approved, shall be deemed as prima facie evidence of compliance with the intent of this code. Nothing herein shall derogate from the authority of the fire code official to determine compliance with codes or standards for those activities or installations within the code official's jurisdiction or responsibility.

❖ This section provides guidance for situations in which no specific standard is designated in the code or otherwise adopted by the jurisdiction. In this instance compliance with the requirements of a standard of the NFPA or other nationally recognized standard can be approved by the fire code official.

102.8 Matters not provided for. Requirements that are essential for the public safety of an existing or proposed activity, building or structure, or for the safety of the occupants thereof, which are not specifically provided for by this code shall be determined by the fire code official.

❖ Evolving technology in our society will sometimes result in a situation or circumstance that the code does not cover. The reasonable application of the code to such hazardous, unforeseen conditions is provided in this section. Clearly, such a section is needed and the fire code official's experience and judgement must be used. The section, however, does not override requirements that may be preferred when the code provides alternative methods. Additionally, the section can be used to implement the general performance-oriented language of the code in specific enforcement situations.

102.9 Conflicting provisions. Where there is a conflict between a general requirement and a specific requirement, the specific requirement shall be applicable.

❖ This section deals with provisions on the same topic that could be different in technical content. In this instance, the specific provision, which is the one having the narrower scope of application, is to govern.

SECTION 103
DEPARTMENT OF FIRE PREVENTION

103.1 General. The department of fire prevention is established within the jurisdiction under the direction of the fire code official. The function of the department shall be the implementation, administration and enforcement of the provisions of this code.

❖ The traditional enforcement agency for the code is the fire department or fire prevention bureau of a state, county or municipal government. Such agencies usually perform administrative functions and provide public safety services related to fire protection; however, a variety of less traditional arrangements have also been used to enforce the code, including private corporations, such as fire districts and fire companies employed by a local government to act as its agent; police and other law enforcement agencies; building, housing or zoning authorities and community and economic development departments. Regardless of who is designated by the legislative or administrative authority to adopt and enforce the code, this section establishes the legal duty of the fire code official to enforce the code.

103.2 Appointment. The fire code official shall be appointed by the chief appointing authority of the jurisdiction; and the fire code official shall not be removed from office except for cause and after full opportunity to be heard on specific and relevant charges by and before the appointing authority.

❖ A fire code official's independence is essential so that public safety decisions are not based on political, economic or social expediencies. This is not to say that social, political and economic considerations should not weigh in deciding some code questions, but the interests of public health, safety and welfare must not be compromised to achieve such objectives. Protection of officials from removal from office without cause helps ensure that reasonable and competent professionals will be willing to serve.

103.3 Deputies. In accordance with the prescribed procedures of this jurisdiction and with the concurrence of the appointing authority, the fire code official shall have the authority to appoint a deputy fire code official, other related technical officers, inspectors and other employees.

❖ Most jurisdictions require more than one official to enforce the code. With the technical and legal demands on code enforcers increasing, additional personnel will certainly be required in this area to serve adequately the public interest. Though the professional qualifications of fire code officials are not detailed in the code, individuals appointed to code enforcement positions should be technically competent, motivated, well-adapted and possess good written and oral communication skills.

Many jurisdictions find it helpful, if not essential, to appoint an individual who is second-in-command and who would assume leadership of the organization in the absence of the chief code enforcement official.

103.4 Liability. The fire code official, officer or employee charged with the enforcement of this code, while acting for the jurisdiction, shall not thereby be rendered liable personally, and is hereby relieved from all personal liability for any damage accruing to persons or property as a result of an act required or permitted in the discharge of official duties.

❖ The fire code official must not be held liable for reasonable and lawful actions taken in accordance with the code. The responsibility of the fire code official in this regard is subject to local, state and federal laws that may supersede this provision. Furthermore, this section establishes that fire code officials or subordinates must not be liable for costs in any legal action in response to the performance of lawful duties. Section 103.4.1 states that these costs must be borne by the state or municipality. The best way to be certain that the fire code official's action is a lawful duty is to always cite the applicable code section supporting the action.

103.4.1 Legal defense. Any suit instituted against any officer or employee because of an act performed by that officer or employee in the lawful discharge of duties and under the provisions of this code shall be defended by the legal representative of the jurisdiction until the final termination of the proceedings. The fire code official or any subordinate shall not be liable for costs in an action, suit or proceeding that is instituted in pursuance of the provisions of this code; and any officer of the department of fire prevention, acting in good faith and without malice, shall be free from liability for acts performed under any of its provisions or by reason of any act or omission in the performance of official duties in connection therewith.

❖ Section 103.4 establishes that fire code officials or subordinates must not be liable for costs in any legal action in response to the performance of lawful duties. This section states that these costs must be borne by the state or municipality. The best way to be certain that the fire code official's action is a lawful duty is to always cite the applicable code section substantiating the action.

SECTION 104 GENERAL AUTHORITY AND RESPONSIBILITIES

104.1 General. The fire code official is hereby authorized to enforce the provisions of this code and shall have the authority to render interpretations of this code, and to adopt policies, procedures, rules and regulations in order to clarify the application of its provisions. Such interpretations, policies, procedures, rules and regulations shall be in compliance with the intent and purpose of this code and shall not have the effect of waiving requirements specifically provided for in this code.

❖ The duty of the fire code official is to enforce the code. Because the fire code official must also act on all questions related to this responsibility, except as specifically exempted by statutory requirements or elsewhere in the code, the fire code official is the "authority having jurisdiction" for all matters relating to the code and its enforcement.

This section also gives the fire code official interpretation authority. Note, however, that the interpretations are to be consistent with the intent and purpose of the code and are not allowed to set aside any specific requirement in the code.

104.2 Applications and permits. The fire code official is authorized to receive applications, review construction documents and issue permits for construction regulated by this code, issue permits for operations regulated by this code, inspect the premises for which such permits have been issued and enforce compliance with the provisions of this code.

❖ The fire code official is obligated to receive, review and act on permit applications required by the code as detailed in Section 105. All permitted premises must be inspected either before or after the permit is issued to determine compliance with the code provisions and terms of the permit.

104.3 Right of entry. Whenever it is necessary to make an inspection to enforce the provisions of this code, or whenever the fire code official has reasonable cause to believe that there exists in a building or upon any premises any conditions or violations of this code which make the building or premises unsafe, dangerous or hazardous, the fire code official shall have the authority to enter the building or premises at all reasonable times to inspect or to perform the duties imposed upon the fire code official by this code. If such building or premises is occupied, the fire code official shall present credentials to the occupant and request entry. If such building or premises is unoccupied, the fire code official shall first make a reasonable effort to locate the owner or other person having charge or control of the building or premises and request entry. If entry is refused, the fire code official has recourse to every remedy provided by law to secure entry.

❖ The first part of this section establishes the right of the fire code official to enter the premises to make the permit inspections required by Section 105.2.2. Permit application forms typically include a statement in the certification signed by the applicant (who is the owner or owner's agent) granting the fire code official the authority to enter areas covered by the permit to enforce code provisions related to the permit.

104.3.1 Warrant. When the fire code official has first obtained a proper inspection warrant or other remedy provided by law to secure entry, an owner or occupant or person having charge, care or control of the building or premises shall not fail or neglect, after proper request is made as herein provided, to permit entry therein by the fire code official for the purpose of inspection and examination pursuant to this code.

❖ Very simply, the requirements in this section specify that when the fire code official has obtained a warrant to inspect the property, the owner or occupant is to allow the fire code official entry to do the inspection (see commentary, Section 104.3).

104.4 Identification. The fire code official shall carry proper identification when inspecting structures or premises in the performance of duties under this code.

❖ This section requires the fire code official (including, by definition, all authorized designees) to carry appropriate official identification in the course of conducting the duties of the position. Such official identification may take the form of a badge, an identification card or both and removes any question as to the purpose and authority of the inspector.

104.5 Notices and orders. The fire code official is authorized to issue such notices or orders as are required to affect compliance with this code in accordance with Sections 109.1 and 109.2.

❖ The fire code official is required to issue orders to abate illegal or hazardous conditions and to pursue correction or abatement of hazardous conditions by issuing legal notices and orders as described by the code. Courts are increasingly ruling that failure to follow up and pursue appropriate legal remedies promptly exposes both the fire code official and the jurisdiction to a liability in tort.

104.6 Official records. The fire code official shall keep official records as required by Sections 104.6.1 through 104.6.4. Such official records shall be retained for not less than five years or for as long as the structure or activity to which such records relate remains in existence, unless otherwise provided by other regulations.

❖ In keeping with the need for an efficiently conducted business practice, the fire code official must keep official records. Such documentation provides a valuable resource of information if questions arise throughout the life of the building and its occupants. The code requires that the construction documents be kept until the project is complete or for at least five years, whichever is longer.

104.6.1 Approvals. A record of approvals shall be maintained by the fire code official and shall be available for public inspection during business hours in accordance with applicable laws.

❖ Records of prior approvals may be needed to determine the status of an existing operation or could be needed for future validation of a specific condition.

104.6.2 Inspections. The fire code official shall keep a record of each inspection made, including notices and orders issued, showing the findings and disposition of each.

❖ Records of inspections are needed to support the issuance of a certificate of occupancy. The inspection records should document any code violations that were subsequently corrected.

104.6.3 Fire records. The fire department shall keep a record of fires occurring within its jurisdiction and of facts concerning the same, including statistics as to the extent of such fires and the damage caused thereby, together with other information as required by the fire code official.

❖ Fire records provide a history of the fire experience of a facility and a cumulative record for all of the facilities of a jurisdiction. Fire records support consideration for construction code requirements based on the need to prevent additional fire occurrences.

104.6.4 Administrative. Application for modification, alternative methods or materials and the final decision of the fire code official shall be in writing and shall be officially recorded in the permanent records of the fire code official.

❖ The written approval of modifications or alternative materials and methods of construction or operation are needed to support the approval of these items in the future. This file could be used to verify that an existing condition had been previously approved.

104.7 Approved materials and equipment. All materials, equipment and devices approved by the fire code official shall be constructed and installed in accordance with such approval.

❖ The code is a compilation of criteria with which materials, equipment, devices and systems must comply to be acceptable for a particular application. The fire code official has a duty to evaluate such materials, equipment, devices and systems for code compliance and, when compliance is determined, approve them for use. As a result of this approval, the material, equipment, device or system must be constructed and installed in compliance with that approval, and with all the conditions and limitations considered as a basis for that approval. For example, the manufacturer's instructions and recommendations are to be followed if the approval of the material was based even in part on those instructions and recommendations.

The approval authority given the fire code official is a significant responsibility and is a key to code compliance. The approval process is first technical and then administrative and must be approached that way. For example, if data to determine code compliance are required, such data should be in the form of test reports or engineering analysis—not simply taken from a sales brochure.

104.7.1 Material and equipment reuse. Materials, equipment and devices shall not be reused or reinstalled unless such elements have been reconditioned, tested and placed in good and proper working condition and approved.

❖ Used materials, equipment and devices are considered to have completed their life span; however, adequate substitutes are occasionally not available for existing items that have become obsolete but still serve a useful and practical purpose. In such cases, existing used equipment should be approved, provided the application is consistent with the purpose for which the equipment was designed, the function is the same as the "new" item, if one were available, and the intended use can be demonstrated as not compromising the public's safety.

104.7.2 Technical assistance. To determine the acceptability of technologies, processes, products, facilities, materials and uses attending the design, operation or use of a building or premises subject to inspection by the fire code official, the fire code official is authorized to require the owner or agent to provide, without charge to the jurisdiction, a technical opinion and report.

The opinion and report shall be prepared by a qualified engineer, specialist, laboratory or fire safety specialty organization acceptable to the fire code official and shall analyze the fire safety properties of the design, operation or use of the building or premises and the facilities and appurtenances situated thereon, to recommend necessary changes. The fire code official is authorized to require design submittals to be prepared by, and bear the stamp of, a registered design professional.

❖ No one person has the technical knowledge to evaluate all of the various operations and uses from a safety standpoint. This section provides the code official the authority to require the owner to provide a technical opinion safety report. The report is to be prepared by parties that have the technical ability to evaluate the design of the facility or the operational process in question. A registered design professional is commonly used for these services. It is critical that the preparer of the report have the proper background and experience for the project since the credibility of the report depends on these qualifications

104.8 Modifications. Whenever there are practical difficulties involved in carrying out the provisions of this code, the fire code official shall have the authority to grant modifications for individual cases, provided the fire code official shall first find that special individual reason makes the strict letter of this code impractical and the modification is in compliance with the intent and purpose of this code and that such modification does not lessen health, life and fire safety requirements. The details of action granting modifications shall be recorded and entered in the files of the department of fire prevention.

❖ The fire code official may amend or make exceptions to the code as needed to respond to "practical difficulties" in work on new or existing buildings. Consideration of a particular difficulty is to be based on the application of the owner and a demonstration that the intent of the code is satisfied. This section is not intended to allow a code provision to be set aside or ignored; rather, it is intended to provide for the acceptance of equivalent protection. Such modifications do not, however, extend to actions that are necessary to correct violations of the code. In other words, a code violation or the expense of correcting a code violation cannot constitute a practical difficulty.

Comprehensive written records are an essential part of an effective administrative system. Unless clearly written records of the considerations and documentation used in the modification process are created and maintained, subsequent enforcement action cannot be supported.

104.9 Alternative materials and methods. The provisions of this code are not intended to prevent the installation of any material or to prohibit any method of construction not specifically prescribed by this code, provided that any such alternative has been approved. The fire code official is authorized to approve an alternative material or method of construction where the fire code official finds that the proposed design is satisfactory and complies with the intent of the provisions of this code, and that the material, method or work offered is, for the purpose intended, at least the equivalent of that prescribed in this code in quality, strength, effectiveness, fire resistance, durability and safety.

❖ Performance requirements have replaced detailed specifications to permit ready substitution and integration of new technologies in the marketplace. The code is not intended to restrict or prevent the development or application of new technologies or applications of existing technologies, provided they meet the intent of the code to protect public health, safety and welfare. When new methods or materials are developed, they should be evaluated.

The fire code official has the authority to recognize alternative and equivalent methods and materials, provided they maintain the level of protection required by the code. One of the most frequent criticisms of codes is that their provisions apply too broadly to classes of occupancies and, therefore, are incapable of recognizing the inherent dissimilarities within occupancy groups. While some criticism may be justified, it is the fire code official's duty to evaluate scrupulously the conditions in each case, as well as judge whether the intent of the code (to provide the minimum acceptable level of protection to life and property) is met. Fire code officials should, therefore, be prepared to use decision aids, the appeal process and outside experts as needed to show that code requirements are met.

104.10 Fire investigations. The fire code official, the fire department or other responsible authority shall have the authority to investigate the cause, origin and circumstances of any fire, explosion or other hazardous condition. Information that could be related to trade secrets or processes shall not be made part of the public record except as directed by a court of law.

❖ The prompt and thorough investigation of fires is important for many reasons, not the least of which is the identification of incendiary fires and prosecution of arsonists. In such cases, the duty of the fire official is clear—evidence must be preserved and leads pursued through criminal prosecution, if possible. However, a more important and frequently overlooked aspect of fire investigation is loss analysis. Whether or not the fire code official has jurisdiction to investigate incendiary fires and prosecute arsonists, it is extremely important that the enforcement agency be involved in the process of determining why fires occur, what can be done to prevent fires, how their effects can be lessened and how persons behave once fires occur. The valuable lessons learned from past tragedies have forged a relationship among the various code organizations across the country.

104.10.1 Assistance from other agencies. Police and other enforcement agencies shall have authority to render necessary assistance in the investigation of fires when requested to do so.

❖ When needed, the fire code official has the authority to ask for assistance from the police department or other enforcement agencies, such as fire code officials in nearby jurisdictions, to investigate fires.

104.11 Authority at fires and other emergencies. The fire chief or officer of the fire department in charge at the scene of a fire or other emergency involving the protection of life or property or any part thereof, shall have the authority to direct such operation as necessary to extinguish or control any fire, perform any rescue operation, investigate the existence of suspected or reported fires, gas leaks or other hazardous conditions or situations, or take any other action necessary in the reasonable performance of duty. In the exercise of such power, the fire chief is authorized to prohibit any person, vehicle, vessel or thing from approaching the scene and is authorized to remove, or cause to be removed or kept away from the scene, any vehicle, vessel or thing which could impede or interfere with the operations of the fire department and, in the judgment of the fire chief, any person not actually and usefully employed in the extinguishing of such fire or in the preservation of property in the vicinity thereof.

❖ This section describes the specific conditions of authority that are granted to the fire code official at a fire or other emergencies. The first half of the paragraph simply describes the fire code official's authority to carry out the fire operation at the site. The fire code official also needs to be able to control who and what is allowed to be at the site so that emergency operations are not hampered.

104.11.1 Barricades. The fire chief or officer of the fire department in charge at the scene of an emergency is authorized to place ropes, guards, barricades or other obstructions across any street, alley, place or private property in the vicinity of such operation so as to prevent accidents or interference with the lawful efforts of the fire department to manage and control the situation and to handle fire apparatus.

❖ This section gives the fire code official the authority to control access to the emergency site so that fire-fighting operations can occur without interference. This authority is also addressed in Section 104.11.

104.11.2 Obstructing operations. No person shall obstruct the operations of the fire department in connection with extinguishment or control of any fire, or actions relative to other emergencies, or disobey any lawful command of the fire chief or officer of the fire department in charge of the emergency, or any part thereof, or any lawful order of a police officer assisting the fire department.

❖ This section requires that the fire department operations not be obstructed and that directions from the fire department official in command at the emergency site be carried out. This is necessary for efficient emergency operations.

104.11.3 Systems and devices. No person shall render a system or device inoperative during an emergency unless by direction of the fire chief or fire department official in charge of the incident.

❖ This section is an extension of the requirements in Section 104.11. The fire department official is in complete charge of the fire-fighting operation at the site. No person is to tamper with the equipment needed for the emergency.

SECTION 105
PERMITS

105.1 General. Permits shall be in accordance with Section 105.

❖ This section includes the regulations covering permits including a comprehensive list of the kinds of activities that require permits.

105.1.1 Permits required. Permits required by this code shall be obtained from the fire code official. Permit fees, if any, shall be paid prior to issuance of the permit. Issued permits shall be kept on the premises designated therein at all times and shall be readily available for inspection by the fire code official.

❖ Establishing permit fees is an important political and economic consideration in the code administration environment. A reasonable fee structure should be both fair and collectible. Permit fees may be grouped with other governmental and nontax revenues under various headings, including "impact fees." To the extent that establishments required to obtain a permit represent a higher-than-normal community hazard, this designation may be appropriate. Nontax charges have become increasingly popular methods of financing governmental programs because of recent tax reforms. Separate from the question of fees, this section also requires the permit holder to keep a copy of the permit on the premises at all times and available for inspection. Unless stipulated in the adopting legislation, a permit need not be conspicuously displayed on the premises to meet this requirement but need only be produced on demand.

105.1.2 Types of permits. There shall be two types of permits as follows:

1. Operational permit. An operational permit allows the applicant to conduct an operation or a business for which a permit is required by Section 105.6 for either:
   - 1.1. A prescribed period.
   - 1.2. Until renewed or revoked.
2. Construction permit. A construction permit allows the applicant to install or modify systems and equipment for which a permit is required by Section 105.7.

❖ The types of activities that require an operational permit are listed in Section 105.6. Construction activities that require a permit are listed in Section 105.7.

105.1.3 Permits for the same location. When more than one permit is required for the same location, the fire code official is authorized to consolidate such permits into a single permit provided that each provision is listed in the permit.

❖ The code allows permits for a number of activities to be included on a single permit in order to decrease the paperwork for all concerned. In this instance, the permit must list in detail the activities that are covered by the combined permit.

105.2 Application. Application for a permit required by this code shall be made to the fire code official in such form and detail as prescribed by the fire code official. Applications for permits shall be accompanied by such plans as prescribed by the fire code official.

❖ Applications provided by the jurisdiction should be complete, concise and relevant. Though the burden of proof is on the applicant to supply all necessary information to determine compliance with the code provisions, it is the fire code official's duty to request sufficient information to make a reasonable and informed judgement prior to approving a permit.

105.2.1 Refusal to issue permit. If the application for a permit describes a use that does not conform to the requirements of this code and other pertinent laws and ordinances, the fire code official shall not issue a permit, but shall return the application to the applicant with the refusal to issue such permit. Such refusal shall, when requested, be in writing and shall contain the reasons for refusal.

❖ This section directs the fire code official not to issue a permit if the application describes a use that does not conform to the requirements of the code. Note that this direction is not advisory. The fire code official would be in violation of the code if a permit were issued in such circumstances.

105.2.2 Inspection authorized. Before a new operational permit is approved, the fire code official is authorized to inspect the receptacles, vehicles, buildings, devices, premises, storage spaces or areas to be used to determine compliance with this code or any operational constraints required.

❖ The inspections described in this section are necessary for the fire code official to determine that the application for an operational permit complies with the code prior to issuing that permit. Operations may not proceed without an operational permit.

105.2.3 Time limitation of application. An application for a permit for any proposed work or operation shall be deemed to have been abandoned six months after the date of filing, unless such application has been diligently prosecuted or a permit shall have been issued; except that the fire code official is authorized to grant one or more extensions of time for additional periods not exceeding 90 days each if there is reasonable cause.

❖ Permit applications lingering indefinitely in an incomplete condition can be an administrative nuisance to the fire code official, while also overburdening the filing system. This section establishes six months as the time limit for the permit applicant to provide the fire code official with sufficient information to evaluate the application and take appropriate action. Six months should normally be more than enough time for an applicant to satisfy code requirements for submittal of construction documents and all other required information.

There may be circumstances, however, that could cause an application to age beyond six months prior to permit issuance, such as awaiting issuance of a report by a quality assurance agency. If the fire code official is satisfied that every effort is being made by the applicant to pursue the application, an extension of time would be acceptable.

105.2.4 Action on application. The fire code official shall examine or cause to be examined applications for permits and amendments thereto within a reasonable time after filing. If the application or the construction documents do not conform to the requirements of pertinent laws, the fire code official shall reject such application in writing, stating the reasons therefor. If the fire code official is satisfied that the proposed work or operation conforms to the requirements of this code and laws and ordinances applicable thereto, the fire code official shall issue a permit therefore as soon as practicable.

❖ While the fire code official has the duty to take all necessary and prudent actions to determine the applicant's compliance with the code, the evaluation must be completed promptly. Once the fire code official's review of the application is complete, either a permit will be issued or a written disapproval notice will be given. The disapproval notice must outline the reasons for rejection and should include a list of applicable code sections with which the applicant must comply to obtain approval.

105.3 Conditions of a permit. A permit shall constitute permission to maintain, store or handle materials; or to conduct processes which produce conditions hazardous to life or property; or to install equipment utilized in connection with such activities; or to install or modify any fire protection system or equipment or any other construction, equipment installation or modification in accordance with the provisions of this code where a permit is required by Section 105.6 or 105.7. Such permission shall not be construed as authority to violate, cancel or set aside any of the provisions of this code or other applicable regulations or laws of the jurisdiction.

❖ In effect, a permit is a contract or covenant between the jurisdiction and the applicant, allowing the applicant to operate, perform, conduct or direct a hazardous operation, process or occupancy. As with all contracts, the terms remain binding for a finite period. This process allows continual review of the applicant's compliance with the contract's terms. Failure to meet the terms of the contract may result in the applicant's forfeiture of the right to conduct or operate the process, operation or occupancy, and subsequently the fire code official may revoke the permit without further notice.

105.3.1 Expiration. An operational permit shall remain in effect until reissued, renewed, or revoked or for such a period of time as specified in the permit. Construction permits shall automatically become invalid unless the work authorized by such permit is commenced within 180 days after its issuance, or if the work authorized by such permit is suspended or abandoned for a period of 180 days after the time the work is commenced. Before such work recommences, a new permit shall be first obtained and the fee to recommence work, if any, shall be one-half the amount required for a new permit for such work, provided no changes have been made or will be made in the original construction documents for such work, and provided further that such suspension or abandonment has not exceeded one year. Permits are not transferable and any change in occupancy, operation, tenancy or ownership shall require that a new permit be issued.

❖ A construction permit is invalid when 180 days go by without any of the authorized work being done. The permit holder should be notified in writing that the permit is invalid, including the reasons why.

Permits are neither transferable nor assignable because they are agreements between two specific parties: the fire code official, who is acting for the jurisdiction, and the applicant. Any changes amending the application or terms of the original agreement will require a new application and permit approval.

105.3.2 Extensions. A permittee holding an unexpired permit shall have the right to apply for an extension of the time within which the permittee will commence work under that permit when work is unable to be commenced within the time required by this section for good and satisfactory reasons. The fire code official is authorized to grant, in writing, one or more extensions of the time period of a permit for periods of not more than 90 days each. Such extensions shall be requested by the permit holder in writing and justifiable cause demonstrated.

❖ The significant issue in this section is that an extension of time is to be granted when justifiable cause is demonstrated by the permit applicant. For example, a construction permit might be granted for certain equipment installation, but the equipment might not be received at the site until after the installation permit expired. To get a time extension, the applicant is to submit a request in writing to the fire code official, including a written explanation of why the work did not proceed within the permit time frame.

105.3.3 Occupancy prohibited before approval. The building or structure shall not be occupied prior to the fire code official issuing a permit that indicates that applicable provisions of this code have been met.

❖ The owner of an existing structure may request that the fire code official issue a certificate of occupancy for a structure, provided that there are no pending violations. A final inspection is usually done to verify that the work covered by the permit has been completed in accordance with the code.

105.3.4 Conditional permits. Where permits are required and upon the request of a permit applicant, the fire code official is authorized to issue a conditional permit to occupy the premises or portion thereof before the entire work or operations on the premises is completed, provided that such portion or portions will be occupied safely prior to full completion or installation of equipment and operations without endangering life or public welfare. The fire code official shall notify the permit applicant in writing of any limitations or restrictions necessary to keep the permit area safe. The holder of a conditional permit shall proceed only to the point for which approval has been given, at the permit holder's own risk and without assurance that approval for the occupancy or the utilization of the entire premises, equipment or operations will be granted.

❖ The fire code official is allowed to issue a conditional permit prior to the completion of all work. Such a permit is to be issued only when the building or structure is available for safe occupancy prior to full completion. The permit is intended to acknowledge that some building features may not be completed even though the building is safe for occupancy.

105.3.5 Posting the permit. Issued permits shall be kept on the premises designated therein at all times and shall be readily available for inspection by the fire code official.

❖ Note that this section does not require that the permit be posted, but it is to be kept on the site at all times for inspection by the fire code official.

105.3.6 Compliance with code. The issuance or granting of a permit shall not be construed to be a permit for, or an approval of, any violation of any of the provisions of this code or of any other ordinance of the jurisdiction. Permits presuming to give authority to violate or cancel the provisions of this code or other ordinances of the jurisdiction shall not be valid. The issuance of a permit based on construction documents and other data shall not prevent the fire code official from requiring the correction of errors in the construction documents and other data. Any addition to or alteration of approved construction documents shall be approved in advance by the fire code official, as evidenced by the issuance of a new or amended permit.

❖ This section includes an important principle regarding construction documents. The fire code official has the authority to require that errors in construction be corrected, even if the construction is based on documents that were part of the applicant's submittal for a construction permit. Thus, the code requirements are not set aside by approved drawings that may include noncomplying items of construction. Any changes amending the application or construction of the original agreement will require a new application and permit approval.

105.3.7 Information on the permit. The fire code official shall issue all permits required by this code on an approved form furnished for that purpose. The permit shall contain a general description of the operation or occupancy and its location and any other information required by the fire code official. Issued permits shall bear the signature of the fire code official or other approved legal authorization.

❖ This section describes the form of the permit and requires that it be either signed by the fire code official or otherwise reflect the legal authorization of the jurisdiction. In many jurisdictions, permits are electronically generated and do not require a traditional signature.

105.4 Construction documents. Construction documents shall be in accordance with this section.

❖ This section states the scope of the sections covering construction documents.

105.4.1 Submittals. Construction documents shall be submitted in one or more sets and in such form and detail as required by the fire code official. The construction documents shall be prepared by a registered design professional where required by the statutes of the jurisdiction in which the project is to be constructed.

❖ A detailed description of the work for which an application is made must be submitted in the form and detail required by the fire code official.

Construction documents are to be prepared by a registered design professional when required by state laws that are in effect in the jurisdiction. States have professional registration laws that specify the type of construction work that is to be designed by a registered design professional. The code relies on these state laws to determine when a design professional is required.

The requirement for the preparation of construction documents and the submittal of calculations is specified by the code in several chapters. For example, Section 901.2 specifies that construction documents and calculations are to be submitted for fire protection systems when required by the fire code official.

105.4.2 Information on construction documents. Construction documents shall be drawn to scale upon suitable material. Electronic media documents are allowed to be submitted when approved by the fire code official. Construction documents shall be of sufficient clarity to indicate the location, nature and extent of the work proposed and show in detail that it will conform to the provisions of this code and relevant laws, ordinances, rules and regulations as determined by the fire code official.

❖ Construction documents are not sketches. They are comprehensive drawings, drawn to scale, that provide the details to verify that the work will comply with the code. The permit applicant must be familiar with the code requirements to prepare construction documents that meet code requirements. If the applicant is not familiar with the code, the construction documents will most likely not have sufficient detail to determine compliance and, thus, not be satisfactory as the basis for a permit.

105.4.3 Applicant responsibility. It shall be the responsibility of the applicant to ensure that the construction documents include all of the fire protection requirements and the shop drawings are complete and in compliance with the applicable codes and standards.

❖ This requirement is similar to the one in Section 901.2 regarding construction documents for fire protection systems.

The requirement in this section regarding shop drawings applies to all types of shop drawings, not just those for fire protection systems. The permit applicant is responsible for the review of the shop drawings, not the fire code official. The permit applicant is also responsible for seeing that the work on the job site complies with the code. Since a lot of the construction work is done in accordance with shop drawings, the applicant should review those drawings for code compliance to make sure field construction complies with the code.

105.4.4 Approved documents. Construction documents approved by the fire code official are approved with the intent that such construction documents comply in all respects with this code. Review and approval by the fire code official shall not relieve the applicant of the responsibility of compliance with this code.

❖ The applicant is responsible for making sure that construction complies with the code. If approved drawings include errors that do not comply with the code, the fire code official still has the authority to require that the errors be corrected. Thus, it is important that the permit applicant be familiar with the code requirements to prevent preparation of construction documents that do not meet the code.

105.4.5 Corrected documents. Where field conditions necessitate any substantial change from the approved construction documents, the fire code official shall have the authority to require the corrected construction documents to be submitted for approval.

❖ It is important that the construction documents include a record of revisions to the construction so that they truly represent the as-built condition. These records are also useful to the permit applicant for future alterations or additions to the facility.

105.4.6 Retention of construction documents. One set of construction documents shall be retained by the fire code official until final approval of the work covered therein. One set of approved construction documents shall be returned to the applicant, and said set shall be kept on the site of the building or work at all times during which the work authorized thereby is in progress.

❖ It is important that a complete, current set of construction documents be kept on the job site at all times. Another set of construction documents is to be kept by the fire code official until final approval of the completed work. The construction documents are part of the official records of the department and should be kept in accordance with Section 104.6.

105.5 Revocation. The fire code official is authorized to revoke a permit issued under the provisions of this code when it is found by inspection or otherwise that there has been a false statement or misrepresentation as to the material facts in the application or construction documents on which the permit or approval was based including, but not limited to, any one of the following:

1. The permit is used for a location or establishment other than that for which it was issued.
2. The permit is used for a condition or activity other than that listed in the permit.
3. Conditions and limitations set forth in the permit have been violated.
4. There have been any false statements or misrepresentations as to the material fact in the application for permit or plans submitted or a condition of the permit.
5. The permit is used by a different person or firm than the name for which it was issued.
6. The permittee failed, refused or neglected to comply with orders or notices duly served in accordance with the provisions of this code within the time provided therein.
7. The permit was issued in error or in violation of an ordinance, regulation or this code.

❖ The fire code official must revoke all permits shown to be based, all or in part, on any false statement or misinterpretation of fact. An applicant may subsequently reapply for a permit.

The code specifies seven specific conditions that allow the fire code official to revoke a permit.

105.6 Required operational permits. The fire code official is authorized to issue operational permits for the operations set forth in Sections 105.6.1 through 105.6.47.

❖ Sections 105.6.1 through 105.6.47 list the conditions requiring operational permits. Many of the items are stated in general terms. The fire code official is to determine whether a specific operation in question is a significant hazard that requires a permit.

The referenced sections of the code indicated in the commentary to Sections 105.6.1 through 105.6.47 are not intended to be all of the requirements that would apply to the operation, but only the unique requirements for that particular operation.

105.6.1 Aerosol products. An operational permit is required to manufacture, store or handle an aggregate quantity of Level 2 or Level 3 aerosol products in excess of 500 pounds (227 kg) net weight.

❖ See Chapter 28 for code requirements covering aerosol products (see commentary, Section 105.6).

105.6.2 Amusement buildings. An operational permit is required to operate a special amusement building.

❖ For requirements that apply to special amusement buildings see Sections 202, 907.2.11 and 903.2.15 and Section 411 of the IBC. (see commentary, Section 105.6).

105.6.3 Aviation facilities. An operational permit is required to use a Group H or Group S occupancy for aircraft servicing or repair and aircraft fuel-servicing vehicles. Additional permits required by other sections of this code include, but are not limited to, hot work, hazardous materials and flammable or combustible finishes.

❖ See Chapter 11 for aviation facility requirements (see commentary, Section 105.6).

105.6.4 Carnivals and fairs. An operational permit is required to conduct a carnival or fair.

❖ See Section 2401.3 for carnival requirements (see commentary, Section 105.6).

105.6.5 Battery systems. A permit is required to install stationary lead-acid battery systems having a liquid capacity of more than 50 gallons (189 L).

❖ See Sections 608 and 602.1 for battery system requirements (see commentary, Section 105.6).

105.6.6 Cellulose nitrate film. An operational permit is required to store, handle or use cellulose nitrate film in a Group A occupancy.

❖ Although cellulose nitrate film is no longer in general use, there are a small number of locations in which this type of film is archived or restored for historical purposes. This section applies to those few locations (see Section 306 for cellulose nitrate film requirements).

105.6.7 Combustible dust-producing operations. An operational permit is required to operate a grain elevator, flour starch mill, feed mill, or a plant pulverizing aluminum, coal, cocoa, magnesium, spices or sugar, or other operations producing combustible dusts as defined in Chapter 2.

❖ See Chapter 13 for combustible dust-producing operations (see commentary, Section 105.6).

105.6.8 Combustible fibers. An operational permit is required for the storage and handling of combustible fibers in quantities greater than 100 cubic feet (2.8 m^3).

Exception: A permit is not required for agricultural storage.

❖ See Chapter 29 for combustible fiber requirements. The exception is for agricultural storage facilities where the hazard to persons is minimal (see Section 105.6).

105.6.9 Compressed gases. An operational permit is required for the storage, use or handling at normal temperature and pressure (NTP) of compressed gases in excess of the amounts listed in Table 105.6.9.

Exception: Vehicles equipped for and using compressed gas as a fuel for propelling the vehicle.

❖ See Chapter 30 for compressed gas requirements. The exception exempts vehicles equipped for compressed gas, since the code requirements for compressed gases do not apply to them.

TABLE 105.6.9
PERMIT AMOUNTS FOR COMPRESSED GASES

TYPE OF GAS	AMOUNT (cubic feet at NTP)
Corrosive	200
Flammable (except cryogenic fluids and liquefied petroleum gases)	200
Highly toxic	Any Amount
Inert and simple asphyxiant	6,000
Oxidizing (including oxygen)	504
Toxic	Any Amount

For SI: 1 cubic foot = 0.02832 m^3.

❖ When the use of indicated compressed gases exceed the amounts indicated in Table 105.6.9, an operational permit is required. The quantities in the table are at normal temperature and pressure (NTP) (see Chapter 30 for compressed gas requirements).

105.6.10 Covered mall buildings. An operational permit is required for:

1. The placement of retail fixtures and displays, concession equipment, displays of highly combustible goods and similar items in the mall.
2. The display of liquid- or gas-fired equipment in the mall.
3. The use of open-flame or flame-producing equipment in the mall.

❖ The listed operations in a covered mall building require an operational permit, since they involve a significant hazard to the occupants. See Section 308 for open-flame regulations (see Section 105.6).

105.6.11 Cryogenic fluids. An operational permit is required to produce, store, transport on site, use, handle or dispense cryogenic fluids in excess of the amounts listed in Table 105.6.11.

Exception: Permits are not required for vehicles equipped for and using cryogenic fluids as a fuel for propelling the vehicle or for refrigerating the lading.

❖ See Chapter 32 for requirements regarding cryogenic fluids. The exception exempts vehicles using cryogenic fluids, since the code requirements do not apply to them.

TABLE 105.6.11
PERMIT AMOUNTS FOR CRYOGENIC FLUIDS

TYPE OF CRYOGENIC FLUID	INSIDE BUILDING (gallons)	OUTSIDE BUILDING (gallons)
Flammable	More than 1	60
Inert	60	500
Oxidizing (includes oxygen)	10	50
Physical or health hazard not indicated above	Any Amount	Any Amount

For SI: 1 gallon = 3.785 L.

❖ Where cryogenic fluids are used in excess of the amounts shown in the table, an operational permit is required. The listed amounts are significantly different inside or outside of a building, since the hazard is greatly reduced if a leak occurs outdoors.

105.6.12 Cutting and welding. An operational permit is required to conduct cutting or welding operations within the jurisdiction.

❖ See Chapter 26 for welding requirements (see commentary, Section 105.6).

105.6.13 Dry cleaning plants. An operational permit is required to engage in the business of dry cleaning or to change to a more hazardous cleaning solvent used in existing dry cleaning equipment.

❖ See Chapter 12 for dry-cleaning regulations (see commentary, Section 105.6).

105.6.14 Exhibits and trade shows. An operational permit is required to operate exhibits and trade shows.

❖ The primary concern is to identify hazardous materials and highly flammable materials that could be involved in an exhibit or booth (see commentary, Section 105.6).

105.6.15 Explosives. An operational permit is required for the manufacture, storage, handling, sale or use of any quantity of explosive, explosive material, fireworks, or pyrotechnic special effects within the scope of Chapter 33.

❖ See Chapter 33 for requirements for explosives and fireworks (see commentary, Section 105.6).

105.6.16 Fire hydrants and valves. An operational permit is required to use or operate fire hydrants or valves intended for fire suppression purposes which are installed on water systems and accessible to a fire apparatus access road that is open to or generally used by the public.

Exception: A permit is not required for authorized employees of the water company that supplies the system or the fire department to use or operate fire hydrants or valves.

❖ An operational permit is required for persons other than authorized employees of the water company or the fire department to operate fire hydrants or valves. This restriction is intended to make sure that the use will not result in a lack of water supply and pressure that may be needed for fire-fighting purposes. The exception allows water company employees or the fire department to use fire hydrants or valves without a permit. Such use is common in order to flush out the piping periodically. When fire departments or fire districts interact with water districts, they should communicate the need for the fire department to use the hydrants and valves for nonemergency situations, such as training. A notification procedure is needed to let the water district know of this planned use.

105.6.17 Flammable and combustible liquids. An operational permit is required:

1. To use or operate a pipeline for the transportation within facilities of flammable or combustible liquids. This requirement shall not apply to the off-site transportation in pipelines regulated by the Department of Transportation (DOTn) nor does it apply to piping systems.
2. To store, handle or use Class I liquids in excess of 5 gallons (19 L) in a building or in excess of 10 gallons (37.9 L) outside of a building, except that a permit is not required for the following:
   - 2.1. The storage or use of Class I liquids in the fuel tank of a motor vehicle, aircraft, motorboat, mobile power plant or mobile heating plant, unless such storage, in the opinion of the code official, would cause an unsafe condition.
   - 2.2. The storage or use of paints, oils, varnishes or similar flammable mixtures when such liquids are stored for maintenance, painting or similar purposes for a period of not more than 30 days.
3. To store, handle or use Class II or Class IIIA liquids in excess of 25 gallons (95 L) in a building or in excess of 60 gallons (227 L) outside a building, except for fuel oil used in connection with oil-burning equipment.
4. To remove Class I or Class II liquids from an underground storage tank used for fueling motor vehicles by any means other than the approved, stationary on-site pumps normally used for dispensing purposes.
5. To operate tank vehicles, equipment, tanks, plants, terminals, wells, fuel-dispensing stations, refineries, distilleries and similar facilities where flammable and combustible liquids are produced, processed, transported, stored, dispensed or used.
6. To place temporarily out of service (for more than 90 days) an underground, protected above-ground or above-ground flammable or combustible liquid tank.
7. To change the type of contents stored in a flammable or combustible liquid tank to a material which poses a greater hazard than that for which the tank was designed and constructed.
8. To manufacture, process, blend or refine flammable or combustible liquids.
9. To engage in the dispensing of liquid fuels into the fuel tanks of motor vehicles at commercial, industrial, governmental or manufacturing establishments.
10. To utilize a site for the dispensing of liquid fuels from tank vehicles into the fuel tanks of motor vehicles at commercial, industrial, governmental or manufacturing establishments.

❖ See Chapter 34 for regulations regarding flammable and combustible liquids (see commentary, Section 105.6).

105.6.18 Floor finishing. An operational permit is required for floor finishing or surfacing operations exceeding 350 square feet (33 m^2) using Class I or Class II liquids.

❖ The concern of this section is the proper use and handling of Class I or Class II liquids that are used in the floor finishing process. If such liquids are not used, an operational permit is not required for floor finishing.

105.6.19 Fruit and crop ripening. An operational permit is required to operate a fruit-, or crop-ripening facility or conduct a fruit-ripening process using ethylene gas.

❖ See Chapter 16 for regulations for fruit and crop ripening processes where ethylene gas is used (see commentary, Section 105.6).

105.6.20 Fumigation and thermal insecticidal fogging. An operational permit is required to operate a business of fumigation or thermal insecticidal fogging and to maintain a room, vault or chamber in which a toxic or flammable fumigant is used.

❖ See Chapter 17 for fumigation and thermal insecticidal fogging regulations within structures (see commentary, Section 105.6).

105.6.21 Hazardous materials. An operational permit is required to store, transport on site, dispense, use or handle hazardous materials in excess of the amounts listed in Table 105.6.21.

❖ See Chapter 27 for the general provisions regarding hazardous materials. Also see Chapters 28 through 44 for regulations regarding a specific hazardous material (see commentary, Section 105.6).

TABLE 105.6.21. See page 1-20.

❖ Where the amounts of hazardous materials in the table are exceeded, an operational permit is required. This applies to the storage, transportation on site, dispensing, use or handling of the hazardous materials that are listed in the table.

105.6.22 HPM facilities. An operational permit is required to store, handle or use hazardous production materials.

❖ See Chapter 18 for the regulations regarding semiconductor fabrication facilities (see commentary, Section 105.6).

105.6.23 High-piled storage. An operational permit is required to use a building or portion thereof as a high-piled storage area exceeding 500 square feet (46 m^2).

❖ See Chapter 23 for high-piled storage provisions (see commentary, Section 105.6).

TABLE 105.6.21
PERMIT AMOUNTS FOR HAZARDOUS MATERIALS

TYPE OF MATERIAL	AMOUNT
Combustible liquids	See Section 105.6.17
Corrosive materials	
Gases	See Section 105.6.9
Liquids	55 gallons
Solids	1000 pounds
Explosive materials	See Section 105.6.15
Flammable materials	
Gases	See Section 105.6.9
Liquids	See Section 105.6.17
Solids	100 pounds
Highly toxic materials	
Gases	See Section 105.6.9
Liquids	Any Amount
Solids	Any Amount
Oxidizing materials	
Gases	See Section 105.6.9
Liquids	
Class 4	Any Amount
Class 3	1 gallon
Class 2	10 gallons
Class 1	55 gallons
Solids	
Class 4	Any Amount
Class 3	10 pounds
Class 2	100 pounds
Class 1	500 pounds
Organic peroxides	
Liquids	
Class I	Any Amount
Class II	Any Amount
Class III	1 gallon
Class IV	2 gallons
Class V	No Permit Required
Solids	
Class I	Any Amount
Class II	Any Amount
Class III	10 pounds
Class IV	20 pounds
Class V	No Permit Required
Pyrophoric materials	
Gases	See Section 105.6.9
Liquids	Any Amount
Solids	Any Amount
Toxic materials	
Gases	See Section 105.6.9
Liquids	10 gallons
Solids	100 pounds

(continued)

TABLE 105.6.21–continued
PERMIT AMOUNTS FOR HAZARDOUS MATERIALS

TYPE OF MATERIAL	AMOUNT
Unstable (reactive) materials	
Liquids	
Class 4	Any Amount
Class 3	Any Amount
Class 2	5 gallons
Class 1	10 gallons
Solids	
Class 4	Any Amount
Class 3	Any Amount
Class 2	50 pounds
Class 1	100 pounds
Water-reactive Materials	
Liquids	
Class 3	Any Amount
Class 2	5 gallons
Class 1	55 gallons
Solids	
Class 3	Any Amount
Class 2	50 pounds
Class 1	500 pounds

For SI: 1 gallon = 3.785 L, 1 pound = 0.454 kg.

105.6.24 Hot work operations. An operational permit is required for hot work including, but not limited to:

1. Public exhibitions and demonstrations where hot work is conducted.
2. Use of portable hot work equipment inside a structure.

   Exception: Work that is conducted under a construction permit.

3. Fixed-site hot work equipment such as welding booths.
4. Hot work conducted within a hazardous fire area.
5. Application of roof coverings with the use of an open-flame device.
6. When approved, the fire code official shall issue a permit to carry out a Hot Work Program. This program allows approved personnel to regulate their facility's hot work operations. The approved personnel shall be trained in the fire safety aspects denoted in this chapter and shall be responsible for issuing permits requiring compliance with the requirements found in Chapter 26. These permits shall be issued only to their employees or hot work operations under their supervision.

❖ See Chapter 26 for hot work regulations. The exception to Item 2 in this section recognizes that work done under a construction permit is already covered by that permit so an operations permit is not required (see commentary Section 105.6).

105.6.25 Industrial ovens. An operational permit is required for operation of industrial ovens regulated by Chapter 21.

❖ See Chapter 21 for regulations regarding industrial ovens (see commentary, Section 105.6).

105.6.26 Lumber yards and woodworking plants. An operational permit is required for the storage or processing of lumber exceeding 100,000 board feet (8,333 ft^3) (236 m^3).

❖ See Chapter 19 for provisions for lumber yards and woodworking plants (see commentary, Section 105.6).

105.6.27 Liquid- or gas-fueled vehicles or equipment in assembly buildings. An operational permit is required to display, operate or demonstrate liquid- or gas-fueled vehicles or equipment in assembly buildings.

❖ See Section 314.4 for requirements regarding liquid- or gas-fueled vehicles inside buildings (see commentary, Section 105.6).

105.6.28 LP-gas. An operational permit is required for:

1. Storage and use of LP-gas.

   Exception: A permit is not required for individual containers with a 500-gallon (1893 L) water capacity or less serving occupancies in Group R-3.

2. Operation of cargo tankers that transport LP-gas.

❖ See Chapter 38 for liquefied petroleum gas regulations. The exception to Item 1 in this section exempts small tanks commonly found for residential service. Item 2 covers cargo tankers, since they transport LP gas onto premises covered by the code and, therefore, represent a potential hazard.

105.6.29 Magnesium. An operational permit is required to melt, cast, heat treat or grind more than 10 pounds (4.54 kg) of magnesium.

❖ See Section 3606 for the code requirements for magnesium (see commentary, Section 105.6).

105.6.30 Miscellaneous combustible storage. An operational permit is required to store in any building or upon any premises in excess of 2,500 cubic feet (71 m^3) gross volume of combustible empty packing cases, boxes, barrels or similar containers, rubber tires, rubber, cork or similar combustible material.

❖ See Section 315 for requirements for miscellaneous combustible material storage (see commentary, Section 105.6).

105.6.31 Open burning. An operational permit is required for the kindling or maintaining of an open fire or a fire on any public street, alley, road, or other public or private ground. Instructions and stipulations of the permit shall be adhered to.

Exception: Recreational fires.

❖ See Section 307 for open burning provisions. Section 302.1 includes the definition of "Open burning." The exception exempts recreational fires, which are defined in Section 302.1.

105.6.32 Open flames and torches. An operational permit is required to remove paint with a torch; or to use a torch or open-flame device in a hazardous fire area.

❖ See Section 308 for regulations regarding open flames (see commentary, Section 105.6).

105.6.33 Open flames and candles. An operational permit is required to use open flames or candles in connection with assembly areas, dining areas of restaurants or drinking establishments.

❖ See Section 308 for regulations regarding open flames (see commentary, Section 105.6).

105.6.34 Organic coatings. An operational permit is required for any organic-coating manufacturing operation producing more than 1 gallon (4 L) of an organic coating in one day.

❖ The manufacture of organic coatings is addressed in Chapter 20 (see commentary, Section 105.6).

105.6.35 Places of assembly. An operational permit is required to operate a place of assembly.

❖ Because of the higher occupant loads found in Group A occupancies, they justify an increased level of scrutiny, such as is provided through the permit process.

105.6.36 Private fire hydrants. An operational permit is required for the removal from service, use or operation of private fire hydrants.

Exception: A permit is not required for private industry with trained maintenance personnel, private fire brigade or fire departments to maintain, test and use private hydrants.

❖ The purpose of an operational permit for the removal of private fire hydrants is to see that adequate fire hydrants are maintained for use during a fire. The exception allows testing and use of private fire hydrants by trained private industry personnel without an operational permit.

105.6.37 Pyrotechnic special effects material. An operational permit is required for use and handling of pyrotechnic special effects material.

❖ See Chapter 33 for fireworks regulations. The definition of "Pyrotechnic special-effect material" is listed in Section 3302 (see commentary, Section 105.6).

105.6.38 Pyroxylin plastics. An operational permit is required for storage or handling of more than 25 pounds (11 kg) of cellulose nitrate (pyroxylin) plastics and for the assembly or manufacture of articles involving pyroxylin plastics.

❖ See Chapter 42 for requirements regarding pyroxylin (cellulose nitrate) plastics (see commentary, Section 105.6).

105.6.39 Refrigeration equipment. An operational permit is required to operate a mechanical refrigeration unit or system regulated by Chapter 6.

❖ See Section 606 for mechanical refrigeration regulations (see commentary, Section 105.6).

105.6.40 Repair garages and motor fuel-dispensing facilities. An operational permit is required for operation of repair garages and automotive, marine and fleet motor fuel-dispensing facilities.

❖ See Chapter 22 for requirements for motor fuel-dispensing facilities and repair garages (see commentary, Section 105.6).

105.6.41 Rooftop heliports. An operational permit is required for the operation of a rooftop heliport.

❖ See Chapter 11 for aviation facility requirements. Section 1107 contains helistop and heliport requirements (see commentary, Section 105.6).

105.6.42 Spraying or dipping. An operational permit is required to conduct a spraying or dipping operation utilizing flammable or combustible liquids or the application of combustible powders regulated by Chapter 15.

❖ See Chapter 15 for flammable finish requirements. Section 1504 contains the spray finishing provisions, Section 1505 addresses dipping operations and Section 1507 includes powder coating regulations (see commentary, Section 105.6).

105.6.43 Storage of scrap tires and tire byproducts. An operational permit is required to establish, conduct or maintain storage of scrap tires and tire byproducts that exceeds 2,500 cubic feet (71 m^3) of total volume of scrap tires and for indoor storage of tires and tire byproducts.

❖ See Chapter 25 for regulations regarding tire rebuilding and tire storage (see Section 105.6).

105.6.44 Temporary membrane structures, tents and canopies. An operational permit is required to operate an air-supported temporary membrane structure or a tent having an area in excess of 200 square feet (19 m^2), or a canopy in excess of 400 square feet (37 m^2).

Exceptions:

1. Tents used exclusively for recreational camping purposes.
2. Fabric canopies open on all sides which comply with all of the following:
   - 2.1. Individual canopies having a maximum size of 700 square feet (65 m^2).
   - 2.2. The aggregate area of multiple canopies placed side by side without a fire break clearance of not less than 12 feet (3658 mm) shall not exceed 700 square feet (65 m^2) total.
   - 2.3. A minimum clearance of 12 feet (3658 mm) to structures and other tents shall be provided.

❖ See Chapter 24 for requirements for tents and other membrane structures. The first exception in this section exempts recreational camping tents, since they are small, temporary and have few occupants. The second exception exempts relatively small tents that are very low hazard, since they are spaced at least 12 feet (3658 mm) apart.

105.6.45 Tire-rebuilding plants. An operational permit is required for the operation and maintenance of a tire-rebuilding plant.

❖ See Chapter 25 for regulations regarding tire rebuilding operations (see commentary, Section 105.6).

105.6.46 Waste handling. An operational permit is required for the operation of wrecking yards, junk yards and waste material-handling facilities.

❖ See Section 315 for miscellaneous combustible materials storage requirements, Section 2704 for provisions regarding the storage of hazardous materials and Section 1908 for provisions regarding yard waste and recycling facilities (see commentary, Section 105.6).

105.6.47 Wood products. An operational permit is required to store chips, hogged material, lumber or plywood in excess of 200 cubic feet (6 m^3).

❖ See Section 1908 for requirements regarding the storage and handling of wood chips, hogged material, fines, compost and raw product in association with yard waste and recycling facilities (see commentary, Section 105.6).

105.7 Required construction permits. The fire code official is authorized to issue construction permits for work as set forth in Sections 105.7.1 through 105.7.12.

❖ This section addresses conditions requiring a construction permit (see Section 105.6). Generally, a construction permit is required when a safety-related system or hazardous material storage is installed or an existing system or facility is modified. Other sections of the code may also apply.

In some cases, the requirements in Sections 105.7.1 through 105.7.12 are stated in only general terms. In these instances, the fire code official is to evaluate the scope of work involved for the modification or installation and determine whether a construction permit is required for the specific project.

105.7.1 Automatic fire-extinguishing systems. A construction permit is required for installation of or modification to an automatic fire-extinguishing system. Maintenance performed in accordance with this code is not considered a modification and does not require a permit.

❖ See Chapter 9 for fire protection system requirements. A construction permit is required for the installation or modification of an automatic fire-extinguishing system so that the work can be verified to meet the code requirements, since the system is obviously safety related (see commentary, Section 105.7).

105.7.2 Compressed gases. When the compressed gases in use or storage exceed the amounts listed in Table 105.6.9, a construction permit is required to install, repair damage to, abandon, remove, place temporarily out of service, or close or substantially modify a compressed gas system.

Exceptions:

1. Routine maintenance.
2. For emergency repair work performed on an emergency basis, application for permit shall be made within two working days of commencement of work.

The permit applicant shall apply for approval to close storage, use or handling facilities at least 30 days prior to the termination of the storage, use or handling of compressed or liquefied gases. Such application shall include any change or alteration of the facility closure plan filed pursuant to Section 2701.6.3. The 30-day period is not applicable when approved based on special circumstances requiring such waiver.

❖ See Chapter 30 for the requirements for compressed gas systems. Where the volume of the compressed gas presents a significant health hazard and the quantity exceeds the allowed amounts in Table 105.6.9, a permit is needed to trigger construction document submittal, document review and inspections of the work on the system. The exceptions address the need for an exemption for maintenance work and to allow emergency work to proceed immediately.

105.7.3 Fire alarm and detection systems and related equipment. A construction permit is required for installation of or modification to fire alarm and detection systems and related equipment. Maintenance performed in accordance with this code is not considered a modification and does not require a permit.

❖ See Section 907 for fire alarm and detection requirements. A construction permit is required for installation or modification of these systems since they are obviously safety related. A permit is not required for maintenance when no modifications are made to the systems (see commentary, Section 105.7).

105.7.4 Fire pumps and related equipment. A construction permit is required for installation of or modification to fire pumps and related fuel tanks, jockey pumps, controllers, and generators. Maintenance performed in accordance with this code is not considered a modification and does not require a permit.

❖ See Section 913 for requirements regarding fire pumps. A construction permit is required for modification or installation of equipment that is necessary to serve the sprinkler or standpipe system. This construction work must be monitored since these are safety related systems (see commentary, Section 105.7).

105.7.5 Flammable and combustible liquids. A construction permit is required:

1. To repair or modify a pipeline for the transportation of flammable or combustible liquids.
2. To install, construct or alter tank vehicles, equipment, tanks, plants, terminals, wells, fuel-dispensing stations, refineries, distilleries and similar facilities where flammable and combustible liquids are produced, processed, transported, stored, dispensed or used.
3. To install, alter, remove, abandon or otherwise dispose of a flammable or combustible liquid tank.

❖ See Chapter 34 for provisions for flammable and combustible liquids. The intent of this section is to require a construction permit for any of the three activities listed, since flammable and combustible liquids are a significant hazard (see commentary, Section 105.7).

105.7.6 Hazardous materials. A construction permit is required to install, repair damage to, abandon, remove, place temporarily out of service, or close or substantially modify a storage facility or other area regulated by Chapter 27 when the hazardous materials in use or storage exceed the amounts listed in Table 105.6.21.

Exceptions:

1. Routine maintenance.
2. For emergency repair work performed on an emergency basis, application for permit shall be made within two working days of commencement of work.

❖ A construction permit is needed for hazardous-material-related construction to ensure submittal of construction documents, document review and inspection of the work for code compliance. The exceptions provide exemptions for maintenance work and allow emergency work to proceed immediately, provided the permit application is submitted within two working days of starting the job.

105.7.7 Industrial ovens. A construction permit is required for installation of industrial ovens covered by Chapter 21.

Exceptions:

1. Routine maintenance.
2. For repair work performed on an emergency basis, application for permit shall be made within two working days of commencement of work.

❖ A construction permit is required for industrial oven installation so that the requirements in Chapter 21 of the code for industrial ovens can be verified. The exceptions provide exemptions for maintenance work and allow emergency work to proceed immediately, provided the permit is applied for within two working days after work begins.

105.7.8 LP-gas. A construction permit is required for installation of or modification to an LP-gas system.

❖ See Chapter 38 for the requirements for liquefied petroleum gas storage, handling and transportation (see commentary, Section 105.7).

105.7.9 Private fire hydrants. A construction permit is required for the installation or modification of private fire hydrants.

❖ A construction permit is needed for the installation or modification of private fire hydrants so that they remain in service for fire protection purposes. The water flow rate and pressure capability need to be maintained.

105.7.10 Spraying or dipping. A construction permit is required to install or modify a spray room, dip tank or booth.

❖ See Chapter 15 for flammable finish requirements. The spray finishing requirements are in Section 1504, while dipping operations regulations are in Section 1505 (see commentary, Section 105.7).

105.7.11 Standpipe systems. A construction permit is required for the installation, modification, or removal from service of a standpipe system. Maintenance performed in accordance with this code is not considered a modification and does not require a permit.

❖ See Section 905 for standpipe system requirements. Construction permits are required for standpipe systems because they are safety-related fire protection systems. Ordinary maintenance that does not involve modifications to the system does not require a construction permit.

105.7.12 Temporary membrane structures, tents and canopies. A construction permit is required to erect an air-supported temporary membrane structure or a tent having an area in excess of 200 square feet (19 m^2), or a canopy in excess of 400 square feet (37 m^2).

Exceptions:

1. Tents used exclusively for recreational camping purposes.

2. Funeral tents and curtains or extensions attached thereto, when used for funeral services.
3. Fabric canopies and awnings open on all sides which comply with all of the following:
   3.1. Individual canopies shall have a maximum size of 700 square feet (65 m^2).
   3.2. The aggregate area of multiple canopies placed side by side without a fire break clearance of not less than 12 feet (3658 mm) shall not exceed 700 square feet (65 m^2) total.
   3.3. A minimum clearance of 12 feet (3658 mm) to structures and other tents shall be maintained.

❖ See Chapter 24 for requirements regarding tents and other membrane structures. The exceptions are for tents where the hazard is very low. They provide needed exemptions for tents used for recreational camping and funerals. Relatively small fabric canopies and awnings that are open on all sides and are located a minimum of 12 feet (3658 mm) apart are also exempt.

SECTION 106
INSPECTIONS

106.1 Inspection authority. The fire code official is authorized to enter and examine any building, structure, marine vessel, vehicle or premises in accordance with Section 104.3 for the purpose of enforcing this code.

❖ The first part of this section establishes the right of the fire code official to enter the premises to make the permit inspections required by Section 104. Permit application forms typically include a statement in the certification signed by the applicant (who is the owner or owner's agent) granting the fire code official the authority to enter areas covered by the permit to enforce code provisions related to the permit.

The right to enter other structures or premises is more limited. First, to protect the right of privacy, the owner or occupant must grant the fire code official permission before the interior of the property can be inspected. Permission is not required for inspections that can be accomplished from within the public right-of-way. Second, such access may be denied by the owner or occupant. Unless the inspector has "reasonable cause" to believe that a violation of the code exists, access may be unattainable. Third, fire code officials must present proper identification (see Section 104.4) and request admittance during reasonable hours—usually the normal business hours of the establishment—to be admitted. Fourth, inspections must be aimed at securing or determining compliance with the provisions and intent of the regulations that are specifically within the established scope of the fire code official's authority. Searches to gather information for the purpose of enforcing other codes, ordinances or regulations are considered unreasonable and are prohibited by the Fourth Amendment to the U.S. Constitution.

Reasonable cause in the context of this section must be distinguished from probable cause, which is required to gain access to property in criminal cases. The burden of proof for establishing reasonable cause may vary among jurisdictions. Usually, an inspector must show that the property is subject to inspection under the provisions of the code (see Section 104); that the interests of the public health, safety and welfare outweigh the individual's right to maintain privacy and that such an inspection is required solely to determine compliance with the provisions of the code. Many jurisdictions do not recognize the concept of an administrative warrant, and may require the fire code official to prove probable cause in order to gain access upon refusal. This burden of proof is usually more substantial, often requiring the fire code official to stipulate in advance why access is needed (usually access is restricted to gathering evidence for seeking an indictment or making an arrest), what specific items or information is sought; its relevance to the case against the individual subject; how knowledge of the relevance of the information or items sought was obtained and how the evidence sought will be used. In all such cases, the right to privacy must always be weighed against the right of the fire code official to conduct an inspection to determine whether the health, safety or welfare of the public is in jeopardy. Such important and complex constitutional issues should be discussed with the jurisdiction's legal counsel. Jurisdictions should establish procedures for securing the necessary court orders when an inspection is considered necessary following a refusal.

106.2 Inspections. The fire code official is authorized to conduct such inspections as are deemed necessary to determine the extent of compliance with the provisions of this code and to approve reports of inspection by approved agencies or individuals. All reports of such inspections shall be prepared and submitted in writing for review and approval. Inspection reports shall be certified by a responsible officer of such approved agency or by the responsible individual. The fire code official is authorized to engage such expert opinion as deemed necessary to report upon unusual, detailed or complex technical issues subject to the approval of the governing body.

❖ This section establishes the fire code official's duty to inspect every building, structure or premises of the jurisdiction to verify that the requirements of the code are met. This section does not, however, establish the frequency of inspections. The code does not presume to interpret each jurisdiction's political, social and economic priorities. Every jurisdiction is likely to establish different inspection priorities and frequencies based on the availability of inspection resources, the level of available fire suppression services, the value of premises to the community or the potential disruption to community services or stability involved if a fire occurs. In brief, each community determines and assumes its own acceptable risk level. This is not to say, however, that every segment of each community will agree on such determination.

106.3 Concealed work. Whenever any installation subject to inspection prior to use is covered or concealed without having first been inspected, the fire code official shall have the authority to require that such work be exposed for inspection.

❖ This section addresses the procedure that is available to the fire code official for inspection of concealed work. In many jurisdictions, the contractor of a construction project is to contact the local inspection department when work is completed but still exposed to allow inspection. If the work that requires inspection is covered up before the inspection takes place, the fire code official has the authority to require removal of the construction that conceals the item to be inspected. Obviously this can be a timely and expensive procedure that can be eliminated by good communication and cooperation between the contractor and the inspection department of the jurisdiction.

SECTION 107 MAINTENANCE

107.1 Maintenance of safeguards. Whenever or wherever any device, equipment, system, condition, arrangement, level of protection, or any other feature is required for compliance with the provisions of this code, or otherwise installed, such device, equipment, system, condition, arrangement, level of protection, or other feature shall thereafter be continuously maintained in accordance with this code and applicable referenced standards.

❖ This section does not identify who is responsible for maintenance because that determination should be made in accordance with the legal documents created between owners and occupants, such as a lease. The owner of a structure or premises, however, is usually the party primarily responsible for its maintenance, since the owner stands to gain the most from a well-maintained property. One of the underlying assumptions is that maintaining a commercial property in good condition allows the owner to recoup a substantial portion of his or her investment in maintenance. There are three factors that may influence owners to comply with code requirements:

- Code compliance requires only a small additional investment in the property;
- The owner has a long-term interest in the property; and
- The owner expects profitability after incurring the additional expense of complying with the code.

While all these factors represent economic incentives, fire code officials should be equally aware of potential disincentives to compliance, such as assessable value, expiring tax credits or historic, architectural or aesthetic criteria. The fire code official need not belabor the justifications for compliance, but should be prepared to acknowledge the owner's rationalizations for failure to comply.

This section also emphasizes that any "otherwise installed" system that currently exists must be maintained. For example, an existing fire protection system cannot be removed from a building just because it is not required in new or existing buildings by current codes.

107.2 Testing and operation. Equipment requiring periodic testing or operation to ensure maintenance shall be tested or operated as specified in this code.

❖ This section addresses periodic testing or operation to verify that the equipment can be expected to operate when needed. For example, see Section 901.6 for inspection and testing requirements for fire protection systems.

107.2.1 Test and inspection records. Required test and inspection records shall be available to the fire code official at all times or such records as the fire code official designates shall be filed with the fire code official.

❖ Test and inspection records must be available to the fire code official for verification that the tests and inspections required by the code and the referenced standards are in compliance.

If the fire code official requests, such records must be filed with the jurisdictional office.

107.2.2 Reinspection and testing. Where any work or installation does not pass an initial test or inspection, the necessary corrections shall be made so as to achieve compliance with this code. The work or installation shall then be resubmitted to the fire code official for inspection and testing.

❖ This section simply requires that an installation be of such quality that it will pass any tests or inspections required by the code. For example, if a fire alarm system did not pass the installation test upon completion of the system, the system is to be reworked until it passes the test.

107.3 Supervision. Maintenance and testing shall be under the supervision of a responsible person who shall ensure that such maintenance and testing are conducted at specified intervals in accordance with this code.

❖ Maintenance supervision is needed to verify that the testing and general supervision is done regularly. Section 901.6 states code requirements regarding testing and maintenance of the fire protection systems.

107.4 Rendering equipment inoperable. Portable or fixed fire-extinguishing systems or devices and fire-warning systems shall not be rendered inoperative or inaccessible except as necessary during emergencies, maintenance, repairs, alterations, drills or prescribed testing.

❖ If fire protection systems are going to be effective when needed, they must be in good operating condition. This section specifies those circumstances when they are allowed to be temporarily out of service. See Section 901.6 for code requirements regarding testing and maintenance of the fire protection systems.

A note from...
Mr. Paul Bechtel

See 107.2

ion and abate-
nsibility of the
ited, hazardous
it shall be held
conditions.

ınd not those
fall within the
ity. Owners,
the unlawful
uisance on a
ey knowingly
of fire, zoning

constitutes an
sponsibility is
equipment in-
e from those
is usually re-
f the building
ances (that is,

nce of any per-
g or a portion
l, upon finding
aisles, passage-
any condition
authorized to
or obstruction

ssfully is good
accurately the
blic assembly.
such events
ze or they are
these difficul-
arge to permit
main clear or
be sought. In
edied by sim-
entering in or-

der to limit the potential hazard to those occupants already inside. If the fire code official determines that preventing further access will be insufficient in itself, he or she is authorized to order the owner or operator to stop the event until the hazardous condition is abated, the approved occupant load is reestablished and resumption of the event is authorized by the fire code official.

SECTION 108
BOARD OF APPEALS

108.1 Board of appeals established. In order to hear and decide appeals of orders, decisions or determinations made by the fire code official relative to the application and interpretation of this code, there shall be and is hereby created a board of appeals. The board of appeals shall be appointed by the governing body and shall hold office at its pleasure. The fire code official shall be an ex officio member of said board but shall have no vote on any matter before the board. The board shall adopt rules of procedure for conducting its business, and shall render all decisions and findings in writing to the appellant with a duplicate copy to the fire code official.

❖ This section provides an objective forum for settling disputes regarding the application or interpretation of the code requirements. The board is required to issue a written decision to the appellant who brought the matter before the board and to the fire code official. Note that the fire code official is a nonvoting member of the board. The board of appeals is an effective decision-making body that is commonly used when the owner or owner's agent and the fire code official do not agree on a matter relating to the application of the code.

108.2 Limitations on authority. An application for appeal shall be based on a claim that the intent of this code or the rules legally adopted hereunder have been incorrectly interpreted, the provisions of this code do not fully apply, or an equivalent method of protection or safety is proposed. The board shall have no authority to waive requirements of this code.

❖ This section states the scope of the issues that are to be addressed by the board of appeals and limits their authority to ruling on these issues. Commonly, the issues relate to the applicability of the code or the interpretation of the code to a given situation. The board listens to both the person who filed the appeal and to the fire code official before ruling on the matter.

This section specifically states that the board does not have the authority to waive code requirements; however, the board has the authority to accept an alternative method of protection or safety if, in its view, it is equivalent to the specific requirement in the code.

108.3 Qualifications. The board of appeals shall consist of members who are qualified by experience and training to pass on matters pertaining to hazards of fire, explosions, hazardous conditions or fire protection systems and are not employees of the jurisdiction.

❖ It is important that the decisions of the appeals board are based purely on the technical merits involved in an appeal; it is not the place for policy or political deliberations. The members of the appeals board are, therefore, expected to have experience in matters within the scope of the code and must be of the highest character, competence and status in their professions and the community at large. Appendix A of the code provides more detailed qualifications for appeals board members and can be adopted by jurisdictions desiring that level of expertise (see commentary, Appendix A).

SECTION 109
VIOLATIONS

109.1 Unlawful acts. It shall be unlawful for a person, firm or corporation to erect, construct, alter, repair, remove, demolish or utilize a building, occupancy, premises or system regulated by this code, or cause same to be done, in conflict with or in violation of any of the provisions of this code.

❖ Section 109 establishes that compliance with the code is required, and what measures are to be taken for noncompliance.

109.2 Notice of violation. When the fire code official finds a building, premises, vehicle, storage facility or outdoor area that is in violation of this code, the fire code official is authorized to prepare a written notice of violation describing the conditions deemed unsafe and, when compliance is not immediate, specifying a time for reinspection.

❖ The fire code official has a duty to supply owners, agents or occupants with a written notice of code violations on the premises under their control. When possible, both the owner and the occupant should be made aware of hazardous conditions. Such notices constitute the first of several steps in the due process procedure. Violation notices must clearly indicate the defect, what must be done to correct the violation and when the work must be completed. Owners, agents or occupants should also be supplied with information regarding penalties, permit applications and appeal procedures. The notice or order must be signed by the fire code official who issued it and he or she should provide a space for the owner, agent or occupant's signature to acknowledge receipt of the document. If possible, duplicate or triplicate copies should be prepared, with the original notice issued to the responsible party. Other copies should be maintained by the inspector and the departmental record keeper.

109.2.1 Service. A notice of violation issued pursuant to this code shall be served upon the owner, operator, occupant, or other person responsible for the condition or violation, either by personal service, mail, or by delivering the same to, and leaving it with, some person of responsibility upon the premises. For unattended or abandoned locations, a copy of such notice of violation shall be posted on the premises in a conspicuous place at or near the entrance to such premises and the notice of violation shall be mailed by certified mail with return receipt requested or a certificate of mailing, to the last known address of the owner, occupant or both.

❖ Service methods are listed by order or preference. Personal service of the owner at the premises cited, followed by the agent and occupant, with a signature acknowledging receipt, is the first and best method of legal service. The next most desirable method is service to these same parties in the order indicated at their place of business when it is not the premises cited.

While post office delivery by ordinary first-class mail is acceptable, most jurisdictions prefer certified mail with return receipt, followed by a certificate of mailing; however, owners familiar with the legal process will often refuse to accept certified mail. As a result, many jurisdictions follow up returned certified mail with a request for a certificate of mailing. A certificate of mailing includes a certification by the mail carrier or post office that the item was physically delivered to the address indicated, but does not verify that the addressee actually took possession of the item. The least desirable method of service is physically posting the premises with the violation notice. When service proves difficult, many jurisdictions pursue the mailing and posting service options simultaneously to exhaust all service methods. Jurisdictions should consult legal counsel about case law regarding legal service in their communities (see Figure 109.2.1).

- Personal to violator.
- Personal to party at premises.
- Certified mail with return receipt.
- First-class mail with certificate of mailing.
- Posting at the premises.

Figure 109.2.1
SERVICE METHODS

109.2.2 Compliance with orders and notices. A notice of violation issued or served as provided by this code shall be complied with by the owner, operator, occupant or other person responsible for the condition or violation to which the notice of violation pertains.

❖ The party responsible for the condition that is in noncompliance is required by this section to bring the property into code compliance. See the remainder of Section 109 for what is to be done if this does not occur.

109.2.3 Prosecution of violations. If the notice of violation is not complied with promptly, the fire code official is authorized to request the legal counsel of the jurisdiction to institute the appropriate legal proceedings at law or in equity to restrain, correct or abate such violation or to require removal or termination of the unlawful occupancy of the structure in violation of the provisions of this code or of the order or direction made pursuant hereto.

❖ The duty to pursue legal remedies through judicial due process is established by this section. Local prosecutors and fire code officials should establish policies covering the following issues regarding judicial due process proceedings:

- Length of compliance period for representative violations;
- Quality or quantity of progress toward compliance warranting an extension or representing reasonable intent to comply;
- Whether court filings should be sought during the appeal application period;
- Rules for obtaining arrest warrants for code violations; and
- Rules for obtaining administrative and criminal search warrants.

The cooperation of the police department and other law enforcement agencies should be coordinated in advance. When necessary to enforce the code provisions, arrangements should be made to have police or other law enforcement personnel make arrests for code violations or ignoring the lawful orders of the fire code official.

109.2.4 Unauthorized tampering. Signs, tags or seals posted or affixed by the fire code official shall not be mutilated, destroyed or tampered with or removed without authorization from the fire code official.

❖ This section states that tampering with signs, seals or tags posted at the property is a violation of the code. The safety of the occupants may depend on the warning signs posted by the fire code official remaining in place.

109.3 Violation penalties. Persons who shall violate a provision of this code or shall fail to comply with any of the requirements thereof or who shall erect, install, alter, repair or do work in violation of the approved construction documents or directive of the fire code official, or of a permit or certificate used under provisions of this code, shall be guilty of a [SPECIFY OFFENSE], punishable by a fine of not more than [AMOUNT] dollars or by imprisonment not exceeding [NUMBER OF DAYS], or both such fine and imprisonment. Each day that a violation continues after due notice has been served shall be deemed a separate offense.

❖ Penalties for violating code provisions must be established in adopting legislation. The specification of the offense, the dollar amount for the fine and the maximum number of days of imprisonment are to be specific in the adopting ordinance of the jurisdiction. See page v in the front of the code for details.

The code does not establish penalties for violations. The jurisdiction's judicial and legislative bodies should work with the fire code official to establish reasonable and equitable penalties for violators. The penalties set for individual violations should be representative of the severity of the act committed and the culpability of the violator. Once served with a violation notice, the violator becomes guilty of a separate offense for each day the violation continues to exist; however, most prosecutors and courts are reluctant to impose this penalty for days during the compliance period.

Many violators wrongly assume that the Seventh Amendment of the U.S. Constitution, which offers protection against double jeopardy, exempts them from compliance once they have paid or served their sentence for a previous fire code violation. This is certainly not the case. Most courts reinforce the compliance requirement in such cases by making compliance a condition for completing the sentence. Failure to comply with the judge's order mandating compliance may result in a contempt of court charge.

109.3.1 Abatement of violation. In addition to the imposition of the penalties herein described, the fire code official is authorized to institute appropriate action to prevent unlawful construction or to restrain, correct or abate a violation; or to prevent illegal occupancy of a structure or premises; or to stop an illegal act, conduct of business or occupancy of a structure on or about any premises.

❖ Even though the person who violated the code has paid a fine and whatever other sentence that may be imposed for the jurisdiction under Section 109.3, the fire code official has the right to require that the code violation be removed. If the violation is not abated, the fire code official has the right to prevent occupancy until the violation is addressed. Usually the court will require that the violation be corrected as part of the sentence of noncompliance prior to the occupancy of the building.

SECTION 110
UNSAFE BUILDINGS

110.1 General. If during the inspection of a premises, a building or structure or any building system, in whole or in part, constitutes a clear and inimical threat to human life, safety or health, the fire code official shall issue such notice or orders to remove or remedy the conditions as shall be deemed necessary in accordance with this section and shall refer the building to the building department for any repairs, alterations, remodeling, removing or demolition required.

❖ The fire code official is required to order the correction or abatement of specific hazardous conditions. The conditions listed in Section 110.1.1 represent many of the most common hazardous conditions encountered. Specific requirements supporting each of these objectives are found throughout the code.

110.1.1 Unsafe conditions. Structures or existing equipment that are or hereafter become unsafe or deficient because of inadequate means of egress or which constitute a fire hazard, or are otherwise dangerous to human life or the public welfare, or which involve illegal or improper occupancy or inadequate maintenance, shall be deemed an unsafe condition. A vacant structure which is not secured against unauthorized entry as required by Section 311 shall be deemed unsafe.

❖ The fire code official is required to report unsafe buildings to the building official to secure abatement of unsafe conditions. Courts have continually upheld the right of states and their authorized subdivisions to abate public nuisances, even by demolition, and bill or assess the property owner through a tax lien for their expenses. However, care must be exercised to maintain compliance with the due process and equal protection doctrines of the Fourth and Fourteenth Amendments of the U.S. Constitution. Jurisdictions should consult legal counsel and adopt appropriate guidelines before engaging in a nuisance abatement program.

110.1.2 Structural hazards. When an apparent structural hazard is caused by the faulty installation, operation or malfunction of any of the items or devices governed by this code, the fire code official shall immediately notify the building code official in accordance with Section 110.1.

❖ The fire code official is required to report structurally unsafe buildings to the building official to secure abatement of unsafe conditions. Courts have continually upheld the right of states and their authorized subdivisions to abate public nuisances, even by demolition, and bill or assess the property owner through a tax lien for their expenses. However, care must be exercised to maintain compliance with the due process and equal protection doctrines of the Fourth and Fourteenth Amendments of the U.S. Constitution.

110.2 Evacuation. The fire code official or the fire department official in charge of an incident shall be authorized to order the immediate evacuation of any occupied building deemed unsafe when such building has hazardous conditions that present imminent danger to building occupants. Persons so notified shall immediately leave the structure or premises and shall not enter or re-enter until authorized to do so by the fire code official or the fire department official in charge of the incident.

❖ The fire code official must immediately order the evacuation of any premises posing a clear and imminent threat to life or property. Building occupants who are warned must comply with the evacuation order without delay. Upon leaving the building, occupants may not reenter until authorization is given by the fire code official. Severe and immediate danger anticipated in this section dictates such extreme measures to protect public health, safety and welfare.

110.3 Summary abatement. Where conditions exist that are deemed hazardous to life and property, the fire code official or fire department official in charge of the incident is authorized to abate summarily such hazardous conditions that are in violation of this code.

❖ As indicated in the commentary to Section 110.1.1, the fire code official is authorized to seek abatement action by the building department and bill the owner for the abatement costs. Obviously this is an extreme measure and should be done only when the owner, operator or occupant does not take such measures under the requirements of Section 110.4.

110.4 Abatement. The owner, operator, or occupant of a building or premises deemed unsafe by the fire code official shall abate or cause to be abated or corrected such unsafe conditions either by repair, rehabilitation, demolition or other approved corrective action.

❖ This section describes the usual circumstance in which a building has such critical violations that it is declared unsafe by the fire code official. The owner, operator or occupant should take abatement measures to correct the unsafe condition. If this is not done promptly, however, the fire code official has the authority to directly abate the unsafe conditions and bill the owner for the abatement work in accordance with Sections 110.1.1 and 110.3.

SECTION 111
STOP WORK ORDER

111.1 Order. Whenever the fire code official finds any work regulated by this code being performed in a manner contrary to the provisions of this code or in a dangerous or unsafe manner, the fire code official is authorized to issue a stop work order.

❖ The fire code official is authorized to issue a stop work order when the work does not comply with the code. Obviously this is an extreme and costly measure that should be reserved for situations in which the violation is a serious safety hazard.

111.2 Issuance. A stop work order shall be in writing and shall be given to the owner of the property, or to the owner's agent, or to the person doing the work. Upon issuance of a stop work order, the cited work shall immediately cease. The stop work order shall state the reason for the order, and the conditions under which the cited work is authorized to resume.

❖ The stop work order is to be in writing and must cite the reason for issuing the order.

Upon receipt of a violation notice from the fire code official, all construction activities identified in the notice must immediately cease, except as expressly permitted to correct the violation.

Construction activities that are outside of the scope of the issue involved with the stop work order are not affected and need not stop; thus, the scope of the order must be clearly stated.

111.3 Emergencies. Where an emergency exists, the fire code official shall not be required to give a written notice prior to stopping the work.

❖ This section gives the fire code official the authority to stop the work in dispute immediately when, in his or her opinion, there is an unsafe emergency condition that has been created by the work. The need for the written notice is suspended for this situation so that the work can be stopped immediately. After the work is stopped, immediate measures should be taken to correct the work at issue.

111.4 Failure to comply. Any person who shall continue any work after having been served with a stop work order, except such work as that person is directed to perform to remove a violation or unsafe condition, shall be liable to a fine of not less than [AMOUNT] dollars or more than [AMOUNT] dollars.

❖ The local jurisdiction is to designate the fine that is to apply to any person who continues work that is at issue, other than abatement work. The dollar amounts for the minimum and maximum fines are to be specified in the adopting ordinance. See the sample ordinance for the adoption of the code on page v in the front of the code for details.

Bibliography

The following resource materials are referenced in this chapter or are relevant to the subject matter addressed in this chapter.

Callahan, T., and C.W. Bahme. *Fire Service and the Law, 2nd edition*. Quincy, MA: National Fire Protection Association, 1987.

Groner, N. *Fire Related Human Behavior*. Course Guide. Unit 9 — "Compliance and Codes." Emmitsburg, MD: Executive Office of the President, U.S. Federal Emergency Management Agency, U.S. Fire Administration, National Fire Academy and Open Learning Fire Service Program, 1990.

Government Liability for Negligent Fire Inspection: Hage vs. Stade. 66 Minnesota Law Review, pp. 1164-1180, 1982.

IBC-2003, *International Building Code*. Falls Church, VA: International Code Council, 2003.

IMC-2003, *International Mechanical Code*. Falls Church, VA: International Code Council, 2003.

IPC-2003, *International Plumbing Code*. Falls Church,VA: International Code Council, 2003.

IPMC-2003, *International Property Maintenance Code*. Falls Church, VA: International Code Council, 2003.

Legal Aspects of Code Administration. Country Club Hills, IL: Building Officials and Code Administrators International, Inc., 1996.

Markman, H. M. "How to Manage Your Liability." *The International Fire Chief*, pp. 14-17, 1983.

Municipal Liability for Negligent Building Inspections — Demise of the Public Duty Doctrine? 65 Iowa Law Review, p. 1416, 1980.

Municipal Liability for Negligent Fire Inspections. 23 Loyola Law Review, pp. 458 and 464, 1977.

Negligent Fire Inspections by City or State Employee. 22 Proof of Facts 2-D, pp. 55 and 69, 1980.

NFPA 13-99, *Installation of Sprinkler Systems*. Quincy, MA: National Fire Protection Association, 1999.

Robertson, J. C., and W.E. Koffel, Jr. *Fire Prevention Organization and Management*. Course Guide. Emmitsburg, MD: Executive Office of the President, U.S. Federal Emergency Management Agency, U.S. Fire Administration, National Fire Academy and Open Learning Fire Service Program, 1990.

Rosenbauer, D. L. *Introduction to Fire Protection Law*. Quincy, MA: National Fire Protection Association, 1978.

"State Tort Liability for Negligent Fire Inspections." *Columbia Journal of Legal and Social Problems,* Volume 25, Issue 13, pp. 303 and 341-44, 1977.

Chapter 2: Definitions

General Comments

All terms defined in the code are listed alphabetically in Chapter 2. The actual definitions of the terms are located as follows:

Where a term is used in more than one chapter, its definition appears in Chapter 2.

Where a term is unique or primarily pertains to a single chapter, its definition appears within that chapter. In many chapters, the second section is devoted to definitions. For example, definitions applicable to means of egress are found in Section 1002.

Purpose

Codes by their very nature are technical documents. Literally every word, term and punctuation mark can add to or change the meaning of the intended result. If a definition is specific to a certain chapter of the code, only a reference to the definition will be noted in the alphabetical listing of the term in this chapter.

Furthermore, the code, with its broad scope of applicability, includes terms that have a different meaning than the generally accepted meaning of the term. Additionally, these terms can have multiple meanings depending on the context or discipline in which they are being used.

For these reasons, maintaining a consensus on the specific meaning of terms contained in the code is essential. Chapter 2 performs this function by stating clearly what specific terms mean for the purpose of the code.

SECTION 201 GENERAL

201.1 Scope. Unless otherwise expressly stated, the following words and terms shall, for the purposes of this code, have the meanings shown in this chapter.

❖ This section contains the definitions for application of the code. As noted, when the definition is specific to a certain chapter within the code, the term is listed with a reference to the chapter where it is applied. If a term is used generally within the code, the term is defined in Chapter 2.

201.2 Interchangeability. Words used in the present tense include the future; words stated in the masculine gender include the feminine and neuter; the singular number includes the plural and the plural, the singular.

❖ While the definitions are to be taken literally, gender and tense are to be considered interchangeable.

201.3 Terms defined in other codes. Where terms are not defined in this code and are defined in the *International Building Code*, *International Fuel Gas Code*, *International Mechanical Code* or *International Plumbing Code*, such terms shall have the meanings ascribed to them as in those codes.

❖ This section states that when a term is not defined in the code but is defined in another volume of the *International Code Family*®, the meaning found in those codes can be used. This adds consistency to the application of the codes.

201.4 Terms not defined. Where terms are not defined through the methods authorized by this section, such terms shall have ordinarily accepted meanings such as the context implies. *Webster's Third New International Dictionary of the English Language, Unabridged*, shall be considered as providing ordinarily accepted meanings.

❖ Another resource for defining words or terms not defined within the code or other *International Codes*® is simply their "ordinarily accepted meaning." With some words, a dictionary definition may be sufficient, if the definition is applied within an appropriate context. Not all dictionaries, however, define words the same and not all parts of this country apply the same meanings to all words. The dictionary referenced in this section provides a standardized resource for defining terms and establishing "ordinarily accepted" meanings of words, thus reducing the likelihood of inconsistent enforcement of the code.

Some terms used throughout the code may not be defined in Chapter 2 or in a dictionary. In those cases, the user should first turn to the definitions contained in the referenced standards (see Chapter 45) and then to published textbooks on the subject in question.

SECTION 202
GENERAL DEFINITIONS

❖ This portion of the commentary addresses only those terms whose definitions appear in Chapter 2. The commentary for definitions that are located elsewhere in the code can be found in the indicated sections that contain those definitions.

[B] ACCESSIBLE MEANS OF EGRESS. See Section 1002.1.

AEROSOL. See Section 2802.1.

Level 1 aerosol products. See Section 2802.1.

Level 2 aerosol products. See Section 2802.1.

Level 3 aerosol products. See Section 2802.1.

AEROSOL CONTAINER. See Section 2802.1.

AEROSOL WAREHOUSE. See Section 2802.1.

AGENT. A person who shall have charge, care or control of any structure as owner, or agent of the owner, or as executor, executrix, administrator, administratrix, trustee or guardian of the estate of the owner. Any such person representing the actual owner shall be bound to comply with the provisions of this code to the same extent as if that person was the owner.

❖ An agent, for purposes of the code, is a person who has full authority under the law to act for or represent the owner of a building subject to the provisions of the code. An agent acts by the authority of the person he or she represents and generally has the same powers as the person represented. It is commonplace for building owners to retain the services of management agents to conduct all affairs pertinent to their building, including code compliance.

AIR-SUPPORTED STRUCTURE. See Section 2402.1.

AIRCRAFT OPERATION AREA (AOA). See Section 1102.1.

AIRPORT. See Section 1102.1.

[B] AISLE ACCESSWAY. See Section 1002.1.

ALARM NOTIFICATION APPLIANCE. See Section 902.1.

ALARM SIGNAL. See Section 902.1.

ALARM VERIFICATION FEATURE. See Section 902.1.

[B] ALTERNATING TREAD DEVICE. See Section 1002.1.

AMMONIUM NITRATE. See Section 3302.1.

ANNUNCIATOR. See Section 902.1.

APPROVED. Acceptable to the fire code official.

❖ As related to the process of acceptance of installations of materials, equipment and fire protection systems required or allowed by the code, this definition identifies where ultimate authority rests. Whenever this term is used, it intends that only the enforcing authority can accept a specific installation, component or system as complying with the code.

[B] AREA OF REFUGE. See Section 1002.1.

ARRAY. See Section 2302.1.

ARRAY, CLOSED. See Section 2302.1.

AUDIBLE ALARM NOTIFICATION APPLIANCE. See Section 902.1.

AUTOMATIC. See Section 902.1.

AUTOMATIC FIRE-EXTINGUISHING SYSTEM. See Section 902.1.

AUTOMATIC SPRINKLER SYSTEM. See Section 902.1.

AUTOMOTIVE MOTOR FUEL-DISPENSING FACILITY. See Section 2202.1.

AVERAGE AMBIENT SOUND LEVEL. See Section 902.1.

BARRICADE. See Section 3302.1.

Artificial barricade. See Section 3302.1.

Natural barricade. See Section 3302.1.

BARRICADED. See Section 3302.1.

BATTERY, LEAD ACID. See Section 602.1.

BATTERY SYSTEM, STATIONARY LEAD ACID. See Section 602.1.

BIN BOX. See Section 2302.1.

BLAST AREA. See Section 3302.1.

BLAST SITE. See Section 3302.1.

BLASTER. See Section 3302.1.

BLASTING AGENT. See Section 3302.1.

[B] BLEACHERS. See Section 1002.1.

BOILING POINT. See Section 2702.1.

BONFIRE. See Section 302.1.

BRITISH THERMAL UNIT (BTU). The heat necessary to raise the temperature of 1 pound (0.454 kg) of water by 1°F (0.5565°C).

❖ This definition describes the English unit of heat used throughout the document. A British thermal unit (Btu) is used as a way to describe the heat content of combustibles. The metric equivalent of a Btu is a Joule. This term should not be confused with heat release rate. Heat release rate would be described as Btu per second (joule per second = watt).

BULK OXYGEN SYSTEM. See Section 4002.1.

BULK PLANT OR TERMINAL. See Section 3402.1.

BULK TRANSFER. See Section 3402.1.

BULLET RESISTANT. See Section 3302.1.

CANOPY. See Section 2402.1.

CARBON DIOXIDE EXTINGUISHING SYSTEM. See Section 902.1.

CARTON. A cardboard or fiberboard box enclosing a product.

❖ This term is commonly used when applying the high-piled storage requirements and also for packaging of aerosols. This definition provides a consistent understanding of a word often used to describe packaging.

CEILING LIMIT. See Section 2702.1.

CHEMICAL. See Section 2702.1.

CHEMICAL NAME. See Section 2702.1.

CLEAN AGENT. See Section 902.1.

CLOSED CONTAINER. See Section 2702.1.

CLOSED SYSTEM. The use of a solid or liquid hazardous material involving a closed vessel or system that remains closed during normal operations where vapors emitted by the product are not liberated outside of the vessel or system and the product is not exposed to the atmosphere during normal operations; and all uses of compressed gases. Examples of closed systems for solids and liquids include product conveyed through a piping system into a closed vessel, system or piece of equipment.

❖ This definition is used primarily with regard to hazardous materials. A closed system is inherently less hazardous than an open system. This difference is related to the fact that vapors, dusts or similar materials are not normally released from closed systems. Because closed systems are less hazardous than open systems, credit is typically given to increase the maximum allowable quantities when systems are considered closed.

COLD DECK. See Section 1902.1.

COMBUSTIBLE DUST. See Section 1302.1.

COMBUSTIBLE FIBERS. See Section 2902.1.

COMBUSTIBLE LIQUID. See Section 3402.1.

Class II. See Section 3402.1.

Class IIIA. See Section 3402.1.

Class IIIB. See Section 3402.1.

[M] COMMERCIAL COOKING APPLIANCES. See Section 602.1.

COMMODITY. See Section 2302.1.

[B] COMMON PATH OF EGRESS TRAVEL. See Section 1002.1.

COMPRESSED GAS. See Section 3002.1.

COMPRESSED GAS CONTAINER. See Section 3002.1.

COMPRESSED GAS SYSTEM. See Section 3002.1.

CONSTANTLY ATTENDED LOCATION. See Section 902.1.

CONSTRUCTION DOCUMENTS. The written, graphic and pictorial documents prepared or assembled for describing the design, location and physical characteristics of the elements of the project necessary for obtaining a permit.

❖ To determine whether proposed construction is in compliance with code requirements, sufficient information must be submitted to the fire code official for review. This definition describes in general which items are to be included in that documentation. This typically will include drawings (floor plans, elevations, sections, details, etc.), specifications and product information describing the proposed work. In the past, these documents were referred to as plans and specifications. Those terms are not broad enough to include all information, including calculations or graphs.

CONTAINER. See Section 2702.1.

CONTAINMENT SYSTEM. See Section 3702.1.

CONTAINMENT VESSEL. See Section 3702.1.

CONTINUOUS GAS DETECTION SYSTEM. See Section 1802.1.

CONTROL AREA. See Section 2702.1.

[B] CORRIDOR. See Section 1002.1.

CORROSIVE. See Section 3102.1.

CRYOGENIC CONTAINER. See Section 3202.1.

CRYOGENIC FLUID. See Section 3202.1.

CRYOGENIC VESSEL. See Section 3202.1.

CYLINDER. See Section 2702.1.

DEFLAGRATION. See Section 2702.1.

DELUGE SYSTEM. See Section 902.1.

DESIGN PRESSURE. See Section 2702.1.

DETACHED BUILDING. See Section 2702.1.

DETEARING. See Section 1502.1.

DETECTOR, HEAT. See Section 902.1.

DETONATING CORD. See Section 3302.1.

DETONATION. See Section 3302.1.

DETONATOR. See Section 3302.1.

DIP TANK. See Section 1502.1.

DISCHARGE SITE. See Section 3302.1.

DISPENSING. See Section 2702.1.

DISPENSING DEVICE, OVERHEAD TYPE. See Section 2202.1.

DISPLAY SITE. See Section 3302.1.

[B] DOOR, BALANCED. See Section 1002.1.

DRAFT CURTAIN. See Section 2302.1.

DRY-CHEMICAL EXTINGUISHING AGENT. See Section 902.1.

DRY CLEANING. See Section 1202.1.

DRY CLEANING PLANT. See Section 1202.1.

DRY CLEANING ROOM. See Section 1202.1.

DRY CLEANING SYSTEM. See Section 1202.1.

EARLY SUPPRESSION FAST-RESPONSE (ESFR) SPRINKLER. See Section 2302.1.

[B] EGRESS COURT. See Section 1002.1.

ELECTROSTATIC FLUIDIZED BED. See Section 1502.1.

EMERGENCY ALARM SYSTEM. See Section 902.1.

EMERGENCY CONTROL STATION. See Section 1802.1.

[B] EMERGENCY ESCAPE AND RESCUE OPENING. See Section 1002.1.

EMERGENCY EVACUATION DRILL. See Section 402.1.

EMERGENCY VOICE/ALARM COMMUNICATIONS. See Section 902.1.

EXCESS FLOW CONTROL. See Section 2702.1.

EXCESS FLOW VALVE. See Section 3702.1.

EXHAUSTED ENCLOSURE. See Section 2702.1.

EXISTING. Buildings, facilities or conditions which are already in existence, constructed or officially authorized prior to the adoption of this code.

❖ This term is specifically defined to reduce any confusion regarding the application of the code to existing buildings, facilities and conditions. This definition would include anything that has already been in use, is already constructed or has been approved by the jurisdiction prior to the adoption of the code. If an occupancy changes use significantly, it may be considered new in some cases.

[B] EXIT. See Section 1002.1.

[B] EXIT ACCESS. See Section 1002.1.

[B] EXIT DISCHARGE. See Section 1002.1.

[B] EXIT DISCHARGE, LEVEL OF. See Section 1002.1.

[B] EXIT ENCLOSURE. See Section 1002.1.

[B] EXIT, HORIZONTAL. See Section 1002.1.

[B] EXIT PASSAGEWAY. See Section 1002.1.

EXPANDED PLASTIC. See Section 2302.1.

EXPLOSION. See Section 2702.1.

EXPLOSIVE. See Section 3302.1.

High Explosive. See Section 3302.1.

Low Explosive. See Section 3302.1.

Mass Detonating Explosives. See Section 3302.1.

UN/DOTn Class 1 Explosives. See Section 3302.1.

Division 1.1. See Section 3302.1.

Division 1.2. See Section 3302.1.

Division 1.3. See Section 3302.1.

Division 1.4. See Section 3302.1.

Division 1.5. See Section 3302.1.

Division 1.6. See Section 3302.1.

EXPLOSIVE MATERIAL. See Section 3302.1.

EXTRA-HIGH-RACK COMBUSTIBLE STORAGE. See Section 2302.1.

FABRICATION AREA. See Section 1802.1.

FACILITY. A building or use in a fixed location including exterior storage areas for flammable and combustible substances and hazardous materials, piers, wharves, tank farms and similar uses. This term includes recreational vehicles, mobile home and manufactured housing parks, sales and storage lots.

❖ The scope of the fire code is broader than the scope of a building code, in that it addresses outdoor storage, tank farms, fire department access and similar activities. The term "facility" helps to more clearly define this scope to clarify the application of this code. As noted in the definition, a building would be included in the definition for "Facility." This definition differs from that used in Chapter 11 of the *International Building Code*® (IBC®), which addresses the topic of accessibility for those with disabilities.

FALLOUT AREA. See Section 3302.1.

FALSE ALARM. The willful and knowing initiation or transmission of a signal, message or other notification of an event of fire when no such danger exists.

❖ The term "false alarm" can have several meanings beyond what this particular definition provides. This definition states that for the purposes of this code a false alarm is an unnecessary, intentional activation of a fire alarm system, signal or message. It would not include alarm activation as a result of a malfunctioning detector.

FINES. See Section 1902.1.

FIRE ALARM. The giving, signaling or transmission to any public fire station, or company or to any officer or employee thereof, whether by telephone, spoken word or otherwise, of information to the effect that there is a fire at or near the place indicated by the person giving, signaling, or transmitting such information.

❖ This is a general definition intended to clarify that a fire alarm is not simply a fire alarm system. This definition would allow a person's actions to be considered part of the fire alarm. The key is that a fire alarm notifies the correct persons or group of persons, such as the fire department or a central station, of a fire.

FIRE ALARM BOX, MANUAL. See Section 902.1.

FIRE ALARM CONTROL UNIT. See Section 902.1.

FIRE ALARM SIGNAL. See Section 902.1.

FIRE ALARM SYSTEM. See Section 902.1.

FIRE APPARATUS ACCESS ROAD. See Section 502.1.

FIRE AREA. See Section 902.1.

FIRE CHIEF. The chief officer of the fire department serving the jurisdiction, or a duly authorized representative.

❖ This definition is necessary to note that when the term "fire chief" is used within the text of the code it is specifically referring to the chief officer of a fire department. This position can be delegated as necessary but must be appropriately authorized.

FIRE CODE OFFICIAL. The fire chief or other designated authority charged with the administration and enforcement of the code, or a duly authorized representative.

❖ Whoever holds the statutory power to enforce the fire code is termed the "fire code official." Normally, responsibility for this enforcement is assigned to a fire prevention bureau or related code enforcement department of the state, county or municipality. In the case of a fire department, the role of fire code official is most often given to the fire chief, the fire marshal or the fire inspector. Often with regard to the fire code, the fire code official will be the fire marshal or fire chief. In some cases, direct reference will be made to the fire chief within the code because some situations are specific to the actions of the fire department, such as authority at fire scenes.

FIRE COMMAND CENTER. See Section 502.1.

FIRE DEPARTMENT MASTER KEY. See Section 502.1.

FIRE DETECTOR, AUTOMATIC. See Section 902.1.

[B] FIRE DOOR ASSEMBLY. Any combination of a fire door, frame, hardware, and other accessories that together provide a specific degree of fire protection to the opening.

❖ Often called "opening protectives," fire door assemblies must be fire tested in accordance with NFPA 80. The test must evaluate the door assembly with all appurtenant hardware in place. To achieve the tested ratings, the assembly must be installed in the same manner as tested. It is important to note that doors and frames are sometimes manufactured separately; nevertheless, each must be labeled.

[B] FIRE EXIT HARDWARE. See Section 1002.1.

FIRE LANE. See Section 502.1.

[B] FIRE PARTITION. A vertical assembly of materials designed to restrict the spread of fire in which openings are protected.

❖ Fire partitions are used as wall assemblies to separate adjacent tenant spaces in covered mall buildings, dwelling units and guestrooms and to enclose corridors. Required fire-resistance ratings for fire partitions are given in Section 708.3 of the IBC.

FIRE POINT. See Section 3402.1.

FIRE PROTECTION SYSTEM. See Section 902.1.

FIRE SAFETY FUNCTIONS. See Section 902.1.

FIRE WATCH. A temporary measure intended to ensure continuous and systematic surveillance of a building or portion thereof by one or more qualified individuals for the purposes of identifying and controlling fire hazards, detecting early signs of unwanted fire, raising an alarm of fire and notifying the fire department.

❖ This term is used in several places throughout the code. A fire watch provides temporary fire safety where there are potential hazards, such as during hot work operations, or when fire protection systems are out of service. A fire watch is not simply to watch for a fire but also to prevent fire by identifying and controlling fire hazards, such as the separation of combustibles from areas where welding is to occur. A fire watch also provides a method of notifying the fire department if a fire should occur.

FIREWORKS. See Section 3302.1.

Fireworks, 1.4G. See Section 3302.1.

Fireworks, 1.3G. See Section 3302.1.

FIREWORKS DISPLAY. See Section 3302.1.

FLAMMABLE CRYOGENIC FLUID. See Section 3202.1.

FLAMMABLE FINISHES. See Section 1502.1.

FLAMMABLE GAS. See Section 3502.1.

FLAMMABLE LIQUEFIED GAS. See Section 3502.1.

FLAMMABLE LIQUID. See Section 3402.1.

Class IA. See Section 3402.1.

Class IB. See Section 3402.1.

Class IC. See Section 3402.1.

FLAMMABLE MATERIAL. A material capable of being readily ignited from common sources of heat or at a temperature of 600°F (316°C) or less.

❖ The primary focus of this term is to classify solid materials that are more hazardous than normal combustibles because of their susceptibility to ignition as flammable materials. Additionally, any material that will readily ignite at or below 600°F (316°C) would be considered flammable. The term "flammable materials" should not be confused with combustible materials.

FLAMMABLE SOLID. See Section 3602.1.

FLAMMABLE VAPORS OR FUMES. See Section 2702.1.

FLASH POINT. See Section 3402.1.

FLEET VEHICLE MOTOR FUEL-DISPENSING FACILITY. See Section 2202.1.

[B] FLOOR AREA, GROSS. See Section 1002.1.

[B] FLOOR AREA, NET. See Section 1002.1.

FLUIDIZED BED. See Section 1502.1.

FOAM-EXTINGUISHING SYSTEM. See Section 902.1.

[B] FOLDING AND TELESCOPIC SEATING. See Section 1002.1.

FUEL LIMIT SWITCH. See Section 3402.1.

FUMIGANT. See Section 1702.1.

FUMIGATION. See Section 1702.1.

FURNACE CLASS A. See Section 2102.1.

FURNACE CLASS B. See Section 2102.1.

FURNACE CLASS C. See Section 2102.1.

FURNACE CLASS D. See Section 2102.1.

GAS CABINET. See Section 2702.1.

GAS ROOM. See Section 2702.1.

[B] GRANDSTAND. See Section 1002.1.

[B] GUARD. See Section 1002.1.

HALOGENATED EXTINGUISHING SYSTEM. See Section 902.1.

HANDLING. See Section 2702.1.

[B] HANDRAIL. See Section 1002.1.

HAZARDOUS MATERIAL. See Section 2702.1.

HAZARDOUS PRODUCTION MATERIAL (HPM). See Section 1802.1.

HEALTH HAZARD. See Section 2702.1.

HELIPORT. See Section 1102.1.

HELISTOP. See Section 1102.1.

HI-BOY. See Section 302.1.

HIGH-PILED COMBUSTIBLE STORAGE. See Section 2302.1.

HIGH-PILED STORAGE AREA. See Section 2302.1.

HIGHLY TOXIC. See Section 3702.1.

HIGHLY VOLATILE LIQUID. A liquefied compressed gas with a boiling point of less than 68°F (20°C).

❖ This definition provides criteria for the classification of a material as being highly volatile, and provides correlation with the defined terms "Liquid" and "Cryogenic fluid." Basically, if the boiling point of a material is at room temperature or lower it would be considered volatile. The concern usually associated with highly volatile liquids is the volume of vapors released to the atmosphere. These vapors could be harmless, but many liquids, for example, may be corrosive, toxic or flammable. Additionally, vapors are more susceptible to ignition than liquids.

HIGHWAY. See Section 3302.1.

HOGGED MATERIALS. See Section 1902.1.

[M] HOOD. See Section 602.1.

Type I. See Section 602.1.

HOT WORK. See Section 2602.1.

HOT WORK AREA. See Section 2602.1.

HOT WORK EQUIPMENT. See Section 2602.1.

HOT WORK PERMITS. See Section 2602.1.

HOT WORK PROGRAM. See Section 2602.1.

HPM FLAMMABLE LIQUID. See Section 1802.1.

HPM ROOM. See Section 1802.1.

IMMEDIATELY DANGEROUS TO LIFE AND HEALTH (IDLH). See Section 2702.1.

IMPAIRMENT COORDINATOR. See Section 902.1.

INCOMPATIBLE MATERIALS. See Section 2702.1.

INHABITED BUILDING. See Section 3302.1.

INITIATING DEVICE. See Section 902.1.

IRRITANT. A chemical which is not corrosive, but which causes a reversible inflammatory effect on living tissue by chemical action at the site of contact. A chemical is a skin irritant if, when tested on the intact skin of albino rabbits by the methods of CPSC 16CFR Part 1500.41 for an exposure of four or more hours or by other appropriate techniques, it results in an empirical score of 5 or more. A chemical is classified as an eye irritant if so determined under the procedure listed in CPSC 16CFR Part 1500.42 or other approved techniques.

❖ Materials classified as irritants include a wide range of materials that pose a health hazard with acute effects caused by short-term exposure. Exposure to irritants may result in a minor, troublesome injury at the point of contact; however, the injury usually heals without leaving a scar. In comparison, corrosives can cause permanent destruction of tissue at the point of contact — with a scar the likely result. Many household insecticides and pesticides are common irritants. The definition is derived from DOL 29 CFR; 1910.1200.

KEY BOX. See Section 502.1.

LABELED. Equipment or material to which has been attached a label, symbol or other identifying mark of a nationally recognized testing laboratory, inspection agency or other organization concerned with product evaluation that maintains periodic inspection of production of labeled equipment or materials, and by whose labeling is indicated compliance with nationally recognized standards or tests to determine suitable usage in a specified manner.

❖ The term "labeled" is not to be confused with the term "listed." A label is a marking or other identifying mark that indicates approval from a nationally recognized testing laboratory, inspection agency or other organization that evaluates products. A label is used to delineate materials and assemblies that must bear the identification of the manufacturer, as well as a third-party quality control agency. The quality control agency allows the use of its label based on the results of periodic audits and inspections of the manufacturer's plant. This is one form of quality control. The code often requires labeled equipment and systems (see the definition for "Listed").

LIMITED SPRAYING SPACE. See Section 1502.1.

LIQUEFIED NATURAL GAS (LNG). See Section 2202.1.

LIQUEFIED PETROLEUM GAS (LP-gas). See Section 3802.1.

LIQUID. See Section 2702.1.

LIQUID STORAGE ROOM. See Section 3402.1.

LISTED. Equipment or materials included on a list published by an approved testing laboratory, inspection agency or other organization concerned with current product evaluation that maintains periodic inspection of production of listed equipment or materials, and whose listing states that equipment or materials comply with approved nationally recognized standards and have been tested or evaluated and found suitable for use in a specified manner.

❖ The term "listed," which is not to be confused with "labeled," is a form of quality control. Essentially, a particular product, piece of equipment or system is evaluated or tested and the results are published in a list by agencies, such as approved testing laboratories and inspection agencies. Listed products and equipment are periodically inspected to maintain the listing. The code often requires listed equipment or systems (see the definition for "Labeled").

LONGITUDINAL FLUE SPACE. See Section 2302.1.

LOW-PRESSURE TANK. See Section 3202.1.

LOWER EXPLOSIVE LIMIT (LEL). See Section 2702.1.

LOWER FLAMMABLE LIMIT (LFL). See Section 2702.1.

MAGAZINE. See Section 3302.1.

Indoor. See Section 3302.1.

Type 1. See Section 3302.1.

Type 2. See Section 3302.1.

Type 3. See Section 3302.1.

Type 4. See Section 3302.1.

Type 5. See Section 3302.1.

MAGNESIUM. See Section 3602.1.

MANUAL FIRE ALARM BOX. See Section 902.1.

MANUAL STOCKING METHODS. See Section 2302.1.

MARINE MOTOR FUEL-DISPENSING FACILITY. See Section 2202.1.

MATERIAL SAFETY DATA SHEET (MSDS). See Section 2702.1.

MAXIMUM ALLOWABLE QUANTITY PER CONTROL AREA. See Section 2702.1.

[B] MEANS OF EGRESS. See Section 1002.1.

MECHANICAL STOCKING METHODS. See Section 2302.1.

MEMBRANE STRUCTURE. See Section 2402.1.

MOBILE FUELING. See Section 3402.1.

MORTAR. See Section 3302.1.

MULTIPLE-STATION ALARM DEVICE. See Section 902.1.

MULTIPLE-STATION SMOKE ALARM. See Section 902.1.

NESTING. See Section 3002.1.

NET EXPLOSIVE WEIGHT (net weight). See Section 3302.1.

NORMAL TEMPERATURE AND PRESSURE (NTP). See Section 2702.1.

[B] NOSING. See Section 1002.1.

NUISANCE ALARM. See Section 902.1.

OCCUPANCY CLASSIFICATION. For the purposes of this code, certain occupancies are defined as follows:

❖ Occupancies are defined and explained in Sections 302 through 312 of the IBC and its commentary.

[B] Assembly Group A. Assembly Group A occupancy includes, among others, the use of a building or structure, or a portion thereof, for the gathering together of persons for purposes such as civic, social or religious functions, recreation, food or drink consumption or awaiting transportation. A room or space used for assembly purposes by less than 50 persons and accessory to another occupancy shall be included as a part of that occupancy. Assembly areas with less than 750 square feet (69.7 m^2) and which are accessory to another occupancy according to Section 302.2.1 of the *International Building Code* are not assembly occupancies. Assembly occupancies which are accessory to Group E in accordance with Section 302.2 of the *International Building Code* are not considered assembly occupancies. Religious educational rooms and religious auditoriums which are accessory to churches in accordance with Section 302.2 of the *International Building Code* and which have occupant loads of less than 100 shall be classified as A-3. A building or tenant space used for assembly purposes by less than 50 persons shall be considered a Group B occupancy.

Assembly occupancies shall include the following:

A-1 Assembly uses, usually with fixed seating, intended for the production and viewing of performing arts or motion pictures including but not limited to:

Motion picture theaters
Symphony and concert halls
Televison and radio studios admitting an audience
Theaters

A-2 Assembly uses intended for food and/or drink consumption including, but not limited to:

Banquet halls
Night clubs
Restaurants
Taverns and bars

A-3 Assembly uses intended for worship, recreation or amusement and other assembly uses not classified elsewhere in Group A, including, but not limited to:

Amusement arcades
Art galleries
Bowling alleys
Churches
Community halls
Courtrooms
Dance halls (not including food or drink consumption)
Exhibition halls
Funeral parlors
Gymnasiums (without spectator seating)
Indoor swimming pools (without spectator seating)
Indoor tennis courts (without spectator seating)
Lecture halls
Libraries
Museums
Waiting areas in transportation terminals
Pool and billiard parlors

A-4 Assembly uses intended for viewing of indoor sporting events and activities with spectator seating including, but not limited to:

Arenas
Skating rinks
Swimming pools
Tennis courts

A-5 Assembly uses intended for participation in or viewing outdoor activities including, but not limited to:

Amusement park structures
Bleachers
Grandstands
Stadiums

[B] Business Group B. Business Group B occupancy includes, among others, the use of a building or structure, or a portion thereof, for office, professional or service-type transactions, including storage of records and accounts. Business occupancies shall include, but not be limited to, the following:

Airport traffic control towers
Animal hospitals, kennels, pounds
Banks
Barber and beauty shops
Car wash
Civic administration
Clinic—outpatient
Dry cleaning and laundries; pick-up and delivery stations and self-service
Educational occupancies above the 12th grade
Electronic data processing
Laboratories; testing and research
Motor vehicle showrooms
Post offices
Print shops
Professional services (architects, attorneys, dentists, physicians, engineers, etc.)
Radio and television stations
Telephone exchanges

[B] Educational Group E. Educational Group E occupancy includes, among others, the use of a building or structure, or a portion thereof, by six or more persons at any one time for educational purposes through the 12th grade. Religious educational rooms and religious auditoriums, which are accessory to churches in accordance with Section 302.2 and have occupant loads of less than 100, shall be classified as Group A-3 occupancies.

Day care. The use of a building or structure, or portion thereof, for educational, supervision or personal care services for more than five children older than $2^1/_2$ years of age shall be classified as an E occupancy.

[B] Factory Industrial Group F. Factory Industrial Group F occupancy includes, among others, the use of a building or structure, or a portion thereof, for assembling, disassembling, fabricating, finishing, manufacturing, packaging, repair or processing operations that are not classified as a Group H high-hazard or Group S storage occupancy.

Factory Industrial F-1 Moderate-Hazard Occupancy. Factory Industrial uses which are not classified as Factory Industrial Group F-2 shall be classified as F-1 Moderate Hazard and shall include, but not be limited to, the following:

Aircraft
Appliances
Athletic equipment
Automobiles and other motor vehicles
Bakeries
Beverages; over 12 percent in alcohol content
Bicycles
Boats
Brooms or brushes
Business machines
Cameras and photo equipment
Canvas and similar fabric
Carpet and rugs (includes cleaning)
Disinfectants
Dry cleaning and dyeing
Electric generation plants
Electronics
Engines (including rebuilding)
Food processing
Furniture
Hemp products
Jute products
Laundries
Leather products
Machinery
Metals
Millwork (sash and doors)

Motion picture and television filming (without spectators)
Musical instruments
Optical goods
Paper mills or products
Photographic film
Plastic products
Printing or publishing
Recreational vehicles
Refuse incineration
Shoes
Soaps and detergents
Textiles
Tobacco
Trailers
Upholstering
Wood; distillation
Woodworking (cabinet)

[B] Factory Industrial F-2 Low-Hazard Occupancy. Factory industrial uses involving the fabrication or manufacturing of noncombustible materials which, during finishing, packaging or processing do not involve a significant fire hazard, shall be classified as Group F-2 occupancies and shall include, but not be limited to, the following:

Beverages; up to and including 12 percent alcohol content
Brick and masonry
Ceramic products
Foundries
Glass products
Gypsum
Ice
Metal products (fabrication and assembly)

High-Hazard Group H. High-hazard Group H occupancy includes, among others, the use of a building or structure, or a portion thereof, that involves the manufacturing, processing, generation or storage of materials that constitute a physical or health hazard in quantities in excess of those found in Tables 307.7(1) and 307.7(2) of the *International Building Code.* (See also definition of "Control area)".

Exception: Occupancies as provided for in the *International Building Code* shall not be classified as Group H, but shall be classified in the occupancy which they most nearly resemble.

High-hazard Group H-1. Buildings and structures containing materials that pose a detonation hazard, shall be classified as Group H-1. Such materials shall include, but not be limited to, the following:

Explosives:
Division 1.1
Division 1.2
Division 1.3

Exception: Materials that are used and maintained in a form where either confinement or configuration will not elevate the hazard from a mass fire to mass explosion hazard shall be allowed in Group H-2 occupancies.

Division 1.4

Exception: Articles, including articles packaged for shipment, that are not regulated as an explosive under Bureau of Alcohol, Tobacco and Firearms regulations, or unpackaged articles used in process operations that do not propagate a detonation or deflagration between articles shall be allowed in Group H-3 occupancies.

Division 1.5
Division 1.6
Organic peroxides, unclassified detonable
Oxidizers, Class 4
Unstable (reactive) materials, Class 3 detonable, and Class 4
Detonable pyrophoric materials

High-hazard Group H-2. Buildings and structures containing materials that pose a deflagration hazard or a hazard from accelerated burning, shall be classified as Group H-2. Such materials shall include, but not be limited to, the following:

Class I, or II or IIIA flammable or combustible liquids which are used or stored in normally open containers or systems, or in closed containers or systems pressurized at more than 15 pounds per square inch (103.4 kPa) gauge
Combustible dusts
Cryogenic fluids, flammable
Flammable gases
Organic peroxides, Class I
Oxidizers, Class 3, that are used or stored in normally open containers or systems, or in closed containers or systems pressurized at more than 15 pounds per square inch (103.4 kPa) gauge
Pyrophoric liquids, solids and gases, nondetonable
Unstable (reactive) materials, Class 3, nondetonable
Water-reactive materials, Class 3

High-hazard Group H-3. Buildings and structures containing materials that readily support combustion or that pose a physical hazard shall be classified as Group H-3. Such materials shall include, but not be limited to, the following:

Class I, II or IIIA flammable or combustible liquids which are used or stored in normally closed containers or systems pressurized at less than 15 pounds per square inch (103.4 kPa) gauge
Combustible fibers
Consumer fireworks, 1.4G (Class C, Common)
Cryogenic fluids, oxidizing
Flammable solids
Organic peroxides, Class II and Class III
Oxidizers, Class 2
Oxidizers, Class 3, that are used or stored in normally closed containers or systems pressurized at less than 15 pounds per square inch (103.4 kPa) gauge
Oxidizing gases
Unstable (reactive) materials, Class 2
Water-reactive materials, Class 2

High-hazard Group H-4. Buildings and structures which contain materials that are health hazards shall be classified as Group H-4. Such materials shall include, but not be limited to, the following:

Corrosives
Highly toxic materials
Toxic materials

High-hazard Group H-5. Semiconductor fabrication facilities and comparable research and development areas in which hazardous production materials (HPM) are used and the aggregate quantity of materials is in excess of those listed in Tables 307.7(1) and 307.7(2) of the *International Building Code*. Such facilities and areas shall be designed and constructed in accordance with Section 415.9 of the *International Building Code*.

[B] Institutional Group I. Institutional Group I occupancy includes, among others, the use of a building or structure, or a portion thereof, in which people, cared for or living in a supervised environment and having physical limitations because of health or age, are harbored for medical treatment or other care or treatment, or in which people are detained for penal or correctional purposes or in which the liberty of the occupants is restricted. Institutional occupancies shall be classified as Group I-1, I-2, I-3 or I-4.

Group I-1. This occupancy shall include buildings, structures or parts thereof housing more than 16 persons, on a 24-hour basis, who because of age, mental disability or other reasons, live in a supervised residential environment that provides personal care services. The occupants are capable of responding to an emergency situation without physical assistance from staff. This group shall include, but not be limited to, the following:

Alcohol and drug centers
Assisted living facilities
Congregate care facilities
Convalescent facilities
Group homes
Half-way houses
Residential board and care facilities
Social rehabilitation facilities

A facility such as the above with five or fewer persons shall be classified as Group R-3 or shall comply with the *International Residential Code*. A facility such as above, housing at least six and not more than 16 persons, shall be classified as Group R-4.

Group I-2. This occupancy shall include buildings and structures used for medical, surgical, psychiatric, nursing or custodial care on a 24-hour basis of more than five persons who are not capable of self-preservation. This group shall include, but not be limited to, the following:

Hospitals
Nursing homes (both intermediate care facilities and skilled nursing facilities)
Mental hospitals
Detoxification facilities.

A facility such as the above with five or fewer persons shall be classified as Group R-3 or shall comply with the *International Residential Code*.

A child care facility which provides care on a 24-hour basis to more than five children $2^1/_2$ years of age or less shall be classified as Group I-2.

Group I-3. This occupancy shall include buildings and structures which are inhabited by more than five persons who are under restraint or security. An I-3 facility is occupied by persons who are generally incapable of self-preservation due to security measures not under the occupants' control. This group shall include, but not be limited to, the following:

Correctional centers
Detention centers
Jails
Prerelease centers
Prisons
Reformatories

Buildings of Group I-3 shall be classified as one of the occupancy conditions indicated in Sections 308.4.1 through 308.4.5 (see Section 408.1) of the *International Building Code*.

Condition 1. This occupancy condition shall include buildings in which free movement is allowed from sleeping areas and other spaces where access or occupancy is permitted, to the exterior via means of egress without restraint. A Condition 1 facility is permitted to be constructed as Group R.

Condition 2. This occupancy condition shall include buildings in which free movement is allowed from sleeping areas and any other occupied smoke compartment to one or more other smoke compartments. Egress to the exterior is impeded by locked exits.

Condition 3. This occupancy condition shall include buildings in which free movement is allowed within individual smoke compartments, such as within a residential unit comprised of individual sleeping units and group activity spaces, where egress is impeded by remote-controlled release of means of egress from such smoke compartment to another smoke compartment.

Condition 4. This occupancy condition shall include buildings in which free movement is restricted from an occupied space. Remote-controlled release is provided to permit movement from sleeping units, activity spaces and other occupied areas within the smoke compartment to other smoke compartments.

Condition 5. This occupancy condition shall include buildings in which free movement is restricted from an occupied space. Staff-controlled manual release is provided to permit movement from sleeping units, activity spaces and other occupied areas within the smoke compartment to other smoke compartments.

Group I-4, day care facilities. This group shall include buildings and structures occupied by persons of any age who receive custodial care for less than 24 hours by individuals other than parents or guardians, relatives by blood marriage, or adoption, and in a place other than the home of the person cared for. A facility such as the above with five or fewer persons shall be classified as Group R-3 or shall comply with the *International Residential Code*. Places of worship during religious functions are not included.

Adult care facility. A facility that provides accommodations for less than 24 hours for more than five unrelated adults and provides supervision and personal care services shall be classified as Group I-4.

Exception: Where the occupants are capable of responding to an emergency situation without physical assistance from the staff the facility shall be classified as Group A-3.

Child care facility. A facility that provides supervision and personal care on less than a 24-hour basis for more than five children $2^1/_2$ years of age or less shall be classified as Group I-4.

Exception: A child day care facility which provides care for more than five but no more than 100 children $2^1/_2$ years or less of age, when the rooms where such children are cared for are located on the level of exit discharge and each of these child care rooms has an exit door directly to the exterior, shall be classified as Group E.

[B] Mercantile Group M. Mercantile Group M occupancy includes, among others, buildings and structures or a portion thereof, for the display and sale of merchandise, and involves stocks of goods, wares or merchandise incidental to such purposes and accessible to the public. Mercantile occupancies shall include, but not be limited to, the following.

Department stores
Drug stores
Markets
Motor fuel-dispensing facilities
Retail or wholesale stores
Sales rooms

[B] Residential Group R. Residential Group R includes, among others, the use of a building or structure, or a portion thereof, for sleeping purposes when not classed as Institutional Group I. Residential occupancies shall include the following:

R-1 Residential occupancies where the occupants are primarily transient in nature including:

Boarding houses (transient)
Hotels (transient)
Motels (transient)

R-2 Residential occupancies containing sleeping units or more than two dwelling units where the occupants are primarily permanent in nature, including:

Apartment houses
Boarding houses (not transient)
Convents
Dormitories
Fraternities and sororities
Hotels (nontransient)
Monasteries
Motels (nontransient)
Vacation timeshare properties

R-3 Residential occupancies where the occupancies are primarily permanent in nature and not classified as R-1, R-2, or I and where buildings do not contain more than two dwelling units, or adult and child care facilities that provide accommodations for five or fewer persons of any age for less than 24-hours. Adult and child care facilities that are within a single-family home are permitted to comply with the *International Residential Code*.

R-4 Residential occupancies shall include buildings arranged for occupancy as Residential Care/Assisted Living Facilities including more than five but not more than 16 occupants.

Group R-4 occupancies shall meet the requirements for construction as defined for Group R-3 except for the height and area limitations provided in Section 503 of the *International Building Code* or shall comply with the *International Residential Code*.

[B] Storage Group S. Storage Group S occupancy includes, among others, the use of a building or structure, or a portion thereof, for storage that is not classified as a hazardous occupancy.

Moderate-hazard storage, Group S-1. Buildings occupied for storage uses which are not classified as Group S-2 including, but not limited to, storage of the following:

Aerosols, Level 2 and 3
Aircraft repair hangar
Bags, cloth, burlap and paper
Bamboo and rattan
Baskets
Belting, canvas and leather
Books and paper in rolls or packs
Boots and shoes
Buttons, including cloth covered, pearl or bone
Cardboard and cardboard boxes
Clothing, woolen wearing apparel
Cordage
Furniture
Furs
Glue, mucilage, paste and size
Grain
Horn and combs, other than celluloid
Leather
Linoleum
Lumber
Motor vehicle repair garages (complying with the *International Building Code* and containing less than the maximum allowable quantities of hazardous materials)
Photo engraving
Resilient flooring
Silk
Soap
Sugar
Tires, bulk storage of
Tobacco, cigars, cigarettes and snuff

Upholstering and mattress
Wax candles

Low-hazard storage, Group S-2. Includes, among others, buildings used for the storage of noncombustible materials such as products on wood pallets or in paper cartons with or without single thickness divisions; or in paper wrappings. Such products may have a negligible amount of plastic trim such as knobs, handles, or film wrapping. Storage uses shall include, but not be limited to, storage of the following:

Aircraft hangar
Asbestos
Beverages up to and including12-percent alcohol in metal, glass or ceramic containers
Cement in bags
Chalk and crayons
Dairy products in nonwaxed coated paper containers
Dry cell batteries
Electrical coils
Electrical motors
Empty cans
Food products
Foods in noncombustible containers
Fresh fruits and vegetables in nonplastic trays or containers
Frozen foods
Glass
Glass bottles, empty or filled with noncombustible liquids
Gypsum board
Inert pigments
Ivory
Metal desks with plastic tops and trim
Metal parts
Metals
Mirrors
Oil-filled and other types of distribution transformers
Parking garages (open or enclosed)
Porcelain and pottery
Stoves
Talc and soapstones
Washers and dryers

[B] Miscellaneous Group U. Buildings and structures of an accessory character and miscellaneous structures not classified in any specific occupancy shall be constructed, equipped and maintained to conform to the requirements of this code commensurate with the fire and life hazard incidental to their occupancy. Group U shall include, but not be limited to, the following:

Agricultural buildings
Aircraft hangar, accessory to a one- or two-family residence (see Section 412.3 of the *International Building Code*
Barns
Carports
Fences more than 6 feet (1829 mm) high
Grain silos, accessory to a residential occupancy
Greenhouses
Livestock shelters
Private garages
Retaining walls
Sheds
Stables
Tanks
Towers

[B] OCCUPANT LOAD. See Section 1002.1.

OPEN BURNING. See Section 302.1.

OPEN SYSTEM. The use of a solid or liquid hazardous material involving a vessel or system that is continuously open to the atmosphere during normal operations and where vapors are liberated, or the product is exposed to the atmosphere during normal operations. Examples of open systems for solids and liquids include dispensing from or into open beakers or containers, dip tank and plating tank operations.

❖ This definition is related primarily to hazardous materials use. Generally, an open system is one that will normally be open to the atmosphere; for example, a dip tank or dispensing or mixing of hazardous materials. Open systems are inherently more hazardous than closed systems. When evaluating the maximum allowable quantities of hazardous materials and the associated requirements open systems are more heavily regulated.

OPERATING BUILDING. See Section 3302.1.

OPERATING PRESSURE. The pressure at which a system operates.

❖ This definition clarifies that when the term "operating pressure" is used within the text of the code, it refers to the specific pressure at which the process or system is designed to operate. This definition is not intended to place a limitation on pressure.

ORGANIC COATING. See Section 2002.1.

ORGANIC PEROXIDE. See Section 3902.1.

Class I. See Section 3902.1.
Class II. See Section 3902.1.
Class III. See Section 3902.1.
Class IV. See Section 3902.1.
Class V. See Section 3902.1.
Unclassified detonable. See Section 3902.1.

OUTDOOR CONTROL AREA. See Section 2702.1.

OVERCROWDING. A condition that exists when either there are more people in a building, structure or portion thereof than have been authorized or posted by the fire code official, or when the fire code official determines that a threat exists to the safety of the occupants due to persons sitting and/or standing in locations that may obstruct or impede the use of aisles, passages, corridors, stairways, exits or other components of the means of egress.

❖ This definition notes that an unsafe condition exists when the actual number of people present in a building or a building space exceeds the maximum allowable occupant load of that building or space as determined and posted on the premises by the fire code official. Section 1004 of the code would allow a maximum occupant load of one person per every 5 square feet (.5 m^2) of building area, as long as the egress components provide sufficient capacity for such a load. When that egress capacity is exceeded, then overcrowding exists. The definition also recognizes that, even though the number of occupants in a building or space may not be excessive, the inability of occupants to use the egress elements due to blockage by patrons is also a life safety hazard.

OWNER. A corporation, firm, partnership, association, organization and any other group acting as a unit, or a person who has legal title to any structure or premises with or without accompanying actual possession thereof, and shall include the duly authorized agent or attorney, a purchaser, devisee, fiduciary and any person having a vested or contingent interest in the premises in question.

❖ This term defines who would be potentially considered as an owner. It is not a single individual in many cases. This clarifies who the responsible parties are for code compliance.

OXIDIZER. See Section 4002.1.

Class 4. See Section 4002.1.
Class 3. See Section 4002.1.
Class 2. See Section 4002.1.
Class 1. See Section 4002.1.

OXIDIZING GAS. See Section 4002.1.

OZONE-GAS GENERATOR. See Section 3702.1.

[B] PANIC HARDWARE. See Section 1002.1.

PASS-THROUGH. See Section 1802.1.

PERMISSIBLE EXPOSURE LIMIT (PEL). See Section 2702.1.

PESTICIDE. See Section 2702.1.

PHYSICAL HAZARD. See Section 2702.1.

PLOSOPHORIC MATERIAL. See Section 3302.1.

PLYWOOD and VENEER MILLS. See Section 1902.1.

POWERED INDUSTRIAL TRUCK. See Section 302.1.

PRESSURE VESSEL. See Section 2702.1.

PRIMARY CONTAINMENT. The first level of containment, consisting of the inside portion of that container which comes into immediate contact on its inner surface with the material being contained.

❖ In most cases, this definition pertains to those components of tanks, portable tanks and containers that are the main mechanism for the containment of liquid (the basic walls of the tank, portable tank or container). The term "secondary containment" refers to the containment provided when the primary containment fails.

PROCESS TRANSFER. See Section 3402.1.

PROPELLANT. See Section 2802.1.

PROXIMATE AUDIENCE. See Section 3302.1.

[B] PUBLIC WAY. See Section 1002.1.

PYROPHORIC. See Section 4102.1.

PYROTECHNIC COMPOSITION. See Section 3302.1.

PYROTECHNIC SPECIAL EFFECT. See Section 3302.1.

PYROTECHNIC SPECIAL-EFFECT MATERIAL. See Section 3302.1.

RAILWAY. See Section 3302.1.

[B] RAMP. See Section 1002.1.

RAW PRODUCT. See Section 1902.1.

READY BOX. See Section 3302.1.

RECORD DRAWINGS. See Section 902.1.

RECREATIONAL FIRE. See Section 302.1.

REDUCED FLOW VALVE. See Section 3702.1.

REFINERY. See Section 3402.1.

REFRIGERANT. See Section 602.1.

REFRIGERATION SYSTEM. See Section 602.1.

[B] REGISTERED DESIGN PROFESSIONAL. An architect or engineer, registered or licensed to practice professional architecture or engineering, as defined by the statutory requirements of the professional registration laws of the state in which the project is to be constructed.

❖ This term is used throughout the code where a special level of expertise and knowledge is required. The definition clearly notes that each state defines its own professional registration laws.

REMOTE EMERGENCY SHUTOFF DEVICE. See Section 3402.1.

REMOTE SOLVENT RESERVOIR. See Section 3402.1.

REPAIR GARAGE. See Section 2202.1.

RESIN APPLICATION AREA. See Section 1502.1.

RESPONSIBLE PERSON. See Section 2602.1.

RETAIL DISPLAY AREA. See Section 2802.1.

ROLL COATING. See Section 1502.1.

RUBBISH (TRASH). Combustible and noncombustible waste materials, including residue from the burning of coal, wood, coke or other combustible material, paper, rags, cartons, tin cans, metals, mineral matter, glass crockery, dust and discarded refrigerators, and heating, cooking or incinerator-type appliances.

❖ The term "rubbish" is normally associated with combustible waste. In this code the term is much broader and would include noncombustible waste, such as metals. Generally, the scope of this definition includes anything that has been discarded

SAFETY CAN. See Section 2702.1.

[B] SCISSOR STAIR. See Section 1002.1.

SECONDARY CONTAINMENT. See Section 2702.1.

SEGREGATED. See Section 2702.1.

SELF-SERVICE MOTOR FUEL-DISPENSING FACILITY. See Section 2202.1.

SEMICONDUCTOR FABRICATION FACILITY. See Section 1802.1.

SERVICE CORRIDOR. See Section 1802.1.

SHELF STORAGE. See Section 2302.1.

SINGLE-STATION SMOKE ALARM. See Section 902.1.

[B] SLEEPING UNIT. See Section 902.1.

SMALL ARMS AMMUNITION. See Section 3302.1.

SMALL ARMS PRIMERS. See Section 3302.1.

SMOKE ALARM. See Section 902.1.

SMOKE DETECTOR. See Section 902.1.

[B] SMOKE-PROTECTED ASSEMBLY SEATING. See Section 1002.1.

SMOKELESS PROPELLANTS. See Section 3302.1.

SOLID. See Section 2702.1.

SOLID SHELVING. See Section 2302.1.

SOLVENT DISTILLATION UNIT. See Section 3402.1.

SOLVENT OR LIQUID CLASSIFICATIONS. See Section 1202.1.
Class I solvents. See Section 1202.1.
Class II solvents. See Section 1202.1.
Class IIIA solvents. See Section 1202.1.
Class IIIB solvents. See Section 1202.1.
Class IV solvents. See Section 1202.1.

SPECIAL AMUSEMENT BUILDING. A building that is temporary, permanent or mobile that contains a device or system that conveys passengers or provides a walkway along, around or over a course in any direction as a form of amusement arranged so that the egress path is not readily apparent due to visual or audio distractions or an intentionally confounded egress path, or is not readily available because of the mode of conveyance through the building or structure.

❖ This definition clarifies which buildings are classified as special amusement buildings. More specifically, it is intended to apply to buildings where the occupants are moved through the building by way of a people mover system or are specifically directed through a walkway. Because of the nature of their use, these buildings contain special effects and other features that make it more difficult for occupants to determine when an emergency exists and where exits are located. Theme parks and traveling carnivals will usually have such buildings. Most of the seasonal, so-called "haunted house" attractions easily fall within the scope of this definition.

SPECIAL INDUSTRIAL EXPLOSIVE DEVICE. See Section 3302.1.

SPRAY AREA. See Section 1502.1.

SPRAY BOOTH. See Section 1502.1.

SPRAY ROOM. See Section 1502.1.

[B] STAIR. See Section 1002.1.

[B] STAIRWAY. See Section 1002.1.

[B] STAIRWAY, EXTERIOR. See Section 1002.1.

[B] STAIRWAY, INTERIOR. See Section 1002.1.

[B] STAIRWAY, SPIRAL. See Section 1002.1.

STANDPIPE SYSTEM, CLASSES OF. See Section 902.1.

Class I system. See Section 902.1.
Class II system. See Section 902.1.
Class III system. See Section 902.1.

STANDPIPE, TYPES OF. See Section 902.1.

Automatic dry. See Section 902.1.
Automatic wet. See Section 902.1.
Manual dry. See Section 902.1.
Manual wet. See Section 902.1.
Semiautomatic dry. See Section 902.1.

STATIC PILES. See Section 1902.1.

STEEL. Hot- or cold-rolled as defined by the *International Building Code.*

❖ This is a basic definition that clarifies that steel is either cold- or hot-rolled as defined by the IBC; however, the IBC does not specifically define cold- or hot-rolled steel but describes how it must be used.

STORAGE, HAZARDOUS MATERIALS. See Section 2702.1.

SUPERVISING STATION. See Section 902.1.

SUPERVISORY SERVICE. See Section 902.1.

SUPERVISORY SIGNAL. See Section 902.1.

SUPERVISORY SIGNAL-INITIATING DEVICE. See Section 902.1.

SYSTEM. See Section 2702.1.

TANK. A vessel containing more than 60 gallons (227 L).

❖ This definition establishes the distinction between containers and tanks for purposes of code application, with a container being defined as a vessel of 60 gallons (227 L) or less capacity (see commentary, Section 2702, definition of "Container").

TANK, ATMOSPHERIC. See Section 2702.1.

TANK, PORTABLE. See Section 2702.1.

TANK, PRIMARY. See Section 3402.1.

TANK, PROTECTED ABOVE GROUND. See Section 3402.1.

TANK, STATIONARY. See Section 2702.1.

TANK VEHICLE. See Section 2702.1.

TENT. See Section 2402.1.

THEFT RESISTANT. See Section 3302.1.

THERMAL INSECTICIDAL FOGGING. See Section 1702.1.

TIMBER and LUMBER PRODUCTION FACILITIES. See Section 1902.1.

TIRES, BULK STORAGE OF. See Section 902.1.

TOOL. See Section 1802.1.

TORCH-APPLIED ROOF SYSTEM. See Section 2602.1.

TOXIC. See Section 3702.1.

TRANSVERSE FLUE SPACE. See Section 2302.1.

TRASH. See 'Rubbish."

TROUBLE SIGNAL. See Section 902.1.

UNAUTHORIZED DISCHARGE. See Section 2702.1.

UNSTABLE (REACTIVE) MATERIAL. See Section 4302.1.

Class 4. See Section 4302.1.
Class 3. See Section 4302.1.
Class 2. See Section 4302.1.
Class 1. See Section 4302.1.

UNWANTED FIRE. A fire not used for cooking, heating or recreational purposes or one not incidental to the normal operations of the property.

❖ For the purposes of applying the code, a clarification is provided to note that certain fires present in buildings would be acceptable; for example, the normal operation of a hot water heater, a gas stove or a fireplace. The definition does not address whether a fire is intentional or unintentional (arson versus a welding accident, for example).

USE (MATERIAL). See Section 2702.1.

VALVE-REGULATED LEAD-ACID (VRLA) BATTERY. See Section 602.1.

VAPOR AREA. See Section 1502.1.

VAPOR PRESSURE. See Section 2702.1.

VENTED (FLOODED) LEAD-ACID BATTERY. See Section 602.1.

VISIBLE ALARM NOTIFICATION APPLIANCE. See Section 902.1.

WATER-REACTIVE MATERIAL. See Section 4402.1.

Class 3. See Section 4402.1.
Class 2. See Section 4402.1.
Class 1. See Section 4402.1.

WET-CHEMICAL EXTINGUISHING AGENT. See Section 902.1.

[B] WINDER. See Section 1002.1.

WIRELESS PROTECTION SYSTEM. See Section 902.1.

WORKSTATION. See Section 1802.1.

ZONE. See Section 902.1.

Bibliography

ASTM E 84-01, *Test Method for Surface Burning Characteristics of Building Materials*. West Conshohocken, PA: ASTM International, 2001.

ASTM E 136-99, *Test Method for Behavior or Materials in a Vertical Tube Furnace at 750° C*. West Conshohocken, PA: ASTM International, 1999.

Burklin, R.W., and R.G. Purington. *Fire Terms: A Guide to Their Meaning and Use*. Quincy, MA: National Fire Protection Association, 1980.

DOL 29 CFR; Part 1910.1200, *Hazard Communication, Occupational Safety and Health Standards*. Washington, DC: U.S. Department of Labor, 1999.

Kimball, Warren Y. *Fire Department Terminology*, 4th edition. Quincy, MA: National Fire Protection Association, 1970.

Kite, Susan L. "Playing It Safe with Haunted Houses." *The Code Official*, vol. 33, no. 5, (September/October 1999): 34-41. Country Club Hills, IL: Building Officials and Code Administrators, International, 1999.

Kuvshinoff, B.W., ed. *Fire Science Dictionary*. New York: John Wiley & Sons. 1977.

National Building Code. New York: National Board of Fire Underwriters. 1955.

Nolan, J.R. and Associates. *Blacks Law Dictionary*, 6th edition. St Paul, MN: West Publishing Co., 1990.

Chapter 3: General Precautions Against Fire

General Comments

Fire is always a concern, whether a building is under construction, is occupied for normal use or is undergoing renovation, restoration, expansion or demolition. But careful planning combined with common sense can make buildings and premises much safer, regardless of the occupancy or other activities at the site.

The primary focus of the requirements in this chapter is making sure the three elements necessary for a fire–ignition source, fuel and oxygen–do not come in contact with one another. NFPA 550 describes in great detail the features of fire safety systems and includes a logic tree called "The Fire Safety Concepts Tree" to graphically show all the possible means of achieving user-defined fire safety objectives. A portion of that tree is reproduced here as Figure 3 to show how to avoid fire ignition. Activities on this diagram that follow a plus sign (+) gate may be undertaken independently of each other to arrive at the desired goal. Alternatives following a dot (•) gate must be combined to achieve the desired result.

Figure 3 shows that eliminating any one of the three elements required for a fire to occur will prevent one from happening. If there is no ignition source, a fuel load of any size should not catch fire. If there is no fuel load, there is nothing for an ignition source to ignite. Lastly, if there is little or no air available to sustain combustion, any fire ignited in a fuel load will quickly die.

The requirements and precautions outlined in this chapter, when applied using good judgement and the common sense mentioned above, will help to insure safety for everyone.

Purpose

The requirements and precautions contained in this chapter are intended to make a premises safe for everyone, from construction workers to tenants, to operations and maintenance personnel.

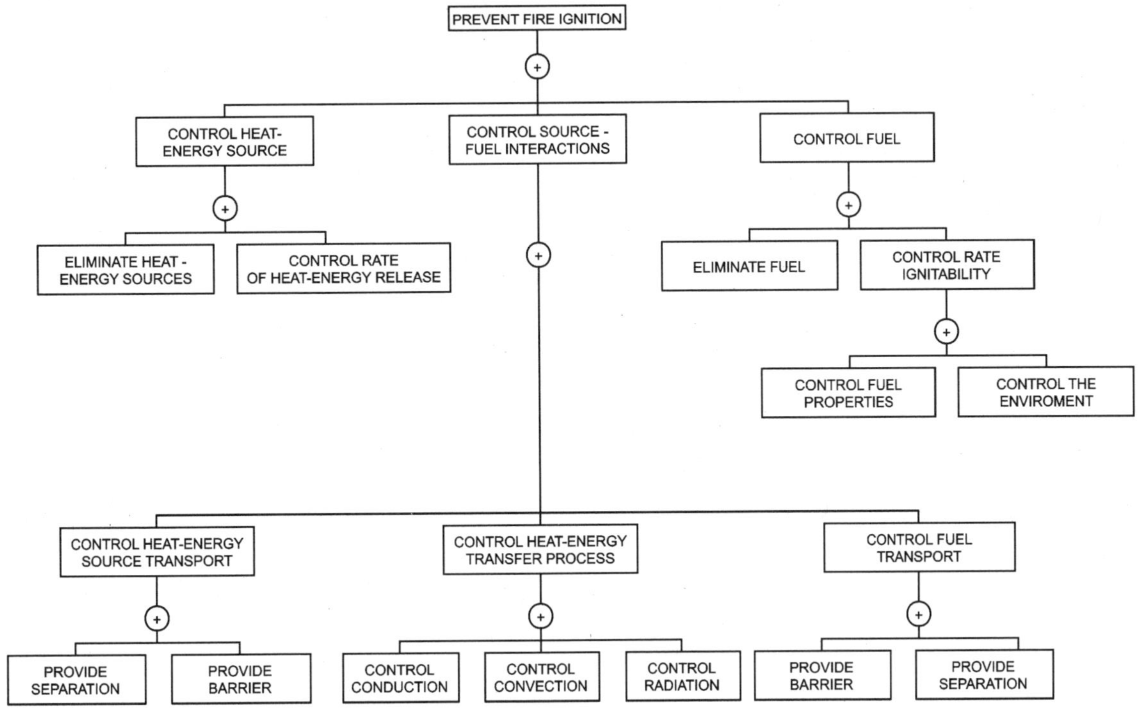

Figure 3
THE FIRE SAFETY CONCEPTS TREE

SECTION 301 GENERAL

301.1 Scope. The provisions of this chapter shall govern the occupancy and maintenance of all structures and premises for precautions against fire and the spread of fire.

❖ The requirements of Chapter 3 prescribe fire safety precautions for conditions that are likely to cause or contribute to the spread of fire in any building or structure or on any premises, regardless of occupancy.

301.2 Permits. Permits shall be required as set forth in Section 105.6 for the activities or uses regulated by Sections 306, 307, 308.3, 308.4, 308.5 and 315.

❖ Issuing permits gives the fire code official an opportunity to carefully evaluate and regulate hazardous operations. Applicants for permits should be required to demonstrate that their operations comply with the intent of the code before the permit is issued. See the commentary to Section 105.6 for a general discussion of operations requiring an operational permit.

SECTION 302 DEFINITIONS

302.1 Definitions. The following words and terms shall, for the purposes of this chapter and as used elsewhere in this code, have the meanings shown herein.

❖ The following words and terms shall, for the purposes of this chapter and as used elsewhere in this code, have the meanings shown herein.

Definitions of terms can help in the understanding and application of the code requirements. The purpose for including those definitions that are associated with the subject matter of this chapter is to provide more convenient access to them without having to refer back to Chapter 2. It is important to emphasize that these terms are not exclusively related to this chapter but are applicable everywhere the term is used in the code. For convenience, these terms are also listed in Chapter 2 with a cross reference to this section. The use and application of all defined terms, including those defined in this section, are set forth in Section 201.

BONFIRE. An outdoor fire utilized for ceremonial purposes.

❖ Bonfires are usually very large and are associated with a crowd activity. Failure to follow good safety practices with these fires can lead to serious injuries and property damage.

HI-BOY. A cart used to transport hot roofing materials on a roof.

❖ A hi-boy, also known as a hot carrier, is a wheeled tank used on the roof deck to move hot asphalt around the work area. Hi-boys are available in either insulated or noninsulated models, and typically hold either 30 or 55 gallons (114 or 208 L).

OPEN BURNING. The burning of materials wherein products of combustion are emitted directly into the ambient air without passing through a stack or chimney from an enclosed chamber. Open burning does not include road flares, smudgepots and similar devices associated with safety or occupational uses typically considered open flames or recreational fires. For the purpose of this definition, a chamber shall be regarded as enclosed when, during the time combustion occurs, only apertures, ducts, stacks, flues or chimneys necessary to provide combustion air and permit the escape of exhaust gas are open.

❖ Open burning is any burning that takes place in an unenclosed space. Examples include burning of leaves or grass clippings, burning construction debris and fires built on the ground for warmth in cold weather.

The burning of wood scraps in a steel drum or in a piece of culvert over which a supply of construction sand can be dumped and kept thawed are common practices on construction sites in cold climates and this could be evaluated by the fire code official as being an allowable "occupational use" as mentioned in the definition.

A recent innovation that is sometimes incorrectly treated as open burning is the patio fireplace. These devices function the same as a masonry or factory-built indoor fireplace except that they are portable, outdoor, solid-fuel-burning patio fireplaces designed to provide ambience and warmth in outdoor settings. They come in many styles and are generally substantially constructed of steel with heavy-duty screening around the firebox, although some types are made of concrete or clay with a small hearth opening and are equipped with a short chimney or a chimney opening in the top. The design also includes a stand to elevate the firebox above the surface upon which it is placed to provide clearance to combustible materials. These devices neither meet the literal definition of "Open burning" nor is their use the type of burning intended to be regulated by Section 307, examples of which could include disposal of brush, construction rubbish or household waste by burning in the open. Stoves, incinerators or other controlled burning devices, which would include patio fireplaces, do not fall under the open-burning regulations.

POWERED INDUSTRIAL TRUCK. A forklift, tractor, platform lift truck or motorized hand truck powered by an electrical motor or internal combustion engine. Powered industrial trucks do not include farm vehicles or automotive vehicles for highway use.

❖ This kind of vehicle includes forklift trucks and other similar vehicles used to move stock in warehouses, industrial buildings, large retail spaces, storage yards and loading docks. These vehicles are not licensed for highway travel and do not include farm machinery.

RECREATIONAL FIRE. An outdoor fire burning materials other than rubbish where the fuel being burned is not contained in an incinerator, outdoor fireplace, barbeque grill or barbeque pit and has a total fuel area of 3 feet (914 mm) or less in diameter and 2 feet (610 mm) or less in height for pleasure, religious, ceremonial, cooking, warmth or similar purposes.

❖ This kind of fire includes ordinary campfires and other small fires used for the activities listed.

SECTION 303
ASPHALT KETTLES

303.1 Transporting. Asphalt (tar) kettles shall not be transported over any highway, road or street when the heat source for the kettle is operating.

Exception: Asphalt (tar) kettles in the process of patching road surfaces.

❖ The hazards of hauling a fired kettle of molten asphalt over public ways are obvious. Most asphalt kettles for roofing, paving and similar uses are currently liquefied petroleum gas (LP-gas) fired. Contractors often wish to keep asphalt in a liquid state to save time between jobs and when work is interrupted. Once asphalt is transformed from a solid to a liquid by heating, it retains much of its heat for some time, and although it becomes increasingly viscous as it cools, it remains fluid for a considerable time. Maintaining a fire under a kettle during transport is usually unnecessary and, therefore, prohibited, since little additional heat is required to return the asphalt to a usable consistency. An accident, flat tire or anything else that could cause the kettle to overturn, spilling the molten asphalt in the presence of an open flame, could lead to a serious fire. Even hitting potholes or other bumps in the road could cause the molten asphalt to splash out of the kettle, causing injury to people nearby or damage to property.

The exception for asphalt being used for road repair is necessary for work crews sealing pavement joints and performing similar roadway repairs for efficient operations.

303.2 Location. Asphalt (tar) kettles shall not be located within 20 feet (6096 mm) of any combustible material, combustible building surface or any building opening and within a controlled area identified by the use of traffic cones, barriers or other approved means. Asphalt (tar) kettles and pots shall not be utilized inside or on the roof of a building or structure. Roofing kettles and operating asphalt (tar) kettles shall not block means of egress, gates, roadways or entrances.

❖ Asphalt kettles sometimes catch fire. Having one located inside a building would present a serious smoke problem as well as the fire hazards of asphalt spills flowing to lower floors or the release of LP-gas inside the building. Having one located next to quantities of combustible materials would also represent a fire hazard as well as the possibility that splashes and splatters could damage construction materials beyond use. Keeping egress pathways and other travel lanes free of obstructions provides a needed immediate exit from an area where an asphalt kettle-related incident might occur and enhances access to such areas for the fire department.

303.3 Location of fuel containers. Fuel containers shall be located at least 10 feet (3048 mm) from the burner.

Exception: Containers properly insulated from heat or flame are allowed to be within 2 feet (610 mm) of the burner.

❖ This section reduces the likelihood that any gas or vapors that might escape from the fuel containers would be ignited by the open flame of the kettle burner and that the heat of the burner would cause overheating of the fuel containers.

The exception acknowledges the greater safety of the insulated containers.

303.4 Attendant. An operating kettle shall be attended by a minimum of one employee knowledgeable of the operations and hazards. The employee shall be within 100 feet (30 480 mm) of the kettle and have the kettle within sight. Ladders or similar obstacles shall not form a part of the route between the attendant and the kettle.

❖ Having a trained attendant watch the kettle helps to create a safe operation. The attendant is usually responsible for making sure the asphalt is at the proper temperature, that the level of liquid in the kettle is maintained at the required level and assuring that the fuel supply for the kettle burner is adequate. The attendant should watch for any change in the kettle that would signal the potential for a safety hazard, and to adjust the burner output or other factors to keep the kettle in safe operating condition. The attendant is also often responsible for keeping the area surrounding the kettle free of combustible materials and other construction debris that could become a safety hazard.

303.5 Fire extinguishers. There shall be a portable fire extinguisher complying with Section 906 and with a minimum 40-B:C rating within 25 feet (7620 mm) of each asphalt (tar) kettle during the period such kettle is being utilized, and one additional portable fire extinguisher with a minimum 40-B:C rating on the roof being covered.

❖ Having fire extinguishers at hand is just good safety practice. This section defines the type and size of extinguisher that must be available for use, both on the ground near the kettle and on the roof level to which the asphalt is being applied. In the event of a kettle fire, water should not be used as an extinguishing agent because it could cause the molten asphalt to froth and possibly overflow the kettle or spatter over the surrounding area and anyone in it.

303.6 Lids. Asphalt (tar) kettles shall be equipped with tight-fitting lids.

❖ A tight-fitting lid on a hot kettle keeps the air supply available to feed a kettle fire to a minimum. Any fire that might start in a closed kettle will quickly burn itself out because of the limited amount of air available for combustion. The lid also helps prevent splashes and splatters that could cause personal injury.

303.7 Hi-boys. Hi-boys shall be constructed of noncombustible materials. Hi-boys shall be limited to a capacity of 55 gallons (208 L). Fuel sources or heating elements shall not be allowed as part of a hi-boy.

❖ Hi-boys are used on the roof of a building to transport hot asphalt from a point of supply near the edge of the roof to the site of the roofing application. Due to the hazards of molten asphalt discussed in Section 303.1, hi-boys are limited in size to control the maximum amount of potential spills on the roof, which could ignite and pose a high-challenge, fire-suppression operation for the fire department. A limited size also enhances the movability and stability of the hi-boy, thus reducing the potential for a tip over. As a further safeguard against a fire incident. hi-boys are prohibited from being fired or equipped with a fuel source for firing. Hi-boys must also be constructed of noncombustible materials to enhance their durability and prevent the container from contributing fuel to a fire. Hi-boys should be well-maintained, including the frame; steering mechanism; tires or wheels; faucets and fill connections (see commentary, Section 302.1 for the definition of "Hi-boy").

303.8 Roofing kettles. Roofing kettles shall be constructed of noncombustible materials.

❖ The requirement for noncombustible materials represents sound safety practice as well as good business practice. Portions of kettles constructed of combustible materials can be easily destroyed and could lead to larger fires. Replacement of destroyed kettles would be expensive. Paying for other fire damage would be even more costly.

Also note that roofing mops soaked in asphalt or pitch must never be left inside a building, near heating equipment or near combustible materials. These mops are subject to spontaneous heating no matter what material they are made of.

303.9 Fuel containers under air pressure. Fuel containers that operate under air pressure shall not exceed 20 gallons (76 L) in capacity and shall be approved.

❖ Limiting the size of pressurized fuel containers limits the probability of a container becoming a major fuel source in case of a kettle fire. Requiring the use of approved containers gives the fire code official more control over the type and suitability of the vessel to be used under pressure.

SECTION 304
COMBUSTIBLE WASTE MATERIAL

304.1 Waste accumulation prohibited. Combustible waste material creating a fire hazard shall not be allowed to accumulate in buildings or structures or upon premises.

❖ Accumulated waste, trash, construction debris and other natural materials, such as grass clippings, leaves and shrubbery cuttings, can become a serious fire hazard. The three subsections that follow this general statement address the most common situations.

304.1.1 Waste material. Accumulations of wastepaper, wood, hay, straw, weeds, litter or combustible or flammable waste or rubbish of any type shall not be permitted to remain on a roof or in any court, yard, vacant lot, alley, parking lot, open space, or beneath a grandstand, bleacher, pier, wharf, manufactured home, recreational vehicle or other similar structure.

❖ This section considers the kind of waste material that is most likely to accumulate during construction, renovation, additions or demolition and is often referred to as "the housekeeping section." It prohibits disorderly, unkempt storage or accumulation of trash; waste rags; wastepaper; scrub brush and weeds; litter and other combustible materials. Litter and trash represent a serious fire hazard because of their ease of ignition and rapid heat release once ignited. The importance of maintaining property and buildings in good order seems obvious, but sloppy housekeeping still occurs and can be the cause of serious fires. In one of the most serious fires in recent years (February 1991), improper storage of linseed-oil-soaked rags used to refinish paneling in a high-rise office building caused a fire that destroyed eight floors of the building and killed three fire fighters.

304.1.2 Vegetation. Weeds, grass, vines or other growth that is capable of being ignited and endangering property, shall be cut down and removed by the owner or occupant of the premises. Vegetation clearance requirements in urban-wildland interface areas shall be in accordance with the *International Urban/Wildland Interface Code.*

❖ Accumulations of natural waste such as grass clippings, weed growth and shrubbery cuttings are not only unsightly, but also represent a serious fire hazard. All too often these accumulations occur at or near fence lines that are adjacent to streets or alleys. This makes accidental ignition by a cigarette butt tossed from a passing vehicle a good possibility. Common sense tells us that removal of this kind of waste is beneficial. The rules of nearly all jurisdictions make waste control and removal the responsibility of the building or property owner, his or her agent, the tenant or the contractor if work is being done on the site. Uncontrolled vegetation growth poses substantial risk to areas designated as urban-wildland interface areas. Accordingly, such areas must comply with the provisions of the referenced code, the *International Urban-Wildland Interface Code*™ (IUWIC™).

304.1.3 Space underneath seats. Spaces underneath grandstand and bleacher seats shall be kept free from combustible and flammable materials. Except where enclosed in not less than 1-hour fire-resistance-rated construction in accordance with the *International Building Code*, spaces underneath grandstand and bleacher seats shall not be occupied or utilized for purposes other than means of egress.

❖ Numerous fires in grandstands and stadiums have shown over the years that the accumulation of flammable or combustible materials under grandstand seating areas can lead to fire disasters. Except as noted in the *International Building Code®* (IBC®), areas under grandstand seating must be kept free of flammable materials, including accumulations of waste or trash. One of the best ways to prevent a fire is to make certain there is no fuel to feed one.

The IBC does allow space under the stands to be used for purposes other than means of egress if that space is separated from the seating area by construction having at least a 1-hour fire-resistance rating. The separation is intended to allow time for occupants in the seating to clear it should a fire occur. The fire code official would usually have to approve plans for use of space under the stands for concession stands, sales areas or storage areas.

304.2 Storage. Storage of combustible rubbish shall not produce conditions that will create a nuisance or a hazard to the public health, safety or welfare.

❖ Storage of combustible rubbish either indoors or outdoors must be approved by the fire code official. Best practice requires combustibles to be accumulated in fire-resistant containers, such as trash cans with tight lids, steel barrels or dumpster bins, that would be removed from the site regularly. This section mentions public health as well as safety and welfare, indicating concern over retention of decomposing organic waste as well as flammable and combustible materials.

304.3 Containers. Combustible rubbish, and waste material kept within a structure shall be stored in accordance with Sections 304.3.1 through 304.3.3.

❖ Proper containers must be used to improve the safety of indoor storage of trash and isolate readily combustible materials. This section introduces the more detailed requirements in Sections 304.3.1 through 304.3.3.

304.3.1 Spontaneous ignition. Materials susceptible to spontaneous ignition, such as oily rags, shall be stored in a listed disposal container. Contents of such containers shall be removed and disposed of daily.

❖ Disposal containers, often called "waste cans" or "oily rag cans," used for storage of materials that might autoignite as a result of the spontaneous combustion process must be tested and listed for that use by a recognized testing laboratory or agency and must bear a label showing that they have been tested, along with the name of the testing agency. Such containers are most commonly round and generally available in sizes ranging from 5 to 40 gallons (19 to 152 L). They are equipped with a manual or foot treadle-operated lid that opens to a maximum angle of 60 degrees (1 rad) and closes by gravity. These containers are designed to prevent continuing combustion of the contents if ignition occurs. Container design includes features that keep the can body containing waste from coming into contact with combustible surfaces of walls or floors (see commentary, Section 202, for the definition of "Listed"). Daily disposal of container contents reduces the amount of time that oily materials will lie dormant, generating internal heat that can lead to ignition. UL 32 provides further information on the construction, testing and listing of these containers.

304.3.2 Capacity exceeding 5.33 cubic feet. Containers with a capacity exceeding 5.33 cubic feet (40 gallons) (0.15 m^3) shall be provided with lids. Containers and lids shall be constructed of noncombustible materials or approved combustible materials.

❖ Requiring larger containers to meet stricter conditions is common sense. The larger volume of waste each can holds represents a larger fire hazard. Isolating the containers from one another with lids helps insure that a fire in one container will not spread to nearby containers. The lid also helps to smother a fire within the container by limiting the oxygen available to feed it. Additionally, closed containers protect flammable and combustible materials from potential ignition sources.

304.3.3 Capacity exceeding 1.5 cubic yards. Dumpsters and containers with an individual capacity of 1.5 cubic yards (40.5 cubic feet) (1.15 m^3) or more shall not be stored in buildings or placed within 5 feet (1524 mm) of combustible walls, openings or combustible roof eave lines.

Exceptions:

1. Dumpsters or containers in areas protected by an approved automatic sprinkler system complying with Chapter 9.
2. Storage in a structure shall not be prohibited where the structure is of Type I or Type IIA construction, located not less than 10 feet (3048 mm) from other buildings and used exclusively for dumpster or container storage.

❖ Although waste containers of this size are nearly always constructed of welded steel because of the weight of the waste load, the very fact that the waste load is large makes the containers a large fire hazard. Keeping these large containers in the open and away from combustible construction is the obvious way to keep the fire hazard low.

Exception 1 permits storage of these large containers indoors if the area is protected by an approved sprinkler system. It would be up to the fire code official to determine the maximum quantities that could be stored under these conditions.

Exception 2 applies only to buildings that are of

fire-resistance-rated construction and are used exclusively for container storage. Such facilities might be found in scrap yards or at recycling centers, but rarely, if ever, in other occupancies.

SECTION 305
IGNITION SOURCES

305.1 Clearance from ignition sources. Clearance between ignition sources, such as light fixtures, heaters and flame-producing devices, and combustible materials shall be maintained in an approved manner.

❖ Establishing safe clearances will usually mean following the requirements of the IBC or other codes adopted by the jurisdiction as well as having the approval of the fire code official.

305.2 Hot ashes and spontaneous ignition sources. Hot ashes, cinders, smoldering coals or greasy or oily materials subject to spontaneous ignition shall not be deposited in a combustible receptacle, within 10 feet (3048 mm) of other combustible material including combustible walls and partitions or within 2 feet (610 mm) of openings to buildings.

Exception: The minimum required separation distance to other combustible materials shall be 2 feet (610 mm) where the material is deposited in a covered, noncombustible receptacle placed on a noncombustible floor, ground surface or stand.

❖ This section covers two different, but equally serious, ignition source problems. First, hot ashes, embers and cinders from fireplaces, stoves or other fireboxes must never be placed in a combustible container. This point seems almost too obvious to be mentioned, but every year fires are started when someone carelessly scoops ashes containing glowing embers into paper bags or cardboard cartons. It is also not uncommon to see construction scrap being burned in steel drums on construction sites in cold weather. Care must be taken when emptying ashes from those containers to make sure no hot coals get dumped on paper waste or other combustible materials.

The second problem, greasy or oily materials subject to spontaneous combustion, is addressed by requiring them to be placed in listed containers (see commentary, Section 304.3.1).

In both cases, safe distances must be maintained from combustible construction and building openings for added protection.

The exception recognizes the added protection of tight-fitting covers on noncombustible trash containers as well as the reduced fire hazard when the containers are placed on a noncombustible surface.

305.3 Open-flame warning devices. Open-flame warning devices shall not be used along an excavation, road, or any place where the dislodgment of such device might permit the device to roll, fall or slide on to any area or land containing combustible material.

❖ Open-flame warning devices other than fusees used to mark road accidents or other short-term emergencies are rarely used today. The old-fashioned kerosene pots used to mark construction hazards in dark areas have been largely replaced by "sawhorse" barriers with battery-powered flashing lights. But, even though use may be limited, the warning in this section is nonetheless real. One fusee not firmly fixed in the ground or on another stable surface can fall into a roadside ditch filled with dry weeds and cause a roadside fire that could spread into dry woodland or cropland, causing enormous fire damage. Likewise, a burned-out hand-held fusee that is carelessly tossed aside while still hot could ignite dry refuse.

305.4 Deliberate or negligent burning. It shall be unlawful to deliberately or through negligence set fire to or cause the burning of combustible material in such a manner as to endanger the safety of persons or property.

❖ The deliberate setting of fires, whether in a structure or in a waste container that is located where it could endanger a structure or its occupants, is normally considered arson, which is a felony that is punishable by a lengthy prison sentenance. Fortunately, arson is not that common. More likely a fire would be caused by carelessness or by someone not considering the consequences.

Regardless of the circumstances, fires must be avoided. Following the requirements in the code as well as those in the IBC, will help to maintain a safe, fire-free site. On construction or demolition sites, secure fencing around the site and its waste containers is good protection. Following good housekeeping practices, including routine disposal of combustible materials, is also an excellent first line of protection against fire. An ignition source cannot cause damage to property or endanger life unless there is a fuel load to be ignited.

SECTION 306
MOTION PICTURE FILM AND SCREENS

306.1 Motion picture projection rooms. Electric arc, xenon or other light source projection equipment which develops hazardous gases, dust or radiation and the projection of ribbon-type cellulose nitrate film, regardless of the light source used in projection, shall be operated within a motion picture projection room complying with Section 409 of the *International Building Code*.

❖ The requirements in this section are specific to spaces housing equipment used to project cellulose acetate film, also called safety film, which is what is in common use today. This film has about the same fire hazard characteristics as paper of the same thickness and form. The equipment used to project the film, however, may also present fire or health hazards that can be minimized by proper construction of the room. Section 409 of the IBC covers these construction requirements in detail.

The older type of motion picture film was made of cellulose nitrate, which is also called pyroxylin, which presents a significantly greater fire hazard and, therefore, calls for stricter construction requirements, including sprinklers. These requirements are contained in Sections 4204 and 903.2.4.3 of the code. The greater hazard of cellulose nitrate film, which today is found mainly in museum collections and other archives or film preservation facilities, comes from the characteristic of the material to begin degrading at temperatures below its ignition temperature, causing a chemical reaction that can lead to spontaneous combustion. The combustion products of cellulose nitrate are both flammable and extremely toxic because they include oxides of nitrogen. Cellulose nitrate film burns at a rate that is as much as 15 times the rate of common combustibles.

306.2 Cellulose nitrate film storage. Storage of cellulose nitrate film shall be in accordance with NFPA 40.

❖ NFPA 40 contains minimum requirements for a reasonable level of protection for the storage and handling of cellulose nitrate film. The standard does not address the manufacture of the film because it has not been made in the United States since 1951.

SECTION 307
OPEN BURNING AND RECREATIONAL FIRES

307.1 General. A person shall not kindle or maintain or authorize to be kindled or maintained any open burning unless conducted and approved in accordance with this section.

❖ To control the hazards associated with it, open burning may not be authorized or undertaken without the approvals specified in Section 307. See the commentary to Section 302.1 for the definition of "Open burning" for a discussion of the types of burning intended to be regulated by this section.

307.2 Permit required. A permit shall be obtained from the fire code official in accordance with Section 105.6 prior to kindling a fire for recognized silvicultural or range or wildlife management practices, prevention or control of disease or pests, or a bonfire. Application for such approval shall only be presented by and permits issued to the owner of the land upon which the fire is to be kindled.

❖ This section defines a rather narrow range of purposes for which permits will be issued. Section 105.6.31 covers open burning permits in general. This section restricts permissible fires to those used for silviculture (the cultivation of forests and shade trees); range or wildlife management; pest control and bonfires as defined in the code. This section further restricts the permitting process to owners of the land on which the fire is to be kindled (see commentary, Section 301.2).

307.2.1 Authorization. Where required by state or local law or regulations, open burning shall only be permitted with prior approval from the state or local air and water quality management authority, provided that all conditions specified in the authorization are followed.

❖ This section requires permit applicants to comply with state and local regulations covering air and water quality as well as safety regulations established by the jurisdiction having authority.

307.2.2 Prohibited open burning. Open burning that will be offensive or objectionable because of smoke or odor emissions when atmospheric conditions or local circumstances make such fires hazardous shall be prohibited. The fire code official is authorized to order the extinguishment by the permit holder or the fire department of open burning which creates or adds to a hazardous or objectionable situation.

❖ This section is intended to protect the public from irresponsible burning and also establishes the authority of the agency granting burning permits to order fires extinguished when they endanger the safety, health or welfare of owners and occupants of property near the burn site or the public in general.

307.3 Location. The location for open burning shall not be less than 50 feet (15 240 mm) from any structure, and provisions shall be made to prevent the fire from spreading to within 50 feet (15 240 mm) of any structure.

Exceptions:

1. Fires in approved containers that are not less than 15 feet (4572 mm) from a structure.
2. The minimum required distance from a structure shall be 25 feet (7620 mm) where the pile size is 3 feet (914 mm) or less in diameter and 2 feet (610 mm) or less in height.

❖ The 50-foot (15 240 mm) restriction applies to large fires in large open areas, such as those defined in Section 307.2. Exception 1 refers to fires that generally would be considerably smaller or would be controlled by the container in which they burn, presenting a reduced exposure risk to nearby buildings. Exception 2 allows a reduction in clearance from buildings based on the lesser hazard of fires that are limited in size.

307.3.1 Bonfires. A bonfire shall not be conducted within 50 feet (15 240 mm) of a structure or combustible material unless the fire is contained in a barbecue pit. Conditions which could cause a fire to spread within 50 feet (15 240 mm) of a structure shall be eliminated prior to ignition.

❖ Bonfires usually are large and are associated with some kind of planned event (for example, school pep rally, holiday celebration or camp celebration). This section restricts the location of these large fires to open areas in which sparks and burning embers would be unlikely to endanger structures and smoke would not be a significant hazard to public health. Allowing a bonfire in a barbeque pit automatically restricts the size of the fire to the fuel load that can be contained within the noncombustible fire pit.

307.3.2 Recreational fires. Recreational fires shall not be conducted within 25 feet (7620 mm) of a structure or combustible material. Conditions which could cause a fire to spread within 25 feet (7620 mm) of a structure shall be eliminated prior to ignition.

❖ Recreational fires are usually fairly small, but can still represent a significant fire hazard if the area in which they are kindled is not kept free of combustible trash and debris. Basic fire safety practices followed by campers make good guidelines. No fire should ever be kindled in a location where it would endanger structures or would be likely to ignite combustible materials close by.

307.4 Attendance. Open burning, bonfires or recreational fires shall be constantly attended until the fire is extinguished. A minimum of one portable fire extinguisher complying with Section 906 with a minimum 4-A rating or other approved on-site fire-extinguishing equipment, such as dirt, sand, water barrel, garden hose or water truck, shall be available for immediate utilization.

❖ This section reiterates basic common sense, but tends to be ignored quite often. Having one or more individuals responsible for keeping watch on a fire, even one of small size, is the first line of fire prevention. All too often news articles tell of wooden decks burning because hot embers from a charcoal grill fell unobserved onto the unprotected wooden surface or of a huge brush or forest fire being caused by careless individuals that did not watch their campfires.

For practical purposes as well as for fire safety, some means of extinguishing a kindled fire should be kept close at hand. For small fires, a shovelful of dirt may be sufficient. For large fires, such as bonfires, large volumes of water may be necessary; however, no matter how much extinguishing equipment is available, it may prove useless unless someone is tending the fire and can sound an alarm.

SECTION 308
OPEN FLAMES

308.1 General. This section shall control open flames, fire and burning on all premises.

❖ This section establishes the scope of the requirements of Section 308 as being applicable to both indoor and outdoor situations involving open flames.

308.2 Where prohibited. A person shall not take or utilize an open flame or light in a structure, vessel, boat or other place where highly flammable, combustible or explosive material is utilized or stored. Lighting appliances shall be well-secured in a glass globe and wire mesh cage or a similar approved device.

❖ This section intends to maintain separation between ignitable materials and ignition sources that involve an open flame in any structure or occupancy.

308.2.1 Throwing or placing sources of ignition. No person shall throw or place, or cause to be thrown or placed, a lighted match, cigar, cigarette, matches, or other flaming or glowing substance or object on any surface or article where it can cause an unwanted fire.

❖ This section recognizes the hazard caused by carelessness in disposing of smoking materials and other flaming or glowing objects. Smoking in bed or in situations where the smoker could forget about lighted smoking materials has caused large numbers of fires and fatalities over the years. Lack of attention to fireplaces and ash pits has caused great property loss as well. As one example, a three-story fraternity house at Iowa State University in Ames, Iowa, was completely destroyed by the fire that resulted from a burning log tumbling from an overflowing ash pit onto a combustible floor surface.

Lighted cigarettes discarded through the windows of moving vehicles each year cause grass, brush and forest fires that consume huge acreage in open country as well as dwellings and other structures.

308.3 Open flame. A person shall not utilize or allow to be utilized, an open flame in connection with a public meeting or gathering for purposes of deliberation, worship, entertainment, amusement, instruction, education, recreation, awaiting transportation or similar purpose in assembly or educational occupancies without first obtaining a permit in accordance with Section 105.6.

❖ This section establishes the authority of the fire code official to control the hazards of using open flame through the permitting process. The restrictions here do not prohibit the use of open-flame devices in the listed activities, but they do allow the fire code official to inspect plans and ongoing activities to make certain they are safe for the occupancy in which they are held.

308.3.1 Open-flame cooking devices. Charcoal burners and other open-flame cooking devices shall not be operated on combustible balconies or within 10 feet (3048 mm) of combustible construction.

Exceptions:

1. One- and two-family dwellings.
2. Where buildings, balconies and decks are protected by an automatic sprinkler system.

❖ This prohibition comes from the potential for hot embers to fall from the firebox of the cooking device and ignite a combustible surface, such as a wooden balcony or deck. The 10-foot (3048 mm) separation also reduces the likelihood that fire starting or cooking flare-ups will come in contact with combustible wall construction that is easily ignited.

Exception 1 exempts one- and two-family dwellings from the requirements of this section. In those occupancies, the level of familiarity and control exercised by the building occupants is recognized as offsetting the hazards of using open-flame cooking devices. There are practical difficulties involved in enforcing such regulations in one- and two-family dwellings as well.

Exception 2 recognizes the added protection of sprinklers.

It should be noted that this section contains a general prohibition on the use of charcoal burners and other open-flame cooking devices in the locations described. Section 308.3.1.1, however, contains a very specific regulation for only LP-gas-fired cooking devices in the described locations and would, therefore, take precedence over the general provisions of this section.

308.3.1.1 Liquefied-petroleum-gas-fueled cooking devices. LP-gas burners having an LP-gas container with a water capacity greater than 2.5 pounds [nominal 1 pound (0.454 kg) LP-gas capacity] shall not be located on combustible balconies or within 10 feet (3048 mm) of combustible construction.

Exception: One- and two-family dwellings.

❖ This section restricts LP-gas burners to small tabletop grills or units that might be used in cooking within residential occupancies. The exception allows the use of LP-gas barbeque grills of any size on balconies of one- and two-family dwellings, but not on balconies or decks of multiple family dwellings where the property and life safety hazard is greater.

308.3.2 Open-flame decorative devices. Open-flame decorative devices shall comply with all of the following restrictions:

1. Class I and Class II liquids and LP-gas shall not be used.
2. Liquid- or solid-fueled lighting devices containing more than 8 ounces (237 ml) of fuel must self-extinguish and not leak fuel at a rate of more than 0.25 teaspoon per minute (1.26 ml per minute) if tipped over.
3. The device or holder shall be constructed to prevent the spillage of liquid fuel or wax at the rate of more than 0.25 teaspoon per minute (1.26 ml per minute) when the device or holder is not in an upright position.
4. The device or holder shall be designed so that it will return to the upright position after being tilted to an angle of 45 degrees from vertical.

   Exception: Devices that self-extinguish if tipped over and do not spill fuel or wax at the rate of more than 0.25 teaspoon per minute (1.26 ml per minute) if tipped over.
5. The flame shall be enclosed except where openings on the side are not more than 0.375 inch (9.5 mm) diameter or where openings are on the top and the distance to the top is such that a piece of tissue paper placed on the top will not ignite in 10 seconds.
6. Chimneys shall be made of noncombustible materials and securely attached to the open-flame device.

   Exception: A chimney is not required to be attached to any open-flame device that will self-extinguish if the device is tipped over.
7. Fuel canisters shall be safely sealed for storage.
8. Storage and handling of combustible liquids shall be in accordance with Chapter 34.
9. Shades, where used, shall be made of noncombustible materials and securely attached to the open-flame device holder or chimney.
10. Candelabras with flame-lighted candles shall be securely fastened in place to prevent overturning, and shall be located away from occupants using the area and away from possible contact with drapes, curtains or other combustibles.

❖ This class of open-flame devices includes items such as wall-mounted candles or torch sconces; bug-repellant candles in glass jars or metal cans; tabletop candles and oil lamps; free-standing torch holders and candelabras. The criteria for the use of this kind of device are all intended to enhance safety.

308.3.3 Location near combustibles. Open flames such as from candles, lanterns, kerosene heaters, and gas-fired heaters shall not be located on or near decorative material or similar combustible materials.

❖ Each year in nearly every county and community in the country at least one house fire occurs that is caused by a gas-fired space heater igniting nearby combustibles. Accidents involving candles and lanterns used in both outdoor and indoor settings are not at all uncommon. In nearly all of these incidents, the exercise of common sense and the practice of keeping ignition sources and fuel packages well separated could have prevented property damage or loss of life.

308.3.4 Aisles and exits. Candles shall be prohibited in areas where occupants stand, or in an aisle or exit.

❖ This prohibition is intended to prevent accidents caused by lighted candles being knocked from their holders onto combustible furniture, carpeting or decorative materials. Candles are commonly found at seasonal religious observances where attendance often exceeds the norm. In case of an emergency, people must be able to move through the aisles toward the exits without risking the ignition of clothing, hair or decorations.

308.3.5 Religious ceremonies. When, in the opinion of the fire code official, adequate safeguards have been taken, participants in religious ceremonies are allowed to carry hand-held candles. Hand-held candles shall not be passed from one person to another while lighted.

❖ This section has a very narrow application. As stated, only religious ceremonies are covered and the judgement of the fire code official is required for final approval. Prohibiting the passing of lighted candles from person to person is intended to minimize the opportunities for the candles to be dropped where they could become an ignition source for flammable or combustible materials or come into contact with clothing or hair. Spiritual significance may be attached to the use of candles in places of worship; therefore, the local fire code official should work closely with religious groups when enforcing this section.

308.3.6 Theatrical performances. Where approved, open-flame devices used in conjunction with theatrical performances are allowed to be used when adequate safety precautions have been taken in accordance with NFPA 160.

❖ Theatrical performances typically occur on stages and involve large quantities of combustible materials. Hazards associated with stages can include: combustible scenery and lighting suspended overhead; scenic elements, contents and acoustical treatment on the back and sides of the stage; workshops, scene docks and dressing rooms located around the stage perimeter and storage areas and property rooms located underneath the stage. Because of the inherent dangers associated with the introduction of open flames into such a fuel-rich environment, the use of open-flame devices in theatrical performances requires review, evaluation and the approval of the fire code official on a case-by-case basis and must be safeguarded in accordance with the provisions of NFPA 160. For further discussion on the special hazard nature of stages, see the commentary for Section 410 of the IBC.

308.3.7 Group A occupancies. Open-flame devices shall not be used in a Group A occupancy.

Exceptions:

1. Open-flame devices are allowed to be used in the following situations, provided approved precautions are taken to prevent ignition of a combustible material or injury to occupants:
   1.1. Where necessary for ceremonial or religious purposes in accordance with Section 308.3.5.
   1.2. On stages and platforms as a necessary part of a performance in accordance with Section 308.3.6.
   1.3. Where candles on tables are securely supported on substantial noncombustible bases and the candle flames are protected.
2. Heat-producing equipment complying with Chapter 6 and the *International Mechanical Code*.
3. Gas lights are allowed to be used provided adequate precautions satisfactory to the fire code official are taken to prevent ignition of combustible materials.

❖ The use of open-flame devices in Group A occupancies where large numbers of people gather for entertainment, instruction, food or drink consumption, deliberation, awaiting transportation or social or religious functions increases the likelihood of the occupants coming into contact with these devices and is, therefore, prohibited. Safe alternatives to open-flame devices should be used where practical, especially in restaurants and other assembly occupancies where the focus is on atmosphere rather than symbolism or religious significance.

Exception 1 refers back to Sections 308.3.4 and 308.3.5 for use in religious and theatrical settings as well as permitting use on tabletops when properly secured and protected.

Exception 2 refers to building service heat-producing equipment that meets other code requirements.

Exception 3 covers gas lights installed with proper flame safeguards and with the approval of the fire code official. This kind of lighting is often a permanent installation that would be covered by additional code requirements.

308.4 Torches for removing paint. Persons utilizing a torch or other flame-producing device for removing paint from a structure shall provide a minimum of one portable fire extinguisher complying with Section 906 and with a minimum 4-A rating, two portable fire extinguishers, each with a minimum 2-A rating, or a water hose connected to the water supply on the premises where such burning is done. The person doing the burning shall remain on the premises 1 hour after the torch or flame-producing device is utilized.

❖ Any time an open flame is used to soften old paint in preparation for removal, there is a risk of fire that must be covered by having approved fire extinguishers or a water source readily available. The requirement for a 1-hour fire watch after discontinuing the use of the open flame covers the possibility that paint fragments could still be hot enough to ignite flammable or combustible materials that might be lying around. It also considers the possibility that the flame used to remove paint from a combustible base material could heat that material to its ignition temperature and leave an almost undetectable smolder that might burst into flame later. Safe and effective means for removing paint at lower temperatures, such as warm-air heat devices capable of generating high-temperature convection air, are readily available for sale or rent and far less likely to result in an ignition of combustible materials.

308.4.1 Permit. A permit in accordance with Section 105.6 shall be secured from the fire code official prior to the utilization of a torch or flame-producing device to remove paint from a structure.

❖ The requirement for a permit issued by the fire code official is intended to strictly control the use of torches and open-flame devices for paint removal. As noted in Section 308.4, the need for allowing such activities is significantly reduced with effective alternatives that are readily available. Accordingly, the number of cases where a torch is necessary should be minimal and permits should be given sparingly (see commentary, Section 301.2).

308.5 Open-flame devices. Torches and other devices, machines or processes liable to start or cause fire shall not be operated or used in or upon hazardous fire areas, except by a permit in accordance with Section 105.6 secured from the fire code official.

Exception: Use within inhabited premises or designated campsites which are a minimum of 30 feet (9144 mm) from grass-, grain-, brush- or forest-covered areas.

❖ This section establishes the fire code official's authority to control through the permitting process the use of open flames in areas susceptible to fires. The term "hazardous fire area" is generally applied to land covered with grass, grain, brush, forest or similar vegetation that, if ignited, could pose a severe fire danger to surrounding areas. The exception recognizes open ground that is free of combustible materials as an acceptable fire barrier (see the IUWIC for further information).

308.5.1 Signals and markers. Flame-employing devices, such as lanterns or kerosene road flares, shall not be operated or used as a signal or marker in or upon hazardous fire areas.

Exception: The proper use of fusees at the scenes of emergencies or as required by standard railroad operating procedures.

❖ This section prohibits the use of flame-producing devices as signal or marker devices except for the use of fusees to mark the scene of an emergency or where routinely employed in railroad procedures, such as when a train is stopped across a roadway not protected by permanent signal lights.

308.5.2 Portable fueled open-flame devices. Portable open-flame devices fueled by flammable or combustible gases or liquids shall be enclosed or installed in such a manner as to prevent the flame from contacting combustible material.

Exceptions:

1. LP-gas-fueled devices used for sweating pipe joints or removing paint in accordance with Chapter 38.
2. Cutting and welding operations in accordance with Chapter 26.
3. Torches or flame-producing devices in accordance with Section 308.4.
4. Candles and open-flame decorative devices in accordance with Section 308.3.

❖ This section prohibits the use of portable devices in situations where they might be placed on unstable platforms or where they could be knocked over by human contact. The exceptions list the types of open flame or heat-producing operations not regulated by Section 308.5.2 but are regulated elsewhere in the code.

308.6 Flaming food and beverage preparation. The preparation of flaming foods or beverages in places of assembly and drinking or dining establishments shall be in accordance with Section 308.6.

❖ The regulations in this section of the code are necessary to give the fire code official guidance in allowing flaming food (also known as "flambé foods") and beverage preparation to be conducted in restaurants in a safe manner. Many restaurants prepare popular selected dishes, such as cherries jubilee, brandied peaches and flaming bananas, in this manner, usually tableside within close proximity of the customers. This process typically includes the use of a small amount of flammable or combustible liquid, such as brandy, rum or other liqueurs or cordials, making regulation of this process appropriate.

308.6.1 Dispensing. Flammable or combustible liquids used in the preparation of flaming foods or beverages shall be dispensed from one of the following:

1. A 1-ounce (29.6 ml) container; or
2. A container not exceeding 1-quart (946.5 ml) capacity with a controlled pouring device that will limit the flow to a 1-ounce (29.6 ml) serving.

❖ These dispensing provisions limit the amount of flammable or combustible liquid being transported around the restaurant for use in flaming food or beverage preparation in order to minimize the fuel potential in a fire incident involving such operations.

308.6.2 Containers not in use. Containers shall be secured to prevent spillage when not in use.

❖ Securing the containers used in flaming food or beverage preparation while not in use reduces the likelihood of an accidental spill.

308.6.3 Serving of flaming food. The serving of flaming foods or beverages shall be done in a safe manner and shall not create high flames. The pouring, ladling or spooning of liquids is restricted to a maximum height of 8 inches (203 mm) above the receiving receptacle.

❖ Limiting the height from which flammable or combustible liquids are poured reduces the likelihood of a spill or overpour that might miss the target dish and be ignited by a table candle, smoking materials or another ignition source.

308.6.4 Location. Flaming foods or beverages shall be prepared only in the immediate vicinity of the table being serviced. They shall not be transported or carried while burning.

❖ This section prohibits movement or transport of "flambé foods" while they are burning in order to reduce the potential for incidents wherein the tray or dish might be dropped or tipped causing a spill of burning liquid, which, while limited in size, could lead to a panic reaction by restaurant patrons or, if spilled on a patron, could cause serious burn injuries.

308.6.5 Fire protection. The person preparing the flaming foods or beverages shall have a wet cloth towel immediately available for use in smothering the flames in the event of an emergency.

❖ Flaming-food preparation is a cooking hazard much the same as stovetop preparation. An efficient method for extinguishing a stovetop fire in a frying pan or similar utensil is to put the cover on the utensil to exclude oxygen. In flambe′ preparation, the food is prepared in a relatively small pan or even in the dish in which it will be served and in which the flames can be easily smothered by a wet towel. Note that this precautionary measure is in addition to the portable fire extinguishers provided in accordance with Section 906.

SECTION 309
POWERED INDUSTRIAL TRUCKS

309.1 General. Powered industrial trucks shall be operated and maintained in accordance with this section.

❖ This statement establishes the fire safety requirement for control of powered industrial trucks. Because these trucks may have either battery-powered electric motors or internal combustion engines using liquid fuel or LP-gas, Sections 309.2 through 309.6 cover the fire safety aspects of both.

309.2 Battery chargers. Battery chargers shall be of an approved type. Combustible storage shall be kept a minimum of 3 feet (915 mm) from battery chargers. Battery charging shall not be conducted in areas accessible to the public.

❖ Battery chargers offer several safety challenges if they are not properly housed and operated. A battery that is connected to the charger with the poles reversed can explode when charging power is turned on, which could result in corrosive liquid being sprayed over the charger room and anyone who happened to be in it. Aside from the possibility of serious bodily injury from flying debris or corrosive spray, eye damage from the spray is a critical consideration. There is also the possibility of energized charging leads shorting, arcing or fusing and causing sparks to become an ignition source for combustibles if they are not kept out of the area.

When a charger is properly connected and energized, the charging process causes generation of hydrogen and oxygen gases as well as acid or alkali fumes. These gases must be vented to prevent them from reaching ignitable or detonable levels (see commentary, Section 309.3).

309.3 Ventilation. Ventilation shall be provided in an approved manner in battery-charging areas to prevent a dangerous accumulation of flammable gases.

❖ Charging lead-acid or nickel iron batteries is a process of electrolysis in which oxides created by operation of the battery are reduced to metal and redeposited on the electrode plates. The process results in the rejuvenation of the electrolyte in the battery and the emission of both oxygen and hydrogen gases as well as corrosive fumes. If these gases are allowed to accumulate in an enclosed space, they could eventually reach an ignitable or detonable level. The charging area must be ventilated in accordance with Section 608.5 and the applicable provisions of the *International Mechanical Code®* (IMC®) to carry off and dilute the concentrations of hazardous gases.

309.4 Fire extinguishers. Battery-charging areas shall be provided with a fire extinguisher complying with Section 906 having a minimum 4-A:20-B:C rating within 20 feet (6096 mm) of the battery charger.

❖ Because of the electrical hazards associated with the battery charging operation; the fuel load presented by the plastic battery cases and other area contents and the potential for the presence of gases in the room, an appropriately sized portable fire extinguisher must be located within the battery charging area. The extinguisher must be accessible with minimum travel.

309.5 Refueling. Powered industrial trucks using liquid fuel or LP-gas shall be refueled outside of buildings or in areas specifically approved for that purpose and in accordance with Chapter 34 or 38.

❖ Because of the hazards associated with liquid fuel spills and gaseous fuel discharges, this section requires that powered industrial trucks be refueled outside where the vapors or gas can be readily dissipated. This section also allows the alternative of fueling inside of buildings where safeguards mitigate the hazards of the operation, the location is specifically approved for the use by the fire code official and the applicable requirements of Chapter 34 (flammable liquids) or 38 (LP-gas) are complied with.

309.6 Repairs. Repairs to fuel systems, electrical systems and repairs utilizing open flame or welding shall be done in approved locations outside of buildings or in areas specifically approved for that purpose.

❖ Repairs that could create ignition sources or a fuel load for an ignition source must be done in indoor or outdoor locations that are designed specifically for vehicle repairs and that have been approved by the fire code official.

SECTION 310
SMOKING

310.1 General. The smoking or carrying of a lighted pipe, cigar, cigarette or any other type of smoking paraphernalia or material is prohibited in the areas indicated in this section.

❖ This section states that smoking is prohibited in those areas designated in Sections 310.2 and 310.8. Before entering an area posted with "no smoking" signs, anyone carrying a lighted smoking product or device must extinguish the smoking material and properly dispose of the ashes and other residue. Suitable ashtrays should be available at the entry to "no smoking" areas for disposal of the prohibited smoking materials.

310.2 Prohibited areas. Smoking shall be prohibited where conditions are such as to make smoking a hazard, and in spaces where flammable or combustible materials are stored or handled.

❖ Smoking can be prohibited wherever it would be a hazard in the judgement of the fire code official.

310.3 "No Smoking" signs. The fire code official is authorized to order the posting of "No Smoking" signs in a conspicuous location in each structure or location in which smoking is prohibited. The content, lettering, size, color and location of required "No Smoking" signs shall be approved.

❖ The fire code official is not only authorized to designate where signs are to be posted, but is also responsible for establishing the specification for all aspects of those signs. A typical sign design is shown in Figure 310.3.

Figure 310.3
NO SMOKING SIGN

310.4 Removal of signs prohibited. A posted "No Smoking" sign shall not be obscured, removed, defaced, mutilated or destroyed.

❖ Posted signs must remain in the locations designated by the fire code official and be readable at all times. Many, if not all, jurisdictions establish penalties for removing, obscuring, defacing or mutilating official signs.

310.5 Compliance with "No Smoking" signs. Smoking shall not be permitted nor shall a person smoke, throw or deposit any lighted or smoldering substance in any place where "No Smoking" signs are posted.

❖ Penalties are usually imposed for violating "no smoking" prohibitions. Smoking must be confined to approved areas, and discarded smoking materials must be deposited only in approved ashtrays or receptacles. Violation of this provision customarily constitutes a misdemeanor. Prosecution of misdemeanor offenses should be coordinated with the jurisdiction's legal counsel. Offense and penalty clauses, such as those included in Section 109, are required in order to prosecute infractions (see commentary, Section 109.3).

310.6 Ash trays. Where smoking is permitted, suitable noncombustible ash trays or match receivers shall be provided on each table and at other appropriate locations.

❖ Where smoking is permitted, smokers must have available a noncombustible ashtray or other receptacle where smoking materials can be safely deposited. These receptacles must be located on each table and other locations throughout the smoking area so that the smoker does not have to look for a place to discard used materials.

310.7 Burning objects. Lighted matches, cigarettes, cigars or other burning object shall not be discarded in such a manner that could cause ignition of other combustible material.

❖ Lighted matches, burning tobacco products and all other burning objects must be deposited in ashtrays or other approved noncombustible containers to separate these ignition sources from any potential fuel loads.

310.8 Hazardous environmental conditions. When the fire code official determines that hazardous environmental conditions necessitate controlled use of smoking materials, the ignition or use of such materials in mountainous, brush-covered or forest-covered areas or other designated areas is prohibited except in approved designated smoking areas.

❖ This section gives the fire code official the authority to establish "no smoking" areas whenever and wherever environmental conditions are considered hazardous. This can include prohibiting burning dry grasses and leaves when lack of rain has rendered the environment dangerously dry or burning of specific materials when the smoke plume from the fire would endanger the health or welfare of a population. The possibilities for determination of hazardous conditions are too numerous to itemize. The judgement of the fire code official, with the advice of other officials being sought as needed, is the determining factor. The fire code official can also exercise the discretion of designating areas in which smoking is permitted.

SECTION 311 VACANT PREMISES

311.1 General. Temporarily unoccupied buildings, structures, premises or portions thereof, including tenant spaces, shall be safeguarded and maintained in accordance with this section.

❖ Vacant buildings or portions of buildings that are open to trespass at doors or windows pose fire safety and criminal trespass hazards to a community and are correctly declared to be unsafe buildings in Section 110.1.1. Such premises are often called an "attractive nuisance" to neighborhood children who may enter them to play or to other persons who may enter seeking shelter from the elements or to engage in potential criminal activities.

311.1.1 Abandoned premises. Buildings, structures and premises for which an owner cannot be identified or located by dispatch of a certificate of mailing to the last known or registered address, which persistently or repeatedly become unprotected or unsecured, which have been occupied by unauthorized persons or for illegal purposes, or which present a danger of structural collapse or fire spread to adjacent properties shall be considered abandoned, declared unsafe and abated by demolition or rehabilitation in accordance with the *International Property Maintenance Code* and the *International Building Code*.

❖ This section establishes the authority to dispose of by demolition or rehabilitation properties that pose a variety of public safety hazards when the owners of the property cannot be located by customary legal means. Because demolition and rehabilitation are regulated by the IBC and *International Property Maintenance Code®* (IPMC®) and applicable state laws, any action taken must be a carefully coordinated effort by all affected code officials in close relationship with the jurisdiction's legal counsel.

311.1.2 Tenant spaces. Storage and lease plans required by this code shall be revised and updated to reflect temporary or partial vacancies.

❖ The intent of this section is to keep storage and lease plans up to date so that the fire service will always have a complete picture of the kinds of hazards they might face in case of a fire, including vacant, unattended spaces.

311.2 Safeguarding vacant premises. Temporarily unoccupied buildings, structures, premises or portions thereof shall be secured and protected in accordance with this section.

❖ This section lists a number of problems that commonly occur when buildings or portions of buildings remain vacant for long periods of time. It covers concerns for security, fire protection and fire separation in vacant spaces.

311.2.1 Security. Exterior openings and interior openings accessible to other tenants or unauthorized persons shall be boarded, locked, blocked or otherwise protected to prevent entry by unauthorized individuals.

❖ Unauthorized or illegal activities in vacant buildings can lead to the presence of unanticipated fire loads susceptible to ready ignition and rapid fire spread, thus increasing the hazard to adjoining properties or spaces and fire department personnel (see commentary, Section 110.1.1). This section requires securing the openings of vacant buildings or spaces against unauthorized entry by any of the methods listed or by other equally effective means approved by the fire code official.

311.2.2 Fire protection. Fire alarm, sprinkler and standpipe systems shall be maintained in an operable condition at all times.

Exceptions:

1. When the premises have been cleared of all combustible materials and debris and, in the opinion of the fire code official, the type of construction, fire separation distance and security of the premises do not create a fire hazard.
2. Where buildings will not be heated and fire protection systems will be exposed to freezing temperatures, fire alarm and sprinkler systems are permitted to be placed out of service and standpipes are permitted to be maintained as dry systems (without an automatic water supply) provided the building has no contents or storage, and windows, doors and other openings are secured to prohibit entry by unauthorized persons.

❖ The basic requirement of this section is clearly stated. The on-site fire protection systems must be maintained whether the property is occupied or vacant. The systems would be subject to the same inspections in either case.

Exception 1 gives the fire code official the authority to determine whether vacant premises pose a significant hazard and lists criteria he or she can use in making that determination.

Exception 2 recognizes that systems located in unheated premises in cold climates could be rendered inoperable by freezing and authorizes the fire code official to grant permission to disable those systems when the security of the premises and fire separation arrangements meet the code requirements stated in Sections 311.2.1 and 311.2.3.

311.2.3 Fire separation. Fire-resistance-rated partitions, fire barriers, and fire walls separating vacant tenant spaces from the remainder of the building shall be maintained. Openings, joints, and penetrations in fire-resistance-rated assemblies shall be protected in accordance with Chapter 7.

❖ Fire-resistance-rated construction separating vacant spaces from the remainder of the building must be maintained to the satisfaction of the fire code official. The requirements for openings, joints and penetrations are covered in Chapter 7.

311.3 Removal of combustibles. Persons owning, or in charge or control of, a vacant building or portion thereof, shall remove therefrom all accumulations of combustible materials, flammable or combustible waste or rubbish and shall securely lock or otherwise secure doors, windows and other openings to prevent entry by unauthorized persons. The premises shall be maintained clear of waste or hazardous materials.

Exceptions:

1. Buildings or portions of buildings undergoing additions, alterations, repairs, or change of occupancy in accordance with the *International Building Code*, where waste is controlled and removed as required by Section 304.
2. Seasonally occupied buildings.

❖ Property owners, their agents and persons leasing vacant spaces are responsible for preventing accumulations of flammable or combustible materials as well as for securing the vacant space against entry by unauthorized persons.

Exception 1 covers building situations in which larger amounts of flammable or combustible waste would reasonably be generated and cites code references to cover requirements in those situations.

Exception 2 allows reasonable accumulations of flammable or combustible materials in spaces that are occupied seasonally. For example, this exception would allow unattended off-season storage of stock for sale.

311.4 Removal of hazardous materials. Persons owning or having charge or control of a vacant building containing hazardous materials regulated by Chapter 27 shall comply with the facility closure requirements of Section 2701.6.

❖ This section gives the fire code official the authority to require property owners, their agents and their tenants to submit a facility closure plan as well as making sure combustible and hazardous materials are removed from the premises. The requirements for the facility closure plan are given in Section 2701.6 (see Section 407.7).

SECTION 312
VEHICLE IMPACT PROTECTION

312.1 General. Vehicle impact protection required by this code shall be provided by posts that comply with Section 312.2 or by other approved physical barriers that comply with Section 312.3.

❖ This section applies to those locations where a moving vehicle could strike a piece of equipment that contains fuel or is fuel fired. These applications include motor fuel-dispensing facilities, above-ground storage tanks and repair garages as well as other locations in which gas- or oil-fired equipment or appliances could be installed where they would be in harm's way. Additional requirements for equipment protection are contained in Chapter 3 of the IMC and Section 305 of the *International Fuel Gas Code®* (IFGC®).

312.2 Posts. Guard posts shall comply with all of the following requirements:

1. Constructed of steel not less than 4 inches (102 mm) in diameter and concrete filled.
2. Spaced not more than 4 feet (1219 mm) between posts on center.
3. Set not less than 3 feet (914 mm) deep in a concrete footing of not less than a 15-inch (381 mm) diameter.
4. Set with the top of the posts not less than 3 feet (914 mm) above ground.
5. Located not less than 3 feet (914 mm) from the protected object.

❖ This section lists five requirements that guard posts must satisfy. Typical installations of posts in service stations and other locations are shown in Figures 312.2(1) and 312.2(2). These guard posts are designed to resist impact from vehicles moving at low speeds, as they would be when pulling up to a fuel pump at a motor fuel-dispensing facility or into a service area in an indoor service facility or repair garage.

312.3 Other barriers. Physical barriers shall be a minimum of 36 inches (914 mm) in height and shall resist a force of 12,000 pounds (53 375 N) applied 36 inches (914 mm) above the adjacent ground surface.

❖ Barriers other than posts could include walls, barricades or elevated locations for equipment (see Figure 312.3). The structural requirements stated in this section are intended to provide protection from moving vehicles traveling at relatively slow approach speeds and are based on U.S. Department of Transportation (DOT) design criteria for a 6,000-pound (2724 kg) impact resistance, plus a safety factor of 2 for a 6-inch-diameter (152 mm) concrete-filled steel pipe set 42 inches (1067 mm) into concrete.

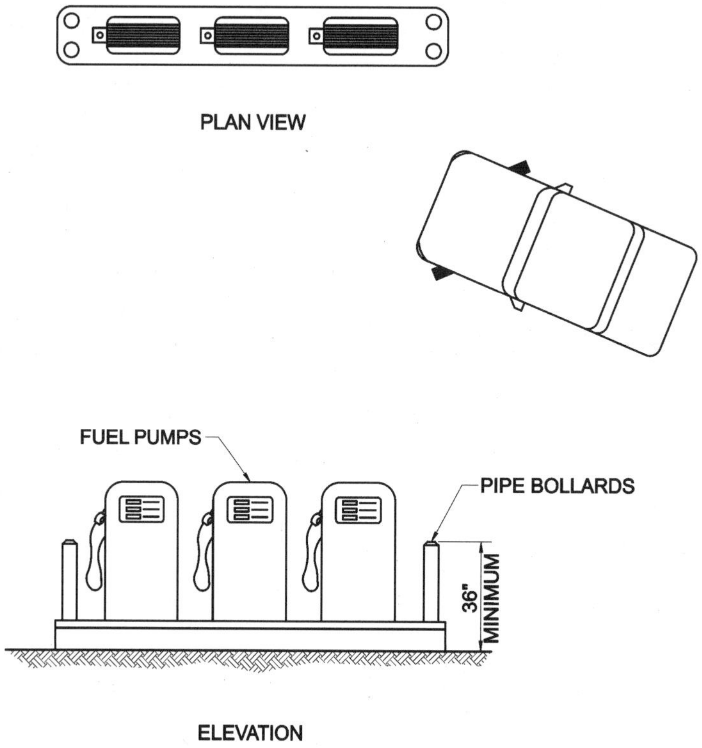

Figure 312.2(1)
PUMP ISLAND PROTECTION

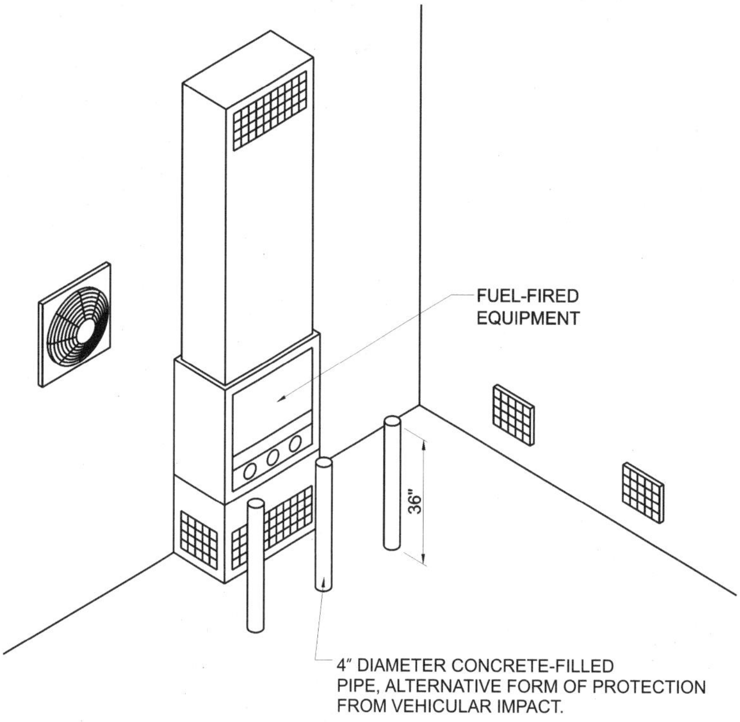

For SI: 1 inch = 25.4 mm.

Figure 312.2(2)
GUARD POST INSTALLATION IN PUBLIC REPAIR GARAGE

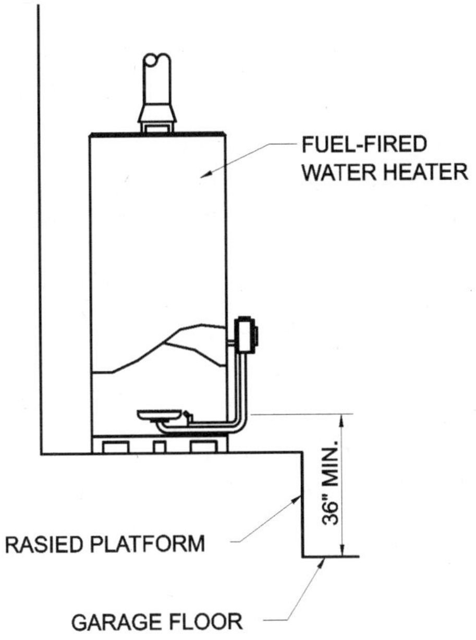

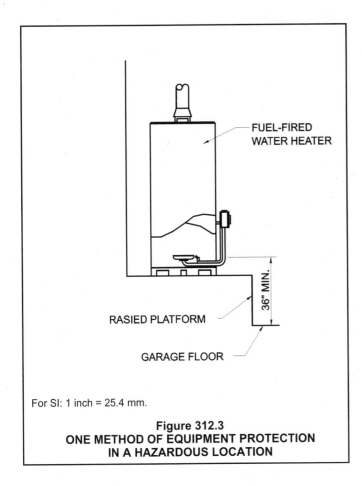

For SI: 1 inch = 25.4 mm.

**Figure 312.3
ONE METHOD OF EQUIPMENT PROTECTION
IN A HAZARDOUS LOCATION**

SECTION 313
FUELED EQUIPMENT

313.1 Fueled equipment. Fueled equipment, including but not limited to motorcycles, mopeds, lawn-care equipment and portable cooking equipment, shall not be stored, operated or repaired within a building.

Exceptions:

1. Buildings or rooms constructed for such use in accordance with the *International Building Code*.
2. Where allowed by Section 313 or 314.

❖ The restrictions in this section are similar to those in Section 314.4, but this section also regulates portable liquid or gas-fueled cooking equipment vehicles and the operation and repair of vehicles and equipment as well as storage (see commentary, Section 314.4).

Exception 1 recognizes the increased safety afforded when these uses are isolated from other parts of the building with fire-resistance rated construction in accordance with the IBC.

Exception 2 defers to the other section of the code that also regulates fueled equipment covered by this section.

313.1.1 Removal. The fire code official is authorized to require removal of fueled equipment from locations where the presence of such equipment is determined by the fire code official to be hazardous.

❖ This section gives the fire code official the authority to conduct inspections for the purpose of determining that vehicle operation, repair and storage in buildings comply with the requirements of this section and, if they do not, the authority to order the removal of fueled equipment from the building as a means of eliminating the hazard.

313.2 Group R occupancies. Vehicles powered by flammable liquids, Class II combustible liquids, or compressed flammable gases shall not be stored within the living space of Group R buildings.

❖ Prohibiting storage of fuel-burning vehicles recognizes the hazards associated with having significant quantities of flammable or combustible liquids or compressed gases in inhabited spaces. Most vehicles that use liquid fuels have tanks that are not pressurized and are not vapor tight. Even a small leak over time can build to dangerous levels in enclosed spaces. For these reasons, this section prohibits the storage of gas- or liquid-fueled vehicles and equipment within the living spaces of Group R buildings.

SECTION 314
INDOOR DISPLAYS

314.1 General. Indoor displays constructed within any occupancy shall comply with Sections 314.2 through 314.4.

❖ Indoor displays of merchandise and the display of all manner of vehicles inside of buildings can create a number of hazards to building occupants, including blocked egress and rapid fire buildup. This section describes reasonable measures to reduce the hazards associated with indoor displays without prohibiting them.

314.2 Fixtures and displays. Fixtures and displays of goods for sale to the public shall be arranged so as to maintain free, immediate and unobstructed access to exits as required by Chapter 10.

❖ The reason for maintaining free and unobstructed access to exits in public shopping spaces is, of course, personal safety in times of emergency. Chapter 10 contains the requirements, criteria and guidelines for this purpose.

314.3 Highly combustible goods. The display of highly combustible goods, including but not limited to fireworks, flammable or combustible liquids, liquefied flammable gases, oxidizing materials, pyroxylin plastics and agricultural goods, in main exit access aisles, corridors, covered malls, or within 5 feet (1524 mm) of entrances to exits and exterior exit doors is prohibited when a fire involving such goods would rapidly prevent or obstruct egress.

❖ As stated in Chapter 10, all elements of the means of egress of any occupancy open to the public must be kept clear of obstructions and other hazards that could prevent the occupants from exiting the premises quickly in an emergency. Displaying the hazardous materials itemized in this section where their involvement in a fire would block exit pathways is prohibited for this reason. The hazards associated with each of the materials mentioned in the section are discussed in Chapters 27 through 44.

314.4 Vehicles. Liquid- or gas-fueled vehicles, boats or other motorcraft shall not be located indoors except as follows:

1. Batteries are disconnected.
2. Fuel in fuel tanks does not exceed one-quarter tank or 5 gallons (19 L) (whichever is least).
3. Fuel tanks and fill openings are closed and sealed to prevent tampering.
4. Vehicles, boats or other motorcraft equipment are not fueled or defueled within the building.

❖ It has become commonplace for covered malls and larger retail stores to have various types of gas- or liquid-fueled vehicles on inside display, such as for promotional events or fire apparatus displays during Fire Prevention Week. Because the hazards of such displays in a public building are similar to those in residential buildings, Section 314.4 parallels Section 313.1, Exception 2 (see commentary, Section 313.1).

SECTION 315 MISCELLANEOUS COMBUSTIBLE MATERIALS STORAGE

315.1 General. Storage, use and handling of miscellaneous combustible materials shall be in accordance with this section. A permit shall be obtained in accordance with Section 105.6.

❖ This section contains regulations for the management of inside and outside miscellaneous combustible materials storage not regulated elsewhere in the code and establishes the requirement for obtaining permits for storage, use and handling of these materials (see commentary, Sections 105.6 and 301.2).

315.2 Storage in buildings. Storage of combustible materials in buildings shall be orderly. Storage shall be separated from heaters or heating devices by distance or shielding so that ignition cannot occur.

❖ Throughout the code, the use of fire-resistance-rated construction and spatial separation distances to minimize fire hazards and fire spread is stated as a requirement for a variety of different materials. This section deals with the requirements for storage of miscellaneous combustible materials inside of buildings. Combustible fibers are discussed in Chapter 29, while control of combustible waste is covered in Sections 304.2 and 304.3. These requirements should be studied in detail.

315.2.1 Ceiling clearance. Storage shall be maintained 2 feet (610 mm) or more below the ceiling in nonsprinklered areas of buildings or a minimum of 18 inches (457 mm) below sprinkler head deflectors in sprinklered areas of buildings.

❖ If the space is not equipped with sprinklers, the clearance between the stored materials and the ceiling must be 2 feet (610 mm) to allow manual hose streams to effectively reach the top of a burning pile as well as to project over and beyond adjacent piles to reach burning materials. Where sprinklers are installed, the 18-inch (457 mm) clearance permits timely activation of the sprinklers and allows unobstructed water distribution over the storage pile. Materials stored too close to sprinklers can not only prevent the heat of a fire from reaching the sprinkler fusible link but also water from reaching the seat of a fire once the sprinklers are activated.

Certain newer types of automatic sprinklers, because of their design or operating characteristics, may require greater clearance distances than the 18-inch (457 mm) minimum prescribed in this section. NFPA 13 and the sprinkler manufacturer's data should be consulted for specific information on the characteristics of the many different types of sprinklers that may be installed in a given building.

315.2.2 Means of egress. Combustible materials shall not be stored in exits or exit enclosures.

❖ As was stated in Section 314.3, all elements of the means of egress must be kept free of obstructions that could block an exit pathway and, thus, jeopardize occupants of the affected space. Chapter 10 offers more guidance on means of egress.

315.2.3 Equipment rooms. Combustible material shall not be stored in boiler rooms, mechanical rooms or electrical equipment rooms.

❖ The intent of this section is to keep the ignition sources inherent in the use of the indicated rooms from coming into contact with combustible materials that might be stored in the rooms and to increase the likelihood that authorized personnel will be able to easily reach critical controls, such as electrical circuit disconnects, in case of an emergency. For additional discussion of requirements applicable to these and other specific occupancy rooms, see Section 302 of the IBC. Further discussion of the hazards of storage in boiler rooms can be found in Section 304 of the IMC.

315.2.4 Attic, under-floor and concealed spaces. Attic, under-floor and concealed spaces used for storage of combustible materials shall be protected on the storage side as required for 1-hour fire-resistance-rated construction. Openings shall be protected by assemblies that are self-closing and are of noncombustible construction or solid wood core not less than 1.75 inches (44.5 mm) in thickness. Storage shall not be placed on exposed joists.

Exceptions:

1. Areas protected by approved automatic sprinkler systems.
2. Group R-3 and Group U occupancies.

❖ This section recognizes the reality that attics, crawl spaces and similar unoccupied concealed spaces in buildings are attractive to building occupants for storage of all kinds of combustible materials. Storage in such unattended and out-of-the-way spaces creates a hazardous condition by introducing a higher fire load to spaces that were neither designed nor intended for such a high-intensity use and in which a fire could rapidly develop unobserved until it had gained a considerable hold on the building. The code provides alternatives to using such spaces for storage and, if they are used for storage, how they can be constructed to isolate the higher fire loads created. Consistent with Section 102.4 of the code, any construction in connection with the concealed spaces regulated by this section must be in accordance with the IBC, especially Section 302.

Placing stored combustibles on exposed joists could hasten collapse of the joists in a fire, which could lead to flaming debris being dropped into the building space below the joists and possible collapse of all or part of the building structure.

Exception 1 recognizes the efficiency and reliability of automatic sprinklers as a trade-off for 1-hour fire-resistance-rated construction.

Exception 2 exempts Group R-3 and Group U occupancies from the requirements of this section. In Group R-3, the level of familiarity and control exercised by the building occupants is recognized as offsetting the hazards of storage in concealed spaces. Because Group U occupancies are generally unoccupied, the hazards of miscellaneous storage are of little or no consequence to the few occupants that might be in such buildings. For further information on Group U occupancies, see Section 312 of the IBC and its commentary.

315.3 Outside storage. Outside storage of combustible materials shall not be located within 10 feet (3048 mm) of a property line.

Exceptions:

1. The separation distance is allowed to be reduced to 3 feet (914 mm) for storage not exceeding 6 feet (1829 mm) in height.
2. The separation distance is allowed to be reduced when the fire code official determines that no hazard to the adjoining property exists.

❖ Outside storage of combustible materials, such as raw materials for production, idle pallets, dunnage and packaging, must be neat and compact. The requirement for a 10-foot (3048 mm) separation is consistent with storage area aisle width requirements throughout the code, often expressed as "one-half the pile height or 10 feet, whichever is greater." The requirement of this section is consistent with that concept as is Section 315.3.2, which limits pile height to 20 feet (6096 mm). The intent of this section is to provide fire suppression access on all sides of storage arrangements and reduce the likelihood of the spread of fire to adjacent properties in the event of a pile collapse. Pile collapses will generally not involve a full-height topple-over of a pile but rather only a partial collapse. Accordingly, Exception 1 allows a reduction in separation distance where the pile height is substantially less than the separation requirement.

Exception 2 allows the fire code official to grant separation reductions when the combustibles are judged to be no threat to adjoining property. Examples of such conditions could include storage where the combustible materials are enclosed in noncombustible containers, the presence of an impervious property line barrier or the provision of fixed fire protection equipment, such as deluge monitors, especially designed for rapid fire suppression and exposure protection.

315.3.1 Storage beneath overhead projections from buildings. Combustible materials stored or displayed outside of buildings that are protected by automatic sprinklers shall not be stored or displayed under nonsprinklered eaves, canopies or other projections or overhangs.

❖ The storage or display of combustible materials beneath unsprinklered canopies or other building projections attached to an otherwise fully sprinklered building could lead to a rapidly developing fire in the stored material to gain sufficient headway beyond the capability of the building sprinkler system to suppress it should it spread into the building's interior. This section reinforces the requirements of NFPA 13 concerning the use of areas beneath building projections, such as eaves or canopies, for the storage or display of combustible materials where those locations are exempt from sprinkler protection as allowed in NFPA 13, Section 4-13.7.1. NFPA 13 mandates that the scope of required sprinkler protection include canopies or roofed-over areas attached to sprinklered buildings unless these projections are constructed of noncombustible materials and the areas are not used for the storage, handling or display of combustible materials. Because NFPA 13 is a design standard and cannot be enforced as a maintenance document, Section 315.3.1 essentially restates the NFPA 13 design requirement exception conditions in enforceable terms. Also note that, in the event that Appendix B of the code is adopted by a jurisdiction, areas used for the storage of combustible materials beneath a building's horizontally projecting elements must be included in the building area for purposes of determining the required fire flow.

315.3.2 Height. Storage in the open shall not exceed 20 feet (6096 mm) in height.

❖ Storage pile height limitations are a means of controlling the size of potential fires and reducing the tip-over potential as well as a way to facilitate the manual fire suppression process by keeping the top of the pile within reach of conventional fire fighting and overhaul tools, such as the ground ladders carried by an engine company or the long pike poles carried by ladder companies. The 20-foot (6096 mm) storage pile height limitation also correlates with Section 315.3 and helps reduce the likelihood that a fire would jeopardize adjacent properties in the event of a pile collapse.

Bibliography

The following resource materials are referenced in this chapter or are relevant to the subject matter addressed in this chapter.

IBC-2003, *International Building Code.* Falls Church, VA: International Code Council, 2003.

IFGC-2003, *International Fuel Gas Code.* Falls Church, VA: International Code Council, 2003.

IMC-2003, *International Mechanical Code.* Falls Church, VA: International Code Council, 2003.

IPMC-2003, *International Property Maintenance Code.* Falls Church, VA: International Code Council, 2003.

IUWIC-2003, *International Urban-Wildland Interface Code.* Falls Church, VA: International Code Council, 2003.

NFPA 40–97, *Storage and Handling of Cellulose Nitrate Motion Picture Film.* Quincy, MA: National Fire Protection Association, 1997.

NFPA 160–01, *Flame Effects Before an Audience.* Quincy, MA: National Fire Protection Association, 2001.

UL 32-94, *Metal Waste Cans.* Northbrook, IL: Underwriters Laboratories, Inc., 1994.

Chapter 4: Emergency Planning and Preparedness

General Comments

This chapter is an expansion of the provisions found in the source fire codes used to develop the *International Fire Code*® (IFC®). The overall approach has been to place all similar requirements into general sections. The unique occupancy and use-specific requirements are provided at the end of Chapter 4.

This chapter first provides general scope and requirements for the reporting of emergencies and the prevention of interference with fire department activities in Section 401.

Section 402 defines "emergency evacuation drill." Section 403 provides the authority for jurisdictions to address hazards associated with public assemblages, regardless of the use or occupancy. Section 404 provides the detailed requirements for fire safety plans and evacuation as they generally apply to the following occupancies:

1. Group A, other than Group A occupancies used exclusively for purposes of religious worship that have an occupant load of less than 2,000 people.
2. Group E.
3. Group H.
4. Group I.
5. Group R-1.
6. Group R-4.
7. High-rise buildings.
8. Group M buildings having an occupant load of 500 or more persons or more than 100 persons above or below the lowest level of exit discharge.
9. Covered malls exceeding 50,000 square feet (4645 m^2) in aggregate floor area.
10. Underground buildings.
11. Buildings with an atrium and having an occupancy in Group A, E or M.

The frequency and required documentation related to evacuation drills are addressed in Section 405. Minimum criteria for the training of occupants for emergency situations are found in Section 406. Section 407 provides requirements that apply to occupancies that contain hazardous materials. Some of the key elements are Hazardous Materials Inventory Statements (HMIS) and Hazardous Materials Management Plans (HMMP). Finally, the occupancy-specific requirements, such as seating plans for Group A occupancies, are included in Section 408.

Purpose

In addition to the requirements found throughout the building and fire codes, the requirements found in this chapter focus on the actions of the occupants. These additional requirements are warranted based on higher levels of care related to the concentration of people; physical and mental capabilities of the occupants; lack of familiarity with a building or simply because of the complexity and size of the building. These requirements are intended to improve the effectiveness of other measures required by the fire and building codes.

Basically, this chapter addresses the human contribution to life safety in buildings when a fire or other emergency occurs. The requirements for continuous training and scheduled evacuation drills can be as important as the required periodic inspections and maintenance of built-in fire protection features. The level of preparation by the occupants also improves the emergency responders' abilities during an emergency.

The *International Building Code*® (IBC®) focuses on built-in fire protection features, such as fire sprinkler systems, fire resistive construction and properly designed egress systems. The human element is only indirectly addressed in the IBC, whereas this chapter fully addresses the human element. Generally, fire codes address the human element more directly in their role in the long-term maintenance of buildings and systems. These issues have traditionally been associated with fire codes.

Chapters 3 and 6 through 10 of the code and Chapters 7 through 10 of the IBC set forth provisions for how and when buildings are to be properly equipped and maintained to prevent damage and loss of life in the event of a fire. These requirements are based on two complementary fire safety strategies: managing the fire and managing the occupants. These strategies are discussed in more detail in Chapter 1.

Managing Fire

A fire can be either prevented or managed. This chapter of the code focuses on training and preparedness, while also emphasizing prevention. In some cases, moving occupants to minimize their exposure to a hazard is difficult or impractical. In these situations, controlling or eliminating the hazard is preferable, especially while it is still manageable. In fact, this is the concept underlying all fire suppression requirements. Successful fire control depends on building occupants recognizing the fire threat, deciding to respond, choosing how to respond and, in the case of choosing fire control, identifying, locating and using the correct method. All of these functions must promptly take place in that order. Failure to perform promptly may preclude alternative strategies; therefore, location and identification of fire extinguishers and occupant standpipe hose lines are provided so that incipient fire-fighting equipment is readily accessible to occupants. These appliances, however, are often difficult to operate, and regardless of experience, fighting fire is a difficult and dangerous task. This chapter prescribes training requirements that assist occupants who are expected to respond to incipient fires to be adequately prepared and trained.

Managing Occupants

The management of occupants is primarily moving them away from the hazard. Verifying that enough exits have ample capacity, are immediately accessible, adequately arranged, appropriately identified and suitably protected are only the first steps toward achieving functional life safety. Occupants must know not only where exits are, but also when and how to use them. For instance, studies have shown that people have a "learned irrelevance" to emergency exits. Learned irrelevance is a psychological phenomenon that occurs when a person is exposed to a stimulus but a majority of the time does not need to respond to it. Because of this phenomenon, most occupants are likely to exit the way they have entered, whether it is the correct way or not; therefore, beyond designing the building with an adequate number of exits, a method of encouraging the use of the best exits must be developed. Identifying dangerous conditions, deciding how to act and responding appropriately and promptly are essential. Various factors and situations can make evacuation not only difficult but potentially impractical. All of these factors involve the interaction between the building; its systems and features; occupants and the fire. The code concentrates on the last two factors, while the IBC regulates the first two factors. Planning for life safety requires a response to these factors by defining the life safety strategies that must be implemented as well as the means to achieve them. Life safety factors, such as buildings, fire and people, are important in managing exposed occupants.

This chapter concentrates on planning and practicing the desired actions of building occupants when a fire occurs. The remainder of the code focuses on the behaviors and procedures that must be practiced or observed to prevent or control a fire. The best way to create a safe building environment is through fire prevention. No system can ensure complete protection of building occupants. Fires are not the only emergencies necessitating the implementation of life safety strategies; therefore, this chapter describes requirements for preparing and implementing life safety plans and programs in occupancies with special life safety problems. These include occupancies in which the number of occupants or the arrangement or complexity of the building may make evacuation or removal from hazardous conditions difficult or impractical.

Not all occupants of each building are equally capable of performing tasks essential to their safety. A growing awareness not only of people with physical disabilities but also of what constitutes a disability has focused life safety on everyone. Federal health care policies and funding criteria have spurred the deinstitutionalization of people who were formerly confined to nursing homes and other traditional health care institutions. This has created a new category of occupancies—board and care homes (institutional, residential care; I-1)—while the number of beds provided in nursing homes and hospitals continues to grow. Similarly, technological advances have promoted the creation of larger and more complex buildings, including high rises, open malls, domed stadiums, mixed-use complexes and convention centers. All of these situations create special life safety problems that physical features alone cannot remedy. Additionally, these situations require not only that adequate physical accommodations be provided but also that building occupants be trained to respond to emergencies in these facilities.

Life Safety

Life safety strategies involve the development of an explicit statement of a desired life safety outcome. This statement, once designed to the capabilities of the building occupants and the physical arrangement of the exits, becomes a life safety strategy. Such approaches stress defining a specific strategy or strategies for protecting occupants. Protection may include moving them (assisting), causing them to move (directing), defending them in place or a combination of these measures. An effective strategy must consider the number and capabilities of building occupants; the type, location and arrangement of building exits; the fire and its effects on the people and the building and the number, training and capability of staff to direct or perform fire evacuation or incipient fire-fighting duties. Each strategy, combined with effective planning and practice, becomes the means for achieving the desired life safety outcome.

The life safety strategies for a health care facility, a high-rise office building and a multiplex theater could vary considerably based on the specific characteristics of the use. First, while the number of occupants will be significant in each case, the actual number occupying the building may be varied. Similarly, the occupant density and location of people in the building will vary, as well as the physical arrangement of the building, which in the first example that follows may be assumed as primarily horizontal and in the second as principally vertical. The most profound difference will be the capabilities of the occupants. In high rises and theaters, building occupants will be expected to perform life safety behaviors themselves, while patients in a health care facility may require substantial assistance; however, high-rise and theater occupants will differ from each other in their levels of familiarity with the building design. Furthermore, in a theater, lighting conditions may interfere with the occupants' ability to discern the path of egress travel.

In the first two examples, health care and high rise, removing all occupants from the building in the event of fire is impractical. In a high-rise building, occupants located above a fire are in greater danger than occupants located below the fire, since combustion products naturally rise. In a health care occupancy, the risk to most occupants is compounded by their weakened or disabled condition prior to the fire. In both of these examples, a life safety strategy should first address the needs of those at greatest risk by removing them from harm. Secondly, the life-safety strategy should stress separating endangered occupants and their immediate neighbors from danger until the hazard can be controlled or confined. As seen, in each example, life safety strategies should also incorporate both partial relocation and defend-in-place concepts. In the health care facility, however, occupants will be moved horizontally to achieve this objective, while in the high rise, occupants will be expected to move downward or upward to separate themselves from danger. In the case of a multiplex theater, occupants will usually be directed to the nearest exit; however, its location may not be known to all or some of them. Furthermore, because employees in assembly occupancies must be trained in the proper use of portable fire extinguishers, the life safety strategy should include instruction in using these appliances to minimize occupant exposure to fire effects.

The resulting life safety strategies for these occupancies may resemble the following:

Example 1 – Health Care Facility: Upon notice of fire, direct or assist evacuation and relocate occupants from the area of fire origin to an adjacent smoke compartment through horizontal exits. Remove the most critically ill patients and those with special needs to an area providing the most appropriate level of care.

Example 2 – High-Rise Building: Direct occupants in the area or floor of fire origin to the nearest exits. Occupants on the fire floor, the floor above and the floor below will relocate sequentially up or down at least two floors. Occupants located two floors above and one floor below the fire floor will be sequentially relocated following movement of fire floor occupants.

Example 3 – Multiplex Theater: Announce exit locations and evacuation instructions prior to each movie. Over voice/alarm systems, direct occupants to nearest exit. Employees in the immediate vicinity of an incipient fire may attempt to control or extinguish it using a portable fire extinguisher after activating the fire alarm system.

Once the appropriate strategy has been defined, a plan can be expanded with little additional effort to form the backbone of a comprehensive life safety protocol.

The following statements provide additional instructions for the aforementioned examples:

Example 1 – Health Care Facility: Monitor or reinforce fire barriers so that they provide adequate defense against fire until it is controlled and extinguished. Staff will report progress of the fire and relocation operation to the Private Branch Exchange (PBX) operator through the nurse call station in adjacent smoke compartments.

Example 2 – High-Rise Building: Building fire manager will meet fire department personnel at the central control station located off the main lobby.

Example 3 – Multiplex Theater: Projectionist will stop films so that the alarm and evacuation instructions are heard and followed. Upon activation, instructions are heard and followed. Upon activation of the alarm, the on-duty manager will telephone the fire department to confirm that the fire was reported. Ushers will follow occupants out of each auditorium as conditions permit, closing exit doors and preventing reentry. Upon completing assigned duties, all staff will report to the manager located in front of the lobby entrance.

These expanded statements certainly do not constitute a fully developed plan; however, with these elements of the life safety plan defined, details can be added to send the plan from preincident preparation through post-incident follow-up. A well-developed plan should include all or most of the following elements:

- Assignment of roles and responsibilities;
- Description of fire protection systems, including operating instructions, if appropriate;
- Building floor plans and sections;
- Seating diagrams and occupant load;
- Number, location and path of travel to exits;
- Emergency notification lists and procedures;
- Post-incident follow-up procedures, including salvage and insurance information; and
- Plan revision and evaluation procedures.

Once a plan is developed, reviewed and approved, it must be distributed, practiced and periodically revised.

SECTION 401
GENERAL

401.1 Scope. Reporting of emergencies, coordination with emergency response forces, emergency plans, and procedures for managing or responding to emergencies shall comply with the provisions of this section.

Exception: Firms that have approved on-premises fire-fighting organizations and that are in compliance with approved procedures for fire reporting.

❖ This section describes the overall scope of Chapter 4, which notes that all procedures relating to reporting and managing fire and other emergencies be in accordance with this chapter. There is one exception that recognizes organizations, such as large industrial sites, that have on-site fire brigades. The fire brigades and the associated reporting procedures must be approved by the authority having jurisdiction.

401.2 Approval. Where required by this code, fire safety plans, emergency procedures, and employee training programs shall be approved by the fire code official.

❖ To verify that emergency procedures, training and fire safety plans have taken all essential factors into account, the plans and procedures must be approved by the fire code official.

401.3 Emergency forces notification. In the event an unwanted fire occurs on a property, the owner or occupant shall immediately report such condition to the fire department. Building employees and tenants shall implement the appropriate emergency plans and procedures. No person shall, by verbal or written directive, require any delay in the reporting of a fire to the fire department.

❖ This section requires prompt notification of the fire department in the event of an emergency. Employees/occupants are prohibited from delaying in any way the notification of the fire department.

401.3.1 Making false report. It shall be unlawful for a person to give, signal, or transmit a false alarm.

❖ Chapter 2 of the code defines a false alarm as an intentional activation of an alarm or notification of a fire or other emergency when no emergency exists. This would not include a malfunctioning alarm system. False alarms have the potential for causing confusion or panic among occupants of the affected premises, a situation that could lead to property damage, personal injury or death. False alarms also place fire fighters and other emergency personnel in potential danger during the unnecessary emergency response. This can jeopardize other lives and property in the community by committing emergency forces to a false situation when they might be needed at a true emergency elsewhere.

401.3.2 Alarm activations. Upon activation of a fire alarm signal, employees or staff shall immediately notify the fire department.

❖ This section specifically requires immediate notification of the fire department or other emergency response groups when an alarm signal is activated. Emergency plans and procedures must not include the requirement that employees report an alarm to a supervisor or similar person first. Quick response is the key to efficient and effective fire fighting.

401.3.3 Emergency evacuation drills. Nothing in this section shall prohibit the sounding of a fire alarm signal for the carrying out of an emergency evacuation drill in accordance with the provisions of Section 405.

❖ This section specifically allows fire alarm signals to be utilized as part of emergency evacuation drills. Without this provision, Section 401.3.1 would not allow the activation of the alarm signal during an emergency evacuation drill.

401.4 Interference with fire department operations. It shall be unlawful to interfere with, attempt to interfere with, conspire to interfere with, obstruct or restrict the mobility of or block the path of travel of a fire department emergency vehicle in any way, or to interfere with, attempt to interfere with, conspire to interfere with, obstruct or hamper any fire department operation.

❖ A potential hazard when fire departments are responding to an emergency is the inability to perform their operations because of physical obstructions, restricted mobility or human interference. This section prohibits any type of interference with emergency response operations. The delay of even a few minutes can cause serious property damage, injuries or fatalities.

401.5 Security device. Any security device or system that emits any medium that could obscure a means of egress in any building, structure or premise shall be prohibited.

❖ Security devices that, when activated, emit into a building a medium such as smoke or other aerosols that could obscure exits or confuse the occupants create an inherently dangerous situation for the public and responding emergency personnel. In cases of activation of these devices, armed criminal perpetrators could be trapped inside buildings. Law enforcement personnel arriving on the scene could easily believe that a building is on fire and responding fire fighters could enter and be confronted by the perpetrator. Another danger is that false fire alarms could be transmitted automatically or by passersby because of the appearance of smoke in the building.

SECTION 402
DEFINITIONS

402.1 Definition. The following word and term shall, for the purposes of this chapter and as used elsewhere in this code, have the meaning shown herein.

❖ The following word and term shall, for the purposes of this chapter and as used elsewhere in this code, have the meaning shown herein.

Definitions of terms can help in the understanding and application of the code requirements. The purpose for including those definitions that are associated with the subject matter of this chapter is to provide more convenient access to them without having to refer back to Chapter 2. It is important to emphasize that this term is not exclusively related to this chapter but is applicable everywhere the term is used in the code. For convenience, the term is also listed in Chapter 2 with a cross reference to this section. The use and application of all defined terms, including those defined in this section, are set forth in Section 201.

EMERGENCY EVACUATION DRILL. An exercise performed to train staff and occupants and to evaluate their efficiency and effectiveness in carrying out emergency evacuation procedures.

❖ This definition provides a consistent explanation of the purpose and extent of such activities. Without drilling the staff and occupants in the emergency procedures for which they have been trained, neither management, staff or the occupants can adequately gauge their readiness to perform in a crisis mode.

SECTION 403
PUBLIC ASSEMBLAGES AND EVENTS

403.1 General. When, in the opinion of the fire code official, it is essential for public safety in a place of assembly or any other place where people congregate, because of the number of persons, or the nature of the performance, exhibition, display, contest or activity, the owner, agent or lessee shall provide one or more fire watch personnel, as required and approved, to remain on duty during the times such places are open to the public, or when such activity is being conducted. The fire watch personnel shall keep diligent watch for fires, obstructions to means of egress and other hazards during the time such place is open to the public or such activity is being conducted and take prompt measures for remediation of hazards, extinguishment of fires that occur and assist in the evacuation of the public from the structures.

❖ Even though Chapter 24 requires standby personnel in tents and membrane structures because of the inherently higher life safety risks associated with such occupancies, this section gives the fire code official the authority to require fire-watch personnel in indoor or outdoor Group A occupancies or other venues where people congregate when the nature of the performance, exhibition, display, contest or activity is such that the presence of fire-watch personnel are essential to public safety (see commentary, Section 202, to the definition of "Fire watch").

403.1.1 Public safety plan. In other than Group A or E occupancies, where the fire code official determines that an indoor or outdoor gathering of persons has an adverse impact on public safety through diminished access to buildings, structures, fire hydrants and fire apparatus access roads or where such gatherings adversely affect public safety services of any kind, the fire code official shall have the authority to order the development of, or prescribe a plan for, the provision of an approved level of public safety.

❖ This section notes that these provisions are not for Group A and E occupancies, which are dealt with more specifically elsewhere in this chapter. This section is important because it provides the fire code official the authority to require the development of or to prescribe a specific plan for large gatherings. Such gatherings could include outdoor festivals, demonstrations or receptions. If such assemblies include the use of tents and canopies, Chapter 24 would also apply.

Again, the primary aim of this section is to address the fact that these large gatherings may hamper the ability of the fire department and other emergency responders to access and protect buildings and building occupants.

403.1.2 Contents. The public safety plan, where required by Section 403.1.1, shall address such items as emergency vehicle ingress and egress, fire protection, emergency medical services, public assembly areas and the directing of both attendees and vehicles (including the parking of vehicles), vendor and food concession distribution, and the need for the presence of law enforcement, and fire and emergency medical services personnel at the event.

❖ As further guidance, this section provides some specific issues to be addressed that include items such as the direction of traffic; vendor and food concession distribution and the need for law enforcement and medical services.

SECTION 404
FIRE SAFETY AND EVACUATION PLANS

404.1 General. Fire safety and evacuation plans shall comply with the requirements of this section.

❖ This section simply states that all fire safety and evacuation plans must comply with Section 404.

404.2 Where required. An approved fire safety and evacuation plan shall be prepared and maintained for the following occupancies and buildings.

1. Group A, other than Group A occupancies used exclusively for purposes of religious worship that have an occupant load less than 2,000.
2. Group E.

3. Group H.
4. Group I.
5. Group R-1.
6. Group R-4.
7. High-rise buildings.
8. Group M buildings having an occupant load of 500 or more persons or more than 100 persons above or below the lowest level of exit discharge.
9. Covered malls exceeding 50,000 square feet (4645 m^2) in aggregate floor area.
10. Underground buildings.
11. Buildings with an atrium and having an occupancy in Group A, E or M.

❖ The list provided notes when plans and procedures need to be developed. As discussed earlier, the occupancies and uses addressed by this chapter were chosen based on the density and location of occupants, the layout of the building or simply the limitations of the occupants during an emergency.

404.3 Contents. Fire safety and evacuation plan contents shall be in accordance with Sections 404.3.1 and 404.3.2.

❖ The two primary plans required by Section 404.3 are a fire evacuation plan and a fire safety plan. The fire evacuation plan focuses primarily on the procedures for the evacuation of the occupants in an emergency. The fire safety plan focuses on the overall understanding of the fire protection package of the building as it pertains to the layout of the building, the contents of the building, the means of egress system, the fire hazards and the identification of key contacts during an emergency.

404.3.1 Fire evacuation plans. Fire evacuation plans shall include the following:

1. Emergency egress or escape routes and whether evacuation of the building is to be complete or, where approved, by selected floors or areas only.
2. Procedures for employees who must remain to operate critical equipment before evacuating.
3. Procedures for accounting for employees and occupants after evacuation has been completed.
4. Identification and assignment of personnel responsible for rescue or emergency medical aid.
5. The preferred and any alternative means of notifying occupants of a fire or emergency.
6. The preferred and any alternative means of reporting fires and other emergencies to the fire department or designated emergency response organization.
7. Identification and assignment of personnel who can be contacted for further information or explanation of duties under the plan.
8. A description of the emergency voice/alarm communication system alert tone and preprogrammed voice messages, where provided.

❖ The primary focus of evacuation plans is to prepare for and define the roles for evacuation and relocation of occupants during an emergency. The fire evacuation plan is important for both the emergency responders and the building or facility occupants. It focuses the occupants' activities on facilitating a smoother evacuation or relocation process and provides the fire department with critical information on the building and the location of the occupants. Keep in mind that these requirements apply to all occupancies listed in Section 404.2. The occupancy- and use-specific requirements are located within Section 408; therefore, the requirements listed here are general and will vary based on many factors, such as the occupants' mobility and familiarity with the building.

Item 1 requires that specific escape routes be defined. This is important because the building is generally designed to facilitate evacuation or relocation in an emergency if the patterns of movement are coordinated. For instance, as noted earlier, a high-rise building will most likely be evacuated in phases. If floors begin evacuating before intended, the evacuation of the occupants in the fire area may be delayed. Also, if everyone tries to use the same exits in a facility, such as a multiplex theater, evacuation of the building will be delayed. As stated earlier, studies have shown that people tend to exit the way they enter a building. Note that the code sometimes requires a certain level of redundancy to account for occupants using the same exits. For example, the IBC requires that the main exit of multiplex theaters be sized for at least half of the occupants even though plenty of egress width may be available elsewhere in the building. The more coordinated the plan, the more evenly the exits will be used.

Item 2 requires that specific procedures for evacuation be provided to those employees who must operate critical equipment before evacuation. These procedures are necessary to ensure a clear understanding to the occupant when evacuation is critical and the operation should be abandoned.

Item 3 simply states that a plan be developed to account for all occupants after evacuation or relocation. This is important not only to the occupants but also to the emergency responders to assess their actions when arriving at the scene.

Item 4 has two roles. First, it provides a designated person for occupants to look to for assistance in an emergency. This will reduce the stress of the situation. Second, when the emergency responders arrive they will have a specific contact to help them assess the situation. These contacts can also be beneficial to emergency responders when preplanning their response to that specific facility.

Item 5 requires that the notification to the occupants of the emergency be standardized. The approach will vary based on the occupancy and use. For instance, all occupants in a multiplex theater would be notified whereas in a correctional facility or hospital only staff will be notified. Also, if the method is standardized, it is easier to differentiate between emergency and nonemergency signals, which facilitates a smoother reaction when an emergency does occur.

Item 6 is focused on the notification of the emergency responders. They are more likely to get the notification of an emergency if a standard protocol exists. This can vary from one occupancy or use to another but as long as a straightforward, consistent method is used, it will facilitate a quicker response. Note that Section 401.3.2 requires direct contact with the fire department once the fire alarm signal is activated; therefore, no intermediate steps, such as investigation, are allowed.

As with Item 4, Item 7 requires a specific contact who is familiar with the plan and how the building operates. This information is helpful for the emergency responders in their preplanning activities. Without a specific contact, the process of getting vital information can become much more difficult for the fire department. In a large building or facility, the safety officer or similar person is most appropriate for such a role.

Item 8 requires documentation of the voice/alarm communications system alert tone and preprogrammed voice messages. This provides emergency responders with a better understanding of the information provided to occupants to better assess the appropriate response. Additionally, if conditions in that building change, the plan can be evaluated to see whether this aspect of the notification system needs to be revised. For instance, if the procedures for evacuation have changed, the voice announcement may need to be revised.

404.3.2 Fire safety plans. Fire safety plans shall include the following:

1. The procedure for reporting a fire or other emergency.
2. The life safety strategy and procedures for notifying, relocating, or evacuating occupants.
3. Site plans indicating the following:
   3.1. The occupancy assembly point.
   3.2. The locations of fire hydrants.
   3.3. The normal routes of fire department vehicle access.
4. Floor plans identifying the locations of the following:
   4.1. Exits.
   4.2. Primary evacuation routes.
   4.3. Secondary evacuation routes.
   4.4. Accessible egress routes.
   4.5. Areas of refuge.
   4.6. Manual fire alarm boxes.
   4.7. Portable fire extinguishers.
   4.8. Occupant-use hose stations.
   4.9. Fire alarm annunciators and controls.
5. A list of major fire hazards associated with the normal use and occupancy of the premises, including maintenance and housekeeping procedures.
6. Identification and assignment of personnel responsible for maintenance of systems and equipment installed to prevent or control fires.
7. Identification and assignment of personnel responsible for maintenance, housekeeping and controlling fuel hazard sources.

❖ This section requires an overall fire safety plan with emphasis on the building and building site layout and hazards. More specifically, information such as the evacuation and relocation aspects of the building layout needs to be clarified, the list of specific hazards associated with normal use of the building needs to be noted and fire department access road locations need to be provided.

This plan also includes identification of the specific personnel who are charged with managing the fire protection systems and equipment and with fire prevention duties, such as controlling combustibles on site. Having specific personnel assigned will work to increase the likelihood of these actions occurring.

The requirements of this plan provide the building owner and occupants a better understanding of how to react in an emergency and how to decrease the likelihood of an emergency occurring. Additionally, this report assists emergency responders during periodic inspections and evaluations of the plans and, more importantly, when responding to an emergency. Generally, buildings that have fairly rigid and well-maintained plans and procedures in place reduce not only the likelihood and magnitude of an incident within the jurisdiction but also the burden to emergency responders.

404.4 Maintenance. Fire safety and evacuation plans shall be reviewed or updated annually or as necessitated by changes in staff assignments, occupancy, or the physical arrangement of the building.

❖ In order to be of optimum value to a facility, plans must accurately reflect building conditions. Plans must be reviewed annually or when building changes affecting the instructions or procedures in the fire safety or emergency evacuation plan occur. Such a review should prompt an immediate revision and redistribution of the plan to all concerned parties, including emergency response personnel.

404.5 Availability. Fire safety and evacuation plans shall be available in the workplace for reference and review by employees, and copies shall be furnished to the fire code official for review upon request.

❖ This essentially requires that these plans be easily accessible to building occupants and the fire code official. If the plans are difficult to access, they are less likely to be updated when necessary and are more likely to be lost or forgotten. This places a burden on the emergency responders when planning methods of response, and puts the occupants of the building at a higher risk during an emergency. Having the documents readily available makes review and use for training occupants more likely.

SECTION 405 EMERGENCY EVACUATION DRILLS

405.1 General. Emergency evacuation drills complying with the provisions of this section shall be conducted in the occupancies listed in Section 404.2 or when required by the fire code official. Drills shall be designed in cooperation with the local authorities.

❖ One of the keys to the success of emergency planning and procedures is the training of the occupants. Section 405 sets out the frequency of drills along with who should be involved and how the drills should be accomplished. These requirements were written generally to apply to all occupancies requiring such drills.

These drills are either as required by the occupancies listed within this chapter or as specifically required by the fire code official. The code has a single section to be used consistently any time an evacuation drill is required. Having a consistent approach provides a level of understanding to the general public on evacuation that would not be obtained if the drill were conducted differently in each occupancy. Depending on the occupancy, certain occupants might not be part of the evacuation drill.

405.2 Frequency. Required emergency evacuation drills shall be held at the intervals specified in Table 405.2 or more frequently where necessary to familiarize all occupants with the drill procedure.

❖ To utilize to their best advantage fire drills and the lessons that they teach, drills should be conducted on a regular basis to familiarize both staff and residents with the evacuation plan. The element of surprise is not necessarily of significant benefit in those occupancies where residents may be prone to maladaptive behavior. Drills should be designed and practiced to reinforce relocation or evacuation behaviors as adaptive planned responses to stressful and potentially dangerous situations. Drills should be scheduled so that all staff members on all shifts have an opportunity to participate in them. Practice makes perfect, and when it comes to effective egress there is no substitute for fire drills at regular intervals so that all occupants are familiar with the plan details and their particular responsibilities in implementing them. Truly effective drills test the plan by varying conditions and force occupants to adapt. Many conditions can conspire to affect available safe egress time, and drills should incorporate some allowance for unanticipated conditions, such as: delayed detection, rapid fire growth, reduced staffing or poor weather conditions, as may be appropriate for the occupancy. Discovering deficiencies in the evacuation plan should be encouraged, and every opportunity should be taken to improve the plans.

TABLE 405.2
FIRE AND EVACUATION DRILL FREQUENCY AND PARTICIPATION

GROUP OR OCCUPANCY	FREQUENCY	PARTICIPATION
Group A	Quarterly	Employees
Group E	Monthly[a]	All occupants
Group I	Quarterly on each shift	Employees[b]
Group R-1	Quarterly on each shift	Employees
Group R-4	Quarterly on each shift	Employees[b]

a. The frequency shall be allowed to be modified in accordance with Section 408.3.2.

b. Fire and evacuation drills in residential care assisted living facilities shall include complete evacuation of the premises in accordance with Section 408.10.5. Where occupants receive habilitation or rehabilitation training, fire prevention and fire safety practices shall be included as part of the training program

❖ Table 405.2 provides the varying frequencies based on occupancy. The table also prescribes who should be involved in these drills. The level of participation is based on the type of occupancy. It is unreasonable to expect that in an occupancy such as Group A the general public would participate. The overall strategy for these occupancies is to provide a package of relevant information to occupants before an emergency to have the staff facilitate and direct occupants during an emergency. Also, in facilities such as hospitals and correctional facilities it is potentially dangerous to involve anyone but the employees in such drills. The necessary participants in the drills are related to the overall emergency strategies for those buildings. Group E, I-1 and R-4 occupancies are the only occupancies requiring that everyone be involved. This is related to the fact that occupants are generally able to evacuate. In the case of educational occupancies, drills serve as a learning tool for children to carry through their lives. Schools have generally stressed these drills because of large losses in fires such as in 1958 at Our Lady of Angels School in Chicago, Illinois. In terms of Group I-1 and R-4 facilities, the occupants generally are able to evacuate with some assistance. It is within their best interest to be familiar with the egress routes. More discussion on Group E, I-1 and R-4 occupancies is found in Section 408.

405.3 Leadership. Responsibility for the planning and conduct of drills shall be assigned to competent persons designated to exercise leadership.

❖ This section requires a focal point in the planning and execution of evacuation drills. Having a single point of contact streamlines the process and provides a necessary leadership role.

405.4 Time. Drills shall be held at unexpected times and under varying conditions to simulate the unusual conditions that occur in case of fire.

❖ If fire and emergency drills are a routine planned occurrence, they will not simulate actual reaction to an emergency but will provide an inaccurate and most likely optimistic outcome; therefore, the drills need to occur at random.

405.5 Record keeping. Records shall be maintained of required emergency evacuation drills and include the following information:

1. Identity of the person conducting the drill.
2. Date and time of the drill.
3. Notification method used.
4. Staff members on duty and participating.
5. Number of occupants evacuated.
6. Special conditions simulated.
7. Problems encountered.
8. Weather conditions when occupants were evacuated.
9. Time required to accomplish complete evacuation.

❖ Documenting the frequency and efficiency of emergency evacuation drills not only aids the fire code official in verifying that drills complying with these provisions have been performed but may also help administrators identify trends in emergency evacuation drill performance. Accurate records help life safety planners determine the adequacy of their plans. Identifying issues such as problems encountered and weather conditions helps to further determine which elements create the largest delays and why.

405.6 Notification. Where required by the fire code official, prior notification of emergency evacuation drills shall be given to the fire code official.

❖ In some cases, the fire code official will want prior notification of evacuation drills because he or she may need to prepare for such an event. This section provides him or her with the authority to require such notification.

405.7 Initiation. Where a fire alarm system is provided, emergency evacuation drills shall be initiated by activating the fire alarm system.

❖ To simulate conditions normally experienced during an emergency, the emergency notification procedures, which would include a fire alarm system in many cases, must be used.

405.8 Accountability. As building occupants arrive at the assembly point, efforts shall be made to determine if all occupants have been successfully evacuated or have been accounted for.

❖ This requirement is key to the success of evacuation plans. If a method is not available to account for the occupants once evacuation or relocation is complete, search and rescue activities will be more difficult for the emergency responders. Also, it would be difficult to measure the success of the plan.

405.9 Recall and reentry. An electrically or mechanically operated signal used to recall occupants after an evacuation shall be separate and distinct from the signal used to initiate the evacuation. The recall signal initiation means shall be manually operated and under the control of the person in charge of the premises or the official in charge of the incident. No one shall reenter the premises until authorized to do so by the official in charge.

❖ This section is primarily aimed at the concern that the occupants will be confused if similar signals are used to notify them of an alarm and also for reentry. This confusion has the consequences of slowing or even halting evacuation during an actual emergency. Additionally, to make sure that occupants do not go back into the building prematurely, any reentry signal must be operated manually to avoid a situation where it automatically sounds. Finally, this section specifically prohibits reentry until authorization is provided by the official in charge at the scene.

SECTION 406 EMPLOYEE TRAINING AND RESPONSE PROCEDURES

406.1 General. Employees in the occupancies listed in Section 404.2 shall be trained in the fire emergency procedures described in their fire evacuation and fire safety plans. Training shall be based on these plans and as described in Section 404.3.

❖ In most cases, the success of an evacuation and fire safety plan hinges on the appropriate reactions of the building occupants. The main activity that building occupants must undertake is removing themselves from the hazards. In some cases, fire safety and evacuation plans involve additional actions by the employees of the facility. For instance, in the case of hospitals, the nurses and other hospital staff must relocate patients; therefore, specific training is required for those activities.

Additionally, employees must be trained based on the specific fire safety and fire evacuation plans.

406.2 Frequency. Employees shall receive training in the contents of fire safety and evacuation plans and their duties as part of new employee orientation and at least annually thereafter. Records shall be kept and made available to the fire code official upon request.

❖ This section requires that employee training occur during new employee indoctrination and annually thereafter. A record of this training must be provided to the fire code official when requested. This section provides a minimum criterion for the training frequency for all occupancies addressed by Chapter 4. Section 408 may require more restrictive training frequencies.

406.3 Employee training program. Employees shall be trained in fire prevention, evacuation and fire safety in accordance with Sections 406.3.1 through 406.3.3.

❖ This is a general section that requires all employees to be trained in fire prevention, evacuation and fire safety in accordance with the subsections that follow. These provisions are primarily intended as a mechanism to ensure that training occurs and not as a requirement for establishing training criteria.

406.3.1 Fire prevention training. Employees shall be apprised of the fire hazards of the materials and processes to which they are exposed. Each employee shall be instructed in the proper procedures for preventing fires in the conduct of their assigned duties.

❖ If a fire can be prevented, evacuation and relocation of the occupants will also be avoided. Employees must be made aware of the potential hazards related to their particular area of the facility and what can be done to avoid a hazardous situation. Having specific procedures increases the likelihood that proper fire prevention techniques will be followed. Generally, people tend to be unaware of many hazards unless they are alerted to them. An example is the use of space heaters.

406.3.2 Evacuation training. Employees shall be familiarized with the fire alarm and evacuation signals, their assigned duties in the event of an alarm or emergency, evacuation routes, areas of refuge, exterior assembly areas, and procedures for evacuation.

❖ In the event that an emergency does occur, the employees must be prepared to assist in the evacuation or relocation of occupants. This training will vary widely from one occupancy type to another. In a high-rise building only some of the occupants will be evacuated at a time, whereas a school will evacuate all occupants at once. This section requires that the training occurs.

406.3.3 Fire safety training. Employees assigned fire-fighting duties shall be trained to know the locations and proper use of portable fire extinguishers or other manual fire-fighting equipment and the protective clothing or equipment required for its safe and proper use.

❖ Any time employees are to take specific action during a fire event, proper training is required. This section holds the building owner or operator responsible for making sure the training occurs.

SECTION 407
HAZARD COMMUNICATION

407.1 General. The provisions of Sections 407.2 through 407.7 shall be applicable where hazardous materials subject to permits under Section 2701.5 are located on the premises or where required by the fire code official.

❖ This section is specific to buildings and facilities that contain hazardous materials over the permitted amounts listed in Section 105.6. Knowledge related to which hazardous materials are on site is critical in several ways. First, it assists emergency responders in knowing what to expect when responding to a scene. Second, it provides emergency responders with an idea of incidents that may occur at a building or facility. Lastly, it provides a better understanding to the occupants of the potential hazards present and how to avoid emergencies.

Facilities that store, use or handle hazardous materials on a large scale generally depend heavily on the actions of the employees to prevent or minimize hazardous materials incidents. Therefore, the occupants play a strong role in the overall protection package for the building.

These types of requirements would, in the past, have been found only within the hazardous materials section of the code. These provisions have also been located within this chapter because of the nature of the information. These requirements are aimed at preparing both the occupants and the emergency responders. Hazardous materials present a wide range of problems because of the significant variation of properties and reactions; therefore, a reference to these requirements has been included in Chapter 4. These specific provisions are found in Section 2701.

This section states that when subject to permits, hazardous materials must also be addressed by this section. This section has requirements for the submittal of Material Safety Data Sheets (MSDS); the labeling or marking of hazardous materials through placarding and related identification; training; the compilation, when required, of both hazardous materials inventory statements and hazardous materials management plans and finally, the submittal of a closure plan when a facility is being shut down.

407.2 Material Safety Data Sheets. Material Safety Data Sheets (MSDS) for all hazardous materials shall be readily available on the premises.

❖ MSDS provide critical information about individual chemicals and their related hazards. This section requires that these data sheets be readily available on the premises. This allows availability for review by both employees and emergency responders. An appropriate location may be the security room at a facility or perhaps the main office. These sheets can potentially play a role in the response to an emergency. For example, if a chemical is noted as being water reactive on the MSDS, depending on the level of water reactivity, applying water to that spill may not be an appropriate response.

407.3 Identification. Individual containers of hazardous materials, cartons or packages shall be marked or labeled in accordance with applicable federal regulations. Buildings, rooms and spaces containing hazardous materials shall be identified by hazard warning signs in accordance with Section 2703.5.

❖ This section requires two activities related to the identification of hazardous materials. First, chemicals must be specifically labeled. Second, rooms or areas where the materials are located must be specifically labeled. In this case, the code essentially requires placarding as defined in Section 2703.5, which references NFPA 704.

407.4 Training. Persons responsible for the operation of areas in which hazardous materials are stored, dispensed, handled or used shall be familiar with the chemical nature of the materials and the appropriate mitigating actions necessary in the event of a fire, leak or spill. Responsible persons shall be designated and trained to be liaison personnel for the Fire Department. These persons shall aid the Fire Department in preplanning emergency responses and identification of the locations where hazardous materials are located, and shall have access to Material Safety Data Sheets and be knowledgeable in the site emergency response procedures.

❖ This section requires training specific to the hazards of the materials located and used at a particular building or facility. As noted, the actions taken will vary based on the hazards associated with the materials. Additionally, this section requires a specific group of persons to be designated as points of contact for the fire department. Having specific points of contact is critical because it eases planning and response procedures. These contacts also provide the fire department with specific persons who are more familiar with the hazards of the building or facility. The fire department is charged with responding to many different businesses within a community. Having specified contacts at facilities containing hazardous materials helps them prepare and respond.

407.5 Hazardous Materials Inventory Statement. Where required by the fire code official, each application for a permit shall include a Hazardous Materials Inventory Statement (HMIS) in accordance with Section 2701.5.2.

❖ A HMIS is further described in Chapter 27 but is essentially a document listing all the hazardous materials found on site. This documentation includes information such as the type of material, amount, specific hazards associated with the material and how it is used. All of this information can be very important for emergency planning and preparedness. An HMIS is required only if the fire code official specifically requires one.

407.6 Hazardous Materials Management Plan. Where required by the fire code official, each application for a permit shall include a Hazardous Materials Management Plan (HMMP) in accordance with Section 2701.5.1. The fire code official is authorized to accept a similar plan required by other regulations.

❖ As with the HMIS, an HMMP is required only when the fire code official specifically requires one. This document is somewhat different from the HMIS in that it is geared to the layout of the building and the location and use of the hazardous materials. This document provides a better understanding of how the facility operates. This information in turn provides more detailed information to the emergency responders. This plan will also include such information as the location of aisles; the type and location of emergency equipment available and location of specific shutoff valves and other operating equipment. The detailed requirements for HMMPs are located in Chapter 27.

407.7 Facility closure plans. The permit holder or applicant shall submit to the fire code official a facility closure plan in accordance with Section 2701.6.3 to terminate storage, dispensing, handling or use of hazardous materials.

❖ It is important for emergency responders to be made aware of the closure of a plant that uses or stores hazardous materials. First, closure means a readjustment in their planning. Second, the extent of the closure must be communicated so that the emergency responders are made aware of hazards that may still be present. Any hazards that are still present may potentially be more dangerous, since the facility is now unattended; therefore, the building owner must develop a plan that is acceptable to the fire code official. In some cases a facility will be only temporarily closed, which would mean maintaining a permit and continuing inspections.

SECTION 408
USE AND OCCUPANCY-RELATED REQUIREMENTS

408.1 General. In addition to the other requirements of this chapter, the provisions of this section are applicable to specific occupancies listed herein.

❖ This section provides specific occupancy and use requirements for emergency planning and preparedness and evacuation procedures. As noted, the occupancies and uses dealt with in this section have unique needs based on the abilities of the occupants; the density and number of occupants; the size and layout of the building and the perceived risk to the occupants.

This section states that these requirements are in addition to the general requirements found in the rest of Chapter 4.

408.2 Group A occupancies. Group A occupancies shall comply with the requirements of Sections 408.2.1 and 408.2.2 and Sections 401 through 406.

❖ Group A occupancies are a special concern because of the high density and number of occupants. Additionally, occupants in Group A occupancies are generally not very familiar with the building, since they tend to consist of the general public.

408.2.1 Seating plan. The fire safety and evacuation plans for assembly occupancies shall include the information required by Section 404.3 and a detailed seating plan, occupant load, and occupant load limit. Deviations from the approved plans shall be allowed provided the occupant load limit for the occupancy is not exceeded and the aisles and exit accessways remain unobstructed.

❖ Proper planning for an assembly occupancy must consider the number, capacity and physical arrangement of exits. In turn, these factors will dictate how seating may be arranged to prevent obstruction of aisles and exits. The number of seats provided may not exceed what is permitted by the number, arrangement and capacity of exits. Floor area factors are only one element in determining whether the exit capacity is adequate. Additionally, the egress plan must be approved by the fire code official, with a copy of the approved plan maintained on the premises for review by employees and inspectors. Many facilities have several approved plans to accommodate various situations and functions. The seating plan selected for any event should reflect the needs of the group and the requirement to keep aisles and exits clear. Deviations from an approved plan may be permitted only if they do not obstruct the complete egress path, including aisles and exits.

408.2.2 Announcements. In theaters, motion picture theaters, auditoriums and similar assembly occupancies in Group A used for noncontinuous programs, an audible announcement shall be made not more than 10 minutes prior to the start of each program to notify the occupants of the location of the exits to be used in the event of a fire or other emergency.

> **Exception:** In motion picture theaters, the announcement is allowed to be projected upon the screen in a manner approved by the fire code official.

❖ Announcements are intended to familiarize occupants with life safety system features that they may need to use if a fire occurs. Information is the most valuable commodity during fires and other emergencies; however, it is often difficult, if not impossible, to override the excitement and confusion caused by a fire or other emergency. Studies have generally shown, however, that occupants do not panic as once thought; therefore, it is imperative that occupants receive information necessary for them to make decisions before an emergency occurs. To convey information and motivate an adaptive response to fires or other emergencies, the message must stimulate interest and speak directly to the topic. Many movie theaters currently use "trailers" or "shorts" to market concession items, as well as fire safety. These messages can be especially effective if they are specific and adequately distinguished from other promotions. Any message should reflect the life safety strategy and must point out specific features of the occupancy. Occupants in most theaters and auditoriums are usually expected to leave the building immediately upon notification of a fire using the nearest available exit. This is largely a reflection upon the building type and arrangement of exits. In these cases, the locations of all exits must be identified. Raising the house lights along the egress path or modulating aisle lighting at appropriate times during the message can reinforce the message.

This section specifically requires an audible announcement, but the exception for motion picture theaters would allow the message to be visually displayed upon the screen as approved. A combination of both audible and visual would most likely be the most effective. This section also has a maximum time from the start of the show to provide such announcements; otherwise, an announcement that comes too early will lose its effectiveness.

408.3 Group E occupancies. Group E occupancies shall comply with the requirements of Sections 408.3.1 through 408.3.4 and Sections 401 through 406.

❖ This a general section noting to the fire code official that compliance with Sections 408.3.1 through 408.3.4 is applicable to Group E occupancies. Group E, Educational, occupancies are one of the primary occupancies that need emergency evacuation drills. It is one of the few occupancies that requires all occupants to participate in evacuation drills. The effectiveness of preemergency planning in educational occupancies has been significant, as evidenced by a remarkable decline in tragic fires in schools over the years. The skills learned by children in school are often carried with them through the rest of their lives.

408.3.1 First emergency evacuation drill. The first emergency evacuation drill of each school year shall be conducted within 10 days of the beginning of classes.

❖ Group E occupants vary from year to year as children enter and leave grade levels. In addition, even though a child may be in the same school for a number of years, location within the building and leaders change. It is important, therefore, that the first evacuation drill occurs within the first 10 days of school. This provides students with nearly immediate training and the school and emergency responders with information about where problems exist. Additionally, in more recent years schools have been immersing students with disabilities into the general classroom. The location and number of these students varies each year. The ability to evacuate them must be assessed early in the school year.

408.3.2 Emergency evacuation drill deferral. In severe climates, the fire code official shall have the authority to modify the emergency evacuation drill frequency specified in Section 405.2.

❖ This section provides the fire code official with the authority to delay required fire drills based on extreme climate conditions. For instance, it would be unreasonable to send students and teachers out during a hurricane or in subzero temperatures; however, all inclement weather should not be avoided since it provides a more realistic evacuation scenario. This section simply recognizes that extreme weather conditions can be considered when scheduling drills.

408.3.3 Time of day. Emergency evacuation drills shall be conducted at different hours of the day or evening, during the changing of classes, when the school is at assembly, during the recess or gymnastic periods, or during other times to avoid distinction between drills and actual fires.

❖ If evacuation drills are done routinely, they will be easily distinguished as drills and not an actual fire. This is potentially dangerous because the behavior patterns will be different and actual preparedness for emergency will be lessened. Time of day, therefore, should be varied whether it is convenient or not. A fire will not differentiate between a convenient and an inconvenient time.

408.3.4 Assembly points. Outdoor assembly areas shall be designated and shall be located a safe distance from the building being evacuated so as to avoid interference with fire department operations. The assembly areas shall be arranged to keep each class separate to provide accountability of all individuals.

❖ A key element in safe evacuation is the exit discharge portion of the evacuation route. Once occupants are outside the building, they need to be located far enough away from it to avoid further hazards. Additionally, there is a potential for the assembly point to interfere with the emergency operations of the fire department, even if they are not present during an evacuation drill. Locations, therefore, need to be designated that both avoid hazards and keep the evacuated occupants away from probable paths of emergency response. Also, to simplify accounting for the occupants once outside, the code requires each class to remain together as a group and separate from other classes.

408.4 Group H-5 occupancies. Group H-5 occupancies shall comply with the requirements of Sections 408.4.1 through 408.4.4 and Sections 401 through 407.

❖ Group H-5 occupancies are semiconductor facilities. These types of facilities are unique in that they have a very large allowable area and house a significant amount of hazardous materials. Essentially, the building is divided into fabrication areas, hazardous production materials (HPM) rooms, which are Group H occupancy storage rooms and networks of service corridors and spaces. The area of the building needs to be large to incorporate all of the operations needed by the semiconductor industry. To facilitate these operations, a special package of requirements and occupancy classification was created. This package is found within the occupancy requirements in the IBC and Chapter 18 of the code. As part of the package, special emergency preparedness and preparation is also required. Sections 408.4.1 through 408.4.4 provide the specific requirements for emergency preparedness and preparation. These requirements are in addition to others pertaining to hazardous materials found in Section 407.

408.4.1 Plans and diagrams. In addition to the requirements of Section 404 and Section 407.6, plans and diagrams shall be maintained in approved locations indicating the approximate plan for each area, the amount and type of HPM stored, handled and used, locations of shutoff valves for HPM supply piping, emergency telephone locations and locations of exits.

❖ The requirements for plans and diagrams are in addition to those required in Sections 404 and 407. More specifically, the additional details required include the approximate plan for each area of the building. This includes the amount of HPM stored, handled and used in both the fabrication areas and the HPM rooms. Additionally, since it is typical that such facilities tend to pipe HPM throughout the building for efficiency, all shutoff valves must be identified. Finally, exits must be clearly marked. This is specifically necessary because many service corridors for the transport of materials are not considered part of the means of egress. These details will assist both the occupants and the emergency responders.

408.4.2 Plan updating. The plans and diagrams required by Section 408.4.1 shall be maintained up to date and the fire code official and fire department shall be informed of all major changes.

❖ Semiconductor facilities are constantly changing because of the needs of new technology; therefore, the types, amounts of materials and their application are constantly changing. These changes must be accounted for within the plans and diagrams required in Sections 408.4.1 and 407; otherwise, the critical information needed by the emergency responders may not be available.

408.4.3 Emergency response team. Responsible persons shall be designated the on-site emergency response team and trained to be liaison personnel for the fire department. These persons shall aid the fire department in preplanning emergency responses, identifying locations where HPM is stored, handled and used, and be familiar with the chemical nature of such material. An adequate number of personnel for each work shift shall be designated.

❖ This section is similar to Section 407.4 but has some specific requirements unique to semiconductor facilities. More specifically, on-site liaisons need to be available who are familiar with the location of HPM and the associated hazards related to those materials. For example, semiconductor facilities make use of silane gas, which is a phyrophoric—it will instantly ignite when exposed to atmospheric conditions. Knowing the characteristics of the material, where it is located and the quantity of the material used is critical information to the responding emergency personnel.

408.4.4 Emergency drills. Emergency drills of the on-site emergency response team shall be conducted on a regular basis but not less than once every three months. Records of drills conducted shall be maintained.

❖ This section specifically requires employees to conduct drills every three months to practice specific emergency procedures for the facilities. Again, this requirement is specific to semiconductor facilities because of their unique layout and contents. The emergency responders rely heavily on the on-site actions of the employees because of their intimate knowledge of the site and the complexity of these buildings. Records must be maintained.

408.5 Group I-1 occupancies. Group I-1 occupancies shall comply with the requirements of Sections 408.5.1 through 408.5.5 and Sections 401 through 406.

❖ A Group I-1 occupancy is one that houses over 16 individuals who live in a supervised residential care facility on a 24-hour basis. This would include: residential board and care homes; congregate care facilities; social rehabilitation facilities; alcohol and drug centers; assisted living facilities and convalescent facilities. Generally, occupants of these facilities are able to respond to an emergency with some assistance from staff.

408.5.1 Fire safety and evacuation plan. The fire safety and evacuation plan required by Section 404 shall include special staff actions including fire protection procedures necessary for residents and shall be amended or revised upon admission of any resident with unusual needs.

❖ This section is in addition to the general requirements found in Section 404. More specifically, any special requirements based on the specific needs of residents must be included as part of the plan. These plans must be reviewed each time a new occupant arrives to assess whether there are any special features that need to be included in the plan to address those particular needs.

408.5.2 Staff training. Employees shall be periodically instructed and kept informed of their duties and responsibilities under the plan. Such instruction shall be reviewed by the staff at least every two months. A copy of the plan shall be readily available at all times within the facility.

❖ These types of facilities are normally occupied by people who have the ability to evacuate or relocate with a certain level of assistance from the staff; therefore, staff training is critical. This section requires that training occur every two months because the needs of the occupants may change over time as certain physical or mental conditions progress or new occupants arrive.

408.5.3 Resident training. Residents capable of assisting in their own evacuation shall be trained in the proper actions to take in the event of a fire. The training shall include actions to take if the primary escape route is blocked. Where the resident is given rehabilitation or habilitation training, training in fire prevention and actions to take in the event of a fire shall be a part of the rehabilitation training program. Residents shall be trained to assist each other in case of fire to the extent their physical and mental abilities permit them to do so without additional personal risk.

❖ As noted, Group I-1 occupants are capable of responding to an emergency but will most likely need direction from staff and perhaps physical assistance to ensure the appropriate response. Unlike other Group I occupancies, I-1 occupancies rely on the abilities of the residents to take some level of responsibility for their own evacuation or relocation, therefore, training residents of these occupancies is critical. One major element that must be communicated to the residents is what to do when the main exit route is blocked.

If residents are receiving rehabilitation or habilitation, fire prevention and appropriate actions to take during a fire should be communicated as part of the sessions. This section also requires the occupants to assist one another as long as a physical or mental condition would not limit their ability to do so.

408.5.4 Drill frequency. Emergency evacuation drills shall be conducted at least six times per year, two times per year on each shift. Twelve drills shall be conducted in the first year of operation. Drills are not required to comply with the time requirements of Section 405.4.

❖ Occupants of Group I-1 facilities need to be reminded often of the procedures because mental capabilities to recall procedures, changes in the resident's abilities over time and the introduction of new residents to the facility may cause memory lapses. Drills, therefore, are to occur six times a year, and of those six drills, two drills are to occur on each shift. To ensure that the staff is fully aware of the procedures for a facility, these drills are required to occur 12 times in the first year of operation.

408.5.5 Resident participation. Emergency evacuation drills shall involve the actual evacuation of all residents to a selected assembly point.

❖ This section clarifies that the drills must include all occupants, not just the staff. In other institutional occupancies where the occupants are not capable or it is not desirable to evacuate them on their own, inclusion of the residents in drills is not necessary. Since dependence is placed on the residents to react in Group I-1 occupancies, their involvement is critical.

408.6 Group I-2 occupancies. Group I-2 occupancies shall comply with the requirements of Sections 408.6.1 and 408.6.2 and Sections 401 through 406. Drills are not required to comply with the time requirements of Section 405.4.

❖ Group I-2 occupancies are medical occupancies where the occupants are not capable of self-preservation and in many cases it is not in their best interest to fully evacuate or even move; therefore, the code generally has a unique set of requirements specific to such occupancies. Some of the elements include quick response sprinklers and the division of the building into smoke zones to assist in the compartmentation of the building. In terms of emergency preparedness and procedures, this section has requirements that address the specific needs of the occupants.

408.6.1 Evacuation not required. During emergency evacuation drills, the movement of patients to safe areas or to the exterior of the building is not required.

❖ As noted above, patients in many cases are not capable of self-preservation and, in most cases, it is not in their best interest to move unless absolutely necessary. Additionally, the relocation and evacuation procedures do not involve the patients in training activities; therefore, the code does not require patients to participate in evacuation and relocation drills.

408.6.2 Coded alarm signal. When emergency evacuation drills are conducted after visiting hours or when patients or residents are expected to be asleep, a coded announcement is allowed instead of audible alarms.

❖ Because sounding an alarm could have serious negative effects, and the staff are the only ones participating in training procedures, the code allows use of a call system or similar approach to notify staff of a drill in place of sounding the regular alarm.

408.7 Group I-3 occupancies. Group I-3 occupancies shall comply with the requirements of Sections 408.7.1 through 408.7.4 and Sections 401 through 406.

❖ Group I-3 occupancies are institutional occupancies where the occupants are under restraint. Typically, the occupants are physically able but are restrained from moving freely. Several levels of restraint in such occupancies are described in more detail in Chapter 2 under the definition of Group I-3 occupancies. As in Group I-2 occupancies, the only participants in drills will be the staff.

408.7.1 Employee training. Employees shall be instructed in the proper use of portable fire extinguishers and other manual fire suppression equipment. Training of new staff shall be provided promptly upon entrance on duty. Refresher training shall be provided at least annually.

❖ Group I-3 facilities are more likely to have incendiary activity; therefore, staff needs to be trained in the use of various fire protection equipment, including fire extinguishers. Additionally, any new employees need to be immediately trained in the use of this equipment. Because of difficulties presented by relocating or evacuating confinees, fires should be managed, when practical, to minimize the threat to occupants. Since combustibles are strictly limited in most of these occupancies, accidental fires generally remain small—at least long enough to be manageable. Incendiary fires often pose greater challenges and generally reflect a breakdown in security discipline. Notwithstanding this problem, fire extinguisher training, and even incipient fire brigades, may be especially effective elements of a fire safety plan in restrained care occupancies.

This training is required of new employees before they can begin their official duties. This training must then be refreshed once a year.

408.7.2 Staffing. Group I-3 occupancies shall be provided with 24-hour staffing. Staff shall be within three floors or 300 feet (91 440 mm) horizontal distance of the access door of each resident housing area. In Use Conditions 3, 4 and 5, as defined in Chapter 2, the arrangement shall be such that the staff involved can start release of locks necessary for emergency evacuation or rescue and initiate other necessary emergency actions within 2 minutes of an alarm.

> **Exception:** Staff shall not be required to be within three floors or 300 feet (9144 mm) in areas in which all locks are unlocked remotely and automatically in accordance with Section 408.4 of the *International Building Code*.

❖ Group I-3 occupancies place a lot of weight on the actions of the staff in emergencies. This particular section provides specific direction as to where the staff is to be located and how they are to react in an emergency. Staff members who are responsible for initiating the relocation or evacuation of confinees must be constantly alert to potential fire hazards and incipient fires. If a fire occurs, two minutes will seem like a long time to confined people. When a remote release locking system is neither required nor provided, the number of locks requiring manual unlocking should be limited with due regard to staff and confinee safety. This may require additional staff to accomplish the unlocking procedure in a timely manner. This section does have a specific exception for systems that utilize a remote locking and unlocking system.

408.7.3 Notification. Provisions shall be made for residents in Use Conditions 3, 4 and 5, as defined in Chapter 2, to readily notify staff of an emergency.

❖ Group I-3 occupancies under Use conditions 3, 4 and 5 where the occupants are very limited in their freedom would be considered moderate- and high-security facilities. Because the occupants are so limited and could be located remotely from guards or other staff members as a result of confinement within a compartment, a method is necessary for staff notification of a fire. In open cell blocks, staff members may be within earshot of occupants but generally this requirement necessitates monitors, intercoms or other communication appliances.

408.7.4 Keys. Keys necessary for unlocking doors installed in a means of egress shall be individually identifiable by both touch and sight.

❖ Keys must be distinctive from one another so they may be promptly and reliably identified under emergency conditions. Fumbling for the right key can cost valuable seconds, and possibly lives, in the event of a fire (see Figure 408.7.4).

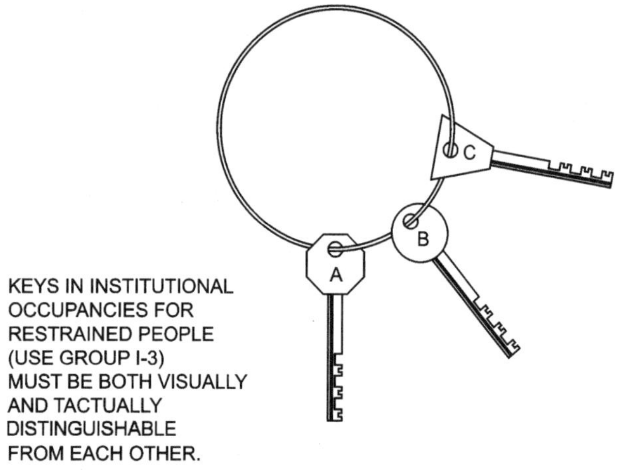

KEYS IN INSTITUTIONAL OCCUPANCIES FOR RESTRAINED PEOPLE (USE GROUP I-3) MUST BE BOTH VISUALLY AND TACTUALLY DISTINGUISHABLE FROM EACH OTHER.
A
B
C

Figure 408.7.4
TACTILE CODING OF INSTITUTIONAL KEYS

408.8 Group R-1 occupancies. Group R-1 occupancies shall comply with the requirements of Sections 408.8.1 through 408.8.3 and Sections 401 through 406.

❖ Group R-1 occupancies are residential occupancies that include hotels and boarding houses. These occupancies contain residents that are temporary in nature; therefore, they are more unfamiliar with their surroundings than Group R-2 and R-3 occupants. Sections 408.8.1 through 408.8.3 provide specific requirements that take into account characteristics of Group R-1 occupancies.

408.8.1 Evacuation diagrams. A diagram depicting two evacuation routes shall be posted on or immediately adjacent to every required egress door from each hotel, motel or dormitory guestroom.

❖ This section requires an evacuation plan drawing to be located in each hotel, motel or dormitory room. This diagram is to display two exit routes in case fire or other obstacles block the main route.

408.8.2 Emergency duties. Upon discovery of a fire or suspected fire, hotel, motel and dormitory employees shall perform the following duties:

1. Activate the fire alarm system, where provided.
2. Notify the public fire department.
3. Take other action as previously instructed.

❖ This section contains specific actions employees are to take if a fire occurs. More specifically, requiring that they first activate the fire alarm and then call the fire department is intended to avoid a situation in which the employee first investigates or calls security. The section requires immediate notification of the occupants to facilitate evacuation or relocation as necessary and notification of the fire department. Notifying the fire department as early as possible will enable them to reach the building at an earlier stage in the fire.

408.8.3 Fire safety and evacuation instructions. Information shall be provided in the fire safety and evacuation plan required by Section 404 to allow guests to decide whether to evacuate to the outside, evacuate to an area of refuge, remain in place, or any combination of the three.

❖ The procedures for isolating occupants from a fire depend on the layout and overall fire protection design of a building. More specifically, a hotel in a high-rise building may be specifically designed to evacuate in phases; therefore, the occupants need to know the procedures to facilitate a smooth and organized reaction to a fire. The appropriate actions that are available should be communicated. Options may include remaining in their rooms, evacuation or relocation.

408.9 Group R-2 occupancies. Group R-2 occupancies shall comply with the requirements of Sections 408.9.1 through 408.9.3 and Sections 401 through 406.

❖ Group R-2 occupancies are permanent residential occupancies that house multiple occupants in multiple dwelling units. Typically, this includes apartment buildings, dormitories and other related residential occupancies. Generally, the occupants tend to be familiar with their surroundings but may be sleeping at times. Sections 408.9.1 through 408.9.3 provide specific requirements in addition to the general requirements of Chapter 4.

408.9.1 Emergency guide. A fire emergency guide shall be provided which describes the location, function and use of fire protection equipment and appliances accessible to residents, including fire alarm systems, smoke alarms, and portable fire extinguishers. The guide shall also include an emergency evacuation plan for each dwelling unit.

❖ This section requires that a guide illustrating the fire safety features of the building be provided to each dwelling unit. This guide must contain the intended evacuation plan for each unit. Providing this information to residents increases the likelihood of a proper response, which in turn, increases resident safety and also makes the fire department's job a bit easier when responding to a scene.

408.9.2 Maintenance. Emergency guides shall be reviewed and approved in accordance with Section 401.2.

❖ This section is a mechanism for the fire code official to make sure that correct and complete information is contained in the emergency guide. This gives the fire code official the ability to ask for changes in the manual if a change in the character of the building or occupants occurs.

408.9.3 Distribution. A copy of the emergency guide shall be given to each tenant prior to initial occupancy.

❖ The guides are effective only when they are properly distributed to the residents. This section provides the authority to make sure the residents are provided with the guide prior to occupancy.

408.10 Group R-4 occupancies. Group R-4 occupancies shall comply with the requirements of Sections 408.10.1 through 408.10.5 and Sections 401 through 406.

❖ A Group R-4 occupancy is a residential care/assisted living facility for more than five but not more than 16 residents. The occupants of these occupancies are similar to those found in a Group I-1 occupancy; therefore, they are capable of self-preservation but in many cases may have mental or physical conditions that could impede their reactions.

408.10.1 Fire safety and evacuation plan. The fire safety and evacuation plan required by Section 404 shall include special staff actions, including fire protection procedures necessary for residents, and shall be amended or revised upon admission of a resident with unusual needs.

❖ This section adds to the general requirements found in Section 404. More specifically, any special requirements based on the specific needs of residents must be included as part of the plan. These plans must be reviewed each time a new occupant arrives to assess whether there are any special features that need to be included in the plan to address those particular needs.

408.10.2 Staff training. Employees shall be periodically instructed and kept informed of their duties and responsibilities under the plan. Such instruction shall be reviewed by the staff at least every two months. A copy of the plan shall be readily available at all times within the facility.

❖ Residents of these types of facilities normally have the ability to evacuate or relocate with a certain level of assistance from staff; therefore it is critical that the staff is properly trained. The requirement that training occur every two months is critical because the needs of the occupants may change over time as certain physical or mental conditions progress or new occupants arrive.

408.10.3 Resident training. Residents capable of assisting in their own evacuation shall be trained in the proper actions to take in the event of a fire. The training shall include actions to take if the primary escape route is blocked. Where the resident is given rehabilitation or habilitation training, training in fire prevention and actions to take in the event of a fire shall be a part of the rehabilitation training program. Residents shall be trained to assist each other in case of fire to the extent their physical and mental abilities permit them to do so without additional personal risk.

❖ Occupants are capable of responding to an emergency but will most likely need direction from staff and perhaps physical assistance to ensure the appropriate response; therefore, training the residents of these occupancies is critical. One major element that must be communicated to the residents is what to do when the main exit route is blocked.

If residents are receiving rehabilitation or habilitation, fire prevention and appropriate actions to take during a fire should be communicated as part of the session.

This section also requires the occupants to assist one another as long as a physical or mental condition would not limit their ability to do so.

408.10.4 Drill frequency. Emergency evacuation drills shall be conducted at least six times per year, two times per year on each shift. Twelve drills shall be conducted in the first year of operation. Drills are not required to comply with the time requirements of Section 405.4.

❖ Occupants of Group R-4 facilities need to be reminded often of the procedures because mental capabilities to recall procedures, changes in the resident's abilities over time and the introduction of new residents to the facility can make retention of details a problem. Dills are to occur, therefore, six times a year, and of those six drills, two drills are to occur on each shift. To ensure that the staff is fully aware of the procedures for a facility, these drills are required to occur 12 times in the first year of operation.

408.10.5 Resident participation. Emergency evacuation drills shall involve the actual evacuation of all residents to a selected assembly point and shall provide residents with experience in exiting through all required exits. All required exits shall be used during emergency evacuation drills.

Exception: Actual exiting from windows shall not be required. Opening the window and signaling for help shall be an acceptable alternative.

❖ Group R-4 occupancies have fewer occupants than Group I-1 occupancies. For that reason some leniency in construction requirements is allowed in such buildings. Also, the level of staffing may not be equivalent; therefore, residents are required to utilize all of the available exits during various drills. This means requiring them to try different routes from one drill to the next. Additionally, in many cases, R-4 occupancies include emergency egress windows as part of the possible exit routes. During a drill, the residents are not required to physically exit through the window; instead, they are allowed to simply open the window and signal for help. Exiting unassisted through windows could lead to serious injuries or death.

408.11 Covered mall buildings. Covered mall buildings shall comply with the provisions of Sections 408.11.1 through 408.11.3.

❖ A covered mall building is a special use as described in Section 402.1 of the IBC and as defined in Section 402.2 of that code. Generally, a covered mall building is a single building housing multiple occupancies, including, but not limited to, retail, assembly, drinking, dining and entertainment in which two or more tenants have a main entrance into a mall area. A mall area is also defined in Section 402.2 of the IBC as "a roofed or covered mall building that serves as access for two or more tenants and does not exceed three levels that are open to each other;" therefore, it presents some unique issues concerning fire department response to an emergency. The requirements of Sections 408.11.1 through 408.11.3 are related primarily to the complexity of the building and provide appropriate information to emergency responders so they can more effectively to respond to a fire or other emergency.

408.11.1 Lease plan. A lease plan shall be prepared for each covered mall building. The plan shall include the following information in addition to that required by Section 404.3.2:

1. Each occupancy, including identification of tenant.
2. Exits from each tenant space.
3. Fire protection features, including the following:
   - 3.1. Fire department connections.
   - 3.2. Fire command center.
   - 3.3. Smoke management system controls.
   - 3.4. Elevators and elevator controls.
   - 3.5. Hose valves outlets.
   - 3.6. Sprinkler and standpipe control valves.
   - 3.7. Automatic fire-extinguishing system areas.
   - 3.8. Automatic fire detector zones.
   - 3.9. Fire barriers.

❖ Item 1 assists the emergency responders by requiring detailed documentation regarding the identification of each tenant, location and occupancy. This will let them know where the highest density of occupants and the types of hazards may be anticipated.

Item 2 is the identification of exits. This will assist the emergency responders in the identification of necessary access routes and how they may interact with the exits.

Item 3 is a report of the available fire protection features. These features, such as identification of the fire detector zones, will help responders quickly assess where a fire is located. If the mall has a smoke control system, access to the controls may be necessary; therefore, the location of those controls is critical. This information gives the fire department a general feel for how the building is intended to perform during a fire. This information is valuable in the sense that they will have more information to promote effective use of the fire protection features installed. If little information is provided, the fire department could actually disrupt the essential activation of a system, such as smoke control.

408.11.1.1 Approval. The lease plan shall be submitted to the fire code official for approval, and shall be maintained on site for immediate reference by responding fire service personnel.

❖ The lease plan must be approved. This allows the fire department to determine whether all necessary information, from their perspective as the responders, is addressed. Also this section provides a requirement to make sure this document is available on site for use by the emergency responders.

408.11.1.2 Revisions. The lease plans shall be revised annually or as often as necessary to keep them current. Modifications or changes in tenants or occupancies shall not be made without prior approval of the fire code official and building official.

❖ This section provides the authority to require a minimum review of the lease plan at least once each year. In addition the fire code official has the authority to ask for more frequent reviews of the plan.

If a change occurs within the building, the lease plan may no longer be valid; therefore, this section requires that no changes to any tenant space be made without approval and review by the fire code official. These changes would have to be documented in the lease plan.

408.11.2 Tenant identification. Each occupied tenant space provided with a secondary exit to the exterior or exit corridor shall be provided with tenant identification by business name and/or address. Letters and numbers shall be posted on the corridor side of the door, be plainly legible and shall contrast with their background.

Exception: Tenant identification is not required for anchor stores.

❖ Identifying secondary exits from tenant spaces that enter into an exit passageway or directly outside is a critical need for emergency responders. This identification can either be the address or the name of the business. Having multiple tenants within a building makes this identification necessary. Anchor stores do not need these labels, since they are fairly recognizable without them.

408.11.3 Maintenance. Unoccupied tenant spaces shall be:

1. Kept free from the storage of any materials.
2. Separated from the remainder of the building by partitions of at least 0.5-inch-thick (12.7 mm) gypsum board or an approved equivalent to the underside of the ceiling of the adjoining tenant spaces.
3. Without doors or other access openings other than one door that shall be kept key locked in the closed position except during that time when opened for inspection.

4. Kept free from combustible waste and be broom-swept clean.

❖ This section is primarily concerned with the hazards posed by a tenant space that is not in use or that is not under the supervision of employees. These spaces are more likely to be targeted by vandalism and possibly incendiary activity. Generally, a fire can grow unnoticed in such spaces as a result of the lack of supervision or activity in the space. To reduce these risks, several requirements are found in Section 408.11.3 that focus on reducing the fire ignition, growth and spread potential by limiting combustibles in the space, securing the space through the use of locks and installing fire separations constructed of $^1/_2$-inch-thick (13 mm) gypsum or similar materials. Note that mall buildings are required to be sprinklered throughout and that the systems in the tenant spaces must be independent of the mall area.

Bibliography

The following resource materials are referenced in this chapter or are relevant to the subject matter addressed in this chapter.

IBC-2003, *International Building Code.* Falls Church, VA: International Code Council, 2003.

Johnson, P. "Shattering the Myths of Fire Protection Engineering." *Fire Protection Engineering,* premier issue, 1998.

Leslie, J.,T. Mclintock, T.J. Shields and A.H. Reinhardt-Rutland "A Behavioural Solution to the Learned Irrelevance of Emergency Exit Signage," *Proceedings, Second International Symposium on Human Behavior in Fire.* Cambridge, MA: Interscience Communications, Massachusetts Institute of Technology, March 2001.

Chapter 5: Fire Service Features

General Comments

The requirements of this chapter apply to all occupancies and pertain to access roads; access to building openings and roofs; premises identification; key boxes; hazards to fire fighters; fire protection water supplies; fire command centers and fire department access to equipment.

Purpose

This chapter contains the requirements for fire service access to the property that is to be protected, including access roads, security devices and access through openings in the building.

The chapter also addresses fire-fighter hazards, the requirements for a fire department command center and fire-fighter access to equipment, such as fire suppression equipment, air-handling equipment, emergency power equipment and access to the roof. In addition this chapter addresses the fire protection water supply.

SECTION 501 GENERAL

501.1 Scope. Fire service features for buildings, structures and premises shall comply with this chapter.

❖ This chapter contains requirements that will enable the fire service to respond to an emergency on the premises of a building or structure.

501.2 Permits. A permit shall be required as set forth in Sections 105.6 and 105.7.

❖ Permits must be obtained from the fire code official. Permit fees, if any, must be paid prior to the issuance of the permit. There are two types of permits: operational and construction. The operational permits required by this section are for the use or operation of fire protection valves and fire hydrants (see Section 105.6.16) or the use or removal from service of a private fire hydrant (see Section 105.6.36). The construction permit (see Section 105.7.9) allows the applicant to install or modify private fir hydrants. See Section 105 for additional information on permits.

501.3 Construction documents. Construction documents for proposed fire apparatus access, location of fire lanes and construction documents and hydraulic calculations for fire hydrant systems shall be submitted to the fire department for review and approval prior to construction.

❖ Construction documents must be drawn to scale to clearly show the details that address the requirements of the code. The jurisdiction adopting the code may require a registered design professional to prepare the construction documents. Each jurisdiction will define the qualifications of a design professional.

501.4 Timing of installation. When fire apparatus access roads or a water supply for fire protection is required to be installed, such protection shall be installed and made serviceable prior to and during the time of construction except when approved alternative methods of protection are provided. Temporary street signs shall be installed at each street intersection when construction of new roadways allows passage by vehicles in accordance with Section 505.2.

❖ Buildings under construction are quite vulnerable to fire and other types of construction incidents, such as injuries from falling objects. Access roads and water for fire protection are essential for fire-fighting purposes. Temporary street signs are also valuable to emergency responders because the streets in new developments will most likely not be familiar to them nor be on their maps.

Marked access roads and an emergency water supply should be in place before any large amount of combustible building materials is placed on site and before any construction is initiated.

SECTION 502 DEFINITIONS

502.1 Definitions. The following words and terms shall, for the purposes of this chapter and as used elsewhere in this code, have the meanings shown herein.

❖ The following words and terms shall, for the purposes of this chapter and as used elsewhere in this code, have the meanings shown herein.

Definitions of terms can help in the understanding and application of the code requirements. The purpose for including those definitions that are associated with the subject matter of this chapter is to provide more convenient access to them without having to refer back to Chapter 2. It is important to emphasize that these terms are not exclusively related to this chapter but are applicable everywhere the term is used in the code. For convenience, these terms are also listed in Chapter 2 with a cross reference to this section. The use and application of all defined terms, including those defined in this section, are set forth in Section 201.

FIRE APPARATUS ACCESS ROAD. A road that provides fire apparatus access from a fire station to a facility, building or portion thereof. This is a general term inclusive of all other terms such as fire lane, public street, private street, parking lot lane and access roadway.

❖ Fire access roads are required to be all-weather surfaced roadways that are designed for the weight and type of emergency vehicle that may use the road. No specific surface material is required for a fire access roadway. It is up to the fire code official to decide whether the surface will support the load of the anticipated emergency vehicles.

FIRE COMMAND CENTER. The principal attended or unattended location where the status of the detection, alarm communications and control systems is displayed, and from which the system(s) can be manually controlled.

❖ Fire command centers are communication centers where dedicated manual and automatic facilities are located for the origination, control and transmission of information and instruction pertaining to a fire emergency to the occupants (including fire department personnel) of the building. Fire command centers must provide facilities for the control and display of the status of all fire protection (detection, signaling, etc.) systems. These stations must be located in secure areas as approved by the fire code official. Often, this is a location near the primary building entrance. Fire command centers also may be combined with other building operations and security facilities when allowed by the fire code official; however, operating controls for use by the fire department must be clearly marked and not subject to tampering by unauthorized persons (see the commentary to Section 509.1 for further discussion).

FIRE DEPARTMENT MASTER KEY. A limited issue key of special or controlled design to be carried by fire department officials in command which will open key boxes on specified properties.

❖ Several companies market emergency entry systems that use master keys. These keys are used to open key boxes and entry gates and turn on/off electronic switches that control electric gates and certain building functions such as smoke control systems, fans and special processes.

FIRE LANE. A road or other passageway developed to allow the passage of fire apparatus. A fire lane is not necessarily intended for vehicular traffic other than fire apparatus.

❖ The term "fire lane" is synonymous with "fire apparatus access road;" however, the driving surface may not be the same as for a public road.

KEY BOX. A secure, tamperproof device with a lock operable only by a fire department master key, and containing building entry keys and other keys that may be required for access in an emergency.

❖ The key box is part of an emergency entry system. The building owner/manager places a key box or key vault on the exterior of the building or at the entrance to the facility, placing keys, access cards or security codes inside the box. The emergency responders can use their special master key to enter the box and gain access to the building or facility.

SECTION 503
FIRE APPARATUS ACCESS ROADS

503.1 Where required. Fire apparatus access roads shall be provided and maintained in accordance with Sections 503.1.1 through 503.1.3.

❖ The code official may require additional access roads to get fire apparatus closer to fire hydrants, fire department connections (FDCs) or emergency access points.

503.1.1 Buildings and facilities. Approved fire apparatus access roads shall be provided for every facility, building or portion of a building hereafter constructed or moved into or within the jurisdiction. The fire apparatus access road shall comply with the requirements of this section and shall extend to within 150 feet (45 720 mm) of all portions of the facility and all portions of the exterior walls of the first story of the building as measured by an approved route around the exterior of the building or facility.

Exception: The fire code official is authorized to increase the dimension of 150 feet (45 720 mm) where:

1. The building is equipped throughout with an approved automatic sprinkler system installed in accordance with Section 903.3.1.1, 903.3.1.2 or 903.3.1.3.
2. Fire apparatus access roads cannot be installed because of location on property, topography, waterways, nonnegotiable grades or other similar conditions, and an approved alternative means of fire protection is provided.
3. There are not more than two Group R-3 or Group U occupancies.

❖ Large area buildings may require a fire apparatus access road on all four sides. An access road is required within 150 feet (45 720 mm) of all portions of the grade level floor of each building [see Figure 503.1.1(1)].

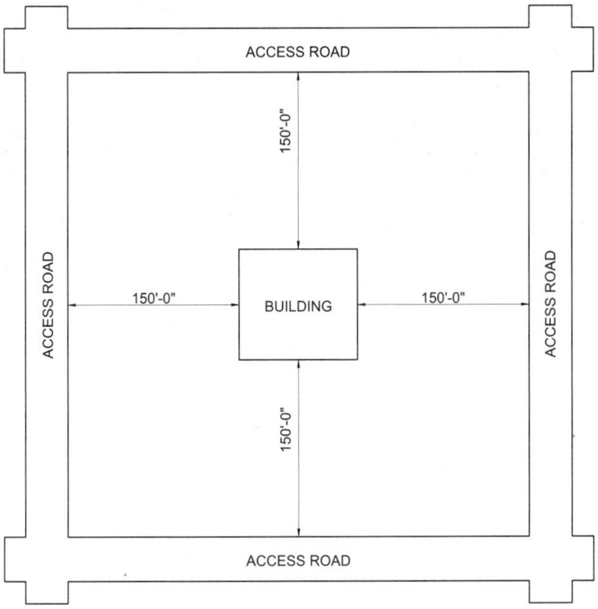

For SI: 1 foot = 304.8 mm.

Figure 503.1.1(1)
FIRE DEPARTMENT ACCESS—LARGE BUILDING

A long narrow building may require fire department access roads on two sides only, if all portions of the grade level floor are within 150 feet (45 720 mm) of the access road [see Figure 503.1.1(2)].

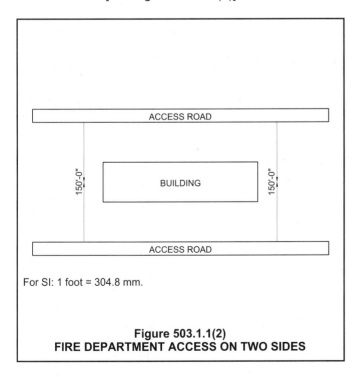

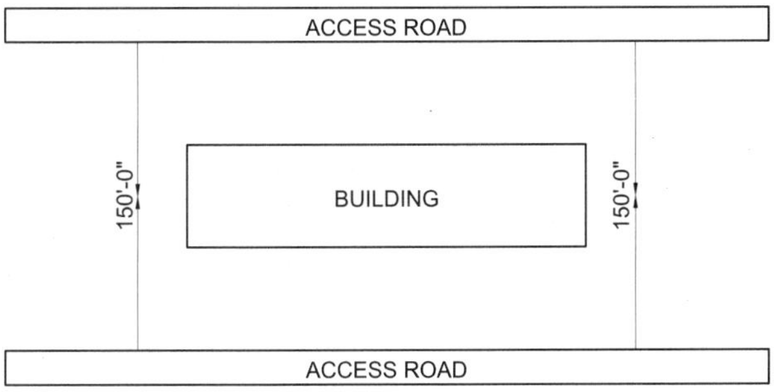

For SI: 1 foot = 304.8 mm.

Figure 503.1.1(2)
FIRE DEPARTMENT ACCESS ON TWO SIDES

Small buildings may require an access road on one side only, if the access road is within 150 feet (45 720 mm) of all portions of the grade level floor [see Figure 503.1.1(3)].

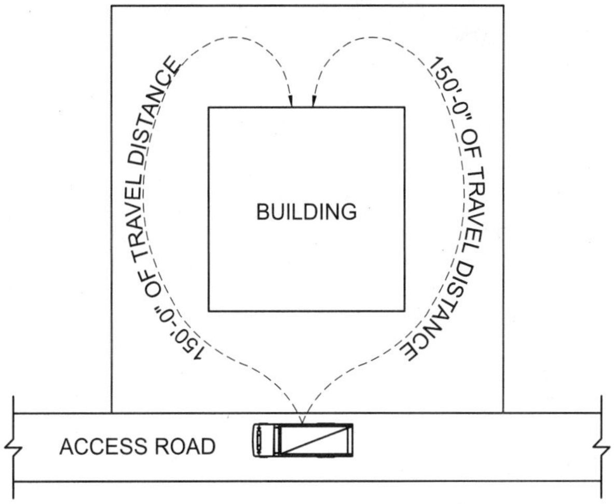

For SI: 1 foot = 304.8 mm.

Figure 503.1.1(3)
SMALL BUILDING ACCESS

Exception 1 states that the 150-foot (45 720 mm) distance may be increased, with the approval of the fire code official, when the building is equipped throughout with an automatic sprinkler system. The code does not give the fire code official guidance on how much over 150 feet (45 720 mm) is reasonable. The fire code official must make the determination based on the response capabilities of his or her emergency response units and the anticipated magnitude of the incident.

The "alternative means" in Exception 2 may include standpipes, automatic sprinklers, remote fire department connections or additional fire hydrants.

The Group R-3 facilities noted in Exception 3 include all detached one- and two-family dwellings and multiple (three or more) single-family dwellings (townhouses) more than three stories in height and all institutional facilities that accommodate five or less people for less than 24 hours per day. Group U occupancies are utility and miscellaneous accessory buildings or structures.

503.1.2 Additional access. The fire code official is authorized to require more than one fire apparatus access road based on the potential for impairment of a single road by vehicle congestion, condition of terrain, climatic conditions or other factors that could limit access.

❖ Additional access roads may be required by the fire code official based on his or her knowledge of traffic patterns, local weather conditions, terrain or anticipated magnitude of a potential incident.

503.1.3 High-piled storage. Fire department vehicle access to buildings used for high-piled combustible storage shall comply with the applicable provisions of Chapter 23.

❖ Chapter 23 has special requirements for building access in occupancies with high-piled storage, but the requirements for fire apparatus access roads are the same as those required in this chapter.

503.2 Specifications. Fire apparatus access roads shall be installed and arranged in accordance with Sections 503.2.1 through 503.2.7.

❖ The dimensions of fire department access roads are based on the size and height of emergency vehicles, their turning radius and the fact that emergency vehicles may be required to pass each other on the access road.

503.2.1 Dimensions. Fire apparatus access roads shall have an unobstructed width of not less than 20 feet (6096 mm), except for approved security gates in accordance with Section 503.6, and an unobstructed vertical clearance of not less than 13 feet 6 inches (4115 mm).

❖ The dimensions in this section are established to give fire apparatus continuous and unobstructed access to buildings and facilities. Section 503.6 requires that security gates be maintained and be capable of operation by the emergency responders at all times. Note that the minimum vertical clearance of 13 feet 6 inches (4115 mm) is the standard clearance used for highway bridges and underpasses. The vertical clearance requirement would apply in cases where a building or portion of a building, such as a canopy or porte-cochere, projects over all or a portion of the required width of the fire apparatus access road. Conversely, if the full required width of the fire apparatus access road is provided outside of the footprint of the projecting building element, the vertical-clearance requirement would not apply. It is not the intent of this section that all projecting elements be constructed with a 13 foot 6 inch (4115 mm) vertical clearance, regardless of whether they encroach upon the required width of a fire apparatus access road. Appendix D contains additional guidance on fire apparatus access road dimensions. It is important to note that the appendices are not considered as part of the code unless specifically adopted (see Section 1 of the sample adopting ordinance on page v of the code).

503.2.2 Authority. The fire code official shall have the authority to require an increase in the minimum access widths where they are inadequate for fire or rescue operations.

❖ The fire code official may require greater dimensions based on the size and maneuverability of the anticipated emergency response apparatus, including mutual-aid apparatus from neighboring communities or agencies.

503.2.3 Surface. Fire apparatus access roads shall be designed and maintained to support the imposed loads of fire apparatus and shall be surfaced so as to provide all-weather driving capabilities.

❖ This provision does not specify a particular type of surface. It is written in performance language; therefore, the surface must carry the load of the anticipated emergency response vehicles and be driveable in all kinds of weather.

503.2.4 Turning radius. The required turning radius of a fire apparatus access road shall be determined by the fire code official.

❖ The turning radius of an access road should be based on the turning radius of the anticipated responding emergency vehicles and must be approved by the fire code official.

503.2.5 Dead ends. Dead-end fire apparatus access roads in excess of 150 feet (45 720 mm) in length shall be provided with an approved area for turning around fire apparatus.

❖ In consideration of the hazards inherent in attempting to back emergency vehicles, especially larger ones such as tower ladders, out of a long dead-end roadway, this section intends to create a safer situation by requiring that dead-end access roads over 150-feet long (45 720 mm) be equipped with an approved turnaround designed for the largest anticipated emergency-response vehicles. Appendix D contains examples of dead-end turnaround configurations. It is important to note that the appendices are not considered as part of the code unless specifically adopted (see Section 1 of the sample adopting ordinance on page v of the code)

503.2.6 Bridges and elevated surfaces. Where a bridge or an elevated surface is part of a fire apparatus access road, the bridge shall be constructed and maintained in accordance with AASHTO *Standard Specification for Highway Bridges*. Bridges and elevated surfaces shall be designed for a live load sufficient to carry the imposed loads of fire apparatus. Vehicle load limits shall be posted at both entrances to bridges when required by the fire code official. Where elevated surfaces designed for emergency vehicle use are adjacent to surfaces which are not designed for such use, approved barriers, approved signs or both shall be installed and maintained when required by the fire code official.

❖ Bridges and elevated surfaces must be capable of carrying the weight of emergency response apparatus and must be marked with signage posting the weight limit of the bridge or elevated surface. Evaluation of bridges should be done in cooperation with the appropriate local or state agency having jurisdiction over private or public roadway bridges.

503.2.7 Grade. The grade of the fire apparatus access road shall be within the limits established by the fire code official based on the fire department's apparatus.

❖ Generally, any grade exceeding 10 percent or a 10-foot (3048 mm) rise in a 100-foot (30 480 mm) length is required to have the approval of the fire code official. See Appendix D for additional guidance on fire apparatus access roads. Note that the appendices are not considered part of the code unless specifically adopted (see Section 1 of the sample adopting ordinance on page v of the code).

503.3 Marking. Where required by the fire code official, approved signs or other approved notices shall be provided for fire apparatus access roads to identify such roads or prohibit the obstruction thereof. Signs or notices shall be maintained in a clean and legible condition at all times and be replaced or repaired when necessary to provide adequate visibility.

❖ Fire department access roads are normally designated on private property to provide fire service access; therefore, maintenance of the access roads, signage and any supplementary markings (pavement marking, curbs markings, etc.) are the responsibility of the owner of the property on which the fire apparatus road is located. Signage and supplemental markings should be in accordance with applicable local or state motor vehicle laws and should be enforced with the cooperation of the local police agency. Appendix D contains examples of signage. It is important to note that the appendices are not considered as part of the code unless specifically adopted (see Section 1 of the sample adopting ordinance on page v of the code).

503.4 Obstruction of fire apparatus access roads. Fire apparatus access roads shall not be obstructed in any manner, including the parking of vehicles. The minimum widths and clearances established in Section 503.2.1 shall be maintained at all times.

❖ To enforce "no parking" in fire apparatus access roads (fire lanes) the roads must be clearly marked. Some jurisdictions cite the building owner if the fire apparatus road is not properly marked and posted and cite the vehicle for parking or blocking the access road if the access road is clearly marked. Other jurisdictions place the responsibility for marking the access roads, as well as the policing of "no parking" zones, on the building owner. In some states, motor vehicle laws may stipulate that fire apparatus access roads/fire lanes posted on private property may only be enforced by a traffic citation where an enforcement contract has been executed between the property owner and the local jurisdiction; that all markings be in accordance with the motor vehicle code and that the designated roadways be described in detail in the local "no parking" ordinances of the jurisdiction.

503.5 Required gates or barricades. The fire code official is authorized to require the installation and maintenance of gates or other approved barricades across fire apparatus access roads, trails or other accessways, not including public streets, alleys or highways.

❖ The fire code official may require the installation and maintenance of gates or barricades across fire apparatus roads to prevent unauthorized vehicles from blocking or parking in the access road. The design and dimensions of the gates or barricade must be approved by the fire code official. Additionally, the gate or barricade must be operable or removable by the responding emergency units.

503.5.1 Secured gates and barricades. When required, gates and barricades shall be secured in an approved manner. Roads, trails and other accessways that have been closed and obstructed in the manner prescribed by Section 503.5 shall not be trespassed on or used unless authorized by the owner and the fire code official.

Exception: The restriction on use shall not apply to public officers acting within the scope of duty.

❖ The owner may secure the fire apparatus access road and restrict its use to emergency vehicles only.

In the exception, the owner is required to maintain the fire apparatus access road in such a way that it will not restrict use by emergency responders.

503.6 Security gates. The installation of security gates across a fire apparatus access road shall be approved by the fire chief. Where security gates are installed, they shall have an approved means of emergency operation. The security gates and the emergency operation shall be maintained operational at all times.

❖ This section does not require that security gates be installed, but since they can affect fire department operations, their installation must be approved by the fire chief. Where installed, security gates must be maintained so as to be operable in an emergency by the emergency response units and the means of operation should be acceptable to the fire chief. Electrically operated gates should include a manual method of operation.

SECTION 504
ACCESS TO BUILDING OPENINGS AND ROOFS

504.1 Required access. Exterior doors and openings required by this code or the *International Building Code* shall be maintained readily accessible for emergency access by the fire department. An approved access walkway leading from fire apparatus access roads to exterior openings shall be provided when required by the fire code official.

❖ The exterior openings referred to in this section are typically exit discharge doors, since such openings provide fire department access directly into the building or to a fire-resistance-rated enclosure from which to operate in multistory buildings. This section also includes access openings for rack or high-piled storage buildings required by Section 2306.6. Under certain circumstances, in order for emergency response personnel to get equipment from the fire or other emergency apparatus to the building, the fire code official is authorized to require approved walkways from the apparatus access road to the building openings on grade level.

504.2 Maintenance of exterior doors and openings. Exterior doors and their function shall not be eliminated without prior approval. Exterior doors that have been rendered nonfunctional and that retain a functional door exterior appearance shall have a sign affixed to the exterior side of the door with the words THIS DOOR BLOCKED. The sign shall consist of letters having a principal stroke of not less than 0.75 inch (19.1 mm) wide and at least 6 inches (152 mm) high on a contrasting background. Required fire department access doors shall not be obstructed or eliminated. Exit and exit access doors shall comply with Chapter 10. Access doors for high-piled combustible storage shall comply with Section 2306.6.1.

❖ This section pertains not only to emergency access openings but also to all exterior doors that have functional appearance from the outside but are not operable. Doors that are part of the required means of egress or that are required by Section 2306.6 must not be rendered unusable or be blocked. Only doors not required by the code or the *International Building Code*® (IBC®) for means of egress may be blocked or made unusable. Even then they must be marked from the outside so that emergency personnel will not attempt to use them.

504.3 Stairway access to roof. New buildings four or more stories in height, except those with a roof slope greater than four units vertical in 12 units horizontal (33.3 percent slope), shall be provided with a stairway to the roof. Stairway access to the roof shall be in accordance with Section 1009.12. Such stairway shall be marked at street and floor levels with a sign indicating that the stairway continues to the roof. Where roofs are used for roof gardens or for other purposes, stairways shall be provided as required for such occupancy classification.

❖ The stairway to the roof required by this section must be a continuation of a rated stair enclosure and must comply with Section 1009.12. If the stair to the roof serves as a means of egress from an occupied roof, then the stair must be equipped with all the components of an exit stairway, such as the required riser and tread dimensions, handrails, etc. If the stair to the roof is not required for roof egress, Section 1009.12 allows the stair segment from the top floor to the roof to be an alternating tread device (see commentary, Section 1009.12). The access to the roof is required for fire-fighter use and not the general public; therefore, the door leading to the roof may be secured in a manner approved by the fire code official with due consideration given to whether the door is an egress element for the roof. These provisions apply only to new buildings four or more stories in height.

SECTION 505
PREMISES IDENTIFICATION

505.1 Address numbers. New and existing buildings shall have approved address numbers, building numbers or approved building identification placed in a position that is plainly legible and visible from the street or road fronting the property. These numbers shall contrast with their background. Address numbers shall be Arabic numerals or alphabet letters. Numbers shall be a minimum of 4 inches (102 mm) high with a minimum stroke width of 0.5 inch (12.7 mm).

❖ Buildings must be easily identified by emergency responders from the emergency response vehicle. This should include the backs of buildings that face alleys or roads, since the emergency response unit may often be directed to the back entrance to a building, such as in a strip shopping center. The back door of each tenant space should have the numerical address and the store name on or above the door.

505.2 Street or road signs. Streets and roads shall be identified with approved signs. Temporary signs shall be installed at each street intersection when construction of new roadways allows passage by vehicles. Signs shall be of an approved size, weather resistant and be maintained until replaced by permanent signs.

❖ The names of streets in new developments may not be on maps, making them hard for emergency responders to find. Temporary street signs must be installed before construction begins and replaced later with permanent signs.

SECTION 506
KEY BOXES

506.1 Where required. Where access to or within a structure or an area is restricted because of secured openings or where immediate access is necessary for life-saving or fire-fighting purposes, the fire code official is authorized to require a key box to be installed in an approved location. The key box shall be of an approved type and shall contain keys to gain necessary access as required by the fire code official.

❖ The fire code official has the authority to require special key vaults when, in his or her opinion, the need for rapid entry into facilities warrants it. The key boxes or vaults are located on the exterior of the building for ready access, and are openable with a special master key in the possession of the emergency responders.

506.1.1 Locks. An approved lock shall be installed on gates or similar barriers when required by the fire code official.

❖ The key-box suppliers also have special padlocks and electronic-key-operated switches that are controlled by the same master key that opens the key vaults. These padlocks can be required by the fire code official for security gates. The key-activated electronic switches may be required for the control of certain equipment in the building, such as smoke control equipment, or to shut down a dangerous process.

506.2 Key box maintenance. The operator of the building shall immediately notify the fire code official and provide the new key when a lock is changed or rekeyed. The key to such lock shall be secured in the key box.

❖ In most cases, the owner of a building cannot open the key vault or box and must call the fire code official to have someone open it to replace keys that have been changed. The building owner is responsible for maintaining the key box as well as keeping the keys inside the box current.

SECTION 507 HAZARDS TO FIRE FIGHTERS

507.1 Trapdoors to be closed. Trapdoors and scuttle covers, other than those that are within a dwelling unit or automatically operated, shall be kept closed at all times except when in use.

❖ Trapdoors and unguarded openings in floors and walkways must remain in the closed position or be designed to automatically close upon activation of the fire alarm system. Openings in floors or walkways can injure emergency responders, especially if vision is obscured.

507.2 Shaftway markings. Vertical shafts shall be identified as required by this section.

❖ This section was developed to prevent fire fighters from falling through shafts when entering buildings off ladders placed on the exterior of the building.

507.2.1 Exterior access to shaftways. Outside openings accessible to the fire department and which open directly on a hoistway or shaftway communicating between two or more floors in a building shall be plainly marked with the word SHAFTWAY in red letters at least 6 inches (152 mm) high on a white background. Such warning signs shall be placed so as to be readily discernible from the outside of the building.

❖ All exterior wall openings that are accessible to fire fighters by way of ladders and aerial equipment and open directly into shafts or hoistways communicating between two or more floors must be clearly marked (see Figure 507.2.1).

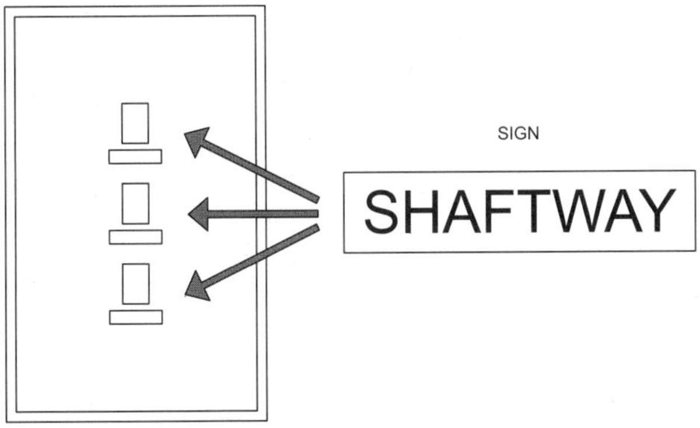

Figure 507.2.1
SHAFTWAY SIGNAGE

507.2.2 Interior access to shaftways. Door or window openings to a hoistway or shaftway from the interior of the building shall be plainly marked with the word SHAFTWAY in red letters at least 6 inches (152 mm) high on a white background. Such warning signs shall be placed so as to be readily discernible.

> **Exception:** Marking shall not be required on shaftway openings which are readily discernible as openings onto a shaftway by the construction or arrangement.

❖ Openings into shaftways from the interior of the building pose a threat to fire fighters when visibility is poor. Interior shaft openings must be marked so that they are plainly visible from the interior of the building.

If fire fighters can readily identify an opening into a shaft by the way the opening is constructed, the shaft opening need not be marked, keeping in mind that the fire fighter may be feeling his or her way in heavy smoke or darkness.

507.3 Pitfalls. The intentional design or alteration of buildings to disable, injure, maim or kill intruders is prohibited. No person shall install and use firearms, sharp or pointed objects, razor wire, explosives, flammable or combustible liquid containers, or dispensers containing highly toxic, toxic, irritant or other hazardous materials in a manner which may passively or actively disable, injure, maim or kill a fire fighter who forcibly enters a building for the purpose of controlling or extinguishing a fire, rescuing trapped occupants or rendering other emergency assistance.

❖ This paragraph prohibits the use of "booby-traps" in buildings, for whatever reason, if they could injure or disable the emergency responder during the performance of his or her duties.

SECTION 508 FIRE PROTECTION WATER SUPPLIES

508.1 Required water supply. An approved water supply capable of supplying the required fire flow for fire protection shall be provided to premises upon which facilities, buildings or portions of buildings are hereafter constructed or moved into or within the jurisdiction.

❖ This section requires that adequate fire protection water be provided to premises upon which new buildings are constructed or onto which a building is moved, from either outside of the jurisdiction or another location within the jurisdiction. See Appendix B for further information on fire flows. It is important to note that the appendices are not considered as part of the code unless specifically adopted (see Section 1 of the sample adopting ordinance on page v of the code).

508.2 Type of water supply. A water supply shall consist of reservoirs, pressure tanks, elevated tanks, water mains or other fixed systems capable of providing the required fire flow.

❖ A good water supply consists of an adequate source of water, a distribution system and proper pressure for delivery. If the water source is not reliable, it should not be considered as an acceptable water supply.

508.2.1 Private fire service mains. Private fire service mains and appurtenances shall be installed in accordance with NFPA 24.

❖ Private fire service mains are often installed when facilities are located well away from municipal water distribution systems. Private hydrants may not be installed on mains less than 6 inches (152 mm) in diameter. When installed, private fire service mains, private hydrants, control valves, hose houses and related equipment must be installed and maintained in accordance with NFPA 24. Private (yard) hydrants are usually painted red to distinguish them from municipal hydrants; however, they may be tested and marked in accordance with NFPA 291 when approved by the fire code official and the fire department.

508.2.2 Water tanks. Water tanks for private fire protection shall be installed in accordance with NFPA 22.

❖ Water tanks for private fire protection may be required where municipal water systems do not exist or are incapable of supplying sprinkler or standpipe demand, or where Section 403.2 of the IBC or Section 903.3.5.2 of the code require secondary water supply for high-rise buildings in Seismic Design Category C, D, E or F. NFPA 22 and Section 1509.3 of the IBC govern the installation of water tanks. Pressure tanks must bear the label of an approved agency and be installed in accordance with the manufacturer's instructions.

508.3 Fire flow. Fire flow requirements for buildings or portions of buildings and facilities shall be determined by an approved method.

❖ Appendix B of the code is a good example of an approved method for determining fire flow and its duration. Table B105.1 bases fire flow on the type of construction and the square footage of the fire area. All calculations in the table are based on a 20 psi (138 kPa) residual pressure. It is important to note that the appendices are not considered as part of the code unless specifically adopted (see Section 1 of the sample adopting ordinance on page v of the code).

508.4 Water supply test. The fire code official shall be notified prior to the water supply test. Water supply tests shall be witnessed by the fire code official or approved documentation of the test shall be provided to the fire code official prior to final approval of the water supply system.

❖ The water supply system must be flushed and pressure tested before the water supply test, and the contractor is required to notify the fire code official prior to performing the flow test on the system. The fire code official will make the final approval by either witnessing the test or accepting the certification documentation. NFPA 24 referenced in Section 508.2.1 contains the testing requirements for private fire mains as well as a test certificate form. It should be noted that the test certificate form has signature blocks only for the building owner's representative and the installing contractor's representative. There is no place on the form for the fire code official's signature nor should he or she expose him or herself to liability of any kind for the installation by signing the form.

508.5 Fire hydrant systems. Fire hydrant systems shall comply with Sections 508.5.1 through 508.5.6.

❖ When fire hydrant systems are part of the approved water supply, the system must comply with this section of the code.

508.5.1 Where required. Where a portion of the facility or building hereafter constructed or moved into or within the jurisdiction is more than 400 feet (122 m) from a hydrant on a fire apparatus access road, as measured by an approved route around the exterior of the facility or building, on-site fire hydrants and mains shall be provided where required by the fire code official.

Exceptions:

1. For Group R-3 and Group U occupancies, the distance requirement shall be 600 feet (183 m).
2. For buildings equipped throughout with an approved automatic sprinkler system installed in accordance with Section 903.3.1.1 or 903.3.1.2, the distance requirement shall be 600 feet (183 m).

❖ Fire fighters should not have to hand lay more than 400 feet (122 m) of hose to reach all portions of the exterior grade level of the building. Each hydrant must be accessible to fire apparatus and the 400-foot (122 m) distance should be measured from the hydrant(s) to all portions of the exterior at ground level [see Figure 508.5.1(1)].

It is important to note that this provision of the code is a subparagraph of 508.5 that states "when fire hydrants are the approved water supply." This paragraph is not intended to prevent development in rural areas when fire hydrants are not available as long as the fire code official has approved an alternate water supply. The alternate water supply could be a fire department water tanker that is approved by the fire code official.

In recognition of the smaller relative size and fire hazard characteristics of one- and two-family dwellings and utility buildings, Exception 1 increases the 400-foot (122 m) distance to 600 feet (183 m) [Figure 508.5.1(2)].

In recognition of the proven efficiency of sprinklers in applying water directly on the seat of the fire for buildings equipped throughout with automatic sprinklers in accordance with NFPA 13 or NFPA 13R, as applicable, Exception 2 increases the 400-foot (122 m) distance to 600 feet (183 m) [see Figure 508.5.1(3)].

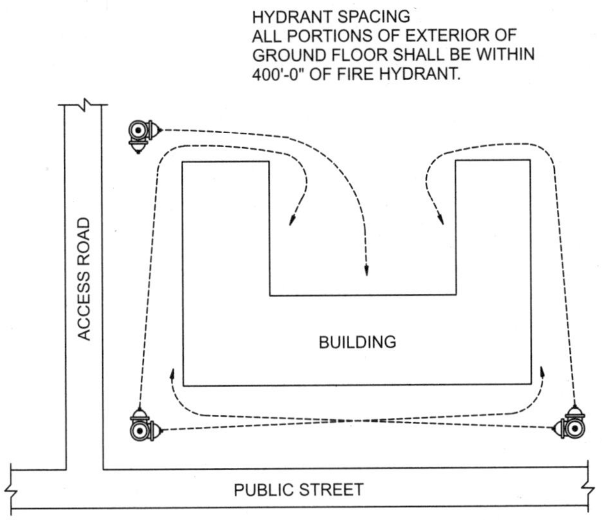

For SI: 1 foot = 304.8

Figure 508.5.1(1)
HYDRANT LAYOUT TO MINIMIZE HOSE RUN

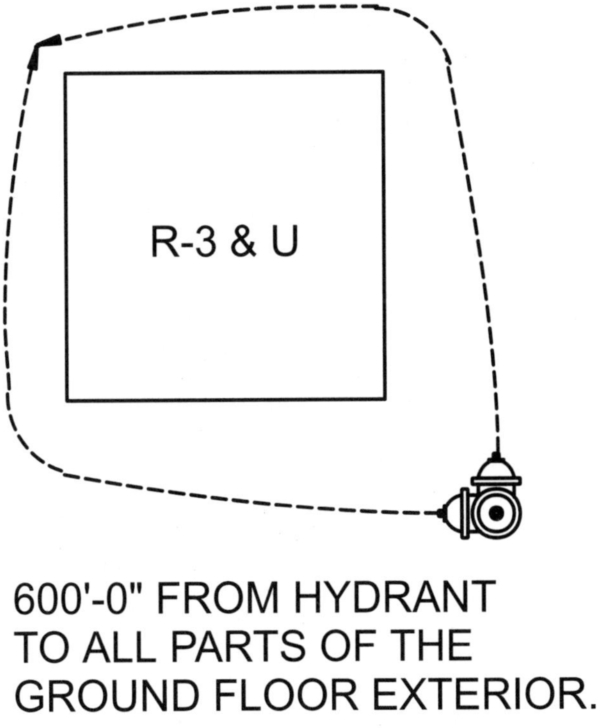

For SI: 1 foot = 304.8

Figure 508.5.1(2)
HYDRANT LOCATION—GROUP R-3 AND U OCCUPANCIES

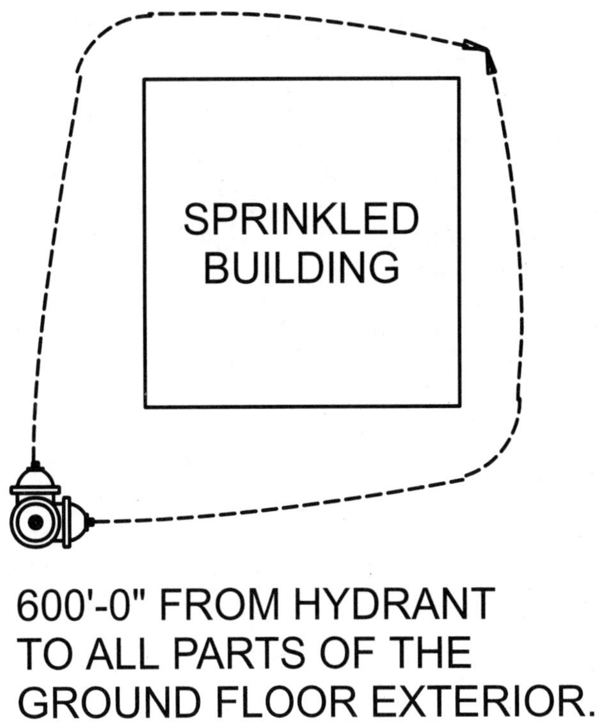

For SI: 1 foot = 304.8 mm

Figure 508.5.1(3)
ACCEPTABLE HYDRANT LOCATION WITH SPRINKLERS

508.5.2 Inspection, testing and maintenance. Fire hydrant systems shall be subject to periodic tests as required by the fire code official. Fire hydrant systems shall be maintained in an operative condition at all times and shall be repaired where defective. Additions, repairs, alterations and servicing shall comply with approved standards.

❖ The fire code official has the authority to require periodic tests and to specify the frequency of such tests. The generally accepted procedure is to inspect hydrants annually for proper operation and drainage by opening and closing the hydrants and lubricating all threads.

508.5.3 Private fire service mains and water tanks. Private fire service mains and water tanks shall be periodically inspected, tested and maintained in accordance with NFPA 25 at the following intervals:

1. Private fire hydrants (all types): Inspection annually and after each operation; flow test and maintenance annually.
2. Fire service main piping: Inspection of exposed, annually; flow test every 5 years.
3. Fire service main piping strainers: Inspection and maintenance after each use.

❖ NFPA 25 is the *Standard for the Inspection, Testing and Maintenance of Water-based Fire Protection Systems.* Chapter 4 of that standard covers private fire service mains and Chapter 6 covers water tanks.

508.5.4 Obstruction. Posts, fences, vehicles, growth, trash, storage and other materials or objects shall not be placed or kept near fire hydrants, fire department inlet connections or fire protection system control valves in a manner that would prevent such equipment or fire hydrants from being immediately discernible. The fire department shall not be deterred or hindered from gaining immediate access to fire protection equipment or fire hydrants.

❖ Nothing should be placed near a fire hydrant, FDC or control valve that would prevent responding fire fighters from immediately recognizing the device. Plants and shrubs on public or private property are probably the most common object that can make fire hydrants, FDCs or fire protection system valves virtually invisible to responding fire apparatus engineers. In residential areas especially, some homeowners don't like "that ugly piece of iron" (i.e., a fire hydrant) in their yard, so they plant all manner of vegetation around it in an effort to hide it—a clear violation of this section. On construction sites, fire hydrants or FDCs are often hidden from view and access by deliveries of construction materials randomly dumped at the most convenient spot on the site without regard for the need of the fire department to gain immediate access to hydrants or FDCs.

508.5.5 Clear space around hydrants. A 3-foot (914 mm) clear space shall be maintained around the circumference of fire hydrants except as otherwise required or approved.

❖ Care must be taken so that fences, utility poles, barricades and other obstructions do not prevent the operation of fire hydrants. A clear space of 3 feet (914 mm) must be maintained around hydrants (Figure 508.5.5). Though not specifically mentioned in this section, it is also important that hydrants be installed with the center of the outlet cap nuts at least 18 inches (457 mm) above adjoining grade to accommodate the free turning of a hydrant wrench when removing the caps (see NFPA 24, Chapter 4, for further information).

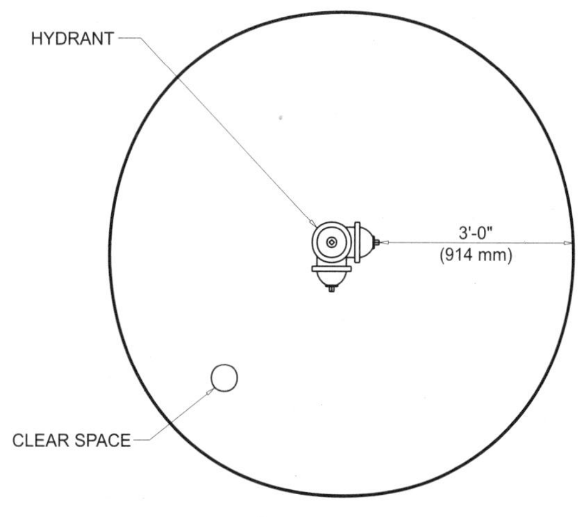

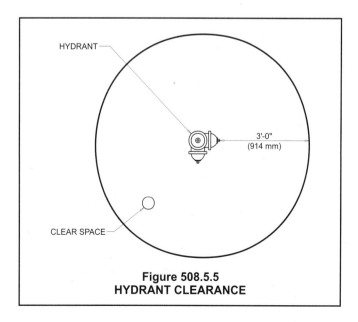

Figure 508.5.5
HYDRANT CLEARANCE

508.5.6 Physical protection. Where fire hydrants are subject to impact by a motor vehicle, guard posts or other approved means shall comply with Section 312.

❖ Section 312 requires vehicle impact protection by placing steel posts filled with concrete around the hydrant (Figure 508.5.6). Section 312 gives the specifications for the posts. Note that the provisions of Section 508.5.5 apply to the installation of posts or other protective features.

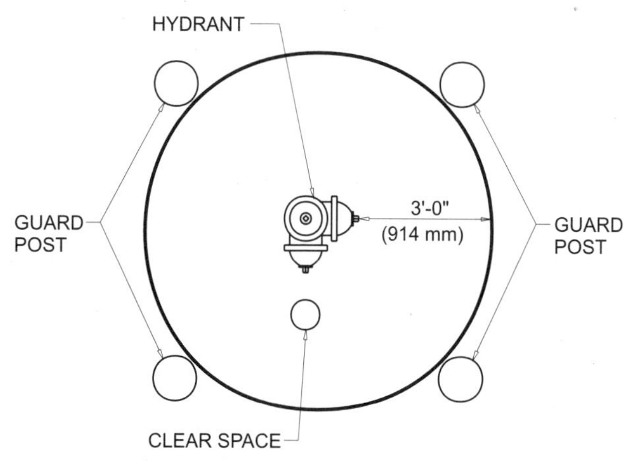

Figure 508.5.6
HYDRANT IMPACT PROTECTION

SECTION 509
FIRE COMMAND CENTER

509.1 Features. Where required by other sections of this code and in all buildings classified as high-rise buildings by the *International Building Code*, a fire command center for fire department operations shall be provided. The location and accessibility of the fire command center shall be approved by the fire department. The fire command center shall be separated from the remainder of the building by not less than a 1-hour fire-resistance-rated fire barrier. The room shall be a minimum of 96 square feet (9 m^2) with a minimum dimension of 8 feet (2438 mm). A layout of the fire command center and all features required by this section to be contained therein shall be submitted for approval prior to installation. The fire command center shall comply with NFPA 72 and shall contain the following features:

1. The emergency voice/alarm communication system unit.
2. The fire department communications system.
3. Fire-detection and alarm system annunciator system.
4. Annunciator visually indicating the location of the elevators and whether they are operational.
5. Status indicators and controls for air-handling systems.
6. The fire-fighter's control panel required by Section 909.16 for smoke control systems installed in the building.
7. Controls for unlocking stairway doors simultaneously.

8. Sprinkler valve and water-flow detector display panels.
9. Emergency and standby power status indicators.
10. A telephone for fire department use with controlled access to the public telephone system.
11. Fire pump status indicators.
12. Schematic building plans indicating the typical floor plan and detailing the building core, means of egress, fire protection systems, fire-fighting equipment and fire department access.
13. Work table.
14. Generator supervision devices, manual start and transfer features.
15. Public address system, where specifically required by other sections of this code.

❖ Fire ground operations usually involve establishing an incident command post where the incident command officer can observe what is happening; control arriving personnel and equipment and direct the resources and fire-fighting operations effectively. Because of the difficulties in controlling a fire in a high-rise building, a protected, readily accessible, separate room within the building must be established to assist the incident command officer. The room must be provided at a location that is acceptable to the fire department, usually at the front of the building or near the main entrance. The room must contain equipment necessary to monitor or control fire protection and other building service systems as listed in this section, including the controls for: voice/alarm signaling systems; fire department communication systems; automatic fire detection and alarm system annunciator panels; an annunciator that visually indicates the floor location of elevators and whether the elevators are operational; air-handling and smoke control system status indicators and controls; controls for unlocking all stairway doors simultaneously; sprinkler valve, water-flow detector and fire-pump status display panels; emergency and standby power; status indicator and a telephone for fire department use to make outside calls (the telephone should not be a pay phone).

SECTION 510
FIRE DEPARTMENT ACCESS TO EQUIPMENT

510.1 Identification. Fire protection equipment shall be identified in an approved manner. Rooms containing controls for air-conditioning systems, sprinkler risers and valves, or other fire detection, suppression or control elements shall be identified for the use of the fire department. Approved signs required to identify fire protection equipment and equipment location, shall be constructed of durable materials, permanently installed and readily visible.

❖ In an emergency, it is vitally important that the fire department and other emergency responders be able to quickly locate and access critical controls for fire protection systems. Obstructed or poorly marked equipment can cause delays in fire-fighting operations while fire fighters locate other hose stations and stretch additional hose, for example. Valves and other controls are often located in rooms or other enclosures and their location must be clearly identified with written or pictographic signs, which must be clearly visible and legible. Signs using the NFPA 170 symbols for fire protection equipment can provide standardized markings throughout a jurisdiction. White reflective symbols on a red reflective background are effective. For exterior signs, heavy-gage, sign-grade aluminum is recommended. Interior signs may be constructed of plastic, light-gage aluminum or other approved, durable, water-resistant material. As a general rule, fire protection piping, cabinets, enclosures, wiring, equipment and accessories are red or are identified by red or red/white markings. The manner of identification is subject to the approval of the fire code official.

Bibliography

The following resource materials are referenced in this chapter or are relevant to the subject matter addressed in this chapter.

HB-16–96, *Standard Specification for Highway Bridges*, 16th edition, with 1997 through 2002 Interim Revisions. Washington, DC: American Association of State Highway and Transportation Officials, 2002.

IBC–2003, *International Building Code.* Falls Church, VA: International Code Council, 2003.

NFPA 22–98, *Water Tanks for Private Fire Protection.* Quincy, MA: National Fire Protection Association, 1998.

NFPA 24–95, *Installation of Private Fire Service Mains and Their Appurtenances.* Quincy, MA: National Fire Protection Association, 1995.

NFPA 25–98, *Inspection, Testing and Maintenance of Water-Based Fire Protection Systems.* Quincy, MA: National Fire Protection Association, 1998.

NFPA 72–99, *National Fire Alarm Code.* Quincy, MA: National Fire Protection Association, 1999.

NFPA 170-02, *Standard for Fire Safety Symbols.* Quincy, MA: National Fire Protection Association, 2002.

NFPA 291-02, *Recommended Practice for Fire Flow Testing and Marking of Hydrants.* Quincy, MA: National Fire Protection Association, 2002.

Chapter 6: Building Services And Systems

General Comments

This chapter is focused on building systems and services as they relate to potential safety hazards and when and how they should be installed. In some cases, many of the provisions are located in other portions of the code. This chapter brings together all building system- and service-related issues for convenience and provides a more systematic view of buildings. The following building services and systems are addressed:

- Fuel-fired appliances (Section 603).
- Emergency and standby power systems (Section 604).
- Electrical equipment, wiring and hazards (Section 605).
- Mechanical refrigeration (Section 606).
- Elevator recall and maintenance (Section 607).
- Stationary lead-acid battery systems (Section 608).
- Valve-regulated lead-acid (VRLA) battery systems (Section 609).
- Commercial kitchen hoods (Section 610).

Some of the sections specifically deal with installation while others deal with reducing the hazards from the use of the services or systems. For example, using too many extension cords on the building electrical system may present a fire hazard (see Section 605). On the other hand, elevator recall and maintenance (see Section 607) simply states when and how recall is required. As with all other chapters of the *International Codes®*, Section 602 contains definitions.

Purpose

As technology progresses and societal expectations increase, building systems and services become more complex and numerous. The use of computers has resulted in a more frequent use of uninterruptible power supplies and emergency power, which are often powered through the use of lead-acid battery systems. In the past, these provisions were simply scattered throughout the code. These various building services and system requirements have been brought together in this chapter to simplify the code requirements and increase the likelihood that these elements are properly addressed.

SECTION 601 GENERAL

601.1 Scope. The provisions of this chapter shall apply to the installation, operation and maintenance of fuel-fired appliances and heating systems, emergency and standby power systems, electrical systems and equipment, mechanical refrigeration systems, elevator recall, stationary lead-acid battery systems and commercial kitchen hoods.

❖ This section provides a laundry list of building systems and services that must comply with Chapter 6 when they are being installed, during their operation and for long-term maintenance.

601.2 Permits. Permits shall be obtained for refrigeration systems and battery systems as set forth in Section 105.6.

❖ Only two systems discussed in Chapter 6 require permits: the operation of refrigeration systems and the installation of lead-acid battery systems. The permit for operation of refrigeration systems is intended to warn emergency responders that a potential hazard exists. This information will better equip them to respond to such a call. The permit for installation of a battery system ensures the proper safety mechanisms, such as ventilation and spill control are installed (see commentary, Section 105).

SECTION 602 DEFINITIONS

602.1 Definitions. The following words and terms shall, for the purposes of this chapter and as used elsewhere in this code, have the meanings shown herein.

❖ The following words and terms shall, for the purposes of this chapter and as used elsewhere in this code, have the meanings shown herein.

Definitions of terms can help in the understanding and application of the code requirements. The purpose for including those definitions that are associated with the subject matter of this chapter is to provide more convenient access to them without having to refer back to Chapter 2. It is important to emphasize that these terms are not exclusively related to this chapter but are applicable everywhere the term is used in the code. For convenience, these terms are also listed in Chapter 2 with a cross reference to this section. The use and application of all defined terms, including those defined in this section, are set forth in Section 201.

BATTERY, LEAD ACID. A group of electrochemical cells interconnected to supply a nominal voltage of DC power to a suitably connected electrical load. The number of cells connected in series determines the nominal voltage rating of the battery. The size of the cells determines the discharge capacity of the entire battery.

❖ The term "lead-acid battery" applies to all wet-cell battery types made up of multiple electrochemical cells interconnected to supply a nominal voltage of direct current (DC) power. Such batteries are typically used for uninterrupted power supply (UPS) systems, emergency power or standby power applications. Lead-acid batteries consist of a positive and a negative lead plate immersed in an electrolyte solution of sulfuric acid, which is considered a corrosive liquid. The definition includes both the vented type (in which hydrogen and oxygen are produced and vented) and the valve-regulated type, which is sealed to prevent the venting of the generated gases .

BATTERY SYSTEM, STATIONARY LEAD ACID. A system which consists of three interconnected subsystems:

1. A lead-acid battery.
2. A battery charger.
3. A collection of rectifiers, inverters, converters, and associated electrical equipment as required for a particular application.

❖ The definition describes the completed battery system, consisting of all the components needed to make a functioning battery power system (see commentary to the definition of "Battery, lead-acid" and Section 608).

[M] COMMERCIAL COOKING APPLIANCES. Appliances used in a commercial food service establishment for heating or cooking food and which produce grease vapors, steam, fumes, smoke or odors that are required to be removed through a local exhaust ventilation system. Such appliances include deep fat fryers; upright broilers; griddles; broilers; steam-jacketed kettles; hot-top ranges; under-fired broilers (charbroilers); ovens; barbecues; rotisseries; and similar appliances. For the purpose of this definition, a food service establishment shall include any building or a portion thereof used for the preparation and serving of food.

❖ This definition is important in the application of Section 610, which requires a commercial kitchen hood above commercial cooking appliances. A definition of "Food service establishment" is included within this definition. "Food service" includes operations such as preparing, handling, cleaning, cooking and packaging food items of any kind.

[M] HOOD. An air-intake device used to capture by entrapment, impingement, adhesion or similar means, grease and similar contaminants before they enter a duct system.

Type I. A kitchen hood for collecting and removing grease vapors and smoke.

❖ A kitchen exhaust system, which includes the hood serving a commercial cooking appliance, is a specialized exhaust system. A commercial cooking appliance can generate large quantities of air contaminants, such as grease vapors, smoke and combustion byproducts. The descriptor "Type I" used in conjunction with the term "hood" refers to an exhaust system that is required for all cooking appliances that are used for commercial purposes and that produce grease-laden vapors or smoke.

REFRIGERANT. The fluid used for heat transfer in a refrigerating system; the refrigerant absorbs heat and transfers it at a higher temperature and a higher pressure, usually with a change of state.

❖ The refrigerant is the working fluid in refrigeration and air-conditioning systems. In vapor refrigeration cycles, refrigerants absorb heat from the load side at the evaporator and reject heat at the condenser. Aside from having suitable thermodynamic properties, the selection of a refrigerant must also take into consideration chemical stability, flammability, toxicity and environmental compatibility. Refrigeration is a result of the physical laws of vaporization (evaporation) of liquids. Basically, evaporation of liquid refrigerant is an endothermic process and condensing of vapors is an exothermic process.

REFRIGERATION SYSTEM. A combination of interconnected refrigerant-containing parts constituting one closed refrigerant circuit in which a refrigerant is circulated for the purpose of extracting heat.

❖ Such systems include, at minimum, a pressure-imposing element or generator, an evaporator, a condenser and interconnecting piping. A single piece of equipment can contain multiple refrigeration systems (circuits).

VALVE-REGULATED LEAD-ACID (VRLA) BATTERY. A lead-acid battery consisting of sealed cells furnished with a valve that opens to vent the battery whenever the internal pressure of the battery exceeds the ambient pressure by a set amount. In VRLA batteries, the liquid electrolyte in the cells is immobilized in an absorptive glass mat (AGM cells or batteries) or by the addition of a gelling agent (gel cells or gelled batteries).

❖ Valve-regulated lead-acid (VRLA) batteries (sometimes referred to as "gel cells") differ substantially from flooded batteries in design, operation and potential hazard. VRLA-type batteries are uniquely different from the traditional liquid electrolyte lead acid batteries in that they have no liquid electrolyte to flow from the container if it were to break. Also, the VRLA batteries do not vent their off-gasses to the atmosphere but rather implement an oxygen recombination cycle that minimizes the emissions of gas from the battery during overcharging.

Though these batteries are considered sealed, their design includes spring-controlled valves that vent gases at a pressure threshold of 2 to 5 pounds per square inch gauge (psig) (14 to 34 kPa). These batteries are sometimes mistakenly called "maintenance free;" however, they should be maintained in accordance with the manufacturer's instructions as with any other building system.

Application of the stationary lead-acid battery provisions of Section 608 to VRLA battery installations would negate the enhanced economic and safety benefits of the new VRLA technology. Accordingly, Section 609 was added to the code to recognize the different characteristics and benefits of these batteries.

VENTED (FLOODED) LEAD-ACID BATTERY. A lead-acid battery consisting of cells that have electrodes immersed in liquid electrolyte. Flooded lead-acid batteries have a provision for the user to add water to the cell and are equipped with a flame-arresting vent which permits the escape of hydrogen and oxygen gas from the cell in a diffused manner such that a spark, or other ignition source, outside the cell will not ignite the gases inside the cell.

❖ There are basically two types of lead-acid storage batteries, which are based on how they are constructed and vented: they are either vented (flooded) or sealed. Vented (flooded) and sealed batteries differ in how they dispose of the hydrogen (explosive in air at 4 percent by volume) and oxygen produced by electrolysis during their recharging (off-gassing). In a vented (flooded) battery, these gases are allowed to escape to the atmosphere. In a sealed battery, the gases are contained within the battery cell(s) and recombined with the electrolyte. Because the gases created during battery charging are vented to the atmosphere, distilled water must be added by the owner periodically to bring the electrolyte level back to that required by the battery specifications. One of the most common types of vented (flooded) lead-acid batteries is the automobile battery.

SECTION 603
FUEL-FIRED APPLIANCES

603.1 Installation. The installation of nonportable fuel gas appliances and systems shall comply the *International Fuel Gas Code*. The installation of all other fuel-fired appliances, other than internal combustion engines, oil lamps and portable devices such as blow torches, melting pots and weed burners, shall comply with this section and the *International Mechanical Code.*

❖ The code regulates the installation of portable gas-fired appliances and portable appliances fueled by methods other than gaseous fuels. The *International Mechanical Code*® (IMC®) and the *International Fuel Gas Code*® (IFGC®) do not cover portable appliances. The code also has provisions that apply to appliances that are not portable and use fuels other than gas.

603.1.1 Manufacturer's instructions. The installation shall be made in accordance with the manufacturer's instructions and applicable federal, state, and local rules and regulations. Where it becomes necessary to change, modify, or alter a manufacturer's instructions in any way, written approval shall first be obtained from the manufacturer.

❖ Compliance with the appliance manufacturer's installation instructions is a fundamental requirement of all *International Codes and those instructions are an enforceable extension of the code. Federal, state, county or municipal laws might supercede part of the installation instructions or could be applied in addition to the requirements in the instructions.*

603.1.2 Approval. The design, construction and installation of fuel-fired appliances shall be in accordance with the *International Fuel Gas Code* and the *International Mechanical Code.*

❖ The code relies on the IMC and the IFGC for the coverage of appliance installations and contains only a limited number of requirements that apply in addition to those of the IMC and IFGC.

603.1.3 Electrical wiring and equipment. Electrical wiring and equipment used in connection with oil-burning equipment shall be installed and maintained in accordance with Section 605 and the ICC *Electrical Code.*

❖ Section 605 contains provisions intended to prevent fire hazards and shock hazards associated with the use of existing appliances. The *ICC Electrical Code*® (ICC EC™) references NFPA 70, which covers the installation of appliances.

603.1.4 Fuel oil. The grade of fuel oil used in a burner shall be that for which the burner is approved and as stipulated by the burner manufacturer. Oil containing gasoline shall not be used. Waste crankcase oil shall be an acceptable fuel in Group F, M and S occupancies, when utilized in equipment listed for use with waste oil and when such equipment is installed in accordance with the manufacturer's instructions and the terms of its listing.

❖ Different grades of fuel oil have different viscosities and chemical makeup. A burner and fuel mismatch could result in poor combustion, sooting and burner component failure. Oil burners are not designed to burn oil contaminated with chemicals of higher volatility. Appliances that consume used engine oil are allowed only in occupancies of low occupant density (those without sleeping rooms) and where it will likely be monitored and maintained by facility personnel. Used engine oil appliances use a specialized type of atomizing oil burner designed to burn dirty waste oil collected from internal combustion engine maintenance operations.

603.1.5 Access. The installation shall be readily accessible for cleaning hot surfaces; removing burners; replacing motors, controls, air filters, chimney connectors, draft regulators, and other working parts; and for adjusting, cleaning and lubricating parts.

❖ The IMC and the IFGC require access for the initial installation as well as for the life of the appliance. Safe operation depends on observation and maintenance, which depend on adequate access to the appliances. In order to be considered readily accessible, the installation should be reachable by personnel without having to remove building elements or obstacles of any kind or use climbing aids to reach it for service.

603.1.6 Testing, diagrams and instructions. After installation of the oil-burning equipment, operation and combustion performance tests shall be conducted to determine that the burner is in proper operating condition and that all accessory equipment, controls, and safety devices function properly.

❖ Testing of an appliance after installation is also required by the IMC and the appliance manufacturer's installation instructions.

603.1.6.1 Diagrams. Contractors installing industrial oil-burning systems shall furnish not less than two copies of diagrams showing the main oil lines and controlling valves, one copy of which shall be posted at the oil-burning equipment and another at an approved location that will be accessible in case of emergency.

❖ For large systems, the piping and control valve layout may be complicated and extensive. In the event of an emergency, facility personnel or fire fighters might need access to control valves to protect piping and to limit the fire hazard.

603.1.6.2 Instructions. After completing the installation, the installer shall instruct the owner or operator in the proper operation of the equipment. The installer shall also furnish the owner or operator with the name and telephone number of persons to contact for technical information or assistance and routine or emergency services.

❖ Appliances are more likely to be properly (safely) operated and maintained if the owner or operator is instructed in the operation of the equipment and given the necessary means to obtain technical and emergency services.

603.1.7 Clearances. Working clearances between oil-fired appliances and electrical panelboards and equipment shall be in accordance with the ICC *Electrical Code*. Clearances between oil-fired equipment and oil supply tanks shall be in accordance with NFPA 31.

❖ The ICC EC requires working clearances around electrical equipment for protection of personnel. NFPA 31 requires clearances between appliances and oil supply tanks to protect the oil tank from excessive heat and to lessen the fire hazard from any oil leakage.

[B, M, FG] 603.2 Chimneys. Masonry chimneys shall be constructed in accordance with the *International Building Code*. Factory-built chimneys shall be installed in accordance with the *International Mechanical Code*. Metal chimneys shall be constructed and installed in accordance with NFPA 211.

❖ The *International Building Code*® (IBC®) regulates masonry chimney construction in Chapter 21. Factory-built chimneys are regulated by Section 805 of the IMC. Metal chimneys are distinct from factory-built chimneys, are industrial occupancy related (e.g., smokestacks) and are regulated by NFPA 211. (see IMC commentary, Section 806.1).

603.3 Fuel oil storage systems. Fuel oil storage systems shall be installed in accordance with this code. Fuel oil piping systems shall be installed in accordance with the *International Mechanical Code*.

❖ The code regulates the storage of fuel oil and the IMC regulates installation of the fuel oil distribution piping system.

603.3.1 Maximum outside fuel oil storage above ground. Where connected to a fuel-oil piping system, the maximum amount of fuel oil storage allowed outside above ground without additional protection shall be 660 gallons (2498 L). The storage of fuel oil above ground in quantities exceeding 660 gallons (2498 L) shall comply with NFPA 31.

❖ To limit the potential fire hazard resulting from oil spillage, the code sets a quantity limitation on unprotected storage. NFPA 31 requires protection, such as spillage containment, for storage in excess of 660 gallons (2498 L). The storage of 660 gallons (2498 L) is allowed to be in any configuration of containers that does not exceed a total of 660 gallons (2498 L).

603.3.2 Maximum inside fuel oil storage. Where connected to a fuel-oil piping system, the maximum amount of fuel oil storage allowed inside any building shall be 660 gallons (2498 L). Where the amount of fuel oil stored inside a building exceeds 660 gallons (2498 L), the storage area shall be in compliance with the *International Building Code*.

❖ This section correlates with Table 2703.1.1(1), Note i; Section 3401.2, Item 3 and IBC Table 307.7(1), Note i and is a specific exception for maximum 660-gallon (2498 L) inside storage of combustible liquids, which are connected to a closed fuel-oil piping system. This exception would apply to most oil-fired stationary equipment, whether in industrial, commercial or residential occupancies. When the aggregate indoor storage quantity of fuel oil connected to a closed fuel-oil piping system exceeds the special amount of 660 gallons (2498 L), then the storage area must comply with the requirements of the IBC for a Group H occupancy.

603.3.3 Underground storage of fuel oil. The storage of fuel oil in underground storage tanks shall comply with NFPA 31.

❖ Section 603.3 does not require that fuel oil tanks be installed underground; however, there may be circumstances under which such an installation is either desirable or advisable. The code user is directed to NFPA 31 for specific requirements applicable to the installation of underground combustible liquids storage tanks.

603.4 Portable unvented heaters. Portable unvented fuel-fired heating equipment shall be prohibited in occupancies in Groups A, E, I, R-1, R-2, R-3 and R-4.

Exception: Listed and approved unvented fuel-fired heaters in one- and two-family dwellings.

❖ Portable unvented fuel-fired heating equipment refers to portable space heaters, such as kerosene-fueled appliances. This section would also apply to gas-fired appliances that connect to gas convenience outlets with gas hose connectors. This section does not apply to permanently installed appliances. Portable space-heating appliances are moved around at will by the occupants and might be placed too close to combustibles or where they are susceptible to being hit, tipped over, etc. Because of potential misuse, such appliances are considered an unacceptable risk in all occupancies except one- and two-family dwellings, business, factory/industrial, high hazard, mercantile, storage and utility. In one- and two-family dwellings, the occupants are assumed to take greater care in the use of such appliances.

603.4.1 Prohibited locations. Unvented fuel-fired heating equipment shall not be located in, or obtain combustion air from, any of the following rooms or spaces: sleeping rooms, bathrooms, toilet rooms or storage closets.

❖ This section is parallel in its intent with Section 303.3 of the IMC and the IFGC. As a subsection to Section 603.4, this section is addressing portable appliances in Groups A, E, I and R. (see IFGC commentary, Section 303.3).

603.5 Heating appliances. Heating appliances shall be listed and shall comply with this section.

❖ The IMC and the IFGC require that all space-heating appliances be listed and labeled.

603.5.1 Guard against contact. The heating element or combustion chamber shall be permanently guarded so as to prevent accidental contact by persons or material.

❖ The injury and ignition protection feature required by this section is typically designed into the appliance.

603.5.2 Heating appliance installation. Heating appliances shall be installed in accordance with the manufacturer's instructions, the *International Building Code*, the *International Mechanical Code*, the *International Fuel Gas Code* and the ICC *Electrical Code*.

❖ Appliance installation is subject to the requirements of multiple codes, including the IMC, the IFGC, the *International Plumbing Code®* (IPC)®, the ICC EC, the *International Energy Conservation Code®* (IECC®) and the IBC.

603.6 Chimneys and appliances. Chimneys, incinerators, smokestacks or similar devices for conveying smoke or hot gases to the outer air and the stoves, furnaces, fireboxes or boilers to which such devices are connected, shall be maintained so as not to create a fire hazard.

❖ A primary function of the code is to reduce or eliminate fire hazards through proper maintenance of appliances and systems that are potential fire and life safety hazards.

603.6.1 Masonry chimneys. Masonry chimneys that, upon inspection, are found to be without a flue liner and that have open mortar joints which will permit smoke or gases to be discharged into the building, or which are cracked as to be dangerous, shall be repaired or relined with a listed chimney liner system installed in accordance with the manufacturer's installation instructions or a flue lining system installed in accordance with the requirements of the *International Building Code* and appropriate for the intended class of chimney service.

❖ See Section 2113.11 of the IBC, Section 801.16 of the IMC and Sections 501.11 and 503.5 of the IFGC for information on masonry chimneys.

603.6.2 Metal chimneys. Metal chimneys which are corroded or improperly supported shall be repaired or replaced.

❖ See the commentary to Section 603.2.

603.6.3 Decorative shrouds. Decorative shrouds installed at the termination of factory-built chimneys shall be removed except where such shrouds are listed and labeled for use with the specific factory-built chimney system and are installed in accordance with the chimney manufacturer's installation instructions.

❖ This section is retroactive in that it requires removal of a previously installed trim item. Section 805.6 of the IMC and Section 503.5.4 of the IFGC prohibit the installation of decorative shrouds not meeting the listing criterion. The code addresses those noncomplying shrouds that were illegally installed (see IMC commentary, Section 805.6, and IFGC commentary, Section 503.5.4).

603.6.4 Factory-built chimneys. Existing factory-built chimneys that are damaged, corroded or improperly supported shall be repaired or replaced.

❖ Defective or inadequately supported chimneys could be leaking flue gas and could fail structurally, resulting in separation, collapse, a fire hazard and a life safety hazard. This section is consistent with the maintenance focus of the code.

603.6.5 Connectors. Existing chimney and vent connectors that are damaged, corroded or improperly supported shall be repaired or replaced.

❖ See the commentary to Section 603.6.4.

603.7 Discontinuing operation of unsafe heating appliances. The fire code official is authorized to order that measures be taken to prevent the operation of any existing stove, oven, furnace, incinerator, boiler or any other heat-producing device or appliance found to be defective or in violation of code requirements for existing appliances after giving notice to this effect to any person, owner, firm or agent or operator in charge of the same. The fire code official is authorized to take measures to prevent the operation of any device or appliance without notice when inspection shows the existence of an immediate fire hazard or when imperiling human life. The defective device shall remain withdrawn from service until all necessary repairs or alterations have been made.

❖ When a heat-producing appliance or system is determined to be unsafe, the fire code official is required to notify the owner or agent of the building as the first step in correcting the difficulty. This notice is to describe the repairs and improvements necessary to correct the deficiency and keep the system in operation or require the unsafe equipment or system to be removed or replaced. The notice must specify a time frame in which the corrective actions must occur. Additionally, the notice should require the immediate response of the owner or agent.

If the owner or agent is not available, public notice of the declaration would be enough to comply with this section. The fire code official may also determine that the system must be disconnected to correct an unsafe condition and must give written notice to that effect; however, an immediate disconnection can be ordered if it is essential for protection of public health and safety.

603.7.1 Unauthorized operation. It shall be a violation of this code for any person, user, firm or agent to continue the utilization of any device or appliance (the operation of which has been discontinued or ordered discontinued in accordance with Section 603.7), unless written authority to resume operation is given by the fire code official. Removing or breaking the means by which operation of the device is prevented shall be a violation of this code.

❖ Appliances or systems removed from service in accordance with Section 603.7 may be sealed or otherwise secured in a manner approved by the fire code official and may only be returned to service upon written authorization of the fire code official.

603.8 Incinerators. Commercial, industrial and residential-type incinerators and chimneys shall be constructed in accordance with the *International Building Code,* the *International Fuel Gas Code* and the *International Mechanical Code.*

❖ See the commentary to Section 907.1 of the IMC, Section 605.1 of the IFGC and Section 2113 of the IBC.

603.8.1 Residential incinerators. Residential incinerators shall be of an approved type.

❖ Residential incinerators have gone out of use today but may still be found in older homes and in rural areas.

603.8.2 Spark arrestor. Incinerators shall be equipped with an effective means for arresting sparks.

❖ Spark arrestor chimney caps are designed with a screened outlet that prevents the escape of burning embers and particles.

603.8.3 Restrictions. Where the fire code official determines that burning in incinerators located within 500 feet (152 m) of mountainous, brush or grass-covered areas will create an undue fire hazard because of atmospheric conditions, such burning shall be prohibited.

❖ The fire code official must determine whether incinerator use would present an unacceptable risk of wild fires in timber, brush and grass-covered areas. For urban-wildland interface areas, see the *International Urban-Wildland Interface Code*™ (IUWIC™). The local air-quality agency may also have restrictions.

603.8.4 Time of burning. Burning shall take place only during approved hours.

❖ The jurisdiction must determine the periods that would be safe for burning and those that would be unsafe. Consideration must be given to daylight, prevailing winds, ambient temperatures, impact on air quality, moisture levels and presence of observers and supervisory personnel.

603.8.5 Discontinuance. The fire code official is authorized to require incinerator use to be discontinued immediately if the fire code official determines that smoke emissions are offensive to occupants of surrounding property or if the use of incinerators is determined by the fire code official to constitute a hazardous condition.

❖ The fire code official can prohibit incinerator use if it would be a nuisance or a health or fire hazard. Coordination with the local air-quality agency may also be necessary.

603.9 Gas meters. Above-ground gas meters, regulators and piping subject to damage shall be protected by a barrier complying with Section 312 or otherwise protected in an approved manner.

❖ Vehicle impact protection is necessary to prevent gas leakage resulting from impact damage to gas service equipment. Protection can be accomplished by location alone or by the construction of barriers as prescribed by Section 312. Barriers would be required only where the gas service equipment is located where vehicle impact is likely to occur.

SECTION 604 EMERGENCY AND STANDBY POWER SYSTEMS

604.1 Installation. Emergency and standby power systems shall be installed in accordance with the ICC *Electrical Code*, NFPA 110 and NFPA 111. Existing installations shall be maintained in accordance with the original approval.

❖ This section is fairly basic in requiring emergency power and standby power in accordance with the ICC EC and NFPA 110 and 111. These two standards are for emergency and standby power systems and stored electrical energy emergency and standby power systems, respectively. NFPA 110 is geared toward power sources, such as diesel-driven generators. NFPA 111 is geared toward power supplies, such as stationary lead-acid battery systems. Section 604 also requires that existing installations continue to comply with the original approval.

A primary difference between emergency and standby power systems is the time in which the power supply activates. More specifically, emergency power is available in 10 seconds whereas standby power is available in 60 seconds. Another type of power system, which operates immediately upon failure of the primary power supply, is called an uninterruptible power supply (UPS), commonly used in data processing operations.

604.1.1 Stationary generators. Stationary emergency and standby power generators required by this code shall be listed in accordance with UL 2200.

❖ This section requires that generators used for emergency and standby power be listed and comply with UL 2200. This is a specific testing and installation standard for generator sets.

604.2 Where required. Emergency and standby power systems shall be provided where required by Sections 604.2.1 through 604.2.18.

❖ Sections 604.2.1 through 604.2.18 is a laundry list of locations throughout buildings and facilities where emergency or standby power is required. In some cases the requirements are occupancy- or use-specific and in others the requirements are system- or equipment-specific. Generally, the requirements are related to life-safety-oriented systems within buildings, such as a fire alarm system or elevators used for egress for those with disabilities.

604.2.1 Group A occupancies. Emergency power shall be provided for emergency voice/alarm communication systems in Group A occupancies in accordance with Section 907.2.1.2.

❖ When a voice communication system is used within a Group A occupancy, the system must have an emergency power supply. Voice communication systems tend to be used in occupancies with high occupant loads and where occupants are generally unfamiliar with their surroundings, making their availability during a power failure extremely important.

604.2.2 Smoke control systems. Standby power shall be provided for smoke control systems in accordance with Section 909.11.

❖ Smoke control systems, as required by Section 909.11, must be equipped with standby power in addition to normal building power. Because smoke control systems are considered a life safety system and generally such systems are more likely to be needed during a building power failure, secondary power is required. In other words, the primary building power may be shut down or lost during a fire in a building when such systems are most critical.

604.2.3 Exit signs. Emergency power shall be provided for exit signs in accordance with Section 1011.5.3

❖ Emergency power is required for exit signs. This requirement applies to all exit signs unless they provide continuous illumination for at least 90 minutes without the need for external power. Examples of signs that can independently remain illuminated are photo-luminescent exit signs. Emergency power is required because exit signs will most likely be needed when the primary power is not available.

604.2.4 Means of egress illumination. Emergency power shall be provided for means of egress illumination in accordance with Section 1006.3.

❖ Illumination of the means-of-egress path is critical during emergencies. Section 1006.3 requires that such illumination be equipped with emergency power. This is similar to the requirement for emergency power for exit sign illumination. As with the exit sign requirements, the most likely time such illumination is needed is when the primary power may not be available.

604.2.5 Accessible means of egress elevators or platform lifts. Standby power shall be provided for elevators or platform lifts that are part of an accessible means of egress in accordance with Section 1007.4 or 1007.5, respectively.

❖ To ensure that those with disabilities are able to egress a building along with other occupants, standby power is required for elevators that are used as part of the accessible means of egress.

604.2.6 Horizontal sliding doors. Standby power shall be provided for horizontal sliding doors in accordance with Section 1008.1.3.3.

❖ Horizontal sliding doors used as part of the means-of-egress system must have an integrated standby power supply as noted in Item 6 of Section 1008.1.3.3. Additionally, Section 1008.1.3.3 has a total of eight specific requirements for horizontal doors. For example, horizontal sliding doors must be openable by a simple method from both sides without any special knowledge or effort.

604.2.7 Semiconductor fabrication facilities. Emergency power shall be provided for semiconductor fabrication facilities in accordance with Section 1803.15.

❖ Chapter 18, specifically Section 1803.15, requires emergency power in semiconductor facilities. The requirements for emergency power extend to several different systems and types of equipment. Some examples include hazardous production material (HPM) exhaust ventilation systems, emergency alarm systems and automatic sprinkler system monitoring and alarms. Semiconductor facilities house various hazardous materials, including silane gas, which is why emergency power is required for certain systems (see Section 1803.15.1).

604.2.8 Membrane structures. Emergency power shall be provided for exit signs in temporary tents and membrane structures in accordance with Section 2403.12.6.1. Standby power shall be provided for auxiliary inflation systems in permanent membrane structures in accordance with the *International Building Code.*

❖ There are two aspects of these requirements. First, exit signs in temporary membrane structures and tents must have an emergency power source. Though temporary in nature, the potential hazards and the loss history associated with such structures warrants the need for exits to be visible. Additionally, the primary power supply is probably less reliable than in a permanent structure. It should be noted that fuel used to generate the secondary power supply may need to meet additional requirements of Chapter 24.

The second requirement found in this section is that permanent air-inflated and air-supported structures must have a standby power source to support the inflation system in case of power failure. The structure would collapse if primary power failed, trapping the occupants under the collapsed top.

604.2.9 Hazardous materials. Emergency or standby power shall be provided in occupancies with hazardous materials in accordance with Sections 2704.7 and 2705.1.5.

❖ This section is a reference to the emergency and standby power requirements in Chapter 27. More specifically, the requirements for emergency and standby power are found in Sections 2704.7 and 2705.1.5; therefore, the requirements will apply only in Group H occupancies. The specific systems and equipment that require such power are as follows:

- Mechanical ventilation.
- Treatment systems.
- Temperature control.
- Alarm systems.
- Detection systems.

Either a standby system or an emergency power supply is allowed. There are several exceptions to these required installations.

604.2.10 Highly toxic and toxic materials. Emergency power shall be provided for occupancies with highly toxic or toxic materials in accordance with Sections 3704.2.2.8 and 3704.3.2.6.

❖ As noted in the commentary for Section 604.2.9, there are several exceptions to the main requirements for emergency and standby power. One of these exceptions is for highly toxic and toxic materials. Essentially, the exception leads to a more specific set of requirements in Sections 3704.2.2.8 and 3704.3.2.6 for emergency power for certain systems and equipment associated with the storage and use of highly toxic and toxic gases. These elements are as follows:

- Exhaust ventilation systems.
- Treatment systems.
- Gas detection systems.
- Smoke detection systems.
- Temperature control systems.
- Fire alarm systems.
- Emergency alarm systems.

These requirements apply to both indoor and outdoor installations.

There is one exception that would not require emergency power for mechanical exhaust ventilation, treatment systems and temperature control systems. This exception is applicable when a fail-safe engineered system is installed. The same exception is also found in Chapter 27 for all types of hazardous materials; therefore, neither emergency nor standby power would be required.

604.2.11 Organic peroxides. Standby power shall be provided for occupancies with organic peroxides in accordance with Section 3904.1.11.

❖ First, Section 3904.1.11 specifically requires standby power in areas where Class I and unclassified detonable organic peroxides are stored. Second, Section 2704.7 requires standby or emergency power as noted in section 604.2.9 for all hazardous materials in storage over the maximum allowable quantities. There is one exception for storage areas containing Class III, IV and V organic peroxides; therefore, the systems listed in Sections 2704.7 and 2705.1.5 where Classes I and II and unclassified detonables above the maximum allowable quantities are stored or used would require either standby or emergency power. Additionally, the systems listed in Section 2705.1.5 where Class III, IV and V organic peroxides over the maximum allowable quantities are being used or handled would require either emergency or standby power unless a fail-safe engineering system is installed.

604.2.12 Pyrophoric materials. Emergency power shall be provided for occupancies with silane gas in accordance with Sections 4106.2.3 and 4106.4.3.

❖ This section references Sections 4106.2.3 and 4106.4.3, which are specific to silane gas. More specifically, ventilation systems associated with indoor storage or use of silane would require emergency power. While silane is not actually considered a pyrophoric, the general characteristic of such gases is ignition in a normal atmosphere. Requiring emergency power for ventilation systems decreases the likelihood of an interaction between the atmosphere and the gas.

604.2.13 Covered mall buildings. Covered mall buildings exceeding 50,000 square feet (4645 m^2) shall be provided with standby power systems which are capable of operating the emergency voice/alarm communication.

❖ The primary requirement for covered mall buildings over 50,000 square feet (4645 m^2) is for a standby power system to provide secondary power to the emergency voice/ alarm communication system. Section 907.2.20 requires emergency/voice alarm communication systems when a covered mall building exceeds 50,000 square feet (4645 m^2); therefore, the emergency voice/alarm communication specified by Section 907.2.20 would require standby power. Such systems are required to have emergency power in Group A occupancies.

604.2.14 High-rise buildings. Standby power, light and emergency systems in high-rise buildings shall comply with the requirements of Sections 604.2.14.1 through 604.2.14.3.

❖ High-rise buildings have a unique package of life safety requirements based on the difficultly of fighting a fire and undertaking rescue operations. Some of these requirements include the need for sprinklers, higher fire-resistance-rated construction, compartmentation and requirements for secondary power in case of failure of the primary power. This section clarifies where standby and emergency power is required throughout these systems and also includes some specific requirements on the location of generators used on site to provide this power.

604.2.14.1 Standby power. A standby power system shall be provided. Where the standby system is a generator set inside a building, the system shall be located in a separate room enclosed with 2-hour fire-resistance-rated fire barrier assemblies. System supervision with manual start and transfer features shall be provided at the fire command center.

❖ This is the general section for standby power requirements. More specifically, it states that if a generator set is within the building it must be enclosed in a 2-hour fire-resistance-rated fire barrier room. Additionally, the controls that allow manual start and transfer features must be available within the fire command center.

604.2.14.1.1 Fuel supply. An on-premises fuel supply, sufficient for not less than 2-hour full-demand operation of the system, shall be provided.

Exception: Where the system is supplied with pipeline natural gas and is approved.

❖ This section contains requirements similar to those in the ICC EC covering the quantity of fuel necessary to operate the generator. Specifically, it states that a 2-hour fuel supply be available for full demand on the system. There is an exception when natural gas is piped directly into the building.

604.2.14.1.2 Capacity. The standby system shall have a capacity and rating that supplies all equipment required to be operational at the same time. The generating capacity is not required to be sized to operate all of the connected electrical equipment simultaneously.

❖ This section requires that the generator be designed to handle the operation of all equipment that must be supplied with secondary power. It then addresses the fact that the standby power does not need to supply power to all equipment and systems in the building at the same time. Instead, only those specific systems that must be in service at the same time, such as the smoke control system and the elevator.

604.2.14.1.3 Connected facilities. Power and lighting facilities for the fire command center and elevators specified in Sections 403.8 and 403.9 of the *International Building Code*, as applicable, and electrically powered fire pumps required to maintain pressure, shall be transferable to the standby source. Standby power shall be provided for at least one elevator to serve all floors and be transferable to any elevator.

❖ This section states which systems and equipment are to be connected to standby power in a high-rise building. The essential equipment and features during an emergency are the fire command center, elevators and fire pumps. Only one elevator would need to operate with standby power during an emergency; however, where there is more than one elevator in the building and standby power cannot support the operation of more than one elevator at the same time, they must connect to standby power in sequence, and return to the designated landing floor one at a time. Once the elevators have been moved to the designated landing, one must be capable of operating under standby power.

604.2.14.2 Separate circuits and fixtures. Separate lighting circuits and fixtures shall be required to provide sufficient light with an intensity of not less than 1 foot-candle (11 lux) measured at floor level in all means of egress corridors, stairways, smokeproof enclosures, elevator cars and lobbies, and other areas which are clearly a part of the escape route.

❖ This section requires a lighting system independent of other building circuits and fixtures to illuminate the means of egress system. This lighting would also include elevator cars and lobbies. This system would be designated for emergency power, which helps to identify it as an emergency system to ensure that the essential lighting is connected to the secondary power system.

604.2.14.2.1 Other circuits. Circuits supplying lighting for the fire command center and mechanical equipment rooms shall be transferable to the standby source.

❖ This section allows the lighting for the fire command center and mechanical equipment rooms to be connected with other circuits within the building during normal operations, but this lighting must automatically switch to standby power in case of the loss of primary power.

604.2.14.3 Emergency systems. Exit signs, exit illumination as required by Chapter 10, and elevator car lighting are classified as emergency systems and shall operate within 10 seconds of failure of the normal power supply and shall be capable of being transferred to the standby source.

Exception: Exit sign, exit and means of egress illumination are permitted to be powered by a standby source in buildings of Group F and S occupancies.

❖ This section requires that exit sign illumination, means of egress illumination and illumination of elevator cars and lobbies be supplied by an emergency power system rather than a standby system. This means that the power must be supplied within 10 seconds and not 60 seconds. These time limits mirror the requirements of the ICC EC for emergency power. There is an exception for Group F and S occupancies that would allow standby power instead of emergency power, since those occupancies have very low occupant loads in comparison to Group B or R-1.

604.2.15 Underground buildings. Emergency and standby power systems in underground buildings covered in Chapter 4 of the *International Building Code* shall comply with Sections 604.2.15.1 and 604.2.15.2.

❖ This section states that standby and emergency power must function as designated in the subsections. Generally, an underground building will present some unique problems because it is located below grade; removing products of combustion during a fire is more complicated. This presents a potentially more hazardous condition for exiting the building and makes it more difficult for fire fighters to perform rescue operations and fight the fire. High-rise buildings pose similar difficulties to fire fighters.

604.2.15.1 Standby power. A standby power system complying with the ICC *Electrical Code* shall be provided for standby power loads as specified in Section 604.2.15.1.1.

❖ This section contains all the relevant requirements for standby power for underground buildings.

[B] 604.2.15.1.1 Standby power loads. The following loads are classified as standby power loads:

1. Smoke control system.
2. Ventilation and automatic fire detection equipment for smokeproof enclosures.
3. Fire pumps.
4. Standby power shall be provided for elevators in accordance with Section 3003 of the *International Building Code.*

❖ This section defines the required capacity of the standby power system by stating which systems and equipment must be connected to standby power.

[B] 604.2.15.1.2 Pickup time. The standby power system shall pick up its connected loads within 60 seconds of failure of the normal power supply.

❖ This section highlights the ICC EC requirement that the secondary power system must pick up the necessary loads within 60 seconds of primary power failure in order to provide continuity of power to the circuits specified in Section 604.2.15.1.1.

604.2.15.2 Emergency power. An emergency power system complying with the ICC *Electrical Code* shall be provided for emergency power loads as specified in Section 604.2.15.2.1.

❖ This section contains the requirements for when and how emergency power is to be supplied to underground buildings. Emergency power must be more quickly available upon loss of power and is, therefore, used for life safety systems, such as fire alarm systems.

604.2.15.2.1 Emergency power loads. The following loads are classified as emergency power loads:

1. Emergency voice/alarm communication systems.

2. Fire alarm systems.
3. Automatic fire detection systems.
4. Elevator car lighting.
5. Means of egress lighting and exit sign illumination as required by Chapter 10.

❖ This section is a laundry list of loads that are to be supplied with emergency power. As noted, because emergency power reacts more quickly than standby power it is generally used for life safety equipment or systems. More specifically, this section requires emergency power for egress lighting, elevator lighting, fire alarm systems and emergency voice/alarm communication systems.

604.2.16 Group I-3 occupancies. Power-operated sliding doors or power-operated locks for swinging doors in Group I-3 occupancies shall be operable by a manual release mechanism at the door, and either emergency power or a remote mechanical operating release shall be provided.

Exception: Emergency power is not required in facilities where provisions for remote locking and unlocking of occupied rooms in Occupancy Condition 4 are not required as set forth in the *International Building Code*.

❖ Group I-3 occupancies are those where the occupants are under restraint or security. The movement of occupants, especially in the higher occupancy conditions such as 4 and 5, are often controlled by a series of locked doors. These locking mechanisms can be power-operated. More specifically, this section is concerned with power-operated horizontal doors and power-operated swing door locks.

Emergency power would be required for such doors and door locks when a remote mechanical means of operating them is not available. In either case, the doors and door locks would be required to contain a manual release mechanism. Basically, the facility must be able to continue to operate the doors remotely, whether through normal activation supported by emergency power or manually through a mechanical release.

604.2.17 Airport traffic control towers. A standby power system shall be provided in airport traffic control towers more than 65 feet (19 812 mm) in height. Power shall be provided to the following equipment:

1. Pressurization equipment, mechanical equipment and lighting.
2. Elevator operating equipment.
3. Fire alarm and smoke detection systems.

❖ This section requires standby power for certain elements of airport traffic control towers over 65-feet (19 812 mm) tall. The requirements primarily relate to the means of egress needs of these towers. Because of the height and configuration of these towers, only one stairway is required and it can be located in the same shaft as the elevator. This is permitted as long as a 4-hour separation between the stairs and the elevators is maintained. As a result of this unique allowance, standby power is required for several elements, including stair pressurization, fire alarms and smoke detection systems. Section 909.20.6.2 of the IBC requires pressurized stairways to have standby power.

604.2.18 Elevators. In buildings and structures where standby power is required or furnished to operate an elevator, the operation shall be in accordance with Sections 604.2.18.1 through 604.2.18.4.

❖ This section states how standby power is to be supplied to elevators when required by other sections, for instance, Section 604.2.17. The requirements from this section are the same as those located in Chapter 30 of the IBC.

604.2.18.1 Manual transfer. Standby power shall be manually transferable to all elevators in each bank.

❖ This section requires that whenever necessary, all elevators in each bank of elevators must be equipped for manual transfer to standby power. All elevators, however, would not need to operate on standby power at the same time; they would have to have manual transfer capability.

604.2.18.2 One elevator. Where only one elevator is installed, the elevator shall automatically transfer to standby power within 60 seconds after failure of normal power.

❖ When a building has a single elevator, it must be automatically transferred to standby power within 60 seconds. The 60 seconds is a reflection of the ICC EC requirements for standby power.

604.2.18.3 Two or more elevators. Where two or more elevators are controlled by a common operating system, all elevators shall automatically transfer to standby power within 60 seconds after failure of normal power where the standby power source is of sufficient capacity to operate all elevators at the same time. Where the standby power source is not of sufficient capacity to operate all elevators at the same time, all elevators shall transfer to standby power in sequence, return to the designated landing and disconnect from the standby power source. After all elevators have been returned to the designated level, at least one elevator shall remain operable from the standby power source.

❖ When there is more than one elevator operating off of a common system, the elevators must all be transferred to standby power within 60 seconds. When only one elevator needs to be available during a loss of power, all elevators must still have the ability to run on standby power. More specifically, all elevators must initially connect to the standby power system and then, in sequence, return to the designated floor where all but one would be disconnected from the standby power.

604.2.18.4 Venting. Where standby power is connected to elevators, the machine room ventilation or air conditioning shall be connected to the standby power source.

❖ This section reduces the likelihood that the equipment running the elevators will not overheat during a loss of power because standby power is also required to power the ventilation or air-conditioning for those areas.

604.3 Maintenance. Emergency and standby power systems shall be maintained such that the system is capable of supplying service within the time specified for the type and duration required.

❖ This section introduces requirements for maintenance of all elements of emergency and standby power systems.

604.3.1 Schedule. Inspection, testing and maintenance of emergency and standby power systems shall be in accordance with an approved schedule established upon completion and approval of the system installation.

❖ Standby power and emergency power are useful only if they continue to work over the life of both the building and its associated equipment; therefore, this section specifically focuses on the maintenance of such systems. The primary specifications of these secondary power supplies are that they be able to supply power within the specified length of time.

This section requires that a specific schedule be created at the completion of the installation of the system to ensure regular maintenance of the power systems.

604.3.2 Written record. Written records of the inspection, testing and maintenance of emergency and standby power systems shall include the date of service, name of the servicing technician, a summary of conditions noted and a detailed description of any conditions requiring correction and what corrective action was taken. Such records shall be kept on the premises served by the emergency or standby power system and be available for inspection by the fire code official.

❖ Documentation of maintenance is key, in that it highlights what specifically was inspected and tested and where potential problems exist. Also, the information about the inspector and inspection agency in the document allows for further reference in the future. This provides a level of accountability. Finally, these documents must be made available to the fire code official upon request.

604.3.3 Switch maintenance. Emergency and standby power system transfer switches shall be included in the inspection, testing and maintenance schedule required by Section 604.3.1. Transfer switches shall be maintained free from accumulated dust and dirt. Inspection shall include examination of the transfer switch contacts for evidence of deterioration. When evidence of contact deterioration is detected, the contacts shall be replaced in accordance with the transfer switch manufacturer's instructions.

❖ One of the most important elements of emergency and standby power systems is the ability for the primary power to be switched to the secondary power supply within the specified time; therefore, this section pays specific attention to the long-term reliability of the transfer switches. This includes inspection for cleanliness and signs of deterioration.

604.4 Operational inspection and testing. Emergency power systems, including all appurtenant components shall be inspected and tested under load in accordance with NFPA 110 and NFPA 111.

> **Exception:** Where the emergency power system is used for standby power or peak load shaving, such use shall be recorded and shall be allowed to be substituted for scheduled testing of the generator set, provided that appropriate records are maintained.

❖ This section requires that emergency power systems be tested and inspected as specified in NFPA 110 and 111. There is an exception to testing emergency power when the emergency power system is either used for peak power periods or for standby power. The use during peak hours means that on a fairly regular basis the power supply will be tested. In terms of the use of emergency power for standby power, the capacity of both types of secondary power systems is the same. The two differ on when they will activate; therefore, testing the system as a standby power supply is adequate when the emergency power system is used as standby power.

604.4.1 Transfer switch test. The test of the transfer switch shall consist of electrically operating the transfer switch from the normal position to the alternate position and then return to the normal position.

❖ This section specifically prescribes the sequence of events for testing transfer switches. The switch must move from the normal position to the alternate position and back again. As noted earlier in this commentary, emergency power must be available within a maximum of 10 seconds upon the loss of primary power, whereas standby power is to be available within 60 seconds. The transfer switches, therefore, must work within the specified time.

604.5 Supervision of maintenance and testing. Routine maintenance, inspection and operational testing shall be overseen by a properly instructed individual.

❖ This section requires a minimum level of qualifications for the testing of emergency and standby power systems. Only trained personnel should do the testing and maintenance of these systems.

SECTION 605
ELECTRICAL EQUIPMENT, WIRING AND HAZARDS

605.1 Abatement of electrical hazards. Identified electrical hazards shall be abated. Identified hazardous electrical conditions in permanent wiring shall be brought to the attention of the code official responsible for enforcement of the ICC *Electrical Code*. Electrical wiring, devices, appliances and other equipment that is modified or damaged and constitutes an electrical shock or fire hazard shall not be used.

❖ Maintenance of electrical systems and services to achieve compliance with the requirements of the ICC EC is required. The leading causes of electrical fires include inadequate or improper maintenance; nonconforming modifications to existing installations; failure to maintain clearances around electrical equipment and devices and improper use of electrical equipment and devices. A detailed analysis of the causes of residential electrical fires by the U.S. Consumer Products Safety Commission suggests that misuse and improper modifications to conforming electrical systems are the leading causes of these fires.

605.2 Illumination. Illumination shall be provided for service equipment areas, motor control centers and electrical panelboards.

❖ Adequate lighting in electrical service distribution areas is required to facilitate the location of the electrical service shutoff during a fire or other emergency and to minimize potential hazards during maintenance or repair work. Although not required, this lighting should be connected to an emergency or standby power source to permit continued illumination when power to service equipment is interrupted during maintenance repair activities or emergencies.

605.3 Working space and clearance. A working space of not less than 30 inches (762 mm) in width, 36 inches (914 mm) in depth and 78 inches (1981 mm) in height shall be provided in front of electrical service equipment. Where the electrical service equipment is wider than 30 inches (762 mm), the working space shall not be less than the width of the equipment. No storage of any materials shall be located within the designated working space.

Exceptions:

1. Where other dimensions are required or allowed by the ICC *Electrical Code*.
2. Access openings into attics or under-floor areas which provide a minimum clear opening of 22 inches (559 mm) by 30 inches (762 mm).

❖ Adequate clearance serves two important purposes: physical separation of combustibles from heat-producing electrical devices and equipment to minimize the possibility of ignition, and providing adequate work space to perform maintenance and repair work safely (see Figure 605.3). The exceptions note that the ICC EC may allow different dimensions than those prescribed in this section. Additionally, the lack of space in areas such as attics and underfloor areas is recognized by also allowing a smaller opening width.

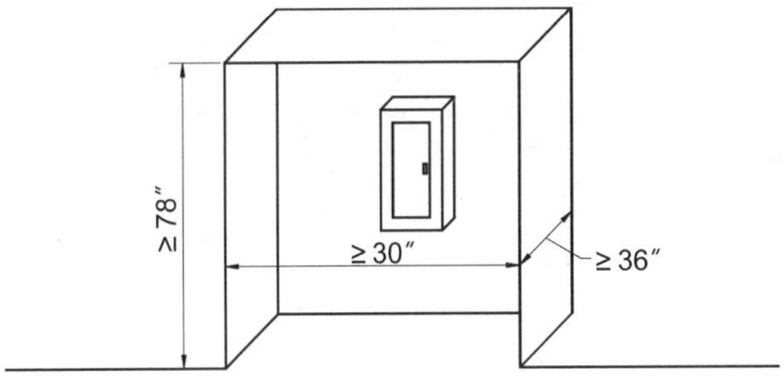

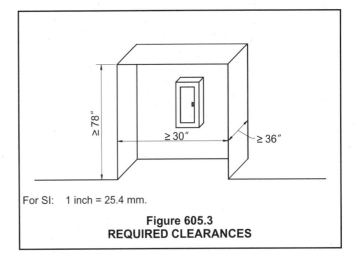

For SI: 1 inch = 25.4 mm.

Figure 605.3
REQUIRED CLEARANCES

605.3.1 Labeling. Doors into electrical control panel rooms shall be marked with a plainly visible and legible sign stating ELECTRICAL ROOM or similar approved wording. The disconnecting means for each service, feeder or branch circuit originating on a switchboard or panelboard shall be legibly and durably marked to indicate its purpose unless such purpose is clearly evident.

❖ In addition to the illumination required in Section 605.2, additional labeling is required for the electrical equipment to assist emergency responders in identifying and then shutting down the electrical service controls during a fire or other emergency.

605.4 Multiplug adapters. Multiplug adaptors, such as cube adaptors, unfused plug strips or any other device not complying with the ICC *Electrical Code* shall be prohibited.

❖ Overcurrent protection interrupts power to an outlet only when connected loads exceed the current rating of the overcurrent device for a specified amount of time. When multiplug adaptors are used for several appliances, the conductor may produce enough heat to ignite nearby combustibles in the time it takes to trip the overcurrent protection. Simultaneous operation of many small loads may cause dangerous localized resistance heating without tripping the overcurrent protection. Additionally, these devices may result in loose electrical connections because of the weight of the connections.

Overuse of these devices may also indicate that the building's electrical service is inadequate for connected loads or current occupancy demands. These devices are intended for temporary use only, not at a fixed location or in place of wiring complying with the ICC EC.

The code does allow for the use of power taps with specific criteria mentioned in Sections 605.4.1 through Section 605.4.3. Power taps are essentially a combination of a multiplug adaptor and an extension cord. The device consists of a plug on one end attached to a flexible cord and two or more receptacles on the opposite end, which contains overcurrent protection. The flexible cord length is dependent on the listing of the particular device, but power taps have been listed for lengths up to 25 feet (7620 mm).

Figure 605.4 shows the difference between a multiplug adapter and a power tap.

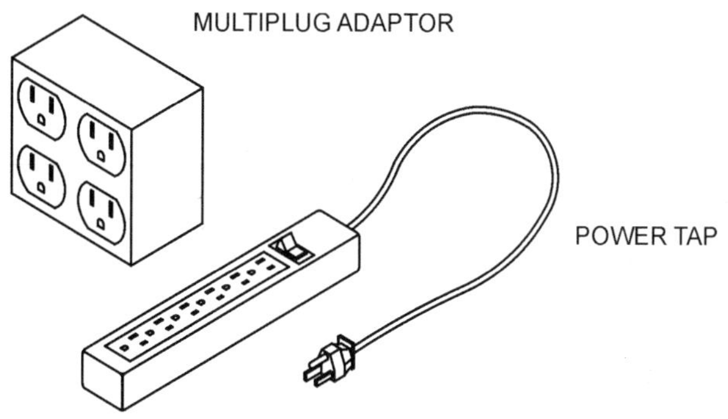

Figure 605.4
TYPICAL MULTIPLUG ADAPTER AND POWER TAP

605.4.1 Power tap design. Relocatable power taps shall be of the polarized or grounded type, equipped with overcurrent protection, and shall be listed.

❖ This section sets out the basic requirements for approved power taps.

605.4.2 Power supply. Relocatable power taps shall be directly connected to a permanently installed receptacle.

❖ The restrictions on power taps are similar to those on extension cords. Power taps should not be used as a substitute for building wiring. Additionally, power taps are to be plugged into a permanently installed receptacle.

605.4.3 Installation. Relocatable power tap cords shall not extend through walls, ceilings, floors, under doors or floor coverings, or be subject to environmental or physical damage.

❖ To prevent use as a substitute for permanent wiring, power taps cannot be placed in locations such as within walls, under doors or where they would be subject to physical damage.

605.5 Extension cords. Extension cords and flexible cords shall not be a substitute for permanent wiring. Extension cords and flexible cords shall not be affixed to structures, extended through walls, ceilings or floors, or under doors or floor coverings, nor shall such cords be subject to environmental damage or physical impact. Extension cords shall be used only with portable appliances.

❖ Frequent or improper use of extension cords in place of permanent fixed wiring is another indication of inadequate electrical service capacity or incompatible demands (see Figure 605.5). Physical damage to extension cords caused by concealment or improper or inadequate maintenance may result in localized resistance heating.

The amount of electrical current that any extension cord can safely conduct is limited by the size of its conductor. This principle is often not understood by the general public. As a result, extension cords are commonly overloaded by connecting either too many appliances or loads in excess of cord capacity.

Overloading of extension cords causes an increase in the conductor's temperature. This increase in temperature can exceed the temperature rating of the conductor insulation, causing it to melt, decompose or burn. The burning insulation can ignite other combustible materi-

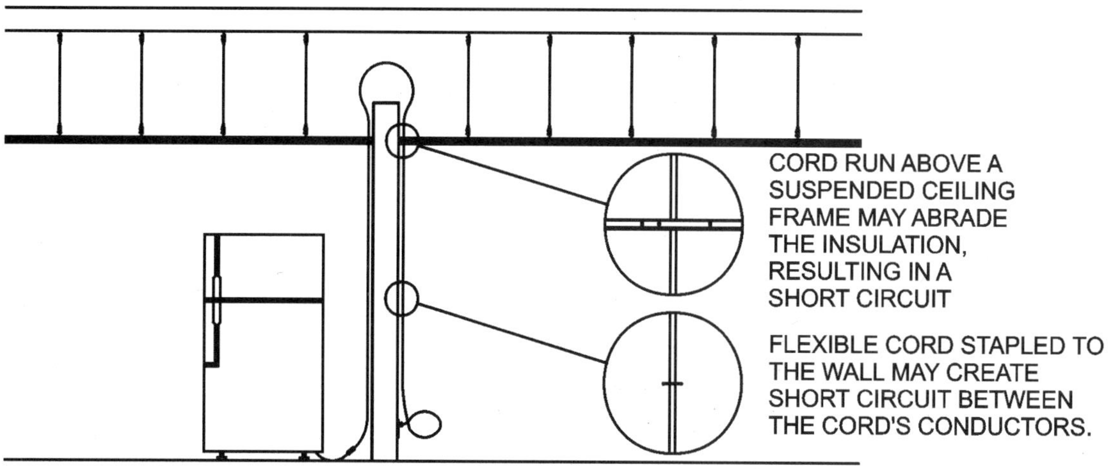

CORD RUN ABOVE A SUSPENDED CEILING FRAME MAY ABRADE THE INSULATION, RESULTING IN A SHORT CIRCUIT
FLEXIBLE CORD STAPLED TO THE WALL MAY CREATE SHORT CIRCUIT BETWEEN THE CORD'S CONDUCTORS.

Figure 605.5
EXAMPLES OF UNACCEPTABLE PRACTICES

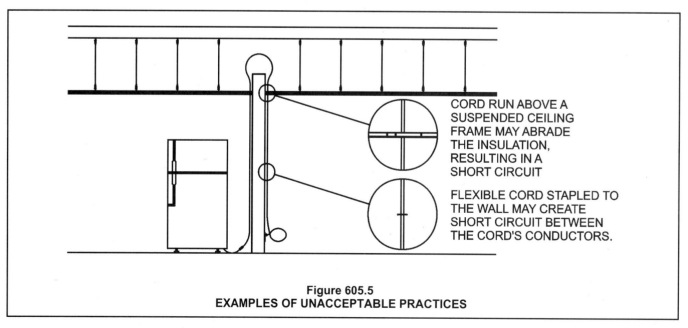

als. The resulting loss of conductor insulation can also cause a short circuit that can act as a source of ignition.

The buildup of heat in an extension cord is often made worse by excessive cord length and by the insulating effect of rugs that often cover extension cords. Extension cords are much more susceptible to physical damage than permanent wiring. Damage to extension cords increases the likelihood of shorts and poor connections, both of which can cause a fire.

In addition to the fire hazard, extension cords pose a tripping hazard to the occupants and when damaged, can pose an electrical shock hazard. Securing flexible cords to a wall baseboard, door jambs, etc., with nails, uninsulated staples or other improper fasteners to eliminate tripping hazards can create another dangerous condition by pinching the cord and causing localized heating of the cord that could lead to ignition.

Additionally, as a way of limiting the use of extension cords, their use is restricted to portable appliances. The reference to "portable" primarily denotes smaller appliances, such as a fan or perhaps an alarm clock. Extension cords should not be used with major appliances, such as refrigerators.

605.5.1 Power supply. Extension cords shall be plugged directly into an approved receptacle, power tap or multiplug adapter and, except for approved multiplug extension cords, shall serve only one portable appliance.

❖ This section allows an extension cord to be plugged into the main electrical supply using an approved receptacle, a power tap or an approved multiplug adapter. This restriction means that multiple extension cords can not be connected to one another for a single appliance. Additionally, extension cords are limited to one appliance unless they are specifically approved multiplug extension cords.

605.5.2 Ampacity. The ampacity of the extension cords shall not be less than the rated capacity of the portable appliance supplied by the cord.

❖ Although most building occupants may have difficulty understanding ampacity, which is the amount of electrical current that a particular electrical device is capable of handling, it still must be addressed. If an appliance requires a higher electrical current than the extension cord is intended to handle, the extension cord will be overloaded and cause potential damage and overheating. A familiarity with the types of extension cords available for sale and the general relationships to common appliances will help with the enforcement of this section.

605.5.3 Maintenance. Extension cords shall be maintained in good condition without splices, deterioration or damage.

❖ When extension cords are damaged, they may become potential shock hazards and a source of ignition because of direct contact with the electrical charge from overheating.

605.5.4 Grounding. Extension cords shall be grounded when serving grounded portable appliances.

❖ If an extension cord serves an appliance that is grounded, to avoid potential shock, that cord must also be grounded.

605.6 Unapproved conditions. Open junction boxes and open-wiring splices shall be prohibited. Approved covers shall be provided for all switch and electrical outlet boxes.

❖ Without covers, connections made in junction boxes may be subject to physical damage. Such damage may loosen electrical connections, resulting in high-resistance arcing. Switches and outlet boxes are subject to arcing from loose connections and reduced clearances between contacts as they age. Accumulation of dirt and debris in open electrical boxes creates an ignitable fuel concentration. Fires in open electrical boxes may spread to wire or cable insulation or other fuels in electrical and mechanical concealed spaces. Furthermore, unprotected electrical connections are an electrical shock hazard to personnel working in concealed spaces.

605.7 Appliances. Electrical appliances and fixtures shall be tested and listed in published reports of inspected electrical equipment by an approved agency and installed in accordance with all instructions included as part of such listing.

❖ The fire code official should look for the listing mark of an approved testing or inspection agency on the appliance and may request the agency's published report showing the listing to verify that the appliance meets an applicable standard for electrical safety. Fire code officials experiencing difficulties interpreting the marking or listing of an agency, as in the case of a laboratory not located within the United States, should consult representatives of the agency or the U. S. Consumer Products Safety Commission for assistance.

605.8 Electrical motors. Electrical motors shall be maintained free from excessive accumulations of oil, dirt, waste and debris.

❖ Internal heating is commonly associated with the operation of electrical motors. Excessive accumulations of dust, oil, grease, dirt or other debris may be easily ignited by the internal frictional heating of electrical motor components.

605.9 Temporary wiring. Temporary wiring for electrical power and lighting installations is allowed for a period not to exceed 90 days. Temporary wiring methods shall meet the applicable provisions of the ICC *Electrical Code.*

Exception: Temporary wiring for electrical power and lighting installations is allowed during periods of construction, remodeling, repair or demolition of buildings, structures, equipment or similar activities.

❖ In some cases, because of a specific need and the temporary nature of the need, temporary wiring is allowed for a period of no more than 90 days. This allowance is primarily aimed at needs such as for holiday lighting. The exception allows temporary wiring to exceed 90 days for certain activities such as remodeling or general construction of a building. Section 305 of the *National Electrical Code* (NEC) contains specific requirements for temporary wiring.

Temporary wiring is not referring to the use of power taps or extension cords. The requirements for temporary wiring are less restrictive than those for permanent wiring but are much more rigorous than the requirements for the use of power taps and extension cords.

605.9.1 Attachment to structures. Temporary wiring attached to a structure shall be attached in an approved manner.

❖ When wiring is specifically attached to a structure, care must be taken to make sure that the attachment will not damage the wiring in a way that would cause resistance heating in localized areas of the wiring.

SECTION 606 MECHANICAL REFRIGERATION

[M] 606.1 Scope. Refrigeration systems shall be installed in accordance with the *International Mechanical Code.*

❖ Chapter 11 of the IMC in conjunction with ASHRAE 15 and IIAR 2, provides complete coverage for the design, installation and maintenance of refrigeration systems.

[M] 606.2 Refrigerants. The use and purity of new, recovered, and reclaimed refrigerants shall be in accordance with the *International Mechanical Code.*

❖ See the commentary to Section 1102.2 of the IMC.

[M] 606.3 Refrigerant classification. Refrigerants shall be classified in accordance with the *International Mechanical Code.*

❖ See the commentary to Section 1103.1 of the IMC.

[M] 606.4 Change in refrigerant type. A change in the type of refrigerant in a refrigeration system shall be in accordance with the *International Mechanical Code.*

❖ See the commentary to Section 1101.8 of the IMC.

606.5 Access. Refrigeration systems having a refrigerant circuit containing more than 220 pounds (100 kg) of Group A1 or 30 pounds (14 kg) of any other group refrigerant shall be accessible to the fire department at all times as required by the fire code official.

❖ Where any one or more refrigeration circuits contain more than the specified quantity limits, the room or building housing the system or systems must be constructed with fire department access for emergency response, inspection and hazard assessment. Refrigerants of other than Group A1 tend to be more flammable or toxic. This section could require that the fire department be given keys to refrigeration machinery rooms and buildings.

606.6 Testing of equipment. Refrigeration equipment and systems having a refrigerant circuit containing more than 220 pounds (100 kg) of Group A1 or 30 pounds (14 kg) of any other group refrigerant shall be subject to periodic testing in accordance with Section 606.6.1. A written record of required testing shall be maintained on the premises. Tests of emergency devices or systems required by this chapter shall be conducted by persons trained and qualified in refrigeration systems.

❖ See the commentary to Section 606.6.

606.6.1 Periodic testing. The following emergency devices or systems shall be periodically tested in accordance with the manufacturer's instructions and as required by the fire code official.

1. Treatment and flaring systems.
2. Valves and appurtenances necessary to the operation of emergency refrigeration control boxes.
3. Fans and associated equipment intended to operate emergency ventilation systems.
4. Detection and alarm systems.

❖ See the commentary to Section 1109.1 of the IMC.

606.7 Emergency signs. Refrigeration units or systems having a refrigerant circuit containing more than 220 pounds (100 kg) of Group A1 or 30 pounds (14 kg) of any other group refrigerant shall be provided with approved emergency signs, charts, and labels in accordance with NFPA 704. Hazard signs shall be in accordance with the *International Mechanical Code* for the classification of refrigerants listed therein.

❖ Signs, charts and labels are necessary to assist emergency response personnel in carrying out their duties to protect building occupants as well as protecting themselves from the hazards associated with refrigerant chemicals (see IMC Table 1103.1).

606.8 Refrigerant detector. Machinery rooms shall contain a refrigerant detector with an audible and visual alarm. The detector, or a sampling tube that draws air to the detector, shall be located in an area where refrigerant from a leak will concentrate. The alarm shall be actuated at a value not greater than the corresponding TLV-TWA values shown in the *International Mechanical Code* for the refrigerant classification. Detectors and alarms shall be placed in approved locations.

Exception: Detectors are not required for ammonia systems where the machinery room complies with Section 1106.3 of the *International Mechanical Code.*

❖ Section 1105.3 of the IMC refers to the code for refrigerant detector requirements. Refrigerant detectors provide early warning of refrigerant leakage. Such leakage could result in a significant fire or health hazard if not discovered and stopped or if occupants are not evacuated from the building. Machinery rooms are required by the IMC where refrigerant quantities exceed specified limits. Refrigerant detectors are used to start the machinery room ventilation systems required by Sections 1105.6.4 and 1106.3 of the IMC (see the definition of "Machinery Room" and Table 1103.1 in the IMC.)

Detector location is critical to early leakage warning and should adhere to the detector manufacturer's instructions. The required detectors must be designed for application with the refrigerant or refrigerants used in the machinery room. The required detectors have the dual role of sending both audible and visual alarms and also starting emergency ventilation/exhaust systems in the machinery rooms.

The exception refers to Section 1106.3 of the IMC, which requires the ventilation/exhaust system in ammonia machinery rooms to operate continuously; however, that section does not state the rate at which the ventilation system must continuously operate. Continuous operation could mean continuous during both occupied and unoccupied conditions at the rate specified in Section 1105.6.3; however, the nature of Exception 1 of Section 1106.3 suggests that the ventilation should be at the emergency rate of Section 1105.6.4, considering that the exception would not require ventilation to start until the ammonia level reaches 1,000 parts per million (ppm)(1000 mg/L) (40 times the threshold for detection required by the code). Exception 1 to Section 1106.3 of the IMC reinstates the detector requirement as an option to continuous ventilation/exhaust operation. Exception 2 to the same section allows the substitution of Class 1, Division 2, hazardous location requirements for continuous ventilation/exhaust operation.

606.9 Remote controls. Remote control of the mechanical equipment and appliances located in the machinery room shall be provided at an approved location immediately outside the machinery room and adjacent to its principal entrance.

❖ Remote controls allow personnel to operate and shut down refrigeration machines and activate emergency systems without requiring personnel to endanger themselves by entering the machinery room.

606.9.1 Refrigeration system. A clearly identified switch of the break-glass type shall provide off-only control of electrically energized equipment and appliances in the machinery room, other than refrigerant leak detectors and machinery room ventilation.

❖ Shutting down compressors and related refrigeration equipment could be necessary to prevent a hazardous condition from worsening and to allow the room to be occupied. The emergency "kill" switch must be a tamper-resistant type (similar to fire alarm pull stations) that requires more than one action to actuate it. To prevent an accidental startup, the switch must be capable of stopping only the controlled machinery. The switch must not affect the operation of life safety systems, such as detectors and exhaust equipment, and should not affect room and egress lighting.

Emergency shutdown controls located outside the machinery room enclosure will allow shutdown of the compressors and related equipment without requiring someone to enter the room and risk being exposed to refrigerant or fire. This arrangement would also permit equipment shutdown by fire-fighting personnel without the risk of fire spreading into or out of the fire-resistance-rated enclosure. The controls must be located near the entrance to the machinery room so that their location is conspicuous. The controls should be labeled and color coded so that their purpose is obvious. Such controls are customarily painted red to make them readily identifiable as emergency devices.

606.9.2 Ventilation system. A clearly identified switch of the break-glass type shall provide on-only control of the machinery room ventilation fans.

❖ For the same reasoning stated in the commentary to Section 606.9.1, a remote switch is required that is not connected to the exhaust system. Although not specifically stated, the intent is for the remote control to activate the emergency mode of operation (see IMC commentary, Section 1105.9.4). To maximize the dependability of the exhaust systems, ASHRAE 15 requires that they be powered from independent dedicated electrical branch circuits.

606.9.3 Emergency control box. Emergency control boxes shall be provided for refrigeration systems required to be equipped with a treatment system, flaring system or ammonia diffusion system.

❖ Emergency control boxes allow emergency response personnel to operate the systems required by Section 606.11.

606.9.3.1 Location. Emergency control boxes shall be located outside of the building at an approved accessible location. All portions of the emergency control box shall be 6 feet (1829 mm) or less above the adjoining grade.

❖ Emergency control boxes must be located where they are readily accessible to emergency personnel. Such boxes should not be obscured by shrubbery, plantings, fences, equipment supported on grade, etc. Thought should also be given to avoiding locations susceptible to vandalism.

606.9.3.2 Construction. Emergency control boxes shall be of iron or steel not less than 0.055 inch (1.4 mm) in thickness and provided with a hinged cover and lock.

❖ Vandalism and unauthorized tampering with emergency controls could create a hazard or cause an unnecessary loss of system operation and refrigerant. The boxes must secure the controls.

606.9.3.3 Operational procedure. Valves and switches shall be identified in an approved manner as to the sequential procedure to be followed in the event of an emergency.

❖ Clear, permanent instructions must be provided to guide emergency personnel in the operation of the emergency controls. It is likely that each system will have unique features and operation sequences necessitating operator instruction for each system.

606.9.3.4 Identification. Emergency control boxes shall be provided with a permanent label on the outside cover reading: FIRE DEPARTMENT USE ONLY— REFRIGERANT CONTROL BOX, and including the name of the refrigerant in the system. Hazard identification in accordance with NFPA 704 shall be posted inside and outside of the control box.

❖ The hazard warning will alert the emergency personnel to the dangers and the precautions that must be considered for the chemicals involved.

606.9.3.5 Instructions. Written instructions and information shall be provided and located in the emergency control box designating the following information:

1. Instructions for suspending operation of the system in the event of an emergency.
2. The name, address and emergency telephone numbers to obtain emergency service.
3. The location and operation of emergency discharge systems.

❖ Clear, permanent instructions must be provided to guide emergency personnel in the operation of the emergency controls. It is likely that each system would have unique features and operation sequences necessitating operator instruction for each system.

606.10 Storage, use and handling. Flammable and combustible materials shall not be stored in machinery rooms for refrigeration systems having a refrigerant circuit containing more than 220 pounds (100 kg) of Group A1 or 30 pounds (14 kg) of any other group refrigerant. Storage, use or handling of extra refrigerant or refrigerant oils shall be as required by Chapters 27, 30, 32 and 34.

Exception: This provision shall not apply to spare parts, tools, and incidental materials necessary for the safe and proper operation and maintenance of the system.

❖ Storage of materials could introduce additional hazards in rooms already considered to be hazardous because of large quantities of refrigerants in the system circuits and machines.

606.11 Termination of relief devices. Pressure relief devices, fusible plugs and purge systems for refrigeration systems containing more than 6.6 pounds (3 kg) of flammable, toxic or highly toxic refrigerants shall be provided with an approved discharge system as required by Sections 606.11.1, 606.11.2 and 606.11.3. Discharge piping and devices connected to the discharge side of a fusible plug or rupture member shall have provisions to prevent plugging the pipe in the event of the fusible plug or rupture member functions.

❖ Discharge systems are intended to treat, incinerate or absorb flammable or toxic refrigerants that would otherwise be released unaltered into the atmosphere. Release would result from the operation of pressure relief devices or intentional dumping of refrigerant in an emergency. This section also intends to increase the likelihood that proper attention will be focused on using discharge piping and devices in such a manner that the emergency pressure relief system will not become obstructed in the event that a fusible plug or rupture member operates and causes debris to be ejected into the relief system.

606.11.1 Flammable refrigerants. Systems containing flammable refrigerants having a density equal to or greater than the density of air shall discharge vapor to the atmosphere only through an approved treatment system in accordance with Section 606.11.4 or a flaring system in accordance with Section 606.11.5. Systems containing flammable refrigerants having a density less than the density of air shall be permitted to discharge vapor to the atmosphere provided that the point of discharge is located outside of the structure at not less than 15 feet (4572 mm) above the adjoining grade level and not less than 20 feet (6096 mm) from any window, ventilation opening or exit.

❖ Where they have a density greater than air (i.e., are heavier than air) and pose the hazard of collecting in low points, which could bring them into contact with ignition sources, flammable refrigerants (A2, B2, A3, B3) must be incinerated in a flaring system (see Section 3704.2.2.7.1). The second sentence of the provision is derived from ANSI/ASHRAE 15, Section 9.7.8 and recognizes the reduced hazard of lighter-than-air flammable refrigerants by allowing them to discharge to the atmosphere without incineration or treatment since they would either dissipate into the air or flare at the point of discharge. Because of their flammability hazard, however, the point of discharge must be located out of reach from grade and well away from building openings.

606.11.2 Toxic and highly toxic refrigerants. Systems containing toxic or highly toxic refrigerants shall discharge vapor to the atmosphere only through an approved treatment system in accordance with Section 606.11.4 or a flaring system in accordance with Section 606.11.5.

❖ Toxic refrigerants, like flammable refrigerants, must be treated to reduce their toxicity or destroyed by incineration (see Section 3704.2.2.7.1).

606.11.3 Ammonia refrigerant. Systems containing ammonia refrigerant shall discharge vapor to the atmosphere through an approved treatment system in accordance with Section 606.11.4, a flaring system in accordance with Section 606.11.5, or through an approved ammonia diffusion system in accordance with Section 606.11.6, or by other approved means.

Exceptions:

1. Ammonia/water absorption systems containing less than 22 pounds (10 kg) of ammonia and for which the ammonia circuit is located entirely outdoors.

2. When the fire code official determines, on review of an engineering analysis prepared in accordance with Section 104.7.2, that a fire, health or environmental hazard would not result from discharging ammonia directly to the atmosphere.

❖ This section parallels Section 1105.8 of the IMC, which references ASHRAE 15. In addition to incineration and absorption in water, ASHRAE 15 allows direct discharge to the atmosphere at prescribed distances above grade and from building openings.

Exception 1 recognizes the reduced hazard in smaller systems where the ammonia circuit is completely outdoors.

Exception 2 recognizes that there are some cases, such as at remote facilities, where atmospheric discharge of ammonia would pose no danger to people or property and provides a basis in the code for permitting relief lines on ammonia refrigeration systems to discharge to the atmosphere when an appropriate analysis has been performed and accepted by the fire code official to show that such discharge can be accomplished safely. In such cases, a flaring or water diffusion system would serve no beneficial purpose, since ammonia is naturally biodegradable. In other cases, engineered designs can be used to activate an alarm and automatically stop the source of a leak rather than relying on manual means to stop the leak and a water tank to treat whatever release may occur before manual intervention can be accomplished. Exception 2 provides correlation with the additional flexibility that is already available in the IMC, while maintaining the fire code official's authority under the IFC to accept or reject any proposal.

606.11.4 Treatment systems. Treatment systems shall be designed to reduce the allowable discharge concentration of the refrigerant gas to not more than 50 percent of the IDLH at the point of exhaust. Treatment systems shall be in accordance with Chapter 37.

❖ See the commentary to Section 3704.2.2.7. "Immediately Dangerous to Life and Health (IDLH)" is defined in Section 2702.1.

606.11.5 Flaring systems. Flaring systems for incineration of flammable refrigerants shall be designed to incinerate the entire discharge. The products of refrigerant incineration shall not pose health or environmental hazards. Incineration shall be automatic upon initiation of discharge, shall be designed to prevent blowback, and shall not expose structures or materials to threat of fire. Standby fuel, such as LP gas, and standby power shall have the capacity to operate for one and one-half the required time for complete incineration of refrigerant in the system.

❖ Destruction of refrigerant by incineration is supposed to render the discharge harmless, so obviously the flames and the combustion byproducts do not pose a hazard themselves. Because most refrigerants would not support combustion unaided, a fuel source is necessary to sustain incineration.

606.11.6 Ammonia diffusion systems. Ammonia diffusion systems shall include a tank containing 1 gallon of water for each pound of ammonia (4 L of water for each 1 kg of ammonia) that will be released in 1 hour from the largest relief device connected to the discharge pipe. The water shall be prevented from freezing. The discharge pipe from the pressure relief device shall distribute ammonia in the bottom of the tank, but no lower than 33 feet (10 058 mm) below the maximum liquid level. The tank shall contain the volume of water and ammonia without overflowing.

❖ Ammonia is readily absorbed by water; therefore, an ammonia discharge into a water tank would be chemically held in the tank. The water tank may have to be heated to prevent it from freezing, which would block the discharge pipe and make the water useless for absorbing ammonia. The deeper the discharge pipe extends below the water surface, the greater the pressure the discharge would have to overcome to escape from the pipe. A 33-foot (10 058 mm) depth of water would exert a pressure approximately equal to sea level atmospheric pressure [14.7 psi (101 kPa)].

606.12 Discharge location for refrigeration machinery room ventilation. Exhaust from mechanical ventilation systems serving refrigeration machinery rooms capable of exceeding 25 percent of the LFL or 50 percent of the IDLH shall be equipped with approved treatment systems to reduce the discharge concentrations of flammable, toxic or highly toxic refrigerants to those values or lower.

❖ Exhaust systems required by Sections 1105 and 1106 of the IMC discharge to the outdoors. Treatment, such as scrubbing or filtering, may be required if it is possible for the exhaust air to contain refrigerant concentrations exceeding the limits set by this section.

606.13 Notification of refrigerant discharges. The fire code official shall be notified immediately when a discharge becomes reportable under state, federal or local regulations in accordance with Section 2703.3.1.

❖ Emergency personnel must be informed of a discharge so that they can respond appropriately. The refrigerant discharge notification requirements of this section parallel those required for all hazardous materials in Section 2703.3.1. There is no reason for different requirements for refrigerants than for other hazardous materials.

606.14 Records. A written record shall be kept of refrigerant quantities brought into and removed from the premises. Such records shall be available to the fire code official.

❖ Emergency personnel must be able to maintain accurate assessments of the potential dangers they may face and the hazards the public may face in and around buildings housing refrigeration systems.

606.15 Electrical equipment. Where refrigerants of Groups A2, A3, B2 and B3, as defined in the *International Mechanical Code*, are used, refrigeration machinery rooms shall conform to the Class I, Division 2 hazardous location classification requirements of the ICC *Electrical Code*.

Exception: Ammonia machinery rooms that are provided with ventilation in accordance with Section 1106.3 of the *International Mechanical Code*.

❖ This section mirrors the text of Section 1106.4 of the IMC and is included in the code because in some jurisdictions, the fire code official is designated to inspect classified electrical equipment. A reference to classified electrical requirements here is consistent with the approach taken in Chapters 27 and 34 for other hazardous materials and flammable liquids, and it provides the fire code official with the provisions that are to be enforced.

The exception for ammonia refrigerant is consistent with requirements of the ICC EC and is included here to avoid possible confusion regarding the need for classified electrical equipment in ammonia machinery rooms. Because ammonia can combust within a limited range of concentrations in the air, the fire code official might be led to believe that classified electrical equipment should be provided in ammonia storage and use areas; however, when such areas are ventilated to maintain ammonia vapor in a concentration that is outside of the flammable range, there is no need for classified electrical equipment.

SECTION 607
ELEVATOR RECALL AND MAINTENANCE

❖ Elevators are often used as a tool by emergency responders when responding to fires and other emergencies. Due to these needs, the elevators must be capable of providing certain functions such as recall and emergency operation. The following sections simply denote when such controls are required and the necessary signage to be used to clarify to building occupants that elevators are not to be used during a fire.

607.1 Required. Existing elevators with a travel distance of 25 feet (7620 mm) or more above or below the main floor or other level of a building and intended to serve the needs of emergency personnel for fire-fighting or rescue purposes shall be provided with emergency operation in accordance with ASME A17.3. New elevators shall be provided with Phase I emergency recall operation and Phase II emergency in-car operation in accordance with ASME A17.1.

❖ This section sets out requirements for both new and existing elevators. Existing elevators that travel 25 feet (635 mm) or more above or below the main level should, as a minimum, be equipped with emergency operation capabilities that comply with ASME A17.3. New elevator installations are held to more restrictive guidelines for increased cost effectiveness and must have both emergency recall and emergency in-car operation to comply with ASME A17.1 for any amount of travel distance. The ASME standards are safety codes for elevators and escalators: ASME A17.3 for existing elevators and ASME A17.1 for new elevator installations.

[B] 607.2 Emergency signs. An approved pictorial sign of a standardized design shall be posted adjacent to each elevator call station on all floors instructing occupants to use the exit stairways and not to use the elevators in case of fire. The sign shall read: IN FIRE EMERGENCY, DO NOT USE ELEVATOR. USE EXIT STAIRS. The emergency sign shall not be required for elevators that are part of an accessible means of egress complying with Section 1007.4.

❖ Because of the needs of fire fighters and the possible risks posed by the use of elevators during a fire, signage is required that prohibits the use of elevators by building occupants during a fire emergency. The need to evacuate all occupants regardless of their physical abilities is becoming a more important issue and, in some cases, elevators are specifically used for such purposes. Elevators used as part of the means of egress must comply with Section 1007.4.

607.3 Elevator keys. Keys for the elevator car doors and fire-fighter service keys shall be kept in an approved location for immediate use by the fire department.

❖ The fire service often responds to elevator emergencies and other emergencies that require the use or operation of the elevators and associated components. Most elevators will have at least five or six key-switched functions, including those for standby power transfer, Phase I emergency elevator recall (hall switch located in the elevator lobby at the designated level), Phase II in-cab emergency operation, inspection function, normal lighting and fan operation. Rule 211.8 of ASME A 17.1 requires that standby power transfer and Phase I and II emergency operation be operated by the same key. Accordingly, three of the elevator keys will be identical but should be clearly labeled as to their respective function, as should the keys for the other key-switched functions. In multiple elevator installations, operating keys must be provided for each elevator. The rule also prohibits the elevator keys from being part of the building master key system. Many elevator installers will provide as part of their package an elevator key box (see Figure 607.3), which will accommodate all of the required operating keys, plus an appropriate hoistway access key or tool for fire department use in accessing the hoistway in case of an elevator emergency.

Experience has shown that this important safety element is often overlooked. Elevator inspection reports should note the presence and locations of the keys. Inspectors may verify the availability of the keys during periodic fire safety inspections and prefire planning surveys.

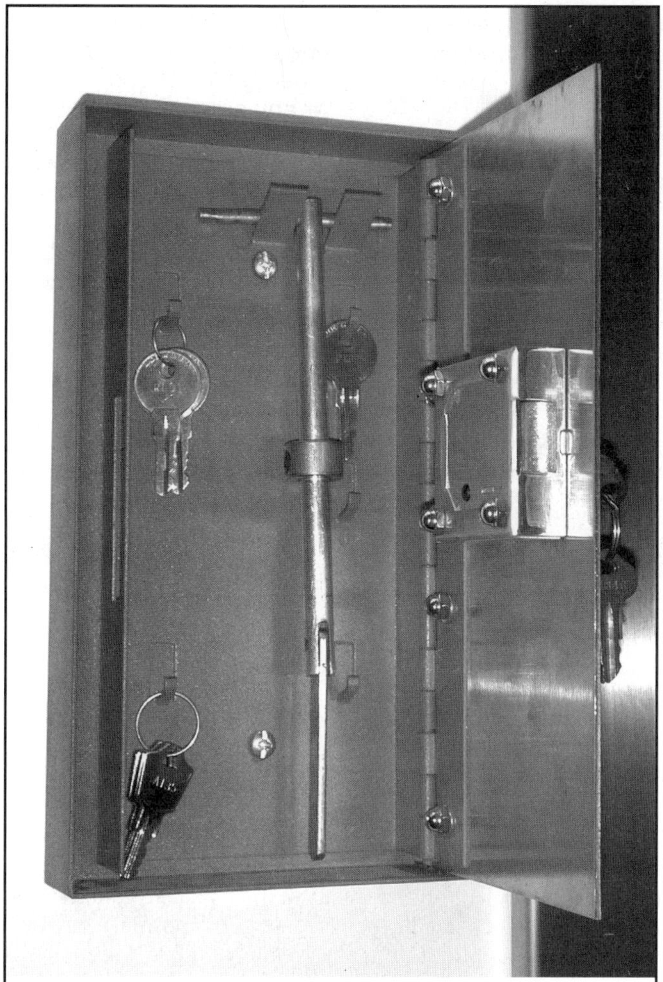

Figure 607.3
ELEVATOR KEY BOX

SECTION 608
STATIONARY LEAD-ACID BATTERY SYSTEMS

608.1 Scope. Stationary lead-acid battery systems using vented (flooded) lead-acid batteries having an electrolyte capacity of more than 50 gallons (189 L) used for facility standby power, emergency power, or uninterrupted power supplies shall comply with this section. Valve-regulated lead-acid batteries are not subject to the requirements of this section, but shall comply with Section 609.

❖ This section applies to lead-acid battery systems used for standby power, emergency power or uninterruptible power systems. These requirements were developed in response to the concern that applying the generic hazardous materials requirements of the code was inappropriate and unnecessary. Lead-acid batteries generally contain sulfuric acid, which is considered a corrosive liquid. In most cases, the battery systems would exceed the maximum allowable quantities prescribed in Chapter 27, which would lead to classification as a Group H occupancy. Generally, these types of systems have had a good safety record and, if the guidelines set out in Section 608 are followed, pose a very low hazard to the building and its occupants. In many cases, such systems are found in buildings with very low occupant loads, such as telephone company exchanges.

This section applies only to lead-acid batteries of the flooded type and then only if a battery system contains over 50 gallons (189 L) of acid. When a system contains less than 50 gallons (189 L), this section does not apply. This section also does not apply to individual lead-acid batteries used in vehicles or other similar applications or to VRLA batteries that are regulated in Section 609. There are some very specific differences in the two types of batteries that should be understood in applying Sections 608 and 609 of the code. First, VRLA batteries are designed so that if punctured or damaged they would not spill because the electrolyte is immobilized in an absorptive glass mat or by a gelling agent. Second, VRLA batteries do not routinely emit hydrogen gas to the atmospheres, as do flooded lead-acid batteries. VRLA batteries vent hydrogen only when the internal pressure exceeds the ambient pressure; therefore, hydrogen release is still an issue and ventilation is still required.

Although the flooded type of lead-acid batteries are more likely to have a spill and routinely emit hydrogen to the atmosphere, they are not as highly subjected to pressure buildup as are the valve-regulated type. Also, VRLA batteries are often used in locations where the flooded type have not traditionally been used. VRLA batteries are more likely to be located in areas such as cabinets in a business occupancy rather than in a phone company exchange with a low occupant load that does use the flooded type. These differences in technologies and applications have been addressed in the code development process by the addition of Section 609 on VRLA batteries to the code (see commentary, Section 609).

608.2 Safety venting. Batteries shall be provided with safety venting caps.

❖ Pressures can build up within batteries as a result of the creation of hydrogen and oxygen, making a method of pressure relief necessary. Both types of batteries, as noted above, vent hydrogen and oxygen to the atmosphere. Flooded batteries vent regularly. Valve-regulated battery systems vent when the internal pressure exceeds ambient pressures. Because of the potential for pressure buildup, especially within valve-regulated batteries, safety-venting caps are required.

608.3 Room design and construction. Enclosure of stationary lead-acid system rooms shall comply with the *International Building Code*. The battery systems are permitted to be in the same room with the equipment they support.

❖ This section states that battery rooms must meet the basic construction requirements of the IBC. No special fire-resistance-rated separation requirements or occupancy classifications are necessary. Additionally, there is no requirement for separating the equipment that the battery system supports from the battery system itself. The package of requirements found in Section 608 is intended to address the hazards associated with potential spills and the presence of hydrogen.

608.4 Spill control and neutralization. An approved method and materials for the control and neutralization of a spill of electrolyte shall be provided. The method and materials shall be capable of controlling and neutralizing a spill from the largest lead-acid battery to a pH between 7.0 and 9.0.

❖ This section is fairly performance based and does not specifically require spill control in the form of containment, nor does it require a specific method of neutralization. Instead, it states that a capability to control and neutralize a spill equal to the liquid content of the largest single battery (to a pH between 7.0 and 9.0) must be available. In the case of flooded lead-acid batteries, this may require initial containment followed by neutralization.

608.5 Ventilation. Ventilation shall be provided in accordance with the *International Mechanical Code* and the following:

1. The ventilation system shall be designed to limit the maximum concentration of hydrogen to 1.0 percent of the total volume of the room; or
2. Continuous ventilation shall be provided at a rate of not less than 1 cubic foot per minute per square foot (1 cfm/ft^2) [(0.0051 m^3/(s · m^2)] of floor area of the room.

❖ Basic ventilation, as with any other building, must comply with the requirements of the IMC. In addition, specific ventilation rates are stated in Chapter 12 of the IBC. Generally with lead-acid battery systems, the main concern is the production of hydrogen and oxygen within an enclosed space. Hydrogen has a wide flammability range and is the lightest element on the periodic chart. To address the concern of hydrogen generation and containment in small areas, a minimum ventilation criterion is set.

There are two methods of compliance. The first is performance based and states that the maximum concentration of hydrogen must be limited to 1.0 percent of the total volume of the room. This method requires an analysis of plausible failure scenarios to justify the ventilation rate ultimately chosen. The second method simply requires continuous ventilation at a rate no less than 1 cubic foot per minute per square foot [0.0051 m^3/(s•m^2)] of room area.

608.6 Signs. Doors into rooms or buildings containing stationary lead-acid battery systems shall be provided with approved signs. The signs shall state that the room contains lead-acid battery systems, that the battery room contains energized electrical circuits, and that the battery electrolyte solutions are corrosive liquids.

❖ Because of the possible hazards associated with exposure to corrosive liquids, energized electrical circuits and the presence of hydrogen and oxygen, specific signage is needed to make building occupants and emergency responders aware of the potential dangers.

608.7 Seismic protection. The battery systems shall be seismically braced in accordance with the *International Building Code*.

❖ Because battery systems pose hazards of corrosive-liquid spills and because they are used for standby power, emergency power and uninterruptible power supplies, this section requires seismic bracing of the systems. Generally, these systems are located in racks that carry a fairly heavy load because of the liquid contained within the batteries. Section 1621 of the IBC contains seismic protection requirements applicable to battery rooms.

608.8 Smoke detection. An approved automatic smoke detection system shall be installed in battery rooms in accordance with Section 907.2.23.

❖ Except for a fire involving hydrogen, battery room fires are likely to be slow-growing and smoldering; therefore a smoke detection system is required for early detection in accordance with Section 907.2.23.

SECTION 609
VALVE-REGULATED LEAD-ACID (VRLA) BATTERY SYSTEMS

609.1 Scope. Valve-regulated lead-acid (VRLA) battery systems having an electrolyte capacity of more than 50 gallons (189 L) used for facility standby power, emergency power or uninterrupted power supplies (UPS) shall comply with this section.

❖ VRLA batteries differ substantially from the flooded batteries that are addressed in Section 608 in design, operation and, more importantly, in potential hazard. VRLA-type batteries are uniquely different from the traditional liquid electrolyte lead acid batteries in that they have no free liquid electrolyte to flow from the container if it were to break. Additionally, the VRLA battery implements an oxygen recombination cycle that minimizes the emissions of gas from the battery during overcharging. This section limits the application of Section 609 to VRLA batteries prevalent in "large" systems such as those commonly used in data centers, customer premise telephone systems and at telecommunication sites, such as central offices, huts, vaults and outdoor cabinet systems. They are selected for these applications specifically due to their lack of a free-flowing liquid electrolyte and, therefore, the lack of any requirement for electrolyte maintenance; flexibility of installation, again due to their lack of a free-flowing liquid electrolyte and economy of installation, since there is no practical need for containment. Smaller, stand-alone personal computer UPS devices, emergency lights, fire alarm control panels and similar installations or devices that use VRLA batteries are specifically intended to be excluded from the provisions of Section 609 (see also Section 602 for the definitions of "Valve-regulated lead-acid battery" and "Vented flooded lead-acid battery" and Section 608.1 for further discussion of the differences between battery types).

609.2 Safety vents. VRLA batteries shall be equipped with self-resealing flame-arresting safety vents.

❖ Self-resealing vents are an integral part of the design of VRLA batteries. Flame-arresting vents are needed to prevent a static spark (or other flame source) outside the battery from propagating to the interior of the battery where oxygen or hydrogen may have accumulated during normal operation of the battery.

609.3 Thermal runaway. VRLA battery systems shall be provided with a listed device or other approved method to preclude, detect and control thermal runaway.

❖ Under certain extreme conditions of high ambient temperature or charging rate, VRLA batteries may experience a phenomenon known as "thermal runaway." Under these conditions, a VRLA battery may generate excessive heat internally that, in rare cases, results in a case rupture or fire. This section requires users to address the potential hazard of thermal runaway, without requiring the use of a specific device.

609.4 Room design and construction. Enclosure of VRLA battery system rooms shall comply with the *International Building Code*. The battery systems are permitted to be in the same room with the equipment they support. When VRLA battery systems are installed in a separate equipment room accessible only to authorized personnel, they shall be allowed to be installed on an open rack for ease of maintenance. When a VRLA battery system is situated in an occupied work center, it shall be housed in a noncombustible cabinet or other enclosure to prevent access by unauthorized personnel.

❖ This section states that battery rooms must meet the basic construction requirements of the IBC. No special fire-resistance-rated separation requirements or occupancy classifications are necessary. Additionally, there is no requirement for separating the equipment that the battery system supports from the battery system itself; however, in order to prevent tampering or accidental contact with the battery system when installed in a work area, this section also requires enclosure of the battery systems in a cabinet under such conditions. VRLA battery systems are often listed under UL 1778, which places similar design and construction requirements on cabinetized systems.

609.5 Neutralization. An approved manual method and materials for the neutralization of a release of electrolyte shall be provided. The method and materials shall be capable of controlling and neutralizing a release of 3 percent of the capacity of the largest VRLA cell or block in the room to a pH between 7.0 and 9.0.

❖ The electrolyte in VRLA batteries is immobilized by either the addition of a gelling agent or by being absorbed in a fiberglass mat (i.e., a sponge). This immobilization creates a situation where a spill of the electrolyte is highly unlikely. A typical accident where a VRLA battery case is broken results in a slight drip or a slow ooze of material out of the battery that cannot be characterized as a spill. Accordingly, spill control, as prescribed in Section 608.4, for flooded lead-acid batteries is not necessary. A means to neutralize any release of the electrolyte must be provided. Typically, either sodium-bicarbonate powder or a liquid buffering solution is provided within the room where the battery system is located for use by trained personnel.

609.6 Room ventilation. Ventilation shall be provided to limit the maximum concentration of hydrogen to 1 percent of the total volume of the room during the worst-case event of simultaneous "boost" charging of all batteries in the room. Where calculations are not provided to substantiate the ventilation rate, continuous ventilation at a rate of not less than 1 cubic foot per minute per square foot (1 $ft^3/min/ft^2$) [(0.0051 $m^3/(s \cdot m^2)$] of floor area of the room shall be provided. The ventilation shall be either mechanically or naturally induced.

❖ Basic ventilation, as with any other building, must comply with the requirements of the IMC. Generally with lead-acid battery systems, the main concern is the production of hydrogen and oxygen within an enclosed space. Hydrogen has a wide flammability range and is the lightest element in the Periodic Table of the Elements. To address the concern of hydrogen generation and containment in small areas, a special minimum ventilation criterion is set.

There are two methods of compliance. The first is performance based and states that the maximum concentration of hydrogen must be limited to 1.0 percent of the total volume of the room under worst-case conditions. This method requires an analysis of plausible failure scenarios to justify the ventilation rate ultimately chosen. The second method simply requires continuous ventilation at a rate not less than 1 cubic foot per minute per square foot [0.0051 $m^3/(s \cdot m^2)$)] of the room area.

609.7 Cabinet ventilation. Where VRLA batteries are installed inside a cabinet, the cabinet shall be vented. The cabinet ventilation shall limit the maximum concentration of hydrogen to 1 percent of the total volume of the cabinet during the worst-case event of simultaneous "boost" charging of all batteries in the cabinet. Where calculations are not provided to substantiate the ventilation rate, continuous ventilation at a rate of not less than 1 cubic foot per minute per square foot (1 $ft^3/min/ft^2$) [0.0051 $m^3/(s \cdot m^2)$] of floor area covered by the cabinet shall be provided. The ventilation shall be either mechanically or naturally induced. The room in which the cabinet is installed shall also be ventilated as required in Section 609.6.

❖ This section addresses the increasingly common practice of placing VRLA battery systems in cabinets. Ventilation of VRLA battery cabinets is treated in the same manner as VRLA battery rooms in Section 609.6 (see commentary, Section 609.6).

609.8 Signs. Doors into electrical equipment rooms containing VRLA battery systems shall be provided with approved signs.

The signs shall state that the room contains lead-acid battery systems and contains energized electrical circuits. Where VRLA batteries are contained in cabinets in occupied work centers, the cabinet enclosures shall be located within 10 feet (3048 mm) of the equipment that they support. The cabinets shall have exterior labels that identify the manufacturer and model number of the system and electrical rating (voltage and current) of the contained battery system. Within the cabinet there shall be signs that indicate the relevant electrical, chemical and fire hazards.

❖ Because of the possible hazards associated with exposure to corrosive electrolytes, energized electrical circuits and the presence of hydrogen and oxygen, specific signage is needed to make building occupants and emergency responders aware of the potential dangers. Additionally, VRLA battery cabinets must be marked with technical data about the system so that emergency responders may properly judge the degree of hazard presented by the system.

609.9 Seismic protection. The battery systems shall be seismically braced in accordance with the *International Building Code.*

❖ Because battery systems are used for essential standby power, emergency power and uninterruptible power supplies, this section requires seismic bracing of the systems. Generally, these systems are located in racks that carry a fairly heavy load because of the liquid contained within the batteries. Section 1621 of the IBC contains seismic protection requirements applicable to battery rooms.

609.10 Smoke detection. An approved automatic smoke detection system shall be installed in rooms containing VRLA battery systems in accordance with Section 907.2.23.

❖ Except for a fire involving hydrogen, battery-room fires are likely to be slow-growing and smoldering; therefore, a smoke detection system is required for early detection in accordance with Section 907.2.23.

SECTION 610
COMMERCIAL KITCHEN HOODS

[M] 610.1 General. Commercial kitchen exhaust hoods shall comply with the requirements of the *International Mechanical Code.*

❖ Whereas the previous edition of the code contained detailed hood requirements duplicated from the IMC, the code now simply references the IMC for hood design. (see IMC commentary, Section 507).

[M] 610.2 Where required. A Type I hood shall be installed at or above all commercial cooking appliances and domestic cooking appliances used for commercial purposes that produce grease vapors.

❖ An exhaust system is required for all appliances used for commercial cooking as defined in Section 602. In addition to the specific cooking appliances identified in the definition, further examples of commercial cooking appliances that require a commercial kitchen exhaust system are griddles (flat or grooved); tilting skillets or woks; braising and frying pans; roasters; pastry ovens; pizza ovens; charbroilers; salamanders and upright broilers; infrared broilers and open-burner stoves and ranges. Furthermore, the definition of "Commercial cooking appliances" defines a food service establishment as "...any building or portion thereof used for the preparation and serving of food."

Bibliography

The following resource materials are referenced in this chapter or are relevant to the subject matter addressed in this chapter.

ASME A17.3, *Safety Code for Existing elevators and Escalators.* New York: American Society of Mechanical Engineers, 1996.

ASME A17.1, *Safety Code for Elevators and Escalators, with A17.1a-97 and A17.1b-98 Addenda.* New York: American Society of Mechanical Engineers, 2000.

IBC-2003, *International Building Code.* Falls Church, VA: International Code Council, 2003.

ICCEC–2003, *ICC Electrical Code.* Falls Church, VA: International Code Council, 2003.

IECC–2003, *International Energy Conservation Code.* Falls Church, VA: International Code Council, 2003.

IFGC-2003, *International Fuel Gas Code.* Falls Church, VA: International Code Council, 2003.

IMC-2003, *International Mechanical Code.* Falls Church, VA: International Code Council, 2003.

IPC-2003, *International Plumbing Code.* Falls Church, VA: International Code Council, 2003.

NFPA 31–01, *Installation of Oil-Burning Equipment.* Quincy, MA: National Fire Prevention Association, 2001.

NFPA 110–00, *Emergency and Standby Power Systems.* Quincy, MA: National Fire Prevention Association, 2000.

NFPA 111–01, *Stored Electrical Energy Emergency and Standby Power Systems.* Quincy, MA: National Fire Prevention Association, 2001.

NFPA 211–00, *Chimneys, Fireplaces, Vents and Solid-Fuel-Burning Appliances.* Quincy, MA: National Fire Prevention Association, 2000.

NFPA 704–96, *Identification of Hazards of Materials for Emergency Response.* Quincy, MA: National Fire Prevention Association, 1996.

UL 1778-03, Uninterruptible Power Systems. Northbrook, IL: Underwriters Laboratories, Inc., 2003.

UL 2200–98, *Stationary Engine Generator Assemblies.* Northbrook, IL: Underwriters Laboratories, Inc., 1998.

Chapter 7: Fire-Resistance-Rated Construction

General Comments

This chapter sets forth general fire safety precautions for buildings and structures. In general, these requirements are intended to maintain required fire-resistance ratings and limit fire spread.

Chapter 7 is divided into four sections. Section 701 gives the general scope of the chapter and provides the basis for enforcement of its provisions. Section 702 refers to Chapter 2 for applicable definitions. The required maintenance of fire-resistance-rated assemblies, opening protectives and fire doors is described in Section 703. Section 704 defines the enclosure requirements for shafts in existing buildings

Purpose

The maintenance of assemblies required to be fire-resistance rated is a key component in a passive fire protection philosophy. This chapter reinforces this component and further regulates floor openings, which are a leading cause of fire spread as well as smoke migration through buildings.

SECTION 701 GENERAL

701.1 Scope. The provisions of this chapter shall specify the requirements for and the maintenance of fire-resistance-rated construction and requirements for enclosing floor openings and shafts in existing buildings. New construction shall comply with the *International Building Code.*

❖ This section states the requirements for maintaining the integrity of fire-resistance-rated assemblies (Section 703) and prescribes the types of floor openings permitted in existing buildings (Section 704).

SECTION 702 DEFINITIONS

702.1 Terms defined in Chapter 2. Words and terms used in this chapter and defined in Chapter 2 shall have the meanings ascribed to them as defined therein.

❖ Words and terms used in this chapter and defined in Chapter 2 shall have the meanings ascribed to them as defined therein.

There are no terms unique to Chapter 7 that warrant inclusion in Chapter 7. Among the definitions in Chapter 2 that are broadly applicable to the entire code, including Chapter 7, are:

- Fire code official.
- Fire door assembly.
- Fire partition.
- Occupancy classification.

SECTION 703 FIRE-RESISTANCE-RATED CONSTRUCTION

703.1 Maintenance. The required fire-resistance rating of fire-resistance-rated construction (including walls, fire stops, shaft enclosures, partitions and floors) shall be maintained. Such elements shall be properly repaired, restored or replaced when damaged, altered, breached or penetrated. Openings made therein for the passage of pipes, electrical conduit, wires, ducts, air transfer openings, and holes made for any reason shall be protected with approved methods capable of resisting the passage of smoke and fire. Openings through fire-resistance-rated assemblies shall be protected by self-closing or automatic-closing doors of approved construction meeting the fire protection requirements for the assembly.

❖ The code requires that all equipment, systems, devices and safeguards required by the current and past codes be maintained in good working order (see Section 102.1). Section 703.1 reiterates this requirement specifically for fire-resistance-rated assemblies.

Once a building is occupied, its component parts are often damaged, altered or penetrated for installation of new piping, wiring and the like. These actions may reduce the effectiveness of assemblies that must be fire-resistance rated. This section requires that any damage to a fire-resistance-rated assembly be repaired in a manner that restores the original required performance characteristics. Similarly, if a fire-resistance-rated assembly is altered or penetrated, the alteration or penetration must comply with the applicable requirements of the *International Building Code*® (IBC®) for that kind of alteration or penetration.

A common violation of fundamental safety principles, as well as the code, is having wooden or rubber floor wedges prop open fire doors or smoke barrier doors.

This renders them totally ineffective as opening protectives. Building maintenance personnel who do not understand the purpose of barrier doors often do this to aid movement of people, equipment or air in a hallway or other passage without realizing the potential hazard if a fire were to occur.

703.1.1 Fireblocking and draftstopping. Required fireblocking and draftstopping in combustible concealed spaces shall be maintained to provide continuity and integrity of the construction.

❖ Fireblocking and draftstopping retard the spread of fire and the products of combustion. To fulfill their intended function, fireblocking and draftstopping must be properly maintained. Most frequently, damage or repairs to other building components, such as mechanical piping, results in fireblocking or draftstopping being removed and not properly replaced. This section specifically requires that when fireblocking and draftstopping required by the IBC are damaged, removed or otherwise altered, they must be replaced or restored.

703.1.2 Smoke barriers. Required smoke barrier partitions shall be maintained to prevent the passage of smoke and all openings protected with approved smoke barrier doors or leakage-rated (smoke) dampers.

❖ Smoke barriers are a key component in a passive fire safety design for institutional occupancies (Groups I-2 and I-3) as part of the defend-in-place strategy for moving occupants to a tenable portion of the building. The IBC includes requirements for fire-resistance ratings (1 hour), continuity and opening and penetration protection. This section of the code reinforces the application of these requirements by specifically citing smoke barriers as a necessary assembly that warrants stringent maintenance.

703.2 Opening protectives. Opening protectives shall be maintained in an operative condition in accordance with NFPA 80. Fire doors and smoke barrier doors shall not be blocked or obstructed or otherwise made inoperable. Fusible links shall be replaced promptly whenever fused or damaged. Fire door assemblies shall not be modified.

❖ Openings in fire-resistance-rated assemblies must be protected to prevent the passage of fire. After opening protectives are installed and approved, they may become damaged, corroded or otherwise less effective than required. This section specifically requires that all opening protectives required by the IBC be maintained in compliance with NFPA 80.

This section requires the maintenance of fire doors and smoke barrier doors so that they can perform their intended function, which is to prevent the passage of smoke, fire or combustion products through openings in fire-resistance-rated walls, ceilings and shafts during a fire emergency. Sections 703.2.2 and 703.2.3 indicate specific points of inspection and enforcement regarding these doors. Prohibited modifications to fire door assemblies include the attachment of materials, cutting, boring holes or other alterations that could affect the performance of the door as a fire protection rated assembly.

703.2.1 Signs. Where required by the fire code official, a sign shall be permanently displayed on or near each fire door in letters not less than 1 inch (25 mm) high to read as follows:

1. For doors designed to be kept normally open: FIRE DOOR—DO NOT BLOCK.
2. For doors designed to be kept normally closed: FIRE DOOR—KEEP CLOSED.

❖ Any door in a fire-resistance-rated wall represents a potential "weak link" in maintaining the degree of compartmentation intended by the code. That is the reason for requiring a rated assembly. The IBC calls for adequate opening protection in the form of a door with a specified fire protection rating. This section allows the fire code official to require signage on or near the rated doors to make the occupants aware of the importance of the door in the passive fire protection philosophy of the code. Also, see the commentary to Section 703.1 for a discussion on the improper use of props to hold doors open.

703.2.2 Hold-open devices and closers. Hold-open devices and automatic door closers, where provided, shall be maintained. During the period that such device is out of service for repairs, the door it operates shall remain in the closed position.

❖ The only devices acceptable for holding fire doors open are fire-detector-activated automatic-closing devices that automatically close the doors (or allow the doors to swing closed using self-closing devices) in the event of a fire. Numerous devices, such as electromagnetic hold-opens, pneumatic systems and systems of pulleys and weights connected to a fusible link, are available.

The detection method for the closing device must be consistent with the purpose of the opening protective; that is, doors in smoke barriers must be activated by smoke detectors. Heat detectors or fusible links are adequate where maintenance of the fire-resistance rating alone is required.

If smoke-detector-activated automatic door closers are used and the detectors are interconnected with a required fire alarm system, the devices and wiring methods must be checked for compatibility with the fire alarm system control panel before installation. Some fire alarm control equipment is compatible only with the manufacturer's automatic smoke detectors.

703.2.3 Door operation. Swinging fire doors shall close from the full-open position and latch automatically. The door closer shall exert enough force to close and latch the door from any partially open position.

❖ Fire doors must be closed to be effective. Swinging fire doors should be frequently checked to make sure they close and latch on their own power from any position.

703.3 Ceilings. The hanging and displaying of salable goods and other decorative materials from acoustical ceiling systems that are part of a fire-resistance-rated floor/ceiling or roof/ceiling assembly, shall be prohibited.

❖ Fire-resistance-rated floor/ceiling and roof/ceiling assemblies must be tested using the methods in ASTM E 119 to demonstrate a fire-resistance rating. Locating a substantial fuel load directly beneath an acoustical ceiling, however, may expose the ceiling to a direct fire source that could breech the ceiling, which is an integral part of the tested assembly. Depending on the contribution of the ceiling to the overall rating, this may result in the assembly not functioning as the code intends.

703.4 Testing. Horizontal and vertical sliding and rolling fire doors shall be inspected and tested annually to confirm proper operation and full closure. A written record shall be maintained and be available to the fire code official.

❖ Annual tests are intended to determine that required fire and smoke-barrier doors operate freely and close completely. Where fusible links are used as the releasing mechanism, the link may be temporarily removed rather than activated during testing. Fusible links in poor condition must be replaced as part of the maintenance of fire-resistance components. Smoke detectors and heat detectors other than fusible links must be tested as required by the manufacturer's instructions (see NFPA 72 for recommended testing procedures for various fire detectors).

Written records must indicate the date, time, test method and person conducting the test for each opening protective. These records must be maintained by the owner and made available to the fire code official for review. This requirement relieves the fire code official of the administrative burden of maintaining test records.

SECTION 704
FLOOR OPENINGS AND SHAFTS

704.1 Enclosure. Interior vertical shafts, including but not limited to stairways, elevator hoistways, service and utility shafts, that connect two or more stories of a building shall be enclosed or protected as specified in Table 704.1. When openings are required to be protected, openings into such shafts shall be maintained self-closing or automatic-closing by smoke detection. Existing fusible-link-type automatic door-closing devices are permitted if the fusible link rating does not exceed 135°F (57°C).

❖ Vertical openings that are not properly protected can act as a chimney for smoke, hot gases and products of combustion. Unprotected floor openings have been a major contributing factor in many large loss-of-life fires. Except as indicated otherwise in Table 704.1, Section 704.1 retroactively requires the enclosure of vertical openings between floors with approved fire barriers. The intent is to increase the level of safety in all buildings by enclosing unprotected floor openings and, thus, removing the avenue for unimpeded fire and smoke spread. In particular, the code intends for the means of egress in existing multistory buildings to be given a reasonable level of protection.

Vertical communication between floors (including stairways, escalator openings, elevator hoistways, trash and laundry chutes and other types of building service and mechanical shafts) can contribute significantly to fire and smoke spread because of stack effect. Except as stated in Table 704.1, this section requires that all interior shafts must be enclosed with fire barriers of the fire-resistance rating required by the table. These assemblies (except for the fire-resistance rating that is specified by the table) must be constructed and protected as required by the IBC.

The listed protective is based on and scaled in accor-

TABLE 704.1
VERTICAL OPENING PROTECTION REQUIRED

OCCUPANCY CLASSIFICATION	CONDITIONS	PROTECTION REQUIRED
Group I	Vertical openings connecting two or more stories	1-hour protection
All, other than Group I	Vertical openings connecting two stories	No protection required[a,b]
All, other than Group I	Vertical openings connecting three to five stories	1-hour protection or automatic sprinklers throughout[a,b]
All, other than Group I	Vertical openings connecting more than five stories	1-hour protection[a,b]
All	Mezzanines open to the floor below	No protection required[a,b]
All, other than Group I	Atriums and covered mall buildings	1-hour protection or automatic sprinklers throughout
All, other than Groups B and M	Escalator openings connecting four or less stories in a sprinklered building. Openings must be protected by a draft curtain and closely spaced sprinklers in accordance with NFPA 13	No protection required
Groups B and M	Escalator openings in a sprinklered building protected by a draft curtain and closely spaced sprinklers in accordance with NFPA 13	No protection required

a. Vertical opening protection is not required for Group R-3 occupancies.
b. Vertical opening protection is not required for open parking garages and ramps.

dance with life safety considerations. The allowance for sprinklers recognizes the value of automatic sprinkler systems conforming to Section 903.3.1.1 (NFPA 13 system) or 903.3.1.2 (NFPA 13R system), as applicable.

Bibliography

The following resource materials are referenced in this chapter or are relevant to the subject matter addressed in this chapter.

ASTM E 119–00, *Standard Test Methods for Fire Tests of Building Construction and Materials.* West Conshohocken, PA: ASTM International, 2000.

IBC-2003, *International Building Code.* Falls Church, VA: International Code Council, 2003.

NFPA 72–99, *National Fire Alarm Code.* Quincy, MA: National Fire Protection Association, 1999.

NFPA 80–99, *Fire Doors and Fire Windows.* Quincy, MA: National Fire Prevention Association, 1999.

Chapter 8: Interior Finish, Decorative Materials and Furnishings

General Comments

This chapter is consistent with the format of the *International Building Code*® (IBC®), which regulates the interior finishes of buildings through the regulation of their flame spread potential. The code goes beyond interior finishes and also regulates furnishings and vegetation within buildings in certain occupancies. Additionally, the code addresses interior finishes and decorative materials in existing buildings.

This chapter is related to fire growth and spread potential in terms of the immediate effect on building occupants. The flame spread characteristics of certain materials will affect the potential fire scenarios within a building. Fire-resistance-rated construction, which is dealt with in Chapter 7 of both the IBC and the code, is more concerned with the spread of fire throughout the structure once the fire has reached a substantial size, with an emphasis on structural failure during a fire.

The regulation of flame spread can be traced back to large life-loss events, such as the Cocoanut Grove nightclub fire that killed 492 people in 1942. This fire was thought to have started when a lightbulb in the basement cocktail lounge came in contact with the cotton cloth that had been applied to the ceiling for decorative purposes. Post-fire testing of the cotton cloth indicated that it had a flame spread rating of 2,500, more than 33 times the maximum allowable flame spread in the IBC. This factor, in addition to a series of problems with the egress system, led to one of the worst fire disasters in history. The need for these regulations was further emphasized after the Station Nightclub fire in West Warwick, Rhode Island, where 100 people died in February 2003. The sound-proofing material in the nightclub was not approved for such use and was a major factor in the fire growth.

In addition to flame spread ratings of surface materials, certain furnishing types and vegetation, such as Christmas trees, pose a large fire hazard because of the potential fire size and intensity. The materials used in furnishings have changed dramatically from past materials and many more plastics are now used for both decoration and furnishings. Plastics not only burn more vigorously than materials such as cotton and wood, but also produce more toxic fire effluents.

Purpose

The overall purpose of Chapter 8 is to ensure that interior finishes, furnishings and vegetation do not significantly add to or create fire hazards within buildings. The provisions tend to aim at occupancies with specific risk characteristics, such as vulnerability of occupants, density of occupants, lack of familiarity with the building and societal expectations of importance. Since this is a fire code, there is an emphasis on both new and existing buildings.

SECTION 801
GENERAL

801.1 Scope. The provisions of this chapter shall govern furniture and furnishings, interior finishes, interior trim, decorative materials and decorative vegetation in buildings. Sections 803, 804 and 805 shall be applicable to new and existing buildings. Section 806 shall be applicable to existing buildings.

❖ This chapter reflects the same scope of issues as Chapter 8 of the IBC but has a slightly different emphasis. Fire codes are intended to address fire hazards of buildings and facilities over their lifespan; therefore, there is greater emphasis on the contents of buildings and on the maintenance of flame spread ratings over time. Section 806 requires the same flame spread ratings as the IBC. Generally, regulating the combustibility of contents is a fairly difficult task once the building is occupied and is considered existing. Because of this difficulty, combustible contents and decorative materials are regulated in a limited number of occupancies. More specifically, the use of combustible furnishings and decorative materials in Group A occupancies is addressed because of the high occupant load and the lack of familiarity of most occupants with the building. The type of furniture allowed in occupancies such as Group I-2 is limited because of the vulnerability of the occupants and the likely fire scenarios that may occur when the building is unsprinklered.

SECTION 802
DEFINITIONS

802.1 Terms defined in Chapter 2. Words and terms used in this chapter and defined in Chapter 2 shall have the meanings ascribed to them as defined therein.

❖ There are no terms that are specific to the application of Chapter 8. The reader is referred to Chapter 2 for general definitions.

SECTION 803
FURNISHINGS

803.1 General requirements. The provisions of Sections 803.1.1 through 803.1.3 shall be applicable to all occupancies covered by Sections 803.2 through 803.7.

❖ Furnishings is an area that codes have not addressed very strongly in the past. Ultimately, the fire hazard potential within a building depends heavily upon what is in the building and where it is placed. First, the burning characteristics of the materials will vary and the location of the material will change the characteristics of a fire. For instance, a couch within a small compartment will create a much different fire event than the same couch burning in a large open atrium (see Figure 803.1). The compartment may be limited by the amount of oxygen available, while the atrium fire will be limited only by the amount of combustibles to burn because oxygen will be plentiful; therefore, the compartment is likely to reach flashover while the atrium is not.

Another key factor is the characteristics of the occupants. Issues such as occupant density, vulnerability, familiarity with the building, whether occupants are sleeping, etc., are all important factors. The codes have made a conscious decision to regulate some of these occupancies where the risk factors are higher. Within Section 803, limitations are placed on the type of building contents for Group A, E, I-4, I-2, I-1 and I-3 occupancies. Additionally, requirements for these particular occupancies are more easily enforced because of the

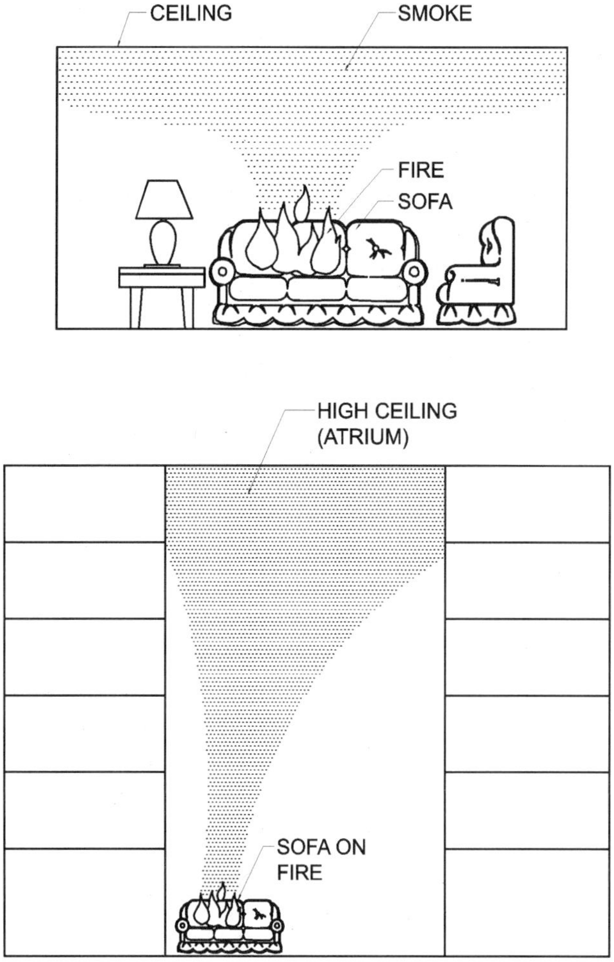

Figure 803.1
EFFECT OF LOCATION OF FIRE

nature of the occupancies. Some of the general requirements in Section 803.1 would apply to all occupancies.

803.1.1 Explosive and highly flammable materials. Furnishings or decorations of an explosive or highly flammable character shall not be used.

❖ This is a general statement that is aimed at the prohibition of furnishings or decorative materials that pose an extreme fire potential. These materials will tend to have a significant impact on the potential fire size in a building.

803.1.2 Fire-retardant coatings. Fire-retardant coatings shall be maintained so as to retain the effectiveness of the treatment under service conditions encountered in actual use.

❖ In many cases, the building and fire codes allow use of certain combustible or flammable materials based on the application of a fire-retardant coating. The use of these coatings must be realistic in that the actual use of the material is taken into account. For example, if a material is likely to be laundered regularly or exposed to extreme climate conditions, tests must demonstrate that these conditions do not eliminate or reduce the effectiveness of the fire-retardant coatings. If the fire-retardant coating is not maintained, the material will not be considered equivalent to a less combustible material. To assist in the maintenance of fire-retardant applications, the manufacturer's instructions must be referred to regularly.

803.1.3 Obstructions. Furnishings or other objects shall not be placed to obstruct exits, access thereto, egress therefrom or visibility thereof.

❖ Although not a flammability or combustibility issue, furnishings can pose a hazard during fires by blocking the means of egress, reducing the visibility of exit signage or simply by placing a fire hazard within the exit. This section prohibits these practices.

803.2 Group A. The requirements in Sections 803.2.1 and 803.2.2 shall apply to occupancies in Group A.

❖ The requirements in Sections 803.2.1 and 803.2.2 are specific to Group A occupancies. These particular requirements are fairly specific to activities and uses that occur in Group A occupancies, such as trade shows and movie theaters. Generally, fire hazards are moderate in Group A occupancies. The concerns are more closely related to the high occupant density and the occupants' lack of familiarity with the building.

803.2.1 Foam plastics. Exposed foam plastic materials and unprotected materials containing foam plastic used for decorative purposes or stage scenery or exhibit booths shall have a maximum heat release rate of 100 kilowatts (kW) when tested in accordance with UL 1975.

Exceptions:

1. Individual foam plastic items or items containing foam plastic where the foam plastic does not exceed 1 pound (0.45 kg) in weight.
2. Cellular or foam plastic shall be allowed for trim not in excess of 10 percent of the wall or ceiling area, provided it is not less than 20 pounds per cubic foot (320 kg per cubic meter) in density, is limited to 0.5 inch (12.7 mm) in thickness and 4 inches (102 mm) in width, and complies with the requirements for Class B interior wall and ceiling finish, except that the smoke-developed index shall not be limited.

❖ Foam plastic burns vigorously and at high heat release rates because the material is petroleum based and contains a large amount of air. Group A occupancies tend to contain uses such as stages, movie theaters and exhibit halls. These uses will likely have highly combustible scenery or exhibit booths; therefore, because the occupants are unlikely to be familiar with the premises, and because of the hazards presented by the combustibles, a heat release rate limit of 100 kW, when tested in accordance with UL1975, is placed upon materials used in Group A occupancies.

In the past, restrictions on the combustibility of materials were based on the heat content of the materials instead of the rate at which the heat content is released. The problem with using it for a limitation is that the heat content, (which can be equated to "potential energy" of the material) does not give us a good understanding of the rate at which the "potential energy" is released. The rate at which the heat is released is a more important characteristic of fire. UL 1975 uses a relationship of the actual heat release rate based on exposure to a specific heat flux as specified in the test.

Some alternatives to testing, as noted in the exceptions, limit the amount of combustibles. The exceptions give two prescriptive approaches with no direct relationship to the actual combustibility of the foam plastic except to limit the amount used. The first is related to plastic display items, such as a small statue, and the second relates specifically to decorative trim. A foam plastic display item would be limited to 1 pound (0.45 kg). Decorative trim is limited to 10 percent of the wall or ceiling area with some additional weight and density limitations. These additional dimension and density limitations are necessary because a small piece of trim that covers the same percentage of wall would be treated the same as a larger (less dense) piece (see Figure 803.2.1). Note also that the trim would have to meet the flame spread rating of a Class B interior but not be subject to the smoke generation limitations (see Section 805.3).

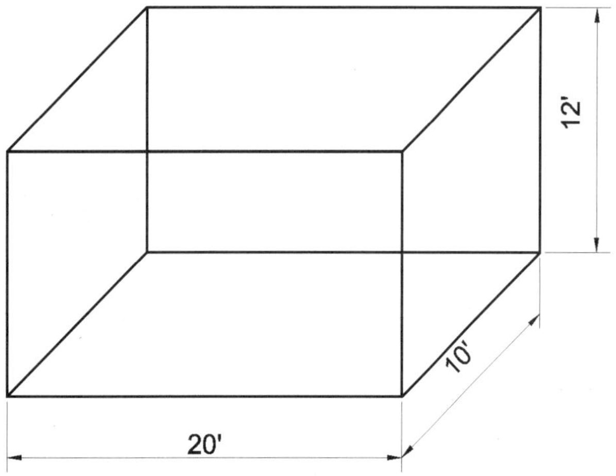

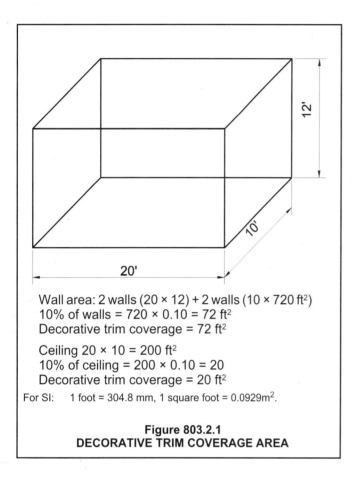

Wall area: 2 walls (20 × 12) + 2 walls (10 × 720 ft^2)
10% of walls = 720 × 0.10 = 72 ft^2
Decorative trim coverage = 72 ft^2

Ceiling 20 × 10 = 200 ft^2
10% of ceiling = 200 × 0.10 = 20
Decorative trim coverage = 20 ft^2

For SI: 1 foot = 304.8 mm, 1 square foot = 0.0929m^2.

Figure 803.2.1
DECORATIVE TRIM COVERAGE AREA

803.2.2 Motion picture screens. The screens upon which motion pictures are projected shall be either flame resistant, as demonstrated by complying with NFPA 701, or shall comply with the requirements for a Class B interior finish.

❖ Movie screens are not considered an interior finish and, therefore, would not be addressed by Chapter 8 of the IBC. These screens consist of a fabric base covered with a thin coating impregnated with reflective glass beads. Movie screens are generally fairly large and typically take up most of the front wall of a movie theater; therefore, the flame spread characteristics must be addressed. Because movie theaters tend to be densely occupied, a material with a high rate of flame spread can pose a significant hazard. Screens can comply in one of two ways. Either the screen must meet the flame-resistance criteria set out in NFPA 701, which focuses on combustibles such as curtains, shades and swages not considered wall coverings, or qualify as a Class B interior finish material.

803.3 Group E. The requirements in Sections 803.3.1 and 803.3.2 shall apply to occupancies in Group E.

❖ Group E occupancies are educational occupancies used by six or more persons for educational purposes through the 12th grade or buildings and structures used for educational, supervision or personal care services for more than five children over the age of $2^{1}/_{2}$ years. This section regulates the amount of combustibles within corridors and lobbies by regulating clothing, personal effects and artwork.

803.3.1 Storage in corridors and lobbies. Clothing and personal effects shall not be stored in corridors and lobbies.

Exceptions:

1. Corridors protected by an approved automatic sprinkler system installed in accordance with Section 903.3.1.1.
2. Corridors protected by an approved smoke detection system installed in accordance with Section 907.
3. Storage in metal lockers provided the minimum required egress width is maintained.

❖ Materials such as clothing, other personal effects and artwork have the potential for creating a fire hazard within the main path of egress travel. This section allows the storage of clothing and other personal effects within these areas only if corridors and lobbies contain one of the following features:

- Sprinkler system,
- Smoke detection system or
- Metal lockers for storage.

The sprinkler system must meet the requirements of NFPA 13. The smoke detection system must be approved by the fire code official. The smoke detection system is intended to specifically focus on the contents of the corridors.

803.3.2 Artwork. Artwork and teaching materials shall be limited on the walls of corridors to not more than 20 percent of the wall area.

❖ Educational occupancies tend to display various artwork and related educational materials on the walls of classrooms and corridors. This section limits the potential combustibility levels of artwork in critical areas of the means of egress system; therefore, decorations or artwork can cover no more than 20 percent of the corridor walls. This requirement applies only to corridors.

803.4 Group I-4, day care facilities. The requirements in Sections 803.4.1 and 803.4.2 shall apply to day care facilities classified in Group I-4.

❖ A Group I-4 day care facility is an occupancy that cares for people of any age for less than 24 hours per day. Most often, these facilities are for small children not yet able to attend school. These occupancies will have features similar to those of Group E occupancies; therefore, the same restrictions on combustibles in corridors and lobbies exist. The key issue here is that the occupants of Group I-4 facilities are often not capable of self-preservation without some level of assistance;

therefore, combustibles must be kept to a minimum, especially in critical portions of the means of egress system.

803.4.1 Storage in corridors and lobbies. Clothing and personal effects shall not be stored in corridors and lobbies.

Exceptions:

1. Corridors protected by an approved automatic sprinkler system installed in accordance with Section 903.3.1.1.
2. Corridors protected by an approved smoke detection system installed in accordance with Section 907.
3. Storage in metal lockers provided the minimum required egress width is maintained.

❖ See the commentary for Section 803.3.1.

803.4.2 Artwork. Artwork and teaching materials shall be limited on walls of corridors to not more than 20 percent of the wall area.

❖ See the commentary for Section 803.3.2.

803.5 Group I-2, nursing homes and hospitals. The requirements in Sections 803.5.1 through 803.5.3 shall apply to nursing homes and hospitals classified in Group I-2.

❖ Occupants of nursing homes and hospitals are considered more vulnerable than the general population. Many of the patients of nursing homes and hospitals are confined because of respirators, IVs and other medical equipment and may not be capable of self-preservation. Hospital employees may have to make several trips into the fire area to assist in evacuating patients; therefore, it is imperative that every effort be made to preserve the integrity of the corridors and minimize the fuel loading caused by furnishings, such as upholstered furniture and mattresses. Historically, fires have been ignited through the use of cigarettes in bed or falling asleep while smoking in a chair. In addition to these hazards, such occupancies usually have additional medical oxygen sources within their rooms. This section, therefore, states several ignitability and combustibility limitations for upholstered furniture and mattresses being introduced into such occupancies.

803.5.1 Upholstered furniture. Newly introduced upholstered furniture shall be shown to resist ignition by cigarettes as determined by tests conducted in accordance with NFPA 261 and shall have a char length not exceeding 1.5 inches (38 mm)

Exceptions:

1. Upholstered furniture belonging to the patient in sleeping rooms of nursing homes (Group I-2), provided that a smoke detector is installed in such rooms. Battery-powered, single-station smoke alarms shall be permitted.
2. Upholstered furniture in rooms or spaces protected by an approved automatic sprinkler system installed in accordance with Section 903.3.1.1.

❖ Most fires in occupancies classified in Group I-2 involve furniture, bedding, linens, draperies or decorative materials. Recent changes in the design philosophy of health care providers and improvements in the performance of synthetic fabrics have resulted in an increase in the quantity of furnishings in these occupancies. Because occupants of health care facilities are especially vulnerable in the event of fire, these requirements are aimed at reducing the contribution of the largest of these fuel packages to any fire hazard. This particular section focuses on the ignitability of this furniture by cigarettes. More specifically, any new upholstered furniture being introduced into nonsprinklered buildings must meet the requirements of NFPA 261, which studies the ability of upholstered furniture to resist ignition by a cigarette in terms of a maximum char length and the overall ignitability of the material. This standard provides the test method, but the code provides the limitation on char length. In this particular section, the maximum char length is set at 1.5 inches (38 mm). Performance in the test is not necessarily representative of real-life conditions. Instead, it serves as a tool for ranking the ignitability of one piece of furniture against another.

As noted, upholstered furniture in sprinklered buildings does not need to meet this criterion. Additionally, if upholstered furniture is located in a nursing home patient's room that is equipped with a smoke detector, the furniture is not required to meet the criteria of NFPA 261 for char length. These exceptions allow patients to bring a piece of furniture from home into their rooms without having to demonstrate compliance through testing.

803.5.2 Upholstered furniture heat release rate. Newly introduced upholstered furniture shall have limited rates of heat release when tested in accordance with ASTM E 1537 or NFPA 266.

1. The peak rate of heat release for the single upholstered furniture item shall not exceed 250 kW.

   Exception: Upholstered furniture in rooms or spaces protected by an approved automatic sprinkler system installed in accordance with Section 903.3.1.1.

2. The total energy released by the single upholstered furniture item during the first 5 minutes of the test shall not exceed 40 megajoules (MJ).

   Exception: Upholstered furniture in rooms or spaces protected by an approved automatic sprinkler system installed in accordance with Section 903.3.1.1.

❖ This section differs from Section 803.5.1 in that it is focusing on the overall fire size and heat release rate potential once furniture is already ignited. The two tests specified in this section measure the overall combustibility of the furniture. Again, if a building is sprinklered, this section does not apply. Upholstered furniture must be tested to either NFPA 266 or ASTM E 1537. Both of these tests use a full-scale furniture calorimeter. A full-scale calorimeter allows a representative piece of furniture to be burned and the products of combustion to be collected and analyzed. This also measures weight loss during burning to help determine the total energy released [see Figure 803.5.2 (1) for a representation of the test].

The acceptance criteria set by the code are as follows:

- Peak heat release is limited to 250 kW and
- Total energy release within the first 5 minutes cannot exceed 40 megajoules.

Limitations are placed on the maximum intensity and fire effluents produced by restricting the peak heat release and the amount of combustibles actually burned.

The total energy release of 40 megajoules could be translated to a steady fire of 133.33 kW (133 kJ/sec) for 5 minutes as follows:

$X \times 5$ minutes = 40 MJ

(5 minutes = 300 sec and 40 MJ = 40,000 kJ)

$X \times 300$ sec = 40,000 kJ

X = 40,000kJ/300 sec = 133.33 kJ/sec

X = 133.33 kJ/sec = 133.33 kW ~ heat release rate

Because fires in more realistic conditions do not burn steadily and vary in their characteristics, the criterion is given in the form of a peak heat release rate and total energy release. To provide a better understanding, if the fire were burning at the maximum peak heat release rate of 250 kW for the first 5 minutes, the total energy output would be 75 MJ, [250 kW (kJ/sec)] × 5 minutes (300 sec) = 250 kJ/sec × 300 sec = 75,000 kJ = 75 MJ, which is well over the criterion of 40 MJ.

A fire burning at a steady rate from start to finish is not realistic because fires must go through an initial growth stage before a peak heat release will be reached, followed by a decay phase; therefore, because a realistic fire will not burn at the peak heat release rate from the start of the fire, it is possible for a piece of furniture to have a peak heat release rate of 250 kW and still stay within the 40 MJ limitation. Figure 803.5.2(2) demonstrates the difference between a steady fire and a more realistic unsteady fire.

Generally, any new facilities will be sprinklered and the restrictions, as stated in the exceptions, will apply only to unsprinklered existing buildings.

803.5.3 Mattresses, heat-release rate. Newly introduced mattresses in Group I-2 occupancies shall have limited rates of heat release when tested in accordance with ASTM E 1590 or NFPA 267.

1. The peak rate of heat release for the mattress shall not exceed 250 kW.

   Exception: Mattresses in rooms or spaces protected by an approved automatic sprinkler system installed in accordance with Section 903.3.1.1.

2. The total energy released by the mattress during the first 5 minutes of the test shall not exceed 40 MJ.

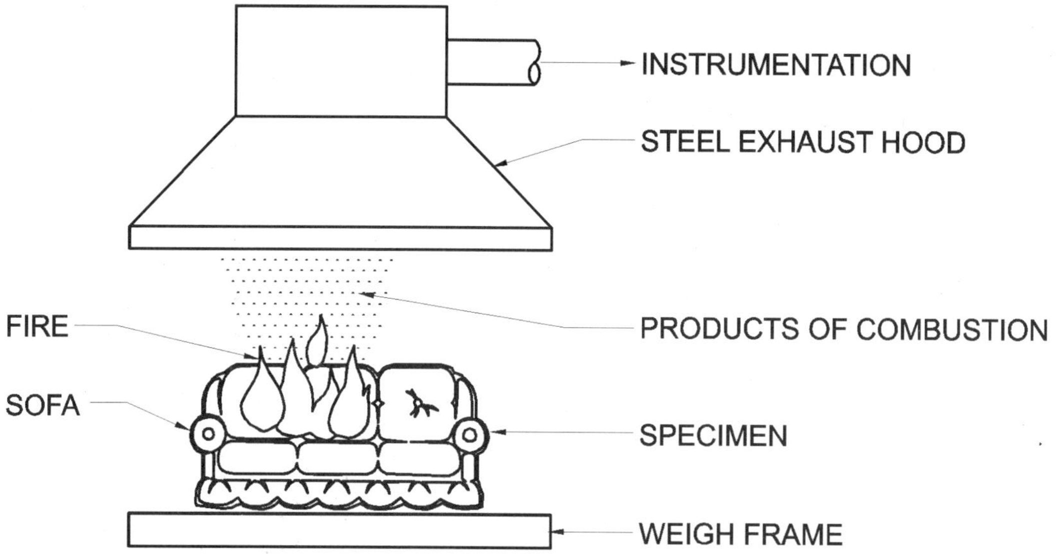

Figure 803.5.2(1)
TYPICAL FIRE TEST SETUP

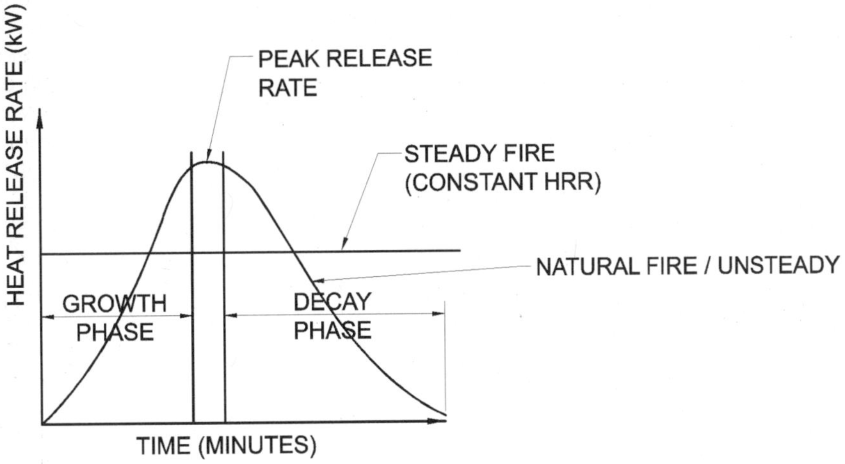

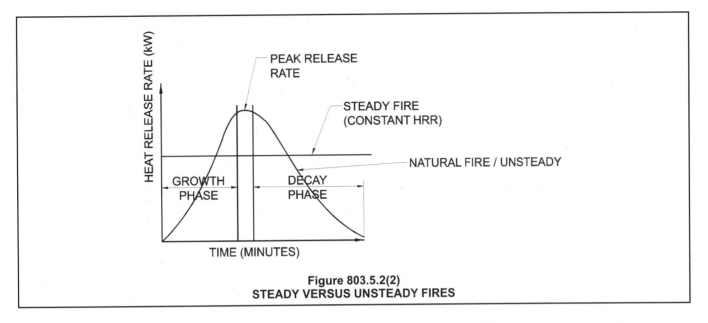

Figure 803.5.2(2)
STEADY VERSUS UNSTEADY FIRES

Exception: Mattresses in rooms or spaces protected by an approved automatic sprinkler system installed in accordance with Section 903.3.1.1.

❖ As noted, an occupant smoking in bed and initiating a mattress fire is a major fire hazard in nursing homes. This section, like Section 803.5.2, limits combustibility through a limitation of the peak heat release rate and on the total energy output in the first 5 minutes of burning. These limitations are the same as those for upholstered furniture, which have a maximum heat release rate of 250 kW and a total energy release of 40 MJ in 5 minutes. As with upholstered furniture, these restrictions are not applicable to sprinklered buildings.

The tests that determine the peak heat release rate and total energy release are specific to mattresses. These tests are detailed in NFPA 267 and ASTM E 1590. These tests, like that referenced in Section 803.5.2, make use of the calorimeter to measure the products of combustion. The major difference is the object being tested (see commentary, Section 803.5.2). Section 803.6.3 contains the same restrictions and testing requirements on combustibility for mattresses in Group I-1 board and care facilities.

803.5.4 Identification. Upholstered furniture shall bear the label of an approved agency, confirming compliance with the requirements of Sections 803.5.1 and 803.5.2.

❖ This section provides the fire code official with a valuable tool in evaluating and approving upholstered furniture in Group 1-2 hospitals and nursing homes. This provision has long had the support of upholstered furniture trade groups.

803.6 Group I-1, board and care facilities. The requirements in Sections 803.6.1 through 803.6.3 shall apply to board and care facilities classified in Group I-1.

❖ Board and care facilities house, in a supervised setting, more than 16 persons on a 24-hour basis because of age, mental disability or other reasons. These occupants are also considered more vulnerable than the general population and have had a history of starting fires in beds or upholstered furniture. There is also more of a concern than with a Group I-2 occupancy of occupants having the ability to purposely start a fire. This is less likely in an assisted living setting but more likely in a halfway house setting; therefore, limitations on combustibility of upholstered furniture and mattresses are also required. Generally, these items are the largest fire hazards in such buildings. Reducing ignitability and combustibility will significantly reduce the level of fire hazard. The requirements are slightly different than for Group I-2 occupancies in that different tests and performance criteria are required because of the nature of the hazards.

803.6.1 Upholstered furniture. Newly introduced upholstered furniture shall meet the requirements for Class I when tested in accordance with NFPA 260.

Exception: Upholstered furniture in rooms or spaces protected by an approved automatic sprinkler system.

❖ Similar to Section 803.5.1, this section is looking at the ignitability of furniture exposed to lighted cigarettes. The test required by this section is NFPA 260. Section 803.5.1 requires compliance with NFPA 261, which specifically looks at ignitability of a mock-up of upholstered furniture. NFPA 261 simply presents a method for study of the ignitability of furniture and a technique to measure the char length. NFPA 260, however, uses an overall classification system that looks at the components that may be found in upholstered furniture. Section 803.6.1 requires a Class I rating for upholstered furniture, meaning that no evidence of ignition occurred and specific limitations on char length must be met based on the specific component of the furniture. For example, the cover material would allow a 1.77 inch (45 mm) char length, whereas the interior fabric would be allowed only a 1.5 inch (38 mm) char length to be considered Class I.

Other components of upholstered furniture tested using NFPA 260 include the welt cords, filling/padding, decking material and barrier materials. Section 803.7.1 also requires compliance with NFPA 260 for Group I-3 detention and correction facilities.

As with Sections 803.5.1 through 803.5.3, when the room or space is sprinkler-protected, this test is not required. The code requires new facilities of this occupancy classification to be sprinklered. This section gives a methodology for existing facilities to follow when introducing furniture to an unsprinklered building.

Also note that Section 803.6 does not state a maximum heat release rate and total energy release limitation for upholstered furniture.

803.6.2 Mattresses. New mattresses shall have a char length not exceeding 2 inches (51 mm) where tested in accordance with DOC 16 CFR Part 1632.

Exception: Mattresses in rooms or spaces protected by an approved automatic sprinkler system.

❖ Sections 803.6.2 and 803.6.3 deal with the combustibility of mattresses. Section 803.6.2 focuses on initial ignition and the ability of a mattress to sustain a fire; Section 803.6.3 is focused primarily on the burning characteristics of mattresses.

This section sets a maximum char length of 2 inches (51 mm) when the mattress is tested under DOC 16 CFR, Part 1632. This test is actually a regulation for all mattresses sold within the United States. More specifically, it is part of the regulations governed by the Consumer Products Safety Commission (CPSC) under the Department of Commerce (DOC). Section 803.7.3 has the same reference and requirements for Group I-3 detention and correction facilities.

803.6.3 Mattresses, heat-release rate. Newly introduced mattresses in Group I-1 occupancies shall have limited rates of heat release when tested in accordance with ASTM E 1590 or NFPA 267.

1. The peak rate of heat release for the mattress shall not exceed 250 kW.

   Exception: Mattresses in rooms or spaces protected by an approved automatic sprinkler system.

2. The total energy released by the mattress during the first 5 minutes of the test shall not exceed 40 MJ.

   Exception: Mattresses in rooms or spaces protected by an approved automatic sprinkler system.

❖ This section is the same as Section 803.5.3 regarding the overall combustibility of mattresses. Combustibility is limited in the maximum peak heat release rate and total energy release over time. The concepts addressed in this section are also the same as those in Section 803.5.2 for upholstered furniture. The test methods are simply specific to mattresses, NFPA 266 and ASTM E 1590. Again, in sprinklered buildings these restrictions do not apply (see commentary, Sections 803.5.2 and 803.5.3).

803.7 Group I-3, detention and correction facilities. The requirements in Sections 803.7.1 through 803.7.6 shall apply to detention and correction facilities classified in Group I-3.

❖ These facilities have a higher likelihood of incendiary activities, and because of the restrictions on movement, the occupants are placed in a more vulnerable position than the general public; therefore, combustibility limitations are placed upon furniture and mattresses.

803.7.1 Upholstered furniture classification. Newly introduced upholstered furniture shall meet the requirements for Class I where tested in accordance with NFPA 260.

Exception: Upholstered furniture in rooms or spaces protected by an approved automatic sprinkler system installed in accordance with Section 903.3.1.1.

❖ Similar to Sections 803.5.1 and 803.6.1, this section considers the ignitability of furniture subjected to cigarette burns in unsprinklered buildings, requires upholstered furniture to meet a Class I rating in accordance with NFPA 260 and requires compliance with NFPA 261, which specifically looks at char length on the furniture. (see commentary, Section 803.5.1 and 803.6.1).

As with Section 803.6.1, when the room or space is sprinklered, compliance with NFPA 260 is not required. This flexibility assists in dealing with the difficulty in controlling these types of furnishings within buildings.

803.7.2 Upholstered furniture heat release rate. Newly introduced upholstered furniture shall have limited rates of heat release, as follows:

1. The peak rate of heat release for the single upholstered furniture item shall not exceed 250 kW.

   Exceptions:

   1. In Use Condition I, II and III occupancies, as defined in the *International Building Code*, upholstered furniture in rooms or spaces protected by approved smoke detectors that initiate, without delay, an alarm that is audible in that room or space.
   2. Upholstered furniture in rooms or spaces protected by an approved automatic sprinkler system installed in accordance with Section 903.3.1.1.

2. The total energy released by the single upholstered furniture item during the first 5 minutes of the test shall not exceed 40 MJ.

   Exception: Upholstered furniture in rooms or spaces protected by an approved automatic sprinkler system installed in accordance with Section 903.3.1.1.

❖ This section is identical to Section 803.5.2 except that there is one exception that allows a smoke detector within the room or space to activate without delay, instead of requiring compliance with one of the tests for the maximum peak heat release and energy output. This exception is specifically for Conditions I, II and III for Group I-3 occupancies. These three conditions represent correctional facilities that have a lower level of security than a Class IV and V and allow reasonably free movement throughout or between smoke compartments. Generally, the vulnerability of occupants in Group I-3 occupancies is related to the level of confinement rather than their physical abilities; therefore, if sufficient area for movement away from a hazard were possible, allowing this exception would be reasonable. Additionally, when the room or space is sprinklered, compliance with these tests is not required (see commentary, Section 803.5.2).

803.7.3 Mattresses, char length. Newly introduced mattresses shall have a char length not exceeding 2 inches (51 mm) when tested in accordance with DOC 16 CFR Part 1632.

Exception: Mattresses in rooms or spaces protected by an approved automatic sprinkler system installed in accordance with Section 903.3.1.1.

❖ This section is the same as Section 803.6.2 and focuses on initial ignition and the ability of a mattress to sustain a fire.

This section sets a maximum char length of 2 inches (51 mm) when tested under DOC 16 CFR Part 1632 (see commentary, Section 803.6.2).

803.7.4 Mattresses, heat release rate. Newly introduced mattresses in detention and correctional occupancies shall have limited rates of heat release when tested in accordance with ASTM E 1590 or NFPA 267, as follows:

1. The peak rate of heat release for the mattress shall not exceed 250 kW.

   Exception: Mattresses in rooms or spaces protected by an approved automatic sprinkler system installed in accordance with Section 903.3.1.1.

2. The total energy released by the mattress during the first 5 minutes of the test shall not exceed 40 mJ.

   Exception: Mattresses in rooms or spaces protected by an approved automatic sprinkler system installed in accordance with Section 903.3.1.1.

❖ This section is the same as Sections 803.5.3 and 803.6.3 and limits the overall combustibility of mattresses. (see commentary, Section 803.5.3).

803.7.5 Wastebaskets. Wastebaskets and other waste containers shall be of noncombustible or other approved materials.

❖ Although residents of Group I-3 occupancies are generally heavily monitored, there is an increased risk of incendiary activity in these facilities. Steps can be taken to reduce the impact of a fire once it has been ignited by limiting wastebaskets to noncombustible or other approved materials. Approved materials may be plastic with a fire-retardant or other method that will contain a fire to the basket and not allow the wastebasket itself to contribute to the fire. Generally, in larger areas of the facility, if a fire can be contained to a wastebasket it will not spread and will simply burn out.

803.7.6 Wastebasket lids. Waste containers with a capacity of more than 20 gallons (76 L) shall be provided with a lid of noncombustible or other approved material.

❖ This section goes one step further than 803.7.5 by requiring a lid on wastebaskets larger than a 20-gallon (76 L) capacity. This helps to ensure that the fire will stay contained to the wastebasket and will possibly reduce the fire size by restricting the combustion process.

SECTION 804
DECORATIVE VEGETATION

804.1 Natural cut trees. Natural cut trees, where permitted by this section, shall have the trunk bottoms cut off at least 0.5 inch (12.7 mm) above the original cut and shall be placed in a support device complying with Section 804.1.2.

❖ This section focuses on the hazards posed by vegetation within a building. The majority of this section deals specifically with fresh Christmas trees. Natural cut trees, such as Christmas trees, within buildings pose a significant fire threat to occupants if they are not properly cared for. Because of the symmetric way in which these trees are normally groomed, the large surface area of the foliage and the amount of airspace throughout the branches, fires have the potential to burn very efficiently and vigorously if the tree is dry.

According to the U. S. CPSC, approximately 400 fires annually are attributed to Christmas trees. These fires account for 10 deaths, 80 injuries and more than $15 million in property damage. In tests performed at the National Institute of Standards and Technology (NIST) in 1999 to better understand the severity of Christmas tree fires, eight Scotch pine Christmas trees were placed in a room at 73°F (23°C) at approximately 50-percent relative humidity for approximately three weeks. Seven of the eight trees were given no additional moisture; the eighth tree was watered according to local rules for trees within business occupancies. This required the tree to be cut in a certain manner and placed in a stand with at least a 2-gallon (7.6 L) capacity. The seven dry trees were ignited and burned intensely; however, the tree that had been watered continuously throughout the three weeks could not be ignited. The peak heat releases from the dry trees ranged from approximately 1,600 kW to 5,000 kW within about 50 to 80 seconds (0.83 min to 1.33 min) of ignition. (see Table 804.1 for a summary of results and Figure 804.1 for a graphical representation of Test No. 3). These test results demonstrate that the proper care for trees makes a significant difference on the fire hazard presented.

This section requires at least a $^1/_2$ inch (13 mm) of tree trunk to be removed above the original cut to make sure the tree has the optimum ability to absorb water to main-

TABLE 804.1
SUMMARY OF TREE PARAMETERS

TEST NO.	WEIGHT (kg)		HEIGHT (m)	WIDTH[a] (m)	MOISTURE CONTENT (%)
	Before Test	After Test			
1	17.2	6.8	2.6	1.7	30
2	15.9	8.2	2.7	1.3	27
3	20.0	6.8	2.3	1.7	30
4	9.5	5.0	2.5	1.2	30
5	19.1	8.6	2.5	1.7	28
6	12.7	7.7	2.5	1.1	32
7	18.6	7.7	3.1	1.5	25
8	28.1	28.1	2.7	1.4	36

From NIST Report FR 4010

a. The width is measured at the wildest point of the tree.

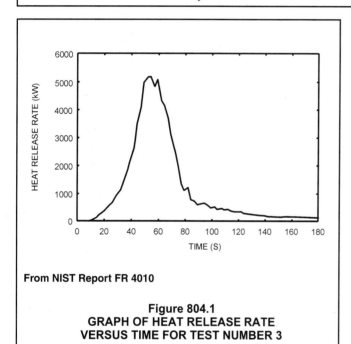

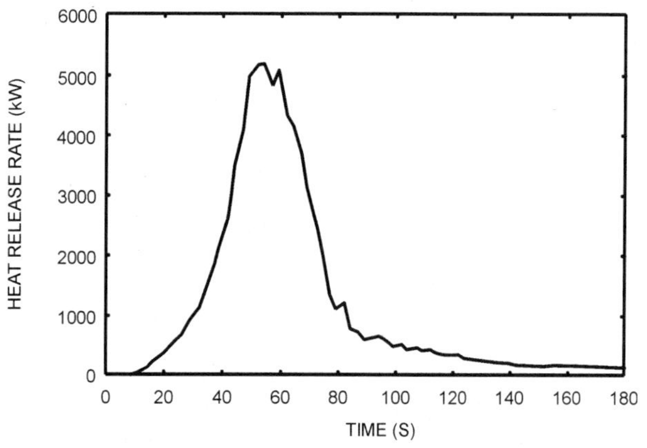

From NIST Report FR 4010

Figure 804.1
GRAPH OF HEAT RELEASE RATE VERSUS TIME FOR TEST NUMBER 3

tain a minimum level of moisture. This section also requires a support device (tree stand or equivalent) that meets the criteria of Section 804.1.2. This device is intended to ensure the correct amount of moisture is in contact with the tree.

The specifics of this section cannot realistically be monitored in all buildings within a jurisdiction. The focus must be on occupancies such as assembly or mercantile. All occupancies benefit, however, because there is now a method that a fire department could use to educate the general public on the treatment of Christmas trees. This tool can be helpful in fire prevention within a jurisdiction.

804.1.1 Restricted occupancies. Natural cut trees shall be prohibited in Group A, E, I-1, I-2, I-3, I-4, M, R-1, R-2 and R-4 occupancies.

Exceptions:

1. Trees located in areas protected by an approved automatic sprinkler system installed in accordance with Section 903.3.1.1 or 903.3.1.2 shall not be prohibited in Groups A, E, M, R-1 and R-2.
2. Trees shall be permitted within dwelling units in Group R-2 occupancies.

❖ Although trees that have been properly watered and handled are less hazardous, there is still a concern that they should not be located in certain occupancies. The occupancies listed in this section are those where the occupant load is high, occupants are vulnerable or the potential hazards that exist if the tree is not properly handled are too great. These occupancies include Groups A, E, I-1, I-2, I-3, I-4, M, R-1 and R-2.

There are exceptions for Group A, E, M, R-1 and R-2 occupancies when the trees are located in areas that are sprinklered in accordance with NFPA 13 or 13R, as applicable, and for trees within individual dwelling units in Group R-2. Essentially, the only occupancies that would be completely prohibited from having cut trees are institutional occupancies because of the vulnerability of the occupants and, in some cases, concerns for incendiary tendencies of some occupants.

804.1.2 Support devices. The support device that holds the tree in an upright position shall be of a type that is stable and that meets all of the following criteria:

1. The device shall hold the tree securely and be of adequate size to avoid tipping over of the tree.
2. The device shall be capable of containing a minimum 2-day supply of water.

3. The water level, when full, shall cover the tree stem at least 2 inches (51 mm). The water level shall be maintained above the fresh cut and checked at least once daily.

❖ This section is intended to ensure that the tree will not tip over, potentially into an ignition source, and that the support device is designed and used to keep the water supply to the tree at a useful level. More specifically, Item 1 requires the device to prevent the tree from tipping; Item 2 states that the device must be capable of holding a two-day water supply and Item 3 requires a minimum coverage of water to 2 inches (51 mm) above the bottom of the stem. The water level must be checked at least once daily to ensure an adequate supply remains.

The restrictions in Section 804.1.1 are in place because the provisions in Section 804.1.2 are generally difficult for a fire department to monitor. Having a sprinkler system or generally prohibiting natural cut trees in the higher risk occupancies provides a redundancy to deal with the potential hazards.

804.1.3 Dryness. The tree shall be removed from the building whenever the needles or leaves fall off readily when a tree branch is shaken or if the needles are brittle and break when bent between the thumb and index finger. The tree shall be checked daily for dryness.

❖ Lack of moisture is the primary problem with natural cut trees within buildings. This section describes a daily test to assist in evaluating whether the tree is considered too dry to remain in the building.

804.2 Obstruction of means of egress. The required width of any portion of a means of egress shall not be obstructed by decorative vegetation.

❖ Decorative vegetation is often placed in a location that does not normally accommodate combustibles. This section restricts locations to ensure that the vegetation does not block the egress width and increase the fire hazard in such areas. For example, this would likely prohibit trees in main lobby areas of a movie theater.

804.3 Open flame. Candles and open flames shall not be used on or near decorative vegetation. Natural cut trees shall be kept a distance from heat vents and any open flame or heat-producing devices at least equal to the height of the tree.

❖ This section addresses the primary ignition hazards associated with decorative vegetation. More specifically, candles should never be placed on or near Christmas trees. Also, heat sources such as heat vents may pose an ignition hazard in addition to being a source of airflow that could dry out a tree, making it more susceptible to ignition.

804.3.1 Electrical fixtures and wiring. The use of unlisted electrical wiring and lighting on decorative vegetation shall be prohibited.

❖ Decorations are quite often used year after year on Christmas trees. Some of this décor consists of lights that are not specifically listed and that tend to become very hot, potentially posing an ignition source. Additionally, this wiring may arc and create an ignition source. This section simply prohibits the use of unlisted wiring or lighting.

804.4 Artificial vegetation. Artificial decorative vegetation shall be flame resistant or flame retardant. Such flame retardance shall be documented and certified by the manufacturer in an approved manner.

❖ Much of the attention of Section 804 has been on natural vegetation, primarily Christmas trees. This particular section addresses artificial vegetation, which generally contains a high concentration of plastic. The goal, therefore, is to reduce the initial ignitability by treating artificial vegetation with flame retardants to increase flame resistance.

804.4.1 Electrical fixtures and wiring. The use of unlisted electrical wiring and lighting on decorative vegetation shall be prohibited. The use of electrical wiring and lighting on metal artificial trees shall be prohibited.

❖ This section is very similar to Section 804.3.1 with a focus upon artificial vegetation. Additionally, because of the shock and ignition potential resulting from a short or direct contact of part of the tree with one of the light sockets, electrical wiring and lighting is not allowed on metal trees.

SECTION 805
DECORATIONS AND TRIM

805.1 General. In occupancies of Groups A, E, I and R-1 and dormitories in Group R-2, curtains, draperies, hangings and other decorative materials suspended from walls or ceilings shall be flame resistant in accordance with Section 805.2 and NFPA 701 or be noncombustible.

In Groups I-1 and I-2, combustible decorations shall be flame retardant unless the decorations, such as photographs and paintings, are of such limited quantities that a hazard of fire development or spread is not present. In Group I-3, combustible decorations are prohibited.

❖ The requirements in this section apply to decorative materials installed in Group A, E, I and R-1 occupancies and dormitories in Group R-2 occupancies. The list of groups is based on the number of persons and the condition or capabilities of the occupants to evacuate quickly in an emergency. Decorative materials must be noncombustible or be maintained flame resistant in accordance with Section 805.2 and NFPA 701.

The requirements for Groups I-1 and I-2 become more stringent because of the nature of the occupants and activities in these types of facilities. Combustible materials in these facilities must be flame retardant, but a limited amount of decorations, such as photographs and paintings, need not be flame retardant. No amount of combustible decorations is permitted in Group I-3 facilities.

805.1.1 Noncombustible materials. The permissible amount of noncombustible decorative material shall not be limited.

❖ Where decorative materials are classified as noncombustible, they are presumed to contribute little, if any, to the growth and spread of fire; therefore, the quantity of noncombustible decorative materials is not limited.

805.1.2 Flame-resistant materials. The permissible amount of flame-resistant decorative materials shall not exceed 10 percent of the aggregate area of walls and ceilings.

Exception: In auditoriums of Group A, the permissible amount of flame-resistant decorative material shall not exceed 50 percent of the aggregate area of walls and ceiling where the building is equipped throughout with an automatic sprinkler system in accordance with Section 903.3.1.1, and where the material is installed in accordance with Section 803.4 of the *International Building Code.*

❖ Flame-resistant materials may contribute to fire growth and spread. More specifically, flame resistance does not mean that the materials will not burn, only that they are going to delay the onset of ignition. These materials are, therefore, limited to a maximum of 10 percent of the total wall and ceiling area of the space under consideration. Unlike incidental trim, decorative materials are not necessarily distributed evenly throughout the room. Additionally, consideration of the long-term maintenance of the materials, including possible periodic retreatment, should be taken into account.

There is an exception for Group A auditoriums that would allow 50-percent coverage of the walls and ceilings (instead of the limit of 10 percent) if the space is sprinklered in accordance with NFPA 13 and the material is applied in accordance with Section 803.3 of the IBC. The décor in Group A auditoriums is treated, therefore, more like wall coverings than decorative materials.

805.2 Acceptance criteria and reports. Where required to be flame resistant, decorative materials shall be tested by an approved agency and pass Test 1 or 2, as described in NFPA 701, or such materials shall be noncombustible. Reports of test results shall be prepared in accordance with NFPA 701 and furnished to the fire code official upon request.

❖ The standard test method to be used to evaluate the flame resistance of a material is NFPA 701, which contains two test methods. Test 1 is a small-scale test and Test 2 is a larger-scale test. NFPA 701 sets out the types of fabrics that should be tested using each method. This particular section only requires compliance with Test 1.

Test 1 is a method to assess the response of fabrics and other multilayer composites to moderate fire exposure. Essentially, Test 1 provides a mechanism to distinguish between materials that allow flames to spread quickly and those that do not.

Test 2 is a larger scale test that is more appropriate for certain materials such as coated fabric black-out linings. Most manufacturers would prefer to test materials under Test 1 because larger-scale tests are more costly.

Materials tested to NFPA 701 are not directly applied to buildings as finish materials, but instead are generally items such as shades, swags, curtains and other similar materials. These tests are used to determine whether flame-resistant materials propagate flame beyond the area exposed to the ignition source. They are not intended to indicate whether the material tested will resist the propagation of flame under fire exposures more extreme than the test conditions.

805.3 Foam plastic. Foam plastic used as interior trim shall comply with Sections 805.3.1 through 805.3.4.

❖ This section regulates foam plastic used as interior trim. Because the requirements in the following four subsections must be taken as a group, the commentary follows Section 805.3.4 and addresses all four requirements.

805.3.1 Density. The minimum density of the interior trim shall be 20 pounds per cubic foot (320 kg/m^3).

❖ See the commentary following Section 805.3.4.

805.3.2 Thickness. The maximum thickness of the interior trim shall be 0.5 inch (12.7 mm) and the maximum width shall be 8 inches (203 mm).

❖ See the commentary following Section 805.3.4.

805.3.3 Area limitation. The interior trim shall not constitute more than 10 percent of the aggregate wall and ceiling area of a room or space.

❖ See the commentary following Section 805.3.4.

805.3.4 Flame spread. The flame spread rating shall not exceed 75 where tested in accordance with ASTM E 84. The smoke-developed index shall not be limited.

❖ This section establishes that foam plastic materials may be used as interior finish materials when in compliance with the criteria set out in Sections 805.3.1 through 805.3.4. These provisions are similar to those found in Exception 2 to Section 803.2.1, which limits the thickness and area of coverage and establishes a minimum density for the foam. Minimum instead of maximum density is specified because the denser the plastic material, the less vigorously it will burn due to the smaller amount of air entrapped within the foam.

The exception to Section 803.5.1 limits the width to 4 inches (102 mm) whereas Section 805.3.2 limits the width to 8 inches (203 mm). Section 803.2.1 is specific to Group A occupancies. This section specifically calls out a numerical flame spread limitation of 75, which is essentially a Class B rating. This criterion alone is not sufficient because, when tested, many foam plastic materials have a flame spread index of less than 25 because they will melt and drop away from the test. This behavior results in an inaccurate indication of the hazard; therefore, the other criteria related to density, as well as dimensional limitations, are necessary. Compliance with Section 2604 of the IBC is also required. The requirements of Section 2604 are essentially identical to those found in this section. Section 2603.4.1.11 of the IBC does not require a thermal barrier for interior trim.

805.4 Pyroxylin plastic. Imitation leather or other material, consisting of or coated with a pyroxylin or similarly hazardous base, shall not be used in Group A occupancies.

❖ Use of pyroxylin plastics, also known as cellulose nitrate plastics, in Group A occupancies as imitation leather or other materials is strictly prohibited because of the normally high occupant loads of such occupancies. This type of plastic is the most hazardous of all plastics and will begin decomposition at temperatures starting at 300°F (149°F). This decomposition generates heat and will eventually raise the material to ignition temperature. In addition, during decomposition the material produces extremely hazardous vapors, such as carbon monoxide and also oxides of nitrogen. The oxides of nitrogen have a delayed effect but can sometimes result in fatalities several hours or days after exposure.

Because cellulose nitrate tends to become somewhat unstable and is easily ignitable, its use has generally declined. Specifically, use of cellulose nitrate for motion picture film was discontinued in 1951.

805.5 Trim. Material used as interior trim shall have a minimum Class C flame spread index and smoke-developed index. Combustible trim, excluding handrails and guardrails, shall not exceed 10 percent of the aggregate wall or ceiling area in which it is located.

❖ In occupancies of any group, unless otherwise noted in the code, the minimum classification of all trim must be at least Class C. Additionally, combustible trim may not exceed 10 percent of the area of the aggregate wall or ceiling area. The 10-percent calculation does not need to include handrails and guardrails. Section 806.2.4 addresses the same issues and provides the same 10-percent limitation. Although a Class C rating may be lower than the rating required for a particular building or facility, this quantity of combustible material will not significantly increase the fuel load (see Figure 803.2.1).

SECTION 806
INTERIOR FINISH AND DECORATIVE MATERIALS

806.1 General. The provisions of this section shall limit the allowable flame spread and smoke development of interior finishes and decorative materials in existing buildings based on location and occupancy classification.

Exceptions:

1. Materials having a thickness less than 0.036 inch (0.9 mm) applied directly to the surface of walls and ceilings.
2. Exposed portions of structural members complying with the requirements of buildings of Type IV construction in accordance with the *International Building Code* shall not be subject to interior finish requirements.

❖ This section specifically addresses existing buildings, whereas Sections 803, 804 and 805 address both new and existing buildings. This section is aimed primarily at the maintenance of the flame spread index, but also provides some guidance on foam plastics used as interior finish or trim.

There are two exceptions that would allow very thin material, less than or equal to 0.036 inch (0.9 mm), when applied directly to the wall and exposed structural members in Type IV construction. These exceptions are the same as those in the IBC.

806.1.1 Requirements based on occupancy. Interior finish and decorative materials shall be restricted by combustibility and flame resistance according to occupancy group in accordance with Table 806.3.

❖ This section refers to Table 806.3 for occupancy-specific flame spread index for walls and ceilings. The ratings are organized by occupancy classification because of the different occupant characteristics, hazards present and general societal expectations of safety. These required classifications are further broken down into groups as follows:

- Vertical exits and exit passageways.
- Exit access corridors and other exit ways.
- Rooms and enclosed spaces.

The way in which these groupings are addressed in the table demonstrates that the intent of the code is to regulate the more critical spaces more heavily. These ratings vary depending on whether or not a sprinkler system has been installed in the building.

806.1.2 Foam plastics. Cellular or foam plastics shall not be used as interior finish or trim.

Exceptions:

1. Cellular or foam plastic materials shall be permitted on the basis of fire tests that substantiate their combustibility characteristics for the use intended under actual fire conditions.
2. Cellular or foam plastic shall be permitted for trim not in excess of 10 percent of the wall or ceiling area, provided such trim is not less than 20 pounds per cubic foot (320 kg/m^3) in density, is limited to 0.5 inch (12.7 mm) in thickness and 8 inches (203 mm) in width, and complies with the requirements for Class A or B interior wall and ceiling finish except that the smoke rating shall not be limited.

❖ This section, like Section 805.3, allows foam plastic decorative trim with certain limitations on minimum density, maximum thickness and width and area of coverage. Additionally, a performance-oriented exception allows cellular or foam plastics as interior trim without limitation if the level of hazard under actual fire conditions can be demonstrated by testing to be low. It appears that this testing is intended to be full-scale testing in realistic applications.

806.1.3 Obstruction of means of egress. No decorations or other objects shall be placed to obstruct exits, access thereto, egress therefrom, or visibility thereof.

❖ Similar to Section 803.1.3, this section prohibits any decorations or other objects from obstructing the means of egress, including visibility.

806.2 Wall and ceiling finish. Interior wall and ceiling finishes shall be classified in accordance with Section 803 of the *International Building Code*. Such interior finishes shall be grouped in the following classes in accordance with their flame spread and smoke-developed index.

Class A: Flame spread index 0-25
Smoke-developed index 0-450

Class B: Flame spread index 26-75
Smoke-developed index 0-450

Class C: Flame spread index 76-200
Smoke-developed index 0-450

Exception: Materials, other than textiles, tested in accordance with Section 806.2.1.

❖ Interior finish and trim materials are required to have flame spread indexes as prescribed in Sections 806.1.1 and 806.3 and smoke-developed indexes as prescribed in Section 806.1.1. ASTM E 84 is the required test method for determining the flame spread and smoke-developed indexes that are required by this and other sections of the code. For example, Section 411.8 of the IBC requires interior finishes within special amusement buildings to be Class A. ASTM E 84 is intended to determine the relative burning behavior of materials on exposed surfaces, such as ceilings and walls, by visually observing the flame spread along the test specimen. Flame spread and smoke density are then reported. The test method may not be appropriate for materials that are not capable of supporting themselves, of being supported in the test tunnel or that drip, melt or delaminate. A distinction is made, therefore, for foam-plastic material and textile materials (see Sections 805.3, 806.1.2 and 806.2.3). ASTM E 84 establishes a flame spread index based on the area under a curve when the actual flame spread distance is plotted as a function of time. The code has divided the acceptable range of flame spread indexes (0-200) into three classes: Class A (0-25), Class B (26-75) and Class C (76-200). An indication of relative flame spread is as follows: an asbestos-cement board has a flame spread of zero, while a red oak species of wood has a flame spread of 100. Not to preclude more detailed information resulting from an ASTM E 84 test report, Figure 806.2 identifies the typical flame spread properties of certain building materials. Textile wall coverings and expanded vinyl wall coverings that have been tested in accordance with Section 806.2.3 (NFPA 265) are exempt from these requirements because they are allowed to be tested for contribution to fire when exposed to radiant conditions.

The exception allows compliance with NFPA 286 in lieu of ASTM E 84. This test is known as a "room-corner" fire test and is similar to that referenced for textile wall coverings in Section 806.2.5 (NFPA 265). In a room-corner fire test, a fire source is set in a corner of a small compartment with the material test sample placed upon the three walls. This test tends to give a more realistic indication of actual field performance. More detail will be provided in Section 806.2.1 regarding the test and the pass/fail criteria.

Material	Flame spread
Glass-fiber sound-absorbing blanks	15 to 30
Mineral-fiber sound-absorbing panels	10 to 25
Shredded wood fiberboard (treated)	20 to 25
Sprayed cellulose fibers (treated)	20
Aluminum (with baked enamel finish on one side)	5 to 10
Asbestos-cement board	0
Brick or concrete block	0
Cork	175
Gypsum board (with paper surface on both sides)	10 to 25
Northern pine (treated)	20
Southern pine (untreated)	130 to 190
Plywood paneling (untreated)	75 to 275
Plywood paneling (treated)	100
Carpeting	10 to 600
Concrete	0

Figure 806.2
TYPICAL FLAME SPREAD OF COMMON MATERIALS

806.2.1 Interior wall and ceiling finishes other than textiles. Interior wall or ceiling finishes, other than textiles, shall be permitted to be tested in accordance with NFPA 286. Finishes tested in accordance with NFPA 286 shall comply with Section 806.2.1.1.

❖ This section allows the use of NFPA 286 in lieu of ASTM E 84. NFPA 286 is known as a "room-corner" fire test. In this test, a fire source consisting of a wood crib is placed in the corner of a compartment. The materials tested are then placed upon the walls of the compartment. This generally provides a more realistic understanding of the hazards involved with the materials. Section 806.2.1.1 provides the pass/fail criteria as accepted by the IBC (see also Section 806.2.3 for more details on the concept of the room-corner test).

806.2.1.1 Acceptance criteria. During the 40 kW exposure, the interior finish shall comply with Item 1. During the 160 kW exposure, the interior finish shall comply with Item 2. During the entire test, the interior finish shall comply with Item 3.

1. During the 40 kW exposure, flames shall not spread to the ceiling.
2. During the 160 kW exposure, the interior finish shall comply with the following:
   2.1. Flame shall not spread to the outer extremity of the sample on any wall or ceiling.
   2.2. Flashover, as defined in NFPA 286, shall not occur.
3. The total smoke released throughout the NFPA 286 test shall not exceed 1,000 m^2.

❖ There are two levels of exposure during an NFPA 286 fire test in order to better represent a growing fire: 40 kW fire size for 5 minutes and 160 kW for 10 minutes. The 40 kW exposure represents the beginning of a fire where the initial spread is critical; therefore, the stated criteria is that the fire cannot spread to the ceiling. The 160 kW exposure is obviously a more intense fire situation and the criteria relates to preventing flashover (as defined by NFPA 286) and the extent of flame spread throughout the entire test assembly. There is also smoke production criteria of 1,000 m^2.

It should be noted that the flashover criteria for NFPA 286 and NFPA 265 is as follows:

- Heat release exceeds 1 MW.
- Heat flux at the floor exceeds 20 kW/m^2.
- Average upper layer temperature exceeds 1112° F (600° C).
- Flames exit the doorway.
- Autoignition of paper target on the floor occurs.

806.2.2 Stability. Interior finish materials regulated by this chapter shall be applied or otherwise fastened in such a manner that such materials will not readily become detached when subjected to a room temperature of 200°F (93°C) for not less than 30 minutes.

❖ Interior finishes are not to become detached for a minimum of 30 minutes under exposure to elevated temperatures [200°F (93°C)]. No standard test method has yet been developed to evaluate this requirement. Some sections of the IBC, however, do offer some additional guidance. For example, the performance of the method of attachment of finish materials in a fire-resistance-rated assembly will usually be adequately ensured by the construction details of the tested assembly. This criterion is necessary because when these materials tend to fall off of walls or the ceiling during a fire they may contribute to the fire and increase the hazard beyond what is typically expected.

806.2.3 Textiles. Textile wall coverings shall have a Class A flame spread rating when tested in accordance with ASTM E 84 and be protected by approved automatic sprinklers installed in accordance with Section 903.3.1.1 or 903.3.1.2 or the covering shall meet the criteria of Section 806.2.3.1 or 806.2.3.2 when tested in accordance with NFPA 265 using the product-mounting system, including adhesive, of actual use.

❖ This section is primarily intended to apply to carpet and carpet-like coverings that include materials having woven or nonwoven, napped, tufted, looped or similar surfaces. If not addressed, these materials can contribute extensively to a fire. This section requires that all such materials on walls and ceilings have a Class A flame spread index and be located in a sprinklered area in accordance with NFPA 13 or 13R. As an alternative to a Class A rating and sprinklers, NFPA 265 may be used.

NFPA 265 is known as a full-scale "room corner" fire test. Past research conducted with this kind of configuration has shown that flame spread indexes produced by ASTM E 84 may not reliably predict the fire behavior of textile wall coverings. A more reliable test procedure was developed at the University of California and involved the use of a room corner fire test with a gas diffusion burner(s). The research findings are described in a report from the University of California Fire Research Laboratory titled, "Room Fire Experiments of Textile Wall Coverings."

Based on the use of test Method A or test Method B from NFPA 265, the sample must meet the criteria of Section 806.2.3.1 or 806.2.3.2, respectively. This test helps to determine the contribution of textile wall coverings to overall fire growth and spread in a compartment fire. This test also exposes the textile to an ignition source while mounted on the wall. The difference between Method A and Method B is that in Method A the materials are placed on only two walls whereas in Method B the sample is placed on three walls. These textiles are exposed to a prescribed heat release of 40 kW for 5 minutes, which is then followed by a heat release rate of 150 kW for 10 minutes.

806.2.3.1 Method A. When using method A, flame shall not spread to the ceiling during a 40 kW exposure. During the 150 kW exposure, all of the following criteria shall be met:

1. Flame shall not spread to the outer extremity of the sample on the 8-foot by 12-foot (2438 mm by 3657 mm) wall.
2. The specimen shall not burn to the outer extremity of the 2-foot-wide (610 mm) samples mounted vertically in the corner of the room.
3. Burning droplets that are judged by the fire code official to be capable of igniting the textile wall covering or that persist in burning for 30 seconds or more shall not be formed and dropped to the floor.

4. Flashover, as defined in NFPA 265, shall not occur.
5. The maximum instantaneous net peak rate of heat release shall not exceed 300 kW.

❖ This section contains a set of pass/fail criteria for test Method A to be used along with NFPA 265 in evaluating performance of a textile. First, the flame cannot spread to the ceiling when the textile is exposed to the 40 kW heat flux. Additionally, at the 150 kW heat flux, the following criteria must be me:

- The fire cannot reach the outer areas of the 8-foot by 12-foot (2438 mm by 3657 mm) wall.
- The specimen cannot burn to the outer portion of the 2-foot (610 mm) wide sample, which is mounted vertically in the room. Burning droplets that are determined to pose further ignition hazards or burn for more than 30 seconds would be considered a failure.
- Flashover, as defined by NFPA 265 (see commentary, Section 806.2.1.1), should not occur.
- A maximum peak instantaneous heat release cannot be greater than 300 kW.

806.2.3.2 Method B. When using method B, flame shall not spread to the ceiling during the 40 kW exposure. During the 150 kW exposure, all of the following criteria shall be met:

1. Flame shall not spread to the outer extremity of the sample on the 8-foot by 12-foot (2438 mm by 3657 mm) wall.
2. Flashover, as defined in NFPA 265, shall not occur.

❖ The criteria for passing test Method B are less restrictive because the sample is mounted on three walls instead of two, which has the effect of increasing the intensity due to the configuration and increased fuel load. First, the flame cannot spread to the ceiling when the textile is exposed to the 40 kW heat flux. Additionally, at the 150 kW heat flux, the following criteria must be met:

- Flashover, as defined by NFPA 265 (see commentary Section 806.2.1.1), must not occur.
- The fire cannot reach the outer areas of the 8-foot by 12-foot (2438 mm by 8657 mm) wall.

806.2.4 Trim and incidental finish. Interior wall and ceiling finish not in excess of 10 percent of the aggregate wall and ceiling areas of any room or space shall be permitted to be Class C materials.

❖ This section, similar to Section 805.5 for trim, allows 10 percent of the wall surface to be covered with Class C-rated materials. Again, because of the small area of coverage, the contribution to the fire load and hazard is minimal.

806.2.5 Expanded vinyl wall coverings. Expanded vinyl wall coverings shall comply with the requirements for textile wall and ceiling materials and their use shall comply with Section 806.2.2.

Exception: Expanded vinyl wall or ceiling coverings complying with Section 806.2.1 shall not be required to comply with Sections 806.2 and 806.3.

❖ Expanded vinyl wall coverings have a woven textile backing and the flame spread indexes produced by ASTM E 84 may not reliably predict its fire behavior; therefore, compliance with the section addressing textiles (Section 806.2.3) is necessary (see commentary, Section 806.2.3). The commentary for Section 806.2.3 also applies to expanded vinyl wall coverings. There is a new exception that allows the use of NFPA 286 for testing (see Section 806.2.1 for discussion of NFPA 286).

806.2.6 Fire-retardant coatings. The required flame spread or smoke-developed classification of surfaces shall be permitted to be achieved by application of approved fire-retardant coatings, paints or solutions to surfaces having a flame spread rating exceeding that permitted. Such applications shall comply with NFPA 703 and the required fire-retardant properties shall be maintained or renewed in accordance with the manufacturer's instructions.

❖ Many times, fire retardants can be used to reduce the flame spread rating of a material that normally has a rating higher than permissible. This section specifically allows this activity but also recognizes that the fire retardants must be reapplied in the future to maintain this equivalent flame spread rating (see commentary, Section 803.1.2).

806.3 Wall and ceiling finish requirements. Interior wall and ceiling finish shall have a flame spread index not greater than that specified in Table 806.3 for the group and location designated. Interior wall and ceiling finish materials, other than textiles, tested in accordance with NFPA 286 and meeting the acceptance criteria of Section 806.2.1.1, shall be permitted to be used where a Class A classification in accordance with ASTM E 84 is required.

❖ The requirements for flame spread indexes for interior finish materials applied to walls and ceilings are contained in Table 806.3. The referenced test for determining flame spread indexes is ASTM E 84, which establishes a relative measurement of flame spread across the surface of the material. The classifications used in Table 806.3 are defined in Section 806.2. (see the commentary to Section 806.2 for additional information on the uses and limitations of the test procedure). This section also allows the use of NFPA 286 as noted in the exception to Sections 806.2 and 806.3. Section 806.3 states that materials that pass this test can be used when a Class A classification is required. In other words, the materials can be used as a replacement for all categories, as Class A is the most restrictive (see commentary on Section 806.2.1 and 806.2.1.1 for more detail on NFPA 286).

TABLE 806.3
INTERIOR WALL AND CEILING FINISH REQUIREMENTS BY OCCUPANCY[k]

GROUP	SPRINKLERED[m]			NONSPRINKLERED		
	Vertical exits and exit passageways[a, b]	Exit access corridors and other exitways	Rooms and enclosed spaces[c]	Vertical exits and exit passageways[a, b]	Exit access corridors and other exitways	Rooms and enclosed spaces[c]
A-1[l] & A-2	B	B	C	A	A[d]	B[e]
A-3[f, l] A-4, A-5	B	B	C	A	A[d]	C
B, E, M, R-1, R-4	B	C	C	A	B	C
F	C	C	C	B	C	C
H	B	B	C[g]	A	A	B
I-1	B	C	C	A	B	B
I-2	B	B	B[h, i]	A	A	B
I-3	A	A[j]	C	A	A	B
I-4	B	B	B[h, i]	A	A	B
R-2	C	C	C	B	B	C
R-3	C	C	C	C	C	C
S	C	C	C	B	B	C
U	No Restrictions			No Restrictions		

For SI: 1 inch = 25.4 mm, 1 square foot = 0.0929 m^2.

a. Class C interior finish materials shall be permitted for wainscotting or paneling of not more than 1,000 square feet of applied surface area in the grade lobby where applied directly to a noncombustible base or over furring strips applied to a noncombustible base and fireblocked as required by Section 803.3 of the *International Building Code.*

b. In vertical exits of buildings less than three stories in height of other than Group I-3, Class B interior finish for unsprinklered buildings and Class C for sprinklered buildings shall be permitted.

c. Requirements for rooms and enclosed spaces shall be based upon spaces enclosed by partitions. Where a fire-resistance rating is required for structural elements, the enclosing partitions shall extend from the floor to the ceiling. Partitions that do not comply with this shall be considered as enclosing spaces and the rooms or spaces on both sides shall be considered as one. In determining the applicable requirements for rooms and enclosed spaces, the specific occupancy thereof shall be the governing factor regardless of the group classification of the building or structure.

d. Lobby areas in Group A-1, A-2 and A-3 occupancies shall not be less than Class B.

e. Class C interior finish materials shall be permitted in Group A occupancies with an occupant load of 300 persons or less.

f. For churches and places or worship, wood used for ornamental purposes, trusses, paneling, or chancel furnishing shall be permitted.

g. Class B required where building exceeds two stories.

h. Class C interior finish materials shall be permitted in administrative spaces.

i. Class C interior finish materials shall be permitted in rooms with a capacity of four persons or less.

j. Class B materials shall be permitted as wainscoting extending not more than 48 inches above the finished floor in exit access corridors.

k. Finish materials as provided for in other sections of this code.

l. Motion picture screens shall comply with Section 803.2.2.

m. Applies when the vertical exits, exit passageways, exit access corridors or exitways, or rooms and spaces are protected by a sprinkler system installed in accordance with Section 903.3.1.1 or 903.3.1.2.

❖ This table prescribes the minimum requirements for interior finishes applied to walls and ceilings; therefore, the use of a Class A material in an area that requires a minimum Class B material is always allowed. Likewise, when the table requires Class C materials, Classes A and B can also be used.

The requirements are based on the use of the space. To determine the applicable criteria, first determine whether the space is an exit passageway or vertical exit (stairway), a corridor, a room or an enclosed space. Interior finishes in spaces that are not separated from a corridor (for example, waiting areas in business or health care facilities) must comply with the requirements for a corridor space. As shown in the table, the code places a higher emphasis on the allowable flame spread index for exits than for enclosed rooms because of the critical nature and relative importance that is placed on maintaining the integrity of exits to evacuate the building.

Numerous notes amend the basic requirements of Table 806.3. Notes a through m apply only where they are specifically referenced in the table.

Note a.

A limited amount of combustible wainscotting or other paneling material does not appreciably reduce the level of safety of an exit element; therefore, up to 1,000 square feet (93 m^2) of Class C wainscotting or paneling is permitted by Note a in a grade-level lobby used as an exit element when applied in accordance with Section 803.4 of the IBC.

Note b.

This note allows the reduction of flame spread ratings in vertical exits in buildings less than three stories high, not including Group I-3 occupancies. The time required to exit a building that is less than three stories is generally shorter than it is for buildings with more than three stories; thus, the class of material allowed in vertical exits is not as critical.

Note c.

Because the intended use of certain rooms in buildings is sometimes more hazardous than the group classification for the overall building, the finish classification must be determined by the group and occupancy classification for the room or area. Additionally, rooms or enclosed spaces not properly separated from one another must be looked at as a single space.

Note d.

This note allows lobbies in Group A-1, A-2 and A-3 occupancies to be Class B instead of Class A. This is likely a result of the low fire load in such areas.

Note e.

Note e recognizes that with a relatively small number of occupants, egress is accomplished more quickly and the activities tend to be more structured and manageable than with a large number of occupants. When the design occupant load is 300 or less in rooms or spaces of Group A-1 and A-2 occupancies, the interior finish materials may be Class C instead of Class B for rooms and enclosed spaces.

Note f.

Churches and places of worship are generally open spaces. The occupants, because of the nature of the activities in these assemblies, are orderly. For this reason, wood, which is a combustible material, is allowed extensively as a finish material without restriction.

Note g.

The time required for exiting increases with the number of stories of the structure. Interior finishes can play a major role in fire spread within a structure. Because of the materials found within a high-hazard occupancy, the rate of fire spread can accelerate much more rapidly than in other occupancies. To abate the hazard of rapid fire growth, Note g places a further restriction of a minimum of a Class B rating on interior finishes of enclosed spaces within sprinklered buildings when the building exceeds two stories.

Note h.

Spaces that are used as offices have a low occupant load and the activity in those spaces is generally not very hazardous; therefore, Class C flame spread ratings are allowed in administrative spaces.

Note i.

Rooms with low occupant loads (four or less) pose a low risk and are quickly evacuated; therefore, Class C finish materials are appropriate.

Note j.

Note j permits interior finish materials, such as wainscotting, that are a maximum height of 48 inches (1219 mm) above the floor in corridors to be Class B. This reduction is based on full-scale fire research that demonstrated that Class B wall finish used on the lower 4 feet (1219 mm) of a corridor wall is not likely to spread fire because it primarily spreads on the ceilings and upper walls before it would affect the lower portion of the walls.

Note k.

This note is generally referenced in the title of Table 806.3 and refers to other sections of the code for material class restrictions for specific groups. More specifically, other tests or restrictions may also apply.

Note l.

This note is specific to the code for motion picture movie screens. They are not considered wall finish but their combustibility and flame spread are still a significant concern. Section 803.2.3 contains specific requirements.

Note m.

Sprinklered facilities provide more protection to the occupants than unsprinklered facilities. In many instances, less restrictive materials may be used as shown in Table 806.3.

Bibliography

The following resource materials are referenced in this chapter or are relevant to the subject matter addressed in this chapter.

ASTM E 84–98, *Test Method for Surface Burning Characteristics of Building Materials.* West Conshohocken, PA: ASTM International, 1998.

ASTM E 1537–99, *Test Method for Fire Testing of Real Scale Upholstered Furniture Items.* West Conshohocken, PA: ASTM International, 1999.

ASTM E 1590–99, *Test Method for Fire Testing of Real Scale Mattresses.* West Conshohocken, PA: ASTM International, 1999.

Babraushas, V., "Burning Rates," Section 3, Chapter 1, *SFPE Handbook of Fire Protection Engineering.* Boston, MA: Society of Fire Protection Engineers, 1995.

Christian, W. J., and T. E. Waterman. "Flame Spread in Corridors: Effects of Location and Area of Wall Finish. *NFPA Fire Journal*, July 1971.

Cohn, B. M., "Plastics and Rubber", Section 3, Chapter 10, *Fire Protection Handbook.* Quincy, MA: National Fire Protection Association, 1991.

"CPSC Releases Holiday Safety Tips for Avoiding Fires and Injuries," Press Release No. 99-029. Washington, DC: Consumer Product Safety Commission, November 30, 1998.

DeLauter, L., J. Lee, G. Roadarmel and D.W. Stroup. "Scotch Pine Christmas Tree Fire Tests," FR4010. Gaithersburg, MD: National Institute of Standards and Technology, US Department of Commerce, 1999.

DOC 16 CFR 1632–90, *Standard for the Flammability of Mattresses and Mattress Pads (FF4–72, Amended).* Washington, DC: Superintendent of Documents, 1990.

Fisher, F. L., B. McCracken and R. B. Williamson. "Room Fire Experiments of Textile Wall Coverings, Final Report of All Materials Tested Between March 1985 and January 1986." *Service to Industry Report.* Berkley, CA: California University, ES-7853, Report 86-2, p.142, March 1986.

"Flammable Decorations, Lack of Exits Create Tragedy at Cocoanut Grove," *Fire Engineering*, August 1977.

IBC-2003, *International Building Code.* Falls Church, VA, International Code Council, 2003.

"Looking Back at the Cocoanut Grove," *Fire Journal.* November, 1982.

NFPA 260–98, *Methods of Tests and Classification System for Cigarette Ignition Resistance of Components of Upholstered Furniture.* Quincy, MA: National Fire Protection Association, 1998.

NFPA 261–98, *Method of Test for Determining Resistance of Mock-up Upholstered Furniture Material Assemblies to Ignition by Smoldering Cigarettes.* Quincy, MA: National Fire Protection Association, 1998.

NFPA 265–98, *Fire Tests for Evaluating Room Fire Growth Contribution of Textile Wall Coverings.* Quincy, MA: National Fire Protection Association, 1998.

NFPA 266–98, *Method of Test for Fire Characteristics of Upholstered Furniture Exposed to Flaming Ignition Source.* Quincy, MA: National Fire Protection Association, 1998.

NFPA 267–98, *Method of Test for Fire Characteristics of Mattress and Bedding Assemblies Exposed to Flaming Ignition Source.* Quincy, MA: National Fire Protection Association, 1998.

NFPA 701–99, *Methods of Fire Tests for Flame-Resistant Textiles and Films.* Quincy, MA: National Fire Protection Association, 1999.

Quintiere, J. "Surface Flame Spread," Section 2, Chapter 14. *SFPE Handbook of Fire Protection Engineering.* Boston, MA: Society of Fire Protection Engineers, 1995.

UL 1975–96, *Fire Tests for Foamed Plastics Used for Decorative Purpose.* Northbrook, IL: Underwriters Laboratories, Inc., 1996.

Chapter 9: Fire Protection Systems

General Comments

The requirements of Chapter 9 are just one aspect of the overall fire protection system of a building or structure. All fire protection requirements contained in the code must be considered as a package or overall system. Noncompliance with any part of the overall system may cause other parts of the system to fail, which may result in an increased loss of life and property from the reduced level of protection.

Every effort must be made to verify the proper design and installation of a given fire protection system, especially those that result in construction alternatives and other code trade-offs.

The requirements in Chapter 9 are active fire safety provisions. They are directed at containing and extinguishing a fire once it has erupted. This chapter parallels and duplicates much of Chapter 9 in the *International Building Code*® (IBC®). The *International Fire Code*® (IFC®), however, contains additional specific provisions that are applicable only to existing buildings. The IFC also contains periodic testing criteria that are not duplicated in the IBC. Proper testing, inspection and maintenance of the various systems are critical to establish the reliability of the system. Additionally, Chapter 9 references and adopts numerous NFPA standards, including the acceptance testing criteria within the standard. The referenced standards will also contain more specific design and installation criteria than are found in this chapter. As noted in Section 102.6, where differences occur between code requirements and the referenced standard, the code provisions apply.

Purpose

Fire protection systems may serve one or more purposes in providing adequate protection from fire. Chapter 9 prescribes the minimum requirements for an active system or systems of fire protection to perform the following functions: to detect a fire; to alert the occupants or fire department of a fire emergency; to control smoke and to control or extinguish the fire. Generally, the requirements are based on the occupancy, the height and the area of the building, because these are the factors that most affect fire-fighting capabilities and the relative hazard of a specific space or area.

SECTION 901 GENERAL

901.1 Scope. The provisions of this chapter shall specify where fire protection systems are required and shall apply to the design, installation, inspection, operation, testing and maintenance of all fire protection systems.

❖ Chapter 9 contains requirements for fire protection systems that may be installed or located in a building. These include automatic suppression systems; standpipe systems; fire alarm and detection systems; smoke control systems; smoke and heat vents and portable fire extinguishers. Besides indicating the conditions under which respective systems are required, this chapter contains the design, installation, maintenance, testing and operational criteria for fire protection systems. While the chapter requires proper maintenance for the reliability of the systems, the actual maintenance provisions (periodic testing, inspections and maintenance) may be contained in one of the referenced NFPA standards.

Chapter 9 is intended to apply to buildings of new construction or when deemed applicable because of a change in occupancy or an addition unless specifically indicated to be applicable to existing buildings only.

901.2 Construction documents. The fire code official shall have the authority to require construction documents and calculations for all fire protection systems and to require permits be issued for the installation, rehabilitation or modification of any fire protection system. Construction documents for fire protection systems shall be submitted for review and approval prior to system installation.

❖ The construction documents and related calculations for all fire protection systems must be reviewed before a permit is issued. The review is done to determine compliance with the code requirements and the referenced standards.

Typical shop drawings for fire protection systems are usually not prepared during the initial submittal for a construction permit. Many jurisdictions require a separate submittal and issue a separate permit to the contractor installing the system (see Section 901.3). Factors such as classification of the hazard, amount of agent or water supply available and the design criteria, including the density or concentration to be achieved by the system, are to be included with the shop drawings. Specific equipment data sheets identifying sprinklers, pipe dimensions, power requirements for smoke detectors, etc., should also be included with the submittal in addition to any required calculations.

901.2.1 Statement of compliance. Before requesting final approval of the installation, where required by the fire code official, the installing contractor shall furnish a written statement to the fire code official that the subject fire protection system has been installed in accordance with approved plans and has been tested in accordance with the manufacturer's specifications and the appropriate installation standard. Any deviations from the design standards shall be noted and copies of the approvals for such deviations shall be attached to the written statement.

❖ A certificate or other approved written statement must be submitted to the fire code official with the proper documentation from the installing contractor that the fire protection system has been installed in accordance with the requirements of the code. The certificate should also indicate that all required inspections and tests of the system have been conducted at the time of application for a certificate of occupancy.

The written statement is to indicate that the system has been installed in accordance with code requirements. As previously stated, contractors may have certificates that specify the criteria of the referenced standards since many of them contain sample certificates. While such certificates may be used, the contractor is required to certify that the system complies with the provisions of the code, which in some instances may vary from the referenced standards.

901.3 Permits. Permits shall be required as set forth in Section 105.6 and 105.7.

❖ Section 105 requires permits of two separate but related types. Section 105.6 requires an operational permit. These permits are required to conduct certain types of businesses or hazardous operations that require a higher level of scrutiny from the fire code official. The second type of permit, required by Section 105.7, is the construction permit, which is required for the installation and modification of all fire protection systems.

901.4 Installation. Fire protection systems shall be maintained in accordance with the original installation standards for that system. Required systems shall be extended, altered, or augmented as necessary to maintain and continue protection whenever the building is altered, remodeled or added to. Alterations to fire protection systems shall be done in accordance with applicable standards.

❖ This section emphasizes the principle that systems installed and maintained in compliance with the codes and standards in effect at the time they were placed in service must remain operational at all times. It is not the intent of the code to require existing systems that are otherwise not being altered to comply with current code and standard requirements.

901.4.1 Required fire protection systems. Fire protection systems required by this code or the *International Building Code* shall be installed, repaired, operated, tested and maintained in accordance with this code.

❖ Fire protection systems that are required by Chapter 9 or by another section of either the IBC or the IFC must be considered as required systems. The fire protection system is an integral component of the protection features of the building and must be properly installed, repaired, operated, tested and maintained in accordance with the code. Improperly installed or maintained systems can fail to provide the anticipated protection and, in fact, create a hazard in itself.

Although the code may not require a fire protection system for a specific building or portion of a building because of its occupancy, the fire protection system would still be considered a required system if some other code trade-off, exception or reduction was taken based on the installation of that fire protection system. For example, a typical small office building may not require an automatic sprinkler system solely because of its Group B occupancy classification. However, if an exit access corridor fire-resistance rating reduction is taken as allowed by Table 1016.1 for buildings equipped throughout with an NFPA 13 sprinkler system, that sprinkler system is now considered a required system.

901.4.2 Nonrequired fire protection systems. Any fire protection system or portion thereof not required by this code or the *International Building Code* shall be allowed to be furnished for partial or complete protection provided such installed system meets the requirements of this code and the *International Building Code*.

❖ A building owner or designer may elect to install a fire protection system that is not required in the code. Even though such a system is not required, it must comply with the applicable requirements of Chapter 9. This requirement is based on the concept that any fire protection system not installed as required by the code is lacking because it could give a false impression of properly installed protection.

For example, if a building owner chooses to install sprinkler protection in a certain area and that protection is not required by any provisions of the code, the system must be installed in accordance with NFPA 13 and other applicable requirements of the code, such as water supply and supervision. The extent of the protection provided would not be regulated.

901.4.3 Additional fire protection systems. In occupancies of a hazardous nature, where special hazards exist in addition to the normal hazards of the occupancy, or where the fire code official determines that access for fire apparatus is unduly difficult, the fire code official shall have the authority to require additional safeguards. Such safeguards include, but shall not be limited to, the following: automatic fire detection systems, fire alarm systems, automatic fire-extinguishing systems, standpipe systems, or portable or fixed extinguishers. Fire protection equipment required under this section shall be installed in accordance with this code and the applicable referenced standards.

❖ This section allows the fire code official to require fire protection safeguards beyond the minimum requirement of Chapter 9 when warranted by potential unsafe conditions. The provisions of the code cannot anticipate every occupancy condition. Hazardous material occupancies or buildings with limited fire department access are potentially a greater hazard to both building occupants and fire fighters. Any additional safeguards should be those needed to abate potential hazards. This section does not give the fire code official the right to require additional systems without cause. All additional safeguards regulated by this section must be considered required systems.

901.4.4 Appearance of equipment. Any device that has the physical appearance of life safety or fire protection equipment but that does not perform that life safety or fire protection function, shall be prohibited.

❖ All required or provided life safety or fire protection-related equipment must continue in use and be maintained to meet the requirements in effect at the time of the original installation. Nonrequired equipment that has been taken out of service or cannot function as intended must be dismantled and removed to prevent creating a false impression of protection.

901.5 Installation acceptance testing. Fire detection and alarm systems, fire-extinguishing systems, fire hydrant systems, fire standpipe systems, fire pump systems, private fire service mains and all other fire protection systems and appurtenances thereto shall be subject to acceptance tests as contained in the installation standards and as approved by the fire code official. The fire code official shall be notified before any required acceptance testing.

❖ Fire protection systems must pass an acceptance test to determine that the system will operate as required by the code. Acceptance tests are usually part of the final inspection procedures. The referenced standards contain specific acceptance test procedures. In most instances, the acceptance test procedures require 100-percent operation of the testable system components to determine that they are operational and functioning as required. Often the design professional may require additional testing that may be beyond the code requirements to verify that the system operates as designed.

The inclusion of the requirement for acceptance tests in the code is not intended to assign responsibility for witnessing the tests. The responsibility to witness the acceptance test is an administrative issue that each municipality must address. Because the acceptance test is critical during design and construction and is a requirement of occupancy, the requirement is located in the code. The section also clarifies that it is the owner's responsibility to conduct the test and the role of the fire code official to witness the test. Typically, the owner will assign the responsibility of conducting the test to the installing contractor.

901.5.1 Occupancy. It shall be unlawful to occupy any portion of a building or structure until the required fire detection, alarm and suppression systems have been tested and approved.

❖ Partial occupancy of any structure should not be permitted unless all fire protection systems for the occupied areas have been tested and approved. Even so, the code assumes that full protection for all areas will be provided as quickly as possible. The installation of many fire protection systems and the associated code trade-offs permitted for a given occupancy assume complete building protection and not just in the occupied areas. All partial occupancy conditions are subject to the final approval of the building official. Section 105.3.4 allows the fire code official to issue conditional occupancy permits.

901.6 Inspection, testing and maintenance. Fire detection, alarm and extinguishing systems shall be maintained in an operative condition at all times, and shall be replaced or repaired where defective. Nonrequired fire protection systems and equipment shall be inspected, tested and maintained or removed.

❖ Adequate maintenance, inspection and periodic testing of all fire protection systems, equipment and devices ensures the systems are ready to perform their intended functions should fire occur.

An inspection consists of a visual check of a system or device to verify that it is in operating condition and free from defects or damage. Indicating valves, gauges and indicator lamps are a few of the features required by the codes to facilitate this activity. Obvious damage and the general condition of the system, particularly the presence of corrosion, should always be noted and recorded. Partially because they are less detailed, inspections are conducted more frequently than tests and maintenance. Because special knowledge and tools are not required, inspections may be performed by any reasonably competent person.

Periodic tests following standardized methods are intended to confirm the results of inspections, determine that all components function properly and that systems meet their original design specifications. Tools, devices or equipment are usually required for these tests.

Because tests are more detailed than inspections, they are usually conducted only once or twice per year in most cases. Some tests, however, may be required as frequently as bimonthly or quarterly (for example, some fire alarm system equipment) or as infrequently as 5-, 6- or 12-year intervals (for example, portable fire extinguisher hydrostatic tests). Since specialized knowledge and equipment are required, testing is usually done by technicians or specialists trained in the proper conduct of the test methods involved.

Periodic maintenance keeps systems in good working order and may be used to repair damage or defects discovered during inspections or testing. Specialized tools and training are required to perform maintenance. Only properly trained technicians or specialists should perform required periodic maintenance. Most maintenance is required only as needed, but many manufacturers suggest or require regular periodic replacement of parts subject to wear or abuse.

Nonrequired fire protection systems, where installed, require the same level of maintenance as required systems. If required maintenance is not being done, there is no way to ensure the system will function as intended. Therefore, inadequately maintained nonrequired systems should be removed to avoid creating a false impression of adequate protection.

901.6.1 Standards. Fire protection systems shall be inspected, tested and maintained in accordance with the referenced standards listed in Table 901.6.1.

❖ Specific requirements related to inspection practices, testing schedules and maintenance procedures are dependent on the type of fire protection system and its corresponding referenced NFPA standard as indicated in Table 901.6.1.

TABLE 901.6.1
FIRE PROTECTION SYSTEM MAINTENANCE STANDARDS

SYSTEM	STANDARD
Portable fire extinguishers	NFPA 10
Carbon dioxide fire-extinguishing system	NFPA 12
Halon 1301 fire-extinguishing systems	NFPA 12A
Dry-chemical extinguishing systems	NFPA 17
Wet-chemical extinguishing systems	NFPA 17A
Water-based fire protection systems	NFPA 25
Fire alarm systems	NFPA 72
Water-mist systems	NFPA 750
Clean-agent extinguishing systems	NFPA 2001

❖ This table lists the NFPA referenced standards that should be used for the inspection, testing and maintenance criteria for various fire protection systems.

901.6.2 Records. Records of all system inspections, tests, and maintenance required by the referenced standards shall be maintained on the premises for a minimum of 3 years and made available to the fire code official upon request.

❖ Accurate, up-to-date records are required to document the history of system inspection, testing and maintenance. Record keeping is not intended simply to prove to the fire code official that required inspection, testing and maintenance are being performed, but to assist the owner or his or her agent in performing these functions. A well-kept log helps an owner or technician determine how the system is performing over time and how changes inside and outside the protected premises are affecting system performance. For example, automatic sprinkler system main drain test results may indicate whether the public water supply is being degraded by development, thereby impairing sprinkler system capabilities. Similarly, a history of accidental alarms at a specific smoke detector may indicate that the device requires cleaning or maintenance.

901.7 Systems out of service. Where a required fire protection system is out of service, the fire department and the fire code official shall be notified immediately and, where required by the fire code official, the building shall either be evacuated or an approved fire watch shall be provided for all occupants left unprotected by the shut down until the fire protection system has been returned to service.

Where utilized, fire watches shall be provided with at least one approved means for notification of the fire department and their only duty shall be to perform constant patrols of the protected premises and keep watch for fires.

❖ Protection must not be diminished in any existing building except for the purpose of conducting tests, maintenance or repairs. The length of service interruptions should be kept to a minimum. The fire department and the fire code official must be notified of any service interruptions and should carefully evaluate the continued operation or occupancy of buildings and structures where protection is interrupted. Whenever possible, all unaffected portions of the system should be kept in service. Until protection is restored, hazardous processes or operations should be suspended and alternative special protection should be considered in addition to an approved fire watch.

901.7.1 Impairment coordinator. The building owner shall assign an impairment coordinator to comply with the requirements of this section. In the absence of a specific designee, the owner shall be considered the impairment coordinator.

❖ The impairment coordinator is the person responsible for maintaining the building fire protection systems. The impairment coordinator may be the building owner or other designee, such as the plant manager or building engineer, if they are trained to comply with the provisions of Section 901.7.

901.7.2 Tag required. A tag shall be used to indicate that a system, or portion thereof, has been removed from service.

❖ When any fire protection system is taken out of service it should be clearly identified with a visible tag that indicates the conditions of the impairment and who to notify. The tag is intended to alert building occupants and fire department personnel that the system in question is impaired. The tag should remain visibly in place until full protection is restored.

901.7.3 Placement of tag. The tag shall be posted at each fire department connection, system control valve, fire alarm control unit, fire alarm annunciator and fire command center, indicating which system, or part thereof, has been removed from service. The fire code official shall specify where the tag is to be placed.

❖ This section specifies some of the impaired locations where a tag should be used. Tagging a fire department connection, for example, is intended to alert the responding fire department that a normal operating condition does not exist. While it is also important to tag system control valves, an impairment tag in the sprinkler riser room may not get noticed until accessed by fire department personnel. The final location of all impairment tags is subject to the approval of the fire code official.

901.7.4 Preplanned impairment programs. Preplanned impairments shall be authorized by the impairment coordinator. Before authorization is given, a designated individual shall be responsible for verifying that all of the following procedures have been implemented:

1. The extent and expected duration of the impairment have been determined.
2. The areas or buildings involved have been inspected and the increased risks determined.
3. Recommendations have been submitted to management or building owner/manager.
4. The fire department has been notified.
5. The insurance carrier, the alarm company, building owner/manager, and other authorities having jurisdiction have been notified.
6. The supervisors in the areas to be affected have been notified.
7. A tag impairment system has been implemented.
8. Necessary tools and materials have been assembled on the impairment site.

❖ This section specifies the procedures that should be followed in a thorough preplanned impairment program. These procedures should be followed whenever systems are purposely impaired, such as for routine sprinkler system alarm testing. Proper notification of responsible parties eliminates the chance of false alarms, reduces disruption of normal business activities and encourages quick resumption of normal operations.

901.7.5 Emergency impairments. When unplanned impairments occur, appropriate emergency action shall be taken to minimize potential injury and damage. The impairment coordinator shall implement the steps outlined in Section 901.7.4.

❖ Unplanned impairments, of course, go beyond typical testing and maintenance procedures but are also not necessarily indicative of a fire event. For example, an unplanned emergency impairment might occur if a sprinkler head or pipe was found leaking or was accidently impacted by a fork-lift truck. To reduce water damage and repair the sprinkler system, the valve controlling the water supply to the affected area would need to be closed, thereby impairing protection to the entire area protected by the sprinkler system. The impairment coordinator should follow the procedures in Section 901.7.4 to restore protection in minimum time.

901.7.6 Restoring systems to service. When impaired equipment is restored to normal working order, the impairment coordinator shall verify that all of the following procedures have been implemented:

1. Necessary inspections and tests have been conducted to verify that affected systems are operational.
2. Supervisors have been advised that protection is restored.
3. The fire department has been advised that protection is restored.
4. The building owner/manager, insurance carrier, alarm company, and other involved parties have been advised that protection is restored.
5. The impairment tag has been removed.

❖ Regardless of whether a system is taken out of service for either a planned impairment or for an emergency, this section specifies the procedures to follow when restoring a system to service. Following these procedures ensures that all responsible parties who were informed of the initial impairment are also fully aware that the system is now fully operational. Restoring the system to service assumes the affected part of the system has been corrected and is in proper working condition.

901.8 Removal of or tampering with equipment. It shall be unlawful for any person to remove, tamper with or otherwise disturb any fire hydrant, fire detection and alarm system, fire suppression system, or other fire appliance required by this code except for the purpose of extinguishing fire, training purposes, recharging or making necessary repairs, or when approved by the fire code official.

❖ Tampering or otherwise unauthorized altering of any fire protection system or component is illegal. A person who unlawfully tampers with equipment could face potential criminal charges. Tampering could include intentionally pulling a manual fire alarm box when no emergency exists, playing with matches to set off a smoke detector or flowing a city fire hydrant. The use of fire protection systems, equipment and other fire appliances is limited to those people authorized to conduct repairs and maintenance unless approved by the fire code official.

901.8.1 Removal of or tampering with appurtenances. Locks, gates, doors, barricades, chains, enclosures, signs, tags or seals which have been installed by or at the direction of the fire code official shall not be removed, unlocked, destroyed, tampered with or otherwise vandalized in any manner.

❖ Tampering with or vandalizing appurtenances that are in place to prevent tampering with the system components is also prohibited. For example, sprinkler system control valves are routinely chained and locked in the open position in addition to being equipped with electronically monitored tamper switches. Any unauthorized removal or tampering with these types of devices is strictly prohibited.

SECTION 902
DEFINITIONS

902.1 Definitions. The following words and terms shall, for the purposes of this chapter and as used elsewhere in this code, have the meanings shown herein.

❖ Definitions of terms can help in the understanding and application of the code requirements. These definitions are included within this chapter to provide more convenient access to them without having to refer back to Chapter 2. For convenience, these terms are also listed in Chapter 2 with a cross reference to this section. The use and application of all defined terms, including those defined here, are set forth in Section 201.

ALARM NOTIFICATION APPLIANCE. A fire alarm system component such as a bell, horn, speaker, light, or text display that provides audible, tactile, or visible outputs, or any combination thereof.

❖ The code requires that fire alarm systems be equipped with approved alarm notification appliances so that in an emergency, the fire alarm system will notify the occupants of the need for evacuation or implementation of the fire emergency plan. Alarm notification devices required by the code are of two general types: visible and audible. Except for emergency voice/alarm communication systems, once the fire alarm system has been activated, all visible and audible communication alarms are required to activate. Emergency voice/alarm communication systems are special signaling systems that are activated selectively in response to specific emergency conditions.

ALARM SIGNAL. A signal indicating an emergency requiring immediate action, such as a signal indicative of fire.

❖ This is a general term for all types of supervisory and trouble signals. An example would be a supervisory (tamper) switch on a sprinkler control valve. The activation of the device does not necessarily indicate that there is a fire; however, the level of protection may have been compromised (see the definition of "Fire alarm signal").

ALARM VERIFICATION FEATURE. A feature of automatic fire detection and alarm systems to reduce unwanted alarms wherein smoke detectors report alarm conditions for a minimum period of time, or confirm alarm conditions within a given time period, after being automatically reset, in order to be accepted as a valid alarm-initiation signal.

❖ False fire (evacuation) alarms are a nuisance. For this reason the code specifies that alarms activated by smoke detectors are not to be sounded until the alarm signal is verified by cross-zoned detectors in a single protected area or by system features that will retard the alarm until the signal is determined to be valid. Valid alarm initiation signals can be determined by detectors that report alarm conditions for a minimum period of time or that, after being reset, continue to report an alarm condition. The alarm verification feature may not retard signal activation for a period of more than 60 seconds and must not apply to alarm-initiating devices other than smoke detectors (which may be connected to the same circuit). Alarm verification should not be confused with presignal features that delay an alarm signal for more than 1 minute and that are allowed only where specifically permitted by the authority having jurisdiction.

ANNUNCIATOR. A unit containing one or more indicator lamps, alphanumeric displays, or other equivalent means in which each indication provides status information about a circuit, condition or location.

❖ This refers to the panel that displays the status of the monitored fire protection systems and devices.

AUDIBLE ALARM NOTIFICATION APPLIANCE. A notification appliance that alerts by the sense of hearing.

❖ Audible alarms that are part of a fire alarm system must be loud enough to be heard in every occupied space of a building. Section 907.10.2 prescribes the minimum sound pressure level for all audible alarm notification appliances depending on the occupancy of the building.

AUTOMATIC. As applied to fire protection devices, is a device or system providing an emergency function without the necessity for human intervention and activated as a result of a predetermined temperature rise, rate of temperature rise, or combustion products.

❖ This term, when used in conjunction with fire protection systems or devices, means that the system or device will perform its intended function without a person being present or performing any task in its control or operation. Automatic devices and systems operate completely without human presence or intervention.

AUTOMATIC FIRE-EXTINGUISHING SYSTEM. An approved system of devices and equipment which automatically detects a fire and discharges an approved fire-extinguishing agent onto or in the area of a fire.

❖ This term is the generic name for all types of automatic fire-extinguishing systems, including the most common type—the automatic sprinkler system. See Section 904 for requirements for particular alternative automatic fire-extinguishing systems, such as wet-chemical, dry-chemical, foam, carbon dioxide, halon and clean-agent systems.

AUTOMATIC SPRINKLER SYSTEM. A sprinkler system, for fire protection purposes, is an integrated system of underground and overhead piping designed in accordance with fire protection engineering standards. The system includes a suitable water supply. The portion of the system above the ground is a network of specially sized or hydraulically designed piping installed in a structure or area, generally overhead, and to which automatic sprinklers are connected in a systematic pattern. The system is usually activated by heat from a fire and discharges water over the fire area.

❖ An automatic sprinkler system is one type of automatic fire-extinguishing system. Automatic sprinkler systems are the most common, and their life safety attributes are widely recognized. The code specifies three types of automatic sprinkler systems: one installed in accordance with NFPA 13, one in accordance with NFPA 13R and the other in accordance with NFPA 13D. To be considered for most code design alternatives, a building automatic sprinkler system must be installed throughout in accordance with NFPA 13 (see Section 903.3.1.1).

In a fire, sprinklers automatically open and discharge water onto the fire in a spray pattern that is designed to contain or extinguish the fire. Originally, automatic sprinkler systems were developed just for the protection of buildings and their contents. Because of the development and improvements in sprinkler head response time and water distribution, however, automatic sprinkler systems are now also considered a life safety system. Proper operation of an automatic sprinkler system requires careful selection of the sprinkler heads so that water in sufficient quantity at adequate pressure and properly distributed will be available to suppress the fire.

AVERAGE AMBIENT SOUND LEVEL. The root mean square, A-weighted sound pressure level measured over a 24-hour period.

❖ The ambient noise that can be expected depends on the occupancy of the building. To attract the attention of the occupants, the audible alarm devices must be heard above the ambient noise in the space. For this reason, the alarm devices must have minimum sound pressure levels above the average ambient sound level. Section 907.10.2 prescribes the minimum sound pressure levels for the audible alarm notification appliances for all occupancy conditions.

CARBON DIOXIDE EXTINGUISHING SYSTEM. A system supplying carbon dioxide (CO_2) from a pressurized vessel through fixed pipes and nozzles. The system includes a manual- or automatic-actuating mechanism.

❖ Carbon dioxide extinguishing systems are useful in extinguishing fires in specific hazards or equipment in occupancies where an inert electrically nonconductive medium is essential or desirable and where cleanup of other extinguishing agents, such as dry-chemical residue, presents a problem. The system works by displacing the oxygen in an enclosed area by flooding the space with carbon dioxide. To effectively flood the enclosure, automatic door and window closers and dampers for the mechancial ventilation system must be installed. NFPA 12 contains minimum requirements for the design, installation, testing, inspection, approval, operation and maintenance of carbon dioxide extinguishing systems.

CLEAN AGENT. Electrically nonconducting, volatile, or gaseous fire extinguishant that does not leave a residue upon evaporation.

❖ The two categories of clean agents are halocarbon compounds and inert gas agents. Halocarbon compounds include bromine, carbon, chloride, fluorine, hydrogen and iodine. Halocarbon compounds suppress the fire through a combination of breaking the chemical chain reaction of the fire, reducing the ambient oxygen supporting the fire and reducing the ambient temperature of the fire origin to reduce the propagation of fire. The clean agents that are inert gas agents contain primary components consisting of helium, neon or argon, or a combination of all three. Inert gases work by reducing the oxygen concentration around the fire origin to a level that does not support combustion (see commentary, Section 904.10).

CONSTANTLY ATTENDED LOCATION. A designated location at a facility staffed by trained personnel on a continuous basis where alarm or supervisory signals are monitored and facilities are provided for notification of the fire department or other emergency services.

❖ These locations are intended to receive trouble, supervisory and alarm signals transmitted by the fire protection equipment installed within a protected facility. It is the intent of this section to have both an approved location and personnel who are acceptable to the authority having jurisdiction responsible for actions taken when the fire protection system requires attention.

DELUGE SYSTEM. A sprinkler system employing open sprinklers attached to a piping system connected to a water supply through a valve that is opened by the operation of a detection system installed in the same area as the sprinklers. When this valve opens, water flows into the piping system and discharges from all sprinklers attached thereto.

❖ A deluge system applies large quantities of water or foam throughout the protected area by means of a system of open sprinklers. In a fire, the system is activated by a fire detection system that makes it possible to apply water to a fire more quickly and to cover a larger area than with a conventional automatic sprinkler system, which depends on sprinklers being activated individually as the fire spreads. Deluge systems are particularly beneficial in hazardous areas where the fuel loads (combustible contents) are of such a nature that fire may flash ahead of the operations of conventional automatic sprinklers.

DETECTOR, HEAT. A fire detector that senses heat produced by burning substances. Heat is the energy produced by combustion that causes substances to rise in temperature.

❖ In a fire, heat is released that causes the temperature in a room or space to increase. Automatic fire detectors that sense abnormally high temperature or rate of temperature rise are known as heat detectors. These include fixed temperature detectors, rate compensation detectors and rate-of-rise detectors. The code requires all automatic fire detectors to be smoke detectors, except that heat detectors tested and approved in accordance with NFPA 72 may be used as an alternative to smoke detectors in rooms and spaces where, during normal operation, products of combustion are present in sufficient quantity to actuate a smoke detector.

DRY-CHEMICAL EXTINGUISHING AGENT. A powder composed of small particles, usually of sodium bicarbonate, potassium bicarbonate, urea-potassium-based bicarbonate, potassium chloride or monoammonium phosphate, with added particulate material supplemented by special treatment to provide resistance to packing, resistance to moisture absorption (caking) and the proper flow capabilities.

❖ A dry-chemical system extinguishes a fire by placing a chemical barrier between the fire and oxygen, which acts to smother a fire. This system is best known for protection for commercial ranges, commercial fryers and exhaust hoods. Wet-chemical extinguishing systems, however, are more commonly used for new installations in commercial cooking equipment.

The type of dry chemical to be used in the extinguishing system is a function of the hazard expected. The type of dry chemical used in a system should not be changed, unless it has been proven changeable by a testing laboratory; is recommended by the manufacturer of the equipment and is acceptable to the fire code official for the hazard expected. Additional guidance on the use of various dry-chemical agents can be found in NFPA 17, which gives minimum requirements for the design, installation, testing, inspection, approval, operation and maintenance of dry-chemical extinguishing systems.

EMERGENCY ALARM SYSTEM. A system to provide indication and warning of emergency situations involving hazardous materials.

❖ Because of the potentially volatile nature of hazardous materials, an emergency alarm system is required outside of interior building rooms or areas containing hazardous materials in excess of the maximum allowable quantities permitted in Tables 2703.1.1(1) and 2703.1.1(2). The intent of the emergency alarm, upon actuation by an alarm-initiating device, such as a pull station, is to alert the occupants to an emergency condition involving hazardous materials.

EMERGENCY VOICE/ALARM COMMUNICATIONS. Dedicated manual or automatic facilities for originating and distributing voice instructions, as well as alert and evacuation signals pertaining to a fire emergency, to the occupants of a building.

❖ An emergency voice/alarm communication system is a special feature of fire alarm systems in buildings with special evacuation considerations, such as a high-rise building. Emergency voice/alarm communication systems automatically communicate a fire emergency message to all occupants of a building on a general or selective basis. Such systems also enable the fire service to manually transmit voice instructions to the building occupants about a fire emergency condition and the action to be taken for evacuation or movement to another area of the building.

FIRE ALARM BOX, MANUAL. See "Manual fire alarm box."

FIRE ALARM CONTROL UNIT. A system component that receives inputs from automatic and manual fire alarm devices and is capable of supplying power to detection devices and transponder(s) of off-premises transmitter(s). The control unit is capable of providing a transfer of power to the notification appliances and transfer of condition to relays of devices.

❖ The fire alarm control unit (panel) acts as a point where all signals initiated within the protected building are received before the signal is transmitted to a constantly attended location.

FIRE ALARM SIGNAL. A signal initiated by a fire alarm-initiating device such as a manual fire alarm box, automatic fire detector, water-flow switch, or other device whose activation is indicative of the presence of a fire or fire signature.

❖ This signal is transmitted to a fire alarm control unit as a warning that requires immediate action. The personnel at the constantly attended location are trained to immediately respond to a fire alarm signal, which indicates the presence of a fire. A fire alarm signal assumes an actual fire has been detected (see the definition of "Alarm signal").

FIRE ALARM SYSTEM. A system or portion of a combination system consisting of components and circuits arranged to monitor and annunciate the status of fire alarm or supervisory signal-initiating devices and to initiate the appropriate response to those signals.

❖ Fire alarm systems are installed in buildings to limit fire casualties and property losses by notifying the occupants of the building, the local fire department or both of an emergency condition. The alarm notification appliances associated with fire alarm systems are intended to be evacuation alarms. All fire alarms must be designed and installed to comply with NFPA 72.

FIRE AREA. The aggregate floor area enclosed and bounded by fire walls, fire barriers, exterior walls, or fire-resistance-rated horizontal assemblies of a building.

❖ This term is used to describe a specific and controlled area within a building that may consist of a portion of the floor area within a single story, one entire story or the combined floor area of several stories, depending on how these areas are enclosed and separated from other floor areas. Where a fire barrier wall with a fire-resistance rating divides the floor area of a one-story building, the floor areas on each side of the wall constitute a separate fire area. If a floor/ceiling assembly separating the two stories in a two-story building is fire-resistance rated, each story is a separate fire area. In cases where mezzanines are present, the floor area of the mezzanine is included in the fire area calculations, even though the area of the mezzanine does not contribute to the building area calculations.

FIRE DETECTOR, AUTOMATIC. A device designed to detect the presence of a fire signature and to initiate action.

❖ Automatic fire detectors include all approved devices designed to detect the presence of a fire and automatically initiate emergency action. These include smoke-sensing fire detectors, heat-sensing fire detectors, flame-sensing fire detectors, gas-sensing fire detectors and other fire detectors that operate on other principles as approved by the fire code official. Automatic fire detectors should be selected based on the type and size of fire to be detected and the response required. Automatic fire detectors must be approved, installed and tested to comply with the code and NFPA 72.

FIRE PROTECTION SYSTEM. Approved devices, equipment and systems or combinations of systems used to detect a fire, activate an alarm, extinguish or control a fire, control or manage smoke and products of a fire or any combination thereof.

❖ A fire protection system is any approved device or equipment used singly or in combination, either manually or automatically, and that is intended to detect a fire, notify the building occupants of a fire or suppress the fire. Fire protection systems include fire suppression systems, standpipe systems, fire alarm systems, fire detection systems, smoke control systems and smoke vents. All fire protection systems must be approved by the fire code official and tested in accordance with the referenced standards and Section 901.6.

FIRE SAFETY FUNCTIONS. Building and fire control functions that are intended to increase the level of life safety for occupants or to control the spread of the harmful effects of fire.

❖ In many cases automatic fire detectors are installed even in buildings not required to have a fire alarm system. These fire detectors perform specific functions such as elevator recall or air distribution system shutdown (see Section 907.11).

FOAM-EXTINGUISHING SYSTEM. A special system discharging a foam made from concentrates, either mechanically or chemically, over the area to be protected.

❖ Foam-extinguishing systems must be of an approved type and installed and tested to comply with NFPA 11, 11A and 16. All foams are intended to exclude oxygen from the fire, cool the fire area and insulate adjoining surfaces from heat caused by fires. Foam systems are commonly used to extinguish flammable or combustible liquid fires (see commentary, Section 904.7).

HALOGENATED EXTINGUISHING SYSTEM. A fire-extinguishing system using one or more atoms of an element from the halogen chemical series: fluorine, chlorine, bromine and iodine.

❖ Halon is a colorless, odorless gas that inhibits the chemical reaction of fire. Halon extinguishing systems are useful in occupancies such as computer rooms where an electrically nonconductive medium is essential or desirable and where cleanup of other extinguishing agents presents a problem. The halon extinguishing system must to be of an approved type and installed and tested to comply with NFPA 12A.

Halon extinguishing agents have been identified as a source of emissions resulting in the depletion of the stratospheric ozone layer. For this reason, production of new supplies of halon has been phased out. Alternative extinguishing agents, such as clean agents, have been developed as alternatives to halon.

IMPAIRMENT COORDINATOR. The person responsible for the maintenance of a particular fire protection system.

❖ To minimize the time a fire protection system is out of service, the building owner or other designee is required to monitor impairment procedures (see commentary, Section 901.7.1).

INITIATING DEVICE. A system component that originates transmission of a change-of-state condition, such as in a smoke detector, manual fire alarm box, or supervisory switch.

❖ All fire protection systems consist of devices, which upon use or actuation, will initiate the intended operation. A manual fire alarm box, for example, upon actuation will activate the building alarm notification appliances and transmit a signal to an approved location.

MANUAL FIRE ALARM BOX. A manually operated device used to initiate an alarm signal.

❖ Manual fire alarm boxes are commonly known as pull stations. Manual fire alarm boxes include all manual devices used to activate a manual fire alarm system and have many configurations, depending on the manufacturer. All manual fire alarm devices, however, must be approved and installed in accordance with NFPA 72 for the particular application. Manual fire alarm boxes may be combined in guard tour boxes.

MULTIPLE-STATION ALARM DEVICE. Two or more single-station alarm devices that can be interconnected such that actuation of one causes all integral or separate audible alarms to operate. It also can consist of one single-station alarm device having connections to other detectors or to a manual fire alarm box.

❖ This definition refers to a combination of similar or different types of alarm devices that could be interconnected. The actuation of any two devices, whether a smoke detector or manual fire alarm box, will activate the required audible alarms.

MULTIPLE-STATION SMOKE ALARM. Two or more single-station alarm devices that are capable of interconnection such that actuation of one causes all integral or separate audible alarms to operate.

❖ In occupancies with sleeping areas, occupants must be notified in a fire so that they can promptly evacuate the premises. In accordance with the requirements of NFPA 72, multiple-station smoke alarms are self-contained, smoke-activated alarm devices built in accordance with UL 217 or UL 268 that can be interconnected with other devices so that all integral or separate alarms will operate when any one device is activated.

NUISANCE ALARM. An alarm caused by mechanical failure, malfunction, improper installation, or lack of proper maintenance, or an alarm activated by a cause that cannot be determined.

❖ A nuisance alarm is essentially any alarm that occurs as a result of a condition that does not arise during the normal operation of the equipment.

RECORD DRAWINGS. Drawings ("as builts") that document the location of all devices, appliances, wiring, sequences, wiring methods, and connections of the components of a fire alarm system as installed.

❖ To verify that the system has been installed to comply with the code and applicable referenced standards, complete as-built drawings of the fire alarm system should be available on-site for review.

SINGLE-STATION SMOKE ALARM. An assembly incorporating the detector, the control equipment, and the alarm-sounding device in one unit, operated from a power supply either in the unit or obtained at the point of installation.

❖ A single-station smoke alarm is a self-contained alarm device that detects visible or invisible particles of combustion. Its function is to detect a fire in the immediate area of the detector location. Single-station smoke alarms are not interconnected with other devices and are not capable of notifying or controlling any other fire protection equipment or systems. They may be battery powered, directly connected to the building power supply or a combination of both. Single-station smoke alarms must be built to comply with UL 217 and are to be installed as required by Section 907.

[B] SLEEPING UNIT. A room or space in which people sleep, which can also include permanent provisions for living, eating, and either sanitation or kitchen facilities but not both. Such rooms and spaces that are also part of a dwelling unit are not sleeping units.

❖ This definition is included to coordinate the *Fair Housing Accessibility Guidelines* (FHAG) with the IBC. The definition for "Sleeping unit" is needed to clarify the differences between sleeping units and dwelling units. Examples of a sleeping unit would be a hotel guestroom, a dormitory, a boarding house, etc. Another example would be a studio apartment with a kitchenette (i.e., microwave oven, sink, refrigerator). Since the cooking arrangements are not permanent, this configuration would be considered a sleeping unit, not a dwelling unit. As already defined in the IBC, a "Dwelling unit" must contain permanent facilities for living, sleeping, eating, cooking and sanitation.

SMOKE ALARM. A single- or multiple-station alarm responsive to smoke and not connected to a system.

❖ This is a general term that applies to both single- and multiple-station smoke alarms that are not part of an automatic fire detection system.

SMOKE DETECTOR. A listed device that senses visible or invisible particles of combustion.

❖ These units are considered early warning devices and have saved many people from smoke inhalation and burns. Smoke detectors have a wide range of uses, from sophisticated fire detection systems for industrial and commercial uses to residential. A smoke detector is a device that activates the fire detection system. This smoke detector system contains only the equipment required to detect the products of combustion and should not be confused with single- and multiple-station smoke alarms.

Smoke detectors typically consist of two types: ionization and photoelectric. An ionization detector contains a small amount of radioactive material that ionizes the air in a sensing chamber and causes a current to flow through the air between two charged electrodes. When smoke enters the chamber, the particles cause a reduction in the current. When the level of conductance decreases to a preset level, the detector responds with an alarm.

A photoelectric smoke detector consists primarily of a light source, a light beam and a photosensitive device. When smoke particles enter the light beam, they reduce the light intensity in the photosensitive device. When obscuration reaches a preset level, the detector initiates an alarm.

STANDPIPE SYSTEM, CLASSES OF. Standpipe classes are as follows:

❖ A standpipe system is typically an arrangement of vertical piping located in exit stairways that allows fire-fighting personnel to connect hand-carried hoses at each level to manually extinguish fires. Section 905 and NFPA 14 recognize three different classes of standpipe systems.

Class I system. A system providing 2.5-inch (64 mm) hose connections to supply water for use by fire departments and those trained in handling heavy fire streams.

❖ A Class I standpipe system is intended for use by trained fire service personnel as a readily available water source for manual fire-fighting operations. A Class I standpipe system is equipped with only 2-inch (51 mm) hose connections to allow the fire service to attach the appropriate hose and nozzles. A Class I standpipe system is not equipped with hose stations, which include a cabinet, hose and nozzle.

Class II system. A system providing 1.5-inch (38 mm) hose stations to supply water for use primarily by the building occupants or by the fire department during initial response.

❖ A Class II standpipe system is intended for use by building occupants for manual suppression. The hose stations defined as part of the Class II standpipe system include a cabinet equipped with a hose and nozzle readily available to the building occupants.

Class III system. A system providing 1.5-inch (38 mm) hose stations to supply water for use by building occupants and 2.5-inch (64 mm) hose connections to supply a larger volume of water for use by fire departments and those trained in handling heavy fire streams.

❖ A Class III standpipe system is intended for use by building occupants as well as trained fire service personnel. The 1-inch (25 mm) hose station is in a cabinet for use by the building occupants for manual fire suppression and the 2-inch (51 mm) hose connection is intended for use by fire service personnel.

STANDPIPE, TYPES OF. Standpipe types are as follows:

❖ Section 905 recognizes five types of standpipe systems. The use of each type of system depends on specific occupancy conditions and the presence of an automatic sprinkler system.

Automatic dry. A dry standpipe system, normally filled with pressurized air, that is arranged through the use of a device, such as a dry pipe valve, to admit water into the system piping automatically upon the opening of a hose valve. The water supply for an automatic dry standpipe system shall be capable of supplying the system demand.

❖ A typical automatic dry standpipe system has an automatic water supply retained by a dry-pipe valve. The dry-pipe-valve clapper is kept in place by air placed in the standpipe system under pressure. Once a standpipe hose valve is opened, the air is released from the system, allowing water to fill the system through the dry-pipe valve. This system is traditionally used in areas where the temperature falls below 40°F (4°C), where a wet system would otherwise freeze.

Automatic wet. A wet standpipe system that has a water supply that is capable of supplying the system demand automatically.

❖ An automatic wet standpipe system is used in locations where the entire system would remain above 40°F (4°C). Because the system is pressurized with water, an immediate release of water occurs when a hose connection valve is opened.

Manual dry. A dry standpipe system that does not have a permanent water supply attached to the system. Manual dry standpipe systems require water from a fire department pumper to be pumped into the system through the fire department connection in order to supply the system demand.

❖ A manual dry standpipe system is filled with water only when the fire service is present. Typically, the fire service connects the discharge from a water source, such as a pumper truck, to the fire department connection of a manual dry standpipe system. When the fire service has suppressed the fire and is preparing to leave, the system is drained of the remaining water. Manual dry standpipe systems are commonly installed in open parking structures.

Manual wet. A wet standpipe system connected to a water supply for the purpose of maintaining water within the system but which does not have a water supply capable of delivering the system demand attached to the system. Manual wet standpipe systems require water from a fire department pumper (or the like) to be pumped into the system in order to supply the system demand.

❖ A manual wet standpipe system is connected to an automatic water supply, but the supply is not capable of providing the system demand. The system demand is met when the fire service provides additional water through the fire department connection from the discharge of a water source such as a pumper truck.

Semiautomatic dry. A dry standpipe system that is arranged through the use of a device, such as a deluge valve, to admit water into the system piping upon activation of a remote control device located at a hose connection. A remote control activation device shall be provided at each hose connection. The water supply for a semiautomatic dry standpipe system shall be capable of supplying the system demand.

❖ This type of dry standpipe is a special design that uses a solenoid-activated valve to retain the automatic water supply. Once the standpipe hose valve is opened, a signal is sent to the deluge valve retaining the automatic water supply to allow water to fill the system. This kind of system is used in areas where the temperature falls below 40°F (4°C), where a wet system would otherwise freeze.

SUPERVISING STATION. A facility that receives signals and at which personnel are in attendance at all times to respond to these signals.

❖ The supervising station is the location where all fire protection-system-related signals are sent and where trained personnel are present to respond to an emergency. The supervising station may be an approved central station, a remote supervising station, a proprietary supervising station or other constantly attended location approved by the fire code official.

SUPERVISORY SERVICE. The service required to monitor performance of guard tours and the operative condition of fixed suppression systems or other systems for the protection of life and property.

❖ The supervisory service is responsible for maintaining the integrity of the fire protection system by notifying the supervising station of a change in protection system status.

SUPERVISORY SIGNAL. A signal indicating the need of action in connection with the supervision of guard tours, the fire suppression systems or equipment, or the maintenance features of related systems.

❖ Activation of a supervisory signal-initiating device transmits a signal indicating that a change in the status of the fire protection system has occurred and that emergency action must be taken. These signals are the basis for the actions taken by the attendant at the supervising station.

SUPERVISORY SIGNAL-INITIATING DEVICE. An initiating device such as a valve supervisory switch, water level indicator, or low-air pressure switch on a dry-pipe sprinkler system whose change of state signals an off-normal condition and its restoration to normal of a fire protection or life safety system; or a need for action in connection with guard tours, fire suppression systems or equipment, or maintenance features of related systems.

❖ The supervisory signal-initiating device detects a change in protection system status. Examples of a supervisory signal-initiating device include a flow switch to detect movement of water through the system and a tamper switch to detect when someone shuts off a water control valve.

TIRES, BULK STORAGE OF. Storage of tires where the area available for storage exceeds 20,000 cubic feet (566 m^3).

❖ This definition describes a storage space that is larger than what would be found in most typical mercantile and storage occupancies. Because of its size and the volume of combustible material it would house, it poses an extraordinary hazard for fire protection. Buildings used for the bulk storage of tires are classified as Group S-1 occupancies in accordance with Section 311.2 of the IBC. All Group S-1 occupancies, regardless of square footage, must be equipped with an automatic sprinkler system if used for the bulk storage of tires. Although Section 903.3.1.1 specifically references NFPA 13, by secondary reference it also requires that bulk tire storage buildings be further protected to comply with NFPA 231D.

TROUBLE SIGNAL. A signal initiated by the fire alarm system or device indicative of a fault in a monitored circuit or component.

❖ This type of signal indicates that there has been a change in status of the fire detection system or devices and that a response is required.

VISIBLE ALARM NOTIFICATION APPLIANCE. A notification appliance that alerts by the sense of sight.

❖ Visible alarm notification appliances are typically located in occupancies where occupants may be hearing impaired and in sleeping accommodations of Group I-1 and R-1 occupancies. The alarm devices should be located and oriented so that they will display alarm signals throughout the required space. Visible alarms, when provided, are typically installed in the public and common areas of buildings.

WET-CHEMICAL EXTINGUISHING AGENT. A solution of water and potassium-carbonate-based chemical, potassium-acetate-based chemical or a combination thereof, forming an extinguishing agent.

❖ This extinguishing agent is a suitable alternative to the use of a dry chemical, especially when protecting commercial kitchen range hoods. There is less cleanup time after system discharge. Wet chemical solutions are considered to be relatively harmless and normally have no lasting effect on the skin or respiratory system. These solutions may produce temporary irritation, which is usually mild and disappears when contact is eliminated. These systems must be preengineered and labeled. NFPA 17A applies to the design, installation, operation, testing and maintenance of wet-chemical extinguishing systems.

WIRELESS PROTECTION SYSTEM. A system or a part of a system that can transmit and receive signals without the aid of wire.

❖ These systems use radio-frequency transmitting devices that comply with the special requirements for supervision of low-power wireless systems in NFPA 72.

ZONE. A defined area within the protected premises. A zone can define an area from which a signal can be received, an area to which a signal can be sent, or an area in which a form of control can be executed.

❖ Zoning a system is important to emergency personnel in locating a fire. When an alarm is designated to a specific zone, it allows the fire service to immediately respond to the area where the fire is in progress instead of searching the entire building for the origin of an alarm.

SECTION 903
AUTOMATIC SPRINKLER SYSTEMS

903.1 General. Automatic sprinkler systems shall comply with this section.

❖ This section identifies the conditions requiring an automatic sprinkler system for all occupancies. The need for an automatic sprinkler system may depend on not only the occupancy but also the occupant load, fuel load, height and area of the building as well as fire-fighting capabilities. Section 903.2 addresses all occupancy conditions requiring an automatic sprinkler system. Section 903.3 contains the installation requirements for all sprinkler systems in addition to the requirements of NFPA 13, NFPA 13R and NFPA 13D. The supervision and alarm requirements for sprinkler systems are contained in Section 903.4, whereas Section 903.5 refers to testing and maintenance requirements for sprinkler systems found in Section 901 and NFPA 25.

The area values contained in this section are intended to apply to fire areas, which are comprised of all floor areas within the fire barriers, fire walls or exterior walls. The minimum required fire-resistance rating of fire barrier assemblies that define a fire area is specified in Table 302.3.2 of the IBC. Because the areas are defined as fire areas, fire barriers, fire walls or exterior walls are the only acceptable means of subdividing a building into smaller areas instead of installing an automatic sprinkler system. Where fire barrier and exterior walls define multiple fire areas within a single building, a fire wall defines separate buildings within one structure. Also note that some of the threshold limitations result in a requirement to install an automatic sprinkler system throughout the building while others may require only specific fire areas to be sprinklered.

Another important point is that one fire area may include floor areas in more than one story of a building (see the commentary to the definition of "Fire area" in Section 902.1).

903.1.1 Alternative protection. Alternative automatic fire-extinguishing systems complying with Section 904 shall be permitted in lieu of automatic sprinkler protection where recognized by the applicable standard and approved by the fire code official.

❖ This section permits the use of an alternative automatic fire-extinguishing system when approved by the fire code official as a means of compliance with the occupancy requirements of Section 903. Although the use of an alternative extinguishing system allowed by Section 904, such as a carbon dioxide system or clean-agent system, would satisfy the requirements of Section 903.2, it would not be considered an acceptable alternative for the purposes of exceptions, reductions or other code trade-offs that would be applicable if an automatic sprinkler system were installed.

903.2 Where required. Approved automatic sprinkler systems in new buildings and structures shall be provided in the locations described in this section.

Exception: Spaces or areas in telecommunications buildings used exclusively for telecommunications equipment, associated electrical power distribution equipment, batteries and standby engines, provided those spaces or areas are equipped throughout with an automatic fire alarm system and are separated from the remainder of the building by a wall with a fire-resistance rating of not less than 1 hour and a floor/ceiling assembly with a fire-resistance rating of not less than 2 hours.

❖ Sections 903.2.1 through 903.2.13 identify the conditions requiring an automatic sprinkler system (see Figure 903.2). The type of sprinkler system must be one that is permitted for the specific occupancy condition. An NFPA 13R sprinkler system, for example, may not be installed to satisfy the sprinkler threshold requirements for a mercantile occupancy (see Section 903.2.6). As indicated in Section 903.3.1.2, the use of an NFPA 13R sprinkler system is limited to Group R occupancies not exceeding four stories in height.

There is one exception for those spaces or areas used exclusively for telecommunications equipment. The telecommunications industry has continually stressed the need for the continuity of telephone service, and the ability to maintain this service is of prime importance. This service is a vital link between the community and the various life safety services, including fire, police and emergency medical services. The integrity of this communications service can be jeopardized not only by fire, but also by water, from whatever the source.

It must be emphasized that the exception applies only to those spaces or areas that are used exclusively for telecommunications equipment. Historically, those spaces have a low incidence of fire events. Fires in telecommunications equipment are difficult to start and, if started, grow slowly, thus permitting early detection. Such fires are typically of the smoldering type, do not spread beyond the immediate area and generally self-extinguish.

Note, however, that this exception requires fire resistive separation from other portions of the building.

903.2.1 Group A. An automatic sprinkler system shall be provided throughout buildings and portions thereof used as Group A occupancies as provided in this section. For Group A-1, A-2, A-3, and A-4 occupancies, the automatic sprinkler system shall be provided throughout the floor area where the Group A-1, A-2, A-3 or A-4 occupancy is located, and in all floors between the Group A occupancy and the level of exit discharge. For group A-5 occupancies, the automatic sprinkler system shall be provided in the spaces indicated in Section 903.2.1.5.

❖ Occupancies of Group A are characterized by a significant number of people who are not familiar with their surroundings. The requirement for a suppression system reflects the additional time needed for egress. The extent of protection is also intended to extend to the occupants of the assembly group from unobserved fires in other building areas located between the floor level containing the assembly occupancy and the level of exit discharge. The only exception to the coverage is for Group A-5 occupancies that are open to the atmosphere. Such occupancies require only certain aspects to be sprinklered, such as concession stands (see commentary, Section 903.2.1.5).

FIGURE 903.1
EXAMPLES OF REQUIREMENTS MODIFIED THROUGH USE OF AUTOMATIC SPRINKLER SYSTEMS

CODE SECTION[a]	MODIFICATION	NFPA 13	NPFA 13R	NFPA 13D
Increases				
504.2	Height increase	yes	yes	no
506.3	Area increase	yes	no	no
Table 1005.1 [b]	Egress width	yes	yes	no
Table 1015.1 [b]	Travel distance	yes	yes	no
Rating Reductions				
302.3.2	Separated uses	yes	no	no
Table 601	Type VA construction	yes	no	no
708.3	Fire partitions (dwelling units, sleeping units)	yes	no	no
Table1016.1 [b]	Corridor walls	yes	yes	no
Miscellaneous				
Tables 307.7(1), 307.7(2)	Hazardous material increase	yes	no	no
403.3	High-rise modification	yes	no	no
404.2	Atriums	yes	no	no
507.2, 507.3, 507.5	Unlimited area buildings	yes	no	no
704.8.1	Allowable area of openings	yes	no	no
704.9	Vertical separation of openings	yes	yes	no
717.3.2	Residential attic draftstopping	yes	yes[c]	no
717.3.3	Non-residential draftstopping	yes	no	no
717.4.2	Group R-1, R-2 draftstopping	yes	yes[c]	no
717.4.3	Other group draftstopping	yes	no	no
Table 803.5	Interior finish	yes	yes	no
804.5.1	Floor finish	yes	no	no
907.2[b]	Fire alarm system	yes (B, F, M)	yes (R-1, R-2)	no
1007.2.1 [b]	Accessible egress	yes	yes	no
1023.1[b]	Exit discharge	yes	yes	no
1025.1[b]	Emergency escape openings	yes	yes	no
1406.3[b]	Balconies	yes	yes[c]	yes[c]

a. Section numbers refer to sections in the *International Building Code*.
b. Section numbers in Chapters 9 and 10 apply to both the *International Building Code* and the code.
c. Sprinkler protection must be extended to the affected areas.

FIGURE 903.2
SUMMARY OF OCCUPANCY-RELATED AUTOMATIC SPRINKLER THRESHOLDS[a]

Occupancy	Threshold	Exception
All occupancies	Buildings with floor level > 55 feet above vehicle access and occupant load > 30.	Airport control towers, open parking structures. F-2, R-3, U
Assembly (A-1, A-3, A-4)	Fire area **>** 12,000 sq ft or fire area occupant load > 300 or fire area above/below level of exit discharge. Multitheater complex (A-1 only)	Participant sport arenas at level of exit discharge. A-3, A-4
Assembly (A-2)	Fire area > 5,000 sq ft or fire area occupant load > 300 or fire area above/below level of exit discharge.	None
Assembly (A-5)	Accessory areas > 1,000 sq ft	None
Educational (E)	Fire area > 20,000 sq ft or below level of exit discharge.	Each classroom has exterior door at grade.
Factory (F-1) Mercantile (M) Storage (S-1)	Fire area > 12,000 sq ft or fire area located > 3 stories above grade, or Combined fire area > 24,000 sq ft Woodworking > 2,500 sq ft (F-1 only) Bulk storage of tires > 20,000 cu ft (S-1 only)	None
High, hazard (H-1, H-2, H-3, H-4, H-5)	Sprinklers required.	None
Institutional (I-1, I-2, I-3, I-4)	Sprinklers required.	None
All Residential (R)	Sprinklers required.	None
Repair garage (S-1)	Fire area > 12,000 sq ft or > 2 stories above grade with fire area 10,000 sq ft or repair garage servicing vehicles in basement.	None
Parking garage (S-2)	Enclosed automobile parking sprinklers required Commercial trucks/buses parking area > 5,000 sq ft	None
Covered malls (402.8)[b]	Sprinklers required	Attached open parking structures
High rises (403.2, 403.3)[b]	> 75 feet above vehicle access	Airport traffic control towers, open garages. A-5
Unlimited area buildings (507)[b]	A-3, A-4, B, E, F, M, S: 1 story. B, F, M, S: 2 story	One story. F-2 or S-2

For SI: 1 foot = 304.8 mm, 1 square foot = 0.0929 m^2.
a. Thresholds located in Section 903.2 unless noted. See also Table 903.2.13 for additional required suppression systems.
b. Numbers refer to sections of the *International Building Code*.

903.2.1.1 Group A-1. An automatic sprinkler system shall be provided for Group A-1 occupancies where one of the following conditions exists:

1. The fire area exceeds 12,000 square feet (1115 m^2);
2. The fire area has an occupant load of 300 or more;
3. The fire area is located on a floor other than the level of exit discharge; or
4. The fire area contains a multitheater complex.

❖ Group A-1 occupancies are identified as assembly occupancies with fixed seating, such as theaters. In addition to the high occupant load associated with these types of facilities, egress is further complicated by the possibility of low lighting levels customary during performances. The fuel load in these buildings is usually of a type and quantity that would support fairly rapid fire development and sustained duration.

Theaters with stages pose a greater hazard. Sections 410.6 and 410.7 of the IBC require stages to be equipped with an automatic sprinkler system and standpipe system, respectively. The proscenium opening must also be protected. These features compensate for the additional hazards associated with stages in Group A-1 occupancies.

This section lists four conditions that require installing

a suppression system in a Group A-1 occupancy. Condition 1 requires that, if any one fire area of Group A-1 exceeds 12,000 square feet (1,115 m^2), the automatic fire suppression system is to be installed throughout the entire story or floor level where a Group A-1 occupancy is located, regardless of whether the building is divided into more than one fire area.

Condition 2 establishes the minimum number of occupants for which a suppression system is considered necessary. The determination of the actual occupant load should be based on Section 1004 of the IBC.

Condition 3 accounts for occupant egress delay when traversing a stairway, requiring a sprinkler system regardless of the size of occupant load.

Condition 4 states that a sprinkler system is required for multitheater complexes to account for the delay associated with the notification of adjacent compartmentalized spaces where the occupants may not be immediately aware of an emergency.

903.2.1.2 Group A-2. An automatic sprinkler system shall be provided for Group A-2 occupancies where one of the following conditions exists:

1. The fire area exceeds 5,000 square feet (464.5 m^2);
2. The fire area has an occupant load of 300 or more; or
3. The fire area is located on a floor other than the level of exit discharge.

❖ Group A-2 assembly occupancies are intended for food or drink consumption, such as banquet halls, nightclubs and restaurants. Occupancies in Group A-2 involve life safety factors such as a high occupant density, flexible fuel loading, movable furnishings and limited lighting; therefore, they must be protected with an automatic sprinkler system under any of the listed conditions.

In the case of an assembly use, the purpose of the automatic sprinkler system is to provide life safety from fire as well as preserving property. By requiring fire suppression in areas through which the occupants may egress, including the level of exit discharge, the possibility of unobserved fire development affecting the occupant egress is minimized.

The 5,000-square-foot (465 m^2) threshold for the automatic sprinkler system reflects the higher degree of life safety hazard associated with Group A-2 occupancies. As alluded to earlier, Group A-2 occupancies could have low lighting levels, loud music, late hours of operation, dense seating with ill-defined aisles and alcoholic beverage service. These factors in combination could delay fire recognition, confuse occupant response and increase egress time. The similar intent of Conditions 2 and 3 is addressed in the commentary to Section 903.2.1.1.

These conditions require sprinklers throughout the fire area containing the Group A-2 occupancy, regardless of the number of fire areas present.

903.2.1.3 Group A-3. An automatic sprinkler system shall be provided for Group A-3 occupancies where one of the following conditions exists:

1. The fire area exceeds 12,000 square feet (1115 m^2);
2. The fire area has an occupant load of 300 or more; or
3. The fire area is located on a floor other than the level of exit discharge.

Exception: Areas used exclusively as participant sports areas where the main floor area is located at the same level as the level of exit discharge of the main entrance and exit.

❖ Group A-3 occupancies are assembly occupancies intended for worship, recreation or amusement and other assembly uses not classified elsewhere in Group A, such as churches, museums and libraries. While Group A-3 occupancies could potentially have a high occupant load, they normally do not have the same potential combination of life safety hazards associated with Group A-2 occupancies. As with most assembly occupancies, however, most of the occupants are typically not completely familiar with their surroundings. When any of the three listed conditions are applicable, an automatic sprinkler system is required throughout the fire area containing the Group A-3 occupancy and in all floors between the Group A occupancy and exit discharge (see commentary, Sections 903.2.1 and 903.2.1.1).

The exception exempts the participant sport area of Group A-3 occupancies from automatic sprinkler system requirements because these areas are typically large open spaces with relatively low fuel loads. The exception includes only the participant sport area, such as an indoor swimming pool or the court area of an indoor tennis court. Note that if the exception is claimed and sprinklers are omitted from the sport area, the building would not be considered completely sprinklered in accordance with Section 903.3.1.1 for purposes of allowing construction alternatives, such as height and area increases, corridor rating reduction and other code trade-offs.

903.2.1.4 Group A-4. An automatic sprinkler system shall be provided for Group A-4 occupancies where one of the following conditions exists:

1. The fire area exceeds 12,000 square feet (1115 m^2);
2. The fire area has an occupant load of 300 or more; or
3. The fire area is located on a floor other than the level of exit discharge.

Exception: Areas used exclusively as participant sports areas where the main floor area is located at the same level as the level of exit discharge of the main entrance and exit.

❖ Group A-4 occupancies are assembly uses intended for viewing of indoor sporting events and activities such as arenas, skating rinks and swimming pools. The occupant load density may be high depending on the extent and style of seating, such as bleachers or fixed seats, and the potential for standing-room viewing.

When any of the three listed conditions are applicable, an automatic sprinkler system is required throughout the fire area containing the Group A-4 occupancy and in all floors between the Group A occupancy and exit discharge (see commentary, Sections 903.2.1 and 903.2.1.1). Similar to Group A-3 occupancies, the participant sport areas on the main floor of Group A-4 occupancies are exempt from the sprinkler system requirement (see commentary, Section 903.2.1.3).

903.2.1.5 Group A-5. An automatic sprinkler system shall be provided in concession stands, retail areas, press boxes, and other accessory use areas in excess of 1,000 square feet (93 m^2).

❖ Group A-5 occupancies are assembly uses intended for viewing of outdoor activities. This occupancy classification could include amusement park structures, grandstands and open stadiums. A sprinkler system is not required in the open area of Group A-5 occupancies because the buildings would not accumulate smoke and hot gases. A fire in open areas would also be obvious to all spectators. Enclosed areas such as retail areas, press boxes and concession stands require sprinklers if they are in excess of 1,000 square feet (93 m^2). The 1,000-square-foot (93 m^2) accessory use area is not intended to be an aggregate condition but rather per space.

903.2.2 Group E. An automatic sprinkler system shall be provided for Group E occupancies as follows:

1. Throughout all Group E fire areas greater than 20,000 square feet (1858 m^2) in area.
2. Throughout every portion of educational buildings below the level of exit discharge.

   Exception: An automatic sprinkler system is not required in any fire area or area below the level of exit discharge where every classroom throughout the building has at least one exterior exit door at ground level.

❖ Group E occupancies are limited to educational purposes through the twelfth grade and day care centers serving children older than 2$^1/_2$ years of age. The 20,000-square-foot (1,852 m^2) fire area threshold for the sprinkler system was established to allow smaller schools and day care centers to be unsprinklered to minimize the economic impact on these facilities.

The exception would allow the omission of the automatic sprinkler system for the Group E fire area if there is a direct exit to the exterior from each classroom at ground level. The students/occupants must be able to go from the classroom directly to the outside without intervening corridors, passageways or exit enclosures.

903.2.3 Group F-1. An automatic sprinkler system shall be provided throughout all buildings containing a Group F-1 occupancy where one of the following conditions exist:

1. Where a Group F-1 fire area exceeds 12,000 square feet (1115 m^2);
2. Where a Group F-1 fire area is located more than three stories above grade; or
3. Where the combined area of all Group F-1 fire areas on all floors, including any mezzanines, exceeds 24,000 square feet (2230 m^2).

❖ Because of the difficulty in manually suppressing a fire involving a large area, occupancies of Group F-1 must be protected throughout with an automatic sprinkler system if the fire area is in excess of 12,000 square feet (1,115 m^2), if the total of all fire areas of Group F-1 in the building is in excess of 24,000 square feet (2,230 m^2) or if the Group F-1 fire area is located more than three stories above grade. This is one of the few locations in the code where the total floor area of the building is aggregated for application of a code requirement. The stipulated conditions for when an automatic sprinkler system is required also apply to Group M (see Section 903.2.6) and Group S-1 (see Section 903.2.8) occupancies.

The following examples illustrate how the criteria should be applied. If a building contains a single fire area of Group F-1 and the fire area is 13,000 square feet (1,208 m^2), an automatic sprinkler system is required throughout the entire building. However, if this fire area is separated into two fire areas and neither is in excess of 12,000 square feet (1,115 m^2), an automatic fire sprinkler system is not required. To be considered separate fire areas, the areas must be separated by fire barrier walls or horizontal assemblies having a fire-resistance rating as required in Table 302.3.2 of the IBC.

If a 30,000-square-foot (2,787 m^2) Group F-1 building was equally divided into separate fire areas of 10,000 square feet (929 m^2) each, an automatic sprinkler system would still be required throughout the entire building. Because the aggregate area of all fire areas exceeds 24,000 square feet (2,230 m^2), additional compartmentation will not eliminate the need for an automatic sprinkler system. However, the use of a fire wall to separate the structure into two buildings would reduce the aggregate area of each building to less than 24,000 square feet (2,230 m^2) and each fire area to less than 12,000 square feet (1,115 m^2), which would offset the need for an automatic sprinkler system.

903.2.3.1 Woodworking operations. An automatic sprinkler system shall be provided throughout all Group F-1 occupancy fire areas that contain woodworking operations in excess of 2,500 square feet in area (232 m^2) which generate finely divided combustible waste or which use finely divided combustible materials.

❖ Because of the potential amount of combustible dust that could be generated during woodworking operations, an automatic sprinkler system is required throughout a fire area when it contains a woodworking operation that exceeds 2,500 square feet (232 m^2) in area. Facilities where woodworking operations take place, such as cabinet making, are considered Group F-1 occupancies. The intent of the phrase "finely divided combustible waste" is to describe particle concentrations that are in the explosive range (see Chapter 13 for discussion of dust-producing operations).

903.2.4 Group H. Automatic sprinkler systems shall be provided in high-hazard occupancies as required in Sections 903.2.4.1 through 903.2.4.3.

❖ Group H occupancies are those intended for the manufacturing, processing or storage of hazardous materials that constitute a physical or health hazard. To be considered a Group H occupancy, the amount of hazardous materials is assumed to be in excess of the maximum allowable quantities permitted by Tables 2703.1.1(1) and 2703.1.1(2).

903.2.4.1 General. An automatic sprinkler system shall be installed in Group H occupancies.

❖ This section requires an automatic sprinkler system in all Group H occupancies. Even though in some instances the hazard associated with the occupancy may be one that is not a fire hazard, an automatic sprinkler system is still required to minimize the potential for fire spreading to the high-hazard use; that is, the sprinklers protect this area from fire outside the area. This section does not prohibit the use of an alternative automatic fire-extinguishing system in accordance with Section 904. When a water-based system is not compatible with the hazardous materials involved and thus creates a dangerous condition, an alternative fire-extinguishing system should be used. For example, combustible metals, such as magnesium and titanium, have a serious record of involvement with fire and are typically not compatible with water (see commentary, Chapter 36).

903.2.4.2 Group H-5 occupancies. An automatic sprinkler system shall be installed throughout buildings containing Group H-5 occupancies. The design of the sprinkler system shall not be less than that required under the *International Building Code* for the occupancy hazard classifications in accordance with Table 903.2.4.2.

Where the design area of the sprinkler system consists of a corridor protected by one row of sprinklers, the maximum number of sprinklers required to be calculated is 13.

❖ Group H-5 occupancies are structures that are typically used as semiconductor fabrication facilities and comparable research laboratory facilities that use hazardous production materials (HPM). Many of the materials used in semiconductor fabrication present unique hazards. Many of the materials are toxic, while some are corrosive, water reactive or pyrophoric. Fire protection for these facilities is aimed at preventing incidents from escalating and producing secondary threats beyond a fire, such as the release of corrosive or toxic materials. Because of the nature of Group H-5 facilities, the overall amount of hazardous materials can far exceed the maximum allowable quantities given in Tables 2703.1.1(1) and 2703.1.1(2). Although the amount of HPM material is restricted in fabrication areas, the quantities of HPM in storage rooms normally will be in excess of those allowed by Tables 2703.1.1(1) and 2703.1.1(2). Additional requirements for Group H-5 facilities are located in Section 415.9 of the IBC and Chapter 18 of the code.

This section also specifies the sprinkler design criteria, based on NFPA 13, for various areas in a Group H-5 occupancy (see commentary, Table 903.2.4.2). When the corridor design area sprinkler option is used, a maximum of 13 sprinklers must be calculated. This exceeds the requirements of NFPA 13 for typical egress corridors, which require a maximum of either five or seven calculated sprinklers, depending on the extent of protected openings in the corridor. The increased number of calculated corridor sprinklers is based on the additional hazard associated with the movement of hazardous materials in corridors of Group H-5 facilities.

TABLE 903.2.4.2
GROUP H-5 SPRINKLER DESIGN CRITERIA

LOCATION	OCCUPANCY HAZARD CLASSIFICATION
Fabrication areas	Ordinary Hazard Group 2
Service corridors	Ordinary Hazard Group 2
Storage rooms without dispensing	Ordinary Hazard Group 2
Storage rooms with dispensing	Extra Hazard Group 2
Corridors	Ordinary Hazard Group 2

❖ Table 903.2.4.2 designates the appropriate occupancy hazard classification for the various areas within a Group H-5 facility. The listed occupancy hazard classifications correspond to specific sprinkler system design criteria in NFPA 13. Ordinary Hazard Group 2 occupancies, for example, require a minimum design density of 0.20 gpm/ft^2 (8.1 L/min/m^2) with a minimum design area of 1,500 square feet (139 m^2). An Extra Hazard Group 2 occupancy, in turn, requires a minimum design density of 0.40 gpm/ft^2 (16.3 L/min/m^2) with a minimum operating area of 2,500 square feet (232 m^2). The increased overall sprinkler demand for Extra Hazard Group 2 occupancies is based on the potential use and handling of substantial amounts of hazardous materials, such as flammable or combustible liquids.

903.2.4.3 Pyroxylin plastics. An automatic sprinkler system shall be provided in buildings, or portions thereof, where cellulose nitrate film or pyroxylin plastics are manufactured, stored or handled in quantities exceeding 100 pounds (45 kg).

❖ Cellulose nitrate (pyroxylin) plastics and motion picture film pose unusual and substantial fire risks. Pyroxylin plastics are the most dangerous and unstable of all plastic compounds. The chemically bound oxygen in their structure permits them to burn vigorously in the absence of atmospheric oxygen. Although these compounds produce approximately the same amount of energy as paper when they burn, pyroxylin plastics burn at a rate as much as 15 times greater than comparable common combustibles. When burning, these materials release highly flammable and toxic combustion byproducts. Consequently, cellulose nitrate fires are very difficult to control. Although this section specifies a sprinkler threshold quantity of 100 pounds (45 kg), the need for additional fire protection should be considered for pyroxylin plastics in any amount.

903.2.5 Group I. An automatic sprinkler system shall be provided throughout buildings with a Group I fire area.

Exception: An automatic sprinkler system installed in accordance with Section 903.3.1.2 or 903.3.1.3 shall be allowed in Group I-1 facilities.

❖ The Group I occupancy is divided into four individual occupancy classifications based on the degree of detention, supervision and physical mobility of the occupants. The evacuation difficulties associated with the building occupants creates the need to incorporate a defend-in-place philosophy of fire protection in occupancies of Group I. For this reason, all such occupancies are to be protected with an automatic sprinkler system.

Of particular note, this section encompasses all Group I-3 occupancies where more than five persons are incarcerated. There has been considerable controversy concerning the use of automatic sprinklers in detention and correctional occupancies. Special design considerations can be taken into account to alleviate the perceived problems with sprinklers in sleeping units. Sprinklers that reduce the likelihood of vandalism as well as the potential to hang oneself are commercially available. Knowledgeable designers can incorporate certain design features to increase reliability and decrease the likelihood of damage to the system.

Group I-4 occupancies would include either adult-only care facilities or occupancies that provide personal care for more than five children $2^1/_2$ years of age or less on a less than 24-hour basis. Because the degree of assistance and the time needed for egress cannot be gauged, an automatic sprinkler system is required. As indicated in Section 202, a Group I-4 child care facility located at the level of exit discharge and accommodating no more than 100 children, with each child care room having an exit directly to the exterior would be classified as a Group E occupancy, and an automatic sprinkler system would not be required unless dictated by the requirements in Section 903.2.2.

The exception permits Group I-1 occupancies to be protected throughout with either an NFPA 13R or 13D sprinkler system instead of a standard NFPA 13 sprinkler system. The exception recognizes the perceived mobility of the occupants in a Group I-1 facility as well as the basic life safety intent to protect the main occupiable areas. However, use of this exception would result in the building not qualifying as a fully sprinklered building in accordance with NFPA 13 for any applicable code trade-offs.

903.2.6 Group M. An automatic sprinkler system shall be provided throughout buildings containing a Group M occupancy where one of the following conditions exists:

1. Where a Group M fire area exceeds 12,000 square feet (1115 m^2);
2. Where a Group M fire area is located more than three stories above grade; or
3. Where the combined area of all Group M fire areas on all floors, including any mezzanines, exceeds 24,000 square feet (2230 m^2).

❖ The sprinkler threshold requirements for Group M occupancies are identical to those of Group F-1 and S-1 occupancies (see commentary, Section 903.2.3). Automatic sprinkler systems for mercantile occupancies are typically designed for an Ordinary Hazard Group 2 classification in accordance with NFPA 13. If high-piled storage (see Section 903.2.6.1) is anticipated, however, additional levels of fire protection may be required. Also, some merchandise in mercantile occupancies, such as aerosols, rubber tires, paints and certain plastic commodities, even at limited storage heights, are considered beyond the scope of NFPA 13 and may require additional fire protection.

903.2.6.1 High-piled storage. An automatic sprinkler system shall be provided as required in Chapter 23 in all buildings of Group M where storage of merchandise is in high-piled or rack storage arrays.

❖ Regardless of the size of the Group M fire area, an automatic sprinkler system may be required in a high-piled storage area. High-piled storage includes piled, palletized, shelf or rack storage of combustibles to a height greater than 12 feet (3,658 mm). Storage of common combustible materials more than 12 feet (3,658 mm) high must meet the requirements of Chapter 23 and referenced standards NFPA 231 and NFPA 231C (Note: The scope of NFPA 13 has been expanded in the 1999 edition to include protection requirements formerly contained in NFPA 231 and 231C).

903.2.7 Group R. An automatic sprinkler system installed in accordance with Section 903.3 shall be provided throughout all buildings with a Group R fire area.

❖ This section requires sprinklers in any building that contains a Group R fire area. This includes uses such as hotels, apartment buildings, group homes and dormitories. There are no minimum criteria and no exceptions.

The International Code Council® (ICC®) Board, in approving the development of the *International Residential Code* (IRC®), indicated that it is to be a stand-alone code for the construction of detached one- and two-family dwellings and multiple single-family dwellings (townhouses) no more than three stories in height with a separate means of egress. That is, all of the provisions for new construction that affect those buildings are to be covered exclusively by the IRC and are not to be covered by another *International Code*®. This section is a new construction requirement and therefore does not apply to buildings covered by the IRC. Buildings that do not fall within the scope of the IRC would be classified in Group R and be subject to these provisions.

With respect to life safety, the need for a sprinkler system is dependent on the occupants' proximity to the fire and the ability to respond to a fire emergency. Group R occupancies could contain occupants who may require assistance to evacuate, such as infants, those with a disability or who may simply be asleep. While the presence of a sprinkler system cannot always protect occupants in residential buildings who are aware of the ignition and either do not respond or respond inappropriately, it can prevent fatalities outside of the area of fire origin regardless of the occupants' response. Section 903.3.2 requires quick-response or residential sprinklers in all Group R occupancies. Full-scale fire tests have demonstrated the ability of quick-response and residential sprinklers to maintain tenability from flaming fires in the room of fire origin.

903.2.8 Group S-1. An automatic sprinkler system shall be provided throughout all buildings containing a Group S-1 occupancy where one of the following conditions exist:

1. Where a Group S-1 fire area exceeds 12,000 square feet (1115 m^2);
2. Where a Group S-1 fire area is located more than three stories above grade; or
3. Where the combined area of all Group S-1 fire areas on all floors, including any mezzanines, exceeds 24,000 square feet (2230 m^2).

❖ An automatic sprinkler system shall be provided throughout all buildings where the fire area containing a Group S-1 occupancy exceeds 12,000 square feet (1,115 m^2), where more than three stories in height or where the combined fire area on all floors, including mezzanines, exceeds 24,000 square feet (2230 m^2).

The sprinkler threshold requirements for Group S-1 occupancies are identical to those of Group F-1 and M (see commentary, Sections 903.2.3 and 903.2.6). Group S-1 occupancies, such as warehouses and self-storage buildings, are assumed to be used for the storage of combustible materials. While high-piled storage does not change the Group S-1 occupancy classification, sprinkler protection, if required, may have to comply with the additional requirements of Chapter 23 and NFPA 231 and 231C (Note: The scope of NFPA 13 has been expanded in the 1999 edition to include protection requirements formerly contained in NFPA 231 and 231C). High-piled stock or rack storage in any occupancy, as indicated in Section 413 of the IBC, must comply with the code.

903.2.8.1 Repair garages. An automatic sprinkler system shall be provided throughout all buildings used as repair garages in accordance with the *International Building Code*, as follows:

1. Buildings two or more stories in height, including basements, with a fire area containing a repair garage exceeding 10,000 square feet (929 m^2).
2. One-story buildings with a fire area containing a repair garage exceeding 12,000 square feet (1115 m^2).
3. Buildings with a repair garage servicing vehicles parked in the basement.

❖ Automatic sprinklers may be required in repair garages, depending on the quantity of combustibles present, their location and floor area. Repair garages may contain significant quantities of flammable liquids and other combustible materials. These occupancies are typically considered Ordinary Hazard Group 2 occupancies as defined in NFPA 13. Portions of repair garages used for parts cleaning using flammable or combustible liquids may require automatic sprinkler protection. In the case of Item 1, the area would require sprinklers based upon whether vehicles are actually being repaired there or if they are simply being parked there.

903.2.8.2 Bulk storage of tires. Buildings and structures where the area for the storage of tires exceeds 20,000 cubic feet (566 m^3) shall be equipped throughout with an automatic sprinkler system in accordance with Section 903.3.1.1.

❖ This section specifies when an automatic sprinkler system is required for the bulk storage of tires based on the volume of the storage area as opposed to a specific number of tires. Even in fully sprinklered buildings, tire fires pose significant problems to local fire departments. Tire fires produce thick smoke and are difficult to extinguish by sprinklers alone. Although this section references Section 903.3.1.1 for an NFPA 13 sprinkler system, by secondary reference through NFPA 13, bulk tire storage buildings must be further protected in accordance with the provisions of NFPA 231D.

903.2.9 Group S-2. An automatic sprinkler system shall be provided throughout buildings classified as an enclosed parking garage in accordance with the *International Building Code* or where located beneath other groups.

Exception: Enclosed parking garages located beneath Group R-3 occupancies.

❖ Fire records have shown that fires in parking structures typically fully involve only a single automobile with minor damage to adjacent vehicles. An enclosed parking garage, however, does not allow the dissipation of smoke and hot gases as readily as an open parking structure, which is also considered a Group S-2 occupancy. If the enclosed parking garage is located beneath another occupancy group, the enclosed parking garage must be protected with an automatic sprinkler system. This requirement is based on the potential for a fire to develop undetected, which would endanger the occupants of the other occupancy. Even though this section does not specify a fire area threshold, enclosed parking garages are considered less hazardous than repair garages, which are classified as Group S-1 occupancies. Enclosed parking garages are not intended to have a more stringent sprinkler threshold than repair garages.

The exception exempts the enclosed garages of buildings that are classified in Group R-3.

903.2.9.1 Commercial parking garages. An automatic sprinkler system shall be provided throughout buildings used for storage of commercial trucks or buses where the fire area exceeds 5,000 square feet (464 m^2).

❖ Because of the larger-sized vehicles involved in commercial parking structures, such as trucks or buses, a more stringent sprinkler threshold is required. Bus garages may also be located adjacent to passenger terminals (Group A-3) that have a substantial occupant load.

903.2.10 All occupancies except Groups R-3 and U. An automatic sprinkler system shall be installed in the locations set forth in Sections 903.2.10.1 through 903.2.10.1.3.

Exception: Group R-3 and Group U.

❖ Sections 903.2.10.1 through 903.2.10.2 specify certain conditions when an automatic sprinkler system is required, even in otherwise nonsprinklered buildings. As indicated in the exception, the listed conditions in the noted sections are applicable to all occupancies except Groups R-3 and U. The exception for Group R-3 occupancies is consistent with other noted sprinkler exceptions for Group R-3 occupancies, such as Section 903.2.11 for enclosed garages. Most structures that qualify as Group U do not typically have the type of conditions stipulated in Sections 903.2.10.1 through 903.2.10.1.3.

903.2.10.1 Stories and basements without openings. An automatic sprinkler system shall be installed in every story or basement of all buildings where the floor area exceeds 1,500 square feet (139.4 m^2) and where there is not provided at least one of the following types of exterior wall openings:

1. Openings below grade that lead directly to ground level by an exterior stairway complying with Section 1009 or an outside ramp complying with Section 1010. Openings shall be located in each 50 linear feet (15 240 mm), or fraction thereof, of exterior wall in the story on at least one side.
2. Openings entirely above the adjoining ground level totaling at least 20 square feet (1.86 m^2) in each 50 linear feet (15 240 mm), or fraction thereof, of exterior wall in the story on at least one side.

❖ Because of both the lack of openings in exterior walls for access by the fire department for fire fighting and rescue and the problems associated with venting the products of combustion during fire suppression operations, all stories and basements of buildings that do not have adequate openings as defined in this section must be equipped with an automatic sprinkler system. This section applies to stories and basements without openings that exceed 1,500 square feet (139 m^2) in a building that is otherwise not required to be fully sprinklered. The requirement for an automatic sprinkler system in this section applies only to the affected area and does not mandate sprinkler protection throughout the entire building.

Stories without openings, as defined in this section, are stories that do not have at least 20 square feet (186 m^2) of opening leading directly to ground level in each 50 lineal feet (15 240 mm) or fraction thereof on at least one side. Since exterior doors will provide openings of 20 square feet (186 m^2), or slightly less in some occupancies, exterior stairways and ramps in each 50 lineal feet (15 240 mm) are considered acceptable.

This section intends that the required openings be distributed such that the lineal distance between adjacent openings does not exceed 50 feet (15 240 mm). If the openings in the exterior wall are located without regard to the location of the adjacent openings, it is possible that segments of the exterior wall will not have the required access to the interior of the building for fire-fighting purposes. Any arrangement of required stairways, ramps or openings that results in a portion of the wall 50 feet (15 240 mm) or more in length with no openings to the exterior does not meet the intent of the code that access be provided in each 50 lineal feet (15 240 mm) (see Figure 903.2.10.1).

903.2.10.1.1 Opening dimensions and access. Openings shall have a minimum dimension of not less than 30 inches (762 mm). Such openings shall be accessible to the fire department from the exterior and shall not be obstructed in a manner that fire fighting or rescue cannot be accomplished from the exterior.

❖ To qualify, an opening must not be less than 30 inches (762 mm) in length or width and must be accessible to the fire department from the exterior. The minimum opening dimension gives fire department personnel access to the interior of the story or basement for fire-fighting and rescue operations and provides openings that are large enough to vent the products of combustion.

903.2.10.1.2 Openings on one side only. Where openings in a story are provided on only one side and the opposite wall of such story is more than 75 feet (22 860 mm) from such openings, the story shall be equipped throughout with an approved automatic sprinkler system or openings as specified above shall be provided on at least two sides the story.

❖ If openings are provided on only one side, an automatic sprinkler system would still be required if the opposite wall of the story is more than 75 feet (22 860 mm) from existing openings. An alternative to providing the automatic sprinkler system would be to design openings on at least two sides of the exterior of the building. In basements, if any portion is more than 75 feet (22 860 mm) from the openings, the entire basement must be equipped with an automatic sprinkler system. Providing openings on more than one wall allows cross ventilation to vent the products of combustion [see Figures 903.2.10.1(1-4)].

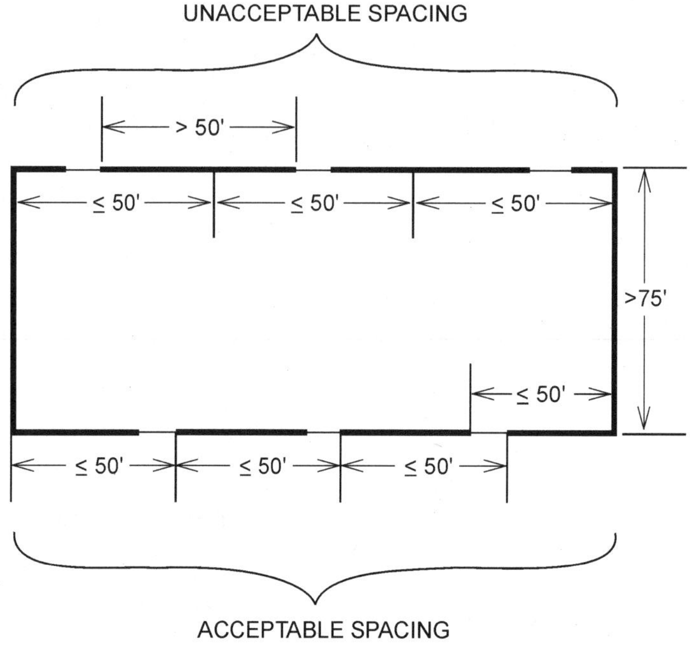

For SI: 1 foot = 304.8 mm.

FIGURE 903.2.10.1
SPACING OF OPENINGS IN STORIES OR BASEMENTS

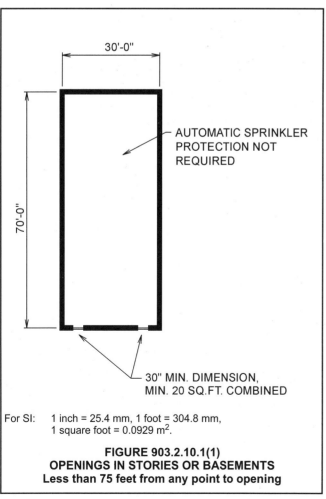

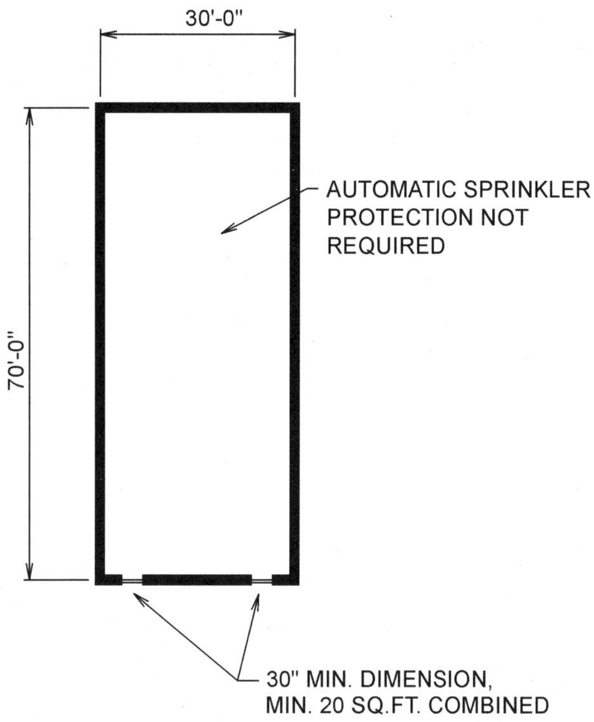

For SI: 1 inch = 25.4 mm, 1 foot = 304.8 mm, 1 square foot = 0.0929 m^2.

FIGURE 903.2.10.1(1)
OPENINGS IN STORIES OR BASEMENTS
Less than 75 feet from any point to opening

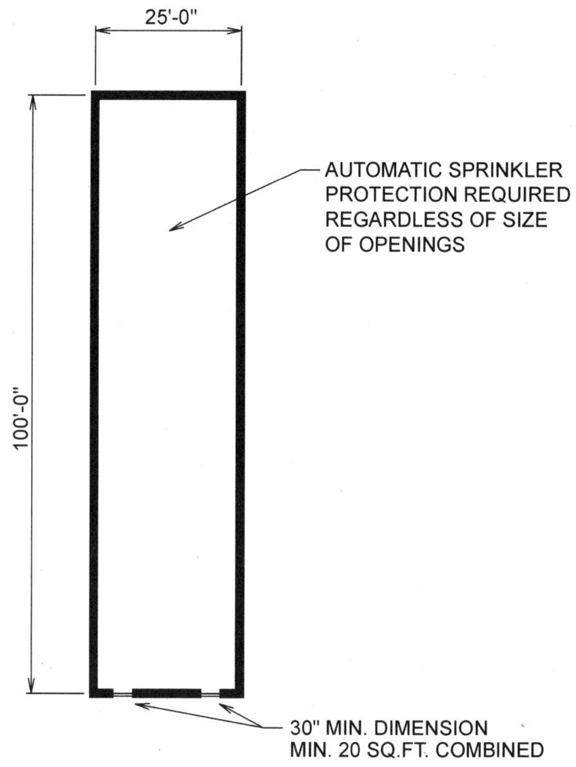

For SI: 1 inch = 25.4 mm, 1 foot = 304.8 mm, 1 square foot = 0.0929 m^2.

FIGURE 903.2.10.1(2)
OPENINGS IN STORIES OR BASEMENTS
Less than 75 feet from any point to opening

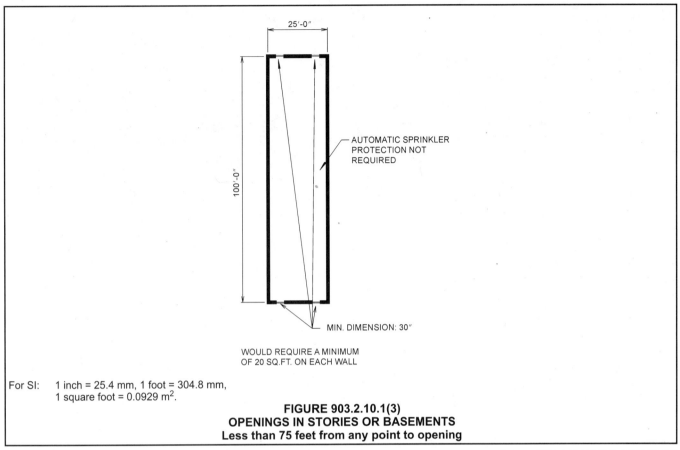

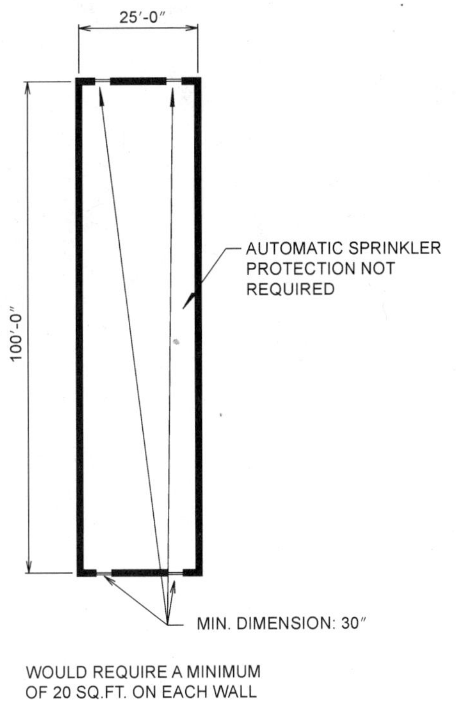

For SI: 1 inch = 25.4 mm, 1 foot = 304.8 mm, 1 square foot = 0.0929 m^2.

FIGURE 903.2.10.1(3)
OPENINGS IN STORIES OR BASEMENTS
Less than 75 feet from any point to opening

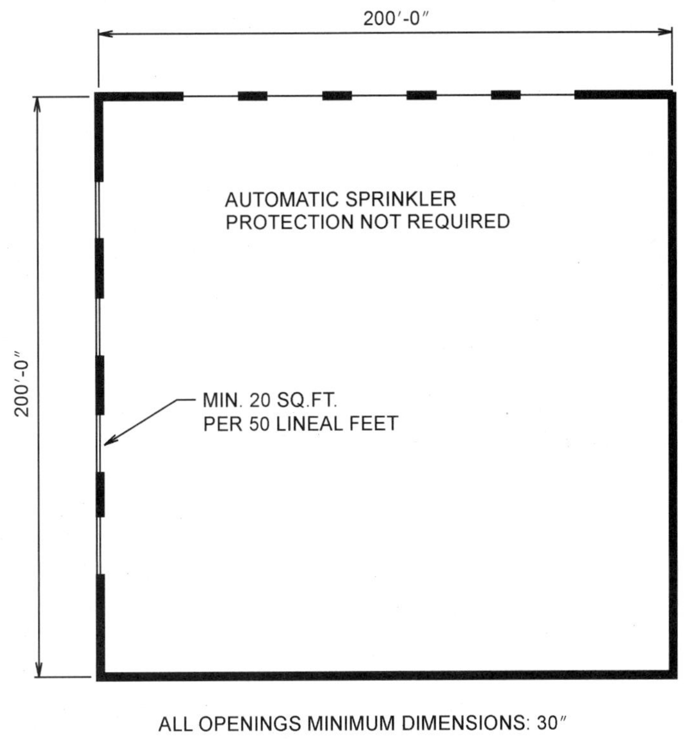

For SI: 1 inch = 25.4 mm, 1 foot = 304.8 mm, 1 square foot = 0.0929 m^2.

FIGURE 903.2.10.1(4)
OPENINGS IN STORIES OR BASEMENTS

903.2.10.1.3 Basements. Where any portion of a basement is located more than 75 feet (22 860 mm) from openings required by Section 903.2.10.1, the basement shall be equipped throughout with an approved automatic sprinkler system.

❖ Where obstructions such as walls or other partitions are present in any given story or basement, the walls and partitions enclosing any room or space must have openings that provide an equivalent degree of fire department access to that provided by the openings prescribed in Section 903.2.10.1 for exterior walls. If an equivalent degree of fire department access to all portions of the floor area is not provided, the story or basement would require an automatic sprinkler system.

903.2.10.2 Rubbish and linen chutes. An automatic sprinkler system shall be installed at the top of rubbish and linen chutes and in their terminal rooms. Chutes extending through three or more floors shall have additional sprinkler heads installed within such chutes at alternate floors. Chute sprinklers shall be accessible for servicing.

❖ This section requires that the chute and termination room associated with a waste or linen system be protected with an automatic sprinkler system. Note that the requirement for suppression is within the chute itself and not within the required shaft that encloses the chute.

903.2.10.3 Buildings more than 55 feet in height. An automatic sprinkler system shall be installed throughout buildings with a floor level having an occupant load of 30 or more that is located 55 feet (16 764 mm) or more above the lowest level of fire department vehicle access.

Exceptions:

1. Airport control towers.
2. Open parking structures.
3. Occupancies in Group F-2.

❖ Because of the difficulties associated with manual suppression of a fire in buildings in excess of 55 feet (16 764 mm) high, an automatic sprinkler system is required throughout the building regardless of occupancy. Buildings that qualify for a sprinkler system under Section 903.2.10.3 are not necessarily high-rise buildings as defined in Section 403.1 of the IBC.

The listed exceptions are occupancies that, based on height only, do not require an automatic sprinkler system. Airport control towers and open parking structures are also exempt from the high-rise provisions of Section 403 of the IBC. Although an automatic sprinkler system is not required in open parking structures, a system may still be needed, depending on the building construction type and the area and number of parking tiers (see Table 406.3.5 of the IBC).

903.2.11 During construction. Automatic sprinkler systems required during construction, alteration and demolition operations shall be provided in accordance with Section 1413.

❖ Chapter 33 of the IBC, as well as Chapter 14 of the code, address fire safety requirements during construction, alteration or demolition work. Working sprinkler systems should remain operative at all times unless it is absolutely necessary to shut down the system because of the proposed work. All sprinkler system impairments should be rectified as quickly as possible unless specific prior approval has been obtained from the fire code official. Buildings with a required sprinkler system should not be occupied unless the sprinkler system has been installed and tested.

903.2.12 Other hazards. Automatic sprinkler protection shall be provided for the hazards indicated in Sections 903.2.12.1 and 903.2.12.2.

❖ This section addresses when an automatic sprinkler system is required for hazardous exhaust ducts and commercial cooking operations. Although Sections 903.2.12.1 and 903.2.12.2 address conditions where an automatic sprinkler system is the extinguishing system, these sections are not intended to exclude approved alternative extinguishing systems.

903.2.12.1 Ducts conveying hazardous exhausts. Where required by the *International Mechanical Code*, automatic sprinklers shall be provided in ducts conveying hazardous exhaust, flammable or combustible materials.

Exception: Ducts where the largest cross-sectional diameter of the duct is less than 10 inches (254 mm).

❖ To protect against the spread of fire within a hazardous exhaust system and to prevent a duct fire from involving the building, an automatic sprinkler system must be installed to protect the exhaust duct system. Where materials conveyed in the ducts are not compatible with water, alternative extinguishing agents should be used. The fire suppression requirement is intended to apply to exhaust systems having an actual fire hazard. An automatic sprinkler system in the duct would be of little value for an exhaust system that conveys only nonflammable or noncombustible materials, fumes, vapors or gases. The exception recognizes the reduced hazard associated with smaller ducts and the impracticality of installing sprinkler protection.

903.2.12.2 Commercial cooking operations. An automatic sprinkler system shall be installed in a commercial kitchen exhaust hood and duct system where an automatic sprinkler system is used to comply with Section 904.

❖ An automatic suppression system is required for commercial kitchen exhaust hood and duct systems where required by Section 610 or the *International Mechanical Code*® (IMC®) to have a Type I hood. Type I hoods are required for commercial cooking equipment that produces grease-laden vapors or smoke. Section 904.11 recognizes that alternative extinguishing systems other than an automatic sprinkler system may be used.

903.2.13 Other required suppression systems. In addition to the requirements of Section 903.2, the provisions indicated in Table 903.2.13 also require the installation of a suppression system for certain buildings and areas.

❖ In addition to Section 903.2, requirements for automatic fire suppression systems are also found elsewhere in the code as indicated in Table 903.2.13.

TABLE 903.2.13
ADDITIONAL REQUIRED FIRE-EXTINGUISHING SYSTEMS

SECTION	SUBJECT
1024.6.2.3	Smoke-protected seating
1208.2	Dry cleaning plants
1208.3	Dry cleaning machines
1504.1	Spray finishing in Group A, E, I or R
1504.6	Spray booths and rooms
1505.1	Dip-tank rooms
1505.6.1	Dip tanks
1505.8.4	Hardening and tempering tanks
1803.10	HPM facilities
1803.10.1.1	HPM work station exhaust
1803.10.2	HPM gas cabinets
1803.10.3	HPM corridors
1803.10.4	HPM exhaust
1803.10.4.1	HPM noncombustible ducts
1803.10.4.2	HPM combustible ducts
1907.3	Lumber production conveyor rooms
1908.7	Recycling facility conveyor rooms
2106.1	Class A and B ovens
2106.2	Class C and D ovens
Table 2306.2	Storage fire protection
2306.4	Storage
2703.8.4.1	Gas rooms
2703.8.5.3	Exhausted enclosures
2704.5	Indoor storage of hazardous materials
2705.1.8	Indoor dispensing of hazardous materials
2804.4.1	Aerosol warehouses

(continued)

TABLE 903.2.13—continued
ADDITIONAL REQUIRED FIRE-EXTINGUISHING SYSTEMS

2904.5	Storage of more than 1,000 cubic feet of loose combustible fibers
3306.5.2.1	Storage of smokeless propellant
3306.5.2.3	Storage of small arms primers
3404.3.7.5.1	Flammable and combustible liquid storage rooms
3404.3.8.4	Flammable and combustible liquid storage warehouses
3405.3.7.3	Flammable and combustible liquid Group H-2 or H-3 areas
3704.1.2	Gas cabinets for highly toxic and toxic gas
3704.1.3	Exhausted enclosures for highly toxic and toxic gas
3704.2.2.6	Gas rooms for highly toxic and toxic gas
3704.3.3	Outdoor storage for highly toxic and toxic gas
4106.2.2	Exhausted enclosures or gas cabinets for silane gas
4204.1.1	Pyroxylin plastic storage cabinets
4204.1.3	Pyroxylin plastic storage vaults
4204.2	Pyroxylin plastic storage and manufacturing
International Building Code	Sprinkler requirements as set forth in Section 903.2.13 of the *International Building Code*

For SI: 1 cubic foot = 0.023 m^3.

❖ Table 903.2.13 identifies other sections of the code that require an automatic fire suppression system based on the specific occupancy, process or operation. The table does not identify the various sections of the code that contain design alternatives based on the use of an automatic fire suppression system, typically an automatic sprinkler system.

903.3 Installation requirements. Automatic sprinkler systems shall be designed and installed in accordance with Sections 903.3.1 through 903.3.7.

❖ Specific design, installation and testing criteria are given for automatic sprinkler systems in the sections and subsections that follow, as well as an indication of the applicability of a nationally recognized standard in the area. The information required to complete a thorough review of an automatic sprinkler system is listed in Figure 903.3.

903.3.1 Standards. Sprinkler systems shall be designed and installed in accordance with Sections 903.3.1.1, 903.3.1.2 or 903.3.1.3.

❖ Automatic sprinkler systems are to be installed to comply with the code and NFPA 13, 13R or 13D. As provided for in Section 102.6, where differences occur between the code and NFPA 13, 13R or 13D, the code applies. The fire code official also has the authority to approve the type of sprinkler system to be installed. See Figure 903.3.1 for typical design parameters for each type of sprinkler system.

1. Information required on shop drawings includes:

__ Name of owner and occupant

__ Location, including street address

__ Point of compass

__ Ceiling construction

__ Full-height cross section

__ Location of fire walls

__ Location of partitions

__ Occupancy of each area or room

__ Location and size of blind spaces and closets

__ Any questionable small enclosures in which no sprinklers are to be installed

__ Size of city main in street, pressure and whether dead end or circulation and, if dead end, direction and distance to nearest circulating main, city main test results

__ Other source of water supply, with pressure or elevation

__ Make, type and orifice size of sprinkler

__ Temperature rating and location of high-temperature sprinklers

__ Number of sprinklers on each riser and on each system by floors and total area by each system on each floor

__ Make, type, model and size of alarm or dry pipe valve

__ Make, type, model and size of preaction or deluge valve

__ Type and location of alarm bells

__ Total number of sprinklers on each dry pipe system or preaction deluge system

__ Approximate capacity in gallons or each dry pipe system

__ Cutting lengths of pipe (or center-to-center dimensions)

__ Type of fittings, riser nipples and size, and all welds and bends

__ Type and location of hangers, inserts and sleeves

__ All control valves, checks, drain pipes and test pipes

__ Small hand-hose equipment

__ Underground pipe size, length, location, weight, material, point of connection to city main; the type of valves, meters and valve pits; and the depth that top of the pipe is laid below grade

__ When the equipment is to be installed as an addition to an old group of sprinklers without additional feed from the yard system, enough of the old system shall be indicated on the plans to show the total number of sprinklers to be supplied and to make all connections clear

__ Name, address and phone number of contractor and sprinkler designer

__ Hydraulic reference points shall be shown by a number and/or letter designation and shall correspond with comparable reference points shown on the hydraulic calculation sheets

__ System design criteria showing the minimum rate of water application (density), the design area of water application and the water required for hose streams both inside and outside

__ Actual calculated requirements showing the total quantity of water and the pressure required at a common reference point for each system

__ Elevation data showing elevations of sprinklers, junction points and supply or reference points

2. Information required on calculations includes:

__ Location

__ Name of owner and occupant

__ Building identification

__ Description of hazard

__ Name and address of contractor and designer

__ Name of approving agency

3. System design requirements include:

__ Design area of water application

__ Minimum rate of water application (density)

__ Area of sprinkler coverage

__ Hazard or commodity classification

__ Building height

__ Storage height

__ Storage method

__ Total water requirements, as calculated, including allowance for hose demand water supply information

__ Location and elevation static and residual test gauge with relation to the riser reference point

__ Flow location

__ Static pressure, psi

__ Residual pressure, psi

__ Flow, gpm

__ Date

__ Time

__ Test conducted by whom

__ Sketch to accompany gridded system calculations to indicate flow quantities and directions for lines with sprinklers operated in the remote area

4. Additional information necessary for complete review includes:

__ Sprinkler description and discharge constant (*K* value)

__ Hydraulic reference points

__ Flow, gpm

__ Pipe diameter (actual internal diameter)

__ Pipe length

__ Equivalent pipe length for fittings and components

__ Friction loss in psi per foot of pipe

__ Total friction loss between reference points

__ Elevation difference between reference points

__ Required pressure in psi at each reference point

__ Velocity pressures and normal pressure if included in calculations

__ Notes to indicate starting points, reference to other sheets or classification of date

5. Included with the submittal must be a graph sheet showing water supply curves and system requirements including:

__ Hose demand plotted on semilogarithmic graph paper so as to present a graphic summary of the complete hydraulic calculations

Figure 903.3
SAMPLE SPRINKLER SYSTEM DRAWING AND DATA SUBMITTALS

Figure 903.3.1
NFPA 13, NFPA 13R, NFPA 13D SYSTEMS

	NFPA 13	NFPA 13R	NFPA 13D
Extent of protection	Equip throughout (Section 903.3.1.1)	Occupied spaces (Section 903.3.1.2)	Occupied spaces (Section 903.3.1.3)
Scope	All occupancies	Low-rise-residential	One- and two-family dwellings
Sprinkler design	Density/area concept	4-head design	2-head design
Sprinklers	All types	Residential only	Residential only
Duration	30 minutes (minimum)	30 minutes	10 minutes
Advantages	Property and life protection	Life safety/tenability	Life safety/tenability

903.3.1.1 NFPA 13 sprinkler systems. Where the provisions of this code require that a building or portion thereof be equipped throughout with an automatic sprinkler system in accordance with this section, sprinklers shall be installed throughout in accordance with NFPA 13 except as provided in Section 903.3.1.1.1.

❖ NFPA 13 contains the minimum requirements for the design and installation of automatic water sprinkler systems and exposure protection sprinkler systems. The requirements contained in the standard include the character and adequacy of the water supply and the selection of sprinklers, piping, valves and all of the materials and accessories. The standard does not include requirements for installation of private fire service mains and their appurtenances; installation of fire pumps or construction and installation of gravity and pressure tanks and towers.

NFPA 13 defines seven classifications or types of water sprinkler systems: wet pipe (see Figure 903.3.1.1); dry pipe; preaction; deluge; combined dry pipe and preaction; sprinkler systems that are designed for a special purpose and outside sprinklers for exposure protection. While numerous variables must be considered in selecting the proper type of sprinkler system, the wet-pipe sprinkler system is recognized as the most effective and efficient. The wet-pipe system is also the most reliable type of sprinkler system, because water under pressure is available at the sprinkler. Therefore, wet-pipe sprinkler systems are recommended wherever possible.

Exceptions for the use of NFPA 13R and NFPA 13D systems are addressed throughout the code when exceptions based upon the use of sprinklers are provided. More specifically, if the use of these other standards is appropriate it will be noted within the exception. For a building to be considered "equipped throughout" with an NFPA 13 sprinkler system, complete protection must be provided in accordance with the referenced standard, subject to the exempt locations indicated in Section 903.3.1.1.1.

903.3.1.1.1 Exempt locations. Automatic sprinklers shall not be required in the following rooms or areas where such rooms or areas are protected with an approved automatic fire detection system in accordance with Section 907.2 that will respond to visible or invisible particles of combustion. Sprinklers shall not be omitted from any room merely because it is damp, of fire-resistance rated construction or contains electrical equipment.

1. Any room where the application of water, or flame and water, constitutes a serious life or fire hazard.
2. Any room or space where sprinklers are considered undesirable because of the nature of the contents, when approved by the fire code official.
3. Generator and transformer rooms separated from the remainder of the building by walls and floor/ceiling or roof/ceiling assemblies having a fire-resistance rating of not less than 2 hours.
4. In rooms or areas that are of noncombustible construction with wholly noncombustible contents.

❖ This section allows the omission of sprinkler protection in certain locations if an approved automatic fire detection system is installed. Buildings in compliance with one of the four listed conditions would still be considered fully sprinklered throughout in compliance with the code and NFPA 13 and thus are eligible for all applicable code trade-offs, exceptions or reductions. Elimination of the sprinkler system in a sensitive area is subject to the approval of the fire code official.

Condition 1 addresses restrictions where the application of water could create a hazardous condition. For example, sprinkler protection should be avoided where it is not compatible with certain stored materials (i.e., some water-reactive hazardous materials). Combustible metals, such as magnesium and aluminum, may burn so intensely that the use of water to attempt fire control will only intensify the reaction.

It is not the intent of Condition 2 to omit sprinklers solely because of a potential for water damage. Also, a desire to not sprinkler a certain area (such as a computer room or operating room) does not fall within the limitations of the exception unless there is something unique about the space that would result in water being incompatible.

Condition 3 recognizes the low fuel load and low occupancy hazards associated with generator and transformer rooms and therefore allows the omission of sprinkler protection if the rooms are separated from adjacent areas by 2-hour fire-resistance-rated construction. This condition assumes the room is not used for any combustible storage.

Condition 4 requires the construction of the room or area, as well as the contents, to be noncombustible. An example would be an area in an unprotected steel frame building (Type IIB construction) used for steel or concrete block storage. Neither involves any significant combustible packaging or sources of ignition, and few combustibles are present (see Figure 903.3.1).

903.3.1.2 NFPA 13R sprinkler systems. Where allowed in buildings of Group R, up to and including four stories in height, automatic sprinkler systems shall be installed throughout in accordance with NFPA 13R.

❖ NFPA 13R contains design and installation requirements for a sprinkler system to aid in the detection and control of fires in low-rise (four stories or less) residential occupancies. Sprinkler systems designed in accordance with NFPA 13R are intended to prevent flashover (total involvement) in the room of fire origin and to improve the chance for occupants to escape or be evacuated. The design criteria in NFPA 13R are similar to those in NFPA 13 except that sprinklers may be omitted from areas in which fatal fires in residential occupancies do not typically originate (bathrooms, closets, attics, porches, garages and concealed spaces).

903.3.1.2.1 Balconies. Sprinkler protection shall be provided for exterior balconies and ground floor patios of dwelling units where the building is of Type V construction. Sidewall sprinklers that are used to protect such areas shall be permitted to be located such that their deflectors are within 1 inch (25 mm) to 6 inches (152 mm) below the structural members, and a maximum distance of 14 inches (356 mm) below the deck of the exterior balconies that are constructed of open wood joist construction.

❖ This section requires additional sprinklers on balconies and patios when Type V construction is used for Group R occupancies. This is in addition to the requirements of NFPA 13R. The intent is to address hazards such as grilling and similar activities. Since NFPA 13R does not require such coverage, there is potential that a fire on a balcony could grow much too large for the system within the building to handle.

903.3.1.3 NFPA 13D sprinkler systems. Where allowed, automatic sprinkler systems installed in one- and two-family dwellings shall be installed throughout in accordance with NFPA 13D.

❖ NFPA 13D contains design and installation requirements for a sprinkler system to aid in the detection and control of fires in one- and two-family dwellings and mobile homes. Similar to NFPA 13R, sprinkler systems designed in accordance with NFPA 13D are intended to

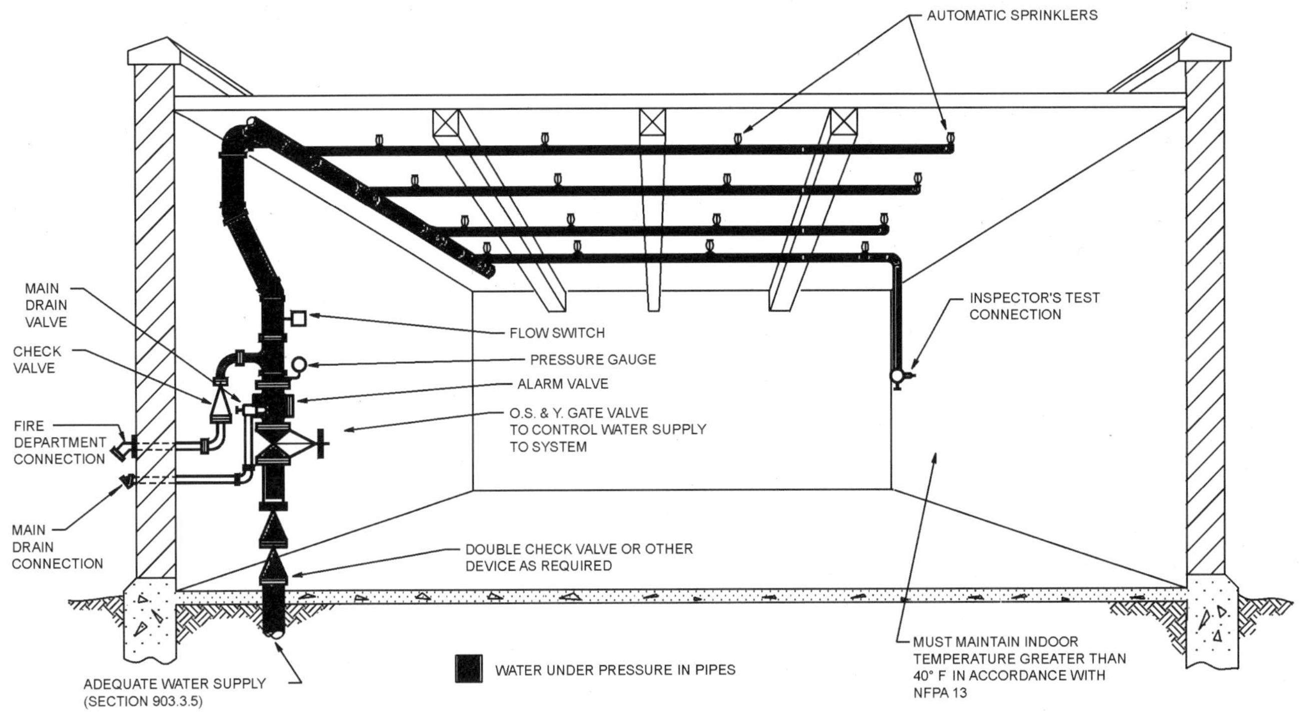

For SI: °C = [(°F) - 32]/1.8

Figure 903.3.1.1
TYPICAL WET PIPE SYSTEM

prevent flashover (total involvement) in the room of fire origin and to improve the chance for occupants to escape or be evacuated. Although the allowable omission of sprinklers in certain areas of the dwelling unit in NFPA 13D is similar to that in NFPA 13R, the water supply requirements are less restrictive. NFPA 13D uses a two-head sprinkler design with a 10-minute duration requirement, while NFPA 13R uses a four-head sprinkler design with a 30-minute duration requirement. The decreased water supply requirement emphasizes the main intent of NFPA 13D to control the fire and maintain tenability during evacuation of the residence.

Since the fire code official has the authority to approve the type of sprinkler system, this section may be used to prevent the use of a specific type of sprinkler system that may be inappropriate for a particular type of occupancy.

903.3.2 Quick-response and residential sprinklers. Where automatic sprinkler systems are required by this code, quick-response or residential automatic sprinklers shall be installed in the following areas in accordance with Section 903.3.1 and their listings:

1. Throughout all spaces within a smoke compartment containing patient sleeping units in Group I-2 in accordance with the *International Building Code.*
2. Dwelling units and sleeping units in Group R and I-1 occupancies.
3. Light-hazard occupancies as defined in NFPA 13.

❖ This section requires the use of either listed quick-response or residential automatic sprinklers depending on the type of sprinkler system required to achieve faster and more effective suppression in certain areas. Residential sprinklers are required in all types of residential buildings that would permit the use of an NFPA 13R or 13D sprinkler system.

Condition 1 reiterates the requirements of Section 407.5 of the IBC to use approved quick-response or residential sprinklers in smoke compartments containing patient sleeping units in Group I-2 occupancies. Even though properly operating standard sprinklers are effective, the extent of fire growth and smoke production that can occur before sprinkler activation creates the need for early warning to enable faster response by staff and initiation of egress that is critical in occupancies containing persons incapable of self-preservation. The faster response time associated with quick-response or residential sprinklers increases the probability that the sprinklers will actuate before the patient's life would be threatened by a fire in his or her room.

Because of the kind of occupants sleeping in Group R and I-1 occupancies, as indicated in Condition 2, a fast-response-type sprinkler is desirable.

Condition 3 recognizes light-hazard occupancies in accordance with NFPA 13. These could include restaurants, schools, office buildings, churches and similar occupancies where the fire load and potential heat release of combustible contents are low.

903.3.3 Obstructed locations. Automatic sprinklers shall be installed with due regard to obstructions that will delay activation or obstruct the water distribution pattern. Automatic sprinklers shall be installed in or under covered kiosks, displays, booths, concession stands, or equipment that exceeds 4 feet (1219 mm) in width. Not less than a 3-foot (914 mm) clearance shall be maintained between automatic sprinklers and the top of piles of combustible fibers.

Exception: Kitchen equipment under exhaust hoods protected with a fire-extinguishing system in accordance with Section 904.

❖ To provide adequate sprinkler coverage, sprinkler protection should be extended under any obstruction that exceeds 4 feet (1,219 mm) in width. Large air ducts are another common obstruction where sprinklers are routinely extended beneath the duct. The 3-foot (914 mm) storage clearance requirement for combustible fibers is caused by their potential high heat release. Most storage conditions require only a minimum 18-inch (457 mm) storage clearance to combustibles, depending on the type of sprinklers used and their actual storage conditions.

The exception recognizes that an alternative extinguishing system is permitted for commercial cooking systems in place of sprinkler protection for exhaust hoods that may be more than 4 feet (1219 mm) wide.

903.3.4 Actuation. Automatic sprinkler systems shall be automatically actuated unless specifically provided for in this code.

❖ The intent of this section is to eliminate the need for occupant intervention during a fire.

Wet-pipe and dry-pipe sprinkler systems, for example, are essentially fail-safe systems in the sense that, if the system is in proper operating condition, it will operate once a sprinkler fuses. Dry systems have an inherent time lag for water to reach the sprinkler; therefore, the response is not as fast as for a wet-pipe system. Other types of sprinkler systems, such as preaction and deluge, rely on the actuation of a detection system to operate the sprinkler valve.

903.3.5 Water supplies. Water supplies for automatic sprinkler systems shall comply with this section and the standards referenced in Section 903.3.1. The potable water supply shall be protected against backflow in accordance with the requirements of this section and the *International Plumbing Code.*

❖ To be effective, all sprinkler systems must have an adequate supply of water. The criteria for an acceptable water supply are contained in the standards referenced in Section 903.3.1. For example, NFPA 13 contains criteria for different types of water supplies as well as the methods to determine the pressure, flow capabilities and capacity necessary to get the intended performance from a sprinkler system. An acceptable water supply could consist of a reliable municipal supply, a gravity tank or a fire pump with a pressure tank or a combination of these.

This section also establishes the requirements for

protecting the potable water system against a nonpotable source, such as stagnant water retained within the sprinkler piping. As stated in Section 608.16.4 of the *International Plumbing Code®* (IPC®), an approved double check valve device or reduced pressure principle backflow preventer is required.

903.3.5.1 Domestic services. Where the domestic service provides the water supply for the automatic sprinkler system, the supply shall be in accordance with this section.

❖ This section establishes the scope of domestic services for limited area sprinkler systems and residential combination services.

903.3.5.1.1 Limited area sprinkler systems. Limited area sprinkler systems serving fewer than 20 sprinklers on any single connection are permitted to be connected to the domestic service where a wet automatic standpipe is not available. Limited area sprinkler systems connected to domestic water supplies shall comply with each of the following requirements:

1. Valves shall not be installed between the domestic water riser control valve and the sprinklers.

   Exception: An approved indicating control valve supervised in the open position in accordance with Section 903.4.

2. The domestic service shall be capable of supplying the simultaneous domestic demand and the sprinkler demand required to be hydraulically calculated by NFPA 13, NFPA 13R or NFPA 13D.

❖ Use of limited area sprinkler systems is primarily limited to fire areas or other areas where the number of sprinklers does not exceed 20.

The use of limited area sprinkler systems is restricted to cases in which the code requires a limited number of sprinklers and not a complete automatic sprinkler system. For example, limited area sprinkler systems may be used to protect stages; storage and workshop areas; stories and basements without openings; painting rooms; trash rooms and chutes; furnace rooms; kitchens and hazardous exhaust systems and incidental use areas as defined in Section 302.1.1 of the IBC. When a wet automatic standpipe is not available, limited-area sprinkler systems may be connected to the domestic water supply.

The water supply to the sprinkler system is to be controlled only by the same valve that controls the domestic water supply to the building; no shutoff valves are permitted in the sprinkler system piping. These restrictions increase the likelihood that the sprinkler system will be operational should a fire occur. Likewise, if the sprinkler system needs restoration after having operated in response to a fire or needs repairs requiring that the water supply be shut off, this section increases the probability that the system will be restored quickly, because having the domestic water supply to the entire building shut off is an inconvenience to occupants that they will not tolerate long.

The exception recognizes the value of standard sprinkler system valve supervision complying with Section 903.4 as providing the level of system reliability contemplated by this section.

Documentation, usually in the form of hydraulic calculations, must be submitted demonstrating that the domestic water system is adequate to supply the sprinkler demand in addition to the peak domestic demand. The domestic demand would normally be determined as stated in the IPC.

903.3.5.1.2 Residential combination services. A single combination water supply shall be permitted provided that the domestic demand is added to the sprinkler demand as required by NFPA 13R.

❖ NFPA 13R permits a common supply main to a building to serve both the sprinkler system and domestic services if the domestic demand is added to the sprinkler demand. NFPA 13R systems do not provide the same level of property protection as NFPA 13 systems.

903.3.5.2 Secondary water supply. A secondary on-site water supply equal to the hydraulically calculated sprinkler demand, including the hose stream requirement, shall be provided for high-rise buildings in Seismic Design Category C, D, E or F as determined by the *International Building Code*. The secondary water supply shall have a duration not less than 30 minutes as determined by the occupancy hazard classification in accordance with NFPA 13.

Exception: Existing buildings.

❖ To increase the reliability of the sprinkler system should an earthquake disable the primary water supply, a secondary on-site water supply is required for high-rise buildings in Seismic Design Category C, D, B or F (see Section 1616.3 of the IBC). The required amount of water is equal to the hydraulically calculated sprinkler demand plus hose stream demand for a 30-minute period related to the appropriate hazard classification in NFPA 13.

The exception recognizes the infeasibility of requiring a secondary water supply in existing high-rise buildings.

903.3.6 Hose threads. Fire hose threads used in connection with automatic sprinkler systems shall be approved and shall be compatible with fire department hose threads.

❖ The threads on connections the fire department will use to connect a hose must be compatible with the fire department threads. The majority of fire departments in the United States use the American National Fire Hose Connection Screw Thread (NH), commonly known as NST and NS. NFPA 1963 gives the screw thread dimensions and the thread size of threaded connections, with nominal sizes ranging from $^{3}/_{4}$ inch (19 mm) to 6 inches (152 mm) for the NH thread. Although efforts to standardize fire hose threads began after the Boston conflagration in 1872, there are still many different screw threads, some of which give the appearance of compatibility with the NH thread. While NFPA 1963 may be used as a guide, the code does not require that any particular standard be adhered to. Design documents should specify the type of thread to be used in order to be compatible with the fire department equipment. The criteria typically apply to fire department connections for sprinkler and standpipe systems, standpipe hose connections, yard hydrants and wall hydrants.

903.3.7 Fire department connections. The location of fire department connections shall be approved by the fire code official.

❖ Fire department connections are required as part of a water-based suppression system as the auxiliary water supply. These connections give the fire department the capability of supplying the necessary water to the automatic sprinkler or standpipe system at a sufficient pressure without pressurizing the underground supply. The fire department connection also serves as an alternative source of water should a valve in the primary water supply be closed. A fire department connection does not, however, constitute an automatic water source.

Section 912 contains additional guidance for the location and accessibility of fire department connections, which should be readily visible from the street and unobstructed. See Figure 903.3.7 for typical fire department connection arrangements. Section 912 also allows for the connections to be locked if appropriate key wrenches will be available (see commentary, Section 912.3.1).

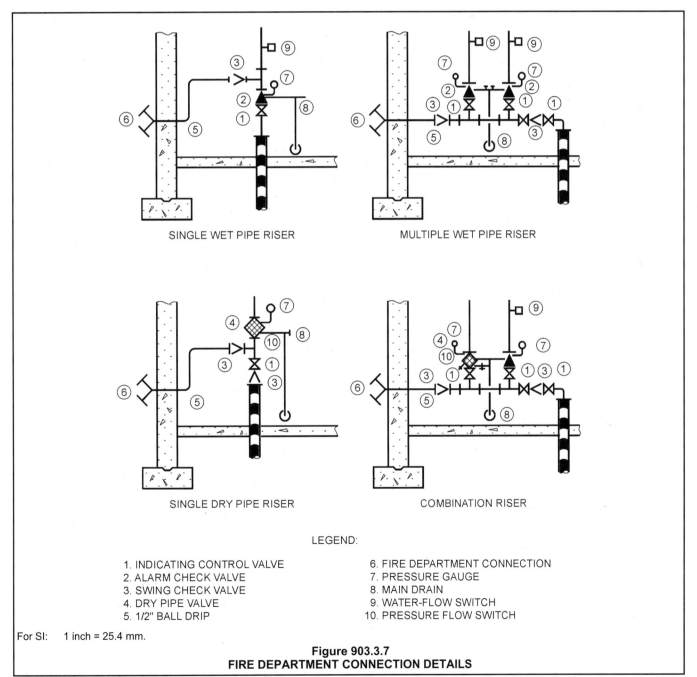

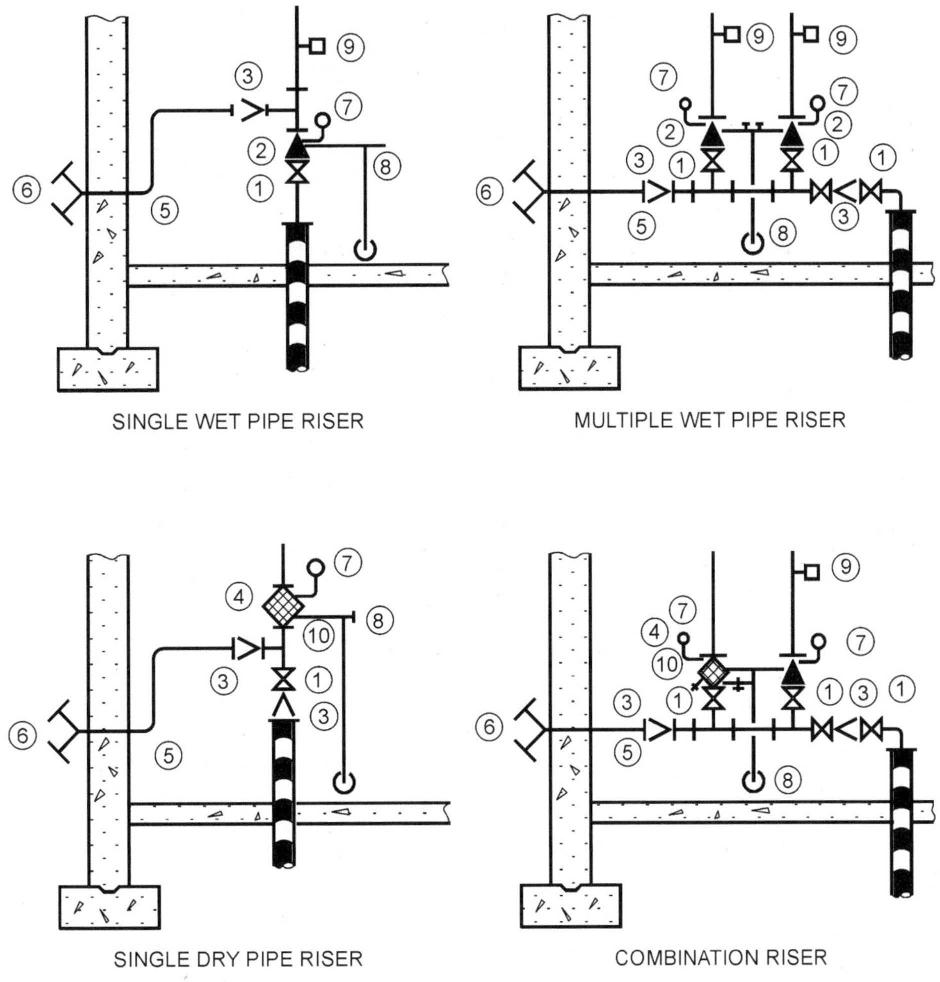

LEGEND:

1. INDICATING CONTROL VALVE
2. ALARM CHECK VALVE
3. SWING CHECK VALVE
4. DRY PIPE VALVE
5. 1/2" BALL DRIP
6. FIRE DEPARTMENT CONNECTION
7. PRESSURE GAUGE
8. MAIN DRAIN
9. WATER-FLOW SWITCH
10. PRESSURE FLOW SWITCH

For SI: 1 inch = 25.4 mm.

Figure 903.3.7
FIRE DEPARTMENT CONNECTION DETAILS

903.4 Sprinkler system monitoring and alarms. All valves controlling the water supply for automatic sprinkler systems, pumps, tanks, water levels and temperatures, critical air pressures, and water-flow switches on all sprinkler systems shall be electrically supervised.

Exceptions:

1. Automatic sprinkler systems protecting one- and two-family dwellings.
2. Limited area systems serving fewer than 20 sprinklers.
3. Automatic sprinkler systems installed in accordance with NFPA 13R where a common supply main is used to supply both domestic water and the automatic sprinkler system, and a separate shutoff valve for the automatic sprinkler system is not provided.
4. Jockey pump control valves that are sealed or locked in the open position.
5. Control valves to commercial kitchen hoods, paint spray booths or dip tanks that are sealed or locked in the open position.
6. Valves controlling the fuel supply to fire pump engines that are sealed or locked in the open position.
7. Trim valves to pressure switches in dry, preaction and deluge sprinkler systems that are sealed or locked in the open position.

❖ The reliability data on automatic sprinkler systems clearly indicate that a closed valve is the leading cause of sprinkler system failure. There are also a number of other critical elements that contribute to successful sprinkler system operation, including, but not limited to, pumps, water tanks and air pressure maintenance devices; therefore, this section requires that the various critical elements that contribute to an available water supply and to the function of the sprinkler system be electrically supervised.

Automatic sprinkler systems in one- and two-family dwellings are typically designed to comply with NFPA 13D, which does not require electrical supervision (see Exception 1).

Limited-area sprinkler systems are generally supervised by their connection to the domestic water service (see Exception 2). Electrical supervision is required only if a control valve is installed between the riser control valve and the sprinkler system piping. Similar to limited-area sprinkler systems, electrical supervision is not required for NFPA 13R residential combination services when a shutoff valve is not installed (see Exception 3). NFPA 13R sprinkler systems are supervised in that the only way to shut off the sprinkler system is to also shut off the domestic water supply. The valves discussed in Exceptions 4 through 7 can be sealed or locked in the open position because they do not control the sprinkler system water supply.

903.4.1 Signals. Alarm, supervisory and trouble signals shall be distinctly different and shall be automatica`lly transmitted to an approved central station, remote supervising station or proprietary supervising station as defined in NFPA 72 or, when approved by the fire code official, shall sound an audible signal at a constantly attended location.

Exceptions:

1. Underground key or hub valves in roadway boxes provided by the municipality or public utility are not required to be monitored.
2. Backflow prevention device test valves, located in limited area sprinkler system supply piping, shall be locked in the open position. In occupancies required to be equipped with a fire alarm system, the backflow preventer valves shall be electrically supervised by a tamper switch installed in accordance with NFPA 72 and separately annunciated.

❖ Automatic sprinkler systems must be supervised as a means of determining that the system is operational. A valve supervisory switch operating as a normally open or normally closed switch is usually used. NFPA 72 does not permit valve supervisory switches to be located on the same zone circuit as the water-flow switch unless it is specifically arranged to actuate a signal that is distinctive from the circuit trouble condition signal.

Required sprinkler systems are to be monitored by an approved supervising service to comply with NFPA 72. Types of supervising stations recognized in NFPA 72 include central station, remote supervising station or proprietary supervising station.

A central station is an independent off-site facility operated and maintained by personnel whose primary business is to furnish, maintain, record and supervise a signaling system. A proprietary system is similar to a central station; however, a proprietary system is typically an on-site facility monitoring a number of buildings on the same site for the same owner. A remote station system has an alarm signal that is transmitted to a remote location acceptable to the authority having jurisdiction that is attended 24 hours a day. The receiving equipment is usually located at a fire station, police station or telephone answering service. Alternatively to the three previous supervising methods, an audible signal can be transmitted to a constantly attended location approved by the fire code official.

Exception 1 recognizes that underground key or hub valves in roadway boxes are not normally supervised or required to be supervised by this section or NFPA 13.

Exception 2 acknowledges that local water utilities and environmental authorities in many instances require, by local ordinances, that backflow prevention devices be installed in limited-area sprinkler system piping. To make the testing and maintenance of backflow prevention devices easier, test valves are installed on each side of the device. These valves are typically indicating-type valves and can function as shutoff valves for the sprinkler system, and therefore require some level of supervision.

Because these infrequently used valves may be the only feature of protection requiring supervision in occupancies not otherwise required to be equipped with a fire alarm system, Exception 2 permits these valves to be locked in the open position. However, if the occupancy is protected by a fire alarm system, these valves must be equipped with approved valve supervisory devices connected to the fire alarm control panel on a separate (supervisory) zone so that the supervisory signal is transmitted to the designated receiving station. Installation and testing of backflow preventers in sprinkler systems are regulated in Sections 312.9 (testing) and 608.16.4 (devices) of the IPC.

903.4.2 Alarms. Approved audible devices shall be connected to every automatic sprinkler system. Such sprinkler water-flow alarm devices shall be activated by water flow equivalent to the flow of a single sprinkler of the smallest orifice size installed in the system. Alarm devices shall be provided on the exterior of the building in an approved location. Where a fire alarm system is installed, actuation of the automatic sprinkler system shall actuate the building fire alarm system.

❖ The audible alarm, sometimes referred to as the "outside ringer" or "water-motor gong," sounds when the sprinkler system has activated. The alarm device may be electrically operated or it may be a true water-motor gong operated by a paddle-wheel-type attachment to the sprinkler system riser that responds to the flow of water in the piping. Though no longer the alarm device of choice, water-motor gongs do have the advantage of not being subject to power failures within or outside the protected building (see Sections 3-10 and 5-15 of NFPA 13 for further information on these devices). The alarm must be installed on the exterior of the building in a location approved by the fire code official. This location is often in close proximity to the fire department connection (FDC), serving a collateral function of helping the responding fire apparatus engineer more promptly locate the FDC. The alarm is not intended to be an evacuation alarm. However, when a fire alarm system is installed, the sprinkler system must be interconnected with the fire alarm system so that when the sprinkler system actuates, it sounds the evacuation alarms required for the fire alarm system.

903.4.3 Floor control valves. Approved supervised indicating control valves shall be provided at the point of connection to the riser on each floor in high-rise buildings.

❖ In high-rise buildings, sprinkler control valves with supervisory initiating devices must be installed at the point of connection to the riser on each floor. Sprinkler control valves on each floor are intended to permit servicing activated systems without impairing the water supply to large portions of the building.

903.5 Testing and maintenance. Sprinkler systems shall be tested and maintained in accordance with Section 901.

❖ Section 901 contains requirements for the testing and maintenance of sprinkler systems. Acceptance tests are necessary to verify that the system performs as intended by design and by the code. Periodic testing and maintenance are essential to verify that the level of protection designed into the building will be operational whenever a fire occurs. Water-based extinguishing systems must be tested and maintained as required by NFPA 25.

903.6 Existing buildings. The provisions of this section are intended to provide a reasonable degree of safety in existing structures not complying with the minimum requirements of the *International Building Code* by requiring installation of an automatic fire-extinguishing system.

❖ This section states the situations in which an automatic fire-extinguishing system must be added to an existing building. This section would not apply to a structure that complies with the requirements for new construction in the IBC.

903.6.1 Pyroxylin plastics. All structures occupied for the manufacture or storage of articles of cellulose nitrate (pyroxylin) plastic shall be equipped with an approved automatic fire-extinguishing system. Vaults located within buildings for the storage of raw pyroxylin shall be protected with an approved automatic sprinkler system capable of discharging 1.66 gallons per minute per square foot (68 L/min/m^2) over the area of the vault.

❖ Although Section 903.6.1 requires an approved automatic fire-extinguishing system in existing buildings used for the manufacture and storage of pyroxylin plastics, the application of this section was not intended to take precedence over the requirements for new buildings housing pyroxylin plastics. Because a quantity limitation is not specified in Section 903.6.1, the sprinkler threshold quantity limitation of 100 pounds (45 kg) in Section 903.2.4.3 should be used. As indicated in the commentary to Section 903.2.4.3, however, the need for additional fire protection should be considered for any amount of pyroxylin plastics because of their flammability and unstable composition.

Even though this section would permit an approved fire-extinguishing system in existing structures containing pyroxylin plastics, vaults must be protected by an approved automatic sprinkler system with a density of 1.66 gpm/ft^2 (68 L/min/m^2) over the entire area of the vault. The high sprinkler density recognizes the need to immerse the pyroxylin plastics in water to counteract the vigorous burn rate of these materials.

SECTION 904 ALTERNATIVE AUTOMATIC FIRE-EXTINGUISHING SYSTEMS

904.1 General. Automatic fire-extinguishing systems, other than automatic sprinkler systems, shall be designed, installed, inspected, tested and maintained in accordance with the provisions of this section and the applicable referenced standards.

❖ Section 904 covers alternative fire-extinguishing systems that use extinguishing agents other than water. Alternative au.tomatic fire-extinguishing systems include wet-chemical, dry-chemical, foam, carbon dioxide, Halon and clean-agent suppression systems. In addition to the provisions of Section 904, the indicated referenced standards include specific installation, maintenance and testing requirements for all systems.

904.2 Where required. Automatic fire-extinguishing systems installed as an alternative to the required automatic sprinkler systems of Section 903 shall be approved by the fire code official. Automatic fire-extinguishing systems shall not be considered alternatives for the purposes of exceptions or reductions permitted by other requirements of this code.

❖ One of the main considerations in selecting an extinguishing agent should be the compatibility of the agent with the hazard. The fire code official is responsible for approving an alternative extinguishing agent. The approval should be based on the compatibility of the agent with the hazard and the potential effectiveness of the agent to suppress a fire involving the hazards present. The code places limitations on alternative systems in that they may not be credited toward a building being equipped throughout with an automatic sprinkler system where the sprinkler system is an alternative or trade-off to a code requirement.

904.2.1 Hood system suppression. Each required commercial kitchen exhaust hood and duct system required by Section 610 to have a Type I hood shall be protected with an approved automatic fire-extinguishing system installed in accordance with this code.

❖ This section requires an effective suppression system to combat fire on the cooking surfaces of grease-producing appliances and within the hood and exhaust system of a commercial kitchen installation. Type I hoods, including the duct system, must be suppressed because they are used for handling grease-laden vapors or smoke whereas Type II hoods handle fumes, steam, heat and odors. Type I hoods are typically required for commercial food heat-processing equipment, such as deep fryers, griddles, charbroilers, broilers and open burner stoves and ranges.

904.3 Installation. Automatic fire-extinguishing systems shall be installed in accordance with this section.

❖ The installation of automatic fire-extinguishing systems must comply with the requirements of Sections 904.3.1 through 904.3.5 in addition to the installation criteria contained in the referenced standard for the proposed type of alternative extinguishing system.

904.3.1 Electrical wiring. Electrical wiring shall be in accordance with the ICC *Electrical Code*.

❖ The ICC *Electrical Code*® (ICC EC™) in turn references NFPA 70 for the design and installation of electrical systems and equipment. All electrical work should also be in compliance with any specific electrical classifications and conditions contained in the referenced standards for each type of system.

904.3.2 Actuation. Automatic fire-extinguishing systems shall be automatically actuated and provided with a manual means of actuation in accordance with Section 904.11.1.

❖ To increase the reliability of the system and to provide the opportunity to initiate the system as a preventive measure, a manual means to activate the system is required (see commentary, Section 904.11.1).

904.3.3 System interlocking. Automatic equipment interlocks with fuel shutoffs, ventilation controls, door closers, window shutters, conveyor openings, smoke and heat vents, and other features necessary for proper operation of the fire-extinguishing system shall be provided as required by the design and installation standard utilized for the hazard.

❖ Shutting off fuel supplies will eliminate potential ignition sources in the protected area. Automatic door and window closers and dampers for forced-air ventilation systems are intended to maintain the desired concentration level of the extinguishing agent in the protected area.

904.3.4 Alarms and warning signs. Where alarms are required to indicate the operation of automatic fire-extinguishing systems, distinctive audible, visible alarms and warning signs shall be provided to warn of pending agent discharge. Where exposure to automatic-extinguishing agents poses a hazard to persons and a delay is required to ensure the evacuation of occupants before agent discharge, a separate warning signal shall be provided to alert occupants once agent discharge has begun. Audible signals shall be in accordance with Section 907.10.2.

❖ Steps and safeguards are necessary to prevent injury or death to personnel in areas where the atmosphere will be made hazardous by agent discharge. Alarms must be installed that will operate on fire detection and agent discharge. A continuous alarm that will sound until the room atmosphere is restored to normal should be installed at entrances to these areas. Warning and instructional signs are to be posted both at the entrances to and within the protected area.

904.3.5 Monitoring. Where a building fire alarm system is installed, automatic fire-extinguishing systems shall be monitored by the building fire alarm system in accordance with NFPA 72.

❖ Automatic fire-extinguishing systems need not be electrically supervised unless the building is equipped with a fire alarm system. This section recognizes the fact that a fire alarm system is not required in all buildings. However, because most alternative fire-extinguishing systems require the space to be evacuated before the system is discharged, they are equipped with evacuation alarms. Interconnection of the fire-extinguishing system evacuation alarm with the building evacuation alarm results in an increased level of hazard notification for the occupants in addition to the electrical supervision of the fire-extinguishing system.

904.4 Inspection and testing. Automatic fire-extinguishing systems shall be inspected and tested in accordance with the provisions of this section prior to acceptance.

❖ The completed installation must be tested and inspected to determine that the system has been installed in compliance with the code and will function as required. Full-scale acceptance tests should be conducted as required by the applicable referenced standard.

904.4.1 Inspection. Prior to conducting final acceptance tests, the following items shall be inspected:

1. Hazard specification for consistency with design hazard.
2. Type, location and spacing of automatic- and manual-initiating devices.
3. Size, placement and position of nozzles or discharge orifices.
4. Location and identification of audible and visible alarm devices.
5. Identification of devices with proper designations.
6. Operating instructions.

❖ This section identifies those items that need to be verified or visually inspected prior to the final acceptance tests. All equipment should be listed, approved and installed in accordance with the manufacturer's recommendations.

904.4.2 Alarm testing. Notification appliances, connections to fire alarm systems, and connections to approved supervising stations shall be tested in accordance with this section and Section 907 to verify proper operation.

❖ Components of fire-extinguishing systems related to alarm devices and their supervision should be tested before the system is approved. Alarm devices should be tested to satisfy the requirements of NFPA 72.

904.4.2.1 Audible and visible signals. The audibility and visibility of notification appliances signaling agent discharge or system operation, where required, shall be verified.

❖ This section requires verification of the audibility and visibility of notification appliances upon installation. For example, Section 907.10.2 prescribes minimum sound pressure levels for audible alarm notification appliances depending upon the occupancy of the building. Audible and visible alarm notification devices must also comply with the applicable requirements in NFPA 72.

904.4.3 Monitor testing. Connections to protected premises and supervising station fire alarm systems shall be tested to verify proper identification and retransmission of alarms from automatic fire-extinguishing systems.

❖ Where monitoring of fire-extinguishing systems is required, such as by Section 904.3.5, all connections related to the supervision of the system must be tested to verify they are in proper working order.

904.5 Wet-chemical systems. Wet-chemical extinguishing systems shall be installed, maintained, periodically inspected and tested in accordance with NFPA 17A and their listing.

❖ NFPA 17A contains minimum requirements for the design, installation, operation, testing and maintenance of wet-chemical preengineered extinguishing systems. Equipment that is typically protected with wet-chemical extinguishing systems includes restaurant, commercial and institutional hoods; plenums; ducts and associated cooking equipment. Strict compliance with the manufacturer's installation instructions is vital for a viable installation.

Wet-chemical solutions used in extinguishing systems are relatively harmless and there is usually no lasting significant effect on a person's skin, respiratory system or clothing. These solutions may produce a mild, temporary irritation but the symptoms will usually disappear when contact is eliminated.

904.5.1 System test. Systems shall be inspected and tested for proper operation at 6-month intervals. Tests shall include a check of the detection system, alarms and releasing devices, including manual stations and other associated equipment. Extinguishing system units shall be weighed and the required amount of agent verified. Stored pressure-type units shall be checked for the required pressure. The cartridge of cartridge-operated units shall be weighed and replaced at intervals indicated by the manufacturer.

❖ This section specifies the frequency for inspection and testing of wet-chemical extinguishing systems. The system and its essential components must be inspected and checked every six months to ensure that the system is in full operating condition.

904.5.2 Fusible link maintenance. Fixed temperature-sensing elements shall be maintained to ensure proper operation of the system.

❖ Wet-chemical extinguishing systems are commonly used to protect commercial cooking equipment. The fusible metal alloy sensing elements are subject to the accumulation of grease or other contaminants that could affect the operation of the fusible link. The sensing elements should be inspected routinely and replaced as needed.

904.6 Dry-chemical systems. Dry-chemical extinguishing systems shall be installed, maintained, periodically inspected and tested in accordance with NFPA 17 and their listing.

❖ NFPA 17 contains the minimum requirements for the design, installation, testing, inspection, approval, operation and maintenance of dry-chemical extinguishing systems.

The fire code official has the authority to approve the type of dry-chemical extinguishing system to be used. NFPA 17 identifies three types of dry-chemical extinguishing systems: total flooding, local application and hand hose-line systems. Only total flooding and local application systems are considered automatic extinguishing systems.

The types of hazards and equipment that can be protected with dry-chemical extinguishing systems include: flammable and combustible liquids and combustible gases; combustible solids, which melt when involved in a fire; electrical hazards, such as transformers or oil circuit breakers; textile operations subject to flash surface fires; ordinary combustibles such as wood, paper or cloth and restaurant and commercial hoods, ducts and associated cooking appliance hazards, such as deep fat fryers and some plastics, depending on the type of material and configuration.

Total flooding dry-chemical extinguishing systems are used only where there is a permanent enclosure about the hazard that is adequate to enable the required concentration to be built up. The total area of unenclosable openings must not exceed 15 percent of the total area of the sides, top and bottom of the enclosure. Consideration must be given to eliminating the probable sources of reignition within the enclosure because the extinguishing action of dry-chemical systems is transient.

Local application of dry-chemical extinguishing systems is to be used for extinguishing fires where the hazard is not enclosed or where the enclosure does not conform to the requirements for total flooding systems. Local application systems have successfully protected hazards involving flammable or combustible liquids, gases and shallow solids, such as paint deposits.

NFPA 17 also discusses preengineered dry-chemical systems consisting of components designed to be installed in accordance with pretested limitations as tested and labeled by a testing agency. Preengineered systems must be installed within the limitations that have been established by the testing agency and may include total flooding, local application or a combination of both types of systems.

The type of dry chemical used in the extinguishing system is a function of the hazard to be protected. The type of dry chemical used in a system should not be changed, unless it has been proven changeable by a testing laboratory, is recommended by the manufacturer of the equipment and is acceptable to the fire code official for the hazard being protected. Additional guidance on the use of various dry-chemical agents can be found in NFPA 17.

904.6.1 System test. Systems shall be inspected and tested for proper operation at 6-month intervals. Tests shall include a check of the detection system, alarms and releasing devices, including manual stations and other associated equipment. Extinguishing system units shall be weighed, and the required amount of agent verified. Stored pressure-type units shall be checked for the required pressure. The cartridge of cartridge-operated units shall be weighed and replaced at intervals indicated by the manufacturer.

❖ This section specifies the frequency for inspection and testing of dry-chemical extinguishing systems. The system and its essential components must be inspected and checked every six months to ensure that the system is in full operating condition.

904.6.2 Fusible link maintenance. Fixed temperature-sensing elements shall be maintained to ensure proper operation of the system.

❖ Wet-chemical extinguishing systems are commonly used to protect commercial cooking systems and other hazardous use conditions. In these applications the fusible metal alloy sensing elements are subject to the accumulation of grease or other contaminants that could affect the operation of the fusible link. The sensing elements should be inspected routinely and replaced as needed.

904.7 Foam systems. Foam-extinguishing systems shall be installed, maintained, periodically inspected and tested in accordance with NFPA 11, NFPA 11A and NFPA 16 and their listing.

❖ NFPA 11 covers the characteristics of foam-producing materials used for fire protection and the requirements for design, installation, operation, testing and maintenance of equipment and systems, including those used in combination with other fire-extinguishing agents. The minimum requirements are covered for flammable and combustible liquid hazards in local areas within buildings, storage tanks and indoor and outdoor processing areas.

Low-expansion foam is defined as an aggregation of air-filled bubbles resulting from the mechanical expansion of a foam solution by air with a foam-to-solution volume ratio of less than 20:1. It is most often used to protect flammable and combustible liquid hazards. Also, low-expansion foam may be used for heat radiation protection. Combined-agent systems involve the application of low-expansion foam to a hazard simultaneously or sequentially with dry-chemical powder.

NFPA 11A gives minimum requirements for the installation, design, operation, testing and maintenance of medium- and high-expansion foam systems. Medium-expansion foam is defined as an aggregation of air-filled bubbles resulting from the mechanical expansion of a foam solution by air or other gases with a foam-to-solution volume ratio of 20:1 to 200:1. High-expansion foam has a foam-to-solution volume ratio of 200:1 to approximately 1,000:1.

Medium-expansion foam may be used on solid fuel and liquid fuel fires where some degree of in-depth coverage is necessary (for example, for the total flooding of small, enclosed or partially enclosed volumes, such as engine test cells, transformer rooms, etc.). High-expansion foam is most suitable for filling volumes in which fires exit at various levels. For example, high-expansion foam can be used effectively against high-rack storage fires in enclosures such as in underground passages, where it may be dangerous to send personnel to control fires involving liquefied natural gas (LNG) and liquefied petroleum gas (LP-gas), and to provide vapor dispersion control for LNG and ammonia spills. High-expansion foam is particularly suited for indoor fires in confined spaces, since it is highly susceptible to wind and lack-of-confinement effects.

NFPA 16 contains the minimum requirements for

open-head deluge-type foam-water sprinkler systems and foam-water spray systems. The systems are especially applicable to the protection of most flammable liquid hazards and have been used successfully to protect aircraft hangars and truck loading racks.

904.7.1 System test. Foam-extinguishing systems shall be inspected and tested at intervals in accordance with NFPA 25.

❖ Although Section 904.7.1 references NFPA 25 as the standard for the inspection and testing of foam-extinguishing systems, NFPA 25 is limited to water-based extinguishing systems. NFPA 25 technically addresses only foam-water sprinkler systems as specified in NFPA 16. NFPA 11 and NFPA 11A should be consulted for inspection and testing intervals for other foam systems.

As with other alternative fire-extinguishing systems, the inspection and testing of foam systems ensure the system is fully operational. In addition to general maintenance of equipment, the condition of the foam concentrate and its storage tanks or containers should be inspected at least once a year to ensure adequate quality. The desired concentration of the foam concentrate in a stagnant storage situation may deteriorate over time.

904.8 Carbon dioxide systems. Carbon dioxide extinguishing systems shall be installed, maintained, periodically inspected and tested in accordance with NFPA 12 and their listing.

❖ NFPA 12 provides minimum requirements for the design, installation, testing, inspection, approval, operation and maintenance of carbon dioxide extinguishing systems.

Carbon dioxide extinguishing systems are useful in extinguishing fires in specific hazards or equipment in occupancies where an inert electrically nonconductive medium is essential or desirable and where cleanup of other extinguishing agents, such as dry-chemical residue, presents a problem. Carbon dioxide systems have satisfactorily protected the following: flammable liquids; electrical hazards, such as transformers, oil switches, rotating equipment and electronic equipment; engines using gasoline and other flammable liquid fuels; ordinary combustibles, such as paper, wood and textiles and hazardous solids.

The fire code official has the authority to approve the type of carbon dioxide system to be installed. NFPA 12 defines four types of carbon dioxide systems: total flooding, local application, hand hose lines and standpipe and mobile supply systems. Only total flooding and local application systems are automatic suppression systems.

Total-flooding systems may be used where there is a permanent enclosure around the hazard that is adequate to allow the required concentration to be built up and maintained for the required period of time, which varies for different hazards. Examples of hazards that have been successfully protected by total flooding systems include rooms, vaults, enclosed machines, ducts, ovens and containers and their contents.

Local application systems may be used for extinguishing surface fires in flammable liquids, gases and shallow solids where the hazard is not enclosed or the enclosure does not conform to the requirements for a total-flooding system. Examples of hazards that have been successfully protected by local application systems include dip tanks, quench tanks, spray booths, oil-filled electric transformers and vapor vents.

904.8.1 System test. Systems shall be inspected and tested for proper operation at 12-month intervals.

❖ To ensure adequate operation, carbon dioxide systems must be inspected and tested at least once a year.

904.8.2 High-pressure cylinders. High-pressure cylinders shall be weighed and the date of the last hydrostatic test shall be verified at 6-month intervals. Where a container shows a loss in original content of more than 10 percent, the cylinder shall be refilled or replaced.

❖ Because of the potential of unobserved leaking high-pressure cylinders, they need to be weighed semi-annually to verify the concentration level is always within at least 10 percent of the original content.

904.8.3 Low-pressure containers. The liquid-level gauges of low-pressure containers shall be observed at one-week intervals. Where a container shows a content loss of more than 10 percent, the container shall be refilled to maintain the minimum gas requirements.

❖ A weekly visual observation of the liquid level gauges of low-pressure containers is required to ensure that there has been no significant leakage.

904.8.4 System hoses. System hoses shall be examined at 12-month intervals for damage. Damaged hoses shall be replaced or tested. At five-year intervals, all hoses shall be tested.

❖ The maintenance of system hoses is essential to ensuring the reliability of their use in an emergency. Although system hoses need to be visually checked on an annual basis only, a complete pressure test as indicated in Section 904.8.4.1 should be done every five years.

904.8.4.1 Test procedure. Hoses shall be tested at not less than 2,500 pounds per square inch (psi) (17 238 kPa) for high-pressure systems and at not less than 900 psi (6206 kPa) for low-pressure systems.

❖ Every five years, system hoses for both high-pressure/low-pressure systems should be pressure tested to verify they are still in proper operating condition. The test typically involves filling the hose with water. The hose is then pressurized at the desired test pressure for at least 1 minute to observe any potential distortions or leakage in the hose. All hose assemblies that do not pass the test should be marked, destroyed and replaced with new hose assemblies. Hose assemblies that pass the test should be marked, dated and returned to service.

904.8.5 Auxiliary equipment. Auxiliary and supplementary components, such as switches, door and window releases, interconnected valves, damper releases and supplementary alarms, shall be manually operated at 12-month intervals to ensure that such components are in proper operating condition.

❖ The effectiveness of the carbon dioxide extinguishing system is also dependent upon the operation of its auxiliary components. These components must also be manually operated at least once a year.

904.9 Halon systems. Halogenated extinguishing systems shall be installed, maintained, periodically inspected and tested in accordance with NFPA 12A and their listing.

❖ NFPA 12A contains minimum requirements for the design, installation, testing, inspection, approval, operation and maintenance of Halon 1301 extinguishing systems. Halon 1301 fire-extinguishing systems are useful in specific hazards, equipment or occupancies where an electrically nonconductive medium is essential or desirable and where cleanup of other extinguishing agents presents a problem.

Halon 1301 systems have satisfactorily protected gaseous and liquid flammable materials; electrical hazards, such as transformers, oil switches and rotating equipment; engines using gasoline and other flammable fuels; ordinary combustibles, such as paper, wood and textiles and hazardous solids. Halon 1301 systems have also satisfactorily protected electronic computers, data processing equipment and control rooms.

The fire code official has the authority to approve the type of halogenated extinguishing system to be installed. NFPA 12A defines two types of halogenated extinguishing system: total flooding and local application. Total-flooding systems may be used where there is a fixed enclosure around the hazard that is adequate to enable the required Halon concentration to be built up and maintained for the required period of time to enable the effective extinguishing of the fire. Total-flooding systems may provide fire protection for rooms, vaults, enclosed machines, ovens, containers, storage tanks and bins.

Local application systems are used where there is not a fixed enclosure around the hazard or where the fixed enclosure around the hazard is not adequate to enable an extinguishing concentration to be built up and maintained in the space. Hazards that may be successfully protected by local application systems include dip tanks, quench tanks, spray booths, oil-filled electric transformers and vapor vents.

Two other considerations in selecting the proper extinguishing system are ambient temperature and the personnel hazards associated with the agent. The ambient temperature of the enclosure for a total-flooding system must be above 70°F (21°C) for Halon 1301 systems. Special consideration must also be given to the use of Halon systems when the temperatures are in excess of 900°F (482°C) because Halon will readily decompose at such temperatures and the products of decomposition can be extremely irritating if inhaled, even in small amounts.

Halon 1301 total-flooding systems must not be used in concentrations greater than 10 percent in normally occupied areas. Where personnel cannot vacate the area within 1 minute, Halon 1301 total-flooding systems must not be used in normally occupied areas with concentrations greater than 7 percent. Halon 1301 total-flooding systems may be used with concentrations of up to 15 percent if the area is not normally occupied and the area can be evacuated within 30 seconds.

The use of halogenated extinguishing systems has become a concern with respect to the potential environmental effects of Halon. Halongenated fire-extinguishing agents have been identified as a source of emissions, resulting in the depletion of the stratospheric ozone layer and, in accordance with the Montreal protocol, the ceasing of its production in January 1994. Therefore, the supply of Halon is limited and new supplies of halogenated extinguishing agents will not be available in the future. Existing supplies of Halon can, however, continue to be used in existing, undischarged systems or to recharge discharged systems. This newfound need for Halon supplies has given rise to new industries geared to the ranking, recycling and reclamation of existing Halon supplies. Alternative extinguishing agents have been developed to replace halogenated agents (see Section 904.10).

904.9.1 System test. Systems shall be inspected and tested for proper operation at 12-month intervals.

❖ To ensure adequate operation, Halon systems must be inspected and tested at least once a year.

904.9.2 Containers. The extinguishing agent quantity and pressure of containers shall be checked at 6-month intervals. Where a container shows a loss in original weight of more than 5 percent or a loss in original pressure (adjusted for temperature) of more than 10 percent, the container shall be refilled or replaced. The weight and pressure of the container shall be recorded on a tag attached to the container.

❖ Because of the potential for unobserved leakage of the Halon containers, they should be checked at least semi-annually to verify the original weight and pressure are within the designated tolerances. When necessary, containers should be refilled or replaced when the desired levels are not maintained. The containers should also be checked for evidence of corrosion or mechanical damage.

904.9.3 System hoses. System hoses shall be examined at 12-month intervals for damage. Damaged hoses shall be replaced or tested. At 5-year intervals, all hoses shall be tested.

❖ Maintenance of system hoses is essential to ensuring the reliability of their use in an emergency. System hoses need to be visually checked annually. A complete pressure test as indicated in Section 904.9.3.1 should be done every five years.

904.9.3.1 Test procedure. For Halon 1301 systems, hoses shall be tested at not less than 1,500 psi (10 343 kPa) for 600 psi (4137 kPa) charging pressure systems and not less than 900 psi (6206 kPa) for 360 psi (2482 kPa) charging pressure systems. For Halon 1211 hand-hose line systems, hoses shall be tested at 2,500 psi (17 238 kPa) for high-pressure systems and 900 psi (6206 kPa) for low-pressure systems.

❖ Every five years, system hoses should be pressure tested to verify they are still in proper operating condition. This section specifies the test pressure for the various types of Halon systems. The pressure test is intended to check for any potential distortion or leaking in the hose (see commentary, Section 904.8.4.1).

904.9.4 Auxiliary equipment. Auxiliary and supplementary components, such as switches, door and window releases, interconnected valves, damper releases and supplementary alarms, shall be manually operated at 12-month intervals to ensure such components are in proper operating condition.

❖ The effectiveness of halogenated extinguishing systems is also dependent upon the operation of its auxiliary components. These components must be manually operated at least once a year.

904.10 Clean-agent systems. Clean-agent fire-extinguishing systems shall be installed, maintained, periodically inspected and tested in accordance with NFPA 2001 and their listing.

❖ NFPA 2001 contains minimum requirements for the design, installation, testing, inspection and operation of clean-agent fire-extinguishing systems. A clean agent is an electrically nonconducting suppression agent that is volatile or gaseous at discharge and does not leave a residue on evaporation. Clean-agent fire-extinguishing systems are installed in locations that are enclosed and have openings in the protected area that can be sealed on activation of the alarm to provide effective clean-agent concentrations. A clean-agent fire-extinguishing system should not be installed in locations that cannot be sealed unless testing has shown that adequate concentrations can be developed and maintained.

The two categories of clean agents are halocarbon compounds and inert gas agents. Halocarbon compounds include bromine, carbon, chlorine, fluorine, hydrogen and iodine. Halocarbon compounds suppress fire by a combination of breaking the chemical chain reaction of the fire, reducing the oxygen supporting the fire and reducing the ambient temperature of the fire origin to reduce the propagation of the fire. Inert gas agents contain primary components consisting of helium, neon, argon or a combination of these. Inert gases work by reducing the oxygen concentration around the fire origin to a level that does not support combustion.

Clean-agent fire-extinguishing systems were developed in response to the demise of Halon as an acceptable fire-extinguishing agent because of its harmful effect on the environment. Although the original hope for a Halon substitute was that these new clean agents could be directly and proportionally substituted for Halon agents in existing systems (drop in replacements), research has shown that clean agents are less efficient in extinguishing fires than are the Halons they were intended to replace and require approximately 60 percent more agent by weight and volume in storage to do the same job. Additionally, the physical and chemical characteristics of clean agents differ sufficiently from Halon to require different nozzles in addition to the need for larger storage vessels. Existing piping systems should be salvaged for use with clean agents only if they are carefully evaluated and determined to be hydraulically compatible with the flow characteristics of the new agent.

This section also relies on strict adherence to the system manufacturer's design and installation instructions for code compliance. As with many of the alternative fire suppression systems covered in this chapter, clean-agent systems are, for the most part, subjected by their manufacturers to a testing and listing program conducted by an approved testing agency. In such testing and listing programs, the clean agent is listed for use with specific equipment and equipment is listed for use with specific clean agents. The resultant listings include reference to the manufacturer's installation manuals, thereby giving the fire code official another valuable resource for reviewing and approving clean-agent systems.

Although clean agents have found a limited market for local application uses, such as a replacement for Halon 1211 in portable fire extinguishers, their primary application is in total flooding systems and they are available in both engineered and preengineered configurations.

Engineered clean-agent systems are specifically designed for protection of a particular hazard, whereas preengineered systems are designed to operate within predetermined limitations up to the noted maximums, thus allowing broader applicability to a variety of hazard applications.

Total flooding systems are used where there is a fixed enclosure around the hazard that is adequate to enable the required clean-agent concentration to build up and be maintained within the space long enough to extinguish the fire. Such applications can include vaults, ovens, containers, tanks, computer rooms, paint lockers or enclosed machinery. In selecting the clean agent to be used in a given application, careful consideration must be given to whether the protected area is a normally occupied space, because different agents have different levels of concentration at which they may be a health hazard to occupants of the area.

The fire code official has the authority to approve the type of clean-agent system to be installed and should become familiar with the unique characteristics and hazards of clean-agent extinguishing systems using all available resources on the subject.

904.10.1 System test. Systems shall be inspected and tested for proper operation at 12-month intervals.

❖ To ensure adequate operation, all clean-agent systems must be inspected and tested at least once a year.

904.10.2 Containers. The extinguishing agent quantity and pressure of the containers shall be checked at 6-month intervals. Where a container shows a loss in original weight of more than 5 percent or a loss in original pressure, adjusted for temperature, of more than 10 percent, the container shall be refilled or replaced. The weight and pressure of the container shall be recorded on a tag attached to the container.

❖ Because of the potential for unobserved leakage of the clean-agent containers, they should be checked at least semi-annually to verify the original weight and pressure are within the designated tolerances. When necessary, containers should be refilled or replaced when the desired levels are not maintained.

904.10.3 System hoses. System hoses shall be examined at 12-month intervals for damage. Damaged hoses shall be replaced or tested. All hoses shall be tested at 5-year intervals.

❖ The maintenance of system hoses is essential to ensuring their reliability in an emergency. System hoses should be visually checked annually. A complete pressure test should be done every five years.

904.11 Commercial cooking systems. The automatic fire-extinguishing system for commercial cooking systems shall be of a type recognized for protection of commercial cooking equipment and exhaust systems of the type and arrangement protected. Preengineered automatic dry- and wet-chemical extinguishing systems shall be tested in accordance with UL 300 and listed and labeled for the intended application. Other types of automatic fire-extinguishing systems shall be listed and labeled for specific use as protection for commercial cooking operations. The system shall be installed in accordance with this code, its listing and the manufacturer's installation instructions. Automatic fire-extinguishing systems of the following types shall be installed in accordance with the referenced standard indicated, as follows:

1. Carbon dioxide extinguishing systems, NFPA 12.
2. Automatic sprinkler systems, NFPA 13.
3. Foam-water sprinkler system or foam-water spray systems, NFPA 16.
4. Dry-chemical extinguishing systems, NFPA 17.
5. Wet-chemical extinguishing systems, NFPA 17A.

Exception: Factory-built commercial cooking recirculating systems that are tested in accordance with UL 197 and listed, labeled and installed in accordance with Section 304.1 of the *International Mechanical Code*.

❖ The history of commercial kitchen exhaust systems shows that the mixture of flammable grease and effluents carried by such systems and the potential for the cooking equipment to act as an ignition source contribute to a higher level of hazard for kitchen exhaust systems than is normally found in many other exhaust systems. Furthermore, fire in a grease exhaust duct can produce temperatures of 2,000°F (1,093°C) or more and heat radiating form the duct can ignite nearby combustibles. As a result, the code requires exhaust systems serving grease-producing equipment to include fire suppression to protect the cooking surfaces, hood, filters and exhaust duct to confine a fire to the hood and duct system, thus reducing the likelihood of it spreading to the structure.

In addition to the general requirements of this section, five industry standards are referenced for the installation of fire-extinguishing systems protecting commercial food heat-processing equipment and kitchen exhaust systems. Design professionals should specify and design fire-extinguishing systems to comply with these referenced standards. Only the installation of fire-extinguishing systems is regulated by these references. Where preengineered automatic dry- and wet-chemical extinguishing systems are installed, they must be listed and labeled for the specific cooking operation and tested in accordance with UL 300. Design and construction requirements for the specific types of fire-extinguishing systems are found in the respective sections of the referenced standards.

Regulatory requirements for the approval and installation of fire-extinguishing systems are the same as the approval required for all mechanical equipment and appliances. This section, therefore, requires extinguishing systems to be listed and labeled by an approved agency and installed in accordance with their listing and the manufacturer's installation instructions.

The exception allows factory-built commercial cooking recirculating systems to be installed if they have been tested and listed in accordance with UL 197. It is important that they be installed in accordance with the manufacturer's installation instructions to ensure the listing requirements are met. An improper installation could result in hazardous vapors being discharged back into the kitchen.

Commercial cooking recirculating systems consist of an electric cooking appliance and an integral or matched packaged hood assembly. The hood assembly consists of a fan, collection hood, grease filter, fire damper, fire-extinguishing system and air filter, such as an electrostatic precipitator. These systems are tested for fire safety and emissions. The grease vapor (condensible particulate matter) in the effluent at the system discharge is not allowed to exceed a concentration of 5.0 mg/m^3. Recirculating systems are not used with fuel-fired appliances because the filtering systems do not remove combustion products. Kitchens require ventilation in accordance with Chapter 4 of the IMC.

904.11.1 Manual system operation. A manual actuation device shall be located at or near a means of egress from the cooking area, a minimum of 10 feet (3048 mm) and a maximum of 20 feet (6096 mm) from the kitchen exhaust system. The manual actuation device shall be located a minimum of 4 feet (1219 mm) and a maximum of 5 feet (1524 mm) above the floor. The manual actuation shall require a maximum force of 40 pounds (178 N) and a maximum movement of 14 inches (356 mm) to actuate the fire suppression system.

Exception: Automatic sprinkler systems shall not be required to be equipped with manual actuation means.

❖ The manual device, usually a pull station, mechanically activates the suppression system. The typical system uses a mechanical circuit of cables under tension to hold the system in the armed (cocked) mode. Melting of a fusible link or actuation of a manual pull station causes the cable to lose tension, which, in turn, starts the discharge of the suppression agent. The manual actuation device must be readily and easily usable by the building occupants; therefore, the device must not require excessive force or range of movement to cause actuation.

Manual actuation is not required for automatic sprinkler systems because the typical system design will employ closed heads and wet system piping. A manual actuation valve would serve no purpose because sprinkler heads are already supplied with pressurized water and will discharge water only when the individual fusible elements open the heads.

904.11.2 System interconnection. The actuation of the fire suppression system shall automatically shut down the fuel or electrical power supply to the cooking equipment. The fuel and electrical supply reset shall be manual.

❖ The actuation of any fire suppression system must automatically shut off all sources of fuel or power to all cooking equipment located beneath the exhaust hood and protected by the suppression system. This requirement is intended to shut off all heat sources that could reignite or intensify a fire. Shutting off a fuel and power supply to cooking appliances will eliminate an ignition source and allow the cooking surfaces to cool down. This shutdown is accomplished with mechanical or electrical interconnections between the suppression system and a shutoff valve or switch located on the fuel or electrical supply.

Common fuel shutoff valves include mechanical-type gas valves and electrical solenoid-type gas valves. Contactor-type switches or shunt-trip circuit breakers can be used for electrically heated appliances. The fuel or electric source must not be automatically restored after the suppression system has been actuated.

Chemical-type fire-extinguishing systems discharge for only a limited time and can discharge only once before recharge and reset; therefore, precautions must be taken to prevent a fire from reigniting. After a fire, the fuel and power supply will be locked out, thereby preventing the operation of the appliances until all systems are again ready for operation. Fuel and power supply shutoff must be manually restored by resetting a mechanical linkage or holding (latching)-type circuit.

904.11.3 Carbon dioxide systems. When carbon dioxide systems are used, there shall be a nozzle at the top of the ventilating duct. Additional nozzles that are symmetrically arranged to give uniform distribution shall be installed within vertical ducts exceeding 20 feet (6096 mm) and horizontal ducts exceeding 50 feet (15 240 mm). Dampers shall be installed at either the top or the bottom of the duct and shall be arranged to operate automatically upon activation of the fire-extinguishing system. When the damper is installed at the top of the duct, the top nozzle shall be immediately below the damper. Automatic carbon dioxide fire-extinguishing systems shall be sufficiently sized to protect all hazards venting through a common duct simultaneously.

❖ This section states specific design requirements for nozzle locations, dampers and ducts for carbon dioxide extinguishing systems that may be used to protect commercial cooking systems. These requirements are intended to supercede similar more general provisions in NFPA 12.

904.11.3.1 Ventilation system. Commercial-type cooking equipment protected by an automatic carbon dioxide extinguishing system shall be arranged to shut off the ventilation system upon activation.

❖ Shutting down the ventilation system upon activation of the carbon dioxide extinguishing system maintains the desired concentration of carbon dioxide to suppress the fire. Leakage of gas from the protected area should be kept to a minimum. Where leakage is anticipated, additional quantities of carbon dioxide should be provided to compensate for any losses.

904.11.4 Special provisions for automatic sprinkler systems. Automatic sprinkler systems protecting commercial-type cooking equipment shall be supplied from a separate, readily accessible, indicating-type control valve that is identified.

❖ This section requires a separate control valve in the water line to the sprinklers protecting the cooking and ventilating system. The additional valve allows the flexibility to shut off the system for repairs or for cleanups after sprinkler discharge without taking the entire system out of service.

904.11.4.1 Listed sprinklers. Sprinklers used for the protection of fryers shall be listed for that application and installed in accordance with their listing.

❖ Sprinklers specifically listed for such use must be used when protecting deep-fat fryers. These specially listed sprinklers use finer water droplets than standard spray sprinklers. The water spray lowers the temperature below a point where the fire can sustain itself and reduces the possibility of expanding the fire.

904.11.5 Commercial cooking equipment. Portable fire extinguishers shall be provided within a 30-foot (9144 mm) travel distance of commercial-type cooking equipment. Cooking equipment involving vegetable or animal oils and fats shall be protected by a Class K rated portable extinguisher.

❖ To combat a fire in its incipient stage, access to a manual means of extinguishment is critical. Although a 30-foot (9144 mm) maximum travel distance is specified, the location of the extinguisher should be a safe distance from the cooking equipment so that it will not become involved in the fire. Only Class K-rated extinguishers that have been tested on commercial cooking appliances should be used (see commentary, Section 906.4).

904.11.6 Operations and maintenance. Commercial cooking systems shall be operated and maintained in accordance with this section.

❖ Most fires in commercial kitchens involve the cooking appliance and exhaust system in some way. Proper operation of the system in accordance with the IMC as well as routine maintenance can reduce the hazards related to the collection and removal of smoke and grease-laden vapors.

904.11.6.1 Ventilation system. The ventilation system in connection with hoods shall be operated at the required rate of air movement, and classified grease filters shall be in place when equipment under a kitchen grease hood is used.

❖ The hood must be designed to adequately collect and exhaust fumes, smoke and vapors from the area over which the hood is installed. To accomplish this, the hood must cause an airflow pattern that will sweep and direct the fumes, smoke and vapors upward from the cooking surfaces into the hood inlet.

The IMC specifies the minimum quantity of exhaust air necessary for effective removal of cooking vapors and the approximate amount of makeup air necessary for proper operation. The quantity of required exhaust is as much a function of the operational characteristics of the cooking equipment as it is a function of the size of the cooking surface or the exhaust hood opening area and the presence of walls and side panels. Manufacturer recommendations should be followed where applicable.

Approved grease filters prevent large amounts of grease from collecting in the hood, in exhaust ducts, on fan blades and at the exhaust system termination. The accumulation of grease can cause blockage in ducts, cause equipment failure and create a fire hazard. It therefore makes sense to have grease filters in place whenever commercial cooking equipment is used.

904.11.6.2 Grease extractors. Where grease extractors are installed, they shall be operated when the commercial-type cooking equipment is used.

❖ As noted in the commentary for Section 904.11.6.1, it is imperative that grease removal devices be operating when commercial cooking equipment is used. Grease removal extractors and similar devices range from simple designs, such as a configuration of baffles, to elaborate hot water scrubbers and electrostatic precipitators. The devices must be installed to comply with the IMC and the manufacturer's installation instructions.

904.11.6.3 Cleaning. Hoods, grease-removal devices, fans, ducts and other appurtenances shall be cleaned at intervals necessary to prevent the accumulation of grease. Cleanings shall be recorded, and records shall state the extent, time and date of cleaning. Such records shall be maintained on the premises.

❖ frequent cleaning schedule must be maintained to remove the accumulation of grease residue within the exhaust system. The hood, grease removal devices, ducts, fans, discharge nozzles and other components must be cleaned regularly to prevent excessive accumulation of grease. The frequency of such cleaning can vary depending on the amount and type of usage; however, the equipment should be inspected daily or weekly or as often as needed to monitor the rate of accumulation. When grease residues are evident, the system should be cleaned. A record of all cleaning must be maintained by the person or party responsible for the system. The records must indicate the method of cleaning and the time between cleanings.

904.11.6.4 Extinguishing system service. Automatic fire-extinguishing systems shall be serviced at least every 6 months and after activation of the system. Inspection shall be by qualified individuals, and a certificate of inspection shall be forwarded to the fire code official upon completion.

❖ Range hood fire-extinguishing systems must be inspected and serviced at regular intervals to determine that they are ready to perform the intended function. The extent of service and maintenance depends on the type of fire-extinguishing system installed. The NFPA standard corresponding to the installed extinguishing agent should be consulted for additional service requirements (see commentary, Section 904.11).

904.11.6.5 Fusible link and sprinkler head replacement. Fusible links and automatic sprinkler heads shall be replaced at least annually, and other protection devices shall be serviced or replaced in accordance with the manufacturer's instructions.

Exception: Frangible bulbs are not required to be replaced annually.

❖ Because of the potential accumulation of grease or other contaminants that could adversely affect proper operation, fusible links and automatic sprinkler heads must be replaced at least annually. The sensing elements of the fusible link devices as well as the sprinkler heads should be routinely visually inspected and replaced as needed, at least annually.

The exception allows frangible bulb-type sprinklers to not be replaced as long as the annual examination shows no accumulation of grease or other contaminants.

SECTION 905
STANDPIPE SYSTEMS

905.1 General. Standpipe systems shall be provided in new buildings and structures in accordance with this section. Fire hose threads used in connection with standpipe systems shall be approved and shall be compatible with fire department hose threads. The location of fire department hose connections shall be approved. In buildings used for high-piled combustible storage, fire protection shall be in accordance with Chapter 23.

❖ Standpipe systems are required in buildings to provide a quick, convenient water source for fire department use where hose lines would otherwise be impractical, such as in high-rise buildings. Standpipe systems can also be used prior to deployment of hose lines from fire department apparatus. The requirements for standpipes are based on practical requirements of typical fire-fighting operations and the nationally recognized standard, NFPA 14.

The threads on connections to which the fire department may connect a hose must be compatible with the fire department hose threads (see commentary, Section 903.3.6). Chapter 23 requires a Class I standpipe system in exit passageways of buildings used for high-piled storage. High-piled storage involves the solid piled, palletized or rack storage of combustible materials over 12 feet (3,658 mm) high.

905.2 Installation standards. Standpipe systems shall be installed in accordance with this section and NFPA 14.

❖ This section requires the installation of standpipe systems to comply with the applicable provisions of NFPA 14 in addition to Section 905. NFPA 14 contains the minimum requirements for the installation of standpipe and hose systems for buildings and structures. The standard addresses additional requirements not addressed in the code, such as pressure limitations, minimum flow rates, piping specifications, hose connection details, valves, fittings, hangers and the testing and inspection of standpipes. The periodic inspection, testing and maintenance of standpipe systems must comply with NFPA 25.

Section 905 and NFPA 14 recognize three classes of standpipe systems: Class I, II or III. The type of system required depends on building height, building area, type of occupancy and the extent of automatic sprinkler protection. Section 905 also recognizes five types of standpipe systems: automatic dry, automatic wet, manual dry, manual wet and semiautomatic dry. The use of each type of system is limited to the building conditions and locations identified in Section 905.3. The classes and types of standpipe systems are defined in Section 902.1.

905.3 Required installations. Standpipe systems shall be installed where required by Sections 905.3.1 through 905.3.6 and in the locations indicated in Sections 905.4, 905.5 and 905.6. Standpipe systems are permitted to be combined with automatic sprinkler systems.

Exception: Standpipe systems are not required in Group R-3 occupancies.

❖ Standpipe systems are installed in buildings based on the occupancy, fire department accessibility and special conditions that may require manual fire suppression exceeding the capacity of a fire extinguisher. Standpipe systems are most commonly required for buildings that exceed the height threshold requirement in Section 905.3.1 or the area threshold requirement in Section 905.3.2. Specific occupancies such as covered mall buildings, stages and underground buildings, because of their use or occupancy, also require a standpipe system.

This section also states that a standpipe system does not have to be separate from an installed sprinkler system. It is common practice in multistory buildings for the standpipe system risers to also serve as risers for the automatic sprinkler systems.

In these instances, precautions need to be taken so that the operation of one system will not interfere with the operation of the other system. Therefore, control valves for the sprinkler system must be installed where the sprinklers are connected to the standpipe riser at each floor level. This allows the standpipe system to remain operational, even if the sprinkler system is shut off at the floor control valve.

The exception recognizes that standpipe systems in Group R-3 occupancies would be of minimal value to the fire department and would send the wrong message to the occupants of a dwelling unit. In the case of multiple single-family dwellings, each dwelling unit has a separate entrance and is separated from the other units by 1-hour fire-resistance-rated construction. These conditions permit ready access to fires and also provide for a degree of fire containment through compartmentation, which is not always present in other occupancies.

905.3.1 Building height. Class III standpipe systems shall be installed throughout buildings where the floor level of the highest story is located more than 30 feet (9144 mm) above the lowest level of the fire department vehicle access, or where the floor level of the lowest story is located more than 30 feet (9144 mm) below the highest level of fire department vehicle access.

Exceptions:

1. Class I standpipes are allowed in buildings equipped throughout with an automatic sprinkler system in accordance with Section 903.3.1.1 or 903.3.1.2.
2. Class I manual standpipes are allowed in open parking garages where the highest floor is located not more than 150 feet (45 720 mm) above the lowest level of fire department vehicle access.
3. Class I manual dry standpipes are allowed in open parking garages that are subject to freezing temperatures, provided that the hose connections are located as required for Class II standpipes in accordance with Section 905.5.
4. Class I standpipes are allowed in basements equipped throughout with an automatic sprinkler system.

❖ Given the available manpower on the fire department vehicle, standard fire-fighting operations and standard hose sizes, a 30-foot (9,144 mm) vertical distance is generally considered the maximum height to which a typical fire department engine company can practically and readily extend its hose lines. Thus, the maximum vertical travel (height) threshold is based on the time it would take a typical fire department engine (pumper) company to manually suppress a fire. The standpipe connection reduces the time needed for the fire department to extend hose lines up or down stairways to advance and apply water to the fire. For this use, a minimum Class III standpipe system is required.

With respect to the height of the building, the threshold is measured from the level at which the fire department can gain access to the building directly from its vehicle. Floor levels above grade are measured from the lowest level of fire department vehicle access to the highest floor level above [see Figure 905.3.1(1)]. If a building contains floor levels below the level of fire department vehicle access, the measurement is made from the highest level of fire department vehicle access to the lowest floor level. In cases where a building has more than one level of fire department vehicle access, the most restrictive measurement is used because it is not known at which level the fire department will access the building. In other words, the vertical distance is to be measured from the more restrictive level of fire department vehicle access to the level of the highest (or lowest, if below) floor [see Figure 905.3.1(2)].

The threshold based on the height of the building is independent of the occupancy of the building, the area of the building or the presence of an automatic sprinkler system. This is based on the universal need to be able to provide a water supply for fire suppression in any building and on the limitations of the physical effort necessary to extend hose lines vertically.

905.3.2 Group A. Class I automatic wet standpipes shall be provided in nonsprinklered Group A buildings having an occupant load exceeding 1,000 persons.

Exceptions:

1. Open-air-seating spaces without enclosed spaces.
2. Class I automatic dry and semiautomatic dry standpipes or manual wet standpipes are allowed in buildings where the highest floor surface used for human occupancy is 75 feet (22 860 mm) or less above the lowest level of fire department vehicle access.

❖ The main concern in assembly occupancies with a high occupant load is evacuation. Many occupants may not be familiar with either their surroundings or the egress arrangement in the building. This section also assumes the building is not sprinklered; therefore, control and suppression of the fire is left to the fire department.

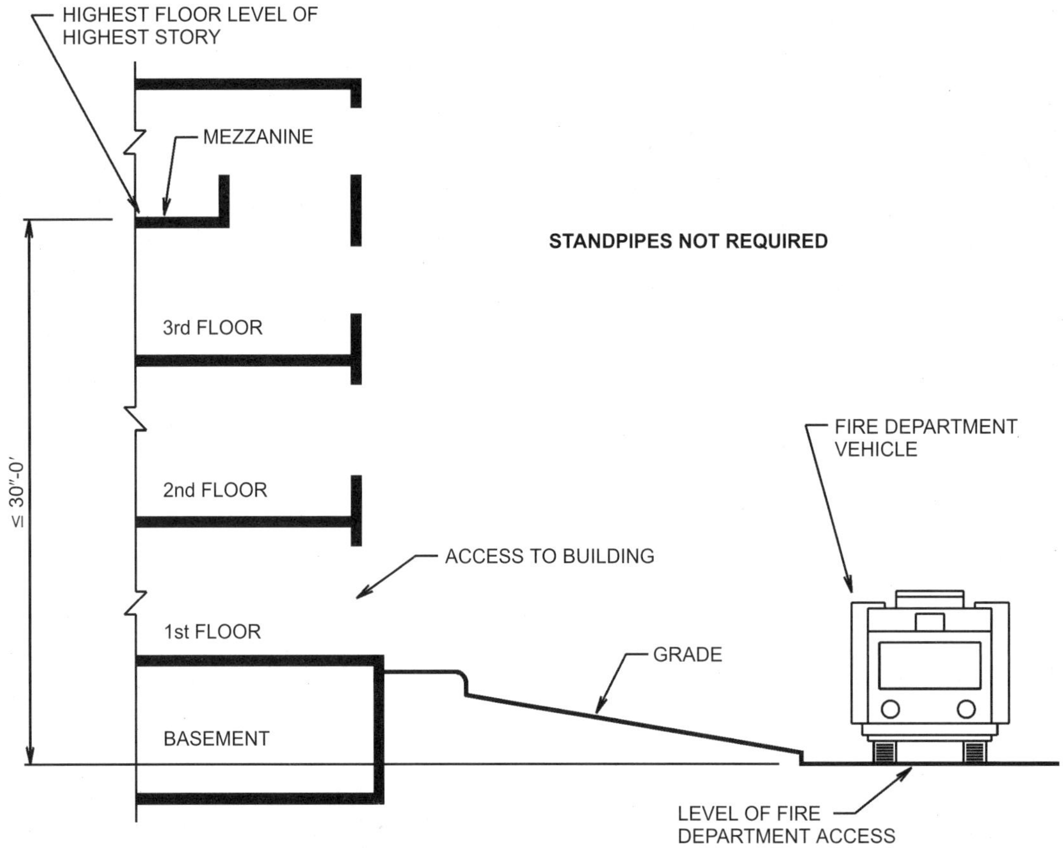

For SI: 1 inch 25.4 mm, 1 foot = 304.8 mm.

Figure 905.3.1(1)
BUILDING HEIGHT THRESHOLD FOR STANDPIPES

Exception 1 exempts open-air seating without enclosed spaces, such as grandstands and bleachers. In such occupancies, a buildup of smoke and hot gases is not possible because these structures are open to the atmosphere.

Exception 2 states that in lieu of a Class I automatic wet standpipe, automatic-dry and semiautomatic dry Class I standpipes are permitted in buildings that are not considered to be a high rise. Class III standpipes may be installed as an alternative to the required Class I standpipes where occupant-use stations are desired.

905.3.3 Covered mall buildings. A covered mall building shall be equipped throughout with a standpipe system where required by Section 905.3. Covered mall buildings not required to be equipped with a standpipe system by Section 905.3 shall be equipped with Class I hose connections connected to a system sized to deliver 250 gallons per minute (946.4 L/min) at the most hydraulically remote outlet. Hose connections shall be provided at each of the following locations:

1. Within the mall at the entrance to each exit passageway or corridor.
2. At each floor-level landing within enclosed stairways opening directly on the mall.
3. At exterior public entrances to the mall.

❖ Covered mall buildings are only required to have a standpipe system if Section 905.3.1 requires such features. If standpipes are not required, Class I hose connections are still required at key locations, such as entrances to exit passageways.

905.3.4 Stages. Stages greater than 1,000 square feet in area (93 m^2) shall be equipped with a Class III wet standpipe system with 1.5-inch and 2.5-inch (38 mm and 64 mm) hose connections on each side of the stage.

Exception: Where the building or area is equipped throughout with an automatic sprinkler system, the hose connections are allowed to be supplied from the automatic sprinkler system and shall have a flow rate of not less than that required by NFPA 14 for Class III standpipes.

❖ Because of the potentially large fuel load and three-dimensional aspect of the fire hazard associated with stages greater than 1,000 square feet (93 m^2) in area, Class III standpipes are required on each side of these large stages. The standpipes must be equipped with a 1-inch (25 mm) hose connection and a 2-inch (51 mm) hose connection. The 1-inch (25 mm) connection is for

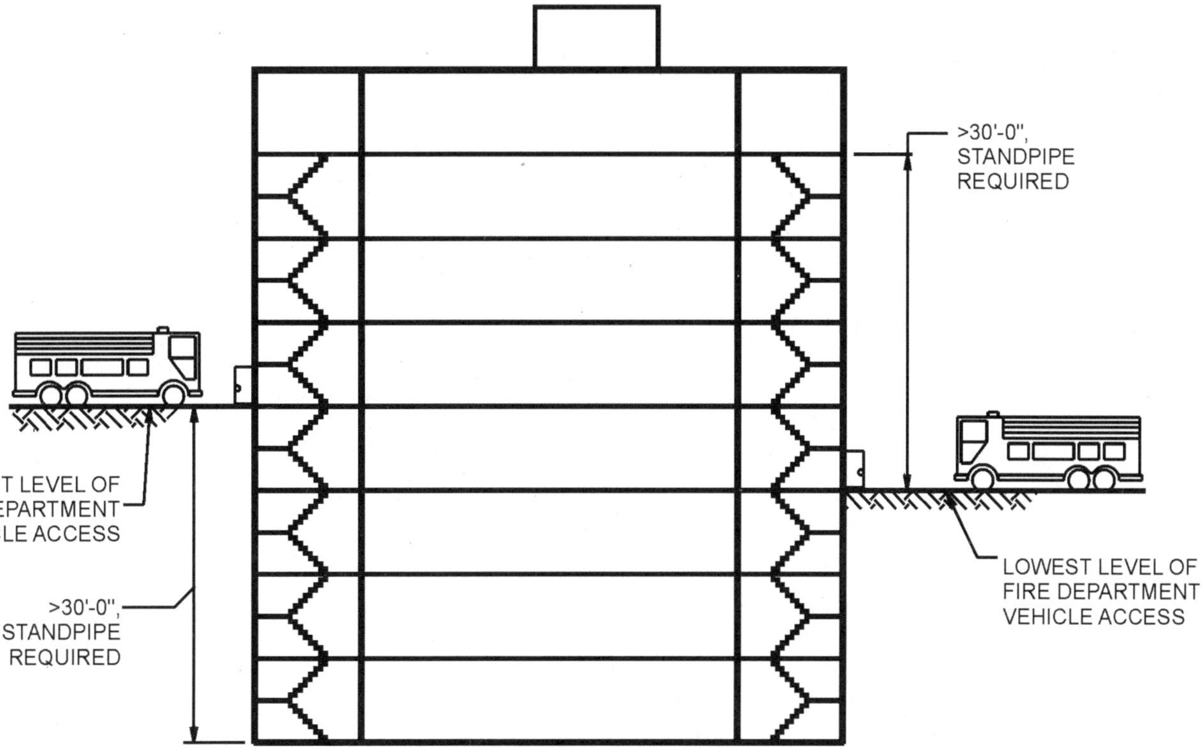

For SI: 1 inch 25.4 mm, 1 foot = 304.8 mm.

Figure 905.3.1(2)
BUILDING HEIGHT REQUIREMENTS FOR STANDPIPES

the hose requirement in Section 905.3.4.1. The 2-$^1/_2$ inch (64 mm) connection is to provide greater flexibility for the fire department in its fire-fighting operations.

The exception recognizes the benefit of the building or area being sprinklered. If the hose connections are supplied through the same standpipe as the sprinklers, the minimum flow must meet the requirements of NFPA 14 for Class III standpipes.

905.3.4.1 Hose and cabinet. The 1.5-inch (38 mm) hose connections shall be equipped with sufficient lengths of 1.5-inch (38 mm) hose to provide fire protection for the stage area. Hose connections shall be equipped with an approved adjustable fog nozzle and be mounted in a cabinet or on a rack.

❖ The 1$^1/_2$ -inch (38 mm) standpipe hose installed for stages greater than 1,000 square feet (93 m^2) in area is intended for use by stage personnel who have been trained to use it. The length of hose provided is a function of the size and configuration of the stage. This includes by definition the entire performance area and adjacent backstage and support areas not fire separated from the performance area. The effective reach of the fire stream from the fog nozzle is a function of the available water supply, and in particular, the pressure. Fog nozzles typically require 100 pounds per square inch (psi) (690 kPa) for optimum performance.

905.3.5 Underground buildings. Underground buildings shall be equipped throughout with a Class I automatic wet or manual wet standpipe system.

❖ Underground buildings present unique hazards to life safety because of their isolation and inaccessibility. Additional fire protection measures are required to ensure safe egress for the occupants and to compensate for the lack of exterior fire suppression and rescue operations. (see Section 405 of the IBC).

905.3.6 Helistops and heliports. Buildings with a helistop or heliport that are equipped with a standpipe shall extend the standpipe to the roof level on which the helistop or heliport is located in accordance with Section 1107.5.

❖ If a building already has a standpipe (required or not) and also contains a helistop or heliport, the standpipe is required to extend to the roof level to make use of the fact that a protection feature is available. Section 1107.5 of the code requires a 2$^1/_2$ -inch (64 mm) standpipe outlet to be within 150 feet (45 675 mm) of all portions of the heliport or helistop area and be either Class I or III.

905.4 Location of Class I standpipe hose connections. Class I standpipe hose connections shall be provided in all of the following locations:

1. In every required stairway, a hose connection shall be provided for each floor level above or below grade. Hose connections shall be located at an intermediate floor level landing between floors, unless otherwise approved by the fire code official.
2. On each side of the wall adjacent to the exit opening of a horizontal exit.
3. In every exit passageway at the entrance from the exit passageway to other areas of a building.
4. In covered mall buildings, adjacent to each exterior public entrance to the mall and adjacent to each entrance from an exit passageway or exit corridor to the mall.
5. Where the roof has a slope less than four units vertical in 12 units horizontal (33.3-percent slope), each standpipe shall be provided with a hose connection located either on the roof or at the highest landing of stairways with stair access to the roof. An additional hose connection shall be provided at the top of the most hydraulically remote standpipe for testing purposes.
6. Where the most remote portion of a nonsprinklered floor or story is more than 150 feet (45 720 mm) from a hose connection or the most remote portion of a sprinklered floor or story is more than 200 feet (60 960 mm) from a hose connection, the fire code official is authorized to require that additional hose connections be provided in approved locations.

❖ Hose connections are required for the fire department to make use of the standpipe system. Since the fire department will typically access the building using the stairways, and most fire departments do not permit entry to the fire floor without an operating hose line, a hose connection must be installed for each floor level of each enclosed stairway.

Item 1 specifies that the hose connections are to be located at intermediate landings between floors. This reduces congestion at the stairway door and may reduce the hose lay distance. The hose connections, however, are still permitted at each floor level of the exit stair instead of at the intermediate landing if this arrangement is approved by the fire code official.

Because horizontal exits are also primary entrances to the fire floor, Item 2 states that hose connections must also be provided at each horizontal exit. The construction of the fire separation assembly used as the horizontal exit will protect the fire fighters while they are connecting to the standpipe system. The hose connections are to be located on each side of the horizontal exit to enable fire fighters to be in a protected area, regardless of the location of the fire.

Item 3 states that an exit passageway in a building required to have a standpipe system is typically used as an extension of a required exit stairway. This allows use of the exit passageway for fire-fighting staging operations in the same way as an exit stair.

In covered mall buildings, Item 4 requires hose connections are required at the entrance to exits and the exterior entrances to allow fire personnel to have a support line as soon as they enter the building.

Depending on the slope of the roof, Item 5 requires a hose connection to aid in the suppression of roof fires either because of the nature of the construction of the roof or the equipment on the roof as well as for exposure protection.

Hose connections in each exit stairway result in hose connections being located based on the travel distances permitted in Table 1015.1, which recognizes that most fire departments carry standpipe hose packs with 150 feet (45 720 mm) of hose or possibly with 100 feet (30 480 mm) of hose and an additional 50-foot (15 240 mm) section that could be easily connected.

With the typical travel distance permitted in nonsprinklered buildings of 200 feet (60 960 mm), reasonable coverage is provided when the effective reach of a fire stream is considered. Depending on the arrangement of the floor, however, all areas may not be effectively protected. Although this situation could easily be corrected by locating additional hose connections on the floor, such connections may rarely be used because of the difficulty in identifying their location during a fire and the fact that most fire departments require an operational hose line before they enter the fire floor. Because longer travel distances are allowed in sprinklered buildings, the problem is increased, but the need for prompt manual suppression is reduced by the presence of the sprinkler system. Item 6 gives the fire code official the authority to require additional hose connections if needed.

905.4.1 Protection. Risers and laterals of Class I standpipe systems not located within an enclosed stairway or pressurized enclosure shall be protected by a degree of fire resistance equal to that required for vertical enclosures in the building in which they are located.

> **Exception:** In buildings equipped throughout with an approved automatic sprinkler system, laterals that are not located within an enclosed stairway or pressurized enclosure are not required to be enclosed within fire-resistance-rated construction.

❖ To minimize the potential for damage to the standpipe systems from a fire, the risers and laterals must be located in an enclosure having the same fire-resistance rating as required for a vertical or shaft enclosure within the building. The required fire-resistance rating for the enclosure can be determined as detailed in Section 707.4 of the IBC.

The enclosure is not required if the building is equipped throughout with an approved automatic sprinkler system. The potential for damage to the standpipe system is minimized by the protection provided by the sprinkler system. The automatic sprinkler system may be either an NFPA 13 or 13R system, depending on what was permitted for the building occupancy.

905.4.2 Interconnection. In buildings where more than one standpipe is provided, the standpipes shall be interconnected in accordance with NFPA 14.

❖ In cases where there are multiple Class I standpipe risers, the risers must be supplied from and interconnected to a common supply line. The required fire department connection should serve all of the sprinklers or standpipes in the building.

905.5 Location of Class II standpipe hose connections. Class II standpipe hose connections shall be accessible and shall be located so that all portions of the building are within 30 feet (9144 mm) of a nozzle attached to 100 feet (30 480 mm) of hose.

❖ Sections 905.5.1 through 905.5.3 specify the requirements for Class II standpipe hose connections. Class II standpipe systems are primarily intended for use by the building occupants.

905.5.1 Groups A-1 and A-2. In Group A-1 and A-2 occupancies with occupant loads of more than 1,000, hose connections shall be located on each side of any stage, on each side of the rear of the auditorium, on each side of the balcony, and on each tier of dressing rooms.

❖ Because of the high occupant load density in Group A-1 and A-2 occupancies, providing additional means for controlling fires in their initial stage is important to enable prompt evacuation of the building. This section is independent of the Class I standpipe requirement for stages based on square footage as indicated in Section 905.3.4.

905.5.2 Protection. Fire-resistance-rated protection of risers and laterals of Class II standpipe systems is not required.

❖ Class II standpipe systems are normally not located in exit stairways; standpipe hose connections are located near the protected area to allow quick access.

905.5.3 Class II system 1-inch hose. A minimum 1-inch (25 mm) hose shall be allowed to be used for hose stations in light-hazard occupancies where investigated and listed for this service and where approved by the fire code official.

❖ This section permits the use of $1^1/_2$-inch (38 mm) listed noncollapsible hose as an alternative to 1-inch (25 mm) hose, subject to the approval of the fire code official. This alternative is limited to light-hazard occupancies, such as office buildings and certain assembly occupancies that tend to have lower fuel loads.

905.6 Location of Class III standpipe hose connections. Class III standpipe systems shall have hose connections located as required for Class I standpipes in Section 905.4 and shall have Class II hose connections as required in Section 905.5.

❖ Class III standpipe systems that have both a $2^1/_2$-inch (64 mm) hose connection and a $1^1/_2$-inch (38 mm) hose connection must comply with the applicable requirements of Sections 905.4, 905.5 and 905.6.

905.6.1 Protection. Risers and laterals of Class III standpipe systems shall be protected as required for Class I systems in accordance with Section 905.4.1.

❖ Because Class III standpipe systems are intended for use by fire-suppression personnel, they must be located in construction that has a fire-resistance rating equivalent to that of the vertical or shaft enclosure requirements of the building (see commentary, Section 905.4.1).

905.6.2 Interconnection. In buildings where more than one Class III standpipe is provided, the standpipes shall be interconnected at the bottom.

❖ As indicated in Section 905.4.2 for Class I standpipe systems, multiple standpipe risers should be interconnected with a common supply line. An indicating valve is typically installed at the base of each riser so that individual risers can be taken out of service without affecting the water supply or the operation of other standpipe risers.

905.7 Cabinets. Cabinets containing fire-fighting equipment, such as standpipes, fire hose, fire extinguishers or fire department valves, shall not be blocked from use or obscured from view.

❖ Cabinets must be readily visible and accessible at all times. Sections 905.7.1 and 905.7.2 contain additional criteria for the construction and identification of the cabinets. Where cabinets are located in fire-resistance-rated assemblies, the integrity of the assembly must be maintained.

905.7.1 Cabinet equipment identification. Cabinets shall be identified in an approved manner by a permanently attached sign with letters not less than 2 inches (51 mm) high in a color that contrasts with the background color, indicating the equipment contained therein.

Exceptions:

1. Doors not large enough to accommodate a written sign shall be marked with a permanently attached pictogram of the equipment contained therein.
2. Doors that have either an approved visual identification clear glass panel or a complete glass door panel are not required to be marked.

❖ This section specifies the minimum criteria to make the signs readily visible. Different color combinations may be approved by the fire code official if the color contrast between the letters and the background is vivid enough to make the sign visible at an approved distance. The exceptions address alternatives to letter signage if the cabinet is still conspicuously identified or the contents are readily visible.

905.7.2 Locking cabinet doors. Cabinets shall be unlocked.

Exceptions:

1. Visual identification panels of glass or other approved transparent frangible material that is easily broken and allows access.
2. Approved locking arrangements.
3. Use Group I-3.

❖ Ready access to all fire-fighting equipment in the cabinet is essential. The exceptions, however, recognize the need to lock cabinets for security reasons and to prevent theft or vandalism.

905.8 Dry standpipes. Dry standpipes shall not be installed.

Exception: Where subject to freezing and in accordance with NFPA 14.

❖ Wet standpipe systems are preferred because they tend to be the most reliable type of standpipe system; therefore, dry standpipes are prohibited unless subject to freezing. For example, Class I manual standpipe systems, which do not have a permanent water supply, are permitted in open parking structures. This recognizes that open parking structures are not heated and that most fires are limited to the vehicle of origin. The use of any dry standpipe system instead of a wet standpipe should take into consideration the added response time and its effect on the occupancy characteristics of the building.

905.9 Valve supervision. Valves controlling water supplies shall be supervised in the open position so that a change in the normal position of the valve will generate a supervisory signal at the supervising station required by Section 903.4. Where a fire alarm system is provided, a signal shall also be transmitted to the control unit.

Exceptions:

1. Valves to underground key or hub valves in roadway boxes provided by the municipality or public utility do not require supervision.
2. Valves locked in the normal position and inspected as provided in this code in buildings not equipped with a fire alarm system.

❖ As with sprinkler systems, water control valves for standpipe systems must be electrically supervised as a means of determining that the system is operational (see commentary, Section 903.4).

Exception 1 recognizes that underground key or hub valves in roadway boxes are not normally supervised or need to be supervised whether the building contains a standpipe system or an automatic sprinkler system.

Exception 2 does not require the control valves for the standpipes to be electrically monitored if they are locked in the normal position and a fire alarm system is not installed in the building. When a fire alarm system is installed, the control valves for the standpipes should be electrically monitored and tied into the supervision required for the fire alarm system.

905.10 During construction. Standpipe systems required during construction and demolition operations shall be provided in accordance with Section 1413.

❖ As stated in Section 1413.1, at least one standpipe is required during construction of buildings four stories or more in height. Standpipe systems must be operable during construction and demolition operations to assist in any potential fire (see commentary, Section 1413).

905.11 Existing buildings. Existing structures with occupied floors located more than 50 feet (15 240 mm) above or below the lowest level of fire department access shall be equipped with standpipes installed in accordance with Section 905. The standpipes shall have an approved fire department connection with hose connections at each floor level above or below the lowest level of fire department access. The fire code official is authorized to approve the installation of manual standpipe systems to achieve compliance with this section where the responding fire department is capable of providing the required hose flow at the highest standpipe outlet.

❖ This section recognizes that some existing buildings do not have a standpipe system. Although it would be inappropriate to require a standpipe system in an existing building based on the requirements of Section 905.3 for new construction, this section establishes a minimum height limitation for existing buildings regardless of occupancy. Many existing buildings may also not be equipped with an automatic sprinkler system. This section will at least ensure that there is some means of manual fire suppression available in buildings where exterior fire department access is limited.

In place of a standpipe system that has a water supply that is capable of supplying the system demand automatically, this section would permit the fire code official to approve the use of a manual standpipe system in an existing building if adequate hose flow is available when the fire department charges the system. The manual standpipe system may be either a dry or a wet system.

SECTION 906 PORTABLE FIRE EXTINGUISHERS

906.1 Where required. Portable fire extinguishers shall be installed in the following locations:

1. In all Group A, B, E, F, H, I, M, R-1, R-2, R-4 and S occupancies.

   Exception: In all Group A, B and E occupancies equipped throughout with quick- response sprinklers, fire extinguishers shall be required only in special-hazard areas.

2. Within 30 feet (9144 mm) of commercial cooking equipment.
3. In areas where flammable or combustible liquids are stored, used or dispensed.
4. On each floor of structures under construction, except Group R-3 occupancies, in accordance with Section 1415.1.
5. Where required by the sections indicated in Table 906.1.
6. Special-hazard areas, including but not limited to laboratories, computer rooms and generator rooms, where required by the fire code official.

❖ Portable fire extinguishers are required in certain instances to give the occupants the means to suppress a fire in its incipient stage. The capability for manual fire suppression can contribute to the protection of the occupants, especially if there are evacuation difficulties associated with the occupancy or the specific hazard in the area. To be effective, personnel must be properly trained in the use of portable fire extinguishers.

Because of the high-hazard nature of building contents, portable fire extinguishers are required in occupancies in Group H.

Portable fire extinguishers are required in occupancies in Groups A, B, E, F, I, M, R-1, R-2, R-4 and S because of the need to control the fire in its early stages and because evacuation can be slowed by the density of the occupant load, the capability of the occupants to evacuate or the overall fuel load in the building.

Portable fire extinguishers are required in areas containing special hazards such as commercial cooking equipment and specific hazardous operations as indicated in Table 906.1. Because of the potential extreme fire hazard associated with such areas or occupancy conditions, prompt extinguishment of the fire is critical.

Portable fire extinguishers are required in all buildings under construction, except in occupancies in Group R-3. The extinguishers are intended for use by construction personnel to suppress a fire in its incipient stages.

Portable fire extinguishers are also required in laboratories, computer rooms and other work spaces in which fire hazards may exist based on the use of the space. Many of these will be addressed by the required occupancy group criteria or by the specific hazard provisions of Table 906.1. Laboratories, for example, may not be considered Group H, but still use limited amounts of hazardous materials that would make manual means of fire extinguishment desirable.

The exception acknowledges the reliable advantages of an automatic sprinkler system designed to comply with NFPA 13. Group A, B and E occupancies are considered light hazard occupancies in NFPA 13. Light hazard occupancies must be protected with quick-response sprinklers (see Section 903.3.2). The faster-acting sprinklers and lower fuel load associated with Group A, B and E occupancies counter the need for portable fire extinguishers. The desire is to have occupants evacuate the building whenever possible rather than fight the fire.

TABLE 906.1
ADDITIONAL REQUIRED PORTABLE FIRE EXTINGUISHERS

SECTION	SUBJECT
303.5	Asphalt kettles
307.4	Open burning
308.4	Open flames
309.4	Powered industrial trucks
1105.2	Aircraft towing vehicles
1105.3	Aircraft welding apparatus
1105.4	Aircraft fuel-servicing tank vehicles
1105.5	Aircraft hydrant fuel-servicing vehicles
1105.6	Aircraft fuel-dispensing stations
1107.7	Heliports and helistops
1208.4	Dry cleaning plants
1415.1	Buildings under construction or demolition
1417.3	Roofing operations
1504.6.4	Spray-finishing operations
1505.5	Dip-tank operations
1904.2	Lumberyards/woodworking facilities
1908.8	Recycling facilities
1909.5	Exterior lumber storage
2003.5	Organic-coating areas
2106.3	Industrial ovens
2205.5	Motor fuel-dispensing facilities
2210.6.4	Marine motor fuel-dispensing facilities
2211.6	Repair garages
2306.10	Rack storage
2404.12	Tents, canopies and membrane structures
2508.2	Tire rebuilding/storage
2604.2.6	Welding and other hot work
2903.6	Combustible fibers
3308.11	Fireworks
3403.2.1	Flammable and combustible liquids, general
3404.3.3.1	Indoor storage of flammable and combustible liquids
3404.3.7.5.2	Liquid storage rooms for flammable and combustible liquids
3405.4.9	Solvent distillation units
3406.2.7	Farms and construction sites—flammable and combustible liquids storage
3406.4.10.1	Bulk plants and terminals for flammable and combustible liquids
3406.5.4.5	Commercial, industrial, governmental or manufacturing establishments—fuel dispensing
3406.6.4	Tank vehicles for flammable and combustible liquids
3606.5.7	Flammable solids
3808.2	LP-gas

❖ Table 906.1 lists those sections of the code that represent specific occupancy conditions requiring portable fire extinguishers for incipient fire control. Wherever the code requires a fire extinguisher because of one of the listed occupancy conditions, it may identify the required rating of the extinguisher that is compatible with the hazard involved in addition to referencing Section 906.

906.2 General requirements. Fire extinguishers shall be selected, installed and maintained in accordance with this section and NFPA 10.

Exception: The travel distance to reach an extinguisher shall not apply to the spectator seating portions of Group A-5 occupancies.

❖ NFPA 10 contains minimum requirements for the selection, installation and maintenance of portable fire extinguishers. Portable fire extinguishers are investigated and rated in conformance to NFPA 10 and listed under a variety of standards. Portable fire extinguishers must be labeled and rated for use on fires of the type, severity and hazard class protected.

The exception recognizes the openness to the atmosphere associated with Group A-5 occupancies. A fire in open areas is more obvious to all spectators. Group A-5 occupancies also do not accumulate smoke and hot gases because they are not enclosed spaces.

906.3 Size and distribution. For occupancies that involve primarily Class A fire hazards, the minimum sizes and distribution shall comply with Table 906.3(1). Fire extinguishers for occupancies involving flammable or combustible liquids with depths of less than or equal to 0.25-inch (6.35 mm) shall be selected and placed in accordance with Table 906.3(2). Fire extinguishers for occupancies involving flammable or combustible liquids with a depth of greater than 0.25-inch (6.35 mm) or involving combustible metals shall be selected and placed in accordance with NFPA 10. Extinguishers for Class C fire hazards shall be selected and placed on the basis of the anticipated Class A or Class B hazard.

❖ Proper selection and distribution of portable fire extinguishers is essential to having adequate protection for the building structure and the occupancy conditions within. Determination of the desired type of portable fire extinguisher depends on the character of the fire anticipated, building occupancy, specific hazards and ambient temperature conditions [see commentary, Tables 906.3(1) and 906.3(2)].

Class A fires generally involve ordinary combustibles, such as wood, cloth, paper, rubber and most plastics, whereas Class B fires involve flammable and combustible liquids, flammable gases, oil-based paints, solvents and similar materials. Class C fires involve energized electrical equipment where the electrical nonconductivity of the extinguishing agent is critical. Class D fires, while not addressed in this section, are fires involving combustible metals, such as magnesium and titanium.

TABLE 906.3(1)
FIRE EXTINGUISHERS FOR CLASS A FIRE HAZARDS

	LIGHT (Low) HAZARD OCCUPANCY	ORDINARY (Moderate) HAZARD OCCUPANCY	EXTRA (High) HAZARD OCCUPANCY
Minimum Rated Single Extinguisher	2-A[c]	2-A	4-A[a]
Maximum Floor Area Per Unit of A	3,000 square feet	1,500 square feet	1,000 square feet
Maximum Floor Area For Extinguisher[b]	11,250 square feet	11,250 square feet	11,250 square feet
Maximum Travel Distance to Extinguisher	75 feet	75 feet	75 feet

For SI: 1 foot = 304.8 mm, 1 square foot = 0.0929 m^2, 1 gallon = 3.785 L.

a. Two 2.5-gallon water-type extinguishers shall be deemed the equivalent of one 4-A rated extinguisher.

b. NFPA 10 Appendix E-3-3 provides more details concerning application of the maximum floor area criteria.

c. Two water-type extinguishers each with a 1-A rating shall be deemed the equivalent of one 2-A rated extinguisher for Light (Low) Hazard Occupancies.

❖ Table 906.3(1) establishes the minimum number and rating of fire extinguishers for Class A fires in any particular occupancy. The occupancy classifications are further defined in NFPA 10. The maximum area that a single fire extinguisher can protect is determined based on the rating of the fire extinguisher. The travel distance limitation of 75 feet (22 860 mm) is intended to be the actual walking distance along a normal path of travel to the extinguisher. For this reason, it is necessary to select fire extinguishers that comply with both the distribution criteria and travel distance limitation for a specific occupancy classification.

TABLE 906.3(2)
FLAMMABLE OR COMBUSTIBLE LIQUIDS WITH DEPTHS OF LESS THAN OR EQUAL TO 0.25-INCH

TYPE OF HAZARD	BASIC MINIMUM EXTINGUISHER RATING	MAXIMUM TRAVEL DISTANCE TO EXTINGUISHERS (feet)
Light (Low)	5-B 10-B	30 50
Ordinary (Moderate)	10-B 20-B	30 50
Extra (High)	40-B 80-B	30 50

For SI: 1 inch = 25.4 mm, 1 foot = 304.8 mm.

NOTE. For requirements on water-soluble flammable liquids and alternative sizing criteria, see NFPA 10, Sections 3-3 and 3-4.

❖ Fires involving flammable or combustible liquids present a severe hazard challenge regardless of occupancy. Table 906.3(2) prescribes the minimum portable fire extinguisher requirements where flammable or combustible liquids are limited in depth [0.25 inch (6 mm) or less]. As can be seen in the table, the size of the extinguisher is directly related to the travel distance to the extinguisher for each given occupancy classification. These fire extinguisher provisions are independent of whether other fixed automatic fire-extinguishing systems are installed. A similar table is included in NFPA 10 for Class B fire hazards.

For occupancy conditions involving flammable or combustible liquids in potential depths greater than 0.25 inch (6 mm), the selection and spacing criteria of NFPA 10 should be used in addition to any applicable requirements in Chapter 34 and NFPA 30.

906.4 Cooking grease fires. Fire extinguishers provided for the protection of cooking grease fires shall be of an approved type compatible with the automatic fire-extinguishing system agent and in accordance with Section 904.11.5.

❖ The combination of high-efficiency cooking appliances and hotter burning cooking media creates a potential severe fire hazard. Although commercial cooking systems must have an approved exhaust hood and be protected by an approved automatic fire-extinguishing system, a manual means of extinguishment is desirable to attack a fire in its incipient stage.

As indicated in Section 904.11.5, a Class K-rated portable fire extinguisher must be located within 30 feet (9144 mm) of travel distance of commercial-type cooking equipment. A Class K-rated extinguisher has been specifically tested on commercial cooking appliances using vegetable or animal oils or fats. The portable fire extinguishers are usually of sodium bicarbonate or potassium bicarbonate dry-chemical type.

906.5 Conspicuous location. Extinguishers shall be located in conspicuous locations where they will be readily accessible and immediately available for use. These locations shall be along normal paths of travel, unless the fire code official determines that the hazard posed indicates the need for placement away from normal paths of travel.

❖ Fire extinguishers should be located in readily accessible locations along normal egress paths. This increases occupants, familiarity with the location of the fire extinguishers.

906.6 Unobstructed and unobscured. Fire extinguishers shall not be obstructed or obscured from view. In rooms or areas in which visual obstruction cannot be completely avoided, means shall be provided to indicate the locations of extinguishers.

❖ Portable fire extinguishers should be located where they are readily visible at all times. If visual obstruction cannot be avoided, the location of the extinguishers must be marked by an approved means of identification. This could include additional signage, lights, arrows or other means approved by the fire code official.

906.7 Hangers and brackets. Hand-held portable fire extinguishers, not housed in cabinets, shall be installed on the hangers or brackets supplied. Hangers or brackets shall be securely anchored to the mounting surface in accordance with the manufacturer's installation instructions.

❖ Portable fire extinguishers not housed in cabinets are usually mounted on walls or columns using securely fastened hangers. Brackets should be used where the fire extinguishers need to be protected from impact or other potential physical damage.

906.8 Cabinets. Cabinets used to house fire extinguishers shall not be locked.

Exceptions:

1. Where fire extinguishers subject to malicious use or damage are provided with a means of ready access.
2. In Group I-3 occupancies and in mental health areas in Group I-2 occupancies, access to portable fire extinguishers shall be permitted to be locked or to be located in staff locations provided the staff has keys.

❖ Cabinets housing fire extinguishers must not be locked in order to provide quick access in an emergency. Exception 1, however, allows the cabinets to be locked in occupancies where vandalism, theft or other malicious behavior is possible. Exception 2 also permits cabinets housing fire extinguishers to be locked or to be located in staff locations in Group I-3 occupancies and mental health areas in Group I-2 occupancies. Occupants in Group I-3 areas of jails, prisons or similar restrained occupancies should not have access to fire extinguishers because they could possibly be used as a weapon or be subject to vandalism. Staff adequately trained in the use of fire extinguishers are assumed to have ready access to the keys for the cabinets at all times.

906.9 Height above floor. Portable fire extinguishers having a gross weight not exceeding 40 pounds (18 kg) shall be installed so that its top is not more than 5 feet (1524 mm) above the floor. Hand-held portable fire extinguishers having a gross weight exceeding 40 pounds (18 kg) shall be installed so that its top is not more than 3.5 feet (1067 mm) above the floor. The clearance between the floor and the bottom of installed hand-held extinguishers shall not be less than 4 inches (102 mm).

❖ Due to the varying height and physical strength levels of persons who might be called upon to operate a portable fire extinguisher, the mounting height of the extinguisher must be commensurate with its weight so that it may be easily retrieved by anyone from its mounting location and placed into use.

906.10 Wheeled units. Wheeled fire extinguishers shall be conspicuously located in a designated location.

❖ Wheeled fire extinguishers consist of a portable fire extinguisher with a carriage and wheels. The wheeled fire extinguishers are constructed so that one able-bodied person could move the unit to the fire area. Wheeled fire extinguishers are capable of delivering greater flow rates and stream range for various extinguishing agents than normal portable fire extinguishers. Wheeled fire extinguishers are generally more effective in high-hazard areas and, as with any extinguisher, should be readily available and stored in an approved location. The wheeled fire extinguisher should be located a safe distance from the hazard area so that it will not become involved in the fire.

SECTION 907
FIRE ALARM AND DETECTION SYSTEMS

907.1 General. This section covers the application, installation, performance and maintenance of fire alarm systems and their components in new and existing buildings and structures. The requirements of Section 907.2 are applicable to new buildings and structures. The requirements of Section 907.3 are applicable to existing buildings and structures.

❖ Fire alarm systems, which typically include manual fire alarm systems and automatic fire detection systems, must be installed in accordance with Section 907 and NFPA 72. As indicated in Section 907.1, only Section 907.3 is intended to be applicable to existing buildings and structures.

Manual fire alarm systems are installed in buildings to limit fire casualties and property losses. Fire alarm systems do this by promptly notifying the occupants of the building of an emergency, which increases the time available for evacuation. Similarly, when fire alarm systems are supervised, the fire department will be promptly notified and its response time relative to the onset of the fire will be reduced.

Automatic fire detection systems are required under certain conditions to increase the likelihood that fire is detected and occupants are given an early warning. The detection system is a system of devices and associated hardware that activates the alarm system. The automatic detecting devices are to be smoke detectors, unless a condition exists that calls for the use of a different type of detector.

907.1.1 Construction documents. Construction documents for fire alarm systems shall be submitted for review and approval prior to system installation. Construction documents shall include, but not be limited to, all of the following:

1. A floor plan which indicates the use of all rooms.
2. Locations of alarm-initiating and notification appliances.
3. Alarm control and trouble signaling equipment.
4. Annunciation.
5. Power connection.
6. Battery calculations.
7. Conductor type and sizes.
8. Voltage drop calculations.
9. Manufacturers, model numbers and listing information for equipment, devices and materials.

10. Details of ceiling height and construction.

11. The interface of fire safety control functions.

❖ Construction documents for fire alarm systems must be submitted for review to determine compliance with the code and NFPA 72. First, the floor plan indicating the use of each space is required. In terms of the actual alarm system, construction documents are to show the location and number of alarm-initiating devices and alarm-notification appliances, and also provide a description of all equipment to be used, proposed zoning, location of the control panel(s) and annunciator(s) and a complete sequence of operation for the system. All of the information required by this section may not be available during the design stage and initial permit process. Submission of more detailed shop drawings may be required. To facilitate the review process, the fire code official may elect to develop a checklist beyond what is listed in Section 907.1.1 stating the information required to complete a thorough review.

907.1.2 Equipment. Systems and their components shall be listed and approved for the purpose for which they are installed.

❖ The components of the fire alarm system must be approved for use in the planned system. NFPA 72 requires all devices, combinations of devices, appliances and equipment to be labeled for their proposed used. The testing agency will test the components for use in various types of systems and stipulate the use of the component on the label. Evidence of labeling of the system components should be submitted with the shop construction documents.

In some instances, the entire system may be labeled. At least one major testing agency, Underwriters Laboratories Inc. (UL), has a program in which alarm installation and service companies are issued a certificate and become listed by the agency as being qualified to design, install and maintain local, auxiliary, remote station or proprietary fire alarm systems. The listed companies may then issue a certificate showing that the system is in compliance with Section 907.

Terms of the company certification by UL include the company being responsible for keeping accurate system documentation, including as-built record drawings, acceptance test records and complete maintenance records on a given system. The company is also responsible for the required periodic inspection and testing of the system under contract with the owner.

A similar program has been available for many years for central station alarm service, whereas the UL program is relatively new to the industry. Even though this company and system listing program is not required by the code or NFPA 72, it can be a valuable tool for the fire code official in determining compliance with the referenced standard.

Another issue that must be considered is the compatibility of the system components as required by NFPA 72. The labeling of system components discussed above should include any compatibility restrictions for components. Compatibility is primarily an issue of the ability of smoke detectors and fire alarm control panels (FACPs) to function properly when interconnected and affects the two-wire type of smoke detectors, which obtain their operating power over the same pair of wires used to transmit signals to the FACP (the control unit initiating device circuits).

Laboratories will test for component compatibility either by actual testing or by reviewing the circuit parameters of both the detector and the FACP. Generally, if both the two-wire detector and the FACP are of the same brand, there should not be a compatibility problem. Nevertheless, the fire code official must be satisfied that the components are listed as being compatible. Failure to comply with the compatibility requirements of NFPA 72 can lead to system malfunction or failure when it may be needed the most.

907.2 Where required—new buildings and structures. An approved manual, automatic, or manual and automatic fire alarm system shall be provided in new buildings and structures in accordance with Sections 907.2.1 through 907.2.23. Where automatic sprinkler protection installed in accordance with Section 903.3.1.1 or 903.3.1.2 is provided and connected to the building fire alarm system, automatic heat detection required by this section shall not be required.

An approved automatic fire detection system shall be installed in accordance with the provisions of this code and NFPA 72. Devices, combinations of devices, appliances and equipment shall comply with Section 907.1.2. The automatic fire detectors shall be smoke detectors, except that an approved alternative type of detector shall be installed in spaces such as boiler rooms where, during normal operation, products of combustion are present in sufficient quantity to actuate a smoke detector.

❖ This section specifies the occupancies or conditions in new buildings or structures that require some form of fire alarm system. The fire alarm system is either a manual fire alarm system (manual fire alarm boxes) or an automatic fire detection system (smoke and heat detectors).

Manual fire alarm systems must be installed in certain occupancies depending on the number of occupants, capabilities of the occupants and height of the building. An automatic fire detection system must be installed in those occupancies and conditions where the need to detect the fire is essential to evacuation or protection of the occupants. The requirements for automatic smoke detection are generally based on the evacuation needs of the occupants and whether the occupancy includes sleeping accommodations.

This section also states that automatic heat detection is not required when buildings are fully sprinklered in accordance with NFPA 13 or 13R. The presence of a sprinkler system exempts areas where a heat detector can be installed in place of a smoke detector, such as in storage or furnace rooms. The sprinkler head in this case essentially acts as a heat detection device.

Automatic fire detection systems must be installed in accordance with the code and NFPA 72. NFPA 72 contains minimum requirements for the performance of automatic fire detectors. The standard covers minimum performance, location, mounting, testing and maintenance requirements for automatic fire detectors. Smoke detectors must be used, except when their use would result in unwanted alarms or when a smoke detector may not provide the desired detection. The manufacturer's literature will identify the limitations on the use of smoke detectors, including environmental conditions such as humidity, temperature and airflow.

Only certain occupancies are required to have either a manual fire alarm or automatic fire detection system installed (see Figure 907.2). The need for either system is determined by the number of occupants and the height of the building. Figure 907.2 contains the conditions that require either system to be installed in a building. Note that the requirements of Section 907.2 do not specifically state that the systems must be installed and maintained throughout buildings, but only in the area that contains the occupancies.

In summary, the fire alarm/detection system installed in mixed occupancy buildings is limited to the fire area that requires the system. Buildings that have more than one occupancy and do not contain more than one fire area are identified as nonseparated mixed-use buildings. A nonseparated mixed-use building that contains one fire area that requires a fire alarm/detection system is required to have the system extended throughout the building in accordance with Section 302.3.1 of the IBC.

FIGURE 907.2. See below.

❖ This figure contains the threshold requirements for when a manual fire alarm system or automatic fire detection system is required based on the occupancy group. Sections 907.2.11 through 907.2.23 contain additional requirements for fire alarm systems depending on special occupancy conditions such as atriums, high-rise buildings or covered mall buildings.

907.2.1 Group A. A manual fire alarm system shall be installed in accordance with NFPA 72 in Group A occupancies having an occupant load of 300 or more. Portions of Group E occupancies

Figure 907.2
SUMMARY OF MANUAL FIRE ALARM AND AUTOMATIC FIRE DETECTION THRESHOLDS

MANUAL FIRE ALARMS	
Occupancy	**Threshold**
Assembly (A-1, A-2, A-3, A-4, A-5)	All with occupant load or > 300
Business (B)	Total occupant load ≥ 500 or occupant load, or > 100 above/below level of exit discharge
Educational (E)	> 50 occupants (several exceptions for manual fire alarm box placement
Factory (F-1, F-2)	≥ 2 stories with occupied load ≥ 500 above or below level of exit discharge
High hazard (H-1, H-2, H-3, H-4, H-5)	H-5 occupancies and occupancies used for manufacture of organic coatings
Institutional (I-1, I-2, I-3, I-4)	All
Mercantile (M)	Total occupant load ≥ 500 or occupant load or > 100 above/below level of exit discharge
Hotels (1)	All unsprinklered buildings except ≤ 2 stories with sleeping units having exit directly to exterior
Apartments (2)	Unsprinklered 3 stories or 1 story below level of exit discharge or > 16 units without exits directly to exterior
AUTOMATIC FIRE DETECTION	
Occupancy	**Threshold**
High hazard (H-1, H-2, H-3, H-4, H-5)	Highly toxic gasses, organic peroxides and oxidizers in accordance with IFC
Institutional (I-1, I-2, I-3, I-4)	All
Hotels (1)	All except buildings without interior corridors and sleeping units having exit directly to exterior

occupied for assembly purposes shall be provided with a fire alarm system as required for the Group E occupancy.

Exception: Manual fire alarm boxes are not required where the building is equipped throughout with an automatic sprinkler system and the alarm notification appliances will activate upon sprinkler water flow.

❖ Group A occupancies are typically occupied by a significant number of people who are not completely familiar with their surroundings. For this reason, a manual fire-alarm system is required in Group A occupancies with an occupant load of 300 or more to aid in the prompt evacuation of the occupants, especially in nonsprinklered buildings.

The exception allows the omission of manual fire alarm boxes in buildings equipped throughout with an automatic sprinkler system, if activation of the sprinkler system will activate the building evacuation alarms associated with the manual fire-alarm system.

This section also permits assembly-type areas in Group E occupancies to comply with Section 907.2.3 instead of the requirements of this section. A typical high school contains many areas used for assembly purposes such as a gymnasium, cafeteria, auditorium or library; however, they all exist to serve as an educational facility as its main function.

907.2.1.1 System initiation in Group A occupancies with an occupant load of 1,000 or more. Activation of the fire alarm in Group A occupancies with an occupant load of 1,000 or more shall initiate a signal using an emergency voice/alarm communications system in accordance with NFPA 72.

Exception: Where approved, the prerecorded announcement is allowed to be manually deactivated for a period of time, not to exceed 3 minutes, for the sole purpose of allowing a live voice announcement from an approved, constantly attended location.

❖ The exception allows the automatic alarm signal to be overridden if the live voice instructions do not exceed 3 minutes. Although this section does not specifically require a fire command center, the location from which the live voice announcement originates should be constantly attended and approved by the fire code official.

907.2.1.2 Emergency power. Emergency voice/alarm communications systems shall be provided with an approved emergency power source.

❖ Because the emergency voice/alarm communication system is a critical aid in evacuating the building, the system must be connected to an approved emergency power source complying with Section 604.

907.2.2 Group B. A manual fire alarm system shall be installed in Group B occupancies having an occupant load of 500 or more persons or more than 100 persons above or below the lowest level of exit discharge.

Exception: Manual fire alarm boxes are not required where the building is equipped throughout with an automatic sprinkler system and the alarm notification appliances will activate upon sprinkler water flow.

❖ Group B occupancies generally involve individuals or groups of people in separate office areas. As a result, the occupants are not necessarily aware of what is going on in other parts of the building. Group B buildings with large occupant loads, even in single-story buildings, or where a substantial number of occupants are above or below the level of exit discharge, increase the difficulty of alerting the occupants of a fire. This is especially true in nonsprinklered buildings with given occupant load thresholds.

The exception does not eliminate the fire alarm system, but rather permits it to be initiated automatically by the sprinkler water flow switch(es) instead of by the manual fire alarm boxes.

907.2.3 Group E. A manual fire alarm system shall be installed in Group E occupancies. When automatic sprinkler systems or smoke detectors are installed, such systems or detectors shall be connected to the building fire alarm system.

Exceptions:

1. Group E occupancies with an occupant load of less than 50.
2. Manual fire alarm boxes are not required in Group E occupancies where all the following apply:
   - 2.1. Interior corridors are protected by smoke detectors with alarm verification.
   - 2.2. Auditoriums, cafeterias, gymnasiums and the like are protected by heat detectors or other approved detection devices.
   - 2.3. Shops and laboratories involving dusts or vapors are protected by heat detectors or other approved detection devices.
   - 2.4. Off-premises monitoring is provided.
   - 2.5. The capability to activate the evacuation signal from a central point is provided.
   - 2.6. In buildings where normally occupied spaces are provided with a two-way communication system between such spaces and a constantly attended receiving station from where a general evacuation alarm can be sounded, except in locations specifically designated by the fire code official.

❖ Group E occupancies involve groups of people distributed throughout a number of classrooms or small rooms. Occupants in one area of the building would not necessarily be aware of an emergency in another part of the building unless a system is provided to alert them. Because of the age and maturity of the occupants, more time may be needed to safely evacuate the building. The requirement for a manual fire alarm system in Group E occupancies is not dependent on the location of the level of exit discharge.

Exception 1 exempts Group E occupancies from requiring a fire alarm system when the occupant load is less than 50. This would exempt small day care centers that serve children older than $2^1/_2$ years of age or a small Sunday school classroom at a church (see Section 202).

Exception 2 exempts manual fire alarm boxes in interior corridors, laboratories, auditoriums, cafeterias, gymnasiums and similar spaces depending on the use of heat/smoke detectors and the extent of supervision. The applicability of Exception 2 is independent of whether an automatic sprinkler system is installed. Section 903.2.2 requires an automatic sprinkler system throughout all Group E fire areas in excess of 20,000 square feet (1,858 m^2). If an automatic sprinkler system or smoke detectors are installed, however, they must be connected to the building fire alarm system.

907.2.4 Group F. A manual fire alarm system shall be installed in Group F occupancies that are two or more stories in height and have an occupant load of 500 or more above or below the lowest level of exit discharge.

> **Exception:** Manual fire alarm boxes are not required where the building is equipped throughout with an automatic sprinkler system and the alarm notification appliances will activate upon sprinkler water flow.

❖ This section is intended to apply only to large multistory manufacturing facilities. For this reason, a manual fire alarm system would be required only if the building were at least two stories in height and had 500 or more occupants above or below the level of exit discharge. An unlimited area two-story Group F occupancy complying with Section 507.3 of the IBC would be indicative of an occupancy requiring a manual fire alarm system.

Buildings in compliance with Section 507.3 of the IBC, and large manufacturing facilities in general, however, must be fully sprinklered and would thus be eligible for the exception. The exception does not eliminate the fire alarm system but rather permits it to be initiated automatically by the sprinkler system water flow switch(es) instead of by the manual fire alarm boxes.

907.2.5 Group H. A manual fire alarm system shall be installed in Group H-5 occupancies and in occupancies used for the manufacture of organic coatings. An automatic smoke detection system shall be installed for highly toxic gases, organic peroxides and oxidizers in accordance with Chapters 37, 39 and 40, respectively.

❖ Because of the nature and potential quantity of hazardous materials in Group H-5 occupancies, a manual means of activating an evacuation alarm is essential for the safety of the occupants. In accordance with Section 1803.12, the activation of the alarm system is to initiate a local alarm and transmit a signal to the emergency control station. The manual fire alarm system requirement for the building is in addition to the emergency alarm requirements in Section 1803.11 (see Section 908.2).

Occupancies involved in the manufacture of organic coatings present special hazardous conditions because of the unstable character of the materials, such as nitrocellulose. Good housekeeping and control of ignition sources is critical. Chapter 20 contains additional requirements for organic coating manufacturing processes.

This section also requires an automatic smoke detection system in certain occupancy conditions involving either highly toxic gases or organic peroxides and oxidizers. The need for the automatic smoke detection system may depend on the class of materials and additional levels of fire protection provided. This requirement also assumes the quantity of materials is in excess of the maximum allowable quantities shown in Tables 2703.1.1(1) and 2703.1.1(2).

907.2.6 Group I. A manual fire alarm system and an automatic fire detection system shall be installed in Group I occupancies. An electrically supervised, automatic smoke detection system shall be provided in waiting areas that are open to corridors.

> **Exception:** Manual fire alarm boxes in patient sleeping areas of Group I-1 and I-2 occupancies shall not be required at exits if located at all nurses' control stations or other constantly attended staff locations, provided such stations are visible and continuously accessible and that travel distances required in Section 907.4.1 are not exceeded.

❖ Because the protection and possible evacuation of the occupants in Group I occupancies are most often dependent on the response by staff, occupancies in Group I must be protected with a manual fire alarm system and an automatic fire detection system. Occupancies in Group I also tend to be compartmentalized into small rooms so that a fire in one area of the building would not easily be noticed by occupants in another part of the building.

To reduce the likelihood that a fire within such a waiting area could develop beyond the incipient stage, thereby jeopardizing the integrity of the corridor, the area must be equipped with an automatic smoke detection system. Whereas the areas also often serve as designated smoking areas, arrangements should be made to minimize the potential for unwanted alarms caused by activating the smoke detection system. The detectors are to be located to provide the required coverage to the waiting area space.

To reduce the potential for unwanted alarms, manual fire alarm boxes may be located at the nurses' control stations or another constantly attended location. The exception assumes the approved location is always accessible by staff and within 200 feet (60 960 mm) of travel distance.

907.2.6.1 Group I-2. Corridors in nursing homes (both intermediate care and skilled nursing facilities), detoxification facilities and spaces open to the corridors shall be equipped with an automatic fire detection system.

Exceptions:

1. Corridor smoke detection is not required in smoke compartments that contain patient sleeping rooms where patient sleeping units are provided with smoke

detectors that comply with UL 268. Such detectors shall provide a visual display on the corridor side of each patient sleeping unit and shall provide an audible and visual alarm at the nursing station attending each unit.

2. Corridor smoke detection is not required in smoke compartments that contain patient sleeping rooms where patient sleeping unit doors are equipped with automatic door-closing devices with integral smoke detectors on the unit sides installed in accordance with their listing, provided that the integral detectors perform the required alerting function.

❖ Automatic fire detection is required in areas open to corridors in occupancies classified as Group I-2 and corridors in nursing homes and detoxification facilities. In recognition of quick-response sprinkler technology and the fact that the sprinkler system is electronically supervised, and because the doors to patient sleeping units are continuously supervised by staff when in the open position, it is now believed that smoke detectors are not required for adequate fire safety in patient sleeping units.

In nursing homes and detoxification facilities, however, some redundance is appropriate because such facilities typically have less control over furnishings and personal items and thereby result in a less predictable and usually higher fire hazard load than other Group I-2 occupancies. Also, there is generally less staff supervision in these facilities than in other health care facilities and thus less control over patient smoking and other fire causes. Therefore, to provide additional protection against fires spreading from the room of origin, automatic fire detection is required in corridors of nursing homes and detoxification facilities.

Fire detection is not required in corridors of other Group I-2 occupancies except where otherwise specifically required in the code (see Section 907.2.6). Similarly, because areas open to the corridor very often are the room of fire origin and because such areas are no longer required by the code to be under visual supervision by staff, some redundance to protection by the sprinkler system is requested. Accordingly, all areas open to corridors must be protected by an automatic fire detection system. This requirement provides an additional level of protection against sprinkler system failures or lapses in staff supervision.

There are two exceptions to the requirement for an automatic fire detection system in corridors of nursing homes and detoxification facilities. In both cases, the required protection serves as an alternative method for redundant protection to the fire protection required in patient sleeping units. For this reason, they provide either a backup to the notification of a fire or containment of fire in the room of origin.

Exception 1 requires smoke detectors in patient sleeping units that activate both a visual display on the corridor side of the patient sleeping unit and a visual and audible alarm at the nurses' station serving or attending the room. Detectors complying with UL 268 are intended for open area protection and for connection to a normal power supply or as part of a fire alarm system.

This exception, however, is specifically designed not to require the detectors to activate the building fire alarm system where approved patient sleeping unit smoke detectors are installed and where visual and audible alarms are provided. This is in response to the concern over unwanted alarms. The required alarm signals will not necessarily indicate to staff that a fire emergency exists because the nursing call system may typically be used to identify numerous conditions within the room.

Exception 2 addresses the situation where smoke detectors are incorporated within automatic door-closing devices. The units are acceptable as long as the required alarm functions are still provided. Such units are usually listed as combination door closer and hold-open devices.

907.2.6.2 Group I-3 occupancies. Group I-3 occupancies shall be equipped with a manual and automatic fire alarm system installed for alerting staff.

❖ Because of the evacuation difficulties associated with Group I-3 occupancies and the dependence on adequate staff response, a manual fire alarm system and an automatic fire detection system are required subject to the special occupancy conditions in Sections 907.2.6.2.1 through 907.2.6.2.3. This section recognizes that the evacuation of Group I-3 occupancies depends on an effective staff response. Section 408.7 contains the requirements for an emergency plan, including employee training, staff availability, the need for occupants to notify staff and the need for the proper keys for unlocking doors for staff in Group I-3 occupancies.

907.2.6.2.1 System initiation. Actuation of an automatic fire-extinguishing system, a manual fire alarm box or a fire detector shall initiate an approved fire alarm signal which automatically notifies staff. Presignal systems shall not be used.

❖ This section specifies the systems that, upon actuation, must initiate the required alarm signal to the staff. So that staff will respond in a timely manner, a presignal system is not permitted. See Section 1-5.4.10 of NFPA 72 for further information on the presignal feature.

907.2.6.2.2 Manual fire alarm boxes. Manual fire alarm boxes are not required to be located in accordance with Section 907.4 where the fire alarm boxes are provided at staff-attended locations having direct supervision over areas where manual fire alarm boxes have been omitted.

Manual fire alarm boxes are allowed to be locked in areas occupied by detainees, provided that staff members are present within the subject area and have keys readily available to operate the manual fire alarm boxes.

❖ Because of the potential for intentional false alarms and the resulting disruption to the facility, manual fire alarm boxes in Group I-3 occupancies may be either locked or made inaccessible to the occupants. The locking of manual fire alarm boxes is permitted only in areas where staff members are present and keys are readily available to them to unlock the boxes, or where the alarm boxes are located in a manned staff location that has direct supervision of the Group I-3 area.

907.2.6.2.3 Smoke detectors. An approved automatic smoke detection system shall be installed throughout resident housing areas, including sleeping areas and contiguous day rooms, group activity spaces and other common spaces normally accessible to residents.

Exceptions:

1. Other approved smoke-detection arrangements providing equivalent protection, including, but not limited to, placing detectors in exhaust ducts from cells or behind protective guards listed for the purpose, are allowed when necessary to prevent damage or tampering.
2. Sleeping units in Use Conditions 2 and 3.
3. Smoke detectors are not required in sleeping units with four or fewer occupants in smoke compartments that are equipped throughout with an approved automatic sprinkler system.

❖ Evacuation of Group I-3 facilities is impractical because of the need to maintain security. An automatic smoke detection system is therefore required to provide early warning of a fire.

As indicated in Exception 1, the installation of automatic smoke detectors must take into account the need to protect the detector from vandalism by residents. As a result, detectors may have to be located in return air ducts or be protected by a substantial physical barrier.

Since occupants in Use Condition II or III are not locked in their sleeping units, Exception 2 reduces the need for smoke detection.

Exception 3 allows smoke detectors to be omitted in sleeping units housing no more than four occupants on the basis that in a building that is protected throughout with an approved automatic sprinkler system, the system will provide both detection and suppression functions. Group I facilities are assumed to be fully sprinklered throughout in accordance with NFPA 13 as required by Section 903.2.5. The limitation of four occupants reduces the potential fuel load (mattresses, clothes, etc.) and the likelihood of involvement over an extended area.

907.2.7 Group M. A manual fire alarm system shall be installed in Group M occupancies, other than covered mall buildings complying with Section 402 of the *International Building Code*, having an occupant load of 500 or more persons or more than 100 persons above or below the lowest level of exit discharge.

Exception: Manual fire alarm boxes are not required where the building is equipped throughout with an automatic sprinkler system and the alarm notification appliances will activate upon sprinkler water flow.

❖ Group M occupancies have the potential for large occupant loads who may not be familiar with their surroundings. The installation of a fire alarm system increases the ability to alert the occupants of a fire. Note that the occupant thresholds are independent. If either the total occupant load is 500 or more persons or more than 100 persons are above or below the level of exit discharge, a manual fire alarm system is required.

The exception does not eliminate the fire alarm system, but rather permits it to be initiated automatically by sprinkler system water flow switch(es) instead of by manual fire alarm boxes. Buildings with a fire area containing a Group M occupancy in excess of 12,000 square feet (1,115 m^2) must be equipped with an automatic sprinkler system complying with Section 903.2.6.

907.2.7.1 Occupant notification. During times that the building is occupied, in lieu of the automatic activation of alarm notification appliances, the manual fire alarm system shall be allowed to activate an alarm signal at a constantly attended location from which evacuation instructions shall be initiated over an emergency voice/alarm communication system installed in accordance with Section 907.2.12.2.

The emergency voice/alarm communication system shall be allowed to be used for other announcements provided the manual fire alarm use takes precedence over any other use.

❖ Occupants in a mercantile occupancy may assume the alarm is a false alarm or act inappropriately and thus delay evacuation of the building. To prevent a panic situation, the manual fire alarm system may be part of an emergency voice/alarm communication system. The signal is to be sent to a constantly attended location on site from which evacuation instructions can be given.

907.2.8 Group R-1. Fire alarm systems shall be installed in Group R-1 occupancies as required in Section 907.2.8.1 through 907.2.8.3.

❖ Because residents of Group R-1 occupancies may be asleep and are usually transients who are unfamiliar with the building, and because such buildings contain numerous small rooms so that the occupants may not notice a fire in another part of the building, occupancies in Group R-1 must have a manual fire alarm system and an automatic fire detection system installed throughout. Requirements for single- or multiple-station smoke alarms in sleeping units are contained in Section 907.2.10.1.1.

907.2.8.1 Manual fire alarm system. A manual fire alarm system shall be installed in Group R-1 occupancies.

Exceptions:

1. A manual fire alarm system is not required in buildings not more than two stories in height where all individual guestrooms and contiguous attic and crawl spaces are

separated from each other and public or common areas by at least 1-hour fire partitions and each individual guestroom has an exit directly to a public way, exit court or yard.

2. Manual fire alarm boxes are not required throughout the building when the following conditions are met:

   2.1. The building is equipped throughout with an automatic sprinkler system installed in accordance with Section 903.3.1.1 or 903.3.1.2;

   2.2. The notification appliances will activate upon sprinkler water flow; and

   2.3. At least one manual fire alarm box is installed at an approved location.

❖ This section is specific to manual fire alarm systems and requires such systems in all Group R-1 occupancies, with two exceptions.

Exception 1 eliminates the requirement for a manual fire alarm system if the sleeping units have an exit discharging directly to a public way, exit court or yard. Even though the building may be two stories in height, the sleeping units on each floor must have access directly to an approved exit at grade level. The use of an exterior exit access balcony with exterior stairs serving the second floor does not constitute an exit directly at grade. The minimum 1-hour fire-resistance rating required for adequate separation of the sleeping units must be maintained.

Exception 2 does not omit the fire alarm system but rather permits it to be initiated automatically by sprinkler system water flow switch(es) in lieu of manual fire alarm boxes. The sprinkler system is to be equipped with local audible alarms that can be heard throughout the building and at least one manual fire alarm box installed at an approved location.

907.2.8.2 Automatic fire alarm system. An automatic fire alarm system shall be installed throughout all interior corridors serving guestrooms.

Exception: An automatic fire detection system is not required in buildings that do not have interior corridors serving guestrooms and each guestroom has a means of egress door opening directly to an exterior exit access that leads directly to an exit.

❖ This section requires an automatic fire alarm system within interior corridors. Such systems make use of smoke detectors for alarm initiation in accordance with Section 907.2, with one exception.

Automatic fire detectors are not required in motels and hotels that do not have interior corridors and in which sleeping units have a door opening directly to an exterior exit access that leads directly to the exits. The intent of Exception 2 is that the exit access from the sleeping unit door be exterior and not require reentering the building prior to entering the exit. Since the exit access is outside, the need for detectors other than in sleeping units is greatly reduced. Unlike Exception 1 to Section 907.2.8.1, exit balconies and exterior stairs are allowed.

907.2.8.3 Smoke alarms. Smoke alarms shall be installed as required by Section 907.2.10. In buildings that are not equipped throughout with an automatic sprinkler system installed in accordance with Section 903.3.1.1 or 903.3.1.2, the smoke alarms in guestrooms shall be connected to an emergency electrical system and shall be annunciated by guestroom at a constantly attended location from which the fire alarm system is capable of being manually activated

❖ The actual requirements for single- and multiple-station smoke alarms are located in Section 907.2.10. This section requires that, in unsprinklered buildings, the single- and multiple-station smoke alarms within sleeping units be connected to the emergency electrical system and that they be annunciated near guestrooms at a location from which the fire alarm system can be manually activated. Automatic activation is avoided to reduce unnecessary alarms within such buildings. There will be some cases where the above conditions exist, but Exception 1 of Section 907.2.8.1 would apply and a fire alarm system would not be required.

907.2.9 Group R-2. A manual fire alarm system shall be installed in Group R-2 occupancies where:

1. Any dwelling unit or sleeping unit is located three or more stories above the lowest level of exit discharge;
2. Any dwelling unit or sleeping unit is located more than one story below the highest level of exit discharge of exits serving the dwelling unit or sleeping unit; or
3. The building contains more than 16 dwelling units or sleeping units.

Exceptions:

1. A fire alarm system is not required in buildings not more than two stories in height where all dwelling units or sleeping units and contiguous attic and crawl spaces are separated from each other and public or common areas by at least 1-hour fire partitions and each dwelling unit or sleeping unit has an exit directly to a public way, exit court or yard.
2. Manual fire alarm boxes are not required throughout the building when the following conditions are met:

   2.1. The building is equipped throughout with an automatic sprinkler system in accordance with Section 903.3.1.1 or Section 903.3.1.2;

   2.2. The notification appliances will activate upon sprinkler flow; and

   2.3. At least one manual fire alarm box is installed at an approved location.

3. A fire alarm system is not required in buildings that do not have interior corridors serving dwelling units and are protected by an approved automatic sprinkler system installed in accordance with Sections 903.3.1.1 or 903.3.1.2, provided that dwelling units either have a means of egress door opening directly to an exterior exit access that leads directly to the exits or are served by open-ended corridors designed in accordance with Section 1022.6, Exception 4.

❖ The occupants of Group R-2 occupancies are not considered to be as transient as those of Group R-1, which increases the probability that residents can more readily notify each other of a fire. Therefore, although Group R-1 occupancies must have a manual fire alarm system subject to the exceptions in Section 907.2.8.1, Group R-2 occupancies are required to have only a manual fire alarm system as stipulated in one of the three listed conditions. The threshold conditions are meant to be applied independent of each other. Exceptions 1 and 2 are essentially identical to Exceptions 1 and 2 in Section 907.2.8.1 (see commentary, Section 907.2.8.1).

Exception 3 allows the omission of a fire alarm system in fully sprinklered buildings (NFPA 13 or 13R) with no interior corridors and that exit directly to an exterior exit access or have open-ended corridors. The important thing to note is that the sprinkler system is not required to activate alarm notification appliances since a fire alarm system would not be required.

907.2.10 Single- and multiple-station smoke alarms. Listed single- and multiple-station smoke alarms shall be installed in accordance with the provisions of this code and the household fire-warning equipment provisions of NFPA 72.

❖ Single- and multiple-station smoke alarms have evolved as one of the most important fire safety features in residential and similar occupancies having sleeping occupants. The value of early fire warning in these occupancies has been repeatedly demonstrated in fires involving both successful and unsuccessful smoke alarm performance.

For successful smoke alarm operation and performance, single- and multiple-station smoke alarms must be installed to comply with the code and NFPA 72, which contains the minimum requirements for the selection, installation, operation and maintenance of fire warning equipment for use in family living units. They are termed “smoke alarms” rather than “smoke detectors” because they are independent of a fire alarm system and include an integral alarm notification device.

907.2.10.1 Where required. Single- or multiple-station smoke alarms shall be installed in the locations described in Sections 907.2.10.1.1 through 907.2.10.1.3.

❖ Section 907.2.10.1 establishes the conditions under which single- or multiple-station smoke alarms are required. Single- and multiple-station smoke alarms are typically required where occupants may be sleeping, and therefore are unaware of a fire in the room or a fire that may affect their egress or escape path, such as in hallways that are adjacent to the bedrooms. Smoke alarms are intended to alert the occupants within the dwelling unit or sleeping unit, not all of the occupants within the building. It is presumed that after the occupant in the unit of origin has safely evacuated the unit, he or she will notify other occupants of the building by either word of mouth or initiation of the manual fire alarm system, if one is installed.

907.2.10.1.1 Group R-1. Single- or multiple-station smoke alarms shall be installed in all of the following locations in Group R-1:

1. In sleeping areas.
2. In every room in the path of the means of egress from the sleeping area to the door leading from the sleeping unit.
3. In each story within the sleeping unit, including basements. For sleeping units with split levels and without an intervening door between the adjacent levels, a smoke alarm installed on the upper level shall suffice for the adjacent lower level provided that the lower level is less than one full story below the upper level.

❖ Because the occupant(s) of a sleeping unit or suite may be asleep and unaware of a fire developing in the room or in the egress path, single- or multiple-station smoke alarms must be provided in the sleeping unit and in any intervening room between the sleeping unit and the exit access door from the room. If the sleeping unit or suite involves more than one level, a smoke alarm must also be installed on every level.

Smoke alarms are required in split-level arrangements, except those that meet the conditions described in Item 3. In accordance with Section 907.2.10.3, all smoke alarms within a sleeping unit or suite must be interconnected so that actuation of one alarm will actuate the alarms in all smoke alarms within the sleeping unit or suite.

Section 907.2.8.3 contains requirements for the single- or multiple-station smoke alarms in the sleeping unit to be connected to an emergency electrical system and be annunciated at a constantly attended location from which manual activation of the fire alarm system can be accomplished (see commentary, Section 907.2.8.3).

907.2.10.1.2 Groups R-2, R-3, R-4 and I-1. Single- or multiple-station smoke alarms shall be installed and maintained in Groups R-2, R-3, R-4 and I-1 regardless of occupant load at all of the following locations:

1. On the ceiling or wall outside of each separate sleeping area in the immediate vicinity of bedrooms.
2. In each room used for sleeping purposes.
3. In each story within a dwelling unit, including basements but not including crawl spaces and uninhabitable attics. In dwellings or dwelling units with split levels and without an intervening door between the adjacent levels, a smoke alarm installed on the upper level shall suffice for the adja-

cent lower level provided that the lower level is less than one full story below the upper level.

❖ Because the occupants of a dwelling unit may be asleep and unaware of a fire developing in the room or in an area within the dwelling unit that will affect their ability to escape, single- or multiple-station smoke alarms must be installed in every bedroom, in the vicinity of all bedrooms (e.g., hallways leading to the bedrooms) and on each story of the dwelling unit (see Figure 907.2.10.1.2). This section also requires smoke alarms in all sleeping areas of Group I-1 occupancies, subject to the automatic fire detection system modification in Section 907.2.10.1.3.

If a sprinkler system was installed throughout the building in accordance with NFPA 13, 13R or 13D, if applicable, smoke alarms would still be required in the bedrooms even if residential sprinklers were used.

Smoke alarms are required in split-level arrangements. As required by Section 907.2.10.3, all smoke alarms within a dwelling unit must be interconnected so that actuation of one alarm will actuate the alarms in all detectors within the dwelling unit.

907.2.10.1.3 Group I-1. Single- or multiple-station smoke alarms shall be installed and maintained in sleeping areas in occupancies in Group I-1. Single- or multiple-station smoke alarms shall not be required where the building is equipped throughout with an automatic fire detection system in accordance with Section 907.2.6.

❖ Even though the occupants of a sleeping unit in a Group I-1 occupancy may be asleep, they are still considered capable of self-preservation. Regardless, smoke alarms are required in sleeping units. Single- or multiple-station smoke alarms need not be installed in the room if an automatic fire detection system that includes room system smoke detectors is installed as required by Section 907.2.6.

907.2.10.2 Power source. In new construction, required smoke alarms shall receive their primary power from the building wiring where such wiring is served from a commercial source and shall be equipped with a battery backup. Smoke alarms shall emit a signal when the batteries are low. Wiring shall be permanent and without a disconnecting switch other than as required for overcurrent protection.

Exception: Smoke alarms are not required to be equipped with battery backup in Group R-1 where they are connected to an emergency electrical system.

❖ Smoke alarms are required to use AC as a primary power source and battery power as a secondary source to improve their reliability. For example, during a power outage, the probability of fire is increased because of the use of candles or lanterns for temporary light. Required backup battery power is intended to ensure continued functioning of the smoke alarms. Smoke alarms are commonly designed to emit a recurring signal when batteries are low and need to be replaced.

Certain Group R-1 occupancies may already have an emergency electrical system in the building to monitor other building system conditions. The emergency electrical system provides a level of reliability equivalent to battery backup.

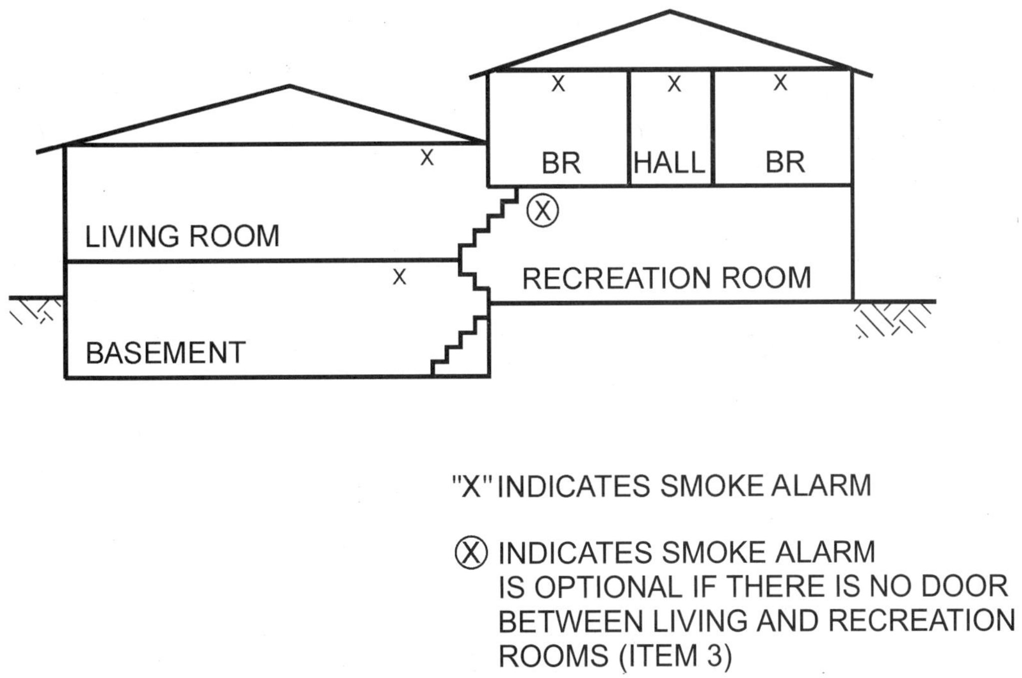

Figure 907.2.10.1.2
LOCATION OF SMOKE ALARMS IN A SINGLE-FAMILY DWELLING

907.2.10.3 Interconnection. Where more than one smoke alarm is required to be installed within an individual dwelling unit or sleeping unit in Group R-2, R-3 or R-4, or within an individual sleeping unit in Group R-1, the smoke alarms shall be interconnected in such a manner that the activation of one alarm will activate all of the alarms in the individual unit. The alarm shall be clearly audible in all bedrooms over background noise levels with all intervening doors closed.

❖ The installation of smoke alarms in areas remote from the sleeping area will be of minimal value if the alarm is not heard by the occupants. Interconnection of multiple smoke alarms within an individual dwelling unit or sleeping unit is required in order to alert a sleeping occupant of a remote fire within the unit before the combustion products reach the smoke alarm in the sleeping area and thus provide additional time for evacuation.

The term interconnection is intended to allow the use of not only hard-wired systems, but also those that use radio signals (wireless systems) (see Section 907.6). UL has listed smoke detectors that use this technology. It is presumed that on safely evacuating the unit or room of fire origin, an occupant will notify other occupants by actuating the manual fire alarm system or using other available means.

907.2.10.4 Acceptance testing. When the installation of the alarm devices is complete, each detector and interconnecting wiring for multiple-station alarm devices shall be tested in accordance with the household fire warning equipment provisions of NFPA 72.

❖ To determine that smoke alarms have been properly installed and are ready to function as intended, they must be actuated during an acceptance test. The test also confirms that interconnected detectors will operate simultaneously as required. The responsibility for conducting the acceptance tests rests with the owner or the owner representative as stated in Section 901.5.

907.2.11 Special amusement buildings. An approved automatic smoke detection system shall be provided in special amusement buildings in accordance with this section.

Exception: In areas where ambient conditions will cause a smoke detection system to alarm, an approved alternative type of automatic detector shall be installed.

❖ Special amusement buildings are buildings in which the means of egress is not readily apparent, is intentionally confounded or is not readily available. Special amusement buildings must also comply with the provisions of Section 411 of the IBC.

The approved automatic smoke detection system is required to provide early warning of a fire. The detection system is required regardless of the presence of staff in the building. The exception recognizes that the ambient conditions in some special amusement buildings may preclude the use of automatic smoke detectors. In those instances, an alternative detection device must be used for early detection of a fire.

907.2.11.1 Alarm. Activation of any single smoke detector, the automatic sprinkler system or any other automatic fire detection device shall immediately sound an alarm at the building at a constantly attended location from which emergency action can be initiated, including the capability of manual initiation of requirements in Section 907.2.11.2.

❖ Upon activation of either a smoke detector or other automatic fire detection device or the automatic sprinkler system, an alarm must sound at a constantly attended location. The staff at the location is expected to be capable of then providing the required egress illumination, stopping the conflicting or confusing sounds and distractions and activating the exit marking required by Section 907.2.11.2. The staff is also expected to be capable of preventing additional people from entering the building.

907.2.11.2 System response. The activation of two or more smoke detectors, a single smoke detector with alarm verification, the automatic sprinkler system or other approved fire detection device shall automatically:

1. Cause illumination of the means of egress with light of not less than 1 foot-candle (11 lux) at the walking surface level;
2. Stop any conflicting or confusing sounds and visual distractions; and
3. Activate an approved directional exit marking that will become apparent in an emergency.

Such system response shall also include activation of a prerecorded message, clearly audible throughout the special amusement building, instructing patrons to proceed to the nearest exit. Alarm signals used in conjunction with the prerecorded message shall produce a sound which is distinctive from other sounds used during normal operation.

The wiring to the auxiliary devices and equipment used to accomplish the above fire safety functions shall be monitored for integrity in accordance with NFPA 72.

❖ Once a fire has been detected, measures must be taken to stop the confusion or distractions. Additionally, the egress path must be illuminated and marked. These measures must occur automatically upon detection of the fire or sprinkler water flow. A prerecorded message that can be heard throughout the building instructing the occupants to proceed to the nearest exit must be automatically activated. The message and alarm signals should be designed to prevent panic. The prerecorded message capability is in addition to the emergency voice/alarm communication system requirement of Section 907.2.11.3. The wiring of all devices must comply with NFPA 72.

907.2.11.3 Emergency voice/alarm communication system. An emergency voice/alarm communication system, which is also allowed to serve as a public address system, shall be installed in accordance with NFPA 72 and be audible throughout the entire special amusement building.

❖ Because of the problem associated with evacuating special amusement buildings, an emergency voice/alarm communication system is required (see Section 907.2.12.2). This section permits the system to also serve as a public address system to have the capability to alert the occupants of a fire and give them evacuation instructions.

907.2.12 High-rise buildings. Buildings having floors used for human occupancy located more than 75 feet (22 860 mm) above the lowest level of fire department vehicle access shall be provided with an automatic fire alarm system and an emergency voice/alarm communication system in accordance with Section 907.2.12.2.

Exceptions:

1. Airport traffic control towers in accordance with Section 907.2.22 and Section 412 of the *International Building Code.*
2. Open parking garages in accordance with Section 406.3 of the *International Building Code.*
3. Buildings with an occupancy in Group A-5 in accordance with Section 303.1 of the *International Building Code.*
4. Low-hazard special occupancies in accordance with Section 503.1.2 of the *International Building Code.*
5. Buildings with an occupancy in Group H-1, H-2 or H-3 in accordance with Section 415 of the *International Building Code.*

❖ High-rise buildings require additional fire protection systems because of the difficulties with smoke movement, egress time and fire department access. As a result, this section requires both an automatic fire alarm system and an emergency voice/alarm communication system (see commentary, Section 907.2.12.2).

The listed exceptions are the same as those in Section 403.1 of the IBC regarding the applicability of the high-rise provisions.

Exception 1 addresses airport traffic control towers and is based on the limited fuel load and the limited number of persons occupying the tower. Open parking garages and places of outdoor assembly (Group A-5) are exempted by Exceptions 2 and 3, respectively, because of the free ventilation to the outside that exists in such structures. In Exception 4, low-hazard special industrial occupancies may be exempted when approved by the fire code official. Such buildings should be evaluated based on the occupant load and the hazards of the occupancy and its contents to determine whether the protection features required by Section 403 of the IBC are necessary. Buildings with occupancies in Groups H-1, H-2 and H-3 are excluded from the requirements of this section by Exception 5 because the fire hazard characteristics of these occupancies have not yet been considered in high-rise buildings.

907.2.12.1 Automatic fire detection. Smoke detectors shall be provided in accordance with this section. Smoke detectors shall be connected to an automatic fire alarm system. The activation of any detector required by this section shall operate the emergency voice/alarm communication system. Smoke detectors shall be located as follows:

1. In each mechanical equipment, electrical, transformer, telephone equipment or similar room which is not provided with sprinkler protection, elevator machine rooms, and in elevator lobbies.
2. In the main return air and exhaust air plenum of each air-conditioning system having a capacity greater than 2,000 cubic feet per minute (cfm) (0.94 m^3/s). Such detectors shall be located in a serviceable area downstream of the last duct inlet.
3. At each connection to a vertical duct or riser serving two or more stories from a return air duct or plenum of an air-conditioning system. In Group R-1 and R-2 occupancies, a listed smoke detector is allowed to be used in each return-air riser carrying not more than 5,000 cfm (2.4 m^3/s) and serving not more than 10 air-inlet openings.

❖ Automatic smoke detectors are required in all high-rise buildings in certain locations so that a fire will be detected in its early stages of development. Smoke detectors must be installed in rooms that are not typically occupied. Spaces specifically identified as requiring automatic fire detection are: mechanical equipment, electrical, transformer telephone equipment and elevator machine rooms. The detectors must be connected to the automatic fire alarm system and be capable of initiating operation of the emergency voice/alarm communication system.

Smoke detectors must be installed in the main return air and exhaust air plenum of each air-conditioning system having a design capacity exceeding 2,000 cubic feet per minute (cfm) (0.94 m^3/s). Systems with design capacities equal to or less than 2,000 cfm (0.94 m^3/s) are exempt from this requirement because their small size limits their capacity for spreading smoke to parts of the building not already involved with fire.

The area that could be served by a 2,000-cfm (0.94 m^3/s) system (approximately 5 tons of cooling capacity) is comparatively small; therefore, the distribution of smoke in a system of that size would be minimal. Smoke detectors must be located so that they monitor the total airflow within the system. If a single detector is unable to sample the total airflow at all times, then multiple detectors are required. The smoke detectors must be made accessible for maintenance and inspection. Many failures and false alarms are caused by a lack of maintenance and cleaning of the smoke detectors.

Consistent with Section 606.2.3 of the IMC, return-air risers serving two or more stories must have smoke detectors installed at each story. Item 3 permits the use of a single listed smoke detector in each return-air riser in a Group R-1 or R-2 occupancy if the capacity of each riser does not exceed 5,000 cfm (2.4 m^3/s) and does not serve more than 10 air-inlet openings. This alternative recognizes that it is not as necessary in buildings dedicated to residential occupancies only to monitor the return air from each story prior to intermixing the return air in the common riser.

907.2.12.2 Emergency voice/alarm communication system. The operation of any automatic fire detector, sprinkler water-flow device or manual fire alarm box shall automatically sound an alert tone followed by voice instructions giving approved information and directions on a general or selective basis to the following terminal areas on a minimum of the alarming floor, the floor above, and the floor below in accordance with the building's fire safety and evacuation plans required by Section 404.

1. Elevator lobbies.
2. Corridors.
3. Rooms and tenant spaces exceeding 1,000 square feet (93 m^2) in area.
4. Dwelling units and sleeping units in Group R-2 occupancies.
5. Sleeping units in Group R-1 occupancies.
6. Areas of refuge as defined in Section 1002.

Exception: In Group I-1 and I-2 occupancies, the alarm shall sound in a constantly attended area and a general occupant notification shall be broadcast over the overhead page.

❖ The section identifies the areas of coverage when an emergency voice/alarm communication system is required. The system may sound a general alarm or be a selective system in which only selected areas of the building receive the alarm indication. This section requires a minimum area of notification that must include the alarming floor and the floors above and below it.

The requirement for alarm notification within dwelling units in Group R-2 occupancies and within sleeping units in Group R-1 occupancies recognizes the need for a sound pressure level that will awaken sleeping occupants. Although Section 907.10.2 requires a minimum sound pressure level of 70 decibel (dBA) in Group R occupancies, a sound pressure level of 75 dBA at the head of the bed is generally considered the minimum to alert sleeping individuals.

This section also indicates that the emergency voice/alarm system is to be initiated as all other fire alarm systems are initiated. The functional operation of the system begins with an alert tone (usually 3 to 10 seconds in duration) followed by the evacuation signal (message).

The exception recognizes the supervised environment typical of institutional uses and the reliance placed on staff to act appropriately in an emergency.

907.2.12.2.1 Manual override. A manual override for emergency voice communication shall be provided for all paging zones.

❖ The intent of this section is to provide the ability to transmit live voice instructions over any previously initiated signals or messages for all zones.

907.2.12.2.2 Live voice messages. The emergency voice/alarm communication system shall also have the capability to broadcast live voice messages through speakers located in elevators, exit stairways, and throughout a selected floor or floors.

❖ The number of speakers installed should be adequate to broadcast the live voice messages over the emergency voice/alarm communication system to all desired areas, including elevators, exit stairs and selected floors. Speakers used for background music should not be used unless specifically listed for fire alarm system use. NFPA 72 has additional requirements for the placement, location and audibility of speakers used as part of an emergency voice/alarm communication system.

907.2.12.2.3 Standard. The emergency voice/alarm communication system shall be designed and installed in accordance with NFPA 72.

❖ NFPA 72 contains the minimum requirements for the design, installation, maintenance and acceptance testing for emergency voice/alarm communication systems. The primary purpose of the emergency voice/alarm communication system is to provide dedicated manual and automatic facilities for the origination, control and transmission of information and instructions pertaining to a fire alarm emergency to the occupants (including fire department personnel) of the building.

907.2.12.3 Fire department communication system. An approved two-way, fire department communication system designed and installed in accordance with NFPA 72 shall be provided for fire department use. It shall operate between a fire command center complying with Section 509 and elevators, elevator lobbies, emergency and standby power rooms, fire pump rooms, areas of refuge and inside enclosed exit stairways. The fire department communication device shall be provided at each floor level within the enclosed exit stairway.

Exception: Fire department radio systems where approved by the fire department.

❖ High-rise buildings have also posed a challenge to the traditional communication systems used by the fire service for fire-to-ground communications. Therefore, a fire department communication system must be installed to assist fire ground officers in communicating with the fire fighters working in various areas of the building. The system must be capable of operating between the fire command center and every elevator, elevator lobby, emergency/standby power room, fire pump room, area of refuge and exit stairway.

The exception permits the use of fire department hand-held radios where approved by the fire department for that specific building. The use of radio systems may not be effective in buildings with a significant amount of structural steel.

907.2.13 Atriums connecting more than two stories. A fire alarm system shall be installed in occupancies with an atrium that connects more than two stories. The system shall be activated in accordance with Section 907.7. Such occupancies in Group A, E or M shall be provided with an emergency voice/alarm communication system complying with the requirements of Section 907.2.12.2.

❖ Buildings containing an atrium that connects more than two stories are to be equipped with a fire alarm system that can be used to notify building occupants to begin evacuating in case of a fire. The alarm system must be initiated in accordance with Section 907.7, which requires that in buildings containing an atrium, the alarm system is to be initiated by the sprinkler system and any automatic or manual fire alarm initiating devices found in the atrium as well as elsewhere in the building. It does not intend to require certain features to be installed within the atrium but rather is simply requiring that any such features present initiate the fire alarm notification system. It would not necessarily be appropriate to also initiate the smoke control system upon activation of the alarm system within a building containing an atrium (see Section 909.12.2).

Groups A, E and M must have an emergency voice/alarm communication system that complies with Section 907.2.12.2 because of the number of persons to be evacuated and the lack of familiarity with the location of exits that is typical of occupants in Groups A and M. The alarm system is intended to warn occupants entering the atrium because smoke is being drawn to the atrium.

907.2.14 High-piled combustible storage areas. An automatic fire detection system shall be installed throughout high-piled combustible storage areas where required by Section 2306.5.

❖ Section 2306.5 requires an automatic fire detection system in high-piled combustible storage areas depending on the commodity class, the size of the high-piled storage area and the presence of an automatic sprinkler system. High-piled storage is the storage of combustible materials in piles, on pallets or in racks more than 12 feet (3,658 mm) high. Chapter 23 and NFPA 231 and 231C contain fire protection requirements for all high-piled storage conditions. (Note: In the 1999 edition of NFPA 13, its scope has been expanded to include protection requirements from NFPA 231 and 231C.)

907.2.15 Delayed egress locks. Where delayed egress locks are installed on means of egress doors in accordance with Section 1008.1.8.6, an automatic smoke or heat detection system shall be installed as required by that section.

❖ This section alerts the code user to additional requirements in Section 1008.1.8.6 that tie the operation of egress doors into the activation of an automatic fire detection system. A smoke or heat detection system is required to unlock delayed egress locks upon activation. The heat detection system can be the sprinkler system. For example, Section 1008.1.3.3 requires horizontal sliding doors used as a component of the means of egress, where required to be rated, to be self-closing or automatic-closing upon smoke detection. Also, access-controlled egress doors in occupancies as required by Section 1008.1.3.4 must be capable of being automatically unlocked by activation of an automatic fire detection system, if one is installed.

907.2.16 Aerosol storage uses. Aerosol storage rooms and general-purpose warehouses containing aerosols shall be provided with an approved manual fire alarm system where required by this code.

❖ Chapter 28 and NFPA 30B contain additional guidance on the storage of and fire protection requirements for aerosol products. The requirements for storing the various levels of aerosol products are dependent on the level of sprinkler protection, the type of storage and the quantity of aerosol products. Although aerosol product fires generally involve property loss as opposed to loss of life, installation of a manual fire alarm system could aid in the prompt evacuation of the occupants. Fires involving aerosol products can spread rapidly through a building that is not properly protected and controlled.

907.2.17 Lumber, plywood and veneer mills. Lumber, plywood and veneer mills shall be provided with a manual fire alarm system.

❖ Any facility using mechanical methods to process wood into finished products, such as waferboard, oriented strandboard, composite wood panels or plywood, produces debris and the potential for combustible dust. Good housekeeping and control of ignition sources are therefore essential. To aid in the quick evacuation of occupants in an emergency, Section 1904.1.1 requires a manual fire alarm system in lumber, plywood and veneer mills that contain product dryers because of their potential as a source of ignition. A manual fire alarm system is not required, however, if the dryers and all other potential sources of ignition are protected by a supervised automatic sprinkler system.

907.2.18 Underground buildings with smoke exhaust systems. Where a smoke exhaust system is installed in an underground building in accordance with the *International Building Code*, automatic fire detectors shall be provided in accordance with this section.

❖ As indicated in Section 405.5.2 of the IBC, each compartment of an underground building must have a smoke control/exhaust system that can be activated both automatically and manually. Floor levels more than 60 feet (18 288 mm) below the lowest level of exit discharge must be compartmented. Compartmentation is a key element in the egress and fire access plan for floor areas in an underground building. The smoke exhaust system must not only facilitate egress during a fire, but also improve fire department access to the fire source by maintaining visibility that is otherwise impossible given the inability of the fire service to manually ventilate the underground portion of the building. To reduce potential involvement of other compartments, the smoke exhaust system for each compartment must be independent (see also Section 909.20).

907.2.18.1 Smoke detectors. A minimum of one smoke detector listed for the intended purpose shall be installed in the following areas:

1. Mechanical equipment, electrical, transformer, telephone equipment, elevator machine or similar rooms.
2. Elevator lobbies.
3. The main return and exhaust air plenum of each air-conditioning system serving more than one story and located in a serviceable area downstream of the last duct inlet.
4. Each connection to a vertical duct or riser serving two or more floors from return air ducts or plenums of heating, ventilating and air-conditioning systems, except that in Group R occupancies, a listed smoke detector is allowed to be used in each return-air riser carrying not more than 5,000 cfm (2.4 m^3/s) and serving not more than 10 air inlet openings.

❖ Automatic smoke detectors are required in certain locations in all underground buildings so that a fire will be detected in its early stages of development. Underground buildings are similar to high-rise buildings in that they present an unusual hazard by being virtually inaccessible to exterior fire department suppression and rescue operations with the increased potential to trap occupants inside the structure. For this reason, the smoke detector location requirements for underground buildings are similar to those in Section 907.2.12.1 for high-rise buildings (see commentary, Section 907.2.12.1).

The requirement for a smoke detector in the main return and exhaust air plenum of an air-conditioning system in an underground building, however, differs from that of a high-rise building in that it is not a function of capacity [2,000 cfm (0.94 m^3/s)] but rather a function of whether the system serves more than one floor level. There is more concern over the threat of smoke movement from floor to floor because the products of combustion cannot be vented directly to the atmosphere.

907.2.18.2 Alarm required. Activation of the smoke exhaust system shall activate an audible alarm at a constantly attended location.

❖ The audible alarm is required to notify qualified personnel immediately that the smoke exhaust system has activated and to put emergency procedures into action quickly.

907.2.19 Underground buildings. Where the lowest level of a structure is more than 60 feet (18 288 mm) below the lowest level of exit discharge, the structure shall be equipped throughout with a manual fire alarm system, including an emergency voice/alarm communication system installed in accordance with Section 907.2.12.2.

❖ The ability to communicate and offer warning of a fire can increase the time available for egress from the building. Underground structures located more than 60 feet (18 288 mm) below the level of exit discharge must therefore have a manual fire alarm system. A voice/alarm communication system is also required as part of this system (see commentary, Section 907.2.12.2).

907.2.19.1 Public address system. Where a fire alarm system is not required by Section 907.2, a public address system shall be provided which shall be capable of transmitting voice communications to the highest level of exit discharge serving the underground portions of the structure and all levels below.

❖ In underground structures where a fire alarm system is not required, a public address system must be installed. This communication between the highest level of discharge and the underground levels can be a vital communication link between emergency personnel and building occupants. This is especially important in an environment in which visibility is likely to be impaired as a means to determine the status of emergency evacuation from a fire scene below grade.

907.2.20 Covered mall buildings. Covered mall buildings exceeding 50,000 square feet (4645 m^2) in total floor area shall be provided with an emergency voice/alarm communication system. An emergency voice/alarm communication system serving a mall, required or otherwise, shall be accessible to the fire department. The system shall be provided in accordance with Section 907.2.12.2.

❖ Because of the potentially large number of occupants and their unfamiliarity with their surroundings, an emergency voice/alarm communication system, accessible by the fire department, is required to aid in evacuation of covered mall buildings exceeding 50,000 square feet (4,645 m^2) in total floor area. Anchor stores are not included as part of the covered mall building.

907.2.21 Residential aircraft hangars. A minimum of one listed smoke alarm shall be installed within a residential aircraft hangar as defined in the *International Building Code* and shall be interconnected into the residential smoke alarm or other sounding device to provide an alarm which will be audible in all sleeping areas of the dwelling.

❖ Residential aircraft hangars are assumed to be on the same property as a one- or two-family dwelling. Section 412.3 of the IBC contains additional requirements for the construction of residential aircraft hangars. The hangar could be located immediately adjacent to the dwelling unit if it is separated by 1-hour fire-resistance-rated construction. Because of the potentially close proximity of the aircraft and its flammability and fuel source, at least one smoke alarm is required in the hangar that is interconnected to the residential smoke alarms. It should be noted, however, that the requirement for a smoke alarm is also applicable to residential aircraft hangars that are detached from the dwelling unit. Because a minimum separation distance is not specified, a fire in the hangar could still present a serious fire hazard to the dwelling unit.

907.2.22 Airport traffic control towers. An automatic fire detection system shall be provided in airport traffic control towers.

❖ Airport traffic control towers must be designed to comply with Section 412 of the IBC. These structures are unique in that they can be built to excessive heights, depending upon construction type, are permitted to have one exit stairway and are typically nonsprinklered. Section 903.2.10.3 specifically exempts airport control towers from the requirements of an automatic sprinkler system. An automatic fire detection system is required, however, for early warning notification of the occupants in an emergency.

907.2.23 Battery rooms. An approved automatic smoke detection system shall be installed in areas containing stationary lead-acid battery systems having a liquid capacity of more than 50 gallons (189 L). The detection system shall be supervised by an approved central, proprietary, or remote station service or a local alarm which will sound an audible signal at a constantly attended location.

❖ Stationary lead-acid battery systems are commonly used for standby power, emergency power or uninterrupted power supplies. The release of hydrogen gas during battery system operation is usually minimal. Adequate ventilation will disperse the small amounts of liberated hydrogen. Because standby power and emergency power systems control many important building emergency systems and functions, a supervised automatic smoke-detection system is required for early warning notification of a hazardous condition. Sections 608 and 609 contain additional requirements, including the need for safety venting; room enclosure requirements; spill control and neutralization provisions; ventilation criteria; signage and seismic protection.

907.3 Where required—retroactive in existing buildings and structures. An approved manual, automatic or manual and automatic fire alarm system shall be installed in existing buildings and structures in accordance with Sections 907.3.1 through 907.3.1.8. Where automatic sprinkler protection is provided in accordance with Section 903.3.1.1 or 903.3.1.2 and connected to the building fire alarm system, automatic heat detection required by this section shall not be required.

An approved automatic fire detection system shall be installed in accordance with the provisions of this code and NFPA 72. Devices, combinations of devices, appliances and equipment shall be approved. The automatic fire detectors shall be smoke detectors, except an approved alternative type of detector shall be installed in spaces such as boiler rooms where, during normal operation, products of combustion are present in sufficient quantity to actuate a smoke detector.

❖ As indicated in Section 907.1, this section is applicable only to existing buildings and structures.

Similar to Section 907.2, this section does not require automatic heat detection in existing buildings that are fully sprinklered in compliance with NFPA 13 or 13R. The sprinkler head, in cases where heat detection is desired, essentially acts as a heat detection device. This provision assumes a heat detector is permitted in place of a smoke detector (see commentary, Section 907.2).

907.3.1 Occupancy requirements. A fire alarm system shall be installed in accordance with Sections 907.3.1.1 through 907.3.1.8.

Exception: Occupancies with an existing, previously approved fire alarm system.

❖ This section specifies the occupancy conditions when an approved fire alarm system is retroactively required in an existing building.

The exception recognizes the infeasibility of requiring existing previously approved fire alarm systems to conform to current code requirements. The existing fire alarm system must be adequately tested, maintained and shown not to create a hazard.

907.3.1.1 Group E. A fire alarm system shall be installed in existing Group E occupancies in accordance with Section 907.2.3.

Exceptions:

1. A building with a maximum area of 1,000 square feet (93 m^2) that contains a single classroom and is located no closer than 50 feet (15 240 mm) from another building.
2. Group E with an occupant load less than 50.

❖ Group E occupancies are limited to educational purposes through the 12th grade. Because of the potentially young age and maturity of the occupants, more time may be needed to safely evacuate the building. The requirement for retroactive installation of at least a fire alarm system recognizes that many existing previously approved unsprinklered educational facilities would most likely require sprinklers under current code provisions (see Section 903.2.2).

Although not limited to this use condition, Exception 1 recognizes the current use of mobile trailer-type facilities on site as additional educational classroom facilities. This exception doesn't exempt the main building, but it would exempt these auxiliary buildings of limited size that do not present an exposure hazard because of the required separation distance.

Exception 2 would exempt small day care centers that serve children older than 2 years of age, a small Sunday school classroom at a church or similar limited educational use areas (see commentary, Section 907.2.3).

907.3.1.2 Group I-1. A fire alarm system shall be installed in existing Group I-1 residential care/assisted living facilities.

Exception: Where each sleeping room has a means of egress door opening directly to an exterior egress balcony that leads directly to the exits in accordance with Section 1013.5, and the building is not more than three stories in height.

❖ Group I-1 facilities are assumed to have more than 16 occupants who because of their age, mental disability or other reasons must live in a supervised environment 24 hours a day. This section would require existing Group I-1 occupancies to have an approved fire alarm system as required by Section 907.2.6. The term "residential care/assisted living facilities" includes, but is not limited to, residential board and care facilities, assisted living facilities, halfway houses, group homes and alcohol and drug abuse centers.

The exception recognizes the increased degree of life safety resulting from having direct access to the exterior from the sleeping rooms. The occupants are not forced to evacuate through the interior of the building during a potential fire. The exterior egress balconies must be sufficiently open to the atmosphere and constructed to minimize the accumulation of smoke and toxic gases.

907.3.1.3 Group I-2. A fire alarm system shall be installed in existing Group I-2 occupancies in accordance with Section 907.2.6.

❖ Because of the potential incapacitation of the occupants and the subsequent reliance on staff, an approved fire alarm system is required in existing Group I-2 occupancies. The system must comply with Section 907.2.6 (see commentary, Section 907.2.6).

907.3.1.4 Group I-3. A fire alarm system shall be installed in existing Group I-3 occupancies in accordance with Section 907.2.6.1.

❖ Because of the potential restraint of the occupants and subsequent evacuation difficulties, an approved fire alarm system is required in existing Group I-3 occupancies. The system must comply with Section 907.2.6.2 (see commentary, Section 907.2.6.2).

907.3.1.5 Group R-1 hotels and motels. A fire alarm system shall be installed in existing Group R-1 hotels and motels more than three stories or with more than 20 guestrooms.

Exception: Buildings less than two stories in height where all guestrooms, attics and crawl spaces are separated by 1-hour fire-resistance-rated construction and each guestroom has direct access to a public way, exit court or yard.

❖ This section specifies the conditions when a fire alarm system is required in existing Group R-1 hotels and motels. The two main criteria are independent of each other in that a fire alarm system is required if the building is more than three stories above grade regardless of the number of guestrooms, or contains 20 guestrooms regardless of the number of stories. Occupants of these types of Group R-1 facilities are assumed to be more transient than the occupants of Group R-1 boarding and rooming houses regulated by Section 907.3.1.6.

The exception recognizes the increased level of life safety afforded by adequate compartmentation using 1-hour fire-resistance-rated construction between guestrooms and direct exterior access for egress.

907.3.1.6 Group R-1 boarding and rooming houses. A fire alarm system shall be installed in existing Group R-1 boarding and rooming houses.

Exception: Buildings that have single-station smoke alarms meeting or exceeding the requirements of Section 907.2.10.1 and where the fire alarm system includes at least one manual fire alarm box per floor arranged to initiate the alarm.

❖ Group R-1 boarding and rooming houses are still assumed to be transient residential occupancies. The functional use of a boarding and rooming house is different from that of a typical hotel/motel. Boarding and rooming houses tend to have more extended living arrangements and border on being classified as Group R-2 facilities. For this reason, this section requires Group R-1 boarding and rooming houses to be equipped with a fire alarm system regardless of the height of the building or number of sleeping rooms.

The exception, however, allows the omission of the fire alarm system if single-station smoke alarms complying with the minimum requirements of Section 907.2.10.1 and at least one manual fire alarm box per floor are installed. The single-station smoke alarms give the desired early warning notification to the occupants and the manual fire alarm box is an additional means to activate the building smoke alarms.

907.3.1.7 Group R-2. A fire alarm system shall be installed in existing Group R-2 occupancies more than three stories in height or with more than 16 dwelling units or sleeping units.

Exceptions:

1. Where each living unit is separated from other contiguous living units by fire barriers having a fire-resistance rating of not less than 0.75 hour, and where each living unit has either its own independent exit or its own independent stairway or ramp discharging at grade.
2. A separate fire alarm system is not required in buildings that are equipped throughout with an approved supervised automatic sprinkler system installed in accordance with Section 903.3.1.1 or 903.3.1.2 and having a local alarm to notify all occupants.
3. A fire alarm system is not required in buildings that do not have interior corridors serving dwelling units and are protected by an approved automatic sprinkler system installed in accordance with Sections 903.3.1.1 or 903.3.1.2, provided that dwelling units either have a means of egress door opening directly to an exterior exit access that leads directly to the exits or are served by

open-ended corridors designed in accordance with Section 1022.6, Exception 4.

❖ This section specifies the conditions when a fire alarm system is required in existing Group R-2 apartment buildings based on height or the number of dwelling units. Occupants of Group R-2 facilities tend to be more permanent than those in Group R-1 facilities.

Exception 1 recognizes the increased degree of life safety afforded by compartmentation using fire-resistance-rated construction and independent means of egress (see Section 907.2.9).

As indicated in Exception 2, existing buildings that are fully sprinklered in accordance with NFPA 13 or 13R do not need a manual fire alarm system if local alarms will sound upon activation of the sprinkler system. The exception essentially eliminates the need for manual fire alarm boxes if evacuation alarms can still be heard throughout the building upon sprinkler system water flow.

Exception 3 mirrors Section 907.2.9, Exception 3 and recognizes the superior fire record of sprinklered multiple-family occupancies by allowing omission of a fire alarm system when a building is fully sprinklered, has no interior egress corridors and provides direct exterior egress from each dwelling unit. Note that in such buildings, rated fire separations are still required between units.

907.3.1.8 Group R-4. A fire alarm system shall be installed in existing Group R-4 residential care/assisted living facilities.

Exceptions:

1. Where there are interconnected smoke alarms meeting the requirements of Section 907.2.10 and there is at least one manual fire alarm box per floor arranged to sound continuously the smoke alarms.
2. Other manually activated, continuously sounding alarms approved by the fire code official.

❖ Existing Group R-4 residential care/assisted living facilities must have a fire alarm system. Group R-4 residential care-assisted living facilities are residential occupancies with more than five but not more than 16 occupants. Although this section requires a fire alarm system in an existing Group R-4 occupancy subject to the listed exceptions, it is not the intent of Section 907.3.1.9 to establish requirements beyond those required for a Group R-4 occupancy of new construction (see Sections 907.2.10.1.2, 907.2.10.3 and 907.3.1).

Exception 1 allows the omission of the fire alarm system if installed interconnected smoke alarms can be manually activated by at least one manual fire alarm box on each floor.

Exception 2 allows the fire code official to approve alternative means of manually activating the alarms in place of a manual fire alarm box.

907.3.2 Single- and multiple-station smoke alarms. Single- and multiple-station smoke alarms shall be installed in existing Group R occupancies in accordance with Sections 907.3.2.1 through 907.3.2.3.

❖ This section establishes the requirements for the installation of smoke alarms in existing Group R occupancies. These requirements recognize the benefit of installing smoke alarms in existing structures, but provide several exceptions for buildings that are not undergoing substantial renovations. Detached single-family dwellings and multiple single-family dwellings constructed under the *International Residential Code*® (IRC®) (see Section R101.2 of the IRC) are not classified in Occupancy Group R and are, therefore, outside the scope of Sections 907.3.2 through 907.3.2.3. See Section 704.2 of the *International Property Maintenance Code*® (IPMC®) for single- or multiple-station smoke alarm requirements for existing dwellings that are not regulated in Occupancy Group R.

907.3.2.1 General. Existing Group R occupancies not already provided with single-station smoke alarms shall be provided with approved single-station smoke alarms. Installation shall be in accordance with Section 907.2.10, except as provided in Sections 907.3.2.2 and 907.3.2.3.

❖ This section requires that Group R occupancies be provided with single-station smoke alarms, if they are not already provided with them. Essentially, a reference is made to Section 907.2.10 for the primary requirements except where Sections 907.3.2.2 and 907.3.2.3 are more specific. Therefore, smoke alarms need to be provided in all locations required by Section 907.2.10, but several exceptions related to interconnection of alarms and power supply are provided, recognizing the practicality of such installations in existing conditions.

907.3.2.2 Interconnection. Where more than one smoke alarm is required to be installed within an individual dwelling unit in Group R-2, R-3 or R-4, or within an individual sleeping unit in Group R-1, the smoke alarms shall be interconnected in such a manner that the activation of one alarm will activate all of the alarms in the individual unit. The alarm shall be clearly audible in all bedrooms over background noise levels with all intervening doors closed.

Exceptions:

1. Interconnection is not required in buildings that are not undergoing alterations, repairs or construction of any kind.
2. Smoke alarms in existing areas are not required to be interconnected where alterations or repairs do not result in the removal of interior wall or ceiling finishes exposing the structure, unless there is an attic, crawl space or basement available which could provide access for interconnection without the removal of interior finishes.

❖ This section, like Section 907.2.10.3, requires that when multiple-station smoke alarms are present, they are to be interconnected and be audible over background noises.

There are two exceptions. Exception 1 does not require interconnection if the building is not undergoing any construction or repairs. Exception 2 clarifies to what extent the building must be undergoing construction before interconnection is required. Generally, the exceptions try to be reasonable based upon the practicality of such installations; therefore, unless areas such as attics or crawl spaces can still be utiltized while the interior finishes are being removed (i.e., drywall removed exposing the studs), interconnection would not be required. Such renovations may only be limited to portions of a structure; therefore, complete interconnection may not be practical or possible. The intent is that additional walls, etc., should not be removed solely to interconnect the smoke alarms.

Battery-powered alarms, as allowed by Section 907.3.2.3, may be required to be interconnected. The exceptions found in Section 907.3.2.3 are similar to this section. The only time battery-powered smoke alarms may need to be interconnected is when a building is not supplied by a commercial power source.

907.3.2.3 Power source. In Group R occupancies, single-station smoke alarms shall receive their primary power from the building wiring provided that such wiring is served from a commercial source and shall be equipped with a battery backup. Smoke alarms shall emit a signal when the batteries are low. Wiring shall be permanent and without a disconnecting switch other than as required for overcurrent protection.

Exception: Smoke alarms are permitted to be solely battery operated: in existing buildings where no construction is taking place; in buildings that are not served from a commercial power source; and in existing areas of buildings undergoing alterations or repairs that do not result in the removal of interior walls or ceiling finishes exposing the structure, unless there is an attic, crawl space or basement available which could provide access for building wiring without the removal of interior finishes.

❖ The section is very similar to Section 907.2.10.2. The primary difference is the exception that allows the use of batteries as the sole power source under several conditions. Two of the conditions are similar to the exceptions to Section 907.3.2.2. If no construction or related repairs are occuring in the building, and if changes are being made that do not expose the structure (removing drywall, etc.), then battery-powered alarms are allowed. There is also the additional stipulation that, in buildings where the wall or ceiling finish are not to be removed exposing the structure but there is adequate ability to use the attic or similar space, the primary power supply should be provided through the building's wiring. The exception that is unique from Section 907.3.2.2 is when a commercial power supply is not available, connection to the building wiring is not required.

907.4 Manual fire alarm boxes. Manual fire alarm boxes shall be installed in accordance with Sections 907.4.1 through 907.4.5.

❖ This section specifies the requirements for manual fire alarm boxes that are part of a manual fire alarm system.

907.4.1 Location. Manual fire alarm boxes shall be located not more than 5 feet (1524 mm) from the entrance to each exit. Additional manual fire alarm boxes shall be located so that travel distance to the nearest box does not exceed 200 feet (60 960 mm).

Exception: Manual fire alarm boxes shall not be required in Group E occupancies where the building is equipped throughout with an approved automatic sprinkler system, the notification appliances will activate on sprinkler water flow and manual activation is provided from a normally occupied location.

❖ Manual fire alarm boxes must be located in the path of egress and be readily accessible to the occupants. They must be located within 5 feet (1,524 mm) of the entrance to each exit on every story of the building. This would include the need to locate manual fire alarm boxes near each horizontal exit, as well as entrances to stairs and exit doors to the exterior.

Manual fire alarm boxes are located near exits to ensure that an adequate number of devices is available in the path of egress to transmit an alarm in a timely manner. These locations also encourage the actuation of a manual fire alarm box on the fire floor prior to entering the stair, resulting in the alarm being received from the actual fire floor and not another floor along the path of egress.

The location also presumes that individuals will be evacuating the area where the fire originated. When evacuation of the fire area is unlikely, consideration could be given to putting manual fire alarm boxes in more convenient places. Examples of such instances would be officer stations in Group I-3 occupancies and nurses' stations in Group I-2 occupancies.

The 200-foot (60 960 mm) travel distance limitation is consistent with the exit access travel distance permitted for most nonsprinklered occupancies. If the 200-foot (60 960 mm) travel distance to a manual fire alarm box is exceeded, even in a fully sprinklered building, additional manual fire alarm boxes would be required.

The exception allows the omission of the manual fire alarm boxes in Group E occupancies equipped throughout with an automatic sprinkler system if the actuation of the sprinkler system will activate the building evacuation alarms associated with the manual fire alarm system.

907.4.2 Height. The height of the manual fire alarm boxes shall be a minimum of 42 inches (1067 mm) and a maximum of 48 inches (1372 mm) measured vertically, from the floor level to the activating handle or lever of the box.

❖ Manual fire alarm boxes must be reachable by the occupants of the building. They must also be mounted high enough to reduce the likelihood of damage or false alarms from something accidentally striking the device. Therefore, manual fire alarm boxes must be mounted a minimum of 42 inches (1,067 mm) and a maximum of 48 inches (1,219 mm) above the floor level. The 48-inch (1,219 mm) measurement corresponds to the maximum unobstructed side reach height by a person in a wheelchair.

907.4.3 Color. Manual fire alarm boxes shall be red in color.

❖ Manual fire alarm boxes are to be painted a distinctive red to help building occupants identify the device.

907.4.4 Signs. Where fire alarm systems are not monitored by a supervising station, an approved permanent sign shall be installed adjacent to each manual fire alarm box that reads: WHEN ALARM SOUNDS—CALL FIRE DEPARTMENT.

Exception: Where the manufacturer has permanently provided this information on the manual fire alarm box.

❖ This section has limited application because, as indicated in Section 907.15, fire alarm systems generally must be monitored by an approved supervising station. When a system is not monitored, such as possibly a fire alarm system that is not required by code, adequate signage must be displayed to tell occupants what response actions must be taken. Most building occupants assume that when an alarm device is activated, the fire department will automatically be notified as well. The sign should be conspicuously located next to the manual fire alarm box unless the signage is mounted on the manual fire alarm box itself by the manufacturer.

907.4.5 Protective covers. The fire code official is authorized to require the installation of listed manual fire alarm box protective covers to prevent malicious false alarms or provide the manual fire alarm box with protection from physical damage. The protective cover shall be transparent or red in color with a transparent face to permit visibility of the manual fire alarm box. Each cover shall include proper operating instructions. A protective cover that emits a local alarm signal shall not be installed unless approved.

❖ Although manual fire alarm boxes should be readily available to all occupants in buildings required to have a manual fire alarm system, this section permits the use of protective covers if they are approved by the fire code official. Protective covers are commonly used to reduce either the potential for intentional false alarms or vandalism.

907.5 Power supply. The primary and secondary power supply for the fire alarm system shall be provided in accordance with NFPA 72.

❖ The operation of fire alarm systems is essential to life safety in buildings and must be reliable in the event the normal power supply fails. To ensure proper operation of fire alarm systems, this section requires that the primary and secondary power supplies comply with NFPA 72. This is in addition to the general requirements for electrical installations in Chapter 27 of the IBC. NFPA 72 offers three alternatives for secondary supply: a 24-hour storage battery; storage batteries with a 4-hour capacity and a generator or multiple generators.

NFPA 72 requires that the primary and secondary power supplies for remotely located control equipment essential to the system operation must conform to the requirements for primary and secondary power supplies for the main system. Also, NFPA 72 contains requirements for monitoring the integrity of primary power supplies and requires a backup power supply.

907.6 Wiring. Wiring shall comply with the requirements of the ICC *Electrical Code* and NFPA 72. Wireless protection systems utilizing radio-frequency transmitting devices shall comply with the special requirements for supervision of low-power wireless systems in NFPA 72.

❖ Wiring for fire alarm systems must be installed so that it is secure and will function reliably in an emergency. The code requires that the wiring for fire alarm systems meet the requirements of NFPA 72. This requirement is in addition to the general requirements for electrical installations set forth in Chapter 27 of the IBC and in the ICC EC. For reliability, systems that use radio-frequency transmitting devices for signal transmission are required to have supervised transmitting and receiving equipment that conforms to the special requirements contained in NFPA 72. This requirement is in addition to the general requirements for supervision in Section 907.15.

907.7 Activation. Where an alarm notification system is required by another section of this code, it shall be activated by:

1. Required automatic fire alarm system.
2. Sprinkler water-flow devices.
3. Required manual fire alarm boxes.

❖ It is not the intent of this section to require that the various initiating devices contained in the list be identified. Rather, the section indicates that when such systems or devices are installed, they must serve as alarm-initiating devices. This section assumes that alarm notification appliances are required by another section of the code that may require either a manual fire alarm or an automatic fire detection system.

907.8 Presignal system. Presignal systems shall not be installed unless approved by the fire code official and the fire department. Where a presignal system is installed, 24-hour personnel supervision shall be provided at a location approved by the fire department, in order that the alarm signal can be actuated in the event of fire or other emergency.

❖ Presignal fire alarm systems have been a contributing factor in several multiple-death fire incidents. In most instances, the staff failed to activate the general alarm quickly and the occupants of the building were unaware of the fire. Therefore, the use of presignal systems is discouraged by the code. Presignal systems may be used only if they are approved by the fire code official and the fire department.

907.9 Zones. Each floor shall be zoned separately and a zone shall not exceed 22,500 square feet (1860 m^2). The length of any zone shall not exceed 300 feet (91 440 mm) in any direction.

Exception: Automatic sprinkler system zones shall not exceed the area permitted by NFPA 13.

❖ Since the fire alarm system also aids emergency personnel in locating the fire, the system must be zoned to shorten response time to the fire area. Zoning is also critical if the fire alarm system initiates certain other fire protection systems or control features, such as smoke control systems.

At a minimum, each floor of a building must constitute one zone of the system. If the floor area exceeds 22,500 square feet (2,090 m^2), additional zones are required. The maximum length of a zone is 300 feet (91 440 mm).

The exception states that NFPA 13 defines the maximum areas to be protected by one sprinkler system and that the sprinkler system need not be designed to meet the 22,500-square- foot area (2,090 m^2) limitations for a fire alarm system zone. For example, NFPA 13 permits a sprinkler system riser in a light-hazard occupancy to protect an area of 52,000 square feet (4,831 m^2) per floor. In accordance with the exception, a single water-flow switch, and consequently a single fire alarm system zone, would be acceptable. If other alarm-initiating devices are present on the floor, they would need to be zoned separately to meet the 22,500-square-foot (2,090 m^2) limitation

907.9.1 Zoning indicator panel. A zoning indicator panel and the associated controls shall be provided in an approved location. The visual zone indication shall lock in until the system is reset and shall not be canceled by the operation of an audible-alarm silencing switch.

❖ The zoning indicator panel must be installed in a location approved by the fire code official. The panel should be located to permit ready access by emergency responders. Once an alarm-initiating device within a zone has been activated, the annunciation of the zone must lock in until the system is reset.

907.9.2 High-rise buildings. In buildings that have floors located more than 75 feet (22 860 mm) above the lowest level of fire department vehicle access that are occupied for human occupancy, a separate zone by floor shall be provided for all of the following types of alarm-initiating devices where provided:

1. Smoke detectors.
2. Sprinkler water-flow devices.
3. Manual fire alarm boxes.
4. Other approved types of automatic fire detection devices or suppression systems.

❖ In addition to at least one zone per floor, high-rise buildings must have a separate zone for each indicated type of alarm-initiating device. Although this feature may be desirable in all buildings, the incremental cost difference is substantially higher in low-rise buildings in which basic fire alarm systems are installed. State-of-the-art fire alarm systems installed in high-rise buildings allow such distinctive zoning at a minimal cost difference.

907.10 Alarm notification appliances. Alarm notification appliances shall be provided and shall be listed for their purpose.

❖ The code requires that fire alarm systems be equipped with approved alarm notification appliances so that in an emergency, the fire alarm system will notify the occupants of the need for evacuation or implementation of the fire emergency plan. Alarm notification devices required by the code are of two general types: visible and audible. Except for voice/alarm signaling systems, once the system has been activated, all visible and audible alarms are required to activate. Voice/alarm signaling systems are special signaling systems that are activated selectively in response to specific emergency conditions.

907.10.1 Visible alarms. Visible alarm notification appliances shall be provided in accordance with Sections 907.10.1.1 through 907.10.1.4.

Exceptions:

1. Visible alarm notification appliances are not required in alterations, except where an existing fire alarm system is upgraded or replaced, or a new fire alarm system is installed.
2. Visible alarm notification appliances shall not be required in exits as defined in Section 1002.1.

❖ This section contains alarm system requirements for occupants who are hearing impaired. Visible alarm notification appliances should be located and oriented so that they will display alarm signals throughout a space.

Exception 1 states that visible alarm devices are not required in previously approved existing fire alarm systems or as part of minor alterations to existing fire alarm systems. Extensive modifications to an existing fire alarm system such as an upgrade or replacement would require the installation of visible alarm devices even if the previous existing system neither had them nor required them.

In Exception 2, visible alarm devices are not required in exit elements because of the potential distraction during evacuation. Exits as defined in Section 1002.1 could include exit enclosures or exit passageways but not exit access corridors.

907.10.1.1 Public and common areas. Visible alarm notification appliances shall be provided in public areas and common areas.

❖ Visible alarm notification appliances must provide coverage in all areas open to the public as well as all shared or common areas (e.g., corridors, public restrooms, shared offices, classrooms, etc). Areas where visible alarm notification appliances are not required include private offices, mechanical rooms or similar spaces.

907.10.1.2 Employee work areas. Where employee work areas have audible alarm coverage, the wiring system shall be designed so that visible alarm notification appliances can be integrated into the alarm system.

❖ This section allows for those with hearing impairments to be accommodated as necessary, but reduces the initial construction cost as such alarms may not be necessary in every situation.

907.10.1.3 Groups I-1 and R-1. Group I-1 and R-1 sleeping units in accordance with Table 907.10.1.3 shall be provided with a visible alarm notification appliance, activated by both the in-room smoke alarm and the building fire alarm system.

❖ Fire alarm systems in Group I-1 and R-1 sleeping accommodations must be equipped with visible alarms to the extent stated in Table 907.10.1.3.

The visible alarm notification devices in these rooms are to be activated by both the required in-room smoke alarm and the building fire alarm system. All visible alarm notification appliances in a building, however, need not be activated by individual room detectors. It is not a requirement that the accessible sleeping units be provided with visible alarm notification appliances even though some elderly patients or residents may be both mobility and hearing impaired.

907.10.1.4 Group R-2. In Group R-2 occupancies required by Section 907 to have a fire alarm system, all dwelling units and sleeping units shall be provided with the capability to support visible alarm notification appliances in accordance with ICC A117.1.

❖ Group R-2 occupancies with a fire alarm system are required to have all dwelling units wired to support visible alarm notification appliances. This includes all dwelling and sleeping units, not just those classified as either Type A or B. By reference to Sections 1004.2 through 1004.4.4 of ICC A117.1, the building alarm system wiring must be extended to the unit smoke detectors so that audible/visible alarm notification appliances may be connected to the building fire alarm system to notify residents with hearing impairments of an emergency situation. Chapter 11 of the IBC contains additional information on the classification criteria and requirements for accessible dwelling units.

TABLE 907.10.1.3
VISIBLE AND AUDIBLE ALARMS

NUMBER OF SLEEPING UNITS	SLEEPING ACCOMMODATIONS WITH VISIBLE AND AUDIBLE ALARMS
6 to 25	2
26 to 50	4
51 to 75	7
76 to 100	9
101 to 150	12
151 to 200	14
201 to 300	17
301 to 400	20
401 to 500	22
501 to 1,000	5% of total
1,001 and over	50 plus 3 for each 100 over 1,000

❖ This table specifies the minimum number of sleeping units that are to be equipped with visible and audible alarms. The numbers are based on the total number of sleeping accommodations in the facility. The requirements in this table are intended to be consistent with the *Americans with Disabilities Act Accessibility Guidelines for Buildings and Facilities* (ADAAG).

907.10.2 Audible alarms. Audible alarm notification appliances shall be provided and sound a distinctive sound that is not to be used for any purpose other than that of a fire alarm. The audible alarm notification appliances shall provide a sound pressure level of 15 decibels (dBA) above the average ambient sound level or 5 dBA above the maximum sound level having a duration of at least 60 seconds, whichever is greater, in every occupied space within the building. The minimum sound pressure levels shall be: 70 dBA in occupancies in Groups R and I-1; 90 dBA in mechanical equipment rooms; and 60 dBA in other occupancies. The maximum sound pressure level for audible alarm notification appliances shall be 120 dBA at the minimum hearing distance from the audible appliance. Where the average ambient noise is greater than 105 dBA, visible alarm notification appliances shall be provided in accordance with NFPA 72 and audible alarm notification appliances shall not be required.

Exception: Visible alarm notification appliances shall be allowed in lieu of audible alarm notification appliances in critical care areas of Group I-2 occupancies.

❖ To attract the attention of building occupants, audible alarms must be distinctive, using a sound that is unique to the signaling system, and be capable of being heard above the ambient noise in the space. In no case may the sound pressure exceed 120 dBA at the minimum hearing distance from the audible appliance. Sound pressures above 120 dBA can cause pain or even permanent hearing loss.

Be aware that in certain work areas, the Occupational Safety and Health Administration (OSHA) requires employees to wear hearing protection, possibly preventing them from hearing an audible alarm. Additionally, the noise factor in these areas is high enough that an audible alarm may not be discernible. In these areas, as well as in others, the primary method of indicating a fire can be by a visible signal. Employees must be capable of identifying such a signal as indicating a fire. The code user is advised that audible alarm notification appliances are not required when visible alarm notification appliances are installed.

The exception recognizes that the occupants in critical care areas of Group I-2 occupancies are usually incapacitated. The audible alarms may have the effect of unnecessarily disrupting the patients who are most likely not capable of self-preservation. Critical care areas are also assumed to be adequately staffed at all times and ready to respond upon activation of a visible alarm device.

907.11 Fire safety functions. Automatic fire detectors utilized for the purpose of performing fire safety functions shall be connected to the building's fire alarm control panel where a fire alarm system is required by Section 907.2. Detectors shall, upon actuation, perform the intended function and activate the alarm notification appliances or activate a visible and audible supervisory signal at a constantly attended location. In buildings not required to be equipped with a fire alarm system, the automatic fire detector shall be powered by normal electrical service and, upon actuation, perform the intended function. The detectors shall be located in accordance with NFPA 72.

❖ When the code requires installation of automatic fire detectors to perform a specific function, such as elevator recall or smokeproof enclosure ventilation, or when detectors are installed to comply with a permitted alternative, such as door-closing devices, these detectors must be connected to the building is automatic fire alarm system if the building is required by the code to have such a system.

In addition to performing its intended function (for example, closing a door), if a detector is activated, it must also activate either the building alarm devices or a supervisory signal at a constantly attended location. This requirement recognizes that these detectors and the devices they control are part of the building fire protection system and are expected to perform as designed. Being connected to the automatic fire alarm system, they will have the supervision necessary to ensure operational reliability.

An exception is provided for fire safety function detectors in buildings not required to have a fire alarm system. The fire safety function detectors must be powered by the building electrical system and be located as required by NFPA 72. Without this exception, these detectors could not be expected to perform as intended because there would be no power supply.

907.12 Duct smoke detectors. Duct smoke detectors shall be connected to the building's fire alarm control panel when a fire alarm system is provided. Activation of a duct smoke detector shall initiate a visible and audible supervisory signal at a constantly attended location. Duct smoke detectors shall not be used as a substitute for required open area detection.

Exceptions:

1. The supervisory signal at a constantly attended location is not required where duct smoke detectors activate the building's alarm notification appliances.
2. In occupancies not required to be equipped with a fire alarm system, actuation of a smoke detector shall activate a visible and an audible signal in an approved location. Smoke detector trouble conditions shall activate a visible or audible signal in an approved location and shall be identified as air duct detector trouble.

❖ It is not the intent of this section to send a signal to the fire department or to activate the alarm notification devices within a building. Instead, this section requires that a supervisory signal be sent to a constantly attended location. Smoke detectors must be connected to a fire alarm system where such systems are installed. Connection to the fire alarm system will activate a visible and audible supervisory signal at a constantly attended location, which will alert building supervisory personnel that a smoke alarm has activated and will also provide electronic supervision of the duct detectors, thereby indicating any problems that may develop in the detector system circuitry or power supply.

Exception 1 allows activation of the building alarm notification appliances in place of a supervisory signal. Causing the building fire alarm system to sound and indicate an alarm would alert the occupants of the building that an alarm condition exists within the air distribution system, thereby performing the same function as a supervisory signal sent to a constantly attended location.

Exception 2 recognizes the fact that not all buildings are required to have a fire alarm system. A visible and audible signal must be activated at an approved location that will alert building supervisory personnel to take action. Additionally, the duct smoke detectors must be electronically supervised to indicate trouble (system fault) in the detector system circuitry or power supply. A trouble condition must activate a distinct visible or audible signal at a location that will alert the responsible personnel.

907.13 Access. Access shall be provided to each detector for periodic inspection, maintenance and testing.

❖ Automatic fire detectors, especially smoke detectors, require periodic cleaning to reduce the likelihood of malfunction. Section 907.20 and NFPA 72 require inspection and testing at regular intervals. Access to perform the required inspections, necessary maintenance and testing is a particularly important consideration for those detectors that are installed within a concealed space, such as an air duct.

907.14 Fire-extinguishing systems. Automatic fire-extinguishing systems shall be connected to the building fire alarm system where a fire alarm system is required by another section of this code or is otherwise installed.

❖ This section requires that alternative automatic fire-extinguishing systems, such as a wet-chemical system for a commercial kitchen exhaust hood and duct system, be monitored by the building fire alarm system, if there is one. Again, it should be noted that this section does not require electrical supervision of all fire-extinguishing systems but only those systems in buildings that contain a fire alarm system.

907.15 Monitoring. Where required by this chapter or by the *International Building Code*, an approved supervising station in accordance with NFPA 72 shall monitor fire alarm systems.

Exception: Supervisory service is not required for:

1. Single- and multiple-station smoke alarms required by Section 907.2.10.
2. Smoke detectors in Group I-3 occupancies.
3. Automatic sprinkler systems in one- and two-family dwellings.

❖ Fire alarm systems required by Section 907 are to be electrically supervised in accordance with NFPA 72.

Exception 1 exempts single- and multiple-station smoke alarms from being supervised due to the potential for unwanted false alarms.

Exception 2 recognizes a similar problem in Group I-3 occupancies. Accordingly, due to the concern over unwanted alarms, smoke detectors in Group I-3 occupancies need only sound an approved alarm signal that automatically notifies staff (see Section 907.2.6.2.1). Smoke detectors in such occupancies are typically subject to misuse and abuse, and frequent unwanted alarms would negate the effectiveness of the system.

Exception 3 clarifies that sprinkler systems in one- and two-family dwellings are not part of a dedicated fire alarm system and are typically designed in accordance with NFPA 13D, which does not require electrical supervision.

907.16 Automatic telephone-dialing devices. Automatic telephone-dialing devices used to transmit an emergency alarm shall not be connected to any fire department telephone number unless approved by the fire chief.

❖ Upon initiation of an alarm, supervisory or trouble signal, an automatic telephone-dialing device takes control of the telephone line for the reliability of transmission of all signals. The device, however, should not be connected to the fire department telephone number because that could disrupt any potential emergency (911) calls. NFPA 72 contains additional guidance on such devices including digital alarm-communicator systems.

907.17 Acceptance tests. Upon completion of the installation of the fire alarm system, alarm notification appliances and circuits, alarm-initiating devices and circuits, supervisory-signal initiating devices and circuits, signaling line circuits, and primary and secondary power supplies shall be tested in accordance with NFPA 72.

❖ A complete performance test of the fire alarm system must be conducted to determine that the system is operating as required by the code. The acceptance test must include a test of each circuit, alarm-initiating device, alarm notification appliance and any supplementary functions, such as activation of closers and dampers. The operation of the primary and secondary (emergency) power supplies must also be tested, as well as the supervisory function of the control panel. Section 901.5 of the IBC assigns responsibility for conducting the acceptance tests to the owner or the owner's representative.

NFPA 72 contains specific acceptance test procedures. Additional guidance on periodic testing and inspection can be also obtained from Section 907.20 and NFPA 72.

907.18 Record of completion. A record of completion in accordance with NFPA 72 verifying that the system has been installed in accordance with the approved plans and specifications shall be provided.

❖ In accordance with NFPA 72, this section requires a written statement from the installing contractor that the fire alarm system has been tested and installed in compliance with the approved plans and the manufacturer's specifications.

907.19 Instructions. Operating, testing and maintenance instructions and record drawings ("as builts") and equipment specifications shall be provided at an approved location.

❖ To permit adequate testing, maintenance and trouble-shooting of the installed fire alarm system, an owner's manual with complete installation instructions should be kept on site or in another approved location. The instructions should include a description of the system, operating procedures and testing and maintenance requirements.

907.20 Inspection, testing and maintenance. The maintenance and testing schedules and procedures for fire alarm and fire detection systems shall be in accordance with this section and Chapter 7 of NFPA 72.

❖ Fire alarms and fire detection systems are to be inspected, tested and maintained in accordance with Sections 907.20.1 through 907.20.5 and the applicable requirements in Chapter 7 of NFPA 72. It is the building owner's responsibility to keep these systems operable at all times.

907.20.1 Maintenance required. Whenever or wherever any device, equipment, system, condition, arrangement, level of protection or any other feature is required for compliance with the provisions of this code, such device, equipment, system, condition, arrangement, level of protection or other feature shall thereafter be continuously maintained in accordance with applicable NFPA requirements or as directed by the fire code official.

❖ Periodic maintenance keeps systems in good working order or allows repair of defects discovered during inspections or testing. Because specialized tools and training are needed, only properly trained technicians or specialists should perform required periodic maintenance. Most maintenance is required only as needed, but many manufacturers suggest or require regular periodic replacement of parts subject to wear or abuse.

907.20.2 Testing. Testing shall be performed in accordance with the schedules in Chapter 7 of NFPA 72 or more frequently where required by the fire code official. Where automatic testing is performed at least weekly by a remotely monitored fire alarm control unit specifically listed for the application, the manual testing frequency shall be permitted to be extended to annual.

Exception: Devices or equipment that are inaccessible for safety considerations shall be tested during scheduled shutdowns where approved by the fire code official, but not less than every 18 months.

❖ Chapter 7 of NFPA 72 includes schedules for testing frequencies of fire alarm and fire detection systems and their components. Periodic tests that follow standardized methods are intended to confirm the results of inspections, determine that all components function properly and that systems meet their original design specifications. Tools, devices or equipment are usually required to perform tests. Because tests are more detailed than inspections, they are usually done only once or twice per year in most cases. Some tests, however, may be required as frequently as bimonthly or quarterly. Because specialized knowledge and equipment are required, tests must usually be performed by technicians or specialists trained in the test methods involved.

Although Section 907.20.2 specifically addresses testing, Chapter 7 of NFPA 72 also contains schedules for visual inspection frequencies. An inspection consists of a visual check of a system or device to verify it is in operating condition and free from visible defects or damage.

Obvious damage and the general condition of the system should always be noted and recorded. Partly because of their cursory nature, inspections are conducted more frequently than tests and maintenance. Because special knowledge and tools are not required, inspections may be done by any reasonably competent person.

The exception recognizes the impracticality of testing every device or piece of equipment related to a fire alarm or fire detection system. Some devices may be inaccessible for safety considerations such as those in continuous process operations. Testing, however, should be done during scheduled shutdowns.

907.20.3 Detector sensitivity. Detector sensitivity shall be checked within 1 year after installation and every alternate year thereafter. After the second calibration test, where sensitivity tests indicate that the detector has remained within its listed and marked sensitivity range (or 4-percent obscuration light grey smoke, if not marked), the length of time between calibration tests shall be permitted to be extended to a maximum of 5 years. Where the frequency is extended, records of detector-caused nuisance alarms and subsequent trends of these alarms shall be maintained. In zones or areas where nuisance alarms show any increase over the previous year, calibration tests shall be performed.

❖ Usually, changes in detector sensitivity are caused by inadequate maintenance. Regular sensitivity testing is intended to determine whether detectors require recalibration or maintenance. This section prescribes the intervals for testing smoke detector sensitivity. Where two successful tests have been conducted, the frequency of the calibration tests can be extended to a maximum of five years. This interval extension recognizes the stability of both the environment and the detector. However, if nuisance alarms occur during this time interval extension, calibration tests may be needed because of potential changes in the environment where the detector is located or in the performance of the detector itself.

907.20.4 Method. To ensure that each smoke detector is within its listed and marked sensitivity range, it shall be tested using either a calibrated test method, the manufacturer's calibrated sensitivity test instrument, listed control equipment arranged for the purpose, a smoke detector/control unit arrangement whereby the detector causes a signal at the control unit where its sensitivity is outside its acceptable sensitivity range or other calibrated sensitivity test method acceptable to the fire code official. Detectors found to have a sensitivity outside the listed and marked sensitivity range shall be cleaned and recalibrated or replaced.

Exceptions:

1. Detectors listed as field adjustable shall be permitted to be either adjusted within the listed and marked sensitivity range and cleaned and recalibrated or they shall be replaced.
2. This requirement shall not apply to single-station smoke alarms.

❖ This section prescribes acceptable test methods to ensure that each smoke detector is within its listed and marked sensitivity range; any of the listed test methods may be used.

With regard to a calibration test method, many manufacturers have designed their devices to be tested by the application of a magnet at a test point on the outside of the detector. This activates a reed switch or pulls a fine wire into the detection chamber to simulate a predetermined level of obscuration.

Another test method may require that a test device such as a key-type tool be inserted in a test port. This either activates a test switch or produces the desired level of obscuration directly.

One detector manufacturer supplies an interface device for connecting a volt-ohm-amp meter to a test port. Pressing a button on the interface device permits a direct reading of detector chamber voltage in an alarm condition.

Other detectors must be removed and inserted in or connected to a device used to calibrate and test the device. The calibrated sensitivity-test instrument must satisfy the manufacturer's recommendation for a specific detector.

Addressable/analog-type detectors produce direct readings of the chamber voltage by the control unit. Many of these systems permit sensitivity adjustments within acceptable limits from the control unit as well. This test method essentially allows remote sensitivity testing.

A system control/detector combination unit detects changes in the environment and in the detector by comparing current readings to previously stored information in the memory of the control unit. Significant changes would indicate that the stability of either the environment or the detector has changed and that further maintenance or recalibration is required.

Any other method or device that permits the user to check the voltage drop across a smoke detection chamber is acceptable subject to the approval of the fire code official. Test devices should be manufactured and supplied by the smoke detector manufacturer.

907.20.4.1 Testing device. Detector sensitivity shall not be tested or measured using a device that administers an unmeasured concentration of smoke or other aerosol into the detector.

❖ Functional testing using smoke or a smoke substitute, such as aerosols, must comply with the manufacturer's recommended test procedures. A precisely measured amount of smoke or other aerosol product must be used to adequately determine detector sensitivity. Some detector manufacturers do not accept testing with aerosol products and void detector warranties when this product is used.

The functional test method selected should not permanently affect detector performance.

907.20.5 Maintenance, inspection and testing. The building owner shall be responsible for ensuring that the fire and life safety systems are maintained in an operable condition at all times. Service personnel shall meet the qualification requirements of NFPA 72 for maintaining, inspecting and testing such systems. A written record shall be maintained and shall be made available to the fire code official.

❖ This section clearly indicates that it is the responsibility of the building owner to maintain all fire alarm systems in proper working order. Often, an outside agency that employs adequately trained personnel, such as a fire alarm contractor, will provide any maintenance and testing that is needed. Chapter 7 of NFPA 72 contains additional guidance on the qualifications for service personnel. Proper maintenance of fire alarm systems is essential so that the systems will perform as intended.

Inspection and test records provide a means for determining compliance with the requirements of the code. Inspectors should be prepared to determine that inspection, test and maintenance logs are accurate and complete. Records must include the nature of the activity or service performed; when the activity occurred; who performed the activity and who witnessed testing or approved the work upon completion. Failure to keep records of tests and maintenance is a code violation.

SECTION 908
EMERGENCY ALARM SYSTEMS

908.1 Group H occupancies. Emergency alarms for the detection and notification of an emergency condition in Group H occupancies shall be provided as required in Chapter 27.

❖ Emergency alarm systems provide indication and warning of emergency situations involving hazardous materials. An emergency alarm system is required in all Group H occupancies as indicated in Sections 2704.9 and 2705.4.4 as well as Group H-5 HPM facilities as indicated in Section 908.2. The Group H occupancy classification assumes the storage or use of hazardous materials exceeds the maximum allowable quantities specified in Tables 2703.1.1(1) and 2703.1.1(2).

An emergency alarm system should include an emergency alarm-initiating device outside each interior door of hazardous material storage areas, a local alarm device and adequate supervision.

Even though ozone gas-generator rooms (Section 908.4), repair garages (Section 908.5) and refrigerant detectors (Section 908.6) are not typically classified as Group H occupancies, the potential hazards associated with these occupancy conditions are great enough to require additional means of early warning detection.

908.2 Group H-5 occupancy. Emergency alarms for notification of an emergency condition in an HPM facility shall be provided as required in Section 1803.12. A continuous gas detection system shall be provided for HPM gases in accordance with Section 1803.13.

❖ In addition to hazardous material storage areas as regulated by Section 2704.9, Section 1803.12.1 also requires emergency alarms for service corridors, exit access corridors and exit enclosures because of the potential transport of hazardous materials through these areas. Section 1803.13 requires a continuous gas detection system for early detection of leaks in areas where HPM gas is used. Gas detection systems are required to initiate a local alarm and transmit a signal to the emergency control station upon detection (see commentary, Sections 1803.12 and 1803.13).

908.3 Highly toxic and toxic materials. Where required by Section 3704.2.2.10, a gas detection system shall be provided for indoor storage and use of highly toxic and toxic compressed gases.

❖ A gas detection system in the room or area used for indoor storage or use of highly toxic or toxic gases gives early notification of a leak that is occurring before the escaping gas reaches hazardous exposure concentration levels. The exception in Section 3704.2.2.10 recognizes that certain toxic compressed gases do not pose a severe exposure hazard. Those toxic gases whose properties under standard conditions are still below the 8-hour weighted average concentration for the permitted exposure limit (PEL) are exempt from the requirement for a gas detection system (see commentary, Section 3704.2.2.10).

908.4 Ozone gas-generator rooms. A gas detection system shall be provided in ozone gas-generator rooms in accordance with Section 3705.3.2.

❖ To monitor the potential buildup of dangerous levels of ozone, a gas detection system is required to, upon actuation, shut off the generator and sound a local alarm. Ozone gas generators are commonly used in water treatment applications. The ozone gas-generator room should not be a normally occupied area or be used for the storage of combustibles or other hazardous materials. Section 3705 contains additional requirements for ozone gas generators.

908.5 Repair garages. A flammable-gas detection system shall be provided in repair garages for vehicles fueled by non-odorized gases in accordance with Section 2211.7.2.

❖ As indicated in Section 2211.7.2, an approved flammable-gas detection system is required for garages used for repair of vehicles fueled by nonodorized gases, such as hydrogen and nonodorized LNG. To prevent a hazardous potential buildup of flammable gas caused by normal leakage and use conditions, the flammable-gas detection system is required to activate when the level of flammable gas exceeds 25 percent of the lower explosive limit (LEL) (see commentary, Section 2211.7.2).

908.6 Refrigeration systems. Refrigeration system machinery rooms shall be provided with a refrigerant detector in accordance with Section 606.8.

❖ A refrigerant-specific detector is required for leak detection, early warning and actuation of emergency exhaust systems. Because most general machinery rooms are unoccupied for long periods of time, a refrigeration leak may go undetected, allowing a buildup of refrigerant that can pose a threat to building occupants and the maintenance personnel who must enter the machinery room. Also, the refrigerants may or may not be detectable by the sense of smell, depending on the chemical nature and concentration in the air of the refrigerant. This can be especially critical when a toxic refrigerant is used in the refrigeration system (see commentary, Section 606.8).

SECTION 909
SMOKE CONTROL SYSTEMS

[B] 909.1 Scope and purpose. This section applies to mechanical or passive smoke control systems when they are required for new buildings or portions thereof by provisions of the *International Building Code* or this code. The purpose of this section is to establish minimum requirements for the design, installation and acceptance testing of smoke control systems that are intended to provide a tenable environment for the evacuation or relocation of occupants. These provisions are not intended for the preservation of contents, the timely restoration of operations, or for assistance in fire suppression or overhaul activities. Smoke control systems regulated by this section serve a different purpose than the smoke- and heat-venting provisions found in Section 910. Mechanical smoke control systems shall not be considered exhaust systems under Chapter 5 of the *International Mechanical Code*.

❖ The intent of this section as well as Section 909 of the IBC and Section 513 of the IMC is to provide a tenable environment to or through that building occupants can evacuate or relocate in case of a fire. These requirements are not intended to protect contents, allow timely restoration of operations or facilitate fire suppression or overhaul activities. These requirements apply only when smoke control is required by other sections of the code. The only place the IBC requires smoke control using these methods is in atriums. A covered mall would require smoke control only when it contains an atrium. Underground buildings require smoke control, but in compliance with Section 909.20 specifically.

In the last several years, smoke control requirements have become increasingly complex because a generic solution of six air changes has repeatedly and scientifically been shown to be inadequate. The problem is related to the fact that six air changes per hour does not take into account factors such as buoyancy; expansion of gases; wind; the geometry of the space and of communicating spaces; the dynamics of the fire; the production and distribution of smoke and the interaction of building systems.

Smoke control systems can be either passive or active. Active systems are sometimes referred to as mechanical. Passive smoke control systems take advantage of smoke barriers surrounding the zone in which the fire occurs or high bay areas that act as reservoirs to control the movement of smoke to other areas of the building. Active systems use either pressure differences to contain smoke within the event zone or exhaust flow rates sufficient to draw off accumulating smoke and prevent its descent below some predetermined position above necessary exit paths through the event zone. On rare occasions, there is also a possibility of controlling the movement of smoke horizontally by opposed airflow, but this method requires a specific architectural geometry to function properly and that does not create an even greater hazard.

Essentially, there are three methods of mechanical or active smoke control that can be used separately or in combination within a design: pressurization, exhaust and, in rare and very special circumstances, opposed airflow.

Of course, all of these active approaches can be used in combination with the passive method.

Typically, the mechanical pressurization method is used in high-rise buildings when pressurizing stairways and for zoned smoke control. Pressurization is not practical in large open spaces such as atriums or malls because it is difficult to develop the required pressure differences in the large volume of the space.

On the other hand, the exhaust method is typically used in large open spaces such as atriums and malls.

The opposed airflow method, which basically uses a velocity of air horizontally to slow the movement of smoke, is typically applied in combination with either a pressurization method or an exhaust method within hallways or openings into atriums and malls.

The application of each of the methods depends on the specifics of the building design. Smoke control within a building is fundamentally an architecturally driven problem. Different architectural geometries first dictate the need for smoke control and then define the solutions to the problem.

Another element addressed in this section is that smoke control systems serve a different purpose than smoke and heat vents. This eliminates any confusion that smoke and heat vents can be used as a substitute for smoke control. Smoke control systems are also not considered an exhaust system that complies with Chapter 5 of the IMC because these systems are unique in their operation and are not necessarily designed to exhaust smoke. Note, however, that Chapter 5 of the IMC also contains smoke control requirements.

[B] 909.2 General design requirements. Buildings, structures, or parts thereof required by the *International Building Code* or this code to have a smoke control system or systems shall have such systems designed in accordance with the applicable requirements of Section 909 and the generally accepted and well-established principles of engineering relevant to the design. The construction documents shall include sufficient information and detail to describe adequately the elements of the design necessary for the proper implementation of the smoke control systems. These documents shall be accompanied with sufficient information and analysis to demonstrate compliance with these provisions.

❖ This section states that when smoke control systems are required by the IBC or the code, the design is required to comply with this section. These designs also need to follow generally accepted and well-established principles of engineering relevant to the design, essentially requiring a certain level of engineering qualifications for their preparation.

Each state within the United States typically requires some sort of minimum qualifications to undertake engineering design. An important reference when designing smoke control systems is the American Society of Heating, Refrigerating and Air-Conditioning Engineers' (ASHRAE) Design of Smoke Management Systems.

A key element discussed by this section is the need for detailed and clear construction documents to ensure the system is installed correctly. In most complex designs the key to success is communication to the contractors of what needs to be installed. Generally, the more complex a design becomes the more likely construction errors become. Most smoke control systems are quite complex, which is why special inspections complying with Section 909.3 and Chapter 17 of the IBC are critical for smoke control systems.

Additionally, for the design to be accepted, the analyses and justifications submitted need to be detailed enough to evaluate for compliance.

[B] 909.3 Special inspection and test requirements. In addition to the ordinary inspection and test requirements to which buildings, structures and parts thereof are required to undergo, smoke control systems subject to the provisions of Section 909 shall undergo special inspections and tests sufficient to verify the proper commissioning of the smoke control design in its final installed condition. The design submission accompanying the construction documents shall clearly detail procedures and methods to be used and the items subject to such inspections and tests. Such commissioning shall be in accordance with generally accepted engineering practice and, where possible, based on published standards for the particular testing involved. The special inspections and tests required by this section shall be conducted under the same terms as in the *International Building Code.*

❖ The complexity and the uniqueness of the design requires specific special inspection and testing. The designer must provide specific recommendations for this special inspection and testing within the submitted documentation. In fact, Chapter 17 of the IBC states that special inspection agencies for smoke control must have expertise in fire protection engineering and mechanical engineering as well as certification as air balancers. Because the designs are unique to each building, a generic approach will probably not be available for inspecting and testing such systems. The designer can and should, however, use any available published standards or guides when developing the special inspection and testing requirements for that particular design.

ASHRAE Guideline 5 is a good starting place, but only as a general outline. Each system will require a unique commissioning plan that can be developed only after careful and thoughtful examination of the final design and all of its components and interrelationships. Generally these requirements can be included in design standards or engineering guides.

[B] 909.4 Analysis. A rational analysis supporting the types of smoke control systems to be employed, the methods of their operations, the systems supporting them, and the methods of construction to be utilized shall accompany the construction documents submission and include, but not be limited to, the items indicated in Sections 909.4.1 through 909.4.6.

❖ This section indicates that simply determining the airflow, exhaust rates and pressures to maintain tenable conditions is not adequate. Many factors could alter the effectiveness of a smoke control system, including stack effect, temperature effect of fire, wind effect, heating, ventilation and air-conditioning (HVAC) system interaction and climate. Additionally, any smoke control system must function for 20 minutes or more from the time the fire is detected. A proper engineering analysis, which takes into account people movement, may determine that a duration longer than 20 minutes is necessary.

There are also occasions when the movement of people can plainly be demonstrated to require significantly less than the time required for the smoke to present a real risk to their safety and, in such situations, the rational analysis can serve as the technical justification for an alternative means of protection. The code cannot reasonably anticipate every conceivable building arrangement or operating condition.

[B] 909.4.1 Stack effect. The system shall be designed such that the maximum probable normal or reverse stack effect will not adversely interfere with the system's capabilities. In determining the maximum probable stack effect, altitude, elevation, weather history and interior temperatures shall be used.

❖ Stack effect is the tendency for air to rise within a heated building when the temperature is colder on the exterior of the building. Reverse stack effect is the tendency for air to flow downward within a building when the interior is cooler than the exterior of the building. This air movement can affect the intended operation of a smoke control system. If the stack effect is great enough, it may overcome the pressures determined during the design analyses and allow smoke to enter into areas outside the zone of origin.

[B] 909.4.2 Temperature effect of fire. Buoyancy and expansion caused by the design fire in accordance with Section 909.9 shall be analyzed. The system shall be designed such that these effects do not adversely interfere with the system's capabilities.

❖ This section requires that the design account for the effect temperature may have on the success of the system. When air or any gases are heated they will expand. This expansion makes the gases lighter and therefore more buoyant. The buoyancy of the hot gases is important when the design is to exhaust gases from a location in or close to the ceiling. Therefore, if sprinklers are part of the design, the gases may be significantly cooler than they would be in an unsprinklered fire, making smoke removal and plume dynamics alterations more difficult. The fact that air expands when heated needs to be accounted for in the design.

When using the pressurization method, the expansion of hot gases must be accounted for because a larger volume of air will be required to create the necessary pressure differences to maintain negative pressure in the area of fire origin. The expansion of the gases has the affect of pushing the hot gases out of the area of fire origin. Since sprinklers will tend to cool the gases, the effect of expansion is lower. The pressure differences required in Section 909.6.1 are specifically based on a sprinklered building. If the building is nonsprinklered, higher pressure differences may be required. The minimum pressure difference for certain ceiling heights in unsprinklered buildings is as follows:

Ceiling height (feet)	Minimum pressure difference (inch water gauge)
9	0.10
15	0.14
21	0.18

This is a complex issue that needs to be part of the design analysis. It must address the type and reaction of the fire protection systems, ceiling heights and the size of the design fire.

[B] 909.4.3 Wind effect. The design shall consider the adverse effects of wind. Such consideration shall be consistent with the wind-loading provisions of the *International Building Code.*

❖ The effect of wind on a smoke control system within a building is very complex. Wind is generally known to exert a load on a building. The loads are looked at as windward (positive pressure) and leeward (negative pressure). The velocity of winds will vary based on the terrain and the height above grade. Therefore, the height of the building and surrounding obstructions will have an effect on these velocities.

These pressures have the effect of altering the operation of fans, especially propeller fans, and altering the pressure differences and airflow direction in the building. There is no easy solution to dealing with these effects. In fact, little research has been done in this area.

In the case of larger buildings, a wind study is normally undertaken for the structural, cladding and roof design. The data from those studies can be used in the analysis of the effects on the pressures and airflows within the building on the performance of the smoke control system.

[B] 909.4.4 Systems. The design shall consider the effects of the heating, ventilating and air-conditioning (HVAC) systems on both smoke and fire transport. The analysis shall include all permutations of systems status. The design shall consider the effects of the fire on the heating, ventilating and air-conditioning systems.

❖ If not properly configured to shut down or be included as part of the design, the HVAC system can alter the smoke control design. More specifically, if dampers are not installed between smoke zones within the HVAC system ducts, smoke could be transported from one smoke zone to another. Additionally, if the HVAC system places more air into the supply air for the smoke control system than expected, the velocity of the air may adversely affect the fire plume or an unwanted positive pressure may be created by having more supply air than exhaust air.

Generally, the smoke control design and the HVAC system must be analyzed in all potential modes and the analysis documented within the design documentation.

[B] 909.4.5 Climate. The design shall consider the effects of low temperatures on systems, property and occupants. Air inlets and exhausts shall be located so as to prevent snow or ice blockage.

❖ This section is focused on protecting equipment from weather conditions that may affect the reliability of the design. For instance, extremely cold or hot air pulled directly from the outside may damage critical equipment within the system. Some listings of duct smoke detectors are for specific ranges of temperatures. Therefore, placing certain detectors within areas exposed to extreme temperatures may void the listing. Also, the equipment and air inlets and outlets should be designed and located so they do not collect snow and ice that could block air from entering or exiting the building.

[B] 909.4.6 Duration of operation. All portions of active or passive smoke control systems shall be capable of continued operation after detection of the fire event for not less than 20 minutes.

❖ The intent of the smoke control requirements is to provide a tenable environment for occupants during either evacuation or relocation to a safe place. Evacuation and relocation activities include notifying occupants, possible investigation time for the occupants, decision time and actual travel time. To achieve this goal, the code has established 20 minutes as a minimum time for evacuation or relocation. A proper engineering analysis of people movement in relationship to smoke development may result in a longer duration of operations. Therefore, systems must be designed to run or be effective for at least 20 minutes after the detection of the fire because the occupants need to be alerted before evacuation can occur.

[B] 909.5 Smoke barrier construction. Smoke barriers shall comply with the *International Building Code*. Smoke barriers shall be constructed and sealed to limit leakage areas exclusive of protected openings. The maximum allowable leakage area shall be the aggregate area calculated using the following leakage area ratios:

1. Walls: $A/A_w = 0.001\ 00$
2. Exit enclosures: $A/A_w = 0.00035$
3. All other shafts: $A/A_w = 0.00150$
4. Floors and roofs: $A/A_F = 0.00050$

where:

A = Total leakage area, square feet (m^2).

A_F = Unit floor or roof area of barrier, square feet (m^2).

A_w = Unit wall area of barrier, square feet (m^2).

The leakage area ratios shown do not include openings due to doors, operable windows or similar gaps. These shall be included in calculating the total leakage area.

❖ Part of the strategy of both passive and mechanical smoke control systems is the use of smoke barriers to divide a building into separate smoke zones. Not all walls, ceilings or floors would be considered smoke barriers. Only walls that designate separate smoke zones within a building need to be constructed as smoke barriers. This section gives the requirements for those walls, floors and ceilings that are used as smoke barriers.

An exhaust-method smoke control system may not need a smoke barrier to divide the building into separate smoke zones. Therefore, the evaluation of barrier construction and leakage area may not be necessary.

The IBC contains specific construction requirements to prevent smoke from traveling from one smoke zone to another. Openings such as doors and windows are dealt with separately in Section 909.5.2 from openings such as cracks or penetrations.

[B] 909.5.1 Leakage area. Total leakage area of the barrier is the product of the smoke barrier gross area monitored by the allowable leakage area ratio, plus the area of other openings such as gaps and operable windows. Compliance shall be determined by achieving the minimum air pressure difference across the barrier with the system in the smoke control mode for mechanical smoke control systems. Passive smoke control systems tested using other approved means such as door fan testing shall be as approved by the fire code official.

❖ It is impossible to construct walls and floors that are completely free from openings that may allow the migration of smoke. Therefore, leakage needs to be considered in the design by calculating the leakage area of walls, ceilings and floors. The factors given in this section, which originate from ASHRAE provisions on leaky buildings, are used to calculate the total leakage area. The total leakage area is then used in the design process to determine the proper amount of air to produce the required pressure differences across the surfaces that form smoke zones. These pressure differences then need to be verified when the system is in smoke control mode.

Additionally, Section 909.5 gives ratios to determine the maximum allowable leakage in walls, exit enclosures, shafts, floors and roofs. These leakage areas are critical in determining whether the proper pressure differences are used in calculations for the pressurization method of smoke control. Pressure differences will decrease as the openings get larger.

[B] 909.5.2 Opening protection. Openings in smoke barriers shall be protected by automatic-closing devices actuated by the required controls for the mechanical smoke control system. Door openings shall be protected by door assemblies complying with the requirements of the *International Building Code* for doors in smoke barriers.

Exceptions:

1. Passive smoke control systems with automatic-closing devices actuated by spot-type smoke detectors listed for releasing service installed in accordance with Section 907.11.
2. Fixed openings between smoke zones which are protected utilizing the airflow method.
3. In Group I-2, where such doors are installed across corridors, a pair of opposite-swinging doors without a center mullion shall be installed having vision panels with approved fire-rated glazing materials in approved fire-rated frames, the area of which shall not exceed that tested. The doors shall be close fitting within operational tolerances, and shall not have undercuts, louvers or grilles. The doors shall have head and jamb stops, astragals or rabbets at meeting edges, and automatic-closing devices. Positive-latching devices are not required.
4. Group I-3.
5. Openings between smoke zones with clear ceiling heights of 14 feet (4267 mm) or greater and bank-down capacity of greater than 20 minutes as determined by the design fire size.

❖ As with concerns of smoke leakage between smoke zones, openings may compromise the necessary pressure differences between smoke zones. Openings in smoke barriers, such as doors and windows, must either be constantly closed or be automatically closed when the smoke control system is operating. This section requires that doors be automatically closed through the activation of an automatic closing device linked to the controls for the smoke control system. Essentially, when the smoke control system is activated, all openings are automatically closed. This most likely would mean that the mechanism that activates the smoke control system would automatically also close all openings. More than likely, the smoke control system will be activated by a specifically zoned smoke detection or sprinkler system as required by Sections 909.12.2 and 909.12.3.

In terms of actual opening protection, Section 909.5.2 refers the user to Section 715.3.3 of the IBC for construction requirements specific to doors located in smoke barriers. Note that smoke barriers are different from a fire-resistance-rated barrier because the intent is somewhat different. One is focused on fire spread from the perspective of heat, and the other from the perspective of smoke passage.

There are several exceptions to this particular section. Exception 1 is specifically for passive systems. Passive systems, as noted, are systems in which there is no use of mechanical systems. Instead, the system operates primarily on the configuration of barriers and layout of the building to control smoke. Passive systems can use spot-type detectors to close doors that are part of a smoke barrier. Essentially, this means a full fire alarm system would not be required. Instead, single-station detectors would be allowed to close the doors. Such doors would need to fail in the closed position if power is lost. NFPA 72 contains the requirements for approved devices.

Exception 2 is based on the fact that some systems take advantage of the opposed-airflow method to prevent smoke from migrating past the doors. Therefore, since the design already accounts for potential smoke migration at these openings through the use of air movement the barrier need not be closed.

Exception 3 is specifically related to the unique requirements for Group I-2 occupancies. Essentially, a very specific alternative, which meets the functional needs of I-2 occupancies, is given. One aspect of the alternative approach is that the doors have vision panels that have approved fire-rated glazing in fire-rated frames of a size that does not exceed the type tested.

Exception 4 allows an exemption from the automatic-closing requirements for all Group I-3 occupancies. Facilities in which occupants are under restraint or that have specific security restrictions have unique requirements as stated in Section 408 of the IBC. These requirements result in reliable barriers between smoke zones because, for the most part, the majority of doors in these facilities will be closed and in a locked position. The staff very closely controls these types of facilities.

Exception 5 relates to the behavior of smoke. The assumption is that smoke rises because of the buoyancy of hot gases and, if the ceiling is sufficiently high, the smoke layer will be contained for a longer time before it begins to move into the next smoke zone. Therefore, it is not as critical that the doors automatically close. This allowance depends on the specific design fire for a building. Different size design fires create different amounts of smoke that, depending on the layout of the building, may migrate in different ways throughout the building. This section states that smoke cannot begin to migrate into the next smoke zone for at least 20 minutes. This is consistent with the 20-minute duration of operation of smoke control systems (see Section 909.4.6). The minimum ceiling height to take advantage of this exception is 14 feet (4,267 mm).

[B] 909.5.2.1 Ducts and air transfer openings. Ducts and air transfer openings are required to be protected with a minimum Class II, 250°F (121°C) smoke damper complying with the *International Building Code.*

❖ Another factor that adds to the reliability of smoke barriers is the protection of ducts and air-transfer openings in smoke barriers. Left open, these openings may allow the transfer of smoke between smoke zones. These ducts and air-transfer openings most often are part of the HVAC system. Damper operation and the reaction with the smoke control system will be evaluated during acceptance testing. Some duct systems used in a smoke control design may be controlled by the smoke control system and should not automatically be closed by a smoke damper upon detection of smoke.

A smoke damper works differently than a fire damper.

Fire dampers react to heat using a fusible link, while smoke dampers activate upon the detection of smoke. The dampers used should be rated as Class II, 250°F (121°C). The class of the smoke damper refers to its leakage level. The temperature rating shows its ability to withstand the heat of smoke resulting from a fire.

Although smoke barriers are required to use only smoke dampers, there may be many instances where a fire damper is also required. For instance, the smoke barrier may also be used as a fire barrier. Also, Section 716.5.3.1 of the IBC would require shaft penetrations to contain both a smoke and a fire damper. There are listings specific to combination smoke and fire dampers.

Chapter 7 of the IBC and Chapter 6 of the IMC contain more specific requirements covering dampers.

[B] 909.6 Pressurization method. The primary mechanical means of controlling smoke shall be by pressure differences across smoke barriers. Maintenance of a tenable environment is not required in the smoke-control zone of fire origin.

❖ There are several methods or strategies that may be used to control smoke. One of these methods is pressurization, wherein the system primarily uses pressure differences across smoke barriers to control the movement of smoke. Basically, if a negative pressure is maintained in the area of fire origin, the smoke will be contained to that smoke zone. A typical approach used to obtain a negative pressure is to exhaust the fire floor. This is a fairly common practice for high-rise buildings. Stairway enclosures also use the concept of pressurization. The pressurization method in large open spaces, such as malls and atriums, is impractical since it would take a large quantity of supply air to create the necessary pressure differences. It should be noted that pressurization is specified as the primary method for mechanical smoke control design; however, airflow and exhaust methods can be used under certain circumstances.

The negative-pressure method of smoke control does not require that tenable conditions be maintained in the smoke zone where the fire originates. Maintaining this area tenable would be impossible because the pressures from the surrounding smoke zones would be holding a negative pressure within the zone of origin to keep the smoke from migrating.

Pressurization is used often with exit stair enclosures. This method results in a positive pressure within the stair enclosure to resist the passage of smoke. Stair pressurization is one method of compliance for stairways in high-rise buildings or underground buildings where the floor surface is located more than 75 feet (22 860 mm) above the lowest level of fire-fighter access or more than 30 feet (9,144 mm) below the level of exit discharge.

The IBC contains two methods that address smoke movement and include a smokeproof enclosure or pressurized stairs. A smokeproof enclosure requires a certain fire-resistance rating along with access through a ventilated vestibule or an exterior balcony. The vestibule can be ventilated using either natural ventilation or mechanical ventilation as outlined in Sections 909.20.3 and 909.20.4 of the IBC. The pressurization method requires a sprinklered building and a minimum pressure difference of 0.15 inch (.037 kPa) of water and a maximum of 0.35 inch (.087 kPa) of water. These pressure differences are to be maintained with all doors closed under maximum stack pressures (see Section 909.20 of the IBC and Section 1019.1.8 for more details).

As noted, the pressurization method uses pressure differences across smoke barriers to achieve control of smoke. Sections 909.6.1 and 909.6.2 contain the design criteria for the smoke control design in terms of minimum and maximum pressure differences.

In summary, the pressurization method is used in two ways. The first is through the use of smoke zones where the zone of origin is exhausted, creating a negative pressure. The second is stair pressurization that creates a positive pressure within the stair to avoid the penetration of smoke. Note that the code allows the use of a smokeproof enclosure instead of pressurization.

[B] 909.6.1 Minimum pressure difference. The minimum pressure difference across a smoke barrier shall be 0.05-inch water gage (0.0124 kPa) in fully sprinklered buildings.

In buildings allowed to be other than fully sprinklered, the smoke control system shall be designed to achieve pressure differences at least two times the maximum calculated pressure difference produced by the design fire.

❖ The minimum pressure difference is established as 0.05-inch (.0124 kPa) water gauge in fully sprinklered buildings. This particular criterion is related to the pressures needed to overcome buoyancy and the pressures generated by the fire, which include expansion. This particular criterion is based on a sprinklered building. The pressure differences would need to be higher in a building that is not sprinklered. Additionally, these pressure differences must consider the possible stack and wind effects present.

[B] 909.6.2 Maximum pressure difference. The maximum air pressure difference across a smoke barrier shall be determined by required door-opening or closing forces. The actual force required to open exit doors when the system is in the smoke control mode shall be in accordance with Section 1008.1.2. Opening and closing forces for other doors shall be determined by standard engineering methods for the resolution of forces and reactions. The calculated force to set a side-hinged, swinging door in motion shall be determined by:

$$F = F_{dc} + K(WA\Delta P)/2(W - d) \qquad \textbf{(Equation 9-1)}$$

where:

A = Door area, square feet (m^2).

d = Distance from door handle to latch edge of door, feet (m).

F = Total door opening force, pounds (N).

F_{dc} = Force required to overcome closing device, pounds (N).

K = Coefficient 5.2 (1.0).

W = Door width, feet (m).

ΔP = Design pressure difference, inches of water (Pa).

❖ Maximum pressure difference is primarily based on the force needed to open and close doors. The IBC establishes maximum opening forces for doors. This maximum opening force cannot be exceeded taking into account the pressure differences across a doorway in a pressurized environment. Essentially, based on the opening force requirements of Section 1008.1.2, the maximum pressure difference can be calculated using Equation 9-1. As stated in Chapter 10, the maximum opening force of a door has three components, including:

- Door latch release:
  Maximum of 15 pounds (67 N)
- Set door in motion:
  Maximum of 30 pounds (134 N)
- Swing to full open position:
  Maximum of 15 pounds (67 N)

Equation 9-1 is used to calculate the total force to set the door into motion when in smoke control mode. Therefore, the limiting criterion would be 30 pounds (134 N). Accessibility requirements would further limit these maximum opening forces.

[B] 909.7 Airflow design method When approved by the fire code official, smoke migration through openings fixed in a permanently open position, which are located between smoke-control zones by the use of the airflow method, shall be permitted. The design airflow shall be in accordance with this section. Airflow shall be directed to limit smoke migration from the fire zone. The geometry of openings shall be considered to prevent flow reversal from turbulent effects.

❖ This method is allowed only when approved by the fire code official. As the title states, this method uses airflow to prevent the migration of smoke across smoke barriers. This has been referred to as opposed airflow. Specifically, this method is suited for the protection against smoke migration through doors and related openings fixed in a permanently open position.

This method consists of maintaining a particular velocity of air based on the temperature of the smoke and the height of the opening. The temperature of the smoke will depend on the design fire that is established for the particular building. The higher the temperature of the smoke and the larger the opening, the higher the velocity necessary to prevent the smoke from migrating between smoke zones

The airflow method seldom works for large openings because the velocity to oppose the smoke becomes too high. This method tends to work better for smaller openings, such as pass-through windows. Use Equation 9-2 to calculate the required velocity.

[B] 909.7.1 Velocity. The minimum average velocity through a fixed opening shall not be less than:

$$v = 217.2\,[h\,(T_f - T_o)/(T_f + 460)]^{1/2} \qquad \textbf{(Equation 9-2)}$$

For SI: $v = 119.9\,[h\,(T_f - T_o)/T_f]^{1/2}$

where:

h = Height of opening, feet (m).

T_f = Temperature of smoke, °F (K).

T_o = Temperature of ambient air, °F (K).

v = Air velocity, feet per minute (m/minute).

❖ This section contains the formula for the minimum average velocity through a fixed opening. The minimum velocity is based on the velocity needed to prevent the smoke from migrating between smoke zones. See the commentary for Section 909.7 for further discussion.

[B] 909.7.2 Prohibited conditions. This method shall not be employed where either the quantity of air or the velocity of the airflow will adversely affect other portions of the smoke control system, unduly intensify the fire, disrupt plume dynamics or interfere with exiting. In no case shall airflow toward the fire exceed 200 feet per minute (1.02 m/s). Where the formula in Section 909.7.1 requires airflows to exceed this limit, the airflow method shall not be used.

❖ The airflow method has a limitation on the maximum velocity that is based on the fact that air may distort the flame and cause additional entrainment and turbulence. Therefore, having a high velocity of air entering the zone of fire origin has the potential of increasing the amount of smoke produced. Finally, the velocity may also interact with other parts of the smoke control design. For instance, the pressure differences in other areas of the building may be altered and exceed the limitations of Sections 909.6.1 and 909.6.2. This section requires that when a velocity of over 200 feet per minute (1.02 m/s) is calculated, the airflow method is not allowed. The solution may require adding a barrier, such as a wall or door.

If the airflow design method is chosen to protect areas communicating with an atrium, the air added to the smoke layer needs to be accounted for in the exhaust rate.

[B] 909.8 Exhaust method. When approved by the fire code official, mechanical smoke control for large enclosed volumes, such as in atriums or malls, shall be permitted to utilize the exhaust method. The design exhaust volumes shall be in accordance with this section.

❖ This method is allowed only when approved by the fire code official. The primary application of the exhaust method is in large spaces, such as atriums and malls. The strategy of this method is to pull the smoke out of the space based on the understanding of the amount of smoke being produced by a particular size fire. Essentially, fires produce different amounts and patterns of smoke based on the material being burned and the location of the fire. Therefore, several equations representing different fire plume configurations are presented to determine the mass loss rate that is then converted to an exhaust rate. The plume calculation that yields the highest exhaust rate will determine the exhaust rate for the space. However, only realistic plume configurations should be considered. The three plume configurations, which will each be analyzed in this section, are: axisymmetric plumes, balcony spill plumes and window plumes.

Additionally, when a fire plume is in contact with a wall, the mass flow rate can be adjusted using Equation 9-7 in Section 909.8.5.

[B] 909.8.1 Exhaust rate. The height of the lowest horizontal surface of the accumulating smoke layer shall be maintained at least 10 feet (3048 mm) above any walking surface which forms a portion of a required egress system within the smoke zone. The required exhaust rate for the zone shall be the largest of the calculated plume mass flow rates for the possible plume configurations. Provisions shall be made for natural or mechanical supply of air from outside or adjacent smoke zones to make up for the air exhausted. Makeup airflow rates, when measured at the potential fire location, shall not exceed 200 feet per minute (60 960 mm per minute) toward the fire. The temperature of the makeup air shall be such that it does not expose temperature-sensitive fire protection systems beyond their limits.

❖ The design criterion of this method is to keep smoke at least 10 feet (3,048 mm) above any walking surface that is considered part of the required egress within the particular smoke zone, such as an atrium, for at least 20 minutes. Chapter 10 of the code considers the majority of occupiable space as part of the means of egress system. Also keep in mind that the criterion of 10 feet (3,048 mm) does not apply just to the floor surface of the mall or atrium but to any level where occupants may be exposed. Again, the required exhaust volume is based on the largest mass loss rate found from each of the plume configurations in Sections 909.8.2 through 909.8.5. However, only realistic plume configurations should be considered.

To exhaust the air from the space, make-up air is required. There are a few considerations related to make-up air. First, as with the airflow method, a 200-foot-per-minute (1.02 m/s) limitation is placed on the speed of the make-up air, to prevent disturbing the plume dynamics and potentially increasing smoke production. The second consideration is the temperature of the incoming air, primarily when the air could be very cool. Fireprotection equipment may be temperature-sensitive and could fail as a result of exposure to extreme temperatures.

There are several considerations when automatic doors are depended on for the make-up air in a smoke control system. First, depending on the climate, the temperature of the air must be accounted for just like any other make-up air entering the building. Additionally, the climate and location of the opening may result in varying air velocities because of the effects of winds. Finally, the owners and users of the building must fully understand the security concerns of a door that is designed to remain open during a fire. If the owner and user cannot tolerate doors being open, other forms of intake must be explored.

[B] 909.8.2 Axisymmetric plumes. The plume mass flow rate (m_p), in pounds per second (kg/s), shall be determined by placing the design fire center on the axis of the space being analyzed. The limiting flame height shall be determined by:

$$z_l = 0.533Q_c^{2/5} \qquad \textbf{(Equation 9-3)}$$

For SI: $z_l = 0.166Q_c^{2/5}$

where:

m_p = Plume mass flow rate, pounds per second (kg/s).

Q = Total heat output.

Q_c = Convective heat output, British thermal units per second (kW). (The value of Q_c shall not be taken as less than 0.70 Q.)

z = Height from top of fuel surface to bottom of smoke layer, feet (m).

z_l = Limiting flame height, feet (m). The z_l value must be greater than the fuel equivalent diameter (see Section 909.9).

for $z > z_l$

$m_p = 0.022Q_c^{1/3}z^{5/3} + 0.0042Q_c$

For SI: $m_p = 0.071\ Q_c^{1/3}z^{5/3} + 0.0018Q_c$

for $z = z_l$

$m_p = 0.011\ Q_c$

For SI: $m_p = 0.035Q_c$

for $z < z_l$

$m_p = 0.0208Q_c^{3/5}z$

For SI: $m_p = 0.032Q_c^{3/5}z$

To convert m_p from pounds per second of mass flow to a volumetric rate, the following formula shall be used:

$$V = 60\ m_p/\rho \qquad \textbf{(Equation 9-4)}$$

where:

V = Volumetric flow rate, cubic feet per minute (m^3/s).

ρ = Density of air at the temperature of the smoke layer, pounds per cubic feet (T: in °F) [kg/m^3 (T: in °C)].

❖ This section presents the methodology to determine the required exhaust rate based on a fire burning away from walls, windows and other factors that may affect it working symmetrically. This type of fire is able to entrain air into the plume from all sides until it reaches the bottom of the smoke layer. See Figure 909.8.2(1) for an illustration of an axisymmetric plume in an atrium.

The user must first determine the limiting flame height through Equation 9-3. This height is then used as the criterion to determine the mass flow rate equation. Figure 909.8.2(1) illustrates the limiting flame height, Z_l. Different behaviors are observed at different flame height ranges because air is entrained differently at different heights and ultimately affects the output of the fire. When the smoke layer is at the same height or lower than the flame height, less air is being entrained into the fire and the hot gases released from the fire are reradiating heat back onto the fire. Therefore, less smoke can be produced because less air is being entrained into the plume when the height of the smoke layer is equal to or less than the flame height. The height, *z*, noted in the equation will be based on the design criterion of 10 feet (3048 mm) above the occupants for 20 minutes. For instance, Figure 909.8.2(2) indicates an atrium 80 feet (24 384 mm) tall having a balcony with a walking surface at 56 feet (17 069 mm), which would require that the smoke layer be at least 66 feet (20 117 mm) above the floor. Therefore z = 66 feet (20 117 mm). The design will then be based on keeping the smoke layer at least 66 feet (20 117 mm) (56 feet + 10 feet) above the fuel source. The fuel source is assumed to be close to the floor.

The actual exhaust volume is determined based on the mass flow rate. The mass flow rate essentially expresses how much mass is being released based on the rate of combustion and air entrainment. As noted above, air entrainment is reduced when the smoke layer height is less than or equal to the flame height and increases as the height differential is increased. When the height differential is positive (i.e., *z* is greater than the limiting flame height), the mass flow rate increases to the $^5/_3$ power. The ability to remove the products of combustion is based on the calculated mass flow rate. An exhaust volume that can pull that mass out of the space is necessary. The exhaust volume relates to the density of the air at the fire temperature and mass flow rate (Density = Mass/Volume). Equation 9-4 takes advantage of this relationship to determine the necessary exhaust volume.

[B] 909.8.3 Balcony spill plumes. The plume mass flow rate (m_p) for spill plumes shall be determined using the geometrically probable width based on architectural elements and projections in the following equation:

$$m_p = 0.124(QW^2)^{1/3}(z_b + 0.25H) \quad \textbf{(Equation 9-5)}$$

For SI: $m_p = 0.36\ (QW^2)^{1/3}(z_b + 0.25H)$

where:

H = Height above fire to underside of balcony, feet (m).

m_p = Plume mass flow rate, pounds per second (kg/s).

Q = Total heat output.

W = Plume width at point of spill, feet (m).

z_b = Height from balcony, feet (m).

❖ The next type of plume for which the mass flow rate is to be determined is a balcony spill plume. This plume occurs when a fire occurs under a balcony and the smoke travels upward and around the balcony. Equation 9-5 gives a method to calculate the mass flow rate for this particular configuration. The mass flow rate calculated

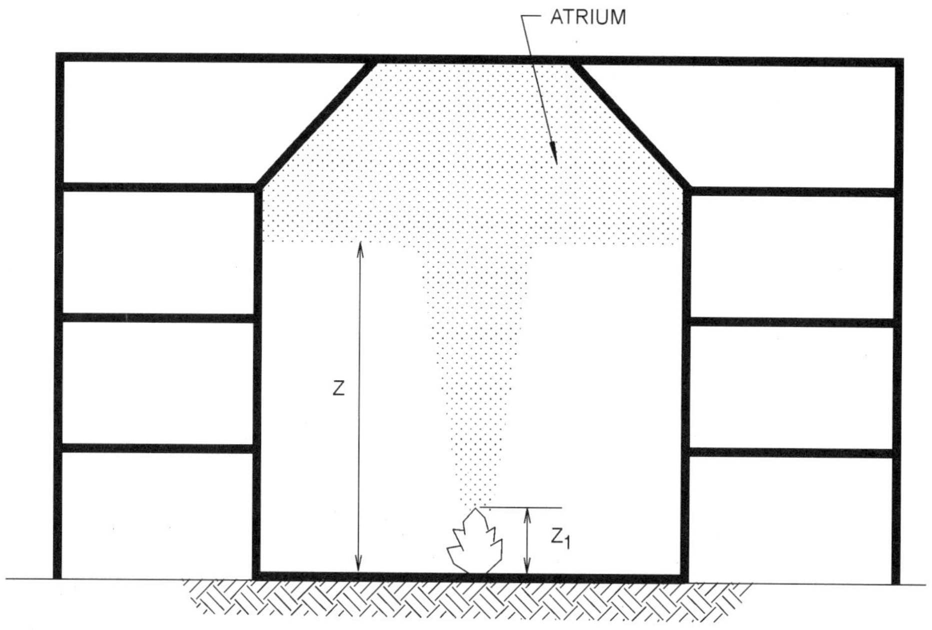

FIGURE 909.8.2(1)
AXISYMMETRIC PLUME AND LIMITING ELEVATION

using this calculation is then applied to Equation 9-4 to determine the necessary exhaust volume. This particular calculation is somewhat difficult because it is difficult to determine the plume width at the point of spill. Additionally, the larger the width used in Equation 9-7, the larger the mass flow rate. Because this width is difficult to determine and because a smaller width yields a lower exhaust rate, architectural features, such as glass or plexiglass draft curtains, are installed or can be used to determine a particular plume width. The mass flow rate also depends on the height above the fire to the underside of the balcony.

[B] 909.8.4 Window plumes. The plume mass flow rate (m_p) shall be determined from:

$$m_p = 0.077(A_w H_w^{1/2})^{1/3}(z_w + a)^{5/3} + 0.18 A_w H_w^{1/2} \quad \textbf{(Equation 9-6)}$$

For SI: $m_p = 0.68(A_w H_w^{1/2})^{1/3}(z_w + a)^{5/3} + 1.5 A_w H_w^{1/2}$

where:

A_w = Area of the opening, square feet (m^2).

H_w = Height of the opening, feet (m).

m_p = Plume mass flow rate, pounds per second (kg/s).

z_w = Height from the top of the window or opening to the bottom of the smoke layer, feet (m).

a = $2.4A_w^{2/5}H_w^{1/5} - 2.1H_w$

❖ Another scenario in terms of possible fire plumes and smoke generation is the window plume. The type of fire this equation addresses could be a fire in a hotel room that is located within an atrium. The fire may break through the window, releasing smoke into the atrium. The key data necessary for this equation beyond the heat release rate (fire size) are the area and the height of the opening. These two elements affect the behavior of the fire and how the smoke is generated. Also, the variable z_w, which is the height from the top of the window or opening to the bottom of the smoke layer, must be determined. Again, the criterion z_w will depend on the design criterion of 10 feet (3,048 mm). This height is important because it helps determine where the plume stops and the smoke layer begins. It is at this point that the mass loss rate of the plume is limited. This calculation is to be used with only one opening.

Again, as with the axisymmetric and balcony spill plumes, the mass loss rate in pounds per second needs to be converted to a volumetric flow rate using Equation 9-4.

[B] 909.8.5 Plume contact with walls. When a plume contacts one or more of the surrounding walls, the mass flow rate shall be adjusted for the reduced entrainment resulting from the contact provided that the contact remains constant. Use of this provision requires calculation of the plume diameter, that shall be calculated by:

$$d = 0.48\,[(T_c + 460)/(T_a + 460)]^{1/2} z \quad \textbf{(Equation 9-7)}$$

For SI: $d = 0.48\,(T_c/T_a)^{1/2} z$

where:

d = Plume diameter, feet (m).

T_a = Ambient air temperature, °F (K).

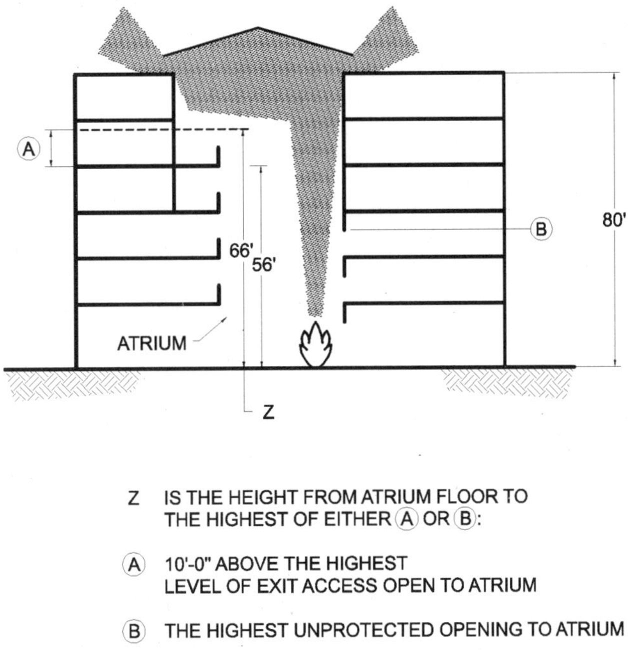

FIGURE 909.8.2(2)
EXAMPLE OF CRITICAL LEVEL OF SMOKE

T_c = Plume centerline temperature, °F (K).

$= 0.60\ (T_a + 460)\ Q_c^{2/3}\ z^{-5/3} + T_a$

z = Height at which T_c is determined, feet (m).

For SI: $T_c = 0.08\ T_a Q_c^{2/3}\ z^{-5/3} + T_a$

❖ When a fire plume is situated either against a wall or in a corner, the plume dynamics are altered and less air is being entrained into the plume. This decrease is 50 percent for walls and 75 percent for corners. The result is a hotter fire that tends to have longer flames. The lack of air entrainment reduces the mass loss rate by 50 percent for walls and 75 percent for corners.

If a plume is large enough in a space such as an atrium, the mass loss rate may be limited by the size of the atrium. When the width of the plume becomes large enough to contact the walls, air can no longer be entrained or is limited in the areas where it can be entrained, thus limiting the mass loss rate and production of smoke. To determine whether the plume makes contact with the wall, the diameter of the plume must be calculated. The calculation will be based on the centerline temperature of the plume at height *z*.

The calculated width obtained with the design criterion height *z* may be larger than the dimensions of the atrium. That is an indication that the plume makes contact with the walls before the design criterion height *z* is reached. Therefore, several different heights may need to be used to determine when the plume makes contact with the wall. Once determined, that information can be used to recalculate the mass loss rate for the various plume calculations, which would then result in a lower exhaust volume requirement.

Generally, basing the necessary exhaust volume on the initial design criterion height *z* is more conservative than using a reduced height. Also, if the plume contacts the walls below the criterion height *z* smoke exposure to the communicating spaces needs to be evaluated.

[B] 909.9 Design fire. The design fire shall be based on a *Q* of not less than 5,000 Btu/s (5275 kW) unless a rational analysis is performed by the registered design professional and approved by the fire code official. The design fire shall be based on the analysis in accordance with Section 909.4 and this section.

❖ The design fire is the critical element in smoke control system design. The fire is what produces the smoke to be controlled by the system. Therefore, it is considered important to establish a minimum fire size of 5,000 Btu per second (5,275 kW) for the application of this section. This particular fire in many cases is fairly conservative. Generally, heat release data is fairly difficult to find for individual combustibles found within buildings. Fire size is sometimes estimated through heat release per square foot with figures such as 44 Btu/ft^2 sec for mercantile occupancies and 20 Btu/ft^2 sec for office buildings (ASHRAE). The 5,000 Btu/sec fire would equate to the following fire area using the above figures:

Mercantile	(5,000 Btu/sec) / (44 Btu/ft^2/sec) = 114 ft^2 (10.7 ft × 10.7 ft)
Office	(5,000 Btu/sec) / (20 Btu/ft^2/sec) = 250 ft^2 (16 ft × 16 ft)

This does not mean that all fires will release 5,000 Btu/sec (5,275 kw). It is simply stating that this is the minimum fire size that should be considered when conducting an analysis to determine the necessary exhaust rate to remove the smoke. The higher the heat release rate, the faster the products of combustion will be released (i.e., that is, the higher the mass flow rate). This section does include the option for establishing a fire size less than 5,000 Btu/sec (5,275 kw) after a rational analysis is conducted. This analysis would be based on the specific hazards associated with a particular facility. For instance, a building may have very sparse furnishings, and a review of fire data and other background information, such as sprinkler activation, shows that the fire size in that particular space would only be 3,500 Btu/sec (3,693 kw). Likewise, after a detailed analysis, the fire size may be reduced based on sprinkler activation.

The analysis to alter the fire size should be conducted by a registered design professional with experience in the area of fire dynamics and possibly in the area of smoke control. Additionally, future use of the building must address reasonably foreseeable changes to the content. For example, holiday displays should be addressed. In addition to the allowance of a reduction in fire size, the designer should also conduct an analysis to determine whether 5,000 Btu/sec (5,275 kw) is adequate.

The fire size determined using the method in this section is a steady fire and not an unsteady fire. A steady fire assumes a constant heat release rate over a period of time where unsteady does not. An unsteady fire includes the growth and decay phases of the fire. An unsteady fire will hit a peak heat release rate when burning in the open, like an axisymmetric fire. To gain an understanding of fire sizes obtained from various combustibles, consider the following data from fire tests. These heat release rates found in Section 3, Chapter 3-1 of the 3rd edition of the *SFPE Handbook of Fire Protection Engineering* are peak heat release rates:

Plastic trash bags/ paper trash	114 332 Btu/sec (120 - 350 kW)
Latex foam pillow	114 Btu/sec approximately (120 kW)
Dry Christmas tree	475-618 BtU/sec (500 - 650 kW)
Sofa	2852 Btu/sec approximately (3000 kW)
Plywood Wardrobe	2947-6084 Btu/sec (3100 - 6400 kW)

Section 909.9 assumes a constant or steady heat release rate of 5,000 Btu/sec (5,275 kw) without consideration of the growth or decay phase. Section 909.4.6 requires that this particular heat release rate fire be countered by a smoke control system for 20 minutes. Therefore, the designer needs to assume a fire of 5,000

Btu/sec (5,275 kw) is burning for a period of 20 minutes. This assumption is fairly conservative in most cases, but accounts for other uncertainties such as ineffectiveness of sprinklers resulting from ceiling height. Also, assuming a steady heat release simplifies the design process.

[B] 909.9.1 Factors considered. The engineering analysis shall include the characteristics of the fuel, fuel load, effects included by the fire, and whether the fire is likely to be steady or unsteady.

❖ The design fire is required to be a minimum of 5,000 Btu/sec (5,275 kw) but it must also be further determined whether 5,000 Btu/sec (5,275 kw) is adequate. To determine the correct fire size, an engineering analysis is necessary, which takes into account the following elements:

- Fuel (potential burning rates),
- Fuel load (how much),
- Effects included by the fire (smoke particulate size and density),
- Steady or unsteady (burn steadily or simply peak and dissipate) and
- Likelihood of sprinkler activation (based on height and distance from the fire).

[B] 909.9.2 Separation distance. Determination of the design fire shall include consideration of the type of fuel, fuel spacing and configuration. The ratio of the separation distance to the fuel equivalent radius shall not be less than 4. The fuel equivalent radius shall be the radius of a circle of equal area to floor area of the fuel package. The design fire shall be increased if other combustibles are within the separation distance as determined by:

$$R = [Q/(12\pi q'')]^{1/2} \qquad \textbf{(Equation 9-8)}$$

where:

q'' = Incident radiant heat flux required for nonpiloted ignition, Btu/ft^2 · s (W/m^2).

Q = Heat release from fire, Btu/s (kW).

R = Separation distance from target to center of fuel package, feet (m).

❖ The design fire size may also be affected by surrounding objects. The surrounding combustibles may have the affect of increasing the fire size. More specifically, there is concern that if sufficient separation is not maintained between other combustibles, a larger design fire is likely. Equation 9-8 calculates the critical separation distance, *R*. This distance is where a nonpiloted ignition of adjacent combustibles will occur.

Nonpiloted ignition means that the radiated heat without direct flame contact or impingement will ignite the adjacent combustibles. This equation is based on the fire size, which in most cases is about 5,000 Btu/sec (5,275 kw), and the incident radiant heat flux required to ignite the specific combustible item in question, such as a polyurethane sofa. This particular data would need to be obtained from sources such as the *SFPE Handbook of Fire Protection Engineering*. The units of the radiant heat flux variable are in Btu/ft^2 /sec (kW/m^2). If the separation distance is less than *R* as determined in Equation 9-8, the fire size needs to be reevaluated to include additional combustibles. The heat release of the combustibles should be added to the fire size.

Equation 9-8 calculates the distance, *R*, from the center of the fuel package. It is possible that because of the area covered by the fuel package itself, the design fire could be much closer to the surrounding combustibles than the distance, *R*. Therefore a minimum distance is set by the following ratio:

$$\text{Minimum distance} = \frac{R(\text{Equation 9} - 8)}{4r^3}$$

Where r = equivalent fuel radius

The equivalent fuel radius is determined by first determining the area of the fuel package, then assuming that area as a circle. For example, the fuel package kiosk for a 4-foot by 3-foot (1,219 mm by 914 mm) area in an atrium is calculated as:

Area = 4 ft × 3 ft = 12 ft^2

Area of circle, A = πr^2

$$r = \sqrt{A / \pi}$$

$$r = \sqrt{12 / \pi} = 1.96 \text{ feet (597 mm)}$$

Therefore: The equivalent fuel radius for this scenario is 1.96 feet (597 mm) ≥ 2 feet (610 mm)

Looking at the ratio established in Section 909.9.2, *R* would be limited as follows in the above example:

$$\frac{R}{r} \geq 4$$

$$R \geq$$

Therefore: *R* should be greater than or equal to:

$$R \geq 4 \times r \geq 4 \times 2 \geq 8 \text{ feet (2,438 mm)}.$$

[B] 909.9.3 Heat-release assumptions. The analysis shall make use of best available data from approved sources and shall not be based on excessively stringent limitations of combustible material.

❖ This section stresses the fact that data obtained for use in a rational analysis needs to come from relevant sources. Data can be obtained from groups such as the National Institute for Standards and Technology (NIST). Data from fire tests is available and is a good resource for such analysis. As noted earlier, such data is not widely available (see also Section 3, Chapter 3-1 of the *SFPE Handbook of Fire Protection Engineering*, 3rd edition).

[B] 909.9.4 Sprinkler effectiveness assumptions. A documented engineering analysis shall be provided for conditions that assume fire growth is halted at the time of sprinkler activation.

❖ This section raises a question concerning an assumption that sprinklered areas will immediately control the fire as soon as they are activated. This assumption may be true in some cases, but for high ceilings, the sprinkler may not activate or may be ineffective, since by the time they are activated the fire is too large to control. In that case, the fire plume may push away and evaporate the water before it actually reaches the seat of the fire. This is a common problem with high-piled storage. Also, if the fire becomes too large before the sprinklers are activated, the available water supply and pressure for the system may be compromised. Additionally, based on the layout of the room and the movement of the fire effluents, the wrong sprinklers could be activated, which leads to a larger fire size and depletion of the available water supply and pressure. Therefore, each scenario must be looked at individually to determine whether sprinklers would effectively halt the growth of a fire. More specifically, the evaluation should include droplet size, density and area of coverage and should also be based on actual test results.

[B] 909.10 Equipment. Equipment such as, but not limited to, fans, ducts, automatic dampers and balance dampers, shall be suitable for their intended use, suitable for the probable exposure temperatures that the rational analysis indicates, and as approved by the fire code official.

❖ Section 909.10 and subsequent sections are primarily related to the reliability of system components to ensure that the smoke control system is working as designed. One of the largest concerns when using smoke control is the overall reliability of the system. These systems have many different components, such as smoke and fire dampers, fans, ducts and controls associated with these components. The more components a system has, the lower the reliability becomes. In fact, one approach to reaching a higher level of reliability is using the normal building systems, such as the HVAC system to be the smoke control system. Basically, systems used every day are more likely to be working properly because they are essentially being tested daily. However, many components, such as exhaust fans in an atrium or the smoke control panel, are specific to the smoke control system.

Also, there is not a generic prescriptive set of requirements for how all smoke control system elements should operate since each design may be fairly unique. The specifics on operation of such a system need to be included in the design and construction documents. Most components used in smoke control systems are elements used in many other applications, such as HVAC systems. Therefore, the basic mechanisms of a fan used in a smoke control system may not be different, although, they may be applied differently.

[B] 909.10.1 Exhaust fans. Components of exhaust fans shall be rated and certified by the manufacturer for the probable temperature rise to which the components will be exposed. This temperature rise shall be computed by:

$$T_s = (Q_c/mc) + (T_a) \qquad \textbf{(Equation 9-9)}$$

where:

c = Specific heat of smoke at smokelayer temperature, Btu/lb°F. (kJ/kg · K).

m = Exhaust rate, pounds per second (kg/s).

Q_c = Convective heat output of fire, Btu/s (kW).

T_a = Ambient temperature, °F (K).

T_s = Smoke temperature, °F (°K).

Exception: Reduced T_s as calculated based on the assurance of adequate dilution air.

❖ The fans used for smoke control systems must be able to tolerate the elevated temperatures to which they will be exposed. Again, like many other factors, this depends on the specifics of the design fire. Equation 9-9 requires calculation of the potential temperature rise. The exhaust fans must be specifically rated and certified by the manufacturer to be able to handle these rises in temperature. There is an exception that allows reduction of the temperature if it can be shown that adequate temperature reduction will occur. In many cases, if the exhaust fans are near the ceiling, the smoke will be much cooler than Equation 9-9 will calculate it to be because the smoke may cool considerably by the time it reaches the ceiling. Also, sprinkler activation will assist in cooling the smoke further.

[B] 909.10.2 Ducts. Duct materials and joints shall be capable of withstanding the probable temperatures and pressures to which they are exposed as determined in accordance with Section 909.10.1. Ducts shall be constructed and supported in accordance with the *International Mechanical Code*. Ducts shall be leak tested to 1.5 times the maximum design pressure in accordance with nationally accepted practices. Measured leakage shall not exceed 5 percent of design flow. Results of such testing shall be a part of the documentation procedure. Ducts shall be supported directly from fire-resistance-rated structural elements of the building by substantial, noncombustible supports.

Exception: Flexible connections (for the purpose of vibration isolation) complying with the *International Mechanical Code* and which are constructed of approved fire-resistance-rated materials.

❖ The next essential component of a smoke control system is the integrity of the ducts to transport supply air and exhaust air. The integrity of ducts is also important for an HVAC system but is more critical in this case because it is not simply a comfort issue but one of life safety. The key concerns with ducts in smoke control systems are that they can withstand elevated temperatures and that there will be minimal leakage. The concern with leakage is the potential for leaking smoke into another smoke zone or not having the right amount of supply air to support the system.

More specifically, all ducts need to be leak tested to

one and one-half times the maximum static design pressure. The leakage resulting should be no more than 5 percent of the design flow. For example, a duct that has design flow of 300 cfm (.141 m^3/s) would be allowed 15 cfm (.007 m^3/s) of leakage when exposed to one and one-half times the design pressure for that duct. The tests should be done using nationally accepted practices. This criterion will often limit ductwork for smoke control systems to lined systems because the amount of leakage in these systems is much less.

As part of the concern for possible exposure to fire and fire products, the ducts must be supported by substantial noncombustible supports connected to the fire-resistance-rated structural elements of the building. As noted, the system must able to run for 20 minutes starting from the detection of the fire. The supports may be other than noncombustible when they are flexible connections installed to reduce the effects of vibration, perhaps as part of a building exposed to seismic loads. The flexible connections still need to be constructed of approved fire-resistance-rated materials.

[B] 909.10.3 Equipment, inlets and outlets. Equipment shall be located so as to not expose uninvolved portions of the building to an additional fire hazard. Outside air inlets shall be located so as to minimize the potential for introducing smoke or flame into the building. Exhaust outlets shall be so located as to minimize reintroduction of smoke into the building and to limit exposure of the building or adjacent buildings to an additional fire hazard.

❖ The intent of this section is to minimize the likelihood of smoke being reintroduced into the building through poorly placed outdoor air inlets and exhaust air outlets. Therefore, placing them right next to each other on the exterior of the building would be poor design. In addition, windy conditions and other adverse conditions should be considered when choosing locations for these inlets and outlets. Particular attention should be paid to introducing exhausted smoke into another smoke zone. Also, exhausted smoke should be exhausted in a direction that will not introduce the smoke into surrounding buildings or facilities. Within the building itself the supply air and exhaust outlets should also be strategically located. The exhaust inlets and supply air should be evenly distributed to reduce the likelihood of a high velocity of air that could disrupt the fire plume and also push smoke back into occupied areas.

[B] 909.10.4 Automatic dampers. Automatic dampers, regardless of the purpose for which they are installed within the smoke control system, shall be listed and conform to the requirements of approved recognized standards.

❖ This section addresses the reliability of any dampers used within a smoke control system. This particular provision requires that the dampers be listed and conform to recognized standards. More specifically, Section 716 of the IBC has more detailed information on the specific requirements for smoke and fire dampers.

Smoke and fire dampers should be listed to comply with UL 555S and 555, respectively. Also, remember that each smoke control design is unique and the sequence and methods used to activate the dampers may vary from design to design. This information should be addressed in the construction documents.

Another factor to take into account when considering the timing of the system is that some dampers react more quickly than others simply because of the particular smoke damper characteristics. Additionally, during the commissioning of the system, the damper is going to be exposed to many repetitions. These repetitions need to be accounted for in the overall reliability of the system.

[B] 909.10.5 Fans. In addition to other requirements, belt-driven fans shall have 1.5 times the number of belts required for the design duty with the minimum number of belts being two. Fans shall be selected for stable performance based on normal temperature and, where applicable, elevated temperature. Calculations and manufacturer's fan curves shall be part of the documentation procedures. Fans shall be supported and restrained by noncombustible devices in accordance with the structural design requirements of Chapter 16 of the *International Building Code*. Motors driving fans shall not be operated beyond their nameplate horsepower (kilowatts) as determined from measurement of actual current draw and shall have a minimum service factor of 1.15.

❖ Part of the overall reliability requires that fans used to provide supply air and exhaust capacity will be functioning when necessary. Therefore, a safety factor of 1.5 is placed on the belts required for fans. Fans used as part of a smoke control system must be equipped with one and one-half times the number of belts required, with a minimum of two belts for each fans.

This section also points out that the fan chosen should fit the specific application. It should be able to withstand the temperature rise as calculated in Section 909.10.1 and generally be able to handle the typical exposure conditions, such as location and wind. For instance, propeller fans are highly sensitive to the effects of wind. When located on the windward side of a building, wall mounted, unhooded propeller fans are not able to compensate for wind effects. Additionally, even hooded propeller fans located on the leeward side of the building may not adequately compensate for the decrease in pressure caused by the wind effects. In general, when designing a system, remember that field conditions might vary from the calculations. Therefore, flexibility should be built into the design to account for things such as variations in wind conditions.

Finally, this section stresses that fan motors are not to be operated beyond their rated horsepower.

[B] 909.11 Power systems. The smoke control system shall be supplied with two sources of power. Primary power shall be the normal building power systems. Secondary power shall be from an approved standby source complying with the ICC *Electrical Code*. The standby power source and its transfer switches shall be in a separate room from the normal power transformers and switch gear and shall be enclosed in a room constructed of not less than 1-hour fire-resistance-rated fire barriers, ventilated directly to and from the exterior. Power distribution from the two sources shall be by independent routes. Transfer to full standby power shall be automatic and within 60 seconds of failure of the primary power. The systems shall comply with the ICC *Electrical Code*.

❖ As with any life safety system, a level of power-supply redundancy is required to ensure the functioning of the system during a fire. The primary source is the normal building power system. The secondary power system will be standby power. One of the key elements is that standby power systems are intended to operate within 60 seconds of loss of primary power. The main difference between standby power and emergency power is that emergency power must operate within 10 seconds of loss of primary power rather than 60 seconds. This section also requires isolation from normal building power systems via a 1-hour fire barrier. This increases the reliability and reduces the likelihood that a single event could remove both power supplies.

[B] 909.11.1 Power sources and power surges. Elements of the smoke management system relying on volatile memories or the like shall be supplied with uninterruptable power sources of sufficient duration to span 15-minute primary power interruption. Elements of the smoke management system susceptible to power surges shall be suitably protected by conditioners, suppressors or other approved means.

❖ Smoke management systems have many components, sometimes highly sensitive electronics, that are adversely affected by any interruption in power or sudden surges of power. Therefore, this section requires that any components of a smoke control system, such as volatile memories, be equipped with an uninterruptible power supply for the first 15 minutes of loss of primary power. Volatile memory components will lose memory upon any loss of power no matter how short the time period. Once the 15 minutes elapses, these elements can be switched over to the already operating standby power supply.

Components sensitive to power surges need to have surge protection in the form of conditioners, suppressors or other approved means.

[B] 909.12 Detection and control systems. Fire detection systems providing control input or output signals to mechanical smoke control systems or elements thereof shall comply with the requirements of Section 907. Such systems shall be equipped with a control unit complying with UL 864 and listed as smoke control equipment.

Control systems for mechanical smoke control systems shall include provisions for verification. Verification shall include positive confirmation of actuation, testing, manual override, the presence of power downstream of all disconnects and, through a preprogrammed weekly test sequence report, abnormal conditions audibly, visually and by printed report.

❖ This section requires that fire detection elements used in mechanical smoke control systems comply with this chapter of the code and NFPA 72, which is the fire alarm standard. Specific to smoke control systems is the requirement that a control unit complying with UL 864 and listed as smoke control equipment be used. UL 864 has a subcategory specific to fire alarm control panels (UUKL).

Additionally, the supervisory system for smoke control systems must verify certain activities of the system using audible and visual signals and a printed report. The following verifications would be included: positive confirmation of actuation, testing, manual override, presence of power downstream of all disconnects, and abnormal conditions obtained by a weekly preprogrammed test sequence.

[B] 909.12.1 Wiring. In addition to meeting requirements of the ICC *Electrical Code*, all wiring, regardless of voltage, shall be fully enclosed within continuous raceways.

❖ The wiring must be placed within continuous raceways. This results in an additional level of reliability for the system. Wiring would not necessarily need to be rated for plenums unless it is located within a plenum.

909.12.2 Activation. Smoke control systems shall be activated in accordance with this section.

❖ The activation of a smoke control system depends on when the system is required. Mechanical smoke control systems, which could include pressurization, airflow or exhaust methods, require an automatic activation mechanism. In a passive system, which simply depends upon compartmentation, spot-type detectors are acceptable for the release of door closers and similar openings.

909.12.2.1 Pressurization, airflow or exhaust method. Mechanical smoke control systems using the pressurization, airflow or exhaust method shall have completely automatic control.

❖ See Sections 909.6 for the pressurization method, 909.7 for the airflow design method and 909.8 for the exhaust method.

909.12.2.2 Passive method. Passive smoke control systems actuated by approved spot-type detectors listed for releasing service shall be permitted.

❖ This section recognizes that a passive system does not require a fire alarm system and would allow single-station detectors to close openings where required by the design.

909.12.3 Automatic control. Where completely automatic control is required or used, the automatic-control sequences shall be initiated from an appropriately zoned automatic sprinkler system complying with Section 903.3.1.1, manual controls that are readily accessible to the fire department, and any smoke detectors required by the engineering analysis.

❖ When automatic activation is required, it must be accomplished by a properly zoned automatic sprinkler system and, if the engineering analysis requires them, smoke detectors. Manual control for the fire department needs to be provided. An important point with this particular requirement is that smoke control systems are engineered systems and a prescribed smoke detection system may not fit the needs of the specific design. Other types of detectors, such as beam detectors (within an atrium), may be used and could be more practical from a maintenance standpoint. The fire alarm system within the building also should not generally activate such systems, as it may alter the effectiveness of the system by pulling smoke through the building versus removing or containing the smoke.

[B] 909.13 Control air tubing. Control air tubing shall be of sufficient size to meet the required response times. Tubing shall be flushed clean and dry prior to final connections and shall be adequately supported and protected from damage. Tubing passing through concrete or masonry shall be sleeved and protected from abrasion and electrolytic action.

❖ Control tubing is used in pneumatic systems that operate components such as the opening and closing of dampers. Because of the sophistication of electronic systems today, pneumatic systems are becoming less common.

These particular requirements provide the criteria for properly designing and installing control tubing. Essentially, it is up to the design professional to determine the size requirements and to properly design supports. This information must be detailed in the construction documents. Additionally, because of the effect of moisture and other contaminants on control tubing, it must be first flushed clean then dried before installation.

[B] 909.13.1 Materials. Control air tubing shall be hard drawn copper, Type L, ACR in accordance with ASTM B 42, ASTM B 43, ASTM B 68, ASTM B 88, ASTM B 251 and ASTM B 280. Fittings shall be wrought copper or brass, solder type, in accordance with ASME B 16.18 or ASME B 16.22. Changes in direction shall be made with appropriate tool bends. Brass compression-type fittings shall be used at final connection to devices; other joints shall be brazed using a BCuP5 brazing alloy with solidus above 1,100°F (593°C) and liquidus below 1,500°F (816°C). Brazing flux shall be used on copper-to-brass joints only.

Exception: Nonmetallic tubing used within control panels and at the final connection to devices, provided all of the following conditions are met:

1. Tubing shall be listed by an approved agency for flame and smoke characteristics.
2. Tubing and the connected device shall be completely enclosed within a galvanized or paint-grade steel enclosure of not less than 0.030 inch (0.76 mm) (No. 22 galvanized sheet gage) thickness. Entry to the enclosure shall be by copper tubing with a protective grommet of neoprene or teflon or by suitable brass compression to male-barbed adapter.
3. Tubing shall be identified by appropriately documented coding.
4. Tubing shall be neatly tied and supported within enclosure. Tubing bridging cabinet and door or moveable device shall be of sufficient length to avoid tension and excessive stress. Tubing shall be protected against abrasion. Tubing serving devices on doors shall be fastened along hinges.

❖ This section addresses the materials allowed for control air tubing along with approved methods of connection. All of this information must be documented, since it will be subject to review by the special inspector.

[B] 909.13.2 Isolation from other functions. Control tubing serving other than smoke control functions shall be isolated by automatic isolation valves or shall be an independent system.

❖ This section requires separation of control tubing used for other functions through the use of isolation valves or by complete separation of the systems. This separation is needed because of the difference in requirements for control tubing used in a smoke control system and other building systems. The isolation of the control air tubing for a smoke control system must be specifically noted on the construction documents.

[B] 909.13.3 Testing. Control air tubing shall be tested at three times the operating pressure for not less than 30 minutes without any noticeable loss in gauge pressure prior to final connection to devices.

❖ As part of the acceptance testing of the smoke control system, the control air tubing will be pressure tested to three times the operating pressure for 30 minutes or more. The failure criterion is whether the control tubing shows a noticeable loss in gauge pressure before final connection of devices during the 30-minute test.

[B] 909.14 Marking and identification. The detection and control systems shall be clearly marked at all junctions, accesses and terminations.

❖ This section requires that all portions of the fire detection system that activate the smoke control system be marked and identified. This includes applicable fire alarm-initiating devices, the respective junction boxes, data-gathering panels and fire alarm control panels.

Additionally, components of the smoke control system that are not considered a fire detection system must be properly identified and marked. This would include applicable junction boxes, control tubing, temperature control modules, relays, damper sensors, automatic door sensors and air movement sensors.

909.15 Control diagrams. Identical control diagrams showing all devices in the system and identifying their location and function shall be maintained current and kept on file with the fire code official, the fire department and in the fire command center in format and manner approved by the fire chief.

❖ Control diagrams provide consistent information on the system in several key locations, including the building department, the fire department and the fire command center. This information is intended to assist emergency response personnel in the use and operation of the smoke control system. The format of the control diagram must be approved by the fire chief because the fire department is the agency that will be using the system during a fire and when the system is tested in the future. The more clearly the information is communicated, the more effective the smoke control system will be.

For the sake of consistency and to simplify training of personnel, the fire department may want all smoke control systems within a jurisdiction to follow a particular protocol for control diagrams.

Generally, the control diagrams should indicate the required reaction of the system in all emergency situations. The status or position of every fan and damper in every situation must be clearly identified.

909.16 Fire-fighter's smoke control panel. A fire-fighter's smoke control panel for fire department emergency response purposes only shall be provided and shall include manual control or override of automatic control for mechanical smoke control systems. The panel shall be located in a fire command center complying with Section 509 and shall comply with Sections 909.16.1 through 909.16.3.

❖ One of the elements that make a smoke control system effective is the successful communication of system activity to the fire department so that the fire department has the information necessary to manually operate the system. The sections that follow describe requirements for a control panel specifically for smoke control systems. This panel must be located within the fire command center. There are two components that include the requirements for the display and also for the controls. This control panel will contain an ability to override any other smoke control system controls, whether manual or automatic.

909.16.1 Smoke control systems. Fans within the building shall be shown on the fire-fighter's control panel. A clear indication of the direction of airflow and the relationship of components shall be displayed. Status indicators shall be provided for all smoke control equipment, annunciated by fan and zone and by pilot-lamp-type indicators as follows:

1. Fans, dampers and other operating equipment in their normal status—WHITE.
2. Fans, dampers and other operating equipment in their off or closed status—RED.
3. Fans, dampers and other operating equipment in their on or open status—GREEN.
4. Fans, dampers and other operating equipment in a fault status—YELLOW/AMBER.

❖ This section details what should be displayed on the control panel. The display is required to include all fans, an indication of the direction of airflow and the relationship of the components. Also, status lights are required and this section sets out specific standardized colors to indicate normal status, closed status, open status and fault status. A standardized approach increases the likelihood that the fire department will be able to more quickly become familiar with a system. Since the fire department has the ability to override the automatic functions of the system this information is critical.

909.16.2 Smoke control panel. The fire-fighter's control panel shall provide control capability over the complete smoke-control system equipment within the building as follows:

1. ON-AUTO-OFF control over each individual piece of operating smoke control equipment that can also be controlled from other sources within the building. This includes stairway pressurization fans; smoke exhaust fans; supply, return and exhaust fans; elevator shaft fans; and other operating equipment used or intended for smoke control purposes.
2. OPEN-AUTO-CLOSE control over individual dampers relating to smoke control and that are also controlled from other sources within the building.
3. ON-OFF or OPEN-CLOSE control over smoke control and other critical equipment associated with a fire or smoke emergency and that can only be controlled from the fire-fighter's control panel.

Exceptions:

1. Complex systems, where approved, where the controls and indicators are combined to control and indicate all elements of a single smoke zone as a unit.
2. Complex systems, where approved, where the control is accomplished by computer interface using approved, plain English commands.

❖ This section sets the requirements for the controls that must be provided for the fire department on the control panel.

There are two aspects to the controls. Essentially, the controls will include on-auto-off and open-auto-close settings or will be strictly on-off or open-close. If the system or component can be set on automatic (auto), it can be controlled from other locations beyond the fire command center. This would include an automatic smoke detection system or one activated manually. If a control contains only on-off or open-close settings, the only way the system component can be controlled is in the fire command center.

Components such as fans are usually associated with on-off-type controls, whereas components such as dampers are associated with open-close-type controls.

909.16.3 Control action and priorities. The fire-fighter's control panel actions shall be as follows:

1. ON-OFF, OPEN-CLOSE control actions shall have the highest priority of any control point within the building. Once issued from the fire-fighter's control panel, no automatic or manual control from any other control point within the building shall contradict the control action. Where automatic means are provided to interrupt normal, nonemergency equipment operation or produce a specific result to safeguard the building or equipment (i.e., duct freezestats, duct smoke detectors, high-temperature cutouts, temperature-actuated linkage and similar devices), such means shall be capable of being overridden by the fire-fighter's control panel. The last control action as indicated by each fire-fighter's control panel switch position shall prevail. In no case shall control actions require the smoke control system to assume more than one configuration at any one time.

   Exception: Power disconnects required by the ICC *Electrical Code*.

2. Only the AUTO position of each three-position fire-fighter's control panel switch shall allow automatic or manual control action from other control points within the building. The AUTO position shall be the NORMAL, nonemergency, building control position. Where a fire-fighter's control panel is in the AUTO position, the actual status of the device (on, off, open, closed) shall continue to be indicated by the status indicator described above. When directed by an automatic signal to assume an emergency condition, the NORMAL position shall become the emergency condition for that device or group of devices within the zone. In no case shall control actions require the smoke control system to assume more than one configuration at any one time.

❖ This section states that when a component of the system is placed in an on-off or open-close configuration, no other control point in the building, whether automatic or manual, can override the action established in the fire command center. If a system component is configured in auto mode, it can be controlled from locations within the building beyond the fire command center. Some controls are specifically designed to allow an action only from the fire command center.

909.17 System response time. Smoke-control system activation shall be initiated immediately after receipt of an appropriate automatic or manual activation command. Smoke control systems shall activate individual components (such as dampers and fans) in the sequence necessary to prevent physical damage to the fans, dampers, ducts and other equipment. For purposes of smoke control, the fire-fighter's control panel response time shall be the same for automatic or manual smoke control action initiated from any other building control point. The total response time, including that necessary for detection, shutdown of operating equipment and smoke control system startup, shall allow for full operational mode to be achieved before the conditions in the space exceed the design smoke condition. The system response time for each component and their sequential relationships shall be detailed in the required rational analysis and verification of their installed condition reported in the required final report.

❖ This particular section gives the criteria for when the smoke control system is required to begin operation. Whether the activation is manual or automatic, this section clarifies that the system must begin operating immediately. Also, this section requires that components activate in a sequence that will not potentially damage the fans, dampers, ducts and other equipment. Unrealistic timing of the system has the potential of creating an unsuccessful system. Delays in the system can be seen in slow dampers, fans that ramp up or down, systems that poll slowly and intentional built-in delays. These factors can add significantly to the reaction time of the system and may hamper achieving the design goals.

The key element is that the system be fully operational before the smoke conditions exceed the design parameters. The sequence of events must be justified in the design analysis and described clearly in the construction documents.

909.18 Acceptance testing. Devices, equipment, components and sequences shall be individually tested. These tests, in addition to those required by other provisions of this code, shall consist of determination of function, sequence and, where applicable, capacity of their installed condition.

❖ To achieve a certain level of performance, the smoke control systems must be thoroughly tested. This section requires that all devices, equipment components and sequences be tested individually.

909.18.1 Detection devices. Smoke or fire detectors that are a part of a smoke control system shall be tested in accordance with Chapter 9 in their installed condition. When applicable, this testing shall include verification of airflow in both minimum and maximum conditions.

❖ Detection devices must be tested in accordance with the fire protection requirements found in Chapter 9 of the code. Also, because such detectors may be subject to higher air velocities than typical detectors, their operation must be verified in the minimum and maximum anticipated airflow conditions.

909.18.2 Ducts. Ducts that are part of a smoke control system shall be traversed using generally accepted practices to determine actual air quantities.

❖ This section requires ducts that are part of the smoke control system to be tested to show that the proper amount of air is flowing. Section 909.10.2 requires that the ducts be leak tested to one and one-half times the maximum design pressure. Leakage at that pressure may not exceed 5 percent of the design flow.

909.18.3 Dampers. Dampers shall be tested for function in their installed condition.

❖ This section notes that dampers must be inspected to make sure they meet the function for which they are installed. For instance, a damper that is supposed to be open when the system is in smoke control mode should be verified to be open during system tests. Also, a damper may have a specific timing associated with its operation that would need to be verified through testing.

909.18.4 Inlets and outlets. Inlets and outlets shall be read using generally accepted practices to determine air quantities.

❖ As with ducts, the amount of air entering or exiting the inlets and outlets, respectively, must be checked.

909.18.5 Fans. Fans shall be examined for correct rotation. Measurements of voltage, amperage, revolutions per minute and belt tension shall be made.

❖ This section requires the testing of fans for correct rotation, voltage, amperage, revolutions per minute and belt tension.

A common problem with fans is that they are often installed in the reversed direction. Also, to verify the reliability of the fans, elements such as correct voltage and belt tension need to be tested.

909.18.6 Smoke barriers. Measurements using inclined manometers or other approved calibrated measuring devices shall be made of the pressure differences across smoke barriers. Such measurements shall be conducted for each possible smoke control condition.

❖ As discussed in Section 909.5.1, the pressure differences across smoke barriers needs to be measured in smoke-control mode. As noted in this section, every possible smoke control condition must be tested and the measurements taken using an inclined manometer or other approved method. Electronic devices are also available. Qualified individuals must calibrate these types of devices. Additionally, an alternative method of testing must be approved by the fire code official before it is used.

909.18.7 Controls. Each smoke zone, equipped with an automatic-initiation device, shall be put into operation by the actuation of one such device. Each additional device within the zone shall be verified to cause the same sequence without requiring the operation of fan motors in order to prevent damage. Control sequences shall be verified throughout the system, including verification of override from the fire-fighter's control panel and simulation of standby power conditions.

❖ This section requires the overall testing of the system. More specifically, each zone needs to individually initiate the smoke control system by the activation of an automatic initiation device. Once that has occurred, all other devices within each zone must be shown to activate the system, but to avoid damage, the fans do not need to be activated.

In addition to determining that all the designated devices initiate the system, all of the controls on the fire-fighter control panel must be shown to initiate the designated components of the smoke control system, including the override capability.

Finally, the initiation and availability of the standby power system must be verified.

909.18.8 Special inspections for smoke control. Smoke control systems shall be tested by a special inspector.

❖ Smoke control systems require special inspection because they are life safety systems.

909.18.8.1 Scope of testing. Special inspections shall be conducted in accordance with the following:

1. During erection of ductwork and prior to concealment for the purposes of leakage testing and recording of device location.
2. Prior to occupancy and after sufficient completion for the purposes of pressure-difference testing, flow measurements, and detection and control verification.

❖ Special inspections must be done at two different stages during construction to facilitate the necessary inspections. The first round of special inspections occurs before concealment of the ductwork or fire protection elements. The special inspector needs to verify the leakage as noted in Section 909.10.2. Additionally, the location of fire protection devices must be verified and documented at this time.

The second round of special inspections occurs just prior to occupancy. The inspections include verification of pressure differences across smoke barriers as required in Sections 909.18.6 and 909.5.1, verification of volumes of airflow as noted in the design and finally, verification of the operation of the detection and control mechanisms as required in Sections 909.18.1 and 909.18.7. These tests must be done just prior to occupancy because the test result will more clearly represent actual conditions.

909.18.8.2 Qualifications. Special inspection agencies for smoke control shall have expertise in fire protection engineering, mechanical engineering and certification as air balancers.

❖ As noted in Section 909.3, special inspections are required for smoke control systems. This means a certain level of qualifications that would include the need for expertise in fire protection engineering, mechanical engineering and certification as air balancers.

909.18.8.3 Reports. A complete report of testing shall be prepared by the special inspector or special inspection agency. The report shall include identification of all devices by manufacturer, nameplate data, design values, measured values and identification tag or mark. The report shall be reviewed by the responsible registered design professional and, when satisfied that the design intent has been achieved, the responsible registered design professional shall seal, sign and date the report.

❖ Once the special inspections are complete, documentation of the activity is required. This documentation is to be prepared in the form of a report that identifies all devices by manufacturer, nameplate data, design values, measured values and identification or mark.

909.18.8.3.1 Report filing. A copy of the final report shall be filed with the fire code official and an identical copy shall be maintained in an approved location at the building.

❖ The report must be reviewed, approved and then signed, sealed and dated. This report is to be given to the fire code official and a copy is also to remain in the building in an approved location. The fire command center is probably the most appropriate location.

909.18.9 Identification and documentation. Charts, drawings and other documents identifying and locating each component of the smoke control system, and describing their proper function and maintenance requirements, shall be maintained on file at the building as an attachment to the report required by Section 909.18.8.3. Devices shall have an approved identifying tag or mark on them consistent with the other required documentation and shall be dated indicating the last time they were successfully tested and by whom.

❖ Additional documentation that must be maintained includes charts, drawings and other related documentation that assists in the identification of each aspect of the smoke control system. This particular documentation is where information such as the last time a device or component was successfully tested and by whom is recorded. This will serve as the main documentation for the system. Again, the fire command center is the most likely location for such information.

909.19 System acceptance. Buildings, or portions thereof, required by this code to comply with this section shall not be issued a certificate of occupancy until such time that the fire code official determines that the provisions of this section have been fully complied with, and that the fire department has received satisfactory instruction on the operation, both automatic and manual, of the system.

Exception: In buildings of phased construction, a temporary certificate of occupancy, as approved by the fire code official, shall be permitted provided that those portions of the building to be occupied meet the requirements of this section and that the remainder does not pose a significant hazard to the safety of the proposed occupants or adjacent buildings.

❖ This section requires that the certificate of occupancy not be issued unless the smoke control system has been accepted. The system must be inspected and approved because it is a life safety system. The exception allows a temporary certificate of occupancy for completed portions of buildings that are constructed in phases.

[B] 909.20 Underground building smoke exhaust system. Where required by the *International Building Code* for underground buildings, a smoke exhaust system shall be provided in accordance with this section.

❖ Buildings that qualify under Section 405 of the IBC as underground buildings require smoke control. This is a prescriptive smoke control approach. The intent of the smoke control provisions is to restrict movement of smoke to the general area of fire origin.

[B] 909.20.1 Exhaust capability. Where compartmentation is required, each compartment shall have an independent, automatically activated smoke exhaust system capable of manual operation. The system shall have an air supply and smoke exhaust capability that will provide a minimum of six air changes per hour.

❖ Section 405.4.1 of the IBC requires compartmentation of an underground building when the building has a floor level more than 60 feet (18 288 mm) below the lowest level of exit discharge. If compartmentation is required, each compartment is required to have a system to exhaust the smoke independent of the other compartments. The exhaust system for each compartment must deliver six air changes per hour.

909.20.2 Operation. The smoke exhaust system shall be operated in the compartment of origin by the following, independently of each other:

1. Two cross-zoned smoke detectors within a single protected area or a single smoke detector monitored by an alarm verification zone or an approved equivalent method.
2. The automatic sprinkler system.
3. Manual controls that are readily accessible to the fire department.

❖ The smoke exhaust system is required to activate the system in the compartment of fire origin in three ways: two cross-zoned smoke detectors within a single protected area, the automatic sprinkler system and manual control. The exhaust systems operate only within the compartment of fire origin.

909.20.3 Alarm required. Activation of the smoke exhaust system shall activate an audible alarm at a constantly attended location.

❖ Regardless of how the system is activated, an audible alarm is required to notify someone in a constantly attended location, such as a security booth, fire command center or central station of the activation of the smoke exhaust system. This will provide notification of a fire or of a malfunction related to the system.

909.21 Maintenance. Smoke control systems shall be maintained to ensure to a reasonable degree that the system is capable of controlling smoke for the duration required. The system shall be maintained in accordance with the manufacturer's instructions and Sections 909.21.1 through 909.21.5.

❖ Routine maintenance and testing of smoke control systems is essential to ensure their performance, as designed, under fire conditions. Maintenance practices must be consistent with the manufacturer's recommendations and as indicated in Sections 909.21.1 through 909.21.5.

909.21.1 Schedule. A routine maintenance and operational testing program shall be initiated immediately after the smoke control system has passed the acceptance tests. A written schedule for routine maintenance and operational testing shall be established.

❖ Operational testing and maintenance must be performed on the smoke control system periodically to verify that it still operates as required by the approved design. A written schedule complying with Section 909.12.2 must be maintained.

909.21.2 Written record. A written record of smoke control system testing and maintenance shall be maintained on the premises. The written record shall include the date of the maintenance, identification of the servicing personnel and notification of any unsatisfactory condition and the corrective action taken, including parts replaced.

❖ This section prescribes the desired content of the written record for the smoke control testing and maintenance program. Test results and maintenance activities should be clearly documented. The written record should be available for inspection and reviewed by the fire code official.

909.21.3 Testing. Operational testing of the smoke control system shall include all equipment such as initiating devices, fans, dampers, controls, doors and windows.

❖ Smoke control systems are made up of components and equipment that are an integral part of other building systems such as fire alarm systems, HVAC equipment and automatic sprinkler systems. For this reason, operational testing of all related system components must ensure that the system as a whole will perform as intended.

909.21.4 Dedicated smoke control systems. Dedicated smoke control systems shall be operated for each control sequence semiannually. The system shall also be tested under standby power conditions.

❖ Because dedicated smoke control systems are designed for smoke control only, the operation of these systems does not adversely affect other building systems or operations. The control sequence for these systems must be tested semiannually to check for system component failures that may not get noticed because dedicated smoke control systems are independent of building HVAC systems.

909.21.5 Nondedicated smoke control systems. Nondedicated smoke control systems shall be operated for each control sequence annually. The system shall also be tested under standby power conditions.

❖ Contrary to dedicated smoke control systems identified in Section 909.21.4, smoke control systems that are not dedicated share system components with other building systems including the HVAC system. Consequently, testing of the control sequence of systems that are not dedicated can be done annually, rather than semiannually, because equipment failures related to other building systems would most likely be noticed and corrected when those other systems were tested or maintained.

SECTION 910
SMOKE AND HEAT VENTS

910.1 General. Where required by this code or otherwise installed, smoke and heat vents, or mechanical smoke exhaust systems, and draft curtains shall conform to the requirements of this section.

Exception: Frozen food warehouses used solely for storage of Class I and Class II commodities where protected by an approved automatic sprinkler system.

❖ Smoke and heat vents must be installed in buildings, structures or portions of them where required by Section 910.2 or as an alternative to another protection feature (see commentary, Section 1015.2). It should be noted that Chapter 23 of the code would also be applicable (see commentary, Section 910.2.3). The systems must be designed, installed, maintained and operated to comply with this section.

The purpose of smoke and heat vents has historically been related to the needs of the fire department. More specifically, smoke and heat vents, when activated, have the potential effect of raising the height of the smoke layer and providing more tenable conditions in which to undertake fire-fighting activities. Other potential benefits are a decrease in property damage and creating more tenable conditions for occupants. The purpose of draft curtains, as addressed in Section 910.3.4, is both to contain the smoke in certain areas and potentially increase the speed of the activation of the smoke and heat vents.

The exception recognizes the building-within-a-building nature of typical frozen food warehouses. Smoke from a fire in a freezer would be contained within the freezer, thus negating the usefulness of smoke and heat vents at the roof level.

910.2 Where required. Approved smoke and heat vents shall be installed in the roofs of one-story buildings or portions thereof occupied for the uses set forth in Sections 910.2.1 through 910.2.4.

❖ Smoke and heat vents are required only in single-story buildings and then only as required by Sections 910.2.1 through 910.2.3 or allowed as a trade-off for increased exit access travel distances in buildings or portions of buildings classified in Group F-1 or S-1.

910.2.1 Groups F-1 and S1. Buildings and portions thereof used as a Group F-1 or S-1 occupancy having more than 50,000 square feet (4645 m^2) of undivided area.

Exception: Group S-1 aircraft repair hangars.

❖ Large-area buildings with moderate to heavy fire loads present special challenges to the fire department in disposing of the smoke generated in a fire. To give the fire department the capability to rapidly and efficiently dispose of smoke in buildings classified in Group F-1 and S-1 occupancies and exceeding 50,000 square feet (4,645 m^2) in area without exposing personnel to the danger associated with cutting ventilation holes in the roof, smoke and heat vents, or mechanical smoke removal facilities must be installed.

910.2.2 Group H. Buildings and portions thereof used as a Group H occupancy as follows:

1. In occupancies classified as Group H-2 or H-3, any of which are more than 15,000 square feet (1394 m^2) in single floor area.

   Exception: Buildings of noncombustible construction containing only noncombustible materials.

2. In areas of buildings in Group H used for storing Class 2, 3 and 4 liquid and solid oxidizers, Class 1 and unclassified detonable organic peroxides, Class 3 and 4 unstable (reactive) materials, or Class 2 or 3 water-reactive materials as required for a high-hazard commodity classification.

Exception: Buildings of noncombustible construction containing only noncombustible materials.

❖ Because of the explosion, deflagration, flammability or combustibility hazard of the contents found in Group H-1, H-2 and H-3 occupancies, smoke and heat vents are required to aid in confinement of the fire and to assist the fire department in ventilation.

910.2.3 High-piled combustible storage. Buildings and portions thereof containing high-piled combustible stock or rack storage in any occupancy group when required by Section 2306.7.

❖ This section alerts the code user to the specific high-piled combustible storage requirements contained in Chapter 23. High-piled storage, whether solid-piled or in racks, in excess of 12 feet (36 758 mm) high requires specific consideration, including fire protection design features and smoke and heat vents to be adequately protected. Not all high-piled storage will require the use of smoke and heat vents and draft curtains. In fact, if the high-piled storage is properly sprinklered (in accordance with Chapter 23 and NFPA 13), draft curtains are not required (see the commentary, Chapter 23).

910.2.4 Exit access travel distance increase. Buildings and portions thereof used as a Group F-1 or S-1 occupancy where the maximum exit access travel distance is increased with Section 1015.2.

❖ This section applies to buildings or portions of buildings in Occupancy Groups F-1 (moderate-hazard indus-trial/factory) and S-1 (moderate-hazard storage) when the exit access travel distance is increased beyond that permitted in Table 1015.1. Table 1015.1 already permits an exit access travel distance of 400 feet (121 920 mm) in sprinklered Group F-2 and S-2 occupancies without smoke and heat vents but protected with an automatic sprinkler system. Note that this section does not require installation of smoke and heat venting but rather allows them to be used as a trade-off to offset the inherent hazards of longer-than-normal travel distances.

The venting system provides a mechanism for removing smoke and products of combustion from the building, thus increasing the amount of time required for smoke to obscure visibility in the path of egress travel and resulting in better visibility and additional time to reach the exits.

910.3 Design and installation. The design and installation of smoke and heat vents and draft curtains shall be as specified in this section and Table 910.3.

❖ Careful design and installation of smoke and heat vents is vital to their efficient operation in case of fire. The design criteria for these fire protection tools are given in Table 910.3 for convenience and ready reference.

TABLE 910.3. See page 9-100.

❖ Table 910.3 identifies the required vent area in terms of ratio of vent area to floor area. The table is essentially divided into two parts. The first part is for Group F-1 occupancies. The second part is for Group S-1 occupancies and high-piled combustible storage. The way in which the table is written makes it appear as if only high-piled storage in Group S-1 occupancies requires smoke and heat venting. If Chapter 23 is referenced, smoke and heat venting requirements are independent of occupancy and simply focus upon the commoditity classification and area of storage. Additionally, Chapter 23 only requires smoke and heat venting and draft curtains for larger storage areas. The required vent areas vary based upon the commodity classification (Group I-IV or high hazard) and height of storage. The higher the storage, the higher the potential for a larger fire. Two options are given for high hazard and Group I-IV commodities. The only significance to these options is that one allows a lower vent-to-floor-area ratio if a deeper draft curtain is chosen, simply providing some credit for the fact that the deeper draft curtains are likely to contain more smoke than a smaller draft curtain. Thus, the area contained by the draft curtains also varies. When in a situation where smoke and heat vents are required and draft curtains are not, the second option would be appropriate.

910.3.1 Vent operation. Smoke and heat vents shall be approved and labeled and shall be capable of being operated by approved automatic and manual means. Automatic operation of smoke and heat vents shall conform to the provisions of this section.

❖ For the purpose of establishing the dependability of the vents, this section requires that they be approved by the fire code official and labeled. Although no standard is referenced in the code for vent labeling, standards such as UL 793, UL 33, UL 521 and FM 4430 may be used.

Because the vents are used as a component of an active venting system, the releasing device must be automatic, such as a fusible link. If large volumes of smoke could be generated before the vent is operated by heat-sensitive devices, approved automatic smoke detectors should be installed to comply with Section 907.11. Special care should be taken when using a method such as smoke detection actuation as opening the vents prior to the sprinkler activation may interrupt the effectiveness of the sprinkler system. In addition to automatic operation of the vents, a manual means of opening them by the fire department during fire suppression operations must also be provided. It should be remembered that one of the main reasons smoke and heat vents were initially introduced was to keep fire fighters from having to get on a roof of a burning building. Therefore the mechanisms for release and the needs of the fire department should be carefully considered.

910.3.1.1 Gravity-operated drop out vents. Automatic smoke and heat vents containing heat-sensitive glazing designed to shrink and drop out of the vent opening when exposed to fire shall fully open within 5 minutes after the vent cavity is exposed to a simulated fire represented by a time-temperature gradient that reaches an air temperature of 500°F (260°C) within 5 minutes.

❖ This section establishes minimum performance criteria for drop-out vents, which include in their design a nonmetallic, clear or opaque glazing element designed to shrink from its frame and fall away when exposed to heat from a fire. This type of vent design must be capable of completely opening the roof vent within 5 minutes of exposure to a simulated fire represented by a time-temperature gradient that reaches an air temperature of 500°F (260°C) within 5 minutes. Drop-out vents tested to comply with UL 793 must begin to operate at a maximum temperature of 286°F (141°C) in order to be labeled.

910.3.1.2 Sprinklered buildings. Where installed in buildings equipped with an approved automatic sprinkler system, smoke and heat vents shall be designed to operate automatically.

❖ Where smoke and heat vents are installed in sprinklered buildings, their operation must be automatic

TABLE 910.3
REQUIREMENTS FOR DRAFT CURTAINS AND SMOKE AND HEAT VENTS[a]

OCCUPANCY GROUP AND COMMODITY CLASSIFICATION	DESIGNATED STORAGE HEIGHT (feet)	MINIMUM DRAFT CURTAIN DEPTH (feet)	MAXIMUM AREA FORMED BY DRAFT CURTAINS (square feet)	VENT AREA TO FLOOR AREA RATIO	MAXIMUM SPACING OF VENT CENTERS (feet)	MAXIMUM DISTANCE TO VENTS FROM WALL OR DRAFT CURTAIN[b] (feet)
Group F-1	—	$0.2 \times H$ but ≥ 4	50,000	1:100	120	60
Group S-1 I-IV (Option 1)	≤ 20	6	10,000	1:100	100	60
	$> 20 \leq 40$	6	8,000	1:75	100	55
Group S-1 I-IV (Option 2)	≤ 20	4	3,000	1:75	100	55
	$> 20 \leq 40$	4	3,000	1:50	100	50
Group S-1 High hazard (Option 1)	≤ 20	6	6,000	1:50	100	50
	$> 20 \leq 30$	6	6,000	1:40	90	45
Group S-1 High hazard (Option 2)	≤ 20	4	4,000	1:50	100	50
	$> 20 \leq 30$	4	2,000	1:30	75	40

For SI: 1 foot = 304.8 mm, 1 square foot = 0.0929 m^2.

a. Requirements for rack storage heights in excess of those indicated shall be in accordance with Chapter 23. For solid-piled storage heights in excess of those indicated, an approved engineered design shall be used.

b. The distance specified is the maximum distance from any vent in a particular draft curtained area to walls or draft curtains which form the perimeter of the draft curtained area.

and coordinated with the operation of the sprinkler system. Caution should be exercised in the design of smoke and heat vents and the required draft curtains so that the draft curtains do not interfere with the operation of the automatic sprinklers. Locating a draft curtain too close to a sprinkler head could prevent proper distribution of water over the fire. In addition, draft curtains contain smoke and hot gases that will direct them away from the area where the fire is actually burning, thus activating sprinklers in the wrong area. This has the potential of overwhelming the sprinkler system. This is especially an issue with specially designed systems such as early suppression fast response (ESFR) sprinklers.

Coordination of fusible link operating temperatures with sprinkler head operating temperatures is also critical to the timely and effective operation of both the vents and the sprinklers. For example, the premature operation of a vent operating mechanism could greatly retard the operation of higher temperature-rated sprinkler heads by dissipating the heat needed to make the fusible link of the sprinkler(s) operate.

Delaying the operation of sprinklers can have the negative effect of causing an excessive number of sprinklers to operate, including some located outside the immediate area of fire danger. Concern over this issue has increased with the introduction of new sprinkler technology, such as the use of ESFR sprinklers, which are designed to act quickly to apply larger volumes of water to extinguish rather than simply control a fire.

910.3.1.3 Nonsprinklered buildings. Where installed in buildings not equipped with an approved automatic sprinkler system, smoke and heat vents shall operate automatically by actuation of a heat-responsive device rated at between 100°F (56°C) and 220°F (122°C) above ambient.

Exception: Gravity-operated drop out vents complying with Section 910.3.1.1.

❖ Where smoke and heat vents are installed in buildings that are not equipped with an automatic sprinkler system, their operation must be automatic, with their operating elements set at between 100 and 220°F (38 and 104°C). The exception indicates that gravity-operated drop-out vents are not subject to this requirement.

910.3.2 Vent dimensions. The effective venting area shall not be less than 16 square feet (1.5 m^2) with no dimension less than 4 feet (1219 mm), excluding ribs or gutters having a total width not exceeding 6 inches (152 mm).

❖ This section prescribes the minimum clear area required for each individual smoke and heat vent, exclusive of any obstructions. The design of the aggregate vent area actually needed is based, in part, on the area defined by the draft curtains and the depth of the curtained area, the objective of the design being to prevent smoke from spilling out of the curtained area or to prevent the smoke interface from interfering with egress visibility. It has also been argued that the draft curtains were intended to speed the operation of the smoke and heat vents by keeping the smoke in a smaller area.

910.3.3 Vent locations. Smoke and heat vents shall be located 20 feet (6096 mm) or more from adjacent lot lines and fire walls and 10 feet (3048 mm) or more from fire barrier walls. Vents shall be uniformly located within the roof area above high-piled storage areas, with consideration given to roof pitch, draft curtain location, sprinkler location and structural members.

❖ This section intends to minimize the effects of the discharge from operating smoke and heat vents on adjacent properties, especially if the smoke being vented contains products of combustion from burning hazardous materials. Because fire walls define separate buildings, and to avoid compromising their integrity by placing a nonfire-resistance-rated roof opening too close to the wall, the clearance between vents and fire walls is the same as for property lines. The integrity of fire barriers is similarly protected; however, being a lesser assembly than a fire wall, the vent clearance requirements are reduced. This section also lists the considerations that must be taken into account when designing roof vent locations for high-piled storage areas.

910.3.4 Draft curtains. Where required, draft curtains shall be provided in accordance with this section.

Exception: Where areas of buildings are equipped with early suppression fast-response (ESFR) sprinklers, draft curtains shall not be provided within these areas. Draft curtains shall only be provided at the separation between the ESFR sprinklers and the conventional sprinklers.

❖ Draft curtains, sometimes termed "curtain boards," must be installed in conjunction with smoke and heat vent installations as stated in Table 910.3 and as required by Chapter 23 for certain high-piled storage conditions. They are installed within and at the perimeter of a protected area to restrict smoke and heat movement beyond the area of fire origin or the protected area and aid in smoke and heat removal through the roof vents. Table 2306.2 does not require draft curtains in sprinklered buildings; instead, only smoke and heat vents are required in certain cases (larger areas of high-piled storage). Also, the extent of the protection is addressed in Chapter 23 (i.e., how much of a building needs smoke and heat vents and draft curtains).

910.3.4.1 Construction. Draft curtains shall be constructed of sheet metal, lath and plaster, gypsum board or other approved materials that provide equivalent performance to resist the passage of smoke. Joints and connections shall be smoke tight.

❖ To avoid contributing to the fire load of a building and to increase the likelihood that draft curtains will remain intact in a fire, they must be constructed of noncombustible materials or an approved equivalent, but are not required to possess a fire-resistance rating. Draft curtains need only be capable of resisting the passage of smoke.

910.3.4.2 Location and depth. The location and minimum depth of draft curtains shall be in accordance with Table 910.3.

❖ The requirements for depth and location of draft curtains are given in Table 910.3 based on the occupancy group (in the case of Group F-1, S-1 and H occupancies) of the building, the commodity classification of the stored materials and the height of the storage.

910.4 Mechanical smoke exhaust. Where approved by the fire code official, engineered mechanical smoke exhaust shall be an acceptable alternative to smoke and heat vents.

❖ This section recognizes that installing a mechanical smoke exhaust system may, under certain circumstances, be more desirable, practical or efficient than installing automatic smoke and heat roof vents. The intent of Sections 910.4.1 through 910.4.6 is to create a mechanical system that is at least as efficient as smoke and heat vents designed to comply with Section 910.3. Installation of an alternative mechanical smoke exhaust system is subject to the specific approval of the fire code official so that the design can be reviewed and the operational sequence and control information can be shared with the fire department. This smoke exhaust system is different from that required in Section 909.

910.4.1 Location. Exhaust fans shall be uniformly spaced within each draft-curtained area and the maximum distance between fans shall not be greater than 100 feet (30 480 mm).

❖ One or more smoke exhaust fans must be installed in each area defined by draft curtains. When more than one fan is installed in a curtained area, the fans must be spaced uniformly within that area, not more that 100 feet (30 480 mm) apart. Locating fans in this manner will improve the uniform removal of smoke from curtained areas and reduce the likelihood of smoke spillage under the draft curtains.

910.4.2 Size. Fans shall have a maximum individual capacity of 30,000 cfm (14.2 m^3/s). The aggregate capacity of smoke exhaust fans shall be determined by the equation:

$$C = A \times 300 \qquad \textbf{(Equation 9-10)}$$

where:

C = Capacity of mechanical ventilation required, in cubic feet per minute (m^3/s).

A = Area of roof vents provided in square feet (m^2) in accordance with Table 910.3.

❖ The intent of the sizing requirements of this section is to obtain a smoke exhaust rate at least equivalent to the venting capacity of roof vents. The exhaust rate required by this section, based on Equation 9-10, is equivalent to 300 cfm per square feet (153 $m^3/s \cdot m^2$) of the roof vent area required by Table 910.3, with no single fan exceeding a 30,000 cfm (14.2 m^3/s) rate. For example, a Group F-1 factory with maximum-sized curtained areas of 50,000 square feet (4,645 m^2) would be required to have a total vent area of 500 square feet (46 m^2) in each curtained area to comply with Table 910.3. The mechanical exhaust rate required based on Equation 9-10 would then be 500 x 300 = 150,000 cfm (70.8 m^3/s), which could be supplied by five 30,000 cfm (14.2 m^3/s) fans spaced as required by Section 910.4.1 in each curtained area.

910.4.3 Operation. Mechanical smoke exhaust fans shall be automatically activated by the automatic sprinkler system or by heat detectors having operating characteristics equivalent to those described in Section 910.3.1. Individual manual controls for each fan unit shall also be provided.

❖ The mechanical smoke exhaust system must be capable of activating upon actuation of the automatic sprinkler system or, in nonsprinklered buildings, by heat detectors with a temperature rating of between 100 and 220°F (38 and 104°C) as required for smoke vents in Section 910.3.1 and manual controls. The manual control is for fire department use to increase the reliability of the system and to allow the fire department to activate the exhaust system to assist in the removal of smoke during or after a fire. Because manual control of the system is primarily for fire department use, the location of the controls should be subject to approval by the fire department. Although not specifically stated in this section, a manual fire alarm system is more prone to false activations; therefore, manual activation of the smoke exhaust system should not be permitted.

910.4.4 Wiring and control. Wiring for operation and control of smoke exhaust fans shall be connected ahead of the main disconnect and protected against exposure to temperatures in excess of 1,000°F (538°C) for a period of not less than 15 minutes. Controls shall be located so as to be immediately accessible to the fire service from the exterior of the building and protected against interior fire exposure by fire barriers having a fire-resistance rating not less than 1 hour.

❖ Unless the mechanical smoke exhaust system also functions as a component of a smoke control system, standby power is not specifically required (see commentary, Section 909.11). To provide an enhanced level of operational reliability, this section requires that the power to smoke exhaust fans must be supplied from a circuit connected ahead of the building main electrical service disconnect. Note that this is one of the sources of power for legally required standby power systems recognized by the ICC EC [NFPA 70, Section 701.11(E)].

This section also requires that the wiring for the smoke exhaust fans be thermally protected in a manner approved by the fire code official that will protect the wiring from heat damage in the event of an interior fire. This protection could be provided by an approved wiring material listed for the temperature application, by physical protection with approved materials or assemblies or by installation outside of the building.

Because smoke exhaust systems are a vital fire-fighting tool, their operating controls must also be protected from interior fire exposure by 1-hour fire-resistance-rated construction. Exterior access to the controls allows fire department personnel to promptly operate the system from a protected area without entering the building. Control identification should be clear, permanent and approved by the fire code official.

910.4.5 Supply air. Supply air for exhaust fans shall be provided at or near the floor level and shall be sized to provide a minimum of 50 percent of required exhaust. Openings for supply air shall be uniformly distributed around the periphery of the area served.

❖ Introduction of makeup air is critical to the proper operation of all exhaust systems. Too little makeup air will cause a negative pressure to develop in the area being exhausted, thereby reducing the exhaust flow.

This section requires that makeup air be introduced to the area equipped with a mechanical smoke exhaust system to maintain the required exhaust flow. Because the system can exhaust only as much air as is introduced into the area and because this section allows mechanical or gravity makeup air openings to provide only 50 percent of the required makeup air, this section allows the designer to rely on infiltration air to make up the remaining 50 percent of the design makeup air required to allow the system to perform. Although not specifically stated in this section, where a mechanical makeup air source is used, it should be electrically interlocked and controlled by a single start switch to ensure that makeup air is always being supplied when the smoke exhaust system is operating. Even distribution of makeup air is important because if too much air is coming from one particular direction it has the potential to vary the dynamics of the fire and the ability of the system to address the smoke.

910.4.6 Interlocks. On combination comfort air-handling/smoke removal systems or independent comfort air-handling systems, fans shall be controlled to shut down in accordance with the approved smoke control sequence.

❖ For HVAC systems and combination HVAC/smoke control systems, this section requires that any fan shutdown requirements, such as those in the IMC, defer to the approved smoke control sequence established to comply with Section 909.

SECTION 911
EXPLOSION CONTROL

911.1 General. Explosion control shall be provided in the following locations:

1. Where a structure, room or space is occupied for purposes involving explosion hazards as identified in Table 911.1.
2. Where quantities of hazardous materials specified in Table 911.1 exceed the maximum allowable quantities in Table 2703.1.1(1).

Such areas shall be provided with explosion (deflagration) venting, explosion (deflagration) prevention systems, or barricades in accordance with this section and NFPA 69, or NFPA 495 as applicable. Deflagration venting shall not be utilized as a means to protect buildings from detonation hazards.

❖ It is usually impractical to design a building to withstand the pressure created by an explosion. Therefore, this section requires an explosion relief system for structures, rooms or spaces with occupancies involving explosion hazards. Explosions may result from the overpressurization of a containing structure, by physical/chemical means or by a chemical reaction. During an explosion, a sudden release of a high-pressure gas occurs and the energy is dissipated in the form of a shock wave.

Structures, rooms or spaces with occupancies involving explosion hazards must be equipped with some method of explosion control as required by the material-specific requirements in the code. Table 911.1 specifies when explosion control is required based on certain materials or occupancies where the quantities of hazardous materials involved exceed the maximum allowable quantities in Table 2703.1.1(1). Section 911 recognizes explosion (deflagration) venting and explosion (deflagration) prevention systems as acceptable methods of explosion control where appropriate. The use of barricades or other explosion protective devices, such as magazines, may be permitted as the means of explosion control where indicated in the code as an acceptable alternative and when approved by the fire code official.

TABLE 911.1. See page 9-104.

❖ This table designates when some methods of explosion control are required for specific material or special use conditions. This table applies when the quantities of hazardous materials involved exceed the maximum allowable quantities in Table 2703.1.1(1). Section 911.2 contains design criteria for explosion (deflagration) venting. Explosion prevention (suppression) systems, where used, must comply with NFPA 69. Barricade construction must be designed and installed to comply with NFPA 495. Chapters 28 through 44 of the code contain additional guidance on the applicability and design criteria for explosion control methods that depend on the specific type of hazardous material involved.

911.2 Required deflagration venting. Areas that are required to be provided with deflagration venting shall comply with the following:

1. Walls, ceilings and roofs exposing surrounding areas shall be designed to resist a minimum internal pressure of 100 pounds per square foot (psf) (4788 Pa). The minimum internal design pressure shall not be less than five times the maximum internal relief pressure specified in Section 911.2, Item 5.

2. Deflagration venting shall be provided only in exterior walls and roofs.

Exception: Where sufficient exterior wall and roof venting cannot be provided because of inadequate exterior wall or roof area, deflagration venting shall be allowed by specially designed shafts vented to the exterior of the building.

3. Deflagration venting shall be designed to prevent unacceptable structural damage. Where relieving a deflagration, vent closures shall not produce projectiles of sufficient velocity and mass to cause life threatening injuries to the occupants or other persons on the property or adjacent public ways.

4. The aggregate clear area of vents and venting devices shall be governed by the pressure resistance of the construction assemblies specified in Item 1 of this section and the maximum internal pressure allowed by Item 5 of this section.

5. Vents shall be designed to withstand loads in accordance with the *International Building Code*. Vents shall consist of any one or any combination of the following to relieve at a maximum internal pressure of 20 pounds per square foot (958 Pa), but not less than the loads required by the *International Building Code*:

   5.1. Exterior walls designed to release outward.

TABLE 911.1
EXPLOSION CONTROL REQUIREMENTS

MATERIAL	CLASS	EXPLOSION CONTROL METHODS	
		Barricade construction	Explosion (deflagration) venting or explosion (deflagration) prevention systems
Hazard Category			
Combustible dusts[a]	—	Not required	Required
Cryogenic fluids	Flammable	Not required	Required
Explosives	Division 1.1 Division 1.2 Division 1.3 Division 1.4 Division 1.5 Division 1.6	Required Required Not required Not required Required Required	Not required Not required Required Required Not required Not required
Flammable gas	Gaseous Liquefied	Not required Not required	Required Required
Flammable liquids	IA[b] IB[c]	Not required Not required	Required Required
Organic peroxides	U I	Required Required	Not permitted Not permitted
Oxidizer liquids and solids	4	Required	Not permitted
Pyrophoric	Gases	Not required	Required
Unstable (reactive)	4 3 detonable 3 nondetonable	Required Required Not required	Not permitted Not permitted Required
Water-reactive liquids and solids	3 2[e]	Not required Not required	Required Required
Special Uses			
Acetylene generator rooms	—	Not required	Required
Grain processing	—	Not required	Required
Liquefied petroleum gas distribution facilities	—	Not required	Required
Where explosion hazards exist[d]	Detonation Deflagration	Required Not required	Not permitted Required

a. Combustible dusts that are generated during manufacturing or processing. See definition of Combustible Dust in Chapter 2.
b. Storage or use.
c. In open use or dispensing.
d. Rooms containing dispensing and use of hazardous materials when an explosive environment can occur because of the characteristics or nature of the hazardous materials or as a result of the dispensing or use process.
e. A method of explosion control shall be provided when Class 2 water-reactive materials can form potentially explosive mixtures.

5.2. Hatch covers.

5.3. Outward swinging doors.

5.4. Roofs designed to uplift.

5.5. Venting devices listed for the purpose.

6. Vents designed to release from the exterior walls or roofs of the building when venting a deflagration shall discharge directly to the exterior of the building where an unoccupied space not less than 50 feet (15 240 mm) in width is provided between the exterior walls of the building and the property line.

   Exception: Vents complying with Item 7 of this section.

7. Vents designed to remain attached to the building when venting a deflagration shall be so located that the discharge opening shall not be less than 10 feet (3048 mm) vertically from window openings and exits in the building and 20 feet (6096 mm) horizontally from exits in the building, from window openings and exits in adjacent buildings on the same property, and from the property line.

8. Discharge from vents shall not be into the interior of the building.

❖ This section prescribes the basic design criteria necessary for deflagration venting. Deflagration venting limits the deflagration pressure in a certain area so that, in case of an explosion, the damage to that enclosed area is minimized or eliminated. Because there are so many variables involved for adequate deflagration venting, the parameters for each design should fit the individual situation. NFPA 68 contains additional guidance on the design and use of deflagration venting systems.

The area of the vent should be adequate to relieve the pressure before it reaches a level in excess of what can be withstood by the weakest building member. The vent area, therefore, is dependent on the actual construction of the enclosed area and the anticipated pressure. The vent panel should be of light-weight construction so that it can easily release at low pressures. Because the light-weight panels have little structural strength, railings may be required along the floor edge to prevent people or objects from falling against the panel.

Item 5 indicates that the vents are to be designed to relieve at a maximum internal pressure of 20 pounds per square foot (958 Pa) but not less than the load design requirements in Chapter 16 of the IBC. In areas commonly subject to high winds, the release pressure has to be increased accordingly to prevent the vents from being actuated by wind forces. Even though the release pressure should be as low as practical, it should always be higher than the external wind pressure.

Venting devices must be located to discharge directly to the open air or to an unoccupied space at least 50 feet (15 240 mm) in width on the same lot. To minimize damage and maintain the integrity of the existing system, window openings and egress facilities are not to be within 10 feet (3,048 mm) vertically or 20 feet (6,096 mm) horizontally of the vent. The spatial distance will permit the pressure to decrease and not to cause additional damage.

911.3 Explosion prevention systems. Explosion prevention systems shall be of an approved type and installed in accordance with the provisions of this code and NFPA 69.

❖ Depending on the conditions of the anticipated explosion hazard, the use of an explosion prevention system may be an effective means of explosion control. An explosion prevention system is most effective in confined spaces or enclosures in which combustible gases, mists or dusts are subject to deflagration in a gas-phase oxidant. Explosion prevention systems are intended to prevent an explosion hazard by combating the process of combustion it its incipient stage.

NFPA 69 contains further information on the installation, operation and design considerations for explosion prevention systems. Explosion prevention systems are commonly used to protect laboratory equipment, such as reactor vessels, mills and dust collectors.

911.4 Barricades. Barricades shall be designed and installed in accordance with NFPA 495.

❖ As indicated in Table 911.1, depending on the type of materials involved, barricade construction may be an acceptable method of explosion control. Barricade construction is an effective method of screening a building containing explosives from other buildings, magazines or public rights-of-way. The barricade could be either natural or artificial, where applicable, as specified in NFPA 495.

SECTION 912
FIRE DEPARTMENT CONNECTIONS

912.1 Installation. Fire department connections shall be installed in accordance with the NFPA standard applicable to the system design.

❖ The requirements for the FDC depend on the type of sprinkler system installed and whether a standpipe system is installed. NFPA 13 and 13R, for example, include design considerations for FDCs that are an auxiliary water supply source for automatic sprinkler systems; NFPA 14 is the design standard to use for FDCs serving standpipe systems. Threads for FDCs to sprinkler systems, standpipes, yard hydrants or any other fire hose connection must be approved (NFPA 1963 may be utitlized as part of the approval or as otherwise approved) and be compatible with the connections used by the local fire department (see commentary, Sections 903.3.6, 903.3.7 and 905.1).

912.2 Location. With respect to hydrants, driveways, buildings and landscaping, fire department connections shall be so located that fire apparatus and hose connected to supply the system will not obstruct access to the buildings for other fire apparatus. The location of fire department connections shall be approved.

❖ This section specifies that the FDC must be located so that vehicles and hose lines will not interfere with access to the building for the use of other fire department apparatus. The location of potential connected hose lines to the FDC and hydrants should be preplanned by the fire department. Many fire departments have a policy restricting the distance that a FDC may be from a fire hydrant. The final location of the FDC is subject to the approval of the fire code official.

912.2.1 Visible location. Fire department connections shall be located on the street side of buildings, fully visible and recognizable from the street or nearest point of fire department vehicle access or as otherwise approved by the fire code official.

❖ FDCs must be readily visible and easily accessed. A local policy constituting what is readily visible and accessible needs to be established. Usually, the policy will address issues such as location on the outside of the building and proximity to fire hydrants.

912.2.2 Existing buildings. On existing buildings, wherever the fire department connection is not visible to approaching fire apparatus, the fire department connection shall be indicated by an approved sign mounted on the street front or on the side of the building. Such sign shall have the letters "FDC" at least 6 inches (152 mm) high and words in letters at least 2 inches (51 mm) high or an arrow to indicate the location. All such signs shall be subject to the approval of the fire code official.

❖ The section acknowledges that FDCs on existing buildings may not always be readily visible from the street or nearest point of fire department vehicle access. In those instances, the location of the connection must be clearly marked with signage.

912.3 Access. Immediate access to fire department connections shall be maintained at all times and without obstruction by fences, bushes, trees, walls or any other object for a minimum of 3 feet (914 mm).

❖ The FDC must be readily accessible to fire fighters. The 3-foot (914 mm) minimum clearance requirement is intended to ensure adequate access is available. Landscaping design should not block a clear view of the FDC.

912.3.1 Locking fire department connection caps. The fire code official is authorized to require locking caps on fire department connections for water-based fire protection systems where the responding fire department carries appropriate key wrenches for removal.

❖ This section allows for the FDC caps to be locked as long as the fire departments that respond to that building or facility have the appropriate key wrenches. This avoids unnecessary vandalism and ensures a more functional FDC when needed.

912.4 Signs. A metal sign with raised letters at least 1 inch (25 mm) in size shall be mounted on all fire department connections serving fire sprinklers, standpipes or fire pump connections. Such signs shall read: AUTOMATIC SPRINKLERS or STANDPIPES or TEST CONNECTION or a combination thereof as applicable.

❖ Signs identify the type of system or zone served by a given FDC. Signs may also distinguish FDCs from fire pump test headers. Usually, FDCs may be distinguished from fire pump test headers by the types of couplings provided. FDCs are customarily equipped with female couplings, while fire pump test headers usually have separately valved male couplings. Furthermore, fire pump test headers are equipped with one $2^1/_2$-inch (64 mm) outlet for each 250 gallon per minute (gpm) (16 L/s) of rated capacity.

[P] 912.5 Backflow protection. The potable water supply to automatic sprinkler and standpipe systems shall be protected against backflow as required by the *International Plumbing Code*.

❖ Section 608.16.4 of the 2000 IPC requires all connections to automatic sprinkler systems and standpipe systems to be equipped with a means to protect the potable water supply. The means of backflow protection can be either a double check-valve assembly or a reduced-pressure-principle back flow preventer. This, in general, assumes a FDC is required. For example, a limited-area sprinkler system off the domestic supply does not necessarily require a FDC and would not require backflow protection.

912.6 Inspection, testing and maintenance. All fire department connections shall be periodically inspected, tested and maintained in accordance with NFPA 25.

❖ Because FDCs are components of a water-based extinguishing system, NFPA 25 is applicable. Inspections must determine that connections are unobstructed, well-protected and in good working order. Plugs or covers must be installed to protect threads and pipe openings, and must be easily removed to permit connection of a fire hose.

Caps or plugs must be kept in place whenever the connection is not in use to discourage the insertion of objects into the connection openings. The interior of piping behind connection clappers must be checked for foreign material and obstructions.

Threads must be compatible with local fire service hose couplings and free of burrs, depressions and other flaws. Couplings must spin freely. Clappers, if installed in the pipe openings, must open easily and automatically return to the closed position.

Exposed piping, fittings, valves and couplings must be free of water where subject to freezing. Defects must be corrected without delay. These and other maintenance features are addressed in NFPA 25.

SECTION 913
FIRE PUMPS

913.1 General. Where provided, fire pumps shall be installed in accordance with this section and NFPA 20.

❖ This section contains specific installation requirements for fire pumps supplying water to fire protection systems. Inspection, testing and maintenance requirements comply with NFPA 20 unless noted otherwise. Applicable maintenance standards are also identified.

Fire pumps are installed in sprinkler and standpipe systems to pressurize the water supply for the minimum required sprinkler and standpipe operation. They are considered a design feature or component of the system. Fire pumps can improve only the pressure of the incoming water supply, not the volume of water available.

When the volume from a water supply is not adequate to supply sprinkler or standpipe demand, water tanks for private fire protection, improvements in the size and capacity of fire mains or water distribution systems or all of these for the installation of a fire pump are needed.

When fire pumps are required to meet the pressure requirements of sprinkler and standpipe systems, they must be installed and tested in accordance with NFPA 20.

913.2 Protection against interruption of service. The fire pump, driver, and controller shall be protected in accordance with NFPA 20 against possible interruption of service through damage caused by explosion, fire, flood, earthquake, rodents, insects, windstorm, freezing, vandalism and other adverse conditions.

❖ This section lists hazards that should be taken into account when determining the extent of protection required for the fire pump and its auxiliary equipment. A pump room in a building that is protected against the listed hazards in compliance with the IBC would be considered in compliance. Because fire pumps are also typically located in separate detached structures, geographical and security issues should be considered.

913.3 Temperature of pump room. Suitable means shall be provided for maintaining the temperature of a pump room or pump house, where required, above 40°F (5°C).

❖ As previously noted for sprinkler systems, standpipe systems and other water-based fire protection systems, pump rooms or pump houses must be maintained at a temperature of 40°F (4°C) or above to prevent the system from freezing.

913.3.1 Engine manufacturer's recommendation. Temperature of the pump room, pump house or area where engines are installed shall never be less than the minimum recommended by the engine manufacturer. The engine manufacturer's recommendations for oil heaters shall be followed.

❖ If the engine manufacturer's recommended minimum temperature is higher than the minimum established in Section 913.3, that recommendation must be complied with. Maintaining the desired engine temperature enhances the startability of the engine. Maintaining water heaters and oil heaters as required for diesel engines, for example, will improve the starting capabilities of the fire pump and reduce engine wear and the drain on batteries.

913.4 Valve supervision. Where provided, the fire pump suction, discharge and bypass valves, and the isolation valves on the backflow prevention device or assembly shall be supervised open by one of the following methods.

1. Central-station, proprietary, or remote-station signaling service.
2. Local signaling service that will cause the sounding of an audible signal at a constantly attended location.
3. Locking valves open.
4. Sealing of valves and approved weekly recorded inspection where valves are located within fenced enclosures under the control of the owner.

❖ As was the case with sprinkler systems, water control valves that are a part of the fire pump installation must be supervised in the open position to ensure the system is operational and also to reduce the chance of a system failure (see commentary, Section 903.4). In most cases the required water-based extinguishing system, which the fire pump is an integral component of, will be electrically supervised. Locking or sealing valves open as the only means of supervision may not be permitted, depending on the type of valve. Section 903.4, for example, specifically exempts jockey pump control valves from being electrically supervised if they are sealed or locked in the open position.

913.4.1 Test outlet valve supervision. Fire pump test outlet valves shall be supervised in the closed position.

❖ Fire pump test outlet valves are for performance testing of the fire pump and do not control the available water supply to either a sprinkler system or a standpipe system. These valves are normally in a closed position and are supervised accordingly.

913.5 Testing and maintenance. Fire pumps shall be inspected, tested and maintained in accordance with the requirements of this section and NFPA 25.

❖ Fire pumps require periodic maintenance to ensure they will perform as required. Monthly maintenance includes running the pump at churn to exercise the pump and driver. Pump packings and relief valve settings must be adjusted as needed. Annually, the pump must be retested to verify its proper performance. Pressure, flow, revolutions per minute, voltage and, for electric motor-driven pumps, voltage and amperage readings must be recorded, plotted and compared with original design criteria. Upon completion of test and maintenance, the pump must be left in the automatic-start condition, ready for service. Because a fire pump is a component of a water-based extinguishing system, NFPA 25 is applicable.

913.5.1 Acceptance test. Acceptance testing shall be done in accordance with the requirements of NFPA 20.

❖ Chapter 11 of NFPA 20 details the procedure for conducting a fire pump acceptance test. This test is run to determine that the installation matches the sprinkler or standpipe design criteria and the manufacturer's performance specifications. When fire pumps are connected to public water mains, the public system should be flushed prior to testing to prevent dirt and debris in the public water supply system from damaging the pump. Acceptance tests for stationary fire pumps are similar to the tests performed on mobile fire department pumping apparatus.

During the test, measurements are taken at varying flow rates, between 0 and 150 percent of the pump rated capacity, in 25-percent increments. At each point, pressure, flow, revolutions per minute and, for electric fire-pump drivers, voltage and amperage readings are taken to assess the pump performance. At the end of the test, the recorded results are plotted and compared to the manufacturer design curve and system demand. A stationary fire pump must be capable of supplying 150 percent of its rated capacity at not less than 65 percent of its rated head or discharge pressure so that the fire pump is capable of exceeding minimum design demands for a period of time without seriously reducing sprinkler or standpipe performance. The intention is to give fire fighters time to augment the water supply system or to modify their tactics to protect fire fighters that are using the systems supplied by the fire pump.

Load tests consisting of a minimum of 10 manual and 10 automatic starts and stops must be performed. Automatic cutout pressures must be set at not less than 140 percent of rated discharge pressure. Pump controllers are also equipped with cutout timers to allow the pump to run for a time between starting and stopping to minimize stresses on working parts. The fire pump must be permitted to run for a minimum of 5 minutes between each start and stop.

To prevent a cavitation condition caused by inadequate supply resulting in potentially damaging air bubbles and excessive turbulence in the impeller and volutes, input pressure for design purposes must be based on 150 percent of the rated discharge capacity. Installation of low-suction cutoff systems to prevent cavitation is not recommended. These systems should be installed only where required by water purveyors, environmental or health authorities to prevent dangerous backflow conditions.

913.5.2 Generator sets. Engine generator sets supplying emergency or standby power to fire pump assemblies shall be periodically tested in accordance with NFPA 110.

❖ This section does not require emergency or standby power for all fire pump installations, but rather requires the testing of on-site generator sets that are used for emergency or standby power to fire pump assemblies. The need for emergency or standby power is typically based on occupancy conditions as indicated in the IBC IBC. Section 403.10 of the IBC, for example, requires standby power for all electrically powered fire pumps in high-rise buildings. A generator set is recognized as a permissible standby power source. NFPA 110 prescribes the operational testing requirements, including load tests, as well as the periodic inspection and maintenance for generator sets.

913.5.3 Transfer switches. Automatic transfer switches shall be periodically tested in accordance with NFPA 110.

❖ Automatic transfer switches are self-acting equipment that is used to transfer power from a normal source of supply to an alternative supply, such as an engine generator set. NFPA 110 requires a test on each automatic transfer switch that simulates failure of the normal power source. Upon failure, the automatic transfer switch should then automatically transfer the load to the emergency power supply. Manual transfer switches are not permitted as the only means to transfer power between the normal supply and the alternative supply to the fire pump controller.

913.5.4 Pump room environmental conditions. Tests of pump room environmental conditions, including heating, ventilation and illumination shall be made to ensure proper manual or automatic operation of the associated equipment.

❖ Maintaining suitable environmental conditions is essential to ensure starting capability, performance and safe operation of fire pumps and associated emergency power supplies, where required. Adequate ventilation, for example, is needed to maintain the ambient temperature in the pump room within the range recommended by the manufacturer for the emergency power supply equipment.

Bibliography

The following resource materials are referenced in this chapter or are relevant to the subject matter addressed in this chapter.

Americans with Disabilities Act Accessibility Guidelines for Buildings and Facilities (ADAAG). Washington, DC: US. Architectural and Transportation Barriers Compliance Board, 1998.

Automatic Sprinkler Systems Handbook, 7th edition. Quincy, MA: National Fire Protection Association, 1996.

Bryan, John L. *Automatic Sprinkler and Standpipe Systems,* 3rd edition. Quincy, MA: National Fire Protection Association, 1997.

Budnick, E.K. *Estimating Effectiveness of State-of-the-Art Detectors and Automatic Sprinklers on Life Safety in Residential Occupancies.* Washington, DC: National Bureau of Standards, NBS IR 84-2819.

Bukowski, R.W. and R.J. O'Laughlin. *Fire Alarm Signaling Systems.* Quincy, MA: National Fire Protection Association, 1994.

Design of Smoke Management Systems. Atlanta, GA: American Society of Heating, Refrigerating and Air-Conditioning Engineers, Inc., 1992.

DOJ 28 CFR, Part 36-91, *Americans With Disabilities Act.* Washington, DC: U.S. Department of Justice, 1991.

DOJ 28 CFR, Part 36 (Appendix A)-91, ADA *Guidelines for Buildings and Facilities.* Washington, DC: U.S. Department of Justice, 1991.

DOTn 49 CFR, Parts 100-178 & 179-199-95, *Specification for Transportation of Explosive and Other Dangerous Articles, Shipping Containers.* Washington, DC: U.S. Department of Transportation, 1995.

Evans, D. and J. Klote. *Smoke Control Provisions of the 2000 IBC.* Falls Church, VA: International Code Council, 2002.

Fire Protection Equipment Directory. Northbrook, IL: Underwriters Laboratories Inc., 2002.

Fire Protection Handbook, 18th edition. Quincy, MA: National Fire Protection Association, 1997.

Fire Pump Handbook. Quincy, MA: National Fire Protection Association, 1998.

FM 4430-88, *Approved Standard for Heat and Smoke Vents.* Norwood, MA: Factory Mutual, 1988.

Grant, Casey Cavanaugh. "Halon and Beyond: Developing New Alternatives". FPA Journal, November/December 1994.

Harrington, Jeff L. "The Halon Phaseout Speeds Up". *NFPA Journal,* March/April 1993.

Health Care Facilities. Falls Church, VA: CABO Board for the Coordination of the Model Codes Report, 1985.

IBC-03, *International Building Code.* Falls Church, VA: International Code Council, 2003.

ICC A117.1-98, *Accessible and Usable Buildings and Facilities.* Falls Church, VA: International Code Council, 1998.

IMC-03, *International Mechanical Code.* Falls Church, VA: International Code Council, 2003.

IPC-03, *International Plumbing Code.* Falls Church, VA: International Code Council, 2003.

IPMC-03, *International Property Maintenance Code.* Falls Church, VA: International Code Council, 2003.

Klote, J. and J. Milke. *Principles of Smoke Management.* Atlanta, GA: American Society of Heating, Refrigerating and Air-Conditioning Engineers, 2002 .

Milke, James A. "Smoke Management for Covered Malls and Atria." *Fire Technology,* August 1990.

National Fire Alarm Code Handbook. Quincy, MA: National Fire Protection Association, 1999.

NFPA 10-98, *Portable Fire Extinguishers.* Quincy, MA: National Fire Protection Association, 1998.

NFPA 11-98, *Low Expansion Foam and Combined Agent Systems.* Quincy, MA: National Fire Protection Association, 1998.

NFPA 11A-99, *Medium and High Expansion Foam Systems.* Quincy, MA: National Fire Protection Association, 1999.

NFPA 12-00, *Carbon Dioxide Extinguishing Systems.* Quincy, MA: National Fire Protection Association, 2000.

NFPA 12A-97, *Halon 1301 Fire Extinguishing Systems.* Quincy, MA: National Fire Protection Association, 1997.

NFPA 13-99, *Installation of Sprinkler Systems.* Quincy, MA: National Fire Protection Association, 1999.

NFPA 13D-99, *Installation of Sprinkler Systems in One- and Two-Family Dwellings and Manufactured Homes.* Quincy, MA: National Fire Protection Association, 1999.

NFPA 13R-99, *Installation of Sprinkler Systems in Residential Occupancies Up to and Including Four Stories in Height.* Quincy, MA: National Fire Protection Association, 1999.

NFPA 14-00, *Standpipe and Hose Systems.* Quincy, MA: National Fire Protection Association, 2000.

NFPA 16-99, *Installation of Deluge Foam-Water Sprinkler and Foam-Water Spray Systems.* Quincy, MA: National Fire Protection Association, 1999.

NFPA 17-98, *Dry Chemical Extinguishing Systems.* Quincy, MA: National Fire Protection Association, 1998.

NFPA 17A-98, *Wet Chemical Extinguishing Systems.* Quincy, MA: National Fire Protection Association, 1998.

NFPA 20-99, *Installation of Centrifugal Fire Pumps.* Quincy, MA: National Fire Protection Association, 1999.

NFPA 24-95, *Installation of Private Fire Service Mains.* Quincy, MA: National Fire Protection Association, 1995.

NFPA 25-98, *Inspection, Testing and Maintenance of Water Based Fire Protection Systems.* Quincy, MA: National Fire Protection Association, 1998.

NFPA 30-00, *Flammable and Combustible Liquids Code.* Quincy, MA: National Fire Protection Association, 2000.

NFPA 30B-00, *Manufacture and Storage of Aerosol Products.* Quincy, MA: National Fire Protection Association, 2000.

NFPA 68-02, *Guide for Venting of Deflagrations.* Quincy, MA: National Fire Protection Association, 2002.

NFPA 69-97, *Explosion Prevention Systems*. Quincy, MA: National Fire Protection Association, 1997.

NFPA 72-99, *National Fire Alarm Code*. Quincy, MA: National Fire Protection Association, 1999.

NFPA 92A-00, *Smoke Control Systems.* Quincy, MA: National Fire Protection Association, 2000.

NFPA 92B-00, *Smoke Management Systems in Malls, Atria and Large Areas*. Quincy, MA: National Fire Protection Association, 2000.

NFPA 204-00, *Standard for Smoke and Heat Venting.* Quincy, MA: National Fire Protection Association, 2000.

NFPA 231-98, *General Storage*. Quincy, MA: National Fire Protection Association, 1998.

NFPA 231C-98, *Rack Storage of Materials*. Quincy, MA: National Fire Protection Association, 1998.

NFPA 231D-98, *Storage of Rubber Tires*. Quincy, MA: National Fire Protection Association, 1998.

NFPA 495-96, *Explosive Materials Code*. Quincy, MA: National Fire Protection Association, 1996.

NFPA 1963-03, *Standard for Screw Threads and Gaskets for Fire Hose Connections*. Quincy, MA: National Fire Protection Association, 2003.

NFPA 2001-00, *Clean Agent Fire Extinguishing Systems.* Quincy, MA: National Fire Protection Association, 2000.

Smoke Control in Fire Safety Design. London, E. & F.N. Spon Ltd., 1979.

The SFPE Handbook of Fire Protection Engineering. Quincy, MA: National Fire Protection Association, 3rd edition, 2002

UL 33-03, *Standard for Heat-Responsive Links*. Northbrook, IL: Underwriters Laboratories Inc., 2003.

UL 197-93, Commercial Electric Cooing Appliances. Northbrook, IL: Underwriters Laboratories Inc., 1993.

UL 217-97, *Single and Multiple Station Smoke Detectors*. Northbrook, IL: Underwriters Laboratories Inc., 1997.

UL 268-96, *Smoke Detectors for Fire Protective Signaling Systems.* Northbrook, IL: Underwriters Laboratories Inc., 1996.

UL 300-96, *Standard for Fire Testing of Fire Extinguishing Systems for Protection of Restaurant Cooking Areas* with Revisions through December 1998. Northbrook, IL: Underwriters Laboratories Inc., 1996.

UL 521-99, *Standard for Heat Detectors for Fire Protective Signaling Systems.* Northbrook, IL: Underwriters Laboratories Inc., 1999.

UL 555-96, *Fire Dampers*. Northbrook, IL: Underwriters Laboratories Inc., 1996.

UL 555S-99, *Smoke Dampers*. Northbrook, IL: Underwriters Laboratories Inc.,1999.

UL 793-03, *Standard for Automatically Operated Roof Vents for Smoke and Heat.* Northbrook, IL: Underwriters Laboratories Inc., 2003.

UL 864-96, *Control Units for Fire Protective Signaling Systems*. Northbrook, IL: Underwriters Laboratories Inc., 1996.

UL 1058-95, *Standard for Halogenated Agent Extinguishing System Units* with Revisions through April 1998. Northbrook, IL: Underwriters Laboratories Inc., 1995.

Chapter 10: Means of Egress

General Comments

The evolution of means of egress requirements has been influenced by lessons learned from real fire incidents. Even though contemporary fires may reinforce some of these lessons, each incident must be viewed as an opportunity to assess critically the safety and reasonableness of current regulations.

Cooperation among the developers of model codes and standards has resulted in agreement on many basic terms and concepts. The text of the code, including Chapter 10, is consistent with these national uniformity efforts.

National uniformity in an area such as means of egress has many benefits for the fire code official and other code users. Not the least important are the lessons to be learned from experiences throughout the nation and the world that can be reported in commonly used terminology and conditions that everyone can relate to and clearly understand.

Chapter 10 includes the minimum requirements for means of egress in all buildings and structures. The requirements detail the size, arrangement, number and protection of the means of egress components. Also specified are the functional and operational characteristics for the components that will permit their safe use without extraordinary knowledge or effort.

Sections 1002 through 1025 are duplicated text from Chapter 10 of the *International Building Code*® (IBC®) and are fully applicable to new buildings constructed after the adoption of the code. The requirements of Sections 1003 through 1025 are also applicable to existing buildings to the extent prescribed in Section 1026. Providing a building with means of egress that are compliant with the code is only the first step in achieving an acceptable minimum level of life safety in a building. To be effective, the means of egress must be continuously maintained to provide the required level of safety. This chapter describes maintenance requirements in Section 1027.

Purpose

A primary purpose of codes in general and building codes in particular is to safeguard life in the presence of a fire. Integral to this purpose is the path of egress travel for occupants to escape and avoid a fire. Means of egress can be considered the lifeline of a building. The principles on which means of egress are based and that form the fundamental criteria for requirements are to provide a means of egress system:

1. That will give occupants alternative paths of travel to a place of safety to avoid fire.
2. That will shelter occupants from fire and the products of combustion.
3. That will accommodate all occupants of a structure.
4. That is clear, unobstructed, well marked and illuminated and in which all components are under control of the user without requiring any tools, keys or special knowledge or effort.

History is marked with the loss of life from fire. Early as well as contemporary multiple fire fatalities can be traced to a compromise of one or more of the above principles.

Life safety from fire is a matter of successfully evacuating or relocating the occupants of a building to a place of safety. As a result, life safety is a function of time: time for detection, time for notification and time for safe egress. The fire growth rate over a period of time is also a critical factor in addressing life safety. Other sections of the code, such as protection of vertical openings (Chapter 7), interior finish (Chapter 8), fire suppression and detection systems (Chapter 9) and numerous others, also have an impact on life safety. Chapter 10 addresses the issues related to the means available to relocate or evacuate building occupants.

SECTION 1001 GENERAL

1001.1 General. Buildings or portions thereof shall be provided with a means of egress system as required by this chapter. The provisions of this chapter shall control the design, construction and arrangement of means of egress components required to provide an approved means of egress from structures and portions thereof. Sections 1003 through 1025 shall apply to new construction. Sections 1026 and 1027 shall apply to existing buildings.

Exception: Detached one- and two-family dwellings and multiple single-family dwellings (townhouses) not more than three stories above grade plane in height with a separate means of egress and their accessory structures shall comply with the *International Residential Code*.

❖ The minimum requirements for means of egress are to be incorporated in all structures as specified in this chapter. Application would be effective on the date the code is adopted and placed into effect.

The means of egress in an existing building that experiences a change of occupancy, such as from Group S-2 (storage) to Group A-3 (assembly), would require reevaluation for code compliance based on the new occupancy as stated in Chapter 34 of the IBC. Similarly, the means of egress in an existing occupancy of Group A-3 in which additional seating is to be provided, thereby increasing the occupant load, would require reevaluation for code compliance based on the increased occupant load.

Fundamental to the level of life safety in any building, whether it is new or many years old, is the provision for an adequate egress system, and it is for that reason that Chapter 10 is also applicable to existing buildings that are not undergoing changes as regulated by Chapter 34 of the IBC. Limitations on and modifications to that applicability are contained in Section 1026. The means of egress in existing buildings must also be properly maintained in accordance with Section 1027 if the intended level of safety is to remain for the life of the building.

Reflecting the correlation and compatibility that is a hallmark of the *International Codes*®, the exception makes it clear that the means of egress in buildings that are within the scope of the *International Residential Code*® (IRC®) are to comply with those requirements instead of Chapter 10.

1001.2 Minimum requirements. It shall be unlawful to alter a building or structure in a manner that will reduce the number of exits or the capacity of the means of egress to less than required by this code.

❖ A fundamental concept in life safety design is that the means of egress system is to be constantly available throughout the life of a building. Any change in the building or its contents, either by physical reconstruction or alteration or by a change of occupancy, is cause to review the resulting egress system. As a minimum, a building's means of egress is to be continued as initially approved. If a building or portion thereof has a change of occupancy, the complete egress system is to be evaluated and approved for compliance with the current code requirements for new occupancies (see Chapter 34 of the IBC).

The means of egress in an existing building that experiences a change of occupancy, such as from Group S-2 (storage) to A-3 (assembly), would require reevaluation for code compliance based on the new occupancy. Similarly, the means of egress in an existing occupancy of Group A-3 in which additional seating is to be provided, thereby increasing the occupant load, would require reevaluation for code compliance based on the increased load.

The temptation is to temporarily remove egress components or other fire protection features from service during an alteration or repair to or temporary occupancy of a building. During such times, a building is frequently more vulnerable to fire and the rapid spread of products of combustion. Either the occupants should not occupy those spaces where the means of egress has been compromised by the construction or compensating fire safety features should be considered, which will provide equivalent safety for the occupants.

It should be noted that occupants in adjacent areas may also require access to the egress facilities in the area under construction.

[B] SECTION 1002 DEFINITIONS

1002.1 Definitions. The following words and terms shall, for the purposes of this chapter and as used elsewhere in this code, have the meanings shown herein.

❖ Definitions of terms can help in the understanding and application of the code requirements. The purpose for including these definitions in this chapter is to provide more convenient access to them without having to refer back to Chapter 2.

For convenience, these terms are also listed in Chapter 2 with a cross reference to this section. The use and application of all defined terms, including those defined herein, are set forth in Section 201.

ACCESSIBLE MEANS OF EGRESS. A continuous and unobstructed way of egress travel from any point in a building or facility that provides an accessible route to an area of refuge, a horizontal exit or a public way.

❖ Accessible means of egress requirements are needed to provide those persons with physical disabilities a means of egress to a safe area in the building or to exit the building.

AISLE ACCESSWAY. That portion of an exit access that leads to an aisle.

❖ As illustrated in Figure 1002.1(1), an aisle accessway is intended for one-way travel or limited two-way travel. The space between tables, seats, displays or other furniture (i.e., aisle accessway) utilized for means of egress will lead to a main aisle.

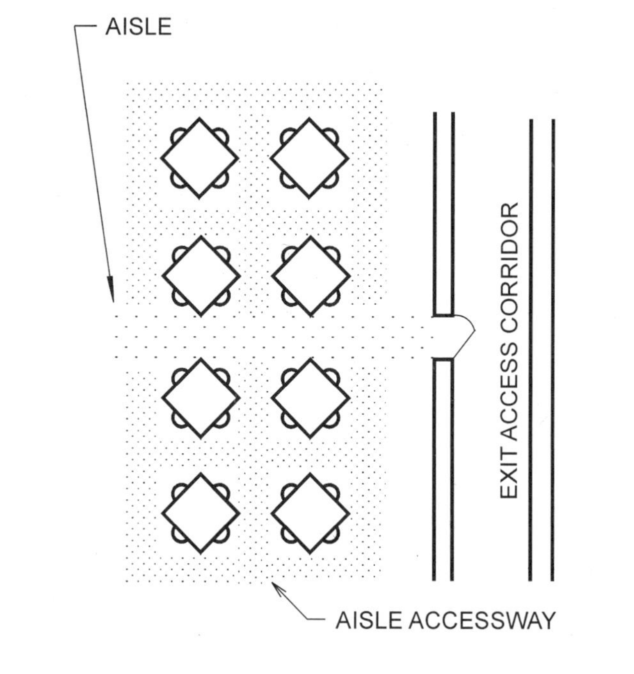

Figure 1002.1(1)
AISLES AND AISLE ACCESSWAYS

ALTERNATING TREAD DEVICE. A device that has a series of steps between 50 and 70 degrees (0.87 and 1.22 rad) from horizontal, usually attached to a center support rail in an alternating manner so that the user does not have both feet on the same level at the same time.

❖ An alternating tread device is commonly used in areas that would otherwise be provided with a ladder. The device is used extensively in industrial facilities for worker access to platforms or equipment.

AREA OF REFUGE. An area where persons unable to use stairways can remain temporarily to await instructions or assistance during emergency evacuation.

❖ The area of refuge provides a safe area in the building during an emergency condition for temporary use by persons who are unable to exit the building using the stairways.

BLEACHERS. Tiered seating facilities.

❖ Bleachers, folding and telescopic seating and grandstands are essentially unique forms of tiered seating. All types are addressed in ICC 300, the new safety standard for these types of seating arrangements. Bleachers typically do not have backrests. The travel path across the bleachers is not restricted to designated rows, aisles and aisle accessways. Without backrests, occupants can traverse from row to row without traveling to the designated egress aisles (see Section 1024.1.1).

COMMON PATH OF EGRESS TRAVEL. That portion of exit access which the occupants are required to traverse before two separate and distinct paths of egress travel to two exits are available. Paths that merge are common paths of travel. Common paths of egress travel shall be included within the permitted travel distance.

❖ The common path of travel is a concept used to refine travel distance criteria. Similar to dead ends in corridors and passageways, a common path of travel is the course an occupant will travel that will cause him or her to pass through the same location regardless of which path is chosen. The length of common path of travel is limited so that the route to a remote means of egress begins to diverge before the occupant has traveled an excessive distance. This reduces the possibility that, although the exits are remote from one another, a single fire condition will render both paths unavailable.

CORRIDOR. An enclosed exit access component that defines and provides a path of egress travel to an exit.

❖ Corridors are regulated in the code because they serve as principal elements of egress travel in the means of egress systems within buildings. Corridors have walls that extend from the floor to at least the ceiling. They need not extend above the ceiling or have doors in their openings unless a fire-resistance rating is required (see Section 1016). The enclosed character of the corridor restricts the sensory perception of the user. A fire located on the other side of the corridor wall, for example, may not be as readily seen, heard or smelled by the occupants traveling through the egress corridor. An egress path bounded by partial-height walls, such as workstation partitions in an office, is not a corridor by definition since it is not enclosed by full-height walls.

DOOR, BALANCED. A door equipped with double-pivoted hardware so designed as to cause a semicounterbalanced swing action when opening.

❖ Balanced doors are commonly used to decrease the force necessary to open the door or to reduce the length of the door swing.

EGRESS COURT. A court or yard which provides access to a public way for one or more exits.

❖ The egress court requirements address situations where the exit discharge portion of the means of egress passes through confined areas near the building and, therefore, faces a hazard not normally found in the exit discharge.

EMERGENCY ESCAPE AND RESCUE OPENING. An operable window, door or other similar device that provides for a means of escape and access for rescue in the event of an emergency.

❖ These are commonly windows that are big enough and located such that they can be used to exit a building directly from a basement or bedroom during an emergency condition. The openings are also used by emergency personnel to rescue the occupants in a building (see Section 1025).

EXIT. That portion of a means of egress system which is separated from other interior spaces of a building or structure by fire-resistance-rated construction and opening protectives as required to provide a protected path of egress travel between the exit access and the exit discharge. Exits include exterior exit doors at ground level, exit enclosures, exit passageways, exterior exit stairs, exterior exit ramps and horizontal exits.

❖ Exits are the critical element of the means of egress system that the building occupants travel through to reach the exterior grade level. Exit stairways from upper and lower stories, as well as horizontal exits, must be separated from adjacent areas with fire-resistance-rated construction. The fire-resistance-rated construction serves as a barrier between the fire and the means of egress and protects the occupants while they travel through the exit. Separation by fire-resistance-rated construction is not required, however, where the exit leads directly to the exterior at the level of exit discharge (e.g., exterior door at grade).

Figure 1002.1(2) illustrates three different types of exits: interior exit stairway, exterior exit stairway and an exterior exit door.

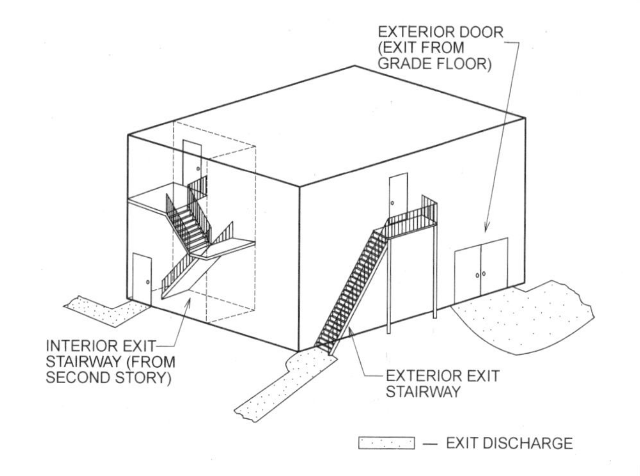

Figure 1002.1(2)
EXITS

EXIT ACCESS. That portion of a means of egress system that leads from any occupied portion of a building or structure to an exit.

❖ The exit access portion of the means of egress consists of all floor areas that lead from usable spaces within the building to the exit or exit(s) serving that floor area. Crawl spaces and concealed attic and roof spaces are not considered to be part of the exit access. As shown in Figure 1002.1(3), the exit access begins at the furthest points within each room or space and ends at the entrance to the exit.

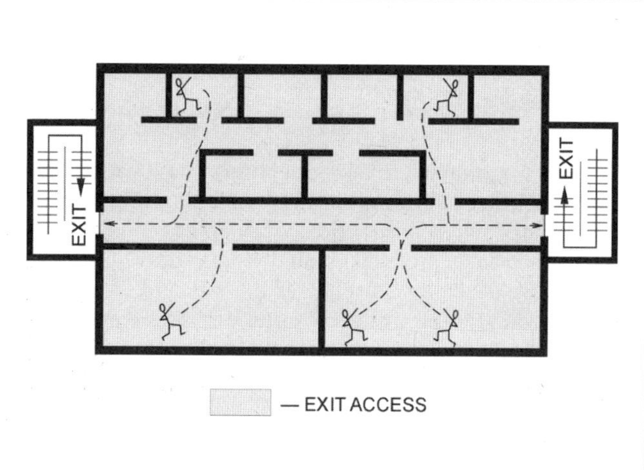

Figure 1002.1(3)
EXIT ACCESS

EXIT DISCHARGE. That portion of a means of egress system between the termination of an exit and a public way.

❖ The exit discharge will typically begin when the building occupants reach the exterior at grade level. It provides occupants with a path of travel away from the building. All components between the building and the public way are considered to be the exit discharge, regardless of the distance. The exit discharge is part of the means of egress and, therefore, its components are subject to the requirements of the code [see Figures 1002.1(2) and 1002.1(4)].

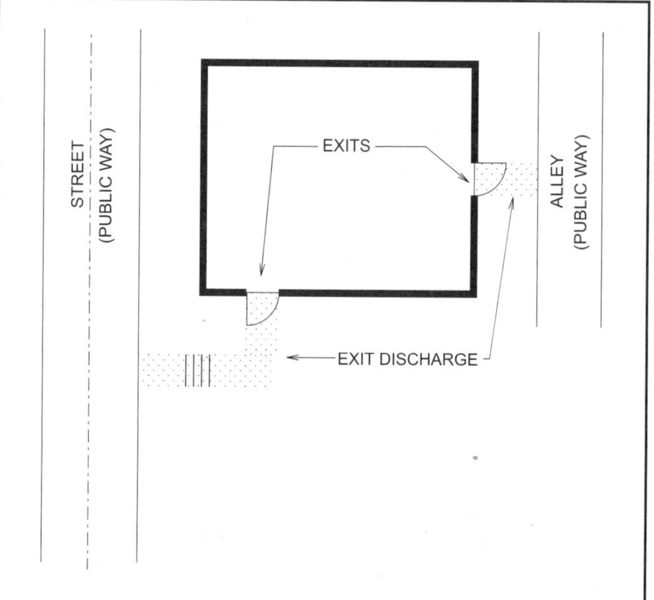

Figure 1002.1(4)
EXIT DISCHARGE

EXIT DISCHARGE, LEVEL OF. The horizontal plane located at the point at which an exit terminates and an exit discharge begins.

❖ The term is intended to describe the level at which the occupants, during egress, leave an exit and are no longer required to ascend or descend to be at the level at which exit discharge begins. At this level, the occupant need only move in a substantially horizontal path to reach the start of exit discharge. Hence, an exit may be composed of both interior and exterior exit stairs that together terminate at the level of exit discharge [see Figure 1002.1(5)].

EXIT ENCLOSURE. An exit component that is separated from other interior spaces of a building or structure by fire-resistance-rated construction and opening protectives, and provides for a protected path of egress travel in a vertical or horizontal direction to the exit discharge or the public way.

❖ This term is used to describe an exit that is within a fire-resistance-rated enclosure for a generally vertical exit path of travel (e.g., a stair) or a generally horizontal path of travel (e.g., a ramp).

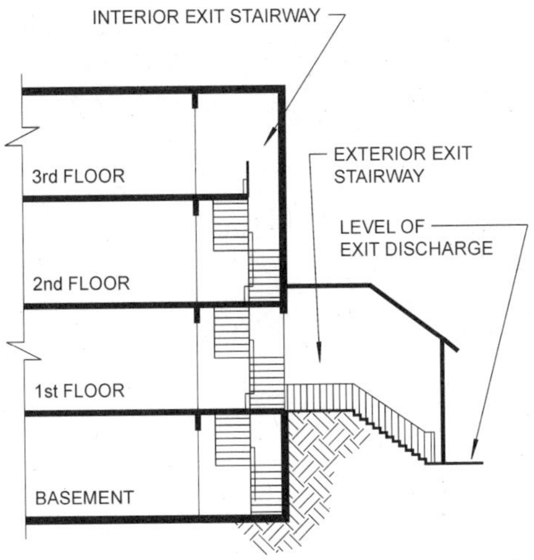

Figure 1002.1(5)
LEVEL EXIT DISCHARGE

EXIT, HORIZONTAL. A path of egress travel from one building to an area in another building on approximately the same level, or a path of egress travel through or around a wall or partition to an area on approximately the same level in the same building, which affords safety from fire and smoke from the area of incidence and areas communicating therewith.

❖ This term refers to a fire-resistance-rated wall that subdivides a building or buildings into multiple compartments and provides an effective barrier to protect occupants from a fire condition within one of the compartments. After occupants pass through a horizontal exit, they must be provided with sufficient space to gather and must also be provided with another exit, such as an exterior door or exit stairway, through which they can exit the building. Figure 1002.1(6) depicts the exits serving a single building that is subdivided with a fire-resistance-rated wall.

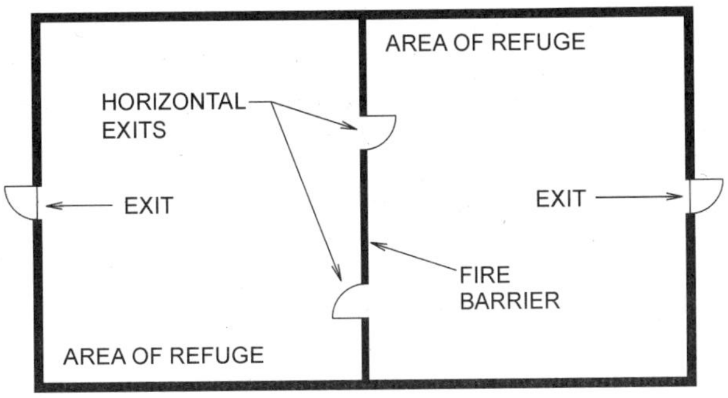

Figure 1002.1(6)
HORIZONTAL EXIT

EXIT PASSAGEWAY. An exit component that is separated from all other interior spaces of a building or structure by fire-resistance-rated construction and opening protectives, and provides for a protected path of egress travel in a horizontal direction to the exit discharge or the public way.

❖ This term refers to the portion of the means of egress that serves as a horizontal exit element and leads to the exit discharge. Since an exit passageway is considered an exit element, it must be protected and separated as required by the code for exits. An exit passageway may be located before or after a vertical exit enclosure. Exit passageways that lead to an exterior exit door are commonly used in malls to satisfy the travel distance in buildings having a large floor area. Exit passageways between a vertical exit enclosure and an exterior exit door are typically found on the level of exit discharge to provide a protected path from a centrally located exit stairway to the exit discharge.

FIRE EXIT HARDWARE. Panic hardware that is listed for use on fire door assemblies.

❖ Where a door that is required to be fire-resistance-rated construction has panic hardware, the hardware is required to be listed for use on the door. Thus, fire door hardware has been demonstrated to function properly when exposed to the effects of a fire.

FLOOR AREA, GROSS. The floor area within the inside perimeter of the exterior walls of the building under consideration, exclusive of vent shafts and courts, without deduction for corridors, stairways, closets, the thickness of interior walls, columns or other features. The floor area of a building, or portion thereof, not provided with surrounding exterior walls shall be the usable area under the horizontal projection of the roof or floor above. The gross floor area shall not include shafts with no openings or interior courts.

❖ Gross floor area is that area measured within the perimeter formed by the inside surface of the exterior walls. The area of all occupiable and nonoccupiable spaces, including mechanical and elevator shafts, toilets, closets, mechanical equipment rooms, etc., is included in the gross floor area. This area could also include any covered porches or other exterior space intended to be used as part of the building's occupiable space. This dimension is primarily used for the determination of occupant load.

FLOOR AREA, NET. The actual occupied area not including unoccupied accessory areas such as corridors, stairways, toilet rooms, mechanical rooms and closets.

❖ This area is intended to be only the room areas that are used for specific occupancy purposes and does not include circulation areas, such as corridors or stairways, and service and utility spaces, such as toilet rooms and mechanical and electrical equipment rooms.

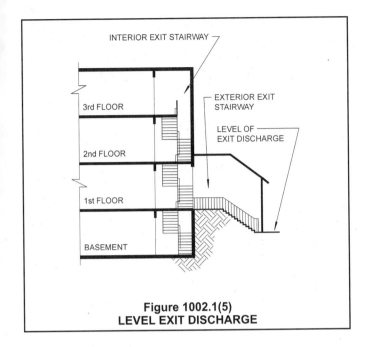

FOLDING AND TELESCOPIC SEATING. Tiered seating facilities having an overall shape and size that are capable of being reduced for purposes of moving or storing.

❖ Bleachers, folding and telescopic seating and grandstands are essentially unique forms of tiered seating. All types are addressed in ICC 300, the new safety standard for these types of seating arrangements. Folding and telescopic seating is commonly used in gymnasiums and sports arenas where the seating can be configured in a variety of ways for various types of events (see Section 1024.1.1).

GRANDSTAND. Tiered seating facilities.

❖ Bleachers, folding and telescopic seating and grandstands are essentially unique forms of tiered seating. All types are addressed in ICC 300, the new safety standard for these types of seating arrangements. Grandstands can be found at a county fairground, along a parade route or within an indoor facility. Examples are sports arenas and public auditoriums, as well as churches and gallery-type lecture halls (see Section 1024.1.1).

GUARD. A building component or a system of building components located at or near the open sides of elevated walking surfaces that minimizes the possibility of a fall from the walking surface to a lower level.

❖ Guards are sometimes mistakenly referred to as "guardrails." Actually, the guard consists of the entire vertical portion of the barrier, not just the top rail (see Section 1012).

HANDRAIL. A horizontal or sloping rail intended for grasping by the hand for guidance or support.

❖ Handrails are provided along walking surfaces that lead from one elevation to another, such as ramps, stairways and landings, and are generally circular in shape. Noncircular shapes could also be acceptable, provided they can be gripped by hand for support and guidance and for checking possible falls on the adjacent walking surface. In addition to being necessary in normal day-to-day use, handrails are especially needed in times of emergency when the pace of egress travel is hurried and the probability for occupant instability while traveling along the sloped or stepped walking surface increases. Handrails are not intended to be used in place of guards to prevent people from falling over the edge.

MEANS OF EGRESS. A continuous and unobstructed path of vertical and horizontal egress travel from any occupied portion of a building or structure to a public way. A means of egress consists of three separate and distinct parts: the exit access, the exit and the exit discharge.

❖ The means of egress is the path traveled by building occupants to leave the building and the site on which it is located. It includes all interior and exterior elements that the occupants must utilize as they make their way from every room and usable space within the building to a public way, such as a street or alley. The elements that make up the means of egress create the lifeline that occupants utilize to travel out of the structure and to a safe distance from the structure. The means of egress provisions of this chapter strive to provide a reasonable level of life safety in every structure. The means of egress provisions are subdivided into three distinct portions (see definitions of "Exit access," "Exit" and "Exit discharge").

NOSING. The leading edge of treads of stairs and of landings at the top of stairway flights.

❖ Limiting the extent of the tread nosings results in a stair that is easy to use. If too large, they are a tripping hazard when walking up a stairway, and reduce the effective tread depth when walking down the stair [see Figures 1009.3(1) and 1009.3(2)].

OCCUPANT LOAD. The number of persons for which the means of egress of a building or portion thereof is designed.

❖ In addition to the code limitation on the maximum occupant load for a space, the code also requires the determination of the occupant load that is to be utilized for the design of the means of egress system. This occupant load is also utilized to determine the required number of plumbing fixtures (see Chapter 29).

PANIC HARDWARE. A door-latching assembly incorporating a device that releases the latch upon the application of a force in the direction of egress travel.

❖ Panic hardware is commonly used in educational and assembly-type spaces where the number of occupants who would use a doorway during a short time frame in an emergency is high in relation to an occupancy with a less dense occupant load, such as an office building. The hardware is required so that the door can be easily opened during an emergency.

PUBLIC WAY. A street, alley or other parcel of land open to the outside air leading to a street, that has been deeded, dedicated or otherwise permanently appropriated to the public for public use and which has a clear width and height of not less than 10 feet (3048 mm).

❖ The public way marks the termination of the exit discharge portion of the means of egress system. It is the final destination for occupants, and is presumed to be safe from the emergency in the structure.

RAMP. A walking surface that has a running slope steeper than one unit vertical in 20 units horizontal (5-percent slope).

❖ This definition is needed to determine the threshold at which the ramp requirements apply to a walking surface. Walking surfaces steeper than specified in the definition are subject to the ramp requirements.

SCISSOR STAIR. Two interlocking stairways providing two separate paths of egress located within one stairwell enclosure.

❖ A scissor or interlocking stair is sometimes used in high-rise buildings or to increase exit capacity of a stairway enclosure. In this configuration, two independent stairways are located within the same exit enclosure and open to one another. If interlocking stairways are separated from each other with appropriate fire barrier assemblies, they are not considered a scissors stairway that may serve as only one exit (see Section 1014.2.1).

SMOKE-PROTECTED ASSEMBLY SEATING. Seating served by means of egress that is not subject to smoke accumulation within or under a structure.

❖ An example of smoke-protected assembly seating is an open outdoor grandstand or an indoor arena with a smoke control system. The code has less stringent requirements for certain aspects of smoke-protected assembly seating than for seating that is not smoke protected, since occupants are subject to less hazard during a fire event. For example, an assembly dead-end aisle is permitted to be longer for a smoke-protected assembly area.

STAIR. A change in elevation, consisting of one or more risers.

❖ All steps, even a single step, are defined as a stair. This makes the stair requirements applicable to all steps unless specifically exempt in the code.

STAIRWAY. One or more flights of stairs, either exterior or interior, with the necessary landings and platforms connecting them, to form a continuous and uninterrupted passage from one level to another.

❖ It is important to note that this definition characterizes a stairway as connecting one level to another. The term "level" is not to be confused with "story." Steps that connect two floor levels, one or both of which are not a "story" of the structure, would be considered a stairway. For example, a set of steps between the basement level in an areaway and the outside ground level would be considered a stairway. A series of steps between the floor of a story and a mezzanine within that story would also be considered a stairway.

STAIRWAY, EXTERIOR. A stairway that is open on at least one side, except for required structural columns, beams, handrails and guards. The adjoining open areas shall be either yards, courts or public ways. The other sides of the exterior stairway need not be open.

❖ This definition is needed since the code requirements for an exterior stairway are different than for an interior stairway (for specific openness requirements, see Section 1022).

STAIRWAY, INTERIOR. A stairway not meeting the definition of an exterior stairway.

❖ This definition is needed since the requirements for an interior stairway are more stringent than those for an exterior stairway (see the definition for "Stairway, exterior").

STAIRWAY, SPIRAL. A stairway having a closed circular form in its plan view with uniform section-shaped treads attached to and radiating from a minimum-diameter supporting column.

❖ Spiral stairways are commonly used where a small number of occupants use the stairway and the floor space for the stair is very limited. Spiral stairways are typically supported by a center pole.

WINDER. A tread with nonparallel edges.

❖ Winders are used as components of stairs that change direction, just as fliers (straight treads) are components in straight stairs. A winder performs the same function as a tread, but its shape allows the additional function of a gradual turning of the stairway direction. The tread depth of a winder at the walk line and the minimum tread depth at the narrow end can control the turn made by each winder.

[B] SECTION 1003 GENERAL MEANS OF EGRESS

1003.1 Applicability. The general requirements specified in Sections 1003 through 1012 shall apply to all three elements of the means of egress system, in addition to those specific requirements for the exit access, the exit and the exit discharge detailed elsewhere in this chapter.

❖ The text of Chapter 10 is subdivided into 27 sections. The requirements in the chapter deal with the three parts of a means of egress system: exit access, exit and exit discharge. Means of egress for existing buildings and means of egress maintenance are also regulated. This section specifies that the requirements of Sections 1003 through 1012 apply to the components of all three parts of the system. For example, the stair tread and riser dimensions in Section 1009 apply to exit access stairs, such as those leading from a small mezzanine, and also apply to enclosed exit stairs according to the vertical exit enclosure requirements in Section 1019.

1003.2 Ceiling height. The means of egress shall have a ceiling height of not less than 7 feet (2134 mm).

Exceptions:

1. Sloped ceilings in accordance with Section 1208.2 of the *International Building Code*.

2. Ceilings of dwelling units and sleeping units within residential occupancies in accordance with Section 1208.2 of the *International Building Code*.
3. Allowable projections in accordance with Section 1003.3.
4. Stair headroom in accordance with Section 1009.2.
5. Door height in accordance with Section 1008.1.1.

❖ Generally, the specified ceiling height is the minimum allowed in any part of the egress path. The exceptions are intended to address conditions where the code allows the ceiling height to be lower than specified in this section. This is also consistent with the headroom requirements at ramps in Section 1010.5.2

The ceiling height for other areas is specified in Section 1208 of the IBC.

1003.3 Protruding objects. Protruding objects shall comply with the requirements of Sections 1003.3.1 through 1003.3.4.

❖ This section identifies the applicable sections that apply to protruding objects and helps to improve awareness of these safety and accessibility-related provisions.

1003.3.1 Headroom. Protruding objects are permitted to extend below the minimum ceiling height required by Section 1003.2 provided a minimum headroom of 80 inches (2032 mm) shall be provided for any walking surface, including walks, corridors, aisles and passageways. Not more than 50 percent of the ceiling area of a means of egress shall be reduced in height by protruding objects.

Exception: Door closers and stops shall not reduce headroom to less than 78 inches (1981 mm).

A barrier shall be provided where the vertical clearance is less than 80 inches (2032 mm) high. The leading edge of such a barrier shall be located 27 inches (686 mm) maximum above the floor.

❖ This provision is applicable to all components of the means of egress. Specifically, the limitations in this section and those in Sections 1003.3.2 and 1003.3.3 provide a reasonable level of safety for those who are preoccupied or not paying attention while walking, as well as for people with impaired vision.

Minimum dimensions for headroom clearance are specified. The minimum headroom clearance over all walking surfaces is required to be maintained at 80 inches (2032 mm). This minimum headroom clearance is consistent with the requirements in Section 1009.2 for stairs and Section 1010.5.2 for ramps. Allowance must be made for door closers and stops, since their design and function necessitates placement within the door opening. The minimum headroom clearance for door closers and stops is allowed to be 78 inches (1981 mm) (see Figure 1003.3.1). The 2-inch (51 mm) projection into the doorway height is reasonable, since these devices are normally mounted away from the center of the door opening, thus minimizing the potential for contact with a person moving through the opening.

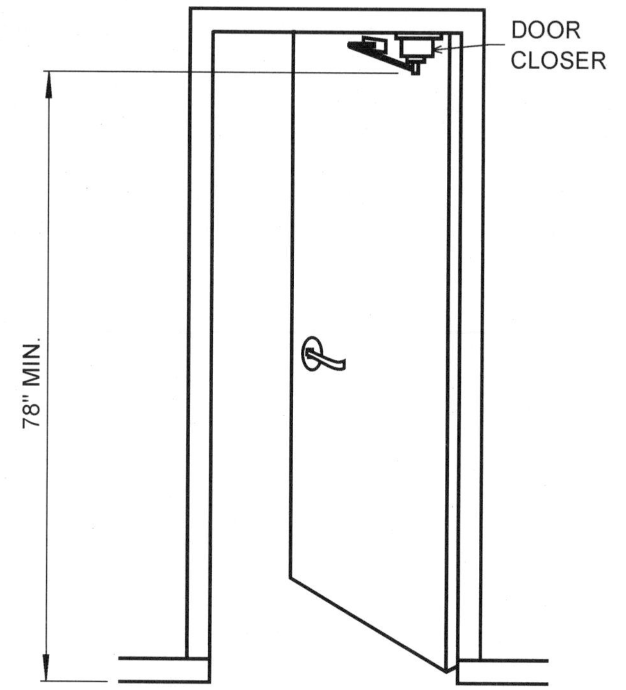

For SI: 1 inch 25.4 mm.

Figure 1003.3.1
DOOR CLOSER HEADROOM PROTRUSIONS FOR WALKING SURFACE

1003.3.2 Free-standing objects. A free-standing object mounted on a post or pylon shall not overhang that post or pylon more than 12 inches (305 mm) where the lowest point of the leading edge is more than 27 inches (686 mm) and less than 80 inches (2032 mm) above the walking surface. Where a sign or other obstruction is mounted between posts or pylons and the clear distance between the posts or pylons is greater than 12 inches (305 mm), the lowest edge of such sign or obstruction shall be 27 inches (685 mm) maximum or 80 inches (2030 mm) minimum above the finish floor or ground.

Exception: This requirement shall not apply to sloping portions of handrails serving stairs and ramps.

❖ Free-standing objects, such as signs mounted on posts, are not permitted to overhang more than 12 inches (305 mm) over the edges of the post where located higher than 27 inches (686 mm) above the walking surface (see Figure 1003.3.2). Since the minimum required height of doorways, stairways and ramps in the means of egress is 80 inches (2032 mm), protruding objects located higher than 80 inches (2032 mm) above the walking surface are not regulated. Protrusions that are located lower than 27 inches (686 mm) above the walking surface are also permitted, since they are more readily detected by a walking cane. The projection of objects located lower than 27 inches (686 mm) is not limited, provided that the minimum required width of the egress element is maintained.

When signs are provided on multiple posts, either the

posts must be located closer then 12 inches (305 mm) apart, or the bottom edge of the sign must be lower than 27 inches (686 mm) so it is within detectable cane range or above 80 inches (2032 mm) so that it is above headroom clearances.

The exception is intended for handrails that occur along a stairway or ramp. The extensions at the top and bottom of stairways and ramps must meet the requirements for protruding objects.

1003.3.3 Horizontal projections. Structural elements, fixtures or furnishings shall not project horizontally from either side more than 4 inches (102 mm) over any walking surface between the heights of 27 inches (686 mm) and 80 inches (2032 mm) above the walking surface.

> **Exception:** Handrails serving stairs and ramps are permitted to protrude 4.5 inches (114 mm) from the wall.

❖ Protruding objects could slow down the egress flow through a passageway and could injure someone hurriedly passing by the protrusion or someone with a visual impairment. Persons with a visual impairment, who use a long cane for guidance, must have sufficient warning of a protruding object. Where protrusions are located higher than 27 inches (686 mm) above the walking surface, the cane will most likely not encounter the protrusion before the person collides with the object.

Additionally, people with poor visual acuity or poor depth perception may have difficulty identifying protruding objects higher than 27 inches (686 mm). Therefore, objects such as lights, signs and door hardware, located between 27 inches (686 mm) and 80 inches (2032 mm) above the walking surface, are not permitted to extend more than 4 inches (102 mm) from each wall (see Figure 1003.3.3).

1003.3.4 Clear width. Protruding objects shall not reduce the minimum clear width of accessible routes as required in Section 1104 of the IBC.

❖ The intent of this section is to limit the projections into an accessible route to those specified in Sections 1003.3 through 1003.3.3. The accessible route requirements are in Section 1104 of the IBC.

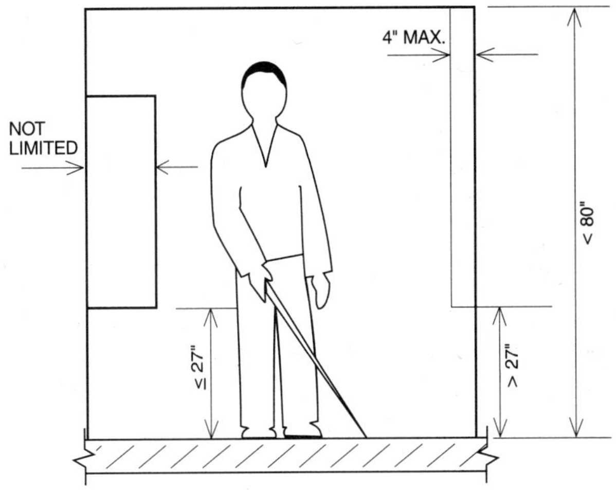

For SI: 1 inch 25.4 mm.

Figure 1003.3.3
HORIZONTAL PROJECTIONS

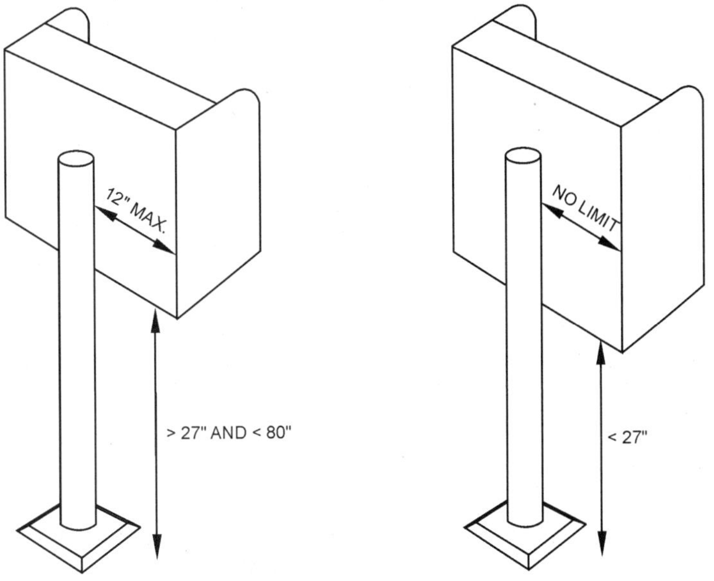

For SI: 1 inch 25.4 mm.

Figure 1003.3.2
FREE-STANDING OBJECT

1003.4 Floor surface. Walking surfaces of the means of egress shall have a slip-resistant surface and be securely attached.

❖ As the pace of exit travel becomes hurried during emergency situations, the probability of slipping on smooth or slick floor surfaces increases. To minimize the hazard, all floor surfaces in the means of egress are required to be slip resistant. The use of hard floor materials with highly polished, glazed, glossy or finely finished surfaces should be avoided.

Field testing and uniform enforcement of the concept of slip resistance are not practical. One method used to establish slip resistance is that the static coefficient of friction between leather [Type 1 (Vegetable Tanned) of Federal Specification KK-L-165C] and the floor surface is greater than 0.5. Laboratory test procedures such as ASTM D 2047 can determine the static coefficient of resistance. Bulletin No. 4 entitled "Surfaces" issued by the U.S. Architectural and Transportation Barriers Compliance Board (ATBCB) contains further information regarding slip resistance.

1003.5 Elevation change. Where changes in elevation of less than 12 inches (305 mm) exist in the means of egress, sloped surfaces shall be used. Where the slope is greater than one unit vertical in 20 units horizontal (5-percent slope), ramps complying with Section 1010 shall be used. Where the difference in elevation is 6 inches (152 mm) or less, the ramp shall be equipped with either handrails or floor finish materials that contrast with adjacent floor finish materials.

Exceptions:

1. A single step with a maximum riser height of 7 inches (178 mm) is permitted for buildings with occupancies in Groups F, H, R-2 and R-3 as applicable in Section 101.2, and Groups S and U at exterior doors not required to be accessible by Chapter 11 of the *International Building Code.*
2. A stair with a single riser or with two risers and a tread is permitted at locations not required to be accessible by Chapter 11 of the *International Building Code,* provided that the risers and treads comply with Section 1009.3, the minimum depth of the tread is 13 inches (330 mm) and at least one handrail complying with Section 1009.11 is provided within 30 inches (762 mm) of the centerline of the normal path of egress travel on the stair.
3. An aisle serving seating that has a difference in elevation less than 12 inches (305 mm) is permitted at locations not required to be accessible by Chapter 11 of the *International Building Code,* provided that the risers and treads comply with Section 1024.11 and the aisle is provided with a handrail complying with Section 1024.13.

Any change in elevation in a corridor serving nonambulatory persons in a Group I-2 occupancy shall be by means of a ramp or sloped walkway.

❖ Minor changes in elevation, such as a single step that is located in any portion of the means of egress (i.e., exit access, exit or exit discharge), may not be readily apparent during normal use or emergency egress and is considered to present a potential tripping hazard. Where the elevation change is less than 12 inches (305 mm), a ramp is specified to make the transition from higher to lower levels. The ramp is intended to reduce accidental falls associated with tripping hazards and must be constructed in accordance with Section 1010.1. The presence of the ramp must be readily apparent from the directions from which it is approached. Handrails are one method of identifying the change in elevation. In lieu of handrails, the surface of the ramp must be finished with materials that contrast with the surrounding floor surfaces. The walking surface of the ramp should contrast both visually and physically.

Exception 1 allows up to a 7-inch (178 mm) step at exterior doors to avoid blocking the outward swing of the door by a buildup of snow or ice in locations that are not used by the public on a regular basis (see Figure 1003.5). This exception supersedes the general provisions of Section 1008.1.4 and is only applicable in occupancies that have relatively low occupant densities, such as factory and industrial structures. This exception is not applicable to exterior doors that are required to serve as an accessible entrance or that are part of a required accessible route.

Exception 2 allows the transition from higher to lower elevations to be accomplished through the construction of stairs with one or two risers. The pitch of the stairway, however, must be shallower than that required for typical stairways (see Section 1009.3). Since the total elevation change is limited to 12 inches (305 mm), each riser must be approximately 6 inches (152 mm) in height. The presence of the elevation change must be readily apparent from the directions from which it is approached. At least one handrail is required, constructed in accordance with Section 1009.11 and located so as to provide a graspable surface from the normal walking path.

Exception 3 is basically a cross reference to the assembly provisions in Section 1024.

None of the exceptions are permitted in a Group I-2 occupancy (e.g., nursing home, hospital) in areas where nonambulatory persons may need access. The mobility impairments of these individuals require additional consideration.

1003.6 Means of egress continuity. The path of egress travel along a means of egress shall not be interrupted by any building element other than a means of egress component as specified in this chapter. Obstructions shall not be placed in the required width of a means of egress except projections permitted by this chapter. The required capacity of a means of egress system shall not be diminished along the path of egress travel.

❖ The purpose of this section is to require that the entire means of egress path is clear of obstructions that would reduce the egress capacity at any point. The egress path is also not allowed to be reduced in width such that the design occupant load would not be served. Note, however, that the egress path could be reduced in width in situations where it is wider than required by the code based on the occupant load. For example, if the required width of a corridor were 52 inches (1321 mm), based on the number of occupants using the corridor, and the corridor provided was 96 inches (2438 mm), in width, the corridor would be allowed to be reduced to the required width of 52 inches (1321 mm) since that width would still serve the number of occupants required by the code.

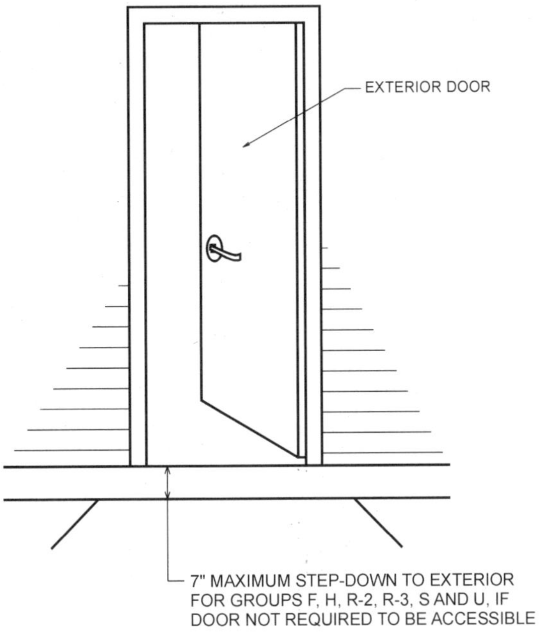

For SI: 1 inch = 25.4 mm.

Figure 1003.5
ELEVATION CHANGE

1003.7 Elevators, escalators and moving walks. Elevators, escalators and moving walks shall not be used as a component of a required means of egress from any other part of the building.

Exception: Elevators used as an accessible means of egress in accordance with Section 1007.4.

❖ Generally, the code does not allow elevators, escalators and moving sidewalks to be used as a required means of egress. Elevators are allowed to be part of an accessible means of egress, provided they comply with the requirements of Section 1007.4. The concern is that escalators and moving sidewalks may not provide a safe and reliable means of egress that is available for use at all times.

[B] SECTION 1004
OCCUPANT LOAD

1004.1 Design occupant load. In determining means of egress requirements, the number of occupants for whom means of egress facilities shall be provided shall be established by the largest number computed in accordance with Sections 1004.1.1 through 1004.1.3.

❖ The design occupant load is the number of people that are intended to occupy a building, or portion thereof, at any one time; consequently, the number for which the means of egress is to be designed. It is the largest number derived by the application of Sections 1004.1 through 1004.1.3. There is a limit to the density of occupants permitted in an area to enable a reasonable amount of freedom of movement (see Section 1004.2). The design occupant load is also utilized to determine the required plumbing fixture count (see the commentary to Chapter 29 of the IBC).

1004.1.1 Actual number. The actual number of occupants for whom each occupied space, floor or building is designed.

❖ The number of occupants that will occupy a space is the actual number and is only limited by Section 1004.2. If the construction documents indicate that the actual occupant load of a space exceeds that determined by Sections 1004.1.2 and 1004.1.3, then the actual number is to be used as the design occupant load of that space. Where the actual number is less than the occupant load determined in accordance with Section 1004.1.2 or 1004.1.3, the largest number must be used in the egress design. For example, if a proposed conference room has a calculated occupant load of 55—using 15 net square (1.4 m^2) feet per person for assembly without fixed seats, unconcentrated tables and chairs (see Table 1004.1.2) —but the owner indicates that the actual number of occupants will not exceed 25, the design occupant load of the room is 55. Therefore, in accordance with Table 1014.1, at least two means of egress must be provided from the conference room. Conversely, if the actual occupant load planned for is 65, the design occupant load is then 65.

1004.1.2 Number by Table 1004.1.2. The number of occupants computed at the rate of one occupant per unit of area as prescribed in Table 1004.1.2.

❖ This number reflects common and traditional occupant density based on empirical data for the density of similar spaces. The number determined using the occupant load rates in Table 1004.1.2 generally establishes the minimum occupant load for which the egress facilities of the rooms, spaces and building must be designed.

It is difficult to predict the many conditions by which a space within a building will be occupied over time. An assembly banquet room in a hotel, for example, could be arranged with rows of chairs to host a business seminar one day and with mixed tables and chairs to host a dinner reception the next day. In some instances, the room will be arranged with no tables and very few chairs to accommodate primarily standing occupants. In such a situation, the egress facilities must safely accommodate the maximum number of persons permitted to occupy the space. When determining the occupant load of this type of occupancy, the various arrangements (e.g., tables and chairs, chairs only, standing space) should be recognized. The worst-case scenario should be utilized to determine the requirements for means of egress elements.

Note that while some of the values in the table utilize the net floor area, most utilize the gross floor area. See the commentary to Table 1004.1.2 for additional discussion and examples.

The occupant load determined in accordance with this section is the minimum occupant load upon which means of egress requirements are to be based. This limitation is true regardless of any indication by the owner that the space will be occupied by fewer people (see commentary, Section 1004.1.1).

Some occupancies may not typically contain an occupant load totally consistent with the occupant load density factors of Table 1004.1.2. Any special considerations for such unique uses must be documented and justified. Additionally, the owner must be aware that such special considerations will impact the future use of the building with respect to the means of egress and other protection features.

TABLE 1004.1.2
MAXIMUM FLOOR AREA ALLOWANCES PER OCCUPANT

OCCUPANCY	FLOOR AREA IN SQ. FT. PER OCCUPANT
Agricultural building	300 gross
Aircraft hangars	500 gross
Airport terminal Baggage claim Baggage handling Concourse Waiting areas	 20 gross 300 gross 100 gross 15 gross
Assembly Gaming floors (keno, slots, etc.)	11 gross
Assembly with fixed seats	See Section 1004.7
Assembly without fixed seats Concentrated (chairs only—not fixed) Standing space Unconcentrated (tables and chairs)	 7 net 5 net 15 net
Bowling centers, allow 5 persons for each lane including 15 feet of runway, and for additional areas	7 net
Business areas	100 gross
Courtrooms—other than fixed seating areas	40 net
Dormitories	50 gross
Educational Classroom area Shops and other vocational room areas	 20 net 50 net
Exercise rooms	50 gross
H-5 Fabrication and manufacturing areas	200 gross
Industrial areas	100 gross

(continued)

TABLE 1004.1.2—continued
MAXIMUM FLOOR AREA ALLOWANCES PER OCCUPANT

OCCUPANCY	FLOOR AREA IN SQ. FT. PER OCCUPANT
Institutional areas Inpatient treatment areas Outpatient areas Sleeping areas	 240 gross 100 gross 120 gross
Kitchens, commercial	200 gross
Library Reading rooms Stack area	 50 net 100 gross
Locker rooms	50 gross
Mercantile Areas on other floors Basement and grade floor areas Storage, stock, shipping areas	 60 gross 30 gross 300 gross
Parking garages	200 gross
Residential	200 gross
Skating rinks, swimming pools Rink and pool Decks	 50 gross 15 gross
Stages and platforms	15 net
Accessory storage areas, mechanical equipment room	300 gross
Warehouses	500 gross

For SI: 1 square foot = 0.0929 m^2.

❖ The table presents the maximum floor area allowance per occupant based on studies and counts of the number of occupants in typical buildings. The use of this table, then, results in the minimum occupant load for which rooms, spaces and the building must be designed. While an assumed normal occupancy may be viewed as somewhat less than that determined by the use of the table factors, such a normal occupant load is not necessarily an appropriate design criterion. The greatest hazard to the occupants occurs when an unusually large crowd is present. The code does not limit the occupant load density of an area, except as provided for in Section 1004.2, but once the occupant load is established, the means of egress must be designed for at least that capacity. If it is intended that the occupant load will exceed that calculated in accordance with the table, the occupant load is to be based on the estimated actual number of people in accordance with Section 1004.1.1. Table 1004.1.2 establishes minimum occupant densities based on the occupancy (not group classification) of the space. Therefore, the occupant load of the office or business areas in a storage warehouse or nightclub is to be determined using the occupant load factor most appropriate to that space—one person for each 100 square feet (9 m^2) of gross floor area.

The use of net and gross floor areas as defined in Section 1002.1 is intended to provide a refinement in the occupant load determination. The gross floor area technique applied to a building only allows the deduction of the plan area of the exterior walls, vent shafts and interior courts from the plan area of the building.

The net floor area permits the exclusion of certain spaces that would be included in the gross floor area. The net floor area is intended to apply to the actual occupied floor areas. The area used for permanent building components, such as shafts, fixed equipment, thicknesses of walls, corridors, stairways, toilet rooms, mechanical rooms and closets, is not included in net floor area. For example, consider a restaurant dining area with dimensions measured from the inside of the enclosing walls of 80 feet by 60 feet (24 384 mm by 18 288 mm) (see Figure 1004.1.2). Within the restaurant area is a 6-inch (152 mm) privacy wall running the length of the room [80 feet by 0.5 feet = 40 square feet (3.7 m^2)], a fireplace [40 square feet (3.7 m^2)] and a cloak room [60 square feet (5.6 m^2)]. Each of these areas is deducted from the restaurant area, resulting in a net floor area of 4,660 square feet (433 m^2). Since the restaurant intends to have unconcentrated seating that involves loose tables and chairs, the resulting occupant load is 311 persons (4,660 divided by 15). As the definition of “Floor area, net” indicates, certain spaces are to be excluded from the gross floor area to derive the net floor area. The key in this definition is that the net floor area is to include the actual occupied area and does not include spaces uncharacteristic of that occupancy.

In determining the occupant load of a building with mixed groups, each floor area of a single occupancy must be separately analyzed, such as required by Section 1004.9. The occupant load of the business portion of an office/warehouse building is determined at a rate of one person for each 100 square feet (9 m^2) of office space, whereas the occupant load of the warehouse portion is determined at the rate of one person for each 300 square feet (28 m^2).

If a specific type of facility is not found in the table, the occupancy it most closely resembles should be utilized. For example, a training room in a business office may utilize the 20 square foot (2 m^2) net established for educational classroom areas.

Table 1004.1.2, in accordance with Section 1004.1.2, presents a method of determining the absolute base minimum occupant load of a space that the means of egress is to accommodate.

In addition to the table, it should be noted that Section 402 of the IBC contains the basis for calculating the occupant load of a covered mall building. However, Table 1004.1.2 should be used for determining the occupant load of each anchor store.

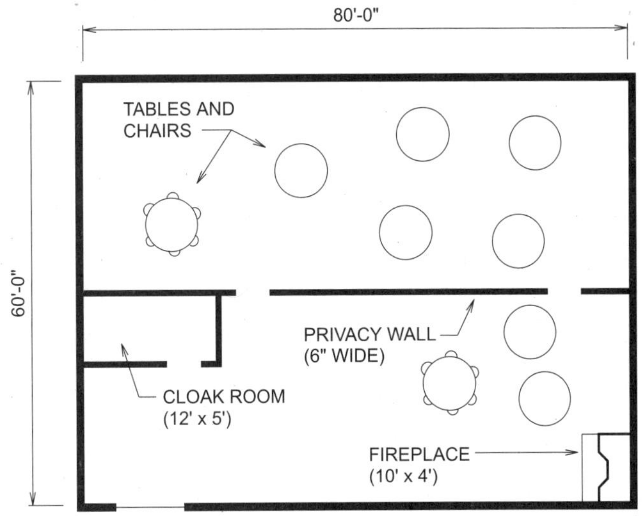

80' x 60' = 4,800 SQ.FT.

PRIVACY WALL:	40 SQ.FT.
FIREPLACE:	40 SQ.FT.
CLOAK ROOM:	60 SQ.FT.
TOTAL:	140 SQ.FT.

(TOTAL AREA WITHIN WALLS) - (EXCLUDED ITEMS) = (NET FLOOR AREA)
4,800 SQ.FT. - 140 SQ.FT. = 4,660 SQ.FT.

(NET FLOOR AREA)/[TABLE 1004.1.2 VALUE] = (OCCUPANT LOAD)
4,660 SQ.FT./15 SQ.FT. PER OCCUPANT = 311 OCCUPANTS

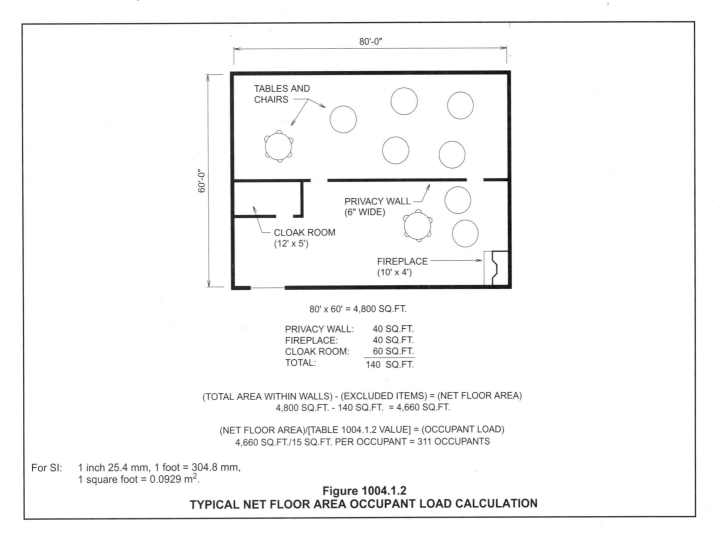

For SI: 1 inch 25.4 mm, 1 foot = 304.8 mm,
1 square foot = 0.0929 m^2.

Figure 1004.1.2
TYPICAL NET FLOOR AREA OCCUPANT LOAD CALCULATION

1004.1.3 Number by combination. Where occupants from accessory spaces egress through a primary area, the calculated occupant load for the primary space shall include the total occupant load of the primary space plus the number of occupants egressing through it from the accessory space.

❖ This section provides a method by which the occupant load of adjacent areas of a building is calculated. The resulting occupant load is what must be considered. For example, the means of egress from a lobby must be sized for the cumulative occupant load of the adjacent office spaces if the occupants must travel through the lobby to reach an exit. Likewise, if an adjacent room has an egress route independent of the lobby, the occupant load of that room would not be combined with the occupant loads of the other rooms that pass through that lobby. If a portion of the adjacent room's occupant load is to travel through the lobby, only that portion would be combined with the lobby occupant load for the design of the means of egress from the lobby (see Figure 1004.1.3). This is particularly important in determining the capacity and the number of means of egress.

1004.2 Increased occupant load. The occupant load permitted in any building or portion thereof is permitted to be increased from that number established for the occupancies in Table 1004.1.2 provided that all other requirements of the code are also met based on such modified number and the occupant load shall not exceed one occupant per 5 square feet (0.47 m^2) of occupiable floor space. Where required by the fire code official, an approved aisle, seating or fixed equipment diagram substantiating any increase in occupant load shall be submitted. Where required by the fire code official, such diagram shall be posted.

❖ An increased occupant load is permitted above that developed by using Table 1004.1.2; for example, utilizing the actual occupant load alternative in Section 1004.1.1. However, if the occupant load exceeds that which is determined in accordance with Section 1004.1.2, the fire code official has the authority to require aisle, seating and equipment diagrams to confirm that all occupants have access to an exit, the exits provide sufficient capacity for all occupants and compliance with this section is attained.

1004.3 Posting of occupant load. Every room or space that is an assembly occupancy shall have the occupant load of the room or space posted in a conspicuous place, near the main exit or exit access doorway from the room or space. Posted signs shall be of an approved legible permanent design and shall be maintained by the owner or authorized agent.

❖ Each room or space used for an assembly occupancy is required to display the approved occupant load. The placard must be posted in a visible location (near the main entrance). See Figure 1004.3 for an example of an occupant load limit sign.

The posting is required to provide a means by which to determine that the maximum approved occupant load is not exceeded. This permanent and readily visible sign provides a constant reminder to building personnel and

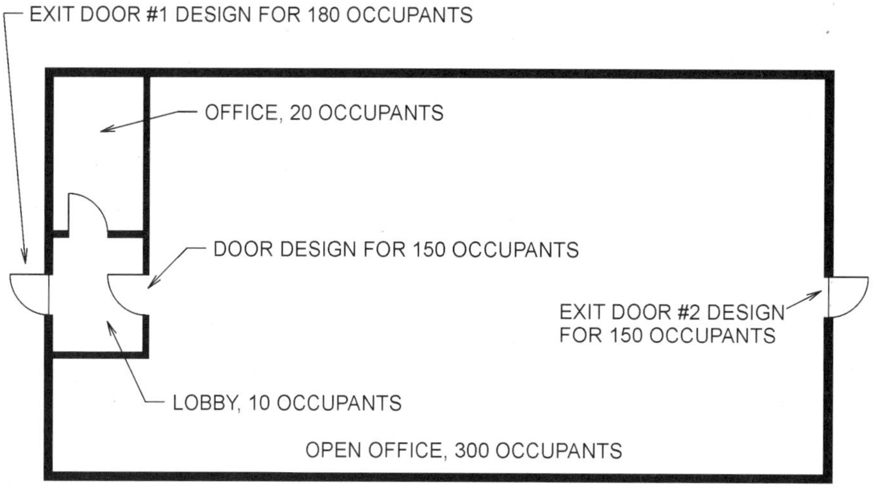

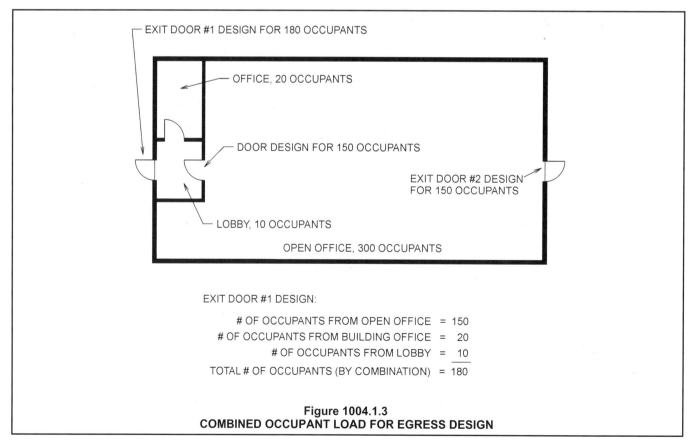

EXIT DOOR #1 DESIGN:

# OF OCCUPANTS FROM OPEN OFFICE	=	150
# OF OCCUPANTS FROM BUILDING OFFICE	=	20
# OF OCCUPANTS FROM LOBBY	=	10
TOTAL # OF OCCUPANTS (BY COMBINATION)	=	180

Figure 1004.1.3
COMBINED OCCUPANT LOAD FOR EGRESS DESIGN

is a reference for fire code officials during periodic inspections.

While the composition and organization of information in the sign are not specified, information must be recorded in a permanent manner. This means that a sign with changeable numbers would not be acceptable.

Figure 1004.3
EXAMPLE OF OCCUPANT LOAD LIMIT SIGN

1004.4 Exiting from multiple levels. Where exits serve more than one floor, only the occupant load of each floor considered individually shall be used in computing the required capacity of the exits at that floor, provided that the exit capacity shall not decrease in the direction of egress travel.

❖ The sum total capacity of the exits that serve a floor is not to be less than the occupant load of the floor as determined by Section 1004.1. If an exit such as a stairway also serves a second floor, and the required capacity of the exit serving the occupants of the second floor is greater than the first floor, the greater capacity would govern the egress components that the occupants of the floors share. For example, if an exit stairway serves two floors, with occupant loads of 300 on the lower floor and 500 on the upper floor, assuming that two stairways serve each floor, the two stairways would be designed for a capacity of 250 people each, using the upper-floor occupant load of 500 as the basis of determination. Note that the doors to the stairways on the lower floor would be designed for a capacity of 150 and the doors to the stairways on the upper floor would be designed for a capacity of 250. Reversing these two floors would result in the portion of the stairways that serves the upper floor to be designed for a capacity of 150 and the stairways that serve the lower floor to be designed for 250. Requiring the egress component to be designed for the largest tributary occupant load accommodates the worst-case situation.

Also note that the capacity of the exits is based on the occupant load of one floor. The occupant loads are not combined with other floors for the exit design. It is assumed that the peak demand or flow of occupants from more than one floor level at a common point in the means of egress will not occur simultaneously, except as provided for in Sections 1004.5 and 1004.6.

1004.5 Egress convergence. Where means of egress from floors above and below converge at an intermediate level, the capacity of the means of egress from the point of convergence shall not be less than the sum of the two floors.

❖ Convergence of occupants can occur whenever the occupants of one floor travel down and occupants of a lower floor travel up and meet at a common, intermediate egress component. The intermediate component may or may not be another occupiable floor and, most often, is an exit discharge door [see Figures 1004.5(1) and 1004.5(2)].

The entire premise of egress convergence is based on the assumption of simultaneous notification (i.e., all occupants of all floors begin moving toward the exits at the same time). As illustrated in Figure 1004.5(3), the occupants of the first floor will have exited the building by the time the occupants of the second floor have reached the exit discharge door. However, as illustrated in Figure 1004.5(1), the occupants of a basement will reach the discharge door simultaneously with the second-floor occupants, thereby creating the need for sizing the components for a larger combined occupant load.

An egress convergence situation can also be created when an intermediate floor level is not present, as illustrated in Figure 1004.5(2). Again, under the assumption of simultaneous notification, occupants of both floors would reach the exit discharge door at approximately the same time, invoking the requirements for a larger egress capacity.

1004.6 Mezzanine levels. The occupant load of a mezzanine level with egress onto a room or area below shall be added to that room or area's occupant load, and the capacity of the exits shall be designed for the total occupant load thus established.

❖ The egress requirements for mezzanines are similar to those addressed in Section 1004.1.3, versus the requirements for exiting from multiple levels in Section 1004.4. That is, the portion of the mezzanine occupant load that discharges to the floor below is to be added to the occupant load of the space on the floor below. The sizing and number of the egress components must reflect this combined occupant load. This does not apply to the means of egress from a mezzanine that does not require travel through another level (i.e., an exit stairway serving the mezzanine). Section 505 contains additional criteria for the means of egress from mezzanines.

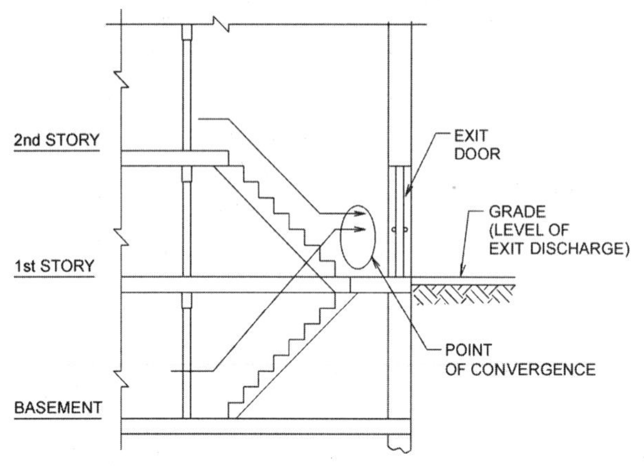

Figure 1004.5(1)
EGRESS CONVERGENCE 1st STORY

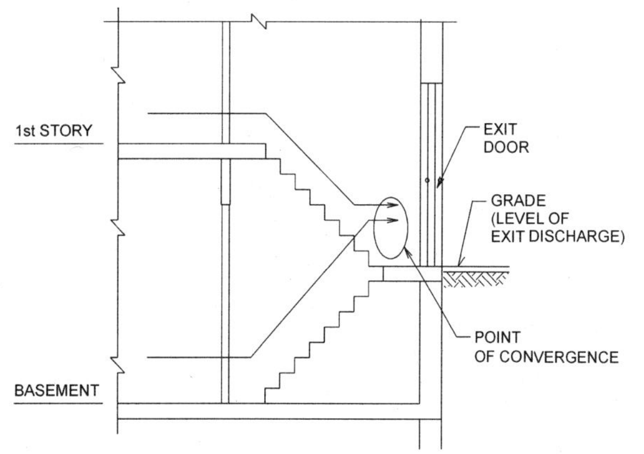

Figure 1004.5(2)
EGRESS CONVERGENCE AT INTERMEDIATE LEVEL

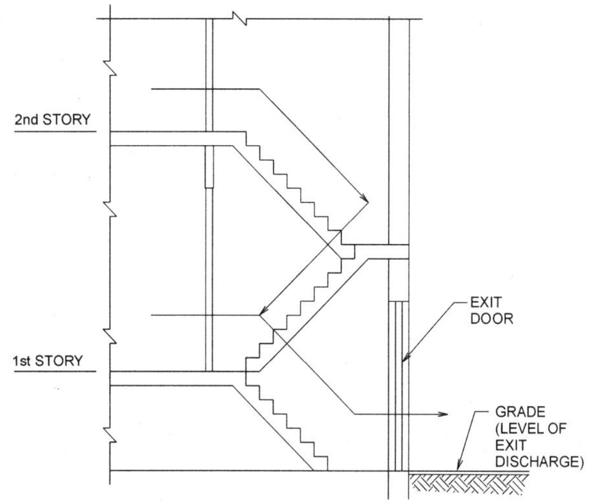

Figure 1004.5(3)
NO EGRESS CONVERGENCE

1004.7 Fixed seating. For areas having fixed seats and aisles, the occupant load shall be determined by the number of fixed seats installed therein.

For areas having fixed seating without dividing arms, the occupant load shall not be less than the number of seats based on one person for each 18 inches (457 mm) of seating length.

The occupant load of seating booths shall be based on one person for each 24 inches (610 mm) of booth seat length measured at the backrest of the seating booth.

❖ The occupant load in an area with fixed seats is readily determined. In spaces with a combination of fixed and loose seating, the occupant load is determined by a combination of the occupant density number from Table 1004.1.2 and a count of the fixed seats.

For bleachers, booths and other seating facilities without dividing arms, the occupant load is simply based on the number of people that can be accommodated in the length of the seat. Measured at the hips, an average person occupies about 18 inches (457 mm) on a bench. In a booth, additional space is necessary for "elbow room" while eating. In a circular or curved booth or bench, the measurement should be taken just a few inches from the back of the seat, which is where a person's hips would be located (see Figure 1004.7).

1004.8 Outdoor areas. Yards, patios, courts and similar outdoor areas accessible to and usable by the building occupants shall be provided with means of egress as required by this chapter. The occupant load of such outdoor areas shall be assigned by the fire code official in accordance with the anticipated use. Where outdoor areas are to be used by persons in addition to the occupants of the building, and the path of egress travel from the outdoor areas passes through the building, means of egress requirements for the building shall be based on the sum of the occupant loads of the building plus the outdoor areas.

Exceptions:

1. Outdoor areas used exclusively for service of the building need only have one means of egress.
2. Both outdoor areas associated with Group R-3 and individual dwelling units of Group R-2, as applicable in Section 1001.1.

❖ This section addresses the means of egress of outdoor areas such as yards, patios and courts. The primary concern is for outdoor areas that are used for functions that would include occupants other than the building occupants. The egress from the outdoor area is back through the building to reach the exit discharge. An example is an interior court of an office building where assembly functions are held during normal business hours for persons other than the building occupants. Where the occupants have to egress from the interior court back through the building, the building's egress system is to be designed for the building occupants plus the assembly occupants from the interior court.

The occupant load is to be assigned by the fire code official based on use. It is suggested that the design occupant load be determined in accordance with Section 1004.1.2.

The exceptions describe conditions where the combination of both occupant loads is not a concern.

1004.9 Multiple occupancies. Where a building contains two or more occupancies, the means of egress requirements shall apply to each portion of the building based on the occupancy of that space. Where two or more occupancies utilize portions of the same means of egress system, those egress components shall meet the more stringent requirements of all occupancies that are served.

❖ Since the means of egress systems are designed for the specific occupancy of a space, the provisions of this chapter are to be applied based on the actual occupancy conditions of the space served.

For example, a hospital is classified as Group I-2 and normally includes the associated administrative or business functions found in the same building. Chapter 3 permits the entire building to be constructed to the more restrictive provisions for Group I-2; however, each area of the building need only have the means of egress designed in accordance with the actual occupancy conditions, such as Groups I-2 and B. If the corridor serves only the occupants in the business use (i.e., administrative staff) and is not intended to serve as a required means of egress for patients, the corridor need only be 36 or 44 inches (914 or 1118 mm) in width, depending on the occupant load.

Where the corridor is used by both Group I-2 and B occupancies, it must meet the most stringent requirement. For example, if a corridor in the business area is also used for the movement of beds (i.e., exit access from a patient care area), it would need to be a minimum of 96 inches (2438 mm) in clear width.

[B] SECTION 1005
EGRESS WIDTH

1005.1 Minimum required egress width. The means of egress width shall not be less than required by this section. The total width of means of egress in inches (mm) shall not be less than the total occupant load served by the means of egress multiplied by the factors in Table 1005.1 and not less than specified elsewhere in this code. Multiple means of egress shall be sized such that the loss of any one means of egress shall not reduce the available capacity to less than 50 percent of the required capacity. The maximum capacity required from any story of a building shall be maintained to the termination of the means of egress.

Exception: Means of egress complying with Section 1024.

❖ The sum of the capacities of the individual means of egress components that serve each space must equal or exceed the occupant load of that space. For example, the two exit access doorways from a room with an occupant load of 300 would each have a required capacity of no less than 150. Likewise, the two exits from a story of a building with a total occupant load of 450 would each have a required capacity of no less than 225. The code does require that when multiple means of egress are required, the loss of any one path would not reduce the available capacity to less than 50 percent. This requirement does not, however, require that the capacities be equally distributed when more than two means of egress are provided. An egress design with a dramatic imbalance of egress component capacities relative to occupant load distribution should be reviewed closely to avoid a needless delay in egressing a floor or area. The balancing of the means of egress

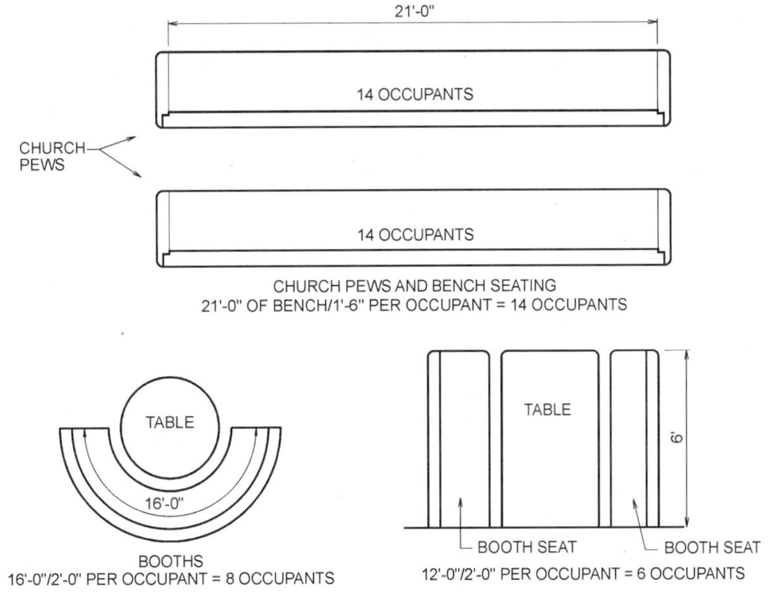

For SI: 1 inch 25.4 mm, 1 foot = 304.8 mm.

Figure 1004.7
TYPICAL FIXED SEATING

components, in accordance with the distribution of the occupant load, is reasonable and, in some cases, necessary for facilities having mixed occupancies with dramatically different occupant loads.

The code requires the utilization of two methods to determine the minimum width of egress components. While this section provides a methodology for determining required widths based on the design occupant load, calculated in accordance with Section 1004.1, other sections provide minimum widths of various components. The actual width that is provided is to be the larger of the two widths. Also note that the width of stairways that are part of the accessible means of egress is further regulated by Section 1007.3. Obviously, an egress door opening with a clear width of 12 inches (305 mm) would be an impossible egress component, although in accordance with Table 1005.1, the capacity of such an opening can be calculated to be 60 persons [60 × 0.2 = 12 inches (305 mm)]. Section 1008.1.1, however, specifies that the minimum required clear width of each door opening is 32 inches (813 mm) with certain exceptions.

The 50 percent minimum of the required capacity results in a fairly uniform distribution of egress paths. For example, the doors for large occupancy rooms are to be located to meet this requirement.

The requirement that the maximum capacity from any floor is to be provided results in an egress width that is adequate to the exit discharge.

The purpose of the exception is to require assembly seating to comply with Section 1024.

TABLE 1005.1
EGRESS WIDTH PER OCCUPANT SERVED

OCCUPANCY	WITHOUT SPRINKLER SYSTEM		WITH SPRINKLER SYSTEM[a]	
	Stairways (inches per occupant)	Other egress components (inches per occupant)	Stairways (inches per occupant)	Other egress components (inches per occupant)
Occupancies other than those listed below	0.3	0.2	0.2	0.15
Hazardous: H-1, H-2, H-3 and H-4	0.7	0.4	0.3	0.2
Institutional: I-2	Not Applicable	Not Applicable	0.3	0.2

For SI: 1 inch = 25.4 mm.

a. Buildings equipped throughout with an automatic sprinkler system in accordance with Section 903.3.1.1or 903.3.1.2.

❖ This table establishes the necessary width of each egress component on a per-occupant basis. When the required occupant capacity of an egress component is determined, multiplication by the appropriate factor from Table 1005.1 results in the required clear width of the component in inches, based on capacity. Similarly, if the clear width of a component is known, division by the appropriate factor from Table 1005.1 results in the permitted capacity of that component.

The following typical calculations illustrate the various uses of Table 1005.1:

1. Determine the minimum required width of a stairway to accommodate 175 occupants in a two-story office building (Group B):
   a. Without an automatic sprinkler system: 175 × 0.3 (from Table 1005.1) = $52^1/_2$ inches (1334 mm), minimum; or
   b. With an automatic sprinkler system: 175 × 0.2 (from Table 1005.1) = 35 inches (889 mm), minimum.

      Section 1009.1, however, prescribes that the width of an interior stairway cannot be less than 44 inches (1118 mm); therefore, 44 inches (1118 mm) is the governing dimension when an automatic sprinkler system is installed. When a sprinkler system is not installed, the capacity criteria are more restrictive and, therefore, the minimum required width is $52^1/_2$ inches (1334 mm).
2. Determine the minimum width of an exit access doorway required to accommodate 240 occupants in a new six-story hotel (Group R-1) (assume the occupant load on the floor is larger and that the capacity of this egress path is being designed to serve 240 of the occupants):
   a. Since Section 903.2.7 requires this building to be equipped throughout with an automatic fire sprinkler system, the proper width per occupant is 0.15 (see Table 1005.1). The required width is 240 × 0.15 = 36 inches (914 mm), minimum.

      Therefore, the egress doorway must be at least 36 inches (914 mm) in clear width [not merely 32 inches (813 mm), as required by Section 1008.1.1]:
   b. It is important to remember that the total width of egress from an area may be distributed among the available means of egress. If Example 2 had a total story occupant load of 240 and two exits available, then two doors that met the 32-inch (813 mm) requirement would be acceptable.

      240 occupants × 0.15 = 36 inches (914 mm) of total width. Since the loss of any one path may not reduce the available capacity below 50 percent, each doorway would need to provide a minimum of 18 inches (457 mm) of that width. However, Section 1008.1.1 would then require each door to be sized to provide 32 inches (813 mm) of clear width.
3. What is the egress capacity of a pair of entrance/exit doors that provide a clear width of 64 inches (1626 mm) in an existing retail store?
   a. If an unsprinklered building or only partially sprinklered: egress capacity = 64 inches × 0.2 = 320 occupants.

b. If a building is sprinklered throughout: egress capacity = 64 inches × 0.15 = 426.67 occupants. (Note that where the clear width provides capacity for a fraction of an occupant, the capacity should be rounded down to the nearest whole number. In this case, the doors can accommodate 426 occupants.)

The examples above demonstrate the use of Table 1005.1 in conjunction with other applicable sections of the code that require minimum widths of various components of the means of egress.

The traditional unit of measurement of egress capacity was based on a unit exit width, which was to simulate the body ellipse with a basic dimensional width of 22 inches (559 mm)—approximately the shoulder width of an average adult male. This unit exit width was combined with assumed egress movement (such as single file or staggered file) to result in an egress capacity per unit exit width for various occupancies, as shown in Table 1005.1. This assumption simplifies the dynamic egress process, since contemporary studies have indicated that people do not egress in such precise and predictable movements. As traditionally used in the codes, the method of determining capacity per unit of clear width implies a higher level of accuracy than can realistically be achieved. The resulting factors in Table 1005.1 preserve the features of past practices, while providing a more straightforward method of determining egress capacity.

The considerations embodied in Table 1005.1 include the nature and conditions of the occupants, the fire hazard of the occupancy, the efficiency of the column of occupants to utilize the component, the physical dimensioning of the components and the presence of an automatic sprinkler system. The presence of an automatic sprinkler system is deemed to control or limit the fire threat to the occupants, thus permitting greater time for the occupants to egress, as well as reducing the need for total and immediate evacuation. The footnote reference to Section 903.3.1.1 or 903.3.1.2 indicates that the sprinkler system is to be designed and installed in accordance with either NFPA 13 or NFPA 13R.

Although an automatic sprinkler system is required for all Group H fire areas, egress width per occupant is shown for such occupancies without an automatic sprinkler system. These factors should be applied in special cases where the use of a sprinkler system is detrimental or of no value to the protection of the hazardous condition and can also be used to establish a recommended minimum safety guideline for existing nonsuppressed Group H occupancies or where the sprinkler system is allowed to be omitted through the appeals process.

The increased factors for buildings of Group I-2 occupancies recognize the decreased mobility of occupants in these buildings. A number is not given for Group I-2 buildings without sprinklers since all Group I-2 facilities are required to be sprinklered. A similar increase is not required in Group I-3 buildings since the population, while not capable of self-preservation, is typically a mobile group and once locks are released, are as capable as the general population with respect to using the egress components.

1005.2 Door encroachment. Doors opening into the path of egress travel shall not reduce the required width to less than one-half during the course of the swing. When fully open, the door shall not project more than 7 inches (178 mm) into the required width.

Exception: The restrictions on a door swing shall not apply to doors within individual dwelling units and sleeping units of Group R-2 and dwelling units of Group R-3.

❖ Projections or restrictions in the required width can impede and restrict occupant travel, causing egress to occur less efficiently than contemplated. The swing of a door, such as from a room into a corridor, is a permitted projection.

However, the arc scribed by the door's outside edge cannot project into more than one-half of the required corridor width. When opened to its fullest extent, the

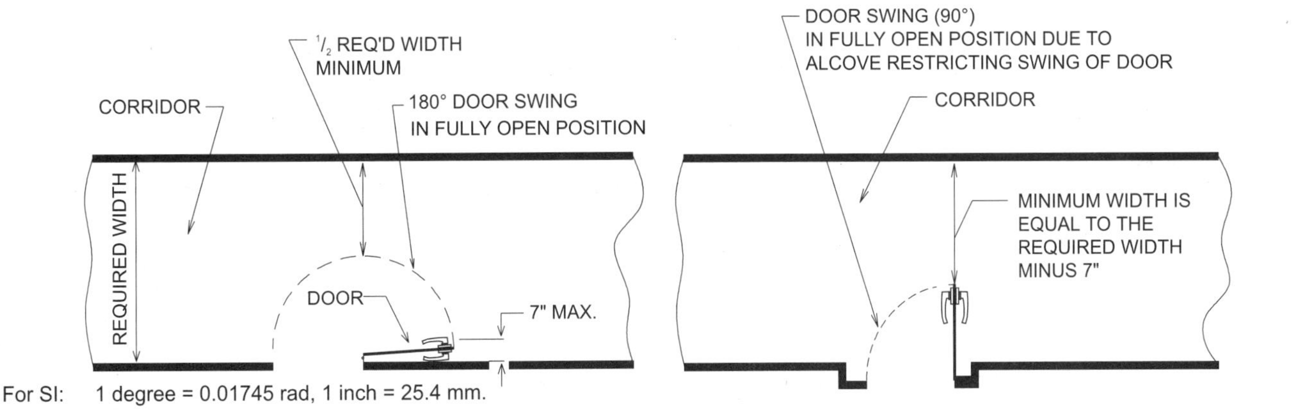

For SI: 1 degree = 0.01745 rad, 1 inch = 25.4 mm.

Figure 1005.2
TYPICAL RESTRICTIONS INTO PASSAGEWAY, AISLE AND CORRIDOR WIDTH

door cannot project more than 7 inches (178 mm) into the required width, which is approximately the sum dimension of a door leaf thickness and protrusion of operational hardware on each side of the leaf as shown in Figure 1005.2. These projections are permitted because they are considered to be temporary and do not significantly impede the flow. Occupants will compensate for the projection by a reduction in the natural cushion they retain between themselves and a boundary, known as the edge effect.

The door swing restrictions do not apply within dwelling units since the occupant load is very low.

[B] SECTION 1006 MEANS OF EGRESS ILLUMINATION

1006.1 Illumination required. The means of egress, including the exit discharge, shall be illuminated at all times the building space served by the means of egress is occupied.

Exceptions:

1. Occupancies in Group U.
2. Aisle accessways in Group A.
3. Dwelling units and sleeping units in Groups R-1, R-2 and R-3.
4. Sleeping units of Group I occupancies.

❖ All means of egress must be continuously illuminated by artificial lighting during the entire time a building is occupied, so that the paths of exit travel are always visible and available for evacuation of the occupants during emergencies. The code makes a special point of noting that the exit discharge must also be provided with adequate illumination so that occupants can safely find the public way should the emergency occur at night.

The exceptions are for occupancies where the constant illumination of the means of egress would interfere with the use of space, such as sleeping areas or theater aisles during a performance.

Bear in mind that means of egress lighting is not emergency lighting. For emergency lighting requirements, see Sections 1006.3 and 1006.4.

1006.2 Illumination level. The means of egress illumination level shall not be less than 1 foot-candle (11 lux) at the floor level.

Exception: For auditoriums, theaters, concert or opera halls and similar assembly occupancies, the illumination at the floor level is permitted to be reduced during performances to not less than 0.2 foot-candle (2.15 lux) provided that the required illumination is automatically restored upon activation of a premise's fire alarm system where such system is provided.

❖ The intensity of floor lighting illuminating the entire means of egress, including open plan spaces, aisles, corridors and exit access passageways, exit stairways, exit doors and places of exit discharge must not be less than 1 foot candle (11 lux). It has been found that this low level of lighting renders enough visibility for the occupants to evacuate the building safely.

It is important to note that this lighting level is measured at the floor in order to make the floor surface visible. Levels of illumination above the floor may be higher or lower, thus allowing lights on the steps to be used rather than general area lights.

The exception addresses occupancies where low light level is needed for the function of the space. The level of intensity of aisle lighting in such spaces may be reduced to 0.2 foot candle (2.15 lux), but only during the time of a performance. This intensity of illumination is sufficient to distinguish the aisles and stairs leading to the egress doors and is not a source of distraction during a performance. It is not the intent of the exception to require a fire alarm system, but to require a connection to the egress lighting where a fire alarm system is provided.

1006.3 Illumination emergency power. The power supply for means of egress illumination shall normally be provided by the premise's electrical supply.

In the event of power supply failure, an emergency electrical system shall automatically illuminate the following areas:

1. Exit access corridors, passageways and aisles in rooms and spaces which require two or more means of egress.
2. Exit access corridors and exit stairways located in buildings required to have two or more exits.
3. Exterior egress components at other than the level of exit discharge until exit discharge is accomplished for buildings required to have two or more exits.
4. Interior exit discharge elements, as permitted in Section 1023.1, in buildings required to have two or more exits.
5. The portion of the exterior exit discharge immediately adjacent to exit discharge doorways in buildings required to have two or more exits.

The emergency power system shall provide power for a duration of not less than 90 minutes and shall consist of storage batteries, unit equipment or an on-site generator. The installation of the emergency power system shall be in accordance with Section 604.

❖ The means of egress must be illuminated, especially in times of emergency when the occupants must have a lighted path of exit travel in order to evacuate the building safely. The code is very specific in the description of the areas that are required to be illuminated by the emergency power system.

The locations include:

1. Dedicated egress routes in larger rooms or spaces. For example, an aisle in an open office

plan but not within individual offices that egress through the open area.

2. Public or common egress elements. Essential portions of the interior egress system, such as stairways and corridors in larger buildings, must be illuminated.
3. Means of egress systems normally used in exit balconies that would provide emergency exit illumination to these components until egress at grade is obtained.
4. Interior exit discharge elements, such as fully sprinklered lobbies and vestibules where stairways discharge into these exit elements.
5. Exterior portions of the exit discharge. Note that only the portion of the exterior discharge that is immediately adjacent to the building exit discharge door is required to have the emergency illumination and not the entire exterior discharge path to the public way.

So that there will be a continuing source of electrical energy for maintaining the illumination of the means of egress when there is a loss of the main power supply, the means of egress lighting system must be connected to an emergency electrical system that consists of storage batteries, unit equipment or an on-site generator. This emergency power-generating facility must be capable of supplying electricity for at least 90 minutes, thereby giving the occupants sufficient time to leave the premises. In most cases, where the loss of the main electrical supply is attributed to a malfunction in the distribution system of the electric power company, experience has shown that such power outages do not usually last longer than 90 minutes.

1006.4 Performance of system. Emergency lighting facilities shall be arranged to provide initial illumination that is at least an average of 1 foot-candle (11 lux) and a minimum at any point of 0.1 foot-candle (1 lux) measured along the path of egress at floor level. Illumination levels shall be permitted to decline to 0.6 foot-candle (6 lux) average and a minimum at any point of 0.06 foot-candle (0.6 lux) at the end of the emergency lighting time duration. A maximum-to-minimum illumination uniformity ratio of 40 to 1 shall not be exceeded.

❖ This section provides the criteria for the illumination levels of the emergency lighting system. The initial average level is the same as for the means of egress illumination in Section 1006.2. The reduction of illumination recognizes the performance characteristics over time of some types of power supplies, such as batteries. The minimum levels are sufficient for the occupants to egress from the building.

The maximum illumination uniformity ratio of 40 means that the variation in the illumination levels is not to exceed 40. For example, a minimum of 0.06 foot candle (0.6 lux) would establish a maximum illumination of 2.4 foot candle (24 lux) in an adjacent area. This is to establish a variation limit such that the means of egress can be seen by a person walking from bright to darker areas along the egress path.

[B] SECTION 1007
ACCESSIBLE MEANS OF EGRESS

1007.1 Accessible means of egress required. Accessible means of egress shall comply with this section. Accessible spaces shall be provided with not less than one accessible means of egress. Where more than one means of egress is required by Section 1014.1 or 1018.1 from any accessible space, each accessible portion of the space shall be served by not less than two accessible means of egress.

Exceptions:

1. Accessible means of egress are not required in alterations to existing buildings.
2. One accessible means of egress is required from an accessible mezzanine level in accordance with Section 1007.3 or 1007.4.
3. In assembly spaces with sloped floors, one accessible means of egress is required from a space where the common path of travel of the accessible route for access to the wheelchair spaces meets the requirements in Section 1024.9.

❖ This section establishes the minimum requirements for means of egress facilities serving all spaces that are required to be accessible to people with physical disabilities. Previously, attention had been focused on response to the civil-rights-based issue of providing adequate access for people with physical disabilities into and throughout buildings. Concerns about life safety and evacuation of people with mobility impairments were frequently cited as reasons for not embracing widespread building accessibility, in the best interest of the disabled community.

The provisions for accessible means of egress are predominantly, though not exclusively, intended to address the safety of persons with a mobility impairment. These requirements reflect the balanced philosophy that accessible means of egress are to be provided for occupants who have gained access into the building but are incapable of independently utilizing the typical means of egress facilities, such as the exit stairways. By making such provisions, the code now addresses means of egress for all building occupants, with and without physical disabilities.

Any space that is not required by the code to be accessible in accordance with Chapter 11 of the IBC is not required to be provided with accessible means of egress. This may include an entire story, a portion of a story or an individual room.

In new construction and additions, the number of required accessible means of egress is the same as the general means of egress, up to a maximum of two. For example, in buildings, stories or spaces required by Section 1014.1 or 1018 to have three or more exits or exit access doors, a minimum of two accessible means of egress is required. The number of exits or exit access doors is based on occupant load; therefore, no matter how large the total occupant load of the space, two fully complying accessible means of egress are considered to provide sufficient capacity for those building occupants with a mobility impairment.

An accessible means of egress is required to provide a continuous path of travel to a public way. This principle is consistent with the general requirements for all means of egress, as reflected in Section 1003.1 and in the definition of the term "Means of egress" in Section 1002. This section also emphasizes the intent that accessible means of egress must be usable by a person with a mobility impairment, such as a person in a wheelchair.

The exceptions address special situations where accessible means of egress requirements need special consideration.

Exception 1 indicates that existing buildings that are undergoing alterations are not required to be provided with accessible means of egress as part of that alteration. In many cases, meeting the requirements for accessible means of egress, especially the 48-inch (1219 mm) clear stair width required in nonsprinklered buildings, would be considered technically infeasible.

Exception 2 is a special consideration for mezzanines. As an example, consider an open mezzanine that does not meet Section 1014.1; therefore, two means of egress are required. In accordance with Section 505.3, open mezzanines are permitted to utilize two open exit access stairways leading to the main level to meet the general means of egress requirements. Stairways are not typically navigable by persons with mobility impairments without assistance. In addition, if a mezzanine is large enough to be required to be accessible (see Section 1104.4 of the IBC), two accessible means of egress are required. The accessible means of egress do not permit exit access stairways. An accessible route must be available to two exits. In this exception, the open mezzanine is permitted to have only one accessible means of egress provided by either an enclosed exit stairway or an elevator with standby power. If the building is not sprinklered throughout, an area of refuge must also be provided.

Exception 3 is in consideration of the practical difficulties of providing accessible routes in assembly areas with sloped floors. Rooms with more than 50 persons are required to have two means of egress; therefore, each accessible seating location is required to have access to two accessible means of egress. Depending on the slope of the seating arrangement, this can be difficult to achieve, especially in small theaters. A maximum travel distance of 30 feet (9144 mm) for ambulatory persons moving from the last seat in dead-end aisles or from box-type seating arrangements to where they have access to a choice of means of egress routes has been established in Section 1024.8. In accordance with this exception, persons using the wheelchair seating spaces have the same maximum 30-foot (9144 mm) travel distance from the accessible seating locations to a cross aisle or out of the room to an adjacent corridor or space where two choices for accessible means of egress are provided. Note that there are increases in travel distance for smoke-protected seating and small spaces, such as boxes, galleries or balconies. For additional information, see Section 1024.8.

1007.2 Continuity and components. Each required accessible means of egress shall be continuous to a public way and shall consist of one or more of the following components:

1. Accessible routes complying with Section 1104 of the *International Building Code.*
2. Stairways within exit enclosures complying with Sections 1007.3 and 1019.1.
3. Elevators complying with Section 1007.4.
4. Platform lifts complying with Section 1007.5.
5. Horizontal exits.
6. Smoke barriers.

Exceptions:

1. Where the exit discharge is not accessible, an exterior area for assisted rescue must be provided in accordance with Section 1007.8.
2. Where the exit stairway is open to the exterior, the accessible means of egress shall include either an area of refuge in accordance with Section 1007.6 or an exterior area for assisted rescue in accordance with Section 1007.8.

❖ This section identifies the various building features that can serve as elements of an accessible means of egress. Accessible routes are readily recognizable as to how they can provide accessible means of egress; however, some nontraditional principles have been established for the total concept of accessible means of egress. This is evident in that stairways and elevators are also identified as elements that can comprise part of an accessible means of egress. Elevators are generally not available for egress during a fire. Stairways are not independently usable by a person in a wheelchair. The concept of accessible means of egress includes the idea that evacuation of people with a mobility impairment may require the assistance of others. In buildings not equipped with an automatic sprinkler system, provisions are also included for the creation of an area of refuge, wherein people can safely await either further instructions or evacuation assistance. Refuge areas can also be established by utilizing either smoke barriers or horizontal exits. All of these elements can be arranged in the manner prescribed in this section to provide accessible means of egress.

Typically, the accessible way into a single-story facility is also one of the accessible means of egress. The exceptions are intended to address problems that most typically arise for the second accessible means of egress. Site constraints or configurations may result in difficulty to provide an accessible route for exit discharge. The alternative for exterior areas for assisted rescue is offered for these types of situations. A change in elevation across a site may result in exit doors leading to exterior steps. These steps are considered exit stairways. For these types of situations, providing either an area of refuge inside the facility or an exterior area of rescue assistance is viable.

1007.2.1 Buildings with four or more stories. In buildings where a required accessible floor is four or more stories above or below a level of exit discharge, at least one required accessible means of egress shall be an elevator complying with Section 1007.4.

Exceptions:

1. In buildings equipped throughout with an automatic sprinkler system installed in accordance with Section 903.3.1.1 or 903.3.1.2, the elevator shall not be required on floors provided with a horizontal exit and located at or above the level of exit discharge.
2. In buildings equipped throughout with an automatic sprinkler system installed in accordance with Section 903.3.1.1 or 903.3.1.2, the elevator shall not be required on floors provided with a ramp conforming to the provisions of Section 1010.

❖ Elevators are the most common and convenient means of providing access in multistory buildings. As such, elevators represent a prime candidate for accessible means of egress from such buildings, especially in light of the difficulties involved in carrying a person in a wheelchair up or down a stairway. The primary consideration for elevators as an accessible means of egress is that the elevator most likely will be available and protected during a fire event.

This section addresses the use of an elevator as part of an accessible means of egress by requiring a standby source of power for the elevator. The standby power requirement establishes a higher degree of reliability that the elevator will be available and usable by reducing the likelihood of power loss caused by fire or other conditions of power failure.

In buildings having four or more stories above or below the level of exit discharge, it is unreasonable to rely solely on exit stairways for all of the required accessible means of egress. This is the point at which complete reliance on assisted evacuation will not be effective or adequate because of the limited availability of either experienced personnel who are trained to handle wheelchair evacuation or special devices, such as self-braking stairway descent equipment or evacuation chairs. In this case, the code requires that at least one elevator, serving all floors of the building, is to serve as one of the required accessible means of egress. This should not represent a hardship, since elevators are typically provided in such buildings for the convenience of the occupants.

The presence of an automatic sprinkler system significantly reduces a potential fire hazard and provides for increased evacuation time. Exception 1 establishes that accessible egress elevator service to floor levels at or above the level of exit discharge is not necessary under specified conditions. These conditions state that the building should be equipped throughout with an automatic sprinkler system in accordance with NFPA 13 or NFPA 13R (see Section 903.3.1.1 or 903.3.1.2) and the floors not serviced by an accessible egress elevator should be provided with a horizontal exit. The combination of automatic sprinklers and a horizontal exit provides adequate protection for the occupants despite their distance to the level of exit discharge. This exception does not apply to floor levels below the level of exit discharge, since such levels are typically below grade and do not have the added advantage of exterior openings that are available for fire-fighting or rescue purposes. Exception 2 specifies that a building sprinklered throughout in accordance with NFPA 13 or NFPA 13R (see Section 903.3.1.1 or 903.3.1.2), with ramp access to each level, such as in a sports stadium, is not required to also have an elevator for accessible means of egress. The reasoning behind this is that the issue of carrying people down stairways does not occur because the ramps may be utilized.

1007.3 Enclosed exit stairways. An enclosed exit stairway, to be considered part of an accessible means of egress, shall have a clear width of 48 inches (1219 mm) minimum between handrails and shall either incorporate an area of refuge within an enlarged floor-level landing or shall be accessed from either an area of refuge complying with Section 1007.6 or a horizontal exit.

Exceptions:

1. Open exit stairways as permitted by Section 1019.1 are permitted to be considered part of an accessible means of egress.
2. The area of refuge is not required at open stairways that are permitted by Section 1019.1 in buildings or facilities that are equipped throughout with an automatic sprinkler system installed in accordance with Section 903.3.1.1.
3. The clear width of 48 inches (1219 mm) between handrails and the area of refuge is not required at exit stairways in buildings or facilities equipped throughout with an automatic sprinkler system installed in accordance with Section 903.3.1.1 or 903.3.1.2.
4. The clear width of 48 inches (1219 mm) between handrails is not required for enclosed exit stairways accessed from a horizontal exit.
5. Areas of refuge are not required at exit stairways serving open parking garages.

❖ The fundamental condition under which a stairway can function as part of an accessible means of egress is that the stairway is sufficiently wide to enable the assisted evacuation.

Sufficient width is accomplished by requiring the stairway to have a minimum clear width between handrails of 48 inches (1219 mm). This dimension is sufficient to enable two or three persons to carry the person in a wheelchair down (or up) to the level of exit discharge from which access to a public way is afforded.

The exit stairway, in combination with an area of refuge, can provide safety from fire in one of two ways. One approach is for the fire-resistance-rated stairway enclosure to afford the necessary safety. To accomplish this, the stairway landing within the stairway enclosure must be able to contain the wheelchair. The concept is that the person in the wheelchair will remain on the stairway landing for a period of time awaiting further instructions or evacuation assistance; therefore, the stairway landing must be able to accommodate the wheelchair without obstructing the use of the stairway by other egressing occupants. An enlarged, story-level landing is required within the stairway enclosure and must be of sufficient size to accommodate the number of wheelchairs required by Section 1007.6.1 [see Figure 1007.3(1)].

The other approach is to utilize a stairway that is accessed from an area of refuge complying with Section 1007.6. Under this approach, the stairway is made safe by virtue of its access being in an area that is separated and protected from the point of fire origin. An area of refuge can be created by constructing a vestibule adjacent and with direct access to the stair enclosure in accordance with Section 1007.6 [see Figure 1007.3(2)]. This is similar in theory to the approach of an enlarged landing within the stairway enclosure.

Note that the exceptions are for areas of refuge or stairway width requirements. They are not intended to indicate that an accessible means of egress is not required.

In the case of a horizontal exit [see Figure 1007.3(3)], each floor area on either side of the exit is considered a refuge area (see commentary, Section 1021.1) by virtue of the construction and separation requirements for horizontal exits. The discharge area is always assumed to be the nonfire side and, therefore, is protected from fire. Any stairway located within the horizontal exit discharge area constitutes an accessible means of egress. Exception 4 relieves the requirement for a minimum 48-inch (1219 mm) clear width for stairways that are accessed from a horizontal exit. This exception considers that an area of refuge created by a horizontal exit will be of considerable size so that movement of occupants from the area of refuge down the stairway can be more deliberate.

Exception 5 for open parking structures is in recognition of the natural ventilation of the products of combustion that will be afforded by the exterior openings required of such structures (see Section 406.3 of the IBC). The most immediate hazard for occupants in a fire incident is exposure to smoke and fumes. Floor areas in open parking structures communicate sufficiently with the outdoors such that the need for protection from smoke is not necessary; therefore, open parking garages are exempted from the requirements for an area of refuge. It is due to this level of natural ventilation that parking garage exit stairways are not required to be enclosed (see Section 1019.1, Exception 5).

Exception 1 is in recognition of the open exit stairways permitted by Section 1019.1. Therefore, the same exit stairway that serves the ambulatory occupants in the building can also serve as part of the accessible means of egress. Note that this is for exit stairways only, not exit access stairways. If the building is not sprinklered, an area of refuge and 48-inch (1219 mm) clear width on the stairway would still be required.

Exception 2 states that for exit stairways permitted to be unenclosed by any exception in Section 1019.1 and in a building sprinklered in accordance with NFPA 13, an area of refuge is not required, but the 48-inch (1219 mm) clear width for the stairway is still required. This exception was not added with the attempt to override any options permitted by a combination of Exceptions 1 and 3.

Exception 3 exempts the 48-inch-width (1219 mm) requirement and the area of refuge or a horizontal exit for any occupancy that is equipped throughout with an automatic sprinkler system in accordance with NFPA 13 or NFPA 13R (see Section 903.3.1.1 or 903.3.1.2). With this exception and Exception 2 to Section 1007.4, the intent of the code is that areas of refuge are not required in fully sprinklered buildings. This exception is in recognition of the increased level of safety and evacuation time that is afforded in a sprinklered occupancy. The expectation is that a supervised system will reduce the threat of fire by reliably controlling and confining the fire to the immediate area of origin. There is also the additional safety of the sprinkler system requirements for automatic notification when the system is activated. This has been substantiated by a study of accessible means of egress conducted for the General Services Administration (GSA). A report issued by the National Institute for Standards and Technology (NIST), NIST IR 4770, concluded that the operation of a properly designed NFPA 13 sprinkler system eliminates the life threat to all building occupants, regardless of their individual physical abilities and is a superior form of protection as compared to areas of refuge. This logic is extended to an NFPA 13R system. One intent of an NFPA 13R system is to provide a minimum level of protection to allow for evacuation of the occupants.

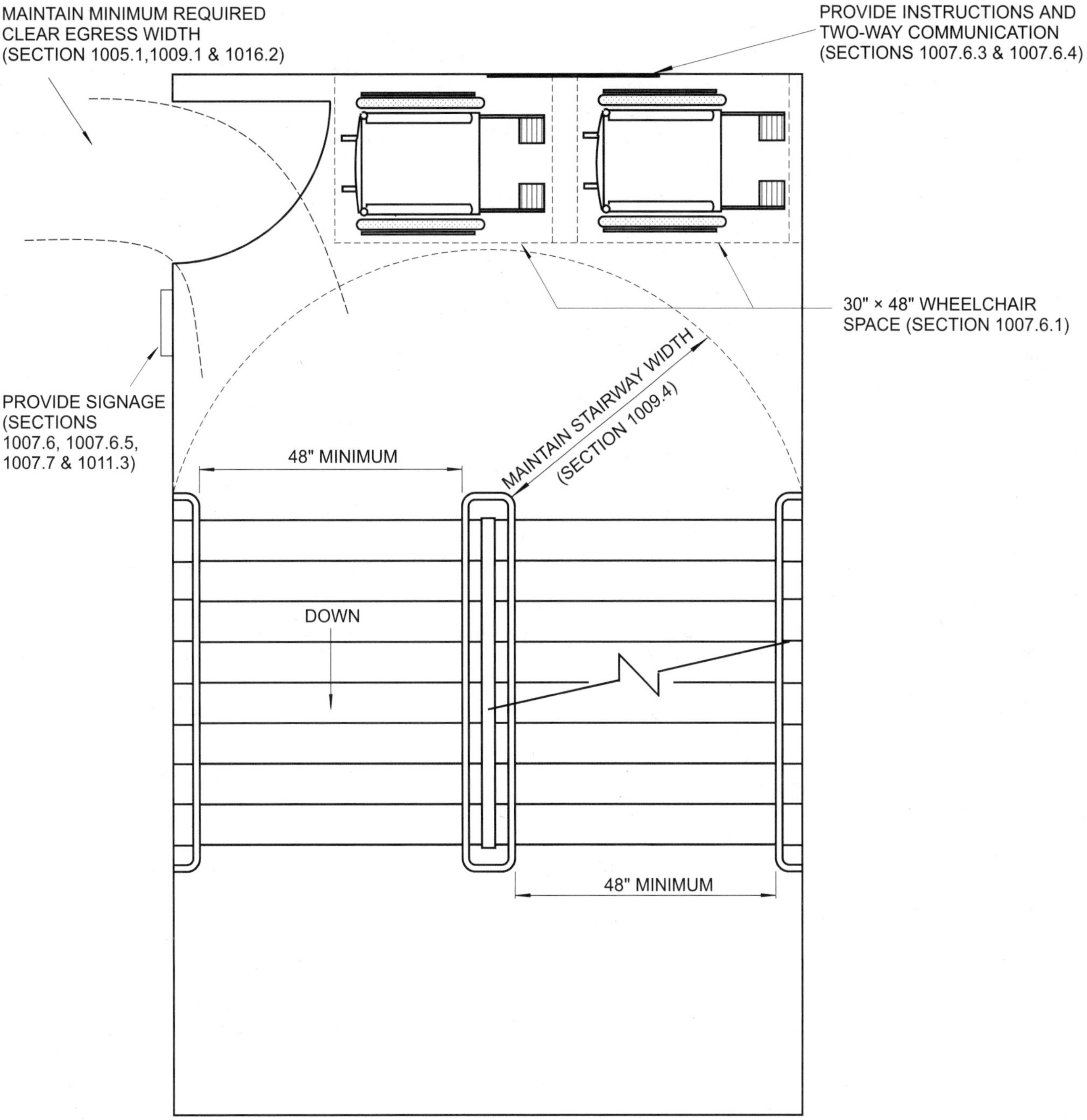

For SI: 1 inch = 25.4 mm.

Figure 1007.3(1)
ACCESSIBLE MEANS OF EGRESS STAIRWAY WITH ENLARGED STORY-LEVEL LANDING

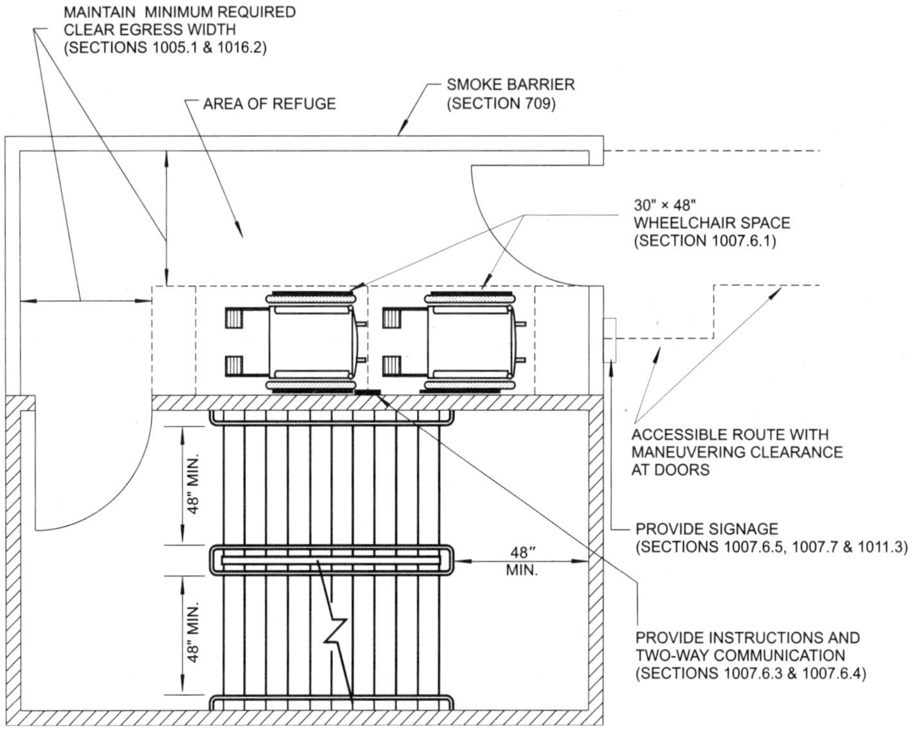

For SI: 1 inch = 25.4 mm.

Figure 1007.3(2)
ACCESSIBLE MEANS OF EGRESS STAIRWAY ACCESSED FROM AN AREA OF REFUGE

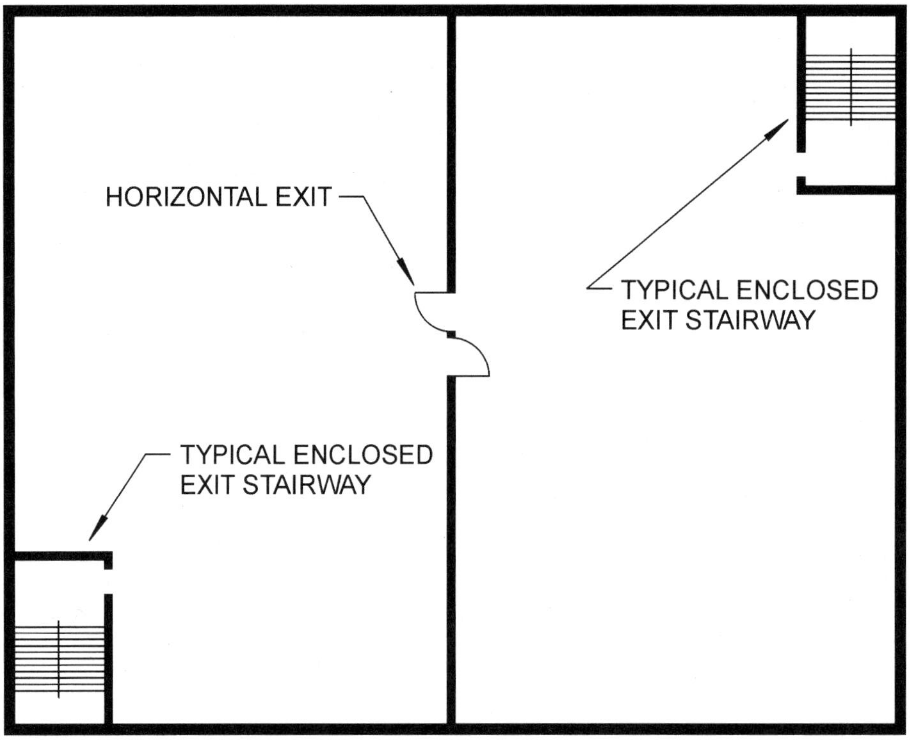

Figure 1007.3(3)
ACCESSIBLE MEANS OF EGRESS STAIRWAY ACCESSED FROM HORIZONTAL EXIT

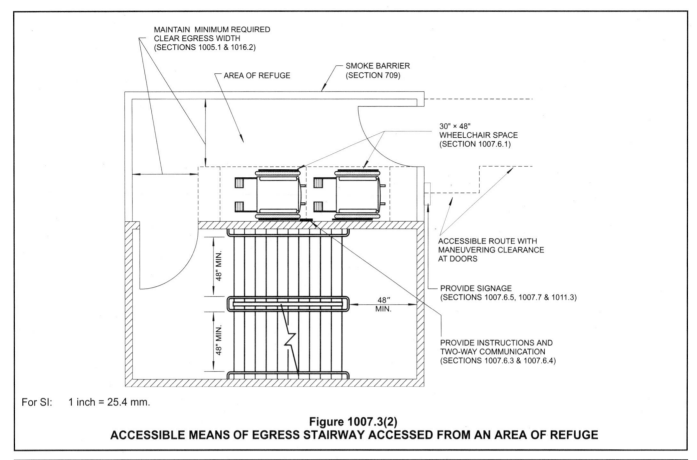

1007.4 Elevators. An elevator to be considered part of an accessible means of egress shall comply with the emergency operation and signaling device requirements of Section 2.27 of ASME A17.1. Standby power shall be provided in accordance with Sections 2702 and 3003 of the *International Building Code*. The elevator shall be accessed from either an area of refuge complying with Section 1007.6 or a horizontal exit.

Exceptions:

1. Elevators are not required to be accessed from an area of refuge or horizontal exit in open parking garages.
2. Elevators are not required to be accessed from an area of refuge or horizontal exit in buildings and facilities equipped throughout with an automatic sprinkler system installed in accordance with Section 903.3.1.1 or 903.3.1.2.

❖ Elevators are the most common and convenient means of providing access in multistory buildings. As such, elevators represent a prime candidate for accessible means of egress from such buildings, especially in light of the difficulties involved in carrying a person in a wheelchair up or down a stairway. The primary consideration for elevators as an accessible means of egress is that the elevator most likely will be available and protected during a fire event.

This section addresses the use of an elevator as part of an accessible means of egress by requiring both a backup source of power for the elevator and access to the elevator from an area of refuge or a horizontal exit. For situations where elevators are required to be part of one of the accessible means of egress, see Section 1007.2.1. The backup power requirement establishes a higher degree of reliability that the elevator will be available and usable by reducing the likelihood of power loss caused by a fire or other conditions. Requiring access from an area of refuge or a horizontal exit affords the same degree of fire safety as described for stairways (see commentary, Section 1007.3). Additionally, the reference to Sections 2702 and 3003 of the IBC clarifies that the elevator will comply with the emergency operation features that relate to elevator operation under fire conditions (see commentary, Sections 2702.2.18 and 3003 of the IBC). Elevators on an accessible route are also required to meet the accessibility provisions of ICC A117.1 (see commentary, Sections 1109.6 and 3001.3 of the IBC).

Note that the exceptions are for areas of refuge requirements only. They are not intended to indicate that an accessible means of egress is not required.

If a level in an open parking garage is accessible, that level is required to have an accessible means of egress. Exception 1 for open parking structures is in recognition of the natural ventilation of the products of combustion that will be afforded by the exterior openings required of such structures (see Section 406.3 of the IBC). The most immediate hazard for occupants in a fire incident is exposure to smoke and fumes. Floor areas in open parking structures communicate sufficiently with the outdoors, thus protection from smoke is not necessary. Therefore, open parking garages are exempt from the requirements for an area of refuge or horizontal exit to access and an elevator that is utilized as part of the accessible means of egress. This is consistent with the exception for stairways in Section 1007.3.

As previously discussed, the presence of an automatic sprinkler system significantly reduces the potential fire hazard and provides for increased evacuation time. Exception 2 indicates that in an occupancy equipped throughout with an automatic sprinkler system in accordance with NFPA 13 or NFPA 13R (see Section 903.3.1.1 or 903.3.1.2), the accessible egress elevator is not required to be accessed from an area of refuge or a horizontal exit. This is consistent with the exception for stairways in Section 1007.3 and recognizes the value of a sprinkler system in limiting fire growth and spread. Note that an elevator lobby that is off a rated corridor must also comply with Section 707.14.1 of the IBC.

1007.5 Platform lifts. Platform (wheelchair) lifts shall not serve as part of an accessible means of egress, except where allowed as part of a required accessible route in Section 1109.7 of the IBC. Platform lifts in accordance with Section 604 shall be installed in accordance with ASME A18.1. Standby power shall be provided in accordance with Section 604.2 for platform lifts permitted to serve as part of a means of egress.

❖ Previously, there have been concerns whether a platform lift will be reliably available at all times. ASME A18.1, the new standard for platform lifts, no longer requires to be key operated. It is important to note that platform lifts are not prohibited by the code. They simply cannot be counted as a required accessible means of egress in other than locations where they are allowed as part of the accessible route into a space (see commentary, Section 1109.7 of the IBC). When platform lifts are utilized as part of an accessible means of egress, they must come equipped with standby power.

In existing buildings undergoing alterations, platform lifts are allowed as part of an accessible route (see commentary, Section 3409.7.3 of the IBC). Note that accessible means of egress are not required in existing buildings undergoing an alteration (see Section 1007.1).

1007.6 Areas of refuge. Every required area of refuge shall be accessible from the space it serves by an accessible means of egress. The maximum travel distance from any accessible space to an area of refuge shall not exceed the travel distance permitted for the occupancy in accordance with Section 1015.1. Every required area of refuge shall have direct access to an enclosed stairway complying with Sections 1007.3 and 1019.1 or an elevator complying with Section 1007.4. Where an elevator lobby is used as an area of refuge, the shaft and lobby shall comply with Section 1019.1.8 for smokeproof enclosures except where the elevators are in an area of refuge formed by a horizontal exit or smoke barrier.

❖ An area of refuge is of no value as part of an accessible means of egress if it is not accessible. The code states an obvious but essential requirement: the path that leads to an area of refuge must qualify as an accessible means of egress. This provision is so that there will be an accessible route leading from every accessible space to each required area of refuge. For consistency in principle with the general means of egress design concepts, the code also limits the travel distance to the area of refuge. The limitation is the same distance as specified in Section 1015.1 for maximum exit access travel distance. This equates the maximum travel distance required to reach an exit with the maximum distance required to reach an area of refuge. It should be noted that an area of refuge is not necessarily an exit in the classic sense. For example, when the area of refuge is an enlarged, story-level landing within an exit stairway, the area of refuge is within the exit and the maximum travel distance for both the conventional exit and the accessible area of refuge is measured to the same point (the entrance to the exit stairway). If, however, the area of refuge is a vestibule immediately adjacent to an enclosed exit stairway, the maximum travel distance for purposes of the required accessible means of egress is measured from the entrance to the area of refuge and, for purposes of the travel distance to the exit, from the entrance to the exit stairway (see Figure 1007.6). In the case of accessible means of egress with an elevator, the maximum travel distance may end up being measured along two different paths, with the only consistency between the conventional means of egress and the accessible means of egress being the maximum travel distance [see Figure 1007.6]. The travel distance within an area of refuge is not directly regulated, but it will be limited by the general provisions for maximum exit access travel distance, which are always applicable.

In summary, the code takes a reasonably consistent approach for both conventional and accessible means of egress by limiting the distance one must travel to reach a safe area from which further egress to a public way is available.

So there is continuity in an accessible means of egress, the code requires that every area of refuge have direct access to either an exit stairway (see Sections 1007.3 and 1019.1) or an elevator (see Section 1007.4). This, again, may be viewed as stating the obvious, but it is necessary so that the egress layout does not involve entering an area of refuge and then having to leave that protected area before gaining access to a stairway or elevator. Once an occupant reaches the safety of an area of refuge, that level of protection must be continuous until the vertical transportation element (the stairway or elevator) is reached.

If one chooses to comply with the accessible means of egress requirements by providing an accessible elevator and an area of refuge in the form of an elevator lobby, the elevator shaft and the lobby are required to be constructed as a smokeproof enclosure in accordance with Section 1019.1.8. Elevator hoistway door assemblies are, by nature, not substantially air tight. Significant quantities of air can move throughout the shaft and adjacent lobby because of stack effect, especially in tall buildings. The requirement for smokeproof enclosure construction provides additional ensurance that the elevator will not be rendered unavailable due to smoke movement into the elevator shaft. If the elevator is in an area of refuge that is formed by the use of a horizontal exit or smoke barrier, which essentially means there is no need for an additional elevator lobby, it is presumed that the area of refuge is free from fire involvement; therefore, smokeproof enclosure construction is not necessary.

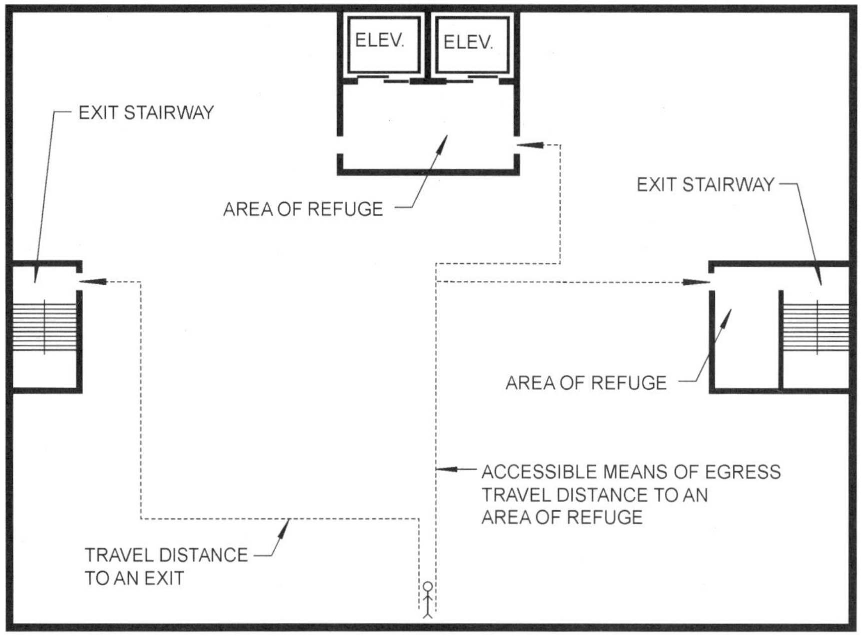

Figure 1007.6
SEPARATE TRAVEL DISTANCES TO AN AREA OF REFUGE

1007.6.1 Size. Each area of refuge shall be sized to accommodate one wheelchair space of 30 inches by 48 inches (762 mm by1219 mm) for each 200 occupants or portion thereof, based on the occupant load of the area of refuge and areas served by the area of refuge. Such wheelchair spaces shall not reduce the required means of egress width. Access to any of the required wheelchair spaces in an area of refuge shall not be obstructed by more than one adjoining wheelchair space.

❖ The number of wheelchair spaces that is required to be provided in an area of refuge is intended to represent broadly the expected population of the average building. As one point of measurement, a 1977 survey conducted by the National Center for Health indicated that one in 333 civilian, noninstitutionalized persons use a wheelchair. Originally, a ratio of one space for each 100 occupants was considered, based on the area served by the area of refuge. The Americans with Disabilities Act (ADA) currently utilizes the criterion of one space for each 200 occupants, based on the area served by the area of refuge. Given the variations and difficulties involved in accurately predicting a representative ratio for application to all occupancies, it was concluded that a requirement for one space for each 200 occupants based on the area of refuge itself, plus the areas served by the area of refuge, represents a reasonable criterion.

Arrangement of the required wheelchair spaces is critical so as not to interfere with the means of egress for ambulatory occupants. Since the design concept is based on the idea that wheelchair occupants will move to the area of refuge and await further instructions or evacuation assistance, the spaces must be located so as not to reduce the required means of egress width of the stairway, door, corridor or other egress path through the area of refuge.

In order to provide for orderly maneuvering of wheelchairs, this section states that access to any of the required wheelchair spaces cannot be obstructed by more than one adjoining wheelchair space. For example, this precludes an arrangement in which three or more wheelchairs could be stacked in a dead-end corridor. This also effectively limits the difficulty any given wheelchair occupant would have in reaching or leaving a given wheelchair space, as well as providing easier access to all wheelchair spaces by persons providing evacuation assistance.

1007.6.2 Separation. Each area of refuge shall be separated from the remainder of the story by a smoke barrier complying with Section 709 of the *International Building Code*. Each area of refuge shall be designed to minimize the intrusion of smoke.

Exceptions:

1. Areas of refuge located within a stairway enclosure.
2. Areas of refuge where the area of refuge and areas served by the area of refuge are equipped throughout with an automatic sprinkler system installed in accordance with Section 903.3.1.1 or 903.3.1.2.

❖ The minimum standard for construction of an area of refuge is a smoke barrier, in accordance with Section 709 of the IBC. This establishes a minimum degree of performance by means of a 1-hour fire-resistance rating, including opening protectives and a minimum degree of performance against the intrusion of smoke into an enclosed area of refuge, as specified in Sections 709.4 and 709.5 of the IBC. This section does not require an area of refuge within an exit stairway to be designed to prevent the intrusion of smoke. This was based on a study of areas of refuge conducted by NIST for the GSA, which concluded that a story-level landing within a fire-resistance-rated exit stairway would provide a satisfactory staging area for evacuation assistance.

This section also exempts the requirement that the area of refuge be designed to prevent intrusion of smoke when the area of refuge and all areas served by it are sprinklered, in accordance with an NFPA 13 or NFPA 13R system (see Section 903.3.1.1 or 903.3.1.2.2). This is based on the value of sprinkler protection. This differs from the exceptions for enclosed stairways and elevators (see Sections 1007.3 and 1007.4), where areas of refuge are not required in buildings equipped throughout with an automatic sprinkler system, whereas this section covers partially sprinklered buildings or buildings that choose to provide areas of refuge in addition to full suppression.

1007.6.3 Two-way communication. Areas of refuge shall be provided with a two-way communication system between the area of refuge and a central control point. If the central control point is not constantly attended, the area of refuge shall also have controlled access to a public telephone system. Location of the central control point shall be approved by the fire department. The two-way communication system shall include both audible and visible signals.

❖ Use of an elevator, stair enclosure or other area of refuge as part of an accessible means of egress requires a disabled person to wait for evacuation assistance or relevant instructions. The two-way communication system allows this person to inform emergency personnel of his or her location and to receive additional instructions or assistance as needed. It is important that both visual and audible signals be provided as part of the communication system so that a person with any disability may utilize the system. The central control point can be where emergency management of the elevators is located, the main desk or lobby or other location where it is likely that trained or experienced personnel will be available to communicate with the occupants of the area of refuge. A suitable central control point is often not available in low-rise buildings or, in a high-rise building, may not be manned on a 24-hour basis. In situations such as these, a public telephone where a caller may reach an appropriate emergency location, such as the fire department, must also be provided in the area of refuge. The fire department must approve the configuration of the system.

1007.6.4 Instructions. In areas of refuge that have a two-way emergency communications system, instructions on the use of the area under emergency conditions shall be posted adjoining the communications system. The instructions shall include all of the following:

1. Directions to find other means of egress.
2. Persons able to use the exit stairway do so as soon as possible, unless they are assisting others.
3. Information on planned availability of assistance in the use of stairs or supervised operation of elevators and how to summon such assistance.
4. Directions for use of the emergency communications system.

❖ The required instructions on the proper use of the area of refuge and the communication system provide a higher degree of ensurance that the communication system will accomplish the intended function and occupants will behave as expected. A two-way communication system will not be of much value if a person in that area does not know how to operate it. Also, since the area of refuge is required by Section 1007.6.5 to be identified as such, ambulatory occupants may mistakenly conclude that they should remain in that area. The instructions provide the opportunity for ambulatory occupants to be reminded that they should continue to egress as soon as possible.

1007.6.5 Identification. Each door providing access to an area of refuge from an adjacent floor area shall be identified by a sign complying with ICC A117.1, stating: AREA OF REFUGE, and including the International Symbol of Accessibility. Where exit sign illumination is required by Section 1011.2, the area of refuge sign shall be illuminated. Additionally, tactile signage complying with ICC A117.1 shall be located at each door to an area of refuge.

❖ Identification of an area of refuge by signage enables an occupant to become aware of the available protected area. The approach that the code takes for identification of the area of refuge is comparable to the general provisions for identification of exits, including the requirement for lighted signage. Tactile signage is also required for the benefit of persons with a visual disability.

1007.7 Signage. At exits and elevators serving a required accessible space but not providing an approved accessible means of egress, signage shall be installed indicating the location of accessible means of egress.

❖ The additional signage required by this section is intended to advise persons of the location of the accessible means of egress. Since not all of the exits will necessarily be accessible means of egress, it is appropriate to provide this information at exit stairways and particularly at elevators that are not the approved accessible means of egress. This section does not require that signage be provided at an accessible exit giving information as to the location of any other accessible means of egress.

1007.8 Exterior area for assisted rescue. The exterior area for assisted rescue must be open to the outside air and meet the requirements of Section 1007.6.1. Separation walls shall comply with the requirements of Section 704 for exterior walls. Where walls or openings are between the area for assisted rescue and the interior of the building, the building exterior walls within 10 feet (3048 mm) horizontally of a nonrated wall or unprotected opening shall be constructed as required for a minimum 1-hour fire-resistance rating with $^3/_4$-hour opening protectives. This construction shall extend vertically from the ground to a point 10 feet (3048 mm) above the floor level of the area for assisted rescue or to the roof line, whichever is lower.

❖ The exterior area of assisted rescue is an alternative for situations where the exit is via an unenclosed exterior stairway or when the exit discharge is not accessible. The protection required would be equivalent to that required for an area of refuge. Note that there is no exception for the exterior area of assisted rescue for buildings that contain sprinkler systems. The separation requirements are similar to exterior exit stairways (see Section 1022.6). The exceptions for exterior exit stairway protection would not be applicable where they would also include an exterior area for assisted rescue. Providing a location that was 10 feet (3048 mm) away from the exterior wall would not serve as a viable alternative for having a rated exterior wall. The persons waiting for assistance must have a minimum level of protection from the fire in the building.

An example where the exterior area for assisted rescue would be a good alternative is along the rear of a strip mall. The front of the facility may be accessible, but if two accessible means of egress are required, than an accessible route must be provided at the rear of each of the tenant spaces. Often in these types of malls, the rear entrance doubles as the loading dock or service entrance. Due to the associated change in elevation, the ramps required to allow accessible exit discharge could be extensive. In addition, the ramps could become damaged over time by trucks maneuvering into the loading dock areas. In this situation, interior areas of refuge are not always a positive alternative. Tenants may tend to use them as convenient storage areas. If persons with a mobility impairment wait for assisted rescue outside of the building, they are already protected from smoke and fumes—the deadliest of the fire hazards. Being visible should also result in a shorter period of time before assisted rescue is effected [see Figures 1007.8(1) and 1007.8(2)].

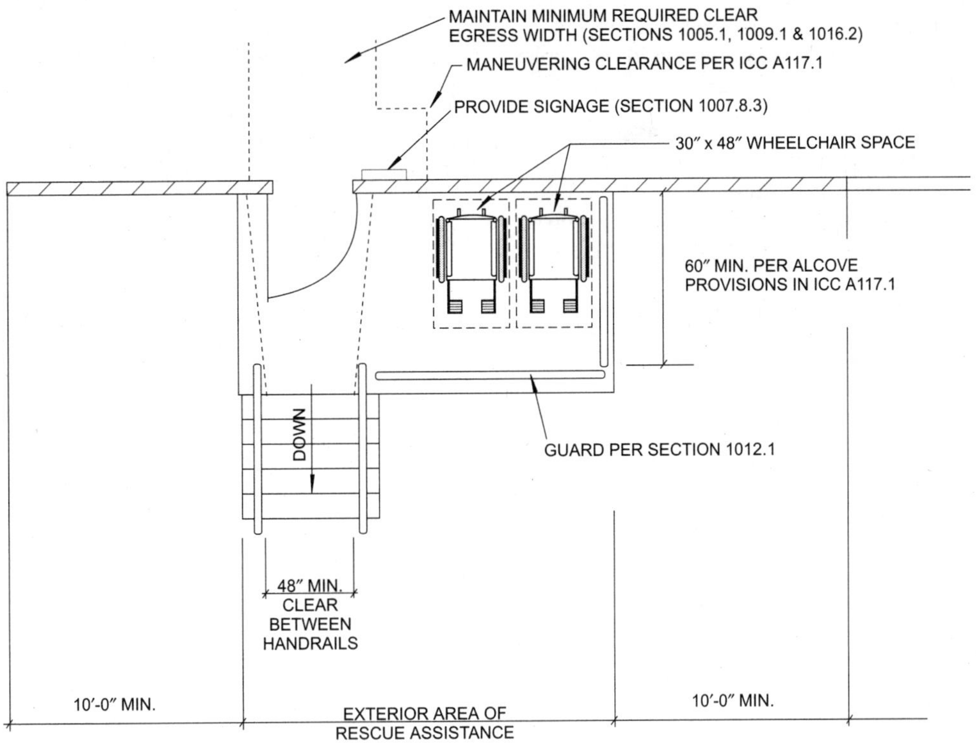

For SI: 1 inch = 25.4 mm, 1 foot = 304.8 mm.

Figure 1007.8(1)
EXTERIOR AREA OF RESCUE ASSISTANCE

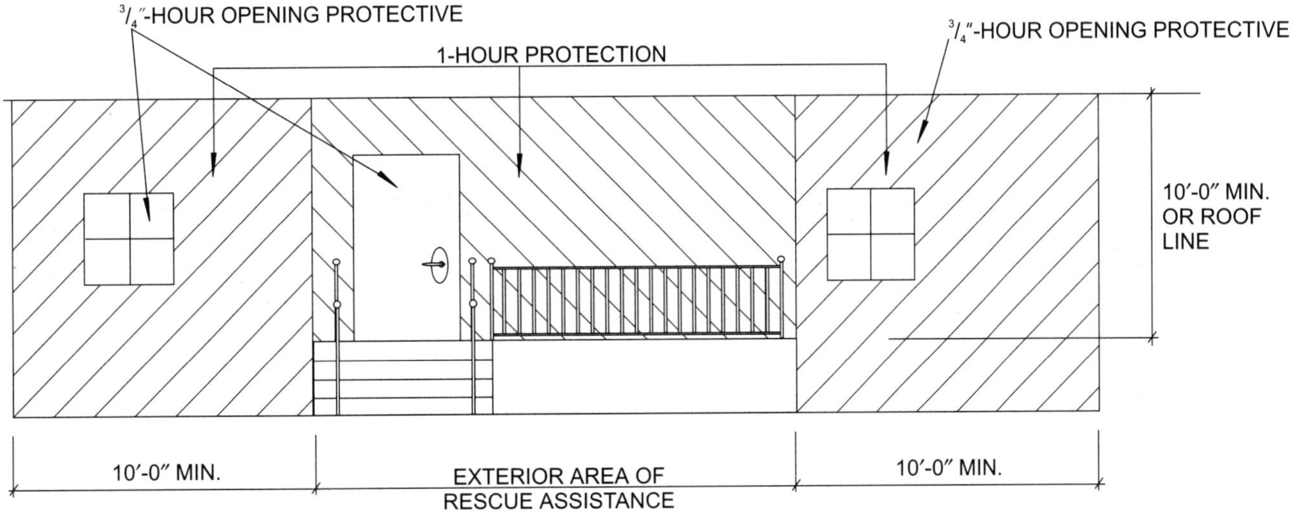

For SI: 1 foot = 304.8 mm.

Figure 1007.8(2)
EXTERIOR AREA OF RESCUE ASSISTANCE

1007.8.1 Openness. The exterior area for assisted rescue shall be at least 50 percent open, and the open area above the guards shall be so distributed as to minimize the accumulation of smoke or toxic gases.

❖ The openness criteria for exterior areas of assisted rescue are similar to the requirements for exterior balconies. The purpose is to ensure that a person at an exterior area of rescue assistance is not in danger from smoke and fumes. The criteria are to address the situation where the area is open to outside air, but a combination of roof overhangs and perimeter walls or guards could still trap enough smoke that the safety of the occupants would be jeopardized.

1007.8.2 Exterior exit stairway. Exterior exit stairways that are part of the means of egress for the exterior area for assisted rescue shall provide a clear width of 48 inches (1219 mm) between handrails.

❖ Any steps that lead from an exterior area for assisted rescue to grade must have a clear width of 48 inches (1219 mm) between handrails. The additional width is to permit adequate room to assist a mobility impaired person down the steps and to move to a safe location.

1007.8.3 Identification. Exterior areas for assisted rescue shall have identification as required for area of refuge that complies with Section 1007.6.5.

❖ The exterior area for assisted rescue must have both visual and tactile signage that states "EXTERIOR AREA FOR ASSISTED RESCUE" and includes the International Symbol of Accessibility. This signage should be illuminated similar to the exit signage. Due to immediate visibility of a person at an exterior area of rescue assistance, two-way communication and instructions are not required.

[B] SECTION 1008 DOORS, GATES AND TURNSTILES

1008.1 Doors. Means of egress doors shall meet the requirements of this section. Doors serving a means of egress system shall meet the requirements of this section and Section 1017.2. Doors provided for egress purposes in numbers greater than required by this code shall meet the requirements of this section.

Means of egress doors shall be readily distinguishable from the adjacent construction and finishes such that the doors are easily recognizable as doors. Mirrors or similar reflecting materials shall not be used on means of egress doors. Means of egress doors shall not be concealed by curtains, drapes, decorations or similar materials.

❖ The general requirements for doors are in this section and the following subsections. Doors need to be easily recognizable for immediate use in an emergency condition. Thus, the code specifies that doors are not to be hidden in such a manner that a person would have trouble seeing where to egress.

1008.1.1 Size of doors. The minimum width of each door opening shall be sufficient for the occupant load thereof and shall provide a clear width of not less than 32 inches (813 mm). Clear openings of doorways with swinging doors shall be measured between the face of the door and the stop, with the door open 90 degrees (1.57 rad). Where this section requires a minimum clear width of 32 inches (813 mm) and a door opening includes two door leaves without a mullion, one leaf shall provide a clear opening width of 32 inches (813 mm). The maximum width of a swinging door leaf shall be 48 inches (1219 mm) nominal. Means of egress doors in an occupancy in Group I-2 used for the movement of beds shall provide a clear width not less than $41^1/_2$ inches (1054 mm). The height of doors shall not be less than 80 inches (2032 mm).

Exceptions:

1. The minimum and maximum width shall not apply to door openings that are not part of the required means of egress in occupancies in Groups R-2 and R-3 as applicable in Section 101.2.
2. Door openings to resident sleeping units in occupancies in Group I-3 shall have a clear width of not less than 28 inches (711 mm).
3. Door openings to storage closets less than 10 square feet (0.93 m^2) in area shall not be limited by the minimum width.
4. Width of door leafs in revolving doors that comply with Section 1008.1.3.1 shall not be limited.
5. Door openings within a dwelling unit or sleeping unit shall not be less than 78 inches (1981 mm) in height.
6. Exterior door openings in dwelling units and sleeping units, other than the required exit door, shall not be less than 76 inches (1930 mm) in height.
7. Interior egress doors within a dwelling unit or sleeping unit which is not required to be adaptable or accessible.
8. Door openings required to be accessible within Type B dwelling units shall have a minimum clear width of $31^3/_4$ inches (806 mm).

❖ The size of a door opening determines its capacity as a component of egress and its ability to fulfill its function in normal use. A door opening must meet certain minimum criteria as to its width and height in order to be used safely and to provide accessibility to people with physical disabilities. Doorways that are not in the means of egress are not limited in size by this section. However, doors that are used for egress purposes, including additional doors over and above the number of means of egress required by the code, are required to meet the requirements of this section unless one of the exceptions applies.

The minimum clear width of an egress doorway for occupant capacity is based on the portion of the occupant load (see Section 1004.1) intended to utilize the doorway for egress purposes multiplied by the egress width per occupant from Table 1005.2. The clear width of a swinging door opening is the horizontal dimension measured between the face of the door and the door stops when the door is in the 90-degree position [see

Figure 1008.1.1(1)].

Using the face of the door as the measurement point is consistent with the provisions of ICC A117.1 and the Americans with Disabilities Act Accessibility Guidelines (ADAAG) Review Advisory Committee. Further, this measurement is not intended to prohibit other projections into the required clear width, such as latching hardware or panic hardware. See the commentary to Section 1008.1.1.1 and Figure 1008.1.1(2) for further discussion on the specific projections allowed in the required clear width. For nonswinging means of egress doors, such as a sliding door, the clear width is to be measured from the face of the door jambs.

The minimum clear-width requirement in a doorway of 32 inches (813 mm) is to allow passage of a wheelchair as well as persons utilizing walking devices or other support apparatus. Similarly, because of the difficulties that a person with physical disabilities would have in opening two doors simultaneously, the 32-inch (813 mm) minimum must be provided by a single door leaf.

Note that in some cases, with standard door construction and hardware, a 36-inch-wide (914 mm) door is the narrowest door that can be used while still providing the minimum clear width of 32 inches (813 mm). A standard 34-inch-wide (864 mm) door has less than a 32-inch (813 mm) clear opening, depending on the thickness of the opposing doorstop and where the door hinges are located. The building designer must verify that the specified swinging door will in fact provide the required clear width. A minimum clear width of 41$^1/_2$ inches (1054 mm) is required for doors in any portion of Group I-2 where bed movement is needed to evacuate patients from the area in the event of a fire.

The maximum width for a means of egress door leaf in a swinging door is 48 inches (1219 mm) because larger doors are difficult to handle and are of sizes that typically are not fire tested. The maximum width only applies to swinging doors and not to horizontal sliding doors.

Minimum door heights are required to provide clear headroom for the users. A minimum height of 80 inches (2032 mm) has been empirically derived as sufficient for most users. Note that although the clear height of a doorway is not specified, typical door-frame dimensions will render an opening very close to 80 inches (2032 mm) clear height.

Exception 1 is very limited in scope and is primarily intended to permit decorative-type doors (e.g., café doors) in dwelling units. This exception addresses spaces that are provided with two or more means of egress when only one is required. These nonrequired egress elements are exempted from the minimum and maximum dimensions.

Exception 2 permits the continued use of doors to resident sleeping rooms (cells) in Group I-3 occupancies according to current practices.

Exception 3 permits doors to storage closets in any group classification less than 10 square feet (0.9 m^2) in area to be less than 32 inches (813 mm). This provision is intended to include those closets that can be reached in an arm's length and thus do not require full passage into the closet to be functional.

Exception 4 permits the door leaves in a revolving door assembly to be of any width when the revolving door passage width complies with Section 1008.1.3.1, which provides for adequate egress width when collapsed into a book-fold position.

Exception 5 permits the doorway within a dwelling or sleeping unit to be a minimum of 78 inches (1981 mm) in clear height. This is deemed acceptable because of the familiarity persons in a dwelling or sleeping unit usually have with the egress system and the lack of adverse injury statistics relating to such doors. Note that this exception does not apply to exterior doors of the dwelling unit.

Exception 6 permits exterior doorways to a dwelling or sleeping unit, except for the required exit door, to be a minimum of 76 inches (1930 mm) in clear height. Accordingly, the required exterior exit door to a dwelling or sleeping unit must be 80 inches (2032 mm) in height (exterior doors are not within the scope of Exception 5), but other exterior doors are allowed to be a height of only 76 inches (1930 mm). This provision allows for the continued use of 76-inch-high (1930 mm) sliding patio doors and swinging doors sized to replace such doors.

Exception 7 allows interior doors in a means of egress in dwelling or sleeping units not required to be accessible to have a clear width less than 32 inches (813 mm). Since the doors specified do not serve a dwelling or sleeping unit that is required to be Accessible, Type A or B units, smaller doors are allowed.

Exception 8 addresses the doors in a Type B dwelling or sleeping unit that are required to be accessible. Refer to Chapter 11 of the IBC for additional information related to Type B dwelling and sleeping units.

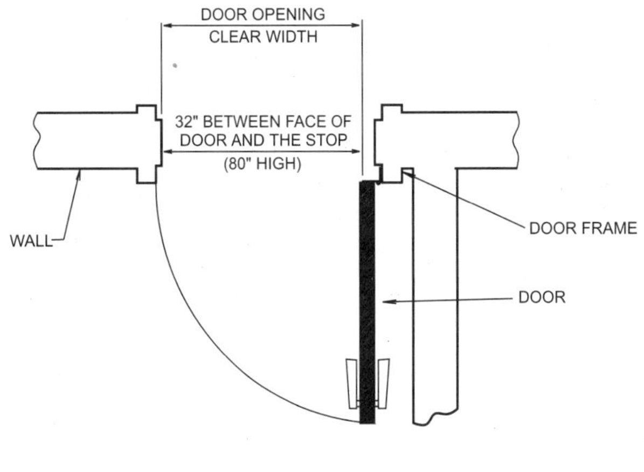

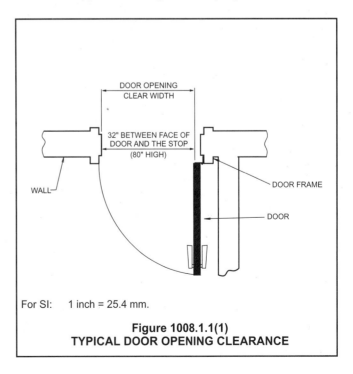

For SI: 1 inch = 25.4 mm.

Figure 1008.1.1(1)
TYPICAL DOOR OPENING CLEARANCE

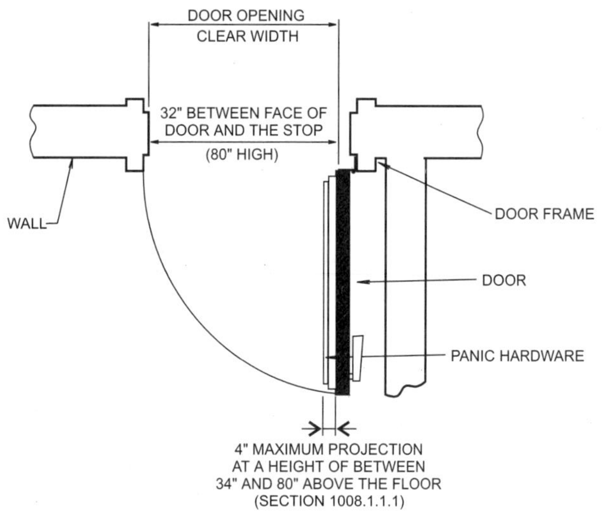

For SI: 1 inch = 25.4 mm.

Figure 1008.1.1(2)
TYPICAL DOOR OPENING CLEARANCE

1008.1.1.1 Projections into clear width. There shall not be projections into the required clear width lower than 34 inches (864 mm) above the floor or ground. Projections into the clear opening width between 34 inches (864 mm) and 80 inches (2032 mm) above the floor or ground shall not exceed 4 inches (102 mm).

❖ This section provides specific allowances for projection into the required clear widths of means of egress doors. These allowances directly correspond with the method of measuring the required clear width of the door as specified in Section 1008.1.1. A reasonable range of projections for door hardware and trim has been established by these requirements. The use of the means of egress door by a wheelchair occupant will not be significantly impacted by small projections located in inconspicuous areas. The key to these allowances is their location. Projections are allowed at a height between 34 inches (864 mm) and 80 inches (2032 mm). Below the 34-inch (864 mm) height, the code does not permit any projections since they would decrease the available width for wheelchair operation. The full 32-inch (813 mm) width must be provided at this location. At 34 inches (864 mm) and higher, projections of up to and including 4 inches (102 mm) are permitted. The 4-inch (102 mm) projection is consistent with the allowances of Section 1003.3.3. This section permits door hardware, such as panic hardware, to extend into the clear width and yet maintain accessibility for persons with physical disabilities [see Figure 1008.1.1(2)].

The protrusion permitted for door closers and stops in Section 1003.3.1 is more specific than this 80-inch (2032 mm) headroom requirement; therefore, a door closer or stop is not limited to a 4-inch (102 mm) protrusion (see Figure 1003.3.1).

1008.1.2 Door swing. Egress doors shall be side-hinged swinging.

Exceptions:

1. Private garages, office areas, factory and storage areas with an occupant load of 10 or less.
2. Group I-3 occupancies used as a place of detention.
3. Doors within or serving a single dwelling unit in Groups R-2 and R-3 as applicable in Section 101.2.
4. In other than Group H occupancies, revolving doors complying with Section 1008.1.3.1.
5. In other than Group H occupancies, horizontal sliding doors complying with Section 1008.1.3.3 are permitted in a means of egress.
6. Power-operated doors in accordance with Section 1008.1.3.1.

Doors shall swing in the direction of egress travel where serving an occupant load of 50 or more persons or a Group H occupancy.

The opening force for interior side-swinging doors without closers shall not exceed a 5-pound (22 N) force. For other side-swinging, sliding and folding doors, the door latch shall release when subjected to a 15-pound (67 N) force. The door shall be set in motion when subjected to a 30-pound (133 N) force. The door shall swing to a full-open position when subjected to a 15-pound (67 N) force. Forces shall be applied to the latch side.

❖ Generally, egress doors are required to be the side-hinged swinging type. Side-hinged swinging doors are familiar to all occupants in the method of operation. Some contemporary door designs have pivots at the top and bottom, which are intended to be permitted by this section, since the door action itself has little difference between the side-hinged-type door.

The code has several conditions where it allows doors that are not the side-hinged swinging type.

Examples of the doors discussed in Exception 1 are an overhead garage door and sliding door. This exception allows doors other than the swinging type for the listed occupancies where the number of occupants is very low.

Exception 2 allows for sliding-type doors that are commonly used in prisons and jails.

Exception 3 allows for sliding-type doors within or serving a residential occupancy. The use of sliding doors on the interior of dwelling units is permitted by Housing and Urban Development (HUD) accessibility requirements and often are needed because of large door requirements of the code to comply with the accessibility provisions.

Exception 4 allows for revolving doors that meet the requirements of Section 1008.1.3.1.

Exception 5 allows for horizontal sliding doors that meet the requirements of Section 1008.1.3.3. This exception is intended to allow wide span openings to be used in a means of egress serving areas of refuge formed by horizontal exits, smoke barriers or elevator lobbies. This is to enhance the movement of people with mobility impairments to areas of safety without obstructions, since the horizontal sliding doors specified in Section 1008.1.3.3 afford simple operation by persons with disabilities.

Exception 6 provides for power-operated doors that comply with the requirements of Section 1008.1.3.2.

A side-hinged door must swing in the direction of egress travel where the required occupant capacity of the room is 50 or more. As such, a room with two doors and an occupant load of 99 would require both doors to swing in the direction of egress travel, even though each door has a calculated occupant usage of less than 50. At this level of occupant load, the possibility exists that, in an emergency situation, a compact line of people could form at a closed door that swings in a direction opposite the egress flow. This could delay or eliminate the first person's ability to open the door inward, with the rest of the queue behind the person.

In a Group H occupancy, the threat of rapid fire buildup or worse is such that any delay in egress caused by door swing may jeopardize the opportunity for all occupants to evacuate the premises. For this reason, all egress doors in Group H occupancies are to swing in the direction of egress.

The ability of all potential users to be physically capable of opening an egress door is a function of the forces required to open the door. The 5-pound (22 N) maximum force for pushing and pulling interior side-swinging doors without closers is based on that which has been deemed appropriate for people with a physical disability that makes it difficult to operate a door. The operating force for other doors is permitted to be higher. This recognizes that doors with closers, particularly fire doors, require greater operating forces in order to close fully in an emergency where combustion gases may be exerting pressure on the door assembly. Similarly, exterior doors are exempted because air pressure differentials and strong winds may prevent doors from being automatically closed. A maximum force of 15 pounds (7 kg) is required for operating the latching mechanism. Once unlatched, a maximum force of 30 pounds (14 kg) is applied to the latch side of the leaf to start the door in motion by overcoming its stationary inertia. Once in motion, it must not take more than 15 pounds (7 kg) of force to keep the door in motion until it reaches its full open position and the required clear width is available. To conform to this requirement on a continuing basis, door closers must be adjusted periodically and door fits must also be checked and adjusted when necessary.

1008.1.3 Special doors. Special doors and security grilles shall comply with the requirements of Sections 1008.1.3.1 through 1008.1.3.5.

❖ This section simply defines the scope of the code requirements for special doors and security grilles.

1008.1.3.1 Revolving doors. Revolving doors shall comply with the following:

1. Each revolving door shall be capable of collapsing into a bookfold position with parallel egress paths providing an aggregate width of 36 inches (914 mm).
2. A revolving door shall not be located within 10 feet (3048 mm) of the foot of or top of stairs or escalators. A dispersal area shall be provided between the stairs or escalators and the revolving doors.
3. The revolutions per minute (rpm) for a revolving door shall not exceed those shown in Table 1008.1.3.1.
4. Each revolving door shall have a side-hinged swinging door which complies with Section 1008.1 in the same wall and within 10 feet (3048 mm) of the revolving door.

❖ Revolving doors must comply with all four provisions.

Item 1: One of the causes contributing to the loss of lives in the 1942 Cocoanut Grove fire in Boston was that the revolving doors at the club's entrance could not collapse for emergency egress and there was not an alternative means of egress adjacent to the revolving doors. Thus, in the panic of the fire, the door became jammed and the club's occupants were trapped.

As a result of this fire experience, all revolving doors, including those for air structures, are now required to be equipped with a collapse feature. A book-fold operation is where all leafs collapse parallel to each other and to the direction of egress [see Figure 1008.1.3.1(1)]. A book-fold operation creates two openings of approximately equal width. The sum of the widths is not to be less than 36 inches (914 mm) so that a stream of pedestrians may use each side of the opening.

Item 2: If a stairway or escalator delivers users to a landing in front of a revolving door at a greater rate than the capacity of the door, a compact line of people will develop. Lines of people formed on a stairway or escalator create an unsafe situation, since stairways and escalators are not intended to be used as standing space for persons who may be waiting to use the revolving doors. Therefore, to avoid congestion at a revolving door, which under normal operation has a maximum delivery capacity of users, a dispersal area is required between the stairways or escalators and the revolving doors to allow for the queuing of people as they enter the door. Accordingly, to create a dispersal area for users of a revolving door, the door is not to be placed closer than 10 feet (3048 mm) from the foot or top of a stairway or escalator.

Item 3: Door speeds also directly relate to the capacity of a revolving door, which is calculated by multiplying the number of leafs (wings) by the revolutions per minute (rpm). For example, if you have a four-leaf door moving at 10 rpm, the door will allow 40 people to move in either direction in 1 minute.

Item 4: In case a revolving door malfunctions or becomes obstructed, the adjacent area is to be equipped with a conventional side-hinged door to provide users with an immediate alternative way to exit a building. The side-hinged door is intended to be used as a relief device for people lined up to use the revolving door or who desire to avoid it because of a physical disability or other reason. It also can be used when the revolving door is obstructed or out of service. The swinging door is to be immediately adjacent to the revolving door so that its availability is obvious [see Figure 1008.1.3.1(2)]. A single swinging door can be located between two revolving doors and comply with this provision.

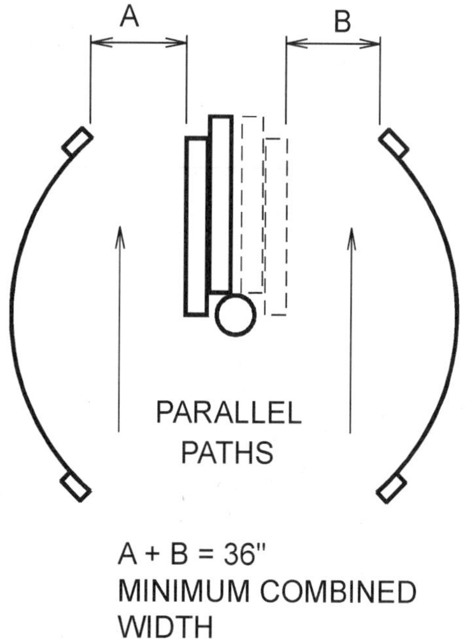

For SI: 1 inch = 25.4 mm.

Figure 1008.1.3.1(1)
REVOLVING DOORS IN BOOK-FOLD POSITION

TABLE 1008.1.3.1
REVOLVING DOOR SPEEDS

INSIDE DIAMETER (feet-inches)	POWER-DRIVEN-TYPE SPEED CONTROL (rpm)	MANUAL-TYPE SPEED CONTROL (rpm)
6-6	11	12
7-0	10	11
7-6	9	11
8-0	9	10
8-6	8	9
9-0	8	9
9-6	7	8
10-0	7	8

For SI: 1 inch = 25.4 mm, 1 foot = 304.8 mm.

❖ Door speeds also directly relate to the capacity of a revolving door, which is calculated by multiplying the number of leafs (wings) by the rpm. For example, if you have a four-leaf door moving at 10 rpm, the door will allow 40 people to move in either direction in 1 minute.

1008.1.3.1.1 Egress component. A revolving door used as a component of a means of egress shall comply with Section 1008.1.3.1 and the following three conditions:

1. Revolving doors shall not be given credit for more than 50 percent of the required egress capacity.
2. Each revolving door shall be credited with no more than a 50-person capacity.

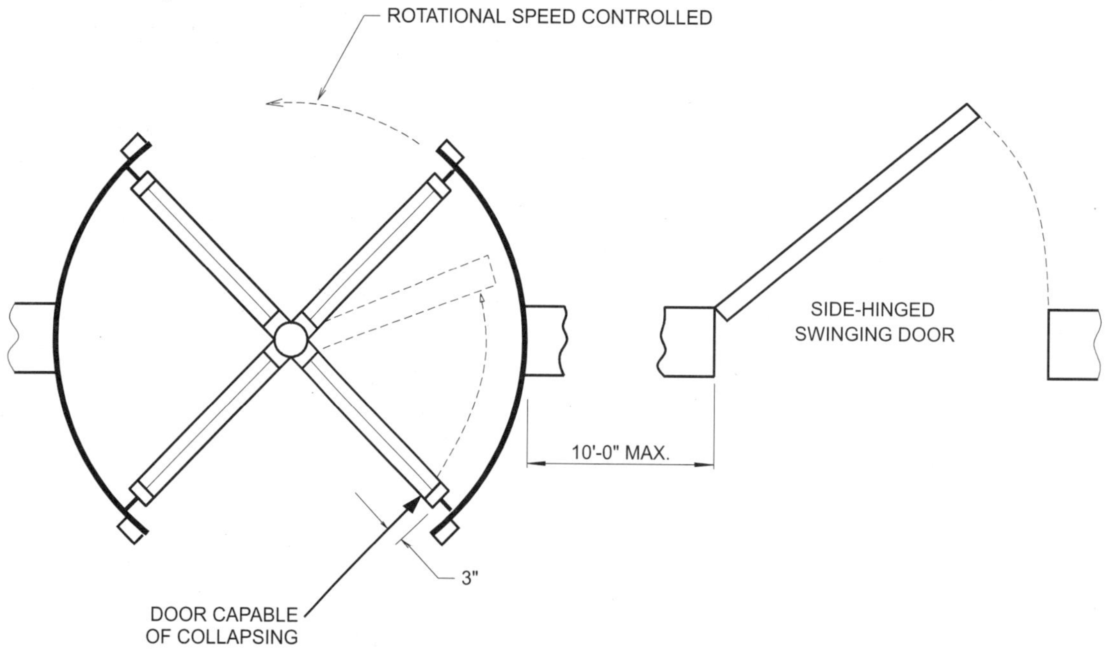

For SI: 1 inch = 25.4 mm.

Figure 1008.1.3.1(2)
TYPICAL REVOLVING DOORS ARRANGEMENT

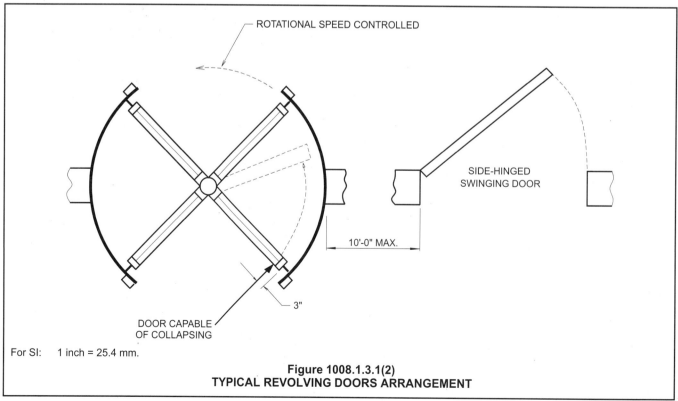

3. Each revolving door shall be capable of being collapsed when a force of not more than 130 pounds (578 N) is applied within 3 inches (76 mm) of the outer edge of a wing.

❖ A revolving door can be incorporated to a very limited extent in a means of egress. Compliance with these three additional conditions is required.

Condition 1 limits the exit capacity that revolving doors can provide in a building. This is so that 50 percent of the capacity has conventional egress components and is not dependent on mechanical devices or fail-safe mechanisms.

Condition 2 limits the capacity of any one revolving door for the same reasons as stated in Condition 1. Each revolving door is, therefore, limited to a 50-person capacity.

Condition 3 limits the collapse force to 130 pounds (59 kg), as opposed to the 180-pound (82 kg) value listed in Section 1008.1.3.1.2. Revolving doors used as means of egress are not permitted to have the collapse force exceed 130 pounds (59 kg) under any circumstances.

1008.1.3.1.2 Other than egress component. A revolving door used as other than a component of a means of egress shall comply with Section 1008.1.3.1. The collapsing force of a revolving door not used as a component of a means of egress shall not be more than 180 pounds (801 N).

Exception: A collapsing force in excess of 180 pounds (801 N) is permitted if the collapsing force is reduced to not more than 130 pounds (578 N) when at least one of the following conditions is satisfied:

1. There is a power failure or power is removed to the device holding the door wings in position.
2. There is an actuation of the automatic sprinkler system where such system is provided.
3. There is an actuation of a smoke detection system which is installed in accordance with Section 907 to provide coverage in areas within the building which are within 75 feet (22 860 mm) of the revolving doors.
4. There is an actuation of a manual control switch, in an approved location and clearly defined, which reduces the holding force to below the 130-pound (578 N) force level.

❖ This section addresses revolving doors that are not used to serve any portion of the occupant egress capacity. For example, where adjacent side-hinged doors have more than the required egress capacity, the revolving door would not be part of the required means of egress.

The maximum collapse force of 180 pounds (82 kg), applied within 3 inches (76 mm) of the outer edge of a wing, is based on industry standards to accommodate normal use conditions and other forces that may act on the leafs, such as that caused by wind or air pressure. An exception for revolving doors that are not a component of a required means of egress allows the collapse force to exceed 180 pounds (82 kg) in normal operating conditions, provided that a force of no more than 130 pounds (59 kg) is required whenever any one of the listed conditions is satisfied.

1008.1.3.2 Power-operated doors. Where means of egress doors are operated by power, such as doors with a photoelectric-actuated mechanism to open the door upon the approach of a person, or doors with power-assisted manual operation, the design shall be such that in the event of power failure, the door is capable of being opened manually to permit means of egress travel or closed where necessary to safeguard means of egress. The forces required to open these doors manually shall not exceed those specified in Section 1008.1.2, except that the force to set the door in motion shall not exceed 50 pounds (220 N). The door shall be capable of swinging from any position to the full width of the opening in which such door is installed when a force is applied to the door on the side from which egress is made. Full-power-operated doors shall comply with BHMA A156.10. Power-assisted and low-energy doors shall comply with BHMA A156.19.

Exceptions:

1. Occupancies in Group I-3.
2. Horizontal sliding doors complying with Section 1008.1.3.3.
3. For a biparting door in the emergency breakout mode, a door leaf located within a multiple-leaf opening shall be exempt from the minimum 32-inch (813 mm) single-leaf requirement of Section 1008.1.1, provided a minimum 32-inch (813 mm) clear opening is provided when the two biparting leaves meeting in the center are broken out.

❖ For convenience purposes, power-operated doors are intended to facilitate the normal nonemergency flow of persons through a doorway. Where a power-operated door is also required to be an egress door, the power-operated or assisted door must conform to the requirements of this section. The essential characteristic is that the door is to be manually openable from any position to its full open position at any time, with or without a power failure or a failure of a door mechanism. Hence, both swinging and horizontal sliding doors, complying with this section, may be used, provided the door can be operated manually from any position as a swinging door and that the minimum required clear width for egress capacity is not less than 32 inches (813 mm). Note that the opening forces of Section 1008.1.2 are applicable, except that the 30-pound (14 kg) force needed to set the door in motion is increased to 50 pounds (23 kg) as an operational tolerance in the design of the power-operated door.

Power-operated doors in detention and correctional occupancies (Group I-3) are not required to be manually operable by the occupants (inmates) for security reasons, but otherwise are required to conform to Section 408. Section 1008.1.3.2 does not apply to horizontal sliding doors that do not meet the provisions of this section (i.e., breakout panels for horizontal power-operated doors) but do comply with Section 1008.1.3.3.

Power-operated doors that meet the requirements of

this section are not required to meet the requirements of Section 1008.1.3.3 for horizontal sliding doors that are not capable of swing operation in the event of power failure.

Exception 3 allows an individual leaf of a four-panel biparting door to be less than 32 inches (813 mm) wide, provided 32 inches (813 mm) of clear space is provided when the two center biparting leaves are broken out.

1008.1.3.3 Horizontal sliding doors. In other than Group H occupancies, horizontal sliding doors permitted to be a component of a means of egress in accordance with Exception 5 to Section 1008.1.2 shall comply with all of the following criteria:

1. The doors shall be power operated and shall be capable of being operated manually in the event of power failure.
2. The doors shall be openable by a simple method from both sides without special knowledge or effort.
3. The force required to operate the door shall not exceed 30 pounds (133 N) to set the door in motion and 15 pounds (67 N) to close the door or open it to the minimum required width.
4. The door shall be openable with a force not to exceed 15 pounds (67 N) when a force of 250 pounds (1100 N) is applied perpendicular to the door adjacent to the operating device.
5. The door assembly shall comply with the applicable fire protection rating and, where rated, shall be self-closing or automatic-closing by smoke detection, shall be installed in accordance with NFPA 80 and shall comply with Section 715 of the *International Building Code.*
6. The door assembly shall have an integrated standby power supply.
7. The door assembly power supply shall be electrically supervised.
8. The door shall open to the minimum required width within 10 seconds after activation of the operating device.

❖ Horizontal sliding doors are permitted in the means of egress, in other than rooms or areas of Group H, under the conditions set forth in this section. Horizontal sliding doors are not permitted to be used in Group H occupancies because of the potential for delaying or impeding egress from those areas and the additional risk to occupants in hazardous occupancies. Note that this section regulates egress doors that do not meet all of the requirements of Section 1008.1.3.2 (e.g., a power-operated horizontal sliding door that does not have "break-out" capabilities to allow the door panels to swing if power is lost).

Such doors permitted in other groups are typically in the open position and close either because of a fire or to provide some degree of separation, often for security purposes to restrict movement to certain parts of a building. When in the closed position, the doors must be easily operated with forces similar to the limitations for swinging doors.

All eight of the criteria listed in this section must be met for a horizontal sliding door since there is a concern that it must be able to be easily opened under all conditions.

Additionally, the door must be openable even if a force of 250 pounds (114 kg) is being applied perpendicular to it, as may occur if a group of people were pushing on the door.

Since the doors are manually operable, they need not automatically open or close during a loss of power. However, a standby power supply must be provided. The primary power supply must be supervised so that an alarm is received at a constantly attended location (such as a security desk) on loss of the primary power. If the doors are also serving as fire doors, they must be automatic or self-closing in accordance with Section 715 of the IBC.

Since the maximum swinging door leaf width limitations of Section 1008.1.1 do not apply, a maximum opening time of 10 seconds is permitted. It should be noted, however, that the door need not open fully within the 10 seconds; rather, it must open to the required width. For example, if the door is protecting an opening that is 10 feet (3048 mm) wide, but the minimum required width of the opening is 32 inches (813 mm) (as determined by Section 1008.1.1), the door need only open 32 inches (813 mm) within the 10-second criterion. In fact, the door may have controls such that the automatic-opening feature only opens the door to a width of 32 inches (813 mm). If additional width is required, it can be accomplished by manual means and, possibly, by an additional activation of the operating device.

1008.1.3.4 Access-controlled egress doors. The entrance doors in a means of egress in buildings with an occupancy in Group A, B, E, M, R-1 or R-2 and entrance doors to tenant spaces in occupancies in Groups A, B, E, M, R-1 and R-2 are permitted to be equipped with an approved entrance and egress access control system which shall be installed in accordance with all of the following criteria:

1. A sensor shall be provided on the egress side arranged to detect an occupant approaching the doors. The doors shall be arranged to unlock by a signal from or loss of power to the sensor.
2. Loss of power to that part of the access control system which locks the doors shall automatically unlock the doors.
3. The doors shall be arranged to unlock from a manual unlocking device located 40 inches to 48 inches (1016 mm to 1219 mm) vertically above the floor and within 5 feet (1524 mm) of the secured doors. Ready access shall be provided to the manual unlocking device and the device shall be clearly identified by a sign that reads: PUSH TO EXIT. When operated, the manual unlocking device shall result in direct interruption of power to the lock—independent of the access control system electronics—and the doors shall remain unlocked for a minimum of 30 seconds.
4. Activation of the building fire alarm system, if provided, shall automatically unlock the doors, and the doors shall

remain unlocked until the fire alarm system has been reset.

5. Activation of the building automatic sprinkler or fire detection system, if provided, shall automatically unlock the doors. The doors shall remain unlocked until the fire alarm system has been reset.

6. Entrance doors in buildings with an occupancy in Group A, B, E or M shall not be secured from the egress side during periods that the building is open to the general public.

❖ Security in buildings is a major concern to owners and occupants from the perspective of property preservation and personal physical safety. Since many occupancies are partially occupied around the clock, after normal business hours or on weekends, it is necessary that some adequate level of security be provided without jeopardizing the egress capabilities of the occupants.

This section permits the building entrance doors in a means of egress and entrance doors to tenant spaces in occupancies of Groups A, B, E, M, R-1 and R-2 to be secured while maintaining them as a means of egress. Items 1 through 6 provide additional life safety measures to permit easier egress during normal and emergency situations. Occupancies in Groups F, S and H are not included here because of their increased potential hazard due to an increase in fuel load and other potentially life-threatening activities. This potential increase in life-threatening circumstances requires an immediate egress capacity without the necessary "waiting period" afforded by the access control of this section. Item 1 requires that such doors be provided with an automatic exit sensor typically operating on an infrared, microwave or sonic principle. This sensor is required to release automatically the lock upon an occupant approaching the door. Item 2 requires that if there is a loss of power to this device or to the access control system itself, the doors must unlock. Item 3 requires that there be a manual exit device, such as a push button, within 5 feet (1524 mm) of the door, mounted 40 to 48 inches (1016 to 1219 mm) above the floor, unobstructed and with a clearly identifiable sign that says "PUSH TO EXIT." When operated, the manual exit device is to interrupt (independent of the access control electronics) the power to the lock directly and cause the doors to remain unlocked for a minimum of 30 seconds. Items 4 and 5 require the building fire alarm system, automatic fire detection system or sprinkler system, if provided, to be interfaced with the access control system to unlock automatically the doors on activation. The doors are to remain unlocked until the fire alarm system is reset. This is so that the building entrance doors with controlled access will remain unlocked and open until fire fighters responding to the alarm have entered.

Item 6 requires that during the hours the building is open to the general public, doors equipped with an access control system will not be secured from the egress side in Group A, B, E and M occupancies. Thus, the building entrance doors in these occupancies are allowed to be secured only during off hours when the building occupant load will generally be reduced. In Groups R-1 and R-2, there is no restriction on the time period that access-controlled egress doors may be used, because the residents of the building are expected to be familiar with the building entrance and locking systems. Hence, with a building access control system, the number of individuals who are not familiar with the building is limited and they are usually accompanied by a resident. Note that access-controlled tenant entrance doors in Group A, B, E and M occupancies may be secured at any time.

1008.1.3.5 Security grilles. In Groups B, F, M and S, horizontal sliding or vertical security grilles are permitted at the main exit and shall be openable from the inside without the use of a key or special knowledge or effort during periods that the space is occupied. The grilles shall remain secured in the full-open position during the period of occupancy by the general public. Where two or more means of egress are required, not more than one-half of the exits or exit access doorways shall be equipped with horizontal sliding or vertical security grilles.

❖ This section really functions as an exception to several sections, including Sections 1008.1.2 and 1008.1.8.3 and permits the use of these security grilles under conditions that are similar to those found in Section 402 for covered mall buildings. These security grilles will be open when the space is occupied and will therefore not obstruct any egress path.

1008.1.4 Floor elevation. There shall be a floor or landing on each side of a door. Such floor or landing shall be at the same elevation on each side of the door. Landings shall be level except for exterior landings, which are permitted to have a slope not to exceed 0.25 unit vertical in 12 units horizontal (2-percent slope).

Exceptions:

1. Doors serving individual dwelling units in Groups R-2 and R-3 as applicable in Section 1001.1 where the following apply:
   1.1. A door is permitted to open at the top step of an interior flight of stairs, provided the door does not swing over the top step.
   1.2. Screen doors and storm doors are permitted to swing over stairs or landings.
2. Exterior doors as provided for in Section 1003.5, Exception 1, and Section 1017.2, which are not on an accessible route.
3. In Group R-3 occupancies, the landing at an exterior doorway shall not be more than 7¾ inches (197 mm) below the top of the threshold, provided the door, other than an exterior storm or screen door, does not swing over the landing.
4. Variations in elevation due to differences in finish materials, but not more than 0.5 inch (12.7 mm).
5. Exterior decks, patios or balconies that are part of Type B dwelling units and have impervious surfaces, and that are not more than 4 inches (102 mm) below the fin-

ished floor level of the adjacent interior space of the dwelling unit.

❖ Changes in floor surface elevation at a door, however small, are often slip or trip hazards. This is because persons passing through a door, including those who are physically disabled, usually do not expect changes in floor surface elevation or are not able to recognize them because of the intervening door leaf. Under emergency conditions, a fall in a doorway could result not only in injury to the falling occupant but could also interrupt orderly egress by other occupants.

Exception 1, which applies to residential occupancies, recognizes that occupants are familiar with the stair and landing arrangements. Note that an interior or exterior door (other than screen or storm doors) is not allowed to swing over a stair.

Except for exterior doors that meet Exceptions 2 and 3, the floor surface elevation of the area, described by the doorway width and length measured from the face of the door on both sides of the door threshold a distance equal to the door width, is to be at the same elevation plus or minus $^1/_2$ inch (12.7 mm) (see Figure 1008.1.4).

Certain exterior doors of Type B dwelling or sleeping units are also exempt from the floor surface requirements of Section 1008.1.4. Please note that this exception is not applicable for the primary entrance door (see Section 1105.1.6 of the IBC). Exterior doors that open out onto an exterior deck, patio or balcony are allowed a 4-inch (102 mm) step. Type B units are established by Chapter 11 of the IBC for residential occupancies containing four or more dwelling or sleeping units. The exterior decks, patios or balconies must be of solid and impervious construction, such as concrete or wood. A 4-inch (102 mm) step from inside the unit down to the exterior surfaces is allowed for weather purposes. This allowance is consistent with the provisions of ICC A117.1 and the Federal Fair Housing Accessibility Guidelines (FHAG).

1008.1.5 Landings at doors. Landings shall have a width not less than the width of the stairway or the door, whichever is the greater. Doors in the fully open position shall not reduce a required dimension by more than 7 inches (178 mm). When a landing serves an occupant load of 50 or more, doors in any position shall not reduce the landing to less than one-half its required width. Landings shall have a length measured in the direction of travel of not less than 44 inches (1118 mm).

Exception: Landing length in the direction of travel in Group R-3 as applicable in Section 1001.1 and Group U and within individual units of Group R-2 as applicable in Section 1001.1, need not exceed 36 inches (914 mm).

❖ This section is intended to address landings at doors (also see Section 1009.4 for stairway landings). The width of a landing at a door in a stairway is to be no less than the width of the stairway or the door, whichever is greater [see Figure 1009.4(4) for an example of these provisions]. Door landings are to have the floor elevation requirements of Section 1008.1.4 extending at least 44 inches (1118 mm) in the direction of egress travel.

The reduction in landing length for certain residential occupancies is a result of the low occupant load of these facilities.

1008.1.6 Thresholds. Thresholds at doorways shall not exceed 0.75 inch (19.1 mm) in height for sliding doors serving dwelling units or 0.5 inch (12.7 mm) for other doors. Raised thresholds and floor level changes greater than 0.25 inch (6.4 mm) at doorways shall be beveled with a slope not greater than one unit vertical in two units horizontal (50-percent slope).

Exception: The threshold height shall be limited to 7 $^3/_4$ inches (197 mm) where the occupancy is Group R-2 or R-3 as applicable in Section 1001.1, the door is an exterior door that is not a component of the required means of egress and the doorway is not on an accessible route.

❖ A threshold is a potential tripping hazard and a barrier to accessibility by people with mobility impairments. For

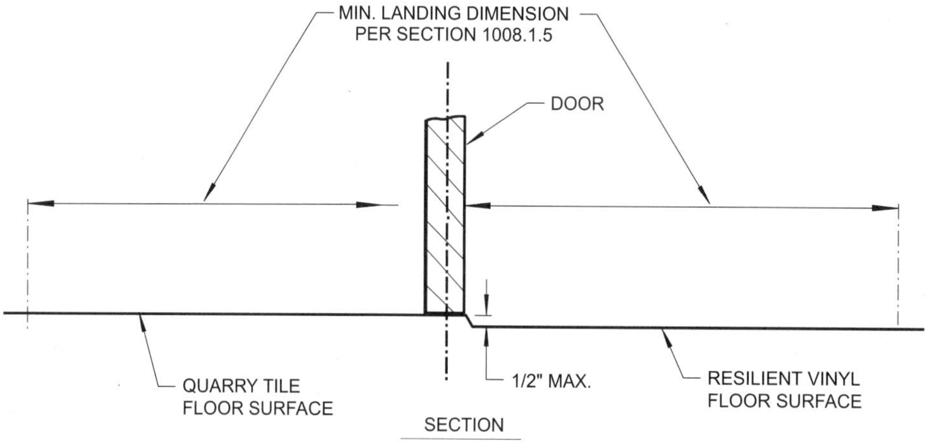

For SI: 1 inch = 25.4 mm.

Figure 1008.1.4
DOOR LANDINGS

these reasons, thresholds for all doorways, except exterior sliding doors serving dwelling units, are to be a maximum of $^{1}/_{2}$ inch (12.7 mm) high. Exterior sliding doors serving dwelling units, however, are permitted to be $^{3}/_{4}$ inch (19.1 mm) high because of practical design considerations, concern for deterioration of the doorway due to snow and ice buildup and lack of adequate drainage in severe climates. Raised threshold and floor level changes at doorways without edge treatment [see Figure 1008.1.6(1)] are permitted to be $^{1}/_{4}$ inch (6.3 mm) high vertically.

Raised threshold and floor level changes with edges beveled with a slope not greater than one unit vertical in two units horizontal (1:2) [see Figure 1008.1.6(2)] are permitted to be $^{1}/_{2}$ inch (12.7 mm) high, with its parts no more than $^{1}/_{4}$ inch (6.3 mm) high and the remainder with beveled edges no more than 1:2 [see Figure 1008.1.6(3)]. This kind of threshold treatment provides for minimum obstructions for wheelchair users and limits the trip hazard for those with other mobility disabilities.

The exception permits a step down at exterior doors for dwelling or sleeping units not required to be Accessible, Type A or Type B units.

1008.1.7 Door arrangement. Space between two doors in series shall be 48 inches (1219 mm) minimum plus the width of a door swinging into the space. Doors in series shall swing either in the same direction or away from the space between doors.

Exceptions:

1. The minimum distance between horizontal sliding power-operated doors in a series shall be 48 inches (1219 mm).
2. Storm and screen doors serving individual dwelling units in Groups R-2 and R-3 as applicable in Section 1001.1 need not be spaced 48 inches (1219 mm) from the other door.
3. Doors within individual dwelling units in Groups R-2 and R-3 as applicable in Section 1001.1 other than within Type A dwelling units.

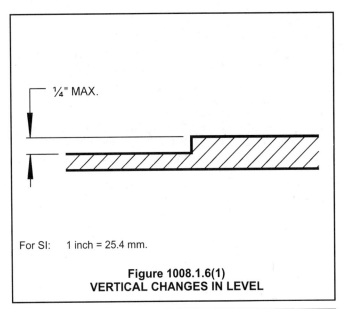

❖ Door arrangement is required to be such that an occupant's use of a means of egress doorway is not hampered by the operation of a preceding door located in the same line of travel so that the occupant flow can be smooth through the openings. Successive doors in a single egress path (i.e., in series) can cause such interference. The 4-foot (1219 mm) clear distance between doors when the first door is open allows an occupant, including one in a wheelchair, to move past one door and its swing before beginning the operation of the next door [see Figure 1008.1.7(1)]. Note that where doors in series are not arranged in a straight line, the intent of the code is to provide sufficient space to enable occupants to negotiate the second door without being encumbered by the first door's swing arc. To facilitate accessibility, the space between doors should provide sufficient clear space for a wheelchair [30 inches by 48 inches (762 mm by 1219 mm)] beyond the arc of the door swing [see Fig-

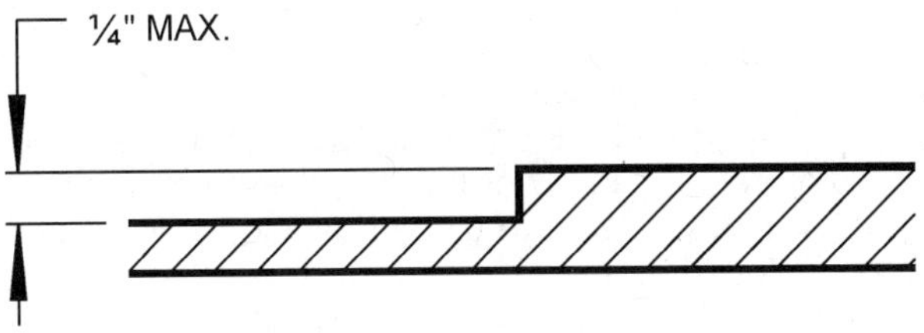

For SI: 1 inch = 25.4 mm.

Figure 1008.1.6(1)
VERTICAL CHANGES IN LEVEL

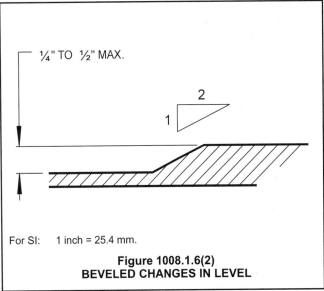

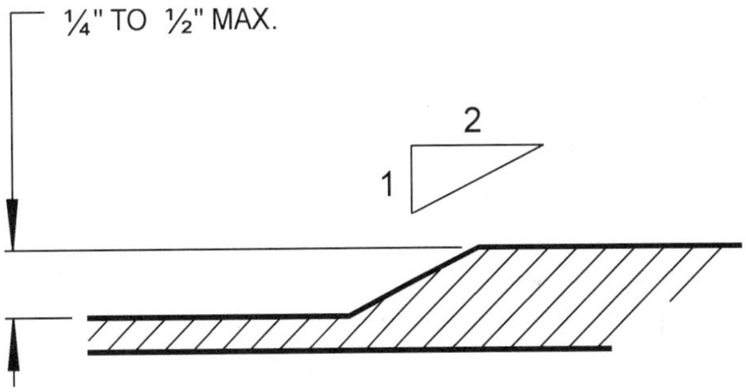

For SI: 1 inch = 25.4 mm.

Figure 1008.1.6(2)
BEVELED CHANGES IN LEVEL

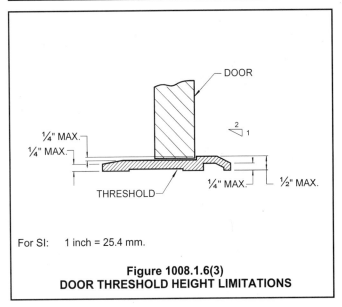

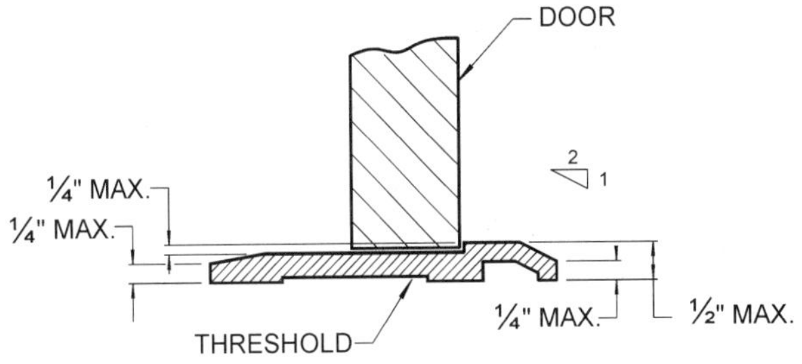

For SI: 1 inch = 25.4 mm.

Figure 1008.1.6(3)
DOOR THRESHOLD HEIGHT LIMITATIONS

ure 1008.1.7(2)]. Additionally, the approach and access provisions of ICC A117.1 should be considered.

The exception is to permit horizontal sliding power-operated doors (see Section 1008.1.3.2) to be designed with a lesser distance between them in a series arrangement, since they are customarily designed to open simultaneously or in sequence such that movement through them is unhampered. Storm and screen doors on residential dwelling units need not be spaced at 48 inches (1219 mm) since it would be impractical. Doors within dwelling units of Group R-2 or R-3 that are not Type A dwelling units (see Chapter 11 of the IBC) are also permitted to have a lesser distance between doors, because the accessibility provisions do not apply.

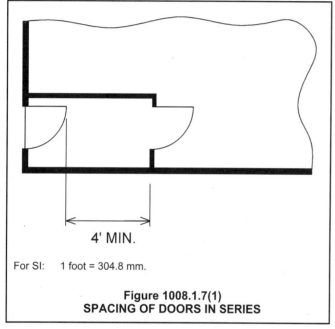

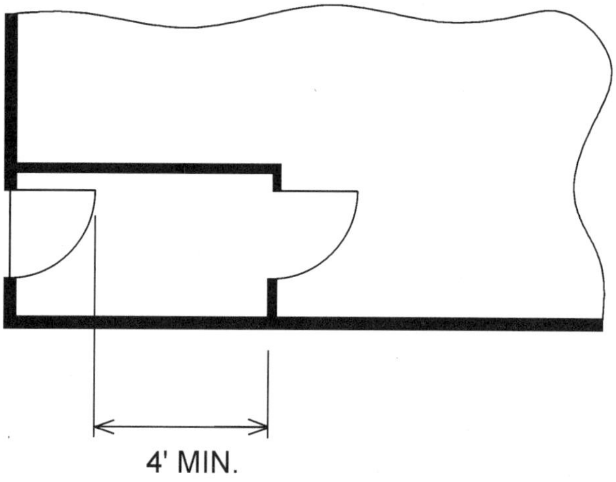

For SI: 1 foot = 304.8 mm.

Figure 1008.1.7(1)
SPACING OF DOORS IN SERIES

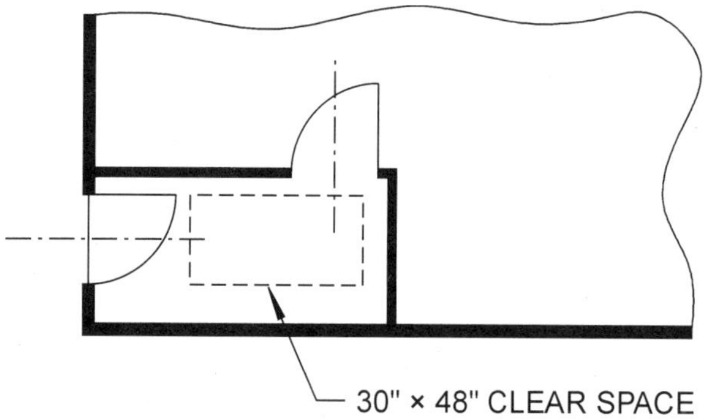

For SI: 1 inch = 25.4 mm.

Figure 1008.1.7(2)
SPACING OF DOORS IN SERIES

1008.1.8 Door operations. Except as specifically permitted by this section egress doors shall be readily openable from the egress side without the use of a key or special knowledge or effort.

❖ When installed for security purposes, locks and latches can intentionally prohibit the use of an egress door, thus interfering with or preventing the egress of occupants at the time of a fire. While the security of property is important for many, the life safety of occupants is essential for everyone. Where security and life safety objectives conflict, alternative measures, such as those permitted by each of the exceptions in Section 1008.1.8.3, may be applicable.

Egress doors are permitted to be locked, but must be capable of being unlocked and readily openable from the side from which egress is to be made. The outside of a door can be key locked as long as the inside (the side from which egress is to be made) can be unlocked without the use of tools, keys or special knowledge or effort. For example, an unlocking operation that is integral with an unlatching operation is acceptable.

Examples of special knowledge would be a combination lock or an unlocking device in an unknown, unexpected or hidden location. Special effort would dictate the need for unusual and unexpected physical ability to unlock or make the door fully available for egress.

Where a pair of egress door leafs is installed, with or without a center mullion, each leaf must in general be provided with its own releasing or unlatching device so as to be readily openable. Door arrangements or devices that depend on the release of one door before the other can be opened are not to be used, except as permitted by Section 1008.1.8.4.

1008.1.8.1 Hardware. Door handles, pulls, latches, locks and other operating devices on doors required to be accessible by Chapter 11 of the *International Building Code* shall not require tight grasping, tight pinching or twisting of the wrist to operate.

❖ Any doors that are located along an accessible route for ingress or egress must have door hardware that is easy to operate by a person with limited mobility. This would include all elements of the door hardware used in typical door operation, such as door levers, locks, security changes, etc. This requirement is also an advantage for persons with arthritis in their hands. Items such as small full twist thumb turns or smooth circular knobs are examples of hardware that is not acceptable.

1008.1.8.2 Hardware height. Door handles, pulls, latches, locks and other operating devices shall be installed 34 inches (864 mm) minimum and 48 inches (1219 mm) maximum above the finished floor. Locks used only for security purposes and not used for normal operation are permitted at any height.

❖ The requirements in this section place the door hardware at a level that is useable by most people, including a person in a wheelchair. The exception allows security locks to be placed at any height. An example would be an unframed glass door at the front door of a tenant space in a mall that has the lock near the floor level. The lock is only used when the store is not open for business. Such locks are not required for the normal operation of the door.

1008.1.8.3 Locks and latches. Locks and latches shall be permitted to prevent operation of doors where any of the following exists:

1. Places of detention or restraint.
2. In buildings in occupancy Group A having an occupant load of 300 or less, Groups B, F, M and S, and in churches, the main exterior door or doors are permitted to be equipped with key-operated locking devices from the egress side provided:
   - 2.1. The locking device is readily distinguishable as locked,
   - 2.2. A readily visible durable sign is posted on the egress side on or adjacent to the door stating: THIS DOOR TO REMAIN UNLOCKED WHEN BUILDING IS OCCUPIED. The sign shall be in letters 1 inch (25 mm) high on a contrasting background,
   - 2.3. The use of the key-operated locking device is revokable by the fire code official for due cause.
3. Where egress doors are used in pairs, approved automatic flush bolts shall be permitted to be used, provided that the door leaf having the automatic flush bolts has no doorknob or surface-mounted hardware.
4. Doors from individual dwelling or sleeping units of Group R occupancies having an occupant load of 10 or less are permitted to be equipped with a night latch, dead bolt or security chain, provided such devices are openable from the inside without the use of a key or tool.

❖ Where security and life safety objectives conflict, alternative measures, such as those permitted by each of the exceptions, may be applicable.

Exception 1 is needed for jails and prisons.

Exception 2 permits a locking device, such as a double-cylinder dead bolt, on the main entrance door. Such locking devices must have an integral indicator that automatically reflects the locked or unlocked status of the device. In addition to being an indicating lock, a sign must be provided that clearly states that the door is to be unlocked when the building is occupied. The sign on the door not only reminds employees to unlock the door, but also advises the public that an unacceptable arrangement exists if one finds the door locked. Ideally, the individual who encounters the locked door will notify management and possibly the fire code official. Note that the use of the key-locking device is revokable by the fire code official. The locking arrangement is not permitted on any door other than the main exit and, therefore, the employees, security and cleaning crews will have access to other exits without requiring the use of a key. This allowance is not limited just to multiple-exit buildings but also to small buildings with one exit.

In Exception 3, an automatic flush bolt device is one that is internal to the inactive leaf of a pair of doors. The device has a small "knuckle" that extends from the inactive leaf into the opening of the active leaf. When the active leaf is opened, the bolt is automatically retracted. When the active leaf is closed, the knuckle is pressed into the inactive leaf by the active leaf, extending the flush bolt(s) in the head or sill of the inactive leaf (see Figure 1008.1.8.3).

Automatic flush bolts on one leaf of a pair of egress doors are acceptable, provided the leaf with the automatic flush bolts is not equipped with a door knob or other hardware that would imply to the user that the door leaf is unlatched independently of the companion leaf.

Exception 4 addresses the need for security in residential units. The occupants are familiar with the operation of the indicated devices, which are intended to be relatively simple to operate without the use of a key or tool. Note that this exception only applies to the door from the dwelling unit.

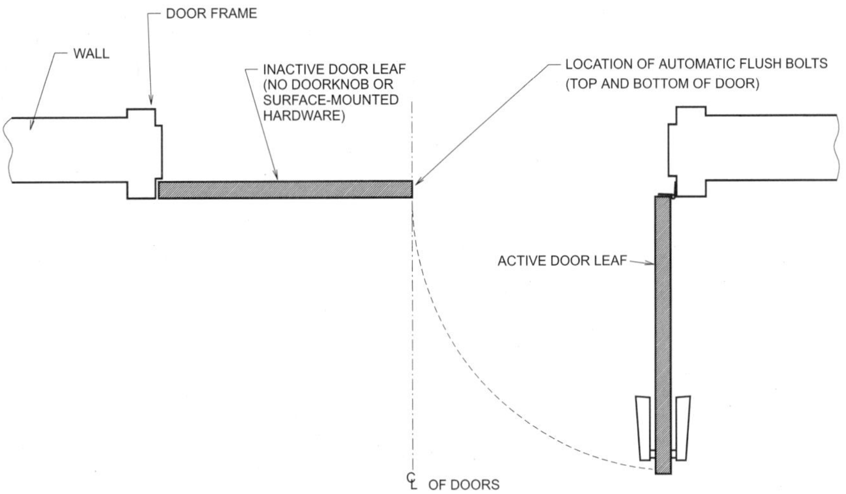

Figure 1008.1.8.3
SURFACE AND FLUSH BOLT LOCATIONS

1008.1.8.4 Bolt locks. Manually operated flush bolts or surface bolts are not permitted.

Exceptions:

1. On doors not required for egress in individual dwelling units or sleeping units.
2. Where a pair of doors serves a storage or equipment room, manually operated edge- or surface-mounted bolts are permitted on the inactive leaf.

❖ This section is applicable to doors that are intended and required to be for means of egress purposes or are identified as a means of egress, such as by an "exit" sign or other device. Doors, as well as a second leaf in a doorway that is provided for a purpose other than means of egress, such as for convenience or building operations, should be arranged or identified so as not to be mistaken as a means of egress.

This section prohibits installation of manually operated flush and surface bolts except in an individual dwelling or sleeping unit. Even then, such bolts may only be used on doors not required for egress. See Section 1008.1.8.3, Exception 4 for security of doors from individual dwelling and sleeping units. Flush and surface bolts represent locking devices that are difficult to operate because of their location and operation (see Figure 1008.1.8.4).

Exception 2 provides for edge-mounted or surface-mounted bolts on the inactive leaf of a pair of door(s) from these limited use areas.

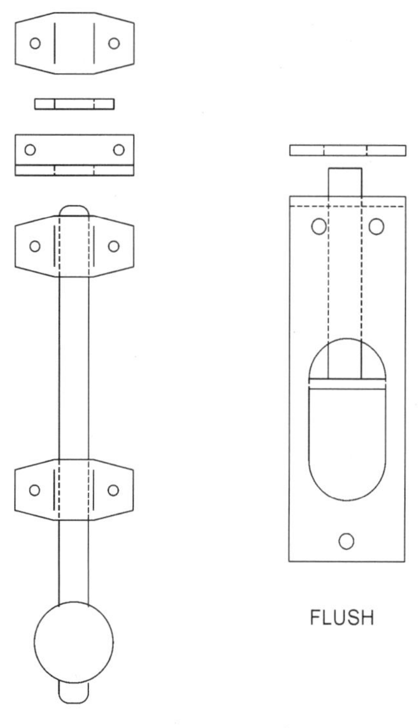

Figure 1008.1.8.4
TYPICAL BOLT HARDWARE

1008.1.8.5 Unlatching. The unlatching of any leaf shall not require more than one operation.

Exception: More than one operation is permitted for unlatching doors in the following locations:

1. Places of detention or restraint.
2. Where manually operated bolt locks are permitted by Section 1008.1.8.4.
3. Doors with automatic flush bolts as permitted by Section 1008.1.8.3, Exception 3.
4. Doors from individual dwelling units and guestrooms of Group R occupancies as permitted by Section 1008.1.8.3, Exception 4.

❖ The code prohibits the use of multiple locks or latching devices on a door. This would be a safety hazard in an emergency situation. The exceptions address locations where multiple locks or latching devices are accessible.

1008.1.8.6 Delayed egress locks. Approved, listed, delayed egress locks shall be permitted to be installed on doors serving any occupancy except Group A, E and H occupancies in buildings that are equipped throughout with an automatic sprinkler system in accordance with Section 903.3.1.1 or an approved automatic smoke or heat detection system installed in accordance with Section 907, provided that the doors unlock in accordance with Items 1 through 6 below. A building occupant shall not be required to pass through more than one door equipped with a delayed egress lock before entering an exit.

1. The doors unlock upon actuation of the automatic sprinkler system or automatic fire detection system.
2. The doors unlock upon loss of power controlling the lock or lock mechanism.
3. The door locks shall have the capability of being unlocked by a signal from the fire command center.
4. The initiation of an irreversible process which will release the latch in not more than 15 seconds when a force of not more than 15 pounds (67 N) is applied for 1 second to the release device. Initiation of the irreversible process shall activate an audible signal in the vicinity of the door. Once the door lock has been released by the application of force to the releasing device, relocking shall be by manual means only.

Exception: Where approved, a delay of not more than 30 seconds is permitted.

5. A sign shall be provided on the door located above and within 12 inches (305 mm) of the release device reading: PUSH UNTIL ALARM SOUNDS. DOOR CAN BE OPENED IN 15 [30] SECONDS.
6. Emergency lighting shall be provided at the door.

❖ For security reasons, special locking arrangements are permitted for doors in a means of egress serving occupancies other than those in Groups A, E and H. The arrangements are not permitted in assembly or educational occupancies because the resulting delay in egress is not acceptable, given the greater possibility of mass panic. Such a delay would also be inconsistent with Section 1008.1.9, which requires the installation of panic hardware on doors in such uses. Also, the delay from Group H would be unreasonable given the potential for rapid fire buildup in such areas.

Because of the possible delay caused by the controlled locking device in the available use of the egress door, the building must be provided throughout with compensating fire protection features to promptly warn occupants of a fire condition. All of the listed conditions must be met in order to permit use of such a locking device.

Condition 1 interconnects the lock with an automatic sprinkler system in accordance with NFPA 13 or, alternatively, an automatic fire detection system in accordance with Section 907, which is required to be installed throughout the building. Such systems are to provide occupants with an early warning of a fire event and thus additional time for egress. Note that the provision for an automatic fire detection system does not include the use of single- or multiple-station detectors. Also note that actuation of the automatic sprinkler or fire detection system is to unlock the control device so as to permit the egress door to be readily and immediately openable.

Condition 2 intends that the control device is fail safe. Since the operation of the device is dependent on electrical power, in the event of electrical power loss to the lock or locking mechanism, the egress doors must be readily and immediately openable from the side from which egress is to be made.

Condition 3 specifies that the door must be capable of being manually unlocked by a signal sent from a fire command center. The personnel at that location are intended to be the first alerted to an emergency event and are expected to take appropriate action to unlock all egress doors equipped with special locking arrangements. This will permit the locks to be deactivated in some cases prior to the sprinklers or smoke detection system detecting the problem or in case of other nonfire emergencies, such as an earthquake.

Condition 4 specifies the operational characteristics of the locking control device, which is similar to a panic device. A user must apply a minimum 15-pound (7 kg) force to the release device for at least 1 second, at which time an audible alarm will sound at the door and the device will automatically start to unlock the door. The 1-second duration is to prevent initiation of the unlocking process because of an inadvertent bump or accidental contact against the device. The unlocking cycle is irreversible; once it is started, it does not stop. Once the cycle starts, the door is required to be unlocked in no more than 15 seconds. When the door is unlocked at the end of the 15-second delay, it stays unlocked until someone comes to the door and manually relocks it. A method of automatically relocking the door from a remote location such as a central control station or security office is not permitted. Therefore, the first users to the door may face a delay, but after that other users would be able to exit immediately.

The exception will permit the fire code official to allow the time delay prior to opening to be increased beyond the basic 15 seconds but never to the point where the delay would exceed 30 seconds.

If a user continues to exert the 15-pound (7 kg) force for more than 1 second, the door is required to be openable after 15 seconds from the start of the force application.

The sign required by Condition 5 informs the user of the type of device and that the door will become available for egress. An undesirable consequence of the door not being immediately unlocked is if the user assumes it will never be available and proceeds to find another exit door.

Condition 6 provides emergency lighting at the door so that the user can read the sign required by Condition 5.

1008.1.8.7 Stairway doors. Interior stairway means of egress doors shall be openable from both sides without the use of a key or special knowledge or effort.

Exceptions:

1. Stairway discharge doors shall be openable from the egress side and shall only be locked from the opposite side.
2. This section shall not apply to doors arranged in accordance with Section 403.12 of the *International Building Code*.

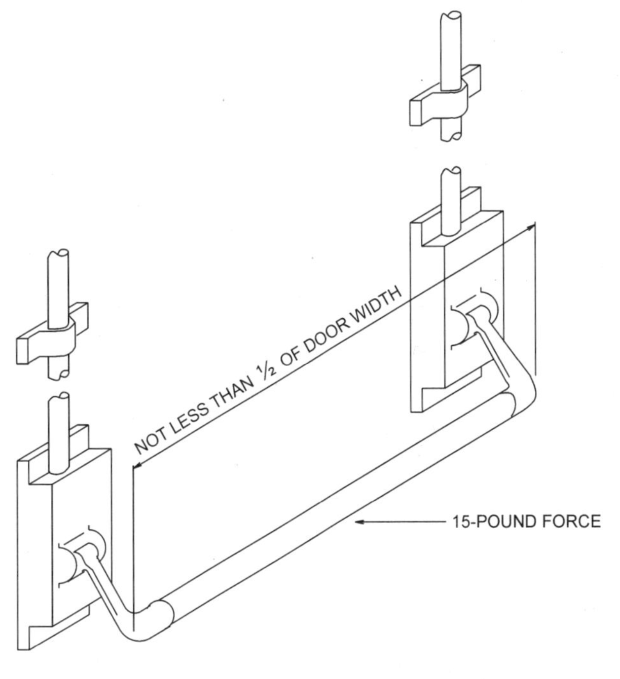

For SI: 1 pound = 4.4N.

Figure 1008.1.9
TYPICAL PANIC HARDWARE

3. In stairways serving not more than four stories, doors are permitted to be locked from the side opposite the egress side, provided they are openable from the egress side.

❖ Based on adverse fire experience where occupants have become trapped in smoke-filled stairway enclosures, generally stairway doors must be arranged to permit reentry into the building without the use of any tools, keys or special knowledge or effort. For security reasons, this restriction does not apply to the discharge door from the stairway enclosure, which is often to the outside. Section 403 for high-rise buildings permits the locking of the doors from the stairway side, provided the doors are capable of being unlocked from a fire command station and there is a communication system within the stairway enclosure that allows contact with the fire command station. It would be reasonable to permit this arrangement in buildings other than high-rise buildings.

Exception 3 addresses the need for security. The exception is limited to four-story buildings to provide a short travel distance to the stairway discharge door.

1008.1.9 Panic and fire exit hardware. Where panic and fire exit hardware is installed, it shall comply with the following:

1. The actuating portion of the releasing device shall extend at least one-half of the door leaf width.
2. A maximum unlatching force of 15 pounds (67 N).

Each door in a means of egress from an occupancy of Group A or E having an occupant load of 100 or more and any occupancy of Group H-1, H-2, H-3 or H-5 shall not be provided with a latch or lock unless it is panic hardware or fire exit hardware.

If balanced doors are used and panic hardware is required, the panic hardware shall be the push-pad type and the pad shall not extend more than one-half the width of the door measured from the latch side.

❖ As its name implies, panic hardware is special unlatching and unlocking hardware that is intended to simplify the unlatching and unlocking operation to a single 15-pound (7 kg) force applied in the direction of egress (see Figure 1008.1.9). In a panic situation with a rush of persons trying to utilize a door, the conventional devices, such as doorknobs or thumb turns, may cause sufficient delay so as to create a crush at the door and prevent or slow the opening operation.

In addition to hazardous occupancies, panic hardware is required on all doors that provide means of egress to rooms and spaces of assembly and educational (Groups A and E) occupancies with an occupant load of 100 or more. These uses are characterized by higher occupant load densities. Whereas doors from an assembly or educational room with an occupant load of less than 100 do not require panic hardware, a door that provides means of egress for two such rooms could require panic hardware when the combination of spaces has a total occupant load of 100 or more.

The locational specifications for the activating panel or bar are based on ready availability and access to the unlatching device. Note that this section requires the activating portion to extend at least one-half the width of the door leaf.

Where a fire door, such as to an exit stairway, is required to be equipped with panic hardware, the hardware must accomplish the dual objectives of panic hardware and continuity of the enclosure in which it is located. In this case, fire exit hardware is to be provided that meets both objectives and requirements, since panic hardware is not tested for use on fire doors.

There are standard test procedures designed to evaluate the performance of panic and fire exit hardware from the panic standpoint as well as from a fire protection standpoint.

The provisions for balanced doors ensure that the occupants push on the latch side of the door, which will swing in the direction of egress travel.

1008.2 Gates. Gates serving the means of egress system shall comply with the requirements of this section. Gates used as a component in a means of egress shall conform to the applicable requirements for doors.

Exception: Horizontal sliding or swinging gates exceeding the 4-foot (1219 mm) maximum leaf width limitation are permitted in fences and walls surrounding a stadium.

❖ This section specifies that all of the requirements for doors also apply to gates, except those that surround a stadium, which are allowed to exceed 4 feet (1219 mm) in width. Usually a large gate is required to adequately serve a stadium crowd for egress purposes.

1008.2.1 Stadiums. Panic hardware is not required on gates surrounding stadiums where such gates are under constant immediate supervision while the public is present, and further provided that safe dispersal areas based on 3 square feet (0.28 m^2) per occupant are located between the fence and enclosed space. Such required safe dispersal areas shall not be located less than 50 feet (15 240 mm) from the enclosed space. See Section 1017 for means of egress from safe dispersal areas.

❖ Panic hardware is impractical for large gates that surround stadiums. Normally these large gates are opened and closed by the stadium grounds crew, which is constantly in attendance during their use. The safe dispersal area requirement provides for the safety of the crowd if for some reason the gate is not open. The safe dispersal area is to be between the stadium enclosure and the surrounding fence and the area to be occupied is not to be closer than 50 feet (15 240 mm) to the stadium enclosure.

1008.3 Turnstiles. Turnstiles or similar devices that restrict travel to one direction shall not be placed so as to obstruct any required means of egress.

Exception: Each turnstile or similar device shall be credited with no more than a 50-person capacity where all of the following provisions are met:

1. Each device shall turn free in the direction of egress travel when primary power is lost, and upon the manual release by an employee in the area.

2. Such devices are not given credit for more than 50 percent of the required egress capacity.
3. Each device is not more than 39 inches (991 mm) high.
4. Each device has at least 16.5 inches (419 mm) clear width at and below a height of 39 inches (991 mm) and at least 22 inches (559 mm) clear width at heights above 39 inches (991 mm).

Where located as part of an accessible route, turnstiles shall have at least 36 inches (914 mm) clear at and below a height of 34 inches (864 mm), at least 32 inches (813 mm) clear width between 34 inches (864 mm) and 80 inches (2032 mm) and shall consist of a mechanism other than a revolving device.

❖ This section provides for a limited use of turnstiles to serve as a means of egress component. The exception to this section limits each turnstile to a maximum of 50 persons for egress capacity. The turnstile must comply with all four listed items to be considered to serve any part of the occupant load means of egress. The turnstiles must rotate freely both when there is a loss of power or when they are manually released. Note that the 50-person limit applies to each individual turnstile.

1008.3.1 High turnstile. Turnstiles more than 39 inches (991 mm) high shall meet the requirements for revolving doors.

❖ Where a turnstile is higher than 39 inches (991 mm), the restriction to egress is much like a revolving door. Thus, the egress limitations for revolving doors in Section 1008.1.3.1 apply to this type of turnstile. If a high turnstile does not meet the revolving door requirements for doors that are an egress component, it is not to be included as serving a portion of the means of egress. It would be necessary to provide doors in these areas for egress.

1008.3.2 Additional door. Where serving an occupant load greater than 300, each turnstile that is not portable shall have a side-hinged swinging door which conforms to Section 1008.1 within 50 feet (15 240 mm).

❖ This section addresses a common egress condition for sports areas where a number of turnstiles are installed for ticket taking. Portable turnstiles are moved from the egress path for proper exiting capacity. Permanent turnstiles are not considered as providing any of the required egress capacity when serving an occupant load greater than 300, no matter how many turnstiles are installed. Doors are required to provide occupants with a path of egress other than through the turnstiles. The doors are to be located within 50 feet (15 240 mm) of the turnstiles.

[B] SECTION 1009 STAIRWAYS AND HANDRAILS

1009.1 Stairway width. The width of stairways shall be determined as specified in Section 1005.1, but such width shall not be less than 44 inches (1118 mm). See Section 1007.3 for accessible means of egress stairways.

Exceptions:

1. Stairways serving an occupant load of 50 or less shall have a width of not less than 36 inches (914 mm).
2. Spiral stairways as provided for in Section 1009.9.
3. Aisle stairs complying with Section 1024.

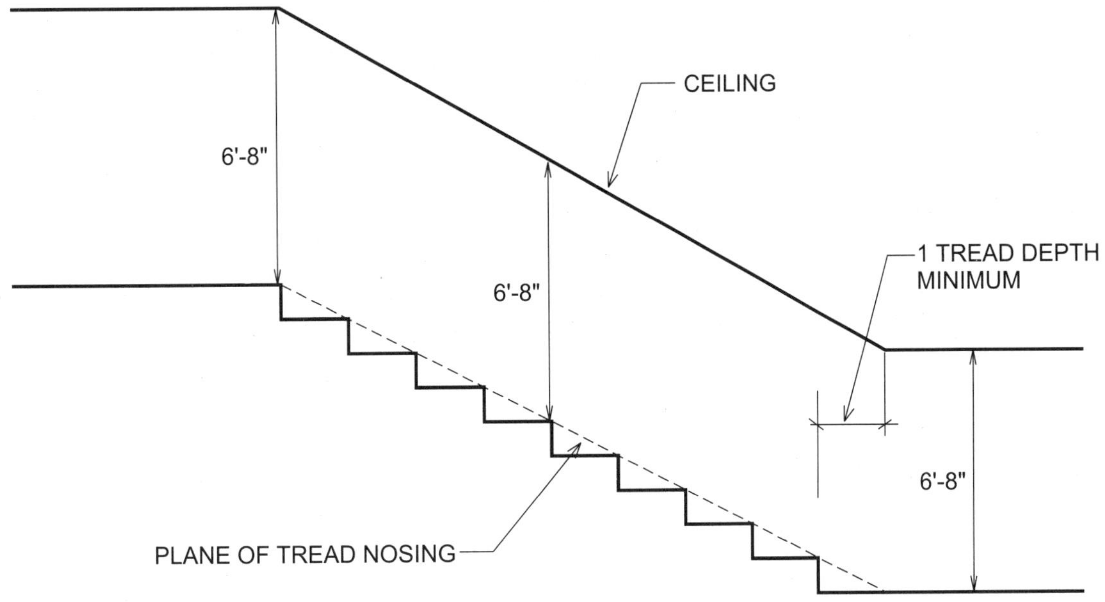

For SI: 1 inch = 25.4 mm, 1 foot = 304.8 mm.

Figure 1009.2
STAIRWAY HEADROOM REQUIREMENTS

4. Where a stairway lift is installed on stairways serving occupancies in Group R-3, or within dwelling units in occupancies in Group R-2, both as applicable in Section 1001.1 a clear passage width not less than 20 inches (508 mm) shall be provided. If the seat and platform can be folded when not in use, the distance shall be measured from the folded position.

❖ To provide adequate space for occupants traveling in opposite directions and to permit the intended full egress capacity to be developed, minimum dimensions are dictated for means of egress stairways. A minimum width of 44 inches (1118 mm) is required for stairway construction to permit two columns of users to travel in the same or opposite directions. The reference to Section 1005.1 is for the determination of stairway width based on occupant load. The larger of the two widths is to be used.

Exception 1 recognizes the relatively small occupant loads of 50 or less that permit a staggered file of users when traveling in the same direction. When traveling in opposite directions, one column of users must stop their ascent (or descent) to permit the opposite column to continue. Again, considering the relatively small occupant loads, any disruption of orderly flow will be infrequent.

Exception 2 permits a spiral stairway to have a minimum width of 26 inches (660 mm) when it conforms to Section 1009.9, on the basis that the configuration of a spiral stairway is such that other than single-file travel is unlikely.

Exception 3 provides for the aisle and aisle stair widths that are specified in Section 1024.

Exception 4 addresses the use of stairway lifts for individual dwelling units. Stairway lifts are typically used by mobility-impaired people in their homes. The code and ASME A 18.1 call for a minimum 20 inches (508 mm) of clear passageway to be maintained on a stairway where a stairway lift is located. If a portion of the stairway lift, such as a seat, can be folded, the minimum clear dimension is to be measured from the folded position. If the lift cannot be folded, then the 20 inches (508 mm) is measured from the fixed position. The track for these lifts typically extends 9 to 12 inches (229 to 305 mm) from the wall, making the 20-inch clear (508 mm) measurement actually 24 to 27 inches (610 to 686 mm) from the edge of the track.

1009.2 Headroom. Stairways shall have a minimum headroom clearance of 80 inches (2032 mm) measured vertically from a line connecting the edge of the nosings. Such headroom shall be continuous above the stairway to the point where the line intersects the landing below, one tread depth beyond the bottom riser. The minimum clearance shall be maintained the full width of the stairway and landing.

Exception: Spiral stairways complying with Section 1009.9 are permitted a 78-inch (1981 mm) headroom clearance.

❖ This requirement is necessary to avoid an obstruction to orderly flow and to provide visibility to the users so that the desired path of travel can be planned and negotiated. Height is a vertical measurement above every point along the stairway stepping and walking surfaces, with minimum height measured vertically from the tread nosing or from the surface of a landing or platform up to the ceiling (see Figure 1009.2). The exception for a clear headroom of 6 feet, 6 inches (1981 mm) for spiral stairs correlates with the provisions of Section 1009.9.

1009.3 Stair treads and risers. Stair riser heights shall be 7 inches (178 mm) maximum and 4 inches (102 mm) minimum. Stair tread depths shall be 11 inches (279 mm) minimum. The riser height shall be measured vertically between the leading edges of adjacent treads. The greatest riser height within any flight of stairs shall not exceed the smallest by more than 0.375 inch (9.5 mm). The tread depth shall be measured horizontally between the vertical planes of the foremost projection of adjacent treads and at right angle to the tread's leading edge. The greatest tread depth within any flight of stairs shall not exceed the smallest by more than 0.375 inch (9.5 mm). Winder treads shall have a minimum tread depth of 11 inches (279 mm) measured at a right angle to the tread's leading edge at a point 12 inches (305 mm) from the side where the treads are narrower and a minimum tread depth of 10 inches (254 mm). The greatest winder tread depth at the 12-inch (305 mm) walk line within any flight of stairs shall not exceed the smallest by more than 0.375 inch (9.5 mm).

Exceptions:

1. Circular stairways in accordance with Section 1009.7.
2. Winders in accordance with Section 1009.8.
3. Spiral stairways in accordance with Section 1009.9.
4. Aisle stairs in assembly seating areas where the stair pitch or slope is set, for sightline reasons, by the slope of the adjacent seating area in accordance with Section 1024.11.2.
5. In occupancies in Group R-3, as applicable in Section 1001.1, within dwelling units in occupancies in Group R-2, as applicable in Section 1001.1, and in occupancies in Group U, which are accessory to an occupancy in Group R-3, as applicable in Section 1001.1, the maximum riser height shall be 7.75 inches (197 mm) and the minimum tread depth shall be 10 inches (254 mm), the minimum winder tread depth at the walk line shall be 10 inches (254 mm), and the minimum winder tread depth shall be 6 inches (152 mm). A nosing not less than 0.75 inch (19.1 mm) but not more than 1.25 inches (32 mm) shall be provided on stairways with solid risers where the tread depth is less than 11 inches (279 mm).
6. See the *International Existing Building Code* for the replacement of existing stairways.

❖ The provisions for treads and risers not only determine the slope of a stairway, but also contribute to the efficient use of the stairway by allowing smooth and orderly travel by users.

The riser height—the vertical dimension from tread surface to tread surface or tread surface to landing surface—is typically limited to not more than 7 inches (178 mm) nor less than 4 inches (102 mm). The minimum tread depth—the horizontal distance from the leading edge (nosing) of one tread to the leading edge (nosing) of the next adjacent tread—is typically limited to not less than 11 inches (279 mm) [see Figure 1009.3(1)]. The minimum tread depth of 11 inches (279 mm) is intended to accommodate the largest shoe size found in 95 percent of the adult population, allowing for an appropriate overhang of the foot beyond the tread nosing while descending a stairway. Tread depths under 11 inches (279 mm) could cause an abnormal overhang (depending on the size of the foot) and could force users to descend a stairway with their feet pointing sideways in a crab-like manner. Such a condition does not promote orderly stairway travel. Based on research of geometrical possibilities of adequate foot placement, the rate of misstep with various step sizes and consideration for user's comfort and energy expenditure, it was found that the 11-inch (279 mm) minimum tread depth and maximum 7-inch (178 mm) riser height resulted in the best proportions for stairway construction.

The size for a winder tread is also considered for proper foot placement along the walking line [see Figure 1009.3(3)]. The dimensional requirements are consistent with the straight tread.

A minimum riser height of 4 inches (102 mm) is necessary to enable the user to identify visually the presence of the riser in ascent or descent.

The precise location of tread depth and riser measurements is to be perpendicular to the tread or riser's leading edge or nosing. This is to duplicate the user's anticipated foot placement in traveling the stairway [see Figure 1009.3(2)].

The maximum variance between the minimum and maximum riser height along a run of a stairway is $^3/_8$ inch (9.5 mm). A cadence is established as a person moves up or down a stairway. A change in more than $^3/_8$ inch (9.5 mm) along a run would become a tripping hazard. A $^3/_8$ inch (9.5 mm) maximum difference between the tread depth of a winder-type stairway is also based on an expected foot placement once a cadence is established.

The exceptions apply only to the extent of the text of each exception. For example, the entire text of Section 1009.3 is set aside for spiral stairways conforming to Section 1009.9 (see Exception 3). However, Exception 5 allows a different maximum riser and minimum tread under limited conditions, but retains the minimum riser height and measurement method of Section 1009.3.

At this time, the requirements for dimensional uniformity are found in both Section 1009.3 and 1009.3.1.

Exceptions 1, 2 and 3 are discussed in Sections 1009.7, 1009.8 and 1009.9, respectively.

Exception 4 provides a practical exception where assembly facilities are designed for viewing. See Sections 1024.11 through 1024.11.3 for assembly aisle stair limiting dimensions.

Exception 5 allows revisions to the standard 7-11 riser/tread requirements for residential occupancies of Group R-3 and within dwelling units of Group R-2 and U buildings where such structures are accessory to a Group R-3 occupancy (such as barns or detached garages). This increase is allowed because of the low occupant load and the high degree of occupant familiarity with the stairways. When this exception is taken for stairways that have solid risers, each tread is required to have a nosing with a minimum dimension of $^3/_4$ inch (19.1 mm) and maximum dimension of $1^1/_4$ inches (32 mm), where the tread depth is less than 11 inches (279 mm). Nosings are not required for residential stairs with open risers and 10-inch (254 mm) treads. The nosing provides a greater stepping surface for those ascending the stairway [see Figure 1009.3(1)].

Exception 6 allows for the replacement of an existing stair. Where a change of occupancy would require compliance with current standards, this exception allows a stairway that may be steeper than that permitted, provided it does not constitute a hazard [see the *International Existing Building Code*® (IEBC®)].

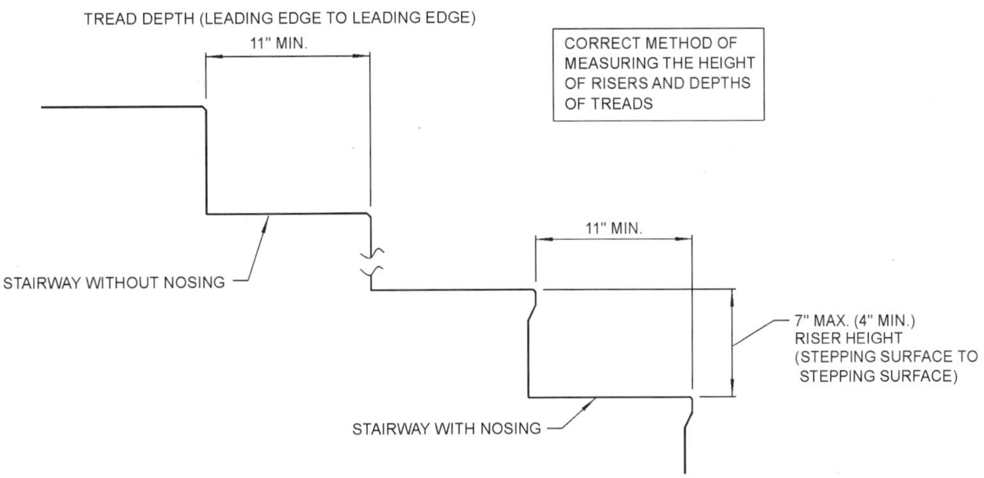

For SI: 1 inch = 25.4 mm.

Figure 1009.3(1)
TREAD/RISER DIMENSIONS

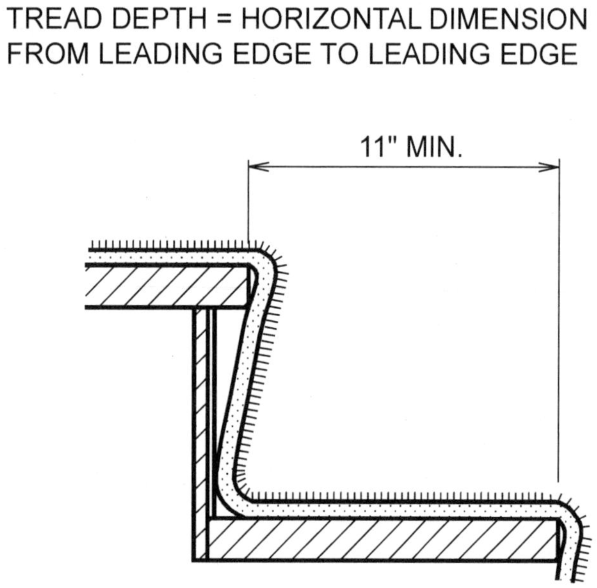

For SI: 1 inch = 25.4 mm.

Figure 1009.3(2)
TREAD DEPTH MEASUREMENT

1009.3.1 Dimensional uniformity. Stair treads and risers shall be of uniform size and shape. The tolerance between the largest and smallest riser or between the largest and smallest tread shall not exceed 0.375 inch (9.5 mm) in any flight of stairs.

Exceptions:

1. Nonuniform riser dimensions of aisle stairs complying with Section 1024.11.2.
2. Consistently shaped winders, complying with Section 1009.8, differing from rectangular treads in the same stairway flight.

Where the bottom or top riser adjoins a sloping public way, walkway or driveway having an established grade and serving as a landing, the bottom or top riser is permitted to be reduced along the slope to less than 4 inches (102 mm) in height with the variation in height of the bottom or top riser not to exceed one unit vertical in 12 units horizontal (8-percent slope) of stairway width. The nosings or leading edges of treads at such nonuniform height risers shall have a distinctive marking stripe, different from any other nosing marking provided on the stair flight. The distinctive marking stripe shall be visible in descent of the stair and shall have a slip-resistant surface. Marking stripes shall have a width of at least 1 inch (25 mm) but not more than 2 inches (51 mm).

❖ Dimensional uniformity in the design and construction of interior stairways contributes to safe stairway use. In

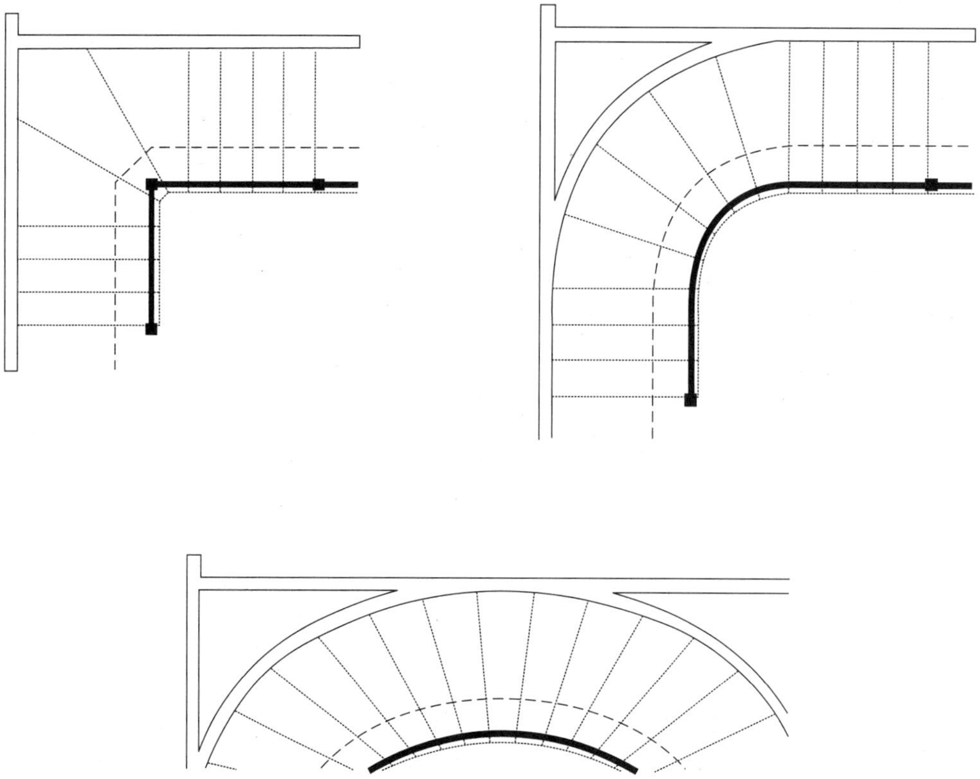

Figure 1009.3(3)
WINDER ALTERNATIVES

ascending or especially when descending a stair, a user sets a natural cadence or rhythmic movement based on the unconscious expectation or "feel" that each step taken will be at the same height and will land in approximately the same balanced position on the tread as the previous steps in the pattern. Any substantial change in tread or riser dimensions in a flight of stairways can break the rhythm and cause a misstep, stumbling or undue physical strain that may result in a fall or serious injury. In emergency situations, building occupants tend to use stairways at a faster pace than under normal conditions, increasing the risk to the user. Therefore, this section limits the dimensional variations to a tolerance of $^3/_8$ inch (9.5 mm) between the largest and smallest riser or tread dimension in a flight of stairs. A "flight" of stairs is commonly defined as the stairs between landings.

For special conditions of construction and as a practical matter, this section allows some greater variations in stairway tread and riser dimensions than the general limitations specified above. Exception 1 addresses conditions where the seating in assembly facilities is on a sloping gradient (for sightline purposes) and the aisle stairs become an integral part of the arrangement. Exception 2 addresses winder treads. They must be consistent along the walking line. The requirements for dimensional uniformity are found in both Section 1009.3 and 1009.3.1. This section also addresses the situation where the bottom riser of a flight of stairways meets a sloped landing such as a public way, walk or driveway.

1009.3.2 Profile. The radius of curvature at the leading edge of the tread shall be not greater than 0.5 inch (12.7 mm). Beveling of nosings shall not exceed 0.5 inch (12.7 mm). Risers shall be solid and vertical or sloped from the underside of the leading edge of the tread above at an angle not more than 30 degrees (0.52 rad) from the vertical. The leading edge (nosings) of treads shall project not more than 1.25 inches (32 mm) beyond the tread below and all projections of the leading edges shall be of uniform size, including the leading edge of the floor at the top of a flight.

Exceptions:

1. Solid risers are not required for stairways that are not required to comply with Section 1007.3, provided that the opening between treads does not permit the passage of a sphere with a diameter of 4 inches (102 mm).
2. Solid risers are not required for occupancies in Group I-3.

❖ The profiles of treads and risers contribute to stairway safety. The maximum radius of curvature at the leading edge of the tread is intended to allow descending foot placement on a surface that does not pitch the foot forward or allow the ball of the foot to slide off the treads. If a stairway design uses a beveled nosing configuration, the bevel is limited to a depth of $^1/_2$ inch (12.7 mm). Solid risers with either no nosing or a slope at the underside of the nosing are required so that the user's foot does not catch while ascending the stairway [see Figure 1009.3.2(1)].

Exception 1 establishes that open risers on stairs are allowed where the accessible means of egress stairway provisions of Section 1007.3 do not apply. The maximum radius for the leading edge, however, is still required. Since the opening limitation in guards is 4 inches (102 mm), even where the riser is allowed to be open, the opening is limited to be consistent with the requirements for guards [see Figure 1009.3.2(2)].

Open risers are not permitted where accessibility or adaptability is required based primarily on the difficulty open risers pose to people with sight impairments who use a long cane and ICC A117.1 specifically prohibits open risers.

Exception 2 recognizes that open risers are commonly used for stairs in occupancies such as detention facilities for practical reasons. Open risers provide a greater degree of security and supervision due to the fact that people cannot effectively conceal themselves behind the stair. The opening limitations of Exception 1 are not applicable to these stairs.

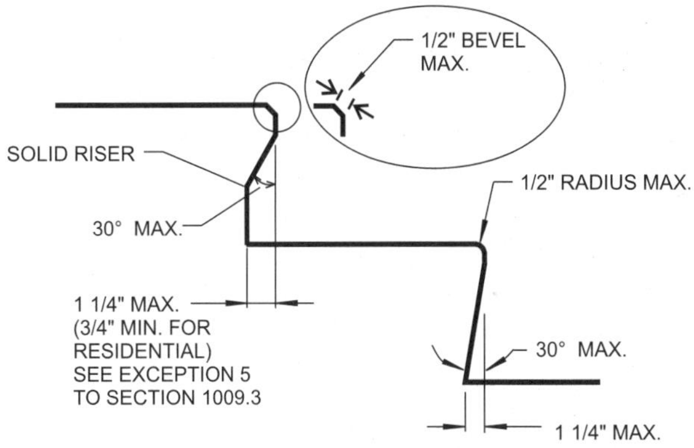

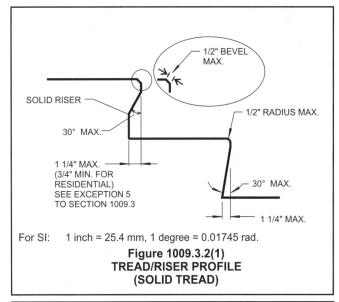

For SI: 1 inch = 25.4 mm, 1 degree = 0.01745 rad.

Figure 1009.3.2(1)
TREAD/RISER PROFILE
(SOLID TREAD)

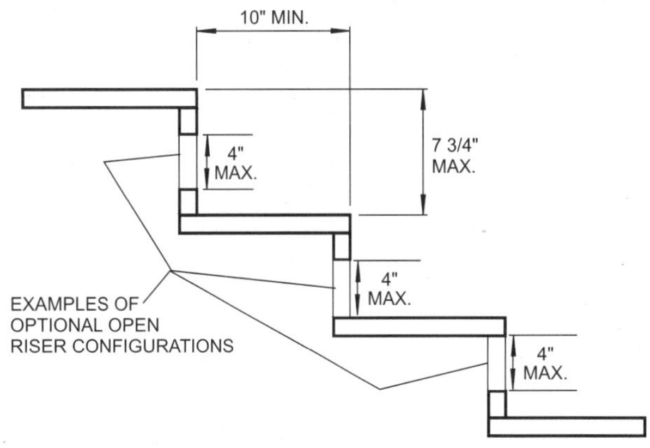

For SI: 1 inch = 25.4 mm.

Figure 1009.3.2(2)
TREAD/RISER PROFILE
(OPEN TREAD)

1009.4 Stairway landings. There shall be a floor or landing at the top and bottom of each stairway. The width of landings shall not be less than the width of stairways they serve. Every landing shall have a minimum dimension measured in the direction of travel equal to the width of the stairway. Such dimension need not exceed 48 inches (1219 mm) where the stairway has a straight run.

Exceptions:

1. Aisle stairs complying with Section 1024.
2. Doors opening onto a landing shall not reduce the landing to less than one-half the required width. When fully open, the door shall not project more than 7 inches (178 mm) into a landing.

❖ A level portion of a stairway provides users with a place to rest in their ascent or descent, to enter a stairway and to adjust their gait before continuing. Landings and platforms also break up the run of a stairway that continues for a considerable distance to arrest falls that may occur (see Section 1009.6).

The minimum size (width and depth) of all landings and platforms in a stairway is determined by the required width of the stairway. If Section 1009.1 requires a stairway to have a width of at least 44 inches (1118 mm), then all landings serving that stairway must be at least 44 inches (1118 mm) wide and 44 inches (1118 mm) deep [see Figure 1009.4(1)]. When a stairway is configured so that it has a straight run, the minimum dimension of the landing between flights in the direction of egress travel is not required to exceed 48 inches (1219 mm), even though the required width of the stair may exceed 48 inches (1219 mm) [see Figure 1009.4(2)]. When successive flights are not in a straight run, the landing must be sized using the required width of the stairway. For example, if Section 1009.1 requires a stairway to have a minimum clear width of 70 inches (1778 mm) because of capacity (see Section 1005.1), then all landings serving that stairway, where there is a change of direction, must be at least 70 inches (1778 mm) wide and 70 inches (1778 mm) deep [see Figure 1009.4(3)]. The measurements for a straight-run stairway are made from nosing to nosing and in clear width.

It is not the intent of this section to require that a stairway landing be shaped as a square or rectangle. A landing shaped as a half-circle would be permitted, as long as the landing provides an area described by an arc with a radius equal to the required stairway width [see Figure 1009.4(3)]. In this case, the minimum required space necessary for means of egress will be available.

Exception 1 provides for aisle stairs where the requirements are in Section 1024 and this section does not apply. Exception 2 limits the arc of the door swing on a landing so that the effect on the means of egress is minimized [see Figure 1009.4(4)].

1009.5 Stairway construction. All stairways shall be built of materials consistent with the types permitted for the type of construction of the building, except that wood handrails shall be permitted for all types of construction.

❖ In keeping with the different levels of fire protection provided by each of the five basic types of construction designated in Chapter 6, the materials used for stairway construction must meet the appropriate combustibility/noncombustibility requirements indicated in Section 602 for the particular type of construction of the building in which the stairway is located. This is required whether or not the stair is part of the required means of egress.

If desired, wood handrails may be used on the basis that the fuel load contributed by this combustible component of stairway construction is insignificant and will not pose a fire hazard.

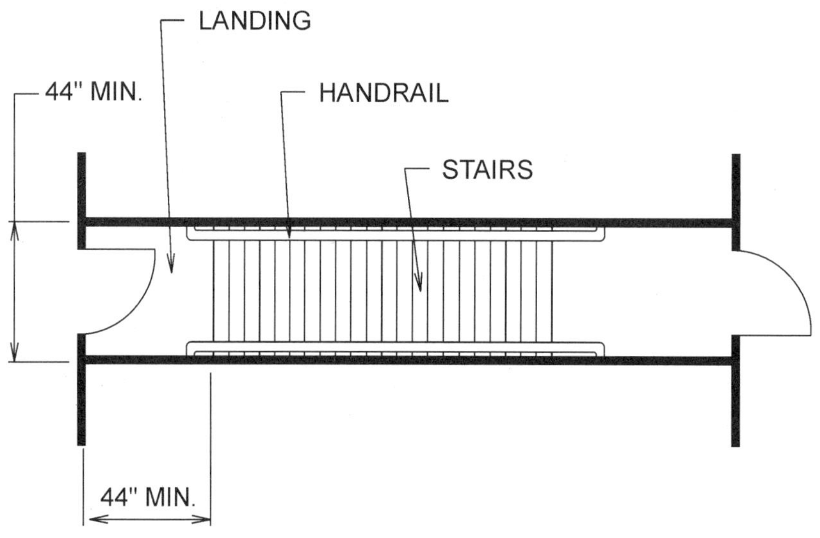

For SI: 1 inch = 25.4 mm.

Figure 1009.4(1)
TYPICAL LANDING DIMENSIONS STRAIGHT-RUN STAIR

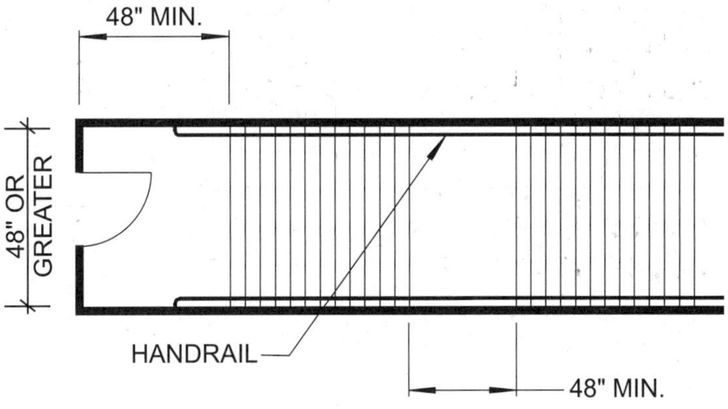

For SI: 1 inch = 25.4 mm.

Figure 1009.4(2)
TYPICAL LANDING DIMENSIONS STRAIGHT-RUN STAIR

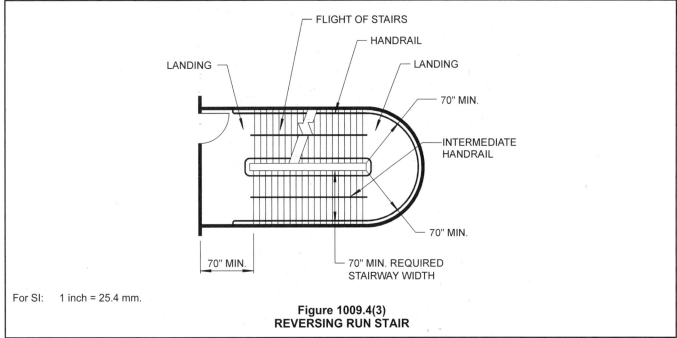

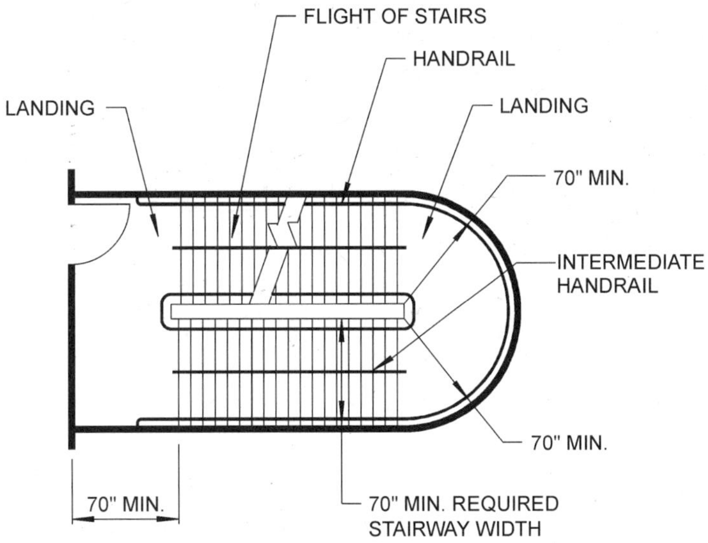

For SI: 1 inch = 25.4 mm.

Figure 1009.4(3)
REVERSING RUN STAIR

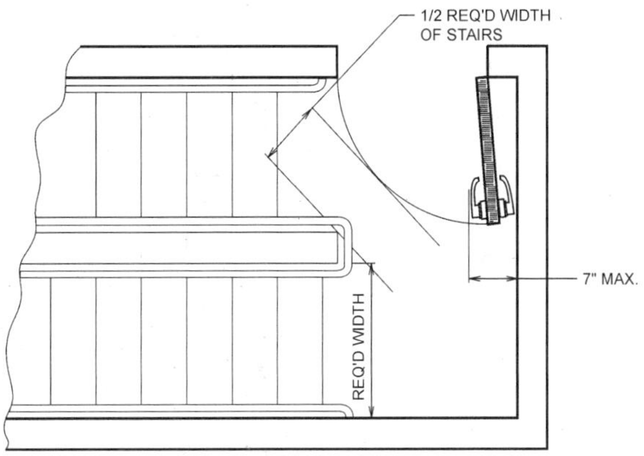

For SI: 1 inch = 25.4 mm.

Figure 1009.4(4)
DOOR CLEARANCE REQUIREMENT IN STAIRWAYS

1009.5.1 Stairway walking surface. The walking surface of treads and landings of a stairway shall not be sloped steeper than one unit vertical in 48 units horizontal (2-percent slope) in any direction. Stairway treads and landings shall have a solid surface. Finish floor surfaces shall be securely attached.

Exception: In Group F, H and S occupancies, other than areas of parking structures accessible to the public, openings in treads and landings shall not be prohibited provided a sphere with a diameter of $1^1/_8$ inches (29 mm) cannot pass through the opening.

❖ It is the intent of this section that both landing and stair treads be solid.

The exception permits the use of open-grate-type material for stairway treads and landings in factory, industrial, storage and high-hazard occupancies. This provision is intended to apply primarily to stairs that provide access to areas not required to be accessible, such as pits, catwalks, tanks, equipment platforms, roofs or mezzanines. Walking surfaces with limited-size openings are typically used because open-grate-type material is less susceptible to accumulation of dirt, debris or moisture as well as being more resistant to corrosion. Most commercially available grate material is manufactured with a maximum nominal 1-inch (25 mm) opening; therefore, the limitation that the opening not allow the passage of a sphere of $1^1/_8$-inch (29 mm) diameter allows the use of most material as well as the ability to account for manufacturing tolerances.

1009.5.2 Outdoor conditions. Outdoor stairways and outdoor approaches to stairways shall be designed so that water will not accumulate on walking surfaces. In other than occupancies in Group R-3, and occupancies in Group U that are accessory to an occupancy in Group R-3, treads, platforms and landings that are part of exterior stairways in climates subject to snow or ice shall be protected to prevent the accumulation of same.

❖ Outdoor stairways and approaches to stairways are to be constructed with a slope that complies with Section 1009.5.1 or are required to be protected such that walking surfaces do not accumulate water.

Where exterior stairways are used in moderate or severe climates, they must be protected from accumulations of snow and ice to provide a safe path of egress travel at all times, including winter. Typical methods for protecting these egress elements include roof overhangs or canopies, heated slab and, when approved by the fire code official, a reliable snow removal maintenance program.

1009.6 Vertical rise. A flight of stairs shall not have a vertical rise greater than 12 feet (3658 mm) between floor levels or landings.

Exception: Aisle stairs complying with Section 1024.

❖ Between landings and platforms, the vertical rise is to be measured from landing walking surface to landing walking surface (see Figure 1009.6). The limited height provides a reasonable interval for users with physical limitations to rest on a level surface and also serves to alleviate potential negative psychological effects of long and uninterrupted stairway flights.

The exception provides for aisle stairs in assembly occupancies, which are regulated by Section 1024 and not by this section.

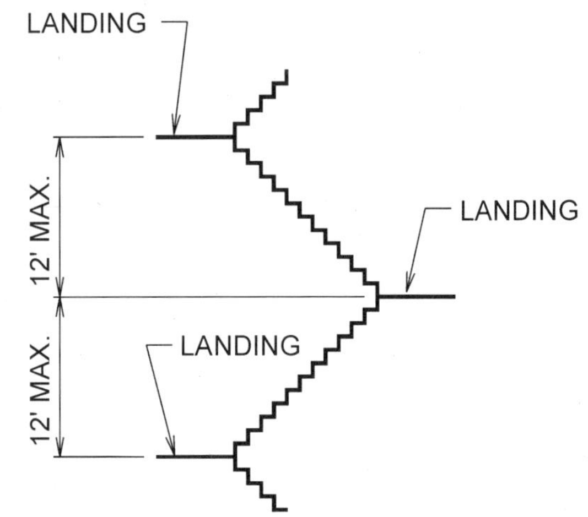

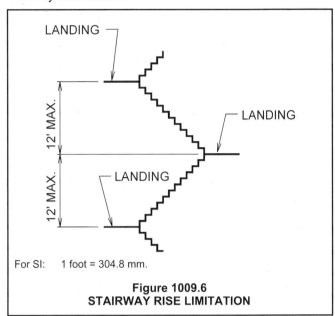

For SI: 1 foot = 304.8 mm.

Figure 1009.6
STAIRWAY RISE LIMITATION

1009.7 Circular stairways. Circular stairways shall have a minimum tread depth and a maximum riser height in accordance with Section 1009.3 and the smaller radius shall not be less than twice the width of the stairway. The minimum tread depth measured 12 inches (305 mm) from the narrower end of the tread shall not be less than 11 inches (279 mm). The minimum tread depth at the narrow end shall not be less than 10 inches (254 mm).

Exception: For occupancies in Group R-3, and within individual dwelling units in occupancies in Group R-2, both as applicable in Section 1001.1.

❖ Circular stairway construction consists of a series of tapered treads that form a stairway with a circular configuration. The commentary to Section 1009.8 regarding the eccentricity of movement on stairways with winders also applies to circular stairways. This type of stairway is only allowed to be used as a component of a means of egress when tread and riser dimensions meet the requirements of Section 1009.3. This means that tapered treads must have a minimum depth of 11 inches (279 mm) measured 12 inches (305 mm) in from the side of the stairway having the shorter radius and that risers are not to exceed 7 inches (178 mm) in height. This section also requires that the shorter radius must be equal to or greater than twice the actual width of the stairway (see Figure 1009.7). If these minimum dimensions are not met, the stairway would be considered a stairway with winders and be subject to the requirements of Section 1009.8.

The exception provides for less stringent requirements for residential units where the occupants are familiar with the circular stair arrangement.

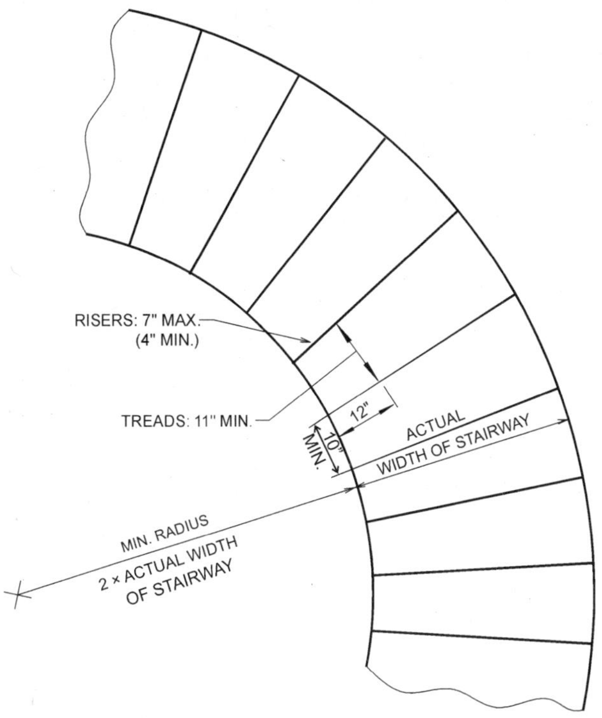

For SI: 1 inch = 25.4 mm.

Figure 1009.7
CIRCULAR STAIRS—MINIMUM DIMENSIONAL REQUIREMENTS

1009.8 Winders. Winders are not permitted in means of egress stairways except within a dwelling unit.

❖ This section specifies the minimum dimensional requirements for the construction of stairway winders. Winders are used to form a bend in a flight of stairs to change the direction of the run.

The risk of injury in the use of stairways constructed with winders is greater than for stairways constructed as straight runs. This is particularly true in emergency situations where the rate of travel up or down a stairway is increased from the pace set under normal conditions of stairway use.

The employment of winders in stairway construction creates a special hazard because of the tapered configuration of the treads. For example, a person descending a straight flight of stairs will set up a natural cadence or rhythmic movement. However, in a stairway constructed with winders, the rhythmic movement of descent is suddenly disturbed when the section of stairway with the winders is reached. Because of the tapered treads, the horizontal distance traveled by each of the footsteps nearest the radial center of the winding section is necessarily shorter than the distance that must be traveled by each if the footsteps are nearest the periphery or outer edge of the stairway (see Figure 1009.8). This condition sets up an eccentric movement. The hazard is further amplified because the inner footsteps (nearest to the radial center of the turn) must land on those portions of the tapered treads that are smaller in depth than the portions receiving the outer footsteps.

Because of the inherent dangers of stairways with winders, this section prohibits winders except for stairways serving a single dwelling unit. This section does not prohibit winders from being used in stairways that are not a required means of egress.

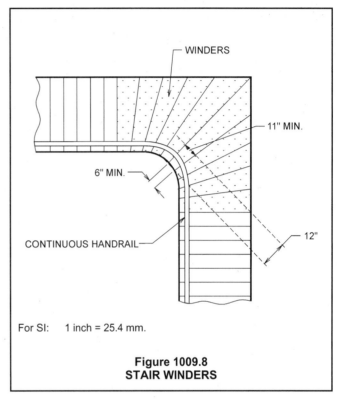

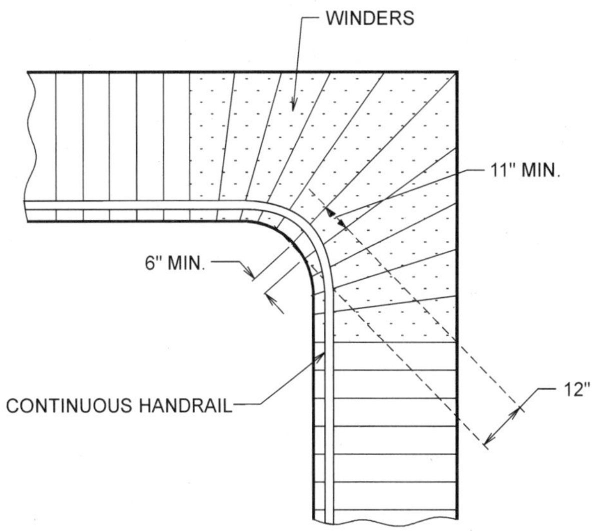

For SI: 1 inch = 25.4 mm.

Figure 1009.8
STAIR WINDERS

1009.9 Spiral stairways. Spiral stairways are permitted to be used as a component in the means of egress only within dwelling units or from a space not more than 250 square feet (23 m^2) in area and serving not more than five occupants, or from galleries, catwalks and gridirons in accordance with Section 1014.6.

A spiral stairway shall have a 7.5-inch (191 mm) minimum clear tread depth at a point 12 inches (305 mm) from the narrow edge. The risers shall be sufficient to provide a headroom of 78 inches (1981 mm) minimum, but riser height shall not be more than 9.5 inches (241 mm). The minimum stairway width shall be 26 inches (660 mm).

❖ Spiral stairways are generally constructed with a fixed center pole that serves as either the primary or the only means of support from which pie-shaped treads radiate to form a winding stairway.

The commentary to Section 1009.8 regarding the eccentricity of movement on stairways with winders also applies to spiral stairways. The nature of stairway construction is such that it does not serve well when used in emergencies that require immediate evacuation, nor does a spiral stairway configuration permit the handling of a large occupant load in an efficient and safe manner. Therefore, this section allows only very limited use of spiral stairways. Like stairways with winders, spiral stairways may be used in any occupancy as long as such stairways are not a component of a required means of egress.

Spiral stairways are required to have dimensional uniformity. Treads must be at least 26 inches (660 mm) wide and the depth of the treads must not be less than $7^1/_2$ inches (191 mm) measured at a point that is 12 inches (305 mm) out from the narrow edge (see Figure 1009.9). Riser heights are required to be the same throughout the stairway, but are not to exceed $9^1/_2$ inches (241 mm). A minimum headroom of 6 feet, 6 inches (1981 mm) is required.

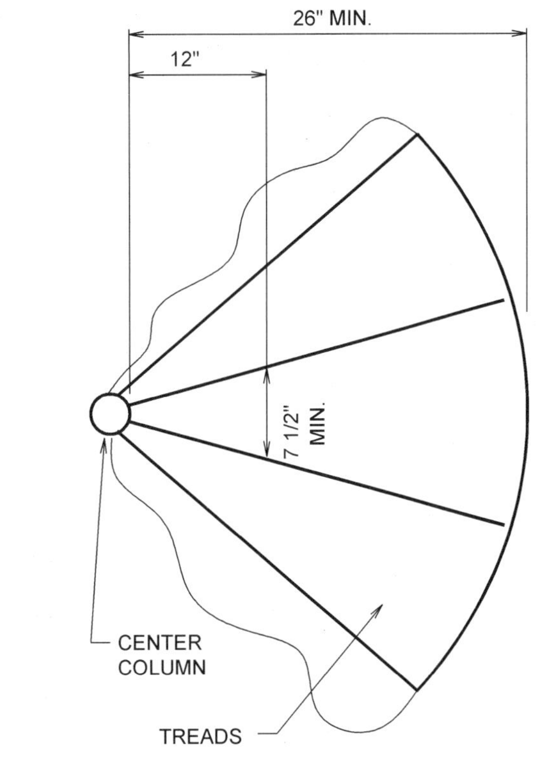

For SI: 1 inch = 25.4 mm.

Figure 1009.9
SPIRAL STAIRS

1009.10 Alternating tread devices. Alternating tread devices are limited to an element of a means of egress in buildings of Groups F, H and S from a mezzanine not more than 250 square feet (23 m^2) in area and which serves not more than five occupants; in buildings of Group I-3 from a guard tower, observation station or control room not more than 250 square feet (23 m^2) in area and for access to unoccupied roofs.

❖ This type of device is constructed in such a way that each tread alternates with each adjacent tread so that the device consists of a system of right-footed and left-footed treads (see Figure 1009.10).

The use of center stringer construction, half-treads and an incline that is considerably steeper than allowed for ordinary stairway construction makes the alternating tread device unique. However, because of its structural features, only single-file use of the device (between handrails) is possible, thus preventing the occupants from passing one another. The pace of occupant travel is set by the slowest user, a condition that could become critical in an emergency situation. Furthermore, it is impossible for fire service personnel to use an alternating tread device at the same time and in a direction opposite that being used by occupants to exit the premises, possibly causing a serious delay in fire-fighting operations. For these reasons, this section greatly restricts the use of alternating tread devices as a means of egress.

Alternating tread devices are considered a modest improvement to ladder construction and, therefore, can be used as an unoccupied roof access in accordance with the requirements of Section 1009.12.

1009.10.1 Handrails of alternating tread devices. Handrails shall be provided on both sides of alternating tread devices and shall conform to Section 1009.11.

❖ For the safety of occupants, this section provides reference to the dimensional requirements for handrail locations to be used in conjunction with the special construction features of alternating tread devices provided in Section 1009.10. Due to the steepness of these devices, additional clearances are required so that handrail movement will not be encumbered by obstructions.

1009.10.2 Treads of alternating tread devices. Alternating tread devices shall have a minimum projected tread of 5 inches (127 mm), a minimum tread depth of 8.5 inches (216 mm), a minimum tread width of 7 inches (178 mm) and a maximum riser height of 9.5 inches (241 mm). The initial tread of the device shall begin at the same elevation as the platform, landing or floor surface.

Exception: Alternating tread devices used as an element of a means of egress in buildings from a mezzanine area not more than 250 square feet (23 m^2) in area which serves not more than five occupants shall have a minimum projected tread of 8.5 inches (216 mm) with a minimum tread depth of 10.5 inches (267 mm). The rise to the next alternating tread surface should not be more than 8 inches (203 mm).

❖ Alternating tread stairways (see Section 1009.10) are required to have tread depths of at least $8^1/_2$ inches (216 mm) and widths of 7 inches (178 mm) or more.

The risers are to be no more than $9^1/_2$ inches (241 mm) when measured from tread to alternating tread (next adjacent surface). Tread projections are not to be less than 5 inches (127 mm) when measured (in a horizontal plane) from tread nosing to the next tread nosing (see Figure 1009.10). Applying the limiting dimensions stated above would result in a device with a very steep incline (approximately 4:1).

For alternating tread devices used as a means of egress from small-area mezzanines as prescribed in the exception, the treads must project at least $8^1/_2$ inches (216 mm) as compared to the 5 inches (127 mm) stated above, treads are to be at least $10^1/_2$ inches (267 mm) in depth [compared to $8^1/_2$ inches (216 mm)] and risers are not to exceed 8 inches (203 mm) in height [compared to $9^1/_2$ inches (341 mm)]. Applying these latter limiting dimensions would result in a device of lesser incline (approximate slope of 2:1) and a more comfortable and safer device to use for egress travel.

1009.11 Handrails. Stairways shall have handrails on each side. Handrails shall be adequate in strength and attachment in accordance with Section 1607.7 of the *International Building Code*. Handrails for ramps, where required by Section 1010.8, shall comply with this section.

Exceptions:

1. Aisle stairs complying with Section 1024 provided with a center handrail need not have additional handrails.
2. Stairways within dwelling units, spiral stairways and aisle stairs serving seating only on one side are permitted to have a handrail on one side only.
3. Decks, patios and walkways that have a single change in elevation where the landing depth on each side of the change of elevation is greater than what is required for a landing do not require handrails.
4. In Group R-3 occupancies, a change in elevation consisting of a single riser at an entrance or egress door does not require handrails.
5. Changes in room elevations of only one riser within dwelling units and sleeping units in Group R-2 and R-3 occupancies do not require handrails.

❖ Changes in room elevations of only one riser within dwelling units and sleeping units in Group R-2 and R-3 occupancies do not require handrails.

Falls are the leading cause of nonfatal injuries in the United States, exceeding even motor vehicle injuries. To protect the user from falls to surfaces below and to aid in the use of the stairway, guards and handrails are to be provided. In cases of fire where vision might be obscured by smoke, handrails serve as guides directing the user along the path of egress travel.

This section requires that handrails be continuous along a stairway and be placed on both sides of a stairway so that a mobility-impaired person can support his or her "strong side" in both ascent and descent [see Figures 1009.4(1) and 1009.4(2)].

This section is referenced when handrails are required on ramps. Handrails on ramps must comply with Sections 1009.11 through 1009.11.7.

The exceptions state conditions where handrails are only required on one side or are not needed at all.

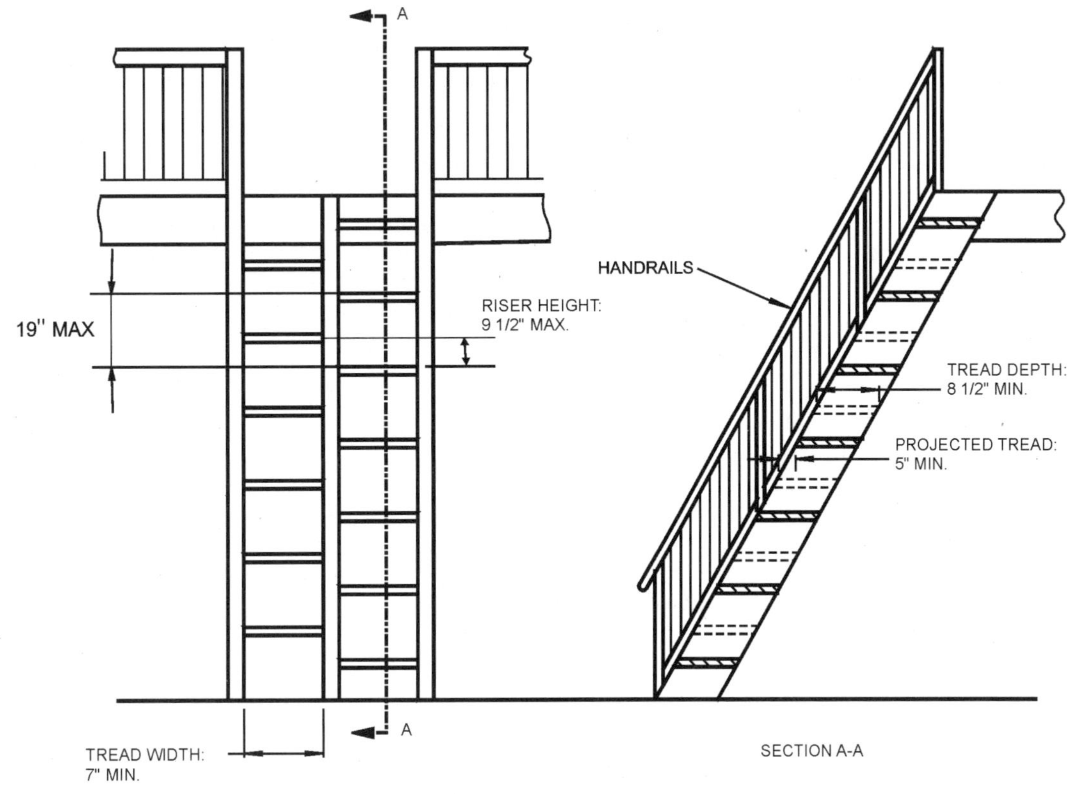

For SI: 1 inch = 25.4 mm.

Figure 1009.10
TYPICAL ALTERNATING TREAD DEVICE

1009.11.1 Height. Handrail height, measured above stair tread nosings, or finish surface of ramp slope, shall be uniform, not less than 34 inches (864 mm) and not more than 38 inches (965 mm).

❖ It has been demonstrated that for safe use, the height of handrails must not be less than 34 inches (864 mm) nor more than 38 inches (965 mm) above the leading edge of stairway treads, landings or other walking surfaces (see Figure 1009.11.1). This requirement is applicable for all uses, including handrails within a dwelling unit.

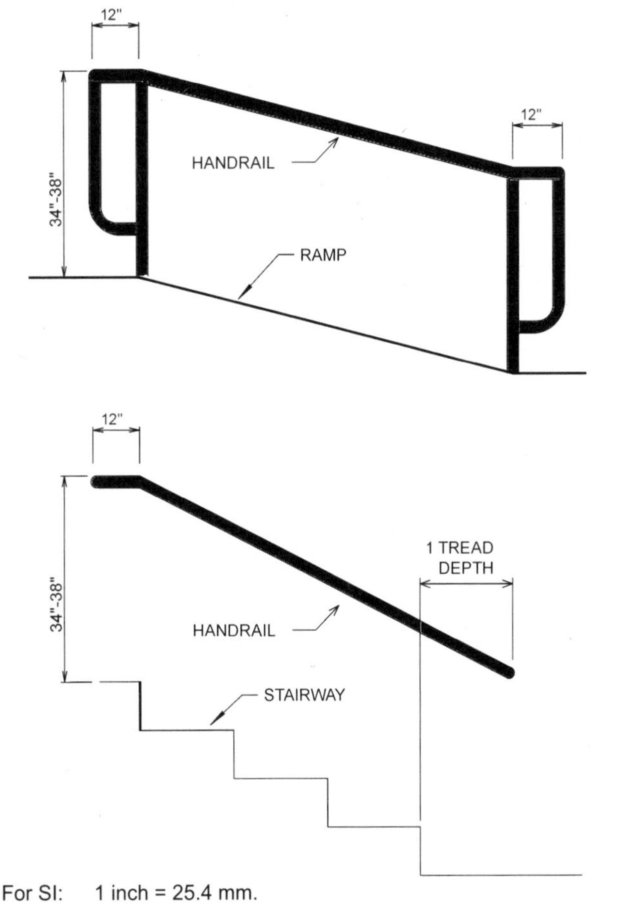

For SI: 1 inch = 25.4 mm.

Figure 1009.11.1
HANDRAIL REQUIREMENTS

1009.11.2 Intermediate handrails. Intermediate handrails are required so that all portions of the stairway width required for egress capacity are within 30 inches (762 mm) of a handrail. On monumental stairs, handrails shall be located along the most direct path of egress travel.

❖ In order to always be available to the user of the stairway, the maximum distance to a handrail from within the required width is to be no more than 30 inches (762 mm). People generally tend to make use of handrails, and if intermediate handrails are not provided for very wide stairways, the center portion of such stairways will normally receive limited use. More importantly, in emergencies, the center portions of wide stairways with handrails would more aptly be used to speed up egress travel rather than be delayed by overcrowding at the sides with the handrails. This would especially be true under panic conditions. Without the requirement for intermediate handrails, the use of wide interior stairways could become particularly hazardous.

It should be noted that the distance to the handrail applies to the required width of the stairway. If a stairway is greater than 60 inches (1524 mm) in width, but only 60 inches (1524 mm) are required based on occupant load (see Section 1005.1), intermediate handrails are not required. Adequate safety is provided since the occupants can use the 30 inches (762 mm) within the handrails provided on each side.

The requirement for monumental stairways deals with the very wide stairway in relation to the required width. While handrails on both sides of the stairway may be sufficient to accommodate the required width, the handrails may not be near the stream of traffic or even apparent to the user. In this case, the handrails are to be placed in a location more reflective of the egress path.

See Figure 1009.11.2 for handrail locations for monumental stairs.

1009.11.3 Handrail graspability. Handrails with a circular cross section shall have an outside diameter of at least 1.25 inches (32 mm) and not greater than 2 inches (51 mm) or shall provide equivalent graspability. If the handrail is not circular, it shall have a perimeter dimension of at least 4 inches (102 mm) and not greater than 6.25 inches (160 mm) with a maximum cross-section dimension of 2.25 inches (57 mm). Edges shall have a minimum radius of 0.01 inch (0.25 mm).

❖ The ability of grasping a handrail firmly and sliding the hand along the rail without meeting obstructions are important factors in the safe use of stairways and ramps. This section requires that handrails have a circular cross section with an outside diameter of at least 1.25 inches (32 mm) but not greater than 2 inches (51 mm). A handrail with either a very narrow or a large cross section is not graspable in a power grip by all able-bodied users and certainly not by those with hand-strength or flexibility deficiencies. Noncircular cross sections can be approved by way of the alternative noncircular criteria in this section, as long as they provide a suitable gripping surface. Edges must be rounded so that there are no sharp edges. An example is shown in Figure 1009.11.3.

1009.11.4 Continuity. Handrail-gripping surfaces shall be continuous, without interruption by newel posts or other obstructions.

Exceptions:

1. Handrails within dwelling units are permitted to be interrupted by a newel post at a stair landing.
2. Within a dwelling unit, the use of a volute, turnout or starting easing is allowed on the lowest tread.

3. Handrail brackets or balusters attached to the bottom surface of the handrail that do not project horizontally beyond the sides of the handrail within 1.5 inches (38 mm) of the bottom of the handrail shall not be considered to be obstructions and provided further that for each 0.5 inch (13 mm) of additional handrail perimeter dimension above 4 inches (102 mm), the vertical clearance dimension of 1.5 inches (38 mm) shall be permitted to be reduced by 0.125 inch (3 mm).

❖ The degree of occupant safety as it relates to handrail use is a function of the features of handrail construction.

Handrails must be usable for their entire length without requiring the users to release their grasp. Typically, in traveling the means of egress, an individual's fingers will trail along the rail. If handrails are to be of service to the occupants, they must be uninterrupted and continuous. Oversize newels or changes in the guard system can cause interruption of the handrail, requiring the occupants to release their grip [see Figure 1009.11.4(1)]. Exceptions 1 and 2 provide for handrail details that have been used for years in dwelling units. Exception 3 provides for conventional methods of handrail support while providing the user with an uninterrupted gripping surface. The larger handrail size permits shorter brackets since geometrically the finger clearance is still maintained. For example, a handrail with a perimeter of 5 inches (127 mm) would be permitted to have a clearance of 1.25 inches (32 mm) to the bottom bracket [see Figures 1009.11.4(2)].

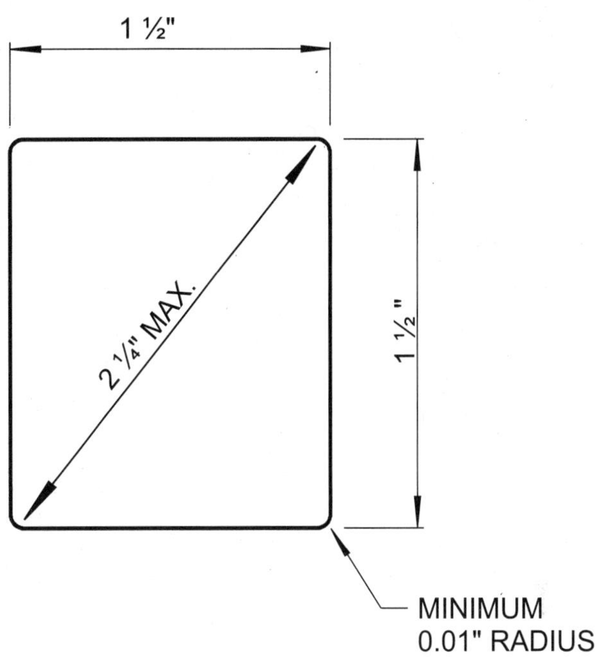

For SI: 1 inch = 25.4 mm.

Figure 1009.11.3
EXAMPLE OF ALTERNATIVE HANDRAIL CONFIGURATION

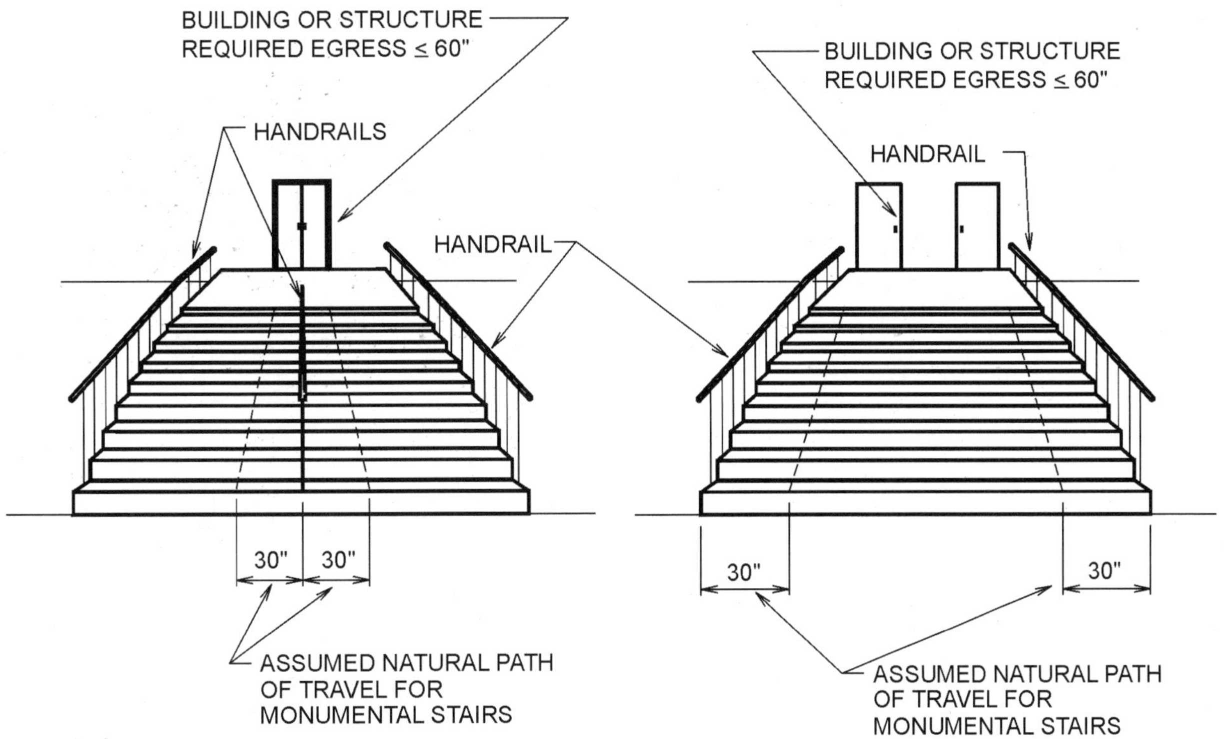

For SI: 1 inch = 25.4 mm.

Figure 1009.11.2
HANDRAILS FOR MONUMENTAL STAIRS

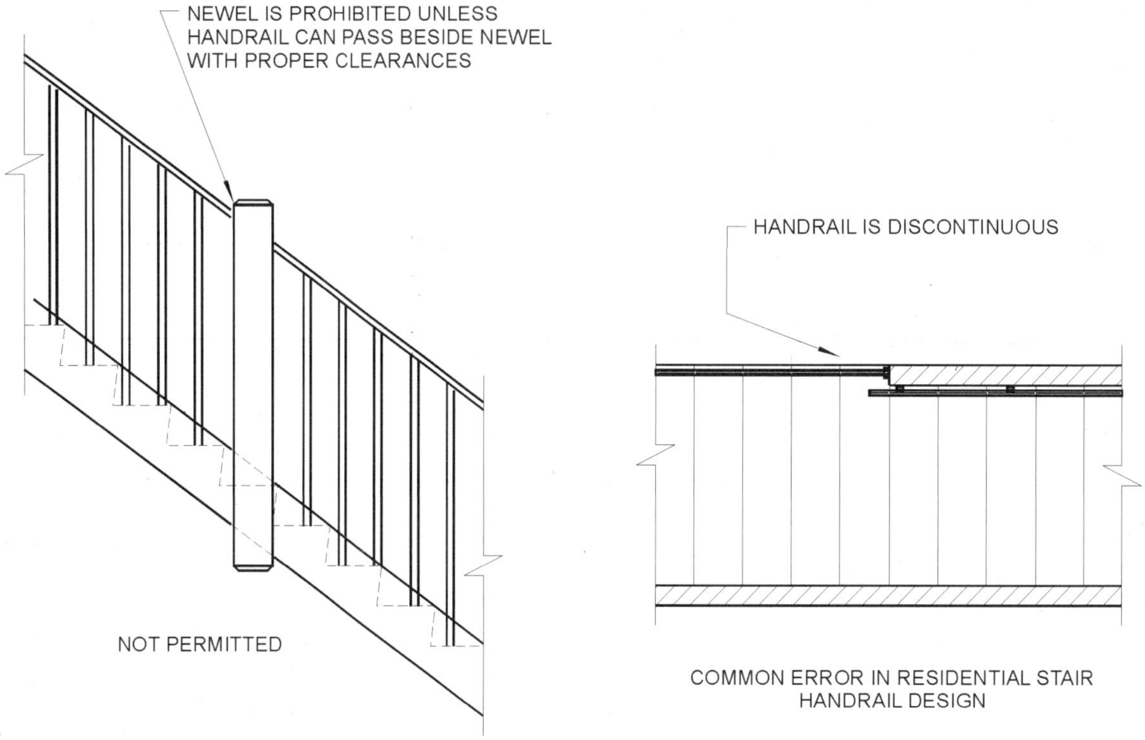

Figure 1009.11.4(1)
EXAMPLES OF PROHIBITED HANDRAIL DESIGNS

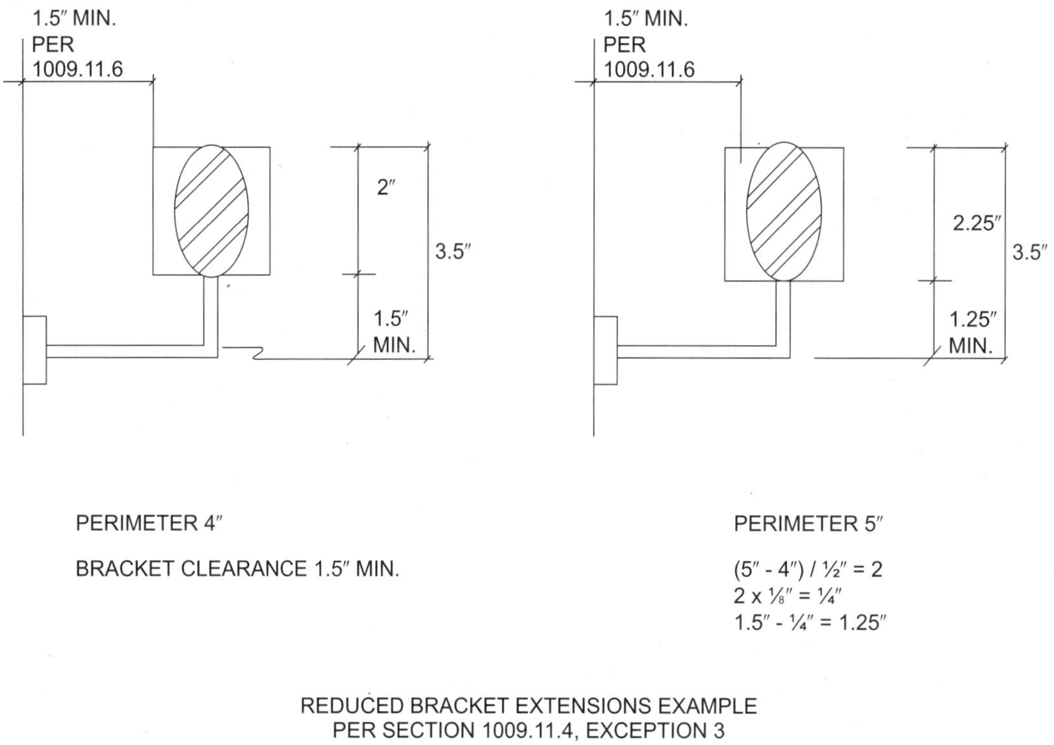

For SI: 1 inch = 25.4 mm.

Figure 1009.11.4(2)
EXAMPLES OF PROHIBITED HANDRAIL DESIGNS

1009.11.5 Handrail extensions. Handrails shall return to a wall, guard or the walking surface or shall be continuous to the handrail of an adjacent stair flight. Where handrails are not continuous between flights, the handrails shall extend horizontally at least 12 inches (305 mm) beyond the top riser and continue to slope for the depth of one tread beyond the bottom riser.

Exceptions:

1. Handrails within a dwelling unit that is not required to be accessible need extend only from the top riser to the bottom riser.
2. Aisle handrails in Group A occupancies in accordance with Section 1024.13.

❖ The purpose of the handrail return requirements is to prevent a person from being injured by falling onto the end of a handrail or catching an article of loose clothing on it.

The length that a handrail extends beyond the top and bottom of a stairway ramp or other location where handrails are otherwise not continuous is an important factor for the safety of the users. An occupant must be able to grasp securely a handrail beyond the last riser of a stairway or the last sloped segment of a ramp.

For stairways, handrails must be extended 12 inches (305 mm) horizontally beyond the top riser and sloped a distance of one tread depth beyond the bottom riser. For ramps, handrails must be extended 12 inches (305 mm) horizontally beyond the last sloped ramp segment at both the top and bottom locations. These handrail extensions are not only required at the top and bottom of stairways and ramps, but also at other places where handrails are not continuous, such as landings and platforms. These changes to previous code requirements are intended to reflect the current provisions of ICC A117.1 (see Figure 1009.11.1). If the handrail extension is at a location that could be considered a protruding object, it must return to the post at a height of less than 27 inches (686 mm) above the floor.

In accordance with Exception 1, the handrail extensions are not needed where a dwelling unit is not required to be accessible since they are not essential for these circumstances. Exception 2 provides for handrails along assembly seating area ramped aisles. It is necessary to have discontinuous handrails for assembly installations to provide for circulation of the occupants from the aisle to the seating areas.

1009.11.6 Clearance. Clear space between a handrail and a wall or other surface shall be a minimum of 1.5 inches (38 mm). A handrail and a wall or other surface adjacent to the handrail shall be free of any sharp or abrasive elements.

❖ See Figures 1009.11.4(2) and 1009.11.7(2) for an illustration of handrail clearance. A clear space is needed between a handrail and the wall or other surface to allow the user to slide his or her hand along the rail with the fingers in the gripping position without contacting the wall surface, which could have an abrasive texture. In climates where persons may be expected to be wearing heavy gloves during the winter, a larger clearance would be desirable at an exterior stairway or a stairway directly inside the entrance to a building.

1009.11.7 Stairway projections. Projections into the required width at each handrail shall not exceed 4.5 inches (114 mm) at or below the handrail height. Projections into the required width shall not be limited above the minimum headroom height required in Section 1009.2.

❖ Handrails may not project more than $4^1/_2$ inches (114 mm) into the required width of a stairway, so that the clear width of the passage will not be seriously reduced [see Figure 1009.11.7(1)]. This projection may exist below the handrail height as well [see Figure 1009.11.7(2)].

1009.12 Stairway to roof. In buildings four or more stories in height above grade, one stairway shall extend to the roof surface, unless the roof has a slope steeper than four units vertical in 12 units horizontal (33-percent slope). In buildings without an occupied roof, access to the roof from the top story shall be permitted to be by an alternating tread device.

❖ Because of safety considerations, roofs used for habitable purposes such as roof gardens, observation decks, sporting facilities (including jogging or walking tracks and tennis courts) or other similar occupancies must be provided with conventional stairways that will serve as required means of egress. Access by ladders or an alternating tread device is not permitted for such uses.

In buildings four or more stories high, roofs that are not used for habitable purposes must be made accessible by conventional stairways or by an alternating tread device (see Section 1009.10). Two purposes of this are access for roof or equipment repair and fire department access during a fire event. Sloping roofs with a rise greater than 4 inches (102 mm) for every 12 inches (305 mm) in horizontal measurement (4:12) are exempt from the requirements of this section because of the steepness of the construction and the inherent dangers to life safety.

While it is not specifically required that this roof access be through an exit stairway enclosure, since part of the intent is for fire department access to the roof, it is advisable. Section 1019.1.7 requires signage at the level of exit discharge indicating if the stairway has roof access.

1009.12.1 Roof access. Where a stairway is provided to a roof, access to the roof shall be provided through a penthouse complying with Section 1509.2 of the *International Building Code*.

Exception: In buildings without an occupied roof, access to the roof shall be permitted to be a roof hatch or trap door not less than 16 square feet (1.5 m^2) in area and having a minimum dimension of 2 feet (610 mm).

❖ The purpose of the penthouse or stairway bulkhead requirement in this section is to protect the walking surface of the stairway to the roof. The exception provides for situations where roof access is only needed for service or maintenance purposes.

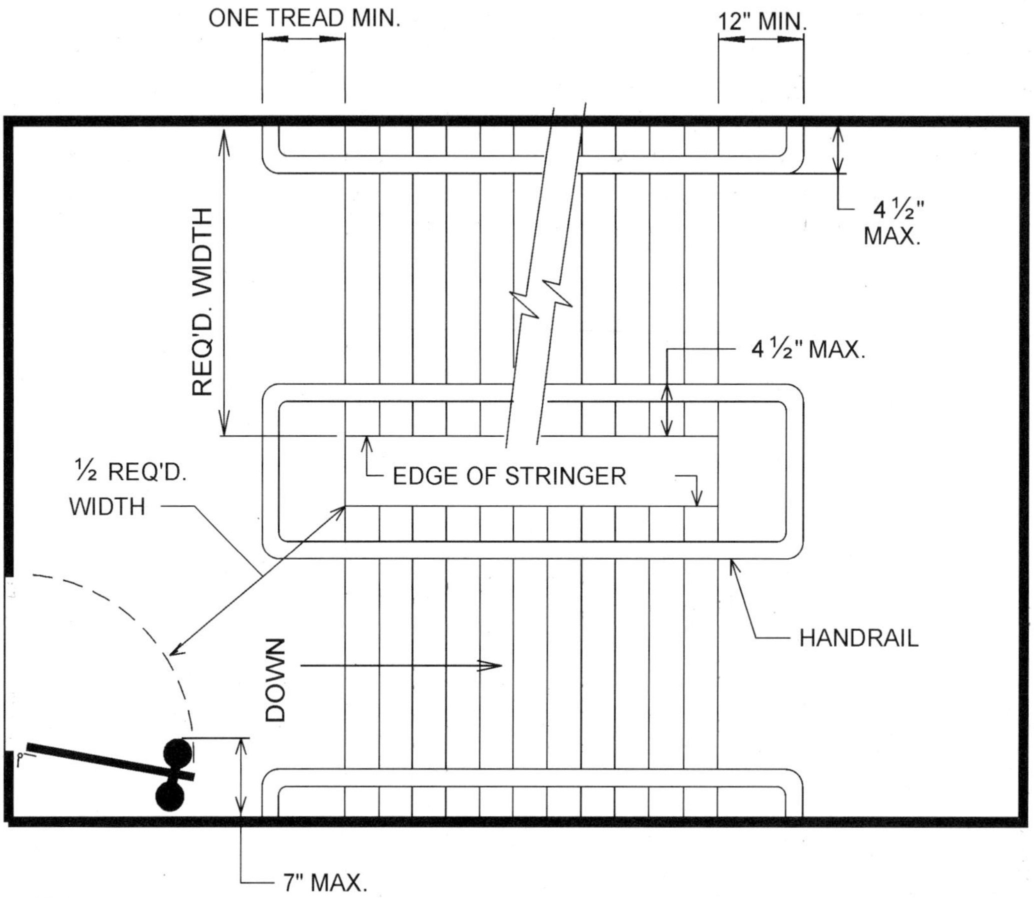

For SI: 1 inch = 25.4 mm.

Figure 1009.11.7(1)
TYPICAL HANDRAIL ARRANGEMENT

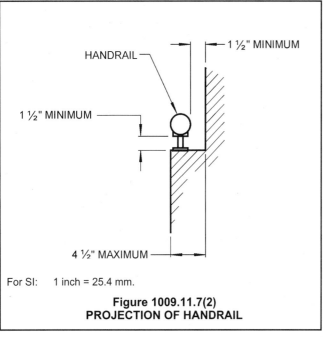

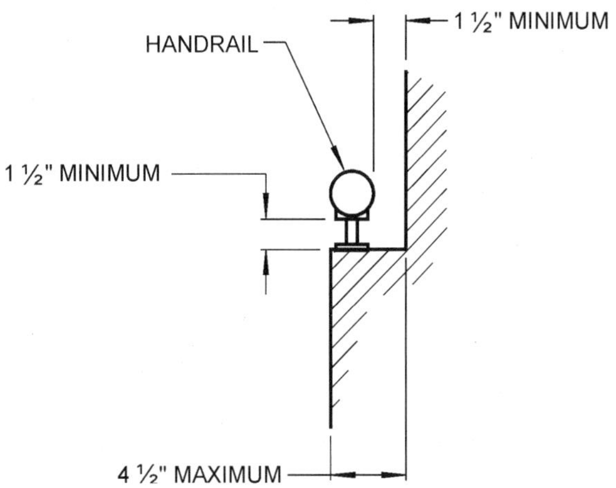

For SI: 1 inch = 25.4 mm.

Figure 1009.11.7(2)
PROJECTION OF HANDRAIL

[B] SECTION 1010
RAMPS

1010.1 Scope. The provisions of this section shall apply to ramps used as a component of a means of egress.

Exceptions:

1. Other than ramps that are part of the accessible routes providing access in accordance with Sections 1108.2.2 through 1108.2.4.1 of the *International Building Code*, ramped aisles within assembly rooms or spaces shall comply with the provisions in Section 1024.11.
2. Curb ramps shall comply with ICC A117.1.
3. Vehicle ramps in parking garages for pedestrian exit access shall not be required to comply with Sections 1010.3 through 1010.9 when they are not an accessible route serving accessible parking spaces, other required accessible elements or part of an accessible means of egress.

❖ Ramps provide an alternative method of vertical means of access to or egress from a building. Ramps are required for access to building areas for mobility-impaired persons (see Chapter 11 of the IBC) and for small changes in floor elevations, which are a safety hazard in themselves. All ramps, whether required or otherwise provided, must comply with the requirements of this section. The code considers any walking surface that has a slope steeper than one unit vertical in 20 units horizontal (5-percent slope) to be a ramp.

Exception 1 clarifies that in assembly rooms with sloped floors, ramps other than the routes utilized to and from accessible wheelchair seating locations (see Section 1108.2 of the IBC) are permitted to be designed in accordance with the sloped aisle provisions in Section 1024.

Exception 2 is a reference to specific curb cut requirements found in ICC A117.1.

Exception 3 is addressing parking garages. An accessible route is required to and from any accessible parking spaces, and all the ramp provisions must be followed. Ramps that provide access to and from the nonaccessible spaces in the remainder of the parking garage, however, need only comply with the provisions for slope and guard requirements. Ramps that are strictly for vehicles, such as jump ramps, are not required to meet any of the ramp provisions.

1010.2 Slope. Ramps used as part of a means of egress shall have a running slope not steeper than one unit vertical in 12 units horizontal (8-percent slope). The slope of other ramps shall not be steeper than one unit vertical in eight units horizontal (12.5-percent slope).

> **Exception:** Aisle ramp slope in occupancies of Group A shall comply with Section 1024.11.

❖ Maximum slope is limited to facilitate the ease of ascent and to control the descent of persons with or without a mobility impairment. The maximum slope of a ramp in the direction of travel is limited to 1 unit vertical in 12 units horizontal (1:12) (see Figure 1010.2). Ramps in existing buildings may be permitted to have a steeper slope at small changes in elevation (see Section 3409.7.5 of the IBC or Section 506.1.4 of the IEBC). An example of a ramp that is not part of a means of egress and, therefore, allowed to be a maximum slope of 1:8, is an industrial access ramp that provides access to a raised floor level around a piece of equipment.

The exception is a reference to the assembly requirements. However, the ramps that are part of an accessible route for ingress or egress are still required to meet the 1:12 maximum slope. Only the ramps used elsewhere in the assembly space may utilize the 1:8 maximum slope permitted in Section 1024.11.

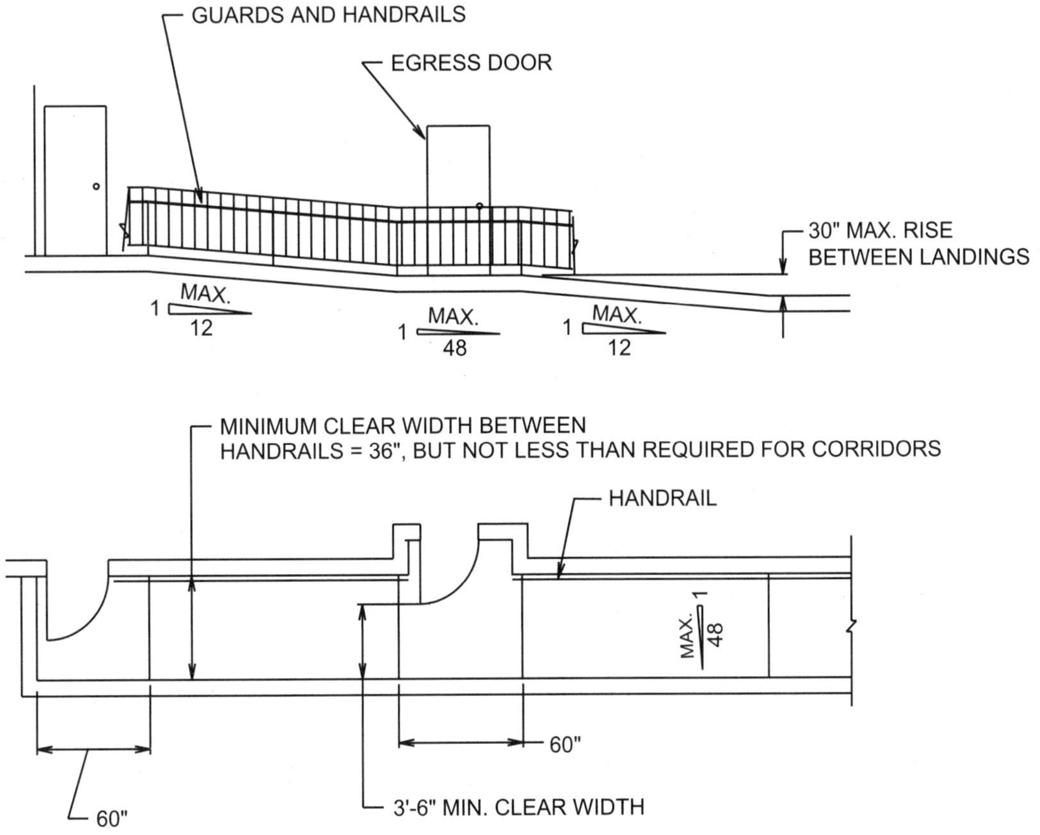

For SI: 1 inch = 25.4 mm, 1 foot = 304.8 mm).

Figure 1010.2
TYPICAL MEANS OF EGRESS RAMP

1010.3 Cross slope. The slope measured perpendicular to the direction of travel of a ramp shall not be steeper than one unit vertical in 48 units horizontal (2-percent slope).

❖ The limitation of 1 unit vertical in 48 units horizontal (1:48) on the slope across the direction of travel is to prevent a severe cross slope that would pitch a user to one side (see Figure 1010.2).

1010.4 Vertical rise. The rise for any ramp run shall be 30 inches (762 mm) maximum.

❖ Because pushing a wheelchair up a ramp requires great energy, landings must be situated so that a person can rest after each 30-inch (762 mm) elevation change (see Figure 1010.2).

1010.5 Minimum dimensions. The minimum dimensions of means of egress ramps shall comply with Sections 1010.5.1 through 1010.5.3.

❖ These minimum dimension requirements allow the ramp to function as a means of egress.

1010.5.1 Width. The minimum width of a means of egress ramp shall not be less than that required for corridors by Section 1016.2. The clear width of a ramp and the clear width between handrails, if provided, shall be 36 inches (914 mm) minimum.

❖ The width requirement of a means of egress ramp is 36 inches (914 mm) minimum, similar to that established by Section 1016.2 for corridors. Note that the clear width of 36 inches (914 mm) is required between the handrails for proper clearance for a person in a wheelchair. This is different from stairways where handrails are permitted to project into the required width. The 36-inch (914 mm) minimum clear width between handrails is consistent with ICC A117.1 and the recommendations of the ADAAG Review Advisory Committee.

1010.5.2 Headroom. The minimum headroom in all parts of the means of egress ramp shall not be less than 80 inches (2032 mm).

❖ The requirement for headroom on any part of an egress ramp is identical to the requirement of a conventional (nonspiral) stairway (see Section 1009.2). The ceiling heights of Section 1003.2 may also be applicable depending on the occupancy of the space.

1010.5.3 Restrictions. Means of egress ramps shall not reduce in width in the direction of egress travel. Projections into the required ramp and landing width are prohibited. Doors opening onto a landing shall not reduce the clear width to less than 42 inches (1067 mm).

❖ The purpose of not allowing ramps to reduce in width in the direction of egress travel is to prevent a restriction of the necessary ramp width that would interfere with the flow of occupants out of a facility. This is not intended to limit a ramp wider than required by the code from being reduced to that required by the code.

Doors that open onto a ramp landing must not reduce the clear width to less than 42 inches (1067 mm). This is a more restrictive provision than for stairways, which would permit the reduction to one-half the required width. Since one of the purposes of a ramp is to accommodate persons with physical disabilities, it must provide the additional clear width for access by those confined to wheelchairs without the interference or potential blockage caused by the swing of a door (see Figure 1010.2 and 1010.5.3).

1010.6 Landings. Ramps shall have landings at the bottom and top of each ramp, points of turning, entrance, exits and at doors. Landings shall comply with Sections 1010.6.1 through 1010.6.5.

❖ Landings must be provided to allow users of a ramp to rest on a level floor surface and to adjust to the change in floor surface pitch.

Landings are required at the top and bottom of each ramp run. In addition, Section 1010.4 requires a landing every 30 inches (762 mm) of vertical rise of the ramp. The requirements for landings allow those occupants of the structure the ability to negotiate all changes in direction, prepare themselves to either ascend or descend the ramp and to rest.

1010.6.1 Slope. Landings shall have a slope not steeper than one unit vertical in 48 units horizontal (2-percent slope) in any direction. Changes in level are not permitted.

❖ Landings must almost be perfectly flat. This allows persons confined to a wheelchair to come to a complete stop without having to activate the brake or hold themselves stationary at the landing. The maximum slope or cross slope of the landing in any direction is 1:48 (see Figure 1010.2).

1010.6.2 Width. The landing shall be at least as wide as the widest ramp run adjoining the landing.

❖ The width of all landings must be consistently as wide as the widths of the ramp runs leading to them. Means of egress ramps cannot be reduced in width in the direction of egress travel. This is also applicable to the landings connecting the ramp runs.

1010.6.3 Length. The landing length shall be 60 inches (1525 mm) minimum.

Exception: Landings in nonaccessible Group R-2 and R-3 individual dwelling units, as applicable in Section 1001.1, are permitted to be 36 inches (914 mm) minimum.

❖ The landings for ramps must be at least 60 inches (1524 mm) long. This allows persons confined to wheelchairs a sufficient distance to stop and rest along with any persons who may be assisting them. This requirement is directly applicable to straight-run ramps, which may require intermediate landing at every 30 inches (762 mm) of vertical rise (see Figure 1010.2). If the landing is also to be used to negotiate a change in the ramp's direction, Section 1010.6.4 is applicable. If a door overlaps the landing, Section 1010.5.3 is also applicable.

The exception provides for smaller landings in dwelling units that are not required to be accessible.

1010.6.4 Change in direction. Where changes in direction of travel occur at landings provided between ramp runs, the landing shall be 60 inches by 60 inches (1524 mm by 1524 mm) minimum.

> **Exception:** Landings in nonaccessible Group R-2 and R-3 individual dwelling units, as applicable in Section 1001.1, are permitted to be 36 inches by 36 inches (914 mm by 914 mm) minimum.

❖ When a change in direction is made in the ramp at a landing, the landing must be at least 60 square inches (0.039 m^2). This allows the person confined to a wheelchair enough room to negotiate the turn with minimal effort. It should be noted that the length of the landing may need to exceed 60 inches (1524 mm) to match the widths of the two ramp runs. In any case, the landing would still need to be 60 inches (1524 mm) wide (see Figure 1010.5.3). If a door overlaps the landing, Section 1010.5.3 is also applicable.

The exception provides for smaller landings in dwelling units that are not required to be accessible.

1010.6.5 Doorways. Where doorways are located adjacent to a ramp landing, maneuvering clearances required by ICC A117.1 are permitted to overlap the required landing area.

❖ This section specifies that the area required for maneuvering to open the door and the area of the landing are allowed to overlap. It is not necessary to provide the sum of the two area requirements (see Figure 1010.5.3).

1010.7 Ramp construction. All ramps shall be built of materials consistent with the types permitted for the type of construction of the building; except that wood handrails shall be permitted for all types of construction. Ramps used as an exit shall conform to the applicable requirements of Sections 1019.1 and 1019.1.1 through 1019.1.3 for vertical exit enclosures.

❖ Material requirements for the type of construction as required by Section 602 of the IBC for floors are also the material requirements for ramp construction. The ramp, if used as an exit, must be enclosed and protected similar to an exit stairway within a vertical exit enclosure.

1010.7.1 Ramp surface. The surface of ramps shall be of slip-resistant materials that are securely attached.

❖ As the pace of exit travel becomes hurried during emergency situations, the probability of slipping on smooth or slick floor surfaces increases. To minimize the hazard, all floor surfaces in the means of egress are required to be slip resistant. The use of hard floor materials with highly polished, glazed, glossy or finely finished surfaces should be avoided.

Field testing and uniform enforcement of the concept of slip resistance is not practical. One method used to establish slip resistance is that the static coefficient of friction between leather [Type 1 (Vegetable Tanned) of Federal Specification KK-L-165C] and the floor surface is greater than 0.5. Laboratory test procedures such as ASTM D 2047 can determine the static coefficient of resistance. Bulletin No. 4 entitled "Surfaces" issued by the

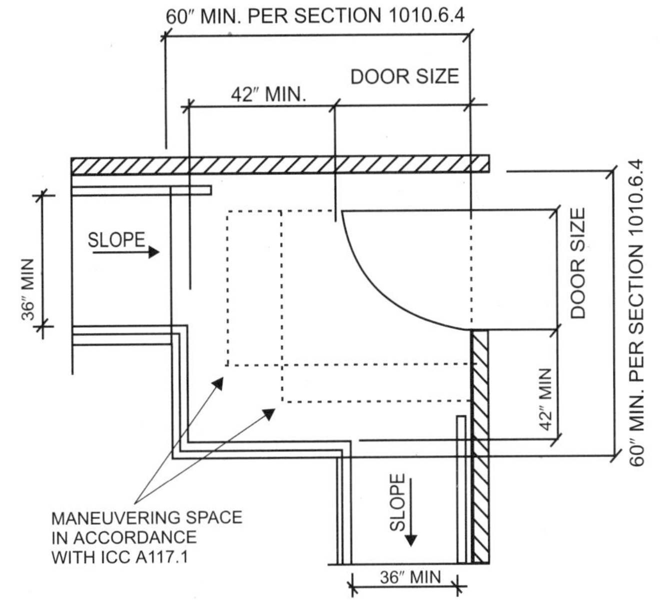

For SI: 1 inch = 25.4 mm.

Figure 1010.5.3
DOORWAY OPENING ONTO LANDING

ATBCB contains further information regarding slip resistance.

1010.7.2 Outdoor conditions. Outdoor ramps and outdoor approaches to ramps shall be designed so that water will not accumulate on walking surfaces. In other than occupancies in Group R-3, and occupancies in Group U that are accessory to an occupancy in Group R-3, surfaces and landings which are part of exterior ramps in climates subject to snow or ice shall be designed to minimize the accumulation of same.

❖ Outdoor ramps must be kept free of water, snow and ice to provide a safe path of egress travel at all times, including winter. Sheltering the ramp can serve to minimize snow and ice accumulation but, in some areas, may not completely eliminate the need to remove the snow or ice by other means. Typical methods for protecting these egress elements include roof overhangs or canopies, heated slabs and, when approved by the fire code official, a reliable snow removal maintenance program. This section does not apply to Group R-3 occupancies or their associated utility structures.

1010.8 Handrails. Ramps with a rise greater than 6 inches (152 mm) shall have handrails on both sides complying with Section 1009.11.

❖ To aid in the use of a ramp, handrails are to be provided. Handrails are intended to provide the user with a graspable surface for guidance and support. All ramps with a vertical rise greater than 6 inches (152 mm) between landings are to be provided with handrails on both sides (see Figures 1010.8 and 1009.11.1). If the handrail extension is at a location that could be considered a protruding object, the handrail must return to the post at a height of less than 27 inches (686 mm) above the floor.

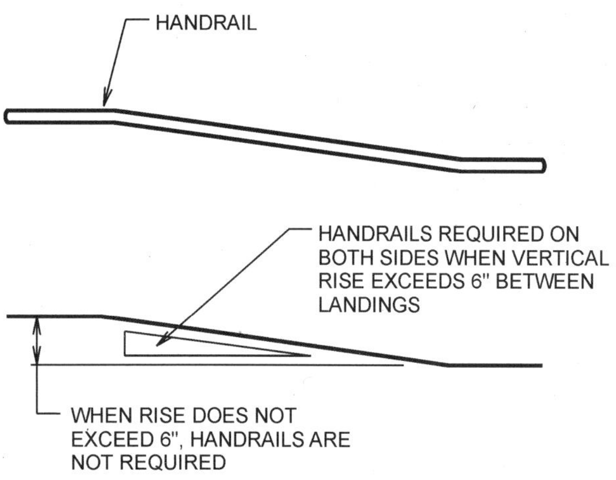
HANDRAIL
HANDRAILS REQUIRED ON BOTH SIDES WHEN VERTICAL RISE EXCEEDS 6" BETWEEN LANDINGS
WHEN RISE DOES NOT EXCEED 6", HANDRAILS ARE NOT REQUIRED

For SI: 1 inch = 25.4 mm.

Figure 1010.8
HANDRAIL ON RAMP

1010.9 Edge protection. Edge protection complying with Section 1010.9.1 or 1010.9.2 shall be provided on each side of ramp runs and at each side of ramp landings.

Exceptions:

1. Edge protection is not required on ramps not required to have handrails, provided they have flared sides that comply with the ICC A117.1 curb ramp provisions.
2. Edge protection is not required on the sides of ramp landings serving an adjoining ramp run or stairway.
3. Edge protection is not required on the sides of ramp landings having a vertical dropoff of not more than 0.5 inch (13 mm) within 10 inches (254 mm) horizontally of the required landing area.

❖ This section of the code addresses the comprehensive requirements for edge protection for all ramps. It must be noted that edge protection is not the same as the requirements for guards. The presence of a guard does not necessarily provide adequate edge protection, and the presence of adequate edge protection does not satisfy the requirements for a guard. Edge protection is necessary to prevent the wheels of a wheelchair from leaving the ramp surface or becoming lodged between the edge of the ramp and any adjacent construction. For example, a ramp may be located relatively adjacent to the exterior wall of a building. However, between the ramp edge and the exterior wall there is a strip of earth for landscape purposes. Without adequate edge protection, persons confined to wheelchairs could possibly have their wheels run off the side of the ramp into the landscape causing them to tip. These requirements are consistent with ICC A117.1 and the ADAAG Review Advisory Committee.

Exception 1 allows a ramp to have minimal edge protection as long as its vertical use is 6 inches (152 mm) or less. The exception is predicated on the ramp not needing any handrails, which is established by the provisions of Section 1010.8. Such a ramp would only need flared sides or returned curbs. For specific details of these types of edge protection, the provisions of ICC A117.1 for curb ramps must be followed.

Exception 2 reiterates that edge protection is not literally required around each side of a ramp landing. Obviously edge protection is not required along that portion of the landing that directly adjoins the next ramp run; it is only required along the unprotected sides of ramp landings.

Exception 3 states that edge protection is not required for those sides of a ramp landing directly adjacent to the ground surface that gently slopes away from the edge of the landing. If the grade adjacent to the ramp landing slopes no more than 1/2:10 (which equals 1:20) away from the landing, additional edge protection is not required. Such a gradual slope would not be detrimental to persons confined to wheelchairs as they negotiate the ramp landing.

1010.9.1 Railings. A rail shall be mounted below the handrail 17 inches to 19 inches (432 mm to 483 mm) above the ramp or landing surface.

❖ The purpose of the railing in this section is to prevent a wheelchair from leaving the ramp surface or becoming lodged between the edge of the ramp and any adjacent construction (see Figure 1010.9.1).

1010.9.2 Curb or barrier. A curb or barrier shall be provided that prevents the passage of a 4-inch-diameter (102 mm) sphere, where any portion of the sphere is within 4 inches (102 mm) of the floor or ground surface.

❖ Edge protection for ramps and ramp landings may be achieved with a builtup curb or other effective barrier. The curb or barrier must be located near the surface of the ramp and landing such that a 4-inch-diameter (102 mm) sphere cannot pass through any openings. An example of an effective barrier would be the bottom rail of a guard system. If the bottom rail is located less than 4 inches (102 mm) above the ramp and landing surface, edge protection has been provided. The curb or barrier prevents the wheel of a wheelchair from running off the edge of the surface and provides people with visual disabilities a toe stop at the edge of the walking surface (see Figure 1010.9.1).

1010.10 Guards. Guards shall be provided where required by Section 1012 and shall be constructed in accordance with Section 1012.

❖ To protect the user from falls to surfaces below, guards are to be provided where the sides of a ramp or landing are more than 30 inches (762 mm) above the adjacent grade. Guards are to be constructed in accordance with Section 1012, including the minimum height of 42 inches (1067 mm) (see Figure 1010.10).

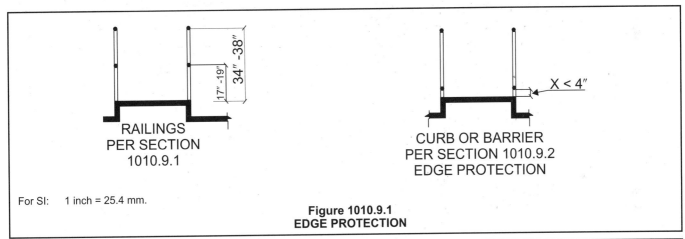

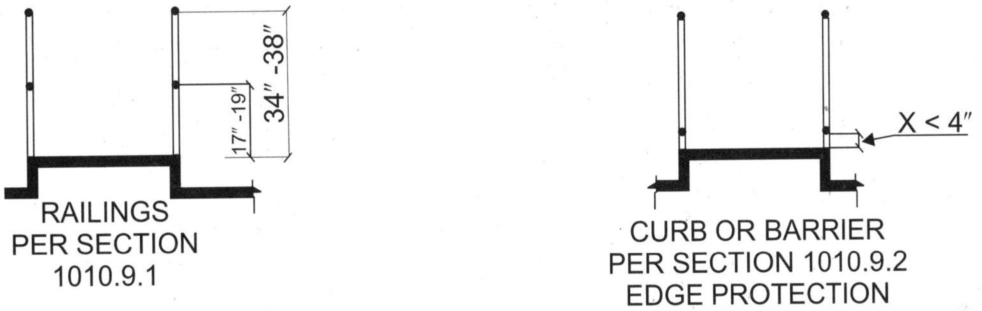

For SI: 1 inch = 25.4 mm.

Figure 1010.9.1
EDGE PROTECTION

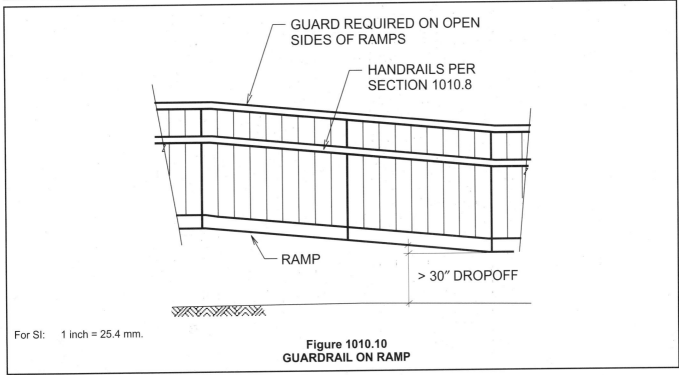

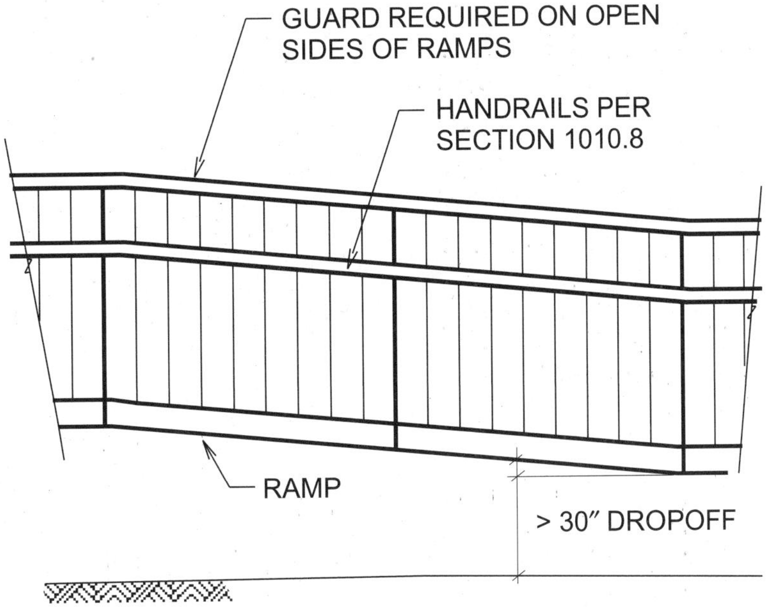

For SI: 1 inch = 25.4 mm.

Figure 1010.10
GUARDRAIL ON RAMP

[B] SECTION 1011
EXIT SIGNS

1011.1 Where required. Exits and exit access doors shall be marked by an approved exit sign readily visible from any direction of egress travel. Access to exits shall be marked by readily visible exit signs in cases where the exit or the path of egress travel is not immediately visible to the occupants. Exit sign placement shall be such that no point in an exit access corridor is more than 100 feet (30 480 mm) or the listed viewing distance for the sign, whichever is less, from the nearest visible exit sign.

Exceptions:

1. Exit signs are not required in rooms or areas which require only one exit or exit access.
2. Main exterior exit doors or gates which obviously and clearly are identifiable as exits need not have exit signs where approved by the fire code official.
3. Exit signs are not required in occupancies in Group U and individual sleeping units or dwelling units in Group R-1, R-2 or R-3.
4. Exit signs are not required in sleeping areas in occupancies in Group I-3.
5. In occupancies in Groups A-4 and A-5, exit signs are not required on the seating side of vomitories or openings into seating areas where exit signs are provided in the concourse that are readily apparent from the vomitories. Egress lighting is provided to identify each vomitory or opening within the seating area in an emergency.

❖ Where an occupancy has two or more required exits or exit accesses, the means of egress must be provided with illuminated signs that readily identify the location of the exits and indicate the path of travel to the exits. The signs must be illuminated with letters reading "exit." The illumination may be internal or external to the sign. The signs should be visible from all directions in the exit access route. In cases where the signs are not visible to the occupants because of turns in the corridor or for other reasons, additional illuminated signs must be provided indicating the direction of egress to an exit. Exit signs must be located so that, where required, the nearest one is within 100 feet (30 480 mm) of the sign's listed viewing distance. While not a referenced standard, UL 924 permits exit signs to be listed with a viewing distance of less than 100 feet (30 480 mm). When a sign is listed for a viewing distance of less than 100 feet (30 480 mm), the label on the sign will indicate the appropriate viewing distance. If such a sign is used, the spacing of the signs should be based on the listed viewing distance.

The exceptions identify conditions where exit signs are not necessary since they would not increase the safety of the egress path.

In accordance with Exception 4, exit signs are not required in sleeping room areas of Group I-3 buildings. In cases of emergency, occupants in Group I-3 are escorted by staff to the exits and to safety. The exit signs also represent potential weapons when they are accessible to the residents.

In accordance with Exception 2, when the exit is identifiable in itself and is the main exterior door through which the occupants would enter the building, exit signs are not required. For example, a two-story Group B building has a main employee/customer entrance. The entrance consists of a storefront arrangement with glass doors and sidelights. The entrance is centrally located within the building. These main exterior exit doors can be quickly observed as being an exit and would not need to be marked with an exit sign.

For Exceptions 1 and 3 the assumption is that the occupants are familiar enough with the space to know the way out. In most cases, the way in is also the way out.

In Group A-4 and A-5 occupancies described in Exception 5, the egress path is obvious; thus, exit signs are not needed. In addition, due to the configuration of the vomitories, the exit signs are not readily visible to the persons immediately adjacent to or above the vomitory.

1011.2 Illumination. Exit signs shall be internally or externally illuminated.

Exception: Tactile signs required by Section 1011.3 need not be provided with illumination.

❖ This section simply provides the scope for illumination of regulated exit signs. Special requirements are established for tactile exit signs in Section 1011.3.

1011.3 Tactile exit signs. A tactile sign stating EXIT and complying with ICC A117.1 shall be provided adjacent to each door to an egress stairway, an exit passageway and the exit discharge.

❖ This signage is needed to provide directions to the exits for those persons with visual impairments. Signs are needed on the required exits in the building, including at doors leading to exit stairway enclosures, at doors leading to exit passageways, doors within the exit enclosures leading to the outside and any exit doors that lead directly to the outside. While not specifically stated, the intent of the code is to also have this provision applicable to ramps within exit enclosures.

Tactile signage in accordance with ICC A117.1 includes both raised lettering and braille. While this sign is not required to be illuminated, it would be advantageous for a person who had partial sight.

1011.4 Internally illuminated exit signs. Internally illuminated exit signs shall be listed and labeled and shall be installed in accordance with the manufacturer's instructions and Section 604. Exit signs shall be illuminated at all times.

❖ While not required to be listed, exit signage may choose to comply with UL 924. Listed internally illuminated exit signs are required by UL 924 to meet the graphics, illumination and power sources defined in Sections 1011.5.1 through 1011.5.3. Exit signs must be illuminated at all times, including when the building may not be fully occupied. If a fire occurs late at night, there may be cleaning crews or persons working overtime in the building who will need to be able to find the exits.

1011.5 Externally illuminated exit signs. Externally illuminated exit signs shall comply with Sections 1011.5.1 through 1011.5.3.

❖ Externally illuminated exit signage must meet the graphic, illumination and emergency power requirements in the referenced sections. The requirements are the same as for internally illuminated signage.

1011.5.1 Graphics. Every exit sign and directional exit sign shall have plainly legible letters not less than 6 inches (152 mm) high with the principal strokes of the letters not less than 0.75 inch (19.1 mm) wide. The word "EXIT" shall have letters having a width not less than 2 inches (51 mm) wide except the letter "I," and the minimum spacing between letters shall not be less than 0.375 inch (9.5 mm). Signs larger than the minimum established in this section shall have letter widths, strokes and spacing in proportion to their height.

The word "EXIT" shall be in high contrast with the background and shall be clearly discernible when the exit sign illumination means is or is not energized. If an arrow is provided as part of the exit sign, the construction shall be such that the arrow direction cannot be readily changed.

❖ Every exit sign and directional sign located in the exit access route is required to have a color contrast vivid enough to make the signs readily visible, even when not illuminated. Letters must be at least 6 inches (152 mm) high and their stroke not less than $^3/_4$-inch (19.1 mm) wide (see Figure 1011.5.1). The sizing of the letters is predicated on the readability of the wording from a distance of 100 feet (30 480 mm).

While red letters are common for exit signs, sometimes green on black is used in auditorium areas with low lighting levels, such as theaters, because that color combination tends not to distract the audience's attention. It is more important that the exit sign be readily visible with respect to the background.

Exit signs may be larger than the minimum size specified. However, the standardized proportion of the letters must be maintained. Externally illuminated signage that are smaller than the specified size could use the requirements in UL 924 for guidance; however, sign spacing would need to be adjusted and alternative approval must be through the fire code official having jurisdiction.

1011.5.2 Exit sign illumination. The face of an exit sign illuminated from an external source shall have an intensity of not less than 5 foot-candles (54 lux).

❖ Every exit sign and directional sign must be continuously illuminated to provide a light intensity at the illuminated surface of at least 5 foot-candles (54 lux). It is not a requirement that the exit signs be internally illuminated. An external illumination source with the power capabilities specified by Section 1011.5.3 is acceptable.

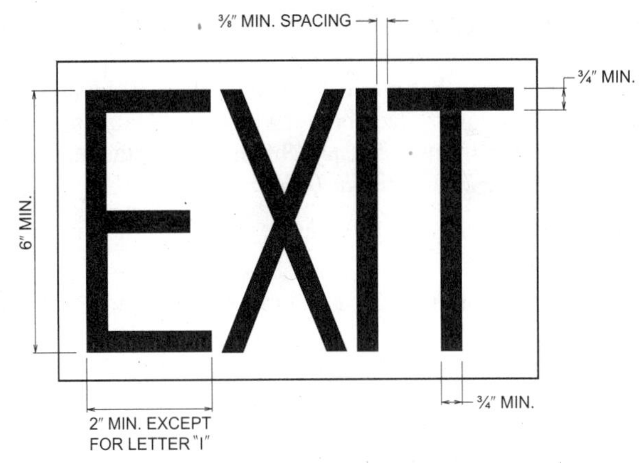

For SI: 1 inch = 25.4 mm.

Figure 1011.5.1
MINIMUM SIZES OF EXIT SIGNS

1011.5.3 Power source. Exit signs shall be illuminated at all times. To ensure continued illumination for a duration of not less than 90 minutes in case of primary power loss, the sign illumination means shall be connected to an emergency power system provided from storage batteries, unit equipment or an on-site generator. The installation of the emergency power system shall be in accordance with Section 604.

Exception: Approved exit sign illumination means that provide continuous illumination independent of external power sources for a duration of not less than 90 minutes, in case of primary power loss, are not required to be connected to an emergency electrical system.

❖ Exit signs must be illuminated on a continuous basis so that when a fire emergency occurs, occupants will be able to identify the locations of the exits. The reliability of the power sources supplying the electrical energy required for maintaining the illumination of exit signs is important. When power interruptions occur, exit sign illumination must be obtained from an emergency power system. This does not imply that the sign must be internally illuminated. Whatever illumination system is used, whether internal or external, it must be connected to a system designed to pick up the power load required by the exit signs after loss of the normal power supply.

Where self-luminous signs are used, connection to the emergency electrical supply system is not required.

[B] SECTION 1012
GUARDS

1012.1 Where required. Guards shall be located along open-sided walking surfaces, mezzanines, industrial equipment platforms, stairways, ramps and landings which are located more than 30 inches (762 mm) above the floor or grade below. Guards shall be adequate in strength and attachment in accordance with Section 1607.7 of the *International Building Code.* Guards shall also be located along glazed sides of stairways, ramps and landings that are located more than 30 inches (762 mm) above the floor or grade below where the glazing provided does not meet the strength and attachment requirements in Section 1607.7 of the *International Building Code.*

Exception: Guards are not required for the following locations:

1. On the loading side of loading docks or piers.
2. On the audience side of stages and raised platforms, including steps leading up to the stage and raised platforms.
3. On raised stage and platform floor areas such as runways, ramps and side stages used for entertainment or presentations.
4. At vertical openings in the performance area of stages and platforms.
5. At elevated walking surfaces appurtenant to stages and platforms for access to and utilization of special lighting or equipment.
6. Along vehicle service pits not accessible to the public.
7. In assembly seating where guards in accordance with Section 1024.14 are permitted and provided.

❖ Where one or more sides of a walking surface are open to the floor level or grade below, a guard system must be provided to minimize the possibility of occupants accidentally falling to the surface below (see Figure 1012.1). A guard is required only where the difference in elevation between the higher walking surface and the surface below is greater than 30 inches (762 mm). The loads for guard design are addressed in Section 1607. Additionally, this section regulates glazing that is installed in a guard on the side of a stairway, ramp or landing where the glazing has not been designed to resist the forces from a fall.

Most of the exceptions identify some typical situations where guards are not practical, such as along loading docks, stages and their approaches and vehicle service pits. Exception 7 references the lower guards permitted at locations where a line of sight for assembly spaces is part of the considerations.

1012.2 Height. Guards shall form a protective barrier not less than 42 inches (1067 mm) high, measured vertically above the leading edge of the tread, adjacent walking surface or adjacent seatboard.

Exceptions:

1. For occupancies in Group R-3, and within individual dwelling units in occupancies in Group R-2, both as applicable in Section 1001.1, guards whose top rail also serves as a handrail shall have a height not less than 34 inches (864 mm) and not more than 38 inches (965 mm) measured vertically from the leading edge of the stair tread nosing.
2. The height in assembly seating areas shall be in accordance with Section 1024.14.

❖ Guards must not be less than 42 inches (1067 mm) in height as measured vertically from the top of the guard

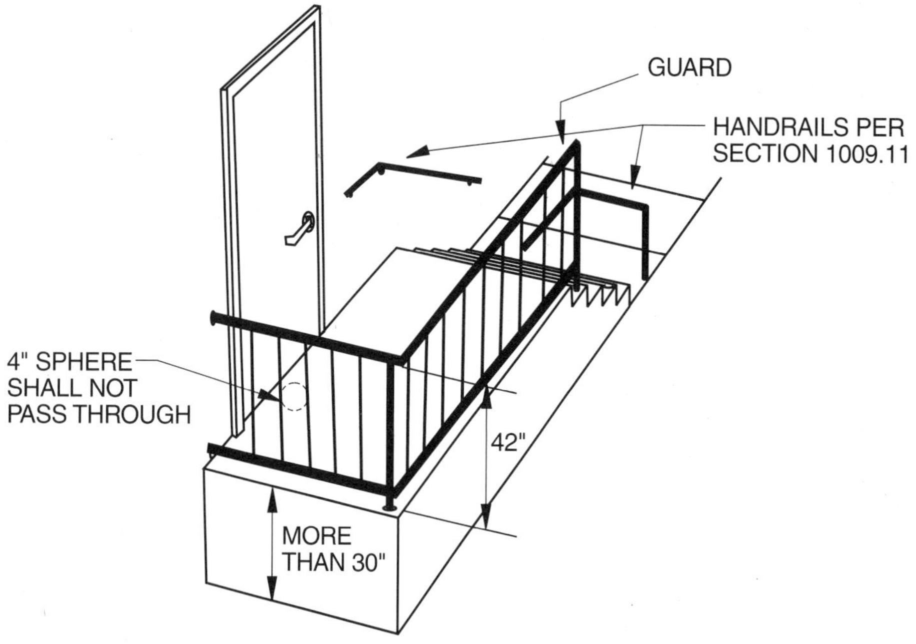

For SI: 1 inch = 25.4 mm.

Figure 1012.1
OPEN-SIDED WALKING AREA

down to the leading edge of the tread or to an adjacent walking surface (see Figures 1012.1 and 1012.2). Experience has shown that 42 inches (1067 mm) or more provides adequate height for protection purposes. This puts the top of the guard above the center of gravity of the average adult. With this height requirement at locations where both a guard and handrail are required, the handrail cannot be at the top of the guard except as permitted in Exception 1.

Due to safety concerns, a designer may want to have a guard where there is a drop-off of less than 30 inches (762 mm). Decorative guards may be utilized to support handrails or serve as part of the edge protection along a ramp. When nonrequired guards are provided, the 42-inch (1067 mm) minimum height is not required.

Exception 1 is for certain residential occupancies and allows for the handrail to be at the top of a lower guard. The reduced allowable guard height is consistent with current construction practice. Exception 2 references the lower guards permitted at locations where a line of sight for assembly spaces is part of the consideration.

1012.3 Opening limitations. Open guards shall have balusters or ornamental patterns such that a 4-inch-diameter (102 mm) sphere cannot pass through any opening up to a height of 34 inches (864 mm). From a height of 34 inches (864 mm) to 42 inches (1067 mm) above the adjacent walking surfaces, a sphere 8 inches (203 mm) in diameter shall not pass.

Exceptions:

1. The triangular openings formed by the riser, tread and bottom rail at the open side of a stairway shall be of a maximum size such that a sphere of 6 inches (152 mm) in diameter cannot pass through the opening.
2. At elevated walking surfaces for access to and use of electrical, mechanical or plumbing systems or equipment, guards shall have balusters or be of solid materials such that a sphere with a diameter of 21 inches (533 mm) cannot pass through any opening.
3. In areas which are not open to the public within occupancies in Group I-3, F, H or S, balusters, horizontal intermediate rails or other construction shall not permit a sphere with a diameter of 21 inches (533 mm) to pass through any opening.
4. In assembly seating areas, guards at the end of aisles where they terminate at a fascia of boxes, balconies and galleries shall have balusters or ornamental patterns such that a 4-inch-diameter (102 mm) sphere cannot pass through any opening up to a height of 26 inches (660 mm). From a height of 26 inches (660 mm) to 42 inches (1067 mm) above the adjacent walking surfaces, a sphere 8 inches (203 mm) in diameter shall not pass.

❖ The basis for limiting openings in a guard to a 4-inch (102 mm) sphere is research that indicates that a 4-inch (102 mm) opening will prevent nearly all children 1 year in age or older from falling through the guard. The allowable opening increases to an 8-inch (203 mm) sphere at heights where falling through the guard is not an issue.

An exception to the 4-inch (102 mm) spacing requirement is that a 6-inch (152 mm) opening is allowed for openings formed by the riser, tread and bottom rail of guards at the open side of a stairway. This is because

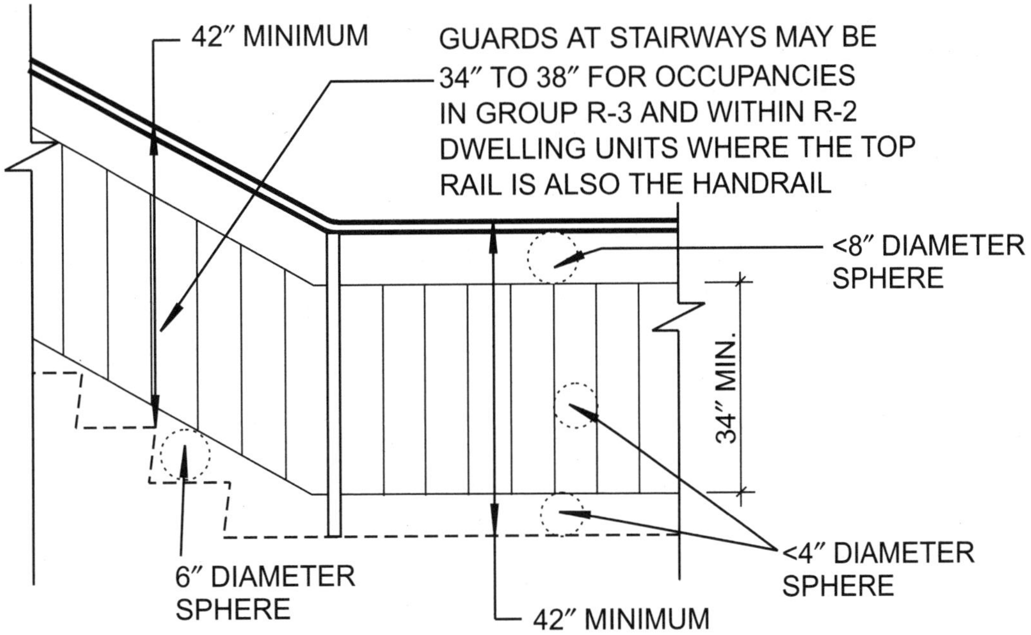

For SI: 1 inch = 25.4 mm.

Figure 1012.2
GUARD HEIGHT

the geometry of the openings is such that the entire body cannot pass through the triangular opening. In the case of a standard stair, limiting such openings to a 4-inch (102 mm) sphere is impractical to achieve with a sloped bottom member in the guard (see Figure 1012.2).

Exceptions 2 and 3 address areas where the presence of small children is unlikely and often prohibited. Guards along walkways leading to electrical, mechanical and plumbing systems or equipment and in occupancies in Groups I-3, F, H and S may be constructed such that a sphere 21 inches (533 mm) in diameter will not pass through any of the openings (see Figure 1012.3). This requirement allows the use of horizontal intermediate members.

The exception for the guard infill near the top of the guard in assembly seating areas is to reduce sightline problems.

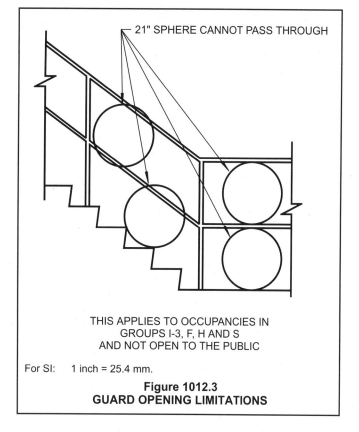

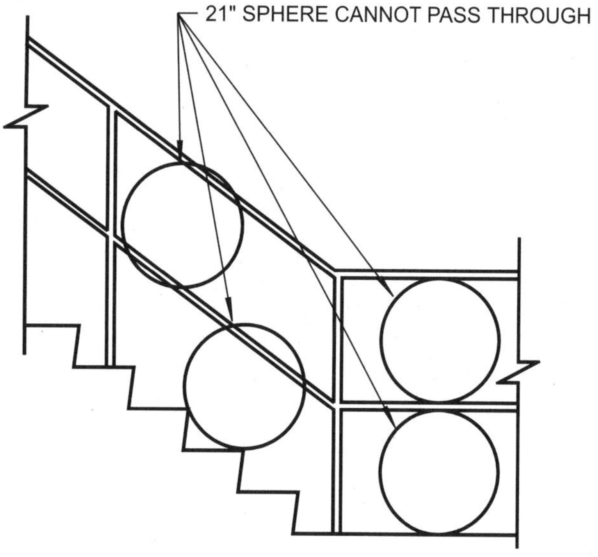
21" SPHERE CANNOT PASS THROUGH
THIS APPLIES TO OCCUPANCIES IN
GROUPS I-3, F, H AND S
AND NOT OPEN TO THE PUBLIC

For SI: 1 inch = 25.4 mm.

Figure 1012.3
GUARD OPENING LIMITATIONS

1012.4 Screen porches. Porches and decks which are enclosed with insect screening shall be provided with guards where the walking surface is located more than 30 inches (762 mm) above the floor or grade below.

❖ Insect screening located on the open sides of porches and decks does not provide an adequate barrier to reasonably protect an occupant from falling to the surface below. Guards are required on the open sides of porches and decks where the floor is located more than 30 inches (762 mm) above the surface below. The guards must comply with the provisions of Section 1012.

1012.5 Mechanical equipment. Guards shall be provided where appliances, equipment, fans or other components that require service are located within 10 feet (3048 mm) of a roof edge or open side of a walking surface and such edge or open side is located more than 30 inches (762 mm) above the floor, roof or grade below. The guard shall be constructed so as to prevent the passage of a 21-inch-diameter (533 mm) sphere.

❖ The purpose of this requirement is to protect workers from falls off of roofs or from open-sided walking surfaces when doing maintenance work on equipment. The guard opening is allowed to be up to 21 inches (533 mm) since children are not likely to be in such areas. This requirement allows the use of horizontal intermediate members.

[B] SECTION 1013
EXIT ACCESS

1013.1 General. The exit access arrangement shall comply with Sections 1013 through 1016 and the applicable provisions of Sections 1003 through 1012.

❖ Sections 1013 through 1016 includes the design requirements for exit access (see Section 1014) and the requirements for exit access components (see Section 1013.4). The general requirements that also apply to the exit access are in Sections 1003 through 1012.

1013.2 Egress through intervening spaces. Egress from a room or space shall not pass through adjoining or intervening rooms or areas, except where such adjoining rooms or areas are accessory to the area served; are not a high-hazard occupancy and provide a discernible path of egress travel to an exit. Egress shall not pass through kitchens, storage rooms, closets or spaces used for similar purposes. An exit access shall not pass through a room that can be locked to prevent egress. Means of egress from dwelling units or sleeping areas shall not lead through other sleeping areas, toilet rooms or bathrooms.

Exceptions:

1. Means of egress are not prohibited through a kitchen area serving adjoining rooms constituting part of the same dwelling unit or sleeping unit.
2. Means of egress are not prohibited through adjoining or intervening rooms or spaces in a Group H occupancy when the adjoining or intervening rooms or spaces are the same or a lesser hazard occupancy group.

❖ This section allows adjoining spaces to be considered as a part of the room or space from which egress originates, provided that there are reasonable assurances that the continuous egress path will always be available. For example, such egress paths must remain unobstructed and must not pass through an extraordinary fire hazard, such as an area of high-hazard use (Group H). Requiring occupants to egress from an area and pass through an adjoining area that can be characterized by rapid fire buildup, or worse, places them in an unreasonable risk situation [see

Figure 1013.2(1)]. An occupant should be provided with an equivalent or increased level of safety as he or she approaches the exit. It should be noted that the code does not limit the number of intervening or adjoining rooms through which egress can be made, provided that all other code requirements (i.e., travel distance, number of doorways, etc.) are met. An exit access route, for example, may be laid out such that an occupant leaves a room or space, passes through an adjoining space, enters an exit access corridor, passes through another room and, finally, into an exit [see Figure 1013.2(2)], as long as all other code requirements are satisfied.

Relying on an egress path through an adjacent dwelling unit to be available at all times is not a reasonable expectation. Egress through an adjacent business tenant space can be unreasonable given the security and privacy measures the adjacent tenant may take to secure such space. However, egress through a reception area that serves a suite of offices of the same tenant is clearly accessory and is permitted.

A common code enforcement problem is the locked door in the egress path. Twenty-five workers perished in September 1991, when they were trapped inside the Imperial Food Processing Plant in Hamlet, North Carolina, due, in part, to locked exit doors. As long as the egress door is readily openable in the direction of egress travel without the use of keys, special knowledge or effort (see Section 1008.1.8.5), the occupants can move unimpeded away from a fire emergency.

Exception 1 allows egress through a kitchen provided it is part of a dwelling unit or guestroom. Egress through a restaurant kitchen is not permitted.

Exception 2 allows egress through high-hazard rooms provided the adjoining rooms are also high-hazard occupancies.

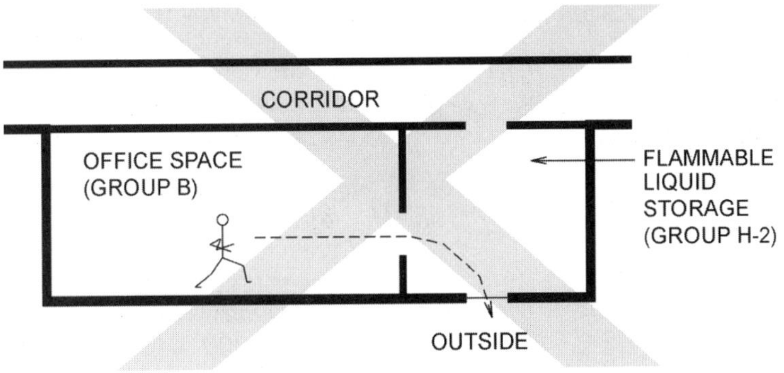

Figure 1013.2(1)
EGRESS PROHIBITED THROUGH GROUP H AREAS

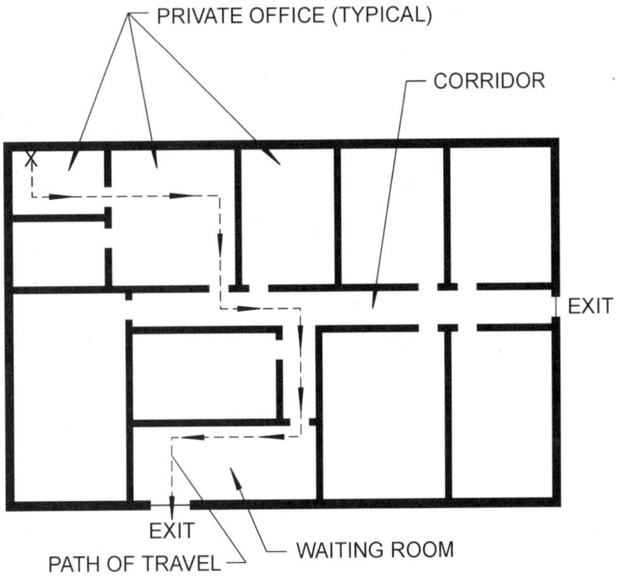

Figure 1013.2(2)
EGRESS THROUGH ADJOINING ROOMS AND SPACES

1013.2.1 Multiple tenants. Where more than one tenant occupies any one floor of a building or structure, each tenant space, dwelling unit and sleeping unit shall be provided with access to the required exits without passing through adjacent tenant spaces, dwelling units and sleeping units.

❖ Where a floor is occupied by multiple tenants, each tenant must be provided with full and direct access to the required exits serving that floor without passing through another tenant space. Tenants frequently lock the doors to their spaces for privacy and security. Should an egress door that is shared by both tenants be locked, occupants in one of the spaces could be trapped and unable to reach an exit. Therefore, an egress layout where occupants from one tenant space travel through another tenant space to gain access to one of the required exits from that floor is prohibited.

This limitation is so that occupants from all tenant spaces will have unrestricted access to the required egress elements while maintaining the security and privacy of the individual tenants. This limitation is based on one of the fundamental principles of egress: to provide a means of egress where all components are capable of being used by the occupants without keys, tools, special knowledge or effort (see Section 1008.1.8.5).

1013.2.2 Group I-2. Habitable rooms or suites in Group I-2 occupancies shall have an exit access door leading directly to an exit access corridor.

Exceptions:

1. Rooms with exit doors opening directly to the outside at ground level.
2. Patient sleeping rooms are permitted to have one intervening room if the intervening room is not used as an exit access for more than eight patient beds.
3. Special nursing suites are permitted to have one intervening room where the arrangement allows for direct and constant visual supervision by nursing personnel.
4. For rooms other than patient sleeping rooms, suites of rooms are permitted to have one intervening room if the travel distance within the suite to the exit access door is not greater than 100 feet (30 480 mm) and are permitted to have two intervening rooms where the travel distance within the suite to the exit access door is not greater than 50 feet (15 240 mm).

Suites of sleeping rooms shall not exceed 5,000 square feet (465 m^2). Suites of rooms, other than patient sleeping rooms, shall not exceed 10,000 square feet (929 m^2). Any patient sleeping room, or any suite that includes patient sleeping rooms, of more than 1,000 square feet (93 m^2) shall have at least two exit access doors remotely located from each other. Any room or suite of rooms, other than patient sleeping rooms, of more than 2,500 square feet (232 m^2) shall have at least two access doors remotely located from each other. The travel distance between any point in a Group I-2 occupancy and an exit access door in the room shall not exceed 50 feet (15 240 mm). The travel distance between any point in a suite of sleeping rooms and an exit access door of that suite shall not exceed 100 feet (30 480 mm).

❖ The purpose of this section is to establish the means of egress requirements that are unique to Group I-2 occupancies. Patient rooms are permitted to be arranged as suites (see Figure 1013.2.2 for an example plan of a patient suite). The figure illustrates a patient suite of more than 1,000 square feet (93 m^2) where two remote egress doors are required. The criteria in this section recognizes the low patient-to-staff ratio of these facilities where the staff is directly responsible for the safety of the patients in the event of a fire.

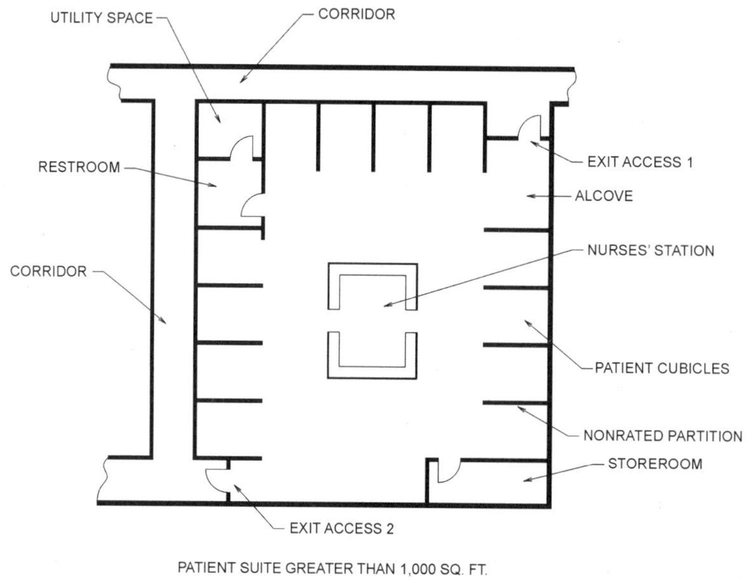

For SI: 1 square foot = 0.0929 m^2.

Figure 1013.2.2
GROUP I-2 MEANS OF EGRESS DOORWAYS

1013.3 Common path of egress travel. In occupancies other than Groups H-1, H-2 and H-3, the common path of egress travel shall not exceed 75 feet (22 860 mm). In occupancies in Groups H-1, H-2, and H-3, the common path of egress travel shall not exceed 25 feet (7620 mm).

Exceptions:

1. The length of a common path of egress travel in an occupancy in Groups B, F and S shall not be more than 100 feet (30 480 mm), provided that the building is equipped throughout with an automatic sprinkler system installed in accordance with Section 903.3.1.1.
2. Where a tenant space in an occupancy in Groups B, S and U has an occupant load of not more than 30, the length of a common path of egress travel shall not be more than 100 feet (30 480 mm).
3. The length of a common path of egress travel in occupancies in Group I-3 shall not be more than 100 feet (30 480 mm).

❖ The common path of travel is the distance measured from the most remote point in a space to the point in the exit path where the occupant has access to two required exits in separate directions. The definition of "Common path of egress travel" is found in Section 1002. The distance limitations are applicable to all paths of travel that lead out of a space or building. An illustration of this distance is found in Figure 1013.3. The illustration reflects two examples of a common path of travel where the occupants at points A and B are able to travel in only one direction before they reach a point at which they have a choice of two paths of travel to the required exits from the building. Note that from point A, the occupants have two available paths, but these merge to form a single path out of the space. This is also considered a common path of travel.

The exceptions increase the allowable length of the common path of travel based on the installation of a sprinkler system, a low occupant load or for a Group I-3 occupancy.

1013.4 Aisles. Aisles serving as a portion of the exit access in the means of egress system shall comply with the requirements of this section. Aisles shall be provided from all occupied portions of the exit access which contain seats, tables, furnishings, displays and similar fixtures or equipment. Aisles serving assembly areas, other than seating at tables, shall comply with Section 1024. Aisles serving reviewing stands, grandstands and bleachers shall also comply with Section 1024.

The required width of aisles shall be unobstructed.

Exception: Doors, when fully opened, and handrails shall not reduce the required width by more than 7 inches (178 mm). Doors in any position shall not reduce the required width by more than one-half. Other nonstructural projections such as trim and similar decorative features are permitted to project into the required width 1.5 inches (38 mm) from each side.

❖ This section addresses aisles for other than assembly areas, which are covered in Section 1024. Aisle accessways for tables and seating are covered by Section 1013.4.2.

The term "aisle accessway" is defined in Section

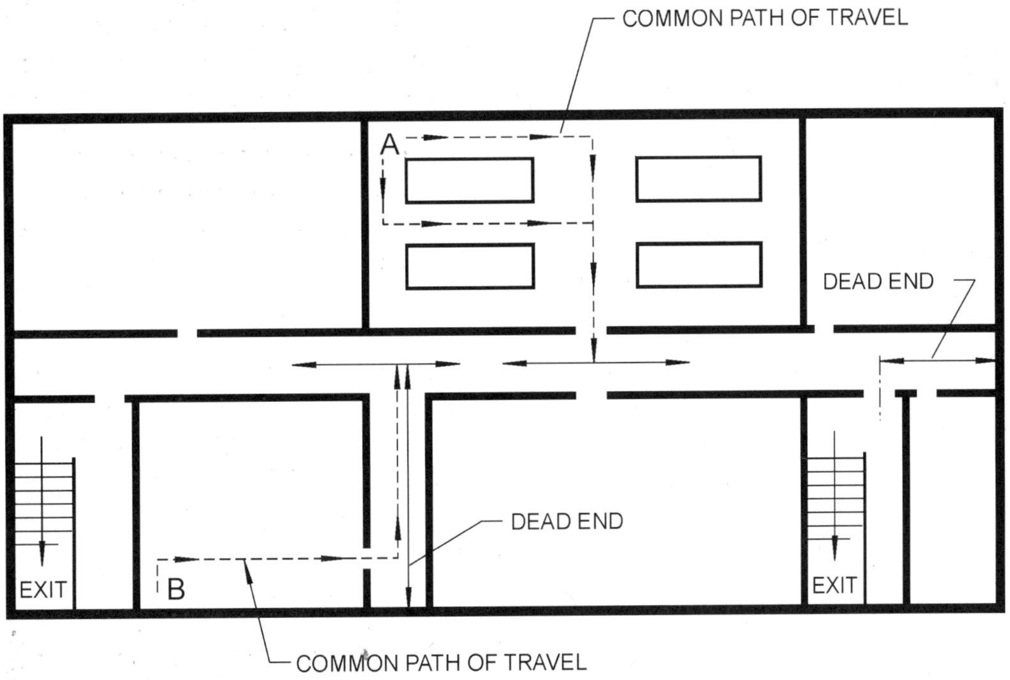

Figure 1013.3
EXAMPLE OF COMMON PATH OF TRAVEL

1002 as "that portion of exit access that leads to an aisle." While the term "aisle" is not defined in the code, the dictionary indicates an aisle to be "a walkway between or along sections of seats, shelves, counters, etc., as in a theater, church or department store." Given the many possible configurations of fixtures and furniture, both permanent and moveable, the determination of where aisle accessways stop and aisles begin is often subject to some interpretation.

Since the aisle serves as a path for means of egress similar to a corridor, the requirements for doors obstructing the aisle are similar. The exception also includes provisions for protrusion of handrails or trim along an aisle.

1013.4.1 Groups B and M. In Group B and M occupancies, the minimum clear aisle width shall be determined by Section 1005.1 for the occupant load served, but shall not be less than 36 inches (914 mm).

> **Exception:** Nonpublic aisles serving less than 50 people, and not required to be accessible by Chapter 11 of the *International Building Code*, need not exceed 28 inches (711 mm) in width.

❖ This requirement establishes aisle width criteria for Group B and M occupancies based on the occupant load served by the aisle. The reference to Section 1005.1 triggers a requirement for aisles wider than 36 inches (917 mm) when the anticipated occupant load that the aisle serves is larger than 180 in nonsprinklered buildings and larger than 240 in sprinklered buildings. The exception addresses aisles that may be found in an archival file room or stock storage racks.

At this time, the code does not include criteria for aisle accessways in these types of facilities. If fixtures are permanent, such as in a typical grocery store, the aisle provisions would be applicable throughout. In a situation where there are groups of displays separated by aisles, the area within the displays may be considered aisle accessways.

Section 105.2, Item 13 exempts movable cases, counters and partitions not over 5 feet, 9 inches (1752 mm) in height from requiring a building permit. It is not practicable to require a means of egress inspection every time a store changes a display or a business moves someone into a new office. At least a minimal amount of space will be provided throughout to allow access to the displays, furniture or equipment. Concerns that should be addressed in the aisle accessways are unobstructed access to an aisle and common path of travel constraints in Section 1013.3. While common path of travel is limited to 30 feet (9144 mm) (see Sections 1013.4.2.3 and 1024.8) in dining and assembly seating, the path of travel for these arrangements is typically limited to one direction. Very few arrangements of fixtures and furniture in Group B and M occupancies have this type of arrangement.

1013.4.2 Seating at tables. Where seating is located at a table or counter and is adjacent to an aisle or aisle accessway, the measurement of required clear width of the aisle or aisle accessway shall be made to a line 19 inches (483 mm) away from and parallel to the edge of the table or counter. The 19-inch (483 mm) distance shall be measured perpendicular to the side of the table or counter. In the case of other side boundaries for aisle or aisle accessways, the clear width shall be measured to walls, edges of seating and tread edges, except that handrail projections are permitted.

> **Exception:** Where tables or counters are served by fixed seats, the width of the aisle accessway shall be measured from the back of the seat.

❖ Seating is often provided adjacent to aisles and aisle accessways. In measuring the width of an aisle or aisle accessway for movable seating, the measurement is taken at a distance of 19 inches (483 mm) perpendicular to the side of the table or counter. This 19-inch (483 mm) space from the edge of the table or counter to the line where the aisle or aisle accessway measurement begins is intended to represent the space occupied by a typical seated occupant. This dimension is also considered to be adequate to accommodate seats with armrests that are too high to fit under the table where fixed seats are used. The aisle width is permitted to be measured from the back of the seat based upon the exception. As indicated in Figure 1013.4.2, where seating abuts an aisle or aisle accessway, 19 inches (483 mm) must be added to the required aisle or aisle accessway width for seating on only one side and 38 inches (965 mm) for seating on both sides. When seating will not be adjacent to the aisles or aisle passageways, as is the case when tables are at an angle to the aisle or aisle accessway, the measurement may be taken to the edge of the seating, table, counter or tread (see Figure 1013.4.2).

1013.4.2.1 Aisle accessway for tables and seating. Aisle accessways serving arrangements of seating at tables or counters shall have sufficient clear width to conform to the capacity requirements of Section 1005.1 but shall not have less than the appropriate minimum clear width specified in Section 1013.4.1.

❖ This section specifies two criteria for the determination of the required width of aisle accessways: the requirements of Sections 1005.1 and 1013.4.1. The aisle accessway width is to be the wider of the two requirements.

The requirements of Sections 1013.4.2.1, 1013.4.2.2 and 1013.4.2.3 are illustrated in Figure 1013.4.2.1.

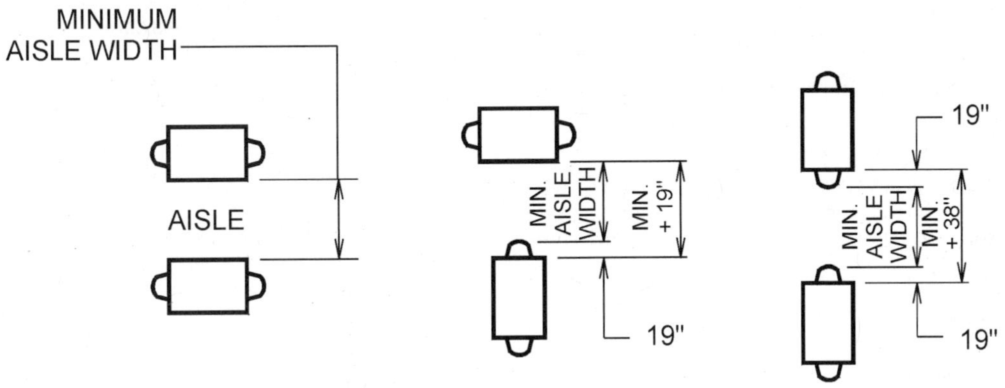

For SI: 1 inch = 25.4 mm.

Figure 1013.4.2
AISLE MEASUREMENT WITH SEATING AND TABLES

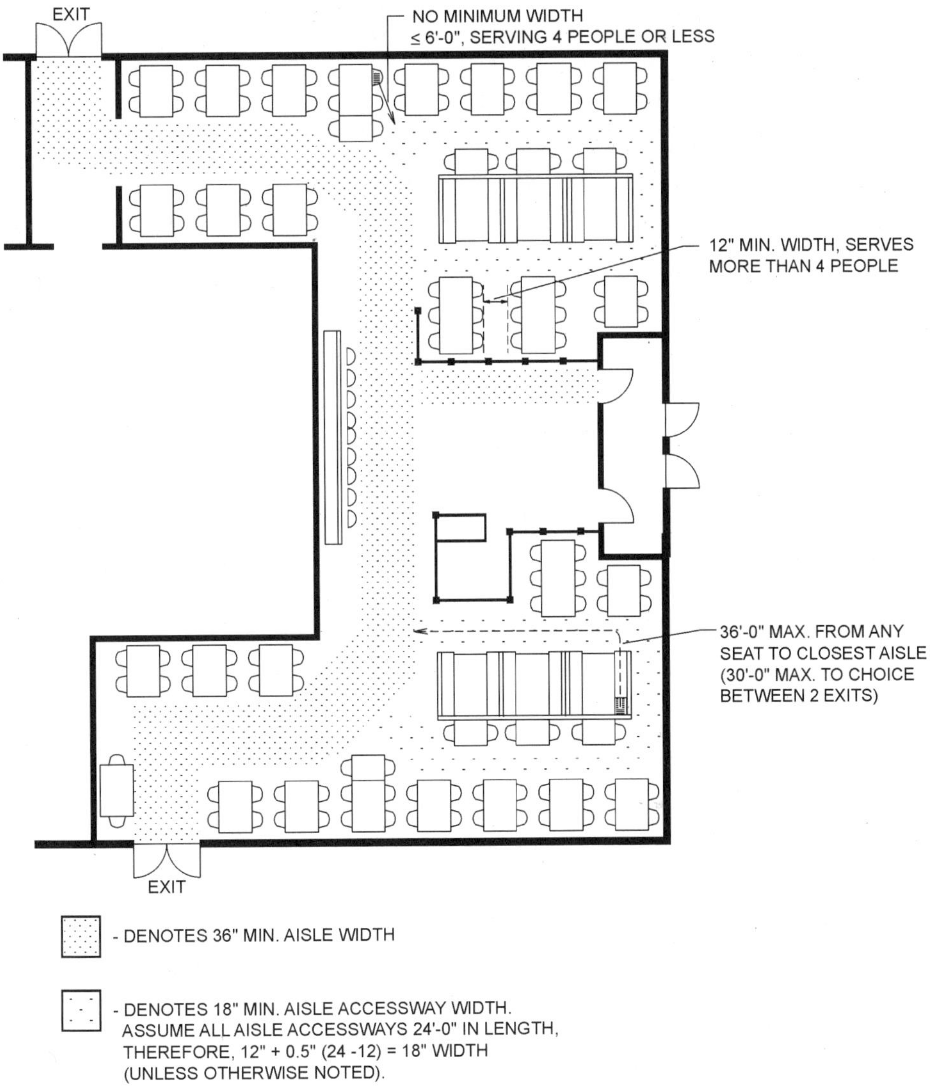

For SI: 1 inch = 25.4 mm, 1 foot = 304.8 mm.

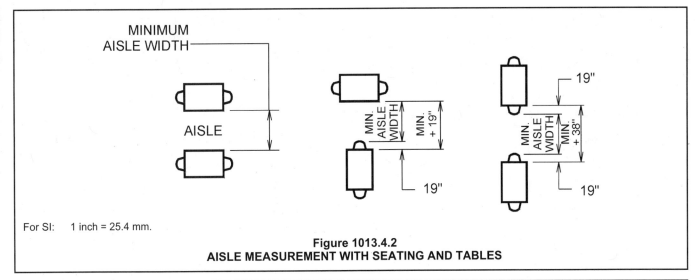

Figure 1013.4.2.1
ACCESS FOR TABLES AND SEATING

1013.4.2.2 Table and seating accessway width. Aisle accessways shall provide a minimum of 12 inches (305 mm) of width plus 0.5 inch (12.7 mm) of width for each additional 1 foot (305 mm), or fraction thereof, beyond 12 feet (3658 mm) of aisle accessway length measured from the center of the seat farthest from an aisle.

Exception: Portions of an aisle accessway having a length not exceeding 6 feet (1829 mm) and used by a total of not more than four persons.

❖ The general relationship of tables and seating results in a situation in which it is more difficult to determine which chairs are served by which aisle accessway. Therefore, the width of the aisle accessway is a function of the distance from the aisle. The same minimum 12 inches (305 mm) is used and is increased 0.5 inch (12.7 mm) for each additional foot of travel beyond 12 feet (3658 mm).

Recognizing that the normal use of table and chair seating will require some clearance for access and service, the exception eliminates the minimum width criteria if the distance to the aisle [or an aisle accessway of at least 12 inches (305 mm)] is less than 6 feet (1829 mm) and the number of people served is not more than four. Therefore, the first 6 feet (1829 mm) are not required to meet any minimum width criteria. After the first 6 feet (1829 mm), the requirements for an aisle accessway will apply. The length of the aisle accessway is then restricted by Section 1013.4.2.3. When the maximum length of the aisle accessway is reached, either an aisle, corridor or exit access door must be provided (see Figure 1013.4.2.1).

1013.4.2.3 Table and seating aisle accessway length. The length of travel along the aisle accessway shall not exceed 30 feet (9144 mm) from any seat to the point where a person has a choice of two or more paths of egress travel to separate exits.

❖ At some point in the exit access travel, it is necessary to reach an aisle complying with the minimum widths of Section 1013.4. An aisle accessway travel distance is not to exceed 30 feet (9144 mm), which may represent a dead-end condition (see Figure 1013.4.2.1).

1013.5 Egress balconies. Balconies used for egress purposes shall conform to the same requirements as corridors for width, headroom, dead ends and projections. Exterior balconies shall be designed to minimize accumulation of snow or ice that impedes the means of egress.

Exception: Exterior balconies and concourses in outdoor stadiums shall be exempt from the design requirement to protect against the accumulation of snow or ice.

❖ This section regulates balconies that are used as an exit access and requires that they meet the same requirements as exit access corridors, except for the enclosure. Exterior exit access balconies must be kept free of snow and ice. Sheltering the exterior balcony can serve to minimize snow and ice accumulation but, in some areas, may not completely eliminate the need to remove the snow or ice by other means. If the exterior side of the balcony is also enclosed to prevent snow and ice accumulation, the balcony essentially creates the same conditions as an interior corridor and must be protected as such.

The exception for protection of snow and ice accumulation for outdoor stadiums is based on the assumption that an adequate snow removal or maintenance policy exists for these facilities.

1013.5.1 Wall separation. Exterior egress balconies shall be separated from the interior of the building by walls and opening protectives as required for corridors.

Exception: Separation is not required where the exterior egress balcony is served by at least two stairs and a dead-end travel condition does not require travel past an unprotected opening to reach a stair.

❖ An exterior exit access balcony has a valuable attribute in that the products of combustion may be freely vented to the open air. In the event of a fire in an adjacent space, the products of combustion would not be expected to build up in the balcony area as would commonly occur in an interior corridor. However, there is still a concern for the egress of occupants who must use the balcony for exit access, and consequently, may have to pass the room or space where the fire is located. Therefore, an exterior exit access balcony is required to be separated from interior spaces by fire partitions, as is required for interior corridors. The other provisions of Section 1016 relative to dead ends and opening protectives also apply.

If there are no dead-end conditions that require travel past an unprotected opening and the balcony is provided with at least two stairways, then the wall separating the balcony from the interior spaces does not need a fire-resistance rating (see Figure 1013.5.1). Such an arrangement reduces the probability that occupants will need to pass the area with the fire to gain access to an exit.

1013.5.2 Openness. The long side of an egress balcony shall be at least 50 percent open, and the open area above the guards shall be so distributed as to minimize the accumulation of smoke or toxic gases.

❖ This section provides an opening requirement that is intended to preclude the rapid buildup of smoke and toxic gases. A minimum of one side of the exterior balcony is required to have a minimum open exterior area of 50 percent of the side area of the balcony. The side openings are to be fairly uniformly distributed along the length of the balcony.

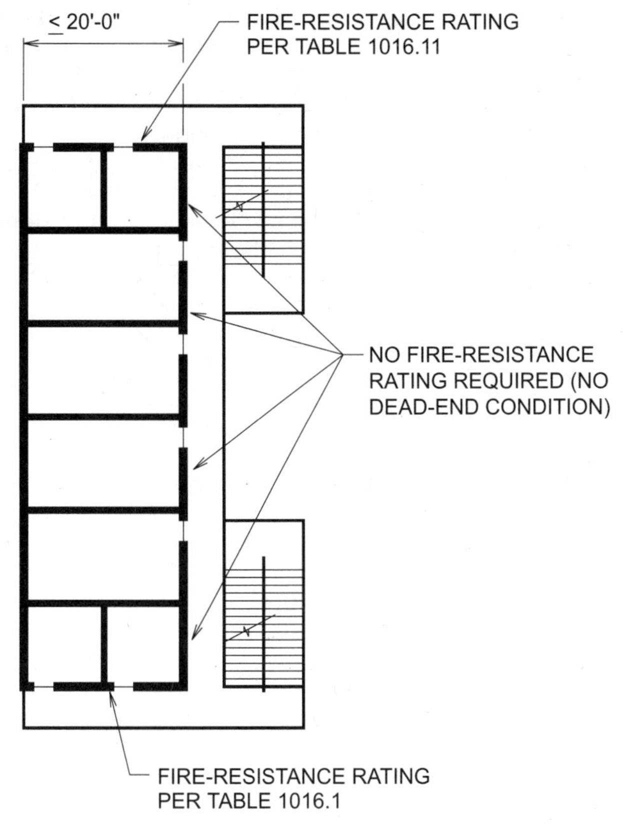

For SI: 1 inch = 25.4 mm, 1 foot = 304.8 mm.

Figure 1013.5.1
EXIT ACCESS BALCONIES

[B] SECTION 1014
EXIT AND EXIT ACCESS DOORWAYS

1014.1 Exit or exit access doorways required. Two exits or exit access doorways from any space shall be provided where one of the following conditions exists:

1. The occupant load of the space exceeds the values in Table 1014.1.
2. The common path of egress travel exceeds the limitations of Section 1013.3.
3. Where required by Sections 1014.3, 1014.4 and 1014.5.

Exception: Group I-2 occupancies shall comply with Section 1013.2.2.

❖ This section dictates the minimum number of paths of travel an occupant is to have available to avoid a fire incident in the room or space occupied. While providing multiple egress doorways from every room is unrealistic, a point does exist where alternative egress paths must be provided based on the number of occupants at risk, the distance any one occupant must travel to reach a doorway and the relative hazards associated with the occupancy of the space. Generally, the number of egress doorways required from any room or space coincides with the occupant load threshold criteria set forth for the minimum number of exits required in a building (see Section 1018.1). The limiting criteria in Table 1014.1 for rooms or spaces permitted to have a single exit access doorway are based on an empirical judgement of the associated risks.

If the occupants of a room are required to egress through another room, as permitted in Section 1013.2, the rooms are to be combined to determine if multiple doorways are required from the combined rooms. For example, if a suite of offices share a common reception area, the entire suite with the reception area must meet both the occupant load and the travel distance criteria.

It should be noted that where two doorways are required, the remoteness requirement of Section 1014.2 is applicable.

Item 2 sets the limits for a single means of egress based on travel distance. Where the common path of travel exceeds the limits in Section 1013.3, two egress paths are required for safe egress from the space.

Item 3 addresses the situation where two means of egress may be required in boiler, incinerator and furnace rooms; refrigerator machinery rooms or refrigerated rooms and spaces.

Group I-2 occupancies are not addressed in Table 1014.1. The exception refers to Section 1013.2.2 for Group I-2 means of egress requirements.

TABLE 1014.1
SPACES WITH ONE MEANS OF EGRESS

OCCUPANCY	MAXIMUM OCCUPANT LOAD
A, B, E, F, M, U	50
H-1, H-2, H-3	3
H-4, H-5, I-1, I-3, I-4, R	10
S	30

❖ The table represents an empirical judgement of the risks associated with a single means of egress from a room or space based on the occupant load in the room, the travel distance to the exit access door and the inherent risks associated with the occupancy (such as occupant mobility, occupant familiarity with the building, occupant response and the fire growth rate).

Since the occupants of Groups I and R may be sleeping and, therefore, may not be able to detect a fire in its early stages without staff supervision or room detectors, the number of occupants in a single egress room or space is limited to 10 persons.

Because of the potential for rapidly developing hazardous conditions, the single egress condition in Groups H-1, H-2 and H-3 is limited to a maximum of three persons. Because the materials contained in Groups H-4 and H-5 do not represent the same fire hazard potential as those found in Groups H-1, H-2 and H-3, the occupant load for spaces with one means of egress is increased.

Because of the reduced occupant density in Group S and the occupants' normal familiarity with the building, the single egress condition is permitted with an occupant load of 30.

1014.1.1 Three or more exits. Access to three or more exits shall be provided from a floor area where required by Section 1018.1.

❖ This section provides a reference to Section 1018.1 for conditions where three or more exits are required. The reference is provided in this section so that the requirements of Section 1018.1 will be obvious.

1014.2 Exit or exit access doorway arrangement. Required exits shall be located in a manner that makes their availability obvious. Exits shall be unobstructed at all times. Exit and exit access doorways shall be arranged in accordance with Sections 1014.2.1 and 1014.2.2.

❖ Exits need to be unobstructed and obvious at all times for the safety of occupants to evacuate the building in an emergency situation. This is consistent with the requirements in Section 1008.1 for exit or exit access doors to not be concealed by curtains, drapes, decorations or mirrors.

1014.2.1 Two exits or exit access doorways. Where two exits or exit access doorways are required from any portion of the exit access, the exit doors or exit access doorways shall be placed a distance apart equal to not less than one-half of the length of the maximum overall diagonal dimension of the building or area to be served measured in a straight line between exit doors or exit access doorways. Interlocking or scissor stairs shall be counted as one exit stairway.

Exceptions:

1. Where exit enclosures are provided as a portion of the required exit and are interconnected by a 1-hour fire-resistance-rated corridor conforming to the requirements of Section 1016, the required exit separation shall be measured along the shortest direct line of travel within the corridor.
2. Where a building is equipped throughout with an automatic sprinkler system in accordance with Section 903.3.1.1 or 903.3.1.2, the separation distance of the exit doors or exit access doorways shall not be less than one-third of the length of the maximum overall diagonal dimension of the area served.

❖ This section provides a method to determine quantitatively remoteness between exits and exit access doors, based on the dimensional characteristics of the space served. This has been a common measuring practice for some years with significant success. Very simply, the method involves determining the maximum dimension between any two points in a floor or a room (e.g., a diagonal between opposite corners in a rectangular room or building or the diameter in a circular room or building). If two doors or exits are required from the room or building (see Sections 1014.1 and 1018.1), the straight-line distance between the center of the thresholds of the doors must be at least one-half of the maximum dimension [see Figure 1014.2.1(1)].

While technical proof is not available to substantiate this method of determining remoteness, it has been found to be realistic and practical for building designs except for the common building with exits in a center core and office spaces around the perimeter.

If a scissor stairway is utilized, regardless of the separation of the two entrances, the scissor stairs may only be counted as one exit. Two stairways within the same enclosure could result in both stairways being unnavigable in an emergency if smoke penetrated the single enclosure (see the definition for "Scissor stairways" in Section 1002).

In Exception 1, a method of permitting the distance between exits to be measured along a complying corridor connecting the exits has served to mitigate the disruption to this design concept [see Figure 1014.2.1(2)].

As reflected in Exception 2, the protection provided by an automatic sprinkler system can reduce the threat of fire buildup so that the reduction in remoteness to one-third of the diagonal dimension is not unreasonable, based on the presumption that it provides the occupants with an acceptable level of safety from fire [see Figure 1014.2.1(3)]. The automatic sprinkler system must be installed throughout the building in accordance with Section 903.3.1.1 or 903.3.1.2 (NFPA 13 or 13R). This reduced separation (one-third diagonal) may also be used when applying the requirements of Exception 1.

In applying the provisions of this section, it is important to recognize any convergence of egress paths that may exist. Figure 1014.2.1(4) illustrates that although the assembly room has remotely located exit access doors, the doors from the entire space do not meet the criteria of this section.

1014.2.2 Three or more exits or exit access doorways. Where access to three or more exits is required, at least two exit doors or exit access doorways shall be placed a distance apart equal to not less than one-half of the length of the maximum overall diagonal dimension of the area served measured in a straight line between such exit doors or exit access doorways. Additional exits or exit access doorways shall be arranged a reasonable distance apart so that if one becomes blocked, the others will be available.

Exception: Where a building is equipped throughout with an automatic sprinkler system in accordance with Section 903.3.1.1 or 903.3.1.2, the separation distance of at least two of the exit doors or exit access doorways shall not be less than one-third of the length of the maximum overall diagonal dimension of the area served.

❖ When there are three or more required exits from a building or exit access doors from a room, they are to be analyzed identically to the method described in Section 1014.2.1. Two of the exits or exit access doors must meet the remoteness test, and any additional exits or doors can be located anywhere within the floor plan that meets the code requirements, including independence, accessibility, capacity and continuity.

The exception is for consistency with the exception for sprinklered buildings in Section 1014.2.1.

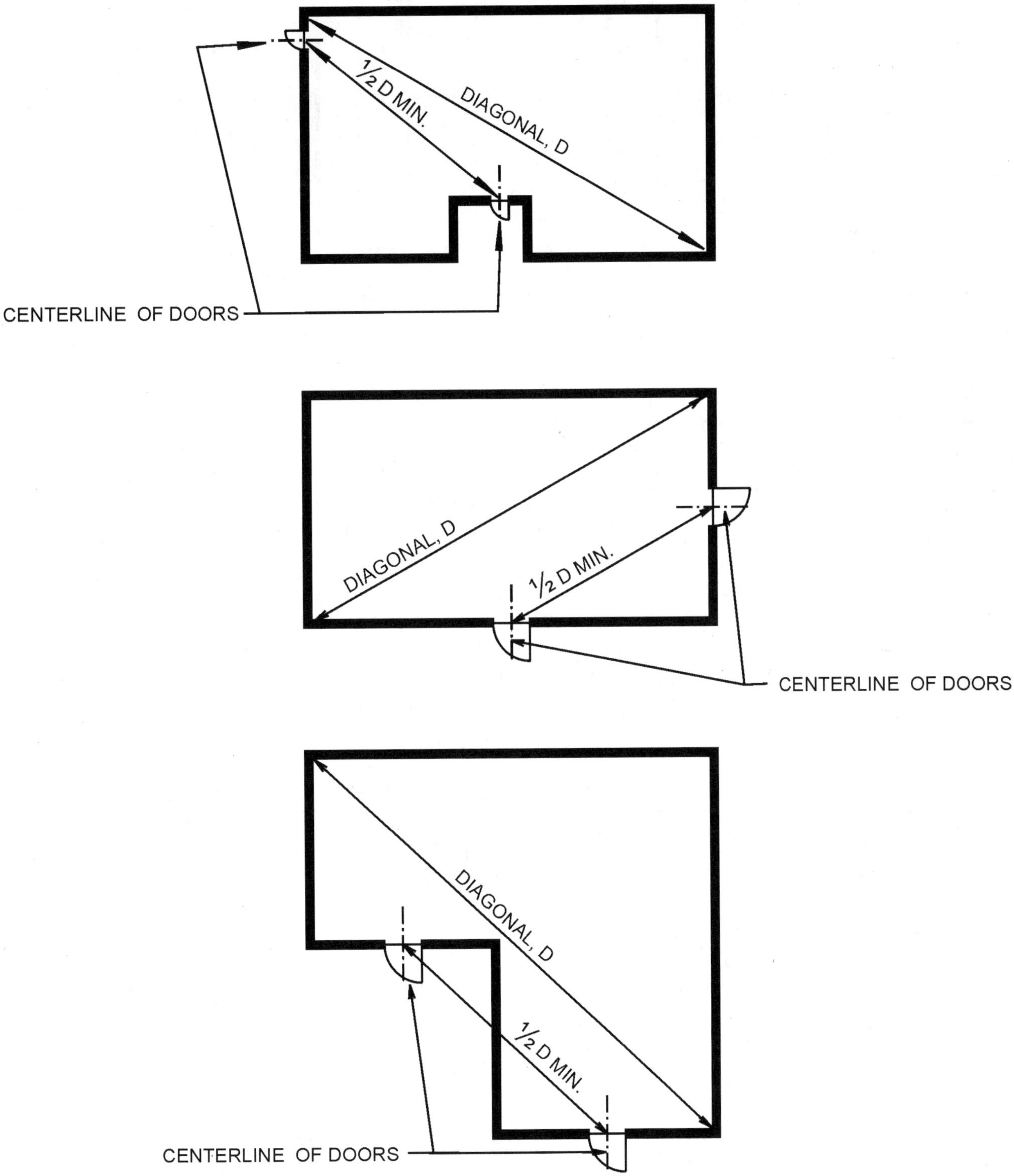

Figure 1014.2.1(1)
EXAMPLES OF REMOTENESS BETWEEN EXITS OR EXIT ACCESS DOORS FROM A ROOM OR BUILDING

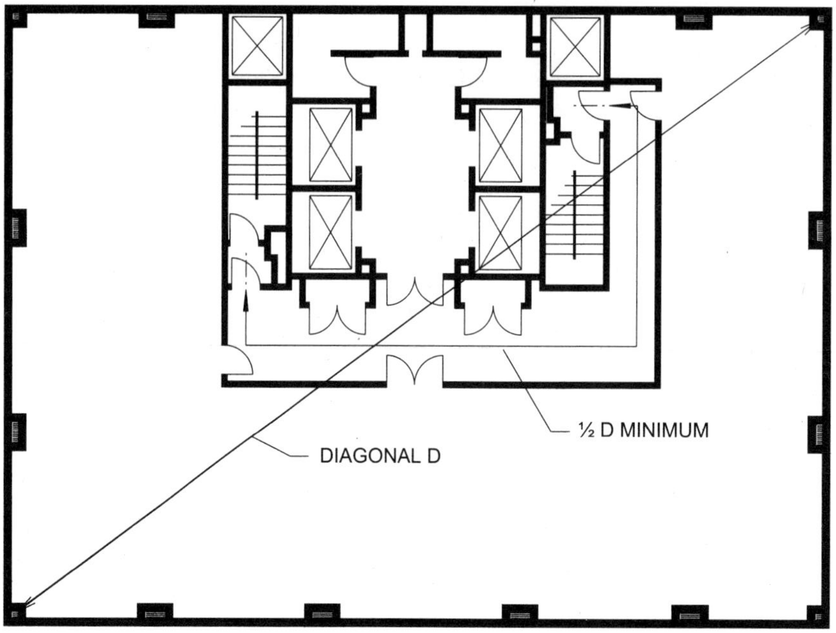

EXAMPLE:
DIAGONAL DIMENSION = 134'-0"
MIN. SEPARATION OF EXITS = 134/2 = 67'-0"

For SI: 1 inch = 25.4 mm, 1 foot = 304.8 mm.

Figure 1014.2.1(2)
REMOTENESS OF EXITS INTERCONNECTED BY A 1-HOUR FIRE-RESISTANCE-RATED CORRIDOR

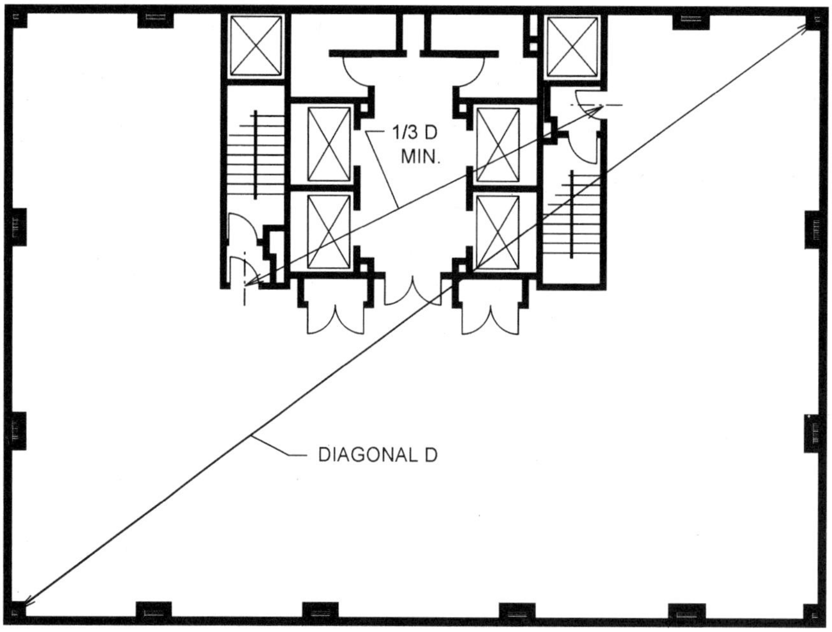

EXAMPLE:
DIAGONAL DIMENSION = 134'-0"
MIN. SEPARATION OF EXITS = 134/3 = 44'-8"

For SI: 1 inch = 25.4 mm, 1 foot = 304.8 mm.

Figure 1014.2.1(3)
REMOTENESS OF EXITS IN A BUILDING WITH AN AUTOMATIC SPRINKLER SYSTEM

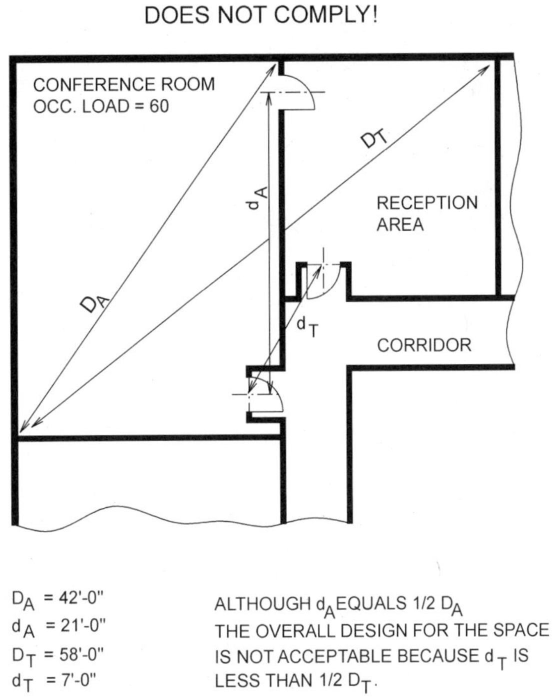

For SI: 1 inch = 25.4 mm, 1 foot = 304.8 mm.

Figure 1014.2.1(4)
REMOTE LOCATION OF EXIT ACCESS DOORS (ADJOINING ROOMS)

1014.3 Boiler, incinerator and furnace rooms. Two exit access doorways are required in boiler, incinerator and furnace rooms where the area is over 500 square feet (46 m^2) and any fuel-fired equipment exceeds 400,000 British thermal units (Btu) (422 000 KJ) input capacity. Where two exit access doorways are required, one is permitted to be a fixed ladder or an alternating tread device. Exit access doorways shall be separated by a horizontal distance equal to one-half the maximum horizontal dimension of the room.

❖ This section requires two exit access doorways for the specified mechanical equipment spaces because of the level of hazards in this type of occupancy. A fixed ladder or an alternating tread device is permitted for the occupants to egress where two doorways are required. The remoteness of the exit access doorways specified in this section is to give the occupants two paths of travel to exit the room so that if one doorway is not available, the alternate path can be used.

1014.4 Refrigeration machinery rooms. Machinery rooms larger than 1,000 square feet (93 m^2) shall have not less than two exits or exit access doors. Where two exit access doorways are required, one such doorway is permitted to be served by a fixed ladder or an alternating tread device. Exit access doorways shall be separated by a horizontal distance equal to one-half the maximum horizontal dimension of room.

All portions of machinery rooms shall be within 150 feet (45 720 mm) of an exit or exit access doorway. An increase in travel distance is permitted in accordance with Section 1015.1.

Doors shall swing in the direction of egress travel, regardless of the occupant load served. Doors shall be tight fitting and self-closing.

❖ The reasons for these requirements are the same as for Section 1014.3. Travel distance is to be limited in accordance with Section 1015.1. The travel distance is to be increased in accordance with Section 1015.1. For example, the travel distance limit for a large refrigeration machinery room classified as Group F-1 that has a sprinkler system throughout the entire building in accordance with Section 903.3.1.1 would be 250 feet (76 200 mm) based on Table 1015.1. The 150-foot (45 720 mm) maximum distance to an exit or exit access doorway that is specified in this section would not apply in this example. The 150-foot (45 720 mm) travel distance is intended to be applied where a sprinkler system is not installed and to shorten the time that occupants would be exposed to the hazards within the machinery room.

1014.5 Refrigerated rooms or spaces. Rooms or spaces having a floor area of 1,000 square feet (93 m^2) or more, containing a refrigerant evaporator and maintained at a temperature below 68°F (20°C), shall have access to not less than two exits or exit access doors.

Travel distance shall be determined as specified in Section 1015.1, but all portions of a refrigerated room or space shall be within 150 feet (45 720 mm) of an exit or exit access door where such rooms are not protected by an approved automatic sprinkler system. Egress is allowed through adjoining refrigerated rooms or spaces.

Exception: Where using refrigerants in quantities limited to the amounts based on the volume set forth in the *International Mechanical Code.*

❖ The commentary to Section 1014.4 also applies to this section. The exception is intended to apply if Chapter 11 of the *International Mechanical Code*® (IMC®) does not require a refrigeration machinery room due to the small amount of refrigerant used. See the commentary for Section 1104 of the IMC for further explanation of the machinery room requirements.

1014.6 Stage means of egress. Where two means of egress are required, based on the stage size or occupant load, one means of egress shall be provided on each side of the stage.

❖ Two means of egress are required from stages in accordance with Section 410.5.4. The stage means of egress paths are to be separate so the two means of egress are independent.

1014.6.1 Gallery, gridiron and catwalk means of egress. The means of egress from lighting and access catwalks, galleries and gridirons shall meet the requirements for occupancies in Group F-2.

Exceptions:

1. A minimum width of 22 inches (559 mm) is permitted for lighting and access catwalks.
2. Spiral stairs are permitted in the means of egress.
3. Stairways required by this subsection need not be enclosed.
4. Stairways with a minimum width of 22 inches (559 mm), ladders, or spiral stairs are permitted in the means of egress.
5. A second means of egress is not required from these areas where a means of escape to a floor or to a roof is provided. Ladders, alternating tread devices or spiral stairs are permitted in the means of escape.
6. Ladders are permitted in the means of egress.

❖ The purpose of this section is to specify the various options that are allowed for means of egress from theater lighting and access catwalks, galleries and gridirons. The requirements are consistent with the use being limited to service personnel.

[B] SECTION 1015 EXIT ACCESS TRAVEL DISTANCE

1015.1 Travel distance limitations. Exits shall be so located on each story such that the maximum length of exit access travel, measured from the most remote point within a story to the entrance to an exit along the natural and unobstructed path of egress travel, shall not exceed the distances given in Table 1015.1.

Where the path of exit access includes unenclosed stairways or ramps within the exit access or includes unenclosed exit ramps or stairways as permitted in Section 1019.1, the distance of travel on such means of egress components shall also be included in the travel distance measurement. The measurement along stairways shall be made on a plane parallel and tangent to the stair tread nosings in the center of the stairway.

Exceptions:

1. Travel distance in open parking garages is permitted to be measured to the closest riser of open stairs.
2. In outdoor facilities with open exit access components and open exterior stairs or ramps, travel distance is permitted to be measured to the closest riser of a stair or the closest slope of the ramp.
3. Where an exit stair is permitted to be unenclosed in accordance with Exception 8 or 9 of Section 1019.1, the travel distance shall be measured from the most remote point within a building to an exit discharge.

❖ The length of travel, as measured from the most remote point within a structure to an exit, is limited to restrict the amount of time that the occupant is exposed to a potential fire condition [see Figure 1015.1(1)]. The route must be assumed to be the natural path of travel without obstruction. This commonly results in a rectilinear path similar to what can be experienced in most occupancies, such as a schoolroom or an office with rows of desks [see Figure 1015.1(2)]. The "arc" method, using an "as the crow flies" linear measurement, must be used with caution, as it seldom represents typical floor design and layout and, in most cases, is not deemed to be the natural, unobstructed path.

The travel distance is measured from each and every occupiable point on a floor to the closest exit. While each occupant may be required to have access to a second or third exit, the travel distance limitation is only applicable to the distance to the nearest exit. In effect, this means that the distance an occupant must travel to the second or third exit is not regulated.

Travel distance is measured along the exit access path. Exit access and travel distance may include travel on a stairway if it is not constructed to meet the definition of an "Exit" (i.e., enclosure, discharge, etc.). An example of this would be an unenclosed stairway from a mezzanine level or steps along the path of travel in a split-floor-level situation. When Section 1019.1 permits an exit stairway to be unenclosed, the travel distance includes travel down the stairway to either an exit to the outside or to a vertical exit enclosure. An example of this would be an open exit stairway within an individual dwelling unit.

The travel distance measurements for very specific open exit stairways, such as in jails in Section 408.4.6 (see Section 1019.1, Exception 6) or for stages in Section 410.5.4 (see Section 1019.1, Exception 7), are examples of where the specific criteria would overrule the general provision.

Exceptions 1 and 2 provide for a travel distance terminating at the top of an open exit stair in an open parking structure or an open exit stair or ramp in outdoor facilities (e.g., stadiums, exterior stairways from open balconies, observation decks or amusement structures). This is appropriate in view of the low hazard in these facilities.

Exception 3 addresses the special concerns for an open exit stairway as permitted in Section 1019.1, Exceptions 8 and 9. The measurement for the travel distance must be from the most remote point down the exit stairway and out of the building to the exit discharge; therefore, when applying Section 1019.1, Exception 8 for open exit stairways between upper levels, this exception would literally require the total travel distance measurement to include any travel distance that was inside an exit enclosure as well as any exterior exit stairways or ramps until the occupants reached grade level. It may be a reasonable interpretation to measure the travel distance to an enclosed exit, horizontal exit or exterior exit stairway.

The distance of travel within an exit and the distance of travel in the exit discharge portion of the means of egress are not regulated.

Section 1018.2 permits certain buildings to be provided with a single exit. In instances where there is a single exit, travel distances less than those permitted in Table 1015.1 apply (see Table 1018.2).

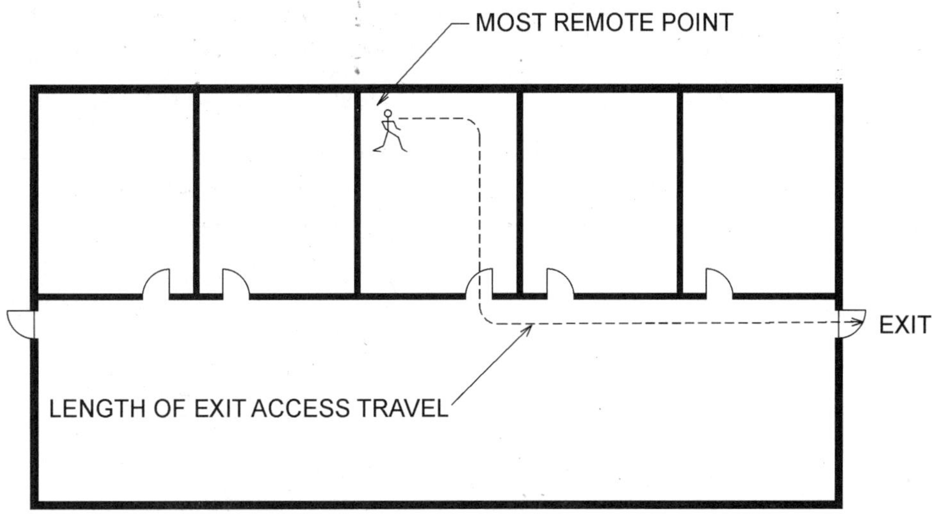

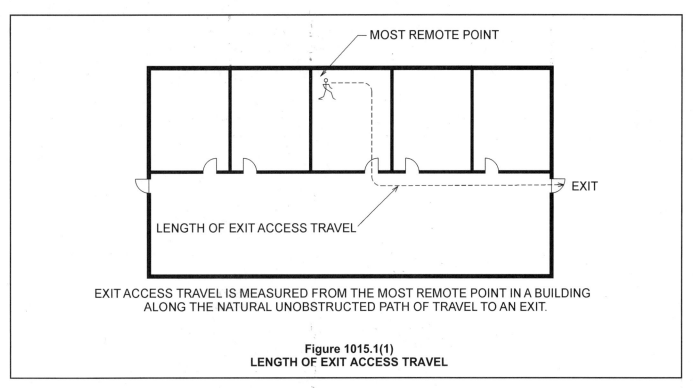

EXIT ACCESS TRAVEL IS MEASURED FROM THE MOST REMOTE POINT IN A BUILDING ALONG THE NATURAL UNOBSTRUCTED PATH OF TRAVEL TO AN EXIT.

Figure 1015.1(1)
LENGTH OF EXIT ACCESS TRAVEL

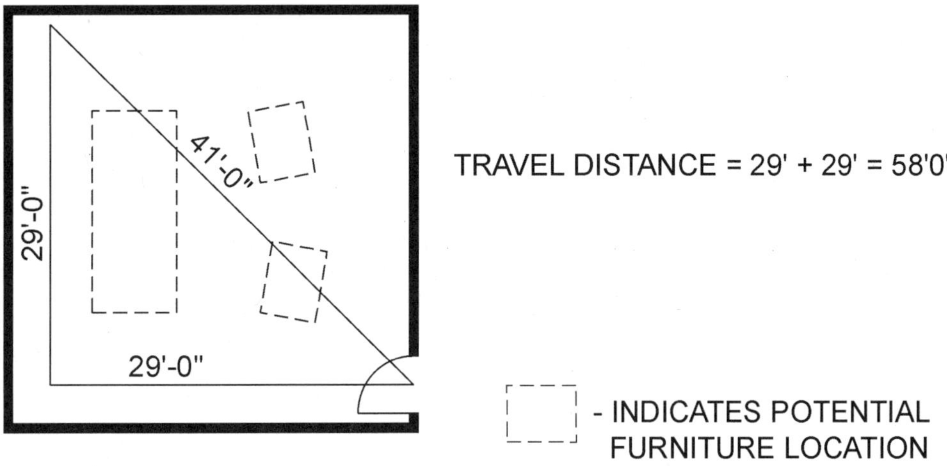

For SI: 1 inch = 25.4 mm, 1 foot = 304.8 mm.

Figure 1015.1(2)
MEASURING TRAVEL DISTANCE USING RECTILINEAR APPROACH

TABLE 1015.1
EXIT ACCESS TRAVEL DISTANCE[a]

OCCUPANCY	WITHOUT SPRINKLER SYSTEM (feet)	WITH SPRINKLER SYSTEM (feet)
A, E, F-1, I-1, M, R, S-1	200	250[b]
B	200	300[c]
F-2, S-2, U	300	400[b]
H-1	Not Permitted	75[c]
H-2	Not Permitted	100[c]
H-3	Not Permitted	150[c]
H-4	Not Permitted	175[c]
H-5	Not Permitted	200[c]
I-2, I-3, I-4	150	200[c]

For SI: 1 foot = 304.8 mm.

a. See the following sections for modifications to exit access travel distance requirements:
Section 402 of the *International Building Code*: For the distance limitation in malls.
Section 404 of the *International Building Code*: For the distance limitation through an atrium space.
Section 1015.2: For increased limitation in Groups F-1 and S-1.
Section 1024.7: For increased limitation in assembly seating.
Section 1024.7: For increased limitation for assembly open-air seating.
Section 1018.2: For buildings with one exit.
Chapter 31 of the *International Building Code*: For the limitation in temporary structures.

b. Buildings equipped throughout with an automatic sprinkler system in accordance with Section 903.3.1.1 or 903.3.1.2. See Section 903 for occupancies where sprinkler systems according to Section 903.3.1.2 are permitted.

c. Buildings equipped throughout with an automatic sprinkler system in accordance with Section 903.3.1.1.

❖ This table reflects the maximum distance a person is allowed to travel from any point in a building floor area to the nearest exit along a natural and unobstructed path. While quantitative determinations or formulas are not available to substantiate the tabular distances, empirical factors are utilized to make relative judgements as to reasonable limitations. Such considerations include the nature and fitness of the occupants; the typical configurations and physical conditions of each group; the level of fire hazard with respect to the specific uses of the facilities, including fire spread, and the potential intensity of a fire. The inclusion of an automatic sprinkler system throughout the building can serve to control, confine or possibly eliminate the fire threat to the occupants so an increased travel distance is permitted. Increased travel distances are permitted when an automatic sprinkler system is installed in accordance with NFPA 13 or 13R (see Section 903.3.1.1 or 903.3.1.2).

When measuring travel distance, it is important to consider the natural path of travel [see Figure 1015.1(1)]. In many cases, the actual layout of furnishings and equipment is not known or is not identified on the plans submitted with the permit application. In such instances, it may be necessary to measure travel distance using the legs of a triangle instead of the hypotenuse [see Figure 1015.1(2)]. Since most people tend to migrate to more open spaces while egressing, measurement of the "natural" path of travel frequently excludes areas of the building within approximately 1 foot (305 mm) of walls, corners, columns and other permanent construction. Where the travel path includes passage through a doorway, the "natural" route is generally measured through the centerline of the door openings.

1015.2 Roof vent increase. In buildings which are one story in height, equipped with automatic heat and smoke roof vents complying with Section 910 and equipped throughout with an automatic sprinkler system in accordance with Section 903.3.1.1, the maximum exit access travel distance shall be 400 feet (122 m) for occupancies in Group F-1 or S.

❖ This section permits an increase in travel distances when roof vents are installed because of the increased visibility provided in a fire event if the roof-venting system is properly designed and installed in accordance with Section 910. While smoke/heat vents and automatic sprinklers serve different life safety functions, they are both active systems of fire protection and can work well together. However, due care must be taken in the design and installation of venting systems so that when they are used in conjunction with automatic sprinklers, they will not cause cooling air drafts to occur at the sprinklers so as to delay activation of the sprinkler system or even render it inoperative. The travel distance increase is limited to single-story buildings of factory (Group F-1) and storage (Group S-1) occupancies because of such characteristics as open areas, fire department accessibility and limited occupant densities. Note that the building must be sprinklered throughout by an automatic sprinkler system designed and installed in accordance with NFPA 13 (see Section 903.3.1.1).

1015.3 Exterior egress balcony increase. Travel distances specified in Section 1015.1 shall be increased up to an additional 100 feet (30 480 mm) provided the last portion of the exit access leading to the exit occurs on an exterior egress balcony constructed in accordance with Section 1013.5. The length of such balcony shall not be less than the amount of the increase taken.

❖ This section allows an additional travel distance on exterior egress balconies since the accumulation of smoke is much less on the balcony. Note that the length of the increase is not to be more than the length of the exterior balcony. For example, if the length of the balcony is 75 feet (22 860 mm), the additional travel distance is limited to 75 feet (22 860). Also note that in order for the increase to apply, the exterior balcony must be located at the end of the path of egress travel and not in some other portion of the egress path.

[B] SECTION 1016
CORRIDORS

1016.1 Construction. Corridors shall be fire-resistance rated in accordance with Table 1016.1. The corridor walls required to be fire-resistance rated shall comply with Section 708 of the *International Building Code* for fire partitions.

Exceptions:

1. A fire-resistance rating is not required for corridors in an occupancy in Group E where each room that is used for instruction has at least one door directly to the exterior and rooms for assembly purposes have at least one-half of the required means of egress doors opening directly to the exterior. Exterior doors specified in this exception are required to be at ground level.
2. A fire-resistance rating is not required for corridors contained within a dwelling or sleeping unit in an occupancy in Group R.
3. A fire-resistance rating is not required for corridors in open parking garages.
4. A fire-resistance rating is not required for corridors in an occupancy in Group B which is a space requiring only a single means of egress complying with Section 1014.1.

❖ The purpose of corridor enclosures is to provide fire protection to occupants as they travel the confined path, perhaps unaware of a fire buildup in an adjacent floor area. The base protection is a fire partition having a 1-hour fire-resistance rating (see Table 1016.1). The table allows a reduction or elimination of the fire-resistance rating depending on the occupant load and the presence of an NFPA 13 or 13R automatic sprinkler system throughout the building.

It should be noted that Section 708 addresses the continuity of fire partitions serving as corridor walls. In addition to allowing the fire partitions to terminate at the underside of a fire-resistance-rated floor/ceiling or roof/ceiling assembly, the supporting construction need not have the same fire-resistance rating in buildings of Type IIB, IIIB and VB construction as specified in Section 708 of the IBC. If such walls were required to be supported by fire-resistance-rated construction, the use of these construction types would be severely restricted when the corridors are required to have a fire-resistance rating. Section 407.3 requires that corridor walls in Group I-2 occupancies that are required to have a fire-resistance rating must be continuous to the underside of the floor or roof deck above or at a smoke-limiting ceiling membrane. Continuity is required because of the defend-in-place protection strategy utilized in such buildings. Requirements for corridor construction within Group I-3 occupancies are found in Section 408.7 of the IBC. For additional requirements for an elevator lobby that is adjacent to or part of a corridor, see the commentary to Section 707.14.1.

Exception 1 indicates that a fire-resistance rating is not required for corridors in Group E when any room adjacent to the corridor that is used for instruction or assembly purposes has a door directly to the outside. Because these rooms are provided with an alternative egress path due to the requirement for exterior exits, the need for a fire-resistance-rated corridor is eliminated.

In accordance with Exception 2, a fire-resistance rating for a corridor contained within a single dwelling unit (e.g., apartment, townhouse) or sleeping unit (e.g., hotel guestroom, assistive living suite) is not required for practical reasons. It is unreasonable to expect fire doors and the associated hardware in homes.

Given the relatively smoke-free environment of open parking structures, Exception 3 does not require fire-resistance-rated corridors in these types of facilities.

If an office suite is small enough so that only one means of egress is required from the suite, Exception 4 indicates that a rated corridor would not be required in that area. The main corridor that connected these suites to the exits would be rated in accordance with Table 1016.1.

TABLE 1016.1. See below.

❖ The required fire-resistance ratings of corridors serving adjacent spaces are provided in Table 1016.1. The fire-resistance rating is based on the group classification (considering characteristics such as occupant mobility, density and knowledge of the building as well as the fire hazard associated with the classification), the total occupant load served by the corridor and the presence of an automatic sprinkler system.

TABLE 1016.1
CORRIDOR FIRE-RESISTANCE RATING

OCCUPANCY	OCCUPANT LOAD SERVED BY CORRIDOR	REQUIRED FIRE-RESISTANCE RATING (hours)	
		Without sprinkler system	With sprinkler system[c]
H-1, H-2, H-3	All	Not Permitted	1
H-4, H-5	Greater than 30	Not Permitted	1
A, B, E, F, M, S, U	Greater than 30	1	0
R	Greater than 10	1	0.5
I-2[a], I-4	All	Not Permitted	0
I-1, I-3	All	Not Permitted	1[b]

a. For requirements for occupancies in Group I-2, see Section 407.3 of the *International Building Code*.
b. For a reduction in the fire-resistance rating for occupancies in Group I-3, see Section 408.7 of the *International Building Code*.
c. Buildings equipped throughout with an automatic sprinkler system in accordance with Section 903.3.1.1 or 903.3.1.2 where allowed.

Where the corridor serves a limited number of people (see the second column in Table 1016.1), the fire-resistance rating is eliminated because of the limited size of the facility and the likelihood that the occupants would become aware of a fire buildup in sufficient time to exit the structure safely. The total occupant load that the corridor serves is used to determine the requirement for a rated corridor enclosure. Corridors serving a total occupant load equal to or less than that indicated in the second column of Table 1016.1 are not required to be enclosed with fire-resistance-rated construction. For example, a corridor serving an occupant load of 30 or less in an unsprinklered Group B occupancy is not required to be enclosed with fire-resistance-rated construction. This example is illustrated in Figure 1016.1.

The purpose of corridor enclosures is to provide fire protection to occupants as they travel the confined path, perhaps unaware of a fire buildup in an adjacent floor area. The base protection is a fire partition having a 1-hour fire-resistance rating. The table allows a reduction or elimination of the fire-resistance rating depending on the occupant load and the presence of an NFPA 13 or 13R (see Section 903.3.1.1 or 903.3.1.2) automatic sprinkler system throughout the building.

A common mistake is to assume that a building is sprinklered throughout and to utilizie the corridor rating reductions, when in fact the requirements in NFPA 13 would not consider the building to be sprinklered throughout. For example, a health club installs a sprinkler system, but chooses to eliminate the sprinklers over the pool in accordance with the exception in Section 903.2.1.3. Any corridors within the building that served more than 30 occupants would still have to be rated because the building would not be considered sprinklered throughout in accordance with NFPA 13 requirements.

Due to the hazardous nature of occupancies in Groups H-1, H-2 and H-3, fire-resistance-rated corridors are required under all conditions. Regardless of the presence of a sprinkler system, a 1-hour-rated corridor enclosure is typically required in high-hazard occupancies. The only exception is for Group H-4 and H-5 occupancies. Occupancies that contain materials or operations constituting a health hazard do not pose the same relative fire or explosion hazard as Group H-1, H-2 or H-3 materials. As such, in Group H-4 or H-5, where the corridor serves a total occupant load of 30 or less, a fire-resistance-rated enclosure is not required. The "not permitted" in the third column is in coordination with Section 903.2.4, which requires all Group H buildings to be fully sprinklered.

The code acknowledges that an automatic sprinkler system can serve to control or eliminate fire development that could threaten the exit access corridor. Most occupancies where sleeping rooms are not present (Groups A, B, E, F, M, S and U) are permitted to have nonfire-resistance-rated corridors if a sprinkler system is installed throughout the building in accordance with NFPA 13 (see Section 903.3.1.1).

In residential facilities, the response time to a fire may be delayed because the residents may be sleeping. With this additional safety concern, the requirements for corridors are more restrictive than nonresidential occupancies. If the corridor serves more than 10 occupants, it is required to be rated. If the building is sprinklered throughout with either an NFPA 13 or 13R system (see Sections 903.3.1.1 and 903.3.1.2), then the rating on the corridor may be reduced to ½ hour. Note the exception for rated corridors within an individual dwelling or sleeping unit in Section 1016.1. Also note that the reduction in the rating of the corridor walls is not permitted when an NFPA 13D (see Section 903.3.1.3) sprinkler system is provided.

While all Group I facilities are supervised environments, the level of supervision in Group I-2 and I-4 occupancies permits assisted evacuation by staff in an emergency; therefore, corridors are not required to be rated. Corridors in Groups I-2 are also regulated by Section 407.3 of the IBC. Due to the lower level of staff to resident ratio in Group I-1 and the limitation on free egress in Group I-3, corridors must have a 1-hour fire-resistance rating. For a reduction in the corridors in Group I-3, see Section 408.7 of the IBC. The "not permitted" in the third column is in coordination with Section 903.2.5, which requires all Group I buildings to be fully sprinklered.

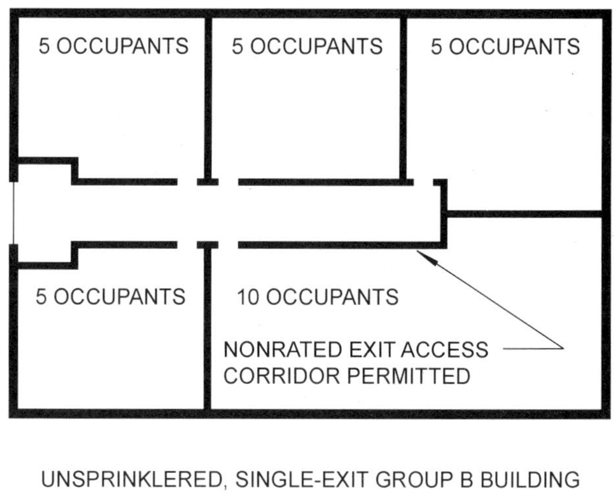

Figure 1016.1
APPLICATION OF TABLE 1016.1

1016.2 Corridor width. The minimum corridor width shall be as determined in Section 1005.1, but not less than 44 inches (1118 mm).

Exceptions:

1. Twenty-four inches (610 mm)—For access to and utilization of electrical, mechanical or plumbing systems or equipment.
2. Thirty-six inches (914 mm)—With a required occupant capacity of 50 or less.
3. Thirty-six inches (914 mm)—Within a dwelling unit.

4. Seventy-two inches (1829 mm)—In Group E with a corridor having a required capacity of 100 or more.
5. Seventy-two inches (1829 mm)—In corridors serving surgical Group I, health care centers for ambulatory patients receiving outpatient medical care, which causes the patient to be not capable of self-preservation.
6. Ninety-six inches (2438 mm)—In Group I-2 in areas where required for bed movement.

❖ The corridor widths specified in this section are minimums. See Section 1005.1 to determine the corridor width based on the number of occupants served by the corridor. Remember, if a corridor goes in two directions, this number can be divided in half. The wider of the two requirements must be utilized for corridor design.

The width of passageways, aisles and corridors is a functional element of building construction that allows the occupants to circulate freely and comfortably throughout the floor area under nonemergency conditions. Under emergency situations, the egress passageways must provide the needed width to accommodate the number of occupants that must utilize the corridor for egress.

When the occupant load of the space exceeds 50, the minimum width of the passageway, aisle or corridor serving that space is required to be 44 inches (1118 mm) to permit two unimpeded parallel columns of users to travel in opposite directions. When the occupant load served is 50 or less, a minimum width of 36 inches (914 mm) is permitted and the users are expected to encounter some intermittent travel interference from fellow users, but the lower occupant load makes those occasions infrequent and tolerable. The 36-inch (914 mm) minimum width is also required within a dwelling unit.

Passageways that lead to building equipment and systems must be at least 24 inches (610 mm) in width to provide a means to access and service the equipment when needed. Due to the frequency of the servicing intervals and the limited number of occupants in these areas, a reduced width is warranted. This minimum width criteria applies to many common situations, such as stage lighting and special effects catwalks, catwalks leading to heating and cooling equipment, as well as passageways providing access to boilers, furnaces, transformers, pumps and other equipment.

Except for small buildings, Group E occupancies are required to have minimum 72-inch-wide (1829 mm) corridors where the corridor serves educational areas. This width is needed not only for proper functional use, but also because of the edge effect caused by student lockers and other boundary attractions and defects. Service and other corridors outside of educational areas, such as an administrative area, would be regulated by the remaining criteria.

In Group I-2 occupancies, where the corridor is utilized during a fire emergency for moving patients confined to beds, the corridor is required to be at least 96 inches (2438 mm) in clear width. This width requirement is only applicable to Group I-2, since occupants in other group classifications are assumed to be ambulatory. This minimum width allows two beds to pass in a corridor and permits the movement of a bed into the corridor through a room door. In areas of Group I-2 where the movement of beds is not anticipated, such as administrative and some outpatient areas of a hospital, the corridor would not be required to provide the 96-inch (2438 mm) width. The minimum width would be determined by one of the remaining applicable criteria. For outpatient medical care where patients may not be capable of self-preservation, such as some outpatient surgery areas or dialysis treatment areas, the 72-inch-wide (1829 mm) corridor is required based on Exception 5.

1016.3 Dead ends. Where more than one exit or exit access doorway is required, the exit access shall be arranged such that there are no dead ends in corridors more than 20 feet (6096 mm) in length.

Exceptions:

1. In occupancies in Group I-3 of Occupancy Condition 2, 3 or 4 (see Section 202, definition of Occupancy Group I-3), the dead end in a corridor shall not exceed 50 feet (15 240 mm).
2. In occupancies in Groups B and F where the building is equipped throughout with an automatic sprinkler system in accordance with Section 903.3.1.1, the length of dead-end corridors shall not exceed 50 feet (15 240 mm).
3. A dead-end corridor shall not be limited in length where the length of the dead-end corridor is less than 2.5 times the least width of the dead-end corridor.

❖ The requirements of this section apply where a space is required to have more than one means of egress according to Section 1014.1.

Dead ends in corridors and passageways can seriously increase the time needed for an occupant to locate the exits. More importantly, dead ends will allow a single fire event to eliminate access to all of the exits by trapping the occupants in the dead-end area. A dead end exists if the occupant of the corridor or passageway has only one direction to travel to reach any of the building exits [see Figure 1016.3(1)]. While a preferred building layout would be one without dead ends, a maximum dead-end length of 20 feet (6096 mm) is permitted and is to be measured from the extreme point in the dead end to the point where the occupants have a choice of two directions to the exits. Having to go back only 20 feet (6096 mm) after coming to a dead end is not such a significant distance as to cause a serious delay in reaching an exit during an emergency situation.

A dead end results whether or not egress elements open into it. A dead end is a hazard for occupants who enter the area from adjacent spaces, travel past an exit into a dead end or enter a dead end with the mistaken assumption that an exit is directly accessible from the dead end.

Note that Section 402.4.5 of the IBC deals with dead-end distances in a covered mall and assumes that, with a sufficiently wide mall in relation to its length, alternative paths of travel will be available in the mall itself to reach an exit (i.e., the mall is not to be confused as being a corridor).

Under special conditions, exceptions to the 20-foot (6096 mm) dead-end limitation apply.

Exception 1 is permitted based on the considerations of the functional needs of Group I-3 Occupancy Conditions 2, 3 or 4, the requirements for smoke compartmentalization in Section 408.6 of the IBC and the requirement for automatic sprinkler protection of the facility in Section 903.2.5.

Exception 2 recognizes the fire protection benefits and performance history of automatic fire sprinkler systems. While the degree of hazard in Group B occupancies does not generally require an automatic fire suppression system, the length of a dead-end corridor or passageway is permitted to be extended to 50 feet (15 240 mm) where an automatic fire sprinkler system in accordance with NFPA 13 (see Section 903.3.1.1) is provided throughout the building. This exception is also permitted in Group F occupancies.

Exception 3 addresses the condition presented by "cul-de-sac" elevator lobbies directly accessible from exit access corridors. In such an elevator lobby, lengths of 20 to 30 feet (6096 to 9144 mm) are common for three- or four-car elevator banks. Typically, the width of this elevator lobby is such that the possibility of confusion with a path of egress is minimized. Below the 2.5:1 ratio, the dead end becomes so wide that it is less likely to be perceived as a corridor leading to an exit. For example, based on the 2.5:1 ratio limitation, a 25-foot-long (7620 mm) dead end over 10 feet (3048 mm) in width would not be considered a dead-end corridor [see Figure 1016.3(2)]. For additional elevator lobby requirements, see the commentary to Section 707.14.1 of the IBC.

1016.4 Air movement in corridors. Exit access corridors shall not serve as supply, return, exhaust, relief or ventilation air ducts or plenums.

Exceptions:

1. Use of a corridor as a source of makeup air for exhaust systems in rooms that open directly onto such corridors, including toilet rooms, bathrooms, dressing rooms, smoking lounges and janitor closets, shall be permitted provided that each such corridor is directly supplied with outdoor air at a rate greater than the rate of makeup air taken from the corridor.
2. Where located within a dwelling unit, the use of corridors for conveying return air shall not be prohibited.
3. Where located within tenant spaces of 1,000 square feet (93 m^2) or less in area, utilization of corridors for conveying return air is permitted.

❖ Two of the most critical elements of the means of egress are the required exit stairways and exit access corridors. Exit stairways serve as protected areas in the building that provide occupants with safe passage to the level of exit discharge. Corridors that provide access to the required exits frequently limit the direction of egress

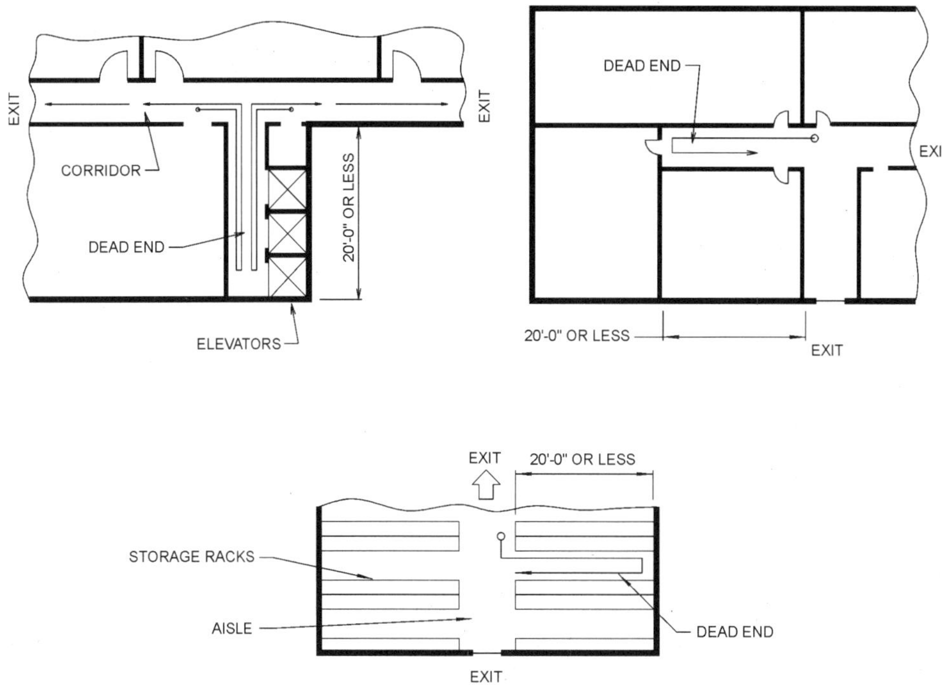

For SI: 1 inch = 25.4 mm, 1 foot = 304.8 mm.

Figure 1016.3(1)
ALLOWABLE DEAD-END CONDITIONS

travel (e.g., travel forward or backward only). Since required exits and exit access corridors are critical elements in the means of egress, the potential spread of smoke and fire into these spaces must be minimized. The scope of this section is exit access corridors. For requirements for the exits, see Section 1019.

The use of these corridors as part of the air distribution system could render those egress elements unusable. Therefore, any air movement condition that could introduce smoke into these vital egress elements is prohibited. It is not the intent of this section to prohibit the air movement necessary for ventilation and space conditioning of exit access corridors, but rather to prevent those spaces from serving as conduits for the distribution of air to, or the collection of air from, adjacent spaces. This restriction also extends to door transoms and door grilles that would allow the spread of smoke into an exit access corridor. This limitation is not, however, intended to restrict slight pressure differences across corridor doors, such as a negative pressure differential maintained in kitchens to prevent odor migration into dining rooms.

The three exceptions to this section identify conditions where an exit access corridor can be utilized as part of the air distribution system. The exceptions do not apply to exits.

Exception 1 addresses the common practice of using air from the corridor as makeup air for small exhaust fans in adjacent rooms. Where the corridor is supplied directly with outdoor air at a rate equal to or greater than the makeup air rate, negative pressure will not be created in the corridor with respect to the adjoining rooms, and smoke would generally not be drawn into the corridor.

Regarding Exception 2, it is common practice to locate return air openings in the corridors of dwelling units and draw return air from adjoining spaces through the corridor. Such use of dwelling unit exit access corridors for conveying return air is not considered to be a significant hazard and is permitted. Dwelling units are permitted to have unprotected openings between floors. Corridors in dwelling units serve small occupant loads, are short in length and are not required to be fire-resistance rated. For these reasons, the use of the corridor or the space above a corridor ceiling for conveying return air does not constitute an unacceptable hazard.

Exception 3 permits exit access corridors located in small tenant spaces to be used for conveying return air based on the relatively low occupant load and the relatively short length of the corridor. These conditions do not pose a significant hazard. In the event of an emergency, the occupants of the space would tend to simply retrace their steps to the entrance.

1016.4.1 Corridor ceiling. Use of the space between the corridor ceiling and the floor or roof structure above as a return air plenum is permitted for one or more of the following conditions:

1. The corridor is not required to be of fire-resistance-rated construction;
2. The corridor is separated from the plenum by fire-resistance-rated construction;

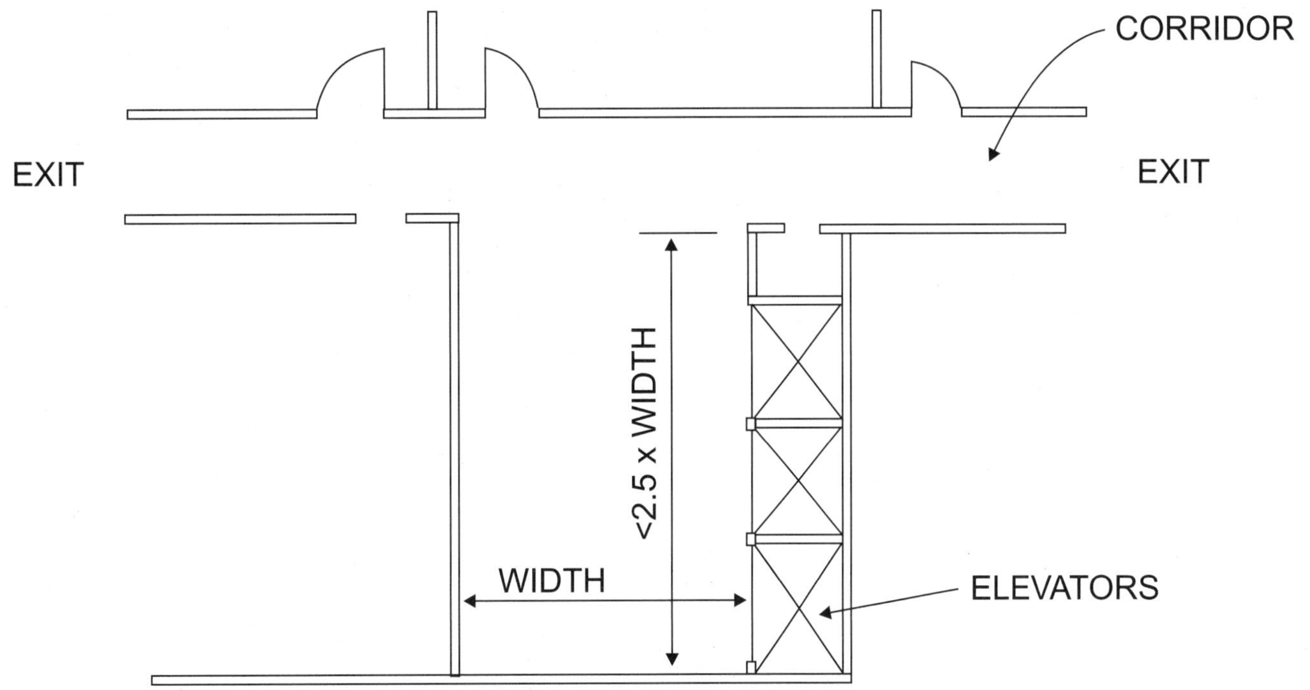

Figure 1016.3(2)
ELEVATOR LOBBIES AND DEAD ENDS

3. The air-handling system serving the corridor is shut down upon activation of the air-handling unit smoke detectors required by the *International Mechanical Code.*
4. The air-handling system serving the corridor is shut down upon detection of sprinkler waterflow where the building is equipped throughout with an automatic sprinkler system; or
5. The space between the corridor ceiling and the floor or roof structure above the corridor is used as a component of an approved engineered smoke control system.

❖ This section identifies five different conditions where the space above the corridor ceiling is permitted to serve as a return air plenum only. Since a return air plenum operates at a negative pressure with respect to the corridor, any smoke and gases within the plenum should be contained within that space. Conversely, a supply plenum operates at a positive pressure with respect to the corridor, thus increasing the likelihood that smoke and gases will infiltrate the corridor enclosure. Where any one of the five conditions is present, the use of the corridor ceiling space as a return air plenum is permitted.

Where the corridor is permitted to be constructed without a fire-resistance rating (see Section 1016.1), Item 1 permits the space above the ceiling to be utilized as a return air plenum.

Item 2 is only applicable to corridors that are required to be enclosed with fire-resistance-rated construction. Compliance with this item requires the plenum to be separated from the corridor by fire-resistance-rated construction equivalent to the rating of the corridor enclosure itself; therefore, the ceiling membrane must provide the fire-resistance rating required of the corridor enclosure. Section 708.4, Exception 3 would be an example of this method of construction.

Items 3 and 4 recognize that the hazard associated with smoke spread through a plenum is minimized if the air movement is stopped.

It is not uncommon for an above-ceiling plenum to be utilized as part of the smoke removal system. This practice is permitted by Item 5. Due to the way these systems are designed, with the higher equipment ratings and power supply provisions, this is considered acceptable.

1016.5 Corridor continuity. Fire-resistance-rated corridors shall be continuous from the point of entry to an exit, and shall not be interrupted by intervening rooms.

Exception: Foyers, lobbies or reception rooms constructed as required for corridors shall not be construed as intervening rooms.

❖ This section requires the fire protection offered by a corridor to be continuous from the point of entry into the corridor to the entrance to an exit. This is to protect occupants from the accumulation of smoke or fire exposure for sufficient time to evacuate the building. For requirements for elevator lobbies that are adjacent to rated corridors, see Section 707.14.1 of the IBC. For requirements for the exit enclosures, see Section 1019. When vertical exit enclosures are not required around an exit stairway or ramp, the "point of entry to the exit" would be the top of the stairway or ramp.

The exception allows a "reception room" to be located on the path of egress from a corridor, provided the room has the same fire-resistance-rated walls and doors as required for the corridor. While the term "reception room" is not defined, it should be viewed as limiting the types of uses that may occur within the protected corridor. Occupied spaces within the corridor should have very limited uses and hazards. Foyers and lobbies are included in this exception based on the low fire hazard of the contents in such rooms.

[B] SECTION 1017
EXITS

1017.1 General. Exits shall comply with Sections 1017 through 1022 and the applicable requirements of Sections 1003 through 1012. An exit shall not be used for any purpose that interferes with its function as a means of egress. Once a given level of exit protection is achieved, such level of protection shall not be reduced until arrival at the exit discharge.

❖ The use of required exterior exit doors, exit stairways, exit passageways and horizontal exits for any purpose other than exiting is prohibited, because it might interfere with use as an exit. For example, the use of an exit stairway landing for storage, vending machines, copy machines or any purpose other than for exiting is not permitted. Such a situation could not only lead to obstruction of the path of exit travel, thereby creating a hazard to life safety, but if the contents consist of combustible materials, then the use of the stairway as a means of egress could be jeopardized by a fire in the exit enclosure. It should be noted, however, that the restriction does not apply solely to combustible contents because of the potential for obstruction and difficulties associated in limiting the materials to noncombustible.

It is recognized that standpipe risers are provided within the stair enclosure and that vertical electrical conduit may be necessary for power or lighting. However, such risers must be located so as not to interfere with the required clear width of the exit. For example, a standpipe riser located in the corner of a stairway will not reduce the required clear width of the landing. Electrical conduit and mechanical equipment are permitted when required for stairway equipment.

It should be noted that Section 1017.1 applies to all exits but does not apply to elements of the means of egress that are not exits, such as exit access corridors and passageways or elements of the exit discharge. For exit discharge options for the enclosure other than a door leading directly to the outside, see the requirements for exit passageways in Section 1020 or the options permitting usage of lobby or vestibule in the exceptions to Section 1023.1.

1017.2 Exterior exit doors. Buildings or structures used for human occupancy shall have at least one exterior door that meets the requirements of Section 1008.1.1.

❖ The purpose of this section is to specify that at least one exterior exit door is required to meet the door size requirements in Section 1008.1.1. It is not the intent of this section to specify the number of exit doors required, which is addressed in Section 1018.1.

1017.2.1 Detailed requirements. Exterior exit doors shall comply with the applicable requirements of Section 1008.1.

❖ The purpose of this section is simply to provide a cross reference from the exit section to all of the detailed requirements for doors that are included in Section 1008.1 and all of its subsections. For example, the requirements for door operation on exterior exit doors are controlled by Section 1008.1.8.

1017.2.2 Arrangement. Exterior exit doors shall lead directly to the exit discharge or the public way.

❖ The exterior exit door is to be the entry point of the exit discharge or lead directly to the public way. When a person reaches the exterior exit door, he or she is directly outside where smoke and toxic gases are not a health hazard. Additionally, this section will keep exterior doors at other locations, such as to an exit balcony, from being viewed as an exit.

[B] SECTION 1018
NUMBER OF EXITS AND CONTINUITY

1018.1 Minimum number of exits. All rooms and spaces within each story shall be provided with and have access to the minimum number of approved independent exits as required by Table 1018.1 based on the occupant load, except as modified in Section 1014.1 or 1018.2. For the purposes of this chapter, occupied roofs shall be provided with exits as required for stories. The required number of exits from any story, basement or individual space shall be maintained until arrival at grade or the public way.

❖ This section requires every floor of a building to be served by at least two exits (see Figure 1018.1). Similarly, every portion of a floor must also be provided with access to at least two exits. Where more than 500 occupants are located on a single floor, additional exits must be provided. For example, if a floor has an occupant load of 750, each occupant of that floor must have access to no less than three exits (see Table 1018.1).

The need for exits to be independent of each other cannot be overstated. Each occupant of each floor must be provided with the required number of exits without having to pass through one exit to gain access to another. Each exit is required to be independent of other exits to prohibit such areas from merging downstream and becoming, in effect, one exit.

As discussed in conjunction with Section 1014.2, the location of two exits is to accomplish certain remoteness, and any additional exits are to be accessible from any point on the floor.

This section also addresses the need for exits from roofs that are occupied by the public, such as roof-top restaurants. It is important that the exit enclosures serve the roof level in addition to the other floor levels.

The text references Sections 1018.2 and 1014.1. See Section 1018.2 for a discussion of buildings that are allowed to have only one exit. Where a building requires more than one exit, the number of required exits, as determined in accordance with Table 1018.1, is required from each floor level. This is true even if the floor level qualifies as a space with one means of egress in accordance with Table 1014.1. Table 1014.1 is intended to be applicable to rooms and spaces but not entire floor levels. One of the main concerns has been that vertical travel takes longer than horizontal travel in emergency exiting situations. This is consistent with Section 1018.2, Item 3.

While not specifically stated in this section, there are two situations where the single means of egress can be used from a multilevel space. Section 505 permits mezzanines to be considered part of the floor below for purposes of means of egress. When a mezzanine meets the occupant load in Table 1014.1 and the common path of travel distance (see Section 1013.3) measured from the most remote point to the bottom of the stairway, it can be considered a space with one means of egress. The second situation is more interpretive, but is based on multiple years of practice within an individual dwelling or sleeping unit. In buildings with multistory dwelling or sleeping units, the means of egress from a dwelling or sleeping unit is typically permitted to be from one level only. If the unit has an occupant load of 10 or less (see Table 1014.1), and the common path of travel from the most remote point on any level to the exit door from the unit itself is a maximum of 75 feet (22 860 mm) (see Section 1013.3), that unit may have only one means of egress. Once the occupants exit the unit itself, the floor level must have access to two or more means of egress for all tenants, depending on the number required for the building as a whole. This would be logically consistent with the residential exception found Section 1018.2, Item 2.

TABLE 1018.1
MINIMUM NUMBER OF EXITS FOR OCCUPANT LOAD

OCCUPANT LOAD	MINIMUM NUMBER OF EXITS
1-500	2
501-1,000	3
More than 1,000	4

❖ Table 1018.1 specifies that the minimum number of exits available to each occupant of a floor is based on the total occupant load of that floor. This is so that at least one exit will be available in case of a fire emergency and to provide increased ensurance that a larger number of occupants can be accommodated by the remaining exits when one exit is not available. While an equal distribution of exit capacity among all of the exits is not required, a proper design would not only balance capacity with the occupant load distribution, but also consider a reasoned distribution of capacity to avoid a severe dependence on one exit.

1018.1.1 Open parking structures. Parking structures shall not have less than two exits from each parking tier, except that only one exit is required where vehicles are mechanically parked. Unenclosed vehicle ramps shall not be considered as required exits unless pedestrian facilities are provided.

❖ At least two exits from each parking tier are required in open parking structures, except where the resulting occupant load is minimal, such as where the vehicles are mechanically parked. A vehicle ramp may be considered as one of the required exits if pedestrian walkways are provided along the ramp. According to Section 1019.1, Exception 5, the exit stairways are not required to be enclosed, since an open parking structure is designed to permit the ready ventilation of the products of combustion to the outside by exterior wall openings (see Section 406.3 of the IBC). Also, parking structures are characterized by open floor areas that allow the occupants to observe a fire condition and choose a travel path that would avoid the fire threat.

1018.1.2 Helistops. The means of egress from helistops shall comply with the provisions of this chapter, provided that landing areas located on buildings or structures shall have two or more exits. For landing platforms or roof areas less than 60 feet (18 288 mm) long, or less than 2,000 square feet (186 m^2) in area, the second means of egress is permitted to be a fire escape or ladder leading to the floor below.

❖ This section provides occupants of helistops with adequate exit facilities. The reduction in the exit requirements for small-area helistops is based on the low number of occupants that are associated with these facilities.

1018.2 Buildings with one exit. Only one exit shall be required in buildings as described below:

1. Buildings described in Table 1018.2, provided that the building has not more than one level below the first story above grade plane.
2. Buildings of Group R-3 occupancy.
3. Single-level buildings with the occupied space at the level of exit discharge provided that the story or space complies with Section 1014.1 as a space with one means of egress.

❖ Buildings with one exit are permitted where the configuration and occupancy meet certain characteristics so as not to present an unacceptable fire risk to the occupants. Buildings that are relatively small in size have a shorter travel distance and fewer occupants; thus, having access to a single exit does not significantly compromise the safety of the occupants since they will also be alerted to and get away from the fire more quickly. It is important to note that the provisions in Section 1018.2 apply to entire buildings only, not individual stories or fire areas.

Occupants of a story of limited size and configuration may have access to a single exit, provided that the building does not have more than one level below the first story above the grade plane (see Table 1018.2 for a list of groups and building characteristics where the occupants are permitted to have access to a single exit). The limitation on the number of levels below the first story is intended to limit the vertical travel an occupant must accomplish to reach the exit discharge in a single-exit building. Taken to its extreme, the code would otherwise allow a single exit in a building with one or two stories above grade and an unlimited number of stories below grade. This would be clearly inadequate for those

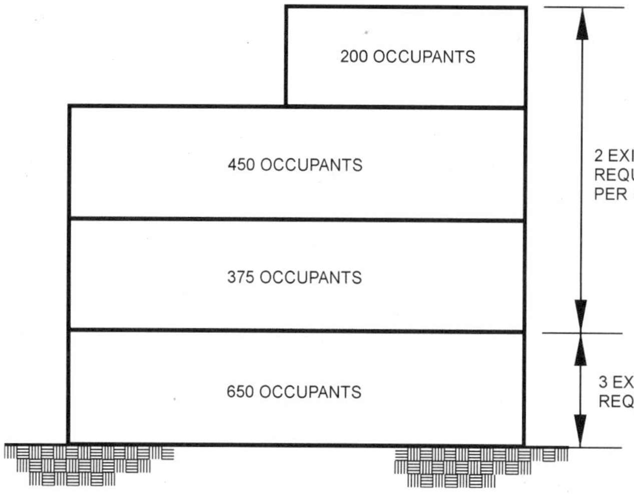

Figure 1018.1
NUMBERS OF EXITS

stories below grade.

Group R-3 building occupancies are permitted to have a single exit since they do not have more than two dwelling units.

Item 3 is a correlation with the requirements of Section 1014.1, which permits a single egress path for certain spaces having a low occupant load. When a single egress doorway is permitted from a given space, it is consistent to permit a single exit to serve a one-story building of the same size as that space.

Since the term "building" limits the area addressed to spaces bordered by exterior walls or fire walls, the common application of Item 3 should be on a tenant-by-tenant basis. For example, a strip mall may not meet the provisions for a "building" with one means of egress; however, assume a tenant meets the provisions for a space with one means of egress in accordance with Section 1014.1. This tenant could exist as a either a stand-alone single-exit building or a single-exit tenant space that exited into an interior corridor. Is it not as safe to permit this tenant to exist as part of a larger building with the door exiting directly to the exterior? Final approval of this alternative approach would be subject to the fire code official having jurisdiction.

TABLE 1018.2
BUILDINGS WITH ONE EXIT

OCCUPANCY	MAXIMUM HEIGHT OF BUILDING ABOVE GRADE PLANE	MAXIMUM OCCUPANTS (OR DWELLING UNITS) PER FLOOR AND TRAVEL DISTANCE
A, B[d], E, F, M, U	1 Story	50 occupants and 75 feet travel distance
H-2, H-3	1 Story	3 occupants and 25 feet travel distance
H-4, H-5, I, R	1 Story	10 occupants and 75 feet travel distance
S[a]	1 Story	30 occupants and 100 feet travel distance
B[b], F, M, S[a]	2 Stories	30 occupants and 75 feet travel distance
R-2	2 Stories[c]	4 dwelling units and 50 feet travel distance

For SI: 1 foot = 304.8 mm.

a. For the required number of exits for open parking structures, see Section 1018.1.1.

b. For the required number of exits for air traffic control towers, see Section 412.1 of the *International Building Code*.

c. Buildings classified as Group R-2 equipped throughout with an automatic sprinkler system in accordance with Section 903.3.1.1 or 903.3.1.2 and provided with emergency escape and rescue openings in accordance with Section 1025 shall have a maximum height of three stories above grade.

d. Buildings equipped throughout with an automatic sprinkler system in accordance with Section 903.3.1.1 with an occupancy in Group B shall have a maximum travel distance of 100 feet.

❖ Table 1018.2 lists the characteristics a building must have to be of single-exit construction, including occupancy, maximum height of building above grade plane, maximum occupants or dwelling units per floor and exit access travel distance per floor. The occupant load of each floor is determined in accordance with the provisions of Section 1004.1. The exit access travel distance is measured along the natural and unobstructed path to the exit, as described in Section 1015.1. If the occupant load is exceeded, as indicated in Table 1018.2, two exits are required from each floor in the building. Likewise, if the travel distance or number of dwelling units is exceeded, as indicated in Table 1018.2, two exits are required from each floor of the building.

The exit enclosure required in a two-story, single-exit building is identical to any other complying exit (i.e., interior stairs, exterior stairs, etc.). Similarly, the fire-resistance rating required for opening protectives is identical to that required by Section 714 of the IBC.

1018.3 Exit continuity. Exits shall be continuous from the point of entry into the exit to the exit discharge.

❖ The intent of this section is to provide safety in all portions of the exit by requiring continuity of the fire protection characteristics of the exit enclosure. This would include but not be limited to the fire-resistance rating of the exit enclosure walls and the opening protection rating of the doors. The code provides no exception for this requirement, which also reinforces the protection issues addressed in Section 1017.1.

1018.4 Exit door arrangement. Exit door arrangement shall meet the requirements of Sections 1014.2 through 1014.2.2.

❖ The intent of this section is to provide a cross reference to the exit door requirements.

[B] SECTION 1019
VERTICAL EXIT ENCLOSURES

1019.1 Enclosures required. Interior exit stairways and interior exit ramps shall be enclosed with fire barriers. Exit enclosures shall have a fire-resistance rating of not less than 2 hours where connecting four stories or more and not less than 1 hour where connecting less than four stories. The number of stories connected by the shaft enclosure shall include any basements but not any mezzanines. An exit enclosure shall not be used for any purpose other than means of egress. Enclosures shall be constructed as fire barriers in accordance with Section 706 of the *International Building Code*.

Exceptions:

1. In other than Group H and I occupancies, a stairway serving an occupant load of less than 10 not more than one story above the level of exit discharge is not required to be enclosed.
2. Exits in buildings of Group A-5 where all portions of the means of egress are essentially open to the outside need not be enclosed.
3. Stairways serving and contained within a single residential dwelling unit or sleeping unit in occupancies in Group R-2 or R-3 and sleeping units in occupancies in Group R-1 are not required to be enclosed.

4. Stairways that are not a required means of egress element are not required to be enclosed where such stairways comply with Section 707.2 of the *International Building Code*.
5. Stairways in open parking structures which serve only the parking structure are not required to be enclosed.
6. Stairways in occupancies in Group I-3 as provided for in Section 408.3.6 of the *International Building Code* are not required to be enclosed.
7. Means of egress stairways as required by Section 410.5.4 of the *International Building Code* are not required to be enclosed.
8. In other than occupancy Groups H and I, a maximum of 50 percent of egress stairways serving one adjacent floor are not required to be enclosed, provided at least two means of egress are provided from both floors served by the unenclosed stairways. Any two such interconnected floors shall not be open to other floors.
9. In other than occupancy Groups H and I, interior egress stairways serving only the first and second stories of a building equipped throughout with an automatic sprinkler system in accordance with Section 903.3.1.1 are not required to be enclosed, provided at least two means of egress are provided from both floors served by the unenclosed stairways. Such interconnected stories shall not be open to other stories.

❖ This section requires that all interior exit stairways or ramps are to be enclosed with fire barriers having a fire-resistance rating of at least 1 hour. The fire-resistance rating of the enclosure must be increased to 2 hours if the stairway or ramp connects four or more stories. Note that the criteria are based on the number of stories connected by the stairway or ramp and not the height of the building. Therefore, a building that has three stories located entirely above the grade plane and a basement would require an enclosure with a 2-hour fire-resistance rating if the stairway or ramp connects all four stories.

The enclosure is needed because an exit stairway or ramp penetrates the floor/ceiling assemblies between the levels, thus creating a vertical opening or shaft. In cases of fire, a vertical opening may act as a chimney, causing smoke, hot gases and light-burning products to flow upward (buoyant force). If an opening is unprotected, these products of combustion will be forced by positive pressure differentials to spread horizontally into the building spaces.

The enclosure of interior stairways or ramp with construction having a fire-resistance rating is intended to prevent the spread of fire from floor to floor. Another important purpose is to provide a safe path of travel for the building occupants and to serve as a protected means of access to the fire floor by fire department personnel. For this reason, this section and Section 1019.1.1 limit the penetrations and openings permitted in a stairway enclosure.

It is important that an exit stairway not be used for any purpose other than as a means of egress. For example, there is a tendency to use stairway landings for storage purposes. Such a situation obstructs the path of exit travel, and if the stored contents consist of combustible materials, the use of the stairway as a means of egress may be jeopardized, creating a hazard to life safety. However, the restriction on the use of an exit stairway is not limited to situations when the contents are combustible.

This section allows exceptions to the requirements stated above:

Exception 1 allows an open exit stairway from the first to the second floor where the stairway does not serve more than 10 occupants. An example of the application of this exception is a retail space that has a small office located on the second story. This exception allows the stair serving the office to be unenclosed based on the small number of occupants served.

In Exception 2, stairways in Group A-5 occupancies in which the means of egress is essentially open to the outside need not be enclosed because of the ability to vent the fire to the outside. The criteria specified in Section 1022.1 should be used to determine if the stairway is "essentially open."

In Exception 3, the exit stairways in the listed Group R occupancies are not required to have enclosures because of the small occupancy load. This exception is limited to stairways located within the individual dwelling or sleeping units.

In Exception 4, stairways that do not serve as a means of egress are not subject to the enclosure requirements (see commentary, Section 707.2 of the IBC). Such stairways may include ornamental stairways within an atrium or any stairways provided in excess of the minimum number required for egress and not designed as an exit. While not specifically stated in this exception, Section 707.2 of the IBC also exempts a ramp that is not part of a required means of egress from enclosure requirements.

In Exception 5, stairways located in open parking structures are exempt from the enclosure requirements because of the ease of accessibility by the fire services, the natural ventilation of such structures, the low level of fire hazard, the small number of people using the structure at any one time and the excellent fire record of such structures (see commentary, Section 1018.1.1).

In Exception 6, because of security needs in detention facilities, one of the exit stairs is permitted to be glazed in a manner similar to atrium enclosures. Specific limitations and requirements are discussed in Section 408.3.6 of the IBC.

Regarding Exception 7, due to the nature of stages and platforms, stage exit stairways from the side of a stage, spaces under the stage, fly galleries and gridirons are not required to be enclosed.

Exception 8 allows certain portions of a required exit stairway to be unenclosed. This exception may be used in either sprinklered or nonsprinklered buildings; however, it may not be used in Group H or I occupancies. Since this exception is limited to 50 percent of the egress stairways, it may not be used in a single-exit building.

There are two instances where this exception has been used. The first is for two-story buildings where one of the required stairways is required to be enclosed but the other is not. The second case is for a multistory building where one of the required exit stairways between levels may be open between two adjacent levels. However, in this situation, after moving down one level, the required number of enclosed exits must be available. Once the enclosed exit is entered, that level of protection for the occupants must be maintained until the exit discharge is reached. The most typical example of the second case would be to provide a second exit for a tenant who did not have access to two enclosed exits on the upper floor because the second stair was only available through another tenant space. Access to exits may not be through another tenant space in accordance with Sections 1013.2 and 1013.2.1.

A stair shaft can not be discontinued at some point in the height of a multistory building and then continued one floor level lower such that a person would have to come out of the stair enclosure then back into it at the next level lower. This arrangement would not be allowed according to Sections 1017.1 and 1018.3, which require the level of protection of an exit to be continuous from the point of entry into the exit to the exit discharge.

Exception 9 is limited to buildings sprinklered throughout by an NFPA 13 system (see Section 903.3.1.1). This exception is not available for Group H or I occupancies. Literally, this exception would permit the second floor of any building to use two, three or four unenclosed stairways to serve as the required exits for that level; however, in most cases in multistory buildings, the second level would utilize the enclosed exit stairways moving down from the upper floors. Sections 1017.1 and 1018.3 require the level of protection of an exit to be continuous from the point of entry to the exit to the exit discharge; therefore, the exit stairways coming down from the upper level could not be open between the second and first levels. This particular exception, therefore, will most typically be utilized in two-story buildings.

For travel distance measurements at the exit stairways in Exceptions 8 and 9, see the commentary to Section 1015.1.

1019.1.1 Openings and penetrations. Exit enclosure opening protectives shall be in accordance with the requirements of Section 715 of the *International Building Code.*

Except as permitted in Section 402.4.6 of the *International Building Code*, openings in exit enclosures other than unexposed exterior openings shall be limited to those necessary for exit access to the enclosure from normally occupied spaces and for egress from the enclosure.

While interior exit enclosures are extended to the exterior of a building by an exit passageway, the door assembly from the exit enclosure to the exit passageway shall be protected by a fire door conforming to the requirements in Section 715.3 of the *International Building Code.* Fire door assemblies in exit enclosures shall comply with Section 715.3.4 of the *International Building Code.*

❖ The only openings that are permitted in fire-resistance-rated exit stairway or ramp enclosures are doors that lead either from normally occupied spaces or from the enclosure. This restriction on openings essentially prohibits the use of windows in an exit enclosure except for those exterior windows that are not exposed to any hazards. This requirement is not intended to prohibit windows or other openings in the exterior walls of the exit enclosure. The verbiage "unexposed exterior openings" includes windows or doors not required to be protected by either Section 704.8 or 1019.1.4. The only exception would be window assemblies that have been tested as wall assemblies in accordance with ASTM E 119. The objective of this provision is to minimize the possibility of fire spreading into an exit enclosure and endangering the occupants or even preventing the use of the exit at a time when it is most needed. The limitation on openings applies regardless of the fire protection rating of the opening protective. The limitation on openings from normally occupied areas is intended to reduce the probability of a fire occurring in an unoccupied area, such as a storage closet, which has an opening into the stairway, thereby resulting in fire spread into the stairway. Other spaces that are not normally occupied include, but are not limited to, toilet rooms, electrical/mechanical equipment rooms and janitorial closets.

The reference to the specific criteria for covered malls in Section 402.4.6 of the IBC is not applicable to exit enclosures between levels. Its application is limited to exit passageways that are discussed in Section 1020.

The last paragraph addresses horizontal exit enclosures of an exit passageway. While the exit passageway is an extension of the protection offered by the vertical exit, there must still be an opening protective at the bottom of the stairway or ramp. This is to prevent any smoke that may migrate into the exit passageway from also moving up the exit stairway or ramp. A door is also required at the bottom of the vertical enclosure for the exit discharge options for lobbies and vestibules (see Section 1023.1).

These opening requirements are very similar to those required for an exit passageway (see Section 1020.4).

1019.1.2 Penetrations. Penetrations into and openings through an exit enclosure are prohibited except for required exit doors, equipment and ductwork necessary for independent pressurization, sprinkler piping, standpipes, electrical raceway for fire department communication and electrical raceway serving the exit enclosure and terminating at a steel box not exceeding 16 square inches (0.010 m^2). Such penetrations shall be protected in accordance with Section 712 of the *International Building Code.* There shall be no penetrations or communication openings, whether protected or not, between adjacent exit enclosures.

❖ This section specifically lists the items that are allowed to penetrate a vertical exit enclosure. This is consistent for all types of exit enclosures, including stair or ramp vertical exit enclosures and exit passageways (see

Section 1020.5). In general, only portions of the building service systems that serve the exit enclosure are allowed to penetrate the exit enclosure. As indicated in the commentary to Section 1017.1, standpipe systems are commonly located in the exit stair enclosures.

1019.1.3 Ventilation. Equipment and ductwork for exit enclosure ventilation shall comply with one of the following items:

1. Such equipment and ductwork shall be located exterior to the building and shall be directly connected to the exit enclosure by ductwork enclosed in construction as required for shafts.
2. Where such equipment and ductwork is located within the exit enclosure, the intake air shall be taken directly from the outdoors and the exhaust air shall be discharged directly to the outdoors, or such air shall be conveyed through ducts enclosed in construction as required for shafts.
3. Where located within the building, such equipment and ductwork shall be separated from the remainder of the building, including other mechanical equipment, with construction as required for shafts.

In each case, openings into the fire-resistance-rated construction shall be limited to those needed for maintenance and operation and shall be protected by self-closing fire-resistance-rated devices in accordance with Chapter 7 of the *International Building Code* for enclosure wall opening protectives.

Exit enclosure ventilation systems shall be independent of other building ventilation systems.

❖ The purpose of the requirements for the ventilation system equipment and ductwork is to maintain the fire resistance of the exit enclosure. The exit enclosure ventilation system is to be independent of other building systems to prevent smoke in the exit enclosure from other areas of the building.

1019.1.4 Vertical enclosure exterior walls. Exterior walls of a vertical exit enclosure shall comply with the requirements of Section 704 of the *International Building Code* for exterior walls. Where nonrated walls or unprotected openings enclose the exterior of the stairway and the walls or openings are exposed by other parts of the building at an angle of less than 180 degrees (3.14 rad), the building exterior walls within 10 feet (3048 mm) horizontally of a nonrated wall or unprotected opening shall be constructed as required for a minimum 1-hour fire-resistance rating with $^3/_4$-hour opening protectives. This construction shall extend vertically from the ground to a point 10 feet (3048 mm) above the topmost landing of the stairway or to the roof line, whichever is lower.

❖ This section does not require exterior walls of a stairway enclosure to have the same fire-resistance rating as interior walls. However, to minimize the potential fire spread into the stairway from the exterior, the issue of the exterior wall of the stairway and adjacent exterior walls of the building is addressed. Essentially, there are two alternatives where an exposure hazard exists: either provide protection to the stairway by having a fire-resistance rating on its exterior wall or provide a fire-resistance rating to the walls adjacent to the stairway for a distance of 10 feet (3048 mm) measured horizontally and vertically from the stairway enclosure where those walls are at an angle of less than 180 degrees (3.14 rad) from the enclosure's exterior wall [see Figure 1019.1.4(1)]. When the adjacent exterior wall is protected in lieu of the stairway enclosure wall, the protection is to extend from the ground to a level of 10 feet (3048 mm) above the highest landing of the stairway. However, the protection is not required to extend beyond the normal roof line of the building.

The 180-degree (3.14 rad) angle criteria is based on the scenario where the exterior wall of the stair enclosure is in the same plane and flush with the exterior wall of the building. In this scenario, a fire would need to travel 180 degrees (3.14 rad) around in order to impinge on the stair. Based on studies of existing buildings, this spread of fire 180 degrees (3.14 rad) does not appear to be a problem. This criteria is only applicable when the angle between the walls is 180 degrees (3.14 rad) or less.

As the fire exposure on the exterior is different than can be expected on the interior, the fire-resistance rating of the exterior wall is not required to exceed 1 hour, regardless of whether it is the stairway enclosure wall or the adjacent exterior wall, unless the exterior wall is required by other sections of the code to have a higher fire-resistance rating. The fire protection rating on any openings in the exterior wall of a stairway enclosure or adjacent exterior wall within 180 degrees (3.1 rad) is to be a minimum of $^3/_4$ hour [see Figure 1019.1.4(2)].

1019.1.5 Enclosures under stairways. The walls and soffits within enclosed usable spaces under enclosed and unenclosed stairways shall be protected by 1-hour fire-resistance-rated construction, or the fire-resistance rating of the stairway enclosure, whichever is greater. Access to the enclosed usable space shall not be directly from within the stair enclosure.

Exception: Spaces under stairways serving and contained within a single residential dwelling unit in Group R-2 or R-3 as applicable in Section 1001.1.

There shall be no enclosed usable space under exterior exit stairways unless the space is completely enclosed in 1-hour fire-resistance-rated construction. The open space under exterior stairways shall not be used for any purpose.

❖ This section addresses the fire hazard of storage under a stairway. The stairway must be protected from a storage area underneath it, even if the stair is not required to be enclosed. The section also requires that the storage area not open into a stair enclosure. This is so that a fire that starts in the storage area would not be mistakenly left open to the stair enclosure. The exception does not require stair protection from storage areas under the stair for the indicated residential occupancies.

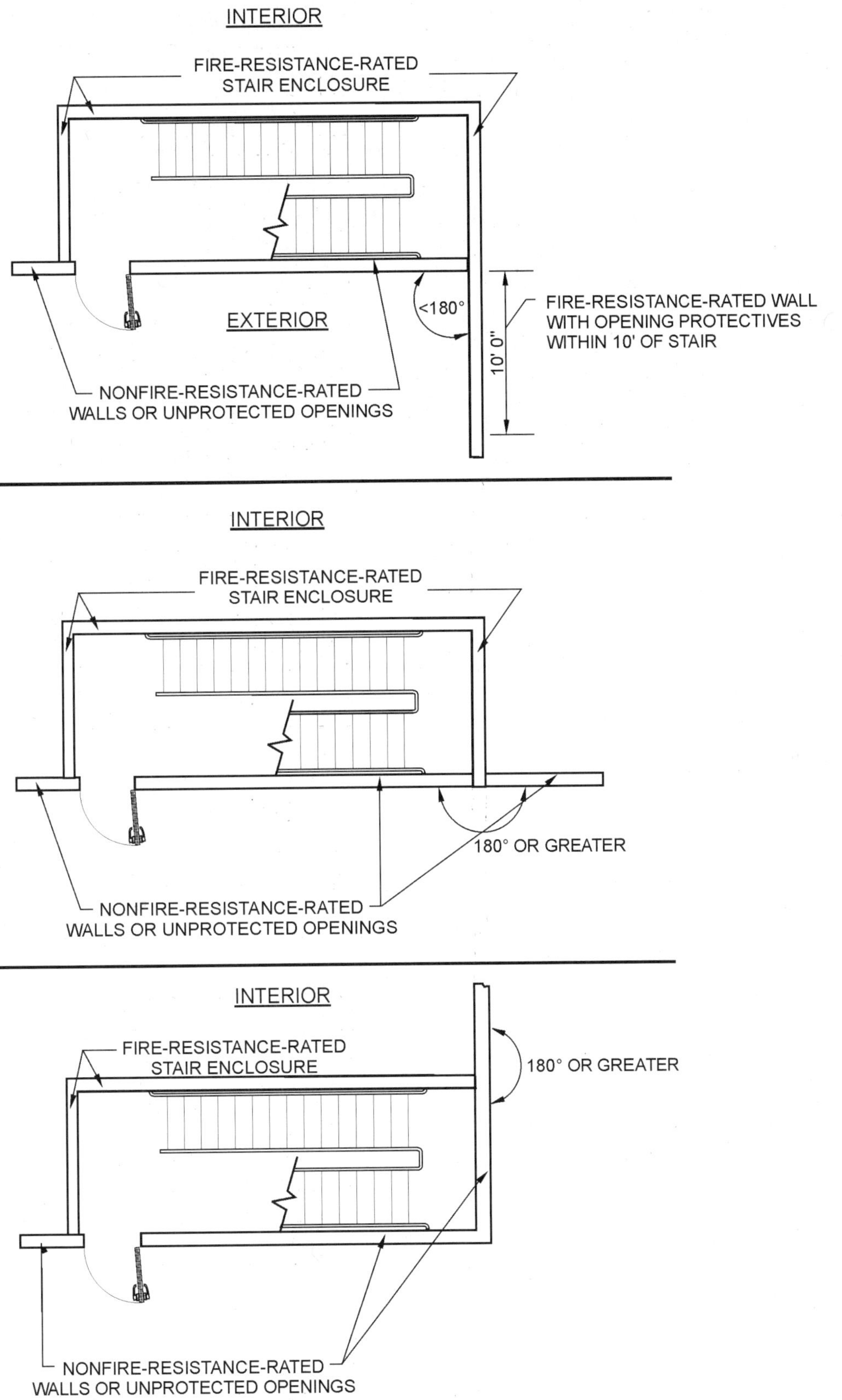

For SI: 1 inch = 25.4 mm, 1 foot = 304.8 mm
1 degree = 0.01745 rad.

Figure 1019.1.4(1)
EXAMPLES OF EXTERIOR WALL PROTECTIVES

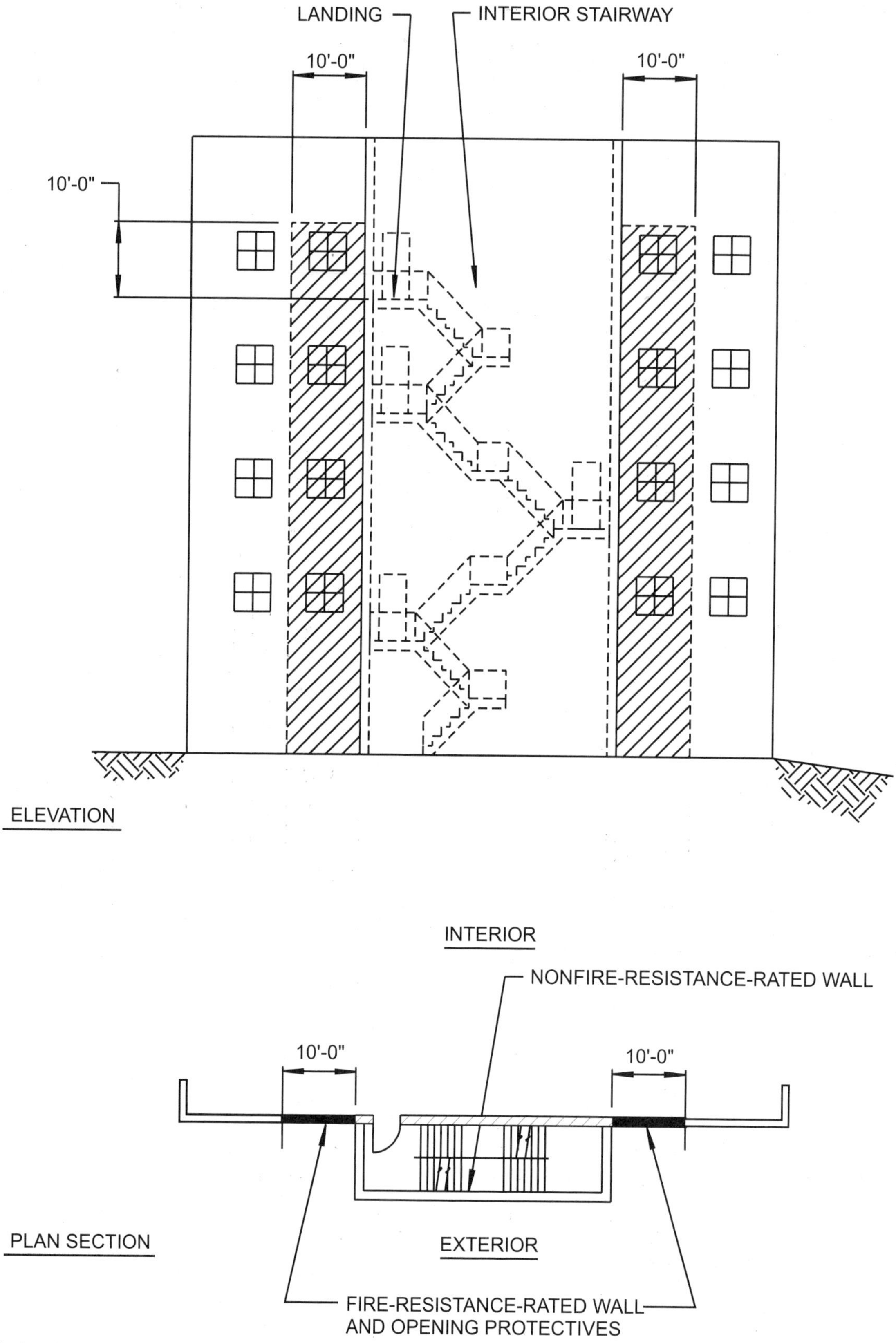

For SI: 1 inch = 25.4 mm, 1 foot = 304.8 mm.

Figure 1019.1.4(2)
EXTERIOR WALL PROTECTION

1019.1.6 Discharge identification. A stairway in an exit enclosure shall not continue below the level of exit discharge unless an approved barrier is provided at the level of exit discharge to prevent persons from unintentionally continuing into levels below. Directional exit signs shall be provided as specified in Section 1011.

❖ So that building occupants using an exit stairway during an emergency situation will be prevented from going past the level of exit discharge, the run of the stairway is to be interrupted by partitions, doors, gates or other approved means. These devices help the users of the stairway to recognize when they have reached the point in the stairway that is the level of exit discharge. Exit signs, including tactile, are to be provided for occupant guidance. Furthermore, signs are to be placed at each floor landing in all interior exit stairways connecting more than three floor levels that designate the level or story of the landings in accordance with Section 1019.1.7.

The code does not specify the type of material or construction of the barrier used to identify the level of exit discharge. The key issues to be considered in the selection and approval of the type of barrier to be used are: (1) will the barrier provide a visible and physical means of alerting occupants who are exiting under emergency conditions that they have reached the level of exit discharge and (2) is the barrier constructed of materials that are permitted by the construction type of the building? In an emergency situation, some occupants are likely to come in contact with the barrier during exiting before realizing that they are at the level of exit discharge. Therefore, the barrier should be constructed in a manner that is substantial enough to withstand the anticipated physical contact, such as pushing or shoving. It would be reasonable, as a minimum, to design the barrier to withstand the structural load requirements of Section 1607.5 for interior walls and partitions. The barrier could be opaque (gypsum wallboard and stud framing) or not (wire grid-type material).

The use of signage only or relatively insubstantial barriers, such as ropes or chains strung across the opening, is typically not sufficient to prevent occupants from attempting to continue past the level of exit discharge during an emergency.

Figure 1019.1.6 is an example of one method of discharge identification.

1019.1.7 Stairway floor number signs. A sign shall be provided at each floor landing in interior vertical exit enclosures connecting more than three stories designating the floor level, the terminus of the top and bottom of the stair enclosure and the identification of the stair. The signage shall also state the story of, and the direction to the exit discharge and the availability of roof access from the stairway for the fire department. The sign shall be located 5 feet (1524 mm) above the floor landing in a position which is readily visible when the doors are in the open and closed positions.

❖ Signs are to be placed at each floor landing in all exit stairways connecting more than three stories. The signs are to designate the level or story of the landings above

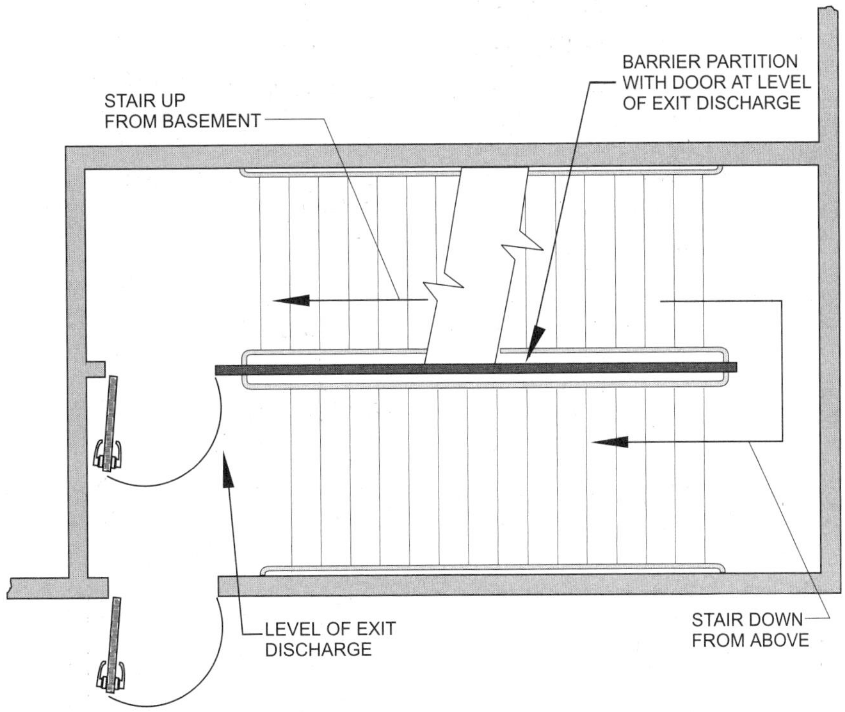

Figure 1019.1.6
EXAMPLE OF DISCHARGE IDENTIFICATION

or below the level of exit discharge. The purpose is to inform the occupants of their location with respect to the level of exit discharge as they use the stairway to leave the building. More importantly, it allows the fire services to locate and gain quick access to the fire floor. At each level, the direction to the exit discharge is required to be indicated. The identification of the level that is the exit discharge also is to be indicated at each level. The identification of the roof access availability is for the fire department; roof access is required by Section 1009.12. For visibility, the signs are required to be located approximately 5 feet (1524 mm) above the floor surface and be visible when the stairway door is open. The need to designate levels remaining to reach the level of exit discharge may mean that the numbering is other than that designation used by building management. For example, a designation of P1, P2, P3, etc., would not be acceptable for stairways in the basement parking garage, since in themselves they do not designate the floor level below the level of exit discharge.

1019.1.8 Smokeproof enclosures. In buildings required to comply with Section 403 or 405, each of the exits of a building that serves stories where the floor surface is located more than 75 feet (22 860 mm) above the lowest level of fire department vehicle access or more than 30 feet (9144 mm) below the level of exit discharge serving such floor levels shall be a smokeproof enclosure or pressurized stairway in accordance with Section 909.20 of the *International Building Code.*

❖ While smokeproof enclosures for exit stairways can, at the designer's option, be used in buildings of any occupancy of any height and area, this section specifically requires smokeproof enclosures to be provided when either of two conditions occur.

The first condition requires all exit stairways in serving buildings with floor levels higher than 75 feet (22 860 mm) above the level of exit discharge to be smokeproof enclosures or pressurized stairways. The reason for this provision is that in very tall buildings, often during fire emergencies, total and immediate evacuation of the occupants cannot be readily accomplished. In such situations, exit stairways become places of safety for the occupants; thus, they must be adequately protected with smokeproof enclosures to provide a safe egress environment. In order to provide this safe environment, the smokeproof enclosure must be constructed to resist the migration of smoke caused by the "stack effect." Stack effect occurs in tall enclosures such as chimneys when a fluid such as smoke, which is less dense than the ambient air, is introduced into the enclosure. The smoke will rise due to the effect of buoyancy and will induce additional flow into the enclosure through openings at the lower levels.

The second condition applies when an occupiable floor level is located more than 30 feet (9144 mm) below the level of exit discharge serving such floor levels. Stairways serving those levels are also required to be protected by smokeproof enclosures because underground portions of a building present unique problems in providing not only for life safety but also access for fire-fighting purposes. The choice of a 30-foot (9144 mm) threshold for this requirement is intended to provide a reasonable limitation on vertical travel distance before the requirement for smokeproof enclosures applies.

Detailed system requirements for a smokeproof enclosure are in Section 909.20.

1019.1.8.1 Enclosure exit. A smokeproof enclosure or pressurized stairway shall exit into a public way or into an exit passageway, yard or open space having direct access to a public way. The exit passageway shall be without other openings and shall be separated from the remainder of the building by 2-hour fire-resistance-rated construction.

Exceptions:

1. Openings in the exit passageway serving a smokeproof enclosure are permitted where the exit passageway is protected and pressurized in the same manner as the smokeproof enclosure, and openings are protected as required for access from other floors.
2. Openings in the exit passageway serving a pressurized stairway are permitted where the exit passageway is protected and pressurized in the same manner as the pressurized stairway.

❖ The walls that comprise the smokeproof enclosure or pressurized stairway, which includes the stairway shaft and the vestibules, must be fire barriers having a fire-resistance rating of at least 2 hours. This level of fire endurance is specified because exit stairways in high-rise buildings serve as principal components of the egress system and as the source of fire service access to the fire floor. This supersedes any allowed reduction of enclosure rating, even if the stair from the level that is more than 30 feet (9144 mm) below exit discharge connects three stories or less. Smokeproof enclosure requirements in this section are more specific; therefore, the exceptions in Section 1023.1 would not apply.

The exceptions apply to openings in the exit passageway that are permitted if the exit passageway is protected and pressurized.

1019.1.8.2 Enclosure access. Access to the stairway within a smokeproof enclosure shall be by way of a vestibule or an open exterior balcony.

Exception: Access is not required by way of a vestibule or exterior balcony for stairways using the pressurization alternative complying with Section 909.20.5 of the *International Building Code.*

❖ See Figures 1019.1.8(1) and 1019.1.8(2) for illustrations of access to the smokeproof stairway by way of a vestibule or an exterior balcony. The purpose of this requirement is to keep the enclosure clear of smoke. Where a pressurized stairway is used, these elements are not necessary.

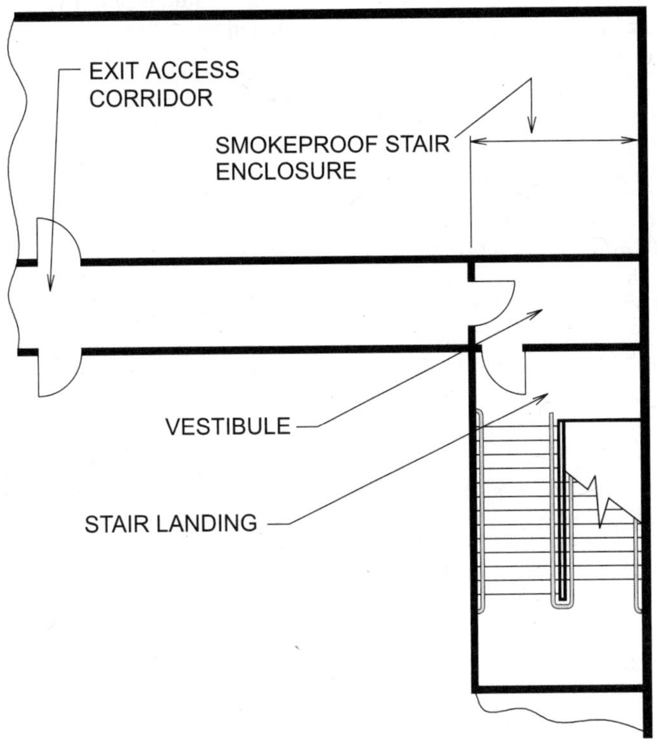

Figure 1019.1.8(1)
TYPICAL SMOKEPROOF ENCLOSURE ENTRY

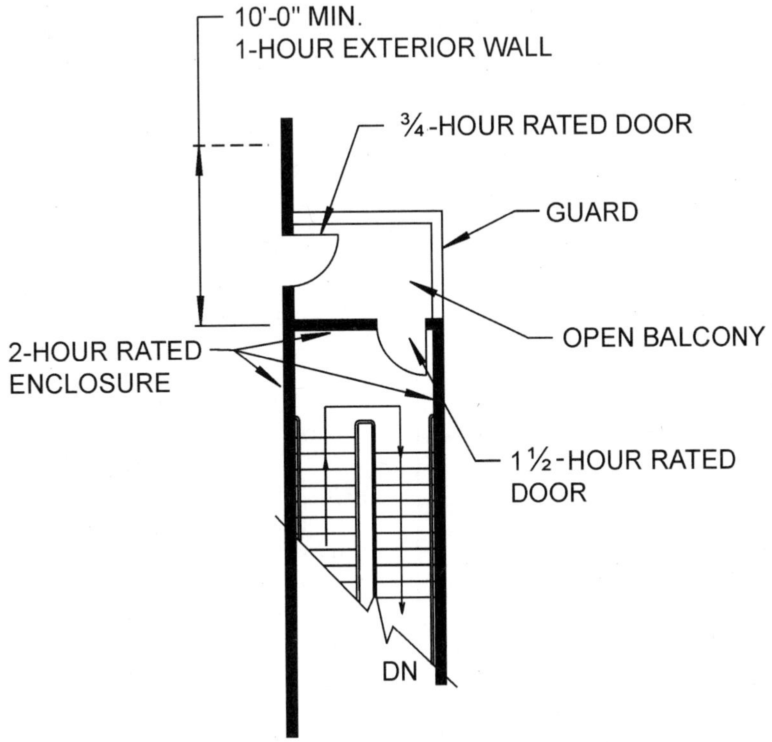

For SI: 1 inch = 25.4 mm, 1 foot = 304.8 mm.

Figure 1019.1.8(2)
TYPICAL SMOKEPROOF ENCLOSURE WITH OPEN BALCONY ARRANGEMENT

[B] SECTION 1020
EXIT PASSAGEWAYS

1020.1 Exit passageway. Exit passageways serving as an exit component in a means of egress system shall comply with the requirements of this section. An exit passageway shall not be used for any purpose other than as a means of egress.

❖ See Figure 1020.1 for an illustration of an exit passageway arrangement. In the case of office buildings or similar types of structures, the exit stairways are often located at the central core or in line with the centrally located exit access corridors. Exit passageways may be used to connect the exit stair to the exterior exit door. Such an arrangement provides great flexibility in the design use of the building. Without the passageway at the grade floor or the level of exit discharge, the occupants of the upper floors or basement levels would have to leave the safety of the exit stairway to travel to the exterior doors. Such a reduction of protection is not acceptable. This section provides acceptable methods of continuing the protected path of travel for building occupants. The building designer/owner is given these different options for achieving this protected path of travel. Exit passageways may also be used on their own in locations not connected with a stair enclosure. Sometimes on large floor plans, an exit passageway may be used to extend an exit into areas that would not otherwise be able to meet the travel distance requirements.

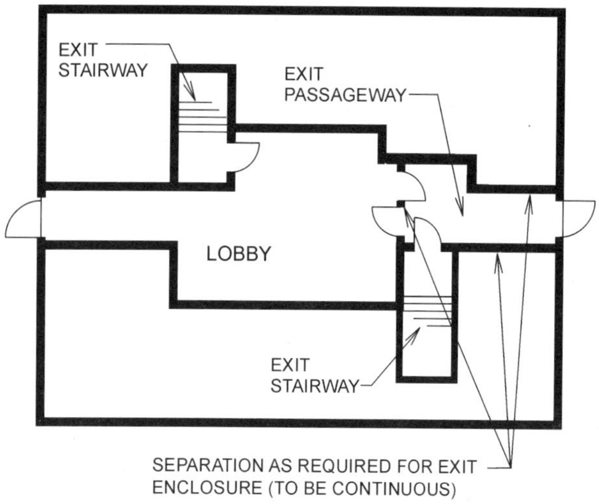

Figure 1020.1
ARRANGEMENT FOR LOBBY AND PASSAGEWAY

1020.2 Width. The width of exit passageways shall be determined as specified in Section 1005.1 but such width shall not be less than 44 inches (1118 mm), except that exit passageways serving an occupant load of less than 50 shall not be less than 36 inches (914 mm) in width.

The required width of exit passageways shall be unobstructed.

Exception: Doors, when fully opened, and handrails, shall not reduce the required width by more than 7 inches (178 mm). Doors in any position shall not reduce the required width by more than one-half. Other nonstructural projections such as trim and similar decorative features are permitted to project into the required width 1.5 inches (38 mm) on each side.

❖ The width of an exit passageway is to be determined in accordance with Section 1005.1 as a function of the number of occupants served in the same manner as for corridors. The greater of the minimum width or the width determined based on occupancy is to be used. In situations where the exit passageway also serves as an exit access corridor for the first floor, the width of the corridor must comply with the stricter requirement.

The limit on the restriction from a door that opens into the exit passageway is the same as for general means of egress requirements for door encroachment (see Section 1005.2) and stairway landings (see Section 1009.4).

1020.3 Construction. Exit passageway enclosures shall have walls, floors and ceilings of not less than 1-hour fire-resistance rating, and not less than that required for any connecting exit enclosure. Exit passageways shall be constructed as fire barriers in accordance with Section 706 of the *International Building Code.*

❖ The entire exit passageway enclosure is to be fire-resistance rated as specified. The floors and ceilings are required to be rated in addition to the walls. When used separately, a minimum 1-hour fire-resistance rating is required. Where extending an exit enclosure, the rating must not be less than the exit enclosure so that the degree of protection is kept at the same level. Remember that if the exit passageway extends over a basement, the continuity requirements for fire barriers and horizontal assemblies require all supporting construction to also have the same fire-resistance rating as the elements supported (see Sections 706.4 and 711.4 of the IBC). The continuity requirements would also be a concern for the rated ceiling of the exit passageway. An alternative for the ceiling of the exit passageway could be a top shaft enclosure (see Section 707.12 of the IBC).

1020.4 Openings and penetrations. Exit passageway opening protectives shall be in accordance with the requirements of Section 715 of the *International Building Code.*

Except as permitted in Section 402.4.6 of the *International Building Code*, openings in exit passageways other than unexposed exterior openings shall be limited to those necessary for exit access to the exit passageway from normally occupied spaces and for egress from the exit passageway.

Where interior exit enclosures are extended to the exterior of a building by an exit passageway, the door assembly from the exit enclosure to the exit passageway shall be protected by a fire door conforming to the requirements in Section 715.3 of the *International Building Code.* Fire door assemblies in exit passageways shall comply with Section 715.3.4 of the *International Building Code.*

Elevators shall not open into an exit passageway.

❖ The requirements for exit passageways are very similar to those required for vertical exit enclosures (see Section 1019.1.1). The only openings that are permitted in fire-resistance-rated exit passageways are doors that lead either from normally occupied spaces or from the vertical exit enclosure. This restriction on openings essentially prohibits the use of windows in an exit passageway except for those exterior windows that are not exposed to any hazards by fire separation distance (see Section 704.8 of the IBC) or adjacent exterior walls similar to what is specified for vertical exit enclosures (see Section 1019.1.4). The only exception would be window assemblies that have been tested as wall assemblies in accordance with ASTM E 119. The objective of this provision is to minimize the possibility of fire spreading into an exit passageway and endangering the occupants or even preventing the use of the exit at a time when it is most needed. The limitation on openings applies regardless of the fire protection rating of the opening protective. The limitation on openings from normally occupied areas is intended to reduce the probability of a fire occurring in an unoccupied area, such as a storage closet, which has an opening into the stairway, thereby resulting in fire spread into the stairway. Other spaces that are not normally occupied include, but are not limited to, toilet rooms, electrical/mechanical equipment rooms and janitorial closets. Exit passageways have an additional requirement that prohibits elevators from opening directly into the passageway. While not specifically stated in Section 1019.1.1, the prohibition for openings from spaces that were not normally occupied could be interpreted as prohibiting elevators from opening into stairway enclosures as well. The are some exceptions for these unoccupied spaces in exit passageways in covered malls (see Section 402.4.6 of the IBC).

The third paragraph addresses horizontal exit enclosures that could move from the vertical enclosure of a stairway or ramp to the horizontal enclosure of an exit passageway. While the exit passageway is an extension of the protection offered by the vertical exit, there must still be an opening protective at the bottom of the stairway or ramp. This is to prevent any smoke that may migrate into the exit passageway from also moving up the exit stairway or ramp.

1020.5 Penetrations. Penetrations into and openings through an exit passageway are prohibited except for required exit doors, equipment and ductwork necessary for independent pressurization, sprinkler piping, standpipes, electrical raceway for fire department communication and electrical raceway serving the exit passageway and terminating at a steel box not exceeding 16 square inches (0.010 m^2). Such penetrations shall be protected in accordance with Section 712 of the *International Building Code.* There shall be no penetrations or communicating openings, whether protected or not, between adjacent exit passageways.

❖ This section specifically lists the items that are allowed to penetrate an exit passageway. This is consistent for all types of exit enclosures, including stair or ramp vertical exit enclosures (see Section 1019.1.2) and exit passageways. In general, only portions of the building service systems that serve the exit enclosure are allowed to penetrate the exit enclosure.

[B] SECTION 1021 HORIZONTAL EXITS

1021.1 Horizontal exits. Horizontal exits serving as an exit in a means of egress system shall comply with the requirements of this section. A horizontal exit shall not serve as the only exit from a portion of a building, and where two or more exits are required, not more than one-half of the total number of exits or total exit width shall be horizontal exits.

Exceptions:

1. Horizontal exits are permitted to comprise two-thirds of the required exits from any building or floor area for occupancies in Group I-2.
2. Horizontal exits are permitted to comprise 100 percent of the exits required for occupancies in Group I-3. At least 6 square feet (0.6 m^2) of accessible space per occupant shall be provided on each side of the horizontal exit for the total number of people in adjoining compartments.

Every fire compartment for which credit is allowed in connection with a horizontal exit shall not be required to have a stairway or door leading directly outside, provided the adjoining fire compartments have stairways or doors leading directly outside and are so arranged that egress shall not require the occupants to return through the compartment from which egress originates.

The area into which a horizontal exit leads shall be provided with exits adequate to meet the occupant requirements of this chapter, but not including the added occupant capacity imposed by persons entering it through horizontal exits from another area. At least one of its exits shall lead directly to the exterior or to an exit enclosure.

❖ See Figure 1021.1 for a typical horizontal exit arrangement. Horizontal exits can comprise up to 50 percent of the exits from an area of a building. The percentage is higher for Group I-2 and I-3 occupancies where the evacuation strategy is defend in place rather than direct egress.

See Section 1002 for the definition of a "Horizontal exit." A horizontal exit may be an element of a means of egress when in compliance with the requirements of this section. The actual horizontal exit is the protected door opening in a wall, open-air balcony or bridge that separates two areas of a building. A horizontal exit is often used in hospitals and in prisons where it is not feasible or desirable that the occupants exit the facility.

Note that the capacity of the exit (such as an exit stairway) from the area that the horizontal exit leads is required to be sufficient for the number of occupants in the area, not including those who come into the space from other areas using the horizontal exit. This is because the adjacent area of refuge is of sufficient safety to house occupants during a fire or until the egress system is available. The occupant capacity of the refuge area is addressed in Section 1021.4.

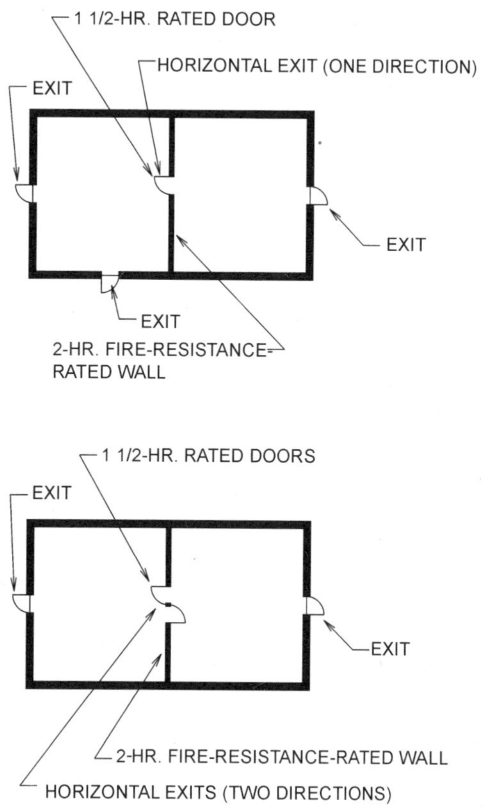

Figure 1021.1
TYPICAL HORIZONTAL EXITS

1021.2 Separation. The separation between buildings or areas of refuge connected by a horizontal exit shall be provided by a fire wall complying with Section 705 of the *International Building Code* or a fire barrier complying with Section 706 of the *International Building Code* and having a fire-resistance rating of not less than 2 hours. Opening protectives in horizontal exit walls shall also comply with Section 715 of the *International Building Code*. The horizontal exit separation shall extend vertically through all levels of the building unless floor assemblies are of 2-hour fire resistance with no unprotected openings.

Exception: A fire-resistance rating is not required at horizontal exits between a building area and an above-grade pedestrian walkway constructed in accordance with Section 3104 of the *International Building Code*, provided that the distance between connected buildings is more than 20 feet (6096 mm).

Horizontal exit walls constructed as fire barriers shall be continuous from exterior wall to exterior wall so as to divide completely the floor served by the horizontal exit.

❖ The basic concept of a horizontal exit is that during a fire emergency, the occupants of a floor will transfer from one side of a fire wall or fire barrier to the other. Separation between areas of a building can be accomplished by either a fire wall (see Section 705 of the IBC or a fire barrier (see Section 706 of the IBC), with a fire-resistance rating not less than 2 hours. Any fire shutters or fire door must have an opening protection of not less than $1^1/_2$ hours (see Table 715.3 of the IBC).

In buildings of Groups I-2 and I-3, it may also be desirable (while not mandatory) for the horizontal exit to serve as a smoke barrier. In such cases, the wall containing the horizontal exit must also comply with the requirements for a smoke barrier (see Section 709 of the IBC).

In order to decrease the amount of smoke that is able to migrate around the edges of a horizontal exit, the horizontal exit must extend from at least the floor to the deck above (i.e., fire barrier), as well as across the floor level from one side of the building to another. To prevent smoke from moving up through the floors there are two choices. One option is that the horizontal exit can extend vertically through all levels of the building (i.e., fire wall or fire barriers). The second option is to utilize fire barriers that are not lined up vertically, but then the floor must have a 2-hour fire-resistance rating and no unprotected openings are permitted between any two refuge areas.

The exception permits a pedestrian walkway or skybridge to act as a horizontal exit when buildings are at least 20 feet (6096 mm) apart.

1021.3 Opening protectives. Fire doors in horizontal exits shall be self-closing or automatic-closing when activated by a smoke detector installed in accordance with Section 907.11. Opening protectives in horizontal exits shall be consistent with the fire-resistance rating of the wall. Such doors where located in a cross-corridor condition shall be automatic-closing by activation of a smoke detector installed in accordance with Section 907.11.

❖ For the safety of occupants using a horizontal exit, it is important that the doors be fire doors that are self-closing or automatic closing by activation of a smoke detector. Smoke detectors that initiate automatic closing should be located at both sides of the doors. See the commentary to Section 907.10 for an additional explanation of the installation requirements. Any openings in the fire barriers or fire walls that are used as horizontal exits must be protected in coordination with the rating of the wall. A reference to Section 715 for opening protectives is found in Section 1021.2.

1021.4 Capacity of refuge area. The refuge area of a horizontal exit shall be spaces occupied by the same tenant or public areas and each such area of refuge shall be adequate to house the original occupant load of the refuge space plus the occupant load anticipated from the adjoining compartment. The anticipated occupant load from the adjoining compartment shall be based on the capacity of the horizontal exit doors entering the area of refuge. The capacity of areas of refuge shall be computed on a net floor area allowance of 3 square feet (0.2787 m^2) for each occupant to be accommodated therein, not including areas of stairways, elevators and other shafts or courts.

Exception: The net floor area allowable per occupant shall be as follows for the indicated occupancies:

1. Six square feet (0.6 m^2) per occupant for occupancies in Group I-3.
2. Fifteen square feet (1.4 m^2) per occupant for ambulatory occupancies in Group I-2.
3. Thirty square feet (2.8 m^2) per occupant for nonambulatory occupancies in Group I-2.

❖ The building area on the discharge side of a horizontal exit must serve as a refuge area for the occupants of both sides of the floor areas connected by the horizontal exit. Therefore, adequate space must be available on each side of the wall to hold the full occupant load of that side, plus the number of occupants from the other side who may be required to use the horizontal exit. These refuge areas are meant to hold temporarily the occupants in a safe place until they can evacuate the premises in an orderly manner or, in the case of hospitals and like facilities, to hold bedridden patients and other nonambulatory occupants in a protected area until the fire emergency has ended. The size of the refuge area is based on the nature of the expected occupants. In the case of Group I-3, the area will be used to hold the occupants until deliberate egress can be accomplished with staff assistance or supervision. In other cases, it is assumed that the occupants simply wait in line to egress through the required exit facilities provided on the discharge side. Although similar language is used in describing the area of refuge for an accessible means of egress, Section 1007.6 specifies area requirements that may be insufficient for use as a refuge area for a horizontal exit. Care must be taken when applying both principles.

The 3-square-feet-(0.28 m^2) per-occupant exception is based on the maximum permitted occupant density at which orderly movement to the exits is reasonable. The 30-square-feet-(2.8 m^2) per-hospital or nursing-home-patient exception is based on the space necessary for a bed or litter. It should be noted that 30 square feet (2.8 m^2) is not based on the total occupant load, as would be determined in accordance with Section 1004.1, but rather on the number of nonambulatory patients. The 15-square-feet (1.4 m^2) requirement for occupancies in Group I-2 facilities is based on each ambulatory patient having a staff attendant.

In a single-tenant facility, any of the spaces that are constantly available (e.g., not lockable) and occupied by the tenant can be used as places of refuge. However, in spaces housing more than one tenant, public areas of refuge such as corridors or passageways must be provided and be accessible at all times. This requirement is necessary because if a horizontal exit connects two areas occupied by different tenants, the tenants could (for privacy and security purposes) render the necessary free access through the horizontal exit ineffective. When the horizontal exit discharges into a public or common space, such as a corridor leading to an exit, each tenant can obtain the desired security.

[B] SECTION 1022
EXTERIOR EXIT RAMPS AND STAIRWAYS

1022.1 Exterior exit ramps and stairways. Exterior exit ramps and stairways serving as an element of a required means of egress shall comply with this section.

Exception: Exterior exit ramps and stairways for outdoor stadiums complying with Section 1019.1, Exception 2.

❖ Exterior exit ramps and stairways are an important element of the means of egress system and must be designed and constructed so that they will serve as a safe path of travel. The general requirements in Section 1009 also apply to exterior stairways. For ramp provisions, see Section 1010.

The exception in this section references Exception 2 in Section 1019.1, which allows exterior ramps and stairways that serve outdoor assembly facilities to be open. The openness criteria in Section 1022.3 would be an appropriate guideline for what is intended.

1022.2 Use in a means of egress. Exterior exit ramps and stairways shall not be used as an element of a required means of egress for occupancies in Group I-2. For occupancies in other than Group I-2, exterior exit ramps and stairways shall be permitted as an element of a required means of egress for buildings not exceeding six stories or 75 feet (22 860 mm) in height.

❖ This section specifies the conditions where an exterior ramp or stairway can be used as a required exit. Exterior exit stairways are not permitted for Group I-2 since quick evacuation of nonambulatory patients from buildings is impractical. Some of the patients may not be capable of self-preservation and may, therefore, require assistance from staff. The period of evacuation of nonambulatory patients could become lengthy.

With the exception of outdoor stadiums (see Section 1022.1), exterior ramps or stairways are not allowed to be a required exit in buildings that exceed six stories or 75 feet (22 860 mm) in height due to the hazard of using such a ramp or stairway in poor weather. Some persons may not use such a stair due to vertigo. When confronted with a view from a great height, vertigo sufferers can become confused, disoriented and dizzy. They could injure themselves, become disabled or refuse to move (freeze). In a fire situation, they could become an obstruction in the path of travel, possibly causing panic and injuries to other users of the exit.

1022.3 Open side. Exterior exit ramps and stairways serving as an element of a required means of egress shall be open on at least one side. An open side shall have a minimum of 35 square feet (3.3 m^2) of aggregate open area adjacent to each floor level and the level of each intermediate landing. The required open area shall be located not less than 42 inches (1067 mm) above the adjacent floor or landing level.

❖ An important factor in considering exterior exit ramps or stairways is that natural ventilation is assumed to occur. This is so that smoke will not be trapped above the ramp or stairway walking surfaces and obscure safe egress.

The exterior ramp or stairway must have at least one of its sides directly facing an outer court, yard or public way. This will allow the products of combustion escaping from the interior of the building to quickly vent to the outdoor atmosphere and let the building occupants egress down the exterior ramp or stairway. Since exterior ramps or stairways are occasionally partially enclosed within the building construction, minimum amounts of exterior openings are specified by the code.

The openings on each and every floor level and landing must total 35 square feet (3.3 m^2) or greater. The opening is to occur higher than 42 inches (1067 mm) from the floor and intermediate landing levels. The high openings dissipate the smoke buildup from the exterior ramp or stairway (see Figure 1022.3). The bottom edge of the opening is consistent with the height requirements for guards (see Section 1012.2).

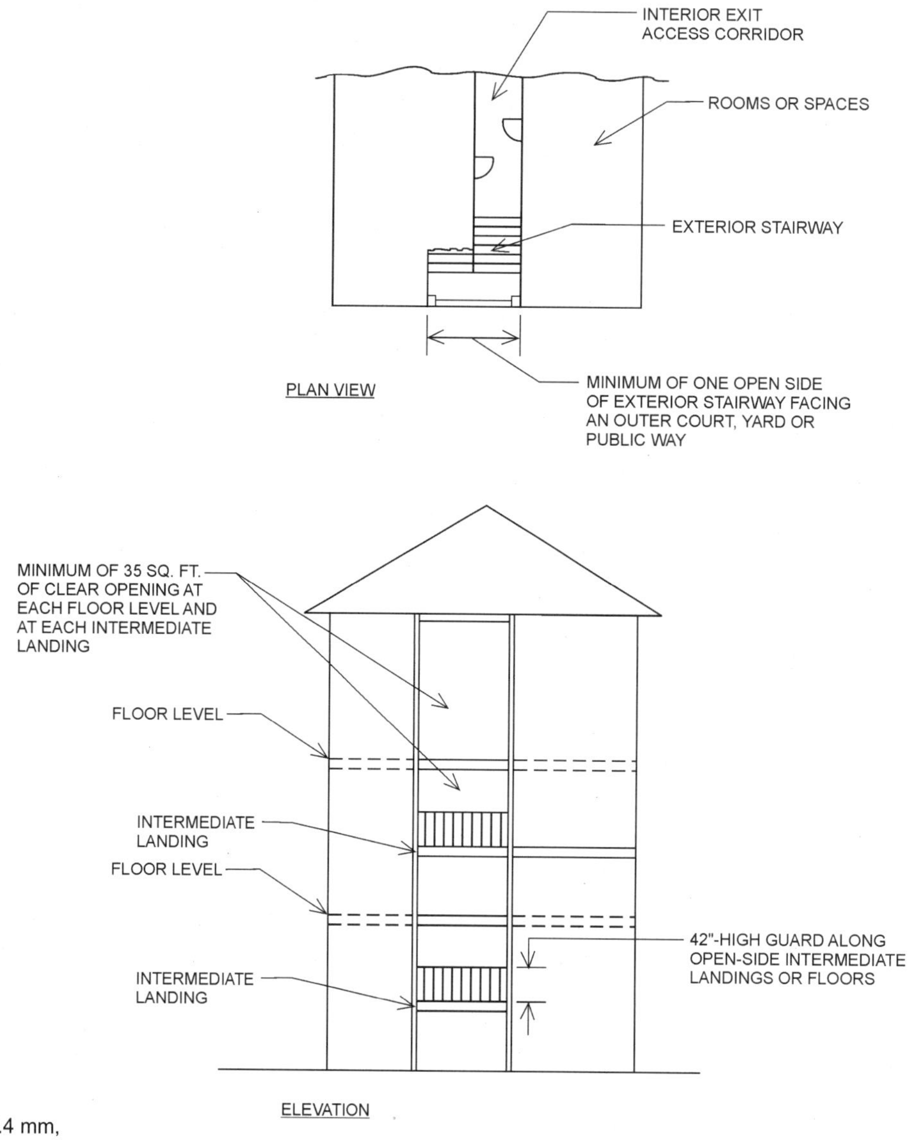

For SI: 1 inch = 25.4 mm,
1 square foot = 0.0929 mm.

Figure 1022.3
EXTERIOR STAIRWAY

1022.4 Side yards. The open areas adjoining exterior exit ramps or stairways shall be either yards, courts or public ways; the remaining sides are permitted to be enclosed by the exterior walls of the building.

❖ See Section 1022.3 for a discussion of the opening requirements. This section simply specifies the type of areas that the exterior opening of the exterior ramp or stair is to adjoin. These open spaces will enable the smoke to dissipate from the exterior ramp or stairway so it will be useable as a required exit.

1022.5 Location. Exterior exit ramps and stairways shall be located in accordance with Section 1023.3.

❖ The location requirements of this section protect the users of the exterior exit ramp or stairway from the effects of a fire in another building on the same lot or an adjacent lot. The separation distance reduces the exposure to heat and smoke. If the stairway is closer than specified, then adjacent buildings' exterior walls and openings are to be protected in accordance with Section 704 of the IBC such that the users of the exterior exit are protected.

1022.6 Exterior ramps and stairway protection. Exterior exit ramps and stairways shall be separated from the interior of the building as required in Section 1019.1. Openings shall be limited to those necessary for egress from normally occupied spaces.

Exceptions:

1. Separation from the interior of the building is not required for occupancies, other than those in Group R-1 or R-2, in buildings that are no more than two stories above grade where the level of exit discharge is the first story above grade.
2. Separation from the interior of the building is not required where the exterior ramp or stairway is served by an exterior ramp and/or balcony that connects two remote exterior stairways or other approved exits, with a perimeter that is not less than 50 percent open. To be considered open, the opening shall be a minimum of 50 percent of the height of the enclosing wall, with the top of the openings no less than 7 feet (2134 mm) above the top of the balcony.
3. Separation from the interior of the building is not required for an exterior ramp or stairway located in a building or structure that is permitted to have unenclosed interior stairways in accordance with Section 1019.1.
4. Separation from the interior of the building is not required for exterior ramps or stairways connected to open-ended corridors, provided that Items 4.1 through 4.4 are met:
   - 4.1. The building, including corridors and ramps and/or stairs, shall be equipped throughout with an automatic sprinkler system in accordance with Section 903.3.1.1 or 903.3.1.2.
   - 4.2. The open-ended corridors comply with Section 1016.
   - 4.3. The open-ended corridors are connected on each end to an exterior exit ramp or stairway complying with Section 1022.
   - 4.4. At any location in an open-ended corridor where a change of direction exceeding 45 degrees (0.79 rad) occurs, a clear opening of not less than 35 square feet (3.3 m^2) or an exterior ramp or stairway shall be provided. Where clear openings are provided, they shall be located so as to minimize the accumulation of smoke or toxic gases.

❖ Exterior exit ramps or stairways must be protected from interior fires that may project through windows or other openings adjacent to the ramp or stairway, possibly endangering the occupants using this means of egress to reach grade. The protection of an exterior ramp or stairway is to be obtained by separating the exterior exit from the interior of the building using walls having a fire-resistance rating of at least 1 hour with opening protectives. Consistent with the protection required in Sections 1019.1 and 1019.1.4 for interior exit stairways, the fire-resistance rating must be provided for a distance of 10 feet (3048 mm) horizontally and vertically from the ramp or stairway edges from the ground to a level 10 feet (3048 mm) above the highest landing.

All window and door openings falling inside the 10-foot (3048 mm) horizontal separation distance as well as all window and door openings 10 feet (3048 mm) above the uppermost landing and below the stairway must be protected with minimum $^3/_4$-hour fire-resistance-rated opening protectives [see Figure 1022.6(1)].

Openings within the width of the stairway must only be from normally occupied spaces. This is consistent with the requirements for vertical exit enclosures (see Sections 1019.1.1 and 1019.1.2).

Exception 1 indicates that opening protectives are not required for occupancies other than Groups R-1 and R-2, which are two stories or less above grade when the level of exit discharge is at the lower story. The reason for this exception is that in cases of fire in low buildings, the occupants are usually able to evacuate the premises before the fire can emerge through exterior wall openings and endanger the exit ramp or stairways. In hotels and apartments, however, the occupants' response to a fire emergency could be significantly reduced because they are unfamiliar with the surroundings and may be sleeping.

Exception 2 allows the opening protectives to be omitted when an exterior exit access balcony is served by two exits and when they are remote from each other. Remoteness is regulated by Section 1014.2. This exception is applicable to all groups. In such instances, it is unlikely that the users of the exterior ramp or stairway will become trapped by fire, since they have the option of using the balcony to gain access to either of the two available exits and the products of combustion will be vented directly to the outside (see Section 1013.5 regarding exterior balconies). At least one-half of the total perimeter of the exterior balcony must be permanently open to the outside. The requirement for at least one-half the height of that level to be open allows for columns, solid guards and decorative elements, such as arches. With the top of the opening at least 7 feet (2134 mm) above the walking surface, products of combustion can vent and allow passage below the smoke layer [see Figure 1022.6(2)].

Exception 3 exempts exterior ramps or stairways from protection when Section 1019.1 permits unenclosed interior exit stairways.

Exception 4 deletes the requirement for a separation from the interior of the building for the exterior wall area where an open-ended corridor (breezeway) interfaces with an exterior ramp or stairway. The separation is not needed as a result of the sprinkler system in all areas of the building, including the open-ended corridor and the exterior exit. The other characteristics of the open-ended corridor described in this exception are needed so that the open-ended corridor provides the safety to be used in the event of a fire. The requirements for an exterior ramp or stairway at each end and the opening or an exterior ramp or stairway where the open-ended corridor has a change of direction of greater than 45 degrees (0.79 rad) are for adequate ventilation of the open-ended corridor.

Similar language is used in describing the exterior wall requirements for an exterior area for assisted rescue (see Section 1007.8). If this option is utilized at an exterior stairway or nonaccessible ramp, the exceptions in this section would not be permitted since occupants of the exterior area for assisted rescue could not immediately move away from the building.

[B] SECTION 1023 EXIT DISCHARGE

1023.1 General. Exits shall discharge directly to the exterior of the building. The exit discharge shall be at grade or shall provide direct access to grade. The exit discharge shall not reenter a building.

Exceptions:

1. A maximum of 50 percent of the number and capacity of the exit enclosures is permitted to egress through areas on the level of discharge provided all of the following are met:

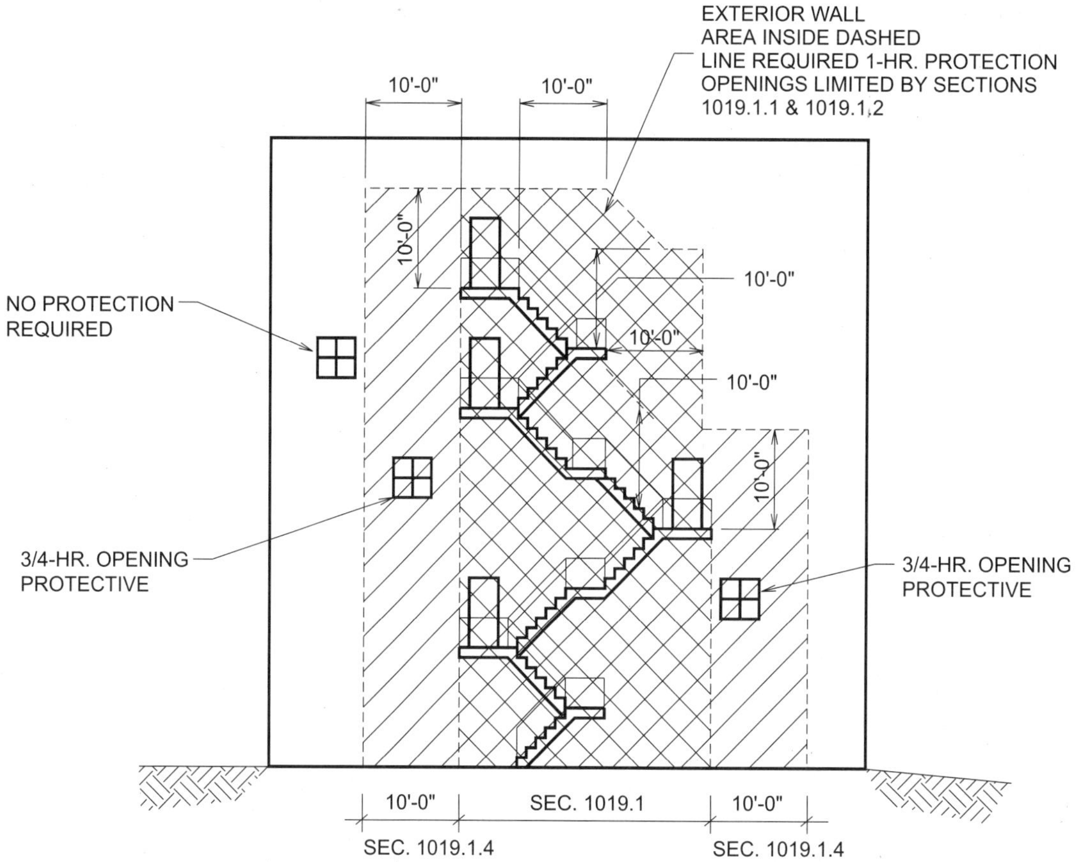

For SI: 1 inch = 25.4 mm, 1 foot = 304.8 mm.

Figure 1022.6(1)
PROTECTION OF EXTERIOR STAIRWAYS

1.1. Such exit enclosures egress to a free and unobstructed way to the exterior of the building, which way is readily visible and identifiable from the point of termination of the exit enclosure.

1.2. The entire area of the level of discharge is separated from areas below by construction conforming to the fire-resistance rating for the exit enclosure.

1.3. The egress path from the exit enclosure on the level of discharge is protected throughout by an approved automatic sprinkler system. All portions of the level of discharge with access to the egress path shall either be protected throughout with an automatic sprinkler system installed in accordance with Section 903.3.1.1 or 903.3.1.2, or separated from the egress path in accordance with the requirements for the enclosure of exits.

2. A maximum of 50 percent of the number and capacity of the exit enclosures is permitted to egress through a vestibule provided all of the following are met:

2.1. The entire area of the vestibule is separated from areas below by construction conforming to the fire-resistance rating for the exit enclosure.

2.2. The depth from the exterior of the building is not greater than 10 feet (3048 mm) and the length is not greater than 30 feet (9144 mm).

2.3. The area is separated from the remainder of the level of exit discharge by construction providing protection at least the equivalent of approved wired glass in steel frames.

2.4. The area is used only for means of egress and exits directly to the outside.

3. Stairways in open parking garages complying with Section 1019.1, Exception 5, are permitted to egress through the open parking garage at the level of exit discharge.

❖ The exit discharge is the third piece of the means of egress system, which consists of exit access, exit and exit discharge. The general provisions for means of egress in Sections 1003 through 1012 are applicable to the exit discharge. The basic provision is that exits shall discharge directly to the outside of the building. The exit discharge is the path from the termination of the exit to the public way. When this is not practical, there are three alternatives: an exit passageway (see Section 1020), an exit discharge lobby (see Section 1023.1, Exception 1) or an exit discharge passageway (see Section 1023.1, Exception 2). Open parking garages are special cases and are addressed in Exception 3.

While Exceptions 1 and 2 could be applicable for exit passageways and exit ramps, most of the real-life application of the exceptions are for exit stairways. For simplicity of discussion, this commentary will be limited to enclosed exit stairways. Up to 50 percent of the exit stairways in a building may use Exception 1 or 2; therefore, neither exception is viable for a single-exit building.

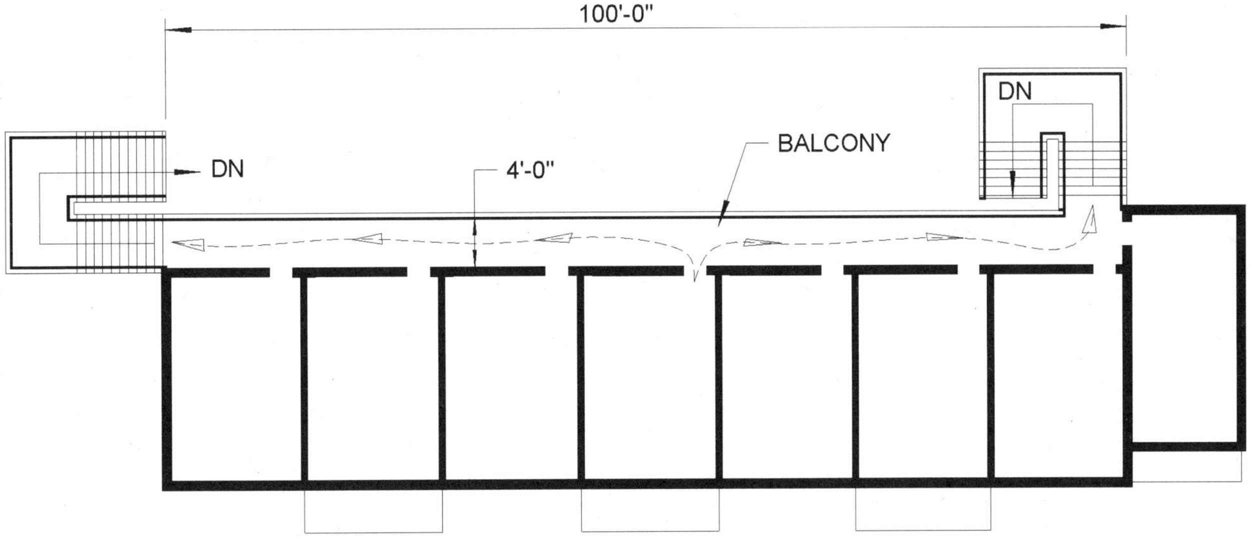

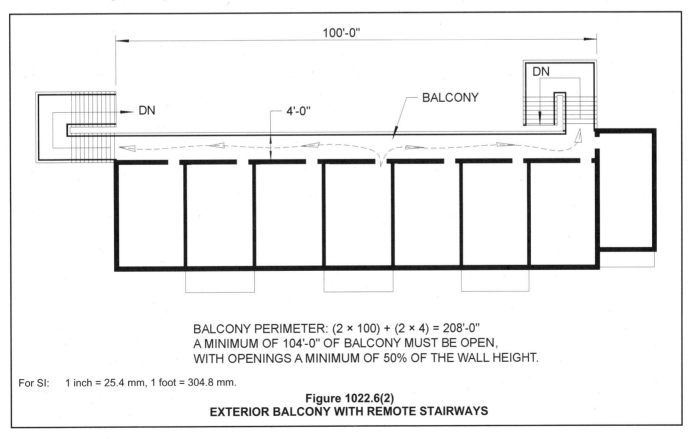

BALCONY PERIMETER: (2 × 100) + (2 × 4) = 208'-0"
A MINIMUM OF 104'-0" OF BALCONY MUST BE OPEN,
WITH OPENINGS A MINIMUM OF 50% OF THE WALL HEIGHT.

For SI: 1 inch = 25.4 mm, 1 foot = 304.8 mm.

Figure 1022.6(2)
EXTERIOR BALCONY WITH REMOTE STAIRWAYS

An interior exit discharge lobby is permitted to receive the discharge from an exit stairway in lieu of the stairway discharging directly to the exterior. A fire door must be provided at the point where the exit stairway discharges into the lobby. Without an opening protective between the stairway and a lobby, it would be possible for the stairway to be directly exposed to smoke movement from a fire in the lobby. The opening protective provides for full continuity of the vertical component of the exit arrangement. Additionally, in buildings where stair towers must be pressurized, pressurization would not be possible without a door at the lobby level.

An exit discharge lobby is the sole location recognized in the code where an exit element can be used for purposes other than pedestrian travel for means of egress. The lobby may contain furniture or decoration and nonoccupiable spaces may open directly into the lobby. The lobby and all other areas on the same level that are not separated from the lobby by fire barriers consistent with the rating of the stair enclosure must be sprinklered in accordance with an NFPA 13 or NFPA 13R system [see Figure 1023.1(1)]. If the entire level is sprinklered, no separation is required. In this case, the automatic sprinkler system is anticipated to control and (perhaps) eliminate the fire threat so as not to jeopardize the path of egress of the occupants. The lobby floor and any supporting construction must be rated the same as the stairway enclosure. If the lobby is slab on grade, this requirement is not applicable. This is consistent with the fundamental concept that an exit enclosure provides the necessary level of protection from adjacent areas. A path of travel through the lobby must be continually clear and available. The exit door leading out of the building must be immediately visible and identifiable when a person leaves the exit. This does not mean the exterior exit door must be directly in front of the door at the bottom of the stairway, but the intent is that it should be within the general range of vision. It should not be required that a person must turn completely around or go around a corner to be able to see the way out.

An exit is also allowed to discharge through a vestibule, provided it complies with the specified requirements of Exception 2. Note that a vestibule is not allowed to have any vending machines within the space or any storage rooms opening into the vestibule. The vestibule floor and any supporting construction must be rated the same as the stairway enclosure. If the vestibule is slab on grade, this requirement is not applicable. The walls of the vestibule itself must have something at least equivalent to wired glass in steel frames. This is typically an opening protection with a fire protection rating of at least 20 minutes. The size of the vestibule is limited so that it cannot be used for other activities, and the travel distance from the exit stairway to the exterior exit doorway is also limited [see Figures 1023.1(2) and 1023.1(3)].

The reason that stairs in open parking garages are allowed to be unenclosed is due to the low hazard and free venting aspects of the open parking garage. When an occupant leaves the stair in an open parking garage, typically he or she can move in several directions to reach the perimeter of the building. This, coupled with the open nature of the parking garage structure, allows occupants a safe path of travel from the bottom of the

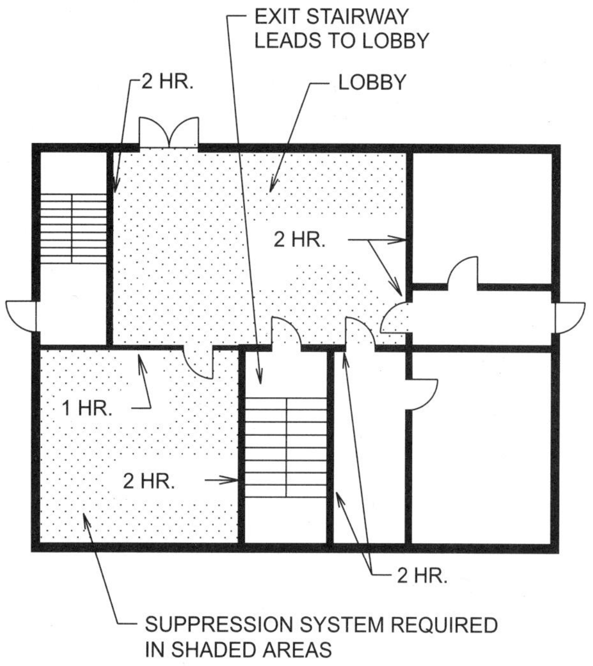

Figure 1023.1(1)
PROTECTION OF LOBBY WITH AN EXIT DISCHARGE

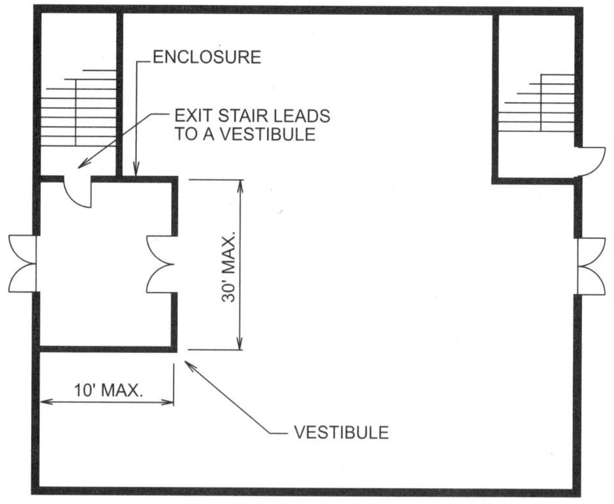

For SI: 1 foot = 304.8 mm.

Figure 1023.1(2)
VESTIBULE EXIT DISCHARGE

stairway across the parking level to the exterior and public way without the additional protection of an exit passageway, exit lobby or exit vestibule. This path must meet the general means of egress requirements and may not be down a driveway unless a pedestrian path is also provided (see Section 1010.1).

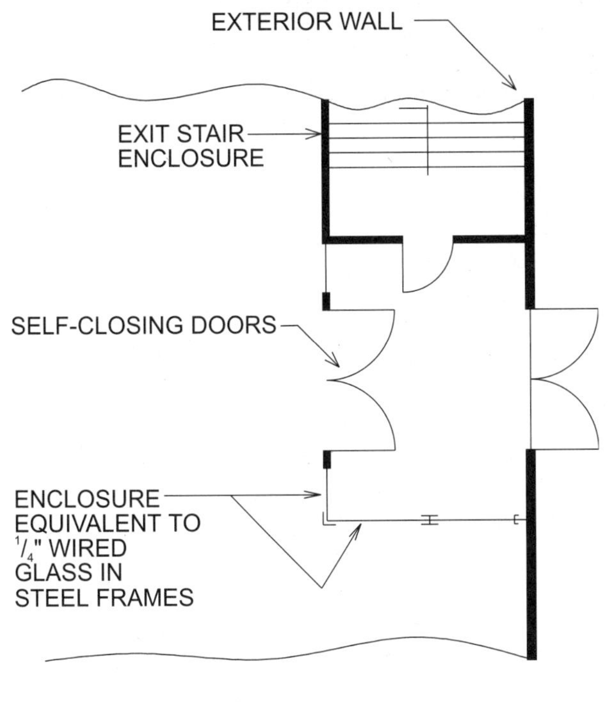

For SI: 1 inch = 25.4 mm.

Figure 1023.1(3)
SEPARATION DETAILS FOR VESTIBULE WITH EXIT DISCHARGE

1023.2 Exit discharge capacity. The capacity of the exit discharge shall be not less than the required discharge capacity of the exits being served.

❖ This section specifies the exit discharge capacity. The exit discharge is required to be designed for the required capacity of all the exits it serves. If the exit discharge serves two exits, it is to be designed for the sum of the occupants served by both exits. Note that the capacity of the exit discharge is not required to be the capacity of both exits, which is higher than the sum of the occupants served by both exits.

1023.3 Exit discharge location. Exterior balconies, stairways and ramps shall be located at least 10 feet (3048 mm) from adjacent lot lines and from other buildings on the same lot unless the adjacent building exterior walls and openings are protected in accordance with Section 704 of the *International Building Code* based on fire separation distance.

❖ The purpose of this section is to protect the users of exterior balconies, stairways and ramps from fire in an adjacent building by requiring the specified separation distance from an adjacent building. The reason for the required distance to a lot line is to provide for a future building that could be built on the adjacent lot.

1023.4 Exit discharge components. Exit discharge components shall be sufficiently open to the exterior so as to minimize the accumulation of smoke and toxic gases.

❖ This section includes exit discharge components that currently only include egress courts. It should be remembered that the general requirements in Sections 1003 through 1012 will still apply when those components are used in the exit discharge.

1023.5 Egress courts. Egress courts serving as a portion of the exit discharge in the means of egress system shall comply with the requirements of Section 1023.

❖ See Figure 1023.5 for an illustration of an exit discharge, including an egress court.

This section and the following subsections address the detailed requirements for egress courts. It is essential that exterior egress courts that serve occupants from an exit to a public way be sufficiently open to prevent the accumulation of smoke and toxic gases in the event of a fire.

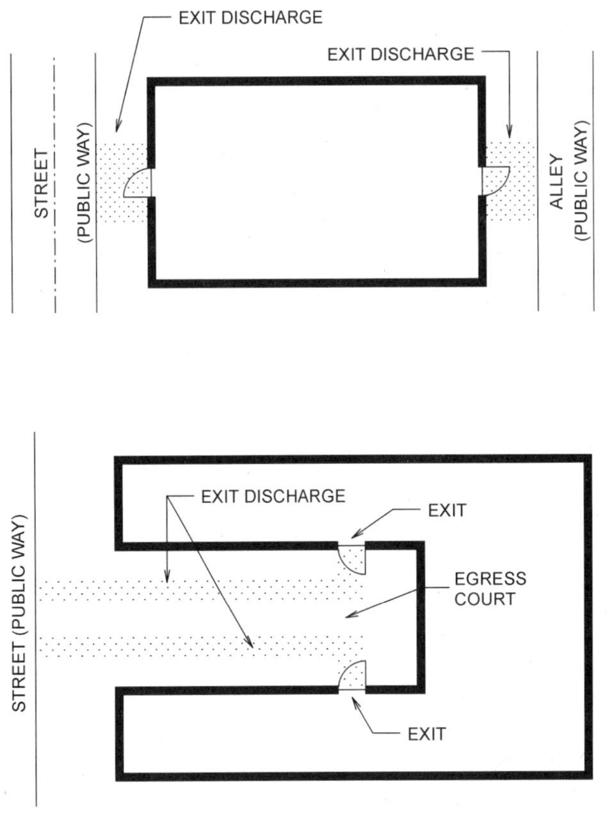

Figure 1023.5
EXAMPLES OF EXIT DISCHARGE

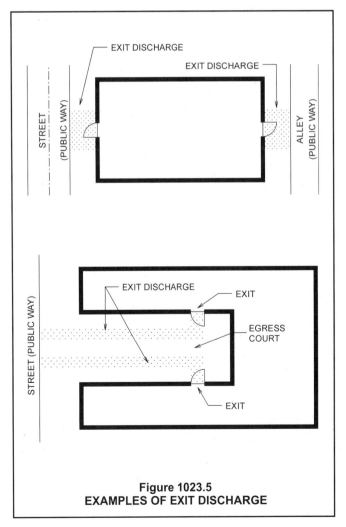

1023.5.1 Width. The width of egress courts shall be determined as specified in Section 1005.1, but such width shall not be less than 44 inches (1118 mm), except as specified herein. Egress courts serving occupancies in Group R-3 applicable in Section 1001.1 and Group U shall not be less than 36 inches (914 mm) in width.

The required width of egress courts shall be unobstructed to a height of 7 feet (2134 mm).

Exception: Doors, when fully opened, and handrails shall not reduce the required width by more than 7 inches (178 mm). Doors in any position shall not reduce the required width by more than one-half. Other nonstructural projections such as trim and similar decorative features are permitted to project into the required width 1.5 inches (38 mm) from each side.

Where an egress court exceeds the minimum required width and the width of such egress court is then reduced along the path of exit travel, the reduction in width shall be gradual. The transition in width shall be affected by a guard not less than 36 inches (914 mm) in height and shall not create an angle of more than 30 degrees (0.52 rad) with respect to the axis of the egress court along the path of egress travel. In no case shall the width of the egress court be less than the required minimum.

❖ The width of an exterior court is to be determined in the same fashion as for an interior corridor. The width is not to be less than required to serve the number of occupants from the exit or exits and also not less than the minimum specified in the section. The protrusion limitations are consistent with the general requirements for door encroachment, aisles and exit discharge (see Sections 1005.2, 1013.4 and 1020.3).

Many egress courts are significantly larger than required. Thus, the code allows such an egress court to decrease in width along the path of travel to the public way. The gradual transition requirement is so the flow of the occupants will be uniform without pockets of congestion. The transition requirements should be applied to egress courts where a reduction results in a width that is near the minimum, based on the number of occupants served. It is this condition where the uniform flow of occupants is essential.

1023.5.2 Construction and openings. Where an egress court serving a building or portion thereof is less than 10 feet (3048 mm) in width, the egress court walls shall be not less than 1-hour fire-resistance-rated exterior walls complying with Section 704 of the *International Building Code* for a distance of 10 feet (3048 mm) above the floor of the court, and openings therein shall be equipped with fixed or self-closing, $^3/_4$-hour opening protective assemblies.

Exceptions:

1. Egress courts serving an occupant load of less than 10.
2. Egress courts serving Group R-3 as applicable in Section 1001.1.

❖ The purpose of this section is to protect the occupants served by the egress court from the building that they are exiting from. If occupants must walk closely by the exterior walls of the exit court, the walls are required to have the specified fire-resistance rating and the openings are required to be protected as specified. This requirement is only for the first 10 feet (3048 mm) above the level of the egress court, since the exposure hazard from walls and openings above this level is reduced.

Exceptions 1 and 2 provide for egress courts that serve a very low number of occupants and the specified residential occupancy where the protection requirement would be excessive.

1023.6 Access to a public way. The exit discharge shall provide a direct and unobstructed access to a public way.

Exception: Where access to a public way cannot be provided, a safe dispersal area shall be provided where all of the following are met:

1. The area shall be of a size to accommodate at least 5 square feet (0.28 m^2) for each person.
2. The area shall be located on the same property at least 50 feet (15 240 mm) away from the building requiring egress.
3. The area shall be permanently maintained and identified as a safe dispersal area.
4. The area shall be provided with a safe and unobstructed path of travel from the building.

❖ There are instances where the path of travel to the public way is not safe or, due to site constraints or security concerns, is not achievable. The provisions in this section specify what would constitute a safe area to allow occupants of a building to assemble in an emergency. The 5 square feet (.5 m^2) would allow adequate space for standing persons as well as some space for persons in wheelchairs or on stretchers.

[B] SECTION 1024 ASSEMBLY

1024.1 General. Occupancies in Group A which contain seats, tables, displays, equipment or other material shall comply with this section.

❖ Assembly spaces that contain elements that would affect the path of travel for the means of egress shall comply with this section. Assembly spaces require special consideration due to the large occupant loads and possible low lighting (e.g., nightclubs or theaters).

1024.1.1 Bleachers. Bleachers, grandstands, and folding and telescopic seating shall comply with the ICC 300.

❖ On February 24, 1999, the Bleacher Safety Act of 1999 was introduced in the U.S. House of Representatives. The bill, which cites the International Code Council® (ICC®) and the IBC, authorized the Consumer Product Safety Commission (CPSC) to issue a standard for bleacher safety. This was in response to concerns relative to accidents on bleacher-type structures. As a result, the CPSC developed and revised the *Guidelines for Retrofitting Bleachers*. The ICC Board of Directors decided that a comprehensive standard dealing with all aspects of both new and existing bleachers was warranted and authorized the formation of the ICC Consensus Committee on Bleacher Safety. The committee is comprised of 12 members, including the requisite balance of general, user interest and producer interest.

The ICC *Standard on Bleachers, Folding and Telescopic Seating and Grandstands* was completed in December 2001 and submitted to ANSI on January 1, 2002. While the term "bleachers" is generic, the standard addresses all aspects of tiered seating associated with bleachers, grandstands and folding telescopic seating. The bleacher standard references Chapter 11 of the code and ICC A117.1 for accessibility requirements.

1024.2 Assembly main exit. Group A occupancies that have an occupant load of greater than 300 shall be provided with a main exit. The main exit shall be of sufficient width to accommodate not less than one-half of the occupant load, but such width shall not be less than the total required width of all means of egress leading to the exit. Where the building is classified as a Group A occupancy, the main exit shall front on at least one street or an unoccupied space of not less than 10 feet (3048 mm) in width that adjoins a street or public way.

Exception: In assembly occupancies where there is no well-defined main exit or where multiple main exits are provided, exits shall be permitted to be distributed around the perimeter of the building provided that the total width of egress is not less than 100 percent of the required width.

❖ Assembly buildings present an unusual life safety problem, which includes frequent high occupant densities, and therefore, large occupant loads and the opportunity for irrational mass response to a perceived emergency (i.e., panic). For this reason, the code requires a specific arrangement of the exits. Studies have indicated that in any emergency, occupants will tend to egress using the same path of travel used to enter the room and building. Therefore, a main entrance to the building must also be designed as the main exit to accommodate this behavior, even if the required exit capacity might be more easily accommodated elsewhere. The main entrance (and exit) must be sized to accommodate at least 50 percent of the total occupant load of the structure and must front on a large, open space, such as a street, for rapid dispersal of the occupants outside the building. The remaining exits must also accommodate at least 50 percent of the total occupant load from each level (see Figure 1024.2). The total occupant load includes the occupants within the theater seating area, the foyer and of any other space (e.g., ticket booth, concession stand, offices, storage and the like).

The required width of the means of egress in places of assembly is more often determined by the occupant load than in most other occupancies. In other occupancies, the minimum required widths and the travel distances will often determine the required widths of the exits.

It should be noted that this section only requires the main exit to accommodate 50 percent of the occupant load when there is a single main entrance. Therefore, a large stadium or civic center, in which there are numerous entrances (and exits), need not comply with the main entrance criteria. This condition is addressed in the exception.

1024.3 Assembly other exits. In addition to having access to a main exit, each level of an occupancy in Group A having an occupant load of greater than 300 shall be provided with additional exits that shall provide an egress capacity for at least one-half of the total occupant load served by that level and comply with Section 1014.2.

Exception: In assembly occupancies where there is no well-defined main exit or where multiple main exits are provided, exits shall be permitted to be distributed around the perimeter of the building provided that the total width of egress is not less than 100 percent of the required width.

❖ This section provides for the egress of one-half of the total occupant load by way of exits other than the main exit that is described in Section 1024.2. The exception addresses a large stadium or civic center in which there are numerous entrances (and exits), none of which are a main entrance or main exit.

1024.4 Foyers and lobbies. In Group A-1 occupancies, where persons are admitted to the building at times when seats are not available and are allowed to wait in a lobby or similar space, such use of lobby or similar space shall not encroach upon the required clear width of the means of egress. Such waiting areas shall be separated from the required means of egress by substantial permanent partitions or by fixed rigid railings not less than 42 inches (1067 mm) high. Such foyer, if not directly connected to a public street by all the main entrances or exits, shall have a straight and unobstructed corridor or path of travel to every such main entrance or exit.

❖ In every case, the main entrance (which can also serve as an exit) and all other exits are to be constantly available for the entire building occupant load.

For example, because of the queuing of large crowds, particularly in theaters where a performance may be in progress and people must wait to attend the next one, standing space is often provided. For reasons of safety, such spaces cannot be located in or interfere with established paths of egress from the assembly areas. It is required to designate these areas using partitions or railings so that the means of egress remains clear (see Figure 1024.2).

1024.5 Interior balcony and gallery means of egress. For balconies or galleries having a seating capacity of over 50 located in Group A occupancies, at least two means of egress shall be provided, one from each side of every balcony or gallery, with at least one leading directly to an exit.

❖ This section states the threshold where two means of egress are required based on the occupant load of the interior balcony. Note that one of the means of egress must lead directly to an exit. It is not allowed to go through any other spaces such as an adjoining room prior to reaching an exit such as an exterior exit doorway. However, Section 1024.5.1 does not require the stairways serving the balcony to be enclosed in an exit enclosure. These requirements will ensure that at least one path of travel is always available and faces a minimum number of hazards.

For balconies with 50 or fewer occupants, see Section 1024.8.

1024.5.1 Enclosure of balcony openings. Interior stairways and other vertical openings shall be enclosed in a vertical exit enclosure as provided in Section 1019.1, except that stairways are permitted to be open between the balcony and the main assembly floor in occupancies such as theaters, churches and auditoriums. At least one accessible means of egress is required from a balcony or gallery level containing accessible seating locations in accordance with Section 1007.3 or 1007.4.

❖ This section allows the stairways that lead from interior balconies to the main floor to be unenclosed where the interior balconies are within theaters, churches and auditoriums. Thus, vertical exit enclosures are not required for interior balconies for these facilities. When balconies or galleries contain accessible wheelchair spaces (see Section 1108.2 of the IBC), at least one means of egress must be accessible. While the section references only indicate exit stairways or elevators (see Sections 1007.3 and 1007.4), there are special allowances for the use of platform lifts (see Sections 1007.5 and Section 1109.7, Item 2 of the IBC) in assembly spaces to allow for dispersion of wheelchair spaces to a variety of locations. This is especially important in assembly spaces with sloped or tiered seating. Section 1007.5 states that if a platform lift is permitted as part of an accessible route, it should also be permitted as part of the accessible means of egress.

1024.6 Width of means of egress for assembly. The clear width of aisles and other means of egress shall comply with Section 1024.6.1 where smoke-protected seating is not provided and with Section 1024.6.2 or 1024.6.3 where smoke-protected seating is provided. The clear width shall be measured to walls, edges of seating and tread edges except for permitted projections.

❖ The means of egress width for assembly occupancy is to be in accordance with this section and the referenced sections instead of the criteria specified in Section 1005.1. Different means of egress width criteria are also

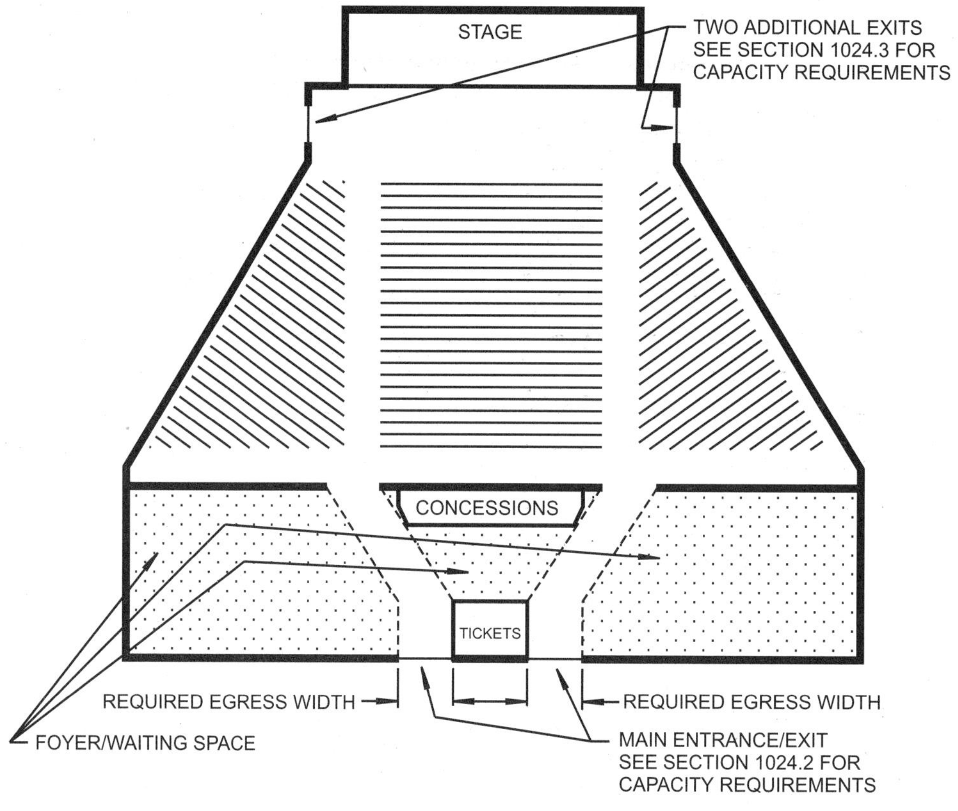

Figure 1024.2
GROUP A-1 WITH WAITING SPACE

specified for assembly seating where smoke protection is provided versus areas it is not provided. The egress width for smoke-protected seating is allowed to be less than for areas where smoke protection is not provided, since the smoke level is required to be maintained at least 6 feet (1829 mm) above the floor of the means of egress according to Section 1024.6.2.1.

1024.6.1 Without smoke protection. The clear width of the means of egress shall provide sufficient capacity in accordance with all of the following, as applicable:

1. At least 0.3 inch (7.6 mm) of width for each occupant served shall be provided on stairs having riser heights 7 inches (178 mm) or less and tread depths 11 inches (279 mm) or greater, measured horizontally between tread nosing.
2. At least 0.005 inch (0.127 mm) of additional stair width for each occupant shall be provided for each 0.10 inch (2.5 mm) of riser height above 7 inches (178 mm).
3. Where egress requires stair descent, at least 0.075 inch (1.9 mm) of additional width for each occupant shall be provided on those portions of stair width having no handrail within a horizontal distance of 30 inches (762 mm).
4. Ramped means of egress, where slopes are steeper than one unit vertical in 12 units horizontal (8-percent slope), shall have at least 0.22 inch (5.6 mm) of clear width for each occupant served. Level or ramped means of egress, where slopes are not steeper than one unit vertical in 12 units horizontal (8-percent slope), shall have at least 0.20 inch (5.1 mm) of clear width for each occupant served.

❖ This section prescribes the criteria needed to calculate the clear widths of aisles and aisle accessways in order to provide sufficient capacity to handle the occupant loads established by the "catchment areas" described in Section 1024.9.2. Clear width is to be measured to walls, edges of seating and tread edges.

The criteria for determining the required widths are based on analytical studies and field tests that used people to model egress situations [see Figures 1024.6.1(1) and 1024.6.1(2)].

Criterion 1 addresses the method for determining the required egress width for aisles and aisle accessways that are stepped. This method corresponds with the requirements of Table 1005.1 for egress width per occupant of stairways in an unsprinklered building.

Criterion 2 addresses the method for determining the additional stair width required for aisle and aisle accessway stairs with risers greater than 7 inches (178 mm).

Criterion 3 addresses the method for determining the additional stair width where a handrail is not located within 30 inches (762 mm).

Criterion 4 addresses the method for determining the required widths for level or ramped means of egress.

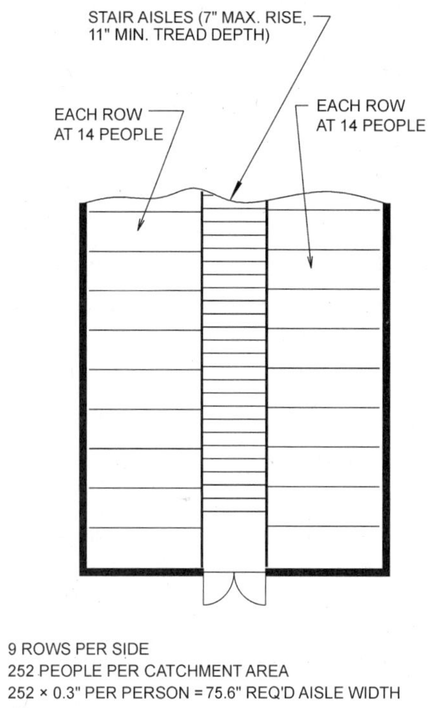

For SI: 1 inch = 25.4 mm.

Figure 1024.6.1(1)
COMPUTATION OF WIDTH FOR STEPPED AISLES

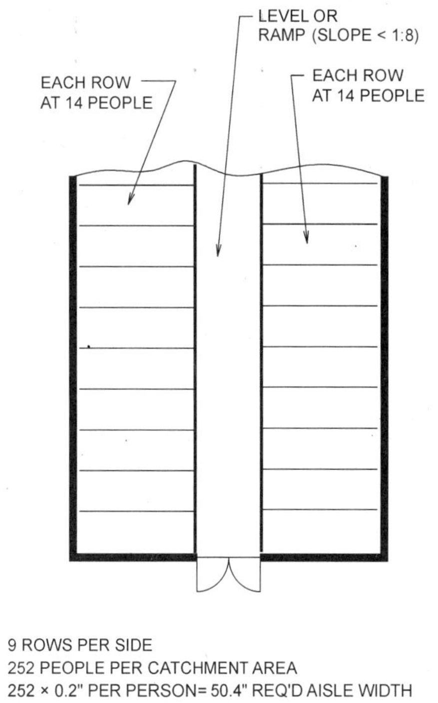

For SI: 1 inch = 25.4 mm.

Figure 1024.6.1(2)
COMPUTATION OF WIDTH FOR LEVEL AISLES OR FOR AISLES WITH RAMPS

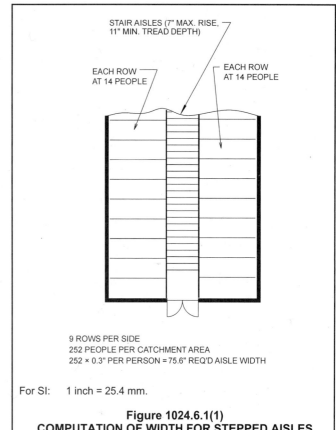

1024.6.2 Smoke-protected seating. The clear width of the means of egress for smoke-protected assembly seating shall be not less than the occupant load served by the egress element multiplied by the appropriate factor in Table 1024.6.2. The total number of seats specified shall be those within a single assembly space and exposed to the same smoke-protected environment. Interpolation is permitted between the specific values shown. A life safety evaluation, complying with NFPA 101, shall be done for a facility utilizing the reduced width requirements of Table 1024.6.2 for smoke-protected assembly seating.

Exception: For an outdoor smoke-protected assembly with an occupant load not greater than 18,000, the clear width shall be determined using the factors in Section 1024.6.3.

❖ Special consideration is given to facilities with features that will prevent the means of egress from being blocked by smoke. Facilities to be considered smoke protected by Sections 1024.6.2.1 through 1024.6.2.3 are permitted increases in travel distance, egress capacity, longer dead-end aisles and increased row lengths. All of these increases result in an increase of allowable egress time. Typically, model codes, based on research by Dr. John Fruin and others, recognize the need for occupants exposed to the fire environment to evacuate to a safe area within 90 seconds of notification and to reach an area of refuge within 5 minutes. With the increases permitted for smoke-protected facilities, these times are effectively doubled, since the time available for safe egress also increases.

The exception is a pointer to the specific criteria for outdoor seating areas. For outdoor stadiums with 18,000 seats or more, use Table 1024.6.2.

TABLE 1024.6.2. See below.

❖ This section requires the egress component to be of adequate size to accommodate the occupant load. The egress width per occupant for nonsmoke-protected seating is to be based on Section 1024.6.1 and is similar to the provisions in Table 1005.1. For smoke-protected seating, the egress width per occupant is based on Table 1024.6.2.

1024.6.2.1 Smoke control. Means of egress serving a smoke-protected assembly seating area shall be provided with a smoke control system complying with Section 909 or natural ventilation designed to maintain the smoke level at least 6 feet (1829 mm) above the floor of the means of egress.

❖ The means of egress and the assembly seating area are required to have some type of smoke control system that will prevent smoke buildup from encroaching on the egress path. This may be a mechanical smoke control system, designed in accordance with Section 909 or a natural ventilation system.

In either type of system, the major consideration is that a smoke-free environment be maintained at least 6 feet (1829 mm) above the floor of the means of egress for a period of at least 20 minutes.

1024.6.2.2 Roof height. A smoke-protected assembly seating area with a roof shall have the lowest portion of the roof deck not less than 15 feet (4572 mm) above the highest aisle or aisle accessway.

Exception: A roof canopy in an outdoor stadium shall be permitted to be less than 15 feet (4572 mm) above the highest aisle or aisle accessway provided that there are no objects less than 80 inches (2032 mm) above the highest aisle or aisle accessway.

❖ One element of a smoke-protected assembly seating facility is that the lowest portion of the roof is required to be at least 15 feet (4572 mm) above the highest aisle or aisle accessway. The objective of this provision is to have a minimum 6-foot (1829 mm) smoke-free height to accommodate safe egress through the area. The additional 9 feet (2743 mm) of height is to provide a volume of space that will act to dissipate smoke. The measurement of the height is shown in Figures 1024.6.2.2(1) and 1024.6.2.2(2).

TABLE 1024.6.2
WIDTH OF AISLES FOR SMOKE-PROTECTED ASSEMBLY

TOTAL NUMBER OF SEATS IN THE SMOKE-PROTECTED ASSEMBLY OCCUPANCY	INCHES OF CLEAR WIDTH PER SEAT SERVED			
	Stairs and aisle steps with handrails within 30 inches	Stairs and aisle steps without handrails within 30 inches	Passageways, doorways and ramps not steeper than 1 in 10 in slope	Ramps steeper than 1 in 10 in slope
Equal to or less than 5,000	0.200	0.250	0.150	0.165
10,000	0.130	0.163	0.100	0.110
15,000	0.096	0.120	0.070	0.077
20,000	0.076	0.095	0.056	0.062
Equal to or greater than 25,000	0.060	0.075	0.044	0.048

For SI: 1 inch = 25.4 mm.

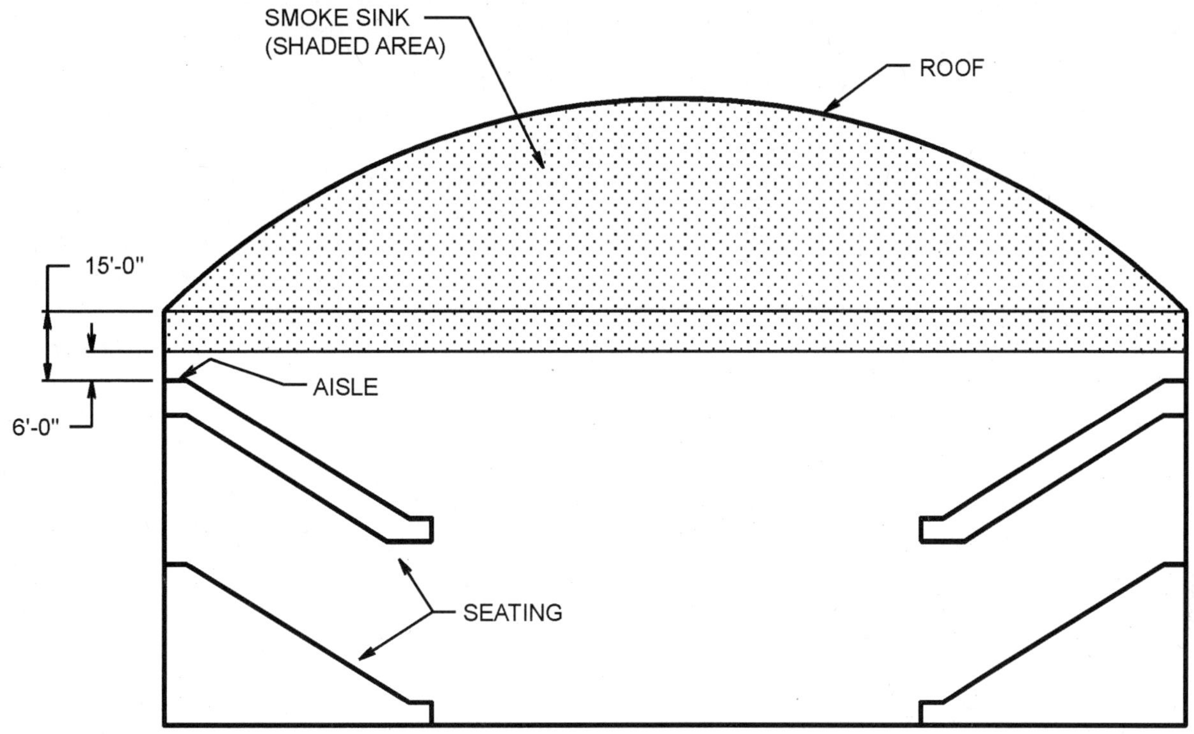

For SI: 1 inch = 25.4 mm, 1 foot = 304.8 mm.

Figure 1024.6.2.2(1)
ROOF HEIGHT (CONVEX DOME ROOF)

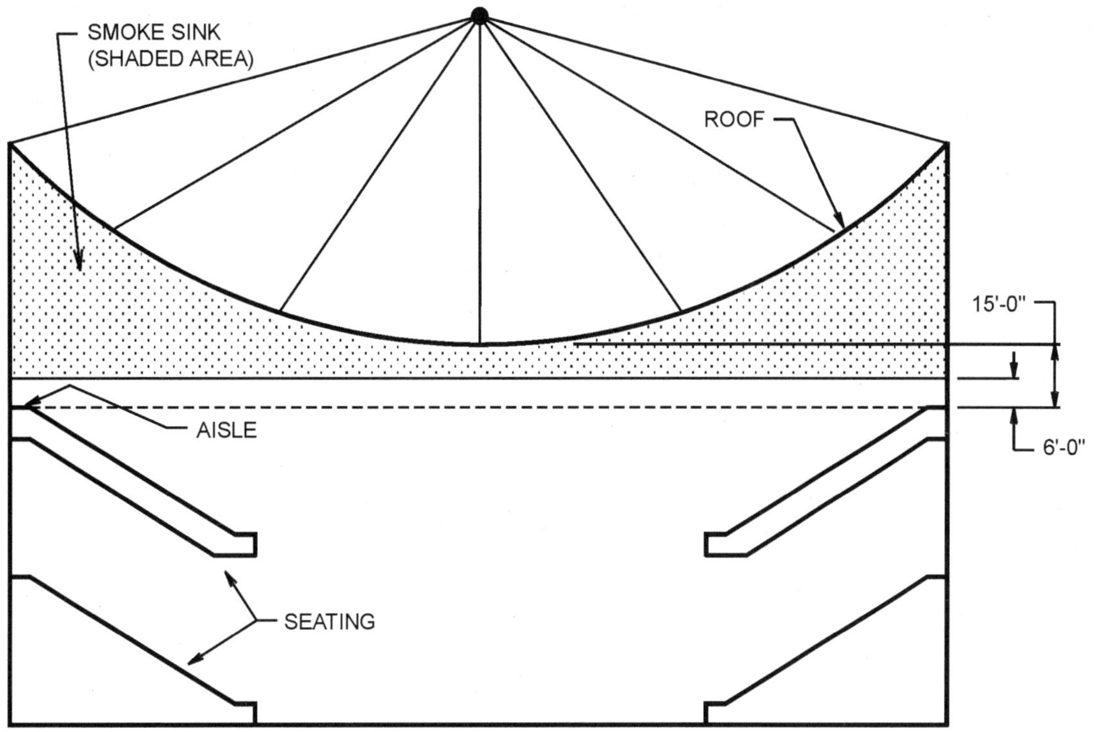

For SI: 1 inch = 25.4 mm, 1 foot = 304.8 mm.

Figure 1024.6.2.2(2)
ROOF HEIGHT (CONCAVE SUSPENDED ROOF)

1024.6.2.3 Automatic sprinklers. Enclosed areas with walls and ceilings in buildings or structures containing smoke-protected assembly seating shall be protected with an approved automatic sprinkler system in accordance with Section 903.3.1.1.

Exceptions:

1. The floor area used for contests, performances or entertainment provided the roof construction is more than 50 feet (15 240 mm) above the floor level and the use is restricted to low fire hazard uses.
2. Press boxes and storage facilities less than 1,000 square feet (93 m^2) in area.
3. Outdoor seating facilities where seating and the means of egress in the seating area are essentially open to the outside.

❖ If there are areas in the smoke-protected assembly seating structure enclosed by walls and ceilings, the entire structure is to be provided with an automatic sprinkler designed to meet the requirements of NFPA 13. NFPA 13R systems are not acceptable for this use.

Exception 1 indicates that the area over the playing field or performance area is not required to be sprinklered if the use of the floor area is restricted. If the facility is used for conventions, trade shows, displays or similar purposes, sprinklers would be required throughout, since the occupancy would no longer be a low-fire-hazard use. A characteristic of a low-fire-hazard occupancy is that the fuel load due to combustibles is approximately 2 pounds per square foot (9.8 kg/m^2) or less.

In order for the contest, performance or entertainment area to be unsprinklered, the roof over that area must be at least 50 feet (15 240 mm) above the floor in addition to the floor area meeting the low-fire-hazard criteria. The 50-foot (15 240 mm) criterion was selected because the response time for sprinklers at this height is extremely slow. It is estimated that the response time for standard sprinklers [50 feet (15 240 mm) above a floor with a fire having a heat release rate of 5 British thermal units (Btu) per square foot per second] exceeds 15 minutes. Therefore, it is not reasonable to install sprinklers at that height with little expectation of timely activation [see Figure 1024.6.2.3(1)]. If this exception is utilized, the trade-offs for a fully sprinklered building, such as increased height and area limitations or decreased corridor ratings, are no longer permitted.

Exception 2 indicates that automatic sprinklers are not required in small spaces in buildings. Sprinklers are required in press box and storage areas of outdoor facilities when the aggregate area exceeds 1,000 square feet (93 m^2). The primary reasons for sprinklers in these areas is that both are anticipated to have a relatively large combustible load when compared to the main seating and participant areas. Additionally, in the case of storage areas, there is an increased potential for an undetected fire condition to occur [see Figure 1024.6.2.3(2)].

Exception 3 provides for outdoor seating facilities where smoke entrapment is not a safety concern.

1024.6.3 Width of means of egress for outdoor smoke-protected assembly. The clear width in inches (mm) of aisles and other means of egress shall be not less than the total occupant load served by the egress element multiplied by 0.08 (2.0 mm) where egress is by aisles and stairs and multiplied by 0.06 (1.52 mm) where egress is by ramps, corridors, tunnels or vomitories.

Exception: The clear width in inches (mm) of aisles and other means of egress shall be permitted to comply with Section 1024.6.2 for the number of seats in the outdoor smoke-protected assembly where Section 1024.6.2 permits less width.

❖ This section has the coefficients for the determination of the width of egress required for outdoor smoke-protected assembly areas. Note that the coefficients are very low compared to the values in Section 1024.6.1 for assembly areas without smoke protection. The coefficients are also lower than those for smoke-protected assembly seating in Table 1024.6.2, except for very large assembly areas. The exception in this section would apply where the coefficients in Table 1024.6.2 are less than those in this section.

Low coefficients are a result of the very low hazard of outdoor smoke-protected assembly areas.

Note that generally an outdoor assembly area meets the smoke control requirements of Section 1024.6.1 by natural ventilation and does not require an automatic sprinkler system according to Section 1024.6.3, Exception 3.

1024.7 Travel distance. Exits and aisles shall be so located that the travel distance to an exit door shall not be greater than 200 feet (60 960 mm) measured along the line of travel in nonsprinklered buildings. Travel distance shall not be more than 250 feet (76 200 mm) in sprinklered buildings. Where aisles are provided for seating, the distance shall be measured along the aisles and aisle accessway without travel over or on the seats.

Exceptions:

1. Smoke-protected assembly seating: The travel distance from each seat to the nearest entrance to a vomitory or concourse shall not exceed 200 feet (60 960 mm). The travel distance from the entrance to the vomitory or concourse to a stair, ramp or walk on the exterior of the building shall not exceed 200 feet (60 960 mm).
2. Open-air seating: The travel distance from each seat to the building exterior shall not exceed 400 feet (122 m). The travel distance shall not be limited in facilities of Type I or II construction.

❖ This section includes the travel distance limits for an assembly occupancy, which are the same as those in Table 1015.1. The travel distance is to be measured in the same path as the occupants would normally take to exit the facility.

Exception 1 provides an extended travel distance for smoke-protected assembly seating that meets the requirements of Sections 1024.6.1 through 1024.6.3. Exception 2 applies to outdoor open-air seating areas where the smoke and fire hazard is very low. The Type I and II construction referred to in this exception is described in Section 602.

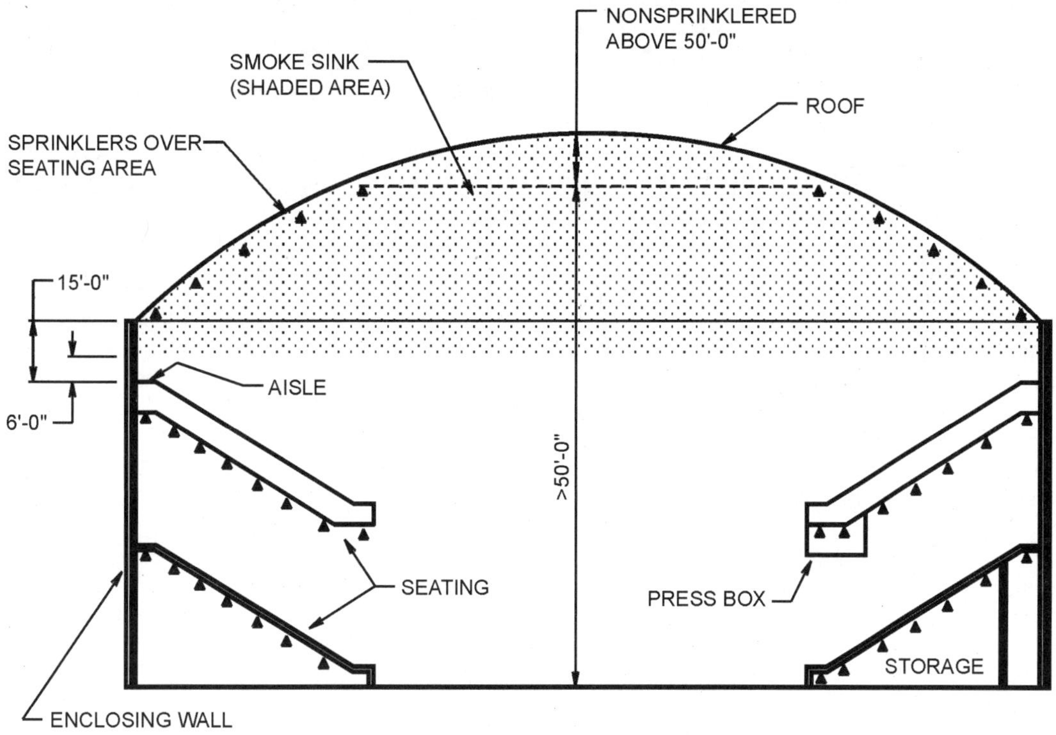

For SI: 1 inch = 25.4 mm, 1 foot = 304.8 mm.

Figure 1024.6.2.3(1)
INDOOR SEATING SPRINKLERED AREAS (CONVEX DOME ROOF)

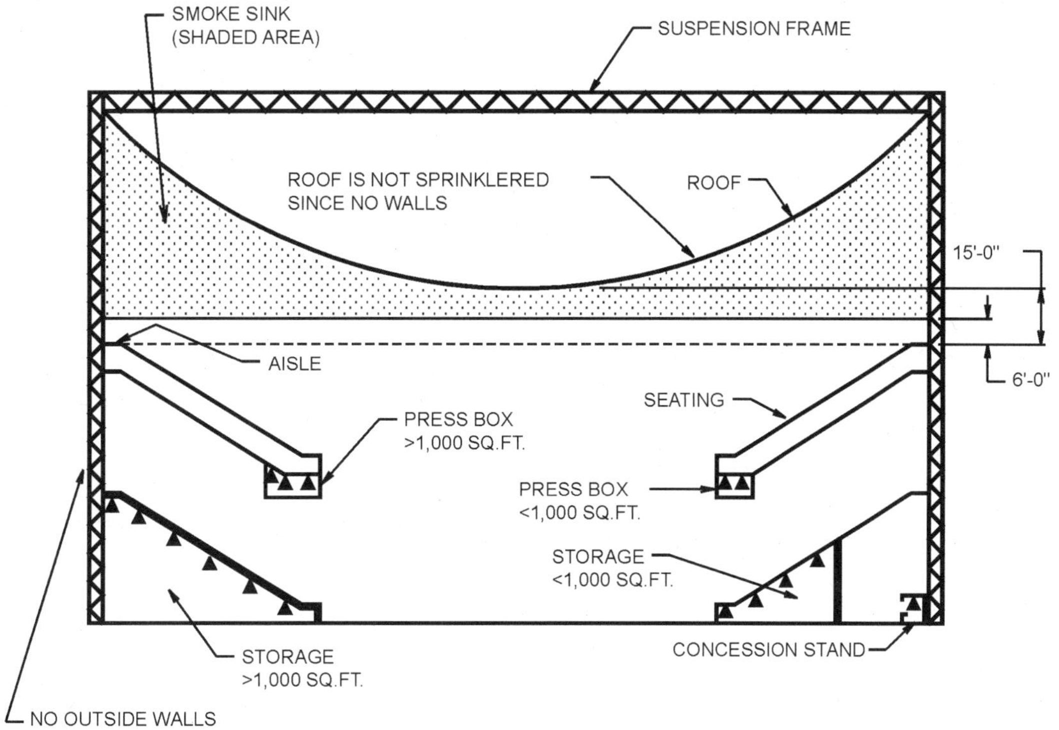

For SI: 1 inch = 25.4 mm, 1 foot = 304.8 mm,
1 square foot = .0929 m^2.

Figure 1024.6.2.3(2)
OUTDOOR SEATING (CONCAVE SUSPENDED ROOF)

1024.8 Common path of travel. The common path of travel shall not exceed 30 feet (9144 mm) from any seat to a point where a person has a choice of two paths of egress travel to two exits.

Exceptions:

1. For areas serving not more than 50 occupants, the common path of travel shall not exceed 75 feet (22 860 mm).
2. For smoke-protected assembly seating, the common path of travel shall not exceed 50 feet (15 240 mm).

❖ The maximum travel distance down a single access row of seating to a location where a patron would have two choices for a way out of the space is 30 feet (9144 mm). In smoke-protected seating, the common path of travel can be up to 50 feet (15 240 mm).

If the room or space (e.g., box, gallery or balcony) has 50 or fewer occupants, the travel distance can be increased to 75 feet (22 860 mm). For example, this allows for a path of travel from a box seat, out of the box and to a main aisle or even a corridor located outside the assembly room itself. When this section is referenced for accessible means of egress (see Section 1007.1, Exception 3), the utilization of Exception 1 would include the entire occupant load of the box, gallery or balcony, not just the number of wheelchair spaces or companion seats.

1024.8.1 Path through adjacent row. Where one of the two paths of travel is across the aisle through a row of seats to another aisle, there shall be not more than 24 seats between the two aisles, and the minimum clear width between rows for the row between the two aisles shall be 12 inches (305 mm) plus 0.6 inch (15.2 mm) for each additional seat above seven in the row between aisles.

Exception: For smoke-protected assembly seating there shall not be more than 40 seats between the two aisles and the minimum clear width shall be 12 inches (305 mm) plus 0.3 inch (7.6 mm) for each additional seat.

❖ In establishing the point where the occupants of a row served by a single access aisle have two distinct paths of travel, the code allows one of those paths to be through the rows of an adjacent seating area or section. This requirement increases the row widths for the single access seating section and the adjacent dual access seating section. This allows the occupants to either travel down the single access aisle or readily traverse the oversized row widths to gain access to a second means of egress. The exception allows a greater number of seats for smoke-protected assembly seating that should be spaced with a minimum clearance of 12 inches (305 mm) in accordance with Sections 1024.6.2.1 through 1024.6.2.3 or Section 1024.6.3. For the base width requirements for single and dual access rows, see the commentary to Sections 1024.10 through 1024.10.2.

1024.9 Assembly aisles are required. Every occupied portion of any occupancy in Group A that contains seats, tables, displays, similar fixtures or equipment shall be provided with aisles leading to exits or exit access doorways in accordance with this section. Aisle accessways for tables and seating shall comply with Section 1013.4.2.

❖ This section requires that each assembly area have designated aisles. For aisle accessway requirements, see Section 1024.10. Assembly area aisle accessways between tables and chairs are to comply with the width requirements in Section 1013.4.2.

1024.9.1 Minimum aisle width. The minimum clear width of aisles shall be as shown:

1. Forty-eight inches (1219 mm) for aisle stairs having seating on each side.

   Exception: Thirty-six inches (914 mm) where the aisle does not serve more than 50 seats.

2. Thirty-six inches (914 mm) for aisle stairs having seating on only one side.
3. Twenty-three inches (584 mm) between an aisle stair handrail or guard and seating where the aisle is subdivided by a handrail.
4. Forty-two inches (1067 mm) for level or ramped aisles having seating on both sides.

   Exceptions:

   1. Thirty-six inches (914 mm) where the aisle does not serve more than 50 seats.
   2. Thirty inches (762 mm) where the aisle does not serve more than 14 seats.

5. Thirty-six inches (914 mm) for level or ramped aisles having seating on only one side.

   Exception: Thirty inches (762 mm) where the aisle does not serve more than 14 seats.

6. Twenty-three inches (584 mm) between an aisle stair handrail and seating where an aisle does not serve more than five rows on one side.

❖ The clear widths of aisles and other means of egress established by the formulas given in Section 1024.6 must not be less than the minimum width requirements of this section. The development of minimum width requirements is based on the association of aisle capacity with the path of exit travel as influenced by the different features of aisle construction. The purpose is to create an aisle system that would provide an even flow of occupant egress.

The minimum width of the aisles is also based on an anticipated movement of people in two directions.

1024.9.2 Aisle width. The aisle width shall provide sufficient egress capacity for the number of persons accommodated by the catchment area served by the aisle. The catchment area served by an aisle is that portion of the total space that is served by that section of the aisle. In establishing catchment areas, the assumption shall be made that there is a balanced use of all means of egress, with the number of persons in proportion to egress capacity.

❖ The determination of required aisle and aisle accessway width is a function of the occupant load. In calculating the required widths, the assumption is that in a system or network of aisles and aisle accessways serving an occupied area, the people will normally exit the area in a way that will distribute the occupant load throughout the system in proportion to the egress capacity of the aisles and aisle accessways. Each aisle and aisle accessway would take its tributary share (catchment area) of the total occupant load (see Figure 1024.9.2).

It should be noted that in addition to the provisions in this section, the requirement for the capacity of the main exit and other exits must also be considered (see Section 1024.2). While this section assumes an equal distribution, Section 1024.2 requires that where the facility has a main exit, the main exit and the access thereto must be capable of handling 50 percent of the occupant load.

1024.9.3 Converging aisles. Where aisles converge to form a single path of egress travel, the required egress capacity of that path shall not be less than the combined required capacity of the converging aisles.

❖ Where one or more aisles or aisle accessways meet to form a single path of egress travel, that path must be sized to handle the combined occupant capacity of the converging aisles and aisle accessways (see Figure 1024.9.3). The reason for this requirement is to maintain the natural pace of travel all the way to the exits and to minimize the queuing of occupants.

It should be noted that this section requires combining the required occupant capacity of converging aisles and aisle accessways, but not necessarily the required widths. For example, if two 48-inch (1219 mm) aisles converge, the result need not be a 96-inch (2438 mm) aisle unless the 48-inch (1219 mm) width of the aisles is required, based on the requirements of Section 1024.6, for the actual occupant load served. However, if the 48-inch (1219 mm) width is not based on the occupant load but is required to comply with the minimum aisle width requirements of Section 1024.9.1, the resulting aisle width must be capable of serving the total occupant load served by the converging aisles, as determined by Section 1024.6, but not less than the minimum widths of Section 1024.9.1.

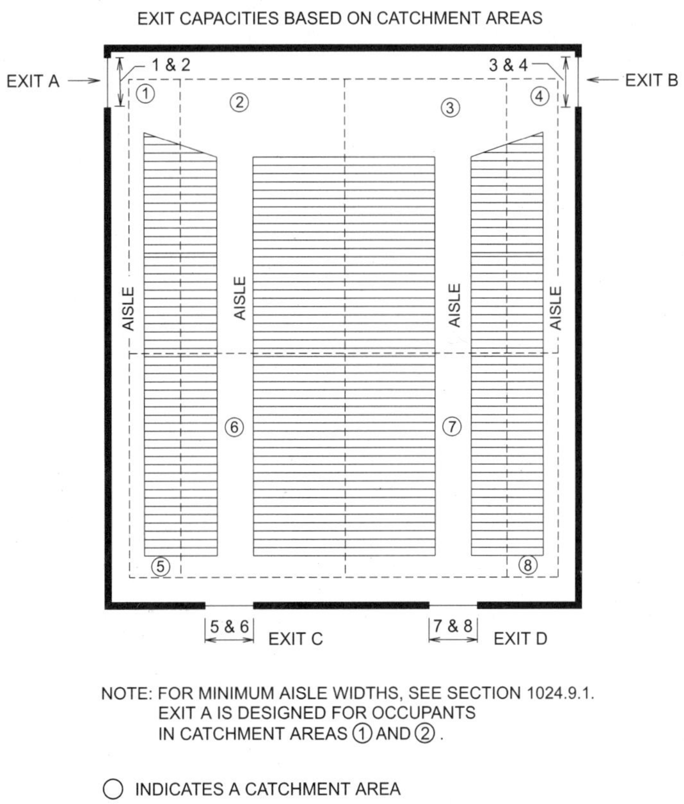

Figure 1024.9.2
TYPICAL AISLE CATCHMENTS

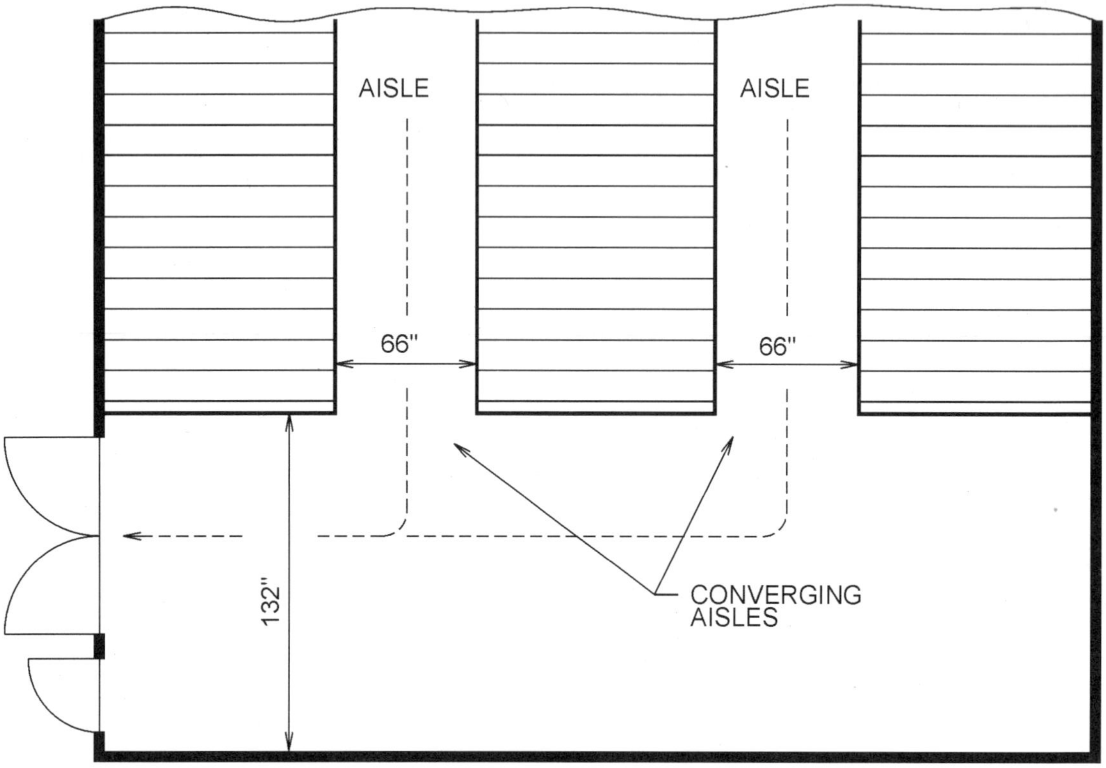

For SI: 1 inch = 25.4 mm.

Figure 1024.9.3
COMPUTATION OF EXIT PASSAGE FOR CONVERGING AISLES

1024.9.4 Uniform width. Those portions of aisles, where egress is possible in either of two directions, shall be uniform in required width.

❖ Aisles that connect or lead to opposite exits must be of uniform width throughout their entire length to allow for exit travel in two directions without creating a traffic bottleneck (see Figure 1024.9.4).

1024.9.5 Assembly aisle termination. Each end of an aisle shall terminate at cross aisle, foyer, doorway, vomitory or concourse having access to an exit.

Exceptions:

1. Dead-end aisles shall not be greater than 20 feet (6096 mm) in length.
2. Dead-end aisles longer than 20 feet (6096 mm) are permitted where seats beyond the 20-foot (6096 mm) dead-end aisle are no more than 24 seats from another aisle, measured along a row of seats having a minimum clear width of 12 inches (305 mm) plus 0.6 inch (15.2 mm) for each additional seat above seven in the row.
3. For smoke-protected assembly seating, the dead-end aisle length of vertical aisles shall not exceed a distance of 21 rows.
4. For smoke-protected assembly seating, a longer dead-end aisle is permitted where seats beyond the 21-row dead-end aisle are not more than 40 seats from another aisle, measured along a row of seats having an aisle accessway with a minimum clear width of 12 inches (305 mm) plus 0.3 inch (7.6 mm) for each additional seat above seven in the row.

❖ Both ends of a cross aisle must terminate at either an intersecting aisle, a foyer, a doorway or a vomitory (lane) that gives access to an exit(s). Dead-end aisles (similar to corridors and passageways) that terminate at one end of a cross aisle or at a foyer, doorway or vomitory must not be more than 20 feet (6096 mm) in length (see Figure 1024.9.5). The intent of the row width requirements in the exceptions is to provide sufficient clear width between rows of seating to allow the occupants in times of emergency to pass quickly from a dead-end aisle to the aisle at the opposite end. In Exception 2, the 0.6-inch (15 mm) increase beyond seven seats is consistent with the minimum width determined in accordance with Section 1024.10.2 for single access rows. The code recognizes that one dead-end aisle may not be usable, thus creating a single access row condition. Exceptions 3 and 4 allow longer dead-end aisles for smoke-protected assembly seating that complies with Sections 1024.6.2.1 through 1024.6.2.3 or Section 1024.6.3.

The overall purpose of this section is to provide aisle/seating arrangements that would allow the occupants to seek safe and rapid passage to exits in case of fire or other emergency.

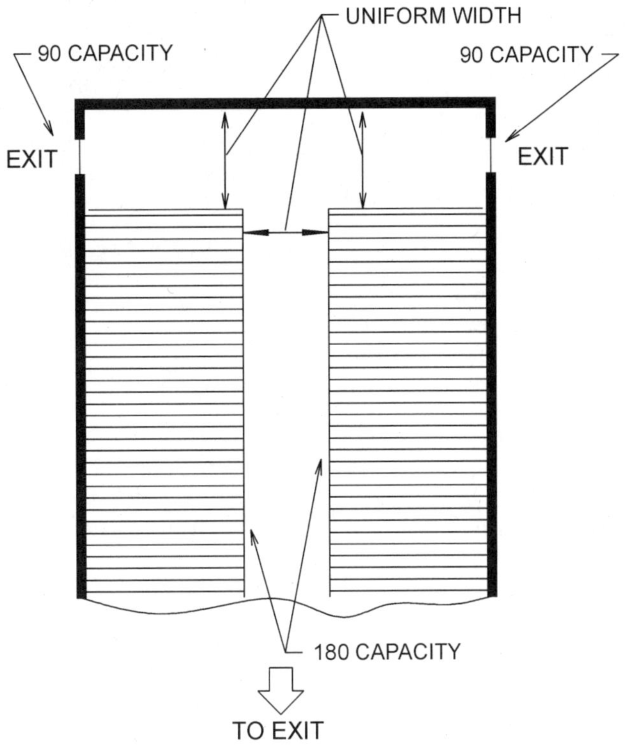

Figure 1024.9.4
AISLE WITH TWO OPPOSITE EXITS

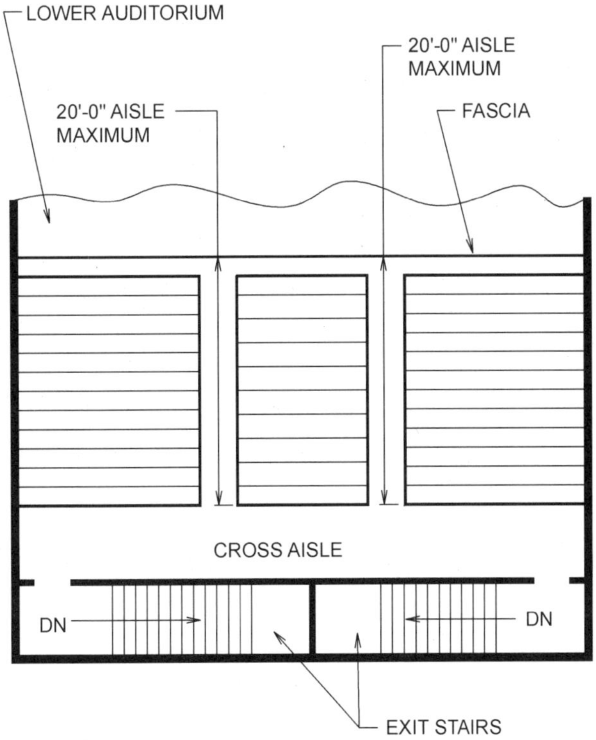

For SI: 1 inch = 25.4 mm, 1 foot = 304.8 mm.

Figure 1024.9.5
TYPICAL BALCONY ARRANGEMENT

1024.9.6 Assembly aisle obstructions. There shall be no obstructions in the required width of aisles except for handrails as provided in Section 1024.13.

❖ Except for handrails, aisles are required to be clear of any obstructions so that the full width is available for egress purposes. Handrails are allowed to project into the required aisle width in the same manner as handrail projections in stairways.

1024.10 Clear width of aisle accessways serving seating. Where seating rows have 14 or fewer seats, the minimum clear aisle accessway width shall not be less than 12 inches (305 mm) measured as the clear horizontal distance from the back of the row ahead and the nearest projection of the row behind. Where chairs have automatic or self-rising seats, the measurement shall be made with seats in the raised position. Where any chair in the row does not have an automatic or self-rising seat, the measurements shall be made with the seat in the down position. For seats with folding tablet arms, row spacing shall be determined with the tablet arm down.

❖ The requirements of this section are applicable to theater-type seating arrangements. This includes both "continental" and "traditional" seating arrangements. Theater-type seating is characterized by a number of seats arranged side by side and in rows. In this type of seating arrangement, the potential exists for a large number of occupants to be present in a confined environment where the ability of the occupants to move is limited. In order to egress, people are required to move within a row before reaching an aisle or aisle accessway and the aisle or aisle accessway also limits movement toward an exit. To provide adequate passage between rows of seats, this section requires that the clear width between the back of a row to the nearest projection of the seating immediately behind must be at least 12 inches (305 mm) (see Figure 1024.10). Where chairs are manufactured with automatic- or self-lifting seats, the minimum width requirement may be measured with the seats in a raised position. When tablet arm chairs are used, the required width is to be determined with the tablet arm in its usable position. It is not acceptable to eliminate the tablet arm from the measurement, even though it is retractable.

With respect to self-rising seats, ASTM F 851 provides one method of determining acceptability.

1024.10.1 Dual access. For rows of seating served by aisles or doorways at both ends, there shall not be more than 100 seats per row. The minimum clear width of 12 inches (305 mm) between rows shall be increased by 0.3 inch (7.6 mm) for every additional seat beyond 14 seats, but the minimum clear width is not required to exceed 22 inches (559 mm).

Exception: For smoke-protected assembly seating, the row length limits for a 12-inch-wide (305 mm) aisle accessway, beyond which the aisle accessway minimum clear width shall be increased, are in Table 1024.10.1.

❖ Where rows of seating are served by aisles or doorways located at both ends of the path of row travel, the number of seats that may be used in a row may be up to but no more than 100 (continental seating). The minimum required clear width aisle accessway of 12 inches (305 mm) between rows of seats must be increased by 0.3 inch (8 mm) for every additional seat beyond 14, but no more than a total of 22 inches (559 mm) (see Figure 1024.10.1). For example, in a row of 24 seats, the minimum clear width would compute to 15 inches (381 mm) [12 + (0.3 by 10)]. For a row of 34 seats, a clear width of 18 inches (457 mm) would be required. Increases in the clear width between rows of seats would occur up to a row of 46 seats. From 47 to 100 seats, a maximum clear width between rows of 22 inches (559 mm) would apply.

Since the row is to provide access to an aisle in both directions, the minimum width applies to the entire length of the row aisle accessway.

The exception allows more seats in a row with the minimum 12-inch (305 mm) seat spacing since safe

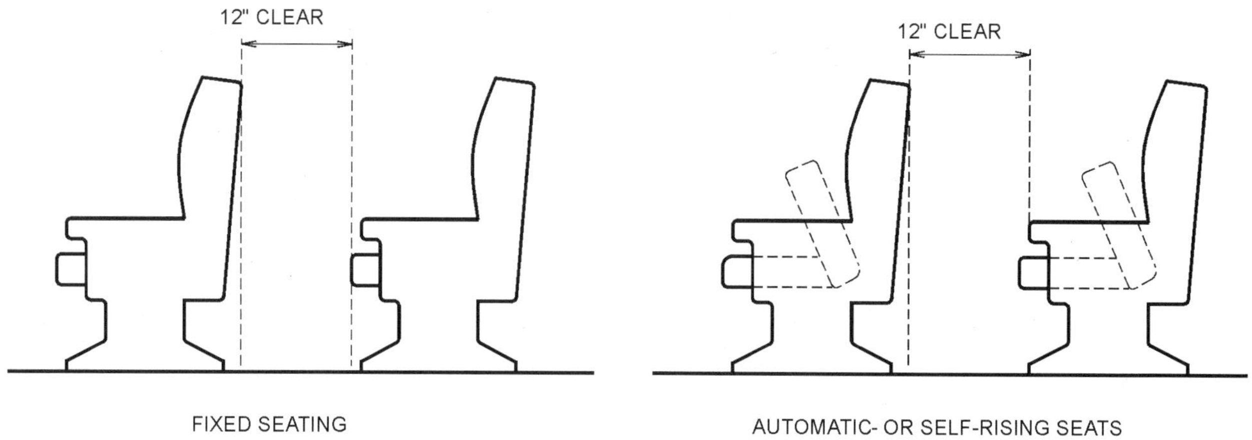

For SI: 1 inch = 25.4 mm.

Figure 1024.10
MINIMUM REQUIRED ROW WIDTH CLEARANCE

egress time is extended for this condition.

For additional aisle accessway width requirements when one of the means of egress at the end of the single access row is through a dual access row, see Section 1024.8.1.

**TABLE 1024.10.1
SMOKE-PROTECTED
ASSEMBLY AISLE ACCESSWAYS**

TOTAL NUMBER OF SEATS IN THE SMOKE-PROTECTED ASSEMBLY OCCUPANCY	MAXIMUM NUMBER OF SEATS PER ROW PERMITTED TO HAVE A MINIMUM 12-INCH CLEAR WIDTH AISLE ACCESSWAY	
	Aisle or doorway at both ends of row	Aisle or doorway at one end of row only
Less than 4,000	14	7
4,000	15	7
7,000	16	8
10,000	17	8
13,000	18	9
16,000	19	9
19,000	20	10
22,000 and greater	21	11

For SI: 1 inch = 25.4 mm.

❖ Table 1024.10.1 recognizes the increased egress time available in smoke-protected assembly seating areas. Therefore, the table permits greater lengths of rows that have the minimum 12 inches (305 mm) of clear width. When a row exceeds the lengths identified in the table, the row width is to be increased in accordance with Section 1024.10.1 [0.3 inch (8 mm) per additional seat] for dual access rows and Section 1024.10.2 [0.6 inch (15 mm) per additional seat] for single access rows. The requirements of this table are based on the total number of seats contained within the assembly space.

1024.10.2 Single access. For rows of seating served by an aisle or doorway at only one end of the row, the minimum clear width of 12 inches (305 mm) between rows shall be increased by 0.6 inch (15.2 mm) for every additional seat beyond seven seats, but the minimum clear width is not required to exceed 22 inches (559 mm).

Exception: For smoke-protected assembly seating, the row length limits for a 12-inch-wide (305 mm) aisle accessway, beyond which the aisle accessway minimum clear width shall be increased, are in Table 1024.10.1.

❖ Where rows of seating are served by an aisle or doorway at only one end of a row, the minimum clear width of 12 inches (305 mm) between rows of seats must be increased by 0.6 inch (15 mm) for every additional seat beyond seven, but not more than a total of 22 inches (559 mm) (see Figure 1024.10.2). While this section does not specify the maximum number of seats permitted in a row, the 30-foot (9144 mm) common path of travel limitation (see Section 1024.8) essentially restricts the single access row to approximately 20 seats, based on an 18-inch (457 mm) width per seat. A row of 12 seats would compute to a required minimum width of 15 inches (381 mm) [12 + (0.5 by 5)]. Similarly, a row of 17 seats would require a clear width of 18 inches (457 mm) and so on. Since dual access is not provided, incremental increases would be permitted in the aisle accessway width as shown in Figure 1024.10.2. Incremental increases in the required width would occur up to the maximum number of seats, which is determined by the 30-foot (9144 mm) dead-end limitation.

The reason for increasing the row accessway widths incrementally, with increases in the number of seats per row, is to provide more efficient passage for the occupants who are using the aisle accessway. As a practical matter, where dual access (see Section 1024.10.1) and

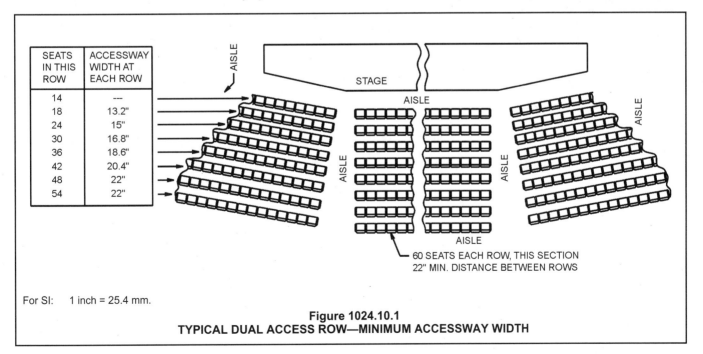

SEATS IN THIS ROW	ACCESSWAY WIDTH AT EACH ROW
14	---
18	13.2"
24	15"
30	16.8"
36	18.6"
42	20.4"
48	22"
54	22"

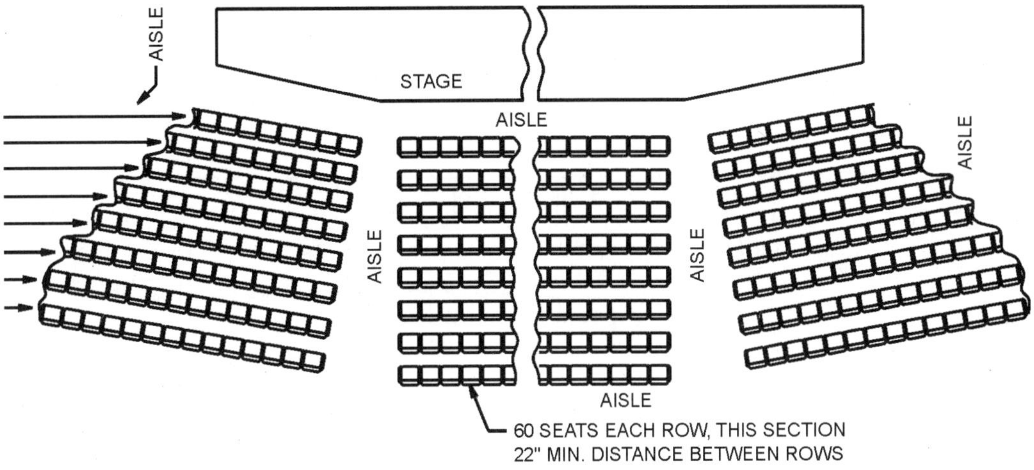

For SI: 1 inch = 25.4 mm.

**Figure 1024.10.1
TYPICAL DUAL ACCESS ROW—MINIMUM ACCESSWAY WIDTH**

single access seating arrangements are used together, the largest computed clear width dimension would normally be applied by the designer to both arrangements so that the rows of seats will be in alignment. For additional aisle accessway width requirements when one of the means of egress at the end of the single access row is through a dual access row, see Section 1024.8.1.

1024.11 Assembly aisle walking surfaces. Aisles with a slope not exceeding one unit vertical in eight units horizontal (12.5-percent slope) shall consist of a ramp having a slip-resistant walking surface. Aisles with a slope exceeding one unit vertical in eight units horizontal (12.5-percent slope) shall consist of a series of risers and treads that extends across the full width of aisles and complies with Sections 1024.11.1 through 1024.11.3.

❖ Assembly facilities such as theaters and auditoriums often require sloping or stepped floors to provide seated occupants with preferred sightlines for viewing presentations. Aisles must, therefore, be designed to accommodate the changing elevations of the floor in such a manner that the path of travel will allow occupants to leave the area at a rapid pace with minimal possibilities for stumbling or falling during times of emergency.

This section requires that aisles with a gradient of one in eight (1:8) or less must consist of a ramp with a slip-resistant surface. Aisles with a gradient exceeding one in eight (1:8) must consist of a series of treads and risers that comply with the requirements of Sections 1024.11.1 through 1024.11.3. Ramps that serve as part of an accessible route to and from accessible wheelchair spaces must comply with the more restrictive requirements for ramps in Section 1010 (see Section 1010.1, Exception 1).

While not specifically indicated for stepped aisles, such floor surfaces must also be slip resistant in accordance with Section 1003.4. Field testing and uniform enforcement of the concept of slip resistance is not practical. One method used to establish slip resistance is that the static coefficient of friction between leather [Type 1 (Vegetable Tanned) of Federal Specification KK-L-165C] and the floor surface is greater than 0.5. Laboratory test procedures can determine the static coefficient of resistance.

What must be recognized here is that stepped aisles are part of the floor construction and are intended to provide horizontal egress. Tread and riser construction for this purpose should not be compared to the requirements for treads and risers in conventional stairways that serve as means of vertical egress. Sometimes, because of design considerations, the gradient of an aisle is required to change from a level floor to a ramp to steps. In cases where there is no uniformity in the path of travel, occupants tend to be considerably more cautious, particularly in the use of stepped aisles, than they would normally be in the use of conventional stairways.

1024.11.1 Treads. Tread depths shall be a minimum of 11 inches (279 mm) and shall have dimensional uniformity.

> **Exception:** The tolerance between adjacent treads shall not exceed 0.188 inch (4.8 mm).

❖ Depths of treads are not to be less than 11 inches (279 mm) and uniform throughout each flight, except that a variance of not more than 0.188 inch (4.8 mm) is permitted between adjacent treads to accommodate variations in construction. While this provision is the same as the limiting dimension for treads in interior stairways (see Section 1009.3), it rarely applies in the construction of stepped aisles. A more common form of stepped aisle construction is to provide a tread depth equal to the back-to-back distance between rows of seats. This way the treads can be extended across the full length of the row and serve as a supporting platform for the seats. Other arrangements might require two treads between rows of seats.

In theaters, for example, the back-to-back distance between rows of fixed seats usually ranges somewhere between 3 and 4 feet (914 and 1219 mm), depending on seat style and seat dimensions as well as the ease of passage between the rows (see Figure 1024.11.1). The

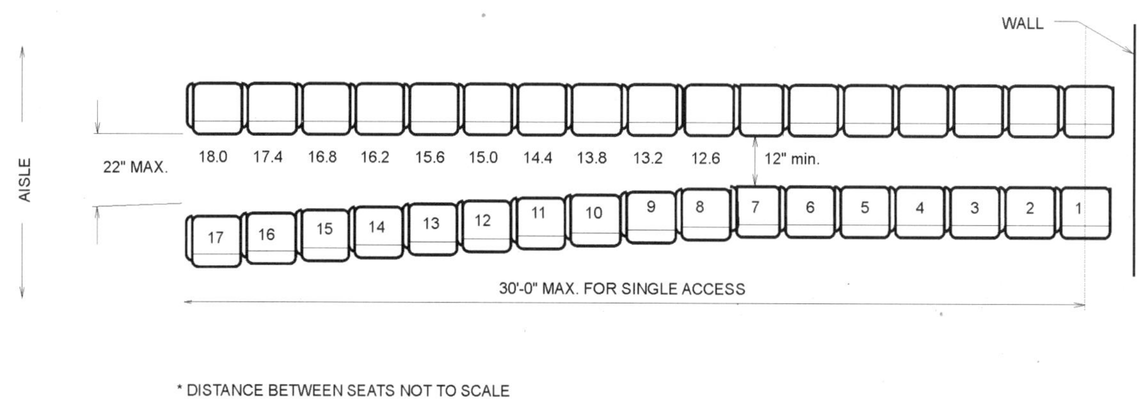

For SI: 1 inch = 25.4 mm.

Figure 1024.10.2
TYPICAL SINGLE ACCESS ROW—MINIMUM ACCESSWAY WIDTH

selection of single-tread or two-tread construction between rows of seats depends on the gradient and suitable riser height (see Section 1024.11.2), as needed for sightlines.

In comparing this section with Section 1024.11.2, it is significant to note the emphasis placed on the tread dimension. While not desirable, the code permits riser heights to deviate; however, tread dimensions must not vary beyond the 0.188-inch (4.8 mm) tolerance.

1024.11.2 Risers. Where the gradient of aisle stairs is to be the same as the gradient of adjoining seating areas, the riser height shall not be less than 4 inches (102 mm) nor more than 8 inches (203 mm) and shall be uniform within each flight.

Exceptions:

1. Riser height nonuniformity shall be limited to the extent necessitated by changes in the gradient of the adjoining seating area to maintain adequate sightlines. Where nonuniformities exceed 0.188 inch (4.8 mm) between adjacent risers, the exact location of such nonuniformities shall be indicated with a distinctive marking stripe on each tread at the nosing or leading edge adjacent to the nonuniform risers. Such stripe shall be a minimum of 1 inch (25 mm), and a maximum of 2 inches (51 mm), wide. The edge marking stripe shall be distinctively different from the contrasting marking stripe.
2. Riser heights not exceeding 9 inches (229 mm) shall be permitted where they are necessitated by the slope of the adjacent seating areas to maintain sightlines.

❖ In stepped aisles where the gradient of the aisle is the same as the gradient of the adjoining seating area, riser heights are not to be less than 4 inches (102 mm) nor more than 8 inches (203 mm) (see Figure 1024.11.2). For the safety of the occupants, risers should have uniform heights, where possible, throughout each flight of steps. However, nonuniformity of riser heights is permitted in cases where changes to the gradient in the adjoining seating area are required because of sightlines and other seating layout considerations.

Where variations in height exceed 0.188 inch (4.8 mm) between adjacent risers, a distinctive marking stripe between 1 and 2 inches (25 to 51 mm) wide is to be located on the nosings of each tread where the variations occur as a visual warning to the occupants to be cautious. Frequently, this is done with "runway" lights. Note that this stripe must be different from the tread contrast marking stripes required in Section 1024.11.3.

In comparing this section with Section 1024.11.1, it is significant to note the emphasis placed on the tread dimension. While not desirable, the code permits riser heights to deviate; however, Section 1024.11.1 does not permit tread dimensions to vary beyond the 0.188-inch (4.8 mm) tolerance.

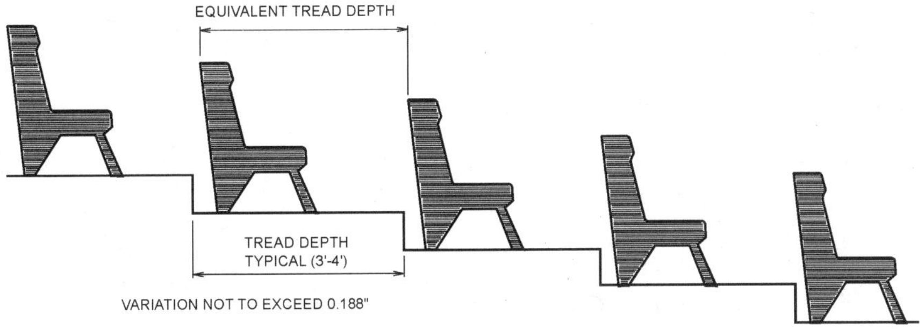

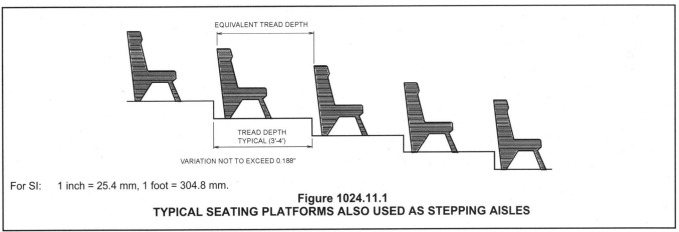

For SI: 1 inch = 25.4 mm, 1 foot = 304.8 mm.

Figure 1024.11.1
TYPICAL SEATING PLATFORMS ALSO USED AS STEPPING AISLES

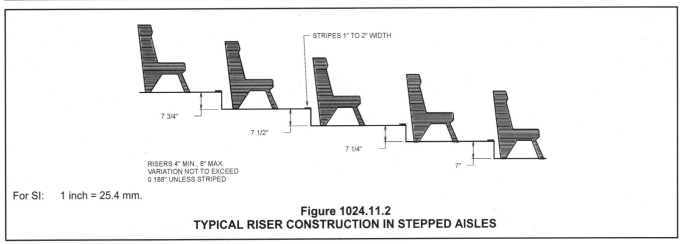

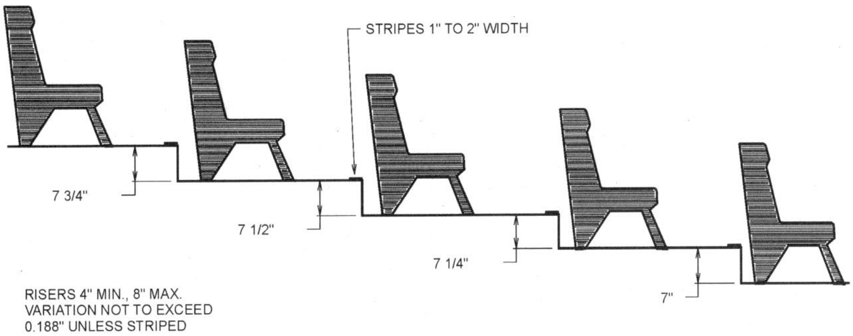

For SI: 1 inch = 25.4 mm.

Figure 1024.11.2
TYPICAL RISER CONSTRUCTION IN STEPPED AISLES

1024.11.3 Tread contrasting marking stripe. A contrasting marking stripe shall be provided on each tread at the nosing or leading edge such that the location of each tread is readily apparent when viewed in descent. Such stripe shall be a minimum of 1 inch (25 mm), and a maximum of 2 inches (51 mm), wide.

Exception: The contrasting marking stripe is permitted to be omitted where tread surfaces are such that the location of each tread is readily apparent when viewed in descent.

❖ The exception provides for the omission of the contrasting marking stripe where the tread is readily apparent, such as when aisle stair treads are provided with a roughened metal nosing strip or where lighted nosings occur. In this situation, the user is aware of the treads without the marking stripe. This stripe must be different from the marking stripe required for nonuniform risers in Section 1024.11.2, Exception 1.

1024.12 Seat stability. In places of assembly, the seats shall be securely fastened to the floor.

Exceptions:

1. In places of assembly or portions thereof without ramped or tiered floors for seating and with 200 or fewer seats, the seats shall not be required to be fastened to the floor.
2. In places of assembly or portions thereof with seating at tables and without ramped or tiered floors for seating, the seats shall not be required to be fastened to the floor.
3. In places of assembly or portions thereof without ramped or tiered floors for seating and with greater than 200 seats, the seats shall be fastened together in groups of not less than three or the seats shall be securely fastened to the floor.
4. In places of assembly where flexibility of the seating arrangement is an integral part of the design and function of the space and seating is on tiered levels, a maximum of 200 seats shall not be required to be fastened to the floor. Plans showing seating, tiers and aisles shall be submitted for approval.
5. Groups of seats within a place of assembly separated from other seating by railings, guards, partial height walls or similar barriers with level floors and having no more than 14 seats per group shall not be required to be fastened to the floor.
6. Seats intended for musicians or other performers and separated by railings, guards, partial height walls or similar barriers shall not be required to be fastened to the floor.

❖ The purpose of this section is to require that assembly seating be fastened to the floor where it would be a significant hazard if they were loose and subject to tipping over. The exceptions allow loose assembly seating for situations where the hazard is lower, such as floors where ramped or tiered seating is not used, where no more than 200 seats are used and for box seating arrangements where a limited number of seats is within railings, guards or partial height walls.

1024.13 Handrails. Ramped aisles having a slope exceeding one unit vertical in 15 units horizontal (6.7-percent slope) and aisle stairs shall be provided with handrails located either at the side or within the aisle width.

Exceptions:

1. Handrails are not required for ramped aisles having a gradient no greater than one unit vertical in eight units horizontal (12.5-percent slope) and seating on both sides.
2. Handrails are not required if, at the side of the aisle, there is a guard that complies with the graspability requirements of handrails.

❖ For the safety of occupants, handrails must be provided in aisles where ramps exceed a gradient of one in 15 (1:15) (see Figure 1024.13).

Exception 1 omits the handrail requirements where ramped aisles are not steep and seats are on both sides to reduce the fall hazard.

Exception 2 allows handrails to be omitted where there is a guard at the side of the aisle with a top rail that complies with the requirements for handrail graspability (see Section 1009.11.3). The guard must meet the height and opening requirements specified in Section 1012 or 1024.14 as applicable.

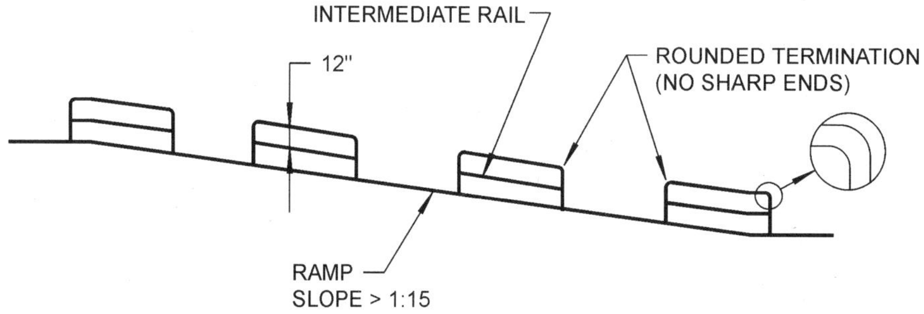

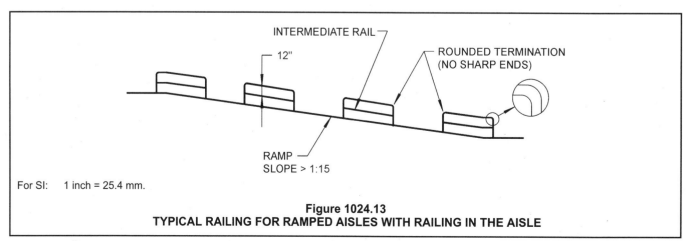

For SI: 1 inch = 25.4 mm.

Figure 1024.13
TYPICAL RAILING FOR RAMPED AISLES WITH RAILING IN THE AISLE

1024.13.1 Discontinuous handrails. Where there is seating on both sides of the aisle, the handrails shall be discontinuous with gaps or breaks at intervals not exceeding five rows to facilitate access to seating and to permit crossing from one side of the aisle to the other. These gaps or breaks shall have a clear width of at least 22 inches (559 mm) and not greater than 36 inches (914 mm), measured horizontally, and the handrail shall have rounded terminations or bends.

❖ Where aisles have seating on both sides, handrails may be located at the sides of the aisles, but are typically located in the center of the aisle. The width of each section of the subdivided aisle between the handrail and the edge of the seating is not to be less than 23 inches (584 mm) (see Section 1024.9.1, Item 3).

For reasons of life safety in fire situations and also as a practical matter in the efficient use of the facility, a handrail down the middle of an aisle should not be continuous along its entire length. Crossovers must be provided by means of gaps or breaks in the handrail installation. Such openings must not be less than 22 inches (559 mm) nor more than 36 inches (914 mm) wide and must be provided at intervals not exceeding the distance of five rows of seats (see Figure 1024.13.1). All handrail terminations should be designed to have rounded ends or bends to avoid possible injury to the occupants (see Figure 1024.13).

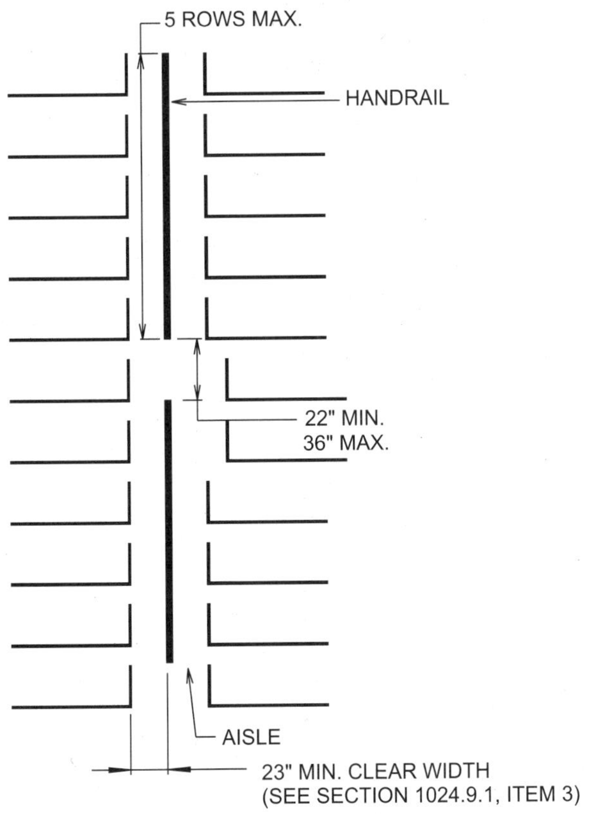

For SI: 1 inch = 25.4 mm.

Figure 1024.13.1
DISCONTINUOUS AISLE HANDRAILS

1024.13.2 Intermediate handrails. Where handrails are provided in the middle of aisle stairs, there shall be an additional intermediate handrail located approximately 12 inches (305 mm) below the main handrail.

❖ Handrail installations down the middle of an aisle must be constructed with intermediate rails located 12 inches (305 mm) below and parallel to main handrails. This is to provide handholds for children and to prevent people from using the handrail like a gym apparatus and possibly injuring themselves and from ducking to get under the rail (see Figure 1024.13).

1024.14 Assembly guards. Assembly guards shall comply with Sections 1024.14.1 through 1024.14.3.

❖ This section establishes the scope of the assembly guard provisions.

1024.14.1 Cross aisles. Cross aisles located more than 30 inches (762 mm) above the floor or grade below shall have guards in accordance with Section 1012.

Where an elevation change of 30 inches (762 mm) or less occurs between a cross aisle and the adjacent floor or grade below, guards not less than 26 inches (660 mm) above the aisle floor shall be provided.

Exception: Where the backs of seats on the front of the cross aisle project 24 inches (610 mm) or more above the adjacent floor of the aisle, a guard need not be provided.

❖ The purpose of this section is to provide for the occupants' safety with guards along elevated cross aisles. The minimum height of the guard is a function of the cross-aisle elevation above the adjacent floor or grade below [i.e., 42 inches (1067 mm) high with more than a 30-inch (762 mm) drop off, and 26 inches (660 mm) high with a 30-inch (762 mm) or less drop off]. When the back of the seats adjacent to the cross aisle are a minimum of 24 inches (610 mm) above the floor level of the cross aisle, they will serve as the guard.

See Figure 1024.14.2 for an illustration of the requirements in this section.

1024.14.2 Sightline-constrained guard heights. Unless subject to the requirements of Section 1024.14.3, a fascia or railing system in accordance with the guard requirements of Section 1012 and having a minimum height of 26 inches (660 mm) shall be provided where the floor or footboard elevation is more than 30 inches (762 mm) above the floor or grade below and the fascia or railing would otherwise interfere with the sightlines of immediately adjacent seating. At bleachers, a guard must be provided where the floor or footboard elevation is more than 24 inches (610 mm) above the floor or grade below and the fascia or railing would otherwise interfere with the sightlines of the immediately adjacent seating.

❖ This section specifies a guard height of 26 inches (660 mm) for guards within assembly seating areas other than at the end of aisles where a vertical 36-inch (914 mm) height is required. This is to provide a reasonable degree of safety while providing for sightlines within the viewing area. The guard opening configuration must comply with Section 1012.3. See Figure 1024.14.2 for an illustration of the requirements in this section.

At bleachers, the maximum required drop off before a guard is 24 inches (610 mm), rather than 30 inches (762 mm). This is consistent with the requirements for guards found in the bleacher standard referenced in Section 1024.1.1; the guard heights and opening requirement is the same. With the lower drop off in bleacher configurations, a guard is required at the locations where cut-ins were made in the lower rows to accommodate wheelchair seating spaces.

1024.14.3 Guards at the end of aisles. A fascia or railing system complying with the guard requirements of Section 1012 shall be provided for the full width of the aisle where the foot of the aisle is more than 30 inches (762 mm) above the floor or grade below. The fascia or railing shall be a minimum of 36 inches (914 mm) high and shall provide a minimum 42 inches (1067 mm) measured diagonally between the top of the rail and the nosing of the nearest tread.

❖ This section applies only at the end of aisles where the foot (the lower end) of the aisle is greater than 30 inches (762 mm) above the adjacent floor or grade below. The guard must satisfy both of the specified height requirements to provide safety for persons at the end of the aisle. The 36-inch (914 mm) minimum height is measured from the floor vertically to the top of the guard. The minimum 42-inch (1067 mm) diagonal dimension from the nosing of the nearest stair tread to the top of the fascia or guard is to provide sufficient height for a fall from the nearest stair tread. See Figure 1024.14.2 for an illustration of the requirements in this section.

1024.15 Bench seating. Where bench seating is used, the number of persons shall be based on one person for each 18 inches (457 mm) of length of the bench.

❖ The purpose of this section is to specify the length of bench for each occupant for bench and bleacher seating. This is commonly used to calculate the occupant load of bench or bleacher seating for egress purposes and is not intended to limit any individual to an 18-inch (457 mm) area. This is consistent with the fixed seating occupant loads indicated in Section 1004.7.

[B] SECTION 1025
EMERGENCY ESCAPE AND RESCUE

1025.1 General. In addition to the means of egress required by this chapter, provisions shall be made for emergency escape and rescue in Group R as applicable in Section 1001.1 and Group I-1 occupancies. Basements and sleeping rooms below the fourth story above grade plane shall have at least one exterior emergency escape and rescue opening in accordance with this section. Where basements contain one or more sleeping rooms, emergency egress and rescue openings shall be required in each

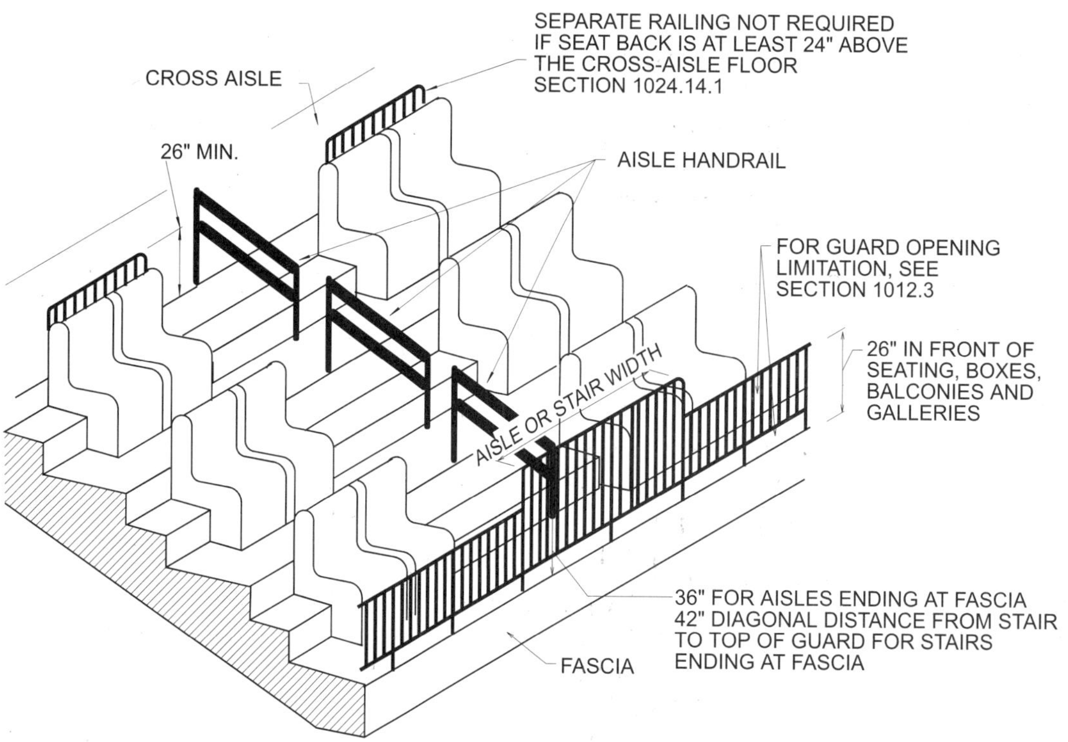

For SI: 1 inch = 25.4 mm.

Figure 1024.14.2
TYPICAL BALCONY GUARDS AT FASCIA

sleeping room, but shall not be required in adjoining areas of the basement. Such opening shall open directly into a public street, public alley, yard or court.

Exceptions:

1. In other than Group R-3 occupancies as applicable in Section 1001.1, buildings equipped throughout with an approved automatic sprinkler system in accordance with Section 903.3.1.1 or 903.3.1.2.
2. In other than Group R-3 occupancies as applicable in Section 1001.1, sleeping rooms provided with a door to a fire-resistance-rated corridor having access to two remote exits in opposite directions.
3. The emergency escape and rescue opening is permitted to open onto a balcony within an atrium in accordance with the requirements of Section 404 of the *International Building Code*, provided the balcony provides access to an exit and the dwelling unit or sleeping unit has a means of egress that is not open to the atrium.
4. Basements with a ceiling height of less than 80 inches (2032 mm) shall not be required to have emergency escape and rescue windows.
5. High-rise buildings in accordance with Section 403 of the *International Building Code*.
6. Emergency escape and rescue openings are not required from basements or sleeping rooms which have an exit door or exit access door that opens directly into a public street, public alley, yard, egress court or to an exterior exit balcony that opens to a public street, public alley, yard or egress court.
7. Basements without habitable spaces and having no more than 200 square feet (18.6 square meters) in floor area shall not be required to have emergency escape windows.

❖ This section requires emergency escape and rescue provisions in groups where occupants may be sleeping during a potential fire buildup but are capable of self-preservation (Groups R and I-1). A basement and each sleeping room are to be provided with an exterior window or door that meets the minimum size requirements and is operable for emergency escape by methods that are obvious and clearly understood by all users. Sleeping rooms four stories or more above grade are not required to be so equipped, since fire service access at that height, as well as escape through such an opening, may not be practical or reliable. In accordance with Chapter 9, such buildings will also be equipped throughout with an automatic fire suppression system. The provision for basements is in recognition that such types of spaces typically only have a single path of egress and often have no alternative routes available as other levels do.

It is important to note that this window is an element of escape and does not comprise any part of the means of egress unless it is a door with appropriate egress component characteristics.

Exception 1 assumes that the automatic sprinkler system can control fire buildup and reduce, if not eliminate, the need for an occupant to use an emergency escape window. The exception applies to buildings equipped throughout with an NFPA 13 or 13R sprinkler system (see Section 903.3.1.1 or 903.3.1.2).

Exception 2 allows another acceptable means of escape; that is, a door directly from the sleeping room to a corridor with exits in opposite directions, to substitute for the escape window.

Exception 3 provides for dwelling and sleeping units that have egress windows to a balcony that is within an atrium. Note that the exception specifies that the dwelling or sleeping unit is to have another means of egress that does not pass through the atrium so that an independent route of egress is provided.

Exceptions 4 and 7 are intended to exempt basements that would not be likely to have sleeping rooms in them from the requirement to have emergency escape and rescue openings.

Exception 5 is in correlation with the exception for emergency escape windows in high-rise buildings found in Section 403.4 of the IBC.

Exception 6 does not require sleeping rooms with a direct access to an exterior-type environment, such as a street or exit balcony, to have an emergency escape window. The open atmosphere of the escape route increases the likelihood of the means of egress being available even with the delayed response time for sleeping residents.

1025.2 Minimum size. Emergency escape and rescue openings shall have a minimum net clear opening of 5.7 square feet (0.53 m^2).

Exception: The minimum net clear opening for emergency escape and rescue grade-floor openings shall be 5 square feet (0.46 m^2).

❖ The dimensional criteria of the opening are intended to permit fire service personnel (in full protective clothing with a breathing apparatus) to enter from a ladder, as well as permit occupants to escape. The net clear opening area and minimum dimensions are intended to provide a clear opening through which an occupant can pass to escape the building or a fire fighter can pass to enter the building for rescue or fire suppression activities. Since the emergency escape windows must be usable to all occupants, including children and guests, the required opening dimensions must be achieved by the normal operation of the window from the inside (e.g., sliding, swinging or lifting the sash). It is impractical to assume that all occupants can operate a window that requires a special sequence of operations to achieve the required opening size. While most occupants are familiar with the normal operation by which to open the window, children and guests are frequently unfamiliar with special procedures necessary to remove the sashes. The time spent in comprehending the special operation unnecessarily delays egress from the bedroom and could lead to panic and further confusion. Thus, windows that achieve the required opening dimensions only by performing operations such as the removal of sashes or mullions are not permitted. It should be noted that the minimum area cannot be achieved by using both the minimum height and minimum width specified in Section 1025.2.1 (see Figure 1025.2).

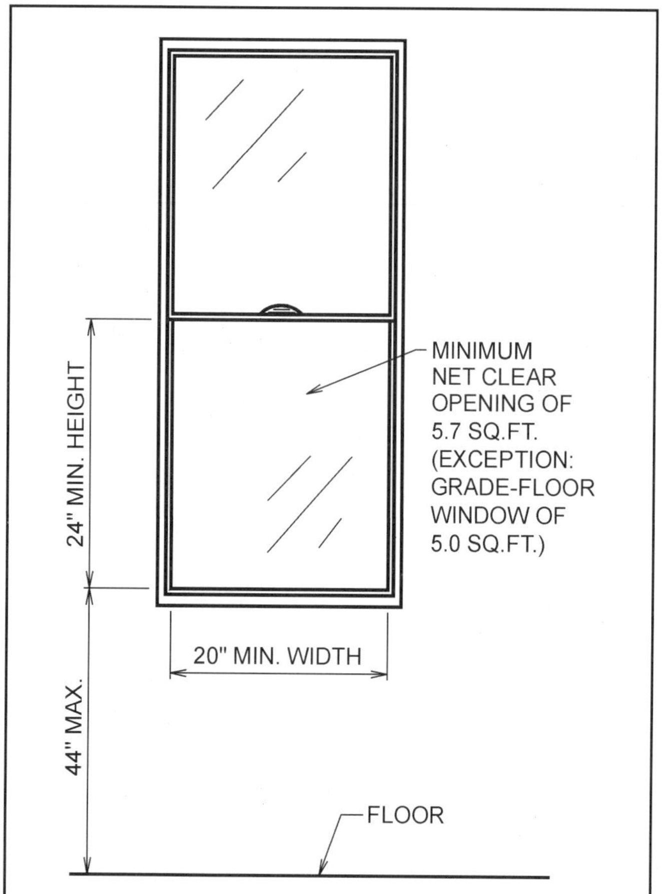

For SI: 1 inch = 25.4 mm, 1 square foot = 0.0929 m^2.

Figure 1025.2
EMERGENCY ESCAPE WINDOW

1025.2.1 Minimum dimensions. The minimum net clear opening height dimension shall be 24 inches (610 mm). The minimum net clear opening width dimension shall be 20 inches (508 mm). The net clear opening dimensions shall be the result of normal operation of the opening.

❖ Note that the minimum dimensions in this section and the minimum area requirements in Section 1025.2 both apply. Thus, a grade-floor window that is only 24 inches (610 mm) in height must be 30 inches (762 mm) wide to meet the 5-square-foot (0.46 m^2) area requirement of Section 1025.2 for grade-floor windows (see Figure 1025.2).

1025.3 Maximum height from floor. Emergency escape and rescue openings shall have the bottom of the clear opening not greater than 44 inches (1118 mm) measured from the floor.

❖ This section limits the height of the bottom of the clear opening to 44 inches (1118 mm) or less such that it can be used effectively as an emergency escape (see Figure 1025.2).

1025.4 Operational constraints. Emergency escape and rescue openings shall be operational from the inside of the room without the use of keys or tools. Bars, grilles, grates or similar devices are permitted to be placed over emergency escape and rescue openings provided the minimum net clear opening size complies with Section 1025.2 and such devices shall be releasable or removable from the inside without the use of a key, tool or force greater than that which is required for normal operation of the escape and rescue opening. Where such bars, grilles, grates or similar devices are installed in existing buildings, smoke alarms shall be installed in accordance with Section 907.2.10 regardless of the valuation of the alteration.

❖ If security grilles, decorations or similar devices are installed on escape windows, such items must be readily removable to permit occupant escape without the use of any tools, keys or a force greater than that required for the normal operation of the window.

Where bars, grilles or grates are placed over the emergency escape and rescue opening, it is important that they are easily removable; thus, the requirements for ease of operation are the same for windows.

The smoke alarms that are required for existing buildings where such items are installed provide advance warning of a fire for safety purposes.

1025.5 Window wells. An emergency escape and rescue opening with a finished sill height below the adjacent ground level shall be provided with a window well in accordance with Sections 1025.5.1 and 1025.5.2.

❖ Emergency escape and rescue openings that are partially or completely below grade need to have window wells so that they can be used effectively (see Figure 1025.5).

1025.5.1 Minimum size. The minimum horizontal area of the window well shall be 9 square feet (0.84 m^2), with a minimum dimension of 36 inches (914 mm). The area of the window well shall allow the emergency escape and rescue opening to be fully opened.

❖ This section specifies the size of the window well that is needed for a rescue person in full protective clothing and breathing apparatus to use the rescue opening. The 9 square feet (0.84 m^2) required is the horizontal cross-sectional area of the window well. Thus, if the window well projects 3 feet (914 mm) away from the plane of the window, the required dimension in the plane of the window along the wall is also 3 feet (914 mm) (see Figure 1025.5).

1025.5.2 Ladders or steps. Window wells with a vertical depth of more than 44 inches (1118 mm) shall be equipped with an approved permanently affixed ladder or steps. Ladders or rungs shall have an inside width of at least 12 inches (305 mm), shall project at least 3 inches (76 mm) from the wall and shall be spaced not more than 18 inches (457 mm) on center (o.c.) vertically for the full height of the window well. The ladder or steps shall not encroach into the required dimensions of the window well by more than 6 inches (152 mm). The ladder or steps shall not be obstructed by the emergency escape and rescue opening. Ladders or steps required by this section are exempt from the stairway requirements of Section 1009.

❖ This section specifies a ladder or steps be provided for ease of getting into and out of window wells that are more than 44 inches (1118 mm) deep.

Usually ladder rungs are embedded in the wall of the window well. The 44-inch (1118 mm) dimension is the depth of the window well, not the distance from the bottom of the window well to grade. Thus, if the floor of a window well is 40 inches (1016 mm) below grade, but the wall of the window well projects above grade by 6 inches (152 mm), steps or a ladder are required since the vertical depth is 46 inches (1168 mm).

It is important that the ladder not obstruct the operation of the emergency escape window (see Figure 1025.5).

SECTION 1026
MEANS OF EGRESS FOR EXISTING BUILDINGS

1026.1 General. Means of egress in existing buildings shall comply with Sections 1003 through 1025, except as amended in Section 1026.

Exception: Mean of egress conforming to the requirements of the building code under which they were constructed shall be considered as complying means of egress if, in the opinion of the fire code official, they do not constitute a distinct hazard to life.

❖ The primary concept of this section is to require existing buildings to comply with the specific means of egress requirements for new buildings as modified by this section. Where an item is specifically addressed by Section 1026, the requirements of this section are intended to override the requirement for new buildings in Sections 1003 through 1025.

For example, the guard height requirements in Sec-

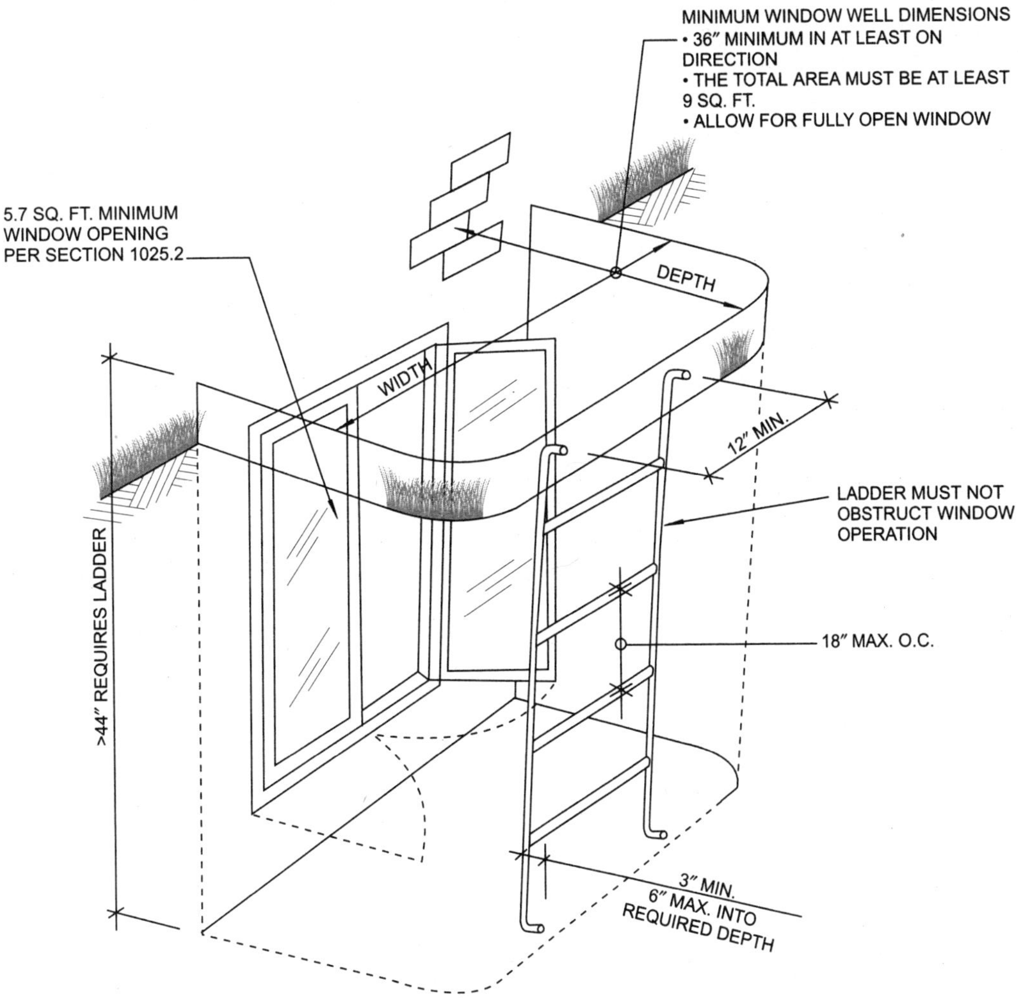

For SI: 1 inch = 25.4 mm, 1 square foot = 0.0929 m^2.

Figure 1025.5
WINDOW WELLS AT EMERGENCY ESCAPE WINDOW

tion 1026.6 supersede the guard height requirement for new buildings in Section 1012.2. In most cases, the requirements for existing buildings in Section 1026 are less stringent than those for new buildings in Sections 1003 through 1025.

The exception in this section applies to most existing buildings. Where a building has been built to meet the requirements of the building code in effect at the time of construction and, in the opinion of the fire code official, the means of egress is not hazardous, the building meets the intent of this section. In this case none of the specific requirements in Section 1026 apply.

1026.2 Elevators, escalators and moving walks. Elevators, escalators and moving walks shall not be used as a component of a required means of egress.

Exceptions:

1. Elevators used as an accessible means of egress where allowed by Section 1007.4.
2. Previously approved escalators and moving walks in existing buildings.

❖ This section is the same as Section 1003.7 except Exception 2 is added. Thus, an escalator or moving walk could be used as part of the required means of egress in an existing building if it had been previously approved by the fire code official.

1026.3 Exit sign illumination. Exit signs shall be internally or externally illuminated. The face of an exit sign illuminated from an external source, shall have an intensity of not less than 5 foot-candles (54 lux). Internally illuminated signs shall provide equivalent luminance and be listed for the purpose.

Exception: Approved self-luminous signs that provide evenly illuminated letters shall have a minimum luminance of 0.06 foot-lamberts (0.21 cd/m^2).

❖ This section is the same as Section 1011.2 for new buildings except that Section 1011.2 includes an exception to illumination for tactile signs. The same exception should apply for existing buildings.

1026.4 Power source. Where emergency illumination is required in Section 1026.5, exit signs shall be visible under emergency illumination conditions.

Exception: Approved signs that provide continuous illumination independent of external power sources are not required to be connected to an emergency electrical system.

❖ This section requires that exit signs serving the occupancies listed in Section 1026.5 be illuminated during the use of emergency power for the means of egress.

The comparable section for new buildings is Section 1011. Exit signs for all new building occupancies must be illuminated at all times and have an emergency power source.

1026.5 Illumination emergency power. The power supply for means of egress illumination shall normally be provided by the premises' electrical supply. In the event of power supply failure, illumination shall be automatically provided from an emergency system for the following occupancies where such occupancies require two or more means of egress:

1. Group A having more than 50 occupants.

   Exception: Assembly occupancies used exclusively as a place of worship and having an occupant load of less than 300.

2. Group B buildings three or more stories in height, buildings with 100 or more occupants above or below the level of exit discharge, or buildings with 1,000 or more total occupants.
3. Group E in interior stairs, corridors, windowless areas with student occupancy, shops and laboratories.
4. Group F having more than 100 occupants.

   Exception: Buildings used only during daylight hours which are provided with windows for natural light in accordance with the International Building Code.

5. Group I.
6. Group M.

   Exception: Buildings less than 3,000 square feet (279 m^2) in gross sales area on one story only, excluding mezzanines.

7. Group R-1.

   Exception: Where each guestroom has direct access to the outside of the building at grade.

8. Group R-2 as applicable in Section 1001.1.

   Exception: Where each living unit has direct access to the outside of the building at grade.

9. Group R-4.

   Exception: Where each sleeping room has direct access to the outside of the building at ground level.

The emergency power system shall provide power for not less than 60 minutes and consist of storage batteries, unit equipment or an on-site generator. The installation of the emergency power system shall be in accordance with Section 604.

❖ This section requires emergency power only for the listed occupancy conditions for existing buildings. The emergency power system is to provide power for at least 60 minutes.

Section 1006.3 requires emergency power for all new buildings with a few exceptions (primarily for utility and sleeping areas). A 90-minute emergency power duration is required by that section.

1026.6 Guards. Guards complying with this section shall be provided at the open sides of means of egress that are more than 30 inches (762 mm) above the floor or grade below.

❖ The guard requirements in this section are the same as in Section 1012 for new buildings (see commentary, Section 1012.1).

1026.6.1 Height of guards. Guards shall form a protective barrier not less than 42 inches (1067 mm) high.

Exceptions:

1. Existing guards on the open side of stairs shall be not less than 30 inches (760 mm) high.
2. Existing guards within dwelling units shall be not less than 36 inches (910 mm) high.
3. Existing guards in assembly seating areas.

❖ This section includes a less restrictive set of rules for existing guards than those for new construction in Sections 1012.2 and for assembly areas in Section 1024.14.

1026.6.2 Opening limitations. Open guards shall have balusters or ornamental patterns such that a 6-inch diameter (152 mm) sphere cannot pass through any opening up to a height of 34 inches (864 mm).

Exceptions:

1. At elevated walking surfaces for access to, and use of electrical, mechanical or plumbing systems or equipment, guards shall have balusters or be of solid materials such that a sphere with a diameter of 21 inches (533 mm) cannot pass through any opening.
2. In occupancies in Group I-3, F, H or S, the clear distance between intermediate rails measured at right angles to the rails shall not exceed 21 inches (533 mm).
3. Approved existing open guards.

❖ Generally, this section allows maximum openings of 6 inches (152 mm) versus the maximum openings of 4 inches (102 mm) allowed by Section 1012.3.

1026.7 Size of doors. The minimum width of each door opening shall be sufficient for the occupant load thereof and shall provide a clear width of not less than 28 inches (711 mm). Where this section requires a minimum clear width of 28 inches (711 mm) and a door opening includes two door leaves without a mullion, one leaf shall provide a clear opening width of 28 inches (711 mm). The maximum width of a swinging door leaf shall be 48 inches (1219 mm) nominal. Means of egress doors in an occupancy in Group I-2 used for the movement of beds shall provide a clear width not less than 41.5 inches (1054 mm). The height of doors shall not be less than 80 inches (2032 mm).

Exceptions:

1. The minimum and maximum width shall not apply to door openings that are not part of the required means of egress in occupancies in Groups R-2 and R-3 as applicable in Section 1001.1.
2. Door openings to storage closets less than 10 square feet (0.93 m^2) in area shall not be limited by the minimum width.
3. Width of door leafs in revolving doors that comply with Section 1003.3.1.3.1 shall not be limited.
4. Door openings within a dwelling unit shall not be less than 78 inches (1981 mm) in height.
5. Exterior door openings in dwelling units, other than the required exit door, shall not be less than 76 inches (1930 mm) in height.
6. Exit access doors serving a room not larger than 70 square feet (6.5 m^2) shall be not less than 24 inches (610 mm) in door width.

❖ Generally, this section requires doors to be at least 28 inches (711 mm) wide versus the minimum width of 32 inches (813 mm) for new construction in Section 1008.1.1. The other provisions and exceptions in this section are the same as those in Section 1008.1.1 for new construction. Exceptions 2 and 8 of Section 1008.1.1 are not included in this section. Exception 2 is not needed since these provisions for existing buildings only require a 28-inch-wide (711 mm) door. Exception 8 relates to accessibility requirements that are not intended to be retroactively applied except within the context of the IBC.

1026.8 Opening force for doors. The opening force for interior side-swinging doors without closers shall not exceed a 5-pound (22 N) force. For other side-swinging, sliding and folding doors, the door latch shall release when subjected to a force of not more than 15 pounds (66 N). The door shall be set in motion when subjected to a force not exceeding a 30-pound (133 N) force. The door shall swing to a full-open position when subjected to a force of not more than 50 pounds (222 N). Forces shall be applied to the latch side.

❖ This section is similar to the third paragraph of Section 1008.1.2. The maximum door operating force in this section is less restrictive than that required in Section 1008.1.2 for new construction.

1026.9 Revolving doors. Revolving doors shall comply with the following:

1. A revolving door shall not be located within 10 feet (3048 mm) of the foot or top of stairs or escalators. A dispersal area shall be provided between the stairs or escalators and the revolving doors.
2. The revolutions per minute for a revolving door shall not exceed those shown in Table 1026.9.
3. Each revolving door shall have a conforming side-hinged swinging door in the same wall as the revolving door and within 10 feet (3048 mm).

Exceptions:

1. A revolving door is permitted to be used without an adjacent swinging door for street floor elevator lobbies provided a stairway, escalator or door from other parts of the building does not discharge through the lobby

and the lobby does not have any occupancy or use other than as a means of travel between elevators and a street.

2. Existing revolving doors where the number of revolving doors does not exceed the number of swinging doors within 20 feet (6096 mm).

❖ This section is comparable with Section 1008.1.3.1 for new buildings. The capability of collapsing to a width of 36 inches (914 mm) in Condition 1 of Section 1008.1.3.1 for new construction does not apply to existing revolving doors. The two exceptions in this section are not allowed for new construction.

TABLE 1026.9
REVOLVING DOOR SPEEDS

INSIDE DIAMETER	POWER-DRIVEN-TYPE SPEED CONTROL (RPM)	MANUAL-TYPE SPEED CONTROL (RPM)
6′6″	11	12
7′0″	10	11
7′6″	9	11
8′0″	9	10
8′6″	8	9
9′0″	8	9
9′6″	7	8
10′0″	7	8

For SI: 1 inch = 25.4 mm, 1 foot = 304.8 mm.

❖ This table is the same as Table 1008.1.3.1 for new construction.

1026.9.1 Egress component. A revolving door used as a component of a means of egress shall comply with Section 1026.9 and all of the following conditions:

1. Revolving doors shall not be given credit for more than 50 percent of the required egress capacity.
2. Each revolving door shall be credited with not more than a 50-person capacity.
3. Revolving doors shall be capable of being collapsed when a force of not more than 130 pounds (578 N) is applied within 3 inches (76 mm) of the outer edge of a wing.

❖ This section is the same as Section 1008.1.3.1.1 for new construction.

1026.10 Stair dimensions for existing stairs. Existing stairs in buildings shall be permitted to remain if the rise does not exceed 8.25 inches (210 mm) and the run is not less than 9 inches (229 mm). Existing stairs can be rebuilt.

Exception: Other stairs approved by the fire code official.

❖ This section includes a much less stringent criterion for stair tread and risers than that in Section 1009.3. This section also allows existing stairways to be rebuilt. The tread and riser dimensions of the rebuilt stair are regulated by Section 1026.10.1.

1026.10.1 Stair dimensions for replacement stairs. The replacement of an existing stairway in a structure shall not be required to comply with the new stairway requirements of Section 1009 where the existing space and construction will not allow a reduction in pitch or slope.

❖ This section is the same as Section 3402.4 for existing buildings in the IBC. The intent of this section is that where existing stairways are replaced, the tread and risers are to conform to the requirements of Section 1009.3 unless the existing space and existing construction will not allow a reduction in stair pitch.

1026.11 Winders. Existing winders shall be allowed to remain in use if they have a minimum tread depth of 6 inches (152 mm) and a minimum tread depth of 9 inches (229 mm) at a point 12 inches (305 mm) from the narrowest edge.

❖ Section 1009.8 for new construction requires that the winder depth at 12 inches (305 mm) from the end be at least 11 inches (279 mm), as compared to the 9-inch (229 mm) minimum dimension in this section.

1026.12 Circular stairways. Existing circular stairs shall be allowed to continue in use provided the minimum depth of tread is 10 inches (254 mm) and the smallest radius shall not be less than twice the width of the stairway.

❖ For other than certain residential occupancies, this section has fewer requirements than Section 1009.7 for new construction.

Section 1009.7 has an exception that is more liberal for certain residential occupancies than this section for existing construction. The relaxed requirements of the exception in Section 1009.7 for Group R-3 occupancies and within individual dwelling units in Group R-2 occupancies, as applicable in Section 1001.1, should also be allowed for existing construction.

1026.13 Stairway handrails. Stairways shall have handrails on at least one side. Handrails shall be located so that all portions of the stairway width required for egress capacity are within 44 inches (1118 mm) of a handrail.

Exception: Aisle stairs provided with a center handrail are not required to have additional handrails.

❖ See Section 1009.11 for handrail requirements for new construction. Handrails are required on both sides, with certain exceptions, for new construction compared to the requirement for one side in this section. The 44-inch (1118 mm) dimension in this section is an increase from the 30-inch (762 mm) dimension in Section 1009.11.2 for new construction. Handrails for assembly occupancies in new construction are covered in Section 1024.13.

1026.13.1 Height. Handrail height, measured above stair tread nosings, shall be uniform, not less than 30 inches (762 mm) and not more than 42 inches (1067 mm).

❖ The allowable range of handrail height installation in this section is broader than that allowed for new construction in Section 1009.11.1. The range for new construction is 34 inches (834 mm) up to and including 38 inches (965 mm).

1026.14 Slope of ramps. Ramp runs utilized as part of a means of egress shall have a running slope not steeper than one unit vertical in ten units horizontal (10-percent slope). The slope of other ramps shall not be steeper than one unit vertical in eight units horizontal (12.5-percent slope).

❖ The slope of means of egress ramps for new construction in Section 1010.2 is a maximum of 1:12. The steeper sloped ramps of 1 in 8 slope in assembly aisles as permitted by Section 1024.11 for new construction should also apply to existing construction.

1026.15 Width of ramps. Existing ramps are permitted to have a minimum width of 30 inches (762 mm) but not less than the width required for the number of occupants served as determined by Section 1005.1.

❖ Ramps for new construction in Section 1010.5.1 are required to have a minimum clear width of 36 inches (914 mm) compared to the 30-inch (762 mm) minimum in this section (see commentary, Section 1010.5.1).

1026.16 Fire escape stairs. Fire escape stairs shall comply with Sections 1026.16.1 through 1026.16.7.

❖ This section and the referenced sections provide the detailed requirements for fire escape stairs. Fire escape stairs are allowed for a required means of egress only in an existing building according to Section 3403 in the IBC. Sections 1026.16.1 through 1026.16.7 of the code are similar to Sections 3403.1 through 3403.5 of the IBC.

1026.16.1 Existing means of egress. Fire escape stairs shall be permitted in existing buildings but shall not constitute more than 50 percent of the required exit capacity.

❖ Fire escapes are limited to serving 50 percent of the required exit capacity because they have minimum usability and are not appropriate for use by persons with limited physical capability.

1026.16.2 Protection of openings. Openings within 10 feet (3048 mm) of fire escape stairs shall be protected by fire assemblies having a minimum 3/4-hour fire-resistance rating.

Exception: In buildings equipped throughout with an approved automatic sprinkler system, opening protection is not required.

❖ It is important that openings in the vicinity of a fire escape are protected so that the fire escape will be usable during a fire. The exception for a sprinkler system takes into account the control of the fire by the sprinkler system.

1026.16.3 Dimensions. Fire escape stairs shall meet the minimum width, capacity, riser height and tread depth as specified in Section 1026.10.

❖ The riser height and tread depth requirement is the same as for existing stairs. The limits are specified in Section 1026.10; however, the minimum width and occupant capacity are not specified in Section 1026.10. The minimum width and occupant capacity should be based on the anticipated occupant load that is likely to use the fire escape. The minimum width could then be determined based on the occupant load by using Table 1005.1.

1026.16.4 Access. Access to a fire escape from a corridor shall not be through an intervening room. Access to a fire escape stair shall be from a door or window meeting the criteria of Table 1005.1. Access to a fire escape stair shall be directly to a balcony, landing or platform. These shall be no higher than the floor or window sill level and no lower than 8 inches (203 mm) below the floor level or 18 inches (457 mm) below the window sill.

❖ This section establishes the arrangement of the fire escape to the building so that the fire escape can be easily reached. Access is not permitted through a room because the room could pose an unacceptable hazard or be locked. The elevation limits of the exterior landing, balcony or platform are to make the fire escape easily accessible.

1026.16.5 Materials and strength. Components of fire escape stairs shall be constructed of noncombustible materials.

Fire escape stairs and balconies shall support the dead load plus a live load of not less than 100 pounds per square foot (4.78 kN/m^2). Fire escape stairs and balconies shall be provided with a top and intermediate handrail on each side.

The fire code official is authorized to require testing or other satisfactory evidence that an existing fire escape stair meets the requirements of this section.

❖ The noncombustible construction requirement is so that the fire escape will be available for use during a fire. The loading of 100 pounds per square foot (psf) (4.78 kN/m^2) anticipates that a number of persons will be using the fire escape at the same time. If the condition of the fire escape or some other factor should result in a questionable loading capacity, the fire code official may require tests or field surveys to validate the capacity.

1026.16.6 Termination. The lowest balcony shall not be more than 18 feet (5486 mm) from the ground. Fire escape stairs shall extend to the ground or be provided with counterbalanced stairs reaching the ground.

Exception: For fire escape stairs serving 10 or fewer occupants, an approved fire escape ladder is allowed to serve as the termination for a fire escape stairs.

❖ This section controls the elevation of the lowest fire escape balcony. For fire escapes that serve 10 or fewer people, a ladder from the lowest balcony to the ground is permitted. In all other cases, the lower balcony is to be served by fire escape stairs. The counter balance keeps the fire escape stairway up off ground level when not in use.

1026.16.7 Maintenance. Fire escapes shall be kept clear and unobstructed at all times and shall be maintained in good working order.

❖ This section prohibits the use of the fire escape for outdoor storage or other use that blocks its use. Fire escapes must be kept in good condition so that they will be available for use.

1026.17 Corridors. Corridors serving an occupant load greater than 30 and the openings therein shall provide an effective barrier to resist the movement of smoke. Transoms, louvers, doors and other openings shall be closed or be self-closing.

Exceptions:

1. Corridors in occupancies other than in Group H, which are equipped throughout with an approved automatic sprinkler system.
2. Patient room doors in corridors in occupancies in Group I-2 where smoke barriers are provided in accordance with the *International Building Code*.
3. Corridors in occupancies in Group E where each room utilized for instruction or assembly has at least one-half of the required means of egress doors opening directly to the exterior of the building at ground level.
4. Corridors that are in accordance with the *International Building Code*.

❖ This section relates to Section 1016.1 for new construction. This section requires the corridors to be an effective barrier to resist the movement of smoke. Section 1016.1 for new construction requires that the corridor has a fire-resistance rating. Generally, solid walls and doors that would resist smoke migration are the intent of this section. The walls or doors need not have a fire-resistance rating.

See Section 407.3.1 of the IBC for the requirements for corridor doors for new construction of Group I-2 facilities.

1026.17.1 Corridor openings. Openings in corridor walls shall comply with the requirements of the *International Building Code*.

Exceptions:

1. Where 20-minute fire assemblies are required, solid wood doors at least 1.75 inches (44 mm) thick or insulated steel doors are permitted.
2. Openings protected with fixed wire glass set in steel frames.
3. Openings covered with 0.5-inch (12.7 mm) gypsum wallboard or 0.75-inch (19.1 mm) plywood on the room side.
4. Opening protection is not required if the building is equipped throughout with an approved automatic sprinkler system.

❖ This section makes reference to the IBC requirements for corridor openings. Section 1016.1 specifies that corridors that are required to be fire-resistance rated are to be constructed according to Section 708 of the IBC for fire partitions. Section 708.6 of the IBC refers to Section 715 of the IBC for protected openings in smoke partitions. Thus, Section 715 of the IBC applies to existing buildings where a fire-resistance-rated corridor would be required for new construction except where one of the exceptions in this section applies. The exceptions in this section provide a number of practical alternatives for existing buildings.

1026.17.2 Dead ends. Where more than one exit or exit access doorway is required, the exit access shall be arranged such that dead ends do not exceed the limits specified in Table 1026.17.2.

Exception: A dead-end passageway or corridor shall not be limited in length where the length of the dead-end passageway or corridor is less than 2.5 times the least width of the dead-end passageway or corridor.

❖ This section references Table 1026.17.2 for the dead-end travel distance limits for existing buildings. The dead-end limit for new construction is in Section 1016.3. That section includes the same exception that is in Section 1026.17.2 for existing buildings. Dead-end travel distance is not a concern for conditions where the exception is satisfied because travel is not limited to a single path. Generally, the dead-end travel limits in Table 1026.17.2 are more liberal than those in Section 1016.3 for new construction.

1026.17.3 Exit access travel distance. Exits shall be located so that the maximum length of exit access travel, measured from the most remote point to an approved exit along the natural and unobstructed path of egress travel, does not exceed the distances given in Table 1026.17.2.

❖ This section references Table 1026.17.2 for the exit access travel distance limits for existing buildings. The travel distance limits for new construction are in Section 1015.1. Generally, the travel distance limits in Table 1026.17.2 for existing construction are the same as those in Table 1015.1 for new construction except for Groups I and H.

1026.17.4 Common path of egress travel. The common path of egress travel shall not exceed the distances given Table 1026.17.2.

❖ This section refers to Table 1026.17.2 for the common path of egress travel for existing buildings. The common path of egress travel limits for new construction are in Section 1013.3. Generally, the limits in Table 1026.17.2 for existing construction are the same as in Section 1013.3 for new construction.

1026.18 Stairway discharge identification. A stairway in an exit enclosure which continues below the level of exit discharge shall be arranged and marked to make the direction of egress to a public way readily identifiable.

Exception: Stairs that continue one-half story beyond the level of exit discharge need not be provided with barriers where the exit discharge is obvious.

❖ The requirements of this section are less stringent than the discharge identification requirements for new construction in Section 1019.1.6. The new construction provisions require that a barrier be placed within the stairway to prevent persons from unintentionally continuing into the levels below the exit discharge. The exception in this section is also not included in the requirements for new construction.

1026.19 Exterior stairway protection. Exterior exit stairs shall be separated from the interior of the building as required in Section 1022.6. Openings shall be limited to those necessary for egress from normally occupied spaces.

Exceptions:

1. Separation from the interior of the building is not required for buildings that are two stories or less above grade where the level of exit discharge is the first story above grade.
2. Separation from the interior of the building is not required where the exterior stairway is served by an exterior balcony that connects two remote exterior stairways or other approved exits, with a perimeter that is not less than 50 percent open. To be considered open, the opening shall be a minimum of 50 percent of the height of the enclosing wall, with the top of the opening not less than 7 feet (2134 mm) above the top of the balcony.
3. Separation from the interior of the building is not required for an exterior stairway located in a building or structure that is permitted to have unenclosed interior stairways in accordance with Section 1019.1.
4. Separation from the interior of the building is not required for exterior stairways connected to open-ended corridors, provided that:
   - 4.1. The building, including corridors and stairs, is equipped throughout with an automatic sprinkler system in accordance with Section 903.3.1.1 or 903.3.1.2.
   - 4.2. The open-ended corridors comply with Section 1016.
   - 4.3. The open-ended corridors are connected on each end to an exterior exit stairway complying with Section 1022.1.
   - 4.4. At any location in an open-ended corridor where a change of direction exceeding 45 degrees occurs, a clear opening of not less than 35 square feet (3 m^2) or an exterior stairway shall be provided. Where clear openings are provided, they shall be located so as to minimize the accumulation of smoke or toxic gases.

❖ The exterior stairway requirements of this section are identical to those for new construction in Section 1022.1, except for Exception 1. While the exception in this section applies to all two-story buildings where the level of the exit discharge is the first story above grade, the exception does not apply to new buildings in Group R-1 and R-2 occupancies.

TABLE 1026.17.2. See page 10-142.

❖ This table contains the existing building limits for common path of travel, dead-end limit and travel distance limits. See the commentary to Sections 1026.17.2, 1026.17.3 and 1026.17.4 for a discussion regarding the table.

1026.20 Minimum aisles width. The minimum clear width of aisles shall be:

1. Forty-two inches (1067 mm) for aisle stairs having seating on each side.

   Exception: Thirty-six inches (914 mm) where the aisle does not serve more than 50 seats.
2. Thirty-six inches (914 mm) for stepped aisles having seating on only one side.

   Exception: Thirty inches (760 mm) for catchment areas serving not more than 60 seats.
3. Twenty inches (508 mm) between a stepped aisle handrail or guard and seating when the aisle is subdivided by the handrail.
4. Forty-two inches (1067 mm) for level or ramped aisles having seating on both sides.

   Exception: Thirty-six inches (914 mm) where the aisle does not serve more than 50 seats.
5. Thirty-six inches (914 mm) for level or ramped aisles having seating on only one side.

   Exception: Thirty inches (760 mm) for catchment areas serving not more than 60 seats.
6. Twenty-three inches (584 mm) between a stepped stair handrail and seating where an aisle does not serve more than five rows on one side.

❖ This section provides the minimum aisle widths for assembly occupancies. Similar requirements for new construction are in Section 1024.9.1. Several of the exceptions in this section are not included in Section 1024.9.1. Thus, the requirements for existing buildings are less stringent than those for new construction.

1026.21 Stairway floor number signs. Existing stairs shall be marked in accordance with Section 1019.1.7.

❖ This section requires that existing stairs be marked in the same manner as new stairs (see commentary, Section 1019.1.7).

TABLE 1026.17.2
COMMON PATH, DEAD-END AND TRAVEL DISTANCE LIMITS (by occupancy)

OCCUPANCY	COMMON PATH LIMIT		DEAD-END LIMIT		TRAVEL DISTANCE LIMIT	
	Unsprinklered (feet)	Sprinklered (feet)	Unsprinklered (feet)	Sprinklered (feet)	Unsprinklered (feet)	Sprinklered (feet)
Group A	20/75[a]	20/75[a]	20[b]	20[b]	200	250
Group B	75	100	50	50	200	250
Group E	75	75	20	20	200	250
Groups F-1, S-1[d]	75	100	50	50	200	250
Groups F-2, S-2[d]	75	100	50	50	300	400
Group H-1	25	25	0	0	75	75
Group H-2	50	100	0	0	75	100
Group H-3	50	100	20	20	100	150
Group H-4	75	75	20	20	150	175
Group H-5	75	75	20	50	150	200
Group I-1	75	75	20	20	200	250
Group I-2 (Health Care)	NR	NR	NR	NR	150	200[c]
Group I-3 (Detention and Correctional—Use Conditions II, III, IV, V	100	100	NR	NR	150[c]	200[c]
Group I-4 (Day Care Centers)	NR	NR	20	20	200	250
Group M (Covered Mall)	75	100	50	50	200	400
Group M (Mercantile)	75	100	50	50	200	250
Group R-1 (Hotels)	75	75	50	50	200	250
Group R-2[e] (Apartments)	75	75	50	50	200	250
Group R-3[e] (One- and Two-Family); Group R-4 (Residential Care/Assisted Living)	NR	NR	NR	NR	NR	NR
Group U	75	75	20	20	200	250

For SI: 1 foot = 304.8 mm.

a. 20 feet for common path serving more than 50 persons; 75 feet for common path serving 50 or fewer persons.

b. See Section 1024.9.5 for dead-end aisles in Group A occupancies.

c. This dimension is for the total travel distance, assuming incremental portions have fully utilized their allowable maximums. For travel distance within the room, and from the room exit access door to the exit, see the appropriate occupancy chapter.

d. See the *International Building Code* for special requirements on spacing of doors in aircraft hangars.

e. As applicable in Section 1001.1.

NR = No requirements.

SECTION 1027
MAINTENANCE OF THE MEANS OF EGRESS

1027.1 General. The means of egress for buildings or portions thereof shall be maintained in accordance with this section.

❖ This section includes the provisions for the maintenance of the means of egress in all buildings and structures.

1027.2 Reliability. Required exit accesses, exits or exit discharges shall be continuously maintained free from obstructions or impediments to full instant use in the case of fire or other emergency. Security devices affecting means of egress shall be subject to approval of the fire code official.

❖ It is important for safety that the pathway from any point in the building to the exit discharge be kept clear so that the occupants can exit the building at any time without obstructions in the egress path. Similarly, handrails, stair treads, flooring materials, door hardware and other fixtures and finishes must be maintained in such a way as to prevent them from becoming hazards themselves. Also, in our society, security is an ever-growing concern, and often the solutions to enhancing the security of buildings conflict with the life safety concerns of building and fire codes. This section provides the fire code official with an important measure of control over the installation or modification of security devices that could have an adverse effect upon the egress system of a building.

1027.3 Obstructions. A means of egress shall be free from obstructions that would prevent its use, including the accumulation of snow and ice.

❖ Blocked exits are one of the most common of all egress problems. Obstructions, impediments and storage or placement of articles in a manner that prevents access to or reduces the effective width of egress elements are prohibited. Such impediments may be movable or fixed. Maintenance of safe egress conditions implies keeping exits free of storage, decorations or debris that obstruct access or visibility. Complex egress paths may also violate the intent of this requirement by confusing or obscuring the path of travel. When inspecting, take extra care to follow the egress path that occupants must take in the event of emergency from a number of points in the same manner. Holding open an egress door may pose an even greater hazard than blocking an exit by giving the fire unimpeded access to the path of egress.

Accumulations of snow and ice could prevent timely exiting from the building in the case of fire or other emergency. Generally, if the exit is used regularly, the exit doorway area is kept free of ice and snow. Where a required exit is not used regularly, it may be necessary to protect the exit doorway area from the accumulation of ice and snow, either by construction of overhangs or enclosures, heated slabs or, when approved by the fire code official, by a reliable snow removal program that is aggressively enforced. For similar requirements for new construction see Sections 1009.5.2, 1010.7.2 and 1013.5.

1027.4 Furnishings and decorations. Furnishings, decorations or other objects shall not be placed so as to obstruct exits, access thereto, egress therefrom, or visibility thereof. Hangings and draperies shall not be placed over exit doors or otherwise be located to conceal or obstruct an exit. Mirrors shall not be placed on exit doors. Mirrors shall not be placed in or adjacent to any exit in such a manner as to confuse the direction of exit.

❖ Displays, furnishings and decorations are frequently intended to create a mood or atmosphere; however, they often make familiar places confusing even to those who normally occupy them. Bright lights, vivid colors, mirrors and hanging material may significantly reduce contrast, impede visibility or confuse direction. Tables, chairs, display cases, coat racks and similar movable objects placed in corridors or aisles may reduce required egress capacity or may require substantial effort to remove or negotiate quickly. Anything that slows egress may also impede access, particularly to fire fighters who may be called to rescue occupants or fight the fire. Similarly, anything that obscures the visibility of exit signs is prohibited.

1027.5 Emergency escape openings. Required emergency escape openings shall be maintained in accordance with the code in effect at the time of construction, and the following: Required emergency escape and rescue openings shall be operational from the inside of the room without the use of keys or tools. Bars, grilles, grates or similar devices are allowed to be placed over emergency escape and rescue openings provided the minimum net clear opening size complies with the code that was in effect at the time of construction and such devices shall be releasable or removable from the inside without the use of a key, tool or force greater than that which is required for normal operation of the escape and rescue opening.

❖ In new construction, Section 1025.1 requires that every sleeping room below the fourth story in Group R and I-1 occupancies have at least one operable window or exterior door approved for emergency escape or rescue. This section provides the necessary maintenance and utilization requirements for emergency escape openings. Maintenance requirements are provided to increase the reliability of the emergency escape elements. Included are allowances to provide the emergency escape elements with security devices that retain the usability of the window or door while allowing a measure of security for the building occupants from intruders. Note that these provisions are applicable only to emergency escape openings that were required to be provided by the applicable building code that was in force when the building was constructed, in accordance with Section 1026.1.

The same blocking, locking and maintenance restrictions that apply to egress doors apply to emergency escape windows and doors. These egress components are frequently found blocked by furniture, fans and window air conditioners, especially in residential sleeping rooms. Inspections of other means of egress elements should include a look at basement escape windows and doors as well. Mechanical parts of exterior doors and windows are particularly susceptible to wear and tear and should be the subject of frequent inspections and periodic preventive maintenance. Hinges must swing and handles must turn at all times so that they are ready to function when needed.

Since the emergency escape windows must be usable to all occupants, including children and guests, the required opening dimensions must be achieved by the normal operation of the window from the inside (e.g., sliding, swinging or lifting the sash). It is impractical to assume that all occupants can operate a window that requires a special sequence of operations to achieve the required opening size. While most occupants are familiar with the normal operation by which to open the window, children and guests are frequently unfamiliar with special procedures necessary to remove the sashes. The time spent in comprehending the special operation unnecessarily delays egress from the bedroom and could lead to panic and further confusion; thus, windows that achieve the required opening dimensions only by performing operations such as the removal of sashes or mullions are not permitted.

Not only are security fixtures frequently added to emergency escape windows to prevent entry, but too often the windows themselves are replaced with others that provide inadequate clearance or improper locking arrangements. Security grille designs are available to deter entry without compromising egress. Bars, grilles or screens placed over emergency escape windows must also be releasable or removable from the inside without the use of a key, tool or force greater than what is required for normal operation of the window.

Bibliography

The following resource materials are referenced in this chapter or are relevant to the subject matter addressed in this chapter.

ASCE 7-02, *Minimum Design Loads for Buildings and Other Structures.* American Society of Civil Engineers, 2002.

ASME A17.1-00, *Safety Code for Elevators and Escalator.* New York: American Society of Mechanical Engineers, 2000.

ASME A18.1-99, *Safety Standard for Platform Lifts and Stairway Chairlifts - with Addenda A18.1a-2001.* New York: American Society of Mechanical Engineers, 2001.

ASTM C 1028-96, *Test Method for Determining the Static Coefficient of Friction of Ceramic Tile and Other Like Surfaces by the Horizontal Dynamometer Pull Meter Method.* West Conshohocken, PA: ASTM International, 1996.

ASTM D 2047-99, *Test Method for Static Coefficient of Friction Polish-Coated Floor Surfaces as Measured by the James Machine.* West Conshohocken, PA: ASTM International, 1999.

ASTM D 2394-83 (Reapproved 1999), *Methods for Simulated Service Testing of Wood and Wood-Base Finish Flooring.* West Conshohocken, PA: ASTM International, 1999.

ASTM E 119-00, *Test Method for Fire Tests of Building Construction and Materials.* West Conshohocken, PA: ASTM International, 2000.

ASTM F 851-87 (Reapproved 2000), *Test Method for Self-Rising Seat Mechanisms.* West Conshohocken, PA: ASTM International, 2000.

Cote, Ron, ed. *Life Safety Code Handbook, 8th ed.* Quincy, MA: National Fire Protection Association, Inc., 2000.

DOJ 28 CFR, Part 36 (Appendix A)-91, ADA *Accessibility Guidelines for Buildings and Facilities.* Washington, DC: U.S. Department of Justice, 1991.

DOJ 28 CFR, Part 36-91, *Americans with Disabilities Act (ADA).* Washington, DC: U.S. Department of Justice, 1991.

DOL 29 CFR, Part 1910-92, *Occupational Safety and Heath Standards.* Washington, DC: U.S. Department of Labor; Occupational Safety and Health Administration, 1992.

Fire Protection Handbook, 19th ed. Quincy, MA: National Fire Protection Association, 2003.

Fruin, John J. *Pedestrian Planning and Design.* Mobile: Elevator World, Inc., 1987.

IBC-03, *International Building Code.* Falls Church, VA: International Code Council, 2003.

ICC 300-02, *ICC Standards on Bleachers, Folding and Telescopic Seating and Grandstands.* Falls Church, VA: International Code Council, 2002.

ICC A117.1-98, *Accessible and Usable Buildings and Facilities.* Falls Church, VA: International Code Council, 1998.

ICC EC-03, *ICC Electrical Code.* Falls Church, VA: International Code Council, 2003.

IMC-03, *International Mechanical Code.* Falls Church, VA: International Code Council, 2003.

IPMC-03, *International Property Maintenance Code.* Falls Church, VA: International Code Council, 2003.

IRC-03, *International Residential Code.* Falls Church, VA: International Code Council, 2003.

"Means of Egress." CABO Board of the Coordination of the Model Codes Report, 1985.

NFPA 13-99, *Installation of Sprinkler Systems.* Quincy, MA: National Fire Protection Association, 1999.

NFPA 13D-99, *Installation of Sprinkler Systems in One- and Two-Family Dwellings and Manufactured Homes.* Quincy, MA: National Protection Association, 1999.

NFPA 13R-99, *Installation of Sprinkler Systems in Residential Occupancies Up to Four Stories in Height.* Quincy, MA: National Fire Protection Association, 1999.

NFPA 92A-00, *Recommended Practice for Smoke Control Systems.* Quincy, MA: National Fire Protection Association, 2000.

NFPA 92B-00, *Guide for Smoke Management Systems in Malls, Atria and Large Areas.* Quincy, MA: National Fire Protection Association, 2000.

NFPA 204-02, *Standard for Smoke and Heat Venting.* Quincy, MA: National Fire Protection Association, 2002.

NFPA 252-99, *Methods of Fire Tests of Door Assemblies.* Quincy, MA: National Fire Protection Association, 1999.

NIST IR 4770-92, *Report on Staging Areas for Persons with Mobility Limitations.* Washington, DC: National Institute of Standards and Technology, 1992.

SFPE Fire Protection Engineering Handbook, 3rd ed. Quincy, MA: National Fire Protection Association, 2002.

"Surfaces." Bulletin No. 4. Washington, DC: U.S. Architectural and Transportation Barriers Compliance Board, April 1994.

UL 199-97, *Automatic Sprinklers for Fire-Protection Service.* Northbrook, IL: Underwriters Laboratories Inc., 1997.

UL 924-95 *Emergency Lighting and Power Equipment.* Northbrook, IL: Underwriters Laboratories Inc., 1995.

UL 1626-01, *Residential Sprinklers for Fire-Protection Service - with Revisions thru March 1997.* Northbrook, IL: Underwriters Laboratories Inc., 2001.

UL 1784-95, *Air Leakage Tests of Door Assemblies.* Northbrook, IL: Underwriters Laboratories Inc., 1995.

Chapter 11: Aviation Facilities

General Comments

Safe and efficient operation of airports, heliports and aircraft service facilities requires a comprehensive understanding of fire safety and aviation activities. The principal nonflight operational hazards associated with aviation involve fuel, facilities and operations. Conflicts have developed in recent years between airport security and life safety requirements because of an increased concern about air piracy and terrorism. The Federal Aviation Administration (FAA) regulates airport and air carrier security operations. These regulations strictly limit access to the air operations area. Unauthorized individuals must be prevented from entering air operations areas during all operating conditions, including emergencies in the terminal building. Concurrently, airport designs have traditionally included large unconfined areas for the movement of people and their belongings. Because most contemporary passenger terminal buildings resemble covered malls, Section 402 of the *International Building Code®* (IBC®) permits passenger transportation terminals to comply with the requirements for a covered mall building and, in fact, includes them in the definition of "Covered mall building."

Purpose

Chapter 11 specifies minimum requirements for the fire-safe operation of airports, heliports and helistops. Safe use of flammable and combustible liquids during fueling and maintenance operations is emphasized. Availability of portable Class B:C-rated fire extinguishers for prompt control or suppression of incipient fires is required.

SECTION 1101 GENERAL

1101.1 Scope. Airports, heliports, helistops and aircraft hangars shall be in accordance with this chapter.

❖ This chapter discusses fire and life safety in the ground environment modes. These modes include aircraft maintenance, aircraft refueling, aircraft hangars, helistops and heliports.

1101.2 Regulations not covered. Regulations not specifically contained herein pertaining to airports, aircraft maintenance, aircraft hangars and appurtenant operations shall be in accordance with nationally recognized standards.

❖ If a regulation is not addressed in this chapter, one must go to a recognized standard for the regulation. Ground operations must be conducted in accordance with recognized standards.

1101.3 Permits. For permits to operate aircraft-refueling vehicles, application of flammable or combustible finishes, and hot work, see Section 105.6.

❖ The process of issuing permits gives the fire code official an opportunity to carefully evaluate and regulate hazardous operations. Permit applicants should be required to demonstrate that their operations comply with the intent of the code before the permit is issued (see the commentary to Section 105.6 for a general discussion of operations requiring a permit). The three operations listed in this section pose a possible fire hazard because an ignition source close to them would create a hazardous situation. The operations must be reviewed for safety concerns and requirements.

SECTION 1102 DEFINITIONS

1102.1 Definitions. The following words and terms shall, for the purposes of this chapter and as used elsewhere in this code, have the meanings shown herein.

❖ Definitions of terms can help in the understanding and application of the code requirements. The purpose for including those definitions that are associated with the subject matter of this chapter is to provide more convenient access to them without having to refer back to Chapter 2. It is important to emphasize that these terms are not exclusively related to this chapter but are applicable everywhere the term is used in the code. For convenience, these terms are also listed in Chapter 2 with a cross reference to this section. The use and application of all defined terms, including those defined in this section, are set forth in Section 201.

AIRCRAFT OPERATION AREA (AOA). Any area used or intended for use for the parking, taxiing, takeoff, landing or other ground-based aircraft activity.

❖ Any area involving aircraft, whether moving or stationary, is known as the aircraft operating area. This area has special operating procedures and hazards.

AIRPORT. An area of land or structural surface that is used, or intended for use, for the landing and taking off of aircraft with an overall length greater than 39 feet (11 887 mm) and an overall exterior fuselage width greater than 6.6 feet (2012 mm), and any appurtenant areas that are used or intended for use for airport buildings and other airport facilities.

❖ Those structures and operations included in an airport are terminal buildings; maintenance hangars; runways; taxiways; loading and unloading of passengers, luggage and freight; and control towers with all the land involved in performing these functions.

HELIPORT. An area of land or water or a structural surface that is used, or intended for use, for the landing and taking off of helicopters, and any appurtenant areas which are used, or intended for use, for heliport buildings and other heliport facilities.

❖ Heliports present special problems because they are frequently located in congested areas of cities, on roofs of buildings, near hospitals and on piers adjacent to water.

HELISTOP. The same as "Heliport," except that no fueling, defueling, maintenance, repairs or storage of helicopters is permitted.

❖ A helistop is a place for landing and taking off for helicopters with no procedures or operations occurring other than loading or off-loading of passengers or freight.

SECTION 1103 GENERAL PRECAUTIONS

1103.1 Sources of ignition. Open flames, flame-producing devices and other sources of ignition shall not be permitted in a hangar, except in approved locations or in any location within 50 feet (15 240 mm) of an aircraft-fueling operation.

❖ Smoking and carrying any open-flame device within 50 feet (15 240 mm) of any fueling operation is prohibited because of flammable vapors that are likely to be present during fueling operations. Electrical equipment on aircraft is usually not approved for use in hazardous (classified) locations, and disconnection of electrical devices often produces sparks, possibly igniting flammable vapor-air mixtures. Consequently, fueling operations must be discontinued before connecting or disconnecting these devices.

1103.2 Smoking. Smoking shall be prohibited in aircraft-refueling vehicles, aircraft hangars and aircraft operation areas used for cleaning, paint removal, painting operations or fueling. "No Smoking" signs shall be provided in accordance with Section 310.

Exception: Designated and approved smoking areas.

❖ An aircraft hangar is simply a building that provides weather protection and shop space during aircraft maintenance and storage. In the maintenance process many hazards are present. Smoking is prohibited in all areas where an aircraft is located because of the potential presence of fuel vapors. The exception gives the fire code official the authority to evaluate and approve designated smoking areas.

1103.3 Housekeeping. The aircraft operation area (AOA) and related areas shall be kept free from combustible debris at all times.

❖ Housekeeping should be a daily practice. The level of fire safety is greatly improved when areas are kept clean and neat. Make sure that all generated waste is removed from the building and safely disposed of each day. Additionally, keeping the areas in which aircraft operate free from debris reduces the likelihood of foreign object damage (FOD) to aircraft engines that could result in engine damage or catastrophic engine failure.

1103.4 Fire department access. Fire apparatus access roads shall be provided and maintained in accordance with Chapter 5. Fire apparatus access roads and aircraft parking positions shall be designed in a manner so as to preclude the possibility of fire vehicles traveling under any portion of a parked aircraft.

❖ Access by emergency vehicles is an important factor. Fire apparatus access roads must be wide enough, well marked and unobstructed in accordance with Section 503. The access roads should also be arranged so there is no confusion over where emergency vehicles are to go in the event of a fire or rescue emergency in a building, on a runway or in an aircraft. Space between aircraft must be large enough to allow emergency response equipment access to buildings and aircraft.

1103.5 Dispensing of flammable and combustible liquids. The dispensing, transferring and storage of flammable and combustible liquids shall be in accordance with this chapter and Chapter 34. Aircraft motor vehicle fuel-dispensing stations shall be in accordance with Chapter 22.

❖ Section 1106 gives guidelines for dispensing fuel into an aircraft. Section 3406 also gives guidelines for dispensing flammable and combustible liquids. Chapter 22 applies when aircraft and airport service vehicles are brought to a fueling station instead of being fueled from a vehicle.

1103.6 Combustible storage. Combustible materials stored in aircraft hangars shall be stored in approved locations and containers.

❖ Combustible materials storage is to be confined to cut-off rooms or approved metal containers with tight-fitting, self-closing or automatic-closing lids to limit the fuel load readily exposed within the hangar. Approved containers provided for oily rags and similar wastes are to be supplied throughout service areas and emptied every day. Combustible materials should be removed from the building as soon as possible.

1103.7 Hazardous material storage. Hazardous materials shall be stored in accordance with Chapter 27.

❖ Chapter 27 contains the requirements for storing hazardous materials. Requirements in Chapters 28 through 44 also apply to specific materials.

SECTION 1104 AIRCRAFT MAINTENANCE

1104.1 Transferring flammable and combustible liquids. Flammable and combustible liquids shall not be dispensed into or removed from a container, tank, vehicle or aircraft except in approved locations.

❖ Due to hazards presented by aviation fuels as discussed in Section 1106, it is necessary that all storage, transfer or dispensing of flammable and combustible liquids be completed outside of and away from structures. This includes the emptying of fuel tanks and the rooftop refueling of helicopters. The large, undivided areas of aircraft hangars coupled with the dollar value of a single aircraft present an unusually high value at risk of loss due to a single fire incident, thus reinforcing the need for the strict regulation of fuels. Dispensing systems generally involve the transfer of liquid from fixed piping systems, drums or 5-gallon (19 L) cans into smaller end-use containers. Because the release of some vapor is practically unavoidable, dispensing must take place in designated areas.

1104.2 Application of flammable and combustible liquid finishes. The application of flammable or Class II combustible liquid finishes is prohibited unless both of the following conditions are met:

1. The application of the liquid finish is accomplished in an approved location.
2. The application methods and procedures are in accordance with Chapter 15.

❖ Application of flammable finishes must comply with Chapter 15. Most exterior aircraft painting is performed using spray apparatus, frequently in aircraft hangars. Usually, control of ignition sources, ventilation and the considerable volume of aircraft hangars is relied on to minimize the hazards typically associated with spraying in more confined areas where vapor-air mixtures can rapidly create an explosive mixture. Small parts and subassemblies should be removed and painted in approved spray booths or areas complying with the requirements of Chapter 15. Exterior painting should not be performed in aircraft hangars not protected throughout by approved automatic fire suppression systems. If systems are inoperable, exterior spray painting is not permitted; only interior painting using water-based products is permitted. Like small exterior parts and subassemblies, application of flammable finishes to removable interior components should be limited to approved spray booths or spray rooms.

1104.3 Cleaning parts. Class IA flammable liquids shall not be used to clean aircraft, aircraft parts or aircraft engines. Cleaning with other flammable and combustible liquids shall be in accordance with Section 3405.3.6.

❖ Class I flammable liquids with flash points below 100° F (38°C) must not be used for cleaning that typically liberates large quantities of flammable vapor and may leave a flammable residue that is easily ignited. Removable parts should be cleaned in approved parts-cleaning machines already tested and labeled for such a purpose. The hazards associated with cleaning an aircraft, aircraft parts and aircraft engines with Class IA flammable liquids are ignition of fires, possible explosions, damaging property and loss of life. Section 3405.3.6 gives requirements for cleaning with Class I, II and IIIA liquids.

1104.4 Spills. This section shall apply to spills of flammable and combustible liquids and other hazardous materials. Fuel spill control shall also comply with Section 1106.11.

❖ The following procedures should be adhered to in the event a spill occurs. The specific requirements for fuel spill prevention are found in Section 1106.11.

1104.4.1 Cessation of work. Activities in the affected area not related to the mitigation of the spill shall cease until the spilled material has been removed or the hazard has been mitigated.

❖ All ongoing activity must stop in the area of a spill until the spill has been cleaned up and removed, since those ongoing activities may cause an ignition to occur. The area should be clear of all hazards before work is resumed.

1104.4.2 Vehicle movement. Aircraft or other vehicles shall not be moved through the spill area until the spilled material has been removed or the hazard has been mitigated.

❖ The movement of vehicles may create an ignition source for the flammable liquids that spilled. Stopping all movement of vehicles reduces the possibility of a fire or an explosion significantly.

1104.4.3 Mitigation. Spills shall be reported, documented and mitigated in accordance with the provisions of this chapter and Section 2703.3.

❖ Any fuel spill, whatever the amount, must be reported to the proper authorities and documented to record the spill details and what was done to clean up the spill. Chapter 27 gives specific procedures for handling a spill.

1104.5 Running engines. Aircraft engines shall not be run in aircraft hangars except in approved engine test areas.

❖ An approved engine test area should have proper ventilation, engine noise control and be separated from other areas of operation. Running engines could create ignition sources that could cause fire or explosions as well as ventilation and noise hazards for the surrounding employees.

1104.6 Open flame. Repairing of aircraft requiring the use of open flames, spark-producing devices or the heating of parts above 500°F (260°C) shall only be done outdoors or in an area complying with the provisions of the *International Building Code* for a Group F-1 occupancy.

❖ No heat-producing, welding, cutting or blow-torch devices should be used inside hangars. Their use is restricted to areas that meet the requirements of a Group F-1 occupancy in the *International Mechanical Code®* (IMC®). Flare pots and other open-flame lights are also included in this category.

SECTION 1105
PORTABLE FIRE EXTINGUISHERS

1105.1 General. Portable fire extinguishers suitable for flammable or combustible liquid and electrical-type fires shall be provided as specified in Sections 1105.2 through 1105.6 and Section 906. Extinguishers required by this section shall be inspected and maintained in accordance with Section 906.

❖ Fire extinguishers must be approved for use on Class B and C fires. Placement and distribution of fire extinguishers should conform to NFPA 10 and 407 and Section 906 of the code. Generally, portable fire extinguishers are required in the immediate vicinity of all flammable and combustible liquid storage; use and dispensing; welding and cutting; spray finishing and other maintenance operations, as well as on aircraft fueler and service vehicles.

1105.2 On towing vehicles. Vehicles used for towing aircraft shall be equipped with a minimum of one listed portable fire extinguisher complying with Section 906 and having a minimum rating of 20-B:C.

❖ Tow motors and other towing vehicles must be equipped with a fire extinguisher that is readily available if a fire occurs away from a service, maintenance or boarding area.

1105.3 On welding apparatus. Welding apparatus shall be equipped with a minimum of one listed portable fire extinguisher complying with Section 906 and having a minimum rating of 2-A:20-B:C.

❖ Consistent with Section 2604.2.6, a fire extinguisher is required on all welding apparatus so that it is readily available during welding or cutting operations outside a welding or cutting shop area.

1105.4 On aircraft fuel-servicing tank vehicles. Aircraft fuel-servicing tank vehicles shall be equipped with a minimum of two listed portable fire extinguishers complying with Section 906, each having a minimum rating of 20-B:C. A portable fire extinguisher shall be readily accessible from either side of the vehicle.

❖ Fuel-servicing tank vehicles for aircraft must have a portable fire extinguisher on each side of the vehicle. Both extinguishers must be easily accessible and not be obstructed. Each fire extinguisher must be effective for the extinguishment of a flammable liquid fire and also be effective for energized electrical components.

1105.5 On hydrant fuel-servicing vehicles. Hydrant fuel-servicing vehicles shall be equipped with a minimum of one listed portable fire extinguisher complying with Section 906, and having a minimum rating of 20-B:C.

❖ Hydrant fuel-servicing vehicles must be equipped with one portable extinguisher that is effective for the extinguishment of a flammable liquid fire and is also effective for energized electrical components.

1105.6 At fuel-dispensing stations. Portable fire extinguishers at fuel-dispensing stations shall be located such that pumps or dispensers are not more than 75 feet (22 860 mm) from one such extinguisher. Fire extinguishers shall be provided as follows:

1. Where the open-hose discharge capacity of the fueling system is not more than 200 gallons per minute (13 L/s), a minimum of two listed portable fire extinguishers complying with Section 906 and having a minimum rating of 20-B:C shall be provided.
2. Where the open-hose discharge capacity of the fueling system is more than 200 gallons per minute (13 L/s) but not more than 350 gallons per minute (22 L/s), a minimum of one listed wheeled extinguisher complying with Section 906 and having a minimum extinguishing rating of 80-B:C, and a minimum agent capacity of 125 pounds (57 kg), shall be provided.
3. Where the open-hose discharge capacity of the fueling system is more than 350 gallons per minute (22 L/s), a minimum of two listed wheeled extinguishers complying with Section 906 and having a minimum rating of 80-B:C each, and a minimum capacity agent of 125 pounds (57 kg) of each, shall be provided.

❖ This section requires portable fire extinguishers with ratings based on the anticipated discharge rate of a broken or ruptured fuel hose.

NFPA 407 contains requirements for the inspection

and maintenance of an aircraft fueling hose, including daily preuse inspection and removal from service of obviously defective hoses. Despite these inspections, however, hoses and fittings can and do fail for a variety of reasons (e.g., unnoticed physical damage, coupling and fitting failure, overpressure rupture, etc.), resulting in a flailing hose "open butt" discharge of fuel under the full pressure of the fueling system. Such uncontrolled fuel discharge could flow under the aircraft; fueling vehicles; passenger stairs or ramps; baggage-handling equipment or in close proximity to building openings. If a hose were to rupture on top of an aircraft wing or a flailing hose were to spray fuel on vehicles, baggage carts, etc., the resulting hazard would increase beyond a simple spill fire. The large amount of property damage and the potential loss of life requires that sufficient portable extinguishers of an adequate size be located in the fueling area. (Note: The extinguishers required by this section are in addition to others required on vehicles that may be present in the fuel area.)

Considerations in locating extinguishers during fueling operations include placing them out and upwind of the fuel-dispensing site and potential spill area, as well as within the access travel distance specified in NFPA 10 for extra-hazard locations. When two fire extinguishers are required, they should be located close enough to each other so NFPA 10 access travel distances are not exceeded and a spill incident does not prevent access to or use of both appliances.

FAA regulations and NFPA 407 require refueling personnel to receive fire extinguisher and fire safety training. Appendix A of NFPA 407 recommends that such training include live-fire exercises. Training should be adequately detailed so that supervisors are capable of properly indoctrinating their subordinates in fire safety essentials. Topics covered in the training program should include electrical bonding and grounding; maintenance of aircraft egress; emergency shutdown of fuel servicing equipment; notification of emergency forces and supporting emergency operations.

The low flow rate in Item 1 requires that at least two 20-B:C hand-held fire extinguishers are provided. It also assumes that the trained personnel available will be able to handle the relatively small anticipated spill. Such extinguishers typically discharge for up to ± 25 seconds for distances up to ± 20 feet (± 6096 mm).

Item 2 states that a ruptured hose discharging up to 350 gallons per minute (gpm) (1325 L/m) creates a potentially larger spill area and a more challenging fire for first-aid appliances. The higher required "B" rating requires a quantity of extinguishing agent, usually dry chemical, that likely exceeds 50 pounds (28 Kg), depending on the agent. Accordingly, a wheeled extinguisher will enable a single operator more mobility in moving the extinguisher for fire attack. The size of the wheeled unit, in addition to its longer discharge hose, allows the operator a greater agent discharge time, a higher agent flow rate, a greater agent discharge distance and more mobility in the hazardous area. The potential for such large fuel discharges increases the extinguishing requirements. For the same reasons discussed under Item 2, a minimum of two wheeled units are required to allow a more aggressive fire attack.

1105.7 Fire extinguisher access. Portable fire extinguishers required by this chapter shall be accessible at all times. Where necessary, provisions shall be made to clear accumulations of snow, ice and other forms of weather-induced obstructions.

❖ Unobstructed access to portable fire extinguishers is essential. In colder climates, snow and ice may block access and must be removed because fire can occur at any time.

1105.7.1 Cabinets. Cabinets and enclosed compartments used to house portable fire extinguishers shall be clearly marked with the words FIRE EXTINGUISHER in letters at least 2 inches (51 mm) high. Cabinets and compartments shall be readily accessible at all times.

❖ In an emergency, people can panic and become confused. Labeling cabinets where portable fire extinguishers are housed with letters 2 inches (51 mm) (often in red) high makes the extinguishers easier to locate.

1105.8 Reporting use. Use of a fire extinguisher under any circumstances shall be reported to the manager of the airport and the fire code official immediately after use.

❖ The fire code official is responsible for the investigation of fires within the jurisdiction and for maintaining records thereof. Likewise, the airport manager is responsible for all activities and events within the airport. Both persons must be notified of extinguisher use so the circumstances of the event can be investigated and appropriate follow-up procedures initiated to mitigate the hazard that resulted in the incident. Discharged extinguishers must be promptly replaced with serviceable units.

SECTION 1106
AIRCRAFT FUELING

1106.1 Aircraft motor vehicle fuel-dispensing stations. Aircraft motor vehicle fuel-dispensing stations shall be in accordance with Chapter 22.

❖ Requirements for fuel-dispensing stations for aircraft motor vehicles are found in Chapter 22. This provision addresses the dispensing of fuel into small general aviation-type aircraft and airport vehicles at stationary fuel-dispensing facilities that use equipment similar to that used at automotive service stations rather than fuel hydrants or fuel tanker trucks that are used on larger aircraft.

1106.2 Airport fuel systems. Airport fuel systems shall be designed and constructed in accordance with NFPA 407.

❖ Aviation fuels present a wide range of hazards. The fuel, ambient temperature, control of ignition sources, drainage, availability of fire protection equipment and the experience and training of fuel-service personnel have the greatest influence over the outcome of fueling accidents. Consequently, this section references NFPA 407, specifying requirements for the design and operation of fueling installations, vehicles and procedures.

At normal ambient temperatures, the kerosene-grade fuels are not readily ignitable, which may explain their popularity. When spilled on a warm aircraft apron, however, kerosene-grade fuels can be readily heated above their flash points. Once ignited, most aviation fuels exhibit relatively similar burning characteristics (see Figure 1106.2 for information on common aviation fuels).

Fuel spills are relatively uncommon occurrences compared to the daily number of refuelings that occur. Most fuel spills occur as the result of a slow or faulty internal shutoff valve that causes overfilling of the tank, resulting in fuel escaping through the tank vent point. To prevent or minimize such accidents, fuel shutoffs and fail-safe, self-closing valves should be exercised and inspected regularly. In addition faulty valves and equipment should be removed from service and repaired or replaced immediately. All spills must be promptly reported to airport fire-fighting personnel and investigated to determine their cause.

The most common ignition source in liquid fuel spills is static electricity. The kerosene and kerosene-gasoline blends are more electrostatically active than AVGAS, and transfer operations may generate considerable amounts of static electricity; therefore, prior to most fueling operations both the aircraft and the refueler shall be independently grounded and then bonded to one another either by the filling hose or a separate bonding line (see the commentary to Section 1106.5.2 and its subsections for further information).

1106.3 Construction of aircraft-fueling vehicles and accessories. Aircraft-fueling vehicles shall comply with this section and shall be designed and constructed in accordance with NFPA 407.

❖ The following sections apply to vehicles operated for refueling aircraft. The sections address transfer apparatus, pumps, dispensing, electrical equipment, venting and smoking. The design and construction of the vehicle tanks, trailers, piping, exhaust system, lighting, venting and safe operating procedures parallel those in NFPA 407, Chapter 2.

1106.3.1 Transfer apparatus. Aircraft-fueling vehicles shall be equipped and maintained with an approved transfer apparatus.

❖ The transfer apparatus installed on the fuel-servicing vehicle must be approved and tested for transferring fuel into an aircraft. All sections involved in a flammable

	Gasoline	Kerosene Grades	Blends of Kerosene and Gasoline
Commercial Designation	AVGAS	JET A, JET A-1	JET B
Military Designation		JP-5, JP-6, JP-8	JP-4
Characteristics			
Freezing Point	-76° F	-49 F to -58° F	-60° F
Vapor Pressure	5.5 to 7.0 psi	0.1 psi	2.0 to 3.0 psi
Flash Point (Closed Cup Method @ MSL)	-50° F	95 to 145° F	-10 to +30° F
Flash Point (Air Saturation Method)	-75 F to -85° F	None	-60° F
Flammable Range Lower Upper Temperature Range	1.5% 7.6% -50 to +30° F	0.74% 5.32% 95 to 165° F	1.16% 7.63% -10 to +100° F
Auto-ignition Temperature	825 to 960° F	440 to 475° F	470 to 480° F
Boiling Points Initial End	110° F 325° F	325° F 450° F	135° F 485° F
Pool Rate of Flame Spread	700 to 800 feet per minute	≤ 100 feet per minute	700 to 800 feet per minute

Source: Brenneman, J.J. *Industrial Fire Hazards Handbook.*

Figure 1106.2
PHYSICAL AND FLAMMABILITY CHARACTERISTICS OF AVIATION FUELS

liquid feed system should be constructed and located to minimize a fire hazard.

1106.3.1.1 Internal combustion type. Where such transfer apparatus is operated by an individual unit of the internal-combustion-motor type, such power unit shall be located as remotely as practicable from pumps, piping, meters, air eliminators, water separators, hose reels, and similar equipment, and shall be housed in a separate compartment from any of the aforementioned items. The fuel tank in connection therewith shall be suitably designed and installed, and the maximum fuel capacity shall not exceed 5 gallons (19 L) where the tank is installed on the engine. The exhaust pipe, muffler and tail pipe shall be shielded.

❖ Isolation of an internal combustion engine from the fuel transfer system helps to control possible ignition sources. The ignition sources need to be shielded and equipment should be housed in separate compartments.

1106.3.1.2 Gear operated. Where operated by gears or chains, the gears, chains, shafts, bearings, housing and all parts thereof shall be of an approved design and shall be installed and maintained in an approved manner.

❖ The gears and other associated parts should be covered and protected from damage, whether the damage comes from environmental or mechanical sources. The design must be approved for the function. Maintenance should be scheduled to provide consistent and proper operation.

1106.3.1.3 Vibration isolation. Flexible connections for the purpose of eliminating vibration are allowed if the material used therein is designed, installed and maintained in an approved manner, provided such connections do not exceed 24 inches (610 mm) in length.

❖ Because hoses are the weak point in any system and they are necessary for flexible connections, they should be kept as short as possible. Hose length must not exceed 24 inches (610 mm).

1106.3.2 Pumps. Pumps of a positive-displacement type shall be provided with a bypass relief valve set at a pressure of not more than 35 percent in excess of the normal working pressure of such unit. Such units shall be equipped and maintained with a pressure gauge on the discharge side of the pump.

❖ A relief valve is needed on positive-displacement pumps to prevent high-pressure damage or an explosion within the fuel feed system. At 35 percent over the normal operating pressure the relief valve will allow the excess pressure to escape, thus preventing a dangerous situation. A pressure gauge also allows monitoring of pump operation to detect overpressure conditions.

1106.3.3 Dispensing hoses and nozzles. Hoses shall be designed for the transferring of hydrocarbon liquids and shall not be any longer than necessary to provide efficient fuel transfer operations. Hoses shall be equipped with an approved shutoff nozzle. Fuel-transfer nozzles shall be self-closing and designed to be actuated by hand pressure only. Notches and other devices shall not be used for holding a nozzle valve handle in the open position. Nozzles shall be equipped with a bonding cable complete with proper attachment for aircraft to be serviced.

❖ Nozzles having notches for holding a nozzle valve handle in the open position creates a risk of an overflow/spill situation if the handle sticks in the open position. The shutoff nozzle also aids in the prevention of an overflow/spill situation. Bonding the nozzle to the aircraft helps to dissipate static electricity that is generated in the fueling operation.

1106.3.4 Protection of electrical equipment. Electric wiring, switches, lights and other sources of ignition, when located in a compartment housing piping, pumps, air eliminators, water separators, hose reels or similar equipment, shall be enclosed in a vapor-tight housing. Electrical motors located in such a compartment shall be of a type approved for use as specified in ICC *Electrical Code.*

❖ Because electrical equipment can be a serious sparking source, precautions must be taken for the elimination of this potential ignition source and possible fire hazard.

1106.3.5 Venting of equipment compartments. Compartments housing piping, pumps, air eliminators, water separators, hose reels and similar equipment shall be adequately ventilated at floor level or within the floor itself.

❖ Venting compartments housing this equipment provides both dilution air to keep possible air-vapor mixtures below the flammable range and airflow for dissipation of any vapors present.

1106.3.6 Accessory equipment. Ladders, hose reels and similar accessory equipment shall be of an approved type and constructed substantially as follows:

1. Ladders constructed of noncombustible material are allowed to be used with or attached to aircraft-fueling vehicles, provided the manner of attachment or use of such ladders is approved and does not constitute an additional fire or accident hazard in the operation of such fueling vehicles.
2. Hose reels used in connection with fueling vehicles shall be constructed of noncombustible materials and shall be provided with a packing gland or other device which will preclude fuel leakage between reels and fuel manifolds.

❖ Ladders made of noncombustible materials and securely attached help to eliminate a potential spark ignition source as well as removing the possibility of falling equipment breaking a component of the fuel system.

Hose reels constructed of noncombustible material help to eliminate the possibility of an ignition source causing a fire hazard. The installation of a packing gland or other device helps to eliminate the possibility of a fuel leak causing a fire hazard.

1106.3.7 Electrical bonding provisions. Transfer apparatus shall be metallically interconnected with tanks, chassis, axles and springs of aircraft-fueling vehicles.

❖ Fuel transfer vehicles must be bonded together as a unit. This helps to eliminate the electrostatic charge created by flowing fuel through pipes and hoses.

1106.3.7.1 Bonding cables. Aircraft-fueling vehicles shall be provided and maintained with a substantial heavy-duty electrical cable of sufficient length to be bonded to the aircraft to be serviced. Such cable shall be metallically connected to the transfer apparatus or chassis of the aircraft-fueling vehicle on one end and shall be provided with a suitable metal clamp on the other end, to be fixed to the aircraft.

❖ The fueling vehicle and the aircraft must be bonded together with a cable before making any fueling connection to the aircraft. This provides a conductive path to equalize the electrostatic charge between the fueling vehicle and the aircraft.

1106.3.7.2 Bonding cable protection. The bonding cable shall be bare or have a transparent protective sleeve and be stored on a reel or in a compartment provided for no other purpose. It shall be carried in such a manner that it will not be subjected to sharp kinks or accidental breakage under conditions of general use.

❖ Bonding cable that is worn, frayed or kinked cannot make good contact or provide a continuous path to eliminate the electrostatic charge. Bonding cables require special care in storage and handling.

1106.3.8 Smoking. Smoking in aircraft-fueling vehicles is prohibited. Signs to this effect shall be conspicuously posted in the driver's compartment of all fueling vehicles.

❖ Because flammable vapors being discharged during refueling can accumulate in closed areas, such as the driver's compartment of the fueling vehicle, smoking in that compartment is strictly prohibited. Signs must be posted conspicuously in the driver's compartment as a warning.

1106.3.9 Smoking equipment. Smoking equipment such as cigarette lighters and ash trays shall not be provided in aircraft-fueling vehicles.

❖ Because smoking is prohibited in fueling vehicles the elimination of cigarette lighters and ashtrays reinforces this regulation.

1106.4 Operation, maintenance and use of aircraft-fueling vehicles. The operation, maintenance and use of aircraft-fueling vehicles shall be in accordance with Sections 1106.4 through 1106.4.4 and other applicable provisions of this chapter.

❖ Fuel servicing equipment must comply with the requirements of the following sections and must be maintained in safe operating condition. Leaking or malfunctioning equipment must be removed from service.

1106.4.1 Proper maintenance. Aircraft-fueling vehicles and all related equipment shall be properly maintained and kept in good repair. Accumulations of oil, grease, fuel and other flammable or combustible materials is prohibited. Maintenance and servicing of such equipment shall be accomplished in approved areas.

❖ Fueling vehicles should be inspected every day. The accumulation of oil, grease or any other material creates a fire hazard if it is ignited by an ignition source. A clean, well-maintained fueling vehicle is a high priority in the line of fire hazard prevention. The cleaning should take place in an area that is constructed and approved for such a task.

1106.4.2 Vehicle integrity. Tanks, pipes, hoses, valves and other fuel delivery equipment shall be maintained leak free at all times.

❖ All fuel delivery equipment should be checked every day for detection of leaks. If a leak occurs, the probability of a fire occuring increases.

1106.4.3 Removal from service. Aircraft-fueling vehicles and related equipment which are in violation of Section 1106.4.1 or 1106.4.2 shall be immediately defueled and removed from service and shall not be returned to service until proper repairs have been made.

❖ Whenever a problem is discovered with the vehicle, tank, pipes, hoses, valves or any equipment related to aircraft fueling, the vehicle must be taken out of service immediately and repaired. Having a vehicle in service that is not working properly increases the potential for a fire.

1106.4.4 Operators. Aircraft-fueling vehicles that are operated by a person, firm or corporation other than the permittee or the permittee's authorized employee shall be provided with a legible sign visible from outside the vehicle showing the name of the person, firm or corporation operating such unit.

❖ A change in the operator, firm or corporation from the permit holder of a fueling vehicle requires notification. In addition to verbal or written communication, a sign that can be read from outside the vehicle must be posted on or in the vehicle notifying observers that someone different is performing the procedure.

1106.5 Fueling and defueling. Aircraft-fueling and defueling operations shall be in accordance with Section 1106.5.

❖ This section covers the requirements for positioning and bonding the fueling vehicle. It also covers the requirements for training of fuel transfer personnel and their responsibilities.

1106.5.1 Positioning of aircraft fuel-servicing vehicles. Aircraft-fueling vehicles shall not be located, parked or permitted to stand in a position where such unit would obstruct egress from an aircraft should a fire occur during fuel-transfer operations. Tank vehicles shall not be located, parked or permitted to stand under any portion of an aircraft.

❖ Safety of crew and passengers is always the highest priority. This section instructs operators to always position their fuel service vehicles away from all parts of an aircraft and always in a location that leaves aircraft egress paths free of obstruction.

1106.5.1.1 Fueling vehicle egress. A clear path shall be maintained for aircraft-fueling vehicles to provide for prompt and timely egress from the fueling area.

❖ The requirement for a clear and unobstructed path into and out of the fueling area calls for cooperation between refueling crews and other ground support personnel to make sure the path exists. This allows for a safe, timely and efficient fueling operation.

1106.5.1.2 Aircraft vent openings. A clear space of at least 10 feet (3048 mm) shall be maintained between aircraft fuel-system vent openings and any part or portion of an aircraft-fueling vehicle.

❖ Fuel vapors are released from fuel-system vent openings. The 10-foot (3048 mm) radius gives a distance sufficient for vapor dissipation.

1106.5.1.3 Parking. Prior to leaving the cab, the aircraft-fueling vehicle operator shall ensure that the parking brake has been set. At least two chock blocks not less than 5 inches by 5 inches by 12 inches (127 mm by 127 mm by 305 mm) in size and dished to fit the contour of the tires shall be utilized and positioned in such a manner as to preclude movement of the vehicle in any direction.

❖ This requirement for securing the fueling vehicle is an important safety consideration. Even a slight grade or dishing of the pavement could cause an unsecured vehicle to roll far enough to strain the fueling link into leaking, disconnecting or breaking, which would create the possibility of a spill and a hazardous situation.

1106.5.2 Electrical bonding. Aircraft-fueling vehicles shall be electrically bonded to the aircraft being fueled or defueled. Bonding connections shall be made prior to making fueling connections and shall not be disconnected until the fuel-transfer operations are completed and the fueling connections have been removed.

Where a hydrant service vehicle or cart is used for fueling, the hydrant coupler shall be connected to the hydrant system prior to bonding the fueling equipment to the aircraft.

❖ Bonding increases the likelihood that the receiving tank (aircraft or vehicle) and the fueling/defueling equipment have the same electrical potential and provides a path for the charges in all parts of the fueling system to neutralize. The bonding of the fuel vehicle and aircraft must be completed before any fuel is moved and must remain in place until after the fueling operation is complete.

1106.5.2.1 Conductive hose. In addition to the bonding cable required by Section 1106.5.2, conductive hose shall be used for all fueling operations.

❖ The use of conductive hose helps to prevent electrostatic discharge. It is not a substitute for required bonding.

1106.5.2.2 Bonding conductors on transfer nozzles. Transfer nozzles shall be equipped with approved bonding conductors which shall be clipped or otherwise positively engaged with the bonding attachment provided on the aircraft adjacent to the fuel tank cap prior to removal of the cap.

Exception: In the case of overwing fueling where no appropriate bonding attachment adjacent to the fuel fill port has been provided on the aircraft, the fueling operator shall touch the fuel tank cap with the nozzle spout prior to removal of the cap. The nozzle shall be kept in contact with the fill port until fueling is completed.

❖ This procedure provides a conductive path for equalizing the potential electrostatic charge between the aircraft and the transfer nozzle. Even if a clip or engaging mechanism is not provided, contact with the fill port must be maintained until the fueling process is complete.

1106.5.2.3 Funnels. Where required, metal funnels are allowed to be used during fueling operations. Direct contact between the fueling receptacle, the funnel and the fueling nozzle shall be maintained during the fueling operation.

❖ Prevention of potential sparks that produce ignition sources for fires is the objective for keeping metal funnels in direct contact with fueling nozzles and receptacles at all times.

1106.5.3 Training. Aircraft-fueling vehicles shall be attended and operated only by persons instructed in methods of proper use and operation and who are qualified to use such fueling vehicles in accordance with minimum safety requirements.

❖ Only personnel trained in the proper and safe operation of the equipment, emergency controls and emergency procedures are allowed to fuel or defuel aircraft.

1106.5.3.1 Fueling hazards. Fuel-servicing personnel shall know and understand the hazards associated with each type of fuel dispensed by the airport fueling-system operator.

❖ Personnel performing fueling or defueling procedures must be familiar with every hazard that is in involved with each and every fuel that is dispensed.

1106.5.3.2 Fire safety training. Employees of fuel agents who fuel aircraft, accept fuel shipments or otherwise handle fuel shall receive approved fire safety training.

❖ Employees must be trained in and be aware of fire safety rules and precautions involving fuel servicing processes. Training must be in areas involving vehicle integrity and placement; bonding of vehicle; aircraft;

hose connections; valves and fuel process equipment. The training must be an approved in-depth program that is recognized by airport authorities.

1106.5.3.2.1 Fire extinguisher training. Fuel-servicing personnel shall receive approved training in the operation of fire-extinguishing equipment.

❖ Employees involved in fuel servicing must have adequate training with extinguishers to use them effectively in the event of an emergency (see commentary, Section 1105.6 for further information).

1106.5.3.2.2 Documentation. The airport fueling-system operator shall maintain records of all training administered to its employees. These records shall be made available to the fire code official on request.

❖ Training must be documented and records kept on premises for verification, if requested, that the personnel have received the proper training.

1106.5.4 Transfer personnel. During fuel-transfer operations, a qualified person shall be in control of each transfer nozzle and another qualified person shall be in immediate control of the fuel-pumping equipment to shut off or otherwise control the flow of fuel from the time fueling operations are begun until they are completed.

Exceptions:

1. For underwing refueling, the person stationed at the point of fuel intake is not required.
2. For overwing refueling, the person stationed at the fuel pumping equipment shall not be required where the person at the fuel dispensing device is within 75 feet (22 800 mm) of the emergency shutoff device, is not on the wing of the aircraft and has a clear and unencumbered path to the fuel pumping equipment; and, the fuel dispensing line does not exceed 50 feet (15 240 mm) in length.

The fueling operator shall monitor the panel of the fueling equipment and the aircraft control panel during pressure fueling or shall monitor the fill port during overwing fueling.

❖ Two qualified trained individuals must perform fuel transfer procedures. One individual controls the nozzles. The other individual must be in the control area for flow monitoring and control and immediate shutdown of fuel flow in case of an emergency (overfill, spill, fire, etc.). The exception for underwing fueling is appropriate because underwing operations require a liquid-tight connection that does not require constant monitoring to ensure freedom from spills and overflows.

The four items that must be met in the overwing refueling process are meant for quick response in case a problem occurs. The individual must be completely alert to the fueling process and free from distractions if alone.

1106.5.5 Fuel flow control. Fuel flow-control valves shall be operable only by the direct hand pressure of the operator. Removal of the operator's hand pressure shall cause an immediate cessation of the flow of fuel.

❖ Deadman controls should be designed so that the operator, while wearing gloves, can hold them for the time required to complete the operation.

1106.6 Emergency fuel shutoff. Emergency fuel shutoff controls and procedures shall comply with Sections 1106.6.1 through 1106.6.4.

❖ The following sections address emergency fuel shutoff accessibility, fire department notification, determining cause if shutoff activates and testing procedures.

1106.6.1 Accessibility. Emergency fuel shutoff controls shall be readily accessible at all times when the fueling system is being operated.

❖ The emergency controls must be unobstructed whenever the fueling system is operating. Time should not be wasted having to move objects or material blocking access to the controls before fueling operations can be shut down in an emergency.

1106.6.2 Notification of the fire department. The fueling-system operator shall establish a procedure by which the fire department will be notified in the event of an activation of an emergency fuel shutoff control.

❖ A plan must be developed for notifying the fire department. The plan should be in a written format and reviewed with trained and certified employees in advance of fueling operations. Notification of the fire department will have emergency responders on their way immediately in case fire or rescue operations are required.

1106.6.3 Determining cause. Prior to reestablishment of normal fuel flow, the cause of fuel shutoff conditions shall be determined and corrected.

❖ If for any reason the flow of fuel is discontinued by the emergency shutoff systems during fueling operations, the cause of the activation must be determined and the problem fixed before fuel flow is started again. Fuel flow may not be restarted if the cause of the shutdown is not determined. If fuel flow were restarted, the problem may occur again and more serious problems could develop.

1106.6.4 Testing. Emergency fuel shutoff devices shall be operationally tested at intervals not exceeding three months. The fueling-system operator shall maintain suitable records of these tests.

❖ Testing of the emergency fuel shutoff devices increases the likelihood that the devices will function properly. Although three-month intervals are the maximum allowed, the test may be performed more frequently if desired for more ensurance. Records are required for documentation that the tests were performed. If an accident occurs, these records will be a great asset for verification that the devices were tested and operational.

1106.7 Protection of hoses. Before an aircraft-fueling vehicle is moved, fuel transfer hoses shall be properly placed on the approved reel or in the compartment provided, or stored on the top decking of the fueling vehicle if proper height rail is provided for security and protection of such equipment. Fuel-transfer hose shall not be looped or draped over any part of the fueling vehicle, except as herein provided. Fuel-transfer hose shall not be dragged when such fueling vehicle is moved from one fueling position to another.

❖ Accidents or fires may occur if hoses are not properly stored and handled. Sparks caused by metal nozzles or hose ends/couplings striking pavement when hoses are dragged from one area to another are a serious potential ignition source. Also the hoses are least likely to be damaged if stored in proper locations.

1106.8 Loading and unloading. Aircraft-fueling vehicles shall be loaded only at an approved loading rack. Such loading racks shall be in accordance with Section 3406.5.1.12.

Exceptions:

1. Aircraft-refueling units may be loaded from the fuel tanks of an aircraft during defueling operations.
2. Fuel transfer between tank vehicles is allowed to be performed in accordance with Section 3406.6 when the operation is at least 200 feet (60 960 mm) from an aircraft.

The fuel cargo of such units shall be unloaded only by approved transfer apparatus into the fuel tanks of aircraft, underground storage tanks or approved gravity storage tanks.

❖ Section 3406.5.1.12 addresses requirements for loading racks. Loading racks, platforms and stairs must be constructed of noncombustible materials. Buildings for pumps and buildings that shelter personnel can be part of the loading rack. The area within 25 feet (7620 mm) of a loading platform shall be electrically classified in accordance with Table 3403.1.1.

Transfer of fuel from one tank vehicle to another and from aircraft to vehicle is allowable using approved methods and procedures. At times, loading and unloading fuel from aircraft and refueling units is necessary when taking units out of service and when defueling an aircraft while preparing for maintenance or repair.

An aircraft refueling unit is normally loaded either for refueling an aircraft or for storage. Fuel must not be unloaded into any other type tank, structure or vehicle.

1106.9 Passengers. Passenger traffic is allowed during the time fuel transfer operations are in progress, provided the following provisions are strictly enforced by the owner of the aircraft or the owner's authorized employee:

1. Smoking and producing an open flame in the cabin of the aircraft or the outside thereof within 50 feet (15 240 mm) of such aircraft shall be prohibited.

A qualified employee of the aircraft owner shall be responsible for seeing that the passengers are not allowed to smoke when remaining aboard the aircraft or while going across the ramp from the gate to such aircraft, or vice versa.

2. Passengers shall not be permitted to linger about the plane, but shall proceed directly between the loading gate and the aircraft.
3. Passenger loading stands or walkways shall be left in loading position until all fuel transfer operations are completed.
4. Fuel transfer operations shall not be performed on the main exit side of any aircraft containing passengers except when the owner of such aircraft or a capable and qualified employee of such owner remains inside the aircraft to direct and assist the escape of such passengers through regular and emergency exits in the event fire should occur during fuel transfer operations.

❖ Special precautions must be taken to protect passengers during fueling operations. If passengers are permitted on or around the aircraft during refueling operations, an employee of the airline or a representative of the airport authority must remain on board and the specified precautions must be taken.

Aircraft fueling; servicing; baggage and cargo loading; movement and passenger boarding are generally conducted in and around the passenger terminal building on the aircraft apron. Aircraft refueling is often accomplished by special refueling vehicles acting like mobile gas stations, transferring fuel directly from their tanks to aircraft fuel tanks. Aircraft fuel tanks are usually topped off after each flight, unless they are making scheduled stops on a multistop flight.

Terminal building discharge to the aircraft apron is not permitted by FAA regulations; however, egress from the aircraft to the apron in the event of a fire or other emergency is permitted. Passengers and unauthorized personnel must be immediately accounted for and escorted from the air operations area.

The terminal building is not required to be separated from the aircraft apron by fire-resistance-rated construction; however, it is considered good practice to take reasonable steps to prevent fire from fueling accidents or other mishaps. Portable fire extinguishers are required on the aircraft apron in accordance with Section 1105.

In Item 1, smoking and open flames are prohibited within and around the aircraft during refueling operations because of the potential for fuel vapors being present. The "no-smoking" signs located in the aircraft cabin must remain illuminated while the aircraft is parked at the gate during refueling operations.

Item 2 requires that deplaning passengers must proceed directly to the terminal and may not remain on the aircraft apron during refueling operations, cargo loading or other similar service functions in order to avoid potential injury.

In Item 3, egress stairways and passageways must remain connected or intact during refueling operations. These stairways and ramps are the primary exit if an emergency occurs during aircraft refueling operations.

Item 4 recognizes that maintenance and fueling operations represent the most serious fire safety concerns at airports and heliports. Fueling represents perhaps the greatest risk because it is commonly conducted near passenger and cargo handling areas and buildings, while maintenance operations are confined to hangars or designated outside maintenance areas. Fueling operations are to be conducted in a manner that avoids locating the refueler on the primary exit side of the aircraft. Airline employees or airport authorities are responsible for directing and assisting egress in the event of an emergency and must direct passengers to exits away from the fueling operation.

1106.10 Sources of ignition. Smoking and producing open flames within 50 feet (15 240 mm) of a point where fuel is being transferred shall be prohibited. Electrical and motor-driven devices shall not be connected to or disconnected from an aircraft at any time fueling operations are in progress on such aircraft.

❖ The 50-foot (15 240 mm) distance is considered a safe distance from the fuel transfer operation for dissipation of any escaping vapors. Prohibiting open flames and smoking eliminates a prime ignition source (examples of open flame devices are lighted cigarettes, cigars, pipes, cigarette lighters, etc.). Because disconnection or connection of any electrical or motor-driven device could create a spark, these activities are not allowed during refueling.

1106.11 Fuel spill prevention and procedures. Fuel spill prevention and the procedures for handling spills shall comply with Sections 1106.11.1 through 1106.11.7.

❖ The following sections address equipment maintenance, fuel nozzles, fueling drums, fuel spill procedures, fire department notification and prohibited procedures for multiple fuel deliveries.

1106.11.1 Fuel-service equipment maintenance. Aircraft fuel-servicing equipment shall be maintained and kept free from leaks. Fuel-servicing equipment that malfunctions or leaks shall not be continued in service.

❖ Fuel-service equipment must not be used unless it is in proper repair and free of possible cracks, frays or breaks that could cause leaks. Defective equipment must be taken out of service and repaired.

1106.11.2 Transporting fuel nozzles. Fuel nozzles shall be carried utilizing appropriate handles. Dragging fuel nozzles along the ground shall be prohibited.

❖ An ignition source may be produced by dragging the nozzle. Using the appropriate handling devices eliminates both sparking and nozzle damage.

1106.11.3 Drum fueling. Fueling from drums or other containers having a capacity greater than 5 gallons (19 L) shall be accomplished with the use of an approved pump.

❖ The use of an approved pump for transferring fuel from drums with capacity in excess of 5 gallons (19 L) greatly reduces the possibility of spills.

1106.11.4 Fuel spill procedures. The fueling-system operator shall establish procedures to follow in the event of a fuel spill. These procedures shall be comprehensive and shall provide for at least all of the following:

1. Upon observation of a fuel spill, the aircraft-fueling operator shall immediately stop the delivery of fuel by releasing hand pressure from the fuel flow-control valve.
2. Failure of the fuel control valve to stop the continued spillage of fuel shall be cause for the activation of the appropriate emergency fuel shutoff device.
3. A supervisor for the fueling-system operator shall respond to the fuel spill area immediately.

❖ Safety and emergency procedures must be in place, and personnel must be trained on these procedures. Should a spill occur, the person operating the fuel delivery hose must immediately release hand pressure to stop fuel flow. If this action does not shut off fuel flow, the emergency shutoff device must be activated.

Fuel system supervisors and operators must be trained in the procedures that they must follow in case of a fuel spill. Supervisors must respond if a spill occurs and assist in containment and cleanup if necessary.

1106.11.5 Notification of the fire department. The fire department shall be notified of any fuel spill which is considered a hazard to people or property or which meets one or more of the following criteria:

1. Any dimension of the spill is greater than 10 feet (3048 mm).
2. The spill area is greater than 50 square feet (4.65 m^2).
3. The fuel flow is continuous in nature.

❖ Three criteria are given for determining when the fire department must be notified. Any amount of spill that exceeds that criteria causes great concern and must be considered a fire hazard. The fire department must be notified because people and property are at stake if a spill ignites and causes explosion and fire.

1106.11.6 Investigation required. An investigation shall be conducted by the fueling-system operator of all spills requiring notification of the fire department. The investigation shall provide conclusive proof of the cause and verification of the appropriate use of emergency procedures. Where it is determined that corrective measures are necessary to prevent future incidents of the same nature, they shall be implemented immediately.

❖ Any spill must be investigated to determine the cause and whether the emergency action plan was properly carried out. If corrective measures are needed, they must be implemented immediately. This investigation must be looked upon as positive for a more efficient emergency action plan that could avoid possible future property damage and loss of life.

1106.11.7 Multiple fuel delivery vehicles. Simultaneous delivery of fuel from more than one aircraft-fueling vehicle to a single aircraft-fueling manifold is prohibited unless proper backflow prevention devices are installed to prevent fuel flow into the tank vehicles.

❖ Only one tank vehicle may be connected to any one aircraft fueling manifold unless a means is provided that prevents fuel from flowing back into the tank vehicle because of a difference in the pumping pressure. Backflow from one tanker to another could result in fuel overflow from the receiving vehicle that would cause a major fuel spill.

1106.12 Aircraft engines and heaters. Operation of aircraft onboard engines and combustion heaters shall be terminated prior to commencing fuel service operations and shall remain off until the fuel-servicing operation is completed.

Exception: In an emergency, a single jet engine is allowed to be operated during fuel servicing where all of the following conditions are met:

1. The emergency shall have resulted from an onboard failure of the aircraft's auxiliary power unit.
2. Restoration of auxiliary power to the aircraft by ground support services is not available.
3. The engine to be operated is either at the rear of the aircraft or on the opposite side of the aircraft from the fuel service operation.
4. The emergency operation is in accordance with a written procedure approved by the fire code official.

❖ Operation of any aircraft engines or combustion heaters allows the possibility of an ignition source being produced. Even in the exceptions allowed in an emergency, care must be taken in preventing a possible fire hazard.

1106.13 Vehicle and equipment restrictions. During aircraft-fueling operations, only the equipment actively involved in the fueling operation is allowed within 50 feet (15 240 mm) of the aircraft being fueled. Other equipment shall be prohibited in this area until the fueling operation is complete.

Exception: Aircraft-fueling operations utilizing single-point refueling with a sealed, mechanically locked fuel line connection and the fuel is not a Class I flammable liquid.

A clear space of at least 10 feet (3048 mm) shall be maintained between aircraft fuel-system vent openings and any part or portion of aircraft-servicing vehicles or equipment.

❖ To control the number and type of ignition sources to which the fueling operation might be exposed, no equipment, other than the equipment performing aircraft servicing functions, is permitted within a 50-foot (15 240 mm) radius of the aircraft during the fueling of the aircraft. The space between service vehicles and equipment and the aircraft fuel vent must be no less than 10 feet (3048 mm) when using a single-point sealed mechanically locked fuel line connector.

1106.13.1 Overwing fueling. Vehicles or equipment shall not be allowed beneath the trailing edge of the wing when aircraft fueling takes place over the wing and the aircraft fuel-system vents are located on the upper surface of the wing.

❖ Equipment under the trailing edge of the wing could cause a fire hazard if a leak or spill were to occur. The fuel would run off the edge of the wing and the equipment could have an ignition source and cause a fire.

1106.14 Electrical equipment. Electrical equipment, including but not limited to, battery chargers, ground or auxiliary power units, fans, compressors or tools, shall not be operated, nor shall they be connected or disconnected from their power source, during fuel service operations.

❖ Even hand lamps used in the fuel-servicing operation must be approved for the proper hazardous location classification. The equipment could produce sparks that could be an ignition source.

1106.14.1 Other equipment. Electrical or other spark-producing equipment shall not be used within 10 feet (3048 mm) of fueling equipment, aircraft fill or vent points, or spill areas unless that equipment is intrinsically safe and approved for use in an explosive atmosphere.

❖ No equipment that produces a spark during its operation can be used within a 10-foot (3048 mm) radius of the fueling operational equipment.

1106.15 Open flames. Open flames and open-flame devices are prohibited within 50 feet (15 240 mm) of any aircraft fuel-servicing operation or fueling equipment.

❖ Blow torches; welding and cutting equipment and flare pots are a few examples of equipment that is not allowed within 50 feet (15 240 mm) of the fueling equipment or the fueling operation.

1106.15.1 Other areas. The fire code official is authorized to establish other locations where open flames and open-flame devices are prohibited.

❖ Depending on the situation, the fire code official may find other locations in which open flames and open-flame devices must be prohibited. Examples could be exposed flame heaters, portable gasoline or kerosene heaters.

1106.15.2 Matches and lighters. Personnel assigned to and engaged in fuel-servicing operations shall not carry matches or lighters on or about their person. Matches or lighters shall be prohibited in, on or about aircraft-fueling equipment.

❖ Because matches, lighters and smoking materials are an ignition hazard, these items are prohibited everywhere in the vicinity of aircraft fueling operations.

1106.16 Lightning procedures. The fire code official is authorized to require the airport authority and the fueling-system operator to establish written procedures to follow when lightning flashes are detected on or near the airport. These procedures shall establish criteria for the suspension and resumption of aircraft-fueling operations.

❖ Fuel-servicing procedures must cease when lightning flashes occur at or in the vicinity of the airport. The fueling supervisor and airport authority must establish a written procedure that sets the criteria for stopping and restarting fueling operations.

1106.17 Fuel-transfer locations. Aircraft fuel-transfer operations shall be prohibited indoors.

Exception: In aircraft hangars built in accordance with the provisions of the *International Building Code* for Group F-1 occupancies, aircraft fuel-transfer operations are allowed where:

1. Necessary to accomplish aircraft fuel-system maintenance operations. Such operations shall be performed in accordance with nationally recognized standards; or
2. The fuel being used has a flash point greater than 100°F (37.8°C).

❖ Aircraft fueling operations must be performed outdoors to minimize the accumulation of flammable vapors. The exception allows indoor fueling only under very specific and limited conditions.

1106.17.1 Position of aircraft. Aircraft being fueled shall be positioned such that any fuel system vents and other fuel tank openings are a minimum of:

1. Twenty-five feet (7620 mm) from buildings or structures other than jet bridges; and
2. Fifty feet (15 240 mm) from air intake vents for boiler, heater or incinerator rooms.

❖ Maintaining minimum distances between fueling operations and other aircraft or buildings allows any flammable vapors released during fueling to dissipate before encountering any possible ignition sources. These distances also provide protection against ignitable vapor concentrations getting to an ignition source in the event of a fuel spill.

1106.17.2 Fire equipment access. Access for fire service equipment to aircraft shall be maintained during fuel-servicing operations.

❖ A clear and unobstructed path is required for quick response for fire personnel. A quick response is essential in an emergency; however, an obstacle may create a delay in emergency operations.

1106.18 Defueling operations. The requirements for fueling operations contained in this section shall also apply to aircraft defueling operations. Additional procedures shall be established by the fueling-system operator to prevent overfilling of the tank vehicle used in the defueling operation.

❖ Transferring of fuel from an aircraft through a hose to a tank vehicle is generally the same process as fueling. The same requirements apply to defueling as to fuel servicing an aircraft. Operators must establish procedures and safeguards for prevention of overfilling the tank vehicle, which is a hazard when defueling.

1106.19 Maintenance of aircraft-fueling hose. Aircraft-fueling hoses shall be maintained in accordance with Sections 1106.19.1 through 1106.19.4.

❖ The following sections address the maintenance of aircraft fueling hose in terms of the frequency of inspections, recognizing damaged hose and repairing the hose before placing it back in service.

1106.19.1 Inspections. Hoses used to fuel or defuel aircraft shall be inspected periodically to ensure their serviceability and suitability for continued service. The fuel-service operator shall maintain records of all tests and inspections performed on fueling hoses. Hoses found to be defective or otherwise damaged shall be immediately removed from service.

❖ Regular inspections of fueling equipment are essential to maintaining serviceable systems. Records of these inspections are required to document inspection results. Damaged or defective items must be removed from service immediately for repair or replacement.

1106.19.1.1 Daily inspection. Each hose shall be inspected daily. This inspection shall include a complete visual scan of the exterior for evidence of damage, blistering or leakage. Each coupling shall be inspected for evidence of leaks, slippage or misalignment.

❖ Daily hose inspections are essential for prevention of leaks. Even minor damage to a hose, coupling, nozzle or other system part can lead to safety hazards.

1106.19.1.2 Monthly inspection. A more thorough inspection, including pressure testing, shall be accomplished for each hose on a monthly basis. This inspection shall include examination of the fuel delivery inlet screen for rubber particles, which indicates problems with the hose lining.

❖ During the monthly inspection, the hose again must be extended to its full length of operation. The area within 12 inches (305 mm) of the coupling must be examined for structural weakness by pressing and observing soft spots. Hoses must be pressure tested at operating pressure and observed for ballooning or twisting that would indicate a weakening of the hose carcass. Examination of the fuel inlet screen for rubber particles is a means of checking the condition of hose linings.

1106.19.2 Damaged hose. Hose that has been subjected to severe abuse shall be immediately removed from service. Such hoses shall be hydrostatically tested prior to being returned to service.

❖ If hose has been subjected to unusual abuse, it must be removed from service immediately as a safety precaution. Such hoses must be hydrostatically tested and inspected. If any soft spots or weakening of the hose is observed, the hose may not be returned to service.

1106.19.3 Repairing hose. Hoses are allowed to be repaired by removing the damaged portion and recoupling the undamaged end. When recoupling hoses, only couplings designed and approved for the size and type of hose in question shall be used. Hoses repaired in this manner shall be visually inspected and hydrostatically tested prior to being placed back in service.

❖ Damaged hoses can be repaired. The damaged portion must be cut off and the undamaged part recoupled. Two lengths from separate hoses may not be joined. To increase the likelihood that the repaired hose is leak tight, only couplings specifically designed for that type and size of hose can be used. Hydrostatic testing of the repaired hose increases efficiency and safety for fueling operations.

1106.19.4 New hose. New hose shall be visually inspected prior to being placed into service.

❖ Before any new hose assembly can be placed into service, it must be inspected visually for evidence of damage or wear and tear of any kind.

1106.20 Aircraft fuel-servicing vehicles parking. Unattended aircraft fuel-servicing vehicles shall be parked in areas that provide for both the unencumbered dispersal of vehicles in the event of an emergency and the control of leakage such that adjacent buildings and storm drains are not contaminated by leaking fuel.

❖ The fuel-servicing vehicles must be positioned so that they can be moved quickly if an emergency occurs. This not only may save the vehicles but also assist in a rapid response to an emergency situation and saving lives. Fuel leaking into storm drains causes not only environmental problems, but vapor accumulating in the confines of a drain line presents a serious explosion hazard.

1106.20.1 Parking area design. Parking areas for tank vehicles shall be designed and utilized such that a clearance of 10 feet (3048 mm) is maintained between each parked vehicle for fire department access. In addition, a minimum clearance of 50 feet (15 240 mm) shall be maintained between tank vehicles and parked aircraft and structures other than those used for the maintenance and/or garaging of aircraft fuel-servicing vehicles.

❖ The 10-foot (3048 mm) requirement permits ready access by fire department equipment in case of an emergency. The fire department may need the area to deploy hose in case of fire or to walk through en route to an investigation of an emergency call. The 50-foot (15 240 mm) requirement is to allow dispersion of flammable vapors in case a fire or other emergency occurs.

1106.21 Radar equipment. Aircraft fuel-servicing operations shall be prohibited while the weather-mapping radar of that aircraft is operating.

Aircraft fuel-servicing or other operations in which flammable liquids, vapors or mists may be present shall not be conducted within 300 feet (91 440 mm) of an operating aircraft surveillance radar.

Aircraft fuel-servicing operations shall not be conducted within 300 feet (91 440 mm) of airport flight traffic surveillance radar equipment.

Aircraft fuel-servicing or other operations in which flammable liquids, vapors or mists may be present shall not be conducted within 100 feet (30 480 mm) of airport ground traffic surveillance radar equipment.

❖ The beam of radar equipment has been known to cause flammable vapor-air mixtures to ignite from electrical arcs or sparks from chance resonant (continuing) conditions. This ability of an arc to ignite flammable vapor-air mixtures depends on the total energy of the arc and the time lapse involved. The key factor in establishing safe distances between fueling operations, storage areas and radar antennas is the peak power output of the radar unit.

1106.21.1 Direction of radar beams. The beam from ground radar equipment shall not be directed toward fuel storage or loading racks.

Exceptions:

1. Fuel storage and loading racks in excess of 300 feet (91 440 mm) from airport flight traffic surveillance equipment.
2. Fuel storage and loading racks in excess of 100 feet (30 480 mm) from airport ground traffic surveillance equipment.

❖ Ground radar for approach control or traffic pattern surveillance is considered the most hazardous type of radar normally operated at an airport. The beam must not be directed toward fuel storage or loading racks because of the possible ignition sources (sparks, arcs) igniting flammable vapor-air mixtures. The exceptions are based on the arc energy emissions of the radars listed.

SECTION 1107
HELISTOPS AND HELIPORTS

1107.1 General. Helistops and heliports shall be maintained in accordance with Section 1107. Helistops and heliports on buildings shall be constructed in accordance with the *International Building Code.*

❖ The following sections address clearances of the touch-down areas, spillage of liquid fuels, exits, fire protection systems and approval of the FAA. Information for construction and design of heliports can be found in FAA A/C 150/5390-2A as well as in the IBC.

1107.2 Clearances. The touchdown area shall be surrounded on all sides by a clear area having minimum average width at roof level of 15 feet (4572 mm) but no width less than 5 feet (1524 mm). The clear area shall be maintained.

❖ Most accidents involving helicopters occur during takeoff and landing procedures, with a significant number occurring at night or in bad weather. The clearance requirements are intended to limit the likelihood of the aircraft colliding with permanent features at the landing site.

1107.3 Flammable and Class II combustible liquid spillage. Landing areas on structures shall be maintained so as to confine flammable or Class II combustible liquid spillage to the landing area itself, and provisions shall be made to drain such spillage away from exits or stairways serving the helicopter landing area or from a structure housing such exit or stairway.

❖ Rooftop drainage must be designed to prevent fuel runoff from the landing area into adjacent or lower building spaces that could contain ignition sources for the fuel vapors. To accomplish this task, most rooftop heliport structures drain the pad to the center and provide some provision for separating fuel runoff from storm water. Additionally, drip pans should be provided below the helicopter engine when the aircraft is parked with the engine off.

1107.4 Exits. Exits and stairways shall be maintained in accordance with Section 412.5 of the *International Building Code*.

❖ Section 412.5 of the IBC refers to Chapter 10 for exit requirements. The section also addresses landing platforms and roof area size and the number of exits required. At least two means of egress must be provided off the landing pad.

1107.5 Standpipe systems. Where a building with a rooftop helistop or heliport is equipped with a standpipe system, the system shall be extended to the roof level on which the helistop or heliport is located. All portions of the helistop and heliport area shall be within 150 feet (45 720 mm) of a 2.5-inch (63.5 mm) outlet on a Class I or III standpipe.

❖ Class I standpipes serve 2.5-inch (63.5 mm) fire hose valves for fire department use. Class III standpipes have both 2.5-inch (63.5 mm) valves for fire department use and 1.5-inch (38 mm) fire hose cabinets for use by building occupants. The extension from either class of standpipe must be the 2.5-inch (63.5 mm) size valve for fire-fighting purposes.

1107.6 Foam protection. Foam fire-protection capabilities shall be provided for rooftop heliports. Such systems shall be designed, installed and maintained in accordance with the applicable provisions of Sections 903, 904 and 905.

❖ A foam fire-extinguishing system must be designed and installed for protection of the rooftop landing pad. Factors considered in the design of the system are aircraft size, effectiveness of agent selected, time required to achieve control and time required to maintain control. Foam discharge rates must also be considered according to the type of foam selected.

1107.7 Fire extinguishers. A minimum of one portable fire extinguisher having a minimum 80-B:C rating shall be provided for each permanent takeoff and landing area and for the aircraft parking areas. Installation, inspection and maintenance of these extinguishers shall be in accordance with Section 906.

❖ The portable fire extinguisher for the takeoff/landing area must be effective for the extinguishment of a flammable liquid fire and for energized electrical components.

1107.8 Federal approval. Before operating helicopters from helistops and heliports, approval shall be obtained from the Federal Aviation Administration.

❖ The FAA approves all installations of heliports and helistops. Design criteria can be found in FAA A/C 150/5390-2A.

Bibliography

The following resource materials are referenced in this chapter or are relevant to the subject matter addressed in this chapter.

"Heliport Design." *Federal Aviation Advisory Circular 150/5390-2A*. Washington, DC: U.S. Department of Transportation, Federal Aviation Administration, 1994.

IBC-2003, *International Building Code*. Falls Church, VA: International Code Council, 2003.

ICC EC-2003, *ICC Electrical Code*. Falls Church, VA: International Code Council, 2003.

IMC-2003, *International Mechanical Code*. Falls Church, VA: International Code Council, 2003.

NFPA 10-98, *Portable Fire Extinguishers*. Quincy, MA: National Fire Protection Association, 1998.

NFPA 407-96, *Aircraft Fuel Services*. Quincy, MA: National Fire Protection Association, 1996.

NFPA 409-01 *Standard on Aircraft Hangars*. Quincy, MA: National Fire Protection Association, 2001.

NFPA 415-02, *Standard on Airport Terminal Buildings, Fueling Ramp Drainage, and Loading Walkways*. Quincy, MA: National Fire Protection Association, 2002.

NFPA 418-01, *Standard for Heliports*. Quincy, MA: National Fire Protection Association, 2001.

Chapter 12: Dry Cleaning

General Comments

Dry cleaning operations remove dirt, grease and other foreign substances from clothing, rugs, textiles and fabrics with solvents that are not water based. These methods may involve several techniques, including immersion and agitation, brushing or scouring or dual-phase processing, where a dry cleaning operation follows a laundering (soap and water) process using the same equipment. The solvents employed may be flammable or nonflammable and may possess certain health hazards. Both the fire code official and fire service personnel should be aware of the health and flammability hazards.

Recently, concern over flammability hazards and hazardous waste disposal have radically altered the nature of the dry cleaning business in the United States. Most plants currently use nonflammable solvents like perchloroethylene or tetrachloroethylene rather than Stoddard solvent, naphtha and specially compounded oils, which were favored in previous years.

Storage of flammable and combustible liquids must comply with the requirements of Chapter 34. Flammable solvent containers and processing equipment must be bonded and grounded during storage, handling and use to prevent buildup of static charges. Solvents must be used only in equipment approved for use with that specific class of solvent.

Purpose

The provisions of Chapter 12 are intended to reduce hazards associated with the use of flammable and combustible dry cleaning solvents. These materials, like all volatile organic chemicals, generate significant quantities of static electricity and are thus readily ignitable. Many flammable and nonflammable dry cleaning solvents also possess health hazards when involved in a fire.

SECTION 1201 GENERAL

1201.1 Scope. Dry cleaning plants and their operations shall comply with the requirements of this chapter.

❖ Hazards associated with dry cleaning operations are addressed by the provisions of this chapter. Solvent storage hazards are addressed by Chapter 34 of the code.

1201.2 Permit required. Permits shall be required as set forth in Section 105.6.

❖ The process of issuing permits gives the fire code official an opportunity to carefully evaluate and regulate hazardous operations. Permit applicants should be required to demonstrate that their operations comply with the intent of the code before the permit is issued. See the commentary to Section 105.6 for a general discussion of operations requiring an operational permit and Section 105.6.13 for discussion of specific operational permits for dry cleaning plants and their operation.

SECTION 1202 DEFINITIONS

1202.1 Definitions. The following words and terms shall, for the purposes of this chapter and as used elsewhere in this code, have the meanings shown herein.

❖ Definitions can help in the understanding and application of the code requirements. These definitions are included in this chapter to provide more convenient access to them when applying the provisions of this chapter. For convenience, the defined terms are also listed in Chapter 2 with a cross reference to this section. The use and application of all defined terms, including those described here, are set forth in Section 201.

DRY CLEANING. The process of removing dirt, grease, paints and other stains from such items as wearing apparel, textiles, fabrics and rugs by use of nonaqueous liquids (solvents).

❖ Dry cleaning is the process of cleaning textile-based items in a closed machine using solvents that are not water based; hence the term "dry." This process generally consists of a cleaning cycle in which the item or items to be cleaned are placed in the machine with the dry cleaning solvent, tumbled for a predetermined length of time, an extraction cycle in which the solvent is centrifugally removed from the clean items (not unlike the "spin" cycle of a washing machine) and, finally, a drying cycle. The process also includes the manual application of solvent and spotting compounds prior to the main cleaning process.

DRY CLEANING PLANT. A facility in which dry cleaning and associated operations are conducted, including the office, receiving area and storage rooms.

❖ The type of construction, occupancy group classification and other building requirements are determined by the type of solvent and machinery used in the cleaning process. The main hazards in a dry cleaning plant are the fire hazards of the flammable and combustible solvents and the health hazards of the chlorinated hydrocarbon solvents.

DRY CLEANING ROOM. An occupiable space within a building used for performing dry cleaning operations, the installation of solvent-handling equipment or the storage of dry cleaning solvents.

❖ The *International Building Code*® (IBC®) defines an occupiable space as "a room or enclosed space designed for human occupancy . . . in which occupants are engaged at labor . . ." Within the context of this chapter, a dry cleaning room is primarily the room or space in which the actual dry cleaning process is conducted. Depending on the scope of the dry cleaning establishment, it could be a room or space within a large plant or could encompass the entire plant in smaller operations. Note that the focus of the term is on the presence of the dry cleaning solvent, whether it is in the cleaning process, being stored for future use within the room or space or being transferred from storage containers to the dry cleaning machines.

DRY CLEANING SYSTEM. Machinery or equipment in which textiles are immersed or agitated in solvent or in which dry cleaning solvent is extracted from textiles.

❖ This term focuses on the actual dry cleaning machines and solvent extractors that use or recover dry cleaning solvent. Systems are classified according to the hazards of the solvent they use. See the commentary to Sections 1203.2 and 1205 for additional discussion of dry cleaning system classification and operation, respectively.

SOLVENT OR LIQUID CLASSIFICATIONS. A method for classifying solvents or liquids according to the following classes:

❖ These dry cleaning solvent classifications parallel the flammable and combustible liquid classifications defined in Chapter 34, with a notable exception that, while flammable liquids (Class I) are divided into three subclasses (Class IA, IB and IC) based on flash point and boiling point, Class I solvents are not. Solvent classifications, on the other hand, include a classification (Class IV) for those solvents considered to be nonflammable whereas the flammable and combustible liquid classifications have no comparable category.

Class I solvents. Liquids having a flash point below 100°F (38°C).

Class II solvents. Liquids having a flash point at or above 100°F (38°C) and below 140°F (60°C).

Class IIIA solvents. Liquids having a flash point at or above 140°F (60°C) and below 200°F (93°C).

Class IIIB solvents. Liquids having a flash point at or above 200°F (93°C).

Class IV solvents. Liquids classified as nonflammable.

SECTION 1203
CLASSIFICATIONS

1203.1 Solvent classification. Dry cleaning solvents shall be classified according to their flash points as follows:

1. Class I solvents are liquids having a flash point below 100°F (38°C).
2. Class II solvents are liquids having a flash point at or above 100°F (38°C) and below 140°F (60°C).
3. Class IIIA solvents are liquids having a flash point at or above 140°F (60°C) and below 200°F (93°C).
4. Class IIIB solvents are liquids having a flash point at or above 200°F (93°C).
5. Class IV solvents are liquids classified as nonflammable.

❖ To establish regulations for mitigating the hazards of dry cleaning, the hazards of dry cleaning solvent must be identified and established. Because the liquid solvents used in dry cleaning operations, other than Class IV solvents, are either flammable or combustible, they are classified using the same criteria as those used to classify flammable and combustible liquids in Chapter 34.

Class I solvents are, by the definition in Chapter 34, flammable liquids because they have a closed-cup flash point of less than 100°F (38°C). The 100°F (38°C) flash point for flammable solvents assumes possible indoor ambient temperature conditions of 100°F (38°C), which means that the solvent could be used at or below its flash point under normal operating conditions in the dry cleaning plant. This solvent classification, however, includes no further subclassifications based on flash point and boiling point as does Chapter 34 for Class I flammable liquids.

Combustible solvents (Classes II and III) differ from flammable solvents in that the closed-cup flash point of all combustible liquids is at or above 100°F (38°C) (see commentary, definition of "Flash point" in Chapter 34). The range of flash point dictates the class of liquid solvent. The flash point range of 100°F (38°C) to 140°F (60°C) for Class II solvents is based on a possible indoor ambient temperature exceeding 100°F (38°C). Only a moderate degree of heating is required to bring the solvent to its flash point in this situation. Class III solvents, which have flash points higher than 140°F (38°C), require a significant heat source in addition to ambient temperature conditions to reach their flash point (see commentary, definition of "Flammable liquid" in Chapter 34).

1203.2 Classification of dry cleaning plants and systems. Dry cleaning plants and systems shall be classified based on the solvents used as follows:

1. Type I—systems using Class I solvents.
2. Type II—systems using Class II solvents.
3. Type III-A—systems using Class IIIA solvents.
4. Type III-B—systems using Class IIIB solvents.
5. Type IV—systems using Class IV solvents in which dry cleaning is not conducted by the public.
6. Type V—systems using Class IV solvents in which dry cleaning is conducted by the public.

Spotting and pretreating operations conducted in accordance with Section 1206 shall not change the type of the dry cleaning plant.

❖ Dry cleaning plants are classified based on the solvents they use and the hazards presented by them.

Type I dry cleaning systems are those systems using a Class I flammable liquid [flash point less than 100°F (38°C)] as a cleaning solvent, such as low-flash-point [less than 50°F (10°C)] naphtha. Such plants are prohibited by Section 1204.1.

Type II dry cleaning systems are those systems using a Class II combustible liquid [flash points 100°F (38°C) to 140°F (60°C)] as a cleaning solvent, such as Stoddard solvent. NFPA 32 classifies the entire dry cleaning room of a Type II plant as a Class I, Division 2 hazardous (classified) location.

Type III-A dry cleaning systems are those systems using a Class IIIA combustible liquid as a cleaning solvent [flash point 140°F (60°C) to 200°F (93°C)].

Type III-B dry cleaning systems are those systems using a Class IIIB combustible liquid as a cleaning solvent [flash point greater than 200°F (93°C)], usually specially compounded oils.

Type IV dry cleaning systems are those systems using nonflammable solvents, and dry cleaning is done only by trained operators.

Type V dry cleaning systems are those systems using nonflammable solvent, typically perchloroethylene, and dry cleaning is done by the general public (e.g., at coin-operated laundries).

This section recognizes the practical need to have limited amounts of higher hazard solvents on hand in the plant for spotting and pretreating by not classifying the plant based on the highest hazard solvent that might be present in any quantity. See the commentary to Section 1206 for further information on spotting and pretreating operations.

1203.2.1 Multiple solvents. Dry cleaning plants using more than one class of solvent for dry cleaning shall be classified based on the numerically lowest solvent class.

❖ Ventilation, and electrical requirements in particular, vary considerably among various types of dry cleaning plants. Additionally, each type of dry cleaning equipment is designed for the hazards specific to the solvent used. The greatest hazard determines the appropriate level of protection where a danger of accidental misuse exists.

The more volatile compounds, such as naphtha, Stoddard solvent and oils, may still be found in some specialized industrial dry cleaning plants to remove specific materials from work clothes. Although perchloroethylene and similar safety solvents are not flammable, they do present certain health hazards, especially under fire conditions. Table 1203.2.1 lists some of the more common dry cleaning solvents with their physical, flammability and toxicity characteristics.

Perchloroethylene concentrations as low as 300 to 1,100 parts per million (ppm) under normal conditions may cause loss of coordination and impairment, while dizziness, drowsiness or loss of consciousness may result at higher concentrations. Perchloroethylene may also emit highly toxic and irritating fumes in a fire.

1203.3 Design. The occupancy classification, design and construction of dry cleaning plants shall comply with the applicable requirements of the *International Building Code.*

❖ Dry cleaning plants are required to comply with the provisions of Section 415.7.4 of the IBC, which, in turn, requires compliance with the applicable provisions of the

TABLE 1203.2.1
CHARACTERISTICS OF DRY CLEANING SOLVENTS

Chemical Name	Flash Point	Ignition Temperature	Health[a]	Flammability[a]	Reactivity[a]
Naptha	28° to 50°F	450°F	1	3	0
Perchloroethylene	None	None	2	0	0
Stoddard solvent	Above 100°F	444°F	0	2	0

a. Based on NFPA 704.

International Plumbing Code® (IPC®) and the *International Mechanical Code*® (IMC®) and references NFPA 32 for plant construction and system installation. Both NFPA 32 and the IMC specify mechanical exhaust ventilation rates for dry cleaning rooms.

The IPC specifies requirements for sanitary sewers and drains in laundries and where hazardous materials are used or stored. Section 415.7.4 of the IBC also references the code for solvent and system classification.

SECTION 1204 GENERAL REQUIREMENTS

1204.1 Prohibited use. Type I dry cleaning plants shall be prohibited. Limited quantities of Class I solvents stored and used in accordance with this section shall not be prohibited in dry cleaning plants.

❖ This section flatly prohibits Class I dry cleaning plants (i.e., those that use Class I flammable liquids) because of the extreme flammability of such liquids. This section also recognizes the practical need to have limited amounts of Class I solvent on hand in the plant for spotting and pretreating, provided that it is stored and used in accordance with Chapter 34. See the commentary to Section 1206 for further information on spotting and pretreating operations.

1204.2 Building services. Building services and systems shall be designed, installed and maintained in accordance with this section and Chapter 6.

❖ Electrical and mechanical systems serving the dry cleaning plant must comply with the provisions of Sections 1204.2.1 through 1204.2.4 and the applicable provisions of Chapter 6.

1204.2.1 Ventilation. Ventilation shall be provided in accordance with Section 502 of the *International Mechanical Code* and DOL 29 CFR Part 1910.1000, where applicable.

❖ The intent of this section is to provide ventilation in dry cleaning plants that is adequate to protect the plant employees and the public from the hazards associated with dry cleaning operations using any of the various classes of dry cleaning solvents, both in approved dry cleaning machines and in spotting and pretreating operations in the open. To achieve this objective, this section requires that mechanical ventilation systems comply with the provisions of Section 502 of the IMC. Compliance with the applicable provisions of the Occupational Safety and Health Administration (OSHA) workplace regulations pertaining to air contaminants contained in DOL 29 CFR 1910.1000 is also required.

1204.2.2 Heating. In Type II dry cleaning plants, heating shall be by indirect means using steam, hot water, or hot oil only.

❖ Open-flame heating appliances, such as unit heaters, must not be located in Type II dry cleaning plants because they could create conditions conducive to the ignition of any fugitive vapors from the Class II dry cleaning solvent used in these plants. Indirect steam, hot water or hot oil heat supplied from appliances located in rooms separated from the Type II dry cleaning plant or in separate buildings are examples of indirect heat methods.

1204.2.3 Electrical wiring and equipment. Electrical wiring and equipment in dry cleaning rooms or other locations subject to flammable vapors shall be installed in accordance with the ICC *Electrical Code.*

❖ To reduce the likelihood of a vapor ignition, electrical wiring and equipment located where flammable vapors might be released by malfunctioning equipment, rupture or breakage is typically classified as Class I, Division 2 equipment under the provisions of the ICC *Electrical Code*® (ICC EC™). This would include rooms or spaces containing closed-system dry cleaning equipment that, under normal conditions, keeps the solvent vapors confined within the equipment. For the same reason, in areas where flammable vapors are present under normal operating conditions, such as in spotting or pretreating areas using Class I solvents in the open, in tubs or on soaking tables, electrical wiring and equipment must be classified as Class I, Division I equipment to meet the requirements of the ICC EC.

1204.2.4 Bonding and grounding. Storage tanks, treatment tanks, filters, pumps, piping, ducts, dry cleaning units, stills, tumblers, drying cabinets and other such equipment, where not inherently electrically conductive, shall be bonded together and grounded. Isolated equipment shall be grounded.

❖ Solvent containers and processing equipment during storage, handling and use must be bonded and grounded to prevent buildup of static charges, thus eliminating a common source of ignition for vapors emitted by flammable and combustible liquid solvents.

SECTION 1205 OPERATING REQUIREMENTS

1205.1 General. The operation of dry cleaning systems shall comply with the requirements of this section.

❖ Section 1205 establishes requirements for the safe operation and maintenance of dry cleaning plants. The fire hazard associated with these plants has diminished significantly in recent years as fire protection, environmental and hazardous waste disposal concerns have caused the industry to move away from the use of flammable and combustible solvents. Poor housekeeping is the primary fire prevention concern these days.

1205.1.1 Written instructions. Written instructions covering the proper installation and safe operation and use of equipment and solvent shall be given to the buyer.

❖ This section applies to persons or firms that engage in wholesale or retail sales of dry cleaning equipment.

Each piece of equipment sold must be accompanied by written instructions for the installation and operation of the equipment and for the safe storage, handling and use of the dry cleaning solvent for which it is designed.

1205.1.1.1 Type II, III-A, III-B and IV systems. In Type II, III-A, III-B and IV dry cleaning systems, machines shall be operated in accordance with the operating instructions furnished by the machinery manufacturer. Employees shall be instructed as to the hazards involved in their departments and in the work they perform.

❖ Safety in dry cleaning operations depends on employees being familiar with not only the dry cleaning machines and their operation as described in the written instructions required by Section 1205.1.1, but also understanding the hazards associated with and the proper methods of storing, handling, dispensing and using dry cleaning solvents within their respective work spaces. This section requires that employees be trained in all aspects of dry cleaning operations.

1205.1.1.2 Type V systems. Operating instructions for customer use of Type V dry cleaning systems shall be conspicuously posted in a location near the dry cleaning unit. A telephone number shall be provided for emergency assistance.

❖ Type V dry cleaning systems are operated by the general public rather than by trained employees as are Type II, III-A and III-B systems. Under ideal conditions, such operations should have an attendant to supervise cleaning activities and be available to answer questions on the proper operation of the machines. Most Type V systems, however, are not attended. For this reason, machine operating instructions must be posted prominently in the machine area where they can be seen readily by patrons. In addition, an emergency phone number must be posted prominently for patron use. Note that this section does not require a public telephone on the premises. With the proliferation of cellular technology, much of the public carries a cell phone that can be used in case of an emergency.

1205.1.2 Equipment identification. The manufacturer shall provide nameplates on dry cleaning machines indicating the class of solvent for which each machine is designed.

❖ Because dry cleaning machines are designed to be used only with a specific class of solvent, this information must be clearly displayed on a permanent nameplate affixed to the machine so that there can be no question of what solvent may be used in a given machine (see also commentary, Section 1205.1.4).

1205.1.3 Open systems prohibited. Dry cleaning by immersion and agitation in open vessels shall be prohibited.

❖ Dry cleaning is intended to be a closed-system process, with the cleaning solvent contained within the machinery. Pouring large quantities of combustible solvent into an open container to soak or agitate items being cleaned allows and increases the escape of combustible vapors or, in the case of Type IV systems, noncombustible but toxic vapors, and is therefore prohibited. Also see the commentary to Section 1206 for a discussion of spotting and pretreating in the open and special handling provisions that allow limited hand agitation in the open.

1205.1.4 Prohibited use of solvent. The use of solvents with a flash point below that for which a machine is designed or listed shall be prohibited.

❖ Dry cleaning equipment is designed and tested for use with specific classifications of solvents. This section mandates that equipment be used only with the class of solvent for which it has been designed. Equipment approved for higher-flash-point or nonflammable solvents may not be equipped with many of the inherent safety features required for lower-flash-point solvents.

1205.1.5 Equipment maintenance and housekeeping. Proper maintenance and operating practices shall be observed in order to prevent the leakage of solvent or the accumulation of lint. The handling of waste material generated by dry cleaning operations and the maintenance of facilities shall comply with the provisions of this section.

❖ Safe dry cleaning depends on safe, well-maintained equipment and good housekeeping practices. Because dry cleaning is intended to be a closed-system use, machines must be checked regularly for loose fittings and connections that could allow leakage of a solvent and its vapors. Also, the lint that accumulates in the cleaning of textile materials must be regularly removed from lint traps and disposed of properly. This section requires that dry cleaning plants comply with the provisions of Sections 1205.1.5.1 through 1205.1.5.4.

1205.1.5.1 Floors. Class I and II liquids shall not be used for cleaning floors.

❖ This section prohibits the use of flammable and Class II combustible liquids for cleaning floors in dry cleaning plants for basically the same reasons that open soaking and agitation with solvents are prohibited by Section 1205.1.3: open use will allow flammable vapors to escape and accumulate. This effect is aggravated by the fact that in many instances, dry cleaning plants operate at high ambient temperatures that may be at or above the flash point of Class I or II liquids, thus forming an immediately ignitable mixture in air.

1205.1.5.2 Filters. Filter residue and other residues containing solvent shall be handled and disposed of in covered metal containers.

❖ The Environmental Protection Agency (EPA) has raised health and safety concerns about dry cleaning workers and people living near dry cleaners. Improper handling and disposal of dry cleaning solvents can pollute outdoor air, soil and water. Dry cleaning and laundry plants that might generate hazardous waste and be subject to Resource Conservation and Recovery Act

(RCRA) requirements covering the generation, transportation and management of hazardous waste include: retail dry cleaning stores; industrial and linen supply plants with dry cleaning operations; self-service laundromats with dry cleaning equipment or other facilities with dry cleaning operations. The volume of hazardous waste produced in these facilities often places them in the EPA category of "small quantity generator." Proper storage and disposal of spent filters and other solvent-containing wastes (empty solvent containers, still residues from solvent distillation and water contaminated with cleaning solvent) as required by this section reduces the potential for the production of solvent vapors and environmental pollution.

1205.1.5.3 Lint. Lint and refuse shall be removed from traps daily, deposited in approved waste cans, removed from the premises, and disposed of safely. At all other times, traps shall be held securely in place.

❖ Lint is finely divided textile fiber sloughed off of articles being cleaned and is generated in large quantities in the dry cleaning process. As with any combustible material, the more finely it is divided, the more surface area of the material is available for ignition and the more readily it will ignite. For this reason, this hazard must be reduced by collection and removal of lint, as well as other combustible waste materials, at least daily, followed by proper disposal.

1205.1.5.4 Customer areas. In Type V dry cleaning systems, customer areas shall be kept clean.

❖ Coin-operated laundries and dry cleaning establishments are often unattended, which can lead to poor housekeeping and hazardous conditions that could contribute to the start or spread of a fire. This section requires that reasonable housekeeping procedures be established and executed regularly to reduce this hazard.

1205.2 Type II systems. Special operating requirements for Type II dry cleaning systems shall comply with the provisions of this section.

❖ Sections 1205.2.1 through 1205.2.3 contain operating procedures uniquely applicable to Type II dry cleaning systems because of their use of Class II solvents and the higher relative hazards associated with that use.

1205.2.1 Inspection of materials. Materials to be dry cleaned shall be searched thoroughly and foreign materials, including matches and metallic substances, shall be removed.

❖ To prevent damage to dry cleaning equipment and eliminate potential sources of ignition, such as matches being ignited or metallic objects striking a spark with dry cleaning machine parts, articles to be dry cleaned must be carefully screened for foreign objects that could create a hazardous condition.

1205.2.2 Material transfer. In removing materials from the washer, provisions shall be made for minimizing the dripping of solvent on the floor. Where materials are transferred from a washer to a drain tub, a nonferrous metal drip apron shall be placed so that the apron rests on the drain tub and the cylinder of the washer.

❖ To reduce the amount of Class II solvent lost during removal of cleaned articles from the dry cleaning machine, means must be provided to minimize the amount of solvent that drips on the plant floor and evaporates into hazardous vapors in the work space. This section describes one method of preventing solvent drips by placing a drain board between the dry cleaning machine and drain tubs. The drain board must be nonferrous to prevent potential metal-to-metal sparks during the transfer process.

1205.2.3 Ventilation. A mechanical ventilation system which is designed to exhaust 1 cubic foot of air per minute for each square foot of floor area [0.0058 $m^3/(s \cdot m^2)$] shall be installed in dry cleaning rooms and in drying rooms. The ventilation system shall operate automatically when the dry cleaning equipment is in operation and shall have manual controls at an approved location.

❖ This section intends to prevent solvent vapors from Type II dry cleaning systems from accumulating to an ignitable concentration in the room in which those appliances are located by providing adequate mechanical ventilation. The prescribed rate of mechanical ventilation will also help prevent the ambient temperature in the room from exceeding the flash point of the solvent being used in the cleaning process. It is not the intent that this system be classified as a "hazardous exhaust system" as addressed in Section 510 of the IMC; however, consistent with Section 502 of that code, this system must be independent of all other exhaust and ventilation systems.

The required ventilation rate of 1 cubic foot per minute (cfm) per square foot [0.0058 $m^3/(s \bullet m^2)$] of floor area is typical of that required in hazardous materials-related areas where fugitive flammable vapors need dilution or removal. Although its exact technical origins remain obscure, it is thought to have been derived from a rule of thumb (possibly from the insurance industry) that established a ventilation requirement in flammable and combustible liquid use areas of six air changes per hour, which has proven to be effective over the years. Although this method provides effective ventilation and hazard reduction, moving that volume of air in industrial or storage buildings with larger-than-average floor-to-floor and floor-to-roof dimensions requires large, costly mechanical equipment installations and creates concerns over energy conservation. To deal with those concerns, a design ceiling height of 10 feet (3048 mm) was assumed in recognition of the fact that solvent vapors will gather at the lowest point in the room. Each square foot of building area, then, would represent 10 cubic feet (.3 m^3) and, at the rate of 6 air changes per hour, 60 cubic feet (1.7 m^3) of air per hour would be moved, which yields 1 cfm per square foot [0.0058 $m^3/(s \cdot m^2)$] of room area.

To be effective, the exhaust equipment is required by this section to be interlocked with the dry cleaning equipment so that it operates whenever the dry cleaning equipment operates. Manual controls are also required for additional flexibility and reliability in the event that the mechanical system runs on after the dry cleaning system is shut down. The location of the controls is to be approved by the fire code official.

1205.3 Type IV and V systems. Type IV and V dry cleaning systems shall be provided with an automatically activated exhaust ventilation system to maintain a minimum of 100 feet per minute (51 m/s) air velocity through the loading door when the door is opened. Such systems for dry cleaning equipment shall comply with the *International Mechanical Code*.

Exception: Dry cleaning units are not required to be provided with exhaust ventilation where an exhaust hood is installed immediately outside of and above the loading door which operates at an airflow rate as follows:

$$Q = 100 \times A_{LD} \qquad \textbf{(Equation 12-1)}$$

where:

Q = flow rate exhausted through the hood, cubic feet per minute (m^3/s).

A_{LD} = area of the loading door, square feet (m^2).

❖ This section intends to prevent solvent vapors from Type IV and Type V dry cleaning systems from escaping into the room where the appliances are located by drawing the required exhaust airflow into the unit through the open door at the minimum velocity of 100 feet per minute (51 m/s). In this way, exposure of employees (in Type IV systems) or the public (in Type V systems) to potentially harmful solvent vapors is minimized. The exhaust capability contemplated by this section is integral with the dry cleaning unit, and its operation must be interlocked with the unit door to automatically start the required exhaust airflow as soon as the unit door is opened. It is not the intent that this system be classified as a "hazardous exhaust system" as addressed in Section 510 of the IMC; however, consistent with Section 502 of that code, this system must be independent of all other exhaust and ventilation systems.

The exception is a design alternative to achieve the goal of solvent vapor capture at the door opening through installation of what is often referred to as an "eyebrow hood" located immediately above the dry cleaning unit door opening. This enables the exhaust airflow to sweep across the loading door opening of the machine and capture any escaping solvent vapors. This kind of hood may be integral with the unit or may be an after-market-installed accessory to the unit. In either case, it must be either interlocked with the unit door to automatically start the required exhaust airflow as soon as the unit door is opened or operate continuously whenever the dry cleaning machine is in operation.

Since exhaust hoods are not designed for a constant air velocity in all parts of the hood, such as can be achieved by drawing air into a machine through a dry cleaning machine loading door opening, exhaust capability for this alternative design calculated in accordance with Equation 12-1 will be expressed as a flow rate rather than a fixed velocity. The fan will be drawing in a large volume of environmental air to achieve the prescribed exhaust airflow across the entire loading door opening. For example, applying the formula to a dry cleaning unit that has a 2-foot-diameter (610 mm) circular loading/unloading door, the exhaust flow rate (Q) would be calculated as follows:

$Q = 100 \text{ x } A_{LD}$

$A_{LD} = \pi r^2$; where $\pi = 3.14$ and $r = 1$, therefore

$Q = 100 \text{ x } (3.14 \text{ x } 1^2)$

$Q = 100 \text{ x } 3.14$

$Q = 314$ cubic feet per minute

It is not the intent that this system be classified as a "hazardous exhaust system" as addressed in Section 510 of the IMC; however, consistent with Section 502.1.3 of that code, this system must be independent of all other exhaust and ventilation systems.

SECTION 1206
SPOTTING AND PRETREATING

1206.1 General. Spotting and pretreating operations and equipment shall comply with the provisions of this section.

❖ Spotting and pretreating operations consist of soaking or direct local application of cleaning solvents to articles about to be dry cleaned. These operations focus the solvent's cleaning power on the removal of more stubborn stains, such as those from grease, oils, makeup, paint, dirt or petroleum products, such as tar, asphalt sealer, etc. The operations typically take place on spotting tables or in scrubbing tubs in the open in the dry cleaning plant. Compliance with the requirements of Section 1206 will result in an acceptable level of safety for spotting and pretreating operations.

1206.2 Type I solvents. The maximum quantity of Type I solvents permitted at any work station shall be 1 gallon (4 L). Class I solvents shall be stored in approved safety cans or in sealed DOTn-approved metal shipping containers of not more than 1-gallon (4 L) capacity. Dispensing shall be from approved safety cans.

❖ This section allows up to 1 gallon (4 L) of Class I (flammable) solvent to be stored in its original shipping container or stored in and dispensed from a safety can complying with UL 30 (see commentary, Sections 2705.1.10 and 3405.2.4).

1206.3 Type II and III solvents. Scouring, brushing, and spotting and pretreating shall be conducted with Class II or III solvents. The maximum quantity of Type II or III solvents permitted at any work station shall be 1 gallon (4 L). In other than a Group H-2 occupancy, the aggregate quantities of solvents shall not exceed the maximum allowable quantity per control area for use-open system.

❖ This section limits solvent use for spotting or pretreating to Class II or III solvents and further limits the quantity of solvent that can be in use at a spotting or pretreating workstation to 1 gallon (4 L) to reduce the likelihood of an unmanageable solvent spill. Aggregate quantities of spotting or pretreating solvent must not exceed the maximum allowable quantity per control area for open system use established by Table 2703.1.1(1), including the increases allowed for the sprinkler system required by Section 1208.2.

1206.3.1 Spotting tables. Scouring, brushing or spotting tables on which articles are soaked in solvent shall have a liquid-tight top with a curb on all sides not less than 1 inch (25 mm) high. The top of the table shall be pitched to ensure thorough draining to a 1.5-inch (38 mm) drain connected to an approved container.

❖ To reduce spills of spotting solvents during their open use, surfaces upon which the spotting is done must be designed to catch and route solvent to an approved container (preferably a closed one) for disposal or recycling.

1206.3.2 Special handling. When approved, articles that cannot be washed in the usual washing machines are allowed to be cleaned in scrubbing tubs. Scrubbing tubs shall comply with the following:

1. Only Class II or III liquids shall be used.
2. The total amount of solvent used in such open containers shall not exceed 3 gallons (11 L).
3. Scrubbing tubs shall be secured to the floor.
4. Scrubbing tubs shall be provided with permanent 1.5-inch (38 mm) drains. Such drain shall be provided with a trap and shall be connected to an approved container.

❖ Although dry cleaning by immersion and agitation in open vessels is prohibited by Section 1205.1.3, there are isolated circumstances in which an article that needs dry cleaning cannot be cleaned in the conventional manner in an approved dry cleaning machine. This section establishes minimum safeguards intended to reduce the hazards associated with open cleaning in a scrubbing tub.

The first step in cleaning an article in the open is to secure the approval of the fire code official for the operation. This enables him or her to inspect the area where the open cleaning is to be done to verify that the housekeeping complies with this chapter, that all the safeguards required by this section are in place, that the ventilation requirements of Section 1206.3.3 are met and that the bonding and grounding required by Section 1206.3.4 are in place.

Although limited use of Class I spotting solvents is allowed by Section 1206.3, this section makes it clear that dry cleaning in the open is limited to only Class II or III solvents because of their higher flash points. To avoid an unmanageable spill while still allowing a workable quantity of solvent in the scrubbing tub, Class II or III solvents are limited to only 3 gallons (11 L) in process, and then only in a tub that is firmly fixed in place by securing it to the floor in an approved manner. Solvents used in open dry cleaning processes must be captured in an approved (preferably closed) container for recycling or proper disposal.

1206.3.3 Ventilation. Scrubbing tubs, scouring, brushing or spotting operations shall be located such that solvent vapors are captured and exhausted by the ventilating system.

❖ When locating the spotting or pretreating operation within the plant, care must be taken to choose a location that will allow adequate airflow on all sides of the spotting or pretreating equipment to maximize the effectiveness of the exhaust ventilation system required by Section 1205.2.3 in preventing the accumulation of hazardous vapor concentrations.

1206.3.4 Bonding and grounding. Metal scouring, brushing and spotting tables and scrubbing tubs shall be permanently and effectively bonded and grounded.

❖ Solvent containers and processing equipment during storage, handling and use must be grounded to prevent buildup of static charges, thus eliminating a common source of ignition for vapors from flammable and combustible liquid solvents.

1206.4 Type IV systems. Flammable and combustible liquids used for spotting operations shall be stored in approved safety cans or in sealed DOTn-approved metal shipping containers of not more than 1 gallon (4 L) in capacity. Dispensing shall be from approved safety cans. Aggregate amounts shall not exceed 10 gallons (38 L).

❖ Type IV dry cleaning systems are those systems using nonflammable solvents where dry cleaning is done only by trained operators who must be familiar with solvent hazards. Up to an aggregate quantity of 10 gallons (38 L) of Class I, II or III spotting or pretreating solvents is allowed if the solvents comply with the minimum storage safeguards of: (a) being stored in their original shipping container or (b) stored in and dispensed from safety cans complying with UL 30 (see commentary, Sections 2705.1.10 and 3405.2.4).

1206.5 Type V systems. Spotting operations using flammable or combustible liquids are prohibited in Type V dry cleaning systems.

❖ Type V dry cleaning systems are those systems using nonflammable solvents where dry cleaning is done by the general public, such as in coin-operated laundries that may or may not be attended by qualified staff. To prevent exposing the general public to the hazards of spotting or pretreating operations using solvents more hazardous than the nonflammable solvents used in the

Type V system itself, and to preclude any action that might violate Section 1205.1.4, this section prohibits spotting and pretreating operations altogether.

SECTION 1207 DRY CLEANING SYSTEMS

1207.1 General equipment requirements. Dry cleaning systems, including dry cleaning units, washing machines, stills, drying cabinets, tumblers, and their appurtenances, including pumps, piping, valves, filters and solvent coolers, shall be installed and maintained in accordance with NFPA 32. The construction of buildings in which such systems are located shall comply with the requirements of this section and the *International Building Code.* B:C portable fire extinguishers shall be provided near the doors inside dry cleaning rooms containing Type II, Type III-A and Type III-B dry cleaning systems.

❖ NFPA 32 contains provisions for the prevention and control of fire and explosion hazards incidental to dry cleaning operations for the protection of the employees and the public. Likewise, the IBC regulates the construction of dry cleaning plant buildings based on the relative hazards of the occupancies. Chapter 3 of the IBC classifies dry cleaning pick-up/drop-off stations and Type V dry cleaning systems in occupancy Group B (business) because of their low-hazard nature. Type II, III and IV dry cleaning plants are classified in occupancy Group F-1 (moderate-hazard factory-industrial) because they include or could include the processing of combustible textiles using combustible solvents and, on a limited basis, flammable liquids for spotting. Certain dry cleaning operations could be classified by the IBC in Group H, depending on the quantity of solvent present and the type of machines used. See the commentary to Section 307.9 of the IBC for further discussion of exceptions to Group H classifications. The fire extinguisher requirements in this section duplicate those in Section 1208.4. Placing the extinguishers at the doors leading out of rooms containing Type II, III-A and III-B dry cleaning systems will enhance personnel safety since it will require them to travel toward the means-of-egress door to gain access to an extinguisher in case of a fire.

1207.2 Type II systems. Type II dry cleaning and solvent tank storage rooms shall not be located below grade or above the lowest floor level of the building and shall comply with Sections 1207.2.1 through 1207.2.3.

> **Exception:** Solvent storage tanks installed underground, in vaults or in special enclosures in accordance with Chapter 34.

❖ Type II dry cleaning systems are those systems using a Class II combustible liquid [flash point from 100°F (38° C) to 140°F (60° C)] as a cleaning solvent, sometimes at or above its lowest flash point under normal plant operating conditions. For this reason, this section prohibits the location of these systems and solvent storage at other than the level of fire department access or exit discharge; that is, other than at grade. Because of the higher relative hazards of these systems and their solvents, locating them above or below the lowest grade floor level could affect rapid egress by employees as well as rapid access to the system by fire fighters in case of an emergency.

1207.2.1 Fire-fighting access. Type II dry cleaning plants shall be located so that access is provided and maintained from one side for fire-fighting and fire control purposes in accordance with Section 503.

❖ The general design of fire apparatus access to buildings is regulated by Section 503. To give fire fighters timely access to the building, this section requires that the building containing the Type II system be accessible on at least one side. This is consistent with the provisions of Section 503.1.1, which allows access on only one side when a building is equipped throughout with an automatic sprinkler system as is required by Section 1208.2 for dry cleaning plants. Fire apparatus access must be kept unobstructed at all times and comply with the provisions of Section 503.

1207.2.2 Number of means of egress. Type II dry cleaning rooms shall have not less than two means of egress doors located at opposite ends of the room, at least one of which shall lead directly to the outside.

❖ The number of exit or exit access doorways required from a room or space is regulated by Section 1014.2. That general section would not require two doors out of a Type II dry cleaning room classified in Occupancy Group F-1 unless it had an occupant load of more than 50 persons; that is, greater than 5,000 square feet (465 m^2) in area based on 100 square feet (9 m^2) per person in accordance with Table 1004.1.2. However, this section is specific to Type II dry cleaning rooms and supercedes the general Group F requirements of Section 1014.1 by requiring two doors out of the room, regardless of occupant load. Note that at least one of the doors must be an exit door that opens directly to the exit discharge or public way outside of the building as required by Section 1017.2.2. All doors must also comply with Section 1017.2.

This section also supercedes the general exit or exit access remoteness requirements of Section 1014.2.1 by mandating that the two means of egress doors from a Type II dry cleaning room be located at opposite ends of the room, rather than a distance apart equal to one-half the overall diagonal dimension of the room.

1207.2.3 Spill control and secondary containment. Curbs, drains, or other provisions for spill control and secondary containment shall be provided in accordance with Section 2704.2 to collect solvent leakage and fire protection water and direct it to a safe location.

❖ To prevent the flow of Class II solvents to rooms or spaces adjoining the dry cleaning room, spill control complying with Section 2704.2 is required by this section. Care must be taken in the design of drainage and secondary containment systems to accurately calculate automatic sprinkler design discharge flow rates and fire suppression hand line [typically $1^1/_2$- or $1^3/_4$-inch (38 mm or 44 mm) hose] flows. Note that secondary containment provisions do not include control of the flammable, irritating or toxic vapors given off by the solvent. Care must be taken to minimize exposure to hazardous vapors. Runoff from spills or manual fire suppression activities may result in environmental contamination if not properly controlled (see also commentary, Section 2704.2).

1207.3 Solvent storage tanks. Solvent storage tanks for Class II, IIIA and IIIB liquids shall conform to the requirements of Chapter 34 and be located underground or outside, above ground.

Exception: As provided in NFPA 32 for inside storage or treatment tanks.

❖ Safe and proper storage of solvents is a paramount concern in the safe operation of a dry cleaning plant or facility. This section requires that combustible solvents be stored in approved tanks located either underground or above ground outside the building, and references Chapter 34 for detailed requirements.

The exception recognizes the provisions contained in NFPA 32 governing the inside storage of solvents. While Chapter 2 of NFPA 32 allows the unenclosed indoor installation of up to three 1,500-gallon (5678 L) storage or treatment tanks of Class II solvent, it should be noted that, in accordance with the IBC, indoor storage of Class II solvents in excess of the maximum allowable quantity per control area indicated in Table 307.7(1) of that code would cause the dry cleaning facility to be classified in Group H-3.

SECTION 1208 FIRE PROTECTION

1208.1 General. Where required by this section, fire protection systems, devices and equipment shall be installed, inspected, tested and maintained in accordance with Chapter 9.

❖ To control the relative hazards posed by dry cleaning operations, this section requires the installation of a fire protection system and references Chapter 9 for installation and maintenance.

1208.2 Automatic sprinkler system. An automatic sprinkler system shall be installed in accordance with Section 903.3.1.1 throughout dry cleaning plants containing Type II, Type III-A or Type III-B dry cleaning systems.

❖ Section 903.2 of the code contains the general "where required" criteria for the installation of automatic sprinklers in buildings. Although the general provisions of that section require automatic sprinkler protection in Group F-1 fire areas based on height and area, Section 1208.2 supercedes those general provisions by requiring that all dry cleaning plants, regardless of height or area, be equipped throughout with an automatic sprinkler system designed and installed in accordance with NFPA 13. This is consistent with the requirements of NFPA 32, and recognizes the hazards of a relatively large contents fire load coupled with the presence of combustible (and, to a lesser degree, flammable) liquids used in both open and closed system operations.

1208.3 Automatic fire-extinguishing systems. Type II dry cleaning units, washer-extractors, and drying tumblers in Type II dry cleaning plants shall be provided with an approved automatic fire-extinguishing system installed and maintained in accordance with Chapter 9.

Exception: Where approved, a manual steam jet not less than 0.75 inch (19 mm) with a continuously available steam supply at a pressure not less than 15 pounds per square inch gauge (psig) (103 kPa) is allowed to be substituted for the automatic fire-extinguishing system.

❖ Section 1208.2 focuses on the protection of the dry cleaning plant occupants and building. This section requires additional fire protection for dry cleaning equipment using Class II solvent because of its lower flash point. Automatic fire-extinguishing systems for local, direct application of suppression media within the dry cleaning machinery must be installed in accordance with Section 904. The type of automatic fire-extinguishing system typically used (and required by NFPA 32) in this application is carbon dioxide, installed in accordance with Section 904.8, although approved dry chemical or gaseous-agent systems could also be used.

The exception recognizes the fire smothering capabilities of steam. Although there is no recognized standard for the design of steam smothering systems, the fact that steam is usually available in large quantities in dry cleaning plants makes it an inexpensive alternative to the more state-of-the-art fire-extinguishing systems. The design and installation of the steam jet(s) that must be available for personnel to manually use must be reviewed and approved by the fire code official. Because steam poses a serious burn hazard, the system design should minimize personnel exposure to this hazard.

1208.4 Portable fire extinguishers. Portable fire extinguishers shall be selected, installed and maintained in accordance with this section and Section 906. A minimum of two 2-A:10-B:C portable fire extinguishers shall be provided near the doors inside dry cleaning rooms containing Type II, Type III-A and Type III-B dry cleaning systems.

❖ Portable fire extinguishers are intended only for fighting incipient fires. Employees should be trained in the proper selection and use of portable fire extinguishers. Both the extinguisher rating and the travel distance must be consistent with Section 906 and NFPA 10 for the moderate hazards expected in dry cleaning plants. Placing the extinguishers at the doors leading out of rooms containing Type II, III-A and III-B dry cleaning systems will enhance personnel safety by requiring them to travel toward the means of egress door to gain access to an extinguisher in case of a fire. The required size of the extinguisher(s) should give the operator sufficient agent capacity and discharge time to handle the magnitude of incipient fires expected.

Bibliography

The following resource materials are referenced in this chapter or are relevant to the subject matter addressed in this chapter.

IBC-2003, *International Building Code.* Falls Church, VA: International Code Council, 2003.

ICC EC-2003, ICC *Electrical Code*. Falls Church, VA: International Code Council, 2003.

IMC-2003, *International Mechanical Code.* Falls Church, VA: International Code Council, 2003.

IPC-2003, *International Plumbing Code.* Falls Church, VA: International Code Council, 2003.

NFPA 10-98, *Portable Fire Extinguishers.* Quincy, MA: National Fire Protection Association, 1998.

NFPA 13-99, *Installation of Sprinkler Systems*. Quincy, MA: National Fire Protection Association, 1999.

NFPA 32-00, *Dry Cleaning Plants*. Quincy, MA: National Fire Protection Association, 2000.

UL 30-95, *Metal Waste Cans.* Northbrook, IL: Underwriters Laboratories Inc., 1995.

Chapter 13:
Combustible Dust-Producing Operations

General Comments

Problems associated with the production and handling of dusts were not widely recognized and understood until the twentieth century. Likewise, the need for adequate protection from dust explosions did not become widely accepted until the latter half of the twentieth century. A relatively incomplete understanding of the theoretical underpinnings of the explosion hazards of dusts is largely responsible for this lag.

The following factors affect the explosion hazards of dusts:

- Chemical composition of the dust;
- Geometry of the dust particles;
- Concentration of dust present in suspension (distance between dust particles or the mass of dust particles in a given volume);
- Nature and concentration of the oxidant, usually air;
- Moisture content of both the atmosphere and the dust and
- Minimum required ignition temperature or ignition energy of the dust and amount of time the two are in contact with each other.

Curiously, not all combustible dusts are explosible. The reasons for this are not very well understood. Explosion hazards of dusts are often defined by the lower and upper explosive limits, explosion pressures and explosion rates of various dusts. These values cannot be derived from the thermal properties of the dusts but rather must be determined by measurement. Combustible dusts may produce explosion pressures as great as 150 pounds per square inch gauge (psig) (103 kPa) and explosion pressure rise rates as high as 15,000 pounds per square inch (psi) per second (103 425 kPa).

Dust explosions usually produce flame fronts traveling at rates less than the speed of sound, and are, therefore, classified as deflagrations. High oxidizer concentrations or the presence of flammable gases may, however, produce detonations with shockwaves traveling in excess of the speed of sound preceding the flame front.

Unlike flammable gases and vapors, concentrations of explosive dusts exceeding stoichiometric quantities (the amount of fuel necessary to consume all the available oxidizer) produce more violent explosions up to the point at which dust concentrations approach several times the stoichiometric concentration.

In 1973, K.N. Palmer listed the following explosion protection measures. Formation of explosive dust suspensions is inevitable, and complete elimination of ignition sources cannot be relied on; therefore, implementing a combination of explosion protection strategies is strongly recommended. Such explosion protection measures include ignition prevention; suppression or containment of the primary explosion flame and allowing the explosion to take its full course in a safe manner.

Fire fighters are especially aware of the hazards of dust explosions in grain elevators in rural areas, but may not recognize the potential for dust explosions in a wide variety of other manufacturing operations. Dusts are produced and used in industrial applications ranging from water treatment to semiconductor manufacturing. The explosive force generated by a dust explosion is comparable to flammable vapor and gas explosions.

The following industries are involved in the production or handling of explosible dusts:

- Agricultural;
- Chemical, including dye;
- Coal mining and use;
- Food (human and animal);
- Metals;
- Pharmaceuticals;
- Plastics and
- Woodworking.

Conditions under which a combustible dust will produce an explosion are quite specific, though they are difficult to quantify. As a result, precautions to prevent dust explosions are often determined by assigning a particular dust to a hazard class. A sample of material is ignited under controlled conditions in a special test apparatus to determine the hazard class of a dust. The results of such tests are listed for common dusts in Table 13(1). The last column of the table describes the hazard class of the material. Table 13(2) describes the appropriate range of values for the critical characteristic used to determine the hazard class of various dusts not listed in Table 13(1).

Table 13(1) describes explosion data for selected combustible dusts. In each column, values of the table are derived experimentally using a special explosion apparatus. For each property, values are presented to fully describe the explosion characteristics of the given material. Take particular note of the units for each value that are presented in parentheses: P_{MAX} is the maximum blast pressure produced inside the test apparatus, while K_{ST} is a coefficient integrating the maximum blast pressure over distance and time. (Note: A "bar" is the pressure exerted by a force of 1,000,000 dynes on a square centimeter of surface.) In turn, these figures may be used to determine explosion venting requirements for a given material or process. The value of K_{ST} is used to determine the dust explosion hazard class for the material. Dust hazard classes are expressed as ST-1 through ST-3, with ST-1 being the highest hazard. Table 13(2) lists the ranges of K_{ST} associated with each hazard class.

TABLE 13(1)
EXPLOSION DATA FOR REPRESENTATIVE POWDERS AND DUSTS

Material	Median Particle Size (μm)	Minimum Explosive Concentration (g/m^3)	P_{MAX}(bar)	K_{ST}(bar m/s)[a]	Dust Hazard Category
Acrylonitrile	25	—	8.5	121	1
Aluminum	29	30	12.4	415	3
Calcium stearate	12	30	9.1	132	1
Coat, bituminous	24	60	9.2	129	1
Corn starch	7	—	10.3	202	2
Magnesium	28	30	17.5	508	3
Milo, powdered	83	60	5.8	28	1
Phenolic resin	Less than 10	15	9.3	129	1
Polyenthylene	Less than 10	30	8	156	1
Polymethyl-methacry late	21	30	9.4	269	2
Polypropylene	25	30	8.4	101	1
Polyvinylchloride (PVC)	107	200	7.6	46	1
Soy flour	20	200	9.2	110	1
Sugar	30	200	8.5	138	1
Sulfur	20	30	6.8	151	1
Wheat starch	22	20	9.9	115	1
Zinc	10	250	6.7	125	1
Zinc	Less than 10	125	7.3	176	1

Note a. $K_{st} = max^{v^{1/3}}$; 1 bar = 29.5 inches of mercury 1 btu = 14.5 psi, 1 psi = 6.894 kPa
Source: Field, P. *Dust Explosions*. New York, Elsevier, 1992.

TABLE 13(2)
CLASSIFICATION OF DUST HAZARDS

Dust Category	K_{ST} (bar m/s)
ST-1	Less than 200
ST-2	200-300
ST-3	Greater than 300

Source: Zalosh, R.G., *The SFPE Handbook of Fire Protection Engineering*. Quincy, MA, National Fire Protection Association, 1988.

Purpose

Awareness and knowledge of the hazards of dusts and powders are less common than of flammable liquids and gases. However, explosions and fires involving dusts and powders are just as hazardous in many industrial settings. The requirements of this chapter seek to reduce the likelihood of dust explosions by managing the hazards of suspensions of ignitable dusts. Ignition source control and good housekeeping practices in occupancies containing dust-producing operations are emphasized.

SECTION 1301
GENERAL

1301.1 Scope. The equipment, processes and operations involving dust explosion hazards shall comply with the provisions of this chapter.

❖ This chapter details general requirements for the protection of properties and processes from explosions and deflagrations involving combustible dust residues and suspensions.

1301.2 Permits. Permits shall be required for combustible dust-producing operations as set forth in Section 105.6.

❖ The process of issuing permits gives the fire code official an opportunity to carefully evaluate and regulate hazardous operations. Permit applicants should be required to demonstrate that their operations comply with the intent of the code before the permit is issued. See the commentary to Section 105.6.7 for discussion of specific combustible dust-producing operations requiring an operational permit.

SECTION 1302
DEFINITIONS

1302.1 Definition. The following word and term shall, for the purposes of this chapter and as used elsewhere in this code, have the meaning shown herein.

❖ Definitions can help in the understanding and application of the code requirements. This definition is given here for convenient access to it when applying the requirements of this chapter. For convenience, the defined term is also listed in Chapter 2 with a cross reference to this section. The use and application of all defined terms, including the one here, are set forth in Section 201.

COMBUSTIBLE DUST. Finely divided solid material which is 420 microns or less in diameter and which, when dispersed in air in the proper proportions, could be ignited by a flame, spark or other source of ignition. Combustible dust will pass through a U.S. No. 40 standard sieve.

❖ Combustible dusts are combustible solids in a finely divided state that are suspended in the air. A hazard exists when the concentration of the combustible dust is within the explosive limits and exposed to an ignition source of sufficient energy and duration to initiate self-sustained combustion. Occupancies that include combustible dust-producing operations are classified in Group H-2 by the *International Building Code*® (IBC®) because of the severe deflagration hazard of combustible dusts. Ways to reduce the explosion potential of these materials must also be addressed.

SECTION 1303
PRECAUTIONS

1303.1 Sources of ignition. Smoking or the use of heating or other devices employing an open flame, or the use of spark-producing equipment is prohibited in areas where combustible dust is generated, stored, manufactured, processed or handled.

❖ Smoking is prohibited in areas where dust explosion hazards exist. Welding and cutting should be confined to approved areas and subject to a hot work permit system and prior approval for each operation (see commentary, Chapter 26). Open-flame heating and spark-producing equipment, as well as other heat-producing devices, are prohibited. Torches and other open-flame devices are never to be used to remove accumulations of dust or dust residues. Electrical wiring, lighting and equipment where dust explosion hazards exist should be approved for use in Class II hazardous (classified) locations as regulated in the ICC *Electrical Code*® (ICC EC™).

1303.2 Housekeeping. Accumulation of combustible dust shall be kept to a minimum in the interior of buildings. Accumulated combustible dust shall be collected by vacuum cleaning or other means that will not place combustible dust into suspension in air. Forced air or similar methods shall not be used to remove dust from surfaces.

❖ Good housekeeping practices are extremely important in occupancies where combustible dust-producing operations are located. Ideally, minimizing the amount of fugitive combustible dust that accumulates in buildings should be accomplished by fixed equipment, such as physical enclosures for dust-producing machinery, to prevent escape of dust from its source and by approved dust collection systems designed to capture the dust at the point of generation. In its simplest form, a dust collection system can consist of a shop vacuum cleaner connected to a woodworking machine, such as a table saw. In larger industrial applications, a ducted dust collection system can serve multiple machines or entire production lines. Such systems are considered to be hazardous exhaust systems and should be designed and installed in accordance with the *International Mechanical Code*® (IMC®) and NFPA 650.

It is important when removing accumulated combustible dust not to place it into suspension in the air, thus creating the potential for a dust explosion. Accordingly, this section requires that dust be collected by vacuum cleaning equipment or other approved means that will not disturb the accumulated dust. Dust should never be brushed from dust-loaded surfaces or be blown off with compressed air.

SECTION 1304
EXPLOSION PROTECTION

1304.1 Standards. The fire code official is authorized to enforce applicable provisions of the codes and standards listed in Table 1304.1 to prevent and control dust explosions.

❖ Because the pressure exerted by a combustible dust explosion typically ranges from 13 psi (90 kPa) to 89 psi (614 kPa), it is impractical to construct a building that will withstand such pressures. Therefore, a means of explosion relief or venting must be provided in accordance with Section 911 and the referenced standards. Additional guidance on the relative fire risk associated with various combustible dusts can be found in the referenced standards and the bibliographic material.

TABLE 1304.1
EXPLOSION PROTECTION STANDARDS

STANDARD	SUBJECT
NFPA 61	Agricultural and Food Products
NFPA 69	Explosion Prevention
NFPA 85	Boiler and Combustion Systems Hazards
NFPA 120	Coal Preparation Plants
NFPA 480	Magnesium Solids and Powders
NFPA 481	Titanium
NFPA 482	Zirconium
NFPA 650	Conveying Combustible Particulate Solids
NFPA 651	Aluminum Powder
NFPA 654	Manufacturing, Processing and Handling of Combustible Particulate Solids
NFPA 655	Sulfur
NFPA 664	Prevention of Fires and Explosions in Wood Processing and Woodworking Facilities
ICC *Electrical Code*	Electrical Installations

❖ Table 1304.1 references 12 NFPA standards and the ICC EC, which detail specific precautions for a wide variety of dust explosion situations covering the broad spectrum of industries involved in dust-producing operations. These standards include the most common dust explosion hazards and regulations for their prevention. Essentially, each of the referenced standards prescribes reasonable requirements for safety to life and property from fire and explosion. The standards also minimize the resulting damage should a fire or explosion occur. More specifically, the standards contain requirements for construction, ventilation, explosion venting, equipment, heating devices, dust control, fire protection and supplemental requirements related to electrical wiring and equipment; provisions concerning protection from sparks; cutting and welding and smoking and signage regulations.

Unusual situations, especially those involving plastics, resins, pharmaceuticals and semiconductor dusts, should be carefully evaluated.

Bibliography

The following resource materials are referenced in this chapter or are relevant to the subject matter addressed in this chapter.

Combustible Dusts. Factory Mutual Loss Prevention Data Sheet 7, Norwood, MA: Factory Mutual, August 1976.

Field, P. *Dust Explosions*. New York: Elsevier, 1982.

Gray, T. A. "National Grain and Feed Association Fire and Explosion Research Program." *SFPE Bulletin*. Fire Record Department, compiler. Boston, MA: Society of Fire Protection Engineers (SFPE), September/October, 1991.

IBC-2003, *International Building Code*. Falls Church, VA: International Code Council, 2003.

ICC EC-2003, *ICC Electrical Code*. Falls Church, VA: International Code Council, 2003.

IMC-2003, *International Mechanical Code*. Falls Church, VA: International Code Council, 2003.

Industrial Fire Hazards Handbook, 3rd ed. Quincy, MA: National Fire Protection Association, 1990.

NFPA Inspection Manual, 7th ed. Quincy, MA: National Fire Protection Association, 1994.

NFPA 61-99, *Standard for the Prevention of Fires and Dust Explosions in Agricultural and Food Products Facilities*. Quincy, MA: National Fire Protection Association, 1999.

NFPA 68-02, *Guide for Venting of Deflagrations*. Quincy, MA: National Fire Protection Association, 2002.

NFPA 69-97, *Explosion Prevention Systems*. Quincy, MA: National Fire Protection Association, 1997.

NFPA 70-02, *National Electrical Code*. Quincy, MA: National Fire Protection Association, 2002.

NFPA 77-00, *Recommended Practice on Static Electricity*. Quincy, MA: National Fire Protection Association, 2000.

NFPA 85-01, *Boiler and Combustion Systems Hazards Code*. Quincy, MA: National Fire Protection Association, 2001.

NFPA 120-99, *Standard for Coal Preparation Plants*. Quincy, MA: National Fire Protection Association, 1999.

NFPA 480-98, *Standard for the Storage, Handling and Processing of Magnesium Solids and Powders*. Quincy, MA: National Fire Protection Association, 1998.

NFPA 481-00, *Standard for the Production, Processing, Handling and Storage of Titanium*. Quincy, MA: National Fire Protection Association, 2000.

NFPA 482-96, *Standard for the Production, Processing, Handling and Storage of Zirconium*. Quincy, MA: National Fire Protection Association, 1996.

NFPA 499-97, *Recommended Practice for the Classification of Combustible Dusts and of Hazardous (Classified) Locations for Electrical Installations in Chemical Process Areas.* Quincy, MA: National Fire Protection Association, 1997.

NFPA 650-98, *Standard for Pneumatic Conveying Systems for Handling Combustible Particulate Solids.* Quincy, MA: National Fire Protection Association, 1998.

NFPA 651-98, *Standard for the Machining and Finishing of Aluminum and the Production and Handling of Aluminum Powders.* Quincy, MA: National Fire Protection Association, 1998.

NFPA 654-00, *Standard for the Prevention of Fire and Dust Explosions from the Manufacturing, Processing and Handling of Combustible Particulate Solids.* Quincy, MA: National Fire Protection Association, 2000.

NFPA 655-93, *Standard for Prevention of Sulfur Fires and Explosions.* Quincy, MA: National Fire Protection Association, 1993.

NFPA 664-98, *Standard for the Prevention of Fires and Explosions in Wood Processing and Woodworking Facilities.* Quincy, MA: National Fire Protection Association, 1998.

Palmer, K. N. *Dust Explosions and Fires.* London, England: Chapman and Hall, 1973.

Report of Important Dust Explosions: A Record of Dust Explosions in the United States and Canada since 1860. Fire Record Department, compiler. Boston, MA: National Fire Protection Association, 1957.

Schram, P.J. and M.W. Earley. *Electrical Installations in Hazardous Locations,* 2nd ed. Quincy, MA: National Fire Protection Association, 1998.

Schwab, R.F. "Section 3/Chapter 12: Dusts." In Cote, A. E., ed., *NFPA Fire Protection Handbook*, 17th ed. Quincy, MA: National Fire Protection Association, 1991.

"Fire Protection in Agricultural Facilities: A Review of Research, Resources and Practices." Journal of Fire Protection Engineering, Volume 13, No 3. Bethesda, MD: Society of Fire Protection Engineers (SFPE), August 2003.

Zalosh, R.G. "Section 2/Chapter 5: Explosion Protection." *The SFPE Handbook of Fire Protection Engineering.* Quincy, MA: National Fire Protection Association, 1988.

Chapter 14: Fire Safety During Construction and Demolition

General Comments

This chapter outlines general fire safety precautions for all structures and all occupancies during construction and demolition operations. In general, these requirements seek to maintain required fire protection, limit fire spread, establish the appropriate operation of equipment and promote prompt response to fire emergencies.

There are 16 sections in Chapter 14. Section 1401 gives the general scope of the chapter and provides the basis for enforcement of its provisions. Section 1402 defines terms specifically relevant to the chapter. The listing, arrangement, fueling and suppression of temporary heating equipment is described in Section 1403. Section 1404 deals with precautions against fire that involve the control of smoking, waste disposal, open burning, spontaneous ignition and temporary electrical wiring. The storage, handling and classification of flammable and combustible liquids, flammable gases and explosive materials are addressed in Sections 1405, 1406 and 1407.

Sections 1408 and 1409 regulate the need for prefire planning, training and maintenance and supervision of fire protection and alarm systems. Access for fire fighting is discussed in Section 1410. Escape by on-site personnel is covered under means of egress in Section 1411. The provision and maintenance of specific fire protection devices such as standpipes, automatic sprinkler systems and portable fire extinguishers is explained in Sections 1413, 1414 and 1415, respectively. The need to regulate heat sources, such as internal combustion engines and fuel-fired asphalt and tar kettles, is addressed in Sections 1416 and 1417.

Purpose

This chapter contains code language that will safeguard people from injury or illness and protect property from damage during the construction or demolition processes.

SECTION 1401 GENERAL

1401.1 Scope. This chapter shall apply to structures in the course of construction, alteration, or demolition, including those in underground locations. Compliance with NFPA 241 is required for items not specifically addressed herein.

❖ Buildings are most vulnerable to fire when undergoing construction, demolition or alteration. Special measures are required to either minimize the potential for a fire or aid in fire control and suppression. These requirements amplify those of other sections of the code and prescribe maintenance of adequate means of egress and on-site incipient fire-fighting equipment. Temporary heating appliances are regulated to prevent ignition of combustible debris and structural elements. Fire apparatus access and maintenance of standpipes are also addressed (see also Chapter 33 of the *International Building Code*® (IBC®), NFPA 241 and Section 6, Chapter 14 of the NFPA *Fire Protection Handbook*).

1401.2 Purpose. This chapter prescribes minimum safeguards for construction, alteration, and demolition operations to provide reasonable safety to life and property from fire during such operations.

❖ This chapter is intended to regulate access by the responding fire department, fire protection systems, operations and maintenance of structures for precautions against fire and spread of fire during construction and demolition.

SECTION 1402 DEFINITIONS

1402.1 Terms defined in Chapter 2. Words and terms used in this chapter and defined in Chapter 2 shall have the meanings ascribed to them as defined therein.

❖ This section directs the code user to Chapter 2 for the proper application of the terms used in this chapter. These terms may be defined in Chapter 2, in another *International Code*® as indicated in Section 201.3 or the dictionary meaning may be all that is needed (see also the commentary to Sections 201 through 201.4).

SECTION 1403
TEMPORARY HEATING EQUIPMENT

1403.1 Listed. Temporary heating devices shall be listed and labeled in accordance with the *International Mechanical Code* or the *International Fuel Gas Code*. Installation, maintenance and use of temporary heating devices shall be in accordance with the terms of the listing.

❖ Listing and labeling are used to identify materials, assemblies and devices that are required to bear the identification of the manufacturer, as well as a third-party quality control agency. The quality control agency allows the use of its listing or label based on periodic audits and inspections of the manufacturer's facility. Not all testing laboratories, inspection agencies and other organizations concerned with product or program evaluation use the same means for identifying listed equipment, materials or agencies. Some do not recognize equipment or materials as listed unless they are also labeled. The fire code official must use the same system as the listing organization to identify listed equipment, materials or agencies.

1403.2 Oil-fired heaters. Oil-fired heaters shall comply with Section 603.

❖ The regulations for the devices that are likely to be used for temporary heat are delineated in Section 603.

1403.3 LP-gas heaters. Fuel supplies for liquefied- petroleum gas-fired heaters shall comply with Chapter 38 and the *International Fuel Gas Code*.

❖ Because propane gas is heaver than air, special attention must be given to the way the fuel tank is arranged and connected to the heating device. Tank location and protecting the tanks from damage are just two of the concerns addressed in the *International Fuel Gas Code*® (IFGC®).

1403.4 Refueling. Refueling operations shall be conducted in accordance with Section 3405. The appliance shall be allowed to cool prior to refueling.

❖ This section address the refueling of liquid-fueled equipment. Section 3405 addresses proper liquid transfer, container filling operations, filling locations, quantity limits and more. Because hot surfaces can cause ignition of flammable vapors and spills, the appliance must be allowed to cool before refueling.

1403.5 Installation. Clearance to combustibles from temporary heating devices shall be maintained in accordance with the labeled equipment. When in operation, temporary heating devices shall be fixed in place and protected from damage, dislodgement or overturning in accordance with the manufacturer's instructions.

❖ Because conditions change during construction or demolition, temporary heating devices must be monitored and maintained. Materials are constantly being moved, which may reduce the clearances to the device. It may be advantageous for the owner to hire a fire watch to check operating conditions while work is in progress and at the end of work shifts. The fire code official may need to decide what is required.

1403.6 Supervision. The use of temporary heating devices shall be supervised and maintained only by competent personnel.

❖ Temporary installations must be as safe as permanent ones. Having qualified people do these installations is important.

SECTION 1404
PRECAUTIONS AGAINST FIRE

1404.1 Smoking. Smoking shall be prohibited except in approved areas. Signs shall be posted in accordance with Section 310. In approved areas where smoking is permitted, approved ashtrays shall be provided in accordance with Section 310.

❖ This smoking prohibition is for fire safety concerns, not health or environmental ones. Every effort must be made to keep hot smoking materials from igniting building materials or debris. Smoking in occupancies subject to ignition hazards from smoking materials should be approved by the fire code official and confined to spaces without significant amounts of combustibles and where approved ash trays or receptacles are provided. Signage indicating "smoking permitted in this area" will encourage the use of the limited area. "No smoking" signs similar to Figure 1404.1 may be used.

1404.2 Waste disposal. Combustible debris shall not be accumulated within buildings. Combustible debris, rubbish and waste material shall be removed from buildings at the end of each shift of work. Combustible debris, rubbish and waste material shall not be disposed of by burning on the site unless approved.

❖ Construction sites must be kept reasonably free of accumulations of combustible waste, debris and rubbish. Accumulations are to be removed at the end of each work shift. Combustibles must not be burned unless local environmental authorities are consulted on local open burning regulations. The local fire authority should also be informed of any open burning.

1404.3 Open burning. Open burning shall comply with Section 307.

❖ Section 307 requires that a permit be obtained for open burning. Other areas of concern are burning location and monitoring. The owner should make documentation of the event available to the fire code official.

Figure 1404.1
NO SMOKING SIGN

1404.4 Spontaneous ignition. Materials susceptible to spontaneous ignition, such as oily rags, shall be stored in a listed disposal container.

❖ Spontaneous ignition, also known as autoignition or self-ignition, is defined by Burklin and Purington as "ignition due to chemical reaction or bacterial action in which there is slow oxidation of organic compounds until the material ignites; usually there is sufficient air for oxidation but insufficient ventilation to carry heat away as it is generated." A detailed treatment of the subject appears in the NFPA *Fire Protection Handbook*. The One Meridian Plaza office building fire in 1991 was allegedly started by spontaneous ignition of oil-soaked rags that were improperly stored during a remodeling operation. This high-rise building in the heart of Philadelphia was so seriously damaged in the fire that it was razed in 1999. A listed container for the storage of the oily rags was not used. The fire code official should determine the kinds of oils or solvents used and research their potential for spontaneous ignition (see commentary, Section 304.3.2).

1404.5 Fire watch. When required by the fire code official for building demolition that is hazardous in nature, qualified personnel shall be provided to serve as an on-site fire watch. Fire watch personnel shall be provided with at least one approved means for notification of the fire department and their sole duty shall be to perform constant patrols and watch for the occurrence of fire.

❖ Hazardous demolition operations may need the services of qualified emergency response personnel, such as hazmat technicians or fire fighters to stand by or actually patrol the area. When such persons are needed, it is essential that they focus on that task only and have no other assignments. A lay person should not be used and it is within the authority of the fire code official to require that professionals be on site. It is critical that such watch personnel be able to contact the fire department immediately in case of an emergency through a reliable means of communication approved by the fire code official.

1404.6 Cutting and welding. Operations involving the use of cutting and welding shall be done in accordance with Chapter 26.

❖ Cutting and welding operations account for 9 percent of fires in industrial operations. With accumulations of combustible materials that are common during construction and demolition, additional precautions must be taken. One of the most effective ways to prevent or promptly respond to fires caused by cutting and welding is to have a vigorous "hot work" permit system. Chapter 26 presents these precautions.

1404.7 Electrical. Temporary wiring for electrical power and lighting installations used in connection with the construction, alteration or demolition of buildings, structures, equipment or similar activities shall comply with the ICC *Electrical Code.*

❖ Storage and use of flammable and combustible liquids require approval of the fire code official for control of hazards and to provide the fire department with vital hazard data for preplanning for incidents involving such materials. A permit is required for storage. See Section 105 and the accompanying commentary for more information on permit requirements.

SECTION 1405
FLAMMABLE AND COMBUSTIBLE LIQUIDS

1405.1 Storage of flammable and combustible liquids. Storage of flammable and combustible liquids shall be in accordance with Section 3404.

❖ Storage and use of flammable and combustible liquids require approval of the fire code official for control of hazards and to provide the fire department with vital hazard data for preplanning for incidents involving such materials. A permit is required for storage. See Section 105 and the accompanying commentary for more information on permit requirements.

1405.2 Class I and Class II liquids. The storage, use and handling of flammable and combustible liquids at construction sites shall be in accordance with Section 3406.2. Ventilation shall be provided for operations involving the application of materials containing flammable solvents.

❖ Section 3406.2 contains comprehensive regulations on the proper storage, use and handling of Class I and II liquids. Areas of particular concern are: signage, storage location, ventilation, sources of ignition and dispensing. Class I liquids are more hazardous than Class II because of their lower flash points [< 100° F (38°C)] (see commentary, Chapter 34).

1405.3 Housekeeping. Flammable and combustible liquid storage areas shall be maintained clear of combustible vegetation and waste materials. Such storage areas shall not be used for the storage of combustible materials.

❖ Housekeeping in this case concerns ignition sources and added fuel load in the storage area. Easily ignited dry weeds, grass and paper are prohibited in the area. Access to the area by fire fighters can be hampered when combustibles in the storage area ignite.

1405.4 Precautions against fire. Sources of ignition and smoking shall be prohibited in flammable and combustible liquid storage areas. Signs shall be posted in accordance with Section 310.

❖ Sources of ignition such as electric arcing, open-flame heating devices and static electricity must be controlled. Smoking must also be controlled by posting "no smoking" signs, providing safe smoking areas and promoting on-the-job awareness of the smoking prohibition as stipulated in Section 310.

1405.5 Handling at point of final use. Class I and II liquids shall be kept in approved safety containers.

❖ This section intends that only approved safety cans (as defined in Section 2702.1) of no more than 5-gallon (19 L) capacity with a spring-loaded, self-closing lid and spout covers, designed to safely relieve internal pressure under fire conditions, be used for the storage of Class I and II liquids at construction and demolition sites.

The key to the proper storage and handling of flammable and combustible liquids is to keep liquids and vapors away from ignition sources. Restrictions on flammable liquid container sizes, separation distances and active and passive fire protection are based on the extent of hazard presented should an uncontrolled release occur.

The key to the proper storage and handling of flammable and combustible liquids is to keep liquids and vapors away from ignition sources. Restrictions on flammable liquid container sizes, separation distances and active and passive fire protection are based on the extent of hazard presented should an uncontrolled release occur.

1405.6 Leakage and spills. Leaking vessels shall be immediately repaired or taken out of service and spills shall be cleaned up and disposed of properly.

❖ Accidental liquid spills create a vapor release that can quickly travel from the spill point to an ignition source. Because spills need immediate attention to neutralize and remove the hazard, the local fire department should be notified; it is the agency best prepared to deal with the immediate hazards of a spill.

Leaks can indicate a developing problem with equipment or piping needing immediate repair. Until repairs are made, the equipment or piping must be taken out of service. If the leak or spill is expected to spread beyond the property lines or contaminate water or air, environmental authorities should also be notified.

SECTION 1406
FLAMMABLE GASES

1406.1 Storage and handling. The storage, use and handling of flammable gases shall comply with Chapter 35.

❖ Chapter 35 deals with the maximum allowable quantities, limits for indoor storage, storage containers, ignition sources and limits for outdoor storage. Also refer to Chapter 30 for requirements on compressed gases.

SECTION 1407
EXPLOSIVE MATERIALS

1407.1 Storage and handling. Explosive materials shall be stored, used and handled in accordance with Chapter 33.

❖ Chapter 33 prescribes minimum requirements for the safe storage, handling and use of explosives, ammunition and blasting agents for commercial and industrial occupancies. Its provisions are intended to protect the general public, emergency responders and individuals who handle explosives in connection with construction or demolition operations.

1407.2 Supervision. Blasting operations shall be conducted in accordance with Chapter 33.

❖ Specific requirements dealing with local physical and governmental controls, blasting area security and post-blast procedures are found in Section 3307.

Security precautions for explosive materials must conform to the requirements of this chapter and the referenced standards. Any discrepancy that suggests the loss or theft of explosives must be reported to local law enforcement authorities and the Bureau of Alcohol, Tobacco and Firearms (ATF) within 24 hours of discovery (see DOTy 27 CFR;55.30). The BATF may be contacted 24 hours a day at (800) 800-3855. Abandoned explosives, including those not claimed by the consignee within 48 hours of their arrival at a terminal, should be returned to the control of the last licensee (manufacturer or distributor) to possess them before they were abandoned. Local law enforcement authorities and BATF should be contacted if this is not possible.

1407.3 Demolition using explosives. Approved fire hoses for use by demolition personnel shall be maintained at the demolition site whenever explosives are used for demolition. Such fire hoses shall be connected to an approved water supply and shall be capable of being brought to bear on post-detonation fires anywhere on the site of the demolition operation.

❖ The code text does not stipulate the number, size or length of hoses needed; therefore, the involvement of the fire code official is essential to anticipate an incident. The competence of the demolition crew to properly use the hose for fire fighting is not addressed, which suggests that fire brigade training may be in order. Periodic inspections by the fire code official are imperative. As is the case throughout the code, fire protection methods and procedures must be acceptable to the fire code official.

SECTION 1408
OWNER'S RESPONSIBILITY FOR FIRE PROTECTION

1408.1 Program superintendent. The owner shall designate a person to be the Fire Prevention Program Superintendent who shall be responsible for the fire prevention program and ensure that it is carried out through completion of the project. The fire prevention program superintendent shall have the authority to enforce the provisions of this chapter and other provisions as necessary to secure the intent of this chapter. Where guard service is provided, the superintendent shall be responsible for the guard service.

❖ Each project must have a fire prevention program superintendent who is in charge of all fire safety efforts such as prefire planning, on-the-job training of personnel, guard service and the other areas covered in Sections 1408.1 through 1408.7. This person acts on behalf of the fire code official and can enforce the provisions of Chapter 14.

1408.2 Prefire plans. The fire prevention program superintendent shall develop and maintain an approved prefire plan in cooperation with the fire chief. The fire chief and the fire code official shall be notified of changes affecting the utilization of information contained in such prefire plans.

❖ Prefire plans are developed by the fire prevention program superintendent to assist the site personnel reacting to a fire. This plan must be coordinated with the local fire chief and the fire code official. Changes in building operations or equipment that could affect or change the fire department's ground attack of a fire must be reported to the fire department responder immediately. For example, if an additional 1,000 gallon (3785 L) propane tank is located alongside an existing tank, the responder needs to know about this situation.

1408.3 Training. Training of responsible personnel in the use of fire protection equipment shall be the responsibility of the fire prevention program superintendent.

❖ The fire responder is expected to know what fire-fighting and fire protection equipment is on the site and how to operate it. The fire prevention program superintendent is responsible for training the job site personnel in the proper use of hand-held fire extinguishers, hose lines, fire alarms and sprinkler systems.

1408.4 Fire protection devices. The fire prevention program superintendent shall determine that all fire protection equipment is maintained and serviced in accordance with this code. The quantity and type of fire protection equipment shall be approved.

❖ Fire protection and detection equipment must be maintained during construction and demolition. The fire prevention program superintendent must decide what is required to enforce maintenance as required by the code. Approval of the equipment and its maintenance is not, however, transferred to the fire prevention program superintendent, but remains with the fire code official.

1408.5 Hot work operations. The superintendent shall be responsible for supervising the permit system for hot work operations in accordance with Chapter 26.

❖ This issue is also discussed in the commentary to Section 1404.6. Chapter 26 contains an in-depth treatment of hot work, especially Section 2603.3, which deals with hot work permits. The fire prevention program superintendent issues the permits to coordinate a response if a fire should occur in the known hot work permit area.

1408.6 Impairment of fire protection systems. Impairments to any fire protection system shall be in accordance with Section 901.

❖ Section 901.7 specifically deals with systems out of service because of planned, emergency or accidental impairment. During demolition, portions of the equipment must be kept in service as long as possible. Likewise, equipment on a construction site must keep pace with the new work and be kept in service as much as possible, especially at the end of the work day.

1408.7 Temporary covering of fire protection devices. Coverings placed on or over fire protection devices to protect them from damage during construction processes shall be immediately removed upon the completion of the construction processes in the room or area in which the devices are installed.

❖ Fire protection devices must be kept in service as much as possible during construction. An example would be that paper or plastic bags must be removed from the sprinkler heads as soon as the painting of the sprinkler piping or the adjacent ceiling is completed. Additional information can be obtained from standards such as NFPA 13 and NFPA 72.

SECTION 1409
FIRE ALARM REPORTING

1409.1 Emergency telephone. Readily accessible emergency telephone facilities shall be provided in an approved location at the construction site. The street address of the construction site and the emergency telephone number of the fire department shall be posted adjacent to the telephone.

❖ The construction site must have an emergency phone located in an approved location. Workers on the site are not expected to know the street address of the site or the fire department emergency number, if it is a number other than 911. Therefore, the site address and fire department emergency number must be prominently posted. Typical customized signs are available from safety equipment suppliers or sign companies.

SECTION 1410
ACCESS FOR FIRE FIGHTING

1410.1 Required access. Approved vehicle access for fire fighting shall be provided to all construction or demolition sites. Vehicle access shall be provided to within 100 feet (30 480 mm) of temporary or permanent fire department connections. Vehicle access shall be provided by either temporary or permanent roads, capable of supporting vehicle loading under all weather conditions. Vehicle access shall be maintained until permanent fire apparatus access roads are available.

❖ Fire-fighting vehicle access is the means by which fire fighters gain access to the construction or demolition site and building for fire suppression and rescue operations until the permanent fire apparatus access roads are constructed. Such access is an integral component of the fire prevention program. The site superintendent or other person responsible for construction and demolition operations is responsible for maintaining and policing fire-fighter access routes, as provided in Section 1408. Fire apparatus must be able to get within 100 feet (30 480 mm) of any installed fire department connection supplying water to temporary or permanent fire protection systems over roads that will support the weight of the heaviest vehicle that might respond. The weight requirements are available from the local fire department. All-weather surfaces are required because the responding fire department should not waste time moving snow or trying to get out of mud (see also commentary, Section 503).

1410.2 Key boxes. Key boxes shall be provided as required by Chapter 5.

❖ As construction nears completion, some areas may not be accessible to the fire department without the use of a key. In those cases, the fire code official may require a key box as stipulated in Section 506.

SECTION 1411
MEANS OF EGRESS

[B] 1411.1 Stairways required. Where a building has been constructed to a height greater than 50 feet (15 240 mm) or four stories, or where an existing building exceeding 50 feet (15 240 mm) in height is altered, at least one temporary lighted stairway shall be provided unless one or more of the permanent stairways are erected as the construction progresses.

❖ Work crews will necessarily be in buildings under construction or demolition at the same time that the means of egress elements are being built or destroyed. This situation requires diligence on the part of the fire code official and the construction managers to make sure a means of escape is available to workers at all times, and that construction of occupiable areas does not unnecessarily extend beyond the construction of the means of egress. By the time the building is substantially enclosed, all required means of egress should be fully constructed and functional. For a building under construction, these precautions are triggered when the building exceeds 50 feet (15 240 mm) in height or four stories. Any temporary stairways must be lighted.

Similarly, the destruction of means of egress should not precede the demolition of areas occupied by workers.

1411.2 Maintenance. Required means of egress shall be maintained during construction and demolition, remodeling or alterations and additions to any building.

Exception: Approved temporary means of egress systems and facilities.

❖ As in any building where people must egress, the required means of egress must be kept clear of construction materials and demolition debris so occupants can exit in an emergency.

Temporary means of egress may be provided when the permanent egress system cannot be maintained in accordance with this section.

SECTION 1412
WATER SUPPLY FOR FIRE PROTECTION

1412.1 When required. An approved water supply for fire protection, either temporary or permanent, shall be made available as soon as combustible material arrives on the site.

❖ A water supply must be connected to the wet standpipe, and underground water supply and hydrants must be available for the dry standpipes as soon as combustible materials are on the job site. As previously stated in this commentary, the unfinished building is most vulnerable to fire and must be protected as much as possible.

SECTION 1413
STANDPIPES

1413.1 Where required. Buildings four or more stories in height shall be provided with not less than one standpipe for use during construction. Such standpipes shall be installed when the progress of construction is not more than 40 feet (12 192 mm) in height above the lowest level of fire department access. Such standpipe shall be provided with fire department hose connections at accessible locations adjacent to usable stairs. Such standpipes shall be extended as construction progresses to within one floor of the highest point of construction having secured decking or flooring.

❖ This requirement complements the installation requirements in Chapter 9 of the IBC and also in the code for new standpipe systems. The requirement for standpipe and hose connections is triggered when construction progresses to a height of no more than 40 feet (12 192 mm) above the lowest level of fire department access. While hoses need not be provided, the hose connection must be located adjacent to a stairway. Although thread requirements are not stated, the hose threads must be compatible with those of the responding fire department. Standpipe hose connections must be ready for use on each floor before installation of the floor deck on the story or level above to provide fire fighters with a means of bringing hose lines to bear on a fire on the highest accessible floor level.

1413.2 Buildings being demolished. Where a building is being demolished and a standpipe is existing within such a building, such standpipe shall be maintained in an operable condition so as to be available for use by the fire department. Such standpipe shall be demolished with the building but shall not be demolished more than one floor below the floor being demolished.

❖ When a structure is being demolished and a standpipe system exists within that structure, the standpipe system must be maintained operable and be available for use by the fire department. When a structure or a floor is to be demolished, its standpipe system must also be demolished with the structure, but the system may not be demolished more than one floor below the floor being demolished.

The availability and abundance of accessible avenues for vertical fire spread make buildings undergoing partial or total demolition highly susceptible to damage from fire. Even more so than construction sites, demolition projects attract vandals and vagrants who may set fires on the property for warmth or criminal purposes. Furthermore, cutting equipment and portable heating appliances may easily ignite combustible debris created during demolition. Once started, fire will spread rapidly through voids and vertical openings created to remove building service equipment. Standpipes provide fire fighters with a means of deploying hose lines quickly against these rapidly spreading fires.

1413.3 Detailed requirements. Standpipes shall be installed in accordance with the provisions of Section 905.

Exception: Standpipes shall be either temporary or permanent in nature, and with or without a water supply, provided that such standpipes comply with the requirements of Section 905 as to capacity, outlets and materials.

❖ Section 905 deals with installation; maintenance and supervision; building height and area; special occupancy application and type as related to standpipe systems in buildings under construction (see commentary, Section 905).

Temporary standpipes, whether dry or wet, are subject to the same requirements of Section 905 as permanent standpipes.

SECTION 1414
AUTOMATIC SPRINKLER SYSTEM

1414.1 Completion before occupancy. In buildings where an automatic sprinkler system is required by this code or the *International Building Code*, it shall be unlawful to occupy any portion of a building or structure until the automatic sprinkler system installation has been tested and approved, except as provided in Section 105.3.3.

❖ Section 105.3.3 is very clear in that when the sprinkler system is not tested and approved, the building cannot be occupied by the owner or tenants. In other words "no protection; no people" as one fire code official puts it. A building that has been given construction alternatives (increased travel distance, increased height and area, and reduced fire-resistance ratings) because of a sprinkler system is not safe for people when the system is not functional.

1414.2 Operation of valves. Operation of sprinkler control valves shall be allowed only by properly authorized personnel and shall be accompanied by notification of duly designated parties. When the sprinkler protection is being regularly turned off and on to facilitate connection of newly completed segments, the sprinkler control valves shall be checked at the end of each work period to ascertain that protection is in service.

❖ Closed control valves at the time of a fire are a major cause of sprinkler system failure. Only properly trained personnel should be operating these valves under the supervision of the fire prevention program superintendent, if available. The systems must be kept in service as much as possible and especially overnight or at the time of shift changes. If a fire watch is employed, duties should include monitoring these valves. The fire department must be notified when the system is out of service for an extended period of time because it responds to a sprinklered building differently than it does to an unsprinklered building.

SECTION 1415 PORTABLE FIRE EXTINGUISHERS

1415.1 Where required. Structures under construction, alteration or demolition shall be provided with not less than one approved portable fire extinguisher in accordance with Section 906 and sized for not less than ordinary hazard as follows:

1. At each stairway on all floor levels where combustible materials have accumulated.
2. In every storage and construction shed.
3. Additional portable fire extinguishers shall be provided where special hazards exist including, but not limited to, the storage and use of flammable and combustible liquids.

❖ Portable extinguishers must be rated for the hazards protected. Section 906 and NFPA 10, the applicable standard for portable fire extinguishers, contain information on fire extinguisher ratings. Other circumstances under which the fire code official may require additional extinguishers include: workers using open-flame devices, flammable or combustible liquids, welding or cutting equipment or painting equipment for applying flammable or combustible finishes during both construction and demolition.

SECTION 1416 MOTORIZED EQUIPMENT

1416.1 Conditions of use. Internal-combustion-powered construction equipment shall be used in accordance with all of the following conditions:

1. Equipment shall be located so that exhausts do not discharge against combustible material.
2. Exhausts shall be piped to the outside of the building.
3. Equipment shall not be refueled while in operation.
4. Fuel for equipment shall be stored in an approved area outside of the building.

❖ Motorized equipment, particularly equipment powered by an internal-combustion engine, must be kept clear of combustibles, must have exhaust arranged so as not to create an environmental hazard, must not be fueled while hot and have fuel stored properly. These issues are similarly addressed in Sections 1403, 1404 and 1405.

SECTION 1417 SAFEGUARDING ROOFING OPERATIONS

1417.1 General. Roofing operations utilizing heat-producing systems or other ignition sources shall be performed by a contractor licensed and bonded for the type of roofing process to be performed.

❖ Licensed and bonded contractors are required for roofing operations using hot materials. The jurisdiction must establish who authorizes the license the fire department or another unit of government. Although there is no specific permit for roofing operations in Section 105, permits for the use of combustible liquids and compressed gas may be appropriate. Roofing permits are typically required by the IBC.

1417.2 Asphalt and tar kettles. Asphalt and tar kettles shall be operated in accordance with Section 303.

❖ Section 303 regulates transportation, location, fueling, supervision, construction and fire protection of asphalt and tar kettles (see commentary, Section 303).

1417.3 Fire extinguishers for roofing operations. Fire extinguishers shall be installed in accordance with Section 906. There shall be not less than one multi-purpose portable fire extinguisher with a minimum 3-A 40-B:C rating on the roof being covered or repaired.

❖ Section 906 generally covers the location and requirements of portable fire extinguishers with a particular reference to asphalt kettles in Table 906.1. Section 303 also has requirements for extinguishers on the kettle, in the proximity of the kettle and on the roof. Fire extinguishers are to be fully charged and ready for service. Many construction sites are littered with building materials and debris. The kettle operator is responsible for maintaining an appropriate distance between the hot kettle and combustible materials.

Bibliography

The following resource materials are referenced in this chapter or are relevant to the subject matter addressed in this chapter.

Burklin, Ralph W. and Robert G. Purington. *Fire Terms: A Guide to Their Meaning and Use*. Quincy, MA: National Fire Protection Association, 1980.

Cote, A.E., ed. *Fire Protection Handbook,* 19th ed. Quincy, MA: National Fire Protection Association, 2002.

DOTy 27 CFR; 55-98, *Commerce in Explosives, as amended through April 1, 1998.* Washington, DC: U.S. Department of Treasury, 1998.

IBC -2003, *International Building Code*. Falls Church, VA: International Code Council, 2003.

IFGC-2003, *International Fuel Gas Code*. Falls Church, VA: International Code Council, 2003.

IFSTA, *Essentials of Fire Fighting,* 4th ed. Oklahoma State University, 1997.

IMC-2003, *International Mechanical Code*. Falls Church, VA: International Code Council, 2003.

NFPA 10-98, *Portable Fire Extinguishers.* Quincy, MA: National Fire Protection Association, 1998.

NFPA 13-99, *Installation of Sprinkler Systems.* Quincy, MA: National Fire Protection Association, 1999.

NFPA 30-99, *Flammable and Combustible Liquids Code.* Quincy, MA: National Fire Protection Association, 1999.

NFPA 31-01, *Installation of Oil Burning Equipment.* Quincy, MA: National Fire Protection Association, 2001.

NFPA 58-01, *Liquefied Petroleum Gas Code.* Quincy, MA: National Fire Protection Association, 2001.

NFPA 72-99, *National Fire Alarm Code.* Quincy, MA: National Fire Protection Association, 1999.

NFPA 241-00, *Safeguarding Construction, Alteration, and Demolition Operations.* Quincy, MA: National Fire Protection Association, 2000.

NFPA 386-90, *Portable Shipping Tanks for Flammable and Combustible Liquids.* Quincy, MA: National Fire Protection Association, 1990.

UL 1313-93, *Nonmetallic Safety Cans for Petroleum Products.* Northbrook, IL: Underwriters Laboratories Inc., 1993.

Chapter 15: Flammable Finishes

General Comments

Roughly one of every six fires occurring in industrial occupancies involves the ignition of a flammable or combustible liquid. This extraordinary statistic underscores the importance of proper fire prevention and protection practices where flammable and combustible coatings are applied using spraying, dipping or flow-coating methods.

Chapter 15 requirements govern operations where flammable or combustible finishes are applied by spraying, dipping, powder coating or flow-coating processes. Like all operations involving flammable or combustible liquids and combustible dusts, controlling ignition sources and methods of reducing or controlling flammable vapors or combustible dusts at or near these operations is emphasized. Open flames and smoking are prohibited near spray areas and dip tanks. Electrical installations must comply with the requirements of the ICC *Electrical Code*® (ICC EC™), especially requirements for electrical equipment and wiring in areas classified as spray areas and vapor areas, as defined in this chapter.

Good housekeeping and maintenance practices will significantly reduce fuel supply hazards. Additionally, care must be taken to control the accumulation and dissipation of static electricity. Managing the quantity of flammable and combustible liquids used and stored in and around spraying and dipping operations also discourages ignition. Because ample fuel and adequate oxygen are available in spraying and dipping operations, safeguards must be installed and maintained to manage the impact of a fire, should it occur. For example, if a match is dropped into a closed jar full of flammable liquids, the match is typically quenched because the fuel-to-air ratio does not allow flammable vapors to ignite; there is not enough oxygen. If the same liquid is poured onto a large surface, such as the floor, the vapors in the air are mixed in ratios that allow ignition to take place much more easily. The same phenomenon applies to atomization of flammable liquids where the liquid droplets have a greater surface area. Please note that in addition to this chapter, other chapters such as 27 (Hazardous Materials - General Provisions), 34 (Flammable and Combustible Liquids) and 39 (Organic Peroxides) are applicable for additional requirements for storage and handling of hazardous materials.

Purpose

Compliance with the requirements of this chapter is intended to reduce the likelihood of fires involving the application of flammable or combustible liquids/powders through spraying, dipping or flow-coating operations. Additionally, compliance with the requirements of this chapter will reduce the impact of a fire, should one occur. Emphasis is placed on controlling ignition sources, managing the quantity and location of fuels/vapors/dust and maintaining fire protection features.

SECTION 1501 GENERAL

1501.1 Scope. This chapter shall apply to locations or areas where any of the following activities are conducted:

1. The application of flammable or combustible paint, varnish, lacquer, stain, fiberglass resins or other flammable or combustible liquid applied by means of spray apparatus in continuous or intermittent processes.
2. Dip-tank operations in which articles or materials are passed through contents of tanks, vats or containers of flammable or combustible liquids, including coating, finishing, treatment and similar processes.
3. The application of combustible powders when applied by powder spray guns, electrostatic powder spray guns, fluidized beds or electrostatic fluidized beds.
4. Floor surfacing or finishing operations in areas exceeding 350 square feet (32.5 m^2).
5. The application of dual-component coatings or Class I or II liquids when applied by brush or roller in quantities exceeding 1 gallon (4 L).
6. Spraying and dipping operations.

❖ This section is very specific in scope. It establishes locations, areas and activities this chapter can be applied to. In addition to the requirements of this chapter, compliance with the requirements of NFPA 33 for spraying operations is established. Although specific conditions or uses are prohibited by this chapter [floor finishing more than 350 square feet (33 m^2) in area or dual component coatings using more than 1 gallon (4 L) when applied by brush or roller], these applications are not completely exempt from code requirements. Requirements within the code, such as Chapters 27 and 34, as well as requirements in other codes, such as the *International Mechanical Code®* (IMC®) for ventilation requirements, would still apply. It is important to note that spray finishing activities do not specifically relate to the maximum allowable quantity per control area as used in Chapters 27 and 34. Chapter 15 primarily addresses the hazard of atomizing flammable and combustible liquids and use of combustible dusts. Storage related to spray finishing would be regulated by Chapters 27 and 34, which would require applying the maximum allowable quantities per control area.

1501.2 Permits. Permits shall be required as set forth in Sections 105.6 and 105.7.

❖ The process of issuing permits gives the fire code official an opportunity to carefully evaluate and regulate the hazardous operations. Permit applicants should be required to demonstrate that their operations comply with the intent of the code before the permit is issued. See the commentary to Section 105.6 for a general discussion of operations requiring an operational permit and Section 105.7 for construction permits. The permit process also notifies the fire department of the need for prefire planning for the hazardous property. Because of the extremely dangerous processes described in this chapter, operational permits are required by Sections 105.6.17 and 105.6.18, and construction permits are required by Sections 105.7.5, 105.7.6 and 105.7.10 for their conduct.

SECTION 1502 DEFINITIONS

1502.1 Definitions. The following words and terms shall, for the purposes of this chapter and as used elsewhere in this code, have the meanings shown herein.

❖ Definitions of terms can help in the understanding and application of the code requirements. The purpose for including those definitions that are associated with the subject matter of this chapter is to provide more convenient access to them without having to refer back to Chapter 2. It is important to emphasize that these terms are not exclusively related to this chapter but are applicable everywhere the term is used in the code. For convenience, these terms are also listed in Chapter 2 with a cross reference to this section. The use and application of all defined terms, including those defined in this section, are set forth in Section 201.

DETEARING. A process for rapidly removing excess wet coating material from a dipped or coated object or material by passing it through an electrostatic field.

❖ Detearing applies to dip tank operations. It is the process of rapidly removing excess wet coating materials from an object.

DIP TANK. A tank, vat or container of flammable or combustible liquid in which articles or materials are immersed for the purpose of coating, finishing, treating and similar processes.

❖ Dip tanks can be almost any size, depending on the size of the work pieces to be immersed. They may be equipped with a fire suppression system, as well as overflow and drain pipes, all of which are based on the tank size. Regardless of size, dip tanks must be constructed of noncombustible materials, equipped with an approved, self-closing cover and properly ventilated [see Figure 1502.1(1)].

ELECTROSTATIC FLUIDIZED BED. A container holding powder coating material that is aerated from below so as to form an air-supported expanded cloud of such material which is electrically charged with a charge opposite to the charge of the object to be coated. Such object is transported through the container immediately above the charged and aerated materials in order to be coated.

❖ Because the powder used in the electrostatic fluidized bed is not as finely divided as that used in electrostatic spray methods, it is not likely to create a dust explosion potential. Electrostatic fluidized beds allow for the coating of materials that then must be cured in a baking oven (see also Chapter 21). This method of coating is typically used for small pieces [dimensions less than 4 inches (102 mm)].

FLAMMABLE FINISHES. Material coatings in which the material being applied is a flammable liquid, combustible liquid, combustible powder or flammable or combustible gel coatings.

❖ This general definition is used to describe all the operations regulated in this chapter including spray applications, dip tank operations and powder coating operations. Refer to definitions in Sections 1302 and 3402 for definitions of "Flammable/combustible liquids" and "Combustible dust." Please note that although "combustible powder" is not defined in this code, the terms "combustible dust" and "combustible powder" have been used interchangeably. Also, the terms "gel" and "combustible gel" are not defined by the code. Gels or pastes and liquids are classified as liquids when classifying hazardous materials. For example, the flash point of a gel, paste or liquid determines the flammable or combustible classification of the product.

FLUIDIZED BED. A container holding powder coating material that is aerated from below so as to form an air-supported expanded cloud of such material through which the preheated object to be coated is immersed and transported.

❖ A fluidized bed differs from an electrostatic fluidized bed in that the fluidized bed uses an air stream instead of electrostatic charge. The air stream behaves like a fluid as the object is passed through it. Additionally, the object is preheated. In an electrostatic fluidized bed, the object is heated/cured in an oven after the coating process.

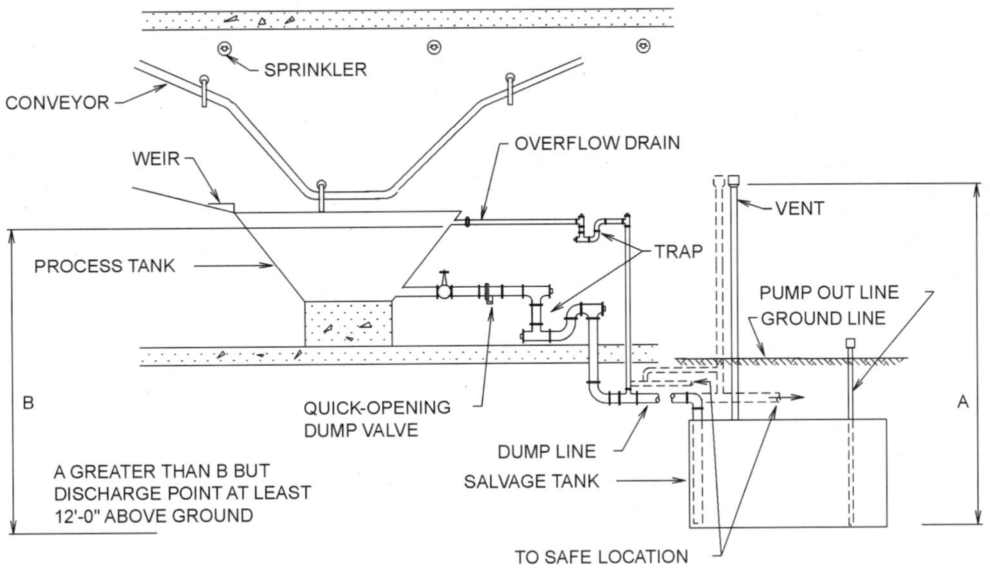

For SI: 1 foot = 304.8 mm.

Figure 1502.1(1)
TYPICAL TANK INSTALLATION

LIMITED SPRAYING SPACE. An area in which spraying operations for touch-up or spot painting of a surface area of 9 square feet (0.84 m^2) or less are conducted.

❖ This definition is applicable only to small "touch-up-" type operations. The requirements in this chapter for limited spraying operations are for an occasional user of flammable/combustible liquids, as opposed to an area used continuously for spraying. An example of such an operation is a furniture distributor that uses a limited spraying space to touch up scratches on products.

RESIN APPLICATION AREA. An area where reinforced plastics are used to manufacture products by hand lay-up or spray-fabrication methods.

❖ Glass fiber is used for reinforcement of plastics or polymers, and the resulting products are typically called glass fiber reinforced plastics or polymers (GFRP). It is common to use the acronym FRP (fiber reinforced polymers), which is applicable to all fiber-reinforced plastics or polymers.

ROLL COATING. The process of coating, spreading and impregnating fabrics, paper or other materials as they are passed directly through a tank or trough containing flammable or combustible liquids, or over the surface of a roller revolving partially submerged in a flammable or combustible liquid.

❖ Roll-coating methods apply material to flat work pieces, usually paper, cardboard, cloth or thin metals using liquid-coated cylinders or rollers. Coating material may be applied to the rollers by rotating them in an open trough or pan or applying liquid to the space between two rollers. Please note that for this term to apply anywhere in the code, the tank or trough must contain flammable or combustible liquids. The requirements in this chapter are for protection against, and mitigation in case of, a fire. In this case, the flammable vapors are typically heavier than air and may travel and spread a long distance unnoticed before reaching a potential ignition source and causing a vapor explosion or fire. In case of a fire within the tank or trough, there are additional concerns. The liquids are typically not water miscible and may overflow when the sprinkler system is activated or during the manual fire-fighting stages. This would spread the fire and liquids even farther.

SPRAY AREA. An area in which dangerous quantities of flammable vapors or combustible residues, dusts or deposits are present because of the operation of spraying processes. It shall include the interior of spray booths, the interior of ducts exhausting from spraying processes, or any area in the direct path of spray or any area containing dangerous quantities of air-suspended powder, combustible residue, dust, deposits, vapor or mists as a result of spraying operations. The fire code official is authorized to define the spray area in any specific case.

❖ Spray areas generally occur in one or a combination of three forms. The least desirable form is open floor area spraying, where the spraying area consists of an entire floor of a building without isolating the spraying operation. A better form is the spray room that isolates the spray operation by construction to less than an entire floor of the facility. The optimum form is a specially designed spray booth that isolates the operational hazards of spraying to an appropriately regulated space. Regardless of the form, all require special safeguards to address hazards, including adequate ventilation, fire suppression and management of overspraying.

SPRAY BOOTH. A mechanically ventilated appliance of varying dimensions and construction provided to enclose or accommodate a spraying operation and to confine and limit the escape of spray vapor and residue and to exhaust it safely.

❖ Spray booths vary in construction, size and design. The definition is clear in that it can be a fully enclosed structure or it can be designed to contain the flammable or combustible vapors. An example of a fully enclosed structure is a spray booth where products are carried/carted into the booth for spraying operation and carted out once the operation is complete. This definition, however, also allows for what is typically described as "open-face" booths, where the spray booth is enclosed on three sides and ventilated on the open side to confine the vapors to the spray booth area [see Figure 1502.1(2)]. An example of this is wood furniture finishing, where products such as wood tables are sprayed with stains and coating.

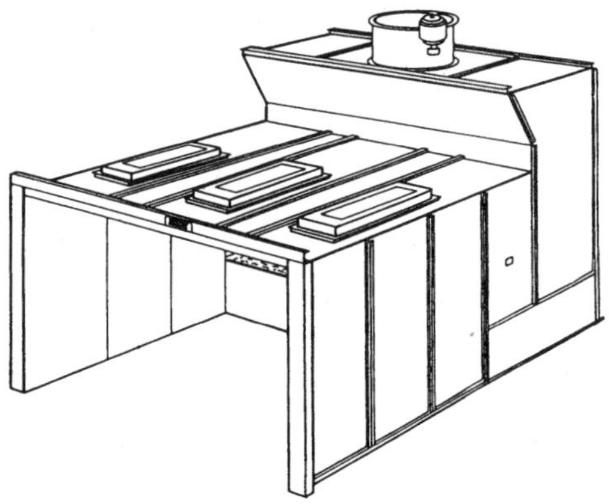

Figure 1502.1(2)
PAINT SPRAY BOOTH

SPRAY ROOM. A room designed to accommodate spraying operations constructed in accordance with the *International Building Code* and separated from the remainder of the building by a minimum 1-hour fire barrier.

❖ When spray booths cannot accommodate the spraying processes, because of size or for economic reasons, an entire room is dedicated to the process. In many cases, the oversprayed flammable/combustible liquids are allowed to remain on the floor until cleaned. These rooms have specific ventilation requirements to prevent the accumulation of vapors at the floor.

VAPOR AREA. An area containing flammable vapors in the vicinity of dip tanks, drain boards or associated drying, conveying or other equipment during operation or shutdown periods. The fire code official is authorized to determine the extent of the vapor area, taking into consideration the characteristics of the liquid, the degree of sustained ventilation and the nature of the operations.

❖ Vapor areas are the interface between fresh air and the surface of a flammable liquid. They are created by the exposed surface of the liquid when its temperature is at or above flash point. The extent of the vapor area should be determined by the fire code official based on an evaluation of the coating process, the liquid being used, the ventilation rate in the area and other variables that might increase the hazard.

SECTION 1503
PROTECTION OF OPERATIONS

1503.1 General. Operations covered by this chapter shall be protected as required by this section.

❖ This section states that the general protection requirements that apply to all processes (spraying, dipping and powder coating) are contained in this section. It states the regulation and mitigation of hazards that all processes described in Section 1501.1 typically have in common.

1503.2 Sources of ignition. Protection against sources of ignition shall be provided in accordance with Sections 1503.2.1 through 1503.2.8.

❖ Protection against sources of ignition is one of the most critical aspects of fire prevention in flammable finish operations. Flammable finish materials are finely divided or atomized during spraying operations, making them much more volatile and subject to ignition. Controlling sources of ignition is a preventive measure in a hazardous area where vapors can exist and potential ignition sources are readily available in the operations (ovens, electrical outlets, etc.).

1503.2.1 Electrical wiring and equipment. Electrical wiring and equipment shall comply with this chapter and the ICC *Electrical Code.*

❖ Electrical wiring must meet the requirements of the ICC EC. That code references NFPA 70 for its technical requirements except as they are modified in Chapter 12 of the ICC EC. Unlike referenced standards, in the event of a conflict between this section and the ICC EC, both requirements must be met. These requirements are found in Article 516 of NFPA 70 for areas in and around the spraying equipment, which place restrictions on the use of portable electric lamps. The classification of an area is based on the amount of flammable vapors, combustible mists, residues, dust or deposits that are present.

1503.2.1.1 Spray spaces and vapor areas. Electrical wiring and equipment in spray spaces and vapor areas shall be of an explosion-proof type approved for use in such hazardous locations. Such areas shall be considered to be Class I, Division 1 or Class II, Division 1 hazardous locations in accordance with the ICC *Electrical Code.*

❖ The areas where flammable vapors [defined as flammable constituents in air that exceed 25 percent of the lower flammable limit (LFL)] are present must meet the requirements for Class I, Division 1. The areas where combustible residues (such as dusts or deposits) are present must meet the requirements for the Class II, Division 1 (for dusts and residue) electrical classification. This is a very critical aspect of prevention of sources of ignition. Based on the definition of "Spray area," the fire code official is authorized to define the spray area in any specific case. Otherwise the areas described in the definition of "Spray area," such as the interior of spray booths, are used for electrical classification. See Figures 1503.2.1.1(1) and 1503.2.1.1(2) for examples of locations classified according to NFPA 70, Section 516.

1503.2.1.2 Electrical wiring and equipment in resin application areas. Electrical wiring and equipment located in resin application areas shall be in accordance with the ICC *Electrical Code*.

❖ See the commentary to Section 1503.2.1.

1503.2.1.3 Areas subject to deposits of residues. Electrical equipment in the vicinity of spray areas and dip tanks or associated drain boards or drying operations which are subject to splashing or dripping of liquids shall be specifically approved for locations containing deposits of readily ignitable residue and explosive vapors.

Exceptions:

1. This provision shall not apply to wiring in rigid conduit, threaded boxes or fittings not containing taps, splices or terminal connections.
2. This provision shall not apply to electrostatic equipment allowed by Section 1506.

In resin application areas, electrical wiring and equipment that is subject to deposits of combustible residues shall be listed for such exposure and shall be installed as required for hazardous (classified) locations. Electrical wiring and equipment not subject to deposits of combustible residues shall be installed as required for ordinary hazard locations.

❖ It is critical that only specifically approved electrical equipment be allowed in spraying areas where deposits of combustible residues may readily accumulate. This section limits the use of electrical equipment within the above mentioned area subject to deposit of residue for two primary reasons:

1. Sparks from such equipment could cause ignition of flammable vapors or overspray residue, and
2. Buildup of combustible overspray or splashed residue accumulating on the surfaces of electrical equipment subject to heating may cause equipment to overheat and ignite. See Figures 1503.2.1.1(1) and 1503.2.1.1(2) for examples of classified locations.

Taps, splices or terminal connections (similar to those of connections within flammable-liquid transfer systems) are the areas where accidents are most likely to occur. By removing risks associated with such fittings and installing rigid conduit to protect against splashing or dripping of liquids, the hazards associated

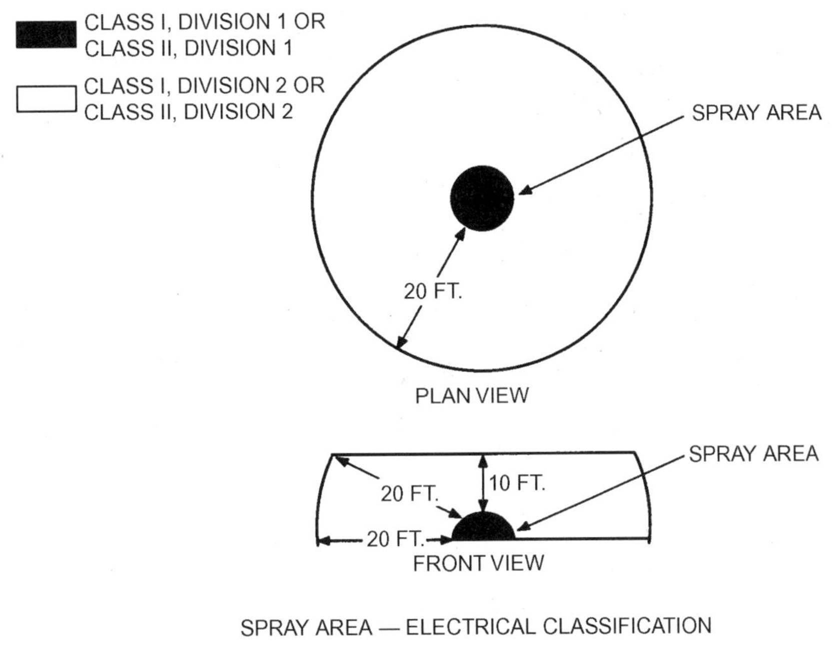

Figure 1503.2.1.1(1)
SPRAY AREA—ELECTRICAL CLASSIFICATION

with such electrical equipment are reduced substantially.

Because the electrostatic equipment is part of many flammable finish processes, Section 1506 requires that the equipment be approved. Approved is defined in Chapter 2 as "accepted by the fire code official" (see commentary, Chapter 2 for this definition).

1503.2.1.4 Areas adjacent to spray booths. Electrical wiring and equipment located outside of, but within 5 feet (1524 mm) horizontally and 3 feet (914 mm) vertically of openings in a spray booth or a spray room shall be approved for Class I, Division 2 or Class II, Division 2 hazardous locations, whichever is applicable.

❖ Class I, Division 2 and Class II, Division 2 typically apply to areas where accumulation of flammable vapors and combustible dust are prevented through ventilation and dust collection systems. However, these flammable vapors and combustible dusts may be present in these areas if the ventilation system fails.

1503.2.1.5 Areas subject to overspray deposits. Electrical equipment in spraying areas located such that deposits of combustible residues could readily accumulate thereon shall be specifically approved for locations containing deposits of readily ignitable residue and explosive vapors in accordance with the ICC *Electrical Code.*

Exceptions:

1. Wiring in rigid conduit.
2. Boxes or fittings not containing taps, splices or terminal connections.
3. Equipment allowed by Sections 1504 and 1506 and Chapter 21.

❖ Hazards associated with overspray deposit are similar to "areas subject to deposits of residues" specified in

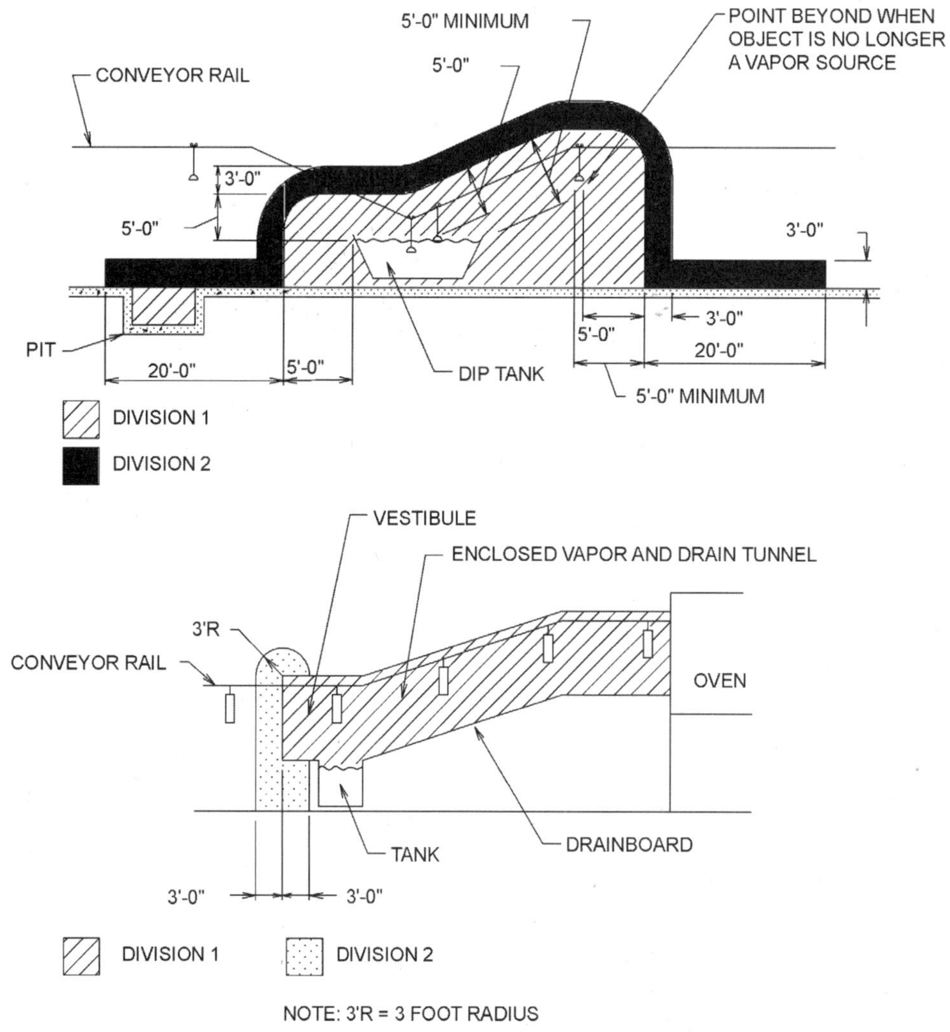

For SI: 1 foot = 304.8 mm.

Figure 1503.2.1.1(2)
EXTENT OF CLASS I, DIVISION I AND DIVISION 2 LOCATIONS IN ENCLOSED COATING PROCESS

Section 1503.2.1.3 (see commentary, Section 1503.2.1.3).

Exception 1 allows the use of rigid conduit to protect against splashing or dripping of liquids, which substantially reduces the hazards associated with electrical equipment.

Exception 2 acknowledges that where boxes do not contain taps, splices or terminal connections (similar to those of connections within flammable-liquid transfer systems) the hazards associated with such electrical equipment are reduced substantially.

Exception 3 exempts equipment allowed by the code (approved or listed) for hazardous uses, such as equipment allowed by Sections 1504 and 1506 and Chapter 21 because those items are designed to protect against ignition/explosion hazards.

1503.2.1.6 Flexible power cords. The use of flexible power cords shall be in accordance with the ICC *Electrical Code.*

❖ The ICC EC has specific electrical requirements for hazardous areas and locations, such as Class I and II, Division 1 and Class I and II, Division 2 areas where extension cords or flexible cords would not be permitted (see commentary, Section 605.5).

1503.2.2 Open flames and sparks. Open flames and spark-producing devices shall not be located in spray spaces or vapor areas and shall not be located within 20 feet (6096 mm) of such areas unless separated by a permanent partition.

> **Exception:** Drying and baking apparatus complying with Section 1504.7.2.

❖ For obvious prevention reasons, open flames and spark-producing devices are not allowed in areas considered as vapor areas and areas that could contain flammable vapors should the ventilation system fail. Open flames are prohibited where flammable or combustible liquids are stored, dispensed or applied. Heaters and all types of open-flame appliances are prohibited in and within 20 feet (6096 mm) of spray spaces and vapor areas, such as dip-tank and spray-finishing areas.

The exception notes that there are additional safety requirements in Section 1504.7.2 and Chapter 21 that reduce the probability of an ignition associated with open flames and sparks from drying and baking apparatus.

1503.2.3 Hot surfaces. Heated surfaces having a temperature sufficient to ignite vapors shall not be located in vapor areas. Space-heating appliances, steam pipes or hot surfaces in a spraying area or a resin application area shall be located such that they are not subject to accumulation of deposits of combustible residues.

> **Exception:** Drying apparatus complying with Section 1504.7.2.

❖ Where practical, auxiliary heating appliances (regardless of it being tested or listed) should not be installed inside a spray booth, room or area. Heating equipment and appliances approved for use only in Class I, Division 1 (flammable spray) or Class II, Division 1 (powder coating) hazardous locations are to be installed in spray booths, rooms or areas. Even the hot surfaces of indirect heating appliances may ignite combustible dusts or flammable or combustible vapors produced by spray or resin application operations.

The exception, as mentioned previously, notes that there are additional safety requirements in Section 1504.7.2 and Chapter 21 that prevent/reduce the probability of an ignition associated with open flames and sparks.

1503.2.4 Equipment enclosures. Equipment or apparatus that is capable of producing sparks or particles of hot metal that would fall into a spray space or vapor area shall be totally enclosed.

❖ Equipment in areas considered as vapor areas and areas that could contain flammable vapors and combustible residue, such as dusts or deposits, is restricted. The same reasoning as for open flames required by Section 1503.2.2 applies. An example of this equipment is metal grinding machines that produce sparks. Also see Section 1503.2.7 for welding requirements and signage.

1503.2.5 Grounding. Metal parts of spray booths, exhaust ducts and piping systems conveying Class I or II liquids shall be electrically grounded in accordance with the ICC *Electrical Code.* Metallic parts located in resin application areas, including but not limited to exhaust ducts, ventilation fans, spray application equipment, workpieces and piping, shall be electrically grounded.

❖ Static electricity is likely the most insidious and most common of all ignition sources. According to D.R. Scarborough in 1990, static sparks are the most common ignition sources involving spray-finishing operations. Additionally, humans are conductors of electricity, meaning that operators must be considered when grounding systems are determined.

1503.2.6 Smoking prohibited. Smoking shall be prohibited in spray spaces or vapor areas. "No Smoking" signs complying with Section 310 shall be conspicuously posted in such areas.

❖ Smoking is prohibited where flammable or combustible liquids or combustible dusts are stored, dispensed or applied. "No smoking" signs should be conspicuously located throughout the work area. Designated smoking areas should be located well outside the spray-finishing area and preferably in a separate room. The requirements of Sections 2703.7.1 and 3406.4.8 also may apply in facilities where flammable finish processes occur (see also commentary, Section 310).

1503.2.7 Welding warning signs. Welding, cutting and similar spark-producing operations shall not be conducted in or adjacent to spray areas or dipping or coating operations unless precautions have been taken to provide safety. Conspicuous signs with the following warning shall be posted in the vicinity of spraying areas, dipping operations and paint storage rooms:

NO WELDING
THE USE OF WELDING OR CUTTING
EQUIPMENT IN OR NEAR THIS AREA
IS DANGEROUS BECAUSE OF FIRE
AND EXPLOSION HAZARDS. WELDING
AND CUTTING SHALL BE DONE ONLY
UNDER THE SUPERVISION OF THE
PERSON IN CHARGE.

❖ For obvious reasons, open flames and spark-producing devices are not allowed in areas considered as vapor areas (and areas that could potentially contain flammable vapors should the prevention system components, such as ventilation, fail). Open flames are prohibited where flammable or combustible liquids are stored, dispensed or applied. Although no specific separation distance between the welding area and spray area is required, the 20 feet (6096 mm) beyond the spray area or to a permanent partition as described Section 1503.2.2 may be a good general separation guideline for these incompatible operations. Because welding and cutting processes can typically be found in buildings that house flammable finish processes, they create a potentially hazardous environment, especially for personnel unaware of such hazards. As with other hazardous conditions such as smoking, warning signs and placards must be posted to give appropriate notice to warn personnel of the hazards of welding near spray areas.

1503.2.8 Powered industrial trucks. Powered industrial trucks used in electrically classified areas shall be listed for such use.

❖ Similar language is used in Section 2703.7.3. Again, because these types of industrial trucks are powered by an electrical motor or internal combustion engine, the ignition of flammable vapors or combustible dusts is likely unless they are listed for such use (see also commentary, Section 2703.7.3).

1503.3 Storage, use and handling of flammable and combustible liquids. The storage, use and handling of flammable and combustible liquids shall be in accordance with this section and Chapter 34.

❖ Provisions of Chapter 34 and this section govern the storage and handling of flammable and combustible liquids for flammable finishes. As mentioned previously, Section 1503 applies to all flammable finishing operations within the scope of this chapter; therefore, Section 1503.3 applies to flammable finish operations that use flammable and combustible liquids.

1503.3.1 Use. Containers supplying spray nozzles shall be of a closed type or provided with metal covers which are kept closed. Containers not resting on floors shall be on noncombustible supports or suspended by wire cables. Containers supplying spray nozzles by gravity flow shall not exceed 10 gallons (37.9 L) in capacity.

❖ Requirements in this section are intended to prevent spills or the release of flammable vapors from flammable liquid containers. Tight-fitting metal lids or covers must be installed on all containers during use. Only those containers used to supply spray apparatus should be in the spray area or spray-finishing enclosure during spray-finishing operations. Containers supplying spray apparatus must rest on a floor or noncombustible stand when in use or be suspended from the ceiling by wire cables. Gravity dispensing of flammable liquids to spray nozzles is permitted for quantities not exceeding 10 gallons (38 L) to control the size of an uncontrolled gravity-fed leak.

1503.3.2 Valves. Containers and piping to which a hose or flexible connection is attached shall be provided with a shutoff valve at the connection. Such valves shall be kept shut when hoses are not in use.

❖ Shutoff valves must be installed at the juncture between flexible hoses or tubing and fixed piping used to dispense flammable liquids to spray-finishing apparatus. These shutoff valves permit the stoppage of liquid flow if a hose or tubing failure occurs. Shutoff valves must be closed when spray apparatus is not in use.

1503.3.3 Pumped liquid supplies. Where flammable or combustible liquids are supplied to spray nozzles by positive displacement pumps, pump discharge lines shall be provided with an approved relief valve discharging to pump suction or a safe detached location.

❖ To prevent excess flows or line ruptures from positive-pressure pumps, pressure-relief valves or other devices must be installed on the discharge side of positive-displacement pumps supplying flammable liquids to spray apparatus. These devices must operate before the discharge pressure exceeds the safe operating pressure of the connected valves, piping and equipment. Any discharge from the devices must be controlled to prevent ignition or environmental damage. See Section 3405.2 for general flammable/combustible liquid transfer requirements.

1503.3.4 Liquid transfer. Where a flammable mixture is transferred from one portable container to another, a bond shall be provided between the two containers. At least one container shall be grounded. Piping systems for Class I and Class II liquids shall be permanently grounded.

❖ The uncontrolled discharge of static electricity is a common ignition source during flammable liquid transfer. Proper bonding and grounding precautions must be followed for the safe discharge of static charges produced during flammable liquid transfer. See Section 3405.2 for general flammable/combustible liquid transfer requirements.

1503.3.5 Class I liquids as solvents. Class I liquids used as solvents shall be used in spray gun and equipment cleaning machines which have been listed and approved for the purpose or shall be used in spray booths or spray rooms in accordance with Sections 1503.3.5.1 and 1503.3.5.2.

❖ Another hazardous aspect of spray finishing is the use of solvents to clean spray guns and related equipment. When spray booths require cleaning, the use of solvents may often pose a greater hazard than the normally sprayed finishing material.

Usually an integral part of the labeling process, manufacturer's installation, operation and maintenance instructions must be carefully followed. The type of solvent to be used in any given machine must be as recommended by the machine manufacturer.

1503.3.5.1 Listed devices. Cleaning machines for spray guns and equipment shall not be located in areas open to the public and shall be separated from ignition sources in accordance with their listings or by a distance of 3 feet (914 mm), whichever is greater. The quantity of solvent used in a machine shall not exceed the design capacity of the machine.

❖ Cleaning machines for spray guns and equipment are commonly used in the industry. See Figure 1503.3.5.1 for an illustration of spray gun cleaning machines. Such machines use solvents powerful enough to dissolve paint residues in the thin tubing and small orifices of the spray equipment. They are very similar to the automotive parts cleaners that have been in widespread use since the early 1970s, and consist of a metal sink-like bowl set on a base unit that houses a storage container for spent solvent after the cleaning process. The solvent is circulated in the machine by a pneumatic pump. Vapors are captured in the sink bowl and are vented to the outdoors. A safety interlock prevents the solvent pump from operating if the exhaust system fails. Installation of such machines must be restricted to areas not accessible to the general public to avoid exposure to the potential hazards of the cleaning operations and prevent the inadvertent introduction of ignition sources to the cleaning area.

1503.3.5.2 Within spray booths and spray rooms. When solvents are used for cleaning spray nozzles and auxiliary equipment within spray booths and spray rooms, the ventilating equipment shall be operated during cleaning.

❖ The safety systems that are an integral part of operating spray booths and spray rooms can be used when solvents are used for equipment cleaning. These systems help, as they would during application of flammable finishes, in reducing the chances of ignition or the severity of a fire, should an ignition occur.

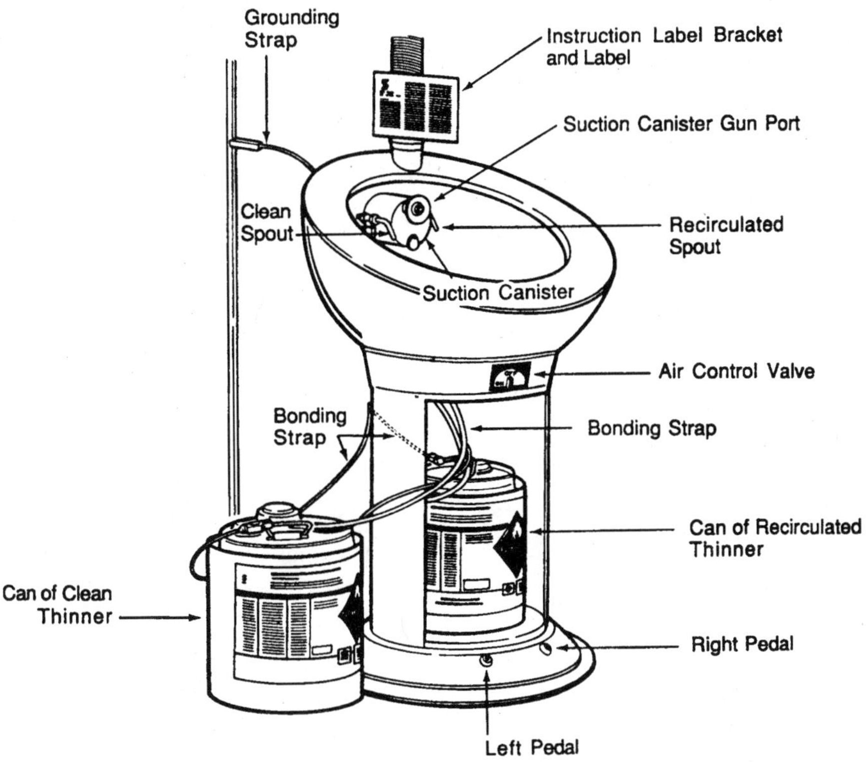

Figure 1503.3.5.1
SPRAY GUN CLEANING MACHINE

1503.3.6 Class II and Class III liquids. Solvents used outside of spray booths, spray rooms or listed and approved spray gun and equipment cleaning machines shall be restricted to Class II and Class III liquids.

❖ The cleaning of spray equipment inside of a spray booth or room designed to accommodate the hazards not only of spray painting but also spray equipment maintenance should be the primary choice of operators. The code recognizes that this is not always a feasible option and thus allows for the cleaning of spray equipment outside of an approved spray booth or room, if a labeled spray gun and equipment cleaning machine are used and the liquids used are combustible liquids having a flashpoint of above 100°F (38°C). This reduces the volatility of the liquids that can be used in cleaning machines outside of spray booths and spray areas.

1503.4 Operations and maintenance. Spraying areas, exhaust fan blades and exhaust ducts shall be kept free from the accumulation of deposits of combustible residues. Where excessive residue accumulates in booths, ducts, or discharge points or other spraying areas, spraying operations shall be discontinued until conditions are corrected.

❖ A regular cleaning schedule should be adopted for spray areas and spray-finishing enclosures. Accumulations of spray residue in interior spray booths or rooms or on exhaust duct surfaces should be removed at least daily or more frequently when accumulations become excessive. Any time overspray residue accumulations become excessive, the operator must suspend spray operations until the spray booth or area is thoroughly and properly cleaned. Water-wash nozzles, strainers and eliminator packs must be checked or cleaned daily or at the end of each shift and tank sludge removed and discarded in a safe manner. Interior surfaces of ductwork and fan blades should be inspected regularly for accumulations of overspray residue caused by fouled nozzles, strainers or eliminator packs. The use of soap-based, water-soluble coatings makes it easier to strip surfaces using high-pressure water spray without greatly increasing the fire hazard.

1503.4.1 Tools. Scrapers, spuds and other tools used for cleaning purposes shall be constructed of nonsparking materials.

❖ Because of the potential ignition problem that exists, the cleaning tools need to be of the nonsparking type. The term "nonsparking" is somewhat inaccurate. Tools made of brass and similar nonsparking materials produce sparks with ignition energies too low to ignite flammable vapors. Nonetheless, such tools should be used carefully to avoid producing sufficient frictional heat to cause an ignition.

1503.4.2 Residue. Residues removed during cleaning and debris contaminated with residue shall be immediately removed from the premises and properly disposed.

❖ Paint and solvent residue, cleaning rags and protective coverings may be susceptible to spontaneous ignition. Residue and cleaning debris must be removed from the building and stored in approved containers far away from the building. Removal of residue and cleaning debris on a regular schedule is highly recommended.

1503.4.3 Waste cans. Approved metal waste cans equipped with self-closing lids shall be provided wherever rags or waste are impregnated with finishing material. Such rags and waste shall be deposited therein immediately after being utilized. The contents of waste cans shall be properly disposed of at least once daily and at the end of each shift.

❖ Waste cans (sometimes called "oily rag cans") used for storage of materials that might auto-ignite as a result of the spontaneous combustion process must be tested and listed for that use by a recognized testing laboratory or agency and must bear a label showing that they have been tested and with the name of the testing agency (see Section 304.3). Such containers are most commonly round and generally available in sizes ranging from 5 to 40 U.S. gallons (19 to 152 L). They are equipped with a manual or foot treadle-operated lid that opens to a maximum angle of 60 degrees (1 rad) and closes by gravity. These containers are designed to prevent continuing combustion of the contents if ignition occurs. The container design includes features that keep the can body from coming into contact with combustible surfaces of walls or floors. Daily disposal of container contents reduces the amount of time that oily materials will lie dormant, generating internal heat that can lead to ignition. UL 32 provides further information on the construction, testing and listing of these containers (see commentary, Section 202, for the definition of "Listed").

1503.4.4 Solvent recycling. Solvent distillation equipment used to recycle and clean dirty solvents shall comply with Section 3405.4.

❖ Section 3405.4 includes detailed requirements for solvent distillation processes and equipment, such as labeling, unit capacity, location and prohibited processes. Note that the terms "solvent distillation unit," "appliance" and "equipment" are used interchangeably in the code.

SECTION 1504
SPRAY FINISHING

1504.1 Location of spray-finishing operations. Spray-finishing operations conducted in buildings used for Group A, E, I or R occupancies shall be located in a spray room protected with an approved automatic sprinkler system installed in accordance with Section 903.3.1.1 and separated vertically and horizontally from other areas in accordance with the *International Building Code*. In other occupancies, spray-finishing operations shall be conducted in a spray room, spray booth or limited spraying space approved for such use.

❖ Separation and protection from hazards to other portions of a building of occupancy groups listed in this section is critical for several reasons. Spray finishing operations are too hazardous to be conducted in occupancies with a high-life or property exposure. The number of occupants in such buildings and physical conditions of those occupants vary. People exposed to the danger of a fire involving spray finishing must be able to evacuate or protect themselves promptly and effectively to avoid the risk of serious injury or death. Also, the psychology of people in a Group A, E, I or R occupancy is very different from those in a Group F manufacturing occupancy. A person walking in a manufacturing environment where flammable finish processes take place in a large area where spray booths and signs are in plain view is more aware of the physical dangers associated with these processes than someone in an R occupancy who may be merely relaxing with a cigarette.

Also note that the protection provided in these occupancies is in many cases less than would be found in manufacturing occupancies. Separation and protection in the form of fire-resistance-rated construction and automatic fire suppression must be provided where such operations must be conducted in the same building. Three-hour fire barrier walls, horizontal assemblies or both having a fire-resistance rating of three hours is required between Groups F-1 and A or I and R occupancies. Except for Group I-2, this requirement can be reduced in sprinklered buildings by one hour. If the quantities of hazardous materials exceed the maximum allowable quantities per control area indicated in Chapter 27, there are more restrictive separation requirements. The separation requirements between Groups H and A or I and R vary depending on the specific hazardous material that is stored [see Section 302.3 of the *International Building Code®* (IBC®)].

1504.1.1 Spray rooms. Spray rooms shall be constructed and designed in accordance with this section and the *International Building Code,* and shall comply with Sections 1504.2, 1504.3, 1504.4, 1504.5 and 1504.6.

❖ The occupancy of buildings or portions of buildings housing these coating operations is typically classified as Group H-2 [because of flammable liquids typically in open system use or under pressure greater than 15 psi (103 kPa)] where the aggregate quantity of flammable liquid stored or used in a single control area exceeds the maximum allowable quantity per control area listed in Table 2703.1.1. However, the occupancy is to be classified as Group F-1 for processes where the quantities do not exceed the maximum allowable quantity per control area. NFPA 33 contains additional technical details regarding hazards associated with these processes.

1504.1.1.1 Floor. Combustible floor construction in spray rooms shall be covered by approved, noncombustible, nonsparking material, except where combustible coverings, such as thin paper or plastic and strippable coatings are utilized over noncombustible materials to facilitate cleaning operations in spray rooms.

❖ Sections 2704 and 2705 require noncombustible floor construction if maximum allowable quantities per control area in storage or use are exceeded. Noncombustible floors, for obvious reasons, are the preferred floor construction. If the floor is not noncombustible, noncombustible and nonsparking material should be used to cover the floor.

Kraft paper and similar coverings are commonly used in spray areas to protect against overspray. Such coverings must be removed and discarded when accumulation becomes excessive. Spraying must be discontinued until the paper or other covering is removed when it becomes saturated. Operations must be discontinued until excess accumulations are stripped and the surfaces are cleaned to reduce the hazard when residues become excessive on any surface in or near the spray booth.

1504.1.2 Spray booths. The design and construction of spray booths shall be in accordance with Sections 1504.1.2.1 through 1504.1.2.6, Sections 1504.2 through 1504.6, and NFPA 33.

❖ Sections 1504.1.2.1 through 1504.1.2.6, 1504.2 through 1504.6 and NFPA 33 contain requirements for controlling hazards associated with spray booths. A lot of the hazard prevention and mitigation of flammable finish fires in a spray booth depend on the proper construction and design of the spray booth.

1504.1.2.1 Construction. Spray booths shall be constructed of approved noncombustible materials. Aluminum shall not be used.

Where walls or ceiling assemblies are constructed of sheet metal, single-skin assemblies shall be no thinner than 0.0478 inch (18 gage) (1.2 mm) and each sheet of double-skin assemblies shall be no thinner than 0.0359 inch (20 gage) (0.9 mm).

Structural sections of spray booths are allowed to be sealed with latex-based or similar caulks and sealants.

❖ Even though spray booths are not required to meet fire-resistance-rated construction requirements of the IBC, the requirements of this section recognize the need to minimize the spread of a fire and prevent a spray booth from contributing to a fire should one start within the booth. Aluminum is not suitable for structurally sound construction enclosures because of its low melting point; it would be likely to melt in case of a substantial fire within the booth. For a spray booth to maintain its structural integrity in a large fire, it should be constructed of steel, masonry or equivalent noncombustible materials. Section 2703.8 also requires hazardous materials storage cabinets to be constructed of 18 gage steel. Both booths and cabinets are viewed as equipment used to control and contain spills or fires.

1504.1.2.2 Surfaces. The interior surfaces of spray booths shall be smooth and shall be constructed so as to permit the free passage of exhaust air from all parts of the interior and to facilitate washing and cleaning, and shall be designed to confine residues within the booth. Aluminum shall not be used.

❖ Rough, corrugated or uneven surfaces are difficult to clean. Periodic cleaning of the interior surfaces reduces the fire hazard posed by the accumulation of flammable or combustible coatings. Because flammable or combustible vapors and dusts are typically heavier than air, design considerations should include the passage of exhaust air and proper air circulation to all parts of the spray booth, especially at or near the floor level. See the commentary to Section 1504.1.2.1 for comments on use of aluminum in spray booth construction.

1504.1.2.3 Floor. Combustible floor construction in spray booths shall be covered by approved, noncombustible, nonsparking material, except where combustible coverings, such as thin paper or plastic and strippable coatings are utilized over noncombustible materials to facilitate cleaning operations in spray booths.

❖ See the commentary to Section 1504.1.1.1.

1504.1.2.4 Means of egress. Means of egress shall be provided in accordance with Chapter 10.

Exception: Means of egress doors from premanufactured spray booths shall not be less than 30 inches (762 mm) in width by 80 inches (2032 mm) in height.

❖ Spray booths are required to comply with exiting requirements in Chapter 10.

Unlike a spraying room, premanufactured spray booths are considered to be an approved appliance or equipment, not a separate occupancy. Therefore, an enclosed spray booth is considered an exception to the number of exit door requirements of Chapter 10 (two exit doors for most Group H occupancies).

1504.1.2.5 Clear space. Spray booths shall be installed so that all parts of the booth are readily accessible for cleaning. A clear space of not less than 3 feet (914 mm) shall be maintained on all sides of the spray booth. This clear space shall be kept free of any storage or combustible construction.

Exceptions:

1. This requirement shall not prohibit locating a spray booth closer than 3 feet (914 mm) to or directly against an interior partition, wall or floor/ceiling assembly, that has a fire-resistance- rating of not less than 1 hour, provided the spray booth can be adequately maintained and cleaned.
2. This requirement shall not prohibit locating a spray booth closer than 3 feet (914 mm) to an exterior wall or a roof assembly provided the wall or roof is constructed of noncombustible material and provided the spray booth can be adequately maintained and cleaned.

❖ Spray booths can at times be viewed as equipment, where accessibility is essential to cleaning and maintenance. This section is also included for housekeeping reasons and to keep combustible debris and materials away from the spray booth to reduce the chances of an ignition on its exterior. This clearance also serves to prevent heat from a fire within the walls of a booth from being transferred to adjacent combustible material. It also will prevent a fire near a booth (from combustible storage) from acting as an ignition source for the booth (see Figure 1504.1.2.5).

Under Exception 1, if the construction materials of an interior portion further inhibit the spread of a fire and if there are proper means of cleaning the spray booth, the intent of Section 1504.1.2.5 is met. This exception, however, still does not allow any storage to within 3 feet (914 mm) of the spray booth.

Exception 2 recognizes that if an exterior wall or roof assembly is of noncombustible construction, it will not contribute to the fuel load or spread of the fire to other interior parts of the building. This meets the intent of this section as long as there are proper means of cleaning the spray booth. This exception still does not allow any storage to within 3 feet (914 mm) of the spray booth.

1504.1.2.6 Size. The aggregate area of spray booths in a building shall not exceed the lesser of 10 percent of the area of any floor of a building or the basic area allowed for a Group H-2 occupancy without area increases, as set forth in the *International Building Code*.

The area of an individual spray booth in a building shall not

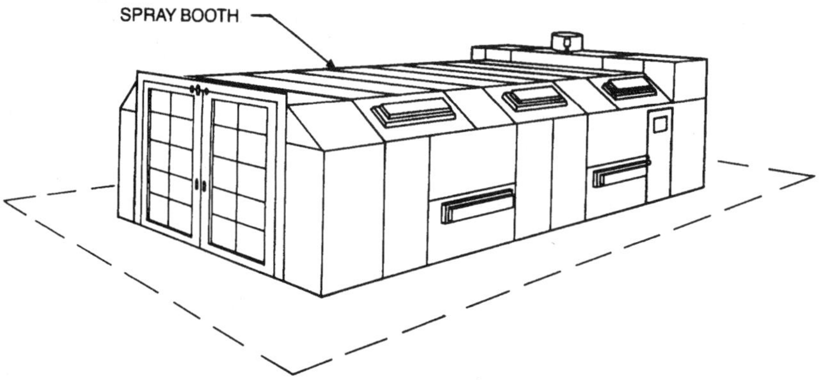

Figure 1504.1.2.5
EXAMPLE OF SPRAY BOOTH

exceed the lesser of the aggregate size limit or 1,500 square feet (139 m^2).

Exception: One individual booth not exceeding 500 square feet (46 m^2).

❖ If the total area of an occupancy is less than 10 percent of the area of any floor of a building, the occupancy is typically viewed as an accessory occupancy group. In these cases, the allowable area for the overall building is not affected (see Section 302.2 of the IBC) as long as the occupancy aggregate square footage meets the allowable requirements of Table 503 in the IBC.

The intent of limiting the size of a spray booth is to compartmentalize or provide passive fire protection. This limits the size of a fire and the processes to a more manageable condition.

The exception references the first paragraph of the section. In smaller businesses, where the building is small [typically less than 5,000 square feet (465 m^2)], it may be impractical to design for a spray booth that is less than 10 percent of the overall building area. In case of a fire, smaller buildings are easier to control from a passive fire protection standpoint than are larger operations. This exception also recognizes that spray booths smaller than 500 square feet (46 m^2) are not practical.

1504.1.3 Spraying spaces. Spraying spaces shall be designed and constructed in accordance with the *International Building Code* and Sections 1504.1.3.1, 1504.2, 1504.3, 1504.4, 1504.5 and 1504.6 of this code.

❖ Spray area, spray space and spraying spaces are terms used in this chapter interchangeably. Because the spraying spaces or areas are not enclosed, the occupancy of the room the spray area is in depends on whether the aggregate quantities of hazardous materials in the control area are exceeded. The occupancy of buildings or portions of buildings housing these coating operations is typically classified as Group H-2 [because flammable liquids are typically in open system use or under pressure greater than 15 psi (103 kPa)] where the aggregate quantity of flammable/combustible liquids or dusts stored or used in a single control area exceeds the maximum allowable quantity per control area listed in Table 2703.1.1. However, the occupancy is to be classified as Group F-1 or mixed use (such as Group F-1/S-1) for processes where the quantities do not exceed the maximum allowable quantity per control area.

1504.1.3.1 Floor. Combustible floor construction in spraying spaces shall be covered by approved, non combustible, nonsparking material, except where combustible coverings, such as thin paper or plastic and strippable coatings are utilized over noncombustible materials to facilitate cleaning operations in spraying spaces.

❖ See the commentary to Section 1504.1.1.1.

1504.1.4 Limited spraying spaces. Limited spraying spaces shall comply with Sections 1504.1.4.1 through 1504.1.4.4.

❖ Limited spraying spaces are to accommodate uses that are limited in their frequency and amounts of hazardous materials used. As mentioned previously, this operation is more for small touch-up work found in a maintenance shop.

1504.1.4.1 Job size. The aggregate surface area to be sprayed shall not exceed 9 square feet (0.84 m^2).

❖ Limiting the workpiece size limits the amount of flammable vapors that are produced in such an area. This is again to accommodate small "incidental-type" spraying processes found in typical manufacturing occupancies.

1504.1.4.2 Frequency. Spraying operations shall not be of a continuous nature.

❖ As mentioned previously, the limited spraying setup is to accommodate small incidental-type processes. If the processes are continuous, a spray booth or spray area should be used. Spray booth requirements are more stringent and are better regulated from a fire prevention and fire protection standpoint.

1504.1.4.3 Ventilation. Positive mechanical ventilation providing a minimum of six complete air changes per hour shall be installed. Such system shall meet the requirements of this code for handling flammable vapors. Explosion venting is not required.

❖ Six air changes per hour is the requirement for limited spraying areas. However, if the quantity of hazardous material in the room or control area in which the limited spray area is located exceeds the allowable quantities for hazardous materials given in Chapter 27 or 34, both exhaust ventilation requirements in those chapters for flammable vapors and the requirement in this section have to be met. A typical example would be a plating shop where allowable quantities of corrosives and toxics are exceeded (Group H-4) and where exhaust ventilation at a rate of 1 cubic foot per minute (cfm) per square foot (.00508 m^3/(s · m^2) of floor area is required. If a limited spray area is located in such a room, the most restrictive requirement has to be met; that is, the greater of 1 cfm per square foot (0.00508 m^3/(s·m^2) or six air changes per hour. The vapor density of the material should be taken into account (whether vapors are heavier or lighter than air) when considering the location of exhaust inlets. Additionally, because Section 502.7.2 of the IMC is identical to this section, conformance to the general requirements of Section 502 of the IMC, especially Section 502.1.1 ("the inlet to an exhaust system shall be located in the area of heaviest concentration of contaminants"), is recommended.

1504.1.4.4 Electrical wiring. Electrical wiring within 10 feet (3048 mm) of the floor and 20 feet (6096 mm) horizontally of the limited spraying space shall be designed for Class I, Division 2 locations in accordance with the ICC *Electrical Code.*

❖ Processes in limited spray areas may still result in some vapor generation. Because they are generally heavier than air, vapors in spray processes typically accumulate near the floor. These vapors may travel long distances undetected before causing a flash fire. To limit the ignition sources, wiring must be installed to meet Class I, Division 2 requirements within 10 feet (3048 mm) of the floor and 20 feet (6096 mm) horizontally.

1504.2 Ventilation. Mechanical ventilation of spraying areas shall be provided in accordance with Section 510 of the *International Mechanical Code.*

❖ Spraying areas as defined in Section 1502 include, but are not limited to, the interior of spray booths and spray rooms. The proper design and installation of exhaust ventilation systems in spray areas is critical because of the potential production of large amounts of flammable vapors in the processes. Exhaust systems in spray finishing areas must also comply with Section 510 of the IMC, which regulates hazardous exhaust systems. Additionally, because Section 502.7.3 of the IMC is identical to this section, compliance with the general section o Section 502 of the IMC, especially Section 502.1.1 ("the inlet to an exhaust system shall be located in the area of heaviest concentration of contaminants"), is recommended. The location of heaviest concentration resulting from the physical characteristics of a material (for example, vapor density) or the process (such as atomization of the material) must be considered when inlets to exhaust systems are designed.

1504.2.1 Operation. Mechanical ventilation shall be kept in operation at all times while spraying operations are being conducted and for a sufficient time thereafter to allow vapors from drying coated articles and finishing material residue to be exhausted. Spraying equipment shall be interlocked with the ventilation of the spraying area such that spraying operations cannot be conducted unless the ventilation system is in operation.

❖ Ventilation must be functioning during the spraying operation phase and the drying phase of a process when vapors are generated. The interlock between the ventilation and spraying equipment will ensure against human error, such as operator failure to activate the ventilation system prior to the use of the spray equipment.

1504.2.2 Recirculation. Air exhausted from spraying operations shall not be recirculated.

Exceptions:

1. Air exhausted from spraying operations is allowed to be recirculated as makeup air for unmanned spray operations provided that:
   - 1.1. The solid particulate has been removed.
   - 1.2. The vapor concentration is less than 25 percent of the LFL.
   - 1.3. Approved equipment is used to monitor the vapor concentration.
   - 1.4. When the vapor concentration exceeds 25 percent of the LFL, the following shall occur:
     - a. An alarm shall sound; and
     - b. Spray operations shall automatically shut down.
   - 1.5. In the event of shutdown of the vapor concentration monitor, 100 percent of the air volume specified in Section 1504.2 is automatically exhausted.
2. Air exhausted from spraying operations is allowed to be recirculated as makeup air to manned spraying operations where all of the conditions provided in Exception 1 are included in the installation and documents have been prepared to show that the installation does not pose a life safety hazard to personnel inside the spray booth, spray space or spray room.

❖ Recirculation of exhausted air containing hazardous materials, such as flammable vapors, is poor design practice for obvious reasons. The recirculation of exhausted air containing flammable vapors will help spread the hazard of flash fires from the area of vapor generation to other parts of the building. At best, it would adversely affect only the area of vapor generation in that it would render the exhaust ventilation useless. It may also give the operators of the facility a false sense that the level of ventilation is safe for the system.

Exception 1 lists five conditions that must be met for recirculation of exhausted air. If solid particulates (dusts) are removed from the exhausted air, the exhausted air is no longer considered a fire hazard because the potential for a dust explosion is eliminated. Flammable vapors are defined by Section 2702.1 as flammable constituents in air that exceed 25 percent of the LFL. By reducing the flammable vapor concentrations to less than 25 percent of the LFL, the exhausted air is no longer considered a flammable vapor. To further ensure that the concentrations remain at less than 25 percent of the LFL, flammable vapor detection systems must automatically shut down the operations, set off an alarm and exhaust 100 percent of the air. This would allow for energy conservation in unmanned operations, without compromising the safety features typically associated with such ventilation systems.

The only difference between Exceptions 1 and 2 is that in a manned operation, additional documentation is need to ensure that personnel, such as people operating the spray equipment, are not at risk from an injury or life safety standpoint. This documentation may include a risk analysis of fire and health hazards associated with the operation of this equipment when some portion of the exhausted air is recirculated.

1504.2.3 Air velocity. Ventilation systems shall be designed, installed and maintained such that the average air velocity over the open face of the booth, or booth cross sectional in the direction of airflow during spraying operations, shall not be less than 100 linear feet per minute (51 m/s).

❖ To ensure that flammable vapors are kept within a designated spray area and that the amount of overspray is limited, the code requires that the exhaust system be adequately sized to maintain an average velocity over the open face of the booth or booth cross section of no less than 100 feet per minute (0.51 m/s), which is the minimum velocity to capture particulate spray material. Velocities exceeding 200 lineal feet per minute have been determined to be too great for this purpose.

To determine the minimum ventilation/exhaust capacity in cfm, multiply the booth width (feet) by booth height (feet) by 100 (lineal per feet).

1504.2.4 Ventilation obstruction. Articles being sprayed shall be positioned in a manner that does not obstruct collection of overspray.

❖ This section is part of the safe operating practices. Anything that would prevent the collection of flammable vapors, such as obstruction of an exhaust inlet, would compromise the effectiveness of the ventilation system.

1504.2.5 Independent ducts. Each spray booth and spray room shall have an independent exhaust duct system discharging to the outside.

Exceptions:

1. Multiple spray booths having a combined frontal area of 18 square feet (1.67 m^2) or less are allowed to have a common exhaust when identical spray-finishing material is used in each booth. If more than one fan serves one booth, fans shall be interconnected such that all fans operate simultaneously.
2. Where treatment of exhaust is necessary for air pollution control or for energy conservation, ducts shall be allowed to be manifolded if all of the following conditions are met:
   - 2.1. The sprayed materials used are compatible and will not react or cause ignition of the residue in the ducts.
   - 2.2. Nitrocellulose-based finishing material shall not be used.
   - 2.3. A filtering system shall be provided to reduce the amount of overspray carried into the duct manifold.
   - 2.4. Automatic sprinkler protection shall be provided at the junction of each booth exhaust with the manifold, in addition to the protection required by this chapter.

❖ This section requires independent duct exhaust of residue from spray-finishing operations. These ducts must be routed directly to the exterior of the building. A similar language is used in Section 510.4 of the IMC. Ducts may not penetrate fire-resistance-rated assemblies [see Figure 1504.2.5(1)].

Exception 1 applies to very small spray booths where the vapor area is very small compared to the area of standard spray booths. Because these individual smaller spray booths are considered as one fire area from a ventilation standpoint, all identical materials are to be used when this exception applies. This will ensure that incompatible materials are not used in booths with a common exhaust system [see Figure 1504.2.5(2)].

Exception 2 notes that because the exhausted air is at times treated, it can be manifolded. However, special hazards must be avoided to ensure fire protection

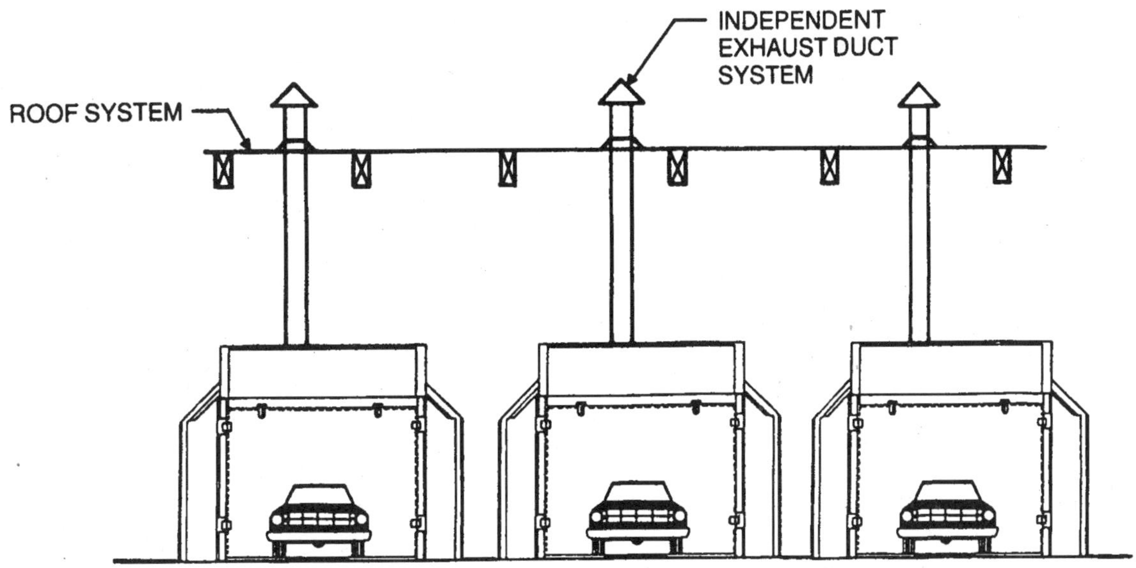

Figure 1504.2.5(1)
INDEPENDENT DUCTS FOR SPRAY BOOTHS

safety. Incompatible materials must be separated in case of a reaction within the ducts. A similar requirement is found in Section 510.4 of the IMC.

Nitrocellulose and nitrocellulose-based products are unstable materials that can easily be ignited, and once ignited, need large quantities of water for suppression. Additionally, nitrocellulose is incompatible with many materials (alkalis, amines, etc.). The cleaning products used in other booths may even ignite the nitrocellulose-based products in a manifolded exhaust system. Therefore, the exhaust of nitrocellulose-based products is considered an exception to this section. Additional protection, such as filtering and sprinklers at the junction booth exhaust, is also required.

1504.2.6 Termination point. The termination point for exhaust ducts discharging to the atmosphere shall not be less than the following distances:

1. Ducts conveying explosive or flammable vapors, fumes or dusts: 30 feet (9144 mm) from the property line; 10 feet (3048 mm) from openings into the building; 6 feet (1829 mm) from exterior walls and roofs; 30 feet (9144 mm) from combustible walls or openings into the building which are in the direction of the exhaust discharge; 10 feet (3048 mm) above adjoining grade.
2. Other product-conveying outlets: 10 feet (3048 mm) from the property line; 3 feet (914 mm) from exterior walls and roofs; 10 feet (3048 mm) from openings into the building; 10 feet (3048 mm) above adjoining grade.
3. Environmental air duct exhaust: 3 feet (914 mm) from the property line; 3 feet (914 mm) from openings into the building.

❖ This section details the requirements of safe outlets/termination points of exhaust ducts to reduce exposure from the dangerous vapors in the exhaust. This is done to:

1. Protect other parts of the building,
2. Protect other buildings,
3. Reduce a potential reaction from materials that may be incompatible and
4. Reduce the severity of a fire, in case of an ignition.

Because this chapter mainly focuses on flammable finishes, vapors that are considered flammable, including dusts, are more restricted than other vapors. To avoid recirculation of flammable vapor fumes or dusts back into the building, the duct must be designed to reduce such exposures. This may be achieved by separating the exhaust outlet from openings in the building, walls and roof, where sources of ignition or incompatible materials may be present [see Figures 1504.2.6(1) and 1504.2.6(2)].

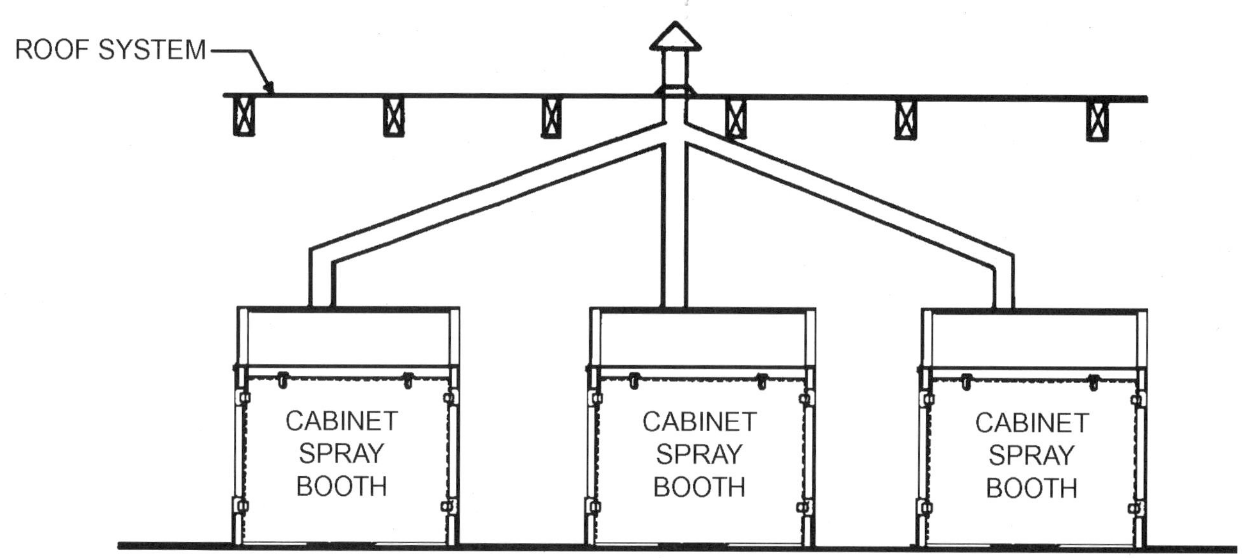

Figure 1504.2.5(2)
COMMON DUCTING SPRAY BOOTHS

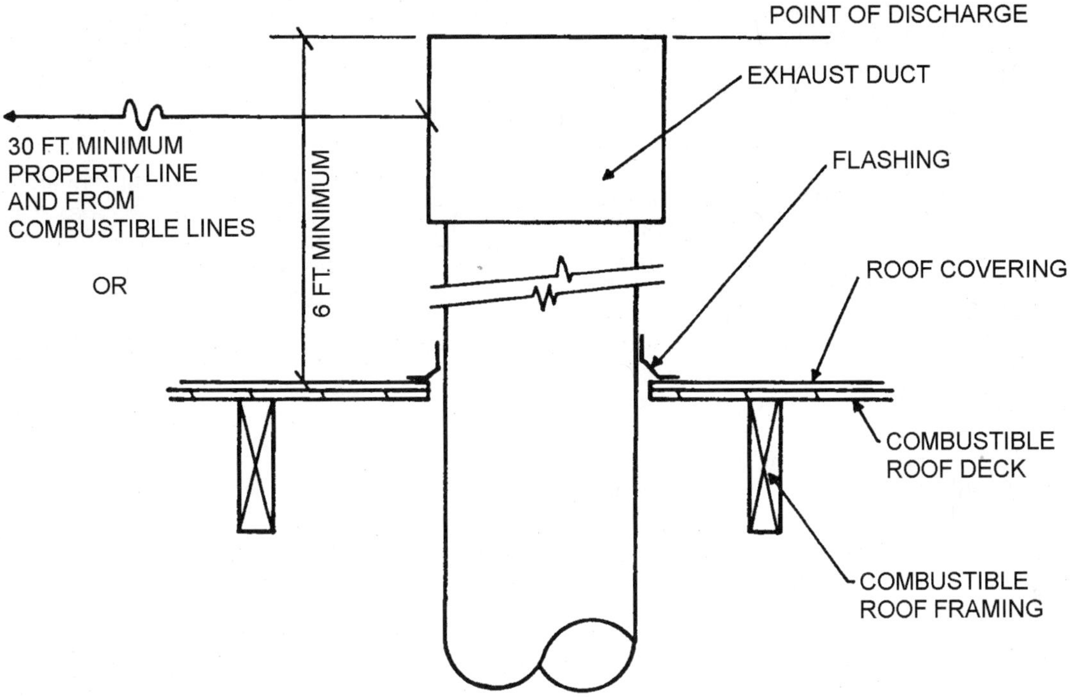

For SI: 1 foot - 304.8 mm.

Figure 1504.2.6(1)
SEPARATION REQUIREMENTS FOR ROOF VENTS

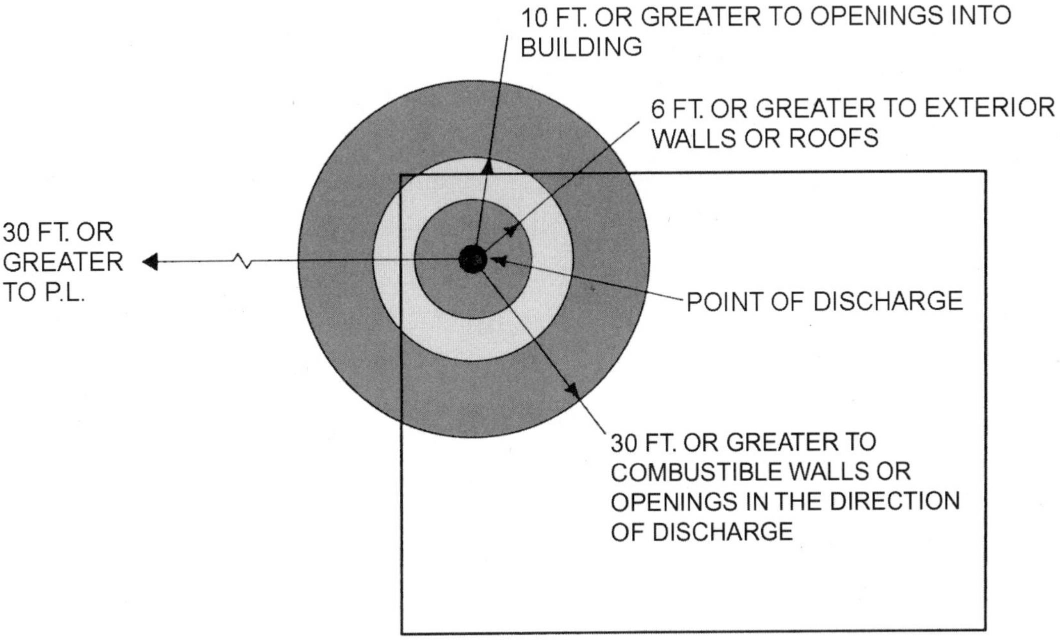

For SI: 1 foot - 304.8 mm.

Figure 1504.2.6(2)
SEPARATION REQUIREMENTS

1504.2.7 Fan motors and belts. Electric motors driving exhaust fans shall not be placed inside booths or ducts. Fan rotating elements shall be nonferrous or nonsparking or the casing shall consist of, or be lined with, such material. Belts shall not enter the duct or booth unless the belt and pulley within the duct are tightly enclosed.

❖ This requirement within the ventilation section is intended to reduce sources of ignition from spark-producing elements. This would increase the likelihood that overspray in the booth or duct cannot accumulate on the motor housing, which could ultimately cause the motor to overheat. Products that are subject to sparking should be avoided within spray areas. Belts that drive exhaust fans are not permitted within the spraying area unless the belts and pulleys are tightly enclosed to prevent solvents in exhaust air from degrading the belt materials and causing a failure of the ventilation system.

As mentioned in the commentary to Section 1503.4.1, the term "nonsparking" is somewhat inaccurate. Parts made of brass and similar nonsparking materials produce sparks with ignition energies too low to ignite flammable vapors. Nevertheless, such parts should be designed carefully to avoid producing sufficient frictional heat to cause an ignition.

1504.3 Filters. Air intake filters that are part of a wall or ceiling assembly shall be listed as Class I or Class II in accordance with UL 900. Exhaust filters shall be required.

❖ Spray booth and ventilation system design should effectively enclose spray operations. To prevent exhausting contaminated vapors into the atmosphere and prevent accumulation of overspray and residue on duct surfaces and at the duct discharge location, filters must be installed ahead of the exhaust ventilation systems from spray areas.

UL 900 requirements cover tests to determine combustibility and the amount of smoke generated for air filter units of both washable and throwaway types used for removal of dust and other airborne particles from air circulated mechanically in equipment and systems. Because the combustibility and smoke generation of an air filter unit once used depends on the chemicals or materials it is impregnated with, the filter test requirements of UL 900 are for the clean condition only.

1504.3.1 Supports. Supports and holders for filters shall be constructed of noncombustible materials.

❖ This section is again intended to minimize the combustible materials within a spray booth. The area near a filter that is used or partially used is very susceptible to ignition.

1504.3.2 Attachment. Overspray collection filters shall be readily removable and accessible for cleaning or replacement.

❖ Dry-type overspray collectors or filters of paper or fiberglass construction are more efficient than baffle plates. They are frequently used in spray booths containing moderate volumes of work. Replaceable flat or cartridge filters are intended to be discarded once they are significantly loaded with finish residue.

1504.3.3 Maintaining air velocity. Visible gauges, audible alarms or pressure-activated devices shall be installed to indicate or ensure that the required air velocity is maintained.

❖ Maintaining air velocities is critical in maintaining a safe environment outside the spray area as well as proper collection of flammable vapors and dusts within the spray area. If air velocities the exhaust system is designed for are not maintained, the spray booth operator must be made aware of this malfunction. To reduce the possibility of a fire or injury caused by human error, an automatic shutdown of the system is recommended when the designed air velocities are not maintained.

1504.3.4 Filter rolls. Spray booths equipped with a filter roll that is automatically advanced when the air velocity is reduced to less than 100 linear feet per minute (51 m/s) shall be arranged to shut down the spraying operation if the filter roll fails to advance automatically.

❖ In the case of roll-type filters, fresh filter material is advanced into the air stream when the air velocity is reduced to less than 100 linear feet per second (0.51 m/s). The impregnated filter is wound on a take-up reel. When the entire filter roll is consumed, it is discarded and replaced with a fresh roll of filter material. Shutting down the spray booth when the filter cannot automatically advance (either because of system failure or the need for cartridge replacement) ensures that the operator of the spray booth does not fail in his or her task of replacing the filter.

1504.3.5 Filter disposal. Discarded filter pads shall be immediately removed to a safe, detached location or placed in a noncombustible container with a tight-fitting lid and disposed of properly.

❖ To reduce the possibility of ignition, filters should be disposed of in approved metal containers with tight-fitting, self-closing lids. Waste containers should be removed from the building when full and at the end of each work shift.

1504.3.6 Spontaneous ignition. Spray booths using dry filters shall not be used for spraying materials that are highly susceptible to spontaneous heating and ignition. Filters shall be changed prior to spraying materials that could react with other materials previously collected. Examples of potentially reactive combinations include lacquer when combined with varnishes, stains or primers.

❖ Ventilation filters must be noncombustible. Once fouled with paint or coating residue, the filters become highly combustible. Moreover, they are more susceptible to spontaneous heating. Dry-type overspray collectors or filters of paper or fiberglass construction are more efficient than baffle plates and are frequently used in spray booths containing moderate volumes of work. Replaceable flat or cartridge filters are intended to be discarded once they are significantly loaded with finish residue.

1504.3.7 Waterwash spray booths. Waterwash spray booths shall be of an approved design so as to prevent excessive accumulation of deposits in ducts and residue at duct outlets. Such booths shall be arranged so that air and overspray are drawn through a continuously flowing water curtain before entering an exhaust duct to the building exterior.

❖ Water-wash booths are typically used for high-volume paint and lacquer usage. Many paints and lacquers are susceptible to spontaneous heating and combustion when left in a poorly ventilated or enclosed area. Materials containing linseed oil are especially prone to this type of reaction. Spray-finishing operations involving these materials should be confined to water-wash booths when possible because water is used as the filtration medium instead of dry filters.

1504.4 Different coatings. Spray booths, spray rooms and spray spaces shall not be alternately utilized for different types of coating materials where the combination of materials is conducive to spontaneous ignition, unless all deposits of one material are removed from the booth, room or space and exhaust ducts prior to spraying with a different material.

❖ Spray operations involving potentially reactive coating materials must be confined to separate spray booths or purged after use. Using the same spray booth for separate operations involving materials that react with each other may produce a reaction between the overspray residues of such materials once they are captured by the filters. Therefore, the entire system, including the ducts/filters, must be purged.

1504.5 Illumination. Where spraying spaces, spray rooms or spray booths are illuminated through glass panels or other transparent materials, only fixed lighting units shall be utilized as a source of illumination.

❖ Fixed lighting units can be designed to minimize the possibility of ignition caused by accidental heating of vapors above their autoignition temperature. Radiative heat from the source of illumination on the flammable finishes may cause ignition. Additionally, the speed and intensity of a fire depends on the temperature of the flammable finish.

1504.5.1 Glass panels. Panels for light fixtures or for observation shall be of heat-treated glass, wired glass or hammered-wire glass and shall be sealed to confine vapors, mists, residues, dusts and deposits to the spraying area. Panels for light fixtures shall be separated from the fixture to prevent the surface temperature of the panel from exceeding 200°F (93°C).

❖ The surface of incandescent bulbs, halogen lamps and other lights often exceeds the ignition temperature of common flammable and combustible liquids. Separation of light fixtures will help reduce the surface temperature on the unexposed side of the glass. Glass panels must be designed, arranged and protected to ease cleaning and prevent breakage.

1504.5.2 Exterior fixtures. Light fixtures attached to the walls or ceilings of a spraying area, but which are outside of any classified area and are separated from the spraying area by vapor-tight glass panels, shall be suitable for use in ordinary hazard locations. Such fixtures shall be serviced from outside the spraying area.

❖ Safety features such as ventilation and separation using vapor-tight glass panels allow for use of ordinary light fixtures that are not electrically classified for hazardous locations. To maintain the integrity of the separation (that is, vapor-tight glass panel and outside classified area), these light fixtures must be serviced from outside the spraying area.

1504.5.3 Integral fixtures. Light fixtures that are an integral part of the walls or ceiling of a spraying area are allowed to be separated from the spraying area by glass panels that are an integral part of the fixture. Such fixtures shall be listed for use in Class I, Division 2 or Class II, Division 2 locations, whichever is applicable, and also shall be suitable for accumulations of deposits of combustible residues. Such fixtures are allowed to be serviced from inside the spraying area.

❖ Light fixtures that are within the spraying area must be Class I, Division 2 because flammable vapors are present. Class II, Division 2 light fixtures are required where combustible residue, such as dust, is present.

1504.5.4 Portable electric lamps. Portable electric lamps shall not be used in spraying areas during spraying operations. Portable electric lamps used during cleaning or repairing operations shall be of a type approved for hazardous locations.

❖ Portable electric lamps are unsuitable for use in spraying areas during spraying operations. Portable electric lamps vary in type and electrical classification. It would be a difficult judgement call for an operator to determine the appropriate type and location of a portable lamp in spray areas. Although portable electric lamps, if approved for hazardous locations, are permitted during cleaning or repairing operations, they should be avoided if possible.

1504.6 Fire protection. Spray booths and spray rooms shall be protected by an approved automatic fire-extinguishing system complying with Chapter 9 which shall also protect exhaust plenums, exhaust ducts and both sides of dry filters when such filters are used.

❖ Automatic sprinkler protection is the preferred method of protection. Other approved automatic fire suppression systems (clean agents, carbon dioxide, etc.) may be installed if approved by the fire code official. Fire protection systems, equipment and devices must be installed in accordance with Chapter 9 and maintained. Because the flammable finish operations are subject to accumulation of residue, failure of the system as a result of poor maintenance is more likely in these areas than in most other types of facilities. Maintenance of such systems cannot be emphasized enough.

1504.6.1 Protection of sprinklers. Automatic sprinklers installed in spraying areas shall be protected from accumulation of residue from spraying operations in an approved manner. Bags used as a protective covering shall be 0.003-inch-thick (0.076 mm) polyethylene or cellophane or shall be thin paper. Automatic sprinklers contaminated by overspray particles shall be replaced with new automatic sprinklers.

❖ Automatic and manual fire protection equipment must be protected from accumulations of spray residue to reduce the likelihood of a malfunction of the equipment in an emergency. For example, protective lightweight paper or plastic bags may be installed over sprinkler heads but must be replaced at regular intervals or when they become heavily coated (see Section 7.5 of NFPA 33).

1504.6.2 Automated spray application operations. Where protecting automated spray application operations, automatic fire-extinguishing systems shall be equipped with an approved interlock feature that will, upon discharge of the system, automatically stop the operation of spraying operations and workpiece conveyors into and out of the spraying area. Where the building is equipped with a fire alarm system, discharge of the automatic fire-extinguishing system shall also activate the building alarm notification appliances.

❖ An interlock is a practical solution to avoid a fire situation in which the automated system would allow for a "moving fire" as the pieces within the automated assembly move through the spray area. The interlock would stop the spray application, which would further contribute to the flammable fuel/vapors, and prevent a moving fire from spreading faster and farther than expected. This is especially significant in larger conveyor and multiple spray area/booth systems, where fire could spread quickly beyond the designed sprinkler area. The interlock between the suppression system and the alarm system would notify the occupants of a building to evacuate.

1504.6.2.1 Alarm station. A manual fire alarm and emergency system shutdown station shall be installed to serve each spraying area. When activated, the station shall accomplish the functions indicated in Section 1504.6.2. At least one such station shall be readily accessible to operating personnel. Where access to this station is likely to involve exposure to danger, an additional station shall be located adjacent to an exit from the area.

❖ In case the operator of a spraying area is aware of a fire prior to the actuation of the extinguishing system, or in case there is an emergency not related to a fire where there is a risk to the occupants of the building, a manual fire alarm station is needed. The manual station is to provide the same functions as the automated interlock in Section 1504.6.2, such as stopping the spread of fire and activating occupant notification. Because areas near the spray area are dangerous, to reduce the risk of injury to an operator who may be attempting to activate the manual alarm, an additional alarm is required adjacent to an exit for the operator to activate while exiting from a hazardous situation.

1504.6.3 Ventilation interlock prohibited. Air makeup and spraying area exhaust systems shall not be interlocked with the fire alarm system and shall remain in operation during a fire alarm condition.

Exception: Where the type of fire-extinguishing system used requires that ventilation be discontinued, air makeup and exhaust systems shall shut down and dampers shall close.

❖ Exhaust ventilation of flammable vapors and smoke during a fire may help reduce the fire severity and increase visibility. Therefore, an interlock system to shut down the spray operations during a fire should not include the safety systems, such as ventilation.

In an emergency, the success of the fire extinguishing system takes precedence over the function of the ventilation system. Therefore, if the ventilation system in any way compromises the fire-extinguishing system, the exception allows these systems to be interlocked to shut down in the event of a fire. Examples of systems that may be compromised if the ventilation system is on are many of the extinguishing systems that are not water based, such as carbon dioxide, halon, clean agent, etc.

1504.6.4 Fire extinguishers. Portable fire extinguishers complying with Section 906 shall be provided for spraying areas in accordance with the requirements for an extra (high) hazard occupancy.

❖ Portable fire extinguishers for incipient fire fighting must be installed for ready access by the spray booth operator. Additionally, they should be selected on the basis of extra-hazard criteria contained in NFPA 10 to provide sufficient extinguishing agents and discharge time for the hazard to be protected. Because a spray area fire would involve ordinary combustibles, as well as flammable/combustible liquids and dusts, the selection of a fire extinguisher will include Class A and B ratings. Chapter 3 of NFPA 10 requires no less than a 4A, 40B extinguisher when the maximum travel distance to the extinguisher does not exceed 30 feet (9144 mm). A 4A, 80B extinguisher is also acceptable when the maximum travel distance does not exceed 50 feet (15 240 mm) (see Figure 1504.6.4).

Employees who are expected to fight incipient fires should receive instruction in the operation of installed fire protection equipment.

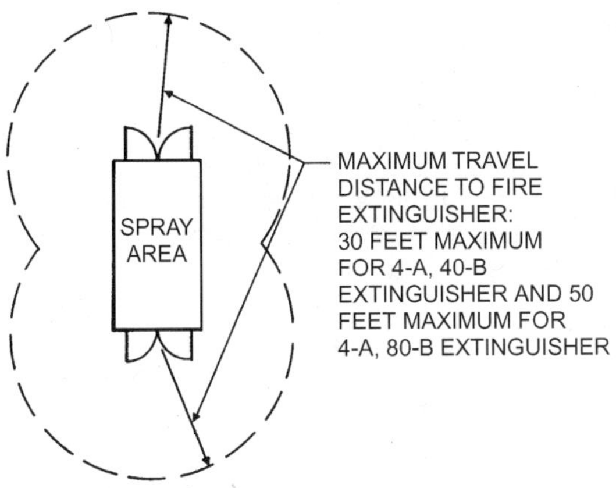

For SI: 1 foot = 304.8 mm.

Figure 1504.6.4
MAXIMUM TRAVEL TO EXTINGUISHERS

1504.7 Drying operations. Spray booths and spray rooms shall not be alternately used for the purpose of drying by arrangements which could cause an increase in the surface temperature of the spray booth or spray room except in accordance with Sections 1504.7.1 through 1504.7.2.3.

❖ Drying apparatus causing significant heating of the workpiece or spray booth surfaces may ignite finishing vapors or overspray residue located inside the spray booth. Separate drying apparatus or enclosures free of overspray residue should be installed when a drying apparatus is used.

1504.7.1 Spraying procedure. The spraying procedure shall use low-volume spray application.

❖ This section would apply only to smaller applications/processes where smaller amounts of flammable finishes are used. This is intended for spray booths and spray rooms that are used only occasionally for spraying and drying operations.

1504.7.2 Drying apparatus. Fixed drying apparatus shall comply with this chapter and the applicable provisions of Chapter 21. When recirculation ventilation is provided in accordance with Section 1504.2.2, the heating system shall not be within the recirculation air path.

❖ The practice of drying finished work pieces in the same enclosure where spray finishing is in progress is common in the automobile refinishing industry. This practice must be confined to spray-finishing enclosures designed for this purpose in compliance with the requirements of this section. Although the recirculated air should not be flammable (less than 25 percent LFL or dust residue is removed), as required by Section 1504.2.2, to reduce the chance of a fire, the heating system should be positioned to avoid possible ignition as a result of any malfunctions or unforeseen incidents.

Although this section appears to require "fixed drying apparatus," the intent of the section is to require both fixed and portable infrared apparatus. This is further evident in Section 1504.7.2.2, in which the additional requirements for portable infrared apparatus are given.

1504.7.2.1 Interlocks. The spraying apparatus, drying apparatus and ventilating system for the spray booth or spray room shall be equipped with interlocks arranged to:

1. Prevent operation of spraying apparatus while drying operations are in progress.
2. Purge spray vapors from the spray booth or spray room for a period of not less than 3 minutes before drying apparatus is rendered operable.
3. Have the ventilating system maintain a safe atmosphere within the spray booth or spray room during the drying process and automatically shut off drying apparatus in the event of a failure of the ventilating system.
4. Shut off the drying apparatus automatically if the air temperature within the booth exceeds 200°F (93°C).

❖ This section prescribes requirements for the safe use of infrared drying units in spray-finishing enclosures used for automobile refinishing and other similar applications. The interlocks listed above are required to reduce the likelihood that a potential source of ignition (drying apparatus) and the fuel (flammable vapors) are not present in the spray booth or spray room simultaneously. Also, the hazard of the ease with which heated flammable vapors are ignited is further mitigated by controlling the temperature of the room. Without these features, and a three-minute purge for the spray area, conditions for ignition of vapors or residue could easily develop.

1504.7.2.2 Portable infrared apparatus. When portable infrared drying apparatus is used, electrical wiring and portable infrared drying equipment shall comply with the ICC *Electrical Code.* Electrical equipment located within 18 inches (457 mm) of floor level shall be approved for Class I, Division 2 hazardous locations. Metallic parts of drying apparatus shall be electrically bonded and grounded. During spraying operations, portable drying apparatus and electrical connections and wiring thereto shall not be located within spray booths, spray rooms or other areas where spray residue would be deposited thereon.

❖ In addition to the drying apparatus requirements stated throughout Section 1504.7, portable infrared apparatus is required to meet the provisions of this section. These requirements, such as bonding and grounding of apparatus, are intended to prevent ignition of vapors in the transport of the apparatus.

1504.7.2.3 Sources of ignition. Except as specifically provided in this section, drying or baking units utilizing a heating system having open flames or which are capable of producing sparks, shall not be installed in a spraying area.

❖ Introduction of additional sources of ignition, including drying and baking units that do not meet the requirements of Section 1504.7.2, is not permitted in spraying areas.

SECTION 1505 DIPPING OPERATIONS

1505.1 Location of dip-tank operations. Dip-tank operations conducted in buildings used for Group A, I or R occupancies shall be located in a room designed for that purpose, equipped with an approved automatic sprinkler system, and separated vertically and horizontally from other areas in accordance with the *International Building Code.*

❖ Separation and protection from hazards of other portions of a building housing the occupancy groups listed in this section is critical for several reasons. Dipping and coating processes are too hazardous to be conducted in occupancies with a high life or property exposure. The number of occupants in such buildings and their physical conditions vary. People exposed to the danger of a fire involving dip tank operations must be able to evacuate or protect themselves promptly and effectively to avoid the risk of serious injury or death. Also, the psychology of people in a Group A, I or R is very different from those in a Group F manufacturing occupancy. A person walking in a manufacturing facility where dipping operations/processes take place in a large area where the dip tanks and signs are in plain view is more aware of the physical dangers associated with these processes than someone in a Group R occupancy, who would more likely be relaxing and thinking of things other than physical hazards.

The protection designed into these occupancies is, in many cases, less than would be found in manufacturing occupancies. Separation and protection in the form of fire-resistance-rated construction and automatic fire suppression must be provided where dipping operations must be conducted in the same building. Fire barrier walls, horizontal assemblies or both having a fire-resistance rating of three hours are required between Group F-1 and A or I and R occupancies. Except for Group I-2, this requirement can be reduced in a sprinklered building by one hour. If the amounts of hazardous materials exceed the exempt quantities in Chapter 27, the occupancy separation between Group H and A or I and R occupancies varies depending on what material(s) are involved (see Section 302.3 of the IBC).

1505.2 Ventilation of vapor areas. Vapor areas shall be provided with mechanical ventilation adequate to prevent the dangerous accumulation of vapors. Required ventilation systems shall be arranged such that the failure of any ventilating fan shall automatically stop the dipping conveyor system.

❖ This section includes performance-based language that requires ventilation to prevent the dangerous accumulation of vapors. Additionally, conveyor systems used with dipping operations need to be interlocked with the ventilation system to avoid continuation of operations in the event of a ventilation system failure.

The IMC contains requirements for hazardous exhaust systems and requires independent duct exhaust of vapor releases from dipping and coating processing tanks, reservoirs, trench drains, drain boards, conveyor tunnels and any other space where flammable vapors may be liberated. Ducts are not to penetrate fire-resistance-rated assemblies. The plenum and ductwork must be protected by automatic sprinklers or other approved fire suppression systems installed in accordance with Chapter 9. Only fans and mechanical equipment approved for use in Class I, Division 1 hazardous locations are to be installed in the exhaust air stream, and ductwork must be of materials, thicknesses and construction methods specified in the IMC.

1505.3 Construction of dip tanks. Dip tanks shall be constructed in accordance with this section and NFPA 34. Dip tanks, including drain boards, shall be constructed of noncombustible material and their supports shall be of heavy metal, reinforced concrete or masonry.

❖ The selection of materials and design of dipping and coating processes must consider the physical properties of the liquid and the processing environment. The corrosivity, density, viscosity, vapor pressure and flash point of the liquid influences the selection of materials and arrangement of processing equipment. Additionally, mechanical hazards, such as impacts and collisions involving conveyor equipment, should be considered. Protected steel or concrete tank supports reduce the likelihood of a collapse in the event of a fire caused by a spill or release. Figure 1502.1(1) shows the arrangement of a typical dip tank, including overflow and emergency release drains.

1505.3.1 Overflow. Dip tanks greater than 150 gallons (568 L) in capacity or 10 square feet (0.93 m^2) in liquid surface area shall be equipped with a trapped overflow pipe leading to an approved location outside the building. The bottom of the overflow connection shall not be less than 6 inches (152 mm) below the top of the tank.

❖ An overflow drain is required for tanks with capacities greater than 150 gallons (568 L) or having a surface area greater than 10 square feet (.93 m^2) to confine spills or uncontrolled releases caused by overfilling or overflowing when parts are immersed. The capacity of overflow drains should exceed the capacity of the tank and discharge through an approved trap and separator or to an approved salvage tank.

1505.3.2 Bottom drains. Dip tanks greater than 500 gallons (1893 L) in liquid capacity shall be equipped with bottom drains that are arranged to automatically and manually drain the tank quickly in the event of a fire unless the viscosity of the liquid at normal atmospheric temperature makes this impractical. Manual operation shall be from a safe, accessible location. Where gravity flow is not practicable, automatic pumps shall be provided. Such drains shall be trapped and discharge to a closed, vented salvage tank or to an approved outside location.

Exception: Dip tanks containing Class IIIB combustible liquids where the liquids are not heated above room temperature, and the process area is protected by automatic sprinklers.

❖ Emergency release drains permit flammable liquids in dip tanks to be safely discharged if a fire occurs. A trap between the drain opening and the separator or salvage tank reduces the likelihood that the fire will flash back into the tank. The arrangement of emergency release drains must permit both automatic and manual operation. Moreover, viscous liquids are exempt from these requirements. These liquids, however, generally have higher flash points and are more difficult to ignite than other, less viscous materials. If a highly viscous combustible liquid is heated for use as a coating material, having a drain remains a good practice. If installed, the drain should be sized for the volume and viscosity of the liquid under normal usage conditions.

When practical, tanks should be arranged to drain by gravity. Pumps must be used if gravity discharge is not practical. When a tank is equipped with a pump, the pump should be installed on the discharge side of the trap. When a pump is installed to aid in emergency release, flammable and combustible liquid piping must comply with Chapter 34.

The exception covers Class IIIB combustible liquids, which have a flashpoint above 200°F (93°C). These liquids are not viewed by the code as extremely hazardous, especially when they are not heated above room temperature. For example, Table 2703.1.1(1), Note f, allows for unlimited quantities of Class IIIB liquids in storage and use in a sprinklered building.

1505.3.3 Dipping liquid temperature control. Protection against the accumulation of vapors, self-ignition and excessively high temperatures shall be provided for dipping liquids that are heated directly or heated by the surfaces of the object being dipped.

❖ The evaporation rate of a liquid increases as the liquid temperature increases. By controlling the maximum liquid temperature, in case of a malfunction or unplanned event, the temperature of the liquid will not increase to a point where the exhaust ventilation cannot accommodate the exhaust of the vapors. Moreover, controlling the maximum liquid temperature would prevent the liquid from reaching its self-ignition (or auto-ignition) temperature.

The temperature control for the liquid must be specifically designed for the liquid. Self-ignition temperatures of liquids can be found in many references. The *SFPE* (Society of Fire Protection Engineers) *Handbook of Fire Protection Engineering* includes tables listing autoignition temperatures for a variety of products.

Although it is a more general application, similar temperature control requirements apply when exempt use and storage amounts of heated hazardous materials are exceeded (see Sections 2704.8.1 and 2705.1.4.3).

1505.4 Conveyors. Dip tanks utilizing a conveyor system shall be arranged such that in the event of fire, the conveyor system shall automatically cease motion and the required tank bottom drains shall open.

❖ Mechanical or electrical interlocks must interrupt conveyor motion and cause required emergency release drains to open in the event of a fire. Stopping the conveyor prevents newly coated parts from transporting a fire beyond the immediate vicinity of the dipping or coating processing area and its protection systems.

1505.5 Portable fire extinguishers. Areas in the vicinity of dip tanks shall be provided with portable fire extinguishers complying with Section 906 and suitable for flammable and combustible liquid fires as specified for extra (high) hazard occupancies.

❖ Readily accessible portable fire extinguishers for incipient fire fighting must be installed for use by employees working around dipping and coating processing equipment. Both the size and distribution of portable fire extinguishers must conform to this section, Section 906.1 and the applicable sections of NFPA 10 for extra (high) hazards. Two units of Class B extinguishing capabilities are required for each square foot of dip tank area if either dry-chemical or carbon dioxide portable extinguishers are installed. Only one unit of Class B rating is required per square foot if aqueous film-forming foam (AFFF) portable extinguishers are provided.

For example, a 40-square-foot (4 m^2) dip tank would require an 80-B-rated dry-chemical or carbon dioxide extinguisher or a 40-B AFFF extinguisher. The maximum travel distance to the nearest required portable fire extinguisher is 30 feet (9144 mm). Employees who are expected to fight incipient fires should receive instructions in the operation of installed fire protection equipment. Fire protection systems, equipment and devices must be maintained in accordance with Section 901.6.

1505.6 Fixed fire-extinguishing equipment. An approved automatic fire-extinguishing system or dip tank covers in accordance with Section 1505.7 shall be provided for the following dip tanks:

1. Dip tanks less than 150 gallons (568 L) in capacity or 10 square feet (0.93 m^2) in liquid surface area.
2. Dip tanks containing a liquid with a flash point below 110°F (43°C), used in such manner that the liquid temperature could equal or be greater than its flash point from artificial or natural causes, and having both a capacity of more than 10 gallons (37.9 L) and a liquid surface area of more than 4 square feet (0.37 m^2).

❖ This section applies to fire-extinguishing equipment for dip tanks. It offers the option of dip tank covers instead of a fixed fire-extinguishing system. Tanks equipped with noncombustible automatic-closing covers conforming to Section 1505.7 need not be protected by a fire suppression system on the basis that the cover can be closed, thus containing a dip tank fire and further restricting oxygen supply to the fire. Additionally, the suppression system will not be effective in controlling a fire when the cover is in the sloped position because the sprinkler water would simply slide off the cover and away from the fire.

Automatic sprinkler protection is recommended for plant areas located around dip tanks and other similar equipment. However, the primary protection for the dip tank should be one of the following types of systems listed in order of desirability and effectiveness:

1. Water-spray fixed system (NFPA 15 or 16);
2. Foam extinguishing system (NFPA 11, 11A or 16);
3. Carbon dioxide system (NFPA 12);
4. Halogenated system (NFPA 12A) or
5. Dry-chemical system (NFPA 17).

Water-spray fixed systems are more effective at controlling flammable liquid pool fires, like those involving dip tanks, because they produce finer water droplets, which are more effective at absorbing heat from these rapidly burning, intense fires. Most water droplets are quickly vaporized by the fire and rarely pose a significant danger of boil-over, froth-over or slop-over from penetrating the surface of the flammable or combustible liquid (see NFPA 30 for further information on these phenomena unique to flammable and combustible liquids in tanks).

Unless dip-tank covers are installed, fixed fire-extinguishing equipment is required for dip tanks less than 150 gallons (568 L) or 10 square feet (.93 m^2).

Unless dip-tank covers are installed, dip tanks containing (1) Class I flammable liquids and Class II combustible liquids with flash points below 110°F (43°C) used at ambient temperatures above their flash points, (2) having tank capacities greater than 10 gallons (38 L) and (3) exposed tank surface areas greater than 4 square feet (.37 m^2) must be protected in accordance with Section 1505.6.1 because of their ease of ignition.

1505.6.1 Fire-extinguishing system. An approved automatic fire-extinguishing system shall be provided for dip tanks with a 150-gallon (568 L) or more capacity, or 10 square feet (0.93 m^2) or larger in a liquid surface area. Fire-extinguishing system design shall be in accordance with NFPA 34.

❖ Automatic fire suppression must be installed at dip tanks with a capacity greater than 150 gallons (568 L) or larger than 10 square feet (.93 m^2) in surface area. See the commentary to Section 1505.6 for the preferred type of extinguishing system.

1505.7 Dip tank covers. Dip tank covers allowed by Section 1505.6 shall be capable of manual operation and shall be automatic-closing by approved automatic-closing devices designed to operate in the event of fire.

❖ Dip-tank covers installed as a method of automatic suppression conforming to Section 1505.6 must be capable of manual actuation. Automatic closure must be initiated by a fusible link or another heat-sensitive device. Covers that cannot close under the force of gravity must be assisted by hydraulic activators, springs, counterweights or other methods.

1505.7.1 Construction. Covers shall be constructed of noncombustible material or be of a tin-clad type with enclosing metal applied with locked joints.

❖ Dip-tank covers must be constructed of noncombustible material to avoid adding to the fire load, as well as to maintain the integrity of the system as a passive containment fire protection system. Tin-clad covers should be designed the same as metal-clad or tin-clad fire doors (see NFPA 80). Materials used for automatic closing mechanisms such as hydraulic actuators, springs, etc., should also be constructed of noncombustible materials.

1505.7.2 Supports. Chain or wire rope shall be utilized for cover supports or operating mechanisms.

❖ Devices for holding the cover open must be designed to permit it to close freely and seal tightly. If noncombustible materials are used, a fusible link or another heat-sensitive device must be used to initiate closure.

1505.7.3 Closed covers. Covers shall be kept closed when tanks are not in use.

❖ Keeping the cover closed when the tank is not in use reduces the release of flammable vapors or the accidental introduction of an ignition source into the vapor area located below the rim of the dip tank.

1505.8 Hardening and tempering tanks. Hardening and tempering tanks shall comply with Sections 1505.3 through 1505.5 but shall be exempt from other provisions of Section 1505.

❖ Requirements of Sections 1505.8.1 through 1505.8.5 apply to oil-quenching tanks used in hardening and tempering processes. Individually, hardening and tempering are usually accomplished by immersing parts in tanks containing certain metal salts heated to temperatures between 1,400 and 2,350°F (760 and 1287°C). This process often requires several steps, with parts cooled by oil quenching between steps. Such oil-quenching baths are the subject of these requirements. Animal, vegetable and mineral oils and various mixtures of each are used for oil quenching, but mineral oils are most commonly used. Recently, polymers have replaced oils on a small scale in some oil-quenching processes. For most oil-quenching processes, oils have flash points above 300°F (149°C) (Class III-B) and many have flash points exceeding 500°F (260°C). Oil-quenching baths are usually maintained 100 to 200°F (38 to 93°C) below their flash points. The same design, construction, operation, maintenance and fire protection requirements applying to dip tanks pertain to hardening and tempering tanks, as do the requirements of Sections 1505.8.1 through 1505.8.5.

1505.8.1 Location. Tanks shall be located as far as practical from furnaces and shall not be located on or near combustible floors.

❖ High temperatures produced by hardening and tempering processes require the maintenance of separation between tanks and combustible materials (see Figure 1505.8.1).

1505.8.2 Hoods. Tanks shall be provided with a noncombustible hood and vent or other approved venting means, terminating outside of the structure to serve as a vent in case of a fire. Such vent ducts shall be treated as flues, and proper clearances shall be maintained from combustible materials.

❖ Though the materials in hardening and tempering baths are usually not considered flammable, the elevated temperatures may represent an ignition source to other combustibles. The contents of oil-quenching tanks used for high-flash-point liquids may be difficult to ignite, but once ignited, the rate of heat release is comparable to other flammable and combustible liquids. Hoods over hardening, tempering and oil-quenching tanks will help maintain tenable conditions near the tanks and help control a fire should it occur by providing a controlled avenue of vertical spread. Therefore, exhaust hood materials must be noncombustible and separated from combustible structural components.

1505.8.3 Alarms. Tanks shall be equipped with a high-temperature limit switch arranged to sound an alarm when the temperature of the quenching medium reaches 50°F (10°C) below the flash point.

❖ Thermostats may be used to limit the heat input to the oil bath for temperature maintenance. However, many systems require cooling systems to maintain oil baths within specified temperature limitations. Water leaking from these cooling systems can pose a special hazard

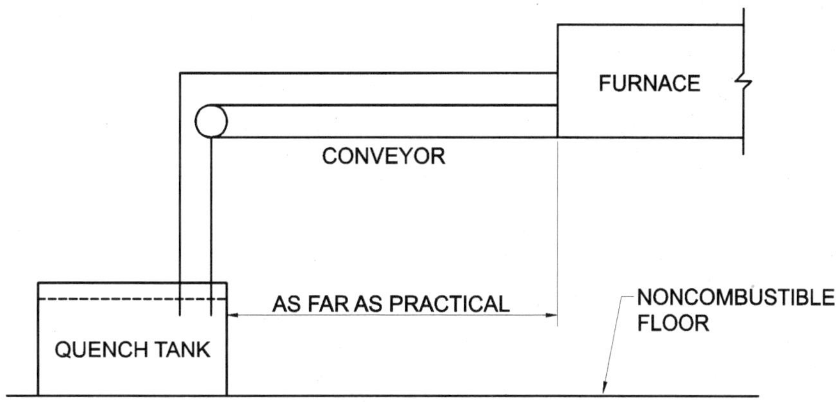

Figure 1505.8.1
SEPARATION REQUIREMENTS FOR QUENCHING TANKS

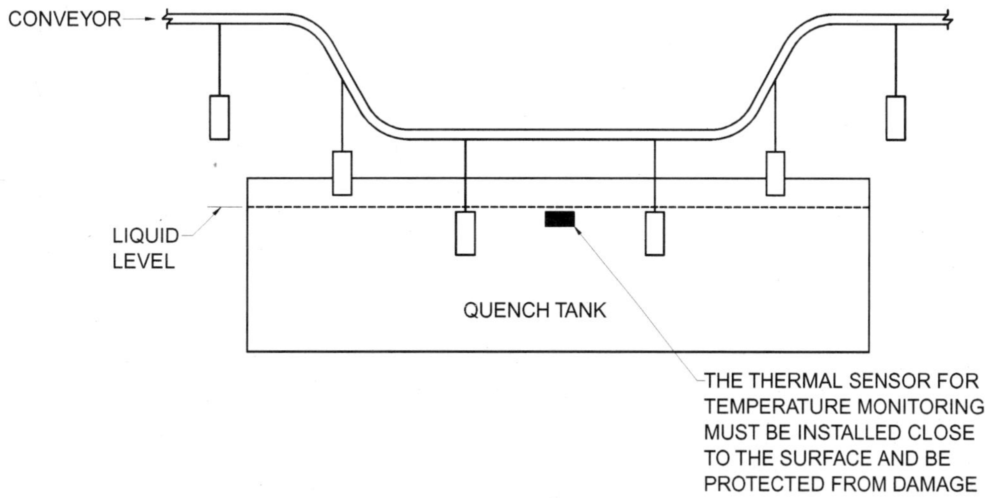

Figure 1505.8.3
TYPICAL QUENCH TANK ALARM LOCATION

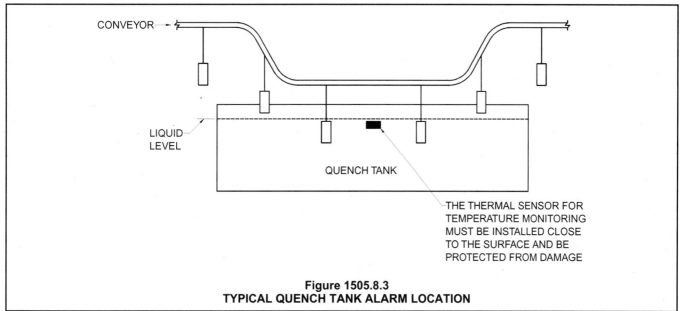

from boil-over as the water is converted to steam (see Figure 1505.8.3).

1505.8.4 Fire protection. Hardening and tempering tanks greater than 500 gallons (1893 L) in capacity or 25 square feet (2.3 m^2) in liquid surface area shall be protected by an approved automatic fire-extinguishing system complying with Chapter 9.

❖ Tanks considered larger than 500 gallons (1893 L) or 25 square feet (2 m^2) in surface area must be protected by an automatic fire suppression system, unless they are equipped with a cover conforming to Section 1505.7 (see commentary, Section 1505.6.1).

1505.8.5 Use of air pressure. Air under pressure shall not be used to fill or agitate oil in tanks.

❖ Heated oil can cause air injected below the surface to expand, allowing oil to spill from the tank. Air containing moisture can contaminate the oil bath, leading to boil-over (see Figure 1505.8.5).

1505.9 Flow-coating operations. Flow-coating operations shall comply with the requirements for dip tanks. The area of the sump and any areas on which paint flows shall be considered to be the area of a dip tank.

❖ In flow-coat operations, a product is applied to the workpiece in an unatomized stream through fixed or oscillating nozzles. Excess product is collected in a trough or sump below the workpiece and recirculated through a reservoir. The principal hazard from flow coating is the liberation of flammable liquid vapor from the surface of excess liquid. The area of the trough or sump for collecting overspray defines the scope of the hazard. Protection of flow-coat operations must be based on the combined area of the trough or sump and any surfaces in which paint or coating material flows en route to the trough or sump (see Figure 1505.9). In large operations, drain tunnels extend outside the enclosure. Moving the object through the drain tunnels reduces solvent evaporation. Using the tunnels also improves the film coating on the object.

Curtain coating operates on a similar principle, except a trough is filled above the workpiece and allowed to overflow in a thin flat stream. This process is often used to coat flat or slightly curved workpieces.

1505.9.1 Paint supply. Paint shall be supplied by a gravity tank not exceeding 10 gallons (37.9 L) in capacity or by direct low-pressure pumps arranged to shut down automatically in case of fire by means of approved heat-actuated devices.

❖ Positive displacement pumps are most commonly used to recirculate paint and coating material from the reservoir to the nozzles. Gravity tanks not exceeding 10 gallons (38 L) in capacity are also permitted. When a pump is used, the pump power supply must be interlocked with heat detectors to shut down if a fire occurs.

1505.10 Roll-coating operations. Roll-coating operations shall comply with Section 1505.9. In roll-coating operations utilizing flammable or combustible liquids, sparks from static electricity shall be prevented by electrically bonding and grounding all metallic rotating and other parts of machinery and equipment and by the installation of static collectors or by maintaining a conductive atmosphere such as a high relative humidity.

❖ Roll-coating methods apply material to flat work pieces, usually paper, cardboard, cloth or thin metals, using liquid-coated cylinders or rollers. Coating material may be

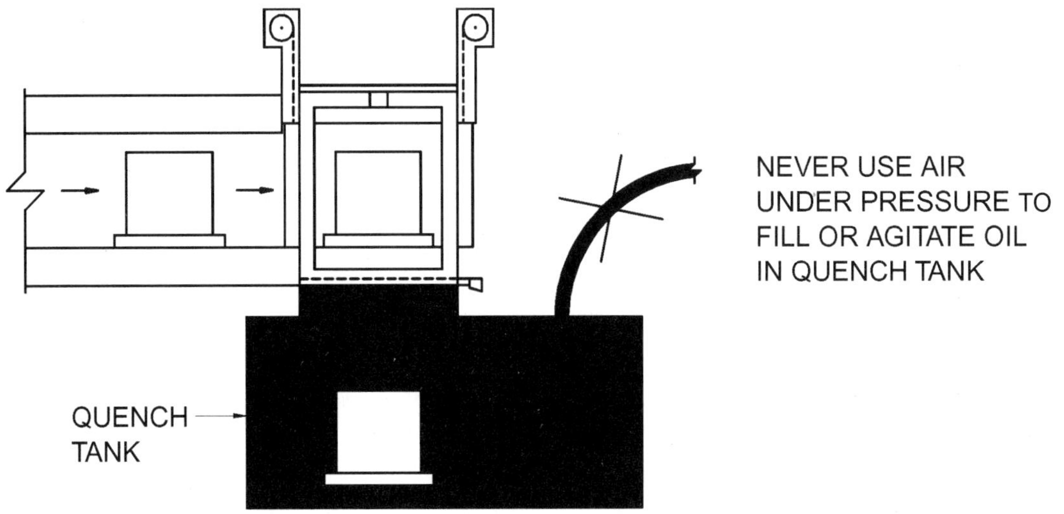
NEVER USE AIR
UNDER PRESSURE TO
FILL OR AGITATE OIL
IN QUENCH TANK
QUENCH
TANK

Figure 1505.8.5
AIR FEED PROHIBITED

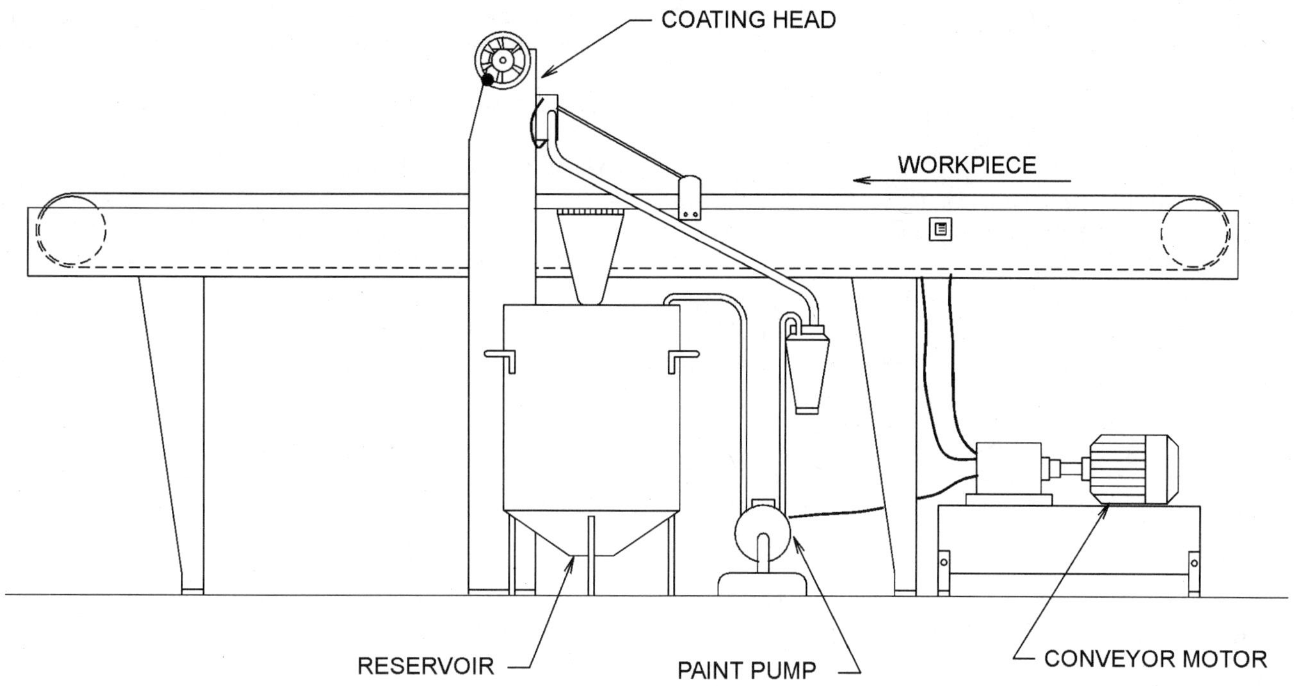

Figure 1505.9
TYPICAL FLOW-COAT SETUP

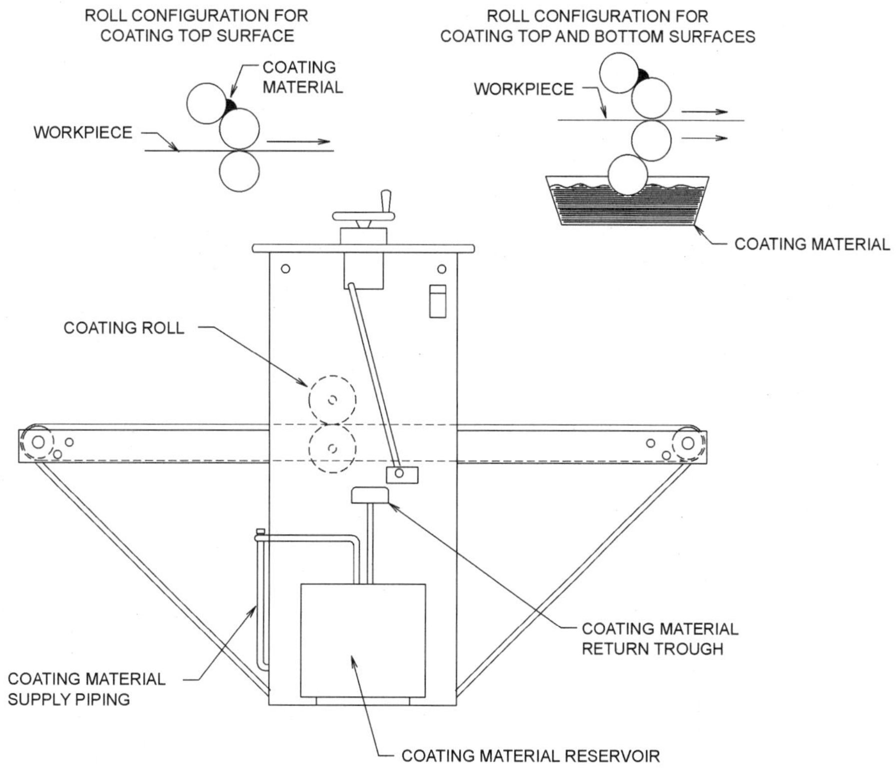

Figure 1505.10
ROLL-COATING SETUP

applied to the rollers by rotating them in an open trough or pan or applying liquid to the space between two rollers (see Figure 1505.10).

SECTION 1506 ELECTROSTATIC APPARATUS

1506.1 General. Electrostatic apparatus and devices used in connection with paint-spraying and paint-detearing operations shall be of an approved type.

❖ Electrostatic spraying and paint detearing equipment, as well as electrostatic devices, such as generators, motors, transformers and electrodes, etc., should bear the label of an independent testing laboratory. Evidence of satisfactory performance is indicated by use or display of the appropriate label or seal of the laboratory. Testing laboratories generally publish directories or lists containing important information about labeled products.

1506.2 Location. Transformers, power packs, control apparatus and all other electrical portions of the equipment, except high-voltage grids and electrostatic atomizing heads and connections, shall be located outside of the spraying or vapor areas, or shall comply with Section 1503.2.

❖ Devices that are not classified must be located outside the vapor area to prevent ignition of flammable vapors and overspray residue. The vapor area is defined as that area where flammable vapors exceed 25 percent of the material's LFL. Equipment must be tested and labeled for use in Class I, Division 1 hazardous locations, as defined by Article 516 of NFPA 70, when locating equipment outside the spraying or vapor area is impractical.

1506.3 Construction of equipment. Electrodes and electrostatic atomizing heads shall be of approved construction, rigidly supported in permanent locations and effectively insulated from ground. Insulators shall be nonporous and noncombustible.

❖ Electrostatic atomizing heads are connected to both an air source and flammable coating products. Additionally, this particular piece of equipment is connected to high-voltage electricity. Oxygen, fuel and sources of ignition are readily available. Therefore, special care and consideration must be given in the construction and installation of such equipment. To avoid any sparking of the equipment, electrostatic atomizing heads must be insulated from grounded objects or parts.

1506.4 Clear space. A space of at least twice the sparking distance shall be maintained between goods being painted or deteared and electrodes, electrostatic atomizing heads or conductors. A sign stating the sparking distance shall be conspicuously posted near the assembly.

❖ Consult the equipment manufacturer's operating instructions to determine the sparking distance of the equipment involved. Once the sparking distance is determined, it must be posted conspicuously in the work area. The sign should be clear, concise and of durable construction. Maintaining the required separation distance prevents sparks generated by the properly maintained equipment from igniting vapors near the surface of the newly coated workpiece before it dries.

1506.5 Emergency shutdown. Electrostatic apparatus shall be equipped with automatic controls operating without time delay to disconnect the power supply to the high-voltage transformer and signal the operator under any of the following conditions:

1. Stoppage of ventilating fans or failure of ventilating equipment from any cause.
2. Stoppage of the conveyor carrying articles past the high-voltage grid.
3. Occurrence of a ground or an imminent ground at any point of the high-voltage system.
4. Reduction of clearance below that required in Section 1506.4.

❖ The required interlocks reduce the likelihood of the apparatus igniting flammable vapors in the event of any of the specified conditions. Conditions 1 and 2 may lead to an increase in the concentration of flammable vapors in the atmosphere. Condition 3 may lead to the release of a spark or arc capable of igniting flammable vapors. Additionally, occurrence of Condition 4 could bring those parts of the system capable of producing an ignition into an area containing an ignitable vapor concentration.

1506.6 Ventilation interlock. Hand electrostatic equipment shall be interlocked with the ventilation system for the spraying area so that the equipment cannot be operated unless the ventilating system is in operation.

❖ The required interlock is intended to prevent the use of hand sprayers without ventilation equipment in operation. Failure to operate exhaust ventilation may lead to the creation of ignitable vapor concentration in the spray area.

1506.7 Protection for automated liquid electrostatic spray application equipment. Automated liquid electrostatic spray application equipment shall be protected by the installation of an approved, supervised flame detection apparatus that shall, in the event of ignition, react to the presence of flame within 0.5 second and shall accomplish all of the following:

1. Activation of a local alarm in the vicinity of the spraying operation and activation of the building alarm system, if such system is provided.
2. Shutting down of the coating material delivery system.
3. Termination of all spray application operations.
4. Stopping of conveyors into and out of the spraying area.
5. Disconnection of power to the high-voltage elements in the spraying area and disconnection of power to the system.

❖ Automated liquid electrostatic spray application may or may not have operators that would take steps to reduce the severity of the fire, such as shutting down the supply of coating material, warning occupants for evacuation purposes, etc. The flame detection system and its associated interlocks are part of the protection system, in case ignition has already occurred.

Systems that are designed to reduce the severity of the accident in the event of a fire need to be initiated automatically. Supervised flame detection systems are required by this section to stop the flow of additional fuel into the system, stop the spread of fire by shutting down the conveyor, activate an alarm for evacuation, terminate spray application operations that would further add to the fuel and disconnect power to the system. The power disconnection referred to in this section is for the spray booth equipment, but does not include power to emergency systems.

1506.8 Barriers. Booths, fencing, railings or guards shall be placed about the equipment such that either by their location or character, or both, isolation of the process is maintained from plant storage and personnel. Railings, fencing and guards shall be of conductive material, adequately grounded, and shall be at least 5 feet (1524 mm) from processing equipment.

❖ The required guards prevent materials with an opposite charge from being placed within the range of the electrostatic spraying apparatus. Guards or railings are grounded so that any charge accumulation or deficit on people or materials will be safely neutralized. The separation distance requirement allows the charge dissipation to occur at a safe distance from ignitable vapors.

1506.9 Signs. Signs shall be posted to provide the following information:

1. Designate the process zone as dangerous with respect to fire and accident.
2. Identify the grounding requirements for all electrically conductive objects in the spray area, including persons.
3. Restrict access to qualified personnel only.

❖ Signs should warn of the smoking, open-flame, grounding areas and high-voltage equipment hazards. Additionally, signs should warn against entrance of unqualified personnel into these areas to avoid accidents caused by people or employees who are not trained for the surrounding hazards.

1506.10 Ventilation. The spraying area shall be ventilated in accordance with Section 1504.2.

❖ See the commentary to Section 1504.2.

1506.11 Maintenance. Insulators shall be kept clean and dry. Drip plates and screens subject to paint deposits shall be removable and taken to a safe place for cleaning.

❖ Accumulation of dirt, oil, moisture or debris may compromise the effectiveness of insulators. Overspray accumulations are less severe with electrostatic processes but still require attention. Equipment and fixtures collecting overspray residue must be cleaned periodically (see Section 1503.4).

1506.12 Fire protection. Areas used for electrostatic spray finishing with fixed equipment shall be protected with an approved automatic fire-extinguishing system complying with Chapter 9.

❖ Although automatic sprinkler protection is the most common method of protection in facilities, other approved automatic fire suppression systems may be installed if approved by the fire code official. Fire protection systems, equipment and devices must be installed in accordance with Chapter 9 and maintained.

SECTION 1507
POWDER COATING

1507.1 General. Operations using finely ground particles of protective finishing material applied in dry powder form by fluidized bed, electrostatic fluidized bed, powder spray guns or electrostatic powder spray guns shall comply with this section.

❖ This section applies to powder coating operations, which involve the application of finely ground particles of protective finishing material. The finish from powder coating is very strong and more durable than conventional finishes. This is a benefit, from an environmental and fire protection standpoint, because there are no liquid finishes or solvents to produce flammable vapors or volatile organic compounds (VOCs). Additionally less solid waste is created with this method of finishing. The foremost hazard associated with such application(s) is fire or explosion as a result of the airborne dust. The hazard associated with powder coating operations is considered less than that of a similar operation using flammable or combustible liquids. The energy required to ignite a cloud of air-suspended coating powder is from 100 to 1,000 times higher than that required to ignite flammable vapors associated with fluid coating processes. Nevertheless, these operations pose a significant explosion hazard when organic powder is suspended in air, forming dust clouds.

1507.2 Location and construction of powder coating rooms and booths. Powder coating operations shall be conducted in enclosed rooms constructed of noncombustible materials, enclosed powder coating facilities which are ventilated, or ventilated spray booths complying with Section 1504.1.2.

Exception: Listed spray-booth assemblies that are constructed of other materials shall be allowed.

❖ The majority of powder coating operations are conducted in spray booths designed specifically to accommodate the airborne dust. Listed spray booth assemblies should be specifically listed for powder coating operations to ensure that the differences in medium between flammable spray finish and powder coating are addressed in the protection systems. An example of such differences is the type of electrical classification between powder coating operations and flammable finish operations.

The exception applies to only listed (not approved) spray booths (see definition of "Listed" in Section 202).

1507.3 Sources of ignition. When parts are heated prior to coating, the temperature of the parts shall not exceed the ignition temperature of the powder to be used.

Precautions shall be taken to minimize the possibility of ignition by static electrical sparks through static bonding and grounding, where possible, of powder transport, application and recovery equipment.

❖ During the heating process and prior to coating, safety controls should be implemented to prevent the ignition of the powder on an overheated piece as a result of a system malfunction (temperature control failures, when a piece is overheated on a conveyor that has stopped, etc.).

1507.4 Ventilation. Exhaust ventilation shall be sufficient to maintain the atmosphere below one-half the minimum explosive concentration for the material being applied. Nondeposited, air-suspended powders shall be removed through exhaust ducts to the powder recovery system.

❖ Powder coating creates explosive atmospheres because of the large surface areas of the particles when dispersed in the air. The explosive limit or concentration, just as with flammable vapors, will depend on the type of material being used. For example, Factory Mutual tested several powder coating materials and found a range of LEL of 0.026 to 0.097 oz/ft^2 and autoignition temperatures between 790 to 1,039°F (412 to 559°C). Therefore, ventilation system requirements may vary from one type of coating to another. Also, a collection system is required to collect any unused powder.

This requirement is similar to ventilation requirements for flammable vapors in its intent to limit the amount of vapors or dusts to a concentration that would not support ignition. Therefore "one-half the minimum explosive concentration" for powder coating is similar in intent to the "25 percent of the Lower Flammable Limit (LFL)" in Section 1504.2.2. The safety factor in the combustible dust is 2, while the safety factor for the LFL is 4. This is partly the result of the lower ignitibility of a dust compared to a vapor (see commentary, Section 1507.1).

1507.5 Drying, curing and fusion equipment. Drying, curing and fusion equipment shall comply with Chapter 21.

❖ Chapter 21 details the construction, operation, maintenance and fire protection of, as well as the equipment/piping associated with, industrial ovens (see commentary, Chapter 21).

1507.6 Operation and maintenance. Powder coating areas shall be kept free from the accumulation of powder coating dusts, including horizontal surfaces such as ledges, beams, pipes, hoods, booths and floors.

❖ Regularly scheduled cleaning of accumulation of powder on surfaces is an important and often most ignored good housekeeping practice. This is especially critical when pieces are bonded to the conveyor. If too much residue is accumulated on the conveyor, the pieces may no longer be bonded. When pieces are preheated, the additional dust accumulation may contribute to air-borne dust and escalate a small fire to a much larger and more severe fire.

1507.6.1 Cleaning. Surfaces shall be cleaned in such a manner so as to avoid scattering dusts to other places or creating dust clouds. Vacuum sweeping equipment shall be of a type approved for use in hazardous location.

❖ Additional air-borne dust must be avoided during cleaning operations. Any type of dust-agitating process can create a potential for flash fires or dust explosions. Removal of dusts from any horizontal surfaces such as ledges, beams, pipes, hoods, booths and floors is important in reducing the excess dust concentrations within powder coating areas. Cleaning should minimize the scattering of dust or creation of a dust cloud that can easily cause a fire or explosion. Vacuum sweeping equipment should be approved for the type of use and be electrically classified in accordance with the ICC EC and Section 1503.2.1 (see commentary, Section 1503.2.1).

1507.6.2 Spark-producing metals. Iron or spark-producing metals shall be prevented from being introduced into the powders being applied by magnetic separators, filter-type separators, or by other approved means.

❖ Iron or spark-producing metals, such as small workpieces or tools, have caused ignitions or explosions of the air-borne combustible dusts in powder coating operations. Although magnetic or filter-type separators are the simplest and most commonly used systems for reducing such a hazard, other approved means of removing metals are acceptable.

1507.6.3 Smoking. "No Smoking" signs complying with Section 310 shall be conspicuously posted at all powder coating areas and powder storage rooms.

❖ Smoking is prohibited at powder coating areas and in powder storage rooms. "No smoking" signs should be conspicuously located throughout the work area. Designated smoking areas should be located well outside the powder coating area and storage rooms, preferably in a separate room. The requirements of Section 2703.7.1 must be met where flammable finish processes occur (see also commentary, Section 310).

1507.7 Fixed electrostatic-spraying equipment. In addition to Section 1507, Section 1506 shall apply to fixed electrostatic equipment used in powder coating operations.

❖ See the commentary to Section 1506.

1507.8 Fire protection. Areas used for powder coating shall be protected by an approved automatic fire-extinguishing system complying with Chapter 9.

❖ Although automatic sprinkler protection is the most common method of protection in facilities, other approved automatic fire suppression systems may be installed where approved by the fire code official. Fire protection systems, equipment and devices must be installed in accordance with Chapter 9 and maintained.

1507.9 Additional protection for fixed systems. Automated powder application equipment shall be protected by the installation of an approved, supervised flame detection apparatus that shall react to the presence of flame within 0.5 second and shall accomplish all of the following:

1. Shutting down of energy supplies (electrical and compressed air) to conveyor, ventilation, application, transfer and powder collection equipment.
2. Closing of segregation dampers in associated ductwork to interrupt airflows from application equipment to powder collectors.
3. Activation of an alarm that is audible throughout the powder coating room or booth.

❖ Automated powder application may or may not have operators who would take steps to reduce the severity of the fire by shutting down the supply of powder, turning off energy supplies, warning occupants for evacuation, etc. The flame detection system and its associated interlocks are part of the protection system, in case ignition has already occurred.

To reduce the severity of a fire or dust cloud explosion, the powder coating application must be shut down immediately upon fire detection. Because automated powder application equipment may at times be unsupervised, flame detection that automatically can shut down the system is required. Shutting down systems such as conveyors that can further spread the fire or an application that further augments the existing fuel load/air-borne dust is critical to reducing the severity of a fire. Furthermore, an audible alarm throughout the powder coating room or area must notify the occupants to evacuate immediately to avoid injury. For example, injury to occupants of the booth may be avoided or reduced if they evacuate when a small incipient fire is detected before it reaches areas within the booth where dust cloud explosions are sustainable.

1507.10 Fire extinguishers. Portable fire extinguishers complying with Section 906 shall be provided for areas used for powder coating in accordance with the requirements for extra hazard occupancy.

❖ Section 906 gives the requirements for portable fire extinguishers. Areas may be classified as an extra-hazard occupancy in accordance with NFPA 10 because of the higher hazard of powder coating operations.

SECTION 1508
AUTOMOBILE UNDERCOATING

1508.1 General. Automobile undercoating spray operations conducted in areas with approved natural or mechanical ventilation shall be exempt from the provisions of Section 1504 when approved and where utilizing Class IIIA or IIIB combustible liquids.

❖ Most automobile undercoating and corrosion inhibitors use combustible liquids with flashpoints greater than 140°F (60°C). Precautions must be taken to prevent these materials from being heated above their flash points or ignited in finely divided or atomized form. Ventilation to reduce the accumulation of hazardous vapors and mists must comply with the requirements of the IMC and Chapter 34 of the code.

SECTION 1509
ORGANIC PEROXIDES AND DUAL-COMPONENT COATINGS

1509.1 Contamination prevention. Organic peroxide initiators shall not be contaminated with foreign substances.

❖ See the commentary to Section 1509.4.

1509.2 Equipment. Spray guns and related handling equipment used with organic peroxides shall be of a type manufactured for such use.

❖ This process involves the discharging of two different components through the same spray gun. The sensitivity of the materials used results in a system that is extremely sensitive to shock, friction, temperature, contaminants, etc. The design and installation of the system cannot be taken lightly.

1509.3 Pressure tanks. Separate pressure vessels and inserts specifically for the application shall be used for the resin and for the organic peroxide, and shall not be interchanged. Organic peroxide pressure tank inserts shall be constructed of stainless steel or polyethylene.

❖ Special consideration must be given to the materials used within organic peroxide systems. Materials that do not react with organic peroxides, such as stainless steel or polyethylene, ensure system component integrity.

1509.4 Residue control. Materials shall not be contaminated by dusts and overspray residues resulting from the sanding or spraying of finishing materials containing organic peroxides.

❖ Because of the hazards associated with organic peroxides and the chemical reactions within the system, issues such as contamination control are very important. To avoid a violent reaction with contaminants that may be accidentally introduced to the organic peroxide initiators, careful consideration must be given to the overall system design.

1509.5 Spilled material. Spilled organic peroxides shall be promptly removed so there are no residues. Spilled material absorbed by using a noncombustible absorbent shall be promptly disposed of in accordance with the manufacturer's recommendation.

❖ To avoid contact with materials and conditions that would cause a reaction, it is good housekeeping practice to promptly clean the spilled organic peroxide. Additionally, to reduce the possibility of a reaction, prompt disposal is recommended. Because the organic peroxides vary in nature and volatility, the manufacturer's recommendations or Material Safety Data Sheets (MSDS) must be followed for disposal.

1509.6 Use of organic peroxide coatings. Spraying operations involving the use of organic peroxides and other dual-component coatings shall be conducted in approved sprinklered spray booths complying with Section 1504.1.2.

❖ Because of the reactivity of the material, the use of organic peroxide and other dual-component coating systems is limited to spray booths only. For example, organic peroxide coatings cannot be applied in the limited spraying spaces described in Section 1504.1.4. This section is more specific and, therefore, supersedes the more general requirements. Storage of organic peroxides shall comply with Chapter 39.

1509.7 Storage. The storage of organic peroxides shall comply with Chapter 39.

❖ Chapter 39 is specific to the storage and use of organic peroxides (see commentary, Chapter 39). The manufacturer's information on handling and care (typically "Special Precautions" and "Reactivity" sections of MSDS) should be referred to for proper storage of the product.

1509.8 Handling. Handling of organic peroxides shall be conducted in a manner that avoids shock and friction that produces decomposition and violent reaction hazards.

❖ Many organic peroxides are unstable reactive materials that are decomposed by heat, shock or friction. The rate of decomposition varies depending on the material and the condition it is exposed to. Organic peroxides vary in reactions from plain decomposition (without a fire or explosion hazard) to deflagration or detonation. Special handling care must be given to these materials. The manufacturer's information on handling and care (typically "Special Precautions" and "Reactivity" sections of MSDS) should be referred to for proper storage of the product.

1509.9 Mixing. Organic peroxides shall not be mixed directly with accelerators or promoters.

❖ To avoid any unexpected reaction, organic peroxides should not be mixed directly. Dual-coating systems specifically designed and approved for such use must be used.

1509.10 Sources of ignition. Smoking shall be prohibited and "No Smoking" signs complying with Section 310 shall be prominently displayed. Only nonsparking tools shall be used in areas where organic peroxides are stored, mixed or applied.

❖ Many organic peroxides are sensitive to heat, shock and friction. Additionally, many of these materials are dissolved in flammable or combustible solvents. Because of the volatility of the material, sources of ignition must be avoided completely. As mentioned in other sections of this chapter, the term "nonsparking" is somewhat inaccurate. Tools made of brass and similar nonsparking materials produce sparks with ignition energies too low to ignite flammable vapors. Nonetheless, such tools should be used carefully to avoid producing sufficient frictional heat to cause an ignition.

Smoking is prohibited in organic peroxide coating areas. "No smoking" signs should be conspicuously located throughout the work area. Designated smoking areas should be located well outside the organic peroxide and dual-coating areas and storage rooms, preferably in a separate room (see also commentary, Section 310).

1509.11 Personnel qualifications. Personnel working with organic peroxides and dual-component coatings shall be specifically trained to work with these materials.

❖ Because of the sensitive nature of organic peroxides, only trained personnel should work with them. Qualified personnel should be trained in material handling to avoid accidents, as well as procedures (such as system shutdown, notification, evacuation, etc.) in case of an accident. Qualified personnel should also be familiar with the MSDS of the flammable finishes.

SECTION 1510
FLOOR SURFACING AND FINISHING OPERATIONS

1510.1 Scope. Floor surfacing and finishing operations exceeding 350 square feet (33 m^2) and using Class I or Class II liquids shall comply with Sections 1510.2 through 1510.5.

❖ Floor surfacing and refinishing using Class I or II liquids in facilities such as bowling alleys pose a high risk because of the spread of liquid over large surface areas. The intent of this section is to provide additional protection and hazard reduction for large surface fires resulting from floor surfacing and finishing.

1510.2 Business operation. Floor surfacing and finishing operations shall not be conducted while an establishment is open to the public.

❖ In case of a fire, the types and numbers of injuries are reduced by allowing surfacing and finishing operations only when the business is not open to the public. This would limit the exposure to fewer people (nighttime employees, etc.).

1510.3 Ventilation. To prevent the accumulation of flammable vapors, mechanical ventilation at a minimum rate of 1 cubic foot per minute per square foot [0.00508 $m^3/(s \times m^2)$] of area being finished shall be provided. Such exhaust shall be by approved temporary or portable means. Vapors shall be exhausted to the exterior of the building.

❖ The air exhausted over the entire surface must be at minimum 1 cfm per square foot [0.0058 $m^3/(s \cdot m^2)$], even in the most remote areas or corners where finishing and surfacing takes place. Please note that floor surfacing and finishing do not occur often in a building. Because outside companies are typically hired to do such surfacing, protection systems such as exhaust are temporary and can therefore, be portable. To prevent recirculation of exhaust air that may contain flammable vapors, the exhausted air must be discharged to the exterior of the building.

1510.4 Mechanical system operation. Heating, ventilation and air-conditioning systems shall not be operated during resurfacing or refinishing operations or within 4 hours of the application of flammable or combustible liquids.

❖ To avoid circulation of potentially flammable vapors into other parts of the building, mechanical systems that are not part of the exhaust system required in Section 1510.3 must be shut down during the application of flammable or combustible liquids and for 4 hours after the last of the flammable or combustible liquids are applied. For example, if a flammable surface is applied and the entire area application takes 2 hours, the number of hours that the mechanical systems must be shut down is 6 (4 hours after the application).

1510.5 Ignition sources. The power to all electrical devices shall be shut down to all electrical sources of ignition within the vapor area, unless those devices are classified for use in Class I, Division 1 hazardous locations.

❖ Because floor surfacing and finishing is a temporary process, the protection system is also temporary. Therefore, unlike processes such as spray booths, the building or area is not required to meet electrical requirements for hazardous areas. These systems that cannot meet the hazard conditions at the time of surfacing and finishing must be shut down to reduce the potential for unrated electrical sources to cause ignition of vapors.

SECTION 1511
INDOOR MANUFACTURING OF REINFORCED PLASTICS

1511.1 General. Indoor manufacturing processes involving spray or hand application of reinforced plastics and using more than 5 gallons (19 L) of resin in a 24-hour period shall be in accordance with this section.

❖ The applicability of this section is limited to operations involving the use of more than 5 gallons (19 L) of resin in a 24-hour period. It is intended to allow very limited amounts of resin to be removed from storage containers and applied in the manufacturing process during that period without further regulation other than those applicable to any hazardous material.

1511.2 Resin application equipment. Equipment used for spray application of resin shall be installed and used in accordance with Sections 1509 and 1511.

❖ The equipment used for resin application should be manufactured for that use. Listed equipment is recommended.

1511.3 Fire protection. Resin application areas shall be protected by an automatic sprinkler system. The sprinkler system design shall not be less than that required for Ordinary Hazard, Group 2, with a minimum design area of 3,000 square feet (279 m^2). Where the materials or storage arrangements are required by other regulations to be provided with a higher level of sprinkler system protection, the higher level of sprinkler system protection shall be provided.

❖ Similar language and requirements can be found in Sections 2705.1.8 and 2704.5 when storage and use or dispensing of hazardous materials exceed the allowable quantities per control area given in Section 2703. If design density required by other sections results in a higher sprinkler demand, the more stringent requirement applies.

1511.4 Sources of ignition in resin application areas. Sources of ignition in resin application areas shall comply with Section 1503.2.

❖ As with other flammable finishes, the sources of ignition must be controlled to a manageable level in hazardous environments.

1511.5 Ventilation. Mechanical ventilation shall be provided throughout resin application areas in accordance with Sections 1504.2 and 1504.3. The ventilation rate shall be adequate to maintain the concentration of flammable vapors in the resin application area at or below 25 percent of the lower flammable limit (LFL).

> **Exception:** Mechanical ventilation is not required for buildings that are unenclosed for at least 75 percent of the perimeter.

❖ In many cases, acetone is used for cleanup of resin accumulations. Therefore, the LFL for both resin and the cleaning solvent must be considered when designing the ventilation for the resin application areas.

The exception recognizes that natural ventilation is considered adequate if the perimeter of the building is 75 percent open.

1511.5.1 Local ventilation. Local ventilation shall be provided inside of workpieces where personnel will be under or inside of the workpiece.

❖ Ventilation in areas where personnel are present is required for fire prevention as well as for reduction of health risks.

1511.6 Storage and use of hazardous materials. Storage and use of organic peroxides shall be in accordance with Section 1509 and Chapter 39. Storage and use of flammable and combustible liquids shall be in accordance with Chapter 34. Storage and use of unstable (reactive) materials shall be in accordance with Chapter 43.

❖ Because a variety of materials are used in the GFRP process, chapters specific to the materials are referenced. In addition to the referenced chapters, the MSDS, as well as the manufacturer's recommendation for safe practices, should be considered. Although the codes contain information for the general hazardous materials category, the MSDS are specific to the material being used. For example, the reactivity section of a MSDS is specific in the types of materials that may be incompatible with the resin used.

1511.7 Handling of excess catalyzed resin. A noncombustible, open-top container shall be provided for disposal of excess catalyzed resin. Excess catalyzed resin shall be drained into the container while still in the liquid state. Enough water shall be provided in the container to maintain a minimum 2-inch (51 mm) water layer over contained resin.

❖ Some catalyzed resin products produce heat as they cure (exothermic reaction). This may be a fire hazard if the container is not open-top. In a confined container, enough heat may be generated to ignite the resin. Additionally, the required water in the container would help cool down the material and the container until the product is cured or safely disposed of.

1511.8 Control of overchop. In areas where chopper guns are used, exposed wall and floor surfaces shall be covered with paper, polyethylene film, or other approved material to allow for removal of overchop. Overchop shall be allowed to cure for not less than 4 hours prior to removal.

❖ To ease removal of excess materials, coverings are required in the areas where chopper guns are used. Again, the materials are best left until they are cured to reduce the likelihood of any reactions or ignition from the heat produced during curing (exothermic reaction).

1511.8.1 Disposal. Following removal, used wall and floor covering materials required by Section 1511.8 shall be placed in a noncombustible container and removed from the facility.

❖ Once the material is cured for 4 hours, immediate disposal is important to reduce further chances of reaction. Additional heat may be produced if the materials are not fully cured. Regular disposal of hazardous materials is an important part of housekeeping in fire prevention.

Bibliography

The following resource materials are referenced in this chapter or are relevant to the subject matter addressed in this chapter.

Fire Protection Handbook, 18th edition. A.E. Cote, ed. Quincy, MA: National Fire Protection Association, 1997.

IBC-2003, *International Building Code*. Falls Church, VA: International Code Council, 2003.

ICC EC-2003, *ICC Electrical Code*. Falls Church VA: International Code Council, 2003.

IMC-2003, *International Mechanical Code*. Falls Church VA: International Code Council, 2003.

NFPA 10-98, *Portable Fire Extinguishers*. Quincy, MA: National Fire Protection Association, 1998.

NFPA 11—98, *Low Expansion Foam*. Quincy, MA: National Fire Protection Association, 1998.

NFPA 11A—99, *Medium- and High-Expansion Foam Systems*. Quincy, MA: National Fire Protection Association, 1999.

NFPA 12—00, *Carbon Dioxide Extinguishing Systems*. Quincy, MA: National Fire Protection Association, 2000.

NFPA 12A—97, *Halon 1301 Fire Extinguishing Systems*. Quincy, MA: National Fire Protection Association, 1997.

NFPA 15—96, *Water Spray Fixed Systems for Fire Protection*. Quincy, MA: National Fire Protection Association, 1996.

NFPA 16—99, *Installation of Foam-Water Sprinkler and Foam-Water Spray Systems*. Quincy, MA: National Fire Protection Association, 1999

NFPA 17—98, *Dry Chemical Extinguishing Systems*. Quincy, MA: National Fire Protection Association, 1998.

NFPA 33-00, *Standard for Spray Application Using Flammable or Combustible Materials*. Quincy, MA: National Fire Protection Association, 2003.

NFPA 34-00, *Standard for Dipping and Coating Processes Using Flammable or Combustible Materials*.

Quincy, MA: National Fire Protection Association, 2003.

NFPA 70-02, *National Electrical Code*. Quincy, MA: National Fire Protection Association, 2002.

Ontario Fire Code Section 5.14 Special Processes Involving Flammable & Combustible Liquids: Illustrated Commentary. Ontario, CA: Office of the Ontario Fire Marshal.

Ostrowski, R. "Oil Quenching and Molten Salt Baths." *Fire Protection Handbook*, 18th ed. Quincy, MA: National Fire Protection Association, 1997.

Scarborough, D.R. "Spray Finishing and Powder Coating." *Industrial Fire Hazard Handbook*, 3rd ed. Quincy, MA: National Fire Protection Association, 1990.

Scarborough, D.R. "Spray Finishing and Powder Coating." *Fire Protection Handbook*, 18th ed. Quincy, MA: National Fire Protection Association, 1997.

Severson, Roger. *Application Guide to Flammable Finishes*. Whittier, CA: International Conference of Building Officials, 2000.

SFPE Handbook of Fire Protection Engineering, 3rd Edition. Quincy, MA: National Fire Protection Association, 2002.

Sheppard, J.W. "Dipping and Coating Processes." *Fire Protection Handbook*, 18th ed. Quincy, MA: National Fire Protection Association, 1997.

Talbot, N.L. and P.H. Dobson. "Dipping and Coating Processes." *Industrial Fire Hazard Handbook*, 3rd ed. Quincy, MA: National Fire Protection Association, 1990.

UL 32-94, *Metal Waste Cans.* Northbrook, IL: Underwriters Laboratories Inc., 1994.

UL 900-94, *Standard for Safety for Air Filter Units.* Northbrook, IL: Underwriters Laboratories Inc., 1995.

Chapter 16: Fruit and Crop Ripening

General Comments

American consumers have become accustomed to having a wide variety of the ripest and most attractive fruits and vegetables available all year. To supply this demand, horticulturists, growers and distributors have devised a means of growing and transporting these commodities to minimize damage and spoilage. Two fruits in greatest demand, tomatoes and bananas, are now shipped hard and green off the vine or tree, minimizing bruising but contributing to poor sales. To make the fruit attractive and edible, it must be ripened. Ethylene, a naturally occurring hormone in many fruit-bearing plants, is used to complete the ripening process after the fruit arrives at the distributor. This material has a wide explosive range and, when inhaled, is a medical anesthetic. Though only small concentrations of the gas are required in the ripening process, explosions have occurred. These incidents are usually attributed to the use of excessive concentrations of gas well above the lower flammable limit (LFL). Placed in a relatively vapor-tight room filled with ethylene gas at small concentrations—100 to 150 parts per million (ppm) (100 to 500 mg/L)—for varying durations, quantities of fruit are ripened slowly, bringing them to just the right point before transporting them to the local retail market.

Purpose

Chapter 16 provides guidance that is intended to reduce the likelihood of explosions resulting from improper use or handling of ethylene gas used for crop-ripening and coloring processes. This is accomplished by regulating ethylene gas generation; storage and distribution systems and controlling ignition sources. Design and construction of facilities for this use are regulated by the *International Building Code®* (IBC®) to reduce the impact of potential accidents on people and buildings.

SECTION 1601 GENERAL

1601.1 Scope. Ripening processes where ethylene gas is introduced into a room to promote the ripening of fruits, vegetables and other crops shall comply with this chapter.

> **Exception:** Mixtures of ethylene and one or more inert gases in concentrations which prevent the gas from reaching greater than 25 percent of the lower explosive limit (LEL) when released to the atmosphere.

❖ This section establishes that this chapter is applicable to fruit- and crop-ripening processes that use ethylene gas as a ripening agent. While still green, many fruits and vegetables are picked and shipped to their point of distribution. Prior to shipping or upon arrival at the distribution warehouse near the retail market, fruits and vegetables are transferred to containers (trailers) or rooms filled with low concentrations of ethylene gas to facilitate ripening. As they ripen, some fruits and vegetables, including bananas, tomatoes, pears, apples and honeydew melons, produce ethylene gas. If kept in tightly enclosed rooms or containers, gas accumulates, expediting the ripening process as concentrations increase.

This process would be relatively safe, simple and effective except for one problem — ethylene is a highly flammable gas (see Table 1601.1). A constituent of liquefied petroleum gas, ethylene is colorless with a sweet odor and taste. It is explosive in sunlight when mixed with chlorine. Moderate concentrations of ethylene in air are a medical anesthetic; thus, care must be taken to prevent inhalation. Atmospheres containing ethylene are classified as Class I, Division 1, Group C hazardous locations by the ICC *Electrical Code®* (ICC EC™). Additionally, ethylene (R-1150) is sometimes used as a refrigerant gas and classified among Group 3 (highly flammable) refrigerants. Explosions involving trucks and warehouses where ethylene gas is used as a fruit-ripening agent are usually attributed to the use of excessive concentrations of gas well above the lower explosive limit (LEL).

The exception recognizes the reduced hazard of ethylene-inert gas mixtures when the mixtures keep the ethylene concentration at or below 25 percent of its LEL/LFL.

**TABLE 1601.1
FLAMMABILITY CHARACTERISTICS
AND ETHYLENE HAZARDS**

Ignition Temperature[a]	842°F to 914°F
Vapor Density	1.0
Flammable Range Lower Upper	 2.7 percent 36 percent
NFPA 704 Hazard Classification Health Flammability Reactivity Other	 1 4 2 —

For SI: °C = (°F-32)/1.8.

Note a. In the *NFPA Fire Protection Guide to Hazardous Materials, 13th edition*, NFPA 325 reports the lower value, while NFPA 49 reports the higher value.

1601.2 Permits. Permits shall be required as set forth in Section 105.6.

❖ The process of issuing permits gives the fire code official an opportunity to carefully evaluate and regulate hazardous operations. Permit applicants should be required to demonstrate that their operations comply with the intent of the code before the permit is issued. Because of the extremely flammable nature of ethylene gas, an operational permit is required by Section 105.6.19 of the code for ethylene-based ripening processes.

1601.3 Ethylene generators. Approved ethylene generators shall be operated and maintained in accordance with Section 1606.

❖ Chapter 16 recognizes the use of listed ethylene generators as a means of safely producing ethylene concentrations needed to ripen crops. This section requires that those devices be properly operated and maintained (see commentary, Section 1606.1).

SECTION 1602
DEFINITIONS

1602.1 Terms defined in Chapter 2. Words and terms used in this chapter and defined in Chapter 2 shall have the meanings ascribed to them as defined therein.

❖ Words and terms used in this chapter and defined in Chapter 2 shall have the meanings ascribed to them as defined therein.

This section directs the code user to Chapter 2 for the proper application of the terms used in this chapter. Such terms may be defined in Chapter 2, in another *International Code*® as indicated in Section 201.3 or the use of their ordinary (dictionary) meaning may be all that is needed (see also commentary, Sections 201.1 through 201.4).

SECTION 1603
ETHYLENE GAS

1603.1 Location. Ethylene gas shall be discharged only into approved rooms or enclosures designed and constructed for this purpose.

❖ Rooms or spaces occupied for the ripening of fruits and crops using ethylene gas must be constructed to accommodate the process without exposing the rest of the building to process hazards. The design and construction of facilities for this use are regulated by the applicable provisions of the IBC to reduce the impact of potential accidents on people and buildings. In particular, buildings and portions of buildings used for crop-ripening or coloring processes must conform to the requirements of Section 417 of the IBC.

Because ethylene is a flammable gas, the occupancy group classification of the building depends on the quantities of gas present and the use of the control area concept (see Section 414.2 of the IBC). If the amount of gas exceeds the maximum allowable quantity per control area [see Table 307.7(1) of the IBC], the applicable provisions of Section 415 of the IBC would also apply. Additionally, Chapter 30 of the code regulates the physical hazards of compressed gases and Chapter 35 regulates the material hazards of flammable gases. Both would apply to ethylene used in fruit-ripening processes as would the general hazardous materials provisions of Chapter 27.

1603.2 Dispensing. Valves controlling discharge of ethylene shall provide positive and fail-closed control of flow and shall be set to limit the concentration of gas in air below 1,000 parts per million (ppm).

❖ Though the hazards of using ethylene should be of considerable concern, technological innovation has produced a safe alternative to the old method of distributing ethylene to the ripening rooms. In the past, compressed ethylene gas in cylinders was piped through regulating equipment into rooms. Regulating equipment was calibrated to restrict gas quantity to a volume that would produce concentrations below the lower explosive gas limit. However, excessive concentrations occasionally escaped and uncontrolled ignition sources produced serious explosions.

Ideally, gas sensors should be used to stop gas flow as it approaches the desired level. Manual overrides must be provided to permit the gas flow to be interrupted if a danger exists. A 1:1000 concentration is only 0.1 percent in air. The LFL of ethylene is 2.6 percent. When used properly, this gas poses little hazard.

SECTION 1604
SOURCES OF IGNITION

1604.1 Ignition prevention. Sources of ignition shall be controlled or protected in accordance with this section and Chapter 3.

❖ Lighted smoking materials and other open-flame sources are prohibited in crop-ripening and color processing rooms. This prohibition should also be extended to the immediate vicinity of the entrance to such rooms. "No smoking" signs are to be conspicuously posted and stringently enforced (see also commentary, Sections 305, 308 and 310).

1604.2 Electrical wiring and equipment. Electrical wiring and equipment, including lighting fixtures, shall be approved for use in Class I, Division 2, Group C hazardous (classified) locations.

❖ Atmospheres containing ethylene are classified as Class I, Division 1, Group C hazardous locations by the ICC EC. All electrical wiring and equipment must be suitably listed and labeled for use in such atmospheres.

1604.3 Static electricity. Containers, piping and equipment used to dispense ethylene shall be bonded and grounded to prevent the discharge of static sparks or arcs.

❖ Suitable grounds must be provided on all ethylene handling systems and equipment to permit the dissipation of static electricity, thus preventing the generation of static sparks. Electrically isolated sections of systems and equipment must be independently grounded. NFPA 77 provides valuable guidance on this topic. Grounding and bonding should be in accordance with the ICC EC.

1604.4 Lighting. Lighting shall be by approved electric lamps or fixtures only.

❖ Open-flame lighting is prohibited in crop-ripening rooms and spaces because ethylene gas is flammable. Only approved electrical lighting equipment is allowed and it must comply with the provisions of Section 1604.2 of the code (see commentary, Section 1604.2).

1604.5 Heating. Heating shall be by indirect means utilizing low-pressure steam, hot water, or warm air.

> **Exception:** Electric or fuel-fired heaters approved for use in hazardous (classified) locations which are installed and operated in accordance with the applicable provisions of the ICC *Electrical Code*, the *International Mechanical Code* or the *International Fuel Gas Code*.

❖ Wherever practical, heat for crop-ripening and coloring rooms should be indirect (steam, hot water or warm air) to eliminate sparks or open-flame ignition hazards. Piped steam and hot-water heating systems operate at temperatures of 250°F to 430°F (121°C to 221°C) and are considered the most appropriate method when providing indirect heat to ripening rooms.

The exception provides that electric heaters may be used when approved by the fire code official. The type of electric heater required must be approved and listed for use in Class I, Division I, Group C electrically classified locations because of the presence of ethylene gas. Such heaters must also be designed to keep all exposed surfaces at temperatures of 800°F (427°C) or below because ethylene gas has an ignition temperature of 842°F (450°C). Fuel-fired heaters using gas or liquid fuel may also be used, provided they are approved and comply with the *International Mechanical Code®* (IMC®) or the *International Fuel Gas Code®* (IFGC®), as applicable. These appliances must also have a sealed combustion chamber to prevent open-flame ignition of the ethylene gas and be installed in accordance with the manufacturer's instructions. All heaters must be protected from physical damage to prevent fuel spills or heater malfunctions.

SECTION 1605
COMBUSTIBLE WASTE

1605.1 Housekeeping. Empty boxes, cartons, pallets and other combustible waste shall be removed from ripening rooms or enclosures and disposed of at regular intervals in accordance with Chapter 3.

❖ Waste accumulation, especially packing materials and crop debris, is to be removed regularly from the premises and disposed of in an approved manner to reduce the volume of nonessential combustible materials susceptible to ignition and contributing to the fire load. Section 304 of the code provides specific regulations for waste disposal (see commentary, Section 304).

SECTION 1606
ETHYLENE GENERATORS

1606.1 Ethylene generators. Ethylene generators shall be listed and labeled by an approved testing laboratory, approved by the fire code official and used only in approved rooms in accordance with the ethylene generator manufacturer's instructions. The listing evaluation shall include documentation that the concentration of ethylene gas does not exceed 25 percent of the lower explosive limit (LEL).

❖ Several companies produce ethylene generators especially for fruit-ripening processes. These devices convert a liquid similar to ethanol into ethylene gas. This process produces ethylene in concentrations far below the LEL, even in very small compartments, an important feature since these generators are portable and could be operating in ripening rooms of different sizes. During a test conducted by an independent testing laboratory and witnessed by representatives of the Los Angeles City Fire Marshal's office, a small [4,000 cubic foot (113 m^3)] fruit-ripening room was filled with 400 cases of green bananas with the generator set at the maximum setting. During the 8-hour test, the device did not produce a concentration greater than 0.1 percent by volume. The LEL for ethylene is 2.6 percent by volume.

While this may not be a worst-case scenario, it certainly suggests a significant safety margin—a factor of 26.

This section mandates that ethylene generators be listed and labeled by a third-party testing laboratory. The basic standard used in the investigation and listing of these devices is UL 499, which, while it covers the most significant electrical hazards of ethylene generators, does not evaluate the flammable vapor-air mixtures produced when the devices are operated in closed rooms. This section intends to fill that gap by requiring that the listing evaluation attests to the safety of the devices by documenting in the report results that the maximum ethylene concentration capable of being generated by the device does not exceed 25 percent of the LEL. This section also requires adherence to the ethylene generator manufacturer's installation instructions, which are considered part of the listing.

1606.2 Ethylene generator rooms. Ethylene generators shall be used in rooms having a volume of not less than 1,000 cubic feet (28 m^3). Rooms shall have air circulation to ensure even distribution of ethylene gas and shall be free from sparks, open flames or other ignition sources.

❖ Because ethylene generators are portable, the level of hazard associated with their use in closed rooms can change each time one is moved to a new ripening room. The intent of this section is to reduce the hazard potential by requiring that ethylene generators be used only in ripening rooms having a minimum volume of 1,000 cubic feet (28 m^3) [10 feet x 10 feet x 10 feet (3048 mm x 3048 mm x 30488 mm)], that are free from ignition hazards and have air circulation to distribute the ethylene gas evenly throughout the space to avoid the potential for pockets of gas having concentrations approaching or exceeding 25 percent of the LEL.

SECTION 1607 WARNING SIGNS

1607.1 When required. Approved warning signs indicating the danger involved and necessary precautions shall be posted on all doors and entrances to the premises.

❖ Given the flammability and the anesthetic quality of ethylene gas, it poses potential dangers to personnel who might accidentally enter a ripening room, building or area unaware of the hazards. As with other hazardous materials, warning signs and placards must be posted to give appropriate notice to warn personnel of the hazards of the ripening process and discourage entry by unauthorized and unprotected people. Warning signs must comply with Sections 2703.5 and 2703.5.1 of the code and include NFPA 704 hazard warning system markings to warn of fire- and health-related hazards of these materials (see commentary, Sections 2703.5 and 2703.5.1).

Bibliography

The following resource materials are referenced in this chapter or are relevant to the subject matter addressed in this chapter.

"Catalytic Generators." *Easy-Ripe® Generator Brochure and Operating Instructions*. Norfolk, VA: Catalytic Generators, Inc., 1990.

"Catalytic Generators." *Material Safety Data Sheet for Ripe-Rite® Liquid*. Norfolk, VA: Catalytic Generators, Inc.

Catalytic Generators, Inc. — Report of Test #177373. Los Angeles, CA: U.S. Testing Company.

"Ethylene Gas Generator-Fruit Ripening." *Field Service Bulletin F-2-81*. Long Grove, IL: Kemper Group of Insurance Companies, 1981.

Fire Protection Guide to Hazardous Materials, 13th edition. Quincy, MA: National Fire Protection Association, 2001.

IBC-2003, *International Building Code*. Falls Church, VA: International Code Council, 2003.

ICC EC-2003, *ICC Electrical Code*. Falls Church, VA: International Code Council, 2003.

IFGC-2003, *International Fuel Gas Code*. Falls Church, VA: International Code Council, 2003.

IMC-2003, *International Mechanical Code*. Falls Church, VA: International Code Council, 2003.

Industrial Fire Hazards Handbook, 3rd ed. Quincy, MA: National Fire Protection Association, 1990.

NFPA 77-00, *Recommended Practice on Static Electricity*. Quincy, MA: National Fire Protection Association, 2000.

NFPA 704-96, *Identification of the Hazards of Materials for Emergency Response*. Quincy, MA: National Fire Protection Association, 1996.

UL 499-97, *Electric Heating Appliances*. Northbrook, IL: Underwriters Laboratories, Inc, 1997.

Chapter 17:
Fumigation And Thermal Insecticidal Fogging

General Comments

Fumigation is the use of toxic pesticide chemicals to kill insects, rodents and other vermin. In addition, agricultural fumigation is used to kill plant and animal parasites, weed seeds and various types of fungi that adversely affect agricultural products.

Fumigants are available as liquids that will vaporize readily at ambient temperatures, solids that can release a toxic gas on reacting with water or acid, or gases. Fumigants and thermal insecticidal fogging agents pose little hazard if properly applied; however, the inherent toxicity of all these agents and the potential flammability of some makes special precautions necessary when they are used. Requirements of this chapter are intended to protect both the public and fire fighters from hazards associated with these products.

The use of fumigants poses the following two distinct hazards to both fire fighters and the general public: some fumigants are flammable or burn under certain circumstances and all fumigants are poisonous or toxic. Compounding these hazards, the fumigation fog may easily be confused with smoke from a fire if it is not properly contained within a building or compartment. Though these concerns seem quite serious, the proper use of fumigants poses little fire hazard. With the exception of allyl alcohol, aluminum phosphide, dichloropropane-dichloropropene mixtures and formaldehyde, most fumigants are relatively difficult to ignite. However, all fumigants are intended to kill something, whether involved in a fire or not. In fact, many of these agents are quite toxic if involved in a fire.

Special Fire Protection Problems

In particular, aluminum phosphide poses a special fire protection problem. This agent readily decomposes in water to form phosphine (PH_3), a pyrophoric toxic gas. In 1990, M.R. Spencer noted that although the lower flammable limit (LFL) of phosphine is only 1.79 percent—this is more than 10 times the effective concentration for fumigation. However, he also noted that poorly distributed agent and storage canisters can easily produce ignitable concentrations.

The other agents—allyl alcohol, dichloropropane-dichloropropene mixtures and formaldehyde—are either flammable or combustible liquids commonly used in diluted form. However, all flammable or combustible liquids are easier to ignite when finely dispersed in air; therefore, special precautions must be taken to eliminate ignition sources during fumigation and fogging operations. Table 17-1 lists hazard classifications of various common fumigants.

Information Sources

Understanding and regulating fumigants and thermal insecticidal fogging agents requires specialized knowledge of their hazards beyond the basic classifications of health, flammability and reactivity. Several references provide additional details necessary to apply code requirements. The most helpful of these references is the product's Material Safety Data Sheet (MSDS). A copy of this document is often shipped with a product in transit (see commentary, Section 407 and Chapter 27). Anyone handling or using these products should also have access to this information in compliance with the provisions of the hazard communication requirements of the Occupational Safety and Health Act (OSH Act).

Reference books, such as the *Farm Chemicals Handbook* and the *Crop Protection Chemical Reference*, are also comprehensive sources for handling and storage information. The former contains a list of common agricultural chemicals based on their use, a list of suppliers or distributors and a chemical dictionary containing each product listed.

Every manufacturer of a pesticide or fumigant also produces a reference known as a label book containing facsimiles of the warning labels for each of its products. This reference is often helpful because Environmental Protection Agency (EPA) regulations require that labels contain breakdowns of the contents of environmentally hazardous substances and signal words providing clues about their toxicity.

Storage and Handling

Other code chapters should be consulted for specific storage and handling requirements for certain fumigants. Chapters 27, 34, 37 and 40 prescribe safeguards for the storage and handling of materials based on their specific hazardous properties. Requirements of this chapter apply only to the use of these products as fumigants or thermal insecticidal fogging agents.

Purpose

Some of these products are also flammable and their involvement in fire may result in serious health or environmental hazards. The key to the safe use of these products is knowledge of the hazards, elimination of ignition sources and isolation of the premises during application.

TABLE 17-1
HAZARDOUS PROPERTIES OF COMMON FUMIGANTS[a,b]

PRODUCT	DOT CLASSIFICATION	NFPA 704 HAZARD CLASSIFICATION			
		Health	Flammability	Reactivity	Other
Allyl Alcohol	Flammable Liquid, Poisonous	3	3	0	
Aluminum Phosphide	Flammable Solid, Poisonous, Water Reactive	4	4	2	~~W~~
Calcium Cyanide	Poison B	3	0	0	
Carbon Tetrachloride	ORM-A	3	0	0	
Chloropicrin	Poison B	4	0	3	
Paradichlorobenzene	ORM-A	2	2	0	
Dichloropropane-Dichloropropene Mixtures	Flammable Liquid	3	3	0	
Ethylene Dibromide (EDB)	ORM-A	3	0	0	
EDB/Chloropicrin mixtures		4	0	3	
Formaldehyde	Combustible Liquid	2	4	0	
Furfural	Combustible Liquid	2	2	0	
Methyl Bromide	Poison B	3	1	0	
Methyl Bromide/Chloropicrin mixtures	Poison B	4	1	3	
Methyl Bromide/EDB mixtures	Poison B	3	1	0	
Methylene Chloride		2	2	0	
Napthalene	ORM-B	2	2	0	
Phosphine	Poison A, Flammable	4	4	4	

Note a. This list was compiled from chemicals listed under the heading "Fumigants" in the *Farm Chemicals Handbook*.

Note b. Other common fumigants include: streptomycin, dibromo-chloropropane, metam ammonium and metan sodium [consult Material Safety Data Sheets (MSDS) or these products before issuing permits.]

SECTION 1701 GENERAL

1701.1 Scope. Fumigation and thermal insecticidal fogging operations within structures shall comply with this chapter.

❖ The requirements of this chapter govern fumigation operations using thermal insecticidal fogging agents and other airborne pesticides used to control insects, rodents, vermin and other similar pests inside structures.

1701.2 Permits. Permits shall be required as set forth in Section 105.6.

❖ Issuing an operational permit as prescribed in Section 105.6.20 gives the fire code official a method for identifying and controlling the hazards of fumigation and thermal insecticidal fogging operations. The process of issuing permits also gives the fire code official a reason or method to advise fire fighters of the potential hazards at fumigation sites. This will allow fire fighters to take special note of the hazards present and will give them the information needed to determine whether the operation is mistaken for a fire.

This chapter states the requirements for approval and special warning signs because of the toxicity and flammability hazards of fumigants. Upon notification of the hazardous potential, the fire code official should notify the fire department responsible for protecting the work site.

A guard or fire watch must be posted at the work site and must have a means available to report a fire or other emergency. If possible, a means should also be provided for the fire department to contact the guards or fire watches to verify the occurrence of a fire and obtain additional information en route to the scene.

SECTION 1702
DEFINITIONS

1702.1 Definitions. The following words and terms shall, for the purposes of this chapter and as used elsewhere in this code, have the meanings shown herein.

❖ Definitions of terms can help in the understanding and application of code requirements. This section contains definitions of terms that are associated with the subject matter of this chapter. These definitions are located here for convenient access to them without having to refer back to Chapter 2. For convenience, these terms are also listed in Chapter 2 with a cross reference to this section. The use and application of all defined terms, including those defined here, are set forth in Section 201.

FUMIGANT. A substance which by itself or in combination with any other substance emits or liberates a gas, fume or vapor utilized for the destruction or control of insects, fungi, vermin, germs, rats or other pests, and shall be distinguished from insecticides and disinfectants which are essentially effective in the solid or liquid phases. Examples are methyl bromide, ethylene dibromide, hydrogen cyanide, carbon disulfide and sulfuryl fluoride.

❖ Pesticides that are applied in gaseous form within a closed space and kill by inhalation are termed "fumigants." The basic fumigant material may be a volatile solid, liquid or gas. An example of a volatile solid fumigant is paradichlorobenzene. This substance, by the process of sublimation, fills a closed space with gas. It is marketed either as moth balls or moth cakes. Sublimation refers to the changing state from a solid to a gas without entering a liquid state. Many fumigants are flammable or combustible and all are toxic, posing health hazards that range from simply hazardous to deadly with minimal exposure.

FUMIGATION. The utilization within an enclosed space of a fumigant in concentrations that are hazardous or acutely toxic to humans.

❖ Fumigation sites are usually contained within a building but may also be at an outdoor location, aboard a vessel or in a vehicle. Soil fumigation can also be successful. The key to any fumigation is applying the appropriate fumigant for the correct duration, usually at least 8 hours for proper soaking to occur. The fumigant gas must penetrate every nook and cranny of the fumigation space or area; therefore, proper enclosure and sealing of the space or area is critical. Rooms can be sealed easily by caulking or taping doors, windows and ventilation openings. Entire structures can be sealed by encapsulation in plastic sheeting. Even trees can be isolated with an air-tight tent, and soils can be fumigated by covering the area with plastic covers. Other sections of this commentary discuss operational safeguards for fumigation operations.

THERMAL INSECTICIDAL FOGGING. The utilization of insecticidal liquids passed through thermal fog-generating units where, by means of heat, pressure and turbulence, such liquids are transformed and discharged in the form of fog or mist blown into an area to be treated.

❖ Thermal insecticidal fogging as a means of pest control was developed during World War II as part of a program to use smoke-screen techniques for control of malaria-bearing mosquitos. The product of the fog-generating process is an aerosol, which is a suspension of liquid particles in air. Many insecticides used in the thermal-fogging process are flammable and pose a fire hazard requiring safeguards, such as the securing of all ignition sources during fogging and for up to 24 hours afterward.

Aerosol insecticides also pose a toxicity hazard; therefore, proper entry precautions for the fogged area should be observed.

SECTION 1703
FIRE SAFETY REQUIREMENTS

1703.1 General. Structures in which fumigation and thermal insecticidal fogging operations are conducted shall comply with the fire protection and safety requirements of Sections 1703.2 through 1703.7.

❖ The provisions in Sections 1703.2 through 1703.7 apply to safe fumigation and thermal insecticidal fogging in structures.

1703.2 Sources of ignition. Fires, open flames and similar sources of ignition shall be eliminated from the space under fumigation or thermal insecticidal fogging. Heating, where needed, shall be of an approved type.

❖ Some insecticides and fumigants are flammable. Others that are considered nonflammable may ignite readily when suspended in air or dispersed as a vapor during fogging or fumigation operations. See the commentary to Chapters 13, 27, 34 and 37 and Section 307 of the *International Building Code*® (IBC®) for further information on the safe storage and handling of dusts and other materials used as pesticides or fogging agents.

1703.2.1 Electricity. Electricity shall be shut off.

Exception: Circulating fans that have been specifically designed for utilization in hazardous atmospheres and installed in accordance with the ICC *Electrical Code*.

❖ The structure's electrical service must be disconnected at the main service disconnect during fumigation or thermal insecticidal fogging operations. While not specifically required, a physical means for preventing the inadvertent, premature restoration of power by anyone other than the fumigation supervisor should be provided, such as by locking the service disconnect switch, if possible, or by equivalent means.

The exception allows the use of fans to assist in the distribution of fumigant throughout the building undergoing fumigation. The power supply for circulating fans must be located outside the building undergoing fumigation, or the fans may be self-powered, either by battery or an integral internal combustion engine. In any case, fans exposed to the hazardous atmosphere must be appropriately classified when an insecticide or fogging agent is susceptible to ignition during application or use, so as not to become an ignition source.

Equipment and devices used in the presence of flammable liquid vapors must be classified for use in Class I, Division 1 hazardous locations. When used where a dust explosion hazard exists, the fans must be classified for Class II hazardous locations.

1703.3 Notification. The fire code official and fire chief shall be notified in writing at least 24 hours before the structure is to be closed in connection with the utilization of any toxic or flammable fumigant. Notification shall give the location of the enclosed space to be fumigated or fogged, the occupancy, the fumigants or insecticides to be utilized, the person or persons responsible for the operation, and the date and time at which the operation will begin. Notice of any fumigation or thermal insecticidal fogging shall be served with sufficient advance notice to the occupants of the enclosed space involved to enable the occupants to evacuate the premises.

❖ Written notice is required at least 24 hours before a structure is closed for fumigation or fogging with flammable or toxic agents. All pesticides must be considered toxic to some degree because they are intended to kill something at some concentration. Occupants of the structure must be given sufficient advance notice to permit evacuation and removal of any belongings or equipment that may be endangered by fumigation or fogging operations. The fire code official should relay the notification information and the nature of the fumigation or fogging operation to fire service personnel of the location upon receipt.

Fire fighters must be aware of the nature of hazards to encourage the use of protective gear and precautions when investigating or operating at sites involving structures undergoing fumigation. Fire code officials may consider the required written notification a prima facie application for a permit under Section 1701.2; however, the fire code official must inspect the premises and equipment to verify compliance with the code before a permit is issued (see commentary, Sections 105.2.2 and 105.6.20).

1703.3.1 Warning signs. Approved warning signs indicating the danger, type of chemical involved and necessary precautions shall be posted on all doors and entrances to the premises and upon all gangplanks and ladders from the deck, pier or land to the ship. Such notices shall be printed in red ink on a white background. Letters in the headlines shall be at least 2 inches (51 mm) in height and shall state the date and time of the operation, the name and address of the person, the name of the operator in charge, and a warning stating that the occupied premises shall be vacated at least 1 hour before the operation begins and shall not be reentered until the danger signs have been removed by the proper authorities.

❖ Warning signs and placards must be posted to give notice for evacuation, warn of the hazards of the fumigant or fogging agent and discourage entry by unauthorized and unprotected people. Warning signs must include the name and address of the party responsible for conducting fumigation operations. Both names and hazards associated with the chemical fumigant must appear on the warning signs.

The NFPA 704 hazard warning system is especially useful for warning of fire-related hazards of these materials. Hazard-warning signal words also provide useful qualitative information about the relative toxicity of pesticides and fumigants (see Table 1703.3.1 and Figure 1703.3.1).

TABLE 1703.3.1
TOXIC HAZARD SIGNAL WORDS

SIGNAL WORDS	TOXICITY LEVEL
Danger, Poison (with Skull and Crossbones Symbol)	High
Warning	Moderate
Caution	Low

Source: Noll, G. G., M.S. Hildenbrand, and J.G. Yvorra. *Hazardous Materials, Managing the Incident* (Figure 6-17, p. 79).

DANGER
FUMIGATION IN PROGRESS
On Dec 14, 2001 at 1:00 pm, this building, 3232 78th St., will be fumigated by the ACME Pest Control Co. All occupants must evacuate the premises no later than Noon that day.
It is UNLAWFUL and DANGEROUS to reenter this building while this notice is posted.
4
2
HAZARD INFORMATION
ALUMINUM PHOSPHIDE
Flammable solid, poison dangerous when wet. Produces Phosphine gas in water. Stay upwind, wear special protective clothing. Contain runoff.
see MSDS

Figure 1703.3.1
FUMIGATION WARNING SIGN

1703.3.2 Breathing apparatus. Persons engaged in the business of fumigation or thermal insecticidal fogging shall maintain and have available approved protective breathing apparatus.

❖ Respiratory protective equipment must be appropriate for the hazards of the material used. Stored-air self-contained breathing apparatus (SCBA) is not necessarily the only appropriate form of respiratory protection. Canister respirators with suitable filters may provide employees greater flexibility to perform required tasks without posing a danger of exposure or contamination to the employee. Employees required to wear respiratory protective equipment should be fitted for the correct size mask and enrolled in a medical surveillance program. The required protective equipment must be available at all times during the fumigation or fogging process.

1703.3.3 Watch personnel. During the period fumigation is in progress, except when fumigation is conducted in a gas-tight vault or tank, a capable, alert watcher shall remain on duty at the entrance or entrances to the enclosed fumigated space until after the fumigation is completed and the premises properly ventilated and safe for occupancy. Sufficient watchers shall be provided to prevent persons from entering the enclosed space under fumigation without being observed.

❖ A minimum of one watch must be posted at each entrance to the structure or space being fumigated. Watches must discourage or prevent entry by unauthorized and unprotected individuals until fumigation is completed; the building, structure or premises has been ventilated and the building, structure or premises is again ready for occupancy. Watches must also have a means available to report emergencies without leaving their posts unattended. If practical, a means should be provided for the fire department to contact the watch while responding to determine whether a fire exists and, if so, obtain additional information before arrival.

1703.4 Thermal insecticidal fogging liquids. Thermal insecticidal fogging liquids with a flash point below 100°F (38°C) shall not be utilized.

❖ The use of Class I flammable liquids as fumigants is prohibited. Many fumigants that are otherwise nonflammable or only combustible may be easily ignited when dispersed or suspended in air as a vapor or fine mist. This prohibition does not apply to flammable active ingredients in nonflammable concentrations when in a solution with other ingredients.

1703.5 Sealing of buildings. Paper and other similar materials that are not flame resistant shall not be used to wrap or cover a building in excess of that required for the sealing of cracks, casements and similar openings.

❖ Fundamental to a successful fumigation operation is achieving the proper concentration of fumigant for the proper length of time to allow the treatment to permeate all portions of the structure, including interstitial spaces and even some building materials to a cellular level. To achieve this level of permeation, the structure must be made as gas tight as possible.

Small holes, cracks and openings can easily be sealed with tape; heavy-weight paper secured with tape or water-soluble or peelable paste; caulking compound; foam plastic sealant or similar, readily available materials. Larger openings can be sealed using plastic sheets held in place with tacks or staples and sealed with tape; heavy-weight paper secured with tape or water-soluble or peelable paste or a spray-on vinyl sealant. The use of these and similar readily combustible materials should be limited to small cracks, casements and similar openings.

The intent of this section is that more extensive sealing procedures than the simple crack and hole sealing described above, such as wrapping the building or tenting it, be done using materials that are, at a minimum, flame resistant. The use of such materials reduces the susceptibility of the enclosing membrane to easy ignition that could quickly engulf the entire structure in fire.

1703.6 Venting and cleanup. At the end of the exposure period, fumigators shall safely and properly ventilate the premises and contents; properly dispose of fumigant containers, residues, debris and other materials used for such fumigation; and clear obstructions from gas-fired appliance vents.

❖ Following the desired fumigant exposure period, the structure must be completely aired by opening as many doors and windows as possible from the outside. Also at this time, any exterior wrapping and sealing materials should be removed. Complete removal of sealing materials from the gas-fired appliance vent and combustion air intake openings that were sealed during fumigation is especially important to avoid potential backups of products of combustion in the building. Approved ventilators and fans powered from a source, and with their switches accessible from outside the building, should be started as well (see commentary, Section 1703.2.1). Personnel should then withdraw from the immediate vicinity of the fumigated structure and wait a reasonable length of time before entering the building to open more doors and windows to complete the ventilation process.

Throughout the ventilation process, personnel should wear personal breathing protection equipment (respirators). Tests for the presence of gas may be conducted as soon as the fumigator determines that the structure is properly aired. Chemical tests for residual fumigant must be carried out to verify that both the structure and its contents are free of toxic concentrations of fumigant, at which time the building may be reoccupied for normal activities.

Consistent with Section 304, rubbish and debris generated by the fumigation project must be removed from the premises and properly disposed. Because fumigants are hazardous materials, fumigant containers and other materials that might contain flammable or toxic fumigant residues must be disposed of in accordance with the fumigant manufacturer's instructions.

1703.7 Flammable fumigants restricted. The use of carbon disulfide and hydrogen cyanide shall be restricted to agricultural fumigation.

❖ While they are very effective fumigants, carbon disulfide (Formula: CS_2; CAS # 75-15-0) and hydrogen cyanide (Formula: HCN; CAS # 74-90-8) present such severe flammability and toxicity hazards that their use is restricted to agricultural fumigation applications only.

In addition to being toxic (see commentary, Chapter 37), carbon disulfide is a Class IB flammable liquid (see commentary, Chapter 34) having a flash point of -22°F (-30°C), a boiling point of 116°F (47°C) and a flammability range of 1 to 50 percent, thus making any release an extreme fire and deflagration hazard to the extent that its vapors can be ignited by contact with the heated surfaces of steam heating pipes or ordinary incandescent lightbulbs. Given these dangers, carbon disulfide is typically used in mixtures with other materials, such as carbon tetrachloride as a fire suppressant, or stabilizing chemicals, such as sulfur dioxide, to mitigate the hazards of use.

The typical agricultural use of carbon disulfide is as a fumigant for treating raw cereal grains in grain bins and silos.

Hydrogen cyanide (also known as formonitrile, hydrocyanic acid or prussic acid) is colorless, possesses a characteristic almond-like aroma and is an insecticide principally used for fumigation of stored agricultural products, especially grains and flour in mills, warehouses and the holds of ships.

In the fruit-growing regions of the United States, it has also been widely used for destroying scale insects on citrus trees while each tree is covered by a gas tent. Hydrogen cyanide can be a Class IA flammable liquid (see commentary, Chapter 34) or a flammable gas (see commentary, Chapter 35). As a liquid, it may also be classified as a Class 2 unstable reactive material (see commentary, Chapter 43), a Class 1 water reactive material (see commentary, Chapter 44) and a highly toxic material (see commentary, Chapter 37). Hydrogen cyanide may enter the body in toxic amounts by absorption through unbroken skin, inhalation and direct ingestion.

Bibliography

The following resource materials are referenced in this chapter or are relevant to the subject matter addressed in this chapter.

Farm Chemicals Handbook. Willoughby, OH: Meister Publishing Co., 1991.

Keffer, W.J. "Pesticides." In Cote, A.E., ed., *Fire Protection Handbook*, 17th ed. Quincy, MA: National Fire Protection Association, 1991.

NFPA 434-02, *Code for the Storage of Pesticides.* Quincy, MA: National Fire Protection Association, 2002.

NFPA 704-96, *Identification of the Fire Hazards of Materials.* Quincy, MA: National Fire Protection Association, 1996.

Schram, P.J. and M.W. Early. *Electrical Installations in Hazardous Locations, 2nd ed.* Quincy, MA: National Fire Protection Association, 1997.

Spencer, M.R. "Storage and Handling of Grain Mill Products." In Cote, A.E., ed., *Fire Protection Handbook*, 17th ed. Quincy, MA: National Fire Protection Association, 1991.

Wiswesser, W.J., ed. *Pesticide Index*. College Park, MD: Entomological Society of America, 1976.

Chapter 18: Semiconductor Fabrication Facilities

General Comments

The invention, development and exploitation of semiconductor technology has changed the world. Without integrated circuits, microchips or just "chips," the world of high technology would not exist. With the benefits of living in a high-tech age come some unique and pressing challenges. The manufacture of microchips is a complex, hazardous and demanding operation involving state-of-the-art design and manufacturing techniques, specially designed processing centers and a highly trained work force. Despite these rigors, the dangers of the processes can neither be avoided nor ignored.

The manufacture of semiconductors and microprocessors has developed into its own industry during the last 25 years. The proliferation of computer technology has resulted in the incredible expansion of the semiconductor manufacturing industry. These sophisticated products require a special processing environment and new rules to match the new technology. Unlike many other hazardous operations, hazards of the production materials are not manifested in the finished product. Considering the unique and often acute hazards of many materials used in semiconductor processing, this contributes to the relatively good safety record of the industry.

Purpose

The requirements of this chapter are intended to control hazards associated with the manufacture of semiconductors. Though the finished product possesses no unusual hazards, materials commonly associated with semiconductor manufacturing are often quite hazardous and include flammable liquids; pyrophoric and flammable gases; toxic substances and corrosives. The requirements are concerned with both life safety and property protection. However, the fire code official should recognize that the risk of extraordinary property damages is far more common than the risk of personal injuries from fire.

SECTION 1801 GENERAL

1801.1 Scope. Semiconductor fabrication facilities and comparable research and development areas classified as Group H-5 shall comply with this chapter and the *International Building Code*. The use, storage and handling of hazardous materials in Group H-5 shall comply with this chapter, other applicable provisions of this code and the *International Building Code*.

❖ Chapter 18 applies when quantities of hazardous materials exceeding the amounts listed in Chapter 27 are used at a facility for semiconductor manufacture or semiconductor research and development operations. Other code requirements governing the storage, use and handling of hazardous materials apply in situations where quantities of the materials listed in Chapter 27 do not exceed the specified limitations and to conditions not specifically addressed in Chapter 18. Semiconductor fabrication facilities are to be considered Group H-5 occupancies and are to comply with applicable *International Building Code®* (IBC®) requirements.

1801.2 Application. The requirements set forth in this chapter are requirements specific only to Group H-5 and shall be applied as exceptions or additions to applicable requirements set forth elsewhere in this code.

❖ Chapter 18 requirements are specific only to Group H-5 occupancies, with the requirements applied as exceptions or additions to requirements addressed elsewhere in the code. Where Chapter 18 contains a specific requirement for a certain condition and a general requirement for the same condition exists elsewhere in the code, the specific Chapter 18 requirements are to be applied. For example, general requirements for spill control and containment for use conditions involving hazardous materials in amounts exceeding maximum allowable quantities are found in Section 2705.2 with conditions specific to Group H-5 occupancies addressed in Section 1805.2.2. For Group H-5 occupancy conditions, the specific Section 1805.2.2 requirements take precedence over the general requirements addressed in Section 2705.2.

1801.3 Multiple hazards. Where a material poses multiple hazards, all hazards shall be addressed in accordance with Section 2701.1.

❖ All hazard classifications of a material are to be considered. For example, glacial acetic acid is classified as both a Class II combustible liquid and a corrosive liquid.

Thus, for glacial acetic acid, the requirements for both Class II combustible liquids and corrosive liquids must be met. This section restates the conditions found in Section 2701.1.

1801.4 Existing buildings and existing fabrication areas. Existing buildings and existing fabrication areas shall comply with this chapter, except that transportation and handling of HPM in exit access corridors and exit enclosures shall be allowed when in compliance with Section 1805.3.2 and the *International Building Code*.

❖ Although the adoption and enforcement of code requirements specifically addressing semiconductor manufacturing and similar research and development operations have been in place for nearly 20 years, there are still some facilities that predate the adoption of regulations specific to these operations. This section requires modifications to existing facilities to comply with Section 415.9 of the IBC. Additionally, requirements found in IBC Sections 415.9.3 and 1805.3.2 must be met when existing conditions or modifications do not include service corridors and existing exit access corridors are used to transport hazardous production materials (HPM) to fabrication areas.

1801.5 Permits. Permits shall be required as set forth in Section 105.6.

❖ The process of issuing permits gives the fire code official an opportunity to carefully evaluate and regulate hazardous operations. Permit applicants should be required to demonstrate that their operations comply with the intent of the code before the permit is issued. The process also notifies the fire department of the need for prefire planning for the hazardous property. See the commentary to Section 105.6 for a general discussion of operations requiring an operational permit, notably Section 105.6.21 for a discussion of specific quantity-based hazardous materials operational permits and Section 105.6.22 for HPM operational permits.

SECTION 1802 DEFINITIONS

1802.1 Definitions. The following words and terms shall, for the purposes of this chapter and as used elsewhere in this code, have the meanings shown herein.

❖ Definitions of terms can help in the understanding and application of code requirements. Including these definitions within this chapter provides more convenient access to them without having to refer back to Chapter 2. It is important to emphasize that these terms are not exclusively related to this chapter but are applicable everywhere the term is used in the code. For convenience, these terms are also listed in Chapter 2 with a cross reference to this section. The use and application of all defined terms, including those defined in Chapter 18, are set forth in Section 201.

CONTINUOUS GAS DETECTION SYSTEM. A gas detection system where the analytical instrument is maintained in continuous operation and sampling is performed without interruption. Analysis is allowed to be performed on a cyclical basis at intervals not to exceed 30 minutes.

❖ This term refers to a system that is capable of constantly monitoring the presence of highly toxic or toxic compressed gases at or below the permissible exposure limit (PEL) for the gas. A continuous gas detection system will provide notification of a leak or rupture in a compressed gas cylinder or tank in a storage or use condition.

Numerous techniques exist for detecting and measuring gases. Two of the most common types of detection methods are briefly described below:

Colorimetric Detection (Chemcassette Detection): Colorimetric detection involves a circuit that is usually completed by the presence of an electrolyte within the cell itself. A carefully prepared reel of porous paper tape is impregnated with a chemical that will change color in the presence of the target gas—this reaction can be very specific to particular gases of interest. This tape is then positioned so that the sample of air or gas passes through a section of the tape. Color changes are measured optically and converted to a direct concentration value of the target gas.

Electrochemical Detection: Electrochemical cells work on the fuel-cell principle. As the target gas enters the cell, it reacts at the active electrode and generates a very small current, which flows through an external measuring circuit back to the counter electrode.

EMERGENCY CONTROL STATION. An approved location on the premises where signals from emergency equipment are received and which is staffed by trained personnel.

❖ This definition identifies the room or area located in the HPM facility that is used to receive various alarms and signals. The smoke detectors located in the building's recirculation ventilation ducts, the gas-monitoring/detection system and the telephone/fire protective signaling systems located outside of HPM storage rooms must all be connected to the emergency control station. The location of this station must be approved by the fire code official. An approved location should be based on personnel being able to adequately monitor the necessary alarms and signals and on the fire department being able to gain access quickly when responding to emergency situations. Additionally, the room must be occupied by persons who are trained to respond to the various alarms and signals.

FABRICATION AREA. An area within a semiconductor fabrication facility and related research and development areas in which there are processes using hazardous production materials. Such areas are allowed to include ancillary rooms or areas such as dressing rooms and offices that are directly related to the fabrication area processes.

❖ This definition describes the basic component of an HPM facility. The code uses this definition to set certain material limitations based on both quantity and density, and to require enclosure of the fabrication areas with fire-barrier assemblies. The fabrication area of an HPM facility is the area where the hazardous materials are actively handled and processed. The fabrication area includes accessory rooms and spaces, such as workstations and employee dressing rooms.

HAZARDOUS PRODUCTION MATERIAL (HPM). A solid, liquid or gas associated with semiconductor manufacturing that has a degree-of-hazard rating in health, flammability or reactivity of Class 3 or 4 as ranked by NFPA 704 and which is used directly in research, laboratory or production processes which have as their end product materials that are not hazardous.

❖ This definition identifies those specific materials that can be contained within an HPM facility. The restriction in the definition for only hazardous materials with a Class 3 or 4 rating is not intended to exclude materials that are less hazardous, but to clarify that materials of the indicated higher ranking are still permitted in an HPM facility without classifying the building as Group H. NFPA 704 is referenced to establish the degree of hazard ratings for all materials as related to health, flammability and reactivity risks. See Table 1802.1 for a list of commonly used HPM and their hazard classifications.

HPM FLAMMABLE LIQUID. An HPM liquid that is defined as either a Class I flammable liquid or a Class II or Class IIIA combustible liquid.

❖ This definition clarifies that an HPM liquid is essentially all classes of flammable or combustible liquids, except Class IIIB combustible liquids, which have a flash point at or above 200°F (93°C). Class IIIB liquids, therefore, are not considered an HPM.

HPM ROOM. A room used in conjunction with or serving a Group H-5 occupancy, where HPM is stored or used and which is classified as a Group H-2, H-3 or H-4 occupancy.

❖ An HPM room in a Group H-5 facility is used for the storage and use of HPM in excess of the maximum allowable quantities permitted in Table 307.7(1) or 307.7(2) of the IBC. The rooms are therefore considered a Group H-2, H-3 or H-4 occupancy, depending on the type of hazardous material.

PASS-THROUGH. An enclosure installed in a wall with a door on each side that allows chemicals, HPM, equipment, and parts to be transferred from one side of the wall to the other.

❖ A pass-through, such as a storage cabinet, is similar to a sally port and is used to store and receive HPM for the fabrication area. The pass-through must be separated from the exit access corridor by fire-resistance-rated construction, including a fire-resistance-rated, self-closing fire door on each side, and be sprinklered.

SEMICONDUCTOR FABRICATION FACILITY. A building or a portion of a building in which electrical circuits or devices are created on solid crystalline substances having electrical conductivity greater than insulators but less than conductors. These circuits or devices are commonly known as semiconductors.

❖ A semiconductor fabrication facility is a building or a portion of a building where semiconductors are produced. See Figure 1802.1 for an example of a typical floor plan of a semiconductor fabrication facility.

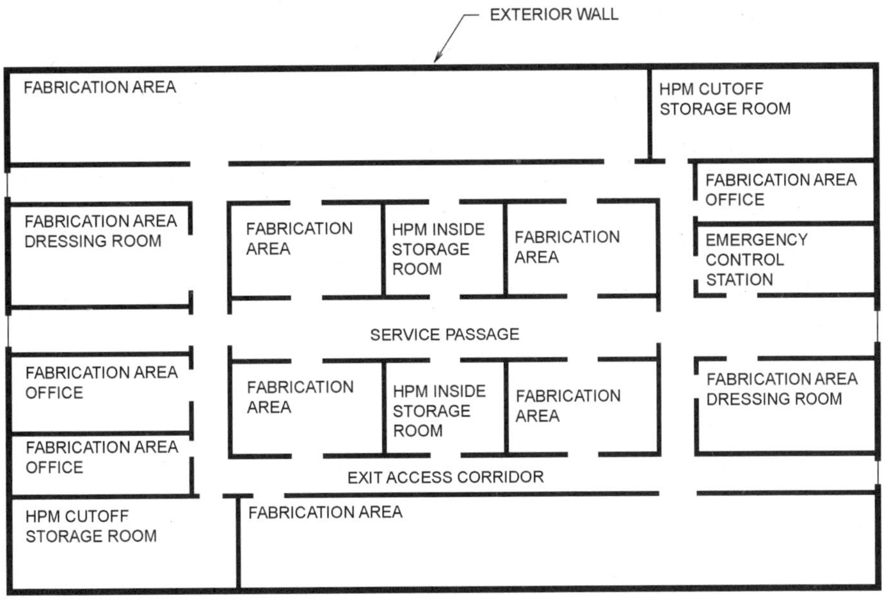

FIGURE 1802.1
TYPICAL COMPONENTS OF AN HPM FACILITY

TABLE 1802.1
HAZARDOUS PRODUCTION MATERIALS (HPM) USED IN THE MANUFACTURE OF SEMICONDUCTORS

Material	Description or Use	NFPA 704 (1990) Hazard Classification			
		Health	Flammability	Reactivity	Other
Acetic acid	Corrosive liquid used for wet etching (metal)	2	2	1	
Acetone	Flammable liquid used for wafer cleaning	1	3	0	
Ammonium fluoride	Corrosive for wet etching (oxide)	3	0	0	
Arsenic trichloride	Diffusion	3	0	1	₩
Arsenic trioxide	Diffusion	4	0	0	
Arsine	Poison flammable gas used for epitaxial growth, diffusion and ion implanation	4	4	3	
Boron tribromide	Corrosive liquid used for diffusion	4	0	3	₩
Boron trichloride	Nonflammable corrosive gas used for diffusion	4	0	1	₩
Chlorine	Poison gas used for dry etching	3	0	0	OXY
Diborane	Highly reactive flammable gas used for diffusion	3	4	3	₩
Dichlorosilane	Flammable liquefied gas used for epitaxial growth	4	4	4	
Gallium	Reactive metal used as a semiconductor crystal material	1	0	3	
Gallium arsenide	Reactive metal salt used as a semiconductor crystal material	3	0	0	
Gallium arsenide phosphide	Reactive metal salt used as a semiconductor crystal material	3	0	0	
Germanium	Reactive metal used as a semiconductor crystal material	0	0	3	
Hydrofluoric acid	Highly corrosive liquid or gas used for wet etching (oxide)	4	0	0	
Hydrogen peroxide[a]	Organic peroxide used for wafer cleaning	2	0	1	OXY
Isopropanol	Flammable liquid used for wafer cleaning	1	3	0	
Methanol	Flammable liquid used for wafer cleaning	1	3	0	
Nitric acid	Corrosive liquid used for wet etching (metal)	3	0	0	OXY
Oxygen (liquid)	Oxidizing gas used for oxidation	3	0	0	OXY
Phosphine	Flammable liquefied poison gas used for diffusion and ion implantation	4	4	4	
Phosphoric acid	Corrosive liquid used for wet etching (metal)	2	0	0	
Phosphorus oxychloride	Corrosive liquid used for diffusion	4	0	3	₩
Phosphorus pentoxide	Corrosive solid sublimed for use in diffusion	4	0	3	₩
Phosphorus tribromide	Corrosive liquid used for diffusion	4	0	3	₩
Silane	Pyrophoric gas used for oxidation	2	4	4	
Silicon	Flammable solid (metal) used as a semiconductor crystal material	2	4	2	₩
1, 1, 1-Trichloroethane	Mildly flammable solvent (difficult to ignite) used or wafer cleaning	2	1	0	
Tetrachlorosilane	Flammable liquid used for epitaxial growth	3	4	2	₩

Note a. NFPA 704 values for 35 to 52 percent by weight (the most concentration) are listed. The reactivity hazard increases to 3 at con centrations above 52 percent.

SERVICE CORRIDOR. A fully enclosed passage used for transporting HPM and purposes other than required means of egress.

❖ Though HPM facility occupants may be exposed to limited HPM quantities during the course of their employment, their means of egress are protected from the HPM hazards by confining the HPM being transferred to its own passageway. A service corridor is required only when the HPM must be carried from a storage room or external area to a fabrication area through a passageway.

TOOL. A device, storage container, workstation, or process machine used in a fabrication area.

❖ A tool is basically any device or piece of equipment, including a workstation, in a fabrication area where hazardous materials are used, stored or handled.

WORKSTATION. A defined space or an independent principal piece of equipment using HPM within a fabrication area where a specific function, laboratory procedure or research activity occurs. Approved or listed hazardous materials storage cabinets, flammable liquid storage cabinets or gas cabinets serving a workstation are included as part of the workstation. A workstation is allowed to contain ventilation equipment, fire protection devices, detection devices, electrical devices and other processing and scientific equipment.

❖ Workstations further subdivide a fabrication area and provide relatively self-contained, specialized areas where HPM processes are conducted. Workstation controls limit the quantity of materials and impose limitations on the design of these processes to include, but not be limited to, protection by local exhaust; sprinklers; automatic and emergency shutoffs; construction materials and HPM compatibility. Excess materials are prohibited and must be contained in storage rooms designed to accommodate such hazards.

SECTION 1803 GENERAL SAFETY PROVISIONS

1803.1 Emergency control station. An emergency control station shall be provided on the premises at an approved location outside of the fabrication area, and shall be continuously staffed by trained personnel. The emergency control station shall receive signals from emergency equipment and alarm and detection systems. Such emergency equipment and alarm and detection systems shall include, but not be limited to, the following where such equipment or systems are required to be provided either in this chapter or elsewhere in this code:

1. Automatic sprinkler system alarm and monitoring systems.
2. Manual fire alarm systems.
3. Emergency alarm systems.
4. Continuous-gas detection systems.
5. Smoke detection systems.
6. Emergency power system.

❖ This section specifies the systems that are to be monitored by an emergency control station. The fire alarm system signals are received at the emergency control station, which must be located in an area approved by the fire code official. The emergency control station should not be located in an area where hazardous materials are stored, used or transported, such as a fabrication area. See the commentary to Section 1802.1 for the definition of "Emergency control station."

1803.2 Systems, equipment and processes. Systems, equipment and processes including, but not limited to, containers, cylinders, tanks, piping, tubing, valves and fittings and shall comply with this section, Section 2703.2, other applicable provisions of this code, the *International Building Code* and the *International Mechanical Code.*

❖ This section contains a general reference to the IBC and the *International Mechanical Code*® (IMC®) as well as to Section 2703.2 and other sections of the code for regulations pertaining to hazardous materials-related systems, equipment and processes. Except as addressed in Section 1803.2.1, the requirements of the code for things such as containers, cylinders, tanks, piping, tubing, valves and fittings in Group H-5 occupancy facilities are basically the same as they are for other occupancies.

1803.2.1 Additional regulations for HPM supply piping and tubing. The requirements set forth in Section 2703.2.2.2 shall apply to supply piping and tubing for HPM gases and liquids. Supply piping and tubing for HPM gases and liquids having a health-hazard ranking of 3 or 4 shall be welded throughout, except for connections located within a ventilated enclosure if the material is a gas, or an approved method of drainage or containment is provided for connections if the material is a liquid.

❖ This section contains specific requirements for supply piping and tubing transferring HPM gases or liquids. The requirements include the general requirements established in Section 2703.2.2.2 as well as more specific requirements for supply piping or tubing containing HPM gases or liquids having a health-hazard ranking of 3 or 4. The primary purpose of the more specific requirements for supply piping or tubing containing a health-hazard ranking of 3 or 4 is to minimize the potential for HPM leakage. The use of mechanical compression-type fittings or other nonwelded joints for these materials is intended to be limited to areas where liquid leaks will be contained or drained and gas leaks will be either contained or exhausted.

1803.3 Construction requirements. Construction of semiconductor fabrication facilities shall be in accordance with Sections 1803.3.1 through 1803.3.8.

❖ This section contains references to the various construction requirements affecting the components of a semiconductor fabrication facility.

1803.3.1 Fabrication areas. Construction and location of fabrication areas shall comply with the *International Building Code*.

❖ Semiconductors are manufactured to specific standards. Current high-speed computers require highly sophisticated microprocessors, which rely on extremely fine electrical pathways to transmit the impulses forming the basis for their operation. A single particle of dust only a fraction of a micron in diameter may destroy a microprocessor by clogging one of these electronic arteries. Specially designed enclosures called "clean rooms" must be used for fabricating these products to protect the finished product during assembly. Table1803.3.1 illustrates various classifications used to describe the degree of cleanliness in a clean room. Environmental particulates inside the clean room are controlled by filtering all air entering the room through special high-efficiency particulate air (HEPA) filters and constantly replacing the air inside the room.

Fabrication areas are portions of an HPM area where semiconductors are actually produced. The circuit pattern is etched on the surface of the semiconductor crystal material during fabrication. Successive layers of etched semiconductor material are assembled to create integrated circuits or microchips. Fabrication is done in clean rooms, which derive their name from the ability of mechanical ventilation systems to maintain an environment nearly free of air-borne particulate contamination. As a result, clean rooms are classified according to the number of particles in a cubic foot of air. For example, in a Class 10 clean room, mechanical ventilation is capable of removing all but 10 particles per cubic foot from the air in the room. The extraordinary precision required in the manufacture of semiconductors demands such cleanliness.

A fire in such an environment represents not only a life and property hazard but also a serious contamination concern. The monetary value of semiconductors is of secondary concern when compared to the value of lost production time during recovery from a fire or other emergency. Such losses may extend into the millions of dollars from even very small fires. Section 415.9 of the IBC specifies requirements for the fabrication of areas to control these hazards. The requirements of this chapter are intended to complement and maintain compliance with the requirements of Section 415.9.

1803.3.2 Pass-throughs in exit access corridors. Pass-throughs in exit access corridors shall be constructed in accordance with the *International Building Code.*

❖ Pass-throughs in exit access corridor walls must be constructed to maintain the integrity of any required fire-resistance rating of such walls in an HPM facility. Pass-through doors must also function to maintain that integrity. The IBC contains construction requirements for both exit access corridor walls and their opening protectives in Chapter 10 (see also the commentary to Section 1802.1 for the definition of "Pass-through").

1803.3.3 Liquid storage rooms. Liquid storage rooms shall comply with Chapter 34 and the *International Building Code.*

❖ Liquid storage rooms are used for the storage of closed containers of flammable or combustible liquids used in the HPM processes. Section 3404.3.7 contains detailed requirements for these rooms (see also the commentary to Section 3402.1 for the definition of "Liquid storage room").

1803.3.4 HPM rooms. HPM rooms shall comply with the *International Building Code.*

❖ Proper storage of hazardous materials can reduce the danger associated with those materials. Section 1804.3 establishes that adequate separation must be provided between materials and limits the quantity of materials stored in HPM rooms. This section requires that the separation of HPM storage rooms must comply with Section 415.9 of the IBC (see also the commentary to Section 1802.1 for the definition of "HPM room").

1803.3.5 Gas cabinets. Gas cabinets shall comply with Section 2703.8.6.

❖ To reduce the hazards presented by gases used in the fabrication of semiconductors, Section 2703.8.5 contains specific construction and ventilation requirements for enclosed gas cabinets used for the storage of HPM gases (see also the commentary to Section 2702.1 for the definition of "Gas cabinet").

TABLE 1803.3.1
CLEAN ROOM CLASSIFICATIONS

	Measured Particle Size (Microns)				
Class	0.1	0.2	0.3	0.5	5.0
1	35	7.5	3	1	—
10	350	75	30	10	—
100	—	750	300	100	—
1,000	—	—	—	1,000	7
10,000	—	—	—	10,000	70
100,000	—	—	—	100,000	700

Source: Pearce, R. J. "Clean Rooms," in Cote, A. E., ed., *Industrial Fire Hazards Handbook.*

1803.3.6 Exhausted enclosures. Exhausted enclosures shall comply with Section 2703.8.5.

❖ To reduce the hazards presented by hazardous materials used in the fabrication of semiconductors, Section 2703.8.4 contains specific construction and ventilation requirements for exhausted enclosures used for the storage of HPM. Whereas gas cabinets must be fully enclosed and equipped with self-closing doors, exhausted enclosures are typically open-fronted and lend themselves to small-scale operations such as fume hoods found in chemical laboratories (see also the commentary to Section 2702.1 for the definition of "Exhausted enclosure").

1803.3.7 Gas rooms. Gas rooms shall comply with Section 2703.8.4.

❖ To reduce the hazards presented by hazardous gases used in the fabrication of semiconductors, Section 2703.8.3 contains specific construction and ventilation requirements for gas rooms used for the storage of HPM gases. These rooms are intended to be restricted to the storage of cylinders and tanks of highly toxic and toxic compressed gases. The storage of other hazardous materials, such as flammable and combustible liquids, is not permitted within a gas room. Similar to gas cabinets, the required exhaust ventilation for the gas room must be operated at negative pressures to prevent leakage of hazardous vapors to adjacent areas.

1803.3.8 Service corridors. Service corridors shall comply with Section 1805.3 and the *International Building Code.*

❖ The likelihood of accidental spillage or contact with HPM is increased during the transportation of such materials. Section 1805.3 regulates materials handling and transportation in passageways used for purposes other than required means of egress.

1803.4 Emergency plan. An emergency plan shall be established as set forth in Section 408.4.

❖ Thorough planning is essential to evacuate personnel effectively and to combat a fire or other emergency. Risks associated with a structure containing HPMs require extraordinary efforts regarding evacuation and fire control. Specific emergency planning procedures are outlined in Section 408.4.

1803.5 Maintenance of equipment, machinery and processes. Maintenance of equipment, machinery and processes shall comply with Section 2703.2.6.

❖ Fundamental to the safe operation of an HPM facility is the care and maintenance of the equipment used in the various fabrication processes. Section 2703.2.6 contains detailed maintenance requirements.

1803.6 Security of areas. Areas shall be secured in accordance with Section 2703.9.2.

❖ Public safety requires that HPM facilities be secure against unauthorized entry in accordance with Section 2703.9.2.

1803.7 Electrical wiring and equipment. Electrical wiring and equipment in HPM facilities shall comply with Sections 1803.7.1 through 1803.7.3.

❖ The safety and integrity of electrical wiring and equipment is a key factor in workplace safety. This section introduces the electrical safety requirements, by area, for HPM facility fabrications areas, workstations and storage rooms.

1803.7.1 Fabrication areas. Electrical wiring and equipment in fabrication areas shall comply with the ICC *Electrical Code.*

❖ The ICC *Electrical Code*® (ICC EC™) contains requirements for the electrical safety of HPM facilities, including fabrication areas.

1803.7.2 Workstations. Electrical equipment and devices within 5 feet (1524 mm) of workstations in which flammable or pyrophoric gases or flammable liquids are used shall comply with the ICC *Electrical Code* for Class I, Division 2 hazardous locations. Workstations shall not be energized without adequate exhaust ventilation in accordance with Section 1803.14.

Exception: Class I, Division 2 hazardous electrical equipment is not required when the air removal from the workstation or dilution will prevent the accumulation of flammable vapors and fumes on a continuous basis.

❖ Areas located in and around workstations are considered Class I, Division 2 hazardous locations as defined by Article 500 of NFPA 70. Incidental exposures to flammable fumes or vapors must be considered horizontally possible within 5 feet (1524 mm) of each workstation where flammable liquids or gases are used. Either a mechanical or electrical interlock must be installed to engage the required exhaust ventilation system before HPM enters the workstation. This reduces the likelihood of gas or vapor exposure.

The exception in this section states that, in some cases, it is both possible and reasonable to assume that a hazardous concentration of flammable gas or vapor is unlikely to occur when ventilation is properly engaged. Class I, Division 2 electrical wiring and equipment is not required where automatic exhaust interlocks are installed.

1803.7.3 Hazardous production material (HPM) rooms, gas rooms and liquid storage rooms. Electrical wiring and equipment in HPM rooms, gas rooms and liquid storage rooms shall comply with the ICC *Electrical Code.*

❖ Electrical wiring, equipment and devices in HPM cutoff rooms used for the storage of flammable liquids or gases must be classified for hazardous locations, in accordance with the ICC EC. The hazardous location requirements in the ICC EC contain special protection features to preclude ignition of flammable vapors, liquids, gases or dusts by sparks or electrical arcing.

1803.8 Exit access corridors and exit enclosures. Hazardous materials shall not be used or stored in exit access corridors or exit access enclosures.

❖ Because exit access corridors and exit enclosures to which they lead are critical to the safe egress and life safety of personnel, this section prohibits the storage or use of HPMs within them. This is consistent with Section 1005.3.2, which prohibits the use of an exit for any purpose other than the egress function.

1803.9 Service corridors. Hazardous materials shall not be used in an open-system use condition in service corridors.

❖ HPM dispensing, the open use or open transfer of HPM from original shipping containers to secondary containers, must not be performed in service corridors in order to reduce the likelihood of liberating dangerous vapors or fumes that might be difficult to dispose of and their resultant hazard exposure to personnel.

1803.10 Automatic sprinkler system. An approved automatic sprinkler system shall be provided in accordance with Sections 1803.10.1 through 1803.10.5 and Chapter 9.

❖ Sections 1803.10.1 through 1803.10.5 contain specific sprinkler protection design requirements for the various parts and components of an HPM facility. Section 903.2.4.2 requires automatic sprinkler protection throughout HPM facilities and states specific design criteria for various areas of the facility. Fabrication areas, service corridors and inside HPM storage rooms are considered Ordinary Hazard Group 2 and HPM storage rooms with dispensing are classified as Extra Hazard Group 2, in accordance with NFPA 13.

1803.10.1 Workstations and tools. The design of the sprinkler system in the area shall take into consideration the spray pattern and the effect on the equipment.

❖ The design of the sprinkler system must consider the potential for collateral damage to equipment from sprinkler discharge that could delay a return to normal operations.

1803.10.1.1 Combustible workstations. A sprinkler head shall be installed within each branch exhaust connection or individual plenums of workstations of combustible construction. The sprinkler head in the exhaust connection or plenum shall be located not more than 2 feet (610 mm) from the point of the duct connection or the connection to the plenum. When necessary to prevent corrosion, the sprinkler head and connecting piping in the duct shall be coated with approved or listed corrosion-resistant materials. The sprinkler head shall be accessible for periodic inspection.

Exceptions:

1. Approved alternative automatic fire-extinguishing systems are allowed. Activation of such systems shall deactivate the related processing equipment.
2. Process equipment which operates at temperatures exceeding 932°F (500°C) and is provided with automatic shutdown capabilities for hazardous materials.
3. Exhaust ducts 10 inches (254 mm) or less in diameter from flammable gas storage cabinets that are part of a workstation.
4. Ducts listed or approved for use without internal automatic sprinkler protection.

❖ Generally, workstations are constructed of noncombustible materials. Wet benches and machines handling certain corrosives may have substantial nonmetallic components or surfaces. Notwithstanding the requirements of Sections 1803.10.4.1 and 1803.10.4.2, an automatic sprinkler must be installed within 2 feet (610 mm) of where the exhaust duct connects to a workstation of combustible construction. It is reasonable to require the installation of a corrosion-resistant sprinkler in accordance with NFPA 13 and the application of a corrosion-resistant pipe coating because combustible (nonmetallic) parts are commonly used as a result of corrosion concerns. The duct must be accessible to permit regular periodic inspection and maintenance. (Note: Corrosion-resistant sprinklers require periodic replacement despite their ability to withstand severe exposures.)

Many fire protection engineers prefer the use of listed special flow control (also called "on-off") sprinklers or cycling (on-off) sprinkler systems to minimize the effects of water damage from automatic sprinklers. Others prefer preaction systems as a backup to an alternative fire suppression system, including clean-agent or carbon dioxide systems.

Exception 1 allows the installation of an alternative automatic fire-extinguishing system in accordance with Section 904 instead of automatic sprinklers. The activation of the alternative system must deactivate or deenergize processing equipment in the protected room or space. Deactivation of processing equipment should not affect the operation of required exhaust ventilation.

Exception 2 states that only the HPM flow must be interrupted in processing equipment operating at temperatures greater than 932°F (500°C).

Exception 3 allows gas cabinet exhaust systems associated with and part of workstations to continue to operate after shutdown.

Exception 4 states that automatic sprinklers are not

required where the risk to people or property is limited, such as when nonmetallic ducts approved for installation without sprinklers are used.

1803.10.1.2 Combustible tools. Where the horizontal surface of a combustible tool is obstructed from ceiling sprinkler discharge, automatic sprinkler protection that covers the horizontal surface of the tool shall be provided.

Exceptions:

1. An automatic gaseous fire-extinguishing local surface application system shall be allowed as an alternative to sprinklers. Gaseous-extinguishing systems shall be actuated by infrared (IR) or ultraviolet/infrared (UVIR) optical detectors.
2. Tools constructed of materials that are listed or approved for use without internal fire extinguishing system protection.

❖ Automatic sprinkler system discharge from ceiling sprinklers that cannot reach the fire because of obstructions above the sprinkler can allow the fire to grow beyond the capability of the system to extinguish or hold the fire's progress in check until help arrives. Where the surface of a combustible tool being used in an HPM process is thus obstructed, a fire could rapidly spread unimpeded in the workstation. This section requires that sprinkler protection be installed to properly cover the entire surface of the tool.

Exception 1 allows surface protection to be by installation of a gaseous fire-extinguishing system, such as a clean agent system, if it is activated by the more sensitive infrared (IR) or ultraviolet/infrared (UVIR) optical detectors (see also the commentary to Section 1802.1 for the definition of "Tool").

Exception 2 recognizes new technology for the fabrication of tools constructed of special materials whose flame propagation characteristics are such that they need not be internally or directly externally sprinklered. Factory Mutual has developed a standard, FM 4910, for the construction of tools (wet benches) that are made from combustible materials that do not propagate fire. Underwriters Laboratories has also developed a similar standard, UL 2360, that classifies the plastic materials used in wet-bench tool construction with a focus on compliance with both NFPA 318 and the code. UL 2360 classifies these materials into three classes based on their flame propagation characteristics: Class 1, nonpropagating; Class 2, limited propagating and Class 3, slow propagating. They are recognized as acceptable materials for tool construction without the installation of internal fire-extinguishing system protection, subject to the approval of the fire code official.

1803.10.2 Gas cabinets and exhausted enclosures. An approved automatic sprinkler system shall be provided in gas cabinets and exhausted enclosures containing HPM compressed gases.

Exception: Gas cabinets located in an HPM room other than those cabinets containing pyrophoric gases.

❖ Automatic sprinkler protection is required inside HPM storage cabinets, including workstation cabinets, except those storing gases other than pyrophoric gases (silane and dichlorosilane) located inside HPM rooms.

The exception recognizes the equivalency between automatic sprinkler protection and the fire-resistance-rated separation provided by an HPM room, except in cases of extremely hazardous pyrophoric gases.

1803.10.3 Pass-throughs in existing exit access corridors. Pass-throughs in existing exit access corridors shall be protected by an approved automatic sprinkler system.

❖ Exit access corridors in new buildings are not to be used for transporting HPM to the fabrication area. In accordance with Section 415.9.2.7 of the IBC, HPM must be transported by service corridors or piping. This section addresses HPM facilities that existed before the adoption and enforcement of those requirements. The intent of this section is to allow a pass-through in existing exit access corridor walls for transporting HPM to a fabrication area. A pass-through, such as a storage cabinet, is used to store and receive HPM for the fabrication area. The pass-through must be separated from the exit access corridor by fire-resistance-rated construction, including rated self-closing fire doors, and be protected by internal automatic sprinklers.

1803.10.4 Exhaust ducts for HPM. An approved automatic sprinkler system shall be provided in exhaust ducts conveying vapors, fumes, mists or dusts generated from HPM in accordance with this section and the *International Mechanical Code*.

❖ Because of the hazardous nature of the materials exhausted in HPM facilities, exhaust systems must comply with the IMC requirements for hazardous exhaust systems, including automatic sprinklers installed within the ducts. Sections 1803.10.4.1 through 1803.10.4.4.5 contain specific requirements for sprinkler installation in these systems.

1803.10.4.1 Metallic and noncombustible nonmetallic exhaust ducts. An approved automatic sprinkler system shall be provided in metallic and noncombustible nonmetallic exhaust ducts when all of the following conditions apply:

1. When the largest cross-sectional diameter is equal to or greater than 10 inches (254 mm).
2. The ducts are within the building.

3. The ducts are conveying flammable vapors or fumes.

❖ Sprinklers are required within each individual duct when all three of the conditions listed in this section exist.

Figure 1803.10.4.1 illustrates how to measure the cross-sectional diameter of various duct shapes. This section requires the square and rounded ducts to be protected by automatic sprinklers. The round or elliptical ducts depicted on the right side of the diagram do not require protection.

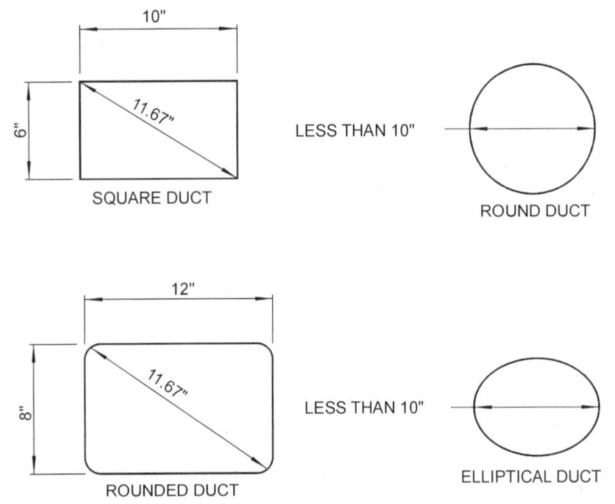

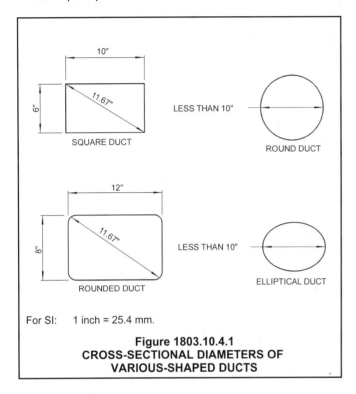

For SI: 1 inch = 25.4 mm.

Figure 1803.10.4.1
CROSS-SECTIONAL DIAMETERS OF VARIOUS-SHAPED DUCTS

1803.10.4.2 Combustible nonmetallic exhaust ducts. An approved automatic sprinkler system shall be provided in combustible nonmetallic exhaust ducts when the largest cross-sectional diameter of the duct is equal to or greater than 10 inches (254 mm).

Exceptions:

1. Ducts listed or approved for applications without automatic sprinkler system protection.
2. Ducts not more than 12 feet (3658 mm) in length installed below ceiling level.

❖ Automatic sprinkler protection is required for all combustible nonmetallic ducts with a cross-sectional diameter equal to or greater than 10 inches (254 mm) because of their ability to add a substantial amount of fuel in case of a fire within the duct system.

Exception 1 states that automatic sprinklers are not required where the risk to people or property is limited, such as when nonmetallic ducts approved for installation without sprinklers are used.

Exception 2 states that when ducts do not exceed 12 feet (3658 mm) in length and are installed exposed below ceiling level, sprinklers may be omitted. A fire within such a system would be readily noticeable and readily accessible for fire attack because of the exposed ductwork.

1803.10.4.3 Exhaust connections and plenums of combustible workstations. Automatic fire-extinguishing system protection for exhaust connections and plenums of combustible workstations shall comply with Section 1803.14.1.

❖ Refer to Section 1803.14.1 for sprinkler protection for combustible workstation exhaust connections and plenums.

1803.10.4.4 Exhaust duct sprinkler system requirements. Automatic sprinklers installed in exhaust duct systems shall be hydraulically designed to provide 0.5 gallons per minute (gpm) (1.9 L/min) over an area derived by multiplying the distance between the sprinklers in a horizontal duct by the width of the duct. Minimum discharge shall be 20 gpm (76 L/min) per sprinkler from the five hydraulically most remote sprinklers.

❖ This section, including Sections 1803.4.4.1 through 1803.4.4.5, addresses the design, installation and maintenance and inspection requirements for sprinklers installed in exhaust duct systems. This section specifies the hydraulic design criteria and minimum flow discharged from sprinklers.

1803.10.4.4.1 Sprinkler head locations. Automatic sprinklers shall be installed at 12-foot (3658 mm) intervals in horizontal ducts and at changes in direction. In vertical runs, automatic sprinklers shall be installed at the top and at alternate floor levels.

❖ This section specifies sprinkler spacing in both horizontal and vertical exhaust ducts.

1803.10.4.4.2 Control valve. A separate indicating control valve shall be provided for sprinklers installed in exhaust ducts.

❖ To isolate the sprinklers installed within the exhaust ducts without depriving the building or other facility components of protection, a separate, approved indicating control valve must be installed to serve the duct system only. This valve should be supervised like any other sprinkler control valve.

1803.10.4.4.3 Drainage. Drainage shall be provided to remove sprinkler water discharged in exhaust ducts.

❖ To prevent contamination of process equipment by sprinkler discharge water that might flow back down the duct, this section requires that the ducts have approved drainage facilities. Prompt drainage of sprinkler discharge water also reduces the likelihood of the duct system collapsing from the weight of retained water. Because the water will likely contain residues of the chemicals being exhausted through the ductwork, it must be disposed of in accordance with applicable environmental laws.

1803.10.4.4.4 Corrosive atmospheres. Where corrosive atmospheres exist, exhaust duct sprinklers and pipe fittings shall be manufactured of corrosion- resistant materials or coated with approved materials.

❖ Corrosion-resistant sprinklers, piping and fittings must be installed in exhaust ducts conveying corrosive vapors, fumes, mists or dusts.

1803.10.4.4.5 Maintenance and inspection. Sprinklers in exhaust ducts shall be accessible for periodic inspection and maintenance.

❖ Access panels must be installed in exhaust ducts for inspection and maintenance of sprinklers.

1803.10.5 Sprinkler alarms and supervision. Automatic sprinkler systems shall be electrically supervised and provided with alarms in accordance with Chapter 9. Automatic sprinkler system alarm and supervisory signals shall be transmitted to the emergency control station.

❖ Automatic sprinkler systems must be electrically supervised and have alarms installed in accordance with Section 903.4. The system alarm and supervisory signals should be transmitted to the on-site emergency control station. See Sections 1802.1 and 1803.1 for conditions and requirements applicable to emergency control stations.

1803.11 Manual fire alarm system. A manual fire alarm system shall be installed throughout buildings containing a Group H-5 occupancy. Activation of the alarm system shall initiate a local alarm and transmit a signal to the emergency control station. Manual fire alarm systems shall be designed and installed in accordance with Section 907.

❖ A manual fire alarm system, designed and installed as specified in Section 907.2, must be installed throughout buildings containing a Group H-5 occupancy. System activation is to set off a local alarm, with the alarm signal transmitted to the on-site emergency control station. Note that the local alarm signal is intended only for the area of alarm origin and is not intended to be a general alarm that sounds throughout the building. Refer to the commentary to Chapter 9 for manual fire alarm system requirements. Note that this section correlates with Section 415.9.8 of the IBC.

1803.12 Emergency alarm system. Emergency alarm systems shall be provided in accordance with this section, Section 2704.9 and Section 2705.4.4. The maximum allowable quantity per control area provisions of Section 2704.1 shall not apply to emergency alarm systems required for HPM.

❖ This section requires an emergency alarm system in all areas where HPM is transported or stored. It also clarifies that the requirement for an emergency alarm system in a Group H-5 facility in the locations identified in Sections 1803.12.1 through 1803.12.1.3 is not dependent on whether the maximum allowable quantities per control area of Section 2704.1 are exceeded. Emergency alarm systems must comply with Section 908.

1803.12.1 Where required. Emergency alarm systems shall be provided in the areas indicated in Sections 1803.12.1.1 through 1803.12.1.3.

❖ This section states that emergency alarm systems must be installed in the locations defined in Sections 1803.12.1.1 through 1803.12.1.3.

1803.12.1.1 Service corridors. An approved emergency alarm system shall be provided in service corridors, with at least one alarm device in the service corridor.

❖ An emergency telephone system or manual alarm pull station that is readily accessible to personnel is required in service corridors. These devices must set off an alarm at the emergency control station as well as activate a local audible signal.

1803.12.1.2 Exit access corridors and exit enclosures. Emergency alarms for exit access corridors and exit enclosures shall comply with Section 2705.4.4.

❖ Because HPM materials would not be in exit access corridors or exit enclosures unless they were being transported to another approved area, the emergency alarm requirements of Section 2705.4.4 for dispensing, use and handling must be complied with.

1803.12.1.3 Liquid storage rooms, HPM rooms and gas rooms. Emergency alarms for liquid storage rooms, HPM rooms and gas rooms shall comply with Section 2704.9.

❖ This section requires compliance with the emergency alarm requirements of Section 2704.9 for hazardous materials in a storage condition. This section addresses storage areas that by their designation contain HPM in quantities greater than those listed in Table 2703.1.1(1) or 2703.1.1(2).

1803.12.2 Alarm-initiating devices. An approved emergency telephone system, local alarm manual pull stations, or other approved alarm-initiating devices are allowed to be used as emergency alarm-initiating devices.

❖ This section classifies what constitutes an approved alarm-initiating device, affording maximum design flexibility.

1803.12.3 Alarm signals. Activation of the emergency alarm system shall sound a local alarm and transmit a signal to the emergency control station.

❖ The alarm signal must be transmitted to the emergency control station to notify trained personnel of an emergency condition. A local alarm is required to alert the occupants of a potential hazardous condition.

1803.13 Continuous gas detection systems. A continuous gas detection system shall be provided for HPM gases when the physiological warning properties of the gas are at a higher level than the accepted permissible exposure limit (PEL) for the gas and for flammable gases in accordance with this section.

❖ A gas detection system in the room or area used for the storage or use of HPM gases provides early notification of a leak that is occurring before the escaping gas reaches hazardous concentration levels.

1803.13.1 Where required. A continuous gas detection system shall be provided in the areas identified in Sections 1803.13.1.1 through 1803.13.1.4.

❖ Sections 1803.13.1.1 through 1803.13.1.4 prescribe the locations in a Group H-5 facility where a gas detection system is required.

1803.13.1.1 Fabrication areas. A continuous gas detection system shall be provided in fabrication areas when gas is used in the fabrication area.

❖ Fabrication areas that use HPM gases must have a gas detection system. It should be noted that gas detection is often installed within workstations as a means of early detection of leaks. Such detection is generally not acceptable as an alternative to gas detection for the fabrication area because a leak may occur at locations that are remote from the workstation.

1803.13.1.2 HPM rooms. A continuous gas detection system shall be provided in HPM rooms when gas is used in the room.

❖ HPM rooms, which by definition contain more than the quantities of hazardous materials per control area permitted by Tables 307.7(1) and 307.7(2) of the IBC, are required to have a gas detection system.

1803.13.1.3 Gas cabinets, exhausted enclosures and gas rooms. A continuous gas detection system shall be provided in gas cabinets and exhausted enclosures. A continuous gas detection system shall be provided in gas rooms when gases are not located in gas cabinets or exhausted enclosures.

❖ In the potential event of a leaking cylinder of a hazardous gas, gas cabinets, exhausted enclosures and gas rooms must have a gas detection system.

1803.13.1.4 Exit access corridors. When gases are transported in piping placed within the space defined by the walls of an exit access corridor and the floor or roof above the exit access corridor, a continuous gas detection system shall be provided where piping is located and in the exit access corridor.

> **Exception:** A continuous gas detection system is not required for occasional transverse crossings of the corridors by supply piping which is enclosed in a ferrous pipe or tube for the width of the corridor.

❖ The installation of HPM piping in the space above an exit access corridor or another occupancy, as well as the cavity of the egress corridor wall, presents a potential source of hazard to the building's occupants; therefore, a gas detection system is required for early notification of a potential leak of an HPM gas. The exception recognizes that when the piping traverses a corridor, the use of a coaxial enclosed pipe around the HPM piping is considered acceptable for the required separation and containment of a potential leak. The assumption is that the open ends of that pipe are in an HPM facility and, therefore, a leak into the outer casing can be monitored. If the adjacent areas that contain the open ends are not in an HPM facility, the outer-jacket method cannot be used.

1803.13.2 Gas detection system operation. The continuous gas detection system shall be capable of monitoring the room, area or equipment in which the gas is located at or below the permissible exposure limit (PEL) or ceiling limit of the gas for which detection is provided. For flammable gases, the monitoring detection threshold level shall be vapor concentrations in excess of 20 percent of the lower flammable limit (LFL). Monitoring for highly toxic and toxic gases shall also comply with Chapter 37.

❖ Gas detection is required based on the potential for health-threatening levels as established by nationally accepted health standards such as those used by the Occupational Safety and Health Administration (OSHA). The permissible exposure limitation (PEL) of a gas is the legal limitation for long-term exposure (8 to 10 hours normally). The American Conference of Government Industrial Hygienists (ACGIH) publishes threshold limit values (TLVs) based on a time-weighted average. State and local laws may also contain limits.

Additionally, gas detection must be installed when dispensing occurs that may result in vapor concentrations in excess of 20 percent of the lower explosive limit (LEL). LELs can be obtained from suppliers or other printed sources, such as the NFPA *Fire Protection Guide to Hazardous Materials*. Chapter 37 contains additional requirements for the monitoring of highly toxic and toxic compressed gases.

1803.13.2.1 Alarms. The gas detection system shall initiate a local alarm and transmit a signal to the emergency control station when a short-term hazard condition is detected. The alarm shall be both visual and audible and shall provide warning both inside and outside the area where the gas is detected. The audible alarm shall be distinct from all other alarms.

❖ The required local alarm is intended to alert occupants to a hazardous condition in the vicinity of where the HPM gases are being stored or used. The alarm is not intended to be an evacuation alarm; however, it must be monitored to hasten emergency personnel response.

1803.13.2.2 Shut off of gas supply. The gas detection system shall automatically close the shutoff valve at the source on gas supply piping and tubing related to the system being monitored for which gas is detected when a short-term hazard condition is detected. Automatic closure of shutoff valves shall comply with the following:

1. Where the gas-detection sampling point initiating the gas detection system alarm is within a gas cabinet or exhausted enclosure, the shutoff valve in the gas cabinet or

exhausted enclosure for the specific gas detected shall automatically close.

2. Where the gas-detection sampling point initiating the gas detection system alarm is within a room and compressed gas containers are not in gas cabinets or exhausted enclosure, the shutoff valves on all gas lines for the specific gas detected shall automatically close.

3. Where the gas-detection sampling point initiating the gas detection system alarm is within a piping distribution manifold enclosure, the shutoff valve supplying the manifold for the compressed gas container of the specific gas detected shall automatically close.

Exception: Where the gas-detection sampling point initiating the gas detection system alarm is at the use location or within a gas valve enclosure of a branch line downstream of a piping distribution manifold, the shutoff valve for the branch line located in the piping distribution manifold enclosure shall automatically close.

❖ Where gas detection systems are required, automatic emergency shutoff valves are required to stop the flow of hazardous materials from possibly deteriorating further in an emergency.

1803.14 Exhaust ventilation systems for HPM. Exhaust ventilation systems and materials for exhaust ducts utilized for the exhaust of HPM shall comply with this section, other applicable provisions of this code, the *International Building Code* and the *International Mechanical Code.*

❖ This section is a summary of the ventilation requirements for semiconductor facilities and references the IBC and IMC for further requirements. Parallel requirements may be found in Section 502.9 of the IMC.

1803.14.1 Where required. Exhaust ventilation systems shall be provided in the following locations in accordance with the requirements of this section and the *International Building Code*:

1. Fabrication areas: Exhaust ventilation for fabrication areas shall comply with the *International Building Code.* The fire code official is authorized to require additional manual control switches.

2. Workstations: A ventilation system shall be provided to capture and exhaust fumes and vapors at workstations.

3. Liquid storage rooms: Exhaust ventilation for liquid storage rooms shall comply with Section 2704.3.1 and the *International Building Code.*

4. HPM rooms: Exhaust ventilation for HPM rooms shall comply with Section 2704.3.1 and the *International Building Code.*

5. Gas cabinets: Exhaust ventilation for gas cabinets shall comply with Section 2703.8.6.2. The gas cabinet ventilation system is allowed to connect to a workstation ventilation system. Exhaust ventilation for gas cabinets containing highly toxic or toxic gases shall also comply with Chapter 37.

6. Exhausted enclosures: Exhaust ventilation for exhausted enclosures shall comply with Section 2703.8.5.2. Exhaust ventilation for exhausted enclosures containing highly toxic or toxic gases shall also comply with Chapter 37.

7. Gas rooms: Exhaust ventilation for gas rooms shall comply with Section 2703.8.4.2. Exhaust ventilation for gas cabinets containing highly toxic or toxic gases shall also comply with Chapter 37.

❖ Items 1 through 7 list the specific ventilation requirements for semiconductor facilities based on the use of the particular area. Many of these requirements are references to other sections of the code. These are essentially the same requirements found in Section 502.9 of the IMC.

1. Fabrication areas must be ventilated in accordance with Section 415.9 of the IBC. This section also gives the fire code official authority to ask for additional manual control switches where facility arrangements make the additional switches necessary.

2. Workstations, typically found within each fabrication area, must have individual exhaust systems to collect any exhaust fumes and vapors as close to the point of generation as possible.

3. Liquid storage rooms within semiconductor facilities are to be treated the same as other liquid storage rooms and should be ventilated to meet the general requirements for hazardous materials found in Section 2704.

4. HPM rooms, which are essentially a Group H-2, H-3 or H-4 occupancy, must also be ventilated in the same way as any other Group H-2, H-3 and H-4 occupancy would be ventilated in accordance with Section 2704.

5. Gas cabinets used in semiconductor facilities should be treated the same as any other use of gas cabinets with hazardous materials. Gas cabinets can share the exhaust system of an individual workstation. Also, semiconductor facilities tend to make use of both toxic and highly toxic gases; therefore, additional requirements are specifically highlighted for these applications.

6. Exhausted enclosures are also regulated just as they are for other applications with hazardous materials. As with gas cabinets, semiconductor facilities tend to make use of both toxic and highly toxic gases; therefore, additional requirements are specifically highlighted for those applications.

7. Gas rooms must also be treated the same as any other gas rooms used with hazardous materials.

1803.14.2 Penetrations. Exhaust ducts penetrating fire barrier assemblies shall be contained in a shaft of equivalent fire-resistance-rated construction. Exhaust ducts shall not penetrate fire walls. Fire dampers shall not be installed in exhaust ducts.

❖ Semiconductor facilities are essentially a combination of many activities involving the storage and use of hazardous materials. These facilities are large and generally complex, and exhaust systems may lead through several areas of a building before the exhaust is processed or released to the atmosphere. This section requires that the protection surrounding an exhaust system is at least equivalent to the fire-resistance rating of the surrounding fire barriers. Ventilation should not be interrupted by a fire damper when a fire or other emergency occurs involving a workstation. This helps reduce the likelihood that hazardous combustion byproducts or hazardous concentrations of HPM will be forced back into the workstation or clean room. Continuous ventilation through a duct enclosed in a fire-resistance-rated shaft is required. Fire walls define separations between buildings. Ducts must never penetrate a barrier common to another building or occupancy. This reduces the likelihood of tampering with or interrupting the duct integrity.

1803.14.3 Treatment systems. Treatment systems for highly toxic and toxic gases shall comply with Chapter 37.

❖ Because of their toxicity hazards, highly toxic and toxic gases should be protected by gas treatment systems. See the commentary to Section 3704 on highly toxic and toxic gases and, in particular, Section 3704.2.2.7 for further discussion of treatment systems.

1803.15 Emergency power system. An emergency power system shall be provided in Group H-5 occupancies where required by Section 604. The emergency power system shall be designed to supply power automatically to required electrical systems when the normal supply system is interrupted.

❖ A backup emergency power source is considered essential for systems monitoring and protecting hazardous materials in a Group H-5 occupancy. Without an emergency power system, all required electrical controls or equipment monitoring hazardous materials would be rendered inoperative if a power failure or other electrical system failure were to occur.

1803.15.1 Required electrical systems. Emergency power shall be provided for electrically operated equipment and connected control circuits for the following systems:

1. HPM exhaust ventilation systems.
2. HPM gas cabinet ventilation systems.
3. HPM exhausted enclosure ventilation systems.
4. HPM gas room ventilation systems.
5. HPM gas detection systems.
6. Emergency alarm systems.
7. Manual fire alarm systems.
8. Automatic sprinkler system monitoring and alarm systems.
9. Electrically operated systems required elsewhere in this code or in the *International Building Code* applicable to the use, storage or handling of HPM.

❖ This section specifies the types of systems within a Group H-5 occupancy that must be connected to an approved emergency power system. As indicated in Section 604, emergency power systems must be installed in accordance with the applicable requirements of the ICC EC and NFPA 110 and 111. Note that the systems included in the list are critical to personnel safety and must remain operable under all conditions of normal power system failure or impairment.

1803.15.2 Exhaust ventilation systems. Exhaust ventilation systems are allowed to be designed to operate at not less than one-half the normal fan speed on the emergency power system when it is demonstrated that the level of exhaust will maintain a safe atmosphere.

❖ Emergency power for exhaust ventilation is required to prevent hazardous concentrations of HPM fumes or vapors in areas such as workstations or fabrication areas. Fans for exhaust ventilation draw a considerable amount of current when operating. Running exhaust fans at a reduced speed may be desirable when it will not endanger the operator or result in a hazardous condition. However, exhaust fans must not be run at a speed less than 50 percent of their rating, even if a slower speed will not produce a serious hazard.

SECTION 1804
STORAGE

1804.1 General. Storage of hazardous materials shall comply with Section 1803 and this section and other applicable provisions of this code.

❖ This section establishes proper storage conditions for hazardous materials in fabrication areas, storage areas and storage in equipment and cabinets.

1804.2 Fabrication areas. Storage of HPM in fabrication areas shall be within approved or listed storage cabinets, gas cabinets or within a workstation.

Flammable and combustible liquid storage cabinets shall comply with Chapter 34.

Hazardous materials storage cabinets shall comply with Section 2703.8.7.

Gas cabinets shall comply with Section 2703.8.6. Gas cabinets for highly toxic or toxic gases shall also comply with Chapter 37.

Workstations shall comply with Section 1805.2.2.

❖ Even though the amount of HPM in a fabrication area is controlled, it still must be within approved cabinets or a workstation. This requirement is intended to limit the exposure to occupants of the fabrication area to only the material in use within that area.

The larger amounts of HPM typically stored in separate areas present a hazard comparable to other Group H facilities. Therefore, storage rooms must meet similar code requirements. Storage rooms containing HPM in quantities greater than permitted by Tables 307.7(1) and 307.7(2) of the IBC must comply with the applicable requirements of Section 415.9.5.2 of that code, depending on the state of the material.

1804.2.1 Maximum aggregate quantities. The aggregate quantities of hazardous materials stored and used in a single fabrication area shall not exceed the quantities set forth in Table 1804.2.1.

Exception: Fabrication areas containing quantities of hazardous materials not exceeding the maximum allowable quantities per control area established by Chapters 27 and 34.

❖ This section regulates the total amount of hazardous materials, whether in use or in storage, within a single fabrication area based on the density/quantity of material specified in Table 1804.2.1. The exception permits a fabrication area to have a total quantity of HPM of either the quantity specified in Table 1804.2.1 or the maximum allowable quantities per control area specified in Table 2703.1.1(1) or 2703.1.1(2), whichever is greater.

For example, a small fabrication area may have a quantity in accordance with Table 2703.1.1(1) or 2703.1.1(2), even though it may exceed the quantities specified in Table1804.2.1. Conversely, a large fabrication area may have an aggregate quantity that exceeds those specified in Table 2703.1.1(1) or 2703.1.1(2) if it does not exceed the quantities specified in Table 1804.2.1.

Again, this section refers to an aggregate quantity of hazardous materials that is in use and storage within a single fabrication area. Section 1804.2.2 further limits the amount of HPM being stored within a fabrication area.

TABLE 1804.2.1. See page 18-16.

❖ The quantity limitations identified in Table 1804.2.1 designate the aggregate amounts of hazardous materials allowed to be both stored and used in each fabrication area. The density basis for managing the maximum quantity of specified hazardous materials controls the hazard distribution within the occupancy. Generally, the lower the permitted density, the greater the required separation between containers.

1804.2.2 Maximum quantities of HPM. The maximum quantities of HPM stored in a single fabrication area shall not exceed the maximum allowable quantities per control area established by Chapters 27 and 34.

❖ This section gives the overall limit of HPM that can be stored within a single fabrication area based on Table 2703.1.1(1) or 2703.1.1(2). Any storage in excess of the permitted maximum allowable quantities would have to be stored in an HPM storage room, liquid storage room or gas room. The aggregate quantities of hazardous materials in both storage and use within a single fabrication area are regulated by Section 1804.2.1.

1804.3 Storage rooms. The storage of HPM in quantities greater than those listed in Chapters 27 and 34 shall be in a room complying with the requirements of the *International Building Code* and this code for a liquid storage room, HPM room or gas room as appropriate for the materials stored. The storage of other hazardous materials shall comply with Chapter 27 and other applicable provisions of this code.

❖ This section deals with construction requirements for two types of storage rooms: those classified as HPM rooms or gas rooms and those used as liquid storage rooms. The size and separation of these rooms is dependent on the type of materials stored.

1804.3.1 Separation of incompatible hazardous materials. Incompatible hazardous materials in storage shall be separated from each other in accordance with Section 2703.9.8.

❖ This section intends to prevent potentially incompatible materials from mixing or reacting with each other. In most instances, a noncombustible partition may suffice as the minimum required level of separation. Compliance with Section 2703.9.8 is required.

SECTION 1805 USE AND HANDLING

1805.1 General. The use and handling of hazardous materials shall comply with this section, Section 1803 and other applicable provisions of this code.

❖ Section 1805 addresses the use of hazardous materials in fabrication areas, and transportation and handling of hazardous materials in buildings containing fabrication areas. Section 1803, Chapter 27 and the chapters of the code specific to hazardous materials are to be applied to the use and handling of other hazardous materials in the building.

1805.2 Fabrication areas. Hazardous production materials located in fabrication areas shall be within approved or listed storage cabinets, gas cabinets or within a workstation.

❖ This section addresses four conditions for hazardous materials located within a fabrication area:

1. Hazardous materials in both use and storage must be within approved or listed cabinets or within a workstation.
2. Section 1804 quantity limits for hazardous materials in use and storage apply.
3. The amount of HPM per individual workstation is limited.
4. Workstation construction, drainage and containment and clearance requirements apply.

TABLE 1804.2.1
QUANTITY LIMITS FOR HAZARDOUS MATERIALS IN A SINGLE FABRICATION AREA IN GROUP H-5[a]

HAZARD CATEGORY	SOLIDS (pounds/square foot)	LIQUIDS (gallons/square foot)	GAS (cubic foot@NTP/square foot)
PHYSICAL-HAZARD MATERIALS			
Combustible dust	Note b	Not Applicable	Not Applicable
Combustible fiber			
Loose	Note b	Not Applicable	Not Applicable
Baled	Note b		
Combustible liquid			
Class II	Not Applicable	0.01	Not Applicable
Class IIIA		0.02	
Class IIIB		Not Limited	
Combination Class I, II and IIIA		0.04	
Cryogenic gas			
Flammable	Not Applicable	Not Applicable	Note c
Oxidizing			1.25
Explosives	Note b	Note b	Note b
Flammable gas			
Gaseous	Not Applicable	Not Applicable	Note c
Liquefied			Note c
Flammable liquid			
Class IA	Not Applicable	0.0025	Not Applicable
Class IB		0.025	
Class IC		0.025	
Combination Class IA, IB and IC		0.025	
Combination Class I, II and IIIA		0.04	
Flammable solid	0.001	Not Applicable	Not Applicable
Organic peroxide			
Unclassified detonable	Note b	Not Applicable	Not Applicable
Class I	Note b		
Class II	0.025		
Class III	0.1		
Class IV	Not Limited		
Class V	Not Limited		
Oxidizing gas			
Gaseous	Not Applicable	Not Applicable	1.25
Liquefied			1.25
Combination of Gaseous and Liquefied			1.25
Oxidizer			
Class 4	Note b	Note b	Not Applicable
Class 3	0.003	0.03	
Class 2	0.003	0.03	
Class 1	0.003	0.03	
Combination oxidizer Class 1, 2, 3	0.003	0.03	
Pyrophoric	Note b	0.00125	Notes c and d
Unstable reactive			
Class 4	Note b	Note b	Note b
Class 3	0.025	0.0025	Note b
Class 2	0.1	0.01	Note b
Class 1	Not Limited	Not Limited	Not Limited
Water reactive			
Class 3	Note b	0.00125	Not Applicable
Class 2	0.25	0.025	
Class 1	Not Limited	Not Limited	

(continued)

TABLE 1804.2.1—continued
QUANTITY LIMITS FOR HAZARDOUS MATERIALS IN A SINGLE FABRICATION AREA IN GROUP H-5[a]

HAZARD CATEGORY	SOLIDS (pounds/square foot)	LIQUIDS (gallons/square foot)	GAS (cubic foot@NTP/square foot)
HEALTH-HAZARD MATERIALS			
Corrosives	Not Limited	Not Limited	Not Limited
Highly toxics	Not Limited	Not Limited	Note c
Toxics	Not Limited	Not Limited	Note c

For SI: 1 pound per square foot = 4.882 kg/m^2, 1 gallon per square foot = 0.025 L/m^2, 1 cubic foot @ NTP/square foot = 0.305 m^3 @NTP/m^2, 1 cubic foot = 0.02832 m^3.

a. Hazardous materials within piping shall not be included in the calculated quantities.

b. Quantity of hazardous materials in a single fabrication shall not exceed the maximum allowable quantities per control area in Tables 2703.1.1(1) and 2703.1.1(2).

c. The aggregate quantity of flammable, pyrophoric, toxic and highly toxic gases shall not exceed 9,000 cubic feet at NTP.

d. The aggregate quantity of pyrophoric gases in the building shall not exceed the amounts set forth in Table 2703.8.2.

1805.2.1 Maximum aggregate quantities. The aggregate quantities of hazardous materials in a single fabrication area shall comply with Sections 1804.2.1 and 1804.2.2, and Table 1804.2.1. The quantity of HPM in use at a workstation shall not exceed the quantities listed in Table 1805.2.1.

❖ This section establishes two quantity limitations for hazardous materials located in a fabrication area:

1. Reemphasizes the Section 1804 hazardous materials quantity limitations allowed in each fabrication area by specific reference to Sections 1804.2.1 and 1804.2.2 and Table 1804.2.1 and

2. Limits the HPM quantity in use at each workstation. It is important to note that hazardous materials that have a hazard ranking of 1 and 2 (materials that are not HPM) are not subject to the same use limitations as HPM.

TABLE 1805.2.1. See page 18-18.

❖ As discussed at the beginning of this chapter, the unique engineering and operational controls inherent in semiconductor fabrication facilities justify what are, in effect, exceptions to an occupancy that would otherwise be classified as Group H-1, H-2, H-3 or H-4. Part of this justification includes limiting the maximum quantities of HPM at an individual workstation, thereby minimizing the hazard potential associated with an accident or equipment failure involving HPM. The table is intended to correlate with the HPM classifications, material conditions and quantity limitations addressed in Table 1804.2.1. It is important to understand that only HPMs are limited to the use quantities addressed in this table.

This table lists maximum quantities of several types of hazardous materials that can be maintained at a workstation, either in storage or in use. From the table it is easy to see that quantities decrease with increased hazards. The footnotes acknowledge the added safety of having approved fire-extinguishing and suppression systems in place as well as the improved conditions with a closed system.

1805.2.2 Workstations. Workstations in fabrication areas shall be constructed of materials compatible with the materials used and stored at the workstation. The portion of the workstation that serves as a cabinet for HPM gases and flammable liquids shall be noncombustible and, if of metal, shall be not less than 0.0478-inch (18 gage) (1.2 mm) steel.

❖ This section addresses five specific requirements for workstations and includes reference to both the IBC and Chapter 30 of the code.

1805.2.2.1 Protection of vessels. Vessels containing HPM located in or connected to a workstation shall be protected from physical damage and shall not project from the workstation. Hazardous gas and liquid vessels located within a workstation shall be protected from seismic forces in an approved manner in accordance with the *International Building Code*. Protection for HPM compressed gases shall also comply with Chapter 30.

❖ HPM vessels located in or connected to a workstation are to be protected from physical damage and seismic forces (see Chapter 30 of the code and Chapter 16 of the IBC). Additionally, these vessels are to be located within the workstation enclosure where they do not project from the workstation.

1805.2.2.2 Drainage and containment for HPM liquids. Each workstation utilizing HPM liquids shall have all of the following:

1. Drainage piping systems connected to a compatible system for disposition of such liquids.

2. The work surface provided with a slope or other means for directing spilled materials to the containment or drainage system.

3. An approved means of containing or directing spilled or leaked liquids to the drainage system.

❖ Federal environmental requirements make it necessary to contain and recover HPM used in the manufacturing process. The design of workstations must facilitate recovery of spilled and spent HPM. The drainage system must be designed so that incompatible materials do not mix and the system is compatible with the materials handled.

TABLE 1805.2.1
MAXIMUM QUANTITIES OF HPM AT A WORKSTATION[e]

HPM CLASSIFICATION	STATE	MAXIMUM QUANTITY
Flammable, highly toxic, pyrophoric and toxic combined	Gas	3 cylinders
Flammable	Liquid Solid	15 gallons[a, b, c] 5 pounds[b, c]
Corrosive	Gas Liquid Solid	3 cylinders Use-Open System 25 gallons[a, c] Use-Closed System: 150 gallons[a,c,f] 20 pounds[b, c]
Highly toxic	Liquid Solid	15 gallons[a, b] 5 pounds[b]
Oxidizer	Gas Liquid Solid	3 cylinders 12 gallons[a, b, c] 20 pounds[b, c]
Pyrophoric	Liquid Solid	0.5 gallon[d] See Table 1804.2.1
Toxic	Liquid Solid	15 gallons[a, b, c] 5 pounds[b, c]
Unstable reactive Class 3	Liquid Solid	0.5 gallon[b, c] 5 pounds[b, c]
Water-reactive Class 3	Liquid Solid	0.5 gallon[d] See Table 1804.2.1

For SI: 1 pound = 0.454 kg, 1 gallon = 3.785 L.

a. DOT shipping containers with capacities of greater than 5.3 gallons shall not be located within a workstation.

b. Maximum allowable quantities shall be increased 100 percent for closed systems operations. When Note c also applies, the increase for both notes shall be allowed.

c. Quantities shall be allowed to be increased 100 percent when workstations are internally protected with an approved automatic fire-extinguishing or suppression system complying with Chapter 9. When Note b also applies, the increase for both notes shall be allowed. When Note f also applies, the maximum increase allowed for both Notes c and f shall not exceed 100 percent.

d. Allowed only in workstations that are internally protected with an approved automatic fire-extinguishing or suppression system complying with Chapter 9.

e. The quantity limits apply only to materials classified as HPM.

f. Quantities shall be allowed to be increased 100 percent for nonflammable, noncombustible corrosive liquids when the materials of construction for workstations are listed or approved for use without internal fire extinguishing or suppression system protection. When Note c also applies, the maximum increase allowed for both Notes c and f shall not exceed 100 percent.

1805.2.2.3 Clearances. Workstations where HPM is used shall be provided with horizontal servicing clearances of not less than 3 feet (914 mm) for electrical equipment, gas cylinder connections and similar hazardous conditions. These clearances shall apply only to normal operational procedures and not to repair or maintenance-related work.

❖ Adequate work space clearances around workstations prevent accidents caused by unsafe adjustments for confined conditions. Operators should be able to conduct all normal operations without having to adopt special maneuvers or unusual positions. Clearance requirements do not apply for the purpose of maintenance or repair work, which should be performed only when the workstation is deactivated and safeguards are in place.

1805.3 Transportation and handling. The transportation and handling of hazardous materials shall comply with this section and other applicable provisions of this code.

❖ Transportation and handling in this section refers to the movement of hazardous materials through means of egress elements, such as exit access corridors and exit enclosures and through service corridors.

1805.3.1 Exit corridors access and exit enclosures. Exit access corridors and exit enclosures in new buildings or serving new fabrication areas shall not contain HPM except as permitted for exit access corridors by Section 415.9.6.3 of the *International Building Code.*

❖ This section is specific to transportation and handling of HPM in both new buildings containing Group H-5 occupancy operations and existing buildings containing new Group H-5 occupancy operations. Transportation or handling of HPM in exit access corridors or exit enclosures is not allowed, except for existing buildings containing existing fabrication areas as specifically addressed in Section 415.9.6.3 of the IBC and Section 1805.3.2 of the code. Note that this section is specific to HPM only and neither prohibits nor restricts the transportation or handling of materials having a health-hazard ranking of 1 or 2 in exit access corridors or exit enclosures. The general handling and transportation requirements addressed in Section 2703.10 apply to materials having a health-hazard ranking of 1 or 2 and to conditions not specifically addressed in this section.

1805.3.2 Transport in existing exit access corridors. When existing fabrication areas are altered or modified in existing buildings, HPM is allowed to be transported in existing exit access corridors when such exit access corridors comply with the *International Building Code*. Transportation in exit access corridors shall comply with Section 2703.10.

❖ Exit access corridors must comply with the requirements of Section 1016 and must be separated from fabrication areas in accordance with Section 415.9.2.2 of the IBC. Additionally, exit access corridors in new buildings are not to be used for transporting HPM to the fabrication area; HPM must be transported in service corridors or piping. This section addresses HPM facilities that existed before the adoption and enforcement of the IBC. It permits the transport of HPM in exit access corridors in existing buildings under the conditions specified in Section 415.9 of the IBC. When alterations are made to a fabrication area, those corridors must be upgraded.

1805.3.3 Service corridors. When a new fabrication area is constructed, a service corridor shall be provided where it is necessary to transport HPM from a liquid storage room, HPM room, gas room or from the outside of a building to the perimeter wall of a fabrication area. Service corridors shall be designed and constructed in accordance with the *International Building Code*.

❖ Under this section, in new buildings, service passageways are separate spaces dedicated to the transportation of HPM between cutoff rooms, the perimeter walls of fabrication areas and the building's exterior. Service passageways must not be used as elements of a means of egress. Requirements of Section 1805.3.4 must govern the number, capacity and type of HPM containers transported in service passageways.

1805.3.4 Carts and trucks. Carts and trucks used to transport HPM in exit acess corridors and exit enclosures in existing buildings shall comply with Section 2703.10.3.

❖ This section identifies the requirements for carts and trucks used to transport HPM in exit corridors and exit enclosures in existing buildings containing existing fabrication areas. Both Sections 1805.3.2 and 1804.3.4 are specific to HPM, and specific only to HPM in glass containers, HPM gas cylinders and to the carts and trucks allowed for transport of HPM through exit access corridors and exit enclosures. The general handling and transportation requirements addressed in Section 2703.10 apply to materials having a hazard ranking of 1 or 2 and to conditions not specifically addressed in Sections 1805.3.2 and 1805.3.4.

1805.3.4.1 Identification. Carts and trucks shall be marked to indicate the contents.

❖ Cart or truck enclosures must bear placards or signs indicating the contents transported. A combination of Department of Transportation (DOT) and NFPA 704 placards is ideal for this purpose. Materials possibly posing a danger of reaction if mixed should never be transported together.

Bibliography

The following resource materials are referenced in this chapter or are relevant to the subject matter addressed in this chapter.

Fire Protection Guide to Hazardous Materials, 13th edition. Quincy, MA: National Fire Protection Association, 2001.

Fluer, L. and A. Goldberg. *H-6 Design Guide to the Uniform Codes for High Tech Facilities*. Codes and Standards Information Company & GRDA Publications, 1st edition, 1986.

IBC-2003, *International Building Code.* Falls Church, VA: International Code Council, 2003.

IMC-2003, *International Mechanical Code*. Falls Church, VA: International Code Council, 2003.

NFPA 13-99, *Installation of Sprinkler Systems.* Quincy, MA: National Fire Protection Association, 1999.

NFPA 70-02, *National Electrical Code.* Quincy, MA: National Fire Protection Association, 2002.

NFPA 110-99, *Emergency and Standby Power Systems.* Quincy, MA: National Fire Protection Association, 1999.

NFPA 111-01, *Stored Electrical Energy Emergency and Standby Power Systems.* Quincy, MA: National Fire Protection Association, 2001.

NFPA 318-02, *Protection of Semiconductor Fabrication Facilities.* Quincy, MA: National Fire Protection Association, 2002.

NFPA 704-96, *Identification of the Fire Hazards of Materials.* Quincy, MA: National Fire Protection Association, 1996.

Triconex, *Improving Fab Performance by Upgrading the Reliability of Your Critical Control and Life Safety Systems.* 2002 Invensys Systems Company, Document Number INDTP01, February 2002.

UL 2360-00, *Test Methods for Determining the Combustibility Characteristics of Plastics Used in Semi-Conductor Tool Construction.* Northbrook, IL: Underwriters Laboratories Inc., 2000.

White, Logan T. *Hazardous Gas Monitoring – A Guide for Semiconductor and Other Hazardous Occupancies.* , Logan T. White Engineering, August 1997.

Chapter 19: Lumber Yards and Woodworking Facilities

General Comments

Woodworking and forest products processing facilities tend to be located close to the source of raw material, but facilities manufacturing finished products for the building trade industries may be found in or near most urban areas. Any facility using mechanical methods to work wood into a more finished form produces dust or debris. The smaller the fuel package, the more easily ignitable it will be. In fact, extremely fine wood dust or wood flour may even produce deflagrations (explosions) under the right conditions. Explosion prevention practices must be implemented to minimize the potential for such hazards. Accumulations of dust must be prevented, controlled mechanically or frequently removed. Automatic fire suppression systems are required by Section 903.2.3.1 of the code. Deflagration venting (see NFPA 69 and NFPA 495) is required where substantial amounts of dust are usually present. Areas where dust suspensions exist are considered Class II hazardous locations by NFPA 70. Manual fire-fighting equipment must be provided in the immediate vicinity of each machine because dust cannot be completely eliminated in any woodworking or wood-processing area.

Methods and procedures used to control hazards associated with wood and forest products storage, sale and processing must recognize not only the fuel properties but the need to control and eliminate ignition sources, minimize the effect of fires and explosions and facilitate fire control efforts by occupants at the incipient phase and fire fighters at more advanced stages. Adherence to these requirements will help secure these objectives.

Purpose

Most everyone is familiar with the hazards of wood as a fuel. However, the scale of wood and forest products storage, mercantile, manufacturing and processing operations stretches the limitations of their understanding. Wood has long been used as a reference fuel for understanding fire, but the sheer volume of materials present and the variety of fuel packages represented in these occupancies make this a hazard requiring special attention.

Sawdust, wood chips, shavings, bark mulch, shorts, finished planks, sheets, posts, poles, timber and raw logs represent a broad continuum of fuels. What they all bear in common is the hazard they represent once ignited, and what is unique about each of these materials is the ease or difficulty with which any single fuel package may be ignited. The regulations recognize both of these concerns.

Provisions of this chapter are intended to prevent fires and explosions, facilitate fire control and reduce exposures to and from facilities storing, selling or processing wood and forest products.

This chapter requires active and passive fire protection features to reduce on- and off-site exposures, limit fire size and development and facilitate fire fighting by employees and the fire service. Design and maintenance of these facilities must provide access to equipment and exposures to facilitate fire control.

SECTION 1901 GENERAL

1901.1 Scope. The storage, manufacturing and processing of timber, lumber, plywood, veneers and byproducts shall be in accordance with this chapter.

❖ These provisions are intended to prevent fires and explosions in lumber yards, woodworking plants, lumber drying and other similar operations. Furthermore, provisions are intended to facilitate fire-fighting operations in these occupancies in the event of a fire.

1901.2 Permit. Permits shall be required as set forth in Section 105.6.

❖ Section 105.6.26 has a permit threshold amount for lumber storage of 100,000 board foot of lumber. The board foot is a measurement used in the lumber and building trades. One board foot (.93 m^2) is the cubic volume of a piece of lumber 1 foot square (.93 m^2) and 1 inch (25 mm) thick or approximately 0.083 cubic foot (.002 m^3). Then, 100,000 board feet is a cubic volume of 8,333 cubic feet (236 m^3) or a solid pile of lumber approximately 20 feet and 5 inches by 20 feet and 5 inches (6223 mm by 6223 mm) by the maximum allowable height of 20 feet (6096 mm). The permit requirement is intended to facilitate the regulation of hazards presented by qualitites of lumber exceeding that amount.

SECTION 1902 DEFINITIONS

1902.1 Definitions. The following words and terms shall, for the purposes of this chapter and as used elsewhere in this code, have the meanings shown herein.

❖ Definitions of terms can help in the understanding and application of the code requirements. The purpose for including those definitions that are associated with the subject matter of this chapter is to provide more convenient access to them without having to refer back to Chapter 2. It is important to emphasize that these terms are not exclusively related to this chapter but are applicable everywhere the term is used in the code. For convenience, these terms are also listed in Chapter 2 with a cross reference to this section. The use and application of all defined terms, including those defined in this section, are set forth in Section 201.

COLD DECK. A pile of unfinished cut logs.

❖ A cold deck is a pile of ranked logs that have different lengths. The lengths are usually greater than 8 feet (2438 mm) and up to 50 feet (15 240 mm) long.

FINES. Small pieces or splinters of wood byproducts that will pass through a 0.25-inch (6.4 mm) screen.

❖ Fines range in size from sawdust to chips and are usually accumulated in piles.

HOGGED MATERIALS. Wood waste materials produced from the lumber production process.

❖ This term refers to mill waste that may include a mixture of bark, chips or dust along with other byproducts of trees. Material designated as hogged fuel is included in this category.

PLYWOOD and VENEER MILLS. Facilities where raw wood products are processed into finished wood products, including waferboard, oriented strandboard, fiberboard, composite wood panels and plywood.

❖ Veneer mills are unique because the milling is done with knives rather than saws. Veneer logs are often air dried for an extended period of time before processing.

RAW PRODUCT. A mixture of natural materials such as tree, brush trimmings, or waste logs and stumps.

❖ A raw product is material brought to the woodworking facility directly from the forest or wood-producing area.

STATIC PILES. Piles in which processed wood product is mounded and is not being turned or moved.

❖ Static piles are long-term bulk storage piles that must be monitored for internal heat buildup.

TIMBER and LUMBER PRODUCTION FACILITIES. Facilities where raw wood products are processed into finished wood products.

❖ Wood is still our most used structural material. With the ever increasing demand for wood products, growth of the forest products industry continues. Some woodworking facilities are a one-man shop while others employ as many as 2,000 people.

SECTION 1903 GENERAL REQUIREMENTS

1903.1 Open yards. Open yards required by the *International Building Code* shall be maintained around structures.

❖ Bulk dimension lumber represents a fuel package of considerable volume in occupancies such as wholesale and retail lumber yards and distribution warehouses. These facilities are often located in urban areas and close to the consumer, thus complicating matters. Though dimension lumber may be harder to ignite than sawdust, scrap or waste, once ignited it could possibly expose more people and property to danger.

1903.2 Dust control. Equipment or machinery located inside buildings which generates or emits combustible dust shall be provided with an approved dust collection and exhaust system installed in accordance with Chapter 13 and the *International Mechanical Code*. Equipment or systems that are used to collect, process or convey combustible dusts shall be provided with an approved explosion control system.

❖ There are two basic designs of dust collection systems. One is a single-stage system that consists of a single dust collector in the form of a cyclone separator or a baghouse. The other is a two-stage system that consists of a cyclone separator followed by a bag-type filter house that is usually called a "bag house."

1903.2.1 Explosion venting. Where a dust explosion hazard exists in equipment rooms, buildings or other enclosures, such areas shall be provided with explosion (deflagration) venting or an approved explosion suppression system complying with Section 911.

❖ When wood dust is ignited while suspended in air in concentrations above the minimum combustible concentration (MCC), it burns violently, producing a deflagration. This likelihood depends on the concentration of fuel per unit volume. Most of the time, dust collection systems remove the larger dust particles, allowing the finer particles to pass through the system and reenter the production area. The finer dust can create an explosion and fire hazard unless an approved suppression system is used.

1903.3 Waste removal. Sawmills, planning mills and other woodworking plants shall be equipped with a waste removal system that will collect and remove sawdust and shavings. Such systems shall be installed in accordance with Chapter 13 and the *International Mechanical Code.*

Exception: Manual waste removal when approved.

❖ Like many other organic materials, sawdust, wood chips and other similar waste materials are susceptible to spontaneous heat. Pilings and debris must be turned over or consumed regularly to minimize the potential for spontaneous combustion. Some waste material is used as fuel, and some might go to pulp or particleboard manufacturers. Finer material is removed by air-moving equipment, cyclones or air bags. Chapter 13 and the *International Mechanical Code*® (IMC®) give the installation requirements for waste removal systems.

1903.3.1 Housekeeping. Provisions shall be made for a systematic and thorough cleaning of the entire plant at sufficient intervals to prevent the accumulations of combustible dust and spilled combustible or flammable liquids.

❖ Good housekeeping should be maintained at all times, including regular and frequent cleaning of material-handling equipment. Combustible waste materials, such as bark, chips and other debris, should not be allowed to accumulate in amounts that will constitute a fire hazard. Air-moving systems should be designed to remove or dilute transient flammable vapors.

1903.3.2 Metal scrap. Provision shall be made for separately collecting and disposing of any metal scrap so that such scrap will not enter the wood handling or processing equipment.

❖ Waste material from rough milling is carried by a belt conveyor to the wood hog, which cuts the waste into small pieces. The conveyor should have a magnetic separator to keep metal from causing fires in the hog, pneumatic conveying and waste-storage silos. Waste material that is swept up and carried to the hog is hazardous because the trash often contains metal objects.

1903.4 Electrical equipment. Electrical wiring and equipment shall comply with the ICC *Electrical Code*.

❖ Arcs, sparks and loose connections often supply the ignition source that a vapor or combustible material needs for fire. Electrical equipment must be listed for the use it is installed for and comply with the requirements of the ICC *Electrical Code*® (ICC EC™).

1903.5 Control of ignition sources. Protection from ignition sources shall be provided in accordance with Sections 1903.5.1 through 1903.5.3.

❖ The two key points that will reduce fire losses in storage areas of forest products are reduction in the sources of ignition and a positive program for detecting incipient fires. The following sections address sources of ignition. Section 1904 addresses programs and systems for early detection of fires.

1903.5.1 Cutting and welding. Cutting and welding shall comply with Chapter 26.

❖ No cutting, welding or other use of open flames and spark-producing equipment are allowed in the storage area without a permit.

1903.5.2 Static electricity. Static electricity shall be prevented from accumulating on machines and equipment subject to static electricity buildup by permanent grounding and bonding wires or other approved means.

❖ Static electricity builds up on operating machines because of the movement of product and machine. Proper grounding and bonding of equipment prevents static electricity from producing an ignition source and igniting a fire.

1903.5.3 Smoking. Where smoking constitutes a fire hazard, the fire code official is authorized to order the owner or occupant to post approved "No Smoking" signs complying with Section 310. The fire code official is authorized to designate specific locations where smoking is allowed.

❖ Smoking must not be allowed in a woodworking facility. Considering the abundance of existing fuel, with much of it in an easily ignitable form, smoking controls are an absolute necessity in lumber yards and forest products facilities. Smoking materials being discarded improperly is an ignition threat in woodworking facilities.

1903.6 Fire apparatus access roads. Fire apparatus access roads shall be provided for buildings and facilities in accordance with Section 503.

❖ Required driveways provide access for fire apparatus and create fire breaks between piles. Access roads must be sized to allow fire emergency equipment to enter and perform emergency operations. Section 503 gives required dimensions for access roads.

1903.7 Access plan. Where storage pile configurations could change because of changes in product operations and processing, the access plan shall be submitted for approval when required by the fire code official.

❖ A site plan showing access to the facilities with consideration given to pile location, access roads, hydrant locations and building locations must be presented to the fire code official for approval prior to construction.

SECTION 1904
FIRE PROTECTION

1904.1 Fire alarms. An approved means for transmitting alarms to the fire department shall be provided in timber and lumber production mills and plywood and veneer mills.

❖ The fire department must be notified if a fire occurs in the facility. How the fire department is notified must be approved by the local fire code official. This may be by manual fire alarm system with pull stations, monitored phone lines or another approved system.

1904.1.1 Manual fire alarms. A manual fire alarm system complying with Section 907.2 shall be installed in areas of timber and lumber production mills and for plywood and veneer mills that contain product dryers.

Exception: Where dryers or other sources of ignition are protected by a supervised automatic sprinkler system complying with Section 903.

❖ Manual fire alarm systems usually have means for local notification, but according to Section 1904.1, alarm systems must have the means to transmit the alarm to fire departments.

1904.2 Portable fire extinguishers and hose. Portable fire extinguishers or standpipes and hose supplied from an approved water system shall be provided within 50 feet (15 240 mm) of travel distance to any machine producing shavings or sawdust. Extinguishers shall be provided in accordance with Section 906 for extra-high hazards.

❖ The degree of protection will vary from facility to facility, but the basic recommendation is for a water system of mains and hydrants capable of supplying at least 1,000 gpm (60 L/s). Standpipes will provide a hose within 50 feet (15 240 mm) of shavings and sanding machines. Portable fire extinguishers are good initial fire knock-down equipment. Section 906 gives these requirements.

1904.3 Automatic sprinkler systems. Automatic sprinkler systems shall be installed in accordance with Section 903.3.1.1.

❖ Woodworking facilities have wood products (raw and finished) stored and laying everywhere. Automatic sprinklers are the considered choice for general fire protection; however, in some older buildings and facilities, addition of sprinklers may be difficult. A general-area protection design may be more feasible. Again, the design must be approved by the local fire code official.

SECTION 1905 PLYWOOD, VENEER AND COMPOSITE BOARD MILLS

1905.1 General. Plant operations of plywood, veneer and composite board mills shall comply with this section.

❖ This section addresses protection of dryers and installation requirements of thermal oil-heating systems.

1905.2 Dryer protection. Dryers shall be protected throughout by an approved, automatic deluge water-spray suppression system complying with Chapter 9. Deluge heads shall be inspected quarterly for pitch buildup. Deluge heads shall be flushed during regular maintenance for functional operation. Manual activation valves shall be located within 75 feet (22 860 mm) of the drying equipment.

❖ A deluge system is a good protective measure for the interior of dryers. This kind of protection is especially important in veneer mills where the temperature of the dryer may approach 392°F (200°C), which is the nominal ignition temperature of wood. Smoke sensors should activate the deluge system and automatically shut down fans, burners and drive machinery. Manual trips should also be installed at each end of the dryer.

1905.3 Thermal oil-heating systems. Facilities that use heat transfer fluids to provide process equipment heat through piped, indirect heating systems shall comply with this code and NFPA 664.

❖ The transfer of heated fluids for process equipment presents a fire hazard. A ruptured pipe line could expose fine particles of wood products to the heated fluid, thus creating the potential for a fire. The code and Chapter 9 of NFPA 664 address the requirements for installation of this equipment.

SECTION 1906 LOG STORAGE AREAS

1906.1 General. Log storage areas shall comply with this section.

❖ The intent of this section is to provide fire protection advice for minimizing fire hazards in log-yard storage areas that contain saw, plywood or pulpwood logs stored in ranked piles, usually referred to as "cold decks." Stacked piles of cordwood are not addressed unless they are stored in ranked piles.

1906.2 Cold decks. Cold decks shall not exceed 500 feet (152.4 m) in length, 300 feet (91 440 mm) in width and 20 feet (6096 mm) in height. Cold decks shall be separated from adjacent cold decks or other exposures by a minimum of 100 feet (30 480 mm).

Exception: The size of cold decks shall be determined by the fire code official where the decks are protected by special fire protection including, but not limited to, additional fire flow, portable turrets and deluge sets, and hydrant hose houses equipped with approved fire-fighting equipment capable of reaching the entire storage area in accordance with Chapter 9.

❖ Because of the fire load that they present, cold decks must be limited in size and isolated from one another as prescribed in this section. All sides of the cold deck should be accessible to fire apparatus over hard-surface fire apparatus access roads. The suggested width of a cold deck fire apparatus access road is one and a half times the height of the pile but a minimum of 20 feet (6096 mm) between alternate rows of two pile groups separated by a clear space of 100 feet (30 480 mm). See the commentary to Section 503 for further discussion of fire apparatus access roads.

1906.3 End stops. Log and pole piles shall be stabilized by approved means.

❖ Stabilizing the log and pole piles minimizes hazards in two ways. In a fire, the piles become weak and can fall on the fire fighters, but the stabilizing method will assist in their protection. The method also minimizes the risk of logs falling on people and causing death or injury when no fire emergency exists.

SECTION 1907 STORAGE OF WOOD CHIPS AND HOGGED MATERIAL ASSOCIATED WITH TIMBER AND LUMBER PRODUCTION FACILITIES

1907.1 General. The storage of wood chips and hogged materials associated with timber and lumber production facilities shall comply with this section.

❖ This section gives fire protection guidance for minimizing fire hazards in yard storage areas containing wood chips and hogged materials. Each individual facility will have its own special conditions for yard use, handling procedures and topography. Fire safety is also affected by climate conditions, wood species and the age of the piles.

1907.2 Size of piles. Piles shall not exceed 60 feet (18 288 mm) in height, 300 feet (91 440 mm) in width and 500 feet (152 m) in length. Piles shall be separated from adjacent piles or other exposures by approved fire apparatus access roads.

Exception: The fire code official is authorized to allow the pile size to be increased when additional fire protection is provided in accordance with Chapter 9. The increase shall be based on the capabilities of the system installed.

❖ Restrictions on pile size are established for practical reasons. When the piles are made low and narrow, fire extinguishment is enhanced. When the piles are extremely high and wide, fire extinguishment is hampered and arrangements must be made for fire-fighting service. Piles should also be constructed with an access roadway to the top of the pile to make all parts of the pile accessible to fire fighters.

1907.3 Pile fire protection. Automatic sprinkler protection shall be provided in conveyor tunnels and combustible enclosures that pass under a pile. Combustible or enclosed conveyor systems shall be equipped with an approved automatic sprinkler system.

❖ Automatic sprinklers are needed because of the difficulty of manual fire suppression operations in concealed, enclosed and elevated areas.

1907.4 Material-handling equipment. Approved material-handling equipment shall be readily available for moving wood chips and hogged material.

❖ Examples of material-handling equipment needed are power-operated shovel and scoop vehicles, dozers and similar types for moving stored material to make fire fighting easier. Pile surfaces can usually be removed with this type of equipment.

1907.5 Emergency plan. The owner or operator shall develop a plan for monitoring, controlling and extinguishing spot fires. The plan shall be submitted to the fire code official for review and approval.

❖ As in all potential hazards, an emergency plan should be designed for monitoring and fighting spot fires. The plan must be submitted to the fire code official for review and approval.

SECTION 1908 STORAGE AND PROCESSING OF WOOD CHIPS, HOGGED MATERIAL, FINES, COMPOST AND RAW PRODUCT ASSOCIATED WITH YARD WASTE AND RECYCLING FACILITIES

1908.1 General. The storage and processing of wood chips, hogged materials, fines, compost and raw product produced from yard waste, debris and recycling facilities shall comply with this section.

❖ This section contains fire protection guidance for minimizing fire hazards in yard storage areas containing wood chips and hogged materials, fines, compost and raw materials.

1908.2 Storage site. Storage sites shall be level and on solid ground or other all-weather surface. Sites shall be thoroughly cleaned before transferring wood products to the site.

❖ The storage site should be reasonably level, solid ground or paved with blacktop, concrete or other hard surface material. The surface must be thoroughly cleaned of scrap and debris before beginning a new pile.

1908.3 Size of piles. Piles shall not exceed 25 feet (7620 mm) in height, 150 feet (45 720 mm) in width and 250 feet (76 200 mm) in length.

Exception: The fire code official is authorized to allow the pile size to be increased when additional fire protection is provided in accordance with Chapter 9. The increase shall be based upon the capabilities of the system installed.

❖ Restrictions on pile sizes are established for good and practical reasons. When the piles are made low and narrow, fire extinguishment is enhanced. When the piles are extremely high and wide, fire extinguishment is hampered and arrangements must be made for fire-fighting service. The fire code official could allow an increase in the pile height if satisfactory fire protection is supplied. Piles should also be constructed with an access roadway to the top of the pile to make all parts of the pile accessible to fire fighters.

1908.4 Pile separation. Piles shall be separated from adjacent piles by approved fire apparatus access roads.

❖ Piles should be divided by fire lanes with at least 20 feet (9144 mm) clear space at the base of the piles.

1908.5 Combustible waste. The storage, accumulation and handling of combustible materials and control of vegetation shall comply with Chapter 3.

❖ Section 304 addresses vegetation and combustible waste requirements for minimizing hazards.

1908.6 Static pile protection. Static piles shall be monitored by an approved means to measure temperatures within the static piles. Internal pile temperatures shall be monitored and recorded weekly. Records shall be kept on file at the facility and made available for inspection. An operational plan indicating procedures and schedules for the inspection, monitoring and restricting of excessive internal temperatures in static piles shall be submitted to the fire code official for review and approval.

❖ Inherent to long-term bulk storage of chips, fines, compost and hogged materials is internal heating, which can progress to spontaneous combustion under certain conditions. Unless procedures are established for measuring internal temperatures, fires could burn undetected for long periods before smoke is seen or produced at the surface. These piles must be monitored on a fixed schedule and records kept of the results of these inspections.

1908.7 Pile fire protection. Automatic sprinkler protection shall be provided in conveyor tunnels and combustible enclosures that pass under a pile. Combustible conveyor systems and enclosed conveyor systems shall be equipped with an approved automatic sprinkler system.

❖ Automatic sprinklers are needed because of the difficulty of manual fire suppression operations in concealed, enclosed and elevated areas (see commentary, Section 1907.3).

1908.8 Fire extinguishers. Portable fire extinguishers complying with Section 906 and with a minimum rating of 4-A:60-B:C shall be provided on all vehicles and equipment operating on piles and at all processing equipment.

❖ Each vehicle operating in the area must be equipped with a portable fire extinguisher that provides the fire extinguishing equivalent of 5 gallons (19 L) of water for use on a substantial Class A fire and is also effective for flammable liquids and energized electrical components.

1908.9 Material-handling equipment. Approved material-handling equipment shall be available for moving wood chips, hogged material, wood fines and raw product during fire-fighting operations.

❖ Examples of material-handling equipment needed are power-operated shovel and scoop vehicles, dozers and similar types for moving stored material to make fire fighting easier. Pile surfaces can usually be removed with this type of equipment.

1908.10 Emergency plan. The owner or operator shall develop a plan for monitoring, controlling and extinguishing spot fires and submit the plan to the fire code official for review and approval.

❖ As in all potential hazards, an emergency plan should be designed for monitoring and fighting spot fires. The plan must be submitted to the fire code official for review and approval.

SECTION 1909
EXTERIOR STORAGE OF FINISHED LUMBER PRODUCTS

1909.1 General. Exterior storage of finished lumber products shall comply with this section.

❖ The following sections address the size and height of piles, access roads and fire protection requirements for exterior lumber storage.

1909.2 Size of piles. Exterior lumber storage shall be arranged to form stable piles with a maximum height of 20 feet (6096 mm). Piles shall not exceed 150,000 cubic feet (4248 m^3) in volume.

❖ To prevent stacks from becoming unstable, height should not exceed 20 feet (6096 mm). Stacks are subject to rapid fire spread through the airspaces and should, therefore, be kept as low as possible.

1909.3 Fire apparatus access roads. Fire apparatus access roads in accordance with Section 503 shall be located so that a maximum grid system unit of 50 feet by 150 feet (15 240 mm by 45 720 mm) is established.

❖ Access roads should be spaced on a minimum grid system not over 50 feet by 150 feet (15 240 mm by 45 720 mm). The access roads should be a minimum of 15 feet (45 720 mm) wide and have an all-weather surface.

1909.4 Security. Permanent lumber storage areas shall be surrounded with an approved fence. Fences shall be a minimum of 6 feet (1829 mm) in height.

> **Exception:** Lumber piles inside of buildings and production mills for lumber, plywood and veneer.

❖ A security fence at least 6 feet (1829 mm) high must be installed around the perimeter of the lumber yard for protection of property and also to prevent anyone from entering the yard without permission and either getting injured or taking items not belonging to them.

1909.5 Fire protection. An approved hydrant and hose system or portable fire-extinguishing equipment suitable for the fire hazard involved shall be provided for open storage yards. Hydrant and hose systems shall be installed in accordance with NFPA 24. Portable fire extinguishers complying with Section 906 shall be located so that the travel distance to the nearest unit does not exceed 75 feet (22 860 mm).

❖ The hydrant system should be capable of supplying four 2.5-inch (64 mm) hose streams simultaneously [1000 gpm (63 L/s)] while maintaining a positive residual pressure of at least 20 psi (138 kPa).

Bibliography

The following resource materials are referenced in this chapter or are relevant to the subject matter addressed in this chapter.

IBC-2003, *International Building Code*. Falls Church, VA: International Code Council, 2003.

ICC EC-2003, *ICC Electrical Code*. Falls Church, VA: International Code Council, 2003.

IMC-2003, *International Mechanical Code*. Falls Church, VA: International Code Council, 2003.

NFPA 24-95, *Installation of Private Fire Service Mains and their Appurtenances*. Quincy, MA: National Fire Protection Association, 1995.

NPFA 69-97, *Explosion Prevention Systems*. Quincy, MA: National Fire Protection Association, 1997.

NPFA 70-02, *National Electrical Code*. Quincy, MA: National Fire Protection Association, 2002.

NPFA 495-96, *Explosives Materials Code*. Quincy, MA: National Fire Protection Association, 1996.

NFPA 664-98, *Prevention of Fires and Explosions in Wood Processing and Woodworking Facilities*. Quincy, MA: National Fire Protection Association, 1998.

Chapter 20: Manufacture of Organic Coatings

General Comments

The term "organic coatings" is used to describe diverse compounds formulated to protect buildings, machines and objects from the effects of weather, corrosion and hostile environmental exposures. Paint for architectural and industrial uses comprises the bulk of organic coating production. Most paints remain solvent-based, though the use of water-based products is becoming more widespread. The most common solvents include: mineral spirits, naphtha (VM&P), xylene, toluene, methyl ethyl ketone (MEK), methyl isobutyl ketone (MIBK), acetone, ethyl acetate, butyl acetate, butanol, isopropanol, ethylene glycol and propylene glycol, most of which are Class I flammable liquids with flash points less than 100°F (38°C). Other coatings are asphaltic and bituminous (for example, roofing tar). Most products in this category are Class II combustible liquids with flash points between 100 and 140°F (38 and 60°C).

The manufacture of organic coatings encompasses operations that produce decorative and protective coatings for architectural uses, industrial products and other specialized purposes. Requirements of this chapter address the hazards associated with the manufacture of solvent-based organic coatings. Water-based materials are exempt from these requirements.

In 1990 Harris and Swartz explained that the manufacture of organic coatings consists of the following six steps:

- Pigment dispersion;
- Mixing of raw materials and intermediates;
- Thinning and tinting;
- Quality-control testing and adjustment;
- Filtering and
- Filling into shipping containers.

Of these six steps, those involving mixing, thinning and container filling are of the most concern. Organic solvents are producers of static electricity, especially when agitated. If dissipation of static electricity is not controlled, agitation of these fluids can release enough energy to ignite flammable vapors. Static sparks are the most common source of ignition in organic coating fires.

By far, the most hazardous organic coating to manufacture is nitrocellulose lacquer because of the release of Class I flammable liquid vapors. Nitrocellulose is commonly shipped and stored alcohol wet. This material becomes increasingly unstable as temperatures increase, thus making it capable of breaking down and burning in the absence of oxygen. Flammable vapors nearly always exist where organic coatings are manufactured. Consequently, the risk of ignition is usually present. Special precautions must be used to protect electrical and mechanical ignition sources.

Requirements of this chapter focus on the separation and control of ignition and fuel sources. Housekeeping is also emphasized to limit the quantity of ignitable material. Storage of flammable and combustible liquids must conform to Chapter 34. Application of organic coatings and other flammable finishes must conform to Chapter 15. Aerosol-charging operations and storage should conform to both Chapter 28 and NFPA 30B. Organic peroxide storage must conform to Chapter 39. Cellulose nitrate storage and handling are regulated by Chapter 42.

Hazards. In declining order, the most hazardous organic coating processes are nitrocellulose lacquer manufacturing and aerosol-charging operations, followed by those involving Class I flammable liquids, Class II combustible liquids and resin manufacturing. Manufacture of water-based products poses little hazard although many water-based coatings exhibit a flash point when tested in a closed-cup apparatus. Research conducted by the National Paint and Coatings Association (NPCA) in 1977, however, indicates that these products pose little fire hazard. Ignition of solvent vapors released from products during the manufacturing process and flammable liquid spills are considered the most serious hazards encountered in daily operations. Concern about both of these scenarios is compounded by the fact that mixing, pumping, agitation and filtering of organic chemicals generate large amounts of static electricity, which is the most common ignition source in paint and coating fires.

Nitrocellulose. Hazards associated with nitrocellulose deserve special attention. Nitrocellulose is a generic term used to describe a group of highly flammable organic fibers and other solid materials, usually cotton or wood in fibrous or finely divided form with nitrogen contents between 10.5 and 12.6 percent. The most dangerous characteristic of these materials is their ability to support combustion in the absence of oxygen. This quality makes fire extinguishment extremely difficult. These materials are usually shipped and stored in 55-gallon (208 L) drums; however, dry product may be found in either fiber or cardboard drums.

Nitrocellulose is usually stored in one of four raw forms: solvent (alcohol) wet, water wet, plasticized or dry. Of these four forms, solvent-wet nitrocellulose (with alcohol as the usual solvent) is most common, while dry is most hazardous. Dry nitrocellulose must be avoided in all operations. Solvent-wet nitrocellulose possesses the same relative hazard as the solvent. Though relatively stable at room temperature, alcohol-wet nitrocellulose decomposes rapidly as temperatures increase; therefore, maintaining room temperature and adequate ventilation is extremely important. Metal drums of the material must be carefully handled to prevent ignition. Drums should never be pushed across the floor or any other surface. Relatively small amounts of frictional heat may ignite the material inside the drum. Burning nitrocellulose produces harmful oxides of nitrogen and carbon monoxide (see commentary, Chapter 42).

Spills. Good housekeeping and prompt response to spills are extremely important fire protection practices. Though not always successful in preventing a disaster, the prompt response of the occupants can often keep the hazard contained and under control until help arrives. Such was the case at a Sherwin-Williams paint warehouse in Dayton, Ohio, in May 1987, where the quick response of employees to a small spill [8 to 10 gallons (30 to 38 L)] saved a worker whose clothing caught on fire. It is very likely that the employees' training and experience prevented their coworker's death.

After their spill control efforts failed, the employees promptly evacuated the building and activated the fire alarm system, resulting in the notification of other employees working in different areas of the building, as well as the local fire department through the central station. Sherwin-Williams employees at the Dayton facility were well trained and well drilled in spill and emergency response procedures. After the incident, employee interviews conducted by the fire department, National Fire Protection Association (NFPA) and Building Officials and Code Administrators® (BOCA) investigators suggested that this program had successfully averted disasters in the past.

Mechanical Ignition Sources. During the Sherwin-Williams warehouse incident, a Type E lift truck was operated in an area that should have been restricted to Type EE or EX lift trucks (see NFPA 505). Many fire investigators believe that the Type E lift truck was the ignition source. However, other possible causes, including static electricity, could not be eliminated. The estimated total loss in damages resulting from this fire topped $49 million, thus requiring an extensive environmental cleanup effort.

Static Electricity. Arcs and sparks from static discharge are the most common ignition sources in organic coating manufacturing plants. As stated earlier, organic chemicals produce a great deal of static electricity. Special measures are required to successfully dissipate this energy without causing ignition. NFPA 77 describes recommended practices for controlling this ignition source.

Ignition Source Control. In this case, fire prevention means preventing ignition. Because the fuel sources (flammable liquids and vapors) usually exist, as do ignition sources (static electricity, electric motors and fixtures), keeping the two apart is nearly impossible. Safeguards must, therefore, be placed to minimize the likelihood that a flammable mixture will be ignited. Some of the most fundamental practices for controlling ignition sources are:

- Installing grounds;
- Observing bonding practices;
- Installing and maintaining electrical equipment classified for hazardous locations and
- Using nonsparking tools.

Purging and inerting systems are often employed where static electricity cannot be adequately dissipated (for example, pebble mixers).

Housekeeping. Good housekeeping practices, including the cleanup of spills and residues, are imperative. Clean-up programs should follow a regular schedule, and every effort should be made to keep the plant as clean and orderly as possible. Nitrocellulose residues should be swept up using a wet-chamber vacuum only. Loose material and scraps must be stored in metal containers with tight-fitting, self-closing or automatic-closing lids.

Installed fire protection systems must conform to requirements of the *International Building Code®* (IBC®) and *International Fire Code®* (IFC®) or another building code in effect at the time of construction, addition or alteration. Automatic sprinklers, foam-water sprinklers or deluge foam-water sprinklers must be installed and maintained in accordance with the code and the appropriate referenced standard (NFPA 11, 11A, 13, 16 and Chapter 9 of the code) when required by the IBC or the code. Process hazards should be protected by special hazard systems designed, installed and maintained in accordance with the code and the appropriate NFPA standard (Table 901.6.1) where flammable liquids exist. A method must be provided to notify employees and plant fire brigade personnel if a fire occurs. Portable fire extinguishers must be provided in the organic coating area and throughout the facility as provided in Section 2003.5 and Chapter 9.

Purpose

Chapter 20 regulates materials and processes associated with the manufacture of paints, as well as bituminous, asphaltic and other organic coatings used for protective or decorative purposes. Painting and processes related to the manufacture of nonflammable and noncombustible or water-based products are exempt from the provisions of this chapter. Application of organic coatings is covered by Chapter 15. Elimination of ignition sources, maintenance of fire protection equipment and isolation or segregation of hazardous operations are emphasized.

SECTION 2001 GENERAL

2001.1 Scope. Organic coating manufacturing processes shall comply with this chapter except that this chapter shall not apply to processes manufacturing nonflammable or water-thinned coatings or to operations applying coating materials.

❖ Manufacture of flammable and combustible paints and other protective or decorative coatings is regulated by this chapter. Materials and processes associated with organic coatings manufacture may present explosion hazards from the ignition of vapors or dusts. Manufacture of nonflammable or water-based products and painting operations are not regulated by this chapter because hazards associated with those operations are insignificant.

2001.2 Permits. Permits shall be required as set forth in Section 105.6.

❖ Any manufacturing operation producing more than 1 gallon (4 L) of an organic coating per day must obtain approval from the fire code official. The process of issuing permits gives the fire code official an opportunity to carefully evaluate and regulate hazardous operations. Permit applicants should be required to demonstrate that their operations comply with the intent of the code before the permit is issued. See the commentary to Section 105.6 for a general discussion of operations requiring an operational permit. The process also notifies the fire department of the need for prefire planning for the hazardous property.

2001.3 Maintenance. Structures and their service equipment shall be maintained in accordance with this code and NFPA 35.

❖ NFPA 35 is referenced to cover conditions not specifically addressed by this chapter, such as closed reactors and thin-down tanks, as well as detailing procedures and safeguards for confined space entry, tank cleaning and piping repair.

SECTION 2002 DEFINITIONS

2002.1 Definition. The following word and term shall, for the purposes of this chapter and as used elsewhere in this code, have the meaning shown herein.

❖ The following word and term shall, for the purposes of this chapter and as used elsewhere in this code, have the meaning shown herein.

Definitions can help in the understanding and application of the code requirements. This definition is given here for convenient access to a key term used in applying the requirements of this chapter. For convenience, the defined term is also listed in Chapter 2 with a cross reference to this section. The use and application of all defined terms, including the one here, are set forth in Section 201.

ORGANIC COATING. A liquid mixture of binders such as alkyd, nitrocellulose, acrylic or oil, and flammable and combustible solvents such as hydrocarbon, ester, ketone or alcohol, which, when spread in a thin film, convert to a durable protective and decorative finish.

❖ Organic coatings are defined as flammable and combustible paints and other protective or decorative coatings.

SECTION 2003 GENERAL PRECAUTIONS

2003.1 Building features. Manufacturing of organic coatings shall be done only in buildings that do not have pits or basements.

❖ The IBC classifies organic coating processes in Group H-2 or H-3, depending on the types and amounts of materials, when the exempt amounts of hazardous materials in Table 307.7(1) of that code are exceeded. Basements, pits or depressed first-floor construction are prohibited because of the tendency for hazardous materials vapors to accumulate in low areas and the difficulty in fighting fires in such areas and occupancies.

2003.2 Location. Organic coating manufacturing operations and operations incidental to or connected with organic coating manufacturing shall not be located in buildings having other occupancies.

❖ Incidental occupancies involve operations and activities closely related to the primary occupancy and are necessary for efficient, continuous and safe organic coatings manufacture. Administration, storage, shipping and receiving, as well as other related but not indispensable operations, should be located in separate buildings. Separations must have a fire-resistance rating, separation distance or a combination of the two as required by the IBC.

2003.3 Fire-fighting access. Organic coating manufacturing operations shall be accessible from at least one side for the purpose of fire control. Approved aisles shall be maintained for the unobstructed movement of personnel and fire suppression equipment.

❖ Access from at least one side conforming to Sections 503 and 504 is required. Fire department connections, fire protection valves, yard hydrants and related fire-fighting equipment should be sited with respect to the provided access and hazards present (see commentary, Chapter 5). Fire department preincident plans should consider operational alternatives if the access is unusable when only one means of access is provided.

Design and layout of equipment and processes must facilitate access for fire control. If provided, standpipes and hose reels should be located at intersections between aisles to facilitate movement of hose lines. Portable fire extinguishers must be located in the path of egress travel.

2003.4 Fire protection systems. Fire protection systems shall be installed, maintained, periodically inspected and tested in accordance with Chapter 9.

❖ Fire protection must be continuously maintained. Prior permission of the fire code official is required for temporary outages for maintenance, repair, testing, alterations or additions. The fire code official may require special protection or precautions during any outage period. Every effort must be made to restore service as quickly as possible (see commentary, Chapter 9).

2003.5 Portable fire extinguishers. A minimum of one portable fire extinguisher complying with Section 906 for extra hazard shall be provided in organic coating areas.

❖ At least one fire extinguisher is required, sized and located for the extra hazardous nature of organic coatings operations. The addition of wheeled fire extinguishers could provide more extinguishing agent, longer discharge time and greater stream reach for the higher hazard in large area operations.

2003.6 Open flames. Open flames and direct-fired heating devices shall be prohibited in areas where flammable vapor-air mixtures exist.

❖ Only indirect heat equipment employing hot water, steam or warm air, or heat equipment approved for use in Class I, Division 2, Group D hazardous locations must be used where flammable vapor-air mixtures exist.

2003.7 Smoking. Smoking shall be prohibited in accordance with Section 310.

❖ Smoking is prohibited in and around organic coating manufacturing areas in accordance with Section 310 to control this ignition source. Approved signs must be posted at entrances and throughout the manufacturing area. Locations approved for smoking should be separated from the manufacturing area by fire-resistance-rated construction and provided with separate ventilation.

2003.8 Power equipment. Power-operated equipment and industrial trucks shall be of a type approved for the location.

❖ Section 309 of the code, Section 5-6 of NFPA 35 and NFPA 505 provide excellent guidance on use, classification of hazards and selection of industrial lift trucks.

2003.9 Tank maintenance. The cleaning of tanks and vessels that have contained flammable or combustible liquids shall be performed under the supervision of persons knowledgeable of the fire and explosion potential.

❖ Trained people well versed in the hazards of the tank cleaning process must supervise the cleaning of tanks containing flammable or combustible vapors. NFPA 326 provides excellent guidance on safety practices. Oxygen deficiency and explosion potential hazards are paramount concerns.

2003.9.1 Repairs. Where necessary to make repairs involving "hot work," the work shall be authorized by the responsible individual before the work begins.

❖ Extreme caution must be exercised when welding and cutting around flammable or combustible liquid tanks. Both Chapter 26 and ANSI Z49.1 specify safety requirements for welding and cutting operations.

2003.9.2 Empty containers. Empty flammable or combustible liquid containers shall be removed to a detached, outside location and, if not cleaned on the premises, the empty containers shall be removed from the plant as soon as practical.

❖ Partially full and empty containers may pose an even greater hazard than full containers. Residual liquid must be completely removed and flammable vapors vented or purged from containers before they can be considered safe. Unused containers must be promptly removed from the premises. A special detached storage facility may be used for containers awaiting cleaning or disposal (see also commentary, Section 2703.2.5).

2003.10 Drainage. Drainage facilities shall be provided to direct flammable and combustible liquid leakage and fire protection water to an approved location away from the building, any other structure, storage area or adjoining premises.

❖ Site drainage must be arranged to minimize hazards to adjacent properties and constructed to keep spills on site and away from significant buildings, means of egress and access routes. See Chapter 34 for further guidance on above-ground diking and drainage.

2003.11 Alarm system. An approved fire alarm system shall be provided in accordance with Section 907.

❖ A manual fire alarm is required in areas where organic coatings are manufactured (see Section 907.2.5).

SECTION 2004
ELECTRICAL EQUIPMENT AND PROTECTION

2004.1 Wiring and equipment. Electrical wiring and equipment shall comply with this chapter and shall be installed in accordance with the ICC *Electrical Code.*

❖ The ICC *Electrical Code*® (ICC EC™) and the requirements of this section apply to electrical wiring and equipment to reduce the fire hazard associated with the presence of such equipment in areas where organic coatings are being manufactured.

2004.2 Hazardous locations. Where Class I liquids are exposed to the air, the design of equipment and ventilation of structures shall be such as to limit the Class I, Division 1, locations to the following:

1. Piping trenches.
2. The interior of equipment.

3. The immediate vicinity of pumps or equipment locations, such as dispensing stations, open centrifuges, plate and frame filters, opened vacuum filters, change cans and the surfaces of open equipment. The immediate vicinity shall include a zone extending from the vapor liberation point 5 feet (1524 mm) horizontally in all directions and vertically from the floor to a level 3 feet (914 mm) above the highest point of vapor liberation.

❖ Class I, Division 1 hazardous locations are areas where flammable or explosive vapors generally exist in quantities sufficient to support an ignition (vapor-air concentrations within the flammable range). Piping trenches and other depressed areas will accumulate vapors because these vapors are usually heavier than air. Vapor production generally creates higher concentrations where flammable liquids are dispensed. Other areas must be adequately ventilated to disperse and dilute vapors to prevent flammable or explosive concentrations. The zone description in Item 3 should be used to define the limitations of "immediate vicinity." Most flammable vapors are susceptible to ignition within relatively narrow concentrations in air so these boundaries provide a relatively good safety margin.

2004.2.1 Other locations. Locations within the confines of the manufacturing room where Class I liquids are handled shall be Class I, Division 2 except locations indicated in Section 2004.2.

❖ Class I, Division 2 locations include all areas where flammable vapors may be liberated in sufficient quantities to present a hazard should an accidental release or spill occur. Areas extending beyond the boundaries defined in Section 2004.2, such as processing areas for Class I liquids, should be considered as such locations. NFPA 497 provides useful guidance for determining the extent of electrically classified boundaries. When Class II liquids are handled and the ambient temperature is cooler than the liquid's flash point, ordinary electrical equipment may be used, provided precautions are taken to prevent hot metal or slag from falling into the liquid.

2004.2.2 Ordinary equipment. Ordinary electrical equipment, including switchgear, shall be prohibited except where installed in a room maintained under positive pressure with respect to the hazardous area. The air or other media utilized for pressurization shall be obtained from a source that will not cause any amount or type of flammable vapor to be introduced into the room.

❖ Electrical equipment not classified for use in hazardous locations must be installed in separate rooms and provided with positive pressure ventilation to prevent infiltration of flammable vapors. Care must be taken to avoid using air for ventilation that is contaminated with flammable vapors. Outside air taken from a source well above the adjacent grade is usually preferred. NFPA 496 provides useful guidance on the design of purged enclosures for electrical equipment.

2004.3 Bonding. Equipment including, but not limited to, tanks, machinery and piping, shall be bonded and connected to a ground where an ignitable mixture is capable of being present.

❖ Both bonding and grounding prevent the accumulation of static charges and the accidental release of electrical energy that could cause an ignition. The bond and ground may be physically applied or be a part of the design of the apparatus.

2004.3.1 Piping. Electrically isolated sections of metallic piping or equipment shall be grounded or bonded to the other grounded portions of the system.

❖ Each electrically isolated section of piping must be grounded or independently bonded to another portion of the system when sections of piping do not form a continuous conductive path.

2004.3.2 Vehicles. Tank vehicles loaded or unloaded through open connections shall be grounded and bonded to the receiving system.

❖ The bond between the tank vehicle and the receiving system may be designed into the dispensing equipment or require physical attachment by the operator (a bond wire is often part of the dispensing hose). The system must also be grounded.

2004.3.3 Containers. Where a flammable mixture is transferred from one portable container to another, a bond shall be provided between the two containers, and one shall be grounded.

❖ Like tank vehicles, the dispensing hose may have the bond designed into the equipment or, if no such bond exists, a separate bond may need to be applied by the operator. A system ground must also be present.

2004.4 Ground. Metal framing of buildings shall be grounded with resistance of not more than 5 ohms.

❖ The building frame of Type 2 buildings or, for that matter, the metal frame of a building of any construction type, must have a ground with a maximum resistance of 5 ohms.

SECTION 2005 PROCESS STRUCTURES

2005.1 Design. Process structures shall be designed and constructed in accordance with the *International Building Code.*

❖ The IBC classifies organic coating processes in Group H-2 or H-3, depending on the type and amounts of materials, when the exempt amounts of hazardous materials in Table 307.7(1) of that code are exceeded.

2005.2 Fire apparatus access. Fire apparatus access complying with Section 503 shall be provided for the purpose of fire control to at least one side of organic coating manufacturing operations.

❖ Access from at least one side that conforms to Section 503 is required. Fire department connections, fire protection valves, yard hydrants and related fire-fighting equipment should be located for ready access to hazards. Fire department preincident plans should consider operational alternatives if the access is unusable when only one means of access is provided.

2005.3 Drainage. Drainage facilities shall be provided in accordance with Section 2003.10 where topographical conditions are such that flammable and combustible liquids are capable of flowing from the organic coating manufacturing operation so as to constitute a fire hazard to other premises.

❖ Facility design should consider both spill containment and control provisions. Drains must discharge to approved containment basins. The location of containment facilities must not endanger adjacent facilities (see also Sections 2003.10 and 2704.2).

2005.4 Explosion control. Explosion control shall be provided in areas subject to potential deflagration hazards as indicated in NFPA 35. Explosion control shall be provided in accordance with Section 911.

❖ Explosion control is required where Class I liquids or flammable dusts create a potential deflagration hazard as defined by NFPA 35. Section 911 of the code and Section 414.5 of the IBC prescribe requirements for explosion control. NFPA 68 also provides guidance on this subject.

2005.5 Ventilation. Enclosed structures in which Class I liquids are processed or handled shall be ventilated at a rate of not less than 1 cubic foot per minute per square foot ($0.00508\ m^3/s \cdot m^2$) of solid floor area. Ventilation shall be accomplished by exhaust fans that take suction at floor levels and discharge to a safe location outside the structure. Noncontaminated intake air shall be introduced in such a manner that all portions of solid floor areas are provided with continuous uniformly distributed air movement.

❖ Ventilation systems must be designed to prevent the accumulation of vapors within the building where Class I flammable liquids are processed. Uncontaminated air, preferably from a source outside the building, should be distributed to dilute and disperse vapors over the entire solid floor area and then discharge them to a safe location outside the building. A ventilation rate of 1 cubic foot per minute per square foot ($0.00508\ m^3/(s \cdot m^2)$ is required [see the *International Mechanical Code*® (IMC®)].

2005.6 Heating. Heating provided in hazardous areas shall be by indirect means. Ignition sources such as open flames or electrical heating elements, except as provided for in Section 2004, shall not be permitted within the structure.

❖ Only indirect-heat appliances are permitted in hazardous areas. Steam or hot water radiators and forced warm air may be used, provided the fans and heat sources are located outside the hazardous area. Appliances and devices using or producing open flames, electrical elements or electric arcs are prohibited.

SECTION 2006
PROCESS MILLS AND KETTLES

2006.1 Mills. Mills, operating with close clearances, which process flammable and heat-sensitive materials, such as nitrocellulose, shall be located in a detached building or in a noncombustible structure without other occupancies. The amount of nitrocellulose or other flammable material brought into the area shall not be more than the amount required for a batch.

❖ Milling of heat-sensitive materials, such as nitrocellulose, is an extraordinary hazard that must be located in single-purpose buildings, away from other uses and high-hazard operations. Pebble mills pose a special vapor ignition hazard caused by static electricity. Both the grinding material and inner lining of these mills are made of materials with good insulating characteristics. Because static electricity is produced during milling, it has nowhere to go. Generally, the atmosphere inside of this type of mill is made either partially or totally inert using nitrogen or carbon dioxide gas to prevent ignition.

2006.2 Mixers. Mixers shall be of the enclosed type or, where of the open type, shall be provided with properly fitted covers. Where flow is by gravity, a shutoff valve shall be installed as close as practical to the mixer, and a control valve shall be provided near the end of the fill pipe.

❖ Like any other part of the manufacturing process, mixing organic chemicals produces large quantities of static electricity. These requirements provide for controlling fires by smothering (closing the cover or lid) and interrupting fuel flow (closing the product valves).

2006.3 Open kettles. Open kettles shall be located in an outside area provided with a protective roof; in a separate structure of noncombustible construction; or separated from other areas by a noncombustible wall having a fire-resistance rating of at least 2 hours.

❖ Kettles and reactors are large warming vessels used to cook various solid, liquid and gaseous materials, including monomers, which are usually solids or liquids, to initiate a controlled chemical reaction to produce resins. Most operations are exothermic and involve controlled polymerization. Finished resins are mixed with solvents, pigments and additives (for example, quality-control agents, texture materials, glass beads, etc.) to form the finished product in organic coating manufacturing. Principal dangers associated with open-fire kettles or reactors are ignition of flammable vapors and uncontrolled polymerization. Vapors are released to the surrounding atmosphere, and uncontrolled reactions can lead to spills or discharges from the reactor vessel in open kettles. These operations must be conducted in an outside area under protective cover, if possible. Otherwise, open-fire kettles must be operated only in a separate building of noncombustible construction. A building separated from the main facility by a 2-hour fire-resistance-rated fire wall complies with this requirement. The area must be positively ventilated, with vapors discharged in a safe location away from the combustion air intake.

2006.4 Closed kettles. Contact-heated kettles containing solvents shall be equipped with safety devices that, in case of a fire, will turn off the process heat, turn on the cooling medium and inject inert gas into the kettle.

❖ Closed reactors confine reaction byproducts within the vessel in a closed loop system. Closed kettles or reactors may be either continuous or batch reactors. In the former, reactants are continuously fed into the system as reactant mass and byproducts are removed. Generally, continuous reactors are safer from a fire protection standpoint.

By comparison, reactants are fed into the batch reactor vessel, which is then sealed before the reaction is initiated by the application of heat. Reaction mass and byproducts are removed upon completion of the reaction, and a new batch is processed with new raw materials. To interrupt the chemical process once the reaction begins, heat must be removed and the reaction mass cooled. This often requires the simultaneous venting of gaseous byproducts. Normal and emergency pressure relief vents must be arranged to discharge flammable vapors to a safe location when necessary. Automatic inerting of the atmosphere further disrupts the reaction process by displacing oxygen and oxidation byproducts. For further information, see NFPA 35.

2006.4.1 Vaporizer location. The vaporizer section of heat-transfer systems that heat closed kettles containing solvents shall be remotely located.

❖ The vaporizer must be located remotely from the process area.

2006.5 Kettle controls. The kettle and thin-down tank shall be instrumented, controlled and interlocked so that any failure of the controls will result in a safe condition. The kettle shall be provided with a pressure-rupture disc in addition to the primary vent. The vent piping from the rupture disc shall be of minimum length and shall discharge to an approved location. The thin-down tank shall be adequately vented. Thinning operations shall be provided with an adequate vapor removal system.

❖ Reactors must be provided with automatic high-temperature limit switches or other approved automatic temperature control methods. Manual and automatic methods must be provided to interrupt the flow of fuel if a flameout or other emergency occurs. The rupture disc is intended to vent excess pressure if the normal vent fails. Ventilation must be by a condenser of adequate size to prevent the accumulation of a vapor fog in the reactor area.

SECTION 2007 PROCESS PIPING

2007.1 Design. All piping, valves and fittings shall be designed for the working pressures and structural stresses to which the piping, valves and fittings will be subjected, and shall be of steel or other material approved for the service intended.

❖ Suitability of piping for the intended purpose is primarily a function of compatibility with the material that it will contain, adequate strength under normal working pressures and durability throughout the intended life of the system. Materials, except cast iron, are permitted by NFPA 35, provided they are approved for the intended service. Good engineering judgment is needed in the design of these systems.

2007.2 Valves. Valves shall be of an indicating type. Terminal valves on remote pumping systems shall be of the dead-man type, shutting off both the pump and the flow of solvent.

❖ NFPA 35 prohibits the use of cast-iron valves and fittings. Indicating valves permit the system operator to determine the position of critical valves at a glance. This aids in prompt control of the product flow if a piping or process failure occurs. Dead-man valves interlocked with product supply pumps cause interruption of product flow if there is an operator error.

2007.3 Support. Piping systems shall be supported adequately and protected against physical damage. Piping shall be pitched to avoid unintentional trapping of liquids, or approved drains shall be provided.

❖ Ideally, product lines should be designed without trapped sections. Cleaning and purging of the system is made easier without releasing vapors when drains are opened and ignition sources are adequately controlled. Adequate support and protection of piping prevent product releases if accidents occur in the processing area.

2007.4 Connectors. Approved flexible connectors shall be installed where vibration exists or frequent movement is necessary. Hose at dispensing stations shall be of an approved type.

❖ Rigid connectors either loosen from repeated vibration or fail from stresses created. Hose should be reinforced and designed with an integral ground wire to permit container bonding.

2007.5 Tests. Before being placed in service, all piping shall be free of leaks when tested for a minimum of 30 minutes at not less than 1.5 times the working pressure or a minimum of 5 pounds per square inch gauge (psig) (35 kPa) at the highest point in the system.

❖ Tests are intended to demonstrate system integrity under both normal and abnormal conditions. The 50-percent margin represents a reasonable safety range if a malfunction occurs. No test provides absolute certainty that a failure will not occur.

SECTION 2008 RAW MATERIALS IN PROCESS AREAS

2008.1 Nitrocellulose quantity. The amount of nitrocellulose brought into the operating area shall not exceed the amount required for a work shift. Nitrocellulose spillage shall be promptly swept up and disposed of properly.

❖ Nitrocellulose lacquer production is among the most hazardous organic coatings manufacturing processes. Nitrocellulose can support combustion in the absence of oxygen and is, therefore, considered very dangerous. As the temperature of alcohol-wet nitrocellulose increases, it becomes unstable. Relatively small amounts of energy are required to initiate combustion. Small amounts of finely divided nitrocellulose fiber should be cleaned up only with a wet-vacuum apparatus. Loose nitrocellulose material, scraps and waste should be stored underwater in a metal container with a tight-fitting, self-closing or automatic-closing lid. The waste container should be removed from the building daily and its contents burned at an approved site. See Chapter 42 for further regulation of pyroxilin plastics.

2008.2 Organic peroxides quantity. Organic peroxides brought into the operating area shall be in the original shipping container. When in the operating area, the organic peroxide shall not be placed in locations exposed to ignition sources, heat or mechanical shocks.

❖ Organic peroxides are both flammable materials and strong oxidizers. Many organic peroxides are principal components in more dangerous compounds like blasting agents. Contamination may sensitize organic peroxides, making them especially sensitive to heat and shock. Organic peroxides should be kept in their original shipping containers to permit easy identification and prevent accidents while transferring contents to other containers.

SECTION 2009 RAW MATERIALS AND FINISHED PRODUCTS

2009.1 General. The storage, handling and use of flammable and combustible liquids in process areas shall be in accordance with Chapter 34.

❖ Chapter 34 details requirements for the storage and handling of flammable liquids. Moreover, it prescribes limitations on the size and location of containers, tanks and piles of flammable liquids in storage. The protection of flammable liquids in bulk storage has been the topic of considerable debate in the aftermath of the 1987 Sherwin-Williams warehouse fire. Fire code officials should carefully consider protection requirements. Appendices D and E of NFPA 30 contain extensive discussions of recommended protection for bulk flammable liquid storage. Also, investigations have been completed of flammable liquid storage in plastic containers. Reports from these tests conducted by the National Fire Protection Research Foundation (NFPRF) are available from NFPA.

2009.2 Tank storage. Tank storage for flammable and combustible liquids located inside of structures shall be limited to storage areas at or above grade which are separated from the processing area in accordance with the *International Building Code*. Processing equipment containing flammable and combustible liquids and storage in quantities essential to the continuity of the operations shall not be prohibited in the processing area.

❖ Tank storage located below grade is prohibited. Basements located under grade-level storage areas should be discouraged. Below-grade flammable liquid fires are extremely difficult to fight. Similarly, above-grade spills will flow to lower floors, possibly resulting in spill fires on more than one building level. Tank storage of raw materials must be confined to locations at or above grade level. Tank storage must be separated from the processing area in accordance with the IBC. If possible, these rooms should be accessible on at least one exterior side of the facility.

2009.3 Tank vehicle. Tank car and tank vehicle loading and unloading stations for Class I liquids shall be separated from the processing area, other plant structures, nearest lot line of property that can be built upon or public thoroughfare by a minimum clear distance of 25 feet (7620 mm).

❖ Like other separation requirements, the 25-foot (7620 mm) clearance is intended to limit accidental fire exposure to process, storage and adjacent buildings, as well as other areas. Additionally, clearance provides the same protection to the tanker or tank car if an incident involving an adjacent facility occurs. Finally, clearance provides access for establishing fire-fighting operations if an incident occurs.

2009.3.1 Loading. Loading and unloading structures and platforms for flammable and combustible liquids shall be designed and installed in accordance with Chapter 34.

❖ Chapter 34 specifies requirements for flammable liquid loading and unloading. Section 3406.5 and Section 5.6 of NFPA 30 specify requirements for tank vehicle and tank car loading and unloading operations, including separate piping and valves for Class I flammable liquids and Class II and III combustible liquids; leak detection on the discharge side of dispensing pumps; provisions for bonding dispensing equipment to the tank vehicle and flow control interlocks.

2009.3.2 Safety. Tank cars for flammable liquids shall be unloaded such that the safety to persons and property is ensured. Tank vehicles for flammable and combustible liquids shall be loaded and unloaded in accordance with Chapter 34.

❖ Section 3406.6 emphasizes the tank vehicle operator's responsibilities. Essentially, the operator must verify that all safety features are properly used and maintained. Before loading or unloading the product, the operator must verify that required bonding is in place, a liquid- and vapor-tight connection has been made and fill or discharge lines are protected from physical damage. The operator must remain with the vehicle during the loading or unloading operation but not in the cab.

2009.4 Nitrocellulose storage. Nitrocellulose storage shall be located on a detached pad or in a separate structure or a room enclosed in accordance with the *International Building Code*. The nitrocellulose storage area shall not be utilized for any other purpose. Electrical wiring and equipment installed in storage areas adjacent to process areas shall comply with Section 2004.2.

❖ Solvent-wet nitrocellulose is the most common type associated with lacquer production. In this case, the solvent is usually an alcohol, which is a Class I flammable liquid. Solvent-wet nitrocellulose possesses the same fire-hazard characteristics as the solvent in which it is stored, with important differences noted. Solvent-wet nitrocellulose becomes increasingly unstable as the temperature rises. Once ignited, nitrocellulose will continue to burn even in the absence of oxygen. Therefore, extra precautions, at a minimum, must be used to prevent ignition and fire spread.

2009.4.1 Containers. Nitrocellulose shall be stored in closed containers. Barrels shall be stored on end and not more than two tiers high. Barrels or other containers of nitrocellulose shall not be opened in the main storage structure but at the point of use or other location intended for that purpose.

❖ Like other materials possessing unusual fire-hazard characteristics that may be susceptible to ignition if improperly handled, nitrocellulose must be kept in its Department of Transportation (DOT) shipping container until used. This reduces the likelihood of accidents associated with unnecessary transfer operations and allows the material to be readily identified by its shipping and DOT hazard labels. Containers must not be stored on their sides because nitrocellulose is commonly shipped solvent wet. One-drum-high storage is preferred, but two-drum-high storage is permitted. Stacking containers poses the risk of toppling or other physical damage to the container with the potential for spillage or ignition.

2009.4.2 Spills. Spilled nitrocellulose shall be promptly wetted with water and disposed of by use or burning in the open at an approved detached location.

❖ Wetting spills with water will help prevent ignition by diluting the flammable solvent, especially in the case of alcohol-wet nitrocellulose. The spilled material must be swept up immediately and placed in a tightly closed metal container and covered with water. Section 6-1 of NFPA 35 requires that waste be burned in an approved and isolated location in accordance with federal, state and local environmental regulations, or denitrated in an agitated 5-percent solution of aqueous sodium hydroxide (NaOH) or lye in a well-ventilated area.

2009.5 Organic peroxide storage. The storage of organic peroxides shall be in accordance with Chapter 39.

❖ Organic peroxides pose a dual hazard of being both flammable and oxidizing materials (see Section 2701.1). Additionally, these compounds are sensitive to friction, heat and shock. Detached storage buildings are required when storage exceeds the exempt amounts of Table 2704.14 to prevent reactions with other materials. These buildings must be constructed in accordance with requirements of the IBC for Group H occupancies, depending on the classification of the organic peroxide involved. Separation distances specified in Table 3904.1.2 are intended to reduce the likelihood of an incident involving materials stored adjacent to hazardous operations that may cause personal injury or property damage. Adequate access must be provided to facilitate defensive fire-fighting operations and support the installed protection. If possible, the fire department connection or connections should be of the free-standing type located far away from the building along the route of fire service access. An organic peroxide may be stored inside a building of another occupancy group in control areas in accordance with Section 3903.1.1 and Table 2703.1.1.

Signs are to be posted at the entrance and within all storage areas. NFPA 704 hazard warning signs and signs indicating "ORGANIC PEROXIDE STORAGE — NO OPEN FLAMES" are recommended for this purpose.

Organic peroxides must be stored in original shipping containers. This practice serves the following three important purposes: reduces the likelihood of accidents caused by unnecessary handling; keeps the material in labeled containers that clearly identify the contents and the hazard and reduces the likelihood of contamination, which may sensitize the material or cause it to decay.

2009.5.1 Size. The size of the package containing organic peroxide shall be selected so that, as nearly as practical, full packages are utilized at one time. Spilled peroxide shall be promptly cleaned up and disposed of as specified by the supplier.

❖ Any leftover organic peroxide poses a storage and handling problem. Most organic peroxides become increasingly unstable if contaminated with foreign materials. Organic peroxides may spontaneously ignite if exposed to sufficient heat or shock because they are both flammables and oxidizers. Extreme care must be taken when cleaning up organic peroxide spills. Hydrocarbons and other volatile organic chemicals may sensitize an organic peroxide. Excessive heat or friction during cleanup operations could result in a deflagration or explosion if the material is contaminated.

2009.6 Finished products. Finished products that are flammable or combustible liquids shall be stored outside of structures, in a separate structure, or in a room separated from the processing area in accordance with the *International Building Code*. The storage of finished products shall be in tanks or closed containers in accordance with Chapter 34.

❖ Flammable liquid warehouses must comply with both the IBC and Chapter 34 requirements. Though flammable liquid warehouses are exempt from storage quantity limitations, pile size and height restrictions still apply. Both height and area limitations specified in the IBC for Group H-3 occupancies and the construction type apply to new construction.

Bibliography

The following resource materials are referenced in this chapter or are relevant to the subject matter addressed in this chapter.

ANSI Z49.1-94, *Safety in Welding and Cutting*. New York: American National Standards Institute, 1994.

Bradford, W.J. "Chemical Processes." *Industrial Fire Hazards Handbook*, 3rd ed. Quincy, MA: National Fire Protection Association, 1990.

Harris, M.V. and A.B. Swartz. *Industrial Fire Hazards Handbook*, 3rd ed. Chapter 10: Paints and Coatings Manufacture. Quincy, MA: National Fire Protection Association, 1990.

IBC-2003, *International Building Code*. Falls Church, VA: International Code Council, 2003.

ICC EC-2003, ICC *Electrical Code*. Falls Church, VA: International Code Council, 2003.

NFPA 30-00, *Flammable and Combustible Liquids Code*. Quincy, MA: National Fire Protection Association, 2000.

NFPA 30B-98, *Manufacture and Storage of Aerosol Products*. Quincy, MA: National Fire Protection Association, 1998.

NFPA 35-95, *Manufacture of Organic Coatings*. Quincy, MA: National Fire Protection Association, 1995.

NFPA 68-02, *Guide for Venting of Deflagrations*. Quincy, MA: National Fire Protection Association, 2002.

NFPA 77-00, *Recommended Practice on Static Electricity*. Quincy, MA: National Fire Protection Association, 2000.

NFPA 326-99, *Safeguarding of Tanks and Containers for Entry, Cleaning, or Repair*. Quincy, MA: National Fire Protection Association, 1999.

NFPA 496-98, *Purged and Pressurized Enclosures for Electrical Equipment*. Quincy, MA: National Fire Protection Association, 1998.

NFPA 497-97, *Classification of Flammable Liquids, Gases or Vapors and of Hazardous (Classified) Locations for Electrical Installations in Chemical Process Areas*. Quincy, MA: National Fire Protection Association, 1997.

NFPA 505-99, *Powered Industrial Trucks, Including Type Designations, Areas of Use, Maintenance, and Operations*. Quincy, MA: National Fire Protection Association, 1999.

NFPA 704-96, *Identification of the Fire Hazards of Materials*. Quincy, MA: National Fire Protection Association, 1996.

Chapter 21: Industrial Ovens

General Comments

Fires in industrial ovens are usually the result of the fuel in use or volatile vapors given off by the materials being heated. This chapter addresses the fuel supply, ventilation, emergency shutdown equipment, fire protection and the operation and maintenance of industrial ovens.

Purpose

Compliance with this chapter is intended to reduce the likelihood of fires involving industrial ovens or to manage the impact if a fire should occur. Industrial ovens are sometimes referred to as industrial heat enclosures or industrial furnaces. Heat may be furnished by gas burners, oil burners, electric heaters, infrared lamps, induction heaters or steam radiation systems.

SECTION 2101 GENERAL

2101.1 Scope. This chapter shall apply to the installation and operation of industrial ovens and furnaces. Industrial ovens and furnaces shall comply with the applicable provisions of NFPA 86, the *International Fuel Gas Code*, *International Mechanical Code* and this chapter. The terms "ovens" and "furnaces" are used interchangeably in this chapter.

❖ This chapter contains provisions for the installation and operation of industrial ovens and furnaces. If a certain provision is not found in this chapter, NFPA 86 and the *International Mechanical Code®* (IMC®) should be used.

2101.2 Permits. Permits shall be required as set forth in Sections 105.6 and 105.7.

❖ The process of issuing permits gives the fire code official an opportunity to carefully evaluate and regulate hazardous operations. Permit applicants should be required to demonstrate that their operations comply with the intent of the code before the permit is issued. See the commentary to Section 105.6 for a general discussion of operations requiring an operational permit. The process also notifies the fire department of the need for prefire planning for the hazardous property.

SECTION 2102 DEFINITIONS

2102.1 Definitions. The following words and terms shall, for the purposes of this chapter and as used elsewhere in this code, have the meanings shown herein.

❖ Definitions of terms can help in the understanding and application of the code requirements. The purpose for including those definitions that are associated with the subject matter of this chapter is to provide more convenient access to them without having to refer back to Chapter 2. It is important to emphasize that these terms are not exclusively related to this chapter but are applicable everywhere the term is used in the code. For convenience, these terms are also listed in Chapter 2 with a cross reference to this section. The use and application of all defined terms, including those defined in this section, are set forth in Section 201.

FURNACE CLASS A. An oven or furnace that has heat utilization equipment operating at approximately atmospheric pressure wherein there is a potential explosion or fire hazard that could be occasioned by the presence of flammable volatiles or combustible materials processed or heated in the furnace.

Note: Such flammable volatiles or combustible materials can, for instance, originate from the following:

1. Paints, powders, inks, and adhesives from finishing processes, such as dipped, coated, sprayed and impregnated materials.
2. The substrate material.
3. Wood, paper and plastic pallets, spacers or packaging materials.
4. Polymerization or other molecular rearrangements.

Potentially flammable materials, such as quench oil, water-borne finishes, cooling oil or cooking oils, that present a hazard are ventilated according to Class A standards.

❖ Ovens may also use a low-oxygen atmosphere to evaporate solvent. This kind of equipment has potential hazards involving the process material and heat generation.

FURNACE CLASS B. An oven or furnace that has heat utilization equipment operating at approximately atmospheric pressure wherein there are no flammable volatiles or combustible materials being heated.

❖ Even though no flammable, volatile or combustible materials are heated in this kind of oven, the process can still be a serious fire and explosion hazard.

FURNACE CLASS C. An oven or furnace that has a potential hazard due to a flammable or other special atmosphere being used for treatment of material in process. This type of furnace can use any type of heating system and includes a special atmosphere supply system. Also included in the Class C classification are integral quench furnaces and molten salt bath furnaces.

❖ These are units in which there is an explosion hazard because a special flammable atmosphere is being used for treatment of material in process. Within this class, an integral quench tank is used, which is a container that holds a quench medium into which a metalwork is immersed for various heat treatment processes. A molten bath furnace is a heated container that holds a melt or fusion into which metalwork is immersed for various heat treatment processes.

FURNACE CLASS D. An oven or furnace that operates at temperatures from above ambient to over 5,000°F (2760°C) and at pressures normally below atmospheric using any type of heating system. These furnaces can include the use of special processing atmospheres.

❖ These are generally referred to as vacuum furnaces because they operate below normal atmospheric pressure. Vacuum furnaces are described as either cold-wall furnaces, hot-wall furnaces or furnaces used for casting or melting of metal at temperatures up to 5,000°F (2760°C) or higher.

SECTION 2103 LOCATION

2103.1 Ventilation. Enclosed rooms or basements containing industrial ovens or furnaces shall be provided with combustion air in accordance with the *International Mechanical Code* and the *International Fuel Gas Code*, and with ventilation air in accordance with the *International Mechanical Code.*

❖ Placing an oven or furnace below grade presents difficulty in ventilation and offers severe obstacles to proper explosion release. When gas or oil is used, the furnace or oven should be vented separately unless the products of discharge deposit directly into the oven.

2103.2 Exposure. When locating ovens, oven heaters and related equipment, the possibility of fire resulting from overheating or from the escape of fuel gas or fuel oil and the possibility of damage to the building and injury to persons resulting from explosion shall be considered.

❖ Ovens and furnaces should be located where they will present the least possible hazard to property and life. To present the least possible hazard, walls or partitions may be needed around the furnaces or ovens.

2103.3 Ignition source. Industrial ovens and furnaces shall be located so as not to pose an ignition hazard to flammable vapors or mists or combustible dusts.

❖ Industrial ovens should be considered an ignition source and should be separated from any materials that may be easily ignited by the oven.

2103.4 Temperatures. Roofs and floors of ovens shall be insulated and ventilated to prevent temperatures at combustible ceilings and floors from exceeding 160°F (71°C).

❖ Furnaces and ovens should be designed to minimize their fire hazard when operating at elevated temperatures. Insulation and ventilation are important in preventing heat transfer from the oven to combustible products surrounding it.

SECTION 2104 FUEL PIPING

2104.1 Fuel-gas piping. Fuel-gas piping serving industrial ovens shall comply with the *International Fuel Gas Code.* Piping for other fuel sources shall comply with this section.

❖ Electrically wired, oil-fired and special-atmosphere furnaces are covered in this section. Fuel-gas piping is covered in the *International Fuel Gas Code*® (IFGC®).

2104.2 Shutoff valves. Each industrial oven or furnace shall be provided with an approved manual fuel shutoff valve in accordance with the *International Mechanical Code* or the *International Fuel Gas Code.*

❖ Individual manual shutoff valves are used for equipment isolation. Requirements are found in either the *International Mechanical Code*® (IMC®) or IFGC.

2104.2.1 Fuel supply lines. Valves for fuel supply lines shall be located within 6 feet (1829 mm) of the appliance served.

Exception: When approved and the valve is located in the same general area as the appliance served.

❖ A fuel supply shutoff valve must be installed in the fuel supply line within 6 feet (1829 mm) of the appliance for easy shutoff of fuel in case of an emergency.

2104.3 Valve position. The design of manual fuel shutoff valves shall incorporate a permanent feature which visually indicates the open or closed position of the valve. Manual fuel shutoff valves shall not be equipped with removable handles or wrenches unless the handle or wrench can only be installed parallel with the fuel line when the valve is in the open position.

❖ Manual shutoff valves should have a permanently affixed visual indicator that is easily recognizable and it should be in the open position. The removable handle requirement prevents fuel supply shutoff for any reason. Accidental shutoff could endanger property and people.

SECTION 2105 INTERLOCKS

2105.1 Shut down. Interlocks shall be provided for Class A ovens so that conveyors or sources of flammable or combustible materials shall shut down if either the exhaust or recirculation air supply fails.

❖ If ventilation or airflow were lost, a safety control circuit would immediately shut down the heating system of the affected section. When necessary, loss of ventilation must shut down the entire system as well as the conveyor.

SECTION 2106 FIRE PROTECTION

2106.1 Required protection. Class A and B ovens which contain, or are utilized for the processing of, combustible materials shall be protected by an approved automatic fire-extinguishing system complying with Chapter 9.

❖ The extent of fire protection required depends on the construction and arrangement of the oven as well as the materials handled. Sprinklers should be located even in the exhaust ducts.

2106.2 Fixed fire-extinguishing systems. Fixed fire-extinguishing systems shall be provided for Class C or D ovens to protect against such hazards as overheating, spillage of molten salts or metals, quench tanks, ignition of hydraulic oil and escape of fuel. It shall be the user's responsibility to consult with the fire code official concerning the necessary requirements for such protection.

❖ Furnaces can present fire hazards to the areas around them. This section is not intended to specify the design of the fire protection system or cover all hazards. The user must collect the requirements from the authority having jurisdiction.

2106.3 Fire extinguishers. Portable fire extinguishers complying with Section 906 shall be provided not closer than 15 feet (4572 mm) or a maximum of 50 feet (15 240 mm) or in accordance with NFPA 10. This shall apply to the oven and related equipment.

❖ Portable fire extinguishers of the proper size and type and using the specified agent must be installed near the furnace and related equipment. The distances mentioned in this section should be used unless requirements in NFPA 10 are more stringent.

SECTION 2107 OPERATION AND MAINTENANCE

2107.1 Furnace system information. An approved, clearly worded, and prominently displayed safety design data form or manufacturer's nameplate shall be provided stating the safe operating condition for which the furnace system was designed, built, altered or extended.

❖ The equipment manufacturer establishes the need for adequate operational checks and maintenance. Specifications, data sheets and procedures provide clear and complete inspection, testing and maintenance instructions.

2107.2 Oven nameplate. Safety data for Class A solvent atmosphere ovens shall be furnished on the manufacturer's nameplate. The nameplate shall provide the following design data:

1. The solvent used.
2. The number of gallons (liters) used per batch or per hour of solvent entering the oven.
3. The required purge time.
4. The oven operating temperature.
5. The exhaust blower rating for the number of gallons (liters) of solvent per hour or batch at the maximum operating temperature.

Exception: For low-oxygen ovens, the maximum allowable oxygen concentration shall be included in place of the exhaust blower ratings.

❖ Safety data must be included on the manufacturer's nameplate for Class A solvent-atmosphere ovens to ensure that safety personnel, operators and maintenance technicians have the information at hand. This same information should also be included in installation, operation and maintenance procedures. The exception in this case is actually an additional nameplate requirement for low-oxygen ovens.

2107.3 Training. Operating, maintenance and supervisory personnel shall be thoroughly instructed and trained in the operation of ovens or furnaces.

❖ Alert and competent operators are essential to safe operations. New operators should be thoroughly trained and tested in the use of the equipment. Regular operators should be reevaluated at regular intervals to make certain their skills and knowledge are current.

2107.4 Equipment maintenance. Equipment shall be maintained in accordance with the manufacturer's instructions.

❖ There should be a program for inspecting and maintaining oven safety controls. The operating and supervisory control equipment should be checked and tested regularly.

Bibliography

The following resource materials are referenced in this chapter or are relevant to the subject matter addressed in this chapter.

IFGC-2003, *International Fuel Gas Code*. Falls Church, VA: International Code Council, 2003.

IMC-2003, *International Mechanical Code*. Falls Church, VA: International Code Council, 2003.

NFPA 10-98, *Portable Fire Extinguishers*. Quincy, MA: National Fire Protection Association, 1998.

NFPA 86-99, *Ovens and Furnaces*. Quincy, MA: National Fire Protection Association, 1999.

Chapter 22: Motor Fuel-Dispensing Facilities and Repair Garages

General Comments

The requirements of this chapter apply to all occupancies that dispense any type of motor fuel and to automotive repair garages.

Purpose

This chapter provides provisions that regulate the dispensing of motor fuels at public and private automotive and marine motor fuel-dispensing facilities, fleet vehicle motor fuel-dispensing facilities, and repair garages. The chapter addresses the dispensing of both liquid and gaseous fuels as well as the storage of those fuels.

SECTION 2201 GENERAL

2201.1 Scope. Automotive motor fuel-dispensing facilities, marine motor fuel-dispensing facilities, fleet vehicle motor fuel-dispensing facilities and repair garages shall be in accordance with this chapter and the *International Building Code, International Fuel Gas Code* and the *International Mechanical Code.* Such operations shall include both operations that are accessible to the public and private operations.

❖ Generally speaking, if fuels, liquid or gas are dispensed from a storage tank, either above or below ground, to the fuel tank of a motor vehicle or marine craft, the operation is within the scope of this chapter.

2201.2 Permits. Permits shall be required as set forth in Section 105.6.

❖ The process of issuing permits gives the fire code official an opportunity to carefully evaluate and regulate hazardous operations. Permit applicants should be required to demonstrate that their operations comply with the intent of the code before the permit is issued. See the commentary for Section 105.6 for a general discussion of operations requiring an operational permit. The process also notifies the fire department of the need for prefire planning for the hazardous property.

2201.3 Construction documents. Construction documents shall be submitted for review and approval prior to the installation or construction of automotive, marine or fleet vehicle motor fuel-dispensing facilities and repair garages in accordance with Section 105.4.

❖ Construction documents, as defined in Section 202, must be drawn to scale with sufficient clarity to be understood by the fire code official, contractors and owners and must address the requirements of the code. State or local laws may require preparation of the construction documents by a registered design professional. See the commentary for Section 106.1 of *International Building Code®* (IBC®).

2201.4 Indoor motor fuel-dispensing facilities. Motor fuel-dispensing facilities located inside buildings shall comply with the *International Building Code* and NFPA 30A.

❖ Generally speaking, this type of motor fuel-dispensing facility is found at parking garages where space is very limited and is subject to the approval of the fire code official. Chapter 6 of NFPA 30A is dedicated to the provisions that govern motor fuel-dispensing facilities inside buildings.

2201.4.1 Protection of floor openings in indoor motor fuel-dispensing facilities. Where motor fuel-dispensing facilities are located inside buildings and the dispensers are located above spaces within the building, openings beneath dispensers shall be sealed to prevent the flow of leaked fuel to lower building spaces.

❖ Floor openings in the dispensing area must be sealed if located over other spaces and drains, and if installed, must be sized to protect areas below from flammable and combustible liquid spills as well as anticipated water from fire hose streams. Drains are to be equipped with separators in accordance with the *International Plumbing Code®* (IPC®).

2201.5 Electrical. Electrical wiring and equipment shall be suitable for the locations in which they are installed and shall comply with Section 605, NFPA 30A and the ICC *Electrical Code.*

❖ Chapter 7 of NFPA 30A addresses electrical equipment at motor fuel-dispensing facilities and is correlated with the ICC *Electrical Code®* (ICC EC™). Electrical equipment must be approved for the particular hazards anticipated at motor fuel-dispensing facilities and the hazardous nature of flammable and combustible liquids.

2201.6 Heat-producing appliances. Heat-producing appliances shall be suitable for the locations in which they are installed and shall comply with NFPA 30A and the *International Fuel Gas Code* or the *International Mechanical Code.*

❖ Heat-producing appliances are ignition sources. Precautions must be taken when using such appliances around flammable and combustible liquids. NFPA 30A, the *International Mechanical Code®* (IMC®) and the *International Fuel Gas Code®* (IFGC®) all provide requirements for the safe use of these appliances.

SECTION 2202 DEFINITIONS

2202.1 Definitions. The following words and terms shall, for the purposes of this chapter and as used elsewhere in this code, have the meanings shown herein.

❖ Definitions of terms can help in the understanding and application of the code requirements. The purpose for including those definitions that are associated with the subject matter of this chapter is to provide more convenient access to them without having to refer back to Chapter 2. It is important to emphasize that these terms are not exclusively related to this chapter but are applicable everywhere the term is used in the code. For convenience, these terms are also listed in Chapter 2 with a cross reference to this section. The use and application of all defined terms, including those defined in this section, are set forth in Section 201.

AUTOMOTIVE MOTOR FUEL-DISPENSING FACILITY. That portion of property where flammable or combustible liquids or gases used as motor fuels are stored and dispensed from fixed equipment into the fuel tanks of motor vehicles.

❖ Automotive motor fuel-dispensing facilities may be attended or unattended and they may take the form of the conventional motor fuel-dispensing facility, convenience store or other location that transfers fuel from a storage tank to the fuel tank of some type of motorized equipment.

DISPENSING DEVICE, OVERHEAD TYPE. A dispensing device that consists of one or more individual units intended for installation in conjunction with each other, mounted above a dispensing area typically within the motor fuel-dispensing facility canopy structure, and characterized by the use of an overhead hose reel.

❖ Dispensing devices are approved pieces of fixed equipment that control the dispensing of fuel through the dispensing hose connected to them. An overhead-type dispensing device is not to be confused with the conventional dispenser, often referred to as a "high-hose" dispenser, that is equipped with a dispensing hose connected at the top of the dispenser frame. The intent of this provision is to identify the overhead hose reel that has special requirements for the classification of electrical equipment in the vicinity of the reel in accordance with Chapter 7 of NFPA 30A.

FLEET VEHICLE MOTOR FUEL-DISPENSING FACILITY. That portion of a commercial, industrial, governmental or manufacturing property where liquids used as fuels are stored and dispensed into the fuel tanks of motor vehicles that are used in connection with such businesses, by persons within the employ of such businesses.

❖ This is sometimes referred to as "you own the tanks, you own the vehicles" motor fuel-dispensing facility. The intent is to allow greater fuel storage tank capacities and reduced separation distances between the dispenser and above-ground tanks when the operator has control of the entire operation, including the vehicles being fueled. In other words, the person dispensing the fuel is an employee of the facility operator.

LIQUEFIED NATURAL GAS (LNG). A fluid in the liquid state composed predominantly of methane and which may contain minor quantities of ethane, propane, nitrogen or other components normally found in natural gas.

❖ Liquefied natural gas (LNG) for vehicles comes from the same source as compressed natural gas (CNG). Unlike liquefied petroleum gas (LP-gas), which changes from vapor to a liquid at room temperature by application of pressure, LNG has to be cooled to liquefy. LNG is usually in a liquid state at the dispensing station.

MARINE MOTOR FUEL-DISPENSING FACILITY. That portion of property where flammable or combustible liquids or gases used as fuel for watercraft are stored and dispensed from fixed equipment on shore, piers, wharves, floats or barges into the fuel tanks of watercraft and shall include all other facilities used in connection therewith.

❖ A marine motor fuel-dispensing facility is not to be confused with a bulk marine terminal, which transfers fuel by way of flange-to-flange connections. A marine motor fuel-dispensing facility uses automotive-type dispensing equipment.

REPAIR GARAGE. A building, structure or portion thereof used for servicing or repairing motor vehicles.

❖ A repair garage may be part of a motor fuel-dispensing facility. The fuel dispensing area will comply with the motor fuel-dispensing facility sections of this chapter and the repair garage area will comply with the repair garage section (Section 2211) of this chapter. A marine pleasure craft dealership with a boat repair area will be classified as a repair garage.

SELF-SERVICE MOTOR FUEL-DISPENSING FACILITY. That portion of motor fuel-dispensing facility where liquid motor fuels are dispensed from fixed approved dispensing equipment into the fuel tanks of motor vehicles by persons other than a motor fuel-dispensing facility attendant.

❖ A self-service motor fuel-dispensing facility may be attended or, with the approval of the fire code official, unattended. A self-service motor fuel-dispensing facility is a facility where the fuel is dispensed by someone other than an employee of the facility operator.

SECTION 2203 LOCATION OF DISPENSING DEVICES

2203.1 Location of dispensing devices. Dispensing devices shall be located as follows:

1. Ten feet (3048 mm) or more from lot lines.
2. Ten feet (3048 mm) or more from buildings having combustible exterior wall surfaces or buildings having noncombustible exterior wall surfaces that are not part of a 1-hour fire-resistance-rated assembly or buildings having combustible overhangs.

   Exception: Canopies constructed in accordance with the *International Building Code* providing weather protection for the fuel islands.
3. Such that all portions of the vehicle being fueled will be on the premises of the motor fuel-dispensing facility.
4. Such that the nozzle, when the hose is fully extended, will not reach within 5 feet (1524 mm) of building openings.
5. Twenty feet (6096 mm) or more from fixed sources of ignition.

❖ In order to reduce the likelihood of motor fuels coming into contact with ignition sources or posing a hazard to persons, adjoining property or on-site buildings, this section provides very specific dispenser location requirements in relation to buildings, lot lines and ignition sources.

Item 1

Figure 2203.1(1) shows the relationship of the dispenser to the lot line (see also commentary, Item 5).

Item 2

Figure 2203.1(2) shows the relationship of the dispenser to a building with a combustible exterior.

Item 3

Figure 2203.1(3) shows the location of the vehicle being fueled in relation to the motor fuel-dispensing facility property lines. Note that this item is a dispenser location requirement or hose length limitation in that neither feature should be such that any portion of a vehicle could be off site while being fueled. In the past, it was not unusual to find dispensing devices installed on the sidewalk or near the curb in front of the motor fuel-dispensing facility so that vehicles could simply pull up to the curb and be fueled. This was especially true for motor fuel-dispensing facilities situated on very small sites. The intent of this requirement is similar to Item 4; that is, to keep the motor fuel from possible contact with off-site ignition sources. It also eliminates the hazard of an off-site vehicle being struck by another vehicle during fueling.

Item 4

Figure 2203.1(4) shows the relationship of the dispenser nozzle to a building opening. The intent of this requirement is similar to Item 3; that is, to keep the motor fuel from possible contact with ignition sources inside the building.

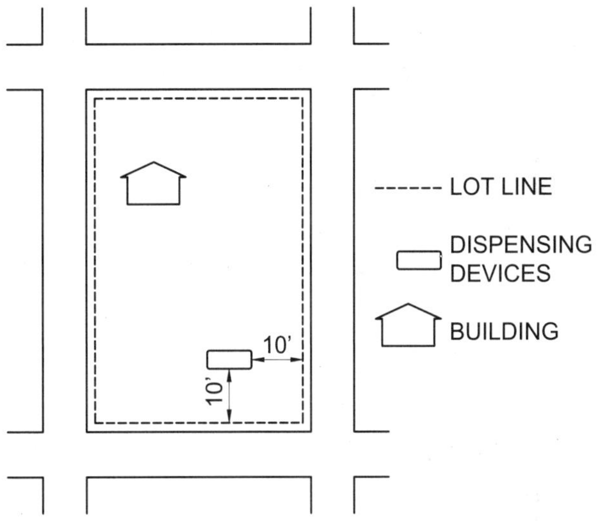

Figure 2203.1(1)
LOCATION OF DISPENSING DEVICES

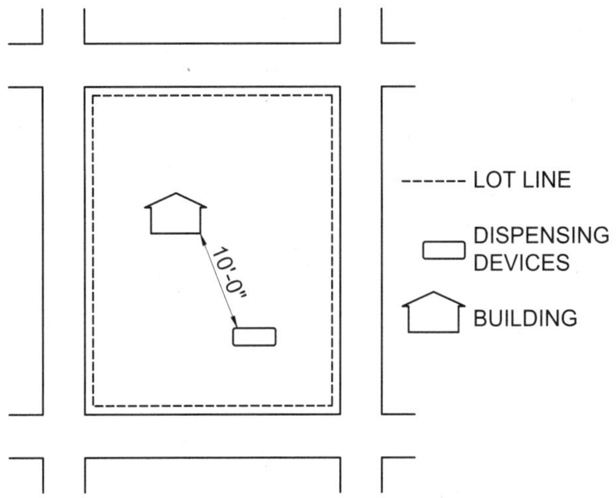

For SI: 1 foot = 304.8 mm.

Figure 2203.1(2)
LOCATION OF DISPENSING DEVICES

Item 5

Figure 2203.1(5) illustrates the area around a dispenser where fixed sources of ignition are prohibited. In planning and reviewing dispenser locations, the 20-foot (6096 mm) dimension should be correlated with Item 1 to prevent any fixed ignition sources beyond the property line from falling within the 20-foot (6096 mm) area required by this item.

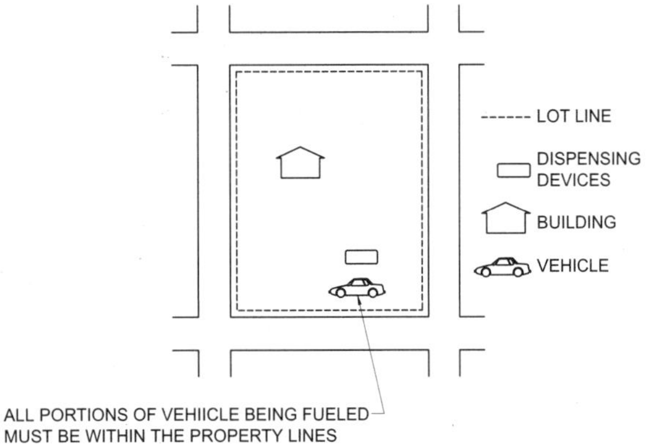

Figure 2203.1(3)
LOCATION OF DISPENSING DEVICES

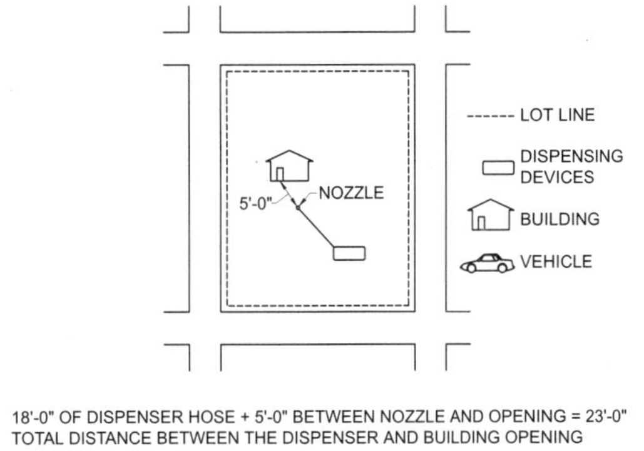

For SI: 1 foot = 304.8 mm.

Figure 2203.1(4)
LOCATION OF DISPENSING DEVICES

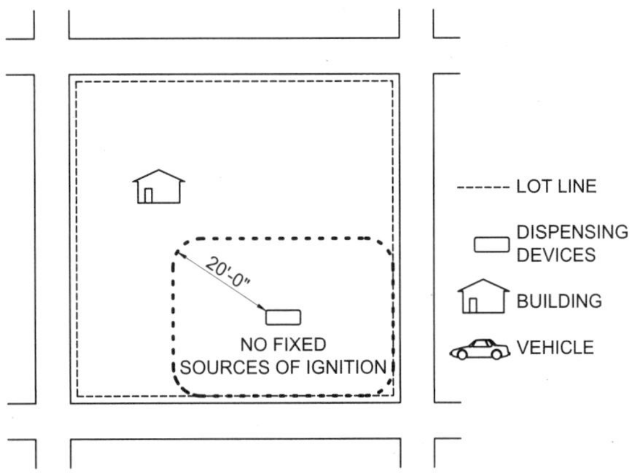

For SI: 1 foot = 304.8 mm.

Figure 2203.1(5)
LOCATION OF DISPENSING DEVICES

2203.2 Emergency disconnect switches. An approved, clearly identified and readily accessible emergency disconnect switch shall be provided at an approved location, to stop the transfer of fuel to the fuel dispensers in the event of a fuel spill or other emergency. An emergency disconnect switch for exterior fuel dispensers shall be located within 100 feet (30 480 mm) of, but not less than 20 feet (6096 mm) from, the fuel dispensers. For interior fuel-dispensing operations, the emergency disconnect switch shall be installed at an approved location. Such devices shall be distinctly labeled as: EMERGENCY FUEL SHUTOFF. Signs shall be provided in approved locations.

❖ This section establishes the requirement for emergency disconnect switches to shut off the flow of fuel in an emergency and specifies where they are to be located for both exterior and interior applications. The emergency disconnect switch must be clearly visible and placed far enough away from the fuel dispenser so that the switch will be easily accessible without entering the fuel spill area, but not so far that it would take too long to get to the switch. The switch location must be prominently indicated by an approved sign and access to the switch must be free of any obstructions, such as displayed merchandise. Figure 2203.2 illustrates the zone in which the disconnect switch must be located.

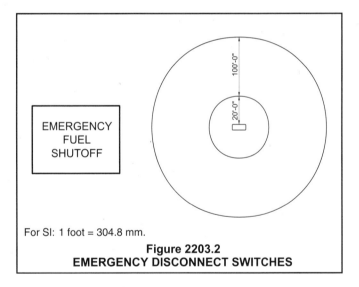

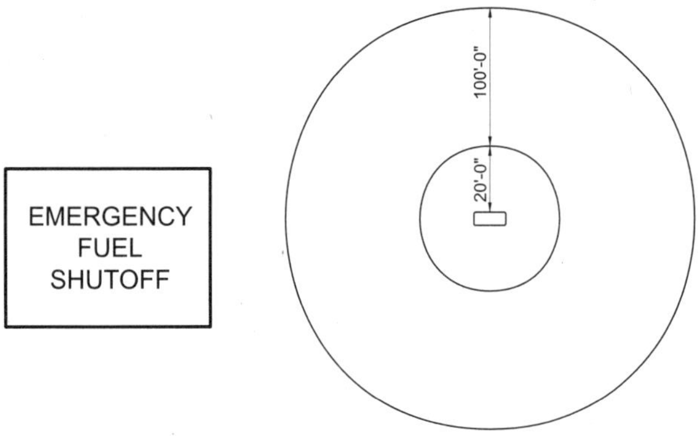

For SI: 1 foot = 304.8 mm.

Figure 2203.2
EMERGENCY DISCONNECT SWITCHES

SECTION 2204
DISPENSING OPERATIONS

2204.1 Supervision of dispensing. The dispensing of fuel at motor fuel-dispensing facilities shall be conducted by a qualified attendant or shall be under the supervision of a qualified attendant at all times or shall be in accordance with Section 2204.3.

❖ Motor fuel-dispensing facilities must have a trained, qualified attendant on duty when the facility is open for business, unless the fire code official specifically approves an unattended location.

2204.2 Attended self-service motor fuel-dispensing facilities. Attended self-service motor fuel-dispensing facilities shall comply with Sections 2204.2.1 through 2204.2.5. Attended self-service motor fuel-dispensing facilities shall have at least one qualified attendant on duty while the facility is open for business. The attendant's primary function shall be to supervise, observe and control the dispensing of fuel. The attendant shall prevent the dispensing of fuel into containers that do not comply with Section 2204.4.1, control sources of ignition, give immediate attention to accidental spills or releases, and be prepared to use fire extinguishers.

❖ An attendant trained in spill control, ignition source control, recognizing approved fuel containers and fire extinguishment is required to be in visual contact with the dispensing operation when the motor fuel-dispensing facility is open for business unless the fire code official has given approval for an unattended self-service facility. The attendant may perform other duties such as those of cashier so long as the attendant can supervise the dispensing operation and has immediate access to emergency shutoff controls. Note that the responsibility of supervision, observation and control of the dispensing operations includes enforcement of the procedures and rules in Sections 2205.6 and 2210.5.

2204.2.1 Special-type dispensers. Approved special-dispensing devices and systems such as, but not limited to, card- or coin-operated and remote-preset types, are allowed at motor fuel-dispensing facilities provided there is at least one qualified attendant on duty while the facility is open to the public. Remote preset-type devices shall be set in the "off" position while not in use so that the dispenser cannot be activated without the knowledge of the attendant.

❖ Special dispensing devices that allow the customer to pay at the dispenser have become very popular. This provision requires that the dispenser be maintained in the off position so that the attendant is alerted before the fuel is dispensed and can, therefore, supervise the dispensing operation, making certain that the customer is placing fuel in an approved container and there are no sources of ignition in the area.

2204.2.2 Emergency controls. Approved emergency controls shall be provided in accordance with Section 2203.2.

❖ A clearly marked emergency fuel shutoff switch must be located no further than 100 feet (30 480 mm) from the dispenser and no closer than 20 feet (6096 mm) to the dispenser. The switch must be readily available to all persons and must cut off power to all dispensers and pumps.

2204.2.3 Operating instructions. Dispenser operating instructions shall be conspicuously posted in approved locations on every dispenser.

❖ Clearly understandable operating instructions for the use of the dispenser must be posted on the dispenser. The location must be approved by the fire code official. These operating instructions are in addition to the warnings required in Section 2205.6.

2204.2.4 Obstructions to view. Dispensing devices shall be in clear view of the attendant at all times. Obstructions shall not be placed between the dispensing area and the attendant.

❖ This provision does not specify a distance the attendant must be from the dispensing operation; however, the attendant must be able to clearly view the entire dispensing area from his or her workstation. In some cases, this is accomplished by closed-circuit television monitoring; however, the fire code official should carefully evaluate the clarity and resolution of the video image to verify that it meets the "clear view" requirement of this section. This section also prohibits the common practice of loading outside areas, including the dispenser islands, with displayed merchandise piled high enough so as to be a visual obstruction to the attendant.

2204.2.5 Communications. The attendant shall be able to communicate with persons in the dispensing area at all times. An approved method of communicating with the fire department shall be provided for the attendant.

❖ This is a two-part requirement. The first part requires that the attendant has the ability to communicate, for example, with an intercom, with the person performing the dispensing. The second part requires the attendant to have some type of communication equipment that will allow the attendant to immediately call the fire department in case of an emergency.

2204.3 Unattended self-service motor fuel-dispensing facilities. Unattended self-service motor fuel-dispensing facilities shall comply with Sections 2204.3.1 through 2204.3.7.

❖ An unattended self-service motor fuel-dispensing facility is allowed only with the specific approval of the fire code official. He or she should consider the location; exposures; the likelihood of vandalism; how emergency equipment such as the emergency shutoff switch, portable fire extinguishers and the means of notifying the fire department in case of spill, fire or other emergency is going to be protected.

2204.3.1 General. Where approved, unattended self-service motor fuel-dispensing facilities are allowed. As a condition of approval, the owner or operator shall provide, and be accountable for, daily site visits, regular equipment inspection and maintenance.

❖ It is imperative that the owner/operator has a responsible person to make daily site visits to the unattended self-service motor fuel-dispensing facility. All emergency equipment must be inspected for proper operation and availability to the customer. Vandalism is a major problem with unattended self-service motor fuel-dispensing facilities. Vandals will trip the emergency fuel shutoff switch and remove or discharge the fire-extinguishing equipment. This equipment must be maintained and be made available to the person dispensing fuel. Also, a method of documenting the daily visits needs to be established and approved by the fire code official.

2204.3.2 Dispensers. Dispensing devices shall comply with Section 2206.7. Dispensing devices operated by the insertion of coins or currency shall not be used unless approved.

❖ The intent of this provision is to allow card or key-operated-type dispensers. Coin- or currency-type dispensers are allowed only with the approval of the fire code official.

2204.3.3 Emergency controls. Approved emergency controls shall be provided in accordance with Section 2203.2. Emergency controls shall be of a type which is only manually resettable.

❖ The emergency controls must be clearly identified and available to the person dispensing fuel. The controls or switch must be located no closer than 20 feet (6096 mm) to the dispenser and no farther than 100 feet (30 480 mm) from the dispenser. The switch must cut off power to all dispensers and pumps and must be manually resettable by the owner/operator. The intent is to prevent anyone from dispensing fuel until the problem has been corrected.

2204.3.4 Operating instructions. Dispenser operating instructions shall be conspicuously posted in approved locations on every dispenser and shall indicate the location of the emergency controls required by Section 2204.3.3.

❖ It is a special requirement for unattended self-service motor fuel-dispensing facilities that the location of the emergency control switch be included with the dispenser instructions and posted on the dispenser in an approved location.

2204.3.5 Emergency procedures. An approved emergency procedures sign, in addition to the signs required by Section 2205.6, shall be posted in a conspicuous location and shall read:

IN CASE OF FIRE, SPILL OR RELEASE

1. USE EMERGENCY PUMP SHUTOFF

2. REPORT THE ACCIDENT!

FIRE DEPARTMENT TELEPHONE NO.______

FACILITY ADDRESS ____________________

❖ Signs must be clearly posted giving the location of the emergency fuel shutoff switch, the fire department's telephone number and the motor fuel-dispensing facility address.

It is imperative that the person dispensing fuel at an unattended motor fuel-dispensing facility knows where the emergency control equipment is located. People not familiar with the area may not know the fire department's telephone number or the address of the motor fuel-dispensing facility; therefore, this information must be included on the sign.

2204.3.6 Communications. A telephone not requiring a coin to operate or other approved, clearly identified means to notify the fire department shall be provided on the site in a location approved by the fire code official.

❖ The intent of this section is to provide a means to call the fire department in an emergency without the use of a coin or phone card. Many jurisdictions with 911 telephone systems have pay phones that do not require the use of a coin or card to dial 911; this type of coin-operated telephone would be allowed.

2204.3.7 Quantity limits. Dispensing equipment used at unsupervised locations shall comply with one of the following:

1. Dispensing devices shall be programmed or set to limit uninterrupted fuel delivery to 25 gallons (95 L) and require a manual action to resume delivery.
2. The amount of fuel being dispensed shall be limited in quantity by a preprogrammed card as approved.

❖ Limiting the amount of uninterrupted fuel delivered before another action is taken reduces the chances of a major fuel spill in the case of vandalism or equipment failure.

2204.4 Dispensing into portable containers. The dispensing of flammable or combustible liquids into portable approved containers shall comply with Sections 2204.4.1 through 2204.4.3.

❖ This section describes an approved container and states requirements for dispensing fuel into an approved container.

2204.4.1 Approved containers required. Class I, II and IIIA liquids shall not be dispensed into a portable container unless such container is of approved material and construction, and has a tight closure with screwed or spring-loaded cover so designed that the contents can be dispensed without spilling. Liquids shall not be dispensed into portable tanks or cargo tanks.

❖ Approved containers must be easily identified as fuel containers and constructed of materials that will maintain structural stability and resist spills. Flammable and combustible liquids should not be dispensed into portable or cargo tanks using conventional automotive dispensing equipment. If the tank is not properly grounded and the dispenser nozzle is not in contact with the tank, static electricity may build up and discharge between the nozzle and the tank.

2204.4.2 Nozzle operation. A hose nozzle valve used for dispensing Class I liquids into a portable container shall be in compliance with Section 2206.7.6 and be manually held open during the dispensing operation.

❖ A listed automatic-closing-type nozzle must be used for dispensing fuel into portable containers. The hold-open device must not be used because the automatic-closing feature may not function properly since it is designed to operate in the neck of an automobile fuel receiver. The fuel tank vent in an automobile terminates in the receiver neck and the blow-back of fuel coming from the vent is what normally triggers the automatic shutoff feature.

2204.4.3 Location of containers being filled. Portable containers shall not be filled while located inside the trunk, passenger compartment or truck bed of a vehicle.

❖ Fuel tanks riding on carpets, mats and bed liners may build up a charge of static electricity. This electrical charge may discharge to the nozzle if not properly grounded. The container must be removed from the vehicle and placed on the ground before fueling. The nozzle must be in contact with the container before discharging the fuel.

SECTION 2205 OPERATIONAL REQUIREMENTS

2205.1 Tank filling operations for Class I, II or IIIA liquids. Delivery operations to tanks for Class I, II or IIIA liquids shall comply with Sections 2205.1.1 through 2205.1.3 and the applicable requirements of Chapter 34.

❖ History has shown that most accidents at motor fuel-dispensing facilities occur during the tank-filling operation. The provisions of this section address those operations.

2205.1.1 Delivery vehicle location. Where liquid delivery to above-ground storage tanks is accomplished by positive-pressure operation, tank vehicles shall be positioned a minimum of 25 feet (7620 mm) from tanks receiving Class I liquids and 15 feet (4572 mm) from tanks receiving Class II and IIIA liquids.

❖ Above-ground tanks are usually above the elevation of the fuel delivery truck, and the fuel must be pumped from the truck into the storage tank. The fuel is delivered in large volumes under pressure and most delivery trucks use a power take-off-driven pump that requires the truck engine to be running. Because of this, the delivery truck and the storage tanks must be separated by the distances indicated so that venting vapors do not find their way to the running engine of the delivery vehicle and ignite.

2205.1.2 Tank capacity calculation. The driver, operator or attendant of a tank vehicle shall, before making delivery to a tank, determine the unfilled, available capacity of such tank by an approved gauging device.

❖ A gauge stick may be used to determine the quantity of fuel in a tank, but is impractical for use on an above-ground tank because the attendant would have to climb on top of the tank each time. In icy, wet or inclement weather this could be dangerous. It is more practical to install a fuel-level gauge accessible to the delivery operator on an above-ground tank.

After the driver places the delivery truck in position, he or she should record the ullage (i.e., the amount of liquid it would take to fill the tank) and set the delivery up for that amount of fuel. This is the first line of defense to prevent fuel spills (see also Section 3406.6.1.5 of the code).

2205.1.3 Tank fill connections. Delivery of flammable liquids to tanks more than 1,000 gallons (3785 L) in capacity shall be made by means of approved liquid- and vapor-tight connections between the delivery hose and tank fill pipe. Where tanks are equipped with any type of vapor recovery system, all connections required to be made for the safe and proper functioning of the particular vapor recovery process shall be made. Such connections shall be made liquid and vapor tight and remain connected throughout the unloading process. Vapors shall not be discharged at grade level during delivery.

❖ The delivery of fuel to either an above-ground or underground storage tank that exceeds 1,000 gallons (3785 L) must be through a liquid-tight fitting to prevent the escape of flammable liquid or vapors at the point of connection. Vapor-tight connections reduce the potential for a release of flammable vapors that can be easily ignited. These include liquid transfer lines and vapor recovery lines, which are designed to prevent the release of polluting, flammable fuel vapor during transfer. These requirements prohibit the extremely dangerous, but not uncommon, practice of delivery tanker operators not connecting vapor return hoses, thus allowing the vapors displaced during delivery to escape at grade level from the unmade connections. Incidents have been reported where the vapors have traveled to nearby buildings, found an ignition source and exploded (see also Section 3406.6.1.10).

2205.2 Equipment maintenance and inspection. Motor fuel-dispensing facility equipment shall be maintained in proper working order at all times in accordance with Sections 2205.2.1 through 2205.2.3.

❖ The provisions of this section address the requirements for maintaining safety equipment at dispensing operations.

2205.2.1 Dispensing devices. Where maintenance to Class I liquid dispensing devices becomes necessary and such maintenance could allow the accidental release or ignition of liquid, the following precautions shall be taken before such maintenance is begun:

1. Only persons knowledgeable in performing the required maintenance shall perform the work.
2. Electrical power to the dispensing device and pump serving the dispenser shall be shut off at the main electrical disconnect panel.
3. The emergency shutoff valve at the dispenser, where installed, shall be closed.
4. Vehicle traffic and unauthorized persons shall be prevented from coming within 12 feet (3658 mm) of the dispensing device.

❖ Dispensers are complex pieces of machinery made up of many listed parts. Certain components of the dispenser are sealed to prevent ignition of fuel vapors.

Therefore, it is imperative that the repair technician be qualified to perform work on the dispenser.

The dispenser and the pump are usually two separate and distinct pieces of equipment. In most cases, the pump is located at the tank, remote from the dispensers, and fuel is supplied under pressure to the dispenser. Therefore, it is imperative that power be disconnected to both the dispenser and the pump before work is begun.

As a safety measure, in case the remote pump kicks on, the dispenser emergency valve must be manually closed. This valve is located below the dispenser in the liquid supply piping. This valve is required in remote pumping systems by Section 2206.7.4.

The separation distance of 12 feet (3658 mm) is intended to keep the public and ignition sources away from possible fuel spills that may occur during the maintenance of dispensing devices.

2205.2.2 Emergency shutoff valves. Automatic-closing emergency shutoff valves required by Section 2206.7.4 shall be checked not less than once per year by manually tripping the hold-open linkage.

❖ The emergency valve is located in the sump below the dispenser and must be installed when fuel is supplied to the dispenser under pressure. This valve consists of a shear section and a fusible link with a spring-loaded valve. This valve must be operated annually to verify that it is operable.

2205.2.3 Leak detectors. Leak detection devices required by Section 2206.7.7.1 shall be checked and tested at least annually in accordance with the manufacturer's specifications to ensure proper installation and operation.

❖ In a remote pumping system (where fuel is supplied under pressure to the dispenser) a leak detection device is required on the discharge/pressure side of the pump. The most common area to leak fuel is the piping between the storage tanks and the dispensers. Therefore, it is imperative that the required leak detection equipment be tested annually and maintained in an operable condition.

2205.3 Spill control. Provisions shall be made to prevent liquids spilled during dispensing operations from flowing into buildings. Acceptable methods include, but shall not be limited to, grading driveways, raising doorsills, or other approved means.

❖ The intent of this provision is to prevent flammable and combustible liquids from entering buildings if a spill occurs. The spill control method must be approved by the fire code official and may be as simple as scoring the concrete pavement adjacent to the dispenser island (similar to highway "rumble strips") to retard the surface flow of spilled fuel.

If the spilled liquid is to be routed to a drain, the drain must be equipped with a sump and an oil-water separator to prevent the fuel from entering the storm drainage system. The oil-water separator must be installed in accordance with the IPC.

2205.4 Sources of ignition. Smoking and open flames shall be prohibited in areas where fuel is dispensed. The engines of vehicles being fueled shall be shut off during fueling. Electrical equipment shall be in accordance with the ICC *Electrical Code*.

❖ The intent of this provision is to control all ignition sources near fuel-dispensing operations. This includes smoking, matches, lighters or any other ignition source. Internal combustion engines should be shut off during the fueling operation. Electrical equipment in close proximity to the fueling operation, including the dispensing equipment, must be in accordance with the ICC EC and Table 3403.1.1 of the code.

2205.5 Fire extinguishers. Approved portable fire extinguishers complying with Section 906 with a minimum rating of 2-A:20-B:C shall be provided and located such that an extinguisher is not more than 75 feet (22 860 mm) from pumps, dispensers or storage tank fill-pipe openings.

❖ A person should not have to travel more than 75 feet (22 860 mm) from a fuel dispenser, pump or a fill opening to reach an extinguisher. If the dispenser, pump and fill opening are in close proximity to each other, one extinguisher may satisfy the requirements.

2205.6 Warning signs. Warning signs shall be conspicuously posted within sight of each dispenser in the fuel-dispensing area and shall state the following:

1. It is illegal and dangerous to fill unapproved containers with fuel.
2. Smoking is prohibited.
3. The engine shall be shut off during the refueling process.
4. Portable containers shall not be filled while located inside the trunk, passenger compartment, or truck bed of a vehicle.

❖ The warning sign must be legible and conspicuously posted in the dispensing area. The intent is to notify the dispenser operator not to use unapproved fuel containers, to remove portable containers from the vehicle before filling (see commentary, Section 2204.4.3) and to eliminate ignition source, such as smoking and operating internal combustion engines. The warnings should be on a sign with a contrasting background that will catch the eye of the person performing the dispensing operation.

2205.7 Control of brush and debris. Fenced and diked areas surrounding above-ground tanks shall be kept free from vegetation, debris and other material that is not necessary to the proper operation of the tank and piping system.

Weeds, grass, brush, trash and other combustible materials

shall be kept not less than 10 feet (3048 mm) from fuel-handling equipment.

❖ Above-ground tanks must to be secured and inaccessible to the public. The secured area around the tanks must be kept clean and free of combustibles. The area around dispensers, remote pumps and fill openings must also be kept clean and free of combustibles for a distance of at least 10 feet (3048 mm). The intent is to prevent accumulation of readily combustible materials near tanks and equipment that could contribute to a fire in the event of an ignition.

SECTION 2206 FLAMMABLE AND COMBUSTIBLE LIQUID MOTOR FUEL-DISPENSING FACILITIES

2206.1 General. Storage of flammable and combustible liquids shall be in accordance with Chapter 34 and this section.

❖ The general requirements for storing flammable and combustible liquids are found in Chapter 34. The provisions specific to motor fuel-dispensing facilities are found in this section. Due to the activities such as convenience stores, vehicle traffic and location of most motor fuel-dispensing facilities, the provisions for tank capacities, locations and dispensing equipment are more stringent in this chapter of the code than the general provisions found in Chapter 34. Accordingly, these specific regulations take precedence over the general provisions of Chapter 34.

2206.2 Method of storage. Approved methods of storage for Class I, II and IIIA liquid fuels at motor fuel-dispensing facilities shall be in accordance with Sections 2206.2.1 through 2206.2.5.

❖ These provisions are specifically for motor fuel-dispensing facilities.

2206.2.1 Underground tanks. Underground tanks for the storage of Class I, II and IIIA liquid fuels shall comply with Chapter 34.

❖ This section relies on Chapter 34, which contains the requirements for the location and installation of underground tanks.

2206.2.1.1 Inventory control for underground tanks. Accurate daily inventory records shall be maintained and reconciled on underground fuel storage tanks for indication of possible leakage from tanks and piping. The records shall be kept at the premises or made available for inspection by the fire code official within 24 hours of a written or verbal request and shall include records for each product showing daily reconciliation between sales, use, receipts and inventory on hand. Where there is more than one system consisting of tanks serving separate pumps or dispensers for a product, the reconciliation shall be ascertained separately for each tank system. A consistent or accidental loss of product shall be immediately reported to the fire code official.

❖ In addition to the leak detection requirements of Section 2205.3, this section requires another level of scrutiny for underground motor fuel storage tanks in the form of daily, documented reconciliation of inventory versus product sold to help detect leaks in underground tanks and piping. Loss in inventory should be reported immediately to the fire code official (see also Section 3404.2.11.5.1).

2206.2.2 Above-ground tanks located inside buildings. Above-ground tanks for the storage of Class I, II and IIIA liquid fuels are allowed to be located in buildings. Such tanks shall be located in special enclosures complying with Section 2206.2.6, in a liquid storage room or a liquid storage warehouse complying with Chapter 34, or shall be listed and labeled as protected above-ground tanks.

❖ The intent of this provision is to establish requirements for placing an above-ground tank inside a building. The options are:

1. A tank in a special enclosure in accordance with Section 2206.2.6.
2. A tank inside a liquid storage room in accordance with Section 3404.3.7.
3. A tank inside a liquid storage warehouse in accordance with Section 3404.3.8.
4. A protected above-ground tank (UL 2085).

2206.2.3 Above-ground tanks located outside, above grade. Above-ground tanks shall not be used for the storage of Class I, II or IIIA liquid fuels except as provided by this section.

1. Above-ground tanks used for outside, above-grade storage of Class I liquids shall be listed and labeled as protected above-ground tanks and be in accordance with Chapter 34. Such tanks shall be located in accordance with Table 2206.2.3.
2. Above-ground tanks used for above-grade storage of Class II or IIIA liquids are allowed to be protected above-ground tanks or, when approved by the fire code official, other above-ground tanks that comply with Chapter 34. Tank locations shall be in accordance with Table 2206.2.3.
3. Tanks containing fuels shall not exceed 12,000 gallons (45 420 L) in individual capacity or 48,000 gallons (181 680 L) in aggregate capacity. Installations with the maximum allowable aggregate capacity shall be separated from other such installations by not less than 100 feet (30 480 mm).
4. Tanks located at farms, construction projects, or rural areas shall comply with Section 3406.2.

❖ The intent of Item 1 is to allow the storage of all classes of flammable and combustible liquids in listed and labeled protected above-ground tanks. A protected above-ground tank is a tank that meets the requirements of Section 3404.2.9.6 and is listed in accordance with UL 2085.

Item 2 essentially states that Class I flammable and combustible liquids must be stored in a UL 2085 protected above-ground tank. Other low-class liquids may be stored in UL 142 above-ground tanks if approved by the fire code official.

In Item 3, the maximum aggregate capacity is the total capacity of all the individual tanks in one area. A separation distance of at least 100 feet (30 480 mm) must be placed between installations of the maximum aggregate capacities [see Figure 2206.2.3(1)]. The 100-foot (30 480 mm) separation distance is intended to protect multiple aggregate capacity installations from one another in the event of a fire or other emergency.

Item 4 provides an exception since the provisions of Section 3406.2 pertain to the permanent and temporary storage and dispensing of Class I (gasoline) and Class II (diesel and kerosene) for private use on farms and rural areas and at construction sites, earth-moving projects and gravel pits. See Section 3406.2 for maximum capacities and other provisions and Section 3406.2.4.4 for locations where above-ground tanks are prohibited.

TABLE 2206.2.3. See below.

❖ It is important to note that Class I liquids can be stored in protected above-ground tanks only. An example of a Class I liquid is gasoline. Class II (e.g., diesel fuel) and III liquids can be stored in any approved above-ground tank for flammable and combustible liquids. Examples of Class II and III liquids are diesel and kerosene. Figure 2206.2.3(2) shows the spacing requirements for a protected (UL 2085) above-ground tank 6,000 gallons (22 710L) or less. Figure 2206.2.3(3) shows the spacing requirements for a protected (UL 2085) above-ground tank greater than 6,000 gallons (22 710 L). Figure 2206.2.3(4) shows the spacing requirements for a listed UL 142 above-ground tank regardless of whether or not it is a double-wall UL 142 tank or a single-wall UL 142 inside of a dike.

TABLE 2206.2.3
MINIMUM SEPARATION REQUIREMENTS FOR ABOVE-GROUND TANKS

CLASS OF LIQUID AND TANK TYPE	INDIVIDUAL TANK CAPACITY (gallons)	MINIMUM DISTANCE FROM NEAREST IMPORTANT BUILDING ON SAME PROPERTY (feet)	MINIMUM DISTANCE FROM NEAREST FUEL DISPENSER (feet)	MINIMUM DISTANCE FROM LOT LINE WHICH IS OR CAN BE BUILT UPON, INCLUDING THE OPPOSITE SIDE OF A PUBLIC WAY (feet)	MINIMUM DISTANCE FROM NEAREST SIDE OF ANY PUBLIC WAY (feet)	MINIMUM DISTANCE BETWEEN TANKS (feet)
Class I protected above-ground tanks or tanks in vaults	Less than or equal to 6,000	5	25[a]	15	5	3
	Greater than 6,000	15	25[a]	25	15	3
Class II and III protected above-ground tanks or tanks in vaults	Same as Class I	Same as Class I	Same as Class I	Same as Class I	Same as Class I	Same as Class I
Other tanks	All	50	50	100	50	3

For SI: 1 foot = 304.8 mm, 1 gallon = 3.785 L.
a. At fleet vehicle motor fuel-dispensing facilities, no minimum separation distance is required.

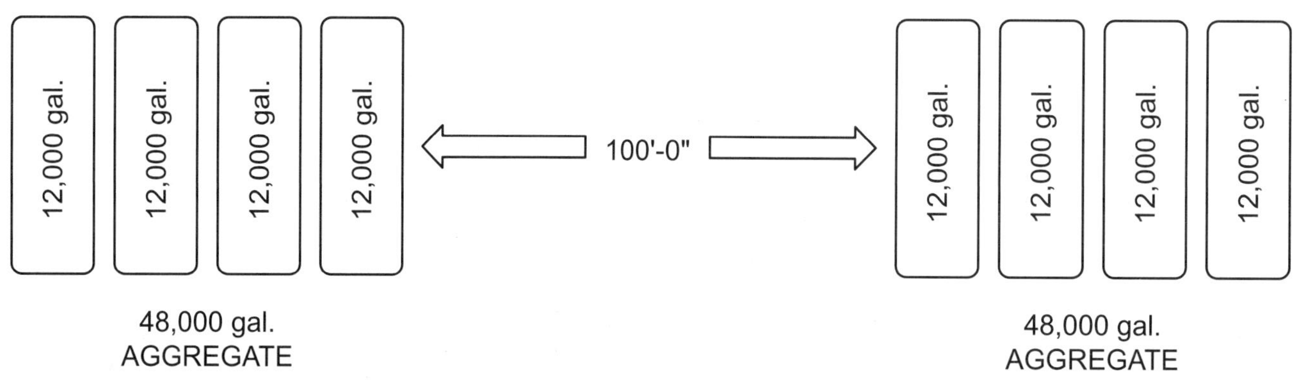

For SI: 1 foot = 304.8 mm, 1 gallon = 3.785 L.

Figure 2206.2.3(1)
ABOVE-GROUND TANKS LOCATED OUTSIDE ABOVE GRADE

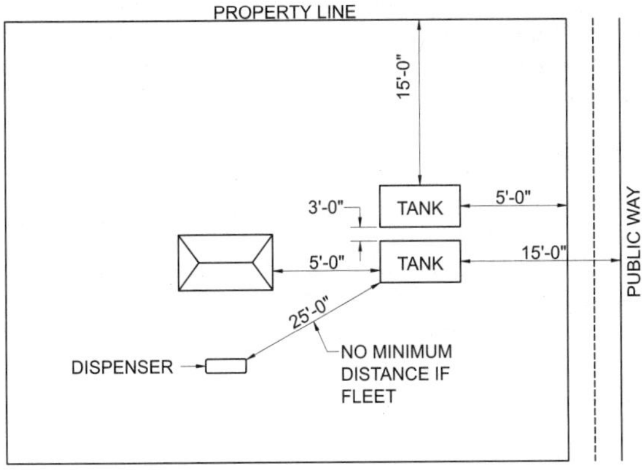

For SI: 1 foot = 304.8 mm.

Figure 2206.2.3(2)
6,000 GALLONS OR LESS PROTECTED ABOVE-GROUND STORAGE TANKS CLASS I, II OR III LIQUIDS

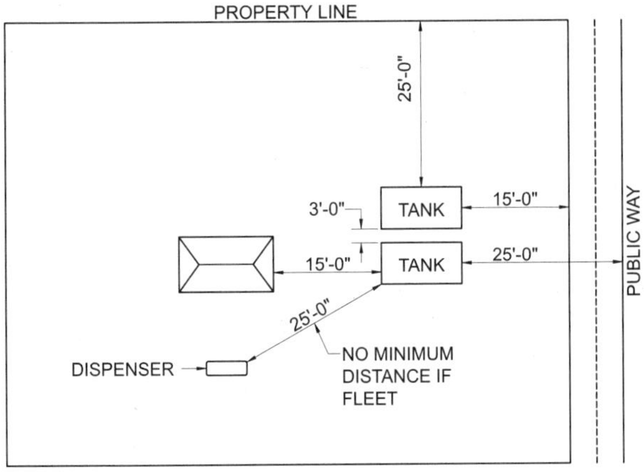

For SI: 1 foot = 304.8 mm.

Figure 2206.2.3(3)
GREATER THAN 6,000 GALLONS PROTECTED AST CLASS I, II OR III LIQUIDS

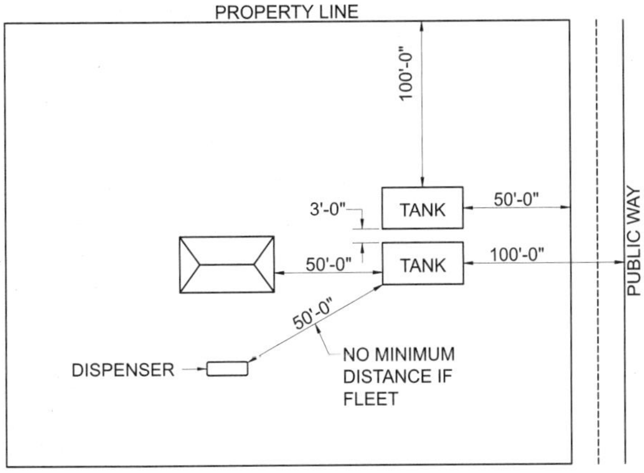

For SI: 1 foot = 304.8 mm.

Figure 2206.2.3(4)
ALL OTHER AST'S CLASS II AND III LIQUIDS ONLY MUST BE APPROVED BY FIRE CODE OFFICIAL

2206.2.4 Above-ground tanks located in above-grade vaults or below-grade vaults. Above-ground tanks used for storage of Class I, II or IIIA liquid fuels are allowed to be installed in vaults located above grade or below grade in accordance with Section 3404.2.8 and shall comply with Sections 2206.2.4.1 and 2206.2.4.2. Tanks in above-grade vaults shall also comply with Table 2206.2.3.

❖ The definition of an above-ground tank is a tank without backfill. An underground tank receives strength from backfill. If the tank is not surrounded by backfill, the tank must be a listed and labeled above-ground tank. Therefore, a tank inside a vault, even if the vault is underground, must be a listed and labeled above-ground tank.

2206.2.4.1 Tank capacity limits. Tanks storing Class I and Class II liquids at an individual site shall be limited to a maximum individual capacity of 15,000 gallons (56 775 L) and an aggregate capacity of 48,000 gallons (181 680 L).

❖ This section places limitations on the capacities of tanks inside vaults. At motor fuel-dispensing facilities, the maximum capacity is 15,000 gallons (56 775 L) for a single tank and 48,000 gallons (181 680 L) aggregate capacity. The individual tank capacity allows the facility to receive the largest single product delivery that may be reasonably expected in a single tank. The aggregate capacity affords the facility the marketing flexibility it needs to carry for sale a wide range of motor fuels and specialty fuels, such as kerosene or diesel fuel.

2206.2.4.2 Fleet vehicle motor fuel-dispensing facilities. Tanks storing Class II and Class IIIA liquids at a fleet vehicle motor fuel-dispensing facility shall be limited to a maximum individual capacity of 20,000 gallons (75 700 L) and an aggregate capacity of 80,000 gallons (302 800 L).

❖ At fleet vehicle motor fuel-dispensing facilities (where the vehicles are used in connection with the fleet operator's business) the maximum storage capacity is 20,000 gallons (75 700 L) for a single tank and 80,000 gallons (302 800 L) aggregate. These higher capacities recognize that the relative hazard of fleet facilities is less than at public motor fuel-dispensing facilities because the fleet fuel of choice is typically diesel fuel, a Class II liquid, rather than gasoline, a Class IB liquid. The higher capacities also reduce the hazard of liquid transfer by reducing the number of times the storage tanks must be refilled. Note that the additional capacity in this section applies only to Class II and IIIA liquid fuels. Class I fuels would remain subject to the provisions of Section 2206.2.4.1.

2206.2.5 Portable tanks. Where approved by the fire code official, portable tanks are allowed to be temporarily used in conjunction with the dispensing of Class I, II or IIIA liquids into the fuel tanks of motor vehicles or motorized equipment on premises not normally accessible to the public. The approval shall include a definite time limit.

❖ This section recognizes the need for allowing portable tanks to be used temporarily for fueling vehicles at locations that, because of topography or security, are not accessible to the general public, such as mining sites, logging camps, well-drilling sites, large rail yards, construction projects, and the like. The key word here is "temporarily." If such sites require a permanent fueling facility, then all applicable provisions of the code apply. The fire code official retains control over such temporary uses since approval is required and a definite time limit on such operations is imposed.

2206.2.6 Special enclosures. Where installation of tanks in accordance with Section 3404.2.11 is impractical, or because of property or building limitations, tanks for liquid fuels are allowed to be installed in buildings in special enclosures in accordance with all of the following:

1. The special enclosure shall be liquid tight and vapor tight.
2. The special enclosure shall not contain backfill.
3. Sides, top and bottom of the special enclosure shall be of reinforced concrete at least 6 inches (152 mm) thick, with openings for inspection through the top only.
4. Tank connections shall be piped or closed such that neither vapors nor liquid can escape into the enclosed space between the special enclosure and any tanks inside the special enclosure.
5. Means shall be provided whereby portable equipment can be employed to discharge to the outside any vapors which might accumulate inside the special enclosure should leakage occur.
6. Tanks containing Class I, II or IIIA liquids inside a special enclosure shall not exceed 6,000 gallons (22 710 L) in individual capacity or 18,000 gallons (68 130 L) in aggregate capacity.
7. Each tank within special enclosures shall be surrounded by a clear space of not less than 3 feet (910 mm) to allow for maintenance and inspection.

❖ Special enclosures (that are essentially a concrete vault) can solve installation difficulties on problematic motor fuel-dispensing facility sites by allowing tanks to be installed inside of the facility building. The intent of this provision is to:

- Protect the building from the tank in case there is a fire involving the tank.
- Protect the tank from the building in case there is a fire involving the building.
- Prevent vapors from accumulating in the space between the tank and the enclosure.
- Limit the capacity of the tank inside the enclosure.
- Provide room for a maintenance technician to enter the space between the tank and the enclosure.

2206.3 Security. Above-ground tanks for the storage of liquid fuels shall be safeguarded from public access or unauthorized entry in an approved manner.

❖ The intent of this section is to protect the above-ground tank from vandalism and at the same time allow enough room for fire fighters to maneuver around the tank.

2206.4 Physical protection. Guard posts complying with Section 312 or other approved means shall be provided to protect above-ground tanks against impact by a motor vehicle unless the tank is listed as a protected above-ground tank with vehicle impact protection.

❖ This provision seeks to prevent leaks and spills caused by vehicle impact. The provision recognizes that vehicle damage can be reduced by placing 4-inch (102 mm) steel posts filled with concrete and spaced on 4-foot (1219 mm) centers around the above-ground tank or by using a listed protected above-ground tank with vehicle impact protection.

2206.5 Secondary containment. Above-ground tanks shall be provided with drainage control or diking in accordance with Chapter 34. Drainage control and diking is not required for listed secondary containment tanks. Secondary containment systems shall be monitored either visually or automatically. Enclosed secondary containment systems shall be provided with emergency venting in accordance with Section 2206.6.2.5.

❖ A single-wall above-ground tank inside a dike or a double-wall secondary containment above-ground tank with emergency relief vents for the inner tank and the interstitial space will provide the level of protection required by this section. Both styles of tanks are listed for above-ground use and both tanks meet the requirements for drainage control and diking.

2206.6 Piping, valves, fittings and ancillary equipment for use with flammable or combustible liquids. The design, fabrication, assembly, testing and inspection of piping, valves, fittings and ancillary equipment for use with flammable or combustible liquids shall be in accordance with Chapter 34 and Sections 2206.6.1 through 2206.6.3.

❖ The majority of leaks come from piping, valves, fittings and ancillary equipment, not from the tank itself. Therefore, the intent of this section is to reduce fuel spills by addressing the requirements for this equipment.

2206.6.1 Protection from damage. Piping shall be located such that it is protected from physical damage.

❖ Piping is easily damaged by vehicles and other equipment. Piping must be installed and located in a way that will minimize damage either by burying the pipe or protecting the pipe by some other physical means. More leaks come from piping than any other source.

2206.6.2 Piping, valves, fittings and ancillary equipment for above-ground tanks for Class I, II and IIIA liquids. Piping, valves, fittings and ancillary equipment for above-ground tanks shall comply with Sections 2206.6.2.1 through 2206.6.2.6.

❖ Because above-ground tanks have a gravity head and some of the piping is always exposed above ground, special provisions apply to piping for above-ground tanks.

2206.6.2.1 Tank openings. Tank openings for above-ground tanks shall be through the top only.

❖ Tank openings must be through the top, above the liquid level, of an above-ground tank to reduce the risk of a liquid spill in the event of a piping failure.

2206.6.2.2 Fill-pipe connections. The fill pipe for above-ground tanks shall be provided with a means for making a direct connection to the tank vehicle's fuel-delivery hose so that the delivery of fuel is not exposed to the open air during the filling operation. Where any portion of the fill pipe exterior to the tank extends below the level of the top of the tank, a check valve shall be installed in the fill pipe not more than 12 inches (305 mm) from the fill-hose connection.

❖ The fill-pipe connection must be a liquid-tight connection such as a cam-lock connection. The connection may be installed on top of the tank, which would require the fuel delivery person to climb on top of the tank, or the connection may be piped down to a lower level to prevent the delivery person from having to climb onto the tank. When the connection is piped down below the fuel level in the tank, a check valve must be installed to prevent the discharge of fuel by siphon flow (see Figure 2206.6.2.2).

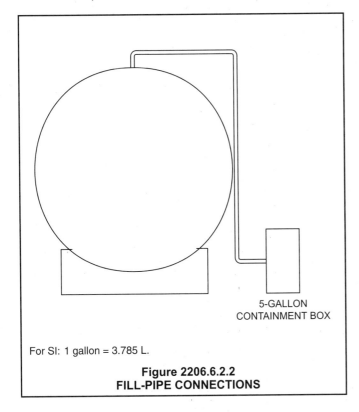

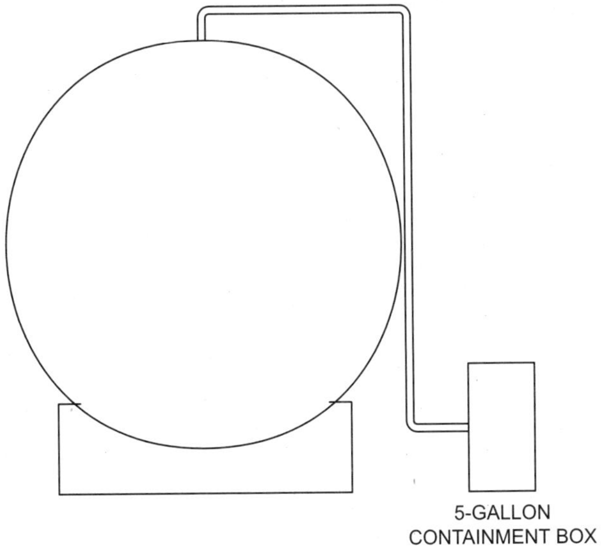

For SI: 1 gallon = 3.785 L.

Figure 2206.6.2.2
FILL-PIPE CONNECTIONS

2206.6.2.3 Overfill protection. Overfill protection shall be provided for above-ground flammable and combustible liquid storage tanks in accordance with Sections 3404.2.7.5.8 and 3404.2.9.6.6.

❖ Overfill prevention is a major concern when dealing with above-ground tanks. The first step in preventing a spill is for the delivery truck operator to check the liquid level in the above-ground tank and set the truck up accordingly to fill the tank to the 90-percent level. The second prevention measure is the audible fill alarm that will sound an alarm when the tank reaches 90-percent capacity. The third level of prevention is for a device that will shut off the flow of liquid at 95-percent capacity.

An above-ground tank should be considered full at 90-percent capacity but under no circumstances should the tank be filled beyond 95-percent capacity.

The first thing the delivery driver should do after he or she properly positions the fuel delivery truck is to check and record the ullage (the amount of liquid required to fill the tank to 90-percent capacity).

The second line of defense to prevent an overfill is an audible alarm that is set to go off when the tank reaches 90-percent capacity.

The third and final line of defense is the tank fill piping must be equipped with a device that will completely shut off the supply of liquid when the tank reaches 95 percent. This device is usually installed inside the tank fill opening and is part of the tank fill tube that extends down into the tank. The delivery operator should never rely on this device to determine when the tank is full; the delivery should have stopped at 90 percent. In the event that the delivery operator overfills the tank and the complete shutoff device closes, the tank delivery driver must have the means to drain the delivery hose without spilling the liquid on the ground.

As an alternative to the complete shutoff device the code allows a device that will slow down the delivery rate when the tank reaches 90 percent. Once the tank reaches 90 percent, the rate of flow should slow down to a point where it will take 30 minutes to fill the tank to 95 percent.

2206.6.2.4 Siphon prevention. An approved antisiphon method shall be provided in the piping system to prevent flow of liquid by siphon action.

❖ Since piping connections to an above-ground tank must enter the tank through the top, this section requires that an approved antisiphon device be installed in the liquid piping near the top of the tank to prevent liquid flow from the tank by siphon action if the supply pipe from the tank to the dispenser is damaged.

2206.6.2.5 Emergency relief venting. Above-ground storage tanks, tank compartments and enclosed secondary containment spaces shall be provided with emergency relief venting in accordance with Chapter 34.

❖ The emergency relief vent is the most important safety device installed on the tank. The emergency relief vent prevents the tank from overpressurizing in the event the tank is exposed to fire. In the case of a secondary containment above-ground tank there must be an emergency relief vent for each tank compartment and an emergency relief vent for the interstitial space.

2206.6.2.6 Spill containers. A spill container having a capacity of not less than 5 gallons (19 L) shall be provided for each fill connection. For tanks with a top fill connection, spill containers shall be noncombustible and shall be fixed to the tank and equipped with a manual drain valve that drains into the primary tank. For tanks with a remote fill connection, a portable spill container is allowed.

❖ The fill connection must be equipped with a spill containment device that will hold at least 5 gallons (19 L). If the tank has a top-fill connection, the container must be noncombustible and attached directly to the tank and equipped with a drain valve that will allow any spilled liquid to be manually drained into the tank. If the tank is equipped with a remote-fill connection, the container must be located at the connection when a delivery connection is made to the fill pipe.

2206.6.3 Piping, valves, fittings and ancillary equipment for underground tanks. Piping, valves, fittings and ancillary equipment for underground tanks shall comply with Chapter 34 and NFPA 30A.

❖ Section 3403.6 of this code and NFPA 30A address the provisions for piping connected to underground tanks:

- Galvanic and corrosion protection.
- Leak detection.
- Special materials.
- Piping supports.
- Backflow protection.
- Flexible connections.
- Testing.

2206.7 Fuel-dispensing systems for flammable or combustible liquids. The design, fabrication and installation of fuel-dispensing systems for flammable or combustible liquid fuels shall be in accordance with this section.

❖ The fuel dispensing system consists of all the equipment required to get the fuel from the tank into the vehicle being fueled. It consists of the pumps, piping, dispensers, hoses, nozzles, break-away devices and any other equipment required for a particular application.

2206.7.1 Listed equipment. Electrical equipment, dispensers, hose, nozzles and submersible or subsurface pumps used in fuel-dispensing systems shall be listed.

❖ The use of listed equipment and devices provides evidence that they have been evaluated by a third-party agency for safe use in the applications for which they are designed. This is especially important where flammable liquids are being transferred or dispensed. See also the commentary to the definition of "Listed" in Chapter 2 for further discussion.

2206.7.2 Fixed pumps required. Class I and Class II liquids shall be transferred from tanks by means of fixed pumps designed and equipped to allow control of the flow and prevent leakage or accidental discharge.

❖ There are two basic types of fixed pumps. One is remote from the dispenser and delivers the fuel under pressure to the dispenser. The most common place to find this type of pump is at the tank. Because the fuel is being delivered to the dispenser under pressure, leak detection devices must be used and a shear valve with a fusible link must be installed in the sump under the dispenser.

The second type of pump is a suction pump mounted in the base of the dispenser. A shear valve and leak detection is not required on this type of pump. Section 2206.7.4 requires an emergency shutoff valve on dispensers equipped with remote pumps.

2206.7.3 Mounting of dispensers. Dispensing devices except those installed on top of a protected above-ground tank that qualifies as vehicle-impact resistant, shall be protected against physical damage by mounting on a concrete island 6 inches (152 mm) or more in height, or shall otherwise be suitably protected in accordance with Section 312. Dispensing devices shall be installed and securely fastened to their mounting surface in accordance with the dispenser manufacturer's instructions. Dispensing devices installed indoors shall be located in an approved position where they cannot be struck by an out-of-control vehicle descending a ramp or other slope.

❖ This provision addresses two alternatives to mounting dispensers so as to provide impact protection. One is to mount the dispenser on an elevated island and the other is to protect the dispenser in accordance with Section 312, which requires the dispenser to be protected by steel posts filled with concrete (see Figure 2206.7.3). Note that these provisions do not apply to dispensers mounted on top of a listed impact-resistant aboveground tank. Regardless of which of the above alternatives is chosen, the stability of an installed dispenser depends on its installation in accordance with the manufacturer's instructions, particularly with respect to ensuring that all bolts required to firmly mount the dispenser to the mounting surface are provided. It is not unusual to find dispensers with only half or fewer of their mounting bolts in place. Such a haphazard installation can make the dispenser piping and electrical conduits more susceptible to damage from a much less serious impact than might otherwise be tolerated if all bolts were in place or even from rocking action in high-wind conditions. This can lead to a liquid leak, an ignition from a damaged electrical circuit or both. NFPA 30A requires dispensers to be bolted to their mounting surface and UL 87 requires that two bolt holes be part of the dispenser base for dispensers not over 6 feet tall (1829 mm), while four bolt holes are needed for dispensers over 6 feet tall (1829 mm). These bolting requirements are for dispenser stability and piping protection and should not be considered part of the physical protection required by this section.

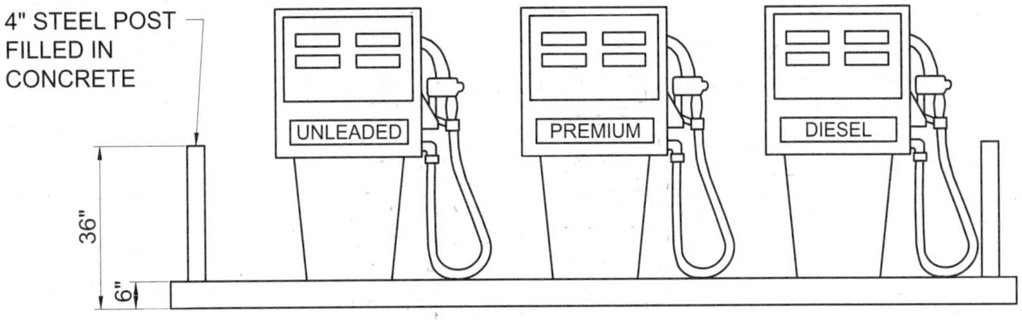

For SI: 1 inch = 25.4 mm.

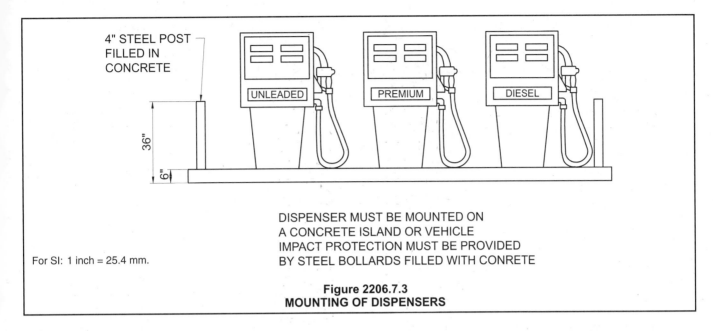

Figure 2206.7.3
MOUNTING OF DISPENSERS

2206.7.4 Dispenser emergency valve. An approved emergency shutoff valve designed to close automatically in the event of a fire or impact shall be properly installed in the liquid supply line at the base of each dispenser supplied by a remote pump. The valve shall be installed so that the shear groove is flush with or within 0.5 inch (12.7 mm) of the top of the concrete dispenser island and there is clearance provided for maintenance purposes around the valve body and operating parts. The valve shall be installed at the liquid supply line inlet of each overhead-type dispenser. Where installed, a vapor return line located inside the dispenser housing shall have a shear section or approved flexible connector for the liquid supply line emergency shutoff valve to function. Emergency shutoff valves shall be installed and maintained in accordance with the manufacturer's instructions, tested at the time of initial installation and tested at least yearly thereafter in accordance with Section 2205.2.2.

❖ The dispenser emergency valve (shear valve) is designed to close automatically when the dispenser is knocked over or if the dispenser is involved in a fire. This so-called "impact" or "breakaway" valve is intended to prevent the free flow of fuel in the event a dispenser is struck by a vehicle. This spring-loaded, fusible-link-operated device may also be tripped by a fire. The relatively low clearance of the shear (breakaway) groove above the top surface of the dispenser island and the manufacturer's requirement for a rigid mounting of the valve body are intended to enable the valve body to fracture at the shear groove upon impact, thus tripping the hold-open linkage and closing the valve. These valves are only required on remote pumping systems (pump is on the tank). Suction system (pump is in dispenser) piping, if broken, will cause the system to lose prime, thus stopping liquid flow. Vapor recovery systems are usually interlocked with the liquid-dispensing system. Consequently, the vapor recovery breakaway design must be compatible with the liquid line shutoff so it functions as intended. Installation, maintenance, acceptance and periodic testing must conform to the manufacturer's instructions. UL 842 describes the tests performed on these valves (see Figure 2206.7.4). The valve must be tested annually by manually tripping the hold- open linkage in accordance with Section 2205.2.2.

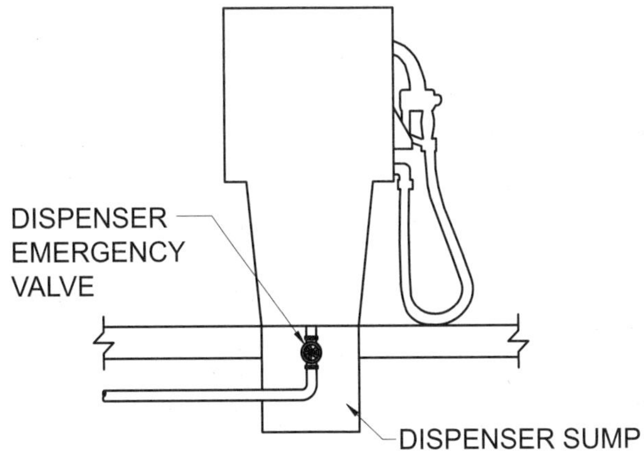

Figure 2206.7.4
DISPENSER EMERGENCY VALVES

2206.7.5 Dispenser hose. Dispenser hoses shall be a maximum of 18 feet (5486 mm) in length unless otherwise approved. Dispenser hoses shall be listed and approved. When not in use, hoses shall be reeled, racked or otherwise protected from damage.

❖ The 18-foot (5486 mm) maximum length for dispenser hose has become a standard, although the original reason for the 18-foot (5486 mm) length had nothing to do with fire safety. It was a weights-and-measures requirement having to do with calibrating the dispenser to accurately give the volume of fuel delivered.

The fire code official should use caution when approving additional lengths of hose because the longer the hose the more susceptible it is to damage by vehicles running over it. Most hose-retrieving mechanisms are designed for 18-foot (5486 mm) hoses. Remember, when the hose is fully extended, the nozzle must not reach within 4 feet (1524 mm) of a building opening.

Means must be provided to protect the hose when it is not in use.

2206.7.5.1 Breakaway devices. Dispenser hoses for Class I and II liquids shall be equipped with a listed emergency breakaway device designed to retain liquid on both sides of a breakaway point. Such devices shall be installed and maintained in accordance with the manufacturer's instructions. Where hoses are attached to hose-retrieving mechanisms, the emergency breakaway device shall be located between the hose nozzle and the point of attachment of the hose-retrieval mechanism to the hose.

❖ This provision requires a breakaway device on dispenser hoses delivering gasoline, diesel and kerosene (Class I and II liquids). These devices are installed to prevent a pull-down of the motor fuel dispenser in the event a car drives away with the hose nozzle valve still in the car's fill pipe. The design of these valves is such that when they operate, the separated sections of the hose are sealed to prevent leakage of liquid from the hose. The placement of the breakaway device between the hose-retrieving mechanism clamp and the nozzle is important to the proper operation of the valve by preventing the pulling force of the driveoff from being transmitted via the retrieving cable directly to the dispenser. Figure 2206.7.5.1 shows a breakaway device installed between the nozzle and the hose-retrieving mechanism.

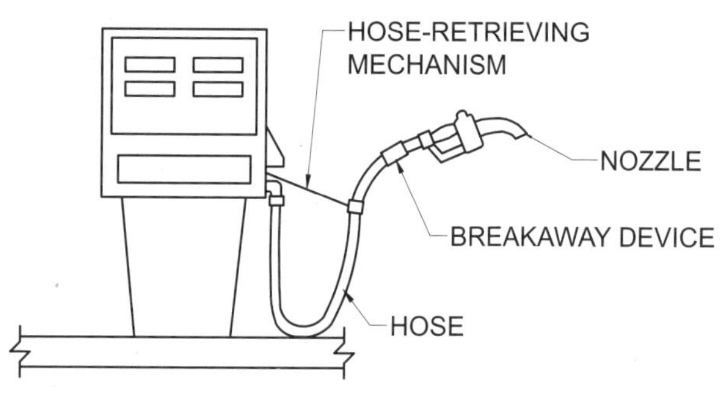

Figure 2206.7.5.1
BREAKAWAY DEVICES

2206.7.6 Fuel delivery nozzles. A listed automatic-closing-type hose nozzle valve with or without a latch-open device shall be provided on island-type dispensers used for dispensing Class I, II or IIIA liquids.

Overhead-type dispensing units shall be provided with a listed automatic-closing-type hose nozzle valve without a latch-open device.

Exception: A listed automatic-closing-type hose nozzle valve with latch-open device is allowed to be used on overhead-type dispensing units where the design of the system is such that the hose nozzle valve will close automatically in the event the valve is released from a fill opening or upon impact with a driveway.

❖ A listed automatic-closing-type hose nozzle with or without a latch-open device is designed to automatically close when the fuel spits back through the vent line and automatically close if the hose nozzle were to fall out of the car fill pipe and strike the ground. When the nozzle is equipped with certain types of vapor recovery equipment, it is designed to shut off when the nozzle is removed from the vehicle. The nozzle on island-type dispensers may or may not be equipped with a latch-open device. The fire code official may want to consider the fact that when a latch-open device is not installed, the dispenser operator may use some unapproved device to hold the nozzle open.

The distinction between an overhead-type dispenser and an island-type dispenser is included in this section even though the overhead dispenser is rarely used. However, the concern was that since the hose is on a retractable reel built into the canopy, the nozzle would not strike the ground if it were to fall out of the vehicle and a latch-open device would fail to work, allowing dangerous, uncontrolled fuel spills.

The exception is that if the overhead-type dispenser hose mechanism is designed to allow the nozzle to strike the ground and automatically shut off or if the nozzle was equipped with vapor return bellows that would automatically shut off the nozzle if it was removed from the vehicle, the latch-open device would be approved.

2206.7.6.1 Special requirements for nozzles. Where dispensing of Class I, II or IIIA liquids is performed, a listed automatic-closing-type hose nozzle valve shall be used incorporating all of the following features:

1. The hose nozzle valve shall be equipped with an integral latch-open device.
2. When the flow of product is normally controlled by devices or equipment other than the hose nozzle valve, the hose nozzle valve shall not be capable of being opened unless the delivery hose is pressurized. If pressure to the hose is lost, the nozzle shall close automatically.

   Exception: Vapor recovery nozzles incorporating insertion interlock devices designed to achieve shutoff on disconnect from the vehicle fill pipe.

3. The hose nozzle shall be designed such that the nozzle is retained in the fill pipe during the filling operation.
4. The system shall include listed equipment with a feature that causes or requires the closing of the hose nozzle valve before the product flow can be resumed or before the hose

nozzle valve can be replaced in its normal position in the dispenser.

❖ **Item 1:**
This provision may seem to be in conflict with Section 2206.7.6, which requires a nozzle with or without a latch-open device. Hose nozzle valves are investigated and labeled by independent testing laboratories and are designed to shut off fuel flow automatically if dropped or jarred, or when fuel flows back into the nozzle spout (tank is full). Many jurisdictions have prohibited the use of latch-open-type hose nozzle valves at retail self-service stations, using the rationale that making customers hold the hose nozzle valve open during fueling would create a safer condition because they will be more alert during the fueling process. Not only does this section prohibit such a ban on a proven safety feature, but the fire record of self-service operations does not support this rationale. In fact, where factory-installed latch-open devices (part of a tested, labeled hose nozzle valve assembly) are removed, it seems to challenge the creativity of customers in seeing how many different makeshift "hold-open" devices can be used to avoid having to hold manually the valve lever open. Such makeshift devices include gas caps, key rings, wallets, magazines, blocks of wood, rubber balls, etc. In one reported incident, a disposable cigarette lighter was used to prop open the hose nozzle valve lever only to strike a spark when the nozzle fell from the fill pipe. The resultant vapor ignition injured the customer and caused substantial damage to the service station and the customer's car. UL 842 describes several types of hose nozzle valves and latch-open features and the tests to which they are subjected.

Item 2:
The statement, "When the flow of product is normally controlled by devices or equipment other than the hose nozzle," is referring to a prepay system; that is, a customer tells the person in the kiosk or convenience store how much fuel he or she wants and the dispenser shuts down and stops the flow, not the nozzle.

The exception addresses the fact that some vapor recovery nozzles are equipped with a bellows. When the nozzle is placed into the vehicle fill pipe, the bellows is compressed and the nozzle will operate. When the nozzle is removed, the bellows expands and the nozzle shuts off. The nozzle does not have to strike something to trigger the automatic shutoff; the nozzle shuts off as soon as the nozzle is removed.

Item 3:
This device may be nothing more than a coil that looks like a spring that is around the dispenser nozzle spout. The spring tends to grip the lip of the vehicle's receiver and helps to prevent the nozzle from falling out of the vehicle receiver.

Item 4:
This safety feature intends to prevent a situation in which a customer could finish fueling without releasing the latch-open device and replace the nozzle in the dispenser boot. The next customer could then remove the nozzle from the dispenser boot, authorize the sale and, before getting the nozzle into the vehicle fill pipe, begin discharging fuel in the open, causing an unacceptable spill situation.

2206.7.7 Remote pumping systems. Remote pumping systems for liquid fuels shall comply with Sections 2206.7.7.1 and 2206.7.7.2.

❖ The intent of this provision is to require special safety devices on pumping systems that deliver fuel by pressure to the dispenser rather than suction.

2206.7.7.1 Leak detection. Where remote pumps are used to supply fuel dispensers, each pump shall have installed on the discharge side a listed leak detection device that will detect a leak in the piping and dispensers and provide an indication at an approved location. A leak detection device is not required if the piping from the pump discharge to under the dispenser is above ground and visible.

❖ In remote pumping systems, the piping between the pump and the dispenser is the number-one cause of fuel spills caused by leaks. This provision addresses both underground and above-ground piping. If the piping is underground, a line leak detector (LLD) must be installed on the discharge side of the remote pump. These devices detect piping leaks by monitoring the pressure in the dispensing system piping. If the LLD detects a pressure loss due to a leak, it will reduce the flow of product in the system to approximately 3 gallons per hour (14.33 L/h) at the hose nozzle (its sensitivity rate) as a visual indicator to the attendant that there is a system problem. The reduced rate of flow indicated at the nozzle fulfills the intent of this section's requirement that detection of a leak be indicated at an approved location. While an electronic leak detection system could fulfill the requirements of this section, it is not the intent of this section to require electronic leak detection.

If the system piping is above ground where a leak would be immediately noticed and the piping can be visually inspected, the leak detection device is not required.

2206.7.7.2 Location. Remote pumps installed above grade, outside of buildings, shall be located not less than 10 feet (3048 mm) from lines of adjoining property that can be built upon and not less than 5 feet (1524 mm) from any building opening. Where an outside pump location is impractical, pumps are permitted to be installed inside buildings as provided for dispensers in Section 2201.4 and Chapter 34. Pumps shall be substantially anchored and protected against physical damage.

❖ Figure 2206.7.7.2 shows the minimum distances between an above-grade remote pump, property lines and openings into a building. Where outside clearances cannot be achieved or are impractical in a given situation, above-grade remote pumps are allowed to be installed inside of buildings on the same basis as motor fuel dispensers.

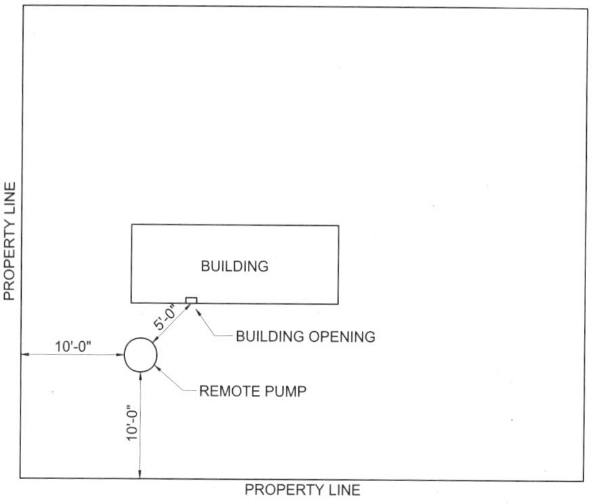

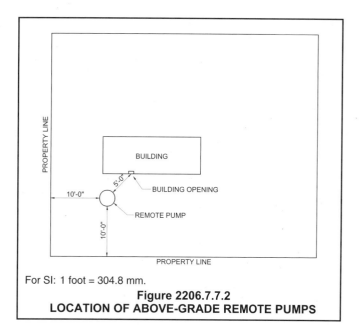

For SI: 1 foot = 304.8 mm.

Figure 2206.7.7.2
LOCATION OF ABOVE-GRADE REMOTE PUMPS

2206.7.8 Gravity and pressure dispensing. Flammable liquids shall not be dispensed by gravity from tanks, drums, barrels or similar containers. Flammable or combustible liquids shall not be dispensed by a device operating through pressure within a storage tank, drum or container.

❖ Delivering fuel by gravity is prohibited because the piping or hose between the tank and the nozzle could rupture or leak, causing the contents of the tank to spill. Tank pressurization is prohibited because the tanks will rupture when excessive pressure is applied. The maximum pressure at which an above-ground tank is tested is 5 psi (34 kPa). Also, the application of air to the tank may lean out the very rich fuel vapors to the point where the atmosphere inside the tank is within the flammable range.

2206.7.9 Vapor-recovery and vapor-processing systems. Vapor-recovery and vapor-processing systems shall be in accordance with Section 2206.7.9.

❖ Some jurisdictions require vapor-recovery or vapor-processing systems. The following provisions address the requirements for these systems. Because of environmental concerns, some jurisdictions prohibit fuel vapors from being discharged into the atmosphere. This section contains the provisions for returning the vapors to the tank or processing the vapors.

2206.7.9.1 Vapor-balance systems. Vapor-balance systems shall comply with Sections 2206.7.9.1.1 through 2206.7.9.1.5.

❖ As fuel enters the fuel tank of the vehicle it displaces fuel vapors. The vapor-balance system simply captures these vapors and returns them to the storage tank.

2206.7.9.1.1 Dispensing devices. Dispensing devices incorporating provisions for vapor recovery shall be listed and labeled. When existing listed or labeled dispensing devices are modified for vapor recovery, such modifications shall be listed by report by a nationally recognized testing laboratory. The listing by report shall contain a description of the component parts used in the modification and recommended method of installation on specific dispensers. Such report shall be made available on request of the fire code official.

Means shall be provided to shut down fuel dispensing in the event the vapor return line becomes blocked.

❖ Motor fuel dispensers equipped with vapor recovery equipment must be listed and labeled by a recognized testing laboratory attesting that they will function properly when installed in accordance with the terms of their listing (see commentary for the definitions of "Listed" and "Labeled" in Chapter 2). When existing dispensing equipment is retrofitted with vapor recovery equipment, the equipment being added must be listed to work with the existing equipment; that is, a person cannot modify an existing dispensing device manufactured by one company with vapor recovery component parts manufactured by the same or a different company unless a report from a recognized testing laboratory is submitted to the fire code official verifying that the retrofit components will work properly with the dispensing device. This is called "listed by report" and is a service provided by major testing laboratories to document equipment compatibility. The report from the testing laboratory must also include the installation instructions.

Means must be provided to shut down fuel dispensing in the event the vapor return line becomes blocked.

This provision is self-explanatory. To prevent vapors from entering the atmosphere, the dispensing operation must be stopped if the vapor return is blocked.

2206.7.9.1.2 Vapor-return line closeoff. An acceptable method shall be provided to close off the vapor return line from dispensers when the product is not being dispensed.

❖ To prevent vapors from escaping after dispensing has stopped, some type of device must be installed that will prevent vapors in the tank from escaping out of the return line. This could be accomplished by a one-way check valve.

2206.7.9.1.3 Piping. Piping in vapor-balance systems shall be in accordance with Sections 3403.6, 3404.2.9 and 3404.2.11. Nonmetallic piping shall be installed in accordance with the manufacturer's installation instructions.

Existing and new vent piping shall be in accordance with Sections 3403.6 and 3404.2. Vapor return piping shall be installed in a manner that drains back to the tank, without sags or traps in which liquid can become trapped. If necessary, because of grade, condensate tanks are allowed in vapor return piping. Condensate tanks shall be designed and installed so that they can be drained without opening.

❖ Section 3403.6 addresses the general requirements for piping. Section 3404.2.9 addresses piping for above-ground tanks and Section 3404.11 lists the requirements for underground tanks. Special piping materials, as described in Section 3403.6.2.1, must be used underground or, if used above ground, must be protected from fire exposure.

The vent piping must comply with the requirements of Section 3403.6 for the general requirements, and Section 3404.2 for the specific vent requirements.

2206.7.9.1.4 Flexible joints and shear joints. Flexible joints shall be installed in accordance with Section 3403.6.9.

An approved shear joint shall be rigidly mounted and connected by a union in the vapor return piping at the base of each dispensing device. The shear joint shall be mounted flush with the top of the surface on which the dispenser is mounted.

❖ Flexible joints must be listed, approved and installed in the following locations:

1. Where piping connects to underground piping.
2. Where piping ends at pump islands and vent risers.
3. At points where differential movement in the piping can occur.

The shear joint mentioned here serves the same function for the vapor recovery lines as the dispenser emergency valve does for the pressurized fuel supply piping to the dispenser (see commentary, Section 2206.7.4). This shear valve must be securely mounted so that the shear groove is located flush or within 0.5 inch (12.7 mm) of the surface of the dispenser island or top of the sump. This shear valve is designed to close in case of a dispenser fire or a dispenser knock down.

2206.7.9.1.5 Testing. Vapor return lines and vent piping shall be tested in accordance with Section 3403.6.3.

❖ Section 3403.6.3 addresses the testing for flammable and combustible liquid piping.

2206.7.9.2 Vapor-processing systems. Vapor-processing systems shall comply with Sections 2206.7.9.2.1 through 2206.7.9.2.4.

❖ Where vapor-recovery systems capture the vapors and return them to the tank, vapor-processing systems do not return the vapors to the tank. Instead, they process the vapors either by refrigeration, absorption or burning them off.

2206.7.9.2.1 Equipment. Equipment in vapor-processing systems, including hose nozzle valves, vapor pumps, flame arresters, fire checks or systems for prevention of flame propagation, controls and vapor-processing equipment, shall be individually listed for the intended use in a specified manner.

Vapor-processing systems that introduce air into the underground piping or storage tanks shall be provided with equipment for prevention of flame propagation that has been tested and listed as suitable for the intended use.

❖ Unlisted vapor processing equipment may not be used. Equipment that is used must be done in accordance with its listing.

Vapor-processing systems that introduce air into the underground piping or storage tanks must be provided with equipment for prevention of flame propagation that has been tested and listed as suitable for the intended use.

When blowers are used to introduce air, equipment such as listed flame arresters must be used to prevent vapors in the flammable range from being ignited.

2206.7.9.2.2 Location. Vapor-processing equipment shall be located at or above grade. Sources of ignition shall be located not less than 50 feet (15 240 mm) from fuel-transfer areas and not less than 18 inches (457 mm) above tank fill openings and tops of dispenser islands. Vapor-processing units shall be located not less than 10 feet (3048 mm) from the nearest building or lot line of a property which can be built upon.

Exception: Where the required distances to buildings, lot lines or fuel-transfer areas cannot be obtained, means shall be provided to protect equipment against fire exposure. Acceptable means shall include but not be limited to:

1. Approved protective enclosures, which extend at least 18 inches (457 mm) above the equipment, constructed of fire-resistant or noncombustible materials; or
2. Fire protection using an approved water-spray system.

Vapor-processing equipment shall be located a minimum of 20 feet (6096 mm) from dispensing devices. Processing equipment shall be protected against physical damage by guardrails, curbs, protective enclosures or fencing. Where approved protective enclosures are used, approved means shall be provided to ventilate the volume within the enclosure to prevent pocketing of flammable vapors.

Where a downslope exists toward the location of the vapor-processing unit from a fuel-transfer area, the fire code official is authorized to require additional separation by distance and height.

❖ The intent of this section is to protect the vapor processing equipment from fire exposure, not to protect other structures or equipment from a fire involving the vapor recovery equipment. To do this, either construct a barrier around the equipment at least 18 inches (457 mm) higher than the equipment or protect the equipment with an approved water spray.

Vapor-processing equipment must be vented to prevent the accumulation of flammable vapors and shall be separated from the dispensers by at least 20 feet (6096 mm). If the vapor-processing equipment is downhill from the dispensers, the fire code official may require greater protection for the vapor-processing equipment.

2206.7.9.2.3 Installation. Vapor-processing units shall be securely mounted on concrete, masonry or structural steel supports on concrete or other noncombustible foundations. Vapor-recovery and vapor-processing equipment is allowed to be installed on roofs when approved.

❖ The intent of this provision is to require the vapor-processing equipment to be mounted on a noncombustible, substantial foundation.

2206.7.9.2.4 Piping. Piping in a mechanical-assist system shall be in accordance with Sections 3403.6.

❖ Section 3403.6 addresses the provisions for flammable and combustible liquid piping.

SECTION 2207 LIQUEFIED PETROLEUM GAS MOTOR FUEL-DISPENSING FACILITIES

2207.1 General. Motor fuel-dispensing facilities for liquefied petroleum gas (LP-gas) fuel shall be in accordance with this section and Chapter 38.

❖ The federal government is mandating that automakers move toward alternative fuels to replace conventional gasoline and diesel. One of the more popular alternative fuels is LP-gas. As more vehicles use alternative fuels, codes will have to be modified to address the safe dispensing of these fuels. This section addresses the dispensing of LP-gas at motor fuel-dispensing facilities. Chapter 38 addresses the general provisions for LP-gas.

2207.2 Approvals. Storage vessels and equipment used for the storage or dispensing of LP-gas shall be approved or listed in accordance with Sections 2207.2.1 and 2207.2.2.

❖ All equipment used in connection with the storage and dispensing of LP-gas motor fuel must be either approved or listed, as specified in the following sections. The terms "approved " and "listed" are defined in Chapter 2. See the commentary to Chapter 2 for these defined terms for further information.

2207.2.1 Approved equipment. Containers, pressure relief devices (including pressure relief valves), pressure regulators and piping for LP-gas shall be approved.

❖ The equipment identified in this section must be approved by the fire code official.

2207.2.2 Listed equipment. Hoses, hose connections, vehicle fuel connections, dispensers, LP-gas pumps and electrical equipment used for LP-gas shall be listed.

❖ The equipment identified in this section must be listed.

2207.3 Attendants. Motor fuel-dispensing operations shall be conducted by qualified attendants or in accordance with Section 2207.6 by persons trained in the proper handling of LP-gas.

❖ LP-gas must be dispensed into a vehicle by trained and qualified persons. Unattended and self-service LP-gas stations that are open to the public are not allowed. Section 2207.6 does allow self-service as long as it is closed to the public, that is, the vehicles belong to the operator of the LP-gas fueling facility and the persons dispensing the fuel are employees of the LP-gas fueling operation and are properly trained.

2207.4 Location of dispensing operations and equipment. In addition to the requirements of Section 2206.7, the point of transfer for dispensing operations shall be 25 feet (7620 mm) or more from buildings having combustible exterior wall surfaces, buildings having noncombustible exterior wall surfaces that are not part of a 1-hour fire-resistance-rated assembly, or buildings having combustible overhangs, lot lines of property which could be built on, public streets, or sidewalks and railroads; and at least 10 feet (3048 mm) from driveways and buildings having noncombustible exterior wall surfaces that are part of a fire-resistance-rated assembly having a rating of 1 hour or more.

> **Exception:** The point of transfer for dispensing operations need not be separated from canopies that are constructed in accordance with the *International Building Code* and which provide weather protection for the dispensing equipment.

LP-gas containers shall be located in accordance with Chapter 38. LP-gas storage and dispensing equipment shall be located outdoors and in accordance with Section 2206.7.

❖ Whereas Section 2203.1 regulates the location of dispensing devices, this section regulates the location of the point of transfer of the LP-gas motor fuel into the vehicle fuel tank. The intent of the term "point of transfer" is to describe the location where LP-gas fueling connections are made and broken and any location where LP-gas is vented during transfer or fueling operations. Whenever LP-gas is dispensed or transferred, there is always at least a "puff" of LP-gas released at the point where the LP-gas dispensing hose connects to and disconnects from the vehicle fuel tank or other LP-gas container, and it is at those locations that a flammable mixture will exist, even if momentarily. This section intends to isolate these "movable locations" by regulating how far away from potential ignition sources or other properties they must remain under in all circumstances.

The IBC has specific construction requirements for canopies located at motor fuel-dispensing facilities. The exception removes any separation requirements between the dispenser area and the canopy when the canopy is constructed in accordance to the special provision for canopies located at motor fuel-dispensing facilities.

Table 3804.3 lists the separation requirements for the containers in relation to buildings, public ways and property lines that are or might be built upon.

2207.5 Installation of LP-gas dispensing devices and equipment. The installation and operation of LP-gas dispensing systems shall be in accordance with Sections 2207.5.1 through 2207.5.3 and Chapter 38. LP-gas dispensers and dispensing stations shall be installed in accordance with the manufacturer's specifications and their listing.

❖ This provision specifically addresses the requirements for valves, hoses and impact protection for LP-gas dispensing equipment.

2207.5.1 Valves. A manual shutoff valve and an excess flow-control check valve shall be located in the liquid line between the pump and the dispenser inlet where the dispensing device is installed at a remote location and is not part of a complete storage and dispensing unit mounted on a common base.

An excess flow-control check valve or an emergency shutoff valve shall be installed in or on the dispenser at the point at which the dispenser hose is connected to the liquid piping. A differential backpressure valve shall be considered equivalent protection.

A listed shutoff valve shall be located at the discharge end of the transfer hose.

❖ When the pump and dispenser are separate components installed remote from one another, the likelihood of pipe failure increases. Safeguards in the form of a manual control valve and an excess flow-control valve are included to control the flow of gas in the event of a leak or rupture of the piping between the pump and the base of the dispenser.

An excess flow-control valve or a differential backpressure valve must be provided to control the flow of gas in the event the dispensing hose is separated. These valves operate by sensing a change in the pressure difference upstream and downstream of the valve and react by shutting off the flow of gas.

The shutoff valve at the end of the dispensing hose must be listed for LP-gas use for dispensing at motor fuel-dispensing facilities.

2207.5.2 Hoses. Hoses and piping for the dispensing of LP-gas shall be provided with hydrostatic relief valves. The hose length shall not exceed 18 feet (5486 mm). An approved method shall be provided to protect the hose against mechanical damage.

❖ To prevent overpressurization of the hoses and piping, a relief valve must be installed in the system. The hose length is limited to 18 feet (5486 mm) to prevent excessive amounts of hose from lying on the ground and being damaged by vehicles. The general public is familiar with the 18-foot (5486 mm) length because of the 18-foot (5486 mm) requirement for liquid fuel-dispensing hoses (Section 2206.7.5) and they know where to position their vehicles for refueling. See the commentary to Section 2206.7.5 for further discussion of hose length.

2207.5.3 Vehicle impact protection. Vehicle impact protection for LP-gas storage containers, pumps and dispensers shall be provided in accordance with Section 2206.4.

❖ This provision requires 4-inch-diameter (102 mm) concrete-filled steel posts on 4-foot (1219 mm) centers, 3 feet (914 mm) high. See Section 312 for more specific construction details, and the commentary to Section 2206.4 for further discussion of such protection.

2207.6 Private fueling of motor vehicles. Self-service LP-gas dispensing systems, including key, code and card lock dispensing systems, shall not be open to the public and shall be limited to the filling of permanently mounted fuel containers on LP-gas powered vehicles.

In addition to the requirements of Sections 2205 and 2206.7, self-service LP-gas dispensing systems shall be in accordance with the following:

1. The system shall be provided with an emergency shutoff switch located within 100 feet (30 480 mm) of, but not less than 20 feet (6096 mm) from, dispensers.
2. The owner of the LP-gas motor fuel-dispensing facility shall provide for the safe operation of the system and the training of users.

❖ If the vehicles being refueled don't belong to the owner of the fuel-dispensing facility and the fuel-dispensing operator is not an employee of the owner, the facility is considered "public."

This provision allows fleet owners to refuel their own vehicles using employees trained by and in the employ of the fleet owner. This provision is not intended to allow a loophole for fuel-dispensing owner/operators to issue fuel cards to other companies or individuals and claim that their facility is not public.

Finally, this section contains provisions for the safe operation of the facility and an emergency shutoff switch similar to the emergency shutoff switch required at motor fuel-dispensing facilities in Section 2203.2.

2207.7 Overfilling. LP-gas containers shall not be filled in excess of the fixed outage installed by the manufacturer or the weight stamped on the tank.

❖ To prevent release of gas by pressure relief due to expansion of the gas, the LP-gas container must not be filled beyond the capacity designated by the manufacturer.

SECTION 2208
COMPRESSED NATURAL GAS MOTOR FUEL-DISPENSING FACILITIES

2208.1 General. Motor fuel-dispensing facilities for compressed natural gas (CNG) fuel shall be in accordance with this section and Chapter 30.

❖ The federal government is mandating that automakers move toward alternative fuels to replace conventional gasoline and diesel. One of the more popular alternative fuels is CNG. As more vehicles use alternative fuels, codes will have to be modified to address the safe dispensing of these fuels. This section addresses the dispensing of CNG at motor fuel-dispensing facilities. Chapter 30 contains the general provisions for CNG.

2208.2 Approvals. Storage vessels and equipment used for the storage, compression or dispensing of CNG shall be approved or listed in accordance with Sections 2208.2.1 and 2208.2.2.

❖ All equipment used in connection with the storage and dispensing of CNG motor fuel must be either approved or listed, as specified in the following sections. The terms "approved " and "listed" are defined in Chapter 2. See the commentary to Chapter 2 for these defined terms for further information.

2208.2.1 Approved equipment. Containers, compressors, pressure relief devices (including pressure relief valves), and pressure regulators and piping used for CNG shall be approved.

❖ The equipment identified in this section must be approved by the fire code official. Listing of the specified components could be the basis for approval, but listing is not mandated. Any such components must be designed for the application and recommended for the application by the manufacturer.

2208.2.2 Listed equipment. Hoses, hose connections, dispensers, gas detection systems and electrical equipment used for CNG shall be listed. Vehicle-fueling connections shall be listed and labeled.

❖ The specified components must be listed and labeled by an approved testing agency as complying with the relevant product standards. As with all listed products, the testing/listing agency will apply its seal or mark to the product.

2208.3 Location of dispensing operations and equipment. Compression, storage and dispensing equipment shall be located above ground, outside.

Exceptions:

1. Compression, storage or dispensing equipment shall be allowed in buildings of noncombustible construction, as set forth in the *International Building Code*, which are unenclosed for three quarters or more of the perimeter.
2. Compression, storage and dispensing equipment shall be allowed indoors in accordance with Chapter 30.

❖ Because of the potential for leakage, the compression, storage and dispensing equipment must be located either outdoors or in a noncombustible, substantially open building, except as allowed by Chapter 30. This section allows compression, storage and dispensing equipment to be located:

1. Outdoors, above ground.
2. Inside noncombustible buildings where at least 75 percent of the wall area is open to the outside atmosphere.
3. Indoors when in accordance with Chapter 30. Chapter 30 refers to NFPA 52 and Section 413 of the IFGC.

2208.3.1 Location on property. In addition to the requirements of Section 2203.1, compression, storage and dispensing equipment shall be installed as follows:

1. Not beneath power lines.
2. Ten feet (3048 mm) or more from the nearest building or lot line which could be built on, public street, sidewalk, or source of ignition.

Exception: Dispensing equipment need not be separated from canopies that are constructed in accordance with the *International Building Code* and which provide weather protection for the dispensing equipment.

3. Twenty-five feet (7620 mm) or more from the nearest rail of any railroad track and 50 feet (15 240 mm) or more from the nearest rail of any railroad main track or any railroad or transit line where power for train propulsion is provided by an outside electrical source such as third rail or overhead catenary.
4. Fifty feet (15 240 mm) or more from the vertical plane below the nearest overhead wire of a trolley bus line.

❖ Natural gas is lighter than air. The requirements for the location of the equipment that compresses and dispenses the gas must take this into consideration. The separation distances from overhead ignition sources should be noted.

2208.4 Private fueling of motor vehicles. Self-service CNG-dispensing systems, including key, code and card lock dispensing systems, shall be limited to the filling of permanently mounted fuel containers on CNG-powered vehicles.

In addition to the requirements in Section 2211, the owner of a self-service CNG motor fuel-dispensing facility shall ensure the safe operation of the system and the training of users.

❖ Unlike LP-gas self-service dispensing stations, CNG self-service dispensing systems can be open to the public. The owner of the system must provide for the training of any users of the system.

2208.5 Pressure regulators. Pressure regulators shall be designed and installed or protected so that their operation will not be affected by the elements (freezing rain, sleet, snow or ice), mud or debris. The protection is allowed to be an integral part of the regulator.

❖ The pressure regulator must be protected from the elements and is intended to prevent the overpressurization of the vehicle fuel tank and the delivery hose. Pressure regulator failure could result in dangerous overpressure and the opening of relief valves; therefore, they must be dependable. The regulator vent is susceptible to blockage by debris and ice.

2208.6 Valves. Gas piping to equipment shall be provided with a remote, readily accessible manual shutoff valve.

❖ Shutoff valves allow isolation of components for service, repair, replacement and emergency shutdown. See the definition of "Ready access" in the IFGC. This valve is independent of the emergency shutdown device required by Section 2208.7.

2208.7 Emergency shutdown device. An emergency shutdown device shall be located within 75 feet (22 860 mm) of, but not less than 25 feet (7620 mm) from, dispensers, and shall also be provided in the compressor area. Upon activation, the emergency shutdown shall automatically shut off the power supply to the compressor and close valves between the main gas supply and the compressor and between the storage containers and dispensers.

❖ CNG systems take natural gas from the utility supply line, compress it to very high pressures and store the compressed gas in vessels from which the dispensers draw the gas for transfer to the vehicle onboard containers. The emergency shutdown device must be located no farther than 75 feet (22 860 mm) from and no closer than 25 feet (7620 mm) to the dispensers. An additional shutdown device must be located near the compressors; the compressors may be remote from the dispensers. The shutdown device must kill the power to the compressors, must actuate automatic valves that isolate the gas supply from the compressors and isolate the storage vessels from the dispenser, thus limiting accidental gas discharge.

2208.8 Discharge of CNG from motor vehicle fuel storage containers. The discharge of CNG from motor vehicle fuel cylinders for the purposes of maintenance, cylinder certification, calibration of dispensers or other activities shall be in accordance with Sections 2208.8.1 through 2208.8.1.2.6.

❖ The equipment referred to in this section is the fixed equipment located at the CNG motor fuel-dispensing facility. Periodically, equipment must be maintained, certified and recalibrated. To do this the CNG must be discharged from the system. This section regulates how this must be done.

2208.8.1 Methods of discharge. The discharge of CNG from motor vehicle fuel cylinders shall be accomplished through a closed transfer system in accordance with Section 2208.8.1.1 or an approved method of atmospheric venting in accordance with Section 2208.8.1.2.

❖ The intentional discharge of gas from vehicle containers is done for several reasons, including vehicle repairs, certification of the container integrity and container replacement. The gas must be discharged in a safe manner as dictated by either Section 2208.8.1.1 or 2208.8.1.2. The vehicle storage container is referred to as a "vessel" and as a "cylinder" in the text to follow.

2208.8.1.1 Closed transfer system. A documented procedure that explains the logical sequence for discharging the cylinder shall be provided to the fire code official for review and approval. The procedure shall include what actions the operator will take in the event of a low-pressure or high-pressure natural gas release during the discharging activity. A drawing illustrating the arrangement of piping, regulators and equipment settings shall be provided to the fire code official for review and approval. The drawing shall illustrate the piping and regulator arrangement and shall be shown in spatial relation to the location of the compressor, storage vessels and emergency shutdown devices.

❖ A closed transfer system uses the same basic components as a dispensing system and withdraws the gas from the vehicle container storing it in vessels. Drawings and a description of the sequence of operation of the transfer system must be provided to the fire code official for approval. This information may then be shared with the fire department for use in case of an emergency where CNG cylinders might need to be discharged.

2208.8.1.2 Atmospheric venting. Atmospheric venting of CNG shall comply with Sections 2208.8.1.2.1 through 2208.8.1.2.6.

❖ The six subsections that follow provide the conditions and requirements under which gas is allowed to be discharged to the atmosphere. Such discharge should be avoided wherever practical because (1) there is an inherent hazard in doing so, (2) methane is an air contaminant and (3) natural resources should never be wasted. A closed transfer system is the preferable way to remove gas from vehicle containers.

2208.8.1.2.1 Plans and specifications. A drawing illustrating the location of the vessel support, piping, the method of grounding and bonding, and other requirements specified herein shall be provided to the fire code official for review and approval.

❖ Plans of the proposed atmospheric venting apparatus and piping system must be reviewed by the fire code official and approved before atmospheric venting can take place.

2208.8.1.2.2 Cylinder stability. A method of rigidly supporting the vessel during the venting of CNG shall be provided. The selected method shall provide not less than two points of support and shall prevent the horizontal and lateral movement of the vessel. The system shall be designed to prevent the movement of the vessel based on the highest gas-release velocity through valve orifices at the vessel's rated pressure and volume. The structure or appurtenance shall be constructed of noncombustible materials.

❖ Vehicle CNG containers (vessels) can hold extremely high pressures, which, if released quickly, can produce large thrust forces that would propel the container like a rocket. Natural gas does not liquefy at normal ambient temperatures; therefore, in order to hold the required amount of fuel on board the vehicle, the gaseous fuel must be compressed to extreme pressures of up to 3,600 psia (24 736 kPa).

2208.8.1.2.3 Separation. The structure or appurtenance used for stabilizing the cylinder shall be separated from the site equipment, features and exposures and shall be located in accordance with Table 2208.8.1.2.3.

❖ The intent of this section is to separate the combustible vapor produced by venting from buildings, building openings, lot lines, public ways, vehicles, CNG compressor and storage vessels and CNG dispensers. This precaution is reasonable considering the potential hazard of working with highly pressurized containers of a flammable gas.

TABLE 2208.8.1.2.3
SEPARATION DISTANCE FOR ATMOSPHERIC VENTING OF CNG

EQUIPMENT OR FEATURE	MINIMUM SEPARATION (feet)
Buildings	25
Building openings	25
Lot lines	15
Public ways	15
Vehicles	25
CNG compressor and storage vessels	25
CNG dispensers	25

For SI: 1 foot = 304.8 mm.

❖ The distances listed in Table 2208.8.1.2.3 are the distances between the equipment or feature and the termination of the vent opening where the gas is entering the atmosphere.

2208.8.1.2.4 Grounding and bonding. The structure or appurtenance used for supporting the cylinder shall be grounded in accordance with the ICC *Electrical Code*. The cylinder valve shall be bonded prior to the commencement of venting operations.

❖ "Grounding" means to intentionally connect to the earth. This could be accomplished by installing a conductor between the cylinder support and the building grounding electrode system. Bonding means to join metallic parts together to form a continuous electrical pathway. Grounding and bonding required by this section are intended to control sparking that could result from current flow produced by voltage differentials across parts of the venting set-up and the building components. Grounding the venting set-up and attaching a bonding jumper to the cylinder valve will put all such components and the building components at the same voltage potential, thereby reducing the possibility of sparks that could ignite flammable vapors or harm the cylinder assembly. The provisions of this section will also help prevent the buildup of static electrical charges that could be a source of ignition.

2208.8.1.2.5 Vent tube. A vent tube that will divert the gas flow to atmosphere shall be installed on the cylinder prior to commencement of the venting and purging operation. The vent tube shall be constructed of pipe or tubing materials approved for use with CNG in accordance with Chapter 30.

The vent tube shall be capable of dispersing the gas a minimum of 10 feet (3048 mm) above grade level. The vent tube shall not be provided with a rain cap or other feature which would limit or obstruct the gas flow.

At the connection fitting of the vent tube and the CNG cylinder, a listed bidirectional detonation flame arrester shall be provided.

❖ The cylinder must be discharged through a vent tube/pipe that is constructed of a material that is compatible with the gas and that has the required strength to withstand the pressure to which it can be exposed. The pressure that the vent can be exposed to must be calculated based on the size of the vent tube, the length of the vent, the friction loss through the vent and the maximum pressure and discharge rate of the cylinder. Vent failure could cause injury resulting from projectile debris and could cause a severe fire/explosion hazard. The intent of this section is to divert the vented vapor upward and away from any potential ignition source.

2208.8.1.2.6 Signage. Approved "No Smoking" signs complying with Section 310 shall be posted within 10 feet (3048 mm) of the cylinder support structure or appurtenance. Approved CYLINDER SHALL BE BONDED signs shall be posted on the cylinder support structure or appurtenance.

❖ The purpose of this section is to prevent ignition sources caused by smoking and stray currents. The provision requires two signs: one is a no smoking sign and the other is a sign reminding the operator that the cylinder must be bonded to divert stray electrical currents.

SECTION 2209
HYDROGEN MOTOR FUEL-DISPENSING AND GENERATION FACILITIES

2209.1 General. Hydrogen motor fuel-dispensing and generation facilities shall be in accordance with this section and Chapter 30. Where a fuel-dispensing facility also includes a repair garage, the repair operation shall comply with Section 2211.

❖ The federal government is mandating that automakers move toward alternative fuels to replace conventional gasoline and diesel. One of the newer alternative fuels is hydrogen and its popularity is sure to increase as its use becomes more widespread.

The United States Department of Energy (DOE), in accordance with the Hydrogen Future Act of 1996, supports a program based on an industry-led cost-sharing approach called the Hydrogen Energy Program. In some markets, agencies of government have mandated that automakers move ahead with production of alternative fueled vehicles to somewhat offset the atmospheric implications of an economy almost entirely driven by petroleum-based fuels, thereby facilitating a shift to the use of renewable energy supplies. Hydrogen is one of those alternative fuels, and the commercial products industry is responding. As more vehicles use alternative fuels, codes will have to be modified to address the safe use, dispensing, storage and generation of hydrogen fuels. Sections 2209 and 2210.8, along with the provisions of Chapter 30, clearly define gaseous hydrogen refueling and generating stations within the scope of the code and provide fire code officials with the necessary tools to create a safe consumer environment as the use of hydrogen as a motor fuel expands.

In many cases, the hydrogen fuel is utilized, with air, within a fuel cell to produce electricity and, in some cases, cogenerate heat. Typically, fire code officials will be faced with two classes of equipment—those that generate hydrogen (for use by other devices) and those that utilize hydrogen as their energy input.

Often, hydrogen will be utilized in a manner similar to the current use of natural gas but there are two important differences that must be noted. First, while both hydrogen and natural gas are lighter than air, hydrogen is lighter than natural gas and is both more diffusive and more buoyant than natural gas. This means that in well-ventilated situations (e.g., outdoors) hydrogen will dissipate more quickly than natural gas, and much more quickly than either propane or gasoline, both of which have fumes that are heavier than air and will linger at an accident scene or release site. However, hydrogen and natural gas can both accumulate in unventilated pockets at the top of indoor structures and could represent a risk in such situations. Similarly, propane and gasoline fumes can accumulate at the floor level in unventilated spaces, posing a different risk. Thus, ignition sources must be regulated at the top of any unventilated spaces for hydrogen and natural gas, while also being regulated near the floor for gasoline or propane vehicles indoors. Second, hydrogen is odorless, colorless and burns with a flame that is not visible to the human eye. This means that it is unlikely that people will be able to detect unsafe conditions (without appropriate instrumentation) if they develop (similar to a carbon monoxide buildup in a structure).

It is important to note that a given volume of natural gas has more than three times the energy of the same volume of hydrogen. Therefore, a given volume of pipe containing natural gas will contain the same energy (potential hazard) as a three times larger volume of hydrogen.

2209.2 Equipment. Equipment used for the generation, compression, storage or dispensing of hydrogen shall be designed for the specific application in accordance with Sections 2209.2.1 through 2209.2.3.

❖ All equipment used in connection with the storage and dispensing of hydrogen motor fuel must be either approved or listed, as specified in the following sections. The terms "approved " and "listed" are defined in Chapter 2. See the commentary to Chapter 2 for these defined terms for further information.

2209.2.1 Approved equipment. Storage vessels, containers, pressure vessels, cylinders, pressure relief devices, including pressure valves, hydrogen vaporizers, pressure regulators and piping used for gaseous hydrogen systems shall be designed and constructed in accordance with Section 2703, NFPA 50A and NFPA 50B.

❖ The equipment identified in this section must be approved by the fire code official. Listing of the specified components could be the basis for approval, but listing is not mandated. Any such components must be designed and recommended for the application by the manufacturer.

Design requirements for storage vessels, containers, pressure vessels, cylinders and pressure relief devices, including pressure valves, hydrogen vaporizers, pressure regulators and piping used for hydrogen, are directly dependent on the type, conditions of use and quantity of material involved. This section is intended to rely on design requirements for this equipment as referenced in Section 2703 and throughout the code.

2209.2.2 Listed equipment. Hoses, hose connections, compressors, hydrogen generators, dispensers, detection systems and electrical equipment used for hydrogen shall be listed for use with hydrogen. Hydrogen motor fueling connections shall be listed and labeled for use with hydrogen.

❖ The specified components must be listed and labeled by an approved testing agency as complying with the relevant product standards. As with all listed products, the testing/listing agency will apply its seal or mark to the product.

Similar to associated piping, hoses, hose connections, compressors, hydrogen generators, dispensers, detection systems and electrical equipment used for hydrogen service must be built to recognized standards and be compatible with the material handled.

ASME B31.3 or CGA G-5.4, which references ASME B31.3, may be appropriate for design and construction of the piping involved in hydrogen service and are examples of common standards employed by industry for piping, tubing and associated distribution equipment involving hazardous materials. Though not specifically referenced here, there are other ASME and industry standards providing further guidance that are considered appropriate for many aspects of gaseous and liquefied hydrogen systems.

2209.2.3 Electrical equipment. Electrical installations shall be in accordance with the ICC *Electrical Code*.

❖ This section addresses electrical equipment at hydrogen motor fuel-dispensing facilities and is correlated with the ICC EC. Electrical equipment must be approved for the particular hazards anticipated at hydrogen motor fuel-dispensing facilities.

2209.3 Location on property. In addition to the requirements of Section 2203.1, generation, compression, storage and dispensing equipment shall be located in accordance with Sections 2209.3.1 through Section 2209.3.4.

❖ In order to reduce the likelihood of hydrogen motor fuels coming into contact with ignition sources or posing a hazard to persons, adjoining property or on-site buildings, this section provides equipment location requirements versus outdoor equipment, canopies and overhead power lines.

2209.3.1 Outdoor exposures. Outdoor exposures shall require separation from other fuels or equivalent risks to life safety and buildings or public areas in accordance with Table 2209.3.1.

Exception: Closed systems with a hydrogen capacity of 3,000 cubic feet or less at NTP (85 m^3).

❖ In order to reduce the likelihood of hydrogen motor fuels coming into contact with ignition sources or posing a hazard to persons, adjoining property or on-site buildings, this section directs the code user to Table 2209.3.1, which establishes very specific equipment location requirements in relation to exposures such as buildings, lot lines and ignition sources.

TABLE 2209.3.1. See this page.

❖ The intent of these requirements is to establish consistency with existing codes and standards wherever possible in the best interest of fire department personnel, fire code officials and other emergency responders. Accordingly, the contents of the table are adaptations of established recognized standard criteria, such as NFPA 50A, NFPA 50B and NFPA 52.

2209.3.2 Location of dispensing operations and equipment. Generation, compression, storage and dispensing equipment shall be located outdoors, above ground.

Exceptions:

1. Generation, compression, storage or dispensing equipment shall be allowed in buildings of Type I and II construction, as defined in the *International Building Code*, which are unenclosed for three quarters or more of the perimeter and constructed in a manner that prevents the accumulation of hydrogen gas.
2. Generation, compression, storage and dispensing equipment shall be allowed indoors in accordance with Chapter 30 and as set forth in the *International Building Code* and *International Fuel Gas Code*.

❖ Because of the physical characteristics of hydrogen and the potential for leakage, the generation, compression, storage and dispensing equipment must be located above ground and outdoors. The exceptions intend to clearly establish provisions already allowed by other sections of the code. For example, the reference to NFPA 50A in Section 3501.1 permits the installation of hydrogen systems indoors, given special considerations for ventilation, type of construction and location of openings. Exception 1 is written in language similar to

TABLE 2209.3.1
OUTDOOR MINIMUM SEPARATION FOR GASEOUS HYDROGEN DISPENSERS, COMPRESSORS, GENERATORS AND STORAGE VESSELS

OUTDOOR EQUIPMENT OR FEATURE	DISTANCE (feet)
Building—Noncombustible walls, sprinklered or nonsprinklered	10
Building—Combustible walls, sprinklered or nonsprinklered	25[b, e]
Building—Noncombustible walls, 2-hour fire barrier interrupts line of sight	5
Offsite sidewalks and on-site/offsite parked vehicles	15[a, b]
Lot line	10[a]
Air intake openings	25[c]
Wall openings located less than 25 feet vertically above	20[c]
Wall openings located greater than 25 feet vertically above	25
Outdoor public assembly	25[a]
Ignition source[d]	10
Flammable or combustible liquid storage—Above ground, diked in accordance with Section 3404.2.9.6.	20
Flammable or combustible liquid storage—Above ground, not diked	50
Flammable or combustible liquid storage—Below ground, vent or fill opening	20
Flammable gas storage (nonhydrogen)—Above ground, with common shutoff	25
Flammable gas storage (nonhydrogen)—Above ground, no common shutoff	50
Combustible waste material (see Section 304.1.1)	50
Liquefied hydrogen storage—Distance to buildings, openings, lot lines, public ways and on-site/off-site parked vehicles	25[a]

For SI: 1 foot = 304.8 mm, 1 cubic foot = 0.02832 m^3.

a. Reduction to 5 feet shall be permitted where a 2-hour fire barrier interrupts the line of sight between the equipment and the exposure. The height of the barrier for vertical tanks shall be no less than one-third of the height of the tank measured vertically, and the length of the wall shall be 1.5 times the maximum diameter of the tank.

b. A reduction to 0 feet shall be permitted for dispensing equipment and vehicles being refueled.

c. Measured along the natural and unobstructed line of travel (e.g., around protective walls, around corners of buildings).

d. Ignition source. A flame, spark or hot surface capable of igniting flammable vapors or fumes. Such sources include appliance burner ignitors and hot work, such as welding and open flames.

e. For storage volume greater than or equal to 15,000 cubic feet at NTP.

that addressing weather-protective canopies for CNG motor fuel-dispensing facilities in Section 2208.3 and Section 406 of the IBC for motor fuel-dispensing facility canopies. Exception 2 relates to the existing provisions within the code that allow up to the maximum allowable quantity per control area of a flammable gas to be stored or used indoors. Additionally, where a maximum allowable quantity per control area threshold in the code is exceeded, the exception would require the construction of the appropriate Group H occupancy to accommodate such indoor generation or refueling operations.

2209.3.3 Canopies. Dispensing equipment need not be separated from canopies that are constructed in accordance with the *International Building Code*, in a manner that prevents the accumulation of hydrogen gas.

❖ The IBC has specific construction requirements for canopies located at motor fuel-dispensing facilities. This section removes any separation requirements between the dispenser area and the canopy when the canopy is constructed in accordance to the special provision for canopies located at motor fuel-dispensing facilities. Note that this section places a condition on the application of IBC canopy construction requirements (i.e., that the canopy be sufficiently open to preclude the accumulation of hydrogen in its construction).

2209.3.4 Overhead lines. The proximity to overhead lines shall be as follows:

1. Not less than 50 feet (15 240 mm) from the vertical plane below the nearest overhead wire of an electric trolley, train or bus line; and
2. Not less than 5 feet (1524 mm) from the vertical plane below the nearest overhead electrical wire.

❖ Hydrogen is much lighter than air. The requirements for the location in relation to overhead power lines of the equipment that compresses and dispenses the gas must take this into consideration. In Item 1, the lateral separation distance from overhead train, trolley car or trolley bus power lines is based on the exposure of the public to the results of an unwanted ignition of fugitive hydrogen and because these types of power lines may spark as part of their normal usage as the trolley poles slide along the wires, showering sparks down several feet out from the wires. Conversely, ordinary power lines do not spark as part of their normal operation and do not pose the same hazard to hydrogen equipment, thus the lesser lateral clearance allowance.

2209.4 Dispensing into motor vehicles at self-service hydrogen motor fuel-dispensing facilities. Self-service hydrogen motor fuel-dispensing systems, including key, code and card lock dispensing systems, shall be limited to the filling of permanently mounted fuel containers on hydrogen-powered vehicles.

In addition to the requirements in Section 2211, the owner of a self-service hydrogen motor fuel-dispensing facility shall provide for the safe operation of the system through the institution of a fire safety plan submitted in accordance with Section 404, the training of employees and operators who use and maintain the system in accordance with Section 406, and provisions for hazard communication in accordance with Section 407.

❖ This section provides regulations for the dispensing of gaseous hydrogen. The dispensing facility owner must demonstrate minimum competency and control of the dispensing of hydrogen including training and supervision for the employees and operators that use and maintain the system.

2209.5 Safety precautions. Safety precautions at hydrogen motor fuel-dispensing and generation facilities shall be in accordance with Sections 2209.5.1 through 2209.5.4.3.

❖ This section establishes a minimum level of safety for hydrogen motor fuel generation and dispensing.

2209.5.1 Valves. Piping to equipment shall be provided with a readily accessible manual shutoff valve that is readily identifiable.

❖ To prevent spillage and to allow servicing of equipment, a remote accessible manual shutoff valve must be installed. This valve is independent of the emergency shutdown equipment required in Section 2209.5.3.

2209.5.2 Protection from vehicles. Guard posts or other approved means shall be provided to protect hydrogen storage systems and use areas subject to vehicular damage in accordance with Section 312.

❖ This provision seeks to prevent leaks caused by vehicle impact. The provision recognizes that vehicle damage can be reduced by placing 4-inch (102 mm) steel posts filled with concrete and spaced on 4-foot (1219 mm) centers around the hydrogen storage system.

2209.5.3 Emergency shutdown. An emergency shutdown device shall be located within 75 feet (22 860 mm) of, but not less than 25 feet (7620 mm) from, dispensers and hydrogen generators, and shall also be provided in the compressor area. On activation, emergency shutdown shall automatically shut off the power supply to all hydrogen storage, compression, dispensing and generating equipment, shut off natural gas or other fuel supply to the hydrogen generator, and close valves between the main supply and the compressor and between the storage containers and dispensing equipment.

❖ Two emergency shutdown devices must be installed, one in the compressor area and the other no closer than 25 feet (7620 mm) nor farther than 75 feet (22 860 mm) from the dispenser. These devices must shut down the power supply to the compressor and close the valves leading to and from the compressor and those between the storage containers and the dispensers in the event of an emergency. In fact, the gaseous hydrogen system may be located more than 300 feet (91 440 mm) from the dispensing operation, but activation of any one emergency shutdown device would activate total shutdown of all generation and dispensing operations on site.

2209.5.4 Emergency venting of hydrogen systems. Hydrogen systems shall be equipped with venting that will relieve excessive internal pressure. Hydrogen systems shall not discharge inside buildings. All portions of the system shall be protected by pressure-relieving devices.

❖ Emergency venting will prevent the excessive buildups of pressure in the system and ensure that the gas will be vented to the outside.

2209.5.4.1 Vent pipe. A vent pipe that will divert the gas flow to atmosphere shall be installed on the vessel for purging operations. The vent pipe shall be designed and constructed as follows:

1. The piping shall be constructed of pipe or tubing materials approved for hydrogen service in accordance with ANSI B31.3 for the rated pressure, volume and temperature. The vent piping shall be designed for the maximum back pressure within the pipe, but not less than 335 pounds per square inch gauge (psig) (2310 kPa).
2. The vent pipe shall be properly supported and shall be provided with a rain cap or other feature which would not limit or obstruct the gas flow from venting vertically upward.
3. A means shall be provided to prevent water, ice and other debris from accumulating inside the vent pipe or obstructing the vent pipe.
4. At the connection fitting of the vent pipe and the hydrogen cylinder, a listed bidirectional detonation flame arrester shall be provided.

❖ In general, four considerations are included in the design of all hydrogen process vent piping: (1) vent to a safe area, (2) assume ignition likely, (3) account for thermal radiation from flame and (4) prevention of unignited flammable mixtures from reaching personnel areas and ignition sources. While these considerations are general in nature and intended for use by designers, fabricators, installers, users and maintainers of hydrogen piping systems, they are also consistent with existing codes and standards wherever possible.

TABLE 2209.5.4.2. See page 22-29.

❖ This table contains the minimum vent height and separation distances based on hydrogen flow rate through the vent pipe and the vent pipe diameter (see also Figure 2209.5.4.2 and commentary, Section 2209.5.4.2).

FIGURE 2209.5.4.2 See page 22-29.

❖ This figure is a companion to Table 2209.5.4.2 and intends to illustrate the table provisions.

2209.5.4.2 Venting of hydrogen gas. Venting of hydrogen gas shall be as follows:

1. The height (*H*) and separation distance (*D*) of the vent pipe shall meet the criteria set forth in Table 2209.5.4.2 for the combinations of maximum hydrogen flow rates and vent stack opening diameters listed;
2. The maximum emergency purging flow rate shall be specified for verification by the authority having jurisdiction. The maximum emergency purging flow rate shall be the pressure relief device release rate in accordance with CGA S-1.3 for a nonengulfing flame or the maximum on-site production rate, whichever is larger; or
3. Where alternative venting arrangements are proposed, an analysis of radiant heat exposures shall be provided showing [in a 30 ft./sec (9.14 m/sec) wind]: exposures to employees are limited to no more than 1,500 Btuh/ft^2 (4732 W/m^2) for a maximum of three minutes, exposures to noncombustible equipment are limited to no more than 8,000 Btuh/ft^2 (25 237 W/m^2), exposures simulated at the property line are limited to no more than 500 Btuh/ft^2 (1577 W/m^2); and that no equipment or personnel within *D* or *H*, or any property line within 1.25 *D* would be exposed to more than one-half of the lower flammable limit (LFL) for hydrogen (2 percent by volume).

❖ The hydrogen storage system must discharge through a vent pipe constructed of a material that is compatible with the gas and has the required strength to withstand the pressure to which it can be exposed. The pressure that the vent can be exposed to must be calculated based on the size of the vent tube, the length of the vent, the friction loss through the vent and the maximum pressure and discharge rate of the system. Vent failure could cause injury resulting from projectile debris and could create a severe fire/explosion hazard. The intent of this section is to divert the vented gas upward and away from any potential ignition source.

The reference in Item 3 to radiant heat exposure is to data as developed by the gas and equipment industry. The exposure to employees criteria are taken from API 521. Table 2209.5.4.2 also reflects conditions where no equipment or personnel within distance (*D*) or any property line within 1¼ *D* would be exposed to more than one-half of the lower flammable limit (LFL) for hydrogen (2 percent by volume).

2209.5.4.2.1 Minimum rate of discharge. The minimum rate of discharge of pressure relief devices on the hydrogen storage tanks shall be in accordance with CGA S-1.3, except for the provision in Section 2209.5.4.3, or the ASME *Boiler and Pressure Vessel Code*, as applicable.

❖ This section states very clearly that pressure relief devises are required for gaseous hydrogen storage tanks. The container standards cover a broad range of container types, from portable to stationary. Pressure relief devices are required only as dictated by the specifications to which the container was fabricated (i.e., ASME or DOTn).

TABLE 2209.5.4.2
VENT PIPE HEIGHT AND SEPARATION DISTANCE
VERSUS HYDROGEN FLOW RATE AND VENT PIPE DIAMETER[a,b,c,d,e,f]

HYDROGEN FLOW RATE	0-500 CFM at NTP		500-1000 CFM at NTP		1,000-2,000 CFM at NTP		2,000-5,000 CFM at NTP			5,000-10,000 CFM at NTP			10,000-20,000 CFM at NTP	
Vent Diameter (inches)	1	2	1	2	1	2	1	2	3	1	2	3	2	3
Height (ft)	8	8	8	8	12	12	17	12	13	25	25	22	36	36
Distance (ft)	13	13	15	17	22	26	39	36	40	53	53	53	81	81

For SI: 1 inch = 25.4 mm, 1 foot = 304.8 mm, 1 Btuh/ft^2 = 3.153 W/m^2, 1 foot/second = 304.8 mm/sec.

a. Minimum distance to lot line is 1.25 times the separation distance.
b. Designs seeking to achieve greater heights with commensurate reductions in separation distances shall be designed in accordance with accepted engineering practice.
c. With this table, personnel on the ground or on the building/equipment are exposed to a maximum of 1,500 Btuh/ ft.^{2}, and are assumed to be provided with a means to escape to a shielded area within 3 minutes, including the case of a 30 ft./sec. wind.
d. Designs seeking to achieve greater radiant exposures to noncombustible equipment shall be designed in accordance with accepted engineering practice.
e. The analysis reflected in this table does not permit hydrogen air mixtures that would exceed one-half of the lower flammable limit (LFL) for hydrogen (2 percent by volume) at the building or equipment, including the case of a 30 ft./sec. wind.
f. See Figure 2209.5.4.2.

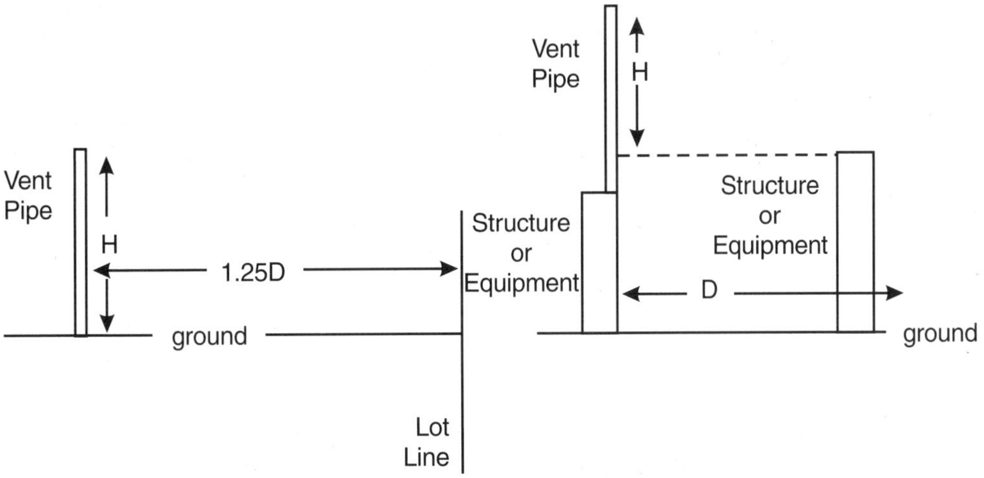

H = Minimum height in feet of vent pipe above the ground or above any structure or equipment within distance (D) where personnel might be present.
D = Distance in feet to adjacent structure or equipment where personnel might be present.

FIGURE 2209.5.4.2
HYDROGEN VENT PIPE HEIGHT (H) VERSUS DISTANCE (D) REQUIREMENTS

2209.5.4.3 Vent pipe flow rates. Where above-ground storage of flammable or combustible liquids occurs and the tanks are diked, or no above-ground storage of flammable or combustible liquids exists, the sizing of the maximum flow for the vent pipe need not include the vent flow as a result of an "engulfing fire" of the hydrogen storage tanks. The pressure relief valve(s) on the gaseous hydrogen storage tanks shall be sized to accommodate a hydrogen compressor that fails to shutdown or unload as a minimum.

❖ This section reflects the industry view that the "engulfing fire case" (exposure fire from flammable or combustible liquid storage) must not be included in the approach to hydrogen safety. Therefore, the intent of this section is effective mitigation of the risk of an engulfing fire through diking of the above-ground flammable or combustible liquid storage tanks rather than address the concept of the maximum hypothetical accident (including hydrogen and other fuels on the site) directly. Typically, the normal sizing of pressure relief devices for other demands (e.g., a runaway hydrogen compressor) is much smaller than the engulfing fire case; hence, the height and distances criteria for the vent stack are easier to accommodate without truly sacrificing safety. Under such circumstances, the pressure relief devices would be sized at or above the maximum compressor flow rate. Accordingly, the minimum vent flow rate to be used in Table 2209.5.4.2 to meet the thermal radiation and unignited vapor criteria would be the nameplate rating of the hydrogen compressor.

SECTION 2210
MARINE MOTOR FUEL-DISPENSING FACILITIES

2210.1 General. The construction of marine motor fuel-dispensing facilities shall be in accordance with the *International Building Code* and NFPA 30A. The storage of Class I, II or IIIA liquids at marine motor fuel-dispensing facilities shall be in accordance with this chapter and Chapter 34.

❖ This section contains code provisions for marine fuel-dispensing facilities that use automotive-type dispensing equipment. This section does not address bulk fuel transfer facilities that use flange-to-flange connections between the shore and the marine vessel.

2210.2 Storage and handling. The storage and handling of Class I, II or IIIA liquids at marine motor fuel-dispensing facilities shall be in accordance with Sections 2210.2.1 through 2210.2.3.

❖ This section introduces provisions for the storage and handling of Class I, II and IIIA flammable and combustible liquids at marine motor fuel-dispensing facilities. Class IIIB combustible liquids do not have specific regulations in the sections that follow because of their relatively low hazard due to their higher flash points.

2210.2.1 Class I, II or IIIA liquid storage. Class I, II or IIIA liquids stored inside of buildings used for marine motor fuel-dispensing facilities shall be stored in approved containers or portable tanks. Storage of Class I liquids shall not exceed 10 gallons (38 L).

Exception: Storage in liquid storage rooms in accordance with Section 3404.3.7.

❖ This section addresses the storage of flammable and combustible liquids inside marine motor fuel-dispensing facility buildings. The provision limits the storage of Class I liquid (e.g., gasoline) to 10 gallons (38 L) unless the building has a flammable and combustible liquid storage room constructed in accordance with the code.

2210.2.2 Class II or IIIA liquid storage and dispensing. Class II or IIIA liquids stored or dispensed inside of buildings used for marine motor fuel-dispensing facilities shall be stored in and dispensed from approved containers or portable tanks. Storage of Class II and IIIA liquids shall not exceed 120 gallons (454 L).

❖ The storage and dispensing inside of marine motor fuel-dispensing facility buildings of Class II (e.g., diesel or kerosene) or IIIA liquids is limited to 120 gallons (454 L) in aggregate. Lubricating oils and gear lubricants (Class III B) are not limited by this provision.

2210.2.3 Heating equipment. Heating equipment installed in Class I, II or IIIA liquid storage or dispensing areas shall comply with Section 2201.6.

❖ When flammable and combustible liquids are stored or dispensed inside of marine motor fuel-dispensing facility buildings, heating equipment must comply with Section 2201.6, which references NFPA 30A, the IFGC and the IMC.

2210.3 Dispensing. The dispensing of liquid fuels at marine motor fuel-dispensing facilities shall comply with Sections 2210.3.1 through 2210.3.5.

❖ The dispensing of fuel at a marine service station requires special consideration because of the location of the dispensing equipment on piers and floating docks and the movement of the pier or dock in relationship to the shore. Dispensing of fuel at marinas generally requires longer dispensing hoses and greater quantities of fuel.

2210.3.1 General. Wharves, piers or floats at marine motor fuel-dispensing facilities shall be used exclusively for the dispensing or transfer of petroleum products to or from marine craft, except that transfer of essential ship stores is allowed.

❖ This section restricts the use of the area adjacent to the fuel-dispensing area to the exclusive use of transferring fuel, with the exception of transferring essential ships' stores. This provision is not intended to restrict the berthing and other uses on the pier or floating structure away from the fuel-transferring docking area. Fuel-dispensing hoses must not be stretched over one vessel to reach another; that is, one vessel cannot be docked parallel and alongside another vessel while taking on fuel.

2210.3.2 Supervision. Marine motor fuel-dispensing facilities shall have an attendant or supervisor who is fully aware of the operation, mechanics and hazards inherent to fueling of boats on duty whenever the facility is open for business. The attendant's primary function shall be to supervise, observe and control the dispensing of Class I, II or IIIA liquids or flammable gases.

❖ Because of the uniqueness of dispensing fuel at marinas and the inherent dangers, the dispensing must be supervised by a trained attendant or supervisor who knows how to control fuel spills, eliminate possible ignition sources, operate emergency shutoff equipment and notify emergency responders.

2210.3.3 Hoses and nozzles. Dispensing of Class I, II or IIIA liquids into the fuel tanks of marine craft shall be by means of an approved-type hose equipped with a listed automatic-closing nozzle without a latch-open device.

Hoses used for dispensing or transferring Class I, II or IIIA liquids, when not in use, shall be reeled, racked or otherwise protected from mechanical damage.

❖ The automatic-closing-type nozzle is designed to operate with automotive-type fuel receivers. Many marine craft are not equipped with automotive-type fuel receivers with the vent line terminating in the receiver tube. It is the fuel returning in the vent line that shuts off the automatic-type nozzle. Therefore, the latch-open device must not be used on marine craft.

Hoses at marinas are not restricted to an 18-foot (5486 mm) maximum length as are those at automotive motor fuel-dispensing facilities. Therefore, the longer hoses are more susceptible to damage and must be properly protected by being placed on a rack or rolled on a reel. It is also wise to mark the nozzle to identify it with a certain dispenser. Because of the length of hose, the operator may attempt to use the wrong nozzle when multiple dispensers are on the pier.

2210.3.4 Portable containers. Class I, II or IIIA liquids shall not be dispensed into a portable container unless such container is approved.

❖ The attendant must supervise the dispensing operation and assure that approved containers are being used. All portable fuel containers must be removed from the marine vessel and placed on the pier or floating dock before being fueled. The nozzle must be in contact with the container before the fuel is dispensed so as to dissipate static electricity.

2210.3.5 Liquefied petroleum gas. Liquefied petroleum gas cylinders shall not be filled at marine motor fuel-dispensing facilities unless approved. Approved storage facilities for LP-gas cylinders shall be provided. See also Section 2207.

❖ Many marine vessels use LP-gas for heating and cooking. LP-gas cylinders may not be refilled at a marina unless specifically approved by the fire code official. Section 2207 has requirements for dispensing LP-gas at motor fuel-dispensing facilities, and this section can be used by the fire code official in making the decision on whether or not to approve the dispensing of LP-gas at marinas.

2210.4 Fueling of marine vehicles at other than approved marine motor fuel-dispensing facilities. Fueling of floating marine craft with Class I fuels at other than a marine motor fuel-dispensing facility is prohibited. Fueling of floating marine craft with Class II or III fuels at other than a marine motor fuel-dispensing facility shall be in accordance with all of the following:

1. The premises and operations shall be approved by the fire code official.
2. Tank vehicles and fueling operations shall comply with Section 3406.6.
3. The dispensing nozzle shall be of the listed automatic-closing type without a latch-open device.
4. Nighttime deliveries shall only be made in lighted areas.
5. The tank vehicle flasher lights shall be in operation while dispensing.
6. Fuel expansion space shall be left in each fuel tank to prevent overflow in the event of temperature increase.

❖ The dispensing of fuel into floating marine craft at locations other than approved marine motor fuel-dispensing facilities is limited to Class II and III liquids (diesel fuel). Gasoline must be dispensed at approved marine motor fuel-dispensing facilities only. Diesel fuel must be dispensed in accordance with all of the six special provisions.

2210.5 Fire prevention regulations. General fire safety regulations for marine motor fuel-dispensing facilities shall comply with Sections 2210.5.1 through 2210.5.7.

❖ This section contains special provisions regarding fire safety at marinas.

2210.5.1 Housekeeping. Marine motor fuel-dispensing facilities shall be maintained in a neat and orderly manner. Accumulations of rubbish or waste oils in excessive amounts shall be prohibited.

❖ This provision limits fuel sources that may readily ignite if subjected to an ignition source or may spontaneously combust.

2210.5.2 Spills. Spills of Class I, II or IIIA liquids at or on the water shall be reported immediately to the fire department and jurisdictional authorities.

❖ The key point in this requirement is to immediately report spills. Petroleum products have a lower specific gravity than water; thus, they will float on the surface of water and a small amount of fuel can be spread over a large area of water.

2210.5.3 Rubbish containers. Metal containers with tight-fitting or self-closing metal lids shall be provided for the temporary storage of combustible trash or rubbish.

❖ The best fire extinguisher for a fire in a metal trash can is the lid. A tight-fitting lid will extinguish the fire by starving the fire of oxygen. Furthermore, if the lid is in place, the combustibles are not likely to ignite in the first place.

2210.5.4 Marine vessels and craft. Vessels or craft shall not be made fast to fuel docks serving other vessels or craft occupying a berth at a marine motor fuel-dispensing facility.

❖ The intent of this provision is to prevent two or more vessels from mooring alongside each other at a fuel dock. If an accident happens at a marine fuel-dispensing facility, the vessels should be able to cast off their docking lines and get underway without having to untie another vessel. Also, the fuel-dispensing hose should never cross one vessel to get to another.

2210.5.5 Sources of ignition. Construction, maintenance, repair and reconditioning work involving the use of open flames, arcs or spark-producing devices shall not be performed at marine motor fuel-dispensing facilities or within 50 feet (15 240 mm) of the dispensing facilities, including piers, wharves or floats, except for emergency repair work approved in writing by the fire code official. Fueling shall not be conducted at the pier, wharf or float during the course of such emergency repairs.

❖ The intent of this requirement is to control possible ignition sources within 50 feet (15 240 mm) of the fuel-dispensing area. In order to control the common ignition hazard of smoking, all hot work within 50 feet (15 240 mm) of a fuel-dispensing area requires a permit from the fire code official in writing and the fuel-dispensing operation must be placed out of service.

2210.5.5.1 Smoking. Smoking or open flames shall be prohibited within 50 feet (15 240 mm) of fueling operations. "No Smoking" signs complying with Section 310 shall be posted conspicuously about the premises. Such signs shall have letters not less than 4 inches (102 mm) in height on a background of contrasting color.

❖ No smoking signs must be conspicuously placed around the fuel-dispensing area.

2210.5.6 Preparation of tanks for fueling. Boat owners and operators shall not offer their craft for fueling unless the tanks being filled are properly vented to dissipate fumes to the outside atmosphere.

❖ Most liquid fuel vapors are heavier than air. Therefore, the vapors, if not properly vented to the atmosphere, will accumulate in the vessel's bilges, causing a very dangerous explosion hazard.

2210.5.7 Warning signs. Warning signs shall be prominently displayed at the face of each wharf, pier or float at such elevation as to be clearly visible from the decks of marine craft being fueled. Such signs shall have letters not less than 3 inches (76 mm) in height on a background of contrasting color bearing the following or approved equivalent wording:

WARNING
NO SMOKING—STOP ENGINE WHILE FUELING,
SHUT OFF ELECTRICITY.

DO NOT START ENGINE UNTIL AFTER BELOW DECK
SPACES ARE VENTILATED.

❖ One of the greatest dangers in dispensing fuel into marine craft is the accumulation of explosive vapors in the vessel's bilges and below-deck spaces. The intent here is to warn the vessel operators to control ignition sources until the bilges and below-deck spaces have been cleared of explosive vapors.

2210.6 Fire protection. Fire protection features for marine motor fuel-dispensing facilities shall comply with Sections 2210.6.1 through 2210.6.4.

❖ This section addresses the requirements for fire protection at marine fuel-dispensing facilities.

2210.6.1 Standpipe hose stations. Fire hose, where provided, shall be enclosed within a cabinet, and hose stations shall be labeled: FIRE HOSE—EMERGENCY USE ONLY.

❖ Section 2210.1 states that the construction requirements for marine motor fuel-dispensing facilities will be found in the IBC and NFPA 30A. NFPA 30A states that marine fuel-dispensing facilities that are located on piers that extend more than a 500-foot (152 400 mm) travel distance from shore are required to have a Class III standpipe installed in accordance with NFPA 14.

2210.6.2 Obstruction of fire protection equipment. Materials shall not be placed on a pier in such a manner as to obstruct access to fire-fighting equipment or piping system control valves.

❖ The intent of this section is to make all fire protection equipment, including means for turning off fuel supply lines in an emergency, readily accessible in an emergency.

2210.6.3 Access. Where the pier is accessible to vehicular traffic, an unobstructed roadway to the shore end of the wharf shall be maintained for access by fire apparatus.

❖ See Chapter 5 and Appendix D for guidance on fire apparatus access roads.

2210.6.4 Portable fire extinguishers. Portable fire extinguishers in accordance with Section 906, each having a minimum rating of 20-B:C, shall be provided as follows:

1. One on each float.
2. One on the pier or wharf within 25 feet (7620 mm) of the head of the gangway to the float, unless the office is within 25 feet (7620 mm) of the gangway or is on the float and an extinguisher is provided thereon.

❖ Section 906 lists the requirements for where and how to mount the portable fire extinguishers. They should be conspicuous and unobstructed.

SECTION 2211
REPAIR GARAGES

2211.1 General. Repair garages shall comply with this section and the *International Building Code*. Repair garages for vehicles that use more than one type of fuel shall comply with the applicable provisions of this section for each type of fuel used.

Where a repair garage also includes a motor fuel-dispensing facility, the fuel-dispensing operation shall comply with the requirements of this chapter for motor fuel-dispensing facilities.

❖ Because of the popularity of alternative fuels and the differences in their properties, repair garages must be designed for the anticipated vehicles and the fuels contained within them. This section includes the provisions for many of the different fuels.

If the repair garage dispenses fuel into vehicles, the repair garage must also meet the requirements for a motor fuel-dispensing facility based on the fuels available.

2211.2 Storage and use of flammable and combustible liquids. The storage and use of flammable and combustible liquids in repair garages shall comply with Chapter 34 and Sections 2211.2.1 through 2211.2.4.

❖ For obvious reasons, the storage of flammable and combustible liquids is always a concern, especially inside buildings. Chapter 34 lists the provisions for the storage and quantity limitations on flammable and combustible liquids. This section addresses specific uses associated with repair garages.

2211.2.1 Cleaning of parts. Cleaning of parts shall be conducted in listed and approved parts-cleaning machines in accordance with Chapter 34.

❖ Section 3405.3.6 refers to the cleaning operation, the equipment and the quantity limits associated with parts cleaning. Caution must be used when cleaning automotive parts because they may contain Class I liquids that may contaminate the cleaning solvents.

2211.2.2 Waste oil, motor oil and other Class IIIB liquids. Waste oil, motor oil and other Class IIIB liquids shall be stored in approved tanks or containers, which are allowed to be stored and dispensed from inside repair garages.

Tanks storing Class IIIB liquids in repair garages are allowed to be located at, below or above grade, provided that adequate drainage or containment is provided.

Crankcase drainings shall be classified as Class IIIB liquids unless otherwise determined by testing.

❖ NFPA 30 will allow storage of Class IIIB liquids (motor oil, crankcase drainings and transmission fluids) in combustible containers inside the building when the building is protected with automatic fire sprinklers. Otherwise, the storage tank inside the building for Class IIIB liquids must be an approved above-ground tank such as one meeting the requirements of UL 142 or UL 80.

The intent is to allow storage of Class IIIB liquids inside below grade when spill protection is provided.

Care must be taken that someone does not pour Class I and II liquids into the crankcase drainings tank.

2211.2.3 Drainage and disposal of liquids and oil-soaked waste. Garage floor drains, where provided, shall drain to approved oil separators or traps discharging to a sewer in accordance with the *International Plumbing Code*. Contents of oil separators, traps and floor drainage systems shall be collected at sufficiently frequent intervals and removed from the premises to prevent oil from being carried into the sewers.

Crankcase drainings and liquids shall not be dumped into sewers, streams or on the ground, but shall be stored in approved tanks or containers in accordance with Chapter 34 until removed from the premises.

Self-closing metal cans shall be used for oily waste.

❖ This section does not require floor drains but, rather, requires oil separators or traps when floor drains are provided. Oil separators must be installed in accordance with the IPC.

The crankcase drainings container must be approved for Class IIIB liquids.

This provision refers to the storage of oily rags, etc. See the commentary to Section 304.3.1 for further discussion of waste cans.

2211.2.4 Spray finishing. Spray finishing with flammable or combustible liquids shall comply with Chapter 15.

❖ Chapter 15 regulates the spraying of flammable and combustible finishes and requires a permit from the fire code official. The spray operation must comply with Section 1504.

2211.3 Sources of ignition. Sources of ignition shall not be located within 18 inches (457 mm) of the floor and shall comply with Chapters 3 and 26.

❖ No open flame device or other sources of ignition may be within 18 inches (457 mm) of the floor in a repair garage. Flammable and combustible vapors found in repair garages are heavier than air and will accumulate below the 18-inch (457 mm) level.

2211.3.1 Equipment. Appliances and equipment installed in a repair garage shall comply with the provisions of the *International Building Code,* the *International Mechanical Code* and the ICC *Electrical Code.*

❖ Special care must be taken in selecting and installing appliances and equipment in repair garages due to the 18-inch (457 mm) hazard zone identified in Section 2211.3. Electrical equipment must be suitable for classified hazardous locations. Appliances, such as furnaces, must be installed with their fire boxes above the 18-inch (457 mm) hazard zone to reduce the likelihood of their becoming an ignition source for flammable vapors that may have escaped and settled to the floor. See the commentary to the applicable sections of the IBC, IMC and ICC EC for further information.

2211.3.2 Smoking. Smoking shall not be permitted in repair garages except in approved locations complying with Section 310.

❖ Smoking is a common ignition hazard that does not occur within 18-inches (457 mm) of the floor but can drop hot ashes into the 18-inch (457 mm) hazard zone near the floor with the potential to ignite any accumulated vapors.

2211.4 Below-grade areas. Pits and below-grade work areas in repair garages shall comply with Sections 2211.4.1 through 2211.4.3.

❖ Note that this section describes this area as a pit or a below-grade work area and not a basement

2211.4.1 Construction. Pits and below-grade work areas shall be constructed in accordance with the *International Building Code*.

❖ Construction requirements are found in the IBC.

2211.4.2 Means of egress. Pits and below-grade work areas shall be provided with means of egress in accordance with Chapter 10.

❖ Section 202 defines a motor vehicle repair garage as a moderate-hazard storage occupancy.

Section 1018.2 allows one means of egress if the building is a storage occupancy with only one level below grade, the occupancy load is less than 30 occupants and the travel distance does not exceed 100 feet (30 500 mm) for a single-story storage occupancy.

2211.4.3 Ventilation. Where Class I liquids or LP-gas are stored or used within a building having a basement or pit wherein flammable vapors could accumulate, the basement or pit shall be provided with mechanical ventilation in accordance with the *International Mechanical Code,* at a minimum rate of 1.5 cubic feet per minute per square foot (cfm/ft^2) [0.008 m^3/(s · m^2)] to prevent the accumulation of flammable vapors.

❖ A ventilation system must be installed in any below-grade area where flammable and combustible vapors might accumulate. The ventilation system must be in operation any time the repair garage is open for business, if a vehicle is left parked over the pit when the garage is closed for business or anytime there is a chance that vapors may accumulate in the below-grade area.

2211.5 Preparation of vehicles for repair. For vehicles powered by gaseous fuels, the fuel shutoff valves shall be closed prior to repairing any portion of the vehicle fuel system.

Vehicles powered by gaseous fuels in which the fuel system has been damaged shall be inspected and evaluated for fuel system integrity prior to being brought into the repair garage. The inspection shall include testing of the entire fuel delivery system for leakage.

❖ Gaseous fuels include such fuels as LP-gas and CNG, among other gases.

If a leak is detected, it must be stopped or the system purged of fuel before the vehicle can be brought into the garage. The intent is to prevent fuel gases from entering the garage area.

2211.6 Fire extinguishers. Fire extinguishers shall be provided in accordance with Section 906.

❖ A motor vehicle repair garage is classified as a moderate-hazard storage occupancy. See Table 906.3(1) for the size and placement of portable fire extinguishers.

2211.7 Repair garages for vehicles fueled by lighter-than-air fuels. Repair garages for the conversion and repair of vehicles which use CNG, liquefied natural gas (LNG), hydrogen or other lighter-than-air motor fuels shall be in accordance with Section 2211.7 in addition to the other requirements of Section 2211.

Exception: Repair garages where work is not performed on the fuel system and is limited to exchange of parts and maintenance requiring no open flame or welding.

❖ Repair garages that install and repair CNG, LNG, hydrogen or other lighter-than-air motor fuels must be equipped with ventilation and gas detection systems in accordance with Sections 2211.7.1 though 2211.7.2.3.

An example of the exception would be a garage that works on automobile mufflers, brakes and shock absorbers and does not repair fuel systems. This type of garage would not have to comply with this section even if the vehicles being repaired are equipped with lighter-than-air fuel systems.

2211.7.1 Ventilation. Repair garages used for the repair of natural gas- or hydrogen-fueled vehicles shall be provided with an approved mechanical ventilation system. The mechanical ventilation system shall be in accordance with the *International Mechanical Code* and Sections 2211.7.1.1 and 2211.7.1.2.

Exception: Repair garages with natural ventilation when approved.

❖ The intent of this section is to prevent the accumulation of lighter-than-air flammable and combustible gases inside the repair garage.

An example of natural ventilation that a fire code official may approve at his or her discretion is a repair garage with at least two opposite sides open all the way to the ceiling. The two opposite sides would allow for cross ventilation. Having the walls open to the ceiling would prevent lighter-than-air gases from accumulating at the ceiling level. The ceiling would have to be sealed to prevent gasses from entering the attic space; otherwise, mechanical ventilation would be required.

2211.7.1.1 Design. Indoor locations shall be ventilated utilizing air supply inlets and exhaust outlets arranged to provide uniform air movement to the extent practical. Inlets shall be uniformly arranged on exterior walls near floor level. Outlets shall be located at the high point of the room in exterior walls or the roof.

Ventilation shall be by a continuous mechanical ventilation system or by a mechanical ventilation system activated by a continuously monitoring natural gas detection system where a gas concentration of not more than 25 percent of the lower flammable limit (LFL) is present. In either case, the system shall shut down the fueling system in the event of failure of the ventilation system.

The ventilation rate shall be at least 1 cubic foot per minute per 12 cubic feet (0.00139 m^3/s · m^3) of room volume.

❖ The intent of this section is to provide uniform ventilation throughout the garage area that will exchange at least 1 cubic foot of air for every 12 cubic feet of room volume every minute (0.00139 m^3/s · m^3). The ventilation must be continuous or be connected to a gas detection system.

2211.7.1.2 Operation. The mechanical ventilation system shall operate continuously.

Exceptions:

1. Mechanical ventilation systems that are interlocked with a gas detection system designed in accordance with Section 2211.7.2.
2. Mechanical ventilation systems in repair garages that are used only for repair of vehicles fueled by liquid fuels or odorized gases, such as CNG, where the ventilation system is electrically interlocked with the lighting circuit.

❖ The intent of this provision is to prevent the accumulation of lighter-than-air gases inside vehicle repair garages.

2211.7.2 Gas detection system. Repair garages used for repair of vehicles fueled by nonodorized gases, such as hydrogen and nonodorized LNG, shall be provided with an approved flammable gas detection system.

❖ Some gases contain additives that produce pungent odors for easy recognition. If the vehicle contains fuel systems that do not use these odorized gases, a gas detection system must be installed.

2211.7.2.1 System design. The flammable gas detection system shall be calibrated to the types of fuels or gases used by vehicles to be repaired. The gas detection system shall be designed to activate when the level of flammable gas exceeds 25 percent of the lower flammable limit (LFL). Gas detection shall also be provided in lubrication or chassis repair pits of repair garages used for repairing nonodorized LNG-fueled vehicles.

❖ This provision will require quick-lube-type facilities that change oil and lubricate vehicles to install gas detection systems in the pit area if they service vehicles that are equipped with LNG fuel systems.

2211.7.2.2 Operation. Activation of the gas detection system shall result in all the following:

1. Initiation of distinct audible and visual alarm signals in the repair garage.
2. Deactivation of all heating systems located in the repair garage.
3. Activation of the mechanical ventilation system, when the system is interlocked with gas detection.

❖ The intent of this section is to identify the equipment that the gas detection system must activate in the event the system detects the presence of gas above 25 percent of the LFL.

2211.7.2.3 Failure of the gas detection system. Failure of the gas detection system shall result in the deactivation of the heating system, activation of the mechanical ventilation system and where the system is interlocked with gas detection and causes a trouble signal to sound in an approved location.

❖ The intent of this section is to require all the equipment that the gas detector would normally activate when it detects gas to also activate if the detector fails.

2211.8 Defueling of hydrogen from motor vehicle fuel storage containers. The discharge or defueling of hydrogen from motor vehicle fuel storage tanks for the purpose of maintenance, cylinder certification, calibration of dispensers or other activities shall be in accordance with Section 2210.8.1.

❖ Because of the emerging use of alternative fuels and the differences in their properties, repair garages must be designed for the anticipated vehicles and the materials fueling them. Section 2211 includes the provisions for many different fuels, including lighter-than-air fuels. Accordingly, if a repair garage makes hydrogen, which is a lighter-than-air fuel, available for dispensing to motor vehicles, the repair garage must also meet the applicable requirements and hazard mitigation criteria for servicing hydrogen-fueled vehicles.

2211.8.1 Methods of discharge. The discharge of hydrogen from motor vehicle fuel storage tanks shall be accomplished through a closed transfer system in accordance with Section 2210.8.1.1 or an approved method of atmospheric venting in accordance with Section 2210.8.1.2.

❖ The intentional discharge of gas from vehicle containers is done for several reasons, including vehicle repairs, certification of the container integrity and container replacement. The gas must be discharged in a safe manner as dictated by either Section 2211.8.1.1 or 2211.8.1.2. The vehicle storage container is referred to as a "vessel" and as a "cylinder" in the text to follow.

2211.8.1.1 Closed transfer system. A documented procedure that explains the logic sequence for discharging the storage tank shall be provided to the code official for review and approval. The procedure shall include what actions the operator is required to take in the event of a low-pressure or high-pressure hydrogen release during discharging activity. Schematic design documents shall be provided illustrating the arrangement of piping, regulators and equipment settings. The construction documents shall illustrate the piping and regulator arrangement and shall be shown in spatial relation to the location of the compressor, storage vessels and emergency shutdown devices.

❖ A closed transfer system uses the same basic components as a dispensing system and withdraws the gas from the vehicle container and stores it in vessels. Drawings and a description of the sequence of operation of the transfer system must be provided to the fire code official for approval. This information may then be shared with the fire department for use in case of an emergency where hydrogen cylinders might need to be discharged.

2211.8.1.2 Atmospheric venting of hydrogen from motor vehicle fuel storage containers. When atmospheric venting is used for the discharge of hydrogen from motor vehicle fuel storage tanks , such venting shall be in accordance with Sections 2210.8.1.2.1 through 2210.8.1.2.4.

❖ The subsections that follow provide the conditions and requirements under which hydrogen is allowed to be discharged to the atmosphere. Such discharge should be avoided wherever practical because there is an inherent hazard in doing so and because resources should never be wasted. A closed transfer system is the preferable way to remove hydrogen from vehicle containers.

2211.8.1.2.1 Defueling equipment required at vehicle maintenance and repair facilities. All facilities for repairing hydrogen systems on hydrogen- fueled vehicles shall have equipment to defuel vehicle storage tanks. Equipment used for defueling shall be listed and labeled for the intended use.

❖ It only makes sense that if a repair garage offers service for hydrogen-powered vehicles that it must have the proper equipment available for that purpose. The hazards of hydrogen are such that using "jerry-rigged" tools and equipment could have disastrous results. As a further precaution, all defueling equipment must be both listed and labeled for that purpose.

2211.8.1.2.1.1 Manufacturer's equipment required. Equipment supplied by the vehicle manufacturer shall be used to connect the vehicle storage tanks to be defueled to the vent pipe system.

❖ Since the vehicle manufacturer provides the necessary equipment and fittings to mate their vehicle to tanks and containers, only that equipment is to be used. The use of incompatible equipment could result in an unexpected gas discharge and possible ignition.

2211.8.1.2.1.2 Vent pipe maximum diameter. Defueling vent pipes shall have a maximum inside diameter of 1 inch (25 mm) and be installed in accordance with Section 2209.5.4.

❖ When discharging hydrogen to the atmosphere, it is important that the height and size of the vent pipe be such that it will provide sufficient discharge velocity to disperse the hydrogen in the air (see also commentary, Section 2209.5.4).

2211.8.1.2.1.3 Maximum flow rate. The maximum rate of hydrogen flow through the vent pipe system shall not exceed 1,000 cfm at NTP (2.5 kg/min) and shall be controlled by means of the manufacturer's equipment, at low pressure and without adjustment.

❖ When gas flows through piping, friction with the pipe wall generates heat. Limiting the flow rate limits the amount of heat generated by the flow of gas.

2211.8.1.2.1.4 Isolated use. The vent pipe used for defueling shall not be connected to another venting system used for any other purpose.

❖ To avoid contamination or the reaction of potentially incompatible materials, this section requires that the defueling vent pipe be dedicated to hydrogen-only use.

2211.8.1.2.2 Construction documents. Construction documents shall be provided illustrating the defueling system to be utilized. Plan details shall be of sufficient detail and clarity to allow for evaluation of the piping and control systems to be utilized and include the method of support for cylinders, containers or tanks to be used as part of a closed transfer system, the method of grounding and bonding, and other requirements specified herein.

❖ Construction documents, as defined in Section 202, must be drawn to scale with sufficient clarity to be understood by the fire code official, contractors and owners and must address the requirements of the code. State or local laws may require preparation of the construction documents by a registered design professional (see the commentary to Section 106.1 of the IBC for further information).

2211.8.1.2.3 Stability of cylinders, containers and tanks. A method of rigidly supporting cylinders, containers or tanks used during the closed transfer system discharge or defueling of hydrogen shall be provided. The method shall provide not less than two points of support and shall be designed to resist lateral movement of the receiving cylinder, container or tank. The system shall be designed to resist movement of the receiver based on the highest gas-release velocity through valve orifices at the receiver's rated service pressure and volume. Supporting structure or appurtenance used to support receivers shall be constructed of noncombustible materials in accordance with the *International Building Code.*

❖ Vehicle hydrogen fuel containers can hold extremely high pressures, which if released quickly can produce large thrust forces that would propel the container like a rocket. Hydrogen does not liquefy at normal ambient temperatures; therefore, in order to hold the required amount of fuel on board the vehicle, the gaseous fuel must be compressed to extreme pressures of from 3,000 to 5,000 psia (20 613 to 34 355 kPa).

2211.8.1.2.4 Grounding and bonding. Cylinders, containers or tanks and piping systems used for defueling shall be bonded and grounded. Structures or appurtenances used for supporting the cylinders, containers or tanks shall be grounded in accordance with the ICC *Electrical Code.* The valve of the vehicle storage tank shall be bonded with the defueling system prior to the commencement of discharge or defueling operations.

❖ "Grounding" means to intentionally connect to the earth. This could be accomplished by installing a conductor between the cylinder support and the building grounding electrode system. Bonding means to join metallic parts together to form a continuous electrical pathway. Grounding and bonding required by this section are intended to control sparking that could result from current flow produced by voltage differentials across parts of the venting set-up and the building components. Grounding the venting set-up and attaching a bonding jumper to the cylinder valve will put all such components and the building components at the same voltage potential, thereby reducing the possibility of sparks that could ignite flammable vapors or harm the cylinder assembly. The provisions of this section will also help prevent the buildup of static electrical charges that could be a source of ignition.

2211.8.2 Repair of hydrogen piping. Piping systems containing hydrogen shall not be opened to the atmosphere for repair without first purging the piping with an inert gas to achieve 1 percent hydrogen or less by volume. Defueling operations and exiting purge flow shall be vented in accordance with Section 2211.8.1.2.

❖ To prevent the release of flammable hydrogen gas to the atmosphere and to prevent an explosive hydrogen-air mixture (above 1-percent hydrogen based on the pipe volume) within system piping, this section requires that piping be purged with an inert gas before being disconnected and repaired.

2211.8.3 Purging. Each individual manufactured component of a hydrogen generating, compression, storage or dispensing system shall have a label affixed as well as a description in the installation and owner's manuals describing the procedure for purging air from the system during startup, regular maintenance and for purging hydrogen from the system prior to disassembly (to admit air).

For the interconnecting piping between the individual manufactured components, the pressure rating must be at least 20 times the absolute pressure present in the piping when any hydrogen meets any air.

❖ This section places the burden of the purging requirement on the equipment manufacturer. Commensurately, these requirements can be verified by the fire code official before, during and after installation. Requirements for rating the interconnecting piping at 20 times the initial pressure reduces the likelihood that a detonation will rupture the vent piping.

2211.8.3.1 System purge required. After installation, repair or maintenance, the hydrogen piping system shall be purged of air in accordance with the manufacturer's procedure for purging air from the system.

❖ Reliable purging procedures are essential to the safe use of hydrogen gas systems. This section requires that the system manufacturer's purge method be used. Such methods could include any of several methodologies in common use in the industry, such as outlined in ASME B31.3 (see commentary, Section 705.3 of the IFGC). The continuous flow method uses a continuous flow of the inert purge gas to remove the hydrogen gas and prevent air or moisture from entering the system. The dilution method uses a sequence of pressurization and venting. This sequence is repeated several times and is very effective in removing gas from dead-end piping, such as pressure gauge lines.

Bibliography

The following resource materials are referenced in this chapter or are relevant to the subject matter addressed in this chapter.

API 521, *Guide for Pressure-Relieving and Depressurizing Systems*, 2nd ed. Washington, DC: American Petroleum Institute, 1982.

ASME B31.3-99, *Process Piping*. New York: American Society of Mechanical Engineers, 1999.

CGA G-5.4-01, *Standard for Hydrogen Piping Systems at Consumer Locations*. Arlington, VA: Compressed Gas Association, 2001.

ICC EC-2003, *ICC Electrical Code*. Falls Church, VA: International Code Council, 2003.

IFGC-2003, *International Fuel Gas Code*. Falls Church, VA: International Code Council, 2003.

IMC-2003, *International Mechanical Code*. Falls Church, VA: International Code Council, 2003.

NFPA 30-00, *Flammable and Combustible Liquids Code*. Quincy, MA: National Fire Protection Association, 2000.

NFPA 30A-00, *Code for Motor Fuel Dispensing Facilities and Repair Garages*. Quincy, MA: National Fire Protection Association, 2000.

NFPA 50A-99, *Gaseous Hydrogen Systems at Consumer Sites*. Quincy, MA: National Fire Protection Association, 1999.

NFPA 50B-99, *Liquefied Hydrogen Systems at Consumer Sites*. Quincy, MA: National Fire Protection Association, 1999.

NFPA 52-98, *Compressed National Gas (CNG) Vehicular Fuel Systems*. Quincy, MA: National Fire Protection Association, 1998.

UL 80-96, *Steel Tanks for Oil-Burner Fuel*. Northbrook, IL: Underwriters Laboratories Inc., 1996.

UL 87-01, *Power-Operated Dispensing Devices for Petroleum Products*. Northbrook, IL: Underwriters Laboratories Inc., 2001.

UL 142-02, *Steel Aboveground Tanks for Flammable and Combustible Liquids*. Northbrook, IL: Underwriters Laboratories Inc., 2002.

UL 842-97, *Valves for Flammable Fluids*. Northbrook, IL: Underwriters Laboratories Inc., 1997.

UL 2085-97, *Protected Aboveground Tanks for Flammable and Combustible Liquids*. Northbrook, IL: Underwriters Laboratories Inc., 1997.

Chapter 23: High-Piled Combustible Storage

General Comments

High-piled combustible storage facilities have presented the fire service with a high challenge in fire incident management. The shear mass of commodities in concentrated form, along with a different configuration for access to those commodities, has led to the development of some unique fire protection and life safety measures. Experience with automatic sprinkler protection has demonstrated over the years that such a system is reliable and effective in suppressing a fire and in supplementing manual fire suppression and extinguishing operations. However, the design and maintenance of any system or measure is essential to ensuring that when needed it will serve its function until such time as the fire service arrives to take command of the incident.

Chapter 23 of the code presents fundamental concepts of fire and life safety protection within high-piled combustible storage areas and buildings. Fire tests and past fire experience have shown that the class of commodity, the quantity of commodities, their relationship to one another and the maintenance and inspection of the systems and building are essential in preventing fires or limiting the spread of fire. This chapter has provisions aimed at verifying the proper design and installation of given fire and life safety protection systems. In some cases the referenced standards will contain more specific design and installation criteria than are found in this chapter.

Other chapters of the code that contain provisions related to Chapter 23 are Chapters 3, 4, 5, 7, 9 and 10. The *International Building Code*® (IBC®) is mentioned within this chapter, which addresses construction classification, occupancy, fire protection features, means of egress, and structural requirements.

This chapter does not specifically cover miscellaneous combustible materials storage (see Section 315) or storage facilities of unusual design. Under those circumstances the provisions of Chapter 1, specifically Section 102.7, would apply.

Unusual storage facilities may include, but are not limited to, storage warehouses located on piers, storage of vehicles, refrigerated storage, underground storage facilities and air-supported storage structures. Each type of specialty storage offers a challenge to the fire department that will require unconventional fire incident management planning and unusual fire-fighting techniques. Access by fire-fighting vehicles and personnel is the greatest of those challenges. As a result, preplanning of the design of the storage facility may require coordination with the fire code official and fire service to determine the means and methods of the fire service to handle an approach by water, addressing floating burning debris from a fire, hazards of other neighboring piers and vessels, etc.

Purpose

Chapter 23 provides guidance for reasonable protection of life from hazards associated with high-piled combustible storage. This chapter provides requirements for identifying various classes of commodities; general fire and life protection safety features and housekeeping and maintenance requirements. This chapter attempts to define the potential fire severity and, in turn, determine fire and life safety protection measures needed to control, and in some cases suppress, a potential fire. The design considerations for determining building construction, occupancy and the fire protection features of the building or structure to accommodate the high-piled combustible storage is the responsibility of the design professional and the building owner. Housekeeping, building maintenance and development and upkeep of an evacuation plan are the responsibility of the building owner, tenant or lessee.

SECTION 2301 GENERAL

2301.1 Scope. High-piled combustible storage shall be in accordance with this chapter. In addition to the requirements of this chapter, the following material-specific requirements shall apply:

1. Aerosols shall be in accordance with Chapter 28.
2. Flammable and combustible liquids shall be in accordance with Chapter 34.
3. Hazardous materials shall be in accordance with Chapter 27.
4. Storage of combustible paper records shall be in accordance with NFPA 231C.
5. Storage of combustible fibers shall be in accordance with Chapter 29.

6. Storage of miscellaneous combustible material shall be in accordance with Chapter 3.

❖ This section is intended as an introduction and references those other chapters within the code that are to be used in conjunction with Chapter 23, when applicable. In many cases, the referenced chapter contains more stringent provisions regarding fire protection and life safety measures and may also cite other referenced standards. For example Chapter 28 references NFPA 30B, Chapter 34 references NFPA 30 and Chapter 27 references NFPA 704.

2301.2 Permits. A permit shall be required as set forth in Section 105.6.

❖ The process of issuing permits gives the fire code official an opportunity to carefully evaluate and regulate hazardous operations. Permit applicants should be required to demonstrate that their operations comply with the intent of the code before the permit is issued. See the commentary to Section 105.6 for a general discussion of operations requiring an operational permit. The permit process also notifies the fire department of the need for prefire planning for the hazardous property.

2301.3 Construction documents. At the time of building permit application for new structures designed to accommodate high-piled storage or for requesting a change of occupancy/use, and at the time of application for a storage permit, plans and specifications shall be submitted for review and approval. In addition to the information required by the *International Building Code*, the storage permit submittal shall include the information specified in this section. Following approval of the plans, a copy of the approved plans shall be maintained on the premises in an approved location. The plans shall include the following:

1. Floor plan of the building showing locations and dimensions of high-piled storage areas.
2. Usable storage height for each storage area.
3. Number of tiers within each rack, if applicable.
4. Commodity clearance between top of storage and the sprinkler deflector for each storage arrangement.
5. Aisle dimensions between each storage array.
6. Maximum pile volume for each storage array.
7. Location and classification of commodities in accordance with Section 2303.
8. Location of commodities which are banded or encapsulated.
9. Location of required fire department access doors.
10. Type of fire suppression and fire detection systems.
11. Location of valves controlling the water supply of ceiling and in-rack sprinklers.
12. Type, location and specifications of smoke removal and curtain board systems.
13. Dimension and location of transverse and longitudinal flue spaces.
14. Additional information regarding required design features, commodities, storage arrangement and fire protection features within the high-piled storage area shall be provided at the time of permit, when required by the fire code official.

❖ Section 2301.3 contains provisions that are unique to the needs of the fire department. These construction document requirements are in addition to those encountered in the IBC and specifically require detailed information that is essential for the permit review process and for future use when conducting fire inspections, in the development of the fire department's prefire emergency plans and when conducting fire incident management during an emergency.

Items 1, 2, 3 and 5 are general requirements that address the general description of work for which the application must be submitted.

Item 4 is significant because the height of any storage in relation to the sprinkler heads will determine sprinkler effectiveness. Sprinklers should be designed for the maximum allowable heights of the stored commodities and the minimum amount of unobstructed space below the sprinkler heads.

Items 6, 7 and 8 recognize that a storage building can contain multiple classes and groups of commodities. As a result, this section requires that the construction documents illustrate the location, size and classification of the various types of commodities.

Items 9, 10, 11 and 12 are important for the development of a prefire emergency plan. Such a plan can be established by the building owner or the tenant in cooperation with the local fire department that will be called to the building during an emergency. Developing and maintaining a prefire emergency plan can help during the incipient stages of a fire by reducing the amount of damage from the fire and containing the fire within a short time.

Item 13 requires a detailed description of the locations of transverse and longitudinal flue spaces. The provisions for minimum flue space dimensions are contained in Section 2308.3.

Item 14 allows the fire code official to use the enforcement powers of that office to collect more detailed information to make a more educated decision.

These provisions recognize that technology is evolving. As a result, the situation presented to the fire code official can require that additional fire protection and life safety measures be applied to address those new designs for storage or new class of commodity.

2301.4 Evacuation plan. When required by the fire code official, an evacuation plan for public accessible areas and a separate set of plans indicating location and width of aisles, location of exits, exit access doors, exit signs, height of storage, and locations of hazardous materials shall be submitted at the time of permit application for review and approval. Following approval of the plans, a copy of the approved plans shall be maintained on the premises in an approved location.

❖ Section 2301.4 contains provisions that are designed as a precautionary mechanism for the fire department. It allows the fire code official to require additional fire and life safety protection features of the building owner or the tenant, in addition to the requirements set forth in the administrative section of this code. This feature serves both the occupants of the building or structure and the fire department by addressing unique conditions. The plans are to be accessible at the site and are for use by the fire code official or fire department during inspections of the building or structure. Maintenance of the plans is the responsibility of the building owner, tenant or lessee.

SECTION 2302 DEFINITIONS

2302.1 Definitions. The following words and terms shall, for the purposes of this chapter and as used elsewhere in this code, have the meanings shown herein.

❖ Definitions of terms can help in the understanding and application of the code requirements. The purpose for including those definitions that are associated with the subject matter of this chapter is to provide more convenient access to them without having to refer back to Chapter 2. It is important to emphasize that these terms are not exclusively related to this chapter but are applicable everywhere the term is used in the code. For convenience, these terms are also listed in Chapter 2 with a cross reference to this section. The use and application of all defined terms, including those defined in this section, are set forth in Section 201.

ARRAY. The configuration of storage. Characteristics considered in defining an array include the type of packaging, flue spaces, height of storage and compactness of storage.

❖ This term defines the configuration of storage, with its essential components being the manner in which the commodity is packaged (stored either in piled formation or on a rack storage system); the amount of flue space between formations of the commodity; height, width and length of any one array and compactness.

ARRAY, CLOSED. A storage configuration having a 6-inch (152 mm) or smaller width vertical flue space that restricts air movement through the stored commodity.

❖ The term "closed array" was effectively defined as the result of a number of fire tests. These tests were conducted with the sample test assembly array employing 6-inch-wide (152 mm) longitudinal flues and no transverse flues.

BIN BOX. A five-sided container with the open side facing an aisle. Bin boxes are self-supporting or supported by a structure designed so that little or no horizontal or vertical space exists around the boxes.

❖ Proprietary storage systems, such as bin box storage, are common in warehousing and manufacturing industries, such as automotive assembly plants and mail-order mercantile operations. Many bin box systems rely on the rigidity of adjacent bins and interlocking design for stability. Provisions for fastening the self-supporting units together to permit higher stacking are typical in most designs. Other bin box systems are designed in conjunction with a supporting rack system to minimize unusable space between bins. Bin boxes can be constructed of combustible materials, such as wood or cardboard, or of noncombustible materials such as metal.

COMMODITY. A combination of products, packing materials and containers.

❖ Commodity is a term used to identify the product being stored, its container or housing and the type of stackable mechanism (with or without pallet). Commodities and their containers are generally identified as classes, with each classification identifying the potential for combustion (heat per unit weight). The quantity and locations of each type of commodity will define the type of general and special fire protection and life safety requirements necessary for the building or structure.

DRAFT CURTAIN. A structure arranged to limit the spread of smoke and heat along the underside of the ceiling or roof.

❖ A draft curtain restricts the passage of smoke at roof or ceiling level. Located at the roof or ceiling, draft curtains are designed to compartmentalize the roof or ceiling area in order to limit the passage of smoke to a defined area and hinder its spread throughout the entire storage area. It has also been argued that such curtains help the activation of smoke and heat vents.

EARLY SUPPRESSION FAST-RESPONSE (ESFR) SPRINKLER. A sprinkler listed for early suppression fast-response performance.

❖ The early suppression fast-response (ESFR) sprinkler head was originally designed to provide fire suppression for heat and fire situations that required a lower thermal response to activate the sprinkler head earlier than the standard sprinkler head. The ESFR sprinkler heads have a Response Time Index (RTI) of 50 (meters-seconds)$^{1/2}$ or less, compared to a standard sprinkler head, which has a RTI of 80 (meters-seconds)$^{1/2}$ or more. ESFR sprinkler design requires the same concepts of sprinkler system design for it to react and suppress a fire when needed. The important difference is that ESFR systems are specifically designed to suppress a high-challange fire.

EXPANDED PLASTIC. A foam or cellular plastic material having a reduced density based on the presence of numerous small cavities or cells dispersed throughout the material.

❖ Expanded plastic is a synthetic or natural organic material that, under high temperatures or pressures, can be shaped, formed and molded into any shape and maintain that shape at ambient temperatures. Plastics are combustible materials, some of which will give off toxic gasses when exposed to fire. The most commonly recognized expanded plastic product is a disposable coffee cup. When plastic is expanded, it becomes much more susceptible to combustion due the larger surface area of the material per unit weight.

EXTRA-HIGH-RACK COMBUSTIBLE STORAGE. Storage on racks of Class I, II, III or IV commodities which exceed 40 feet (12 192 mm) in height and storage on racks of high-hazard commodities which exceed 30 feet (9144 mm) in height.

❖ This term was generated to address those storage areas that were of unusual height, and where that height would have an influence on the building or structure, the contents and the safeguards from the hazards of a fire or explosion. Extra-high-rack combustible storage is common to building occupancies that require a substantial amount of storage of commodities (supply) necessary to ensure that the process remains operational. An example of this type of storage is an automotive assembly plant where the rack storage contains a sufficient amount of material to service the assembly plant operation for several days or weeks.

HIGH-PILED COMBUSTIBLE STORAGE. Storage of combustible materials in closely packed piles or combustible materials on pallets, in racks or on shelves where the top of storage is greater than 12 feet (3658 mm) in height. When required by the fire code official, high-piled combustible storage also includes certain high-hazard commodities, such as rubber tires, Group A plastics, flammable liquids, idle pallets and similar commodities, where the top of storage is greater than 6 feet (1829 mm) in height.

❖ High-piled combustible storage has two distinct features not common to other storage areas: the large quantity of commodities (or products) and storage in a compact arrangement (density). The height values used to distinguish high-piled storage from general or incidental storage were largely based on fire tests [12 feet (3638 mm) for Class I-IV and 6 feet (1829 mm) high hazard]. These tests were conducted to determine the effects of various configurations, quantities and classes of commodities as well as various fire protection features.

HIGH-PILED STORAGE AREA. An area within a building which is designated, intended, proposed or actually used for high-piled combustible storage.

❖ This term defines the area or space where the combustible commodity is actually located. The intent of defining this area is to differentiate such a space from any of the other more traditionally defined occupancies (for example, business, factory, etc.) and the amount of fire protection that is required in these unique spaces.

LONGITUDINAL FLUE SPACE. The flue space between rows of storage perpendicular to the direction of loading.

❖ The longitudinal flue space is a continuous open area between a double row or multiple-row-type rack storage system. The flue space is to be clear for a set dimension, as required in Table 2308.3, from the floor to the top of the highest commodity for the entire length of the rack system. The flue spaces are an important feature for automatic sprinkler systems to effectively suppress and potentially control the fire. Reducing or eliminating such spaces will potentially reduce, if not eliminate, sprinkler effectiveness.

MANUAL STOCKING METHODS. Stocking methods utilizing ladders or other nonmechanical equipment to move stock.

❖ Manual stocking methods do not require the use of mechanical means to stock and retrieve commodities. Commodities may be stocked or retrieved on foot where the commodity is accessible from a floor or landing, or through the use of a portable ladder or portable stair.

MECHANICAL STOCKING METHODS. Stocking methods utilizing motorized vehicles or hydraulic jacks to move stock.

❖ Mechanical stocking methods are generally associated with rack storage. The mass of the commodity, the height of the rack storage or the shear quantity of commodity on any one pallet will normally put stocking or retrieval beyond human reach or strength. Methods commonly employed for this type of operation are manned forklifts or unmanned mechanized storage and retrieval systems.

SHELF STORAGE. Storage on shelves less than 30 inches (762 mm) deep with the distance between shelves not exceeding 3 feet (914 mm) vertically. For other shelving arrangements, see the requirements for rack storage.

❖ To be considered shelf storage, the shelving must be no deeper than 30 inches (762 mm) and must be separated by no more than 3 feet (914 mm) vertically. Further, shelf storage units must be separated horizontally by aisles not less than 30 inches (762 mm) wide to reduce the transfer of fire from one shelf unit across the aisle to another. The aisles also allow convenient access for fire department personnel to combat a fire and for salvage and debris removal after an incident.

SOLID SHELVING. Shelving that is solid, slatted or of other construction located in racks and which obstructs sprinkler discharge down into the racks.

❖ Solid shelving generally consists of nominal wood (lumber), plywood, particleboard or metal shelves that span between the supports of the storage system to support the commodities. Solid shelving creates a condition where the rack storage system is effectively divided into areas by the shelving. The shelving can potentially act as a protective barrier for a fire and from fire service hose streams by preventing the penetration of water into the fire area. As a result, the sprinkler system design will require an in-rack sprinkler system in addition to the ceiling sprinkler system for water distribution to penetrate the shelving barriers.

TRANSVERSE FLUE SPACE. The space between rows of storage parallel to the direction of loading.

❖ The transverse flue space is a continuous open area between commodities in single-row, double-row and multiple-row-type rack storage systems. The flue space is to be clear for a set dimension, as required in Table 2308.3, from the floor to the top of the highest commodity for the entire width of the rack system. The flue spaces are an important feature for automatic sprinkler systems to effectively suppress and potentially control the fire. Reducing or eliminating such spaces will potentially reduce, if not eliminate, sprinkler effectiveness.

SECTION 2303 COMMODITY CLASSIFICATION

2303.1 Classification of commodities. Commodities shall be classified as Class I, II, III, IV or high hazard in accordance with this section. Materials listed within each commodity classification are assumed to be unmodified for improved combustibility characteristics. Use of flame-retarding modifiers or the physical form of the material could change the classification. See Section 2303.7 for classification of Group A, B and C plastics.

❖ The classification of commodities contains information that addresses the materials that make up the commodity and its packaging. As an example, lubricating fluid in metal cans is designated as Class III, whereas lubricating fluid in plastic containers is designated as Class IV. Although this section addresses only the classification, such factors as the location of the commodity, location relative to other commodities of differing classes, flue spaces, fire protection, ventilation, access and egress will affect the storage layout. One of the most important considerations is the packaging of the commodity and the type of support the commodity is resting upon (i.e., pallets).

For rack storage, the distinction between the classes is made with respect to the arrangement of the products; for example, products encapsulated with plastic wrap and those products not encapsulated.

The plastic wrap acts effectively as a container for the entire arrangement of products, and because it encapsulates five of the six sides of the arrangement, it can prevent sprinklers from having an effect on the products if they do ignite and makes it difficult for prewetting of the surrounding commodities. The encapsulated category can be considered nonencapsulated by removing, at a minimum, the plastic wrap from the top of the arranged products, thus allowing some penetration for the sprinklers.

CLASSIFICATION LEVELS	
	Low Fire Hazard
Class I	Essentially noncombustible products on noncombustible pallets
	High Fire Hazard
Class II	Essentially Class I products on wood pallets or crates
Class III	Essentially wood, paper and fiber products and Group C plastics (limited Class A and B plastics)
Class IV	Essentially Class I, II or III containing significant Group A plastics (unlimited Class B Plastics and free-flowing Group A plastics)
	High Hazard
—	Essentially present fire hazard beyond Class I, II, III or IV

2303.2 Class I commodities. Class I commodities are essentially noncombustible products on wooden or nonexpanded polyethylene solid deck pallets, in ordinary corrugated cartons with or without single-thickness dividers, or in ordinary paper wrappings with or without pallets. Class I commodities are allowed to contain a limited amount of Group A plastics in accordance with Section 2303.7.4. Examples of Class I commodities include, but are not limited to, the following:

- Alcoholic beverages not exceeding 20-percent alcohol
- Appliances noncombustible, electrical
- Cement in bags
- Ceramics
- Dairy products in nonwax-coated containers (excluding bottles)
- Dry insecticides
- Foods in noncombustible containers
- Fresh fruits and vegetables in nonplastic trays or containers
- Frozen foods
- Glass
- Glycol in metal cans
- Gypsum board
- Inert materials, bagged
- Insulation, noncombustible
- Noncombustible liquids in plastic containers having less than a 5-gallon (19 L) capacity
- Noncombustible metal products

❖ Class I commodities are basically products that are noncombustible that are packaged in corrugated cardboard cartons. The products may be arranged on combustible pallets. Examples of products that create this hazard are listed in this section.

2303.3 Class II commodities. Class II commodities are Class I products in slatted wooden crates, solid wooden boxes, multiple-thickness paperboard cartons or equivalent combustible packaging material with or without pallets. Class II commodities are allowed to contain a limited amount of Group A plastics in accordance with Section 2303.7.4. Examples of Class II commodities include, but are not limited to, the following:

- Alcoholic beverages not exceeding 20-percent alcohol, in combustible containers
- Foods in combustible containers
- Incandescent or fluorescent light bulbs in cartons
- Thinly coated fine wire on reels or in cartons

❖ Class II commodities consist of noncombustible products that are packaged in slatted wooden crates, solid wooden boxes, multiple-thickness paperboard cartons or equivalent combustible packaging materials that may or may not be arranged on pallets. If the commodity is stored on a rack storage system it will generally be placed on wooden pallets. Examples of products that create this hazard are listed in this section.

2303.4 Class III commodities. Class III commodities are commodities of wood, paper, natural fiber cloth, or Group C plastics or products thereof, with or without pallets. Products are allowed to contain limited amounts of Group A or B plastics, such as metal bicycles with plastic handles, pedals, seats and tires. Group A plastics shall be limited in accordance with Section 2303.7.4. Examples of Class III commodities include, but are not limited to, the following:

- Aerosol, Level 1 (see Chapter 28)
- Combustible fiberboard
- Cork, baled
- Feed, bagged
- Fertilizers, bagged
- Food in plastic containers
- Furniture: wood, natural fiber, upholstered, nonplastic, wood or metal with plastic-padded and covered arm rests
- Glycol in combustible containers not exceeding 25 percent
- Lubricating or hydraulic fluid in metal cans
- Lumber
- Mattresses, excluding foam rubber and foam plastics
- Noncombustible liquids in plastic containers having a capacity of more than 5 gallons (19 L)
- Paints, oil base, in metal cans
- Paper, waste, baled
- Paper and pulp, horizontal storage, or vertical storage that is banded or protected with approved wrap
- Paper in cardboard boxes
- Pillows, excluding foam rubber and foam plastics
- Plastic-coated paper food containers
- Plywood
- Rags, baled
- Rugs, without foam backing
- Sugar, bagged
- Wood, baled
- Wood doors, frames and cabinets
- Yarns of natural fiber and viscose

❖ Class III commodities are products of wood, paper, natural fiber or Group C plastics that may be arranged with or without pallets. The limitations concerning Group A are determined by Section 2303.7.4 and Figure 2303.7.4. Limited Group B plastics are allowed. Examples of products that create this hazard are listed in this section.

2303.5 Class IV commodities. Class IV commodities are Class I, II or III products containing Group A plastics in ordinary corrugated cartons and Class I, II and III products, with Group A plastic packaging, with or without pallets. Group B plastics and free-flowing Group A plastics are also included in this class. The total amount of nonfree-flowing Group A plastics shall be in accordance with Section 2303.7.4. Examples of Class IV commodities include, but are not limited to, the following:

- Aerosol, Level 2 (see Chapter 28)
- Alcoholic beverages, exceeding 20-percent but less than 80-percent alcohol, in cans or bottles in cartons.
- Clothing, synthetic or nonviscose
- Combustible metal products (solid)
- Furniture, plastic upholstered
- Furniture, wood or metal with plastic covering and padding
- Glycol in combustible containers (greater than 25 percent and less than 50 percent)
- Linoleum products
- Paints, oil base in combustible containers
- Pharmaceutical, alcoholic elixirs, tonics, etc.
- Rugs, foam back
- Shingles, asphalt
- Thread or yarn, synthetic or nonviscose

❖ Class IV commodities represent Class I, II or III products that contain an appreciable amount of Group A plastics. These plastics can be a part of the actual product or may be part of the packaging. Packing is in ordinary corrugated cardboard cartons, and arrangement may or may not be on pallets. This class also allows unlimited Group B plastics and free-flowing Group A plastics. Examples of products that create this hazard are listed in this section.

2303.6 High-hazard commodities. High-hazard commodities are high-hazard products presenting special fire hazards beyond those of Class I, II, III or IV. Group A plastics not otherwise classified are included in this class. Examples of high-hazard commodities include, but are not limited to, the following:

- Aerosol, Level 3 (see Chapter 28)
- Alcoholic beverages, exceeding 80-percent alcohol, in bottles or cartons
- Commodities of any class in plastic containers in carousel storage
- Flammable solids (except solid combustible metals)
- Glycol in combustible containers (50 percent or greater)

- Lacquers, which dry by solvent evaporation, in metal cans or cartons
- Lubricating or hydraulic fluid in plastic containers
- Mattresses, foam rubber or foam plastics
- Pallets and flats which are idle combustible
- Paper, asphalt, rolled, horizontal storage
- Paper, asphalt, rolled, vertical storage
- Paper and pulp, rolled, in vertical storage which is unbanded or not protected with an approved wrap
- Pillows, foam rubber and foam plastics
- Pyroxylin
- Rubber tires
- Vegetable oil and butter in plastic containers

❖ High-hazard commodities present a hazard inasmuch as they contain materials that readily support combustion or materials that can support a physical or health hazard. Those products that represent a physical hazard are those presenting a detonation hazard, deflagration hazard or that readily support combustion. The products representing a health hazard are those that present a risk to people from handling or exposure to the product. Examples of products that create this kind of hazard are listed in this section.

2303.7 Classification of plastics. Plastics shall be designated as Group A, B or C in accordance with this section.

❖ The classification of plastics is contained in Sections 2303.7.1 through 2303.7.4. This section recognizes that plastics tend to have high combustion and burn rates. For example, a Class IV commodity could be evaluated as a high hazard because of the quantity of Group A plastic cushioning within the container used to sustain or hold the product.

2303.7.1 Group A plastics. Group A plastics are plastic materials having a heat of combustion that is much higher than that of ordinary combustibles, and a burning rate higher than that of Group B plastics. Examples of Group A plastics include, but are not limited to, the following:

- ABS (acrylonitrile-butadiene-styrene copolymer)
- Acetal (polyformaldehyde)
- Acrylic (polymethyl methacrylate)
- Butyl rubber
- EPDM (ethylene propylene rubber)
- FRP (fiberglass-reinforced polyester)
- Natural rubber (expanded)
- Nitrile rubber (acrylonitrile butadiene rubber)
- PET or PETE (polyethylene terephthalate)
- Polybutadiene
- Polycarbonate
- Polyester elastomer
- Polyethylene
- Polypropylene
- Polystyrene (expanded and unexpanded)
- Polyurethane (expanded and unexpanded)
- PVC (polyvinyl chloride greater than 15 percent plasticized, e.g., coated fabric unsupported film)
- SAN (styrene acrylonitrile)
- SBR (styrene butadiene rubber)

❖ Group A plastics are generally considered a high-hazard commodity because they have the fastest burn rate of the materials in the classes and groups. Examples of products that create this hazard are listed in this section.

2303.7.2 Group B plastics. Group B plastics are plastic materials having a heat of combustion and a burning rate higher than that of ordinary combustibles, but not as high as those of Group A plastics. Examples of Group B plastics include, but are not limited to, the following:

- Cellulosics (cellulose acetate, cellulose acetate butyrate, ethyl cellulose)
- Chloroprene rubber
- Fluoroplastics (ECTFE, ethylene-chlorotrifluoroethylene copolymer; ETFE, ethylene-tetrafluoroethylene copolymer; FEP, fluorinated ethylene-propylene copolymer)
- Natural rubber (nonexpanded)
- Nylon (Nylon 6, Nylon 6/6)
- PVC (polyvinyl chloride greater than 5-percent, but not exceeding 15-percent plasticized)
- Silicone rubber

❖ Group B plastics represent the next level of plastics that have a heat of combustion and burning rate higher than that of ordinary combustibles, but less than Group A plastics. Examples of products that create this hazard are listed in this section.

2303.7.3 Group C plastics. Group C plastics are plastic materials having a heat of combustion and a burning rate similar to those of ordinary combustibles. Examples of Group C plastics include, but are not limited to, the following:

- Fluoroplastics (PCTFE, polychlorotrifluoroethylene; PTFE, polytetrafluoroethylene)
- Melamine (melamine formaldehyde)
- Phenol
- PVC (polyvinyl chloride, rigid or plasticized less than 5 percent, e.g., pipe, pipe fittings)
- PVDC (polyvinylidene chloride)
- PVDF (polyvinylidene fluoride)
- PVF (polyvinyl fluoride)
- Urea (urea formaldehyde)

❖ Group C plastics are those products that can be similar to ordinary combustibles because of their heat of combustion and burning rate, which will be less than the Group A and B plastics. Examples of products that create this hazard are listed in this section.

2303.7.4 Limited quantities of Group A plastics in mixed commodities. Figure 2303.7.4 shall be used to determine the quantity of Group A plastics allowed to be stored in a package or carton or on a pallet without increasing the commodity classification.

❖ Limited quantities of Group A plastics can be allowed within other classes of commodities. Figure 2303.7.4 shows the pertinent requirements, or limits, for determining which class the limited amount of Group A plastic is assigned.

FIGURE 2303.7.4. See page 23-9.

❖ The figure represents an empirical judgement of the risks associated with assigning a limited amount of Group A plastics to another class of commodity. The figure is intended to help determine the commodity classification of a mixed commodity in a package or carton or on a pallet when plastics are involved. Plastics can be involved (mixed) with products through the use of packaging or can be contained in or integrated into the actual product.

SECTION 2304 DESIGNATION OF HIGH-PILED STORAGE AREAS

2304.1 General. High-piled storage areas, and portions of high-piled storage areas intended for storage of a different commodity class than adjacent areas, shall be designed and specifically designated to contain Class I, Class II, Class III, Class IV or high-hazard commodities. The designation of a high-piled combustible storage area, or portion thereof intended for storage of a different commodity class, shall be based on the highest hazard commodity class stored except as provided in Section 2304.2.

❖ This section establishes the designation requirements for the different commodity classes and the designation of the area within the building for use in determining the general fire protection and life safety features.

2304.2 Designation based on engineering analysis. The designation of a high-piled combustible storage area, or portion thereof, is allowed to be based on a lower hazard class than that of the highest class of commodity stored when a limited quantity of the higher hazard commodity has been demonstrated by engineering analysis to be adequately protected by the automatic sprinkler system provided. The engineering analysis shall consider the ability of the sprinkler system to deliver the higher density required by the higher hazard commodity. The higher density shall be based on the actual storage height of the pile or rack and the minimum allowable design area for sprinkler operation as set forth in the density/area figures provided in NFPA 231 and NFPA 231C. The contiguous area occupied by the higher hazard commodity shall not exceed 120 square feet (11 m^2), and additional areas of higher hazard commodity shall be separated from other such areas by 25 feet (7620 mm) or more. The sprinkler system shall be capable of delivering the higher density over a minimum area of 900 square feet (84 m^2) for wet pipe systems and 1,200 square feet (11 m^2) for dry pipe systems. The shape of the design area shall be in accordance with Section 903.

❖ This section allows the designer to make use of the benefits of hydraulic design. In other words, in an area rated for a low-hazard commodity it is likely that in some portions of that space the available water and pressure will be sufficient for a higher-hazard commodity. This is true since sprinklers are designed for the most hydraulically remote areas; therefore, areas closer to the riser will receive more water and higher pressure. There are certain criteria that accompany this allowance that limit the size of the higher-hazard storage areas to 120 square feet (11 m^2) each and require that they be separated from other similar areas by 25 feet (7620 mm). It also provides minimum areas in which this increased water supply needs to be available. It should be noted that there are some potential maintenance limitations. It may be difficult to keep track of these particular designated areas and the separation between the areas; therefore, it is suggested that the areas be marked.

SECTION 2305 HOUSEKEEPING AND MAINTENANCE

2305.1 Rack structures. The structural integrity of racks shall be maintained.

❖ This section's emphasis on "structural integrity" has its roots in the IBC, specifically Chapters 16 and 22. These two chapters govern the structural design of rack storage. The integrity of rack storage includes such items as the primary structural components, the connections, bracing members and relationship with the superstructure of the building. Racks can be a very significant part of the structure and pose structural failure concerns, especially for the fire suppression forces that may be operating in close proximity to the racks.

2305.2 Ignition sources. Clearance from ignition sources shall be provided in accordance with Section 305.

❖ Section 2305.2 directs the reader to Section 305, which addresses housekeeping provisions for such things as open flames, heaters, flame-producing devices, light fixtures and materials subject to spontaneous combustion. The responsibility for the care and maintenance of the building rests with the building owner or the tenant.

2305.3 Smoking. Smoking shall be prohibited. Approved "No Smoking" signs shall be conspicuously posted in accordance with Section 310.

❖ This section prohibits smoking within combustible storage areas to limit the chance for a fire to ignite. Section 310 outlines the limitations for smoking areas, the installation of signage and the proper methods for discarding burning, smoking paraphernalia. Included with the provisions is a requirement that the fire code official determine when a hazardous environment or condition exists, and that the building official make certain that fire prevention measures are in place to reduce the risk. The responsibility for the designation of permissible smoking areas rests with the building owner, tenant or lessee.

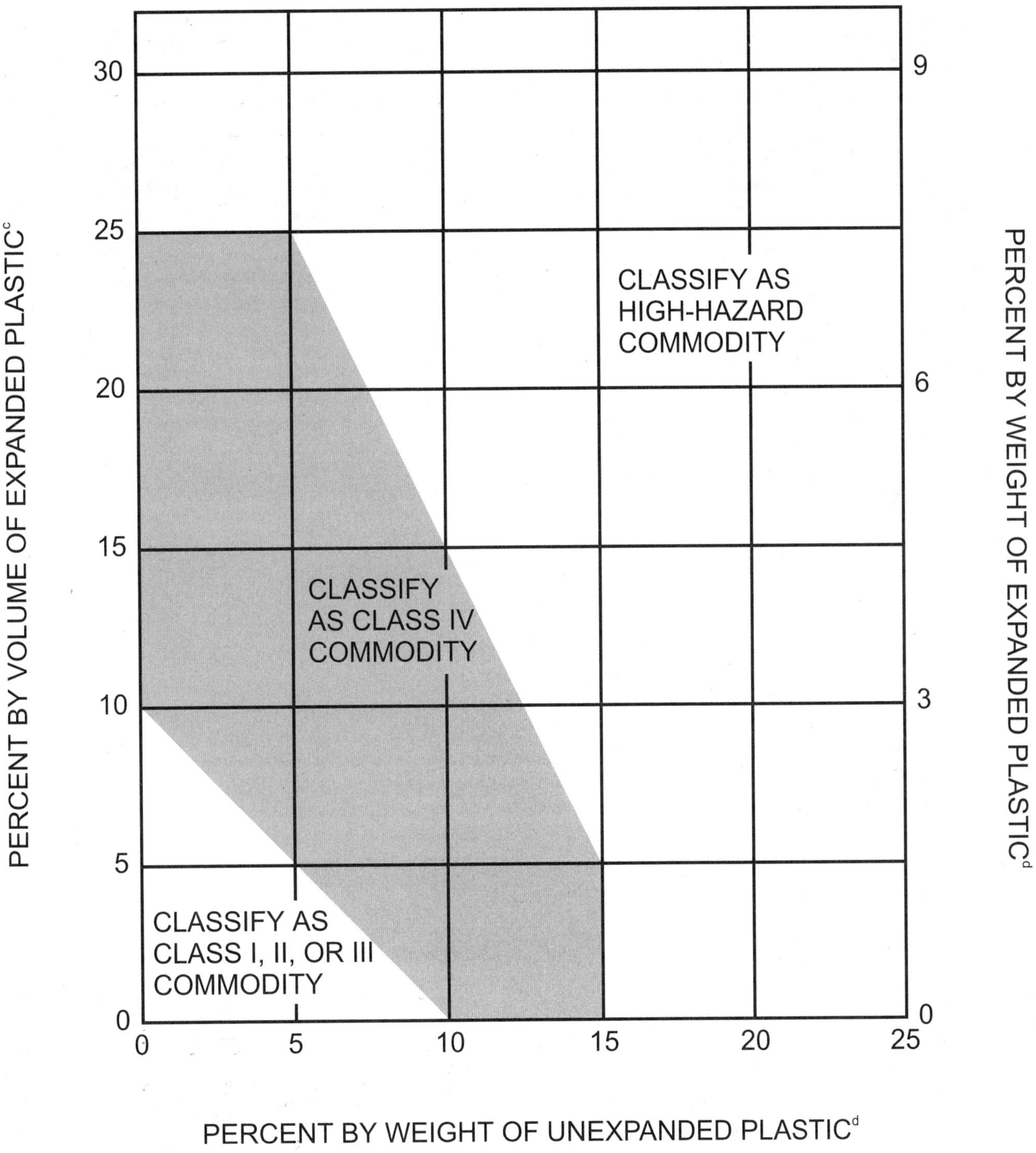

FIGURE 2303.7.4
MIXED COMMODITIES[a, b]

a. This figure is intended to determine the commodity classification of a mixed commodity in a package, carton or on a pallet where plastics are involved.

b. The following is an example of how to apply the figure: A package containing a Class III commodity has 12-percent Group A expanded plastic by volume. The weight of the unexpanded Group A plastic is 10 percent. This commodity is classified as a Class IV commodity. If the weight of the unexpanded plastic is increased to 14 percent, the classification changes to a high-hazard commodity.

c. $\text{Percent by volume} = \dfrac{\text{Volume of plastic in pallet load}}{\text{Total volume of pallet load, including pallet}}$

d. $\text{Percent by weight} = \dfrac{\text{Weight of plastic in pallet load}}{\text{Total weight of pallet load, including pallet}}$

2305.4 Aisle maintenance. When restocking is not being conducted, aisles shall be kept clear of storage, waste material and debris. Fire department access doors, aisles and exit doors shall not be obstructed. During restocking operations using manual stocking methods, a minimum unobstructed aisle width of 24 inches (610 mm) shall be maintained in 48-inch (1219 mm) or smaller aisles, and a minimum unobstructed aisle width of one-half of the required aisle width shall be maintained in aisles greater than 48 inches (1219 mm). During mechanical stocking operations, a minimum unobstructed aisle width of 44 inches (1118 mm) shall be maintained in accordance with Section 2306.9.

❖ This section emphasizes the importance of keeping aisles and access corridors open at all times for both employee egress and fire-fighter access. Also, such aisles serve as fire breaks. Miscellaneous storage, waste and other objects can be the most obtrusive objects anytime during building operation, which is why it is critical that the building owner or the tenant have a rigorous maintenance program. This section is applicable to buildings that are accessible by the public as well as those that are not.

Restocking operations have the potential for being another impediment. Establishing minimum aisle widths coupled with strict enforcement increases the likelihood that clear aisle space will be available.

2305.5 Pile dimension and height limitations. Pile dimensions and height limitations shall comply with Section 2307.3.

❖ Pile dimensions and height limitations are directly regulated by the commodity class and the size of the storage area in accordance with Section 2307.3 and Table 2306.2.

2305.6 Arrays. Arrays shall comply with Section 2307.4.

❖ This section directs the reader to Section 2307.4, which in turn is governed by referenced standard NFPA 231. More specifically, it is requiring that any necessary spacings be provided.

2305.7 Flue spaces. Flue spaces shall comply with Section 2308.3.

❖ This section directs the reader to Section 2308.3, which in turn is governed by Table 2308.3 and provides criteria for flue spaces.

SECTION 2306
GENERAL FIRE PROTECTION AND LIFE SAFETY FEATURES

2306.1 General. Fire protection and life safety features for high-piled storage areas shall be in accordance with this section.

❖ This section establishes the minimum requirements for fire protection and life safety features. These provisions include separation requirements; automatic sprinklers; fire detection; access and egress; smoke and heat removal; hose connections; minimum dimensions for aisles; clear heights; dead ends and fire extinguishers. In addition to the fire protection and life safety requirements, Table 2306.2 contains provisions concerning commodity class and size of storage area requirements that define which fire protection and life safety requirements are to be used in the building design.

Although this section addresses minimum fire protection and life safety features, the provisions that require proper maintenance of those features are included in Section 2305 and Chapter 3. The responsibility for the care and maintenance of the fire protection and life safety features rests with the building owner, tenant or lessee. The responsibility for notifying the fire service (local fire department) when fire protection systems are inoperable is also assigned to the building owner, tenant or lessee. This notification is necessary so that another means of protection is provided and that the fire department response and fire-ground management operations are based on system condition.

2306.2 Extent and type of protection. Where required by Table 2306.2, fire detection systems, smoke and heat removal, draft curtains and automatic sprinkler design densities shall extend the lesser of 15 feet (4572 mm) beyond the high-piled storage area or to a permanent partition. Where portions of high-piled storage areas have different fire protection requirements because of commodity, method of storage or storage height, the fire protection features required by Table 2306.2 within this area shall be based on the most restrictive design requirements.

❖ The fire protection features noted in this section include a requirement for the extension of those features 15 feet (4572 mm) beyond the storage area if no partitions are present. For the fire protection systems to be effective, they must be designed with the knowledge that a fire within a storage area (high piled or rack) has the potential of jumping the aisle into another area of the same building. Note also that where mixed classes are stored in an area, the fire detection and protection system must be designed for the most hazardous commodity except as allowed in Section 2304.2. The size of the high-piled storage area should include the aisles. The 15-foot (4572 mm) extension of protection should include the aisle width when determining how far to extend the protection.

TABLE 2306.2. See page 23-12.

❖ In addition to the provisions of Sections 2306 through 2310, this table defines, through a prescriptive approach, the fire protection requirements for high-piled storage. The table bases the requirements on the commodity class and size of the high-piled storage area to indicate the necessary automatic fire-extinguishing system, fire detection, building access, smoke and heat removal and curtain boards. In addition, the type of commodity and size of high-piled storage area govern the limitations for the size, height and volume of solid piled storage, shelf storage and palletized storage.

Example: A storage area contains 200,000 square feet (18 580 m^2) of high-piled storage classified as commodity Class I. It is a single-story building with floor space only, no rack storage and is protected by standard automatic sprinklers. The subject building would be required to comply with the following:

- Automatic fire-extinguishing system required throughout (Section 2306.4).
- Fire detection system not required.
- Building access required (Section 2306.6).
- Smoke and heat removal required (Section 2306.7).
- Draft curtains not required.
- Maximum pile dimension of 100 feet (30 480 mm) (Section 2306.9).
- Maximum permissible storage height of 40 feet (12 192 mm).
- Maximum pile volume of 400,000 cubic feet (11 328 m^3).

2306.3 Separation of high-piled storage areas. High-piled storage areas shall be separated from other portions of the building where required by Sections 2306.3.1 through 2306.3.2.2.

❖ This section sets forth the requirements for separating the storage area from other uses within the same building and for multiple classifications and heights of storage.

2306.3.1 Separation from other uses. Mixed occupancies shall be separated in accordance with the *International Building Code.*

❖ Section 302.3 of the IBC describes the provisions governing the condition when a building contains more than one of the occupancy groups identified in Sections 301 through 312 of the IBC (see also Section 202). Buildings very often will contain more than one occupancy group classification; for example, an office (Group B business) connected to a storage area (Group S-1, low-hazard storage). Once a building is determined to house more than one occupancy group, the codes require that a combination of potential fire and life safety hazards be addressed. Section 302.3 of the IBC provides for two conditions under which certain dissimilar occupancies are treated: nonseparated uses and separated uses.

A word of caution is needed regarding the design of a mixed occupancy building. When the mixed occupancy building is separated by a fire wall it technically becomes two buildings (see the definition of "Building area" in Section 502.1 of the IBC). As a result, the provisions of Sections 503 and 507 of the IBC will need to be reviewed for applicability and the evaluation and design adjusted for that difference.

2306.3.2 Multiple high-piled storage areas. Multiple high-piled storage areas shall be in accordance with Section 2306.3.2.1 or 2306.3.2.2.

❖ This section sets forth the requirements for determining the fire and life safety requirements for a single building or area containing multiple classes of commodities.

2306.3.2.1 Aggregate area. The aggregate of all high-piled storage areas within a building shall be used for application of Table 2306.2 unless such areas are separated from each other by 1-hour fire-resistance-rated fire barrier walls constructed in accordance with the *International Building Code.* Openings in such walls shall be protected by opening protective assemblies having a 1-hour fire protection rating.

❖ The intent of this section is to recognize that, in actual storage circumstances, storage arrays do not always correspond to any one commodity classification. It is not unusual for arrays to consist of thousands of products. As a result, this section recognizes that, in storage areas, multiples of commodity classifications may be contained in one area and must be treated as an aggregate when applying Table 2306.2.

As an alternative, this provision allows for each class of commodity to be separated within the same building by 1-hour fire-resistance-rated fire barriers constructed in accordance with Section 706 of the IBC, which applies to fire barriers used for separating fire areas. To maintain integrity of the separation of fire areas, fire barriers must be continuous from the top of the floor below to the underside of the roof slab or deck above and be securely attached thereto.

As with any fire-resistance-rated assembly, consideration must be given to openings and penetrations. The number and size of openings in the fire barrier must comply with the provisions of Section 706 of the IBC, and the opening protectives must have a minimum fire protection rating of 1 hour. Note that Section 715.3 of the IBC allows an opening protective fire protection rating of $^3/_4$-hour for nonexit-enclosure fire barriers; however, this section, with its specific requirement of a 1-hour opening protective assembly rating, would supercede that requirement. Penetrations of fire barriers must comply with Section 712 of the IBC. Maintenance of the fire-resistance-rated construction rests with the building owner or the tenant as described in Section 703 of the code.

The main purpose for this section is to determine what area is required to be used when addressing Table 2306.2. Section 2306.3.2.2 is focused upon the need to separate different hazard levels of storage in order to increase or reduce the fire protection requirements (see also commentary, Section 2306.3.2.2).

TABLE 2306.2
GENERAL FIRE PROTECTION AND LIFE SAFETY REQUIREMENTS

COMMODITY CLASS	SIZE OF HIGH-PILED STORAGE AREA[a] (square feet) (see Sections 2306.2 and 2306.4)	ALL STORAGE AREAS (See Sections 2306, 2307 and 2308)[b]					SOLID-PILED STORAGE, SHELF STORAGE AND PALLETIZED STORAGE (see Section 2307.3)		
		Automatic fire-extinguishing system (see Section 2306.4)	Fire detection system (see Section 2306.5)	Building access (see Section 2306.6)	Smoke and heat removal (see Section 2306.7)	Draft curtains (see Section 2306.7)	Maximum pile dimension[c] (feet)	Maximum permissible storage height[d] (feet)	Maximum pile volume (cubic feet)
I-IV	0-500	Not Required[a]	Not Required	Not Required[e]	Not Required	Not Required	Not Required	Not Required	Not Required
I-IV	501-2,500	Not Required[a]	Yes[i]	Not Required[e]	Not Required	Not Required	100	40	100,000
I-IV	2,501-12,000 Public accessible	Yes	Not Required	Not Required[e]	Not Required	Not Required	100	40	400,000
I-IV	2,501-12,000 Nonpublic accessible (Option 1)	Yes	Not Required	Not Required[e]	Not Required	Not Required	100	40	400,000
I-IV	2,501-12,000 Nonpublic accessible (Option 2)	Not Required[a]	Yes	Yes	Yes[j]	Yes[j]	100	30[f]	200,000
I-IV	12,001-20,000	Yes	Not Required	Yes	Yes[j]	Not Required	100	40	400,000
I-IV	20,001-500,000	Yes	Not Required	Yes	Yes[j]	Not Required	100	40	400,000
I-IV	Greater than 500,000[g]	Yes	Not Required	Yes	Yes[j]	Not Required	100	40	400,000
High hazard	0-500	Not Required[a]	Not Required	Not Required[e]	Not Required	Not Required	50	Not Required	Not Required
High hazard	501-2,500 Public accessible	Yes	Not Required	Not Required[e]	Not Required	Not Required	50	30	75,000
High hazard	501-2,500 Nonpublic accessible (Option 1)	Yes	Not Required	Not Required[e]	Not Required	Not Required	50	30	75,000
High hazard	501-2,500 Nonpublic accessible (Option 2)	Not Required[a]	Yes	Yes	Yes[j]	Yes[j]	50	20	50,000
High hazard	2,501-300,000	Yes	Not Required	Yes	Yes[j]	Not Required	50	30	75,000
High hazard	300,001-500,000[g, h]	Yes	Not Required	Yes	Yes[j]	Not Required	50	30	75,000

For SI: 1 foot = 304.8 mm, 1 cubic foot = 0.02832 m^3, 1 square foot = 0.0929 m^2.

a. When automatic sprinklers are required for reasons other than those in Chapter 23, the portion of the sprinkler system protecting the high-piled storage area shall be designed and installed in accordance with Sections 2307 and 2308.
b. For aisles, see Section 2306.9.
c. Piles shall be separated by aisles complying with Section 2306.9.
d. For storage in excess of the height indicated, special fire protection shall be provided in accordance with Note g when required by the fire code official. See also Chapters 28 and 34 for special limitations for aerosols and flammable and combustible liquids.
e. Section 503 shall apply for fire apparatus access.
f. For storage exceeding 30 feet in height, Option 1 shall be used.
g. Special fire protection provisions including, but not limited to, fire protection of exposed steel columns; increased sprinkler density; additional in-rack sprinklers, without associated reductions in ceiling sprinkler density; or additional fire department hose connections shall be provided when required by the fire code official.
h. High-piled storage areas shall not exceed 500,000 square feet. A 2-hour fire wall constructed in accordance with the *International Building Code* shall be used to divide high-piled storage exceeding 500,000 square feet in area.
i. Not required when an automatic fire-extinguishing system is designed and installed to protect the high-piled storage area in accordance with Sections 2307 and 2308.
j. Not required when storage areas are protected by early suppression fast-response (ESFR) sprinkler systems installed in accordance with NFPA 13.

2306.3.2.2 Multiclass high-piled storage areas. High-piled storage areas classified as Class I through Class IV not separated from high-piled storage areas classified as high hazard shall utilize the aggregate of all high-piled storage areas as high hazard for purposes of application of Table 2306.2. To be considered as separated, 1-hour fire-resistance-rated fire barrier walls shall be constructed in accordance with the *International Building Code*. Openings in such walls shall be protected by opening protective assemblies having a 1-hour fire protection rating.

Exception: As provided for in Section 2304.2.

❖ The intent of this section is to recognize that in actual circumstances storage arrays may contain a combination of Class I through IV commodities alongside high-hazard commodities.

It is not unusual for arrays to consist of thousands of products. As in Section 2306.3.2.1, this section recognizes that in storage areas multiples of commodity classifications may be contained in one area and must be treated as an aggregate under the highest hazard commodity class category found in that area when applying Table 2306.2.

As an alternative, this provision allows for each class of commodity to be separated within the same building by 1-hour fire-resistance-rated fire barriers constructed in accordance with Section 706 of the IBC, which applies to fire barriers used for separating fire areas. To maintain integrity of the separation of fire areas, fire barriers must be continuous from the top of the floor below to the underside of the roof slab or deck above and be securely attached thereto. As with any fire-resistive barriers, consideration must be given to openings and penetrations. The number and size of openings in the fire barrier must comply with the provisions of Section 706 of the IBC, and the opening protective assemblies must have a minimum fire-resistance rating of 1 hour. Note that Section 715.3 of the IBC allows an opening protection rating of $^3/_4$-hour for nonexit-enclosure fire barriers; however, this section, with its specific requirement of a 1-hour opening protective assembly rating, would supercede that requirement. Penetrations of fire barriers must comply with Section 712 of the IBC. Maintenance of the fire-resistance-rated construction rests with the building owner or the tenant as described in Section 703 of the code.

The use of separation allows the application of less restrictive requirements in some areas. The exception permits the use of an accepted engineering analysis in accordance with Section 2304.2, which allows one to take advantage of the benefits of hydraulic design (see Section 2304.2).

2306.4 Automatic sprinklers. Automatic sprinkler systems shall be provided in accordance with Sections 2307, 2308 and 2309.

❖ This section refers the reader to Sections 2307, 2308 and 2309. Sprinkler systems are discussed in Sections 2307.2, 2308.2 and 2309.2 for the various ways in which high-piled storage is configured.

2306.5 Fire detection. Where fire detection is required by Table 2306.2, an approved automatic fire detection system shall be installed throughout the high-piled storage area. The system shall be monitored and be in accordance with Section 907.

❖ Automatic fire detection systems are designed to increase the likelihood that the fire is detected and the occupants are given an early warning to allow enough time to exit the area or building. The detectors are smoke detectors approved for the particular use. Table 2306.2 shows when a fire detection system will be installed in a storage area, and Section 907 covers the application, installation, performance and maintenance of the fire alarm system and their components. Section 907.9 provides the minimum requirements for zoning of the system as well as requirements that will reduce the frequency of unwanted alarms. Automatic detection systems are to consist of either cross-zoned detectors or detectors provided with alarm verification. Such systems are only required in some cases in Table 2306.2, primarily when the building is not sprinklered.

2306.6 Building access. Where building access is required by Table 2306.2, fire apparatus access roads in accordance with Section 503 shall be provided within 150 feet (45 720 mm) of all portions of the exterior walls of buildings used for high-piled storage.

Exception: Where fire apparatus access roads cannot be installed because of topography, railways, waterways, non-negotiable grades or other similar conditions, the fire code official is authorized to require additional fire protection.

❖ This section contains the requirements for fire department access to a building. Section 503 discusses the minimum requirements for the location of such a roadway, additional access when required, minimum dimensions, minimum surface load tolerances, turning radii, grades, marking and overall access by the fire department. Additional measures are necessary when a road cannot be built on the property surrounding the building, including additional fire sprinkler protection, detection, building fire resistance, etc. This reliance on fire protection systems is necessary to compensate for the limited access available to fire department equipment.

2306.6.1 Access doors. Where building access is required by Table 2306.2, fire department access doors shall be provided in accordance with this section. Access doors shall be accessible without the use of a ladder.

❖ Building access requirements are summarized in Table 2306.2. The access requirements include the application of Sections 2306.6.6.1.1 through 2306.6.1.3. In addition, the provisions of Chapter 10, which address the minimum clear width dimensions, allowable projections into the clear width, floor landings, door hardware, door identification, etc., must be considered. Section 504 states that an approved walkway must be provided from the fire apparatus access road, or from other approved fire apparatus access roadways, to an exterior door opening required for fire department access.

2306.6.1.1 Number of doors required. A minimum of one access door shall be provided in each 100 lineal feet (30 480 mm), or fraction thereof, of the exterior walls which face required fire apparatus access roads.

❖ Door openings in exterior walls are located in each 100 lineal feet (30 480 mm) to provide the fire department access within short runs for quick access to assess, combat and control a fire within the building. These doors are only required around areas of the building housing high-piled storage. Table 2306.2 tends to require such doors for larger sized high-piled storage areas where fire fighting may be more difficult.

2306.6.1.2 Door size and type. Access doors shall not be less than 3 feet (914 mm) in width and 6 feet 8 inches (2032 mm) in height. Roll-up doors shall not be used unless approved.

❖ Door sizes are the same as those in the means of egress door provisions of Chapter 10. The emphasis on the side-hinged type of door is to ensure that other doors, such as roll-up and sliding, are not used. The swing should be outward and be no less than 90 degrees (1.75 rad). Section 1003.5, Exception 1 permits the door sill elevation to be above the outside grade if the door is not a required accessible route in accordance with the IBC. In some regions the raised sill is preferred because of the potential buildup of snow and ice.

2306.6.1.3 Locking devices. Only approved locking devices shall be used.

❖ Locking mechanisms must be approved by the fire code official. The fire service must be able to open the doors from the exterior side during an emergency. The locking mechanism must be designed to maintain the security of the building, to be readily openable from the egress side and to be openable by fire department personnel from the exterior.

2306.7 Smoke and heat removal. Where smoke and heat removal are required by Table 2306.2, smoke and heat vents shall be provided in accordance with Section 910. Where draft curtains are required by Table 2306.2, they shall be provided in accordance with Section 910.3.4.

❖ Table 2306.2 identifies the requirement for smoke and heat removal and draft curtains in terms of high-piled storage area and whether the storage is accessible by the public. If required, the smoke and heat vents and draft curtains are to be designed and installed in accordance with Section 910, which provides the criteria that are to be used in determining the vent-to-area ratio and spacing of vents, as well as the construction dimensions and spacing of draft curtains. If sprinklers are used, draft curtains are not required.

Caution should be exercised when designing smoke and heat removal systems and locating draft curtains to make certain the system and draft curtains do not interfere with required automatic sprinkler systems. Locating a draft curtain too close to a sprinkler head could prevent the head from properly discharging. In addition, vent locations must take into consideration the temperature requirements of both the fusible link of the vent and the temperature rating range of the sprinkler heads. A vent opening could keep the temperature around nearby sprinkler heads low enough to prevent them from activating because of the cooler air passing to the opened vent; therefore, smoke and heat vents should have a higher temperature rating than the sprinklers.

2306.8 Fire department hose connections. Where exit passageways are required by the *International Building Code* for egress, a Class I standpipe system shall be provided in accordance with Section 905.

❖ An exit passageway is a means of continuing the exit enclosure protection horizontally to the exit discharge. Accordingly, an exit passageway may be used to connect an interior exit stairway to the exit discharge. Another use of an exit passageway is to bring an exit entrance within the allowable limit of exit access travel distance.

This section states that when the IBC requires the installation of an exit passageway, a Class I standpipe system must be provided in accordance with Section 905 of the code. The intent is to recognize that the standpipe will provide a quick and convenient water source for fire department use where fire hose lines would otherwise be impractical to use because of travel distance. The intent is to allow the fire department to initiate an attack on the fire more quickly, thus reducing the possibility of loss of life or property.

2306.9 Aisles. Aisles providing access to exits and fire department access doors shall be provided in high-piled storage areas exceeding 500 square feet (46 m^2), in accordance with Sections 2306.9.1 through 2306.9.3. Aisles separating storage piles or racks shall comply with NFPA 13. Aisles shall also comply with Chapter 10.

Exception: Where aisles are precluded by rack storage systems, alternate methods of access and protection are allowed when approved.

❖ The width of aisles is a functional element of the building occupancy, which allows the occupants to circulate freely throughout the floor area under normal conditions, is a functional element for the automatic fire sprinklers to successfully operate and is a functional element of the operations of the storage area (see Figure 2306.9). Under normal conditions, the aisles primarily serve the persons working in the building, and the design for minimum aisle width is based on the maneuvering capabilities of the commodity-handling methods and equipment. Under emergency situations, the aisle automatically becomes an egress pathway that provides the needed width to accommodate the number of occupants that must use the aisles and also serves to allow the water discharge from the sprinklers to penetrate the high-piled storage. Aisles also provide access for fire department personnel and serve as fire breaks between piles and racks.

2306.9.1 Width. Aisle width shall be in accordance with Sections 2306.9.1.1 and 2306.9.1.2.

Exceptions:

1. Cross aisles used only for employee access between aisles shall be a minimum of 24 inches (610 mm) wide.
2. Aisles separating shelves classified as shelf storage shall be a minimum of 30 inches (762 mm) wide.

❖ This section alerts the code user to the various minimum aisle widths based on the commodity, commodity area and whether or not the building contains automatic fire sprinklers. There are two exceptions to the aisle requirements. Exception 1 allows cross aisles to be 24 inches (610 mm), the logic being that those aisles are not required and are simply a convenience for building owners and users. Exception 2 is for shelving that allows a 30-inch (762 mm) aisle. This is likely related to the outcome of fire tests on this configuration of storage.

2306.9.1.1 Sprinklered buildings. Aisles in sprinklered buildings shall be a minimum of 44 inches (1118 mm) wide. Aisles shall be a minimum of 96 inches (2438 mm) wide in high-piled storage areas exceeding 2,500 square feet (232 m^2) in area, that are accessible to the public and designated to contain high-hazard commodities.

Exception: Aisles in high-piled storage areas exceeding 2,500 square feet (232 m^2) in area, that are accessible to the public and designated to contain high-hazard commodities, are protected by a sprinkler system designed for multiple-row racks of high-hazard commodities shall be a minimum of 44 inches (1118 mm) wide.

Aisles shall be a minimum of 96 inches (2438 mm) wide in areas accessible to the public where mechanical stocking methods are used.

❖ The 44-inch (1118 mm) minimum aisle width allows two unimpeded parallel columns of users to travel in opposite directions. The 96-inch (2438 mm) minimum width is designed to allow the two unimpeded parallel columns of users in addition to anticipating other obstructions occurring during an emergency situation. These obstructions could be in the form of a lift truck, idle pallets, mercantile displays and even the results of unstable commodities falling into the aisle. In addition, the 96-inch (2438 mm) width reduces the hazard of "aisle jumps" (radiant heat from a burning pile/rack causing the pile/rack across the aisle to ignite) in case of a fire and provides the fire department with greater maneuvering capabilities during a fire.

The exception recognizes the increased safety provided by a higher level of sprinkler protection, such as an ESFR system or additional levels of in-rack sprinklers, by allowing for smaller aisle widths for high-hazard commodity areas exceeding 2,500 square feet (232 m^2). If the sprinkler protection can protect multiple-row racks of

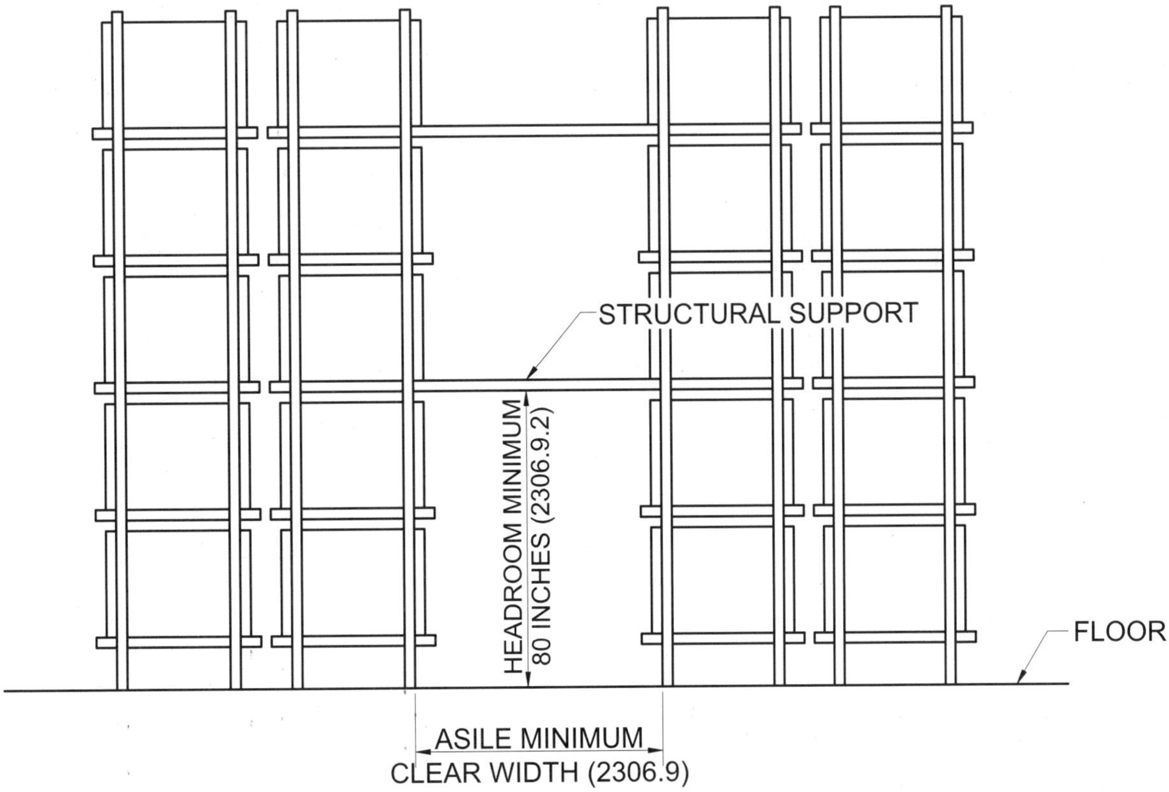

For SI: 1 inch = 25.4 mm.

Figure 2306.9
RACK STORAGE (DOUBLE ROW)

high-hazard commodities, it should adequately protect small aisle displays between piles or racks.

2306.9.1.2 Nonsprinklered buildings. Aisles in nonsprinklered buildings shall be a minimum of 96 inches (2438 mm) wide.

❖ The minimum 96-inch (2438 mm) aisle width for all aisles in nonsprinklered buildings recognizes the greater hazard caused by the lack of an automatic fire sprinkler system. This minimum width allows the building users a greater area for anticipating obstructions as well as greater flexibility for fire department personnel to assess, combat and control the fire.

2306.9.2 Clear height. The required aisle width shall extend from floor to ceiling. Rack structural supports and catwalks are allowed to cross aisles at a minimum height of 6 feet 8 inches (2032 mm) above the finished floor level, provided that such supports do not interfere with fire department hose stream trajectory.

❖ The clear ceiling height, or head room minimum (see Figure 2306.9), is necessary for the occupants to avoid an obstruction, to provide visibility to the occupants so that the path of travel can be planned and negotiated and to allow sufficient area for effective use of the fire department's hose streams. The height is the vertical measurement above every point along the finished floor to a ceiling, to the underside of a rack storage catwalk or to the underside of a structural member.

2306.9.3 Dead ends. Dead-end aisles shall be in accordance with Chapter 10.

❖ Dead ends in corridors and passageways can seriously increase the time needed for an occupant to locate exits. More importantly, dead ends can allow a single fire event to eliminate access to all the exits by trapping the occupants in the dead-end area. While a preferred building layout would be one without dead ends, this section requires the provisions of Chapter 10, specifically Sections 1016.3 and 1026.17.2, to be used in the building design. In this case, a Group S-1 or S-2 occupancy is allowed a 20-foot (6096 mm) maximum dead-end condition for new construction and a 50-foot (15 240 mm) maximum dead-end condition in existing buildings.

2306.10 Portable fire extinguishers. Portable fire extinguishers shall be provided in accordance with Section 906.

❖ Portable fire extinguishers provide the building occupants with an opportunity to suppress a fire in its incipient stage. In storage facilities, the fire extinguisher can contribute to the protection of occupants when there are evacuation difficulties or a specific hazard within that occupancy. For portable extinguishers to be effective, personnel must be properly trained in their use and maintenance (see Sections 2301.3 and 2301.4). Section 906 provides criteria for the location, installation, inspection and testing and maintenance of portable fire extinguishers. Section 906 also contains criteria for cabinets in which portable fire extinguishers may be stored, as well as the designated locations for wheeled portable fire extinguishers.

SECTION 2307
SOLID-PILED AND SHELF STORAGE

2307.1 General. Shelf storage and storage in solid piles, solid piles on pallets and bin box storage in bin boxes not exceeding 5 feet (1524 mm) in any dimension, shall be in accordance with Sections 2306 and this section.

❖ Solid piling in piles or on pallets generally consists of commodities in cartons, boxes, bales or bags [see Figure 2307.1(1)] or which may be encapsulated in a plastic wrap for containment [see Figures 2307.1(2) and (3)]. These commodities are usually stacked manually or with lift trucks. Palletized storage consists of commodities set onto pallets that are generally made of wood or plastic, are square in shape, measure 3 to 4 feet (914 to 1219 mm) on each side and have an appearance to that of a cavity wall. In this case, the cavity portion of the pallet is designed to accept the fork prongs of the lifting device for the commodity and pallet to be transported. The height limitations for stacking palletized storage are directly proportional to stackability of the commodity. Heights of palletized storage can reach 30 feet (9144 mm) in some cases.

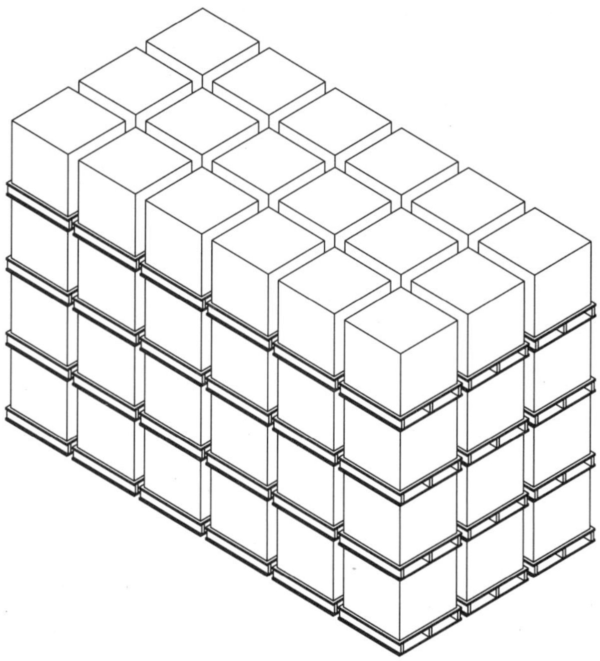

Figure 2307.1(1)
SOLID-PILED ARRAY
(COMMODITY CONFIGURED WITHOUT FLUE SPACES)

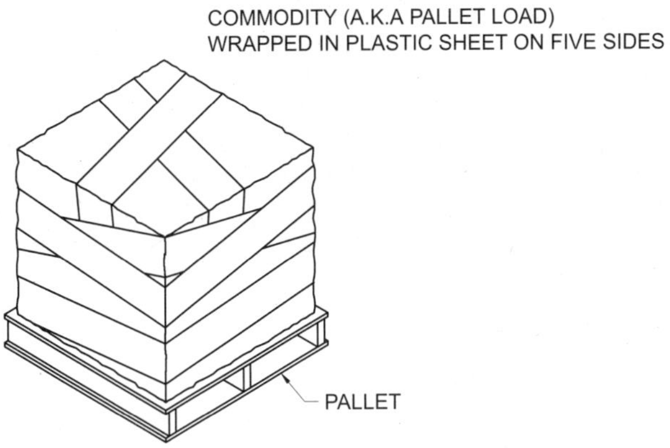

Figure 2307.1(2)
ENCAPSULATED COMMODITY

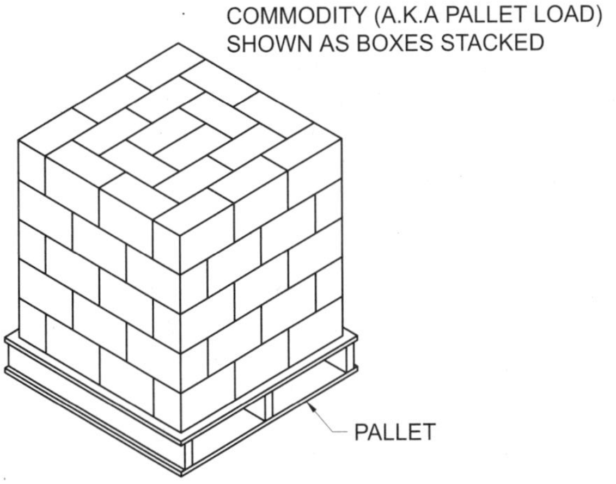

Figure 2307.1(3)
UNENCAPSULATED COMMODITY

2307.2 Fire protection. Where automatic sprinklers are required by Table 2306.2, an approved automatic sprinkler system shall be installed throughout the building or to 1-hour fire-resistance-rated fire barrier walls constructed in accordance with the *International Building Code*. Openings in such walls shall be protected by opening protective assemblies having 1-hour fire protection ratings. The design and installation of the automatic sprinkler system and other applicable fire protection shall be in accordance with the *International Building Code* and NFPA 231.

❖ The provision for approved automatic sprinklers is based on the commodity class and the size of the storage area. Automatic sprinkler protection, when required, is to be provided throughout the area containing the storage. For example, in a mixed-use occupancy, the area designated for storage would require an automatic sprinkler protection system throughout the storage area up to the fire barrier(s), whereas the other areas of the mixed-use occupancy may or may not require automatic sprinkler protection.

NFPA 231 is referenced because the development of this standard's provisions was based on a number of large-scale fire tests conducted to determine the advantages and limitations of automatic sprinklers in various storage array configurations. However, test information was not available for all configurations. As a result, some of the protection criteria are based on extrapolation of the test information. The automatic sprinkler system is expected to remain in operation as the fire fighters are attempting to control and extinguish the fire. As a result, design considerations for water supply must ensure that the water supply is not compromised by other operations. In no case should manual fire-fighting operations be substituted for automatic sprinklers.

The sprinkler protection is to extend throughout the building containing the rack storage or to a fire barrier wall. The fire-resistance-rated fire barrier wall is to comply with the provisions of Section 706 of the IBC which applies to fire barriers used for separating fire areas. To maintain integrity of the separation of fire areas, fire barriers must be continuous from the top of the floor below to the underside of the roof slab or deck above and be securely attached thereto. As with any fire-resistive barriers, consideration must be given to openings and penetrations. The number and size of openings in the fire barrier must comply with the provisions of Section 706 of the IBC and the opening protective assemblies must have a minimum fire-resistance rating of 1 hour. Note that Section 715.3 of the IBC allows an opening protection rating of $^{3}/_{4}$-hour for nonexit-enclosure fire barriers; however, this section, with its specific requirement of a 1-hour rating, would supercede that requirement. Penetrations of fire barriers must comply with Section 712 of the IBC. Maintenance of the fire-resistance-rated construction rests with the building owner or the tenant as described in Section 703 of the code.

2307.2.1 Shelf storage. Shelf storage greater than 12 feet (3658 mm) but less than 15 feet (4572 mm) in height shall be in accordance with the fire protection requirements set forth in NFPA 231. Shelf storage 15 feet (4572 mm) or more in height shall be protected in an approved manner with special fire protection, such as in-rack sprinklers.

❖ This section recognizes the provisions of NFPA 231 that apply to the height range in question because those provisions were developed based on full-scale tests of both unencapsulated and encapsulated products up to the 15-foot (4572 mm) height. As a result, the provisions of NFPA 231C should be used when addressing shelf storage of commodities above height. The standard states that sprinklers are required both at the ceiling and at the underside of each solid shelf (in-rack) to address the obstruction of the flue spaces created by the shelving.

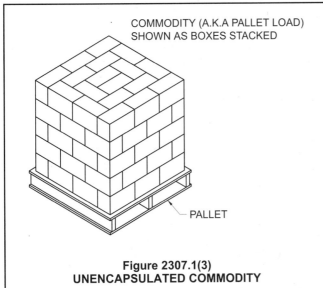

2307.3 Pile dimension and height limitations. Pile dimensions, the maximum permissible storage height and pile volume shall be in accordance with Table 2306.2.

❖ The commodity class and the size of the storage area directly regulate pile dimensions and height limitations. Table 2306.2 contains the categories for those limitations, and associated with those categories are the minimum requirements for an automatic sprinkler system, detection, building access, smoke and heat removal and curtain boards. The storage area column also designates which types of storage areas are public accessible and which are not (for example, bulk storage mercantile retail facility versus storage facility for automotive assembly plant). More options are provided when the facility is not open to the public.

2307.4 Array. Where an automatic sprinkler system design utilizes protection based on a closed array, array clearances shall be provided and maintained as specified by the standard used.

❖ The closed array is defined in NFPA 231 as "a storage arrangement where air movement through a pile is restricted due to vertical flues six inches in width or narrower." In this case, the provisions allow a departure from Table 2308.3 by allowing the use of NFPA 231 or NFPA 231C.

SECTION 2308 RACK STORAGE

2308.1 General. Rack storage shall be in accordance with Section 2306 and this section. Bin boxes exceeding 5 feet (1524 mm) in any dimension shall be regulated as rack storage.

❖ Storage racks are commonly steel frames (see Figure 2308.1) able to store commodities in various configurations. There are numerous variations of rack storage, such as single-row, double-row and multiple-row racks. The rack structure may or may not be connected to the building superstructure, which is governed by Chapters 16 and 22 of the IBC. The commodity is generally set on pallets for easy and economical transport using lift trucks or automated materials handling systems (automatic/unmanned). Manual rack storage operations require reasonably wide aisles to accommodate the lift trucks, whereas automated systems can operate in much narrower aisles.

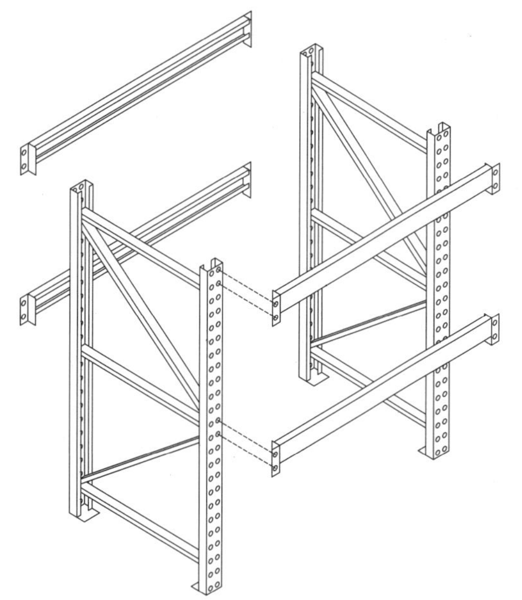

Figure 2308.1
TYPICAL RACK SYSTEM

2308.2 Fire protection. Where automatic sprinklers are required by Table 2306.2, an approved automatic sprinkler system shall be installed throughout the building or to 1-hour fire barrier walls constructed in accordance with the *International Building Code*. Openings in such walls shall be protected by opening protective assemblies having 1-hour fire protection ratings. The design and installation of the automatic sprinkler system and other applicable fire protection shall be in accordance with Section 903.3.1.1, the *International Building Code* and NFPA 231C.

❖ The provision for approved automatic sprinklers is based on the commodity class and the size of the storage area. The automatic sprinkler protection, when required, is to be installed throughout the area containing the storage. For example, in a mixed-use occupancy, the area designated for storage would require an automatic sprinkler protection system throughout the storage area up to the fire barrier(s), whereas the other areas of the mixed-use occupancy may or may not require automatic sprinkler protection. It should be noted that 15 feet (4572 mm) beyond the high-piled storage area a different type of sprinkler system can be used. It may be necessary to provide draft curtains to divide the two types of systems. An example is dividing an ESFR sprinkler system from a standard sprinkler system.

When the rack storage generally exceeds 25 feet (7620 mm) in height, in-rack automatic sprinkler systems should be incorporated into the design because ceiling sprinklers alone may not be able to provide protection for those commodities in the lower levels of the rack storage. In addition, the protection of the columns of a building by that same sprinkler may be necessary to satisfy the provisions of Section 2308.4 (see NFPA 231C).

NFPA 231C is referenced because development of this standard was based on a number of large-scale fire tests conducted to determine the advantages and limitations of automatic sprinklers in various storage array configurations. However, test information was not available for all configurations. As a result, some of the protection criteria are based on extrapolation of the test information. The automatic sprinkler system is expected to remain in operation as the fire fighters are attempting to control and extinguish the fire. As a result, design considerations for water supply must ensure that the water supply is not compromised by other operations. In no case should manual fire-fighting operations be substituted for automatic sprinklers.

The sprinkler protection is to extend throughout the building containing the rack storage or to a fire barrier wall. The fire-resistance-rated fire barrier wall is to comply with the provisions of Section 706 of the IBC which applies to fire barriers used for separating fire areas. To maintain integrity of the separation of fire areas, fire barriers must be continuous from the top of the floor below to the underside of the roof slab or deck above and be securely attached thereto. As with any fire-resistance-rated barriers, consideration must be given to openings and penetrations. The number and size of openings in the fire barrier must comply with the provisions of Section 706 of the IBC, and the opening protective assemblies must have a minimum fire-resistance rating of 1 hour. Note that Section 715.3 of the IBC allows an opening protection rating of $^3/_4$-hour for nonexit-enclosure fire barriers; however, this section, with its specific requirement of a 1-hour rating, would supercede that requirement. Penetrations of fire barriers must comply with Section 712 of the IBC. Maintenance of the fire-resistance-rated construction rests with the building owner or the tenant as described in Section 703 of the code.

2308.2.1 Plastic pallets and shelves. Storage on plastic pallets or plastic shelves shall be protected by approved specially engineered fire protection systems.

❖ Plastic pallets and plastic shelving, when used, increase the fuel load to the commodity that is being stored on top of the pallet or shelving. NFPA 231C states, "When plastic pallets are used, the classification of commodity unit shall be increased by one class (e.g., Class II becomes Class IV...)." In addition, this section states an additional requirement for a specially engineered fire protection system(s) to address this additional fuel load.

2308.2.2 Racks with solid shelving. Racks with solid shelving having an area greater than 32 square feet (3 m^2), measured between approved flue spaces at all four edges of the shelf, shall be in accordance with this section.

Exceptions:

1. Racks with mesh, grated, slatted or similar shelves having uniform openings not more than 6 inches (152 mm) apart, comprising at least 50 percent of overall shelf area, and with approved flue spaces, are allowed to be treated as racks without solid shelves.
2. Racks used for the storage of combustible paper records, with solid shelving, shall be in accordance with NFPA 231C.

❖ Racks with solid shelving provide a challenge to the fire protection system design. Ceiling-only automatic sprinkler protection will not provide sufficient suppression and protection of the neighboring commodities or the building. Additional sprinkler protection will be necessary to control and potentially suppress a fire. In this case, in-rack sprinklers have demonstrated that they can supplement the ceiling sprinklers for acceptable protection overall by reducing the fire's ability to develop a high rate of heat release. Designs for in-rack protection must consider necessary clearances, distance from structural components, flue space clearances and protection of the heads to allow the sprinklers to function properly.

2308.2.2.1 Fire protection. Fire protection for racks with solid shelving shall be in accordance with NFPA 231C.

❖ The provisions of NFPA 231C stipulate the minimum requirements for the design and construction of an automatic fire suppression system within a building containing rack storage with solid shelving. Rack storage with solid shelving requires the installation of an in-rack automatic fire suppression system. Since most rack storage systems are permanent, they can support the piping for the in-rack system.

The in-rack fire suppression system provides an additional layer of protection that ceiling-only systems could not provide because the shelving prevents the water discharge from entering the individually shelved commodities. The in-rack sprinklers can contain the fire to a small area of the rack structure and wet the commodities adjacent to the fire, thus preventing the fire from spreading to other commodities located horizontally and vertically from the commodities involved in the fire. Important design considerations are the location and protection of supply lines from commodity storage and handling operations, protection of individual sprinkler heads and the separation of water supply to the in-rack fire suppression system from other systems, such as the ceiling-mounted fire suppression system.

2308.3 Flue spaces. Flue spaces shall be provided in accordance with Table 2308.3. Required flue spaces shall be maintained.

❖ This section refers to the requirements in Table 2308.3 for the minimum longitudinal and transverse flue dimensions and design for the automatic fire suppression system, in association with Table 2306.2. The responsibility for the care and maintenance of the building rests with the building owner, tenant or lessee. Generally, the purpose of flue spaces is to enable sprinkler water to penetrate the storage commodities; therefore, when in-rack sprinklers are provided or if the building is not sprinklered, flue spaces are not required.

TABLE 2308.3. See page 23-20.

❖ Table 2308.3 prescribes the necessary flue spacing for rack storage systems (see Figure 2308.3). The table is organized with requirements based on rack configuration.

TABLE 2308.3
REQUIRED FLUE SPACES FOR RACK STORAGE

RACK CONFIGURATION	AUTOMATIC SPRINKLER PROTECTION		SPRINKLER AT THE CEILING WITH OR WITHOUT MINIMUM IN-RACK SPRINKLERS			IN-RACK SPRINKLERS AT EVERY TIER	NONSPRINKLERED
			≤ 25 feet				
	Storage height		Option 1	Option 2	≥ 25 feet	Any height	Any height
Single-row rack	Transverse flue space	Size[b]	3 inches	Not Applicable	3 inches	Not Required	Not Required
		Vertically aligned	Not Required	Not Applicable	Yes	Not Applicable	Not Required
	Longitudinal flue space		Not Required	Not Applicable	Not Required	Not Required	Not Required
Double-row rack	Transverse flue space	Size[b]	6 inches[a]	3 inches	3 inches	Not Required	Not Required
		Vertically aligned	Not Required	Not Required	Yes	Not Applicable	Not Required
	Longitudinal flue space		Not Required	6 inches	6 inches	Not Required	Not Required
Multi-row rack	Transverse flue space	Size[b]	6 inches	Not Applicable	6 inches	Not Required	Not Required
		Vertically aligned	Not Required	Not Applicable	Yes	Not Applicable	Not Required
	Longitudinal flue space		Not Required	Not Applicable	Not Required	Not Required	Not Required

For SI: 1 inch = 25.4 mm, 1 foot = 304.8 mm.

a. Three-inch transverse flue spaces shall be provided at least every 10 feet where ESFR sprinkler protection is provided.

b. Random variations are allowed, provided that the configuration does not obstruct water penetration.

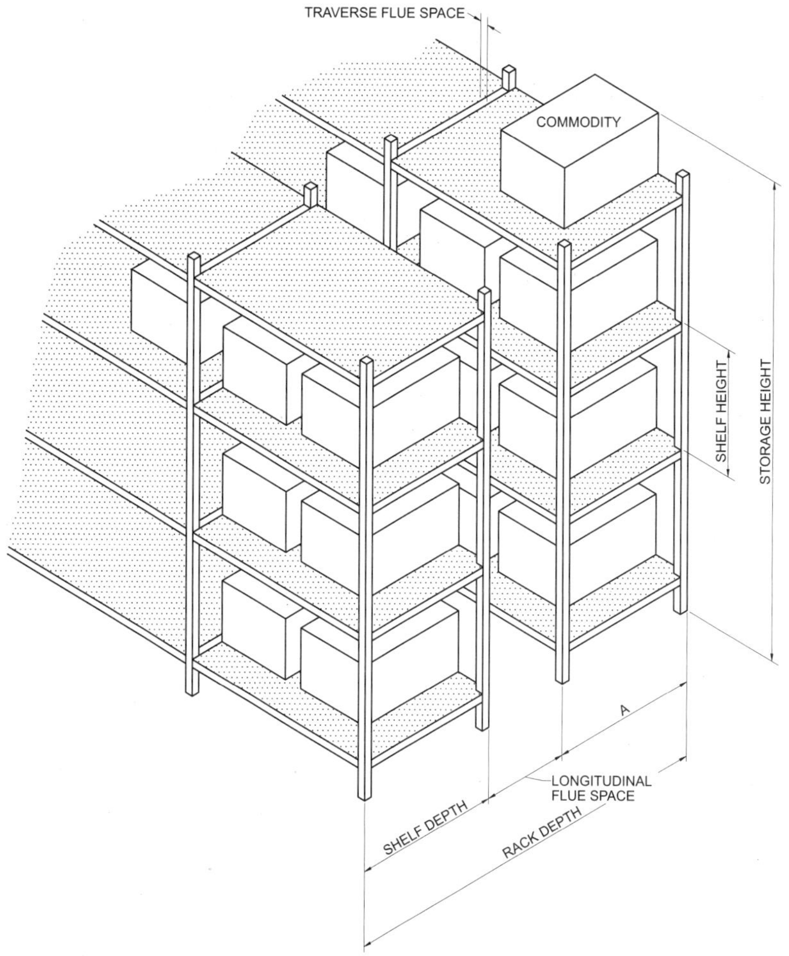

Figure 2308.3
DOUBLE-ROW RACKS WITH SOLID SHELVES

2308.4 Column protection. Steel building columns shall be protected in accordance with NFPA 231C.

❖ NFPA 231C provisions describe the requirements for the protection of steel columns as being a minimum of a 1-hour enclosure, protection using sidewall sprinklers at each column or protection through the design of a greater sprinkler density and flow. These provisions are required when the rack storage system achieves a height at which the design of the ordinary ceiling sprinkler system will not sufficiently protect the columns.

2308.5 Extra-high-rack storage systems. Approval of the fire code official shall be obtained prior to installing extra-high-rack combustible storage.

❖ In addition to the permit requirements contained in Section 105, the building owner, tenant or lessee is required to seek permission from the fire code official to construct an extra-high-rack storage system.

2308.5.1 Fire protection. Buildings with extra-high-rack combustible storage shall be protected with a specially engineered automatic sprinkler system. Extra-high-rack combustible storage shall be provided with additional special fire protection, such as separation from other buildings and additional built-in fire protection features and fire department access, when required by the fire code official.

❖ Extra-high rack storage is considered to have additional risks that warrant the requirement for an automatic sprinkler system engineered with additional protection features. These features could include additional heads along the vertical axis, change in the type of sprinkler heads and the use of high-expansion foam systems.

To further protect the building and neighboring buildings, the fire code official may require that additional fire prevention measures be taken to ensure that, if a fire does occur, it will be contained to that building. These measures can include building separation requirements greater than required in Table 602 of the IBC, greater than the number of fire apparatus access roads and proximity of roads to the building, greater number of hydrants and greater than the number of building openings. All additional fire prevention and life safety systems require the approval of the fire code official.

SECTION 2309
AUTOMATED STORAGE

2309.1 General. Automated storage shall be in accordance with this section.

❖ Automated storage is generally designed for large operations in which the commodity storage necessary to serve the operation is large and where that facility may operate 24 hours a day, seven days a week. This type of storage is generally programmed for commodity stocking and retrieving, and contains a sufficient amount of commodity for the operation to continue working even if restocking may experience a slow period. Operations that commonly employ such facilities are automotive parts manufacturers, electronic parts manufacturers and even the food industry.

One of the primary fire concerns with automated storage is the potential for the fire to be carried by the mechanism that transports the commodity. If such a fire were to occur while the commodity is in motion, it could reduce, if not overwhelm, the effectiveness of the automatic fire suppression system. As a result, additional measures are required in an attempt to compensate for that potential fire hazard.

2309.2 Automatic sprinklers. Where automatic sprinklers are required by Table 2306.2, an approved automatic sprinkler system shall be installed throughout the building. The design and installation of the automatic sprinkler system shall be in accordance with Section 903.

❖ The provisions requiring an automatic sprinkler system throughout the building housing the automatic storage system is an additional fire protection feature based on the concept that the area may be difficult for the fire service to suppress a fire manually. The reference to the provisions of Chapter 9 covers just one aspect of the automatic sprinkler system; Chapter 23 and Chapter 9 combined effectively require such systems to be designed in accordance with NFPA 13, NFPA 231 and NFPA 231C.

2309.3 Carousel storage. High-piled storage areas having greater than 500 square feet (46 m^2) of carousel storage shall be provided with automatic shutdown in accordance with one of the following:

1. An automatic smoke detection system installed in accordance with Section 907, with coverage extending 15 feet (4575 mm) in all directions beyond unenclosed carousel storage systems and which sounds a local alarm at the operator's station and stops the carousel storage system upon the activation of a single detector.
2. An automatic smoke detection system installed in accordance with Section 907 and within enclosed carousel storage systems, which sounds a local alarm at the operator's station and stops the carousel storage system upon the activation of a single detector.
3. A single dead-man-type control switch that allows the operation of the carousel storage system only when the operator is present. The switch shall be in the same room as the carousel storage system and located to provide for observation of the carousel system.

❖ Carousel storage is used in large storage operations when the commodity must be protected (security), when the commodity requires strict environmental control or both. In addition, carousel storage is selected because it may provide an economic advantage to an operation or business by having the commodity come directly to the employee rather than the employee stocking and retrieving the commodity. The carousel area is normally not occupied. An exception may be when the carousel system is undergoing maintenance or repair or during an annual inspection of the fire protection features. Examples of commodities within carousel storage may include distribution centers that distribute pharmaceutical or refrigerated/frozen foods and beverages.

The additional fire safety requirements are a precautionary measure as a direct result of the limited access to these types of storage areas and to address the potential of a moving fire, courtesy of the carousel mechanism. Because the area of the carousel is controlled by a mechanized, generally computer-controlled system, immediate access to the actual storage area is normally limited. These limits serve to protect both personnel and the commodity. The additional fire safety measures include provisions for an elaborate automatic detection system, which upon activation stops the carousel from operating (moving). This precaution prevents the fire from spreading throughout the entire storage area.

SECTION 2310 SPECIALTY STORAGE

2310.1 General. Records storage facilities used for the rack or shelf storage of combustible paper records greater than 12 feet (3658 mm) in height shall be in accordance with Sections 2306 and 2308 and NFPA 231C. Palletized storage of records shall be in accordance with Section 2307.

❖ Recording storage can have a relatively high intrinsic value. This is where protection of the actual documents is essential to any business to retrieve information or to reconstruct records. The provisions of this section specifically identify the fire and life safety requirements for records stored on pallets and in-rack or shelf-storage systems. Although the provisions of this section set the minimum safeguards from the hazards of fire and explosion, it is up to the owners of the records to determine the actual protection that must be provided. NFPA 231 and 231C are the minimum referenced standards.

Bibliography

The following resource materials are referenced in this chapter or are relevant to the subject matter addressed in this chapter.

"Commodity Classification." Data sheet 8-1. Norwood, MA: Factory Mutual Research Corporation.

Fire Protection Handbook, A.E. Cote and J.L. Linville, ed. Quincy, MA: National Fire Protection Association, 1997.

FMRC, *Large Scale Fire Tests of Rack Storage Group A Plastics in Retail Operation Scenarios Protected by Extra Large Orifice (ELO) Sprinklers.* Report FMRCJ.I.0X1R0.RR. Norwood, MA: Factory Mutual Research Corporation, 1994.

IBC-2003, *International Building Code.* Falls Church, VA: International Code Council, 2003.

Klausbruckner, Elley, P.E. *Application Guide to High-piled Storage.* Whittier, CA: International Conference of Building Officials, 2000.

Kolodner, Herbert J., Ph.D., P.E., "Emergency Operations in High-Rack Warehouses," *Fire Journal*, pages 63-65 May, 1982.

NFPA 10-98, *Standard for Portable Fire Extinguishers.* Quincy, MA: National Fire Protection Association, 1998.

NFPA 13-99, *Installation of Sprinkler Systems.* Quincy, MA: National Fire Protection Association, 1999.

NFPA 231-98, *Standard for General Storage.* Quincy, MA: National Fire Protection Association, 1998.

NFPA 231C-98, *Standard for Rack Storage of Materials.* Quincy, MA: National Fire Protection Association, 1998.

Chapter 24: Tents, Canopies and Other Membrane Structures

General Comments

Though they don't happen often, fires occurring in tents and air-supported structures have historically caused significant loss of life. Perhaps the most notable of these tragedies occurred on July 6, 1944, in Hartford, Connecticut, where a Ringling Brothers, Barnum and Bailey circus tent caught fire during a matinee performance, killing 167 people and injuring 487. Since then, protection of tents and air-supported structures has focused on construction methods and materials, as well as limiting use and occupancy.

Construction Methods and Materials

Tents and air-supported structures are constructed of diverse materials, usually fabrics, textiles and films. Section 3102 of the *International Building Code*® (IBC®) requires these membrane materials to be either noncombustible as defined in Section 703.4 of the IBC or fire resistant as defined by NFPA 701. The use of lightweight high-tensile-strength membrane coverings is perhaps the most significant similarity between tents and air-supported structures. Beyond this similarity, an increasingly wide variety of structural configurations is becoming common. Tents include all structures using rigid structural frames or supports for lateral and compressive stability. In the case of air-supported structures, a positive pressure differential between the inside and the outside of the structure, coupled with the favorable tensile properties of the membrane, yields these structural properties.

In every case, anchors and cables are used for either additional structural stability or to act as fail-safe devices against extreme wind, rain or snow loading. The more commonplace membrane coverings include cotton and plastic canvas fabrics. Exotic new materials, such as high-tensile-strength plastic films, have spawned a new generation of air-supported structures, including the spectacularly covered and domed stadiums built in recent years in many large metropolitan areas. Protecting the structure from collapse and fire remains the most significant fire and life safety concern.

Occupancy

Fire poses a dual threat to a tent or air-supported structure. First, the fire presents a danger to the occupants by exposing them to heat, smoke and toxic combustion products. Just as important, the fire represents an imminent threat to the structure. Even the best flame-resistant fabrics may ignite or fail under extreme conditions posed by a fire, and the fire's demand for air may compromise the structural support of the air inside the building, if not the integrity of the membrane in the case of an air-supported structure. Full or partial collapse of the membrane covering of a tent or air-supported structure may occur earlier and with less warning than in any other structural type.

Egress may become difficult, if not nearly impossible, if a collapse occurs. Therefore, this chapter limits the storage or handling of combustible materials inside tents and air-supported structures because of their contribution to fuel loading. Similarly, heat energy sources that may ignite the membrane fabric or other combustibles are prohibited or restricted. Even spot lighting must be used with caution to prevent heat energy from igniting the covering. Portable fire extinguishers must be readily available for incipient fire fighting as an additional safeguard against fire.

Purpose

The requirements in this chapter are intended to protect tents, canopies and air-supported structures from fire by requiring regular inspections, certifying continued compliance with fire safety regulations, as well as the requirements of the IBC regulating their use and occupancy.

SECTION 2401
GENERAL

2401.1 Scope. Tents, canopies and membrane structures shall comply with this chapter. The provisions of Section 2403 are applicable only to temporary membrane structures. The provisions of Section 2404 are applicable to temporary and permanent membrane structures.

❖ This section defines the kinds of structures covered by this chapter and designates which sections apply to temporary structures and which apply to permanent structures. Structures can range from 10-foot by 10-foot (3048 mm by 3048 mm) canvas shelters to major indoor sports arenas. The common feature of all types of tents, whether they are air-supported, air-inflated or tensioned membrane structures, is the nature of the structural skin. In all cases, a textile material is used to create an indoor or protected space by separating the area under the covering from wind, precipitation and temperature extremes. Although most membrane structures are intended for temporary or seasonal use, elegant all-weather permanent structures are becoming increasingly common.

SECTION 2402
DEFINITIONS

2402.1 Definitions. The following words and terms shall, for the purposes of this chapter and as used elsewhere in this code, have the meanings shown herein.

❖ Definitions of terms can help in the understanding and application of the code requirements. The purpose for including those definitions that are associated with the subject matter of this chapter is to provide more convenient access to them without having to refer back to Chapter 2. It is important to emphasize that these terms are not exclusively related to this chapter but are applicable everywhere the term is used in the code. For convenience, these terms are also listed in Chapter 2 with a cross reference to this section. The use and application of all defined terms, including those defined in this section, are set forth in Section 201.

Although "air-inflated structure" is not a defined term in this section, it is defined in Section 3102.2 of the IBC and is used in Sections 2403.5, 2403.7, 2403.10 and 2403.10.2 of the code. This type of structure is generally much smaller than an air-supported structure and differs in that it depends for support on the inflation of balloon-like sections over, under or around the occupants. The occupants normally are found within a surrounding structure consisting of these inflated sections. Note that the occupants of the structure are not subjected to the pressurized areas, as they are in air-supported membrane structures. Possibly the most common example of this kind of structure is the "Moon Walk" children's entertainment structure, which has an inflated floor structure for children to play on and also inflated columns that support an overhead canopy and plastic mesh walls. Figure 2402.1(2) illustrates an air-inflated structure.

AIR-SUPPORTED STRUCTURE. A structure wherein the shape of the structure is attained by air pressure, and occupants of the structure are within the elevated pressure area.

❖ The term air-supported structure identifies those membrane structures that are completely pressurized for the purposes of supporting the membrane covering. Most domed sports arenas use air pressure within the structure to support the membrane covering. The membrane covering can consist of one layer or multiple layers. Figure 2402.1(1) illustrates an air-supported structure.

CANOPY. A structure, enclosure or shelter constructed of fabric or pliable materials supported by any manner, except by air or the contents it protects, and is open without sidewalls or drops on 75 percent or more of the perimeter.

❖ A canopy is an architectural projection or structure comprised of a rigid structure over which a membrane covering is typically attached, providing overhead weather protection, a means of identity or decoration. It may be supported by both the building to which it is attached and stanchions at the outer end or may be free-standing and completely supported by stanchions. Note the definition requires that the canopy be overhead only, except for a drop on one side or 25 percent of the perimeter. Structures with more than 25 percent of their perimeter enclosed by drops fall under the definition of a "Tent." Figure 2402.1(3) illustrates a canopy.

MEMBRANE STRUCTURE. An air-inflated, air-supported, cable or frame-covered structure as defined by the *International Building Code* and not otherwise defined as a tent or canopy. See Chapter 31 of the *International Building Code.*

❖ This definition is broadly inclusive of all types of membrane structures, regardless of the supporting mechanism or structure, and intends to include the structures defined in Section 3102 of the IBC.

TENT. A structure, enclosure or shelter constructed of fabric or pliable material supported by any manner except by air or the contents that it protects.

❖ Tents can be temporary or permanent structures. When permanent, they are regulated by Section 2404. When erected as temporary enclosures, they are regulated by Sections 2403 and 2304 (see also Chapter 31 of the IBC). Figure 2402.1(4) illustrates a tent.

Figure 2402.1(1)
AIR-SUPPORTED MEMBRANE STRUCTURE—SWIMMING POOL COVER

Figure 2402.1(2)
EXAMPLE OF AIR-INFLATED MEMBRANE STRUCTURE

Figure 2402.1(3)
EXAMPLE OF CANOPY

Figure 2402.1(4)
EXAMPLE OF TENT STRUCTURE

SECTION 2403 TEMPORARY TENTS, CANOPIES AND MEMBRANE STRUCTURES

2403.1 General. All temporary tents, canopies and membrane structures shall comply with this section.

❖ This section addresses tents, canopies and membrane structures that are considered temporary in terms of the duration of their erection and use (see Section 2403.5 for the definition of "Temporary" as it applies to membrane structures). The intent of this section is that if a canopy or awning is used, it should be soundly designed so as not to present a hazard to its users, emergency responders or the public during the time it is in place.

2403.2 Approval required. Tents and membrane structures having an area in excess of 200 square feet (19 m^2) and canopies in excess of 400 square feet (37 m^2) shall not be erected, operated or maintained for any purpose without first obtaining a permit and approval from the fire code official.

Exceptions:

1. Tents used exclusively for recreational camping purposes.
2. Fabric canopies open on all sides which comply with all of the following:
   - 2.1. Individual canopies having a maximum size of 700 square feet (65 m^2).
   - 2.2. The aggregate area of multiple canopies placed side by side without a fire break clearance of 12 feet (3658 mm), not exceeding 700 square feet (65 m^2) total.
   - 2.3. A minimum clearance of 12 feet (3658 mm) to all structures and other tents.

❖ Use of membrane structures results in great flexibility and a large volume of weather-protected space. However, these benefits are balanced by the sensitivity of these structures to strict maintenance requirements. The approval process allows the fire code official to exercise strict control, assuring compliance with the requirements of this chapter.

This section sets the minimum size structure that requires approval.

Exception 1 covers tents that are normally used by families or very small groups for short periods under widely varying conditions that would be difficult or impossible for a fire code official to police.

2403.3 Place of assembly. For the purposes of this chapter, a place of assembly shall include a circus, carnival, tent show, theater, skating rink, dance hall or other place of assembly in or under which persons gather for any purpose.

❖ This section gives examples of common types of assembly use for membrane structures, but leaves the issue open so that the fire code official has the discretion to determine whether some other intended use should be considered a place of assembly that would require permits and approvals.

2403.4 Permits. Permits shall be required as set forth in Sections 105.6 and 105.7.

❖ The process of issuing permits gives the fire code official an opportunity to carefully evaluate and regulate hazardous operations or special structures. Permit applicants should be required to demonstrate that their operation or construction complies with the intent of the code before the permit is issued. See the commentary to Section 105.7 for a general discussion of operations requiring a construction permit. The process also notifies the fire department of the need for prefire planning for the hazardous property.

2403.5 Use period. Temporary tents, air-supported, air-inflated or tensioned membrane structures and canopies shall be used for a period of not more than 180 days within a 12-month period on a single premise.

❖ This section defines the term "temporary." Any membrane structure used for more than 180 days in any 12-month period at a single location must be considered permanent and would be subject to all requirements for permanent structures as set forth in Section 2404.

2403.6 Construction documents. A detailed site and floor plan for tents, canopies or membrane structures with an occupant load of 50 or more shall be provided with each application for approval. The tent, canopy or membrane structure floor plan shall indicate details of the means of egress facilities, seating capacity, arrangement of the seating and location and type of heating and electrical equipment.

❖ The requirement for a floor plan showing means of egress facilities and seating locations eliminates possible conflicts at the time of field inspection. Evaluating means of egress for hastily arranged or moveable seating is a challenge to the inspector. The detailed means of egress plan allows the inspector to verify that the actual configuration matches the approved plan. The applicant and designer also benefit by having the plans reviewed in detail before construction begins.

2403.7 Inspections. The entire tent, air-supported, air-inflated or tensioned membrane structure system shall be inspected at regular intervals, but not less than two times per permit use period, by the permittee, owner or agent to determine that the installation is maintained in accordance with this chapter.

Exception: Permit use periods of less than 30 days.

❖ The periodic inspections required here are conducted by the permit holder, the owner or his or her agent to make certain the structure continues to meet code requirements and is being properly maintained.

The exception states that structures used for less than 30 days do not need to be reinspected.

2403.7.1 Inspection report. When required by the fire code official, an inspection report shall be provided and shall consist of maintenance, anchors and fabric inspections.

❖ This section allows the fire code official to require submittal of a complete inspection report, including maintenance work done, for each finished inspection by the permit holder, owner or agent.

2403.8 Access, location and parking. Access location and parking for temporary tents, canopies and membrane structures shall be in accordance with this section.

❖ This section addresses the issues of fire apparatus access, separation from vehicles and other structures and the need for fire breaks for tents, canopies and membrane structures that are considered temporary in terms of the duration of their erection and use.

2403.8.1 Access. Fire apparatus access roads shall be provided in accordance with Section 503.

❖ The same access rules apply to membrane structures as apply to structures erected using conventional construction materials. Because membrane structures can become serious fire hazards, depending on membrane material and structure contents, maintaining code-required fire access roads and lanes is especially important.

2403.8.2 Location. Tents, canopies or membrane structures shall not be located within 20 feet (6096 mm) of lot lines, buildings, other tents, canopies or membrane structures, parked vehicles or internal combustion engines. For the purpose of determining required distances, support ropes and guy wires shall be considered as part of the temporary membrane structure, tent or canopy.

Exceptions:

1. Separation distance between membrane structures, tents and canopies not used for cooking, is not required when the aggregate floor area does not exceed 15,000 square feet (1394 m^2).
2. Membrane structures, tents or canopies need not be separated from buildings when all of the following conditions are met:
   - 2.1. The aggregate floor area of the membrane structure, tent or canopy shall not exceed 10,000 square feet (929 m^2).
   - 2.2. The aggregate floor area of the building and membrane structure, tent or canopy shall not exceed the allowable floor area including increases as indicated in the *International Building Code*.
   - 2.3. Required means of egress provisions are provided for both the building and the membrane structure, tent or canopy, including travel distances.
   - 2.4. Fire apparatus access roads are provided in accordance with Section 503.

❖ The 20-foot (6096 mm) separation distance is consistent with requirements for conventional structures, especially those that could represent an above-average fire hazard. A fire of any size within a membrane structure would almost certainly involve the membrane itself. Because the support ropes and guy wires are under tension, particularly with large membrane structures, a membrane weakened by fire would fail, causing the ropes or wires to recoil and possibly pull portions of the burning membrane well clear of its original position. Requiring 20 feet (6096 mm) of clear ground around the structure helps reduce the likelihood that burning membrane sections and flying embers would not endanger other structures or public trafficways.

Exception 1 acknowledges the reduced hazard of small membrane structures that do not house cooking appliances.

Exception 2 lists four criteria for determining whether a membrane structure must conform to the separation criteria of this section. To be exempt, the structure would have to meet all four criteria.

2403.8.3 Location of structures in excess of 15,000 square feet in area. Membrane structures having an area of 15,000 square feet (1394 m^2) or more shall be located not less than 50 feet (15 240 mm) from any other tent or structure as measured from the sidewall of the tent or membrane structure unless joined together by a corridor.

❖ The larger separation distance required by this section is consistent with the hazards presented by the larger structures. The exception for structures connected by corridors considers the smaller hazard posed by the covering of the corridor. Corridors are not likely to contain significant amounts of combustible materials other than the membrane itself.

2403.8.4 Connecting corridors. Tents or membrane structures are allowed to be joined together by means of corridors. Exit doors shall be provided at each end of such corridor. On each side of such corridor and approximately opposite each other, there shall be provided openings not less than 12 feet (3658 mm) wide.

❖ Corridors connecting membrane structures to each other or to permanent structures must have openings in their side walls for convenient egress in case of an emergency. These openings could also be entry points for emergency response personnel.

2403.8.5 Fire break. An unobstructed fire break passageway or fire road not less than 12 feet (3658 mm) wide and free from guy ropes or other obstructions shall be maintained on all sides of all tents, canopies and membrane structures unless otherwise approved by the fire code official.

❖ This requirement for a clear path makes the membrane structures more accessible to emergency response personnel. It also results in an open space into which flaming embers or other debris can fall without endangering other structures or public trafficways.

2403.9 Anchorage required. Tents, canopies or membrane structures and their appurtenances shall be adequately roped, braced and anchored to withstand the elements of weather and prevent against collapsing. Documentation of structural stability shall be furnished to the fire code official on request.

❖ Having secure anchorage to prevent damage or loss caused by wind or precipitation makes good economic sense. This section also gives the fire code official the authority to review and approve both plans and actual installations to ensure that the structures have been designed and erected using good engineering practices.

2403.10 Temporary air-supported and air-inflated membrane structures. Temporary air-supported and air-inflated membrane structures shall be in accordance with this section.

❖ This section addresses air-supported and air-inflated membrane structures that are considered temporary in terms of the duration of their erection and use (see Section 2403.5 for the definition of "Temporary" as it applies to these structures). The intent of this section is that if air-supported or air-inflated membrane structures are used, they should be soundly designed so as not to present a hazard to their users, emergency responders or the public during the time they are in place.

2403.10.1 Door operation. During high winds exceeding 50 miles per hour (80 kph) or in snow conditions, the use of doors in air-supported structures shall be controlled to avoid excessive air loss. Doors shall not be left open.

❖ Because the design pressure is critical in maintaining the structural integrity of an air-supported structure, doors must not be kept open for extended periods. Controls on door usage help prevent excessive losses of internal pressure. When large openings are necessary, such as for vehicular traffic, vestibules help avoid excessive pressure loss.

2403.10.2 Fabric envelope design and construction. Air-supported and air-inflated structures shall have the design and construction of the fabric envelope and the method of anchoring in accordance with Architectural Fabric Structures Institute ASI 77.

❖ The referenced document is both a design manual that contains the engineering formulas needed to calculate stresses and other parameters associated with structural stability and a design standard for air-supported structures.

2403.10.3 Blowers. An air-supported structure used as a place of assembly shall be furnished with not less than two blowers, each of which has adequate capacity to maintain full inflation pressure with normal leakage. The design of the blower shall be so as to provide integral limiting pressure at the design pressure specified by the manufacturer.

❖ Requiring two blowers that have the capacity to keep the structure fully inflated builds redundancy into the system. Should one blower fail, the other would be capable of maintaining full inflation while the failed unit was repaired or replaced. The pressure-limiting device is required to prevent overpressurizing the structure and thus causing failure.

2403.10.4 Auxiliary power. Places of public assembly for more than 200 persons shall be furnished with either a fully automatic auxiliary engine-generator set capable of powering one blower continuously for 4 hours, or a supplementary blower powered by an internal combustion engine which shall be automatic in operation.

❖ Because an air-supported structure must be maintained at full design pressure, a power failure that would disable all of the blowers used to maintain pressure would result in a gradual collapse of the structure as pressure dropped as a result of normal leakage. Having an auxiliary power source on site is intended to prevent this situation. This section allows either an auxiliary generator that will supply power to the blowers or a self-contained supplementary blower unit that has an internal combustion engine to drive it.

2403.11 Seating arrangements. Seating in tents, canopies or membrane structures shall be in accordance with Chapter 10.

❖ See the commentary for Section 2404.14.

2403.12 Means of egress. Means of egress for temporary tents, canopies and membrane structures shall be in accordance with this section.

❖ Because of the unique nature of the structures regulated by this chapter, it is vital that their means of egress systems be carefully designed and constructed. This section regulates the access to, number, location, marking and illumination of the means of egress for temporary membrane structures.

2403.12.1 Distribution. Exits shall be spaced at approximately equal intervals around the perimeter of the tent, canopy or membrane structure, and shall be located such that all points are 100 feet (30 480 mm) or less from an exit.

❖ The requirement for exits to be equally spaced around the perimeter of the membrane structure considers the probability that any one structure could be used for several different purposes over its life and that each use could represent a different seating arrangement. Because travel distances to an exit must not exceed 100 feet (30 480 mm), exits must be located to minimize travel distances no matter what the seating arrangement is.

2403.12.2 Number. Tents, canopies or membrane structures or a usable portion thereof shall have at least one exit and not less than the number of exits required by Table 2403.12.2. The widths of means of egress required by Table 2403.12.2 shall be divided approximately equally among the separate means of egress. The total width of means of egress in inches (mm) shall not be less than the total occupant load served by a means of egress multiplied by 0.2 inches (5 mm) per person.

❖ This section specifies the use of the exit requirements contained in Table 2403.12.2 to determine the number and size of exits for all membrane structures. The formula given in the last sentence of the section allows calculation of total exit requirements for any given number of occupants.

TABLE 2403.12.2
MINIMUM NUMBER OF MEANS OF EGRESS AND MEANS OF EGRESS WIDTHS FROM TEMPORARY MEMBRANE STRUCTURES, TENTS AND CANOPIES

OCCUPANT LOAD	MINIMUM NUMBER OF MEANS OF EGRESS	MINIMUM WIDTH OF EACH MEANS OF EGRESS (inches)	MINIMUM WIDTH OF EACH MEANS OF EGRESS (inches)
		Tent or Canopy	Membrane Structure
10 to 199	2	72	36
200 to 499	3	72	72
500 to 999	4	96	72
1,000 to 1,999	5	120	96
2,000 to 2,999	6	120	96
Over 3,000[a]	7	120	96

For SI: 1 inch = 25.4 mm.

a. When the occupant load exceeds 3,000, the total width of means of egress (in inches) shall not be less than the total occupant load multiplied by 0.2 inches per person.

❖ The minimum means of egress widths shown in this table were established, in all likelihood, on two very different bases. For tents, the widths are possibly conservative because so many in the fire service remember the Ringling Brothers, Barnum and Bailey circus tent fire that happened in Hartford, Connecticut, in July 1944. In that fire, 167 people died and 487 were injured because they were unable to escape from the burning tent. Also, because the side panels of a tent can be constructed to varying widths without too much difficulty, the wider openings are practical.

On the other hand, means of egress widths required for air-supported structures seem to be much more optimistic. The narrower means of egress widths are necessary to prevent excess loss of internal air pressure in the air-supported structure and can be linked to the exit requirements in ASI 77. Excessively large pressure losses could lead to the collapse of the structure onto the very occupants the exits are serving.

2403.12.3 Exit openings from tents. Exit openings from tents shall remain open unless covered by a flame-resistant curtain. The curtain shall comply with the following requirements:

1. Curtains shall be free sliding on a metal support. The support shall be a minimum of 80 inches (2032 mm) above the floor level at the exit. The curtains shall be so arranged that, when open, no part of the curtain obstructs the exit.
2. Curtains shall be of a color, or colors, that contrasts with the color of the tent.

❖ The requirement for exit openings to remain free and clear is meant to ensure that the exit is not blocked or hidden from view of the occupants attempting to exit. Allowing the opening to be covered by an easily moveable curtain is meant to add to the comfort of occupants by protecting them from wind, precipitation and temperature extremes. The requirement for the curtains to be of a contrasting color makes the exits readily identifiable even at a distance.

2403.12.4 Doors. Exit doors shall swing in the direction of exit travel. To avoid hazardous air and pressure loss in air-supported membrane structures, such doors shall be automatic closing against operating pressures. Opening force at the door edge shall not exceed 15 pounds (7 kg).

❖ Pressure loss in an air-supported structure is always a concern. The requirement that the door be capable of closing automatically against the internal pressure supporting the membrane structure is intended to protect against this kind of loss. The 15-pound (66 N) force required for opening acknowledges that the door must be held closed against the interior air pressure. This level of opening force is, however, consistent with the requirements of the IBC for exterior doors in conventional buildings.

2403.12.5 Aisle. The width of aisles without fixed seating shall be in accordance with the following:

1. In areas serving employees only, the minimum aisle width shall be 24 inches (610 mm) but not less than the width required by the number of employees served.
2. In public areas, smooth-surfaced, unobstructed aisles having a minimum width of not less than 44 inches (1118 mm) shall be provided from seating areas, and aisles shall be progressively increased in width to provide, at all points, not less than 1 foot (305 mm) of aisle width for each 50 persons served by such aisle at that point.

❖ The first requirement for areas serving only employees is indefinite because the configuration of the area could vary widely from one use to another. Required width beyond 24 inches (610 mm) would have to be determined based on seating arrangements, if any, and traffic patterns in the affected area. Guidance for assembly areas of fixed conventional structures is given in Chapter 10 of the code as well as in Chapter 10 of the IBC. The judgement of the fire code official would also have to be considered in the final decision.

The second requirement is quite specific for public-use areas. The wide aisles are intended to make egress quicker and easier because a fire in an air-supported structure could easily burn through the membrane covering and cause the structure to collapse on the occupants.

2403.12.5.1 Arrangement and maintenance. The arrangement of aisles shall be subject to approval by the fire code official and shall be maintained clear at all times during occupancy.

❖ This section gives the fire code official the authority to approve seating arrangements in air-supported structures and to inspect those arrangements periodically to verify that they have not been changed to an unacceptable configuration.

2403.12.6 Exit signs. Exits shall be clearly marked. Exit signs shall be installed at required exit doorways and where otherwise necessary to indicate clearly the direction of egress when the exit serves an occupant load of 50 or more.

❖ This charging statement establishes the requirement for exit signs. See the commentary to Section 1011.2 for further discussion of exit sign requirements.

2403.12.6.1 Exit sign illumination. Exit signs shall be of an approved self-luminous type or shall be internally or externally illuminated by fixtures supplied in the following manner:

1. Two separate circuits, one of which shall be separate from all other circuits, for occupant loads of 300 or less; or
2. Two separate sources of power, one of which shall be an approved emergency system, shall be provided when the occupant load exceeds 300. Emergency systems shall be supplied from storage batteries or from the on-site generator set, and the system shall be installed in accordance with the ICC *Electrical Code*.

❖ Because there is always the possibility of power failure in a fire or other emergency, exit signs must have a power source—batteries, backup power or auxiliary power—that will keep them illuminated if the primary power source to the structure lighting system fails. The requirement for this power source is established in Section 604.2.3 and is further defined in Section 1003.2.10.5.

2403.12.7 Means of egress illumination. Means of egress shall be illuminated with light having an intensity of not less than 1 foot-candle (11 lux) at floor level while the structure is occupied. Fixtures required for means of egress illumination shall be supplied from a separate circuit or source of power.

❖ All means of egress must be continuously illuminated by artificial lighting during the entire time the air-supported structure is occupied so that the paths of exit travel are always visible and available for evacuation of the occupants during emergencies. The code makes a special point of noting that the exit discharge must also be provided with adequate illumination so that occupants can safely find the public way should the emergency occur at night.

The intensity of floor lighting illuminating the entire means of egress, including open plan spaces, aisles, corridors and exit access passageways and exit doors, must not be less than 1 foot-candle. It has been found that this low level of lighting renders enough visibility for the occupants to evacuate the building safely.

This lighting level is measured at the floor to make the floor surface visible. Levels of illumination above the floor may be higher or lower.

The means of egress must be illuminated, especially in times of emergency when the occupants must have a lighted path of exit travel to evacuate the building safely.

So that there will be a continuing source of electrical energy for maintaining the illumination of the means of egress when there is a loss of the main power supply, the means of egress lighting system must be connected to an emergency electrical system that consists of storage batteries, unit equipment or an on-site generator. This emergency power-generating facility must be capable of supplying electricity for at least 90 minutes, thereby giving the occupants sufficient time to leave the premises. In most cases, where the loss of the main electrical supply is attributed to a malfunction in the distribution system of the electric power company, experience has shown that such power outages do not usually last as long as 90 minutes.

2403.12.8 Maintenance of means of egress. The required width of exits, aisles and passageways shall be maintained at all times to a public way. Guy wires, guy ropes and other support members shall not cross a means of egress at a height of less than 8 feet (2438 mm). The surface of means of egress shall be maintained in an approved manner.

❖ The requirement for keeping the means of egress clear applies to all structures. It is especially important for air-supported structures and tents because of the added hazards they present.

SECTION 2404
TEMPORARY AND PERMANENT TENTS, CANOPIES AND MEMBRANE STRUCTURES

2404.1 General. All tents, canopies and membrane structures, both temporary and permanent, shall be in accordance with this section. Permanent tents, canopies and membrane structures shall also comply with the *International Building Code*.

❖ This section addresses tents, canopies and membrane structures that are considered either temporary or permanent in terms of the duration of their erection and use (see Section 2403.5 for the definition of "Temporary" and, by default, "Permanent" as they apply to membrane structures). The intent of this section is that if a canopy or awning is used, it should be soundly designed so as not to present a hazard to its users, emergency personnel or the public during the time it is in place, whether that is more or less than 180 days per calendar year.

2404.2 Flame-resistant treatment. Before a permit is granted, the owner or agent shall file with the fire code official a certificate executed by an approved testing laboratory, certifying that the tents, canopies and membrane structures and their appurtenances, sidewalls, drops and tarpaulins, floor coverings, bunting, combustible decorative materials and effects, including sawdust when used on floors or passageways, shall be composed of flame-resistant material or shall be treated with a flame retardant in an approved manner and meet the requirements for flame resistance as determined in accordance with NFPA 701, and that such flame resistance is effective for the period specified by the permit.

❖ The reference to NFPA 701 in this section is consistent with the requirements of Section 3102 of the IBC or the structural materials, and Sections 802 and 805 of the IBC and Sections 803 and 805 of the code for finishes and trim materials. Making certain the structure and as much of its contents as possible are either noncombustible or are treated to make them fire resistant is an important first step in fire safety.

This section also gives the fire code official the authority to inspect the facility, its contents and all documentation of flame-retardant treatment to ensure permit requirements have been met.

2404.3 Label. Membrane structures, tents or canopies shall have a permanently affixed label bearing the identification of size and fabric or material type.

❖ This required label gives the fire code official important information about the membrane and the designed size of the structure. This information will help him or her determine whether the structure and its covering meet code requirements.

2404.4 Certification. An affidavit or affirmation shall be submitted to the fire code official and a copy retained on the premises on which the tent or air-supported structure is located. The affidavit shall attest to the following information relative to the flame resistance of the fabric:

1. Names and address of the owners of the tent, canopy or air-supported structure.
2. Date the fabric was last treated with flame-resistant solution.
3. Trade name or kind of chemical used in treatment.
4. Name of person or firm treating the material.
5. Name of testing agency and test standard by which the fabric was tested.

❖ This certificate is another piece of information the fire code official must have to make a decision on code compliance. This certificate will indicate whether the membrane fabric needs retreatment to maintain its fire-retarding properties.

2404.5 Combustible materials. Hay, straw, shavings or similar combustible materials shall not be located within any tent, canopy or membrane structure containing an assembly occupancy, except the materials necessary for the daily feeding and care of animals. Sawdust and shavings utilized for a public performance or exhibit shall not be prohibited provided the sawdust and shavings are kept damp. Combustible materials shall not be permitted under stands or seats at any time. The areas within and adjacent to the tent or air-supported structure shall be maintained clear of all combustible materials or vegetation that could create a fire hazard within 20 feet (6096 mm) from the structure. Combustible trash shall be removed at least once a day from the structure during the period the structure is occupied by the public.

❖ This section contains a "laundry list" of unacceptable practices. Hay, straw, shavings and sawdust are all readily ignitable materials. Minimizing their use is important to fire safety. A carelessly discarded match or other smoking material could easily cause any of these materials to burst into flame.

Requiring sawdust and shavings to be maintained damp reduces their ignition potential to almost zero.

Keeping areas under seating areas free of combustible materials removes a significant fire hazard from these areas of high occupant density.

Maintaining a clear buffer zone around the membrane structure minimizes the possibility of a wind-blown fire causing the structure to ignite. The 20-foot (6096 mm) clearing is consistent with separation requirements elsewhere in the code.

Keeping trash cleared both inside and outside of the structure also removes a potential fuel source from the area. This requirement is consistent with requirements for permanent structures used as places of assembly.

2404.6 Smoking. Smoking shall not be permitted in tents, canopies or membrane structures. Approved "No Smoking" signs shall be conspicuously posted in accordance with Section 310.

❖ Because smoking is prohibited in membrane structures, the requirement for posting signage is obvious. Because no specification exists for standard signs, each jurisdiction having authority is responsible for establishing its own criteria. To be approved, signs must be large enough to be read from a distance and be worded simply and clearly.

2404.7 Open or exposed flame. Open flame or other devices emitting flame, fire or heat or any flammable or combustible liquids, gas, charcoal or other cooking device or any other unapproved devices shall not be permitted inside or located within 20 feet (6096 mm) of the tent, canopy or membrane structures while open to the public unless approved by the fire code official.

❖ Prohibiting open flames and high-heat appliances fueled by flammable or combustible gases, liquids and solids inside or within 20 feet (6096 mm) of a membrane structure is just common sense. Fires cannot start if there is no source of ignition. This section, however, gives the fire code official the authority to approve an open flame or exposed flame appliance considered to be a minimal fire hazard. These might include catalytic heaters that are located well away from the membrane walls and any combustible materials or limited cooking facilities using charcoal or LP-gas for fuel and located under a canopy that is open on at least three sides.

2404.8 Fireworks. Fireworks shall not be used within 100 feet (30 480 mm) of tents, canopies or membrane structures.

❖ Fireworks, whether designed to explode at ground level or as aerial displays, result in hot embers that could become a source of ignition if they contact the membrane of the structure. Establishing a 100-foot (30 480 mm) clear zone both minimizes the possibility of hot embers contacting the membrane and also gives air-borne embers time to cool before they could reach membrane level.

2404.9 Spot lighting. Spot or effect lighting shall only be by electricity, and all combustible construction located within 6 feet (1829 mm) of such equipment shall be protected with approved noncombustible insulation not less than 9.25 inches (235 mm) thick.

❖ Spotlights can generate considerable heat. The metal housings of these lights generally become too hot to touch if they are on for more than a minute or two. Requiring that they be positioned well away from combustible materials and that the nearby combustibles be insulated helps ensure that the heat from the lights does not become an ignition source for either the membrane structure or combustibles inside the structure. Noncombustible insulation must conform to the requirements of Section 718 of the IBC.

2404.10 Safety film. Motion pictures shall not be displayed in tents, canopies or membrane structures unless the motion picture film is safety film.

❖ The display of motion pictures on cellulose nitrate film is prohibited because of the extreme hazards associated with this type of material. However, such motion pictures are extremely rare because production of the raw film was suspended in 1951. Most known motion pictures printed on cellulose nitrate film are in the possession of special film repositories, such as the National Archives and the Smithsonian Institution, or are being restored by conservators.

2404.11 Clearance. There shall be a minimum clearance of at least 3 feet (914 mm) between the fabric envelope and all contents located inside the tent or membrane structure.

❖ The 3-foot (914 mm) clearance is intended not only to give fire fighters access to fires, but also to prevent or minimize exposure of the membrane envelope if stored material becomes involved in a fire.

2404.12 Portable fire extinguishers. Portable fire extinguishers shall be provided as required by Section 906.

❖ Section 906 states that portable extinguishers are required in Group A occupancies and in special-hazards areas as designated by the fire code official. This section also refers to NFPA 10 for guidance on selection and placement of the extinguishers. Employees and staff who will be manning the membrane structure must be trained to use the extinguishers because they are likely to become the first line of emergency response in case of a fire.

2404.13 Fire protection equipment. Fire hose lines, water supplies and other auxiliary fire equipment shall be maintained at the site in such numbers and sizes as required by the fire code official.

❖ This section gives the fire code official the authority to establish reasonable equipment requirements for membrane structures. Because the size, construction and intended use of membrane structures vary so widely, each installation must be evaluated individually.

2404.14 Occupant load factors. The occupant load allowed in an assembly structure, or portion thereof, shall be determined in accordance with Chapter 10.

❖ This reference to Chapter 10 tells us that a membrane structure used as a place of assembly is treated in the same way as a permanent structure would be. The same means of egress requirements apply whether the structure is conventional construction or a membrane structure.

2404.15 Heating and cooking equipment. Heating and cooking equipment shall be in accordance with this section.

❖ Because of the unique nature of the structures regulated by this chapter, it is vital that ignition sources be carefully regulated. This section regulates the dual hazards of heat-producing appliances and cooking equipment in membrane structures.

2404.15.1 Installation. Heating or cooking equipment, tanks, piping, hoses, fittings, valves, tubing and other related components shall be installed as specified in the *International Mechanical Code* and the *International Fuel Gas Code,* and shall be approved by the fire code official.

❖ This section refers to the installation requirements contained in the *International Fuel Gas Code*® (IFGC®) and *International Mechanical Code*® (IMC®) and gives the fire code official the authority to inspect and approve completed installations. Because of the special hazards that exist in membrane structures of all kinds, the importance of proper installation cannot be stressed too strongly.

2404.15.2 Venting. Gas, liquid and solid fuel-burning equipment designed to be vented shall be vented to the outside air as specified in the *International Fuel Gas Code* and the *International Mechanical Code.* Such vents shall be equipped with approved spark arresters when required. Where vents or flues are used, all portions of the tent, canopy or membrane structure shall be not less than 12 inches (305 mm) from the flue or vent.

❖ As with installation requirements in Section 2404.15.1, venting must comply with the applicable provisions of the IFGC and IMC and be approved by the fire code official. The 12-inch (305 mm) separation between the vent or flue stack and the membrane fabric and support structures is intended to prevent heating of those elements to possible points of ignition.

2404.15.3 Location. Cooking and heating equipment shall not be located within 10 feet (3048 mm) of exits or combustible materials.

❖ This location requirement is intended to help make certain a fire or other emergency involving one of the equipment items would not be likely to block a means of egress. Equipment locations would be subject to the approval of the fire code official in all cases.

2404.15.4 Operations. Operations such as warming of foods, cooking demonstrations and similar operations that use solid flammables, butane or other similar devices which do not pose an ignition hazard, shall be approved.

❖ This section authorizes the listed operations, but at the same time authorizes the fire code official to make certain the operations meet the requirements of the other subsections within this section.

2404.15.5 Cooking tents. Tents where cooking is performed shall be separated from other tents, canopies or membrane structures by a minimum of 20 feet (6096 mm).

❖ This separation requirement is consistent with the overall requirement covering open or exposed flames that is contained in Section 2404.7. The 20-foot (6096 mm) separation is protection against hot embers from a fire reaching the main membrane structure. The requirements for keeping the open area free of combustible materials or debris that could limit access to emergency response personnel also apply.

2404.15.6 Outdoor cooking. Outdoor cooking that produces sparks or grease-laden vapors shall not be performed within 20 feet (6096 mm) from a tent, canopy or membrane structure.

❖ As with cooking in tents, the 20-foot (6096 mm) separation is intended to separate this potential source of ignition from the membrane structure. Requirements for keeping the open area free of combustible materials or other debris that could hinder emergency response efforts also apply.

2404.15.7 Electrical heating and cooking equipment. Electrical cooking and heating equipment shall comply with the ICC *Electrical Code.*

❖ This reference to the ICC *Electrical Code*® (ICC EC™) establishes the requirements for the equipment and its installation. It also establishes the authority for the fire code official to approve the equipment and the installation.

2404.16 LP-gas. The storage, handling and use of LP-gas and LP-gas equipment shall be in accordance with this section.

❖ Because of the unique nature of the structures regulated by this chapter, it is vital that the storage, handling and use of hazardous materials in or around such structures be carefully regulated. In general, Chapter 38 regulates LP-gas. This section specifically regulates its storage, handling and use in connection with membrane structures.

2404.16.1 General. LP-gas equipment such as tanks, piping, hoses, fittings, valves, tubing and other related components shall be approved and in accordance with Chapter 38 and with the *International Fuel Gas Code.*

❖ The requirements stated in this section mean that LP-gas containers and all associated equipment used in or around membrane structures of any kind will be inspected and must meet basically the same requirements as similar installations and equipment in conventional structures.

2404.16.2 Location of containers. LP-gas containers shall be located outside. Safety release valves shall be pointed away from the tent, canopy or membrane structure.

❖ Containers must be located outdoors to prevent vapors from safety-release valves, inadvertent spills during filling or from any other source from accumulating inside the membrane structure to flammable levels. Requiring the safety-release valves to be pointed away from the structure also helps to ensure that vapors do not infiltrate the structure.

2404.16.2.1 Containers 500 gallons or less. Portable LP-gas containers with a capacity of 500 gallons (1893 L) or less shall have a minimum separation between the container and structure not less than 10 feet (3048 mm).

❖ The 10-foot (3048 mm) separation distance is consistent with the requirements of Table 3804.3. See the commentary for Section 3804.3 and Table 3804.3 for further discussion.

2404.16.2.2 Containers more than 500 gallons. Portable LP-gas containers with a capacity of more than 500 gallons (1893 L) shall have a minimum separation between the container and structures not less than 25 feet (7620 mm).

❖ The 25-foot (7620 mm) separation distance is consistent with the requirements of Table 3804.3. See the commentary for Section 3804.3 and Table 3804.3 for further discussion.

2404.16.3 Protection and security. Portable LP-gas containers, piping, valves and fittings which are located outside and are being used to fuel equipment inside a tent, canopy or membrane structure shall be adequately protected to prevent tampering, damage by vehicles or other hazards and shall be located in an approved location. Portable LP-gas containers shall be securely fastened in place to prevent unauthorized movement.

❖ LP-gas containers and associated equipment must be protected from impacts by vehicles or other objects that could cause damage and leakage. They must also be protected from tampering and vandalism as well as from theft. See Section 3807 for additional guidance on protection and security.

2404.17 Flammable and combustible liquids. The storage of flammable and combustible liquids and the use of flammable-liquid-fueled equipment shall be in accordance with this section.

❖ Because of the unique nature of the structures regulated by this chapter, it is vital that the storage, handling and use of hazardous materials in or around such structures be carefully regulated. In general, Chapter 34 regulates flammable and combustible liquids. This section specifically regulates their storage, handling and use in connection with membrane structures.

2404.17.1 Use. Flammable-liquid-fueled equipment shall not be used in tents, canopies or membrane structures.

❖ Equipment of any kind that uses a flammable liquid for fuel must not be used inside a membrane structure because of the possibility of fuel leakage as well as the risk of accumulation of noxious exhaust fumes. Fluid leaks could vaporize and reach a flammable concentration. Exhaust fumes normally contain carbon monoxide and other gases that are harmful to occupants. Membrane structures, particularly air-supported and air-inflated structures, are not ventilated to the same extent that conventional structures are. Harmful vapors and fumes, therefore, are not diluted or dispersed as quickly or as efficiently as they are in those structures.

2404.17.2 Flammable and combustible liquid storage. Flammable and combustible liquids shall be stored outside in an approved manner not less than 50 feet (15 240 mm) from tents, canopies or membrane structures. Storage shall be in accordance with Chapter 34.

❖ This section states a general separation requirement for containers used to store flammable or combustible liquids that is consistent with the requirements in Table 2206.2.3 for other tanks. The more detailed requirements are contained in Chapter 34 and depend, for the most part, on whether the liquid is Class I, II or III and the quantity of the liquid being stored. Storage arrangements must comply with Chapter 34 requirements. They should, however, be discussed in advance with the fire code official to make certain all affected parties are in agreement.

2404.17.3 Refueling. Refueling shall be performed in an approved location not less than 20 feet (6096 mm) from tents, canopies or membrane structures.

❖ This section gives only a general guideline for location of a refueling station and gives the fire code official the authority to approve the selected location. When planning a station of this kind, the requirements of Chapter 34 must be considered. Some of the requirements of Sections 2204, 2205 and 2206 might also apply if vehicles are being refueled.

2404.18 Display of motor vehicles. Liquid- and gas-fueled vehicles and equipment used for display within tents, canopies or membrane structures shall be in accordance with this section.

❖ This statement is an introduction to the subsections that follow. It contains only the kinds of structures and the kind of vehicles and equipment covered by the overall section.

2404.18.1 Batteries. Batteries shall be disconnected in an appropriate manner.

❖ Requiring that batteries be disconnected serves two purposes. First, it prevents unauthorized persons from starting and running engines that exhaust noxious fumes. Second, it removes a potential ignition source from the site by opening the battery-powered electrical circuit, thereby removing the possibility that the circuit could give off a stray spark. Equipment/vehicle manufacturers will, in all likelihood, have a recommended procedure in their owner's manuals for disconnecting batteries safely. The fire code official would also want to make sure the procedures used present no hazards.

2404.18.2 Fuel systems. Vehicles or equipment shall not be fueled or defueled within the tent, canopy or membrane structure.

❖ This statement means that all fueling stations must be located outdoors.

2404.18.2.1 Quantity limit. Fuel in the fuel tank shall not exceed one-quarter of the tank capacity or 5 gallons (19 L), whichever is less.

❖ Vehicles and equipment on display are allowed only a limited quantity of fuel in their tanks. Full fuel tanks could leak if temperatures reach a level that would cause the contained fuel to expand beyond the capacity of the tank. This kind of liquid spill would evaporate with the potential for enough flammable vapor to accumulate to reach a flammable level. Full tanks also leave very little room for vapors given off by the contained fuel. Practically all fuel tanks have some kind of pressure-release valve built in to prevent overpressurization of the tank. Leaving three quarters of the tank empty allows space for moderate quantities of vapor to accumulate without being released to the building atmosphere.

2404.18.2.2 Inspection. Fuel systems shall be inspected for leaks.

❖ This statement places an inspection requirement on the party displaying the vehicles or equipment. The requirement should be looked at as being a routine housekeeping chore. The statement also gives the fire code official the authority to inspect the facility to assure compliance with code requirements.

2404.18.2.3 Closure. Fuel tank openings shall be locked and sealed to prevent the escape of vapors.

❖ Locking and sealing fuel tank openings prevents vapors from escaping from the tanks and also explains why fuel tanks are to be left mostly empty.

2404.18.3 Location. The location of vehicles or equipment shall not obstruct means of egress.

❖ If the structure is occupied, blocking a means of egress would interfere with orderly evacuation of the space. Whether the structure is occupied or not, blocking a means of egress could interfere with access for emergency response personnel.

2404.18.4 Places of assembly. When a compressed natural gas (CNG) or liquefied petroleum gas (LP-gas) powered vehicle is parked inside a place of assembly, all the following conditions shall be met:

1. The quarter-turn shutoff valve or other shutoff valve on the outlet of the CNG or LP-gas container shall be closed and the engine shall be operated until it stops. Valves shall remain closed while the vehicle is indoors.
2. The hot lead of the battery shall be disconnected.
3. Dual-fuel vehicles equipped to operate on gasoline and CNG or LP-gas shall comply with this section and Sections 2404.18.1 through 2404.18.5.3 for gasoline-powered vehicles.

❖ The three requirements given in this section are applicable to vehicles powered by CNG or LP-gas, but they are consistent with requirements for gasoline-fueled vehicles that are given elsewhere in the code.

The first requirement is intended to prevent vapor buildup in confined spaces such as inside an engine compartment. Fuel left standing in a warm engine will "cook off" (vaporize) after the engine is turned off. Requiring the engine to be left running after the supply valve is closed means that all fuel left in the engine will be burned off through internal combustion, leaving the engine relatively free of fuel that could vaporize.

The second requirement is a restatement of Section 2404.18.1 (see the commentary for that section).

The third requirement states that specialty vehicles that are designed to run on either gasoline or CNG/LP-gas must comply with the same requirements as vehicles designed to run on gasoline alone.

2404.18.5 Competitions and demonstrations. Liquid- and gas-fueled vehicles and equipment used for competition or demonstration within a tent, canopy or membrane structure shall comply with Sections 2404.18.5.1 through 2404.18.5.3.

❖ This generally states that the vehicles must comply with the requirements in the three subsections that follow.

2404.18.5.1 Fuel storage. Fuel for vehicles or equipment shall be stored in approved containers in an approved location outside of the structure in accordance with Section 2404.17.2.

❖ This section gives the fire code official the authority to approve both the storage containers used and the location of those containers. The site designer, competition manager or other responsible party must have the fire code official's approval before the storage scheme is put into use. See also Section 3804 for further discussion of LP-gas storage and Sections 2203 and 2206 for additional information on storage of CNG.

2404.18.5.2 Fueling. Refueling shall be performed outside of the structure in accordance with Section 2404.17.3.

❖ This section refers back to Section 2413.3, which states a requirement for a 20-foot (6096 mm) separation between the fueling station and the membrane structure. Further information on fueling with LP-gas is contained in Section 3806. Similar information covering CNG is contained in Sections 2204, 2207 and 2208.

2404.18.5.3 Spills. Fuel spills shall be cleaned up immediately.

❖ Controlling spills is important to fire safety. Section 2205.3 gives requirements for spill containment. The requirements for immediate cleanup are intended to prevent buildup of flammable vapors to hazardous levels. Provisions for fuel spill containment and cleanup would require approval of the fire code official.

2404.19 Separation of generators. Generators and other internal combustion power sources shall be separated from tents, canopies or membrane structures by a minimum of 20 feet (6096 mm) and shall be isolated from contact with the public by fencing, enclosure or other approved means.

❖ This section states two requirements. First, consistent with other separation requirements, generators and other equipment driven by internal combustion engines must be kept separate from the membrane structure to minimize fire hazards if the generator or other equipment should fail and catch on fire. The isolation requirement is intended to both keep the public from coming into contact with hazardous equipment and protect the equipment from vandalism or accidental damage by the public.

2404.20 Standby personnel. When, in the opinion of the fire code official, it is essential for public safety in a tent, canopy or membrane structure used as a place of assembly or any other use where people congregate, because of the number of persons, or the nature of the performance, exhibition, display, contest or activity, the owner, agent or lessee shall employ one or more qualified persons, as required and approved, to remain on duty during the times such places are open to the public, or when such activity is being conducted.

Before each performance or the start of such activity, standby personnel shall keep diligent watch for fires during the time such place is open to the public or such activity is being conducted and take prompt measures for extinguishment of fires that occur and assist in the evacuation of the public from the structure.

There shall be trained crowd managers or crowd manager supervisors at a ratio of one crowd manager/supervisor for every 250 occupants, as approved.

❖ The exact duties and responsibilities of the individuals employed as crowd managers are not defined here other than the requirement that they maintain a careful fire watch. They could serve as ushers, tour guides, service supervisors for table seating or in some other capacity related to making sure occupants are moved to or from assigned places in an orderly way. The key to this section is that the crowd managers must be trained in crowd management procedures appropriate to the activity being carried on in the membrane structure and they must be present in the required numbers.

Training of personnel and the duties assigned to them would have to be approved by the fire code official.

2404.21 Vegetation removal. Combustible vegetation shall be removed from the area occupied by a tent, canopy or membrane structure, and from areas within 30 feet (9144 mm) of such structures.

❖ The vegetation removal requirements here are more restrictive than similar restrictions given in Sections 2205.7 and 3303.5.2 because membrane structures are considered a greater hazard than conventional construction. The requirement that combustible vegetation be removed, however, is consistent with the basic rules of good housekeeping. Removing a fuel source helps ensure that destructive fires cannot start.

2404.22 Waste material. The floor surface inside tents, canopies or membrane structures and the grounds outside and within a 30-foot (9144 mm) perimeter shall be kept clear of combustible waste. Such waste shall be stored in approved containers until removed from the premises.

❖ This section extends the requirement for cleanliness and trash removal indoors as well as outdoors. Again, the 30-foot (9144 mm) perimeter recognizes the increased hazard of membrane structures.

Bibliography

The following resource materials are referenced in this chapter or are relevant to the subject matter addressed in this chapter.

ASI-77, *Design and Standards Manual.* Roseville, MN: Industrial Fabric Association, 1977.

Dent, R.N. *Principles of Pneumatic Architecture.* New York, NY: Halsted Press Division, John Wiley & Sons, 1971.

Frei, O. *Tensile Structures: Design, Structure, and Calculation of Buildings, Cables, Nets and Membranes,* 2 vol. Cambridge, MA: MIT Press, 1969.

Herzog, T. *Pneumatic Structures: A Handbook of Inflatable Architecture,* 1976.

IBC-2003, *International Building Code.* Falls Church, VA: International Code Council, 2003.

ICC EC-2003, *ICC Electrical Code.* Falls Church, VA: International Code Council, 2003.

IFGC-2003, *International Fuel Gas Code.* Falls Church, VA: International Code Council, 2003.

IMC-2003, *International Mechanical Code.* Falls Church, VA: International Code Council, 2003.

NFPA 10-98, *Portable Fire Extinguishers.* Quincy, MA: National Fire Protection Association, 1998.

NFPA 701-99, *Methods of Fire Tests for Flame-Resistant Textiles and Films.* Quincy, MA: National Fire Protection Association, 1999.

Tensioned Fabric Structures. R.E. Shaeffer, ed. Reston, VA: American Society of Civil Engineers, 1996.

Chapter 25: Tire Rebuilding and Tire Storage

General Comments

Each year, over 270 million vehicle tires are disposed of in the United States. Recycling, reuse and energy recovery are having a major impact, but tire storage continues to present an environmental and fire safety hazard. The risk associated with tire fires demonstrates the need to address this problem.

This chapter prescribes ways to prevent and control fires in tire rebuilding plants, tire storage and tire byproduct facilities. Fires in storage facilities do not occur often; however, these fires usually result in millions of dollars in direct and indirect losses. Tire storage "dumps" can burn for a long time, overwhelming fire department resources. In 1990, Canadian officials estimated that a fire at a tire storage facility in Hagarsville, Ontario, caused $3 million in damage and cost $1.5 million to extinguish. The environmental results of large tire fires can, at times, be disastrous. The extreme heat of the Hagarsville fire turned the tire rubber into oil. One tire generates about two gallons of oil as it burns and liquefies. About half the tires in Hagarsville liquefied, resulting in over 14 million gallons (52 990 000 L) of toxic oil.

During a fire in a tire storage yard in Rhinehart, Virginia, a plume of smoke 3,000 feet (914 400 mm) high and 50 miles long covered parts of three states. This fire also threatened the potable water supply for Washington, D.C. The cleanup cost was estimated at $1.3 million.

Purpose

The requirements of Chapter 25 are intended to prevent or control fires and explosions associated with the remanufacture of tires. Additionally, these requirements are intended to minimize the impact of tire storage fires by segregating the various operations and controlling ignition sources. Although the finished product, the tire, is not an unusual hazard, once ignition occurs the fire is extremely difficult to extinguish. Facilities designed and constructed for tire remanufacture are regulated by the *International Building Code*® (IBC®) to reduce the impact of potential fires on buildings and the environment.

SECTION 2501 GENERAL

2501.1 Scope. Tire rebuilding plants, tire storage and tire byproduct facilities shall comply with this chapter, other applicable requirements of this code and NFPA 231D. Tire storage in buildings shall also comply with Chapter 23.

❖ In addition to the provisions of this chapter, NFPA 231D is referenced for alternative separation requirements. Spray operations using flammable or combustible solvents must also comply with Chapter 15.

2501.2 Permit required. Permits shall be required as set forth in Section 105.6.

❖ The process of issuing permits gives the fire code official an opportunity to carefully evaluate and regulate hazardous operations. Permit applicants should be required to demonstrate that their operations comply with the intent of the code before the permit is issued. See the commentary to Section 105.6 for a general discussion of operations requiring an operational permit. The process also notifies the fire department of the need for prefire planning for the hazardous property.

SECTION 2502 DEFINITIONS

2502.1 Terms defined in Chapter 2. Words and terms used in this chapter and defined in Chapter 2 shall have the meanings ascribed to them as defined therein.

❖ This section directs the code user to Chapter 2 for the proper application of the terms used in this chapter. Such terms may be defined in Chapter 2, in another *International Code*® as indicated in Section 201.3 or the use of their ordinary (dictionary) meaning may be all that is needed (also see commentary, Sections 201.1 through 201.4).

SECTION 2503 TIRE REBUILDING

2503.1 Construction. Tire rebuilding plants shall comply with the requirements of the *International Building Code*, as to construction, separation from other buildings or other portions of the same building, and protection.

❖ This section requires that construction must comply with the IBC. At a minimum, tire rebuilding plants are classified in Group F-1 (factory/industrial moderate hazard) in accordance with Section 306.2 of the IBC. Depending on an evaluation of the actual hazards presented by a given operation, which could include grinding, buffing or gluing of tires or tire components, the plant or portions of it might be classified in a Group H.

2503.2 Location. Buffing operations shall be located in a room separated from the remainder of the building housing the tire rebuilding or tire recapping operations by a 1-hour fire barrier.

Exception: Buffing operations are not required to be separated where all of the following conditions are met:

1. Buffing operations are equipped with an approved continuous automatic water-spray system directed at the point of cutting action;
2. Buffing machines are connected to particle-collecting systems providing a minimum air movement of 1,500 cubic feet per minute (cfm) (0.71 m^3/s) in volume and 4,500 feet per minute (fpm) (23 m/s) in-line velocity; and
3. The collecting system shall discharge the rubber particles to an approved outdoor noncombustible or fire-resistant container, which is emptied at frequent intervals to prevent overflow.

❖ This section specifies that the buffing operations must be separated from the other operations by a 1-hour fire barrier. The intent is identical to the intent regarding construction: to keep the higher hazard operations separate, thereby reducing the potential for a rapidly spreading fire.

The exception recognizes that meeting the three outlined conditions will afford protection equivalent to the 1-hour fire barrier required by this section.

2503.3 Cleaning. The buffing area shall be cleaned at frequent intervals to prevent the accumulation of rubber particles.

❖ This section recognizes the importance of keeping the buffing area clean to reduce the possibility of igniting waste material. A maintenance schedule should be developed for the removal of particles. Cleaning frequency should be determined by the amount of equipment in operation.

2503.4 Spray rooms and booths. Each spray room or spray booth where flammable or combustible solvents are applied, shall comply with Chapter 15.

❖ When flammable or combustible solvents are used, the room or spray booth must comply with the requirements of Chapter 15. This provision is intended to reduce the likelihood of igniting solvents used in this process and to keep the incident within manageable proportions if ignition should occur.

SECTION 2504
PRECAUTIONS AGAINST FIRE

2504.1 Open burning. Open burning is prohibited in tire storage yards.

❖ Due to the stubborn nature of tire fires once they are ignited, this section prohibits open burning in tire storage areas where ignition of tires is a hazard.

2504.2 Sources of heat. Cutting, welding or heating devices shall not be operated in tire storage yards.

❖ Like the previous section, this section prohibits additional sources of ignition from welding or heating devices.

2504.3 Smoking prohibited. Smoking is prohibited in tire storage yards, except in designated areas.

❖ Except in smoking locations designated by the owner and approved by the fire code official, smoking is prohibited in tire storage yards. "No smoking" signs should be posted prominently in the yards and rigorously enforced.

2504.4 Power lines. Tire storage piles shall not be located beneath electrical power lines having a voltage in excess of 750 volts or that supply power to fire emergency systems.

❖ This requirement prohibits the location of tire storage yards under power lines that are either more than 750 volts or supply power to fire emergency systems. Should a fire occur in a tire storage yard, it could burn for some time before it is extinguished. Because these fires can be quite extensive, this precaution reduces the possibility that large electrical grids or emergency power supplies will be disabled for extended periods.

2504.5 Fire safety plan. The owner or individual in charge of the tire storage yard shall be required to prepare and submit to the fire code official a fire safety plan for review and approval. The fire safety plan shall include provisions for fire department vehicle access. At least one copy of the fire safety plan shall be prominently posted and maintained at the storage yard.

❖ This section requires that the owner develop a fire safety plan and submit it to the fire code official for approval. This plan should be as detailed as necessary, depending on the size and layout of the yard. This plan should also be coordinated with fire department preincident plans. The plan should include the size and composition of the storage material; layout of access and egress routes; the physical infrastructure of the roads and other possible access routes. Maps should include hydrant and water sources, interior access lanes and fuel load configurations. The composition of the storage pile should be indicated because shredded scrap or chip piles require a different fire-fighting approach than a pile of whole tires. The location of utilities on the site should also be included.

2504.6 Telephone number. The telephone number of the fire department and location of the nearest telephone shall be posted conspicuously in attended locations.

❖ This section requires posting of the fire department telephone number along with the location of the telephone in conspicuous locations.

SECTION 2505
OUTDOOR STORAGE

2505.1 Individual piles. Tire storage shall be restricted to individual piles not exceeding 5,000 square feet (464.5 m^2) of continuous area. Piles shall not exceed 50,000 cubic feet (1416 m^3) in volume or 10 feet (3048 mm) in height.

❖ Whole tires, shredded tires or tire chip piles represent a fuel package of considerable volume. To manage this hazard, the code places limitations on pile heights and sizes. Pile limitations are a means of controlling the size of potential fires and facilitate fire-fighting operations. The code limits individual piles to a maximum 5,000 square feet (465 m^2) in area, 50,000 cubic feet (1416 m^3) in volume and a maximum 10 feet (3048 mm) in height.

2505.2 Separation of piles. Individual tire storage piles shall be separated from other piles of salvage by a clear space of at least 40 feet (12 192 mm).

❖ The intent of this section is to create firebreaks between piles of tire storage. The distance required is 40 feet (12 192 mm) in all directions.

2505.3 Distance between piles of other stored products. Tire storage piles shall be separated by a clear space of at least 40 feet (12 192 mm) from piles of other stored product.

❖ This section also requires a 40-foot (12 192 mm) firebreak from piles of other materials, not just other tire storage piles.

2505.4 Distance from lot lines and buildings. Tire storage piles shall be located at least 50 feet (15 240 mm) from lot lines and buildings.

❖ This section requires increasing the firebreak to 50 feet (15 240 mm) from lot lines and buildings.

2505.5 Fire breaks. Storage yards shall be maintained free from combustible ground vegetation for a distance of 40 feet (12 192 mm) from the stored material to grass and weeds; and for a distance of 100 feet (30 480 mm) from the stored product to brush and forested areas.

❖ Because dry grass and weeds represent a significant fuel accumulation and fire hazard, this section requires that all combustible vegetation (grasses and weeds) be kept at least 40 feet (12 192 mm) from storage piles. The distance is increased to 100 feet (30 480 mm) when next to brush or forested areas, not only to protect the brush and forested area from a fire in a storage yard but also to protect the storage yard from a brush or forest fire. For additional information on the protection of such areas, see the *International Urban-Wildland Interface Code®* (IUWIC™). The clearances required by this section require a vegetation control program to reduce the hazards.

2505.6 Volume more than 150,000 cubic feet. Where the bulk volume of stored product is more than 150,000 cubic feet (4248 m^3), storage arrangement shall be in accordance with the following:

1. Individual storage piles shall comply with size and separation requirements in Sections 2505.1 through 2505.5.
2. Adjacent storage piles shall be considered a group, and the aggregate volume of storage piles in a group shall not exceed 150,000 cubic feet (4248 m^3).

Separation between groups shall be at least 75 feet (22 860 m) wide.

❖ This section correlates with the requirements in Section 2505.1, which allow storage piles up to 50,000 cubic feet (1416 m^3) in volume. There is really no reason to limit the number of piles but rather to limit the volume of storage. This section limits the group volume to 150,000 cubic feet (4248 m^3) and clarifies that the piles must still comply with Sections 2505.1 through 2505.5 on pile size and separations. Basically, with the 75-foot (22 860 mm) separation requirement between groups, unlimited amounts of storage would be allowed.

2505.7 Location of storage. Outdoor waste tire storage shall not be located under bridges, elevated trestles, elevated roadways or elevated railroads.

❖ This section states restrictions on where waste tire storage may be located. As a result of past fire experiences that created extended disruption in mass transit, storage is prohibited under elevated roadways and bridges.

SECTION 2506
FIRE DEPARTMENT ACCESS

2506.1 Required access. New and existing tire storage yards shall be provided with fire apparatus access roads in accordance with Section 503 and this section.

❖ Fire apparatus access roads must be arranged to provide clear, unobstructed access to required yard hydrants, if provided, and all points of the yard storage area. Turn-arounds and turning radii must be sized so that the fire apparatus can maneuver to protect exposures as well as fire fighters. Access roads must also comply with Section 503. Appendix D covers design guidelines for access roads. The appendices are not considered as part of the code unless specifically adopted. See Section 1 of the sample adopting ordinance on page v of the code.

2506.2 Location. Fire apparatus access roads shall be located within all pile clearances identified in Sections 2505.4 and within all fire breaks required in Section 2505.5. Access roadways shall be within 150 feet (45 720 mm) of any point in the storage yard where storage piles are located, at least 20 feet (6096 mm) from any storage pile.

❖ In order to provide adequate access for fire suppression operations in tire yard storage fires, fire apparatus access roads must be located within all pile clearances and within all firebreaks. No portion of a pile is to be more than 150 feet (45 720 mm) from an access road and no less than 20 feet (6096 mm) from any storage pile. Maintaining a 20-foot (6096 mm) clearance between storage piles and the edge of the fire apparatus access road reduces the likelihood that the collapse of a pile [limited to 10 feet (3048 mm) in height by Section 2505.1] will totally obstruct access.

SECTION 2507
FENCING

2507.1 Where required. Where the bulk volume of stored material is more than 20,000 cubic feet (566 m^3), a firmly anchored fence or other approved method of security that controls unauthorized access to the storage yard shall surround the storage yard.

❖ Perimeter fences for site security are required to limit access to stored materials and deter theft, vandalism and arson in storage yards where the volume of tire material stored exceeds 20,000 cubic feet (566 m^3).

2507.2 Construction. The fence shall be constructed of approved materials and shall be at least 6 feet (1829 mm) high and provided with gates at least 20 feet (6096 mm) wide.

❖ Owners should be encouraged to construct noncombustible fences. Chain-link fencing is an excellent choice because it can be firmly anchored and requires little maintenance. To accommodate fire apparatus, all gates must be at least 20 feet (6096 mm) wide. Depending on local security concerns, fences may be topped with barbed wire for intruder control to further enhance the level of security provided. Such an installation should be approved by the fire code official and the local administrative authority, since some communities have ordinances or a zoning code that prohibits the use of barbed wire on fences.

2507.3 Locking. All gates to the storage yard shall be locked when the storage yard is not staffed.

❖ This section states that access gates to the storage yard are to be locked when not staffed to deter unauthorized entry. Since the yard entrance is considered part of the fire apparatus road network required for the yard, gates must also comply with the provisions of Section 503.6.

2507.4 Unobstructed. Gateways shall be kept clear of obstructions and be fully openable at all times.

❖ This section requires that gates be accessible and be fully openable at all times, which will allow full and quick access for fire apparatus.

SECTION 2508
FIRE PROTECTION

2508.1 Water supply. A public or private fire protection water supply shall be provided in accordance with Section 508. The water supply shall be arranged such that any part of the storage yard can be reached by using not more than 500 feet (152 m) of hose.

❖ This section requires that a fire protection water supply, in accordance with Section 508, be provided to the storage yard. Open storage yards are often located substantial distances from public fire mains, which may mean that water must be delivered to the scene by fire department tankers. This requirement should be included in both the prefire plan and the facilities fire safety plan. Alternative water supplies such as a stream, lake or other body of water in the vicinity should be explored. Storage of tires totaling no more than 50,000 cubic feet (1416 m^3) should have a water supply of 1,000 gallons per minute (gpm) (3785 L/m) for at least 6 hours (see Appendix C of NFPA 231D).

2508.2 Fire extinguishers. Buildings or structures shall be provided with portable fire extinguishers in accordance with Section 906. Fuel-fired vehicles operating in the storage yard shall be equipped with a minimum 2-A:20-B:C rated portable extinguisher.

❖ This section, as well as Section 906, requires that fire extinguishers be available. Portable fire extinguishers are provided for incipient fire control. These appliances should be located where they are readily available. They are also required on fuel-fired vehicles operating in the yard.

SECTION 2509
INDOOR STORAGE ARRANGEMENT

2509.1 Pile dimensions. Where tires are stored on-tread, the dimension of the pile in the direction of the wheel hole shall not be more than 50 feet (15 240 mm). Tires stored adjacent to or along one wall shall not extend more than 25 feet (7620 mm) from that wall. Other piles shall not be more than 50 feet (15 240 mm) in width.

❖ This section places further restrictions on storage adjacent to a wall to enhance fire-fighting access to the inside of the stored tires in case of a fire. Storage piled against a wall must not extend more than 25 feet (7620 mm) from the wall. When the tires are stored on their tread, the piles shall not be more than 50 feet (15 240 mm), measured in the direction of the wheel hole.

Bibliography

The following resource materials are referenced in this chapter or are relevant to the subject matter addressed in this chapter.

IBC -2003, *International Building Code*. Falls Church, VA: International Code Council, 2003.

IUWIC-2003, *International Urban-Wildland Interface Code*. Falls Church, VA: International Code Council, 2003.

NFPA 231D-98, *Standard for Storage of Rubber Tires*. Quincy, MA: National Fire Protection Association, 1998.

Prevention and Management of Scrap Tire Fires. Washington, DC: Rubber Manufacturers Association.

Chapter 26: Welding and Other Hot Work

General Comments

Welding and other hot work are frequent ignition sources. Statistics from a major property insurance company for a recent five-year period showed 290 hot-work-related ignitions that led to losses of $407 million, or an average of $1.4 million per incident. Of these 290 losses, 42 percent were caused by employees and 58 percent were caused by outside contractors. To compare the magnitude of these losses, this same insurance company saw 395 fires associated with housekeeping and 262 losses associated with smoking. The average poor housekeeping loss was $902,000; the average smoking loss was about $440,000.

Both hot-work operations themselves and the equipment and materials associated with such work can create significant ignition and fire hazards. Hot work creates sparks and slag and gives off heat. Materials such as acetylene and oxygen are used in gas welding and an electrical current is used for arc welding. Additionally, these activities tend to occur in buildings that were not designed for these materials and hazards. Hot work often occurs within buildings undergoing renovations, which are even more susceptible to ignition. Hot work can be either temporary or ongoing. Permanent installations generally have the ability to address ignition hazards more consistently.

Several different types of hot work would fall under the requirements found in Chapter 26, including both gas and electric arc methods and any open-torch operations.

The important factor in avoiding ignition hazards is preparing for and monitoring hot-work activities. Primarily these precautions relate to basic fire prevention and fire control. Chapter 26 details a program that allows a facility to assign an employee to be the administrator of a hot-work program as defined in Section 2602.1. This administrator would be allowed to issue permits for work on site, would be required to perform prework inspections and would be responsible for ensuring that the correct safety measures are taken. The fire code official has the authority to make periodic checks of these records, so they must be made available for at least 48 hours after the work ends. This chapter provides specific requirements for the protection of combustibles and requirements for fire watches.

Personnel undertaking hot work will have varying levels of familiarity with the building or facility where the work is being done. Often the person undertaking hot work is not an employee at the facility and may not be under the direct control of the hot work program manager. The qualifications of the hot work operator are discussed in Section 2603.4.

Purpose

This chapter covers requirements for safety in welding and other types of hot work by reducing the potential for fire ignitions. Many of the activities of this chapter focus on the actions of the occupants. As noted, welding and other hot work are responsible for a large percentage of fire ignitions that usually result in large losses.

SECTION 2601 GENERAL

2601.1 Scope. Welding, cutting, open torches and other hot work operations and equipment shall comply with this chapter.

❖ This section is fairly straightforward and states that any welding, cutting, open torches and other hot work would be regulated by Chapter 26. Generally, this would encompass any kind of welding and cutting operations, whether electric or gas.

2601.2 Permits. Permits shall be required as set forth in Section 105.6.

❖ The process of issuing permits gives the fire code official an opportunity to carefully evaluate and regulate hazardous operations. Permit applicants should be required to demonstrate that their operations comply with the intent of the code before the permit is issued. See the commentary to Section 105.6 for a general discussion of operations requiring an operational permit. The process also notifies the fire department of the need for prefire planning for the hazardous property and helps to verify that proper procedures will be followed.

The actual permit requirements for hot work are in Section 105.6.24. This section lists several specific instances where a permit would be required for hot-work operations. One of the items would allow a single permit to be issued to allow a hot work program. This program will be explained in more detail in this chapter, but essentially it allows a person on site to manage the hot work activities. This program has a number of safety requirements, including fire department review of documentation at the facility for a minimum of 48 hours after the work is completed. Generally, this type of permit provides much needed flexibility for facilities where hot work is a common occurrence.

2601.3 Restricted areas. Hot work shall only be conducted in areas designed or authorized for that purpose by the personnel responsible for a Hot Work Program. Hot work shall not be conducted in the following areas unless approval has been obtained from the fire code official:

1. Areas where the sprinkler system is impaired.
2. Areas where there exists the potential of an explosive atmosphere, such as locations where flammable gases, liquids or vapors are present.
3. Areas with readily ignitable materials, such as storage of large quantities of bulk sulfur, baled paper, cotton, lint, dust or loose combustible materials.
4. On board ships at dock or ships under construction or repair.
5. At other locations as specified by the fire code official.

❖ This section describes restrictions on the areas where hot work can take place. Normally hot work activities are restricted to designated areas, however, there are times when hot work may be needed in specific locations, such as in a building undergoing renovation. This section presents this list as a way to verify that when hot work is needed, notification of the activity is made and special precautions are undertaken. In addition, this section also authorizes the fire code official to add other areas where special approval would be necessary. The code cannot anticipate all potentially hazardous situations. For this reason, this section does not explicitly prohibit hot work in these areas; it simply requires special approval.

2601.4 Cylinders and containers. Compressed gas cylinders and fuel containers shall comply with this chapter and Chapter 30.

❖ This section is focused on any cylinders or containers used to store gases used in hot work operations, primarily oxygen and acetylene. Oxygen is an oxidizing gas and acetylene is a highly flammable gas and an unstable reactive Class 2. This section requires that any specific requirements within Chapter 26 be addressed along with the general requirements found in Chapter 30 regarding compressed gases.

2601.5 Design and installation of oxygen-fuel gas systems. An oxygen-fuel gas system with two or more manifolded cylinders of oxygen shall be in accordance with NFPA 51.

❖ This section references NFPA 51 for any oxygen-fuel gas systems where any number of oxygen containers are manifolded. The scope of this standard specifically states that it addresses only situations where two or more cylinders are manifolded.

SECTION 2602 DEFINITIONS

2602.1 Definitions. The following words and terms shall, for the purposes of this chapter and as used elsewhere in this code, have the meanings shown herein.

❖ Definitions of terms can help in the understanding and application of the code requirements. The purpose for including those definitions that are associated with the subject matter of this chapter is to provide more convenient access to them without having to refer back to Chapter 2. It is important to emphasize that these terms are not exclusively related to this chapter but are applicable everywhere the term is used in the code. For convenience, these terms are also listed in Chapter 2 with a cross reference to this section. The use and application of all defined terms, including those defined in this section, are set forth in Section 201.

HOT WORK. Operations including cutting, welding, Thermit welding, brazing, soldering, grinding, thermal spraying, thawing pipe, installation of torch-applied roof systems or any other similar activity.

❖ This term describes the scope of what would be considered hot work as it is regulated in this chapter. The scope is broad and would include any activity that produces sparks, slag or other waste products. This would include both gas and electric methods. Torch-applied roof systems are also included.

HOT WORK AREA. The area exposed to sparks, hot slag, radiant heat, or convective heat as a result of the hot work.

❖ This definition helps to locate which areas would be considered part of the hot work area to better understand the level of susceptibility to ignition.

HOT WORK EQUIPMENT. Electric or gas welding or cutting equipment use for hot work.

❖ In the past, chapters dealing with hot work focused primarily on gas welding. Electric welding, though it does not deal with oxygen and fuel gases, still presents ignition hazards.

HOT WORK PERMITS. Permits issued by the responsible person at the facility under the hot work permit program permitting welding or other hot work to be done in locations referred to in Section 2603.3 and pre-permitted by the fire code official.

❖ As applied in Chapter 26, this permit differs from a typical permit in that it is not directly issued by the fire code official. Instead, a hot work operations facility is given permission to designate a person, perhaps the safety officer, to issue permits as needed. This results in flexibility for facilities where hot work is a common occurrence. These permits are issued under what is called a hot work program, which is also defined in this section.

HOT WORK PROGRAM. A permitted program, carried out by approved facilities-designated personnel, allowing them to oversee and issue permits for hot work conducted by their personnel or at their facility. The intent is to have trained, on-site, responsible personnel ensure that required hot work safety measures are taken to prevent fires and fire spread.

❖ This kind of program is described in the definition for "hot work permits." This program allows someone on site to control the issuing of permits for hot work. The person who is charged with this responsibility must be trained in hot work operations and have the necessary authority. Having such a program at a facility encourages a better understanding of fire safety and perhaps more incentive to play an active role in the prevention of fires. This program reduces the administrative burden on the fire department and ensures that hot work operations can proceed as needed.

RESPONSIBLE PERSON. A person trained in the safety and fire safety considerations concerned with hot work. Responsible for reviewing the sites prior to issuing permits as part of the hot work permit program and following up as the job progresses.

❖ This is the person designated to administer the hot work program. Without this definition, the term "responsible person" is a vague descriptor. The definition includes the scope of responsibilities for this person.

TORCH-APPLIED ROOF SYSTEM. Bituminous roofing systems using membranes that are adhered by heating with a torch and melting asphalt back coating instead of mopping hot asphalt for adhesion.

❖ This is a very specific operation that relates to hot work in that it uses a torch to adhere the materials. It is not considered welding, but still falls within the definition of "Hot work."

SECTION 2603
GENERAL REQUIREMENTS

2603.1 General. Hot work conditions and operations shall comply with this chapter.

❖ This section is generally applicable to all hot work activities, which would include welding and cutting, but also includes activities such as torch-applied roof system activities. The requirements are primarily related to the hot work permit program, qualifications and general administrative provisions related to fire safety procedures.

2603.2 Temporary and fixed hot work areas. Temporary and fixed hot work areas shall comply with this section.

❖ These provisions are applicable to both temporary and permanent activities because the same fire hazard exists in both cases. Temporary situations generally pose a greater hazard, however, because they typically occur in areas not designed for such ignition hazards. For example, hot work is fairly common in buildings undergoing renovation.

2603.3 Hot work program permit. Hot work permits, issued by an approved responsible person under a hot work program, shall be available for review by the fire code official at the time the work is conducted and for 48 hours after work is complete.

❖ Because individual facilities are allowed to manage the process of issuing permits, permit information must be available to the fire code official for periodic review. As noted in the definition for "Hot work permits," the records must be available for at least 48 hours following completion of work. The 48-hour period gives the fire code official the time necessary to verify that permitting was done according to established procedures if a fire should occur.

This section does not require that the permits be submitted to the fire code official, it asks only that they be available for review. This section, along with Section 2603.5, is a part of the package that allows a periodic random check of the permitting and hot work administrative procedures.

2603.4 Qualifications of operators. A permit for hot work operations shall not be issued unless the individuals in charge of performing such operations are capable of performing such operations safely. Demonstration of a working knowledge of the provisions of this chapter shall constitute acceptable evidence of compliance with this requirement.

❖ The definitions for "Hot work program" and "Responsible person" were specifically aimed at the individual who coordinates issuing and managing permits. This particular section is aimed at the qualifications of the person undertaking the hot work itself. This could be someone unfamiliar with the facility. The person performing the hot work must both understand the particular operation and understand and demonstrate knowledge of the specific provisions of this chapter. The qualifications of the operator and administrative follow-through are likely the most critical aspect in preventing fires.

2603.5 Records. The individual responsible for the hot work area shall maintain "prework check" reports in accordance with Section 2604.3.1. These reports shall be maintained on the premises for a minimum of 48 hours after work is complete.

❖ This section is specific to the hot work program administrator and requires that the prework checks be available for review for at least 48 hours. Again, this allows periodic checks by the fire code official and would allow reasonable time for review of documentation after a fire.

2603.6 Signage. Visible hazard identification signs shall be provided where required by Chapter 27. Where the hot work area is accessible to persons other than the operator of the hot work equipment, conspicuous signs shall be posted to warn others before they enter the hot work area. Such signs shall display the following warning:

CAUTION
HOT WORK IN PROGRESS
STAY CLEAR.

❖ Signage, as with many other code applications, is used as a method to warn of hazards. This is especially critical in areas where people unfamiliar with the hazards may be present. For example, this may be more important during renovations in an occupied office building. The signage requirements apply to both temporary and fixed situations, and the visibility of the signs must be consistent with the requirements of Chapter 27.

SECTION 2604 FIRE SAFETY REQUIREMENTS

2604.1 Protection of combustibles. Protection of combustibles shall be in accordance with Sections 2604.1.1 through 2604.1.9.

❖ This section deals with the basic fire safety activities that should be addressed when undertaking hot work. Some of the requirements may be more applicable to certain types of hot work than others because of the specific hazards presented. The three specific issues addressed include the protection of the area and fire protection systems located in the vicinity of the hot work; requirements for fire watches during and after the hot work is undertaken and the administrative procedures used to confirm that all the applicable safety steps have been taken. This section is focusing primarily on the protection of combustibles through both passive and active means. More specifically, combustibles must either be removed or be protected by fire protection systems that remain operational during hot work activities.

2604.1.1 Combustibles. Hot work areas shall not contain combustibles or shall be provided with appropriate shielding to prevent sparks, slag or heat from igniting exposed combustibles.

❖ This section requires that combustibles either be removed from the area or, if not removed, be properly shielded from sparks or excessive heat that could ignite a fire. A shield may need to be both noncombustible and insulating. In the case of sparks that provide little heat, a metal shield may be allowable. If the hot work operation gives off large amounts of heat, such as the use of a torch, the shield would also need to resist the transmission of heat.

2604.1.2 Openings. Openings or cracks in walls, floors, ducts or shafts within the hot work area shall be tightly covered to prevent the passage of sparks to adjacent combustible areas, or shielded by metal fire-resistant guards, or curtains shall be provided to prevent passage of sparks or slag.

❖ Openings or cracks in walls, floors, ducts or shafts within the hot work area shall be tightly covered to prevent the passage of sparks to adjacent combustible areas; shields constructed of metal fire-resistant guards or curtains shall be provided to prevent passage of sparks or slag.

Openings or cracks in walls are potential travel paths for sparks or slag from hot work operations. It is very common for a partition in an existing building to have penetrations for pipes or other utilities. Sparks that penetrate a partition or wall have the potential of starting a fire in adjoining rooms or areas.

2604.1.3 Housekeeping. Floors shall be kept clean within the hot work area.

❖ This section addresses a basic fire safety issue of keeping the floors in the area of the hot work clean. Accumulations of dust and other high-surface-area materials are highly susceptible to ignition, flash fires and explosions. This kind of housekeeping is doubly important for welding and similar hot work in a building under renovation because it tends to have a higher concentration of dusts and other hazardous combustibles.

2604.1.4 Conveyor systems. Conveyor systems that are capable of carrying sparks to distant combustibles shall be shielded or shut down.

❖ A hazard, particularly in industrial applications, is the use of conveyor systems. More specifically, if a welding operation ignites a fire on a conveyor system, the fire can move throughout the parts of the building served by the conveyor and possibly ignite multiple fires. Generally, most fire protection features are not designed for multiple-fire ignitions, but instead are designed for a single event. For instance, a sprinkler system has a specific design density for a specified area of operation. If the fire demand is higher than what the sprinkler system was designed to handle, the water supply and pressure could be inadequate and the sprinkler system will likely be overcome. Also, because the fire is moving, it is more difficult for the sprinklers to activate because the sprinkler might not heat sufficiently. This results in the fire growing larger before intervention from the fire protection systems. Even if a sprinkler does activate, it may not be addressing the origin of the fire.

2604.1.5 Partitions. Partitions segregating hot work areas from other areas of the building shall be noncombustible. In fixed hot work areas, the partitions shall be securely connected to the floor such that no gap exists between the floor and the partition. Partitions shall prevent the passage of sparks, slag, and heat from the hot work area.

❖ The partitions discussed in this section act as shields, as described in Section 2604.1.1. The requirements state that partitions must be installed so that there is no room for sparks, slag or heat to affect the area surrounding the hot work area. These requirements are primarily geared toward fixed hot work areas where the conditions can be more permanently controlled. Hot work operations in temporary locations will likely not have partitions installed solely for the purpose of protecting combustible materials. Section 2604.1.1 would be more applicable to temporary hot work.

2604.1.6 Floors. Fixed hot work areas shall have floors with noncombustible surfaces.

❖ The requirement for noncombustible floors is specific to fixed hot work operations because a fixed situation can be more easily controlled than a temporary hot work operation. In many cases, program administrators will have little control over the type of floor and the location of combustibles in the areas were temporary hot work occurs.

2604.1.7 Precautions in hot work. Hot work shall not be performed on containers or equipment that contains or has contained flammable liquids, gases or solids until the containers and equipment have been thoroughly cleaned, inerted or purged; except that "hot tapping" shall be allowed on tanks and pipe lines when such work is to be conducted by approved personnel.

❖ The title of this section is somewhat misleading because the section pertains specifically to hot work on containers or equipment that may contain flammable or combustible liquids. Simply because a container is empty does not mean that it is safe to conduct hot work. In fact, the vapors are usually much more susceptible to ignition than the liquid itself. Therefore, before any hot work can occur, the vapors and liquids must be purged. This section does allow hot work on containers or piping without the specific removal of liquids. This is termed "hot tapping" and must be done by an experienced individual.

2604.1.8 Sprinkler protection. Automatic sprinkler protection shall not be shut off while hot work is performed. Where hot work is performed close to automatic sprinklers, noncombustible barriers or damp cloth guards shall shield the individual sprinkler heads and shall be removed when the work is completed. If the work extends over several days, the shields shall be removed at the end of each workday. The fire code official shall approve hot work where sprinkler protection is impaired.

❖ One of the major sources of losses related to hot work occurs either at buildings under construction or under renovation where an installed sprinkler system has been shut off. In many cases, there is a concern that the sprinkler system will be damaged or will accidentally activate during construction so the system is shut off. Unfortunately, this is the most likely time for an ignition to occur.

As a result of the concerns related to accidental operation of sprinklers, this section does include some methods of protecting the sprinkler without shutting the system down. It is important to note that any shields placed onto the sprinklers are to be removed at the end of each workday.

2604.1.9 Fire detection systems. Approved special precautions shall be taken to avoid accidental operation of automatic fire detection systems.

❖ As with sprinkler systems, there is a concern for false alarms. This section does not give a methodology for protecting against false alarms but does state that precautions must be taken. There are many different technologies and approaches for fire detection systems in addition to the building-specific applications. Each situation should be looked at individually. One possible scenario may be that smoke detectors in the area where hot work is being done be shut down and a fire watch put in place. As soon as the hot work is complete, those detectors are placed back on line.

2604.2 Fire watch. Fire watches shall be established and conducted in accordance with Sections 2604.2.1 through 2604.2.6.

❖ This section is critical to avoiding ignition as a result of hot work operations. The six subsections list criteria for setting a fire watch and how it should be undertaken.

2604.2.1 When required. A fire watch shall be provided during hot work activities and shall continue for a minimum of 30 minutes after the conclusion of the work. The fire code official, or the responsible manager under a hot work program, is authorized to extend the fire watch based on the hazards or work being performed.

Exception: Where the hot work area has no fire hazards or combustible exposures.

❖ Fire watches are required any time hot work is undertaken and are to extend a minimum of 30 minutes beyond completion of the work. The time may need to be extended, depending on the specific hazards present, such as a large amount of combustibles or the facility being open to the public. There is an exception for those situations when combustibles are simply not present. The combustibility of the floor should also be considered.

2604.2.2 Location. The fire watch shall include the entire hot work area. Hot work conducted in areas with vertical or horizontal fire exposures that are not observable by a single individual shall have additional personnel assigned to fire watches to ensure that exposed areas are monitored.

❖ This section states that a fire watch is required in all hot work areas. The term "hot work area" is defined but is necessarily a general definition because many things will affect the extent of the area. These factors include the type and application of hot work, the configuration and layout of the space and the types of materials in the area. Also, in some situations the fire watch may need to consist of more than one person because the layout of the area may prevent one person from watching an entire area, such as when many pieces of equipment act as obstructions or the shape of the room or placement of partitions blocks a line of sight.

2604.2.3 Duties. Individuals designated to fire watch duty shall have fire-extinguishing equipment readily available and shall be trained in the use of such equipment. Individuals assigned to fire watch duty shall be responsible for extinguishing spot fires and communicating an alarm.

❖ The individuals who undertake a fire watch have specific duties. They are not only to watch for and notify of an ignition of combustibles, they are also to be prepared to extinguish spot fires with portable extinguishers. Intervention when fires are small is the best line of defense in extinguishing and controlling fires. Waiting until the fire department or fire brigade arrives will allow a fire to increase dramatically in size and intensity.

2604.2.4 Fire training. The individuals responsible for performing the hot work and individuals responsible for providing the fire watch shall be trained in the use of portable fire extinguishers.

❖ A person conducting a fire watch must be trained to operate fire extinguishers located in the watch area. As noted previously, intervention in the incipient stages of a fire is extremely effective.

2604.2.5 Fire hoses. Where hoselines are required, they shall be connected, charged and ready for operation.

❖ This section requires that when a hoseline is required, it should be properly charged and ready for use during a fire watch. Otherwise, the effectiveness is much lower. As noted already, fires are more likely to be extinguished or controlled when intervention occurs early. An uncharged hoseline will defeat the purpose of the equipment.

2604.2.6 Fire extinguisher. A minimum of one portable fire extinguisher complying with Section 906 and with a minimum 2-A:20-B:C rating shall be readily accessible within 30 feet (9144 mm) of the location where hot work is performed.

❖ This section specifies that the fire extinguishers required for a fire watch must be an all-purpose extinguisher for all fire types; the potential fire type will vary with the type of hot work and the surrounding combustibles. The 30-foot (9144 mm) travel distance specified here is more restrictive than what is required for similar ratings of extinguishers in Section 906. Table 906.3(2) would allow a maximum travel distance of 50 feet (15 240 mm) for other applications with the same rating of extinguisher.

2604.3 Area reviews. Before hot work is permitted and at least once per day while the permit is in effect, the area shall be inspected by the individual responsible for authorizing hot work operations to ensure that it is a fire safe area. Information shown on the permit shall be verified prior to signing the permit in accordance with Section 105.6.

❖ This section is part of the hot work program. It requires the person administering the program to check the area where hot work has been permitted. A specific checklist is contained within Section 2604.3.1 to provide guidance as to what is to be inspected. Because the authority is given to those other than the fire department to issue permits and manage the safety of hot work operations, documentation of the inspections is a mechanism for review by the fire department to verify that operations are proceeding safely.

2604.3.1 Pre-hot-work check. A pre-hot-work check shall be conducted prior to work to ensure that all equipment is safe and hazards are recognized and protected. A report of the check shall be kept at the work site during the work and available upon request. The pre-hot-work check shall determine all of the following:

1. Hot work equipment to be used shall be in satisfactory operating condition and in good repair.
2. Hot work site is clear of combustibles or combustibles are protected.
3. Exposed construction is of noncombustible materials or, if combustible, then protected.
4. Openings are protected.
5. Floors are kept clean.
6. No exposed combustibles are located on the opposite side of partitions, walls, ceilings or floors.
7. Fire watches, where required, are assigned.
8. Approved actions have been taken to prevent accidental activation of suppression and detection equipment in accordance with Sections 2604.1.8 and 2604.1.9.
9. Fire extinguishers and fire hoses (where provided) are operable and available.

❖ As noted in Section 2604.3, this section includes a list of items to be reviewed in hot work areas. These checks confirm that the requirements in Sections 2604.1 and 2604.2 are actually being met.

SECTION 2605 GAS WELDING AND CUTTING

2605.1 General. Devices or attachments mixing air or oxygen with combustible gases prior to consumption, except at the burner or in a standard torch or blow pipe, shall not be allowed unless approved.

❖ The materials used for gas cutting and welding are generally materials such as acetylene and oxygen. Acetylene is a flammable gas and oxygen is an oxidizer. Together, these gases produce an intense high-temperature flame. The high temperature of combustion is excellent for welding and cutting and, at the same time, is a significant ignition hazard. Acetylene is the preferred gas because it burns at high temperatures, but other flammable gases, such as methyl acetylene-propadiene and propylene, are also sometimes used. The other major type of welding is electric arc welding, which is discussed in Section 2606.

Because of the enriched combustion process created by pure oxygen, this section prohibits mixing oxygen and air prior to use with the flammable gas of choice for welding or cutting.

2605.2 Cylinder and container storage, handling and use. Storage, handling and use of compressed gas cylinders, containers and tanks shall be in accordance with this section and Chapter 30.

❖ A major hazard with gas welding and cutting is not necessarily the hot work operation itself, but instead the materials used in the operation. These gases are stored in cylinders and manifolded into systems for use. The storage, handling and use of such gases is specifically regulated by Chapter 30, which addresses issues such as security, valve protection, separation from hazardous materials and container and cylinder marking (see commentary, Chapter 30). Note that oxygen-fuel gas systems where two or more cylinders of oxygen are manifolded must meet the requirements of NFPA 51.

2605.3 Precautions. Cylinders, valves, regulators, hose and other apparatus and fittings for oxygen shall be kept free from oil or grease. Oxygen cylinders, apparatus and fittings shall not be handled with oily hands, oily gloves, or greasy tools or equipment.

❖ Oxygen, which is an oxidizer, will increase the intensity of combustion. Therefore, traces of combustibles, such as oils or greases, that normally pose a moderate hazard will be a higher hazard when found on cylinders of oxygen. Keeping gas cylinders free of grease and oil is especially important because gas-welding operations use flammable gases in combination with oxygen.

2605.4 Acetylene gas. Acetylene gas shall not be piped except in approved cylinder manifolds and cylinder manifold connections, or utilized at a pressure exceeding 15 pounds per square inch gauge (psig) (103 kPa) unless dissolved in a suitable solvent in cylinders manufactured in accordance with DOTn 49 CFR. Acetylene gas shall not be brought in contact with unalloyed copper, except in a blowpipe or torch.

❖ Acetylene is the preferred gas for welding because it burns at very high temperatures, which is more conducive to welding and cutting operations. This also creates a high fire hazard. Therefore, this section limits the piping of acetylene to approved cylinder manifolds and manifold connections, which can be accomplished in a variety of ways. The intent is simply to stress that these connections are critical and must be approved. Also, the pressure is limited to 15 psig (103 kPa) to avoid large releases of acetylene, especially in pressurized spaces.

There is an exception when acetylene is dissolved in a solvent, probably because the overall flammability is reduced.

This section also prohibits contact with unalloyed copper, unless used with a blowpipe and torch. Unalloyed copper is one of a number of metals that is subject to chemical conversion to a metallic acetylide. Copper acetylide is an extremely shock-sensitive explosive. If detonated, even in very small quantities, it can initiate acetylene decomposition, cause hose or tank rupture and potentially cause catastrophic detonations or deflagrations, especially when tanks are manifolded.

2605.5 Remote locations. Oxygen and fuel-gas cylinders and acetylene generators shall be located away from the hot work area to prevent such cylinders or generators from being heated by radiation from heated materials, sparks or slag, or misdirection of the torch flame.

❖ Because of the hazards of using gases for welding and cutting, the cylinders and generators must be located away from the hot work operation itself. This means that the connection must be a sufficient length to allow locating the cylinders a safe distance from the hot work. Acetylene is both a compressed gas and a flammable gas. Heating could cause overpressures in the cylinder and ignition of the gas.

2605.6 Cylinders shutoff. The torch valve shall be closed and the gas supply to the torch completely shut off when gas welding or cutting operations are discontinued for a period of 1 hour or more.

❖ Although the torch has a valve controlled by the operator, when hot work is discontinued for more than an hour, the valves that supply the gas to the torch should also be closed. This is a fairly basic concept for prevention. This limits the amount of flammable or oxidizing gases that can be emitted to the atmosphere if the torch or hose is damaged in any way or if a valve at the torch head is not tightly closed.

2605.7 Prohibited operation. Welding or cutting work shall not be held or supported on compressed gas cylinders or containers.

❖ Section 2605.5 addresses physical separation of the welding and cutting gas cylinders from the hot work. This section goes a step further and prohibits welding operations on top of compressed gas containers or cylinders. This requirement would apply to any compressed gas cylinders or containers, whether they contain flammable gases or not. The concern is explosion caused by overpressure in the compressed gas cylinders when they are heated.

2605.8 Tests. Tests for leaks in piping systems and equipment shall be made with soapy water. The use of flames shall be prohibited for leak testing.

❖ This section is mandating that piping and equipment subjected to hot work be tested using soapy water to recognize any leaks rather than using a flame to indicate leaks. This is important because the piping may contain some residual flammable or oxidizing materials.

SECTION 2606 ELECTRIC ARC HOT WORK

2606.1 General. The frame or case of electric hot work machines, except internal-combustion-engine-driven machines, shall be grounded. Ground connections shall be mechanically strong and electrically adequate for the required current.

❖ The following sections are specific to welding and cutting using an electrical arc. This process produces enough heat to join metals together or cut metals. There are several variations of arc welding and cutting. More specifically, in some arc welding processes a shielding gas is used to protect the weld from contaminants and prevent metal oxidation. Also, the electrode is either nonconsumable or consumable, which may dictate whether additional materials may need to be entered into the process.

Electric arc hot work poses some potential hazards because the process is working with electric power and current. Therefore, unless the welder is powered using an internal combustion engine, the frame or case of the welding piece must be grounded.

2606.2 Return circuits. Welding current return circuits from the work to the machine shall have proper electrical contact at joints. The electrical contact shall be periodically inspected.

❖ The current for welding must be able to loop back to the power source from the item being welded or cut; therefore, the item must be grounded to avoid shock and potential ignition hazards. Basically the electric charge needs to be directed back to the power source, or it will find another direction to go if the connections are not reliable.

2606.3 Disconnecting. Electrodes shall be removed from the holders when electric arc welding or cutting is discontinued for any period of 1 hour or more. The holders shall be located to prevent accidental contact and the machines shall be disconnected from the power source.

❖ This section is trying to prevent materials and machinery used as part of the electric arc welding process from being a source of ignition after the hot work operations have been discontinued for an hour through the disconnection of the electrodes and disconnection from the power supply. This is a similar to the intent of Section 2605.6 for gas hot work, which requires the supply gas valves to be shut off when an extended interruption occurs.

2606.4 Emergency disconnect. A switch or circuit breaker shall be provided so that fixed electric welders and control equipment can be disconnected from the supply circuit. The disconnect shall be installed in accordance with the ICC *Electrical Code*.

❖ When an electric arc welding or cutting operation is in a permanent location, it is important that the particular operation be isolated onto a single disconnect switch. This makes it easier to verify that operations are in fact disconnected from the power supply during an emergency. The ICC *Electrical Code*® (ICC EC™) will have the specific requirements for this disconnect.

2606.5 Damaged cable. Damaged cable shall be removed from service until properly repaired or replaced.

❖ Damaged cable increases the likelihood of ignition hazards because wires may be exposed, resulting in resistance heating or potential electric shock hazards.

SECTION 2607 CALCIUM CARBIDE SYSTEMS

2607.1 Calcium carbide storage. Storage and handling of calcium carbide shall comply with Chapter 27 of this code and Chapter 7 of NFPA 51.

❖ Calcium carbide gas is used for the creation of acetylene gas used in gas welding and cutting. Acetylene gas as discussed earlier in this commentary is extremely flammable and burns at very high temperatures. Calcium carbide itself is considered water reactive and will form acetylene upon contact with moisture. This section increases the likelihood that the gas is handled to avoid a hazardous situation. Compliance with both Chapter 27 of the code and Chapter 7 of NFPA 51A is required. Chapter 7 specifically relates to storage requirements for calcium carbide gas. These requirements focus on keeping the calcium carbide dry and separated from other areas when it is stored in large quantities.

SECTION 2608 ACETYLENE GENERATORS

2608.1 Use of acetylene generators. The use of acetylene generators shall comply with this section and Chapter 4 of NFPA 51A.

❖ Acetylene generators essentially use calcium carbide gas and moisture to generate acetylene. Because acetylene is extremely flammable, the equipment must be handled with care.

Detailed requirements for acetylene generators are left to Chapter 6 of NFPA 51A, which is specific to such generators. These provisions deal with issues such as the construction requirements of generator rooms or houses and detailed installation requirements for the generators themselves.

There are several types of generators, including those with automatic water feed and manual water feed.

2608.2 Portable generators. The minimum volume of rooms containing portable generators shall be 35 times the total gas-generating capacity per charge of all generators in the room. The gas-generating capacity in cubic feet per charge shall be assumed to be 4.5 times the weight of carbide per charge in pounds. The minimum ceiling height of rooms containing generators shall be 10 feet (3048 mm). An acetylene generator shall not be moved by derrick, crane or hoist while charged.

❖ When generators are portable, they are generally at a greater risk for damage and are more likely to be in a location that may not have been specifically designed for this equipment. Therefore, this section provides some minimum volume requirements to make sure that a particular space can handle the volume of gases produced. The 35-times requirement is likely related to the lower flammability limit for acetylene.

To make it easier for the user of the code to determine how much acetylene could be generated, this section includes a conservative approach to calculate the generation capacity. Additionally, the ceiling height of 10 feet (3048 mm) or greater is to allow the gas, which is lighter than air, to have more room to dissipate. The high ceiling is important because acetylene is considered an asphyxiant. The last portion of this section simply prohibits the movement of charged generators by means such as a crane or hoist, which are likely to cause damage.

2608.3 Protection against freezing. Generators shall be located where water will not freeze. Common salt such as sodium chloride or other corrosive chemicals shall not be utilized for protection against freezing.

❖ Because acetylene generators use calcium carbide and moisture to create acetylene, temperatures should not drop below freezing. Additionally, corrosive materials such as salt used to thaw ice on walkways, should not be used near the generators. Essentially, there is a concern that corrosion could damage the machinery and potentially lead to failure.

SECTION 2609 PIPING MANIFOLDS AND HOSE SYSTEMS FOR FUEL GASES AND OXYGEN

2609.1 General. The use of piping manifolds and hose systems shall be in accordance with Section 2609, Chapter 30 and Chapter 3 of NFPA 51.

❖ Quite often, manifolding of several cylinders is desirable to decrease the amount of time taken to disconnect spent cylinders and reconnect new ones. When cylinders are manifolded, the potential hazard increases because the amount of material that can be released in one event is much larger. Special care and specific requirements are necessary to address these hazards. Compliance with Chapter 3 of NFPA 51 and Chapter 30 for compressed gases is referenced. Although manifolded cylinders may create a higher risk for a large release, numerous disconnections and connections using a single cylinder may increase the likelihood of a failure.

2609.2 Protection. Piping shall be protected against physical damage.

❖ This section requires that piping associated with manifolded cylinders be protected from damage. Again, when cylinders are manifolded, the potential size of the release increases. Manifold piping is more likely to be damaged than the cylinders themselves.

2609.3 Signage. Signage shall be provided for piping and hose systems as follows:

1. Above-ground piping systems shall be marked in accordance with ANSI A13.1.
2. Station outlets shall be marked to indicate their intended usage.
3. Signs shall be posted, indicating clearly the location and identity of section shutoff valves.

❖ Signage for the piping and hosing associated with manifolded cylinders must indicate the content of the system, gas flow direction, intended use of the gas outlets and the location of shutoff valves. This information helps to quickly assess emergencies and take corrective action.

2609.4 Manifolding of cylinders. Oxygen manifolds shall not be located in an acetylene generator room. Oxygen manifolds shall be located at least 20 feet (6096 mm) away from combustible material such as oil or grease, and gas cylinders containing flammable gases, unless the gas cylinders are separated by a fire partition.

❖ This section focuses on the hazard of oxygen as an oxidizer. The main requirements are that manifolded oxygen be stored in a room separate from acetylene gas generators and that a distance of 20 feet (6096 mm) be maintained from combustible residues, such as oil and grease, unless a fire partition is installed. As discussed earlier, oxidizers have the potential to intensify burning.

2609.5 Identification of manifolds. Signs shall be posted for oxygen manifolds with service pressures not exceeding 200 psig (1379 kPa). Such signs shall include the words:

LOW-PRESSURE MANIFOLD
DO NOT CONNECT HIGH-PRESSURE CYLINDERS
MAXIMUM PRESSURE 250 PSIG

❖ The signage required by this section is intended to prevent high-pressure oxygen cylinders from being connected to a low-pressure manifold. A high-pressure cylinder may cause a rupture of the system, resulting in an instantaneous release of large volumes of oxygen.

2609.6 Clamps. Hose connections shall be clamped or otherwise securely fastened.

❖ Hoses are potentially less reliable than a pipe if not properly cared for. This section requires clamps or other methods of securing to provide stability.

2609.7 Inspection. Hoses shall be inspected frequently for leaks, burns, wear, loose connections or other defects rendering the hose unfit for service.

❖ Because hoses are subject to aging, wear and damage, they must be inspected regularly.

Bibliography

The following resource materials are referenced in this chapter or are relevant to the subject matter addressed in this chapter.

ANSI A13.1-96, *Scheme for the Identification of Piping Systems*. New York, NY: American National Standards Institute, 1991.

Manz, A., "Welding and Cutting," Section 3, Chapter 14, *Fire Protection Handbook*, 18th Edition. Quincy, MA: National Fire Protection Association, 1997.

NFPA 51-97, *Design and Installation of Oxygen-Fuel Gas Systems for Welding, Cutting, and Allied Processes.* Quincy, MA: National Fire Protection Association, 1997.

NFPA 51A-01, *Acetylene Cylinder Charging Plants.* Quincy, MA: National Fire Protection Association, 2001.

Chapter 27: Hazardous Materials—General Provisions

General Comments

The requirements of this chapter apply to all hazardous chemicals. Hazardous chemicals are defined as those that pose an unreasonable risk to the health and safety of operating or emergency personnel, the public and the environment if not properly controlled during handling, storage, manufacture, processing, packaging, use, disposal or transportation. The requirements of this chapter and the other associated chapters are considered the minimum safety requirements for the use, production and storage of hazardous chemicals.

Purpose

This chapter contains the general requirements for hazardous chemicals in all occupancies. The general provisions of this chapter are intended to be companion provisions with the specific requirements of Chapters 28 through 44 regarding a given hazardous material.

SECTION 2701 GENERAL

2701.1 Scope. Prevention, control and mitigation of dangerous conditions related to storage, dispensing, use and handling of hazardous materials shall be in accordance with this chapter.

This chapter shall apply to all hazardous materials, including those materials regulated elsewhere in this code, except that when specific requirements are provided in other chapters, those specific requirements shall apply in accordance with the applicable chapter. Where a material has multiple hazards, all hazards shall be addressed.

Exceptions:

1. The quantities of alcoholic beverages, medicines, foodstuffs, cosmetics, and consumer or industrial products containing not more than 50 percent by volume of water-miscible liquids and with the remainder of the solutions not being flammable, in retail or wholesale sales occupancies, are unlimited when packaged in individual containers not exceeding 1.3 gallons (5 L).
2. Application and release of pesticide and agricultural products and materials intended for use in weed abatement, erosion control, soil amendment or similar applications when applied in accordance with the manufacturer's instructions and label directions.
3. The off-site transportation of hazardous materials when in accordance with DOT regulations.
4. Building materials not otherwise regulated by this code.
5. Refrigeration systems (see Section 606).
6. Stationary lead-acid batteries regulated by Section 608.
7. The display, storage, sale or use of fireworks and explosives in accordance with Chapter 33.
8. Corrosives utilized in personal and household products in the manufacturer's original consumer packaging in Group M occupancies.
9. The storage of distilled spirits and wines in wooden barrels and casks.

❖ This chapter contains the general requirements for controlling the hazardous materials regulated by Chapter 4 of the *International Building Code*® (IBC®) and Chapter 7 of the code. It is important to understand that not all buildings in which chemicals are used are classified as Group H occupancies. To be classified as a Group H occupancy the amount of hazardous materials storage or use for a control area must exceed the maximum allowable quantity per control areas as found in Tables 2703.1.1(1), 2703.1.1(2), 2703.1.1(3), 2703.1.1(4) and 2703.8.3.2. If the amount of a chemical is below the maximum allowable quantity per control area, the use area would need to meet the requirements of the most appropriate occupancy group for the activity in the area. For hazardous chemicals, the general provisions of this chapter are applicable. In addition, Chapters 28 through 43 may also be applicable. Any specific provisions in these chapters will take precedence over the general requirements of this chapter when both chapters contain a requirement on the same subject.

The exceptions cover storage of amounts over the required limits. They allow for the amounts to exceed those called for in Tables 2703.1.1(1) through

2703.1.1(4) and the area or building would still not be considered Group H.

Exception 1 deals with the amount of storage allowed when individual containers do not exceed 1.3 gallons (5 L) in wholesale and retail stores. The thought is that in the event of a leak or broken package, the amount of chemical lost would be small and present a smaller spill cleanup problem. This is limited to products containing no more than 50 percent by volume of watermiscible liquids and with the remaining content being nonflammable.

Exception 2 states that the application of pesticides and agricultural products is not considered a hazardous process if they are applied in accordance with manufacturer instructions and label directions as they have been tested in these application methods.

Exception 3 states that any time a hazardous chemical is transported over public highways or by public transportation, the Department of Transportation (DOT) requirements must be enforced.

Exception 4 deals with building products that may exhibit some hazardous properties but do not fall into the realm of a hazardous materials definition in either the code or the IBC.

Exception 5 states that refrigeration systems are governed by Section 606, which has specific requirements that would be enforced over any general requirements of the hazardous material chapters of the code.

Exception 6 notes that stationary lead-acid battery systems are governed by Section 608, which has specific requirements that would be enforced over any general requirements of the hazardous material chapter of the code. Note that the IBC considers this an incidental storage area if the battery capacity is over 100 gallons (379 L). See Table 302.1.1 of the IBC for incidental storage requirements.

Exception 7 refers to Chapter 33 for specific requirements concerning fireworks and explosives. Those specific requirements would be enforced over any general requirements of the hazardous material chapters of the code.

Exception 8 states that corrosives used in personal and household products are not considered hazardous when maintained in their original packaging in Group M use areas. Note that this is specific to Group M use areas only.

Exception 9 covers the storage of distilled spirits and wines in wooden barrels and casks. This statement may appear to exempt all requirements for these products from being a Group H occupancy. However, the IBC will still classify the storage area as a Group H occupancy if the amounts exceed the amounts listed in Tables 307.7(1) of that code for flammable or combustible liquids. All requirements for a Group H occupancy in the IBC are still applicable; however, any requirements from the code are not.

2701.1.1 Waiver. The provisions of this chapter are waived when the fire code official determines that such enforcement is preempted by other codes, statutes or ordinances. The details of any action granting such a waiver shall be recorded and entered in the files of the code enforcement agency.

❖ If the fire code official feels that the requirements of this chapter are preempted by other codes, statutes or ordinances, he or she has the authority to waive the requirements of this chapter. Documentation must be placed into the files of the code enforcement agency stating the reasons for the waiver.

2701.2 Material classification. Hazardous materials are those chemicals or substances defined as such in this code. Definitions of hazardous materials shall apply to all hazardous materials, including those materials regulated elsewhere in this code.

❖ Hazardous materials are chemicals or substances defined in the code as such. Be sure to view the Material Safety Data Sheets (MSDS) for information on the properties of the material.

2701.2.1 Mixtures. Mixtures shall be classified in accordance with hazards of the mixture as a whole. Mixtures of hazardous materials shall be classified in accordance with nationally recognized reference standards; by an approved qualified organization, individual, or Material Safety Data Sheet (MSDS); or by other approved methods.

❖ Mixtures are classified to their specific blend of chemicals. The MSDS are a required source of information on the properties of the chemical mixtures.

2701.2.2 Hazard categories. Hazardous materials shall be classified according to hazard categories. The categories include materials regulated by this chapter and materials regulated elsewhere in this code.

❖ The properties of the chemical will mandate the hazard categories that the chemical is listed under. These properties are found in the MSDS.

2701.2.2.1 Physical hazards. The material categories listed in this section are classified as physical hazards. A material with a primary classification as a physical hazard can also pose a health hazard.

1. Explosives and blasting agents.
2. Flammable and combustible liquids.
3. Flammable solids and gases.
4. Organic peroxide materials.
5. Oxidizer materials.
6. Pyrophoric materials.
7. Unstable (reactive) materials.
8. Water-reactive solids and liquids.
9. Cryogenic fluids.

❖ This section lists those hazardous materials regulated in Chapters 24 through 43 that are classified as a physical hazard. The definition for the term "Physical hazard" is located in Section 2702. Because of their potential detonation or fire hazard, buildings containing more than the maximum allowable quantity per control area of hazardous materials listed in this section are classified as Group H1, H2 or H3. Materials posing multiple hazards must be classified as both a health hazard and a physical hazard and meet the requirements of both classifications.

2701.2.2.2 Health hazards. The material categories listed in this section are classified as health hazards. A material with a primary classification as a health hazard can also pose a physical hazard.

1. Highly toxic and toxic materials.
2. Corrosive materials.

❖ This section lists those hazardous materials regulated in Chapters 27 through 44 that are classified as a health hazard. When hazardous chemicals pose more of a health problem than a fire, explosion or reactivity hazard, buildings containing more than the maximum allowable quantity per control area of hazardous materials listed in this section are classified as Group H4. This section also notes that a hazardous material may pose multiple hazards and thus be considered both a physical hazard and a health hazard. For example, a material classified as toxic or corrosive may also be a Class 2 or 3 oxidizer. Requirements for each material classification are applicable where a multiple hazard exists.

2701.3 Performance-based design alternative. When approved by the fire code official, buildings and facilities where hazardous materials are stored, used or handled shall be permitted to comply with this section as an alternative to compliance with the other requirements set forth in this chapter and Chapters 28 through 45.

❖ The purpose of performance-based design criteria is to promote innovative, flexible and responsive solutions that optimize the expenditure and consumption of resources while preserving social and economic value. The model codes, including the *International Codes®*, have traditionally incorporated alternative materials, designs and methods of construction provisions, such as those found in Section 104.9. This section provides a framework and opportunity to use new materials and methods when design equivalence to the prescriptive requirements of the code is demonstrated to and approved by the fire code official.

Large chemical manufacturing and production facilities are typically required to comply with federal risk management plans (RMP) and process safety management (PSM) requirements, which require a level of safety that generally exceeds current hazardous maerials regulations in the code. This section intends to be compatible with federal RMP and PSM programs, so it would reduce the burden on businesses in having to comply with duplicative or conflicting local and federal regulations. Yet, compliance with these provisions will yield a level of safety that should equal or exceed the prescriptive code requirements.

Based on the fact that performance-based approaches are already in widespread use in federal laws regulating chemical storage and handling facilities, the topic of hazardous materials regulation stands out as a good place to start phasing performance-based concepts into the code. While a casual glance through the text of Section 2701.3 might lead one to conclude that the proposed section represents a huge loophole for the industry to jump through to avoid compliance with the code's prescriptive requirements, a closer investigation will reveal that this is not the case. This text, duplicated from Chapter 22 of the *ICC Performance Code™ for Buildings and Facilities* (ICC PC™), was developed by trying to incorporate federal RMP and PSM regulations and then looking through the code's prescriptive hazardous materials regulations to pick up any topics that appeared to have been overlooked.

The text of this section was developed through a cooperative effort of fire officials and industry representatives in the IFC Performance Code Development Forum, and those who worked on the text believed that compliance with the proposed provisions would yield a facility that would be at least as safe, if not safer, than a facility constructed in accordance with the code's prescriptive requirements.

It is absolutely true that use of this approach would require a tremendous amount of time, money and effort on the part of a permit applicant to demonstrate code compliance to local fire code officials, and clearly, the approach will not be for everyone; however, even for those who might not use the approach in its entirety, the performance objectives would greatly assist code users and enforcers when dealing with alternate methods by better defining the intent of the code.

See the ICC PC and its "User's Guide" for a more comprehensive discussion of performance-based design concepts and philosophy.

2701.3.1 Objective. The objective of Section 2701.3 is to protect people and property from the consequences of unauthorized discharge, fires or explosions involving hazardous materials.

❖ The intent and scope of this section is to protect the occupants of the building, people in the surrounding area, emergency response personnel and property from acute consequences associated with unintended or unauthorized releases of hazardous materials. These performance-based design requirements encourage the use of both accident prevention and control measures to reduce risk.

It is not the intent of this section or the prescriptive requirements of the code to regulate all hazardous materials. Within the scopes of building and fire codes, hazardous materials are generally defined as those materials that are acutely dangerous to people or property. Building and fire codes usually defer regulation of materials that present only a risk of chronic or environmental effects to other regulatory agencies, such as the Occupational Safety and Health Administration (OSHA) or the Environmental Protection Agency (EPA) in the United States. Exposure of workers to hazardous materials in the normal course of their jobs is also beyond the scope of building and fire codes. Such workplace safety issues are instead regulated by occupational safety and health codes, which in the United States fall under the jurisdiction of OSHA.

When developing a performance–based design involving hazardous materials concerns, consideration should be given not only to the hazardous materials categories in Section 2701.2.2 but also to the quantity, state, situation (storage/use), arrangement and location of materials and processes.

2701.3.2 Functional statements. Performance-based design alternatives are based on the following functional statements:

1. Provide safeguards to minimize the risk of unwanted releases, fires or explosions involving hazardous materials.
2. Provide safeguards to minimize the consequences of an unsafe condition involving hazardous materials during normal operations and in the event of an abnormal condition.

❖ This section includes two functional statements that serve the overall objective of Section 2701.3. These two statements focus on reducing the probability of unsafe conditions involving hazardous materials and minimizing the consequences of an unsafe condition, if one occurs. The concepts can be summarized as prevention and control. Specific means by which these functional statements can be accommodated are listed in Section 2701.3.3.

2701.3.3 Performance requirements. When safeguards, systems, documentation, written plans or procedures, audits, process hazards analysis, mitigation measures, engineering controls or construction features are required by Sections 2701.3.3.1 through 2701.3.3.18, the details of the design alternative shall be subject to approval by the code official. The details of actions granting the use of the design alternatives shall be recorded and entered in the files of the jurisdiction.

❖ Section 2701.3 allows the use of this section of the code based on the approval of the fire code official; however, the specifics of the design alternatives selected by designers, evaluators and operators should also be subject to review by a third party representing the public. The fire code official has the responsibility to verify that the performance alternatives provided by Section 2701.3 will protect from conditions hazardous to life, property or public welfare as required by Section 101.2. When acceptance for the use of the design is granted, a record of the approval should be made in the public record to document acceptance of the design alternative.

2701.3.3.1 Properties of hazardous materials. The physical and health-hazard properties of hazardous materials on site shall be known and shall be made readily available to employees, neighbors and the fire code official.

❖ This section correlates with the reporting requirements set forth in the Superfund Amendments and Reauthorization Act of 1986 (SARA) Title III and to some degree with the prescriptive reporting requirements set forth in the code. Compliance with these reporting requirements can be accomplished through the use of MSDS, inventory reports, SARA Title III reporting documents, which are typically mandatory under federal law, and Section 2701.5.2. This section ensures that interested parties will have access to information about the characteristics of hazardous materials that are located on site.

2701.3.3.2 Reliability of equipment and operations. Equipment and operations involving hazardous materials shall be designed, installed and maintained to ensure that they reliably operate as intended.

❖ Equipment and operations at facilities regulated by federal PSM rules should have little difficulty demonstrating compliance with the requirements of this section. The PSM rules generally address this topic area.

At facilities that are not required to comply with PSM rules, the selection of equipment and design of operations would have to go through a great deal of scrutiny by qualified individuals. In addition, equipment manuals and operational protocols would need to be developed and followed, as applicable.

2701.3.3.3 Prevention of unintentional reaction or release. Safeguards shall be provided to minimize the risk of an unintentional reaction or release that could endanger people or property.

❖ Facilities regulated by federal RMP rules are required to evaluate the potential consequences of various release scenarios on the surrounding area; therefore, many such facilities provide safety systems to reduce these potential consequences, recognizing that the consequence analysis information must be made available to the public.

Depending on the classification and state (solid, liquid or gas) of hazardous materials stored or used at a given site, a variety of mitigation measures may be provided to comply with this provision. Such measures might include process controls, spill control and containment systems and ventilation controls.

2701.3.3.4 Spill mitigation. Spill containment systems or means to render a spill harmless to people or property shall be provided where a spill is determined to be a plausible event and where such an event would endanger people or property.

❖ This requirement is primarily derived from the prescriptive provisions in the code. As a general rule, storage facilities are regarded as less likely candidates for dangerous spills than facilities that involve dispensing or processing operations. In addition, dangerous spill conditions are probably more likely to occur in facilities with large quantity vessels or systems than those with only small containers. Information that may be useful in determining whether a spill is plausible and whether dangerous conditions would result includes the following:

- Specific material and process hazards involved.

- A block flow diagram for the facility.
- Piping and instrument drawings.
- A list of all safety devices showing their location, design basis and capacity, date of installation, etc.
- Equipment manufacturers' operational instructions, including safe operating limits for the equipment.
- Equipment drawings and specifications that reflect built and installed equipment.

2701.3.3.5 Ignition hazards. Safeguards shall be provided to minimize the risk of exposing combustible hazardous materials to unintended sources of ignition.

❖ The primary design and operating intent is to ensure that flammable and combustible materials are always completely controlled in accordance with process design parameters; however, where flammable and combustible hazardous materials are present, a degree of redundancy is sometimes necessary to provide an additional level of safety. Where there is a plausible risk of spills or leaks, such as in loading and unloading or packaging operations, additional measures such as ignition source controls are prudent. To that end, process design and operation should ensure to the greatest degree possible that ignition sources are kept away from areas where flammable or combustible hazardous materials are present. Where separation is not feasible, ignition source controls may be warranted. Such controls may involve the following:

- Electrical classification of areas where flammable hazardous materials might be present.
- Classification of mobile equipment that might operate in areas where flammable hazardous materials might be present.
- The use of grounding systems and equipment to minimize the potential for sparking in areas where flammable hazardous materials might be present.

2701.3.3.6 Protection of hazardous materials. Safeguards shall be provided to minimize the risk of exposing hazardous materials to a fire or physical damage whereby such exposure could endanger or lead to the endangerment of people or property.

❖ This section directs the designer and the operator to review and ensure that vessels or systems containing hazardous materials are not exposed to or are protected from damage by external fire. The design should focus, first, on reducing the possibility for fire or other hazards, such as vehicular impact and, second, on isolating hazardous materials from exposure to unsafe conditions, such as a fire.

All storage areas and systems should be formally reviewed to find and correct any sources of exposure to fire, including the following:

- Nearby storage of combustibles.
- Nearby hot work operation.
- Nearby vehicular operation.

All systems subject to fire exposure should be formally reviewed to ensure adequate protection, including the following:

- Sprinkler installation.
- Insulation of equipment.
- Fire-resistance-rated barriers.

2701.3.3.7 Exposure hazards. Safeguards shall be provided to minimize the risk of and limit damage from a fire or explosion involving explosive hazardous materials whereby such fire or explosion could endanger or lead to the endangerment of people or property.

❖ This section directs the designer and the operator to review and ensure that vessels or systems containing hazardous materials are not subject to damage from internal fire, chemical reaction or explosion. The design criteria should be, first, to reduce the risk of an internal fire or explosion and, second, where the first is not feasible, to design vessels and systems in such a manner that loss of integrity will not occur in an overpressure situation.

All systems should be formally reviewed to identify and correct any sources of internal fire, explosion or overpressure. The review should include the following:

- The potential for inadvertent or improper mixing of reactive components.
- The potential for overheating of unstable materials.
- The potential for inadequate venting of unstable reaction byproducts.
- The potential for inadequate dilutent material supply.

Where overpressure or explosion conditions cannot be reasonably ruled out, the design should consider overpressure protection, containment and explosion control systems.

2701.3.3.8 Detection of gas or vapor release. Where a release of hazardous materials gas or vapor would cause immediate harm to persons or property, means of mitigating the dangerous effects of a release shall be provided.

❖ This section increases the likelihood that hazardous vapor releases are detected and mitigated before they can harm individuals or property. In occupied areas, detection of a vapor release may be by sight, smell or an automatic detection system. For many hazardous materials, such as chlorine or ammonia, vapor releases are readily evident before concentrations are truly hazardous based on the presence of vapor fog or a noxious odor. Where this is not the case, automatic detection systems and alarms may be warranted. Sensors can take the form of ambient sampling devices at strategic area locations, sampling devices in key vent streams or specially designed leak-detection systems, such as acoustic emission systems. The performance measurement is based on the ability of the sensing equipment or operators to provide adequate warning so that safety precautions can be taken before unsafe conditions are present.

Mitigation-based solutions can range from special process equipment designs to elaborate ventilation and air-scrubbing systems. Where practical, the simplest mitigation consists of overdesign of the process system so that the likelihood of release is extremely low. The performance measurement of a ventilation or treatment system is based on the reduction of the concentration of the hazardous materials in the workplace and nearby environment to levels that are not acutely hazardous.

2701.3.3.9 Reliable power source. Where a power supply is relied upon to prevent or control an emergency condition that could endanger people or property, the power supply shall be from a reliable source.

❖ This section is derived from the prescriptive requirements of the code. It is essential to ensure that a reliable power supply is provided for systems that are critical to safety. Some examples of systems that may require a reliable power supply include mechanical ventilation systems; treatment systems; gas detection and alarm systems and emergency shutdown systems. The reliability needs of the system are related to the potential risks associated with system failure.

A reliable power source does not necessarily equate to a generator or a battery system. The type of system to be used depends on the relative level of hazard that might result in the event of a power failure, and in some cases, such as those where hazardous processes shut down upon loss of power, a connection ahead of the building's main disconnect switch may be adequate to qualify as a reliable source. Guidance on the selection and performance requirements for power supply systems providing an alternate source of electrical power can be found in the ICC *Electrical Code®* (ICC EC™) and NFPA 110.

2701.3.3.10 Ventilation. Where ventilation is necessary to limit the risk of creating an emergency condition resulting from normal or abnormal operations, means of ventilation shall be provided.

❖ In many cases involving hazardous materials, ventilation must be provided to limit the risk of creating an emergency condition. Ventilation might be necessary during both normal and abnormal operating conditions. Some examples of operations that may require ventilation are storage or processing of flammable and combustible liquids or gases inside buildings; drum filling operations inside buildings; laboratory use of chemicals and dust-handling systems. Ventilation may also be used as a means for reducing vapor concentrations below lower flammable limits in areas where ignition sources are present or for pressurization of areas to isolate hazardous vapors.

Guidance on the performance requirements for ventilation systems can be found in a number of sources, including the OSHA PSM regulations, 29 CFR 1922.226, NFPA 30, NFPA 45, NFPA 69, NFPA 497 and the ICC EC.

2701.3.3.11 Process hazard analyses. Process hazard analyses shall be conducted to ensure reasonably the protection of people and property from dangerous conditions involving hazardous materials.

❖ This section establishes an administrative safety control plan addressing process hazard analysis. Guidance on process hazard analysis techniques can be found in the OSHA PSM regulations, 29 CFR Part 1910.119. The process hazard analysis must be appropriate to the complexity of the process and must identify, evaluate and control the hazards involved in the process. The analysis can be accomplished through various methods. Some of these are "what-if" scenarios, process hazard analysis, fault tree, etc. A person trained in these and other hazard evaluation techniques should be employed to complete this analysis.

2701.3.3.12 Pre-startup safety review. Written documentation of pre-startup safety review procedures shall be developed and enforced to ensure that operations are initiated in a safe manner. The process of developing and updating such procedures shall involve the participation of affected employees.

❖ This section establishes an administrative safety control plan addressing prestartup safety review procedures. Guidance on techniques for written documentation of prestartup safety review procedures can be found in the OSHA PSM regulations, 29 CFR Part 1910.119. Prestartup safety reviews are typically necessary when new facilities are prepared for operation and where existing facilities are modified to a degree that is significant enough to require a change in the process safety information.

A prestartup safety review should confirm that, prior to the introduction of highly hazardous chemicals to a process, the following verifications have been accomplished at a minimum:

- Construction and equipment is in accordance with design specifications.
- Safety, operating, maintenance and emergency procedures are in place and are adequate.
- For new facilities, a process hazard analysis has been performed and recommendations have been resolved or implemented before startup; for modified facilities, requirements contained in management of change documents have been met.
- Training of each employee involved in operating a process has been completed.

2701.3.3.13 Operating and emergency procedures. Written documentation of operating procedures and procedures for emergency shut down shall be developed and enforced to ensure that operations are conducted in a safe manner. The process of developing and updating such procedures shall involve the participation of affected employees.

❖ This section establishes an administrative safety control plan addressing written documentation of operating and emergency shutdown procedures. Guidance on developing written documentation for operating procedures and emergency shutdown techniques can be found in the OSHA PSM regulations, 29 CFR Part 1910.119. Overall, there are 14 elements that employers covered by PSM are required to complete to meet the federal PSM regulations. Two elements that relate to this section are as follows:

- 29 CFR 1910.119 (c): This element requires that employees and their representatives be consulted on the development and conduct of hazard assessments and the development of chemical accident prevention plans and provide access to these and other records required under the federal law.
- 29 CFR 1910.119 (f): This element requires that written operating procedures for the chemical process, including procedures for each operating phase, operating limitations and safety and health considerations, must be developed and implemented.

2701.3.3.14 Management of change. A written plan for management of change shall be developed and enforced. The process of developing and updating the plan shall involve the participation of affected employees.

❖ This section establishes an administrative safety control plan addressing management of change. Guidance on developing written documentation for management of change can be found in the OSHA PSM regulations, 29 CFR Part 1910.119. The PSM element that relates to this section is 29 CFR 1910.119 (l), which states that this element requires a review of the technical basis for the proposed change; the impact of change on safety and health; possible modifications to operating procedures and process safety information; the necessary time period for the change and authorization requirements for the proposed change.

Employees involved in operating a process, and maintenance and contract employees whose job tasks will be affected by a change in the process, should be informed of and trained in the change prior to startup of the process or the affected part of the process.

2701.3.3.15 Emergency response plan. A written emergency plan shall be developed to ensure that proper actions are taken in the event of an emergency, and the plan shall be followed if an emergency condition occurs. The process of developing and updating the plan shall involve the participation of affected employees.

❖ This section establishes an administrative safety control plan addressing emergency response planning. Guidance on developing written documentation for an emergency response plan can be found in the OSHA PSM regulations, 29 CFR Part 1910.119. The PSM element that relates to this section is 29 CFR 1910.119 (n), which references other portions of the federal regulations. Such plans may include identification of actions to be taken by employees in the event of an emergency and the assignment of a staff liaison who can assist emergency response personnel.

2701.3.3.16 Accident procedures. Written procedures for investigation and documentation of accidents shall be developed, and accidents shall be investigated and documented in accordance with these procedures.

❖ This section establishes an administrative safety control plan addressing accident investigation and reporting. Guidance on accident investigation and reporting can be found in the OSHA PSM regulations, 29 CFR Part 1910.119. The PSM element that relates to this section is 29 CFR 1910.119 (m).

Some of the guidelines specified in the federal regulations include the following:

- The need for an incident investigation team to be established, consisting of at least one person knowledgeable in the process involved, a contract employee if the incident involved contractor work and other persons with appropriate knowledge and experience to thoroughly investigate and analyze the incident.
- The need for a report to be prepared at the conclusion of each investigation, including at a minimum the date of the incident and when the investigation began; description of the incident; factors that contributed to the incident and recommendations resulting from the investigation.
- The need for the establishment of a system to promptly address and resolve the incident report findings and recommendations and to document resolutions and corrective actions.
- The need for accident investigation reports to be reviewed by all affected persons whose job tasks are relevant to the incident findings, including contract employees where applicable.

2701.3.3.17 Consequence analysis. Where an accidental release of hazardous materials could endanger people or property, either on or off-site, an analysis of the expected consequences of a plausible release shall be performed and utilized in the analysis and selection of active and passive hazard mitigation controls.

❖ This section establishes an administrative safety control plan addressing an analysis of off-site consequences. Guidance on accident investigation and reporting can be found in the EPA RMP regulations, 40 CFR Part 68. These regulations amend the accident release prevention requirements under Section 112 (r) of the Clean Air Act.

EPA's RMP rules are a good source of examples for

alternate release scenarios for a particular site, and, through the identification and analysis of plausible release scenarios, changes can be implemented to minimize the probability and consequences of a release. A plausible release is one that has occurred in the past or could occur under reasonable single-system failures.

Devices that normally use some kind of motion or energy to prevent or minimize the release represent active mitigation controls. Active mitigation controls might include valves, switches, pumps and blowers. Passive mitigation controls include devices that are permanently in place and have an inherently safe design that allows them to be used at all times. Passive mitigation controls might include dikes, walls, ponds and sumps.

The off-site consequence analysis can be accomplished through various methods. Some of these are "what-if" scenarios, process hazard analysis, HAZOP and fault tree. A person trained in these and other hazard evaluation techniques should be used to complete this analysis.

2701.3.3.18 Safety audits. Safety audits shall be conducted on a periodic basis to verify compliance with the requirements of this section.

❖ This section establishes an administrative safety control plan addressing safety compliance audits. Guidance on safety audits can be found in the OSHA PSM regulations, 29 CFR Part 1910.119. The PSM element that relates to this section is 29 CFR 1910.119 (o). On a routine basis, each facility must review its continuing compliance with each of the subsections in Section 2701.3 and other related provisions of the code. The word "periodic" reflects a need for adequate frequency to check that safety programs, features and systems will perform as intended. Recognizing that many code sections contain issues that change very little over time, compliance audit frequencies will not be the same for all programs, features and systems and, depending on the particular safety element, the audit frequency may range to as much as five-year intervals under the PSM regulations.

2701.4 Retail and wholesale storage and display. For retail and wholesale storage and display of nonflammable solid and nonflammable or noncombustible liquid hazardous materials in Group M occupancies and storage in Group S occupancies, see Section 2703.11.

❖ This section deals only with nonflammable solid and nonflammable or noncombustible liquid hazardous materials in Group M and S occupancies. For flammable or combustible liquids see Section 3404.3.4 of the code.

2701.5 Permits. Permits shall be required as set forth in Sections 105.6 and 105.7.

When required by the fire code official, permittees shall apply for approval to permanently close a storage, use or handling facility. Such application shall be submitted at least 30 days prior to the termination of the storage, use or handling of hazardous materials. The fire code official is authorized to require that the application be accompanied by an approved facility closure plan in accordance with Section 2701.5.3.

❖ The process of issuing permits gives the fire code official an opportunity to carefully evaluate and regulate hazardous operations. Permit applicants should be required to demonstrate that their operations comply with the intent of the code before the permit is issued. See the commentary to Section 105.6 for a general discussion of operations requiring an operational permit and Section 105.7 for a general discussion of activities requiring a construction permit. The permit process also notifies the fire department of the need for prefire planning for the hazardous property.

2701.5.1 Hazardous Materials Management Plan. Where required by the fire code official, each application for a permit shall include a Hazardous Materials Management Plan (HMMP). The HMMP shall include a facility site plan designating the following:

1. Storage and use areas.
2. Maximum amount of each material stored or used in each area.
3. Range of container sizes.
4. Locations of emergency isolation and mitigation valves and devices.
5. Product conveying piping containing liquids or gases, other than utility-owned fuel gas lines and low-pressure fuel gas lines.
6. On and off positions of valves for valves that are of the self-indicating type.
7. Storage plan showing the intended storage arrangement, including the location and dimensions of aisles.
8. The location and type of emergency equipment.The plans shall be legible and drawn approximately to scale. Separate distribution systems are allowed to be shown on separate pages.

❖ A Hazardous Materials Management Plan (HMMP) or other approved plan is required for facilities using or storing quantities of hazardous materials exceeding permit quantities indicated in Chapters 27 through 44. An HMMP is designed to aid fire department personnel in the building design's preplanning phase. The plan should also be updated when a change in occupancy or other relevant condition occurs in existing structures.

The plan must include a site plan that shows details on drainage, storm sewer location, underground utilities, retaining ponds, streams and underground water sources, fire hydrants, fire lanes and any other information that might help in fire fighting.

Stating the maximum amount of each material stored or used in each area satisfies two requirements: It allows for a good review of the plans to ensure that Chapter 27 through 44 requirements have been met, and it provides information that will be vital to a fire department preplan, such as the size of secondary containment that could be required in a spill, the amounts and types of extinguishing materials required for each area and exposures that could add to the fire load of the incident.

Knowing the range of container sizes is important because the methods for handling spills from a 10,000-gallon (37 850 L) tank are different from those of handling spills from 1-pint containers. Having the information on the size of containers will aid in the preplan for the area.

Emergency isolation and mitigation may be a remote tank or impounding area that is hooked to the emergency drain system; control valves that are used to stop additional flow from broken pipes and tanks or include dikes and containment areas. This information is vital to the responding units so that they can reduce the flow of liquids or isolate and control any spill or fire that arises. This information must be established prior to the incident, and proper procedures to fight or control the spill or emergency must be put in place.

Piping conveying liquids or gases, other than utility-owned and low-pressure fuel gas lines, must be shown in the HMMP. The location of any underground or above-ground piping needs to be noted. Any monitoring wells should also be noted on the drawing.

On and off positions of valves that are of the self-indicating type must be designated. A method should be developed for the indication of on and off positions. This method should take into account the limited visibility and functions of the responding fire fighters in their protective equipment.

Storage plans must show the intended storage arrangements, including the location and dimensions of aisles; the height of the storage; whether it is in racks, tanks or on pallets; whether the pallets are plastic or wood and the depth of the storage. All this information will affect the fire-containment plans of the incident commander.

Also required are the locations of self-contained breathing apparatus spare air tanks and their compatibility with the tanks of the responding agencies; whether there are additional containers of foam and the type and quantity of the foam. Because all chemicals cannot be protected with the same foam, it will be important to have a record of the various types and amounts available.

Fire fighters will also want to know whether nonsparking tools are available and where the MSDS are kept.

These are only a few of the questions that need to be asked. A list should be developed based on the code user's needs and command style.

All plans shall be legible and drawn to scale or dimensioned. These drawings must be kept updated.

The fire code official may waive the requirement for an HMMP. By themselves, approvals of the building or site may suffice for the HMMP when dealing with small quantities of hazardous materials not posing a complex storage, use or maintenance problem.

2701.5.2 Hazardous Materials Inventory Statement (HMIS). Where required by the fire code official, an application for a permit shall include an HMIS, such as SARA (Superfund Amendments and Reauthorization Act of 1986) Title III, Tier II Report, or other approved statement. The HMIS shall include the following information:

1. Manufacturer's name.
2. Chemical name, trade names, hazardous ingredients.
3. Hazard classification.
4. MSDS or equivalent.
5. United Nations (UN), North America (NA) or the Chemical Abstract Service (CAS) identification number.
6. Maximum quantity stored or used on-site at one time.
7. Storage conditions related to the storage type, temperature and pressure.

❖ A Hazardous Material Inventory Statement (HMIS) or other approved statement is required for all facilities with hazardous materials exceeding permit quantities indicated in Chapters 27 through 44. As the name suggests, HMIS represents a hazardous materials inventory and is designed to alert the fire code official to the presence of hazardous materials by specific name and hazard class. Both descriptive information and maximum quantity estimates are required for all hazardous materials that may be present at any given time.

SARA Title III, known as the Emergency Planning and Right to Know Act, established requirements for emergency planning and reporting on hazardous materials. SARA Title II is representative of a typical HMIS as required by this section. The exception, which is listed in SARA and is similar to Section 2701.4.1, leaves the requirement for submission of the HMIS up to the discretion of the fire code official for circumstances in which hazards can be managed without the need for HMIS.

2701.6 Facility closure. Facilities shall be placed out of service in accordance with Sections 2701.6.1 through 2701.6.3.

❖ This section contains requirements for handling and disposing of hazardous materials prior to a facility's closure. The 30-day notice prior to the facility's closure is intended to enable proper measures to be taken for disposing of or eliminating all hazardous materials before the building owner vacates the premises.

2701.6.1 Temporarily out-of-service facilities. Facilities that are temporarily out of service shall continue to maintain a permit and be monitored and inspected.

❖ Facilities warranting a closure plan are considered to be either temporarily or permanently out of service. Although a facility may not be closed permanently, this section recognizes the need to regulate hazardous materials in buildings that may be temporarily out of service. Because the out-of-service condition may be temporary, the storage or presence of any hazardous materials in a facility must be monitored and inspected as required by the fire code official.

2701.6.2 Permanently out-of-service facilities. Facilities for which a permit is not kept current or is not monitored and inspected on a regular basis shall be deemed to be permanently out of service and shall be closed in an approved manner. When required by the fire code official, permittees shall apply for approval to close permanently storage, use or handling facilities. The fire code official is authorized to require that such application be accompanied by an approved facility closure plan in accordance with Section 2701.5.3.

❖ Facilities warranting a closure plan are considered to be either temporarily or permanently out of service. A facility is to be classified as permanently out of service if approval is not kept current as required for a temporarily out-of-service facility or it is not properly monitored or inspected as required by the fire code official.

2701.6.3 Facility closure plan. When a facility closure plan is required in accordance with Section 2701.4 to terminate storage, dispensing, handling or use of hazardous materials, it shall be submitted to the fire code official at least 30 days prior to facility closure. The plan shall demonstrate that hazardous materials which are stored, dispensed, handled or used in the facility will be transported, disposed of or reused in a manner that eliminates the need for further maintenance and any threat to public health and safety.

❖ This plan is used to document the timetable for the proper transportation, disposal or other approved disposal method of all chemicals that are on site. It is important to note that this could include any contaminated soils or dike facilities in the area.

SECTION 2702 DEFINITIONS

2702.1 Definitions. The following words and terms shall, for the purposes of this chapter, Chapters 28 through 44, and as used elsewhere in this code, have the meanings shown herein.

❖ Definitions can help in the understanding and application of the code requirements. These definitions are included here to give the code user convenient access to key terms used in this chapter. For convenience, the defined terms are also listed in Chapter 2 with a cross reference to this section. The use and application of all defined terms are set forth in Section 201 of the code.

BOILING POINT. The temperature at which the vapor pressure of a liquid equals the atmospheric pressure of 14.7 pounds per square inch (psia) (101 kPa) or 760 mm of mercury. Where an accurate boiling point is unavailable for the material in question, or for mixtures which do not have a constant boiling point, for the purposes of this classification, the 20-percent evaporated point of a distillation performed in accordance with ASTM D 86 shall be used as the boiling point of the liquid.

❖ The purpose of the boiling point is to assist in classifying flammable liquids. The classification of flammable liquids is based on the flash point and the boiling point. When one looks at a Class IA liquid and a Class IB liquid the only difference in the definition is the boiling point. This information will be found in the MSDS and must be evaluated by the fire code official.

CEILING LIMIT. The maximum concentration of an air-borne contaminant to which one may be exposed. The ceiling limits utilized are those published in DOL 29 CFR Part 1910.1000. The ceiling Recommended Exposure Limit (REL-C) concentrations published by the U.S. National Institute for Occupational Safety and Health (NIOSH), Threshold Limit Value — Ceiling (TLV-C) concentrations published by the American Conference of Governmental Industrial Hygenists (ACGIH), ceiling Workplace Environmental Exposure Level (WEEL-Ceiling) Guides published by the American Industrial Hygiene Association (AIHA), and other approved, consistent measures are allowed as surrogates for hazardous substances not listed in DOL 29 CFR Part 1910.1000.

❖ This limit is the concentration at which immediate irritation to skin, respiratory system or both will occur. This is an important level for emergency personnel and workers to be aware of. Mechanical ventilation may be used to assist in keeping the working environment below these levels.

CHEMICAL. An element, chemical compound or mixture of elements or compounds or both.

❖ Chemicals may be hazardous, in which case the buildings, processes and storage that use them are regulated, or they may be nonhazardous, in which case their storage is regulated. These regulations are typically found in the IBC, the *International Mechanical Code®* (IMC®), the *International Plumbing Code®* (IPC®), the *International Fuel Gas Code®* (IFGC®) and the IFC.

CHEMICAL NAME. The scientific designation of a chemical in accordance with the nomenclature system developed by the International Union of Pure and Applied Chemistry, the Chemical Abstracts Service rules of nomenclature, or a name which will clearly identify a chemical for the purpose of conducting an evaluation.

❖ This name can be used to research the product in resource manuals to determine the hazardous characteristics of the product. Some examples of resources are: National Toxicology Program (NTP), Chemical Hazards Response Information System (CHRIS) manual, emergency response guidebook (ERG), MSDS, etc.

CLOSED CONTAINER. A container sealed by means of a lid or other device such that liquid, vapor or dusts will not escape from it under ordinary conditions of use or handling.

❖ A closed container is one that is sealed so that no vapors or dust can escape. The important difference in dealing with an open container and a closed container is that the open container is more dangerous and calls for more safety requirements. The fire tetrahedron is made up of fuel, heat, oxygen and chain reaction. In a closed container you have only the fuel part of the tetrahedron; in an open container you have the fuel and oxygen sides of the tetrahedron. In this case, a closed container is less hazardous than an open container unless there is a leak.

CONTAINER. A vessel of 60 gallons (227 L) or less in capacity used for transporting or storing hazardous materials. Pipes, piping systems, engines and engine fuel tanks are not considered to be containers.

❖ This definition establishes the intended capacity of the container to avoid confusion with portable or stationary tanks. A container could include typical 55-gallon (208 L) drums or 2-ounce (57 g) cans. It is important to note the size difference between a drum and a barrel. A drum has a capacity of 55 U.S. gallons (208 L) and a barrel has a capacity of 42 U.S. gallons (158 L). These terms are sometimes reported incorrectly when determining the amount of storage in a facility.

CONTROL AREA. Spaces within a building which are enclosed and bounded by exterior walls, fire walls, fire barriers and roofs, or a combination thereof, where quantities of hazardous materials not exceeding the maximum allowable quantities per control area are stored, dispensed, used or handled.

❖ The use of control areas allows for the use and storage of hazardous materials without classifying the building or structure as a high-hazard occupancy (Group H) when the total quantity of hazardous materials in the entire building might exceed the maximum allowable quantity per control area. This concept is based on regulating the allowable quantities of hazardous materials in each control area by giving credit for further compartmentation through the use of fire separation assemblies having a minimum fire-resistance rating of 1 hour. Maximum quantities of hazardous materials within each control area cannot exceed the maximum allowable quantity per control area for a given material. Thus, the quantities in each control area will be less than the maximum allowable quantity per control area, while the overall quantity in the entire building could exceed the maximum allowable quantity per control area (see commentary, Section 2703.8.2).

CYLINDER. A pressure vessel designed for pressures higher than 40 psia (275.6 kPa) and having a circular cross section. It does not include a portable tank, multi-unit tank car tank, cargo tank or tank car.

❖ As referenced in the code, cylinders are vessels containing flammable or nonflammable compressed gases. Gas cylinders are fabricated to comply with regulations specified by DOT and are generally limited to a capacity equivalent to the volume of 1,000 pounds (454 kg) of water.

DEFLAGRATION. An exothermic reaction, such as the extremely rapid oxidation of a flammable dust or vapor in air, in which the reaction progresses through the unburned material at a rate less than the velocity of sound. A deflagration can have an explosive effect.

❖ Materials posing a deflagration hazard usually burn very quickly with an energy release from a chemical reaction in the form of intense heat. Confined deflagration hazards under pressure can result in an explosion. Most hazardous materials posing a severe deflagration hazard are classified as Group H2.

DESIGN PRESSURE. The maximum gauge pressure that a pressure vessel, device, component or system is designed to withstand safely under the temperature and conditions of use expected.

❖ A container that is subjected to an internal pressure higher than its design pressure faces a high-failure rate. If the container is used to store flammable liquids or gases, a boiling liquid expanding vapor explosion (BLEVE), which is a form of pressure-releasing explosion, may occur.

DETACHED BUILDING. A separate single-story building, without a basement or crawl space, used for the storage or use of hazardous materials and located an approved distance from all structures.

❖ This term refers to the type of structure that the code recognizes for using and storing Group H materials exceeding maximum allowable quantities per control area. A detached storage building is required only for Group H, as indicated in Table 2703.8.2 of the code or Table 415.3.2 of the IBC and as may be required in Chapters 28 through 44 of the code.

DISPENSING. The pouring or transferring of any material from a container, tank or similar vessel, whereby vapors, dusts, fumes, mists or gases are liberated to the atmosphere.

❖ This term refers to any transfer of a hazardous material from one container to another that is open to the atmosphere.

EXCESS FLOW CONTROL. A fail-safe system or other approved means designed to shut off flow caused by a rupture in pressurized piping systems.

❖ This refers to a fail-safe valve or other approved device that is required on pressurized piping systems to prevent the uncontrolled release of excess quantities of hazardous liquids and gases if piping or valve failure occurs. To safeguard as much piping as practical, the device must be installed as close to the supply container as possible.

EXHAUSTED ENCLOSURE. An appliance or piece of equipment which consists of a top, a back and two sides providing a means of local exhaust for capturing gases, fumes, vapors and mists. Such enclosures include laboratory hoods, exhaust fume hoods and similar appliances and equipment used to retain and exhaust locally the gases, fumes, vapors and mists that could be released. Rooms or areas provided with general ventilation, in themselves, are not exhausted enclosures.

❖ When a hazardous chemical is being dispensed or used, an exhausted enclosure can be used to reduce the exposure of personnel to a toxic or hazardous atmosphere. These enclosures have special requirements for protection (see Section 2703.8.4).

EXPLOSION. An effect produced by the sudden violent expansion of gases, which may be accompanied by a shock wave or disruption, or both, of enclosing materials or structures. An explosion could result from any of the following:

1. Chemical changes such as rapid oxidation, deflagration or detonation, decomposition of molecules and runaway polymerization (usually detonations).
2. Physical changes such as pressure tank ruptures.
3. Atomic changes (nuclear fission or fusion).

❖ Materials that pose a threat of explosion are classified as H1 when present in the use area in quantities above the maximum allowable quantity per control area and are required to be in a detached storage building meeting the requirements of Section 415.4 of the IBC and Chapter 33 of the code.

FLAMMABLE VAPORS OR FUMES. The concentration of flammable constituents in air that exceeds 25 percent of their lower flammable limit (LFL).

❖ Vapors or fumes are only considered flammable when there is a high enough concentration for an ignition to occur if exposed to an ignition source. The code specifically defines flammable as being greater than 25 percent of the lower flammable limit (LFL).

GAS CABINET. A fully enclosed, noncombustible enclosure used to provide an isolated environment for compressed gas cylinders in storage or use. Doors and access ports for exchanging cylinders and accessing pressure-regulating controls are allowed to be included.

❖ An assembly constructed and designed to protect compressed gas cylinders and associated equipment.

GAS ROOM. A separately ventilated, fully enclosed room in which only compressed gases and associated equipment and supplies are stored or used.

❖ A gas room is an on-site built room that meets the construction requirements of the IBC. This room will require separation based on the amount of gases stored in the room.

HANDLING. The deliberate transport by any means to a point of storage or use.

❖ Handling is concerned with transporting hazardous materials within a building's means of egress. A hazardous material that is not in either storage or use is essentially being handled.

HAZARDOUS MATERIALS. Those chemicals or substances which are physical hazards or health hazards as defined and classified in this chapter, whether the materials are in usable or waste condition.

❖ The term "hazardous materials" refers to materials posing either a physical or health hazard.

HEALTH HAZARD. A classification of a chemical for which there is statistically significant evidence that acute or chronic health effects are capable of occurring in exposed persons. The term "health hazard" includes chemicals that are toxic, highly toxic and corrosive.

❖ Materials that pose risks to people from handling or exposure are considered health hazards. Even though the materials may also be flammable, those classified as health hazards either will not burn or will not pose a fire hazard similar to that of ordinary combustible materials. Materials that pose a health hazard may also pose a physical hazard and must comply with the requirements of the code applicable to both hazards.

Toxins that attack specific organs are indicative of the other health-hazard materials regulated by this chapter. Hepatotoxins, such as carbon tetrachloride, are capable of causing liver damage and nephrotoxins, such as halogenated hydrocarbons, can cause kidney damage. Neurotoxins include mercury and calcium disulfide, which may produce toxic effects on the nervous system. Although the definition of "Health hazard" includes a reference to carcinogens, it is not the intent of this chapter to regulate carcinogens that are not otherwise classified as an irritant, sensitizer or other known health hazard, such as a targetorgan toxin. Federal regulations address the permitted workplace exposure conditions to known carcinogens.

IMMEDIATELY DANGEROUS TO LIFE AND HEALTH (IDLH). The concentration of air-borne contaminants which poses a threat of death, immediate or delayed permanent adverse health effects, or effects that could prevent escape from such an environment. This contaminant concentration level is established by the National Institute of Occupational Safety and Health (NIOSH) based on both toxicity and flammability. It generally is expressed in parts per million by volume (ppm v/v) or milligrams per cubic meter (mg/m^3). If adequate data do not exist for precise establishment of IDLH concentrations, an independent certified industrial hygienist, industrial toxicologist, appropriate regulatory agency or other source approved by the fire code official shall make such determination.

❖ There are three general atmospheres that make up an IDLH toxic condition. These are toxic, flammable and oxygen deficient. In the absence of an IDLH value, the fire code official may consider using an estimated IDLH of 10 times the lower explosive limit (LEL) while an IDLH oxygen-deficient atmosphere is 19.5-percent oxygen or lower. The EPA has determined that 10 percent of the IDLH value is an acceptable level of concern for evaluating hazmat release concentrations and public protective options.

INCOMPATIBLE MATERIALS. Materials that, when mixed, have the potential to react in a manner which generates heat, fumes, gases or byproducts which are hazardous to life or property.

❖ Incompatible materials constitute a dangerous chemical combination whether in storage or in use. Determining which chemicals in combination pose a hazard is not always easy. MSDS may not provide all of the necessary information. When in doubt, the fire code official should seek additional information from the manufacturer of the chemicals involved, the building owner or experts who are knowledgeable in industrial hygiene or chemistry. NFPA 491 also contains useful information on hazardous chemical reactions.

LIQUID. A material having a melting point that is equal to or less than 68°F (20°C) and a boiling point which is greater than 68°F (20°C) at 14.7 psia (101 kPa). When not otherwise identified, the term "liquid" includes both flammable and combustible liquids.

❖ In dealing with liquids, two areas are important to check on the MSDS sheet:

1. What is the specific gravity of the liquid? The specific gravity is the chemical's weight compared to the weight of an equal volume of water. If the specific gravity is lower than 1.0 (which is the specific gravity of water) the chemical will float. If it is higher than 1.0, it will sink. A flammable liquid that has a specific gravity lower than 1.0 will float on top of any fire-fighting water that is applied. It can then become a running fire as it floats on top of the water that is running off from the scene.

2. The other area is water solubility: will the chemical mix with water? If a chemical will mix with water, it limits the fire-fighting ability of water and another method of extinguishment should be considered.

LOWER EXPLOSIVE LIMIT (LEL). See "Lower flammable limit."

❖ See the commentary for the definition of "Lower flammable limit (LFL)."

LOWER FLAMMABLE LIMIT (LFL). The minimum concentration of vapor in air at which propagation of flame will occur in the presence of an ignition source. The LFL is sometimes referred to as LEL or lower explosive limit.

❖ LFL or LEL is the bottom limit on a flammability range, which is the range in which a flammable vapor is mixed with air in just the right percentages to allow combustion. This is an important concept because the requirement for ventilation is based on keeping the vapor concentrations outside of the flammability range. The upper portion of the range is called the UEL, or upper explosive limit, another term that carries the same meaning is UFL, or upper flammable limit. As long as flammable vapors are not within the range between UFL and LFL, combustion is unlikely.

MATERIAL SAFETY DATA SHEET (MSDS). Information concerning a hazardous material which is prepared in accordance with the provisions of DOL 29 CFR Part 1910.1200 or in accordance with the provisions of a federally approved state OSHA plan.

❖ To comply with right-to-know legislation, building owners are required to prepare a MSDS for all hazardous materials that may be on the premises. The MSDS is the single best source of information in dealing with the requirements of the IBC and the code.

MAXIMUM ALLOWABLE QUANTITY PER CONTROL AREA. The maximum amount of a hazardous material allowed to be stored or used within a control area inside a building or an outdoor control area. The maximum allowable quantity per control area is based on the material state (solid, liquid or gas) and the material storage or use conditions.

❖ Exceeding this amount will place the use area or building into a hazardous occupancy. See Tables 2703.1.1(1), 2703.1.1(2), 2703.1.1(3), 2703.1.1(4) and 2703.8.3.2.

NORMAL TEMPERATURE AND PRESSURE (NTP). A temperature of 70°F (21°C) and a pressure of 1 atmosphere [14.7 psia (101 kPa)].

❖ Reaction by some chemicals is based on temperature and pressure. Understanding this relationship to normal room temperature and pressure (elevation) can provide information on the hazards for the chemical being considered.

OUTDOOR CONTROL AREA. An outdoor area that contains hazardous materials in amounts not exceeding the maximum allowable quantities of Table 2703.1.1(3) or 2703.1.1(4).

❖ A storage area that is exposed to the elements (wind, rain, snow, etc., that cannot exceed the maximum allowable quantity per control area listed in the code and the IBC. See Tables 2703.1.1(3), 2703.1.1(4) and Sections 2701 and 2703 in the code and Tables 307.7(1), 307.7(2) and 414.6 in the IBC.

PERMISSIBLE EXPOSURE LIMIT (PEL). The maximum permitted 8-hour time-weighted-average concentration of an air-borne contaminant. The exposure limits to be utilized are those published in DOL 29 CFR Part 1910.1000. The Recommended Exposure Limit (REL) concentrations published by the U.S. National Institute for Occupational Safety and Health (NIOSH), Threshold Limit Value-Time Weighted Average (TLV-TWA) concentrations published by the American Conference of Governmental Industrial Hygienists (ACGIH), Workplace Environmental Exposure Level (WEEL) Guides published by the American Industrial Hygiene Association (AIHA), and other approved, consistent measures are allowed as surrogates for hazardous substances not listed in DOL 29 CFR Part 1910.1000.

❖ The PEL is a maximum time-weighted concentration at which 95 percent of exposed, healthy adults suffer no adverse effects over a 40-hour work week. The lower the PEL, the more toxic the substance.

PESTICIDE. A substance or mixture of substances, including fungicides, intended for preventing, destroying, repelling or mitigating pests and substances or a mixture of substances intended for use as a plant regulator, defoliant or desiccant. Products defined as drugs in the Federal Food, Drug and Cosmetic Act are not pesticides.

❖ Typically, pesticides are ranked in the toxic category of health hazards. They are primarily used to control a variety of pests.

PHYSICAL HAZARD. A chemical for which there is evidence that it is a combustible liquid, compressed gas, cryogenic, explosive, flammable gas, flammable liquid, flammable solid, organic peroxide, oxidizer, pyrophoric or unstable (reactive) or water-reactive material.

❖ Materials posing a detonation or deflagration hazard, or materials that readily support combustion, are considered physical hazards. Structures containing materials posing a physical hazard exceeding maximum allowable quantities per control area are classified in Group H1, H2 or H3. Materials posing a physical hazard may also present a health hazard.

PRESSURE VESSEL. A closed vessel designed to operate at pressures above 15 psig (103 kPa).

❖ A pressure vessel used to contain flammable liquids is classified as Group H2 when the pressure is over 15 pounds per square inch gauge (psig) (103 kPa). A pressure vessel when exposed to fire has a greater tendency to cause a BLEVE because of the increased pressure in the container and the more rapid increase of internal pressure when exposed to fire.

SAFETY CAN. An approved container of not more than 5-gallon (19 L) capacity having a spring-closing lid and spout cover so designed that it will relieve internal pressure when subjected to fire exposure.

❖ The use of safety cans is one of the methods for storage in a facility. Limited quantities of combustible and flammable liquids are allowed in a building or work area when they are stored in safety cans, and the area would not be classified as a Group H.

SECONDARY CONTAINMENT. That level of containment that is external to and separate from primary containment.

❖ If a spill or leak from the primary means of containment occurs, an additional level of containment may be necessary to isolate the hazardous materials from adjoining areas or the environment.

SEGREGATED. Storage in the same room or inside area, but physically separated by distance from incompatible materials.

❖ The mixture of two or more chemicals can create a toxic or explosive chemical that is more dangerous than any of the individual chemicals. When chemicals are stored in the same room or storage area it is important to segregate them either by distance or by curbs that will prevent the chemicals from mixing in the case of a discharge.

SOLID. A material that has a melting point and decomposes or sublimes at a temperature greater than 68°F (20°C).

❖ One of the three states of matter, solids must decompose (pyrolysis) before they can produce vapors that will support combustion. The surface area of a solid in relation to the heat source is a concern for fire fighters; the greater the surface being subjected to a heat source, the more rapid the pyrolysis.

STORAGE, HAZARDOUS MATERIALS. The keeping, retention or leaving of hazardous materials in closed containers, tanks, cylinders, or similar vessels; or vessels supplying operations through closed connections to the vessel.

❖ Storage of hazardous materials in a structure is governed either by the occupancy group that it is accessory to or by the hazard group that it is associated with. The determination of the use is based on the maximum allowable quantities per control area and the exceptions found in the IBC and the code.

SYSTEM. An assembly of equipment consisting of a container or containers, appurtenances, pumps, compressors and connecting piping.

❖ As with containers, a system can be either open or closed. The type of system being used with flammable or combustible liquids will determine whether it is Group H2 (open) or H3 (closed) when the amounts of chemicals are over the maximum allowable quantity per control area.

TANK, ATMOSPHERIC. A storage tank designed to operate at pressures from atmospheric through 1.0 pound per square inch gauge (760 mm Hg through 812 mm Hg) measured at the top of the tank.

❖ This tank is not designed for an internal pressure that exceeds atmospheric pressure. Most require some type of emergency venting system to assist in relief of pressure in a fire.

TANK, PORTABLE. A packaging of more than 60-gallon (227 L) capacity and designed primarily to be loaded into or on or temporarily attached to a transport vehicle or ship and equipped with skids, mountings or accessories to facilitate handling of the tank by mechanical means. It does not include any cylinder having less than a 1,000-pound (454 kg) water capacity, cargo tank, tank car tank or trailers carrying cylinders of more than 1,000-pound (454 kg) water capacity.

❖ The tank must be movable without having to detach permanently mounted electrical controls for the pumping or dispensing systems.

TANK, STATIONARY. Packaging designed primarily for stationary installations not intended for loading, unloading or attachment to a transport vehicle as part of its normal operation in the process of use. It does not include cylinders having less than a 1,000-pound (454 kg) water capacity.

❖ The tank is placed in a permanent location and typically has electrically mounted controls attached to a permanent power source.

TANK VEHICLE. A vehicle other than a railroad tank car or boat, with a cargo tank mounted thereon or built as an integral part thereof, used for the transportation of flammable or combustible liquids, LP-gas or hazardous chemicals. Tank vehicles include self-propelled vehicles and full trailers and semitrailers, with or without motive power, and carrying part or all of the load.

❖ Tank vehicles used for storage and transportation of hazardous chemicals over public roadways are governed by the DOT.

UNAUTHORIZED DISCHARGE. A release or emission of materials in a manner which does not conform to the provisions of this code or applicable public health and safety regulations.

❖ Remember that hazardous chemicals pose a threat to life and property only when release is not controlled or properly protected.

USE (MATERIAL). Placing a material into action, including solids, liquids and gases.

❖ This term refers to when a chemical or material is used in a process that forms another substance, whether it is hazardous or not, and when the chemical is used independently.

VAPOR PRESSURE. The pressure exerted by a volatile fluid as determined in accordance with ASTM D 323.

❖ Vapor pressure is a characteristic property of liquids and varies with their temperature. As the temperature of the liquid increases, more and more of the liquid enters the vapor stage. This increased pressure can cause an emergency vent to release, or, if the conditions are serious enough, can result in a BLEVE.

SECTION 2703
GENERAL REQUIREMENTS

2703.1 Scope. The storage, use and handling of all hazardous materials shall be in accordance with this section.

❖ Once a chemical is considered a hazardous chemical that is not exempt under the exceptions of Section 2701.1, it is expected to meet the general requirements listed in Section 2703 and any of the referenced codes or standards.

2703.1.1 Maximum allowable quantity per control area. The maximum allowable quantity per control area shall be as specified in Tables 2703.1.1(1) through 2703.1.1(4).

For retail and wholesale storage and display in Group M occupancies and Group S storage, see Section 2703.11.

❖ This section references Tables 2703.1.1(1) through 2703.1.1(4) for the maximum quantities allowed per a control area (see Section 2703.8.2 for design and protection requirements). If the amount of chemicals used in a building does not exceed the maximum allowed per control area and the number of control areas does not exceed the number and percentage of chemicals allowed by Table 2703.8.3.2, an area can be considered accessory to the use area and does not have to meet all of the requirements for a hazardous occupancy.

TABLE 2703.1.1(1). See page 27-18.

❖ Table 2703.1.1(1) is subdivided based on whether the material is in storage or in use in a closed or open system. Definitions of both closed and open systems are found in Section 202. Within these subdivisions, the appropriate maximum allowable quantity per control area is listed in accordance with the physical state (solid, liquid or gas) of the material. A column for gas in open systems is not indicated because hazardous gaseous materials should not be allowed in a system that is continuously open to the atmosphere.

Note b clearly indicates that the aggregate quantity of hazardous materials in use and storage, within a given control area, cannot exceed the quantity listed in the table for storage.

Without Note c, many common alcoholic beverages and household products containing a negligible amount of a hazardous material could result in a Group M occupancy being classified as a high hazard. Note c recognizes the reduced hazard of the materials based on their water miscibility and limited container size. A similar exception is indicated in Table 2307.1.1(2) and Tables 307.7(1) and 307.7(2) of the IBC.

Notes d and e of the table are significant in that, for certain materials, the maximum allowable quantity may be increased due to the use of approved special hazardous material storage methods (cabinets, exhausted enclosures or safety containers), sprinklers or both. The notes are intended to be cumulative in that up to four times the base maximum allowable quantity may be allowed per control area, if the building is sprinklered and approved special storage methods are used, without classifying the building as Group H. While the use of approved special storage methods is not always a feasible or practical method of storage, they do provide sufficient additional protection to warrant an increase if provided.

While classified as a hazardous material, the code recognizes the relative lower hazard of Class IIIB liquids as compared to that of other flammable and combustible liquids by classifying them as Group H-3 instead of Group H-2 and by establishing a base maximum allowable quantity per control area of 13,200 gallons (49 962 L). As indicated in Note f, the quantity of Class I oxidizers and Class IIIB liquids, without classifying the occupancy as Group H-3, would not be limited, provided the building is fully sprinklered in accordance with NFPA 13. The hazard presented by Class I oxidizers is that they slightly increase the burning rate of combustible materials that they may come into contact with during a fire. Class IIIB combustible liquids have flash points at or above 200°F (93°C). Motor oil is a typical example of a Class IIIB combustible liquid.

Note g recognizes that the hazard presented by certain materials is such that they may be stored or used only inside buildings that are fully sprinklered.

Note h clarifies for the user that while there is a combination maximum allowable quantity for flammable liquids, no individual class of liquid (Class IA, IB or IC) may exceed its own individual maximum allowable quantity.

Note i is a specific exception to the maximum allowable quantity per control area for maximum 660-gallon (2498 L) inside storage tanks of combustible liquids that are connected to a fuel-oil piping system. This exception would apply to most oil-fired stationary equipment, whether in industrial, commercial or residential occupancies. Oil-fired heating equipment and diesel engine-driven generator sets and their fuel supplies are indicative of the types of fuel-oil piping systems to which this note would apply. Note i also provides consistency with Sections 603.3.2 and 3401.2 of the code for inside storage.

Note k permits a larger amount of Class III oxidizers in a building when used for maintenance and health purposes. The quantities proposed are reasonable for occupancies such as the health care industry where Class III oxidizers are used for maintenance purposes, sterilization and sanitation of equipment and operation sanitation. The method used to store the oxidizers is subject to the evaluation and approval of the fire code official.

Note l clarifies that the 125 pounds (57 kg) of storage permitted for consumer fireworks represents the net weight of the pyrotechnic composition of the fireworks in a nonsprinklered building. This amount represents approximately $12^1/_2$ shipping cases (less than one-and-a-half pallet loads) of fireworks in a nonsprinklered storage condition. In cases where the net weight of the pyrotechnic composition of the fireworks is unknown, 25 percent of the gross weight of the fireworks is to be used. The gross weight is to include the weight of the packaging.

Note n provides an exception when the amount of hazardous material in storage and display in Group M and S occupancies meets the requirements of Section 2703.11.

TABLE 2703.1.1(2). See page 27-20.

❖ Table 2703.1.1(2), similar to Table 2703.1.1(1), specifies the maximum allowable quantities of hazardous materials, liquids or chemicals allowed per control area before having to classify a part of the (or the entire) building as a Group H occupancy and is subdivided based on whether the material is in storage or in use in a closed or open system. Definitions of both "Closed" and "Open systems" are found in Section 202. Within these subdivisions, the appropriate maximum allowable quantity per control area is listed in accordance with the physical state (solid, liquid or gas) of the material. A column for gas in open systems is not indicated because hazardous gaseous materials should not be allowed in a system that is continuously open to the atmosphere. This table contains health-hazard materials classified as Group H-4 in accordance with Section 307.6 of the IBC. While the materials listed in this table are considered health hazards, some of the materials may also possess physical hazard characteristics more indicative of materials classified as Group H-1, H-2 or H-3 (see Section 2701.1).The maximum allowable quantities per control area listed in the table are indicative of industry practice and assume the materials are properly stored and handled in accordance with the code. Group H-4 materials, while indeed hazardous, are primarily considered a handling problem and do not possess the same fire, explosion or reactivity hazard associated with other hazardous materials. The base maximum allowable quantity per control area of 810 cubic feet (23 m^3) for gases that are either corrosive or toxic is based on a standard-size chlorine cylinder.

Without Note b, many common household products containing a negligible amount of a hazardous material could result in a Group M occupancy being classified as a high hazard. Note b recognizes the reduced hazard of the materials based on their water miscibility and limited container size. A similar exception is indicated in Table 2703.1.1(1), and Tables 307.7(1) and 307.7(2) of the IBC.

Note c provides an exception when the amount of hazardous material in storage and display in Group M and S occupancies meets the requirements of Section 2703.11.

Note d clearly indicates that the aggregate quantity of hazardous materials in use and storage, within a given control area, cannot exceed the quantity listed in the table for storage.

Where applicable, Notes e and f provide an increase in the base maximum allowable amount similar to that in Table 2703.1.1(1) [see commentary, Table 2703.1.1(1)].

Note g exempts a building from a Group H-4 classification that contains no more than a single 150-pound (68 kg) cylinder per control area of anhydrous ammonia in a nonsprinklered building and no more than two cylinders each containing 150 pounds (68 kg) or less in a single control area of a fully sprinklered building. Anhydrous ammonia, which has a very low density when unconfined, is approximately one-fifth the density of chlorine. Based on their respective LC_{50} values, chlorine is also approximately 80 times more toxic than anhydrous ammonia. The use of cabinets is not recognized as a realistic method of storing cylinders of anhydrous ammonia.

Note h of the table is significant in that, for certain materials, their hazard is so great that their maximum allowable quantity per control area may be stored in the building only when approved exhausted enclosures or gas cabinets complying with Sections 2703.8.5 and 2703.8.6, respectively, are utilized.

TABLE 2703.1.1(3). See page 27-21.

❖ Table 2703.1.1(3) specifies the maximum allowable quantities of hazardous materials, liquids or chemicals allowed per outdoor control area before being subject to additional regulations contained in Chapters 28 through 44 and is subdivided based on whether the material is in storage or in use in a closed or open system. Definitions of both "Closed" and "Open systems" are found in Section 202. Within these subdivisions, the appropriate maximum allowable quantity per control area is listed in accordance with the physical state (solid, liquid or gas) of the material. A column for gas in open systems is not indicated because hazardous gaseous materials should not be allowed in a system that is continuously open to the atmosphere. This table contains physical-hazard materials as defined in Section 2702.1. While the materials listed in this table are considered physical hazards, some of the materials may also possess health-hazard characteristics as defined in Section 2702.1. The maximum allowable quantities per outdoor control area listed in the table are indicative of industry practice and assume the materials are properly stored and handled in accordance with the code. The base maximum allowable quantity per outdoor control area of 810 cubic feet (23 m^3) for gases that are either corrosive or toxic is based on a standard-size chlorine cylinder.

While hazardous materials within a closed or open system are considered to be "in use," Note b clearly indicates that the aggregate quantity of hazardous materials in use and storage within a given outdoor control area cannot exceed the quantity listed in the table for storage.

Note c provides an exception when the amount of hazardous materials in outdoor storage in conjunction with a retail or wholesale Group M occupancy meets the requirements of Section 2703.11. The outside storage area must be under the same ownership as the Group M occupancy.

TABLE 2703.1.1(4). See page 27-22.

❖ Table 2703.1.1(4), similar to Table 2703.1.1(3), specifies the maximum allowable quantities of hazardous materials, liquids or chemicals allowed per outdoor control area before being subject to additional regulations contained in Chapters 28 through 44. It also is subdivided based on whether the material is in storage or in use in a closed or open system. Definitions of both "Closed" and "Open systems" are found in Section 202. Within these subdivisions, the appropriate maximum allowable quantity per control area is listed in accordance with the physical state (solid, liquid or gas) of the material. A column for gas in open systems is not indicated because hazardous gaseous materials should not be allowed in a system that is continuously open to the atmosphere. This table contains health-hazard materials as defined in Section 2702.1. While the materials listed in this table are considered health hazards, some of the materials may also possess physical-hazard characteristics as defined in Section 2702.1. The maximum allowable quantities per outdoor control area listed in the table are indicative of industry practice and assume the materials are properly stored and handled in accordance with the code. These materials, while indeed hazardous, are primarily considered a handling problem and do not possess the same fire, explosion or reactivity hazard associated with other hazardous materials. The base maximum allowable quantity per outdoor control area of 810 cubic feet (23 m^3) for gases that are either corrosive or toxic is based on a standard-size chlorine cylinder.

Note b clearly indicates that the aggregate quantity of hazardous materials in use and storage within a given outdoor control area cannot exceed the quantity listed in the table for storage.

Note c provides an exception when the amount of hazardous material in outdoor storage in conjunction with a retail or wholesale Group M occupancy meets the requirements of Section 2703.11. The outside storage area must be under the same ownership as the Group M occupancy.

Note d of the table is significant in that, for certain materials, their hazard is so great that their maximum allowable quantity in an outdoor control area may be stored only when approved exhausted enclosures or gas cabinets complying with Sections 2703.8.5 and 2703.8.6, respectively, or laboratory fume hoods are utilized.

Note e states that when toxic liquids with a vapor pressure in excess of 1 pound per square inch absolute (psia) (7 kPa) at 77°F (25°C) are stored, the maximum allowable quantity per control area is limited to the amount listed for highly toxic materials.

Note g establishes a special maximum allowable quantity of no more than two cylinders each containing 150 pounds (68 kg) or less in a single outdoor control. Anhydrous ammonia, which has a very low density when unconfined, is approximately one-fifth the density of chlorine. Based on their respective median lethal (LC_{50}) values, chlorine is also approximately 80 times more toxic than anhydrous ammonia. The use of cabinets is not recognized as a realistic method of storing cylinders of anhydrous ammonia.

TABLE 2703.1.1(1)
MAXIMUM ALLOWABLE QUANTITY PER CONTROL AREA OF HAZARDOUS MATERIALS POSING A PHYSICAL HAZARD[a, j, m, n]

MATERIAL	CLASS	GROUP WHEN THE MAXIMUM ALLOWABLE QUANTITY IS EXCEEDED	STORAGE[b]			USE-CLOSED SYSTEMS[b]			USE-OPEN SYSTEMS[b]	
			Solid pounds (cubic feet)	Liquid gallons (pounds)	Gas cubic feet at NTP	Solid pounds (cubic feet)	Liquid gallons (pounds)	Gas cubic feet at NTP	Solid pounds (cubic feet)	Liquid gallons (pounds)
Combustible liquid[c, i]	II IIIA IIIB	H-2 or H-3 H-2 or H-3 Not Applicable	Not Applicable	120[d, e] 330[d, e] 13,200[e, f]	Not Applicable	Not Applicable	120[d] 330[d] 13,200[f]	Not Applicable	Not Applicable	30[d] 80[d] 3,300[f]
Combustible fiber	Loose Baled	H-3	(100) (1,000)	Not Applicable	Not Applicable	(100) (1,000)	Not Applicable	Not Applicable	(20) (200)	Not Applicable
Cryogenic Flammable	Not Applicable	H-2	Not Applicable	45[d]	Not Applicable	Not Applicable	45[d]	Not Applicable	Not Applicable	10[d]
Consumer fireworks (Class C Common)	1.4G	H-3	125[d, e, l]	Not Applicable	Not Applicable	Not Applicable	Not Applicable	Not Applicable	Not Applicable	Not Applicable
Cryogenic Oxidizing	Not Applicable	H-3	Not Applicable	45[d]	Not Applicable	Not Applicable	45[d]	Not Applicable	Not Applicable	10[d]
Explosives	Division 1.1 Division 1.2 Division 1.3 Division 1.4 Division 1.4G Division 1.5 Division 1.6	H-1 H-1 H-1 or H-2 H-3 H-3 H-1 H-1	1[e, g] 1[e, g] 5[e, g] 50[e, g] 125[d, e, l] 1[e, g] 1[d, e, g]	(1)[e, g] (1)[e, g] (5)[e, g] (50)[e, g] Not Applicable (1)[e, g] Not Applicable	Not Applicable	0.25[g] 0.25[g] 1[g] 50[g] Not Applicable 0.25[g] Not Applicable	(0.25)[g] (0.25)[g] (1)[g] (50)[g] Not Applicable (0.25)[g] Not Applicable	Not Applicable	0.25[g] 0.25[g] 1[g] Not Applicable Not Applicable 0.25[g] Not Applicable	(0.25)[g] (0.25)[g] (1)[g] Not Applicable Not Applicable (0.25)[g] Not Applicable
Flammable gas	Gaseous Liquefied	H-2	Not Applicable	Not Applicable 30[d, e]	1,000[d, e] Not Applicable	Not Applicable	Not Applicable 30[d, e]	1,000[d, e] Not Applicable	Not Applicable	Not Applicable
Flammable liquids[c]	IA IB and IC	H-2 or H-3	Not Applicable	30[d, e] 120[d, e]	Not Applicable	Not Applicable	30[d] 120[d]	Not Applicable	Not Applicable	10[d] 30[d]
Combination Flammable liquid (IA, IB, IC)	Not Applicable	H-2 or H-3	Not Applicable	120[d, e, h]	Not Applicable	Not Applicable	120[d, h]	Not Applicable	Not Applicable	30[d, h]
Flammable solid	Not Applicable	H-3	125[d, e]	Not Applicable	Not Applicable	125[d]	Not Applicable	Not Applicable	25[d]	Not Applicable
Organic peroxide	UD I II III IV V	H-1 H-2 H-3 H-3 Not Applicable Not Applicable	1[e, g] 5[d, e] 50[d, e] 125[d, e] Not Limited Not Limited	(1)[e, g] (5)[d, e] (50)[d, e] (125)[d, e] Not Limited Not Limited	Not Applicable	0.25[g] 1[d] 50[d] 125[d] Not Limited Not Limited	(0.25)[g] (1)[d] (50)[d] (125)[d] Not Limited Not Limited	Not Applicable	0.25[g] 1[d] 10[d] 25[d] Not Limited Not Limited	(0.25)[g] (1)[d] (10)[d] (25)[d] Not Limited Not Limited

(continued)

TABLE 2703.1.1(1)—(continued)
MAXIMUM ALLOWABLE QUANTITY PER CONTROL AREA OF HAZARDOUS MATERIALS POSING A PHYSICAL HAZARD[a, j, m, n]

MATERIAL	CLASS	GROUP WHEN THE MAXIMUM ALLOWABLE QUANTITY IS EXCEEDED	STORAGE[b]			USE-CLOSED SYSTEMS[b]			USE-OPEN SYSTEMS[b]	
			Solid pounds (cubic feet)	Liquid gallons (pounds)	Gas cubic feet at NTP	Solid pounds (cubic feet)	Liquid gallons (pounds)	Gas cubic feet at NTP	Solid pounds (cubic feet)	Liquid gallons (pounds)
Oxidizer	4	H-1	1[g]	(1)[e, g]	Not Applicable	0.25[g]	(0.25)[g]	Not Applicable	0.25[g]	(0.25)[g]
	3[k]	H-2	10[d, e]	(10)[d, e]		2[d]	(2)[d]		2[d]	(2)[d]
	2	H-3	250[d, e]	(250)[d, e]		250[d]	(250)[d]		50[d]	(50)[d]
	1	H-3	4,000[e,f]	(4,000)[e,f]		4,000[f]	(4,000)[f]		1,000[f]	(1,000)[f]
Oxidizing gas	Gaseous	H-3	Not Applicable	Not Applicable	1,500[d, e]	Not Applicable	Not Applicable	1,500[d, e]	Not Applicable	Not Applicable
	Liquefied			15[d, e]	Not Applicable		15[d, e]	Not Applicable		
Pyrophoric	Not Applicable	H-2	4[e, g]	(4)[e, g]	50[e, g]	1[g]	(1)[g]	10[e, g]	0	0
Unstable (reactive)	4	H-1	1[e, g]	(1)[e, g]	10[e, g]	0.25[g]	(0.25)[g]	2[e, g]	0.25[g]	(0.25)[g]
	3	H-1 or H-2	5[d, e]	(5)[d, e]	50[d, e]	1[d]	(1)[d]	10[d, e]	1[d]	(1)[d]
	2	H-3	50[d, e]	(50)[d, e]	250[d, e]	50[d]	(50)[d]	250[d, e]	10[d]	(10)[d]
	1	Not Applicable	Not Limited	Not Limited	Not Limited	Not Limited	Not Limited	Not Limited	Not Limited	Not Limited
Water reactive	3	H-2	5[d,e]	(5)[d, e]	Not Applicable	5[d]	(5)[d]	Not Applicable	1[d]	(1)[d]
	2	H-3	50[d, e]	(50)[d, e]		50[d]	(50)[d]		10[d]	(10)[d]
	1	Not Applicable	Not Limited	Not Limited		Not Limited	Not Limited		Not Limited	Not Limited

For SI: 1 cubic foot = 0.023 m^3, 1 pound = 0.454 kg, 1 gallon = 3.785 L.

a. For use of control areas, see Section 2703.8.3.
b. The aggregate quantity in use and storage shall not exceed the quantity listed for storage.
c. The quantities of alcoholic beverages in retail and wholesale sales occupancies shall not be limited providing the liquids are packaged in individual containers not exceeding 1.3 gallons. In retail and wholesale sales occupancies, the quantities of medicines, foodstuffs, consumer or industrial products, and cosmetics containing not more than 50 percent by volume of water-miscible liquids with the remainder of the solutions not being flammable shall not be limited, provided that such materials are packaged in individual containers not exceeding 1.3 gallons.
d. Maximum allowable quantities shall be increased 100 percent in buildings equipped throughout with an approved automatic sprinkler system in accordance with Section 903.3.1.1. Where Note e also applies, the increase for both notes shall be applied accumulatively.
e. Maximum allowable quantities shall be increased 100 percent when stored in approved storage cabinets, gas cabinets, exhausted enclosures or safety cans. Where Note d also applies, the increase for both notes shall be applied accumulatively.
f. Quantities shall not be limited in a building equipped throughout with an approved automatic sprinkler system.
g. Allowed only in buildings equipped throughout with an approved automatic sprinkler system.
h. Containing not more than the maximum allowable quantity per control area of Class IA, Class IB or Class IC flammable liquids.
i. Inside a building, the maximum capacity of a combustible liquid storage system that is connected to a fuel-oil piping system shall be 660 gallons provided such system conforms to this code.
j. Quantities in parenthesis indicate quantity units in parenthesis at the head of each column.
k. A maximum quantity of 200 pounds of solid or 20 gallons of liquid Class 3 oxidizers is allowed when such materials are necessary for maintenance purposes, operation or sanitation of equipment when the storage containers and the manner of storage are approved.
l. Net weight of pyrotechnic composition of the fireworks. Where the net weight of the pyrotechnic composition of the fireworks is not known, 25 percent of the gross weight of the fireworks including packaging shall be used.
m. For gallons of liquids, divide the amount in pounds by 10 in accordance with Section 2703.1.2.
n. For storage and display squantities in Group M and storage quantities in Group S occupancies complying with Section 2703.11, see Table 2703.11.1.

TABLE 2703.1.1(2)
MAXIMUM ALLOWABLE QUANTITY PER CONTROL AREA OF HAZARDOUS MATERIAL POSING A HEALTH HAZARD[a,b,c,j]

MATERIAL	STORAGE[d]			USE-CLOSED SYSTEMS[d]			USE-OPEN SYSTEMS[d]	
	Solid pounds[e, f]	Liquid gallons (pounds)[e, f]	Gas cubic feet[e]	Solid pounds[e]	Liquid gallons (pounds)[e]	Gas cubic feet[e]	Solid pounds[e]	Liquid gallons (pounds)[e]
Corrosive	5,000	500	810[f, g]	5,000	500	810[f, g]	1,000	100
Highly toxic	10	(10)[i]	20[h]	10	(10)[i]	20[h]	3	(3)[i]
Toxic	500	(500)[i]	810[f]	500	(500)[i]	810[f]	125	(125)[i]

For SI: 1 cubic foot = 0.028 m^3, 1 pound = 0.454 kg, 1 gallon = 3.785 L.

a. For use of control areas, see Section 2703.8.3.

b. In retail and wholesale sales occupancies, the quantities of medicines, foodstuffs consumer or industrial products, and cosmetics, containing not more than 50 percent by volume of water-miscible liquids and with the remainder of the solutions not being flammable, shall not be limited, provided that such materials are packaged in individual containers not exceeding 1.3 gallons.

c. For storage and display quantities in Group M and storage quantities in Group S occupancies complying with Section 2703.11, see Table 2703.11.1.

d. The aggregate quantity in use and storage shall not exceed the quantity listed for storage.

e. Maximum allowable quantities shall be increased 100 percent in buildings equipped throughout with an approved automatic sprinkler system in accordance with Section 903.3.1.1. Where Note f also applies, the increase for both notes shall be applied accumulatively.

f. Maximum allowable quantities shall be increased 100 percent when stored in approved storage cabinets, gas cabinets, or exhausted enclosures. Where Note e also applies, the increase for both notes shall be applied accumulatively.

g. A single cylinder containing 150 pounds or less of anhydrous ammonia in a single control area in a nonsprinklered building shall be considered a maximum allowable quantity. Two cylinders, each containing 150 pounds or less in a single control area shall be considered a maximum allowable quantity provided the building is equipped throughout with an automatic sprinkler system in accordance with Section 903.3.1.1.

h. Allowed only when stored in approved exhausted gas cabinets or exhausted enclosures.

i. Quantities in parenthesis indicate quantity units in parenthesis at the head of each column.

j. For gallons of liquids, divide the amount in pounds by 10 in accordance with Section 2703.1.2.

TABLE 2703.1.1(3) MAXIMUM ALLOWABLE QUANTITY PER CONTROL AREA OF HAZARDOUS MATERIALS POSING A PHYSICAL HAZARD IN AN OUTDOOR CONTROL AREA[a, b, c]

MATERIAL	CLASS	STORAGE[b]			USE-CLOSED SYSTEMS[b]			USE-OPEN SYSTEMS[b]	
		Solid pounds	Liquid gallons (pounds)	Gas cubic feet at NTP	Solid pounds	Liquid gallons (pounds)	Gas cubic feet at NTP	Solid pounds	Liquid gallons (pounds)
Flammable gas	Gaseous Liquefied	Not Applicable	Not Applicable 30	3,000 Not Applicable	Not Applicable	Not Applicable 15	1,500 Not Applicable	Not Applicable	Not Applicable
Flammable solid	Not Applicable	500	Not Applicable	Not Applicable	250	Not Applicable	Not Applicable	50	Not Applicable
Organic peroxide	Unclassified Detonable	1	(1)	Not Applicable	0.25	(0.25)[d]	Not Applicable	0.25	(0.25)[d]
Organic peroxide	I II III IV V	20 200 500 1,000 Not Limited	(20)[d] (200)[d] (500)[d] (1,000)[d] Not Limited	Not Applicable	10 100 250 500 Not Limited	(10)[d] (100)[d] (250)[d] (500)[d] Not Limited	Not Applicable	2 20 50 100 Not Limited	(2)[d] (20)[d] (50)[d] (100)[d] Not Limited
Oxidizer	4 3 2 1	2 40 1,000 Not Limited	(2)[d] (40)[d] (1,000)[d] Not Limited	Not Applicable	1 20 500 Not Limited	(1)[d] (20)[d] (500)[d] Not Limited	Not Applicable	0.25 4 100 Not Limited	(0.25)[d] (4)[d] (100)[d] Not Limited
Oxidizing gas	Gaseous Liquefied	Not Applicable	Not Applicable 60	6,000 Not Applicable	Not Applicable	Not Applicable 30	3,000 Not Applicable	Not Applicable	Not Applicable
Pyrophoric materials	Not Applicable	8	(8)[d]	100	4	(4)[d]	10	0	0
Unstable (reactive)	4 3 2 1	2 20 200 Not Limited	(2)[d] (20)[d] (200)[d] Not Limited	20 200 1,000 1,500	1 10 100 Not Limited	(1)[d] (10)[d] (100)[d] Not Limited	2 10 250 Not Limited	0.25 1 10 Not Limited	(0.25)[d] 1 10 Not Limited
Water reactive	3 2 1	20 200 Not Limited	(20)[d] (200)[d] Not Limited	Not Applicable	10 100 Not Limited	(10)[d] (100)[d] Not Limited	Not Applicable	1 10 Not Limited	(1)[d] (10)[d] Not Limited

For SI: 1 pound = 0.454 kg, 1 gallon = 3.785 L, 1 cubic foot = 0.02832 m^3.

a. For gallons of liquids, divide the amount in pounds by 10 in accordance with Section 2703.1.2.

b. The aggregate quantities in storage and use shall not exceed the quantity listed for storage.

c. The aggregate quantity of nonflammable solid and nonflammable or noncombustible liquid hazardous materials allowed in outdoor storage per single property under the same ownership or control used for retail or wholesale sales is allowed to exceed the maximum allowable quantity per control area when such storage is in accordance with Section 2703.11.

d. Quantities in parentheses indicate quantity units in parentheses at the head of each column.

TABLE 2703.1.1(4)
MAXIMUM ALLOWABLE QUANTITY PER CONTROL AREA OF HAZARDOUS MATERIALS POSING A HEALTH HAZARD IN AN OUTDOOR CONTROL AREA[a,b,c]

MATERIAL	STORAGE			USE-CLOSED SYSTEMS			USE-OPEN SYSTEMS	
	Solid pounds	Liquid gallons (pounds)	Gas cubic feet at NTP	Solid pounds	Liquid gallons (pounds)	Gas cubic feet at NTP	Solid pounds	Liquid gallons (pounds)
Corrosives	20,000	2,000	1,620[g]	10,000	1,000	810[g]	1,000	100
Highly toxics	20	(20)[f]	40[d]	10	(10)[f]	20[d]	3	(3)[f]
Toxics	1,000	(1,000)[e, f]	1,620	500	50[e]	810	25	(25)[e, f]

For SI: 1 cubic foot = 0.02832 m^3, 1 pound = 0.454 kg, 1 gallon = 3.785 L, 1 pound per square inch absolute = 6.895 kPa, °C = [(°F)-32/1.8].

a. For gallons of liquids, divide the amount in pounds by 10 in accordance with Section 2703.1.2.

b. The aggregate quantities in storage and use shall not exceed the quantity listed for storage.

c. The aggregate quantity of nonflammable solid and nonflammable or noncombustible liquid hazardous materials allowed in outdoor storage per single property under the same ownership or control used for retail or wholesale sales is allowed to exceed the maximum allowable quantity per control area when such storage is in accordance with Section 2703.11.

d. Allowed only when used in approved exhausted gas cabinets, exhausted enclosures or under fume hoods.

e. The maximum allowable quantity per control area for toxic liquids with vapor pressures in excess of 1 psia at 77°F shall be the maximum allowable quantity per control area listed for highly toxic liquids.

f. Quantities in parentheses indicate quantity units in parentheses at the head of each column.

g. Two cylinders, each cylinder containing 150 pounds or less of anhydrous ammonia, shall be considered a maximum allowable quantity in an outdoor control area.

2703.1.2 Conversion. Where quantities are indicated in pounds and when the weight per gallon of the liquid is not provided to the fire code official, a conversion of 10 pounds per gallon (1.2 kg/L) shall be used.

❖ If the weight of a liquid is not given either in the MSDS or in other documentation, the fire code official is allowed to use a conversion rate of 10 gallons (38 L) of liquid to 1 pound (.45 kg) of material (Note: 1 U.S. gallon = 0.1336805 cubic foot).

2703.1.3 Quantities not exceeding the maximum allowable quantity per control area. The storage, use and handling of hazardous materials in quantities not exceeding the maximum allowable quantity per control area indicated in Tables 2703.1.1(1) through 2703.1.1(4) shall be in accordance with Sections 2701 and 2703.

❖ See the commentary for Section 2703.1.1.

2703.1.4 Quantities exceeding the maximum allowable quantity per control area. The storage and use of hazardous materials in quantities exceeding the maximum allowable quantity per control area indicated in Tables 2703.1.1(1) through 2703.1.1(4) shall be in accordance with this chapter.

❖ Once the amounts of chemicals exceed the maximum allowed in Tables 2703.1.1(1) through 2703.1.1(4) per control area, the area or building is considered a hazardous occupancy and must meet the general requirements and specific requirements in both the code and the IBC, based on the chemicals being used.

2703.2 Systems, equipment and processes. Systems, equipment and processes utilized for storage, dispensing, use or handling of hazardous materials shall be in accordance with Sections 2703.2.1 through 2703.2.8.

❖ Devices used in the process and storage or dispensing of hazardous materials are to meet the requirements of Sections 2703.1 to 2703.2.8.

2703.2.1 Design and construction of containers, cylinders and tanks. Containers, cylinders and tanks shall be designed and constructed in accordance with approved standards. Containers, cylinders, tanks and other means used for containment of hazardous materials shall be of an approved type.

❖ The design and construction of containers must meet the requirements listed in Chapters 27 to 44 and the referenced standards in Chapter 45.

2703.2.2 Piping, tubing, valves and fittings. Piping, tubing, valves and fittings conveying hazardous materials shall be designed and installed in accordance with approved standards and shall be in accordance with Sections 2703.2.2.1 and 2703.2.2.2.

❖ The design and construction of piping, tubing, valves and fittings shall meet the requirements listed in Chapters 27 to 44 and the referenced standards in Chapter 45.

2703.2.2.1 Design and construction. Piping, tubing, valves, fittings and related components used for hazardous materials shall be in accordance with the following:

1. Piping, tubing, valves, fittings and related components shall be designed and fabricated from materials compatible with the material to be contained and shall be of adequate strength and durability to withstand the pressure, structural and seismic stress, and exposure to which they are subject.
2. Piping and tubing shall be identified in accordance with ANSI A13.1 to indicate the material conveyed.
3. Readily accessible manual valves, or automatic remotely activated fail-safe emergency shutoff valves shall be installed on supply piping and tubing at the following locations:
   - 3.1. The point of use.
   - 3.2. The tank, cylinder or bulk source.
4. Emergency shutoff valves shall be identified and the location shall be clearly visible and accessible and indicated by means of a sign.
5. Backflow prevention or check valves shall be provided when the backflow of hazardous materials could create a hazardous condition or cause the unauthorized discharge of hazardous materials.
6. Where gases or liquids having a hazard ranking of:

Health hazard Class 3 or 4
Flammability Class 4
Reactivity Class 3 or 4

in accordance with NFPA 704 are carried in pressurized piping above 15 pounds per square inch gauge (psig) (103 kPa), an approved means of leak detection and emergency shutoff or excess flow control shall be provided. Where the piping originates from within a hazardous material storage room or area, the excess flow control shall be located within the storage room or area. Where the piping originates from a bulk source, the excess flow control shall be located as close to the bulk source as practical.

Exceptions:

1. Piping for inlet connections designed to prevent backflow.
2. Piping for pressure relief devices.

❖ This section specifies minimum design requirements for any piping system handling hazardous materials. All associated piping, valves and fittings also should be compatible with the material to be used. As may be required by industry standards and OSHA regulations, piping and tubing conveying hazardous materials must be properly identified. Identification could include color-coded piping and permanent labeling on the piping and tubing. The method of operation of accessible manual shutoff valves must be identified as to what they control as well. Backflow protection must be provided as necessary to protect potable water supplies and the environment as required by local health departments, environmental agencies and fire code officials.

2703.2.2.2 Additional regulations for supply piping for health-hazard materials. Supply piping and tubing for gases and liquids having a health-hazard ranking of 3 or 4 in accordance with NFPA 704 shall be in accordance with ANSI B31.3 and the following:

1. Piping and tubing utilized for the transmission of highly toxic, toxic or highly volatile corrosive liquids and gases shall have welded, threaded or flanged connections throughout except for connections located within a ventilated enclosure if the material is a gas, or an approved method of drainage or containment is provided for connections if the material is a liquid.
2. Piping and tubing shall not be located within corridors, within any portion of a means of egress required to be enclosed in fire-resistance-rated construction or in concealed spaces in areas not classified as Group H occupancies.

Exception: Piping and tubing within the space defined by the walls of corridors and the floor or roof above or in concealed spaces above other occupancies when installed in accordance with Section 415.9.6.3 of the *International Building Code* for Group H-5 occupancies.

❖ Requirements of this section are dependent on whether the hazardous material involved has a health-hazard ranking of 3 or 4 based on the degree of hazard classification system in NFPA 704. This section is not applicable to materials with a flammability or reactivity rating of 3 or 4; rather, it addresses piping systems handling highly toxic or toxic liquids and gases.

Leaks or piping failures are most common around threaded connections in the piping or tubing. Therefore, a means of containment is required around connections unless metallic piping or tubing with welded connections is used.

Hazardous material piping may not be located in an exit access corridor, exit or areas not classified as high hazard so that the required means of egress are available. The exception in this section recognizes specific design provisions for existing hazardous production materials facilities, which commonly transport hazardous materials through service passages or corridors. Also required are excess-flow control valves (to regulate the rate of flow of hazardous materials within the piping system) and emergency shutoff valves (to stop the flow of hazardous materials from possibly deteriorating further in an emergency scenario).

2703.2.3 Equipment, machinery and alarms. Equipment, machinery and required detection and alarm systems associated with the use, storage or handling of hazardous materials shall be listed or approved.

❖ Equipment associated with the use or storage of hazardous materials must be listed or approved by a third-party testing agency.

2703.2.4 Installation of tanks. Installation of tanks shall be in accordance with Sections 2703.2.4.1 through 2703.2.4.2.1.

❖ The installation of tanks shall be in accordance with Chapter 34, the referenced standards in Chapter 45 and Sections 2703.2.4.1 through 2703.2.4.2.1.

2703.2.4.1 Underground tanks. Underground tanks used for the storage of liquid hazardous materials shall be provided with secondary containment. In lieu of providing secondary containment for an underground tank, an above-ground tank in an underground vault complying with Section 3404.2.8 shall be permitted. Underground vaults shall be otherwise regulated as underground tank installations.

❖ All underground tanks used to store liquid hazardous materials must have a secondary containment that meets the requirements of Section 3404.2.7 or, alternatively, must be installed in an underground vault in compliance with Section 3404.2.8.

2703.2.4.2 Above-ground tanks. Above-ground stationary tanks used for the storage of hazardous materials shall be located and protected in accordance with the requirements for outdoor storage of the particular material involved.

Exception: Above-ground tanks that are installed in vaults complying with Section 3404.2.8 shall not be required to comply with location and protection requirements for outdoor storage.

❖ Requirements for above-ground tanks depend on the hazards associated with the material being stored. Tank requirements can be found in Sections 2206, 3104 and 3404.2. The exception allows the use of an above-ground tank in a below-grade tank vault in lieu of an underground tank. Installation of a tank in a vault is regarded as equivalent in safety to an underground tank by Chapter 34, and it is a superior method with regard to environmental safety. When such an installation is provided, requirements for location on site and similar provisions should be regarded as an underground installation.

2703.2.4.2.1 Marking. Above-ground stationary tanks shall be marked as required by Section 2703.5.

❖ This section requires that NFPA 704 hazard identification signs in accordance with Section 2703.5 be provided for above-ground tanks to assist emergency responders in identifying the hazards of the tanks' contents in case of a spill or fire incident.

2703.2.5 Empty containers and tanks. Empty containers and tanks previously used for the storage of hazardous materials shall be free from residual material and vapor as defined by DOTn, the Resource Conservation and Recovery Act (RCRA) or other regulating authority or maintained as specified for the storage of hazardous material.

❖ Tanks must be purged and cleaned of all residual hazardous chemicals before reuse for storage. Containers and tanks that have not been cleaned must be stored in areas meeting the requirements for a hazardous use area. Under the Resource Conservation and Recovery Act, the EPA issues "cradle-to-grave" regulations for storing, using and disposing of hazardous waste. The act, enacted by Congress in 1976, established a uniform national policy for hazardous and solid waste disposal. This section requires that containers, cylinders and tanks, if still in use, either be properly maintained as required or comply with the EPA or other state and local environmental regulations.

2703.2.6 Maintenance. In addition to the requirements of Section 2703.2.3, equipment, machinery and required detection and alarm systems associated with hazardous materials shall be maintained in an operable condition. Defective containers, cylinders and tanks shall be removed from service, repaired or disposed of in an approved manner. Defective equipment or machinery shall be removed from service and repaired or replaced. Required detection and alarm systems shall be replaced or repaired where defective.

❖ Equipment, machinery and required detection and alarm equipment must be maintained in an operable condition at all times.

2703.2.6.1 Tanks out of service for 90 days. Stationary tanks not used for a period of 90 days shall be properly safeguarded or removed in an approved manner. Such tanks shall have the fill line, gauge opening and pump connection secured against tampering. Vent lines shall be properly maintained.

❖ This section places a time limitation on out-of-service storage tanks of hazardous materials. Within 90 days, proper steps should be taken to address any potential hazard with the storage tank. Without a time limitation for safeguarding the out-of-service storage tank, it is likely that the tank will not be properly monitored or inspected as necessary and, consequently, the risk of the tank becoming defective increases. Care shall be taken to prevent tampering with all associated equipment. Vent lines must be kept clear and be properly maintained during this time.

2703.2.6.1.1 Return to service. Tanks that are to be placed back in service shall be tested in an approved manner.

❖ Tanks must be tested as a new installation before being placed back in service.

2703.2.6.2 Defective containers and tanks. Defective containers and tanks shall be removed from service, repaired in accordance with approved standards or disposed of in an approved manner.

❖ Damaged containers, cylinders and tanks pose the potential hazard of content release. Care must be taken to determine that disposal does not present a greater hazard than the damaged container, cylinder or tank.

Small containers usually cannot be repaired easily. Large containers and fixed tank installations often cannot be easily replaced and must be repaired. Temporary storage of hazardous materials during container repair must comply with code requirements.

2703.2.7 Liquid-level limit control. Atmospheric tanks having a capacity greater than 500 gallons (1893 L) and which contain hazardous material liquids shall be equipped with a liquid-level limit control or other approved means to prevent overfilling of the tank.

❖ Overfilling of tanks has been a problem over the years. This section requires that atmospheric tanks with a capacity greater than 500 gallons (1893 L) containing hazardous materials be equipped with an approved method of fill control.

2703.2.8 Seismic protection. Machinery and equipment utilizing hazardous materials shall be braced and anchored in accordance with the seismic design requirements of the *International Building Code* for the seismic design category in which the machinery or equipment is classified.

❖ In areas that are listed by Sections 1614 through 1623 of the IBC as requiring seismic protection, machinery and equipment containing hazardous materials must be protected again seismic activity. Table 1604.5 of the IBC assigns importance factors in classifying buildings. Buildings containing sufficient quantities of toxic or explosive substances to be dangerous to the public if release occurs carry an importance factor of 1.25 and structures containing highly toxic materials where the storage or use exceeds the maximum allowable quantity per control area have an importance factor of 1.50.

2703.3 Release of hazardous materials. Hazardous materials in any quantity shall not be released into a sewer, storm drain, ditch, drainage canal, creek, stream, river, lake or tidal waterway or on the ground, sidewalk, street, highway or into the atmosphere.

Exceptions:

1. The release or emission of hazardous materials is allowed when in compliance with federal, state, or local governmental agencies, regulations or permits.
2. The release of pesticides is allowed when used in accordance with registered label directions.
3. The release of fertilizer and soil amendments is allowed when used in accordance with manufacturer's specifications.

❖ Because of the toxic and hazardous nature of chemicals governed by this section, no amount of release is allowed, unless it is in compliance with federal, state or local regulations. The release of pesticides is allowed when they are used in compliance with the manufacturer's instructions. Fertilizer and soil amendments are also allowed when they are used as the manufacturer specifies.

2703.3.1 Unauthorized discharges. When hazardous materials are released in quantities reportable under state, federal or local regulations, the fire code official shall be notified and the following procedures required in accordance with Sections 2703.3.1.1 through 2703.3.1.4.

❖ As stated above, release of hazardous chemicals is prohibited; however, when a release does occur in quantities that exceed the requirements of federal, state and local regulations, the fire code official must be notified.

2703.3.1.1 Records. Accurate records shall be kept of the unauthorized discharge of hazardous materials by the permittee.

❖ The amounts released, the cause of the release, containment efforts, cleanup efforts and environmental impact are items that should be included in the records submitted and kept on the incident.

2703.3.1.2 Preparation. Provisions shall be made for controlling and mitigating unauthorized discharges.

❖ A facility preplan for controlling and mitigating a release is recommended. After a release, this plan should be immediately put into action.

2703.3.1.3 Control. When an unauthorized discharge caused by primary container failure is discovered, the involved primary container shall be repaired or removed from service.

❖ Once the cause of the release has been determined, repair of defective equipment or changes in operating procedures must begin immediately.

2703.3.1.4 Responsibility for cleanup. The person, firm or corporation responsible for an unauthorized discharge shall institute and complete all actions necessary to remedy the effects of such unauthorized discharge, whether sudden or gradual, at no cost to the jurisdiction. When deemed necessary by the fire code official, cleanup may be initiated by the fire department or by an authorized individual or firm. Costs associated with such cleanup shall be borne by the owner, operator or other person responsible for the unauthorized discharge.

❖ Cleanup is the responsibility of the person, firm or corporation responsible for an unauthorized release. Cleanup must begin immediately once the incident is stable.

2703.4 Material Safety Data Sheets. Material Safety Data Sheets (MSDS) shall be readily available on the premises for hazardous materials regulated by this chapter. When a hazardous substance is developed in a laboratory, available information shall be documented.

Exception: Designated hazardous waste.

❖ Both the number and the diversity of industrial chemicals are constantly increasing. At the same time, the inventory of chemicals at modern industrial operations is sometimes quite variable. These factors make accurate and timely information more important to emergency responders. This section specifies requirements for submitting MSDS and other emergency response information. Regardless of quantity, MSDS are required for all hazardous materials regulated by the code, even if the intended quantities do not require a permit or exceed the maximum quantities allowed per control area relative to a high-hazard occupancy classification. Preincident planning is essential for buildings containing hazardous materials, regardless of quantity.

The exception recognizes that the information contained in the *Uniform Hazardous Waste Manifest* (EPA Form 870022A) required by DOT 49 CFR regulations contains sufficient information on the hazards of the waste material. Because EPA shipping rules for hazardous waste require that the manifest be with the waste material at all times, the manifest serves the same purpose as a MSDS. The following is the minimum information needed from the MSDS to assist in determining the hazardous occupancy. The special consideration area will tell when items such as emergency showers, eye wash centers, acid piping and special ventilation are required. The MSDS will have the following information recorded on it.

- The chemical name;
- The boiling point;
- The flash point;
- The UFL;
- The LFL;
- The solubility of the chemical;
- The IDLH;
- The LD_{50};
- The NFPA 704 classification and
- Any special consideration with the chemical.

2703.5 Hazard identification signs. Unless otherwise exempted by the fire code official, visible hazard identification signs as specified in NFPA 704 for the specific material contained shall be placed on stationary containers and above-ground tanks and at entrances to locations where hazardous materials are stored, dispensed, used or handled in quantities requiring a permit and at specific entrances and locations designated by the fire code official.

❖ This section contains requirements for identification signage and labeling of containers with hazardous materials. Signs are required to alert occupants who may unknowingly enter an area containing hazardous materials. The exception in this section permits a listing of hazardous materials at the room entrance when approved by the fire code official as an alternative to the signage system required by NFPA 704.

The hazard identification symbol (see Figure 2703.5) is a color-coded array of four numbers or letters arranged in a diamond shape. This symbol appears on the label of many chemicals acquired from commercial vendors.

The **blue diamond**, appearing on the left side of the label, conveys **health hazard** information. A number from 0 to 4 appears in the blue diamond indicating the degree of the hazard. The higher the number, the higher the hazard, as follows:

0—No hazard.

1—Can cause irritation if not treated.

2—Can cause injury. Requires prompt treatment.

3—Can cause serious injury despite medical treatment.

4—Can cause death or major injury despite medical treatment.

The **red diamond**, appearing at the top of the label, conveys **flammability hazard** information. Again, the numbers 0 to 4 are used to rate the flammability hazard as follows:

0—No hazard.

1—Ignites after considerable heating.

2—Ignites if moderately heated.

3—Can be ignited at all normal temperatures.

4—Very flammable gases or very volatile flammable liquid.

The **yellow diamond**, appearing on the right side of the label, conveys **reactivity hazard** information. The numbers 0 to 4 are used to rank reactivity hazards as follows:

0—Normally stable. Not reactive with water.

1—Normally stable. Unstable at high temperatures and pressure. Reacts with water.

2—Normally unstable but will not detonate.

3—Can detonate or explode but requires strong initiating force or heating under confinement.

4—Readily detonates or explodes.

The **white diamond**, appearing at the bottom of the label, conveys **special hazard** information. This information is conveyed by the use of symbols that represent the special hazard.Two of the common symbols are:

~~W~~—Denotes the material is water reactive.

OX—Denotes an oxidizing agent.

Some labels use the white diamond to convey personal protective equipment or engineering controls required to work with the material safely. You may see a picture of gloves, safety glasses or a fume hood in the white diamond.

NFPA hazard ratings can be found on the MSDS for a given chemical. Also, this symbol, or a form of this symbol, often appears on the label of commercial chemical products.

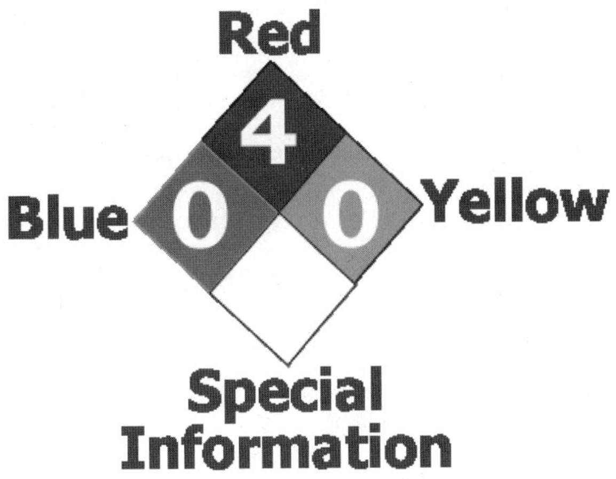

Figure 2703.5
HAZARD IDENTIFICATION SYMBOLS

2703.5.1 Markings. Individual containers, cartons or packages shall be conspicuously marked or labeled in an approved manner. Rooms or cabinets containing compressed gases shall be conspicuously labeled: COMPRESSED GAS.

❖ DOT 49 CFR requires labels on all containers, cartons and packages of hazardous materials during transportation. Many manufacturers also post comprehensive labels on all containers and packages. These labels often include hazard information beyond that required by DOT 49 CFR. The international pictorial symbols likely to be found on these labels are shown in Figure 2703.5.1.

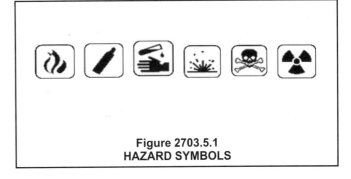

Figure 2703.5.1
HAZARD SYMBOLS

2703.6 Signs. Signs and markings required by Sections 2703.5 and 2703.5.1 shall not be obscured or removed, shall be in English as a primary language or in symbols allowed by this code, shall be durable, and the size, color and lettering shall be approved.

❖ Signs must be in English as the primary language, or in symbols allowed by the code, and be made of a durable material with the size, color and lettering approved by the fire code official or other sections of the *International Codes*.

2703.7 Sources of ignition. Sources of ignition shall comply with Sections 2703.7.1 through 2703.7.3.

❖ A form of heat is required to ignite flammable and combustible chemicals. By limiting the sources of ignition in the storage or use area one can reduce the possibility of a fire.

2703.7.1 Smoking. Smoking shall be prohibited and "No Smoking" signs provided as follows:

1. In rooms or areas where hazardous materials are stored or dispensed or used in open systems in amounts requiring a permit in accordance with Section 2701.2.1.
2. Within 25 feet (7620 mm) of outdoor storage, dispensing or open use areas.
3. Facilities or areas within facilities that have been designated as totally "no smoking" shall have "No Smoking" signs placed at all entrances to the facility or area. Designated areas within such facilities where smoking is permitted either permanently or temporarily, shall be identified with signs designating that smoking is permitted in these areas only.
4. In rooms or areas where flammable or combustible hazardous materials are stored, dispensed or used.

Signs required by this section shall be in English as a primary language or in symbols allowed by this code and shall comply with Section 310.

❖ The four areas listed in this section, because of their hazard characteristics, have been designated to receive "no smoking" signs in accordance with Section 310.

2703.7.2 Open flames. Open flames and high-temperature devices shall not be used in a manner which creates a hazardous condition and shall be listed for use with the hazardous materials stored or used.

❖ Processes that use open flames must be installed and safeguarded in accordance with the manufacturer's specifications. It should be noted that a chemical that is preheated can change from a nonhazardous classification to a hazardous classification or move from a lower hazard to a higher hazard. Safety is further enhanced by the requirement for devices to be listed for use in proximity to hazardous materials.

2703.7.3 Industrial trucks. Powered industrial trucks used in areas designated as hazardous (classified) locations in accordance with the ICC *Electrical Code* shall be listed and labeled for use in the environment intended in accordance with NFPA 505.

❖ Powered industrial trucks used in hazardous locations in accordance with the ICC EC shall meet the requirements of NFPA 505 (see also Section 309).

2703.8 Construction requirements. Buildings, control areas, enclosures and cabinets for hazardous materials shall be in accordance with Sections 2703.8.1 through 2703.8.6.2.

❖ Both the code and the IBC have requirements for construction of hazardous use areas. Each of these codes must be consulted for the specific and general requirements based on the chemicals being protected.

2703.8.1 Buildings. Buildings, or portions thereof, in which hazardous materials are stored, handled or used shall be constructed in accordance with the *International Building Code*.

❖ Buildings must meet the construction requirements of the IBC. Note that Section 414 of the IBC is a good starting point for this review. Figure 2703.8.1 shows check sheets that can be used for review of the building code restrictions and requirements for hazardous occupancies.

2703.8.2 Required detached buildings. Group H occupancies containing quantities of hazardous materials in excess of those set forth in Table 2703.8.2 shall be in detached buildings.

❖ The definition of a "Detached storage building" is found in Section 2702 and in Section 307.2 in the IBC. As defined, a detached storage building is a separate, single-story building, without a basement or crawl space that is used for the storage of hazardous materials and located an approved distance from all other structures.

The manufacture and storage of hazardous materials are frequently integrated into a single building. Additionally, it is not unusual for certain manufacturing buildings to contain multiple Group H uses where the threshold quantities are exceeded. Limiting the uses of identified hazardous materials to separate buildings containing only Group H uses maintains the intent of the code, which is to isolate large quantities of certain physical hazardous materials from uses other than those in Group H.

TABLE 2703.8.2. See page 27-32.

❖ Table 2703.8.2 gives the maximum allowable quantities of materials that can be stored in Group H occupancies without detached storage.

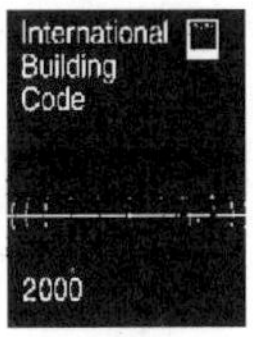

PROJECT________________________________PLAN REVIEW NO.____________________

HAZARDOUS OCCUPANCY REQUIREMENTS
2000 International Building Code

CODE SECTION	ITEM	O.K., COMMENT OR N/A
1603.3	Posted Floor Live Loads	
307.1	Occupancy Classification	
415	Storage of Hazardous Materials in Excess of the Maximum Allowable Quantities in Table 307.7 (1) or 307.7 (2)	
307.3	Subclassification H-1 (detonation hazard)	
307.4	Subclassification H-2 (deflagration hazard or a hazard from accelerated burning)	
307.5	Subclassification H-3 (materials support combustion or present a physical hazard)	
307.6	Subclassification H-4 (health hazards)	
307.7	Subclassification H-5 (semiconductor fabrication facilities)	
307.9	Exception to Group H Classification	
506	Area Modification	
Table 503	Building Height and Area	
504.2	Height Modifications for Group H-4	
Table 302.3.2	HAZARDOUS OCCUPANCY SEPARATION	
903.2.4, 903.2.12.1 903.3	Automatic Sprinkler Systems	
903.4	Valves Controlling Water Supply	
414.1.3	Information Required	
414.2	Control Areas	
414.3	Ventilation	
414.5	Inside Storage, Dispensing and Use of Hazardous Materials in Excess of the Maximum Allowable Quantities	
907.2.5	Manual Fire Alarm System and Smoke Detection	
414.7	Emergency Alarm	
414.5.4	Standby Power or Emergency Power	

CODE SECTION	ITEM	O.K., COMMENT OR N/A
414.5.5	Spill Control, Drainage and Containment	
414.6	Outdoor Storage, Dispensing and Use	
415.3	Location on Property	
415.3.1	Minimum Distance to Lot Lines	
2702.2.9, 2702.2.10, 2702.2.11, 2702.2.12	Emergency and Standby Power for Highly Toxic, Organic Peroxides, and Pyrophoric Materials	
	H-1 OCCUPANCY	
415.4	Maximum of 1 Story, no basement	
415.4.1	Floors in Storage Rooms	
	H-2 & H-3 OCCUPANCIES	
415.5	Maximum of 1 Story, no basement	
415.5.1	Floors in Storage Rooms	
415.6	Smoke & Heat Venting	
	H-2 OCCUPANCY	
415.7.1	Combustible Dusts, Grain Processing and Storage	
415.7.1.1	Type of Construction and Height Exceptions	
415.7.1.2	Grinding Rooms	
415.7.1.5	Grain Elevators	
415.7.1.6	Coal Pockets	
415.7.2	Flammable & Combustible Liquids	
415.7.3	Liquid Petroleum Gas Distribution Facilities	
415.7.4	Dry Cleaning Plants	
	H-3 & H-4 OCCUPANCY	
415.8.1	Gas Rooms	
415.8.2	Floors in Storage Rooms	
415.8.3	Separation - Highly Toxic Solids and Liquids	

HAZARDOUS

Figure 2703.8.1
PLAN REVIEW FORMS

International Building Code 2000

PROJECT____________________ PLAN REVIEW NO.____________________

HAZARDOUS OCCUPANCY REQUIREMENTS
2000 International Building Code

CODE SECTION	ITEM	O.K., COMMENT OR N/A
	INTERNATIONAL FIRE CODE	
IFC 2705	Use, Dispensing and Handling of Hazardous Materials	
IFC 2704.2	Spill Control and Secondary Containment for Liquid and Solid Hazardous Materials	
IFC 3001	Storage, Use and Handling of Compressed Gases	
IFC 3304	Explosive Materials Storage and Handling	
IFC 3404	Storage of Flammable and Combustible Liquids	
IFC 3604	Storage of Flammable Solids	
IFC 4004	Storage of Oxidizers	
IFC 3804	Storage and Handling of Liquid Petroleum Gases	
IFC 3904	Storage of Organic Peroxides	
IFC 4104	Storage of Pyrophoric Materials	
IFC 4304	Storage of Unstable (Reactive) Materials	
IFC 4404	Storage of Water-Reactive Materials	
IFC 3204	Storage of Cryogenic Fluids	
IFC 3704	Storage and Use of Highly Toxic Materials	
IFC 3104	Storage of Corrosives	
IFC Chapter 12	Dry Cleaning	
415.9	**H-5 OCCUPANCY**	
415.9.2	Fabrication Areas	
Table 503	Building Height and Area	
415.9.2.2	Separation	
415.9.2.4	Floors	
415.9.2.6	Ventilation	
415.9.3	Exit Access Corridors	
415.9.4	Service Corridors	

CODE SECTION	ITEM	O.K., COMMENT OR N/A
415.9.4.2	Separation of Service Corridors From Exit Access	
415.9.4.3	Ventilation	
415.9.4.5	Minimum Width of Service Corridor	
415.9.4.4	Maximum Travel Distance	
415.9.4.6	Emergency Alarm Systems	
415.9.5	Storage of Hazardous Production Materials	
415.9.5.1	General	
415.9.5.3	Location Within Building	
415.9.5.5	Exits from HPM Rooms	
415.9.5.7	Ventilation	
415.9.5.8	Emergency Alarm Systems	
415.9.5.9	Separation of HPM	
415.9.6	Piping and Tubing	
415.9.6.3	Installations in Exit Corridors and Above Other Occupancies	
415.9.5.2	Construction of HPM Rooms and Gas Rooms	
415.9.7	Continuous Gas Detection Systems	
415.9.5.4, 415.5.1	Explosion Control	
415.9.8	Manual Fire Alarm System	
415.9.9	Emergency Control Station	
415.9.10	Emergency Power System	
415.9.11	Fire Sprinkler System Protection in Exhaust Ducts for HPM	

Figure 2703.8.1—continued
PLAN REVIEW FORMS

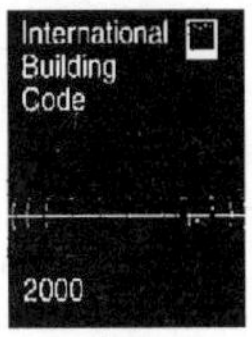

PROJECT__________________ PLAN REVIEW NO.__________________

HAZARDOUS OCCUPANCY REQUIREMENTS
2000 International Building Code

EGRESS REQUIREMENTS - CHAPTER 10

CODE SECTION	ITEM	O.K., COMMENT OR N/A
1003	General	
1005.1	Exit Capacity ? Occupant Content	
1004.5	Converging Exit	
1015	Travel Distance & Arrangement of Exits	
Table 1014.1	Spaces with One Means of Egress	
1013.3	Common Path of Travel	
1014.2.1	Exits Separation (½ Diagonal Dimension of Building)	
1018.1	Minimum Number of Exits	
1016.2	Minimum Width of Access Corridor	
1009.1	Minimum Stair Width	
1008.1.1	Minimum Exit Door Width	
1016.3	Dead-End Corridor	
1019.1.8	Smokeproof Enclosures	
1019.1	Exit Stairs	
1022.1	EXTERIOR EXIT STAIRS	
1009.5	STAIRWAY CONSTRUCTION	G N/A
1009.3.2	Closed/Open Risers	
1019.1.5	Separation of Closets Below Stairways	

CODE SECTION	ITEM	O.K., COMMENT OR N/A
1019.1.6	Stairway Identification for Stairs that Continue Below the Level of Discharge	
1009.3	Stair Treads and Risers	
1009.3.1	Uniform Treads & Risers	
1009.10	Alternating Tread Stairways	
1009.12	Access to Roof	
1021.1	Horizontal Exits	
1023	Exit Discharge	
1008.1	Doors	
1008.1.2	Door Swing	
1010.1	Ramp	
1013.5	Egress Balconies	
1012.1	Guards	
1012.3	Opening Limitations	
1011.2	Exit Illumination and Signs	
1019.1	Stairway Floor Number Signs	

Notes:____

HAZARDOUS

Figure 2703.8.1—continued
PLAN REVIEW FORMS

TABLE 2703.8.2
REQUIRED DETACHED STORAGE

DETACHED STORAGE IS REQUIRED WHEN THE QUANTITY OF MATERIAL EXCEEDS THAT LISTED HEREIN			
Material	**Class**	**Solids and liquids (tons)[a, b]**	**Gases (cubic feet)[a, b]**
Explosives	Division 1.1 Division 1.2 Division 1.3 Division 1.4 Division 1.4[c] Division 1.5 Division 1.6	Maximum Allowable Quantity Maximum Allowable Quantity Maximum Allowable Quantity Maximum Allowable Quantity 1 Maximum Allowable Quantity Maximum Allowable Quantity	Not Applicable
Oxidizers	Class 4	Maximum Allowable Quantity	Maximum Allowable Quantity
Unstable (reactives) detonable	Class 3 or 4	Maximum Allowable Quantity	Maximum Allowable Quantity
Oxidizer, liquids and solids	Class 3 Class 2	1,200 2,000	Not Applicable
Organic peroxides	Detonable Class I Class II Class III	Maximum Allowable Quantity Maximum Allowable Quantity 25 50	Not Applicable
Unstable (reactives) nondetonable	Class 3 Class 2	1 25	2,000 10,000
Water reactives	Class 3 Class 2	1 25	Not Applicable
Pyrophoric gases	Not Applicable	Not Applicable	2,000

For SI: 1 pound = 0.454 kg, 1 cubic foot = 0.02832 m^3.

a. For materials which are detonable, the distance to other buildings or lot lines shall be as specified in the *International Building Code*. For materials classified as explosives, the required separation distances shall be as specified in Chapter 33.

b. "Maximum Allowable Quantity" means the maximum allowable quantity per control area set forth in Table 2703.1.1(1).

c. Limited to Division 1.4 materials and articles, including articles packaged for shipment, that are not regulated as an explosive under Bureau of Alcohol, Tobacco and Firearms regulations, or unpackaged articles used in process operations that do not propagate a detonation or deflagration between articles, providing the net explosive weight of individual articles does not exceed 1 pound.

2703.8.3 Control areas. Control areas shall be those spaces within a building and outdoor areas where quantities of hazardous materials not exceeding the maximum quantities allowed by this code are stored, dispensed, used or handled.

❖ Control areas allow a building to be built without having to classify it or the use area as hazardous. The requirements for the construction, number and separation requirements follow. A control area may be an entire building or any portion of a building. Where a building is not compartmented as required by the code for control areas, the entire building would be considered a single control area. In that case, the maximum quantity is that which can be present in the entire building. By using multiple control areas, the overall quantity of hazardous materials in the building can be increased because the allowable quantity can be present in each control area and the building would not be classified as Group H.

It is not the code's intent to require multiple control areas. As previously indicated, in a building that is entirely one control area, if the total quantity of hazardous materials does not exceed the maximum quantities, the building would not be classified as Group H. Similarly, if the owner is satisfied with an occupancy classification of Group H, multiple control areas would not be required. Therefore, control areas are characterized as an alternative means by which a building can be classified as other than Group H. Again, the provisions of this section are applicable only when control areas are chosen as a design alternative to classification of the occupancy as Group H.

Requiring fire-resistance-rated compartmentation of control areas minimizes the possibility of simultaneous involvement of multiple control areas in a single fire. A fire in a single control area would involve only the amount of hazardous materials in that control area, which would not exceed the maximum allowable quantity per control area.

Application of the control area provisions is intended only as a means for a building to avoid classification as Group H. It is not intended in all cases to result in adequate separation of certain hazardous materials storage. For example, a control area for the storage of flammable liquids not exceeding the maximum allowable quantity per control area indicated in Table 2703.1.1(1) and constructed in accordance with this section may require a fire-resistance rating of 1 hour for a building to avoid classification in Group H. However, the provisions of the referenced standard (in Chapter 34) on flammable liquids, NFPA 30, may require a higher fire-resistance rating to provide proper separation based on the hazards of the flammable liquid being stored. In such a case, the provisions requiring the higher rating would apply.

[B] 2703.8.3.1 Construction requirements. Control areas shall be separated from each other by not less than a 1-hour fire barrier constructed in accordance with the *International Building Code*.

❖ It is important to note that walls of the control area must be fire barriers (see Section 706 of the IBC).

2703.8.3.2 Number. The maximum number of control areas within a building shall be in accordance with Table 2703.8.3.2.

❖ See the commentary to Table 2703.8.3.2.

[B] 2703.8.3.3 Separation. The required fire-resistance rating for fire barrier assemblies shall be in accordance with Table 2703.8.3.2. The floor construction of the control area and construction supporting the floor of the control area shall have a minimum 2-hour fire-resistance rating.

❖ Note that the fire-resistance requirements for the fire barrier walls increase to 2 hours beginning at the fourth floor. The floor assembly and all supporting construction for the control area would require a minimum 2-hour fire-resistance rating if the control area is located above the third floor. The increased fire-resistance rating and reduced quantities are intended to aid fire department and emergency response personnel.

TABLE 2703.8.3.2. See below.

❖ Table 2703.8.3.2 establishes the maximum quantity of hazardous materials permitted in a building and the required fire-resistance-rated separation for control areas. The overall maximum quantity of hazardous materials that can be present in the entire building is established based on the number of permitted control areas in Table 2703.8.3.2 and the maximum quantity of materials allowed in each control area.

Based on Table 2703.8.3.2, the first floor level could contain four control areas with up to 100 percent of the maximum allowable quantity per control area of hazardous materials per control area. For example, a single control area in a one-story, nonsprinklered building could contain up to 30 gallons (114 L) of Class 1A flammable liquids [see Table 2703.1.1(1)], 125 pounds (57 kg) of Class III organic peroxides [see Table 2703.1.1(1)], 250 pounds (114 kg) of Class 2 oxidizers [see Table 2703.1.1(1)] and 500 gallons (1892 L) of corrosive liquids [see Table2703.1.1.(2)]. These quantities could be contained in each of four different control areas (each chemical in a different control area) if they are separated from each by a minimum 1-hour fire-resistance-rated wall assembly.

The amount of hazardous materials per control area and the number of control areas per floor are reduced if hazardous materials are stored or used above the first floor because there is a relatively greater difficulty in gaining access for fire fighting or other emergency response purposes and a relatively greater potential hazard to building occupants who must egress that area. The use of control areas on upper floors can be advantageous for multistory research and laboratory-type facilities that often have a functional need to use limited amounts of hazardous materials throughout various portions of the building. Without control areas, the maximum allowable quantity for a hazardous material would be limited to an entire building area regardless of the overall size or height of the building. For example, if control areas are not used, a 50,000-square-foot (4645 m^2) multistory building would be limited to the same maximum allowable quantity per control area of hazardous materials as a single-story building of 5,000 square feet (465 m^2).

TABLE 2703.8.3.2
DESIGN AND NUMBER OF CONTROL AREAS

FLOOR LEVEL		PERCENTAGE OF THE MAXIMUM ALLOWABLE QUANTITY PER CONTROL AREA[a]	NUMBER OF CONTROL AREAS PER FLOOR[b]	FIRE-RESISTANCE RATING FOR FIRE BARRIERS IN HOURS[c]
Above grade	Higher than 9	5	1	2
	7-9	5	2	2
	6	12.5	2	2
	5	12.5	2	2
	4	12.5	2	2
	3	50	2	1
	2	75	3	1
	1	100	4	1
Below grade	1	75	3	1
	2	50	2	1
	Lower than 2	Not Applicable	Not Applicable	Not Applicable

a. Percentages shall be of the maximum allowable quantity per control area shown in Tables 2703.1.1(1) and 2703.1.1(2), with all increases allowed in the footnotes to those tables.

b. There shall be a maximum of two control areas per floor in Group M occupancies and in buildings or portions of buildings having Group S occupancies with storage conditions and quantities in accordance with Section 2703.11.

c. Fire barriers shall include walls and floors as necessary to provide separation from other portions of the building.

2703.8.3.4 Hazardous materials in Group M and S occupancies. The aggregate quantity of nonflammable solid and nonflammable or noncombustible liquid hazardous materials allowed within a single control area of a Group M or S occupancy is allowed to exceed the maximum allowable quantities specified in Tables 2703.1.1(1) and 2703.1.1(2) without classifying the building or use as a Group H occupancy, provided that the materials are stored in accordance with Section 2703.11.

❖ This statement refers to Section 2703.11 for the methods and amounts of storage of nonflammable solids and nonflammable or noncombustible liquids within a single control area.

2703.8.4 Gas rooms. Where a gas room is provided to comply with the provisions of Chapter 37, the gas room shall be in accordance with Sections 2703.8.4.1 and 2703.8.4.2.

❖ When gas rooms are provided as required by Chapter 37 they must meet the requirements stated in the subsections that follow.

2703.8.4.1 Construction. Gas rooms shall be protected with an automatic sprinkler system. Gas rooms shall be separated from the remainder of the building in accordance with the requirements of the *International Building Code* based on the occupancy group into which it has been classified.

❖ Construction shall be based on the occupancy separation requirements of Table 302.3.3 in the IBC. A gas room would typically be a Group H-4 occupancy; therefore, the separation requirements would be from a Group H-4 occupancy to the other occupancy (i.e., business, institutional, storage, etc.).

2703.8.4.2 Ventilation system. The ventilation system for gas rooms shall be designed to operate at a negative pressure in relation to the surrounding area. Highly toxic and toxic gases shall also comply with Section 3704.2.2.6. The ventilation system shall be installed in accordance with the *International Mechanical Code.*

❖ The area must be designed to operate at a negative pressure to maintain the atmosphere inside the room in the event of a leak. Section 3704.2.2.6 also has specific ventilation requirements for highly toxic and toxic gases. Ventilation systems must comply with the IMC.

2703.8.5 Exhausted enclosures. Where an exhausted enclosure is used to increase maximum allowable quantity per control area or when the location of hazardous materials in exhausted enclosures is provided to comply with the provisions of Chapter 37, the exhausted enclosure shall be in accordance with Sections 2703.8.5.1 through 2703.8.5.3.

❖ When exhausted enclosures are provided as required by Chapter 37, they must meet the following requirements.

2703.8.5.1 Construction. Exhausted enclosures shall be of noncombustible construction.

❖ See Section 703.4 of the IBC for the definition of "Noncombustible construction."

2703.8.5.2 Ventilation. The ventilation system for exhausted enclosures shall be designed to operate at a negative pressure in relation to the surrounding area. Ventilation systems used for highly toxic and toxic gases shall also comply with Items 1, 2 and 3 of Section 3704.1.2. The ventilation system shall be installed in accordance with the *International Mechanical Code.*

❖ The area must be designed to operate at a negative pressure to maintain the atmosphere inside the room in the event of a leak. Section 3704.2.2.6 also has specific ventilation requirements for highly toxic and toxic gases. Ventilation systems must comply with the IMC.

2703.8.5.3 Fire-extinguishing system. Exhausted enclosures where flammable materials are used shall be protected by an approved automatic fire-extinguishing system in accordance with Chapter 9.

❖ Special care is required in selecting the correct type of automatic fire suppression system. The MSDS should be consulted for compatibility of the suppression system and the chemicals being used.

2703.8.6 Gas cabinets. Where a gas cabinet is used to increase the maximum allowable quantity per control area or when the location of compressed gases in gas cabinets is provided to comply with the provisions of Chapter 37, the gas cabinet shall be in accordance with Sections 2703.8.6.1 through 2703.8.6.3.

❖ When gas cabinets are provided as required by Chapter 37, they must meet the following requirements.

2703.8.6.1 Construction. Gas cabinets shall be constructed in accordance with the following:

1. Constructed of not less than 0.097-inch (2.5 mm) (No. 12 gage) steel.
2. Be provided with self-closing limited access ports or noncombustible windows to give access to equipment controls.
3. Be provided with self-closing doors.
4. Gas cabinet interiors shall be treated, coated or constructed of materials that are compatible with the hazardous materials stored. Such treatment, coating or construction shall include the entire interior of the cabinet.

❖ This section itemizes four requirements for the construction of gas cabinets. Listed and labeled gas cabinets can be assumed to meet these minimum requirements when they have been tested by an approved third-party testing agency.

2703.8.6.2 Ventilation. The ventilation system for gas cabinets shall be designed to operate at a negative pressure in relation to the surrounding area. Ventilation systems used for highly toxic and toxic gases shall also comply with Items 1, 2 and 3 of Section 3704.1.2. The ventilation system shall be installed in accordance with the *International Mechanical Code.*

❖ The area must be designed to operate at a negative pressure to maintain the atmosphere inside the room in the event of a leak. Section 3704.2.2.6 also has specific ventilation requirements for highly toxic and toxic gases. Ventilation systems must comply with the IMC.

2703.8.6.3 Maximum number of cylinders per gas cabinet. The number of cylinders contained in a single gas cabinet shall not exceed three.

❖ This section limits the number of cylinders of compressed gases within a gas cabinet for quantity control purposes in an effort to reduce the potential involvement of other gas cylinders in a fire.

2703.8.7 Hazardous materials storage cabinets. Where storage cabinets are used to increase maximum allowable quantity per control area or to comply with this chapter, such cabinets shall be in accordance with Sections 2703.8.7.1 and 2703.8.7.2.

❖ This section recognizes that the use of approved storage cabinets is an acceptable alternative to increase the maximum allowable quantity per control area of certain hazardous materials. For example, the base maximum allowable quantity per control area in each control area for Class IB flammable liquids is 60 gallons (227 L) as indicated in Table 3201.4. However, if Class IB flammable liquids are stored in an approved storage cabinet, the base maximum allowable quantity in each control area could be increased to 120 gallons (454 L). Therefore, a building could contain 120 gallons (454 L) of Class IB liquids in each control area without classifying the building as a high-hazard occupancy if all flammable liquids are stored in an approved storage cabinet. These types of storage cabinets are sometimes known as flammable liquid cabinets or acid storage cabinets.

2703.8.7.1 Construction. The interior of cabinets shall be treated, coated or constructed of materials that are nonreactive with the hazardous material stored. Such treatment, coating or construction shall include the entire interior of the cabinet. Cabinets shall either be listed in accordance with UL 1275 as suitable for the intended storage or constructed in accordance with the following:

1. Cabinets shall be of steel having a thickness of not less than 0.0478 inch (1.2 mm) (No. 18 gage). The cabinet, including the door, shall be double walled with a 1.5-inch (38 mm) airspace between the walls. Joints shall be riveted or welded and shall be tight fitting. Doors shall be well fitted, self-closing and equipped with a self-latching device.
2. The bottoms of cabinets utilized for the storage of liquids shall be liquid tight to a minimum height of 2 inches (51 mm).

Electrical equipment and devices within cabinets used for the storage of hazardous gases or liquids shall be in accordance with the ICC *Electrical Code.*

❖ This section gives two methods of acceptance. Either the cabinet is listed in accordance with UL 1275 or construction meets the requirements of this section.

This section specifies minimum construction requirements for hazardous material storage cabinets similar to those cabinets required for flammable and combustible liquids in NFPA 30. While both this section and NFPA 30 require the door of the cabinet to be equipped with a latching device, this section does not specifically require a three-point latch arrangement on all cabinet doors. The three-point latch arrangement is recommended to enhance the integrity of the cabinet in a fire. The door sill of hazardous material storage cabinets used for storing liquids must be raised 2 inches (51 mm) above the bottom of the cabinet. The 2-inch (51 mm) raised sill is intended to retain any spilled liquid within the cabinet. The surface of the cabinets must be compatible with the material stored in the cabinets. Any electrical equipment must comply with the ICC EC and Article 500 of NFPA 70.

2703.8.7.2 Warning markings. Cabinets shall be clearly identified in an approved manner with red letters on a contrasting background to read:

HAZARDOUS — KEEP FIRE AWAY.

❖ To warn the general public and employees of potential exposure to hazardous materials within storage cabinets, appropriate warning labels are required. Cabinets shall be clearly identified with the wording given in this section.

2703.9 General safety precautions. General precautions for the safe storage, handling or care of hazardous materials shall be in accordance with Sections 2703.9.1 through 2703.9.9.

❖ The following sections deal with the safe handling and storage of hazardous materials.

2703.9.1 Personnel training and written procedures. Persons responsible for the operation of areas in which hazardous materials are stored, dispensed, handled or used shall be familiar with the chemical nature of the materials and the appropriate mitigating actions necessary in the event of fire, leak or spill.

❖ Each tenant or owner should develop a mitigation plan. This plan should be posted and be familiar to all workers. In the event of a release, this plan should be immediately placed into effect.

2703.9.1.1 Fire department liaison. Responsible persons shall be designated and trained to be liaison personnel to the fire department. These persons shall aid the fire department in preplanning emergency responses and identifying the locations where hazardous materials are located, and shall have access to Material Safety Data Sheets and be knowledgeable in the site's emergency response procedures.

❖ As we know, an emergency is not the time to wonder who, what, when or where. The code section recommends that a working relationship be established with the emergency services prior to the emergency. A plant engineer or chemist who knows the type of materials, methods of storage, location of storage in the structure and chemical makeup would typically be the best persons for this liaison work.

2703.9.2 Security. Storage, dispensing, use and handling areas shall be secured against unauthorized entry and safeguarded in a manner approved by the fire code official.

❖ Safeguards must be in place to assist in the prevention of unauthorized entry into the building or the removal of hazardous chemicals.

2703.9.3 Protection from vehicles. Guard posts or other approved means shall be provided to protect storage tanks and connected piping, valves and fittings; dispensing areas; and use areas subject to vehicular damage in accordance with Section 312.

❖ Storage, piping or other process equipment that could be damaged by vehicular traffic must be protected with barriers designed to meet the requirements of Section 312.

2703.9.4 Electrical wiring and equipment. Electrical wiring and equipment shall be installed and maintained in accordance with the ICC *Electrical Code*.

❖ Article 500 of NFPA 70 will also provide guidance, as referenced in the ICC EC.

2703.9.5 Static accumulation. When processes or conditions exist where a flammable mixture could be ignited by static electricity, means shall be provided to prevent the accumulation of a static charge.

❖ NFPA 77 can give some guidance into the methods of protection against static electricity. The ICC EC also provides guidance on the requirements.

2703.9.6 Protection from light. Materials that are sensitive to light shall be stored in containers designed to protect them from such exposure.

❖ Light-sensitive chemicals must be protected from UV lights or other spectrum patterns that may be damaging. See the MSDS for additional information and requirements for chemicals.

2703.9.7 Shock padding. Materials that are shock sensitive shall be padded, suspended or otherwise protected against accidental dislodgement and dislodgement during seismic activity.

❖ Several explosive chemicals are shock sensitive and must be protected against accidents that could knock them from their shelves. If the building in which they are stored or used is in a seismic zone, they must also be protected.

2703.9.8 Separation of incompatible materials. Incompatible materials in storage and storage of materials that are incompatible with materials in use shall be separated when the stored materials are in containers having a capacity of more than 5 pounds (2 kg) or 0.5 gallon (2 L). Separation shall be accomplished by:

1. Segregating incompatible materials in storage by a distance of not less than 20 feet (6096 mm).
2. Isolating incompatible materials in storage by a noncombustible partition extending not less than 18 inches (457 mm) above and to the sides of the stored material.
3. Storing liquid and solid materials in hazardous material storage cabinets.
4. Storing compressed gases in gas cabinets or exhausted enclosures in accordance with Sections 2703.8.5 and 2703.8.6. Materials that are incompatible shall not be stored within the same cabinet or exhausted enclosure.

❖ Materials that are incompatible with each other must be separated. This section gives four methods of accomplishing this separation. If cabinets or exhausted enclosures are being used, only compatible chemicals can be stored in any one cabinet or enclosure.

2703.9.9 Shelf storage. Shelving shall be of substantial construction, and shall be braced and anchored in accordance with the seismic design requirements of the *International Building Code* for the seismic zone in which the material is located. Shelving shall be treated, coated or constructed of materials that are compatible with the hazardous materials stored. Shelves shall be provided with a lip or guard when used for the storage of individual containers.

Exceptions:

1. Storage in hazardous material storage cabinets or laboratory furniture specifically designed for such use.
2. Storage of hazardous materials in amounts not requiring a permit in accordance with Section 2701.5.

Shelf storage of hazardous materials shall be maintained in an orderly manner.

❖ Where hazardous chemicals are stored on shelves, the shelves must have a lip or guard at the edges. The shelving must be treated or otherwise protected to be compatible with the chemicals stored.

2703.10 Handling and transportation. In addition to the requirements of Section 2703.2, the handling and transportation of hazardous materials in corridors or exit enclosures shall be in accordance with Sections 2703.10.1 through 2703.10.3.6.

❖ This section deals with the handling and transportation of hazardous chemicals in corridors or exit enclosures.

2703.10.1 Valve protection. Hazardous material gas containers, cylinders and tanks in transit shall have their protective caps in place. Containers, cylinders and tanks of highly toxic or toxic compressed gases shall have their valve outlets capped or plugged with an approved closure device in accordance with Chapter 30.

❖ Whenever a cylinder or tank of hazardous gases is transported, it can have a protective cover on the outlet valves or be plugged with an approved closure valve as stated in Section 3003.4.

2703.10.2 Carts and trucks required. Liquids in containers exceeding 5 gallons (19 L) in a corridor or exit enclosure shall be transported on a cart or truck. Containers of hazardous materials having a hazard ranking of 3 or 4 in accordance with NFPA 704 and transported within corridors or exit enclosures, shall be on a cart or truck. Where carts and trucks are required for transporting hazardous materials, they shall be in accordance with Section 2703.10.3.

Exceptions:

1. Two hazardous material liquid containers, which are hand carried in acceptable safety carriers.
2. Not more than four drums not exceeding 55 gallons (208 L) each, which are transported by suitable drum trucks.
3. Containers and cylinders of compressed gases, which are transported by approved hand trucks, and containers and cylinders not exceeding 25 pounds (11 kg), which are hand carried.
4. Solid hazardous materials not exceeding 100 pounds (45 kg), which are transported by approved hand trucks, and a single container not exceeding 50 pounds (23 kg), which is hand carried.

❖ This section limits the amount of hazardous liquid that can be transported by hand in approved containers to 5 gallons (19 L) or less. Containers with hazardous liquids exceeding 5 gallons (19 L) must be transported using approved carts or trucks that meet the construction requirements of Section 2703.10.3. This section addresses not only hand push-type carts and trucks, but also gas carts, motorized hand trucks and specialized industrial trucks. Additional guidance on the approved use and construction of motorized hand trucks and electrical industrial trucks can be found in NFPA 505.

This section prohibits the transportation of more than the maximum allowable quantity of any material within an exit. The movement of hazardous materials through exit enclosures within a building is undesirable. However, it may also be unavoidable in multistory buildings. This section is intended to limit the amount of hazardous materials within an exit at any time.

Exception 1 recognizes that two safety carriers can be hand carried.

Exception 2 recognizes that a maximum of four drums with a capacity of 55 gallons (208 L) or less can be transported using an approved drum truck.

Exception 3 recognizes that some containers and cylinders of compressed gases can be transported if they are secured on approved hand trucks. Also, the hand carrying of containers and cylinders is approved when they weigh no more than 25 pounds (11 kg).

Exception 4 recognizes that solid hazardous materials not exceeding 100 pounds (45 kg) can be transported using approved hand trucks. Single containers weighing no more than 50 pounds (23 kg) may be hand carried.

2703.10.3 Carts and trucks. Carts and trucks required by Section 2703.10.2 to be used to transport hazardous materials shall be in accordance with Sections 2703.10.3.1 through 2703.10.3.6.

❖ Where carts and trucks are required, they must be designed in accordance with Sections 2703.10.3.1 through 2703.10.3.6.

2703.10.3.1 Design. Carts and trucks used to transport hazardous materials shall be designed to provide a stable base for the commodities to be transported and shall have a means of restraining containers to prevent accidental dislodgement. Compressed gas cylinders placed on carts and trucks shall be individually restrained.

❖ Carts and trucks must be of an inherently stable design to minimize tipping, rolling or other uncontrolled movement. They must also be equipped with a means to restrain cylinders or other containers from falling, tipping or rolling.

2703.10.3.2 Speed-control devices. Carts and trucks shall be provided with a device that will enable the operator to control safely movement by providing stops or speed-reduction devices.

❖ Carts and trucks must be equipped with speed control devices, brakes, steering stops or other controls for use when uncontrolled movements occur.

2703.10.3.3 Construction. Construction materials for hazardous material carts or trucks shall be compatible with the material transported. The cart or truck shall be of substantial construction.

❖ The surface of the cart or truck must be compatible with the material transported. Although this section only clarifies that the cart or truck be made of substantial construction, trucks are built of noncombustible materials. The cart or truck should not be a contributing factor in a fire.

2703.10.3.4 Spill control. Carts and trucks transporting liquids shall be capable of containing a spill from the largest single container transported.

❖ Each cart or hand truck used must be able to control the spill of the largest container it transports.

2703.10.3.5 Attendance. Carts and trucks used to transport materials shall not obstruct or be left unattended within any part of a means of egress.

❖ When transporting material through a corridor or exit enclosure, the material must not be left unattended.

2703.10.3.6 Incompatible materials. Incompatible materials shall not be transported on the same cart or truck.

❖ As has been stated throughout the code, incompatible materials must not be mixed during transport.

2703.11 Group M storage and display and Group S storage. The aggregate quantity of nonflammable solid and nonflammable or noncombustible liquid hazardous materials stored and displayed within a single control area of a Group M occupancy, or an outdoor control area, or stored in a single control area of a Group S occupancy, is allowed to exceed the maximum allowable quantity per control area indicated in Section 2703.1 when in accordance with Sections 2703.11.1 through 2703.11.3.10.

❖ This section names three specific use or storage areas in which limits on nonflammable solid and nonflammable or noncombustible liquid chemicals may be exceeded:

1. When they are being stored or displayed in a Group M occupancy.
2. When they are being stored in a single control area of a Group S occupancy.
3. When they are in an outdoor control area.

Maximum allowable quantities can be exceeded when the area is protected as required by Sections 2703.11.1 through 2703.11.3.10.

2703.11.1 Maximum allowable quantity per control area in Group M or S occupancies. The aggregate amount of nonflammable solid and nonflammable or noncombustible liquid hazardous materials stored and displayed within a single control area of a Group M occupancy or stored in a single control area of a Group S occupancy shall not exceed the amounts set forth in Table 2703.11.1.

❖ The amounts of nonflammable solid and nonflammable or noncombustible liquid hazardous material stored and displayed in inside areas may not exceed the maximum allowable quantities shown in Table 2703.11.1 unless a footnote allows an increase.

TABLE 2703.11.1. See page 27-39.

❖ Table 2703.11.1 lists the hazardous materials eligible for the mercantile and storage occupancy option and the corresponding maximum allowable quantities per control area depending on the extent of protection provided. The permitted quantities of each listed material are independent of each other as well as the various classes or physical state of a specific material. For example, a given control area could contain up to the permitted maximum quantity of Class 2 solid oxidixers, Class 3 solid oxidizers and Class 2 liquid oxidizers, in addition to the permitted quantities of corrosive materials.

Notes b and c would allow the listed maximum quantity in Table 2703.11.1 to be increased due to the use of sprinklers or approved hazardous materials storage cabinets or both. The notes are intended to be cumulative in that up to four times the listed amount may be allowed per control area, if the building is fully sprinklered and approved cabinets are utilized, without classifying the building as Group H.

Note d simply refers to Table 2703.8.3.2 for the design and permitted number of control areas. Note b of Table 2703.8.3.2 limits mercantile and storage occupancies utilizing this option to two control areas.

The 100-percent increase in maximum allowable quantities for outdoor control areas permitted by Note f is based on the reduced exposure hazard to the building and its occupants. The increase encourages exterior storage applications without mandating sprinkler protection or approved hazardous material storage cabinets.

Notes g and h recognize that Class 2 and 3 solid oxidizers include several disinfectants that are commonly used in recreational, potable and wastewater treatment. Without these exceptions, the tabular maximum allowable quantities allowed in Group M and S occupancies would not be sufficient to sustain trade demand during times of peak usage. Because small containers of these materials have not been involved in losses, the exceptions permit additional containers of 10 pounds (5 kg) or less. Note that Section 2703.11.3.6 limits the tabular quantities to individual containers of 100 pounds (45 kg) or less, whereas these exceptions give the retailer/wholesaler the option of increasing quantities on the shelves when the packaging sizes are reduced and limited to 10 pounds (5 kg) or less.

Note i recognizes the inherently higher level of protection and safety afforded by a sprinkler system and that, by definition, the only hazard presented by Class 1 oxidizers is that they slightly increase the burning rate of combustible materials with which they may come into contact during a fire. Materials with such properties present nowhere near the level of hazard of many ordinary commodities that might be found in a Group M or S occupancy, such as foam plastics. To put this matter into perspective, Class 1 oxidizers are materials with a degree of hazard similar to that of toilet bowl cleaner crystals. Note i also provides correlation with Note f of Table 2703.1.1(1).

Note j further recognizes the lesser hazard of Class 1 oxidizers and the inherent safety of storing hazardous materials outdoors by allowing quantities to be unlimited in outdoor control areas. Note j also provides correlation with Table 2703.1.1(3).

TABLE 2703.11.1
MAXIMUM ALLOWABLE QUANTITY PER INDOOR AND OUTDOOR CONTROL AREA IN GROUP M AND S OCCUPANCIES NONFLAMMABLE SOLIDS, NONFLAMMABLE AND NONCOMBUSTIBLE LIQUIDS [d, e, f]

CONDITION		MAXIMUM ALLOWABLE QUANTITY PER CONTROL AREA	
Material[a]	Class	Solids pounds	Liquids gallons
A. HEALTH-HAZARD MATERIALS—NONFLAMMABLE AND NONCOMBUSTIBLE SOLIDS AND LIQUIDS			
1. Corrosives[b, c]	Not Applicable	9,750	975
2. Highly Toxics	Not Applicable	20[b, c]	2[b, c]
3. Toxics[b, c]	Not Applicable	1,000	100
B. PHYSICAL-HAZARD MATERIALS —NONFLAMMABLE AND NONCOMBUSTIBLE SOLIDS AND LIQUIDS			
1. Oxidizers[b, c]	4	Not Allowed	Not Allowed
	3	1,150[g]	115
	2	2,250[h]	225
	1	18,000[i, j]	1,800[i, j]
2. Unstable (Reactives)[b, c]	4	Not Allowed	Not Allowed
	3	550	55
	2	1,150	115
	1	Not Limited	Not Limited
3. Water (Reactives)	3[b, c]	550	55
	2[b, c]	1,150	115
	1	Not Limited	Not Limited

For SI: 1 pound = 0.454 kg, 1 gallon = 3.785 L, 1 cubic foot = 0.02832 m^3.

a. Hazard categories are as specified in Section 2701.2.2.

b. Maximum allowable quantities shall be increased 100 percent in buildings equipped throughout with an approved automatic sprinkler system in accordance with Section 903.3.1.1. When Note c also applies, the increase for both notes shall be applied accumulatively.

c. Maximum allowable quantities shall be increased 100 percent when stored in approved storage cabinets in accordance with Section 2703.8. When Note b also applies, the increase for both notes shall be applied accumulatively.

d. See Table 2703.8.3.2 for design and number of control areas.

e. Allowable quantities for other hazardous material categories shall be in accordance with Section 2703.1.

f. Maximum quantities shall be increased 100 percent in outdoor control areas.

g. Maximum amounts are permitted to be increased to 2,250 pounds when individual packages are in the original sealed containers from the manufacturer or packager and do not exceed 10 pounds each.

h. Maximum amounts are permitted to be increased to 4,500 pounds when individual packages are in the original sealed containers from the manufacturer or packager and do not exceed 10 pounds each.

i. Quantities are unlimited where protected by an automatic sprinkler system.

j. Quantities are unlimited in an outdoor control area.

2703.11.2 Maximum allowable quantity per outdoor control area in Group M or S occupancies. The aggregate amount of nonflammable solid and nonflammable or noncombustible liquid hazardous materials stored and displayed within a single outdoor control area of a Group M occupancy shall not exceed the amounts set forth in Table 2703.11.1.

❖ The amounts of storage of nonflammable solid and nonflammable or noncombustible liquid hazardous material stored and displayed in outside areas may not exceed the maximum allowable quantities given in Table 2703.11.1.

2703.11.3 Storage and display. Storage and display shall be in accordance with Sections 2703.11.3.1 through 2703.11.3.10.

❖ Storage and display of nonflammable solid and nonflammable or noncombustible liquid chemicals must comply with Sections 2703.11.3.1 through 2703.11.3.10.

2703.11.3.1 Density. Storage and display of solids shall not exceed 200 pounds per square foot (976 kg/m^2) of floor area actually occupied by solid merchandise. Storage and display of liquids shall not exceed 20 gallons per square foot (0.50 L/m^2) of floor area actually occupied by liquid merchandise.

❖ The key element to this section is the statement "floor area actually occupied by solid or liquid merchandise." As an example, if 10 square feet (.9 m^2) of floor area is to be used for storage or display, the limit allowed would be 2,000 pounds (908 kg) (200 pounds x 10 square feet) of solid material or 200 gallons (757 L) (20 gallons x 10 square feet) of liquid material.

2703.11.3.2 Storage and display height. Display height shall not exceed 6 feet (1829 mm) above the finished floor in display areas of Group M occupancies. Storage height shall not exceed 8 feet (2438 mm) above the finished floor in storage areas of Group M and Group S occupancies.

❖ This section limits Group M display height to 6 feet (1829 mm). In storage areas of Group M occupancies and in Group S occupancies, the storage height is increased to 8 feet (2438 mm) in recognition of the fact that these areas are not normally open to the public. These areas are also subject to the density requirements of Section 2703.11.3.1.

2703.11.3.3 Container location. Individual containers less than 5 gallons (19 L) or less than 25 pounds (11 kg) shall be stored or displayed on pallets, racks or shelves.

❖ When the capacity of individual containers is less than either 5 gallons (19 L) or 25 pounds (11 kg), they must be displayed on pallets, racks or shelves.

2703.11.3.4 Racks and shelves. Racks and shelves used for storage or display shall be in accordance with Section 2703.9.9.

❖ The design of racks and shelves must meet the requirements of Section 2703.9.9.

2703.11.3.5 Container type. Containers shall be approved for the intended use and identified as to their content.

❖ Containers must be approved for the storage or display conditions.

2703.11.3.6 Container size. Individual containers shall not exceed 100 pounds (45 kg) for solids or 10 gallons (38 L) for liquids in storage and display areas.

❖ The individual containers may not exceed 10 gallons (38 L) for liquids or 100 pounds (45 kg) for solids.

2703.11.3.7 Incompatible materials. Incompatible materials shall be separated in accordance with Section 2703.9.8.

❖ Incompatible materials may not be stored together unless separated in accordance with Section 2703.9.8.

2703.11.3.8 Floors. Floors shall be in accordance with Section 2704.12.

❖ Floors in storage and display areas must meet the requirements of Section 2704.12.

2703.11.3.9 Aisles. Aisles 4 feet (1219 mm) in width shall be maintained on three sides of the storage or display area.

❖ The storage or display area must be surrounded on at least three sides with an aisle that is at least 4 feet (1219 mm) wide.

2703.11.3.10 Signs. Hazard identification signs shall be provided in accordance with Section 2703.5.

❖ Signs meeting the requirements of NFPA 704 and Section 2703.5 must be installed.

2703.12 Outdoor control areas. Outdoor control areas for hazardous materials in amounts not exceeding the maximum allowable quantity per outdoor control area shall be in accordance with the following:

1. Outdoor control area shall be kept free from weeds, debris and common combustible materials not necessary to the storage. The area surrounding an outdoor control area shall be kept clear of such materials for a minimum of 15 feet (4572 mm).
2. Outdoor control areas shall be located not closer than 20 feet (6096 mm) from a lot line that can be built upon, public street, public alley or public way. A 2-hour fire-resistance-rated wall without openings extending not less than 30 inches (762 mm) above and to the sides of the storage area is allowed in lieu of such distance.
3. Where a property exceeds 10,000 square feet (929 m^2), a group of two outdoor control areas is allowed when approved and when each control area is separated by a minimum distance of 50 feet (15 240 mm).
4. Where a property exceeds 35,000 square feet (3252 m^2), additional groups of outdoor control areas are allowed when approved and when each group is separated by a minimum distance of 300 feet (91 440 mm).

❖ This section lists four requirements for an outdoor control area.

Item 1 requires that outdoor control areas be kept clear of combustible materials for a minimum of 15 feet (4572 mm) around the control area.

Item 2 requires the control area to either be 20 feet (6096 mm) from the lot lines, street, alley or public way or have a 2-hour blank wall extending a minimum of 30 inches (762 mm) above and to either side of the storage area, if the distance requirement cannot be met.

Item 3 allows two control areas on properties larger than 10,000 square feet (929 m^2) when they are separated by at least 50 feet (15 240 mm). The 20-foot (6096 mm) distance from the lot lines, street, alley or public way is still in effect.

Item 4 allows for more than two control areas when the property exceeds 35,000 square feet (3252 m^2). Each group must be separated by a minimum distance of 300 feet (91 440 mm).

SECTION 2704
STORAGE

2704.1 Scope. Storage of hazardous materials in amounts exceeding the maximum allowable quantity per control area as set forth in Section 2703.1 shall be in accordance with Sections 2701, 2703 and 2704. Storage of hazardous materials in amounts not exceeding the maximum allowable quantity per control area as set forth in Section 2703.1 shall be in accordance with Sections 2701 and 2703. Retail and wholesale storage and display of nonflammable solid and nonflammable and noncombustible liquid hazardous materials in Group M occupancies and Group S storage shall be in accordance with Section 2703.11.

❖ This scope paragraph considers three storage situations: storage of quantities exceeding the maximum, storage of quantities within the allowable limits and retail and wholesale storage and display in Group M occupancies and storage areas in Group S occupancies.

Section references are given for each of the three storage and use situations.

2704.2 Spill control and secondary containment for liquid and solid hazardous materials. Rooms, buildings or areas used for the storage of liquid or solid hazardous materials shall be provided with spill control and secondary containment in accordance with Sections 2704.2.1 through 2704.2.3.

Exception: Outdoor storage of containers on approved containment pallets in accordance with Section 2704.2.3.

❖ The requirement for spill control in a room or area is based on two items. The first is that the storage container(s) have a capacity of more than 55 gallons (208 L). The second is that the aggregate capacity of multiple vessels be more than 1,000 gallons (3785 L). The area, once determined to require spill control, must be protected so that the containment area will handle the release from the largest container in the area. This section recommends four methods of containment.

2704.2.1 Spill control for hazardous material liquids. Rooms, buildings or areas used for the storage of hazardous material liquids in individual vessels having a capacity of more than 55 gallons (208 L), or in which the aggregate capacity of multiple vessels exceeds 1,000 gallons (3785 L), shall be provided with spill control to prevent the flow of liquids to adjoining areas. Floors in indoor locations and similar surfaces in outdoor locations shall be constructed to contain a spill from the largest single vessel by one of the following methods:

1. Liquid-tight sloped or recessed floors in indoor locations or similar areas in outdoor locations.
2. Liquid-tight floors in indoor locations or similar areas in outdoor locations provided with liquid-tight raised or recessed sills or dikes.
3. Sumps and collection systems.
4. Other approved engineered systems.

Except for surfacing, the floors, sills, dikes, sumps and collection systems shall be constructed of noncombustible material, and the liquid-tight seal shall be compatible with the material stored. When liquid-tight sills or dikes are provided, they are not required at perimeter openings having an open-grate trench across the opening that connects to an approved collection system.

❖ The requirement for spill control in a room or area is based on two items. The first is that the storage container(s) have a capacity of more than 55 gallons (208 L). The second is that the aggregate capacity of multiple vessels be more than 1,000 gallons (3785 L). The area, once determined to require spill control, must be protected so that the containment area will handle the release from the largest container in the area. This section recommends four methods of containment.

1. *Liquid-tight sloped or recessed floors.* The chemicals being stored or used in the area must be evaluated to ensure that the method of making the floor liquid tight will not cause a reaction with the chemicals.
2. *Liquid-tight floors with a containment sill or trench around the area.* To determine the volume of the containment sill or dike, the following procedure is recommended:
   - Determine the greatest amount of liquid that can be released from the largest tank or container within the containment area.
   - If more than one tank or group of containers are in the containment area, the volume of the tank or group of containers below the height of the containment sill or dike can be subtracted from the volume of the containment sill or dike.
   - The following is the equation for determining the volumetric capacity of a tank:

$$V = \frac{3.1416(d)^2 h}{4}$$

where:

V = Tank capacity, in gallons.
d = Diameter of tank, in feet.
h = Height of tank, in feet.

Determine whether the containment sill or dike is of sufficient size to control the spill.

The following is the equation for determining the volumetric capacity of the containment sill or dike:

$$V = a^* h$$

where:

V = Volumetric capacity of dike, in gallons.
a^* = Area of storage floor, in square feet.
h = Height of sill, in feet.

Example:

You have a sprinklered first floor inside storage room that is 20 feet by 15 feet (6096 mm by 4572 mm). Storage consists of 32 drums [55 gallon (208 L)] of a Class II combustible liquid stacked two pallets high. The containment sill is 4 inches (102 mm) high. Is this height adequate for spill containment?

The largest tank is 55 gallons (208 L). It is 36 inches (914 mm) tall and 22 inches (559 mm) wide, and has a volumetric capacity of:

$$V = \frac{3.1416(22/12)^2\,(36/12)}{4}$$

(Convert inches to feet)

V = 7.91 cubic feet (.22 m^3) of spill from a 55-gallon (208 L) drum.

You have four pallets that are 54 inches by 54 inches square (1372 mm by 1372 mm) and 4 inches (102 mm) tall.

V = (54/12) × (54/12) × 4/12 inches

V = 6.68 cubic feet (.2 m^3) per pallet

6.68 × 4 (pallets) = 26.73 cubic feet (.76 m^3) of space taken up in the containment area.

The room has a containment area of 20 feet by 15 feet by 4/12 or 99 cubic feet (3 m^3).

When you subtract the volume of the pallets, you have 72.27 cubic feet (.2 m^3) of containment. The largest container [55-gallon (208 L) drum] would produce a spill volume of 7.91 cubic feet (.2 m^3). Thus you see that the secondary containment of a 20-foot by 15-foot (6096 mm by 4572 mm) room with a 4-inch (102 mm) curb would contain the spill from the largest container.

3. *Sumps and collection systems that can consist of floor drains to a remote collection tank.* Oil/water separators would also be required by Section 1003.4 of the IPC for floor drains that discharge into the building drainage system or other point of disposal.

4. *Any other approved engineered systems.* Any system that has been engineered and evaluated for the hazards present can be accepted by the fire code official.

2704.2.2 Secondary containment for hazardous material liquids and solids. Where required by Table 2704.2.2 buildings, rooms or areas used for the storage of hazardous materials liquids or solids shall be provided with secondary containment in accordance with this section when the capacity of an individual vessel or the aggregate capacity of multiple vessels exceeds the following:

1. Liquids: Capacity of an individual vessel exceeds 55 gallons (208 L) or the aggregate capacity of multiple vessels exceeds 1,000 gallons (3785 L); and
2. Solids: Capacity of an individual vessel exceeds 550 pounds (250 kg) or the aggregate capacity of multiple vessels exceeds 10,000 pounds (4540 kg).

❖ Table 2704.2.2 contains requirements for secondary containment according to the type of material and method of storage. In addition, the capacity of an individual container must exceed 55 gallons (208 L) of liquid or multiple containers exceed 1,000 gallons (3785 L) of liquid. If the chemical is in solid form and the capacity of an individual container exceeds 550 pounds (250 kg) or multiple containers exceed a cumulative of 10,000 pounds (4540 kg), secondary containment is required when called for in Table 2704.2.2.

TABLE 2704.2.2. See page 27-44.

❖ This table is divided into three main columns: materials, indoor storage and outdoor storage. Indoor and outdoor storage is further divided into two subcolumns: solids and liquids. This table indicates when secondary containment is required.

2704.2.2.1 Containment and drainage methods. The building, room or area shall contain or drain the hazardous materials and fire protection water through the use of one of the following methods:

1. Liquid-tight sloped or recessed floors in indoor locations or similar areas in outdoor locations.
2. Liquid-tight floors in indoor locations or similar areas in outdoor locations provided with liquid-tight raised or recessed sills or dikes.
3. Sumps and collection systems.
4. Drainage systems leading to an approved location.
5. Other approved engineered systems.

❖ The five methods of containment and drainage for secondary containment in this section are similar to those of spill control.

Method 1 consists of liquid-tight sloped or recessed floors. The chemicals being stored or used in the area must be evaluated to ensure that the method of making the floor liquid tight will not cause a reaction with the chemicals.

Method 2 consists of liquid-tight floors with a containment sill or trench around the area.

To determine the height of the containment sill or dike, the following procedure is recommended:

- Determine the greatest amount of liquid that can be released from the largest tank or container within the containment area.
- If more than one tank or group of containers are in the containment area, the volume of the tank or group of containers below the height of the containment sill or dike must be subtracted from the volume of the containment sill or dike.
- The following is the equation for determining the volumetric capacity of a tank:

$$V = \frac{3.1416(d)^2 h}{4}$$

- Determine whether the containment sill or dike is of sufficient size to control a spill and the fire-fighting water, as required.

  The following is the equation for determining the volumetric capacity of a containment sill or dike:

$$V = (a)\,(h)$$

where

V = Volume of the containment, in cubic feet.

a = Area of the containment, in square feet.

h = Height of the sill or dike, in feet.

Example:

You have a sprinklered first-floor inside storage room that is 20 feet by 15 feet (6096 mm by 4572 mm). Stored in this area are 32 drums [55 gallons (208 L)] of a Class II combustible liquid stacked two pallets high. The containment sill is 4 inches (102 mm) high. Will this provide secondary containment?

The largest tank is 55 gallons (208 L). It has dimensions of 36 inches (914 mm) tall and 22 inches (539 mm) wide, and has a volumetric capacity of:

$$V = \frac{3.1416(22/12)^2\ (36/12)}{4}$$

(Convert inches to feet)

$V = 7.91$ cubic feet (.2 m^3) of spill from a 55-gallon drum

You have four pallets with a size of 54 inches by 54 inches (1372 mm by 1372 mm) and 4 inches (102 mm) tall.

$V = (54/12)(54/12)\ (4/12)$

$V = 6.68$ cubic feet (.2 m^3) per pallet

6.68 × 4 (pallets) = 26.73 cubic feet (.76 m^3) of space taken up in the containment area.

The room has a containment area of 20 feet by 15 feet by 4 inches (6096 mm by 4572 mm by 102 mm) or 99 cubic feet (3 m^3).

When you subtract the volume of the pallets, we have 72.27 cubic feet (2 m^3) of containment. The largest container [55-gallon (208 L) drum] would produce a spill of 7.91 cubic feet (.2 m^3). Thus you see that the secondary containment of a 20-foot by 15-foot (6096 mm by 4572 mm) room with a 4-inch (102 mm) curb would contain the spill from the largest container. Now you must factor in the requirements of the fire-fighting water from the sprinkler system as required by Section 2704.2.2.3. From NFPA 13 Figure 7-2.3.1.2 (Area Density Curves) the design density is 0.2. This design density of 0.2 times the fire area of 300 square feet (28 m^2) equals 60 gallons (227 L) per minute discharge. The sprinkler discharge will be 60 gallons (227 L) × 20 minutes (as required by Section 2704.2.2.3) or 1,200 gallons (4542 L) of fire-fighting water.

The conversion factor from gallon to cubic foot is 0.1335805. Therefore, 1,200 gallons (4542 L) of fire-fighting water × 0.1335805 = 167.79 cubic feet (5 m^3) of containment area required.

The largest tank volume is 7.91 cubic feet (.2 m^3)

The volume of the pallets is 26.73 cubic feet (.8 m^3)

The volume for fire-fighting water is 167.79 cubic feet (5 m^3)

Total cubic feet required for secondary containment is 202.43 cubic feet (6 m^3)

In this case, the secondary containment does not have enough capacity. Although the 4-inch (102 mm) sill does give enough capacity for spill containment, with the requirement of fire-fighting water you now do not have proper secondary containment.

Method 3 recommends sumps and collection systems, which can consist of floor drains to remote collection tanks. It is important to remember that oil/water separators are required by Section 1003.4 of the IPC for floor drains that discharge into the building drainage system or other point of disposal.

Method 4 is a drainage system leading to an approved location. If water is being used as the automatic extinguishing agent for the area, a large containment area will be required to hold the runoff of the sprinkler water and the spill.

Method 5 is any other approved engineered system. Any system that has been engineered and evaluated for the hazards present can be accepted by the fire code official.

2704.2.2.2 Incompatible materials. Incompatible materials used in open systems shall be separated from each other in the secondary containment system.

❖ Different containment areas cannot be manifolded if the products in the different areas are incompatible. Even in containment of a spill in a single area, incompatible materials must be kept separated.

2704.2.2.3 Indoor design. Secondary containment for indoor storage areas shall be designed to contain a spill from the largest vessel plus the design flow volume of fire protection water calculated to discharge from the fire-extinguishing system over the minimum required system design area or area of the room or area in which the storage is located, whichever is smaller. The containment capacity shall be designed to contain the flow for a period of 20 minutes.

❖ Secondary containment must be sized to hold the release of the largest container in the area plus the design flow volume of the fire protection for a 20-minute period.

For example, you have a 500-square-foot (46 m^2) flammable liquid storage area that is sprinklered. Section 2704.5 states that the minimum design of the sprinkler system is an ordinary hazard Group 2 with the minimum design area of 3,000 square feet (279 m^2). From NFPA 13 Figure 7-2.3.1.2 (Area Density Curves) the design density is 0.2. This design density of 0.2 times the fire area of 500 square feet (46 m^2) equals 100-gallon-per-minute (6 L/s) discharge.

The largest tank in the fire area is 100 gallons (379 L). The secondary containment must be sized to hold the 100 gallons (379 L) plus the sprinkler discharge of 100 gallons (379 L) for 20 minutes, or 2,100 gallons (7949 L). This must be the minimum capacity of the secondary containment.

TABLE 2704.2.2
REQUIRED SECONDARY CONTAINMENT—HAZARDOUS MATERIAL SOLIDS AND LIQUIDS STORAGE

MATERIAL		INDOOR STORAGE		OUTDOOR STORAGE	
		Solids	Liquids	Solids	Liquids
1. Physical-hazard materials					
Combustible liquids	Class II	Not Applicable	See Chapter 34	Not Applicable	See Chapter 34
	Class IIIA		See Chapter 34		See Chapter 34
	Class IIIB		See Chapter 34		See Chapter 34
Cryogenic fluids			See Chapter 32		See Chapter 32
Explosives		See Chapter 33		See Chapter 32	
Flammable liquids	Class IA	Not Applicable	See Chapter 34	Not Applicable	See Chapter 34
	Class IB		See Chapter 34		See Chapter 34
	Class IC		See Chapter 34		See Chapter 34
Flammable solids		Not Required	Not Applicable	Not Required	Not Applicable
Organic peroxides	Unclassified Detonable	Required	Required	Not Required	Not Required
	Class I				
	Class II				
	Class III				
	Class IV				
	Class V	Not Required	Not Required	Not Required	Not Required
Oxidizers	Class 4	Required	Required	Not Required	Not Required
	Class 3				
	Class 2				
	Class 1	Not Required	Not Required	Not Required	Not Required
Pyrophorics		Not Required	Required	Not Required	Required
Unstable (reactives)	Class 4	Required	Required	Required	Required
	Class 3				
	Class 2				
	Class 1	Not Required	Not Required	Not Required	Not Required
Water reactives	Class 3	Required	Required	Required	Required
	Class 2				
	Class 1	Not Required	Not Required	Not Required	Not Required
2. Health-hazard materials					
Corrosives		Not Required	Required	Not Required	Required
Highly toxics		Required	Required	Required	Required
Toxics					

❖ This table is divided into three main columns: materials, indoor storage and outdoor storage. Indoor and outdoor storage is further divided into two subcolumns: solids and liquids. This table indicates when secondary containment is required.

2704.2.2.4 Outdoor design. Secondary containment for outdoor storage areas shall be designed to contain a spill from the largest individual vessel. If the area is open to rainfall, secondary containment shall be designed to include the volume of a 24-hour rainfall as determined by a 25-year storm and provisions shall be made to drain accumulations of ground water and rainwater.

❖ In addition to holding the volume of the largest container, if the area is open to rainfall, the secondary containment must include the volume of a 24-hour rainfall as determined by a 25-year storm. Drains must be sized to carry off accumulations of ground water and rainwater. See Section 1106.1 of the IPC for information on rainfall maps.

2704.2.2.5 Monitoring. An approved monitoring method shall be provided to detect hazardous materials in the secondary containment system. The monitoring method is allowed to be visual inspection of the primary or secondary containment, or other approved means. Where secondary containment is subject to the intrusion of water, a monitoring method for detecting water shall be provided. Where monitoring devices are provided, they shall be connected to approved visual or audible alarms.

❖ Visual inspection of the primary or secondary containment system is permitted; otherwise, an electronic monitoring system must be installed. These electronic systems must be connected to both audible and visual alarms.

2704.2.2.6 Drainage system design. Drainage systems shall be in accordance with the *International Plumbing Code* and all of the following:

1. The slope of floors to drains in indoor locations, or similar areas in outdoor locations shall not be less than 1 percent.
2. Drains from indoor storage areas shall be sized to carry the volume of the fire protection water as determined by the design density discharged from the automatic fire-extinguishing system over the minimum required system design area or area of the room or area in which the storage is located, whichever is smaller.
3. Drains from outdoor storage areas shall be sized to carry the volume of the fire flow and the volume of a 24-hour rainfall as determined by a 25-year storm.
4. Materials of construction for drainage systems shall be compatible with the materials stored.
5. Incompatible materials used in open systems shall be separated from each other in the drainage system.
6. Drains shall terminate in an approved location away from buildings, valves, means of egress, fire access roadways, adjoining property and storm drains.

❖ Drainage systems for hazardous materials must meet the requirements of Section 1003 of the IPC in addition to the six elements listed in this section:

1. The slope of the floor may not be more than 1 percent.
2. The drains for an indoor storage area must be sized to carry the automatic fire-extinguishing agent from the room. If an agent other than a liquid is used, no additional volume will be needed in your calculations.
3. The drains for outdoor storage must be sized to carry the volume of fire flow and the volume of a 24-hour rainfall.
4. The piping and other elements of the system must be chosen to be compatible with the chemicals and extinguishing agents that will be flowing through the drainage system.
5. Incompatible materials pose a great hazard both in use and when an emergency occurs. This section requires that incompatible materials, when used in open systems, be separated into different drainage systems.
6. Drains must terminate in safe locations so that they do not pose additional threats to lives and property.

2704.2.3 Containment pallets. When used as an alternative to spill control and secondary containment for outdoor storage in accordance with the exception in Section 2704.2, containment pallets shall comply with all of the following:

1. A liquid-tight sump accessible for visual inspection shall be provided.
2. The sump shall be designed to contain not less than 66 gallons (250 L).
3. Exposed surfaces shall be compatible with material stored.
4. Containment pallets shall be protected to prevent collection of rainwater within the sump.

❖ Another option that is now available is the use of containment pallets. This type of pallet is designed to contain a leak should one occur. This section specifies the minimum design for the pallet.

2704.3 Ventilation. Indoor storage areas and storage buildings shall be provided with mechanical exhaust ventilation or natural ventilation where natural ventilation can be shown to be acceptable for the materials as stored.

Exception: Storage areas for flammable solids complying with Chapter 36.

❖ Indoor storage areas and buildings must be ventilated either mechanically or naturally so that the level of vapors is maintained below the LFL or the permissible exposure limits (PEL). Keeping the area/building at these levels maintains a level of safety for the area. Storage areas for flammable solids that meet the requirements of Chapter 13 are exempted from this requirement.

2704.3.1 System requirements. Exhaust ventilation systems shall comply with all of the following:

1. Installation shall be in accordance with the *International Mechanical Code.*

2. Mechanical ventilation shall be at a rate of not less than 1 cubic foot per minute per square foot [0.00508 $m^3/(s \cdot m^2)$] of floor area over the storage area.
3. Systems shall operate continuously unless alternative designs are approved.
4. A manual shutoff control shall be provided outside of the room in a position adjacent to the access door to the room or in an approved location. The switch shall be of the break-glass type and shall be labeled: VENTILATION SYSTEM EMERGENCY SHUTOFF.
5. Exhaust ventilation shall be designed to consider the density of the potential fumes or vapors released. For fumes or vapors that are heavier than air, exhaust shall be taken from a point within 12 inches (305 mm) of the floor.
6. The location of both the exhaust and inlet air openings shall be designed to provide air movement across all portions of the floor or room to prevent the accumulation of vapors.
7. Exhaust ventilation shall not be recirculated within the room or building if the materials stored are capable of emitting hazardous vapors.

❖ The exhaust ventilation system must comply with all of the seven requirements of this section:

1. Systems must be installed as required by Section 414.4 of the IMC.
2. The minimum rate for mechanical ventilation is listed here; however, the MSDS must also be reviewed. The ventilation needed to maintain a safe environment could require a much higher flow rate.
3. The exhaust system must provide continuous ventilation in the area.
4. The entry door into the area must be equipped with an emergency shutoff that can be used to disable the ventilation system in case of a fire.
5. The system design must consider the vapor density of the chemicals being stored. The vapor density for the chemicals can be found in the MSDS.
6. The system must provide air movement across the entire area being protected.
7. Hazardous atmospheres may not be recirculated.

2704.4 Separation of incompatible hazardous materials. Incompatible materials shall be separated in accordance with Section 2703.9.8.

❖ Incompatible materials must be kept separated so that accidental mixing of chemicals in an emergency does not create a more dangerous incident.

2704.5 Automatic sprinkler systems. Indoor storage areas and storage buildings shall be equipped throughout with an approved automatic sprinkler system in accordance with Section 903.3.1.1. The design of the sprinkler system shall not be less than that required for Ordinary Hazard Group 2 with a minimum design area of 3,000 square feet (279 m^2). Where the materials or storage arrangement are required by other regulations to be provided with a higher level of sprinkler system protection, the higher level of sprinkler system protection shall be provided.

❖ This section requires a sprinkler system for indoor storage areas and buildings. The system must comply with NFPA 13 and provide a minimum density as required for an Ordinary Hazard Group 2. The MSDS should be consulted for compatibility with water. If the materials are incompatible, other methods of automatic extinguishing should be used.

2704.6 Explosion control. Indoor storage rooms, areas and buildings shall be provided with explosion control in accordance with Section 911.

❖ Storage rooms and use areas must be provided with explosion control in accordance with Table 911.1. The design of the explosion control must met the requirements of Section 911.

2704.7 Standby or emergency power. Where mechanical ventilation, treatment systems, temperature control, alarm, detection or other electrically operated systems are required, such systems shall be provided with an emergency or standby power system in accordance with the ICC *Electrical Code* and Section 604.

Exceptions:

1. Storage areas for Class 1 and 2 oxidizers.
2. Storage areas for Class III, IV and V organic peroxides.
3. For storage areas for highly toxic or toxic materials, see Sections 3704.2.2.8 and 3704.3.2.6.
4. Standby power for mechanical ventilation, treatment systems and temperature control systems shall not be required where an approved fail-safe engineered system is installed.

❖ Based on the required level of protection, an emergency electrical system or standby power system is required. Exceptions include Class 1 and 2 oxidizers; Class III, IV and V organic peroxides and highly toxic or toxic materials as noted in Sections 3704.2.2.8 and 3704.3.2.6, if a fail-safe engineered system is installed.

2704.8 Limit controls. Limit controls shall be provided in accordance with Sections 2704.8.1 and 2704.8.2.

❖ Based on the MSDS, limit controls may be required to protect the chemicals.

2704.8.1 Temperature control. Materials that must be kept at temperatures other than normal ambient temperatures to prevent a hazardous reaction shall be provided with an approved means to maintain the temperature within a safe range. Redundant temperature control equipment that will operate on failure of the primary temperature control system shall be provided. Where approved, alternative means that prevent a hazardous reaction are allowed.

❖ When a chemical is temperature sensitive, a temperature control system with a redundant backup is required.

2704.8.2 Pressure control. Stationary tanks and equipment containing hazardous material liquids that can generate pressures exceeding design limits because of exposure fires or internal reaction, shall have some form of construction or other approved means that will relieve excessive internal pressure. The means of pressure relief shall vent to an approved location or to an exhaust scrubber or treatment system where required by Chapter 37.

❖ Emergency vents must be installed when the vapor density of a chemical could cause a BLEVE during a fire.

2704.9 Emergency alarm. An approved manual emergency alarm system shall be provided in buildings, rooms or areas used for storage of hazardous materials. Emergency alarm-initiating devices shall be installed outside of each interior exit or exit access door of storage buildings, rooms or areas. Activation of an emergency alarm-initiating device shall sound a local alarm to alert occupants of an emergency situation involving hazardous materials.

❖ A manual emergency alarm system that meets the requirements of Section 907 must be installed.

2704.10 Supervision. Emergency alarm, detection and automatic fire-extinguishing systems required by Section 2704 shall be supervised by an approved central, proprietary or remote station service or shall initiate an audible and visual signal at a constantly attended on-site location.

❖ Manual alarm systems, detection systems and automatic extinguishing systems must be supervised by an approved central, proprietary or remote system that will also cause both audible and visual signals at a constantly attended on-site location.

2704.11 Clearance from combustibles. The area surrounding an outdoor storage area or tank shall be kept clear of combustible materials and vegetation for a minimum distance of 25 feet (7620 mm).

❖ Vegetation and other combustible materials must be cut back for a distance of at least 25 feet (7620 mm) from outdoor storage or tanks.

2704.12 Noncombustible floor. Except for surfacing, floors of storage areas shall be of noncombustible construction.

❖ The floor of hazardous occupancies must be of noncombustible construction.

2704.13 Weather protection. Where overhead noncombustible construction is provided for sheltering outdoor hazardous material storage areas, such storage shall not be considered indoor storage when the area is constructed in accordance with the requirements for weather protection as required by the *International Building Code.*

Exception: Storage of explosive materials shall be considered as indoor storage.

❖ This section allows for the construction of noncombustible covers over outdoor storage areas for weather protection. This area is not considered inside storage, except when explosive materials are stored under the noncombustible roof. Section 414.6.1 of the IBC contains the specific requirements for weather protection construction.

SECTION 2705
USE, DISPENSING AND HANDLING

2705.1 General. Use, dispensing and handling of hazardous materials in amounts exceeding the maximum allowable quantity per control area set forth in Section 2703.1 shall be in accordance with Sections 2701, 2703 and 2705. Use, dispensing and handling of hazardous materials in amounts not exceeding the maximum allowable quantity per control area set forth in Section 2703.1 shall be in accordance with Sections 2701 and 2703.

❖ The section covers two areas: (1) when the amount of chemicals exceeds the maximum allowable quantity per control area, and (2) when the amount of chemicals is within the maximum allowable quantity per control area.

2705.1.1 Separation of incompatible materials. Separation of incompatible materials shall be in accordance with Section 2703.9.8.

❖ When using, dispensing or handling incompatible materials, care must be taken to ensure that the chemicals do not mix.

2705.1.2 Noncombustible floor. Except for surfacing, floors of areas where liquid or solid hazardous materials are dispensed or used in open systems shall be of noncombustible, liquid-tight construction.

❖ When chemicals are used in an open system, the floors must be both liquid tight and of noncombustible construction.

2705.1.3 Spill control and secondary containment for hazardous material liquids. Where required by other provisions of Section 2705, spill control and secondary containment shall be provided for hazardous material liquids in accordance with Section 2704.2.

❖ If required by Section 2705, the spill control and secondary containment must meet the requirements of Section 2704.2.

2705.1.4 Limit controls. Limit controls shall be provided in accordance with Sections 2705.1.4.1 through 2705.1.4.4.

❖ Limit controls required in use, dispensing and handling areas are described in the four subsections to this section.

2705.1.4.1 High-liquid-level control. Open tanks in which liquid hazardous materials are used shall be equipped with a liquid-level limit control or other means to prevent overfilling of the tank.

❖ When there is a danger of overfilling a tank, especially in open systems, liquid level controls are required.

2705.1.4.2 Low-liquid-level control. Approved safeguards shall be provided to prevent a low-liquid level in a tank from creating a hazardous condition, including but not limited to, overheating of a tank or its contents.

❖ If the method of storage can allow for the collapse of the tank or other types of failure as a result of a low level of chemicals, low-liquid-level controls are required.

2705.1.4.3 Temperature control. Temperature control shall be provided in accordance with Section 2704.8.1.

❖ See Section 2704.8.1 for requirements for temperature controls.

2705.1.4.4 Pressure control. Pressure control shall be provided in accordance with Section 2704.8.2.

❖ See Section 2704.8.2 for requirements for pressure control systems.

2705.1.5 Standby or emergency power. Where mechanical ventilation, treatment systems, temperature control, manual alarm, detection or other electrically operated systems are required, such systems shall be provided with an emergency or standby power system in accordance with the ICC *Electrical Code* and Section 604.

Exceptions:

1. Standby power for mechanical ventilation, treatment systems and temperature control systems shall not be required where an approved fail-safe engineered system is installed.
2. Systems for highly toxic or toxic gases shall be provided with emergency power in accordance with Sections 3704.2.2.8 and 3704.3.2.6.

❖ Mechanical ventilation, treatment systems, temperature controls or other important safety controls must be connected to an emergency electrical system or standby power as required by Section 604. Exceptions include installations having approved fail-safe engineered systems and systems handling highly toxic or toxic gases that meet the installation requirements of Sections 3704.2.2.8 and 3704.3.2.6.

2705.1.6 Supervision. Manual alarm, detection and automatic fire-extinguishing systems required by other provisions of Section 2705 shall be supervised by an approved central, proprietary or remote station service or shall initiate an audible and visual signal at a constantly attended on-site location.

❖ Automatic fire suppression systems, detection systems and manual alarm systems meeting the requirements of Section 2705 must be supervised by an approved central, proprietary or remote station that meets the requirements of NFPA 72.

2705.1.7 Lighting. Adequate lighting by natural or artificial means shall be provided.

❖ All areas must be adequately lighted.

2705.1.8 Fire-extinguishing systems. Indoor rooms or areas in which hazardous materials are dispensed or used shall be protected by an automatic fire-extinguishing system in accordance with Chapter 9. Sprinkler system design shall not be less than that required for Ordinary Hazard, Group 2, with a minimum design area of 3,000 square feet (279 m^2). Where the materials or storage arrangement are required by other regulations to be provided with a higher level of sprinkler system protection, the higher level of sprinkler system protection shall be provided.

❖ An automatic sprinkler system is required in use, dispensing and handling areas. Special attention must be paid to chemical/water compatibility.

2705.1.9 Ventilation. Indoor dispensing and use areas shall be provided with exhaust ventilation in accordance with Section 2704.3.

Exception: Ventilation is not required for dispensing and use of flammable solids other than finely divided particles.

❖ Exhaust ventilation meeting the requirements of Section 2704.3 must be installed in indoor use and dispensing areas so that the level of vapors is maintained below the LFL or the PEL. Keeping the area/building at these levels provides a level of safety for the area.

2705.1.10 Liquid transfer. Liquids having a hazard ranking of 3 or 4 in accordance with NFPA 704 shall be transferred by one of the following methods:

1. From safety cans complying with UL 30.
2. Through an approved closed piping system.
3. From containers or tanks by an approved pump taking suction through an opening in the top of the container or tank.
4. From containers or tanks by gravity through an approved self-closing or automatic-closing valve when the container or tank and dispensing operations are provided with spill control and secondary containment in accordance with Section 2704.2. Highly toxic liquids shall not be dispensed by gravity from tanks.

5. Approved engineered liquid transfer systems.

Exceptions:

1. Liquids having a hazard ranking of 4 when dispensed from approved containers not exceeding 1.3 gallons (5 L).
2. Liquids having a hazard ranking of 3 when dispensed from approved containers not exceeding 5.3 gallons (20 L).

❖ Liquids having a hazard rating of 3 or 4 in NFPA 704 must be transferred using one of the five methods listed in this section. Exceptions include liquids having a hazard ranking of 4 being dispensed from approved containers not exceeding 1.3 gallons (5 L) and liquids having a hazard ranking of 3 when dispensed from approved containers not exceeding 5.3 gallons (20 L).

2705.2 Indoor dispensing and use. Indoor dispensing and use of hazardous materials shall be in buildings complying with the *International Building Code* and in accordance with Section 2705.1 and Sections 2705.2.1 through 2705.2.2.5.

❖ Indoor dispensing and use areas must be constructed to meet the requirements of the IBC along with Sections 2705.1 and Sections 2705.2.1 through 2705.2.2.5 of the code.

2705.2.1 Open systems. Dispensing and use of hazardous materials in open containers or systems shall be in accordance with Sections 2705.2.1.1 through 2705.2.1.4.

❖ See the definition for "Open systems" in Chapter 2.

2705.2.1.1 Ventilation. Where gases, liquids or solids having a hazard ranking of 3 or 4 in accordance with NFPA 704 are dispensed or used, mechanical exhaust ventilation shall be provided to capture fumes, mists or vapors at the point of generation.

Exception: Gases, liquids or solids which can be demonstrated not to create harmful fumes, mists or vapors.

❖ Areas where gases, liquids or solids having a ranking of 3 or 4 in accordance with NFPA 704 are used or dispensed must have a mechanical exhaust system to capture vapors at the point of generation unless the chemicals do not produce harmful fumes, mists or vapors.

2705.2.1.2 Explosion control. Explosion control shall be provided in accordance with Section 2704.6 when an explosive environment can occur because of the characteristics or nature of the hazardous materials dispensed or used, or as a result of the dispensing or use process.

❖ Based on the requirements of Section 2704.6, explosion control must be provided when the dispensing or use of a chemical could cause an explosion.

2705.2.1.3 Spill control for hazardous material liquids. Buildings, rooms or areas where hazardous material liquids are dispensed into vessels exceeding a 1.3-gallon (5 L) capacity or used in open systems exceeding a 5.3-gallon (20 L) capacity shall be provided with spill control in accordance with Section 2704.2.1.

❖ When dispensing hazardous materials into containers larger than 1.3 gallons (5 L) or in an open system with a capacity of 5.3 gallons (20 L), spill control must be provided.

2705.2.1.4 Secondary containment for hazardous material liquids. Where required by Table 2705.2.1.4, buildings, rooms or areas where hazardous material liquids are dispensed or used in open systems shall be provided with secondary containment in accordance with Section 2704.2.2 when the capacity of an individual vessel or system or the capacity of multiple vessels or systems exceeds the following:

1. Individual vessel or system: greater than 1.3 gallons (5 L).
2. Multiple vessels or systems: greater than 5.3 gallons (20 L).

❖ Table 2705.2.1.4 lists conditions when secondary spill containment is required. The containment must meet the requirements of Section 2704.2.2 when the capacity of a single vessel exceeds 1.3 gallons (5 L) or multiple vessels exceed 5.3 gallons (20 L).

TABLE 2705.2.1.4. See page 49-50.

❖ This table specifies when secondary containment is required. The table is divided into three columns: material, indoor use and outdoor use. Subcolumns of solid and liquids further divide the table. By finding the type of material in question, the physical state and the method of use, one can determine whether secondary containment is required.

2705.2.2 Closed systems. Use of hazardous materials in closed containers or systems shall be in accordance with Sections 2705.2.2.1 through 2705.2.2.5.

❖ See the definition of "Closed systems" in Chapter 2.

2705.2.2.1 Design. Systems shall be suitable for the use intended and shall be designed by persons competent in such design. Controls shall be designed to prevent materials from entering or leaving the process or reaction systems at other than the intended time, rate or path. Where automatic controls are provided, they shall be designed to be fail safe.

❖ A UL-listed or other third-party tested and designed system is recommended. This design should incorporate a fail-safe method of controlling the release of chemicals.

2705.2.2.2 Ventilation. Where closed systems are designed to be opened as part of normal operations, ventilation shall be provided in accordance with Section 2705.2.1.1.

❖ See Section 2705.2.1.1 for requirements on ventilation of areas with closed systems.

TABLE 2705.2.1.4
REQUIRED SECONDARY CONTAINMENT—HAZARDOUS MATERIAL SOLIDS AND LIQUIDS USE

<table>
<tr><th rowspan="2" colspan="2">MATERIAL</th><th colspan="2">INDOOR USE</th><th colspan="2">OUTDOOR USE</th></tr>
<tr><th>Solids</th><th>Liquids</th><th>Solids</th><th>Liquids</th></tr>
<tr><td colspan="6">1. Physical-hazard materials</td></tr>
<tr><td rowspan="3">Combustible liquids</td><td>Class II</td><td rowspan="3">Not Applicable</td><td>See Chapter 34</td><td rowspan="3">Not Applicable</td><td>See Chapter 34</td></tr>
<tr><td>Class IIIA</td><td>See Chapter 34</td><td>See Chapter 34</td></tr>
<tr><td>Class IIIB</td><td>See Chapter 34</td><td>See Chapter 34</td></tr>
<tr><td colspan="2">Cryogenic fluids</td><td>Not Applicable</td><td>See Chapter 32</td><td>Not Applicable</td><td>See Chapter 32</td></tr>
<tr><td colspan="2">Explosives</td><td colspan="2">See Chapter 33</td><td colspan="2">See Chapter 33</td></tr>
<tr><td rowspan="3">Flammable liquids</td><td>Class IA</td><td rowspan="3">Not Applicable</td><td>See Chapter 34</td><td rowspan="3">Not Applicable</td><td>See Chapter 34</td></tr>
<tr><td>Class IB</td><td>See Chapter 34</td><td>See Chapter 34</td></tr>
<tr><td>Class IC</td><td>See Chapter 34</td><td>See Chapter 34</td></tr>
<tr><td colspan="2">Flammable solids</td><td>Not Required</td><td>Not Applicable</td><td>Not Required</td><td>Not Applicable</td></tr>
<tr><td rowspan="6">Organic peroxides</td><td>Unclassified Detonable</td><td rowspan="10">Not Required</td><td>Required</td><td>Not Required</td><td>Required</td></tr>
<tr><td>Class I</td><td rowspan="4">Required</td><td rowspan="4">Not Required</td><td rowspan="4">Required</td></tr>
<tr><td>Class II</td></tr>
<tr><td>Class III</td></tr>
<tr><td>Class IV</td></tr>
<tr><td>Class V</td><td>Not Required</td><td>Not Required</td><td>Not Required</td></tr>
<tr><td rowspan="4">Oxidizers</td><td>Class 4</td><td rowspan="4">Required</td><td rowspan="4">Not Required</td><td rowspan="4">Required</td></tr>
<tr><td>Class 3</td></tr>
<tr><td>Class 2</td></tr>
<tr><td>Class 1</td></tr>
<tr><td colspan="2">Pyrophorics</td><td>Not Required</td><td>Required</td><td>Not Required</td><td>Required</td></tr>
<tr><td rowspan="4">Unstable (reactives)</td><td>Class 4</td><td rowspan="3">Not Required</td><td rowspan="3">Required</td><td rowspan="3">Required</td><td rowspan="3">Required</td></tr>
<tr><td>Class 3</td></tr>
<tr><td>Class 2</td></tr>
<tr><td>Class 1</td><td>Not Required</td><td>Not Required</td><td>Required</td><td>Required</td></tr>
<tr><td rowspan="3">Water reactives</td><td>Class 3</td><td rowspan="2">Not Required</td><td rowspan="2">Required</td><td rowspan="2">Required</td><td rowspan="2">Required</td></tr>
<tr><td>Class 2</td></tr>
<tr><td>Class 1</td><td>Not Required</td><td>Not Required</td><td>Required</td><td>Required</td></tr>
<tr><td colspan="6">2. Health-hazard materials</td></tr>
<tr><td colspan="2">Corrosives</td><td>Not Required</td><td rowspan="3">Required</td><td rowspan="3">Not Required</td><td rowspan="3">Required</td></tr>
<tr><td colspan="2">Highly toxics</td><td>Required</td></tr>
<tr><td colspan="2">Toxics</td><td>Not Required</td></tr>
</table>

2705.2.2.3 Explosion control. Explosion control shall be provided in accordance with Section 2704.6 where an explosive environment exists because of the hazardous materials dispensed or used, or as a result of the dispensing or use process.

Exception: Where process vessels are designed to contain fully the worst-case explosion anticipated within the vessel under process conditions based on the most likely failure.

❖ If the use or process could produce an explosion and the vessels are not rated as explosionproof, explosion control must meet the requirements of Section 2704.6. The exception covers process vessels designed to contain potential explosions.

2705.2.2.4 Spill control for hazardous material liquids. Buildings, rooms or areas where hazardous material liquids are used in individual vessels exceeding a 55-gallon (208 L) capacity shall be provided with spill control in accordance with Section 2704.2.1.

❖ In addition to the other requirements that have been covered, if an individual vessel exceeds 55 gallons (208 L), spill containment is required.

2705.2.2.5 Secondary containment for hazardous material liquids. Where required by Table 2705.2.1.4, buildings, rooms or areas where hazardous material liquids are used in vessels or systems shall be provided with secondary containment in accordance with Section 2704.2.2 when the capacity of an individual vessel or system or the capacity of multiple vessels or systems exceeds the following:

1. Individual vessel or system: greater than 55 gallons (208 L).
2. Multiple vessels or systems: greater than 1,000 gallons (3785 L).

❖ Secondary containment must be installed where required by Table 2705.2.1.4, where the capacity of an individual vessel is over 55 gallons (208 L) or where multiple vessels have a combined capacity greater than 1,000 gallons (3785 L).

2705.3 Outdoor dispensing and use. Dispensing and use of hazardous materials outdoors shall be in accordance with Sections 2705.3.1 through 2705.3.9.

❖ The following sections refer to storage that is outdoors. It is important to note that a noncombustible shed without sides is also considered outdoor storage.

2705.3.1 Quantities exceeding the maximum allowable quantity per control area. Outdoor dispensing or use of hazardous materials, in either closed or open containers or systems, in amounts exceeding the maximum allowable quantity per control area indicated in Tables 2703.1.1(3) and 2703.1.1(4) shall be in accordance with Sections 2701, 2703, 2705.1 and 2705.3.

❖ Once the quantities in storage exceed the maximum allowable quantities of Table 2703.1.1(3) and Section 2703.11(4), the storage must meet the requirements of Sections 2701, 2703, 2705.1 and 2705.3.

2705.3.2 Quantities not exceeding the maximum allowable quantity per control area. Outdoor dispensing or use of hazardous materials, in either closed or open containers or systems, in amounts not exceeding the maximum allowable quantity per control area indicated in Tables 2703.1.1(3) and 2703.1.1(4) shall be in accordance with Sections 2701 and 2703.

❖ If the quantities in storage do not exceed the maximum allowable quantities of Tables 2703.1.1(3) and 2703.1.1(4), the storage must meet the requirements of Sections 2701 and 2703.

2705.3.3 Location. Outdoor dispensing and use areas for hazardous materials shall be located as required for outdoor storage in accordance with Section 2704.

❖ See Section 2704 for location of storage.

2705.3.4 Spill control for hazardous material liquids in open systems. Outdoor areas where hazardous material liquids are dispensed in vessels exceeding a 1.3-gallon (5 L) capacity or used in open systems exceeding a 5.3-gallon (20 L) capacity shall be provided with spill control in accordance with Section 2704.2.1.

❖ The dispensing and use areas of open systems located outdoors also require spill controls when the individual containers being filled exceed 1.3 gallons (5 L) or the combined vessels or systems exceed 5.3 gallons (20 L).

2705.3.5 Secondary containment for hazardous material liquids in open systems. Where required by Table 2705.2.1.4, outdoor areas where hazardous material liquids are dispensed or used in open systems shall be provided with secondary containment in accordance with Section 2704.2.2 when the capacity of an individual vessel or system or the capacity of multiple vessels or systems exceeds the following:

1. Individual vessel or system: greater than 1.3 gallons (5 L).
2. Multiple vessels or systems: greater than 5.3 gallons (20 L).

❖ Dispensing and use areas of open systems located outdoors also require secondary spill controls when the requirements of Table 2705.2.1.4 are met and the individual containers being filled exceed 1.3 gallons (5 L) or the combined vessels or systems exceed 5.3 gallons (208 L).

2705.3.6 Spill control for hazardous material liquids in closed systems. Outdoor areas where hazardous material liquids are used in closed systems exceeding 55 gallons (208 L) shall be provided with spill control in accordance with Section 2704.2.1.

❖ Outdoor closed systems with a liquid capacity of over 55 gallons (208 L) must have spill control.

2705.3.7 Secondary containment for hazardous material liquids in closed systems. Where required by Table 2705.2.1.4, outdoor areas where hazardous material liquids are dispensed or used in closed systems shall be provided with secondary containment in accordance with Section 2704.2.2 when the capacity of an individual vessel or system or the capacity of multiple vessels or systems exceeds the following:

1. Individual vessel or system: greater than 55 gallons (208 L).
2. Multiple vessels or systems: greater than 1,000 gallons (3785 L).

❖ Dispensing and use areas of closed systems located outdoors also require secondary spill controls when the individual containers being filled exceed 55 gallons (208 L) or the combined vessels or systems exceed 1,000 gallons (3785 L).

2705.3.8 Clearance from combustibles. The area surrounding an outdoor dispensing or use area shall be kept clear of combustible materials and vegetation for a minimum distance of 30 feet (9144 mm).

❖ The area around an outdoor dispensing or use area must be kept clean for a distance of at least 30 feet (9144 mm).

2705.3.9 Weather protection. Where overhead noncombustible construction is provided for sheltering outdoor hazardous material use areas, such use shall not be considered indoor use when the area is constructed in accordance with the requirements for weather protection as required in the *International Building Code.*

Exception: Use of explosive materials shall be considered as indoor use.

❖ This section allows for the construction of noncombustible covers over outdoor dispensing and use areas for weather protection. This area is not considered inside dispensing or use, except when explosive materials are dispensed or used under the noncombustible roof. Section 414.6.1 of the IBC contains the specific requirements for weather protection construction.

2705.4 Handling. Handling of hazardous materials shall be in accordance with Sections 2705.4.1 through 2705.4.4.

❖ The handling of hazardous materials must meet the following requirements.

2705.4.1 Quantities exceeding the maximum allowable quantity per control area. Handling of hazardous materials in indoor and outdoor locations in amounts exceeding the maximum allowable quantity per control area indicated in Tables 2703.1.1(1) through 2703.1.1(4) shall be in accordance with Sections 2701, 2703, 2705.1 and 2705.4.

❖ When the indoor and outdoor storage amounts of chemicals are below the maximum allowable quantities per control area, the area must meet the requirements of Sections 2701, 2703, 2705.1 and 2705.4.

2705.4.2 Quantities not exceeding the maximum allowable quantity per control area. Handling of hazardous materials in indoor locations in amounts not exceeding the maximum allowable quantity per control area indicated in Tables 2703.1.1(1) and 2703.1.1(2) shall be in accordance with Sections 2701, 2703 and 2705.1. Handling of hazardous materials in outdoor locations in amounts not exceeding the maximum allowable quantity per control area indicated in Tables 2703.1.1(3) and 2703.1.1(4) shall be in accordance with Sections 2701 and 2703.

❖ When the indoor storage amounts of chemicals exceed the maximum allowable quantity per control area, the area must meet the requirements of Sections 2701, 2703 and 2705.1. When outdoor storage exceeds the maximum allowable quantities per control area listed in Tables 2703.1.1(3) and 2703.1.1(4), the area must meet the requirements of Section 2701 and 2703.

2705.4.3 Location. Outdoor handling areas for hazardous materials shall be located as required for outdoor storage in accordance with Section 2704.

❖ Outdoor handling areas must meet the requirements of Section 2704.

2705.4.4 Emergency alarm. Where hazardous materials having a hazard ranking of 3 or 4 in accordance with NFPA 704 are transported through corridors or exit enclosures, there shall be an emergency telephone system, a local manual alarm station or an approved alarm-initiating device at not more than 150-foot (45 720 mm) intervals and at each exit and exit access doorway throughout the transport route. The signal shall be relayed to an approved central station, proprietary supervising station or remote supervising station or a constantly attended on-site location and shall also initiate a local audible alarm.

❖ When exit access corridors or exit enclosures are used to transport hazardous materials with a ranking of 3 or 4, a supervised emergency telephone system, local manual alarm or approved alarm-initiating device must be installed at intervals of no more than 150 feet (45 720 mm). In addition, these devices must be located at each exit and exit access doorway throughout the transport route.

Bibliography

The following resource materials are referenced in this chapter or are relevant to the subject matter addressed in this chapter.

IBC-2003, *International Building Code.* Falls Church, VA: International Code Council, 2003.

ICC EC-2003, *ICC Electrical Code.* Falls Church, VA: International Code Council, 2003.

ICC PC-2003, *ICC Performance Code for Buildings and Facilities.* Falls Church, VA: International Code Council, 2003.

IMC-2003, *International Mechanical Code.* Falls Church, VA: International Code Council, 2003.

NFPA 13-99, *Installation of Sprinkler Systems.* Quincy, MA: National Fire Protection Association, 1998.

NFPA 30-00, *Flammable and Combustible Liquids Code*. Quincy, MA: National Fire Protection Association, 2000.

NFPA 45-00, *Laboratories Using Chemicals*. Quincy, MA: National Fire Protection Association, 2000

NFPA 69-97, *Explosion Prevention Systems*. Quincy, MA: National Fire Protection Association, 1997.

NFPA 70-02, *National Electrical Code*. Quincy, MA: National Fire Protection Association, 2002.

NFPA 72-99, *National Fire Alarm Code*. Quincy, MA: National Fire Protection Association, 1999.

NFPA 110-99, *Emergency and Standby Power Systems*. Quincy, MA: National Fire Protection Association, 1999.

NFPA 497-97, *Recommended Practice for Classification of Flammable Liquids, Gases or Vapors and of Hazardous (Classified) Locations for Electrical Installations in Chemical Process Areas*. Quincy, MA: National Fire Protection Association, 1997

NFPA 505-99, *Powered Industrial Trucks*. Quincy, MA: National Fire Protection Association, 1997

NFPA 704-96. *Identification of the Fire Hazards of Materials*. Quincy, MA: National Fire Protection Association, 1996.

UL 1275-94, *Flammable Liquid Storage Cabinets*. Northbrook, IL: Underwriters Laboratories, Inc. 1994.

Chapter 28: Aerosols

General Comments

The adequacy of aerosol storage protection became a major concern in the late 1970s and early 1980s because of a few major warehouse fires in which the involvement of aerosol products was a primary factor. These fire-loss incidents showed that, though these buildings were fully sprinklered, the level of sprinkler protection was inadequate and strict storage limitations for aerosol products were necessary.

The fire losses prompted extensive large-scale fire tests to develop protection requirements for aerosol product storage. The results of these tests led to the development of NFPA 30B, which is the basis for the provisions of this chapter. NFPA 30B recognizes that aerosol products represent a wide range of flammability and that classification criteria are necessary for all aerosol products to determine the desired level of protection. Prior to the development of NFPA 30B, flammable aerosols were classified as Class 1A flammable liquids in accordance with NFPA 30; however, this classification was based primarily on a flame extension test designed to assess the in-use aerosol flammability hazard, but had little relevance to warehouse storage conditions.

Increased sprinkler water density and sprinkler head sensitivity, along with adequate separation and quantity control, are the most important factors in controlling aerosol product fires. Increased water density requirements improve the capability of the system to suppress the fire, as opposed to just controlling its spread, by applying more water to the base of the fire. Increased sprinkler activation speed also provides additional benefits. Early suppression fast response (ESFR) sprinklers are specifically listed for high-challenge fire hazards. Increased sensitivity of ESFR sprinklers allows them to activate in response to a cardboard-carton packaging fire before aerosol cans rupture.

Purpose

These requirements address the prevention, control and extinguishment of fires and explosions in facilities where retail aerosol products are displayed or stored. They are concerned with both life safety and property protection from a fire; however, historically, aerosol product fires have caused property loss more frequently than loss of life. Requirements for storing aerosol products are dependent on the level of sprinkler protection, type of storage condition and quantity of aerosol products.

SECTION 2801
GENERAL

2801.1 Scope. The provisions of this chapter, the *International Building Code* and NFPA 30B shall apply to the manufacturing, storage and display of aerosol products in addition to the requirements of Chapter 27.

❖ In addition to the provisions of this chapter, NFPA 30B is referenced for storing and displaying aerosol products because it was the original source for this chapter. The requirements in Chapter 23 must also be complied with where applicable.

Although this chapter deals primarily with the storage and retail display of aerosol products, NFPA 30B must be consulted for guidance on facilities involved in the manufacture of aerosol-containing products. For example, aerosol-charging operations that use flammable aerosols require special design considerations and operating procedures. An aerosol-charging room is where the aerosol containers are filled with the propellant. These rooms are either located in a separate building or are separated within a building and in spaces that have explosion venting because of the deflagration potential of the products being handled.

2801.2 Permit required. Permits shall be required as set forth in Section 105.6.

❖ The process of issuing permits gives the fire code official an opportunity to carefully evaluate and regulate hazardous operations. Permit applicants should be required to demonstrate that their operations comply with the intent of the code before a permit is issued. See the commentary to Section 105.6 for a general discussion of operations requiring an operational permit, Section 105.6.1 for discussion of specific quantity-based operational permits for the materials regulated in this chapter and Section 105.7 for a general discussion of activities requiring a construction permit. The permit process also notifies the fire department of the need for prefire planning for hazardous properties.

2801.3 Material Safety Data Sheets. Material Safety Data Sheet (MSDS) information for aerosol products displayed shall be kept on the premises at an approved location.

❖ This section gives the fire code official the authority to designate or approve the location where the Material Safety Data Sheets (MSDS) covering aerosol materials are kept.

SECTION 2802 DEFINITIONS

2802.1 Definitions. The following words and terms shall, for the purposes of this chapter and as used elsewhere in this code, have the meanings shown herein.

❖ Definitions of terms can help in the understanding and application of the code requirements. The purpose for including those definitions that are associated with the subject matter of this chapter is to provide more convenient access to them without having to refer back to Chapter 2. It is important to emphasize that these terms are not exclusively related to this chapter but are applicable everywhere the term is used in the code. For convenience, these terms are also listed in Chapter 2 with a cross reference to this section. The use and application of all defined terms, including those defined in this section, are set forth in Section 201.

AEROSOL. A product that is dispensed from an aerosol container by a propellant.

Aerosol products shall be classified by means of the calculation of their chemical heats of combustion and shall be designated Level 1, Level 2 or Level 3.

Level 1 aerosol products. Those with a total chemical heat of combustion that is less than or equal to 8,600 British thermal units per pound (Btu/lb) (20 kJ/g).

Level 2 aerosol products. Those with a total chemical heat of combustion that is greater than 8,600 Btu/lb (20 kJ/g), but less than or equal to 13,000 Btu/lb (30 kJ/g).

Level 3 aerosol products. Those with a total chemical heat of combustion that is greater than 13,000 Btu/lb (30 kJ/g).

❖ The intent of the code is to regulate those aerosols that contain a flammable propellant such as butane, isobutane or propane. An aerosol product such as whipped cream is a water-based material with a nonflammable propellant (nitrous oxide) and would, therefore, not be regulated as a hazardous material. The contents of the aerosol container may be dispensed in the form of a mist spray, foam, gel or aerated powder.

AEROSOL CONTAINER. A metal can, or a glass or plastic bottle designed to dispense an aerosol. Metal cans shall be limited to a maximum size of 33.8 fluid ounces (1000 ml). Glass or plastic bottles shall be limited to a maximum size of 4 fluid ounces (118 ml).

❖ All design criteria for the aerosol container, including the maximum size and minimum strength, are set by the U.S. Department of Transportation (DOTn 49 CFR). These container regulations are necessary for the safe transportation of aerosol products.

AEROSOL WAREHOUSE. A building used for warehousing aerosol products.

❖ Any building used for storing large quantities of aerosol products would be considered an aerosol warehouse and would be subject to fire safety requirements that are consistent with the type of aerosol stored and its known hazards.

PROPELLANT. The liquefied or compressed gas in an aerosol container that expels the contents from an aerosol container when the valve is actuated. A propellant is considered flammable if it forms a flammable mixture with air, or if a flame is self-propagating in a mixture with air.

❖ The amount of flammable propellant content is important in properly classifying aerosol products, since it can affect the overall chemical heat of combustion value. Common flammable propellants are hydrocarbons, such as butane, propane, isobutane or a combination of these.

RETAIL DISPLAY AREA. The area of a Group M occupancy open for the purpose of viewing or purchasing merchandise offered for sale. Individuals in such establishments are free to circulate among the items offered for sale which are typically displayed on shelves, racks or the floor.

❖ Products containing aerosol propellants range from hairspray to paint to pesticides, lubricants and adhesives. Most often, products of this kind are grouped on shelves, but may also be set out in aisle displays of stacked cartons with the top carton open to display the product.

SECTION 2803 CLASSIFICATION OF AEROSOL PRODUCTS

2803.1 Classification levels. Aerosol products shall be classified as Level 1, 2 or 3 in accordance with Table 2803.1 and NFPA 30B. Aerosol products in cartons which are not identified in accordance with this section shall be classified as Level 3.

❖ Because of the wide range of flammability of aerosol products, a classification system was established to determine the required level of fire protection. Categories are defined according to the aerosol's chemical heat of combustion expressed in Btus per pound (Btu/lb) (see the commentary to Table 2803.1 for additional discussion of factors affecting classification).

Three categories determine the level of fire protection required (see Chapter 4 of NFPA 30B for fire protection requirements). Aerosol category classifications of Levels 1, 2 and 3 are used to avoid confusion with flammable liquid classifications (Classes I, II and III). Table 2803.1 shows the aerosol classifications. Appendixes A and C of NFPA 30B contain additional background information on the development of the classification system.

This section also recognizes the importance of identifying the level of aerosols in storage cartons. Unless cartons are marked to identify their contents, it is difficult to determine the permitted quantity and fire protection requirements. Where cartons are not marked, there is no alternative but to consider the aerosol storage as Level 3 and apply applicable code requirements for that level (see commentary, Section 2806.1).

TABLE 2803.1
CLASSIFICATION OF AEROSOL PRODUCTS

CHEMICAL HEAT OF COMBUSTION		AEROSOL CLASSIFICATION
Greater than (Btu/lb)	Less than or equal to (Btu/lb)	
0	8,600	1
8,600	13,000	2
13,000	—	3

For SI: 1 British thermal unit per pound = 0.002326 KJ/g.

❖ Table 2803.1 shows how to classify aerosol products based on their chemical heat of combustion. NFPA 30B contains tables giving the heat of combustion (expressed in kJ/gram) for representative materials and gives examples for calculating the chemical heat of combustion for aerosol products. Heat of combustion is determined using the method in ASTM D 240. Factory Mutual Research Corporation (FMRC) correlated the chemical heat of combustion with the results of full-scale pallet tests to provide a classification method based on chemical heat of combustion that is more consistent than the current classification based on weight percentages. Examples are given in NFPA 30B for calculating the heat of combustion for aerosol products that contain a number of components so that they can be classified using Table 2803.1.

Examples of Level 1 aerosol products are shaving gel, whipped cream and air fresheners. Fire tests involving Level 1 aerosols have demonstrated that they pose a fire hazard no greater than that of Class III commodities as defined in NFPA 231. Consequently, Level 1 aerosols are not regulated as a hazardous material and are essentially exempt from the requirements of this chapter.

Examples of Level 2 aerosol products include some hair sprays and insect repellents based on the values in Table 2803.1.

Level 3 aerosol products, such as carburetor cleaner and other petroleum based aerosols, require the highest level of fire protection.

2803.2 Identification. Cartons shall be identified on at least one side with the classification level of the aerosol products contained within the carton as follows:

LEVEL ________ AEROSOLS

❖ Aerosol products are generally stored and transported in cardboard cartons that must be clearly marked with the level of aerosol products they contain so that the proper storage arrangements and fire protection are provided. The product label on the aerosol container may define whether the product is flammable or extremely flammable but does not need to explain the actual classification level. If cartons are discarded or not marked, classification may be obtained from the submitted MSDS (see Section 2801.3). Fire protection requirements for the highest level of aerosols must be used where storage is mixed or unknown (see Section 2803.1).

SECTION 2804
INSIDE STORAGE OF AEROSOL PRODUCTS

2804.1 General. The inside storage of Level 2 and 3 aerosol products shall comply with Sections 2804.2 through 2804.7 and NFPA 30B. Level I aerosol products shall be considered equivalent to a Class III commodity and shall comply with the requirements for palletized or rack storage in NFPA 13.

❖ Sections 2804.2 through 2804.7 and applicable requirements in NFPA 30B regulate the inside storage of aerosol products. Permissible quantities, separation and fire protection depend on the occupancy type (see Sections 2804.2 and 2804.7) and storage condition, such as a general purpose warehouse (see Section 2804.3), aerosol warehouse (see Section 2804.4), inside flammable liquid storage room (see Section 2804.5) or liquid warehouse (see Section 2804.6).

This section also specifies the level of protection required for Level 1 aerosols. The fire hazard associated with Level 1 aerosols has been demonstrated to be equivalent to Class III commodities as defined in NFPA 13. Therefore, Sections 2804.2 through 2804.7 apply only to Level 2 and 3 aerosol products.

2804.2 Storage in Groups A, B, E, F, I and R. Storage of Level 2 and 3 aerosol products in occupancies in Groups A, B, E, F, I and R shall be limited to the following maximum quantities:

1. A net weight of 1,000 pounds (454 kg) of Level 2 aerosol products.
2. A net weight of 500 pounds (227 kg) of Level 3 aerosol products.
3. A combined net weight of 1,000 pounds (454 kg) of Level 2 and 3 aerosol products.

The maximum quantity shall be increased 100 percent where the excess quantity is stored in storage cabinets in accordance with Section 3404.3.2.

❖ This section strictly limits quantities of aerosols in buildings of Groups A, B, E, F, I and R because these occupancy types will have higher occupant loads and many activities not related to the storage of aerosol products. Storage exceeding the maximum permitted quantities would result in the building being classified as a high-hazard occupancy in accordance with the *International Building Code*® (IBC®). The indicated maximum quantities are applicable per building area and not per control area (see commentary, Section 2703).

This section would also allow the maximum quantities to be increased by 100 percent, if the quantities exceeding those indicated for Level 2 and 3 aerosol products are stored in approved storage cabinets in accordance with Section 3404.3.2 which contains specific requirements for the design, construction and capacity of storage cabinets.

2804.2.1 Excess storage. Storage of quantities exceeding the maximum quantities indicated in Section 2804.2 shall be stored in separate inside flammable liquid storage rooms in accordance with Section 2804.5.

❖ This section recognizes that in certain occupancies, the maximum quantities in Section 2804.2 may be exceeded. Any Level 2 or 3 aerosol product exceeding the maximum quantities specified in Section 2804.2, including the exception for approved storage cabinets, must be stored in a separate inside flammable liquid storage room to maintain a Group A, B, E, F, I or R classification for the building. For example, 1,000 pounds (454 kg) of Level 2 aerosols is permitted in a building of Group B. An additional 1,000 pounds (454 kg) of Level 2 aerosols is also permitted, if the product is stored in approved storage cabinets (see Section 2804.2). If additional storage is desired, 1,000 pounds (454 kg) of Level 2 aerosols would be permitted in an inside flammable liquid storage room having a floor area 500 square feet (46 m^2) or less, in accordance with Section 2804.5.1. Therefore, up to 3,000 pounds (1362 kg) of Level 2 aerosols could be stored in a building of Group B, if all excess storage is in approved storage cabinets and in an inside flammable liquid storage room having a floor area of no more than 500 square feet (46 m^2).

If an inside flammable liquid storage room exceeding 500 square feet (46 m^2) were used for the excess storage, the permissible quantity of Level 2 aerosols in the room could be increased to 2,500 pounds (1135 kg). The increased permitted quantities in the larger storage room results from the presence of an automatic sprinkler system. Compliance with these requirements would maintain the Group B classification for the building (see commentary, Section 2804.5).

2804.3 Storage in general purpose warehouses. Aerosol storage in general purpose warehouses utilized only for warehousing-type operations involving mixed commodities shall comply with Section 2804.3.1 or 2804.3.2.

❖ General purpose warehouses are used for storing general commodities and aerosol products. Aerosol product storage located in general purpose warehouses is classified as either nonsegregated storage (Section 2804.3.1) or segregated storage (Section 2804.3.2).

Distribution warehouses for major department store chains containing large amounts of aerosols and various common mercantile commodities are typical of buildings classified as general purpose warehouses. Proper protection and separation of aerosol product storage areas are essential to protect ordinary storage commodities from the hazards of aerosol products in these types of buildings.

2804.3.1 Nonsegregated storage. Storage consisting of solid pile, palletized or rack storage of Level 2 and 3 aerosol products not segregated into areas utilized exclusively for the storage of aerosols shall comply with Table 2804.3.1.

❖ Nonsegregated storage is located in a general purpose warehouse in which Level 2 and 3 aerosol products are not physically or spatially separated from other commodities. The maximum quantity of Level 2 and 3 aerosol products that can be located in nonsegregated storage areas is specified in Table 2804.3.1.

TABLE 2804.3.1
NONSEGREGATED STORAGE OF LEVEL 2 AND 3 AEROSOL PRODUCTS IN GENERAL PURPOSE WAREHOUSES[b]

AEROSOL LEVEL	MAXIMUM NET WEIGHT PER FLOOR (pounds)[b]			
	Palletized or solid-pile storage		Rack storage	
	Unprotected	Protected[a]	Unprotected	Protected[a]
2	2,500	12,000	2,500	24,000
3	1,000	12,000	1,000	24,000
Combination 2 and 3	2,500	12,000	2,500	24,000

For SI: 1 foot = 304.8 mm, 1 pound = 0.454 kg,
1 square foot = 0.0929 m^2.

a. Approved automatic sprinkler system protection and storage arrangements shall comply with NFPA 30B. Sprinkler system protection shall extend 20 feet beyond the storage area containing the aerosol products.

b. Storage quantities indicated are the maximum permitted in any 50,000-square-foot area.

❖ Table 2403.3.1 lists maximum quantities of Level 2 and 3 aerosols permitted in general purpose warehouses with nonsegregated storage. Maximum quantities of aerosol products depend on the type of storage condition and the level of sprinkler protection. Unprotected storage refers to areas that have no sprinkler protection or in which the sprinkler system does not meet specific NFPA 30B design requirements; therefore, protected storage refers only to areas in which sprinkler protection meets NFPA 30B design parameters for solid pile, palletized and rack structure storage conditions.

2804.3.2 Segregated storage. Storage of Level 2 and 3 aerosol products segregated into areas utilized exclusively for the storage of aerosols shall comply with Table 2804.3.2 and Sections 2804.3.2.1 and 2804.3.2.2.

❖ Special requirements for storing aerosol products to prevent mixing with other commodities recognizes that without proper sprinkler protection and separation fires involving Level 2 and 3 aerosols are difficult to control. Storage areas can be segregated by fire-resistance-rated interior walls, a chain-link fence enclosure or the establishment of a separation area. All aerosol product storage areas in segregated storage must have sprinkler protection in accordance with NFPA 30B. The maximum quantity of Level 2 and 3 aerosol products that can be placed in segregated storage is specified in Table 2804.3.2.

TABLE 2804.3.2
SEGREGATED STORAGE OF LEVEL 2 AND 3 AEROSOL PRODUCTS IN GENERAL PURPOSE WAREHOUSES

STORAGE SEPARATION	MAXIMUM SEGREGATED STORAGE AREA[a]		SPRINKLER REQUIREMENTS
	Percentage of building area (percent)	Area limitation (square feet)	
Separation area[e, f]	15	20,000	Notes b, c
Chain-link fence enclosure [d]	20	20,000	Notes b, c
1-hour fire-resistance-rated interior walls	20	30,000	Note b
2-hour fire-resistance-rated interior walls	25	40,000	Note b
3-hour fire-resistance-rated interior walls	30	50,000	Note b

For SI: 1 foot = 304.8 mm, 1 square foot = 0.0929 m^2.

a. The maximum segregated storage area shall be limited to the smaller of the two areas resulting from the percentage of building area limitation and the area limitation.

b. Automatic sprinkler system protection in aerosol product storage areas shall comply with NFPA 30B and be approved. Building areas not containing aerosol product storage shall be equipped throughout with an approved automatic sprinkler system in accordance with Section 903.3.1.1.

c. Automatic sprinkler system protection in aerosol product storage areas shall comply with NFPA 30B and be approved. Sprinkler system protection shall extend a minimum 20 feet beyond the aerosol storage area.

d. Chain-link fence enclosures shall comply with Section 2804.3.2.1.

e. A separation area shall be defined as an area extending outward from the periphery of the segregated aerosol product storage area as follows.

1. The limits of the aerosol product storage shall be clearly marked on the floor.
2. The separation distance shall be a minimum of 25 feet and maintained clear of all materials with a commodity classification greater than Class III in accordance with Section 903.3.1.1.

f. Separation areas shall only be permitted where approved.

❖ Table 2403.3.2 lists maximum quantities of Level 2 and 3 aerosols permitted in general purpose warehouses with segregated storage. Storage areas can be segregated using either fire-resistance-rated interior walls, a chain-link fence enclosure (see Section 2804.3.2.1) or a separation area. As indicated in Note e, a minimum separation distance of 25 feet (7620 mm) from commodity classifications greater than Class III must be maintained when using the separation area option (for additional guidance on commodity classifications, see NFPA 231).

Depending on the separation type provided, the segregated storage area is limited by both a percentage of the building area and a maximum area limitation, as indicated in Note a. The allowable area of segregated storage increases with the increased degree of separation. Segregated storage areas of Level 2 and 3 aerosols must have sprinkler protection in accordance with NFPA 30B.

2804.3.2.1 Chain-link fence enclosures. Chain-link fence enclosures required by Table 2804.3.2 shall comply with the following:

1. The fence shall not be less than No. 9 gage steel wire, woven into a maximum 2-inch (51 mm) diamond mesh.
2. The fence shall be installed from the floor to the underside of the roof or ceiling above.
3. Class III, IV and high-hazard commodities shall be stored outside of the aerosol storage area and a minimum of 8 feet (2438 mm) from the fence.
4. Access openings in the fence shall be provided with either self-closing or automatic-closing devices or a labyrinth opening arrangement preventing aerosol containers from rocketing through the access openings.
5. Not less than two means of egress shall be provided from the fenced enclosure.

❖ Table 2804.3.2 establishes limits on the use of a chain-link fence as a means of segregated storage of Level 2 and 3 aerosol products. A chain-link fence enclosure is intended to reduce the potential hazard of "rocketing" aerosol cans in a fire. Both fire tests and loss history have shown that ruptured aerosol cans in fires will rocket through the warehouse storage area, resulting in multiple ignition locations and overtaxing of the fire protection systems. Chain-link fence construction details, as indicated in this section, in combination with proper sprinkler protection as specified in Notes b and c of Table 2804.3.2, provide a measure of additional safety, permitting increased storage quantities over that permitted where no physical barrier exists.

The No. 9 gage steel requirement for the chain-link fence is similar to a standard industrial-grade chain link gage that is commonly used for fencing property. This chain-link gage is considered the lightest acceptable fencing and is capable of restraining flying aerosol containers.

A labyrinth opening arrangement in the fencing essentially works as an entryway maze. In place of a door, the access area to the chain-link fence enclosure would be constructed of other fencing, which is usually located at right angles, and through its arrangement, would obstruct the opening to prevent the aerosol containers from rocketing through the access opening.

2804.3.2.2 Aisles. The minimum aisle requirements for segregated storage in general purpose warehouses shall comply with Table 2804.3.2.2.

❖ Fires involving aerosol products can spread across aisles located between two rows of racks or palletized and solid pile storage. If the fire is severe enough, the radiant energy may ignite combustible cartons across an aisle, or a fireball from a rupturing container may be large enough to engulf adjacent storage. Table 2804.3.2.2 gives aisle requirements for segregated storage of Level 2 and 3 aerosol products located in general purpose warehouses.

TABLE 2804.3.2.2
SEGREGATED STORAGE AISLE WIDTHS AND DISTANCE TO AISLES IN GENERAL PURPOSE WAREHOUSES

STORAGE CONDITION	MINIMUM AISLE WIDTH (feet)	MAXIMUM DISTANCE FROM STORAGE TO AISLE (feet)
Solid pile or palletized[a]	4 feet between piles	25
Racks with ESFR sprinklers[a]	4 feet between racks and adjacent Level 2 and 3 aerosol product storage	25
Racks without ESFR sprinklers[a]	8 feet between racks and adjacent Level 2 and 3 aerosol product storage	25

For SI: 1 foot = 304.8 mm.
a. Sprinklers shall comply with NFPA 30B.

❖ Table 2804.3.2.2 shows a minimum aisle width and a 25-foot (7620 mm) maximum separation distance from aerosol storage to aisles where aerosols cannot be stored for segregated storage of Level 2 and 3 aerosol products located in general purpose warehouses. The minimum aisle width depends on the storage condition and the level of sprinkler protection. The 25-foot (7620 mm) distance between the storage and aisle not only isolates aerosols but also limits travel distances within the storage area.

Note a in this table is intended to require design of the automatic sprinkler system to meet the requirements of NFPA 30B. If the automatic sprinkler system is not designed in accordance with the conditions for ESFR sprinklers in NFPA 30B, aisles must be 8 feet (2438 mm) wide when storing Level 2 or 3 aerosol products in rack structures. The storage area must be configured so that storage is no more than 25 feet (7620 mm) from the aisle, thus providing a storage footprint 50 feet (15240 mm) on a side.

2804.4 Storage in aerosol warehouses. The total quantity of Level 2 and 3 aerosol products in a warehouse utilized for the storage, shipping and receiving of aerosol products shall not be restricted in structures complying with Sections 2804.4.1 through 2804.4.4.

❖ Buildings classified as aerosol warehouses are detached buildings or a separate portion of a building used exclusively for storing and handling aerosol products. This section does not limit the quantity of Level 2 and 3 aerosol products in a building classified as an aerosol warehouse complying with the provisions of Sections 2804.4.1 through 2804.4.5.

2804.4.1 Automatic sprinkler system. Aerosol warehouses shall be protected by an approved wet-pipe automatic sprinkler system in accordance with NFPA 30B. Sprinkler protection shall be designed based on the highest classification level of aerosol product present.

❖ The automatic sprinkler system for aerosol warehouses must be designed in accordance with NFPA 30B. The sprinkler protection must be designed for Level 3 aerosol products if both Level 2 and 3 aerosols are stored in the aerosol warehouse.

With the approval of the fire code official, sprinkler protection can be omitted from an aerosol warehouse where a minimum separation of 100 feet (30 480 mm) from a lot line or structure is maintained and exposure protection is provided. Exposure protection would include either a public fire department or plant fire brigade that could apply cooling water streams on adjacent property or structures. Exposure protection is not required where the distance is 200 feet (60 960 mm) or more to other buildings or lot lines.

2804.4.2 Pile and palletized storage aisles. Solid pile and palletized storage shall be arranged so the maximum travel distance to an aisle is 25 feet (7620 mm). Aisles shall have a minimum width of 4 feet (1219 mm).

❖ The maximum travel distance of 25 feet (7620 mm) to an aisle provides an adequate means of egress if an emergency occurs. This results in an allowable storage area of 50 feet by 50 feet (15 240 mm by 15 240 mm) surrounded by 4-foot (1219 mm) aisles for palletized and pile storage. The 4-foot (1219 mm) aisle width, in conjunction with a sprinkler system and in accordance with NFPA 30B, reduces the chance of fire spread across aisles involving pile and palletized storage.

2804.4.3 Rack storage aisles. Rack storage shall be arranged with a minimum aisle width of 8 feet (2438 mm) between rows of racks and 8 feet (2438 mm) between racks and adjacent solid pile or palletized storage. Where early suppression fast-response (ESFR) sprinklers provide automatic sprinkler protection, the minimum aisle width shall be 4 feet (1219 mm).

❖ The requirements in this section are similar to those in Table 2804.3.2.2 for segregated storage in general purpose warehouses. This section requires 8-foot (2438 mm) aisles between rows of racks or between racks and adjacent piles or palletized storage when the area is protected with sprinkler systems designed in accordance with NFPA 30B. The exception in this section allows the aisle width to be reduced to 4 feet (1219 mm) when ESFR sprinklers are used and the system is designed in accordance with NFPA 30B. The maximum travel distance to an aisle is intended to be 25-feet (7620 mm) as indicated in Table 2403.3.2.2.

2804.4.4 Combustible commodities. Combustible commodities other than flammable and combustible liquids shall be permitted to be stored in an aerosol warehouse.

Exception: Flammable and combustible liquids in 1-quart (0.95 L) metal containers and smaller shall be permitted to be stored in an aerosol warehouse.

❖ This section (and its exception) correlates with NFPA 30B, by allowing ordinary combustible commodities and small 1 quart (0.95 L) containers of flammable liquid to be stored in aerosol warehouses. Compliance with fire protection design tables in NFPA 30B is required.

2804.5 Storage in inside flammable liquid storage rooms. Inside flammable liquid storage rooms shall comply with Section 3404.3.7. The maximum quantities of aerosol products shall comply with Section 2804.5.1 or 2804.5.2.

❖ This section recognizes another option for storing flammable aerosols. Depending on the anticipated quantities, storage in an inside flammable liquid storage room meeting the requirements of Section 3404.3.7 may be more economical. Both Sections 2804.5.1 and 2804.5.2 limit maximum quantities of Level 2 and 3 aerosols that can be stored within an inside flammable liquid storage room depending on the size of the room and the degree of fire-resistance-rated separation and sprinkler protection installed.

Indicated maximum quantities of aerosol products are the same quantities listed in Section 2804.2. Storage of aerosols exceeding the quantities indicated in Sections 2804.5.1 and 2804.5.2 and the storage of flammable and combustible liquids exceeding maximum allowable quantities should be classified as high-hazard occupancy groups.

2804.5.1 Storage rooms of 500 square feet or less. The storage of aerosol products in flammable liquid storage rooms less than or equal to 500 square feet (46 m^2) in area shall not exceed the following quantities:

1. A net weight of 1,000 pounds (454 kg) of Level 2 aerosol products.
2. A net weight of 500 pounds (227 kg) of Level 3 aerosol products.
3. A combined net weight of 1,000 pounds (454 kg) of Level 2 and 3 aerosol products.

❖ This section limits the amount of Level 2 and 3 aerosols that can be stored in an inside flammable liquid storage room 500 square feet (46 m^2) or less in area that complies with NFPA 30. The degree of fire-resistance-rated construction depends on room size. Sprinkler protection may or may not be required, depending on the quantity of other flammable and combustible liquids in the inside storage room (see commentary, Section 2804.2.1).

2804.5.2 Storage rooms greater than 500 square feet. The storage of aerosol products in flammable liquid storage rooms greater than 500 square feet (46 m^2) in area shall not exceed the following quantities:

1. A net weight of 2,500 pounds (1135 kg) of Level 2 aerosol products.
2. A net weight of 1,000 pounds (454 kg) of Level 3 aerosol products.
3. A combined net weight of 2,500 pounds (1135 kg) of Level 2 and 3 aerosol products.

The maximum aggregate storage quantity of Level 2 and 3 aerosol products permitted in separate inside storage rooms protected by an approved automatic sprinkler system in accordance with NFPA 30B shall be 5,000 pounds (2270 kg).

❖ This section allows an increase in the quantity of Level 2 and 3 aerosols for inside flammable liquid storage rooms exceeding 500 square feet (46 m^2) in area. These provisions assume that the storage room complies with fire-resistance-rated construction and the fire protection requirements of NFPA 30. Increased quantities compared to Section 2804.5.1 for rooms less than 500 square feet (46 m^2) in area are based on having an automatic sprinkler system throughout the building in accordance with NFPA 13.

The additional quantities allowed by the last paragraph of this section require the sprinkler system to comply with NFPA 30B (see commentary, Section 2804.2.1).

2804.6 Storage in liquid warehouses. The storage of Level 2 and 3 aerosol products in liquid warehouses shall comply with NFPA 30B. The storage shall be located within segregated storage areas in accordance with Section 2804.3.2 and Sections 2804.6.1 through 2804.6.3.

❖ This section contains requirements for storing Level 2 and 3 aerosol products in warehouses used primarily for storing flammable and combustible liquids. The aerosol product storage area is required to meet the segregated storage requirements of Section 2804.3.2, Sections 2804.6.1 through 2804.6.3 and NFPA 30B. This section is intended to address only protection for the segregated aerosol product storage area. NFPA 30 addresses protection for the flammable and combustible liquids in the liquid warehouse.

2804.6.1 Containment. Spill control or drainage shall be provided to prevent the flow of liquid to within 8 feet (2438 mm) of the segregated storage area.

❖ To reduce the potential for the segregated aerosol storage area to become involved in an emergency created by the flammable and combustible liquids stored elsewhere in the building, an approved means of spill control and drainage must be provided. The spill control and drainage system is intended to keep the liquids away from the aerosols. NFPA 30 contains additional guidance on the means of spill control and drainage for flammable and combustible liquids (see also Section 2704.2).

2804.6.2 Sprinkler design. Sprinkler protection shall be designed based on the highest level of aerosol product present.

❖ Design requirements for the sprinkler system are based on the highest level of aerosol products stored and the anticipated storage condition. The sprinkler system must be designed for Level 3 aerosol products where both Level 2 and 3 aerosol products are stored.

2804.6.3 Opening protection into segregated storage areas. Fire doors or gates opening into the segregated storage area shall either be self-closing or provided with automatic-closing devices activated by sprinkler water flow or an approved fire detection system.

❖ Openings for access to and from segregated storage areas must provide the intended degree of protection. Chain-link fence enclosures can have either self-closing or automatic-closing gates. If the segregated storage opening is protected with automatic-closing fire doors or gates, the closing device must be activated by water flow from the sprinkler system or by an approved fire detection system.

2804.7 Storage in Group M occupancies. Storage of Level 2 and 3 aerosol products in occupancies in Group M shall comply with Table 2804.7. Retail display shall comply with Section 2806.

❖ Both this section and Table 2804.7 contain storage limitations for Level 2 and 3 aerosol products in the storage area of a mercantile occupancy that is physically separated from the sales area and not accessible to the public. These areas are generally referred to as "back-stock" storage areas. The storage limitation for aerosol products located in the retail sales area of mercantile occupancies is specified in Section 2806.

TABLE 2804.7
MAXIMUM QUANTITIES OF LEVEL 2 AND 3 AEROSOL PRODUCTS IN RETAIL STORAGE AREAS

MAXIMUM NET WEIGHT PER FLOOR (pounds)			
		Segregated storage	
Floor	Nonsegregated storage[a, b]	Storage cabinets[b]	Separated from retail area[c]
Basement	Not permitted	Not permitted	Not permitted
Ground floor	2,500	5,000	Note d
Upper floors	500	1,000	Note d

For SI: 1 pound = 0.454 kg, 1 square foot = 0.0929 m^2.

a. The total aggregate quantity on display and in storage shall not exceed the maximum retail display quantity indicated in Section 2806.3.

b. Storage quantities indicated are the maximum permitted in any 50,000-square-foot area.

c. The storage area shall be separated from the retail area with a 1-hour fire-resistance-rated assembly.

d. See Table 2804.3.2.

❖ Table 2804.7 shows the maximum quantity limitations for Level 2 and 3 aerosol products in the back-stock storage areas of mercantile occupancies in both nonsegregated and segregated conditions.

In nonsegregated storage, the aerosol product storage is neither separated from the retail display by a fire-resistance-rated barrier nor stored in approved flammable liquid storage cabinets. When the storage is nonsegregated, the quantity of Level 2 and 3 aerosols is the total quantity in the back-stock storage and display areas combined. Level 2 and 3 aerosol products would be limited to 2,500 pounds (1135 kg) and 500 pounds (227 kg) on the ground and upper floors, respectively, in any 50,000-square-foot (4645 m^2) area.

In the case of segregated back-stock storage using approved flammable liquid storage cabinets, the total aerosol product storage is independent of the displayed quantities. The back-stock storage area quantities can be increased to those specified in Table 2403.3.2.2 for segregated storage in general purpose warehouses where the storage has been separated from the retail sales area by no less than 1-hour fire-resistance-rated construction.

SECTION 2805
OUTSIDE STORAGE

2805.1 General. The outside storage of Level 2 and 3 aerosol products, including storage in temporary storage trailers, shall be separated from exposures in accordance with Table 2805.1.

❖ Outside storage of Level 2 and 3 aerosol products must be separated from buildings, lot lines, public ways and other outside combustible storage as shown in Table 2805.1 to minimize potential exposure hazards. Temporary storage trailers must also be located a minimum of 50 feet (15 240 mm) from buildings and other outside storage. Minimum separation distances are consistent with NFPA 30B and 80A. NFPA 231 also contains additional guidance on the outside storage of combustible commodities.

TABLE 2805.1
DISTANCE TO EXPOSURES FOR OUTSIDE STORAGE OF LEVEL 2 AND 3 AEROSOL PRODUCTS

EXPOSURE	MINIMUM DISTANCE FROM AEROSOL STORAGE (feet) [a]
Public alleys, public ways, public streets	20
Buildings	50
Exit discharge to a public way	50
Lot lines	20
Other outside storage	50

For SI: 1 inch = 25.4 mm, 1 foot = 304.8 mm.

a. The minimum separation distance indicated is not required where exterior walls having a 2-hour fire-resistance rating without penetrations separate the storage from the exposure. The walls shall extend not less than 30 inches above and to the sides of Level 2 and 3 aerosol products.

❖ Table 2805.1 specifies minimum separation requirements for the outside storage of Level 2 and 3 aerosol products for various exposure conditions. In place of the minimum separation distance requirement, Note a permits the use of a minimum 2-hour fire-resistance-rated wall without openings or penetrations as the means of exposure protection. The wall with the required extensions beyond the aerosol product storage area is intended to act as an effective shield to protect the exposed area from the Level 2 and 3 aerosol products.

SECTION 2806
RETAIL DISPLAY

2806.1 General. This section shall apply to the retail display of 500 pounds (227 kg) or more of Level 2 and 3 aerosol products.

❖ Storage of Level 2 and 3 aerosol products with a net weight exceeding 500 pounds (227 kg) must comply with Sections 2806.2 and 2806.3. Quantities of Level 1 aerosol products are not limited in mercantile display areas because of their low content of flammable product or propellant.

2806.2 Maximum quantities in retail display areas. Aerosol products in retail display areas shall not exceed quantities needed for display and normal merchandising and shall not exceed the quantities in Table 2806.2.

❖ The intent of this section is to restrict the quantities of hazardous materials in display areas where the public would be exposed to them and where they might not be protected by fire barriers or sprinkler systems, as they would be in approved storage areas.

TABLE 2806.2
MAXIMUM QUANTITIES OF LEVEL 2 AND 3 AEROSOL PRODUCTS IN RETAIL DISPLAY AREAS

FLOOR	MAXIMUM NET WEIGHT PER FLOOR (pounds)[a, b]	
	Unprotected[c]	Protected[c,d]
Basement	Not allowed	500
Ground	2,500	10,000
Upper	500	2,000

For SI: 1 pound = 0.454 kg, 1 square foot = 0.0929 m^2.

a. The total quantity shall not exceed 1,000 pounds net weight in any one 100-square-foot retail display area.

b. When packaged, stored and protected in accordance with NFPA 30B, quantity limits shall be limited to those specified in NFPA 30B.

c. Per 25,000-square-foot retail display area.

d. Minimum Ordinary Hazard Group 2 wet-pipe automatic sprinkler system through the retail sales occupancy.

❖ This table establishes quantity limits for display areas that are protected by sprinklers as well as those that are not. The footnote reference to NFPA 30B is consistent with references throughout this chapter where sprinklers are installed for protection. Although Note a allows 1,000 pounds (454 kg) of Level 2 and 3 aerosols per 100 square feet (9 m^2) of display space, the reference to NFPA 30B in Note b could alter this restriction. The quantity is further restricted by the limits imposed by Note c.

2806.3 Maximum quantities in storage areas. Aerosol products in storage areas adjacent to retail display areas shall not exceed the quantities in Table 2806.3.

❖ This section establishes quantity limits on aerosol products held in storage areas adjacent to retail sales floors. The limits are given in Table 2806.3.

TABLE 2806.3
MAXIMUM STORAGE QUANTITIES FOR STORAGE AREAS ADJACENT TO RETAIL DISPLAY OF LEVEL 2 AND LEVEL 3 AEROSOLS

MAXIMUM NET WEIGHT PER FLOOR (pounds)			
		Separated	
Floor	Unseparated[a, b]	Storage cabinets[b]	1-hour occupancy separation
Basement	Not allowed	Not allowed	Not allowed
Ground	2,500	5,000	In accordance with NFPA 30B, Sections 4-3.4.2 and 4-3.4.3
Upper	500	1,000	In accordance with NFPA 30B, Sections 4-3.4.2 and 4-3.4.3

For SI: 1 pound = 0.454 kg, 1 square foot = 0.0929 m^2.

a. The aggregate quantity in storage and retail display shall not exceed the quantity limits for retail display.

b. In any 50,000-square-foot area.

❖ The quantities stated in this table are the same as those in Table 2804.7 except for those in the right-hand column. Table 2804.7 designates this column as applying to areas that are separated from the retail sales area. The requirement for a barrier having a 1-hour fire-resistance rating is given in the footnotes to Table 2804.7. Applicable quantities for that storage condition are given in Table 2804.3.2.

In the case of Table 2806.3, the separation is called a 1-hour occupancy separation and the quantity limits are as found in NFPA 30B.

2806.4 Display of containers. Level 2 and 3 aerosol containers shall not be stacked more than 6 feet (1829 mm) high from the base of the aerosol array to the top of the aerosol array unless the containers are placed on fixed shelving or otherwise secured in an approved manner. When storage or retail display is on shelves, the height of such storage or retail display to the top of aerosol containers shall not exceed 8 feet (2438 mm).

Exception: Storage or display protected in accordance with Sections 2806.2 and 2806.3.

❖ Restricting display heights of Level 2 and 3 aerosol containers serves three purposes. The first is, obviously, to limit the quantity of aerosol that can be housed in a given area; the second is to make certain the displays are stable and not likely to topple and the third is to make certain the piles are not so high that they would be likely to interfere with the efficient operation of area sprinklers.

The exception clarifies that the height limitation does not apply to materials that are protected in accordance with the provisions of NFPA 30B.

2806.5 Combustible cartons. Aerosol products located in retail display areas shall be removed from combustible cartons.

Exceptions:

1. Display areas that use a portion of combustible cartons, which consist of only the bottom panel and not more than 2 inches (51 mm) of side panel is allowed.

2. When the display area is protected in accordance with Table 4-3 of NFPA 30B, storage of aerosol products in combustible cartons is allowed.

❖ Removing combustible cartons from display areas reduces the potential fuel supply in case of a fire and also removes the combustible material that would surround the aerosol containers and heat them to the point that they could ignite, "rocket" or explode. It also keeps the display more "open" so that sprinklers can reach the individual containers more readily to help keep them below dangerous temperatures as well as controlling or extinguishing fire in other surrounding combustibles.

Exception 1 allows cut-down cartons in displays because they represent a somewhat reduced fuel load for a fire and also leave the sides of the aerosol containers exposed so that sprinkler water can reach them. The carton bottoms also help stabilize stacks in displays to keep them from toppling.

Exception 2 allows use of cartons in displays under strictly controlled conditions established by the referenced NFPA 30B table.

2806.6 Aisles. Aisles not less than 4 feet (1219 mm) in width shall be maintained on three sides of a retail display area containing aerosol products.

❖ The requirement for aisles on three sides of a display area serves three purposes. First, the aisles are the escape path in case of a fire; second, they give ready access to emergency response personnel and third, they help prevent the spread of a fire by providing space for a display to topple without falling into an adjacent display area as well as some space to accommodate bursting or rocketing containers. This requirement is consistent with the aisle width requirement in Table 2804.3.2.2.

2806.7 Retail display automatic sprinkler system. When an automatic sprinkler system is required for the protected retail display of aerosol products, the wet-pipe automatic sprinkler system shall be in accordance with Section 903.3.1.1. The minimum system design shall be for an Ordinary Hazard Group 2 occupancy. The system shall be provided throughout the retail display area.

❖ The reference to Section 903.3.1.1 indicates that the sprinkler system must meet the requirements of NFPA 13, which classifies mercantile groups used for retail display as an Ordinary Hazard, Group 2 occupancy. This would require a minimum sprinkler design density of 0.20 gallon per minute per square foot [8.14(L/min/m^2)] applied over the hydraulically most remote 1,500 square feet (139 m^2) of system coverage, which is common for most mercantile occupancies. In areas that are equipped with a sprinkler system conforming to NFPA 13, this section limits the quantity of aerosols on display to 2 pounds per square foot (10 kg/m^2) of the gross sales floor area with a maximum weight of 1000 pounds (454 kg) of aerosols in any 10-foot by 10-foot (3048 mm by 3048 mm) area of the sales floor. If more aerosol product storage is desired or anticipated, therefore, the level of sprinkler protection must comply with NFPA 30B and quantities must not exceed those given in Table 2804.3.1.

2806.8 Storage automatic fire-extinguishing system. When the height of storage or display exceeds the limits in Section 2806.4, the design of the automatic sprinkler system shall be in accordance with NFPA 30B.

❖ This section permits increased quantities of Level 2 and 3 aerosol products in storage areas adjacent to retail display areas when the storage area is protected by an automatic sprinkler system conforming to NFPA 30B. The increased level of sprinkler protection may include a combination of higher water densities, larger-orifice sprinkler heads and ESFR sprinklers, depending on the storage condition.

The increased level of sprinkler protection justifies allowing the maximum quantity in retail display areas to be regulated by Table 2804.3.1.

SECTION 2807 MANUFACTURING FACILITIES

2807.1 General. Manufacturing facilities shall be in accordance with NFPA 30B.

❖ This brief statement requires manufacturing areas to be protected by sprinklers that have a capacity consistent with the aerosol loads in the manufacturing area. The fire code official will have to approve sprinkler arrangements on a case-by-case basis.

Bibliography

The following resource materials are referenced in this chapter or are relevant to the subject matter addressed in this chapter.

ASTM D 240-92, *Standard Test Method for Heat of Combustion of Liquid Hydrocarbon Fuels by Bomb Calorimeter*. West Conshohocken, PA: ASTM International, 1992.

DOTn 49 CFR; 100-178 & 179-199-94, *Specification for Transportation of Explosive and Other Dangerous Articles, Shipping Containers.* Washington, DC: U.S. Department of Transportation, 1994.

Fire Protection Handbook, 18th ed. Quincy, MA: National Fire Protection Association, 1997.

IBC-03, *International Building Code*. Falls Church, VA: International Code Council, 2003.

NFPA 13-99, *Installation of Sprinkler Systems*. Quincy, MA: National Fire Protection Association, 1999.

NFPA 30-00, *Flammable and Combustible Liquids Code*. Quincy, MA: National Fire Protection Association, 2000.

NFPA 30B-98, *Manufacture and Storage of Aerosol Products*. Quincy, MA: National Fire Protection Association, 1998.

NFPA 58-01, *Storage and Handling of Liquefied Petroleum Gases*. Quincy, MA: National Fire Protection Association, 2001.

NFPA 80A-01, *Protection of Buildings from Exterior Fire Exposures*. Quincy, MA: National Fire Protection Association, 2001.

NFPA 231-98, *General Storage*. Quincy, MA: National Fire Protection Association, 1998.

NFPA 231C-98, *Rack Storage of Materials*. Quincy, MA: National Fire Protection Association, 1998.

NFPA 704-96, *Identification of the Fire Hazards of Materials*. Quincy, MA: National Fire Protection Association, 1996.

Chapter 29:
Combustible Fibers

General Comments

Operations involving combustible fibers are typically associated with salvage, paper milling, recycling, cloth manufacturing, carpet and textile mills and agricultural operations, among others.

The primary hazard associated with these operations is the abundance of materials and their ready ignitability. These so-called "Rag Districts," where cloth scrap and clippings are collected and separated for reuse in paper manufacturing, have been associated with catastrophic conflagrations as recently as the 1970s.

Because of these hazards, occupancies storing or handling more than 100 cubic feet (3 m^3) of loose, or 1,000 cubic feet (28 m^3) of baled, combustible fibers are classified as Group H-3 (high hazard) by the *International Building Code*® (IBC®).

Purpose

Chapter 29 establishes the requirements for storage and handling of combustible fibers, including animal, vegetable and synthetic fibers, whether woven into textiles, baled, packaged or loose.

SECTION 2901
GENERAL

2901.1 Scope. The equipment, processes and operations involving combustible fibers shall comply with this chapter.

❖ Fibers and textiles are an integral part of our daily lives. Almost all fibers are combustible. This characteristic is a problem when the fibers are manufactured, made into fabrics and collected as waste.

2901.2 Applicability. Storage of combustible fibers in any quantity shall comply with this section.

❖ Fibers stored as raw material or finished product must comply with this section of the code.

2901.3 Permits. Permits shall be required as set forth in Section 105.6.

❖ The process of issuing permits gives the fire code official an opportunity to carefully evaluate and regulate hazardous operations. Permit applicants should be required to demonstrate that their operations comply with the intent of the code before the permit is issued. See the commentary to Section 105.6 for a general discussion of operations requiring an operational permit, Section 105.6.8 for discussion of specific quantity-based operational permits for the materials regulated in this chapter and Section 105.7 for a general discussion of activities requiring a construction permit. The permit process also notifies the fire department of the need for prefire planning for hazardous property.

SECTION 2902
DEFINITIONS

2902.1 Definition. The following word and term shall, for the purposes of this chapter and as used elsewhere in this code, have the meaning shown herein.

❖ Definitions of terms can help in the understanding and application of the code requirements. The purpose for including here the definition that is associated with the subject matter of this chapter is to provide more convenient access to it without having to refer back to Chapter 2. It is important to emphasize that this term is not exclusively related to this chapter but is applicable everywhere the term is used in the code. For convenience, the term is also listed in Chapter 2 with a cross reference to this section. The use and application of all defined terms, including the one defined in this section, are set forth in Section 201.

COMBUSTIBLE FIBERS. Readily ignitable and free-burning fibers, such as cocoa fiber, cloth, cotton, excelsior, hay, hemp, henequen, istle, jute, kapok, oakum, rags, sisal, Spanish moss, straw, tow, wastepaper, certain synthetic fibers or other like materials.

❖ The basic component of all textiles is fibers. Fibers may be either natural or man-made. Cellulosic, protein and mineral fibers are considered natural fibers. Fibers produced by chemical processes (nylon, rayon, Orlon, etc.) are considered man-made.

SECTION 2903 GENERAL PRECAUTIONS

2903.1 Use of combustible receptacles. Ashes, waste, rubbish or sweepings shall not be placed in wood or other combustible receptacles and shall be removed daily from the structure.

❖ The regular and proper disposal of ashes, waste, rubbish and sweepings is of the utmost importance. Not giving a fire a place of origin is the main objective. The waste containers cannot be combustible. Routine and regular handling and disposal of waste products is an integral part of housekeeping.

2903.2 Vegetation. Grass or weeds shall not be allowed to accumulate at any point on the premises.

❖ Tall grass, dry weeds and bushes around buildings, on and along side roadways, highways and streets present a definite fire hazard. Controlling or clearing grass, weeds and bushes reduces this hazard.

2903.3 Clearances. A minimum clearance of 3 feet (914 mm) shall be maintained between automatic sprinklers and the top of piles.

❖ Although a 3-foot (914 mm) clearance is required by this section, other sections of the code and NFPA 13 allow less clearance. Sprinkler clearances are intended to provide room for the fire plume to develop, thus ensuring sprinkler activation and preventing obstruction of the sprinkler spray pattern.

2903.4 Agricultural products. Hay, straw or similar agricultural products shall not be stored adjacent to structures or combustible materials unless a clear horizontal distance equal to the height of a pile is maintained between such storage and structures or combustible materials. Storage shall be limited to stacks of 100 tons (91 metric tons) each. Stacks shall be separated by a minimum of 20 feet (6096 mm) of clear space. Quantities of hay, straw and other agricultural products shall not be limited where stored in or near farm structures located outside closely built areas. A permit shall not be required for agricultural storage.

❖ Farm and other agricultural buildings are exempt from the requirements of this section. Agricultural outbuildings probably represent the single largest class of buildings used for combustible fiber storage; however, despite the hazards associated with spontaneous combustion of green hay, these occupancies generally pose little threat to life or adjacent property.

2903.5 Dust collection. Where located within a building, equipment or machinery which generates or emits combustible fibers shall be provided with an approved dust-collecting and exhaust system. Such systems shall comply with Chapter 13 and Section 511 of the *International Mechanical Code.*

❖ Dust may create a considerable charge when being displaced from a surface on which it rests and it may also be the material ignited by static sparks. For this reason, a dust-collecting hood and exhaust system is required.

2903.6 Portable fire extinguishers. Portable fire extinguishers shall be provided in accordance with Section 906 as required for extra-hazard occupancy protection as indicated in Table 906.3(1).

❖ Section 906 gives the requirements for portable fire extinguishers. Areas may be classified as an extra-hazard occupancy because of the high possibility of ignition of combustible fibers or dust.

SECTION 2904 LOOSE FIBER STORAGE

2904.1 General. Loose combustible fibers, not in suitable bales or packages and whether housed or in the open, shall not be stored within 100 feet (30 480 mm) of any structure, except as indicated in this chapter.

❖ Piles of loose free-flowing materials may be considered bulk storage. The materials are typically stored in large piles on the floor of a building or out in the open. Because of the greater exposed surface area, loose (unbaled or unpackaged) fibers present a greater risk of ignition and may burn with greater intensity. Consequently, limiting exposures by maintaining operations at a specific distance from occupied buildings and important structures is required.

2904.2 Storage of 100 cubic feet or less. Loose combustible fibers in quantities of not more than 100 cubic feet (3 m^3) located in a structure shall be stored in a metal or metal-lined bin equipped with a self-closing cover.

❖ Small amounts of material may be stored inside a building in approved containers. Acceptable receptacles are metal or metal-lined containers with self-closing lids. This may be accomplished by equipping container covers with heat-activated, automatic-closing covers. Storing loose fibers in a metal or metal-lined container with a self-closing lid prevents the loose fibers from being ignited by outside sources such as cigarettes, candles, etc.

2904.3 Storage of more than 100 cubic feet to 500 cubic feet. Loose combustible fibers in quantities exceeding 100 cubic feet (3 m^3) but not exceeding 500 cubic feet (14 m^3) shall be stored in rooms enclosed with 1-hour fire-resistance-rated fire barriers, with openings protected by an approved opening protective assembly having a fire protection rating of $^3/_4$-hour, constructed in accordance with the *International Building Code.*

❖ Moderate quantities of material must be stored in 1-hour fire-resistance-rated rooms. Openings in floors, walls and ceilings must be protected with self-closing or automatic-closing $^3/_4$-hour fire doors or windows in approved frames. Section 706 of the IBC deals with the construction of and fire-resistance ratings for fire barriers. The section deals with fire barriers used for separation of incidental use areas.

2904.4 Storage of more than 500 cubic feet to 1,000 cubic feet. Loose combustible fibers in quantities exceeding 500 cubic feet (14 m^3) but not exceeding 1,000 cubic feet (28 m^3) shall be stored in rooms enclosed with 2-hour fire-resistance-rated fire barriers, with openings protected by an approved opening protective assembly having a fire protection rating of $1^1/_2$ hours, and constructed in accordance with the *International Building Code.*

❖ Large amounts of material require a 2-hour fire-resistance-rated enclosure. Openings in floors, walls and ceilings require self-closing or automatic-closing $1^1/_2$-hour fire doors or windows in approved frames. Again, the IBC establishes the requirements for the construction of fire barriers and required opening protectives.

2904.5 Storage of more than 1,000 cubic feet. Loose combustible fibers in quantities exceeding 1,000 cubic feet (28 m^3) shall be stored in rooms enclosed with 2-hour fire-resistance-rated fire barriers, with openings protected by an approved opening protective assembly having a fire protection rating of $1^1/_2$ hours, and constructed in accordance with the *International Building Code*. The storage room shall be protected by an automatic sprinkler system installed in accordance with Section 903.3.1.1.

❖ The enclosure requirements in this section are the same as in Section 2904.4; however, if the storage quantity exceeds 1,000 cubic feet (28 m^3), the storage room must be sprinklered for extra protection.

2904.6 Detached storage structure. A maximum of 2,500 cubic feet (70 m^3) of loose combustible fibers shall be stored in a detached structure suitably located, with openings protected against entrance of sparks. The structure shall not be occupied for any other purpose.

❖ Separate, special-purpose structures are required for the storage of extremely large volumes of loose material. Only loose fiber material may be stored in the structure, and openings must restrict the entrance of sparks. These occupancies must be classified in Group H-3 and must be constructed and protected in accordance with the IBC.

SECTION 2905
BALED STORAGE

2905.1 Bale size and separation. Baled combustible fibers shall be limited to single blocks or piles not more than 25,000 cubic feet (700 m^3) in volume, not including aisles or clearances. Blocks or piles of baled fiber shall be separated from adjacent storage by aisles not less than 5 feet (1524 mm) wide, or by flash-fire barriers constructed of continuous sheets of noncombustible material extending from the floor to a minimum height of 1 foot (305 mm) above the highest point of the piles and projecting not less than 1 foot (305 mm) beyond the sides of the piles.

❖ Bulk restrictions and aisle requirements for piles or blocks of combustible fibers are established in this section. These restrictions provide access for fire fighting and limit the fuel load of piles and blocks. Aisles or flash-fire barriers must be provided between piles and blocks. Aisles must be a minimum of 5 feet (1524 mm) wide to provide fire fighters access to fires. In place of aisles, equivalent protection in the form of flash-fire barriers constructed of noncombustible materials may be used as specified in this section.

2905.2 Special baling conditions. Sisal and other fibers in bales bound with combustible tie ropes, jute and other fibers that swell when wet, shall be stored to allow for expansion in any direction without affecting building walls, ceilings or columns. A minimum clearance of 3 feet (914 mm) shall be required between walls and sides of piles, except that where the storage compartment is not more than 30 feet (9144 mm) wide, the minimum clearance at side walls shall be 1 foot (305 mm), provided that a center aisle not less than 5 feet (1524 mm) wide is maintained.

❖ Allowances or clearances must be provided for the expansion of combustible fibers susceptible to swelling when wet. In addition to the lateral forces that these piles may cause if placed against structural elements, their contribution to the building live load should be considered. This section prescribes clearances from building elements. Structural loading concerns should be referred to the building official or to a qualified structural engineer. As an example of special baling conditions, rolled paper is commonly stored on its side rather than on its end to promote water runoff. Structural damage has resulted from the expansion of wetted paper rolls or bales of rags when minimum clearances were not provided.

Bibliography

The following resource materials are referenced in this chapter or are relevant to the subject matter addressed in this chapter.

IBC-2003, *International Building Code*. Falls Church, VA: International Code Council, 2003.

IMC-2003, *International Mechanical Code*. Falls Church, VA: International Code Council, 2003.

NFPA 13-99, *Installation of Sprinkler Systems*. Quincy, MA: National Fire Protection Association, 1999.

Chapter 30: Compressed Gases

General Comments

This chapter regulates the storage, use and handling of all flammable and nonflammable compressed gases, such as those that are used in medical facilities, air separation plants, industrial plants, agricultural equipment and similar occupancies. Standards for the design, construction and marking of compressed gas cylinders and pressure vessels are referenced. Compressed gases used in welding and cutting, cryogenic liquids and liquefied petroleum gases are also regulated under Chapters 26, 32 and 38, respectively. Compressed gases that are classified as hazardous materials are also regulated in Chapter 27, which includes general requirements.

Purpose

Chapter 30 sets requirements for the storage, handling and use of all compressed gases, whether flammable or nonflammable.

Flammable compressed gases.

The principal hazard posed by flammable compressed gas is its ready ignitability, or even explosivity, when mixed with air in the proper proportions. The question in a flammable gas release usually is not if the mixture will ignite, but rather when or how it will ignite if not controlled. Consequently, occupancies storing or handling more than 1,000 cubic feet (28 m^3) of flammable compressed gases or 30 gallons (114 L) of liquefied flammable gases per control area are classified as Group H-2 (high hazard) by the *International Building Code*® (IBC®).

Nonflammable compressed gases.

The principal danger in the case of nonflammable compressed gas is its toxicity, reactivity or the ability to support combustion. Many gases do not fall into any of these categories; however, oxygen, perhaps the most abundant compressed gas, obviously does. Gaseous commodities containing oxygen, such as oxygen-helium and oxygen-nitrogen mixtures, oxides (e.g., nitrous oxide) and peroxides, all support combustion. Fluorine (F_2) and chlorine (Cl_2) may also support combustion of certain materials. Saacke and Associates (1990) discussed some of the properties of these two gases. Since these commodities are stored under pressure, their release from a cylinder causes them to expand to many times their initial volume. In confined and poorly ventilated spaces, this will cause certain gases to create an oxygen-enriched atmosphere that will accelerate burning. The release of other gases displaces atmospheric oxygen, thereby creating a dangerous oxygen-deficient atmosphere. Many materials that do not burn in atmospheric concentrations of oxygen (19 to 21 percent) will support combustion at elevated levels. Chapter 37 contains specific provisions for highly toxic and toxic compressed gases.

Container supports.

The one common danger presented by all compressed gases is the enormous amount of energy released by container or fitting failures. For this reason, Section 3003.3.2 requires all compressed gas cylinders to be properly secured as a means of protection against physical or mechanical damage. Containers should always be checked for obvious physical damage. Dented, bulging, gouged or corroded cylinders should be returned to the gas supplier for inspection and, if necessary, retesting.

Overpressure protection.

All compressed gas storage vessels, except those containing highly toxic materials, are equipped with pressure relief devices as a measure of protection against catastrophic container failure. These devices operate when compressed gas pressure, temperature or both exceed safe limits. Fusible discs and plugs, bimetallic "snap" discs, spring-operated valves or a combination of these methods are used to vent excess gas to the atmosphere. Some of these devices self-restore and stop the release of container contents; others do not. The fire code official should check pressure relief devices to see that they have not been painted, removed, damaged, contaminated, obstructed or otherwise impaired. Additionally, temperature extremes in the operating area [greater than 120°F (49° C) and less than 20°F (-7° C)] should be avoided.

Housekeeping.

Check valves, filters, flash arrestors and other gas system apparatus must be maintained in good operating condition. Dirt is the primary enemy of any gas system: debris can clog filters and block valves; damaged valves may permit the gradual release of tank contents and accumulations of grease and other organic materials become a fire hazard in the presence of oxidizing gases. Combustibles should be kept clear of all compressed gas installations, especially oxidizing gases. External fires involving accumulated combustibles can cause containers to fail.

Separation.

Separating gas system installations and incompatible gases (flammables and oxidizers) to minimize exposures is one of the simplest safeguards to implement. If possible, a suitable fire barrier should be constructed around container installations. The best location for compressed gas container installations is outside of buildings. In such cases, adequate access and weather protection should be provided to facilitate maintenance and emergency response.

Ignition sources.

Controlling ignition sources around both flammable compressed gas and oxidizing gas installations is the most important and most difficult safeguard to implement. Sparks, frictional heat, electrical arcs, static electricity, smoking materials and hot surfaces are only a few of the potential ignition sources. Every reasonable effort must be made to eliminate these sources or limit the adverse impact of an unintentional ignition. Smoking, welding, cutting and other obvious sources of ignition must be prohibited in the vicinity of compressed gases, and steps must be taken to prevent the accumulation of static electricity. Compressed gas cylinders should never be permitted to come into contact with energized electrical equipment.

SECTION 3001 GENERAL

3001.1 Scope. Storage, use and handling of compressed gases in compressed gas containers, cylinders, tanks and systems shall comply with this chapter, including those gases regulated elsewhere in this code. Partially full compressed gas containers, cylinders or tanks containing residual gases shall be considered as full for the purposes of the controls required.

Exceptions:

1. Gases used as refrigerants in refrigeration systems (see Section 606).
2. Compressed natural gas (CNG) for use as a vehicular fuel shall comply with Chapter 22, NFPA 52 and the *International Fuel Gas Code.*

Cutting and welding gases shall also comply with Chapter 27.

Cryogenic fluids shall also comply with Chapter 32. Liquefied natural gas for use as a vehicular fuel shall also comply with NFPA 57 and NFPA 59A.

Compressed gases classified as hazardous materials shall also comply with Chapter 27 for general requirements and chapters addressing specific hazards, including Chapters 35 (Flammable Gases), 37 (Highly Toxic and Toxic Materials), 40 (Oxidizers) and 41 (Pyrophoric).

LP-gas shall also comply with Chapter 38 and the *International Fuel Gas Code.*

❖ This section establishes the scope of Chapter 30 with respect to the storage, use and handling of compressed gases, containers, cylinders, tanks and systems. The requirements of this chapter and the referenced standards are applicable, in addition to the general storage requirements of Chapter 27 for hazardous materials.

3001.2 Permits. Permits shall be required as set forth in Section 105.6.

❖ The process of issuing permits gives the fire code official an opportunity to carefully evaluate and regulate hazardous operations. Permit applicants should be required to demonstrate that their operations comply with the intent of the code before the permit is issued. See the commentary to Section 105.6 for a general discussion of operations requiring an operational permit, Section 105.6.9 for discussion of specific quantity-based operational permits for the materials regulated in this chapter and Section 105.7 for a general discussion of activities requiring a construction permit. The permit process also notifies the fire department of the need for prefire planning for hazardous property.

SECTION 3002 DEFINITIONS

3002.1 Definitions. The following words and terms shall, for the purposes of this chapter and as used elsewhere in this code, have the meanings shown herein.

❖ Definitions of terms can help in the understanding and application of the code requirements. The purpose for including here those definitions that are associated with the subject matter of this chapter is to provide more convenient access to them without having to refer back to Chapter 2. It is important to emphasize that these terms are not exclusively related to this chapter but are applicable everywhere the term is used in the code. For convenience, these terms are also listed in Chapter 2 with a cross reference to this section. The use and application of all defined terms, including those defined in this section, are set forth in Section 201.

COMPRESSED GAS. A material, or mixture of materials which:

1. Is a gas at 68°F (20°C) or less at 14.7 psia (101 kPa) of pressure; and
2. Has a boiling point of 68°F (20°C) or less at 14.7 psia (101 kPa) which is either liquefied, nonliquefied or in solution, except those gases which have no other health- or physical-hazard properties are not considered to be compressed until the pressure in the packaging exceeds 41 psia (28 kPa) at 68°F (20°C).

The states of a compressed gas are categorized as follows:

1. Nonliquefied compressed gases are gases, other than those in solution, which are in a packaging under the charged pressure and are entirely gaseous at a temperature of 68°F (20°C).

2. Liquefied compressed gases are gases that, in a packaging under the charged pressure, are partially liquid at a temperature of 68°F (20°C).

3. Compressed gases in solution are nonliquefied gases that are dissolved in a solvent.

4. Compressed gas mixtures consist of a mixture of two or more compressed gases contained in a packaging, the hazard properties of which are represented by the properties of the mixture as a whole.

❖ This term refers to all types of gases that are under pressure at normal room or outdoor temperatures inside their containers, including, but not limited to, flammable, nonflammable, highly toxic, toxic, cryogenic and liquefied gases. The vapor pressure limitations provide the distinction between a liquid and a gas.

COMPRESSED GAS CONTAINER. A pressure vessel designed to hold compressed gases at pressures greater than one atmosphere at 68°F (20°C) and includes cylinders, containers and tanks.

❖ Containers covered by this definition range from the small compressed air tanks carried by some road service trucks to the very large storage tanks mounted on permanent bases at industrial plants. Designs vary considerably depending on the gases they are intended to hold, whether they are designed for upright or horizontal mounting and the environment they will be used in.

COMPRESSED GAS SYSTEM. An assembly of equipment designed to contain, distribute or transport compressed gases. It can consist of a compressed gas container or containers, reactors and appurtenances, including pumps, compressors and connecting piping and tubing.

❖ This definition is intended to include every component used to convey the gas to or from manufacturing, storage and use facilities. System designs will vary widely depending on their intended use.

NESTING. A method of securing flat-bottomed compressed gas cylinders upright in a tight mass using a contiguous three-point contact system whereby all cylinders within a group have a minimum of three points of contact with other cylinders, walls or bracing.

❖ Nesting is most often used when multiple cylinders are required to supply needed quantities of gases. An example of nesting is the grouping of three or more cylinders of acetylene and/or oxygen in use at a given location in which the cylinders are strapped together. This configuration is more stable than single cylinders and is, therefore, much less likely to be knocked over. Nesting improves safety because falling cylinders can cause valve damage that could result in leakage of hazardous gases.

SECTION 3003
GENERAL REQUIREMENTS

3003.1 Containers, cylinders and tanks. Compressed gas containers, cylinders and tanks shall comply with this section. Compressed gas containers, cylinders or tanks that are not designed for refillable use shall not be refilled after use of the original contents.

❖ This section addresses protection from physical damage and prevents the gas containers, cylinders and tanks from being tampered with and otherwise damaged. It also requires marking for identification of the contents and the hazard degree. Only compressed gas containers, cylinders and tanks designed to be refillable may be done so after the original contents have been used.

3003.2 Marking. Stationary and portable compressed gas containers, cylinders, tanks and systems shall be marked in accordance with Sections 3003.2.1, 3003.2.2 and 3003.2.3.

❖ The referenced standard details the marking requirements for specific systems or gas containers, cylinders and tanks, including color and labeling of the name of the gas it contains.

3003.2.1 Stationary compressed gas containers, cylinders and tanks. Stationary compressed gas containers, cylinders and tanks shall be marked with the name of the gas and in accordance with Sections 2703.5 and 2703.6. Markings shall be visible from any direction of approach.

❖ These units are to be marked in accordance with Section 2703.5 (which references NFPA 704) and 2703.6. The markings need to be visible from any direction of approach (see Figure 3003.2.1).

3003.2.2 Portable containers, cylinders and tanks. Portable compressed gas containers, cylinders and tanks shall be marked in accordance with CGA C-7.

❖ These portable units are to be marked in accordance with CGA C-7. Markings need to include the contents and the direction of flow.

3003.2.3 Piping systems. Piping systems shall be marked in accordance with ANSI A13.1. Markings used for piping systems shall consist of the content's name and include a direction-of-flow arrow. Markings shall be provided at each valve; at wall, floor or ceiling penetrations; at each change of direction; and at a minimum of every 20 feet (6096 mm) or fraction thereof throughout the piping run.

Exceptions:

1. Piping that is designed or intended to carry more than one gas at various times shall have appropriate signs or markings posted at the manifold, along the piping and

at each point of use to provide clear identification and warning.

2. Piping within gas-manufacturing plants, gas-processing plants, refineries and similar occupancies shall be marked in an approved manner.

❖ Piping systems are to be marked in accordance with ANSI A13.1.

Exception 1 addresses piping systems that carry more than one gas. In these cases, signs are posted at the manifold and at each point of use. This will provide a clear identification and warning.

Exception 2 recognizes that gas-manufacturing plants need to be evaluated individually to provide clear and adequate protection.

3003.3 Security. Compressed gas containers, cylinders, tanks and systems shall be secured against accidental dislodgement and against access by unauthorized personnel in accordance with Sections 3003.3.1 through 3003.3.3.

❖ Compressed gas containers, cylinders and tanks must be adequately safeguarded. A fence or other approved protection should be provided around the storage area to minimize the likelihood of someone unknowingly entering the area, as well as to deter vandalism or theft.

3003.3.1 Security of areas. Areas used for the storage, use and handling of compressed gas containers, cylinders, tanks and systems shall be secured against unauthorized entry and safeguarded in an approved manner.

❖ Areas such as yards, loading platforms and any area where gas containers, cylinders and tanks are used, handled or stored are to be secured and safeguarded against unauthorized access.

3003.3.2 Physical protection. Compressed gas containers, cylinders, tanks and systems which could be exposed to physical damage shall be protected. Guard posts or other approved means shall be provided to protect compressed gas containers, cylinders, tanks and systems indoors and outdoors from vehicular damage and shall comply with Section 312.

❖ Compressed gas containers, cylinders and tanks must be adequately protected against physical or mechanical damage to the container and any valves or fittings. Where damage from vehicles may occur, guard posts or other kinds of guards are required to afford protection that complies with Section 312.

3003.3.3 Securing compressed gas containers, cylinders and tanks. Compressed gas containers, cylinders and tanks shall be secured to prevent falling caused by contact, vibration or seismic activity. Securing of compressed gas containers, cylinders and tanks shall be by one of the following methods:

1. Securing containers, cylinders and tanks to a fixed object with one or more restraints.
2. Securing containers, cylinders and tanks on a cart or other mobile device designed for the movement of compressed gas containers, cylinders or tanks.
3. Nesting of compressed gas containers, cylinders and tanks at container filling or servicing facilities or in seller's warehouses not accessible to the public. Nesting shall be allowed provided the nested containers, cylinders or tanks, if dislodged, do not obstruct the required means of egress.

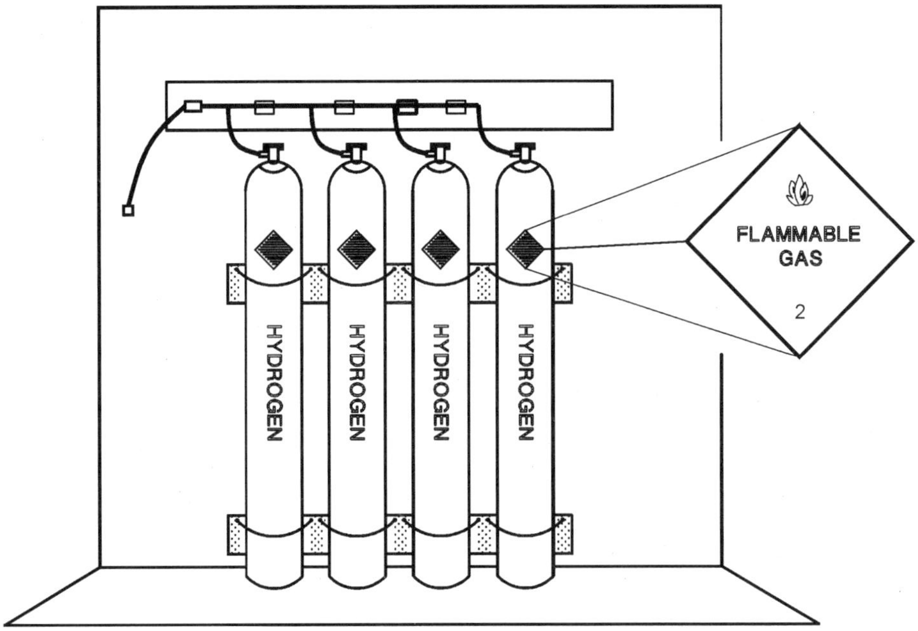

Figure 3003.2.1
STORAGE AND LABELING OF CYLINDERS

4. Securing of compressed gas containers, cylinders and tanks to or within a rack, framework, cabinet or similar assembly designed for such use.

Exception: Compressed gas containers, cylinders and tanks in the process of examination, filling, transport or servicing.

❖ Compressed gas cylinders must be secured to provide protection against physical and mechanical damage to the container and any valves or fittings. Several methods that are outlined in this section ensure the adequacy of securing containers, cylinders and tanks. For an example of restrained cylinders in a compressed gas system, see Figure 3003.2.1.

The exception recognizes that the container will not be secure when it is being filled, examined, transported or serviced. The container should be attended and monitored by adequately trained personnel during these operations.

3003.4 Valve protection. Compressed gas container, cylinder and tank valves shall be protected from physical damage by means of protective caps, collars or similar devices in accordance with Sections 3003.4.1 and 3003.4.2.

❖ This section states that all compressed gas containers, cylinders and tank valves must be protected from physical damage in accordance with Sections 3003.4.1 and 3003.4.2. Avoiding damage to the cylinder valves is extremely important because it could result in the cylinder being propelled by the sudden release of its contents, which are under high pressure.

3003.4.1 Compressed gas container, cylinder or tank protective caps or collars. Compressed gas containers, cylinders and tanks designed for protective caps, collars or other protective devices shall have the caps or devices in place except when the containers, cylinders or tanks are in use or are being serviced or filled.

❖ When the design of gas containers, cylinders or tanks includes protective caps, collars or other protective devices, these devices must be in place except when the container, cylinders or tanks are in use or being serviced or filled.

3003.4.2 Caps and plugs. Compressed gas containers, cylinders and tanks designed for valve protection caps or other protective devices shall have the caps or devices attached. When outlet caps or plugs are installed, they shall be in place.

Exception: Compressed gas containers, cylinders or tanks in use, being serviced or being filled.

❖ In gas containers, cylinders and tanks designed to include valve protection caps or other protective devices, these devices are to be tightly in place except when the container, cylinder or tank is in use or connected for use.

3003.5 Separation from hazardous conditions. Compressed gas containers, cylinders and tanks and systems in storage or use shall be separated from materials and conditions which pose exposure hazards to or from each other. Compressed gas containers, cylinders, tanks and systems in storage or use shall be separated in accordance with Sections 3003.5.1 through 3003.5.10.

❖ This section recognizes the danger of improperly storing materials, which can pose an exposure hazard to surrounding materials. Compressed gas containers, cylinders and tanks must be separated in accordance with Sections 3003.5.1 through 3003.5.8.

3003.5.1 Incompatible materials. Compressed gas containers, cylinders and tanks shall be separated from each other based on the hazard class of their contents. Compressed gas containers, cylinders and tanks shall be separated from incompatible materials in accordance with Section 2703.9.8.

❖ Separating gas systems to minimize exposures is one of the simplest safeguards to implement. If possible, a suitable fire barrier should be constructed around container installations. Gas containers, cylinders and tanks are to be separated from incompatible materials in accordance with Section 2703.9.8.

3003.5.2 Combustible waste, vegetation and similar materials. Combustible waste, vegetation and similar materials shall be kept a minimum of 10 feet (3048 mm) from compressed gas containers, cylinders, tanks and systems. A noncombustible partition, without openings or penetrations and extending not less than 18 inches (457 mm) above and to the sides of the storage area is allowed in lieu of such distance. The wall shall either be an independent structure, or the exterior wall of the building adjacent to the storage area.

❖ This section indicates that combustible waste, vegetation and similar materials must be kept at least 10 feet (3048 mm) from compressed gas containers, cylinders and tanks because external fires involving accumulated combustibles can cause them to fail. This section does permit a noncombustible partition instead of the 10-foot (3048 mm) requirement, provided the partition has no openings or penetrations and it extends at least 18 inches (457 mm) above the sides of the storage area. If the partition is part of a structure, it has to be an independent structure or an exterior wall of a building adjacent to the storage area.

3003.5.3 Ledges, platforms and elevators. Compressed gas containers, cylinders and tanks shall not be placed near elevators, unprotected platform ledges or other areas where falling would result in compressed gas containers, cylinders or tanks being allowed to drop distances exceeding one-half the height of the container, cylinder or tank.

❖ This section addresses concerns about gas containers, cylinders and tanks being damaged by a fall. Gas containers, cylinders and tanks cannot be placed on platform ledges, elevators and other areas where the container, if dropped, would fall more than one-half the height of the container.

3003.5.4 Temperature extremes. Compressed gas containers, cylinders and tanks, whether full or partially full, shall not be exposed to artificially created high temperatures exceeding 125°F (52°C) or subambient (low) temperatures unless designed for use under the exposed conditions.

❖ This section identifies the temperature extremes to which gas containers, cylinders, and tanks can be exposed. The section also provides an exception for gases that are designed for use under the exposure conditions.

3003.5.5 Falling objects. Compressed gas containers, cylinders, tanks and systems shall not be placed in areas where they are capable of being damaged by falling objects.

❖ This section is similar in intent to Section 3003.3.2 for physical protection of gas containers, cylinders and tanks, except this section addresses falling objects, which could cause physical damage.

3003.5.6 Heating. Compressed gas containers, cylinders and tanks, whether full or partially full, shall not be heated by devices which could raise the surface temperature of the container, cylinder or tank to above 125°F (52°C). Heating devices shall comply with the *International Mechanical Code* and the ICC *Electrical Code* Approved heating methods involving temperatures of less than 125°F (52°C) are allowed to be used by trained personnel. Devices designed to maintain individual compressed gas containers, cylinders or tanks at constant temperature shall be approved and shall be designed to be fail safe.

❖ This section requires that any heating devices comply with the *International Mechanical Code®* (IMC®) and the ICC *Electrical Code®* (ICC EC™). It also requires that heating devices be used only by trained personnel and be designed to be fail safe.

3003.5.7 Sources of ignition. Open flames and high-temperature devices shall not be used in a manner which creates a hazardous condition.

❖ Controlling ignition sources around compressed gas is both very important and somewhat difficult. This section addresses open flames and high-temperature devices. These ignition sources should be used only when necessary and by trained personnel, provided adequate safeguards are in place.

3003.5.8 Exposure to chemicals. Compressed gas containers, cylinders, tanks and systems shall not be exposed to corrosive chemicals or fumes which could damage containers, cylinders, tanks, valves or valve-protective caps.

❖ This section provides protection against corrosive chemicals or fumes that could cause damage to the container or valves and valve caps.

3003.5.9 Exhausted enclosures. When exhausted enclosures are provided as a means to segregate compressed gas containers, cylinders and tanks from exposure hazards, such enclosures shall comply with the requirements of Section 2703.8.5.

❖ Exhausted enclosures are used as a means to isolate compressed gas containers from hazardous conditions. When such equipment is utilized, performance requirements are needed. The requirements for exhausted enclosures are found in Section 2703.8.5. By providing a cross reference to the base provisions in Chapter 27, the fundamental requirements for these enclosures can be consistently applied.

3003.5.10 Gas cabinets. When gas cabinets are provided as a means to separate compressed gas containers, cylinders and tanks from exposure hazards, such gas cabinets shall comply with the requirements of Section 2703.8.6.

❖ Gas cabinets are used as a means to isolate compressed gas containers from hazardous conditions. When such equipment is utilized, performance requirements are needed. The requirements for gas cabinets are found in Section 2703.8.6. By providing a cross reference to the base provisions in Chapter 27, the fundamental requirements for gas cabinets can be consistently applied.

3003.6 Wiring and equipment. Electrical wiring and equipment shall comply with the ICC *Electrical Code.* Compressed gas containers, cylinders, tanks and systems shall not be located where they could become part of an electrical circuit. Compressed gas containers, cylinders, tanks and systems shall not be used for electrical grounding.

❖ This section requires that any electrical wiring and equipment comply with the ICC EC. It further requires that no compressed gas containers, cylinders and tanks be placed where they could become part of an electrical circuit. Cylinders and systems cannot be part of any electrical grounding system.

3003.7 Service and repair. Service, repair, modification or removal of valves, pressure-relief devices or other compressed gas container, cylinder or tank appurtenances shall be performed by trained personnel.

❖ This section requires individuals who service, repair or modify gas containers, cylinders and tanks or tank appurtenances to be fully trained in the particular function they are performing.

3003.8 Unauthorized use. Compressed gas containers, cylinders, tanks and systems shall not be used for any purpose other than to serve as a vessel for containing the product which it is designed to contain.

❖ This section requires that gas containers, cylinders and tanks hold only the product they were designed to contain. These containers, cylinders and tanks may not be used for any other purpose.

3003.9 Exposure to fire. Compressed gas containers, cylinders and tanks which have been exposed to fire shall be removed from service. Containers, cylinders and tanks so removed shall be handled by approved qualified persons.

❖ This section requires that any gas containers, cylinders and tanks that have been exposed to fire be removed from service and be handled only by qualified persons. This is especially important because the container may still be under pressure.

3003.10 Leaks, damage or corrosion. Leaking, damaged or corroded compressed gas containers, cylinders and tanks shall be removed from service. Leaking, damaged or corroded compressed gas systems shall be replaced or repaired in accordance with the following:

1. Compressed gas containers, cylinders and tanks which have been removed from service shall be handled in an approved manner.
2. Compressed gas systems which are determined to be leaking, damaged or corroded shall be repaired to a serviceable condition or removed from service.

❖ This section addresses concerns related to gas containers, cylinders and tanks leaking, being corroded or damaged. Leaking, damaged and corroded gas containers, cylinders and tanks must be removed from service and be replaced or repaired in accordance with Items 1 and 2 of this section. Damaged gas containers, cylinders and tanks pose a potential content-release hazard. Care must be taken to determine that disposal does not present a greater hazard than the damaged container. Small gas containers, cylinders and tanks usually cannot be repaired easily. Large gas containers, cylinders and tanks and fixed tank installations often cannot be easily replaced and must be repaired. Temporary storage of materials during repair of these gas containers, cylinders and tanks must comply with code requirements.

3003.11 Surface of unprotected storage or use areas. Unless otherwise specified in Section 3003.12, compressed gas containers, cylinders and tanks are allowed to be stored or used without being placed under overhead cover. To prevent bottom corrosion, containers, cylinders and tanks shall be protected from direct contact with soil or unimproved surfaces. The surface of the area on which the containers are placed shall be graded to prevent accumulation of water.

❖ This section requires that adequate protection be provided to prevent bottom corrosion of the container from direct contact with soil or unimproved surfaces. Grading is also required to prevent the accumulation of water in the container storage area. Overhead cover is not required unless extreme temperatures are present. Section 3003.12 addresses overhead protection.

3003.12 Overhead cover. Compressed gas containers, cylinders and tanks are allowed to be stored or used in the sun except in locations where extreme temperatures prevail. When extreme temperatures prevail, overhead covers shall be provided.

❖ This section requires that overhead cover be provided when extreme temperatures exist. Many of these structures are constructed as an attached canopy to provide ready access to materials.

3003.13 Lighting. Approved lighting by natural or artificial means shall be provided.

❖ Lighting sufficient for good visibility must be maintained at all times in both inside and outside areas. Natural light for outdoor areas may be sufficient where operations are limited to daylight hours.

SECTION 3004
STORAGE OF COMPRESSED GASES

3004.1 Upright storage. Compressed gas containers, cylinders and tanks, except those designed for use in a horizontal position, and all compressed gas containers, cylinders and tanks containing nonliquefied gases, shall be stored in an upright position with the valve end up. An upright position shall include conditions where the container, cylinder or tank axis is inclined as much as 45 degrees (0.80 rad) from the vertical.

Exceptions:

1. Compressed gas containers with a water volume less than 1.3 gallons (5 L) are allowed to be stored in a horizontal position.
2. Cylinders, containers and tanks containing nonflammable gases or cylinders, containers and tanks containing nonliquefied flammable gases, which have been secured to a pallet for transportation purposes.

❖ Containers, cylinders or tanks designed to be used in the horizontal position with the valve end up and accessible are to be stored in an upright position unless the conditions of one of the two listed exceptions are met. Exception 1 allows gas containers, cylinders and tanks with a water volume less than 1.3 gallons (5 L) to be stored in the horizontal position. Exception 2 permits gas containers, cylinders and tanks containing nonflammable gases to be in the horizontal position when secured to a pallet for transportation.

3004.2 Material-specific regulations. In addition to the requirements of this section, indoor and outdoor storage of compressed gases shall comply with the material-specific provisions of Chapters 31, 35 and 37 through 44.

❖ In addition to meeting the requirements of this section, storage of compressed gases needs to comply with the material-specific provisions of Chapters 31, 35 and 37 through 44. For example, the storage and use of flammable gases must be in accordance with Chapter 35 in addition to the requirements of this section.

SECTION 3005
USE AND HANDLING OF COMPRESSED GASES

3005.1 Compressed gas systems. Compressed gas systems shall be suitable for the use intended and shall be designed by persons competent in such design. Compressed gas equipment, machinery and processes shall be listed or approved.

❖ This section contains requirements for gas systems. A system includes stationary or movable gas containers, cylinders and tanks; pressure regulators; safety relief devices; manifolds; interconnecting piping and controls. It is important that persons designing these systems be competent in this area. All equipment, machinery and processes need to be listed or approved for that particular use.

3005.2 Controls. Compressed gas system controls shall be designed to prevent materials from entering or leaving process or reaction systems at other than the intended time, rate or path. Automatic controls shall be designed to be fail safe.

❖ In designing the system controls, provisions must be made to prevent materials from escaping the system other than at the intended time. Automatic controls must be designed so that they are fail safe to provide safe conditions when personnel are not present.

3005.3 Piping systems. Piping, including tubing, valves, fittings and pressure regulators, shall comply with this section and Chapter 27. Piping, tubing, pressure regulators, valves and other apparatus shall be kept gas tight to prevent leakage.

❖ All apparatus, connected and part of the system, shall comply with this section and Chapter 27. Proper maintenance of all the components ensures that gas leaks are prevented.

3005.4 Valves. Valves utilized on compressed gas systems shall be suitable for the use intended and shall be accessible. Valve handles or operators for required shutoff valves shall not be removed or otherwise altered to prevent access.

❖ This section is intended to provide adequate safeguards for valves. Valves are to be employed solely for the use for which they were designed. They must be accessible and must not be removed or altered to prevent access.

3005.5 Venting. Venting of gases shall be directed to an approved location. Venting shall comply with the *International Mechanical Code.*

❖ This section indicates that gases must be vented to an approved location. It refers the user to the IMC to ensure compliance with venting requirements.

3005.6 Upright use. Compressed gas containers, cylinders and tanks, except those designed for use in a horizontal position, and all compressed gas containers, cylinders and tanks containing nonliquefied gases, shall be used in an upright position with the valve end up. An upright position shall include conditions where the container, cylinder or tank axis is inclined as much as 45 degrees (0.80 rad) from the vertical. Use of nonflammable liquefied gases in the inverted position when the liquid phase is used shall not be prohibited provided that the container, cylinder or tank is properly secured and the dispensing apparatus is designed for liquefied gas use.

Exception: Compressed gas containers, cylinders and tanks with a water volume less than 1.3 gallons (5 L) are allowed to be used in a horizontal position.

❖ This section has some of the general requirements found in Section 3004.1. Unless the container is designed to be used in the horizontal position, it must be in the upright position with the valve end up and accessible. In a gas system, the container may be in the inverted position when using nonflammable liquefied gases, provided that the container is secured and the dispensing apparatus is designed for liquefied gas use. Exception 1 to Section 3004.1 also applies here: if the container has a water volume less than 1.3 gallons (5 L) it may be used in the horizontal position.

3005.7 Transfer. Transfer of gases between containers, cylinders and tanks shall be performed by qualified personnel using equipment and operating procedures in accordance with CGA P-1.

Exception: Fueling of vehicles with compressed natural gas (CNG).

❖ This section requires that qualified personnel, using equipment and procedures in accordance with CGA P1, execute the transfer of gases between gas containers, cylinders and tanks. The exception to this occurs when fueling vehicles with compressed natural gas (CNG). In such a case, fueling must comply with the code requirements for CNG.

3005.8 Use of compressed gas for inflation. Inflatable equipment, devices or balloons shall only be pressurized or filled with compressed air or inert gases.

❖ This section permits inflatable items, such as balloons, to be pressurized with compressed air or an inert gas only.

3005.9 Material-specific regulations. In addition to the requirements of this section, indoor and outdoor use of compressed gases shall comply with the material-specific provisions of Chapters 31, 35 and 37 through 44.

❖ In addition to the requirements of this section, storage of compressed gases must comply with the material-specific provisions of Chapters 31, 35 and 37 through 44. For example, the storage and use of a flammable gas system must comply with Chapter 35 in addition to the requirements of this section.

3005.10 Handling. The handling of compressed gas containers, cylinders and tanks shall comply with Sections 3005.10.1 and 3005.10.2.

❖ The requirements of Sections 3003.10.1 and 3005.10.2 are specific for handling gas containers, cylinders and tanks.

3005.10.1 Carts and trucks. Containers, cylinders and tanks shall be moved using an approved method. Where containers, cylinders or tanks are moved by hand cart, hand truck or other mobile device, such carts, trucks or devices shall be designed for the secure movement of containers, cylinders or tanks. Carts and trucks utilized for transport of compressed gas containers, cylinders and tanks within buildings shall comply with Section 2703.10. Carts and trucks utilized for transport of compressed gas containers, cylinders and tanks exterior to buildings shall be designed so that the containers, cylinders and tanks will be secured against dropping or otherwise striking against each other or other surfaces.

❖ This section contains regulations for moving gas containers, cylinders and tanks with carts and trucks. When carts or trucks are used inside buildings, they must also comply with Section 2703.10. The potential release of gas during the handling process is greatly increased. A leak caused by improper handling, such as dropping a container, striking containers against each other or striking a container on another surface, poses a potential hazard to persons and property. Proper handling of the containers greatly reduces this risk.

3005.10.2 Lifting devices. Ropes, chains or slings shall not be used to suspend compressed gas containers, cylinders and tanks unless provisions at time of manufacture have been made on the container, cylinder or tank for appropriate lifting attachments, such as lugs.

❖ This section states that compressed gas containers, cylinders and tanks must be fitted with lifting attachments at the time of manufacture before ropes, slings or chains may be used to suspend them.

SECTION 3006
MEDICAL GAS SYSTEMS

3006.1 General. Compressed gases at hospitals and similar facilities intended for inhalation or sedation including, but not limited to, analgesia systems for dentistry, podiatry, veterinary and similar uses shall comply with this section in addition to other requirements of this chapter.

❖ Compressed gases used in medical facilities include oxygen, nitrogen oxides and other gases that pose a variety of hazards if not handled and stored properly. Oxygen, for example, represents both a fire and an explosion hazard. Nitrogen oxides are useful for sedation, but in uncontrolled amounts in a confined space can asphyxiate an unknowing occupant. The requirements of the remainder of these sections are intended to minimize hazards by setting requirements for storage and supply locations.

3006.2 Interior supply location. Medical gases shall be stored in areas dedicated to the storage of such gases without other storage or uses. Where containers of medical gases in quantities greater than the permit amount are located inside buildings, they shall be in a 1-hour exterior room, a 1-hour interior room or a gas cabinet in accordance with Section 3006.2.1, 3006.2.2 or 3006.2.3.

❖ This section states that medical gases must be stored in dedicated areas not used for other storage or any other purpose. When medical gases are stored in quantities greater than the permitted amount, which is located in Table 105.6.9, and are located inside buildings, they are required to be in a 1-hour exterior room, in a 1-hour interior room or a gas cabinet in accordance with Section 3006.2.1, 3006.2.2 or 3006.2.3.

3006.2.1 One-hour exterior rooms. A 1-hour exterior room shall be a room or enclosure separated from the remainder of the building by fire barriers with a fire-resistance rating of not less than 1 hour. Openings between the room or enclosure and interior spaces shall be self-closing smoke- and draft-control assemblies having a fire protection rating of not less than 1 hour. Rooms shall have at least one exterior wall that is provided with at least two vents. Each vent shall not be less than 36 square inches (0.023 m^2) in area. One vent shall be within 6 inches (152 mm) of the floor and one shall be within 6 inches (152 mm) of the ceiling. Rooms shall be provided with at least one automatic sprinkler to provide container cooling in case of fire.

❖ This section contains specific construction, ventilation and suppression requirements for exterior rooms that contain medical gases in quantities greater than the permitted amount.

3006.2.2 One-hour interior room. When an exterior wall cannot be provided for the room, automatic sprinklers shall be installed within the room. The room shall be exhausted through a duct to the exterior. Supply and exhaust ducts shall be enclosed in a 1-hour-rated shaft enclosure from the room to the exterior. Approved mechanical ventilation shall comply with the *International Mechanical Code* and be provided at a minimum rate of 1 cubic foot per minute per square foot [0.00508 $m^3/(s \cdot m^2)$] of the area of the room.

❖ This section contains specific construction, ventilation and suppression requirements for interior rooms that contain medical gases in quantities greater than the permitted amount.

3006.2.3 Gas cabinets. Gas cabinets shall be constructed in accordance with Section 2703.8.6 and the following:

1. The average velocity of ventilation at the face of access ports or windows shall not be less than 200 feet per minute (61 m/s) with a minimum of 150 feet per minute (46 m/s) at any point of the access port or window.
2. Connected to an exhaust system.
3. Internally sprinklered.

❖ This section contains specific requirements for cabinet construction, which are in addition to requirements contained in Section 2703.8.5.

3006.3 Exterior supply locations. Oxidizer medical gas systems located on the exterior of a building with quantities greater than the permit amount shall be located in accordance with Section 4004.2.1.

❖ This section requires compliance with Section 4004.2.1 for oxidizer medical gas systems located on the exterior of the building in amounts greater than permitted in Table 105.6.9.

3006.4 Medical gas systems. Medical gas systems including, but not limited to, distribution piping, supply manifolds, connections, pressure regulators, and relief devices and valves, shall comply with NFPA 99 and the general provisions of this chapter.

❖ Compressed gas piping systems in hospitals and similar institutions may not be used to distribute flammable gases. Nonflammable compressed gas piping systems installed and used as detailed in NFPA 99 are permitted. Systems are to comply with the general provisions of this chapter and NFPA 99.

SECTION 3007 COMPRESSED GASES NOT OTHERWISE REGULATED

3007.1 General. Compressed gases in storage or use not regulated by the material-specific provisions of Chapters 6, 31, 35 and 37 through 45, including asphyxiant, irritant and radioactive gases, shall comply with this section in addition to other requirements of this chapter.

❖ Statistics show that there are more deaths every year from the use of inert and asphyxiant gases than from toxic gases. The indoor storage or use of compressed gases falling into categories such as inerts, asphyxiants, irritants and radioactive can be hazardous regardless of the type of gas involved simply due to the hazards related to asphyxiation. Sections 3001 through 3005 have been designed as generic provisions that apply to all compressed gases. Section 3006 is unique in that it establishes additional requirements for specific uses. Section 3007 intends to fill a void in the regulation of compressed gases by addressing compressed gases posing material hazards not otherwise regulated while not creating unnecessary additional material-specific chapters in the code.

3007.2 Ventilation. Indoor storage and use areas and storage buildings shall be provided with mechanical exhaust ventilation or natural ventilation in accordance with the requirements of Section 2704.3 or 2705.1.9. When mechanical ventilation is provided, the systems shall be operational during such time as the building or space is occupied.

❖ The provisions of this section are limited to basic ventilation requirements when these materials are stored or used indoors. Section 2704.3 provides the fundamental requirements for ventilation systems. The basic provisions allow natural ventilation to be used as a means to address the concern, and mechanical ventilation is not required when it can be shown that natural ventilation is adequate. By including basic provisions for ventilation for the specified material hazards, a reasonable level of safety for building occupants is maintained.

Bibliography

The following resource materials are referenced in this chapter or are relevant to the subject matter addressed in this chapter.

Barlen, W., F.C. Saacke and G.R. Spies. *Industrial Fire Hazards Handbook*, 3rd ed. Quincy, MA: National Fire Protection Association, 1990.

CGA, *Handbook of Compressed Gases*, 4th ed. Arlington, VA: Compressed Gas Association.

Fire Protection Handbook, 18th ed. Quincy, MA: National Fire Protection Association, 1997.

Handbook of Compressed Gases, 3rd ed. Arlington, VA: Compressed Gas Association, 1990.

IMC-03, *International Mechanical Code*. Falls Church, VA: International Code Council, 2003.

Klein, B.R., ed. *Health Care Facilities Handbook*, 5th ed. Quincy, MA: National Fire Protection Association.

NFPA 50-01, *Bulk Oxygen Systems at Consumer Sites*. Quincy, MA: National Fire Protection Association, 2001.

NFPA 70-02, *National Electrical Code*. Quincy, MA: National Fire Protection Association, 2002.

NFPA 99-99, *Health Care Facilities*. Quincy, MA: National Fire Protection Association, 1999.

NFPA 704-96, *Identification of the Fire Hazards of Material*. Quincy, MA: National Fire Protection Association, 1996.

Chapter 31: Corrosive Materials

General Comments

This chapter regulates the storage of corrosive materials. These materials pose unusual risks to fire fighters, emergency personnel and the general public when they are involved in a spill or in a fire or explosion.

Section 3101 establishes the scope of Chapter 31 and the quantities requiring a permit. This section also lists the exempt amounts per control area for corrosive materials to aid in establishing the appropriate occupancy group classification for the building.

Section 3102 provides the definition of corrosive, which is derived from DOT 49 CFR; Part 173.

Section 3103 provides the requirements for structures used for the storage of corrosive materials. These requirements are applicable to storage in excess of the exempt amounts per control area.

Section 3104 provides the maintenance provisions for both indoor and outdoor storage conditions, regardless of quantities.

Section 3105 provides the requirements for both indoor and outdoor use of corrosives, regardless of the quantities.

Corrosive materials may be found in solid, liquid or gaseous states, although the most frequently encountered corrosive materials are liquids and solids. They may be found in all types of occupancies, including research laboratories, hospitals, industrial facilities, warehouses and retail stores.

Corrosive materials may pose multiple hazards, such as radioactivity, toxicity, flammability or detonability. The focus of this chapter is on materials whose primary hazard is corrosivity; that is, the ability to destroy or irreparably damage living tissue on contact.

These regulations are intended to minimize the exposure of the public, fire fighters and other emergency responders to the harmful vapors, liquid matter or splashes resulting from fire or accidental releases. Advance knowledge of the materials in structures through the permit process and the submittal of Material Safety Data Sheets (MSDS) are essential for effective prefire planning by the fire department. NFPA 471 provides useful guidance for managing hazardous material incidents.

Purpose

Chapter 31 addresses the hazards of corrosive materials that have a destructive effect on living tissues. Though corrosive gases exist, most corrosive materials are solid and classified as either acids or bases (alkalis). These materials may pose a wide range of hazards other than corrosivity, such as combustibility, reactivity or oxidizing hazards. This chapter, however, addresses the hazards associated with the storage of materials based on their corrosivity only. Materials posing multiple hazards must conform to the requirements of the code with respect to all their known hazards.

SECTION 3101 GENERAL

3101.1 Scope. The storage and use of corrosive materials shall be in accordance with this chapter. Compressed gases shall also comply with Chapter 30.

Exceptions:

1. Display and storage in Group M and storage in Group S occupancies complying with Section 2703.11.
2. Stationary lead-acid battery systems in accordance with Section 608.
3. This chapter shall not apply to R-717 (ammonia) where used as a refrigerant in a refrigeration system (see Section 606).

❖ Chapter 31 details specific requirements for the storage and use of corrosive materials. The requirements of this chapter are intended to complement the hazardous materials general storage requirements of Chapter 27.

Exception 1 addresses mercantile display to which this chapter is not applicable. Section 2703.11 addresses requirements for the storage of hazardous material in mercantile occupancies. Exceptions 2 and 3 state that stationary lead-acid battery systems and R-717 refrigerants are not addressed in this chapter.

3101.2 Permits. Permits shall be required as set forth in Section 105.6.

❖ The process of issuing permits gives the fire code official an opportunity to carefully evaluate and regulate hazardous operations. Permit applicants should be required to demonstrate that their operations comply with the intent of the code before the permit is issued. See the commentary to Section 105.6 for a general discussion of operations requiring an operational permit, Section 105.6.21 for discussion of specific quantity-based operational permits for the materials regulated in this chapter and Section 105.7 for a general discussion of activities requiring a construction permit. The permit process also notifies the fire department of the need for prefire planning for hazardous property.

SECTION 3102 DEFINITIONS

3102.1 Definition. The following word and term shall, for the purposes of this chapter and as used elsewhere in this code, have the meaning shown herein.

❖ Definitions of terms can help in the understanding and application of the code requirements. The purpose for including here the definition that is associated with the subject matter of this chapter is to provide more convenient access to it without having to refer back to Chapter 2. It is important to emphasize that this term is not exclusively related to this chapter but is applicable everywhere the term is used in the code. For convenience, the term is also listed in Chapter 2 with a cross reference to this section. The use and application of all defined terms, including the one defined in this section, are set forth in Section 201.

CORROSIVE. A chemical that causes visible destruction of, or irreversible alterations in, living tissue by chemical action at the point of contact. A chemical shall be considered corrosive if, when tested on the intact skin of albino rabbits by the method described in DOTn 49 CFR 173.137, such chemical destroys or changes irreversibly the structure of the tissue at the point of contact following an exposure period of 4 hours. This term does not refer to action on inanimate surfaces.

❖ This definition is derived from DOT 49 CFR; Part 173. While corrosive materials may not pose a fire, explosion or reactivity hazard, they do pose a handling and storage problem. Corrosive materials, therefore, are primarily considered a health hazard, and an occupancy containing such materials in excess of the maximum allowable quantity per control area is classified in Group H-4. It should be noted that many corrosive chemicals are also strong oxidizing agents that would require review as a multiple-hazard material in accordance with Section 2701.1.

SECTION 3103 GENERAL REQUIREMENTS

3103.1 Quantities not exceeding the maximum allowable quantity per control area. The storage and use of corrosive materials in amounts not exceeding the maximum allowable quantity per control area indicated in Section 2703.1 shall be in accordance with Sections 2701, 2703 and 3101.

❖ The stored quantities per control area are addressed in Section 2701 if the quantities do not exceed the amounts in Table 2703.1.1(2) or 2703.1.1(4). The tables address quantities posing a health hazard for control areas located inside and outside.

3103.2 Quantities exceeding the maximum allowable quantity per control area. The storage and use of corrosive materials in amounts exceeding the maximum allowable quantity per control area indicated in Section 2703.1 shall be in accordance with this chapter and Chapter 27.

❖ Tables 2703.1.1(2) and 2703.1.1(4) give the maximum allowable quantities for a control area. If the amounts exceed the allowable quantities, the occupancy classifications in Chapter 2 and this chapter must be used for requirements. Section 2703.1.4 states that Chapter 27 is also required.

SECTION 3104 STORAGE

3104.1 Indoor storage. Indoor storage of corrosive materials in amounts exceeding the maximum allowable quantity per control area indicated in Table 2703.1.1(2), shall be in accordance with Sections 2701, 2703, 2704 and this chapter.

❖ Table 2703.1.1(2) gives the maximum quantities for an indoor storage control area. Section 2704 addresses storage requirements of hazardous materials in the amounts that exceed the maximum allowed per control area.

3104.1.1 Liquid-tight floor. In addition to the provisions of Section 2704.12, floors in storage areas for corrosive liquids shall be of liquid-tight construction.

❖ Section 2704.12 requires that floors of storage areas must be of noncombustible construction except for surfacing. This section requires them to be liquid tight as well.

3104.2 Outdoor storage. Outdoor storage of corrosive materials in amounts exceeding the maximum allowable quantity per control area indicated in Table 2703.1.1(4) shall be in accordance with Sections 2701, 2703, 2704 and this chapter.

❖ Table 2703.1.1(4) gives the maximum quantities for an outdoor storage control area. Section 2704 addresses storage of hazardous materials in amounts that exceed the maximum allowed per control area.

3104.2.1 Above-ground outside storage tanks. Above-ground outside storage tanks exceeding an aggregate quantity of 1,000 gallons (3785 L) of corrosive liquids shall be provided with secondary containment in accordance with Section 2704.2.2.

❖ In order to help confine a leak or spill from an outside above-ground storage tank containing corrosive liquids, secondary containment in accordance with Section 2704.2 must be provided. This section applies to all outside storage areas with an aggregate tank capacity in excess of 1,000 gallons (3785 L).

3104.2.2 Distance from storage to exposures. Outdoor storage of corrosive materials shall not be within 20 feet (6096 mm) of buildings not associated with the manufacturing or distribution of such materials, lot lines, public streets, public alleys, public ways or means of egress. A 2-hour fire barrier wall without openings or penetrations, and extending not less than 30 inches (762 mm) above and to the sides of the storage area, is allowed in lieu of such distance. The wall shall either be an independent structure, or the exterior wall of the building adjacent to the storage area.

❖ The required separation distances for corrosive materials are intended to reduce the hazard of radiant heat transfer from nearby exposures on or off the property. This section also recognizes that a 2-hour fire barrier constructed in accordance with Section 706 of the *International Building Code*® (IBC®) will provide physical barrier protection as an alternative to the required separation distance. Note that the wall must be positioned to restrict the spread of fire around or over it.

SECTION 3105 USE

3105.1 Indoor use. The indoor use of corrosive materials in amounts exceeding the maximum allowable quantity per control area indicated in Table 2703.1.1(2) shall be in accordance with Sections 2701, 2703, 2705 and this chapter.

❖ Table 2703.1.1(2) gives the maximum allowable quantities per control area for use in either a closed system or an open system. Section 2705 addresses the use, dispensing and handling of hazardous materials in amounts that exceed the maximum allowable quantity per control area.

3105.1.1 Liquid transfer. Corrosive liquids shall be transferred in accordance with Section 2705.1.10.

❖ Section 2705.1.10 lists five methods that can be used for the transfer of corrosive liquids. These methods are for liquids having a hazard ranking of 3 or 4 in accordance with NFPA 704.

3105.1.2 Ventilation. When corrosive materials are dispensed or used, mechanical exhaust ventilation in accordance with Section 2705.2.1.1 shall be provided.

❖ This section requires mechanical exhaust ventilation for all storage areas containing corrosive liquids with positive vapor pressures, which, if exposed under standard room temperature and atmospheric pressure, give off hazardous fumes and vapors. Adequate mechanical ventilation will reduce the chance for accumulation of hazardous concentration levels of toxic fumes and vapors. Corrosive liquids without a positive vapor pressure do not readily give off vapors at hazardous concentration levels under normal conditions and, therefore, do not require mechanical exhaust ventilation.

3105.2 Outdoor use. The outdoor use of corrosive materials in amounts exceeding the maximum allowable quantity per control area indicated in Table 2703.1.1(4) shall be in accordance with Sections 2701, 2703, 2705 and this chapter.

❖ Table 2703.1.1(4) gives the maximum allowable quantities for an outdoor control area. Section 2705 addresses the use, dispensing and handling of hazardous materials in the amounts that exceed the maximum allowable quantity per control area.

3105.2.1 Distance from use to exposures. Outdoor use of corrosive materials shall be located in accordance with Section 3104.2.2.

❖ The required separation distances for corrosive materials are intended to reduce the hazard of radiant heat transfer to nearby structures or public areas. The distances also help protect the property in question from heat exposure from incidents on or off the property. The exception recognizes that a 2-hour fire-separation wall will provide physical-barrier protection as an alternative to the required separation distance.

Bibliography

The following resource materials are referenced in this chapter or are relevant to the subject matter addressed in this chapter.

DOL 29 CFR; Part 1910.1200-99, *Occupational Safety and Health Standards*. Washington, DC: U.S. Department of Labor, 1999.

DOL 49 CFR; Part 171-177-98. *Hazardous Materials Regulations*. Washington, DC: U.S. Department of Labor, 1998.

IBC-2003, *International Building Code*. Falls Church, VA: International Code Council, 2003.

NFPA 471-02, *Responding to Hazardous Materials Incidents*. Quincy, MA: National Fire Protection Association, 2002.

NFPA 704-96, *Identification of the Fire Hazards of Materials*: Quincy, MA: National Fire Protection Association, 1996.

Chapter 32: Cryogenic Fluids

General Comments

The *International Fire Code*® (IFC®) regulates the hazards associated with materials that are considered to be cryogenic fluids. These requirements are in addition to the other code requirements that address hazards such as flammability and toxicity. Cryogenics are hazardous because they are held at extremely low temperatures and high pressures. Many cryogenic fluids, however, are actually inert gases and would not be regulated elsewhere in the fire code.

Cryogenics pose several hazards to humans who come in close contact with them. The first is the potential for severe freeze burns and tissue damage that can result from direct contact with cryogenic liquids, uninsulated cryogenic pipes or uninsulated cryogenic equipment. Also, a jet of cryogen can freeze the skin or the eyes faster than liquid contact. Eyes are especially susceptible to cryogen exposure.

In addition to freeze hazards, cryogens pose an asphyxiation hazard because they rapidly boil and convert from a liquid to a gas. As will be discussed in the definitions, cryogens have extremely low boiling points. When converting from a liquid to a gas, there is a great deal of expansion that displaces breathable air. This hazard is compounded when the gases are also toxic or flammable.

There is also the potential for air surrounding a cryogen containment system to condense. This is especially the case when transferring liquid nitrogen through uninsulated metal pipes. Because nitrogen has a lower boiling point than oxygen, it will evaporate first and leave an oxygen-enriched atmosphere, which has the potential of improving the conditions for combustion.

Finally, a release of cryogens at extremely low temperatures has the effect of making some materials such as rubber, carbon steel and plastic so brittle that failure of those materials can occur very easily. Materials like stainless steel, copper, brass and most alloys of aluminum must be used when handling cryogenic fluids.

The extreme low temperatures also have the tendency to create thermal stresses in badly designed cryogen containment systems. Uneven temperature distributions can create stress in some piping or related equipment.

The more common cryogens include helium, hydrogen, nitrogen, argon, oxygen and methane. Note that some of these gases are inert but, based on the specific concerns related to cryogens, are still a potential hazard. Cryogens are used for many applications but specifically have had widespread use in the biomedical field and in space programs.

Purpose

Hazards created by cryogenic materials are sometimes compounded by additional hazard characteristics, such as flammability or toxicity. These other characteristics are dealt with in Chapter 27 and other chapters, such as Chapter 35 dealing with flammable gases. This chapter covers the storage, use and handling of cryogenic fluids through regulation of such things as pressure relief mechanisms and proper container storage.

SECTION 3201 GENERAL

3201.1 Scope. Storage, use and handling of cryogenic fluids shall comply with this chapter. Cryogenic fluids classified as hazardous materials shall also comply with Chapter 27 for general requirements. Partially full containers having residual cryogenic fluids shall be considered as full for the purposes of the controls required.

Exceptions:

1. Fluids used as refrigerants in refrigeration systems (see Section 606).
2. Liquified natural gas (LNG). Liquified natural gas shall comply with NFPA 59A.

Oxidizing cryogenic fluids, including oxygen, shall comply with NFPA 50.

Flammable cryogenic fluids, including hydrogen, methane and carbon monoxide, shall comply with NFPA 50B.

Inert cryogenic fluids, including argon, helium and nitrogen, shall comply with CGA P-18.

❖ This section states that this chapter focuses upon the hazards related to cryogenic fluids. If the materials are also classified as a hazardous material, additional requirements, as found in Chapter 27, must be reviewed as well. An inert cryogenic liquid, therefore, such as nitrogen, would need to comply with only this chapter.

Additionally, there are two exceptions. Exception 1 is for cryogens used in refrigerant systems. Exception 2 is for liquefied natural gas since there are more specific and appropriate sections or standards that address those hazards. For refrigerants, refer to Section 606, and for liquefied natural gas refer to NFPA 59A.

Following the exceptions are three references to associated standards dealing with oxidizing cryogenic liquids, flammable cryogenic fluids and inert cryogenic fluids that must be adhered to in addition to complying with Chapter 32. See also Appendix G of the code for weight and volume equivalents of cryogenic fluids. It is important to note that the appendices are not considered part of the code unless specifically adopted (see Section 1 of the sample adopting ordinance on page v of the code).

3201.2 Permits. Permits shall be required as set forth in Section 105.6.

❖ The process of issuing permits gives the fire code official an opportunity to carefully evaluate and regulate hazardous operations. Permit applicants should be required to demonstrate that their operations comply with the intent of the code before the permit is issued. See the commentary to Section 105.6 for a general discussion of operations requiring an operational permit, Section 105.6.11 for discussion of specific quantity-based operational permits for the materials regulated in this chapter and Section 105.7 for a general discussion of activities requiring a construction permit. The permit process also notifies the fire department of the need for prefire planning for hazardous property.

SECTION 3202
DEFINITIONS

3202.1 Definitions. The following words and terms shall, for the purposes of this chapter and as used elsewhere in this code, have the meanings shown herein.

❖ Definitions of terms can help in the understanding and application of the code requirements. The purpose for including those definitions that are associated with the subject matter of this chapter is to provide more convenient access to them without having to refer back to Chapter 2. It is important to emphasize that these terms are not exclusively related to this chapter but are applicable everywhere the term is used in the code. For convenience, these terms are also listed in Chapter 2 with a cross reference to this section. The use and application of all defined terms, including those defined in this section, are set forth in Section 201.

CRYOGENIC CONTAINER. A cryogenic vessel of any size used for the transportation, handling or storage of cryogenic fluids.

❖ This definition is necessary to differentiate containers that hold ordinary compressed gases from those that are used for flammable and combustible liquids as addressed in Chapter 34. Because of the extreme pressures and temperatures, these containers tend to be unique in construction.

CRYOGENIC FLUID. A fluid having a boiling point lower than -130 °F. (-89.9 °C.) at 14.7 pounds per square inch atmosphere (psia) (an absolute pressure of 101.3 kPa).

❖ This definition contains the criteria for determining whether this chapter is applicable. If a fluid falls outside of these criteria, it would likely be treated as a compressed gas and addressed within Chapter 30. Chapter 30 contains a specific definition for compressed gases as well.

CRYOGENIC VESSEL. A pressure vessel, low-pressure tank or atmospheric tank designed to contain a cryogenic fluid on which venting, insulation, refrigeration or a combination of these is used in order to maintain the operating pressure within the design pressure and the contents in a liquid phase.

❖ Cryogenic vessels differ from basic containers in that they are designed for use as either a pressure vessel, a low pressure tank or an atmospheric tank that regulates the operating pressure to maintain the fluid as a liquid. Such vessels not only contain the fluids but also play an active role in regulating the state of the fluids.

FLAMMABLE CRYOGENIC FLUID. A cryogenic fluid that is flammable in its vapor state.

❖ These fluids are flammable in a vapor stage or are to be considered as flammable. It may be possible for a fluid to be nonflammable in the liquid phase but flammable in the vapor stage. The vapor phase would be the more hazardous form of the material. Again, similar to the definition of "Cryogenic fluids," this describes the applicability of the code requirements. Flammability is dealt with primarily in Chapter 35 and NFPA 50B.

LOW-PRESSURE TANK. A storage tank designed to withstand an internal pressure greater than 0.5 pounds per square inch gauge (psig) (3.4 kPa) but not greater than 15 psig (103.4 kPa).

❖ This definition makes a differentiation between what is considered a low-pressure tank and what is considered a high-pressure tank. Low-pressure tanks are generally less hazardous than high-pressure tanks because the rate of release of cryogenic fluids is much lower.

SECTION 3203
GENERAL REQUIREMENTS

3203.1 Containers. Containers employed for storage or use of cryogenic fluids shall comply with Sections 3203.1.1 through 3203.1.3.2 and Chapter 27.

❖ The focus of this particular section is the proper use of containers when storing, using or handling cryogenic fluids to ensure that they can handle concerns such as pressures, thermal stresses and embrittlement. Chapter 27 would also apply in cases where issues are not specifically dealt with in Chapter 32.

3203.1.1 Nonstandard containers. Containers, equipment and devices which are not in compliance with recognized standards for design and construction shall be approved upon presentation of satisfactory evidence that they are designed and constructed for safe operation.

❖ Various standards are available for the construction of containers for use with cryogenic materials. In some cases, applications using cryogenics can facilitate the use of a uniquely designed container that does not meet the specifics of the available standards. This section allows such containers with proper technical justification; criteria are listed. Material properties of cryogenic fluids vary based on differences in the boiling point and critical point. For instance, a material with a lower boiling point would require colder temperatures and perhaps higher pressures; therefore, design calculations or test results must show that container designs can meet the specific pressure and temperature criteria for the cryogenic fluid to be stored.

3203.1.1.1 Data submitted for approval. The following data shall be submitted to the fire code official with reference to the deviation from the recognized standard with the application for approval.

1. Type and use of container, equipment or device.
2. Material to be stored, used or transported.
3. Description showing dimensions and materials used in construction.
4. Design pressure, maximum operating pressure and test pressure.
5. Type, size and setting of pressure relief devices.
6. Other data requested by the fire code official.

❖ This section lists approval criteria for nonstandard containers to be used with cryogenic fluids. Material properties of cryogenic fluids vary based on differences in the boiling point and critical point. For instance, a material with a lower boiling point would require colder temperatures and perhaps higher pressures; therefore, design calculations or test results must show that container designs can meet the specific pressure and temperature criteria for the cryogenic fluid to be stored. Approval applications must include specific references to the portions of recognized standards from which they deviate.

3203.1.2 Concrete containers. Concrete containers shall be built in accordance with the *International Building Code*. Barrier materials and membranes used in connection with concrete, but not functioning structurally, shall be compatible with the materials contained.

❖ Because concrete containers are often built on-site, this section requires that they be built in accordance with the *International Building Code*® (IBC®) requirements for strength and other related issues. Additionally, this section acknowledges the fact that any material in contact with cryogens must be compatible. This will vary based on what cryogenic fluids are stored and the type of barrier materials chosen.

3203.1.3 Foundations and supports. Containers shall be provided with substantial concrete or masonry foundations, or structural steel supports on firm concrete or masonry foundations. Containers shall be supported to prevent the concentration of excessive loads on the supporting portion of the shell. Foundations for horizontal containers shall be constructed to accommodate expansion and contraction of the container. Foundations shall be provided to support the weight of vaporizers or heat exchangers.

❖ Critical to avoiding catastrophic failure of a container is the proper construction of the elements that support the container. The design must anticipate the equipment loads that may be used in conjunction with the container, such as vaporizers or heat exchangers. The foundations and supports need to be strong and firmly set in place by either placing the containers directly on a concrete or masonry foundation or on steel supports that are set into a concrete or masonry foundation.

3203.1.3.1 Temperature effects. When container foundations or supports are subject to exposure to temperatures below -150°F (-101°C), the foundations or supports shall be constructed of materials to withstand the low-temperature effects of cryogenic fluid spillage.

❖ As was mentioned in the general comments at the beginning of this chapter, exposure to extremely low temperatures will cause some materials, such as carbon steel, to become brittle and lose structural strength. For this reason, any portion of a foundation or structural support for cryogenic fluid tanks that might be exposed to the fluid in case of a spill must be constructed of materials that will not be affected by the exposure.

3203.1.3.2 Corrosion protection. Portions of containers in contact with foundations or saddles shall be painted to protect against corrosion.

❖ Corrosion can occur for a number of reasons. Because tanks and their supports will probably be constructed of dissimilar materials, exposure to moisture can cause galvanic corrosion. That same exposure to moisture can cause oxidation of metals, otherwise known as rust.

If the corrosion occurs in the metal skin of the tank, leakage of the cryogenic fluid in the tank could occur. If the corrosion is extensive, tank failure and spillage of pressurized tank contents could occur.

If the corrosion occurs in the saddle or other part of the supporting foundation structure, collapse of the structure could cause the tank to fall and rupture, resulting in spillage of the cryogenic fluid. If the cryogenic fluid is flammable, tank failure could lead to a fire or explosion.

Prevention of corrosion is, therefore, important. The most common way of protecting metal structures from corrosion is to paint them. This painting becomes even more important where dissimilar metals contact each other. This section specifies that the containers must be painted where they are in contact with foundation structures. Good maintenance practice suggests that both tanks and their supporting structures might benefit from painting.

3203.2 Pressure relief devices. Pressure relief devices shall be provided in accordance with Sections 3203.2.1 through 3203.2.7 to protect containers and systems containing cryogenic fluids from rupture in the event of overpressure. Pressure relief devices shall be designed in accordance with CGA S-1.1, CGA S-1.2 and CGA S-1.3.

❖ Pressure relief devices are essential for cryogenics because of the high pressures and low temperatures at which cryogenics are maintained. Although storage tanks, other containers and transfer piping are normally well insulated, some heating of the contents will occur over time, causing internal pressures to increase. Pressure relief mechanisms provide a method of relieving these overpressures and avoiding a hazardous situation. Three Compressed Gas Association (CGA) standards that cover the full range of container types, from portable to stationary, are referenced at the end of this section.

Sections 3203.2.1 through 3203.2.7 contain requirements related to accessibility for maintenance, general sizing, installation requirements and device integrity.

3203.2.1 Containers. Containers shall be provided with pressure relief devices.

❖ This section states very clearly that pressure relief devices are required for cryogenic containers. Section 3203.2.2 addresses associated equipment. Generally, the container is the main area of concern because it is the primary storage area for cryogenics and has the largest potential for overpressurization.

3203.2.2 Vessels or equipment other than containers. Heat exchangers, vaporizers, insulation casings surrounding containers, vessels and coaxial piping systems in which liquefied cryogenic fluids could be trapped because of leakage from the primary container shall be provided with a pressure relief device.

❖ Just as Section 3203.2.1 requires pressure relief devices on the container itself, this section mentions other areas where overpressures occur if there is a leak in the primary container. This is a smaller hazard potential but requirements ensure that all potential overpressures are addressed.

3203.2.3 Sizing. Pressure relief devices shall be sized in accordance with the specifications to which the container was fabricated. The relief device shall have sufficient capacity to prevent the maximum design pressure of the container or system from being exceeded.

❖ This section contains only general language that ensures that the pressure relief device is properly designed to fit the needs of the particular container. In most cases, the manufacturer will already have the devices installed on the container. There are cases, however, when a cryogenic system or container may be constructed for a specific purpose and user. In those cases, the relief valves must be sized and installed by the user.

3203.2.4 Accessibility. Pressure relief devices shall be located such that they are provided with ready access for inspection and repair.

❖ This section addresses the long-term reliability of pressure relief devices by making maintenance and repair more convenient through accessibility.

3203.2.5 Arrangement. Pressure relief devices shall be arranged to discharge unobstructed to the open air in such a manner as to prevent impingement of escaping gas on personnel, containers, equipment and adjacent structures or to enter enclosed spaces.

Exception: DOTn-specified containers with an internal volume of 2 cubic feet ($0.057m^3$) or less.

❖ Pressure relief devices must be located to direct vented vapors away from personnel, containers and structures as well as enclosed spaces to prevent personal injury and property damage. The arrangement will vary from one installation to another based on the use and location of the container and the system.

The exception for DOTn containers with an internal volume less than or equal to 2 cubic feet (.06 m^3) recognizes that the amount of vapor released from these containers is small enough to make this section impractical and unnecessary.

3203.2.6 Shutoffs between pressure relief devices and containers. Shutoff valves shall not be installed between pressure relief devices and containers.

Exception: A shutoff valve is allowed on containers equipped with multiple pressure-relief device installations where the arrangement of the valves provides the full required flow through the minimum number of required relief devices at all times.

❖ This section prohibits shutoff valves from being installed between the pressure relief device and the container because the closing of a valve in that position would allow overpressure in the tank to build without relief. There is an exception if a container is designed with multiple pressure relief devices that can still handle the pressure relief even when a valve is shut off.

3203.2.7 Temperature limits. Pressure relief devices shall not be subjected to cryogenic fluid temperatures except when operating.

❖ One concern related to the integrity of the relief valves is deterioration of the valves as a result of extended exposure to extremely low temperatures, which often have the tendency to make materials brittle and more susceptible to failure. This requirement ensures that pressure relief devices are not subject to these extreme temperatures. Instead, the materials can directly interact with the pressure relief valve only when the valve is operating.

3203.3 Pressure-relief vent piping. Pressure-relief vent-piping systems shall be constructed and arranged so as to remain functional and direct the flow of gas to a safe location in accordance with Sections 3203.3.1 and 3203.3.2.

❖ Because of tank locations, piping must sometimes be used to extend the discharge location of the pressure relief vent. This section generally states that piping can be used along with the pressure relief device when needed for safety. The requirements that follow are similar to those found in the section on pressure relief devices.

3203.3.1 Sizing. Pressure-relief-device vent piping shall have a cross-sectional area not less than that of the pressure-relief-device vent opening and shall be arranged so as not to restrict the flow of escaping gas.

❖ Though somewhat obvious, this section points out in the most generic way the need for piping to have at least the same diameter as the pressure relief device itself, if not larger. This section also states that piping must be free of any bends or other features that could potentially restrict the flow of gases.

3203.3.2 Arrangement. Pressure-relief-device vent piping and drains in vent lines shall be arranged so that escaping gas will discharge unobstructed to the open air and not impinge on personnel, containers, equipment and adjacent structures or enter enclosed spaces. Pressure-relief-device vent lines shall be installed in such a manner to exclude or remove moisture and condensation and prevent malfunction of the pressure relief device because of freezing or ice accumulation.

❖ This section is similar to Section 3203.2.5 in that it requires the vent piping to be arranged so that the discharge itself does not cause additional damage simply because of its location. As discussed in Section 3203.2.5, cryogenic liquids have the potential for damage to materials as well as asphyxiation and burn hazards to people in the surrounding areas.

This section also addresses the fact that moisture accumulated in piping has the potential to freeze and cause an obstruction that could interfere with the proper operation of the pressure relief device.

3203.4 Marking. Cryogenic containers and systems shall be marked in accordance with Sections 3203.4.1 through 3203.4.6.

❖ This section is important for the safe and effective use of cryogenics and is also a necessary tool for emergency response teams. The marking requirements apply to a variety of locations, from the area in which these materials are stored, to the marking of the piping and emergency shutoff valves.

3203.4.1 Identification signs. Visible hazard identification signs in accordance with NFPA 704 shall be provided at entrances to buildings or areas in which cryogenic fluids are stored, handled or used.

❖ This section refers to the placarding requirements in NFPA 704, which are an important tool for emergency response teams when entering a building or room containing hazardous materials. This section requires all rooms and buildings containing cryogenic fluids, regardless of the amount, to be properly marked.

3203.4.2 Identification of contents. Stationary and portable containers shall be marked with the name of the gas contained. Stationary above-ground containers shall be placarded in accordance with Sections 2703.5 and 2703.6. Portable containers shall be identified in accordance with CGA C-7.

❖ Because different cryogens present different hazards, the contents of the container must be specifically identified. This section refers to other sections or standards that contain more detailed identification requirements.

3203.4.3 Identification of containers. Stationary containers shall be identified with the manufacturing specification and maximum allowable working pressure with a permanent nameplate. The nameplate shall be installed on the container in an accessible location. The nameplate shall be marked in accordance with the ASME *Boiler and Pressure Vessel Code* or DOTn 49 CFR Part 1.

❖ This section establishes a requirement for the manufacturer of permanently installed tanks to label the tanks with a nameplate meeting the requirements of the ASME *Boiler and Pressure Vessel Code* or DOTn 49 CFR Part 1. The nameplate must be clearly visible to anyone entering the tank location. The manufacturer's specifications and maximum allowable tank working pressure must be included on the nameplate.

3203.4.4 Identification of container connections. Container inlet and outlet connections, liquid-level limit controls, valves and pressure gauges shall be identified in accordance with one of the following: marked with a permanent tag or label identifying their function, or identified by a schematic drawing which portrays their function and designates whether they are connected to the vapor or liquid space of the container. Where a schematic drawing is provided, it shall be attached to the container and maintained in a legible condition.

❖ Cryogens are often used as part of a system or process. Failure of parts of the system or process can be as hazardous as failure of the main container. Also, the better the components of a process or system are understood, the less likely a hazardous situation will be. This section requires that all components, including shutoff valves, gauges, inlet and outlet connections and others, be labeled or that a schematic drawing of the process or system be attached to the container. The labeling and drawings will help personnel responding to an emergency understand both what has gone wrong and how to most effectively handle the situation.

3203.4.5 Identification of piping systems. Piping systems shall be identified in accordance with ANSI A 13.1.

❖ Piping systems are to be labeled using the industry standard requirements found in ANSI A13.1. This standard is specific to the identification of piping systems for all types of materials. Use of this standard results in consistency from one type of industrial process to another and again decreases the likelihood of large-scale failures.

3203.4.6 Identification of emergency shutoff valves. Emergency shutoff valves shall be identified and the location shall be clearly visible and indicated by means of a sign.

❖ Emergency valves can play the most important role during an emergency. For this reason, they must be clearly visible and identified using signs. This is particularly important because the emergency responders may not be as familiar with the facility as are the daily occupants.

3203.5 Security. Cryogenic containers and systems shall be secured against accidental dislodgement and against access by unauthorized personnel in accordance with Sections 3203.5.1 through 3203.5.4.

❖ The security measures stated in the four subsections that follow this general statement are intended to safeguard cryogenic containers and systems from both accidental and intentional damage. Entry of unauthorized personnel into a cryogenic storage or use area can result in personal injury as well as physical damage.

3203.5.1 Security of areas. Containers and systems shall be secured against unauthorized entry and safeguarded in an approved manner.

❖ This section addresses the restriction of access to the cryogenic containers and associated equipment. The requirements are generic because each facility has unique characteristics. The fire code official is responsible for reviewing and approving security plans.

3203.5.2 Securing of containers. Stationary containers shall be secured to foundations in accordance with the *International Building Code*. Portable containers subject to shifting or upset shall be secured. Nesting shall be an acceptable means of securing containers.

❖ This section focuses on the protection of containers from unintentional physical damage by increasing the integrity of the storage arrangement. Stationary containers are addressed through a reference to the IBC for requirements on foundations and similar support systems because they are permanent structures.

Portable containers are addressed only generally because container sizes and locations vary so widely. Nesting is noted as being an acceptable method of securing, but other methods may also be acceptable. Fire code official approval is required for proposed security plans.

3203.5.3 Securing of vaporizers. Vaporizers, heat exchangers and similar equipment shall be anchored to a suitable foundation and its connecting piping shall be sufficiently flexible to provide for the effects of expansion and contraction due to temperature changes.

❖ The stability of vaporizers and other equipment can be as important as the containers themselves because they regulate the conditions of the fluids. Equipment must be anchored securely and protected from damage. Piping must be able to expand and contract with temperature changes either from the process itself or climatic conditions.

3203.5.4 Physical protection. Containers, piping, valves, pressure relief devices, regulating equipment and other appurtenances shall be protected against physical damage and tampering.

❖ The wording of this section is broad and general because the equipment involved can vary so widely. The intent of the section is to ensure that all associated equipment, components and piping are protected from impacts that are fairly likely to occur within a facility and also to reduce the chances for someone to tamper with the components. This could mean restricting access to certain critical areas of the system.

3203.6 Separation from hazardous conditions. Cryogenic containers and systems in storage or use shall be separated from materials and conditions which pose exposure hazards to or from each other in accordance with Sections 3203.6.1 through 3203.6.2.1.

❖ This introductory statement emphasizes that cryogenic containers and systems, whether for storage or for use, must be separated from other materials and conditions that could pose a hazard or that they could be hazardous to. The two subsections that follow address stationary containers and portable containers.

3203.6.1 Stationary containers. Stationary containers shall be separated from exposure hazards in accordance with the provisions applicable to the type of fluid contained and the minimum separation distances indicated in Table 3203.6.1.

❖ This section refers to Table 3203.6.1 and also notes that there may be other separation requirements associated with the fluids that may increase this distance. For example, Table 3504.2.1 for flammable gases has more restrictive distance requirements in certain cases. This table states that the minimum distance to buildings for 0 to 4225 cubic feet (0 to 120 m^3) of flammable gases is 5 feet (1524 mm), whereas Table 3203.6.1 requires only a 1-foot (305 mm) separation.

TABLE 3203.6.1
SEPARATION OF STATIONARY CONTAINERS FROM EXPOSURE HAZARDS

EXPOSURE	MINIMUM DISTANCE (feet)
Buildings, regardless of construction type	1
Wall openings	1
Air intakes	10
Lot lines	5
Places of public assembly	50
Nonambulatory patient areas	50
Combustible materials such as paper, leaves, weeds, dry grass or debris	15
Other hazardous materials	In accordance with Chapter 27

For SI: 1 foot = 304.8 mm.

❖ This table has several different separation distance criteria for stationary tanks. Generally, the distances are larger when exposure to people is involved, such as places of public assembly and areas where nonambulatory patients are housed. These restrictions are related more to the hazards associated with their low temperatures, high pressures and asphyxiation hazards than they are to flammability or toxicity, which are dealt with in other chapters. Depending on these additional characteristics, larger distances to buildings and lot lines are sometimes required. Also, the restrictions in Table 3203.6.1 are for any amounts of cryogens where the requirements for other hazards typically vary based on the volume of gas.

3203.6.1.1 Point-of-fill connections. Remote transfer points and fill connection points shall not be positioned closer to exposures than the minimum distances required for stationary containers.

❖ Because there is a potential for a large release of cryogens from faulty connections, the same separation requirements found in Table 3203.6.1 would apply to the point-of-fill connection.

3203.6.1.2 Surfaces beneath containers. The surface of the area on which stationary containers are placed, including the surface of the area located below the point where connections are made for the purpose of filling such containers, shall be compatible with the fluid in the container.

❖ To ensure that a reaction does not occur with the surface below the container in the event of a cryogen spill, the surface beneath the container and its point of fill must be compatible with the fluid stored. The surface below the container must never be made of materials that may become brittle when exposed to extremely low temperatures or be corroded by the stored fluid. Either embrittlement or corrosion could weaken the supporting structure and affect the stability of the container.

3203.6.2 Portable containers. Portable containers shall be separated from exposure hazards in accordance with Table 3203.6.2.

❖ Concerns with portable containers are similar to those with stationary containers because of low temperatures and high pressures. This section refers to Table 3203.6.2 for the required separation distances.

TABLE 3203.6.2
SEPARATION OF PORTABLE CONTAINERS FROM EXPOSURE HAZARDS

EXPOSURE	MINIMUM DISTANCE (feet)
Building exits	10
Wall openings	1
Air intakes	10
Lot lines	5
Room or area exits	3
Combustible materials such as paper, leaves, weeds, dry grass or debris	15
Other hazardous materials	In accordance with Chapter 27

For SI: 1 foot = 304.8 mm.

❖ Because portable tanks are generally much smaller than stationary containers and are more easily moved, the separation distances do not address areas of assembly or buildings housing nonambulatory patients. Instead, the focus is on exits and areas where portable tanks are more likely to cause an immediate threat. For those locations that correlate with Table 3203.6.1, the distances are the same. For instance, the distance to wall openings must be at least 1 foot (305 mm) for both stationary and portable containers. Again, these separation requirements are for all amounts of cryogens stored in portable containers. Concerns over flammability or toxicity are covered in other chapters.

3203.6.2.1 Surfaces beneath containers. Containers shall be placed on surfaces that are compatible with the fluid in the container.

❖ This section is identical in intent to Section 3203.6.1.2; however, because portable containers are not permanently attached to foundations or similar supports, they can be moved to various locations with different surfaces. Personnel responsible for handling and placing the portable containers must be educated and trained on the interaction of cryogenics with certain materials.

3203.7 Electrical wiring and equipment. Electrical wiring and equipment shall comply with the ICC *Electrical Code* and Sections 3203.7.1 and 3203.7.2.

❖ This section addresses compliance with the ICC *Electrical Code®* (ICC EC™) requirements.

3203.7.1 Location. Containers and systems shall not be located where they could become part of an electrical circuit.

❖ Containers and systems accidentally becoming part of an electrical circuit simply because of where they are placed could overheat, which could lead to expansion of the fluids and overpressure situations.

3203.7.2 Electrical grounding and bonding. Containers and systems shall not be used for electrical grounding. When electrical grounding and bonding is required, the system shall comply with the ICC *Electrical Code*. The grounding system shall be protected against corrosion, including corrosion caused by stray electric currents.

❖ Containers and systems should not be used to ground or bond because this may cause a rise in temperature, which, similar to being part of a circuit, may cause heating (see also Section 3203.7.1).

3203.8 Service and repair. Service, repair, modification or removal of valves, pressure relief devices or other container appurtenances, shall comply with Sections 3203.8.1 and 3203.8.2 and the ASME *Boiler and Pressure Vessel Code*, Section VIII or DOTn 49 CFR Part 1.

❖ To ensure that the initial reliability of the containers and systems is maintained, minimum requirements for the qualifications and procedures for repair are mandated. The general requirements in Sections 3202.8.1 and 3202.8.2 are supplemented by a reference to the ASME *Boiler and Pressure Vessel Code* and DOTn requirements.

3203.8.1 Containers. Containers that have been removed from service shall be handled in an approved manner.

❖ This section requires that containers out of service must be handled according to procedures previously approved by the authority having jurisdiction. These procedures have to be written to accommodate all types and uses of containers including stationary and portable containers.

3203.8.2 Systems. Service and repair of systems shall be performed by trained personnel.

❖ This section gives the jurisdiction having authority the right to enforce training requirements that are consistent with the fluids being handled and the physical plant being operated.

3203.9 Unauthorized use. Containers shall not be used for any purpose other than to serve as a vessel for containing the product which it is designed to contain.

❖ To ensure that the container can actually handle the pressures and temperatures for which it was designed, containers are limited to the materials they were specifically designed to contain. This can be a material compatibility concern as well. A container designed to be compatible with one cryogenic fluid may not be compatible with other fluids.

3203.10 Leaks, damage and corrosion. Leaking, damaged or corroded containers shall be removed from service. Leaking, damaged or corroded systems shall be replaced, repaired or removed in accordance with Section 3203.8.

❖ This requirement states that faulty containers must be removed from service before a failure can occur. The section allows repair of the faulty container either before or after it is removed from service as well as the option for replacement.

3203.11 Lighting. When required, lighting, including emergency lighting, shall be provided for fire appliances and operating facilities such as walkways, control valves and gates ancillary to stationary containers.

❖ This section gives the fire code official the authority to require lighting around essential features of the facility both for routine use and during emergency operations. This might include paths to such features. This requirement also allows the fire code official to ask for additional lighting that can be used as emergency lighting when it is appropriate. Typically, the phrase "where required" in other places within the code means that another section would enact the requirement. In this case, however, it appears that the intent was to provide authority to the fire code official to ask for specific lighting to address emergency response in individual situations.

SECTION 3204
STORAGE

3204.1 General. Storage of containers shall comply with this section.

❖ As with all the chapters associated with hazardous materials, there is a section specific to storage and one specific to use. This section covers storage aspects, which tend to be less hazardous than use. The requirements are split into indoor and outdoor storage locations. Whether cryogens are indoors or outdoors will affect the number and type of safeguards required.

3204.2 Indoor storage. Indoor storage of containers shall be in accordance with Sections 3204.2.1 through 3204.2.2.3.

❖ The indoor storage requirements are divided into stationary and portable containers.

3204.2.1 Stationary containers. Stationary containers shall be installed in accordance with the provisions applicable to the type of fluid stored and this section.

❖ This section acknowledges that requirements will vary depending on the type of fluid being stored.

3204.2.1.1 Containers. Stationary containers shall comply with Section 3203.1.

❖ See the commentary for Section 3203.1.

3204.2.1.2 Construction of indoor areas. Cryogenic fluids in stationary containers stored indoors shall be located in buildings, rooms or areas constructed in accordance with the *International Building Code.*

❖ This section requires that indoor storage areas be constructed in accordance with the IBC. This addresses occupancy requirements in terms of maximum allowable quantities and other relevant issues. Again, note that all cryogens do not necessarily have a characteristic that would classify them as hazardous in the context of maximum allowable quantities.

3204.2.1.3 Ventilation. Storage areas for stationary containers shall be ventilated in accordance with the *International Mechanical Code.*

❖ Ventilation is more critical for indoor areas than for outdoor areas because the fluids cannot disperse and be removed as easily. This section refers to the *International Mechanical Code®* (IMC®) for the ventilation requirements. The IMC ventilation requirements are based on the use of the space. Chapter 27 of the code also addresses ventilation requirements in Sections 2703.8 and 2705.2. These particular requirements apply only when the maximum allowable quantities have been exceeded. Further requirements for ventilation and treatment systems in some cases are found in the hazard-specific chapters of the code, such as Chapter 37, which addresses highly toxic and toxic materials.

3204.2.2 Portable containers. Indoor storage of portable containers shall comply with the provisions applicable to the type of fluid stored and Sections 3204.2.2.1 through 32042.2.3.

❖ This section is a reminder that storage requirements will vary with the fluid being stored.

3204.2.2.1 Containers. Portable containers shall comply with Section 3203.1.

❖ See the commentary for Section 3203.1.

3204.2.2.2 Construction of indoor areas. Cryogenic fluids in portable containers stored indoors shall be stored in buildings, rooms or areas constructed in accordance with the *International Building Code.*

❖ This section is the same as Section 3204.2.1.2 for stationary containers and requires compliance with the IBC for the construction of the building, which means the occupancy classification and related construction requirements must be considered. The use of cryogenics alone may not drive the need for a Group H occupancy. The additional hazard characteristics and amounts of fluids will drive those requirements.

3204.2.2.3 Ventilation. Storage areas shall be ventilated in accordance with the *International Mechanical Code.*

❖ These requirements are the same as those for stationary containers (see commentary, Section 3204.2.1.3).

3204.3 Outdoor storage. Outdoor storage of containers shall be in accordance with Sections 3204.3.1 through 3204.3.2.2.

❖ This section is specific to outdoor areas, which are generally less hazardous because the fluids can more easily disperse if released, thereby reducing the potential for harm to both people in the surrounding area and property. Outdoor storage areas are, however, more susceptible to damage from both weather and people. As with the indoor area section, the requirements are divided into portable and stationary containers.

3204.3.1 Stationary containers. The outdoor storage of stationary containers shall comply with Section 3203 and this section.

❖ The requirements for stationary containers are more detailed than for portable containers because they are more permanent.

3204.3.1.1 Location. Stationary containers shall be located in accordance with Section 3203.6. Containers of cryogenic fluids shall not be located within diked areas containing other hazardous materials.

Storage of flammable cryogenic fluids in stationary containers outside of buildings is prohibited within the limits established by law as the limits of districts in which such storage is prohibited (see Section 3 of the Sample Ordinance for Adoption of the *International Fire Code* on page v).

❖ Section 3203.6 refers to Table 3203.6.1 for minimum distance requirements specific to stationary containers. Cryogenic fluids cannot share diked areas with other hazardous materials because of the risk that one type of hazardous material could create an exposure hazard for the other. Cryogens, if released, will be at very low temperatures and high pressures that could potentially compromise the integrity of the other containers and tanks.

The second paragraph of this section reminds users that there may be an ordinance within the jurisdiction that would limit the amount of cryogen that could be stored. Generally, such restrictions will be based on locations within densely populated areas or similar factors. This section does not limit the amount that can be stored; it is simply a note to the users to review any local rules or ordinances for restrictions before investing time and money in a facility that may not be useable.

3204.3.1.2 Areas subject to flooding. Stationary containers located in areas subject to flooding shall be securely anchored or elevated to prevent the containers from separating from foundations or supports.

❖ Because stationary containers are considered permanent, issues such as flooding and the effect that flooding will have on the stability of the container are important. Flooding has the tendency to cause containers to pull away from their foundations because of their potential buoyancy. This section requires that the container either be anchored or be located in an area where hazards are minimized.

3204.3.1.3 Drainage. The area surrounding stationary containers shall be provided with a means to prevent accidental discharge of fluids from endangering personnel, containers, equipment and adjacent structures or to enter enclosed spaces. The stationary container shall not be placed where spilled or discharged fluids will be retained around the container.

> **Exception:** These provisions shall not apply when it is determined by the fire code official that the container does not constitute a hazard, after consideration of special features such as crushed rock utilized as a heat sink, topographical conditions, nature of occupancy, proximity to structures on the same or adjacent property, and the capacity and construction of containers and character of fluids to be stored.

❖ This section deals with both exposure hazards and the need to avoid a buildup of spilled fluids. People and property must be protected from the exposure hazards of the initial spill. How this is to be done is not specified because the needs for installation vary considerably.

In terms of the drainage, if the fluids are released, they cannot be held in the immediate vicinity of the container; therefore, diking may be a practical method of containment, but the impounded spill would have to be diverted away from the container once it has been released.

There is an exception for areas where the exposure hazards are very low, such as a container in a very remote location. Several factors need to be addressed before such allowances can be given. In some cases, a smaller container may be able to take advantage of this exception whereas a larger container may not, based on the potential size of the discharge.

3204.3.2 Portable containers. Outdoor storage of portable containers shall comply with Section 3203 and this section.

❖ Portable containers are typically smaller than stationary containers; therefore, the requirements are less detailed and are generally less substantial.

3204.3.2.1 Location. Portable containers shall be located in accordance with Section 3203.6.

❖ See the commentary for Section 3203.6.

3204.3.2.2 Drainage. The area surrounding portable containers shall be provided with a means to prevent accidental discharge of fluids from endangering adjacent containers, buildings, equipment or adjoining property.

> **Exception:** These provisions shall not apply when it is determined by the fire code official that the container does not constitute a hazard.

❖ This section requires placement of the containers where they will not further endanger buildings and other exposures if a spill occurs. If impounding is used to contain a spill, the spilled fluid must be diverted to a location where it would not cause personal injury or property damage. This requirement is in addition to the distances specified in Section 3203.6. As in Section 3204.3.1.3, there is an exception that allows the fire code official to review container locations and evaluate the potential hazard of a spill. If the arrangement of the container does not create a hazard, compliance with Section 3204.3.2.2 is not necessary. A good example is a well-isolated storage area.

SECTION 3205
USE AND HANDLING

3205.1 Applicability. Use and handling of containers and systems shall comply with this section.

❖ These provisions are not quantity-specific but apply generally any time cryogens are being used or handled. This section addresses system components and the integrity of the components. The container itself is addressed primarily in Section 3204.

3205.1.1 Cryogenic fluid systems. Cryogenic fluid systems shall be suitable for the use intended and designed by persons competent in such design. Equipment, machinery and processes shall be listed or approved.

❖ This section contains general requirements for the design of systems, including the requirement for the competency of the designer. Because the use of cryogens is varied, a single listed system is not available. There is, however, a requirement that all components of the system either be listed or specifically approved.

The nature of the listing should be understood because it may be unrelated to the performance of the system. Use of listed parts and components alone does not guarantee that a system as a whole operates as intended.

3205.1.2 Piping systems. Piping, tubing, valves and joints and fittings conveying cryogenic fluids shall be installed in accordance with the material-specific provisions of Sections 3201.1 and 3205.1.2.1 through 3205.1.2.6.

❖ See the commentary for Section 3201.1 and for the subsections that follow.

3205.1.2.1 Design and construction. Piping systems shall be suitable for the use intended through the full range of pressure and temperature to which they will be subjected. Piping systems shall be designed and constructed to provide adequate allowance for expansion, contraction, vibration, settlement and fire exposure.

❖ This section sets out the basic elements that need to be addressed in an acceptable design, including temperature and pressures and the type of events or climatic exposures that the system must withstand. A competent designer must show that all of these elements have been satisfied.

3205.1.2.2 Joints. Joints on container piping and tubing shall be threaded, welded, silver brazed or flanged.

❖ This section lists the acceptable types of joints allowed with cryogenic fluids. Joints represent potential weak points in a system if they are not properly designed and constructed. Friction joints would not be allowed. A high level of reliability is needed from joints to decrease the likelihood of joint failure.

3205.1.2.3 Valves and accessory equipment. Valves and accessory equipment shall be suitable for the intended use at the temperatures of the application and shall be designed and constructed to withstand the maximum pressure at the minimum temperature to which they will be subjected.

❖ This section sets the basic criterion for approval, which requires valves and any accessory equipment to be able to withstand the minimum temperature at the highest operating pressure. This will ensure that the valves can withstand the most critical forces to which they may be subject during normal operation.

3205.1.2.3.1 Shutoff valves on containers. Shutoff valves shall be provided on all container connections except for pressure relief devices. Shutoff valves shall be provided with access thereto and located as close as practical to the container.

❖ This section requires shutoff valves on all container connections. This allows isolation of the container to prevent a large release. It is critical that shutoff valves not be placed on pressure relief devices.

3205.1.2.3.2 Shutoff valves on piping. Shutoff valves shall be installed in piping containing cryogenic fluids where needed to limit the volume of liquid discharged in the event of piping or equipment failure. Pressure relief valves shall be installed where liquid is capable of being trapped between shutoff-valves in the piping system (see Section 3203.2).

❖ Further shutoff valves are required at strategic locations within the piping systems where the likelihood of a large release may be possible. The containers themselves must have shutoff valves in accordance with Section 3205.1.2.3.1. This section is quite general. Because piping systems vary so much, there is no single solution based on either use or process layouts. There is an additional requirement for pressure relief devices to be installed where shutoff of the flow could potentially cause a buildup of pressure between the source and the shutoff valve.

3205.1.2.4 Physical protection and support. Above-ground piping systems shall be supported and protected from physical damage. Piping passing through walls shall be protected from mechanical damage.

❖ This is a general section that requires a level of protection from physical and other damage when piping is either exposed or passes through elements such as walls and floors that are subject to movement. These requirements, along with the others, are intended to increase the reliability of the piping and associated processes.

3205.1.2.5 Corrosion protection. Above-ground piping that is subject to corrosion because of exposure to corrosive atmospheres, shall be constructed of materials to resist the corrosive environment or otherwise protected against corrosion. Below-ground piping shall be protected against corrosion.

❖ Another possible mode of failure is the weakening of the piping system caused by corrosion. The extent of corrosion protection required will depend on the piping material used, the particular climate or, in the case of the underground installations, the soil condition and content of the soil. In some areas, this may be a significant problem whereas in others it may not. Corrosion-resistant construction materials are the preferred means of compliance with this section.

3205.1.2.6 Testing. Piping systems shall be tested and proven free of leaks after installation as required by the standards to which they were designed and constructed. Test pressures shall not be less than 150 percent of the maximum allowable working pressure when hydraulic testing is conducted or 110 percent when testing is conducted pneumatically.

❖ To increase the reliability of the system to work as designed without the occurrence of leaks or other more substantial failures, proof testing of the piping is required. This section sets the testing criteria for either a hydraulic test or a pneumatic test.

3205.2 Indoor use. Indoor use of cryogenic fluids shall comply with the material-specific provisions of Section 3201.1.

❖ The code does not provide detail on the specifics of indoor use. That is contained in the standards listed in Section 3201.1.

3205.3 Outdoor use. Outdoor use of cryogenic fluids shall comply with the material specific provisions of Sections 3201.1, 3205.3.1 and 3205.3.2.

❖ This section refers back to the standards in Section 3201.1 for the specific details on the use of cryogens, but does emphasize the separation requirements and placement of necessary shutoff valves.

3205.3.1 Separation. Distances from property lines, buildings and exposure hazards shall comply with Section 3203.6 and the material specific provisions of Section 3201.1.

❖ Because use is more hazardous than storage, the basic separation requirements found in Tables 3203.6.1 and 3203.6.2 would apply in addition to any of the restrictions found in the referenced standards in Section 3201.1.

3205.3.2 Emergency shutoff valves. Readily available shutoff valves shall be provided to shut off the cryogenic fluid supply in case of emergency. A shutoff valve shall be located at the source of supply and at the point where the system enters the building.

❖ This section establishes the requirement for shutoff valves to be readily accessible during an emergency, either by on-site personnel or by the local emergency response personnel. Essentially, the ability must exist to both shut down the supply of cryogens and also to shut off the supply at the location where it enters the building.

Pressure relief devices may be necessary where liquid could be trapped between the source and the emergency shutoff valve at the entrance to the building (see also commentary, Section 3205.1.2.3.2).

3205.4 Filling and dispensing. Filling and dispensing of cryogenic fluids shall comply with Sections 3205.4.1 through 3205.4.3.

❖ Filling and dispensing is an activity that could result in fluid release because of the connections and disconnections involved with such processes. This section addresses the location and construction of dispensing areas, the approval of loading and unloading activities and controls on the amount of fluids dispensed into stationary tanks. All of these requirements focus on preventing an unwanted fluid release and possible physical damage or personal injury.

3205.4.1 Dispensing areas. Dispensing of cryogenic fluids with physical or health hazards shall be conducted in approved locations. Dispensing indoors shall be conducted in areas constructed in accordance with the *International Building Code.*

❖ This section focuses on dispensing of fluids with a physical or health hazard rating, such as flammability or toxicity. This section also refers back to the IBC for construction requirements, primarily for the determination of whether a Group H occupancy would be required. This specifically relates to whether the maximum allowable quantities of hazardous materials have been exceeded.

3205.4.1.1 Ventilation. Indoor areas where cryogenic fluids are dispensed shall be ventilated in accordance with the requirements of the *International Mechanical Code* in a manner that captures any vapor at the point of generation.

Exception: Cryogenic fluids that can be demonstrated not to create harmful vapors.

❖ In addition to the basic ventilation requirements for the indoor storage of cryogens, this section requires that vapors be captured at the source; therefore, special ventilation would be required at the point of fill for the dispensing operation. These ventilation requirements appear to be for normal operation rather than emergency operation.

There is an exception if the amount and type of gas would not be sufficient enough to create a hazardous situation.

3205.4.1.2 Piping systems. Piping systems utilized for filling or dispensing of cryogenic fluids shall be designed and constructed in accordance with Section 3205.1.2.

❖ This is simply a reference back to Section 3205.1.2 for the general piping requirements to ensure correct design and construction (see commentary, Section 3205.1.2).

3205.4.2 Vehicle loading and unloading areas. Loading or unloading areas shall be conducted in an approved manner in accordance with the standards referenced in Section 3201.1.

❖ Loading, unloading, dispensing and filling operations, in general, have a high potential for creating a hazardous situation if not properly conducted. This section ensures that the areas are constructed and operated as required by approved standards.

3205.4.3 Limit controls. Limit controls shall be provided to prevent overfilling of stationary containers during filling operations.

❖ This is a mechanism to prevent a release of fluids caused by an overflow.

3205.5 Handling. Handling of cryogenic containers shall comply with this section.

❖ This section is focused on portable containers and the prevention of a fluid release.

3205.5.1 Carts and trucks. Cryogenic containers shall be moved using an approved method. Where cryogenic containers are moved by hand cart, hand truck or other mobile device, such carts, trucks or devices shall be designed for the secure movement of the container.

Carts and trucks used to transport cryogenic containers shall be designed to provide a stable base for the commodities to be transported and shall have a means of restraining containers to prevent accidental dislodgement.

❖ This section does not specify a particular handling method, but gives the fire code official the authority to determine whether the method chosen by the facility is acceptable. If carts, hand trucks or other similar methods are chosen, they must be able to transport the containers safely and securely. This includes restraining containers while they are being moved.

3205.5.2 Closed containers. Pressurized containers shall be transported in a closed condition. Containers designed for use at atmospheric conditions shall be transported with appropriate loose fitting covers in place to prevent spillage.

❖ This section recognizes that containers designed to hold pressure can be moved safely if they are properly closed. Containers designed for use at normal atmospheric pressure need only have an appropriate cover to ensure that fluid does not escape.

Bibliography

The following resource materials are referenced in this chapter or are relevant to the subject matter addressed in this chapter.

ANSI A13.1-96, *Scheme for the Identification of Piping Systems.* New York: American National Standards Institute, 1996.

ASME-98, *Boiler and Pressure Vessel Code, Section VIII, Divisions 1, 2 & 3*. New York: The American Society of Mechanical Engineers, 1998.

CGA S1.2-95, *Pressure Relief Device Standards—Part 2—Cargo and Portable Tanks for Compressed Gases*. Arlington, VA: Compressed Gas Association, 1995.

CGA S1.3-94, *Pressure Relief Device Standards—Part 3—Compressed Gas Stationary Storage Containers*. Arlington, VA: Compressed Gas Association, 1994.

DOTn 49 CFR 100-178 and 179-199. *Specification for Transportation of Explosive and Other Dangerous Articles, Shipping Containers.* Washington, DC: U. S. Department of Transportation, 1994.

Fire Protection Handbook. 18th ed. Quincy, MA: National Fire Protection Association, 1997.

IBC-2003, *International Building Code*. Falls Church, VA: International Code Council, 2003.

ICC EC-2003, *ICC Electrical Code*. Falls Church, VA: International Code Council, 2003.

IMC-2003, *International Mechanical Code.* Falls Church, VA: International Code Council, 2003.

NFPA 50-96, *Bulk Oxygen Systems at Consumer Sites.* Quincy, MA: National Fire Protection Association, 1996.

NFPA 50A-99, *Gaseous Hydrogen Systems at Consumer Sites.* Quincy, MA: National Fire Protection Association, 1999.

NFPA 50B-99, *Liquefied Hydrogen Systems at Consumer Sites.* Quincy, MA: National Fire Protection Association, 1999.

NFPA 59A-96, *Production, Storage and Handling of Liquefied Natural Gas (LNG).* Quincy, MA: National Fire Protection Association, 1996.

NFPA 704-96, *Identification of the Fire Hazards of Materials.* Quincy, MA: National Fire Protection Association, 1996.

NFPA Inspection Manual. 7th ed. Quincy, MA: National Fire Protection Association, 1994.

Saacke, F. C., G. R. Spies and W. Barlen. *Industrial Fire Hazards Handbook.* 3rd ed. Quincy, MA: National Fire Protection Association, 1990.

Chapter 33: Explosives and Fireworks

General Comments

The safe handling of explosives in transportation, storage and use requires preventing ignitions and reducing the hazard to exposures. These exposures include people who may be harmed and property that may be damaged by an accidental detonation and other circumstances involving accidental or malicious detonation of the explosive materials. All requirements applicable to explosives are based on these principles. To apply the requirements of this chapter, the fire code official must recognize conditions that are liable to cause ignition or create exposures.

To varying degrees, all explosives, ammunition and blasting agents are susceptible to ignition from heat, sparks and, in certain cases, shock or pressure. Explosives are considered the most susceptible to ignition from these stimuli and may detonate when exposed to any of them, even under controlled conditions. Blasting agents are generally more stable and less susceptible to detonation, but such agents may become sensitized or unstable when exposed to heat or contaminated by certain organic materials. Sources of ignition include: sparks from tools, friction, static electricity, electrical devices, hot surfaces, open flames and open-flame devices, smoking materials, chemical reactions, electric currents, pressure and shock from explosions or impact.

Exposure protection involves protecting people, buildings and public rights-of-way from detonations, and protecting explosives, ammunition and blasting agents from fires or explosions occurring outside the magazine or blasting area. Such protection also includes security—protecting the magazine or blasting area from entry by unauthorized personnel. Two techniques—separation distance requirements and security precautions—are used to reduce exposure hazards.

An excellent discussion of the origin and rationale behind the separation distances specified in Table 3304.5.2(2) is presented in Appendix B of NFPA 495. Required magazine construction features include: weather resistance, bullet resistance, spark resistance, fire resistance, theft resistance and ventilation to prevent excessive heating or dampening of explosives.

Regulation of explosives is a complex enforcement issue. Local fire code officials must understand that many state and federal agencies have concurrent jurisdiction over the manufacture, transportation, storage, sale, handling and use of explosive materials. The U.S. Department of Treasury (DOT), which oversees the Bureau of Alcohol, Tobacco and Firearms (ATF) division, is the federal agency responsible (see 18 USC, Chapter 40) for regulating the manufacture, sale, distribution and storage of explosives. In addition to enforcing explosives laws and regulations, ATF investigates incidents of theft or misuse of explosives and has extensive resources for investigating bombings and incendiary acts. Recently, ATF's inspection of explosives manufacturing plants has been credited with improving industry safety.

Many states, especially where mining is an important industry, also regulate explosives. States often pick up where federal authority ends, especially in the areas of blasting and siting of explosive material facilities.

Fireworks regulation is one of the most controversial and hotly debated topics in American fire protection. The National Fire Protection Association (NFPA) and U.S. Consumer Product Safety Commission (CPSC) have called for stricter rules to govern the sale and use of common fireworks and trick and novelty items. To understand the tenor of the debate, it is helpful to appreciate the scope of the problem and look back at the origins of this chapter and other standards that were drafted in response to fireworks misuse and injuries.

According to statistics quoted by NFPA and gathered by the CPSC, more than 10,000 citizens are injured each year by the misuse of common and illegally manufactured fireworks. These injuries result in millions of dollars in medical and legal expenses, and untold suffering. The injuries include burns and the loss of fingers, limbs, vision or hearing; most injuries leave permanent scarring. The overwhelming majority of persons injured are younger than 20 years old. Public displays of fireworks have also resulted in several serious accidents. Many of these accidents have involved local fire departments and untrained operators performing public fireworks displays.

Parts of Chapter 33 had their origin in former NFPA 1121L, *Model Fireworks Law*, which was first published in 1938. According to NFPA, this is the most widely adopted fireworks regulation in the United States. Many states have enacted this model as part of state law and prohibit all fireworks except toy paper or plastic caps and authorized public displays. Other states have adopted modified versions that prohibit all but trick and novelty items, toy paper or plastic caps and permitted public displays. The NFPA Standards Council withdrew NFPA 1121L as an NFPA standard and in 1988 transferred control of the document to the International Fire Marshals Association (IFMA), a membership section of NFPA from whom copies of the Model Fireworks Law may be obtained.

This chapter requires the display of fireworks to comply with NFPA 1123. The standard includes criteria for the firing and on-site storage of fireworks, display site location, fallout area and operator qualifications. The chapter also references NFPA 1124 for the regulation of fireworks manufacture and storage of fireworks at manufacturing plants. Federal regulations also provide useful guidance.

Purpose

Chapter 33 prescribes minimum requirements for the safe manufacture, storage, handling and use of explosives, ammunition and blasting agents for commercial and industrial occupancies. These provisions are intended to protect the general public, emergency responders and individuals who handle explosives.

This chapter regulates the manufacturing, retail sale, display and wholesale distribution of fireworks, establishing the requirements for obtaining approval to manufacture, store, sell, discharge or conduct a public display, and references national standards for regulations governing manufacture, storage and public displays.

SECTION 3301
GENERAL

3301.1 Scope. The provisions of this chapter shall govern the possession, manufacture, storage, handling, sale and use of explosives, explosive materials, fireworks and small arms ammunition.

Exceptions:

1. The Armed Forces of the United States, Coast Guard or National Guard.
2. Explosives in forms prescribed by the official United States Pharmacopoeia.
3. The possession, storage and use of small arms ammunition when packaged in accordance with DOTn packaging requirements.
4. The possession, storage, and use of not more than 1 pound (0.454 kg) of commercially manufactured sporting black powder, 20 pounds (9 kg) of smokeless powder and 10,000 small arms primers for hand loading of small arms ammunition for personal consumption.
5. The use of explosive materials by federal, state and local regulatory, law enforcement and fire agencies acting in their official capacities.
6. Special industrial explosive devices which in the aggregate contain less than 50 pounds (23 kg) of explosive materials.
7. The possession, storage and use of blank industrial-power load cartridges when packaged in accordance with DOTn packaging regulations.
8. Transportation in accordance with DOTn 49 CFR Parts 100-178.
9. Items preempted by federal regulations.

❖ This chapter contains specific requirements for the manufacture, transportation, handling, storage and use of explosives, ammunition and blasting agents by nonmilitary and nongovernmental agencies and individuals.

Exceptions in this section detail situations in which compliance with the requirements of this chapter is not required. Generally, these situations are governed by more stringent federal requirements or another chapter of the code or they may represent a low hazard.

Governmental agencies responsible for national defense or public safety, including local law enforcement and fire suppression agencies, are exempt from these requirements. State and local agencies should adhere to these requirements for liability reasons.

The U.S. military promulgates its own regulations governing the manufacture and storage of explosives, ammunition and blasting agents. These regulations are similar to those that apply to civilian explosives with a few exceptions for identifying and marking them in transportation and storage.

Nitroglycerin tablets and transdermal patches used for the treatment of angina pectoris and other pharmaceuticals containing explosive materials are not dangerous in the form dispensed and thus are exempt from these requirements.

Pyrotechnic devices used in transportation are exempt because they pose a minimal mass fire hazard.

3301.1.1 Explosive material standard. In addition to the requirements of this chapter, NFPA 495 shall govern the manufacture, transportation, storage, sale, handling and use of explosive materials.

❖ The requirements in NFPA 495 apply to situations not specifically addressed by this chapter.

3301.1.2 Explosive material terminals. In addition to the requirements of this chapter, the operation of explosive material terminals shall conform to the provisions of NFPA 498.

❖ The requirements of NFPA 498 apply to situations not specifically addressed by this chapter.

3301.1.3 Fireworks. The possession, manufacture, storage, sale, handling and use of fireworks are prohibited.

Exceptions:

1. Storage and handling of fireworks as permitted in Section 3304.
2. Manufacture, assembly and testing of fireworks as permitted in Section 3305.
3. The use of fireworks for display as permitted in Section 3308.
4. The possession, storage, sale, handling and use of specific types of Division 1.4G fireworks where allowed by applicable local or state laws, ordinances and regu-

lations provided such fireworks comply with CPSC 16 CFR, Parts 1500 and 1507, and DOTn 49 CFR, Parts 100-178, for consumer fireworks.

❖ The possession, manufacture, storage, sale and unauthorized use of fireworks is prohibited by this section. The prohibition of retail sales allows communities to have direct control over the hazards associated with small amounts of storage typical of retail sales.

Exception 1 allows storage and handling of fireworks within the limitations of Section 3304. Exception 2 allows the manufacture, assembly and testing of fireworks within the limitations of Section 3305. Exception 3 allows those displays to be specifically approved by the fire code official in accordance with Section 3308. Exception 4 recognizes that in some instances, the possession, storage, sale, handling and use of certain types of fireworks may be allowed by provisions of the preemptive laws of superior jurisdictions, most typically the state. Even if such laws preempt the local jurisdiction, the exception stipulates that the fireworks allowed by such laws must still meet the minimum requirements of the referenced standards.

3301.1.4 Rocketry. The storage, handling and use of model and high-power rockets shall comply with the requirements of NFPA 1122, NFPA 1125, and NFPA 1127.

❖ NFPA 1122 contains instructional guidelines and specific standards for the design, construction, limitation of charge and power, and reliability of rocket motors manufactured for sale to the general public; for the design and construction of rockets propelled by these motors and for tests, launchings and other operations involving such rockets in order to minimize hazards. NFPA 1125 applies to the manufacture of model rocket motors designed, sold and used for the purpose of propelling recoverable aero models. NFPA 1127 contains instructional guidelines and specific standards for the design, construction, limitation of charge and power and reliability of high-power rocket motors manufactured for sale to users; for the qualification and certification of users; for the design and construction of high-power rockets propelled by these motors and for tests, launchings and other operations involving rockets so that hazards are minimized.

3301.1.5 Ammonium nitrate. The storage and handling of ammonium nitrate shall comply with the requirements of NFPA 490 and Chapter 40.

Exception: Storage of ammonium nitrate in magazines with blasting agents shall comply with the requirements of NFPA 495.

❖ NFPA 490 addresses storage and Chapter 40 addresses oxidizers and separation distances for ammonium nitrate. Ammonium nitrate can be sensitized by both heat and contaminants, causing it to become a greater explosive danger (commercial blasting agents are made from a mixture of diesel oil and ammonium nitrate).

It cannot be determined when or if contamination will occur. No matter how many times it has failed to explode in fires, the important point is that ammonium nitrate might explode and has the potential to explode in any fire.

3301.2 Permit required. Permits shall be required as set forth in Section 105.6 and regulated in accordance with this section.

❖ The process of issuing permits gives the fire code official an opportunity to carefully evaluate and regulate hazardous operations. Permit applicants should be required to demonstrate that their operations comply with the intent of the code before the permit is issued. See the commentary to Section 105.6 for a general discussion of operations requiring an operational permit, Section 105.6.15 for a discussion of specific quantity-based operational permits for the materials regulated in this chapter and Section 105.7 for a general discussion of activities requiring a construction permit. The permit process also notifies the fire department of the need for prefire planning for the hazardous property.

3301.2.1 Residential uses. No person shall keep or store, nor shall any permit be issued to keep or store, any explosives at any place of habitation, or within 100 feet (30 480 mm) thereof.

Exception: Storage of smokeless propellant, black powder, and small arms primers for personal use and not for resale in accordance with Section 3306.

❖ Small amounts of materials for making small arms ammunition for personal use are exempt based on the limited potential hazard of the small quantities.

3301.2.2 Sale and retail display. No person shall construct a retail display nor offer for sale explosives, explosive materials, or fireworks upon highways, sidewalks, public property, or in assembly or educational occupancies.

❖ Public display and sale of explosives, including Class B and C fireworks, are prohibited in public rights-of-way and in assembly and educational buildings. This reduces the likelihood of theft and personal injury if a fire or explosion occurs.

3301.2.3 Permit restrictions. The fire code official is authorized to limit the quantity of explosives, explosive materials, or fireworks permitted at a given location. No person, possessing a permit for storage of explosives at any place, shall keep or store an amount greater than authorized in such permit. Only the kind of explosive specified in such a permit shall be kept or stored.

❖ The fire code official may set limits on the quantity of explosive materials or blasting agents stored at any site as a means of maintaining control over the degree of hazard posed by explosive storage. Limits should be based on the severity of the exposure if an explosion or fire occurs in the magazine. This section is not intended to give the fire code official authority to prohibit the storage of explosives or blasting agents on any site.

3301.2.4 Financial responsibility. Before a permit is issued, as required by Section 3301.2, the applicant shall file with the jurisdiction a corporate surety bond in the principal sum of $100,000 or a public liability insurance policy for the same amount, for the purpose of the payment of all damages to persons or property which arise from, or are caused by, the conduct of any act authorized by the permit upon which any judicial judgment results. The fire code official is authorized to specify a greater or lesser amount when, in his or her opinion, conditions at the location of use indicate a greater or lesser amount is required. Government entities shall be exempt from this bond requirement.

❖ The fire code official should understand that some insurance coverages are invalidated by violations of federal, state and local regulations. Insurance coverages obtained by an owner or operator provide no protection from liability for the fire code official who is responsible for issuing approvals or conducting inspections. Moreover, third-party insurance may conflict with other coverages obtained by the jurisdiction, as well as governmental immunity or tort claims protections under state or local statutes.

3301.2.4.1 Blasting. Before approval to do blasting is issued, the applicant for approval shall file a bond or submit a certificate of insurance in such form, amount and coverage as determined by the legal department of the jurisdiction to be adequate in each case to indemnify the jurisdiction against any and all damages arising from permitted blasting.

❖ Insurance coverage is required in an amount specified by a jurisdiction's legal department. This coverage is intended to indemnify the operator or individual who is responsible for blasting operations involving explosives or blasting agents from damages arising from accidents involving these operations.

3301.2.4.2 Fireworks display. The permit holder shall furnish a bond or certificate of insurance in an amount deemed adequate by the fire code official for the payment of all potential damages to a person or persons or to property by reason of the permitted display, and arising from any acts of the permit holder, the agent, employees or subcontractors.

❖ The bonding requirement is intended to indemnify the display operator and, if required by the fire code official, the jurisdiction in the event of an accident. Jurisdictions desiring coverage under the display operator's policy should require that the jurisdiction be named on the policy as an additional insured or a named insured. (Before making this requirement, check with legal counsel to determine the fire code official's and the jurisdiction's liability. Many tort claims acts exempt the government from certain claims, while others limit the amount of liability. The standard of care that must be exercised by the fire code official when reviewing conditions for a permit varies widely.)

The fire code official must exercise great care when establishing bonding requirements. Insurance companies underwriting fireworks displays often issue a large number of policies at the same time of year. Often the face value of these policies far exceeds the total of the companies' assets and reserves. Insurance evaluation services should be consulted to evaluate the companies' ratings. (Like credit bureaus, these evaluation services rate the companies' financial health on a letter scale—AAA being the highest rating, B and C the lowest.) Most policies include some coverage restrictions. Losses within 150 feet (45 720 mm) of the discharge site are often excluded from coverage, and claims within 600 feet (182 880 mm) of the discharge site are frequently severely limited. When included in the policy, these separation distances usually conform to the separation requirements of NFPA 1123. This is done purposely to encourage display operators to follow nationally recognized standards.

The fire code official may require the approval holder to submit an original copy of the certificate of insurance verifying indemnification of the display. When an original copy cannot be obtained, a facsimile of the original from the issuing broker is a good alternative. Most insurance companies authorize only highly trusted, specially trained, bonded employees to issue these certificates.

The fire code official should examine the certificate carefully and never accept a photocopy of this document unless it can be thoroughly authenticated. Unscrupulous operators have been known to alter old certificates or produce and submit counterfeit certificates. The fire code official may contact the broker, underwriter or other insurance company representative to verify coverage, although these agents may be reluctant to confirm coverage if the jurisdiction is not named as an additional insured on the policy.

3301.3 Prohibited explosives. Permits shall not be issued or renewed for possession, manufacture, storage, handling, sale or use of the following materials and such materials currently in storage or use shall be disposed of in an approved manner.

1. Liquid nitroglycerin.
2. Dynamite containing more than 60-percent liquid explosive ingredient.
3. Dynamite having an unsatisfactory absorbent or one that permits leakage of a liquid explosive ingredient under any conditions liable to exist during storage.
4. Nitrocellulose in a dry and uncompressed condition in a quantity greater than 10 pounds (4.54 kg) of net weight in one package.
5. Fulminate of mercury in a dry condition and fulminate of all other metals in any condition except as a component of manufactured articles not hereinafter forbidden.
6. Explosive compositions that ignite spontaneously or undergo marked decomposition, rendering the products of their use more hazardous, when subjected for 48 consecutive hours or less to a temperature of 167°F (75°C).
7. New explosive materials until approved by DOTn, except that permits are allowed to be issued to educational, governmental or industrial laboratories for instructional or research purposes.
8. Explosive materials condemned by DOTn.

9. Explosive materials containing an ammonium salt and a chlorate.

10. Explosives not packed or marked as required by DOTn 49 CFR, Parts 100-178.

Exception: Gelatin dynamite.

❖ The fire code official is not authorized to issue approval for manufacture, transportation, storage, sale or use because of extreme or unusual hazards presented by the listed materials.

3301.4 Qualifications. Persons in charge of magazines, blasting, fireworks display, or pyrotechnic special effect operations shall not be under the influence of alcohol or drugs which impair sensory or motor skills, shall be at least 21 years of age, and shall demonstrate knowledge of all safety precautions related to the storage, handling or use of explosives, explosive materials or fireworks.

❖ The only discharge of fireworks permissible under the code is a public display conducted by competent pyrotechnicians in accordance with the requirements of NFPA 1123 and authorized by the fire code official.

The competence of the display operator is first among the important safeguards that must be observed for a safe and enjoyable public display. Chapter 6 of NFPA 1123 details the qualifications of competent fireworks display operators. Many jurisdictions also require display operators to possess a license or certificate of fitness. To obtain such a certificate, the operator must be bonded or indemnified, pass a written examination and serve an apprenticeship under another licensed or certified pyrotechnician.

The best sites are free of overhead obstructions and are well isolated, with clear viewing paths and landing areas. Fallout areas should be large, open areas, clear of spectators, vehicles and combustible materials. Generally, the discharge site must have a minimum radius of 70 feet (21 336 mm) for each inch of aerial shell diameter. Table 31.3 of NFPA 1123 specifies the separation distances.

Fireworks discharge sites must be separated from institutional and high-hazard occupancies by at least twice the distance specified in the table [140 feet (42 672 mm) per inch of shell diameter]. When mortars are positioned vertically (zero degrees), they must be located at the center of the display area. When mortars or shells stored at the discharge site are angled, they must be aimed away from principal spectator viewing and shell storage areas. When angled, mortars may be placed up to one-third the distance from the center of the display area to the principal spectator viewing area. Aerial shell trajectories must not come within 25 feet (7,620 mm) of overhead obstructions, such as power lines and trees. Tents and canvas structures must be at least 100 feet (30 480 mm) from the discharge site.

High winds, precipitation or extremely hot, dry conditions should be avoided. Moisture-damaged shells must not be fired. If, in the opinion of the fire code official or the display operator, weather conditions present a danger, the display must be postponed or canceled.

3301.5 Supervision. The fire code official is authorized to require operations permitted under the provisions of Section 3301.2 to be supervised at any time by the fire code official in order to determine compliance with all safety and fire regulations.

❖ Only supervised public displays of fireworks, approved in advance by the fire code official, and wholesale sales in accordance with Section 3301.1.3, Exception 4 are permitted under the requirements of this chapter. Written application for approval of public displays must be made at least 15 days prior to the display. Before approval can be issued, the fire code official must review the qualifications and determine the competence of the display operator, verify the operator's proof of insurance or indemnification, inspect the proposed discharge site and viewing area and review the operator's fire protection and crowd control plans.

3301.6 Notification. Whenever a new explosive material storage or manufacturing site is established, including a temporary job site, the local law enforcement agency, fire department, and local emergency planning committee shall be notified 48 hours in advance, not including Saturdays, Sundays and holidays, of the type, quantity and location of explosive materials at the site.

❖ Local law enforcement, fire department and emergency planning officials must be notified of all new explosive materials storage and handling sites at least 48 hours, excluding weekends and holidays, prior to operations beginning. This period is intended to give officials time to prepare response plans for the new site, communicate the hazards incident to the operations to first responders and train site personnel and emergency responders on the management of emergencies at the new site.

In practice, this notice should be submitted well in advance of the required 48 hours. Approval should not be granted or should be suspended if local law enforcement, fire department or emergency planning officials believe that adequate and proper preparations cannot be made to safeguard the public and emergency responders prior to the planned start of such operations.

3301.7 Seizure. The fire code official is authorized to remove or cause to be removed or disposed of in an approved manner, at the expense of the owner, explosives, explosive materials or fireworks offered or exposed for sale, stored, possessed or used in violation of this chapter.

❖ The seizure and disposal of controlled articles, in this case fireworks, at the owner's expense is usually considered a lawful taking of private property under the U.S. Constitution. However, before taking such measures, fire code officials should consult with legal counsel regarding due process requirements.

Seizure requires probable cause that the article or device being seized is unlawful itself, is being used in an unlawful manner, has been used in conjunction with an unlawful activity or poses an imminent danger to life and property. Recovery of expenses may require the filing of civil or criminal charges against the owner or his or her agents.

Before taking possession of fireworks, it is also prudent for the jurisdiction to verify that it has adequate facilities for safely handling, transporting, storing and disposing of the articles. The jurisdiction may require assistance to dispose of large quantities of special fireworks. Transportation of special fireworks is governed by DOTn 49 CFR. A commercial driver's license with a hazardous materials endorsement is required, as well as a vehicle with special equipment and inspections.

3301.8 Establishment of quantity of explosives and distances. The quantity of explosives and distances shall be in accordance with Sections 3301.8.1 and 3301.8.1.1.

❖ This section provides a methodology for establishing the explosive quantities and distances based on the class of explosives.

3301.8.1 Quantity of explosives. The quantity-distance tables in Sections 3304.5 and 3305.3 shall be used to provide appropriate distances from potential explosion sites. The classification of the explosives and the weight of the explosives are primary characteristics governing the use of these tables. The net explosive weight shall be determined in accordance with Sections 3301.8.1.1 through 3301.8.1.4.

❖ The hazards presented by explosive materials ranges across a spectrum identified by hazard Divisions 1.1 through 1.6. When all of the explosive materials are in a single hazard division, the quantity is easily determined; however, when the materials are mixed among multiple hazard divisions, a method is needed to determine the level of hazard in order to apply the quantity-distance tables contained in Sections 3304 and 3305. The concepts in the following sections have been drawn from applicable federal regulations governing the use of these materials (i.e., the *DOD Contractor's Safety Manual For Ammunition and Explosives*).

3301.8.1.1 Mass-detonating explosives. The total net explosive weight of Division 1.1, 1.2 or 1.5 explosives shall be used. See Table 3304.5.2 (2) or Table 3305.3 as appropriate.

Exception: When the TNT equivalence of the explosive material has been determined, the equivalence is allowed to be used to establish the net explosive weight.

❖ Mass-detonating explosives are typically classified as Group H-1 by the *International Building Code*® (IBC®), and present a detonation hazard and a greater threat to adjacent objects and structures. The code, therefore, contains provisions in Table 3305.3 to deal with the separation distances for mass explosion hazards. This section establishes the weights to be used when applying that table as the total net weight of all mass-detonating explosive hazard divisions.

3301.8.1.2 Non-mass-detonating explosives (excluding Division 1.4). Non-mass-detonating explosives shall be as follows:

1. Division 1.3 propellants. The total weight of the propellants alone shall be the net explosive weight. The net weight of propellant shall be used. See Table 3304.5.2(3).
2. Combinations of bulk metal powder and pyrotechnic compositions. The sum of the net weights of metal powders and pyrotechnic compositions in the containers shall be the net explosive weight. See Table 3304.5.2(3).

❖ This section establishes the explosive material weight for entering Table 3304.5.2(3) for explosives classified as nonmass detonating.

3301.8.1.3 Combinations of mass-detonating and non-mass-detonating explosives (excluding Division 1.4). Combination of mass-detonating and non-mass-detonating explosives shall be as follows:

1. When Division 1.1 and 1.2 explosives are located in the same site, determine the distance for the total quantity considered first as 1.1 and then as 1.2. The required distance is the greater of the two. When the Division 1.1 requirements are controlling and the TNT equivalence of the 1.2 is known, the TNT equivalent weight of the 1.2 items shall be allowed to be added to the total explosive weight of Division 1.1 items to determine the net explosive weight for Division 1.1 distance determination. See Table 3304.5.2(3) or Table 3305.3 as appropriate.
2. When Division 1.1 and 1.3 explosives are located in the same site, determine the distances for the total quantity considered first as 1.1 and then as 1.3. The required distance is the greater of the two. When the Division 1.1 requirements are controlling and the TNT equivalence of the 1.3 is known, the TNT equivalent weight of the 1.3 items shall be allowed to be added to the total explosive weight of Division 1.1 items to determine the net explosive weight for Division 1.1 distance determination. See Table 3304.5.2(2), 3304.5.2 (3) or 3305.3, as appropriate.
3. When Division 1.1, 1.2 and 1.3 explosives are located in the same site, determine the distances for the total quantity considered first as 1.1, next as 1.2 and finally as 1.3. The required distance is the greatest of the three. As permitted by paragraphs 1 and 2 above, TNT equivalent weights for 1.2 and 1.3 items are allowed to be used to determine the net weight of explosives for Division 1.1 distance determination. Table 3304.5.2 (2) or 3305.3 shall be used when TNT equivalency is used to establish the net explosive weight.
4. For composite pyrotechnic items Division 1.1 and Division 1.3, the sum of the net weights of the pyrotechnic composition and the explosives involved shall be used. See Tables 3304.5.2 (2) and 3304.5.2 (3).

❖ This section establishes the explosive material weights to be used when applying the various distance tables in Sections 3304 and 3305 for mixed storage combinations of explosives classified as both mass detonating and nonmass detonating. The most hazardous explosives will likely drive the distance requirements.

3301.8.1.4 Moderate fire — no blast hazards. Division 1.4 explosives. The total weight of the explosive material alone is the net weight. The net weight of the explosive material shall be used. See Table 3304.5.2 (4).

❖ This section establishes the explosive material weight to be used when applying Table 3304.5.2(4) for explosives classified as a moderate fire hazard without a blast hazard.

SECTION 3302 DEFINITIONS

3302.1 Definitions. The following words and terms shall, for the purposes of this chapter and as used elsewhere in this code, have the meanings shown herein.

❖ Definitions of terms can help in the understanding and application of the code requirements. The purpose for including those definitions that are associated with the subject matter of this chapter is to provide more convenient access to them without having to refer back to Chapter 2. It is important to emphasize that these terms are not exclusively related to this chapter but are applicable everywhere the term is used in the code. For convenience, these terms are also listed in Chapter 2 with a cross reference to this section. The use and application of all defined terms, including those defined in this section, are set forth in Section 201.

These definitions are extracted from federal regulations and the Institute of Makers of Explosives (IME) Safety Library Publication No. 12, *Glossary of Commercial Explosives Industry Terms*. These definitions are intended to complement those found in DOTy 27 CFR; 55.11 and NFPA 495.

AMMONIUM NITRATE. A chemical compound represented by the formula NH_4NO_3.

❖ Ammonium nitrate is a fairly simple compound (NH_4NO_3) but one that has an extremely complex set of reactions in a fire. The history of ammonium nitrate has been marked by major fires and explosions. Certain characteristics have caused or contributed to these fires and explosions. The following properties of ammonium nitrate are similar to those of other nitrates: (1) it is an oxidizing agent—it contains available oxygen that can make ignition easier and cause a fire to burn with surprising intensity; (2) it is hygroscopic—absorbing moisture from the air or from substances it touches and (3) it is deliquescent—as it absorbs moisture, a portion will liquify and nearby combustibles can become impregnated with an oxidizing salt.

BARRICADE. A structure that consists of a combination of walls, floor and roof, which is designed to withstand the rapid release of energy in an explosion and which is fully confined, partially vented or fully vented; or other effective method of shielding from explosive materials by a natural or artificial barrier.

❖ Barricade means effectively screening a building containing explosives by means of a natural or artificial barrier from a magazine, another building, a railway or a highway.

Artificial barricade. An artificial mound or revetment a minimum thickness of 3 feet (914 mm).

❖ When suitable natural features do not exist to protect adjacent people and structures from flying debris and blast effects if an explosion in a magazine occurs, an artificial barrier must be constructed.

Natural barricade. Natural features of the ground, such as hills, or timber of sufficient density that the surrounding exposures that require protection cannot be seen from the magazine or building containing explosives when the trees are bare of leaves.

❖ Native land features capable of protecting adjacent buildings, people and property from blast effects if an explosion in a magazine occurs may qualify as reduced separation distances permitted by Table 3304.5.2(2). Trees and other ground cover must be thick enough for effective visual screening for the magazine when the branches are bare of leaves or other seasonal foliage.

BARRICADED. The effective screening of a building containing explosive materials from the magazine or other building, railway, or highway by a natural or an artificial barrier. A straight line from the top of any sidewall of the building containing explosive materials to the eave line of any magazine or other building or to a point 12 feet (3658 mm) above the center of a railway or highway shall pass through such barrier.

❖ Natural or artificial barriers must exist or be constructed to qualify for the separation distance reductions allowed in Table 3304.5.2(2). These barriers are intended to provide protection for people and property equivalent to the larger separation distances by absorbing or deflecting blast effects and debris from explosions involving magazines. Figure 3302.1 illustrates the arrangement of natural and artificial barricades for protection from two 2,000-pound (908 kg) Type 1 magazines.

BLAST AREA. The area including the blast site and the immediate adjacent area within the influence of flying rock, missiles and concussion.

❖ The area of a blast is affected by flying rock missiles, gases and concussion and also includes the blast site and the immediately adjacent area that is owned, leased or controlled by the blast operator.

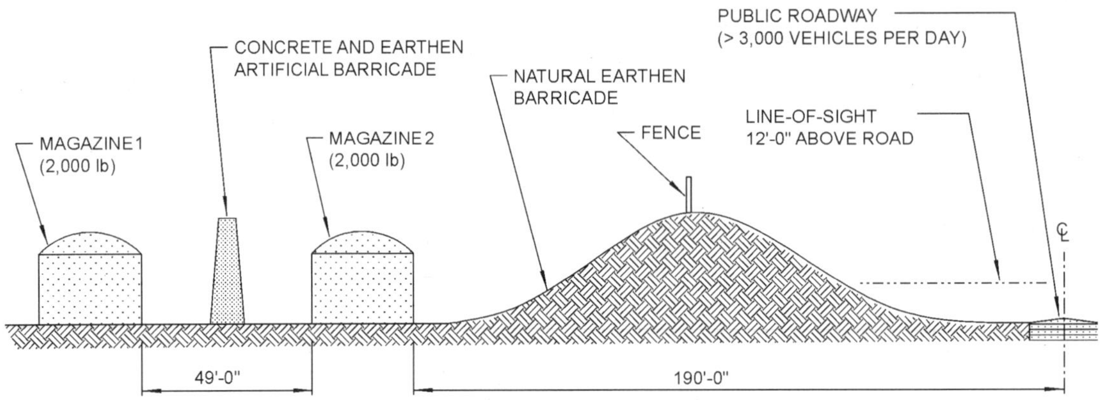

For SI: 1 inch = 25.4 mm, 1 foot = 304.8 mm, 1 pound = .454 kg.

Figure 3302.1
ARRANGEMENT OF BARRICADES

BLAST SITE. The area in which explosive materials are being or have been loaded and which includes all holes loaded or to be loaded for the same blast and a distance of 50 feet (15 240 mm) in all directions.

❖ The area where explosive material is handled during loading, which includes 50 feet (15 240 mm) in all directions from loaded blast holes or holes to be loaded.

BLASTER. A person qualified in accordance with Section 3301.4 to be in charge of and responsible for the loading and firing of a blast.

❖ In general, the blaster should be qualified, experienced and of sound judgement when performing the duties of blasting. This person is authorized to use explosives for blasting purposes. The blaster is trained and experienced in the use of explosives and licensed by the department.

BLASTING AGENT. A material or mixture consisting of fuel and oxidizer, intended for blasting provided that the finished product, as mixed for use or shipment, cannot be detonated by means of a No. 8 test detonator when unconfined. Blasting agents are labeled and placarded as Class 1.5 material by US DOTn.

❖ This definition is derived from federal explosive regulations and is intended to distinguish blasting agents from more dangerous explosive materials on the basis of their propensity to mass detonate by initiation using a standard device.

BULLET RESISTANT. Constructed so as to resist penetration of a bullet of 150-grain M2 ball ammunition having a nominal muzzle velocity of 2,700 feet per second (fps) (824 mps) when fired from a 30-caliber rifle at a distance of 100 feet (30 480 mm), measured perpendicular to the target.

❖ Tests to determine bullet resistance are to be conducted on test panels or empty magazines. The panels or magazines are to resist a penetration of five out of five shots placed independently of each other in an area at least 3 feet by 3 feet (0.9 m by 0.9 m). If hardwood or softwood is used, its water content is not to exceed 15 percent.

Where a magazine roof or ceiling is required to be bullet resistant, it must be constructed of materials comparable to the sidewalls or of other materials that can withstand the penetration of bullets fired at an angle of 45 degrees (.79 rad) from perpendicular.

DETONATING CORD. A flexible cord containing a center core of high explosive used to initiate other explosives.

❖ Outwardly, detonating cord, or primacord, is somewhat similar to a safety fuse in size and appearance. But instead of being filled with black powder, detonating cord generally contains pentaerythritoltetranitrate (PETN). Like other explosives, primacord is generally set off by a blasting cap. It propagates a detonation at a rate of 9,000 yards per second, a little over 5 miles. A length of detonating cord stretching from San Francisco to New York will explode over its entire length of 3,000 miles in about 10 minutes.

DETONATION. An exothermic reaction characterized by the presence of a shock wave in the material which establishes and maintains the reaction. The reaction zone progresses through the material at a rate greater than the velocity of sound. The principal heating mechanism is one of shock compression. Detonations have an explosive effect.

❖ Detonations are distinguished from deflagrations, which are produced by explosive gases, dusts, vapors and mists, by the speed with which they propagate a blast effect. Both detonations and deflagrations may produce explosive results when they occur in a confined space.

DETONATOR. A device containing any initiating or primary explosive that is used for initiating detonation. A detonator shall not contain more than 154.32 grains (10 grams) of total explosives by weight, excluding ignition or delay charges. The term includes, but is not limited to, electric blasting caps of instantaneous and delay types, blasting caps for use with safety fuses, detonating cord delay connectors, and noninstantaneous and delay blasting caps which use detonating cord, shock tube or any other replacement for electric leg wires. All types of detonators in strengths through No. 8 cap should be rated at 1.5 pounds (0.68 kg) of explosives per 1,000 caps. For strengths higher than No. 8 cap, consult the manufacturer.

❖ These devices contain a primary explosive charge and are used to initiate other explosives. Detonators include, but are not limited to:

1. Electric detonators of instantaneous and delay types.
2. Detonators for use with safety fuses, detonating cord delay connectors and nonelectric instantaneous delay detonators that use detonating cord, shock tube or any other replacement for electric leg wires.

DISCHARGE SITE. The immediate area surrounding the fireworks mortars used for an outdoor fireworks display.

❖ The area selected for the discharge of aerial shells must be located so that the trajectory of the shells does not come within 25 feet (7.7 m) of any overhead object. Ground display pieces must be located a minimum distance of 75 feet (22 860 m) from spectator viewing areas and parking areas.

DISPLAY SITE. The immediate area where a fireworks display is conducted. The display area includes the discharge site, the fallout area, and the required separation distance from the mortars to spectator viewing areas. The display area does not include spectator viewing areas or vehicle parking areas.

❖ Where added safety precautions have been taken, or particularly favorable conditions exist, the fire code official can decrease the required separation distances upon demonstration that the hazard has been reduced or the risk has been properly protected. Where unusual or safety-threatening conditions exist, he or she can also increase the required separation distances as he or she deems necessary.

EXPLOSIVE. A chemical compound, mixture or device, the primary or common purpose of which is to function by explosion. The term includes, but is not limited to, dynamite, black powder, pellet powder, initiating explosives, detonators, safety fuses, squibs, detonating cord, igniter cord, igniters and display fireworks, 1.3G (Class B, Special).

The term "explosive" includes any material determined to be within the scope of USC Title 18: Chapter 40 and also includes any material classified as an explosive other than consumer fireworks, 1.4G (Class C, Common) by the hazardous materials regulations of DOTn 49 CFR.

❖ Explosives either detonate or deflagrate, rather than burn, when initiated by either heat, shock or electric current. Although these materials are normally designed and intended to be initiated by detonators under controlled conditions, heat, shock and electric current from uncontrolled sources may initiate these materials to produce an explosion.

High explosive. Explosive material, such as dynamite, which can be caused to detonate by means of a No. 8 test blasting cap when unconfined.

❖ High explosives can also deflagrate but they are capable of even more. Before any appreciable amount of gas can escape and reduce the pressure, the body of the explosive is completely vaporized. The strength of a particular explosive depends on the amount of gas and heat it produces. At the very instant of their formation, the gases in a high-explosive state occupy only the original volume of the explosive material. Gas pressures can soar above 1 million pounds per square inch (psi); temperatures of 6,000°F (3316°C) and higher are common. Rapid expansion must occur and pressure waves move out at enormous velocities, often supersonic, doing tremendous damage.

Low explosive. Explosive material that will burn or deflagrate when ignited. It is characterized by a rate of reaction that is less than the speed of sound. Examples of low explosives include, but are not limited to, black powder, safety fuse, igniters, igniter cord, fuse lighters, fireworks, 1.3G (Class B special) and propellants, 1.3C.

❖ In low-explosive materials the burning rate is rapid but the created gases still have time to escape as the explosive is consumed. Gunpowder, for example, burns extremely rapidly and the gases can be put to work projecting an object out of a gun barrel. Pressures remain relatively low when the explosive is unconfined. Low explosives, like black powder, will deflagrate (burn intensely), producing a visible flame, such as a muzzle flash. But they do not ordinarily detonate.

Mass-detonating explosives. Division 1.1, 1.2 and 1.5 explosives alone or in combination, or loaded into various types of ammunition or containers, most of which can be expected to explode virtually instantaneously when a small portion is subjected to fire, severe concussion, impact, the impulse of an initiating agent, or the effect of a considerable discharge of energy from without. Materials that react in this manner represent a mass explosion hazard. Such an explosive will normally cause severe structural damage to adjacent objects. Explosive propagation could occur immediately to other items of ammunition and explosives stored sufficiently close to and not adequately protected from the initially exploding pile with a time interval short enough so that two or more quantities must be considered as one for quantity-distance purposes.

❖ Mass-detonating explosives (typically stored in occupancies classified as Group H-1 by the IBC) present a detonation hazard, as do high explosives, and present a greater threat to adjacent objects and structures. The code, therefore, contains provisions in Table 3305.3 to deal with the separation distances for mass-explosion hazards.

UN/DOTn Class 1 explosives. The former classification system used by DOTn included the terms "high" and "low" explosives as defined herein. The following terms further define explosives under the current system applied by DOTn for all explosive materials defined as hazard Class 1 materials. Compatibility group letters are used in concert with the Division to specify further limitations on each division noted, (i.e., the letter G identifies the material as a pyrotechnic substance or article containing a pyrotechnic substance and similar materials).

❖ The DOT issued new regulations in 1991 based on the United Nations Recommendations on the Transportation of Dangerous Goods for the transportation of hazardous materials. DOT's hazardous materials requirements with respect to hazard communications, classifications and packaging of explosives were revised by the new regulations.

Division 1.1. Explosives that have a mass explosion hazard. A mass explosion is one which affects almost the entire load instantaneously.

Division 1.2. Explosives that have a projection hazard but not a mass explosion hazard.

Division 1.3. Explosives that have a fire hazard and either a minor blast hazard or a minor projection hazard or both, but not a mass explosion hazard.

Division 1.4. Explosives that pose a minor explosion hazard. The explosive effects are largely confined to the package and no projection of fragments of appreciable size or range is to be expected. An external fire must not cause virtually instantaneous explosion of almost the entire contents of the package.

Division 1.5. Very insensitive explosives. This division is comprised of substances that have a mass explosion hazard but which are so insensitive that there is very little probability of initiation or of transition from burning to detonation under normal conditions of transport.

Division 1.6. Extremely insensitive articles which do not have a mass explosion hazard. This division is comprised of articles that contain only extremely insensitive detonating substances and which demonstrate a negligible probability of accidental initiation or propagation.

❖ DOT puts explosives in six classes according to the degree of hazard posed by the material (see Table 3302.1). The most dangerous of these materials is capable of almost simultaneous detonation of all of the material in a single load or store. The least sensitive explosives produce blasts limited to the packages in which they are transported. This definition of "Explosive" includes materials such as: detonators, blasting agents and water gels. Examples of these materials are listed in DOTy 27 CFR 55.23.

EXPLOSIVE MATERIAL. The term "explosive" material means explosives, blasting agents, and detonators.

❖ A list of explosive materials is maintained by the ATF pursuant to 18 USC, Chapter 40. This list (ATF Publication 5400.8) is available at no cost from the ATF Distribution Center, 7943 Angus Court, Springfield, VA 22153.

FALLOUT AREA. The area over which aerial shells are fired. The shells burst over the area, and unsafe debris and malfunctioning aerial shells fall into this area. The fallout area is the location where a typical aerial shell dud falls to the ground depending on the wind and the angle or mortar placement.

❖ The area described by this definition is a large, unoccupied open space that functions as a "safety zone" between the fireworks display launch area and the spectator areas and is included in the display site definition in this section. The layout of the fallout area is usually centered around the location of, and takes into account the launch angle of, the mortars used to launch the aerial displays. Local weather wind direction and speed forecasts may also affect the layout of the area. The fallout area provides a relatively safe area in which display debris and dud or malfunctioned shells may fall and is the starting point for the post-display inspection required by Section 3308.9. Within the fallout area, no spectators or vehicles should be allowed and the area should be free of combustible materials that could be ignited by falling

TABLE 3302.1
COMPARISON OF CLASSIFICATION SYSTEMS FOR EXPLOSIVE MATERIALS

BATF	OLD DOTn[a]	NEW DOTn[b, c]	Description
High Explosives	Class A Explosives	Division 1.1	Mass Explosion Potential
		Divisions 1.2	Projectile Hazard
Low Explosives	Class B Explosives	Division 1.3	Predominantly a Fire Hazard
	Class C Explosives	Divisions 1.4	No Significant Blast Hazard
Blasting Agents	Blasting Agents	Division 1.5	Very Sensitive Explosives
	Dangerous	Division 1.6	Extremely Insensitive Detonating Substances

Note a. Prior to September 30, 1991
Note b. Effective October 1, 1991
Note c. *International Fire Code* and new DOT classifications correlate.

debris. In reviewing the fallout area site plan, the presence and location of trees, brush and other vegetation that could "hide" undetonated shells or be a fire ignition hazard should be considered. In some cases, it may be necessary to wet-down the fallout area to further reduce the ignition hazard to vegetation by falling hot debris.

FIREWORKS. Any composition or device for the purpose of producing a visible or an audible effect for entertainment purposes by combustion, deflagration or detonation that meets the definition of 1.4G fireworks or 1.3G fireworks as set forth herein.

❖ This term refers to any device, other than a novelty or theatrical pyrotechnic article, intended to produce visible and/or audible effects by combustion, deflagration or detonation and any chemical compound or mechanically mixed preparation of an explosive or inflammable nature that is used for the purpose of making any manufactured fireworks and is not included in any other class of explosives.

Fireworks, 1.4G. (Formerly known as Class C, Common Fireworks.) Small fireworks devices containing restricted amounts of pyrotechnic composition designed primarily to produce visible or audible effects by combustion. Such 1.4G fireworks which comply with the construction, chemical composition and labeling regulations of the DOTn for Fireworks, UN 0336, and the U.S. Consumer Product Safety Commission as set forth in CPSC 16 CFR: Parts 1500 and 1507, are not explosive materials for the purpose of this code.

❖ The requirements for storage, display and labeling depend on the correct application of this definition. This definition reflects the construction, chemical composition and labeling requirements of the CPSC, found in Title 16, Code of Federal Regulations, Parts 1500 and 1507. Consumer 1.4G fireworks are not considered to be explosives subject to the provisions of Chapter 33.

Fireworks, 1.3G. (Formerly Class B, Special Fireworks.) Large fireworks devices, which are explosive materials, intended for use in fireworks displays and designed to produce audible or visible effects by combustion, deflagration or detonation. Such 1.3G fireworks include, but are not limited to, firecrackers containing more than 130 milligrams (2 grains) of explosive composition, aerial shells containing more than 40 grams of pyrotechnic composition, and other display pieces which exceed the limits for classification as 1.4G fireworks. Such 1.3G fireworks, are also described as Fireworks, UN0335 by the DOTn.

❖ This category of fireworks represents considerable life safety hazard in comparison to the consumer 1.4G fireworks. This definition reflects the construction, chemical composition and labeling requirements of the CPSC, found in Title 16, Code of Federal Regulations, Parts 1500 and 1507.

FIREWORKS DISPLAY. A presentation of fireworks for a public or private gathering.

❖ The areas selected for the discharge site, spectator viewing area, parking areas and the fallout area must be inspected and approved by the authority having jurisdiction.

HIGHWAY. A public street, public alley or public road.

❖ Roads, alleys and similar thoroughfares or vehicular accessways on private property are not included in this definition. However, when private roadways adjacent to an explosives magazine or blasting site are commonly used for vehicular traffic by the public, the separation required should be the same as that required for public rights-of-way or public access should be suspended.

INHABITED BUILDING. A building regularly occupied in whole or in part as a habitation for human beings, or any church, schoolhouse, railroad station, store or other structure where people are accustomed to assemble, except any building or structure occupied in connection with the manufacture, transportation, storage or use of explosive materials.

❖ Building use need not conform to regular intervals or schedules to be considered inhabited if people routinely occupy the building. Those buildings at a storage or blasting site used for the manufacture, transportation or storage of explosive materials must be considered as magazines rather than inhabited buildings for the purpose of applying the provisions of Section 3304.

MAGAZINE. A building, structure or container, other than an operating building, approved for storage of explosive materials.

❖ Structures for the storage of explosive materials are not considered inhabited buildings for the purpose of applying Section 3304. Explosives, a necessary part of our industrialized society, must be transported from places of manufacture to a location of use. Somewhere along the line, they will probably have to be stored. This type of storage place is called a magazine.

Indoor. A portable structure, such as a box, bin or other container, constructed as required for Type 2, 4 or 5 magazines in accordance with NFPA 495, NFPA 1124 or DOTy 27 CFR Part 55 so as to be fire resistant and theft resistant.

❖ These magazines are sometimes found inside warehouses, wholesale houses and retail establishments on wheels or casters to allow easy movement when needed.

Type 1. A permanent structure, such as a building or igloo, that is bullet resistant, fire resistant, theft resistant, weather resistant and ventilated in accordance with the requirements of NFPA 495, NFPA 1124, or DOTy 27 CFR Part 55.

❖ The walls of a Type I magazine should be constructed of either masonry, metal or wood. The foundation should consist of brick, concrete, cement, block, stone or wood post. The floors should be constructed of nonsparking material and be strong enough to bear the weight of the maximum quantity to be stored. The listed standards give the sizes of material for each type of construction material.

Type 2. A portable or mobile structure, such as a box, skid-magazine, trailer or semitrailer, constructed in accordance with the requirements of NFPA 495, NFPA 1124 or DOTy 27 CFR, Part 55 that is fire resistant, theft resistant, weather resistant and ventilated. If used outdoors, a Type 2 magazine is also bullet resistant.

❖ Construction may be either:

1. Wood [having sides, bottoms, and covers or doors constructed of 2-inch (51 mm) hardwood, well braced at corners and covered with sheet metal (not less than 26 gauge) with exposed nails countersunk].
2. Metal [having sides, bottoms and covers or doors constructed of 12-gauge metal and lined inside with a nonsparking material. Edges of metal shall overlap sides at least 1 inch (25 mm)].

Type 3. A fire-resistant, theft-resistant and weather-resistant "day box" or portable structure constructed in accordance with NFPA 495, NFPA 1124, or DOTy 27 CFR Part 55 used for the temporary storage of explosive materials.

❖ Construction may be of not less than 12-gauge (0.1046 inch) steel, lined with $^1/_2$-inch (12.7 mm) plywood or $^1/_2$-inch (12.7 mm) hardboard. The door or lid must overlap the door opening by at least 1 inch (25 mm).

Type 4. A permanent, portable or mobile structure such as a building, igloo, box, semitrailer or other mobile container that is fire resistant, theft resistant and weather resistant and constructed in accordance with NFPA 495, NFPA 1124, or DOTy 27 CFR, Part 55.

❖ Construction may be of masonry, metal-covered wood, fabricated metal or a combination of these materials.

Type 5. A permanent, portable or mobile structure such as a building, igloo, box, bin, tank, semitrailer, bulk trailer, tank trailer, bulk truck, tank truck or other mobile container that is theft resistant, which is constructed in accordance with NFPA 495, NFPA 1124, or DOTy 27 CFR, Part 55.

❖ Restrictions on Type 5 outdoor storage facilities:

Ground around the storage facility must slope away for drainage. Wheels must be removed from vehicles used as unattended storage facilities or they must be immobilized by kingpin locking devices.

Restrictions on Type 5 indoor storage facilities:

Blasting agents are not to be stored in any kind of indoor storage facility in a residence or dwelling.

MORTAR. A tube from which fireworks shells are fired into the air.

❖ Mortars must be inspected carefully for defects, such as dents, bent ends, damaged interiors and damaged plugs, prior to placement and use. Defective mortars must not be used. Careful inspection of mortars is of particular importance for paper mortars that can sustain undetected damage to their interiors that can result in serious malfunctions.

NET EXPLOSIVE WEIGHT (net weight). The weight of explosive material expressed in pounds. The net explosive weight is the aggregate amount of explosive material contained within buildings, magazines, structures or portions thereof, used to establish quantity-distance relationships.

❖ This definition is included to correlate the use of the provisions of this chapter. The definition is based on Department of Defense (DOD) concepts; however, prescriptive requirements have been removed from the DOD definition and placed into the body of the code. The net explosive weight may vary depending on building construction, for example, in cases where appropriate barrier walls or appropriate distances to avoid propagation have been employed.

OPERATING BUILDING. A building occupied in conjunction with the manufacture, transportation, storage, or use of explosive materials. Operating buildings are separated from one another with the use of intraplant or intraline distances.

❖ Magazines are used for storage of explosive materials. Manufacturing or operating buildings used for the storage of explosives are not magazines and are not intended to be used for storage, although at times there may be storage incidental to the manufacturing function. This definition is included here to clarify this difference.

PLOSOPHORIC MATERIAL. Two or more unmixed, commercially manufactured, prepackaged chemical substances including oxidizers, flammable liquids or solids, or similar substances that are not independently classified as explosives but which, when mixed or combined, form an explosive that is intended for blasting.

❖ Plosophoric materials, or plosophors, also are known as two-component or binary explosives. When plosophoric materials are mixed or combined at the point of use, the procedures recommended by the manufacturer should be strictly enforced. Mixed or combined plosophoric materials must be transported, stored and used in the same manner as explosives.

PROXIMATE AUDIENCE. An audience closer to pyrotechnic devices than permitted by NFPA 1123.

❖ The separation distance of the pyrotechnic devices and the audience permitted by NFPA 1123 is in relationship to the shell size of the mortar (see Table 3-1.3 of NFPA

1123). NFPA 1126 addresses the requirements for proximate audiences.

PYROTECHNIC COMPOSITION. A chemical mixture that produces visible light displays or sounds through a self-propagating, heat-releasing chemical reaction which is initiated by ignition.

❖ This definition parallels the definition in NFPA 1124, referenced in Section 3305.1. The term is used in the definition of "Display, 1.3G fireworks," and is the basis for establishing the allowable amounts in Table 3304.3.

PYROTECHNIC SPECIAL EFFECT. A visible or audible effect for entertainment created through the use of pyrotechnic materials and devices.

❖ Pyrotechnic special effects are widely used in motion-picture production to create all types of effects involving explosions, fires, light, smoke and sound concussions. The types of pyrotechnic materials used include flash powder, flash paper, gun cotton, black powder (gunpowder), smokeless powder, detonator explosives and many more. They are used in bullet hits (squibs), blank cartridges, flash pots, fuses, mortars, smoke pots, sparkle pots, etc.

The main problems of pyrotechnics include prematurely triggering the pyrotechnic effect; use of larger quantities or more dangerous materials than needed; causing a fire; lack of adequate fire-extinguishing capabilities and, of course, inadequately trained and experienced pyrotechnic operators. As a result of these risks, all pyrotechnic special effects are regulated at the federal, state and local level.

PYROTECHNIC SPECIAL-EFFECT MATERIAL. A chemical mixture used in the entertainment industry, to produce visible or audible effects by combustion, deflagration or detonation. Such a chemical mixture predominantly consists of solids capable of producing a controlled, self-sustaining and self-contained exothermic chemical reaction that results in heat, gas sound, light or a combination of these effects. The chemical reaction functions without external oxygen.

❖ Pyrotechnic special effects materials are Division 1.3 explosives. They will burn but not explode, unless confined. Examples are black powder and pellet powder, safety fuses, igniters, igniter cord, fuse lighters, Division 1.3 special fireworks and Division 1.3 composite solid propellants.

RAILWAY. A steam, electric or other railroad or railway that carriers passengers for hire.

❖ The definition of "Railway" is intended to minimize the exposure of passengers to danger if an explosion occurs involving an explosives magazine.

READY BOX. A weather-resistant container with a self-closing or automatic-closing cover that protects fireworks shells from burning debris. Tarpaulins shall not be considered as ready boxes.

❖ After delivery and prior to the display, shells must be separated according to size and their designation as salutes. Any display fireworks that will be temporarily stored at the display site during the fireworks display must be stored in ready boxes separated according to size and their designation as salutes.

SMALL ARMS AMMUNITION. A shotgun, rifle or pistol cartridge and any cartridge for propellant-actuated devices. This definition does not include military ammunition containing bursting charges or incendiary, trace, spotting or pyrotechnic projectiles.

❖ Small arms ammunition consists of cylindrical casings containing a small amount of smokeless powder. These items are usually designed to propel a missile or bullet at a target, but they may also be used in explosive actuated devices, such as nail guns and riveters. Individual cartridges are typically packed in paperboard boxes shipped in cardboard cartons. In a fire, these articles pose no mass detonation and only a moderate (low-velocity) projectile hazard. The ATF definition of "ammunition" also includes percussion caps and $^{3}/_{32}$ inch (2.4 mm) and other external burning pyrotechnic hobby fuses. Black powder ammunition and black powder are not included in this definition.

SMALL ARMS PRIMERS. Small percussion-sensitive explosive charges, encased in a cap, used to ignite propellant powder.

❖ "Small arm primers" and "percussion caps" mean primers used for small arms ammunition.

SMOKELESS PROPELLANTS. Solid propellants, commonly referred to as smokeless powders, used in small arms ammunition, cannons, rockets, propellant-actuated devices and similar articles.

❖ This term refers to a propellant explosive from which there is little or no smoke when fired, including smokeless powder for cannons and smokeless powder for small arms.

SPECIAL INDUSTRIAL EXPLOSIVE DEVICE. An explosive power pack containing an explosive charge in the form of a cartridge or construction device. The term includes but is not limited to explosive rivets, explosive bolts, explosive charges for driving pins or studs, cartridges for explosive-actuated power tools and charges of explosives used in automotive air bag inflators, jet tapping of open hearth furnaces and jet perforation of oil well casings.

❖ Special industrial explosive devices means explosive actuated power devices and propellant actuated power devices. "Explosive actuated device" means a tool or special mechanized device that is actuated by explosives. Examples of explosive actuated power devices are jet tappers and jet perforators. "Propellant actuated device" means a tool or special mechanized device or gas generator system that is actuated by a smokeless propellant or that releases and directs work through a smokeless propellant charge.

THEFT RESISTANT. Construction designed to deter illegal entry into facilities for the storage of explosive materials.

❖ Theft-resistant designs are intended to provide security against illegal or unauthorized entry into magazines containing explosives. Security measures specified in this and subsequent sections require special tools, keys or excessive force to compromise the security measure.

SECTION 3303 RECORD KEEPING AND REPORTING

3303.1 General. Records of the receipt, handling, use or disposal of explosive materials, and reports of any accidents, thefts, or unauthorized activities involving explosive materials shall conform to the requirements of this section.

❖ An accumulation of invoices, sales slips, delivery tickets or receipts or similar records representing individual transactions will satisfy the requirements for record keeping if they include the signature of the receiver of the explosive materials.

3303.2 Transaction record. The permittee shall maintain a record of all transactions involving receipt, removal, use or disposal of explosive materials. Such a record shall be maintained for a period of 5 years, and shall be furnished to the fire code official for inspection upon request.

Exception: Where only Division 1.4G (consumer fireworks) are handled, records need only be maintained for a period of 3 years.

❖ A permit holder must keep a record of all transactions or operations involving explosive materials for five years and should be made available to the issuing authority upon request.

3303.3 Loss, theft or unauthorized removal. The loss, theft or unauthorized removal of explosive materials from a magazine or permitted facility shall be reported to the fire code official, local law enforcement authorities, and the U.S. Department of Treasury, Bureau of Alcohol, Tobacco and Firearms within 24 hours.

Exception: Loss of Division 1.4G (consumer fireworks) need not be reported to the Bureau of Alcohol, Tobacco and Firearms.

❖ The loss, theft or unlawful removal of explosive materials must be reported within 24 hours to the ATF, to the permit-issuing authority, and to the local law enforcement agency.

3303.4 Accidents. Accidents involving the use of explosives, explosive materials and fireworks, which result in injuries or property damage, shall be reported to the fire code official immediately.

❖ Accidents involving explosive material that cause a lost-time injury or property damage must be reported immediately to the authority having jurisdiction for its records.

3303.5 Misfires. The pyrotechnic display operator or blaster in charge shall keep a record of all aerial shells that fail to fire or charges that fail to detonate.

❖ A record of all misfires must be kept, eliminating the possibility of using the misfires in other displays and shows.

3303.6 Hazard communication. Manufacturers of explosive materials and fireworks shall maintain records of chemicals, chemical compounds and mixtures required by DOL 29 CFR, Part 1910.1200, and Section 407.

❖ Manufacturers of explosive materials and fireworks are required to assess the hazards of chemicals that they produce or import. Employers must inform their employees of the hazardous chemicals to which they are exposed, using a hazard communication program, labels and other forms of warning, Material Safety Data Sheets (MSDS) and information and training. In addition, this section requires distributors to transmit the required information to employers.

3303.7 Safety rules. Current safety rules covering the operation of magazines, as described in Section 3304.7, shall be posted on the interior of the magazine in a visible location.

❖ Magazine safety rules must be posted conspicuously on the interior of the magazine as a reminder to employees of the requirements for the safe operation of the magazine. These safety rules must cover security; open flames and lights; area around the magazine; separation distance of combustibles; instructions for packing and unpacking; acceptable tools and equipment.

SECTION 3304 EXPLOSIVE MATERIALS STORAGE AND HANDLING

3304.1 General. Storage of explosives and explosive materials, small arms ammunition, small arms primers, propellant-actuated cartridges and smokeless propellants in magazines, shall conform to the provisions of this section.

❖ This section establishes requirements for various magazines associated with explosive materials storage.

3304.2 Magazine required. Explosives and explosive materials, and Division 1.3G fireworks shall be stored in magazines constructed, located, operated and maintained in accordance with the provisions of Section 3304 and NFPA 495 or NFPA 1124.

Exceptions:

1. Storage of fireworks at display sites in accordance with Section 3308.5 and NFPA 1123 or NFPA 1126.
2. Portable or mobile magazines not exceeding 120 square feet (11 m^2) in area shall not be required to comply with the requirements of the *International Building Code.*

❖ Explosives and blasting agents must be kept in magazines when not in use to reduce the exposure hazard.

3304.3 Magazines. The storage of explosives and explosive materials in magazines shall comply with Table 3304.3.

❖ This section gives the requirements for magazine storage of explosives and explosive materials. The section covers high explosives, low explosives and detonating cord.

TABLE 3304.3. See below.

❖ This table gives both the new and the old classification of explosives. It also states whether the explosive is a low or high type of explosive. The table specifies the maximum amounts allowed in the control areas, whether indoors or outdoors. The last columns specify the type of magazine required for storage of various explosives. Then Table 3304.5.2(1) refers to the proper table for separation distances.

3304.3.1 High explosives. Explosive materials classified as Division 1.1 or 1.2 or formerly classified as Class A by the U.S. Department of Transportation shall be stored in Type 1, 2 or 3 magazines.

Exceptions:

1. Black powder shall be stored in a Type 1, 2, 3 or 4 magazine.
2. Cap-sensitive explosive material that is demonstrated not to be bullet sensitive, shall be stored in a Type 1, 2, 3, 4 or 5 magazine.

❖ Type 1, 2 and 3 magazines are constructed with stricter requirements for the prevention of possible fire, water damage or bullet hazards.

3304.3.2 Low explosives. Explosive materials that are not cap sensitive shall be stored in a Type 1, 2, 3, 4 or 5 magazine.

❖ There are fewer restrictions on the storage of low explosives that are not cap sensitive. Protection against theft is a primary concern.

3304.3.3 Detonating cord. For quantity and distance purposes, detonating cord of 50 grains per foot shall be calculated as equivalent to 8 pounds (4 kg) of high explosives per 1,000 feet (305 m). Heavier or lighter core loads shall be rated proportionally.

❖ Standard detonating cord averaging about 50 grains per foot, which is considered the equivalent of 8 pounds (4 kg) of high explosives per 1,000 feet (305 m), is a benchmark for detonating cord. Heavier cord would be 60 grains per foot equaling 9 pounds (4 kg) of high explosive per 1,000 feet (305 m).

3304.4 Prohibited storage. Detonators shall be stored in a separate magazine for blasting supplies and shall not be stored in a magazine with other explosive materials.

❖ Storage of detonators with explosives may result in an accidental mass detonation. Detonators may be stored

TABLE 3304.3
STORAGE AMOUNTS AND MAGAZINE REQUIREMENTS FOR EXPLOSIVES, EXPLOSIVE MATERIALS AND FIREWORKS, 1.3G MAXIMUM ALLOWABLE QUANTITY PER CONTROL AREA

NEW UN/ DOTn DIVISION	OLD DOTn CLASS	ATF/OSHA CLASS	INDOOR[a] (pounds)				OUTDOOR (pounds)	MAGAZINE TYPE REQUIRED				
			Unprotected	Cabinet	Sprinklers	Sprinklers & cabinet		1	2	3	4	5
1.1[b]	A	High	0	0	1	2	1	X	X	X	—	—
1.2	A	High	0	0	1	2	1	X	X	X	—	—
1.2	B	Low	0	0	1	2	1	X	X	X	X	—
1.3	B	Low	0	0	5	10	1	X	X	X	X	—
1.4	B	Low	0	0	50	100	1	X	X	X	X	—
1.5	C	Low	0	0	1	2	1	X	X	X	X	—
1.5	Blasting Agent	Blasting Agent	0	0	1	2	1	X	X	X	X	X
1.6	N/A	N/A	0	0	1	2	1	X	X	X	X	X

For SI: 1 pound = 0.454 kg, 1 pound per gallon = 0.12 kg per liter, 1 ounce = 28.35 g.
a. A factor of 10 pounds per gallon shall be used for converting pounds (solid) to gallons (liquid) in accordance with Section 2703.1.2.
b. Black powder shall be stored in a Type 1, 2, 3 or 4 magazine as provided for in Section 3304.3.1.

with other explosives only if that storage is approved by the fire code official and is under the following conditions [see DOTy 27 CFR 55.213(b)]:

1. Nonmass-detonating detonators may be stored with electric squibs, safety fuses, ignitors and ignitor cord in a Type 4 magazine.
2. Detonators may be stored with delay devices, electric squibs, safety fuses, ignitors and ignitor cord in Type 1 and 2 magazines.

3304.5 Location. The use of magazines for storage of explosives and explosive materials shall comply with Sections 3304.5.1 through 3304.5.3.3.

❖ Explosives and blasting agents must be kept in magazines when not in use to reduce the exposure hazard.

3304.5.1 Indoor magazines. The use of indoor magazines for storage of explosives and explosive materials shall comply with the requirements of this section.

❖ The following sections address the requirements for indoor magazines. These sections cover the use, construction, quantity limit, prohibited use, number and separation distance of indoor magazines.

3304.5.1.1 Use. The use of indoor magazines for storage of explosives and explosive materials shall be limited to occupancies of Group F, H, M or S, and research and development laboratories.

❖ Section 3304.2 requires all explosives in storage to be located in magazines. This section, however, recognizes that certain operations engaged in manufacturing or research processes require the use of magazines (commonly referred to as "day boxes") inside the building for the purposes of storing small quantities of material awaiting use in the process operations.

3304.5.1.2 Construction. Indoor magazines shall comply with the following construction requirements:

1. Construction shall be fire resistant and theft resistant.
2. Exterior shall be painted red.
3. Base shall be fitted with wheels, casters or rollers to facilitate removal from the building in an emergency.
4. Lid or door shall be marked with conspicuous white lettering not less than 3 inches (76 mm) high and minimum 0.5 inch (12.7 mm) stroke, reading EXPLOSIVES — KEEP FIRE AWAY.
5. The least horizontal dimension shall not exceed the clear width of the entrance door.

❖ Indoor magazines need not be bullet resistant if the building in which they are located gives protection from bullet penetration. Not more than 50 pounds (23 kg) of high explosives may be stored in an indoor magazine, and not more than one indoor magazine may be stored inside a single building if the 50-pound (23 kg) limit has not been exceeded. The following comments clarify the intent of special requirements applying to magazines used indoors:

1. Construction specifications in this section state the minimum acceptable degree of protection from physical hazards and an exterior fire exposure when wood is used as a construction material for the magazine. Interior nails and screws must be countersunk to prevent damage to explosives and friction or sparks when the magazine is in transit or explosives are moved within the magazine.
2. The required color serves as a reminder of flammable, explosive and detonation hazards.
3. If a fire or an emergency occurs in the building, wheels or casters will permit the magazine to be removed from the building or relocated to a safer area.
4. The required sign serves as a reminder to keep open flames and ignition sources away from the magazine.
5. In case of emergency, the magazine should clear the opening of the door for transfer of the magazine to the outdoors.

3304.5.1.3 Quantity limit. Not more than 50 pounds (23 kg) of explosives or explosive materials shall be stored within an indoor magazine.

❖ The quantity of materials kept on display should be the minimum required to illustrate available wares and should not exceed the quantity usually sold on an average business day.

3304.5.1.4 Prohibited use. Indoor magazines shall not be used within buildings containing Group R occupancies.

❖ Indoor magazines are not permitted inside residences or dwellings because of the life hazard present and the difficulty of maintaining control over their use through routine inspections.

3304.5.1.5 Location. Indoor magazines shall be located within 10 feet (3048 mm) of an entrance and only on floors at or having ramp access to the exterior grade level.

❖ The location of these magazines must be chosen to minimize exposure of means of egress but maintain ready access from the exterior to facilitate fire fighting.

3304.5.1.6 Number. Not more than two indoor magazines shall be located in the same building. Where two such magazines are located in the same building, one magazine shall be used solely for the storage of not more than 5,000 detonators.

❖ The location of magazines in wholesale and retail establishments should be coordinated with and approved by the fire department's prefire plans. Any changes in magazine locations require prior approval of the fire code official and should be noted on the permit.

3304.5.1.7 Separation distance. When two magazines are located in the same building, they shall be separated by a distance of not less than 10 feet (3048 mm).

❖ Explosives and blasting agents that are offered for sale must be stored in separate magazines to prevent an explosion in one magazine from triggering an explosion in another magazine or mass detonation.

3304.5.2 Outdoor magazines. All outdoor magazines other than Type 3 shall be located so as to comply with Table 3304.5.2(2), Table 3304.5.2 (3) or Table 3304.5.2(4) as set forth in Table 3304.5.2(1).

❖ Specified separation distances are intended to minimize the potential damage to life and property that may result from a blast inside a magazine. Additionally, these separation distances help control the hazard to a given magazine from exposure fires involving ground cover or a fire occurring in another magazine and thus provide clear access for emergency response personnel (see Figure 3304.5.2). This section also recognizes that Type 3 magazines (day boxes) are intended for short-term storage of explosives at a blasting site. Separation and the rules governing blasting for those magazines are addressed by Section 3307.

TABLE 3304.5.2(1). See below.

❖ This table refers to the proper table for separation of explosives according to which division the explosive is classified in. Note that Table 3304.5.2(3) is basically for low explosives. If magazines contain different explosives, the most restrictive separation distances will apply.

TABLE 3304.5.2(2). See page 33-19.

❖ Table 3304.5.2(2) establishes minimum separation distances for the permanent storage of explosives from selected property classes. The intent of the table is to place explosive storage far enough away from occupied buildings, highways, railways and other magazines to reduce exposure of those properties to damage if a detonation of a magazine occurs. Storage exceeding tabular amounts in a single magazine is rare and thus requires special approval by the fire code official.

TABLE 3304.5.2(3). See page 33-21.

❖ Magazines containing special fireworks, other than special salutes, and black powder or other low explosives must be separated from each other and from inhabited buildings, public highways and passenger railways. Amounts of low explosives are located in the left two columns. The separation distances are located on the row of the amount of low explosives in reference to inhabited buildings, railways or other magazines.

TABLE 3304.5.2(4). See page 33-22.

❖ Items in Division 1.4 present a fire hazard with no blast hazard and virtually no fragmentation hazard beyond the fire-hazard clearance ordinarily specified for high-risk materials. Separate facilities for storage and handling of this division should not be less than 100 feet (30 480 mm) from other facilities, except those of noncombustible construction, which can be 50 feet (15 240 mm) from each other, provided both are noncombustible.

Division 1.4 materials are finished goods or devices that contain energetic materials. A fire or explosive event with Division 1.4 materials is limited to the individual devices involved and, unlike the materials in the other divisions, there is no mass reaction. The table of distances has been based on federal regulations as published by the DOD.

3304.5.3 Special requirements for Type 3 magazines. Type 3 magazines shall comply with Sections 3304.5.3.1 through 3304.5.3.3.

❖ These requirements are intended to provide a reasonable degree of safety for Type 3 magazines at blasting sites.

TABLE 3304.5.2(1)
APPLICATION OF SEPARATION DISTANCE TABLE[a]

DOTn DIVISION	AMERICAN TABLE OF DISTANCES FOR STORAGE OF EXPLOSIVE MATERIALS 3304.5.2(2) (DOTy 27 CFR, Part 55.218)	TABLE OF SEPARATION DISTANCES FOR LOW EXPLOSIVES 3304.5.2(3) (DOTy 27 CFR, Part 55.219)	TABLE OF DISTANCES FOR BUILDINGS CONTAINING EXPLOSIVES DIVISION 1.4 3304.5.2(4)
1.1	X	—	—
1.2	X	—	—
1.3	—	X	—
1.4G or 1.4S fireworks	—	—	X
1.4B or 1.4S detonators	—	—	X
1.5	X	—	—
1.6	Not Applicable	Not Applicable	Not Applicable

a. Where adjacent magazines contain different classes of explosive materials, the separation between magazines shall be as prescribed by Table 3304.5.2(2).

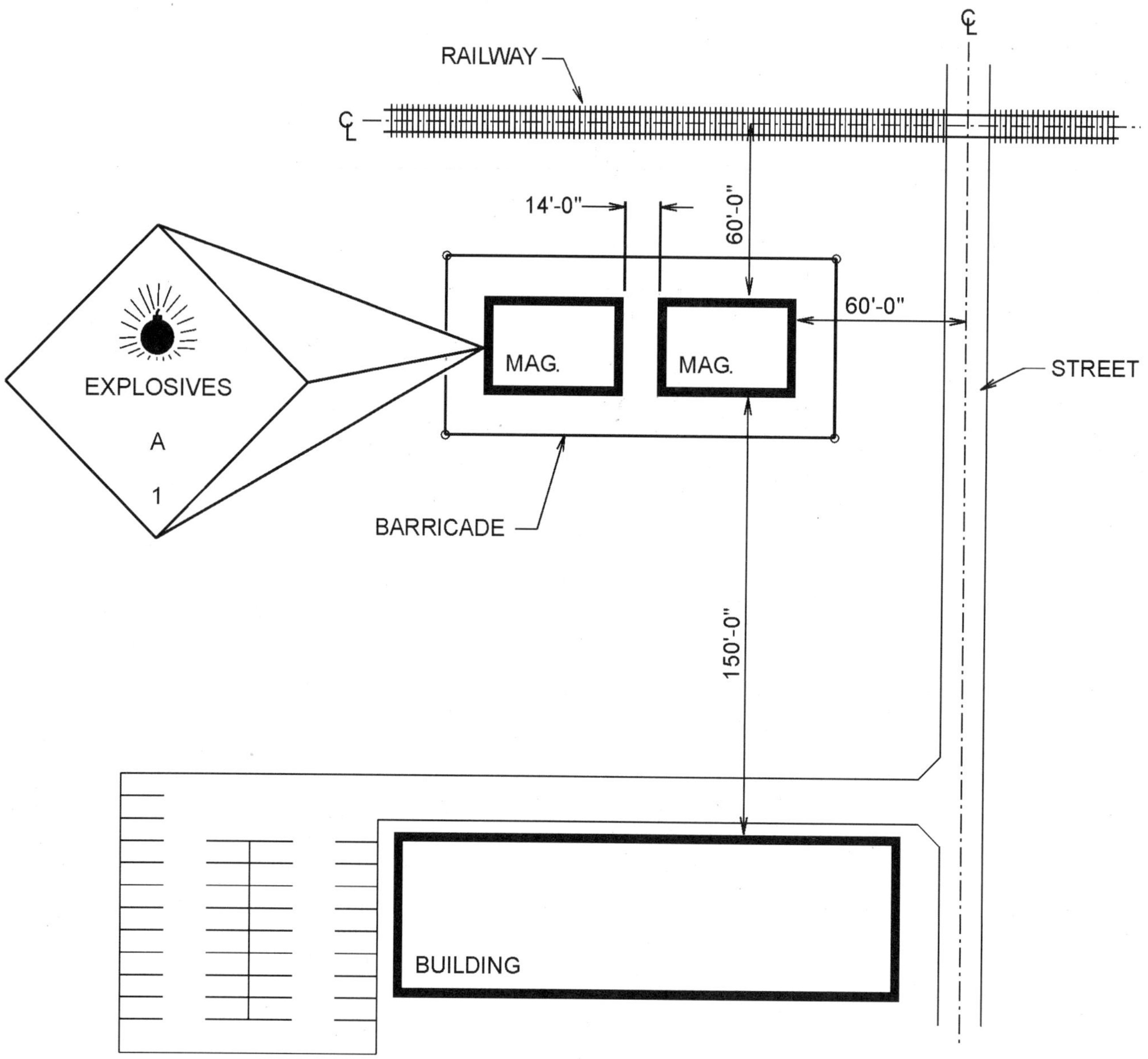

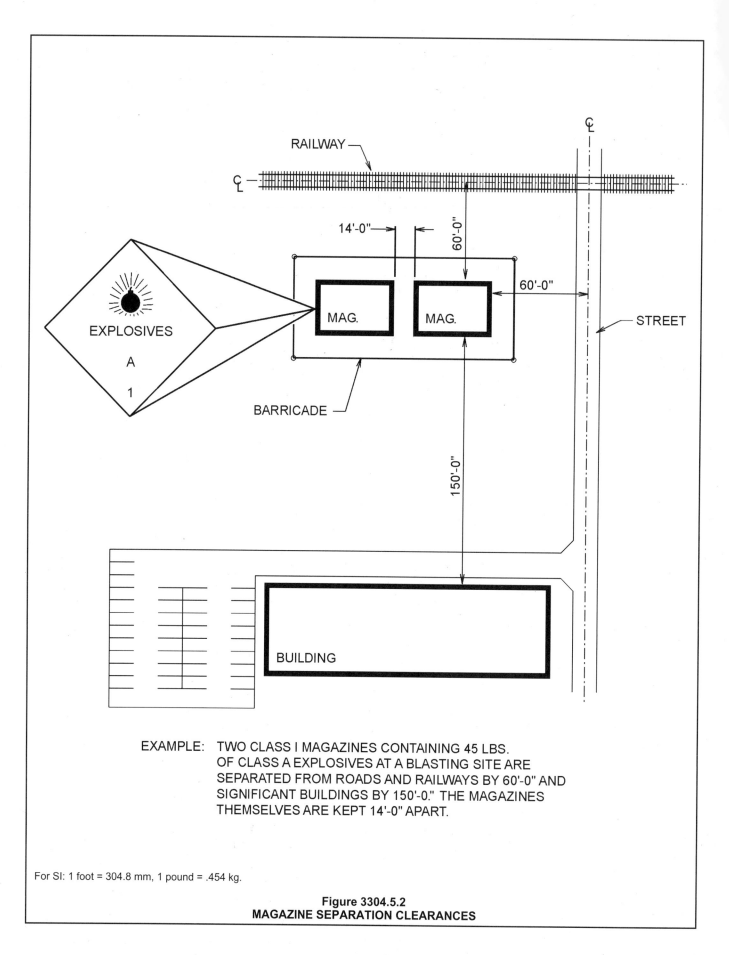

EXAMPLE: TWO CLASS I MAGAZINES CONTAINING 45 LBS. OF CLASS A EXPLOSIVES AT A BLASTING SITE ARE SEPARATED FROM ROADS AND RAILWAYS BY 60'-0" AND SIGNIFICANT BUILDINGS BY 150'-0." THE MAGAZINES THEMSELVES ARE KEPT 14'-0" APART.

For SI: 1 foot = 304.8 mm, 1 pound = .454 kg.

Figure 3304.5.2
MAGAZINE SEPARATION CLEARANCES

TABLE 3304.5.2(2) AMERICAN TABLE OF DISTANCES FOR STORAGE OF EXPLOSIVES AS APPROVED BY THE INSTITUTE OF MAKERS OF EXPLOSIVES AND REVISED JUNE 1991[a]

QUANTITY OF EXPLOSIVE MATERIALS[c]		DISTANCES IN FEET							
		Inhabited buildings		Public highways with traffic volume less than 3,000 vehicles per day		Public highways with traffic volume greater than 3,000 vehicles per day and passenger railways		Separation of magazines[d]	
Pounds over	Pounds not over	Barricaded	Unbarricaded	Barricaded	Unbarricaded	Barricaded	Unbarricaded	Barricaded	Unbarricaded
0	5	70	140	30	60	51	102	6	12
5	10	90	180	35	70	64	128	8	16
10	20	110	220	45	90	81	162	10	20
20	30	125	250	50	100	93	186	11	22
30	40	140	280	55	110	103	206	12	24
40	50	150	300	60	120	110	220	14	28
50	75	170	340	70	140	127	254	15	30
75	100	190	380	75	150	139	278	16	32
100	125	200	400	80	160	150	300	18	36
125	150	215	430	85	170	159	318	19	38
150	200	235	470	95	190	175	350	21	42
200	250	255	510	105	210	189	378	23	46
250	300	270	540	110	220	201	402	24	48
300	400	295	590	120	240	221	442	27	54
400	500	320	640	130	260	238	476	29	58
500	600	240	480	135	270	253	506	31	62
600	700	355	710	145	290	266	532	32	64
700	800	375	750	150	300	278	556	33	66
800	900	390	780	155	310	289	578	35	70
900	1,000	400	800	160	320	300	600	36	72
1,000	1,200	425	850	165	330	318	636	39	78
1,200	1,400	450	900	170	340	336	672	41	82
1,400	1,600	470	940	175	350	351	702	43	86
1,600	1,800	490	980	180	360	366	732	44	88
1,800	2,000	505	1,010	185	370	378	756	45	90
2,000	2,500	545	1,090	190	380	408	816	49	98
2,500	3,000	580	1,160	195	390	432	864	52	104
3,000	4,000	635	1,270	210	420	474	948	58	116
4,000	5,000	685	1,370	225	450	513	1,026	61	122
5,000	6,000	730	1,460	235	470	546	1,092	65	130
6,000	7,000	770	1,540	245	490	573	1,146	68	136
7,000	8,000	800	1,600	250	500	600	1,200	72	144
8,000	9,000	835	1,670	255	510	624	1,248	75	150
9,000	10,000	865	1,730	260	520	645	1,290	78	156
10,000	12,000	875	1,750	270	540	687	1,374	82	164

(continued)

TABLE 3304.5.2(2)–continued
AMERICAN TABLE OF DISTANCES FOR STORAGE OF EXPLOSIVES AS APPROVED BY THE INSTITUTE OF MAKERS OF EXPLOSIVES AND REVISED JUNE 1991[a]

QUANTITY OF EXPLOSIVE MATERIALS[c]		DISTANCES IN FEET							
		Inhabited buildings		Public highways with traffic volume less than 3,000 vehicles per day		Public highways with traffic volume greater than 3,000 vehicles per day and passenger railways		Separation of magazines[d]	
Pounds over	Pounds not over	Barricaded	Unbarricaded	Barricaded	Unbarricaded	Barricaded	Unbarricaded	Barricaded	Unbarricaded
12,000	14,000	885	1,770	275	550	723	1,446	87	174
14,000	16,000	900	1,800	280	560	756	1,512	90	180
16,000	18,000	940	1,880	285	570	786	1,572	94	188
18,000	20,000	975	1,950	290	580	813	1,626	98	196
20,000	25,000	1,055	2,000	315	630	876	1,752	105	210
25,000	30,000	1,130	2,000	340	680	933	1,866	112	224
30,000	35,000	1,205	2,000	360	720	981	1,962	119	238
35,000	40,000	1,275	2,000	380	760	1,026	2,000	124	248
40,000	45,000	1,340	2,000	400	800	1,068	2,000	129	258
45,000	50,000	1,400	2,000	420	840	1,104	2,000	135	270
50,000	55,000	1,460	2,000	44	88	1,140	2,000	140	280
55,000	60,000	1,515	2,000	455	910	1,173	2,000	145	290
60,000	65,000	1,565	2,000	470	940	1,206	2,000	150	300
65,000	70,000	1,610	2,000	485	970	1,236	2,000	155	310
70,000	75,000	1,655	2,000	500	1,000	1,263	2,000	160	320
75,000	80,000	1,695	2,000	510	1,020	1,293	2,000	165	330
80,000	85,000	1,730	2,000	520	1,040	1,317	2,000	170	340
85,000	90,000	1,760	2,000	530	1,060	1,344	2,000	175	350
90,000	95,000	1,790	2,000	540	1,080	1,368	2,000	180	360
95,000	100,000	1,815	2,000	545	1,090	1,392	2,000	185	370
100,000	110,000	1,835	2,000	550	1,100	1,437	2,000	195	390
110,000	120,000	1,855	2,000	555	1,110	1,479	2,000	205	410
120,000	130,000	1,875	2,000	560	1,120	1,521	2,000	215	430
130,000	140,000	1,890	2,000	565	1,130	1,557	2,000	225	450
140,000	150,000	1,900	2,000	570	1,140	1,593	2,000	235	470
150,000	160,000	1,935	2,000	580	1,160	1,629	2,000	245	490
160,000	170,000	1,965	2,000	590	1,180	1,662	2,000	255	510
170,000	180,000	1,990	2,000	600	1,200	1,695	2,000	265	530
180,000	190,000	2,010	2,010	605	1,210	1,725	2,000	275	550
190,000	200,000	2,030	2,030	610	1,220	1,755	2,000	285	570

(continued)

TABLE 3304.5.2(2)—continued
AMERICAN TABLE OF DISTANCES FOR STORAGE OF EXPLOSIVES AS APPROVED BY THE INSTITUTE OF MAKERS OF EXPLOSIVES AND REVISED JUNE 1991[a]

QUANTITY OF EXPLOSIVE MATERIALS[c]		DISTANCES IN FEET							
		Inhabited buildings		Public highways with traffic volume less than 3,000 vehicles per day		Public highways with traffic volume greater than 3,000 vehicles per day and passenger railways		Separation of magazines[d]	
Pounds over	Pounds not over	Barricaded	Unbarricaded	Barricaded	Unbarricaded	Barricaded	Unbarricaded	Barricaded	Unbarricaded
200,000	210,000	2,055	2,055	620	1,240	1,782	2,000	295	590
210,000	230,000	2,100	2,100	635	1,270	1,836	2,000	315	630
230,000	250,000	2,155	2,155	650	1,300	1,890	2,000	335	670
250,000	275,000	2,215	2,215	670	1,340	1,950	2,000	360	720
275,000	300,000[b]	2,275	2,275	690	1,380	2,000	2,000	385	770

For SI: 1 foot = 304.8 mm, 1 pound = 0.454 kg.

a. This table applies only to the manufacture and permanent storage of commercial explosive materials. It is not applicable to transportation of explosives or any handling or temporary storage necessary or incident thereto. It is not intended to apply to bombs, projectiles or other heavily encased explosives.

b. Storage in excess of 300,000 pounds of explosive materials in one magazine is not allowed.

c. Where a manufacturing building on an explosive materials plant site is designed to contain explosive materials, such building shall be located with respect to its proximity to inhabited buildings, public highways and passenger railways based on the maximum quantity of explosive materials permitted to be in the building at one time.

d. Where two or more storage magazines are located on the same property, each magazine shall comply with the minimum distances specified from inhabited buildings, railways and highways, and, in addition, they should be separated from each other by not less than the distances shown for separation of magazines, except that the quantity of explosives in detonator magazines shall govern in regard to the spacing of said detonator magazines from magazines containing other explosive materials. Where any two or more magazines are separated from each other by less than the specified separation of magazines distances, then two or more such magazines, as a group, shall be considered as one magazine, and the total quantity of explosive materials stored in such group shall be treated as if stored in a single magazine located on the site of any magazine in the group and shall comply with the minimum distances specified from other magazines, inhabited buildings, railways and highways.

TABLE 3304.5.2(3)
TABLE OF DISTANCES FOR BUILDINGS CONTAINING EXPLOSIVES—DIVISION 1.3—MASS-FIRE HAZARD[a,b,c]

QUANTITY OF DIVISION 1.3 EXPLOSIVES		DISTANCES IN FEET		
Pounds over	Pounds not over	Inhabited buildings	Passenger railways and public highways	Magazines and operating buildings
0	1,000	75	75	50
1,000	5,000	115	115	75
5,000	10,000	150	150	100
10,000	20,000	190	190	125
20,000	30,000	215	215	145
30,000	40,000	235	235	155
40,000	50,000	250	250	165
50,000	60,000	260	260	175
60,000	70,000	270	270	185
70,000	80,000	280	280	190
80,000	90,000	295	295	195
90,000	100,000	300	300	200
100,000	200,000	375	375	250
200,000	300,000	450	450	300

For SI: 1 foot = 304.8 mm, 1 pound = 0.454 kg.

a. Black powder, when stored in magazines, is defined as low explosive by the Bureau of Alcohol, Tobacco and Firearms (BATF).

b. For quantities less than 1,000 pounds, the required distances are those specified for 1,000 pounds. The use of lesser distances is permitted when supported by approved test data and/or analysis.

c. Linear interpolation of explosive quantities between table entries is permitted.

TABLE 3304.5.2(4)
TABLE OF DISTANCES FOR BUILDINGS CONTAINING EXPLOSIVES—DIVISION 1.4[c]

QUANTITY OF DIVISION 1.4 EXPLOSIVES		DISTANCES IN FEET		
Pounds Over	Pounds Not Over	From Inhabited Building	From Public Railroad and Highway	From Above- ground Magazine and Operating Buildings[a, b]
50	Not Limited	100	100	50

For SI: 1 foot = 304.8 mm, 1 pound = 0.454 kg.

a. A separation distance of 100 feet is required for buildings of other than Type I or Type II construction as defined in the *International Building Code.*

b. For earth-covered magazines, no specified separation is required.

(1) Earth cover material used for magazines shall be relatively cohesive. Solid or wet clay and similar types of soil are to cohesive and shall not be used. Soil shall be free from unsanitary organic matter, trash, debris and stones heavier than 10 pounds or larger than 6 inches in diameter. Compaction and surface preparation shall be provided, as necessary, to maintain structural integrity and avoid erosion. Where cohesive material cannot be used, as in sandy soil, the earth cover over magazines shall be finished with a suitable material to ensure structural integrity.

(2) The earth fill or earth cover between earth-covered magazines shall be either solid or sloped, in accordance with the requirements of other construction features, but a minimum of 2 feet of earth cover shall be maintained over the top of each magazines. To reduce erosion and facilitate maintenance operations, the cover shall have a slope of 2 horizontal to 1 vertical.

c. Restricted to articles, including articles packaged for shipment, that are not regulated as an explosive under Bureau of Alcohol, Tobacco and Firearms regulations, or unpacked articles used in process operations that do not propagate a detonation of deflagration between articles.

3304.5.3.1 Location. Wherever practicable, Type 3 magazines shall be located away from neighboring inhabited buildings, railways, highways, and other magazines in accordance with Table 3304.5.2(2), 3304.5.2(3) or 3304.5.2(4) as applicable.

❖ Type 3 magazines must be located as remotely as practical from public rights-of-way, buildings and other magazines to reduce exposures between them.

3304.5.3.2 Supervision. Type 3 magazines shall be attended when explosive materials are stored within. Explosive materials shall be removed to appropriate storage magazines for unattended storage at the end of the work day.

❖ The magazine must be constantly supervised by a trained attendant when explosives are being stored to prevent unauthorized removal of the contents and to enforce safety rules.

3304.5.3.3 Use. Not more than two Type 3 magazines shall be located at the same blasting site. Where two Type 3 magazines are located at the same blasting site, one magazine shall used solely for the storage of detonators.

❖ Detonators must be stored in separate magazines from explosives to prevent an accidental mass detonation.

3304.6 Construction. Magazines shall be constructed in accordance with Sections 3304.6.1 through 3304.6.5.2.

❖ Magazine construction requirements are based on requirements in DOTy 27 CFR 55 and NFPA 495. Magazines are identified by designation as one of five types: Types 1 through 5.

3304.6.1 Drainage. The ground around a magazine shall be graded so that water drains away from the magazine.

❖ Grading land away from the magazine allows storm water to run off and not accumulate in, under or around a magazine. Water may damage explosives. Contaminated water may conduct an electric current, possibly resulting in the detonation of explosives.

3304.6.2 Heating. Magazines requiring heat shall be heated as prescribed in NFPA 495 by either hot water radiant heating within the magazine or by indirect warm air heating.

❖ Indirect heating systems can prevent freezing and avoid excessive heating of magazine contents. These systems also do not constitute an ignition hazard.

3304.6.3 Lighting. When lighting is necessary within a magazine, electric safety flashlights or electric safety lanterns shall be used, except as provided in NFPA 495.

❖ Safety flashlights and safety lanterns used inside magazines must be designed for use in hazardous locations. Fixed electric lighting installed as specified in Section 65.5.1 of NFPA 495 is acceptable. All electrical work must conform to NFPA 70 requirements. DOTy 27 CFR; 55.217(c) requires that copies of documents indicating that the electrical installation conforms to *ICC Electrical Code®* (ICC EC™) be maintained on site and always available for inspections. Requirements of ICC EC and NFPA 495 are intended to minimize the possibility that sparks generated by electrical arcing during normal operations or through equipment failures will ignite explosives within the magazine.

3304.6.4 Nonsparking materials. In other than Type 5 magazines, there shall be no exposed ferrous metal on the interior of a magazine containing packages of explosives.

❖ Ferrous metal must not be exposed in the interior of a Type 1, 2, 3 or 4 magazine where the metal could come in contact with packages of explosives.

3304.6.5 Signs and placards. Property upon which Type 1 magazines and outdoor magazines of Types 2, 4 and 5 are located shall be posted with signs stating: EXPLOSIVES — KEEP OFF. These signs shall be of contrasting colors with a minimum letter height of 3 inches (76 mm) with a minimum brush stroke of 0.5 inch (12.7 mm). The signs shall be located to minimize the possibility of a bullet shot at the sign hitting the magazine.

❖ Signs are intended to identify the public hazard and provide warning to fire fighters. Signs should also be prominently displayed at the entrance to the property and at regular intervals around the perimeter of the property. Contrasting letters and backgrounds on the signs (red on white preferred) should be used with letters that are legible from a distance. A minimum 3-inch (76 mm) height is recommended for the letters. The IME recommends the following wording in addition to that required by this section:

DANGER!
NEVER FIGHT EXPLOSIVE FIRES.
EXPLOSIVES ARE STORED AT THIS SITE.
CALL: _________.

Military and DOT explosives warning signs use black letters, numerals and symbols on orange backgrounds. Nonreflective sign materials make these types of signs difficult to identify and read. Signs must always be conspicuously posted and maintained in a clear, clean and legible condition.

Signs are attractive targets for irresponsible firearms bearers. Therefore, signs must be located in anticipation of an individual shooting a rifle or handgun at the sign so the bullet will not hit the magazine.

The absence of clear warning signs was cited as a contributing factor in the November 1988 deaths of six Kansas City, Missouri, fire fighters who were approaching a Type 4 magazine at a remote highway construction site when it exploded. A similar incident in March 1989 in Peterborough, England, killed one fire fighter and injured 76 other persons, including six fire fighters. The Peterborough, England, incident occurred while fire fighters were trying to control a fire in a motor carrier hauling explosives.

3304.6.5.1 Access road signs. At the entrance to explosive material manufacturing and storage sites, all access roads shall be posted with the following warning sign or other approved sign:

DANGER!
NEVER FIGHT EXPLOSIVE FIRES.
EXPLOSIVES ARE STORED ON THIS SITE
CALL _______.

The sign shall be weather resistant with a reflective surface and have lettering at least 2 inches (51 mm) high.

❖ Signs are intended to identify the public hazard and warn fire fighters. Signs should also be prominently displayed at the entrance to the property and at regular intervals around the perimeter of the property. Contrasting letters and backgrounds on the signs (red on white preferred) should be used with letters that are legible from a distance. A minimum 2-inch (51 mm) height is recommended for the letters.

3304.6.5.2 Placards. Type 5 magazines containing Division 1.5 blasting agents shall be prominently placarded as required during transportation by DOTn 49 CFR, Part 172 and DOTy 27 CFR, Part 55.

❖ Requirements for placarding explosives shipments are found in DOTn 49 CFR; 172, Subpart F. Figure 3304.6.5.2 depicts the format of a typical explosives placard (black graphics on an orange background). Placards allow emergency response personnel to readily identify the explosive materials being transported if a fire or accident involving the transport vehicle occurs.

Figure 3304.6.5.2
AN EXAMPLE OF AN EXPLOSIVES PLACARD

3304.7 Operation. Magazines shall be operated in accordance with Sections 3304.7.1 through 3304.7.9.

❖ The following sections address the operations of magazines. Items covered include security; open flames; brush; combustible storage; unpacking and packing; tools and equipment.

3304.7.1 Security. Magazines shall be kept locked in the manner prescribed in NFPA 495 at all times except during placement or removal of explosives or inspection.

❖ Security precautions must be taken to minimize the possibility of theft, tampering or misuse of explosives.

3304.7.2 Open flames and lights. Smoking, matches, flame-producing devices, open flames, firearms and firearms cartridges shall not be permitted inside of or within 50 feet (15 240 mm) of magazines.

❖ The carrying or use of open flames, spark-producing devices, matches and firearms is prohibited to prevent accidental ignition of explosives within the magazine.

3304.7.3 Brush. The area located around a magazine shall be kept clear of brush, dried grass, leaves, trash, debris, and similar combustible materials for a distance of 25 feet (7620 mm).

❖ Brush and dry grass, leaves, trash and other easily ignitable debris must be kept clear for a distance of 25 feet (7620 mm) around magazines containing explosives to minimize the threat to magazine contents from fire exposure.

3304.7.4 Combustible storage. Combustible materials shall not be stored within 50 feet (15 240 mm) of magazines.

❖ Elimination of combustible materials within 50 feet (15 240 mm) of magazines eliminates both a fuel source and a possible source of ignition for stored explosives.

3304.7.5 Unpacking and repacking explosive materials. Containers of explosive materials, except fiberboard containers, and packages of damaged or deteriorated explosive materials or fireworks shall not be unpacked or repacked inside or within 50 feet (15 240 mm) of a magazine or in close proximity to other explosive materials.

❖ The packing or unpacking of boxes, crates, drums or other containers with metal staples, bands, nails or other ferrous parts may generate sparks, which could possibly ignite explosives stored within a magazine. The separation distance is intended to reduce the possibility of secondary explosions if an accident involving the explosives being handled occurs.

3304.7.5.1 Storage of opened packages. Packages of explosive materials that have been opened shall be closed before being placed in a magazine.

❖ The closing of the opened packages eliminates the hazard of mixing with other explosives, being available in case of an ignition source and causing an explosion or fire.

3304.7.5.2 Nonsparking tools. Tools used for the opening and closing of packages of explosive materials, other than metal slitters for opening paper, plastic or fiberboard containers, shall be made of nonsparking materials.

❖ The nonsparking tools are used to avoid producing sparks and igniting the explosives. These tools must either be made from nonsparking materials or be painted to prevent them from producing sparks.

3304.7.5.3 Disposal of packaging. Empty containers and paper and fiber packaging materials that previously contained explosive materials shall be disposed of or reused in a approved manner.

❖ Debris from explosives shipments or use, including discarded packages, bags and cartons, must be disposed in accordance with the explosive manufacturer's instructions.

3304.7.6 Tools and equipment. Metal tools, other than nonferrous transfer conveyors and ferrous metal conveyor stands protected by a coat of paint, shall not be stored in a magazine containing explosive materials or detonators.

❖ Tools that can be stored in a magazine are limited to those needed to maintain the magazine and facilitate the transfer of explosives. These tools must be nonsparking or painted to prevent them from producing sparks. Ferrous tools with chipped or worn protective coatings must be removed from service and repainted before being placed back in use.

3304.7.7 Contents. Magazines shall be used exclusively for the storage of explosive materials, blasting materials and blasting accessories.

❖ Additional materials likely to pose a fire hazard or produce sparks when used cannot be stored inside magazines.

3304.7.8 Compatibility. Corresponding grades and brands of explosive materials shall be stored together and in such a manner that the grade and brand marks are visible. Stocks shall be stored so as to be easily counted and checked. Packages of explosive materials shall be stacked in a stable manner not exceeding 8 feet (2438 mm) in height.

❖ Identical grades and brands of explosives must be stored together, with the brands and grade marks showing. Explosive materials must be stored so they can be easily checked and counted.

3304.7.9 Stock rotation. When explosive material is removed from a magazine for use, the oldest usable stocks shall be removed first.

❖ Use of the oldest useable stock first assists in elimination of explosive material before it can deteriorate and become hazardous.

3304.8 Maintenance. Maintenance of magazines shall comply with Sections 3304.8.1 through 3304.8.3.

❖ This section addresses housekeeping, repairs and the floors on which the magazines are maintained.

3304.8.1 Housekeeping. Magazine floors shall be regularly swept and be kept clean, dry and free of grit, paper, empty packages and rubbish. Brooms and other cleaning utensils shall not have any spark-producing metal parts. Sweepings from magazine floors shall be disposed of in accordance with the manufacturers' approved instructions.

❖ Dirt and debris may contain materials that produce sparks or friction when explosives are moved within the magazine. Some blasting agents become sensitized or unstable when contaminated. Therefore, regular cleaning and a maintenance schedule should be established and maintained.

Cleaning implements with ferrous parts may produce

sparks when used improperly. Only nonsparking tools may be used inside the magazine.

Finely divided residue on packages may pose an extreme fire hazard when improperly discarded. Most manufacturers recommend burning packages and debris from explosives at an approved remote location. The IME Safety Library Publication No. 21 contains general guidance on the subject of disposing packages of explosives. Poor housekeeping often indicates other safety hazards.

3304.8.2 Repairs. Explosive materials shall be removed from the magazine before making repairs to the interior of a magazine. Explosive materials shall be removed from the magazine before making repairs to the exterior of the magazine where there is a possibility of causing a fire. Explosive materials removed from a magazine under repair shall either be placed in another magazine or placed a safe distance from the magazine, where they shall be properly guarded and protected until repairs have been completed. Upon completion of repairs, the explosive materials shall be promptly returned to the magazine. Floors shall be cleaned before and after repairs.

❖ Explosives must be removed from the magazine under repair and placed in another magazine or a safe distance away before beginning repair activities that could cause sparks or fire, and the floor must be cleaned before beginning repairs inside a magazine. Explosives must be properly guarded until they are returned to the magazine.

3304.8.3 Floors. Magazine floors stained with liquid shall be dealt with according to instructions obtained from the manufacturer of the explosive material stored in the magazine.

❖ Floors stained with nitroglycerin must be cleaned according to the manufacturer's instructions. If the manufacturer cannot be identified or reached, another manufacturer, the nearest U.S. military installation, the ATF regional office or a local law enforcement agency must be contacted to request an explosive ordnance disposal (EOD) team.

3304.9 Inspection. Magazines containing explosive materials shall be opened and inspected at maximum 7-day intervals. The inspection shall determine whether there has been an unauthorized or attempted entry into a magazine or an unauthorized removal of a magazine or its contents.

❖ The importance of magazine security and explosives accountability cannot be overestimated, especially in light of the upsurge in incidents of international terrorists using bombs. In most cases of illicit use of explosives, the source of the explosive materials has been legitimate operating explosive material users. Magazines must be inspected at least weekly to determine that no theft or attempted theft has occurred. Thefts or tampering should be reported immediately to the fire code official and local and federal law enforcement officials.

3304.10 Disposal of explosive materials. Explosive materials shall be disposed of in accordance with Sections 3304.10.1 through 3304.10.7.

❖ Sections 3304.10.1 through 3304.10.7 cover the procedures for proper disposal of explosive materials. Areas covered are notification, deteriorated materials, qualified persons, storage of misfires, disposal sites, reuse of sites and personnel safeguards.

3304.10.1 Notification. The fire code official shall be notified immediately when deteriorated or leaking explosive materials are determined to be dangerous or unstable and in need of disposal.

❖ The notification must include the following for each site where explosive material is stored: type of explosives, magazine capacity and location.

3304.10.2 Deteriorated materials. When an explosive material has deteriorated to an extent that it is in an unstable or dangerous condition, or when a liquid has leaked from an explosive material, the person in possession of such material shall immediately contact the material's manufacturer to obtain disposal and handling instructions.

❖ Damaged or deteriorated explosives may become sensitized and unpredictable. Extreme caution must be used in handling such materials. Some explosives, such as nitroglycerin, become shock sensitive when contaminated or deteriorated because of age, exposure to ultraviolet light or excessive heating. Emergency forces should be summoned for standby while cleanup procedures are underway in situations involving sensitized or unstable explosives. If an accident occurs, fire fighters should attempt to protect exposures only from a distance and rescue individuals who are in immediate danger and safely reachable. Explosives manufacturers do not recommend fighting fires involving explosives because of the extreme danger associated with such operations. Manufacturers can provide advice and assistance in identifying a safe means of disposing of damaged explosives.

3304.10.3 Qualified person. The work of destroying explosive materials shall be directed by persons experienced in the destruction of explosive materials.

❖ Only an approved and experienced explosives technician is permitted to dispose of damaged, dangerous or unstable explosives. IME Safety Library Publication No. 21 contains general guidance on the safe destruction of damaged explosives and explosive waste. However, the manufacturer's specific guidance should always be followed first.

A military or law enforcement EOD team should be contacted if the licensee or permittee cannot provide a suitable explosives technician. Member companies of the IME have entered into a cooperative agreement to give advice and consultation to law enforcement agencies, fire departments and regulatory officials regarding the destruction or disposal of dangerous, damaged or unstable commercial explosives. Another manufacturer can lend assistance if the manufacturer of the damaged or dangerous goods cannot be identified or located.

3304.10.4 Storage of misfires. Explosive materials and fireworks recovered from blasting or display misfires shall be placed in a magazine until an experienced person has determined the proper method for disposal.

❖ Explosives recovered from misfires must be placed in a separate licensed magazine until they can be disposed of according to the manufacturer's recommendations.

3304.10.5 Disposal sites. Sites for the destruction of explosive materials and fireworks shall be approved and located at the maximum practicable safe distance from inhabited buildings, public highways, operating buildings, and all other exposures to ensure keeping air blast and ground vibration to a minimum. The location of disposal sites shall be no closer to magazines, inhabited buildings, railways, highways and other rights-of-way than is permitted by Tables 3304.5.2(1), 3304.5.2(2) and 3304.5.2(3). When possible, barricades shall be utilized between the destruction site and inhabited buildings. Areas where explosives are detonated or burned shall be posted with adequate warning signs.

❖ A blasting shelter should be located near the burn area for protection from the blast and for emergency use of personnel.

3304.10.6 Reuse of site. Unless an approved burning site has been thoroughly saturated with water and has passed a safety inspection, 48 hours shall elapse between the completion of a burn and the placement of scrap explosive materials for a subsequent burn.

❖ Procedures and safeguards must be in place to prevent scrap explosive material from being placed in any burn location until at least 48 hours has passed since the last fires have gone out.

3304.10.7 Personnel safeguards. Once an explosive burn operation has been started, personnel shall relocate to a safe location where adequate protection from air blast and flying debris is provided. Personnel shall not return to the burn area until the person in charge has inspected the burn site and determined that it is safe for personnel to return.

❖ A warning device must be used when explosives and fireworks materials are being destroyed. This warning device may also be used to advise personnel when the area is safe for reentry.

SECTION 3305 MANUFACTURE, ASSEMBLY AND TESTING OF EXPLOSIVES, EXPLOSIVE MATERIALS AND FIREWORKS

3305.1 General. The manufacture, assembly and testing of explosives, ammunition, blasting agents and fireworks shall comply with the requirements of this section and NFPA 495 or NFPA 1124.

Exceptions:

1. The hand loading of small arms ammunition prepared for personal use and not offered for resale.
2. The mixing and loading of blasting agents at blasting sites in accordance with NFPA 495.
3. The use of binary explosives or plosophoric materials in blasting or pyrotechnic special effects applications in accordance with NFPA 495 or NFPA 1126.

❖ This section covers emergency planning; separation of operating and manufacturing buildings; control areas; operations and maintenance for the manufacturing, assembling, and testing of explosives, blasting agents and fireworks. Section 3306 covers small arms ammunition. Exceptions 2 and 3 are addressed in NFPA standards.

Fireworks manufacturing is a relatively localized or concentrated industry. Most fireworks plants have settled and remain in predominantly rural locations for several reasons. Substantial sites are required to allow for isolation of manufacturing and storage buildings from one another to limit hazard exposures. These sites permit insulation from regulatory supervision. Skilled workers are required to assemble the fireworks; as a result, a loyal workforce usually develops in the regions where plants are located. Consequently, the local economy usually becomes heavily dependent on the industry's presence. Furthermore, public concern and outcry is minimized by these factors, since occasional accidents are more or less expected. Increased regulatory attention and the prohibition of common fireworks in many states has, however, led to increased foreign competition that has further impacted the industry. Today, fewer than 100 American firms are engaged in this activity and they employ less than 3,000 persons. The U.S. Department of Treasury enforces regulations (DOTy 27 CFR; 55 and 181) restricting the import, manufacture, storage and use of fireworks and explosives.

3305.2 Emergency planning and preparedness. Emergency plans, emergency drills, employee training and hazard communication shall conform to the provisions of this section and Sections 404, 405, 406 and 407.

❖ Sections 404, 405 and 407 are also included for assistance with this section.

3305.2.1 Hazardous Materials Management Plans and Inventory Statements required. Detailed Hazardous Materials Management Plans (HMMP) and Hazardous Materials Inventory Statements (HMIS) complying with the requirements of Section 407 shall be prepared and submitted to the local emergency planning committee, the fire code official, and the local fire department.

❖ Section 407.6 requires that Hazardous Materials Management Plans (HMMP) be in accordance with Section 2701.4.1, which gives the requirements for a facility site plan. Section 407.5 requires that the Hazardous Materials Inventory Statements (HMIS) be in accordance with Section 2701.4.2, which gives the items required in an HMIS.

3305.2.2 Maintenance of plans. A copy of the required HMMP and HMIS shall be maintained on-site and furnished to the fire code official on request.

❖ A copy of the plant's plans must be kept in the office on the premises of each explosives, ammunition, blasting agent or fireworks manufacturing plant and must be made available to the fire code official or an authorized representative upon request.

3305.2.3 Employee training. Workers who handle explosives or explosive charges or dispose of explosives shall be trained in the hazards of the materials and processes in which they are to be engaged and with the safety rules governing such materials and processes.

❖ Employees who are required to handle the explosive materials need instructions concerning the hazards of the products. MSDS are a great source for the hazards of the materials. The employees must go through an orientation on the rules and regulations concerning the processes.

3305.2.4 Emergency procedures. Approved emergency procedures shall be formulated for each plant which will include personal instruction in any emergency that may be anticipated. All personnel shall be made aware of an emergency warning signal.

❖ Emergency procedures must be developed for each plant and building. Fire and disaster drills must be conducted and records of those drills must be kept at the plant office.

3305.3 Intraplant separation of operating buildings. Explosives and fireworks manufacturing buildings, including those where explosive charges are assembled, manufactured, prepared or loaded utilizing Division 1.1, 1.2, 1.3, 1.4 or 1.5 explosives, shall be separated from all other buildings, including magazines, within the confines of the manufacturing plant at a distance not less than those shown in Table 3305.3, 3304.5.2 (3), or Table 3304.5.2 (4), as appropriate.

The quantity of explosives in an operating building shall be the net weight of all explosives contained therein. Distances shall be based on the hazard division requiring the greatest separation, unless the aggregate explosive weight is divided by approved walls or shields designed for that purpose. When dividing a quantity of explosives into smaller stacks, a suitable barrier or adequate separation distance shall be provided to prevent propagation from one stack to another.

When distance is used as the sole means of separation within a building, such distance shall be established by testing. Testing shall demonstrate that propagation between stacks will not result. Barriers provided to protect against explosive effects shall be designed and installed in accordance with approved standards.

Exception: Fireworks-manufacturing buildings separated in accordance with NFPA 1124.

❖ Manufacturing operations frequently contain explosive materials that may fall into more than one division. When such is the case, quantity/distance relationships are established on the most severe case. The separation of explosives into piles by distance or the construction of substantial dividing walls (barrier wall) is a recognized means of mitigating the effects of conflagration by propagation. Propagation distances are established by testing. The construction of barriers can be accomplished by various means; however, the design of such structures must be specific to address the nature (impulse pressures) and time factors involved with the energy produced by explosive materials. The primary reference is U.S. Army Technical Manual TM 51300, Air Force Manual (AFM) 8822 and Navy NAVFAC P397, *Structures to Resist the Effects of Accidental Explosions*. The exception recognizes that NFPA 1124 requirements cover site security, separation distances for manufacturing facilities; storage of salute and black powder; building construction and transportation of fireworks.

TABLE 3305.3. See page 33-28.

❖ Table 3305.3 has been designed to recognize explosives that either detonate or present a mass-explosion hazard (Divisions 1.1, 1.2 and 1.5). Under the classification system used by the code there is a need to recognize the various hazards of explosive materials when classified in other divisions, specifically Division 1.3 and 1.4 materials. The hazards of Division 1.3 materials are mass fire hazards where fire can involve the aggregate amount of undivided material. As finished goods, the hazards of Division 1.4 materials are limited to events with the individual articles. Division 1.5 materials are capable of detonation, and including them in Table 3305.3 represents the most conservative case. This table applies to the separation distances of buildings within an explosive manufacturing plant. Of course, the distance is related to the amount of explosives in storage. Distance and barricades are means of reducing damage to property and injury to personnel. With a barricade, the distance is sufficient, but without a barricade, the distance must be doubled to be the equivalent of a barricade. The quantity is calculated in pounds in each structure or building.

3305.4 Separation of manufacturing buildings from inhabited buildings, rights-of-way, and magazines. When a manufacturing building on an explosive materials plant site is designed to contain explosive materials, such building shall be located away from inhabited buildings, public highways, and passenger railways in accordance with Table 3304.5.2(2), 3304.5.2(3) or 3304.5.2(4) as appropriate, based on the maximum quantity of explosive materials permitted to be in the building at one time.

Exception: Fireworks-manufacturing buildings constructed and operated in accordance with NFPA 1124.

❖ Buildings and other facilities used for mixing of blasting agents at a fixed location must comply with the separation requirements of Table 3304.5.2(2). The buildings must be isolated for the protection of life and nearby property. Suitable exterior barricades, natural or artificial, are to be provided on pressure-relief sides of buildings. The exception recognizes that NFPA 1124 requirements cover site security; separation distances for manufacturing facilities; storage of salute and black powder; building construction and transportation of fireworks.

3305.5 Buildings and equipment. Buildings or rooms that exceed the maximum allowable quantity per control area of explosive materials shall be operated in accordance with this section and constructed in accordance with the requirements of the *International Building Code* for Group H occupancies.

Exception: Fireworks-manufacturing buildings constructed and operated in accordance with NFPA 1124.

❖ The IBC addresses the issue of buildings that exceed the maximum allowable quantities per control area found in Table 307.7(1) of the IBC as well as in Chapter 27 of the code. The exception recognizes that NFPA 1124 requirements cover site security; separation distances for manufacturing facilities; storage of salute and black powder; building construction and transportation of fireworks.

TABLE 3305.3
MINIMUM INTRAPLANT SEPARATION DISTANCES BETWEEN BARRICADED OPERATING BUILDINGS CONTAINING EXPLOSIVES—DIVISION 1.1, 1.2 OR 1.5—MASS EXPLOSION HAZARD[a]

EXPLOSIVES			EXPLOSIVES		
Pounds over	Pounds not over	Minimum distance (feet)	Pounds over	Pounds not over	Minimum distance (feet)
0	50	30	20,000	25,000	265
50	100	40	25,000	30,000	280
100	200	50	30,000	35,000	295
200	300	60	35,000	40,000	310
300	400	65	40,000	45,000	320
400	500	70	45,000	50,000	330
500	600	75	50,000	55,000	340
600	700	80	55,000	60,000	350
700	800	85	60,000	65,000	360
800	900	90	65,000	70,000	370
900	1,000	95	70,000	75,000	385
1,000	1,500	105	75,000	80,000	390
1,500	2,000	115	80,000	85,000	395
2,000	3,000	130	85,000	90,000	400
3,000	4,000	140	90,000	95,000	410
4,000	5,000	150	95,000	100,000	415
5,000	6,000	160	100,000	125,000	450
6,000	7,000	170	125,000	150,000	475
7,000	8,000	180	150,000	175,000	500
8,000	9,000	190	175,000	200,000	525
9,000	10,000	200	200,000	225,000	550
10,000	15,000	225	225,000	250,000	575
15,000	20,000	245	250,000	275,000	600
—	—	—	275,000	300,000	635

For SI: 1 foot = 304.8 mm, 1 pound = 0.454 kg.

a. Where a building or magazine containing explosives is not barricaded, the intraline distances shown in this table shall be doubled.

3305.5.1 Explosives dust. Explosives dust shall not be exhausted to the atmosphere.

❖ Exhausting explosive dust into the atmosphere creates an explosive hazard. When the concentration of air and dust reaches the ignition stage, explosions can occur and cause both physical and structural damage.

3305.5.1.1 Wet collector. When collecting explosives dust, a wet collector system shall be used. Wetting agents shall be compatible with the explosives. Collector systems shall be interlocked with process power supplies so that the process cannot continue without the collector systems also operating.

❖ Wet processes will be used wherever practicable in mixing explosive materials. Wet processing eliminates the probability of dust being an explosive hazard.

3305.5.1.2 Waste disposal and maintenance. Explosives dust shall be removed from the collection chamber as often as necessary to prevent overloading. The entire system shall be cleaned at a frequency that will eliminate hazardous concentrations of explosives dust in pipes, tubing and ducts.

❖ Ducts are to be grounded and be as short and straight as possible with no caps, outlets, pockets or other dead-end spaces where explosives might accumulate.

3305.5.2 Exhaust fans. Squirrel cage blowers shall not be used for exhausting hazardous fumes, vapors or gases. Only nonferrous fan blades shall be used for fans located within the ductwork and through which hazardous materials are exhausted. Motors shall be located outside the duct.

❖ Ventilation and dust control equipment must be of such a type and installed and operated to not endanger employees by possible ignition of explosives.

3305.5.3 Work stations. Work stations shall be separated by distance, barrier or other approved alternatives so that fire in one station will not ignite material in another work station. Where necessary, the operator shall be protected by a personnel shield located between the operator and the explosive device or explosive material being processed. This shield and its support shall be capable of withstanding a blast from the maximum amount of explosives allowed behind it.

❖ Using safety precautions should eliminate most problems in a work station; however, sometimes accidents happen beyond our control. The separation requirements in this section are intended to protect workstations from accidents in another station and their operators from all accidental blasts. Protection of the worker must always be of high priority.

3305.6 Operations. Operations involving explosives shall comply with Sections 3305.6.1 through 3305.6.10.

❖ The following sections address the operations that involve explosives. These sections apply to isolating the operations, controlling static electricity, using approved containers for bulk materials, having quantity limits, properly disposing of waste, abiding by safety rules, posting limits and not storing explosive materials near a heat source.

3305.6.1 Isolation of operations. When the type of material and processing warrants, mechanical operations involving explosives in excess of 1 pound (0.454 kg) shall be carried on at isolated stations or at intraplant distances, and machinery shall be controlled from remote locations behind barricades or at separations so that workers will be at a safe distance while machinery is operating.

❖ The machinery used in processing operations on explosive items weighing more than 1 pound (0.454 kg) must be controlled from remote locations and behind barricades for the safety of personnel. These extra safety controls are not required when no unit or separate article of any manufactured or assembled explosive device contains more than 1 pound (.45 kg) of explosive material.

3305.6.2 Static controls. The work area where the screening, grinding, blending and other processing of static-sensitive explosives or pyrotechnic materials is done shall be provided with approved static controls.

❖ The working area must be maintained above 20-percent relative humidity. If the relative humidity drops below 20 percent, the above operations must be stopped and secured until the relative humidity can be raised above 20 percent. It is desirable to keep the relative humidity above 20 to 30 percent, except where metal powders are involved, then the relative humidity should be between 50 and 60 percent. Means must be provided and used to discharge static electricity from hand trucks, buggies and similar equipment before they enter buildings containing static-sensitive explosives. Conductive wheels, including metal wheels that could not cause sparks, are recommended for this equipment.

3305.6.3 Approved containers. Bulk explosives shall be kept in approved, nonsparking containers when not being used or processed. Explosives shall not be stored or transported in open containers.

❖ Explosive materials must be kept in covered containers except when being used or processed. The closed containers eliminate the possibility of an ignition source falling into the containers and causing an explosion.

3305.6.4 Quantity limits. The quantity of explosives at any particular work station shall be limited to that posted on the load limit signs for the individual work station. The total quantity of explosives for multiple workstations shall not exceed that established by the intraplant distances in Table 3305.3, 3304.5.2 (3) or 3304.5.2 (4), as appropriate.

❖ The maximum permissible quantities of explosive materials allowed by Table 3305.3 must be clearly indicated with suitable signs, usually with letters not less that 3 inches (76 mm) high.

3305.6.4.1 Magazines. Magazines used for storage in processing areas shall be in accordance with the requirements of Section 3304.5.1. All explosive materials shall be removed to appropriate storage magazines for unattended storage at the end of the work day. The contents of indoor magazines shall be added to the quantity of explosives contained at individual workstations and the total quantity of material stored, processed or used shall be utilized to establish the intraplant separation distances indicated by Table 3305.3, 3304.5.2(3) or 3304.5.2(4), as appropriate.

❖ This section makes it clear that indoor magazines can be used in manufacturing areas when in accordance with the requirements of the code, including quantity limits, separation distances and limitations on the number of magazines. Materials contained in magazines add to the net explosive weight (NEW) for the building, and intraplant separations are based on NEW.

This section also recognizes that there may be more than one workstation in a process area, with the control on the location of process areas established by intraplant distances. Removal of any residual stored contents at the end of the day is consistent with requirements of NFPA 495 and military standards governing process operations.

3305.6.5 Waste disposal. Approved receptacles with covers shall be provided for each location for disposing of waste material and debris. These waste receptacles shall be emptied and cleaned as often as necessary but not less than once each day or at the end of each shift.

❖ Waste receptacles with tight-fitting covers must be supplied at or just outside of each working area for the disposal of waste material, cleaning rags and other combustible waste and debris. The containers must be emptied daily or at the end of each shift.

3305.6.6 Safety rules. General safety rules and operating instructions governing the particular operation or process conducted at that location shall be available at each location.

❖ Operating instructions must be posted at or near each working area. These instructions must include the technical steps for the process and the safety precautions the operators must follow for their personal safety.

3305.6.7 Personnel limits. The number of occupants in each process building and in each magazine shall not exceed the number necessary for proper conduct of production operations.

❖ The number of persons involved in performing the process must be held to a minimum to make sure there is enough room for those performing the operation. Too many occupants reduces operating space, thus causing a possible accident that could result in personnel injury or building damage.

3305.6.8 Pyrotechnic and explosive composition quantity limits. Not more than 500 pounds (227 kg) of pyrotechnic or explosive composition, including not more than 10 pounds (5 kg) of salute powder shall be allowed at one time in any process building or area. All compositions not in current use shall be kept in covered nonferrous containers.

Exception: Composition that has been loaded or pressed into tubes or other containers as consumer fireworks.

❖ The maximum amount of pyrotechnic or explosive materials contained at any one time in any building used for the manufacture or assembly of products using or containing explosive materials must not exceed the amounts listed above. This action will assist in eliminating the possibility of a catastrophic explosion.

3305.6.9 Posting limits. The maximum number of occupants and maximum weight of pyrotechnic and explosive composition permitted in each process building shall be posted in a conspicuous location in each process building or magazine.

❖ Only persons essential to the operation can be allowed in the mixing and packaging area. Usually no more than one day's production of explosive material should be allowed in the mixing and packaging area. These limitations must be posted in an obvious location.

3305.6.10 Heat sources. Fireworks, explosives or explosive charges in explosive materials manufacturing, assembly or testing shall not be stored near any source of heat.

Exception: Approved drying or curing operations.

❖ Storing the explosive materials near heat sources obviously will be an explosion hazard. Stored explosive products must be separated an adequate distance from any heat source. The building heat source should be located outside the building.

3305.7 Maintenance. Maintenance and repair of explosives-manufacturing facilities and areas shall comply with Section 3304.8.

❖ Section 3304.8 covers the subjects of housekeeping, repairs and floors of buildings and facilities. Having a periodic maintenance program in place will keep hazards to a minimum.

3305.8 Explosive materials testing sites. Detonation of explosive materials or ignition of fireworks for testing purposes shall be done only in isolated areas at sites where distance, protection from missiles, shrapnel or flyrock, and other safeguards provides protection against injury to personnel or damage to property.

❖ Operations or activities on a site where explosive materials are used, stored or handled could be subject to additional or more restrictive requirements or conditions at the discretion of the local fire code official.

3305.8.1 Protective clothing and equipment. Protective clothing and equipment shall be provided to protect persons engaged in the testing, ignition or detonation of explosive materials.

❖ Personnel must be dressed in proper personal protective equipment to eliminate the hazard of physical injury.

3305.8.2 Site security. When tests are being conducted or explosives are being detonated, only authorized persons shall be present. Areas where explosives are regularly or frequently detonated or burned shall be approved and posted with adequate warning signs. Warning devices shall be activated before burning or detonating explosives to alert persons approaching from any direction that they are approaching a danger zone.

❖ Only authorized personnel can be allowed in test area enclosures. Audible and visible warning devices must be used to warn nearby personnel before detonating or burning any explosive material.

3305.9 Waste disposal. Disposal of explosive materials waste from manufacturing, assembly or testing operations shall be in accordance with Section 3304.10.

❖ Section 3304.10 addresses the issues of notification, deteriorated materials, qualified people, storage of misfires, sites for disposal and reuse and safety for personnel.

SECTION 3306 SMALL ARMS AMMUNITION

3306.1 General. Indoor storage and display of black powder, smokeless propellants and small arms ammunition shall comply with this section and NFPA 495.

❖ This section addresses storage that is prohibited, proper packages and residential and commercial storage of black powder; smokeless propellants and small arms ammunition. NFPA 495 contains requirements for these explosives.

3306.2 Prohibited storage. Small arms ammunition shall not be stored together with Division 1.1, Division 1.2 or Division 1.3 explosives unless the storage facility is suitable for the storage of explosive materials.

❖ Small arms ammunition must not be stored with explosives that have mass explosion, projection or fire hazards. If the storage facility meets the requirements of the local fire code official, however, this storage is allowed. Separation and explosive wave prevention are two requirements that should be met.

3306.3 Packages. Smokeless propellants shall be stored in approved shipping containers conforming to DOTn 49 CFR, Part 173.

❖ Smokeless propellants must not be stored in their original factory containers. DOTn 49 CFR, Part 173 addresses precautions such as not allowing metal (nails, staples, etc.) to penetrate the interior of a container, interior packaging not allowing contents to shift and become loose and packaging being made compatible with the explosives kept in the container.

3306.3.1 Repackaging. The bulk repackaging of smokeless propellants, black powder, and small arms primers shall not be performed in retail establishments.

❖ Bulk repackaging must be performed by qualified personnel. The product should be handled by trained explosives personnel and in the proper location for human and property safety.

3306.3.2 Damaged packages. Damaged containers shall not be repackaged.

Exception: Approved repackaging of damaged containers of smokeless propellant into containers of the same type and size as the original container.

❖ Damaged containers must not be repackaged. Smokeless propellants must not be transferred from the approved container into one that is not approved.

3306.4 Storage in residences. The storage of small arms ammunition shall comply with Sections 3306.4.1 and 3306.4.2.

❖ The following sections cover storage requirements for black powder, smokeless propellants and small arms primers in private residences.

3306.4.1 Black powder and smokeless propellants. Propellants for personal use in quantities not exceeding 20 pounds (9 kg) of black powder or 20 pounds (9 kg) of smokeless powder shall be stored in original containers in occupancies limited to Group R-3. Smokeless powder in quantities exceeding 20 pounds (9 kg) but not exceeding 50 pounds (23 kg) kept in a wooden box or cabinet having walls of at least 1 inch (25 mm) nominal thickness shall be allowed to be stored in occupancies limited to Group R-3. Quantities exceeding these amounts shall not be stored in any Group R occupancy.

❖ Small arms smokeless propellant intended for personal use in quantities not exceeding 20 pounds (9 kg) may be stored without restriction in residences; quantities over 25 pounds (11 kg) but not exceeding 50 pounds (23 kg) must be stored in a strong box or cabinet constructed with 1-inch (25 mm) wood (minimum), or equivalent, on all sides, top and bottom. Black powder as used in muzzle-loading firearms may be transported in a private vehicle or stored without restriction in private residences in quantities not to exceed 20 pounds (9 kg).

3306.4.2 Small arms primers. No more than 10,000 small arms primers shall be stored in occupancies limited to Group R-3.

❖ NFPA 495 allows storage of no more than 10,000 primers in a private residence. This recommendation is law in some communities. Even a deeply involved reloader should have no reason to store in excess of 1,000 of each large rifle, large rifle magnum, small rifle, large pistol, large pistol magnum, small pistol, small pistol magnum and shotgun primers—that's 8,000 primers.

3306.5 Display and storage in Group M occupancies. The display and storage of small arms ammunition in Group M occupancies shall comply with this section.

❖ Ignited stored powder can raise pressure within the storage area; therefore, there are specific requirements for keeping powder on hand. The first of these is never to transfer any propellant from its original container, which is designed to burst or partially open at very low pressures. Some container caps will simply just push off if the contents are ignited.

3306.5.1 Display. Display of small arms ammunition in Group M occupancies shall comply with Sections 3306.5.1.1 through 3306.5.1.3.

❖ The display quantities of smokeless propellants, black powder and small arms primers are given in the following sections.

3306.5.1.1 Smokeless propellant. No more than 20 pounds (9 kg) of smokeless propellants, each in containers of 1 pound (0.454 kg) or less capacity, shall be displayed in Group M occupancies.

❖ Displayed smokeless propellant must be limited to 1 pound (0.454 kg) of each type and cannot be accessible to the public. The total quantity of smokeless propellant not in an approved magazine is limited to 20 pounds (9 kg).

3306.5.1.2 Black powder. No more than 1 pound (0.454 kg) of black powder shall be displayed in Group M occupancies.

❖ Safety must be the first priority when black powder is accessible to the public. The 1-pound (.45 kg) limitation holds to the objective of keeping the public from having access to enough black powder to cause serious damage if mishandled.

3306.5.1.3 Small arms primers. No more than 10,000 small arms primers shall be displayed in Group M occupancies.

❖ Because of their explosive nature, only the absolute minimum should be kept on display. With care in replacing exhausted supplies, it is not difficult to adhere to the 10,000 primer limit.

3306.5.2 Storage. Storage of small arms ammunition shall comply with Sections 3306.5.2.1 through 3306.5.2.3.

❖ Generally, propellants used in sporting arms cartridges will not explode or detonate when ignited. Burning smokeless propellants do not generate the shock wave produced by an explosive. Unlike smokeless propellants, shock waves produced by ignition of an explosive cannot be adequately vented, even when ignition occurs in the open.

3306.5.2.1 Smokeless propellant. Commercial stocks of smokeless propellants shall be stored as follows:

1. Quantities exceeding 20 pounds (9 kg), but not exceeding 100 pounds (45 kg) shall be stored in portable wooden boxes having walls of at least 1 inch (25 mm) nominal thickness.
2. Quantities exceeding 100 pounds (45 kg), but not exceeding 800 pounds (363 kg), shall be stored in nonportable storage cabinets having walls at least 1 inch (25 mm) nominal thickness. Not more than 400 pounds (182 kg) shall be stored in any one cabinet, and cabinets shall be separated by a distance of at least 25 feet (7620 mm) or by a fire partition having a fire-resistance rating of at least 1 hour.
3. Storage of quantities exceeding 800 pounds (363 kg), but not exceeding 5,000 pounds (2270 kg) in a building shall comply with all of the following:
   - 3.1. The warehouse or storage room is unaccessible to unauthorized personnel.
   - 3.2. Smokeless propellant shall be stored in nonportable storage cabinets having wood walls at least 1 inch (25 mm) nominal thickness and having shelves with no more than 3 feet (914 mm) of separation between shelves.
   - 3.3. No more than 400 pounds (182 kg) is stored in any one cabinet.
   - 3.4. Cabinets shall be located against walls of the storage room or warehouse with at least 40 feet (12 192 mm) between cabinets.
   - 3.5. The minimum required separation between cabinets shall be 20 feet (6096 mm) provided that barricades twice the height of the cabinets are attached to the wall, midway between each cabinet. The barricades must extend a minimum of 10 feet (3048 mm) outward, be firmly attached to the wall, and be constructed of steel not less than 0.25 inch thick (6.4 mm), 2-inch (51 mm) nominal thickness wood, brick, or concrete block.
   - 3.6. Smokeless propellant shall be separated from materials classified as combustible liquids, flammable liquids, flammable solids, or oxidizing materials by a distance of 25 feet (7620 mm) or by a fire partition having a fire-resistance rating of 1 hour.
   - 3.7. The building shall be equipped throughout with an automatic sprinkler system installed in accordance with Section 903.3.1.1.
4. Smokeless propellants not stored according to Item 1, 2, or 3 above shall be stored in a Type 2 or 4 magazine in accordance with Section 3304 and NFPA 495.

❖ Storage cabinets are often made of 1-inch-thick (25 mm) wood with one or more walls designed to open outward or blow free at very low pressures. They should be many times larger than necessary to store the minimum quantities of powder. Propellants must not be stored in the same area with solvents, flammable gases, primers or highly combustible materials. Smoking should never be allowed in the storage areas or while handling or using powder.

3306.5.2.2 Black powder. Commercial stocks of black powder in quantities less than 50 pounds (23 kg) shall be allowed to be stored in Type 2 or 4 indoor or outdoor magazines. Quantities greater than 50 pounds (23 kg) shall be stored in outdoor Type 2 or 4 magazines. When black powder and smokeless propellants are stored together in the same magazine, the total quantity shall not exceed that permitted for black powder.

❖ Black powder must be stored in a magazine no matter the quantity. The guideline quantity of 50 pounds (23 kg) is the dividing line between storage in an indoor or outdoor magazine. The code does allow storage of smokeless propellants along with black powder as long as the total quantity given in this section for black powder is not exceeded. The maximum quantity is given to limit the explosive potential in considering life safety and property damage.

3306.5.2.3 Small arms primers. Commercial stocks of small arms primers shall be stored as follows.

1. Quantities not to exceed 750,000 small arms primers stored in a building shall be arranged such that not more than 100,000 small arms primers are stored in any one pile and piles are at least 15 feet (4572 mm) apart.
2. Quantities exceeding 750,000 small arms primers stored in a building shall comply with all of the following:
   - 2.1. The warehouse or storage building shall not be accessible to unauthorized personnel.
   - 2.2. Small arms primers shall be stored in cabinets. No more than 200,000 small arms primers shall be stored in any one cabinet.
   - 2.3. Shelves in cabinets shall have vertical separation of at least 2 feet (610 mm).
   - 2.4. Cabinets shall be located against walls of the warehouse or storage room with at least 40 feet (12 192 mm) between cabinets.
   - 2.5. The minimum required separation between cabinets shall be 20 feet (6096 mm) provided that barricades twice the height of the cabinets are attached to the wall, midway between each cabinet. The barricades shall be firmly attached to the wall, and shall be constructed of steel not less than 0.25 inch thick (6.4 mm), 2-inch (51 mm) nominal thickness wood, brick, or concrete block.
   - 2.6. Small arms primers shall be separated from materials classified as combustible liquids, flammable liquids, flammable solids, or oxidizing materials by a distance of 25 feet (7620 mm) or by a fire partition having a fire-resistance rating of 1 hour.
   - 2.7. The building shall be protected throughout with an automatic sprinkler system installed in accordance with Section 903.3.1.1.
3. Small arms primers not stored in accordance with Item 1 or 2 of this section shall be stored in a magazine meeting the requirements of Section 3304 and NFPA 495.

❖ Primers must be stored in a remote area away from any possible source of ignition, including bullet impact. Primers are by nature explosive. As a result, they may explode if subjected to friction, percussion, crushing or excessive heat from any cause, whether open flame or not. Static electricity and many other abuses can cause primers to explode; therefore, public access to storage containers must not be allowed.

Primers must be stored away from oxidizing agents and flammable liquids and solids. Naturally, smoking should be prohibited around primers. The storage cabinet is strongly recommended. The cabinet should be constructed of 1-inch-thick (25 mm) lumber to delay the transfer of heat to contents in the event of a fire or other mishap.

SECTION 3307
BLASTING

3307.1 General. Blasting operations shall be conducted only by approved, competent operators familiar with the required safety precautions and the hazards involved and in accordance with the provisions of NFPA 495.

❖ Personnel using explosives must be at least 21 years old and possess all required federal (see 18 USC, Chapter 40), state and local approvals or permits. Employees under 21 years old but at least 18 years old may serve as apprentices or assistants under direct supervision of a permittee. In addition to these requirements, individuals who handle explosives must not:

1. Be under the influence of alcohol or drugs;
2. Smoke or carry matches or firearms; or
3. Use or carry an open flame or open flame-producing device.

Many states require individuals who handle explosives to obtain a certificate of fitness, which is dubbed a "blaster's permit," before being authorized to handle or use explosives. The individual must be bonded or show acceptable proof of insurance, pass a written examination demonstrating an adequate understanding of the applicable codes and ordinances and demonstrate adequate experience (usually a minimum of three years) in handling explosives safely in an approved apprenticeship program to obtain a fitness certificate.

3307.2 Manufacturer's instructions. Blasting operations shall be performed in accordance with the instructions of the manufacturer of the explosive materials being used.

❖ No blasting operation will be performed in a manner contrary to the instructions of the manufacturer of the explosive materials being used. Each manufacturer has its own instructions for discharging its explosives.

3307.3 Blasting in congested areas. When blasting is done in a congested area or in close proximity to a structure, railway or highway, or any other installation, precautions shall be taken to minimize earth vibrations and air blast effects. Blasting mats or other protective means shall be used to prevent fragments from being thrown.

❖ Precautions must be taken to minimize damage from blasting operations where those operations are likely to generate projectiles, scatter debris or produce significant blast-wave effects.

3307.4 Restricted hours. Surface-blasting operations shall only be conducted during daylight hours. Other blasting shall be performed during daylight hours unless otherwise approved by the fire code official.

❖ Darkness impairs the effectiveness of the most important sense involved in setting explosive charges—sight. Care must be taken throughout blasting operations to inspect all elements of the explosive setup. Nighttime blasting operations require approved artificial illumination. The lighting could pose a hazard if improper equipment or procedures are used. Only the fire code official may authorize nighttime blasting operations.

3307.5 Utility notification. Whenever blasting is being conducted in the vicinity of utility lines or rights-of-way, the blaster shall notify the appropriate representatives of the utilities at least 24 hours in advance of blasting, specifying the location and intended time of such blasting. Verbal notices shall be confirmed with written notice.

Exception: In an emergency situation, the time limit shall not apply when approved.

❖ Steps must be taken to prevent disruption of public utility services. Location of underground utilities in many areas must be designated by a utility locator service before blasting or construction begins. Notification may not relieve the blaster of liability if operations disrupt utility service. Therefore, great care must be exercised when blasting in the vicinity of utilities. Blasting must not proceed without the approval of utility officials and the fire code official.

The exception recognizes that there may be emergency conditions under which it is not possible to strictly comply with the full 24-hour advance notice requirement. Conditions represented as an emergency must be approved as such by the fire code official. (Note: Both verbal and written notification must still be made in as timely a manner as the emergency will permit prior to blasting operations.)

3307.6 Electric detonator precautions. Precautions shall be taken to prevent accidental discharge of electric detonators from currents induced by radar and radio transmitters, lightning, adjacent power lines, dust and snow storms, or other sources of extraneous electricity.

❖ Wires connecting elements of the explosive train may act as an antenna or conduct electricity from sources outside of those controlled by the blaster. Therefore, special precautions are required to prevent inadvertent detonation of explosive materials.

1. The 350-foot (106 680 mm) separation distance provides a significant safety margin. Recommended minimum clearances for citizens band, VHF, UHF and cellular telephone transmitters with up to 50 watts vary from 5 to 180 feet (1524 to 54 864 mm). Nearly all mobile transmitters operate between 3 and 30 watts.
2. Power lines may also emit radio frequency energy (see IME Safety Library Publication No. 20 for guidance on separation from other radio frequency sources).
3. Lightning produces more than sufficient energy to detonate explosive materials. Charges should not be placed when there has been adequate warning of an approaching storm. If charges have been set before warning of an approaching storm is received, blast site workers should abandon their operations and take cover until the storm passes.

3307.7 Nonelectric detonator precautions. Precautions shall be taken to prevent accidental initiation of nonelectric detonators from stray currents induced by lightning or static electricity.

❖ All elements of pneumatic loading devices must be electrically bonded and a positive grounding device for the equipment must be used to prevent the accumulation of static electricity. Water lines, air lines, rails or permanent electric grounding systems for other equipment must not be used to ground pneumatic loading equipment.

3307.8 Blasting area security. During the time that holes are being loaded or are loaded with explosive materials, blasting agents or detonators, only authorized persons engaged in drilling and loading operations or otherwise authorized to enter the site shall be allowed at the blast site. The blast site shall be guarded or barricaded and posted. Blast site security shall be maintained until after the post-blast inspection has been completed.

❖ Loading and firing operations are the most hazardous steps in the blasting process. To prevent inadvertent detonation while electrically fired charges are being set, only the individual making lead wire connections must fire the shot [see DOL 29 CFR 1910.109 (e)(4)(viii) and 1926.906(s)]. Similar precautions must be taken when firing explosives using other methods. Individuals performing or supervising loading and firing operations must hold all required federal (see USC Chapter 40 and DOTy 27 CFR 55, Subpart D), state and local approvals or permits.

3307.9 Drill holes. Holes drilled for the loading of explosive charges shall made and loaded in accordance with NFPA 495.

❖ Drill holes must be large enough to allow free insertion of cartridges of explosive materials and cannot be collared in bootlegs or in holes that previously contained explosive materials. Holes must not be drilled where there is a danger of intersecting another hole containing explosive material. Of utmost importance, all drill holes must be inspected and cleared of any obstruction before loading. NFPA 495 gives other requirements for drilling the holes.

3307.10 Removal of excess explosive materials. After loading for a blast is completed and before firing, excess explosive materials shall be removed from the area and returned to the proper storage facilities.

❖ After loading is completed, surplus explosive materials must be returned to an approved magazine before firing.

3307.11 Initiation means. The initiation of blasts shall be by means conforming to the provisions of NFPA 495.

❖ Cap and fuse cannot be used to initiate blasts on or adjacent to highways open to traffic or in congested areas. NFPA 495 gives the burning rate and safety fuse length in units of time.

3307.12 Connections. The blaster shall supervise the connecting of the blastholes and the connection of the loadline to the power source or initiation point. Connections shall be made progressively from the blasthole back to the initiation point.

Blasting lead lines shall remain shunted (shorted) and shall not be connected to the blasting machine or other source of current until the blast is to be fired.

❖ Connections must be made progressively from the blast holes to the source of firing current. The shorted lead wire must remain unconnected until the blast is ready for firing. Only insulated leading wire of adequate current-carrying capacity can be used.

3307.13 Firing control. No blast shall be fired until the blaster has made certain that all surplus explosive materials are in a safe place in accordance with Section 3307.10, all persons and equipment are at a safe distance or under sufficient cover, and that an adequate warning signal has been given.

❖ The blaster must conduct a complete survey of the blast site before blasting begins. Surplus charges must be placed in a magazine or day box and individuals and vehicles must be cleared from the blast site. A loud warning signal (claxon horn or whistle) must be sounded before charges are fired to alert workers that blasting is about to begin (see also IME Safety Library Publication Nos. 3, 4 and 17).

3307.14 Post-blast procedures. After the blast, the following procedures shall be observed.

1. No person shall return to the blast area until allowed to do so by the blaster in charge.
2. The blaster shall allow sufficient time for smoke and fumes to dissipate and for dust to settle before returning to or approaching the blast area.
3. The blaster shall inspect the entire blast site for misfires before allowing other personnel to return to the blast area.

❖ Blast areas must not be reentered after firing until concentrations of smoke, dust and fumes have been reduced to safe limits as determined by the blaster in charge. The blaster must determine the length-of-waiting period before any person is permitted in the blast area. When a misfire is known or suspected, no person should enter the area for at least 1 hour.

3307.15 Misfires. Where a misfire is suspected, all initiating circuits shall be traced and a search made for unexploded charges. Where a misfire is found, the blaster shall provide proper safeguards for excluding all personnel from the blast area. Misfires shall be reported to the blasting supervisor immediately. Misfires shall be handled under the direction of the person in charge of the blasting operation in accordance with NFPA 495.

❖ All misfires are potential accidents. A misfire may be caused by many different factors. Only qualified and experienced personnel should attempt to handle disposal of a misfire. DOL 29 CFR 1910.109(e)(4)(v) and 1926.911(c) prohibit the extraction of explosives from a blasthole unless it is impossible to detonate the unexploded charge by insertion of a fresh primer.

No one should enter the blasting area to handle a misfire or investigate the cause until it is safe to do so. Recommended waiting periods for investigating misfires are listed in Table 3307.15. The IME waiting periods correspond with the recommendation in IME Safety Library Publication No. 17. Occupational Safety and Health Administration (OSHA) waiting periods are those specified in DOL 29 CFR 1910.109 and 1926.911(d).

The safest way to dispose of most misfires is by detonation. However, even this method has distinct hazards. Misfired explosives may be unstable or sensitive, so any procedure to dispose of a misfire must be carried out with extreme caution. Removing a misfired explosive charge should be attempted only if all other alternatives are unsuccessful or unadvisable. After a misfired charge has been removed, it must be transported immediately to a storage magazine and disposed of as soon as possible.

TABLE 3307.15
RECOMMENDED WAITING PERIODS FOR INVESTIGATING MISFIRES

Description	IME	OSHA
Blasting cap and fuse	30 minutes	1 hour
Electric blasting cap, shock Tube, gas tube or detonating cord	15 minutes	30 minutes
Refiring of a misfired charge	1 hour	Not rated

Source: DOL 29 CFR 1910 and IME Safety Library Publication No. 17.

SECTION 3308
FIREWORKS DISPLAY

3308.1 General. The display of fireworks, including proximate audience displays and pyrotechnic special effects in motion picture, television, theatrical, and group entertainment productions, shall comply with this chapter and NFPA 1123 or NFPA 1126.

❖ The only discharge of fireworks permissible under the code is a public display conducted by competent pyrotechnicians in accordance with the requirements of NFPA 1123 and NFPA 1126 and authorized by the fire code official.

The use of pyrotechnic special effects in theatrical performances has become a topic of some concern. In December 1990, a pyrotechnic effect exploded during a band concert in a Florida hotel lounge, injuring seven people, one critically. NFPA 1126 was published in response to this event, although it should be noted that it is not a referenced standard in the code nor are indoor displays allowed by this section.

3308.2 Permit application. Prior to issuing permits for fireworks display, plans for the display, inspections of the display site, and demonstrations of the display operations shall be approved.

❖ This section addresses approval for fireworks displays. A 15-day time limitation is suggested to give the fire code official reasonable time to thoroughly verify the completeness and accuracy of the information in the application and to allow a thorough inspection of the display site. Customarily, only the display operator or pyrotechnician may apply for approval. Issuing approvals to corporations, associations, boards or other corporate entities may hamper accountability.

Because only a single individual can be assigned overall control of the discharge of fireworks, only that individual responsible for operating the display should be issued an approval. However, all other organizations, institutions or individuals acting together or alone to contract the display should also be named in the approval application and, if considered appropriate by the fire code official, named as additional insureds in the bond indemnifying the display. The strict prohibitions against transfers and extensions are needed for the continuing supervision of fireworks operations.

3308.2.1 Outdoor displays. In addition to the requirements of Section 403, permit applications for outdoor fireworks displays using Division 1.3G fireworks shall include a diagram of the location at which the display will be conducted, including the site from which fireworks will be discharged; the location of buildings, highways, overhead obstructions and utilities; and the lines behind which the audience will be restrained.

❖ The best sites are free of overhead obstructions and are well isolated, with clear viewing paths and landing areas. Fallout areas should be large, open areas, clear of spectators, vehicles and combustible materials. Generally, the discharge site must have a minimum radius of 70 feet (21 336 mm) for each inch of aerial shell diameter.

3308.2.2 Proximate audience displays. Where the separation distances required by Section 3308.4 and NFPA 1123 are unavailable or cannot be secured, only proximate audience displays conducted in accordance with NFPA 1126 shall be allowed. Applications for proximate audience displays shall include plans indicating the required clearances for spectators and combustibles, crowd control measures, smoke control measures, and requirements for standby personnel and equipment when provision of such personnel or equipment is required by the fire code official.

❖ This section applies to any outdoor use of pyrotechnics at distances less than those required by NFPA 1123. The use of pyrotechnics before a proximate audience is not a display of fireworks as regulated by NFPA 1123.

The separation distance between the audience and where the pyrotechnic device is fired during a performance must be at least 15 feet (4572 mm) but not less than twice the fallout radius of the device. The audience must be separated from concussion mortars by a minimum of 25 feet (7620 mm) and there must be no glowing or flaming particles within 10 feet (3048 mm) of the audience.

3308.3 Approved displays. Approved displays shall include only the approved Division 1.3G, Division 1.4G, and Division 1.4S fireworks, shall be handled by an approved competent operator, and the fireworks shall be arranged, located, discharged and fired in a manner that will not pose a hazard to property or endanger any person.

❖ The competence of the display operator is first among the important safeguards that must be observed for a safe and enjoyable public display. Chapter 6 of NFPA 1123 details the qualifications of competent fireworks display operators. Many jurisdictions require display operators to possess a license or certificate of fitness. To obtain such a certificate, the operator must be bonded or indemnified, pass a written examination and serve an apprenticeship under another licensed or certified pyrotechnician.

Fireworks discharge sites must be separated from institutional and high-hazard occupancies by at least twice the distance specified in the table [140 feet (42 672 mm) per inch of shell diameter]. When mortars are positioned vertically (zero degrees), they must be located at the center of the display area. When mortars or shells stored at the discharge site are angled, they must be aimed away from principal spectator viewing and shell storage areas. When angled, mortars may be placed up to one-third the distance from the center of the display area to the principal spectator viewing area. Aerial shell trajectories must not come within 25 feet (7620 mm) of overhead obstructions, such as power lines and trees. Tents and canvas structures must be at least 100 feet (30 480 mm) from the discharge site.

High winds, precipitation or extremely hot, dry conditions should be avoided. Moisture-damaged shells must not be fired. If, in the opinion of the fire code official or the display operator, weather conditions present a danger, the display must be postponed or canceled.

3308.4 Clearance. Spectators, spectator parking areas, and dwellings, buildings or structures shall not be located within the display site.

Exceptions:

1. This provision shall not apply to pyrotechnic special effects and displays using Division 1.4G materials before a proximate audience in accordance with NFPA 1126.
2. This provision shall not apply to unoccupied dwellings, buildings and structures with the approval of the building owner and the fire code official.

❖ Aerial displays must meet the requirements of NFPA 1123. The site for the outdoor display should have at least a 70-foot (21 336 mm) radius per inch of the internal mortar diameter of the largest aerial shell being fired, except as noted in NFPA 1123. Some jurisdictions require 100 feet (30 480 mm) of radius per inch of diameter of aerial shell. No spectators, dwellings or spectator parking areas can be located within the display site.

3308.5 Storage of fireworks at display site. The storage of fireworks at the display site shall comply with the requirements of this section and NFPA 1123 or NFPA 1126.

❖ NFPA 1123 addresses weather protection, inspection, sorting and ready boxes for fireworks. The standard also addresses construction of display fireworks aerial shells.

3308.5.1 Supervision and weather protection. Beginning as soon as fireworks have been delivered to the display site, they shall not be left unattended.

❖ Never leave fireworks unattended. Too many events can cause serious problems during the show. Someone could replace good fireworks with faulty ones. Constant supervision is necessary both after and before inspection of the fireworks.

3308.5.2 Weather protection. Fireworks shall be kept dry after delivery to the display site.

❖ Protecting the fireworks from inclement weather is the responsibility of everyone involved in the display. If the fireworks get wet from rain, they will not fire properly, thus possibly causing harm to personnel and spectators. A tarpaulin will serve nicely for protection from rain.

3308.5.3 Inspection. Shells shall be inspected by the operator or assistants after delivery to the display site. Shells having tears, leaks, broken fuses or signs of having been wet shall be set aside and shall not be fired. Aerial shells shall be checked for proper fit in mortars prior to discharge. Aerial shells that do not fit properly shall not be fired. After the display, damaged, deteriorated or dud shells shall either be returned to the supplier or destroyed in accordance with the supplier's instructions and Section 3304.10.

Exception: Minor repairs to fuses shall be allowed. For electrically ignited displays, attachment of electric matches and similar tasks shall be allowed.

❖ Prior to acceptance of display fireworks from a wholesaler, the permit holder or designated agent must confirm that the outside of all cartons, containers or cases is in good condition and all documentation is in order. Shells can be damaged during transport from the factory. It is good safety practice to examine all shells before placing them into the mortars for firing. If there is any sign of damage, the shells should be set aside and not fired, reducing the risk of injury to personnel.

The exception does allow minor repairs that are safe in the judgement of the pyrotechnician.

3308.5.4 Sorting and separation. After delivery to the display site and prior to the display, all shells shall be separated according to size and their designation as salutes.

Exception: For electrically fired displays, or displays where all shells are loaded into mortars prior to the show, there is no requirement for separation of shells according to size or their designation as salutes.

❖ Where aerial shells are to be stored at the discharge site for subsequent loading into mortars during the display, the mortars must be placed usually at one-sixteenth, but not more than one-third, the distance from the center of the display site toward the main spectator area.

The exception allows preloading of mortars without respect to shell size.

3308.5.5 Ready boxes. Display fireworks (Division 1.3G) that will be temporarily stored at the site during the fireworks display shall be stored in ready boxes located upwind and at least 25 feet (7620 mm) from the mortar placement and separated according to size and their designation as salutes.

Exception: For electrically fired displays, or displays where all shells are loaded into mortars prior to the show, there is no requirement for separation of shells according to size, their designation as salutes, or for the use of ready boxes.

❖ A ready box should be a weather-resistant container that protects contents from burning debris with a self closing cover or equivalent means of closure required. If the wind shifts during a display, the ready boxes must be relocated to again be upwind from the discharge site. Tarpaulins can be used as weather protection for ready boxes but not considered as ready boxes.

3308.6 Installation of mortars. Mortars for firing fireworks shells shall be installed in accordance with NFPA 1123 and shall be positioned so that shells are propelled away from spectators and over the fallout area. Under no circumstances shall mortars be angled toward the spectator viewing area. Prior to placement, mortars shall be inspected for defects, such as dents, bent ends, damaged interiors and damaged plugs. Defective mortars shall not be used.

❖ Mortars can be buried to a depth of at least two-thirds to three-quarters of their length, either in the ground or in above-ground troughs or drums or however the local fire code official considers necessary. Eliminating as much of a risk as possible from spectator injury is a good reason for angling away from the viewing area. See also the commentary to the definition of "Fallout area" for further information.

3308.7 Handling. Aerial shells shall be carried to mortars by the shell body. For the purpose of loading mortars, aerial shells shall be held by the thick portion of the fuse and carefully loaded into mortars.

❖ During the firing of the display, personnel in the discharge site should wear head protection, eye protection, hearing protection, foot protection and cotton, wool or similarly flame-resistant, long-sleeved, long-legged clothing. Personal-protective equipment, as necessary, should be worn by the operator and assistants during the setup and cleanup of the display. Shells must be carried from the storage area to the discharge site only by their bodies and shall never be carried by their fuses.

3308.8 Display supervision. Whenever in the opinion of the fire code official or the operator a hazardous condition exists, the fireworks display shall be discontinued immediately until such time as the dangerous situation is corrected.

❖ All displays must be set up using methods that allow an interruption in firing in case an unforseen danger becomes evident. The judgement of the display operator and the fire code official will determine whether an ongoing display or one that has been set up and is ready to begin must be stopped or delayed because of hazardous conditions.

3308.9 Post-display inspection. After the display, the firing crew shall conduct an inspection of the fallout area for the purpose of locating unexploded aerial shells or live components. This inspection shall be conducted before public access to the site shall be allowed. Where fireworks are displayed at night and it is not possible to inspect the site thoroughly, the operator or designated assistant shall inspect the entire site at first light.

❖ The firing crew will consist of the operator and assistants. For the public's own safety, no one should be allowed entrance into the fallout site until inspection is completed. Mortar inspection and removal should be conducted within 10 minutes after show completion. When fireworks are displayed at night and it is impossible to thoroughly inspect the site, the crew must ensure that the entire site is reinspected very early the following morning.

3308.10 Disposal. Any shells found during the inspection required in Section 3308.9 shall not be handled until at least 15 minutes have elapsed from the time the shells were fired. The fireworks shall then be doused with water and allowed to remain for at least 5 additional minutes before being placed in a plastic bucket or fiberboard box. The disposal instructions of the manufacturer as provided by the fireworks supplier shall then be followed in disposing of the fireworks in accordance with Section 3304.10.

❖ In addition to the above requirements, any aerial shell that misfires in a mortar should be left alone for a minimum of 30 minutes, carefully loaded into a bucket of water and left for a minimum of 15 minutes and then properly disposed of.

3308.11 Retail display and sale. Fireworks displayed for retail sale shall not be made readily accessible to the public. A minimum of one pressurized-water portable fire extinguisher complying with Section 906 shall be located not more than 15 feet (4572 mm) and not less than10 feet (3048 mm) from the hazard. "No Smoking" signs complying with Section 310 shall be conspicuously posted in areas where fireworks are stored or displayed for retail sale.

❖ The location of an outdoor retail sales outlet must not present a significant risk of fire or injury to those individuals conducting sales of retail fireworks, individual members of the general public and any surrounding property. A good rule of thumb for extinguishers is to have one 2A:10B:C-rated fire extinguisher in place for every 100 feet (30 480 mm) of sales floor area. No smoking or open flames can be tolerated in the floor sales area. Signs to this effect should be conspicuously posted approximately every 10 feet (3048 mm) of sales area.

Bibliography

The following resource materials are referenced in this chapter or are relevant to the subject matter addressed in this chapter.

18 USC, Chapter 40, *Importation, Distribution and Storage of Explosive Materials*. Washington, DC: U.S. Department of Treasury, Bureau of Alcohol, Tobacco and Firearms.

APA 871-87, *Standards for Construction and Approval for Transportation of Fireworks and Novelties.*

Chestertown, MD: American Pyrotechnics Association, 1987.

ATF, *Explosives Laws and Regulations* (ATF P 5400.7). Washington, DC: U.S. Department of Treasury, Bureau of Alcohol, Tobacco and Firearms.

Blaster's Handbook, 16th ed. Wilmington, DE: E. I. DuPont de Nemours & Co., Inc., 1978.

Callahan, T. *Fire Service and the Law*, 2nd ed. Quincy, MA: National Fire Protection Association, 1987.

DOD 6055.9STD, *Ammunition and Explosives Safety*. Washington, DC: U.S. Department of Defense, Explosives Safety Board.

DOD Contractor's Safety Manual For Ammunition and Explosives, Publication DOD 4145.26-M. Washington, DC: U.S. Department of Defense, July 1997.

DOL 29 CFR 1910.109-92, *Explosives and Blasting Agents*. Washington, DC: U.S. Department of Labor, Occupational Safety and Health Administration, 1992.

DOL 29 CFR 1926 Subpart U, *Blasting and Use of Explosives*. Washington, DC: U.S. Department of Labor, Occupational Safety and Health Administration, 1992.

DOTn 49 CFR 100-177 & 178-199-94, *Transportation*. Washington, DC: U.S. Department of Transportation, Research and Special Programs Administration, 1994.

DOTn 49 CFR Chapter II, Subchapter B, *Federal Motor Carrier Safety Regulations.* Washington, DC: U.S. Department of Transportation, Federal Highway Administration, 1992.

DOTy 27 CFR 55-92, *Alcohol, Tobacco Products and Firearms: Commerce in Explosives*. Washington, DC: U.S. Department of Treasury, Bureau of Alcohol, Tobacco and Firearms, 1992.

DOTy 27 CFR 178-92, *Alcohol, Tobacco Products and Firearms: Commerce in Firearms and Ammunition*. Washington, DC: U.S. Department of the Treasury, Bureau of Alcohol, Tobacco and Firearms. 1992.

IBC-2003, *International Building Code*. Falls Church, VA: International Code Council, 2003.

ICC EC-2003, *ICC Electrical Code.* Falls Church, VA: International Code Council, 2003.

IME Safety Library Publication No. 1, *Construction Guide for Storage Magazines*. Washington, DC: Institute of Makers of Explosives, 1986.

IME Safety Library Publication No. 2, *American Table of Distances*. Washington, DC: Institute of Makers of Explosives, 1991.

IME Safety Library Publication No. 3, *Suggested Code of Regulations for the Manufacture, Transportation, Storage, Sale, Possession, and Use of Explosive Materials*. Washington, DC: Institute of Makers of Explosives, 1985.

IME Safety Library Publication No. 12, *Glossary of Commercial Explosives Industry Terms*. Washington, DC: Institute of Makers of Explosives, 1985.

IME Safety Library Publication No. 14, *Handbook for the Transportation and Distribution of Explosive Materials*. Washington, DC: Institute of Makers of Explosives, 1986.

IME Safety Library Publication No. 17, *Safety in the Transportation, Storage, Handling and Use of Explosive Materials*. Washington, DC: Institute of Makers of Explosives, 1987.

IME Safety Library Publication No. 20, *Safety Guide for the Prevention of Radio Frequency Radiation Hazards in the Use of Blasting Caps*. Washington, DC: Institute of Makers of Explosives, 1988.

IME Safety Library Publication No. 21, *Destruction of Commercial Explosive Materials.* Washington, DC: Institute of Makers of Explosives, 1987.

IME Safety Library Publication No. 22, *Recommendations for the Safe Transportation of Detonators in a Vehicle with Certain Other Explosive Materials*. Washington, DC: Institute of Makers of Explosives, 1985.

NFPA 490-98, *Storage of Ammonium Nitrate*. Quincy, MA: National Fire Protection Association, 1998.

NFPA 495-96, *Explosive Materials Code*. Quincy, MA: National Fire Protection Association, 1996.

NFPA 498-96, *Safe Havens and Interchange Lots for Vehicles Transporting Explosives*. Quincy, MA: National Fire Protection Association, 1996.

NFPA *Inspection Manual*, 7th ed. Quincy, MA: National Fire Protection Association, 1994.

Structures to Resist the Effects of Accidental Explosions. U.S. Army Technical Manual TM 51300, Air Force Manual (AFM) 8822 and Navy NAVFAC P397. Washington, DC: U.S. Government Printing Office.

Chapter 34: Flammable and Combustible Liquids

General Comments

Flammable and combustible liquids are essential in our modern lifestyles. These liquids are used for fuel, lubricants, cleaners, solvents, medicine and even drinking. The danger associated with flammable and combustible liquids is that the vapors from these liquids, when combined with air in their flammable range, will burn or explode at temperatures near our normal living and working environment.

The use of these liquids is accepted in all occupancies if the liquids are in appropriate containers and the quantity is very limited. When the quantities exceed these limited amounts or the use of the flammable or combustible liquids increases the potential danger, the code requires that measures be taken to control the potential danger. These measures are to prevent the possibility of flammable and combustible liquids igniting.

Although the dangers of flammable liquids are well known, accidents involving flammable liquids remain one of the most common fire scenarios in the United States. Statistically, the more common of the flammable or combustible liquids (gasoline) rather than the most dangerous flammable or combustible liquids account for the most fires. There are five factors that account for the involvement of flammable liquids in these fires: (1) personnel inadequately trained in safe operating procedures; (2) hazardous operations not isolated from other operations; (3) equipment and flammable or combustible liquids improperly used; (4) poor property maintenance and supervision and (5) inadequate control systems.

These five factors suggest that a holistic approach to flammable and combustible liquid fire safety is required. By beginning with people, a safety system has a better chance of working successfully and consistently. Trained personnel recognize the importance of safe practices to their personal safety, and are more likely to demand that necessary safeguards be installed in their homes and workplaces. Because all of us tend to become complacent as time passes, inspectors, owners, operators, managers and employees must work together to maintain vigilance over the system's continued operation.

Hazardous characteristics

Although the classification boundaries are somewhat arbitrary, flammable and combustible liquids are distinguished by their flash points. The flash point is that temperature at which the liquid produces sufficient vapor to form an ignitable vapor-air mixture above its surface. Because Class I flammable liquids all have flash points below 100°F (38°C), it is prudent to assume that they may be capable of igniting when unconfined under normal environmental conditions. On the other hand, combustible liquids [those materials with flash points above 100°F (38°C)] must usually be heated above their flash points, or in the case of extremely high flash-point liquids, above their boiling points, before they will ignite.

Physical characteristics

Flammable liquids possess other characteristics besides their low flash points. Significant characteristics when evaluating relative fire hazards include ignition temperature, autoignition temperature, flammable (explosive) range, viscosity, vapor density, vapor pressure, boiling point, evaporation rate, specific gravity and water solubility. Once the liquid is ignited, these variables have little influence over the material's heat release rate; however, factors such as evaporation rate, viscosity and water solubility may profoundly affect how these fires are extinguished.

Hazards

In general, flammable and combustible liquids have low specific gravities, high vapor densities and narrow flammable (explosive) ranges. These characteristics mean that the liquids will usually float on water, the vapors will usually hug the ground and ignitable vapor-air mixtures will be confined to a range between 6 and 15 percent in air. Thus, smothering is difficult and ignition sources near the ground are more likely to pose a hazard. Low and high concentrations within the flammable range are likely to produce deflagrations, while concentrations near the middle of the flammable (explosive) range are more likely to produce detonations.

Protection

The protection provided by the code is to prevent the flammable and combustible liquids from becoming ignited. This is accomplished by one or more of the following procedures:

1. Preventing the flammable and combustible liquids from vaporizing. The use of equipment and devices to safely store, transport, dispense, mix or use the flammable and combustible liquids so that the liquid does not have the opportunity to evaporate except where intended.

2. Preventing the concentration of vapors of flammable and combustible liquids from reaching the vapor-air mixture between the lower flammable limit (LFL) and the upper flammable limit (UFL). Ventilation of the area is used to dilute and disperse the

vapors before the vapor-air mixture reaches the LFL.

3. Preventing an accumulation of the vapor-air mixture of flammable and combustible liquids. The ventilation of the area as well as the design of the area is used to prevent the vapors from collecting.
4. Removing ignition sources from the area. These ignition sources are from people, equipment or static electricity sources.
5. Precautions against spontaneous ignition.
6. Removing other combustibles from the area. Other combustibles may become involved in a fire, creating the ignition source for the flammable and combustible liquid.
7. Designing the equipment (piping, vessels, containers, etc.) and facilities to prevent the loss of flammable and combustible liquids and to protect the flammable and combustible liquids from exterior fire exposure. This protection is also used to protect other property and flammable and combustible liquids from fire exposure.
8. Controlling the volume of flammable and combustible liquids to reduce the size of a potential fire.
9. Providing fire-fighting equipment to control fire from combustibles other than flammable and combustible liquids as well as fire from flammable and combustible liquids.
10. Security to ensure that only trained personnel have access to flammable and combustible liquids.
11. Training of personnel to ensure they are aware of the potential dangers and know the operating procedures and safety procedures.

Purpose

The requirements of this chapter are intended to reduce the likelihood of fires involving the storage, handling, use or transportation of flammable and combustible liquids. Adherence to these practices may also limit damage in the event of an accidental fire involving these materials.

SECTION 3401 GENERAL

3401.1 Scope and application. Prevention, control and mitigation of dangerous conditions related to storage, use, dispensing, mixing and handling of flammable and combustible liquids shall be in accordance with Chapter 27 and this chapter.

❖ This chapter regulates the storage, handling and use of flammable and combustible liquids. Although U.S. Department of Transportation (DOT) regulations govern the construction of tank vehicles and the interstate transportation of flammable and combustible liquids, this chapter regulates the parking, garaging, filling and discharging of tank vehicles. This chapter is used in conjunction with Chapter 27 to regulate flammable and combustible liquids. Related operations, materials and processes are regulated elsewhere in the code.

3401.2 Nonapplicability. This chapter shall not apply to liquids as otherwise provided in other laws or regulations or chapters of this code, including:

1. Specific provisions for flammable liquids in motor fuel-dispensing facilities, repair garages, airports and marinas in Chapter 22.
2. Medicines, foodstuffs, cosmetics, and commercial, institutional and industrial products in the same concentration and packaging containing not more than 50 percent by volume of water-miscible liquids and with the remainder of the solution not being flammable, and alcoholic beverages in retail or wholesale sales or storage uses when packaged in individual containers not exceeding 1.3 gallons (5 L).
3. Storage and use of fuel oil tanks and containers connected to oil-burning equipment. Such storage and use shall be in accordance with Section 603. For abandonment of fuel oil tanks, this chapter applies.
4. Refrigerant liquids and oils in refrigeration systems (see Section 606).
5. Storage and display of aerosol products complying with Chapter 28.
6. Storage and use of liquids that have no fire point when tested in accordance with ASTM D 92.
7. Liquids with a flashpoint greater than 95°F (35°C) in a water-miscible solution or dispersion with a water and inert (noncombustible) solids content of more than 80 percent by weight, which do not sustain combustion.
8. Liquids without flash points that can be flammable under some conditions, such as certain halogenated hydrocarbons and mixtures containing halogenated hydrocarbons.
9. The storage of distilled spirits and wines in wooden barrels and casks.

❖ This section recognizes that some flammable and combustible liquids are regulated by other laws or by other chapters of the code. The application of these other laws or code chapters takes precedence over the requirements in this chapter.

Exception 1 covers the unique use of flammable liquids at motor fuel-dispensing facilities, airports and marinas, which require specific regulations that are addressed in Chapter 22.

Individually packaged consumer products consisting of medicines, foodstuffs, cosmetics and commercial, industrial and institutional products with a limited quantity of flammable liquid used in the product are exempt from the requirements of this chapter. Alcoholic beverages in retail or wholesale mercantile occupancies or in storage occupancies are not regulated by this chapter if the individual containers do not exceed 1.3 gallons (5 L).

Exception 3 states that fuel oil tanks and containers connected to oil-burning equipment are regulated by Chapter 6. This chapter regulates abandoned fuel tanks.

Exception 4 notes that refrigerant liquids and oils in refrigeration systems are also regulated by Chapter 6.

Exception 5 refers the reader to Chapter 28 for requirements covering the storage and display of aerosol products. If the storage or display of aerosols does not comply with Chapter 28, this chapter will regulate the storage and display of aerosols in those specific circumstances.

Exception 6 notes that the storage and use of a petroleum product whose vapor cannot be ignited and sustain burning for a minimum of 5 seconds (fire point) is not regulated by this chapter. The test procedure for determining the fire point of a petroleum product is ASTM D 92. The fire point is based on a specific barometric pressure established by ASTM International.

Exception 7 exempts certain liquids that, although they have a flash point, will not sustain combustion.

Exception 8 notes that some liquids do not have a flash point except in specific circumstances; for example, halogenated hydrocarbons that may become explosive when exposed to aluminum. These liquids are not regulated by the code.

Exception 9 makes the storage of distilled spirits and wines in wooden barrels and casks exempt from this chapter. Although their contents are classified as flammable liquids, the containers do not pose the rupture hazard as do other containers. Barrels and casks will leak their contents and contribute to the fire as the metal bands that secure the staves expand and loosen. Even this hazard feature is generally mitigated by the operation of automatic sprinklers that prevent the fire from progressing to the point where the metal bands get hot enough to expand.

3401.3 Referenced documents. The applicable requirements of Chapter 27, other chapters of this code, the *International Building Code* and the *International Mechanical Code* pertaining to flammable liquids shall apply.

❖ The requirements to regulate the design, construction and maintenance of facilities using flammable and combustible liquids are contained in more than one document. The code contains several chapters that address requirements for the storage, handling, dispensing, processing, transportation and use of flammable and combustible liquids. The *International Building Code*® (IBC®) covers the construction requirements for the structure and the *International Mechanical Code*® (IMC®) covers the construction requirements for mechanical systems.

3401.4 Permits. Permits shall be required as set forth in Sections 105.6 and 105.7.

❖ The process of issuing permits gives the fire code official an opportunity to carefully evaluate and regulate hazardous operations. Permit applicants should be required to demonstrate that their operations comply with the intent of the code before the permit is issued. See the commentary to Section 105.6 for a general discussion of operations requiring an operational permit, Section 105.6.17 for discussion of specific quantity-based operational permits for the materials regulated in this chapter and Section 105.7.5 for a discussion of activities requiring a construction permit. The permit process also notifies the fire department of the need for prefire planning for the hazardous property.

3401.5 Material classification. Flammable and combustible liquids shall be classified in accordance with the definitions in Section 3402.1.

When mixed with lower flash-point liquids, Class II or III liquids are capable of assuming the characteristics of the lower flash-point liquids. Under such conditions the appropriate provisions of this chapter for the actual flash point of the mixed liquid shall apply. When heated above their flash points, Class II and III liquids assume the characteristics of Class I liquids. Under such conditions, the appropriate provisions of this chapter for flammable liquids shall apply.

❖ Flammable and combustible liquids are defined in Chapter 2 and Section 3402. Mixing or heating the liquid may modify flammable and combustible liquids. This process will change the flash point of the liquid. The mixed liquid is to be classified and handled based on the flash point determined by the appropriate test procedure and apparatus as specified in ASTM D 56, ASTM D 93 or ASTM D 3278. A Class II or III liquid that is heated above its flash point during handling or processing must be treated as a Class I liquid.

SECTION 3402
DEFINITIONS

3402.1 Definitions. The following words and terms shall, for the purposes of this chapter and as used elsewhere in this code, have the meanings shown herein.

❖ Definitions of terms can help in the understanding and application of the code requirements. The purpose for including those definitions that are associated with the subject matter of this chapter is to provide more convenient access to them without having to refer back to Chapter 2. It is important to emphasize that these terms are not exclusively related to this chapter but are applicable everywhere the term is used in the code. For convenience, these terms are also listed in Chapter 2 with a cross reference to this section. The use and application of all defined terms, including those defined in this section, are set forth in Section 201.

BULK PLANT OR TERMINAL. That portion of a property where flammable or combustible liquids are received by tank vessel, pipelines, tank car or tank vehicle and are stored or blended in bulk for the purpose of distributing such liquids by tank vessel, pipeline, tank car, tank vehicle, portable tank or container.

❖ All of us have seen the large storage tanks surrounded by containment berms that constitute what is commonly called a "tank farm." A facility of this kind may consist of one tank or several tanks and typically receives, stores and dispenses anywhere from thousands of gallons to hundreds of thousands of gallons of flammable or combustible liquids daily.

BULK TRANSFER. The loading or unloading of flammable or combustible liquids from or between tank vehicles, tank cars or storage tanks.

❖ This term refers to the loading or unloading of a flammable or combustible liquid from or between tanks. This transfer is for the storage or transportation of flammable or combustible liquids.

COMBUSTIBLE LIQUID. A liquid having a closed cup flash point at or above 100°F (38°C). Combustible liquids shall be subdivided as follows:

Class II. Liquids having a closed cup flash point at or above 100°F (38°C) and below 140°F (60°C).

Class IIIA. Liquids having a closed cup flash point at or above 140°F (60°C) and below 200°F (93°C).

Class IIIB. Liquids having closed cup flash points at or above 200°F (93°C).

The category of combustible liquids does not include compressed gases or cryogenic fluids.

❖ Combustible liquids differ from flammable liquids in that the closed-cup flash point of all combustible liquids is at or above 100°F (38°C) (see the definition of "Flash point"). There are three categories of combustible liquids. The range of their closed-cup flash point dictates the class of combustible liquid. The flash point range of 100°F (38°C) to 140°F (60°C) for Class II liquids was based on a possible indoor ambient temperature exceeding 100°F (38°C). Only a moderate degree of heating would be required to bring the liquid to its flash point at this temperature. Class III liquids, which have flash points higher than 140°F (60°C), would require a significant heat source above ambient temperature to reach their flash point (see the definition of "Flammable liquids"). Class IIIA has a closed-cup flash point range of 140°F (60°C) to 200°F (93°C). Class IIIB has a closed-cup flash point at or above 200°F (93°C). All combustible liquids are considered Group H-2 materials except for Class IIIB liquids, which are considered Group H-3 because of their relatively high flash points. Motor oil is a typical example of a Class IIIB combustible liquid. Combustible liquids do not include compressed gases or cryogenic fluids. Compressed gases are regulated in Chapter 30 and cryogenic fluids are regulated in Chapter 32.

FIRE POINT. The lowest temperature at which a liquid will ignite and achieve sustained burning when exposed to a test flame in accordance with ASTM D 92.

❖ The fire point is the lowest temperature at which a liquid will ignite and sustain burning for a minimum of 5 seconds (fire point) when exposed to the test flame under a specific barometric pressure according to ASTM D 92.

FLAMMABLE LIQUID. A liquid having a closed cup flash point below 100°F (38°C). Flammable liquids are further categorized into a group known as Class I liquids. The Class I category is subdivided as follows:

Class IA. Liquids having a flash point below 73°F (23°C) and having a boiling point below 100°F (38°C).

Class IB. Liquids having a flash point below 73°F (23°C) and having a boiling point at or above 100°F (38°C).

Class IC. Liquids having a flash point at or above 73°F (23°C) and below 100°F (38°C).

The category of flammable liquids does not include compressed gases or cryogenic fluids.

❖ Flammable liquids have a closed-cup flash point less than 100°F (38°C); the classification of Class I liquid into three classes is dependent on their flash point. The 100°F (38°C) flash point limitation for flammable liquids assumes possible indoor ambient temperature conditions of 100°F (38°C). The vapor pressure limitation of 40 pound per square inch absolute (psia) (276 kPa) at 100°F (38°C) is the threshold for the definition of what constitutes a liquid for the purposes of classifying the material as a flammable or combustible liquid. Flammable liquids are classified into three classes based on a combination of their flash point and boiling point. Class IA has a flash point below 73°F (23°C) and a boiling point below 100°F (38°C). Class IB has a flash point below 73°F (23°C) and a boiling point at or above 100°F (38°C). Class IC has a flash point above 73°F (23°C) and below 100°F (38°C). Flammable liquids do not include compressed gases or cryogenic fluids. Compressed gases are regulated in Chapter 30 and cryogenic fluids are regulated in Chapter 32.

FLASH POINT. The minimum temperature in degrees Fahrenheit at which a liquid will give off sufficient vapors to form an ignitable mixture with air near the surface or in the container, but will not sustain combustion. The flash point of a liquid shall be determined by appropriate test procedure and apparatus as specified in ASTM D 56, ASTM D 93 or ASTM D 3278.

❖ The flash point is the characteristic used in the classification of flammable and combustible liquids. The flash point is the minimum temperature of a liquid at which it gives off sufficient vapor to form an ignitable mixture with air above its surface. The Tag Closed Tester (ASTM D 56), the Pensky-Martens Closed Tester (ASTM D 93) and the Small Scale Closed-Cup Apparatus (ASTM D 3278) are the referenced test procedures for determining the flash points of liquids. The applicability of the three test methods depends on the viscosity of the test liquid and the expected flash point.

FUEL LIMIT SWITCH. A mechanism, located on a tank vehicle, that limits the quantity of product dispensed at one time.

❖ This definition pertains to mobile fueling operations regulated by Section 3406.5.4.5 and describes a limit control that prevents more than a specified amount of liquid fuel from being dispensed at one time from a tank vehicle used in the mobile fueling operation (see commentary, Section 3406.5.4.5).

LIQUID STORAGE ROOM. A room classified as a Group H-3 occupancy used for the storage of flammable or combustible liquids in a closed condition.

❖ Group H-3 occupancy for storage of flammable or combustible liquids in closed containers recognizes the hazardous nature of these materials.

MOBILE FUELING. The operation of dispensing liquid fuels from tank vehicles into the fuel tanks of motor vehicles. Mobile fueling may also be known by the terms "Mobile Fleet Fueling," "Wet Fueling" and "Wet Hosing."

❖ This definition pertains to the fueling process regulated by Section 3406.5.4.5, wherein fuel is dispensed from the tank vehicle directly to the fuel tank of a vehicle (see commentary, Section 3406.5.4.5).

PROCESS TRANSFER. The transfer of flammable or combustible liquids between tank vehicles or tank cars and process operations. Process operations may include containers, tanks, piping and equipment.

❖ The transfer of flammable or combustible liquids during any process operation may include the introduction of the flammable or combustible liquids into or within the process operation.

REFINERY. A plant in which flammable or combustible liquids are produced on a commercial scale from crude petroleum, natural gasoline or other hydrocarbon sources.

❖ A refinery is the facility that produces flammable or combustible liquids from raw materials.

REMOTE EMERGENCY SHUTOFF DEVICE. The combination of an operator-carried signaling device and a mechanism on the tank vehicle. Activation of the remote emergency shutoff device sends a signal to the tanker-mounted mechanism and causes fuel flow to cease.

❖ This definition describes an important safety device used in the mobile fueling operation regulated by Section 3406.5.4.5 and describes a portable device that a tank vehicle driver may use to prevent an overfill spill during mobile fueling when the driver is out of immediate reach of the tanker shutoff controls (see commentary, Section 3406.5.4.5).

REMOTE SOLVENT RESERVOIR. A liquid solvent container enclosed against evaporative losses to the atmosphere during periods when the container is not being utilized, except for a solvent return opening not larger than 16 square inches (10 322 mm^2). Such return allows pump-cycled used solvent to drain back into the reservoir from a separate solvent sink or work area.

❖ A remote solvent reservoir is the storage of flammable or combustible liquid in a container that is not in the same control area as the machine using the flammable or combustible liquid. The remote solvent reservoir is connected to the machine by piping or tubing.

SOLVENT DISTILLATION UNIT. An appliance that receives contaminated flammable or combustible liquids and which distills the contents to remove contaminants and recover the solvents.

❖ A solvent distillation unit recycles flammable and combustible liquids by the condensation and collection of the vapors that are produced as the mixture is heated. The solvent distillation unit processes waste solvents in a separate, stand-alone batch, on-line batch or continuous systems. The distillation units heat the waste solvent to its boiling point. This causes the solvent to evaporate and the solvent vapors are then condensed in a separate container. The basic components of a distillation unit are the process chamber or boiler, the encapsulated heaters, a water-cooled chamber and associated piping and instrumentation. Temperature sensors monitor the temperature and help maintain the required distillation temperature. Disposable vessel liners can be used for simple collection and disposal of still bottoms. Vacuum pumps that can distill high-boiling solvents at lower temperatures are also available. Solvent distillation units having a distillation chamber capacity of 60 gallons (227 L) or less are listed under UL 2208. Solvent distillation units having a distillation chamber capacity greater than 60 gallons (227 L) must comply with Section 3405.4.2.

TANK, PRIMARY. A listed atmospheric tank used to store liquid. See "Primary containment."

❖ The primary tank is the principal storage vessel for flammable and combustible liquids. The tank may use a secondary containment system or be installed in a dike area to control leaks and spills.

TANK, PROTECTED ABOVE GROUND. A tank listed in accordance with UL 2085 consisting of a primary tank provided with protection from physical damage and fire-resistive protection from a high-intensity liquid pool fire exposure. The tank may provide protection elements as a unit or may be an assembly of components, or a combination thereof.

❖ A protected above-ground tank is a tank that is listed under UL 2085. The listing for these tanks indicates that the primary tank is protected from physical damage and also has fire-resistive protection from exposure to a high-intensity liquid pool fire. The listing states that this primary tank complies with the code as a "protected above-ground tank."

SECTION 3403 GENERAL REQUIREMENTS

3403.1 Electrical. Electrical wiring and equipment shall be installed and maintained in accordance with the ICC *Electrical Code.*

❖ The installation of electrical wiring and equipment is regulated by the ICC *Electrical Code*® (ICC EC™).

3403.1.1 Classified locations for flammable liquids. Areas where flammable liquids are stored, handled, dispensed or mixed shall be in accordance with Table 3403.1.1. A classified area shall not extend beyond an unpierced floor, roof or other solid partition.

The extent of the classified area is allowed to be reduced, or eliminated, where sufficient technical justification is provided to the fire code official that a concentration in the area in excess of 25 percent of the lower flammable limit (LFL) cannot be generated.

❖ Electrical systems can create sparks that can be a source of ignition. Areas of a building where flammable liquids are present are to have the electrical wiring and equipment installed to prevent the electrical system from becoming an ignition source. This type of electrical installation is not required in other areas of the building where the construction prevents the spread of flammable liquids and their vapors. This construction can be accomplished only by eliminating penetrations. Class I equipment locations, which require special installation of wiring and equipment, are listed in Table 3403.1.1.

TABLE 3403.1.1. See page 34-7.

❖ The table lists the Class I Group D locations that require installation of the electrical system or its components to prevent them from becoming an ignition source. Class I is defined as a location where flammable liquids or gases may be present in sufficient quantities to produce an explosive or ignitable mixture. Group D is defined as an atmosphere containing flammable or combustible liquids or gases. Group D is divided into two divisions. Division 1 is for locations where flammable or combustible vapors are present. Division 2 is for locations where flammable or combustible vapors may be present.

3403.1.2 Classified locations for combustible liquids. Areas where Class II or III liquids are heated above their flash points shall have electrical installations in accordance with Section 3403.1.1.

Exception: Solvent distillation units in accordance with Section 3405.4.

❖ Electrical systems can create sparks that can be a source of ignition. The sparks may be caused by electrical current or static electricity. Areas of a building where combustible liquids are heated to or above their flash points are treated the same as areas containing flammable liquids. Class I equipment locations, which require special installation of the wiring and equipment, are listed in Table 3403.1.1.

The exception indicates that use of a solvent distillation unit for the recovery of combustible liquids is not considered a Class I, Group D location. Solvent distillation units are listed, which will include installation instructions.

3403.1.3 Other applications. The fire code official is authorized to determine the extent of the Class I electrical equipment and wiring location when a condition is not specifically covered by these requirements or the ICC *Electrical Code.*

❖ There may be situations when the code does not specifically cover the hazardous location that should have Class I electrical equipment and wiring. The fire code official has the authority to require Class I electrical equipment and wiring in a location not specifically identified by the code for Division 1 or Division 2 locations.

3403.2 Fire protection. Fire protection for the storage, use, dispensing, mixing, handling and on-site transportation of flammable and combustible liquids shall be in accordance with this chapter and applicable sections of Chapter 9.

❖ The requirements of Chapter 9 apply to flammable and combustible liquids. Fire protection is a principal means of preventing and controlling the spread of fire.

3403.2.1 Portable fire extinguishers and hose lines. Portable fire extinguishers shall be provided in accordance with Section 906. Hose lines shall be provided in accordance with Section 905.

❖ Portable fire extinguishers and hose lines are to be installed where flammable and combustible liquids are stored, used or dispensed. These fire protection devices, operated by trained personnel, are to handle small emergencies. They are not an alternative to fire protection systems mandated by this section, Chapter 9 or the IBC.

TABLE 3403.1.1
CLASS I ELECTRICAL EQUIPMENT LOCATIONS[a]

LOCATION	GROUP D DIVISION	EXTENT OF CLASSIFIED AREA
Underground tank fill opening	1	Pits, boxes or spaces below grade level, any part of which is within the Division 1 or 2 classified area.
	2	Up to 18 inches above grade level within a horizontal radius of 10 feet from a loose-fill connection and within a horizontal radius of 5 feet from a tight-fill connection.
Vent—Discharging upward	1	Within 3 feet of open end of vent, extending in all directions.
	2	Area between 3 feet and 5 feet of open end of vent, extending in all directions.
Drum and container filling		
Outdoor or indoor with adequate ventilation	1	Within 3 feet of vent and fill opening, extending in all directions.
	2	Area between 3 feet and 5 feet from vent of fill opening, extending in all directions. Also up to 18 inches above floor or grade level within a horizontal radius of 10 feet from vent or fill opening.
Pumps, bleeders, withdrawal fittings, meters and similar devices		
Indoor	2	Within 5 feet of any edge of such devices, extending in all directions. Also up to 3 feet above floor or grade level within 25 feet horizontally from any edge of such devices.
Outdoor	2	Within 3 feet of any edge of such devices, extending in all directions. Also up to 18 inches horizontally from an edge of such devices.
Pits		
Without mechanical ventilation	1	Entire area within pit if any part is within a Division 1 or 2 classified area.
With mechanical ventilation	2	Entire area within pit if any part is within a Division 1 or 2 classified area.
Containing valves, fittings or piping, and not within a Division 1 or 2 classified area	2	Entire pit.
Drainage ditches, separators, impounding basins		
Indoor	1 or 2	Same as pits.
Outdoor	2	Area up to 18 inches above ditch, separator or basin. Also up to 18 inches above grade within 15 feet horizontal from any edge.
Tank vehicle and tank car[b]		
Loading through open dome	1	Within 3 feet of edge of dome, extending in all directions.
	2	Area between 3 feet and 15 feet from edge of dome, extending in all directions.
Loading through bottom connections with atmospheric venting	1	Within 3 feet of point of venting to atmosphere, extending in all directions.
	2	Area between 3 feet and 15 feet from point of venting to atmosphere, extending in all directions. Also up to 18 inches above grade within a horizontal radius of 10 feet from point of loading connection.
Office and restrooms	Ordinary	Where there is an opening to these rooms within the extent of an indoor classified location, the room shall be classified the same as if the wall, curb or partition did not exist.

(continued)

TABLE 3403.1.1—continued
CLASS I ELECTRICAL EQUIPMENT LOCATIONS[a]

LOCATION	GROUP D DIVISION	EXTENT OF CLASSIFIED AREA
Tank vehicle and tank car[b]–continued		
Loading through closed dome with atmospheric venting	1	Within 3 feet of open end of vent, extending in all directions.
	2	Area between 3 feet and 15 feet from open end of vent, extending in all directions. Also within 3 feet of edge of dome, extending in all directions.
Loading through closed dome with vapor control	2	Within 3 feet of point of connection of both fill and vapor lines, extending in all directions.
Bottom loading with vapor control or any bottom unloading	2	Within 3 feet of point of connection, extending in all directions. Also up to 18 inches above grade within a horizontal radius of 10 feet from point of connection.
Storage and repair garage for tank vehicles	1	Pits or spaces below floor level.
	2	Area up to 18 inches above floor or grade level for entire storage or repair garage.
Garages for other than tank vehicles	Ordinary	Where there is an opening to these rooms within the extent of an outdoor classified area, the entire room shall be classified the same as the area classification at the point of the opening.
Outdoor drum storage	Ordinary	
Indoor warehousing where there is no flammable liquid transfer	Ordinary	Where there is an opening to these rooms within the extent of an indoor classified area, the room shall be classified the same as if the wall, curb or partition did not exist.
Indoor equipment where flammable vapor/air mixtures could exist under normal operations	1	Area within 5 feet of any edge of such equipment, extending in all directions.
	2	Area between 5 feet and 8 feet of any edge of such equipment, extending in all directions. Also, area up to 3 feet above floor or grade level within 5 feet to 25 feet horizontally from any edge of such equipment.[c]
Outdoor equipment where flammable vapor/air mixtures could exist under normal operations	1	Area within 3 feet of any edge of such equipment, extending in all directions.
	2	Area between 3 feet and 8 feet of any edge of such equipment extending in all directions. Also, area up to 3 feet above floor or grade level within 3 feet to 10 feet horizontally from any edge of such equipment.
Tank—Above ground		
Shell, ends or roof and dike area	1	Area inside dike where dike height is greater than the distance from the tank to the dike for more than 50 percent of the tank circumference.
	2	Area within 10 feet from shell, ends or roof of tank. Area inside dikes to level of top of dike.
Vent	1	Area within 5 feet of open end of vent, extending in all directions.
	2	Area between 5 feet and 10 feet from open end of vent, extending in all directions.
Floating roof	1	Area above the roof and within the shell.

For SI: 1 inch = 25.4 mm, 1 foot = 304.8 mm.

a. Locations as classified in the ICC *Electrical Code*.

b. When classifying extent of area, consideration shall be given to the fact that tank cars or tank vehicles can be spotted at varying points. Therefore, the extremities of the loading or unloading positions shall be used.

c. The release of Class I liquids can generate vapors to the extent that the entire building, and possibly a zone surrounding it, are considered a Class I, Division 2 location.

3403.3 Site assessment. In the event of a spill, leak or discharge from a tank system, a site assessment shall be completed by the owner or operator of such tank system if the fire code official determines that a potential fire or explosion hazard exists. Such site assessments shall be conducted to ascertain potential fire hazards and shall be completed and submitted to the fire department within a time period established by the fire code official, not to exceed 60 days.

❖ Site assessment is to ensure that a spill, leak or discharge of flammable or combustible liquid is investigated and corrective action is taken. The corrective action may involve only cleaning the area and covering the cause of the problem during personnel safety meetings or it may require a revision to the equipment or operation procedures. The fire code official is to establish the deadline for completion of the site assessment. The maximum time for completing a site assessment is 60 days. The corrective actions may take longer than the time to complete the site assessment.

3403.4 Spill control and secondary containment. Where the maximum allowable quantity per control area is exceeded, and when required by Section 2704.2, rooms, buildings or areas used for storage, dispensing, use, mixing or handling of Class I, II and III-A liquids shall be provided with spill control and secondary containment in accordance with Section 2704.2.

❖ Where the maximum allowable quantity per control area of flammable or Class II or IIIA combustible liquids is exceeded, spills must be controlled to prevent the spread of liquid and vapors. Section 2704.2 discusses the use of liquid-tight floors, curbs, dikes and drainage systems to divert the liquid to a location where it can be contained and safely handled. Section 2704.3 discusses mechanical and natural exhaust systems. The exhaust system is to remove the vapors to prevent them from accumulating in concentrations in the flammable range of the vapor.

3403.5 Labeling and signage. The fire code official is authorized to require warning signs for the purpose of identifying the hazards of storing or using flammable liquids. Signage for identification and warning such as for the inherent hazard of flammable liquids or smoking shall be provided in accordance with this chapter and Sections 2703.5 and 2703.6.

❖ Signs are used to identify the flammable or combustible liquid being stored or used and to provide any warning or information necessary for its storage or use. Sections 2703.5 and 2703.6 include a reference to NFPA 704, which details the locations and construction of the signs. NFPA 704 uses a diamond with each of its four points colored either red, blue, yellow or white. Each diamond point represents a different hazard. For flammable and combustible liquids, the red diamond point (uppermost point) represents flammability. The number in this diamond point will vary from 0 (will not burn) to 4 (rapidly burn). For flammable and combustible reference, numbers 2 to 4 usually represent flammable and combustible liquids. These signs are permanent durable signs that are to be readily visible and are not to be covered or removed.

3403.5.1 Style. Warning signs shall be of a durable material. Signs warning of the hazard of flammable liquids shall have white lettering on a red background and shall read: DANGER—FLAMMABLE LIQUIDS. Letters shall not be less than 3 inches (76 mm) in height and 0.5 inch (12.7 mm) in stroke.

❖ The signs used with flammable liquids are to read "DANGER – FLAMMABLE LIQUIDS." The code defines color and size of lettering to ensure uniformity.

3403.5.2 Location. Signs shall be posted in locations as required by the fire code official. Piping containing flammable liquids shall be identified in accordance with ANSI A13.1.

❖ The location of signs is discussed in Section 2703.5 for containers, tanks, entrances, etc., by reference to NFPA 704. Signage for piping containing flammable liquids is covered under ANSI A13.1. The signage is in English text and requires arrows to indicate flow direction.

3403.5.3 Warning labels. Individual containers, packages and cartons shall be identified, marked, labeled and placarded in accordance with federal regulations and applicable state laws.

❖ The warning labels on individual containers, packages and cartons may be different than what is required under NFPA 704. Other federal and state laws may address the warning label for these individual containers, packages and cartons.

3403.5.4 Identification. Color coding or other approved identification means shall be provided on each loading and unloading riser for flammable or combustible liquids to identify the contents of the tank served by the riser.

❖ More than one flammable or combustible liquid may be present at a facility in addition to other liquids. The loading and unloading risers are to be color coded or identified by other approved identification means to ensure that operators know which material is being used. If color coding is not used, the fire code official is to approve any other identification means.

3403.6 Piping systems. Piping systems, and their component parts, for flammable and combustible liquids shall be in accordance with this section.

❖ Piping must be designed to provide protection against overpressure or other conditions that could create leaks at joints or rupture the pipes.

3403.6.1 Nonapplicability. The provisions of Section 3403.6 shall not apply to gas or oil well installations; piping that is integral to stationary or portable engines, including aircraft, watercraft and motor vehicles; and piping in connection with boilers and pressure vessels regulated by the *International Mechanical Code.*

❖ Piping of some applications of flammable and combustible liquids is addressed under other code documents or other laws. Other state and federal agencies have laws and authority to regulate the piping for oil and gas wells. Stationary and portable engines are manufactured under other standards that are enforced by other state and federal agencies. Because the piping associated with heating equipment (fuel oil and pressure systems) is addressed by the IMC, those requirements are not repeated.

3403.6.2 Design and fabrication of system components. Piping system components shall be designed and fabricated in accordance with NFPA 30, Chapter 3, except as modified by this section.

❖ NFPA 30 is a comprehensive document that sets forth requirements for the handling, storage and use of flammable and combustible liquids in a variety of situations. NFPA 30 presents a systematic approach to the design and operation of flammable and combustible liquid storage and handling facilities. The appendixes of NFPA 30 contain helpful guidance for meeting the intent of the code, including sizing data for above-ground tank emergency relief vents, recommended procedures for the abandonment or removal of underground tanks and recommended sprinkler design densities for protecting flammable liquid in bulk piles and rack storage.

3403.6.2.1 Special materials. Low-melting-point materials (such as aluminum, copper or brass), materials that soften on fire exposure (such as nonmetallic materials) and nonductile material (such as cast iron) shall be acceptable for use underground in accordance with ANSI B31.9. When such materials are used outdoors in above-ground piping systems or within buildings, they shall be in accordance with ANSI B31.9 and one of the following:

1. Suitably protected against fire exposure.
2. Located where leakage from failure would not unduly expose people or structures.
3. Located where leakage can be readily controlled by operation of accessible remotely located valves.

In all cases, nonmetallic piping shall be used in accordance with Section 3.3.6 of NFPA 30.

❖ Piping that may fail under fire exposure as a result of heat reducing its material strength is limited to specific installations and locations. Pipes that have "low melting points" can fail by sagging when exposed to fire. Sagging can cause a joint to separate or the pipes to burst if piping is under pressure. These failures would result in the flammable and combustible liquids being exposed to the fire and air. The limitations on the use of these piping materials reduces their exposure to a fire, or where a failure occurs, does not increase the potential danger. This kind of piping can also be used where the piping system can be isolated in an emergency.

3403.6.3 Testing. Unless tested in accordance with the applicable section of ANSI B31.9, piping, before being covered, enclosed or placed in use, shall be hydrostatically tested to 150 percent of the maximum anticipated pressure of the system, or pneumatically tested to 110 percent of the maximum anticipated pressure of the system, but not less than 5 pounds per square gauge (psig) (34.47 kPa) at the highest point of the system. This test shall be maintained for a sufficient time period to complete visual inspection of joints and connections. For a minimum of 10 minutes, there shall be no leakage or permanent distortion. Care shall be exercised to ensure that these pressures are not applied to vented storage tanks. Such storage tanks shall be tested independently from the piping.

❖ Piping must be tested to a minimum pressure of 5 pounds per square inch gauge (psig) (34.47 kPa) or to a pressure greater than the anticipated pressure of the system. The pipe joints and connections are to be visually inspected. The test pressure is to be maintained for a minimum of 10 minutes, but not less than the time required for a visual inspection.

3403.6.3.1 Existing piping. Existing piping shall be tested in accordance with this section when the fire code official has reasonable cause to believe that a leak exists. Piping that could contain flammable or combustible liquids shall not be tested pneumatically. Such tests shall be at the expense of the owner or operator.

Exception: Vapor-recovery piping is allowed to be tested using an inert gas.

❖ The fire code official may require testing of existing piping. Existing piping is to be tested to the same criteria as new piping, except that piping containing flammable or combustible liquids is not to be pneumatically tested. The introduction of air into these pipes can create a vapor and air mixture that reaches the flammable range.

The exception allows pneumatic testing of a vapor recovery system with an inert gas (such as nitrogen, carbon dioxide, etc.). Because vapor-recovery systems are designed to remove the flammable or combustible vapors and recycle the liquid, these vapors could be removed from the piping during the recovery process; however, the inert gas is still required to prevent the vapor and air mixture from reaching the flammable range before or during the recovery process.

3403.6.4 Protection from vehicles. Guard posts or other approved means shall be provided to protect piping, valves or fittings subject to vehicular damage in accordance with Section 312.

❖ Protection from vehicle impact is provided by guard posts or other approved barriers. Section 312 states the specifications for guard posts or the design forces required for an approved barrier should comply with the code.

3403.6.5 Protection from corrosion and galvanic action. Where subject to external corrosion, piping, related fluid-handling components and supports for both underground and above-ground applications shall be fabricated from noncorrosive materials, and coated or provided with corrosion protection. Dissimilar metallic parts that promote galvanic action shall not be joined.

❖ Deterioration of piping and components can cause leaks and spillage of flammable and combustible liquids. Using uncorrodible materials, protective coatings, galvanic protection or a combination of these methods can protect the piping and components. Dissimilar metals are prohibited because of the localized galvanic action that could occur between them. This localized galvanic action could cause one of the metals to corrode so that the other metal is protected from corrosion.

3403.6.6 Valves. Piping systems shall contain a sufficient number of manual control valves and check valves to operate the system properly and to protect the plant under both normal and emergency conditions. Piping systems in connection with pumps shall contain a sufficient number of such valves to control properly the flow of liquids in normal operation and in the event of physical damage or fire exposure.

❖ Valves are essential to proper operation. Check valves prevent the backflow or siphonage of flammable and combustible liquids. Other valves are used to isolate piping sections and equipment for maintenance. Valves are also used to stop the flow of flammable and combustible liquids.

3403.6.6.1 Backflow protections. Connections to pipelines or piping by which equipment (such as tank cars, tank vehicles or marine vessels) discharges liquids into storage tanks shall be provided with check valves or block valves for automatic protection against backflow where the piping arrangement is such that backflow from the system is possible. Where loading and unloading is done through a common pipe system, a check valve is not required. However, a block valve shall be provided which is located so as to be readily accessible or remotely operable.

❖ Check valves prevent the backflow or siphonage of flammable and combustible liquids. A check valve cannot be used for a common pipe used to both load and unload flammable and combustible liquids because check valves are designed to allow flow in only one direction. This type of valve would prohibit a common pipe for both loading and unloading. A block valve is to be used for this common pipe. The block valve control mechanism must be readily accessible or remotely operable in the event that the valve is needed to stop a spill or accidental discharge.

3403.6.6.2 Manual drainage. Manual drainage-control valves shall be located at approved locations remote from the tanks, diked area, drainage system and impounding basin to ensure their operation in a fire condition.

❖ In case of a fire, it may be necessary to drain the piping system. This is to be accomplished by a manual drainage-control valve. The number of manual drainage-control valves will depend on the facility. The manual drainage-control valves are to be located to isolate sections of the piping and equipment for maintenance, repair, replacement and control of flammable and combustible liquids during an emergency.

3403.6.7 Connections. Above-ground tanks with connections located below normal liquid level shall be provided with internal or external isolation valves located as close as practical to the shell of the tank. Except for liquids whose chemical characteristics are incompatible with steel, such valves, when external, and their connections to the tank shall be of steel.

❖ The isolation valve is located as near as practical to above-ground tanks to control the flow of flammable and combustible liquids. This location is to reduce the quantity of flammable and combustible liquids that may be discharged during an emergency. The loss of a portion of the piping system between an above-ground tank and the isolation valve may allow the discharge of the flammable and combustible liquids under gravity flow.

The isolation valve is to be of steel unless the flammable and combustible liquids are not compatible with steel. Other valve materials may be damaged and fail under the heat from a fire. If the isolation valve fails, the flammable and combustible liquids may be discharged under gravity flow and increase the fire potential.

3403.6.8 Piping supports. Piping systems shall be substantially supported and protected against physical damage and excessive stresses arising from settlement, vibration, expansion, contraction or exposure to fire. The supports shall be protected against exposure to fire by one of the following:

1. Draining liquid away from the piping system at a minimum slope of not less than 1 percent.
2. Providing protection with a fire-resistance rating of not less than 2 hours.
3. Other approved methods.

❖ Pipe supports are necessary to reduce stress on the pipe from both external and internal sources. Personnel are an external source of potential damage to piping, as is unattached equipment hitting the piping system. The pipe supports are to absorb these impact loads to protect the pipe from excess deflection.

Internal forces are caused by the positive and negative pressures created by the operation of pumps and valves. Pumping of flammable and combustible liquids will generate positive pressure in the pipe. This pressure, combined with any pressure required to support the dead weight of the piping system and flammable and combustible liquids in the piping system, can cause the pipe wall to rupture.

The pumping action and the operation of the valve can cause shock waves to travel through the flammable and combustible liquids in a pipe, which can create internal pressures several times larger than normal operating pressures. One type of shock wave is caused by the fast opening and closing of a valve. This pressure wave can place high internal pressures on the piping system.

This section lists three methods of protecting pipe supports from a fire:

1. The piping is required to have a minimum slope of 1 percent to allow for drainage. The viscosity of the flammable and combustible liquids may mandate a greater drainage slope.
2. The pipe supports must support the piping in a fire for a minimum of 2 hours based on ASTM E 119 test criteria. This fire protection of the supports is to keep the pipe with any flammable and combustible liquids in the pipe from adding to an existing fire.
3. The code always recognizes that there may be other methods available for accomplishing the intent. The fire code official has the responsibility to review these alternative methods and the authority to approve an alternative method, but only if it has been demonstrated or documented to comply with the intent of the code requirement.

3403.6.9 Flexible joints. Flexible joints shall be listed and approved and shall be installed on underground liquid, vapor and vent piping at all of the following locations:

1. Where piping connects to underground tanks.
2. Where piping ends at pump islands and vent risers.
3. At points where differential movement in the piping can occur.

❖ Flexible joints are necessary to handle expansion and contraction of the piping system and for vibration control. Expansion and contraction of the piping system will create stresses in the pipe and pipe joints because of the increase or decrease in pipe length. These changes in pipe length may cause the pipe to buckle, or pull or push a joint apart. The vibration of a pipe may cause a fatigue failure. Fatigue failures result from the reversal of stresses in a material. The flexing of the pipe wall of a section of pipe or pipe joint from a positive pressure to a negative pressure over time will create a fatigue failure in the pipe wall. This failure can cause the pipe or pipe joint to fail because of internal operating pressures or external loading, such as dead weight or an impact load.

This section lists three locations where flexible joints must be used:

1. *Piping connected to an underground tank.* An underground tank is unmovable so that the flexible joint connecting the piping to the tank must handle the expansion and contraction of the pipe system.
2. *Piping ending at pump islands and vent risers.* The mechanical equipment vibrations make it necessary to isolate the piping with flexible joints at a pumping island. Equipment vibrations can cause failure of vent risers or joints in vent risers. Vent risers may vibrate more than other piping because there is no liquid in the pipe to help dampen the vibrations.
3. *Points of differential movement.* Any location where differential settlement may occur will need to be isolated by a flexible joint. Differential settlement can cause the piping to become the support or to have inadequate support. Either case can cause the pipe to buckle or fail.

3403.6.9.1 Fiberglass-reinforced plastic piping. Fiberglass-reinforced plastic (FRP) piping is not required to be provided with flexible joints in locations where both of the following conditions are present:

1. Piping does not exceed 4 inches (102 mm) in diameter.
2. Piping has a straight run of not less than 4 feet (1219 mm) on one side of the connection when such connections result in a change of direction.

In lieu of the minimum 4-foot (1219 mm) straight run length, approved and listed flexible joints are allowed to be used under dispensers and suction pumps, at submerged pumps and tanks, and where vents extend above-ground.

❖ Fiberglass-reinforced plastic piping is more flexible than metal piping. This flexibility of fiberglass-reinforced plastic piping can be used to handle expansion and contraction of piping, or vibrations that would be handled by a flexible joint under a set of conditions of pipe diameter and minimum straight section of fiberglass-reinforced plastic piping. This section lists two conditions that must be met:

1. *Maximum 4 inches (102 mm) in diameter.* Fiberglass-reinforced plastic piping greater than 4 inches (102 mm) in diameter can be too stiff to have the flexibility necessary to be an alternative to a flexible joint. The greater the diameter of the pipe, the greater the stiffness of the pipe.
2. *Minimum 4-foot (1219 mm) straight run.* Straight runs of less than 4 feet (1219 mm) are too stiff to have the flexibility necessary to be an alternative to a flexible joint. The longer a pipe run is, the more the pipe can flex without causing stresses that will damage the pipe.

These two conditions are required before fiberglass-reinforced plastic piping can be used as the flexible joint. Flexible joints can be used with fiberglass-reinforced plastic piping. There are locations where there is not sufficient space to meet the two requirements for fiberglass-reinforced plastic piping as a flexible joint or where the use of fiberglass-reinforced plastic piping is not desired. These locations can use a flexible joint.

3403.6.10 Pipe joints. Joints shall be liquid tight and shall be welded, flanged or threaded except that listed flexible connectors are allowed in accordance with Section 3403.6.9. Threaded or flanged joints shall fit tightly by using approved methods and materials for the type of joint. Joints in piping systems used for

Class I liquids shall be welded when located in concealed spaces within buildings.

Nonmetallic joints shall be approved and shall be installed in accordance with the manufacturer's instructions.

Pipe joints that are dependent on the friction characteristics or resiliency of combustible materials for liquid tightness of piping shall not be used in buildings. Piping shall be secured to prevent disengagement at the fitting.

❖ Pipe joints are to be liquid tight. The code recognizes only three types of generic mechanical joints as being adequate for pipes carrying flammable and combustible liquids. Welded joints, flanged joints and threaded joints provide a liquid-tight joint. ANSI B31.3 contains criteria for the welding of piping. This reference is not in this chapter, but is cited in Chapter 27.

Flanged joints are to be made with materials that are compatible with the piping system and the flammable and combustible liquids in the pipe.

Threaded joints are to be fabricated by methods that ensure a liquid-tight joint by the selection of thread pitch and length of the threaded connection. Listed flexible joints are to be approved by the fire code official.

Pipe joints in a building's concealed space that carry Class I flammable liquids are limited to welded joints. Because Class I flammable liquids can become vapor at ambient temperature, a joint leak in a concealed space could go unnoticed. There could be no liquid escaping the concealed space for personnel to notice. A welded joint that has passed the test requirements of Section 3403.6.3 would be a liquid-tight joint that meets code requirements.

3403.6.11 Bends. Pipe and tubing shall be bent in accordance with ANSI B31.9.

❖ Pipe direction can be changed using either fittings or bends. Bends are to be done according to ANSI B31.9. Bending a pipe can damage the pipe. An improper bend may kink the interior portion of the pipe wall and that could cause increased pipe stresses resulting from the kink restricting flow. An improper bend could stretch the outer portion of the pipe wall. The stretched portion of pipe would have a thinner wall thickness. This thinner wall could develop pinhole leaks or even cause the pipe to rupture.

SECTION 3404
STORAGE

3404.1 General. The storage of flammable and combustible liquids in containers and tanks shall be in accordance with this section and the applicable sections of Chapter 27.

❖ This section and Chapter 27 cover the storage of flammable and combustible liquids in containers and tanks.

3404.2 Tank storage. The provisions of this section shall apply to:

1. The storage of flammable and combustible liquids in fixed above-ground and underground tanks.
2. The storage of flammable and combustible liquids in fixed above-ground tanks inside of buildings.
3. The storage of flammable and combustible liquids in portable tanks whose capacity exceeds 660 gallons (2498 L).
4. The installation of such tanks and portable tanks.

❖ The scope of this section is limited to the storage of flammable and combustible liquids in above-ground tanks, underground tanks, above-ground tanks in buildings and portable tanks exceeding 660 gallons (2498 L) and the installation of these tanks.

3404.2.1 Change of tank contents. Tanks subject to change in contents shall be in accordance with Section 3404.2.7. Prior to a change in contents, the fire code official is authorized to require testing of a tank.

Tanks that have previously contained Class I liquids shall not be loaded with Class II or Class III liquids until such tanks and all piping, pumps, hoses and meters connected thereto have been completely drained and flushed.

❖ The type of flammable or combustible liquid stored in a tank can change. When the type of flammable or combustible liquid is changed, the design and construction of the tank is to comply with NFPA 30 for the type of flammable or combustible liquid to be stored. A change in the tank contents can effect safety by altering the flashpoint of the contents through contamination. Accordingly, changing the LFL, the fire code official has the authority to require the tank to be tested before placing the tank in operation with the new type of flammable and combustible liquid.

If the change of flammable and combustible liquid is from a Class I to a Class II or III liquid, the tank and accessory piping and equipment are to be drained and cleaned of the Class I liquid. The Class I liquid left in the tank and accessory piping and equipment can generate vapors that could create a hazard not associated with the Class II or III liquid that has replaced the Class I liquid.

3404.2.2 Use of tank vehicles and tank cars as storage tanks. Tank cars and tank vehicles shall not be used as storage tanks.

❖ The code does not regulate tank vehicles and tank cars that contain flammable and combustible liquids. Monitoring the use of these portable tanks would place a major burden on the fire code official to verify that the tank vehicles and tank cars comply with federal and state regulations. Because these tanks are portable, their location on the facility could require extensive monitoring to be certain that the locations do not violate code requirements. The problems in verifying code compliance and controlling the safe use and operation of portable tanks at a facility forces the code to prohibit their use as permanent storage.

3404.2.3 Labeling and signs. Labeling and signs for storage tanks and storage tank areas shall comply with Sections 3404.2.3.1 and 3404.2.3.2.

❖ Above-ground tanks, underground tanks, above-ground tanks in buildings and portable tanks exceeding 660 gallons (2,498 L) are to be provided with warning and identification signs.

3404.2.3.1 Smoking and open flame. Signs shall be posted in storage areas prohibiting open flames and smoking. Signs shall comply with Section 3403.5.

❖ Warning signs prohibiting smoking and open flames are to comply with NFPA 704.

3404.2.3.2 Label or placard. Tanks more than 100 gallons (379 L) in capacity, which are permanently installed or mounted and used for the storage of Class I, II or IIIA liquids, shall bear a label and placard identifying the material therein. Placards shall be in accordance with NFPA 704.

Exceptions:

1. Tanks of 300-gallon (1136 L) capacity or less located on private property and used for heating and cooking fuels in single-family dwellings.
2. Tanks located underground.

❖ Tanks in excess of 100 gallons (379 L) used to stored Class I, II or IIIA liquids are to be permanently labeled to identify the flammable or combustible liquid.

There are two exceptions to this labeling of storage tanks based on the use of the storage tank and its location:

1. A storage tank for use by a single-family dwelling does not need to be labeled. This exception is based on the tank's sole use for the storage of heating or cooking fuel used by the occupants of the single-family dwelling. This storage tank is limited to a maximum capacity of 300 gallons (1136 L).
2. Underground tanks do not need to be labeled.

3404.2.4 Sources of ignition. Smoking and open flames are prohibited in storage areas in accordance with Section 2703.7.

Exception: Areas designated as smoking and hot work areas, and areas where hot work permits have been issued in accordance with this code.

❖ Smoking and open flames are strictly limited around above-ground tanks, underground tanks, above-ground tanks in buildings and portable tanks exceeding 660 gallons (2498 L). Section 2703.7 prohibits smoking or open flames within 25 feet (7620 mm) of outdoor storage of flammable and combustible liquids and indoors where there are flammable and combustible liquids or where vapors from flammable and combustible liquids may occur.

The exception allows establishment of designated smoking areas and hot work areas. Hot work may be done in areas not designated as a hot work area when a hot work permit is obtained.

3404.2.5 Explosion control. Explosion control shall be provided in accordance with Section 911.

❖ The vapor from flammable and combustible liquids can cause an explosion when the vapor-air mixture is in an explosive ratio. Explosion control is required for a facility that is storing or using Class IA liquids or for a facility that has open use or dispensing of Class IIB liquids. Section 911 requires deflagration venting to direct the force of an explosion out of the structure and into an unoccupied area. This section also references NFPA 69, which may require monitoring of gases and other methods to suppress factors affecting an explosion.

3404.2.6 Separation from incompatible materials. Storage of flammable and combustible liquids shall be separated from incompatible materials in accordance with Section 2703.9.8.

Grass, weeds, combustible materials and waste Class I, II or IIIA liquids shall not be accumulated in an unsafe manner at a storage site.

❖ Materials that could create a fire or explosive hazard when in contact with flammable or combustible liquids are to be separated from flammable or combustible liquids by either distance or physical barriers.

Materials that may not be incompatible with flammable or combustible liquids but may be an ignition source or a fuel source are to be removed.

3404.2.7 Design, construction and general installation requirements for tanks. The design, fabrication and construction of tanks shall comply with NFPA 30. Each tank shall bear a permanent nameplate or marking indicating the standard used as the basis of design.

❖ Tanks are to be designed according to NFPA 30 and labeled to indicate the design standard.

3404.2.7.1 Materials used in tank construction. The materials used in tank construction shall be in accordance with NFPA 30.

❖ The preferred material for tank construction is steel and concrete. These materials have high levels of resistance to heat.

3404.2.7.2 Pressure limitations for tanks. Tanks shall be designed for the pressures to which they will be subjected in accordance with NFPA 30.

❖ Tanks may be designed for operation under atmospheric pressure, low pressure or high pressure. The design criteria for pressure are in NFPA 30.

3404.2.7.3 Tank vents for normal venting. Tank vents for normal venting shall be installed and maintained in accordance with Sections 3404.2.7.3.1 through 3404.2.7.3.6.

❖ Tanks are vented to maintain the internal tank pressure within the design operating range. A low pressure can increase the generation of vapors while a high pressure can damage the tank or piping system. Any pressure outside of the design pressure range can have an adverse affect on the operation of the system as well as the piping and equipment.

3404.2.7.3.1 Vent lines. Vent lines from tanks shall not be used for purposes other than venting unless approved.

❖ Vent lines are to be used only as vents unless an additional use is approved by the fire code official.

3404.2.7.3.2 Vent-line flame arresters and venting devices. Vent-line flame arresters and venting devices shall be installed in accordance with their listings. Use of flame arresters in piping systems shall be in accordance with API 2028.

❖ The vapors from a vent will be mixing with air and will become an ignitable mixture. This condition requires that measures be taken to either suppress an ignition source or to disperse the ignitable mixture with additional air to drop the vapor air mixture below the LFL.

3404.2.7.3.3 Vent pipe outlets. Vent pipe outlets for tanks storing Class I, II or IIIA liquids shall be located such that the vapors are released at a safe point outside of buildings and not less than 12 feet (3658 mm) above the adjacent ground level. Vapors shall be discharged upward or horizontally away from adjacent walls to assist in vapor dispersion. Vent outlets shall be located such that flammable vapors will not be trapped by eaves or other obstructions and shall be at least 5 feet (1524 mm) from building openings or lot lines of properties that can be built upon. Vent outlets on atmospheric tanks storing Class IIIB liquids are allowed to discharge inside a building if the vent is a normally closed vent.

❖ Vent pipes must be terminated to direct vapors away from the building. Vapors from flammable liquids are normally heavier than air so that the vapor will settle to lower levels. The termination of a vent pipe a minimum of 12 feet (3658 mm) above grade will provide space for the vapors to disperse to below the LFL. This high termination elevation also reduces the potential for the termination being close to grade-level ignition sources. Because flammable liquid vapors are heavier than air, attention needs to be placed on the building design near the termination. Building design and features that may allow the flammable liquid vapors to reenter the building or to collect on the building are to be eliminated. Because Class IIIB liquids have a relatively high boiling point, the vent termination for atmospheric tanks containing these combustible liquids may terminate inside the building, if the vent is normally closed.

3404.2.7.3.4 Installation of vent piping. Vent piping shall be designed, sized, constructed and installed in accordance with Section 3403.6. Vent pipes shall be installed such that they will drain toward the tank without sags or traps in which liquid can collect. Vent pipes shall be installed in such a manner so as not to be subject to physical damage or vibration.

❖ Section 3403.6 covers the design, installation, testing and protection of vent pipes. Vent pipes are to drain to the tank and not accumulate condensation in the vent pipe.

3404.2.7.3.5 Manifolding. Tank vent piping shall not be manifolded unless required for special purposes such as vapor recovery, vapor conservation or air pollution control.

❖ The combining of vent pipes into a manifold is not permitted as a function for venting unless it is required for the purposes stated in this section. Manifolding of several vents can cause pressure problems during tank filling operation or in case of a fire in or near the tanks. When vapor recovery, vapor conservation or pollution control is necessary, the vents can be manifolded so that the same equipment can be used for multiple vents.

3404.2.7.3.5.1 Above-ground tanks. For above-ground tanks, manifolded vent pipes shall be adequately sized to prevent system pressure limits from being exceeded when manifolded tanks are subject to the same fire exposure.

❖ Manifolding of above-ground tanks requires the manifold to be designed to handle the additional pressure generated by the heating of the flammable or combustible liquid. This additional pressure could cause pressure to build up in other tanks or in the piping system. This additional pressure could create leaks or failure of other tanks or piping.

3404.2.7.3.5.2 Underground tanks. For underground tanks, manifolded vent pipes shall be sized to prevent system pressure limits from being exceeded when manifolded tanks are filled simultaneously.

❖ Manifolding of underground tanks must consider the buildup of pressure when the underground tanks are filled simultaneously. The introduction of flammable or combustible liquids into several tanks at the same time will require the vent system to release the vapor in the tank being displaced by the flammable or combustible liquids. This displaced vapor could exceed the capacity of the vent, causing a buildup of pressure.

3404.2.7.3.5.3 Tanks storing Class I liquids. Vent piping for tanks storing Class I liquids shall not be manifolded with vent piping for tanks storing Class II and III liquids unless positive means are provided to prevent the vapors from Class I liquids from entering tanks storing Class II and III liquids, to prevent contamination and possible change in classification of less volatile liquid.

❖ Vapor from a Class I liquid is not to be vented with Class II or III liquid vapors unless a positive means is provided to prevent the Class I vapors from entering tanks storing Class II or III liquids. The flash point of the Class I liquid vapor may affect the flash point of the vapors in the tanks storing the Class II or III liquids.

3404.2.7.3.6 Tank venting for tanks and pressure vessels storing Class IB and IC liquids. Tanks and pressure vessels storing Class IB or IC liquids shall be equipped with venting devices which shall be normally closed except when venting under pressure or vacuum conditions, or with listed flame arresters. The vents shall be installed and maintained in accordance with Section 2.2.5.1 of NFPA 30 or API 2000.

❖ Class IB and IC liquids are to be vented to relieve excess pressure or to relieve a vacuum. Normally these vents will be closed to prevent the mixing of air with the vapors from Class IB and IC liquids. The vapor-air mixture for Class IB and IC liquids has a low ignition point.

3404.2.7.4 Emergency venting. Stationary, above-ground tanks shall be equipped with additional venting that will relieve excessive internal pressure caused by exposure to fires. Emergency vents for Class I, II and IIIA liquids shall not discharge inside buildings. The venting shall be installed and maintained in accordance with Section 2.2.5.2 of NFPA 30.

Exception: Tanks larger than 12,000 gallons (45 420 L) in capacity storing Class IIIB liquids which are not within the diked area or the drainage path of Class I or II liquids do not require emergency relief venting.

❖ Stationary above-ground tanks can be exposed to an external fire. This fire hazard will heat the tank, generating a greater volume of flammable or combustible liquid vapors in the tank. These vapors can create pressures that could damage the tank or piping system. These tanks are vented to relieve this additional pressure.

Stationary above-ground tanks storing more than 12,000 gallons (45 420 L) of Class IIIB liquids do not need emergency venting if the tank is not in the same containment area or drainage path as tanks with Class I or II liquids. The high boiling point of Class IIIB liquids and the high volume provide a degree of safety before the buildup of internal pressure. With the tank safety features, the volume of Class IIIB liquid and having the tank located so that it does not affect tanks storing Class I or II liquids, emergency venting is not required.

3404.2.7.5 Tank openings other than vents. Tank openings for other than vents shall comply with Sections 3404.2.7.5.1 through 3404.2.7.5.8.

❖ Tanks will have openings other than a vent opening. It will be necessary to have openings for the transfer of flammable and combustible liquids, to monitor the contents, sampling, etc. These openings will need to be controlled to avoid the escape of vapors of flammable and combustible liquids or the entrance of air.

3404.2.7.5.1 Connections below liquid level. Connections for tank openings below the liquid level shall be liquid tight.

❖ Connections below the liquid level for flammable and combustible liquids are to be liquid tight. A connection that is not liquid tight may allow the flammable or combustible liquid to leak from the connection. The leakage could cause a vapor-air mixture that is between the LFL and the UFL.

3404.2.7.5.2 Filling, emptying and vapor recovery connections. Filling, emptying and vapor recovery connections to tanks containing Class I, II or IIIA liquids shall be located outside of buildings at a location free from sources of ignition and not less than 5 feet (1524 mm) away from building openings or lot lines of property that can be built on. Such openings shall be provided with a liquid-tight cap which shall be closed when not in use and properly identified.

❖ Tanks for Class I, II and IIIA liquids are to have openings for filling, emptying and vapor recovery connections outside of the building and away from property lines and ignition sources. The low flash point of the liquids creates a hazard that is not acceptable inside a building where vapors could accumulate.

3404.2.7.5.3 Piping, connections and fittings. Piping, connections, fittings and other appurtenances shall be installed in accordance with Section 3403.6.

❖ Section 3403.6 covers the design, installation, testing and protection of vent pipes.

3404.2.7.5.4 Manual gauging. Openings for manual gauging, if independent of the fill pipe, shall be provided with a liquid-tight cap or cover. Covers shall be kept closed when not gauging. If inside a building, such openings shall be protected against liquid overflow and possible vapor release by means of a spring- loaded check valve or other approved device.

❖ A manual gauge opening that has a liquid-tight cap and protection against overfill and vapor release is acceptable in tanks located inside a building. The manual gauge opening is permitted if it meets the safety requirements to prevent spillage, leakage and vapor release as well as the operation feature that the opening has to be closed except when used to check the contents of the tank.

3404.2.7.5.5 Fill pipes and discharge lines. For top-loaded tanks, a metallic fill pipe shall be designed and installed to minimize the generation of static electricity by terminating the pipe within 6 inches (152 mm) of the bottom of the tank, and it shall be installed in a manner which avoids excessive vibration.

❖ The filling of a tank with flammable or combustible liquids can generate static electricity. To reduce the generation of static electricity for a top-loaded tank, the fill pipe is to be metallic and extend to within 6 inches (152 mm) of the tank bottom.

3404.2.7.5.5.1 Class I liquids. For Class I liquids other than crude oil, gasoline and asphalt, the fill pipe shall be designed and installed in a manner which will minimize the possibility of generating static electricity by terminating within 6 inches (152 mm) of the bottom of the tank.

❖ Class I liquids other than crude oil, gasoline and asphalt are to comply with Section 3404.2.7.5.5.

3404.2.7.5.5.2 Underground tanks. For underground tanks, fill pipe and discharge lines shall enter only through the top. Fill lines shall be sloped toward the tank. Underground tanks for Class I liquids having a capacity greater than 1,000 gallons (3785 L) shall be equipped with a tight fill device for connecting the fill hose to the tank.

❖ Underground tanks are to have the fill pipe and discharge pipe through the top. These fill and discharge pipes are to slope to the top to prevent the accumulation of flammable and combustible liquid. The tight fill device for tanks having a capacity of 1,000 gallons (3785 L) of Class I liquid is to ensure a liquid-tight mechanical connection of the fill hose to the tank. This connection will reduce the potential for spills or leakage and the mixing of Class I vapors with air.

3404.2.7.5.6 Location of connections that are made or broken. Filling, withdrawal and vapor-recovery connections for Class I, II and IIIA liquids which are made and broken shall be located outside of buildings at a location away from sources of ignition and not less than 5 feet (1524 mm) away from building openings. Such connections shall be closed and liquid tight when not in use and shall be properly identified.

❖ Connections for Class I, II and IIIA liquids that are made or broken for filling, emptying and vapor recovery are to be outside of the building and away from property lines and ignition sources. The low flash point of the liquids creates a hazard that is not acceptable inside of a building where vapors could accumulate.

3404.2.7.5.7 Protection against vapor release. Tank openings provided for purposes of vapor recovery shall be protected against possible vapor release by means of a spring-loaded check valve or dry-break connections, or other approved device, unless the opening is a pipe connected to a vapor processing system. Openings designed for combined fill and vapor recovery shall also be protected against vapor release unless connection of the liquid delivery line to the fill pipe simultaneously connects the vapor recovery line. Connections shall be vapor tight.

❖ Connections are to be vapor tight to prevent the release of vapor from flammable or combustible liquids into the area surrounding the tank. The release of vapor from flammable or combustible liquids could create a vapor-air mixture that exceeds the LFL. The code does permit connections that are not vapor tight if the connection is part of the vapor recovery system. This exception is permitted because the vapor recovery system should be operating at a pressure lower than atmospheric pressure. This lower pressure should prevent vapor from escaping the vapor recovery system.

3404.2.7.5.8 Overfill prevention. An approved means or method in accordance with Section 3404.2.9.6.6 shall be provided to prevent the overfill of all Class I, II and IIIA liquid storage tanks.

❖ Overfill protection for tanks containing Class I, II or IIIA liquids is covered in Section 3404.2.9.6.6.

3404.2.7.6 Repair, alteration or reconstruction of tanks and piping. The repair, alteration or reconstruction, including welding, cutting and hot tapping of storage tanks and piping that have been placed in service, shall be in accordance with NFPA 30.

❖ Tanks and piping for flammable and combustible liquids can be repaired, altered or reconstructed under the criteria of NFPA 30.

3404.2.7.7 Design of supports. The design of the supporting structure for tanks shall be in accordance with the *International Building Code* and NFPA 30.

❖ Footings, foundations and structural supports for tanks must comply with the IBC and NFPA 30.

3404.2.7.8 Locations subject to flooding. Where a tank is located in an area where it is subject to buoyancy because of a rise in the water table, flooding or accumulation of water from fire suppression operations, uplift protection shall be provided in accordance with Sections 2.3.2.6 and 2.3.3.5 of NFPA 30.

❖ The tank, with its content of flammable or combustible liquid, may weigh less than an equivalent volume of water. If this occurs and the tank is subjected to flooding, the tank will float. This will place stresses on piping systems that could fail, causing the flammable or combustible liquid to spill. The flooding may be from natural causes or from fire suppression operations. The application of water by a fire department could cause flooding in the area of the tank. Flood zones may change over time as a result of upstream development or flood control projects. Section 26.6 of NFPA 30 contains the criteria for designing anchorage for both above-ground and underground tanks.

3404.2.7.9 Corrosion protection. Where subject to external corrosion, tanks shall be fabricated from corrosion-resistant materials, coated or provided with corrosion protection in accordance with Section 2.2.6.1 of NFPA 30.

❖ Soil conditions and environmental conditions can cause tanks to deteriorate. Corrosion can weaken the tank, creating a potential for leakage. The tank is to be protected from corrosion by use of corrosion-resistant material, coatings, cathodic protection or methods described in Section 24.3 of NFPA 30.

3404.2.7.10 Leak reporting. A consistent or accidental loss of liquid, or other indication of a leak from a tank system, shall be reported immediately to the fire department, the fire code official and other authorities having jurisdiction.

❖ The leakage of flammable or combustible liquids is a serious fire hazard. This hazard is to be immediately reported to the fire department, the fire code official and other authorities having jurisdiction.

The leaking tank will have to be repaired or taken out of service. The fire code official will have to review plans and issue a permit for the repair.

3404.2.7.10.1 Leaking tank disposition. Leaking tanks shall be promptly emptied, repaired and returned to service, abandoned or removed in accordance with Section 3404.2.13 or 3404.2.14.

❖ Leaking tanks must be repaired, taken out of service or removed. The fire code official must review plans and issue a permit for these activities.

3404.2.7.11 Tank lining. Steel tanks are allowed to be lined only for the purpose of protecting the interior from corrosion or providing compatibility with a material to be stored. Only those liquids tested for compatibility with the lining material are allowed to be stored in lined tanks.

❖ Tanks are lined to prevent corrosion from attacking the interior surface. Water vapor can condense inside a tank or the flammable or combustible liquid stored can be corrosive. The type of lining will limit the use of the tank to flammable and combustible liquids that are compatible with the lining.

3404.2.8 Vaults. Vaults shall be allowed to be either above or below grade and shall comply with Sections 3404.2.8.1 through 3404.2.8.18.

❖ Vaults are designed and constructed for the protection of tanks and as a secondary containment for flammable and combustible liquids. Above-ground tanks may be installed in vaults.

3404.2.8.1 Listing required. Vaults shall be listed in accordance with UL 2245.

Exception: Where approved by the fire code official, below-grade vaults are allowed to be constructed on site, provided that the design is in accordance with the *International Building Code* and that special inspections are conducted to verify structural strength and compliance of the installation with the approved design in accordance with the *International Building Code,* Section 1707. Installation plans for below-grade vaults that are constructed on site shall be prepared by, and the design shall bear the stamp of, a professional engineer. Consideration shall be given to soil and hydrostatic loading on the floors, walls and lid; anticipated seismic forces; uplifting by ground water or flooding; and to loads imposed from above such as traffic and equipment loading on the vault lid.

❖ Vaults must be listed to UL 2245. The fire code official can approve below-grade vaults that are constructed on site. These below-grade vaults must be designed by a professional engineer to comply with the IBC. The construction is to be inspected by a design professional.

3404.2.8.2 Design and construction. The vault shall completely enclose each tank. There shall be no openings in the vault enclosure except those necessary for access to, inspection of, and filling, emptying and venting of the tank. The walls and floor of the vault shall be constructed of reinforced concrete at least 6 inches (152 mm) thick. The top of an above-grade vault shall be constructed of noncombustible material and shall be designed to be weaker than the walls of the vault, to ensure that the thrust of an explosion occurring inside the vault is directed upward before significantly high pressure can develop within the vault.

The top of an at-grade or below-grade vault shall be designed to relieve safely or contain the force of an explosion occurring inside the vault. The top and floor of the vault and the tank foundation shall be designed to withstand the anticipated loading, including loading from vehicular traffic, where applicable. The walls and floor of a vault installed below grade shall be designed to withstand anticipated soil and hydrostatic loading.

Vaults shall be designed to be wind and earthquake resistant, in accordance with the *International Building Code.*

❖ The vault is to be of noncombustible materials with openings required only for the operation and maintenance of the enclosed tank. Reinforced concrete is to be used for the walls and floor and the top is designed to vent an explosion.

3404.2.8.3 Secondary containment. Vaults shall be substantially liquid tight and there shall be no backfill around the tank or within the vault. The vault floor shall drain to a sump. For premanufactured vaults, liquid tightness shall be certified as part of the listing provided by a nationally recognized testing laboratory. For field-erected vaults, liquid tightness shall be certified in an approved manner.

❖ The vault is to function as a secondary containment for the tank in the event of a leak or spillage. Any leakage or spillage must be removed by draining to a sump in the vault floor. Premanufactured vaults are required to be listed in accordance with UL 2245 by Section 3404.2.8.1, and part of the analysis program performed is an evaluation during the listing process of the effectiveness of secondary containment provided by a vault. It is much more difficult to accomplish a liquid-tight installation on a field-erected vault, as evidenced by the proliferation of leaky basements found throughout the country; therefore, to enhance the quality of field-erected vaults, such units must be certified for liquid tightness to the satisfaction of the fire code official.

Methods that might be used include third-party inspection/evaluation and full-scale liquid retention testing.

3404.2.8.4 Internal clearance. There shall be sufficient clearance between the tank and the vault to allow for visual inspection and maintenance of the tank and its appurtenances. Dispensing devices are allowed to be installed on tops of vaults.

❖ Maintenance and inspection of the tank requires that the clearance between the tank and the vault be sufficient for personnel to perform these functions.

3404.2.8.5 Anchoring. Vaults and their tanks shall be suitably anchored to withstand uplifting by ground water or flooding, including when the tank is empty.

❖ The tank and the vault may float if the water table is high or flooding occurs. The tank is to be anchored to the vault to prevent the tank from floating, while the vault is also to be anchored to prevent it from floating. Anchoring is also required under Section 3404.2.7.8.

3404.2.8.6 Vehicle impact protection. Vaults shall be resistant to damage from the impact of a motor vehicle, or vehicle impact protection shall be provided in accordance with Section 312.

❖ Protection from impact by vehicles is provided by guard posts or other approved barriers. Section 312 contains the specifications for guard posts or the design forces required for an approved barrier to comply with the code. Protection from vehicle impact is also required under Section 3403.6.4.

3404.2.8.7 Arrangement. Tanks shall be listed for above-ground use, and each tank shall be in its own vault. Compartmentalized tanks shall be allowed and shall be considered as a single tank. Adjacent vaults shall be allowed to share a common wall. The common wall shall be liquid and vapor tight and shall be designed to withstand the load imposed when the vault on either side of the wall is filled with water.

❖ Above-ground tanks installed in vaults are to be independent of each other. A separate vault is to be constructed for each tank. The individual vaults can use a common separation wall if it does not allow the flammable and combustible liquids or their vapors from one vault to enter another vault. This common wall must be able to resist the hydrostatic loads if the adjacent vault is flooded.

3404.2.8.8 Connections. Connections shall be provided to permit venting of each vault to dilute, disperse and remove vapors prior to personnel entering the vault.

❖ Because flammable and combustible vapors are normally heavier than air, the vault must have connections for venting these vapors.

3404.2.8.9 Ventilation. Vaults that contain tanks of Class I liquids shall be provided with an exhaust ventilation system installed in accordance with Section 2704.3. The ventilation system shall operate continuously or be designed to operate upon activation of the vapor or liquid detection system. The system shall provide ventilation at a rate of not less than 1 cubic foot per minute (cfm) per square foot of floor area [0.00508 $m^3/(s \times m^2)$], but not less than 150 cfm [0.071 $m^3/(s \times m^2)$]. The exhaust system shall be designed to provide air movement across all parts of the vault floor. Supply and exhaust ducts shall extend to within 3 inches (76 mm), but not more than 12 inches (305 mm), of the floor. The exhaust system shall be installed in accordance with the *International Mechanical Code.*

❖ The removal of Class I vapors from a vault requires that a ventilation system be installed according to Section 2704.3 and the IMC. The ventilation system must have supply and exhaust ducts within 3 inches (76 mm) to 12 inches (305 mm) of the vault floor. These ducts are to provide ventilation across the entire vault floor to remove the vapors and provide breathable air.

3404.2.8.10 Liquid detection. Vaults shall be equipped with a detection system capable of detecting liquids, including water, and activating an alarm.

❖ The liquid detection system is to identify any leakage or spillage of flammable or combustible liquids. There are two concerns for liquid detection: flammable or combustible liquids leaking into the vault and water leaking into the vault.

3404.2.8.11 Monitoring and detection. Vaults shall be provided with approved vapor and liquid detection systems and equipped with on-site audible and visual warning devices with battery backup. Vapor detection systems shall sound an alarm when the system detects vapors that reach or exceed 25 percent of the lower explosive limit (LEL) of the liquid stored. Vapor detectors shall be located no higher than 12 inches (305 mm) above the lowest point in the vault. Liquid detection systems shall sound an alarm upon detection of any liquid, including water. Liquid detectors shall be located in accordance with the manufacturer's instructions. Activation of either vapor or liquid detection systems shall cause a signal to be sounded at an approved, constantly attended location within the facility serving the tanks or at an approved location. Activation of vapor detection systems shall also shut off dispenser pumps.

❖ The vault is to have liquid and vapor detection systems that sound an alarm when either is present. The systems are to have battery backup in the event of a power failure.

The liquid and vapor detection systems are to sound an alarm at a constantly attended location that is approved by the fire code official.

The vapor detection system is to sound an alarm and shut off dispensing pumps when the detection system senses a vapor concentration of 25 percent or greater of the flammable or combustible LFL.

3404.2.8.12 Liquid removal. Means shall be provided to recover liquid from the vault. Where a pump is used to meet this requirement, the pump shall not be permanently installed in the vault. Electric-powered portable pumps shall be suitable for use in Class I, Division 1 locations, as defined in the ICC *Electrical Code*.

❖ A method for removing liquid from the vault must be provided. This can be a gravity drain, if the site is appropriate, or manual or portable electric pumps.

3404.2.8.13 Normal vents. Vent pipes that are provided for normal tank venting shall terminate at least 12 feet (3658 mm) above ground level.

❖ Vault vent pipe termination must comply with Section 3404.2.7.3.3.

3404.2.8.14 Emergency vents. Emergency vents shall be vapor tight and shall be allowed to discharge inside the vault. Long-bolt manhole covers shall not be permitted for this purpose.

❖ An emergency vent is necessary to release any pressure that develops in a tank when it is exposed to fire. The fire will cause a release of vapor from the flammable or combustible liquid that greatly exceeds that which is expected during normal operation. The emergency vent on a tank is to prevent the tank from rupturing, which would expose a greater volume of flammable or combustible liquids to the fire.

3404.2.8.15 Accessway. Vaults shall be provided with an approved personnel accessway with a minimum dimension of 30 inches (762 mm) and with a permanently affixed, nonferrous ladder. Accessways shall be designed to be nonsparking. Travel distance from any point inside a vault to an accessway shall not exceed 20 feet (6096 mm). At each entry point, a warning sign indicating the need for procedures for safe entry into confined spaces shall be posted. Entry points shall be secured against unauthorized entry and vandalism.

❖ Access to the vault must have a minimum dimension of 30 inches (762 mm) for ease of personnel passage. Because there is the possibility of vapors from flammable and combustible liquids in the vault, the accessway and ladder are to be nonsparking. Vapor from flammable and combustible liquids can replace the air in a vault. For personnel safety, the travel distance is limited and warning signs must be posted to remind personnel of potential hazards.

3404.2.8.16 Fire protection. Vaults shall be provided with a suitable means to admit a fire suppression agent.

❖ The fire suppression agent used to control a flammable or combustible liquid fire in a vault may be water, foam or some combination. The vault must be equipped with a suitable means for applying the fire suppression agent.

3404.2.8.17 Classified area. The interior of a vault containing a tank that stores a Class I liquid shall be designated a Class I, Division 1 location, as defined in the ICC *Electrical Code*.

❖ The interior of a vault storing Class I liquid is a Class I Division I location for determining the type of electrical system to be installed.

3404.2.8.18 Overfill protection. Overfill protection shall be provided in accordance with Section 3404.2.9.6.6. The use of a float vent valve shall be prohibited.

❖ Overfill protection is covered in Section 3404.2.9.6.6 and applies to all flammable and combustible liquids.

3404.2.9 Above-ground tanks. Above-ground storage of flammable and combustible liquids in tanks shall comply with Section 3404.2 and Sections 3404.2.9.1 through 3404.2.9.6.10.

❖ The storage of flammable and combustible liquids is permitted in above-ground tanks if the tanks are equipped to prevent the flammable and combustible liquid from escaping and becoming a vapor-air mixture in the flammable range. Operation of these above-ground tanks, sources of ignition, locations, security, etc., are covered to ensure a safe facility and operation.

3404.2.9.1 Fire protection. Fire protection for above-ground tanks shall comply with Sections 3404.2.9.1.1 through 3404.2.9.1.4.

❖ Above-ground tanks are equipped with fire protection to control exterior fire exposure; control fire from flammable and combustible liquid; protect the tank structure and protect the facility.

3404.2.9.1.1 Required foam fire protection systems. When required by the fire code official, foam fire protection shall be provided for above-ground tanks, other than pressure tanks operating at or above 1 pound per square inch gauge (psig) (6.89 kPa) when such tank, or group of tanks spaced less than 50 feet (15 240 mm) apart measured shell to shell, has a liquid surface area in excess of 1,500 square feet (139 m^2), and is in accordance with one of the following:

1. Used for the storage of Class I or II liquids.
2. Used for the storage of crude oil.
3. Used for in-process products and is located within 100 feet (30 480 mm) of a fired still, heater, related fractioning or processing apparatus or similar device at a processing plant or petroleum refinery as herein defined.
4. Considered by the fire code official as posing an unusual exposure hazard because of topographical conditions; nature of occupancy, proximity on the same or adjoining property, and height and character of liquids to be stored; degree of private fire protection to be provided; and facilities of the fire department to cope with flammable liquid fires.

❖ The fire code official has the authority to require a foam fire protection system for above-ground tanks when the tanks satisfy the criteria in this section. These limits are based on the capacity of the foam fire protection system.

3404.2.9.1.2 Foam fire protection system installation. Where foam fire protection is required, it shall be installed in accordance with NFPA 11 and NFPA 11A.

❖ Foam fire protection systems are to comply with NFPA 11 or NFPA 11A. Systems conforming to NFPA 11 are for use on flammable and combustible liquid hazards in local areas of a building, for storage tanks and for indoor or outdoor processing areas. NFPA 11A covers systems for use on liquid fuel fires in small enclosed or partially enclosed spaces. Foam can provide quick and effective coverage for flammable liquid spill fires where rapid vapor suppression is essential. High-expansion foams are for use on liquid fuel fires where depth of coverage is important to fill volumes where fire exists at various levels. High-expansion foams are more effective indoors than outdoors.

3404.2.9.1.2.1 Foam storage. Where foam fire protection is required, foam-producing materials shall be stored on the premises.

Exception: Storage of foam-producing materials off the premises is allowed as follows:

1. Such materials stored off the premises shall be of the proper type suitable for use with the equipment at the installation where required.
2. Such materials shall be readily available at the storage location at all times.
3. Adequate loading and transportation facilities shall be provided.
4. The time required to deliver such materials to the required location in the event of fire shall be consistent with the hazards and fire scenarios for which the foam supply is intended.
5. At the time of a fire, these off-premises supplies shall be accumulated in sufficient quantities before placing the equipment in operation to ensure foam production at an adequate rate without interruption until extinguishment is accomplished.

❖ When the fire code official requires a foam fire protection system, the foam must be stored on site.

The exception allows the fire code official to authorize off-site foam storage if the off-site storage does not hamper the foam fire protection. Consideration must be given to type of foam, availability, time needed to get the foam material to the site and the quantity of foam material required before authorization for off-site foam storage is approved.

3404.2.9.1.3 Fire protection of supports. Supports or pilings for above-ground tanks storing Class I, II or IIIA liquids elevated more than 12 inches (305 mm) above grade shall have a fire-resistance rating of not less than 2 hours in accordance with the fire exposure criteria specified in ASTM E 1529.

Exceptions:

1. Structural supports tested as part of a protected above-ground tank in accordance with UL 2085.
2. Stationary tanks located outside of buildings when protected by an approved water-spray system designed in accordance with Chapter 9 and NFPA 15.
3. Stationary tanks located inside of buildings equipped throughout with an approved automatic sprinkler system designed in accordance with Section 903.3.1.1.

❖ Above-ground tank supports extending more than 12 inches (305 mm) above grade must have no less than 2-hour fire protection. This fire rating is based on ASTM E 1529, which is based on a hydrocarbon pool fire instead of the ASTM E 119 estimate of fuel content in a building.

The exceptions to this fire rating for the structural support are based on the fire protection provided by other means listed under UL 2085, protection provided by a waterspray system under NFPA 15 or for a tank inside a sprinklered building.

3404.2.9.1.4 Inerting of tanks with boilover liquids. Liquids with boilover characteristics shall not be stored in fixed roof tanks larger than 150 feet (45 720 mm) in diameter unless an approved gas enrichment or inerting system is provided on the tank.

Exception: Crude oil storage tanks in production fields with no other exposures adjacent to the storage tank.

❖ The application of water to a liquid with boilover characteristics can cause a rapid increase in the fire. The boilover liquid may be hot enough to vaporize water that has been added to the fire or water that may be under the boilover liquid. The sudden release of steam can cause the fire to greatly increase in intensity or cause an explosion. Using an inert system to suppress a fire in a large diameter tank will reduce the potential danger from boilover.

The exception notes that where the hazard is limited to an isolated crude oil tank in a production field, the inert system is not required.

3404.2.9.2 Supports, foundations and anchorage. Supports, foundations and anchorages for above-ground tanks shall be designed and constructed in accordance with NFPA 30 and the *International Building Code*.

❖ Footings, foundations and structural supports for above-ground tanks must comply with the IBC and NFPA 30.

3404.2.9.3 Stairs, platforms and walkways. Stairs, platforms and walkways shall be of noncombustible construction and shall be designed and constructed in accordance with NFPA 30 and the *International Building Code*.

❖ Stairs, platforms and walkways for above-ground tanks must be of noncombustible construction and comply with the IBC and NFPA 30.

3404.2.9.4 Above-ground tanks inside of buildings. Tanks storing Class I, II and IIIA liquids inside buildings shall be equipped with a device or other means to prevent overflow into the building including, but not limited to: a float valve; a preset meter on the fill line; a valve actuated by the weight of the tanks contents; a low head pump which is incapable of producing overflow; or a liquid-tight overflow pipe at least one pipe size larger than the fill pipe and discharging by gravity back to the outside source of liquid or to an approved location.

❖ Overfilling of tanks containing Class I, II and IIIA liquids inside a building can release vapors that could reach concentrations at or above the LFL. Devices and equipment for filling of these tanks must be designed to prevent spillage and damage to the storage tank.

3404.2.9.5 Above-ground tanks outside of buildings. Above-ground tanks outside of buildings shall comply with Sections 3404.2.9.5.1 through 3404.2.9.5.3.

❖ Above-ground tanks outside of a building are separated from other tanks and facilities to minimize the exposure to fire and transfer of fire from tank to tank or to another facility.

3404.2.9.5.1 Locations where above-ground tanks are prohibited. Storage of Class I and II liquids in above-ground tanks outside of buildings is prohibited within the limits established by law as the limits of districts in which such storage is prohibited (see Section 3 of the Sample Ordinance for Adoption of the *International Fire Code* on page v).

❖ This section enables the adopting jurisdiction to prohibit the installation of above-ground tanks in certain geographic areas of the jurisdiction by enumerating them in the adopting legislation. The code includes a sample adopting ordinance that contains a blank space for the jurisdiction to fill in describing the particular areas where above-ground tank installations are to be prohibited.

3404.2.9.5.1.1 Location of tanks with pressures 2.5 psig or less. Above-ground tanks operating at pressures not exceeding 2.5 psig (17.2 kPa) for storage of Class I, II or IIIA liquids, which are designed with a floating roof, a weak roof-to-shell seam or equipped with emergency venting devices limiting pressure to 2.5 psig (17.2 kPa), shall be located in accordance with Table 2.3.2.1.1(a) of NFPA 30.

Exceptions:

1. Vertical tanks having a weak roof-to-shell seam and storing Class IIIA liquids are allowed to be located at one-half the distances specified in Table 2.3.2.1.1(a) of NFPA 30, provided the tanks are not within a diked area or drainage path for a tank storing Class I or II liquids.
2. Liquids with boilover characteristics and unstable liquids in accordance with Sections 3404.2.9.5.1.3 and 3404.2.9.5.1.4.
3. For protected above-ground tanks in accordance with Section 3404.2.9.6 and tanks in at-grade or above-grade vaults in accordance with Section 3404.2.8, the distances in Table 2.3.2.1.1(b) of NFPA 30 shall apply and shall be reduced by one-half, but not to less than 5 feet (1524 mm).

❖ Above-ground tanks for Class I, II and IIIA liquids are to be located at clearances specified by Table 2.3.2.1.1(a) of NFPA 30.

Exception 1 permits locating above-ground tanks for Class IIIA liquids at half the clearances of Table 2.3.2.1.1(a) of NFPA 30 if the tanks are isolated from tanks containing Class I and II liquids.

Exception 2 recognizes the special hazards of boilover and unstable liquids by deferring to Section 3904.2.9.5.1.3 or 3904.2.9.5.1.4, respectively.

Exception 3 recognizes the additional fire protection provided by protected above-ground tanks and tanks in vaults by allowing reduced separation distances.

3404.2.9.5.1.2 Location of tanks with pressures exceeding 2.5 psig. Above-ground tanks for the storage of Class I, II or IIIA liquids operating at pressures exceeding 2.5 psig (17.2 kPa) or equipped with emergency venting allowing pressures to exceed 2.5 psig (17.2 kPa) shall be located in accordance with Table 2.3.2.1.2 of NFPA 30.

Exception: Liquids with boilover characteristics and unstable liquids in accordance with Sections 3404.2.9.5.1.4 and 3404.2.9.5.1.5.

❖ Above-ground tanks for Class I, II and IIIA liquids are to be located at clearances specified by Table 2.3.2.1.2 of NFPA 30.

The exception prohibits this tank spacing if the tanks contain liquids with boilover characteristic and unstable liquids (see commentary, Sections 3404.2.9.5.1.3 and 3404.2.9.5.1.4).

3404.2.9.5.1.3 Location of tanks for boilover liquids. Above-ground tanks for storage of liquids with boilover characteristics shall be located in accordance with Table 2.3.2.1.3 of NFPA 30.

❖ Above-ground tanks containing boilover liquids are to be located at clearances specified by Table 2.3.2.1.3 of NFPA 30.

3404.2.9.5.1.4 Location of tanks for unstable liquids. Above-ground tanks for the storage of unstable liquids shall be located in accordance with Table 2.3.2.1.4 of NFPA 30.

❖ Above-ground tanks for unstable liquids are to be located at clearances specified by Table 2.3.2.1.4 of NFPA 30.

3404.2.9.5.1.5 Location of tanks for Class IIIB liquids. Above-ground tanks for the storage of Class IIIB liquids, excluding unstable liquids, shall be located in accordance with Table 2.3.2.1.5 of NFPA 30, except when located within a diked area or drainage path for a tank or tanks storing Class I or II liquids. Where a Class IIIB liquid storage tank is within the diked area or drainage path for a Class I or II liquid, distances required by Section 3404.2.9.5.1.2 shall apply.

❖ Aboveground tanks for Class IIIB liquids, excluding unstable liquids,are to be located at clearances specified by Table 2.3.2.1.5 of NFPA 30. Because of the hazards of Class I and Class II liquids, when an aboveground tank for Class IIIB liquids is not isolated from these liquids, the clearances specified by Table 2.3.2.1.2 of NFPA 30 are to apply.

3404.2.9.5.1.6 Reduction of separation distances to adjacent property. Where two tank properties of diverse ownership have a common boundary, the code fire official is authorized to, with the written consent of the owners of the two properties, apply the distances in Sections 3404.2.9.5.1.2 through 3404.2.9.5.1.6 assuming a single property.

❖ The fire code official has the authority to consider two independent properties as a single property for tank farms for determining the clearances. This authority is limited to having the written consent from both property owners. This does not reduce the clearances between above-ground tanks.

3404.2.9.5.2 Separation between adjacent stable or unstable liquid tanks. The separation between tanks containing stable liquids shall be in accordance with Table 2.3.2.2.1 of NFPA 30. Where tanks are in a diked area containing Class I or II liquids, or in the drainage path of Class I or II liquids, and are compacted in three or more rows or in an irregular pattern, the fire code official is authorized to require greater separation than specified in Table 2.3.2.2.1 of NFPA 30 or other means to make tanks in the interior of the pattern accessible for fire-fighting purposes.

Exception: Tanks used for storing Class IIIB liquids are allowed to be spaced 3 feet (914 mm) apart unless within a diked area or drainage path for a tank storing Class I or II liquids.

The separation between tanks containing unstable liquids shall not be less than one-half the sum of their diameters.

❖ Chapter 2 of NFPA 30 specifies clearances between above-ground tanks containing stable and unstable liquids. Because the placement of tanks can create an access problem for fire department equipment, the fire code official has the authority to require a greater separation than required by Table 2.3.2.1.1 of NFPA 30 to ensure fire equipment access.

The exception allows a 3-foot (914 mm) clearance between above-ground tanks containing Class IIIB liquid if they meet the requirements stated.

The minimum clearance for above-ground tanks containing unstable liquids must be greater because of the greater hazard they pose.

3404.2.9.5.3 Separation between adjacent tanks containing flammable or combustible liquids and LP-gas. The minimum horizontal separation between an LP-gas container and a Class I, II or IIIA liquid storage tank shall be 20 feet (6096 mm) except in the case of Class I, II or IIIA liquid tanks operating at pressures exceeding 2.5 psig (17.2 kPa) or equipped with emergency venting allowing pressures to exceed 2.5 psig (17.2 kPa), in which case the provisions of Section 3404.2.9.5.2 shall apply.

An approved means shall be provided to prevent the accumulation of Class I, II or IIIA liquids under adjacent LP-gas containers such as by dikes, diversion curbs or grading. When flammable or combustible liquid storage tanks are within a diked area, the LP-gas containers shall be outside the diked area and at least 10 feet (3048 mm) away from the centerline of the wall of the diked area.

Exceptions:

1. Liquefied petroleum gas containers of 125 gallons (473 L) or less in capacity installed adjacent to fuel-oil supply tanks of 660 gallons (2498 L) or less in capacity.
2. Horizontal separation is not required between above-ground LP-gas containers and underground flammable and combustible liquid tanks.

❖ LP-gas above-ground tanks need a minimum clearance of 20 feet (6069 mm) from above-ground tanks containing Class I, II and IIIA liquids. When the above-ground tanks containing Class I, II and IIIA liquids are under pressure exceeding 2.5 psig (17.2 kPa), the minimum clearance is according to Section 3404.2.9.5.2.

LP-gas tanks require additional protection from Class I, II and IIIA liquids. Spills and leakage of the flammable and combustible liquids are to be kept away from the LP-gas tank by not placing the tanks in a common diked area. Drainage from the above-ground tanks containing Class I, II and IIIA liquids is to be directed away from the LP-gas tank.

The exception states that small tanks of LP-gas and small fuel oil tanks need not be separated. There is no separation required between LP-gas tanks and underground tanks of flammable and combustible liquids.

3404.2.9.6 Additional requirements for protected above-ground tanks. In addition to the requirements of this chapter for above-ground tanks, the installation of protected above-ground tanks shall be in accordance with Sections 3404.2.9.6.1 through 3404.2.9.6.10.

❖ Besides being located to have the required clearance between above-ground tanks and between above-ground tanks and property lines or other structures, the above-ground tanks are to be designed and constructed with safety features to protect and control flammable and combustible liquids. These features include venting, flame arresters, secondary containment, impact protection, overfill protection and antisiphon devices.

3404.2.9.6.1 Tank construction. The construction of a protected above-ground tank and its primary tank shall be in accordance with Section 3404.2.7.

❖ Tanks are to be designed according to NFPA 30 and labeled to indicate the design standard according to Section 3404.2.7.

3404.2.9.6.2 Normal and emergency venting. Normal and emergency venting for protected above-ground tanks shall be provided in accordance with Sections 3404.2.7.3 and 3404.2.7.4. The vent capacity reduction factor shall not be allowed.

❖ Tanks must be vented to maintain the internal tank pressure within the design operating range. A low pressure can increase the generation of vapors. A high pressure can damage the tank or piping system. Any pressure outside of the design pressure range can have an adverse effect on the operation of the system as well as on the piping and equipment. Vent lines are to be used solely as vents unless other uses are approved by the fire code official. The termination of a vent pipe is to direct the vapors away from the building. Vapors from flammable liquids are normally heavier than air so that the vapor will settle to lower levels. The termination of a vent pipe a minimum of 12 feet (3658 mm) above grade will provide space for the vapors to disperse to concentrations below the LFL. This high termination elevation also reduces the potential for the termination to be close to grade-level ignition sources. Because flammable liquid vapors are heavier than air, attention needs to be given to the building design near the termination. Building design and features that may allow the flammable liquid vapors to reenter the building or to collect on the building must be eliminated. Because Class IIIB has a relatively high boiling point, the vent termination for this combustible liquid may terminate inside the building if the vent is normally closed. Section 3403.6 covers the design, installation, testing and protection of vent pipes. Vent pipes must drain to the tank and not accumulate condensation in the vent pipe.

3404.2.9.6.3 Flame arresters. Approved flame arresters or pressure vacuum breather valves shall be installed in normal vents.

❖ The vapors from a vent will be mixing with air and could become an ignitable mixture. This condition requires that measures be taken to either suppress an ignition source or to disperse the ignitable mixture with additional air to drop the vapor air mixture to below the LFL.

3404.2.9.6.4 Secondary containment. Protected above-ground tanks shall be provided with secondary containment, drainage control or diking in accordance with Section 2704.2. A means shall be provided to establish the integrity of the secondary containment in accordance with NFPA 30.

❖ Above-ground tanks are to be located within secondary containment according to Section 2704.2 and NFPA 30. Section 2704.2 discusses the use of liquid-tight floors, curbs, dikes and drainage systems to divert the liquid to a location where it can be contained and safely handled.

3404.2.9.6.5 Vehicle impact protection. Where protected above-ground tanks, piping, electrical conduit or dispensers are subject to vehicular impact, they shall be protected therefrom, either by having the impact protection incorporated into the system design in compliance with the impact test protocol of UL 2085, or by meeting the provisions of Section 312, or where necessary, a combination of both. Where guard posts or other approved barriers are provided, they shall be independent of each above-ground tank.

❖ Above-ground tanks can be protected from vehicle impact in one of two ways. Either the above-ground tank system can be designed to satisfy the impact criteria of UL 2085 or guard posts or other approved barriers complying with Section 312 can be used for protection from vehicle impact. Section 312 contains the specifications for guard posts or the design forces required for an approved barrier to comply with the code.

3404.2.9.6.6 Overfill prevention. Protected above-ground tanks shall not be filled in excess of 95 percent of their capacity. An overfill prevention system shall be provided for each tank. During tank-filling operations, the system shall:

1. Provide an independent means of notifying the person filling the tank that the fluid level has reached 90 percent of tank capacity by providing an audible or visual alarm signal, providing a tank level gauge marked at 90 percent of tank capacity, or other approved means.
2. Automatically shut off the flow of fuel to the tank when the quantity of liquid in the tank reaches 95 percent of tank capacity. For rigid hose fuel-delivery systems, an approved means shall be provided to empty the fill hose into the tank after the automatic shutoff device is activated.
3. Reduce the flow rate to not more than 15 gallons per minute (0.95 L/sec) so that at the reduced flow rate, the tank will not overfill for 30 minutes, and automatically shut off flow into the tank so that none of the fittings on the top of the tank are exposed to product because of overfilling.

A permanent sign shall be provided at the fill point for the tank, documenting the filling procedure and the tank calibration chart.

Exception: Where climatic conditions are such that the sign may be obscured by ice or snow, or weathered beyond readability or otherwise impaired, said procedures and chart shall be located in the office window, lock box or other area accessible to the person filling the tank.

The filling procedure shall require the person filling the tank to determine the gallonage (literage) required to fill it to 90 percent of capacity before commencing the fill operation.

❖ To prevent spillage during filling, above-ground tanks are limited to 95 percent of their capacity by an overfill protection system. Several methods are acceptable that provide either an audible or visible alarm. The protection system must include draining the fill hose into the above-ground tank without exceeding the 95-percent capacity. As an additional safety measure, the filling procedure is to include a requirement that the operator determine the gallonage (literage) required to fill the tank to 90-percent capacity before commencing the operation.

A permanent sign displaying filling instructions and a tank calibration chart must be located at the fill point.

If weather conditions exist that make an outdoor permanent sign impractical, the exception allows storing the instructions and calibration charts at another location that is accessible to the individual filling the tank.

3404.2.9.6.7 Fill pipe connections. The fill pipe shall be provided with a means for making a direct connection to the tank vehicle's fuel delivery hose so that the delivery of fuel is not exposed to the open air during the filling operation. Where any portion of the fill pipe exterior to the tank extends below the level of the top of the tank, a check valve shall be installed in the fill pipe not more than 12 inches (305 mm) from the fill hose connection.

❖ The filling operation could create a vapor-air mixture above the LFL. The fill hose is to be tight fitting to reduce the potential for the flammable or combustible liquid to be exposed to air.

3404.2.9.6.8 Spill containers. A spill container having a capacity of not less than 5 gallons (19 L) shall be provided for each fill connection. For tanks with a top fill connection, spill containers shall be noncombustible and shall be fixed to the tank and equipped with a manual drain valve that drains into the primary tank. For tanks with a remote fill connection, a portable spill container shall be allowed.

❖ To control spillage during filling, a spill container is to be provided for each above-ground tank. The spill container must have a top fill connection, be noncombustible and be permanently fixed to the tank. This spill container is to drain directly into the above-ground tank. A remote fill connection can use a portable spill container.

3404.2.9.6.9 Tank openings. Tank openings in protected above-ground tanks shall be through the top only.

❖ Above-ground tanks can have an opening only through the tank top.

3404.2.9.6.10 Antisiphon devices. Approved antisiphon devices shall be installed in each external pipe connected to the protected above-ground tank when the pipe extends below the level of the top of the tank.

❖ To prevent spillage by siphoning through the fill connection, an approved antisiphon device must be installed on the fill connection.

3404.2.10 Drainage and diking. The area surrounding a tank or group of tanks shall be provided with drainage control or shall be diked to prevent accidental discharge of liquid from endangering adjacent tanks, adjoining property or reaching waterways.

Exceptions:

1. The fire code official is authorized to alter or waive these requirements based on a technical report which demonstrates that such tank or group of tanks does not constitute a hazard to other tanks, waterways or adjoining property, after consideration of special features such as topographical conditions, nature of occupancy and proximity to buildings on the same or adjacent property, capacity, and construction of proposed tanks and character of liquids to be stored, and nature and quantity of private and public fire protection provided.
2. Drainage control and diking is not required for listed secondary containment tanks.

❖ Leaks and spills of flammable and combustible liquids must be controlled by dikes and drainage. The flammable and combustible liquids are to be collected in a manner that will not endanger other tanks, properties or waterways.

Exception 1 gives the fire code official the authority to consider installations where the use of drainage and dikes may not be necessary if technical documentation exists to show that this is practical.

Exception 2 states that listed secondary containment tanks do not require drainage control or dikes in the surrounding area. The secondary containment system in these tanks is considered equivalent to drainage and diking.

3404.2.10.1 Volumetric capacity. The volumetric capacity of the diked area shall not be less than the greatest amount of liquid that can be released from the largest tank within the diked area. The capacity of the diked area enclosing more than one tank shall be calculated by deducting the volume of the tanks other than the largest tank below the height of the dike.

❖ The diked area must have sufficient capacity to contain the spillage of flammable and combustible liquids. The volume of flammable and combustible liquid to be held in the diked area is the greatest amount that can be released from the largest tank. This is the volume of the largest tank that is above the lowest elevation of the dike. For diked areas containing more than one tank, the volume is determined using the volume of the largest tank. This determination is based on the assumption that major leakage from more than one tank at any given time is very unlikely.

Because dike storage capacity is critical to spill control, designers must make certain they design their impoundment based on tank capacity. The fire code official, the plan reviewers and the inspectors responsible for approving the design must also be able to determine whether the design is adequate. Below are two examples to aid in understanding the calculations.

Example 1:

Determine the dike storage volume for a single tank.

Tank:

Diameter = 100 feet (30 480 mm)

Height = 30 feet (9144 mm)

Dike:

Length	=	140feet (42 672 mm)
Width	=	140feet (42 672 mm)
Height	=	5feet (1524 mm)

Maximum dike storage volume:
(140 feet) (140 feet) (5 feet) = 98,000 ft^3 (2775 m^3)

Maximum potential volume of spill:
(100 feet/2)^2 (π) (30 feet – 5 feet) = 196,350 ft^3 (5560 m^3)

Tank volume below dike height:
(100 feet/2)^2 (π) (5 feet) = 39,270 ft^3 (1112 m^3)

Available dike storage volume for spill containment:
98,000 ft^3 – 39,270 ft^3 = 58,730 ft^3 (1663 m^3)

Available dike storage volume for spill containment is *not* sufficient to contain for the maximum potential spill.

58,730 ft^3(1,663 m^3) < 196,350 ft^3 (5560 m^3)

Determine the minimum size dike based on a square dike area.

$$L=\sqrt{\frac{(100\text{ feet}/2)^2\ (30\text{ feet})}{5}} = 218\text{ feet (66 446 mm)}$$

Maximum dike storage volume:
(218 feet) (218 feet) (5 feet) = 237,620 ft^3 (6729 m^3)

Available dike storage volume for spill containment:
237,620 ft^3 – 39,270 ft^3 = 198,350 ft^3 (5617 m^3)

Available dike storage volume for spill containment is sufficient to contain for the maximum potential spill.

198,350 ft^3(5617 m^3) > 196,350 ft^3 (5560 m^3)

Example 2:

Determine the dike storage volume for a dike with three tanks.

Tank 1:

Diameter	=	100feet (30 480 mm)
Height	=	30 feet (9144 mm)

Tank 2:

Diameter	=	50feet (15 240 mm)
Height	=	20 feet (6096 mm)

Tank 3:

Diameter	=	50feet (15 240 mm)
Height	=	20 feet (6096 mm)

Dike:

Length	=	300feet (91 440 mm)
Width	=	150feet (45 720 mm)
Height	=	5 feet (1524 mm)

Maximum dike storage volume:
300 feet (150 feet) (5 feet) = 225,000 ft^3 (6371 m^3)

Maximum potential volume of spill:

Tank 1:
(100 feet/2)^2 (π) (30 feet – 5 feet) = 196,350 ft^3 (5560 m^3)

Tank 1 volume below dike height:
(100 feet/2)^2 (π) (5 feet) = 39,270 ft^3 (1112 m^3)

Tank 2 volume below dike height:
(50 feet/2)^2 (π) (5 feet) = 9,817 ft^3 (278 m^3)

Tank 3 volume below dike height:
(50 feet/2)^2 (π) (5 feet) = 9,817 ft^3 (278 m^3)

Available dike storage volume for spill containment:
225,000 ft^3 – 39,270 ft^3 – 9,817 ft^3 – 9,817 ft^3
= 166,095 ft^3 (4703 m^3)

Available dike storage volume for spill containment is not sufficient to contain for the maximum potential spill.

166,095 ft^3 (4703 m^3) < 196,350 ft^3 (5560 m^3)

Revise the dike design by increasing the dike height to 6 feet (1829 mm).

Maximum dike storage volume:
300 feet (150 feet) (6 feet) = 270,000 ft^3 (7646 m^3)

Maximum potential volume of spill:
Tank 1:
(100 feet/2)^2 (π) (30 feet – 6 feet) = 188,496 ft^3 (5338 m^3)

Tank 1 volume below dike height:
(100 feet/2)^2 (π) (6 feet) = 47,124 ft^3 (1334 m^3)

Tank 2 volume below dike height:
(50 feet/2)^2 (π) (6 feet) = 11,781 ft^3 (334 m^3)

Tank 3 volume below dike height:
(50 feet/2)^2 (π) (6 feet) = 11,781 ft^3 (334 m^3)

Available dike storage volume for spill containment:
270,000 ft^3 – 47,124 ft^3 – 11,781 ft^3 – 11,781 ft^3
= 199,314 ft^3 (5644 m^3)

Available dike storage volume for spill containment is sufficient to contain for the maximum potential spill.

199,314 ft^3(5644 m^3) > 188,496 ft^3 (5338 m^3)

3404.2.10.2 Diked areas containing two or more tanks. Diked areas containing two or more tanks shall be subdivided in accordance with NFPA 30.

❖ Diked areas are to be subdivided according to NFPA 30 to control the flow of flammable and combustible liquids.

3404.2.10.3 Protection of piping from exposure fires. Piping shall not pass through adjacent diked areas or impounding basins, unless provided with a sealed sleeve or otherwise protected from exposure to fire.

❖ Piping can be damaged and fail as a result of fire exposure. A failed pipe could add fuel to a fire. To prevent a pipe from an adjacent diked area or impoundment basin from providing additional fuel, the pipe is to be protected with a sealed sleeve or otherwise protected from exposure to fire. The best method is to install piping so that it does not enter an adjacent diked area or impoundment basin whenever possible. The sealed pipe sleeve provides some protection from direct fire exposure and is also a secondary containment system.

3404.2.10.4 Combustible materials in diked areas. Diked areas shall be kept free from combustible materials, drums and barrels.

❖ Good housekeeping must be practiced around flammable and combustible materials, whether they are solid or liquid. Combustible materials that accumulate in diked areas are a source of ignition and fuel. These materials must be removed.

3404.2.10.5 Equipment, controls and piping in diked areas. Pumps, manifolds and fire protection equipment or controls shall not be located within diked areas or drainage basins or in a location where such equipment and controls would be endangered by fire in the diked area or drainage basin. Piping above ground shall be minimized and located as close as practical to the shell of the tank in diked areas or drainage basins.

Exceptions:

1. Pumps, manifolds and piping integral to the tanks or equipment being served which is protected by intermediate diking, berms, drainage or fire protection such as water spray, monitors or resistive coating.
2. Fire protection equipment or controls which are appurtenances to the tanks or equipment being protected, such as foam chambers or foam piping and water or foam monitors and hydrants, or hand and wheeled extinguishers.

❖ Equipment and controls must be located outside of the diked areas for fire protection. Piping must be in the diked areas to operate the facility, but the piping must be underground except at the tank. This provides as much fire protection for the piping as is practical while still allowing the facility to function.

Exception 1 allows locating equipment that is integral to the tank inside the diked area. Other service equipment that is provided with fire protection may also be in diked areas. Separation (immediate diking), water spray or coatings can accomplish this fire protection.

Exception 2 allows fire protection equipment that is part of the tank system and for the fire protection of the tank to be in the diked area.

3404.2.11 Underground tanks. Underground storage of flammable and combustible liquids in tanks shall comply with Section 3404.2 and Sections 3404.2.11.1 through 3404.2.11.5.2.

❖ Underground tanks are exposed to conditions not associated with above-ground tanks. Leakage of flammable and combustible liquids is harder to detect. Protection from loads being placed on top of or adjacent to the underground tank is needed to prevent damage to the underground tank. The underground tank must be protected from flooding and from floating in areas having a groundwater table that may be above the bottom of the underground tank.

3404.2.11.1 Contents. Underground tanks shall not contain petroleum products containing mixtures of a nonpetroleum nature, such as ethanol or methanol blends, without evidence of compatibility.

❖ Petroleum products containing mixtures that are not petroleum based may be incompatible with components in the underground tank. The nonpetroleum products could attack liners, gaskets, etc. The loss of or damage to these components may cause the underground tank or piping system to develop leaks.

3404.2.11.2 Location. Flammable and combustible liquid storage tanks located underground, either outside or under buildings, shall be in accordance with all of the following:

1. Tanks shall be located with respect to existing foundations and supports such that the loads carried by the latter cannot be transmitted to the tank.
2. The distance from any part of a tank storing liquids to the nearest wall of a basement, pit, cellar, or lot line shall not be less than 3 feet (914 mm).
3. A minimum distance of 1 foot (305 mm), shell to shell, shall be maintained between underground tanks.

❖ Underground tanks must be located away from a building or structure so that the tank does not support the building or structure. The loads from a building or structure can cause the underground tank to rupture.

Underground tanks are to be a minimum 3 feet (914 mm) from the nearest below-grade wall or property line. Any leakage from an underground tank may migrate through the soil and enter a basement, cellar or pit. The accumulation of flammable and combustible liquids in this location could develop into a hazardous condition. Because the owner of the underground tank does not own the adjacent property, the location of the underground tank should not affect the adjacent property or the use of this property.

To ensure that underground tanks are independent of each other, a minimum clearance of 1 foot (305 mm) is required between the shells of adjacent underground tanks.

3404.2.11.3 Depth and cover. Excavation for underground storage tanks shall be made with due care to avoid undermining of foundations of existing structures. Underground tanks shall be set on firm foundations and surrounded with at least 6 inches (152 mm) of noncorrosive inert material, such as clean sand.

❖ Excavation for underground tanks must not damage existing structures and must provide a sound foundation for the underground tank. The use of 6 inches (152 mm) of a noncorrosive material (sand) around an underground tank prevents concentrated loads from being applied to the tank. These concentrated loads can be caused during backfilling when a hard solid object, such as a rock, comes in contact with the underground tank. Soil pressures on the rock can be concentrated into a small contact area with the underground tank.

3404.2.11.4 Overfill protection and prevention systems. Fill pipes shall be equipped with a spill container and an overfill prevention system in accordance with NFPA 30.

❖ To control spillage during filling of underground tanks, each tank must have a spill container that is noncombustible and permanently fixed to the tank. This spill container is to drain directly into the underground tank.

3404.2.11.5 Leak prevention. Leak prevention for underground tanks shall comply with Sections 3404.2.11.5.1 and 3404.2.11.5.2.

❖ Leakage from underground tanks is to be detected by either monitoring or an approved leak detection system.

3404.2.11.5.1 Inventory control. Daily inventory records shall be maintained for underground storage tank systems.

❖ Leakage detection by inventory control requires accurate records of the volume of flammable and combustible liquid dispensed into the underground tank and records of the volume of flammable and combustible liquid removed. A discrepancy between these two volumes is used to identify a leak.

3404.2.11.5.2 Leak detection. Underground storage tank systems shall be provided with an approved method of leak detection from any component of the system that is designed and installed in accordance with NFPA 30.

❖ Leakage detection systems must comply with NFPA 30.

3404.2.12 Testing. Tank testing shall comply with Sections 3404.2.12.1 and 3404.2.12.2.

❖ Tanks must be tested before being placed in service.

3404.2.12.1 Acceptance testing. Prior to being placed into service, tanks shall be tested in accordance with Section 2.4 of NFPA 30.

❖ The acceptance test for tanks is described in Section 2.4 of NFPA 30.

3404.2.12.2 Testing of underground tanks. Before being covered or placed in use, tanks and piping connected to underground tanks shall be tested for tightness in the presence of the fire code official. Piping shall be tested in accordance with Section 3403.6.3. The system shall not be covered until it has been approved.

❖ The fire code official is to be present for testing of underground tanks. The tank, connections and piping are to be tested for tightness. Piping is to be tested under Section 3403.6.3.

3404.2.13 Abandonment and status of tanks. Tanks taken out of service shall be removed in accordance with Section 3404.2.14, or safeguarded in accordance with Sections 3404.2.13.1 through 3404.2.13.2.3 and API 1604.

❖ Tanks that are no longer in service are to be removed or secured. Abandoned tanks can develop leaks or be damaged when new construction occurs near the tank.

3404.2.13.1 Underground tanks. Underground tanks taken out of service shall comply with Sections 3404.2.13.1.1 through 3404.2.13.1.5.

❖ The procedure for safeguarding an underground tank will depend on the period that the tank is to be out of service.

3404.2.13.1.1 Temporarily out of service. Underground tanks temporarily out of service shall have the fill line, gauge opening, vapor return and pump connection secure against tampering. Vent lines shall remain open and be maintained in accordance with Sections 3404.2.7.3 and 3404.2.7.4.

❖ An underground tank that will be temporarily out of service for a period of less than 90 days is to be secured from tampering. The vents are to remain in operation and are to be maintained to allow for continuous venting of the tank. The flammable or combustible liquids may remain in the underground tank.

3404.2.13.1.2 Out of service for 90 days. Underground tanks not used for a period of 90 days shall be safeguarded in accordance with all the following or be removed in accordance with Section 3404.2.14:

1. Flammable or combustible liquids shall be removed from the tank.
2. All piping, including fill line, gauge opening, vapor return and pump connection, shall be capped or plugged and secured from tampering.
3. Vent lines shall remain open and be maintained in accordance with Sections 3404.2.7.3 and 3404.2.7.4.

❖ Flammable or combustible liquid must be removed from underground tanks that are out of service for 90 days but less than a year. The tank is to be secured and vents are to be opened and maintained as required by Section 3404.2.13.1.1.

The owner also has the option to remove the underground tank according to Section 3404.2.14. The code does not provide the option for abandoning the underground tank in place because the tank could not be returned to service.

3404.2.13.1.3 Out of service for 1 year. Underground tanks that have been out of service for a period of 1 year shall be removed from the ground in accordance with Section 3404.2.14 or abandoned in place in accordance with Section 3404.2.13.1.4.

❖ An underground tank that is out of service for one year is to be removed or abandoned in place.

3404.2.13.1.4 Tanks abandoned in place. Tanks abandoned in place shall be abandoned as follows:

1. Flammable and combustible liquids shall be removed from the tank and connected piping.
2. The suction, inlet, gauge, vapor return and vapor lines shall be disconnected.
3. The tank shall be filled completely with an approved, inert solid material.

   Exception: Residential heating oil tanks of 1,100 gallons (4164 L) or less, provided the fill line is permanently capped or plugged, below grade, to prevent refilling of the tank.
4. Remaining underground piping shall be capped or plugged.
5. A record of tank size, location and date of abandonment shall be retained.

❖ An underground tank that is abandoned in place is to have the flammable or combustible liquid removed. The tank is to be stripped of any appurtenances and filled with an inert material. Piping from the underground tank is to be capped or plugged. The owner is to retain a record of the tank. This action should remove the flammable or combustible liquid from the underground tank and prevent any residue from readily mixing with air.

The exception allows capping an underground residential fuel tank of 1,000 gallons (4164 L) or less to prevent refilling of the tank.

3404.2.13.1.5 Reinstallation of underground tanks. Tanks which are to be reinstalled for flammable or combustible liquid service shall be in accordance with this chapter, ASME *Boiler and Pressure Vessel Code* (Section VIII), API 12-P, API 1615, UL 58 and UL 1316.

❖ An underground tank that is reinstalled for flammable or combustible liquids has to comply as a new installation.

3404.2.13.2 Above-ground tanks. Above-ground tanks taken out of service shall comply with Sections 3404.2.13.2.1 through 3404.2.13.2.3.

❖ The procedure for safeguarding an above-ground tank will depend on the period that the tank is to be out of service.

3404.2.13.2.1 Temporarily out of service. Above-ground tanks temporarily out of service shall have all connecting lines isolated from the tank and be secured against tampering.

Exception: In-place fire protection (foam) system lines.

❖ An above-ground tank that is temporarily out of service for a period of less than 90 days is to be secured from tampering, and all lines to the tank are to be disconnected. The flammable or combustible liquids may remain in the above-ground tank.

The exception requires keeping fire protection lines intact. Disconnecting the fire protection lines would reduce the fire safety of the above-ground tank.

3404.2.13.2.2 Out of service for 90 days. Above-ground tanks not used for a period of 90 days shall be safeguarded in accordance with Section 3404.2.13.1.2 or removed in accordance with Section 3404.2.14.

Exceptions:

1. Tanks and containers connected to oil burners that are not in use during the warm season of the year or are used as a backup heating system to gas.
2. In-place, active fire protection (foam) system lines.

❖ Flammable or combustible liquid must be removed from an above-ground tank that is out of service for at least 90 days but less than one year. The tank is to be secured and vents are to be opened and maintained as required by Section 3404.2.13.1.1.

The owner has the option to remove the above-ground tank according to Section 3404.2.14.

Exception 1 exempts an above-ground tank used for seasonal heating or as a backup for a gas heater from being considered out of service.

Exception 2 requires fire protection lines to remain intact. To disconnect the fire protection lines would reduce the fire safety of the above-ground tank.

3404.2.13.2.3 Out of service for 1 year. Above-ground tanks that have been out of service for a period of 1 year shall be removed in accordance with Section 3404.2.14.

Exception: Tanks within operating facilities.

❖ An above-ground tank that is out of service for one year is to be removed.

3404.2.14 Removal and disposal of tanks. Removal and disposal of tanks shall comply with Sections 3404.2.14.1 and 3404.2.14.2.

❖ Removal and disposal of above-ground and underground tanks requires care because the vapors in the tank may be above the LFL for the flammable or combustible liquid. Operations to remove equipment or piping may involve heat that could ignite the vapor-air mixture or force the vapor-air mixture out of the tank where it may be ignited.

3404.2.14.1 Removal. Removal of above-ground and underground tanks shall be in accordance with all of the following:

1. Flammable and combustible liquids shall be removed from the tank and connecting piping.
2. Piping at tank openings which is not to be used further shall be disconnected.
3. Piping shall be removed from the ground.

   Exception: Piping is allowed to be abandoned in place where the fire code official determines that removal is not practical. Abandoned piping shall be capped and safeguarded as required by the fire code official.
4. Tank openings shall be capped or plugged, leaving a 0.125-inch to 0.25-inch-diameter (3.2 mm to 6.4 mm) opening for pressure equalization.
5. Tanks shall be purged of vapor and inserted prior to removal.

❖ The piping to the tank is to be disconnected and capped. The flammable or combustible liquid is to be removed from the tank and piping and the tank purged. An inert gas such as nitrogen or carbon dioxide should be used to purge the tank, not air. Tank openings are to be used as a vent for equalization of internal and atmospheric pressure. The vent opening is to be between 0.125 inch and 0.25 inch (3.2 mm to 6.4 mm) in diameter. Abandoned piping in the ground is to be removed to the maximum practical extent. Any piping approved by the fire code official to remain in the ground is to be capped.

3404.2.14.2 Disposal. Tanks shall be disposed of in accordance with federal, state and local regulations.

❖ Other federal, state and local regulations address the disposal of above-ground and underground tanks.

3404.3 Container and portable tank storage. Storage of flammable and combustible liquids in closed containers that do not exceed 60 gallons (227 L) in individual capacity and portable tanks that do not exceed 660 gallons (2498 L) in individual capacity, and limited transfers incidental thereto, shall comply with this section.

❖ Storage containers not exceeding 60 gallons (227 L) and portable tanks not exceeding 660 gallons (2798 L) are regulated by this section. The use of these containers and portable tanks is limited to incidental transfers of flammable or combustible liquids.

3404.3.1 Design, construction and capacity of containers and portable tanks. The design, construction and capacity of containers for the storage of flammable and combustible liquids shall be in accordance with this section and Section 4.2 of NFPA 30.

❖ Design, construction and capacity of containers and portable tanks are addressed in Section 4.2 of NFPA 30.

3404.3.1.1 Approved containers. Only approved containers and portable tanks shall be used.

❖ It is impossible to determine by examination that a container complies with Section 4.2 of NFPA 30. The only practical method is to use approved, listed containers.

3404.3.2 Liquid storage cabinets. Where other sections of this code require that liquid containers be stored in storage cabinets, such cabinets and storage shall be in accordance with Sections 3404.3.2.1 through 3404.3.2.3.

❖ Containers and portable storage tanks are to be stored in liquid storage cabinets.

3404.3.2.1 Design and construction of storage cabinets. Design and construction of liquid storage cabinets shall be in accordance with this section.

❖ Liquid storage cabinets are designed to protect containers and portable storage tanks and their contents from damage and ignition sources.

Liquid storage cabinets can be constructed of metal or wood. Cabinets listed under UL 1275 and cabinets constructed according to this section are approved.

3404.3.2.1.1 Materials. Cabinets shall be listed in accordance with UL 1275, or constructed of approved wood or metal in accordance with the following:

1. Unlisted metal cabinets shall be constructed of steel having a thickness of not less than 0.044 inch (1.12 mm) (18 gage). The cabinet, including the door, shall be double walled with 1.5-inch (38 mm) airspace between the walls. Joints shall be riveted or welded and shall be tight fitting.
2. Unlisted wooden cabinets, including doors, shall be constructed of not less than 1-inch (25 mm) exterior grade plywood. Joints shall be rabbeted and shall be fastened in two directions with wood screws. Door hinges shall be of steel or brass. Cabinets shall be painted with an intumescent-type paint.

❖ Unlisted liquid storage cabinets and cabinet doors made of steel are to be double-wall cabinets with tight-fitting joints. Minimum steel thickness is 18 gage [0.044 inch (1.12 mm)].

Unlisted liquid storage cabinets and cabinet doors made of wood are to use rabbet joints fastened with wood screws in two directions to develop a tight joint. The minimum wood is to be 1-inch (25 mm) exterior-grade plywood. The cabinet is to be painted with an intumescent-type paint. The plywood grade will increase the cabinet's durability and the intumescent-type paint will reduce the ignition properties of the plywood.

3404.3.2.1.2 Labeling. Cabinets shall be provided with a conspicuous label in red letters on contrasting background which reads: FLAMMABLE—KEEP FIRE AWAY.

❖ Label the liquid storage cabinet to restrict ignition sources from the immediate area.

3404.3.2.1.3 Doors. Doors shall be well fitted, self-closing and equipped with a three-point latch.

❖ The door is to be self-closing and tight fitting. This prevents flammable or combustible liquid that has leaked or spilled in the storage cabinet from easily escaping the storage cabinet.

3404.3.2.1.4 Bottom. The bottom of the cabinet shall be liquid tight to a height of at least 2 inches (51 mm).

❖ To control the flow of any spills, the cabinet is to be liquid tight for at least 2 inches (51 mm) from the bottom. This prevents flammable or combustible liquid that has leaked or spilled in the storage cabinet from easily escaping the storage cabinet.

3404.3.2.2 Capacity. The combined total quantity of liquids in a cabinet shall not exceed 120 gallons (454 L).

❖ The quantity of flammable and combustible liquids in a cabinet is not to exceed 120 gallons (454 L). Controlling the quantity of flammable and combustible liquids in a liquid storage cabinet controls the fire hazard by limiting the amount of liquid that can be involved in a single incident.

3404.3.2.3 Number of storage cabinets. Not more than three storage cabinets shall be located in a single fire area, except that in a Group F occupancy, additional cabinets are allowed to be located in the same fire area if the additional cabinets (or groups of up to three cabinets) are separated from other cabinets or groups of cabinets by at least 100 feet (30 480 mm).

❖ Controlling the number of liquid storage cabinets to a maximum of three in a fire area regulates the fire potential. The number of cabinets in a fire area in a Group F occupancy can be increased if the cabinets or groups of cabinets are separated by at least 100 feet (30 480 mm). This separation requirement recognizes that the larger number of cabinets in the area is an increased fire hazard.

3404.3.3 Indoor storage. Storage of flammable and combustible liquids inside buildings in containers and portable tanks shall be in accordance with this section.

Exceptions:

1. Liquids in the fuel tanks of motor vehicles, aircraft, boats or portable or stationary engines.
2. The storage of distilled spirits and wines in wooden barrels or casks.

❖ Indoor storage in containers or portable tanks is governed by this section with two exceptions.

The requirements of this section are not applicable to fuel in vehicles and portable engines. The small quantity of fuel and the protection provided by the fuel containers provide sufficient safety for these uses to be indoors.

Distilled spirits and wines in wooden barrels or casks are allowed indoors. This exception is covered in Section 3401.1.

3404.3.3.1 Portable fire extinguishers. Approved portable fire extinguishers shall be provided in accordance with specific sections of this chapter and Section 906.

❖ Portable fire extinguishers are useful for controlling small fires. Section 906 contains the size and spacing for portable fire extinguishers to use on fire of flammable or combustible liquids that have a liquid depth of 0.25 inches (6.4 mm) or less.

3404.3.3.2 Incompatible materials. Materials that will react with water or other liquids to produce a hazard shall not be stored in the same room with flammable and combustible liquids in accordance with Section 2703.9.8.

❖ Materials that generate heat or become combustible when exposed to water or other liquids are not to be in the same room as flammable or combustible liquids. This is to remove a potential ignition source.

3404.3.3.3 Clear means of egress. Storage of any liquids, including stock for sale, shall not be stored near or be allowed to obstruct physically the route of egress.

❖ The means of egress must be usable to be effective. The placement of flammable or combustible liquids near or in the route used to exit the room or building produces a risk that is not acceptable.

3404.3.3.4 Empty containers or portable tank storage. The storage of empty tanks and containers previously used for the storage of flammable or combustible liquids, unless free from explosive vapors, shall be stored as required for filled containers and portable tanks. Portable tanks and containers, when emptied, shall have the covers or plugs immediately replaced in openings.

❖ An empty container or portable tank is as dangerous and possibly more dangerous than a full container or portable storage tank. There is a possibility that the vapor air mixture in the container or portable storage tank could reach the LFL. This potential danger requires that empty containers and portable tanks be handled and stored as if full of flammable or combustible liquid.

3404.3.3.5 Shelf storage. Shelving shall be of approved construction, adequately braced and anchored. Seismic requirements shall be in accordance with the *International Building Code*.

❖ Shelving for containers and portable tanks is to be adequate to support the container and portable tank under normal loads and seismic loads. Failure of shelving could cause damage to containers and portable tanks or leakage.

3404.3.3.5.1 Use of wood. Wood of at least 1 inch (25 mm) nominal thickness is allowed to be used as shelving, racks, dunnage, scuffboards, floor overlay and similar installations.

❖ The minimum thickness of wood is to be 1 inch (25 mm).

3404.3.3.5.2 Displacement protection. Shelves shall be of sufficient depth and provided with a lip or guard to prevent individual containers from being displaced.

Exception: Shelves in storage cabinets or on laboratory furniture specifically designed for such use.

❖ Shelving must be designed and constructed to prevent containers or portable tanks from sliding off the shelving. A container or portable tank that falls from a shelf is subject to damage or leakage.

The exception covers shelving in storage cabinets, which may have locked doors or doors fitting snugly against the front of shelves to prevent containers from falling. Shelving that is part of laboratory furniture is not required to have a lip or guard because typically this kind of shelving would hold only small containers. Large containers or tanks would be floor mounted in nearly all laboratory settings. These kinds of shelves may also have other features to prevent containers or portable tanks from sliding or being knocked off the shelf.

3404.3.3.5.3 Orderly storage. Shelf storage of flammable and combustible liquids shall be maintained in an orderly manner.

❖ The handling of containers and portable tanks increases the possibility that an accident can occur. Containers and portable tanks arranged on shelves in an orderly manner make moving one container or portable tank to get to another unnecessary.

3404.3.3.6 Rack storage. Where storage on racks is allowed elsewhere in this code, a minimum 4-foot-wide (1219 mm) aisle shall be provided between adjacent rack sections and any adjacent storage of liquids. Main aisles shall be a minimum of 8 feet (2438 mm) wide.

❖ Rack storage indicates that a larger quantity of flammable and combustible liquids is available for use. Requiring a minimum aisle width of 4 feet (1219 mm) between racks and a minimum main aisle width of 8 feet (2438 mm) provides room to access the flammable and combustible liquids while reducing interference with other racks or other containers and portable tanks. In case of a fire, these aisles are wide enough to give emergency response personnel ready access to the fire and also serve as fire breaks that help prevent fire spread from rack to rack.

3404.3.3.7 Pile or palletized storage. Solid pile and palletized storage in liquid warehouses shall be arranged so that piles are separated from each other by at least 4 feet (1219 mm). Aisles shall be provided and arranged so that no container or portable tank is more than 20 feet (6096 mm) from an aisle. Main aisles shall be a minimum of 8 feet (2438 mm) wide.

❖ The rationale for aisle widths is similar to that for rack storage. The 20-foot (6069 mm) restriction is intended to place containers and portable tanks within easy range of on-site fire-fighting equipment as well as minimize the number of items that must be moved to get to the desired container or portable tank.

3404.3.3.8 Limited combustible storage. Limited quantities of combustible commodities are allowed to be stored in liquid storage areas where the ordinary combustibles, other than those used for packaging the liquids, are separated from the liquids in storage by a minimum of 8 feet (2438 mm) horizontally, either by open aisles or by open racks, and where protection is provided in accordance with Chapter 9.

❖ Combustible products may be stored in a storage facility with flammable or combustible liquids because having a separate storage facility for these combustible materials may be impractical. The storage of these combustible materials with flammable and combustible liquids results in a source of ignition that must be regulated. The code recognizes a clearance of 8 feet (2438 mm) between the combustible materials and the flammable and combustible liquids, with the building fire protection systems required by Chapter 9, as being acceptable.

3404.3.3.9 Idle combustible pallets. Storage of empty or idle combustible pallets inside an unprotected liquid storage area shall be limited to a maximum pile size of 2,500 square feet (232 m^2) and to a maximum storage height of 6 feet (1829 mm). Storage of empty or idle combustible pallets inside a protected liquid storage area shall comply with NFPA 231. Pallet storage shall be separated from liquid storage by aisles that are at least 8 feet (2438 mm) in width.

❖ Combustible pallets are both a fuel source and an ignition source; however, combustible pallets are often necessary for the storage of flammable and combustible liquids. When combustible pallets are no longer in use, their quantity is controlled to reduce the potential fire hazard that is caused by them. The separation distance required is intended to reduce the possibility of fire spread from the pallet storage pile to stored flammable or combustible liquids.

3404.3.3.10 Containers in piles. Containers in piles shall be stacked in such a manner as to provide stability and to prevent excessive stress on container walls. Portable tanks stored more than one tier high shall be designed to nest securely, without dunnage. Material-handling equipment shall be suitable to handle containers and tanks safely at the upper tier level.

❖ The requirements in this section for stacking containers reduce the possibility of damaging the containers. Piling containers can damage them by placing loading on sides that were not intended to be loaded. Piles are subject to collapse if stress loads on containers become excessive. Collapse could cause the containers to rupture or rub against one another and bump one another, resulting in liquid spills and possibly sparking that could cause ignition. By controlling container stacks, this potential for damage is reduced.

Unless tanks are specifically designed to allow stacking, they must all be stored at ground level.

Requiring materials-handling equipment to be designed specifically to handle the stored items is intended to increase safety and reduce the potential for serious accidents.

3404.3.4 Quantity limits for storage. Liquid storage quantity limitations shall comply with Sections 3404.3.4.1 through 3404.3.4.4.

❖ The quantity of flammable and combustible liquids in an area is limited to reduce the potential fire hazard.

3404.3.4.1 Maximum allowable quantity per control area. For occupancies other than Group M wholesale and retail sales uses, indoor storage of flammable and combustible liquids shall not exceed the maximum allowable quantities per control area indicated in Table 2703.1.1(1) and shall not exceed the additional limitations set forth in this section.

For Group M occupancy wholesale and retail sales uses, indoor storage of flammable and combustible liquids shall not exceed the maximum allowable quantities per control area indicated in Table 3404.3.4.1.

Storage of hazardous production material flammable and combustible liquids in Group H-5 occupancies shall be in accordance with Chapter 18.

❖ Table 2703.1.1(1) lists the quantity permitted per control area. Limiting the quantity per control area reduces the flammable and combustible liquid hazard to a level that the fire protection and egress requirements are designed to accept. These limits result in a building that can function with a reasonable degree of safety.

TABLE 3404.3.4.1. See below.

❖ Group M (mercantile) occupancies used for wholesale and retail sales of flammable and combustible liquids are not designed and constructed for flammable and combustible liquids. Businesses in Group M occupancies must be able to display flammable and combustible liquids for sale to the public. To control the potential hazard, the quantities of flammable and combustible liquids are limited in a control area. The limitations are based on the type of flammable and combustible liquid in the control area, the type of storage of the flammable and combustible liquids and the fire sprinkler system installed in the mercantile occupancy.

The easier the flammable or combustible liquid is to ignite, the smaller the quantity of the liquid allowed. This table shows that the quantity of Class IA liquids is smaller than any of the flammable and combustible liquids, and Class III liquids are unlimited in a building with a fire sprinkler system.

The fire sprinkler system must be an automatic sprinkler system complying with NFPA 13. Flammable and combustible liquids displayed on shelves of 6 feet (1829 mm) or less are treated as Ordinary Hazard Group 2, which would require a minimum sprinkler density of 0.19 gallon per minute (0.72 L/min) per square foot over the most remote 1,500-square-foot (139.4 m^2) area.

Because the flammable and combustible liquid is more exposed in individual packaging, the sprinkler system can provide better fire control. If the flammable and combustible liquids are displayed or stored in cartons, pallets or racks, the minimum sprinkler density is

TABLE 3404.3.4.1
MAXIMUM ALLOWABLE QUANTITY OF FLAMMABLE AND COMBUSTIBLE LIQUIDS IN WHOLESALE AND RETAIL SALES USES PER CONTROL AREA[a]

TYPE OF LIQUID	MAXIMUM ALLOWABLE QUANTITY PER CONTROL AREA (gallons)		
	Sprinklered[b] per footnote densities and arrangements	Sprinklered per Tables 3404.3.6.3(4) through 3404.3.6.3(8) and Table 3404.3.7.5.1	Nonsprinklered
Class IA	60	60	30
Class IB, IC, II and IIIA	7,500[c]	15,000[c]	1,600
Class IIIB	Unlimited	Unlimited	13,200

For SI: 1 foot = 304.8 mm, 1 square foot = 0.0929m^2, 1 gallon = 3.785 L, 1 gallon per minute per square foot = 40.75 L/min/m^2.

a. Control areas shall be separated from each other by not less than a 1-hour fire barrier wall.

b. To be considered as sprinklered, a building shall be equipped throughout with an approved automatic sprinkler system with a design providing minimum densities as follows:
  1. For uncartoned commodities on shelves 6 feet or less in height where the ceiling height does not exceed 18 feet, quantities are those permitted with a minimum sprinkler design density of Ordinary Hazard Group 2.
  2. For cartoned, palletized or racked commodities where storage is 4 feet 6 inches or less in height and where the ceiling height does not exceed 18 feet, quantities are those permitted with a minimum sprinkler design density of 0.21 gallon per minute per square foot over the most remote 1,500-square-foot area.

c. Where wholesale and retail sales or storage areas exceed 50,000 square feet in area, the maximum allowable quantities are allowed to be increased by 2 percent for each 1,000 square feet of area in excess of 50,000 square feet, up to a maximum of 100 percent of the table amounts. A control area separation is not required. The cumulative amounts, including amounts attained by having an additional control area, shall not exceed 30,000 gallons.

0.21 gallon per minute (0.79 L/min) per square foot over the most remote 1,500-square-foot (139.4 m^2) area. This type of display or storage is limited to a maximum height of 4 feet, 6 inches (1372 mm). This type of packaging is more difficult for the fire sprinkler system to handle so a greater density is required. To allow a larger quantity of flammable and combustible liquids, the mercantile occupancy can use a fire sprinkler system that complies with either Table 3404.3.6.3(4), Table 3404.3.6.3(9) or Table 3404.3.7.5.1. These tables require the fire sprinkler system to have a greater capacity for fire fighting.

3404.3.4.2 Occupancy quantity limits. The following limits for quantities of stored flammable or combustible liquids shall not be exceeded:

1. Group A occupancies: Quantities in Group A occupancies shall not exceed that necessary for demonstration, treatment, laboratory work, maintenance purposes and operation of equipment, and shall not exceed quantities set forth in Table 2703.1.1(1).
2. Group B occupancies: Quantities in drinking, dining, office and school uses within Group B occupancies shall not exceed that necessary for demonstration, treatment, laboratory work, maintenance purposes and operation of equipment, and shall not exceed quantities set forth in Table 2703.1.1(1).
3. Group E occupancies: Quantities in Group E occupancies shall not exceed that necessary for demonstration, treatment, laboratory work, maintenance purposes and operation of equipment, and shall not exceed quantities set forth in Table 2703.1.1(1).
4. Group F occupancies: Quantities in dining, office, and school uses within Group F occupancies shall not exceed that necessary for demonstration, laboratory work, maintenance purposes and operation of equipment, and shall not exceed quantities set forth in Table 2703.1.1(1).
5. Group I occupancies: Quantities in Group I occupancies shall not exceed that necessary for demonstration, laboratory work, maintenance purposes and operation of equipment, and shall not exceed quantities set forth in Table 2703.1.1(1).
6. Group M occupancies: Quantities in dining, office, and school uses within Group M occupancies shall not exceed that necessary for demonstration, laboratory work, maintenance purposes and operation of equipment, and shall not exceed quantities set forth in Table 2703.1.1(1). The maximum allowable quantities for storage in wholesale and retail sales areas shall be in accordance with Section 3404.3.4.1.
7. Group R occupancies: Quantities in Group R occupancies shall not exceed that necessary for maintenance purposes and operation of equipment, and shall not exceed quantities set forth in Table 2703.1.1(1).
8. Group S occupancies: Quantities in dining and office uses within Group S occupancies shall not exceed that necessary for demonstration, laboratory work, maintenance purposes and operation of equipment, and shall not exceed quantities set forth in Table 2703.1.1(1).

❖ Flammable and combustible liquids may be used in occupancies other than Group H. These other occupancies are not designed specifically for flammable and combustible liquids, so there are limitations on the quantities that can be in use or stored in these occupancies. The use of flammable and combustible liquids must be consistent with the function of the occupancy. This prevents these other occupancies from becoming a Group H occupancy. The quantities of flammable and combustible liquid are listed in Table 2703.1.1(1). The occupancies covered by these limitations are Group A, B, E, F, I, M, R and S.

3404.3.4.3 Quantities exceeding limits for control areas. Quantities exceeding those allowed in control areas set forth in Section 3404.3.4.1 shall be in liquid storage rooms or liquid storage warehouses in accordance with Sections 3404.3.7 and 3404.3.8.

❖ It is possible to have a quantity of flammable and combustible liquids greater than allowed in Section 3404.3.4.1. To control the potential hazard, these quantities must be given additional protection by being in specifically designed liquid storage rooms.

3404.3.4.4 Liquids for maintenance and operation of equipment. In all occupancies, quantities of flammable and combustible liquids in excess of 10 gallons (38 L) used for maintenance purposes and the operation of equipment shall be stored in liquid storage cabinets in accordance with Section 3404.3.2. Quantities not exceeding 10 gallons (38 L) are allowed to be stored outside of a cabinet when in approved containers located in private garages or other approved locations.

❖ The operation of a building will require the use of flammable and combustible liquids for maintenance. A quantity of 10 gallons (38 L) or less of flammable and combustible liquids is recognized as practical to allow the use of such liquids for maintenance without requiring fire protection.

3404.3.5 Storage in control areas. Storage of flammable and combustible liquids in control areas shall be in accordance with Sections 3404.3.5.1 through 3404.3.5.4.

❖ The location of flammable and combustible liquids in a control area can increase the fire hazard.

3404.3.5.1 Basement storage. Class I liquids shall not be permitted in basement areas. Class II and IIIA liquids shall be allowed to be stored in basements provided that automatic suppression and other fire protection is provided in accordance with Chapter 9.

❖ The storage of Class I liquids is prohibited in a basement because of the extreme fire hazard they pose. A Class I liquid fire in a basement could quickly spread to the upper floors. Class II and IIIA liquids are permitted in a basement that has an approved sprinkler system.

3404.3.5.2 Storage pile heights. Containers having less than a 30-gallon (114 L) capacity which contain Class I or II liquids shall not be stacked more than 3 feet (914.4 mm) or two containers high, whichever is greater, unless stacked on fixed shelving or otherwise satisfactorily secured. Containers of Class I or II liquids having a capacity of 30 gallons (114 L) or more shall not be stored more than one container high. Containers shall be stored in an upright position.

❖ The storage height of containers of flammable and combustible liquids increases the risk by having a larger quantity per floor area and possible damage to the container. The stacking operation and the weight of containers can damage the containers, creating leaks.

3404.3.5.3 Storage distance from ceilings and roofs. Piles of containers or portable tanks shall not be stored closer than 3 feet (914 mm) to the nearest beam, chord, girder or other obstruction, and shall be 3 feet (914 mm) below sprinkler deflectors or discharge orifices of water spray or other overhead fire protection system.

❖ The storage of containers or portable tanks near a ceiling or roof can create several problems. The high fire load that close to a structural member could overpower the fire protection of the structure or membrane, allowing the fire to spread to another level. The height could reduce the effectiveness of the automatic sprinkler system. If the flammable and combustible liquid containers and portable tanks are stacked too high, the sprinkler heads may not be able to apply water to the fire. The containers and portable tanks may be above the sprinkler spray or the containers and portable tanks may shield areas from the spray.

3404.3.5.4 Combustible materials. In areas that are inaccessible to the public, Class I, II and IIIA liquids shall not be stored in the same pile or rack section as ordinary combustible commodities unless such materials are packaged together as kits.

❖ The mixing of flammable and combustible liquid containers and portable tanks with other combustibles can place the flammable and combustible liquid in the vicinity of a possible ignition source and a fuel source.

3404.3.6 Wholesale and retail sales uses. Flammable and combustible liquids in Group M occupancy wholesale and retail sales uses shall be in accordance with Sections 3404.3.6.1 through 3404.3.6.5, or NFPA 30 Sections 4.4.3.3, 4.5.6.7, 4.8.2, Tables 4.8.2(a) through (f), and Figures 4.8.2(a) through (d).

❖ Mercantile occupancy is not designed to address the fire hazards associated with flammable and combustible liquids, so the limited quantities, specific packing requirements and handling are used to control the fire hazard.

3404.3.6.1 Container type. Containers for Class I liquids shall be metal.

> **Exception:** In sprinklered buildings, an aggregate quantity of 120 gallons (454 L) of water-miscible Class IB and Class IC liquids is allowed in nonmetallic containers, each having a capacity of 16 ounces (0.473 L) or less.

❖ Metal containers are required for Class I liquids. Class II and III liquids may be stored in any container designed for that specific liquid.

The exception recognizes that the additional fire protection provided by an automatic sprinkler system will allow an aggregate quantity of 120 gallons (454 L) of Classes IB and IC to be in individual noncombustible containers of 16 ounces (0.473 L) or less. The lower flash point of these Class I liquids, combined with the individual packaging in an automatic sprinkler building, is the rationale for this exception.

3404.3.6.2 Container capacity. Containers for Class I liquids shall not exceed a capacity of 5 gallons (19 L).

> **Exception:** Metal containers not exceeding 55 gallons (208 L) are permitted to store up to 240 gallons (908 L) of the maximum allowable quantity per control area of Class IB and IC liquids in a control area. The building shall be equipped throughout with an approved automatic sprinkler system in accordance with Table 3404.3.4.1. The containers shall be provided with plastic caps without cap seals and shall be stored upright. Containers shall not be stacked or stored in racks and shall not be located in areas accessible to the public.

❖ Limiting the capacity of containers in a wholesale or retail establishment reduces the size of a potential spill or fire.

The lengthy and detailed exception recognizes that sprinklers provide additional protection so that a greater quantity of flammable and combustible liquids can be placed in the wholesale or retail establishment. Even with this greater fire protection, the flammable and combustible liquids must have the physical property that would allow them to be mixed with water. This ability of the flammable or combustible liquid to be miscible will allow the sprinkler system to dilute the flammable and combustible liquid. Additional restrictions on the size of the individual container control the volume of Class IB and IC liquids that can be exposed at one time.

3404.3.6.3 Fire protection and storage arrangements. Fire protection and container storage arrangements shall be in accordance with Table 3404.3.6.3(1) or the following:

1. Storage on shelves shall not exceed 6 feet (1829 mm) in height, and shelving shall be metal.
2. Storage on pallets or in piles greater than 4 feet 6 inches (1372 mm) in height, or where the ceiling exceeds 18 feet (5486 mm) in height, shall be protected in accordance with Table 3404.3.6.3(4), and the storage heights and ar-

rangements shall be limited to those specified in Table 3404.3.6.3(2).

3. Storage on racks greater than 4 feet 6 inches (1372 mm) in height, or where the ceiling exceeds 18 feet (5486 mm) in height shall be protected in accordance with Tables 3404.3.6.3(5), 3404.3.6.3(6), and 3404.3.6.3(7) as appropriate, and the storage heights and arrangements shall be limited to those specified in Table 3404.3.6.3(3).

Combustible commodities shall not be stored above flammable and combustible liquids.

❖ The methods and the maximum height of storage are intended to control the flammable and combustible liquids so that damage to containers is limited and fire protection equipment is adequate for the protection of flammable and combustible liquids as well as to control a potential fire.

TABLE 3404.3.6.3(1)
MAXIMUM STORAGE HEIGHT IN CONTROL AREA

TYPE OF LIQUID	NONSPRINKLERED AREA (feet)	SPRINKLERED AREA (feet)	SPRINKLERED[a] WITH IN-RACK PROTECTION (feet)
Flammable liquids:			
Class IA	4	4	4
Class IB	4	8	12
Class IC	4	8	12
Combustible liquids:			
Class II	6	8	12
Class IIIA	8	12	16
Class IIIB	8	12	20

For SI: 1 foot = 304.8 mm.

a. In-rack protection shall be in accordance with Table 3404.3.6.3(5), 3404.3.6.3(6) or 3404.3.6.3(7).

❖ The storage height of flammable and combustible liquids has an effect on the ability of the fire sprinkler system to control a fire. The first consideration is that the higher the storage, the greater the quantity. By limiting the height of storage, the code limits the volume of flammable and combustible liquids under a sprinkler head. Because the fire sprinkler system is designed to deliver a minimum density of water on a square foot area, the less flammable and combustible the liquid is, the easier it will be for the fire sprinkler system to control the fire. The second function of limiting the height of storage is to make certain the fire sprinkler system will cover the flammable and combustible liquids. The upper containers of flammable and combustible liquids could act as a shield for lower containers. This shielding could divert the water spray from the sprinkler head away from the fire. With the fire sprinkler system shielded during early development of the fire, the fire could develop to a stage that the fire sprinkler system would not be effective.

TABLE 3404.3.6.3(2). See page 34-38.

❖ The palletized or solid-pile storage of flammable and combustible liquids in liquid storage rooms and warehouses is regulated by floor, type of container, maximum storage height, maximum quantity per pile and maximum quantity per room. The table restricts the storage by floor to ensure that the basement is not used for the storage of Class I liquids. The accumulation of vapors from Class I liquids in a basement with a low ignition point for these vapors is not an acceptable situation.

The type of container is regulated to control the problem of leakage and damage to the container caused by handling.

The storage height is used to control the quantity of flammable and combustible liquids and to provide access to the flammable and combustible liquids for water spray from the fire sprinkler system.

The maximum quantity per pile is to limit the volume of flammable and combustible liquids so that the fire sprinkler system can control a fire based on the design density of the fire sprinkler system.

The maximum quantity per room of flammable and combustible liquids is to control the volume of flammable and combustible liquids with the automatic sprinkler system.

The sprinkler system is designed for a minimum flow rate. If volume of flammable and combustible liquids in a room is not regulated, the room could contain more flammable and combustible liquids than the design flow rate of the fire sprinkler system could possibly handle.

TABLE 3404.3.6.3(3). See page 34-38.

❖ The storage of flammable and combustible liquids in liquid storage rooms and warehouses is regulated by floor, type of rack, maximum storage height and maximum quantity per room. The table restricts the storage by floor to ensure that the basement is not used for the storage of Class I liquids. The accumulation of vapors from Class I liquids in a basement with a low ignition point for these vapors is not an acceptable situation.

The storage height is used to control the quantity of flammable and combustible liquids and to provide access to the flammable and combustible liquids for water spray from the automatic sprinkler system.

The maximum quantity per room of flammable and combustible liquids is to control the volume of flammable and combustible liquids with the automatic sprinkler system.

The automatic sprinkler system is designed for a minimum flow rate. If volume of flammable and combustible liquids in a room is not regulated, the room could contain more flammable and combustible liquids than the design flow rate of the fire sprinkler system could possibly handle.

TABLE 3404.3.6.3(4). See page 34-39.

❖ The quantity of flammable and combustible liquids can be increased if they are in containers and arranged as described in this table and the sprinkler system criteria are increased to comply with this table. The containers and arrangements control the quantity of flammable and combustible liquid that is exposed. The sprinkler system criteria have been increased to provide a higher density, larger water demand and longer duration than normally required under NFPA 13. These additional sprinkler requirements permit an increase in quantity of flammable and combustible liquids that are permitted per control area in wholesale and retail uses.

TABLE 3404.3.6.3(5). See page 34-40.

❖ Rack storage is designed for easy access by personnel and to contain huge quantities of goods that are stored to great heights. The rack storage arrangement maximizes the fuel surface area accessible to flames. Fire can rapidly spread up between containers. Heat from flames burning on one surface augments the heat transfer from the flames burning on the opposing surface. Rack storage geometry can be the most hazardous of all fire geometries. High rack storage is often protected with in-rack sprinklers inside or on the face of the high rack storage. The storage of flammable and combustible liquids can be as high as 25 feet (7620 mm) because the containers have been limited in size to control volume in individual containers and the sprinkler system requirements have been increased to provide more fire protection.

TABLE 3404.3.6.3(6). See page 34-41.

❖ Rack storage is designed for easy access by personnel and to contain huge quantities of goods that are stored to great heights. The rack storage arrangement maximizes the fuel surface area accessible to flames. Fire can rapidly spread up between containers. Heat from flames burning on one surface augments the heat transfer from the flames burning on the opposing surface. Rack storage geometry can be the most hazardous of all fire geometries. High rack storage is often protected with in-rack sprinklers inside or on the face of the high rack storage. The storage of Class I and II liquids up to 25 feet (7620 mm) and for Class III liquids up to 40 feet (12 192 mm) in containers larger than 5-gallon (19 L) capacity is similar to Table 3404.3.6.3(5), except the sprinkler system requirements have been increased to provide more fire protection.

TABLE 3404.3.6.3(7). See page 34-42.

❖ Rack storage is designed for easy access by personnel and to contain huge quantities of goods that are stored to great heights. The rack storage arrangement maximizes the fuel surface area accessible to flames. Fire can rapidly spread up between containers. Heat from flames burning on one surface augments the heat transfer from the flames burning on the opposing surface. Rack storage geometry can be the most hazardous of all fire geometries. High rack storage is often protected with in-rack sprinklers inside or on the face of the high rack storage. Aqueous film-forming foam (AFFF) is effective because the film forms a barrier that starves the fire for oxygen, cools the hydrocarbon and suppresses the release of flammable vapors. The increased fire protection provided by the additional requirements on the AFFF above that required under NFPA 231C permits the storage of flammable and combustible liquids in rack storage.

TABLE 3404.3.6.3(8). See page 34-42.

❖ Group M (mercantile) occupancies used for wholesale and retail sales of flammable and combustible liquids are not designed and constructed for flammable and combustible liquids. Table 3404.3.4.1 permits the display and storage of Class I liquid if the automatic fire protection system complies with this table. The limited container size, the limited volume under Table 3404.3.4.1 and the sprinkler protection provided under this table permit display and storage in racks up to 6 feet, 6 inches (1981 mm) high.

Rack storage is designed for easy access by personnel and to contain huge quantities of goods that are stored to great heights. The rack storage arrangement maximizes the fuel surface area accessible to flames. Fire can rapidly spread up between containers. Heat from flames burning on one surface augments the heat transfer from the flames burning on the opposing surface. Rack storage geometry can be the most hazardous of all fire geometries. High rack storage is often protected with in-rack sprinklers inside or on the face of the high rack storage. Double-row racks are rack storage that has two racks placed back to back. The double-row rack storage requires additional sprinkler heads to control fire between containers within the rack storage. This table provides the requirements for the additional sprinkler heads and increases the sprinkler density.

3404.3.6.4 Warning for containers. All cans, containers and vessels containing flammable liquids or flammable liquid compounds or mixtures offered for sale shall be provided with a warning indicator, painted or printed on the container and stating that the liquid is flammable, and shall be kept away from heat and an open flame.

❖ Individual containers and their packaging must bear labels warning handling personnel and the public of flammable and combustible liquids. This warning is to prevent the containers from being accidently or deliberately exposed to heat or open flame.

3404.3.6.5 Storage plan. When required by fire the code official, aisle and storage plans shall be submitted in accordance with Chapter 27.

❖ The storage plan would be needed if there are conditions that the fire code official believes would lead to the mishandling of flammable and combustible liquids. This could be a result of the quantity of flammable and combustible liquids, the turnover rate of flammable and combustible liquids, the turnover rate of personnel or other facts that could lead to the mishandling of flammable and combustible liquids.

TABLE 3404.3.6.3(2)
STORAGE ARRANGEMENTS FOR PALLETIZED OR SOLID-PILE STORAGE IN LIQUID STORAGE ROOMS AND WAREHOUSES

CLASS	STORAGE LEVEL	MAXIMUM STORAGE HEIGHT			MAXIMUM QUANTITY PER PILE (gallons)		MAXIMUM QUANTITY PER ROOM[a] (gallons)	
		Drums	Containers[b] (feet)	Portable tanks (feet)	Containers	Portable tanks	Containers	Portable tanks
IA	Ground floor	1	5	Not Allowed	3,000	Not Allowed	12,000	Not Allowed
	Upper floors	1	5	Not Allowed	2,000	Not Allowed	8,000	Not Allowed
	Basements	0	Not Allowed	Not Allowed	Not Allowed	Not Allowed	Not Allowed	Not Allowed
IB	Ground floor	1	6.5	7	5,000	20,000	15,000	40,000
	Upper floors	1	6.5	7	3,000	10,000	12,000	20,000
	Basements	0	Not Allowed	Not Allowed	Not Allowed	Not Allowed	Not Allowed	Not Allowed
IC	Ground floor	1	6.5[c]	7	5,000	20,000	15,000	40,000
	Upper floors	1	6.5[c]	7	3,000	10,000	12,000	20,000
	Basements	0	Not Allowed	Not Allowed	Not Allowed	Not Allowed	Not Allowed	Not Allowed
II	Ground floor	3	10	14	10,000	40,000	25,000	80,000
	Upper floors	3	10	14	10,000	40,000	25,000	80,000
	Basements	1	5	7	7,500	20,000	7,500	20,000
III	Ground floor	5	20	14	15,000	60,000	50,000	100,000
	Upper floors	5	20	14	15,000	60,000	50,000	100,000
	Basements	3	10	7	10,000	20,000	25,000	40,000

For SI: 1 foot = 304.8 mm, 1 gallon = 3.785 L.

a. See Section 3404.3.8.1 for unlimited quantities in liquid storage warehouses.

b. Storage heights are allowed to be increased for Class IB, IC, II and III liquids in metal containers having a capacity of 5 gallons or less where an automatic AFFF-water protection system is provided in accordance with Table 3404.3.7.5.1.

c. These height limitations are allowed to be increased to 10 feet for containers having a capacity of 5 gallons or less.

TABLE 3404.3.6.3(3)
STORAGE ARRANGEMENTS FOR RACK STORAGE IN LIQUID STORAGE ROOMS AND WAREHOUSES

CLASS	TYPE RACK	STORAGE LEVEL	MAXIMUM STORAGE HEIGHT (feet)	MAXIMUM QUANTITY PER ROOM (gallons)
			Containers	Containers
IA	Double row or Single row	Ground floor	25	7,500
		Upper floors	15	4,500
		Basements	Not Allowed	Not Allowed
IB IC	Double row or Single row	Ground floor	25	15,000
		Upper floors	15	9,000
		Basements	Not Allowed	Not Allowed
II	Double row or Single row	Ground floor	25	24,000
		Upper floors	25	24,000
		Basements	15	9,000
III	Multirow	Ground floor	40	48,000
	Double room	Upper floors	20	48,000
	Single row	Basements	20	24,000

For SI: 1 foot = 304.8 mm, 1 gallon = 3.785 L.

TABLE 3404.3.6.3(4)
AUTOMATIC SPRINKLER PROTECTION FOR SOLID-PILE AND PALLETIZED STORAGE OF LIQUIDS IN CONTAINERS AND PORTABLE TANKS[a]

STORAGE CONDITIONS		CEILING SPRINKLER DESIGN AND DEMAND				MINIMUM HOSE STREAM DEMAND (gpm)	MINIMUM DURATION SPRINKLERS AND HOSE STREAMS (hours)
			Area (square feet)				
Class liquid	Container size and arrangement	Density (gpm/ft^2)	High-temperature sprinklers	Ordinary temperature sprinklers	Maximum spacing (square feet)		
IA	5 gallons or less, with or without cartons, palletized or solid pile[b]	0.30	3,000	5,000	100	750	2
	Containers greater than 5 gallons, on end or side, palletized or solid pile	0.60	5,000	8,000	80	750	
IB, IC and II	5 gallons or less, with or without cartons, palletized or solid pile[b]	0.30	3,000	5,000	100	500	2
	Containers greater than 5 gallons on pallets or solid pile, one high	0.25	5,000	8,000	100		
II	Containers greater than 5 gallons on pallets or solid pile, more than one high, on end or side	0.60	5,000	8,000	80	750	2
IB, IC and II	Portable tanks, one high	0.30	3,000	5,000	100	500	2
II	Portable tanks, two high	0.60	5,000	8,000	80	750	2
III	5 gallons or less, with or without cartons, palletized or solid pile	0.25	3,000	5,000	120	500	1
	Containers greater than 5 gallons on pallets or solid pile, on end or sides, up to three high	0.25	3,000	5,000	120	500	1
III	Containers greater than 5 gallons, on pallets or solid pile, on end or sides, up to 18 feet high	0.35	3,000	5,000	100	750	2
	Portable tanks, one high	0.25	3,000	5,000	120	500	1
	Portable tanks, two high	0.50	3,000	5,000	80	750	2

For SI: 1 foot = 304.8 mm, 1 gallon = 3.785 L, 1 square foot = 0.0929 m^2, 1 gallon per minute = 3.785 L/m, 1 gallon per minute per square foot = 40.75 L/min/m^2.

a. The design area contemplates the use of Class II standpipe systems. Where Class I standpipe systems are used, the area of application shall be increased by 30 percent without revising density.

b. For storage heights above 4 feet or ceiling heights greater than 18 feet, an approved engineering design shall be provided in accordance with Section 104.7.2.

TABLE 3404.3.6.3(5)
AUTOMATIC SPRINKLER PROTECTION REQUIREMENTS FOR RACK STORAGE OF LIQUIDS IN CONTAINERS OF 5-GALLON CAPACITY OR LESS WITH OR WITHOUT CARTONS ON CONVENTIONAL WOOD PALLETS[a]

CLASS LIQUID	CEILING SPRINKLER DESIGN AND DEMAND			IN-RACK SPRINKLER ARRANGEMENT AND DEMAND					MINIMUM HOSE STREAM DEMAND (gpm)	MINIMUM DURATION SPRINKLER AND HOSE STREAM (hours)
	Density (gpm/ft^2)	Area (square feet) High-temperature sprinklers	Area (square feet) Ordinary temperature sprinklers	Maximum spacing	Racks up to 9 feet deep	Racks more than 9 feet to 12 feet deep	30 psi (standard orifice) 14 psi (large orifice)	Number of sprinklers operating		
I (maximum 25-foot height) Option 1	0.40	3,000	5,000	80 ft^2/head	1. Ordinary temperature, quick-response sprinklers, maximum 8 feet 3 inches horizontal spacing 2. One line sprinklers above each level of storage 3. Locate in longitudinal flue space, staggered vertical 4. Shields required where multi-level	1. Ordinary temperature, quick-response sprinklers, maximum 8 feet 3 inches horizontal spacing 2. One line sprinklers above each level of storage 3. Locate in transverse flue spaces, staggered vertical and within 20 inches of aisle 4. Shields required where multi-level	30 psi (0.5-inch orifice)	1. Eight sprinklers if only one level 2. Six sprinklers each on two levels if only two levels 3. Six sprinklers each on top three levels, if three or more levels 4. Hydraulically most remote	750	2
I (maximum 25-foot height) Option 2	0.55	2,000[b]	Not Applicable	100 ft^2/head	1. Ordinary temperature, quick-response sprinklers, maximum 8 feet 3 inches horizontal spacing 2. See 2 above 3. See 3 above 4. See 4 above	1. Ordinary temperature, quick-response sprinklers, maximum 8 feet 3 inches horizontal spacing 2. See 2 above 3. See 3 above 4. See 4 above	14 psi (0.53-inch orifice)	See 1 through 4 above	500	2
I and II (maximum 14-foot storage height) (maximum three tiers)	0.55[c]	2,000[b, d]	Not Applicable	100 ft^2/head	Not Applicable None for maximum 6-foot-deep racks	Not Applicable	Not Applicable	Not Applicable	500	2
II (maximum 25-foot height)	0.30	3,000	5,000	100 ft^2/head	1. Ordinary temperature sprinklers 8 feet apart horizontally 2. One line sprinklers between levels at nearest 10-foot vertical intervals 3. Locate in longitudinal flue space, staggered vertical 4. Shields required where multi-level	1. Ordinary temperature sprinklers 8 feet apart horizontally 2. Two lines between levels at nearest 10-foot vertical intervals 3. Locate in transverse flue spaces, staggered vertical and within 20 inches of aisle 4. Shields required where multi-level	30 psi	Hydraulically most remote—six sprinklers at each level, up to a maximum of three levels	750	2
III (40-foot height)	0.25	3,000	5,000	120 ft^2/head	Same as for Class II liquids	Same as for Class II liquids	30 psi	Same as for Class II liquids	500	2

For SI: 1 inch = 25.4 mm, 1 foot = 304.8 mm, 1 square foot = 0.0929 m^2, 1 pound per square inch = 6.895 kPa, 1 gallon = 3.785 L, 1 gallon per minute = 3.785 L/m, 1 gallon per minute per square foot = 40.75 L/min/m^2.

a. The design area contemplates the use of Class II standpipe systems. Where Class I standpipe systems are used, the area of application shall be increased by 30 percent without revising density.

b. Using listed or approved extra-large orifices, high-temperature quick-response or standard element sprinklers under a maximum 30-foot ceiling with minimum 7.5-foot aisles.

c. For friction lid cans and other metal containers equipped with plastic nozzles or caps, the density shall be increased to 0.65 gpm per square foot using listed or approved extra-large orifice, high-temperature quick-response sprinklers.

d. Using listed or approved extra-large orifice, high-temperature quick-response or standard element sprinklers under a maximum 18-foot ceiling with minimum 7.5-foot aisles and metal containers.

TABLE 3404.3.6.3(6)
AUTOMATIC SPRINKLER PROTECTION REQUIREMENTS FOR RACK STORAGE OF LIQUIDS IN CONTAINERS GREATER THAN 5-GALLON CAPACITY[a]

CLASS LIQUID	CEILING SPRINKLER DESIGN AND DEMAND			IN-RACK SPRINKLER ARRANGEMENT AND DEMAND					MINIMUM HOSE STREAM DEMAND (gpm)	MINIMUM DURATION SPRINKLER AND HOSE STREAM (hours)
	Density (gpm/ft^2)	Area (square feet)		Maximum spacing	On-side storage racks up to 9-foot-deep racks	On-end storage (on pallets) up to 9-foot-deep racks	Minimum nozzle pressure	Number of sprinklers operating		
		High-temperature sprinklers	Ordinary temperature sprinklers							
IA (maximum 25-foot height)	0.60	3,000	5,000	80 ft^2/head	1. Ordinary temperature sprinklers 8 feet apart horizontally 2. One line sprinklers above each tier of storage 3. Locate in longitudinal flue space, staggered vertical 4. Shields required where multi-level	1. Ordinary temperature sprinklers 8 feet apart horizontally 2. One line sprinklers above each tier of storage 3. Locate in longitudinal flue space, staggered vertical 4. Shields required where multi-level	30 psi	Hydraulically most remote—six sprinklers at each level	1,000	2
IB, IC and II (maximum 25-foot height)	0.60	3,000	5,000	100 ft^2/head	1. See 1 above 2. One line sprinklers every three tiers of storage 3. See 3 above 4. See 4 above	1. See 1 above 2. See 2 above 3. See 3 above 4. See 4 above	30 psi	Hydraulically most remote—six sprinklers at each level	750	2
III (maximum 40-foot height)	0.25	3,000	5,000	120 ft^2/head	1. See 1 above 2. One line sprinklers every sixth level (maximum) 3. See 3 above 4. See 4 above	1. See 1 above 2. One line sprinklers every third level (maximum) 3. See 3 above 4. See 4 above	15 psi	Hydraulically most remote—six sprinklers at each level	500	1

For SI: 1 foot = 304.8 mm, 1 square foot = 0.0929 m^2, 1 pound per square inch = 6.895 kPa, 1 gallon = 3.785 L, 1 gallon per minute = 3.785 L/m, 1 gallon per minute per square foot = 40.75 L/min/m^2.

a. The design assumes the use of Class II standpipe systems. Where a Class I standpipe system is used, the area of application shall be increased by 30 percent without revising density.

TABLE 3404.3.6.3(7)
AUTOMATIC AFFF WATER PROTECTION REQUIREMENTS FOR RACK STORAGE OF LIQUIDS IN CONTAINERS GREATER THAN 5-GALLON CAPACITY[a,b]

CLASS LIQUID	CEILING SPRINKLER DESIGN AND DEMAND			IN-RACK SPRINKLER ARRANGEMENT AND DEMAND[c]				DURATION AFFF SUPPLY (minimum)	DURATION WATER SUPPLY (hours)
	Density (gpm/ft^2)	Area (square feet) High-temperature sprinklers	Area (square feet) Ordinary temperature sprinklers	On-end storage of drums on pallets, up to 25 feet	Minimum nozzle pressure (psi)	Number of sprinklers operating	Hose stream demand[d] (gpm)		
IA, IB, IC and II	0.30	1,500	2,500	1. Ordinary temperature sprinkler up to 10 feet apart horizontally 2. One line sprinklers above each level of storage 3. Locate in longitudinal flue space, staggered vertically 4. Shields required for multilevel	30	Three sprinklers per level	500	15	2

For SI: 1 inch = 25.4 mm, 1 foot = 304.8 mm, 1 square foot = 0.0929 m^2, 1 pound per square inch = 6.895 kPa, 1 gallon = 3.785 L, 1 gallon per minute = 3.785 L/m, 1 gallon per minute per square foot = 40.75 L/min/m^2.

a. System shall be a closed-head wet system with approved devices for proportioning aqueous film-forming foam.
b. Except as modified herein, in-rack sprinklers shall be installed in accordance with NFPA 231C.
c. The height of storage shall not exceed 25 feet.
d. Hose stream demand includes 1.5-inch inside hand hose, when required.

TABLE 3404.3.6.3(8)
AUTOMATIC SPRINKLER PROTECTION REQUIREMENTS FOR CLASS I LIQUID STORAGE OF 1-GALLON CAPACITY OR LESS WITH UNCARTONED OR CASE-CUT SHELF DISPLAY UP TO 6.5 FEET, AND PALLETIZED STORAGE ABOVE IN A DOUBLE-ROW RACK ARRAY[a]

STORAGE HEIGHT	CEILING SPRINKLER DESIGN AND DEMAND				IN-RACK SPRINKLER ARRANGEMENT AND DEMAND				MINIMUM HOSE STREAM DEMAND (gpm)	MINIMUM DURATION SPRINKLERS AND HOSE STREAM (hours)
	Density (gpm/ft^2)	Area (square feet) High temperature	Area (square feet) Ordinary temperature	Maximum spacing	Racks up to 9 feet deep	Racks 9 to 12 feet	Minimum nozzle pressure	Number of sprinklers operating		
Maximum 20-foot storage height	0.60	2,000[b]	Not Applicable	100 ft^2/head	1. Ordinary temperature, quick- response sprinklers, maximum 8 feet 3 inches horizontal spacing 2. One line of sprinklers at the 6-foot level and the 11.5-foot level of storage 3. Locate in longitudinal flue space, staggered vertical 4. Shields required where multilevel	Not Applicable	30 psi (standard orifice) or 14 psi (large orifice)	1. Six sprinklers each on two levels 2. Hydraulically most remote 12 sprinklers	500	2

For SI: 1 inch = 25.4 mm, 1 foot = 304.8 mm, 1 square foot = 0.0929 m^2, 1 pound per square inch = 6.895 kPa, 1 gallon = 3.785 L, 1 gallon per minute = 3.785 L/m, 1 gallon per minute per square foot = 40.75 L/min/m^2.
a. This table shall not apply to racks with solid shelves.
b. Using extra-large orifice sprinklers under a ceiling 30 feet or less in height. Minimum aisle width is 7.5 feet.

3404.3.7 Liquid storage rooms. Liquid storage rooms shall comply with Sections 3404.3.7.1 through 3404.3.7.5.2.

❖ Liquid storage rooms are a protected location for the storage of flammable and combustible liquids in occupancies that normally are not associated with the use and storage of flammable and combustible liquids. Liquid storage rooms are intended to store flammable and combustible liquids that exceed the quantities permitted in a control area.

3404.3.7.1 General. Quantities of liquids exceeding those set forth in Section 3404.3.4.1 for storage in control areas shall be stored in a liquid storage room complying with this section and constructed and separated as required by the *International Building Code*.

❖ Liquid storage rooms are intended to store flammable and combustible liquids that exceed the quantities permitted in a control area.

3404.3.7.2 Quantities and arrangement of storage. The quantity limits and storage arrangements in liquid storage rooms shall be in accordance with Tables 3404.3.6.3(2) and 3404.3.6.3(3) and Sections 3404.3.7.2.1 through 3404.3.7.2.3.

❖ Liquid storage rooms are limited in the quantity of flammable and combustible liquids and the method of storage of flammable and combustible liquids. The maximum height and maximum volume per pile are limited by Table 3404.3.6.3 (2) and Table 3404.3.6.3 (3).

3404.3.7.2.1 Mixed storage. Where two or more classes of liquids are stored in a pile or rack section:

1. The quantity in that pile or rack shall not exceed the smallest of the maximum quantities for the classes of liquids stored in accordance with Table 3404.3.6.3(2) or 3404.3.6.3(3); and
2. The height of storage in that pile or rack shall not exceed the smallest of the maximum heights for the classes of liquids stored in accordance with Table 3404.3.6.3(2) or 3404.3.6.3(3).

❖ A liquid storage room is intended for the storage of a maximum quantity of flammable and combustible liquid. When more than one class of flammable and combustible liquid is stored in the same room, the aggregate quantity of flammable and combustible liquids cannot exceed the smallest quantity allowed in Table 3404.3.6.3(2) or 3404.3.6.3(3) for the flammable and combustible liquids being stored.

1. Because the liquid storage room is adequate for a maximum volume of a flammable or combustible liquid, the liquid storage room will be adequate for the aggregate quantity based on the allowable volume of the most hazardous flammable and combustible liquid being stored.
2. Because the liquid storage room is adequate for a maximum height of a flammable or combustible liquid, the liquid storage room will be adequate for the height based on the allowable height of the most hazardous flammable and combustible liquid being stored.

3404.3.7.2.2 Separation and aisles. Piles shall be separated from each other by at least 4-foot (1219 mm) aisles. Aisles shall be provided so that all containers are 20 feet (6096 mm) or less from an aisle. Where the storage of liquids is on racks, a minimum 4-foot-wide (1219 mm) aisle shall be provided between adjacent rows of racks and adjacent storage of liquids. Main aisles shall be a minimum of 8 feet (2438 mm) wide.

Additional aisles shall be provided for access to doors, required windows and ventilation openings, standpipe connections, mechanical equipment and switches. Such aisles shall be at least 3 feet (914 mm) in width, unless greater widths are required for separation of piles or racks, in which case the greater width shall be provided.

❖ Aisles in a liquid storage room are designed to the same aisle criteria as for rack storage. A minimum aisle width of 4 feet (1219 mm) between piles and a minimum main aisle of 8 feet (2438 mm) is to provide clearance for personnel to access the flammable and combustible liquids without interference with other piles.

Egress from the room for evacuation and access to fire protection and other facilities within the room must be maintained.

3404.3.7.2.3 Stabilizing and supports. Containers and piles shall be separated by pallets or dunnage to provide stability and to prevent excessive stress to container walls. Portable tanks stored over one tier shall be designed to nest securely without dunnage.

Requirements for portable tank design shall be in accordance with Chapter 4 of NFPA 30. Shelving, racks, dunnage, scuffboards, floor overlay and similar installations shall be of noncombustible construction or of wood not less than a 1-inch (25 mm) nominal thickness. Adequate material-handling equipment shall be available to handle tanks safely at upper tier levels.

❖ The requirements of this section are intended to prevent piles from collapsing and causing container damage or liquid spills. Requiring pallets or dunnage spreads stress loads over broad surfaces, reducing stresses in container walls that could lead to damage or leakage.

Requirements for portable tank design are intended to result in stability that will prevent tanks from toppling. Having materials-handling equipment that is designed to handle the tanks also helps to ensure safe storage and handling.

3404.3.7.3 Spill control and secondary containment. Liquid storage rooms shall be provided with spill control and secondary containment in accordance with Section 2704.2.

❖ Spills must be contained and controlled to prevent the spread of the liquid and the vapors. Section 2704.2 discusses the use of liquid-tight floors, curbs, dikes and drainage systems to divert the liquid to a location where it can be contained and safely handled. Storage rooms shall be ventilated in accordance with Section 2704.3.

Section 2704.3 discusses mechanical and natural exhaust systems. The exhaust system is to remove the vapors to prevent them from accumulating in concentrations in the flammable range of the vapor.

3404.3.7.4 Ventilation. Liquid storage rooms shall be ventilated in accordance with Section 2704.3.

❖ See the commentary to Section 2704.3 for discussions of ventilation requirements.

3404.3.7.5 Fire protection. Fire protection for liquid storage rooms shall comply with Sections 3404.3.7.5.1 and 3404.3.7.5.2.

❖ Fire protection is a principal means of preventing and controlling fires.

3404.3.7.5.1 Fire-extinguishing systems. Liquid storage rooms shall be protected by automatic sprinkler systems installed in accordance with Chapter 9 and Tables 3404.3.6.3(4) through 3404.3.6.3(7) and Table 3404.3.7.5.1. In-rack sprinklers shall also comply with NFPA 13 and NFPA 231C.

Automatic foam-water systems and automatic aqueous film-forming foam (AFFF) water sprinkler systems shall not be used except when approved.

Protection criteria developed from fire modeling or full-scale fire testing conducted at an approved testing laboratory are allowed in lieu of the protection as shown in Tables 3404.3.6.3(2) through 3404.3.6.3(7) and Table 3404.3.7.5.1 when approved.

❖ The fire protection system must be designed to handle anticipated fires. For a liquid storage room that contains a variety of flammable and combustible liquids, the fire protection needs to be selected from Table 3404.3.6.3(4) through 3404.3.6.3(7) and Table 3404.3.7.5.1. These tables address the classification of flammable and combustible liquid, container size and the method of storing the liquids. The installation of this fire protection is to comply with Chapter 9. For liquid storage rooms that use rack storage not covered in the tables, use NFPA 13 and NFPA 231C for the fire protection requirements.

Automatic foam-water systems and AFFF are to be used if approved by the fire code official. These fire protection systems starve the fire for oxygen by forming a barrier between the hydrocarbon and the air, by cooling the hydrocarbon and by suppressing the release of flammable vapors.

These fire protection systems, however, can be harmful to the environment. Containment of the foam solution should be considered when using these systems.

The code always recognizes that there may be other methods available for accomplishing the intent. The fire code official is responsible for reviewing proposed alternative methods and has the authority to approve them when they have been demonstrated or documented to comply with the intent of the code requirement.

TABLE 3404.3.7.5.1. See below.

❖ Solid-pile or palletized storage of flammable and combustible liquids, except for Class IA liquids, in metal containers is allowed when AFFF is installed. The storage system permits these flammable and combustible liquids to either be in cartons or uncartoned. Flammable and combustible liquids in carton packaging can be piled 11 feet (3353 mm) high while uncartoned containers can be piled 12 feet (3658 mm) high. The higher storage is allowed for containers out of cartons because the combustible cartons have been removed, reducing the fire load.

The fire protection for this storage is AFFF, which is effective because it forms a barrier that starves the fire for oxygen, cools the hydrocarbon and suppresses the release of flammable vapors. The 5-gallon (19 L) metal containers provide some protection from ignition as well as limit the quantity of flammable and combustible liquids that can be exposed. The cartons and pallets are a source of combustibles.

The AFFF provides fire prevention and fire protection in case of a liquid spill or pool fire. The AFFF will also provide protection for the metal containers if the cartons or pallets become involved in a fire. The increased fire protection provided by using 5-gallon (19 L) containers or less with the additional requirements on the AFFF above that required under NFPA 231C permit the storage of flammable and combustible liquids in solid-pile and palletized storage.

TABLE 3404.3.7.5.1
AUTOMATIC AFFF-WATER PROTECTION REQUIREMENTS FOR SOLID-PILE AND PALLETIZED STORAGE OF LIQUIDS IN METAL CONTAINERS OF 5-GALLON CAPACITY OR LESS[a,b]

PACKAGE TYPE	CLASS LIQUID	CEILING SPRINKLER DESIGN AND DEMAND					STORAGE HEIGHT (feet)	HOSE DEMAND (gpm)[c]	DURATION AFFF SUPPLY (minimum)	DURATION WATER SUPPLY (hours)
		Density (gpm/ft^2)	Area (square feet)	Temperature rating	Maximum spacing	Orifice size (inch)				
Cartoned	IB, IC, II and III	0.40	2,000	286°F	100 ft^2/head	0.531	11	500	15	2
Uncartoned	IB, IC, II and III	0.30	2,000	286°F	100 ft^2/head	0.5 or 0.531	12	500	15	2

For SI: 1 inch = 25.4 mm, 1 foot = 304.8 mm, 1 square foot = 0.0929 m^2, 1 gallon per minute = 3.785 L/m, 1 gallon per minute per square foot = 40.75 L/min/m^2, °C. = [(°F)-32]/1.8.

a. System shall be a closed-head wet system with approved devices for proportioning aqueous film-forming foam.

b. Maximum ceiling height of 30 feet.

c. Hose stream demand includes 1.5-inch inside hand hose, when required.

3404.3.7.5.2 Portable fire extinguishers. A minimum of one approved portable fire extinguisher complying with Section 906 and having a rating of not less than 20-B shall be located not less than 10 feet (3048 mm) or more than 50 feet (15 240 mm) from any Class I or II liquid storage area located outside of a liquid storage room.

A minimum of one portable fire extinguisher having a rating of not less than 20-B shall be located outside of, but not more than 10 feet (3048 mm) from, the door opening into a liquid storage room.

❖ Portable fire extinguishers are to be available outside of the liquid storage room because a fire in the liquid storage room could prevent personnel from getting to portable fire extinguishers in the room. The 20B portable fire extinguisher is for the control of small flammable liquid fires.

3404.3.8 Liquid storage warehouses. Buildings used for storage of flammable or combustible liquids in quantities exceeding those set forth in Section 3404.3.4 for control areas and Section 3404.3.7 for liquid storage rooms shall comply with Sections 3404.3.8.1 through 3404.3.8.5 and shall be constructed and separated as required by the *International Building Code*.

❖ Liquid storage warehouses provide a protected location for the storage of flammable and combustible liquids in occupancies that normally are not associated with the use and storage of flammable and combustible liquids. Liquid storage warehouses are intended to store flammable and combustible liquids that exceed the quantities permitted in a control area or a liquid storage room. A liquid storage warehouse must be separated by occupancy separation from the remainder of the building according to the IBC.

3404.3.8.1 Quantities and storage arrangement. The total quantities of liquids in a liquid storage warehouse shall not be limited. The arrangement of storage shall be in accordance with Table 3404.3.6.3(2) or 3404.3.6.3(3).

❖ Liquid storage warehouses are not restricted in the quantity of flammable and combustible liquids that may be stored; however, the maximum height and maximum volume per pile is limited by Tables 3404.3.6.3(2) and 3404.3.6.3(3), which effectively limits the quantity of flammable or combustible liquids that can be stored in a given space. Aisle width requirements further restrict quantities.

3404.3.8.1.1 Mixed storage. Mixed storage shall be in accordance with Section 3404.3.7.2.1.

❖ Liquid storage warehouses have to meet the same criteria as a liquid storage room. Liquid storage warehouses are intended for the storage of a maximum quantity of flammable and combustible liquid. When more than one class of flammable and combustible liquid is stored in the same liquid storage warehouse, the aggregate quantity of flammable and combustible liquids cannot exceed the smallest quantity allowed in Table 3404.3.6.3(2) or Table 3404.3.6.3(3) for the flammable and combustible liquids being stored.

Because the liquid storage warehouse is adequate for a maximum volume of a flammable or combustible liquid, the liquid storage warehouse will be adequate for the aggregate quantity based on the allowable volume of the most hazardous flammable and combustible liquid being stored.

Because the liquid storage warehouse is adequate for a maximum height of a flammable or combustible liquid, the liquid storage warehouse will be adequate for the height based on the allowable height of the most hazardous flammable and combustible liquid being stored.

3404.3.8.1.2 Separation and aisles. Separation and aisles shall be in accordance with Section 3404.3.7.2.2.

❖ Liquid storage warehouses have to meet the same criteria as a liquid storage room. A liquid storage warehouse is designed to the same aisle criteria as rack storage. A minimum aisle width of 4 feet (1219 mm) between piles and a minimum main aisle of 8 feet (2438 mm) provide clearance for personnel and materials-handling equipment to access the flammable and combustible liquids without interference with other piles. The aisles are also intended for ready access by emergency responders.

Egress from the warehouse for evacuation and access to fire protection and other facilities within the warehouse needs to be maintained.

3404.3.8.2 Spill control and secondary containment. Liquid storage warehouses shall be provided with spill control and secondary containment as set forth in Section 2704.2.

❖ The spill must be controlled to prevent the spread of liquid. Section 2704.2 discusses the use of liquid-tight floors, curbs, dikes and drainage systems to divert the liquid to a location where it can be contained and safely handled.

3404.3.8.3 Ventilation. Liquid storage warehouses storing containers greater than 5 gallons (19 L) in capacity shall be ventilated at a rate of not less than 0.25 cfm/sq. ft. (0.075 m^3/min per m^2) of floor area over the storage area.

❖ This section specifies the ventilation rate for liquid warehouses storing containers whose capacity exceeds 5 gallons (19 L). Note that the provisions of Section 2704.3 do not apply. The exhaust system is to remove the vapors to prevent them from accumulating in concentrations in the flammable range of the vapor.

3404.3.8.4 Fire-extinguishing systems. Liquid storage warehouses shall be protected by automatic sprinkler systems installed in accordance with Chapter 9 and Tables 3404.3.6.3(4) through 3404.3.6.3(7) and Table 3404.3.7.5.1, or Section 4.8.2 and Tables 4.8.2(a) through (f) of NFPA 30. In-rack sprinklers shall also comply with NFPA 13 and NFPA 231C.

Automatic foam water systems and automatic aqueous film-forming foam water sprinkler systems shall not be used except when approved.

Protection criteria developed from fire modeling or full-scale fire testing conducted at an approved testing laboratory are allowed in lieu of the protection as shown in Tables 3404.3.6.3(2) through 3404.3.6.3(7) and Table 3404.3.7.5.1 when approved.

❖ The fire protection system must be designed to cope with anticipated fires. For a liquid storage warehouse that contains a variety of flammable and combustible liquids, the fire protection needs to be selected from Tables 3404.3.6.3(4) through 3404.3.6.3(7) and Table 3404.3.7.5.1. These tables address the classification of flammable and combustible liquid, container size and the method of storing the liquids. The installation of this fire protection is to comply with Chapter 9. For liquid storage warehouses that use rack storage not covered in the tables, use NFPA 13 and NFPA 231C for the fire protection requirements.

Automatic foam-water systems and AFFF are to be used if approved by the fire code official. These fire protection systems starve the fire for oxygen by forming a barrier between the hydrocarbon and the air, by cooling the hydrocarbon and by suppressing the release of flammable vapors.

These fire protection systems, however, can be harmful to the environment. Containment of the foam solution should be considered when using these systems.

The code always recognizes that there may be other methods available for accomplishing the intent. The fire code official is responsible for reviewing these alternative methods and has the authority to approve them if they have been demonstrated or documented to comply with the intent of the code requirement.

3404.3.8.5 Warehouse hose lines. In liquid storage warehouses, either 1.5-inch (38 mm) lined or 1-inch (25 mm) hard rubber hand hose lines shall be provided in sufficient number to reach all liquid storage areas and shall be in accordance with Section 903 or Section 905.

❖ The requirements for hose lines stated in this section are consistent with the requirements of Chapter 9. They may be supplied by either the automatic sprinkler system in accordance with Section 903 or a standpipe system in accordance with Section 905.

3404.4 Outdoor storage of containers and portable tanks. Storage of flammable and combustible liquids in closed containers and portable tanks outside of buildings shall be in accordance with Section 3403 and Sections 3404.4.1 through 3404.4.8. Capacity limits for containers and portable tanks shall be in accordance with Section 3404.3.

❖ Outdoor storage in containers and portable tanks is regulated to prevent ignition sources from coming in contact with containers and portable tanks, to prevent damage to containers and portable tanks and to protect property and buildings.

3404.4.1 Plans. Storage shall be in accordance with approved plans.

❖ The fire code official must approve the layout of the outdoor storage. The clearances between tanks, buildings and property lines are to ensure that ignition sources are not in the immediate vicinity of the outdoor storage area.

3404.4.2 Location on property. Outdoor storage of liquids in containers and portable tanks shall be in accordance with Table 3404.4.2. Storage of liquids near buildings located on the same property shall be in accordance with this section.

❖ Restrictions on pile size, required pile separation distances and required distances from lot lines, structures and public throughways are intended to result in improved fire safety. Limiting quantities that can be stored in a single pile limits the potential fuel load in any fire. Keeping piles away from structures, lot lines and public ways protects the piles from stray ignition sources as well as keeping buildings, personnel and the general public safe from the hazards of a possible fire.

TABLE 3404.4.2. See page 34-47.

❖ The outdoor storage of flammable and combustible liquids in containers and portable tanks is limited to control the size and location of a fire. The volume of flammable and combustible liquids is limited based on the type of liquid. The lower the ignition point of the liquid, the smaller the volume of the liquid permitted in a pile. The same approach is taken in determining the maximum height of a pile. These two features control the size of a fire.

Separation is used to control the spread of fire from pile to pile and from a pile to a structure or other property. The distance between piles is the same for all classes of flammable and combustible liquids. The distance between a pile and a structure or property line serves several functions. The clearance is greater for flammable or combustible liquids with a lower ignition point to protect the flammable and combustible liquid from an ignition source as well as protecting the surrounding area from a pool fire of flammable or combustible liquid.

3404.4.2.1 Mixed liquid piles. Where two or more classes of liquids are stored in a single pile, the quantity in the pile shall not exceed the smallest of maximum quantities for the classes of material stored.

❖ When more than one class of flammable and combustible liquid is stored in the same pile, the aggregate quantity of flammable and combustible liquids cannot exceed the smallest quantity allowed in Table 3404.4.2 for the flammable and combustible liquids being stored.

Because the pile is adequate for a maximum volume of a flammable or combustible liquid, the pile will be adequate for the aggregate quantity based on the allowable volume of the most hazardous flammable and combustible liquid being stored.

Because the pile is adequate for a maximum height of a flammable or combustible liquid, the pile will be adequate for the height based on the allowable height of the most hazardous flammable and combustible liquid being stored.

3404.4.2.2 Access. Storage of containers or portable tanks shall be provided with fire apparatus access roads in accordance with Chapter 5.

❖ Access roads for fire department apparatus must be maintained in the outdoor storage so that fire department apparatus can gain access to any pile storage. Access may involve more than one access point. Consideration must be given to the turning radius of fire apparatus, dead-end lanes and entrance through security gates or locked gates.

3404.4.2.3 Security. The storage area shall be protected against tampering or trespassers where necessary and shall be kept free from weeds, debris and other combustible materials not necessary to the storage.

❖ Unauthorized personnel may not know the potential dangers associated with flammable and combustible liquids. By controlling access to the outdoor storage facility, the owner will be able to monitor sources of heat and open flame. Security will be able to monitor the accumulation of combustibles and take appropriate action to remove these fuel sources.

3404.4.2.4 Storage adjacent to buildings. A maximum of 1,100 gallons (4163 L) of liquids stored in closed containers and portable tanks is allowed adjacent to a building located on the same premises and under the same management, provided that:

1. The building does not exceed one story in height. Such building shall be of fire-resistance-rated construction with noncombustible exterior surfaces or noncombustible construction and shall be used principally for the storage of liquids; or
2. The exterior building wall adjacent to the storage area shall have a fire-resistance rating of not less than 2 hours, having no openings to above-grade areas within 10 feet (3048 mm) horizontally of such storage and no openings to below-grade areas within 50 feet (15 240 mm) horizontally of such storage.

The quantity of liquids stored adjacent to a building protected in accordance with Item 2 is allowed to exceed 1,100 gallons (4163 L), provided that the maximum quantity per pile does not exceed 1,100 gallons (4163 L) and each pile is separated by a 10-foot-minimum (3048 mm) clear space along the common wall.

Where the quantity stored exceeds 1,100 gallons (4163 L) adjacent to a building complying with Item 1, or the provisions of Item 1 cannot be met, a minimum distance in accordance with the column for distance to a lot line that can be built on in Table 3404.4.2 shall be maintained between buildings and the nearest container or portable tank.

❖ Storage of flammable and combustible liquid next to a building can expose the flammable and combustible liquid or the building to a fire; however, placing flammable and combustible liquids next to a building may be necessary for the operation of the facility. To control the fire exposure, both the flammable and combustible liquid next to the building and the building itself must be on the same property and under the same management. Two types of buildings may have flammable and combustible liquid stored adjacent to them.

A building used for the storage of flammable and combustible liquid that is of noncombustible construction and not more than one story in height can have outdoor storage next to it. This building has fire protection features installed to protect the building and its contents of flammable and combustible liquid from an interior fire, but is not designed for the exterior fire exposure. Limiting the quantity of flammable and combustible liquid to

TABLE 3404.4.2
OUTDOOR LIQUID STORAGE IN CONTAINERS AND PORTABLE TANKS

CLASS OF LIQUID	CONTAINER STORAGE—MAXIMUM PER PILE		PORTABLE TANK STORAGE—MAXIMUM PER PILE		MINIMUM DISTANCE BETWEEN PILES OR RACKS (feet)	MINIMUM DISTANCE TO LOT LINE OF PROPERTY THAT CAN BE BUILT UPON[c,d] (feet)	MINIMUM DISTANCE TO PUBLIC STREET, PUBLIC ALLEY OR PUBLIC WAY[d] (feet)
	Quantity[a, b] (gallons)	Height (feet)	Quantity[a, b] (gallons)	Height (feet)			
IA	1,100	10	2,200	7	5	50	10
IB	2,200	12	4,400	14	5	50	10
IC	4,400	12	8,800	14	5	50	10
II	8,800	12	17,600	14	5	25	5
III	22,000	18	44,000	14	5	10	5

For SI: 1 foot = 304.8 mm, 1 gallon 3.785 L.

a. For mixed class storage, see Section 3404.4.2.

b. For storage in racks, the quantity limits per pile do not apply, but the rack arrangement shall be limited to a maximum of 50 feet in length and two rows or 9 feet in depth.

c. If protection by a public fire department or private fire brigade capable of providing cooling water streams is not available, the distance shall be doubled.

d. When the total quantity stored does not exceed 50 percent of the maximum allowed per pile, the distances are allowed to be reduced 50 percent, but not less than 3 feet.

1100 gallons (4136 L) in closed containers reduces the exterior fire exposure. If the exterior quantity of flammable and combustible liquid exceeds 1100 gallons (4136 L) in closed containers, or this building does not meet the criteria for construction, height or use as storage for flammable and combustible liquids, the fire protection must be increased by using distance between the building and the pile. The clearance required between the class of flammable and combustible liquid and the property line is to be used.

A building can have outdoor storage next to it when the building exterior wall adjacent to the storage of flammable and combustible liquid has a minimum 2-hour fire-resistance rating and protected openings. The restriction on openings above grade is to prevent an interior or exterior fire from bypassing the 2-hour fire-resistance-rated wall construction. The restriction on an opening below grade is to prevent flammable or combustible liquid or vapors from entering the building. The maximum quantity of flammable and combustible liquid is limited to 1100 gallons (4136 L) in closed containers per pile. More than one pile can be located adjacent to this fire-resistant construction if the piles are located a minimum of 10 feet (3480 mm) from the building.

3404.4.3 Spill control and secondary containment. Storage areas shall be provided with spill control and secondary containment in accordance with Section 3403.4.

Exception: Containers stored on approved containment pallets in accordance with Section 2704.2.3 and containers stored in cabinets and lockers with integral spill containment.

❖ Outdoor storage of flammable or Class II combustible liquids must be controlled to prevent the spread of liquid. Section 2704.2 discusses the use of liquid-tight floors, curbs, dikes and drainage systems to divert the liquid to a location where it can be contained and safely handled.

The exception allows use of approved containment pallets as an acceptable alternative to an integral spill containment for containers. Section 2704.2.3 requires containment pallets to have a light-tight sump that is accessible for visible inspection with a minimum capacity of 66 gallons (250 L). The containment pallet must be designed to prevent the collection of rainwater in the sump.

3404.4.4 Security. Storage areas shall be protected against tampering or trespassers by fencing or other approved control measures.

❖ Unauthorized personnel may not know the potential dangers associated with flammable and combustible liquids. By controlling access to the outdoor storage facility, the owner will be able to monitor sources of heat and open flame.

3404.4.5 Protection from vehicles. Guard posts or other means shall be provided to protect exterior storage tanks from vehicular damage. When guard posts are installed, the posts shall be installed in accordance with Section 312.

❖ Protection from vehicle impact is provided by guard posts or other approved barriers. Section 312 contains the specifications for guard posts or the design forces required for an approved barrier to comply with the code.

3404.4.6 Clearance from combustibles. The storage area shall be kept free from weeds, debris and combustible materials not necessary to the storage. The area surrounding an exterior storage area shall be kept clear of such materials for a minimum distance of 15 feet (4572 mm).

❖ This is the same requirement as in Section 3404.4.2.3 with the additional focus that combustible materials be cleared for a minimum of 15 feet (4572 mm) around the storage area. Keeping combustibles a minimum of 15 feet (4572 mm) from each pile removes a fuel source from the immediate area.

3404.4.7 Weather protection. Weather protection for outdoor storage shall be in accordance with Section 2704.13.

❖ Weather protection is not required for outdoor storage; however, where overhead structure is erected, it must conform to the requirements of Section 2704.13, which refers to the requirements of the IBC. The code reference notes that an open structure consisting of only a noncombustible roof structure does not change the storage facility to indoor storage. Natural airflow through a structure with no walls and only a noncombustible roof has sufficient ventilation for the storage area to be treated as outdoor storage for flammable and combustible liquids.

3404.4.8 Empty containers and tank storage. The storage of empty tanks and containers previously used for the storage of flammable or combustible liquids, unless free from explosive vapors, shall be stored as required for filled containers and tanks. Tanks and containers when emptied shall have the covers or plugs immediately replaced in openings.

❖ An empty container or portable tank is at least as and possibly more dangerous than a full container or portable storage tank. There is a possibility that a vapor-air mixture in the container or portable storage tank could reach the LFL. This potential danger requires that empty containers and portable tanks be handled and stored as if full of flammable or combustible liquid.

SECTION 3405
DISPENSING, USE, MIXING AND HANDLING

3405.1 Scope. Dispensing, use, mixing and handling of flammable liquids shall be in accordance with Section 3403 and this section. Tank vehicle and tank car loading and unloading and other special operations shall be in accordance with Section 3406.

> **Exception:** Containers of organic coatings having no fire point and which are opened for pigmentation are not required to comply with this section.

❖ The dispensing, use, mixing and handling of flammable and combustible liquids provide opportunities for flammable and combustible liquids to become mixed with air. These operations can create a vapor-air mixture between the LFL and the UFL.

This section does not cover organic coating with no fire point used for pigmentation. This exception is also covered in Section 3401.1.

3405.2 Liquid transfer. Liquid transfer equipment and methods for transfer of Class I, II and IIIA liquids shall be approved and be in accordance with Sections 3405.2.1 through 3405.2.6.

❖ The low flash points for Class I, II and IIIA liquids require restrictions on transferring of the flammable and combustible liquids.

3405.2.1 Pumps. Positive-displacement pumps shall be provided with pressure relief discharging back to the tank, pump suction or other approved location, or shall be provided with interlocks to prevent over-pressure.

❖ Positive-displacement pumps are used because the pumping action is forward only and is a good pump for viscous fluids. The positive-displacement pump will create pressure waves in the fluid. A buildup of pressure can damage the pump and the piping system. This pressure buildup must be relieved by discharging the excess pressure into the tank or the pump suction (intake) or other approved location.

3405.2.2 Pressured systems. Where gases are introduced to provide for transfer of Class I liquids, or Class II and III liquids transferred at temperatures at or above their flash points by pressure, only inert gases shall be used. Controls, including pressure relief devices, shall be provided to limit the pressure so that the maximum working pressure of tanks, containers and piping systems cannot be exceeded. Where devices operating through pressure within a tank or container are used, the tank or container shall be a pressure vessel approved for the intended use. Air or oxygen shall not be used for pressurization.

> **Exception:** Air transfer of Class II and Class III liquids at temperatures below their flash points.

❖ Compressed inert gas can be used to transfer flammable and combustible liquids. The inert gas prevents a vapor-air mixture from entering the LFL. The pressure system must be designed to prevent overloading any component of the system. An overload could create a leak or failure in the system.

The use of air as the compressed gas is permitted for Class II and III liquids under limited conditions. If the Class II and III liquids are to be dispensed at a temperature below their flash points, there is no ignition source.

3405.2.3 Piping, hoses and valves. Piping, hoses and valves used in liquid transfer operations shall be approved or listed for the intended use.

❖ Piping, hoses and valves must be designed to function with the flammable or combustible liquid being transferred and at the temperatures and pressures of the dispensing system.

3405.2.4 Class I, II and III liquids. Class I and II liquids or Class III liquids that are heated up to or above their flash points shall be transferred by one of the following methods:

> **Exception:** Liquids in containers not exceeding a 5.3-gallon (20 L) capacity.

1. From safety cans complying with UL 30.
2. Through an approved closed piping system.
3. From containers or tanks by an approved pump taking suction through an opening in the top of the container or tank.
4. For Class IB, IC, II and III liquids, from containers or tanks by gravity through an approved self-closing or automatic-closing valve when the container or tank and dispensing operations are provided with spill control and secondary containment in accordance with Section 3403.4. Class IA liquids shall not be dispensed by gravity from tanks.
5. Approved engineered liquid transfer systems.

❖ The lower flash points for Class I and II liquids require that dispensing be done by an approved procedure to avoid the development of a vapor-air mixture above the LFL. Class III liquids, when heated up to or above their flash point, become as readily ignitable as Class I liquids. This section correlates with Section 3401.5 and similar provisions in Chapter 5 of NFPA 30.

The exception recognizes the use of Class I and II liquids as fuel for small engines or for some maintenance procedures, and exempts containers no larger than 5.3 gallons (20 L) from the limitations of this section.

3405.2.5 Manual container filling operations for Class I liquids. Class I liquids and Class II or III liquids heated to or above their flash points shall not be transferred into containers unless the nozzle and containers are electrically interconnected. Acceptable methods of electrical interconnection include:

1. Metallic floor plates on which containers stand while filling, when such floor plates are electrically connected to the fill stem; or
2. Where the fill stem is bonded to the container during filling by means of a bond wire.

❖ Flammable and combustible liquids being transferred near or above their flash points will be generating vapors and are near the point of ignition. Transferring these liquids to a container must be done in a manner to prevent a static spark. Grounding the container and the nozzle will remove the potential for an electrical hazard and the possibility for a spark to jump.

3405.2.6 Automatic container-filling operations for Class I liquids. Container-filling operations for Class I liquids involving conveyor belts or other automatic-feeding operations shall be designed to prevent static accumulations.

❖ The conveyor belts and other automatic-feeding equipment can develop a static charge. This static charge must be discharged by grounding the equipment so that no electrical potential can develop.

3405.3 Use, dispensing and mixing inside of buildings. Indoor use, dispensing and mixing of flammable and combustible liquids shall be in accordance with Sections 3405.2 and 3405.3.1 through 3405.3.5.3.

❖ The use, dispensing and mixing of flammable and combustible liquids creates an environment where vapor may accumulate. Theses vapors must be controlled to keep the vapor-air mixture from reaching the LFL.

3405.3.1 Closure of mixing or blending vessels. Vessels used for mixing or blending of Class I liquids and Class II or III liquids heated up to or above their flash points shall be provided with self-closing, tight-fitting, noncombustible lids that will control a fire within such vessel.

Exception: Where such devices are impractical, approved automatic or manually controlled fire-extinguishing devices shall be provided.

❖ Flammable and combustible liquids being mixed or blended near or above their flash points will generate vapors and are near the point of ignition. Using a vessel with a self-closing, tight-fitting, noncombustible lid will allow a fire to be contained within the vessel and for the fire to be cut off from any source of additional air.

The exception covers situations in which the use of a vessel to contain and extinguish the fire may be impractical. These processes can use a fire-extinguishing device that operates automatically or manually.

3405.3.2 Bonding of vessels. Where differences of potential could be created, vessels containing Class I liquids or liquids handled at or above their flash points shall be electrically connected by bond wires, ground cables, piping or similar means to a static grounding system to maintain equipment at the same electrical potential to prevent sparking.

❖ Static electricity will spark between surfaces with different electrical potential. This spark is an ignition source for Class I liquids. Vessels used to contain Class I liquids near or above their flash points are to be bonded using wires, ground cables, metal piping or other similar means that will carry current.

3405.3.3 Heating, lighting and cooking appliances. Heating, lighting and cooking appliances which utilize Class I liquids shall not be operated within a building or structure.

Exception: Operation in single-family dwellings.

❖ The use of Class I liquids for the operation of heating, lighting and cooking appliances is not permitted inside of a building or structure except for a single-family dwelling.

3405.3.4 Location of processing vessels. Processing vessels shall be located with respect to distances to lot lines of adjoining property which can be built on, in accordance with Tables 3405.3.4(1) and 3405.3.4(2).

Exception: Where the exterior wall facing the adjoining lot line is a blank wall having a fire-resistance rating of not less than 4 hours, the fire code official is authorized to modify the distances. The distance shall not be less than that set forth in the *International Building Code*, and when Class IA or unstable liquids are involved, explosion control shall be provided in accordance with Section 911.

❖ Adjacent property is protected from processing vessels by separation. The three factors to be considered in determining the minimum clearance to protect property are: (1) the processing vessel operating pressure, (2) the processing vessel capacity and (3) the stability of the flammable or combustible liquid. The larger these factors become, the greater the clearance needed for safety. The processing of an unstable flammable or combustible liquid will require a greater clearance than a similar operation using a stable flammable or combustible liquid. Tables 3405.3.4(1) and 3405.3.4(2) are used together to determine the minimum clearance required to protect adjacent property or important buildings on the same property.

The exception allows fire-resistant construction to be substituted for clearance to a property. A 4-hour fire-resistant exterior wall without any openings may be used to reduce the clearance too as little as 30 feet (9144 mm). The fire code official can use the requirements in Section 415.3.1 of IBC, along with a 4-hour fire-resistant exterior wall without any openings to approve smaller clearance to a property line. The clearances in Section 415.3.1 establish the minimum clearance.

For process vessels using Class I liquids, this exception requires that the building be designed for explosion control as well as having a 4-hour fire-resistant exterior wall without any openings to take advantage of the smaller clearance.

Here are some sample problems to determine the separation for processing vessels:

Example 1:

Processing vessel:
Tank capacity = 800 gallons (3028 L)
Emergency relief venting = 2 psig (14 kPa)

Flammable liquid is stable.

Factors from Table 3405.3.4(1):
Emergency relief venting under 2.5 psig (17 kPa) with stable liquid = 1

Minimum distance from Table 3405.3.4(2):
Distance from lot line or opposite side of public way = 15 feet (4572 mm)

Distance to nearest side of public way or important building = 5 feet (1524 mm)

Adjustment to the minimum distance resulting from emergency relief venting and stability of flammable liquid:

Distance from lot line or opposite side of public way: 1 (15 feet) = 15 feet (4572 mm)
Distance to nearest side of public way or important building:
1 (5 feet) = 5 feet (1524 mm)

Example 2:
Processing vessel:
Tank capacity = 800 gallons (3028 L)
Emergency relief venting = 3 psig (21 kPa)

Flammable liquid is stable.

Factors from Table 3405.3.4(1):
Emergency relief venting over 2.5 psig with stable liquid = 2.5

Minimum distance from Table 3405.3.4(2):
Distance from lot line or opposite side of public way = 15 feet (4572 mm)
Distance to nearest side of public way or important building = 5 feet (1524 mm)

Adjustment to the minimum distance resulting from emergency relief venting and stability of flammable liquid:

Distance from lot line or opposite side of public way: 2.5 (15 feet) = 37.5 feet (11430 mm)
Distance to nearest side of public way or important building: 2.5 (5 feet) = 12.5 feet (3810 mm)

Example 3:
Processing vessel:
Tank capacity = 800 gallons (3028 L)
Emergency relief venting = 2 psig (14 kPa)

Flammable liquid is stable.

Factors from Table 3405.3.4(1):
Emergency relief venting under 2.5 psig (17 kPa) with stable
Liquid = 1.5

Minimum distance from Table 3405.3.4(2):
Distance from lot line or opposite side of public way = 15 feet (4572 mm)
Distance to nearest side of public way or important building = 5 feet (1524 mm)

Adjustment to the minimum distance resulting from emergency relief venting and stability of flammable liquid:

Distance from lot line or opposite side of public way: 1.5 (15 feet) = 22.5 feet (6858 mm)
Distance to nearest side of public way or important building: 1.5 (5 feet) = 7.5 feet (2286 mm)

Example 4:
Processing vessel:
Tank capacity = 800 gallons (3028 L)
Emergency relief venting = 3 psig (17 kPa)

Flammable liquid is unstable.

Factors from Table 3405.3.4(1):
Emergency relief venting over 2.5 psig (17 kPa) with unstable liquid = 4

Minimum distance from Table 3405.3.4(2):
Distance from lot line or opposite side of public way = 15 feet (4572 mm)
Distance to nearest side of public way or important building = 5 feet (1524 mm)

Adjustment to the minimum distance resulting from emergency relief venting and stability of flammable liquid:

Distance from lot line or opposite side of public way: 4 (15 feet) = 60 feet (18288 mm)
Distance to nearest side of public way or important building: 4 (5 feet) = 20 feet (6096 mm)

TABLE 3405.3.4(1). See below.

❖ Processing vessels containing flammable and combustible liquids must be located a minimum distance from property lines. The clearance is to protect the flammable and combustible liquids from ignition sources that are not under the control of the facility operators. This clearance also protects the property adjacent to the facility.

The separation varies depending on the type of flammable and combustible liquid and operating pressure of the processing vessel. The table divides flammable and combustible liquids into two categories: stable and unstable. An unstable liquid can self-react when exposed to heat. This autoignition property requires that the separation distance for an unstable flammable or combustible liquid be several times greater than for a stable flammable or combustible liquid.

The operating pressure of the processing vessel is divided into two categories: 2.5 psig (17.2 kPa) or less and over 2.5 psig (17.2 kPa). Pressure can be the cause of damage to the processing system, and this damage could result in a vapor or liquid leak of flammable or combustible liquid. A higher pressure would also force more vapor or liquid out of the system during a leak. To provide protection from a processing vessel with an operating pressure greater than 2.5 psig (17.2 kPa), the separation distance is several times greater than for a processing vessel with an operating pressure 2.5 psig (17.2 kPa) or less. These two factors are cumulative.

One or both of these two factors may increase the separation distance. This table provides the factor to be used to calculate the separation distance listed in Table 3405.3.4(2).

TABLE 3405.3.4(2). See below.

❖ The separation distance from the processing vessel is affected by the tank capacity of the processing vessel and the object. The larger the tank capacity, the greater the separation distance. This greater separation distance is to handle the potential pool fire with the tank capacity. The separation distance is greater to a property line to protect the processing vessel from ignition sources that are not under control of the facility operators, as well as to protect the adjacent property from a potential fire. The separation distance to an important building on the property or to a public way is the same or less than for a property line. Because buildings on the property are under the control and operation of the facility operators, these buildings should be maintained and operated to prevent an ignition source. The separation from a public way is handled the same as from an important building on the property. The public way does not have buildings that have ignition sources so that the potential is not as great as from an adjacent property.

3405.3.5 Quantity limits for use. Liquid use quantity limitations shall comply with Sections 3405.3.5.1 through 3405.3.5.3.

❖ The volume of flammable and combustible liquid is controlled to limit the potential fire. The smaller the quantity, the smaller the potential fire. The dispensing and mixing of flammable and combustible liquids indoors can generate a vapor-air mixture. If this vapor-air mixture is ignited, the size of the fire could increase because of the availability of flammable and combustible liquid in the area. Limiting the volume will help control the overall size of any fire.

3405.3.5.1 Maximum allowable quantity per control area. Indoor use, dispensing and mixing of flammable and combustible liquids shall not exceed the maximum allowable quantity per control area indicated in Table 2703.1.1(1) and shall not exceed the additional limitations set forth in Section 3405.3.5.

Exception: Cleaning with Class I, II and IIIA liquids shall be in accordance with Section 3405.3.6.

Use of hazardous production material flammable and combustible liquids in Group H-5 occupancies shall be in accordance with Chapter 18.

❖ The maximum quantity per control area for use, dispensing and mixing of flammable and combustible liq-

TABLE 3405.3.4(1)
SEPARATION OF PROCESSING VESSELS FROM LOT LINES

PROCESSING VESSELS WITH EMERGENCY RELIEF VENTING	LOCATION[a]	
	Stable liquids	Unstable liquids
Not in excess of 2.5 psig	Table 3405.3.4(2)	2.5 times Table 3405.3.4(2)
Over 2.5 psig	1.5 times Table 3405.3.4(2)	4 times Table 3405.3.4(2)

For SI: 1 pound per square inch gauge = 6.895 kPa.

a. Where protection of exposures by a public fire department or private fire brigade capable of providing cooling water streams on structures is not provided, distances shall be doubled.

TABLE 3405.3.4(2)
REFERENCE TABLE FOR USE WITH TABLE 3405.3.4(1)

TANK CAPACITY (gallons)	MINIMUM DISTANCE FROM LOT LINE OF A LOT WHICH IS OR CAN BE BUILT UPON, INCLUDING THE OPPOSITE SIDE OF A PUBLIC WAY (feet)	MINIMUM DISTANCE FROM NEAREST SIDE OF ANY PUBLIC WAY OR FROM NEAREST IMPORTANT BUILDING ON THE SAME PROPERTY (feet)
275 or less	5	5
276 to 750	10	5
751 to 12,000	15	5
12,001 to 30,000	20	5
30,001 to 50,000	30	10
50,001 to 100,000	50	15
100,001 to 500,000	80	25
500,001 to 1,000,000	100	35
1,000,001 to 2,000,000	135	45
2,000,001 to 3,000,000	165	55
3,000,001 or more	175	60

For SI: 1 foot = 304.8 mm, 1 gallon = 3.785 L.

uids indoors is identical to that for indoor storage (see Section 3404.3.3) with some additional limits caused by the processing operation. Table 2703.1.1(1) contains the quantity permitted per control area. By limiting the quantity per control area, the flammable and combustible liquid hazard is reduced to a level that the fire protection can handle and that will not interfere with egress requirements, which results in a building that can function and provide a reasonable degree of safety.

Cleaning with Class I, II, and IIIA liquids is covered in Section 3405.3.6.

The use of flammable and combustible liquids in a semiconductor fabrication facility is covered by Chapter 18.

3405.3.5.2 Occupancy quantity limits. The following limits for quantities of flammable and combustible liquids used, dispensed or mixed based on occupancy classification shall not be exceeded.

Exception: Cleaning with Class I, II, or IIIA liquids shall be in accordance with Section 3405.3.6.

1. Group A occupancies: Quantities in Group A occupancies shall not exceed that necessary for demonstration, treatment, laboratory work, maintenance purposes and operation of equipment, and shall not exceed quantities set forth in Table 2703.1.1(1).
2. Group B occupancies: Quantities in drinking, dining, office and school uses within Group B occupancies shall not exceed that necessary for demonstration, treatment, laboratory work, maintenance purposes and operation of equipment, and shall not exceed quantities set forth in Table 2703.1.1(1).
3. Group E occupancies: Quantities in Group E occupancies shall not exceed that necessary for demonstration, treatment, laboratory work, maintenance purposes and operation of equipment and shall not exceed quantities set forth in Table 2703.1.1(1).
4. Group F occupancies: Quantities in dining, office and school uses within Group F occupancies shall not exceed that necessary for demonstration, laboratory work, maintenance purposes and operation of equipment, and shall not exceed quantities set forth in Table 2703.1.1(1).
5. Group I occupancies: Quantities in Group I occupancies shall not exceed that necessary for demonstration, laboratory work, maintenance purposes and operation of equipment, and shall not exceed quantities set forth in Table 2703.1.1(1).
6. Group M occupancies: Quantities in dining, office and school uses within Group M occupancies shall not exceed that necessary for demonstration, laboratory work, maintenance purposes and operation of equipment, and shall not exceed quantities set forth in Table 2703.1.1(1).
7. Group R occupancies: Quantities in Group R occupancies shall not exceed that necessary for maintenance purposes and operation of equipment, and shall not exceed quantities set forth in Table 2703.1.1(1).
8. Group S occupancies: Quantities in dining and office uses within Group S occupancies shall not exceed that necessary for demonstration, laboratory work, maintenance purposes and operation of equipment and shall not exceed quantities set forth in Table 2703.1.1(1).

❖ Flammable and combustible liquids may be used in occupancies other than Group H. The control of these liquids is based on the quantity in that occupancy.

The limitations for occupancies are based on the quantity limits for the storage of flammable and combustible liquids in containers and portable tanks in other occupancies. These other occupancies are not designed for flammable and combustible liquids so there are limitations on the quantity of flammable and combustible liquids that can be stored or used in these occupancies.

The use of flammable and combustible liquids must be consistent with the function of the occupancy. This prevents these other occupancies from being converted into Group H. The quantities of flammable and combustible liquids are listed in Table 2703.1.1(1). The occupancies covered by these limitations are Groups A, B, E, F, I, M, R and S.

3405.3.5.3 Quantities exceeding limits for control areas. Quantities exceeding the maximum allowable quantity per control area indicated in Sections 3405.3.5.1 and 3405.3.5.2 shall be in accordance with the following:

1. For open systems, indoor use, dispensing and mixing of flammable and combustible liquids shall be within a room or building complying with the *International Building Code* and Sections 3405.3.7.1 through 3405.3.7.5.
2. For closed systems, indoor use, dispensing and mixing of flammable and combustible liquids shall be within a room or building complying with the *International Building Code* and Sections 3405.3.7 through 3405.3.7.4 and 3405.3.7.6.

❖ It is possible to have a quantity of flammable and combustible liquids greater than allowed in Sections 3404.3.5 and 3405.3.5.2. To use, dispense and mix flammable and combustible liquids indoors in quantities larger than approved in this section, the building or room must be designed and constructed to provide additional protection. The design and construction requirements are based on the potential for the use, dispensing and mixing to generate vapors.

3405.3.6 Cleaning with flammable and combustible liquids. Cleaning with Class I, II and IIIA liquids shall be in accordance with this section.

Exceptions:

1. Dry cleaning shall be in accordance with Chapter 12.
2. Spray-nozzle cleaning shall be in accordance with Section 1503.5.

❖ Cleaning machines using flammable or combustible liquids may generate a vapor-air mixture in the flammable range. The use of cleaning machines using flammable or combustible liquids is regulated to control the vapor-air mixture, ignition sources and quantity of flammable or combustible liquids in the machines and the work area.

The exception directs the reader to other chapters for detailed discussion of dry cleaning operations and spray nozzle cleaning.

3405.3.6.1 Cleaning operations. Class IA liquids shall not be used for cleaning. Cleaning with Class IB, IC or II liquids shall be conducted as follows:

1. In a room or building in accordance with Section 3405.3.7; or
2. In a machine listed and approved for the purpose in accordance with Section 3405.3.6.2.

Exception: Materials used in commercial and industrial process-related cleaning operations in accordance with other provisions of this code and not involving facilities maintenance cleaning operations.

❖ The lower flash point for a Class IA liquid is the reason for prohibiting its use as a cleaning solution. Class IB, IC and II liquids can be used in rooms designed and constructed with adequate fire protection, where there is control of spillage and adequate ventilation or in a machine that is designed to protect and control the liquid. The exception is provided to make it clear that process-related cleaning operations that are not considered part of facility maintenance cleaning and have a good safety history are not included in the operations regulated by this section.

3405.3.6.2 Listed and approved machines. Parts cleaning and degreasing conducted in listed and approved machines in accordance with Section 3405.3.6.1 shall be in accordance with Sections 3405.3.6.2.1 through 3405.3.6.2.7.

❖ Prior to the development of parts washing machines, the typical method of washing automotive or other machinery parts was primitive and dangerous and typically consisted of allowing parts to soak in an open bucket of gasoline or low-flash-point solvent. Labeled machines used for the washing of parts are similar to the spray equipment cleaning machines discussed in Section 1503.3.5.1. Such machines consist of a sink-like open container set upon a solvent reservoir or connected to it by approved hoses. A noncombustible lid to the sink is typically held in the open position by a fusible element that will melt and allow the lid to close in the event of a fire in the sink [see Figure 3405.3.6.2(1)]. Some labeled parts washing machines do not include a remote solvent reservoir but are simply a solvent-filled rinsing tank in which parts may be soaked, manually agitated or scrubbed with a brush, but in which circulation of the solvent does not occur [see Figure 3405.3.6.2(2)]. The manufacturers' installation instructions are evaluated as part of the labeling process and, thus, must be carefully followed when setting up and using these machines.

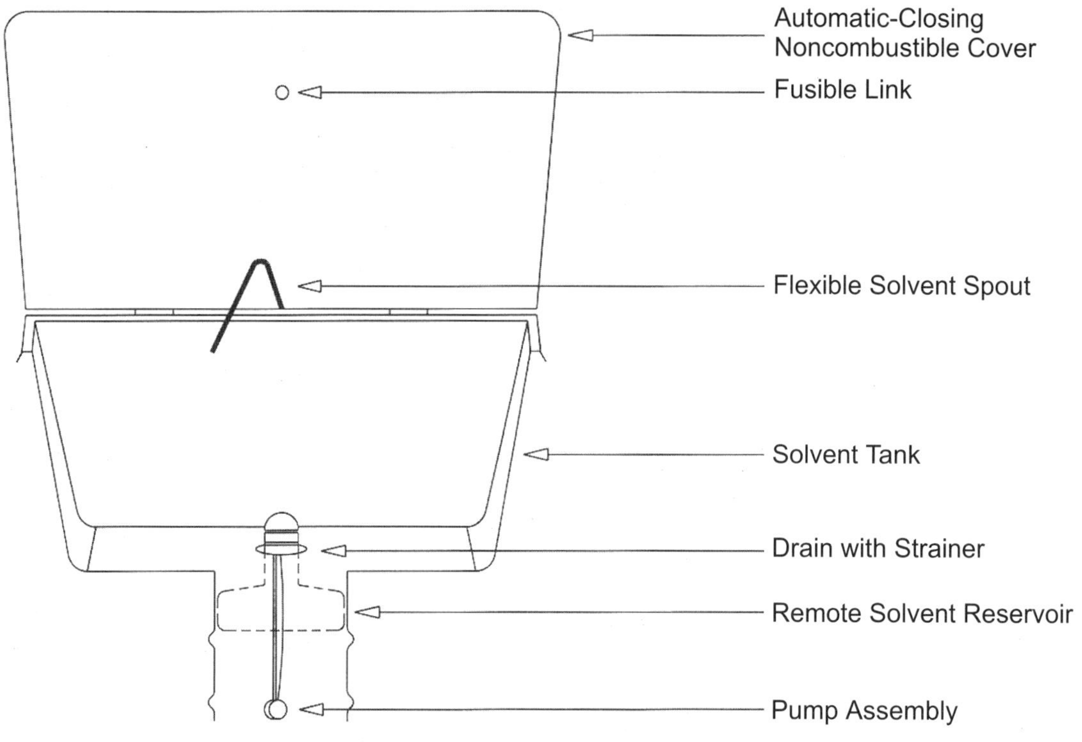

Figure 3405.3.6.2(1)
PARTS CLEANING MACHINE WITH REMOTE SOLVENT RESERVOIR

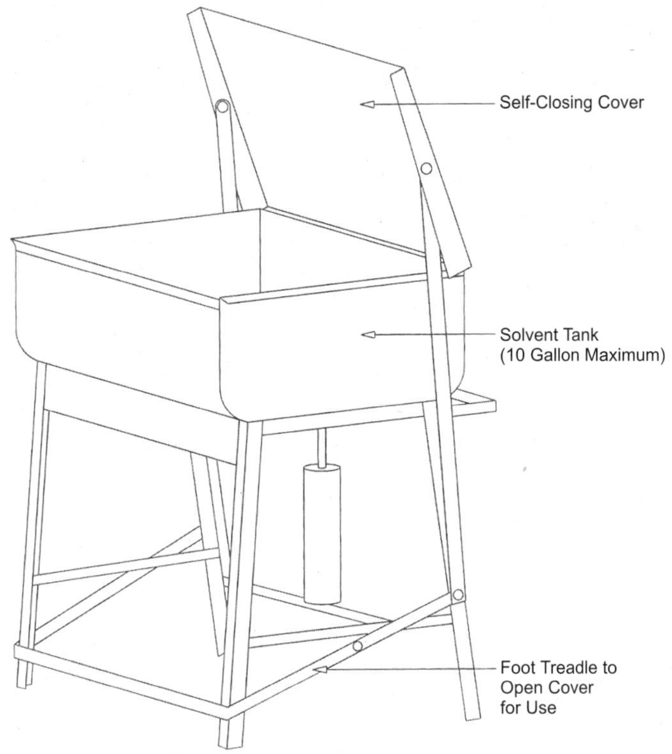

For SI: 1 gallon = 3.785 L.

Figure 3405.3.6.2(2)
PARTS CLEANING MACHINE WITH REMOTE SOLVENT RESERVOIR

3405.3.6.2.1 Solvents. Solvents shall be classified and shall be compatible with the machines within which they are used.

❖ Manufacturers of labeled parts washing machines either manufacture and market or recommend specific solvents that may be used in their machines in order to maintain the level of safety contemplated in the design of the machine. The testing and labeling process is based on only the manufacturer-recommended solvents being used in the machines. Typically, these are Class II or III solvents.

3405.3.6.2.2 Machine capacities. The quantity of solvent shall not exceed the listed design capacity of the machine for the solvent being used with the machine.

❖ The machine listing will state the quantity of solvent permitted. The listing agency tests the machine to standards to determine the maximum capacity of flammable or combustible liquid. This quantity is based on the testing required to get the listing. The listing indicates that the machine is safe for use based on the limitations of the listing.

3405.3.6.2.3 Solvent quantity limits. Solvent quantities shall be limited as follows:

1. Machines without remote solvent reservoirs shall be limited to quantities set forth in Section 3405.3.5.
2. Machines with remote solvent reservoirs using Class I liquids shall be limited to quantities set forth in Section 3405.3.5.
3. Machines with remote solvent reservoirs using Class II liquids shall be limited to 35 gallons (132 L) per machine. The total quantities shall not exceed an aggregate of 240 gallons (908 L) per control area in buildings not equipped throughout with an approved automatic sprinkler system and an aggregate of 480 gallons (1817 L) per control area in buildings equipped throughout with an approved automatic sprinkler system in accordance with Section 903.3.1.1.
4. Machines with remote solvent reservoirs using Class IIIA liquids shall be limited to 80 gallons (303 L) per machine.

❖ This section lists four restrictions on solvent quantities in machines:

1. A machine without a remote solvent reservoir is under the quantity limitations for a control area. The quantity of solvent in the machine and any other flammable and combustible liquids in the control area cannot exceed the maximum quantity permitted under Table 2703.1.1(1). These quantities can be increased by application of Section 3405.3.5.3.
2. A machine with a remote solvent reservoir using Class IB or IC liquids is treated the same as a machine without a remote solvent reservoir.

3. A machine with a remote solvent reservoir using Class II liquid is permitted to have larger quantities than Table 2703.1.1(1) allows. The allowable quantity is increased even more if the control area has an automatic sprinkler system. The safety features built into the machine permit the increase in Class II solvent quantities for each machine and for the control area.
4. A machine with a remote solvent reservoir using Class IIIA liquid is permitted to have 80 gallons (303 L) per machine.

3405.3.6.2.4 Immersion soaking of parts. Work areas of machines with remote solvent reservoirs shall not be used for immersion soaking of parts.

❖ The action of immersion soaking of parts for a machine with a remote reservoir would have the effect of changing the machine from having a remote reservoir to a machine without a remote reservoir. This function of immersion soaking of parts changes the justification for permitting machines with a remote reservoir to have greater quantities of solvent than machines without a remote reservoir.

3405.3.6.2.5 Separation. Multiple machines shall be separated from each other by a distance of not less than 30 feet (9144 mm) or by a fire barrier with a minimum 1-hour fire-resistance rating.

❖ Machines are isolated from one another to reduce the potential of a fire from one machine affecting an adjacent machine. The isolation can be by a clearance of 30 feet (9144 mm) or by a 1-hour fire-resistant barrier.

3405.3.6.2.6 Ventilation. Machines shall be located in areas adequately ventilated to prevent accumulation of vapors.

❖ The machines are to be located in areas having mechanical or natural ventilation complying with Section 2704.3 to remove the vapors in order to prevent the vapor-air mixture from accumulating in concentrations above the LFL.

3405.3.6.2.7 Installation. Machines shall be installed in accordance with their listings.

❖ Listed machines normally have written installation instructions that must be followed to make certain the machine is installed correctly. A machine installed incorrectly could generate vapors, leaks or sparks.

3405.3.7 Rooms or buildings for quantities exceeding the maximum allowable quantity per control area. Where required by Section 3405.3.5.3 or 3405.3.6.1, rooms or buildings used for use, dispensing or mixing of flammable and combustible liquids shall be in accordance with Sections 3405.3.7.1 through 3405.3.7.6.3.

❖ Some uses of flammable and combustible liquids may require a quantity that is greater than maximum allowable quantity permitted in a control area. These sections state requirements for a room or building so that a greater quantity of flammable and combustible liquid may be used.

3405.3.7.1 Construction, location and fire protection. Rooms or buildings classified in accordance with the *International Building Code* as Group H-2 or H-3 occupancies based on use, dispensing or mixing of flammable or combustible liquids shall be constructed in accordance with the *International Building Code.*

❖ Group H-2 occupancy is the classification for a facility that uses or stores Class I, II or III liquids in open containers and systems, or in closed containers or systems under a pressure greater than 15 psig (103 kPa). Group H-3 occupancy is the classification for a facility that uses or stores Class I, II or III liquids in closed containers or systems under a pressure of 15 psig (103 kPa) or less. Sections 414 and 415 of the IBC give guidance for determining the construction requirements for these two occupancies.

3405.3.7.2 Basements. In rooms or buildings classified in accordance with the *International Building Code* as Group H-2 or H-3, dispensing or mixing of flammable or combustible liquids shall not be conducted in basements.

❖ The mixing and dispensing of flammable and combustible liquids may generate vapors. The accumulation of flammable and combustible liquids or vapors in a basement could develop into a hazardous condition. These operations are prohibited in a basement of a Group H-2 or H-3 occupancy.

3405.3.7.3 Fire protection. Rooms or buildings classified in accordance with the *International Building Code* as Group H-2 or H-3 occupancies shall be equipped with an approved automatic fire-extinguishing system in accordance with Chapter 9.

❖ Group H-2 and H-3 occupancies must have an automatic sprinkler system according to Section 903.4 of the IBC. The same requirement is in Section 903.2.4 of the code. Any building that can be classified as a Group H-2 or H-3 occupancy that is not equipped with an automatic sprinkler system is in violation of the code.

3405.3.7.4 Doors. Interior doors to rooms or portions of such buildings shall be self-closing fire doors in accordance with the *International Building Code.*

❖ Interior fire-resistance-rated wall construction used to separate or isolate portions of a building must have fire-resistance-rated doors in the openings in the wall. A fire-resistance-rated interior door consists of the door, door frame, latches, locks, hinges, closers and any other hardware that is necessary for the door to function. The door is to be self-closing to ensure that the door is not open during a fire. Section 714 of the IBC contains the criteria for the fire-resistance-rated door. Any obstruction to the operation or damage to this door that prevents the door from forming a barrier with the fire-resistance-rated wall must be removed or repaired.

3405.3.7.5 Open systems. Use, dispensing and mixing of flammable and combustible liquids in open systems shall be in accordance with Sections 3405.3.7.5.1 through 3405.3.7.5.3.

❖ Rooms and buildings where flammable and combustible liquids are dispensed, mixed or used in open containers or open systems must be designed and constructed to control the potential fire hazard created by the liquid and its vapor.

3405.3.7.5.1 Ventilation. Continuous mechanical ventilation shall be provided at a rate of not less than 1 cubic foot per minute per square foot [0.00508 $m^3/(s \times m^2)$] of floor area over the design area. Provisions shall be made for introduction of makeup air in such a manner to include all floor areas or pits where vapors can collect. Local or spot ventilation shall be provided when needed to prevent the accumulation of hazardous vapors. Ventilation system design shall comply with the *International Building Code* and *International Mechanical Code*.

Exception: Where natural ventilation can be shown to be effective for the materials used, dispensed or mixed.

❖ The requirements of this section for a mechanical ventilation system are identical to those in Section 2704.3. The mechanical ventilation system must remove the vapors to prevent accumulation in concentrations in the flammable range for the flammable and combustible liquids being used.

The exception allows use of natural ventilation complying with Section 2704.3 that can be demonstrated to be effective for the flammable and combustible liquids being used.

3405.3.7.5.2 Explosion control. Explosion control shall be provided in accordance with Section 911.

❖ Refer to Section 911 for deflagration venting to direct the force of an explosion out of the structure into an unoccupied area. That section also references NFPA 69, which may require monitoring of gases and other methods to suppress factors affecting an explosion.

3405.3.7.5.3 Spill control and secondary containment. Spill control shall be provided in accordance with Section 3403.4 where Class I, II or IIIA liquids are dispensed into containers exceeding a 1.3-gallon (5 L) capacity or mixed or used in open containers or systems exceeding a 5.3-gallon (20 L) capacity. Spill control and secondary containment shall be provided in accordance with Section 3403.4 when the capacity of an individual container exceeds 55 gallons (208 L) or the aggregate capacity of multiple containers or tanks exceeds 100 gallons (378.5 L).

❖ Spill control complying with Section 3403.4 must be built into a room or a building where Class I, II or IIIA liquids are used in open containers or open systems that dispense the liquid into containers greater than 1.3 gallons (5 L), mixed or used in quantities greater than 5.3 gallons (20 L), in containers greater than 55 gallons (208 L) or exceed an aggregate quantity of 100 gallons (378.5 L). This section references Section 2704.2, which discusses the use of liquid-tight floors, curbs, dikes and drainage systems to divert the liquid to a location where it can be contained and safely handled.

3405.3.7.6 Closed systems. Use or mixing of flammable or combustible liquids in closed systems shall be in accordance with Sections 3405.3.7.6.1 through 3405.3.7.6.3.

❖ Rooms and buildings where flammable and combustible liquids are dispensed, mixed or used in closed systems must be designed and constructed to control the potential fire hazard created by the liquid and its vapor.

3405.3.7.6.1 Ventilation. Closed systems designed to be opened as part of normal operations shall be provided with ventilation in accordance with Section 3405.3.7.5.1.

❖ The requirements of this section for a mechanical ventilation system are identical to those in Section 2704.3. The mechanical ventilation system must remove the vapors to prevent accumulation in concentrations in the flammable range for the flammable and combustible liquids being used. Natural ventilation complying with Section 2704.3 that can be demonstrated to be effective for the flammable and combustible liquids being used is an acceptable alternative.

3405.3.7.6.2 Explosion control. Explosion control shall be provided when an explosive environment can occur as a result of the mixing or use process. Explosion control shall be designed in accordance with Section 911.

Exception: When process vessels are designed to contain fully the worst-case explosion anticipated within the vessel under process conditions considering the most likely failure.

❖ Refer to Section 911 for deflagration venting to direct the force of an explosion out of the structure into an unoccupied area. This section also references NFPA 69, which may require monitoring of gases and other methods to suppress factors affecting an explosion.

The exception recognizes that a closed system can be designed to absorb the forces from an internal explosion. If the closed system is designed to resist the worst case explosion, the room or building does not have to be designed to comply with Section 911.

3405.3.7.6.3 Spill control and secondary containment. Spill control shall be provided in accordance with Section 3403.4 when flammable or combustible liquids are dispensed into containers exceeding a 1.3-gallon (5 L) capacity or mixed or used in open containers or systems exceeding a 5.3-gallon (20 L) capacity. Spill control and secondary containment shall be provided in accordance with Section 3403.4 when the capacity of an individual container exceeds 55 gallons (208 L) or the aggregate capacity of multiple containers or tanks exceeds 1,000 gallons (3785 L).

❖ Spill control complying with Section 3403.4 must be designed into a room or a building where flammable or combustible liquids are used in closed systems that dispense the liquid into containers greater than 1.3 gallons (5 L), mixed or used in quantities greater than 5.3 gallons (20 L), in containers greater than 55 gallons (208 L) or exceed an aggregate quantity of 1,000 gallons (3785 L). This section references Section 2704.2, which discusses the use of liquid-tight floors, curbs, dikes and drainage systems to divert the liquid to a location where it can be contained and safely handled. This is identical to the requirement for an open system except that the aggregate quantity for a closed system is 1,000 gallons (3785 L) and only 100 gallons (378.5 L) for an open system.

3405.3.8 Use, dispensing and handling outside of buildings. Outside use, dispensing and handling shall be in accordance with Sections 3405.3.8.1 through 3405.3.8.3.

Dispensing of liquids into motor vehicle fuel tanks at motor fuel-dispensing facilities shall be in accordance with Chapter 22.

❖ Dispensing and handling of flammable and combustible liquids outside of a building requires that the location be protected from ignition sources by being separated from structures, property lines, streets, etc. Spill control and drainage control must be designed into the outside location to prevent flammable and combustible liquids from affecting other areas

Dispensing into motor vehicle fuel tanks is covered in Chapter 22.

3405.3.8.1 Spill control and drainage control. Outside use, dispensing and handling areas shall be provided with spill control as set forth in Section 3403.4.

❖ Outside use, dispensing and handling areas must have spill control according to Section 3403.4. Section 2704.2 discusses the use of liquid-tight floors, curbs, dikes and drainage systems to divert the liquid to a location where it can be contained and safely handled.

3405.3.8.2 Location on property. Dispensing activities which exceed the quantities set forth in Table 3405.3.8.2 shall not be conducted within 15 feet (4572 mm) of buildings or combustible materials or within 25 feet (7620 mm) of building openings, lot lines, public streets, public alleys or public ways. Dispensing activities that exceed the quantities set forth in Table 3405.3.8.2 shall not be conducted within 15 feet (4572 mm) of storage of Class I, II or III liquids unless such liquids are stored in tanks which are listed and labeled as 2-hour protected tank assemblies in accordance with UL 2085.

Exceptions:

1. The requirements shall not apply to areas where only the following are dispensed: Class III liquids; liquids that are heavier than water; water-miscible liquids; and liquids with viscosities greater than 10,000 centipoise (cp).
2. Flammable and combustible liquid dispensing in refineries, chemical plants, process facilities, gas and crude oil production facilities and oil blending and packaging facilities, terminals and bulk plants.

❖ The dispensing of flammable and combustible liquids can generate vapors. Flammable and combustible liquid exceeding the quantities in Table 3405.3.8.2 must be separated from combustibles, buildings, other property, public property, etc. The separation allows the vapor to disperse to a vapor-air mixture below its LFL. The clearance isolates the dispensing area from ignition sources and other fuel sources. When the quantity of flammable and combustible liquid exceeds the quantity in Table 3405.3.8.2, the storage tank for the flammable or combustible liquid must be protected by separation from the dispensing area. The clearance between the storage tank and the dispensing area can be reduced if the storage tank is a 2-hour protected tank in accordance with UL 2085.

Exception 1 is based on the viscosity of the hazardous liquid. Class III liquids liquids heavier than water, water-miscible liquids and liquids with a viscosity greater than 10,000 cp need not be isolated. The isolation is not necessary for liquids with higher flash points, liquids that can be mixed with water, liquids that can be covered by a barrier of water or liquids that are slow moving. These liquids can be controlled by other procedures rather than mandate a clearance.

Exception 2 exempts facilities that are designed specifically for processing, manufacturing, storage or transfer of large quantities of flammable and combustible liquids.

TABLE 3405.3.8.2. See page 34-59.

❖ The quantities of flammable and combustible liquids dispensed outdoors are limited when the dispensing is near a building. This table contains the maximum volume of flammable and combustible liquids that can be dispensed in an outdoor control area. An outdoor control area is defined as the area within 15 feet (4572 mm) of a building or combustible materials or a Class I, II or III liquid storage tank unless the storage tanks have 2-hour fire protection, or within 25 feet (7620 mm) of a building opening, a property line, a street, an alley or a public way. This quantity limit protects the vapor of flammable or combustible liquids from ignition sources. It also protects building and adjacent property from a fire involving flammable or combustible liquids.

3405.3.8.3 Location of processing vessels. Processing vessels shall be located with respect to distances to lot lines which can be built on in accordance with Table 3405.3.4(1).

Exception: In refineries and distilleries.

❖ Processing vessels outside of a building must be treated the same as a processing vessel inside a building. The exception covers refineries and distilleries, which are designed specifically for the processing of large quantities of hazardous liquids.

3405.4 Solvent distillation units. Solvent distillation units shall comply with Sections 3405.4.1 through 3405.4.9.

❖ Distillation units generate vapors by heating the flammable and combustible liquids. This process must be controlled to prevent the vapor-air mixture from reaching the LFL or an ignition source from coming into contact with the mixture.

3405.4.1 Unit with a capacity of 60 gallons or less. Solvent distillation units used to recycle Class I, II or IIIA liquids having a distillation chamber capacity of 60 gallons (227 L) or less shall be listed, labeled and installed in accordance with Section 3405.4 and UL 2208.

Exceptions:

1. Solvent distillation units installed in dry cleaning plants in accordance with Chapter 12.
2. Solvent distillation units used in continuous through-put industrial processes where the source of heat is remotely supplied using steam, hot water, oil or other heat transfer fluids, the temperature of which is below the auto-ignition point of the solvent.
3. Solvent distillation units listed for and used in laboratories.
4. Approved research, testing and experimental processes.

❖ Distillation equipment of this size can be manufactured so that the design, construction and operation can be tested for safety. These units are tested, listed and labeled according to this section and UL 2208. There are four exceptions to these requirements.

Exception 1 notes that the recovery of dry cleaning solution in a dry cleaning plant is covered in Chapter 12.

Exception 2 exempts solvent distillation units in continuous operation that do not use heat from fluid transfer and the heat transfer fluid temperature is below the autoignition point.

Exception 3 exempts solvent distillation units in and used in a laboratory, which are listed under a different standard.

Exception 4 exempts solvent distillation units that are approved for research or experimental processes from complying with this section or UL 2208.

3405.4.2 Units with a capacity exceeding 60 gallons. Solvent distillation units used to recycle Class I, II or IIIA liquids, having a distillation chamber capacity exceeding 60 gallons (227 L) shall be used in locations that comply with the use and mixing requirements of Section 3405 and other applicable provisions in this chapter.

❖ Solvent distillation units must be treated as if the unit was for mixing and blending. The operation involves vapor concentrations from flammable or combustible liquid that may be in the flammable range.

3405.4.3 Prohibited processing. Class I, II and IIIA liquids also classified as unstable (reactive) shall not be processed in solvent distillation units.

Exception: Appliances listed for the distillation of unstable (reactive) solvents.

❖ Flammable and combustible liquids that are also unstable are prohibited from being processed by a solvent distillation unit. An unstable liquid can self-react when exposed to heat. By removing the heat source, the code is reducing the potential danger.

The exception recognizes that some equipment is designed for the safe distillation of unstable flammable and combustible liquids and is listed for that function.

3405.4.4 Labeling. A permanent label shall be affixed to the unit by the manufacturer. The label shall indicate the capacity of the distillation chamber, and the distance the unit shall be placed away from sources of ignition. The label shall indicate the products for which the unit has been listed for use or refer to the instruction manual for a list of the products.

❖ The distillation unit is to be permanently labeled to provide information for its safe operation. The minimum information is capacity, clearance from ignition sources and information on products that unit is designed to process.

TABLE 3405.3.8.2
MAXIMUM ALLOWABLE QUANTITIES FOR DISPENSING OF FLAMMABLE AND COMBUSTIBLE LIQUIDS IN OUTDOOR CONTROL AREAS[a,b]

CLASS OF LIQUID	QUANTITY (gallons)
Flammable	
Class IA	10
Class IB	15
Class IC	20
Combination Class IA, IB and IC	30[c]
Combustible	
Class II	30
Class IIIA	80
Class IIIB	3,300

For SI: 1 gallon = 3.785 L.

a. For definition of "Outdoor Control Area," see Section 2702.1.

b. The fire code official is authorized to impose special conditions regarding locations, types of containers, dispensing units, fire control measures and other factors involving fire safety.

c. Containing not more than the maximum allowable quantity per control area of each individual class.

3405.4.5 Manufacturer's instruction manual. An instruction manual shall be provided. The manual shall be readily available for the user and the fire code official. The manual shall include installation, use and servicing instructions. It shall identify the liquids for which the unit has been listed for distillation purposes along with each liquid's flash point and auto-ignition temperature. For units with adjustable controls, the manual shall include directions for setting the heater temperature for each liquid to be instilled.

❖ Manufacturers of listed and labeled equipment normally prepare an instruction manual for their equipment. The instruction manual is to be available for use by installation personnel and the fire code official. The instruction manual should include not only installation instructions, but also operating procedures and the products that it can distill.

3405.4.6 Location. Solvent distillation units shall be used in locations in accordance with the listing. Solvent distillation units shall not be used in basements.

❖ The label and instruction manual will include information on acceptable locations, separation from ignition sources, light sources, ventilation, etc. Regardless of the instruction manual, the distillation unit cannot be located in a basement. Vapors that may escape during distillation will normally be heavier than air and settle to the lower levels, making their removal from a basement location difficult.

3405.4.7 Storage of liquids. Distilled liquids and liquids awaiting distillation shall be stored in accordance with Section 3404.

❖ Distilled liquids governed by this section will always present storage hazards. It is therefore important that all the liquids, both before and after distillation, be handled as prescribed in Section 3404.

3405.4.8 Storage of residues. Hazardous residue from the distillation process shall be stored in accordance with Section 3404 and Chapter 27.

❖ The residue from the distillation process may be a flammable or combustible material, or some other hazardous material. The instruction manual must include information on the products that the distillation unit is designed to process and identify the residue. If the residue is hazardous, it must be handled according to Section 3404 and Chapter 27.

3405.4.9 Portable fire extinguishers. Approved portable fire extinguishers shall be provided in accordance with Section 906. At least one portable fire extinguisher having a rating of not less than 40-B shall be located not less than 10 feet (3048 mm) or more than 30 feet (9144 mm) from any solvent distillation unit.

❖ Portable fire extinguishers must be located in clear view and within 30 feet (9144 mm) of the distillation unit. The 40B portable fire extinguisher is for the control of small flammable liquid fires.

SECTION 3406 SPECIAL OPERATIONS

3406.1 General. This section shall cover the provisions for special operations which include, but are not limited to, storage, use, dispensing, mixing or handling of flammable and combustible liquids. The following special operations shall be in accordance with Sections 3401, 3403, 3404 and 3405, except as provided in Section 3406.

1. Storage and dispensing of flammable and combustible liquids on farms and construction sites.
2. Well drilling and operating.
3. Bulk plants or terminals
4. Bulk transfer and process transfer operations utilizing tank vehicles and tank cars.
5. Tank vehicles and tank vehicle operation.
6. Refineries.
7. Vapor recovery and vapor-processing systems.

❖ This section covers uses of flammable and combustible liquids for specific occupancies. The requirements of Sections 3401, 3403, 3404 and 3405 apply to these occupancies except as specifically directed by Section 3406.

3406.2 Storage and dispensing of flammable and combustible liquids on farms and construction sites. Permanent and temporary storage and dispensing of Class I and II liquids for private use on farms and rural areas and at construction sites, earth-moving projects, gravel pits or borrow pits shall be in accordance with Sections 3406.2.1 through 3406.2.8.1.

Exception: Storage and use of fuel oil and containers connected with oil-burning equipment regulated by Section 603 and the *International Mechanical Code*.

❖ These provisions cover both temporary and permanent storage and dispensing of Class I and II liquids, primarily in outdoor locations.

The exception covers storage and use of fuel oil for building service equipment, which is governed by other sections of the code as well as other codes.

3406.2.1 Combustibles and open flames near tanks. Storage areas shall be kept free from weeds and extraneous combustible material. Open flames and smoking are prohibited in flammable or combustible liquid storage areas.

❖ The area around storage tanks is to be free of combustibles and ignition sources.

3406.2.2 Marking of tanks and containers. Tanks and containers for the storage of liquids above ground shall be conspicuously marked with the name of the product which they contain and the words: FLAMMABLE—KEEP FIRE AND FLAME AWAY. Tanks shall bear the additional marking: KEEP 50 FEET FROM BUILDINGS.

❖ When hazardous liquids are stored in above-ground storage tanks, warning signs to keep ignition sources from the tanks and safety information are to be placed on the tanks.

3406.2.3 Containers for storage and use. Metal containers used for storage of Class I or II liquids shall be in accordance with DOTn requirements or shall be of an approved design.

Discharge devices shall be of a type that do not develop an internal pressure on the container. Pumping devices or approved self-closing faucets used for dispensing liquids shall not leak and shall be well-maintained. Individual containers shall not be interconnected and shall be kept closed when not in use.

Containers stored outside of buildings shall be in accordance with Section 3404 and the *International Building Code*.

❖ Individual containers for the storage of Class I and II liquids are to be DOT approved or approved for use with these liquids. To prevent leaks and spills, the discharge devices must be designed for use with Class I and II liquids. The storage of individual containers outside of a building is to be treated as outside storage under Section 3404.

3406.2.4 Permanent and temporary tanks. The capacity of permanent above-ground tanks containing Class I or II liquids shall not exceed 1,100 gallons (4164 L). The capacity of temporary above-ground tanks containing Class I or II liquids shall not exceed 10,000 gallons (37 854 L). Tanks shall be of the single-compartment design.

Exception: Permanent above-ground tanks of greater capacity which meet the requirements of Section 3404.2.

❖ Above-ground single-compartment storage tanks must be used for permanent or temporary storage of Class I and II liquids. Permanent storage is limited to 1,100 gallons (3785 L) where temporary storage is limited to 10,000 gallons (37 854 L).

The exception covers permanent storage of quantities larger than 1,100 gallons (3785 L). Requirements for permanent storage of larger quantities are contained in Section 3404.2.

3406.2.4.1 Fill-opening security. Fill openings shall be equipped with a locking closure device. Fill openings shall be separate from vent openings.

❖ Security to prevent the dispensing of flammable and combustible liquids is provided by a locking device on the fill opening. Fill openings are separated from vent openings so that locking does not affect the operation of the vent openings.

3406.2.4.2 Vents. Tanks shall be provided with a method of normal and emergency venting. Normal vents shall also be in accordance with Section 3404.2.7.3.

Emergency vents shall be in accordance with Section 3404.2.7.4. Emergency vents shall be arranged to discharge in a manner which prevents localized overheating or flame impingement on any part of the tank in the event that vapors from such vents are ignited.

❖ The need for normal venting and emergency venting is no different than required on other storage tanks. The requirements of Sections 3402.7.3 and 3402.7.4 apply to venting the storage tanks.

3406.2.4.3 Location. Tanks containing Class I or II liquids shall be kept outside and at least 50 feet (15 240 mm) from buildings and combustible storage. Additional distance shall be provided when necessary to ensure that vehicles, equipment and containers being filled directly from such tanks will not be less than 50 feet (15 240 mm) from structures, haystacks or other combustible storage.

❖ The clearance required between above-ground storage tanks containing Class I and II liquids, combustible storage and other structures is 50 feet (15 240 mm). Vehicles, equipment or containers using the storage tank are to have a clearance of 50 feet (15 240 mm) from these same ignition sources and fuel sources.

3406.2.4.4 Locations where above-ground tanks are prohibited. The storage of Class I and II liquids in above-ground tanks is prohibited within the limits established by law as the limits of districts in which such storage is prohibited (see Section 3 of the Sample Ordinance for Adoption of the *International Fire Code* on page v).

❖ This section enables the adopting jurisdiction to prohibit the installation of above-ground tanks in certain geographic areas of the jurisdiction by enumerating them in the adopting legislation. The code includes a sample adopting ordinance that contains a blank space for the jurisdiction to fill in describing the particular areas where above-ground tank installations are to be prohibited.

3406.2.5 Type of tank. Tanks shall be provided with top openings only or shall be elevated for gravity discharge.

❖ Above-ground storage tanks are to be top-opening or gravity-discharge tanks.

3406.2.5.1 Tanks with top openings only. Tanks with top openings shall be mounted as follows:

1. On well-constructed metal legs connected to shoes or runners designed so that the tank is stabilized and the entire tank and its supports can be moved as a unit; or
2. For stationary tanks, on a stable base of timbers or blocks approximately 6 inches (152 mm) in height which prevents the tank from contacting the ground.

❖ Above-ground storage tanks with top openings must be on stable supports of metal or wood. The above-ground storage tank shell is not to be in contact with the ground. Stability is extremely important because movement of the tank could damage the tank shell or connected piping, creating a leak.

3406.2.5.1.1 Pumps and fittings. Tanks with top openings only shall be equipped with a tightly and permanently attached, approved pumping device having an approved hose of sufficient length for filling vehicles, equipment or containers to be served from the tank. Either the pump or the hose shall be equipped with a padlock to its hanger to prevent tampering. An effective antisiphoning device shall be included in the pump discharge unless a self-closing nozzle is provided. Siphons or internal pressure discharge devices shall not be used.

❖ Above-ground storage tanks with top openings must be equipped with an approved permanently attached tight-fitting pump and hose. Pumps must be secured against unauthorized use and designed to prevent siphoning through the pump.

3406.2.5.2 Tanks for gravity discharge. Tanks with a connection in the bottom or the end for gravity-dispensing liquids shall be mounted and equipped as follows:

1. Supports to elevate the tank for gravity discharge shall be designed to carry all required loads and provide stability.
2. Bottom or end openings for gravity discharge shall be equipped with a valve located adjacent to the tank shell which will close automatically in the event of fire through the operation of an effective heat-activated releasing device. Where this valve cannot be operated manually, it shall be supplemented by a second, manually operated valve.

The gravity discharge outlet shall be provided with an approved hose equipped with a self-closing valve at the discharge end of a type that can be padlocked to its hanger.

❖ Above-ground storage tanks for gravity discharge must be elevated on stable supports capable of supporting all required loads. Valves for bottom and end openings must be located adjacent to the tank shell. These valves are to be heat-activated valves that must close in case of a fire. The valve is to prevent the tank from dispensing its content through one of the tank openings. If the heat-activated valve cannot be manually operated, a second manually operable valve must be installed. The hose must be secured to prevent unauthorized operation.

3406.2.6 Spill control drainage control and diking. Indoor storage and dispensing areas shall be provided with spill control and drainage control as set forth in Section 3403.4. Outdoor storage areas shall be provided with drainage control or diking as set forth in Section 3404.2.10.

❖ Indoor above-ground storage tanks must be equipped to control leaks and spills to prevent the spread of liquid and vapors. Section 2704.2 discusses the use of liquid-tight floors, curbs, dikes and drainage systems to divert the liquid to a location where it can be contained and safely handled. Section 2704.3 discusses mechanical and natural exhaust systems. The exhaust system is to remove the vapors to prevent them from accumulating in concentrations in the flammable range of the vapor.

Outdoor above-ground storage tanks must use dikes and drainage to control leaks or spills from a tank. The flammable and combustible liquids must be collected in a manner that will not endanger other tanks, properties or waterways.

The fire code official has the authority to consider installations where the use of drainage and dikes may not be necessary, such as listed secondary containment tanks in lieu of drainage control or dikes. The secondary containment system in these tanks is considered equivalent to drainage and diking.

3406.2.7 Portable fire extinguishers. Portable fire extinguishers with a minimum rating of 20-B:C and complying with Section 906 shall be provided where required by the fire code official.

❖ Portable fire extinguishers must be in clear view and within 50 feet (15 240 mm) of the storage tank. The 20BC portable fire extinguisher is for the control of small flammable liquid fires and electrical fires.

3406.2.8 Dispensing from tank vehicles. Where approved, liquids used as fuels are allowed to be transferred from tank vehicles into the tanks of motor vehicles or special equipment, provided:

1. The tank vehicle's specific function is that of supplying fuel to motor vehicle fuel tanks.
2. The dispensing hose does not exceed 100 feet (30 480 mm) in length.
3. The dispensing nozzle is an approved type.
4. The dispensing hose is properly placed on an approved reel or in a compartment provided before the tank vehicle is moved.
5. Signs prohibiting smoking or open flames within 25 feet (7620 mm) of the vehicle or the point of refueling are prominently posted on the tank vehicle.
6. Electrical devices and wiring in areas where fuel dispensing is conducted are in accordance with the ICC *Electrical Code.*
7. Tank vehicle-dispensing equipment is operated only by designated personnel who are trained to handle and dispense motor fuels.
8. Provisions are made for controlling and mitigating unauthorized discharges.

❖ The dispensing of fuel at a farm or construction site may be from a tanker to a vehicle or equipment. Dispensing fuel using a tanker requires the tanker to be designed and equipped specifically for fueling other vehicles and equipment. The dispensing line cannot exceed 100 feet (30 480 mm). The tanker operator is responsible for the dispensing line and the tanker, so these two items need to be in the same area. The dispensing cannot be done around an ignition source, such as smoking, or electrical equipment that is not classified for use in hazardous locations.

3406.2.8.1 Location. Dispensing from tank vehicles shall be conducted at least 50 feet (15 240 mm) from structures or combustible storage.

❖ Fuel must be dispensed at least 50 feet (15 240 mm) from a structure or combustible storage to prevent the accumulation of vapor-air mixtures, control ignition sources and minimize the potential loss of property.

3406.3 Well drilling and operating. Wells for oil and natural gas shall be drilled and operated in accordance with Sections 3406.3.1 through 3406.3.8.

❖ The drilling and operation of a well may allow the escape of vapors from petroleum products. Well drilling and operation are essential for the extraction of petroleum products. The requirements are to prevent the vapor-air mixture from reaching the LFL, to prevent ignition sources from being in the immediate vicinity and other functions to control the hazard.

3406.3.1 Location. The location of wells shall comply with Sections 3406.3.1.1 through 3406.3.1.3.2.

❖ The location of wells is regulated to permit the dilution of any vapor-air mixtures that may escape the well head or facilities, prevent ignition sources from being in close proximity and other safety and security to control the hazards created by the well drilling and production operation.

3406.3.1.1 Storage tanks and sources of ignition. Storage tanks or boilers, fired heaters, open-flame devices or other sources of ignition shall not be located within 25 feet (7620 mm) of well heads. Smoking is prohibited at wells or tank locations except as designated and in approved posted areas.

Exception: Engines used in the drilling, production and serving of wells.

❖ Sources of ignition from equipment and open flame must be kept a minimum of 25 feet (7620 mm) from a well head. The distance provides an area for any vapors leaking from the well head to dilute to a vapor-air mixture below the LFL.

Only equipment necessary for the drilling, installation and production of the well is permitted. These engines are designed for this use.

3406.3.1.2 Streets and railways. Wells shall not be drilled within 75 feet (22 860 mm) of any dedicated public street, highway or nearest rail of an operating railway.

❖ Sources of ignition from vehicle and rail traffic are to be kept a minimum of 75 feet (22 860 mm) from a well head. Vehicle traffic can produce sources of ignition from either the vehicle or the passengers. The rail traffic can produce sources of ignition by engine, brakes or personnel. The distance provides an area for any vapors leaking from the well head to dilute to a vapor-air mixture below the LFL.

3406.3.1.3 Buildings. Wells shall not be drilled within 100 feet (30 480 mm) of buildings not necessary to the operation of the well.

❖ Sources of ignition from buildings are to be kept a minimum of 100 feet (30 480 mm) from a well head. Building equipment and occupancy can become sources of ignition. The building also represents a potential for loss of life and property. The distance provides an area for any vapors leaking from the well head to dilute to a vapor-air mixture below the LFL.

The separation distance also protects buildings and their occupants from hazards associated with the well and its operation.

A building necessary for operation of a well need not be remote from the well head, but it should be constructed and maintained to prevent it from becoming an ignition source.

3406.3.1.3.1 Group A, E or I buildings. Wells shall not be drilled within 300 feet (91 440 mm) of buildings with an occupancy in Group A, E or I.

❖ Sources of ignition from a Group A, E, or I occupancy must be kept a minimum of 300 feet (91 440 mm) from a well head. Building equipment and occupancy can become sources of ignition. Because these occupancies also represent a potential for loss of life and property, the clearance has been increased over that required for other occupancies. The distance provides an area for any vapors leaking from the well head to dilute to a vapor-air mixture below the LFL.

3406.3.1.3.2 Existing wells. Where wells are existing, buildings shall not be constructed within the distances set forth in Section 3406.3.1 for separation of wells or buildings.

❖ The same problems exist regardless of whether the well head is installed first or the building is constructed first. The code regulates whichever is built second. The well head clearance to the building is to be maintained.

3406.3.2 Waste control. Control of waste materials associated with wells shall comply with Sections 3406.3.2.1 and 3406.3.2.2.

❖ Liquids containing petroleum or its products are to be disposed of according to federal and state laws and any local ordinances.

3406.3.2.1 Discharge on a street or water channel. Liquids containing crude petroleum or its products shall not be discharged into or on streets, highways, drainage canals or ditches, storm drains or flood control channels.

❖ The discharge of liquids containing petroleum or its products is prohibited. The unregulated discharge of liquids containing petroleum or its products can place flammable or combustible liquids and gases into contact with ignition sources. The liquids and their vapors also can become environmental problems and can cause health concerns.

3406.3.2.2 Discharge and combustible materials on ground. The surface of the ground under, around or near wells, pumps, boilers, oil storage tanks or buildings shall be kept free from oil, waste oil, refuse or waste material.

❖ The discharge of liquids containing petroleum or its products is prohibited. The unregulated discharge of liquids containing petroleum or its products can place flammable or combustible liquids and gases into contact with ignition sources as well as becoming a source of environmental contamination and health hazards.

3406.3.3 Sumps. Sumps associated with wells shall comply with Sections 3406.3.3.1 through 3406.3.3.3.

❖ Sumps and basins may be necessary to temporarily store drilling materials that may contain petroleum products. The use of sumps and basins is regulated to control their size, service life and security.

3406.3.3.1 Maximum width. Sumps or other basins for the retention of oil or petroleum products shall not exceed 12 feet (3658 mm) in width.

❖ The maximum width of 12 feet (3658 mm) for a sump or basin keeps the surface area of the sump or basin at a size that will help in diluting the vapor-air mixture to below the LFL and that is readily accessible to fire-fighting equipment. The narrow width allows the vapors of the petroleum products to diffuse with a greater volume of air than if the sump or basin were allowed to be larger.

3406.3.3.2 Backfilling. Sumps or other basins for the retention of oil or petroleum products larger than 6 feet by 6 feet by 6 feet (1829 mm by 1829 mm by 1829 mm) shall not be maintained longer than 60 days after the cessation of drilling operations.

❖ The temporary storage of oil or petroleum products in sumps or basins larger than 6 feet by 6 feet by 6 feet (1829 mm by 1829 mm by 1829 mm) is restricted to no more than 60 days. Drilling operations may require that materials containing petroleum products be stored in a sump or basin. This open storage may allow the vapors from the petroleum products to mix with air, creating a hazard. To control the hazard, the use of a sump or basin is restricted to a maximum of 60 days. The storage of oil or petroleum products in volumes greater than the capacity of these sumps or basins needs to be equipped with safety features to control the hazard.

3406.3.3.3 Security. Sumps, diversion ditches and depressions used as sumps shall be securely fenced or covered.

❖ Security for the open storage of oil or petroleum products in sumps, basins and ditches is needed to prevent unauthorized access. Individuals may not realize the hazard presented by this open storage and accidentally introduce ignition sources.

3406.3.4 Prevention of blowouts. Protection shall be provided to control and prevent the blowout of a well. Protection equipment shall meet federal, state and other applicable jurisdiction requirements.

❖ The blowout of a well can release flammable and combustible liquids and vapors into the atmosphere. Federal and state regulations govern the protective equipment required to prevent blowouts.

3406.3.5 Storage tanks. Storage of flammable or combustible liquids in tanks shall be in accordance with Section 3404. Oil storage tanks or groups of tanks shall have posted in a conspicuous place, on or near such tank or tanks, an approved sign with the name of the owner or operator, or the lease number and the telephone number where a responsible person can be reached at any time.

❖ The storage requirements for flammable and combustible liquids at a well head are identical to those for any other storage of flammable and combustible liquids. Clearances, security, corrosion protection, drainage, diking, signing, etc., are to control ignition sources, combustibles, leaks, spills, etc.

3406.3.6 Soundproofing. Where soundproofing material is required during oil field operations, such material shall be noncombustible.

❖ The operation of drilling, pumping and dispensing equipment may create noise that needs to be reduced. Soundproofing material must not add a fuel source around the flammable or combustible liquids. Combustible soundproofing is to be used only at locations that are beyond the distances for the clearance between flammable and combustible liquids and combustibles.

3406.3.7 Signs. Well locations shall have posted in a conspicuous place on or near such tank or tanks an approved sign with the name of the owner or operator, name of the leasee or the lease number, the well number and the telephone number where a responsible person can be reached at any time. Such signs shall be maintained on the premises from the time materials are delivered for drilling purposes until the well is abandoned.

❖ Well heads may or may not have personnel on site so that it may be necessary to contact the operator/owner. The signs need to provide sufficient information so that the operator/owner can be notified of any problems with the site.

3406.3.8 Field-loading racks. Field-loading racks shall be in accordance with Section 3406.5.

❖ See the commentary for Section 3406.5.

3406.4 Bulk plants or terminals. Portions of properties where flammable and combustible liquids are received by tank vessels, pipelines, tank cars or tank vehicles and which are stored or blended in bulk for the purpose of distributing such liquids by tank vessels, pipelines, tanks cars, tank vehicles or containers shall be in accordance with Sections 3406.4.1 through 3406.4.10.4.

❖ Bulk plants and terminals are used for the blending or transfer of large volumes of flammable and combustible liquids. Besides normally involving large quantities of flammable and combustible liquids, these facilities can create numerous opportunities for leaks, spills and the escape of vapors.

3406.4.1 Building construction. Buildings shall be constructed in accordance with the *International Building Code*.

❖ These facilities are hazardous occupancies divided into Group H-2 or H-3. Sections 414 and 415 of the IBC contain guidance for determining the construction requirements for these two occupancies.

3406.4.2 Means of egress. Rooms in which liquids are stored, used or transferred by pumps shall have means of egress arranged to prevent occupants from being trapped in the event of fire.

❖ The means of egress from rooms where flammable or combustible liquids are present must be an escape route for personnel that cannot be blocked in case of a fire. This may require having more than one independent escape route or having a short travel distance to a door exiting the room. Doors exiting rooms should swing in the direction of exit travel.

3406.4.3 Heating. Rooms in which Class I liquids are stored or used shall be heated only by means not constituting a source of ignition, such as steam or hot water. Rooms containing heating appliances involving sources of ignition shall be located and arranged to prevent entry of flammable vapors.

❖ Heating units that use open flames and other ignition sources must be separated from areas where flammable or combustible liquids are present or where vapors from these liquids can migrate. The low flash point for Class I liquids requires that where Class I liquid is present, either steam or hot water must be used as the heat source.

3406.4.4 Ventilation. Ventilation shall be provided for rooms, buildings and enclosures in which Class I liquids are pumped, used or transferred. Design of ventilation systems shall consider the relatively high specific gravity of the vapors. When natural ventilation is used, adequate openings in outside walls at floor level, unobstructed except by louvers or coarse screens, shall be provided. When natural ventilation is inadequate, mechanical ventilation shall be provided in accordance with the *International Mechanical Code*.

❖ The low flash point for Class I liquids requires that where Class I liquid is present the area be ventilated to prevent the accumulation of a vapor-air mixture above the LFL. Ventilation can be either mechanical or natural. Because the vapors from Class I liquids are heavier than air, the ventilation must remove the air at the floor level. Natural ventilation must be to the outdoors.

3406.4.4.1 Basements and pits. Class I liquids shall not be stored or used within a building having a basement or pit into which flammable vapors can travel, unless such area is provided with ventilation designed to prevent the accumulation of flammable vapors therein.

❖ The vapors from Class I liquids are heavier than air, so these vapors can settle from the upper floors and accumulate in a basement or pit. In a building with a basement or pit that has any connection between the basement or pit and upper floors, the basement or pit is not to be used for the storage or mixing of Class I liquids. The vapors from Class I liquids may settle in the basement or pit through openings for mechanical systems or that are part of an egress path, creating a hazard. Unless there is a mechanical or natural ventilation system that is adequate for removing accumulated vapors from Class I liquids from the basement or pit, the building cannot be used for the storage or mixing of Class I liquids.

3406.4.4.2 Dispensing of Class I liquids. Containers of Class I liquids shall not be drawn from or filled within buildings unless a provision is made to prevent the accumulation of flammable vapors in hazardous concentrations. Where mechanical ventilation is required, it shall be kept in operation while flammable vapors could be present.

❖ Containers of Class I liquids cannot be dispensed or filled in a building unless ventilation is installed to remove accumulated vapors. To make certain that vapors do not accumulate, any mechanical ventilation system must operate continuously or whenever containers of Class I liquids are open.

3406.4.5 Storage. Storage of Class I, II and IIIA liquids in bulk plants shall be in accordance with the applicable provisions of Section 3404.

❖ The potential hazards of storing Class I, II and IIIA liquids are the same for bulk plants and terminals as for any other facility. The requirements for storage are covered in Section 3404.

3406.4.6 Overfill protection of Class I liquids. Manual and automatic systems shall be provided to prevent overfill during the transfer of Class I liquids from mainline pipelines and marine vessels in accordance with API 2350.

❖ Overfill protection must comply with API 2350. This standard is limited to above-ground tanks that receive Class I liquids from main pipelines or marine vessels. The standard accomplishes overfill protection by awareness of available tank capacity and inventory and careful monitoring and control of product movement.

3406.4.7 Wharves. This section shall apply to all wharves, piers, bulkheads and other structures over or contiguous to navigable water having a primary function of transferring liquid cargo in bulk between shore installations and tank vessels, ships, barges, lighter boats or other mobile floating craft.

Exception: Marine motor fuel-dispensing facilities in accordance with Chapter 22.

❖ This section regulates the transfer of flammable and combustible liquids in bulk quantities between a shore installation and any marine vessel or other floating craft, except for marine motor fuel-dispensing facilities, which are covered in Chapter 22.

3406.4.7.1 Transferring approvals. Handling packaged cargo of liquids, including full and empty drums, bulk fuel and stores, over a wharf during cargo transfer shall be subject to the approval of the wharf supervisor and the senior deck officer on duty.

❖ The handling and transferring of flammable and combustible liquids, including any container that contains or has contained these liquids, must be approved by the personnel responsible for the wharf. This will usually be the wharf supervisor or senior deck officer. Because these materials represent a potential hazard if mishandled, the person in authority must be aware of the potential hazard and approve the process being used.

3406.4.7.2 Transferring location. Wharves at which liquid cargoes are to be transferred in bulk quantities to or from tank vessels shall be at least 100 feet (30 480 mm) from any bridge over a navigable waterway; or from an entrance to, or superstructure of, any vehicular or railroad tunnel under a waterway. The termination of the fixed piping used for loading or unloading at a wharf shall be at least 200 feet (60 960 mm) from a bridge or from an entrance to, or superstructures of, a tunnel.

❖ Transportation facilities on a wharf must be protected by separation from public throughways. The separation distance required for vessel loading or unloading is 100 feet (30 480 mm). Fixed piping must terminate 200 feet (60 960 mm) from the same throughways.

3406.4.7.3 Superstructure and decking material. Superstructure and decking shall be designed for the intended use. Decking shall be constructed of materials that will afford the desired combination of flexibility, resistance to shock, durability, strength and fire resistance.

❖ The wharf superstructure and decking materials have to be selected to withstand normal loads, impact loads, fire exposure, environmental factors, etc.

3406.4.7.4 Tanks allowed. Tanks used exclusively for ballast water or Class II or III liquids are allowed to be installed on suitably designed wharves.

❖ Tanks other than tanks used for Class I liquids can be installed on a wharf if the wharf is designed for this purpose.

3406.4.7.5 Transferring equipment. Loading pumps capable of building up pressures in excess of the safe working pressure of cargo hose or loading arms shall be provided with bypasses, relief valves or other arrangements to protect the loading facilities against excessive pressure. Relief devices shall be tested at least annually to determine that they function satisfactorily at their set pressure.

❖ Pump systems must be designed so that excess pressure cannot be created in the piping and storage system. If excess pressure develops and is not relieved, the equipment or piping and storage system can be damaged. A damaged system could develop a leak.

3406.4.7.6 Piping, valves and fittings. Piping valves and fittings shall be in accordance with Section 3403.6 except as modified by the following:

1. Flexibility of piping shall be ensured by appropriate layout and arrangement of piping supports so that motion of the wharf structure resulting from wave action, currents, tides or the mooring of vessels will not subject the pipe to repeated excessive strain.
2. Pipe joints that depend on the friction characteristics of combustible materials or on the grooving of pipe ends for mechanical continuity of piping shall not be used.
3. Swivel joints are allowed in piping to which hoses are connected and for articulated, swivel-joint transfer systems, provided the design is such that the mechanical strength of the joint will not be impaired if the packing materials fail such as by exposure to fire.
4. Each line conveying Class I or II liquids leading to a wharf shall be provided with a readily accessible block valve located on shore near the approach to the wharf and outside of any diked area. Where more than one line is involved, the valves shall be grouped in one location.
5. Means shall be provided for easy access to cargo line valves located below the wharf deck.
6. Piping systems shall contain a sufficient number of valves to operate the system properly and to control the flow of liquid in normal operation and in the event of physical damage.
7. Piping on wharves shall be bonded and grounded where Class I and II liquids are transported. Where excessive stray currents are encountered, insulating joints shall be installed. Bonding and grounding connections on piping shall be located on the wharf side of hose riser insulating

flanges, where used, and shall be accessible for inspection.

8. Hose or articulated swivel-joint pipe connections used for cargo transfer shall be capable of accommodating the combined effects of change in draft and maximum tidal range, and mooring lines shall be kept adjusted to prevent surge of the vessel from placing stress on the cargo transfer system.
9. Hoses shall be supported to avoid kinking and damage from chafing.

❖ The requirements of Section 3403.6 apply to the piping on a wharf, with the exception of specific requirements that are unique to the design and operation of a wharf:

1. Wharves, especially floating wharves, need flexible piping that can withstand its wharf's movement.
2. Slip joint pipe connections and threaded connections are prohibited. The flexible piping required to handle the wharf's movement may move enough to cause these types of joints to loosen. A loose joint can create a leak or a spill.
3. Swivel joints are necessary for connections between wharves; between the wharf and land and between the wharf and marine vessels. These joints are to be designed to withstand loads placed on them even if the packing material used to seal the joint fails.
4. Lines carrying Class I and II liquids require a block valve on each line feeding the wharf. The block valve(s) is to be near the wharf and outside of the dike area. A block valve is used to isolate the line from the wharf.
5. Piping under a wharf used to transfer flammable and combustible liquids must be accessible. If these lines are not accessible, it may be difficult or impossible to inspect, maintain or provide fire protection to these lines.
6. Valves must be installed in the piping system so that sections of the system can be isolated for maintenance or repairs or to stop the transfer of flammable and combustible liquids in case of leaks, spills or fire.
7. Piping must be bonded to prevent static sparks or sparks from the electrical equipment.
8. Swivel joints must be protected from loads caused by movement of the wharf relative to movement of the marine vessel.
9. Hoses must be maintained to prevent damage.

3406.4.7.7 Loading and unloading. Loading or discharging shall not commence until the wharf superintendent and officer in charge of the tank vessel agree that the tank vessel is properly moored and connections are properly made.

❖ Section 3406.4.7.1 requires approval from the individual responsible for the wharf for transferring flammable and combustible liquids over the wharf. This section requires that this same individual approve the mooring of the marine vessel before transfer begins. Failure of the moorings or excessive movement of the marine vessel can damage the piping.

3406.4.7.8 Mechanical work. Mechanical work shall not be performed on the wharf during cargo transfer, except under special authorization by the fire code official based on a review of the area involved, methods to be employed and precautions necessary.

❖ Work on the wharf or transfer equipment and piping are prohibited during the transfer of flammable or combustible liquids. Work can introduce an ignition source or possible damage to the equipment or piping system.

3406.4.8 Sources of ignition. Class I, II or IIIA liquids shall not be used, drawn or dispensed where flammable vapors can reach a source of ignition. Smoking shall be prohibited except in designated locations. "No Smoking" signs complying with Section 310 shall be conspicuously posted where a hazard from flammable vapors is normally present.

❖ One of the principal safety measures for using flammable and combustible liquids is to keep sources of ignition away from the liquids.

3406.4.9 Drainage control. Loading and unloading areas shall be provided with drainage control in accordance with Section 3404.2.10.

❖ Flammable and combustible liquids that leak or spill from a tank must be controlled by dikes and drainage. The flammable and combustible liquids are to be collected in a manner that will not endanger other tanks, properties or waterways. The fire code official has the authority to consider installations where the use of drainage and dikes may not be necessary. Listed secondary containment tanks need not be surrounded by drainage control or dikes. The secondary containment system in these tanks is considered equivalent to drainage and diking.

3406.4.10 Fire protection. Fire protection shall be in accordance with Chapter 9 and Sections 3406.4.10.1 through 3406.4.10.4.

❖ Fire protection complying with Chapter 9 is required for wharves.

3406.4.10.1 Portable fire extinguishers. Portable fire extinguishers with a rating of not less than 20-B and complying with Section 906 shall be located within 75 feet (22 860 mm) of hose connections, pumps and separator tanks.

❖ Portable fire extinguishers must be located within 75 feet (22 860 mm) of hose connections, pumps and separator tanks. These locations are where a possible leak can occur and a portable fire extinguisher may be adequate in preventing a fire or controlling a small flammable liquid fire. The 20B portable fire extinguisher is for the control of a small flammable liquid fire.

3406.4.10.2 Fire hoses. Where piped water is available, ready-connected fire hose in a size appropriate for the water supply shall be provided in accordance with Section 905 so that manifolds where connections are made and broken can be reached by at least one hose stream.

❖ If the fire control system uses piped water, a fire hose of adequate size for the piped water that complies with Section 905 must be available. The fire hose must be capable of washing down manifolds in case of a leak or spill or for fighting a fire at the manifold.

3406.4.10.3 Obstruction of equipment. Material shall not be placed on wharves in such a manner that would obstruct access to fire-fighting equipment or important pipeline control valves.

❖ Access for fire equipment and to valves for isolating and controlling flammable and combustible liquids must be kept clear. Obstruction would prevent or slow any response to a leak, spill or fire.

3406.4.10.4 Fire apparatus access. Where the wharf is accessible to vehicular traffic, an unobstructed fire apparatus access road to the shore end of the wharf shall be maintained in accordance with Chapter 5.

❖ Wharves that are accessible to vehicle traffic are to be made accessible for fire department equipment. The access requirements are in Chapter 5. If vehicle access to the wharf does not exist, fire department equipment access is not required.

3406.5 Bulk transfer and process transfer operations. Bulk transfer and process transfer operations shall be approved and be in accordance with Sections 3406.5.1 through 3406.5.4.4. Motor fuel-dispensing facilities shall comply with Chapter 22.

❖ This section gives the fire code official the authority to require approval of bulk transfer and process transfer operations.

3406.5.1 General. The provisions of Sections 3406.5.1.1 through 3406.5.1.18 shall apply to bulk transfer and process transfer operations; Sections 3406.5.2 and 3406.5.2.1 shall apply to bulk transfer operations; Sections 3406.5.3 through 3406.5.3.3 shall apply to process transfer operations and Sections 3406.5.4 through 3406.5.4.4 shall apply to dispensing from tank vehicles and tank cars.

❖ Facilities that are used for bulk transfer, bulk processing and bulk dispensing to tank vehicles and tank cars are covered by these sections.

3406.5.1.1 Location. Bulk transfer and process transfer operations shall be conducted in approved locations. Tank cars shall be unloaded only on private sidings or railroad-siding facilities equipped for transferring flammable or combustible liquids. Tank vehicle and tank car transfer facilities shall be separated from buildings, above-ground tanks, combustible materials, lot lines, public streets, public alleys or public ways by a distance of 25 feet (7620 mm) for Class I liquids and 15 feet (4572 mm) for Class II and III liquids measured from the nearest position of any loading or unloading valve. Buildings for pumps or shelters for personnel shall be considered part of the transfer facility.

❖ The volume of flammable and combustible liquids being transferred and processed requires that the location be protected from ignition sources. This is accomplished by maintaining minimum clearance between the transfer and process operation and buildings, property lines, streets, etc.

3406.5.1.2 Weather protection canopies. Where weather protection canopies are provided, they shall be constructed in accordance with Section 2704.13. Weather protection canopies shall not be located within 15 feet (4572 mm) of a building or combustible material or within 25 feet (7620 mm) of building openings, lot lines, public streets, public alleys or public ways.

❖ This section makes clear that an open structure consisting of only a noncombustible roof structure is considered outdoor storage. A structure with no walls and only a noncombustible roof provides sufficient ventilation for the storage area to be treated as outdoor storage. Separation from buildings and openings is needed to prevent the vapors from flammable and combustible liquids from being exposed to an ignition source. These separation requirements also limit potential property and life losses.

3406.5.1.3 Ventilation. Ventilation shall be provided to prevent accumulation of vapors in accordance with Section 3405.3.7.5.1.

❖ The requirements of this section for a mechanical ventilation system are identical to those in Section 2704.3. The mechanical ventilation system must remove vapors to prevent accumulation in concentrations in the flammable range for the flammable and combustible liquids being used. Natural ventilation complying with Section 2704.3 that can be demonstrated to be effective for the flammable and combustible liquids being used is an acceptable alternative.

3406.5.1.4 Sources of ignition. Sources of ignition shall be controlled or eliminated in accordance with Section 2703.7.

❖ Smoking and open flames are strictly limited around flammable and combustible liquids. Section 2703.7 prohibits smoking or open flames within 25 feet (7620 mm) of outdoor storage of flammable and combustible liquids and anywhere indoors where there are flammable and combustible liquids or where vapors from flammable and combustible liquids may occur.

3406.5.1.5 Spill control and secondary containment. Areas where transfer operations are located shall be provided with spill control and secondary containment in accordance with Section 3403.4. The spill control and secondary containment system shall have a design capacity capable of containing the capacity of the largest tank compartment located in the area where transfer operations are conducted. Containment of the rainfall volume specified in Section 2704.2.2.6 is not required.

❖ Where flammable or combustible liquids are present, spills must be controlled to prevent the spread of liquid and vapors. Section 2704.2 discusses the use of liquid-tight floors, curbs, dikes and drainage systems to divert the liquid to a location where it can be contained and safely handled. Section 2704.3 discusses mechanical and natural exhaust systems for indoor facilities. The exhaust system must remove the vapors to prevent them from accumulating in concentrations in the flammable range. The diked area must have sufficient capacity to contain the largest spillage of flammable and combustible liquids that can be released from the largest tank. This would be the volume of the largest tank that extends above the top of the dike.

3406.5.1.6 Fire protection. Fire protection shall be in accordance with Section 3403.2.

❖ The requirements of Chapter 9 apply to flammable and combustible liquids. Fire protection is a principal means of preventing and controlling the spread of a fire.

3406.5.1.7 Static protection. Static protection shall be provided to prevent the accumulation of static charges during transfer operations. Bonding facilities shall be provided during the transfer through open domes where Class I liquids are transferred, or where Class II and III liquids are transferred into tank vehicles or tank cars which could contain vapors from previous cargoes of Class I liquids.

Protection shall consist of a metallic bond wire permanently electrically connected to the fill stem. The fill pipe assembly shall form a continuous electrically conductive path downstream from the point of bonding. The free end of such bond wire shall be provided with a clamp or equivalent device for convenient attachment to a metallic part in electrical contact with the cargo tank of the tank vehicle or tank car. For tank vehicles, protection shall consist of a flexible bond wire of adequate strength for the intended service and the electrical resistance shall not exceed 1 megohm. For tank cars, bonding shall be provided where the resistance of a tank car to ground through the rails is 25 ohms or greater.

Such bonding connection shall be fastened to the vehicle, car or tank before dome covers are raised and shall remain in place until filling is complete and all dome covers have been closed and secured.

Exceptions:

1. Where vehicles and cars are loaded exclusively with products not having a static-accumulating tendency, such as asphalt, cutback asphalt, most crude oils, residual oils and water-miscible liquids.
2. When Class I liquids are not handled at the transfer facility and the tank vehicles are used exclusively for Class II and III liquids.
3. Where vehicles and cars are loaded or unloaded through closed top or bottom connections whether the hose is conductive or nonconductive.

Filling through open domes into the tanks of tank vehicles or tank cars that contain vapor-air mixtures within the flammable range, or where the liquid being filled can form such a mixture, shall be by means of a downspout which extends to near the bottom of the tank.

❖ Static electricity will spark between surfaces with different electrical potential. Vehicles that are or may have been used to carry Class I liquids are to be bonded. Vehicles with open domes are to be bonded before the dome is opened. Bonding to control static electricity can be done using wires, ground cables, metal piping or other similar means that will carry current. Open dome tanks must use a downspout that extends to near the bottom of the tank. This downspout will reduce the mixing of the flammable or combustible liquid with air.

Bonding to control static electricity is not necessary where static electricity is not an issue or a static spark would not act as an ignition source. Three exceptions are given to this section:

1. Combustible liquids whose viscosity prevents the accumulation of static electricity.
2. Facilities and vehicles that do not handle Class I liquids.
3. Vehicles that use a closed system for loading and unloading flammable and combustible liquids. These closed systems prevent the vapors of flammable and combustible liquids from mixing with air to form a vapor-air mixture that can reach the LFL.

3406.5.1.8 Stray current protection. Tank car loading facilities where Class I, II or IIIA liquids are transferred through open domes shall be protected against stray currents by permanently bonding the pipe to at least one rail and to the transfer apparatus. Multiple pipes entering the transfer areas shall be permanently electrically bonded together. In areas where excessive stray currents are known to exist, all pipes entering the transfer area shall be provided with insulating sections to isolate electrically the transfer apparatus from the pipelines.

❖ Bonding of piping to the transfer apparatus and to a rail prevents any stray current from arcing. Any electrical equipment can develop stray current. Grounding of electrical equipment, dispensing piping and vehicles will prevent the accumulation of electrical potential, which will prevent the stray current from arcing and forming an ignition source.

3406.5.1.9 Top loading. When top loading a tank vehicle with Class I and II liquids without vapor control, valves used for the final control of flow shall be of the self-closing type and shall be manually held open except where automatic means are provided for shutting off the flow when the tank is full. When used, automatic shutoff systems shall be provided with a manual shutoff valve located at a safe distance from the loading nozzle to stop the flow if the automatic system fails.

When top loading a tank vehicle with vapor control, flow control shall be in accordance with Section 3406.5.1.10. Self-closing valves shall not be tied or locked in the open position.

❖ A top-loading tank vehicle for use with Class I and II flammable and combustible liquids must have either a self-closing valve, a control valve that is manually held open or an automatic shutoff control valve to prevent overfilling and spillage. The automatic shutoff control valve is to have a manual backup control valve located a safe distance from the loading nozzle.

If the dispensing system includes vapor control, the top loading is to comply with Section 3406.5.1.10.

3406.5.1.10 Bottom loading. When bottom loading a tank vehicle or tank car with or without vapor control, a positive means shall be provided for loading a predetermined quantity of liquid, together with an automatic secondary shutoff control to prevent overfill. The connecting components between the transfer equipment and the tank vehicle or tank car required to operate the secondary control shall be functionally compatible.

❖ Spillage is prevented when filling a bottom-loading tank vehicle by using a loading system that loads a predetermined quantity of flammable or combustible liquid. An automatic secondary shutoff control valve is required for a bottom-loading tank vehicle.

3406.5.1.10.1 Dry disconnect coupling. When bottom loading a tank vehicle, the coupling between the liquid loading hose or pipe and the truck piping shall be a dry disconnect coupling.

❖ To prevent spillage when disconnecting the dispensing system from the bottom-loading vehicle, the system is to use a dry connection.

3406.5.1.10.2 Venting. When bottom loading a tank vehicle or tank car that is equipped for vapor control and vapor control is not used, the tank shall be vented to the atmosphere to prevent pressurization of the tank. Such venting shall be at a height equal to or greater than the top of the cargo tank.

❖ Pressure is relieved in a bottom-loading tank during the loading process by venting the tank to the atmosphere. The tank vent is to be above the top of the upper limit to which flammable and combustible liquid can be loaded. This prevents the vent from spilling flammable or combustible liquid.

3406.5.1.10.3 Vapor-tight connection. Connections to the plant vapor control system shall be designed to prevent the escape of vapor to the atmosphere when not connected to a tank vehicle or tank car.

❖ Vapor recovery systems must be designed so that when they are not in use, vapors cannot escape the connection to the tank vehicle.

3406.5.1.10.4 Vapor-processing equipment. Vapor-processing equipment shall be separated from above-ground tanks, warehouses, other plant buildings, transfer facilities or nearest lot line of adjoining property that can be built on by a distance of at least 25 feet (7620 mm). Vapor-processing equipment shall be protected from physical damage by remote location, guardrails, curbs or fencing.

❖ Vapor recovery is the most dangerous component of flammable or combustible liquid transfer. To protect property and life, the vapor recovery equipment is to be separated from other buildings, equipment, property lines, etc. The vapor recovery equipment must be protected from physical damage either by separation or by physical barriers. Any damage to the vapor recovery equipment could create a vapor-air mixture that is flammable. The same damage could generate an ignition source.

3406.5.1.11 Switch loading. Tank vehicles or tank cars which have previously contained Class I liquids shall not be loaded with Class II or III liquids until such vehicles and all piping, pumps, hoses and meters connected thereto have been completely drained and flushed.

❖ Class I liquids and their vapors must be cleaned from a tank vehicle before loading it with Class II or III combustible liquids. The vapors from Class I can affect the response to Class II or III liquids by being the source of ignition and the initial fuel source to start a Class II or III liquid fire.

3406.5.1.12 Loading racks. Where provided, loading racks, stairs or platforms shall be constructed of noncombustible materials. Buildings for pumps or for shelter of loading personnel are allowed to be part of the loading rack. Wiring and electrical equipment located within 25 feet (7620 mm) of any portion of the loading rack shall be in accordance with Section 3403.1.1.

❖ Loading racks and buildings used to shelter pumps and loading personnel that are part of the loading rack are to be noncombustible. This reduces the fuel sources in the vicinity of the flammable and combustible liquids. Electrical equipment within 25 feet (7620 mm) of the loading racks is to be classified as a hazardous location. This removes an ignition source from the immediate vicinity.

3406.5.1.13 Transfer apparatus. Bulk and process transfer apparatus shall be of an approved type.

❖ This section gives the fire code official the authority to approve any transfer apparatus.

3406.5.1.14 Inside buildings. Tank vehicles and tank cars shall not be located inside a building while transferring Class I, II or IIIA liquids, unless approved by the fire code official.

Exception: Tank vehicles are allowed under weather protection canopies and canopies of automobile motor vehicle fuel-dispensing stations.

❖ Tank vehicles are to be loaded outside of a building unless approved by the fire code official. The potential for the accumulation of vapors in a building where the bulk loading of tank vehicles takes place produces an unacceptable hazard.

The exception allows loading tank vehicles under a canopy because a canopy structure is considered to be outdoor usage.

3406.5.1.15 Tank vehicle and tank car certification. Certification shall be maintained for tank vehicles and tank cars in accordance with DOTn 49 CFR, Parts 100-178.

❖ Tank vehicles for use with flammable and combustible liquids are regulated under DOTn 49 CFR, Parts 100-178. Tank vehicles and tank cars not having a current certification are not to be used.

3406.5.1.16 Tank vehicle and tank car stability. Tank vehicles and tank cars shall be stabilized against movement during loading and unloading in accordance with Sections 3406.5.1.16.1 through 3406.5.1.16.3.

❖ The stability of tank vehicles and tank cars is to be maintained during loading and unloading. The movement of a tank vehicle or tank car during loading or unloading could cause a spill.

3406.5.1.16.1 Tank vehicles. When the vehicle is parked for loading or unloading, the cargo trailer portion of the tank vehicle shall be secured in a manner that will prevent unintentional movement.

❖ Cargo trailers must be secured to prevent movement during loading and unloading.

3406.5.1.16.2 Chock blocks. At least two chock blocks not less than 5 inches by 5 inches by 12 inches (127 mm by 127 mm by 305 mm) in size and dished to fit the contour of the tires shall be used during transfer operations of tank vehicles.

❖ Tank vehicles are to use chock blocks that conform to the wheels to prevent movement. Movement of the tank vehicle could cause a spill.

3406.5.1.16.3 Tank cars. Brakes shall be set and the wheels shall be blocked to prevent rolling.

❖ Tank cars are to use both the tank car brakes and blocks to prevent movement. Movement of the tank car could cause a spill.

3406.5.1.17 Monitoring. Transfer operations shall be monitored by an approved monitoring system or by an attendant. When monitoring is by an attendant, the operator or other competent person shall be present at all times.

❖ To prevent spillage, the transfer operations are to be continually monitored by an individual at the site.

3406.5.1.18 Security. Transfer operations shall be surrounded by a noncombustible fence not less than 5 feet (1524 mm) in height. Tank vehicles and tank cars shall not be loaded or unloaded unless such vehicles are entirely within the fenced area.

Exceptions:

1. Motor fuel-dispensing facilities complying with Chapter 22.
2. Installations where adequate public safety exists because of isolation, natural barriers or other factors as determined appropriate by the fire code official.
3. Facilities or properties that are entirely enclosed or protected from entry.

❖ Transfer operations must take place in a fenced-in area to prevent access by unauthorized personnel and the possibility of introduction of ignition sources around the transfer operation.

The three exceptions cover special sets of circumstances:

1. Motor fuel-dispensing facilities that comply with Chapter 22.
2. Installation approved by the fire code official that provides security that complies with the intent of the code.
3. Facilities or properties that are completely enclosed or protected from entry by unauthorized persons.

3406.5.2 Bulk transfer. Bulk transfer shall be in accordance with Sections 3406.5.1 and 3406.5.2.1.

❖ The operation of a motor vehicle is prohibited during bulk transfer to prevent the presence of an ignition source.

3406.5.2.1 Vehicle motor. Motors of tank vehicles or tank cars shall be shut off during the making and breaking of hose connections and during the unloading operation.

Exception: Where unloading is performed with a pump deriving its power from the tank vehicle motor.

❖ The engine of a motor vehicle is a source of ignition. This source of ignition is to be turned off during transfer of flammable or combustible liquids.

The exception allows the engine of a motor vehicle to operate during transfer of flammable or combustible liquids only if the pump to move the flammable or combustible liquids is powered by the vehicle engine.

3406.5.3 Process transfer. Process transfer shall be in accordance with Section 3406.5.1 and Sections 3406.5.3.1 through 3406.5.3.3.

❖ Safety features are installed in the processing system to protect the system from failure. The failure may be caused by operational procedures or equipment. The process transfer is to be designed and operated to prevent leaks, spills, overpressure buildup, siphonage, accumulation of vapors, exposure to ignition sources, etc.

3406.5.3.1 Piping, valves, hoses and fittings. Piping, valves, hoses and fittings which are not a part of the tank vehicle or tank car shall be in accordance with Section 3403.6. Caps or plugs which prevent leakage or spillage shall be provided at all points of connection to transfer piping.

❖ Piping complying with Section 3403.6 provides protection against leaks and overpressures that may create leaks at joints or rupture the pipes. To prevent any residue from leaking or spilling from the piping system, caps or plugs are to be available for all openings.

3406.5.3.1.1 Shutoff valves. Approved automatically or manually activated shutoff valves shall be provided where the transfer hose connects to the process piping, and on both sides of any exterior fire-resistance-rated wall through which the piping passes. Manual shutoff valves shall be arranged so that they are accessible from grade. Valves shall not be locked in the open position.

❖ Shutoff valves must be installed to isolate portions of the piping system. The valves must be located where the flammable or combustible liquids connect to the process equipment. Valves are to be located on both sides of exterior fire-resistance rated walls. Valves are not to be locked in the open position.

3406.5.3.1.2 Hydrostatic relief. Hydrostatic pressure-limiting or relief devices shall be provided where pressure buildup in trapped sections of the system could exceed the design pressure of the components of the system.

Devices shall relieve to other portions of the system or to another approved location.

❖ Pressure relief valves must be installed to prevent damage to the system by releasing the pressure either into another portion of the system or to an approved location. This prevents any vapors from the flammable and combustible liquids from being released into the atmosphere of the building or facility.

3406.5.3.1.3 Antisiphon valves. Antisiphon valves shall be provided when the system design would allow siphonage.

❖ To prevent leakage or spillage by siphoning flammable or combustible liquids from one portion of the process system to another, an approved antisiphon device must be installed.

3406.5.3.2 Vents. Normal and emergency vents shall be maintained operable at all times.

❖ Vents are installed to help prevent overpressurization of tanks and transfer piping. If these vents fail to operate as designed, serious damage to the system could result.

3406.5.3.3 Motive power. Motors of tank vehicles or tank cars shall be shut off during the making and breaking of hose connections and during the unloading operation.

Exception: When unloading is performed with a pump deriving its power from the tank vehicle motor.

❖ Because there is always a possibility that vapors or minor liquid spillage could present a fire hazard, it is important to remove all possible sources of ignition from the vicinity. Requiring internal combustion engines to be shut off before connections are made or broken removes an ignition source from the vicinity.

The exception acknowledges that some pumps are powered by the engine of the tank vehicle involved in the liquid transfer operation.

3406.5.4 Dispensing from tank vehicles and tank cars. Dispensing from tank vehicles and tank cars into the fuel tanks of motor vehicles shall be prohibited unless allowed by and conducted in accordance with Sections 3406.5.4.1 through 3406.5.4.5.

❖ The dispensing of flammable or combustible liquids from a tank vehicle or tank car into the fuel tank of a motor vehicle is covered by this section.

3406.5.4.1 Marine craft and special equipment. Liquids intended for use as motor fuels are allowed to be transferred from tank vehicles into the fuel tanks of marine craft and special equipment when approved by the fire code official, and when:

1. The tank vehicle's specific function is that of supplying fuel to fuel tanks.
2. The operation is not performed where the public has access or where there is unusual exposure to life and property.
3. The dispensing line does not exceed 50 feet (15 240 mm) in length.
4. The dispensing nozzle is approved.

❖ Transfer of motor fuel into marine craft and special equipment is covered by this section.

The tank vehicle or tank car must be designed and equipped specifically for fueling other vehicles and equipment and the dispensing equipment must be approved for this use. The dispensing line cannot be more than 50 feet (15 240 mm) long. The dispensing must be done in a controlled area where there is minimal exposure to life or property.

3406.5.4.2 Emergency refueling. When approved by the fire code official, dispensing of motor vehicle fuel from tank vehicles into the fuel tanks of motor vehicles is allowed during emergencies. Dispensing from tank vehicles shall be in accordance with Sections 3406.2.8 and 3406.6.

❖ Emergency refueling of vehicles is permitted only when approved by the fire code official.

3406.5.4.3 Aircraft fueling. Transfer of liquids from tank vehicles to the fuel tanks of aircraft shall be in accordance with Chapter 11.

❖ Chapter 11 covers the requirements for aircraft fueling.

3406.5.4.4 Fueling of vehicles at farms, construction sites and similar areas. Transfer of liquid from tank vehicles to motor vehicles for private use on farms and rural areas and at construction sites, earth-moving projects, gravel pits and borrow pits is allowed in accordance with Section 3406.2.8.

❖ Fueling of vehicles at farms, construction sites and similar areas is covered by Section 3406.2.8. The dispensing of fuel at a farm or construction site may be from a tanker to a vehicle or equipment. Dispensing using a tanker requires that the tanker be designed and equipped specifically for fueling other vehicles and equipment. The dispensing equipment must be approved for this use. The dispensing line cannot be more than 100 feet (30 480 mm) long.

The tanker operator is responsible for the dispensing line and the tanker, so these two items need to be in the same area. The dispensing cannot be done around an ignition source, such as smoking, or electrical equipment that is not classified for use in hazardous locations.

3406.5.4.5 Commercial, industrial, governmental or manufacturing. Dispensing of Class II and III motor vehicle fuel from tank vehicles into the fuel tanks of motor vehicles located at commercial, industrial, governmental or manufacturing establishments is allowed where permitted, provided such dispensing operations are conducted in accordance with the following:

1. Dispensing shall occur only at sites that have been permitted to conduct mobile fueling.
2. The owner of a mobile fueling operation shall provide to the jurisdiction a written response plan which demonstrates readiness to respond to a fuel spill and carry out appropriate mitigation measures, and describes the process to dispose properly of contaminated materials.
3. A detailed site plan shall be submitted with each application for a permit. The site plan shall indicate: all buildings, structures and appurtenances on site and their use or function; all uses adjacent to the property lines of the site; the locations of all storm drain openings, adjacent waterways or wetlands; information regarding slope, natural drainage, curbing, impounding and how a spill will be retained upon the site property; and the scale of the site plan.

   Provisions shall be made to prevent liquids spilled during dispensing operations from flowing into buildings or off-site. Acceptable methods include, but shall not be limited to, grading driveways, raising doorsills or other approved means.
4. The fire code official is allowed to impose limits on the times and/or days during which mobile fueling operations may take place, and specific locations on a site where fueling is permitted.
5. Mobile fueling operations shall be conducted in areas not accessible to the public or shall be limited to times when the public is not present.
6. Mobile fueling shall not take place within 15 feet (4572 mm) of buildings, property lines or combustible storage.
7. The tank vehicle shall comply with the requirements of NFPA 385 and local, state and federal requirements. The tank vehicle's specific functions shall include that of supplying fuel to motor vehicle fuel tanks. The vehicle and all its equipment shall be maintained in good repair.
8. Signs prohibiting smoking or open flames within 25 feet (7620 mm) of the tank vehicle or the point of fueling shall be prominently posted on three sides of the vehicle including the back and both sides.
9. A portable fire extinguisher with a minimum rating of 40:BC shall be provided on the vehicle with signage clearly indicating its location.
10. The dispensing nozzles and hoses shall be of an approved and listed type.
11. The dispensing hose shall not be extended from the reel more than 100 feet (30 480 mm) in length.
12. Absorbent materials, nonwater-absorbent pads, a 10-foot-long (3048 mm) containment boom, an approved container with lid and a nonmetallic shovel shall be provided to mitigate a minimum 5-gallon (19 L) fuel spill.
13. Tank vehicles shall be equipped with a "fuel limit" switch such as a count-back switch, to limit the amount of a single fueling operation to a maximum of 500 gallons (1893 L) before resetting the limit switch.

    Exception: Tank vehicles where the operator carries and can utilize a remote emergency shutoff device which, when activated, immediately causes flow of fuel from the tank vehicle to cease.
14. Persons responsible for dispensing operations shall be trained in the appropriate mitigating actions in the event of a fire, leak or spill. Training records shall be maintained by the dispensing company and shall be made available to the fire code official upon request.
15. Operators of tank vehicles used for mobile fueling operations shall have in their possession at all times an emergency communications device to notify the proper authorities in the event of an emergency.
16. The tank vehicle dispensing equipment shall be constantly attended and operated only by designated personnel who are trained to handle and dispense motor fuels.
17. Prior to beginning dispensing operations, precautions shall be taken to ensure ignition sources are not present.

18. The engines of vehicles being fueled shall be shut off during dispensing operations.
19. Nighttime fueling operations shall only take place in adequately lighted areas.
20. The tank vehicle shall be positioned with respect to vehicles being fueled to prevent traffic from driving over the delivery hose.
21. During fueling operations, tank vehicle brakes shall be set, chock blocks shall be in place and warning lights shall be in operation.
22. Motor vehicle fuel tanks shall not be topped off.
23. The dispensing hose shall be properly placed on an approved reel or in an approved compartment prior to moving the tank vehicle.
24. The fire code official and other appropriate authorities shall be notified when a reportable spill or unauthorized discharge occurs.

❖ This section codifies minimum safety requirements for the regulation of certain mobile fueling operations and provides administrative controls over fueling sites, specifies the types of tank vehicles required in such operations and specifies training and licensing requirements for persons engaged in mobile fueling operations.

Regardless of previous allowances or restrictions by the legacy model fire codes (i.e., national, standard and uniform), mobile fueling is flourishing nationally and is welcomed in many jurisdictions. In fact, a number of local jurisdictions have developed regulations specifically allowing for expanded mobile fueling operations. The regulations in this section are consistent with those local regulations as well as those contained in NFPA 30A and provide an international model for jurisdictions to follow.

The greatest operational uses of mobile fueling are in conjunction with fueling of fleets, such as trucking companies, bus companies, delivery companies, municipal fleets, the U.S. Postal Service fleets and similar operations. Nationwide, the annual mobile fueling volume is conservatively estimated to exceed 500,000,000 gallons (18,992,500,000 L) and it is increasing at a rate exceeding 30 percent per year. It is a business carried out by both public and private entities. One private company alone delivers over 120,000,000 gallons (454,200,000 L) per year. There are several other companies that deliver from 500,000 to 3,000,000 gallons (1,892,500 to 11,355,000 L) per month. These operations are occurring in all regions of the United States since the early 1990s and in some areas, even longer.

Mobile fueling thrives because it meets several needs:

1. It is more cost efficient to bring delivery of fuel to high-volume consuming vehicles versus having each of those vehicles go to the fuel source.
2. Fueling is a safer operation when carried out by a trained, focused specialist who is familiar with the dispensing equipment, the safety regulations, is trained in spill control and mitigation and knows what steps to take should an incident occur.
3. Fuel-consuming fleet operators recognize both safety and cost benefits by not involving a multitude of employees in a fueling operation for which they are not trained and are not likely to develop an overriding concern for safety.
4. Additional safety controls are evolving through the use of technology systems. Some systems are not economically feasible unless they include additional services. For example, some mobile fueling operations bring cost-saving advantages to their clients through the provision of a data capturing system. In addition to having instant remote fuel-flow stopping capabilities, the system frees the operator to concentrate solely on the task of safe dispensing.
5. On-site mobile fueling eliminates the traffic hazards and air pollution problems associated with driving fleets of vehicles to a fixed fueling facility.

There is no known adverse fire incident history associated with mobile fueling. Given the length of time that mobile fueling has been occurring, the data points to an outstanding safety record. There is no safety-related reason that mobile fueling operations should not be allowed, provided such operations are carried out in keeping with reasonable safety requirements, including those designed to protect water supply and environment, as provided in this section.

3406.6 Tank vehicles and vehicle operation. Tank vehicles shall be designed, constructed, equipped and maintained in accordance with NFPA 385 and Sections 3406.6.1 through 3406.6.4.

❖ This section sets forth rules for the operation of tank vehicles used for the transportation of flammable and combustible liquids. Included are parking and garaging regulations, fire extinguisher requirements and regulations for the discharge of flammable and combustible liquids into underground storage tanks. These requirements are intended to apply to the extent that they do not conflict with federal requirements detailed in DOTn 49 CFR and other federal transportation regulations.

NFPA 385 details requirements for materials and methods of construction for flammable and combustible liquid tank vehicles to provide protection for their cargo under highway conditions. Design and construction of these vehicles must also conform to the requirements set forth in DOTn 49 CFR.

3406.6.1 Operation of tank vehicles. Tank vehicles shall be utilized and operated in accordance with NFPA 385 and Sections 3406.6.1.1 through 3406.6.1.11.

❖ Chapter 6 of NFPA 385 outlines the requirements for the operation of tank vehicles. These requirements detail required practices during transit; parking; loading and unloading; maintenance and repair designed to reduce the potential for accidental release of the vehicle cargo.

3406.6.1.1 Vehicle maintenance. Tank vehicles shall not be operated unless they are in proper state of repair and free from accumulation of grease, oil or other flammable substance, and leaks.

❖ Good housekeeping practices extend to tank vehicles just as much as they do to permanent facilities. Keeping vehicles clean and in good repair reduces the possibility for accumulations of flammable materials to become a fuel source for a fire.

3406.6.1.2 Leaving vehicle unattended. The driver, operator or attendant of a tank vehicle shall not remain in the vehicle cab and shall not leave the vehicle while it is being filled or discharged. The delivery hose, when attached to a tank vehicle, shall be considered to be a part of the tank vehicle.

❖ The tank vehicle driver, operator or attendant must be prepared to immediately suspend dispensing; loading or unloading operations; control a spill or extinguish a fire should an incident occur. This section and NFPA 385 emphasize that the driver, operator or attendant must be outside the cab of the vehicle in order to be in compliance. It further makes it clear that the driver, operator or attendant may be no further away from the vehicle than the length of the delivery hose connected to the vehicle. These restrictions are especially important in climates subject to extreme weather when drivers are often tempted to sit in the truck cab or go into a building to keep cool or warm, depending on the season. Keeping the person most qualified to take prompt emergency action in the best possible location to do so reduces the likelihood of a major spill in the event of an equipment failure or accident.

3406.6.1.3 Vehicle motor shutdown. Motors of tank vehicles or tractors shall be shut down during the making or breaking of hose connections. If loading or unloading is performed without the use of a power pump, the tank vehicle or tractor motor shall be shut down throughout such operations.

❖ Motors must be shut down when not required to operate unloading equipment. When the vehicle motor is required to operate transfer pumps, it must be shut off when loading or unloading is commencing or is completed and before hose connections are made or broken in order to control it as a source of ignition.

3406.6.1.4 Outage. A cargo tank or compartment thereof used for the transportation of flammable or combustible liquids shall not be loaded to absolute capacity. The vacant space in a cargo tank or compartment thereof used in the transportation of flammable or combustible liquids shall not be less than 1 percent. Sufficient space shall be left vacant to prevent leakage from or distortion of such tank or compartment by expansion of the contents caused by rise in temperature in transit.

❖ To allow for the thermal expansion of flammable or combustible liquids, the maximum allowable capacity of a cargo tank or compartment is 99 percent of the actual capacity. The expansion of flammable or combustible liquids in a confined space will create internal pressures, which may damage the cargo tank or compartment, possibly creating a leak.

3406.6.1.5 Overfill protection. The driver, operator or attendant of a tank vehicle shall, before making delivery to a tank, determine the unfilled capacity of such tank by a suitable gauging device. To prevent overfilling, the driver, operator or attendant shall not deliver in excess of that amount.

❖ The individual responsible for the operation of the tank vehicle is also responsible for overfill protection. This individual is to calculate the fill quantity. This is the maximum quantity that the individual responsible for the operation of the tank vehicle is to deliver.

3406.6.1.6 Securing hatches. During loading, hatch covers shall be secured on all but the receiving compartment.

❖ To ensure that the minimum amount of flammable and combustible liquids is exposed to air or to an ignition source, only hatches in actual use during dispensing are to be opened.

3406.6.1.7 Liquid temperature. Materials shall not be loaded into or transported in a tank vehicle at a temperature above the material's ignition temperature unless safeguarded in an approved manner.

❖ Loading and transporting of flammable and combustible liquids above their ignition temperature are prohibited unless the loading operation and the transporting process are protected from the hazards involved in this process. Risks are increased significantly when the liquid is at an elevated temperature.

3406.6.1.8 Bonding to underground tanks. An external bond-wire connection or bond-wire integral with a hose shall be provided for the transferring of flammable liquids through open connections into underground tanks.

❖ Static electricity will spark between surfaces with different electrical potential. This spark is an ignition source for flammable and combustible liquids. Bonding of the dispensing hose to the underground tank is required, if the transfer is through an open connection. Bonding can be done using wires, ground cables, metal piping or other similar means that will carry current.

3406.6.1.9 Smoking. Smoking by tank vehicle drivers, helpers or other personnel is prohibited while they are driving, making deliveries, filling or making repairs to tank vehicles.

❖ Smoking is prohibited in the vicinity of the tank vehicle. This eliminates a source of ignition.

3406.6.1.10 Hose connections. Delivery of flammable liquids to underground tanks with a capacity of more than 1,000 gallons (3785 L) shall be made by means of approved liquid and vapor-tight connections between the delivery hose and fill tank pipe. Where underground tanks are equipped with any type of vapor recovery system, all connections required to be made for the safe and proper functioning of the particular vapor recovery process shall be made. Such connections shall be made liquid and vapor tight and remain connected throughout the unloading process. Vapors shall not be discharged at grade level during delivery.

❖ Vapor-tight connections reduce the potential release of flammable vapors that can be easily ignited. These include liquid transfer lines and vapor recovery lines that are designed to prevent the release of polluted, flammable fuel vapor during transfer. These requirements prohibit the extremely dangerous but not uncommon practice of not connecting vapor return hoses, thus allowing the vapors displaced during delivery to escape at grade level from the unmade connections. Incidents have been reported where the vapors have traveled to nearby buildings, found an ignition source and exploded (see also Section 2205.1.3).

3406.6.1.10.1 Simultaneous delivery. Simultaneous delivery to underground tanks of any capacity from two or more discharge hoses shall be made by means of mechanically tight connections between the hose and fill pipe.

❖ The simultaneous delivery of flammable and combustible liquids to an underground tank using two or more hoses requires a mechanically tight connection for each delivery hose to the tank fill pipe. The pressure generated by the delivery could cause one or more of the delivery hoses to leak or to disconnect if the connections are not tight.

3406.6.1.11 Hose protection. Upon arrival at a point of delivery and prior to discharging any flammable or combustible liquids into underground tanks, the driver, operator or attendant of the tank vehicle shall ensure that all hoses utilized for liquid delivery and vapor recovery, where required, will be protected from physical damage by motor vehicles. Such protection shall be provided by positioning the tank vehicle to prevent motor vehicles from passing through the area or areas occupied by hoses, or by other approved equivalent means.

❖ Whenever possible, fill connections for new underground storage tanks should be located where tank vehicle hose lines will be out of the way of traffic during liquid transfer operations. Otherwise, the tank vehicle should be positioned to minimize the likelihood of damage to the discharge line (which could result in a large spill and potentially a large fire), or traffic cones or pylons should be placed to warn approaching vehicles (see Figure 3406.6.1.11).

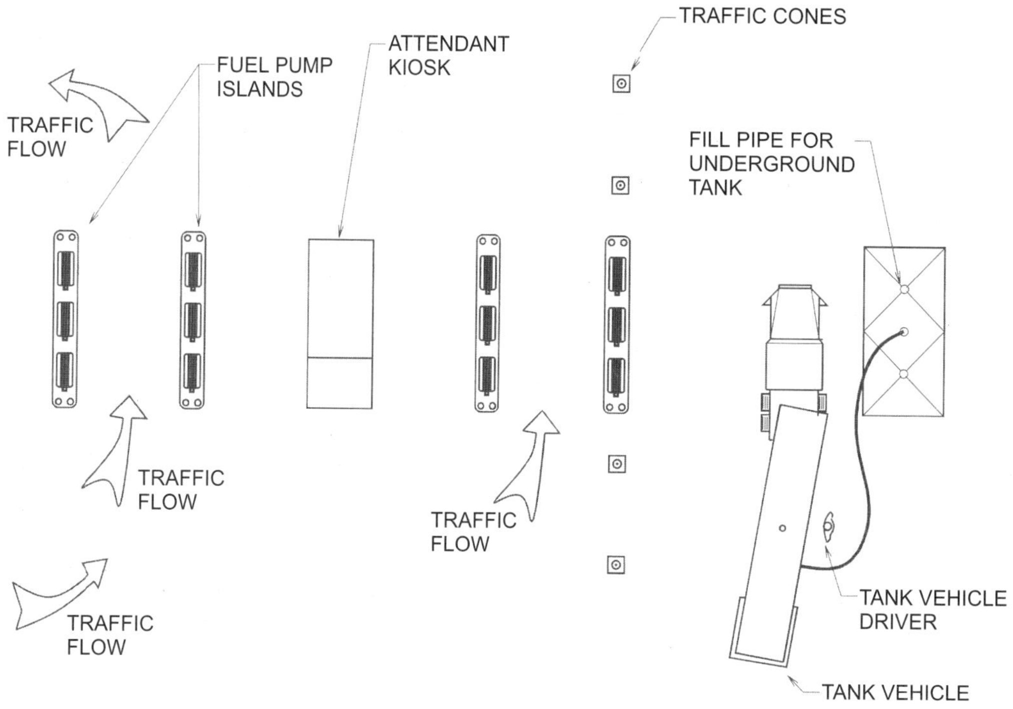

Figure 3406.6.1.11
TANK VEHICLE POSITIONING DURING UNLOADING

3406.6.2 Parking. Parking of tank vehicles shall be in accordance with Sections 3406.6.2.1 through 3406.6.2.3.

Exception: In cases of accident, breakdown or other emergencies, tank vehicles are allowed to be parked and left unattended at any location while the operator is obtaining assistance.

❖ The parking of tank vehicles is as important as aboveground tank storage. The quantity of flammable and combustible liquids can be substantial in a tank vehicle and exposure to ignition sources is more problematic than in a secured area. The facilities adjacent to the tank vehicle may represent a potential for large loss of life or large property damage. The spread of fire from a tank vehicle is not contained by dikes or a similar containment and drainage system. The fire can spread to areas outside of the tank vehicle location and fire-fighting operations may not be effective if it goes into an underground drainage system.

The restrictions on parking do not apply in case of a vehicle breakdown. Because it is impossible to predict or prevent all vehicle failures, the code does not make temporary parking resulting from mechanical failure of the tank vehicle a violation.

3406.6.2.1 Parking near residential, educational and institutional occupancies and other high-risk areas. Tank vehicles shall not be left unattended at any time on residential streets, or within 500 feet (152 m) of a residential area, apartment or hotel complex, educational facility, hospital or care facility. Tank vehicles shall not be left unattended at any other place that would, in the opinion of the fire chief, pose an extreme life hazard.

❖ Tank vehicles must not be parked or left unattended in congested districts or near schools, nursing homes, hospitals or similar buildings where the hazards to life may be high. Maintaining a distance of 500 feet (152 400 mm) to the listed premises provides a spatial buffer between the tank vehicle and the high life-risk property in case of a spill, accident or ignition. This section also allows the fire chief to declare certain other areas of the jurisdiction as being off limits to parking, based on the hazard exposures present. Some jurisdictions have, by local ordinance, specified certain road routes to be used by hazardous cargo vehicles to even further reduce the likelihood of a major hazardous materials incident or fire.

3406.6.2.2 Parking on thoroughfares. Tank vehicles shall not be left unattended on a public street, highway, public avenue or public alley.

Exceptions:

1. The necessary absence in connection with loading or unloading the vehicle. During actual fuel transfer, Section 3406.6.1.2 shall apply. The vehicle location shall be in accordance with Section 3406.6.2.1.
2. Stops for meals during the day or night, if the street is well lighted at the point of parking. The vehicle location shall be in accordance with Section 3406.6.2.1.

❖ Leaving a tank vehicle parked or unattended in a public way invites malicious vandalism to the vehicle or an accidental collision by another vehicle, either of which could cause a release of the cargo creating a large-scale spill or possible large-area fire if ignition were to occur. The exceptions in this section are very specific in describing the limited conditions under which the vehicle operator may be absent from the vehicle. The locations where such absence is permitted may be expected to be generally off limits to the public (i.e., unauthorized personnel) or, in the case of meal breaks, sufficiently lighted so as to discourage vandalism and reduce the likelihood of accidental collision.

3406.6.2.3 Duration exceeding 1 hour. Tank vehicles parked at one point for longer than 1 hour shall be located off of public streets, highways, public avenues or alleys, and:

1. Inside of a bulk plant and either 25 feet (7620 mm) or more from the nearest lot line or within a building approved for such use; or
2. At other approved locations not less than 50 feet (15 240 mm) from the buildings other than those approved for the storage or servicing of such vehicles.

❖ Tank vehicles parked for more than 1 hour must be in a secured location. This prevents the tank vehicle from from being subject to accidents and vandalism. This section contains two specific requirements for vehicle parking beyond 1 hour:

1. The tank vehicle must be inside a bulk plant facility at least 25 feet (7620 mm) from a property line or within a building designed for the storage or service of tank vehicles.
2. Any approved location must be at least 50 feet (15 240 mm) from a property line or within a building designed for the storage or service of tank vehicles.

3406.6.3 Garaging. Tank vehicles shall not be parked or garaged in buildings other than those specifically approved for such use by the fire code official.

❖ Vehicles must be unloaded, if necessary, cleaned and purged of vapors or made inert before parking inside any building or structure, unless the building or structure is specially designed for that purpose and approved by the fire code official. Buildings used for the storage of flammable liquid tank vehicles have safety features for control of leaks, spills, ignition sources, fire and fire containment that are not in other buildings and are usually classified in Occupancy Group H-2, as defined in the IBC and Section 202 of the code.

3406.6.4 Fire protection. Tank vehicles shall be equipped with a fire extinguisher complying with Section 906 and having a minimum rating of 2-A:20-B:C.

During unloading of the tank vehicle, the fire extinguisher shall be out of the carrying device on the vehicle and shall be 15 feet (4572 mm) or more from the unloading valves.

❖ A fire extinguisher (2A:20BC) must be available to control a small fire. The fire extinguisher is designed for use on ordinary combustible fires, flammable liquid fires and electrical fires. The extinguisher is to be at least 15 feet (4572 mm) from the unloading valve. This location is convenient for the attendant to close the valve, control the flow of the flammable or combustible liquid and have access to the fire extinguisher to control a fire.

3406.7 Refineries. Plants and portions of plants in which flammable liquids are produced on a scale from crude petroleum, natural gasoline or other hydrocarbon sources shall be in accordance with Sections 3406.7.1 through 3406.7.3. Petroleum-processing plants and facilities or portions of plants or facilities in which flammable or combustible liquids are handled, treated or produced on a commercial scale from crude petroleum, natural gasoline, or other hydrocarbon sources shall also be in accordance with API 651, API 653, API 752, API 1615, API 2001, API 2003, API 2009, API 2015, API RP2023, API 2201 and API 2350.

❖ Refineries process crude petroleum or other hydrocarbon sources into flammable or combustible liquids on a commercial scale. These facilities are covered by American Petroleum Institute (API) standards:

API 651: The standard describes corrosion problems characteristic to above-ground storage tanks and associated piping systems. The standard covers the two current methods used to provide cathodic protection against corrosion.

API 653: The standard covers the inspection, repair, alteration and reconstruction of steel above-ground storage tanks and includes the minimum requirements for maintaining the integrity of welded or riveted, nonrefrigerated, atmospheric pressure, aboveground storage tanks that have been placed in service.

API 752: The guide describes a methodology for assessing and evaluating the hazards associated with the location of process plant buildings.

API 1615: The guide contains the procedures and lists equipment needed for the proper installation of underground petroleum storage systems. The guide applies to underground storage tank systems that store petroleum products for retail and commercial facilities.

API 2001: The publication provides an understanding of the fire protection problems and the steps needed to ensure the safe storage, handling and processing of petroleum products in refineries and the safe shipment of petroleum products.

API 2003: This publication describes some of the conditions that have resulted in fires caused by electrical sparks and arcs from natural causes. The publication describes methods currently used to prevent ignition from these sources.

API 2009: The publication contains suggested precautions for the protection of personnel from injury and protection of property from damage by fire that may arise during the operation of gas and electric cutting and welding equipment in and around petroleum operations.

API 2015: This standard contains safety procedures for preparing, emptying, isolating, ventilating, atmospheric testing, cleaning, entry, hot work and recommissioning activities in, on and around atmospheric and low pressure [15 psig (103 kPa) or less] above-ground storage tanks. The standard applies to stationary tanks used in petroleum and petrochemical plants and terminals.

APIRP 2023: The publication is a recommended practice that describes the phenomena that can occur and precautions to be taken in the storage of asphalt products and residue derived from crude oil petroleum. The recommendations apply when these materials are stored in heated tanks at refineries and bulk storage facilities and transported in tank vehicles.

API 2201: The publication covers the safety aspects to be considered when hot tapping or welding without hot tapping on in-service piping or equipment.

API 2350: The publication is a recommended practice to prevent petroleum storage tanks from being overfilled. Tank overfill can be effectively reduced by developing and implementing practical and safe operating procedures for storage facilities and by providing for careful selection and application of equipment, scheduled maintenance programs and personnel training. The publication covers overfill protection for aboveground storage tanks in petroleum facilities, including refineries, terminals, bulk plants and pipeline terminals for Class I liquids from mainline pipelines or marine vessels.

3406.7.1 Corrosion protection. Above-ground tanks and piping systems shall be protected against corrosion in accordance with API 651.

❖ Above-ground tanks must have corrosion protection complying with API 651. Soil and environmental conditions can cause the deterioration of tanks. Corrosion can weaken the tank, creating a potential for leakage.

Cathodic protection uses a sacrificial metal to prevent the corrosion of the metal storage tank. The sacrificial metal is bonded to the tank so the electrochemical reaction of the sacrificial metal prevents the corrosion of the steel storage tank. The sacrificial metal is the anode, the steel storage tank is the cathode and a power supply connects the two. Replacement of the anode will depend on the rate of corrosion.

3406.7.2 Cleaning of tanks. The safe entry and cleaning of petroleum storage tanks shall be conducted in accordance with API 2015.

❖ The safe entry and cleaning of tanks must comply with API 2015. The environment in a storage tank can be hostile for personnel. The air may not be breathable or may contain a vapor-air mixture within the flammable range.

3406.7.3 Storage of heated petroleum products. Where petroleum-derived asphalts and residues are stored in heated tanks at refineries and bulk storage facilities or in tank vehicles, such products shall be in accordance with API 2023.

❖ The storage of heated petroleum products must comply with API 2023. The guide describes the phenomena that can occur and the precautions to be taken in the storage and handling of asphalt products and residue derived from crude-oil petroleum. The heating of these flammable and combustible liquids can result in the release of vapors as well as increase the potential for ignition.

3406.8 Vapor recovery and vapor-processing systems. Vapor-processing systems in which the vapor source operates at pressures from vacuum, up to and including 1 psig (6.9 kPa) or in which a potential exists for vapor mixtures in the flammable range, shall comply with Sections 3406.8.1 through 3406.8.5.

Exceptions:

1. Marine systems complying with federal transportation waterway regulations such as DOTn 33 CFR, Parts 154 through 156, and CGR 46 CFR, Parts 30, 32, 35 and 39.
2. Motor fuel-dispensing facility systems complying with Chapter 22.

❖ Vapor recovery systems must function to prevent the release of vapors and protect the vapor recovery system during operation in the flammable range.

Exception 1 covers marine systems complying with federal regulations. A marine system not complying with federal regulations is in violation of the code and should be reported to the appropriate federal authority for enforcement of federal regulations. Exception 2 addresses service station systems complying with Chapter 22. A service station system not complying with Chapter 22 is in violation of the code. The motor fuel-dispensing facility system is to be brought into compliance with Chapter 22.

3406.8.1 Over-pressure/vacuum protection. Tanks and equipment shall have independent venting for over-pressure or vacuum conditions that might occur from malfunction of the vapor recovery or processing system.

Exception: For tanks, venting shall comply with Section 3404.2.7.3.

❖ A vapor recovery system must be equipped with an independent system for venting overpressure or a vacuum. An overpressure condition can damage the vapor recovery system, creating leaks or other damage. The leak or damage could allow the vapor from a flammable or combustible liquid to escape and generate a vapor-air mixture in the flammable range.

A vacuum condition can result in the introduction of air into the vapor recovery system. The introduction of air could generate a vapor-air mixture in the flammable range within the vapor recovery system.

The exception recognizes that tank venting complying with Section 3404.2.7.3 is exempt from this requirement. Tanks are vented to maintain the internal tank pressure within the design operating range. A low pressure can increase the generation of vapors. A high pressure can damage the tank or piping system. Any pressure outside of the design pressure range can have an adverse affect on the operation of the system as well as the piping and equipment.

3406.8.2 Vent location. Vents on vapor-processing equipment shall be not less than 12 feet (3658 mm) from adjacent ground level, with outlets located and directed so that flammable vapors will disperse to below the lower flammable limit (LFL) before reaching locations containing potential ignition sources.

❖ The termination of a vent pipe must direct vapors away from the building. Vapors from flammable liquids are normally heavier than air so that the vapor will settle to lower levels. The termination of a vent pipe a minimum of 12 feet (3658 mm) above grade will provide space for the vapors to disperse to below the LFL. This high termination elevation also reduces the potential for the termination being in close proximity to grade-level ignition sources. Because flammable liquid vapor is heavier than air, attention must be given to the building design near the termination.

3406.8.3 Vapor collection systems and overfill protection. The design and operation of the vapor collection system and overfill protection shall be in accordance with this section and Section 5.10 of NFPA 30.

❖ The collection system and overfill protection are to comply with Section 5.10 of NFPA 30. This section addresses the same topics as Section 3406.8. Section 5.10 of NFPA 30 contains additional requirements not specifically listed in Section 3406.8. These requirements are in other sections of this chapter, but not specifically contained in this section. These additional requirements are: (1) vapor collection system is to be designed not to trap liquid in the vapor collection piping; (2) unless the vapor recovery system is designed to handle flammable or combustible liquid, the system is to be designed to eliminate any liquid from the vapor recovery system; (3) provide protection from ignition sources; (4) classification of locations for electrical system; (5) protection from static electricity; (6) precautions for spontaneous ignition; (7) prevent friction heat or sparks from mechanical equipment; (8) prevent the propagation of flame through the vapor recovery system; (9) when necessary, provisions for explosion prevention and (10) requirement for an emergency shutdown system.

3406.8.4 Liquid-level monitoring. A liquid knock-out vessel used in the vapor collection system shall have means to verify the liquid level and a high-liquid-level sensor that activates an alarm. For unpopulated facilities, the high-liquid-level sensor shall initiate the shutdown of liquid transfer into the vessel and shutdown of vapor recovery or vapor-processing systems.

❖ Liquid monitoring is necessary to prevent the liquid knockout vessel from overfilling. The liquid-level monitor must sound an alarm or shut down the vapor recovery process before the liquid knockout vessel is overfilled. Overfill of the liquid knockout vessel could cause a spill of the flammable or combustible liquid.

3406.8.5 Overfill protection. Storage tanks served by vapor recovery or processing systems shall be equipped with overfill protection in accordance with Section 3404.2.7.5.8.

❖ To prevent spillage of storage tanks used with a vapor recovery or processing system, the tanks are limited to 95 percent of their capacity by an overfill protection system. Several audible or visible alarm methods are acceptable. The protection system must include provisions for draining the vapor recovery or processing system into the storage tank without exceeding the 95-percent capacity.

Bibliography

The following resource materials are referenced in this chapter or are relevant to the subject matter addressed in this chapter.

ANSI A13.1-96, *Scheme for Identification of Piping Systems*. New York: American National Standards Institute, 1996.

ANSI B31.3-99, *Process Piping*. New York: American National Standards Institute, 1996.

ANSI B31.9-96, *Building Service Piping*. New York: American National Standards Institute, 1996.

API 12P-95, *Specification for Fiberglass Reinforced Plastic Tanks*. Washington, DC: American Petroleum Institute, 1995.

API RP 651-97, *Cathodic Protection of Aboveground Petroleum Storage Tanks*. Washington, DC: American Petroleum Institute, 1997.

API RP 752-95, *Management of Hazards Associated with Location of Process Plant Buildings, CMA Manager's Guide*. Washington, DC: American Petroleum Institute, 1995.

API RP 1604-96, *Closure of Underground Petroleum Storage Tanks*. Washington, DC: American Petroleum Institute, 1996.

API RP 1615-96, *Installation of Underground Petroleum Storage Systems*. Washington, DC: American Petroleum Institute, 1996.

API RP 2001-98, *Fire Protection in Refineries*. Washington, DC; American Petroleum Institute, 1998.

API RP 2003-98, *Protection Against Ignition Arising Out of Static, Lightning and Stray Currents*. Washington, DC: American Petroleum Institute, 1998.

API RP 2009-95, *Safe Welding, Cutting, and Other Hot Work Practices in Petroleum and Petrochemical Industries*. Washington, DC: American Petroleum Institute, 1995.

API RP 2023-88, *Guide for Entry and Cleaning of Petroleum Storage Tanks*. Washington, DC: American Petroleum Institute, 1988.

API RP 2028-91, *Flame Arrestors in Piping Systems*. Washington, DC, American Petroleum Institute, 1991.

API RP 2201-95, *Procedures for Welding or Hot Tapping on Equipment in Service*. Washington, DC: American Petroleum Institute, 1995.

API RP 2350-96, *Overfill Protection for Storage Tanks in Petroleum Facilities*. Washington, DC: American Petroleum Institute, 1996.

API STD 653-98, *Tank Inspection, Repair, Alteration, and Reconstruction*. Washington, DC: American Petroleum Institute, 1998.

API STD 2000-98, *Venting Atmospheric and Low Pressure Storage Tanks Nonrefrigerated and Refrigerated*. Washington, DC: American Petroleum Institute, 1998.

API STD 2015-94, *Safe Entry and Clearing of Petroleum Storage Tanks, Planning and Managing Tank Entry from Decommissioning Through Recommissioning*. Washington, DC: American Petroleum Institute, 1994.

ASTM D 56-01, *Standard Test Method for Flash Point by Tag Closed Tester*. West Conshohocken, PA: ASTM International, 2001.

ASTM D 92-01, *Standard Test Method for Flash and Fire Points by Cleveland Open Cup Tester*. West Conshohocken, PA: ASTM International, 2001.

ASTM D 93-00, *Standard Test Methods for Flash Point by Pensky-Martens Closed Cup Tester*. West Conshohocken, PA: ASTM International, 2000.

ASTM D 3278-96e1, *Standard Test Methods for Flash Point of Liquids by Small Scale Closed-Cup Apparatus*. West Conshohocken, PA: ASTM International, 1996.

ASTM E 1529-00, *Standard Test Methods for Determining Effects of Large Hydrocarbon Pool Fires on Structural Members and Assemblies*. West Conshohocken, PA: ASTM International, 2000.

BPVC-2001, *ASME Boiler and Pressure Vessel Code*. New York: American Society of Mechanical Engineers, 2001.

CGR 46 CFR 30, 32, 35 & 39, *Shipping*. Washington, DC: U.S. Government Printing Office, 1999.

DOTn 33 CFR 154-156, *Navigation and Navigable Waters*. Washington, DC: U.S. Department of Transportation, 1998.

IBC-03 *International Building Code*. Falls Church, VA: International Code Council, 2003.

ICC EC-03, *ICC Electrical Code*. Falls Church, VA: International Code Council, 2003.

IFGC-03, *International Fuel Gas Code*. Falls Church, VA: International Code Council, 2003.

IMC-03, *International Mechanical Code*. Falls Church, VA: International Code Council, 2003.

NFPA 11-98, *Standard for Low-Expansion Foam Systems*. Quincy, MA: National Fire Protection Association, 1998.

NFPA 11A-99, *Standard for Medium, and High-Expansion Foam Systems*. Quincy, MA: National Fire Protection Association, 1999.

NFPA 13-96, *Standard For the Installation of Sprinkler Systems*. Quincy, MA: National Fire Protection Association, 1996.

NFPA 15-96, *Standard for Water-Spray Fixed System for Fire Protection*. Quincy, MA: National Fire Protection Association, 1996.

NFPA 30-00, *Flammable and Combustible Liquids Code*. Quincy, MA: National Fire Protection Association, 2000.

NFPA 69-97, *Explosion Prevention Systems*. Quincy, MA: National Fire Protection Association, 1997.

NFPA 231C-98, *Rack Storage of Materials*. Quincy, MA: National Fire Protection Association, 1998.

NFPA 385-00, *Standard for Tank Vehicles for Storage of Flammable and Combustible Liquids*. Quincy, MA: National Fire Protection Association, 2000.

UL 58-96, *Standard for Steel Underground Tanks for Flammable and Combustible Liquids*. Northbrook, IL: Underwriters Laboratories, 1996.

UL 1275-94, *Standard for Flammable Liquid Storage Cabinets*. Northbrook, IL: Underwriters Laboratories, 1994.

UL 1316-94, *Glass-Fiber-Underground Storage Tanks for Petroleum Product, Alcohols, and Alcohol-Gasoline Mixtures*. Northbrook, IL: Underwriters Laboratories, 1994.

UL 2085-97, *Standard for Protected Aboveground Storage Tanks for Flammable and Combustible Liquids*. Northbrook, IL: Underwriters Laboratories, 1997.

UL 2208-96, *Solvent distillation Units*. Northbrook, IL:Underwriters Laboratories, 1996.

UL 2245-99 *Standard for Below-Grade Vaults for Flammable Liquid Storage Tanks*. Northbrook, IL:Underwriters Laboratories, 1999.

Chapter 35: Flammable Gases

General Comments

Of the three physical states of matter, only gas will ignite without preheating. In order to burn, all other materials must be transformed into a vaporous or gaseous state, regardless of their initial phase.

The principal hazard posed by flammable gas is its ready ignitability, or even explosivity, when mixed with air in the proper proportions. The question in a flammable gas release usually is not whether the mixture will ignite, but rather when or how it will ignite if not controlled. Consequently, occupancies storing or handling more than 1,000 cubic feet (28 m^3) of flammable gas or 30 gallons (114 L) of liquefied flammable gas per control area are classified as Group H-2 (high hazard) by the *International Building Code*® (IBC®).

Purpose

Chapter 35 sets requirements for the storage and use of flammable gases. For safety purposes, there is a limit on the quantities of flammable gas allowed per control area. Exceeding these limitations increases the possibility of damage to both property and individuals.

SECTION 3501 GENERAL

3501.1 Scope. The storage and use of flammable gases shall be in accordance with this chapter. Compressed gases shall also comply with Chapter 30. Gaseous hydrogen systems at consumer sites shall also comply with NFPA 50A.

Exceptions:

1. Gases used as refrigerants in refrigeration systems (see Section 606).
2. Liquefied petroleum gases and natural gases regulated by Chapter 38.
3. Fuel-gas systems and appliances regulated under the *International Fuel Gas Code.*
4. Hydrogen motor fuel-dispensing facilities designed and constructed in accordance with Chapter 22.

❖ This section establishes the scope of Chapter 35 with respect to the storage and use of flammable and nonflammable gases. The requirements of this chapter and the referenced standards are applicable in addition to the general storage requirements of Chapter 27 for hazardous materials.

3501.2 Permits. Permits shall be required as set forth in Section 105.6.

❖ The process of issuing permits gives the fire code official an opportunity to carefully evaluate and regulate hazardous operations. Permit applicants should be required to demonstrate that their operations comply with the intent of the code before the permit is issued. See the commentary to Section 105.6 for a general discussion of operations requiring an operational permit, Section 105.6.21 for discussion of specific quantity-based operational permits for the materials regulated in this chapter and Section 105.7 for a general discussion of activities requiring a construction permit. The permit process also notifies the fire department of the need for prefire planning for hazardous property.

SECTION 3502 DEFINITIONS

3502.1 Definitions. The following words and terms shall, for the purposes of this chapter and as used elsewhere in this code, have the meanings shown herein.

❖ Definitions of terms can help in the understanding and application of the code requirements. The purpose for including those definitions that are associated with the subject matter of this chapter is to provide more convenient access to them without having to refer back to Chapter 2. It is important to emphasize that these terms are not exclusively related to this chapter but are applicable everywhere the term is used in the code. For convenience, these terms are also listed in Chapter 2 with a cross reference to this section. The use and application of all defined terms, including those defined in this section, are set forth in Section 201.

FLAMMABLE GAS. A material which is a gas at 68°F (20°C) or less at 14.7 pounds per square inch atmosphere (psia) (101 kPa) of pressure [a material that has a boiling point of 68°F (20°C) or less at 14.7 psia (101 kPa)] which:

1. Is ignitable at 14.7 psia (101 kPa) when in a mixture of 13 percent or less by volume with air; or

2. Has a flammable range at 14.7 psia (101 kPa) with air of at least 12 percent, regardless of the lower limit.

The limits specified shall be determined at 14.7 psi (101 kPa) of pressure and a temperature of 68°F (20°C) in accordance with ASTM E 681.

❖ The ASTM E 681 test method covers the determination of the lower and upper concentration limits of chemicals having sufficient vapor pressure to form flammable mixtures in air at atmospheric pressure at the test temperature. The flammability limits depend on the test temperature and pressure. This test method is limited to an initial pressure of the local ambient or less, with a practical lower pressure limit of approximately 13 kPa (100 mm Hg). The maximum practical operating temperature of this equipment is approximately 150°C (302°F).

FLAMMABLE LIQUEFIED GAS. A liquefied compressed gas which, under a charged pressure, is partially liquid at a temperature of 68°F (20°C) and which is flammable.

❖ Flammable liquefied gases are widely useful because of their properties, including high heat output in combustion for most gases, high reactivity in chemical processing with other gases, extremely low temperatures available from some gases and the economy of handling them all in a compact form at high pressure or low temperature.

SECTION 3503 GENERAL REQUIREMENTS

3503.1 Quantities not exceeding the maximum allowable quantity per control area. The storage and use of flammable gases in amounts not exceeding the maximum allowable quantity per control area indicated in Section 2703.1 shall be in accordance with Sections 2701, 2703, 3501 and 3503.

❖ Basically, when the amounts stored or used do not exceed the amounts in Tables 2703.1(1) through 2703.1(4), certain factors that must be adhered to include systems and processes, release of hazardous materials into the air, Material Safety Data Sheets (MSDS), hazard identification signs, sources of ignition and construction requirements.

3503.1.1 Special limitations for indoor storage and use. Flammable gases shall not be stored or used in Group A, B, E, I or R occupancies.

Exceptions:

1. Cylinders not exceeding a capacity of 250 cubic feet (7.08m^3) each at normal temperature and pressure (NTP) used for maintenance purposes, patient care or operation of equipment.
2. Food service operations in accordance with Section 3803.2.1.7.

❖ Flammable gases must not be stored or used where an accident could cause a large loss of life. They must be stored and used in accordance with preventive guidelines for safety purposes. Cylinders under low pressure are allowed for maintenance of buildings, taking care of patients and equipment operation. No systems are allowed for these purposes because the large volume would create the potential for a catastrophic event with large loss of life and property.

3503.1.1.1 Medical gases. Medical gas system supply cylinders shall be located in medical gas storage rooms or gas cabinets as set forth in Section 3006.

❖ Section 3006 gives requirements for storage rooms and gas cabinets. Storage rooms are to be of 1-hour rated construction and be either interior or exterior rooms. Gas cabinets must be connected to an exhaust system, sprinklered internally and meet certain air velocity ventilation requirements.

3503.1.1.2 Aggregate quantity. The aggregate quantities of flammable gases used for maintenance purposes and operation of equipment shall not exceed the maximum allowable quantity per control area indicated in Table 2703.1.1(1).

❖ Table 2703.1.1(1) contains categories of storage, use-closed systems and use-open systems. Flammable gas is categorized only in the storage and use-closed system. Because gas is in closed containers and systems, use-open systems do not apply. The maximum amount for either storage or use-closed system is 1,000 cubic feet (28 m^3).

3503.1.2 Storage containers. Cylinders and pressure vessels for flammable gases shall be designed, constructed, installed, tested and maintained in accordance with Chapter 30.

❖ Sections 3003 and 3004 give requirements for storage containers, such as markings on tanks, securing the tanks to prevent dislodging, protection of the valves, separation from hazards, electrical wiring, exposure to fire, unauthorized use, leak prevention, overhead protection and grounding to properly prevent a lightning hazard.

3503.1.3 Emergency shutoff. Compressed gas systems conveying flammable gases shall be provided with approved emergency shutoff valves that can be activated at each point of use and each source.

❖ If a leak, fire or other hazardous situation occurs, emergency shutoff valves at each point of use and each source will allow isolation of any line affected by the emergency situation without time being taken to close off all valves.

3503.1.4 Ignition source control. Ignition sources in areas containing flammable gases shall be controlled in accordance with Section 2703.7.

Static-producing equipment located in flammable gas storage areas shall be grounded.

"No Smoking" signs shall be posted in areas containing flammable gases in accordance with Section 2703.6.

❖ As discussed in the commentary to Chapter 3, an ignition source is needed to ignite a fuel load and cause a fire. Controlling ignition sources by making sure electrical equipment located in gas storage areas is properly grounded is one of the primary ways to achieve control; establishing "no-smoking" zones around storage and use areas is another. Section 2703.7 addresses smoking and open flames, as well as industrial trucks. No open flames are allowed because of explosion hazard. Industrial trucks must meet the requirements of NFPA 505 and be listed and labeled for use in areas designated as hazardous.

3503.1.5 Liquefied flammable gases and flammable gases in solution. Containers of liquefied flammable gases and flammable gases in solution shall be positioned in the upright position or positioned so that the pressure relief valve is in direct contact with the vapor space of the container.

Exceptions:

1. Containers of flammable gases in solution with a capacity of 1.3 gallons (5 L) or less.
2. Containers of flammable liquefied gases, with a capacity not exceeding 1.3 gallons (5 L), designed to preclude the discharge of liquid from safety relief devices.

❖ Proper storage and handling of containers of liquified flammable gases and flammable gases avoids many possible incidents. Careful handling and securing the cylinders at all times helps protect against hazards resulting from the containers falling.

3503.2 Quantities exceeding the maximum allowable quantity per control area. The storage and use of flammable gases in amounts exceeding the maximum allowable quantity per control area indicated in Section 2703.1 shall be in accordance with Chapter 27 and this chapter.

❖ Section 2703.1.4 addresses quantities that exceed the maximum allowable per control area. The limits on the quantities are located in Tables 2703.1.1(1) through 2703.1.1(4).

SECTION 3504 STORAGE

3504.1 Indoor storage. Indoor storage of flammable gases in amounts exceeding the maximum allowable quantity per control area indicated in Table 2703.1.1(1), shall be in accordance with Sections 2701, 2703 and 2704, and this chapter.

❖ Sections 2703.1 through 2703.11 give the requirements for the indoor storage of flammable gases, including design and construction; equipment; maintenance; markings and signs.

3504.1.1 Explosion control. Buildings or portions thereof containing flammable gases shall be provided with explosion control in accordance with Section 911.

❖ Table 911.1 gives the barricade and explosion control requirements for flammable gases. Section 911 gives deflagration, explosion venting and barricade requirements.

3504.2 Outdoor storage. Outdoor storage of flammable gases in amounts exceeding the maximum allowable quantity per control area indicated in Table 2703.1.1(3) shall be in accordance with Sections 2701, 2703 and 2704, and this chapter.

❖ Section 2701 covers the classification of the hazard of flammable gases. The maximum quantities per control area of the gases are given in Table 2703.1.1(3). Section 2704 deals with the storage of flammable gases. The section covers ventilation, separation, fire-extinguishing systems, emergency power, limit controls, supervision and weather protection.

3504.2.1 Outdoor storage areas. Outdoor storage areas for flammable gases shall be located in accordance with Table 3504.2.1.

❖ All flammable gases must be kept in a dry location within the distances specified in Table 3504.2.1. The cylinders must also be protected against direct rays of the sun, accumulation of ice and snow and access by unauthorized persons.

TABLE 3504.2.1
FLAMMABLE GASES DISTANCE FROM OUTDOOR STORAGE AREAS TO EXPOSURES[a]

AGGREGATE QUANTITY PER STORAGE AREA (cubic feet)	MINIMUM DISTANCE TO BUILDINGS, PUBLIC STREETS, PUBLIC ALLEYS, PUBLIC WAYS OR LOT LINES (feet)	MINIMUM DISTANCE BETWEEN STORAGE AREAS (feet)
0-4,225	5	5
4,226-21,125	10	10
21,126-50,700	15	10
50,701-84,500	20	10
84,501 or greater	25	20

For SI: 1 foot = 304.8 mm, 1 cubic foot = 0.02832 m^3.

a. The minimum required distances shall be reduced to 5 feet when protective structures having a minimum fire-resistance rating of 2 hours interrupt the line of sight between the container and the exposure. The protective structure shall be at least 5 feet from the exposure. The configuration of the protective structure shall be designed to allow natural ventilation to prevent the accumulation of hazardous gas concentrations.

❖ The table gives the separation distance, in feet, of aggregate quantities of flammable gases. The first column has the aggregate quantities in six different ranges. According to the amount for the location, the second column gives the distance from public ways, lot lines, alleys and buildings. The third column gives the separation distance between storage areas.

SECTION 3505
USE

3505.1 General. The use of flammable gases in amounts exceeding the maximum allowable quantity per control area indicated in Table 2703.1.1(1) or 2703.1.1(3) shall be in accordance with Sections 2701, 2703 and 2705, and this chapter.

❖ Section 2705 addresses separation, spill control, limit controls, lighting, fire-extinguishing systems, open systems, ventilation, outdoor dispensing and handling of flammable gases.

Bibliography

The following resource materials are referenced in this chapter or are relevant to the subject matter addressed in this chapter.

ASTM E 681-98, *Standard Test Method for Concentration Limits of Flammability of Chemicals* (vapors and gases). West Conshohocken, PA: ASTM International, 1998.

Barlen W., F.C. Saacke and G.R. Spies. *Industrial Fire Hazards Handbook*, 3rd ed. Quincy, MA: National Fire Protection Association, 1990.

DOT 29 CFR; 1910-74, *Occupational Safety and Health Standards*. Washington, DC: U.S. Department of Labor, 1974.

Fire Protection Guide to Hazardous Materials, 13th ed. Quincy MA: National Fire Protection Association, 2001.

Fire Protection Handbook, 18th ed. Quincy MA: National Fire Protection Association, 1997.

Handbook of Compressed Gases, 3rd ed. Arlington, VA: Compressed Gas Association, 1999.

IBC-2003, *International Building Code*. Falls Church, VA: International Code Council, 2003.

IFGC-2003, *International Fuel Gas Code*. Falls Church, VA: International Code Council, 2003.

IMC-2003, *International Mechanical Code*. Falls Church, VA: International Code Council, 2003.

Klein, B. R., ed. *Health Care Facilities Handbook*, 5th ed. Quincy, MA: National Fire Protection Association.

NFPA 30-00, *Flammable and Combustible Liquids Code*. Quincy, MA: National Fire Protection Association, 2000.

NFPA 50-01, *Bulk Oxygen Systems at Consumer Sites*. Quincy, MA: National Fire Protection Association, 2001.

NFPA 50A-99, *Gaseous Hydrogen at Consumer Sites*. Quincy, MA: National Fire Protection Association, 1999.

NFPA 68-02, *Deflagration Venting*. Quincy, MA: National Fire Protection Association, 2002.

NFPA 69-97, *Explosion Prevention Systems*. Quincy, MA: National Fire Protection Association, 1997.

NFPA 70-02, *National Electrical Code*. Quincy, MA: National Fire Protection Association, 2002.

NFPA 99-99, *Health Care Facilities*. Quincy, MA: National Fire Protection Association, 1999.

NFPA 505-99, *Fire Safety Standard for Powered Industrial Trucks*. Quincy, MA: National Fire Protection Association, 1999.

NFPA 704-96, *Identification of Fire Hazards of Material*. Quincy, MA: National Fire Protection Association, 1996.

Chapter 36: Flammable Solids

General Comments

Although this chapter addresses magnesium almost to the exclusion of all other flammable solids, it is important to know that several other solid materials, primarily metals, are also flammable and under the right conditions can be explosion hazards.

The list of other metals that can become fire hazards consists of titanium, zirconium, hafnium, calcium, zinc, sodium, lithium, potassium, sodium/potassium alloys, aluminum, iron and steel, uranium, thorium and plutonium. Some of these metals have few highly specialized commercial uses; they are almost exclusively laboratory materials. But where they are used, both plant and fire service personnel must be trained to handle emergency situations. Because uranium, thorium and plutonium are also radioactive materials, they present still more specialized problems for plant fire brigades and local fire service personnel.

The form of the material being used (powder, sheets, castings or billets) also is critical to the way fire services respond to an incident. Fine powders of any of the materials listed can ignite or even explode under various atmospheres, including nitrogen. Some molten metals can ignite or explode under certain conditions. Castings of some of these metals can ignite or detonate if they are not handled properly. Even bulky billets can be ignited if there is sufficient heat to bring the metal to its ignition temperature, resulting in self-sustained burning.

Conventional fire-extinguishing agents may only increase the intensity of the fire being fought. Magnesium, for example, burns fiercely in a steam atmosphere. Likewise, carbon dioxide, foam and dry chemical extinguishers are not effective on titanium fires. Additionally, water, foam and vaporizing liquids should never be used on lithium, sodium and potassium fires. Each material is different and requires different extinguishing treatment.

The National Fire Protection Association (NFPA) has developed standards for handling several of the listed materials, which are included in the bibliography at the end of this chapter. Most industry associations and companies manufacturing primary metals and alloys also have available recommended practices for storage, handling, use and scrap disposal that are based on extensive testing and usage history. The fire code official should require any person or business storing, handling or processing any of these materials to demonstrate a thorough knowledge of safe practices in both facility design and operating procedures.

Purpose

This chapter sets general requirements for storage and handling of flammable solids in the first five sections before addressing the subject of magnesium in Section 3606.

A word of caution is necessary when applying the general requirements stated in Sections 3601 through 3605. Each of the flammable metals mentioned in the "General Comments" section requires special precautions. What works with one material may cause a major disaster with another. It is always best to make certain the building, facility and operating conditions proposed by the owner, architect or builder are based on the latest safety information available from the suppliers of the materials to be used or stored. The fire code official must ensure that the emergency plan prepared for use by the owner or tenant is workable and compatible with the emergency response personnel and equipment available to the jurisdiction.

SECTION 3601 GENERAL

3601.1 Scope. The storage and use of flammable solids shall be in accordance with this chapter.

❖ This section establishes the applicability of this chapter to all materials meeting the definition of and classified as flammable solids.

3601.2 Permits. Permits shall be required as set forth in Section 105.6.

❖ The process of issuing permits gives the fire code official an opportunity to carefully evaluate and regulate hazardous operations. Permit applicants should be required to demonstrate that their operations comply with the intent of the code before the permit is issued. See the commentary to Section 105.6 for a general discussion of operations requiring an operational permit, Section 105.6.21 for a discussion of specific quantity-based operational permits for the materials regulated in this chapter and Section 105.7 for a general discussion of activities requiring a construction permit. The permit process also notifies the fire department of the need for prefire planning for hazardous property.

SECTION 3602 DEFINITIONS

3602.1 Definitions. The following words and terms shall, for the purposes of this chapter and as used elsewhere in this code, have the meanings shown herein.

❖ Definitions of terms can help in the understanding and application of the code requirements. The purpose for including those definitions that are associated with the subject matter of this chapter is to provide more convenient access to them without having to refer back to Chapter 2. It is important to emphasize that these terms are not exclusively related to this chapter but are applicable everywhere the term is used in the code. For convenience, these terms are also listed in Chapter 2 with a cross reference to this section. The use and application of all defined terms, including those defined in this section, are set forth in Section 201. Additional hazardous material definitions are given in Chapter 27.

FLAMMABLE SOLID. A solid, other than a blasting agent or explosive, that is capable of causing fire through friction, absorption or moisture, spontaneous chemical change, or retained heat from manufacturing or processing, or which has an ignition temperature below 212°F (100°C) or which burns so vigorously and persistently when ignited as to create a serious hazard. A chemical shall be considered a flammable solid as determined in accordance with the test method of CPSC 16 CFR; Part 1500.44, if it ignites and burns with a self-sustained flame at a rate greater than 0.1 inch (2.5 mm) per second along its major axis.

❖ Flammable solids include various materials that either ignite readily, burn vigorously or are difficult to extinguish. Materials that may not ignite easily or burn vigorously in bulk form may do so in finely divided form. This is especially true of most flammable metals. Ignition sources for flammable solids include frictional heat from machining or cutting operations; absorption of moisture from air (as opposed to water-reactive materials forming flammable vapors when mixed with water) spontaneous chemical changes such as sublimation (the chemical process through which solids emit vapors without first changing phase to liquids); and heat absorbed during manufacturing processes like oil quenching or heat treating. Solid materials with ignition temperatures below 212°F (100°C) that ignite before melting are also considered flammable solids, as are materials burning robustly and persistently when ignited, including magnesium and coal.

The Consumer Product Safety Commission (CPSC) has developed a standard test method (CPSC 16 CFR; 1500.44) that is referenced for determining when a material complies with the definition. Figure 3602.1(1) depicts the equipment and test method. A material burning at a rate greater than 0.1 inch (2.5 mm) per second is considered a flammable solid for the purpose of apply-

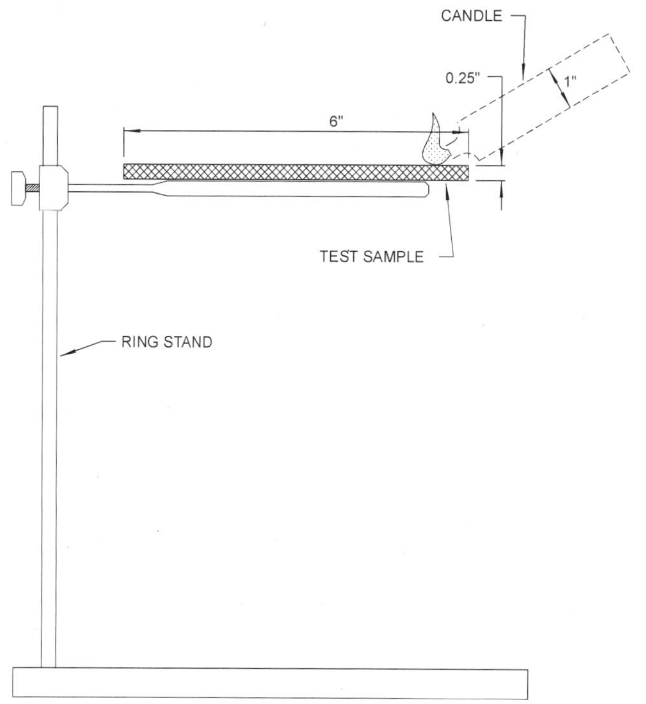

For SI: 1 inch = 25.4 mm.

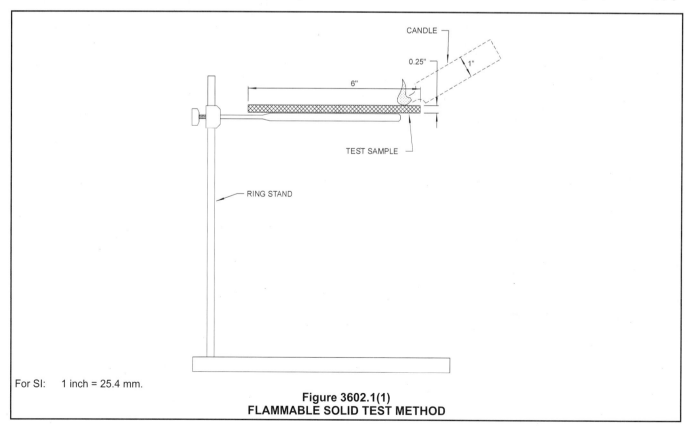

Figure 3602.1(1)
FLAMMABLE SOLID TEST METHOD

ing the requirements of this chapter.

Table 3602.1(1) lists both the melting points and ignition temperatures for several pure metals in bulk form. These materials may ignite at much lower temperatures when finely divided. Moreover, many metals, such as calcium, hafnium, plutonium, sodium, thorium and zirconium, will ignite in air under certain conditions. Thorium and plutonium will release radiation when they burn. Likewise, many metals react with each other in finely divided form. For example, iron or steel filings and fine magnesium particles, combined with frictional heat or cutting oil, can ignite in a thermite reaction. Aluminum, iron and steel are not usually recognized as combustible metals; however, aluminum, iron and steel can be ignited in powdered form. Rather than producing an open flame, both iron and steel produce a vigorous sparking reaction when ignited. On the other hand, aluminum may burn with explosive force.

Table 3602.1(2) describes fire hazards of selected common flammable solids that are regulated by this chapter.

MAGNESIUM. The pure metal and alloys, of which the major part is magnesium.

❖ Magnesium is a silvery-white combustible metal weighing only two-thirds as much as aluminum and having good structural properties when suitably alloyed. For this reason, magnesium alloys are used to a great extent in the construction of aircraft; automobiles and trucks; household appliances; furniture; office equipment, machine parts and numerous other applications. Powdered magnesium is used in signal flares and other fireworks to produce an intense white light.

The melting point of pure magnesium is 1,202°F (650°C). The ignition temperature is generally considered to be very close to the melting point, but ignition of magnesium in certain forms may occur at lower air temperatures. Magnesium ribbon and fine magnesium shavings can be ignited under some conditions at temperatures of 950°F (510°C), and very finely divided magnesium powder has been ignited at an air temperature below 900°F (482°C).

The ease of ignition of magnesium depends on a large extent upon the size and shape of the material as well as the intensity of the ignition source. The flame of a match may be sufficient to ignite magnesium ribbon, shavings or chips with thin feather edges, and a spark will ignite fine dust such as is produced in grinding operations. Heavier pieces, such as ingots or thick-walled castings, are difficult to ignite because heat is rapidly conducted away from a localized ignition source. If the entire piece of metal can be raised to the ignition temperature, however, self-sustained burning will occur.

Because the melting point of magnesium is low, the metal melts as it burns and, after some minutes of burning, produces puddles of molten magnesium. The production of molten metal will depend to a considerable extent on the physical condition of the material. Finely divided magnesium, such as shavings, dust and small scraps, will burn more rapidly and produce less molten metal than will an equal quantity of magnesium in the more solid form of ingots or castings.

Metal products marketed under a variety of trade names and designations and commonly referred to as "magnesium" may, in fact, be one of a large number of alloys containing widely differing percentages of mag-

TABLE 3602.1(1)
MELTING AND IGNITION TEMPERATURES OF SELECTED PURE METALS IN SOLID FORM

Material	Melting point °F	Ignition temperature °F
Aluminum[a]	1,220	1,832
Barium	1,337	347
Calcium	1,548	1,300
Hafnium	4,032	—
Iron[b]	2,795	1,706
Lithium	367	356
Magnesium	1,202	1,153
Plutonium	1,184	1,112
Potassium[c]	144	156
Sodium	208	239
Strontium	1,425	1,328
Thorium	3,353	932
Titanium	3,140	2,900
Uranium[d]	2,070	6,900
Zinc	786	1,652
Zinconium	3,326	2,552

Note a. Above indicated temperature.
Note b. Ignition in oxygen.
Note c. Spontaneous ignition in moist air.
Note d. Below indicated temperature.

nesium, aluminum, zinc and manganese. Some of these alloys have melting points and ignition temperatures considerably lower than that of pure magnesium.

SECTION 3603 GENERAL REQUIREMENTS

3603.1 Quantities not exceeding the maximum allowable quantity per control area. The storage and use of flammable solids in amounts not exceeding the maximum allowable quantity per control area as indicated in Section 2703.1 shall be in accordance with Sections 2701, 2703 and 3601.

❖ This section complements the requirements of Chapter 27 in structures occupied for the storage, handling or use of flammable solids. The regulations assume that the quantity of flammable solids in a given building is limited to the maximum allowable quantity per control area as established in Section 2703.1; thus, the building is not classified in Occupancy Group H. The general requirements of Sections 2701 and 2703 are fully applicable to the storage and use of flammable solids, in addition to the requirements of this chapter.

3603.2 Quantities exceeding the maximum allowable quantity per control area. The storage and use of flammable solids exceeding the maximum allowable quantity per control area as indicated in Section 2703.1 shall be in accordance with Chapter 27 and this chapter.

❖ This section complements the requirements of Chapter 27 for structures occupied for the storage, handling or

TABLE 3602.1(2)
COMMON FLAMMABLE SOLIDS AND THEIR PROPERTIES

Material	Description
Carbon	
Carbon black	Carbon black is formed by combustion of certain gaseous hydrocarbons and hydrocarbon cracking. It is most hazardous after manufacture when particles may still be hot. Carbon black absorbs oxygen while cooling and slow may develop. After cooling, the material is not subject to spontaneous heating. A Mixture of carbon black and oxidizable oils may produce heating.
Lamp black	Lamp black is formed by incomplete burning of carbonaceous oils. It absorbs gases to some degree and has a strong affinity for liquids. Heats when in contact with drying oils and may ignite spontaneously soon after bagging begins.
Lead sulfocyanate	Burns slowly and decomposes to form flammable and toxic hydrogen disulfide and toxic carbon disulfide when heated.
Nitroaniline	Melts at 295°F with a flash point of 390°F. When in contact with organic materials it may produce spontaneous ignition.
Nitrochlorobenzene	A solid material giving off flammable vapors when heated (sublimation).
Sulfides	
Antimony pentasulfide	Antimony pentasulfide is readily ignited when in contact with oxidizing materials and yields flammable and toxic hydrogen sulfide when in contact with strong acids.
Phosphorus pentasulfide	Phosphorus pentasulfide ignites readily and is subject to spontaneous heating in the presence of moisture. The ignition temperature is 287°F. Phosphorus pentasulfide produces toxic sulfur dioxide and phosphorus pentoxide when it burns, as well as flammable and toxic hydrogen sulfide when in contact with water.
Phosphorus sesquisulfide	Phosphorus sesquisulfide is highly flammable and ignites at 212°F to produce toxic sulfur dioxide.
Potassium and sodium sulfides	Both potassium and sodium sulfide are moderately flammable and they produce sulfur dioxide when burning and hydrogen sulfide comes in contact with acids.
Sulfur	The melting point is 234°F and the boiling point is 832°F with a flash point of 405°F. Sulfur vapors are highly flammable in air. Sulfur dust is a severe explosion hazard with ignition temperatures in the range of 274°F.
Naphthalene	Combustible in both solid and liquid form. Vapors and dusts form explosive mixtures in air.

For SI: °C = [(°F) - 32]/1.8.

use of flammable solids. The regulations contained in this section assume that the quantity of flammable solids in a given building is in excess of the maximum allowable quantity per control area as established in Section 2703.1; thus, the building is classified in Occupancy Group H. The requirements of Chapter 27 apply to the storage and use of flammable solids, in addition to the requirements of this chapter.

SECTION 3604
STORAGE

3604.1 Indoor storage. Indoor storage of flammable solids in amounts exceeding the maximum allowable quantity per control area indicated in Table 2703.1.1(1) shall be in accordance with Sections 2701, 2703, 2704 and this chapter.

❖ This section regulates the indoor storage of flammable solids when in excess of the maximum allowable quantity per control area in buildings or portions of buildings classified in Occupancy Group H. The general and storage requirements of Chapter 27 are applicable in addition to the requirements of this section. Storage of flammable solids inside of structures must comply with Sections 3604.1.1 through 3604.1.3 to prevent exposure to conditions that may result in a fire or explosion.

3604.1.1 Pile size limits and location. Flammable solids stored in quantities greater than 1,000 cubic feet (28 m^3) shall be separated into piles each not larger than 1,000 cubic feet (28 m^3).

❖ Storage piles are restricted to 1,000 cubic feet (28 m^3) in size to limit the quantity of flammable solids exposed to a single fire and to facilitate fire-fighting operations. Aisles must be provided on all sides to permit access for fire fighting and reduce the likelihood of the spread of fire to adjacent piles if a pile collapse occurs.

3604.1.2 Aisles. Aisle widths between piles shall not be less than the height of the piles or 4 feet (1219 mm), whichever is greater.

❖ As with all other storage, aisles allow fire response personnel ready access to the immediate area of the fire. The requirement that aisle width depends on pile height acknowledges the effectiveness of physical separation in preventing fire spread as well as making room for more or larger fire-fighting equipment that may be needed to fight fires in large storage piles of flammable solids.

3604.1.3 Basement storage. Flammable solids shall not be stored in basements.

❖ Basement storage is prohibited because of the limited access for fire fighters in most basements and the hazards associated with the vigorous and persistent fires that can occur in flammable solids.

3604.2 Outdoor storage. Outdoor storage of flammable solids in amounts exceeding the maximum allowable quantities per control area indicated in Table 2703.1.1(3) shall be in accordance with Sections 2701, 2703, 2704 and this chapter. Outdoor storage of magnesium shall be in accordance with Section 3606.

❖ This section regulates the outdoor storage of flammable solids when in excess of the maximum allowable quantity per outdoor control area established by Table 2703.1.1(3). The general and storage requirements of Chapter 27 are applicable in addition to the requirements of this section. Storage of flammable solids in outdoor control areas must comply with Sections 3604.2.1 and 3404.2.2 to reduce the likelihood of uncontrolled release or exposure to conditions that may result in a fire or explosion.

3604.2.1 Distance from storage to exposures. Outdoor storage of flammable solids shall not be located within 20 feet (6096 mm) of a building, lot line, public street, public alley, public way or means of egress. A 2-hour fire barrier without openings or penetrations and extending 30 inches (762 mm) above and to the sides of the storage area is allowed in lieu of such distance. The wall shall either be an independent structure, or the exterior wall of the building adjacent to the storage area.

❖ The required separation distance is intended to minimize radiant heat transfer between exposures. Separation distances provide a measure of protection against the possibility of fire spread if a fire occurs involving either the stored material or another exposure, such as a building located on the same or an adjacent lot or a vehicle in the public right-of-way.

This section also recognizes the concept that a 2-hour fire barrier is an equivalent means of achieving this objective. Where a separation assembly is installed in place of separation distance, the wall must extend vertically at least 30 inches (762 mm) beyond the roof or wall opening of the larger structure on each side (top and sides) to prevent a fire from lapping over or extending around the wall. Fire barriers must be constructed in accordance with the *International Building Code®* (IBC®).

3604.2.2 Pile size limits. Outdoor storage of flammable solids shall be separated into piles not larger than 5,000 cubic feet (141 m^3) each. Piles shall be separated by aisles with a minimum width of not less than one-half the pile height or 10 feet (3048 mm), whichever is greater.

❖ Outside storage piles may be increased in size (over inside storage) to 5,000 cubic feet (141 m^3) based on the reduced danger to people and property associated with outdoor storage. Required aisles are intended to facilitate fire-fighting access and prevent the spread of fire to adjacent piles if a pile collapse occurs. Aisles should permit unobstructed access to the pile on all sides, as well as permit approach from more than one direction.

SECTION 3605
USE

3605.1 General. The use of flammable solids in amounts exceeding the maximum allowable quantity per control area indicated in Table 2703.1.1(1) or 2703.1.1(3) shall be in accordance with Sections 2701, 2703, 2705 and this chapter.The use of magnesium shall be in accordance with Section 3606.

❖ This section applies to all indoor and outdoor use and handling operations involving flammable solids, except magnesium, when the amounts being used or handled are in excess of the maximum allowable quantities per indoor control area (buildings or portions of buildings classified in Occupancy Group H) and outdoor control area indicated in Tables 2703.1.1(1) and 2703.1.1(3), respectively. The administrative, general use and handling provisions of Chapter 27 are applicable, in addition to the requirements of this chapter.

SECTION 3606
MAGNESIUM

3606.1 General. Storage, use, handling and processing of magnesium, including the pure metal and alloys of which the major part is magnesium, shall be in accordance with Chapter 27 and this section.

❖ This section establishes the applicability of Section 3606 and Chapter 27 to the production, storage, processing and disposal of magnesium products. These requirements are applicable not only to pure magnesium but also to alloys having a magnesium content in excess of 50 percent.

3606.2 Storage of magnesium articles. The storage of magnesium shall comply with Sections 3606.2.1 through 3606.4.3.

❖ The requirements of Sections 3606.2.1 through 3606.4.3 apply to the storage of any quantity of magnesium based on the various physical forms of the material and the fire hazard and extinguishing problems it poses.

3606.2.1 Storage of greater than 50 cubic feet. Magnesium storage in quantities greater than 50 cubic feet (1.4 m^3) shall be separated from storage of other materials that are either combustible or in combustible containers by aisles. Piles shall be separated by aisles with a minimum width of not less than the pile height.

❖ This is a general criterion applicable to all forms of magnesium. As is noted in the following sections, storage requirements and allowable quantities vary with the form of the product being stored.

Isolation of magnesium from other combustible materials of any kind helps reduce the amount of magnesium exposed to a single fire originating outside of the magnesium pile and protects materials outside the magnesium pile from exposure to it. Properly established and maintained aisles provide fire suppression personnel ready access to the immediate area of the fire as well as proper egress circulation within the storage area. The requirement that aisle width depends on pile height acknowledges the effectiveness of physical separation in preventing fire spread.

3606.2.2 Storage of greater than 1,000 cubic feet. Magnesium storage in quantities greater than 1,000 cubic feet (28 m^3) shall be separated into piles not larger than 1,000 cubic feet (28 m^3) each. Piles shall be separated by aisles with a minimum width of not less than the pile height. Such storage shall not be located in nonsprinklered buildings of Type III, IV or V construction, as defined in the *International Building Code*.

❖ Again, this is a general criterion applicable to all forms of magnesium and establishes the maximum pile size at 1,000 cubic feet (28 m^3) of material (see also the discussion in the commentary to Section 3606.2.1). In this scenario, a pile could be approximately 10 feet long by 10 feet wide by 10 feet high (3048 mm by 3048 mm by 3048 mm) with an established aisle width between piles of 10 feet (3048 mm). Note that when the quantity of stored magnesium exceeds 1,000 cubic feet (28 m^3), this section requires that the storage building be of Type I or II construction or, when equipped throughout with an automatic sprinkler system, of Type III, IV or V construction.

3606.2.3 Storage in combustible containers or within 30 feet of other combustibles. Where in nonsprinklered buildings of Type III, IV or V construction, as defined in the *International Building Code*, magnesium shall not be stored in combustible containers or within 30 feet (9144 mm) of other combustibles.

❖ This section recognizes the increased hazard of storing magnesium in unsprinklered buildings of Type III, IV and V construction by requiring that any containers used for the storage of magnesium be constructed of noncombustible material to provide a layer of shielding to the magnesium in the event of a fire. To further isolate magnesium piles, reduce the likelihood of fire spread among adjacent piles and enhance the effectiveness of the shielding provided by noncombustible storage containers where used, this section requires a substantial increase in the spatial separation between the magnesium and any other combustible materials. Note that this increased separation is not related to pile height.

3606.2.4 Storage in foundries and processing plants. The size of storage piles of magnesium articles in foundries and processing plants shall not exceed 1,250 cubic feet (25 m^3). Piles shall be separated by aisles with a minimum width of not less than one-half the pile height.

❖ Allowing storage of an increased quantity of flammable solids in foundries and processing plants acknowledges that ignition of pigs, ingots and billets is unlikely under conditions normally found in foundries and other processing buildings where ordinary combustible materials are rarely found in any significant quantity.

3606.3 Storage of pigs, ingots and billets. The storage of magnesium pigs, ingots and billets shall comply with Sections 3606.3.1 and 3606.3.2.

❖ The requirements of Sections 3606.3.1 and 3606.3.2 apply to the storage of any quantity of magnesium in the forms of pigs (a mass of magnesium that has been run into a mold while molten), ingots (a mass of magnesium cast in a convenient form for further processing) and billets (a bar of magnesium forged from an ingot) and the hazards they pose. The quantities allowed again indicate that ignition of these large shapes is unlikely under the conditions required.

3606.3.1 Indoor storage. Indoor storage of pigs, ingots and billets shall only be on floors of noncombustible construction. Piles shall not be larger than 500,000 pounds (226.8 metric tons) each. Piles shall be separated by aisles with a minimum width of not less than one-half the pile height.

❖ This section establishes maximum quantity for single piles, but does not restrict the number of piles that can be located in one building or structure. The requirement for aisles that are at least half the pile height allows adequate clearance for both materials-handling equipment and emergency response equipment and personnel. Where pile heights are kept low, the aisles must still be maintained at a width that will allow for equipment travel between piles. Requiring storage on a noncombustible surface reduces the likelihood that the floor would contribute any fuel or contribute to the spread of a fire involving magnesium.

3606.3.2 Outdoor storage. Outdoor storage of magnesium pigs, ingots and billets shall be in piles not exceeding 1,000,000 pounds (453.6 metric tons) each. Piles shall be separated by aisles with a minimum width of not less than one-half the pile height. Piles shall be separated from combustible materials or buildings on the same or adjoining property by a distance of not less than the height of the nearest pile.

❖ This section recognizes the inherently higher level of safety provided by the storage of materials outdoors by doubling the amount of magnesium per pile that can be stored. The required separation distance is intended to minimize radiant heat transfer between exposures. Separation distances provide a measure of protection against the possibility of fire spread if a fire occurs either in the stored material or in another exposure, such as a building located on the same or an adjacent lot. Separation also serves the purpose of allowing passage of materials-handling equipment and emergency response equipment.

3606.4 Storage of fine magnesium scrap. The storage of scrap magnesium shall comply with Sections 3606.4.1 through 3606.4.3.

❖ The requirements of Sections 3606.4.1 through 3606.4.3 apply to the storage of any quantity of magnesium in the form of scrap chips, fines and dust and the hazards they pose. These less dense forms of magnesium, typically produced in machine processing, are recovered for subsequent reuse and present a substantial fire and explosion risk, which is addressed in the following sections.

3606.4.1 Separation. Magnesium fines shall be kept separate from other combustible materials.

❖ Separation of stored magnesium fines from ordinary combustible materials is required because fines are extremely combustible and easily ignitable. A small pile of fines can be ignited by a common match flame. Section 3606.2.1 contains minimum separation criteria for all types of magnesium storage.

3606.4.2 Storage of 50 to 1,000 cubic feet. Storage of fine magnesium scrap in quantities greater than 50 cubic feet (1.4 m^3) [six 55-gallon (208 L) steel drums] shall be separated from other occupancies by an open space of at least 50 feet (15 240 mm) or by a fire barrier constructed in accordance with the *International Building Code.*

❖ Because fines are usually wet with coolant from the processing operation, there is a possibility of hydrogen generation. For this reason, fines must be stored in approved steel containers with vented lids to prevent hydrogen buildup. Because there is also a possibility of spontaneous heating of fines, they must be stored separate from combustible materials, including other storage piles of magnesium. The required separation distance is intended to minimize radiant heat transfer between the stored magnesium and other materials and provide a measure of protection against the possibility of spread if a fire occurs.

This section also recognizes the concept that a 2-hour fire barrier is an equivalent means of achieving this objective. Where a separation assembly is installed in place of a separation distance, the wall must extend vertically at least 30 inches (762 mm) beyond the roof or wall opening of the larger structure on each side (top and sides) to prevent a fire from lapping over or extending around the wall. Fire barriers must be constructed in accordance with the IBC.

3606.4.3 Storage of greater than 1,000 cubic feet. Storage of fine magnesium scrap in quantities greater than 1,000 cubic feet (28 m^3) shall be separated from all buildings other than those used for magnesium scrap recovery operations by a distance of not less than 100 feet (30 480 mm).

❖ This section recognizes the increased risk associated with the storage of significant quantities of magnesium fines by establishing a blanket 100-foot (30 480 mm) separation distance to any building that is not specifically part of a magnesium recovery operation.

3606.5 Use of magnesium. The use of magnesium shall comply with Sections 3606.5.1 through 3606.5.8.

❖ The requirements of Sections 3606.5.1 through 3606.5.8 apply to the use, handling and processing of any quantity of magnesium and the processing hazards it poses.

3606.5.1 Melting pots. Floors under and around melting pots shall be of noncombustible construction.

❖ This requirement should seem obvious for any operation that involves melting any quantity of any metal. It is particularly important with flammable metals, however, because of the possibility of fire from an overheated melting pot or an explosion caused by the addition of alloying metal to the molten magnesium that has not been thoroughly dried before being added to the pot. Any quantity of absorbed moisture on the added metal will turn instantly to steam and cause a violent eruption of molten metal from the pot.

Eruptions of molten metal present a serious burn hazard to personnel as well as the potential for the hot metal to ignite combustible construction under or near the pot.

When molten magnesium is handled according to standard industry safety practices, it is not a serious hazard. Magnesium has been used for well over 50 years in both military and commercial applications with very few serious fires. This long history of safe operations suggests that safety rules and precautions developed by both suppliers and users are effective. The fire code official must require that a user of molten magnesium have safe operating procedures in place as well as an emergency response plan that includes both plant fire brigades and jurisdiction emergency response units.

3606.5.2 Heat-treating ovens. Approved means shall be provided for control of magnesium fires in heat-treating ovens.

❖ Heat-treating ovens present a significant potential for fires. Some of the heat treatments required to obtain the physical properties necessary for certain applications are done at temperatures very close to the ignition temperature of magnesium. Accordingly, this section requires that fire control means approved by the fire code official be readily available to the heat-treating process. Large castings that have thick sections are not as likely to ignite, but fine fins or very thin sections can ignite if overheated, as can dust or fine chips. For this reason, oven controls are critical to safe operation. Making sure items to be heat treated are free of dust and chips is also essential. Fire control can also be achieved in higher temperature ovens by operating them with an inert gas atmosphere to reduce the risk of magnesium ignition.

Although not stated specifically in this section, the requirement for keeping the heat-treating area clear of scrap and other combustibles should be obvious.

3606.5.3 Dust collection. Magnesium grinding, buffing and wire-brushing operations, other than rough finishing of castings, shall be provided with approved hoods or enclosures for dust collection which are connected to a liquid-precipitation type of separator that converts dust to sludge without contact (in a dry state) with any high-speed moving parts.

❖ The potential for fires and explosions involving magnesium dust and other fines has already been stated. The dust from grinding, buffing and wirebrushing operations must be collected and contained in a closed system that is equipped with a waterspray dust precipitator and an exhaust blower. The exhaust system is required because wet magnesium fines will generate hydrogen gas that must be diluted below its lower flammability level.

Equipment used for grinding, buffing and wirebrushing must be dedicated for use on magnesium only. Figure 3606.5.3 shows a typical grinding machine setup.

3606.5.3.1 Duct construction. Connecting ducts or suction tubes shall be completely grounded, as short as possible, and without bends. Ducts shall be fabricated and assembled with a smooth interior, with internal lap joints pointing in the direction of airflow and without unused capped side outlets, pockets or other dead-end spaces which allow an accumulation of dust.

❖ Although the requirements stated in this section are applicable to magnesium manufacturing operations, they are virtually the same as requirements for handling any other kind of flammable or detonable dust and fine particles. The construction specifications are intended to prevent fines from accumulating in sufficient quantities to become a hazard. Grounding is required to prevent sparking that could become a source of ignition. See the commentary to Sections 510 and 511 of the *International Mechanical Code*® (IMC®) for a discussion of hazardous exhaust systems and dust conveying systems.

3606.5.3.2 Independent dust separators. Each machine shall be equipped with an individual dust-separating unit.

Exceptions:

1. One separator is allowed to serve two dust-producing units on multiunit machines.
2. One separator is allowed to serve not more than four portable dust-producing units in a single enclosure or stand.

❖ This requirement complements the one in Section 3606.5.3.1 that ducts be straight and of minimum length in order to reduce the likelihood of static buildup and dust buildup within the duct.

Exception 1 does allow two units working in tandem to be served by one collector; however, the ducts must still meet the specifications in Section 3606.5.3.1.

Exception 2 allows one collector to serve up to four small units if they are operated in an enclosure that would prevent dust from escaping into the general work area. The collector would function in approximately the same way as a vacuum system in a paint booth. The requirements for exhausting, water spray and ductwork would still apply.

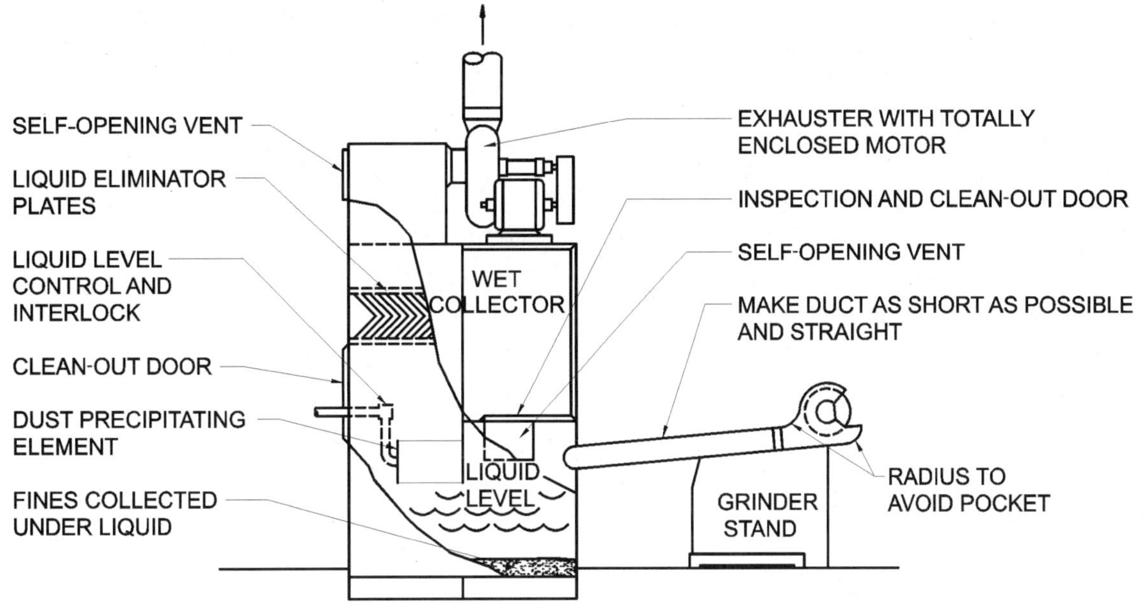

Figure 3606.5.3
A SCHEMATIC DIAGRAM OF WATER PRECIPITATION-TYPE COLLECTOR

3606.5.4 Power supply interlock. Power supply to machines shall be interlocked with exhaust airflow, and liquid pressure level or flow. The interlock shall be designed to shut down the machine it serves when the dust removal or separator system is not operating properly.

❖ The power supply to the machines used for grinding, buffing, drilling or brushing magnesium must be interlocked with the exhaust equipment and dust collector to prevent or automatically shut down operations when either it is not operating as designed or not operating at all.

3606.5.5 Electrical equipment. Electric wiring, fixtures and equipment in the immediate vicinity of and attached to dust-producing machines, including those used in connection with separator equipment, shall be of approved types and shall be approved for use in Class II, Division 1 hazardous locations in accordance with the ICC *Electrical Code.*

❖ This requirement for classified electrical equipment is intended to reduce the likelihood that any fixed or portable electrical wiring and electrical equipment attached to, located or used near dust-producing and dust-collecting machines will become an ignition source.

3606.5.6 Grounding. Equipment shall be securely grounded by permanent ground wires in accordance with the ICC *Electrical Code.*

❖ As with Section 3606.5.5, this requirement is intended to reduce the likelihood of a buildup of static electricity that could produce a spark that would be an ignition source for flammable solids dust. This requirement is similar to grounding requirements for electrical equipment in other dust-producing environments.

3606.5.7 Fire-extinguishing materials. Fire-extinguishing materials shall be provided for every operator performing machining, grinding or other processing operation on magnesium as follows:

1. Within 30 feet (9144 mm), a supply of extinguishing materials in an approved container with a hand scoop or shovel for applying the material; or
2. Within 75 feet (22860 mm), a portable fire extinguisher complying with Section 906.

All extinguishing materials shall be approved for use on magnesium fires. Where extinguishing materials are stored in cabinets or other enclosed areas, the enclosures shall be openable without the use of a key or special knowledge.

❖ Magnesium fires present unusual fire suppression challenges because none of the common extinguishing materials can be used safely. The problem of using water has already been mentioned, but it must be emphasized again. Water sprayed on a magnesium fire will do two things. First, if the fire involves small pieces such as chips, fines or dust, the reaction of the burning metal with water can be explosive, causing burning brands to fly onto surrounding materials and equipment. Second, the extraordinary heat generated by burning magnesium can break the chemical bonds between hydrogen and oxygen atoms in the water molecule. Once this occurs, the hydrogen will be burned and the oxygen will support continued combustion, making the fire that much more intense; however, the danger of a hydrogen explosion under fire conditions is generally slight. If magnesium burns in the open, an excess of oxygen will be available to burn any hydrogen as rapidly as it is generated, thus preventing formation of an explosive accumulation of gas. Despite the noted dangers, however, magnesium fires can be extinguished by cooling the metal below its melting point using relatively large amounts of water carefully applied from a safe distance.

The use of sand can produce a similar reaction to the use of water. Sand is composed of silicon dioxide (SiO_2) molecules that will break down, allowing the oxygen to support the combustion of the magnesium.

Carbon dioxide cannot be used on magnesium fires for the same reason. Magnesium reacts so strongly with oxygen that the carbon dioxide will decompose, giving the fire additional oxygen and making it burn more intensely.

In extreme cases, magnesium will even burn in a nitrogen atmosphere, forming magnesium nitride.

Halon cannot be used because magnesium reacts violently with the chlorine molecules in the gas to form magnesium chloride.

Consequently, special fire-fighting agents are required for fighting magnesium and other flammable metal fires. Because flammable metal fires are considered Class D fires, it is important to verify that portable fire extinguishers provided in accordance with this section are labeled as being effective on such fires. Similarly, any installed fire-extinguishing system must be chosen carefully with the focus on extinguishing agent compatibility with the protected content.

The form of the material involved in the fire will dictate what can be used to extinguish it. The most difficult fire is one involving fines. Normally, a dry extinguishing agent is manually spread over the fire to smother the burning material. Care must be taken, however, to make sure the application does not raise a dust cloud that could cause an explosion.

The long history of magnesium use has resulted in a well-documented fire-fighting strategy for each form of magnesium that could be involved in a fire. The fire code official must verify that the user facility is equipped to fight potential fires and that emergency response personnel are trained in the hazards they may face as well as in the use of the available extinguishers.

Prompt control of magnesium fires requires rapid and ready access to fire-extinguishing materials and equipment by every operator at every workstation processing the metal. The fire extinguisher access travel distance of 75 feet (22 860 mm) correlates with Section 906. The 30-foot (9144 mm) travel distance to a stockpiled supply of dry extinguishing agent correlates with the travel distance for an extra-hazard occupancy and recognizes the potential need to make repeated trips to the stockpile to bring a sufficient amount of agent to accomplish extinguishment.

3606.5.8 Collection of chips, turnings and fines. Chips, turnings and other fine magnesium scrap shall be collected from the pans or spaces under machines and from other places where they collect at least once each working day. Such material shall be placed in a covered, vented steel container and removed to an approved location.

❖ This requirement for collection of highly flammable material is really nothing more than good housekeeping. Keeping the working area free of accumulations of fire fuel means that the fuel load for any potential fire is minimized. As indicated in the commentary to Section 3606.4.2, because magnesium chips, turnings and fines are usually wet with coolant from the processing operation, there is a possibility of hydrogen generation. For this reason, fines must be stored in accordance with Section 3606.4 and in approved steel containers with vented lids to prevent hydrogen buildup.

As was discussed in Chapter 3, the three required elements for a fire are fuel, oxygen and an ignition source. If any one of these three is absent, there will be no fire. Depriving a potential fire of its fuel source is one of the primary ways to prevent fires. Following the other requirements in this chapter will also minimize the oxygen supply and prevent fuel from coming in contact with ignition sources.

Bibliography

The following resource materials are referenced in this chapter or are relevant to the subject matter addressed in this chapter.

Fire Protection Handbook, 18th Ed. Quincy, MA: National Fire Protection Association, 1997.

IBC-2003, *International Building Code*. Falls Church, VA: International Code Council, 2003.

ICC EC-2003, *ICC Electrical Code*. Falls Church, VA: International Code Council, 2003.

NFPA 70-02, *National Electrical Code*. Quincy, MA: National Fire Protection Association, 2002.

NFPA 86C-99, *Standard for Industrial Furnaces Using a Special Processing Atmosphere.* Quincy, MA: National Fire Protection Association, 1999.

NFPA 480-98, *Standard for Storage, Handling and Processing of Magnesium Solids and Powder.* Quincy, MA: National Fire Protection Association, 1998.

NFPA 481-00, *Standard for the Production, Processing, Handling and Storage of Titanium.* Quincy, MA: National Fire Protection Association, 2000.

NFPA 482-96, *Standard for Production, Processing, Handling and Storage of Zirconium.* Quincy, MA: National Fire Protection Association, 1996.

NFPA 484-02, *Standard for Combustible Metals, Metal Powders, and Metal Dusts.* Quincy, MA: National Fire Protection Association, 2002.

NFPA 651-98, *Standard for Machining and Finishing of Aluminum and Production and Handling of Aluminum Powder.* Quincy, MA: National Fire Protection Association, 1998.

Chapter 37: Highly Toxic and Toxic Materials

General Comments

Toxic and highly toxic materials are addressed in the code because of the immediate threat they pose to occupants, others in the vicinity of a building and facility and emergency responders. As with other health hazard materials, the solid state is usually the least hazardous, while the gaseous form is the most hazardous.

Materials are often listed as being toxic or highly toxic on Material Safety Data Sheets (MSDS). These descriptors do not necessarily mean that the materials would be considered toxic or highly toxic according to the specific definitions found in Section 3702. Those definitions provide specific criteria that will be discussed in more detail. Generally, the requirements for toxic and highly toxic materials are the most regulated health hazards in the code.

This chapter deals with all three states of toxic and highly toxic materials: solids, liquids and gases. As will be discussed, gases will generally require treatment systems and related ventilation systems.

Purpose

The main purpose of this chapter is, as noted, to protect occupants, emergency responders and those in the immediate area of the building and facility from short-term, acute hazards associated with a release or general exposure to toxic and highly toxic materials. The code does not address long-term exposure effects. Such issues are addressed by agencies Environmental Protection Agency (EPA) and Occupational Safety and Health Administration (OSHA).

SECTION 3701 GENERAL

3701.1 Scope. The storage and use of highly toxic and toxic materials shall comply with this chapter. Compressed gases shall also comply with Chapter 30.

Exceptions:

1. Display and storage in Group M and storage in Group S occupancies complying with Section 2703.11.
2. Conditions involving pesticides or agricultural products as follows:
   - 2.1. Application and release of pesticide, agricultural products and materials intended for use in weed abatement, erosion control, soil amendment or similar applications when applied in accordance with the manufacturer's instruction and label directions.
   - 2.2. Transportation of pesticides in compliance with the Federal Hazardous Materials Transportation Act and regulations thereunder.
   - 2.3. Storage in dwellings or private garages of pesticides registered by the U.S. Environmental Protection Agency to be utilized in and around the home, garden, pool, spa and patio.

❖ This section states that highly toxic materials must be stored and used in accordance with this chapter. Additionally, it notes that gases are subject to the requirements in Chapter 30, which focuses on the hazards associated with the fact that the material is a compressed gas.

There are various exceptions to this chapter. Exception 1 is related to the increased amounts allowed for storage and display in Group M and S occupancies. These increased amounts apply only to solids and liquids. For this exception to apply, all requirements found in Section 2703.11 are applicable. Exception 2 is specific to pesticides or other agriculture-related products. Essentially, only storage in a building or facility would be regulated. The application, release or transportation of such materials would be exempt because federal standards would preempt a local jurisdiction from enforcement. In terms of application and release, regulations are specific to activities such as weed control, erosion control and soil amendment. Additionally, pesticides approved for use around homes, gardens, pools, spas and patios can be stored without regulation in private garages and within dwellings.

3701.2 Permits. Permits shall be required as set forth in Section 105.6.

❖ The process of issuing permits gives the fire code official an opportunity to carefully evaluate and regulate hazardous operations. Permit applicants should be required to demonstrate that their operations comply with the intent of the code before the permit is issued. See the commentary to Section 105.6 for a general discussion of operations requiring an operational permit, Section 105.6.21 for a discussion of specific quantity-based operational permits for the materials regulated in this chapter and Section 105.7 for a general discussion of activities requiring a construction permit. The permit process also notifies the fire department of the need for prefire planning for hazardous property.

SECTION 3702 DEFINITIONS

3702.1 Definitions. The following words and terms shall, for the purposes of this chapter and as used elsewhere in this code, have the meanings shown herein.

❖ Definitions of terms can help in the understanding and application of the code requirements. The purpose for including those definitions that are associated with the subject matter of this chapter is to provide more convenient access to them without having to refer back to Chapter 2. It is important to emphasize that these terms are not exclusively related to this chapter but are applicable everywhere the term is used in the code. For convenience, these terms are also listed in Chapter 2 with a cross reference to this section. The use and application of all defined terms, including those defined in this section, are set forth in Section 201.

CONTAINMENT SYSTEM. A gas-tight recovery system comprised of equipment or devices which can be placed over a leak in a compressed gas container, thereby stopping or controlling the escape of gas from the leaking container.

❖ A containment system consists of various components that will capture gases from a leaking container by being placed at the source of the leak.

CONTAINMENT VESSEL. A gas-tight recovery vessel designed so that a leaking compressed gas container can be placed within its confines thereby, encapsulating the leaking container.

❖ A containment vessel is a closed unit that a leaking container can be placed in that will fully contain any unwanted release.

EXCESS FLOW VALVE. A valve inserted into a compressed gas cylinder, portable tank or stationary tank that is designed to positively shut off the flow of gas in the event that its predetermined flow is exceeded.

❖ Such a valve has the ability to shut down flow when the intended flow rate has been exceeded. Quite often, when the predetermined flow has been exceeded, a failure will occur.

HIGHLY TOXIC. A material which produces a lethal dose or lethal concentration which falls within any of the following categories:

1. A chemical that has a median lethal dose (LD_{50}) of 50 milligrams or less per kilogram of body weight when administered orally to albino rats weighing between 200 and 300 grams each.
2. A chemical that has a median lethal dose (LD_{50}) of 200 milligrams or less per kilogram of body weight when administered by continuous contact for 24 hours (or less if death occurs within 24 hours) with the bare skin of albino rabbits weighing between 2 and 3 kilograms each.
3. A chemical that has a median lethal concentration (LC_{50}) in air of 200 parts per million by volume or less of gas or vapor, or 2 milligrams per liter or less of mist, fume or dust, when administered by continuous inhalation for one hour (or less if death occurs within 1 hour) to albino rats weighing between 200 and 300 grams each.

Mixtures of these materials with ordinary materials, such as water, might not warrant classification as highly toxic. While this system is basically simple in application, any hazard evaluation that is required for the precise categorization of this type of material shall be performed by experienced, technically competent persons.

❖ This definition, as does the definition of "Toxic," gives very specific criteria in the form of lethal doses and lethal concentrations as administered to albino rats and albino rabbits. The lethal dosages are related to the ingestion and skin contact with materials, generally liquids and solids. The lethal concentrations are related to vapors, dusts, gases or mists as inhaled by albino rats. Inhalation can occur from either a gas, vapor or mist that is generated from highly toxic or toxic liquids. In some cases, a liquid may be considered highly toxic or toxic if ingested or if skin contact occurs, but vapors are not an inhalation hazard according to the criteria. These definitions give criteria to help determine what materials are regulated by this chapter and Chapter 27. Often, materials are listed as toxic or highly toxic on MSDS but may not necessarily meet these criteria. Instead, the terminology may be used to describe irritant characteristics of the material.

OZONE-GAS GENERATOR. Equipment which causes the production of ozone.

❖ Ozone is considered a highly toxic gas. Ozone generators are addressed separately in Section 3705 because the code has traditionally dealt with the storage and use of hazardous materials, but not the generation.

REDUCED FLOW VALVE. A valve equipped with a restricted flow orifice and inserted into a compressed gas cylinder, portable tank or stationary tank that is designed to reduce the maximum flow from the valve under full-flow conditions. The maximum flow rate from the valve is determined with the valve allowed to flow to atmosphere with no other piping or fittings attached.

❖ This is a valve that allows the maximum flow rate from a container to be reduced. For the reduction to be accurate, the maximum flow rate of a container must be known. The maximum flow rate must be determined without any piping or fittings attached to the container.

This ensures that the reduction valve can actually achieve what is intended.

TOXIC. A chemical falling within any of the following categories:

1. A chemical that has a median lethal dose (LD_{50}) of more than 50 milligrams per kilogram, but not more than 500 milligrams per kilogram of body weight when administered orally to albino rats weighing between 200 and 300 grams each.
2. A chemical that has a median lethal dose (LD_{50}) of more than 200 milligrams per kilogram but not more than 1,000 milligrams per kilogram of body weight when administered by continuous contact for 24 hours (or less if death occurs within 24 hours) with the bare skin of albino rabbits weighing between 2 and 3 kilograms each.
3. A chemical that has a median lethal concentration (LC_{50}) in air of more than 200 parts per million but not more than 2,000 parts per million by volume of gas or vapor, or more than 2 milligrams per liter but not more than 20 milligrams per liter of mist, fume or dust, when administered by continuous inhalation for 1 hour (or less if death occurs within 1 hour) to albino rats weighing between 200 and 300 grams each.

❖ See the commentary for the definition of "Highly toxic."

SECTION 3703 HIGHLY TOXIC AND TOXIC SOLIDS AND LIQUIDS

3703.1 Indoor storage and use. The indoor storage and use of highly toxic and toxic materials shall comply with Sections 3703.1.1 through 3703.1.5.3.

❖ As noted in the "General Comments" section, liquids and solids are dealt with in one section and gases are dealt with in another. Generally, gases pose a greater hazard because they are more difficult to contain and can have a much more immediate effect. Section 3703.1 contains requirements for indoor storage and use. Outdoor storage and use are discussed in Section 3703.2.

3703.1.1 Quantities not exceeding the maximum allowable quantity per control area. The indoor storage or use of highly toxic and toxic solids or liquids in amounts not exceeding the maximum allowable quantity per control area indicated in Table 2703.1.1(2) shall be in accordance with Sections 2701, 2703 and 3701.

❖ This section sends the code user to the appropriate sections when the maximum allowable quantities have not been exceeded. As with other materials, when the maximum allowable quantities have not been exceeded, the requirements are less restrictive. The code user must comply with Section 3701 as well as Sections 2701 and 2703, which are the general requirements for hazardous materials related to permits; material classification and management plans; hazard identification and other basic requirements. Section 3703 would not apply.

3703.1.2 Quantities exceeding the maximum allowable quantity per control area. The indoor storage or use of highly toxic and toxic solids or liquids in amounts exceeding the maximum allowable quantity per control area set forth in Table 2703.1.1(2) shall be in accordance with Sections 3701 through 3703.1.3 and Chapter 27.

❖ When the maximum allowable quantities have been exceeded, the requirements in Sections 3701 and 3703 and all of Chapter 27 are applicable.

3703.1.3 Treatment system—highly toxic liquids. Exhaust scrubbers or other systems for processing vapors of highly toxic liquids shall be provided where a spill or accidental release of such liquids can be expected to release highly toxic vapors at normal temperature and pressure. Treatment systems and other processing systems shall be installed in accordance with the *International Mechanical Code.*

❖ This requirement is specific to highly toxic liquids and would require a treatment system to collect and process any vapors that might escape if a spill should occur at "normal temperature and pressure." In other words, if at normal temperature and pressure conditions vapors would not be highly toxic, a treatment system would not be required. The focus of this section is on the inhalation hazards associated with highly toxic materials. A material may be considered highly toxic by skin contact or ingestion, but not create an inhalation hazard because of the low volatility of the liquid.

3703.1.4 Indoor storage. Indoor storage of highly toxic and toxic solids and liquids shall comply with Sections 3703.1.4.1 and 3703.1.4.2.

❖ This section is specific to indoor storage and focuses on floor surfaces and separation requirements.

3703.1.4.1 Floors. In addition to the requirements set forth in Section 2704.12, floors of storage areas shall be of liquid-tight construction.

❖ This section ensures that if a highly toxic or toxic liquid comes in contact with the floor, it will not soak and be difficult to remove. If concrete is not properly treated, spills could seep into the floor and give off vapors over time. The use of liquid-tight floors is one of the possible methods mentioned in Section 2704 for drainage control and secondary containment. The requirements of this section would apply in any case.

The reference to Section 2704.12 requires the floor to be noncombustible except for the surfacing; therefore, the method used to make the floor liquid tight does not need to be noncombustible. The same requirement is stated in Section 2704.2.1.

3703.1.4.2 Separation—highly toxic solids and liquids. In addition to the requirements set forth in Section 2703.9.8, highly toxic solids and liquids in storage shall be located in approved hazardous material storage cabinets or isolated from other hazardous material storage by construction in accordance with the *International Building Code.*

❖ To ensure that highly toxic and toxic materials do not react with incompatible materials and to generally reduce the likelihood of a release, this section specifically requires that highly toxic liquids and solids be contained in either a hazardous materials cabinet or separated by construction. This would require a completely separate storage room for such materials if a cabinet is not used. This section would not allow a 20-foot (6096 mm) physical separation or partition that extended only 18 inches (457 mm) above and to the sides of the material, as would be possible with other hazardous materials.

3703.1.5 Indoor use. Indoor use of highly toxic and toxic solids and liquids shall comply with Sections 3703.1.5.1 through 3703.1.5.3.

❖ Use is more hazardous than storage because the materials are more susceptible to release. The focus of this particular section is on the transfer of highly toxic liquids and requirements for exhaust ventilation systems where highly toxic and toxic materials are being used.

3703.1.5.1 Liquid transfer. Highly toxic and toxic liquids shall be transferred in accordance with Section 2705.1.10.

❖ This section refers the code user back to Section 2705.1.10 for requirements on the transfer of liquids with a hazard ranking of 3 or 4. These liquids can be transferred using several different methods, including safety cans, closed piping, approved pump arrangements or an approved engineered liquid transfer system, as well as by gravity under certain conditions. It should be noted that Section 2705.1.10 prohibits highly toxic liquids from being transferred where gravity feed is involved regardless of the safeguards. There are exceptions for small amounts of liquids [1.3 gallons (5 L) for a hazard ranking of 4 and 5.3 gallons (20 L) for a hazard ranking of 3]. Section 2705.1.10 is also referenced in Section 3703.2.6 for outdoor liquid transfer.

3703.1.5.2 Exhaust ventilation for open systems. Mechanical exhaust ventilation shall be provided for highly toxic and toxic liquids used in open systems in accordance with Section 2705.2.1.1.

Exception: Liquids or solids that do not generate highly toxic or toxic fumes, mists or vapors.

❖ This section requires any open use of highly toxic or toxic liquids to be properly ventilated and refers to Section 2705.2.1.1. That section contains the general ventilation requirements for open systems using gases, liquids or solids with a hazard ranking of 3 or 4. Essentially, it requires that vapors be captured at the point of generation.

The exception to both Sections 3703.1.5.2 and 2705.2.1.1 states that liquids that do not produce hazardous vapors, mists or fumes do not require compliance with these ventilation requirements. Much of this will depend on the volatility of the liquid, the degree of hazard of the liquid and how it is used.

3703.1.5.3 Exhaust ventilation for closed systems. Mechanical exhaust ventilation shall be provided for highly toxic and toxic liquids used in closed systems in accordance with Section 2705.2.2.2.

Exception: Liquids or solids that do not generate highly toxic or toxic fumes, mists or vapors.

❖ Section 2705.2.2.2 requires ventilation in accordance with Section 2705.1.1 if the closed system is designed to be opened during normal operations. Section 2705.1.1 is for open systems using materials with a hazard ranking of 3 or 4. The same exception is stated in Section 3703.1.5.2, which exempts liquids that do not produce highly toxic or toxic fumes, vapors or mists (see commentary, Section 3703.1.5.2).

3703.2 Outdoor storage and use. Outdoor storage and use of highly toxic and toxic materials shall comply with Sections 3703.2.1 through 3703.2.6.

❖ Outdoor storage and use is generally less hazardous than indoor storage and use because the vapors can disperse more easily to the atmosphere, posing less of a hazard to occupants and those in the vicinity of the building. Because the materials are located outside, however, there are other exposure concerns such as weather and location of storage and use.

3703.2.1 Quantities not exceeding the maximum allowable quantity per control area. The outdoor storage or use of highly toxic and toxic solids or liquids in amounts not exceeding the maximum allowable quantity per control area indicated in Table 2703.1.1(4) shall be in accordance with Sections 2701, 2703 and 3701.

❖ As with indoor storage and use, when the maximum allowable quantities have not been exceeded, the only requirements are those found in Section 3701 and the general requirements of Chapter 27. These requirements address issues such as permits, hazardous materials plans, pipe connections (especially with health-hazard ranking materials of 3 or 4), facility closures and hazard identification and signage.

3703.2.2 Quantities exceeding the maximum allowable quantity per control area. The outdoor storage or use of highly toxic and toxic solids or liquids in amounts exceeding the maximum allowable quantity per control area set forth in Table 2703.1.1(4) shall be in accordance with Sections 3701 and 3703.2 and Chapter 27.

❖ When the maximum allowable quantities have been exceeded for the outdoor control areas, the requirements become more extensive. More specifically, compliance with Section 3703.2 is required in addition to Section 3701 and all of Chapter 27, as applicable.

3703.2.3 General outdoor requirements. The general requirements applicable to the outdoor storage of highly toxic or toxic solids and liquids shall be in accordance with Sections 3703.2.3.1 and 3703.2.3.2.

❖ This section sets general requirements for the location and the need for treatment systems to collect vapors from highly toxic liquids.

3703.2.3.1 Location. Outdoor storage or use of highly toxic or toxic solids and liquids shall not be located within 20 feet (6096 mm) of lot lines, public streets, public alleys, public ways, exit discharges or exterior wall openings. A 2-hour fire barrier wall without openings or penetrations extending not less than 30 inches (762 mm) above and to the sides of the storage is allowed in lieu of such distance. The wall shall either be an independent structure, or the exterior wall of the building adjacent to the storage area.

❖ This section requires that toxic and highly toxic solids and liquids be at least 20 feet (6096 mm) from possible exposure hazards such as an exit discharge. The concern with highly toxic and toxic materials is the health hazards for building occupants, emergency responders and others in the immediate area.

An alternative to the 20-foot (6096 mm) distance is offered in this section. Essentially, it would allow the use of a 2-hour fire barrier, without openings, that extends 30 inches (763 mm) above and to the sides. The wall can be either freestanding or a wall of the building that fits the requirements as a 2-hour fire barrier without openings. Generally, this wall and the distance requirements are for the protection of the toxic and highly toxic liquids and solids from fire exposures that could lead to the release of materials. Additionally, both methods allowed in this section result in a separation (by either distance or construction) from people to reduce the likelihood of contact with hazards.

3703.2.3.2 Treatment system—highly toxic liquids. Exhaust scrubbers or other systems for processing vapors of highly toxic liquid shall be provided where a spill or accidental release of such liquids can be expected to release highly toxic vapors at normal temperature and pressure (NTP). Treatment systems and other processing systems shall be installed in accordance with the *International Mechanical Code*.

❖ This section is the same as Section 3703.1.3 for indoor storage and the use of highly toxic liquids (see commentary, Section 3703.1.3).

3703.2.4 Outdoor storage piles. Outdoor storage piles of highly toxic and toxic solids and liquids shall be separated into piles not larger than 2,500 cubic feet (71 m^3). Aisle widths between piles shall not be less than one-half the height of the pile or 10 feet (3048 mm), whichever is greater.

❖ The requirement in this section seeks to reduce the hazard level of a release of highly toxic and toxic liquids and solids by reducing the amount allowed in a single pile. Additionally, minimum separations between piles are required. These separations serve both as a fire barrier and as access for emergency responders. If the piles become too large, they become difficult to manage if a release should happen or a fire should occur.

3703.2.5 Weather protection for highly toxic liquids and solids — outdoor storage or use. Where overhead weather protection is provided for outdoor storage or use of highly toxic liquids or solids, and the weather protection is attached to a building, the storage or use area shall either be equipped throughout with an approved automatic sprinkler system in accordance with Section 903.3.1.1, or storage or use vessels shall be fire-resistive rated. Weather protection shall be provided in accordance with Section 2704.13 for storage and Section 2705.3.9 for use.

❖ This section is specific to outdoor storage or use of highly toxic liquids and solids when they are located in an area with weather protection attached to a building. The storage or use must be either sprinklered or placed within fire-resistive containers. Because the storage or use area is next to the building, the concern for a fire and the potential release of liquids and solids is greater. This poses a hazard to both building occupants and emergency responders. Section 2704.13 contains the general requirements for weather-protected storage.

3703.2.6 Outdoor liquid transfer. Highly toxic and toxic liquids shall be transferred in accordance with Section 2705.1.10.

❖ This section refers to Section 2705.1.10 for the general requirements for liquid transfer. This is the same reference used in Section 3703.1.5.1 for indoor liquid transfer (see commentary, Section 3703.1.5.1).

SECTION 3704
HIGHLY TOXIC AND TOXIC COMPRESSED GASES

3704.1 General. The storage and use of highly toxic and toxic compressed gases shall comply with this section.

❖ This section requires all highly toxic and toxic gases to comply with the following subsections.

3704.1.1 Special limitations for indoor storage and use by occupancy. The indoor storage and use of highly toxic and toxic compressed gases in certain occupancies shall be subject to the limitations contained in Sections 3704.1.1.1 through 3704.1.1.3.

❖ This section places additional limitations on the storage and use of toxic and highly toxic gases in several occupancy types and uses. These are further restrictions on the maximum allowable quantities given in Chapter 27.

3704.1.1.1 Group A, E, I or U occupancies. Toxic and highly toxic compressed gases shall not be stored or used within Group A, E, I or U occupancies.

Exception: Cylinders not exceeding 20 cubic feet (0.566 m^3) at normal temperature and pressure (NTP) are allowed within gas cabinets or fume hoods.

❖ This section prohibits having any large quantities of highly toxic and toxic gases in these occupancies. With the exception of Group U, these are occupancies that have typically high occupant densities or a vulnerable population. The prohibition for Group U occupancies is likely related to the potential lack of supervision and the types of materials that are likely to be stored along with such gases.

The exception to this section allows small amounts of gases if they are stored or used within gas cabinets or fume hoods. The allowance for small cylinders results from the small potential for release when they are stored in a cabinet or used within a fume hood and the probability that the volume of the release would be low.

3704.1.1.2 Group R occupancies. Toxic and highly toxic compressed gases shall not be stored or used in Group R occupancies.

❖ This section prohibits the storage and use of highly toxic and toxic gases in all Group R occupancies without exception. Group R occupancies cover a wide variety of dwelling-type occupancies, such as one- and two-family dwellings, apartment buildings, hotels, motels, etc. The hazards posed by the storage and use of toxic or highly toxic gases would be much higher than would reasonably be anticipated by the occupants.

3704.1.1.3 Offices, retail sales and classrooms. Toxic and highly toxic compressed gases shall not be stored or used in offices, retail sales or classroom portions of Group B, F, M or S occupancies.

Exception: In classrooms of Group B occupancies, cylinders with a capacity not exceeding 20 cubic feet (0.566 m^3) at NTP are allowed in gas cabinets or fume hoods.

❖ This section does not completely prohibit the storage and use of highly toxic liquids in these particular occupancies, but instead focuses on certain portions of occupancies. In particular, this section addresses the offices, retail sales areas and classrooms of Group B, F, M or S occupancies. A normal storage area with no public access, therefore, could be used to store highly toxic or toxic gases, but the office incidental to that storage facility could not.

The exception allows small cylinders in classrooms in Group B occupancies when they are used within gas cabinets or fume hoods. This acknowledges special needs at university laboratories and similar facilities.

3704.1.2 Gas cabinets. Gas cabinets containing highly toxic or toxic compressed gases shall comply with Section 2703.8.5 and the following requirements:

1. The average ventilation velocity at the face of gas cabinet access ports or windows shall not be less than 200 feet per minute (1.02 m/s) with a minimum of 150 feet per minute (0.76 m/s) at any point of the access port or window.
2. Gas cabinets shall be connected to an exhaust system.
3. Gas cabinets shall not be used as the sole means of exhaust for any room or area.
4. The maximum number of cylinders located in a single gas cabinet shall not exceed three, except that cabinets containing cylinders not over 1 pound (0.454 kg) net contents are allowed to contain up to 100 cylinders.
5. Gas cabinets required by Section 3704.2 or 3704.3 shall be equipped with an approved automatic sprinkler system in accordance with Section 903.3.1.1. Alternative fire-extinguishing systems shall not be used.

❖ This section sets additional requirements for gas cabinets used specifically for highly toxic and toxic gases. Section 2703.8.5 contains the general requirements for all gas cabinets. More specifically, Section 2703.8.5 sets out construction specifications, requires negative pressure for ventilation and restricts the number of cylinders to three. Section 3704.1.2 is more restrictive, requiring an air velocity of at least 200 feet per minute (1.02 m/s) at the face of the cabinet as well as a connection to an exhaust system and an area ventilation system in addition to the ventilation system in the gas cabinet. A sprinkler system is also required.

Section 3704.2.2.7 would require the exhaust to be connected to a treatment system. Item 4 in Section 3704.1.2 does allow up to 100 small cylinders [under 1 pound (0.454 kg) each] instead of the restriction of three larger cylinders. This allowance recognizes the reduced potential for a large release and increases flexibility to meet the needs of the facilities such as laboratories.

3704.1.3 Exhausted enclosures. Exhausted enclosures containing highly toxic or toxic compressed gases shall comply with Section 2703.8.5 and the following requirements:

1. The average ventilation velocity at the face of the enclosure shall not be less than 200 feet per minute (1.02 m/s) with a minimum of 150 feet per minute (0.76 m/s).
2. Exhausted enclosures shall be connected to an exhaust system.
3. Exhausted enclosures shall not be used as the sole means of exhaust for any room or area.
4. Exhausted enclosures required by Section 3704.2 or 3704.3 shall be equipped with an approved automatic sprinkler system in accordance with Section 903.3.1.1. Alternative fire-extinguishing systems shall not be used.

❖ This section, like Section 3704.1.2, requires compliance with the general exhausted enclosure requirements in Section 2703.8.4. Section 2703.8.4, like Section

2703.8.5 for gas cabinets, contains basic construction specifications and requires that the enclosure be at a negative pressure. Additionally, there is a requirement in Section 2703.8.4 that a fire-extinguishing system be installed when the materials stored or used within the enclosure are flammable.

Section 3704.1.2 additionally requires an air velocity of at least 200 feet per minute (1.02 m/s) at the face of the enclosure and that the enclosure be connected to an exhaust system that is not the sole source of ventilation for that area. In addition, this section requires a sprinkler system within the enclosure. This is independent of whether the material is considered flammable.

3704.2 Indoor storage and use. The indoor storage and use of highly toxic or toxic compressed gases shall be in accordance with Sections 3704.2.1 through 3704.2.2.10.3.

❖ This section is specific to the indoor storage and use of toxic and highly toxic gases. When Section 3704.2 applies, it focuses on the location of cylinders and the removal of unwanted releases of gases. Treatment systems are required to process any gases collected when ventilating results in the release of toxic and highly toxic gases.

3704.2.1 Applicability. The applicability of regulations governing the indoor storage and use of highly toxic and toxic compressed gases shall be as set forth in Sections 3704.2.1.1 through 3704.2.1.3.

❖ This section clarifies which requirements apply based on the amount of material in storage or in use.

3704.2.1.1 Quantities not exceeding the maximum allowable quantity per control area. The indoor storage or use of highly toxic and toxic gases in amounts not exceeding the maximum allowable quantity per control area set forth in Table 2703.1.1(2) shall be in accordance with Sections 2701, 2703, 3701 and 3704.1.

❖ When the maximum allowable quantities have not been exceeded, only the more general requirements would apply. These include restrictions on the storage and use in certain occupancies, piping connection requirements based on the level of health hazards, permits and other similar requirements. Tables 2703.1.1(2) and 2703.1.1(4) would require putting highly toxic gases in a gas cabinet or exhausted enclosure regardless of the amount of gases stored or used. These gas cabinets and exhausted enclosures need be in accordance only with the basic requirements of Chapter 27 and do not need to be connected to a treatment system.

3704.2.1.2 Quantities exceeding the maximum allowable quantity per control area. The indoor storage or use of highly toxic and toxic gases in amounts exceeding the maximum allowable quantity per control area set forth in Table 2703.1.1(2) shall be in accordance with Sections 3701, 3704.1, 3704.2 and Chapter 27.

❖ If the maximum allowable quantities have been exceeded, the requirements become much more extensive. This requires compliance with all applicable sections of Chapter 27 and also Section 3704.2, which has requirements for treatment systems and gas detection systems.

3704.2.1.3 Ozone gas generators. The indoor use of ozone gas-generating equipment shall be in accordance with Section 3705.

❖ This section is a specific reference to Section 3705, which deals with the process of ozone generation. The operation is unique and, therefore, has unique requirements. The requirements in Section 3705 apply when the ozone-generating capacity exceeds 0.5 pound (.227 kg) in a 24-hour period.

3704.2.2 General indoor requirements. The general requirements applicable to the indoor storage and use of highly toxic and toxic compressed gases shall be in accordance with Sections 3704.2.2.1 through 3704.2.2.10.3.

❖ The requirements in this section are for both storage and use of toxic and highly toxic gases when the maximum allowable quantities have been exceeded.

3704.2.2.1 Cylinder and tank location. Cylinders shall be located within gas cabinets, exhausted enclosures or gas rooms. Portable and stationary tanks shall be located within gas rooms or exhausted enclosures.

❖ Toxic and highly toxic gases pose a high threat to occupants and emergency responders if released to the atmosphere; therefore, this section places restrictions on where cylinders and tanks can be located. More specifically, cylinders must be in a gas cabinet or within gas rooms or exhausted enclosures. Tanks, both portable and stationary, must be either in a gas room or an exhausted enclosure. Gas cabinets are not an option because of the size of portable and stationary tanks. Gas cabinets, exhausted enclosures and gas rooms have specific requirements in Sections 3704.1.2, 3704.1.3 and 3704.2.2.6 in addition to the general requirements in Chapter 27.

3704.2.2.2 Ventilated areas. The room or area in which gas cabinets or exhausted enclosures are located shall be provided with exhaust ventilation. Gas cabinets or exhausted enclosures shall not be used as the sole means of exhaust for any room or area.

❖ This section requires that gas cabinets and exhausted enclosures not be the only ventilation provided when toxic or highly toxic gases are stored or used. The room or area must have additional ventilation. The exhaust ventilation for the room need not be processed through a treatment system.

3704.2.2.3 Leaking cylinders and tanks. One or more gas cabinets or exhausted enclosures shall be provided to handle leaking cylinders, containers or tanks.

Exceptions:

1. Where cylinders, containers or tanks are located within gas cabinets or exhausted enclosures.
2. Where approved containment vessels or containment systems are provided in accordance with all of the following:
   - 2.1. Containment vessels or containment systems shall be capable of fully containing or terminating a release.
   - 2.2. Trained personnel shall be available at an approved location.
   - 2.3. Containment vessels or containment systems shall be capable of being transported to the leaking cylinder, container or tank.

❖ Section 3704.2.2.1 requires the use of gas cabinets, exhausted enclosures or gas rooms for the storage of cylinders and tanks. This section takes the requirements one step further and requires that one or more additional gas cabinets or exhausted enclosures be provided and ready to receive leaking cylinders or tanks.

Exception 1 is for cylinders and tanks that are already contained within gas cabinets or exhausted enclosures.

Exception 2 allows the use of containment vessels and containment systems in place of a gas cabinet or exhausted enclosure to address the release of gases. There are several conditions, which include that the vessel or system actually contain the potential release, that a trained person be available and that the containment vessel or system be transportable to the leaking cylinder or tank.

3704.2.2.3.1 Location. Gas cabinets and exhausted enclosures shall be located in gas rooms and connected to an exhaust system.

❖ When gas cabinets and exhausted enclosures are used with leaking tanks, they must be contained within gas rooms. Containment vessels and systems would not have to be located within a gas room.

3704.2.2.4 Local exhaust for portable tanks. A means of local exhaust shall be provided to capture leaks from portable tanks. The local exhaust shall consist of portable ducts or collection systems designed to be applied to the site of a leak in a valve or fitting on the tank. The local exhaust system shall be located in a gas room. Exhaust shall be directed to a treatment system in accordance with Section 3704.2.2.7.

❖ This section requires portable tanks located in gas rooms to have an additional local exhaust mechanism. More specifically, Section 3704.2.2.1 requires locating portable tanks in a gas room or exhausted enclosure as a minimum. Since exhausted enclosures would be considered local, therefore, an additional local exhaust system would not be required. Only when the sole exhaust mechanism is a gas room would a local exhaust be required. The exhaust from the gas room and the local exhaust system must be processed through a treatment system in accordance with Section 3704.2.2.7. The focus of a local exhaust system should be on the valves or valve fittings where a leak is more likely.

3704.2.2.5 Piping and controls—stationary tanks. In addition to the requirements of Section 2703.2.2, piping and controls on stationary tanks shall comply with the following requirements:

1. Pressure relief devices shall be vented to a treatment system designed in accordance with Section 3704.2.2.7.

   Exception: Pressure relief devices on outdoor tanks provided exclusively for relieving pressure due to fire exposure are not required to be vented to a treatment system provided that:

   1. The material in the tank is not flammable.
   2. The tank is not located in a diked area with other tanks containing combustible materials.
   3. The tank is located not less than 30 feet (9144 mm) from combustible materials or structures or is shielded by a fire barrier complying with Section 3704.3.2.1.1.

2. Filling or dispensing connections shall be provided with a means of local exhaust. Such exhaust shall be designed to capture fumes and vapors. The exhaust shall be directed to a treatment system in accordance with Section 3704.2.2.7.
3. Stationary tanks shall be provided with a means of excess flow control on all tank inlet or outlet connections.

   Exceptions:

   1. Inlet connections designed to prevent backflow.
   2. Pressure relief devices.

❖ This section is simply in addition to Section 2703.2.2, which addresses piping, tubing, valves and fittings in general for all hazardous materials. This section focuses on exhaust ventilation for potential leaks and releases associated with piping, filling and dispensing connections on stationary tanks. Section 2703.2.2 contains general requirements covering issues such as compatibility, shutoff valves, backflow prevention and leak detection. This section is specific to stationary tanks because they are more permanent and piping and other controls are more likely on stationary tanks versus portable tanks and cylinders.

The specific requirements found in this section are as follows:

1. If a pressure relief valve is used, the potential release must be vented directly to a treatment system. There is an exception that pertains to out-

door tanks with pressure valves specifically for pressure relief in a fire. There are several conditions for the exception that address exposure hazards, such as neighboring tanks or combustible hazards. Also, the material in the tank itself cannot be flammable. Although this exception for outdoor tanks is found within the indoor storage and use requirements, Section 3704.3 for outdoor storage and use has a specific requirement within Section 3704.3.2.3 that refers back to this section for piping and controls for outdoor stationary tanks.

2. This item requires a local exhaust system connected to a treatment system for filling and dispensing connections.

3. Excess flow control is required at every tank inlet and outlet to reduce the size of the release and to avoid dangerous reactions and overpressures in other areas of a process designed for a particular flow rate and pressure. There are two exceptions that relate to devices serving a specific purpose, which include an inlet connection designed to address back-flow or a pressure relief valve. A pressure relief valve is specifically designed to allow excessive flow beyond the design pressures of the tank to avoid overpressures in the tank.

3704.2.2.6 Gas rooms. Gas rooms shall comply with Section 2703.8.4 and both of the following requirements:

1. The exhaust ventilation from gas rooms shall be directed to an exhaust system.
2. Gas rooms shall be equipped with an approved automatic sprinkler system. Alternative fire-extinguishing systems shall not be used.

❖ The requirements in Chapter 27 address construction and basic ventilation. General gas room requirements are found in Section 2703.8.3. More specifically, Section 2703.8.3 requires an automatic sprinkler system, separation as required in the IBC and maintaining negative pressure in the room. This section takes the requirements one step further and requires that exhaust ventilation be directed to an exhaust system. Section 3704.2.2.7 also requires directing this exhaust system to a treatment system. The second criterion is a restriction that does not allow any alternative fire-extinguishing systems in place of an automatic sprinkler system.

3704.2.2.7 Treatment systems. The exhaust ventilation from gas cabinets, exhausted enclosures and gas rooms, and local exhaust systems required in Sections 3704.2.2.4 and 3704.2.2.5 shall be directed to a treatment system. The treatment system shall be utilized to handle the accidental release of gas and to process exhaust ventilation. The treatment system shall be designed in accordance with Sections 3704.2.2.7.1 through 3704.2.2.7.5 and Section 510 of the *International Mechanical Code*.

Exceptions:

1. Highly toxic and toxic gases—storage. A treatment system is not required for cylinders, containers and tanks in storage when all of the following controls are provided:
   - 1.1. Valve outlets are equipped with gas-tight outlet plugs or caps.
   - 1.2. Handwheel-operated valves have handles secured to prevent movement.
   - 1.3. Approved containment vessels or containment systems are provided in accordance with Section 3704.2.2.3.
2. Toxic gases—use. Treatment systems are not required for toxic gases supplied by cylinders or portable tanks not exceeding 660 gallons (2 498 L) liquid capacity when the following are provided:
   - 2.1. A gas detection system with a sensing interval not exceeding 5 minutes.
   - 2.2. An approved automatic-closing fail-safe valve located immediately adjacent to cylinder valves. The fail-safe valve shall close when gas is detected at the permissible exposure limit (PEL) by a gas detection system monitoring the exhaust system at the point of discharge from the gas cabinet, exhausted enclosure, ventilated enclosure or gas room. The gas detection shall comply with Section 3704.2.2.10.

❖ Treatment systems are required for all exhaust ventilation and accidental releases of toxic and highly toxic gases. A treatment system essentially treats the exhaust through methods such as diluting, absorbing, burning and other various methods, which are discussed in Section 3704.2.2.7.1. Generally, Section 3704.2.2.7 and related subsections contain the design criteria for such systems.

There are two overall exceptions where treatment systems would not be required. They are broken into storage and use, respectively. Treatment systems can be very costly and the exceptions give credit to situations where the hazard is low and the benefit of a costly treatment system cannot be justified.

Exception 1 deals with cylinders, containers and tanks in storage. A treatment system would not be required if all three criteria are met. The criteria include caps or plugs on any valves to provide a level of redundancy in case of an accidental release, handwheel-operated valves that are secured in place and the installation of containment vessels or containment systems.

Exception 2 is for toxic gases supplied by cylinders. To avoid the use of a treatment system, a gas detection system must be accompanied by a fail-safe valve adjacent to the cylinder valve. The fail-safe valve must operate when gas is detected at the point of discharge of the location of the cylinder. Section 3704.2.2.1 requires cylinders to be located in a gas cabinet, an exhausted enclosure or a gas room.

3704.2.2.7.1 Design. Treatment systems shall be capable of diluting, adsorbing, absorbing, containing, neutralizing, burning or otherwise processing the contents of the largest single vessel of compressed gas. Where a total containment system is used, the system shall be designed to handle the maximum anticipated pressure of release to the system when it reaches equilibrium.

❖ This section states that a treatment system must process the exhaust ventilation or an accidental release. Various methods are listed but the section is written to allow exploration of other methods. Additionally, this section sets important capacity criteria for the treatment system, which would require it to either be capable of processing the largest vessel or handle the maximum pressure of release at equilibrium when a total containment system is used.

3704.2.2.7.2 Performance. Treatment systems shall be designed to reduce the maximum allowable discharge concentrations of the gas to one-half immediate dangerous to life and health (IDLH) at the point of discharge to the atmosphere. Where more than one gas is emitted to the treatment system, the treatment system shall be designed to handle the worst-case release based on the release rate, the quantity and the IDLH for all compressed gases stored or used.

❖ Now that the method and the capacity have been established in Section 3704.2.2.7.1, the actual treatment capabilities are described in this section. More specifically, this section describes how well the gases need to be treated. Once treated, the output from the treatment system must not exceed one-half of the Immediately Dangerous to Life and Health (IDLH) concentration. If the treatment system is used for a variety of different stored gases that have various levels of toxicity, the treatment system must be able to accommodate the worst-case situation. As an example, the least toxic gas may have the largest release potential, but the treatment system does not have to work as hard to reduce the IDLH; therefore, both the level of hazard and the amount of the gas must be addressed.

3704.2.2.7.3 Sizing. Treatment systems shall be sized to process the maximum worst-case release of gas based on the maximum flow rate of release from the largest vessel utilized. The entire contents of the largest compressed gas vessel shall be considered.

❖ This section reemphasizes that the treatment system must be capable of treating the largest single vessel. In addition, this section requires that the maximum flow rates be considered; therefore, it is not simply the capacity of the largest single vessel but also how fast that gas is released. Treatment systems need to account for only a single failure of a vessel, but at the highest flow rate.

3704.2.2.7.4 Stationary tanks. Stationary tanks shall be labeled with the maximum rate of release for the compressed gas contained based on valves or fittings that are inserted directly into the tank. Where multiple valves or fittings are provided, the maximum flow rate of release for valves or fittings with the highest flow rate shall be indicated. Where liquefied compressed gases are in contact with valves or fittings, the liquid flow rate shall be utilized for computation purposes. Flow rates indicated on the label shall be converted to cubic feet per minute (ft^3/min) (m^3/s) of gas at normal temperature and pressure (NTP).

❖ The potential release rates of all valves directly connected to the tank that have the potential for a release of gases must be properly labeled. In addition, the valves that have the highest potential release rates need to be specifically identified. This is especially important when there are multiple valves.

Another requirement of this section is for situations in which the valve is located where it is interacting with the liquid form of the compressed gas. Such interaction will alter the release rate; therefore, the liquid rate of release must be used. When a gas is in liquid form it is denser than the gas form. If released, the liquid vaporizes, expanding into a much larger volume of gas.

Finally, to more readily assess the amount of gas that can be released into the room or area, the tank liquid flow rate must be shown on the label as the equivalent gas volume in cubic feet per minute at normal temperature and pressure (NTP). In other words, the flow rate is not as important as the gas volume it will produce when released into the room or area. NTP provides a comparable base value.

3704.2.2.7.5 Portable tanks and cylinders. The maximum flow rate of release for portable tanks and cylinders shall be calculated based on the total release from the cylinder or tank within the time specified in Table 3704.2.2.7.5. When portable tanks or cylinders are equipped with approved excess flow or reduced flow valves, the worst-case release shall be determined by the maximum achievable flow from the valve as determined by the valve manufacturer or compressed gas supplier. Reduced flow and excess flow valves shall be permanently marked by the valve manufacturer to indicate the maximum design flow rate. Such markings shall indicate the flow rate for air under normal temperature and pressure.

❖ Because the flow rates of different portable tanks and cylinders tend to vary because of their portability, this information would be difficult to track; therefore, this section simply requires that the maximum flow rate be determined based on the capacity of the tank and the time prescribed in Table 3704.2.2.7.5. For example, say a portable tank had a capacity (at NTP) of 1,000 cubic feet (28 m^3) and was not under liquefied conditions. The maximum rate of release based on the criteria given in Table 3704.2.2.7.5 would be as follows:

1,000 cubic feet/40 minutes =
25 cubic feet per minute at NTP

In some cases portable tanks and cylinders may be equipped with valves that either alter the rate of flow or will stop in the case of flow over the designed rate. In those cases, credit is given to the actual amount of gas released at NTP in determining the maximum flow rate. Valves that are used in this manner must be properly labeled to note the specific flow rate at NTP or function.

TABLE 3704.2.2.7.5
RATE OF RELEASE FOR CYLINDERS AND PORTABLE TANKS

VESSEL TYPE	NONLIQUEFIED (minutes)	LIQUEFIED (minutes)
Containers	5	30
Portable tanks	40	240

3704.2.2.8 Emergency power. Emergency power in accordance with the ICC *Electrical Code* shall be provided in lieu of standby power where any of the following systems are required:

1. Exhaust ventilation system.
2. Treatment system.
3. Gas detection system.
4. Smoke detection system.
5. Temperature control system.
6. Fire alarm system.
7. Emergency alarm system.

Exception: Emergency power is not required for mechanical exhaust ventilation, treatment systems and temperature control systems where approved fail-safe engineered systems are installed.

❖ Because of the immediate health hazard posed by the release of toxic or highly toxic gases, emergency, rather than standby, power is required. The major difference between them is that standby power activates within 60 seconds whereas emergency power activates within 10 seconds. This section provides a fairly specific list of systems that would require emergency power. The most critical is likely the treatment system.

The exception allows standby power only if a fail-safe engineered system is installed that will shut down in a manner that will contain the gases.

3704.2.2.9 Automatic fire detection system—highly toxic compressed gases. An approved automatic fire detection system shall be installed in rooms or areas where highly toxic compressed gases are stored or used. Activation of the detection system shall sound a local alarm. The fire detection system shall comply with Section 907.

❖ This section requires a fire detection system in rooms or areas where highly toxic gases are stored or used. The intent is that a fire within the area could lead to the release of the highly toxic gases. Fires can heat gases stored and cause expansion, leading to overpressures and releases; therefore, warning of a fire is critical to avoiding such releases. The alarm system needs to provide a local alarm at the building, but the detection is required only in the room or area where the highly toxic gas is stored.

3704.2.2.10 Gas detection system. A gas detection system shall be provided to detect the presence of gas at or below the permissible exposure limit (PEL) or ceiling limit of the gas for which detection is provided. The system shall be capable of monitoring the discharge from the treatment system at or below one-half the IDLH limit.

Exception: A gas detection system is not required for toxic gases when the physiological warning properties for the gas are at a level below the accepted PEL for the gas.

❖ This section requires a system to detect the presence of gas in a large enough concentration to exceed the permissible exposure limit (PEL). The system must also be designed to be capable of detecting whether one half the IDLH has been exceeded at the discharge from the treatment system. This is to ensure that the treatment system is working to capacity or to indicate that there may be other problems, such as a release larger than the treatment system has been designed to handle. It is important to stress the terminology of "capable," which would not require monitoring, but would require an installation that would enable such monitoring if desired. In many cases, monitoring is done intermittently with portable monitors. Such monitoring may become necessary as a result of environmental restrictions, which are beyond the scope of the code.

There is an exception to the requirement of a gas detection system where the odor of the gas or its physical effects are noticeable far before the PEL is reached. Those indicators should be sufficient to notify people to leave the area.

3704.2.2.10.1 Alarms. The gas detection system shall initiate a local alarm and transmit a signal to a constantly attended control station when a short-term hazard condition is detected. The alarm shall be both visual and audible and shall provide warning both inside and outside the area where gas is detected. The audible alarm shall be distinct from all other alarms.

Exception: Signal transmission to a constantly attended control station is not required where not more than one cylinder of highly toxic or toxic gas is stored.

❖ Once the gas is detected at the levels noted in Section 3704.2.2.10, a local alarm must be initiated and a signal at a constantly attended control station (such as a security room or fire command center) must be transmitted. The alarm is intended to alert those both inside the particular area of detection and in the immediate vicinity. This is to prevent any gases that might escape from causing harm to those outside the area of release because they were not aware of the activities within the building.

The notification to the control station provides information to those who must take a role in emergency response, whereas the local alarm is a warning for those in the vicinity of the release.

A signal need not be sent to the control station if the amount of gas stored or used is a maximum of one cylinder. In that case, a local alarm is sufficient to notify people of the immediate hazard.

3704.2.2.10.2 Shut off of gas supply. The gas-detection system shall automatically close the shutoff valve at the source on gas supply piping and tubing related to the system being monitored for whichever gas is detected.

Exception: Automatic shutdown is not required for reactors utilized for the production of highly toxic or toxic compressed gases where such reactors are:

1. Operated at pressures less than 15 pounds per square inch gauge (psig) (103.4 kPa).
2. Constantly attended.
3. Provided with readily accessible emergency shutoff valves.

❖ The gas detection system also initiates the shutdown of gases at the source for gas supply piping and tubing, making this requirement applicable to the use of the gas moreso than to storage. The exception applies only to equipment used to make toxic or highly toxic gases, and only when all three of the stated conditions are met. This exception recognizes that pressure is a critical element in how much and how fast a gas is released. A low operating pressure normally means a smaller, more easily controlled release. Having an operator monitoring the equipment at all times is considered an adequate safeguard when shutoff valves are easy to reach in case of an emergency. Under these circumstances, notification by an alarm system and a signal to a constantly attended control station is sufficient to deal with the particular hazard. Automatic shutoff would probably be overly restrictive.

3704.2.2.10.3 Valve closure. Automatic closure of shutoff valves shall be in accordance with the following:

1. When the gas-detection sampling point initiating the gas detection system alarm is within a gas cabinet or exhausted enclosure, the shutoff valve in the gas cabinet or exhausted enclosure for the specific gas detected shall automatically close.
2. Where the gas-detection sampling point initiating the gas detection system alarm is within a gas room and compressed gas containers are not in gas cabinets or exhausted enclosures, the shutoff valves on all gas lines for the specific gas detected shall automatically close.
3. Where the gas-detection sampling point initiating the gas detection system alarm is within a piping distribution manifold enclosure, the shutoff valve for the compressed container of specific gas detected supplying the manifold shall automatically close.

Exception: When the gas-detection sampling point initiating the gas-detection system alarm is at a use location or within a gas valve enclosure of a branch line downstream of a piping distribution manifold, the shutoff valve in the gas valve enclosure for the branch line located in the piping distribution manifold enclosure shall automatically close.

❖ This section describes three common situations in which gas lines need to automatically close when gas is detected. If the gas is detected within a gas cabinet or exhausted enclosure, only the gas line related to the gas cylinder or container within the cabinet or enclosure must be shut down. If the gas is detected within a gas room, all gas lines containing that particular gas must be shut down because it is difficult to determine where the leak originates when the storage and use are in a larger area. The next criterion is related to situations where the gas sampling occurs within a piping manifold enclosure. Gas sampling in such locations is much more localized and the code requires shutting down only the cylinder supplying the manifold.

The exception is for situations where the gas is clearly being released downstream from the piping distribution manifold enclosure. The problem is with the piping or perhaps at a point of use and can be isolated by simply shutting off that particular branch line from the piping distribution manifold enclosure. This would be allowed only if the gas detection system was sampling at the location of use or within a gas valve enclosure downstream from the distribution piping. Otherwise, it would be difficult to determine where the leak originated and the supply at the cylinder or tank would have to be shut down.

3704.3 Outdoor storage and use. The outdoor storage and use of highly toxic and toxic compressed gases shall be in accordance with Sections 3704.3.1 through 3704.3.9.

❖ Outdoor storage and use is generally less hazardous than indoor storage and use because the toxic and highly toxic gases, if released, are more easily diluted and, thus, the hazard to the people in the surrounding area is reduced. When gases are released outdoors, however, it is more difficult to control where the released gases will go and what or who is being exposed to the hazards; therefore, this section contains restrictions on locations and distance to exposures. This section refers back to many of the requirements found in Section 3704.2 such as for piping and controls for stationary tanks and gas detection.

3704.3.1 Applicability. The applicability of regulations governing the outdoor storage and use of highly toxic and toxic compressed gases shall be as set forth in Sections 3704.3.1.1 through 3704.3.1.3.

❖ This section defines which requirements of the code apply based on the amount of gases stored and used.

3704.3.1.1 Quantities not exceeding the maximum allowable quantity per control area. The outdoor storage or use of highly toxic and toxic gases in amounts not exceeding the maximum allowable quantity per control area set forth in Table 2703.1.1(4) shall be in accordance with Sections 2701, 2703 and 3701.

❖ This section states that when the maximum allowable quantities have not been exceeded, the general requirements found in Sections 2701 and 2703 would apply, as appropriate. Section 3701, primarily Section 3701.2, would also apply; therefore, the requirements would be fairly limited for amounts under the maximum allowable quantities.

3704.3.1.2 Quantities exceeding the maximum allowable quantity per control area. The outdoor storage or use of highly toxic and toxic gases in amounts exceeding the maximum allowable quantity per control area set forth in Table 2703.1.1(4) shall be in accordance with Sections 3701 and 3704.3 and Chapter 27.

❖ When the maximum allowable quantities have been exceeded, the requirements become more restrictive. All of Chapter 27 would apply, as appropriate to the outdoor storage and use of gases, and Section 3704.3 would apply in its entirety.

3704.3.1.3 Ozone gas generators. The outdoor use of ozone gas-generating equipment shall be in accordance with Section 3705.

❖ This section refers to Section 3705 for requirements specific to ozone gas generators (see also commentary, Section 3704.2.1.3).

3704.3.2 General outdoor requirements. The general requirements applicable to the outdoor storage and use of highly toxic and toxic compressed gases shall be in accordance with Sections 3704.3.2.1 through 3704.3.2.7.

❖ This section contains the bulk of the requirements for outdoor storage and the use of highly toxic and toxic gases. In several places the text refers back to various sections within Section 3704.2 to avoid repetitive wording when the same requirements apply to both indoor and outdoor storage and use. The sections that are unique to outdoor storage and use concern location and exposures.

3704.3.2.1 Location. Outdoor storage or use of highly toxic or toxic compressed gases shall be located in accordance with Sections 3704.3.2.1.1 through 3704.3.2.1.3.

Exception: Compressed gases located in gas cabinets complying with Sections 2703.8.5 and 3704.1.2 and located 5 feet (1524 mm) or more from buildings and 25 feet (7620 mm) or more from an exit discharge.

❖ As noted, outdoor storage and the use of highly toxic and toxic gases is more difficult to contain because an accidental release could disperse in various directions based on the configuration of the storage, the properties of the gases stored and used and climatic conditions. Additionally, there is less control of exposure hazards, which may affect the storage and use of such gases. The subsections that follow address distance to exposures, openings in exposed buildings and hazards associated with air intakes into buildings.

There is an exception when cylinders are contained within a gas cabinet that complies with both the general requirements of Chapter 27 and the specific requirements of Section 3704.1.2 for use of highly toxic and toxic gases. In addition, the cabinet must be sufficiently separated from buildings and the exit discharge.

3704.3.2.1.1 Distance limitation to exposures. Outdoor storage or use of highly toxic or toxic compressed gases shall not be located within 75 feet (22 860 mm) of a lot line, public street, public alley, public way, exit discharge or building not associated with the manufacture or distribution of such gases, unless all of the following conditions are met:

1. Storage is shielded by a 2-hour fire barrier which interrupts the line of sight between the storage and the exposure.
2. The 2-hour fire barrier shall be located at least 5 feet (1524 mm) from any exposure.
3. The 2-hour fire barrier shall not have more than two sides at approximately 90-degree (1.57 rad) directions, or three sides with connecting angles of approximately 135 degrees (2.36 rad).

❖ A minimum distance of 75 feet (22 860 mm) is required to exposures such as an exit discharge, lot line or public way. The distance to exposures is in terms of buildings not associated with the manufacture or distribution of highly toxic and toxic gases because buildings associated with those activities would not have separation requirements. It would be impractical for such facilities to be laid out with this kind of separation, and the level of hazard is known and accepted at these facilities. An alternative to the 75-foot (22 860 mm) separation is allowed when certain conditions are met. A 2-hour fire barrier can be substituted when the conditions stated in this section are met [see Figure 3704.3.2.1.1(1)]. The code criteria do not set a limit on the height or width of the barrier as long as it interrupts the line of sight between the storage and the exposure, although areas located above and to the sides of the barrier will likely have more restrictive limitations on openings (see commentary, Section 3704.3.2.1.2).

Limitations are placed on barriers in terms of the number of walls creating the separation based on the angle of the walls to one another. These are shown in Figure 3704.3.2.1.1(2). Generally, the limitations are related to the fact that when more than one wall is used there is the potential that some of the natural ventilation of the outdoor storage location would be lost. For instance, allowing three walls at 90 degrees (1.57 rad) versus two walls practically encloses the storage. The allowance of three walls at 135-degree (2.36 rad) angles still results in a level of openness that allows the benefits of natural ventilation.

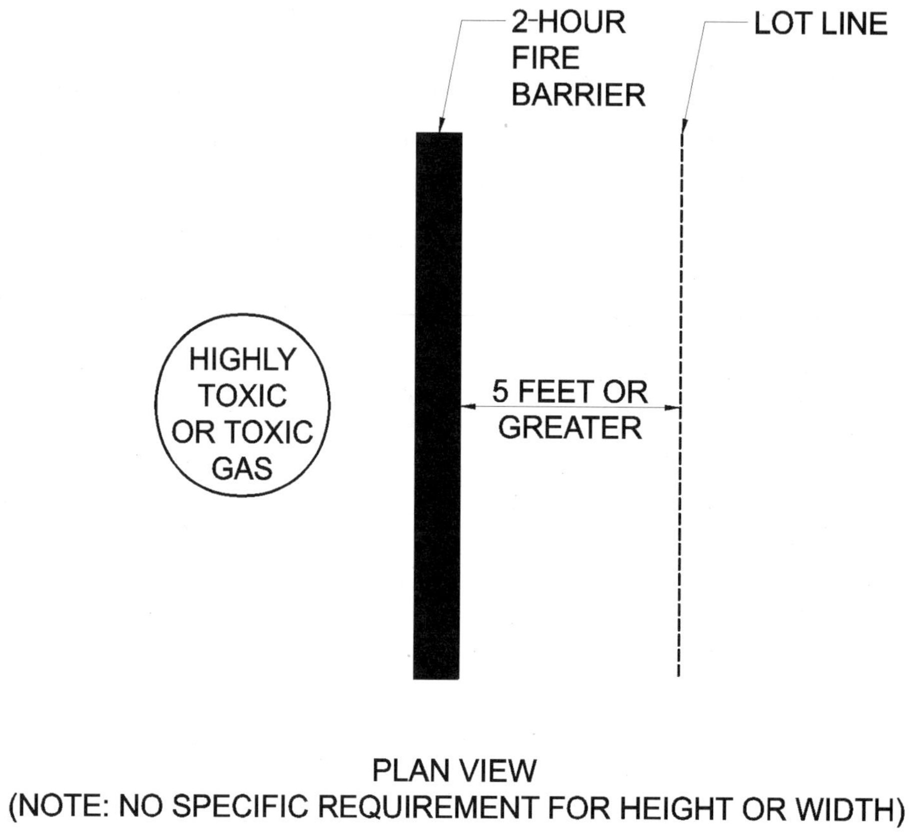

For SI: 1 foot = 304.8 mm.

Figure 3704.3.2.1.1(1)
FIRE BARRIER

3704.3.2.1.2 Openings in exposed buildings. Where the storage or use area is located closer than 75 feet (22 860 mm) to a building not associated with the manufacture or distribution of highly toxic or toxic compressed gases, openings into a building other than for piping are not allowed above the height of the top of the 2-hour fire barrier or within 50 feet (15 240 mm) horizontally from the storage area whether or not shielded by a fire barrier.

❖ When a fire barrier is used in place of the 75-foot (22 860 mm) separation, there are further limitations on openings. Openings other than pipe penetrations are allowed only when the exposure is 50 feet (15 240 mm) or more from the barrier, and then only below the height of the barrier. Piping penetrations are allowed because they are small enough to be considered an insignificant hazard (see Figure 3704.3.2.1.2).

The primary intent of this section is to protect the highly toxic and toxic gases from a release caused by a fire in the building. Openings such as windows increase the likelihood of a fire heating up a cylinder or tank and causing overpressures that could result in a release. A solid wall decreases the possibility of a fire penetrating or the radiation from the fire from contacting the storage area. Also, openings are a potential method for highly toxic and toxic gases to enter a building. A fire barrier will not stop gases from traveling upward if they are lighter than air.

3704.3.2.1.3 Air intakes. The storage or use area shall not be located within 75 feet (22 860 mm) of air intakes.

❖ This section recognizes the threat of a release of toxic or highly toxic gases getting into a building's HVAC system. The 75-foot (22 860 mm) separation is consistent with the distances previously laid out. The important distinction with this section is that exposure of the building's exterior is not a concern. Instead, the concern is for the health risks of a gas released into the building's HVAC system.

This section does not differentiate between a building associated with the storage and manufacture of highly toxic and toxic gases and one that is not. The risk to building occupants and emergency responders and the potential hazards with such gases entering into main sources of air for facilities, occupants would be the same in either case. Also, gases entering other openings do not have the same potential for circulation around the building as those entering an air intake with the specific purpose of circulation.

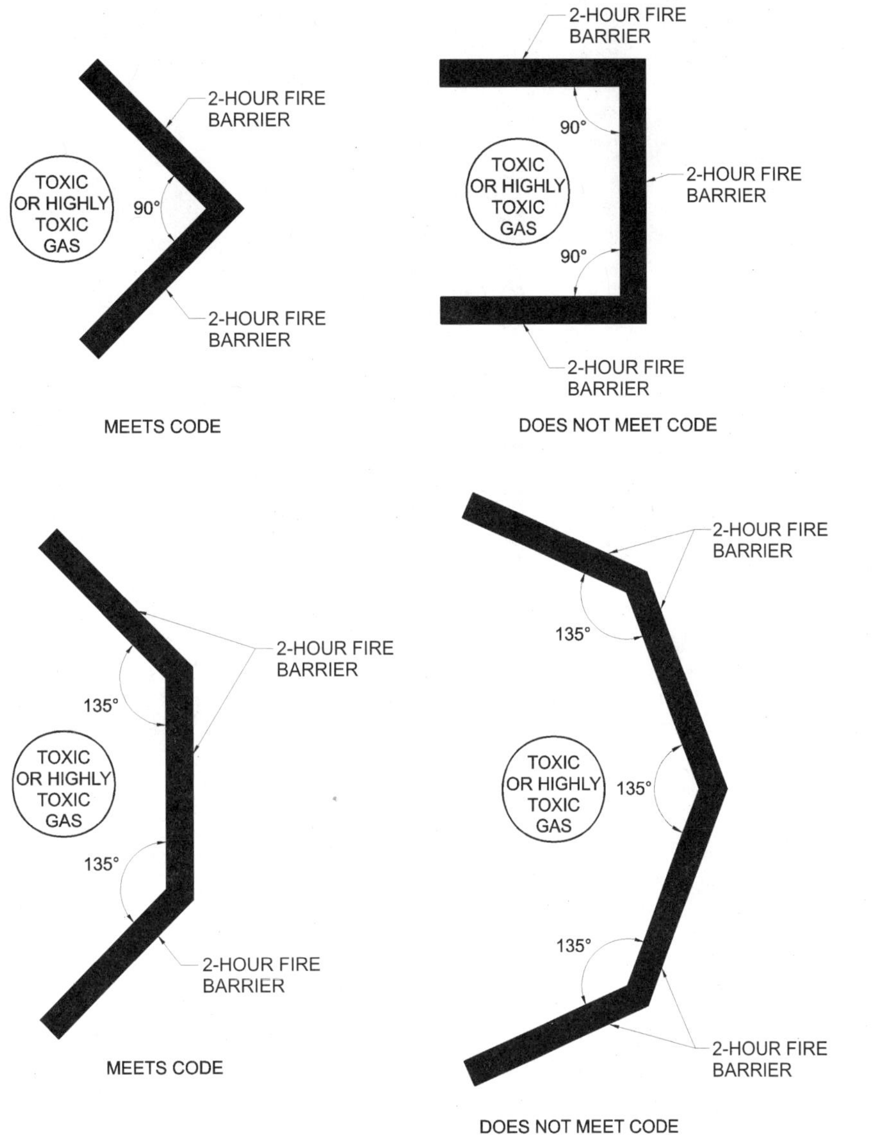

For SI: 1 degree = 0.01745 rad.

Figure 3704.3.2.1.1(2)
FIRE BARRIER LIMITATIONS

3704.3.2.2 Leaking cylinders and tanks. The requirements of Section 3704.2.2.3 shall apply to outdoor cylinders and tanks. Gas cabinets and exhausted enclosures shall be located within or immediately adjacent to outdoor storage or use areas.

❖ This section refers back to Section 3704.2.2.3 for requirements for handling leaking cylinders and tanks. Essentially, a gas cabinet or exhausted enclosure is required to contain leaking cylinders and tanks. Section 3704.2.2.3.1 requires that gas cabinets and exhausted enclosures used with leaking cylinders and tanks be located within a gas room. Although this section refers back to these requirements, it was not the intent for the code to require a gas room in outdoor storage and use areas. If a gas room with either a gas cabinet or exhausted enclosure is available nearby, it can be used, but to require an outdoor storage and use area to go indoors would be impractical. Regardless, gas rooms, gas cabinets and exhausted enclosures would have to be connected to an exhaust system that would then be connected to a treatment system. The major difference between the indoor requirements and the outdoor requirements is that cylinders, portable tanks and stationary tanks stored outside of buildings need not be placed within either gas cabinets, exhausted enclosures or gas rooms except when they are leaking.

A containment vessel or containment system could also be used as long as several criteria are met (see definitions for “Containment vessel” and “Containment system” in Section 3702 and the commentary to Section 3704.2.2.3).

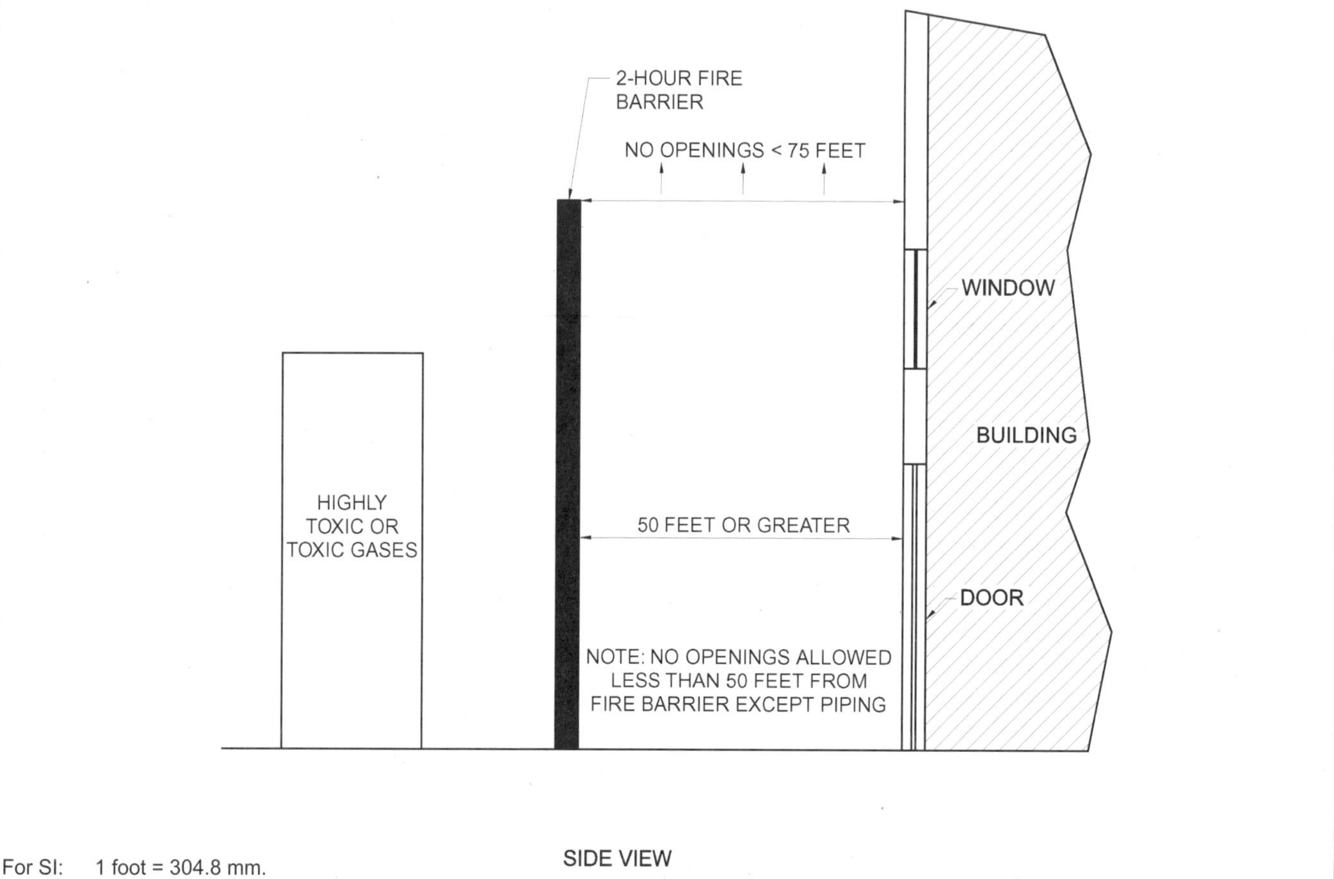

Figure 3704.3.2.1.2
OPENINGS

3704.3.2.3 Local exhaust for portable tanks. Local exhaust for outdoor portable tanks shall be provided in accordance with the requirements set forth in Section 3704.2.2.4.

❖ This section refers to Section 3704.2.2.4, which would require local exhaust systems for portable tanks to address areas on a tank where leaks are likely to occur. These systems need to be flexible to address the types of tanks likely to be stored or used. Although Section 3704.2.2.4 would require the exhaust system to be located within a gas room, the intent of this section would warrant the elimination of a gas room where the exhaust system is within or immediately adjacent to the outdoor storage area. Again, the intent is not to require outdoor storage to be placed indoors within a gas room. Outdoor storage has slightly different concerns, such as exposure hazards (see commentary, Section 3704.2.2.4).

3704.3.2.4 Piping and controls—stationary tanks. Piping and controls for outdoor stationary tanks shall be in accordance with the requirements set forth in Section 3704.2.2.5.

❖ This section refers to Section 3704.2.2.5 for requirements. The exception to Section 3704.2.2.5(1) is specific to outdoor tanks. This exception allows for the elimination of a treatment system to collect a release from a pressure relief device that is specifically installed for overpressure caused by fire. To take advantage of this exception several criteria must be met. These criteria address the reduction of possible fire hazards by ensuring both that the gas stored is not flammable and the exposures to other fire hazards are reduced. This exception recognizes that gases are more readily dispersed outside, which reduces the need for a treatment system to process a release (see commentary, Section 3704.2.2.5).

3704.3.2.5 Treatment systems. The treatment system requirements set forth in Section 3704.2.2.7 shall apply to highly toxic or toxic gases located outdoors.

❖ This section refers to the design criteria for treatment systems in Section 3704.2. As noted in Section 3704.2.2.7, a treatment system is required for exhaust ventilation coming from gas cabinets, exhausted enclosures, gas rooms and local exhaust systems. There are exceptions in Section 3704.2.2.7 for both storage and use. The storage exception focuses on securing valve outlets and using containment vessels and systems.

The use exception requires gas detection and fail-safe valve operation. Note that the exception for use is specific only to toxic gases.

For outdoor areas there are only a few occasions where a gas cabinet, exhausted enclosure, gas room or local exhaust would be required. Again, this is related to the fact that gases stored and used outdoors can be more easily dispersed if a release should occur, generally posing a lower hazard to those in the area. The requirements generally are based on the location of the gases. Treatment systems would come into play only for gas cabinets, exhausted enclosures and gas rooms associated with leaking cylinders or tanks and local exhaust for portable containers.

3704.3.2.6 Emergency power. The requirements for emergency power set forth in Section 3704.2.2.8 shall apply to highly toxic or toxic gases located outdoors.

❖ See the commentary to Section 3704.2.2.8.

3704.3.2.7 Gas detection system. The gas detection system requirements set forth in Section 3704.2.2.10 shall apply to highly toxic or toxic gases located outdoors.

❖ Because this section pertains to outdoor storage and use of highly toxic and toxic gases, the requirement for gas detection systems is less restrictive than it is for indoor storage and use. Essentially, detection systems would be required for small enclosures such as gas cabinets and exhausted enclosures, and for local enclosures such as piping manifold enclosures. Otherwise, gas detection is fairly ineffective in the outdoors. The detailed requirements for gas detection are found in Section 3704.2.2.10.

3704.3.3 Outdoor storage weather protection for portable tanks and cylinders. Weather protection in accordance with Section 2704.13 shall be provided for portable tanks and cylinders located outdoors and not within gas cabinets or exhausted enclosures. The storage area shall be equipped with an approved automatic sprinkler system in accordance with Section 903.3.1.1.

Exception: An automatic sprinkler system is not required when:

1. All materials under the weather protection structure, including hazardous materials and the containers in which they are stored, are noncombustible.
2. The weather protection structure is located not less than 30 feet (9144 mm) from combustible materials or structures or is separated from such materials or structures using a fire barrier complying with Section 3704.3.2.1.1.

❖ Cylinders and portable tanks that are used strictly for the outdoor storage of highly toxic and toxic gases must be placed in gas cabinets and exhausted enclosures only if they are leaking. Otherwise, outdoor storage simply requires protection from the elements.

Sprinklers are required unless every material under the weather protection is noncombustible and the weather protection structure is properly separated either by distance or by construction from combustible exposures such as other storage or buildings. The exception refers back to the exposure requirements in Section 3704.3.2.1.1 for construction requirements.

Protecting materials from the elements reduces the likelihood of the tanks and cylinders being damaged and possibly leading to a release through corrosion. Additionally, the sprinklers protect the storage from a large release if a fire were to occur that resulted in overpressures in tanks and cylinders and ultimately releases of gases. This protection is especially important because these particular gases are not being processed through a treatment system should a release occur.

3704.3.4 Outdoor use of cylinders, containers and portable tanks. Cylinders, containers and portable tanks in outdoor use shall be located in gas cabinets or exhausted enclosures.

❖ Because use is more likely to lead to a release of gas than storage, the cylinders, containers or portable tanks that supply the gases need to be located within a gas cabinet or exhausted enclosure. These cabinets and enclosures would require connection to an exhaust system and ultimately a treatment system.

SECTION 3705
OZONE GAS GENERATORS

3705.1 Scope. Ozone gas generators having a maximum ozone-generating capacity of 0.5 pound (0.23 kg) or more over a 24-hour period shall be in accordance with this section.

Exception: Ozone-generating equipment used in Group R-3 occupancies.

❖ This section differs from most of the rest of this chapter because it deals with the generation (i.e., creation) of a highly toxic gas and not simply its storage or use. The requirements center on the design, location, piping integrity and methods of ozone generator shutdown.

Ozone is a molecule composed of three atoms of oxygen. Two atoms of oxygen form the basic oxygen molecule, which is the oxygen we breathe. The third oxygen atom can detach from the ozone molecule and reattach to molecules of other substances, thereby altering their chemical composition. Generally, this ability to reattach is the reason ozone is generated and used as a method of purification. The primary uses are for the purification of water and air. The two uses are very different in their applications and the success of the air purification is a controversial one because it appears to be ineffective unless the ozone is present in the atmosphere at concentrations that would be harmful to people. Ozone is actually considered a pollutant when in the atmosphere we breathe.

The ability of ozone to react with organic materials is why it is used in purification. It is this same ability that causes harm to humans. The following table is taken from documentation from the Environmental Protection Agency (EPA) regarding the health effects, risk factors

Figure 3705.1

HEALTH EFFECTS	RISK FACTORS	HEALTH STANDARDS*
Potential risk of experiencing: Decreases in lung function Aggravation of asthma Throat irritation and cough Chest pain and shortness of breath Inflammation of lung tissue Higher susceptibility to respiratory infection	**Factors expected to increase risk and severity of health effects are:** Increase in ozone concentration in air Greater duration of exposure for some health effects Activities that raise the breathing rate (e.g., exercise) Certain preexisting lung diseases (e.g., asthma)	**Food and Drug Administration (FDA)** requires ozone output of indoor medical devices to be no more than 0.05 ppm. **Occupational Safety and Health Administration (OSHA)** requires that worker exposure be limited to an average concentration of no more than 0.10 ppm for a maximum of 8 hours. **National Institute of Occupational Safety and Health (NIOSH)** recommends an upper limit of 0.10 ppm, not to be exceeded at any time. **Environmental Protection Agency (EPA)'s** National Ambient Air Quality Standard for Ozone is a maximum of 8-hour average outdoor concentration of 0.08 ppm.

(*ppm = parts per million)

and health standards of various federal agencies.

Because of the nature of fire codes, the main goal of the requirements in this section is to avoid a situation where building occupants and emergency responders are exposed to unsafe levels of ozone on a more immediate basis, which is slightly different than the objectives of the federal agencies listed in Figure 3705.1.

These requirements apply only to generators that produce 0.5 pound (0.23 kg) or more over a 24-hour period. These tend to be the larger commercial and industrial application generators.

The exception for ozone generators that are found in one- and two-family dwellings (Group R-3) is based on the assumption that these generators would not produce the quantity noted within this section. This exception avoids impractical enforcement.

3705.2 Design. Ozone gas generators shall be designed, fabricated and tested in accordance with NEMA 250.

❖ Ozone is generated by applying an electrical current to at least one ozone-generating plate that then charges oxygen in the incoming air and produces ozone; therefore, the generator must comply with NEMA 250 for design, fabrication and testing.

3705.3 Location. Ozone generators shall be located in approved cabinets or ozone generator rooms in accordance with Section 3705.3.1 or 3705.3.2.

> **Exception:** An ozone gas generator within an approved pressure vessel when located outside of buildings.

❖ Because of the potential health hazards related to the generation of ozone, this section states limitations on the location of generating equipment. Ozone generators must be either in an approved cabinet or in an ozone gas generator room. Details of each are discussed in the subsections that follow.

The exception for ozone gas generators located outside of buildings is applicable only if the generators are contained within an approved pressure vessel. Generally, as with storage and use of gases outdoors, the gas is more likely to be dispersed to the atmosphere without harming people in the surrounding area.

3705.3.1 Cabinets. Ozone cabinets shall be constructed of approved materials and compatible with ozone. Cabinets shall display an approved sign stating:

OZONE GAS GENERATOR—HIGHLY TOXIC—OXIDIZER.

Cabinets shall be braced for seismic activity in accordance with the *International Building Code*.

Cabinets shall be mechanically ventilated in accordance with the *International Mechanical Code* with a minimum of six air changes per hour.

The average velocity of ventilation at makeup air openings with cabinet doors closed shall not be less than 200 feet per minute (1.02 m/s).

❖ This section lists criteria for the design and construction of cabinets intended for the ozone gas generators. This includes labeling, ventilation and seismic bracing requirements. For ventilation, it both refers to the IMC for requirements and states the following criteria:

- Six air changes per hour.
- Average velocity of 200 feet per minute (1.02 m/s) across the opening of the cabinet.

These criteria result in a negative pressure cabinet with appropriate intake air sizing.

3705.3.2 Ozone gas generator rooms. Ozone gas generator rooms shall be mechanically ventilated in accordance with the *International Mechanical Code* with a minimum of six air changes per hour. Ozone gas generator rooms shall be equipped with a continuous gas detection system which will shut off the generator and sound a local alarm when concentrations above the permissible exposure limit occur.

Ozone gas-generator rooms shall not be normally occupied, and such rooms shall be kept free of combustible and hazardous material storage. Room access doors shall display an approved sign stating:

OZONE GAS GENERATOR—HIGHLY TOXIC—OXIDIZER.

❖ For larger generators and particular processes, an ozone gas generator room may be more practical than cabinets. This section gives the criteria for design and construction for these rooms. The key construction element is the ventilation requirement of six air changes per hour, which is the same requirement as for cabinets. A gas detection system is also required that will activate a local alarm when the permissible exposure limit has been reached. This limit appears to be 10 parts per million (ppm) in accordance with the information provided by the EPA.

Because ozone is also an oxidizer, combustibles would not be allowed within the room to reduce the potential for an intense fire. In addition, because of the health effects, the rooms are not intended to be normally occupied. These two limitations essentially necessitate a mechanical room-type arrangement. The labeling of the generator room is essential to emergency responders and also to those unfamiliar with the facility.

3705.4 Piping, valves and fittings. Piping, valves, fittings and related components used to convey ozone shall be in accordance with Sections 3705.4.1 through 3705.4.3.

❖ The requirements found within this section are intended to reduce the weak links in the system that can be found with improper piping connections or fittings.

3705.4.1 Piping. Piping shall be welded stainless steel piping or tubing.

Exceptions:

1. Double-walled piping.
2. Piping, valves, fittings and related components located in exhausted enclosures.

❖ A welded connection is required for piping because it is more reliable than a friction connection or a threaded connection. Additionally, this section requires the piping to be stainless steel to reduce the likelihood of corrosion that other piping may be susceptible to.

There are two exceptions to the welded connection and stainless steel requirement that include double wall piping or when all portions of the piping, valves and fittings are located within an exhausted enclosure. Both of these methods will reduce the likelihood that a failure in the piping, valves or fittings would result in a release of highly toxic ozone to the atmosphere.

3705.4.2 Materials. Materials shall be compatible with ozone and shall be rated for the design operating pressures.

❖ It is important that valves or fittings used be compatible with ozone. Section 3705.4.1 requires the piping to be stainless steel unless one of the two exceptions applies. Regardless, the materials used need to be compatible. Essentially, the intent is to avoid leaks or major failures resulting from corrosion.

Also, this section discusses the need for piping to withstand the pressures that are expected, because that could be another potential mode of failure.

3705.4.3 Identification. Piping shall be identified with the following:

OZONE GAS—HIGHLY TOXIC—OXIDIZER.

❖ To alert occupants and emergency responders to the potential dangers of ozone, the piping carrying the ozone must be labeled to list the hazards of this highly toxic and oxidizing gas.

3705.5 Automatic shutdown. Ozone gas generators shall be designed to shut down automatically under the following conditions:

1. When the dissolved ozone concentration in the water being treated is above saturation when measured at the point where the water is exposed to the atmosphere.
2. When the process using generated ozone is shut down.
3. When the gas detection system detects ozone.
4. Failure of the ventilation system for the cabinet or ozone-generator room.
5. Failure of the gas-detection system.

There are several conditions where the ozone generator needs to shut down automatically. Primarily, when the ozone being generated is not placed directly into use, as in the case of water purification or when ozone is either detected or the ventilation and gas detection systems fail, the generator must automatically shutdown. This adds to the levels of redundancies available to assure personnel safety.

❖ There are several conditions where the ozone generator needs to shut down automatically. Primarily, when the ozone being generated is not placed directly into use, as in the case of water purification or when ozone is either detected or the ventilation and gas detection systems fail, the generator must automatically shut down. This adds to the levels of redundancies available to enhance personnel safety.

3705.6 Manual shutdown. Manual shutdown controls shall be provided at the generator and, where in a room, within 10 feet (3048 mm) of the main exit or exit access door.

❖ A manual shutdown control on the ozone generator is a backup to the automated shutdown system that uses sensors to detect leaks, sound alarms and shut down the system. Because an ozone generator located in a room would almost always be a large-capacity machine, the manual shutdown controls are needed near the main access door to minimize the possibility that any significant quantity of escaping ozone would reach building occupants. Having the controls near the door also minimizes the chance that the person who shuts down the generator in an emergency would be exposed to harmful amounts of ozone. With smaller generators that would more likely be located in a ventilated cabinet or exhausted enclosure, there is no requirement for manual shutdown controls other than those on the generator itself.

Bibliography

The following resource materials are referenced in this chapter or are relevant to the subject matter addressed in this chapter.

EPA-600/R-95-154, *Ozone Generators in Indoor Air Settings*. Report prepared for the Office of Research and Development by Raymond Steiber. Research Triangle Park, NC: National Risk Management Research Laboratory, U.S. Environmental Protection Agency (U.S. EPA), 1995.

EPA-452/R-96-007, *Review of National Ambient Air Quality Standards for Ozone: Assessment of Scientific and Technical Information*. OAQPS Staff Paper. Research Triangle Park, NC: Office of Air Quality Planning and Standards, U.S. Environmental Protection Agency (U.S. EPA), 1996.

ICC EC-2003, *International Electrical Code*, Falls Church, VA: International Code Council, 2003.

IBC-2003, *International Building Code*, Falls Church, VA: International Code Council, 2003.

IMC-2003, *International Mechanical Code*. Falls Church, VA: International Code Council, 2003.

NEMA 250-97, *Enclosures for Electrical Equipment (1,000 volt maximum)*. Rosslyn, VA: National Electrical Manufacturer's Association, 1997.

Chapter 38: Liquefied Petroleum Gases

General Comments

The use and popularity of liquefied petroleum gas (LP-gas) for domestic purposes varies depending on location. In the United States, propane is the most widely used (LP-gas), with butane a distant runner-up. Rural communities, especially those located in cooler climates, tend to use propane as a principal fuel source for domestic cooking, clothes drying, water heating and space heating. This practice tends to be less common in urban and suburban communities served by public utility companies supplying piped natural gas. On the other hand, LP-gas is widely used by industries as a motor fuel for industrial lift trucks; aerosol charging; welding and cutting; auxiliary heating and lighting and, in the agricultural community, as a fuel for drying crops, operating pumps and heating livestock shelters.

Propane is well known as a camping fuel for cooking, lighting, heating and refrigerating. LP-gas also remains a popular standby fuel supply for auxiliary generators. Additionally, utility companies use propane as a substitute for natural gas as a "peak shaving" alternative when supplies run low or prices are high; in this application, propane is usually mixed with air to achieve the same energy content per cubic foot as natural gas.

Propane is widely used as an alternative motor vehicle fuel, and its characteristic as a clean burning fuel has resulted in the addition of propane dispensers to service stations throughout the country.

Another use for propane, although less popular because it is flammable, is as the working fluid in a refrigeration cycle. In some applications it is a suitable alternative to R-22 and other CFC-rich refrigerants.

Future uses for LP-gas, and propane in particular, may arise out of an intense interest in independent electrical power generation units such as fuel cell and microturbine generators. Propane is a leading candidate for these units because of its portability and moderate storage pressures. In addition, ongoing research on absorption refrigeration cycles may lead to greater interest in propane-powered air conditioners.

Purpose

Chapter 38 includes requirements for handling, storing and using LP-gas, principally propane, to reduce the possibility of damage to containers, accidental releases of LP-gas and exposure of flammable concentrations of LP-gas to ignition sources.

SECTION 3801
GENERAL

3801.1 Scope. Storage, handling and transportation of LP-gas and the installation of LP-gas equipment pertinent to systems for such uses shall comply with this chapter and NFPA 58. Properties of LP-gases shall be determined in accordance with Appendix B of NFPA 58.

❖ Some of the principal characteristics of LP-gases are shown in Table 3801.1(1). They include the limits of flammability, the specific gravity of the gas vapor at 60°F (16°C) and the ignition temperatures of the flammable mixtures.

It is important to recognize that LP-gases exist in both liquid and vapor states at ambient temperatures. When stored in a pressure vessel, the liquid and vapor states of LP-gases will be in equilibrium until the system is called upon to deliver vapor to the appliance it is serving. When this happens, the pressure in the container drops and the liquid begins to boil to produce more vapor and once again reach equilibrium. As vapor is produced through the boiling process, the liquid in the container begins to refrigerate itself and the entire system cools down. Table 3801.1(2) shows the vapor pressure of propane at various temperatures.

LP-gases are heavier than air, and although public perception holds that they will sink to the lowest level of a space, these gases disperse according to the laws of physics just as other gases do. It is not safe to assume that all LP-gases, such as propane, immediately sink to the floor and "pool" there. LP-gases can and will, under ambient conditions, disperse to all parts of a room or space.

LP-gases, such as propane, must be odorized with a warning agent that is detectable at a minimum concentration of one-fifth the lower limit of flammability. The most frequently used odorant for propane is ethyl mercaptan, which is usually added at a pipeline terminal or other supply point before the gas is shipped to a retail bulk storage facility.

LP-gas containers pose the danger of a Boiling-Liquid-Expanding-Vapor-Explosion (BLEVE) if they are exposed to a fire. Prolonged flame impingement above the liquid-gas interface, where the heat-absorbing properties of the liquid cannot protect the container shell from thermal stress, may generate catastrophic failures, expelling tank contents in a fireball and propelling container fragments from the site of the accident.

Such violent container failures are extremely dangerous and were much more common in the past, before safety improvements in railcars and cargo transport trucks were introduced. Recently, the propane industry, through the National Propane Gas Association and the Propane Education and Research Council, has developed a program for training emergency response personnel to respond effectively and knowledgeably to threatening incidents involving the transportation, storage or use of propane. The curriculum was made available free of charge to every fire department in the United States.

The requirements of this chapter are intended to address hazards associated with the storage, handling and use of LP-gas. NFPA 58 is supplemented by NFPA 59, which specifically addresses the use of LP-gas at utility gas plants.

The U.S. Department of Transportation (DOT) classifies LP-gas as a hazardous material, and the transportation of LP-gases "in commerce" is regulated by Title 49 of the Code of Federal Regulations. The term "in commerce" is defined to be transportation by a commercial entity. In other words, transportation of a propane grill cylinder in the passenger compartment of a car is not regulated by the DOT but is by NFPA 58.

3801.2 Permits. Permits shall be required as set forth in Sections 105.6 and 105.7.

Distributors shall not fill an LP-gas container for which a permit is required unless a permit for installation has been issued for that location by the fire code official.

❖ The process of issuing permits gives the fire code official an opportunity to carefully evaluate and regulate hazardous operations. Permit applicants should be required to demonstrate that their operations comply with the intent of the code before the permit is issued. See the commentary to Section 105.6 for a general discussion of operations requiring an operational permit, Section 105.6.28 for a discussion of specific quantity-based operational permits for the materials regulated in this chapter and Section 105.7 for a general discussion of activities requiring a construction permit. The permit process also notifies the fire department of the need for prefire planning for hazardous property.

An operational permit for installations using or storing LP-gas is required except when the installation is a container of 500-gallon (1893 L) water capacity or less and serving a Group R-3 occupancy. In addition, Section 105.7.8 requires that a construction permit be obtained for any installation of or modification to an LP-gas system.

The last sentence of this section is addressing stationary installations of LP-gas and states that for those installations requiring a permit, the container must not be filled unless that permit has been issued.

TABLE 3801.1(1)
PROPERTIES OF COMMON LP-GASES

GAS	IGNITION TEMPERATURE (°F)	LOWER FLAMMABLE LIMITATIONS[a]	UPPER FLAMMABLE LIMITATIONS[a]	SPECIFIC GRAVITY OF VAPOR @ 60°F
Butane	550	1.9	8.5	2.0
1-Butene	725	1.6	10.0	1.9
2-Butene (cis)	617	1.7	9.0	1.9
2-Butene (trans)	615	1.8	9.7	1.9
Propane	920	2.1	9.5	1.5
Propylene	851	2.0	11.1	1.5

Note a. Lower and upper flammable limitations are percentages of gas in gas-air mixtures. Specific gravity describes the relative weight of a unit of gas compared to air (air = 1.0).

Source: NFPA 325M - Properties of Flammable Solids, Liquids and Gases (1991).

TABLE 3801.1(2)
VAPOR PRESSURES OF PROPANE

TEMP (°F)	PRESS. (psig)	TEMP (°F)	PRESS. (psig)	TEMP (°F)	PRESS. (psig)	TEMP (°F)	PRESS. (psig)
130	257	70	109	20	40	-20	10
120	225	65	100	10	31	-25	8
110	197	60	92	0	23	-30	5
100	172	50	77	-5	20	-35	3
90	149	40	63	-10	16	-40	1
80	128	30	51	-15	13	-44	0

3801.3 Construction documents. Where a single container is more than 2,000 gallons (7570 L) in water capacity or the aggregate capacity of containers is more than 4,000 gallons (15 140 L) in water capacity, the installer shall submit construction documents for such installation.

❖ This section applies to those installations for which a construction permit is required, namely when a new system is being installed or an existing system is being modified. When container size exceeds the thresholds specified, applicants must submit plans indicating compliance with requirements before approval can be authorized. Plans should be clear and concise, legible, prepared on standard-sized paper and include the following:

- Location and legal identification (address and lot or parcel number) of the lot or site;
- Legal boundaries of the site, including reference to source or survey;
- Location of significant buildings on the lot or site and adjacent lots or sites;
- Location of nearest public roadways and site access;
- Location of all underground and overhead utilities;
- North arrow;
- Topographical features;
- Proposed container location with respect to buildings, building openings, lot lines, public roadways and underground or overhead utilities;
- Container dimensions and capacity;
- Container compliance markings (e.g., ASME, DOTn, API);
- Details of container foundation and supports;
- Section through container showing supports and anchors and, if an underground tank, backfill and corrosion protection;
- Arrangement of valves and piping;
- Specifications for containers, valves, piping, tank mounts and pads and other related equipment and appliances;
- Means for protecting valves from tampering;
- Name, address and telephone number of property owner; and
- Name, address and telephone number of installer and servicing contractor.

SECTION 3802 DEFINITIONS

3802.1 Definition. The following word and term shall, for the purposes of this chapter and as used elsewhere in this code, have the meaning shown herein.

❖ Definitions of terms can help in the understanding and application of the code requirements. The purpose for including the definition that is associated with the subject matter of this chapter is to provide more convenient access to it without having to refer back to Chapter 2. It is important to emphasize that this term is not exclusively related to this chapter but is applicable everywhere the term is used in the code. For convenience, the term is also listed in Chapter 2 with a cross reference to this section. The use and application of all defined terms, including the one defined in this section, are set forth in Section 201.

LIQUEFIED PETROLEUM GAS (LP-gas). A material which is composed predominantly of the following hydrocarbons or mixtures of them: propane, propylene, butane (normal butane or isobutane) and butylenes.

❖ The definition of LP-gas is consistent with that found in NFPA 58, with one exception: NFPA 58 requires that the vapor pressure of an LP-gas mixture be less than or equal to that of commercial propane.

SECTION 3803 INSTALLATION OF EQUIPMENT

3803.1 General. Liquefied petroleum gas equipment shall be installed in accordance with the *International Fuel Gas Code* and NFPA 58, except as otherwise provided in this chapter.

❖ The *International Fuel Gas Code*® (IFGC®) addresses the installation of equipment downstream from the final pressure regulator in the system, which typically is the second-stage regulator. The IFGC would, therefore, regulate LP-gas piping typically installed indoors, as well as appliance installations.

The remainder of Chapter 38 is devoted primarily to energy-consuming installations at which fuel-burning appliances are installed. NFPA 58 should be referred to for propane bulk plant installations; refrigerated storage installations; marine shipping and receiving installations and equipment-specific requirements such as for installing vaporizers and dispensers. The more detailed provisions in NFPA 58 should be reviewed by the fire code official before approving any LP-gas installation.

3803.2 Use of LP-gas containers in buildings. The use of LP-gas containers in buildings shall be in accordance with Sections 3803.2.1 and 3803.2.2.

❖ In general, storing and using LP-gas inside buildings is limited to relatively small quantities. There are some exceptions, such as the use of LP-gas for temporary heating in buildings under construction.

LP-gas containers are either fabricated to the requirements of the *ASME Boiler and Pressure Vessel Code* or the U.S. Department of Transportation rules contained in Title 49 of the Code of Federal Regulations. A typical ASME tank is shown in Figure 3803.2.

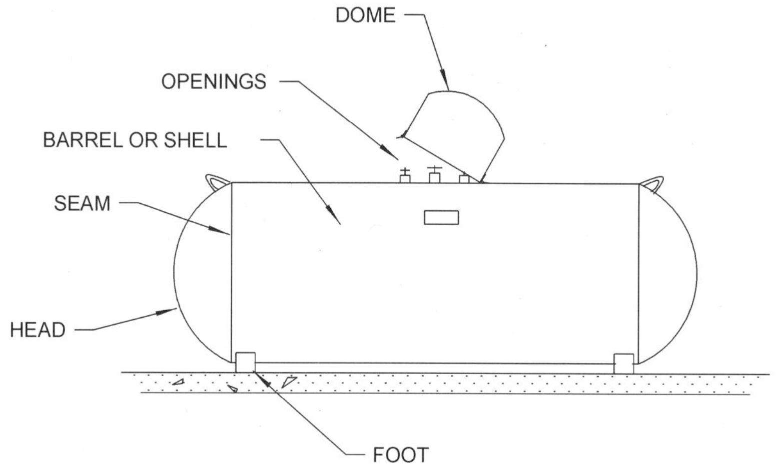

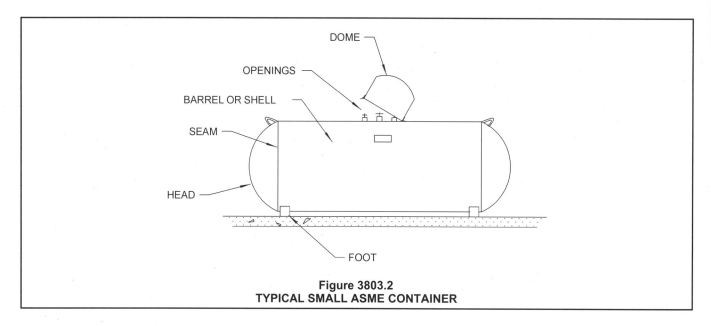

Figure 3803.2
TYPICAL SMALL ASME CONTAINER

3803.2.1 Portable containers. Portable LP-gas containers, as defined in NFPA 58, shall not be used in buildings except as specified in NFPA 58 and Sections 3803.2.1.1 through 3803.2.1.7.

❖ Portable containers are typically DOT cylinders, which are used in the vertical position. This is not always the case though; some recreational vehicles use DOT cylinders in a horizontal position (see Figure 3803.2.1).

One of the most common violations involving LP-gas is the storage or use of it inside buildings, particularly residential structures. This is a long-standing requirement that relates to the potential for releasing propane within a building.

In the past, any propane cylinder would release propane simply by opening the service valve; however, advances have been made in the technologies used in the propane industry. Since 1995, propane cylinders shipped with new outdoor cooking equipment (grills, smokers, etc.) must be fitted with valves that will not release propane even if the valve handle is opened, unless a positive, leak-tight connection is made between the cylinder and the cooking equipment. This technology, as well as two others that would stop the flow of gas under fire conditions or if there were a hose line break, are required by ANSI Z21.58.

3803.2.1.1 Use in basement, pit or similar location. LP-gas containers shall not be used in a basement, pit or similar location where heavier-than-air gas might collect. LP-gas containers shall not be used in an above-grade underfloor space or basement unless such location is provided with an approved means of ventilation.

Exception: Use with self-contained torch assemblies in accordance with Section 3803.2.1.6.

❖ Because propane has a specific gravity of 1.52 at 60°F (16°C) (which means it is 1.52 times heavier than air at that temperature), propane has been assumed to automatically sink to the ground when it is released to the atmosphere. Even though the laws of physics have proven this concept to be untrue, concerns persist about the ability for propane to disperse in locations where air circulation is restricted. Any ventilation system design should take into account the expected temperatures at the location and an expected leak rate from the piping system.

Only LP-gas containers are prohibited from being used in basements, pits and similar locations. Hardpiped systems in which the container remains outside the building can serve systems installed in basements, pits and similar locations. The hazard associated with containers in buildings relates to the fact that the pressure in the container is dependent on the temperature of the propane within it and can exceed 100 pounds per square inch gauge (psig) at 70°F (21°C), whereas the pressure in the building piping of a propane service is usually 0.5 psig (3 kPa).

In the past, the model codes prohibited the use of any propane system in spaces below grade. The U.S. Consumer Product Safety Commission staff concluded, however, that the rate of incidents occurring in propane systems used in basements and underground locations is no different than that occurring in above-ground locations.

The exception to this requirement acknowledges the use of self-contained torch assemblies, which are limited to about 1 pound (.454 kg) (2.5 pounds water) of propane per container.

3803.2.1.2 Construction and temporary heating. Portable containers are allowed to be used in buildings or areas of buildings undergoing construction or for temporary heating as set forth in Sections 3.4.3, 3.4.4, and 3.4.7 of NFPA 58.

❖ The referenced sections in NFPA 58 address the use of temporary heating in buildings undergoing construction or major or minor renovations, or industrial buildings under certain conditions.

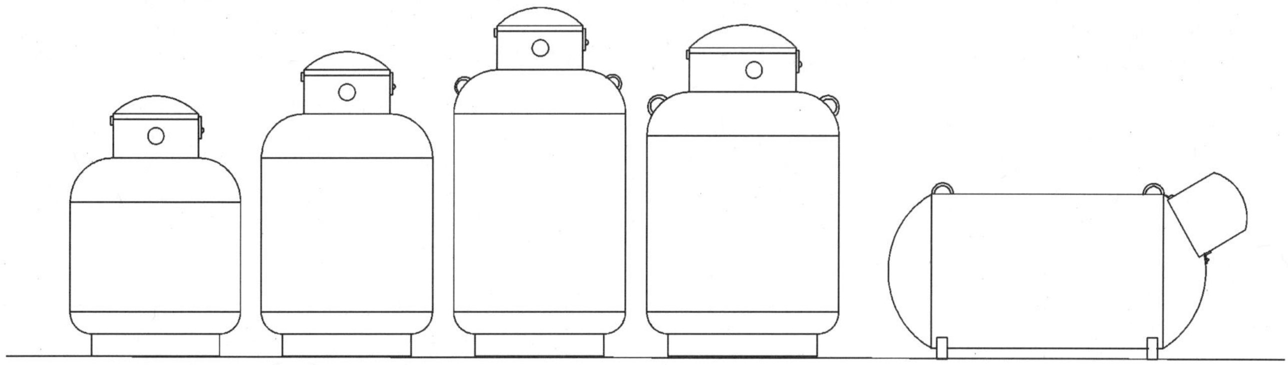

Figure 3803.2.1
TYPICAL DOT STATIONARY CYLINDERS

3803.2.1.3 Group F occupancies. In Group F occupancies, portable LP-gas containers are allowed to be used to supply quantities necessary for processing, research or experimentation. Where manifolded, the aggregate water capacity of such containers shall not exceed 735 pounds (334 kg) per manifold. Where multiple manifolds of such containers are present in the same room, each manifold shall be separated from other manifolds by a distance of not less than 20 feet (6096 mm).

❖ The use of manifolds to connect containers increases the overall capacity of the system by increasing its vaporization capacity. The vaporization rate is directly related to the total surface area of the containers. Using manifolds, however, also increases the number of connections made on a system, and, because the connections in a piping system under pressure are where leaks are likely to originate, the total volume of propane in manifolded systems is limited to that shown. The amount of propane in pounds can be calculated from the water capacity in pounds by multiplying by 0.42. In this instance, the amount of propane permitted is about 310 pounds (141 kg).

3803.2.1.4 Group E and I occupancies. In Group E and I occupancies, portable LP-gas containers are allowed to be used for research and experimentation. Such containers shall not be used in classrooms. Such containers shall not exceed a 50-pound (23 kg) water capacity in occupancies used for educational purposes and shall not exceed a 12-pound (5 kg) water capacity in occupancies used for institutional purposes. Where more than one such container is present in the same room, each container shall be separated from other containers by a distance of not less than 20 feet (6096 mm).

❖ Further restrictions on the use of propane in educational (schools) and institutional occupancies reflects the likelihood that the occupants of those spaces will be limited in their ability to respond to an emergency.

The minimum distance of 20 feet (6096 mm) is intended to provide a factor of safety so that if one installation becomes a source of an ignited jet plume, any other similar installations would not be directly exposed to the plume.

3803.2.1.5 Demonstration uses. Portable LP-gas containers are allowed to be used temporarily for demonstrations and public exhibitions. Such containers shall not exceed a water capacity of 12 pounds (5 kg). Where more than one such container is present in the same room, each container shall be separated from other containers by a distance of not less than 20 feet (6096 mm).

❖ This section restricts the size of LP-gas containers used for demonstrations to roughly 5 pounds (2.3 kg) of propane capacity. There is no restriction on the number of containers, however; only a limitation that they be separated by at least 20 feet (6096 mm) for the same reason as specified in the previous section.

3803.2.1.6 Use with self-contained torch assemblies. Portable LP-gas containers are allowed to be used to supply approved self-contained torch assemblies or similar appliances. Such containers shall not exceed a water capacity of 2.5 pounds (1 kg).

❖ The containers used with torches of this type are limited to 1 pound (.454 kg) (3 pounds water) nominal propane capacity. NFPA 58 also requires that these containers comply with UL 147A.

3803.2.1.7 Use for food preparation. Where approved, listed LP-gas commercial food service appliances are allowed to be used for food-preparation within restaurants and in attended commercial food-catering operations in accordance with the *International Fuel Gas Code,* the *International Mechanical Code* and NFPA 58.

❖ Although not stated explicitly, this section intends to permit the use of LP-gas containers in buildings with commercial food service cooking appliances. The code requires that those appliances be approved. Typically, this type of appliance will be labeled in accordance with one of the ANSI Z83 standards for commercial gas cooking equipment.

3803.2.2 Industrial vehicles and floor maintenance machines. Containers on industrial vehicles and floor maintenance machines shall comply with NFPA 58, Section 8.3 and 8.4.

❖ This section merely refers to the correct section reference in NFPA 58 for containers serving engines mounted on industrial vehicles and floor maintenance machines.

3803.3 Location of equipment and piping. Equipment and piping shall not be installed in locations where such equipment and piping is prohibited by the *International Fuel Gas Code.*

❖ With respect to LP-gas systems, the IFGC addresses equipment and piping downstream of the second-stage regulator, which would typically include all piping and appliances within the building.

SECTION 3804 LOCATION OF CONTAINERS

3804.1 General. The storage and handling of LP-gas and the installation and maintenance of related equipment shall comply with NFPA 58 and be subject to the approval of the fire code official, except as provided in this chapter.

❖ This section reinforces the need to review NFPA 58 for detailed provisions before any installation of LP-gas is approved.

3804.2 Maximum capacity within established limits. Within the limits established by law restricting the storage of liquefied petroleum gas for the protection of heavily populated or congested areas, the aggregate capacity of any one installation shall not exceed a water capacity of 2,000 gallons (7570 L) (see Section 3 of the Sample Ordinance for Adoption of the *International Fire Code* on page v).

> **Exception:** In particular installations, this capacity limit shall be determined by the fire code official, after consideration of special features such as topographical conditions, nature of occupancy, and proximity to buildings, capacity of proposed containers, degree of fire protection to be provided and capabilities of the local fire department.

❖ This section originated in NFPA 58 and is intended to establish a limit on ordinary above-ground containers in densely populated areas. The requirement establishes a level of acceptable risk to the general public; however, the scientific basis for the 2,000 gallon (908 kg) limitation is not clear.

The exception recognizes that consideration of the specific features of the installation, such as fire protection systems, mounding or burying of containers or the capabilities of the local fire-fighting service, should be taken into account before establishing any limit on the system's capacity.

3804.3 Container location. Containers shall be located with respect to buildings, public ways, and lot lines of adjoining property that can be built upon, in accordance with Table 3804.3.

❖ This paragraph references Table 3804.3, which contains the requirements for siting LP-gas containers on a piece of property.

Separating and protecting containers from each other and from buildings and heat-producing appliances are the principal means for preventing container failures and accidental ignitions. Separation serves at least two purposes. Outside a building, separating tanks from structures allows escaping gas to disperse or dilute before it can enter a building or come in contact with an ignition source. Similarly, the container is protected to an extent from a hazardous exposure if escaping gas is ignited, and the tank is protected if the building becomes involved in a fire.

Separating containers from each other or from other groups of containers minimizes the scale of an accident if a tank or group of containers becomes involved in a fire. Though the separation distances do not result in absolute protection from fire exposures and leaks, they do provide a measure of access for establishing fire-fighting positions to protect tanks and exposures.

TABLE 3804.3. See page 38-7.

❖ This table has its origins in NFPA 58 and has existed for many years. This establishes reasonable distances that can be used to determine separation requirements between an LP-gas container installation and buildings, lot lines and public ways, as well as between multiple containers installed at the same site. The important factors to consider when establishing suitable distances include the container capacity and whether it is installed above ground, mounded or below ground. See Figure 3804.3(1) for a graphic depiction of the requirements.

As one would expect, the required distances are reduced for containers that are mounded or installed underground because of the reduced exposure of the container to external sources of heat. Although not defined in the code, the difference between a mounded container and one installed underground is that a mounded container is an ASME container labeled for underground installation but installed above the minimum depth required for underground service. NFPA 58 contains specific provisions for mounding tanks.

Note a to Table 3804.3 defines the endpoint for measuring the separation distance for underground containers as being the relief device, which will release propane to the atmosphere only under abnormal conditions, and the filling connection and liquid-level gauge, both of which release propane during a normal filling operation.

Note b contains a requirement for building overhangs in which the overhang is less than 50 feet (15 240 mm) above the relief device discharge outlet. This dimension represents what is considered a safe distance to prevent the accumulation of LP-gas should the relief device discharge [see Figure 3804.3(2)].

Note c requires that clearances be provided to allow installation and maintenance, if necessary, for underground tanks, which must be done using a backhoe or some other large piece of equipment.

Note d anticipates an increase in the hazard of those installations made up of at least four containers because of the greater number of joints and connections needed as well as perhaps a greater risk of involvement of the storage system in a building fire.

Note e addresses installing smaller containers and focuses on requirements for locating relief devices and other appurtenances an acceptable distance from potential sources of ignition.

Note f permits a reduction in the required distances for a 1,200-gallon (4,542 L) water capacity container, which is typically the maximum size for a container that would be used for any residential service.

3804.3.1 Special hazards. Containers shall also be located with respect to special hazards such as above-ground flammable or combustible liquid tanks, oxygen or gaseous hydrogen containers, flooding or electric power lines as specified in NFPA 58, Section 3.2.2.6.

❖ Additional requirements may apply when siting an LP-gas container. Other products stored in the vicinity may present specific hazards that must be addressed.

TABLE 3804.3
LOCATION OF LP-GAS CONTAINERS

CONTAINER CAPACITY (water gallons)	MINIMUM SEPARATION BETWEEN CONTAINERS AND BUILDINGS, PUBLIC WAYS OR LOT LINES OF ADJOINING PROPERTY THAT CAN BE BUILT UPON		MINIMUM SEPARATION BETWEEN CONTAINERS[b, c] (feet)
	Mounded or underground containers[a] (feet)	Above-ground containers[b] (feet)	
Less than 125[c, d]	10	5[e]	None
125 to 250	10	10	None
251 to 500	10	10	3
501 to 2,000	10	25[e, f]	3
2,001 to 30,000	50	50	5
30,001 to 70,000	50	75	(0.25 of sum of diameters of adjacent containers)
70,001 to 90,000	50	100	
90,001 to 120,000	50	125	

For SI: 1 foot = 304.8 mm, 1 gallon = 3.785 L.

a. Minimum distance for underground containers shall be measured from the pressure relief device and the filling or liquid-level gauge vent connection at the container, except that all parts of an underground container shall be 10 feet or more from a building or lot line of adjoining property which can be built upon.

b. For other than installations in which the overhanging structure is 50 feet or more above the relief-valve discharge outlet. In applying the distance between buildings and ASME containers with a water capacity of 125 gallons or more, a minimum of 50 percent of this horizontal distance shall also apply to all portions of the building which project more than 5 feet from the building wall and which are higher than the relief valve discharge outlet. This horizontal distance shall be measured from a point determined by projecting the outside edge of such overhanging structure vertically downward to grade or other level upon which the container is installed. Distances to the building wall shall not be less than those prescribed in this table.

c. When underground multicontainer installations are comprised of individual containers having a water capacity of 125 gallons or more, such containers shall be installed so as to provide access at their ends or sides to facilitate working with cranes or hoists.

d. At a consumer site, if the aggregate water capacity of a multicontainer installation, comprised of individual containers having a water capacity of less than 125 gallons, is 500 gallons or more, the minimum distance shall comply with the appropriate portion of Table 3804.3, applying the aggregate capacity rather than the capacity per container. If more than one such installation is made, each installation shall be separated from other installations by at least 25 feet. Minimum distances between containers need not be applied.

e. The following shall apply to above-ground containers installed alongside buildings:
   1. Containers of less than a 125-gallon water capacity are allowed next to the building they serve when in compliance with Items 2, 3 and 4.
   2. Department of Transportation (DOTn) specification containers shall be located and installed so that the discharge from the container pressure relief device is at least 3 feet horizontally from building openings below the level of such discharge and shall not be beneath buildings unless the space is well ventilated to the outside and is not enclosed for more than 50 percent of its perimeter. The discharge from container pressure relief devices shall be located not less than 5 feet from exterior sources of ignition, openings into direct-vent (sealed combustion system) appliances or mechanical ventilation air intakes.
   3. ASME containers of less than a 125-gallon water capacity shall be located and installed such that the discharge from pressure relief devices shall not terminate in or beneath buildings and shall be located at least 5 feet horizontally from building openings below the level of such discharge and not less than 5 feet from exterior sources of ignition, openings into direct vent (sealed combustion system) appliances, or mechanical ventilation air intakes.
   4. The filling connection and the vent from liquid-level gauges on either DOTn or ASME containers filled at the point of installation shall not be less than 10 feet from exterior sources of ignition, openings into direct vent (sealed combustion system) appliances or mechanical ventilation air intakes.

f. This distance is allowed to be reduced to not less than 10 feet for a single container of 1,200-gallon water capacity or less, provided such container is at least 25 feet from other LP-gas containers of more than 125-gallon water capacity.

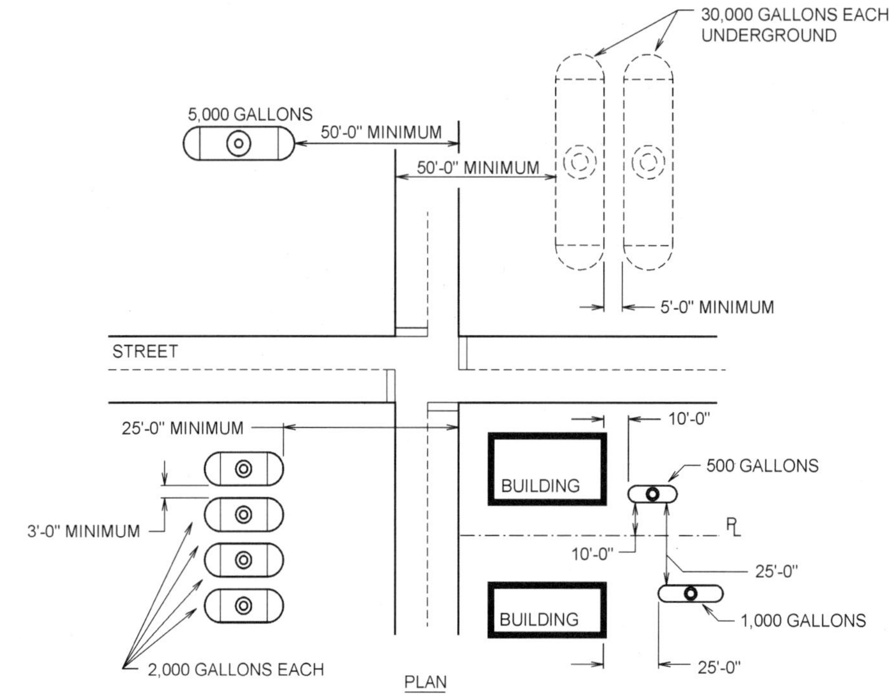

For SI: 1 inch = 25.4 mm,
1 foot = 304.8 mm,
1 gallon = 3.785 L.

Figure 3804.3(1)
LOCATION OF LPG CYLINDER/CONTAINER

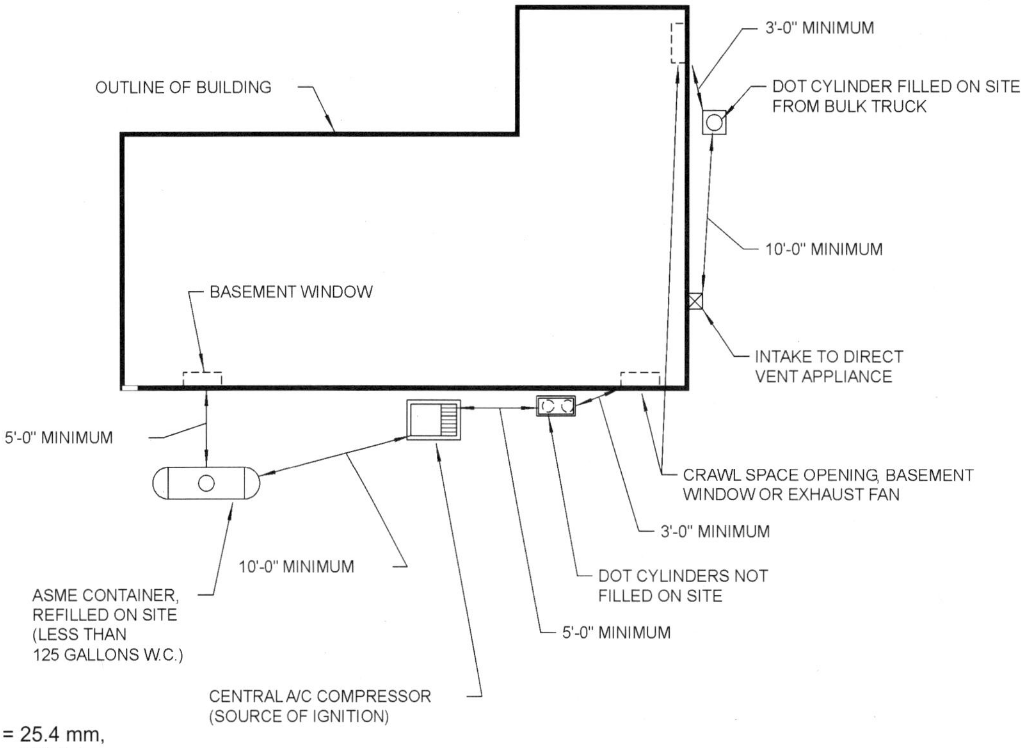

For SI: 1 inch = 25.4 mm,
1 foot = 304.8 mm,
1 gallon = 3.785 L.

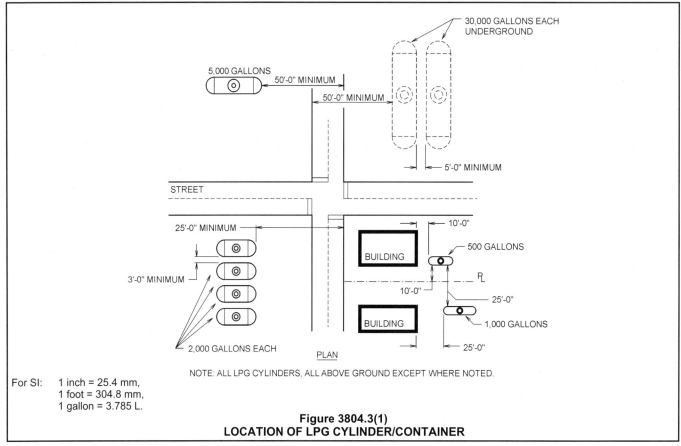

Figure 3804.3(2)
LOCATION OF DOT AND ASME CONTAINERS

3804.4 Multiple container installation. Multiple container installations with a total water storage capacity of more than 180,000 gallons (681 300 L) [150,000-gallon (567 750 L) LP-gas capacity] shall be subdivided into groups containing not more than 180,000 gallons (681 300 L) in each group. Such groups shall be separated by a distance of not less than 50 feet (15 240 mm), unless the containers are protected in accordance with one of the following:

1. Mounded in an approved manner.
2. Protected with approved insulation on areas that are subject to impingement of ignited gas from pipelines or other leakage.
3. Protected by firewalls of approved construction.
4. Protected by an approved system for application of water as specified in NFPA 58, Table 3.2.2.4.
5. Protected by other approved means.

Where one of these forms of protection is provided, the separation shall not be less than 25 feet (7620 mm) between container groups.

❖ These separation requirements are intended to minimize the exposure of adjacent containers following ignition of a release and to establish access for fire fighting. When used, insulation must be capable of preventing container temperatures from exceeding 800°F (427°C) for at least 50 minutes. Appendix H of NFPA 58 details the performance test for LP-gas tank insulation systems.

SECTION 3805 PROHIBITED USE OF LP-GAS

3805.1 Nonapproved equipment. Liquefied petroleum gas shall not be used for the purpose of operating devices or equipment unless such device or equipment is approved for use with LP-gas.

❖ Appliances and equipment must be approved for use with LP-gas. For example, appliances should be listed for use with LP-gas; hoses used to serve LP-gas installations should be listed and labeled for that use.

3805.2 Release to the atmosphere. Liquefied petroleum gas shall not be released to the atmosphere, except through an approved liquid-level gauge or other approved device.

❖ The release of LP-gas to the atmosphere is not recommended, but sometimes cannot be avoided. One such instance is when filling a container at a residence. The delivery person must rely on the "fixed liquid level gauge" to determine when the maximum amount of LP-gas has been reached. This gauge is a very small circular orifice that is connected to a tube that extends into the container to a level consistent with the height of the liquid when the container would be 80-percent full. When the liquid level in the container being filled reaches the bottom of the tube opening, a small amount of liquid escapes from the gauge to the atmosphere, causing it to vaporize. The vaporizing gas cools the surrounding air to the point that the moisture in the air condenses and forms a white cloud. When the delivery person observes this cloud, the filling operation is stopped.

Another instance in which the release of gas to the atmosphere is unavoidable is when the filling hose is disconnected from the container fill valve. The liquid remaining in the hose is released to the atmosphere and vaporizes.

SECTION 3806 DISPENSING AND OVERFILLING

3806.1 Attendants. Dispensing of LP-gas shall be performed by a qualified attendant.

❖ This section requires that any person transferring LP-gas must be qualified, which means that person must be trained. A number of training materials are available, most notably the *Certified Employees Training Program* published by the National Propane Gas Association.

3806.2 Overfilling. Liquefied petroleum gas containers shall not be filled or maintained with LP-gas in excess of either the volume determined using the fixed liquid-level gauge installed by the manufacturer, or the weight determined by the required percentage of the water capacity marked on the container.

❖ Overfilling a container can result in a catastrophic release of LP-gas because of the relatively high coefficient of expansion of liquid LP-gas. Depending on how much a container has been overfilled, a temperature difference of just a few degrees can result in the container becoming "liquid full," which increases the pressure in the container dramatically and leads to the pressure relief device opening, releasing LP-gas to the atmosphere.

Because of the hazards of overfilling, it is considered to be a violation of the code and can be cited by a fire code official as such.

Detailed requirements for filling containers are given in Chapter 4 of NFPA 58. As stated there, the maximum amount of LP-gas permitted in a container may vary between containers filled by weight or by volume, because the density of liquid LP-gas varies greatly with different temperatures.

3806.3 Dispensing locations. The point of transfer of LP-gas from one container to another shall be separated from exposures as specified in NFPA 58.

❖ Because the release of LP-gas to the atmosphere in small quantities may be unavoidable when filling a container, all possible sources of ignition in the vicinity of the point of transfer must be accounted for. NFPA 58 includes detailed requirements for establishing clearances from those sources of ignition.

SECTION 3807 SAFETY PRECAUTIONS AND DEVICES

3807.1 Safety devices. Safety devices on LP-gas containers, equipment and systems shall not be tampered with or made ineffective.

❖ The safety devices found on all propane containers include the service valve, the filler valve (part of the service valve on some containers) and the pressure-relief device. In addition, small cylinders with a propane capacity between 4 pounds and 40 pounds (1.8 kg and 18 kg) now may have a separate device called an "overfilling prevention device" attached to the service valve inside the container. This is a float-operated mechanism that is designed to act as a backup to the normal filling procedures (using either a scale or the fixed liquid level gauge) and will automatically close the filler valve when the liquid in the container reaches 80 percent of the container volume. The "OPD," as it is referred to, cannot be seen, but new containers having this safety device are identifiable by the three lobes on the handwheel on the valve (see Figure 3807.1).

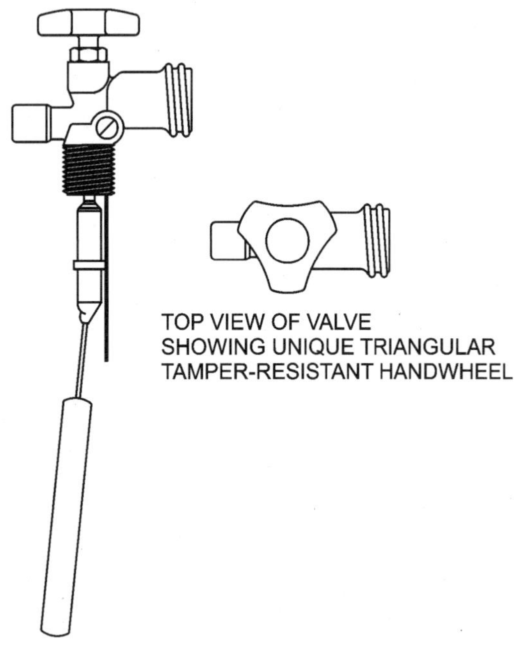

Figure 3807.1
CYLINDER VALVE WITH OVERFILL PREVENTION DEVICE

3807.2 Smoking and other sources of ignition. "No Smoking" signs complying with Section 310 shall be posted when required by the fire code official. Smoking within 25 feet (7620 mm) of a point of transfer, while filling operations are in progress at containers or vehicles, shall be prohibited.

Control of other sources of ignition shall comply with Chapter 3 and NFPA 58, Section 3.7.

❖ Because of the nature of the filling operation, it may be inevitable that some quantity of gas will be released to the atmosphere; therefore, strict observance of the requirement for prohibiting sources of ignition within the limits prescribed here is essential.

3807.3 Clearance to combustibles. Weeds, grass, brush, trash and other combustible materials shall be kept a minimum of 10 feet (3048 mm) from LP-gas tanks or containers.

❖ The concern addressed here is the potential threat to a container of LP-gas from the ignition of nearby combustible material. It has been common practice to exempt from this requirement landscaping materials, such as grass and shrubs, that are regularly maintained.

3807.4 Protecting containers from vehicles. Where exposed to vehicular damage due to proximity to alleys, driveways or parking areas, LP-gas containers, regulators and piping shall be protected in accordance with Section 312.

❖ See the commentary to Section 312 for detailed information on the protection of containers from vehicular impact.

SECTION 3808 FIRE PROTECTION

3808.1 General. Fire protection shall be provided for installations having storage containers with a water capacity of more than 4,000 gallons (15 140 L), as required by Section 3-10 of NFPA 58.

❖ This section refers to NFPA 58 for the criteria to establish where fire protection should be installed at any location where the aggregate water capacity of LP-gas containers exceeds 4,000 gallons (15 140 L). The objective is to provide protection to those containers that are exposed to the threat of fire. NFPA 58 should be referred to for exceptions to this requirement.

3808.2 Fire extinguishers. Fire extinguishers complying with Section 906 shall be provided as specified in NFPA 58.

❖ Refer to NFPA 58 to determine where fire extinguishers must be provided.

SECTION 3809 STORAGE OF PORTABLE LP-GAS CONTAINERS AWAITING USE OR RESALE

3809.1 General. Storage of portable containers of 1,000 pounds (454 kg) or less, whether filled, partially filled or empty, at consumer sites or distributing points, and for resale by dealers or resellers shall comply with Sections 3809.2 through 3809.15.

Exceptions:

1. Containers that have not previously been in LP-gas service.
2. Containers at distributing plants.

3. Containers at consumer sites or distributing points, which are connected for use.

❖ This section is intended to apply to DOTn cylinders, which are distinguishable from ASME containers by their vertical orientation (ASME containers are usually horizontally oriented) and markings on the collar.

The container size threshold of 1,000 pounds (454 kg) is the water capacity of the container; therefore, the maximum propane capacity would be 420 pounds (191 kg).

The exceptions to Section 3809.1 include those containers already in use and those at LP-gas distribution plants, at which employees are trained in the hazards of LP-gas and whose everyday job functions include the handling of the containers.

3809.2 Exposure hazards. Containers in storage shall be located in a manner which minimizes exposure to excessive temperature rise, physical damage or tampering.

❖ Because of the relatively high coefficient of expansion for liquid LP-gas, precautions must be taken to prevent overheating of containers that could lead to releases of LP-gas to the atmosphere.

3809.3 Position. Containers in storage having individual water capacity greater than 2.5 pounds (1 kg) [nominal 1-pound (0.454 kg) LP-gas capacity] shall be positioned with the pressure relief valve in direct communication with the vapor space of the container.

❖ Liquid propane released to the atmosphere through the relief valve will expand to 270 times its original volume after it vaporizes, whereas propane vapor will expand much less. To avoid the release of liquid propane, it is important to keep the pressure relief device in contact with the vapor space. Most cylinders are designed to be oriented vertically, so the relief device is at the top of the cylinder; however, some cylinders used in the recreational vehicle industry are designed for horizontal orientation only, and must be kept in that position for the relief valve to be in contact with the vapor space. Some cylinders are designed for use in either the horizontal or vertical orientation.

3809.4 Separation from means of egress. Containers stored in buildings in accordance with Sections 3809.9 and 3809.11 shall not be located near exit access doors, exits, stairways, or in areas normally used, or intended to be used, as a means of egress.

❖ This section states that LP-gas containers may not be stored where they would block a means of egress in a building. The required means of egress widths and capacities must not be infringed upon by either LP-gas containers or the cabinets in which they may be stored.

3809.5 Quantity. Empty containers that have been in LP-gas service shall be considered as full containers for the purpose of determining the maximum quantities of LP-gas allowed in Sections 3809.9 and 3809.11.

❖ Because the actual quantity of LP-gas in a container cannot always be determined, every container stored at a location must be treated as a full container.

3809.6 Storage on roofs. Containers which are not connected for use shall not be stored on roofs.

❖ The roof is a location that is likely to become warmer than atmospheric conditions on a sunny day; therefore, LP-gas containers must not be stored on roofs when they are not connected for use.

3809.7 Storage in basement, pit or similar location. Liquefied petroleum gas containers shall not be stored in a basement, pit or similar location where heavier-than-air gas might collect. Liquefied petroleum gas containers shall not be stored in above-grade underfloor spaces or basements unless such location is provided with an approved means of ventilation.

Exception: Department of Transportation (DOTn) specification cylinders with a maximum water capacity of 2.5 pounds (1 kg) for use in completely self-contained hand torches and similar applications. The quantity of LP-gas shall not exceed 20 pounds (9 kg).

❖ Because propane has a specific gravity of 1.52 at 60°F (16°C) (which means it is 1.52 times heavier than air at that temperature), propane has been assumed to automatically sink to the ground when it is released to the atmosphere. Even though the laws of physics have proven this concept to be untrue, concerns persist about propane dispersing in locations where air circulation is restricted. Any proposed ventilation system should take into account the expected temperatures at the location.

The exception to this section would permit very small, disposable cylinders to be stored in these locations.

3809.8 Protection of valves on containers in storage. Container valves shall be protected by screw-on-type caps or collars which shall be securely in place on all containers stored regardless of whether they are full, partially full or empty. Container outlet valves shall be closed or plugged.

❖ Valve assemblies must be protected from physical impact. Cylinders having propane capacities up to 60 pounds (27 kg) will usually have collars that extend above the height of the valves. Larger cylinders will have screw-on caps or domes that serve the same function.

3809.9 Storage within buildings accessible to the public. Department of Transportation (DOTn) specification cylinders with maximum water capacity of 2.5 pounds (1 kg) used in completely self-contained hand torches and similar applications are allowed to be stored or displayed in a building accessible to the public. The quantity of LP-gas shall not exceed 200 pounds (91 kg) except as provided in Section 3809.11.

❖ This section recognizes that retail stores require inventory that is accessible to the public where handheld torch assemblies are sold.

3809.10 Storage within buildings not accessible to the public. The maximum quantity allowed in one storage location in buildings not accessible to the public, such as industrial buildings, shall not exceed a water capacity of 735 pounds (334 kg) [nominal 300 pounds (136 kg) of LP-gas]. Where additional storage locations are required on the same floor within the same building, they shall be separated by a minimum of 300 feet (91 440 mm). Storage beyond these limitations shall comply with Section 3809.11.

❖ This requirement is applicable to industrial or storage facilities where LP-gas-powered forklift trucks are frequently operated. The requirement permits up to 300 pounds (136 kg) of propane to be stored at a single location. Forklift cylinders, which typically have a 33-pound (15 kg) propane capacity, can be stored in small quantities for convenient exchange during a working shift.

The reference to Section 3809.11 applies where more than 300 pounds (136 kg) must be stored at a location without the required 300-foot (91 440 mm) separation distance.

3809.10.1 Quantities on equipment and vehicles. Containers carried as part of service equipment on highway mobile vehicles need not be considered in the total storage capacity in Section 3809.10, provided such vehicles are stored in private garages and do not carry more than three LP-gas containers with a total aggregate LP-gas capacity not exceeding 100 pounds (45.4 kg) per vehicle. Container valves shall be closed.

❖ This section addresses vehicles carrying LP-gas onboard that is ancillary to service equipment on the vehicle. Typically, LP-gas may be found on fabrication trucks and mechanics' trucks for use with welding equipment, roofing tar trucks for torches used on roofs and plumbers' trucks for use with lead melting pots.

3809.11 Storage within rooms used for gas manufacturing. Storage within buildings or rooms used for gas manufacturing, gas storage, gas-air mixing and vaporization, and compressors not associated with liquid transfer shall comply with Sections 3809.11.1 and 3809.11.2.

❖ The requirements of this section might typically apply to the storage of LP-gas containers used for "standby" gas systems in which LP-gas is mixed with air to achieve an energy content similar to that of natural gas, thereby permitting gas appliances and equipment to continue operating without a change in burner controls. These systems are commonly found in larger industrial applications.

3809.11.1 Quantity limits. The maximum quantity of LP-gas shall be 10,000 pounds (4540 kg).

❖ Limiting LP-gas to 10,000 pounds (4,540 kg) [roughly 3,000-gallon (11 355 L) water capacity container] is considered to be an acceptable threshold for the safe and efficient operation of standby gas systems.

3809.11.2 Construction. The construction of such buildings and rooms shall comply with requirements for Group H occupancies in the *International Building Code*; NFPA 58, Chapter 7; and both of the following:

1. Adequate vents shall be provided to the outside at both top and bottom, located at least 5 feet (1524 mm) from building openings.
2. The entire area shall be classified for the purposes of ignition source control in accordance with NFPA 58, Section 3.7.

❖ Chapter 7 of the *International Building Code*® (IBC®) contains specific requirements to address the perceived hazard of storing up to 10,000 pounds (4,540 kg) of LP-gas in a room or a building. In addition to those requirements, Item 1 of this section requires adequate ventilation openings to assist in the dilution of gas should a leak occur. Chapter 7 of NFPA 58 contains information on venting as well as other aspects of the design of rooms of this nature. Item 2 refers to NFPA 58, which contains extensive information on establishing the proper hazard classification that would govern the requirements for electrical wiring systems and limit ignition sources.

3809.12 Location of storage outside of buildings. Storage outside of buildings, for containers awaiting use, resale or part of a cylinder exchange program shall be located not less than 20 feet (6096 mm) from openings into buildings, 20 feet (6096 mm) from any motor vehicle fuel dispenser and 10 feet (3048 mm) from any combustible material and in accordance with Table 3809.12.

❖ This section regulates the placement of cylinders awaiting use or resale and cylinder exchange cabinets, which are increasingly being used to distribute the "20-pound" (9 kg) gas grill cylinders to the public. Exchange cabinets are typically constructed of steel and are ventilated to permit the dispersal of gas should a leak occur.

In order to not compromise the means of egress in a potential fire scenario and to prevent LP-gas from entering the building and finding an ignition source in case of a leak, the LP-gas container storage must be a minimum of 20 feet (6,096 mm) from any door opening. Additionally, motor fuel-dispensing facilities are considered Group M occupancies. These types of facilities frequently include convenience stores in addition to motor fueling operations. This section specifies a minimum fire separation distance between the LP-gas container area and any flammable or combustible liquid dispensing operation or the outdoor storage of combustible materials to reduce the potential fire hazard exposure.

TABLE 3809.12
LOCATION OF CONTAINERS AWAITING USE OR RESALE STORED OUTSIDE OF BUILDINGS

QUANTITY OF LP-GAS STORED	DISTANCES TO A BUILDING OR GROUP OF BUILDINGS, PUBLIC WAY OR LOT LINE OF PROPERTY THAT CAN BE BUILT UPON (feet)
500 pounds or less	0
501 to 2,500 pounds	10[a]
2,501 to 6,000 pounds	15
6,001 to 10,000 pounds	20
Over 10,000 pounds	25

For SI: 1 foot = 304.8 mm, 1 pound = 0.454 kg.
a. Containers are allowed to be located a lesser distance.

❖ This table addresses the minimum distances from any cylinder awaiting use or resale from a building or group of buildings, public way or lot line of a property that can be built upon. As in Section 3804, this table is based on the quantity of LP-gas stored (not the size of the container) and is intended to limit the two-way exposure between containers and the built environment.

3809.13 Protection of containers. Containers shall be stored within a suitable enclosure or otherwise protected against tampering. Vehicular protection shall be provided as required by the fire code official.

❖ At public facilities, tampering with LP-gas containers may be a problem. For that reason, locked metal cabinets are used to provide not only tamper protection but also substantial protection from vehicular impact (see Figure 3809.13). At locations where the cabinet is installed close to the direct path of motor vehicles, the fire code official can require additional protection.

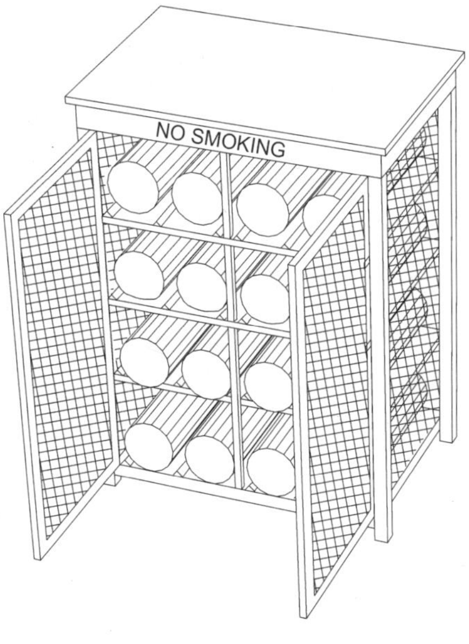

Figure 3809.13
TYPICAL INSTALLATION OF LP-GAS CONTAINERS FOR RETAIL SALE

3809.14 Separation from means of egress for containers located outside of buildings. Containers located outside of buildings shall not be located within 20 feet (6096 mm) of any exit access doors, exits, stairways or in areas normally used, or intended to be used, as a means of egress.

❖ This section intends to keep LP-gas containers from obstructing the means of egress and, if the containers are involved in a fire, to prevent the container from constituting a hazard to people using the means of egress.

3809.15 Alternative location and protection of storage. Where the provisions of Sections 3809.12 and 3809.13 are impractical at construction sites, or at buildings or structures undergoing major renovation or repairs, the storage of containers shall be as required by the fire code official.

❖ These requirements permit the fire code official to approve storage sites for LP-gas even if compliance with Sections 3809.12 and 3809.13 is not practical. This would especially be the case at construction sites and other transient installations in which LP-gas containers may be moved frequently.

SECTION 3810
CONTAINERS NOT IN SERVICE

3810.1 Temporarily out of service. Containers whose use has been temporarily discontinued shall comply with all of the following:

1. Be disconnected from appliance piping.
2. Have container outlets, except relief valves, closed or plugged.
3. Be positioned with the relief valve in direct communication with container vapor space.

❖ When a new LP-gas supplier installs a tank at a customer's facility, he or she is responsible for ensuring that the previous supplier's tank is removed from service and safely stored until it can be retrieved.

3810.2 Permanently out of service. Containers to be placed permanently out of service shall be removed from the site.

❖ This provision is an extension of Section 3810.1 and requires that any container placed out of service must be retrieved by its owner.

SECTION 3811
PARKING AND GARAGING

3811.1 General. Parking of LP-gas tank vehicles shall comply with Sections 3811.2 and 3811.3.

> **Exception:** In cases of accident, breakdown or other emergencies, tank vehicles are allowed to be parked and left unattended at any location while the operator is obtaining assistance.

❖ In this context, LP-gas tank vehicles include bulk cargo tank vehicles, either transports (semitruck trailers) or bobtails [usually up to 5,000-gallon (18 925 L) water capacity]. The exception to this section recognizes the possibility of mechanical breakdown of a vehicle while in service.

3811.2 Unattended parking. The unattended parking of LP-gas tank vehicle shall be in accordance with Sections 3811.2.1 and 3811.2.2.

❖ Unattended parking may be defined as the parking of a cargo tank vehicle used to transport LP-gas where either the driver or other responsible person is not able to respond to any situation that may occur that involves the vehicle.

3811.2.1 Near residential, educational and institutional occupancies and other high-risk areas. Liquefied petroleum gas tank vehicles shall not be left unattended at any time on residential streets or within 500 feet (152 m) of a residential area, apartment or hotel complex, educational facility, hospital or care facility. Tank vehicles shall not be left unattended at any other place that would, in the opinion of the fire code official, pose an extreme life hazard.

❖ Because of the hazardous cargo involved, operator control must be maintained over cargo tank vehicles used to transport LP-gas. In other words, LP-gas transports must not be left unattended. This is especially true in areas where high occupancy levels exist and the possibility of tampering with the vehicle increases.

3811.2.2 Durations exceeding 1 hour. Liquefied petroleum gas tank vehicles parked at any one point for longer than1 hour shall be located as follows:

1. Off public streets, highways, public avenues or public alleys.
2. Inside of a bulk plant.
3. At other approved locations not less than 50 feet (15 240 mm) from buildings other than those approved for the storage or servicing of such vehicles.

❖ When a vehicle will be parked for more than 1 hour, it will probably be unattended and, therefore, special conditions are imposed.

The first condition specifies that the vehicle must not be parked on a public way. The second condition permits the vehicle to be parked within an LP-gas bulk plant, which is a plant with containers used for storing LP-gas, until the gas is delivered to the end user. The third condition permits the fire code official to approve alternate locations for parking LP-gas cargo tank vehicles. The fire code official might base approval on the proximity to occupied buildings and spaces as well as the potential for tampering with the vehicle.

3811.3 Garaging. Garaging of LP-gas tank vehicles shall be as specified in NFPA 58. Vehicles with LP-gas fuel systems are allowed to be stored or serviced in garages as specified in NFPA 58, Section 8.6.

❖ This section applies not only to cargo tank vehicles used to transport and deliver LP-gas, but also to motor vehicles that are fueled by LP-gas. NFPA 58 contains extensive provisions for garaging these vehicles.

Bibliography

The following resource materials are referenced in this chapter or are relevant to the subject matter addressed in this chapter.

ANSI Z21.5895, *Outdoor Cooking Gas Appliances with Addendum Z21.58a1998.* New York: American National Standards Institute, 1995.

BPVC01, *ASME Boiler and Pressure Vessel Code, Section VIII, Divisions 1, 2 & 3.* New York: The American Society of Mechanical Engineers, 2001.

Certified Employee Training Program, Basic Principles and Practices. Lisle, IL: National Propane Gas Association, 1994.

DOTn 49 CFR; 100-78 & 179-199-94, *Specification for Transportation of Explosive and Other Dangerous Articles, Shipping Containers.* Washington, DC: U.S. Department of Transportation, 1994.

Handbook of Compressed Gases, 3rd ed. Arlington, VA: Compressed Gas Association, Inc. 1990.

IBC-03, *International Building Code.* Falls Church, VA: International Code Council, 2003.

IFGC-03, *International Fuel Gas Code.* Falls Church, VA: International Code Council, 2003.

IMC-03, *International Mechanical Code.* Falls Church, VA: International Code Council, 2003.

Lemoff, T.C., ed. *LP-gas Code Handbook*, 5th ed. Quincy, MA: National Fire Protection Association, 1998.

NFPA 58-01, *Storage and Handling of Liquefied Petroleum Gases.* Quincy, MA: National Fire Protection Association, 2001.

NFPA 59-01, *Storage and Handling of Liquefied Petroleum Gases at Utility Gas Plants.* Quincy, MA: National Fire Protection Association, 2001.

UL 147A-96, *Standard for Nonrefillable (Disposable) Type Fuel Gas Cylinder Assemblies.* Northbrook, IL: Underwriters Laboratories Inc, 1996.

Chapter 39: Organic Peroxides

General Comments

This chapter addresses the hazards associated with the storage, handling and use of organic peroxides. These chemicals possess the characteristics of flammable or combustible liquids and are also strong oxidizers. Class V organic peroxides pose little fire hazard; therefore, these materials are not regulated by specific storage or use requirements. Some organic peroxides are unstable and become increasingly reactive with age or heating.

Organic peroxides pose the dual hazards of being both oxidizers and flammable or explosive compounds. This unusual combination of properties requires special storage and handling precautions to prevent uncontrolled release, contamination, hazardous chemical reactions, fires or explosions. In addition to these properties, organic peroxides are unusually sensitive to temperature. Heat, whether by fire exposure or environmental, is a major factor in the decomposition of peroxide compounds. Some organic peroxides will decompose uneventfully when subject to a gradual temperature increase but may explode if they undergo the thermal shock of a rapid, uncontrolled temperature rise.

J.S. Townsend (1993) notes that some organic peroxides are just as dangerous when they become too cold as when they are too hot. For example, acetyl peroxide becomes unstable above 122°F (50°C) — its self-accelerating decomposition temperature (SADT). When cooled below 17°F (-8.3°F), acetyl peroxide forms crystals that are shock sensitive. Consequently, special precautions must be taken to transport and store acetyl peroxide between 32°F (0°C) and 90°F (32°C).

Organic peroxides are commonly used in the plastics industry to initiate polymerization. Although the requirements of this chapter pertain to industrial applications in which significant quantities of organic peroxides are stored or used, smaller quantities of organic peroxides still pose a significant hazard. These materials, therefore, must be stored and used in accordance with the applicable provisions of this chapter and Chapter 27.

Purpose

The provisions of this chapter are intended to manage the fire and oxidation hazards of organic peroxides by preventing their uncontrolled release.

SECTION 3901 GENERAL

3901.1 Scope. The storage and use of organic peroxides shall be in accordance with this chapter and Chapter 27.

Unclassified detonable organic peroxides that are capable of detonation in their normal shipping containers under conditions of fire exposure shall be stored in accordance with Chapter 33.

❖ The specific requirements for the storage, handling and use of organic peroxides in this chapter are intended to complement the general hazardous materials requirements of Chapter 27. Because of the hazards to people and property, organic peroxides capable of being detonated in their usual shipping containers or packages under fire conditions must be stored in accordance with the provisions of Chapter 33 for explosives. Examples of unclassified detonable (UD) materials are organic peroxides classified as Type A by DOTn 49 CFR; 173.128(b)(1) and materials classified as explosives by DOTn 49 CFR; 173, Subpart C. DOTn Type A organic peroxides have an SADT of 122°F (50°C) or less.

3901.2 Permits. Permits shall be required for organic peroxides as set forth in Section 105.6.

❖ The process of issuing permits gives the fire code official an opportunity to carefully evaluate and regulate hazardous operations. Permit applicants should be required to demonstrate that their operations comply with the intent of the code before the permit is issued. See the commentary to Section 105.6 for a general discussion of operations requiring an operational permit, Section 105.6.21 for a discussion of specific quantity-based operational permits for the materials regulated in this chapter and Section 105.7 for a general discussion of activities requiring a construction permit. The permit process also notifies the fire department of the need for prefire planning for hazardous property.

SECTION 3902 DEFINITIONS

3902.1 Definition. The following word and term shall, for the purposes of this chapter and as used elsewhere in this code, have the meanings shown herein.

❖ Definitions of terms can help in the understanding and application of the code requirements. The purpose for including those definitions that are associated with the subject matter of this chapter is to provide more convenient access to them without having to refer back to Chapter 2. It is important to emphasize that these terms are not exclusively related to this chapter but are applicable everywhere the term is used in the code. For convenience, these terms are also listed in Chapter 2 with a cross reference to this section. The use and application of all defined terms, including those defined in this section, are set forth in Section 201.

ORGANIC PEROXIDE. An organic compound that contains the bivalent -O-O- structure and which may be considered to be a structural derivative of hydrogen peroxide where one or both of the hydrogen atoms have been replaced by an organic radical. Organic peroxides can present an explosion hazard (detonation or deflagration) or they can be shock sensitive. They can also decompose into various unstable compounds over an extended period of time.

Class I. Describes those formulations that are capable of deflagration but not detonation.

Class II. Describes those formulations that burn very rapidly and that pose a moderate reactivity hazard.

Class III. Describes those formulations that burn rapidly and that pose a moderate reactivity hazard.

Class IV. Describes those formulations that burn in the same manner as ordinary combustibles and that pose a minimal reactivity hazard.

Class V. Describes those formulations that burn with less intensity than ordinary combustibles or do not sustain combustion and that pose no reactivity hazard.

Unclassified detonable. Organic peroxides that are capable of detonation. These peroxides pose an extremely high-explosion hazard through rapid explosive decomposition.

❖ The chemical structure of organic peroxides differs from that of hydrogen peroxide (an oxidizer) in that an organic radical replaces the hydrogen atoms. Figure 3902.1 shows an example of this chemical structure in which a benzoyl radical (C_6H_5CO) in the widely used Class I organic peroxide benzoyl peroxide replaces the hydrogen atoms in hydrogen peroxide (H_2O_2). Organic chemicals are all carbon based. As a result, organic peroxides pose varying degrees of fire or explosion hazards in addition to their oxidizing properties. The classification system in this chapter (see Table 3902.1) is derived from a system developed by the Society of the Plastics Industry (Bulletin 19A).

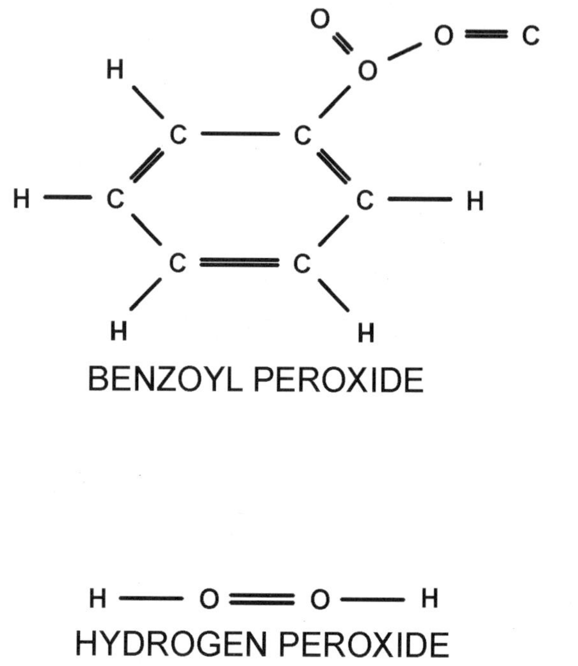

Figure 3902.1
COMPARISON OF HYDROGEN PEROXIDE STRUCTURE WITH BENZOYL PEROXIDE

SECTION 3903
GENERAL REQUIREMENTS

3903.1 Quantities not exceeding the maximum allowable quantity per control area. The storage and use of organic peroxides in amounts not exceeding the maximum allowable quantity per control area indicated in Section 2703.1 shall be in accordance with Sections 2701, 2703, 3901 and 3903.

❖ The provisions of this section complement the requirements of Chapter 27 in structures occupied for the storage, handling or use of organic peroxides. The regulations contained in Sections 3903.1.1 through 3903.1.1.4 assume that the quantity of organic peroxides in a given building is limited to the maximum allowable quantities per control area as established in Sec-

TABLE 3902.1
COMPARISON OF ORGANIC PEROXIDE CLASSIFICATION SYSTEMS

IFC	DOTn 49 CFR; PART 173.128(b)	HAZARD DESCRIPTION
Unclassified detonatable	Type A	Detonation hazard when confined
Class I	Type B	Deflagration hazard when confined
Class II	Type C and Type D	Mass fire hazard similar to flammable liquid[a]
Class III	Type E	Fire hazard similar to combustible liquid
Class IV	Type F	Little or no fire or reactivity hazard
Class V	Type G	No fire or reactivity hazard

Note a. Moderate detonation or deflagration hazard when heated under confinement.

tion 2703.1; thus, the building is not classified in Occupancy Group H. The general requirements of Sections 2701 and 2703 are fully applicable to the storage and use of organic peroxides, in addition to the provisions of this chapter.

3903.1.1 Special limitations for indoor storage and use by occupancy. The indoor storage and use of organic peroxides shall be in accordance with Sections 3903.1.1.1 through 3903.1.1.4.

❖ Because certain occupancies may need to have organic peroxides on hand, Sections 3903.1.1.1 through 3903.1.1.4 provide regulations that are specific to occupancy group classifications and that recognize the relative hazards of both the occupancy and the organic peroxide.

3903.1.1.1 Group A, E, I or U occupancies. In Group A, E, I or U occupancies, any amount of unclassified detonable and Class I organic peroxides shall be stored in accordance with the following:

1. Unclassified detonable and Class I organic peroxides shall be stored in hazardous materials storage cabinets complying with Section 2703.8.7.
2. The hazardous materials storage cabinets shall not contain other storage.

❖ Because of their respective explosive or higher deflagration hazard characteristics, even the smallest quantity of unclassified detonable or Class I organic peroxides present in Group A, E, I or U occupancies must be stored in an approved hazardous material storage cabinet constructed and placarded in accordance with Section 2703.8.6 to reduce the exposure of the materials to hazards from the surrounding environment.

In accordance with Table 2703.1.1(1), Note g, storage of unclassified detonable organic peroxide in any amount is allowed only in buildings equipped throughout with an approved automatic sprinkler system in accordance with Section 903.3.1. Also, based on Table 2703.1.1(1), Note e, where an approved storage cabinet is used, the maximum allowable quantity per control area that could be kept in the occupancies can be doubled.

To reduce the likelihood of contamination of the organic peroxide materials or damage to their packaging, Item 2 prohibits the storage of other materials in the approved organic peroxide storage cabinet.

In accordance with Section 3901.1, storage of unclassified detonable organic peroxides must also comply with the applicable provisions of Chapter 33 for explosives.

3903.1.1.2 Group R occupancies. Unclassified detonable and Class I organic peroxides shall not be stored or used within Group R occupancies.

❖ Because of their explosive or higher deflagration hazard characteristics, unclassified detonable and Class I organic peroxide cannot be stored in any residential occupancy within the scope of the code's regulations. Storage would result in an increased danger to the occupants as well as exposure of the peroxides to the otherwise unregulated environment.

3903.1.1.3 Group B, F, M or S occupancies. Unclassified detonable and Class I organic peroxides shall not be stored or used in offices, or retail sales areas of Group B, F, M or S occupancies.

❖ Because of their explosive or higher deflagration hazard characteristics, unclassified detonable and Class I organic peroxide cannot be stored in occupancies in Group B, F, M or S. Storage would result in an increased danger to the occupants as well as exposure of the peroxides to the higher relative fire loads typically encountered in these occupancies.

3903.1.1.4 Classrooms. In classrooms in Group B, F or M occupancies, any amount of unclassified detonable and Class 1 organic peroxides shall be stored in accordance with the following.

1. Unclassified detonable and Class 1 organic peroxides shall be stored in hazardous materials storage cabinets complying with Section 2703.8.7.
2. The hazardous materials storage cabinets shall not contain other storage.

❖ It is the intent of this section to allow for the occasional use of limited amounts of organic peroxides in certain scientific, experimental or demonstration settings; however, this section does not allow storage of any quantity of these materials for any length of time. The fire code official may limit the amount of organic peroxides brought into a structure for these uses. The quantity actually needed for the experiment should determine the amount allowed into a structure.

Because of their explosive or higher deflagration hazard characteristics, even the smallest quantity of unclassified detonable or Class I organic peroxides must be stored in an approved hazardous material storage cabinet constructed and placarded in accordance with Section 2703.8.6 to reduce the exposure of the materials to hazards from the surrounding environment.

In accordance with Table 2703.1.1(1), Note g, storage of unclassified detonable organic peroxide in any amount is allowed only in buildings equipped throughout with an approved automatic sprinkler system in accordance with Section 903.3.1. Also, based on Table 2703.1.1(1), Note e, where an approved storage cabinet is used, the maximum allowable quantity per control area that could be kept in these occupancies can be doubled.

To reduce the likelihood of contaminating organic peroxide materials or damaging their packaging, Item 2 prohibits the storage of other materials in the approved organic peroxide storage cabinet.

Note also that in accordance with Section 3901.1, storage of unclassified detonable organic peroxides must also comply with the applicable provisions of Chapter 33 for explosives.

3903.2 Quantities exceeding the maximum allowable quantity per control area. The storage and use of organic peroxides in amounts exceeding the maximum allowable quantity per control area indicated in Section 2703.1 shall be in accordance with Chapter 27 and this chapter.

❖ The provisions of this section complement the requirements of Chapter 27 in structures occupied for the storage, handling or use of organic peroxides. The regulations in this section assume that the quantity of organic peroxides in a given building is in excess of the maximum allowable quantities per control area as established in Section 2703.1; thus, that the building is classified in Occupancy Group H. The requirements of Chapter 27 apply to the storage and use of organic peroxides in addition to the provisions of this chapter.

SECTION 3904 STORAGE

3904.1 Indoor storage. Indoor storage of organic peroxides in amounts exceeding the maximum allowable quantity per control area indicated in Table 2703.1.1(1) shall be in accordance with Sections 2701, 2703, 2704 and this chapter.

Indoor storage of unclassified detonable organic peroxides that are capable of detonation in their normal shipping containers under conditions of fire exposure shall be stored in accordance with Chapter 33.

❖ This section regulates the indoor storage of organic peroxides when in excess of the maximum allowable quantity per control area in buildings or portions of buildings classified in Occupancy Group H. The general and storage provisions of Chapter 27 are applicable in addition to the requirements of this section. Storage of organic peroxides inside of structures must comply with Sections 3904.1.1 through 3904.1.12 to prevent uncontrolled release or exposure to conditions that may result in a fire or explosion.

Because of the explosion hazard, unclassified detonable organic peroxides must be stored in accordance with the provisions of this section and Chapter 33 for explosives.

3904.1.1 Detached storage. Storage of organic peroxides shall be in detached buildings when required by Section 2703.8.2.

❖ Detached structures designed and constructed for the sole purpose of organic peroxide storage provide the best protection for people and property from fire and explosions. Detached storage structures should be constructed of noncombustible materials to prevent them from becoming involved in either an outside fire that may endanger their contents or in a fire themselves, should their contents be ignited.

Section 2703.8.2 and Table 2703.8.2 require detached storage when the indoor storage quantities are more than 2 tons (4 metric tons) of Class I, more than 25 tons (50 metric tons) of Class II and more than 50 tons (100 metric tons) of Class III organic peroxides. See the commentary to Section 2703.8.2 and Table 2703.8.2 for further discussion of detached storage requirements. Figure 3904.1.1 shows a cross-sectional diagram of a typical detached storage building.

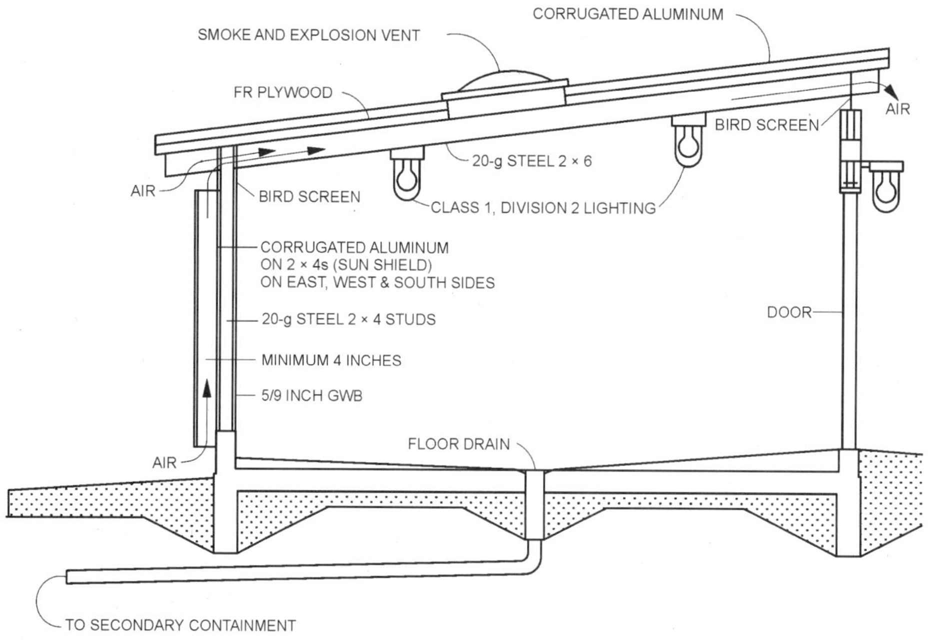

For SI: 1 inch = 25.4 mm.

Figure 3904.1.1
TYPICAL DETACHED STORAGE STRUCTURE

3904.1.2 Distance from detached storage buildings to exposures. In addition to the requirements of the *International Building Code*, detached storage buildings shall be located in accordance with Table 3904.1.2.

❖ The *International Building Code*® (IBC®) prescribes minimum fire-resistance ratings for exterior walls of buildings based on clearances from adjacent structures and lot lines. The separation distances prescribed by this section are minimums notwithstanding the inherent fire-resistance ratings of the exterior wall assemblies. These clearances are intended to reduce the hazard to nearby structures and people in the event of a fire or explosion in a detached storage structure used for the storage of organic peroxides.

TABLE 3904.1.2. See below.

❖ The separation distances specified are intended to reduce the effects of radiant heat exposure in the event of a fire in an adjacent storage structure. Separation distances for unclassified detonable organic peroxides must conform to the requirements of Table 3304.5.2(2)

The exception indicates that automatic fire detection is not required in detached storage buildings when an automatic fire suppression system is provided. Automatic sprinklers or other approved fire suppression systems reduce the danger to people and property from fire by detecting the fire, sounding an alarm, transmitting the alarm to the fire department and containing or suppressing the fire. Sprinkler systems in detached storage buildings must be supervised by a connection to a central station, a remote supervising station or a proprietary supervising station or, when approved by the fire code official, be connected to a system that produces an audible and visual signal monitored at a constantly attended on-site location. See the commentary to Section 903.4 for further discussion of sprinkler system supervision and Sections 904.3.5 and 907.14 for further discussion of alternative automatic fire-extinguishing system supervision.

3904.1.3 Liquid-tight floor. In addition to the requirements of Section 2704.12, floors of storage areas shall be of liquid-tight construction.

❖ In addition to the requirement of Section 2704.12 that floors be constructed of noncombustible materials, the requirement for floors to be liquid tight is intended to result in floor construction that has the ability to stop the passage of liquids to adjacent spaces.

3904.1.4 Electrical wiring and equipment. In addition to the requirements of Section 2703.9.4, electrical wiring and equipment in storage areas for Class I or II organic peroxides shall comply with the requirements for electrical Class I, Division 2 locations.

❖ Because of the danger of ignition by arcs or sparks in the event of an accidental spill or leak, electrical equipment and devices in storage areas for organic peroxides must be classified for use in Class I, Division 2, hazardous locations as described in the ICC *Electrical Code*® (ICC EC™).

3904.1.5 Smoke detection. An approved supervised smoke detection system in accordance with Section 907 shall be provided in rooms or areas where Class I, II or III organic peroxides are stored. Activation of the smoke detection system shall sound a local alarm.

> **Exception:** A smoke detection system shall not be required in detached storage buildings equipped throughout with an approved automatic fire-extinguishing system complying with Chapter 9.

❖ An automatic fire detection system with central, remote or proprietary station supervision and a local alarm signal is required to provide early warning of fire to building occupants and fire suppression personnel. Detectors and installation methods in organic peroxide storage areas should be selected for compatibility with the materials stored. Fire alarm equipment and installation methods must comply with Section 907. Materials that give off strong vapors may interfere with fire detection sys-

TABLE 3904.1.2
ORGANIC PEROXIDES—DISTANCE TO EXPOSURES FROM DETACHED STORAGE BUILDINGS OR OUTDOOR STORAGE AREAS

ORGANIC PEROXIDE CLASS	MAXIMUM STORAGE QUANTITY (POUNDS) AT MINIMUM SEPARATION DISTANCE					
	Distance to buildings, lot lines, public streets, public alleys, public ways or means of egress			Distance between individual detached storage buildings or individual outdoor storage areas		
	50 feet	100 feet	150 feet	20 feet	75 feet	100 feet
I	2,000	20,000	175,000	2,000	20,000	175,000
II	100,000	200,000	No Limit	100,000[a]	No Limit	No Limit
III	200,000	No Limit	No Limit	200,000[a]	No Limit	No Limit
IV	No Limit	No Limit	No Limit	No Limit	No Limit	No Limit
V	No Limit	No Limit	No Limit	No Limit	No Limit	No Limit

For SI: 1 foot = 304.8 mm, 1 pound = 0.454 kg.
a. When the amount of organic peroxide stored exceeds this amount, the minimum separation shall be 50 feet.

tem components or trigger nuisance alarm signals. The required local alarm is intended to alert the occupants in the immediate vicinity of the storage area to a potential hazardous condition. The local alarm is not intended to be part of an evacuation alarm system for the entire structure.

3904.1.6 Maximum quantities. Maximum allowable quantities per building in a mixed occupancy building shall not exceed the amounts set forth in Table 2703.8.2. Maximum allowable quantities per building in a detached storage building shall not exceed the amounts specified in Table 3904.1.2.

❖ This section establishes the maximum allowable indoor storage quantities of organic peroxide on a per-building basis, as opposed to a per-control area basis, and regulates both mixed occupancy buildings and detached organic peroxide storage buildings.

3904.1.7 Storage arrangement. Storage arrangements for organic peroxides shall be in accordance with Table 3904.2.4 and shall comply with all of the following:

1. Containers and packages in storage areas shall be closed.
2. Bulk storage shall not be in piles or bins.
3. A minimum 2-foot (610 mm) clear space shall be maintained between storage and uninsulated metal walls.
4. Fifty-five-gallon (208 L) drums shall not be stored more than one drum high.

❖ These provisions detail storage requirements related to the hazards of release or ignition of liquids and vapors of organic peroxide stored in structures. Some of the factors that are considered are as follows:

1. Open containers or packages may permit the release of flammable or oxidizing materials or vapors.
2. Containers in bulk piles or bins may be susceptible to physical damage from stacking or product movement, which could damage the container and lead to an uncontrolled release of product.
3. The required separation distance is intended to minimize the effects of radiant heat exposures to stored materials in the event of a fire outside the structure.
4. The stacking of drums may result in a container being damaged during product movement if a drum is dropped or tipped over.

3904.1.8 Location in building. The storage of Class I or II organic peroxides shall be on the ground floor. Class III organic peroxides shall not be stored in basements.

❖ Class I and II organic peroxides may not be stored above or below the ground floor because of concerns for fire-fighter access. Because of their lower relative hazard, Class III organic peroxides may be stored on floors above grade; however, they may not be stored on floors below grade because of the difficulty of manual fire suppression operations in below-grade areas.

3904.1.9 Contamination. Organic peroxides shall be stored in their original DOTn shipping containers. Organic peroxides shall be stored in a manner to prevent contamination.

❖ Organic peroxides must be kept in original Department of Transportation (DOTn)-approved shipping containers to facilitate identification and to minimize the possibility of accidental spills or ignition. The stability of organic peroxides can be markedly reduced through contamination by various materials such as strong acids or alkalis, sulfur-based compounds or reducing agents of any type. Storage practices must prevent contamination and the hazards associated with it.

3904.1.10 Explosion control. Indoor storage rooms, areas and buildings containing unclassified detonable and Class I organic peroxides shall be provided with explosion control in accordance with Section 911.

❖ Because of the possibility of a deflagration or detonation in the event of ignition of Class I organic peroxides, explosion relief venting must be provided to protect the storage building or structure from collapse. Explosion venting must conform to the requirements in Section 911.

3904.1.11 Standby power. Standby power in accordance with Section 604 shall be provided for storage areas of Class I and unclassified detonable organic 2703.1.1(3).

❖ A standby power system complying with Section 604 is required as a backup power supply for mechanical and electrical systems, such as refrigeration equipment, neutralizer systems associated with secondary containment and mechanical ventilation equipment for vapor control. These systems may also be connected to an approved emergency power system instead of a separate standby power system. Automatic fire detection systems should also be connected to approved emergency power supplies (see also commentary, Section 604).

3904.2 Outdoor storage. Outdoor storage of organic peroxides in amounts exceeding the maximum allowable quantities per control area indicated in Table 2703.1.1(3) shall be in accordance with Sections 2701, 2703, 2704 and this chapter.

❖ This section regulates the outdoor storage of organic peroxides when in excess of the maximum allowable quantity per outdoor control area established by Table 2703.1.1(3). The general and storage provisions of Chapter 27 are applicable in addition to the requirements of this section. Storage of organic peroxides in outdoor control areas must comply with Sections 3904.2.1 through 3904.2.5 to prevent uncontrolled release or exposure to conditions that may result in a fire or explosion.

3904.2.1 Distance from storage to exposures. Outdoor storage areas for organic peroxides shall be located in accordance with Table 3904.1.2

❖ The IBC prescribes minimum fire-resistance ratings for exterior walls of buildings based on clearances from adjacent structures and lot lines. The separation distances prescribed by this section are minimums notwithstanding the inherent fire-resistance ratings of the exterior wall assemblies. These clearances are intended to reduce the hazard to nearby structures and people in the event of a fire or explosion in a detached storage structure or outdoor area used for the storage of organic peroxides.

3904.2.2 Electrical wiring and equipment. In addition to the requirements of Section 2703.9.4, electrical wiring and equipment in outdoor storage areas containing unclassified detonable, Class I or II organic peroxides shall comply with the requirements for electrical Class I, Division 2 locations.

❖ Because of the danger of ignition by arcs or sparks in the event of an accidental spill or leak, electrical equipment and devices in outdoor organic peroxide storage areas must be classified for use in Class I, Division 2, hazardous locations as described in the ICC EC.

3904.2.3 Maximum quantities. Maximum quantities of organic peroxides in outdoor storage shall be in accordance with Table 3904.1.2.

❖ Outdoor storage of organic peroxides must comply with the storage requirements specified in Table 3904.1.2, which establishes the minimum separation distances required between the outdoor storage area and exposures, including other outdoor storage areas, and the maximum allowable quantity of organic peroxides being stored.

3904.2.4 Storage arrangement. Storage arrangements shall be in accordance with Table 3904.2.4.

❖ Similar to Section 3904.1.8 for indoor storage, this section references the provisions of Table 3904.2.4 for organic peroxide storage arrangements. Pile limitations specified in the table are based on the relative hazard of the material when involved in fire. These limits apply to Class I, II, III and IV organic peroxides in combustible and noncombustible containers and packages.

3904.2.5 Separation. In addition to the requirements of Section 2703.9.8, outdoor storage areas for organic peroxides in amounts exceeding those specified in Table 2703.8.2 shall be located a minimum distance of 50 feet (15 240 mm) from other hazardous material storage.

❖ The required minimum 50-foot (15 240 mm) separation distance from other hazardous materials storage is intended to reduce the hazard of dangerous chemical reactions with other incompatible materials in the event of a spill, fire or explosion.

SECTION 3905
USE

3905.1 General. The use of organic peroxides in amounts exceeding the maximum allowable quantity per control area indicated in Table 2703.1.1(1) or 2703.1.1(3) shall be in accordance with Sections 2701, 2703, 2705 and this chapter.

❖ This section applies to indoor and outdoor dispensing, use and handling of organic peroxides when the amounts being dispensed, used or handled are in excess of the maximum allowable quantities per indoor or outdoor control area indicated in Table 2703.1.1(1) or 2703.1.1(3), respectively. The administrative, general, use, dispensing and handling provisions of Chapter 27

TABLE 3904.2.4
STORAGE OF ORGANIC PEROXIDES

ORGANIC PEROXIDE CLASS	PILE CONFIGURATION				MAXIMUM QUANTITY PER BUILDING
	Maximum width (feet)	Maximum height (feet)	Minimum distance to next pile (feet)	Minimum distance to walls (feet)	
I	6	8	4[a]	4[b]	Note c
II	10	8	4[a]	4[b]	Note c
III	10	8	4[a]	4[b]	Note c
IV	16	10	3[a,d]	4[b]	No Requirement
V	No Requirement	No Requirement	No Requirement	No Requirement	No Requirement

For SI: 1 foot = 304.8 mm.
a. At least one main aisle with a minimum width of 8 feet shall divide the storage area.
b. Distance to noncombustible walls is allowed to be reduced to 2 feet.
c. See Table 3904.1.2 for maximum quantities.
d. The distance shall not be less than one-half the pile height.

are applicable, in addition to the requirements of this chapter.

Once the maximum allowable quantity of organic peroxide per control area has been exceeded, indoor areas where materials are being dispensed, used or handled must be located in a building or portion of a building complying with the IBC for a Group H occupancy because of the increased hazards associated with quantity.

Although no occupancy group is assigned to them, outside organic peroxide use areas require an increased level of regulation when quantities exceed the maximum allowable quantities per outdoor control area. The maximum allowable quantities per control area listed in Tables 2703.1.1(1) and 2703.1.1(3) have been divided into closed use and open use systems. Corresponding maximum allowable quantities per control area recognize that an open-use condition is generally more hazardous than a closed-use condition because the organic peroxide is more directly exposed to the surrounding environment and can become more readily involved in an incident than if it is totally confined. The maximum allowable quantities per control area for use are based on the aggregate quantity in both use and storage, not exceeding the exempt amount listed for storage.

Bibliography

The following resource materials are referenced in this chapter or are relevant to the subject matter addressed in this chapter.

Bulletin 19A, Suggested Relative Hazard Classification of Organic Peroxides. Washington, DC: Society of the Plastics Industry, 1975.

DOTn 49 CFR; 100-178-94, *Specification for Transportation of Explosive and Other Dangerous Articles, Shipping Containers*. Washington, DC: U.S. Department of Transportation, 1994.

ICC EC-2003, *ICC Electrical Code*. Falls Church, VA: International Code Council, 2003.

IBC-2003, *International Building Code*. Falls Church, VA: International Code Council, 2003.

NFPA 432-02, *Code for the Storage of Organic Peroxide Formulations*. Quincy, MA: National Fire Protection Association, 2002.

Recommendations on the Transport of Dangerous Goods, Test and Criteria, Parts I, II and III, 10th rev. ed. New York: United Nations, 1997.

Publication AS109, Safety and Handling of Organic Peroxides: A Guide. Washington, DC: Society of the Plastics Industry, 1999.

Townsend, J.S. "Heat: The Forgotten Reactant." *Industrial Fire World*, May/June 1993, p. 1416.

Chapter 40: Oxidizers

General Comments

Solid and liquid oxidizers are common industrial chemicals. These compounds are often used because of their reactive properties. Common oxidizers include bromates, chlorates, chlorites, dichromates, hypochlorites, nitrates, nitrites, permanganates, inorganic peroxides and inorganic superoxides.

Although oxidizers themselves do not burn, they pose unique fire hazards because of their ability to support combustion by breaking down and giving off oxygen. The hazard classification system for these materials described in the definition of "Oxidizers" in Section 4002.1 reflects the varying reactivity of oxidizers when they come in contact with combustible materials. The classification system originated in NFPA 430, upon which the provisions of this chapter are based but that is not referenced by the code except for sprinkler design criteria in Section 4004.1.4.

The oxidizer hazard classifications defined in this chapter do not correspond to those given in NFPA 704.

The NFPA 704 system indicates whether a given material will support combustion by using the "OXY" symbol in the lower quadrant of the placard. No numerical hazard value is assigned to a material's oxidizing ability in the NFPA 704 system.

When the specific hazard classification of a material is not known, the fire code official must use judgement in approving the assignment of materials to specific hazard classes. The best source of guidance is the information supplied by the Material Safety Data Sheets (MSDS) (see commentary, Section 407).

Purpose

Chapter 40 addresses the hazards associated with oxidizing materials and establishes criteria for their safe storage and protection in indoor and outdoor storage facilities, minimizing the potential for uncontrolled releases and contact with fuel sources.

SECTION 4001 GENERAL

4001.1 Scope. The storage and use of oxidizers shall be in accordance with this chapter and Chapter 27. Compressed gases shall also comply with Chapter 30.

> **Exception:** Display and storage in Group M and storage in Group S occupancies complying with Section 2703.11.

Bulk oxygen systems at industrial and institutional consumer sites shall be in accordance with NFPA 50.

❖ This chapter is based on NFPA 430, addressing the hazards presented by the storage and use of oxidizers. In addition to the requirements of this chapter, oxidizing compressed gases are subject to Chapter 30. Bulk oxygen systems, as defined in Section 4002.1, are required to comply with the provisions of NFPA 50.

The exception states that this chapter does not apply to oxidizers where stored and displayed in Group M occupancies or where stored in Group S occupancies. Instead, they are governed by Section 2703.11 for those occupancy groups. In that section, Class 4 oxidizers are prohibited in Group M mercantile occupancies because of their detonable hazard potential. Quantities of all other oxidizer classes are limited on a per-control-area basis to limit exposure to people and property (see commentary, Section 2703.11). This exception would permit an increase above the exempt amounts per control area indicated in Table 2703.1.1(1) for oxidizers while still maintaining a mercantile or storage occupancy group classification.

4001.2 Permits. Permits shall be required as set forth in Section 105.6.

❖ The process of issuing permits gives the fire code official an opportunity to carefully evaluate and regulate hazardous operations. Permit applicants should be required to demonstrate that their operations comply with the intent of the code before the permit is issued. See the commentary to Section 105.6 for a general discussion of operations requiring an operational permit, Section 105.6.21 for a discussion of specific quantity-based operational permits for the materials regulated in this chapter and Section 105.7 for a general discussion of activities requiring a construction permit. The permit process also notifies the fire department of the need for prefire planning for hazardous property.

SECTION 4002 DEFINITIONS

4002.1 Definitions. The following words and terms shall, for the purposes of this chapter and as used elsewhere in this code, have the meanings shown herein.

❖ Definitions of terms can help in the understanding and application of the code requirements. The purpose for including those definitions that are associated with the subject matter of this chapter is to provide more convenient access to them without having to refer back to Chapter 2. It is important to emphasize that these terms are not exclusively related to this chapter but are applicable everywhere the term is used in the code. For convenience, these terms are also listed in Chapter 2 with a cross reference to this section. The use and application of all defined terms, including those defined in this section, are set forth in Section 201.

BULK OXYGEN SYSTEM. An assembly of equipment, such as oxygen storage containers, pressure regulators, safety devices, vaporizers, manifolds and interconnecting piping, that has a storage capacity of more than 20,000 cubic feet (566 m^3) of oxygen at normal temperature and pressure (NTP) including unconnected reserves on hand at the site. The bulk oxygen system terminates at the point where oxygen at service pressure first enters the supply line. The oxygen containers can be stationary or movable, and the oxygen can be stored as a gas or liquid.

❖ As indicated in Section 4001.1, NFPA 50 contains installation and maintenance requirements for bulk oxygen systems. NFPA 50 also has the requirements for the protection of bulk oxygen systems from potential fire exposures. Oxygen storage systems with less than the capacities indicated in the definition are not required to comply with NFPA 50.

OXIDIZER. A material that readily yields oxygen or other oxidizing gas, or that readily reacts to promote or initiate combustion of combustible materials. Examples of other oxidizing gases include bromine, chlorine and fluorine.

Class 4. An oxidizer that can undergo an explosive reaction due to contamination or exposure to thermal or physical shock. In addition, the oxidizer will enhance the burning rate and can cause spontaneous ignition of combustibles.

Class 3. An oxidizer that will cause a severe increase in the burning rate of combustible materials with which it comes in contact or that will undergo vigorous self-sustained decomposition caused by contamination or exposure to heat.

Class 2. An oxidizer that will cause a moderate increase in the burning rate or that causes spontaneous ignition of combustible materials with which it comes in contact.

Class 1. An oxidizer whose primary hazard is that it slightly increases the burning rate but which does not cause spontaneous ignition when it comes in contact with combustible materials.

❖ The classification of oxidizers had its origins in the provisions of NFPA 430 and is consistent with Department of Transportation (DOTn) hazardous materials regulations. Oxidizers, whether a solid, liquid or gas, yield oxygen or another oxidizing gas during a chemical reaction or readily react to oxidize combustibles and increase their burning rate. This characteristic is a result of the enrichment of the air to more than 21-percent oxygen content.

This enrichment is a hazard because an ordinary combustible material that will burn freely at the atmospheric oxygen level of 21 percent will burn more rapidly at higher concentrations of oxygen. The rate of reaction varies with the class of oxidizer. Specific classification of oxidizers is important because of the varying degree of hazard. Examples of oxidizers include liquid hydrogen peroxide, nitric acid, sulfuric acid, and solids, such as sodium chlorite, chromic acid and calcium hypochlorite. Many commercially available swimming pool chemicals are examples of Class 2 or 3 oxidizers.

OXIDIZING GAS. A gas that can support and accelerate combustion of other materials.

❖ Oxidizing gases present essentially the same hazard characteristics as solids and liquids. Examples of oxidizing gases are oxygen, ozone and the oxides of nitrogen, fluorine and chlorine.

SECTION 4003 GENERAL REQUIREMENTS

4003.1 Quantities not exceeding the maximum allowable quantity per control area. The storage and use of oxidizers in amounts not exceeding the maximum allowable quantity per control area indicated in Section 2703.1 shall be in accordance with Sections 2701, 2703, 4001 and 4003. Oxidizing gases shall also comply with Chapter 30.

❖ The provisions of this section complement the requirements of Chapter 27 in structures used for the storage, handling or use of oxidizing materials. Unless otherwise indicated in a particular section, the regulations contained in Sections 4003.1.1 through 4003.1.3 assume that the quantity of oxidizers in a given building is limited to the maximum allowable quantities per control area established in Section 2703.1; thus, the building is not classified in Occupancy Group H. The general requirements of Sections 2701 and 2703 are fully applicable to the storage and use of organic peroxides in addition to the provisions of this chapter. In the event that the oxidizer being stored is a gas, it will also be subject to Chapter 30.

4003.1.1 Special limitations for indoor storage and use by occupancy. The indoor storage and use of oxidizers shall be in accordance with Sections 4003.1.1.1 through 4003.1.1.3.

❖ Because certain occupancies may need to have oxidizers on hand, Sections 4003.1.1.1 through 4003.1.1.3 provide regulations that are specific to occupancy group classifications and recognize the relative hazards of both the occupancy and the oxidizer.

4003.1.1.1 Class 4 liquid and solid oxidizers. The storage and use of Class 4 liquid and solid oxidizers shall comply with Sections 4003.1.1.1.1 through 4003.1.1.1.4.

❖ Because of their explosive or higher deflagration hazard characteristics, Class 4 liquid and solid oxidizers warrant special consideration and limitations when stored or used within certain occupancies, as indicated in Sections 4003.1.1.1.1 through 4003.1.1.1.4.

4003.1.1.1.1 Group A, E, I or U occupancies. In Group A, E, I or U occupancies, any amount of Class 4 liquid and solid oxidizers shall be stored in accordance with the following:

1. Class 4 liquid and solid oxidizers shall be stored in hazardous materials storage cabinets complying with Section 2703.8.7.
2. The hazardous materials storage cabinets shall not contain other storage.

❖ Because of their explosive or higher deflagration hazard characteristics, even the smallest quantity of Class 4 liquid and solid oxidizers present in Group A, E, I or U occupancies must be stored in an approved hazardous materials storage cabinet constructed and placarded in accordance with Section 2703.8.6 to reduce the exposure of the materials to hazards from the surrounding environment.

Note also that Note g to Table 2703.1.1(1) limits storage of Class 4 liquid and solid oxidizers in any amount to buildings equipped throughout with an approved automatic sprinkler system in accordance with Section 903.3.1. Also, based on Note e to Table 2703.1.1(1), if an approved storage cabinet is used, the maximum allowable quantity per control area that could be kept in occupancies equipped with sprinklers can be doubled.

4003.1.1.1.2 Group R occupancies. Class 4 liquid and solid oxidizers shall not be stored or used within Group R occupancies.

❖ Because Class 4 liquid and solid oxidizers are a great explosive hazard, they cannot be stored in any residential occupancy within the scope of the code's regulations because of the increased danger to the occupants and the otherwise unregulated environment to which the materials could be exposed.

4003.1.1.1.3 Offices, and retail sales areas. Class 4 liquid and solid oxidizers shall not be stored or used in offices, or retail sales areas of Group B, F, M or S occupancies.

❖ Because Class 4 liquid and solid oxidizers are a great explosive hazard, they cannot be stored in offices or retail sales areas of occupancies in Group B, F, M or S because of the increased danger to the occupants and the higher relative fire loads typically encountered in these occupancies.

4003.1.1.1.4 Classrooms. In classrooms of Group B, F or M occupancies, any amount of Class 4 liquid and solid oxidizers shall be stored in accordance with the following:

1. Class 4 liquid and solid oxidizers shall be stored in hazardous materials storage cabinets complying with Section 2703.8.7.
2. Hazardous materials storage cabinets shall not contain other storage.

❖ This section allows the occasional use of limited amounts of Class 4 liquid and solid oxidizers in certain scientific, experimental or demonstration settings; however, storage of any quantity of these materials for any length of time is not allowed. The fire code official may limit the amount of Class 4 liquid and solid oxidizers brought into a structure for these uses. The quantity actually needed for the experiment should determine the amount allowed into a structure.

Because of their explosive hazard characteristics, even the smallest quantity of Class 4 liquid and solid oxidizers must be stored in an approved hazardous materials storage cabinet, constructed and placarded in accordance with Section 2703.8.6, to reduce the exposure of the materials to hazards from the surrounding environment. Based on Table 2703.1.1(1), Note e, where an approved storage cabinet is used, the maximum allowable quantity per control area that could be kept in such occupancies can be doubled.

Note g to Table 2703.1.1(1) allows storage of Class 4 liquid and solid oxidizers in any amount only in buildings equipped throughout with an approved automatic sprinkler system in accordance with Section 903.3.1.

To reduce the likelihood of contamination of the Class 4 oxidizer materials or damage to their packaging, Item 2 prohibits the storage of other materials within the approved storage cabinet.

4003.1.1.2 Class 3 liquid and solid oxidizers. A maximum of 200 pounds (91 kg) of solid or 20 gallons (76 L) of liquid Class 3 oxidizer is allowed in Group I occupancies when such materials are necessary for maintenance purposes or operation of equipment. The oxidizers shall be stored in approved containers and in an approved manner.

❖ The higher amounts of Class 3 liquid and solid oxidizers permitted by this section in institutional occupancies (Group I) are in recognition of their common use in these occupancies for building maintenance purposes or equipment operation. Proper storage practices and safeguards must still be observed.

4003.1.1.3 Oxidizing gases. Except for cylinders not exceeding a capacity of 250 cubic feet (7 m^3) each used for maintenance purposes, patient care or operation of equipment, oxidizing gases shall not be stored or used in Group A, B, E, I, or R occupancies.

The aggregate quantities of gases used for maintenance purposes and operation of equipment shall not exceed the maximum allowable quantity per control area listed in Table 2703.1.1(1).

Medical gas systems and medical gas supply cylinders shall also be in accordance with Section 3006.

❖ In Occupancy Groups A (assembly), B (business), E (educational), I (institutional) or R (residential), up to the maximum allowable quantity of oxidizing gases per control area established in Section 2703.1.1(1) is allowed for maintenance and critical functions such as patient care (in Group B occupancies) and maintenance of equipment. To limit the amount of gas that could be released in any given incident, no individual cylinder may exceed a capacity of 250 cubic feet (7 m^3). Consistent with Section 4001.1, the provisions of Chapter 30 are also applicable to oxidizer gases in addition to the provisions of this section.

4003.1.2 Emergency shutoff. Compressed gas systems conveying oxidizer gases shall be provided with approved emergency shutoff valves that can be activated at each point of use and each source.

❖ To maintain control over the flow of oxidizer gases under emergency conditions, each supply source as well as each point of use of oxidizer compressed gas must have an approved emergency shutoff valve. These valves must be excess flow control valves to regulate the rate of flow of hazardous materials within the piping system, fail-safe valves or other approved types that will operate automatically or there must be a manual means of operation to give the fire department or other responsible persons the ability to stop the flow of hazardous materials in an emergency. The valves must be readily accessible and comply with Section 2703.2.2.1.

4003.1.3 Ignition source control. Ignition sources in areas containing oxidizing gases shall be controlled in accordance with Section 2703.7.

❖ Given the fact that oxidizers enhance or accelerate the combustion process, ignition sources, such as smoking and open flames, must be strictly controlled in oxidizer storage or use areas in accordance with the provisions of Section 2703.7 to reduce the likelihood of a fire involving the stored material.

4003.2 Quantities exceeding the maximum allowable quantity per control area. The storage and use of oxidizers in amounts exceeding the maximum allowable quantity per control area indicated in Section 2703.1 shall be in accordance with Chapter 27 and this chapter.

❖ When the amount of oxidizer being used or stored exceeds the maximum allowable quantities per control area as established in Section 2703.1, the requirements of Chapter 27 apply (see commentary, Chapter 27).

SECTION 4004
STORAGE

4004.1 Indoor storage. Indoor storage of oxidizers in amounts exceeding the maximum allowable quantity per control area indicated in Table 2703.1.1(1) shall be in accordance with Sections 2701, 2703, 2704 and this chapter.

❖ This section regulates the indoor storage of oxidizers when quantities are in excess of the maximum allowable quantity per control area classified in Occupancy Group H. The general and storage provisions of Chapter 27 are applicable in addition to the requirements of this section. Storage of oxidizers inside structures must comply with Sections 4004.1.1 through 4004.2.4 to prevent uncontrolled release or exposure to conditions that may result in a fire or explosion.

4004.1.1 Detached storage. Storage of liquid and solid oxidizers shall be in detached buildings when required by Section 2703.8.2.

❖ Detached structures designed and constructed for the sole purpose of oxidizer storage provide the best protection for people and property from fire and explosion. Detached storage structures should be constructed of noncombustible materials to prevent them from becoming involved in an outside fire that could endanger their contents or becoming involved in a fire themselves should their contents be ignited.

Section 2703.8.2 and Table 2703.8.2 require detached storage when the indoor storage quantities are more than 2 pounds (.908 kg) of Class 4 liquid and solid oxidizers, 1,200 tons (2400 metric tons) of Class 3 liquid and solid oxidizers, more than 2,000 tons (4000 metric tons) of Class 2 liquid and solid oxidizers or more than the maximum allowable quantity per control area of oxidizer gases. See the commentary to Section 2703.8.2 and Table 2703.8.2 for further discussion of detached storage requirements.

4004.1.2 Distance from detached storage buildings to exposures. In addition to the requirements of the *International Building Code*, detached storage buildings shall be located in accordance with Table 4004.1.2.

❖ The *International Building Code*® (IBC®) prescribes minimum fire-resistance ratings for exterior walls of buildings based on clearances from adjacent structures and lot lines. The separation distances prescribed in this section are minimums notwithstanding the inherent fire-resistance ratings of the exterior wall assemblies. These clearances are intended to reduce the hazard to nearby structures and people in the event of a fire or explosion in a detached structure used for the storage of oxidizers.

TABLE 4004.1.2
OXIDIZER LIQUIDS AND SOLIDS —DISTANCE FROM DETACHED BUILDINGS AND OUTDOOR STORAGE AREAS TO EXPOSURES

OXIDIZER CLASS	WEIGHT (pounds)	MINIMUM DISTANCE TO BUILDINGS, LOT LINES, PUBLIC STREETS, PUBLIC ALLEYS, PUBLIC WAYS OR MEANS OF EGRESS (feet)
1	Note a	Not Required
2	Note a	35
3	Note a	50
4	over 10 to 100	75
	101 to 500	100
	501 to 1,000	125
	1,001 to 3,000	200
	3,001 to 5,000	300
	5,001 to 10,000	400
	over 10,000	As required by the fire code official

For SI: 1 foot = 304.8 mm, 1 pound = 0.454 kg.

a. Any quantity over the amount required for detached storage in accordance with Section 2703.8.2, or over the outdoor maximum allowable quantity for outdoor control areas.

❖ The separation distances specified are intended to reduce the effects of radiant heat exposure in the event of a fire in a detached storage structure.

4004.1.3 Explosion control. Indoor storage rooms, areas and buildings containing Class 4 liquid or solid oxidizers shall be provided with explosion control in accordance with Section 911.

❖ Because of the possibility of a deflagration or detonation in the event of ignition of oxidizers, explosion relief venting meeting the requirements of Section 911 must be installed to protect the storage building or structure from collapse.

4004.1.4 Automatic sprinkler system. The automatic sprinkler system shall be designed in accordance with NFPA 430.

❖ This section, as is all of Section 4004.1, is applicable to oxidizer storage buildings or portions of buildings classified in Occupancy Group H. Section 903.2.4.1 generally requires that sprinklers be installed throughout the Group H occupancy in accordance with Section 903.3.1.1. Because of the unique fire protection challenges presented by the storage of liquid and solid oxidizers, this section supercedes the general sprinkler design criteria of Section 903.3.1.1 (NFPA 13) by mandating that the sprinkler design for oxidizer storage be in accordance with NFPA 430, which includes special sprinkler design densities based on the class of oxidizer being protected and the manner of storage (see Table 4004.1.4). Note that only the sprinkler design criteria contained in NFPA 430 may be used because of the limited reference in this section See the commentary to Section 102.6 for further discussion of the limitations on the use of referenced standards.

4004.1.5 Liquid-tight floor. In addition to Section 2704.12, floors of storage areas for liquid and solid oxidizers shall be of liquid-tight construction.

❖ Floors and sills of rooms or areas used to contain hazardous material spills must be liquid tight to prevent the flow of liquids to adjoining areas (see commentary, Section 2704.2). The floor surface should be compatible with the oxidizer materials to be retained and must be noncombustible, as required by Section 2704.12.

TABLE 4004.1.4
SPRINKLER SYSTEM DESIGN DENSITIES FOR OXIDIZER STORAGE

OXIDIZER CLASS	STORAGE TYPE/MAXIMUM HEIGHT (feet)	DENSITY (gpm/ft^2)	DESIGN AREA (ft^2)	NFPA 430 SECTION
4	All	0.35	Entire storage area[a]	6-4.1
3[d]	Palletized/5; Rack/10[b]	0.35	5,000	5-4.1
3[d]	Bulk/10	0.65	5,000	2-3.3, 3-3.4, 5-4.1
2[d]	Palletized/8; Rack/12[c]	0.20	3,750	4-4.1
2[d]	Rack/16[c]	0.30	2,000	
2[d]	Bulk/12	0.35	3,750	2-3.3, 2-3.4, 5-4.1
1	Based on storage type, storage height and commodity classifications of NFPA 231 and 231C			3-3.1, 3-3.2

Note a. Deluge sprinkler system in accordance with Section 3-3 of NFPA 13.
Note b. One level of in-rack sprinklers at the midpoint of the rack.
Note c. One line of in-rack sprinklers above each storage level except the top.
Note d. 286°F ceiling sprinklers

4004.1.6 Smoke detection. An approved supervised smoke detection system in accordance with Section 907 shall be installed in liquid and solid oxidizer storage areas. Activation of the smoke detection system shall sound a local alarm.

Exception: Detached storage buildings protected by an approved automatic fire-extinguishing system.

❖ An automatic fire detection system with central, remote or proprietary station supervision and a local alarm signal is required for early warning of fire to building occupants and fire suppression personnel. Detectors and installation methods in oxidizer storage areas must be selected for compatibility with the materials stored. Fire alarm equipment and installation methods must comply with Section 907. Materials that emit strong vapors may interfere with fire detection system components or trigger nuisance alarm signals. The required local alarm is intended to alert the occupants in the immediate vicinity of the storage area to a potentially hazardous condition. The alarm is not intended to be part of an evacuation alarm system for the entire structure.

The exception indicates that automatic fire detection is not required in detached storage buildings when there is an automatic fire suppression system. Automatic sprinklers or approved alternative fire-extinguishing systems reduce the danger to people and property from fire by detecting the fire, sounding an alarm, transmitting the alarm to the fire department and suppressing the fire.

Fire-extinguishing systems in detached storage buildings must be supervised by connection to a central station, a remote supervising station or a proprietary supervising station or, when approved by the fire code official, be connected to a system that produces an audible and visual signal monitored at a constantly attended on-site location. See the commentary to Section 903.4 for further discussion of sprinkler system supervision and Sections 904.3.5 and 907.14 for further discussion of alternative automatic fire-extinguishing system supervision.

4004.1.7 Storage conditions. The maximum quantity of oxidizers per building in detached storage buildings shall not exceed those quantities set forth in Tables 4004.1.7(1) through 4004.1.7(4).

The storage configuration for liquid and solid oxidizers shall be as set forth in Tables 4004.1.7(1) through 4004.1.7(4).

Class 2 oxidizers shall not be stored in basements except when such storage is in stationary tanks.

Class 3 and 4 oxidizers in amounts exceeding the maximum allowable quantity per control area set forth in Section 2703.1 shall be stored on the ground floor only.

❖ This section covers storage requirements related to the hazards of oxidizers stored in detached storage structures. Maximum allowable quantities of oxidizers and their arrangement within the building must comply with Tables 4004.1.7(1) through (4).

Because of the moderate hazard of Class 2 oxidizers, they can be stored in basements, but only when contained in approved stationary tanks that comply with the applicable provisions of Section 2703.2.

Because of their greater relative hazard, storage of Class 3 and 4 oxidizers in excess of the maximum allowable quantities per control area above or below the ground floor is prohibited. The greater hazard results in more problems associated with manual fire suppression operations in below-grade areas. Ground-floor storage also facilitates fire-fighter access to the storage area.

TABLE 4004.1.7(1)
STORAGE OF CLASS 1 OXIDIZER LIQUIDS AND SOLIDS IN COMBUSTIBLE CONTAINERS[a]

STORAGE CONFIGURATION	LIMITS (feet)
Piles	
Maximum length	No Limit
Maximum width	50
Maximum height	20
Minimum distance to next pile	3
Minimum distance to walls	2
Maximum quantity per pile	No Limit
Maximum quantity per building	No Limit

For SI: 1 foot = 304.8 mm.

a. Storage in noncombustible containers or in bulk in detached storage buildings is not limited as to quantity or arrangement.

❖ The limitations of Table 4004.1.7(1) apply to Class 1 oxidizers in combustible containers and packages. Combustible containers such as fiber drums; cardboard or plastic boxes; paper or plastic bags and plastic bottles or jugs are common packaging materials for oxidizers. Expanded polystyrene foam and other plastic and cellulosic combustible materials are also commonly used as packing materials. Noncombustible containers, such as glass jugs and bottles, are common containers but are usually shipped in combustible packing materials and boxes to prevent damage. Occasionally, some materials may be shipped or stored in metal drums. Whenever possible, materials should be kept in original shipping containers to prevent spillage or contamination.

Class 1 oxidizers are the least susceptible to spontaneous reactions at elevated temperatures (such as from exposure fires); therefore, eliminating combustible packaging materials significantly reduces the fire hazard, making unlimited quantities reasonable, as indicated in Note a. Likewise, when bulk Class 1 oxidizers are stored in detached buildings, the low relative hazard of such materials is further mitigated, making unlimited quantities reasonable. The term "bulk" intends to mean more than 600 gallons (2 L) of liquid oxidizer in a single portable or stationary tank or more than 6,000 pounds (2,724 kg) of solid oxidizer in a single package (see Section 2-3 of NFPA 430).

**TABLE 4004.1.7(2)
STORAGE OF CLASS 2 OXIDIZER LIQUIDS AND SOLIDS[a,b]**

STORAGE CONFIGURATION	LIMITS		
	Segregated storage	Cutoff storage rooms[c]	Detached building
Piles			
Maximum width	16 feet	25 feet	25 feet
Maximum height	10 feet	12 feet	12 feet
Minimum distance to next pile	Note d	Note d	Note d
Minimum distance to walls	2 feet	2 feet	2 feet
Maximum quantity per pile	20 tons	50 tons	200 tons
Maximum quantity per building	200 tons	500 tons	No Limit

For SI: 1 foot = 204.8 mm, 1 ton = 0.907185 metric ton.

a. Storage in noncombustible containers is not limited as to quantity or arrangement, except that piles shall be at least 2 feet from walls in sprinklered buildings and 4 feet from walls in nonsprinklered buildings; the distance between piles shall not be less than the pile height.

b. Quantity limits shall be reduced by 50 percent in buildings or portions of buildings used for retail sales.

c. Cutoff storage rooms shall be separated from the remainder of the building by 2-hour fire barriers.

d. Aisle width shall not be less than the pile height.

❖ Table 4004.1.7(2) limits the amounts of Class 2 liquid and solid oxidizers that can be stored in combustible containers [see the discussion of packaging materials and methods in the commentary to Table 4004.1.7(1)]. See Note a for guidance on applying these restrictions to Class 2 liquid and solid oxidizers in noncombustible containers and packages.

Regarding Note a, although noncombustible packages themselves may not contribute to the fuel present in a reaction with the oxidizer in the event of an exposure fire, the container will either conduct heat to the material (metal) or provide inadequate insulation (glass) from an exposure fire. If there is no automatic sprinkler protection, interior separation distances from exterior walls must be increased to prevent degradation of the package contents by heat conduction or radiation from potential outside exposure fires.

Note b is a general requirement that limits the quantity of Class 2 liquid and solid oxidizers to one-half the tabular amount in retail sales areas. For specific requirements applicable to storage and display in Group M occupancies, see the commentary to Section 2703.11.

Note c indicates that the separation between cutoff rooms and adjacent spaces must have a 2-hour fire-resistance rating because of the increased reactivity of Class 2 liquid and solid oxidizers.

Aisle widths that are equal to the pile height, as stated in Note d, reduce the fire exposure hazard between piles and, in the event of a fire-related collapse of a pile, reduce the likelihood of a "domino" effect that could not only increase fire intensity but also completely block access to pile areas by fire suppression personnel.

**TABLE 4004.1.7(3)
STORAGE OF CLASS 3 OXIDIZER LIQUIDS AND SOLIDS[a,b]**

STORAGE CONFIGURATION	LIMITS		
	Segregated storage	Cutoff storage rooms[c]	Detached building
Piles			
Maximum width	12 feet	16 feet	20 feet
Maximum height	8 feet	10 feet	10 feet
Minimum distance to next pile	Note d	Note d	Note d
Minimum distance to walls	4 feet	4 feet	4 feet
Maximum quantity per pile	20 tons	30 tons	150 tons
Maximum quantity per building	100 tons	500 tons	No Limit

For SI: 1 foot = 204.8 mm, 1 ton = 0.907185 metric ton.

a. Storage in noncombustible containers is not limited as to quantity or arrangement, except that piles shall be at least 2 feet from walls in sprinklered buildings and 4 feet from walls in nonsprinklered buildings; the distance between piles shall not be less than the pile height.

b. Quantity limits shall be reduced by 50 percent in buildings or portions of buildings used for retail sales.

c. Cutoff storage rooms shall be separated from the remainder of the building by 2-hour fire barriers.

d. Aisle width shall not be less than the pile height.

❖ Table 4004.1.7(3) applies equally to Class 3 oxidizers in combustible and noncombustible containers and packages in segregated piles, cutoff rooms and detached storage structures.

Note a acknowledges that the use of noncombustible containers reduces the fire hazard by eliminating the potential for ignition of the container or packaging; therefore, quantities and pile heights may be unlimited when Class 3 oxidizers are stored in noncombustible containers. The distance required from exterior walls has been adjusted because of the increased level of protection afforded by sprinklers and the enhanced protection of noncombustible containers in the event of an exposure fire [see also the commentary to Note a of Table 4004.1.7(2)].

Note b is a general requirement that limits the quantity of Class 2 liquid and solid oxidizers to one-half the tabular amount in retail sales areas. For specific requirements applicable to storage and display in Group M occupancies, see the commentary to Section 2703.11.

Note c indicates that the separation between cutoff rooms and adjacent spaces must have a 2-hour fire-resistance rating because of the increased reactivity of Class 3 oxidizers.

Aisle widths that are equal to the pile height, as required in Note d, reduce the fire exposure hazard between piles and, in the event of a fire-related collapse of a pile, reduce the likelihood of a "domino" effect that could not only increase fire intensity but also completely block access to pile areas by fire suppression personnel.

TABLE 4004.1.7(4)
STORAGE OF CLASS 4 OXIDIZER LIQUIDS AND SOLIDS

STORAGE CONFIGURATION	LIMITS (feet)
Piles	
Maximum length	10
Maximum width	4
Maximum height	8
Minimum distance to next pile	8
Maximum quantity per building	No Limit

For SI: 1 foot = 304.8 mm.

❖ Table 4004.1.7(4) applies to the storage of Class 4 oxidizers in quantities greater than the exempt amounts specified in Table 2703.1.1(1). Aisle widths that are equal to the pile height, as required in the table, reduce the fire exposure hazard between piles and, in the event of a fire-related collapse of a pile, reduce the likelihood of a "domino" effect that could not only increase fire intensity but also completely block access to pile areas by fire suppression personnel.

4004.1.8 Separation of Class 4 oxidizers from other materials. In addition to the requirements in Section 2703.9.8, Class 4 oxidizer liquids and solids shall be separated from other hazardous materials by not less than a 1-hour fire barrier or stored in hazardous materials storage cabinets.

Detached storage buildings for Class 4 oxidizer liquids and solids shall be located a minimum of 50 feet (15 240 mm) from other hazardous materials storage.

❖ Above all else, oxidizers must be kept away from flammable and combustible liquids and explosive materials. Other chemicals may also react violently with some oxidizers and some oxidizers react with other oxidizers. For example, triazinetriones (chlorinated isocyanurates) are incompatible with hypochlorites. Materials should be checked for compatibility before storage arrangements are approved. See the MSDS for each product for guidance on compatibility.

Class 4 oxidizers must be stored in separate hazardous materials storage cabinets (see commentary, Section 407) or in rooms separated from other storage areas by 1-hour fire-resistance-rated construction. Detached storage of Class 4 oxidizers must be separated from other structures by at least 50 feet (15 240 mm) because of the potentially violent nature of any fuel/oxidizer reaction (see commentary, Section 2703.9.8).

4004.1.9 Contamination. Liquid and solid oxidizers shall not be stored on or against combustible surfaces. Liquid and solid oxidizers shall be stored in a manner to prevent contamination.

❖ Combustible surfaces may be ignited in the event of a spill or uncontrolled release of oxidizing material, including its vapors. Protective floor and wall coverings should be checked for compatibility with the oxidizing materials even if the underlying material is compatible. Noncombustible materials, such as unprotected ferrous metals, may corrode or react violently in the presence of certain oxidizing materials. For example, concentrated hydrogen peroxide (greater than 52-percent solution) may produce a violent decomposition reaction in contact with iron, copper, chromium, brass, bronze, lead, silver and manganese. The corrosive effects of many oxidizers may also damage concrete surfaces.

4004.2 Outdoor storage. Outdoor storage of oxidizers in amounts exceeding the maximum allowable quantities per control area set forth in Table 2703.1.1(3) shall be in accordance with Sections 2701, 2703, 2704 and this chapter. Oxidizing gases shall also comply with Chapter 30.

❖ This section regulates the outdoor storage of oxidizers when in excess of the maximum allowable quantity per outdoor control area established by Table 2703.1.1(3). The general and storage provisions of Chapter 27 are applicable in addition to the requirements of this section. Storage of oxidizers in outdoor control areas must comply with Sections 4004.2.1 through 4004.2.4 to prevent uncontrolled release or exposure to conditions that may result in a fire or explosion. In addition to the requirements of this chapter, oxidizing compressed gases are subject to the requirements of Chapter 30.

4004.2.1 Distance from storage to exposures for liquid and solid oxidizers. Outdoor storage areas for liquid and solid oxidizers shall be located in accordance with Table 4004.1.2.

❖ The separation distances specified in Table 4004.1.2 are intended to reduce the effects of radiant heat exposure in the event of a fire in an outdoor storage area.

4004.2.2 Distance from storage to exposures for oxidizer gases. Outdoor storage areas for oxidizer gases shall be located in accordance with Table 4004.2.2.

❖ Table 4004.2.2 contains separation distances for outdoor storage of oxidizer gases that are intended to minimize radiant heat transfer and fire spread between exposures.

TABLE 4004.2.2
OXIDIZER GASES— DISTANCE ROM STORAGE TO EXPOSURES[a]

QUANTITY OF GAS STORED (cubic feet at NTP)	DISTANCE TO A BUILDING NOT ASSOCIATED WITH THE MANUFACTURE OR DISTRIBUTION OF OXIDIZER GASES OR PUBLIC WAY OR LOT LINE THAT CAN BE BUILT UPON (feet)	DISTANCE BETWEEN STORAGE AREAS (feet)
0-50,000	5	5
50,001-100,000	10	10
100,001 or greater	15	10

For SI: 1 foot = 304.8 mm, 1 cubic foot = 0.02832 m^3.

a. The distances do not apply when protective structures having a minimum fire-resistance rating of 2 hours interrupt the line of sight between the storage container and the exposure. The protective structure shall be at least 5 feet from the exposure. The configuration of the protective structure shall be designed to allow natural ventilation to prevent the accumulation of hazardous gas concentrations.

❖ Table 4004.2.2 establishes spatial separation (or alternative) safeguards required for the outside storage of oxidizing compressed gases when in excess of the outside storage maximum allowable quantity per control area shown in Table 2703.1.1(3).

The required separation distance is intended to minimize radiant heat transfer between exposures. Separation distances provide a measure of protection against the possibility of fire spread in the event of fire involving the stored material or an incident involving another exposure, such as a building on the same or an adjacent lot or vehicle in the public right-of-way. Section 416.6 of the IBC contains additional protection requirements for outside storage conditions in areas that are provided with an overhead roof structure or canopy to protect the materials from the weather.

Note a recognizes that a 2-hour fire barrier wall (constructed as required by Section 706 of the IBC) provides an equivalent means of achieving this objective. Where a separation assembly is used instead of separation distance, the wall must extend vertically beyond the roof or wall opening of the larger structure on each side (top and sides) to prevent a fire from lapping over or extending around the wall and must be arranged to prevent hazardous accumulation of gas.

4004.2.3 Storage configuration for liquid and solid oxidizers. Storage configuration for liquid and solid oxidizers shall be in accordance with Tables 4004.1.7(1) through 4004.1.7(4).

❖ The outside storage arrangements for liquid and solid oxidizers are the same as for inside storage arrangements contained in Section 4004.1.8 and its referenced tables (see commentary, Section 4004.1.8).

4004.2.4 Storage configuration for oxidizer gases. Storage configuration for oxidizer gases shall be in accordance with Table 4004.2.2.

❖ To minimize radiant heat transfer and fire spread between exposures, Table 4004.2.2 contains separation distances for the outdoor storage of oxidizer gases when in excess of the outside storage maximum allowable quantity per control area shown in Table 2701.1.1(3).

SECTION 4005 USE

4005.1 Scope. The use of oxidizers in amounts exceeding the maximum allowable quantity per control area indicated in Table 2703.1.1(1) or 2703.1.1(3) shall be in accordance with Sections 2701, 2703, 2705 and this chapter. Oxidizing gases shall also comply with Chapter 30.

❖ This section applies to all indoor and outdoor dispensing, use and handling of oxidizers when these amounts are in excess of the maximum allowable quantities per indoor or outdoor control area indicated in Table 2703.1.1(1) or 2703.1.1(3), respectively. The administrative, general, use, handling and dispensing provisions of Chapter 27 are applicable in addition to the requirements of this chapter.

Once the maximum allowable quantity of oxidizers per control area has been exceeded, indoor areas where materials are being dispensed, used or handled must be located in a building or portion of a building complying with the IBC for a Group H occupancy as a result of the increased hazards associated with quantity.

Although no occupancy group is assigned to them, outside oxidizer use areas must have an increased level of regulation when quantities exceed the maximum allowable quantities per outdoor control area listed in Table 2703.1.1(3). Corresponding maximum allowable quantities per control area recognize that an open-use condition is generally more hazardous than a closed-use condition because the oxidizer is more directly exposed to the surrounding environment and can become more readily involved in an incident than if totally confined. The maximum allowable quantities per control area for use are based on the aggregate quantity in both use and storage being within the maximum allowable quantities per control area listed for storage.

Bibliography

The following resource materials are referenced in this chapter or are relevant to the subject matter addressed in this chapter.

DOTn 49 CFR; 100-178-94, *Hazardous Materials Regulations*. Washington, DC: U.S. Department of Transportation, 1994.

Holtzclaw, H. F and W. R. Robinson. *General Chemistry*, 8th ed. Lexington, MA: D. C. Heath and Co., 1988.

IBC–2003, *International Building Code*. Falls Church, VA: International Code Council, 2003.

IMC–2003, *International Mechanical Code*. Falls Church, VA: International Code Council, 2003.

NFPA 13-99, *Installation of Sprinkler Systems*. Quincy, MA: National Fire Protection Association, 1999.

NFPA 50-01, *Bulk Oxygen Systems at Consumer Sites*. Quincy, MA: National Fire Protection Association, 2001.

NFPA 72-99 *National Fire Alarm Code*. Quincy, MA: National Fire Protection Association, 1999

NFPA 430-95, *Code for the Storage of Liquid and Solid Oxidizers*. Quincy, MA: National Fire Protection Association, 1995.

NFPA 704-96, *Identification of the Fire Hazards of Materials*. Quincy, MA: National Fire Protection Association, 1996.

Chapter 41:
Pyrophoric Materials

General Comments

Because of their capacity to ignite spontaneously at low temperatures, pyrophoric materials pose unusual deflagration and detonation hazards to building occupants and fire-fighting personnel. Advance knowledge of the materials present in the building through the issuance of permits and the submittal of Material Safety Data Sheets (MSDS), as noted in Section 407 of the code, are essential for effective preplanning by the fire department. The ability to fight fires involving pyrophoric materials may be somewhat limited since many pyrophoric materials are also highly reactive with water. Interior fire fighting may not be an option unless adequate alternative extinguishing agents are available.

Some flammable liquids may also be considered pyrophoric liquids. Diethylzine, for example, is considered a pyrophoric liquid in accordance with NFPA 49, but has a U.S. Department of Transportation (DOT) classification as a flammable liquid. Since this material is known to ignite spontaneously when exposed to air, the multiple hazards of being both pyrophoric and flammable must be addressed.

Alkali metals, such as sodium, potassium and lithium, are examples of pyrophoric solids. Most of these reactive metals are not found free in nature, but are combined with other elements. Sodium, for example, burns violently and ignites spontaneously when exposed to moist air. Sodium that comes in contact with water may be accompanied by hydrogen explosions.

Diborane and phosphine are pyrophoric gases that must comply with the provisions of this chapter as well as those of Chapter 30. This chapter also contains specific detailed regulations for silane [also known as silicon tetrahydride (SiH_4); CAS No. 7803-62-5], a common but dangerous pyrophoric gas that is also classified as unstable (reactive) (see commentary, Chapter 43).

The proper classification of materials as pyrophoric is essential to providing adequate means of hazard mitigation. Pyrophoric materials, as indicated earlier, may also pose multiple hazards similar to flammable liquids, flammable solids and water-reactive materials.

Purpose

This chapter regulates the hazards associated with pyrophoric materials, which are capable of spontaneously igniting in the air at or below a temperature of 130°F (54°C). Many pyrophoric materials also pose severe flammability or reactivity hazards. This chapter addresses only the hazards associated with pyrophoric materials. Other materials that pose multiple hazards must conform to the requirements of the code with respect to all hazards (see Section 2701.1 of the code). Strict compliance with the provisions of this chapter, along with proper housekeeping and storage arrangements, can reduce the exposure hazards associated with the involvement of pyrophoric materials in a fire or other emergency.

SECTION 4101
GENERAL

4101.1 Scope. The storage and use of pyrophoric materials shall be in accordance with this chapter. Compressed gases shall also comply with Chapter 30.

❖ This chapter details specific requirements for the storage of pyrophoric materials. The requirements of this chapter are intended to complement the general requirements for hazardous materials in Chapter 27. Pyrophoric gases are also regulated by Chapter 30. The classification of pyrophoric materials is based on the definition indicated in Section 2702, which is derived from DOL 29 CFR; Part 1910.1200.

4101.2 Permits. Permits shall be required as set forth in Section 105.6.

❖ The process of issuing permits gives the fire code official an opportunity to carefully evaluate and regulate hazardous operations. Permit applicants should be required to demonstrate that their operations comply with the intent of the code before the permit is issued. See the commentary to Section 105.6 for a general discussion of operations requiring an operational permit, Section 105.6.21 for discussion of specific quantity-based operational permits for the materials regulated in this chapter and Section 105.7 for a general discussion of activities requiring a construction permit. The permit process also notifies the fire department of the need for prefire planning for hazardous property.

SECTION 4102 DEFINITIONS

4102.1 Definition. The following word and term shall, for the purposes of this chapter and as used elsewhere in this code, have the meaning shown herein.

❖ Definitions of terms can help in the understanding and application of the code requirements. The purpose for including the definition that is associated with the subject matter of this chapter is to provide more convenient access to it without having to refer back to Chapter 2. It is important to emphasize that these terms are not exclusively related to this chapter but are applicable everywhere the term is used in the code. For convenience, these terms are also listed in Chapter 2 with a cross reference to this section. The use and application of all defined terms, including those defined in this section, are set forth in Section 201.

PYROPHORIC. A chemical with an autoignition temperature in air, at or below a temperature of 130°F (54°C).

❖ The definition is derived from DOL 29 CFR; Part 1910.1200. Pyrophoric materials, whether in a gas, liquid or solid form, are capable of spontaneous ignition at low temperatures. Pyrophoric materials, regardless of their physical state, may spontaneously ignite when exposed to air at normal or slightly elevated temperatures, even in small quantities. Many pyrophoric materials are also highly reactive with water. While even moist air may increase the possibility of ignition, the application of water may cause an explosive reaction (see commentary, Chapter 44).

SECTION 4103 GENERAL REQUIREMENTS

4103.1 Quantities not exceeding the maximum allowable quantity per control area. The storage and use of pyrophoric materials in amounts not exceeding the maximum allowable quantity per control area indicated in Section 2703.1 shall be in accordance with Sections 2701, 2703, 4101 and 4103.

❖ The provisions of this section complement the requirements of Chapter 27 for structures occupied for the storage, handling or use of pyrophoric materials. The general requirements of Sections 2701 and 2703, in addition to the provisions of this chapter, are fully applicable to the storage and use of organic peroxides.

4103.1.1 Emergency shutoff. Compressed gas systems conveying pyrophoric gases shall be provided with approved emergency shutoff valves that can be activated at each point of use and each source.

❖ To provide control over the flow of pyrophoric gases under emergency conditions, each supply source as well as each point of use of pyrophoric compressed gas must be equipped with an approved emergency shutoff valve. These are either excess flow control valves that regulate the rate of flow of hazardous materials within the piping system, fail-safe valves or other approved types that will operate automatically or manually to give the fire department or other responsible persons the ability to stop the flow of hazardous material in an emergency. The valves must be readily accessible and comply with Section 2703.2.2.1.

4103.2 Quantities exceeding the maximum allowable quantity per control area. The storage and use of pyrophoric materials in amounts exceeding the maximum allowable quantity per control area indicated in Section 2703.1 shall be in accordance with Chapter 27 and this chapter.

❖ This section complements the requirements of Chapter 27 for structures used for the storage, handling or use of pyrophoric materials. The regulations contained in this section assume that the quantity of pyrophoric materials in a given building is in excess of the maximum allowable quantities per control area as established in Section 2703.1, thus classifying the building as a Group H occupancy. The requirements of Chapter 27 apply to the storage and use of pyrophoric materials, in addition to the provisions of this chapter.

SECTION 4104 STORAGE

4104.1 Indoor storage. Indoor storage of pyrophoric materials in amounts exceeding the maximum allowable quantity per control area indicated in Table 2703.1.1(1), shall be in accordance with Sections 2701, 2703, 2704 and this chapter.

The storage of silane gas and gas mixtures with a silane concentration of 2 percent or more by volume, shall be in accordance with Section 4106.

❖ This section regulates the indoor storage of pyrophoric materials (other than silane gas) when they exceed the maximum allowable quantity per control area in buildings or portions of buildings classified as a Group H occupancy. The general and storage provisions of Chapter 27 are applicable, in addition to the requirements of this section. Storage of pyrophoric materials inside of structures must comply with Sections 4104.1.1 through 4104.1.4 to prevent uncontrolled release or exposure to conditions that may result in a fire or explosion. Section 4106 covers silane gas.

4104.1.1 Liquid-tight floor. In addition to the requirements of Section 2704.12, floors of storage areas containing pyrophoric liquids shall be of liquid-tight construction.

❖ Floors and sills of rooms or areas used to contain hazardous material spills must be liquid tight to prevent the flow of liquids to adjoining areas (see commentary, Section 2704.2). The floor surface should be compatible with the pyrophoric materials to be retained and must be noncombustible, as required by Section 2704.12.

4104.1.2 Pyrophoric solids and liquids. Storage of pyrophoric solids and liquids shall be limited to a maximum area of 100 square feet (9.3 m^2) per pile. Storage shall not exceed 5 feet (1524 mm) in height. Individual containers shall not be stacked.

Aisles between storage piles shall be a minimum of 10 feet (3048 mm) in width.

Individual tanks or containers shall not exceed 500 gallons (1893 L) in capacity.

❖ Inside storage restrictions, including pile height, container arrangement and aisle width, are intended to reduce the potential involvement of multiple piles, reduce the exposure hazard to occupants and facilitate fire department access to the storage areas. Inside storage of pyrophoric liquids in tanks or containers is limited to an individual capacity of 500 gallons (1893 L) to reduce the exposure hazard in the event of a single container failure. Prohibiting the stacking of individual containers reduces the likelihood of container failure from stacking stresses. Wide aisles reduce the fire exposure hazard between piles and, in the event of a fire-related collapse of a pile, reduce the likelihood of a "domino" effect that could not only increase fire intensity but also completely block access to pile areas for fire suppression personnel. While not specifically mentioned in this section, inside storage must also be protected as required by Chapter 27, which includes provisions for incompatible material storage, security, signage, control of ignition sources and submittal of a storage plan.

4104.1.3 Pyrophoric gases. Storage of pyrophoric gases shall be in detached buildings where required by Section 2703.8.2.

❖ This section is an important cross reference to Section 2704.14 and Table 2704.14. The table requires that storage of more than 2,000 cubic feet (57 m^3) of pyrophoric gas must be in a detached building complying with Section 2704.14.

Detached structures designed and constructed for the sole purpose of pyrophoric material storage provide the best protection for people and property from fire and explosion. Detached storage structures should be constructed of noncombustible materials to prevent them from becoming involved in an outside fire that may endanger their contents or from becoming involved in a fire themselves, should their contents be ignited (see commentary, Section 2704.14 and Table 2704.14 for further discussion of detached storage requirements).

4104.1.4 Separation from incompatible materials. In addition to the requirements of Section 2703.9.8, indoor storage of pyrophoric materials shall be isolated from incompatible hazardous materials by 1-hour fire barriers with openings protected in accordance with the *International Building Code.*

Exception: Storage in approved hazardous materials storage cabinets constructed in accordance with Section 2703.8.7.

❖ The intent of this section is to separate all incompatible as well as flammable, explosive or other highly reactive materials from the inside storage areas of pyrophoric materials. A 1-hour fire barrier assembly or, as allowed by the exception, an approved hazardous materials storage cabinet can reduce the potential involvement of pyrophoric materials in a fire involving other incompatible hazardous materials. Note that the provisions of Section 4104.1.4 are to be applied in addition to the requirements of Section 2703.9.8.

4104.2 Outdoor storage. Outdoor storage of pyrophoric materials in amounts exceeding the maximum allowable quantity per control area indicated in Table 2703.1.1(3) shall be in accordance with Sections 2701, 2703, 2704 and this chapter.

The storage of silane gas, and gas mixtures with a silane concentration of 2 percent or more by volume, shall be in accordance with Section 4106.

❖ Sections 4104.2.1 and 4104.2.2 are applicable for the outdoor storage of pyrophoric materials in excess of the maximum allowable quantity per outdoor control area listed in Table 2703.1.1(3). The provisions of this section, in addition to the provisions of Sections 2701, 2703 and 2704, are applicable to outdoor areas used for the storage of pyrophoric materials because of the severity of the hazards posed by these materials. Regardless of quantity, silane gas and certain silane gas mixtures must also comply with the provisions of Section 4106.

4104.2.1 Distance from storage to exposures. The separation of pyrophoric solids, liquids and gases from buildings, lot lines, public streets, public alleys, public ways or means of egress shall be in accordance with the following:

1. Solids and liquids. Two times the separation required by Chapter 34 for Class IB flammable liquids.
2. Gases. The location and maximum amount of pyrophoric gas per storage area shall be in accordance with Table 4104.2.1.

❖ To minimize radiant heat transfer and fire spread between stored pyrophoric solid and liquid materials and the listed exposures, the values shown for Class IB flammable liquids in Table 3404.3.4.2 are applied and then doubled. Similarly, Table 4104.2.1 is referenced for the outdoor storage of pyrophoric gases.

TABLE 4104.2.1. See page 41-4.

❖ Table 4104.2.1 establishes the spatial separation distance (or alternative safeguards) that is required for the outside storage of pyrophoric compressed gases when in excess of the outside storage maximum allowable quantity per control area shown in Table 2703.1.1(3).

The required separation distance is intended to minimize radiant heat transfer between storage and exposures. Separation distances provide a measure of protection against the possibility of fire spread in the event of a fire involving either the stored material or another exposure, such as a building on the same or an adjacent lot or a vehicle in the public right-of-way. Section 416.6 of the *International Building Code*® (IBC®) contains additional protection requirements for outside storage in areas that have an overhead roof structure or canopy to protect the materials from the weather.

Note a recognizes that a 2-hour fire barrier wall (constructed in accordance with Section 706 of the IBC, provides an equivalent means of achieving the objectives of spatial separation. Where a separation assembly is used instead of the tabular separation distance, the wall must be arranged to prevent a hazardous accumulation of gas.

4104.2.2 Weather protection. When overhead construction is provided for sheltering outdoor storage areas of pyrophoric materials, the storage areas shall be provided with approved automatic fire-extinguishing system protection.

❖ The general requirements for construction of weather protection roofs over outdoor storage areas are contained in Section 2704.13 of the code and Section 414.6.1 of the IBC; however, due to the hazards of pyrophoric materials, this section is a specific provision applicable to pyrophoric materials that requires roofed-over outdoor storage areas to be sprinklered.

SECTION 4105 USE

4105.1 General. The use of pyrophoric materials in amounts exceeding the maximum allowable quantity per control area indicated in Table 2703.1.1(1) or 2703.1.1(3) shall be in accordance with Sections 2701, 2703, 2705 and this chapter.

❖ This section applies to all indoor and outdoor dispensing, use and handling of pyrophoric material when the amounts being dispensed, used or handled are in excess of the maximum allowable quantities per indoor or outdoor control area as indicated in Table 2703.1.1(1) or 2703.1.1(3), respectively. The administrative, general use, handling and dispensing provisions of Chapter 27 are applicable, in addition to the requirements of this chapter.

Once the maximum allowable quantity per control area of pyrophoric material has been exceeded, indoor areas where materials are being dispensed, used or handled must be located in a building or portion of a building complying with the IBC for a Group H occupancy because of the increased hazards associated with quantity.

Although no occupancy group is assigned to them, outside pyrophoric material use areas require an increased level of regulation when quantities exceed the maximum allowable quantities per outdoor control area listed in Table 2703.1.1(3). Corresponding maximum allowable quantities per control area recognize that an open use condition is generally more hazardous than a closed use condition because the pyrophoric material is more directly exposed to the surrounding environment and can become more readily involved in an incident than if it is totally confined. The maximum allowable quantities per control area for use are based on the aggregate quantity in both use and storage not exceeding the maximum allowable quantities per control area listed for storage.

4105.2 Weather protection. When overhead construction is provided for sheltering of outdoor use areas of pyrophoric materials, the use areas shall be provided with approved automatic fire-extinguishing system protection.

❖ The general requirements for the construction of weather protection roofs over outdoor use areas are contained in Section 2705.3.9 of the code and Section 414.6.1 of the IBC; however, due to the hazards of pyrophoric materials, this section is a specific provision applicable to pyrophoric materials that requires roofed-over outdoor use areas to be sprinklered.

4105.3 Silane gas. The use of silane gas, and gas mixtures with a silane concentration of 2 percent or more by volume, shall be in accordance with Section 4106.

❖ Regardless of quantity, the use of silane gas and certain silane gas mixtures must comply with the provisions of Section 4106, in addition to the requirements of Section 4105.1.

TABLE 4104.2.1
PYROPHORIC GASES—DISTANCE FROM STORAGE TO EXPOSURES[a]

MAXIMUM AMOUNT PER STORAGE AREA (cubic feet)	MINIMUM DISTANCE BETWEEN STORAGE AREAS (feet)	MINIMUM DISTANCE TO LOT LINES OF PROPERTY THAT CAN BE BUILT UPON (feet)	MINIMUM DISTANCE TO PUBLIC STREETS, PUBLIC ALLEYS OR PUBLIC WAYS (feet)	MINIMUM DISTANCE TO BUILDINGS ON THE SAME PROPERTY		
				Nonrated construction or openings within 25 feet	Two-hour construction and no openings within 25 feet	Four-hour construction and no openings within 25 feet
250	5	25	5	5	0	0
2,500	10	50	10	10	5	0
7,500	20	100	20	20	10	0

For SI: 1 foot = 304.8 mm, 1 cubic foot = 0.02832 m^3.

a. The minimum required distances shall be reduced to 5 feet when protective structures having a minimum fire resistance of 2 hours interrupt the line of sight between the container and the exposure. The protective structure shall be at least 5 feet from the exposure. The configuration of the protective structure shall allow natural ventilation to prevent the accumulation of hazardous gas concentrations.

SECTION 4106 SILANE GAS

4106.1 General requirements. The storage and use of silane gas and gas mixtures with a silane concentration of 2 percent or more by volume, in amounts exceeding the maximum allowable quantity per control area indicated in Table 2703.1.1(1) or 2703.1.1(3), shall be in accordance with this section.

❖ This section regulates the indoor and outdoor storage and use of silane gas when in excess of the maximum allowable quantity per control area in buildings or portions of buildings classified as a Group H occupancy or in outdoor control areas. The general and storage provisions of Chapter 27 are applicable, in addition to the requirements of this section. Storage of silane gas must comply with Sections 4106.1.1 through 4106.1.3 to prevent uncontrolled release or exposure to conditions that may result in a fire or explosion. NFPA 318 and CGA P-32 provide additional information on safeguarding silane gas and its mixtures.

4106.1.1 Building construction. Indoor storage and use of silane gas shall be within a room or building conforming to the *International Building Code.*

❖ Because of the violent potential of silane compressed gas, the structure or room used for storing the gas must be of noncombustible construction and meet the requirements of Section 307 of the IBC so that it will not be a contributing factor in a fire and will be isolated from other portions of the building. Structures or portions of structures containing silane compressed gas exceeding the maximum allowable quantity per control area as indicated in Table 2703.1.1(1) are classified as a Group H occupancy and may be required to be detached structures in accordance with Section 2704.14.

4106.1.2 Flow control. Compressed gas containers, cylinders and tanks containing silane gas, and gas mixtures with a silane concentration of 2 percent or more by volume, shall be equipped with reduced flow valves equipped with restrictive-flow orifices not exceeding 0.010 inch (0.254 mm) in diameter. The presence of the restrictive flow orifice shall be indicated on the valve and on the container, cylinder or tank by means of a label placed at a prominent location by the manufacturer.

Exceptions:

1. Manufacturing and filling facilities where silane is produced or mixed and stored prior to sale.
2. Outdoor installations consisting of permanently mounted cylinders connected to a manifold, provided that the outlet connection from the manifold is equipped with a restrictive flow orifice not exceeding 0.125 inch (3.175 mm) in diameter and the setback distance to exposures is not less than 40 feet (12 192 mm). Footnote a of Table 4104.2.1 shall not apply.

❖ Reduced-flow valves, as defined in Section 3702.1, must be installed on vessels containing silane gas to provide a positive means of limiting the flow of gas under normal use conditions. This type of valve is used to control the maximum rate of release of gas from cylinders or portable or stationary tanks. The valves should be marked by the manufacturer to indicate the associated maximum design flow rate. It is possible that their use in connection with silane may reduce the likelihood of silane explosions through maintaining the flow speed of the silane molecules from an opening in a silane container at a level lower than the flame propagation rate of silane, although this is still being investigated.

Exception 1 recognizes the higher level of design safety in silane manufacturing and cylinder facilities.

Exception 2 recognizes the decreased hazard of outdoor installations where silane gas can be readily dispersed.

4106.1.3 Valves. Container, cylinder and tank valves shall be constructed of stainless steel or other approved materials. Valves shall be equipped with outlet fittings in accordance with CGA V-1.

❖ To reduce the possibility of rust or corrosion, valves on silane gas cylinders, containers and tanks must be constructed of corrosion-resistant materials, such as stainless steel. Fittings for these valves must comply with CGA V-1.

4106.2 Indoor storage. Indoor storage of silane gas, and gas mixtures with a silane concentration of 2 percent or more by volume, shall be in accordance with Section 4104.1 and Sections 4106.2.1 through 4106.2.3.

❖ Storage of silane gas inside structures must comply with Sections 4104.1 and 4106.2.1 through 4106.2.3 to prevent uncontrolled release or exposure to conditions that may result in a fire or explosion.

4106.2.1 Fire protection. When automatic fire-extinguishing systems are required, automatic sprinkler systems shall be used.

❖ This section makes it clear that the fire protection medium of choice for the protection of indoor silane gas storage areas is water effectively applied through automatic sprinklers. This section does not require the installation of sprinklers but rather states that if an automatic fire-extinguishing system is required by some other code provision, alternative fire-extinguishing systems (such as those regulated by Section 904) may not be used as an alternative to sprinklers.

4106.2.2 Exhausted enclosures or gas cabinets. When provided, exhausted enclosures and gas cabinets shall be constructed as follows:

1. Exhausted enclosures and gas cabinets shall be in accordance with Sections 2703.8.5 and 2703.8.6.
2. Exhausted enclosures and gas cabinets shall be internally sprinklered.
3. The velocity of ventilation across unwelded fittings and connections on the piping system shall not be less than 200 linear feet per minute (102 m/s).

4. The average velocity at the face of the access ports or windows in the gas cabinet shall not be less than 200 linear feet per minute (102 m/s) with a minimum velocity of 150 linear feet per minute (76 m/s) at any point of the access port or window.

❖ This section provides the construction and performance requirements for exhausted enclosures and gas cabinets when used for the storage of more than 50 cubic feet (1 m^3) of silane gas inside a building. Whereas gas cabinets must be fully enclosed and equipped with self-closing doors, exhausted enclosures are typically open fronted and lend themselves to small-scale ventilators, such as fume hoods found in chemical laboratories.

Though silane is not a health hazard like toxic and highly toxic gases, the level of protection afforded by Item 1 of this section is the same as required for those gases in Chapter 37 (see commentary, Chapter 37) because of the unpredictable nature of silane gas. Item 2 intends to reduce the likelihood that an incident within an exhausted enclosure or gas cabinet will spread to the building. Items 3 and 4 require that both exhausted enclosures and gas cabinets must be operated at a negative pressure with respect to the surrounding area to capture and dispose of any fugitive silane gas.

4106.2.3 Emergency power. The ventilation system shall be provided with an automatic emergency power source in accordance with Section 604 and designed to operate at full capacity.

❖ This section requires exhaust ventilation systems used in connection with silane gas storage to be connected to an emergency electrical system in accordance with Section 604. Without emergency power, the exhaust ventilation system would be rendered inoperative if a power failure or other electrical system failure occurred, which could lead to the escape of silane gas.

4106.3 Outdoor storage. Outdoor storage of silane gas, and gas mixtures with a silane concentration of 2 percent or more by volume, shall be in accordance with Section 4104.2 and Sections 4106.3.1 through 4106.3.3.

❖ Storage of silane gas in outdoor control areas must comply with Sections 4106.3.1 through 4106.3.3 to prevent uncontrolled release or exposure to conditions that may result in a fire or explosion. In addition to the requirements of this chapter, silane gas is subject to the provisions of Chapter 30.

4106.3.1 Volume. The maximum volume for each nest shall not exceed 10,000 cubic feet (283.2 m^3) of gas.

❖ To limit the amount of material involved in a single fire, silane gas is limited to cylinder nests containing not more than 10,000 cubic feet (283.2 m^3) of gas (see commentary, Sections 3002 for the definition of "Nesting" and Section 3003.3.3).

4106.3.2 Aisles. Storage nests shall be separated by aisles a minimum of 6 feet (1829 mm) in width.

❖ Aisle widths that are equal to or greater than the height of a silane gas cylinder reduce the fire exposure hazard between cylinder nests and, in the event of a fire-related collapse of a nest, reduce the likelihood of a "domino" effect that could not only increase fire intensity or result in cylinder valve damage but also completely block access to nest areas for fire suppression personnel.

4106.3.3 Separation. Storage shall be located a minimum of 25 feet (7620 mm) from lot lines, public streets, public alleys, public ways, means of egress or buildings.

❖ The clearances required by this section are intended to reduce the hazard to nearby properties and people in the event of a fire or explosion in an outdoor area used for the storage of silane gas.

4106.3.4 Weather protection. The clear height of overhead construction provided for sheltering of outdoor storage shall not be less than 12 feet (3658 mm).

❖ The general requirements for construction of weather protection roofs over outdoor storage areas are contained in Section 2704.13 of the code and Section 414.6.1 of the IBC. Due to the hazards of silane gas, however, this section is a specific provision applicable to silane gas and silane gas mixtures that requires a minimum height of 12 feet (3658 mm) for the weather protection roof structure. The intent of this requirement is to enhance the dissipation of silane gas by natural ventilation in the event of a leak thereby reducing the likelihood of the gas pocketing that could result in delayed and unexpected detonation of the accumulated gas.

4106.4 Indoor use and dispensing. The indoor use and dispensing of silane gas and gas mixtures with a silane concentration of 2 percent or more by volume, in amounts exceeding the maximum allowable quantity per control area indicated in Table 2703.1.1(1) shall be in accordance with Sections 4105 and this section.

❖ This section applies to all indoor dispensing, use and handling of silane gas when the amounts are in excess of the maximum allowable quantities per control area of pyrophoric gas as indicated in Table 2703.1.1(1).

Once the maximum allowable quantity per control area of silane gas has been exceeded, indoor areas where the gas is being dispensed, used or handled must be located in a building or portion of a building complying with the IBC for a Group H occupancy because of the increased hazards associated with quantity. The maximum allowable quantities per control area for use are based on the aggregate quantity in both use and storage not exceeding the exempt amount listed for storage.

4106.4.1 Exhausted enclosures or gas cabinets. When provided, exhausted enclosures and gas cabinets shall be installed in accordance with Section 4106.2.2.

❖ The exhausted enclosure or gas cabinet requirements applicable to the storage of silane gas are also applicable to its use and dispensing (see commentary, Section 4106.2.2).

4106.4.2 Remote manual shutdown. Remote manual shutdown of process gas flow shall be provided outside each gas cabinet.

❖ To provide control over the flow of silane gas under emergency conditions, each process supply source must have a remote manual shutoff valve of an approved type to provide the fire department or other responsible persons with the ability to stop the flow of hazardous material in an emergency. These valves must be installed outside the gas cabinet in an approved, readily accessible location and comply with Section 2703.2.2.1.

4106.4.3 Emergency power. The ventilation system shall be provided with an approved automatic emergency power source in accordance with Section 604 and designed to operate at full capacity.

❖ This section requires that exhaust ventilation systems used in connection with silane gas use and dispensing be connected to an emergency electrical system in accordance with Section 604. Without emergency power, the exhaust ventilation system would not work if a power failure or other electrical system failure occurred, which could lead to the escape of silane gas.

4106.4.4 Purge panels. Automated purge panels shall be provided.

❖ Where silane gas is in use, the gas installation must include means to automatically purge the area between the cylinder valve and the regulator with an inert gas prior to breaking connections for maintenance or cylinder change. Purging silane gas piping and controls with an inert gas is an extremely important factor when changing cylinders because it will: (1) remove air and moisture from the system before process gas is introduced, which preserves the purity of the gas and promotes system reliability; (2) remove process gas from the system before the system is opened to the atmosphere, thereby reducing the risk of personnel exposure to the gas; (3) prevent the release of silane gas during cylinder removal and prevent air from entering the system when the new cylinder is connected; (4) prevent valve lock-up and (5) prevent "regulator creep," which allows full cylinder pressure to be transferred to the low pressure side of the regulator.

4106.4.4.1 Purge gases. Purging of piping and controls located in gas cabinets or exhausted enclosures shall only be performed using a dedicated inert gas supply that is designed to prevent silane from entering the inert gas supply. The use of nondedicated systems or portions of piping systems is allowed on portions of the venting system that are continuously vented to atmosphere. Devices that could interrupt the continuous flow of purge gas to the atmosphere shall be prohibited.

Exception: Manufacturing and filling facilities where silane is produced or mixed.

❖ To aid in the confinement of silane gas, inert gases used to purge systems not continuously vented to the atmosphere must be used solely for that purpose and not connected to other apparatus. A check valve must be installed in the purge gas line to prevent backflow of process gas to the purge gas supply source or the system design must be specifically approved as meeting that performance objective. Where a system is continuously vented to the atmosphere, small quantities of silane that may find their way into purge gas piping do not pose a significant hazard and, therefore, the purge gas supply need not be dedicated and protected from backflow. The exception recognizes the higher level of design safety in silane manufacturing and cylinder facilities.

4106.4.4.2 Venting. Gas vent headers or individual purge panel vent lines shall have a continuous flow of inert gas. The inert gas shall be introduced upstream of the first vent or exhaust connection to the header.

❖ This section requires that, in order to safely and properly dispose of purged silane gas from process piping, inert purge gas must flow constantly in the purge panel vent lines or in the gas vent header where multiple lines vent to a header.

4106.4.4.3 Purging operations. Purging operations shall be performed by means ensuring complete purging of the piping and control system before the system is opened to the atmosphere.

❖ Reliable purging procedures are essential to the safe use of silane gas systems. Although all purging methods must be approved, there are several methodologies in common use in the industry.

The continuous flow method uses a continuous flow of the inert purge gas to remove the silane gas and prevent air or moisture from entering the system. This method is efficient at purging the gas directly in the flow path, but will not efficiently purge dead-end lines, such as pressure gauge lines.

The dilution method uses a sequence of pressurization followed by depressurization and venting. This sequence is repeated several times and is very effective at diluting the silane gas, efficiently replacing it with the inert purge gas and preventing air or moisture from entering the piping. This method is also effective in removing gas from dead-end piping, such as pressure gauge lines.

4106.5 Outdoor use and dispensing. The outdoor use and dispensing of silane gas, and gas mixtures with a silane concentration of 2 percent or more by volume, exceeding the maximum allowable quantity per control area indicated in Table

2703.1.1(3) shall be in accordance with Sections 4105, 4106.4 and 4106.5.1.

❖ Although no occupancy group is assigned to them, increased regulation is required in outside silane gas use areas when quantities exceed the maximum allowable quantities per outdoor control area as listed in Table 2703.1.1(3). The maximum allowable quantities per control area for use are based on the aggregate quantity in both use and storage not exceeding the maximum allowable quantities per control area listed for storage.

4106.5.1 Outdoor use weather protection. When overhead construction is provided for sheltering outdoor use areas containing silane gas, or gas mixtures with a silane concentration of 2 percent or more by volume, the use areas shall be provided with approved automatic fire-extinguishing system protection.

❖ The general requirements for the construction of weather protection roofs over outdoor use areas are contained in Section 2705.3.9 of the code and Section 414.6.1 of the IBC. Due to the hazards of silane gas, however, this section is a specific provision applicable to silane gas and silane gas mixtures that requires roofed-over outdoor use areas to be sprinklered.

Bibliography

The following resource materials are referenced in this chapter or are relevant to the subject matter addressed in this chapter.

CGA P-32-00, *Safe Storage and Handling of Silane and Silane Mixtures*. Arlington, VA: Compressed Gas Association, 2000.

CGA V1-01, *Compressed Gas Cylinder Valve Outlet and Inlet Connections*. Arlington, VA: Compressed Gas Association, 2001.

DOL 29 CFR; Part 1910.1200-99, *Hazard Communication*. Washington, DC: U.S. Department of Labor, 1999.

Fire Protection Guide on Hazardous Materials, 13th ed. Quincy, MA: National Fire Protection Association, 2001.

IBC-2003, *International Building Code*. Falls Church, VA: International Code Council, 2003.

NFPA 318-02, *Protection of Semiconductor Fabrication Facilities*. Quincy, MA: National Fire Protection Association, 2002.

NFPA 704-96, *Identification of the Fire Hazards of Materials*. Quincy, MA: National Fire Protection Association, 1996.

Chapter 42: Pyroxylin (Cellulose Nitrate) Plastics

General Comments

Pyroxylin (cellulose nitrate) plastic is formulated from a combination of cellulose, nitric acid and sulfuric acid. The resulting compound, also known as pyroxylin or nitrocellulose, is an unstable and extremely combustible plastic. Once exposed to elevated temperatures, pyroxylin (cellulose nitrate) plastic is subject to spontaneous ignition. The products of combustion of pyroxylin (cellulose nitrate) plastic are extremely toxic because nitrogen oxides are produced at the elevated temperatures reached when this material burns rapidly.

Pyroxylin (cellulose nitrate) plastic, also called gun cotton, was discovered to be a powerful explosive and replaced common gunpowder as the explosive charge in the ammunition for rifles and artillery in World War I. Other early uses included being a replacement material for elephant tusk ivory in billiard balls (which had a tendency to explode on a hard break), as casino dice and as the laminating adhesive in early forms of automotive safety glass. It has also been used as a film base for photographic film, as well as to encase documents, book leaves, etc.; however, its flammability and the fact that it causes severe deterioration of the materials it supposedly protects has prevented its widespread use in preservation work.

Many of these products are no longer manufactured from pyroxylin (cellulose nitrate) plastic. Although it has been replaced in many applications by safer plastics, it is still used in many industrial applications such as shoe heels, housewares and lacquers.

Raw pyroxylin (cellulose nitrate) plastic is usually shipped in drums and covered with water or another solvent, usually alcohol.

Water-wet pyroxylin plastic presents the least hazard, while alcohol-wet and other types of solvent-wet pyroxylin plastic possess hazards similar to those of the solvent involved. Because nitrocellulose becomes increasingly unstable as temperatures increase and may ignite easily from frictional heat, drums must never be pushed, rolled or dragged across the floor. Pyroxylin (cellulose nitrate) plastic is especially susceptible to ignition, burns vigorously once ignited (at a rate approximately 15 times that of a comparable mass of paper), produces toxic nitrogen oxides and will burn in the absence of oxygen. Additional information is available in NFPA 42.

Purpose

This chapter addresses the significant hazards associated with pyroxylin (cellulose nitrate) plastics, which are the most dangerous and unstable of all plastic compounds. The chemically bound oxygen in their structure permits them to burn vigorously in the absence of atmospheric oxygen. Although these compounds produce approximately the same amount of energy when they burn as paper, pyroxylin (cellulose nitrate) plastics burn at a rate 15 times greater than comparable common combustibles. When burning, these materials release highly flammable and toxic combustion byproducts. Consequently, pyroxylin (cellulose nitrate) plastic fires are very difficult to control and must be virtually flooded to extinguish them. Even storage cabinets, which may be used only for temporary (e.g., overnight) storage of pyroxylin (cellulose nitrate) plastic films, must be sprinklered. Strict compliance with the provisions of this chapter, along with proper housekeeping and storage arrangements, help to reduce the hazards associated with pyroxylin (cellulose nitrate) plastics in a fire or other emergencies.

SECTION 4201 GENERAL

4201.1 Scope. This chapter shall apply to the storage and handling of plastic substances, materials or compounds with cellulose nitrate as a base, by whatever name known, in the form of blocks, sheets, tubes or fabricated shapes.

Cellulose nitrate motion picture film shall comply with the requirements of Section 306.

❖ This section establishes that any raw materials or finished products that contain any amount of pyroxylin (cellulose nitrate) plastic are subject to the regulations of this chapter. This is true even if the materials are called something other than pyroxylin (cellulose nitrate) plastic, such as xyloidin, collodion, photocotton, pyrocollodion, guncotton or smokeless powder. Compliance with the applicable provisions of Chapter 27 and Section 407, which include, but are not limited to, Material Saftey Data Sheets (MSDS) submittal, hazard identification signs and labeling provisions, is also required. Pyroxylin (cellulose nitrate) plastic in the form of motion picture film is regulated by Section 306 as noted in this section.

4201.2 Permits. Permits shall be required as set forth in Section 105.6.

❖ The process of issuing permits gives the fire code official an opportunity to carefully evaluate and regulate hazardous operations. Permit applicants should be required to demonstrate that their operations comply with the intent of the code before the permit is issued. See the commentary to Section 105.6 for a general discussion of operations requiring an operational permit, Section 105.6.38 for discussion of specific quantity-based operational permits for the materials regulated in this chapter and Section 105.7 for a general discussion of activities requiring a construction permit. The permit process also notifies the fire department of the need for prefire planning for hazardous property.

SECTION 4202 DEFINITIONS

4202.1 Terms defined in Chapter 2. Words and terms used in this chapter and defined in Chapter 2 shall have the meanings ascribed to them as defined therein.

❖ This section directs the code user to Chapter 2 of the code for the proper application of the terms used in this chapter. These terms may be defined in Chapter 2 of the code, in another *International Code*® as indicated in Section 201.3 or may retain their ordinary (dictionary) meaning (see also commentary, Sections 201.1 through 201.4).

SECTION 4203 GENERAL REQUIREMENTS

4203.1 Displays. Cellulose nitrate (pyroxylin) plastic articles are allowed to be placed on tables not more than 3 feet (914 mm) wide and 10 feet (3048 mm) long. Tables shall be spaced at least 3 feet (914 mm) apart. Where articles are displayed on counters, they shall be arranged in a like manner.

❖ Because of pyroxylin (cellulose nitrate) plastic's hazardous instability and easily ignitable composition, its exposure to the ambient environment in public venues must be strictly controlled to minimize the amount of material exposed to potential ignition sources and the number of persons exposed to its hazards, should it be ignited. This section places strict limits on the size of an individual display surface and requires that multiple display surfaces be separated from one another by minimum 3-foot (914 mm) aisles to not only limit the number of tables involved in a single incident but also to provide free egress circulation in the event of an emergency.

4203.2 Space under tables. Spaces underneath tables shall be kept free from storage of any kind and accumulation of paper, refuse and other combustible material.

❖ The importance of good housekeeping in reducing the exposure of pyroxylin (cellulose nitrate) plastic display objects to fire exposure cannot be overstated. Prohibiting the storage or accumulation of any combustible material beneath tables on which pyroxylin (cellulose nitrate) plastic display objects are arrayed will reduce the display's exposure to fire in the event such stored materials were to be ignited.

4203.3 Location. Sales or display tables shall be so located that in the event of a fire at the table, the table will not interfere with free means of egress from the room in at least one direction.

❖ Display tables placed in corridors or aisles may unduly expose egressing persons to rapidly burning pyroxylin (cellulose nitrate) plastic objects, reduce required egress capacity or require substantial effort to remove the tables or negotiate the reduced egress path quickly. Anything that slows egress may also impede access, particularly to fire fighters who may be called to rescue occupants or fight the fire.

4203.4 Lighting. Lighting shall not be located directly above cellulose nitrate (pyroxylin) plastic material, unless provided with a suitable guard to prevent heated particles from falling.

❖ Because it is pyroxylin (cellulose nitrate) plastic's nature to rapidly deteriorate and ignite when exposed to heat, the hazard posed by light fixtures located over display tables must be reduced in an approved manner. Ideally, the display tables should be located where hot particles from a broken light bulb cannot fall on the pyroxylin (cellulose nitrate) plastic materials. When that is not possible, fixtures must be enclosed to prevent debris from a broken bulb from falling out of the fixture.

SECTION 4204 STORAGE AND HANDLING

4204.1 Raw material. Raw cellulose nitrate (pyroxylin) plastic material in a Group F building shall be stored and handled in accordance with Sections 4204.1.1 through 4204.1.7.

❖ Raw pyroxylin (cellulose nitrate) plastic materials present increased hazards when they are involved in manufacturing processes. To reduce those increased hazards, the provisions of Sections 4204.1.1 through 4204.1.7 are applicable to buildings used for the fabrication of items containing pyroxylin (cellulose nitrate) plastic.

4204.1.1 Storage of incoming material. Where raw material in excess of 25 pounds (11 kg) is received in a building or fire area, an approved vented cabinet or approved vented vault equipped with an approved automatic sprinkler system shall be provided for the storage of material.

❖ The amount of heat liberated by pyroxylin (cellulose nitrate) plastic when burning and the speed with which it burns make conventional extinguishment virtually impossible. This, coupled with the poisonous combustion gases, produce conditions that require material confinement and rapid fire suppression.

This section requires that quantities of pyroxylin (cellulose nitrate) plastic in excess of 25 pounds (11 kg) be confined to an approved storage cabinet complying with Section 2703.8.7 or an approved vault constructed in accordance with the *International Building Code*® (IBC®), while typical pyroxylin (cellulose nitrate) plastic storage cabinets are limited to a 30-cubic-foot (.8 m^3) volume, while typical pyroxylin (cellulose nitrate) plastic vault construction has a fire-resistance rating of 4 hours because of the violent nature of the stored material.

Cabinets and vaults must be vented to the outdoors to relieve the pressure buildup resulting from the rapid decomposition of the stored material and equipped with automatic sprinklers designed in accordance with Section 903.6.1. For additional cabinet and vault information, see NFPA 42.

4204.1.2 Capacity limitations. Cabinets in any one workroom shall not contain more than 1,000 pounds (454 kg) of raw material. Each cabinet shall not contain more than 500 pounds (227 kg). Each compartment shall not contain more than 250 pounds (114 kg).

❖ To provide a reasonable quantity of material to work with in the manufacturing process while minimizing the quantity of pyroxylin (cellulose nitrate) plastic exposed to a single fire, this section limits the aggregate and per-cabinet and cabinet compartment quantities.

4204.1.3 Storage of additional material. Raw material in excess of that allowed by Section 4204.1.2 shall be kept in vented vaults not exceeding 1,500-cubic-foot capacity (43 m^3) of total vault space, and with approved construction, venting and sprinkler protection.

❖ Quantities of pyroxylin (cellulose nitrate) plastic in excess of the 1,000 pounds (454 kg) allowed by Section 4204.1.2 to be readily accessible in cabinets in work spaces must be stored in approved sprinklered and vented vaults no larger than 1,500 cubic feet (43 m^3) each in volume [approximately 10 feet by 15 feet by 10 feet (3048 mm by 4572 mm by 3048 mm) high] to limit the amount of material involved in a fire.

4204.1.4 Heat sources. Cellulose nitrate (pyroxylin) plastic shall not be stored within 2 feet (610 mm) of heat-producing appliances, steam pipes, radiators or chimneys.

❖ Because pyroxylin (cellulose nitrate) plastic rapidly decomposes and will spontaneously burst into flame when in contact with heated objects in the presence of sufficient oxygen, it is extremely important to reduce the hazard by keeping stored materials well away from the common sources of heat noted in this section. Storing pyroxylin (cellulose nitrate) plastic in an unheated room would be an ideal strategy for this hazard.

4204.1.5 Accumulation of material. In factories manufacturing articles of cellulose nitrate (pyroxylin) plastics, approved sprinklered and vented cabinets, vaults or storage rooms shall be provided to prevent the accumulation in workrooms of raw stock in process or finished articles.

❖ This section reinforces the provisions of Sections 4204.1.1, 4204.1.2 and 4204.1.3 and intends to limit the amount of raw pyroxylin (cellulose nitrate) plastic stock or finished product exposed to the hazards of manufacturing operations by mandating that all pyroxylin (cellulose nitrate) plastic materials, whether raw stock or finished product, be stored in sprinklered, vented vaults or cabinets.

4204.1.6 Operators. In workrooms of cellulose nitrate (pyroxylin) plastic factories, operators shall not be stationed closer together than 3 feet (914 mm), and the amount of material per operator shall not exceed one-shift's supply and shall be limited to the capacity of three tote boxes, including material awaiting removal or use.

❖ To isolate the manufacturing hazard of labor, that is, the hazards involved in the manipulation of materials in the manufacturing, packing or shipping of finished products, as well as the in-process quantity of pyroxylin (cellulose nitrate) plastic (either raw or finished product), this section requires that workstations in the factory have a clearance from one another of no less than 3 feet (914 mm). Depending on the specific articles being manufactured, the quantity of pyroxylin raw material or finished product allowed at any workstation may vary but cannot exceed the amount needed for a single work shift or three tote boxes, whichever is the greater amount. The term "tote box," although not defined in the code, describes portable containers used to transport raw pyroxylin (cellulose nitrate) plastic or finished pyroxylin (cellulose nitrate) plastic products between a central storage vault and a work station. Though not a referenced standard in the code, NFPA 42 contains additional information on the use and construction of tote boxes.

4204.1.7 Waste material. Waste cellulose nitrate (pyroxylin) plastic materials such as shavings, chips, turnings, sawdust, edgings and trimmings shall be kept under water in metal receptacles until removed from the premises.

❖ The manufacturing process may involve shaving, scraping, sanding or cutting of pyroxylin (cellulose nitrate) plastic, all of which produce finely divided scrap material that presents much more surface area that is susceptible to the hazards of ignition than the larger, more dense work pieces from which they came. To reduce the measurable increase in the hazard, this section requires extraordinary collection and storage safeguards for such scrap material in water-filled metal containers. These containers should be removed from the work area at the end of each shift for disposal.

4204.2 Fire protection. The manufacture or storage of articles of cellulose nitrate (pyroxylin) plastic in quantities exceeding 100 pounds (45 kg) shall be located in a building or portion thereof equipped throughout with an approved automatic sprinkler system in accordance with Section 903.3.1.1.

❖ Because pyroxylin (cellulose nitrate) plastics pose unusual and substantial fire risks, burn at a rate 15 times greater than comparable common combustibles and, when burning, release highly flammable and toxic combustion byproducts, fires involving these materials are very difficult to control. This section mirrors the provisions of Section 903.2.4.3 and specifies a sprinkler threshold quantity of 100 pounds (45 kg); however, the need for additional fire protection should be considered for pyroxylin (cellulose nitrate) plastics in any amount.

4204.3 Sources of ignition. Sources of ignition shall not be located in rooms in which cellulose nitrate (pyroxylin) plastic in excess of 25 pounds (11 kg) is handled or stored.

❖ Consistent with the hazards presented by pyroxylin (cellulose nitrate) plastic, this section prohibits the presence of recognized ignition sources in storage and manufacturing areas where more than 25 pounds (11 kg) of material are present to minimize the amount of material exposed.

4204.4 Heating. Rooms in which cellulose nitrate (pyroxylin) plastic is handled or stored shall be heated by low-pressure steam or hot water radiators.

❖ This section reinforces the provisions of Sections 4204.1.4 and 4204.3 by mandating indirect heat only by either low-pressure steam or hot water in rooms where pyroxylin (cellulose nitrate) plastic is stored or processed. In this way, the hazards of open-flame, fuel-fired heat sources do not exist and cannot expose the pyroxylin (cellulose nitrate) plastic to ignition.

Bibliography

The following resource materials are referenced in this chapter or are relevant to the subject matter addressed in this chapter.

Barrow, William J. *Manuscripts and Documents: Their Deterioration and Restoration*. 2nd ed. Charlottesville, VA: University of Virginia Press, 1972.

IBC-2003, *International Building Code*. Falls Church, VA: International Code Council, 2003.

NFPA 40-97, *Storage and Handling of Pyroxylin (cellulose nitrate) Plastic Motion Picture Film*. Quincy, MA: National Fire Protection Association, 1997.

NFPA 42-02, *Storage of Pyroxylin Plastics*. Quincy, MA: National Fire Protection Association, 2002.

"Pyroxylin (cellulose nitrate) plastic." *Special Interest Bulletin No. 49*. New York: American Insurance Association, 1971.

Chapter 43: Unstable (Reactive) Materials

General Comments

This chapter regulates the storage of unstable (reactive) materials. Unstable (reactive) materials may react spontaneously with themselves, other chemicals or when exposed to light, heat, cold, moisture, air or physical shock. These materials may burn, explode, polymerize or decompose to form toxic materials. Unstable (reactive) materials are used in a variety of industrial applications, including food processing and the manufacture of plastics, textiles, fireworks, explosives, rocket propellants, special fuel systems and dyes. These hazardous materials may also be found in the preparation of certain medicines or fumigants. They pose unusual and substantial risks to the general public, fire fighters and emergency response personnel under a variety of conditions.

Advance knowledge of the materials being stored in structures through the permit process and the submittal of Material Safety Data Sheets (MSDS), in accordance with Section 407, is essential for adequate control of the hazard and for prefire planning by the fire department. This is especially important for unstable (reactive) liquid and solid materials because of their unpredictable and violent nature.

During a mishap, a variety of hazards can be created by these materials, including explosions, violent decomposition, toxic products of combustion, toxic vapors, corrosion injuries, poisoning or violent polymerization.

Spills or leaks of unstable materials may be handled by absorbing them with inert materials and them removal or, in cases of larger spills, flushing with large volumes of water if no additional hazard will be created.

Fires in structures or vehicles storing or carrying unstable chemicals should be approached with extreme caution to avoid injury from an explosion or tank BLEVE (Boiling Liquid Expanding Vapor Explosion). Unless and until specific and reliable information on the stored materials is available to fire command personnel, the scene should be treated as though the incident involves explosives, and personnel and apparatus should be kept well away.

Purpose

This chapter addresses the hazards of unstable (reactive) liquid and solid materials as well as unstable (reactive) compressed gas materials. In addition to their unstable reactivity, these materials may pose other hazards, such as toxicity, corrosivity, explosivity, flammability or oxidizing potential. This chapter, however, intends to address those materials whose primary hazard is unstable reactivity. Materials that pose multiple hazards must conform to the requirements of the code with respect to all hazards (see commentary, Section 2701.1). Strict compliance with the provisions of this chapter, along with proper housekeeping and storage arrangements, help to reduce the exposure hazards associated with unstable (reactive) materials in a fire or other emergency.

SECTION 4301 GENERAL

4301.1 Scope. The storage and use of unstable (reactive) materials shall be in accordance with this chapter. Compressed gases shall also comply with Chapter 30.

Exceptions:

1. Display and storage in Group M and storage in Group S occupancies complying with Section 2703.11.
2. Detonable unstable (reactive) materials shall be stored in accordance with Chapter 33.

❖ This chapter details specific requirements for the storage of unstable (reactive) materials. The requirements of this chapter are intended to complement the hazardous materials general storage requirements of Chapter 27. Hazardous gases are also regulated by Chapter 30. The chemicals classified as unstable (reactive) are based on the definition in Section 4302.1, which is derived from NFPA 704.

Exception 1 makes it clear that this chapter does not apply to unstable (reactive) materials stored and displayed in Group M occupancies or stored in Group S occupancies. Section 2703.11 covers those occupancy groups. Also in that section, Class 4 unstable (reactive) materials are prohibited in Group M (mercantile) occupancies because of their detonable hazard potential. Quantities of all other classes are limited on a per-control-area basis to limit exposure to people and property (see commentary, Section 2703.11). This exception would permit an increase above the exempt amounts per control area indicated in Table 2703.1.1(1) for unstable (reactive) materials while still maintaining a mercantile or storage occupancy group classification.

Exception 2 recognizes the violent, explosive nature of detonable unstable (reactive) materials and directs that they be stored in accordance with the requirements for explosives and blasting agents in Chapter 33 to provide the safeguards commensurate with their hazards.

4301.2 Permits. Permits shall be required as set forth in Section 105.6.

❖ The process of issuing permits gives the fire code official an opportunity to carefully evaluate and regulate hazardous operations. Permit applicants should be required to demonstrate that their operations comply with the intent of the code before the permit is issued. See the commentary to Section 105.6 for a general discussion of operations requiring an operational permit, Section 105.6.21 for discussion of specific quantity-based operational permits for the materials regulated in this chapter and Section 105.7 for a general discussion of activities requiring a construction permit. The permit process also notifies the fire department of the need for prefire planning for hazardous property.

SECTION 4302 DEFINITIONS

4302.1 Definition. The following word and term shall, for the purposes of this chapter and as used elsewhere in this code, have the meaning shown herein.

❖ Definitions of terms can help in the understanding and application of the code requirements. The purpose for including here the definition that is associated with the subject matter of this chapter is to provide more convenient access to it without having to refer back to Chapter 2. It is important to emphasize that this term is not exclusively related to this chapter but is applicable everywhere the term is used in the code. For convenience, the term is also listed in Chapter 2 with a cross reference to this section. The use and application of all defined terms, including the one defined in this section, are set forth in Section 201.

UNSTABLE (REACTIVE) MATERIAL. A material, other than an explosive, which in the pure state or as commercially produced, will vigorously polymerize, decompose, condense or become self-reactive and undergo other violent chemical changes, including explosion, when exposed to heat, friction or shock, or in the absence of an inhibitor, or in the presence of contaminants, or in contact with incompatible materials. Unstable (reactive) materials are subdivided as follows:

Class 4. Materials that in themselves are readily capable of detonation or explosive decomposition or explosive reaction at normal temperatures and pressures. This class includes materials that are sensitive to mechanical or localized thermal shock at normal temperatures and pressures.

Class 3. Materials that in themselves are capable of detonation or of explosive decomposition or explosive reaction but which require a strong initiating source or which must be heated under confinement before initiation. This class includes materials that are sensitive to thermal or mechanical shock at elevated temperatures and pressures.

Class 2. Materials that in themselves are normally unstable and readily undergo violent chemical change but do not detonate. This class includes materials that can undergo chemical change with rapid release of energy at normal temperatures and pressures, and that can undergo violent chemical change at elevated temperatures and pressures.

Class 1. Materials that in themselves are normally stable but which can become unstable at elevated temperatures and pressure.

❖ The definition of an "Unstable (reactive) material" is based on NFPA 704. The different classes of unstable (reactive) material reflect the degree of susceptibility of the materials to release energy. Unstable (reactive) materials polymerize, decompose or become self-reactive when exposed to heat, air, moisture, pressure or shock. Separation from incompatible materials is essential to minimizing the hazards. Examples of unstable (reactive) materials include acetaldheyde, ammonium nitrate, ethylene oxide, hydrogen cyanide, nitromethane, perchloric acid, sodium perchlorate, vinyl acetate and acetic acid.

SECTION 4303 GENERAL REQUIREMENTS

4303.1 Quantities not exceeding the maximum allowable quantity per control area. Quantities of unstable (reactive) materials not exceeding the maximum allowable quantity per control area shall be in accordance with Sections 4303.1.1 through 4303.1.2.5.

❖ The regulations contained in Sections 4303.1.1 through 4303.1.2.5 assume that the quantity of unstable (reactive) materials in a given building is limited to the maximum allowable quantities per control area as established in Section 2703.1; thus, the building is not classified in Occupancy Group H.

4303.1.1 General. The storage and use of unstable (reactive) materials in amounts not exceeding the maximum allowable quantity per control area indicated in Section 2703.1 shall be in accordance with Sections 2701, 2703 4301 and 4303.

❖ This section complements the requirements of Chapter 27 in structures occupied for the storage, handling or use of unstable (reactive) materials. Unless otherwise indicated in a particular section, the regulations contained in Sections 4303.1.2 through 4303.1.2.5 assume that the quantity of unstable (reactive) materials in a given building is limited to the maximum allowable quantities per control area as established in Section 2703.1. The general requirements of Sections 2701 and 2703 are fully applicable to the storage and use of unstable (reactive) materials, in addition to the provisions of this chapter.

4303.1.2 Limitations for indoor storage and use by occupancy. The indoor storage of unstable (reactive) materials shall be in accordance with Sections 4303.1.2.1 through 4303.1.2.5.

❖ Because unstable (reactive) materials may be needed in certain occupancies, Sections 4303.1.2.1 through 4303.1.2.5 provide regulations that are specific to occupancy group classifications and that recognize the relative hazards of both the occupancy and the unstable (reactive) materials.

4303.1.2.1 Group A, E, I or U occupancies. In Group A, E, I or U occupancies, any amount of Class 3 and 4 unstable (reactive) materials shall be stored in accordance with the following:

1. Class 3 and 4 unstable (reactive) materials shall be stored in hazardous material storage cabinets complying with Section 2703.8.7.
2. The hazardous material storage cabinets shall not contain other storage.

❖ Because of their explosive or higher deflagration hazard characteristics, even the smallest quantity of Class 3 and 4 unstable (reactive) materials present in Group A, E, I or U occupancies must be stored in an approved hazardous material storage cabinet constructed and placarded in accordance with Section 2703.8.6 to reduce the exposure of the materials to hazards from the surrounding environment.

In accordance with Table 2703.1.1(1), Note g, storage of Class 4 unstable (reactive) materials in any amount is allowed only in buildings equipped throughout with an approved automatic sprinkler system in accordance with Section 903.3.1. Also, based on Table 2703.1.1(1), Note e, where an approved storage cabinet is used, the maximum allowable quantity per control area of unstable (reactive) materials that could be kept in those occupancies can be doubled.

To reduce the likelihood of contamination of the organic peroxide materials or damage to their packaging, Item 2 prohibits the storage of other materials within the approved unstable (reactive) materials storage cabinet.

4303.1.2.2 Group R occupancies. Class 3 and 4 unstable (reactive) materials shall not be stored or used within Group R occupancies.

❖ Because of their respective explosive or higher deflagration hazard characteristics, Class 3 and 4 unstable (reactive) materials storage in any residential occupancy within the scope of the code's regulations is prohibited because of the increased danger to the occupants and the otherwise unregulated environment to which those materials could be exposed.

4303.1.2.3 Group M occupancies. Class 4 unstable (reactive) materials shall not be stored or used in retail sales portions of Group M occupancies.

❖ Because of their explosive hazard characteristics, Class 4 unstable (reactive) materials storage in occupancies in Group M is prohibited because of the increased danger to the occupants and the higher relative fire loads typically encountered in these occupancies to which the materials could be exposed.

4303.1.2.4 Offices. Class 3 and 4 unstable (reactive) materials shall not be stored or used in offices of Group B, F, M or S occupancies.

❖ Because of their explosive hazard characteristics, Class 3 and 4 unstable (reactive) materials storage in office areas of occupancies in Group B, F, M or S is prohibited because of the increased danger to the occupants and the higher relative fire loads typically encountered in these occupancies.

4303.1.2.5 Classrooms. In classrooms in Group B, F or M occupancies, any amount of Class 3 and 4 unstable (reactive) materials shall be stored in accordance with the following:

1. Class 3 and 4 unstable (reactive) materials shall be stored in hazardous material storage cabinets complying with Section 2703.8.7.
2. The hazardous material storage cabinets shall not contain other storage.

❖ It is the intent of this section to allow for the occasional use of limited amounts of Class 3 and 4 unstable (reactive) materials in certain scientific, experimental or demonstration settings; however, it is not the intent to allow storage of any quantity of these materials for any length of time. The fire code official may limit the amount of unstable (reactive) materials brought into a structure for these uses. The quantity actually needed for the experiment or demonstration should determine the amount allowed into a structure.

Because of its respective explosive or higher deflagration hazard characteristics, even the smallest quantity of Class 3 or 4 unstable (reactive) materials must be stored in an approved hazardous material storage cabinet constructed and placarded in accordance with Section 2703.8.6 to reduce the exposure of the materials to hazards from the surrounding environment.

In accordance with Table 2703.1.1(1), Note g, storage of Class 4 unstable (reactive) materials in any amount is allowed only in buildings equipped throughout with an approved automatic sprinkler system in accordance with Section 903.3.1. Also, based on Table 2703.1.1(1), Note e, where an approved storage cabinet is used, the maximum allowable quantity per control area that could be kept in such occupancies can be doubled.

To reduce the likelihood of contamination of the unstable (reactive) materials or damage to their packaging, Item 2 prohibits the storage of other materials within the approved unstable (reactive) material storage cabinet.

4303.2 Quantities exceeding the maximum allowable quantity per control area. The storage and use of unstable (reactive) materials in amounts exceeding the maximum allowable quantity per control area indicated in Section 2703.1 shall be in accordance with Chapter 27 and this chapter.

❖ This section complements the requirements of Chapter 27 in structures occupied for the storage, handling or use of unstable (reactive) materials. The regulations contained in this section assume that the quantity of unstable (reactive) material in a given building or portion of a building is in excess of the maximum allowable quantities per control area as established in Section 2703.1; thus, the building or portion of the building is classified in Occupancy Group H. The requirements of Chapter 27 apply to the storage and use of unstable (reactive) material, in addition to the provisions of this chapter.

SECTION 4304 STORAGE

4304.1 Indoor storage. Indoor storage of unstable (reactive) materials in amounts exceeding the maximum allowable quantity per control area indicated in Table 2703.1.1(1) shall be in accordance with Sections 2701, 2703, 2704 and this chapter.

In addition, Class 3 and 4 unstable (reactive) detonable materials shall be stored in accordance with the *International Building Code* requirements for explosives.

❖ This section regulates the indoor storage of unstable (reactive) materials when in excess of the maximum allowable quantity per control area in buildings or portions of buildings classified in Occupancy Group H. The general and storage provisions of Chapter 27 are applicable in addition to the requirements of this section. Storage of unstable (reactive) materials inside structures must comply with Sections 4304.1.1 through 4304.1.7 to prevent uncontrolled release or exposure to conditions that may result in a fire or explosion.

Because of the explosion hazard, Class 3 and 4 detonable unstable (reactive) materials must be stored as required by this section, Chapter 33 and the *International Building Code*® (IBC®) requirements for explosives.

4304.1.1 Detached storage. Storage of unstable (reactive) materials shall be in detached buildings when required in Section 2703.8.2.

❖ Detached structures designed and constructed for the sole purpose of unstable (reactive) materials storage are the best protection for people and property from fire and explosion. Detached storage structures should be constructed of noncombustible materials to prevent them from becoming involved in an outside fire that may endanger their contents or from becoming involved in a fire due to their own contents igniting.

Section 2704.14 and Table 2704.14 require detached storage when the indoor storage quantities are more than 1 ton (908 kg) of Class 3 unstable (reactive) solid or liquid materials, more than 25 tons (22 700 kg) of Class 2 unstable (reactive) solid or liquid material, more than 2,000 cubic feet (57 m^3) of Class 3 gaseous unstable (reactive) material or more than 10,000 cubic feet (283 m^3) of Class 2 unstable (reactive) gaseous material. See the commentary to Section 2704.14 and Table 2704.14 for further discussion of detached storage requirements.

4304.1.2 Explosion control. Indoor storage rooms, areas and buildings containing Class 3 or 4 unstable (reactive) materials shall be provided with explosion control in accordance with Section 911.

❖ Because of the possibility of a deflagration or detonation in the event of ignition of Class 3 or 4 unstable (reactive) materials, explosion relief venting must be installed to protect the storage building or structure from collapse. Explosion venting must conform to the requirements of Section 911.

4304.1.3 Liquid-tight floor. In addition to Section 2704.12, floors of storage areas for liquids and solids shall be of liquid-tight construction.

❖ Floors and sills of rooms or areas used to contain hazardous material spills must be liquid tight to prevent the flow of liquids to adjoining areas (see commentary, Section 2704.2). The floor surface should be compatible with the unstable (reactive) materials to be retained and must be noncombustible, as required by Section 2704.12.

4304.1.4 Storage configuration. Unstable (reactive) materials stored in quantities greater than 500 cubic feet (14 m^3) shall be separated into piles, each not larger than 500 cubic feet (14 m^3). Aisle width shall not be less than the height of the piles or 4 feet (1219 mm), whichever is greater.

Exception: Materials stored in tanks.

❖ These provisions detail storage requirements related to the hazards of release or ignition of unstable (reactive) materials stored in structures and intend to reduce the amount of material exposed in a single incident by managing pile sizes and their separation. Aisle widths that are equal to the pile height, or 4 feet (3048 mm), whichever is greater reduce the fire exposure hazard between piles and, in the event of a fire-related collapse of a pile, reduce the likelihood of a "domino" effect that could not only increase fire intensity but also completely block access to pile areas by fire suppression personnel.

Although not specifically mentioned in this section, inside storage must also be protected in accordance with Chapter 27, which includes provisions for incompatible material storage, security, signage, control of ignition sources and submittal of a storage plan.

4304.1.5 Location in building. Unstable (reactive) materials shall not be stored in basements.

❖ Storage of unstable (reactive) material below the ground floor is prohibited to facilitate fire-fighter access and also because of the inherent difficulty associated with manual fire suppression operations in below-grade areas.

4304.2 Outdoor storage. Outdoor storage of unstable (reactive) materials in amounts exceeding the maximum allowable quantities per control area indicated in Table 2703.1.1(3) shall be in accordance with Sections 2701, 2703, 2704 and this chapter.

❖ This section regulates the outdoor storage of unstable (reactive) materials when in excess of the maximum allowable quantity per outdoor control area established by Table 2703.1.1(3). The general and storage provisions of Chapter 27 are applicable in addition to the requirements of this section. Storage of unstable (reactive) materials in outdoor control areas must comply with Sections 4304.2.1 and 4304.2.2 to prevent uncontrolled release or exposure to conditions that could result in a fire or explosion.

4304.2.1 Distance from storage to exposures. Outdoor storage of unstable (reactive) material that can deflagrate shall not be within 75 feet (22 860 mm) of buildings, lot lines, public streets, public alleys, public ways or means of egress. Outdoor storage of nondeflagrating unstable (reactive) materials shall not be within 20 feet (6096 mm) of buildings, lot lines, public streets, public alleys, public ways or means of egress. A 2-hour fire barrier wall without openings or penetrations extending not less than 30 inches (762 mm) above and to the sides of the storage is allowed in lieu of such distance. The wall shall either be an independent structure, or the exterior wall of the building adjacent to the storage area.

❖ The clearances required by this section are intended to reduce the hazard to nearby structures and people in the event of a fire or explosion in an outdoor area used for the storage of unstable (reactive) materials capable of deflagration. This section also recognizes that, for nondeflagrating unstable (reactive) materials, a minimum separation distance of 20 feet (6096 mm) is adequate protection and that a solid, 2-hour-rated fire barrier wall (constructed in accordance with Section 706 of the IBC is an equivalent means of achieving the objectives of spatial separation. Where a separation assembly is installed in place of the separation distance, the wall must extend vertically beyond the roof or wall opening of the larger structure on the top and sides to prevent a fire from lapping over or extending around the wall.

4304.2.2 Storage configuration. Piles of unstable (reactive) materials shall not exceed 1,000 cubic feet (28 m^3).

❖ The size of storage piles of unstable (reactive) materials is regulated by this section based on the hazards of the unstable (reactive) materials being stored. To limit the amount of material involved in a single fire, unstable (reactive) materials are limited to piles no larger than 10 feet by 10 feet by 10 feet (3048 mm by 3048 mm by 3048 mm) high.

4304.2.3 Aisle widths. Aisle widths between piles shall not be less than one-half the height of the pile or 10 feet (3048 mm), whichever is greater.

❖ Minimum 10-foot-wide (3048 mm) aisles provide access to the storage area for emergency personnel and reduce the likelihood of a "domino" effect that could not only increase fire intensity but also completely block access to pile areas by fire suppression personnel should a pile topple over.

SECTION 4305
USE

4305.1 General. The use of unstable (reactive) materials in amounts exceeding the maximum allowable quantity per control area indicated in Table 2703.1.1(1) or 2703.1.1(3) shall be in accordance with Sections 2701, 2703, 2705 and this chapter.

❖ This section applies to all indoor and outdoor dispensing, use and handling of unstable (reactive) materials when the amounts being dispensed, used or handled are in excess of the maximum allowable quantities per indoor or outdoor control area indicated in Tables 2703.1.1(1) or 2703.1.1(3), respectively. The administrative; general and use and handling and dispensing provisions of Chapter 27 are applicable, in addition to the requirements of this chapter.

Once the maximum allowable quantity per control area of unstable (reactive) materials has been exceeded, indoor areas where materials are being dispensed, used or handled must be located in a building or portion of a building complying with the IBC for a Group H occupancy because of the increased hazards associated with quantity. Although no occupancy group is assigned to them, outside unstable (reactive) materials use areas must be regulated more closely when quantities exceed the maximum allowable quantities per outdoor control area.

The maximum allowable quantities per control area listed in Table 2703.1.1(1) or 2703.1.1(3) have been divided into closed-use and open-use systems. Corresponding maximum allowable quantities per control area recognize that an open-use system is generally more hazardous than a closed-use system because the unstable (reactive) materials are more directly exposed to the surrounding environment and can become more readily involved in an incident than if they are totally confined. The maximum allowable quantities per control area for use are based on the aggregate quantity in both use and storage not exceeding the exempt amount listed for storage.

Bibliography

The following resource materials are referenced in this chapter or are relevant to the subject matter addressed in this chapter.

DOTn 49 CFR; 100-178 - 94, *Hazardous Materials Regulations*. Washington, DC: U.S. Department of Transportation, 1994.

Fire Protection Guide on Hazardous Materials, 13th ed. Quincy, MA: National Fire Protection Association, 2001.

IBC-2003, *International Building Code*. Falls Church, VA: International Code Council, 2003.

Isman, W.E. and G.P. Carlson. *Hazardous Materials*. Encino, CA: Glencoe Publishing Co., Inc., 1980.

NFPA 704-96, *Identification of the Fire Hazards of Materials*. Quincy, MA: National Fire Protection Association, 1996.

Police and Fire Interest Bulletin No 7, *Unstable (Reactive) Chemicals*. New York: American Insurance Association, 1973.

Chapter 44: Water-Reactive Solids and Liquids

General Comments

Water-reactive materials are used in a variety of industrial applications for the processing of other materials, such as descaling (salt) baths in the metal processing industry, as dehydrating agents in sulfonation processes (the addition of fuming sulfuric acid to a product being treated) and a variety of other complex chemical industrial processes. They also may be found in the manufacture of a variety of consumer products, such as soaps and detergents, rodenticides, fertilizers, silicone rubber and pharmaceutical products.

Advance knowledge of the materials stored in structures through permits and the submittal of Material Safety Data Sheets (MSDS) are essential for effective prefire planning by the fire department. This is especially important in prefire planning for water-reactive materials, because the presence of these materials severely limits the fire department's ability to use water as the primary fire suppression tool.

Water-reactive materials may react to water in a variety of ways, including explosion, violent splattering, production of toxic gases, rapid decomposition with the evolution of large volumes of heat that may ignite nearby combustibles or any combination of these reactions. Alternative fire suppression protocols must be developed well in advance of an incident and the requisite alternative suppression medium obtained, rather than having to go searching for it at the time of an emergency. Depending on the specific water-reactive material involved in an incident, the alternative extinguishing media could include dry graphite, soda ash, sodium chloride, dry sand or specialized dry powder agents such as Na-X or Met-L-X, both manufactured by the Ansul Fire Protection Company

If a manageable spill occurs without a fire, the material can be confined by a dam of dry sand and then covered with an absorbent material, such as vermiculite, dolomite or more dry sand. Once absorbed, the material must be moved outside with care for disposal. Water-reactive materials that are spilled outdoors are sometimes best handled by allowing them to react and hastening that process by applying water. Large volumes of water are required for this procedure and they must be applied from a safe distance by personnel wearing full protective clothing and breathing apparatus. This method will not only accelerate the reaction but also disperse and dilute the fumes generated.

Purpose

This chapter addresses the hazards associated with water-reactive materials that are solid or liquid at normal temperatures and pressures. In addition to their water reactivity, these materials may pose a wide range of other hazards, such as toxicity, flammability, corrosiveness or oxidizing potential. This chapter addresses only those materials whose primary hazard is water reactivity. Materials that pose multiple hazards must conform to the requirements of the code with respect to all hazards (see commentary, Section 2701.1). Strict compliance with the requirements of this chapter, along with proper housekeeping and storage arrangements, helps to reduce the exposure hazards associated with water-reactive materials in a fire or other emergency.

SECTION 4401
GENERAL

4401.1 Scope. The storage and use of water-reactive solids and liquids shall be in accordance with this chapter.

Exceptions:

1. Display and storage in Group M and storage in Group S occupancies complying with Section 2703.11.
2. Detonable water-reactive solids and liquids shall be stored in accordance with Chapter 33.

❖ This chapter contains specific requirements for the storage of water-reactive materials. The requirements of this chapter are intended to complement the hazardous materials general storage requirements of Chapter 27. Classification of solids and liquids as being water reactive is based on their relative degree of hazard as described in the definition in Section 4202.

Exception 1 makes it clear that this chapter does not apply to water-reactive materials stored and displayed in Group M occupancies or stored in Group S occupancies, and defers to the requirements of Section 2703.11 for those occupancy groups. In that section, quantities are limited on a per-control-area basis to limit exposure to people and property (see commentary, Section 2703.11). This exception would permit an increase above the exempt amounts per control area indicated in Table 2703.1.1(1) for water-reactive materials while still maintaining a mercantile or storage occupancy group classification.

Exception 2 recognizes the violent, explosive nature of detonable water-reactive materials and directs that they be stored in accordance with the requirements for explosives and blasting agents (see Chapter 33) to provide the safeguards commensurate with their hazards.

4401.2 Permits. Permits shall be required as set forth in Section 105.6.

❖ The process of issuing permits gives the fire code official an opportunity to carefully evaluate and regulate hazardous operations. Permit applicants should be required to demonstrate that their operations comply with the intent of the code before the permit is issued. See the commentary to Section 105.6 for a general discussion of operations requiring an operational permit, Section 105.6.21 for discussion of specific quantity-based operational permits for the materials regulated in this chapter and Section 105.7 for a general discussion of activities requiring a construction permit. The permit process also notifies the fire department of the need for prefire planning for hazardous property.

SECTION 4402 DEFINITIONS

4402.1 Definition. The following word and term shall, for the purposes of this chapter and as used elsewhere in this code, have the meaning shown herein.

❖ Definitions of terms can help in the understanding and application of the code requirements. The purpose for including here the definition that is associated with the subject matter of this chapter is to provide more convenient access to it without having to refer back to Chapter 2. It is important to emphasize that this term is not exclusively related to this chapter but is applicable everywhere the term is used in the code. For convenience, the term is also listed in Chapter 2 with a cross reference to this section. The use and application of all defined terms, including the one defined in this section, are set forth in Section 201.

WATER-REACTIVE MATERIAL. A material that explodes; violently reacts; produces flammable, toxic or other hazardous gases; or evolves enough heat to cause self-ignition or ignition of nearby combustibles upon exposure to water or moisture. Water-reactive materials are subdivided as follows:

Class 3. Materials that react explosively with water without requiring heat or confinement.

Class 2. Materials that may form potentially explosive mixtures with water.

Class 1. Materials that may react with water with some release of energy, but not violently.

❖ Class 2 and 3 water-reactive materials can liberate significant quantities of heat and hazardous gases when reacting with water. Combustible water-reactive materials are capable of self-ignition. Even noncombustible water-reactive materials pose a hazard because of the heat released during their reaction with water, which may be sufficient to ignite surrounding combustible materials.

SECTION 4403 GENERAL REQUIREMENTS

4403.1 Quantities not exceeding the maximum allowable quantity per control area. The storage and use of water-reactive solids and liquids in amounts not exceeding the maximum allowable quantity per control area indicated in Section 2703.1 shall be in accordance with Sections 2701, 2703, 4401 and 4403.

❖ This section complements the requirements of Chapter 27 in structures occupied for the storage, handling or use of water-reactive materials limited to the maximum allowable quantities per control area as established in Section 2703.1. The general requirements of Sections 2701 and 2703 are fully applicable to the storage and use of water-reactive materials, in addition to the requirements of this chapter.

4403.2 Quantities exceeding the maximum allowable quantity per control area. The storage and use of water-reactive solids and liquids in amounts exceeding the maximum allowable quantity per control area indicated in Section 2703.1 shall be in accordance with Chapter 27 and this chapter.

❖ This section complements the requirements of Chapter 27 in structures occupied for the storage, handling or use of water-reactive materials. The regulations contained in this section assume that the quantity of water-reactive materials in a given building is in excess of the maximum allowable quantities per control area as established in Section 2703.1; thus, the building is classified in Occupancy Group H. The requirements of Chapter 27 apply to the storage and use of water-reactive material, in addition to the requirements of this chapter.

SECTION 4404 STORAGE

4404.1 Indoor storage. Indoor storage of water-reactive solids and liquids in amounts exceeding the maximum allowable quantity per control area indicated in Table 2703.1.1(1), shall be in accordance with Sections 2701, 2703, 2704 and this chapter.

❖ This section regulates the indoor storage of water-reactive materials when in excess of the maximum allowable quantity per control area in buildings or portions of buildings classified in Occupancy Group H. The general and storage requirements of Chapter 27 are applicable in addition to the requirements of this section. Storage of water-reactive materials inside structures must comply with Sections 4401.1 through 4404.1.7 to prevent uncontrolled release or exposure to conditions that may result in a fire or explosion.

4404.1.1 Detached storage. Storage of water-reactive solids and liquids shall be in detached buildings when required by Section 2703.8.2.

❖ Detached structures designed and constructed for the sole purpose of water-reactive materials storage provide the best protection for people and property from fire and explosion. Detached storage structures should be constructed of noncombustible materials to prevent them from becoming involved in an outside fire that may endanger their contents or from becoming involved in a fire due to their own contents igniting.

Section 2703.8.2 and Table 2703.8.2 require detached storage when the indoor storage quantities are more than 1 ton (908 kg) of Class 3 water-reactive solid or liquid materials or more than 25 tons (22 700 kg) of Class 2 water-reactive solid or liquid materials. See the commentary to Section 2703.8.2 and Table 2703.8.2 for further discussion of detached storage requirements.

4404.1.2 Liquid-tight floor. In addition to the provisions of Section 2704.12, floors in storage areas for water-reactive solids and liquids shall be of liquid-tight construction.

❖ Floors and sills of rooms or areas used to contain hazardous material spills must be liquid tight to prevent the flow of liquids to adjoining areas (see commentary, Section 2704.2). The floor surface should be compatible with the water-reactive materials to be retained and must be noncombustible, as required by Section 2704.12.

4404.1.3 Waterproof room. Rooms or areas used for the storage of water-reactive solids and liquids shall be constructed in a manner which resists the penetration of water through the use of waterproof materials. Piping carrying water for other than approved automatic sprinkler systems shall not be within such rooms or areas.

❖ The design and construction of rooms used for storing water-reactive materials must prevent water from coming into contact with the stored materials. The building materials should be noncombustible so they do not contribute fuel to a fire or reaction and should be designed to resist the passage of flowing water.

A major safeguard is planning the design of the room and locating it to minimize water hazards. For example, in a two-story building it would not be appropriate to put a water-reactive-materials storage room directly below a locker room with showers or a bathroom located on the second floor.

Similarly, the enclosure walls of a storage room should not contain plumbing piping. Though water piping may not be run into or through storage rooms, the code recognizes that automatic sprinkler systems are a more regulated type of water piping system and have a low leakage and failure rate when properly installed and maintained.

4404.1.4 Water-tight containers. When Class 3 water-reactive solids and liquids are stored in areas equipped with an automatic sprinkler system, the materials shall be stored in closed water-tight containers.

❖ In the event of a sprinkler discharge in response to a fire, the application of water should not aggravate the fire by causing a violent exothermic reaction with the water-reactive materials in storage that might not otherwise become involved. To reduce this hazard and to complement the provisions of Section 4404.1.4, water-reactive materials must be stored in water-tight containers of less than a 60-gallon (227 L) capacity (see commentary, Section 2702 and definition of "Container") that comply with Section 2703.2.

4404.1.5 Storage configuration. Water-reactive solids and liquids stored in quantities greater than 500 cubic feet (14 m^3) shall be separated into piles, each not larger than 500 cubic feet (14 m^3). Aisle widths between piles shall not be less than the height of the pile or 4 feet (1219 mm), whichever is greater.

Exception: Water-reactive solids and liquids stored in tanks.

Class 2 water-reactive solids and liquids shall not be stored in basements unless such materials are stored in closed water-tight containers or tanks.

Class 3 water-reactive solids and liquids shall not be stored in basements.

Class 2 or 3 water-reactive solids and liquids shall not be stored with flammable liquids.

❖ This section details storage requirements related to the hazards of release or ignition of water-reactive materials stored in structures and is intended to reduce the amount of material exposed in a single incident by managing pile sizes and their separation. Water-reactive materials are limited to piles no larger than 500 cubic feet (14 m^3) in volume [approximately 8-feet by 8-feet by 8-feet high (2438 mm by 2438 mm by 2438 mm)]. Aisle widths that are equal to the pile height or 4 feet (1219 mm), whichever is greater, reduce the fire exposure hazard between piles and, in the event of a fire-related collapse of a pile, reduce the likelihood of a "domino" effect that could not only increase fire intensity but also completely block access to pile areas by fire suppression personnel. Although not specifically mentioned in this section, inside storage must also comply with Chapter 27, which includes provisions for incompatible material storage, security, signage, control of ignition sources and submittal of a storage plan.

The exception exempts tank storage from the requirements of this section; however, the tanks must conform to the requirements of Section 2703.2.

Because Class 2 oxidizers are potentially explosive, they may be stored in basements only when contained in approved stationary tanks or containers that comply with the applicable requirements of Sections 2703.2 and 4404.1.5.

Because Class 3 water-reactive materials are violently reactive, they may not be stored in a basement because of the limited access for fire suppression operations, the potential for increased damage to the structure and exposure of the occupants to danger.

Class 2 and 3 water-reactive materials may not be stored in the same room or area as flammable liquids because of their violent reactivity and their incompatibility. See the commentary to Section 2703.9.8 for a discussion of precautions to be taken with incompatible materials.

4404.1.6 Explosion control. Indoor storage rooms, areas and buildings containing Class 2 or 3 water-reactive solids and liquids shall be provided with explosion control in accordance with Section 911.

❖ The violently reactive nature of Class 2 and 3 water-reactive materials can seriously damage or destroy a storage room or structure. To prevent damage or destruction in storage rooms or structures where Class 2 or 3 water-reactive materials are stored, an explosion control system must be installed as required in Section 911.

4404.2 Outdoor storage. Outdoor storage of water-reactive solids and liquids in quantities exceeding the maximum allowable quantity per control area indicated in Table 2703.1.1(3) shall be in accordance with Sections 2701, 2703, 2704 and this chapter.

❖ This section regulates the outdoor storage of water-reactive material when in excess of the maximum allowable quantity per outdoor control area established by Table 2703.1.1(3). The general and storage provisions of Chapter 27 are applicable in addition to the requirements of this section.

4404.2.1 General. Outdoor storage of water-reactive solids and liquids shall be within tanks or closed water-tight containers and shall be in accordance with Sections 4404.2.2 through 4404.2.5.

❖ Similar to Sections 4404.1.3 and 4404.1.5, which require waterproof rooms and water-tight containers for the storage of water-reactive materials, this section carries the protection of the material one step further by requiring that water-reactive materials stored outdoors be in closed, water-tight containers or tanks to reduce the likelihood that rain or snow will come into contact with them. Storage of water-reactive material in outdoor control areas must comply with Sections 4404.2.1 through 4404.2.5 to prevent uncontrolled release or exposure to conditions that could result in a fire or explosion.

4404.2.2 Class 3 distance to exposures. Outdoor storage of Class 3 water-reactive solids and liquids shall not be within 75 feet (22 860 mm) of buildings, lot lines, public streets, public alleys, public ways or means of egress.

❖ The required separation distances are based on the class of water-reactive material and are intended to reduce the hazard of radiant heat transfer to nearby structures, public streets or alleys or egress elements from buildings. The distances also protect the property in question from heat exposure from incidents on or off the property.

4404.2.3 Class 2 distance to exposures. Outdoor storage of Class 2 water-reactive solids and liquids shall not be within 20 feet (6096 mm) of buildings, lot lines, public streets, public alleys, public ways or means of egress. A 2-hour fire barrier wall without openings or penetrations, and extending not less than 30 inches (762 mm) above and to the sides of the storage area, is allowed in lieu of such distance. The wall shall either be an independent structure, or the exterior wall of the building adjacent to the storage area.

❖ The required separation distances in this section are based on the less violently reactive Class 1 and 2 water-reactive materials and are intended to reduce the hazard of radiant heat transfer to nearby structures, public streets or alleys or egress elements from buildings. The distances also protect the property in question from heat exposure from incidents on or off the property.

This section also recognizes that, for these less violently reactive water-reactive materials, a minimum separation distance of 20 feet (6096 mm) provides adequate protection and that a solid, 2-hour fire barrier wall, constructed in accordance with Section 706 of the *International Building Code®* (IBC®), is an equivalent way to meet the objectives of spatial separation. Where a separation assembly is installed instead of having the separation distance, the wall must extend vertically beyond the roof or wall of the larger structure on each side (top and sides) to prevent a fire from lapping over or extending around the wall.

4404.2.4 Storage conditions. Class 3 water-reactive solids and liquids shall be limited to piles not greater than 500 cubic feet (14 m^3).

Class 2 water-reactive solids and liquids shall be limited to piles not greater than 1,000 cubic feet (28 m^3).

Aisle widths between piles shall not be less than one-half the height of the pile or 10 feet (3048 mm), whichever is greater.

❖ The size of storage piles of water-reactive materials is regulated by this section based on the hazards of the water-reactive materials being stored. To limit the amount of material involved in a single fire, water-reactive materials are limited to piles no larger than approximately 8 feet by 8 feet by 8 feet (2438 mm by 2438 mm by 2438 mm) high for Class 3 or 10 feet by 10 feet by 10 feet (3048 mm by 3048 mm by 3048 mm) high for Class 2 materials. The required aisle width ensures access to the storage area for emergency personnel and reduces the likelihood of a "domino" effect that could not only increase fire intensity but also completely block access to pile areas by fire suppression personnel should a pile topple over.

4404.2.5 Containment. Secondary containment shall be provided in accordance with the provisions of Section 2704.2.2.

❖ To prevent the flow of water-reactive liquids to adjoining rooms or spaces, secondary containment complying with Section 2704.2.2 is required by this section. The design of drainage and secondary containment systems must take into consideration automatic sprinkler design discharge flow rates and fire suppression hand line [typically 1½- or 1¾-inch (38 or 44 mm) hose] flows.

Note that secondary containment requirements do not provide for control of the flammable, irritating or toxic vapors given off by reacted materials, and care must be taken to minimize exposure to hazardous vapors.

Runoff from spills or manual fire suppression activities may result in environmental contamination if not properly controlled (see also commentary, Section 2704.2).

SECTION 4405 USE

4405.1 General. The use of water-reactive solids and liquids in amounts exceeding the maximum allowable quantity per control area indicated in Table 2703.1.1(1) or 2703.1.1(3) shall be in accordance with Sections 2701, 2703, 2705 and this chapter.

❖ This section applies to all indoor and outdoor dispensing, use and handling operations of water-reactive materials when the amounts being dispensed, used or handled are in excess of the maximum allowable quantities per indoor or outdoor area indicated in Table 2703.1.1(1) or 2703.1.1(3), respectively. The administrative; general, use and handling and dispensing provisions of Chapter 27 are applicable, in addition to the requirements of this chapter.

Once the maximum allowable quantity per control area of water-reactive materials has been exceeded, indoor areas where materials are being dispensed, used or handled must be located in a building or portion of a building complying with the IBC for a Group H occupancy because of the increased hazards associated with quantity. Although no occupancy group is assigned to them, outside water-reactive materials use areas must be regulated more heavily when quantities exceed the maximum allowable quantities per outdoor control area.

The maximum allowable quantities per control area listed in Table 2703.1.1(1) or 2703.1.1(3) have been divided into closed-use and open-use systems. Corresponding maximum allowable quantities per control area recognize that an open-use system is generally more hazardous that a closed-use system because the water-reactive materials are more directly exposed to the surrounding environment and can become more readily involved in an incident than if they are totally confined. The maximum allowable quantities per control area for use are based on the aggregate quantity in both use and storage not exceeding the exempt amount listed for storage.

Bibliography

The following resource materials are referenced in this chapter or are relevant to the subject matter addressed in this chapter.

Bradford, W.I. Section 5, Chapter 6, "Chemicals," and Section 11, Chapter 6, "Storage and Handling of Chemicals." In Cote, A.E., ed. *Fire Protection Handbook*, 16th ed. Quincy, MA: National Fire Protection Association, 1986.

"Cellar Fires." *Special Interest Bulletin* No. 67. New York: National Board of Fire Underwriters, 1953.

DOTn 49 CFR; 100-178-94, *Hazardous Materials Regulations*. Washington, DC: U.S. Department of Transportation, 1994.

Fire Inspector Guidebook, 8th ed. Country Club Hills, IL: Building Officials and Code Administrators International, Inc., 2001.

Fire Protection Guide on Hazardous Materials, 12th ed. Quincy, MA: National Fire Protection Association, 1997.

IBC-2003, *International Building Code.* Falls Church, VA: International Code Council, 2003.

Isman, W.E. and G.P. Carlson. *Hazardous Materials*. Enrico, CA: Glencoe Publishing Co., Inc., 1980.

NFPA 704-96, *Identification of the Fire Hazards of Materials.* Quincy, MA: National Fire Protection Association, 1996.

"Sodium." *Special Interest Bulletin* No. 208. New York: National Board of Fire Underwriters, 1956.

"Sodium Hydride Descaling." *Special Interest Bulletin* No. 209. New York: National Board of Fire Underwriters, 1956.

Chapter 45: Referenced Standards

General Comments

Not every document related to fire safety system design, installation and construction is qualified to be a referenced standard. The International Code Council® (ICC®) has adopted a criterion that referenced standards in the *International Codes*® and standards intended for adoption into the *International Codes* must meet to qualify as a referenced standard. The policy is as follows:

Referenced Standards: In order for a standard to be considered for reference or to continue to be referenced by the codes, a standard shall meet the following criteria:

Code References:

1. The standard and the manner in which it is to be utilized shall be specifically referenced in the code text.
2. The need for the standard to be referenced shall be established.

Standard Content:

1. A standard or portions of a standard intended to be enforced shall be written in mandatory language.
2. The standard shall be appropriate for the subject covered.
3. All terms shall be defined when they deviate from an ordinarily accepted meaning or a dictionary definition.
4. The scope or application of a standard shall be clearly described.
5. The standard shall not have the effect of requiring proprietary materials.
6. The standard shall not prescribe a proprietary agency for quality control or testing.
7. The test standard shall describe, in detail, preparation of the test sample, sample selection or both.
8. The test standard shall prescribe the reporting format for the test results. The format shall identify the key performance criteria of the element(s) tested.
9. The measure of performance for which the test is conducted shall be clearly defined in either the test standard or in code text.
10. The standard shall not state that its provisions shall govern whenever the referenced standard is in conflict with the requirements of the referencing code.
11. The preface to the standard shall announce that the standard is promulgated according to a consensus procedure.

Standard Promulgation:

1. The standard shall be readily available.
2. The standard shall be developed and maintained through a consensus process such as ASTM or ANSI. Standards using the ANSI Canvass Method shall comply with *Report of ICC Modifications to the ANSI General Procedures* and to *ANSI Annex B — Procedures for Canvass by an Accredited Sponsor.*

Once a standard is incorporated into the code through the code development process, it becomes an enforceable part of the code, subject to the limitations of the text referenced in accordance with Section 102.6. When the code is adopted by a jurisdiction, the standard also is part of that jurisdiction's adopted code. It is for this reason that the criteria were developed.

Compliance with this policy means that documents or portions of documents incorporated into the code by reference are, among others, developed through the use of a consensus process, written in mandatory language and do not mandate the use of proprietary materials or agencies. The requirement that a standard be developed through a consensus process is vital, because it means that the standard will be a representative of the most current body of available knowledge on the subject as determined by a broad range of interested or affected parties without dominance by any single interest group. A true consensus process has many attributes, including but not limited to:

- An open process that has formal (published) procedures that allow for the consideration of all viewpoints;
- A definitive review period that allows for the standard to be updated or revised;
- A process of notification to all interested parties; and
- An appeals process.

Many available documents related to fire safety system design, installation and construction, though useful, are not "standards" and are not appropriate for reference in the code. Often, these documents are developed or written with the intention of being used for regulatory purposes and are unsuitable for use as a regulation because of extensive use of recommendations, advisory comments and nonmandatory terms. Typical examples of such documents include installation instructions, guidelines and practices.

The ICC's standards policy results in regulations that are clear, concise and enforceable, thus the requirement that standards be written in mandatory language. This requirement is not intended to mean that a standard cannot contain informational or explanatory material that will aid the user of the standard in its application. When it is the desire of the standard's promulgating agency for such material to be included, however, the information must appear in a nonmandatory location, such as an annex or appendix, and be clearly identified as not being part of the standard.

Overall, standards referenced by the code must be authoritative, relevant, up to date and, most important, reasonable and enforceable. Standards that comply with ICC's standards policy fulfill these expectations.

Purpose

As a performance-based code, it contains numerous references to documents that are used to regulate materials and methods of construction. The references to these documents within the code text consist of the promulgating agency's acronym and its publication designation (for example, ASME A17.3) as well as a further indication that the document being referenced is the one that is listed in Chapter 45. Chapter 45 contains all of the information that is necessary to identify the specific referenced document. Included is the following information on a document's promulgating agency (see Figure 45):

- The promulgating agency (the agency's title);
- The promulgating agency's acronym; and
- The promulgating agency's address.

For example, a reference to an ASME standard within the code indicates that the document is promulgated by the American Society of Mechanical Engineers (ASME), which is located in New York City. Chapter 45 lists the standards' agencies alphabetically for ease of identification.

Chapter 45 also includes the following information on the referenced document itself (see Figure 45):

- The document's publication designation;
- The document's edition year;
- The document's title;
- Any addenda or revisions to the document known at the time of the code's publication; and
- Every section of the code in which the document is referenced.

For example, a reference to ASME B16.18 indicates that this document can be found in Chapter 45 under the heading ASME. The specific standards designation is B16.18. For convenience, these designations are listed in alphanumeric order. Chapter 45 identifies that: ASME 16.18 is titled *Cast Copper Alloy Solder Joint Pressure Fittings*, the applicable edition (its year of publication) is 1984, and it is referenced in one specifically identified section of the code.

Chapter 45 will also note when a document has been discontinued or replaced by its promulgating agency. When a document is replaced by a different one, a note will appear to tell the user the designation and title of the new document.

Using the system established for the family of *International Codes,* the specific edition of a specific standard is clearly identified and the requirements necessary for compliance can be readily determined. The basis for code compliance is, therefore, established and available on an equal basis to the code official, builder, designer and owner.

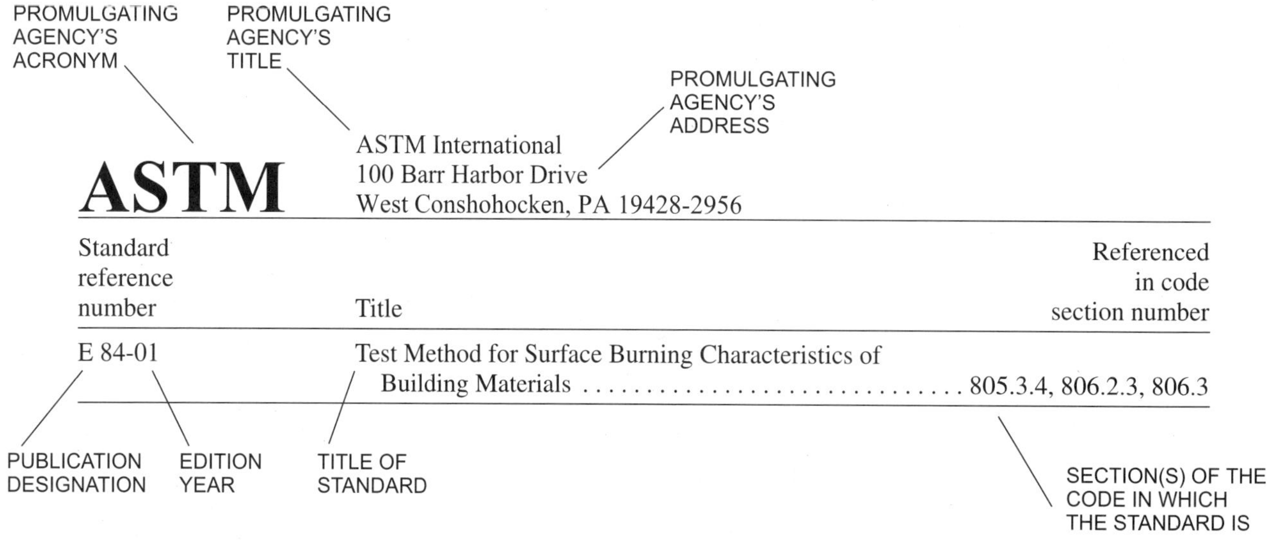

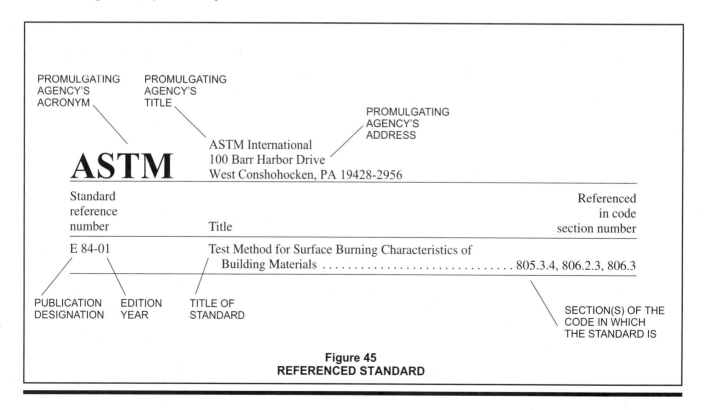

Figure 45
REFERENCED STANDARD

This chapter lists the standards that are referenced in various sections of this document. The standards are listed herein by the promulgating agency of the standard, the standard identification, the effective date and title, and the section or sections of this document that reference the standard. The application of the referenced standards shall be as specified in Section 102.6.

AASHTO

American Association of State Highway
and Transportation Officials
444 North Capitol Street, Northwest, #249
Washington, DC 20001

Standard reference number	Title	Referenced in code section number
HB-16—1996	Standard Specification for Highway Bridges, 16th edition —with 1997 through 2000 Interim Revisions	503.2.6

AFSI

Architectural Fabric Structures Institute
c/o Industrial Fabric Association International
1801 County Road B West
Roseville, MN 55113

Standard reference number	Title	Referenced in code section number
ASI—77	Design and Standard Manual	2403.10.2

ANSI

American National Standards Institute
25 West 43rd Street, Fourth Floor
New York, NY 10036

Standard reference number	Title	Referenced in code section number
A13.1—96	Scheme for the Identification of Piping Systems	2609.3, 2703.2.2.1, 3003.2.3, 3203.4.5, 3403.5.2
B31.3—99	Process Piping, including addendum	2209.5.4.1, 2703.2.2.2
B31.9—96	Building Services Piping Code for Pressure Piping	3403.6.2.1, 3403.6.3, 3403.6.11

API

American Petroleum Institute
1220 L Street, Northwest
Washington, DC 20005

Standard reference number	Title	Referenced in code section number
Spec 12P—1995	Specification for Fiberglass Reinforced Plastic Tanks	3404.2.13.1.5
RP 651—1997	Cathodic Protection of Aboveground Petroleum Storage Tanks	3406.7, 3406.7.1
Std 653—1998	Tank Inspection, Repair, Alteration and Reconstruction	3406.7
RP 752—1995	Management of Hazards Associated with Location of Process Plant Buildings, CMA Managers Guide	3406.7
RP 1604—1996	Closure of Underground Petroleum Storage Tanks	3404.2.13
RP 1615—1996	Installation of Underground Petroleum Storage Systems	3404.2.13.1.5, 3406.7
Std 2000—1998	Venting Atmosphere and Low Pressure Storage Tanks: Nonrefrigerated and Refrigerated	3404.2.7.3.6
RP 2001—1998	Fire Protection in Refineries	3406.7
RP 2003—1998	Protection Against Ignitions Arising out of Static, Lightening, and Stray Currents	3406.7
Publ 2009—1995	Safe Welding and Cutting Practices in Refineries, Gas Plants and Petrochemical Plants	3406.7

API—continued

Std 2015—1994	Safe Entry and Clearing of Petroleum Storage Tanks	3406.7, 3406.7.2
Publ 2028—1991	Flame Arrestors in Piping Systems	3404.2.7.3.2
RP 2023—1988	Guide for Safe Storage and Handling of Heated Petroleum-Derived Asphalt Products and Crude-Oil Residue	3406.7, 3406.7.3
Publ 2201—1995	Procedures for Welding or Hot Tapping on Equipment in Service	3406.7
RP 2350—1996	Overfill Protection for Storage Tanks in Petroleum Facilities	3406.4.6, 3406.7

ASME

The American Society of Mechanical Engineers
Three Park Avenue
New York, NY 10016-5990

Standard reference number	Title	Referenced in code section number
A17.1—2000	Safety Code for Elevators and Escalators	607.1, 1007.4
A17.3—1996	Safety Code for Existing Elevators and Escalators with A17.3a-2000 Addenda	607.1
A18.1-1999	Safety Standard for Platform Lifts and Stairway Chair Lifts — with A18.1a-2001 Addenda	1007.5
B16.18—1984 (Reaffirmed 1994)	Cast Copper Alloy Solder Joint Pressure Fittings	909.13.1
B16.22—1995	Wrought Copper and Copper Alloy Solder-Joint Pressure Fittings—with B16.22a-1998 Addenda	909.13.1
BPVC-2001	ASME Boiler and Pressure Vessel Code, 2001 Edition of	2209.5.4.2.1, 3203.4.3, 3203.8, 3404.2.13.1.5

ASTM

ASTM International
100 Barr Harbor Drive
West Conshohocken, PA 19428-2959

Standard reference number	Title	Referenced in code section number
B 42—98	Specification for Seamless Copper Pipe, Standard Sizes	909.13.1
B 43—98	Specification for Seamless Red Brass Pipe, Standard Sizes	909.13.1
B 68—99	Specification for Seamless Copper Tube, Bright Annealed	909.13.1
B 88—99e1	Specification for Seamless Copper Water Tube	909.13.1
B 251—97	Specification for General Requirements for Wrought Seamless Copper and Copper-Alloy Tube	909.13.1
B 280—99e1	Specification for Seamless Copper Tube for Air Conditioning and Refrigeration Field Service	909.13.1
D 56—01	Test Method for Flash Point by Tag Closed Tester	3402.1
D 86—01e01	Test Method for Distillation of Petroleum Products at Atmospheric Pressure	2702.1
D 92—01	Test Method for Flash and Fire Points by Cleveland Open Cup	3401.2, 3402.1
D 93—00	Test Method for Flash Point by Pensky-Martens Closed Up Tester	3402.1
D 323—99a	Test Method for Vapor Pressure of Petroleum Products (Reid Method)	2702.1
D 3278—96e01	Test Methods for Flash Point of Liquids by Small Scale Closed-Cup Apparatus	3402.1
E 84—01	Test Method for Surface Burning Characteristics of Building Materials	805.3.4, 806.2.3, 806.3
E 681—01	Test Method for Concentration Limits of Flammability of Chemicals (Vapors and Gases)	3502.1
E 1529—00	Test Method for Determining Effects of Large Hydrocarbon Pool Fires on Structural Members and Assemblies	3404.2.9.1.3
E 1537—01	Test Method for Fire Testing of Upholstered Furniture	803.5.2
E 1590—01	Test Method for Fire Testing of Mattresses	803.5.3, 803.6.3, 803.7.4

BHMA

Builders Hardware Manufacturers' Association
355 Lexington Avenue, 17th Floor
New York, NY 10017-6603

Standard reference number	Title	Referenced in code section number
A156.10—99	American National Standard for Power Operated Pedestrian Doors	1008.1.3.2
A156.19—97	American National Standard for Power Assist and Low Energy Power Operated Doors	1008.1.3.2

CGA

Compressed Gas Association
1725 Jefferson Davis Highway
5th Floor
Arlington, VA 22202-4102

Standard reference number	Title	Referenced in code section number
C-7—(2000)	Guide to the Preparation of Precautionary Labeling and Marking of Compressed Gas Containers	3003.2.2, 3203.4.2
P-1—(2000)	Safe Handling of Compressed Gases in Containers	3005.7
P-18—(1992)	Standard for Bulk Inert Gas Systems at Consumer Sites	3201.1
S-1.1—(1994)	Pressure Relief Device Standards—Part 1—Cylinders for Compressed Gases	3203.2
S-1.2—(1995)	Pressure Relief Device Standards—Part 2—Cargo and Portable Tanks for Compressed Gases	3203.2
S-1.3—(1995)	Pressure Relief Device Standards—Part 3—Stationary Storage Containers for Compressed Gases	2209.5.4.1, 2209.5.4.2, 3203.2
V-1—(2001)	Compressed Gas Association Standard Compressed Gas Cylinder Valve Outlet and Inlet Connections	4106.1.3

CGR

Coast Guard Regulations
c/o Superintendent of Documents
U.S. Government Printing Office
Washington, DC 20402-9325

Standard reference number	Title	Referenced in code section number
46 CFR Parts 30, 32, 35 & 39—1999	Shipping	3406.8

CPSC

Consumer Product Safety Commission
4330 East West Highway
Bethesda, MD 20814

Standard reference number	Title	Referenced in code section number
16 CFR Part 1500.41—1984	Method for Testing Primary Irritant Substances	202
16 CFR Part 1500.42—1984	Test for Eye Irritants	202
16 CFR Part 1500.44—2001	Method for Testing Extremely Flammable and Flammable Solids	3602.1
16 CFR Part 1500—1984	Hazardous Substances and Articles; Administration and Enforcement Regulations	3301.1.3, 3302.1
16 CFR Part 1507—2001	Fireworks Devices	3301.1.3, 3302.1

DOC

U.S. Department of Commerce
100 Bureau Drive, Stop 3460
Gaithersburg, MD 20899

Standard reference number	Title	Referenced in code section number
16 CFR Part 1632—1999	Standard for the Flammability of Mattress and Mattress Pads (FF 4—72, Amended)	803.6.2, 803.7.3

DOL

U.S. Department of Labor
c/o Superintendent of Documents
U.S. Government Printing Office
Washington, DC 20402-9325

Standard reference number	Title	Referenced in code section number
29 CFR Part 1910.1000—1974	Air Contaminants	1204.2.1, 2702.1
29 CFR Part 1910.1200—1999	Hazard Communication	2702.1, 3303.6

DOT

U.S. Department of Transportation
Office of Hazardous Material Standards
400 7th Street, Southwest
Washington, DC 20590

Standard reference number	Title	Referenced in code section number
33 CFR Part 154 —1998	Facilities Transferring Oil or Hazardous Material in Bulk	3406.8
33 CFR Part 155 —1998	Oil or Hazardous Material Pollution Prevention Regulations for Vessels	3406.8
33 CFR Part 156 —1998	Oil and Hazardous Material Transfer Operations	3406.8
49 CFR—1998	Transportation	2605.4, 3302.1
49 CFR Part 1—1999	Transportation	3203.4.3, 3203.8
49 CFR Part 172—1999	Hazardous Materials Tables, Special Provisions, Hazardous Materials Communications, Emergency Response Information and Training Requirements	3304.6.5.2
49 CFR Part 173—1999	Shippers — General Requirements for Shipments and Packagings	3306.3
49 CFR Part 173.137—1990	Shippers — General Requirements for Shipments and Packagings: Class 8 — Assignment of Packing Group	3102.1
49 CFR Parts 100-178—1994	Hazardous Materials Regulations	3301.1, 3301.1.3, 3301.3, 3406.5.1.15

DOTy

U.S. Department of Treasury
c/o Superintendent of Documents
U.S. Government Printing Office
Washington, DC 20402-9325

Standard reference number	Title	Referenced in code section number
27 CFR Part 55—1998	Commerce in Explosives, as amended through April 1, 1998	3302.1, Table 3304.5.2(1), 3304.6.5.2

ICC

International Code Council, Inc.
5203 Leesburg Pike, Suite 600
Falls Church, VA 22041

Standard reference number	Title	Referenced in code section number
ICC/ANSI A117.1—98	Accessible and Usable Buildings and Facilities	907.10.1.4, 1007.6.5, 1010.1, 1010.6.5, 1010.9, 1011.3
ICC 300—02	Standard on Bleachers, Folding and Telescopic Seating and Grandstands	1024.1.1
ICC EC—03	ICC Electrical Code®	603.1.3, 603.1.7, 603.5.2, 604.1, 604.2.15.1, 604.2.15.2, 604.2.16, 605.1, 605.3, 605.4, 605.9, 606.15, 904.3.1, 907.6, 909.11, 909.12.1, 909.16.3, 1106.3.4, 1204.2.3, Table 1304.1, 1404.7, 1503.2.1, 1503.2.1.1, 1503.2.1.2, 1503.2.1.5, 1503.2.1.6, 1503.2.5, 1504.1.4.4, 1504.7.2.2, 1604.5, 1703.2.1, 1803.7.1, 1803.7.2, 1803.7.3, 1903.4, 2004.1, 2201.5, 2205.4, 2208.8.1.2.4, 2209.2.3, 2211.3.1, 2403.12.6.1, 2404.15.7, 2606.4, 2703.7.3, 2703.8.7.1, 2703.9.4, 2704.7, 2705.1.5, 3003.5.6, 3003.6, 3203.7, 3203.7.2, 3403.1, Table 3403.1.1, 3403.1.3, 3404.2.8.12, 3404.2.8.17, 3406.2.8, 3606.5.5, 3606.5.6, 3704.2.2.8
IBC—03	International Building Code®	201.3, 202, 304.1.3, 306.1, 311.1.1, 311.3, 313.1, 313.2, 408.7.2, 504.1, 509.1, 603.2, 603.3.2, 603.5.2, 603.6.1, 603.8, 604.2.8, 604.2.14.1.3, 604.2.15, 604.2.15.1.1, 604.2.16, 608.3, 608.7, 609.4, 609.9, 701.1, 803.7.2, 805.1.2, 806.1, 806.2, Table 806.3, 901.4.1, 901.4.2, 903.1.2, 903.2.4.2, 903.2.8.1, 903.2.9, Table 903.2.13, 903.3.2, 903.3.5.2, 903.6, 907.2.7, 907.2.12, 907.2.18, 907.2.21, 907.15, 909.1, 909.2, 909.3, 909.4.3, 909.5, 909.5.2, 909.5.2.1, 909.10.5, 909.20, 911.2, 1003.2, 1003.3.4, 1003.5, 1007.2, 1007.4, 1007.5, 1007.6.2, 1008.1.3.3, 1008.1.8.1, 1008.1.8.7, 1009.11, 1009.12.1, 1010.1, 1012.1, 1013.4.1, Table 1015.1, 1016.1, Table 1016.1, Table 1018.2, 1019.1, 1019.1.1, 1019.1.2, 1019.1.3, 1019.1.4, 1019.1.8.2, 1020.3, 1020.4, 1020.5, 1021.2, 1023.3, 1023.5.2, 1025.1, 1026.5, 1026.17, 1026.17.1, 1104.6, 1106.17, 1107.1, 1107.4, 1203.3, 1207.1, 1414.1, 1502.1, 1504.1, 1504.1.1, 1504.1.2.6, 1504.1.3, 1505.1, 1801.1, 1801.4, 1803.2, 1803.3.1, 1803.3.2, 1803.3.3, 1803.3.4, 1803.3.8, 1803.14, 1803.14.1, 1803.15.1, 1804.3, 1805.2.2.1, 1805.3.1, 1805.3.2, 1805.3.3, 1903.1, 2005.1, 2009.2, 2009.4, 2009.6, 2201.1, 2201.4, 2203.1, 2207.4, 2208.3, 2208.3.1, 2209.3.2, 2209.3.3, 2210.1, 2211.1, 2211.3.1, 2211.4.1, 2211.8.1.2.3, 2301.3, Table 2306.2, 2306.3.1, 2306.3.2.1, 2306.3.2.2, 2306.8, 2307.2, 2402.1, 2403.8.2, 2404.1, 2503.1, 2703.2.2.2, 2703.2.8, 2703.8.1, Table 2703.8.2, 2703.8.3.1, 2703.8.4.1, 2703.9.9, 2704.13, 2705.2, 2705.3.9, 2801.1, 2904.3, 2904.4, 2904.5, 3203.1.2, 3203.5.2, 3204.2.1.2, 3204.2.2.2, 3205.4.1, 3304.2, 3305.5, 3401.3, 3404.2.7.7, 3404.2.8.1, 3404.2.8.2, 3404.2.9.2, 3404.2.9.3, 3404.3.3.5, 3404.3.7.1, 3404.3.8, 3405.3.4, 3405.3.5.3, 3405.3.7.1, 3405.3.7.2, 3405.3.7.3, 3405.3.7.4, 3405.3.7.5.1, 3406.2.3, 3406.4.1, 3606.2.3, 3606.4.2, 3703.1.4.2, 3705.3.1, 3809.11.2, 3904.1.2, 4004.1.2, 4104.1.4, 4106.1.1, 4304.1
IEBC—03	International Existing Building Code®	102.3, 102.4, 102.5, 1009.3
IFGC—03	International Fuel Gas Code®	201.3, 603.1, 603.1.2, 603.5.2, 603.8, 1403.1, 1403.3, 1604.5, 2101.1, 2103.1, 2104.1, 2104.2, 2201.1, 2201.6, 2404.15.1, 2404.15.2, 2404.16.1, 3001.1, 3501.1, 3803.1, 3803.2.1.7, 3803.3

ICC—continued

IMC—03	International Mechanical Code®	201.3, 308.3.7, 603.1, 603.1.2, 603.2, 603.3, 603.5.2, 603.8, 606.1, 606.2, 606.3, 606.4, 606.7, 606.8, 606.15, 608.5, 610.1, 610.1.1, 909.1, 909.10.2, 1014.5, 1016.4.1, 1204.2.1, 1205.3, 1403.1, 1504.2, 1604.5, 1803.2, 1803.10.4, 1803.14, 1903.2, 1903.3, 2101.1, 2103.1, 2104.2, 2201.1, 2201.6, 2209.3.2, 2211.3.1, 2211.4.3, 2211.7.1, 2404.15.1, 2404.15.2, 2703.8.4.2, 2703.8.5.2, 2703.8.6.2, 2704.3.1, 2903.5, 3003.5.6, 3005.5, 3006.2.2, 3204.2.1.3, 3204.2.2.3, 3205.4.1.1, 3401.3, 3403.6.1, 3404.2.8.9, 3405.3.7.5.1, 3406.2, 3406.4.4, 3703.1.3, 3703.2.3.2, 3704.2.2.7, 3705.3.1, 3705.3.2, 3803.2.1.7
IPC—03	International Plumbing Code®	201.3, 903.3.5, 912.5, 2211.2.3, 2704.2.2.6
IPMC—03	International Property Maintenance Code®	311.1.1
IRC—03	International Residential Code®	202, 1001.1
IUWIC—03	International Urban-Wildland Interface Code™	304.1.2

NEMA

National Electrical Manufacturer's Association
1300 N. 17th Street
Suite 1847
Rosslyn, VA 22209

Standard reference number	Title	Referenced in code section number
250—1997	Enclosures for Electrical Equipment (1,000 Volt Maximum)	3705.2

NFPA

National Fire Protection Association
Batterymarch Park
Quincy, MA 02269

Standard reference number	Title	Referenced in code section number
10—98	Portable Fire Extinguishers	Table 901.6.1, 906.2, 906.3, Table 906.3(1), Table 906.3(2), 2106.3
11—98	Low Expansion Foam	904.7, 3404.2.9.1.2
11A—99	Medium- and High-Expansion Foam Systems	904.7, 3404.2.9.1.2
12—00	Carbon Dioxide Extinguishing Systems	Table 901.6.1, 904.8, 904.11
12A—97	Halon 1301 Fire Extinguishing Systems	Table 901.6.1, 904.9
13—99	Installation of Sprinkler Systems	Table 704.1, 903.3.1.1, 903.3.2, 903.3.5.1.1, 903.3.5.2, 904.11, 907.9, Table 2306.2, 2306.9, 2804.1, 3404.3.7.5.1, 3404.3.8.4
13D—99	Installation of Sprinkler Systems in One- and Two-Family Dwellings and Manufactured Homes	903.3.1.3, 903.3.5.1.1
13R—99	Installation of Sprinkler Systems in Residential Occupancies up to and Including Four Stories in Height	903.1.2, 903.3.1.2, 903.3.5.1.1, 903.3.5.1.2, 903.4
14—00	Installation of Standpipe, Private Hydrants and Hose Systems	905.2, 905.3.4, 905.4.2, 905.8
15—96	Water Spray Fixed Systems for Fire Protection	3404.2.9.1.3
16—99	Installation of Foam-Water Sprinkler and Foam-Water Spray Systems	904.7, 904.11
17—98	Dry Chemical Extinguishing Systems	Table 901.6.1, 904.6, 904.11
17A—98	Wet Chemical Extinguishing Systems	Table 901.6.1, 904.5, 904.11
20—99	Installation of Stationary Pumps for Fire Protection	913.1, 913.2, 913.5.1
22—98	Water Tanks for Private Fire Protection	508.2.2
24—95	Installation of Private Fire Service Mains and their Appurtenances	508.2.1, 1909.5
25—98	Inspection, Testing and Maintenance of Water-Based Fire Protection Systems	508.5.3, Table 901.6.1, 904.7.1, 912.6, 913.5,
30—00	Flammable and Combustible Liquids Code	3403.6.2, 3403.6.2.1, 3404.2.7, 3404.2.7.1 3404.2.7.2, 3404.2.7.3.6, 3404.2.7.4, 3404.2.7.6, 3404.2.7.7, 3404.2.7.8, 3404.2.7.9, 3404.2.9.2, 3404.2.9.3, 3404.2.9.5.1.1, 3404.2.9.5.1.2, 3404.2.9.5.1.3, 3404.2.9.5.1.4, 3404.2.9.5.1.5, 3404.2.9.5.2, 3404.2.9.6.4, 3404.2.10.2, 3404.2.11.4, 3404.2.11.5.2, 3404.2.12.1, 3404.3.1, 3404.3.6, 3404.3.7.2.3, 3404.3.8.4, 3406.8.3
30A—00	Code for Motor Fuel-Dispensing Facilities and Repair Garages	2201.4, 2201.5, 2201.6, 2206.6.3, 2210.1
30B—98	Manufacture and Storage of Aerosol Products	2801.1, 2803.1, 2804.1, Table 2804.3.1, Table 2804.3.2, Table 2804.3.2.2, 2804.4.1, 2804.5.2, 2804.6, Table 2806.2, Table 2806.3, 2806.5, 2806.8, 2807.1, Table 2804.3.2, Table 2804.3.2.2, 2804.4.1, 2804.5.2, 2804.6, Table 2806.2, Table 2806.3
31—01	Installation of Oil-Burning Equipment	603.1.7, 603.3.1, 603.3.3
32—00	Drycleaning Plants	1201.1, 1207.1, 1207.3
33—00	Spray Application Using Flammable or Combustible Materials	1504.1.2
34—00	Dipping and Coating Processes Using Flammable or Combustible Liquids	1505.3, 1505.6.1

NFPA—continued

35—99	Manufacture of Organic Coatings	2001.3, 2005.4
40—97	Storage and Handling of Cellulose Nitrate Motion Picture Film	306.2
50—01	Bulk Oxygen Systems at Consumer Sites	3201.1, 4001.1
50A—99	Gaseous Hydrogen Systems at Consumer Sites	2209.2.1, 3501.1
50B—99	Liquefied Hydrogen Systems at Consumer Sites	2209.2.1, 3201.1
51—97	Design and Installation of Oxygen-Fuel Gas Systems for Welding, Cutting, and Allied Processes	2601.5, 2607.1, 2609.1
51A—01	Acetylene Cylinder Charging Plants	2608.1
52—98	Compressed Natural Gas (CNG) Vehicular Fuel Systems	3001.1
57—99	Liquefied Natural Gas (LNG) Vehicular Fuel Systems	3001.1
58—01	Liquefied Petroleum Gas Code	3801.1, 3803.1, 3803.2.1, 3803.2.1.2 3803.2.1.7, 3803.2.2, 3804.1, 3804.3.1, 3804.4, 3806.3, 3807.2, 3808.1, 3808.2, 3809.11.2, 3811.3
59A—01	Production, Storage and Handling of Liquefied Natural Gas (LNG)	3001.1, 3201.1
61—99	Prevention of Fires and Dust Explosions in Agricultural and Food Products Facilities	Table 1304.1
69—97	Explosion Prevention Systems	911.1, 911.3, Table 1304.1
72—99	National Fire Alarm Code	509.1, Table 901.6.1, 903.4.1, 904.3.5, 907.2, 907.2.1, 907.2.1.1, 907.2.10, 907.2.10.4, 907.2.11.2, 907.2.11.3, 907.2.12.2.3, 907.2.12.3, 907.3, 907.5, 907.6, 907.10.2, 907.11, 907.15, 907.17, 907.18, 907.20, 907.20.2, 907.20.5
80—99	Fire Doors and Fire Windows	703.2, 1008.1.3.3
85—01	Boiler and Combustion System Hazards Code (Note: NFPA 8503 has been incorporated into NFPA 85)	Table 1304.1
86—99	Ovens and Furnaces	2101.1
99—99	Health Care Facilities	3006.4
101—00	Life Safety Code	1024.6.2
110—99	Emergency and Standby Power Systems	604.1, 604.4, 913.5.2, 913.5.3
111—01	Stored Electrical Energy Emergency and Standby Power Systems	604.1, 604.4
120—99	Coal Preparation Plants	Table 1304.1
160—01	Flame Effects Before an Audience	308.3.6
211—00	Chimneys, Fireplaces, Vents and Solid Fuel-Burning Appliances	603.2
231—98	General Storage	2304.2, 2307.2, 2307.2.1, 3404.3.3.9
231C—98	Rack Storage of Materials	2301.1, 2304.2, 2308.2, 2308.2.2, 2308.2.2.1, 2308.4, 2310.1, Table 3404.3.6.3(7), 3404.3.7.5.1, 3404.3.8.4
231D—98	Storage of Rubber Tires	2501.1
241—00	Safeguarding Construction, Alteration, and Demolition Operations	1401.1
260—98	Methods of Tests and Classification System for Cigarette Ignition Resistance of Components of Upholstered Furniture	803.6.1, 803.7.1
261—98	Method of Test for Determining Resistance of Mock-Up Upholstered Furniture Material Assemblies to Ignition by Smoldering Cigarettes	803.5.1
265—98	Standard Method of Fire Tests for Evaluating Room Fire Growth Contribution of Textile Wall Coverings	806.2.3, 806.2.3.1, 806.2.3.2
266—98	Method of Test for Fire Characteristics of Upholstered Furniture Exposed to Flaming Ignition Source	803.5.2
267—98	Method of Test for Fire Characteristics of Mattresses and Bedding Assemblies Exposed to Flaming Ignition Source	803.5.3, 803.6.3, 803.7.4
286—00	Standard Method of Fire Tests for Evaluating Contribution of Wall and Ceiling Interior Finish to Room Fire Growth	806.2.1, 806.2.1.1, 806.3
385—00	Tank Vehicles for Flammable and Combustible Liquids	3406.5.4.5, 3406.6, 3406.6.1
407—96	Aircraft Fuel Servicing	1106.2, 1106.3
430—95	Storage of Liquid and Solid Oxidizers	4004.1.4
480—98	Storage, Handling, and Processing of Magnesium Solids and Powders	Table 1304.1
481—00	Production, Processing, Handling and Storage of Titanium	Table 1304.1
482—96	Production, Processing, Handling and Storage of Zirconium	Table 1304.1
490—98	Storage of Ammonium Nitrate	3301.1.5
495—96	Explosive Materials Code	911.1, 911.4, 3301.1.1, 3301.1.5, 3302.1, 3304.2, 3304.6.2, 3304.6.3, 3304.7.1, 3305.1, 3306.1, 3306.5.2.1, 3306.5.2.3, 3307.1, 3307.9, 3307.11, 3307.15
498—96	Safe Havens and Interchange Lots for Vehicles Transporting Explosives	3301.1.2
505—99	Powered Industrial Trucks, Including Type Designations, Areas of Use, Maintenance, and Operation	2703.7.3
650—98	Pneumatic Conveying Systems for Handling Combustible Particulate Solids	Table 1304.1
651—98	Machining and Finishing of Aluminum and the Production and Handling of Aluminum Powders	Table 1304.1
654—00	Prevention of Fire and Dust Explosions from the Manufacturing, Processing and Handling of Combustible Particulate Solids	Table 1304.1
655—93	Prevention of Sulfur Fires and Explosions	Table 1304.1

NFPA—continued

664—98	Prevention of Fires and Explosions in Wood Processing and Woodworking Facilities	Table 1304.1, 1905.3
701—99	Standard Methods of Fire Tests for Flame-Propagation of Textiles and Films	803.2.2, 805.1, 805.2, 2402.2
703—00	Fire Retardant Impregnated Wood and Fire Retardant Coatings for Building Materials	806.2.6
704—96	Identification of the Hazards of Materials for Emergency Response	606.7, 606.9.3.4, 1802.1, 2703.2.2.1, 2703.2.2.2, 2703.5, 2703.10.2, 2705.1.10, 2705.2.1.1, 2705.4.4, 3203.4.1, 3404.2.3.2
750—00	Standard on Water Mist Fire Protection Systems	Table 901.6.1
1122—97	Model Rocketry	3301.1.4
1123—00	Fireworks Display	3302.1, 3304.2, 3308.1, 3308.2.2, 3308.5, 3308.6
1124—98	Manufacture, Transportation, and Storage of Fireworks and Pyrotechnic Articles	3302.1, 3304.2, 3305.1, 3305.3, 3305.4, 3305.5
1125—95	Manufacture of Model Rocket and High Power Rocket Motors	3301.1.4
1126—01	Use of Pyrotechnics Before a Proximate Audience	3304.2, 3305.1, 3308.1, 3308.2.2, 3308.4, 3308.5
1127—98	High Power Rocketry	3301.1.4
2001—00	Clean Agent Fire Extinguishing Systems	Table 901.6.1, 904.10

UL

Underwriters Laboratories Inc.
333 Pfingsten Road
Northbrook, IL 60062

Standard reference number	Title	Referenced in code section number
30—95	Metal Safety Cans—with Revisions through 2000	2705.1.10, 3405.2.4
58—96	Steel Underground Tanks for Flammable and Combustible Liquids—with Revisions through July 1998	3404.2.13.1.5
197—93	Commercial Electric Cooking Appliances—with Revisions through January 2000	904.11
268—96	Smoke Detectors for Fire Protective Signaling Systems—with Revisions through January 1999	907.2.6.1
300—96	Fire Testing of Fire Extinguishing Systems for Protection of Restaurant Cooking Areas —with Revisions through December 1998	904.11
864—96	Control Units for Fire Protective Signaling Systems—with Revisions through March 1999	909.12
900—94	Air Filter Units—with Revisions through October 1999	1504.3
1275—94	Flammable Liquid Storage Cabinets—with Revisions through March 1997	2703.8.7.1, 3404.3.2.1.1
1316—94	Glass Fiber Reinforced Plastic Underground Storage Tanks for Petroleum Products, Alcohols, and Alcohol-Gasoline Mixtures—with Revisions through April 1996	3404.2.13.1.5
1975—96	Fire Tests for Foamed Plastics Used for Decorative Purpose	803.2.1
2085—97	Protected Aboveground Tanks for Flammable and Combustible Liquids—with Revisions through December 1999	3402.1, 3404.2.9.1.3, 3404.2.9.6.5, 3405.3.8.2
2200—98	Stationary Engine Generator Assemblies	604.1.1
2208—96	Solvent Distillation Units—with Revisions through March 1999	3405.4.1
2245—99	Below-Grade Vaults for Flammable Liquid Storage Tanks	3404.2.8.1

USC

United States Code
c/o Superintendent of Documents
U.S. Government Printing Office
Washington, DC 20402-9325

Standard reference number	Title	Referenced in code section number
18 USC Part 1, Chapter 40	Importation, Manufacture, Distribution and Storage of Explosive Materials	3302.1

Appendix A: Board of Appeals

The provisions contained in this appendix are not mandatory unless specifically referenced in the adopting ordinance.

General Comments

When adopted, this appendix provides jurisdictions with detailed appeals board member qualifications and administrative procedures to supplement the basic requirements found in Section 108 of the code.

Purpose

This appendix contains optional criteria for administrative procedures of the board of appeals and board member qualifications. A jurisdiction that wants to make this appendix a mandatory part of the code needs to specifically list this appendix in its adoption ordinance (see page v of the code for a sample ordinance for adoption).

SECTION A101 GENERAL

A101.1 Scope. A board of appeals shall be established within the jurisdiction for the purpose of hearing applications for modification of the requirements of the *International Fire Code* pursuant to the provisions of Section 108. The board shall be established and operated in accordance with this section, and shall be authorized to hear evidence from appellants and the fire code official pertaining to the application and intent of this code for the purpose of issuing orders pursuant to these provisions.

❖ Section 108 establishes a board of appeals, provides a framework for the composition of the board, defines the limits of the board's authority and requires that the board adopt a set of rules of procedure for its operation. This appendix describes a model board of appeals that the jurisdiction may adopt and gives both a recommended board membership and a recommended operating procedure for the conduct of the board. The jurisdiction must adopt this appendix as part of its fire code before the board and its operations, as described in the appendix, can be authorized.

A101.2 Membership. The membership of the board shall consist of five voting members having the qualifications established by this section. Members shall be nominated by the fire code official or the chief administrative officer of the jurisdiction, subject to confirmation by a majority vote of the governing body. Members shall serve without remuneration or compensation, and shall be removed from office prior to the end of their appointed terms only for cause.

❖ This section details the method of appointment and the general administrative policy pertaining to the board. The five subsections, Sections A101.2.1 through A101.2.5, list the recommended makeup of the board and the qualifications each member should have to serve on the board.

It is important that the decisions of the appeals board be based purely on the technical merits involved in an appeal and that only technical people rule on technical matters, with due regard for state-of-the-art fire protection and construction technology. The board should not be expected to engage in policy or political deliberations. Therefore, members of the appeals board are expected to have experience in matters of fire safety and building construction technology as prescribed in Sections A101.2.1 through A101.2.5.

The board of appeals is to consist of five members recommended by the fire code official and appointed by the chief administrative officer of the jurisdiction — typically, the mayor or city manager. To enhance the integrity of the board and its deliberations and preclude any accusations of partiality on the part of a board member, the members are to serve strictly as volunteers in the community interest, without salary, stipend or any other form of compensation.

A101.2.1 Design professional. One member shall be a practicing design professional registered in the practice of engineering or architecture in the state in which the board is established.

❖ The general architectural or engineering design professional serves on the board to give a balanced perspective to board deliberations. His or her role should be to evaluate the general design features of the appeal to determine whether they satisfy the intent of code requirements.

A101.2.2 Fire protection engineering professional. One member shall be a qualified engineer, technologist, technician or safety professional trained in fire protection engineering, fire science or fire technology. Qualified representatives in this category shall include fire protection contractors and certified technicians engaged in fire protection system design.

❖ This board member is expected to evaluate appeals to determine whether they represent good, logical solutions to fire safety questions that satisfy code requirements and are consistent with current fire protection engineering principles. Note that this position does not specifically require professional registration in the state but may be filled by any technical person qualified in fire protection technology, including experienced fire protection system installation contractors and system designers.

A101.2.3 Industrial safety professional. One member shall be a registered industrial or chemical engineer, certified hygienist, certified safety professional, certified hazardous materials manager or comparably qualified specialist experienced in chemical process safety or industrial safety.

❖ The key words in this section are "chemical process safety" or "industrial safety." This board member should contribute specialized knowledge of this field to board deliberations. His or her concerns should be more with industrial processes and inventories than with design or safety features, although his or her knowledge of the processes and the materials involved in them should bear on both. Note that this position also does not necessarily require professional registration in the state but may be filled by any technical person qualified in matters of chemical or industrial safety. This member's input would be especially valuable in appeals involving the application and enforcement of the hazardous material provisions of the code.

A101.2.4 General contractor. One member shall be a contractor regularly engaged in the construction, alteration, maintenance, repair or remodeling of buildings or building services and systems regulated by the code.

❖ The experienced general contractor adds to the board expertise in determining the practicality of an appeal. This member many times may be a counterbalance to the opinion of the design professional. That is, the design may satisfy the definition of good engineering practice but not be practical or economical to construct or be compatible with an existing structure in the case of building modifications or additions.

A101.2.5 General industry or business representative. One member shall be a representative of business or industry not represented by a member from one of the other categories of board members described above.

❖ This board member, as stated in the code text, is expected to present a point of view not represented by the other four board members. This member could, for example, be nominated by the jurisdiction's Chamber of Commerce or Industrial Development Board.

A101.3 Terms of office. Members shall be appointed for terms of four years. No member shall be reappointed to serve more than two consecutive full terms.

❖ Limiting terms of service serves two purposes. First, it ensures a turnover of membership so that professionals not currently on the board will have an opportunity to serve. Second, it tells the prospective board nominee that there is a definite term of service being committed to. In larger jurisdictions where the board may be very busy, these principles can also help reduce the potential for the so-called "burnout" syndrome among board members.

A101.3.1 Initial appointments. Of the members first appointed, two shall be appointed for a term of 1 year, two for a term of 2 years, one for a term of 3 years.

❖ The staggered terms for initial appointees ensures that no more than two board members will come up for reappointment or replacement in any one year unless one or more members resigns for personal reasons or is replaced for cause. This method of staggered appointment also allows for a smooth transition of board of appeals members, providing continuity of board action over the years.

A101.3.2 Vacancies. Vacancies shall be filled for an unexpired term in the manner in which original appointments are required to be made. Members appointed to fill a vacancy in an unexpired term shall be eligible for reappointment to two full terms.

❖ This section authorizes filling vacancies on the board outside of regular term expirations and sets the limits on terms of service for the new appointee. Vacancies are filled in the same manner as outlined in Section A101.2, with persons possessing qualifications equivalent to those of the board member being replaced.

A101.3.3 Removal from office. Members shall be removed from office prior to the end of their terms only for cause. Continued absence of any member from regular meetings of the board shall, at the discretion of the applicable governing body, render any such member liable to immediate removal from office.

❖ No board member can be removed from office without cause. Although there may be many reasons for removing a serving board member, the only one identified here is chronic failure to perform board duties by attending scheduled regular meetings. This does not mean absence is the only reason for removal.

A101.4 Quorum. Three members of the board shall constitute a quorum. In varying the application of any provisions of this code or in modifying an order of the fire code official, affirmative votes of the majority present, but not less than three, shall be required.

❖ This section clearly defines a quorum and also states that no matter how many members are present at the meeting, at least three affirmative votes are required for passage of a proposal for code variance or modification.

A101.5 Secretary of board. The fire code official shall act as secretary of the board and shall keep a detailed record of all its proceedings, which shall set forth the reasons for its decisions, the vote of each member, the absence of a member and any failure of a member to vote.

❖ This section establishes the fire code official as the board secretary and defines the secretary's duties. Because the deliberations and actions of a board of appeals are considered legal proceedings, the secretary is required to record the proceedings in substantial detail. These details may be needed for any future review of the board's decision or for documentation in executing the procedures established by this appendix.

A101.6 Legal counsel. The jurisdiction shall furnish legal counsel to the board to provide members with general legal advice concerning matters before them for consideration. Members shall be represented by legal counsel at the jurisdiction's expense in all matters arising from service within the scope of their duties.

❖ This section requires the jurisdiction to appoint legal counsel to give the board opinions on the legality of proposed variances and also to represent board members at the jurisdiction's expense should a legal action result from their decisions within the scope of their duties. This legal representation for board members would not be considered remuneration or compensation to the board member, which is prohibited by Section A101.2.

A101.7 Meetings. The board shall meet at regular intervals, to be determined by the chairman. In any event, the board shall meet within 10 days after notice of appeal has been received.

❖ This section establishes ground rules for when meetings are to be held and gives the chairperson of the board the responsibility for making sure meetings are held to give timely response to appellants. In order that an appellant's request be heard in a timely manner, the board must meet within 10 days of the filing of an appeal. In large jurisdictions, where there are likely to be more appeals, the board will often set a regular schedule of meeting dates, such as monthly, and the 10-day rule would not apply.

A101.8 Conflict of interest. Members with a material or financial interest in a matter before the board shall declare such interest and refrain from participating in discussions, deliberations, and voting on such matters.

❖ This section defines conflict of interest for board members and states their expected behavior. All members must recuse themselves from any appeal proceeding in which they have a personal, professional or financial interest.

A101.9 Decisions. Every decision shall be promptly filed in writing in the office of the fire code official and shall be open to public inspection. A certified copy shall be sent by mail or otherwise to the appellant, and a copy shall be kept publicly posted in the office of the fire code official for 2 weeks after filing.

❖ This section establishes guidelines for posting of decisions and notification of appellants. Once an appeal is concluded and a board has taken action, the board secretary must prepare one or more certified copies of the written report for record and distribution. The report copies must be certified as required by applicable state laws pertaining to such matters, which can often be accomplished by a commissioned notary public. The office of the municipal clerk or the board's legal counsel should be able to provide guidance on the subject. To ensure that the decisions and actions of the board are publicly broadcast, a copy of all decisions must be posted in a publicly accessible location in the office of the fire code official for at least two weeks from the date of the action. The appellant must also be given a certified copy of the decision by personal service or ordinary mail service, although certified mail with a requested return receipt would be advisable.

A101.10 Procedures. The board shall be operated in accordance with the Administrative Procedures Act of the state in which it is established or shall establish rules and regulations for its own procedure not inconsistent with the provisions of this code and applicable state law.

❖ This section establishes the responsibility of the board to operate using procedures established by the state or jurisdiction having authority. If those administrative procedures do not exist in a given jurisdiction, the board is to write its own procedures consistent with both the code and existing state law.

Hearings before the board must be open to the public as required by state law. The appellant, the appellant's representative, the building official and any person whose interests are affected must be heard.

Appendix B:
Fire-flow Requirements For Buildings

The provisions contained in this appendix are not mandatory unless specifically referenced in the adopting ordinance.

General Comments

The availability of water is essential for fire-fighting operations. The amount of water required to fight a fire depends on many things, including the type of construction, the location of the fire, the contents of the building, response time and the capabilities of the fire department. Fires will increase in size very quickly from the time of ignition to the arrival of the fire department. Couple these unknowns with the fact that the actual water available varies significantly from one jurisdiction to another, in many cases within the same jurisdiction, and it is easy to see that determining the necessary water supply is not an exact science. The fire flow rates given in this appendix are a simplified version of the method previously published by the Insurance Services Office (ISO) titled *Guide for Determination of Required Fire Flow* (ISO 1972). This particular method took several factors into account that included construction type, size and location of the building. The actual equation used with the ISO Guide was as follows:

$$F = 18\,C\,(A)^{0.5}$$

where:

F = Required fire flow (gpm).

C = Coefficient related to the type of construction.

A = Total floor area (included all stories but excluded the basement).

Type of Construction	Coefficient
Wood-frame construction	1.5
For ordinary construction	1.0
Noncombustible construction	0.8
Fire-resistive construction	0.6

This equation came with various increases and decreases as will be discussed throughout this commentary. The simplified version of this method is included here for two reasons. First, the guidelines were difficult to obtain and second, the methodology was considered overly complex for the degree of accuracy it gave.

Fire-flow determination is not an exact science. Several methods beyond the one presented by ISO have been available over the years and none are able to provide a correct answer for all situations. Fires grow quickly during their initial stages and the amount of water necessary increases as the fire grows. The larger the fire the larger the water supply necessary. This is why sprinklers require, comparably, much less water as they can attack the fire at a very early stage. For these reasons, this appendix does not provide a single answer to solve the problem of determining the amount of fire flow required. It is a decision that must involve many factors.

This appendix was developed independent of the sprinkler standards NFPA 13, 13R and 13D. These standards sometimes have requirements for inside and outside hose streams that are independent of the fire-flow requirements.

Purpose

This appendix provides a tool for jurisdictions to establish a policy for fire-flow requirements. The determination of required fire flow is not an exact science, but having some level of information provides a consistent way of choosing the appropriate fire flow for buildings throughout a jurisdiction.

The primary tool used in this appendix is Table B105.1, which presents fire flows based on construction type and building area. This table was based on the correlation of the ISO method and the construction types used in the *International Building Code*® (IBC®). Because of the wide variations in water availability and the application of fire flow in different communities these provisions are presented in this appendix.

The important message sent by this appendix is that some sort of policy should be in place to ensure that requirements are consistent within a jurisdiction. Fire-flow requirements have the tendency to be somewhat controversial for the simple reason that they can be very costly to construct and install and appear to the building owners, in many cases, to yield little benefit.

SECTION B101 GENERAL

B101.1 Scope. The procedure for determining fire-flow requirements for buildings or portions of buildings hereafter constructed shall be in accordance with this appendix. This appendix does not apply to structures other than buildings.

❖ This appendix is clearly intended for buildings only and would not be applicable to outside storage areas or similar hazards. The provisions were drafted based on the ISO method, which focused on buildings and takes into account the construction type and building size. This method may not translate very well to other hazards but could be used as a starting point for other types of facilities. Also, the scope of this appendix is intended for new construction rather than existing buildings. Providing fire flow is generally costly, and requiring it for existing buildings would likely be unreasonable. Again, the appendix applies only if specifically adopted by a jurisdiction.

SECTION B102 DEFINITIONS

B102.1 Definitions. For the purpose of this appendix, certain terms are defined as follows:

❖ Definitions can help in the understanding and application of the code requirements. Having this definition here puts it close to the subject matter it pertains to.

FIRE FLOW. The flow rate of a water supply, measured at 20 pounds per square inch (psi) (138 kPa) residual pressure, that is available for fire fighting.

❖ A set of consistent criteria is used to measure the water available for fire fighting. The criterion is that the fire flow available be measured at a residual pressure of 20 pounds per square inch (psi) (138 kPa). Residual pressure is the pressure measured when the water supply is flowing versus static pressure, which is measured when the water is not flowing. The criterion of 20psi (138 kPa) residual is used because it is the minimum pressure recommended for fire engine use and provides a consistent point from which to measure the available flow. Flow will vary based on the pressure for each system.

FIRE-FLOW CALCULATION AREA. The floor area, in square feet (m^2), used to determine the required fire flow.

❖ This term defines what portion of the building is to be accounted for when applying Table B105.1. This term differs from the IBC definition of "Fire area" in that this definition allows a fire flow calculation area, for the purposes of defining fire flow, to be divided only by a fire wall with no openings. Fire barriers and partitions could not be used to create separate fire-flow calculation areas (see commentary, Section B104.2).

SECTION B103 MODIFICATIONS

B103.1 Decreases. The fire chief is authorized to reduce the fire-flow requirements for isolated buildings or a group of buildings in rural areas or small communities where the development of full fire-flow requirements is impractical.

❖ The purpose of this section is to recognize that many factors may require adjustments to the numbers in Table B105.1. This particular section generally addresses issues such as proximity to exposures, general location, configuration and practicality.

This section gives the fire code official the authority to make adjustments based on the impracticality of fire-flow requirements in rural areas. The text conveys the message that the requirements found here will not be appropriate for all situations. For example, requiring that a fire main be pulled to a house located by itself in the middle of a large open field is impractical. This is especially the case if the fire department has a considerable response time. When a considerable response time exists for buildings, such as one- and two-family dwellings in isolated locations, the effectiveness of fire flow is likely to be low because intervention may not be necessary once the fire department arrives. Although, if this same house is located in an urban-wildland interface area this may be a different issue. The house may be a complete loss but the protection of the wildland from the exposure of this fire may be necessary.

This section is attempting to provide flexibility to better fit the needs of a specific community. In addition to the example above, the following examples describe instances where requiring the full fire flow given in Table B105.1 would be unreasonable:

- A rural area dependent on tanker supplies and on-site water sources;
- A water system for a small town or community is provided for domestic consumption with some incidental fire hydrants, but with no serious intent to provide fire protection water or
- A fire department that does not have the equipment to pump the required fire flow.

Section B103.3 discusses alternative approaches for water supplies.

B103.2 Increases. The fire chief is authorized to increase the fire-flow requirements where conditions indicate an unusual susceptibility to group fires or conflagrations. An increase shall not be more than twice that required for the building under consideration.

❖ As discussed in Section B103.1, the fire flows provided in this appendix are not appropriate for all situations. This section gives the jurisdictions the authority to increase fire flow when necessary. The focus of this section is on densely populated occupancies or simply buildings arranged in a way that makes conflagration more likely. These provisions were based on the ISO Guidelines, and this particular section is primarily focusing on proximity to exposures. The ISO Guidelines included specific increases for close proximity buildings and other exposures. This appendix does not include specific increases but does give the jurisdictions the authority to make adjustments based on these concerns. Adjustments, however, are not to exceed twice the required fire flow. This section and Section B103.1 together suggest that the jurisdiction have a specific policy that can anticipate various scenarios to enable a consistent approach to those undertaking construction in the jurisdiction.

B103.3 Areas without water supply systems. For information regarding water supplies for fire-fighting purposes in rural and suburban areas in which adequate and reliable water supply systems do not exist, the fire code official is authorized to utilize NFPA 1142 or the *International Urban Wildland Interface Code.*

❖ In many cases the infrastructure simply does not exist to provide large amounts of water as required by Table B105.1, but the hazards require that some level of water be available for fire-fighting activities. This section provides an outside resource in either NFPA 1231 or the use of the International *Urban-Wildland Interface Code*™ (IUWIC™). NFPA 1231 gives options for areas where adequate and reliable water supplies are not available. It provides minimum requirements for situations where the water is to come from sources such as a river, canal, stream or pond.

The IUWIC, which is currently undergoing public comment before becoming a code within the family of *International Codes*®, includes some alternative approaches to address the lack of water supplies in areas where providing necessary fire flow is typically difficult. Basically, this document provides a series of alternatives, including more restrictive construction types and providing defensible spaces to compensate for a reduced water supply. Also, water supply methodologies are presented for both natural and man-made sources.

SECTION B104
FIRE-FLOW CALCULATION AREA

B104.1 General. The fire-flow calculation area shall be the total floor area of all floor levels within the exterior walls, and under the horizontal projections of the roof of a building, except as modified in Section B104.3.

❖ This section establishes the area that is to be taken to Table B105.1 to determine the minimum fire flow including how to separate a building into multiple fire-flow calculation areas in order to have lower fire-flow requirements. Fire-flow calculation areas are defined differently here than is the term "Fire area" in the IBC. This difference is described in the commentary for the definition of fire area in Section B102 of this appendix.

The fire-flow calculation area includes all floors of a building and the horizontal projections (see Figure B104.1). The area under horizontal projections is important because either combustibles may be located below those areas or the construction itself is combustible. Both situations add to the fire loading of the building. In some cases, horizontal projections can cover a significant area. There are some exceptions for Type IA and IB construction because they are inherently less combustible structures (see commentary, Section B104.3).

B104.2 Area separation. Portions of buildings which are separated by fire walls without openings, constructed in accordance with the *International Building Code,* are allowed to be considered as separate fire-flow calculation areas.

❖ To reduce the amount of fire flow required, fire walls without openings can be constructed to create separate fire-flow calculation areas. Fire barriers or fire partitions cannot be used to create separate fire-flow calculation areas.

B104.3 Type IA and Type IB construction. The fire-flow calculation area of buildings constructed of Type IA and Type IB construction shall be the area of the three largest successive floors.

> **Exception:** Fire-flow calculation area for open parking garages shall be determined by the area of the largest floor.

❖ Type IA and IB construction are essentially noncombustible and have the tendency to limit fire spread within the buildings more so than other construction types. Therefore, the fire-flow calculation area needs to include only the three largest successive floors. Successive floors are specified because of the logical progression of a fire. The concept of three largest successive floors appears to come from the ISO Guidelines. These guidelines allowed the fire-flow calculation area for fire-resistive construction to only include six successive floors if vertical openings were not protected, and three successive floors if the vertical openings were protected. Taking the three largest floors when they are separated from one another may be overly conservative.

The exception to this section allows open parking garages to count only the largest floor for the fire-flow calculation area. This is probably related to the fact that fires in such facilities tend to be limited to one or two cars as well as to the large openings through which the hot gases and smoke from a fire can dissipate quickly and limit the intensity of the fire.

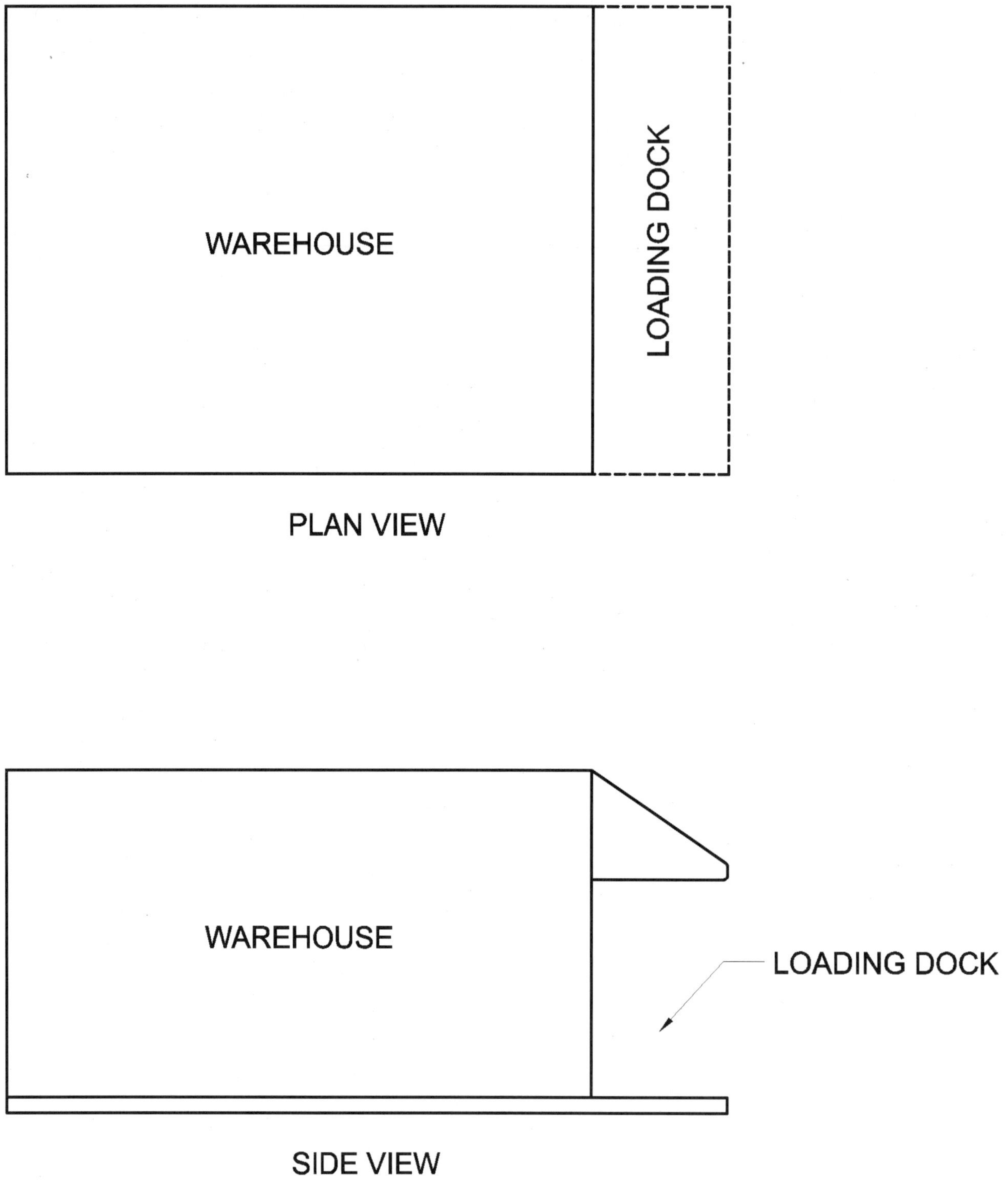

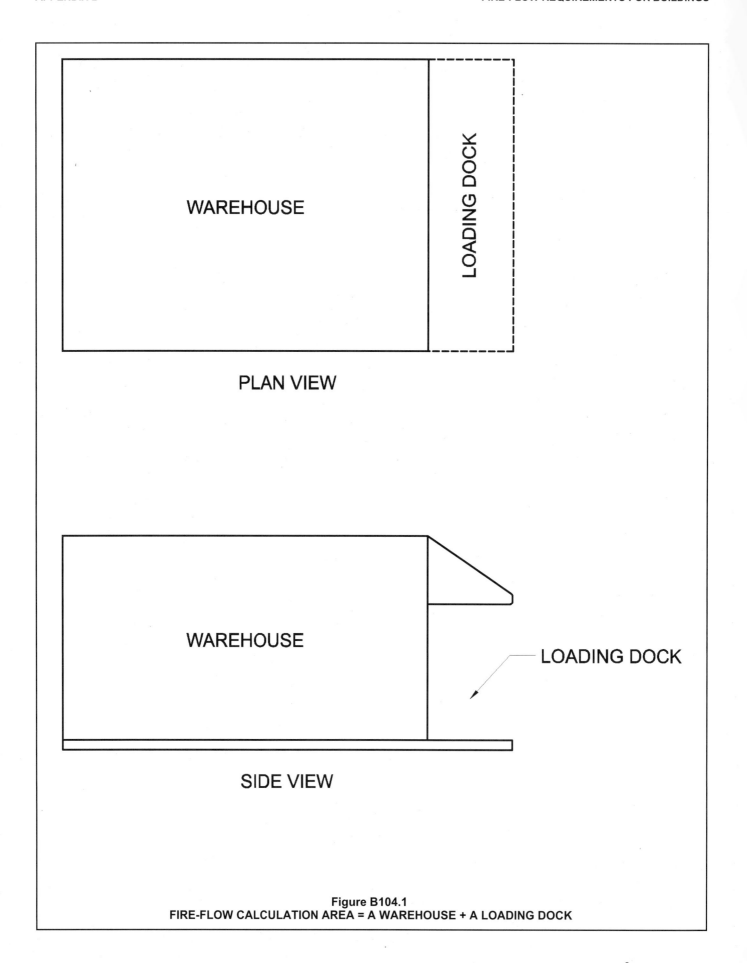

Figure B104.1
FIRE-FLOW CALCULATION AREA = A WAREHOUSE + A LOADING DOCK

SECTION B105
FIRE-FLOW REQUIREMENTS FOR BUILDINGS

B105.1 One- and two-family dwellings. The minimum fire-flow requirements for one- and two-family dwellings having a fire-flow calculation area which does not exceed 3,600 square feet (344.5 m^2) shall be 1,000 gallons per minute (3785.4 L/min). Fire flow and flow duration for dwellings having a fire-flow calculation area in excess of 3,600 square feet (344.5 m^2) shall not be less than that specified in Table B105.1.

Exception: A reduction in required fire flow of 50 percent, as approved, is allowed when the building is provided with an approved automatic sprinkler system.

❖ One- and two-family dwellings under 3,600 square feet (345 m^2) in area require a minimum fire flow of 1,000 gallons per minute (3785 L/min). The original ISO Guidelines provide a simplified approach for one- and two-family dwellings. That approach stated that fire flows should be based on a limitation of two stories and a relationship to proximity to exposures. The fire-flow requirements based on proximity to exposures in the ISO Guidelines were as follows:

Exposure Distance (ft)	Fire-flow Requirement (gpm)
Over 100	500
31-100	750-1000
11-30	1000-1500
10 or less	1500-2000

This appendix uses 1,000 gpm (3785 L/min) , which would be equivalent to a 30-foot (9144 mm) distance from exposures. This was taken as an average to provide a reasonable number for a majority of one- and two-family dwellings. Also, as discussed in the beginning of this appendix, based on the amount of variability involved with fighting fires, taking an average and applying to all one- and two-family dwellings may be the most reasonable approach.

This appendix also does not use the two-story limitation and instead uses an area limitation of 3,600 square feet (345 m^2) for the 1,000-gpm (3785 L/min) requirement. This is a more realistic approach because the ISO Guidelines probably did not anticipate the larger floor area of today's houses and the large number of townhouses. When one- or two-family dwellings exceed 3,600 square feet (345 m^2), Table B105.1 must be used.

There is an exception to this section that allows a reduction in fire flow by 50 percent when sprinklers are installed. A similar exception is found also in Section B105.2 for all other types of buildings.

Application of this concept has been heavily debated. First, this exception is allowed only when approved by the jurisdiction. Also, the exception does not address the type of sprinklers used or whether the system is monitored. More specifically, it does not restrict the type of system to NFPA 13 or NFPA 13R.

A review of the original ISO Guidelines reveals that there was no reduction for sprinklers in one- and two-family dwellings. However, in 1972 sprinklers were extremely uncommon within homes, and since that time sprinkler technology has changed dramatically. Section B105.2 contains more discussion on the application of this concept of reductions for sprinklers. Generally, the exception is intended to encourage the use of sprinklers because it is easier to control a fire that is attacked during the incipient stages.

B105.2 Buildings other than one- and two-family dwellings. The minimum fire flow and flow duration for buildings other than one- and two-family dwellings shall be as specified in Table B105.1.

Exception: A reduction in required fire flow of up to 50 percent, as approved, is allowed when the building is provided with an approved automatic sprinkler system installed in accordance with Section 903.3.1.1 or 903.3.1.2 of the *International Fire Code*. Where buildings are also of Type I or II construction and are a light-hazard occupancy as defined by NFPA 13, the reduction may be up to 75 percent. The resulting fire flow shall not be less than 1,500 gallons per minute (5678 l/min) for the prescribed duration as specified in Table B 105.1.

❖ This section refers all other buildings beyond one- and two-family dwellings to Table B105.1 for the minimum fire flow requirements. As already noted, this section contains an exception that allows a reduction in available fire flow when sprinklers are installed. In this case, the reduction is 75 percent versus 50 percent as allowed for one- and two-family dwellings. As with Section B105.1, this reduction must be approved by the jurisdiction.

Again, the intent of the exception is to encourage the use of sprinklers. This exception does not link back to any other portions of the *International Fire Code®* (IFC®) or the IBC in terms of height and area requirements and limitations. Therefore, it can be used in addition to any trade-offs for sprinklers. Keep in mind that as the area of the building increases so do the fire-flow requirements. Therefore, even though a reduction may be given to a building that has already increased its area based on sprinklers, the overall fire flow will be larger because of this area increase.

This section has historically been applied in many different ways because of concerns with sprinkler reliability and the specific capabilities of a particular fire department. The important thing to note is that the code gives the fire code official the authority to adjust the allowance in the reduction with the use of the term "approved." Therefore, there is no single way to approach this issue, but, regardless of which way is chosen, the jurisdiction must be as consistent as possible. This may require the use of a policy or other methodology.

The original ISO Guidelines allowed a 25-percent reduction for sprinklers. As mentioned in Section B105.1, sprinkler technology has changed dramatically since the guidelines were developed. Also, the ISO Guidelines allowed reduction in fire flow for buildings with light fire loads that this appendix does not.

A common debate that arises is whether or not sprinkler flow and fire flow are cumulative. In other words, if a building had a fire-flow requirement of 1,500 gpm (5678 L/min) and a sprinkler-flow requirement of 500 gpm (1893 L/min), would the total required water supply be 2,000 gpm (7570 L/min) or 1,500 gpm (5678 L/min).

This appendix does not specifically address this issue; however, if the sprinkler water supply and the fire-flow water supply are from completely separate sources, both the 1,500 gpm (5678 L/min) for fire flow and the 500 gpm (1893 L/min) for sprinkler flow should be required. Generally, the intent is that the water supply be cumulative when they are shared, especially if a reduction has already been taken for sprinklers. If a reduction is not taken for sprinklers, sharing the water supply may be possible. An approach typically observed is to allow a 50-percent reduction and not require the sprinkler flow and fire flow to be cumulative. If the sprinkler reduction allowed is 75 percent, the water supply would be cumulative.

Another common question regarding the reductions allowed within the exception is whether the duration required in Table B105.1 can also be reduced. This appendix does not address this issue. The ISO Guidelines did not address duration for any situation within the guidelines.

TABLE B105.1. See page B-7.

❖ Table B105.1 states the fire-flow and duration requirements based on the fire area, as defined by the definition in this appendix and Section B103, and the construction types defined in the IBC. As the construction type becomes more combustible, the fire-flow requirements will increase. Likewise, as the area of the building increases, the fire-flow requirements increase. The last column also specifies a minimum duration of fire flow. The duration of fire flow varies from a minimum of 2 hours to 4 hours. Flow duration may be an issue that each jurisdiction may need to consider when assessing the capabilities of the department, the hazards presented and the realistic availability of water supply.

Applying this table, for example, a 50,000-square-foot (4546 m^2) Type IV building would require a fire flow of 4,000 gpm (15 140 L/min) with a duration of 4 hours. If the building was sprinklered and the full 75-percent reduction was allowed, the required fire flow would be 1,500 gpm (5678 L/min) [75-percent reduction would result in 1,000 gpm (3785 L/min), which is lower than the minimum of 1,500 gpm) (5678 L/min)].

This table does not address use and occupancy classifications. A Type IA construction building housing a Group A occupancy would be treated the same as a Type IA construction building housing a Group H or Group F occupancy. Again, this table was formed based on the approaches presented by the ISO Guidelines which focus on construction types. It should be noted that Group R occupancies are specifically allowed a 25-percent reduction. This reflects the reduction allowed by the ISO Guidelines for residential occupancies.

A common question asked when applying this table is how to deal with a building that incorporates multiple construction types. Such scenarios would be better addressed through a percentage approach. For example, in a building that has two construction types, Types IA and VA, having areas of 25,000 square feet (2323 m^2) and 10,000 square feet (929 m^2), respectively, the fire flow would be calculated as follows:

Total building area

25,000 square feet (Type IA) + 10,000 square feet (Type VA) = 35,000 square feet (3252 m^2)

Fire flow per construction type

Type IA at 35,000 square feet = 2,000 gpm (7370 L/min)

Type VA at 35,000 square feet = 3,250 gpm (12 112 L/min)

Percentage of building

IA = 25,000/35,000 × 100 = 71.4 percent

VA = 10,000/35,000 × 100 = 28.6 percent

Therefore

0.714 (2,000 gpm) + 0.286 (3,250 gpm) = 2,357.5

= Approximately 2,350 gpm (8894 L/min)

SECTION B106 REFERENCED STANDARDS

ICC	IBC	International Building Code	B104.2, Table B105.1
ICC	IFC	International Fire Code	B105.2
ICC	IUWIC	International Urban-Wildland Interface Code	B103.3
NFPA	1142	Standard on Water Supplies for Suburban and Rural Fire Fighting	B103.3

Bibliography

Davis, L. "Rural Fire Fighting Operations." *Fire Service Information*, Iowa State University, February 1984.

IFCI, *UFC Code Applications Manual*. Whittier, CA: International Fire Code Institute, 1998.

IUWIC-03 *International Urban-Wildland Interface Code*. Falls Church, VA: International Code Council, 2003.

"Guide for Determination of Required Fire Flow." New York: Insurance Services Office, 1972.

Linder, K. "Water Supply Requirements for Fire Protection." Section 6, Chapter 5. *NFPA Handbook*, 18th ed. Quincy, MA: National Fire Protection Association, 1997.

NFPA 1142-01, *Water Supplies for Suburban and Rural Fire Fighting*. Quincy, MA: National Fire Protection Association, 2001.

Pumping Apparatus Driver/Operator Handbook, 1st ed. Stillwater, OK: International Fire Service Training Association, 1998.

Smith, P.D. "What Are the Real Fire Flow Requirements?" *Fire Journal*, 1975.

TABLE B105.1
MINIMUM REQUIRED FIRE FLOW AND FLOW DURATION FOR BUILDINGS[a]

FIRE-FLOW CALCULATION AREA (square feet)					FIRE FLOW (gallons per minute)[c]	FLOW DURATION (hours)
Type IA and IB[b]	Type IIA and IIIA[b]	Type IV and V-A[b]	Type IIB and IIIB[b]	Type V-B[b]		
0-22,700	0-12,700	0-8,200	0-5,900	0-3,600	1,500	2
22,701-30,200	12,701-17,000	8,201-10,900	5,901-7,900	3,601-4,800	1,750	
30,201-38,700	17,001-21,800	10,901-12,900	7,901-9,800	4,801-6,200	2,000	
38,701-48,300	21,801-24,200	12,901-17,400	9,801-12,600	6,201-7,700	2,250	
48,301-59,000	24,201-33,200	17,401-21,300	12,601-15,400	7,701-9,400	2,500	
59,001-70,900	33,201-39,700	21,301-25,500	15,401-18,400	9,401-11,300	2,750	
70,901-83,700	39,701-47,100	25,501-30,100	18,401-21,800	11,301-13,400	3,000	3
83,701-97,700	47,101-54,900	30,101-35,200	21,801-25,900	13,401-15,600	3,250	
97,701-112,700	54,901-63,400	35,201-40,600	25,901-29,300	15,601-18,000	3,500	
112,701-128,700	63,401-72,400	40,601-46,400	29,301-33,500	18,001-20,600	3,750	
128,701-145,900	72,401-82,100	46,401-52,500	33,501-37,900	20,601-23,300	4,000	4
145,901-164,200	82,101-92,400	52,501-59,100	37,901-42,700	23,301-26,300	4,250	
164,201-183,400	92,401-103,100	59,101-66,000	42,701-47,700	26,301-29,300	4,500	
183,401-203,700	103,101-114,600	66,001-73,300	47,701-53,000	29,301-32,600	4,750	
203,701-225,200	114,601-126,700	73,301-81,100	53,001-58,600	32,601-36,000	5,000	
225,201-247,700	126,701-139,400	81,101-89,200	58,601-65,400	36,001-39,600	5,250	
247,701-271,200	139,401-152,600	89,201-97,700	65,401-70,600	39,601-43,400	5,500	
271,201-295,900	152,601-166,500	97,701-106,500	70,601-77,000	43,401-47,400	5,750	
295,901-Greater	166,501-Greater	106,501-115,800	77,001-83,700	47,401-51,500	6,000	
—	—	115,801-125,500	83,701-90,600	51,501-55,700	6,250	
—	—	125,501-135,500	90,601-97,900	55,701-60,200	6,500	
—	—	135,501-145,800	97,901-106,800	60,201-64,800	6,750	
—	—	145,801-156,700	106,801-113,200	64,801-69,600	7,000	
—	—	156,701-167,900	113,201-121,300	69,601-74,600	7,250	
—	—	167,901-179,400	121,301-129,600	74,601-79,800	7,500	
—	—	179,401-191,400	129,601-138,300	79,801-85,100	7,750	
—	—	191,401-Greater	138,301-Greater	85,101-Greater	8,000	

For SI: 1 square foot = 0.0929 m^2, 1 gallon per minute = 3.785 L/m, 1 pound per square inch = 6.895 kPa.

a. The minimum required fire flow shall be permitted to be reduced by 25 percent for Group R.

b. Types of construction are based on the *International Building Code.*

c. Measured at 20 psi.

Appendix C:
Fire Hydrant Locations and Distribution

The provisions contained in this appendix are not mandatory unless specifically referenced in the adopting ordinance.

General Comments

Fire hydrants are the one of the primary ways to access water for fighting fires. The location and spacing of hydrants is important to the success of fire-fighting operations. The difficulty with determining the spacing of fire hydrants is that every situation is unique and has unique challenges. Finding one methodology for determining hydrant spacing is difficult. This particular appendix gives one methodology fire departments can work with to set a policy for new buildings and facilities.

This methodology is located in an appendix because, as with fire-flow requirements (Appendix B), many factors affect the need for and location of hydrants. Also, in many jurisdictions, hydrant spacing is prescribed by zoning regulations or by the water authority. This appendix is simply one approach for spacing hydrants when no other guidelines are given.

Purpose

Section 508.5.1 requires that hydrants be within 400 feet of all portions of new buildings and facilities. If a building or facility cannot meet that criterion, on-site fire hydrants and mains are required. This appendix provides some guidance on the spacing of the on-site hydrants based on the required fire flow. These guidelines could also be used to require hydrants on public streets and roads when no other entity is regulating such hydrants.

The general approach is to use fire-flow requirements to determine the number and spacing of hydrants–the higher the fire-flow requirements, the larger the number of hydrants required and the smaller the spacings between hydrants. The spacings given in this appendix are independent of the distance to a building and are simply focused on having the correct number and spacing of hydrants on a fire apparatus access road.

SECTION C101
GENERAL

C101.1 Scope. Fire hydrants shall be provided in accordance with this appendix for the protection of buildings, or portions of buildings, hereafter constructed.

❖ This section states that the appendix is intended to address only new buildings and portions of buildings. This is similar to the scope of Appendix B. Generally, both fire flow and fire hydrants are fairly expensive and requiring them for an existing building would likely be cost prohibitive.

SECTION C102
LOCATION

C102.1 Fire hydrant locations. Fire hydrants shall be provided along required fire apparatus access roads and adjacent public streets.

❖ For fire flow to be effective it should be located along the fire apparatus roads and adjacent public streets. An approved fire apparatus access road is required to have a proper water supply and access to that water supply. Hydrants must be located along the road to be of use to the fire department.

SECTION C103
NUMBER OF FIRE HYDRANTS

C103.1 Fire hydrants available. The minimum number of fire hydrants available to a building shall not be less than that listed in Table C105.1. The number of fire hydrants available to a complex or subdivision shall not be less than that determined by spacing requirements listed in Table C105.1 when applied to fire apparatus access roads and perimeter public streets from which fire operations could be conducted.

❖ This section is focused only on the number of hydrants required. Table C105.1 shows a minimum number of hydrants based on the required fire flow and spacing limitations. In many cases, a particular fire flow may require only one or two hydrants. If there is a complex of buildings, each with a low required fire flow, the spacing requirements would dictate additional hydrants. A fire flow of 2,000 gpm (126 L/s) would require only two hydrants with a maximum average spacing of 450 feet (137 160 mm) but because of the size of the complex, additional hydrants may be necessary to achieve proper spacing along the access road (see Figure C103.1). In this particular case, the number of hydrants, not the spacing, drives the layout in this particular site.

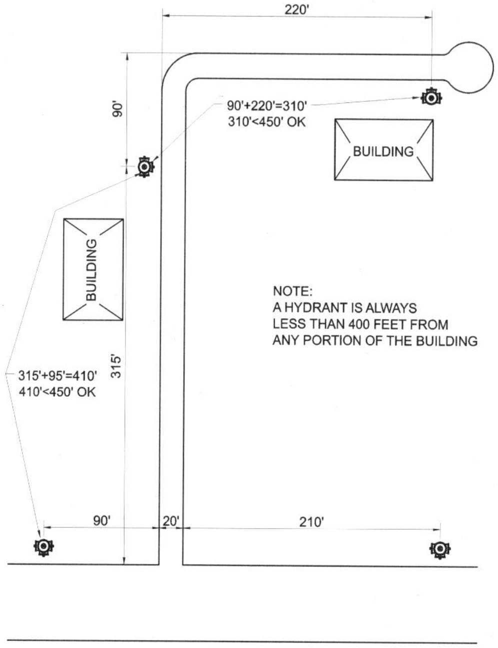

For SI: 1 foot = 304.8 mm.

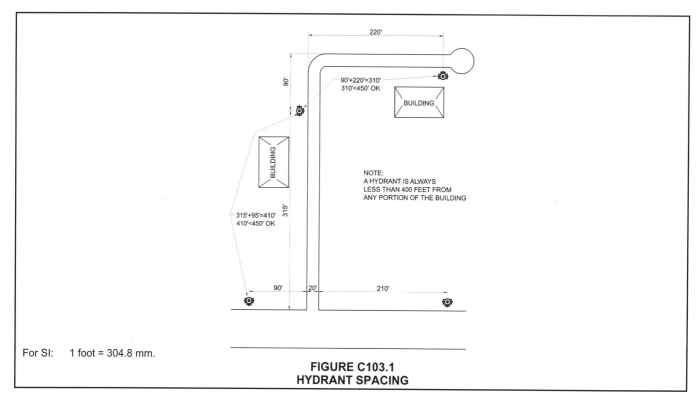

FIGURE C103.1
HYDRANT SPACING

SECTION C104
CONSIDERATION OF EXISTING FIRE HYDRANTS

C104.1 Existing fire hydrants. Existing fire hydrants on public streets are allowed to be considered as available. Existing fire hydrants on adjacent properties shall not be considered available unless fire apparatus access roads extend between properties and easements are established to prevent obstruction of such roads.

❖ To meet the number and spacing requirements of this appendix, existing hydrants may be counted as available. In smaller buildings, this could mean that the fire hydrants available may be sufficient to meet the hydrant spacing needs. Hydrants on adjacent property should not be used unless access is always available to the fire department. This may require an easement on the adjacent property.

SECTION C105
DISTRIBUTION OF FIRE HYDRANTS

C105.1 Hydrant spacing. The average spacing between fire hydrants shall not exceed that listed in Table C105.1.

> **Exception:** The fire chief is authorized to accept a deficiency of up to 10 percent where existing fire hydrants provide all or a portion of the required fire hydrant service.

Regardless of the average spacing, fire hydrants shall be located such that all points on streets and access roads adjacent to a building are within the distances listed in Table C105.1.

❖ This section states that the spacing designated in Table C105.1 is an average. Real-life conditions, however, often result in the need for approximate spacing of hydrants. In some cases, good judgement would dictate deviation from rigid exact spacing so that hydrants could be located where more than one building could access them [see Figure C105.1(1)].

There is an exception that places a limit of a 10-percent deficiency in average hydrant spacings when using existing hydrants. For example, if the average spacing allowed is 350 feet (106 680 mm), the largest average spacing allowed by this exception would be 350 x 1.10 = 385 feet (117 348 mm). The allowance of longer average spacings for existing hydrants is important in enabling their use with new buildings.

The last sentence of this section requires that no point on a street or access road be beyond the distance to a hydrant shown in the last column of Table C105.1. This distance is not an average but a maximum distance [see Figure C105.1(2)].

TABLE C105.1. See page C-3.

❖ This table is referenced throughout this appendix for minimum number of hydrants and spacing limits. These limits are based on the fire flow required for a particular building. In terms of the spacing limitations, there are two criteria. First is the average distance between hydrants. Second is a maximum distance limitation from any point on a street or access way to a hydrant. When the number of required fire hydrants increases, the maximum distance does not always correspond to half of the average distance between hydrants. Instead, this distance is larger than half the distance to compensate for the fact that the average spacing between hydrants as the number of hydrants increases may result in spacings larger than those shown in column three. Otherwise, the maximum distance listed in column four would limit the spacing [see Figure C105.1(3)].

One question that is often asked with these requirements is whether the table or the entire appendix, for that matter, requires minimum flows per hydrant. This particular appendix does not address minimum flow per hydrant or how the flow is to be subdivided. Fire-flow requirements are addressed in Appendix B.

Also, it is important to point out that there is no direct connection between Appendix B and Appendix C. In other words, there is no restriction on using a reduced or base fire flow from Appendix B in determining hydrant requirements, according to Appendix C. The fire flow used to determine hydrant number and layout can be the reduced fire flow determined in Appendix B or whatever method is used within a jurisdiction to determine fire flow for a building. This has been an area of confusion in the past.

Note a

This note is referenced in the third column for the average spacing between hydrants and restricts the spacing between hydrants on dead-end streets or roads to 100 less then the required spacing. This avoids a hydrant being located unnecessarily remote from hazards. Otherwise, a hydrant could literally be placed at the end of street and provide little benefit to buildings on that street.

Note b

This note provides specific guidance for multilane roads, such as freeways, that carry a heavy traffic load each day. These hydrants are typically intended for use only with highway hazards and may be spaced every 500 feet (152 400 mm) on each side in alternating positions [resulting in a hydrant every 250 feet (76 200 mm)] [see Figure C105.1(4)].

Note c

This note requires that fire mains on access roads without structures have hydrants at least every 1,000 feet (394 800 mm) for hazards such as car fires or other traffic-related hazards. This does not require that water mains specifically be placed along roadways for this purpose; only that when such mains exists, hydrants must be installed.

Note d

This note requires, as does Note a, that the distance to a hydrant be reduced by 50 feet (13 240 mm) on dead-end streets and roads. Again, this avoids a hydrant being unnecessarily far from buildings.

Note e

This note applies only to situations where the table requires eight or more hydrants and that hydrants be added for every 1,000 gpm (63 L/s) or fraction thereof of fire flow required beyond 7,500 gpm (473 L/s). This means that one additional hydrant would be required for fire flows up to 8,500 gpm (536 L/s). No minimum fire flow per hydrant is specified.

TABLE C105.1
NUMBER AND DISTRIBUTION OF FIRE HYDRANTS

FIRE-FLOW REQUIREMENT (gpm)	MINIMUM NUMBER OF HYDRANTS	AVERAGE SPACING BETWEEN HYDRANTS[a, b, c] (feet)	MAXIMUM DISTANCE FROM ANY POINT ON STREET OR ROAD FRONTAGE TO A HYDRANT[d]
1,750 or less	1	500	250
2,000-2,250	2	450	225
2,500	3	450	225
3,000	3	400	225
3,500-4,000	4	350	210
4,500-5,000	5	300	180
5,500	6	300	180
6,000	6	250	150
6,500-7,000	7	250	150
7,500 or more	8 or more[e]	200	120

For SI: 1 foot = 304.8 mm, 1 gallon per minute = 3.785 L/m.

a. Reduce by 100 feet for dead-end streets or roads.

b. Where streets are provided with median dividers which can be crossed by fire fighters pulling hose lines, or where arterial streets are provided with four or more traffic lanes and have a traffic count of more than 30,000 vehicles per day, hydrant spacing shall average 500 feet on each side of the street and be arranged on an alternating basis up to a fire-flow requirement of 7,000 gallons per minute and 400 feet for higher fire-flow requirements.

c. Where new water mains are extended along streets where hydrants are not needed for protection of structures or similar fire problems, fire hydrants shall be provided at spacing not to exceed 1,000 feet to provide for transportation hazards.

d. Reduce by 50 feet for dead-end streets or roads.

e. One hydrant for each 1,000 gallons per minute or fraction thereof.

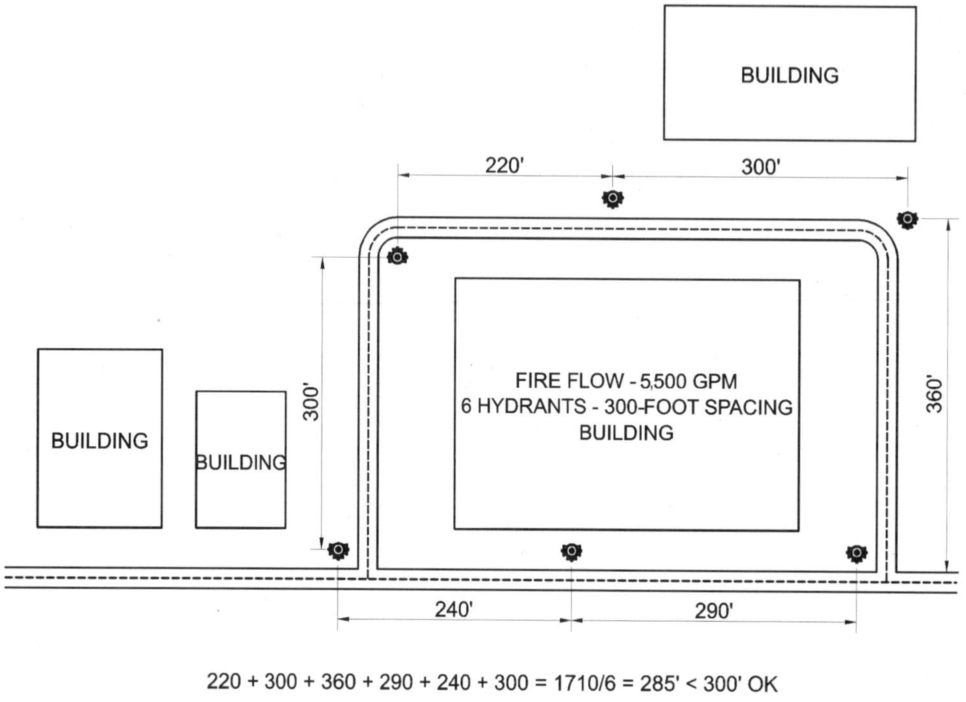

For SI: 1 foot = 304.8 mm, 1 gallon per minute = .06308 L/s.

FIGURE C105.1(1)
AVERAGE HYDRANT SPACING

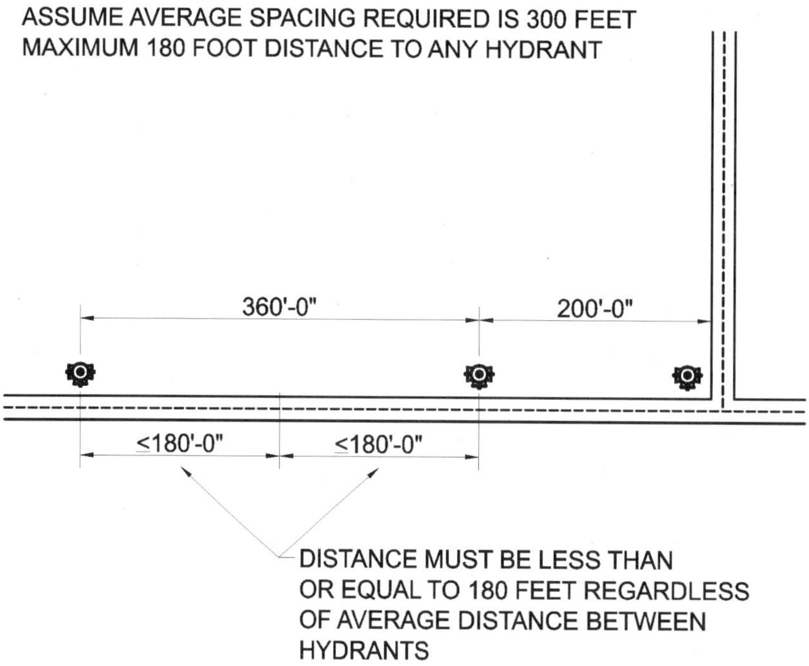

For SI: 1 foot = 304.8 mm.

FIGURE C105.1(2)
TRAVEL DISTANCE TO HYDRANTS

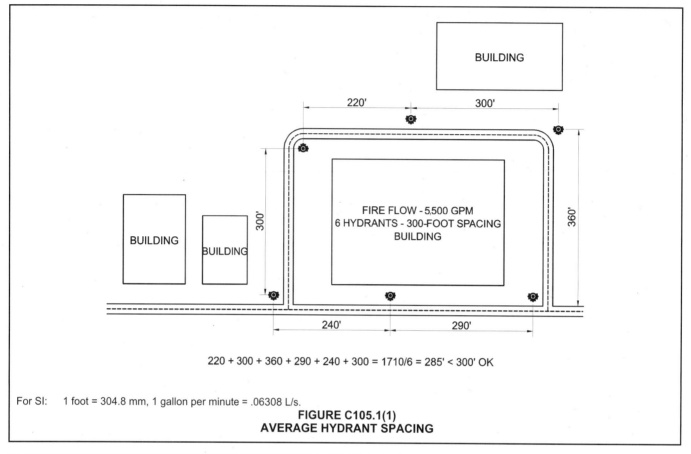

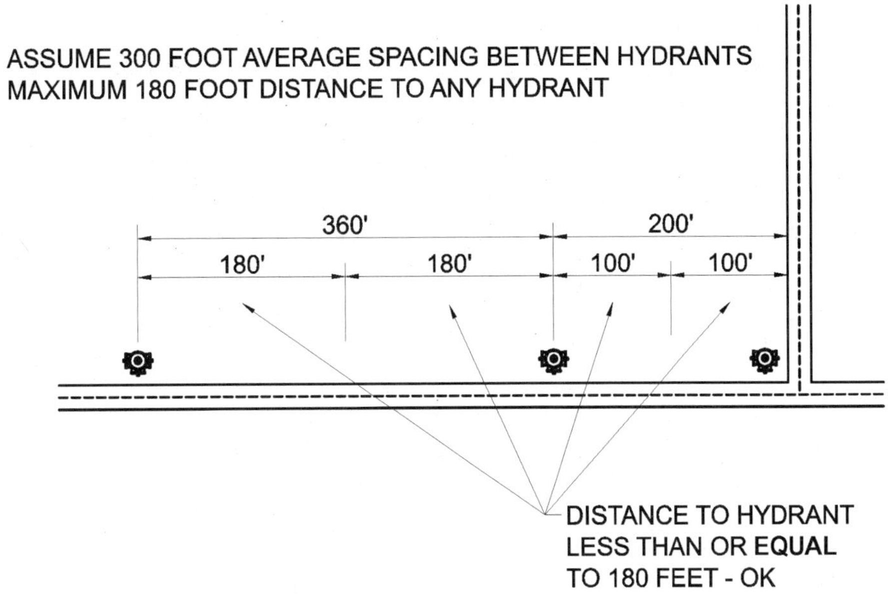

For SI: 1 foot = 304.8 mm.

Figure C105.1(3)
TRAVEL DISTANCE TO HYDRANTS

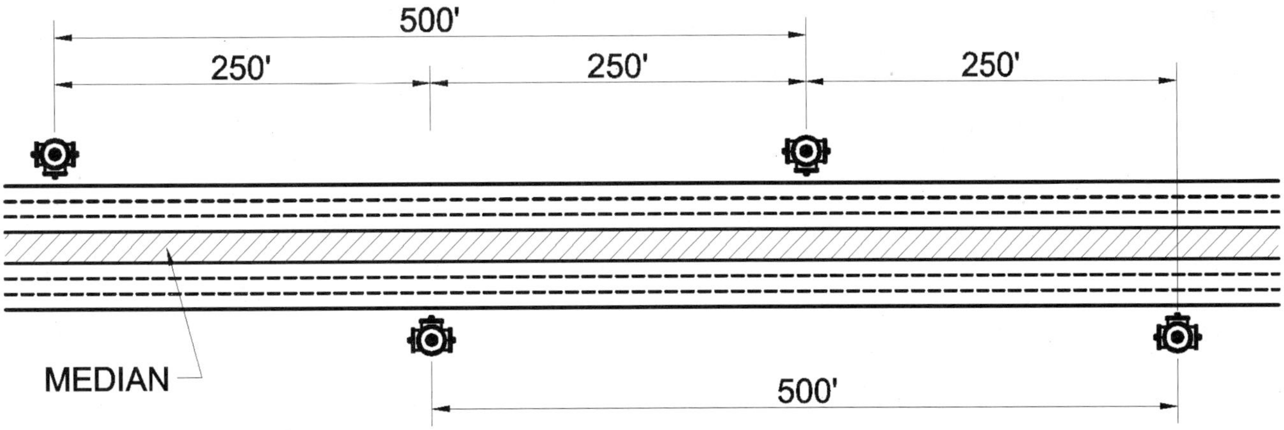

For SI: 1 foot = 304.8 mm.

Figure C105.1(4)
HYDRANT SPACING ALONG FREEWAYS

Appendix D: Fire Apparatus Access Roads

The provisions contained in this appendix are not mandatory unless specifically referenced in the adopting ordinance.

General Comments

Access is a necessity when it comes to fire fighting. Fire department vehicles vary widely in size. All must be able to maneuver into position to properly undertake fire-fighting activities. The needs of each jurisdiction will therefore vary with the equipment used. Access roads must be designed to ensure the fire department has the required access to all structures on a site.

Purpose

This appendix contains more detailed elements for use with the basic access requirements found in Section 503. Section 503 gives some minimum criteria, such as a maximum distance of 150 feet (45 720 mm) and a minimum width of 20 feet (6090 mm), but in many cases Section 503 does not state specific criteria. For example, what specific load should a fire apparatus access road be able to carry and what specific grade is allowed? Section 503 cannot go into this level of detail because the needs vary widely from one jurisdiction to another. This appendix, like Appendixes B and C, is a tool for jurisdictions looking for guidance in establishing access requirements.

Some of the other requirements found in this appendix address access layouts for multiple-family residential developments and large one- and two-family subdivisions. Also, specific examples for various types of turnarounds for fire department apparatus are shown.

SECTION D101 GENERAL

D101.1 Scope. Fire apparatus access roads shall be in accordance with this appendix and all other applicable requirements of the *International Fire Code.*

❖ If this appendix has been adopted by a jurisdiction, this particular section simply states that all fire apparatus access roads must meet the requirements of this appendix and other applicable requirements. More specifically, Section 503 would also apply.

Be aware that essentially all roads leading to a particular building or facility, whether public or private, are fire apparatus access roads. Generally, the requirements of this appendix and Section 503 would be required only for new buildings and facilities. However, in some cases, improvements to existing roads and access ways may be neccessary to meet the needs of the fire department.

SECTION D102 REQUIRED ACCESS

D102.1 Access and loading. Facilities, buildings or portions of buildings hereafter constructed shall be accessible to fire department apparatus by way of an approved fire apparatus access road with an asphalt, concrete or other approved driving surface capable of supporting the imposed load of fire apparatus weighing at least 75,000 pounds (34 050 kg).

❖ This section contains more detailed specifications for the road surface and applied loads. In Section 503, it simply states that the road must be able to withstand the loads and be of "all-weather driving capability." This section states that the surface be of asphalt, concrete or other approved material and be able to withstand a load of 75,000 pounds (330 000 N).

SECTION D103 MINIMUM SPECIFICATIONS

D103.1 Access road width with a hydrant. Where a fire hydrant is located on a fire apparatus access road, the minimum road width shall be 26 feet (7925 mm). See Figure D103.1.

❖ The access road width of 20 feet (6096 mm) stated in Section 503 does not specifically account for the presence of the hydrant. This section specifically requires a minimum width of 26 feet (7925 mm) when a hydrant is located along that access roadway. This provides more room for the fire department vehicle to maneuver and connect to the hydrant. In many cases, a full 26 foot (7925 mm) width may not be possible for a majority of the access road and a possible solution is to simply widen the access road for a short distance to accommodate hydrant use. Section 503 is generic because available water supplies are not always accessed using hydrants. In some cases, the water comes from a tanker or from an on-site water supply.

D103.2 Grade. Fire apparatus access roads shall not exceed 10 percent in grade.

> **Exception:** Grades steeper than 10 percent as approved by the fire chief.

❖ Section 503 discusses grade in generalities and states that the grade be within the limits established by the fire code official. The criteria are generic because the conditions in different jurisdictions will vary. For example, some fire department apparatus is able to handle steeper grades than others, and the likelihood of inclement weather, such as snow, will affect the ability of the vehicles to handle the terrain.

This appendix states a numerical criterion of 10 percent, which is fairly conservative for most situations. This number gives something specific for a jurisdiction to cite without having to determine the actual grade. There is an exception to this section that would allow the fire chief to approve a grade greater than 10 percent. This gives the jurisdiction flexibility for specific situations where terrain might call for a steeper grade.

This figure shows various turnaround configurations, all of which call for a turning radius of 28 feet (8534 mm) (see commentary, Table D103.4).

D103.3 Turning radius. The minimum turning radius shall be determined by the fire code official.

❖ The turning radius is left generic within both Section 503 and this section because of the large variation in the equipment used by fire departments. Each fire department must assess the specific abilities of its vehicles to set a minimum turning radius. The diagrams in Figure D103.1 set the turning radius at 28 feet (8534 mm), which may not be satisfactory for all jurisdictions.

D103.4 Dead ends. Dead-end fire apparatus access roads in excess of 150 feet (45 720 mm) shall be provided with width and turnaround provisions in accordance with Table D103.4.

TABLE D103.4
REQUIREMENTS FOR DEAD-END FIRE APPARATUS ACCESS ROADS

LENGTH (feet)	WIDTH (feet)	TURNAROUNDS REQUIRED
0–150	20	None required
151–500	20	120-foot Hammerhead, 60-foot "Y" or 96-foot-diameter cul-de-sac in accordance with Figure D103.1
501–750	26	120-foot Hammerhead, 60-foot "Y" or 96-foot-diameter cul-de-sac in accordance with Figure D103.1
Over 750	Special approval required	

For SI: 1 foot = 304.8 mm.

❖ Though the widths of the access roadways may be sufficient to move and operate the necessary equipment at a fire scene, they may not be wide enough for the vehicles to turn around. On through streets this is not an is-

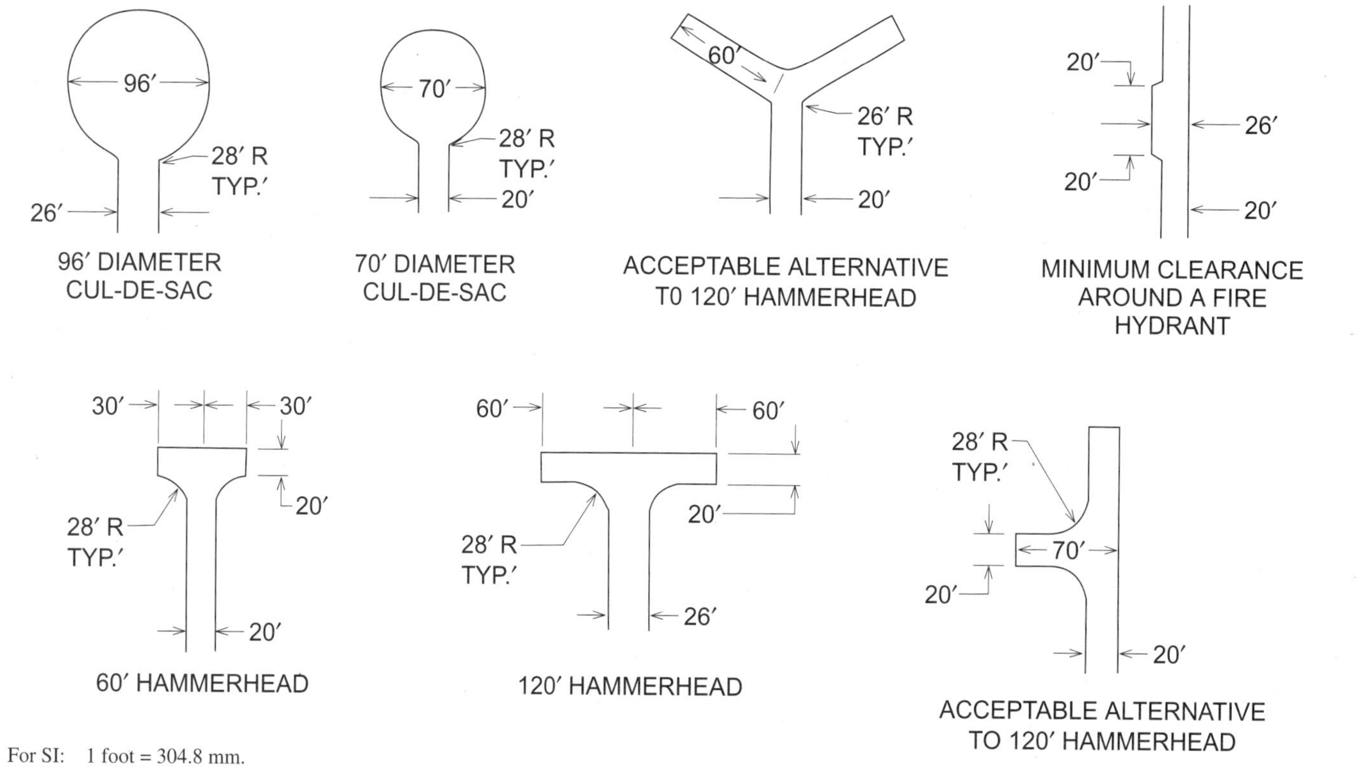

For SI: 1 foot = 304.8 mm.

FIGURE D103.1
DEAD-END FIRE APPARATUS ACCESS ROAD TURNAROUND

sue, but when the road is a dead end and is sufficiently long, some means are necessary to enable fire department vehicles to turn around. The three major methods used are cul-de-sac, hammerhead and "Y." Figure D103.1 shows examples of all three types. Section 503 does not give any specific guidance. Each jurisdiction can choose from a variety of ways to accomplish this.

Dead ends require a fire vehicle turnaround when they exceed 150 feet (45 720 mm). This distance is thought to be reasonable. Backing a large vehicle over 150 feet (45 720 mm) becomes too difficult. Refer to Table D103.4 for more guidance in determining the kind of turning radius required.

This table, which is based on the depth of a dead end, sets minimum widths and recommends which types of turnarounds should be used. The diagrams in Figure D103.1 show the configurations of these turnarounds. Note that Table D103.4 allows only three of these configurations, the 120-foot (29 261 mm) hammerhead, the 96-foot-diameter (36 576 mm) cul-de-sac and the 60-foot (18 288 mm) "Y." Each jurisdiction is free to adopt or modify these requirements as it sees fit.

D103.5 Fire apparatus access road gates. Gates securing the fire apparatus access roads shall comply with all of the following criteria:

1. The minimum gate width shall be 20 feet (6096 mm).
2. Gates shall be of the swinging or sliding type.
3. Construction of gates shall be of materials that allow manual operation by one person.
4. Gate components shall be maintained in an operative condition at all times and replaced or repaired when defective.
5. Electric gates shall be equipped with a means of opening the gate by fire department personnel for emergency access. Emergency opening devices shall be approved by the fire code official.
6. Manual opening gates shall not be locked with a padlock or chain and padlock unless they are capable of being opened by means of forcible entry tools.
7. Locking device specifications shall be submitted for approval by the fire code official.

❖ Gates are sometimes required by the fire code official to limit access to certain hazardous fire areas. They are also often used as a security mechanism for gated communities and complexes. Section 503 discusses the use of gates in general terms. This section gives some specific guidelines for maintaining gates and requirements for emergency access. The seven requirements stated here all must be complied with. They focus on maintaining the required width, ease of use, and ability to open in an emergency. The methods for opening the gates, whether by manual lock or by an electrical mechanism, must be approved by the fire code official. This ensures that the operating procedures of the fire department are taken into account.

D103.6 Signs. Where required by the fire code official, fire apparatus access roads shall be marked with permanent NO PARKING—FIRE LANE signs complying with Figure D103.6. Signs shall have a minimum dimension of 12 inches (305 mm) wide by 18 inches (457 mm) high and have red letters on a white reflective background. Signs shall be posted on one or both sides of the fire apparatus road as required by Section D103.6.1 or D103.6.2.

❖ One of the more challenging aspects of access roads is maintaining the necessary width. Parked cars can reduce this width if parking is not prohibited and the prohibition is posted. Section 503.3 addresses this need by giving the fire code official the authority to require marking of fire access roads. This section and Figure D103.6 add wording and dimension specifications for the signs needed to mark areas where parking is prohibited.

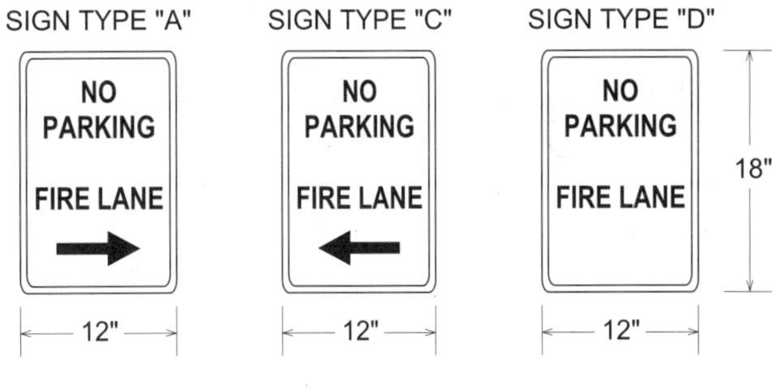

FIGURE D103.6
FIRE LANE SIGNS

D103.6.1 Roads 20 to 26 feet in width. Fire apparatus access roads 20 to 26 feet wide (6096 to 7925 mm) shall be posted on both sides as a fire lane.

❖ This section requires that parking should be prohibited on both sides of the narrower fire access roads. Twenty feet (6096 mm) is the appropriate width needed for two average-size fire trucks to pass one another. If that width is reduced by parking even on one side, it will be potentially difficult for a fire department to undertake emergency operations in that area.

D103.6.2 Roads more than 26 feet in width. Fire apparatus access roads more than 26 feet wide (7925 mm) to 32 feet wide (9754 mm) shall be posted on one side of the road as a fire lane.

❖ Because this width is more than sufficient for maneuvering at least two fire department vehicles by one another, parking would be allowed on one side.

SECTION D104
COMMERCIAL AND INDUSTRIAL DEVELOPMENTS

D104.1 Buildings exceeding three stories or 30 feet in height. Buildings or facilities exceeding 30 feet (9144 mm) or three stories in height shall have at least three means of fire apparatus access for each structure.

❖ This section addresses commercial and industrial buildings that, because of their size and/or height, have the potential of creating a large challenge to a fire department. This section, along with Sections D105, D106 and D107, contains requirements for fire apparatus access roads for specific kinds of buildings or developments. Section 503 gives the fire code official the authority to require more access roads but does not specify when the additional roads are required. The need for additional access roads will depend on so many factors that each situation must be judged individually.

Because of the height of these buildings, various types of vehicles may be needed, and having three means of approaching the site may be necessary to manage and manipulate the vehicles.

D104.2 Buildings exceeding 62,000 square feet in area. Buildings or facilities having a gross building area of more than 62,000 square feet (5760 m^2) shall be provided with two separate and approved fire apparatus access roads.

> **Exception:** Projects having a gross building area of up to 124,000 square feet (11 520 m^2) that have a single approved fire apparatus access road when all buildings are equipped throughout with approved automatic sprinkler systems.

❖ When buildings are simply large in area, two separate fire apparatus access roads are required because a large building may be difficult to access quickly, and if one of the access roads is blocked there is a large potential for loss. The exception acknowledges the ability of sprinklers to prevent most fires from growing quickly.

D104.3 Remoteness. Where two access roads are required, they shall be placed a distance apart equal to not less than one half of the length of the maximum overall diagonal dimension of the property or area to be served, measured in a straight line between accesses.

❖ This concept is similar to the one dealing with the remoteness of exits. One of the primary reasons for multiple access roads is to ensure that if one access road is blocked or otherwise unavailable, another will allow access to the fire department. Therefore, when more than one access road is required, they need to be separated by enough distance to avoid a situation where both would be blocked or unavailable simply because they are too close to one another.

SECTION D105 AERIAL FIRE APPARATUS ACCESS ROADS

D105.1 Where required. Buildings or portions of buildings or facilities exceeding 30 feet (9144 mm) in height above the lowest level of fire department vehicle access shall be provided with approved fire apparatus access roads capable of accommodating fire department aerial apparatus. Overhead utility and power lines shall not be located within the aerial fire apparatus access roadway.

❖ When buildings exceed 30 feet (9144 mm) in height above the lowest level of fire department access, the use of aerial fire apparatus becomes more necessary. This section states in general terms that the access roads must be capable of handling the larger aerial equipment. The requirement for clear overhead space prevents interference with the aerial apparatus and avoids the possibility of personnel injury and equipment damage from electrical shock. These factors must be included in site design to make certain the fire department has the needed access to the buildings.

D105.2 Width. Fire apparatus access roads shall have a minimum unobstructed width of 26 feet (7925 mm) in the immediate vicinity of any building or portion of building more than 30 feet (9144 mm) in height.

❖ This section specifies the minimum road width needed for aerial vehicles. This width allows the aerial vehicle supports to be set solidly on the road surface for safe operation of the aerial equipment.

D105.3 Proximity to building. At least one of the required access routes meeting this condition shall be located within a minimum of 15 feet (4572 mm) and a maximum of 30 feet (9144 mm) from the building, and shall be positioned parallel to one entire side of the building.

❖ This section ensures that the access road is specifically located where aerial equipment will have maximum access to the building. The distance from the building to the road must be established to match the capabilities of the fire department aerial equipment.

SECTION D106 MULTIPLE-FAMILY RESIDENTIAL DEVELOPMENTS

D106.1 Projects having more than 100 dwelling units. Multiple-family residential projects having more than 100 dwelling units shall be equipped throughout with two separate and approved fire apparatus access roads.

> **Exception:** Projects having up to 200 dwelling units may have a single approved fire apparatus access road when all buildings, including nonresidential occupancies, are equipped throughout with approved automatic sprinkler systems installed in accordance with Section 903.3.1.1 or 903.3.1.2 of the *International Fire Code*.

❖ This section is intended to provide some specific guidance to jurisdictions for dealing with larger apartment complexes. Again, Section 503 suggests that more than one access road is needed when there is a potential for an access road to be unavailable. In a large complex there is a large potential for loss. Lack of access should not become a factor in such a loss.

This section requires at least two separate access roads any time the number of dwelling units exceeds 100. The term "approved" is used because the layout of the complex may require some specific considerations when providing the access roads. For example, having two access road leading onto a facility that come together before reaching the actual buildings may not satisfy the criterion of remoteness to be effective in an emergency.

The exception would allow a single access road for

up to 200 welling units if all buildings on the site are fully sprinklered to meet code requirements. This exception acknowledges the effectiveness of sprinklers in slowing the growth of fires; therefore, the risk of having the access road blocked or unusable is a more acceptable.

D106.2 Projects having more than 200 dwelling units. Multiple-family residential projects having more than 200 dwelling units shall be provided with two separate and approved fire apparatus access roads regardless of whether they are equipped with an approved automatic sprinkler system.

❖ Because of the large size of such complexes and the potential for large losses, even where sprinklers are installed, two approved access roads should be required. This section emphasizes that the exception in Section D106.1 is for up to 200 units only.

SECTION D107 ONE- OR TWO-FAMILY RESIDENTIAL DEVELOPMENTS

D107.1 One- or two-family dwelling residential developments. Developments of one- or two-family dwellings where the number of dwelling units exceeds 30 shall be provided with separate and approved fire apparatus access roads, and shall meet the requirements of Section D104.3.

Exceptions:

1. Where there are 30 or fewer dwelling units on a single public or private access way and all dwelling units are protected by approved residential sprinkler systems, access from two directions shall not be required.
2. The number of dwelling units on a single fire apparatus access road shall not be increased unless fire apparatus access roads will connect with future development, as determined by the fire code official.

❖ The presence of this section in an appendix of the *International Fire Code*® (IFC®) raises the question, "Is it the intent of the IFC that its regulations (including duly adopted appendices) concerning fire apparatus access roads and fire protection water supply be applicable to one- and two-family residential development sites upon which buildings are constructed under the provisions of the *International Residential Code*® (IRC®)?" The answer to that question is that the IRC is intended to be a stand-alone code for the construction of detached one- and two-family dwellings and townhouses not more than three stories in height. That is, all of the provisions for the construction of buildings of those descriptions are to be regulated exclusively by the IRC and not by another *International Code*®. Note, however, that the IRC applies only to the construction of the structures of one- and two-family dwellings and not to the development of the site upon which multiple such structures are built. Accordingly, where the IFC is adopted, its fire apparatus access road and water supply provisions contained in Chapter 5 (and, where specifically adopted, the related appendices) would apply because they are dealing only with land development requirements providing fire protection access and fire protection water to the community.

This section requires that subdivisions consisting of more than 30 units have more than one access road into the complex. The second access road is needed in case one access road is unusable. Because the number of units is higher, the potential for loss becomes higher. This section, however, does not consider how close these units are to one another. In some parts of the country, houses are on much smaller lots than they are in others and the potential for a conflagration is greater.

The two access roads must also be remote from one another as required in Section D104.3 to reduce the likelihood that both access roads would be unavailable during a fire or other emergency.

Exception 1 states that when there are 30 or less dwelling units with approved residential sprinkler systems, a second access road is not required. It does not consider any development having more than 30 units regardless of whether or not they are equipped with sprinklers.

Exception 2 requires approval by the fire code official any time a new house is constructed on an existing access road. This gives the fire code official an opportunity to assess whether additional access is required.

Appendix E: Hazard Categories

The provisions contained in this appendix are not mandatory unless specifically referenced in the adopting ordinance.

General Comments

This appendix contains guidance for designers, engineers, architects, building officials, plans reviewers and inspectors in the classifying of hazardous materials so that proposed designs can be evaluated intelligently and accurately. If adopted, it can be an extremely useful tool for all concerned. A thorough understanding of how Material Safety Data Sheets (MSDS) are created and how to read them makes these hazard evaluations much easier.

Purpose

The descriptive materials and explanations of hazardous materials and how to report and evaluate them on the MSDS that are contained in this appendix are intended to be instructional as well as informative. A thorough understanding of how to prepare and interpret the MSDS makes everyone's job easier.

SECTION E101 GENERAL

E101.1 Scope. This appendix provides information, explanations and examples to illustrate and clarify the hazard categories contained in Chapter 27 of the *International Fire Code*. The hazard categories are based upon the DOL 29 CFR. Where numerical classifications are included, they are in accordance with nationally recognized standards.

This appendix should not be used as the sole means of hazardous materials classification.

❖ Although this appendix contains guidance in classifying hazardous materials, there are several other sources available to the fire code official or building official. The actual MSDS should be the only method used in the final classification of products because of the different characteristics that blends of products produce. General classification information can come from NFPA 704, CHRIS manuals, MSDS online and other methods.

It is important to remember that many chemicals have multiple hazards associated with them. A chemical may be a toxic poison (H4) and also have a flammable or combustible base (H2 or H3), which creates additional problems. Section 2701 states, "Where a material has multiple hazards, all hazards shall be addressed." This means that the building must be protected in the above example as an H4 Use Group and an H3 Use Group when the products are in sealed containers and an H2 Use Group if they are open to the air.

SECTION E102 HAZARD CATEGORIES

E102.1 Physical hazards. Materials classified in this section pose a physical hazard.

❖ A chemical presents a physical hazard when there is evidence that it is a combustible liquid, compressed gas, cryogenic, explosive, flammable gas, flammable liquid, flammable solid, organic peroxide, oxidizer, pyrophoric or unstable (reactive) or water-reactive material.

E102.1.1 Explosives and blasting agents. The current UN/DOT classification system recognized by international authorities, the Department of Defense and others classifies all explosives as Class 1 materials. They are then divided into six separate divisions to indicate their relative hazard. There is not a direct correlation between the designations used by the old DOT system and those used by the current system nor is there correlation with the system (high and low) established by the Bureau of Alcohol, Tobacco and Firearms (BATF). Table 3304.3 provides some guidance with regard to the current categories and their relationship to the old categories. Some items may appear in more than one division, depending on factors such as the degree of confinement or separation, by type of packaging, storage configuration or state of assembly.

In order to determine the level of hazard presented by explosive materials, testing to establish quantitatively their explosive nature is required. There are numerous test methods that have been used to establish the character of an explosive material. Standardized tests, required for finished goods containing explosives or explosive materials in a packaged form suitable for shipment or storage, have been established by UN/DOT and BATF. However, these tests do not consider key elements that should be examined in a manufacturing situation. In manufacturing operations, the condition and/or the state of a material may vary within the process. The in-process material classification and classification requirements for materials used in the manufacturing process may be different from the classification of the same material when found in finished goods depending on the stage of the process in which the material is found. A classification methodology must be used that recognizes the hazards commensurate with the application to the variable physical conditions as well as potential variations of physical character and type of explosive under consideration.

Test methods or guidelines for hazard classification of energetic materials used for in-process operations shall be approved by the fire code official. Test methods used shall be DOD, BATF, UN/DOT or other approved criteria. The results of such testing shall become a portion of the files of the jurisdiction and be included as an independent section of any Hazardous Materials Management Plan (HMMP) required by Section 3305.2.1. Also see Section 104.7.2.

Examples of materials in various Divisions are as follows:

1. Division 1.1 (High Explosives). Consists of explosives that have a mass explosion hazard. A mass explosion is one which affects almost the entire pile of material instantaneously. Includes substances that, when tested in accordance with approved methods, can be caused to detonate by means of a blasting cap when unconfined or will transition from deflagration to a detonation when confined or unconfined. Examples: dynamite, TNT, nitroglycerine, C-3, HMX, RDX, encased explosives, military ammunition.
2. Division 1.2 (Low Explosives). Consists of explosives that have a projection hazard, but not a mass explosion hazard. Examples: nondetonating encased explosives, military ammunition and the like.
3. Division 1.3 (Low Explosives). Consists of explosives that have a fire hazard and either a minor blast hazard or a minor projection hazard or both, but not a mass explosion hazard. The major hazard is radiant heat or violent burning, or both. Can be deflagrated when confined. Examples: smokeless powder, propellant explosives, display fireworks.
4. Division 1.4. Consists of explosives that pose a minor explosion hazard. The explosive effects are largely confined to the package and no projection of fragments of appreciable size or range is expected. An internal fire must not cause virtually instantaneous explosion of almost the entire contents of the package. Examples: squibs (nondetonating igniters), explosive actuators, explosive trains (low level detonating cord).
5. Division 1.5 (Blasting Agents). Consists of very insensitive explosives. This division is comprised of substances which have a mass explosion hazard, but are so insensitive that there is very little probability of initiation or of transition from burning to detonation under normal conditions of transport. Materials are not cap sensitive; however, they are mass detonating when provided with sufficient input. Examples: oxidizer and liquid fuel slurry mixtures and gels, ammonium nitrate combined with fuel oil.
6. Division 1.6. Consists of extremely insensitive articles which do not have a mass explosive hazard. This division is comprised of articles which contain only extremely insensitive detonating substances and which demonstrate a negligible probability of accidental initiation or propagation. Although this category of materials has been defined, the primary application is currently limited to military uses. Examples: Low vulnerability military weapons.

Explosives in each division are assigned a compatibility group letter by the Associate Administrator for Hazardous Materials Safety (DOT) based on criteria specified by DOTn 49 CFR. Compatibility group letters are used to specify the controls for the transportation and storage related to various materials to prevent an increase in hazard that might result if certain types of explosives were stored or transported together. Altogether, there are 35 possible classification codes for explosives, e.g., 1.1A, 1.3C, 1.4S, etc.

❖ It is recognized that the hazard classification of goods packaged for release to the Department of Transportation (DOT) system is likely to be different from that of the same material when found in an unpackaged or bulk form. For example, a finished article as manufactured or packaged and classified as Division 1.4 may contain Division 1.1 or 1.3 material. Although the ingredients may in fact be classified differently when removed from the package or article, the hazard level of the finished device may be classified with a different level of hazard from the ingredients alone. Classification, in part, is based on the protection offered by the final package, the article itself, as well as the configuration or nature of the energetic materials as they exist within the article. Recognizing the fact that the hazard level of an energetic material can vary based on the physical character (size and shape) or configuration (physical arrangement) in which the material is found is key to understanding the nature of explosives and proper application of the code.

Scientifically based methods are needed to appraise the hazards of materials as they appear in the manufacturing process. By granting the fire code official authority to approve test methods, flexibility is provided while at the same time maintaining a reasonable level of control. By recording the technical basis of decision within the Hazardous Materials Management Plan (HMMP), a record of the criteria used to establish the hazard level is maintained by the applicant. The reference to Section 104.7.2 refers the fire code official back to the administrative section of the code to exercise authority for outside technical assistance in those cases where such assistance is warranted. The terminology used in the examples reflects the current terminology, and examples accurately portray materials that may be encountered. The explanatory language has been drawn in pertinent part from 49 CFR Sections 173.50 and 173.58. Although the DOT has eliminated the Class A, B and C nomenclature, the Bureau of Alcohol, Tobacco and Firearms (ATF) has retained the use of the term "high and low explosives." These terms are included in parenthesis where appropriate.

By definition, an "Explosive" is a chemical compound, mixture or device, the primary or common purpose of which is to function by explosion. The term includes, but is not limited to, dynamite, black powder, pellet powder, initiating explosives, detonators, safety fuses, squibs, detonating cord, igniter cord, igniters and display fireworks, 1.3G (Class B, Special).

The term "explosive" includes any material determined to be within the scope of USC Title 18: Chapter 40 and also includes any material classified as an explosive other than consumer fireworks, 1.4G (Class C, Common) by the hazardous materials regulations of DOTn 49 CFR.

The former classification system used by DOTn (UN/DOTn Class 1 explosives) included the term "high" and "low" explosives. The following terms further define explosives under the current system used by DOTn for all explosive materials defined as hazard Class 1 materials. Compatibility group letters are used together with the division to specify further limitations on each division noted; that is, the letter "G" identifies the material as a pyrotechnic substance or article containing a pyrotechnic substance and similar materials.

Division 1.1. Explosives that are a mass explosion hazard. A mass explosion is one that affects almost the entire load instantaneously.

Division 1.2. Explosives that have a projection hazard but not a mass explosion hazard.

Division 1.3. Explosives that have a fire hazard and either a minor blast hazard or a minor projection hazard or both, but not a mass explosion hazard.

Division 1.4. Explosives that pose a minor explosion hazard. The explosive effects are largely confined to the package and no projection of fragments of appreciable size or range is to be expected. An external fire must not cause virtually instantaneous explosion of almost the entire contents of the package.

Division 1.5. Very insensitive explosives. This division is comprised of substances that have a mass explosion hazard but that are so insensitive that there is very little probability of initiation or of transition from burning to detonation under normal conditions of transport.

Division 1.6. Extremely insensitive articles that do not have a mass explosion hazard. This division is comprised of articles that contain only extremely insensitive detonating substances and which demonstrate a negligible probability of accidental initiation or propagation.

EXPLOSIVE MATERIAL. The term "explosive" material means explosives, blasting agents and detonators.

Chemicals in this classification must meet the minimum requirements of Sections 2701, 2703, 2704, 2705 and Chapter 33.

E102.1.2 Compressed gases. Examples include:

1. Flammable: acetylene, carbon monoxide, ethane, ethylene, hydrogen, methane. Ammonia will ignite and burn although its flammable range is too narrow for it to fit the definition of flammable gas.
2. Oxidizing: oxygen, ozone, oxides of nitrogen, chlorine and fluorine. Chlorine and fluorine do not contain oxygen but reaction with flammables is similar to that of oxygen.
3. Corrosive: ammonia, hydrogen chloride, fluorine.
4. Highly toxic: arsine, cyanogen, fluorine, germane, hydrogen cyanide, nitric oxide, phosphine, hydrogen selenide, stibine.
5. Toxic: chlorine, hydrogen fluoride, hydrogen sulfide, phosgene, silicon tetrafluoride.
6. Inert (chemically unreactive): argon, helium, krypton, neon, nitrogen, xenon.
7. Pyrophoric: diborane, dichloroborane, phosphine, silane.
8. Unstable (reactive): butadiene (unstabilized), ethylene oxide, vinyl chloride.

❖ Compressed gases by definition are a material, or mixture of materials, that:

1. Is a gas at 68°F (20°C) or less at 14.7 pounds per square inch atmosphere (psia) (101 kPa) of pressure; and
2. Has a boiling point of 68°F (20°C) or less at 14.7 psia (101 kPa), which is either liquefied, nonliquefied or in solution, except those gases that have no other health or physical hazard properties and are not considered to be compressed until the pressure in the packaging exceeds 41 psia (282 kPa) at 68°F (20°C).

The states of a compressed gas are categorized as follows:

1. Nonliquefied compressed gases are gases, other than those in solution, that are in a packaging under the charged pressure and are entirely gaseous at a temperature of 68°F (20°C).
2. Liquefied compressed gases are gases that, in a packaging under the charged pressure, are partially liquid at a temperature of 68°F (20°C).
3. Compressed gases in solution are nonliquefied gases that are dissolved in a solvent.
4. Compressed gas mixtures consist of a mixture of two or more compressed gases contained in a packaging, the hazard properties of which are

represented by the properties of the mixture as a whole.

Chemicals in this classification must meet the minimum requirements of Sections 2701, 2703, 2704, 2705 and 3704 and Chapters 30, 31, 32 and 35.

E102.1.3 Flammable and combustible liquids. Examples include:

1. Flammable liquids.

   Class IA liquids shall include those having flash points below 73°F (23°C) and having a boiling point at or below 100°F (38°C).

   Class IB liquids shall include those having flash points below 73°F (23°C) and having a boiling point at or above 100°F (38°C).

   Class IC liquids shall include those having flash points at or above 73°F (23°C.) and below 100°F (38°C).

2. Combustible liquids.

   Class II liquids shall include those having flash points at or above 100°F (38°C) and below 140°F (60°C).

   Class IIIA liquids shall include those having flash points at or above 140°F (60°C) and below 200°F (93°C).

   Class IIIB liquids shall include those liquids having flash points at or above 200°F (93°C).

❖ Chemicals in this classification must meet the minimum requirements of the Sections 2701, 2703, 2704, 2705 and Chapter 34.

E102.1.4 Flammable solids. Examples include:

1. Organic solids: camphor, cellulose nitrate, naphthalene.
2. Inorganic solids: decaborane, lithium amide, phosphorous heptasulfide, phosphorous sesquisulfide, potassium sulfide, anhydrous sodium sulfide, sulfur.
3. Combustible metals (except dusts and powders): cesium, magnesium, zirconium.

❖ A flammable solid is solid, other than a blasting agent or explosive, that is capable of causing fire through friction; absorption of moisture; spontaneous chemical change or retained heat from manufacturing or processing; has an ignition temperature below 212ºF (100ºC) or which burns so vigorously and persistently when ignited as to create a serious hazard. A chemical shall be considered a flammable solid as determined in accordance with the test method of CPSC 16 CFR; Part 1500.44 if it ignites and burns with a self-sustained flame at a rate greater than 0.1 inch (2.5 mm) per second along its major axis.

Chemicals in this classification must meet the minimum requirements of Sections 2701, 2703, 2704, 2705 and Chapter 36.

E102.1.5 Combustible dusts and powders. Finely divided flammable solids which may be dispersed in air as a dust cloud: wood sawdust, plastics, coal, flour, powdered metals (few exceptions).

❖ Combustible dusts and combustible fibers are listed separately because they are not subcategories of flammable solids in either the maximum allowable quantity tables in Chapter 27 or in their respective material-specific hazard chapters.

E102.1.6 Combustible fibers. See Section 2902.1.

❖ See the commentary to Appendix Section E102.1.5.

E102.1.7 Oxidizers. Examples include:

1. Gases: oxygen, ozone, oxides of nitrogen, fluorine and chlorine (reaction with flammables is similar to that of oxygen).
2. Liquids: bromine, hydrogen peroxide, nitric acid, perchloric acid, sulfuric acid.
3. Solids: chlorates, chromates, chromic acid, iodine, nitrates, nitrites, perchlorates, peroxides.

❖ An oxidizer is material that readily yields oxygen or other oxidizing gas, or that readily reacts to promote or initiate combustion of combustible materials. Examples of other oxidizing gases include bromine, chlorine and fluorine.

OXIDIZING GAS. A gas that can support and accelerate combustion of other materials.

E102.1.7.1 Examples of liquid and solid oxidizers according to hazard.

Class 4: ammonium perchlorate (particle size greater than 15 microns), ammonium permanganate, guanidine nitrate, hydrogen peroxide solutions more than 91 percent by weight, perchloric acid solutions more than 72.5 percent by weight, potassium superoxide, tetranitromethane.

Class 3: ammonium dichromate, calcium hypochlorite (over 50 percent by weight), chloric acid (10 percent maximum concentration), hydrogen peroxide solutions (greater than 52 percent up to 91 percent), mono-(trichloro)-tetra-(monopotassium dichloro)-penta-s-triazinetrione, nitric acid, (fuming —more than 86 percent concentration), perchloric acid solutions (60 percent to 72 percent by weight), potassium bromate, potassium chlorate, potassium dichloro-s-triazinetrione (potassium dichloro- isocyanurate), sodium bromate, sodium chlorate, sodium chlorite (over 40 percent by weight) and sodium dichloro-s-triazinetrione (sodium dichloro- isocyanurate).

Class 2: barium bromate, barium chlorate, barium hypochlorite, barium perchlorate, barium permanganate, 1-bromo-3-chloro-5, 5-dimethylhydantoin, calcium chlorate, calcium chlorite, calcium hypochlorite (50 percent or less by weight), calcium perchlorate, calcium permanganate, chromium trioxide (chromic acid), copper chlorate, halane

(1, 3-dichloro-5, 5-dimethylhydantoin), hydrogen peroxide (greater than 27.5 percent up to 52 percent), lead perchlorate, lithium chlorate, lithium hypochlorite (more than 39 percent available chlorine), lithium perchlorate, magnesium bromate, magnesium chlorate, magnesium perchlorate, mercurous chlorate, nitric acid (more than 40 percent but less than 86 percent), perchloric acid solutions (more than 50 percent but less than 60 percent), potassium perchlorate, potassium permanganate, potassium peroxide, potassium superoxide, silver peroxide, sodium chlorite (40 percent or less by weight), sodium perchlorate, sodium perchlorate monohydrate, sodium permanganate, sodium peroxide, strontium chlorate, strontium perchlorate, thallium chlorate, trichloro-s-triazinetrione (trichloroisocyanuric acid), urea hydrogen peroxide, zinc bromate, zinc chlorate and zinc permanganate.

Class 1: all inorganic nitrates (unless otherwise classified), all inorganic nitrites (unless otherwise classified), ammonium persulfate, barium peroxide, calcium peroxide, hydrogen peroxide solutions (greater than 8 percent up to 27.5 percent), lead dioxide, lithium hypochlorite (39 percent or less available chlorine), lithium peroxide, magnesium peroxide, manganese dioxide, nitric acid (40 percent concentration or less), perchloric acid solutions (less than 50 percent by weight), potassium dichromate, potassium percarbonate, potassium persulfate, sodium carbonate peroxide, sodium dichloro-s-triazinetrione dihydrate, sodium dichromate, sodium perborate (anhydrous), sodium perborate monohydrate, sodium perborate tetrahydrate, sodium percarbonate, sodium persulfate, strontium peroxide and zinc peroxide.

❖ Class 4. An oxidizer that can undergo an explosive reaction as a result of contamination or exposure to thermal or physical shock. In addition, the oxidizer will enhance the burning rate and can cause spontaneous ignition of combustibles.

Class 3. An oxidizer that will cause a severe increase in the burning rate of combustible materials with which it comes in contact or that will undergo vigorous self-sustained decomposition as a result of contamination or exposure to heat.

Class 2. An oxidizer that will cause a moderate increase in the burning rate or that causes spontaneous ignition of combustible materials with which it comes in contact.

Class 1. An oxidizer whose primary hazard is that it slightly increases the burning rate but that does not cause spontaneous ignition when it comes in contact with combustible materials.

Chemicals in this classification must meet the minimum requirements of Sections 2701, 2703, 2704, 2705 and Chapter 40.

E102.1.8 Organic peroxides. Organic peroxides contain the double oxygen or peroxy (-o-o) group. Some are flammable compounds and subject to explosive decomposition. They are available as:

1. Liquids.
2. Pastes.
3. Solids (usually finely divided powers).

❖ An organic peroxide is an organic compound that contains the bivalent structure and that may be considered to be a structural derivative of hydrogen peroxide where one or both of the hydrogen atoms have been replaced by an organic radical. Organic peroxides can pose an explosion hazard (detonation or deflagration) or they can be shock sensitive. They can also decompose into various unstable compounds over an extended period of time.

E102.1.8.1 Classification of organic peroxides according to hazard.

Unclassified: Unclassified organic peroxides are capable of detonation and are regulated in accordance with Chapter 33.

Class I: acetyl cyclohexane sulfonyl 60-65 percent concentration by weight, fulfonyl peroxide, benzoyl peroxide over 98 percent concentration, t-butyl hydroperoxide 90 percent, t-butyl peroxyacetate 75 percent, t-butyl peroxyisopropylcarbonate 92 percent, diisopropyl peroxydicarbonate 100 percent, di-n-propyl peroxydicarbonate 98 percent, and di-n-propyl peroxydicarbonate 85 percent.

Class II: acetyl peroxide 25 percent, t-butyl hydroperoxide 70 percent (with DTBP and t-BuOH diluents), t-butyl peroxybenzoate 98 percent, t-butyl peroxy-2-ethylhexanoate 97 percent, t-butyl peroxyisobutyrate 75 percent, t-butyl peroxyisopropyl-carbonate 75 percent, t-butyl peroxypivalate 75 percent, dybenzoyl peroxydicarbonate 85 percent, di-sec-butyl peroxydicarbonate 98 percent, di-sec-butyl peroxydicarbonate 75 percent, 1,1-di-(t-butylperoxy)-3,5,5-trimethyecyclohexane 95 percent, di-(2-ethythexyl) peroxydicarbonate 97 percent, 2,5-dymethyl-2-5 di (benzoylperoxy) hexane 92 percent, and peroxyacetic acid 43 percent.

Class III: acetyl cyclohexane sulfonal peroxide 29 percent, benzoyl peroxide 78 percent, benzoyl peroxide paste 55 percent, benzoyl peroxide paste 50 percent peroxide/50 percent butylbenzylphthalate diluent, cumene hydroperoxide 86 percent, di-(4-butylcyclohexyl) peroxydicarbonate 98 percent, t-butyl peroxy-2-ethylhexanoate 97 percent, t-butyl peroxyneodecanoate 75 percent, decanoyl peroxide 98.5 percent, di-t-butyl peroxide 99 percent,1,1-di-(t-butylperoxy) 3,5,5-trimethylcyclohexane 75 percent, 2,4-dichlorobenzoyl peroxide 50 percent, diisopropyl peroxydicarbonate 30 percent, 2,-5-dimethyl-2,5-di-(2-ethylhexanolyperoxy)-hexane 90 percent, 2,5-dimethyl-2,5-di-(t-butylperoxy) hexane 90 percent and methyl ethyl ketone peroxide 9 percent active oxygen diluted in dimethyl phthalate.

Class IV: benzoyl peroxide 70 percent, benzoyl peroxide paste 50 percent peroxide/15 percent water/35 percent butylphthalate diluent, benzoyl peroxide slurry 40 percent, benzoyl peroxide powder 35 percent, t-butyl hydroperoxide 70 percent, (with water diluent), t-butyl peroxy-2-ethylhexanoate 50 percent, decumyl peroxide 98 percent, di-(2-ethylhexal) peroxydicarbonate 40 percent, laurel peroxide 98 percent, p-methane hydroperoxide 52.5 percent, methyl ethyl ketone peroxide 5.5 percent active oxygen and methyl ethyl ketone peroxide 9 percent active oxygen diluted in water and glycols.

Class V: benzoyl peroxide 35 percent, 1,1-di-t-butyl peroxy 3,5,5-trimethylcyclohexane 40 percent, 2,5-di-(t-butyl peroxy) hexane 47 percent and 2,4-pentanedione peroxide 4 percent active oxygen.

❖ Unclassified detonable. Organic peroxides that are capable of detonation. These peroxides pose an extremely high explosion hazard through rapid explosive decomposition.

Class I. Those formulations that are capable of deflagration but not detonation.

Class II. Those formulations that burn very rapidly and that pose a moderate reactivity hazard.

Class III. Those formulations that burn rapidly and that pose a moderate reactivity hazard.

Class IV. Those formulations that burn in the same manner as ordinary combustibles and that pose a minimal reactivity hazard.

Class V. Those formulations that burn with less intensity than ordinary combustibles or do not sustain combustion and that pose no reactivity hazard.

Chemicals in this classification must meet the minimum requirements of Sections 2701, 2703, 2704, 2705 and Chapter 39.

E102.1.9 Pyrophoric materials. Examples include:

1. Gases: diborane, phosphine, silane.
2. Liquids: diethylaluminum chloride, diethylberyllium, diethylphosphine, diethylzinc, dimethylarsine, triethylaluminum etherate, triethylbismuthine, triethylboron, trimethylaluminum, trimethylgallium.
3. Solids: cesium, hafnium, lithium, white or yellow phosphorous, plutonium, potassium, rubidium, sodium, thorium.

❖ Pyrophoric materials are two or more unmixed, commercially manufactured, prepackaged chemical substances including oxidizers, flammable liquids or solids or similar substances that are not independently classified as explosives but that, when mixed or combined, form an explosive that is intended for blasting.

Chemicals in this classification must meet the minimum requirements of Sections 2701, 2703, 2704, 2705 and Chapter 41.

E102.1.10 Unstable (reactive) materials. Examples include:

Class 4: acetyl peroxide, dibutyl peroxide, dinitrobenzene, ethyl nitrate, peroxyacetic acid and picric acid (dry) trinitrobenzene.

Class 3: hydrogen peroxide (greater than 52 percent), hydroxylamine, nitromethane, paranitroaniline, perchloric acid and tetrafluoroethylene monomer.

Class 2: acrolein, acrylic acid, hydrazine, methacrylic acid, sodium perchlorate, styrene and vinyl acetate.

Class 1: acetic acid, hydrogen peroxide 35 percent to 52 percent, paraldehyde and tetrahydrofuran.

❖ An unstable (reactive) material is a material, other than an explosive, that in the pure state or as commercially produced, will vigorously polymerize, decompose, condense or become self-reactive and undergo other violent chemical changes, including explosion, when exposed to heat, friction or shock; when in the absence of an inhibitor in the presence of contaminants or in contact with incompatible materials. Unstable (reactive) materials are subdivided as follows:

Class 4. Materials that in themselves are readily capable of detonation or explosive decomposition or reaction at normal temperatures and pressures. This class includes materials that are sensitive to mechanical or localized thermal shock at normal temperatures and pressures.

Class 3. Materials that in themselves are capable of detonation or of explosive decomposition or reaction but that require a strong initiating source or must be heated under confinement before initiation. This class includes materials that are sensitive to thermal or mechanical shock at elevated temperatures and pressures.

Class 2. Materials that in themselves are normally unstable and readily undergo violent chemical change but do not detonate. This class includes materials that can undergo chemical change with rapid release of energy at normal temperatures and pressures, and that can undergo violent chemical change at elevated temperatures and pressures.

Class 1. Materials that in themselves are normally stable but that can become unstable at elevated temperatures and pressure.

Chemicals in this classification must meet the minimum requirements of Sections 2701, 2703, 2704, 2705 and Chapter 43.

E102.1.11 Water-reactive materials. Examples include:

Class 3: aluminum alkyls such as triethylaluminum, isobutylaluminum and trimethylaluminum; bromine pentafluoride, bromine trifluoride, chlorodiethylaluminium and diethylzinc.

Class 2: calcium carbide, calcium metal, cyanogen bromide, lithium hydride, methyldichlorosilane, potassium metal, potassium peroxide, sodium metal, sodium peroxide, sulfuric acid and trichlorosilane.

Class 1: acetic anhydride, sodium hydroxide, sulfur monochloride and titanium tetrachloride.

❖ A water-reactive material explodes; violently reacts; produces flammable, toxic or other hazardous gases or evolves enough heat to cause self-ignition or ignition of nearby combustibles upon exposure to water or moisture. Water-reactive materials are subdivided as follows:

Class 3. Materials that react explosively with water without requiring heat or confinement.

Class 2. Materials that may form potentially explosive mixtures with water.

Class 1. Materials that may react with water with some release of energy, but not violently.

Chemicals in this classification must meet the minimum requirements of Sections 2701, 2703, 2704, 2705 and Chapter 44.

E102.1.12 Cryogenic fluids. The cryogenics listed will exist as compressed gases when they are stored at ambient temperatures.

1. Flammable: carbon monoxide, deuterium (heavy hydrogen), ethylene, hydrogen, methane.
2. Oxidizing: fluorine, nitric oxide, oxygen.
3. Corrosive: fluorine, nitric oxide.
4. Inert (chemically unreactive): argon, helium, krypton, neon, nitrogen, xenon.
5. Highly toxic: fluorine, nitric oxide.

❖ A cryogenic fluid is liquid having a boiling point lower than 150°F (101°C) at 14.7 pounds per square inch atmosphere (psia) (101 kPa).

A flammable cryogenic fluid is cryogenic fluid that is flammable in its vapor state.

Chemicals in this classification must meet the minimum requirements of Sections 2701, 2703, 2704, 2705 and Chapter 32.

E102.2 Health hazards. Materials classified in this section pose a health hazard.

❖ If a chemical is classified as a health hazard, there is statistically significant evidence that it can cause acute or chronic health effects in exposed persons. The term "health hazard" includes chemicals that are toxic, highly toxic and corrosive.

E102.2.1 Highly toxic materials. Examples include:

1. Gases: arsine, cyanogen, diborane, fluorine, germane, hydrogen cyanide, nitric oxide, nitrogen dioxide, ozone, phosphine, hydrogen selenide, stibine.
2. Liquids: acrolein, acrylic acid, 2-chloroethanol (ethylene chlorohydrin), hydrazine, hydrocyanic acid, 2-methylaziridine (propylenimine), 2-methylacetonitrile (acetone cyanohydrin), methyl ester isocyanic acid (methyl isocyanate), nicotine, tetranitromethane and tetraethylstannane (tetraethyltin).
3. Solids: (aceto) phenylmercury (phenyl mercuric acetate), 4-aminopyridine, arsenic pentoxide, arsenic trioxide, calcium cyanide, 2-chloroacetophenone, aflatoxin B, decaborane(14), mercury (II) bromide (mercuric bromide), mercury (II) chloride (corrosive mercury chloride), pentachlorophenol, methyl parathion, phosphorus (white) and sodium azide.

❖ A highly toxic material produces a lethal dose or lethal concentration that falls within any of the following categories:

1. A chemical that has a median lethal dose (LD^{50}) of 50 milligrams or less per kilogram of body weight when administered orally to albino rats weighing between 200 and 300 grams each.
2. A chemical that has a median lethal dose (LD^{50}) of 200 milligrams or less per kilogram of body weight when administered by continuous contact for 24 hours (or less if death occurs within 24 hours) with the bare skin of albino rabbits weighing between 2 and 3 kilograms each.
3. A chemical that has a median lethal concentration (LC^{50}) in air of 200 parts per million by volume or less of gas or vapor, or 2 milligrams per liter or less of mist, fume or dust, when administered by continuous inhalation for 1 hour (or less if death occurs within 1 hour) to albino rats weighing between 200 and 300 grams each.

Mixtures of these materials with ordinary materials, such as water, might not warrant classification as highly toxic. Although this system is basically simple in application, any hazard evaluation that is required for the precise categorization of this type of material must be performed by experienced, technically competent persons.

Chemicals in this classification must meet the minimum requirements of Sections 2701, 2703, 2704, 2705 and Chapter 37.

E102.2.2 Toxic materials. Examples include:

1. Gases: boron trichloride, boron trifluoride, chlorine, chlorine trifluoride, hydrogen fluoride, hydrogen sulfide, phosgene, silicon tetrafluoride.

2. Liquids: acrylonitrile, allyl alcohol, alpha-chlorotoluene, aniline, 1-chloro- 2,3-epoxypropane, chloroformic acid (allyl ester), 3-chloropropene (allyl chloride), o-cresol, crotonaldehyde, dibromomethane, diisopropylamine, diethyl ester sulfuric acid, dimethyl ester sulfuric acid, 2-furaldehyde (furfural), furfural alcohol, phosphorus chloride, phosphoryl chloride (phosphorus oxychloride) and thionyl chloride.

3. Solids: acrylamide, barium chloride, barium (II) nitrate, benzidine, p-benzoquinone, beryllium chloride, cadmium chloride, cadmium oxide, chloroacetic acid, chlorophenylmercury (phenyl mercuric chloride), chromium (VI) oxide (chromic acid, solid), 2,4-dinitrotoluene, hydroquinone, mercury chloride (calomel), mercury (II) sulfate (mercuric sulfate), osmium tetroxide, oxalic acid, phenol, P-phenylenediamine, phenylhydrazine, 4-phenylmorpholine, phosphorus sulfide, potassium fluoride, potassium hydroxide, selenium (IV) disulfide and sodium fluoride.

❖ A toxic chemical falls within any of the following categories:

1. A chemical that has a median lethal dose (LD^{50}) of more than 50 milligrams per kilogram, but not more than 500 milligrams per kilogram of body weight when administered orally to albino rats weighing between 200 and 300 grams each.

2. A chemical that has a median lethal dose (LD^{50}) of more than 200 milligrams per kilogram but not more than 1,000 milligrams per kilogram of body weight when administered by continuous contact for 24 hours (or less if death occurs within 24 hours) with the bare skin of albino rabbits weighing between 2 and 3 kilograms each.

3. In air of more than 200 parts per million but not more than 2,000 parts per million by volume of gas or vapor, or more than 2 milligrams per liter but not more than 20 milligrams per liter of mist, fume or dust, when administered by continuous inhalation for 1 hour (or less if death occurs within 1 hour) to albino rats weighing between 200 and 300 grams each.

Chemicals in this classification must meet the minimum requirements of Sections 2701, 2703, 2704, 2705 and Chapter 37.

E102.2.3 Corrosives. Examples include:

1. Acids: Examples: chromic, formic, hydrochloric (muriatic) greater than 15 percent, hydrofluoric, nitric (greater than 6 percent, perchloric, sulfuric (4 percent or more).

2. Bases (alkalis): hydroxides—ammonium (greater than 10 percent), calcium, potassium (greater than 1 percent), sodium (greater than 1 percent); certain carbonates—potassium.

3. Other corrosives: bromine, chlorine, fluorine, iodine, ammonia.

Note: Corrosives that are oxidizers, e.g., nitric acid, chlorine, fluorine; or are compressed gases, e.g., ammonia, chlorine, fluorine; or are water-reactive, e.g., concentrated sulfuric acid, sodium hydroxide, are physical hazards in addition to being health hazards.

❖ Corrosive chemicals cause visible destruction of, or irreversible alterations in, living tissue by chemical action at the point of contact. A chemical shall be considered corrosive if, when tested on the intact skin of albino rabbits by the method described in DOTn 49 CFR, Part 173, it destroys or changes irreversibly the structure of the tissue at the point of contact following an exposure period of 4 hours. This term does not refer to action on inanimate surfaces.

Chemicals in this classification must meet the minimum requirements of Sections 2701, 2703, 2704, 2705 and Chapter 31.

SECTION E103 EVALUATION OF HAZARDS

E103.1 Degree of hazard. The degree of hazard present depends on many variables which should be considered individually and in combination. Some of these variables are as shown in Sections E103.1.1 through E103.1.5.

❖ How can the degree of hazard presented by the various chemicals that may be in the structure under design or review be determined? The only true method to be assured that the structure is constructed or reviewed to the correct standards is to obtain and review the MSDS on the specific chemicals proposed. Owners can expect their structures to be required to meet a number of federal laws that regulate hazardous materials, including: the Superfund Amendments and Reauthorization Act of 1986 (SARA), the Resource Conservation and Recovery Act of 1976 (RCRA), the Hazardous Materials Transportation Act (HMTA), the Occupational Safety and Health Act (OSHA), the Toxic Substances Control Act (TSCA) and the Clean Air Act and Title III of SARA.

Community building/fire departments must make certain that the properties in their area meet Title III of SARA, which regulates the packaging, labeling, handling, storage and transportation of hazardous materials. The law requires facilities to furnish information about the quantities and health effects of materials used at the facility and to promptly notify local and state officials whenever a significant release of hazardous materials occurs. Title 42, Chapter 116, Subchapter II, Section 11021 of the Emergency Planning and Community Right-to-Know Act contains the following requirements for the submittal of MSDS (excerpt):

Basic requirement

1. Submission of MSDS or list

The owner or operator of any facility which is required to prepare or have available a material safety data sheet for a hazardous chemical under the Occupational Safety and Health Act of 1970 (29 U.S. C. 651 et seq.) and regulations promulgated under that Act shall submit a material safety data sheet for each chemical.

These MSDS are a guide to determining the specific hazards presented to a community. The sections that follow, along with their commentary, should help in determination of whether the MSDS is complete and correct in identifying the properties and characteristics of the hazardous materials expected in the subject occupancy.

E103.1.1 Chemical properties of the material. Chemical properties of the material determine self reactions and reactions which may occur with other materials. Generally, materials within subdivisions of hazard categories will exhibit similar chemical properties. However, materials with similar chemical properties may pose very different hazards. Each individual material should be researched to determine its hazardous properties and then considered in relation to other materials that it might contact and the surrounding environment.

❖ When we think of chemical properties, what should we be looking for in the MSDS in terms of reactivity hazards?

Hypergolic materials are those chemicals that will ignite when they come in contact with each other. Rocket propellants and military munitions are the most common uses of this type of material.

Pyrophoric materials react and ignite on contact with air. Common storage practices have this type of material stored in inert substances or under pressure in sealed containers to prevent the introduction of air.

Water-reactive materials will react on contact with water. It is very important to the emergency personnel that these chemicals be identified and labeled properly. The use of fire hoses and sprinklers with this type of material can be very dangerous for the responding personnel.

Unstable materials can violently decompose with little or no outside stimulus. Several of the materials used in the development of plastics can exhibit these traits.

E103.1.2 Physical properties of the material. Physical properties, such as whether a material is a solid, liquid or gas at ordinary temperatures and pressures, considered along with chemical properties will determine requirements for containment of the material. Specific gravity (weight of a liquid compared to water) and vapor density (weight of a gas compared to air) are both physical properties which are important in evaluating the hazards of a material.

❖ When we think of physical properties, what should we be looking for in the MSDS?

Flammable and combustible liquids classifications are determined by their chemical flash points and boiling points. The only difference between the various classifications of flammable and combustible liquids is the degree difference in these two chemical properties.

Ignition temperature is the temperature that the fuel in air must be heated for self-sustained combustion without help from a heat source. This gives the MSDS user a point to determine how the method of storage can change the properties of the chemical.

Flammability range is a very important property to understand because it will relate to the amount of ventilation that may be required to satisfy the *International Mechanical Code*® (IMC®). If a storage atmosphere is within the flammability range, a dangerous situation exists. In this condition the only element missing to start a fire is an ignition source; thus, an incident could happen at any time. In the reverse, when the storage arrangement is outside the flammability range, a much safer condition exists because the air-to-chemical ratio must be within the range prior to a dangerous incident.

Specific gravity is the weight of the item compared to the weight of an equal volume of water. Water is given a specific gravity of 1.00. If the chemical has a specific gravity less than 1.00, the chemical will float on water. For that reason, water would not be an effective extinguishing tool. Also, the chemical will flow with any runoff of fire-fighting water. Diking or another containment method is needed to contain the fire-fighting water and chemicals. If the chemical has a specific gravity greater that than 1.00, the chemical will sink in water. It is important to note that most flammable liquids have a specific gravity less than 1.00. Review the MSDS for the specific requirements for the chemicals.

Vapor density is similar to specific gravity; however, it is a comparison between the densities of a gas/vapor and air. Again, air is given a density of 1.00. If the chemical has a vapor density less than 1.00, it will rise. In many cases the vapor density being less than 1.00 is a good property because there are far fewer high ignition sources; however, remember that in an incident, the gas/vapor can be spread over a larger area. The spreading can reduce the flammability range enough to greatly lessen the possibility of a fire. If the vapor density is greater than 1.00, the gas/vapor will sink to the lowest point in the storage area. An example of a protection method in the codes for this property is the requirement that gas-fired appliances in garages must be 18 inches (457 mm) above the floor surface. This allows a chemical with a high vapor density to spread across the floor surface without encountering an ignition source.

Water solubility refers to the chemical's ability to mix with water. This information can be used in creating a proper response to an emergency incident.

E103.1.3 Amount and concentration of the material. The amount of material present and its concentration must be considered along with physical and chemical properties to determine the magnitude of the hazard. Hydrogen peroxide, for example, is used as an antiseptic and a hair bleach in low concentrations (approximately 8 percent in water solution). Over 8 percent, hydrogen peroxide is classed as an oxidizer and is toxic. Above 90 percent, it is a Class 4 oxidizer "that can undergo an explosive reaction when catalyzed or exposed to heat, shock or friction," a definition which incidentally also places hydrogen peroxide over 90-percent concentration in the unstable (reactive) category. Small amounts at high concentrations may present a greater hazard than large amounts at low concentrations.

❖ Many chemicals exhibit combined chemical traits, one of which is toxicity. The amount of damage that can be inflicted on the environment is based on the toxicity and the threshold limit values, lethal dosage, lethal concentration or emergency exposure limits.

Review the chemicals that will be housed in the structure or used in the process. Look at their chemical, physical and toxicity properties. The method of storage and the amounts of chemical will also play a major role in determining hazards. Chemical and physical characteristics were covered in Sections E103.1.1 and E103.1.2. This section covers the properties of toxicity.

Threshold limit value, also known as "time weighted average" (TLVTWA), is the maximum amount of chemical that the human body can be exposed to for 8 hours a day or 40 hours a week without dangerous effects. This number is important when checking or reviewing the ventilation for a structure. TLVTWA is normally expressed in parts per million, while others are sometimes expressed as a percentage per billion. This is the number of molecules of chemical per molecules of air. It is important to note that the smaller the ratio, the more toxic the material.

Lethal dosage (LD) is the minimum amount of solid or liquid that, when ingested or absorbed through the skin, may be fatal. MSDS provide this information typically as LD_{50}. This notation represents the amount of chemical that will kill at least 50 percent of the test subjects when exposed. The amount is expressed in milligrams per kilogram of body weight.

Lethal concentration (LC) is the minimum concentration in the gaseous state that, when inhaled, may be fatal. MSDS provide this information typically as LC_{50}. This notation is expressed in milligrams per liter.

Emergency exposure limit or TLV short-term exposure limit (EEL or TLVSTEL) is the maximum amount of chemical that can be tolerated with no permanent toxic effects.

Ceiling limit (TLVC) is the amount of chemical that normally will not cause immediate irritation. This limit should never be reached in normal operation, even for an instant.

E103.1.3.1 Mixtures. Gases—toxic and highly toxic gases include those gases which have an LC_{50} of 2,000 parts per million (ppm) or less when rats are exposed for a period of 1 hour or less. To maintain consistency with the definitions for these materials, exposure data for periods other than 1 hour must be normalized to 1 hour. To classify mixtures of compressed gases that contain one or more toxic or highly toxic components, the LC_{50} of the mixture must be determined. Mixtures that contain only two components are binary mixtures. Those that contain more than two components are multi-component mixtures. When two or more hazardous substances (components) having an LC_{50} below 2,000 ppm are present in a mixture, their combined effect, rather than that of the individual substances (components), must be considered. In the absence of information to the contrary, the effects of the hazards present must be considered as additive. Exceptions to the above rule may be made when there is a good reason to believe that the principal effects of the different harmful substances (components) are not additive.

For binary mixtures where the hazardous component is diluted with a nontoxic gas such as an inert gas, the LC_{50} of the mixture is estimated by use of the following formula:

$$LC_{50m} = \frac{1}{[C_i / LC_{50i}]}$$ **(Equation E-1)**

For multi-component mixtures where more than one component has a listed LC_{50}, the LC_{50} of the mixture is estimated by use of the following formula:

$$C_{50m} = \frac{1}{(C_{il} / LC_{50il}) + (C_{i2} / LC_{50i2}) + (C_{in} / LC_5)}$$

(Equation E-2)

where:

LC_{50m} = LC_{50} of the mixture in parts per million (ppm).

C_i = concentration of component (i) in decimal percent. The concentration of the individual components in a mixture of gases is to be expressed in terms of percent by volume.

LC_{50i} = LC_{50} of component (i). The LC_{50} of the component is based on a 1-hour exposure. LC_{50} data which are for other than 1-hour exposures shall be normal ized to 1-hour by multiplying the LC_{50} for the time determined by the factor indicated in Table E103.1.3.1. The preferred mammalian species for LC_{50} data is the rat, as specified in the definitions of toxic and highly toxic in Chapter 2 of the *Interna tional Fire Code*. If data for rats are unavailable, and in the absence of information to the contrary, data for other species may be utilized. The data shall be taken in the following order of preference: rat, mouse, rabbit, guinea pig, cat, dog, monkey.

i_n = component 1, component 2 and so on to the nth component.

Examples:

a. What is the LC_{50} of a mixture of 15-percent chlorine, 85-percent nitrogen?

The 1-hour (rat) LC_{50} of pure chlorine is 293 ppm.

LC_{50m} =1 / (0.15 / 293) or 1,953 ppm. Therefore, the mixture is toxic.

b. What is the LC_{50} of a mixture of 15-percent chlorine, 15-percent fluorine and 70-percent nitrogen? The 1-hour (rat) LC_{50} of chlorine is 293 ppm. The 1-hour (rat) LC_{50} of fluorine is 185 ppm.

$LC_{50m} = 1 / (0.15 / 293) + (0.15 / 185)$ or 755 ppm. Therefore the mixture is toxic.

c. Is the mixture of 1 percent phosphine in argon toxic or highly toxic? The 1-hour (rat) LC_{50} is 11 ppm.

$LC_{50m} = 1 / [0.01 / (11 \quad 2)]$ or 2,200 ppm. Therefore the mixture is neither toxic nor highly toxic. Note that the 4-hour LC_{50} of 11 ppm was normalized to 1-hour by use of E103.1.3.1.

❖ This information allows evaluation of the methods of calculation submitted by the chemical engineer. It is not designed for the inspector/plans examiner to perform the calculations during inspection or plan review. Mixtures and chemicals should be evaluated by a qualified person such as a chemical engineer, with the information submitted to the inspector/plans examiner. MSDS for the specific mixtures can be submitted in place of separate chemical engineering evaluations.

TABLE E103.1.3.1
NORMALIZATION FACTOR

TIME (hours)	MULTIPLY BY
0.5	0.7
1.0	1.0
1.5	1.2
2.0	1.4
3.0	1.7
4.0	2.0
5.0	2.2
6.0	2.4
7.0	2.6
8.0	2.8

E103.1.4 Actual use, activity or process involving the material. The definition of handling, storage and use in closed systems refers to materials in packages or containers. Dispensing and use in open containers or systems describes situations where a material is exposed to ambient conditions or vapors are liberated to the atmosphere. Dispensing and use in open systems, then, are generally more hazardous situations than handling, storage or use in closed systems. The actual use or process may include heating, electric or other sparks, catalytic or reactive materials and many other factors which could affect the hazard and must therefore be thoroughly analyzed.

❖ Questions must be asked of the owner on how the products will be used in the facility. Tables 2703.1.1(1) and (2) are based on normal storage arrangements, open systems and closed systems. The amount of chemical allowed per control area is based on how the chemical is being used or stored. Correct interpretation of the tables depends on knowing how the chemicals will be used. Remember that any process that changes the chemical traits must be accounted for. These would include preheating and pressurizing the chemicals.

E103.1.5 Surrounding conditions. Conditions such as other materials or processes in the area, type of construction of the structure, fire protection features (e.g., fire walls, sprinkler systems, alarms, etc.), occupancy (use) of adjoining areas, normal temperatures, exposure to weather, etc., must be taken into account in evaluating the hazard.

❖ A final element in our determination of hazards is the conditions of storage and use. Will the chemical be in outdoor storage, indoor storage or in a detached building? What types of protection for the chemicals are proposed? What types of protection are required for the type of hazard? What are the requirements for temperature control for the chemicals? What types of processes will the chemicals be used in? The answers to these questions will determine whether the proposed structure meets the minimum requirements of the code.

E103.2 Evaluation questions. The following are sample evaluation questions:

❖ Evaluation of each chemical proposed to be in a structure and then determining whether the structure will be a hazardous use group is very important. In today's society, even those occupancies that would normally not be thought of as having hazardous materials can have enough involved in production to change the use group. For example, during review of a dentist's office (normally considered a business use group) MSDS were requested for all chemicals, along with the amounts and storage methods. When they were received, the chemicals reported included one that was listed as explosive when at room temperature. The occupancy was to have 3 pounds (1.4 kg) of the product.

The introduction of this one chemical made the structure move from a Group B use to a Group H-1 use, and the product had to be stored in a detached storage building.

These questions must be asked and evaluated to determine whether a building is a Group H use group.

1. What is the material? Correct identification is important; exact spelling is vital. Check labels, MSDS, ask responsible persons, etc.

❖ The properties of chemical change with the mixture, manufacturer and blends. Generic information cannot be used; information recorded on the MSDS must match the exact chemical that will be used. Currently, there are at least six different blends of diesel fuel that, based on the flash point and boiling point, can range from a Class IC to a Class III flammable liquid; thus, different protection and exempt amounts are allowed and required for the chemical. It is important to note that the changing of one letter in the chemical name can change a chemical that is not considered hazardous to being extremely hazardous. Precise information is required to determine whether a hazardous use group designation is needed.

2. What are the concentration and strength?

❖ Based on the blend, concentration and strength of solution, each chemical takes on different traits. One MSDS does not cover all concentrations and blends; a separate MSDS must be submitted for each specific chemical.

3. What is the physical form of the material? Liquids, gases and finely divided solids have differing requirements for spill and leak control and containment.

❖ The physical form of the material is important from the standpoint of hazards and the concepts of fire. Gases are more dangerous than liquids and liquids are more dangerous than solids. Understanding the form and makeup of the chemical will assist in determining the danger to the structure and the community. Each physical form will have its own problems with protection and containment that must be considered in the review.

4. How much material is present? Consider in relation to permit amounts, maximum allowable quantity per control area (from Group H occupancy requirements), amounts which require detached storage and overall magnitude of the hazard.

❖ A review of the chemicals and amounts of product that will be present will raise some questions: Will the chemicals be in an open or closed container? What type of protection will be provided? Will the chemical be placed in control areas in accordance with Section 2703.8.2? Will the amounts be more than the exempt amounts allowed by Tables 2703.1.1(1) and 2703.1.1(2) or be one of the exceptions listed in Section 2701.1? The answers to these questions will govern the types of requirements that must be addressed.

5. What other materials (including furniture, equipment and building components) are close enough to interact with the material?

❖ Will other items be stored close to the chemical storage? Will the storage of other chemicals that are normally not hazardous cause a more hazardous condition when mixed with the hazardous chemicals in an emergency. Is the fire load of the area increased by the presence of the chemicals and general storage? If so, additional protection may be required based on the additional fire loading and hazards created.

6. What are the likely reactions?

❖ Will the mixing of chemical or fire protection features create a dangerous or hazardous condition if the chemicals are not separated into different containment areas? Checking the MSDS for adverse reactions to other products in the storage area is extremely important.

7. What is the activity involving the material?

❖ What process will the chemical be used in? Will the physical conditions of the chemical be changed by heating, pressurizing or other methods during the use, process or manufacturing of the final product? It is important to note that some nonhazardous chemicals, through the processing phases, can create a hazardous atmosphere. As much information and knowledge as possible of the processing system and use of the chemical must be gained for a good review/inspection process.

8. How does the activity impact the hazardous characteristics of the material? Consider vapors released or hazards otherwise exposed.

❖ Is the chemical being used in an open atmosphere? Chemicals in an open system are normally much more dangerous than they would be in a closed system. When dealing with a toxic chemical, what types of filters are required to protect personnel and the environment?

9. What must the material be protected from? Consider other materials, temperature, shock, pressure, etc.

❖ High explosives can be denoted by light to moderate shock. Some chemicals can react to changes in temperatures. Containers under pressure can react violently. The MSDS must be checked for these types of hazards to make certain protective storage and use arrangements fit requirements.

10. What effects of the material must people and the environment be protected from?

❖ What are the dangers to the environment and/or to personnel if there is an emergency release? Dikes, special filters, distance from other structures and other methods of protection may be required to reduce the impact from an emergency release.

11. How can protection be accomplished? Consider:

11.1. Proper containers and equipment.

❖ Plastic pint bottles, glass bottles, safety cans, drums, barrels, above-ground tanks and underground tanks are just a few of the types of containers that may be used for storage. Each type of container has special requirements for size and methods of storage. These requirements can be found in Chapter 27 and other related references.

11.2. Separation by distance or construction.

❖ Group H-1 use areas must be located in a detached building. Outdoor storage must be separated from other buildings and property lines by distances as outlined in Section 2703.12 and other related sections.

11.3. Enclosure in cabinets or rooms.

❖ Third-party tested hazardous materials storage cabinets can be used to increase the exempt amounts allowed by Tables 2703.1.1(1) and (2). These cabinets must meet the minimum requirements of Sections 2703.8.4, 3704.1.2 and other related sections. Gas rooms must meet the requirements of Sections 2703.8.3, 3704.1.2 and other related sections. Flammable liquid storage rooms must meet the requirements of Section 3404.3.7.3 and related codes.

11.4. Spill control, drainage and containment.

❖ If one can minimize the spill area and contain any accidental release, the amount of potential damage will be greatly reduced. The cleanup efforts and reduction of the hazard will be much easier for the fire service and emergency workers. Spill control, drainage and containment must meet the minimum requirements of Section 2704.2 and related sections. Remember that the containment area is, in a lot of cases, more hazardous than the initial storage area.

11.5. Control systems — ventilation, special electrical, detection and alarm, extinguishment, explosion venting, limit controls, exhaust scrubbers and excess flow control.

❖ The methods of control, how much ventilation, special detection and alarms will be obtained from the MSDS. Section 2703 contains guidance into requirements for additional protection.

11.6. Administrative (operational) controls—signs, ignition source control, security, personnel training, established procedures, storage plans and emergency plans.

Evaluation of the hazard is a strongly subjective process; therefore, the person charged with this responsibility must gather as much relevant data as possible so that the decision will be objective and within the limits prescribed in laws, policies and standards.

It may be necessary to cause the responsible persons in charge to have tests made by qualified persons or testing laboratories to support contentions that a particular material or process is or is not hazardous. See Section 104.7.2 of the *International Fire Code.*

❖ The posting of signs as required by Section 2703.5 is an important component to alert emergency response personnel to possible dangers to them and the community. Section 2703.7 contains guidance into the requirements for ignition source controls. Section 2703.9.1 contains information on personnel training. Section 2703.9.2 contains information on security. Storage plans and emergency plans requirements can be found in Sections 409.4 and 2701.4.1 and other related sections. Having an up-to-date emergency plan is very important to both the community and emergency services; however, it is often overlooked.

Appendix F: Hazard Ranking

The provisions contained in this appendix are not mandatory unless specifically referenced in the adopting ordinance.

General Comments

The presence of hazardous materials in all types of occupancies is becoming increasingly more customary. While industrial facilities are constantly developing and expanding their use of hazardous materials, significant quantities of an ever increasing variety of materials can also be found in other occupancies, such as hospitals, research laboratories and even mercantile uses. Knowledge of hazardous materials within a given occupancy is essential not only for proper code enforcement but also for prefire planning purposes. Although submittal of Material Safety Data Sheets (MSDS) and compliance with right-to-know legislation are necessary to determine the proper mitigation means for all hazardous materials, it is also true that in an emergency response situation, responders must have a clear and readily available warning of the material hazards that confront them. The NFPA 704 hazard classification system was developed for that express purpose and it is the intent of this appendix to provide the fire code official with a ready reference tool for approval of the hazard warnings required by Chapter 27, among others, of the code.

Purpose

The code regulates the storage, dispensing and use of all hazardous materials classified as either physical or health hazards. These materials pose diverse hazards, including instability, reactivity, flammability, oxidizing potential, radioactivity or toxicity; therefore, identifying them by hazard ranking is essential. The information in this appendix is intended to be a companion to the specific requirements of Chapters 28 through 44.

The table included in this appendix lists the various hazardous materials categories that are defined in Chapter 2 of the code, along with the NFPA 704 hazard ranking for each. Once a specific material is properly identified and categorized as meeting a given hazardous material definition, the appropriate flammability, health, reactivity, oxidizing and special hazard designations of the NFPA 704 system can be readily found in the table.

SECTION F101 GENERAL

F101.1 Scope. Assignment of levels of hazards to be applied to specific hazard classes as required by NFPA 704 shall be in accordance with this appendix. The appendix is based on application of the degrees of hazard as defined in NFPA 704 arranged by hazard class as for specific categories defined in Chapter 2 of the *International Fire Code* and used throughout.

❖ This paragraph establishes the relationship between the hazardous materials definitions found in Chapter 2 of the code and the methods used in NFPA 704 to establish the degree of hazard for a wide variety of hazardous materials.

F101.2 General. The hazard rankings shown in Table F101.2 have been established by using guidelines found within NFPA 704. As noted in Section 1-5 of NFPA 704, there could be specific reasons to alter the degree of hazard assigned to a specific material; for example, ignition temperature, flammable range or susceptibility of a container to rupture by an internal combustion explosion or to metal failure while under pressure or because of heat from external fire. As a result, the degree of hazard assigned for the same material can vary when assessed by different people of equal competence.

The hazard rankings assigned to each class represent reasonable minimum hazard levels for a given class based on the use of criteria established by NFPA 704. Specific cases of use or storage may dictate the use of higher degrees of hazard in certain cases.

❖ These two paragraphs explain the rationale for the hazard rankings shown in Table F101.2. The word of caution sounded here is that rankings will vary depending on the circumstances of each individual application. In addition, because fire code officials are required to use personal judgement to determine rankings, those rankings may vary from one official to another because of their differing personal backgrounds and experience.

The message given in the last sentence of the first paragraph is very significant. The fire code official will be called upon to use every bit of knowledge gained through years of experience to make judgements of hazards on a case-by-case basis. The rankings shown in Table F101.2 are minimums. The fire code official may decide that circumstances in individual cases require higher rankings than are shown in the table.

As an example of how conditions can cause a change in ranking, consider this. The method of storage and processing of a chemical can greatly change the ranking. If a Class IIIB combustible liquid is preheated to its flashpoint, or pressurized to greater than 15 psi (103 kPa), it will exhibit the same traits as a Class IA flammable liquid. Thus the ranking would change from a 1 (F1) in the flammable diamond to a 4 (F4). This increase in ranking is a result of the change of storage method and process, not any change in the chemical makeup of the combustible liquid.

TABLE F101.2
FIRE FIGHTER WARNING PLACARD DESIGNATIONS BASED ON HAZARD CLASSIFICATION CATEGORIES

HAZARD CATEGORY	DESIGNATION
Combustible liquid II	F2
Combustible liquid IIIA	F2
Combustible liquid IIIB	F1
Combustible dust	F4
Combustible fiber	F3
Cryogenic flammable	F4, H3
Cryogenic oxidizing	OX, H3
Explosive	R4
Flammable solid	F2
Flammable gas (gaseous)	F4
Flammable gas (liquefied)	F4
Flammable liquid IA	F4
Flammable liquid IB	F3
Flammable liquid IC	F3
Organic peroxide, UD	R4
Organic peroxide I	F4, R3
Organic peroxide II	F3, R3
Organic peroxide III	F2, R2
Organic peroxide IV	F1, R1
Organic peroxide V	Nonhazard
Oxidizing gas (gaseous)	OX
Oxidizing gas (liquefied)	OX
Oxidizer 4	OX
Oxidizer 3	OX
Oxidizer 2	OX
Oxidizer 1	F4
Pyrophoric gases	F3
Pyrophoric solids, liquids	R4
Unstable reactive 4D	R4
Unstable reactive 3D	R3
Unstable reactive 3N	R2
Unstable reactive 2	W, R3
Water reactive 3	W, R2
Water reactive 2	H3, COR
Corrosive	H3
Toxic	H4
Highly toxic	

F—Flammable category.
R—Reactive category.
H—Health category.
W—Special hazard: water reactive.
OX—Special hazard: oxidizing properties.
COR—Corrosive.
UD—Unclassified detonable material.
4D—Class 4 detonable material.
3D—Class 3 detonable material.
3N—Class 3 nondetonable material.

❖ This table lists the various hazardous materials classes that are defined in Chapter 2 of the code and includes the NFPA 704 hazard ranking for each. This information will provide a ready reference to assist in the application of certain sections of the code that contain specific safety requirements based on the NFPA 704 ranking of a material, such as Section 2703.2.2.1. The table can also be a useful tool in the preparation of the NFPA 704 hazard identification signs required by Section 2703.5, among others.

SECTION F102 REFERENCED STANDARDS

ICC	IFC	International Fire Code	F101.1
NFPA	704	Identification of the Hazards of Materials for Emergency Response	F101.1, F101.2

Appendix G: Cryogenic Fluids—Weight And Volume Equivalents

The provisions contained in this appendix are not mandatory unless specifically referenced in the adopting ordinance.

General Comments

Chapter 32 of the code regulates cryogenic fluids (liquefied gases) that are used to provide very low temperatures in a variety of scientific and industrial processes. Low temperatures in the cryogenic processes are achieved primarily by the liquefaction of gases, resulting in the family of hazardous materials known as cryogenic fluids (see the commentary to the definition of "Cryogenic fluid" in Section 3202.1). More than 25 such cryogens are currently in use in the cryogenic area; however, this appendix focuses on the six that account for the greatest volume of use and application in research and industry: helium, hydrogen, nitrogen, argon, oxygen and liquefied natural gas (LNG; methane).

Although cryogenic fluids are characterized by extreme low temperatures, ranging from a boiling point of 258.5°F (161.4°C) for LNG to 453.1°F (269.5°C) for helium, they also posses another hazard characteristic: a high volume expansion ratio from liquid to gas, from approximately 696 to 1 for nitrogen, to 860 to 1 for oxygen. And while argon, helium, oxygen and nitrogen in the cryogenic range are not toxic or flammable, they can cause asphyxiation by displacing the air necessary for the support of life when released. Even oxygen, an oxidizer, may have harmful physiological effects if it is breathed over an extended period.

There is also the flammable gas hazard when cryogenic fluids, such as hydrogen and LNG, are stored or used. However, the fire hazard may also be greatly increased when cryogenic fluids normally thought to be nonflammable are used. The presence of oxygen, for example, will greatly increase the combustibility of ordinary combustible materials, and may even cause some noncombustible materials like carbon steel to burn readily under the right conditions. Liquefied inert gases, such as liquid nitrogen or liquid helium, are also capable, under the right conditions, of condensing oxygen from the atmosphere and causing unsuspected oxygen enrichment or entrapment in areas where there may be ignition sources. Reduction of this and the asphyxiation potential hazard of cryogenic fluids (liquefied gases) discussed above may be accomplished, in part, by a properly designed ventilation system.

To properly apply the requirements of Chapter 32 of the code to reduce the hazards discussed here, the volume of gas capable of being generated by a cryogenic fluid installation must be determined. Rather than rely on leaving such determination to case-by-case mathematical calculation, this appendix gives the fire code official and design professional a convenient tool for determining the correct gas volumes for selected cryogenic fluids. For further discussion of cryogenic fluids, see the commentary to Chapter 32.

Purpose

This appendix gives the fire code official a ready reference tool for the conversion of the liquid weight and volume of cryogenic fluid to their corresponding volume of gas and vice versa.

SECTION G101 GENERAL

G101.1 Scope. This appendix is used to convert from liquid to gas for cryogenic fluids.

❖ This appendix gives the liquid volume for the most frequently used cryogenic fluids for a given weight at the normal boiling point of the liquid, and then shows the volume of gas that will expand from that volume of liquid at normal temperature and pressure (NTP). An example showing how to use the table is given in the paragraphs that follow.

Anyone who has followed a truck containing liquid nitrogen down a highway has seen the vapor plume venting from the tank. Anyone who has watched the launch of a liquid-fueled rocket, such as the space shuttle booster or the Saturn V used to put Apollo capsules in moon orbit, has seen a vapor plume coming from the liquid oxygen tanks. In both cases the plume is water vapor that has been condensed from the air by the cold gas escaping from the tanks through overpressure valves.

All of these common cryogenics represent hazards of one kind or another. Liquefied natural gas (LNG) and hydrogen are both flammable and present combustion and explosion hazards at concentrations that are addressed in Chapter 32. Oxygen can also contribute tremendously to the intensity of a fire, its size and its spread. On the other hand, helium, argon and nitrogen are nonflammable, but are also not breathable. Helium, being lighter than air, would rise to the highest levels of the storage space, presenting a minimal hazard unless the concentration became so large that it displaced breathable air at lower levels. Nitrogen and argon are heavier and would concentrate at lower levels, displacing breathable air.

G101.2 Conversion. Table G101.2 shall be used to determine the equivalent amounts of cryogenic fluids in either the liquid or gas phase.

❖ Table G101.2 is a convenient reference to determine the volume of gas that will expand from a given volume of cryogenic fluid. This kind of information is important to the decision of what quantities can be safely stored or used under any set of conditions and also to the determination of what ventilation capacity would be required for those storage or use conditions.

G101.2.1 Use of the table. To use Table G101.2, read horizontally across the line of interest. For example, to determine the number of cubic feet of gas contained in 1.0 gallon (3.785 L) of liquid argon, find 1.000 in the column entitled "Volume of Liquid at Normal Boiling Point." Reading across the line under the column entitled "Volume of Gas at 70°F and 1 atmosphere 14.7 psia," the value of 112.45 cubic feet (3.184 m^3) is found.

❖ This example emphasizes the importance of expansion volume of these cryogenic liquids to determining safe storage and use conditions. Leakage of as little as 10 gallons (38 L) of liquid argon would displace all of the air in a space measuring 10 feet by 10 feet by 10 feet (3048 mm by 3048 mm by 3048mm), unless the ventilation system for this space is sized to remove the expanding gas as it escapes from its container or transport system.

G101.2.2 Other quantities. If other quantities are of interest, the numbers obtained can be multiplied or divided to obtain the quantity of interest. For example, to determine the number of cubic feet of argon gas contained in a volume of 1,000 gallons (3785 L) of liquid argon at its normal boiling point, multiply 112.45 by 1,000 to obtain 112,450 cubic feet (3184 m^3).

❖ This example shows that the basic numbers in the table can be used to calculate the effect of any quantity of cryogenic liquid on any given storage or work space.

The table can also be used to calculate needed information about design requirements. Consider this simplified example.

Assume a cryogenic tank that holds 75 gallons (284 L) of liquid oxygen (LOX) and weighs 200 pounds (91 kg) empty. The tank is to be located in a room in an existing retirement home building to fill portable oxygen bottles for individuals needing breathing assistance. An exhaust hood will be installed over the tank to assist in carrying off any oxygen vapors. To be determined is whether the floor structure of the existing room will carry the added load of the filled tank and also what volume of oxygen gas would have to be exhausted in case of a valve failure.

Table G101.2 shows that 1 gallon (3.785 L) of LOX weighs 9.527 pounds (4.3 kg). The weight of the tank and its contents may then be calculated as follows:

75 gal.× 9.527 lb/gal = 714.525 lb
714.525 lb + 200 lb = 915 lb
Add safety factor of 2 = 1830 lb

The floor structure would have to support the added load of the LOX tank and its contents having a weight of approximately 2,000 pounds (1080 kg) [see the commentary to Chapter 16 of the *International Building Code*® (IBC®) for discussion of structural loads.]

Table G101.2 also shows that 1 gallon (3.785 L) of LOX will expand to 115.05 cubic feet (3.26 m^3) of gas. Assuming the tank is full and a valve failure would result in the tank emptying completely in 2 hours, the exhaust hood would have to be able to remove the following volume of gas:

75 gal. × 115.05 cu. ft./gal. = 8628.75 cu. ft. (244.4 m^3) total
8628.75 cu. ft. ÷ 120 min. = 71.9 cfm (2 m^3/min)

The exhaust hood fan would have to be able to remove 72 cubic feet per minute (cfm) (.03 m^3/s) of gas from the room.

This example is obviously a simplified version of a real problem. In real-world terms, more information would be needed about the volume of the room the tank is in, and additional safety features to make sure no spark sources or flammable materials are added where concentrations of oxygen could intensify a fire would need to be considered. But it does serve to show how Table G101.2 can be used to help designers, contractors, inspectors, plans examiners and building owners and tenants determine whether planned installations meet minimum safety requirements.

TABLE G101.2
WEIGHT AND VOLUME EQUIVALENTS FOR COMMON CRYOGENIC FLUIDS[a]

CRYOGENIC FLUID	WEIGHT OF LIQUID OR GAS		VOLUME OF LIQUID AT NORMAL BOILING POINT		VOLUME OF GAS AT 70°F AND 1 ATMOSPHERE 14.7 PSIA	
	Pounds	Kilograms	Liters	Gallons	Cubic feet	Cubic meters
Argon	1.000	0.454	0.326	0.086	9.67	0.274
	2.205	1.000	0.718	0.190	21.32	0.604
	3.072	1.393	1.000	0.264	29.71	0.841
	11.628	5.274	3.785	1.000	112.45	3.184
	10.340	4.690	3.366	0.889	100.00	2.832
	3.652	1.656	1.189	0.314	35.31	1.000
Helium	1.000	0.454	3.631	0.959	96.72	2.739
	2.205	1.000	8.006	2.115	213.23	6.038
	0.275	0.125	1.000	0.264	26.63	0.754
	1.042	0.473	3.785	1.000	100.82	2.855
	1.034	0.469	3.754	0.992	100.00	2.832
	0.365	0.166	1.326	0.350	35.31	1.000
Hydrogen	1.000	0.454	6.409	1.693	191.96	5.436
	2.205	1.000	14.130	3.733	423.20	11.984
	0.156	0.071	1.000	0.264	29.95	0.848
	0.591	0.268	3.785	1.000	113.37	3.210
	0.521	0.236	3.339	0.882	100.00	2.832
	0.184	0.083	1.179	0.311	35.31	1.000
Oxygen	1.000	0.454	0.397	0.105	12.00	0.342
	2.205	1.000	0.876	0.231	26.62	0.754
	2.517	1.142	1.000	0.264	30.39	0.861
	9.527	4.321	3.785	1.000	115.05	3.250
	8.281	3.756	3.290	0.869	100.00	2.832
	2.924	1.327	1.162	0.307	35.31	1.000
Nitrogen	1.000	0.454	0.561	0.148	13.80	0.391
	2.205	1.000	1.237	0.327	30.43	0.862
	1.782	0.808	1.000	0.264	24.60	0.697
	6.746	3.060	3.785	1.000	93.11	2.637
	7.245	3.286	4.065	1.074	100.00	2.832
	2.558	1.160	1.436	0.379	35.31	1.000
LNG1	1.000	0.454	1.052	0.278	22.968	0.650
	2.205	1.000	2.320	0.613	50.646	1.434
	0.951	0.431	1.000	0.264	21.812	0.618
	3.600	1.633	3.785	1.000	82.62	2.340
	4.356	1.976	4.580	1.210	100.00	2.832
	11.501	5.217	1.616	0.427	35.31	1.000

For SI: 1 pound = 0.454 kg, 1 gallon = 3.785 L, 1 cubic foot = 0.02832 m^3, °C = [(°F)-32]/1.8, 1 pound per square inch atmosphere = 6.895 kPa.

a. The values listed for liquefied natural gas (LNG) are "typical" values. LNG is a mixture of hydrocarbon gases, and no two LNG streams have exactly the same composition.

INDEX

B

C

D

E

F

G

H

I

K

L

P

R

S

T

U

V

W

Z